Climate Change 2007
Mitigation of Climate Change

The Intergovernmental Panel on Climate Change (IPCC) was set up jointly by the World Meteorological Organization and the United Nations Environment Programme to provide an authoritative international statement of scientific understanding of climate change. The IPCC's periodic assessments of the causes, impacts and possible response strategies to climate change are the most comprehensive and up-to-date reports available on the subject, and form the standard reference for all concerned with climate change in academia, government and industry worldwide. Through three working groups, many hundreds of international experts assess climate change in this Fourth Assessment Report. The Report consists of three main volumes under the umbrella title Climate Change 2007, all available from Cambridge University Press:

Climate Change 2007 – The Physical Science Basis
Contribution of Working Group I to the Fourth Assessment Report of the IPCC
(ISBN 978 0521 88009-1 Hardback; 978 0521 70596-7 Paperback)

Climate Change 2007 – Impacts, Adaptation and Vulnerability
Contribution of Working Group II to the Fourth Assessment Report of the IPCC
(978 0521 88010-7 Hardback; 978 0521 70597-4 Paperback)

Climate Change 2007 – Mitigation of Climate Change
Contribution of Working Group III to the Fourth Assessment Report of the IPCC
(978 0521 88011-4 Hardback; 978 0521 70598-1 Paperback)

Climate Change 2007 – Mitigation of Climate Change aims to answer essentially five questions relevant to policymaking worldwide:

- What can we do to reduce or avoid the threats of climate change?

- What are the costs of these actions and how do they relate to the costs of inaction?

- How much time is available to realise the drastic reductions needed to stabilise greenhouse gas concentrations in the atmosphere?

- What are the policy actions that can overcome the barriers to implementation?

- How can climate mitigation policy be aligned with sustainable development policies?

This latest assessment of the IPCC provides a comprehensive, state-of-the-art and worldwide overview of scientific knowledge related to the mitigation of climate change. It includes a detailed assessment of costs and potentials of mitigation technologies and practices, implementation barriers, and policy options for the sectors: energy supply, transport, buildings, industry, agriculture, forestry and waste management. It links sustainable development policies with climate change practices. This volume will again be the standard reference for all those concerned with climate change, including students and researchers, analysts and decision-makers in governments and the private sector.

From reviews of the Third Assessment Report – Climate Change 2001:

'The detail is truly amazing … invaluable works of reference … no reference or science library should be without a set [of the IPCC volumes] … unreservedly recommended to all readers.'
Journal of Meteorology

'This well-edited set of three volumes will surely be the standard reference for nearly all arguments related with global warming and climate change in the next years. It should not be missing in the libraries of atmospheric and climate research institutes and those administrative and political institutions which have to deal with global change and sustainable development.'
Meteorologische Zeitschrift

'… likely to remain a vital reference work until further research renders the details outdated by the time of the next survey … another significant step forward in the understanding of the likely impacts of climate change on a global scale.'
International Journal of Climatology

'The IPCC has conducted what is arguably the largest, most comprehensive and transparent study ever undertaken by mankind … The result is a work of substance and authority, which only the foolish would deride.'
Wind Engineering

'… the weight of evidence presented, the authority that IPCC commands and the breadth of view can hardly fail to impress and earn respect. Each of the volumes is essentially a remarkable work of reference, containing a plethora of information and copious bibliographies. There can be few natural scientists who will not want to have at least one of these volumes to hand on their bookshelves, at least until further research renders the details outdated by the time of the next survey.'
The Holocene

'The subject is explored in great depth and should prove valuable to policy makers, researchers, analysts, and students.'
American Meteorological Society

From reviews of the Second Assessment Report – Climate Change 1995:

' … essential reading for anyone interested in global environmental change, either past, present or future. … These volumes have a deservedly high reputation'
Geological Magazine

'… a tremendous achievement of coordinating the contributons of well over a thousand individuals to produce an authoritative, state-of-the-art review which will be of great value to decision-makers and the scientific community at large … an indispensable reference.'
International Journal of Climatology

'… a wealth of clear, well-organized information that is all in one place ... there is much to applaud.'
Environment International

Climate Change 2007
Mitigation of Climate Change

Edited by

Bert Metz

Co-chair Working Group III
Netherlands Environmental
Assessment Agency

Ogunlade Davidson

Co-chair Working Group III
University of Sierra Leone

Peter Bosch **Rutu Dave** **Leo Meyer**

Technical Support Unit IPCC Working Group III
Netherlands Environmental Assessment Agency

Contribution of Working Group III
to the Fourth Assessment Report of the
Intergovernmental Panel on Climate Change

Published for the Intergovernmental Panel on Climate Change

CAMBRIDGE UNIVERSITY PRESS
Cambridge, New York, Melbourne, Madrid, Cape Town, Singapore, São Paolo, Delhi

Cambridge University Press
32 Avenue of the Americas, New York, NY 10013-2473, USA

www.cambridge.org
Information on this title: www.cambridge.org/9780521880114

First published 2007

Printed in Canada by Friesens

A catalog record for this publication is available from the British Library.

ISBN 978-0-521-88011-4 hardback
ISBN 978-0-521-70598-1 paperback

Please use the following reference to the whole report:

IPCC, 2007: *Climate Change 2007: Mitigation of Climate Change. Contribution of Working Group III to the Fourth Assessment Report of the Intergovernmental Panel on Climate Change* [B. Metz, O.R. Davidson, P.R. Bosch, R. Dave, L.A. Meyer (eds)], Cambridge University Press, Cambridge, United Kingdom and New York, NY, USA., 851 pp.

Cover photo:
Solar PV screen and high voltage power line. © (FREELENS Pool) Tack / Still Pictures

Contents

Foreword

"Climate Change 2007 – Mitigation of Climate Change", the third volume of the Fourth Assessment Report of the Intergovernmental Panel on Climate Change (IPCC), provides an in-depth analysis of the costs and benefits of different approaches to mitigating and avoiding climate change.

In the first two volumes of the "Climate Change 2007" Assessment Report, the IPCC analyses the physical science basis of climate change and the expected consequences for natural and human systems. The third volume of the report presents an analysis of costs, policies and technologies that could be used to limit and/or prevent emissions of greenhouse gases, along with a range of activities to remove these gases from the atmosphere. It recognizes that a portfolio of adaptation and mitigation actions is required to reduce the risks of climate change. It also has broadened the assessment to include the relationship between sustainable development and climate change mitigation.

At regular intervals of five or six years, the IPCC presents comprehensive scientific reports on climate change that assess the existing scientific, technical and socioeconomic literature. The rigorous multi-stage review process of the reports, the broad and geographically-balanced participation of experts from all relevant fields of knowledge and the thousands of comments taken into account guarantee a transparent and unbiased result.

As an intergovernmental body established by the World Meteorological Organization and the United Nations Environment Programme, the IPCC has the responsibility of providing policymakers with objective scientific and technical findings that are policy relevant but not policy prescriptive. This is especially evident in the Mitigation report, which presents tools that governments can consider and implement in their domestic policies and measures in the framework of international agreements.

Hundreds of authors contributed to the preparation of this report. They come from different backgrounds and possess a wide range of expertise, from emissions modelling to economics, from policies to technologies. They all dedicated a large part of their valuable time to the preparation of the report. We would like to thank them all, in particular the 168 Coordinating Lead Authors and Lead Authors most closely engaged in the process.

The preparation of an IPCC Assessment Report is a complex and absorbing process. We would like to express our gratitude to the Technical Support Unit for its massive organizational efforts. We would also like to thank the IPCC Secretariat for its dedication to the efficient completion of the report.

We express our appreciation to the Government of the Netherlands, which hosted the Technical Support Unit; the Government of Thailand, which hosted the plenary session for the approval of the report; the Governments of China, Germany, New Zealand and Peru, which hosted the Lead Authors' meetings; and to all the countries that contributed to IPCC work through financial and logistic support.

We wish to sincerely thank Dr Rajendra K. Pachauri, Chairman of the IPCC, for his steady and discreet guidance and to express our deep gratitude to Drs Ogunlade Davidson and Bert Metz, Co-Chairs of Working Group III, who successfully led their team with positive, efficient and constructive direction.

M. Jarraud
Secretary General
World Meteorological Organization

A. Steiner
Executive Director
United Nations Environment Programme

Preface

The Fourth Assessment Report of IPCC Working Group III, "Mitigation of Climate Change", aims to answer essentially five questions relevant to policymakers worldwide:

- What can we do to reduce or avoid climate change?
- What are the costs of these actions and how do they relate to the costs of inaction?
- How much time is available to realise the drastic reductions needed to stabilise greenhouse gas concentrations in the atmosphere?
- What are the policy actions that can overcome the barriers to implementation?
- How can climate mitigation policy be aligned with sustainable development policies?

A description of mitigation options for the various societal sectors that contribute to emissions forms the core of this report. Seven chapters cover mitigation options in energy supply, transport, buildings, industry, agriculture, forestry and waste management, with one additional chapter dealing with the cross-sectoral issues. The authors have provided the reader with an up-to-date overview of the characteristics of the various sectors, the mitigation measures that could be employed, the costs and specific barriers, and the policy implementation issues. In addition, estimates are given of the overall mitigation potential and costs per sector, and for the world as a whole. The report combines information from bottom-up technological studies with results of top-down modelling exercises. Mitigation measures for the short term are placed in the long-term perspective of realising stabilisation of global average temperatures. This provides policy-relevant information on the relation between the stringency of stabilisation targets and the timing and amount of mitigation necessary. Policies and measures to achieve mitigation action, both at national and international levels, are covered in chapter 13; this is additional to what is included in the sector chapters. The link between climate change mitigation, adaptation and sustainable development has been further elaborated in the relevant chapters of the report, with one chapter presenting an overview of the connections between sustainable development and climate change mitigation.

The process

After two scoping meetings to establish possible content, the formal assessment production process got underway in 2003 with the approval of the report outline by the IPCC at the Panel's 21st session. Soon after this, an author team of 168 lead authors (55 from developing countries, 5 from EIT countries and 108 from OECD countries) and 85 contributing authors was formed by the Working Group III Bureau, based on nominations from governments and international organisations. Thirty-six per cent of the lead authors came from developing countries and countries with economies in transition. The IPCC review procedure was followed, in which drafts produced by the authors were subject to two reviews. Thousands of comments from a total of 485 expert reviewers, and governments and international organisations were processed. The processing into new drafts was overseen by two review editors per chapter, who ensured that all substantive comments received appropriate consideration.

The Summary for Policymakers was approved line by line, and the main report and Technical Summary were accepted at the 9th session of the IPCC Working Group III held in Bangkok, Thailand from 30 April to 4 May 2007.

Acknowledgements

Production of this report was a major enterprise, in which many people all around the world delivered a wide variety of contributions. This input could not have been made without the generous support from the governments and institutions involved, which enabled the authors, review editors and reviewers to participate in this process. To them, our thanks.

We are particularly grateful to the governments of Germany, Peru, China and New Zealand, who, in collaboration with local institutions, hosted the crucial lead author meetings in Leipzig (October 2004), Lima (June 2005), Beijing (February 2006) and Christchurch (October 2006).

Various countries and institutions supported expert meetings and stakeholder consultations that have contributed to the depth and scope of the report, namely:

- Adaptation, mitigation and sustainable development in La Réunion (supported by the government of France)
- Emissions scenarios in Washington DC (supported by the US Government)
- Input by industry representatives in Tokyo (supported by the Japanese government) and Cape Town, South Africa (co-sponsored by ESKOM), and
- Input from environmental NGOs, intergovernmental organisations, research organisations and members of the International Energy Agency and its technology network in Paris (in cooperation with the IEA).

Throughout the process, the Working Group III Bureau – consisting of Ramón Pichs Madruga (Cuba), R.T.M. Sutamihardja (Indonesia), Hans Larsen (Denmark), (up to May 2005), Olav Hohmeyer (Germany, from June 2005), Eduardo Calvo (Peru), Ziad H.Abu-Ghararah (Saudi Arabia, up to September 2005), and Taha M. Zatari (Saudi Arabia, after September 2005), Ismail A.R. Elgizouli (Sudan) – delivered constructive support and continuous encouragement.

The success of this report is, however, fully based on the expertise and enthusiasm of the author team for which we are grateful. We would also like to express our appreciation of the expert reviewer inputs. Without their comments, the report would not have achieved its current quality level. Our review editors had a similar critical role in supporting the author team in dealing with the comments.

The assessment process was supported by the Technical Support Unit, financed by the government of the Netherlands. The following persons provided support, advice and coordination: Leo Meyer, Peter Bosch, Rutu Dave, Monique Hoogwijk, Thelma van den Brink, Anita Meier, Sander Brinkman, Heleen de Coninck, Bertjan Heij, David de

Jager, John Kessels, Eveline Trines, Manuela Loos (editing support), Martin Middelburg (layout), Rob Puijk (webmaster), Ruth de Wijs (coordination of copyediting), Marilyn Anderson (index) and many more from the secretariats of MNP and ECN in the Netherlands.

Finally, we would like to thank the IPCC secretariat in Geneva in the persons of Renate Christ (Secretary of the IPCC), Jian Liu, Carola Saibante, Rudie Bourgeois, Annie Courtin and Joelle Fernandez for their continuous support throughout the process.

Bert Metz
Ogunlade Davidson

Co-chairs, IPCC Working Group III

This report is dedicated to

Gerhard Petschel-Held, Germany

Lead Author in Chapter 12

Gerhard Petschel-Held died unexpectedly on September 9, 2005, at the age of 41 years. He worked at the Potsdam Institute for Climate Impact Research as head of the department Integrated Systems Analysis. He was an excellent scientist and a wonderful person to work with.
Based on his scientific credentials and his capacity to integrate, Gerhard Petschel-Held played a key role in several international research networks. He believed strongly in the communication of scientific knowledge to a wider public, and to improve the world with the help of science.

Contribution of Working Group III to the
Fourth Assessment Report of the
Intergovernmental Panel on Climate Change

Summary for Policymakers

This Summary for Policymakers was formally approved at the 9th Session of Working Group III of the IPCC, Bangkok, Thailand. 30 April - 4 May 2007

Drafting authors:

Terry Barker, Igor Bashmakov, Lenny Bernstein, Jean Bogner, Peter Bosch, Rutu Dave, Ogunlade Davidson, Brian Fisher, Michael Grubb, Sujata Gupta, Kirsten Halsnaes, BertJan Heij, Suzana Kahn Ribeiro, Shigeki Kobayashi, Mark Levine, Daniel Martino, Omar Masera Cerutti, Bert Metz, Leo Meyer, Gert-Jan Nabuurs, Adil Najam, Nebojsa Nakicenovic, Hans Holger Rogner, Joyashree Roy, Jayant Sathaye, Robert Schock, Priyaradshi Shukla, Ralph Sims, Pete Smith, Rob Swart, Dennis Tirpak, Diana Urge-Vorsatz, Zhou Dadi

This Summary for Policymakers should be cited as:

IPCC, 2007: Summary for Policymakers. In: *Climate Change 2007: Mitigation. Contribution of Working Group III to the Fourth Assessment Report of the Intergovernmental Panel on Climate Change* [B. Metz, O.R. Davidson, P.R. Bosch, R. Dave, L.A. Meyer (eds)], Cambridge University Press, Cambridge, United Kingdom and New York, NY, USA.

Table of Contents

A. Introduction

1.   The Working Group III contribution to the IPCC Fourth Assessment Report (AR4) focuses on new literature on the scientific, technological, environmental, economic and social aspects of mitigation of climate change, published since the IPCC Third Assessment Report (TAR) and the Special Reports on CO_2 Capture and Storage (SRCCS) and on Safeguarding the Ozone Layer and the Global Climate System (SROC).

The following summary is organised into six sections after this introduction:
- Greenhouse gas (GHG) emission trends
- Mitigation in the short and medium term, across different economic sectors (until 2030)
- Mitigation in the long-term (beyond 2030)
- Policies, measures and instruments to mitigate climate change
- Sustainable development and climate change mitigation
- Gaps in knowledge.

References to the corresponding chapter sections are indicated at each paragraph in square brackets. An explanation of terms, acronyms and chemical symbols used in this SPM can be found in the glossary to the main report.

B. Greenhouse gas emission trends

2.   **Global greenhouse gas (GHG) emissions have grown since pre-industrial times, with an increase of 70% between 1970 and 2004** *(high agreement, much evidence)*[1].
- Since pre-industrial times, increasing emissions of GHGs due to human activities have led to a marked increase in atmospheric GHG concentrations [1.3; Working Group I SPM].
- Between 1970 and 2004, global emissions of CO_2, CH_4, N_2O, HFCs, PFCs and SF_6, weighted by their global warming potential (GWP), have increased by 70% (24%

between 1990 and 2004), from 28.7 to 49 Gigatonnes of carbon dioxide equivalents (GtCO$_2$-eq)[2] (see Figure SPM.1). The emissions of these gases have increased at different rates. CO_2 emissions have grown between 1970 and 2004 by about 80% (28% between 1990 and 2004) and represented 77% of total anthropogenic GHG emissions in 2004.

- The largest growth in global GHG emissions between 1970 and 2004 has come from the energy supply sector (an increase of 145%). The growth in direct emissions[3] from transport in this period was 120%, industry 65% and land use, land use change, and forestry (LULUCF)[4] 40%[5]. Between 1970 and 1990 direct emissions from agriculture grew by 27% and from buildings by 26%, and the latter remained at approximately at 1990 levels thereafter. However, the buildings sector has a high level of electricity use and hence the total of direct and indirect emissions in this sector is much higher (75%) than direct emissions [1.3, 6.1, 11.3, Figures 1.1 and 1.3].

- The effect on global emissions of the decrease in global energy intensity (-33%) during 1970 to 2004 has been smaller than the combined effect of global per capita income growth (77 %) and global population growth (69%); both drivers of increasing energy-related CO_2 emissions (Figure SPM.2). The long-term trend of a declining carbon intensity of energy supply reversed after 2000. Differences in terms of per capita income, per capita emissions, and energy intensity among countries remain significant. (Figure SPM.3). In 2004 UNFCCC Annex I countries held a 20% share in world population, produced 57% of world Gross Domestic Product based on Purchasing Power Parity (GDP$_{ppp}$)[6], and accounted for 46% of global GHG emissions (Figure SPM.3) [1.3].

- The emissions of ozone depleting substances (ODS) controlled under the Montreal Protocol[7], which are also GHGs, have declined significantly since the 1990s. By 2004 the emissions of these gases were about 20% of their 1990 level [1.3].

- A range of policies, including those on climate change, energy security[8], and sustainable development, have been effective in reducing GHG emissions in different sectors and many countries. The scale of such measures, however, has not yet been large enough to counteract the global growth in emissions [1.3, 12.2].

---

1   Each headline statement has an "agreement/evidence" assessment attached that is supported by the bullets underneath. This does not necessarily mean that this level of "agreement/evidence"applies to each bullet. Endbox 1 provides an explanation of this representation of uncertainty.

2   The definition of carbon dioxide equivalent (CO$_2$-eq) is the amount of CO2 emission that would cause the same radiative forcing as an emitted amount of a well mixed greenhouse gas or a mixture of well mixed greenhouse gases, all multiplied with their respective GWPs to take into account the differing times they remain in the atmosphere [WGI AR4 Glossary].

3   Direct emissions in each sector do not include emissions from the electricity sector for the electricity consumed in the building, industry and agricultural sectors or of the emissions from refinery operations supplying fuel to the transport sector.

4   The term "land use, land use change and forestry" is used here to describe the aggregated emissions of CO$_2$, CH$_4$, N$_2$O from deforestation, biomass and burning, decay of biomass from logging and deforestation, decay of peat and peat fires [1.3.1]. This is broader than emissions from deforestation, which is included as a subset.   The emissions reported here do not include carbon uptake (removals).

5   This trend is for the total LULUCF emissions, of which emissions from deforestation are a subset and, owing to large data uncertainties, is significantly less certain than for other sectors. The rate of deforestation globally was slightly lower in the 2000-2005 period than in the 1990-2000 period [9.2.1].

6   The GDPppp metric is used for illustrative purposes only for this report. For an explanation of PPP and Market Exchange Rate (MER) GDP calculations, see footnote 12.

7   Halons, chlorofluorocarbons (CFCs), hydrochlorofluorocarbons (HCFCs), methyl chloroform (CH$_3$CCl$_3$), carbon tetrachloride (CCl$_4$) and methyl bromide (CH$_3$Br).

8   Energy security refers to security of energy supply.

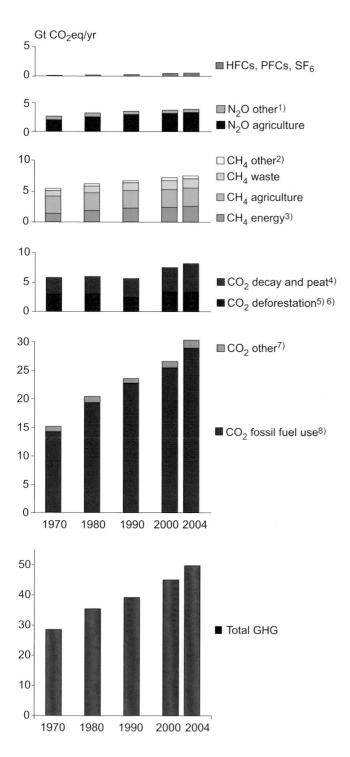

Figure SPM.1: *Global Warming Potential (GWP) weighted global greenhouse gas emissions 1970-2004. 100 year GWPs from IPCC 1996 (SAR) were used to convert emissions to CO_2-eq. (cf. UNFCCC reporting guidelines). CO_2, CH_4, N_2O, HFCs, PFCs and SF_6 from all sources are included.*
The two CO_2 emission categories reflect CO_2 emissions from energy production and use (second from bottom) and from land use changes (third from the bottom) [Figure 1.1a].

Notes:
1. Other N_2O includes industrial processes, deforestation/savannah burning, waste water and waste incineration.
2. Other is CH_4 from industrial processes and savannah burning.
3. Including emissions from bioenergy production and use
4. CO_2 emissions from decay (decomposition) of above ground biomass that remains after logging and deforestation and CO_2 from peat fires and decay of drained peat soils.
5. As well as traditional biomass use at 10% of total, assuming 90% is from sustainable biomass production. Corrected for 10% carbon of biomass that is assumed to remain as charcoal after combustion.
6. For large-scale forest and scrubland biomass burning averaged data for 1997-2002 based on Global Fire Emissions Data base satellite data.
7. Cement production and natural gas flaring.
8. Fossil fuel use includes emissions from feedstocks.

3. **With current climate change mitigation policies and related sustainable development practices, global GHG emissions will continue to grow over the next few decades** *(high agreement, much evidence).*

    - The SRES (non-mitigation) scenarios project an increase of baseline global GHG emissions by a range of 9.7 GtCO$_2$-eq to 36.7 GtCO$_2$-eq (25-90%) between 2000 and 2030[9] (Box SPM.1 and Figure SPM.4). In these scenarios, fossil fuels are projected to maintain their dominant position in the global energy mix to 2030 and beyond. Hence CO$_2$ emissions between 2000 and 2030 from energy use are projected to grow 40 to 110% over that period. Two thirds to three quarters of this increase in energy CO$_2$ emissions is projected to come from non-Annex I regions, with their average per capita energy CO$_2$ emissions being projected to remain substantially lower (2.8-5.1 tCO$_2$/cap) than those in Annex I regions (9.6-15.1 tCO$_2$/cap) by 2030. According to SRES scenarios, their economies are projected to have a lower energy use per unit of GDP (6.2 – 9.9 MJ/US\$ GDP) than that of non-Annex I countries (11.0 – 21.6 MJ/US\$ GDP). [1.3, 3.2]

---

9   The SRES 2000 GHG emissions assumed here are 39.8 GtCO2-eq, i.e. lower than the emissions reported in the EDGAR database for 2000 (45 GtCO2-eq). This is mostly due to differences in LULUCF emissions.

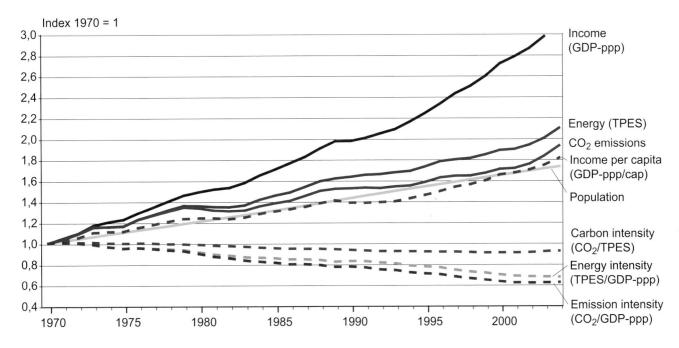

Figure SPM.2: *Relative global development of Gross Domestic Product measured in PPP (GDPppp), Total Primary Energy Supply (TPES), CO2 emissions (from fossil fuel burning, gas flaring and cement manufacturing) and Population (Pop). In addition, in dotted lines, the figure shows Income per capita (GDPppp/Pop), Energy Intensity (TPES/GDPppp), Carbon Intensity of energy supply (CO2/TPES), and Emission Intensity of the economic production process (CO2/GDPppp) for the period 1970-2004. [Figure 1.5]*

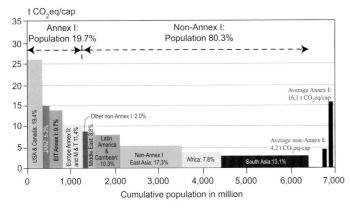

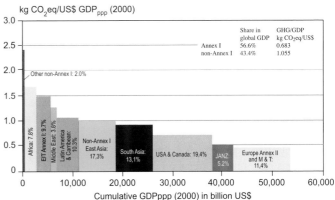

Figure SPM.3a: *Year 2004 distribution of regional per capita GHG emissions (all Kyoto gases, including those from land-use) over the population of different country groupings. The percentages in the bars indicate a regions share in global GHG emissions [Figure 1.4a].*

Figure SPM.3b: *Year 2004 distribution of regional GHG emissions (all Kyoto gases, including those from land-use) per US$ of GDPppp over the GDPppp of different country groupings. The percentages in the bars indicate a regions share in global GHG emissions [Figure 1.4b].*

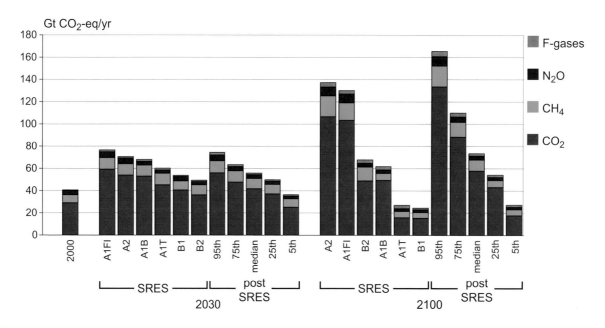

Figure SPM.4: *Global GHG emissions for 2000 and projected baseline emissions[10] for 2030 and 2100 from IPCC SRES and the post-SRES literature. The figure provides the emissions from the six illustrative SRES scenarios. It also provides the frequency distribution of the emissions in the post-SRES scenarios (5th, 25th, median, 75th, 95th percentile), as covered in Chapter 3. F-gases cover HFCs, PFCs and SF6 [1.3, 3.2, Figure 1.7].*

4. **Baseline emissions scenarios published since SRES[10], are comparable in range to those presented in the IPCC Special Report on Emission Scenarios (SRES) (25- 135 GtCO2-eq/yr in 2100, see Figure SPM.4)** *(high agreement, much evidence).*

   • Studies since SRES used lower values for some drivers for emissions, notably population projections. However, for those studies incorporating these new population projections, changes in other drivers, such as economic growth, resulted in little change in overall emission levels. Economic growth projections for Africa, Latin America and the Middle East to 2030 in post-SRES baseline scenarios are lower than in SRES, but this has only minor effects on global economic growth and overall emissions [3.2].

   • Representation of aerosol and aerosol precursor emissions, including sulphur dioxide, black carbon, and organic carbon, which have a net cooling effect[11] has improved. Generally, they are projected to be lower than reported in SRES [3.2].

   • Available studies indicate that the choice of exchange rate for GDP (MER or PPP) does not appreciably affect the projected emissions, when used consistently[12]. The differences, if any, are small compared to the uncertainties caused by assumptions on other parameters in the scenarios, e.g. technological change [3.2].

---

10 Baseline scenarios do not include additional climate policy above current ones; more recent studies differ with respect to UNFCCC and Kyoto Protocol inclusion.
11 See AR4 WG I report, Chapter 10.2.
12 Since TAR, there has been a debate on the use of different exchange rates in emission scenarios. Two metrics are used to compare GDP between countries. Use of MER is preferable for analyses involving internationally traded products. Use of PPP, is preferable for analyses involving comparisons of income between countries at very different stages of development. Most of the monetary units in this report are expressed in MER. This reflects the large majority of emissions mitigation literature that is calibrated in MER. When monetary units are expressed in PPP, this is denoted by GDP_ppp.

Box SPM.1: The emission scenarios of the IPCC Special Report on Emission Scenarios (SRES)

A1. The A1 storyline and scenario family describes a future world of very rapid economic growth, global population that peaks in mid-century and declines thereafter, and the rapid introduction of new and more efficient technologies. Major underlying themes are convergence among regions, capacity building and increased cultural and social interactions, with a substantial reduction in regional differences in per capita income. The A1 scenario family develops into three groups that describe alternative directions of technological change in the energy system. The three A1 groups are distinguished by their technological emphasis: fossil intensive (A1FI), non fossil energy sources (A1T), or a balance across all sources (A1B) (where balanced is defined as not relying too heavily on one particular energy source, on the assumption that similar improvement rates apply to all energy supply and end use technologies).

A2. The A2 storyline and scenario family describes a very heterogeneous world. The underlying theme is self reliance and preservation of local identities. Fertility patterns across regions converge very slowly, which results in continuously increasing population. Economic development is primarily regionally oriented and per capita economic growth and technological change more fragmented and slower than other storylines.

B1. The B1 storyline and scenario family describes a convergent world with the same global population, that peaks in mid-century and declines thereafter, as in the A1 storyline, but with rapid change in economic structures toward a service and information economy, with reductions in material intensity and the introduction of clean and resource efficient technologies. The emphasis is on global solutions to economic, social and environmental sustainability, including improved equity, but without additional climate initiatives.

B2. The B2 storyline and scenario family describes a world in which the emphasis is on local solutions to economic, social and environmental sustainability. It is a world with continuously increasing global population, at a rate lower than A2, intermediate levels of economic development, and less rapid and more diverse technological change than in the B1 and A1 storylines. While the scenario is also oriented towards environmental protection and social equity, it focuses on local and regional levels.

An illustrative scenario was chosen for each of the six scenario groups A1B, A1FI, A1T, A2, B1 and B2. All should be considered equally sound.

The SRES scenarios do not include additional climate initiatives, which means that no scenarios are included that explicitly assume implementation of the United Nations Framework Convention on Climate Change or the emissions targets of the Kyoto Protocol.

This box summarizing the SRES scenarios is taken from the Third Assessment Report and has been subject to prior line by line approval by the Panel.

Box SPM.2: Mitigation potential and analytical approaches

The concept of "mitigation potential" has been developed to assess the scale of GHG reductions that could be made, relative to emission baselines, for a given level of carbon price (expressed in cost per unit of carbon dioxide equivalent emissions avoided or reduced). Mitigation potential is further differentiated in terms of "market potential" and "economic potential".

Market potential is the mitigation potential based on private costs and private discount rates[13], which might be expected to occur under forecast market conditions, including policies and measures currently in place, noting that barriers limit actual uptake [2.4].

---

13 Private costs and discount rates reflect the perspective of private consumers and companies; see Glossary for a fuller description.

(Box SPM.2 Continued)

Economic potential is the mitigation potential, which takes into account social costs and benefits and social discount rates[14], assuming that market efficiency is improved by policies and measures and barriers are removed [2.4].

Studies of market potential can be used to inform policy makers about mitigation potential with existing policies and barriers, while studies of economic potentials show what might be achieved if appropriate new and additional policies were put into place to remove barriers and include social costs and benefits. The economic potential is therefore generally greater than the market potential.

Mitigation potential is estimated using different types of approaches. There are two broad classes – "bottom-up" and "top-down" approaches, which primarily have been used to assess the economic potential.

Bottom-up studies are based on assessment of mitigation options, emphasizing specific technologies and regulations. They are typically sectoral studies taking the macro-economy as unchanged. Sector estimates have been aggregated, as in the TAR, to provide an estimate of global mitigation potential for this assessment.

Top-down studies assess the economy-wide potential of mitigation options. They use globally consistent frameworks and aggregated information about mitigation options and capture macro-economic and market feedbacks.

Bottom-up and top-down models have become more similar since the TAR as top-down models have incorporated more technological mitigation options and bottom-up models have incorporated more macroeconomic and market feedbacks as well as adopting barrier analysis into their model structures. Bottom-up studies in particular are useful for the assessment of specific policy options at sectoral level, e.g. options for improving energy efficiency, while top-down studies are useful for assessing cross-sectoral and economy-wide climate change policies, such as carbon taxes and stabilization policies. However, current bottom-up and top-down studies of economic potential have limitations in considering life-style choices, and in including all externalities such as local air pollution. They have limited representation of some regions, countries, sectors, gases, and barriers. The projected mitigation costs do not take into account potential benefits of avoided climate change.

Box SPM.3: Assumptions in studies on mitigation portfolios and macro-economic costs

Studies on mitigation portfolios and macro-economic costs assessed in this report are based on top-down modelling. Most models use a global least cost approach to mitigation portfolios and with universal emissions trading, assuming transparent markets, no transaction cost, and thus perfect implementation of mitigation measures throughout the 21st century. Costs are given for a specific point in time.

Global modelled costs will increase if some regions, sectors (e.g. land-use), options or gases are excluded. Global modelled costs will decrease with lower baselines, use of revenues from carbon taxes and auctioned permits, and if induced technological learning is included. These models do not consider climate benefits and generally also co-benefits of mitigation measures, or equity issues.

Box SPM.4: Modelling induced technological change

Relevant literature implies that policies and measures may induce technological change. Remarkable progress has been achieved in applying approaches based on induced technological change to stabilisation studies; however, conceptual issues remain. In the models that adopt these approaches, projected costs for a given stabilization level are reduced; the reductions are greater at lower stabilisation levels.

---

14 Social costs and discount rates reflect the perspective of society. Social discount rates are lower than those used by private investors; see Glossary for a fuller description.

C. Mitigation in the short and medium term (until 2030)

5. Both bottom-up and top-down studies indicate that there is substantial economic potential for the mitigation of global GHG emissions over the coming decades, that could offset the projected growth of global emissions or reduce emissions below current levels *(high agreement, much evidence).*

Uncertainties in the estimates are shown as ranges in the tables below to reflect the ranges of baselines, rates of technological change and other factors that are specific to the different approaches. Furthermore, uncertainties also arise from the limited information for global coverage of countries, sectors and gases.

Bottom-up studies:
- In 2030, the economic potential estimated for this assessment from bottom-up approaches (see Box SPM.2) is presented in Table SPM.1 below and Figure SPM.5A. For reference: emissions in 2000 were equal to 43 GtCO$_2$-eq. [11.3]:

- Studies suggest that mitigation opportunities with net negative costs[15] have the potential to reduce emissions by around 6 GtCO$_2$-eq/yr in 2030. Realizing these requires dealing with implementation barriers [11.3].
- No one sector or technology can address the entire mitigation challenge. All assessed sectors contribute to the total (see Figure SPM.6). The key mitigation technologies and practices for the respective sectors are shown in Table SPM 3 [4.3, 4.4, 5.4, 6.5, 7.5, 8.4, 9.4, 10.4].

Top-down studies:
- Top-down studies calculate an emission reduction for 2030 as presented in Table SPM.2 below and Figure SPM.5B. The global economic potentials found in the top-down studies are in line with bottom-up studies (see Box SPM.2), though there are considerable differences at the sectoral level [3.6].
- The estimates in Table SPM.2 were derived from stabilization scenarios, i.e., runs towards long-run stabilization of atmospheric GHG concentration [3.6].

Table SPM.1: *Global economic mitigation potential in 2030 estimated from bottom-up studies.*

Carbon price (US$/tCO$_2$-eq)	Economic potential (GtCO$_2$-eq/yr)	Reduction relative to SRES A1 B (68 GtCO$_2$-eq/yr) (%)	Reduction relative to SRES B2 (49 GtCO$_2$-eq/yr) (%)
0	5-7	7-10	10-14
20	9-17	14-25	19-35
50	13-26	20-38	27-52
100	16-31	23-46	32-63

Table SPM.2: *Global economic mitigation potential in 2030 estimated from top-down studies.*

Carbon price (US$/tCO$_2$-eq)	Economic potential (GtCO$_2$-eq/yr)	Reduction relative to SRES A1 B (68 GtCO$_2$-eq/yr) (%)	Reduction relative to SRES B2 (49 GtCO$_2$-eq/yr) (%)
20	9-18	13-27	18-37
50	14-23	21-34	29-47
100	17-26	25-38	35-53

15 In this report, as in the SAR and the TAR, options with net negative costs (no regrets opportunities) are defined as those options whose benefits such as reduced energy costs and reduced emissions of local/regional pollutants equal or exceed their costs to society, excluding the benefits of avoided climate change (see Box SPM.1).

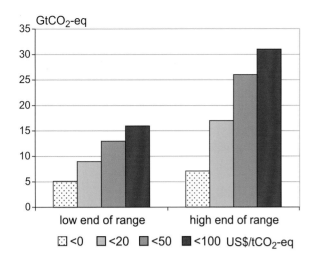

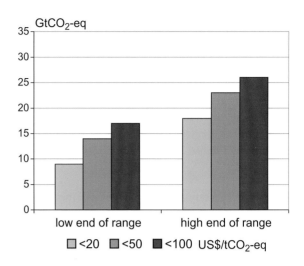

Figure SPM.5A: *Global economic mitigation potential in 2030 estimated from bottom-up studies (data from Table SPM.1)*

Figure SPM.5B: *Global economic mitigation potential in 2030 estimated from top-down studies (data from Table SPM.2)*

Table SPM.3: *Key mitigation technologies and practices by sector. Sectors and technologies are listed in no particular order. Non-technological practices, such as lifestyle changes, which are cross-cutting, are not included in this table (but are addressed in paragraph 7 in this SPM).*

Sector	Key mitigation technologies and practices currently commercially available	Key mitigation technologies and practices projected to be commercialized before 2030
Energy supply [4.3, 4.4]	Improved supply and distribution efficiency; fuel switching from coal to gas; nuclear power; renewable heat and power (hydropower, solar, wind, geothermal and bioenergy); combined heat and power; early applications of Carbon Capture and Storage (CCS, e.g. storage of removed CO_2 from natural gas).	CCS for gas, biomass and coal-fired electricity generating facilities; advanced nuclear power; advanced renewable energy, including tidal and waves energy, concentrating solar, and solar PV.
Transport [5.4]	More fuel efficient vehicles; hybrid vehicles; cleaner diesel vehicles; biofuels; modal shifts from road transport to rail and public transport systems; non-motorised transport (cycling, walking); land-use and transport planning.	Second generation biofuels; higher efficiency aircraft; advanced electric and hybrid vehicles with more powerful and reliable batteries.
Buildings [6.5]	Efficient lighting and daylighting; more efficient electrical appliances and heating and cooling devices; improved cook stoves, improved insulation ; passive and active solar design for heating and cooling; alternative refrigeration fluids, recovery and recycle of fluorinated gases.	Integrated design of commercial buildings including technologies, such as intelligent meters that provide feedback and control; solar PV integrated in buildings.
Industry [7.5]	More efficient end-use electrical equipment; heat and power recovery; material recycling and substitution; control of non-CO_2 gas emissions; and a wide array of process-specific technologies.	Advanced energy efficiency; CCS for cement, ammonia, and iron manufacture; inert electrodes for aluminium manufacture.
Agriculture [8.4]	Improved crop and grazing land management to increase soil carbon storage; restoration of cultivated peaty soils and degraded lands; improved rice cultivation techniques and livestock and manure management to reduce CH_4 emissions; improved nitrogen fertilizer application techniques to reduce N_2O emissions; dedicated energy crops to replace fossil fuel use; improved energy efficiency.	Improvements of crops yields.
Forestry/forests [9.4]	Afforestation; reforestation; forest management; reduced deforestation; harvested wood product management; use of forestry products for bioenergy to replace fossil fuel use.	Tree species improvement to increase biomass productivity and carbon sequestration. Improved remote sensing technologies for analysis of vegetation/ soil carbon sequestration potential and mapping land use change.
Waste management [10.4]	Landfill methane recovery; waste incineration with energy recovery; composting of organic waste; controlled waste water treatment; recycling and waste minimization.	Biocovers and biofilters to optimize CH_4 oxidation.

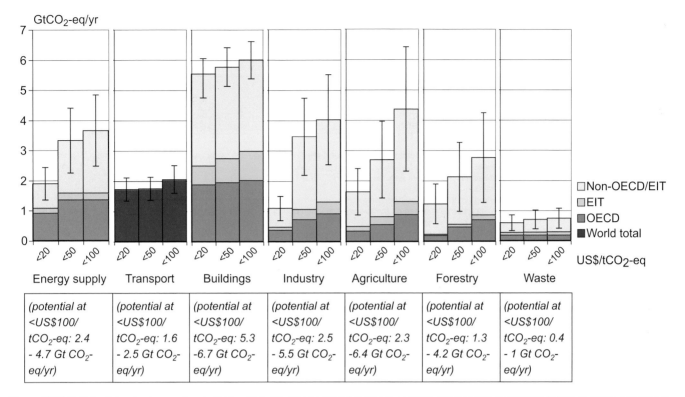

Figure SPM.6: *Estimated sectoral economic potential for global mitigation for different regions as a function of carbon price in 2030 from bottom-up studies, compared to the respective baselines assumed in the sector assessments. A full explanation of the derivation of this figure is found in Section 11.3.*

Notes:
1. The ranges for global economic potentials as assessed in each sector are shown by vertical lines. The ranges are based on end-use allocations of emissions, meaning that emissions of electricity use are counted towards the end-use sectors and not to the energy supply sector.
2. The estimated potentials have been constrained by the availability of studies particularly at high carbon price levels.
3. Sectors used different baselines. For industry the SRES B2 baseline was taken, for energy supply and transport the WEO 2004 baseline was used; the building sector is based on a baseline in between SRES B2 and A1B; for waste, SRES A1B driving forces were used to construct a waste specific baseline, agriculture and forestry used baselines that mostly used B2 driving forces.
4. Only global totals for transport are shown because international aviation is included [5.4].
5. Categories excluded are: non-CO_2 emissions in buildings and transport, part of material efficiency options, heat production and cogeneration in energy supply, heavy duty vehicles, shipping and high-occupancy passenger transport, most high-cost options for buildings, wastewater treatment, emission reduction from coal mines and gas pipelines, fluorinated gases from energy supply and transport. The underestimation of the total economic potential from these emissions is of the order of 10-15%.

6. **In 2030 macro-economic costs for multi-gas mitigation, consistent with emissions trajectories towards stabilization between 445 and 710 ppm CO_2-eq, are estimated at between a 3% decrease of global GDP and a small increase, compared to the baseline (see Table SPM.4). However, regional costs may differ significantly from global averages** (*high agreement, medium evidence*) **(see Box SPM.3 for the methodologies and assumptions of these results).**

- The majority of studies conclude that reduction of GDP relative to the GDP baseline increases with the stringency of the stabilization target.
- Depending on the existing tax system and spending of the revenues, modelling studies indicate that costs may be substantially lower under the assumption that revenues from carbon taxes or auctioned permits under an emission trading system are used to promote low-carbon technologies or reform of existing taxes [11.4].

- Studies that assume the possibility that climate change policy induces enhanced technological change also give lower costs. However, this may require higher upfront investment in order to achieve costs reductions thereafter (see Box SPM.4) [3.3, 3.4, 11.4, 11.5, 11.6].
- Although most models show GDP losses, some show GDP gains because they assume that baselines are non-optimal and mitigation policies improve market efficiencies, or they assume that more technological change may be induced by mitigation policies. Examples of market inefficiencies include unemployed resources, distortionary taxes and/or subsidies [3.3, 11.4].
- A multi-gas approach and inclusion of carbon sinks generally reduces costs substantially compared to CO_2 emission abatement only [3.3].
- Regional costs are largely dependent on the assumed stabilization level and baseline scenario. The allocation regime is also important, but for most countries to a lesser extent than the stabilization level [11.4, 13.3].

Table SPM.4: *Estimated global macro-economic costs in 2030[a] for least-cost trajectories towards different long-term stabilization levels.[b), c)]*

Stabilization levels (ppm CO_2-eq)	Median GDP reduction[d] (%)	Range of GDP reduction[d), e)] (%)	Reduction of average annual GDP growth rates[d), f)] (percentage points)
590-710	0.2	-0.6-1.2	<0.06
535-590	0.6	0.2-2.5	<0.1
445-535[g]	not available	<3	<0.12

Notes:
a) For a given stabilization level, GDP reduction would increase over time in most models after 2030. Long-term costs also become more uncertain. [Figure 3.25]
b) Results based on studies using various baselines.
c) Studies vary in terms of the point in time stabilization is achieved; generally this is in 2100 or later.
d) This is global GDP based market exchange rates.
e) The median and the 10th and 90th percentile range of the analyzed data are given.
f) The calculation of the reduction of the annual growth rate is based on the average reduction during the period till 2030 that would result in the indicated GDP decrease in 2030.
g) The number of studies that report GDP results is relatively small and they generally use low baselines.

7. **Changes in lifestyle and behaviour patterns can contribute to climate change mitigation across all sectors. Management practices can also have a positive role** *(high agreement, medium evidence).*

- Lifestyle changes can reduce GHG emissions. Changes in lifestyles and consumption patterns that emphasize resource conservation can contribute to developing a low-carbon economy that is both equitable and sustainable [4.1, 6.7].
- Education and training programmes can help overcome barriers to the market acceptance of energy efficiency, particularly in combination with other measures [Table 6.6].
- Changes in occupant behaviour, cultural patterns and consumer choice and use of technologies can result in considerable reduction in CO_2 emissions related to energy use in buildings [6.7].
- Transport Demand Management, which includes urban planning (that can reduce the demand for travel) and provision of information and educational techniques (that can reduce car usage and lead to an efficient driving style) can support GHG mitigation [5.1].
- In industry, management tools that include staff training, reward systems, regular feedback, documentation of existing practices can help overcome industrial organization barriers, reduce energy use, and GHG emissions [7.3].

8. **While studies use different methodologies, in all analyzed world regions near-term health co-benefits from reduced air pollution as a result of actions to reduce GHG emissions can be substantial and may offset a substantial fraction of mitigation costs** *(high agreement, much evidence).*

- Including co-benefits other than health, such as increased energy security, and increased agricultural production and reduced pressure on natural ecosystems, due to decreased tropospheric ozone concentrations, would further enhance cost savings [11.8].
- Integrating air pollution abatement and climate change mitigation policies offers potentially large cost reductions compared to treating those policies in isolation [11.8].

9. **Literature since TAR confirms that there may be effects from Annex I countries' action on the global economy and global emissions, although the scale of carbon leakage remains uncertain** *(high agreement, medium evidence).*

- Fossil fuel exporting nations (in both Annex I and non-Annex I countries) may expect, as indicated in TAR[16], lower demand and prices and lower GDP growth due to mitigation policies. The extent of this spill over[17] depends strongly on assumptions related to policy decisions and oil market conditions [11.7].
- Critical uncertainties remain in the assessment of carbon leakage[18]. Most equilibrium modelling support the conclusion in the TAR of economy-wide leakage from Kyoto action in the order of 5-20%, which would be less if competitive low-emissions technologies were effectively diffused [11.7].

10. **New energy infrastructure investments in developing countries, upgrades of energy infrastructure in industrialized countries, and policies that promote energy security, can, in many cases, create opportunities to achieve GHG emission reductions[19] compared to baseline scenarios. Additional co-benefits are country-**

16 See TAR WG III (2001) SPM paragraph 16.
17 Spill over effects of mitigation in a cross-sectoral perspective are the effects of mitigation policies and measures in one country or group of countries on sectors in other countries.
18 Carbon leakage is defined as the increase in CO_2 emissions outside the countries taking domestic mitigation action divided by the reduction in the emissions of these countries.
19 See table SPM.3 and Figure SPM.6

specific but often include air pollution abatement, balance of trade improvement, provision of modern energy services to rural areas and employment *(high agreement, much evidence)*.

- Future energy infrastructure investment decisions, expected to total over 20 trillion US$[20] between now and 2030, will have long term impacts on GHG emissions, because of the long life-times of energy plants and other infrastructure capital stock. The widespread diffusion of low-carbon technologies may take many decades, even if early investments in these technologies are made attractive. Initial estimates show that returning global energy-related CO_2 emissions to 2005 levels by 2030 would require a large shift in the pattern of investment, although the net additional investment required ranges from negligible to 5-10% [4.1, 4.4, 11.6].

- It is often more cost-effective to invest in end-use energy efficiency improvement than in increasing energy supply to satisfy demand for energy services. Efficiency improvement has a positive effect on energy security, local and regional air pollution abatement, and employment [4.2, 4.3, 6.5, 7.7, 11.3, 11.8].

- Renewable energy generally has a positive effect on energy security, employment and on air quality. Given costs relative to other supply options, renewable electricity, which accounted for 18% of the electricity supply in 2005, can have a 30-35% share of the total electricity supply in 2030 at carbon prices up to 50 US$/$tCO_2$-eq [4.3, 4.4, 11.3, 11.6, 11.8].

- The higher the market prices of fossil fuels, the more low-carbon alternatives will be competitive, although price volatility will be a disincentive for investors. Higher priced conventional oil resources, on the other hand, may be replaced by high carbon alternatives such as from oil sands, oil shales, heavy oils, and synthetic fuels from coal and gas, leading to increasing GHG emissions, unless production plants are equipped with CCS [4.2, 4.3, 4.4, 4.5].

- Given costs relative to other supply options, nuclear power, which accounted for 16% of the electricity supply in 2005, can have an 18% share of the total electricity supply in 2030 at carbon prices up to 50 US$/$tCO_2$-eq, but safety, weapons proliferation and waste remain as constraints [4.2, 4.3, 4.4][21].

- CCS in underground geological formations is a new technology with the potential to make an important contribution to mitigation by 2030. Technical, economic and regulatory developments will affect the actual contribution [4.3, 4.4, 7.3].

11. **There are multiple mitigation options in the transport sector[19], but their effect may be counteracted by growth in the sector. Mitigation options are faced with many barriers, such as consumer preferences and lack of policy frameworks** *(medium agreement, medium evidence)*.

- Improved vehicle efficiency measures, leading to fuel savings, in many cases have net benefits (at least for light-duty vehicles), but the market potential is much lower than the economic potential due to the influence of other consumer considerations, such as performance and size. There is not enough information to assess the mitigation potential for heavy-duty vehicles. Market forces alone, including rising fuel costs, are therefore not expected to lead to significant emission reductions [5.3, 5.4].

- Biofuels might play an important role in addressing GHG emissions in the transport sector, depending on their production pathway. Biofuels used as gasoline and diesel fuel additives/substitutes are projected to grow to 3% of total transport energy demand in the baseline in 2030. This could increase to about 5-10%, depending on future oil and carbon prices, improvements in vehicle efficiency and the success of technologies to utilise cellulose biomass [5.3, 5.4].

- Modal shifts from road to rail and to inland and coastal shipping and from low-occupancy to high-occupancy passenger transportation[22], as well as land-use, urban planning and non-motorized transport offer opportunities for GHG mitigation, depending on local conditions and policies [5.3, 5.5].

- Medium term mitigation potential for CO_2 emissions from the aviation sector can come from improved fuel efficiency, which can be achieved through a variety of means, including technology, operations and air traffic management. However, such improvements are expected to only partially offset the growth of aviation emissions. Total mitigation potential in the sector would also need to account for non-CO_2 climate impacts of aviation emissions [5.3, 5.4].

- Realizing emissions reductions in the transport sector is often a co-benefit of addressing traffic congestion, air quality and energy security [5.5].

12. **Energy efficiency options[19] for new and existing buildings could considerably reduce CO_2 emissions with net economic benefit. Many barriers exist against tapping this potential, but there are also large co-benefits** *(high agreement, much evidence)*.

- By 2030, about 30% of the projected GHG emissions in the building sector can be avoided with net economic benefit [6.4, 6.5].

---

20 20 trillion = 20000 billion= $20*10^{12}$.
21 Austria could not agree with this statement.
22 Including rail, road and marine mass transit and carpooling.

- Energy efficient buildings, while limiting the growth of CO_2 emissions, can also improve indoor and outdoor air quality, improve social welfare and enhance energy security [6.6, 6.7].
- Opportunities for realising GHG reductions in the building sector exist worldwide. However, multiple barriers make it difficult to realise this potential. These barriers include availability of technology, financing, poverty, higher costs of reliable information, limitations inherent in building designs and an appropriate portfolio of policies and programs [6.7, 6.8].
- The magnitude of the above barriers is higher in the developing countries and this makes it more difficult for them to achieve the GHG reduction potential of the building sector [6.7].

13. **The economic potential in the industrial sector[19] is predominantly located in energy intensive industries. Full use of available mitigation options is not being made in either industrialized or developing nations** (*high agreement, much evidence*).

- Many industrial facilities in developing countries are new and include the latest technology with the lowest specific emissions. However, many older, inefficient facilities remain in both industrialized and developing countries. Upgrading these facilities can deliver significant emission reductions [7.1, 7.3, 7.4].
- The slow rate of capital stock turnover, lack of financial and technical resources, and limitations in the ability of firms, particularly small and medium-sized enterprises, to access and absorb technological information are key barriers to full use of available mitigation options [7.6].

14. **Agricultural practices collectively can make a significant contribution at low cost[19] to increasing soil carbon sinks, to GHG emission reductions, and by contributing biomass feedstocks for energy use** (*medium agreement, medium evidence*).

- A large proportion of the mitigation potential of agriculture (excluding bioenergy) arises from soil carbon sequestration, which has strong synergies with sustainable agriculture and generally reduces vulnerability to climate change [8.4, 8.5, 8.8].
- Stored soil carbon may be vulnerable to loss through both land management change and climate change [8.10].
- Considerable mitigation potential is also available from reductions in methane and nitrous oxide emissions in some agricultural systems [8.4, 8.5].

- There is no universally applicable list of mitigation practices; practices need to be evaluated for individual agricultural systems and settings [8.4].
- Biomass from agricultural residues and dedicated energy crops can be an important bioenergy feedstock, but its contribution to mitigation depends on demand for bioenergy from transport and energy supply, on water availability, and on requirements of land for food and fibre production. Widespread use of agricultural land for biomass production for energy may compete with other land uses and can have positive and negative environmental impacts and implications for food security [8.4, 8.8].

15. **Forest-related mitigation activities can considerably reduce emissions from sources and increase CO_2 removals by sinks at low costs[19], and can be designed to create synergies with adaptation and sustainable development** (*high agreement, much evidence*)[23].

- About 65% of the total mitigation potential (up to 100 US$/$tCO_2$-eq) is located in the tropics and about 50% of the total could be achieved by reducing emissions from deforestation [9.4].
- Climate change can affect the mitigation potential of the forest sector (i.e., native and planted forests) and is expected to be different for different regions and sub-regions, both in magnitude and direction [9.5].
- Forest-related mitigation options can be designed and implemented to be compatible with adaptation, and can have substantial co-benefits in terms of employment, income generation, biodiversity and watershed conservation, renewable energy supply and poverty alleviation [9.5, 9.6, 9.7].

16. **Post-consumer waste[24] is a small contributor to global GHG emissions[25] (<5%), but the waste sector can positively contribute to GHG mitigation at low cost[19] and promote sustainable development** (*high agreement, much evidence*).

- Existing waste management practices can provide effective mitigation of GHG emissions from this sector: a wide range of mature, environmentally effective technologies are commercially available to mitigate emissions and provide co-benefits for improved public health and safety, soil protection and pollution prevention, and local energy supply [10.3, 10.4, 10.5].
- Waste minimization and recycling provide important indirect mitigation benefits through the conservation of energy and materials [10.4].

---

23 Tuvalu noted difficulties with the reference to "low costs" as Chapter 9, page 15 of the WG III report states that: "the cost of forest mitigation projects rise significantly when opportunity costs of land are taken into account".
24 Industrial waste is covered in the industry sector.
25 GHGs from waste include landfill and wastewater methane, wastewater N_2O, and CO_2 from incineration of fossil carbon.

- Lack of local capital is a key constraint for waste and wastewater management in developing countries and countries with economies in transition. Lack of expertise on sustainable technology is also an important barrier [10.6].

17. **Geo-engineering options, such as ocean fertilization to remove CO_2 directly from the atmosphere, or blocking sunlight by bringing material into the upper atmosphere, remain largely speculative and unproven, and with the risk of unknown side-effects. Reliable cost estimates for these options have not been published** *(medium agreement, limited evidence)* [11.2].

D.   Mitigation in the long term (after 2030)

18. **In order to stabilize the concentration of GHGs in the atmosphere, emissions would need to peak and decline thereafter. The lower the stabilization level, the more quickly this peak and decline would need to occur. Mitigation efforts over the next two to three decades will have a large impact on opportunities to achieve lower stabilization levels (see Table SPM.5, and Figure SPM. 8)[26]** *(high agreement, much evidence)*.

- Recent studies using multi-gas reduction have explored lower stabilization levels than reported in TAR [3.3].
- Assessed studies contain a range of emissions profiles for achieving stabilization of GHG concentrations[27]. Most of these studies used a least cost approach and include both early and delayed emission reductions (Figure SPM.7) [Box SPM.2]. Table SPM.5 summarizes the required emissions levels for different groups of stabilization concentrations and the associated equilibrium global mean temperature increase[28], using the 'best estimate' of climate sensitivity (see also Figure SPM.8 for the likely range of uncertainty)[29]. Stabilization at lower concentration and related equilibrium temperature levels advances the date when emissions need to peak, and requires greater emissions reductions by 2050 [3.3].

Table SPM.5: *Characteristics of post-TAR stabilization scenarios [Table TS 2, 3.10][a)*

Category	Radiative forcing (W/m²)	CO_2 concentration[c)] (ppm)	CO_2-eq concentration[c)] (ppm)	Global mean temperature increase above pre-industrial at equilibrium, using "best estimate" climate sensitivity[b), c)] (°C)	Peaking year for CO_2 emissions[d)]	Change in global CO_2 emissions in 2050 (% of 2000 emissions)[d)]	No. of assessed scenarios
I	2.5-3.0	350-400	445-490	2.0-2.4	2000-2015	-85 to -50	6
II	3.0-3.5	400-440	490-535	2.4-2.8	2000-2020	-60 to -30	18
III	3.5-4.0	440-485	535-590	2.8-3.2	2010-2030	-30 to +5	21
IV	4.0-5.0	485-570	590-710	3.2-4.0	2020-2060	+10 to +60	118
V	5.0-6.0	570-660	710-855	4.0-4.9	2050-2080	+25 to +85	9
VI	6.0-7.5	660-790	855-1130	4.9-6.1	2060-2090	+90 to +140	5
						Total	177

a) The understanding of the climate system response to radiative forcing as well as feedbacks is assessed in detail in the AR4 WGI Report. Feedbacks between the carbon cycle and climate change affect the required mitigation for a particular stabilization level of atmospheric carbon dioxide concentration. These feedbacks are expected to increase the fraction of anthropogenic emissions that remains in the atmosphere as the climate system warms. Therefore, the emission reductions to meet a particular stabilization level reported in the mitigation studies assessed here might be underestimated.

b) The best estimate of climate sensitivity is 3°C [WG 1 SPM].

c) Note that global mean temperature at equilibrium is different from expected global mean temperature at the time of stabilization of GHG concentrations due to the inertia of the climate system. For the majority of scenarios assessed, stabilisation of GHG concentrations occurs between 2100 and 2150.

d) Ranges correspond to the 15th to 85th percentile of the post-TAR scenario distribution. CO_2 emissions are shown so multi-gas scenarios can be compared with CO_2-only scenarios.

---

26 Paragraph 2 addresses historical GHG emissions since pre-industrial times.

27 Studies vary in terms of the point in time stabilization is achieved; generally this is around 2100 or later.

28 The information on global mean temperature is taken from the AR4 WGI report, chapter 10.8. These temperatures are reached well after concentrations are stabilized.

29 The equilibrium climate sensitivity is a measure of the climate system response to sustained radiative forcing. It is not a projection but is defined as the global average surface warming following a doubling of carbon dioxide concentrations [AR4 WGI SPM].

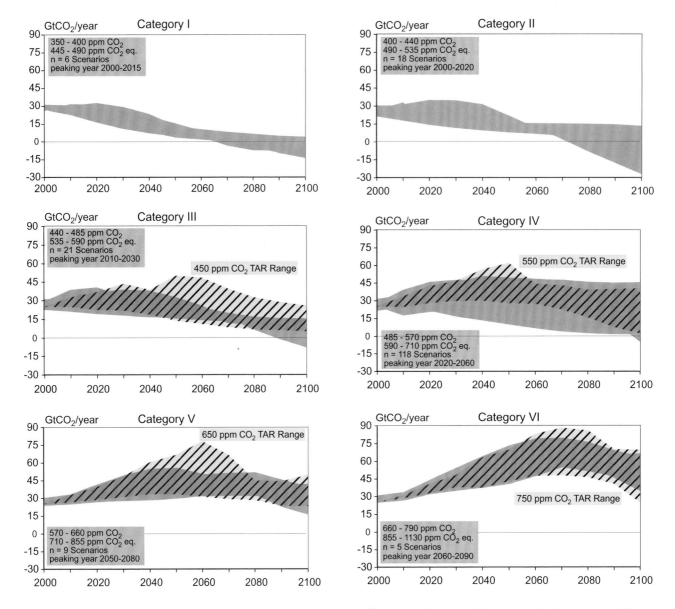

Figure SPM.7: *Emissions pathways of mitigation scenarios for alternative categories of stabilization levels (Category I to VI as defined in the box in each panel). The pathways are for CO_2 emissions only. Light brown shaded areas give the CO_2 emissions for the post-TAR emissions scenarios. Green shaded and hatched areas depict the range of more than 80 TAR stabilization scenarios. Base year emissions may differ between models due to differences in sector and industry coverage. To reach the lower stabilization levels some scenarios deploy removal of CO_2 from the atmosphere (negative emissions) using technologies such as biomass energy production utilizing carbon capture and storage. [Figure 3.17]*

19. **The range of stabilization levels assessed can be achieved by deployment of a portfolio of technologies that are currently available and those that are expected to be commercialised in coming decades. This assumes that appropriate and effective incentives are in place for development, acquisition, deployment and diffusion of technologies and for addressing related barriers** *(high agreement, much evidence).*

   • The contribution of different technologies to emission reductions required for stabilization will vary over time, region and stabilization level.

     o Energy efficiency plays a key role across many scenarios for most regions and timescales.

   o For lower stabilization levels, scenarios put more emphasis on the use of low-carbon energy sources, such as renewable energy and nuclear power, and the use of CO_2 capture and storage (CCS). In these scenarios improvements of carbon intensity of energy supply and the whole economy need to be much faster than in the past.

   o Including non-CO_2 and CO_2 land-use and forestry mitigation options provides greater flexibility and cost-effectiveness for achieving stabilization. Modern bioenergy could contribute substantially to the share of renewable energy in the mitigation portfolio.

Equilibrium global mean temperature increase
above pre-industrial (°C)

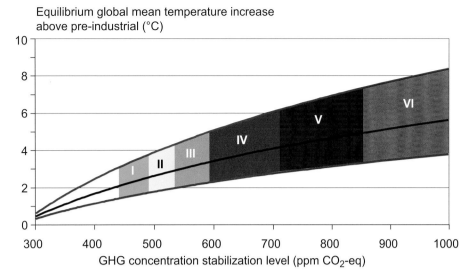

Figure SPM.8: *Stabilization scenario categories as reported in Figure SPM.7 (coloured bands) and their relationship to equilibrium global mean temperature change above pre-industrial, using (i) "best estimate" climate sensitivity of 3°C (black line in middle of shaded area), (ii) upper bound of likely range of climate sensitivity of 4.5°C (red line at top of shaded area) (iii) lower bound of likely range of climate sensitivity of 2°C (blue line at bottom of shaded area). Coloured shading shows the concentration bands for stabilization of greenhouse gases in the atmosphere corresponding to the stabilization scenario categories I to VI as indicated in Figure SPM.7. The data are drawn from AR4 WGI, Chapter 10.8.*

o For illustrative examples of portfolios of mitigation options, see figure SPM.9 [3.3, 3.4].

• Investments in and world-wide deployment of low-GHG emission technologies as well as technology improvements through public and private Research,

Development & Demonstration (RD&D) would be required for achieving stabilization targets as well as cost reduction. The lower the stabilization levels, especially those of 550 ppm CO$_2$-eq or lower, the greater the need for more efficient RD&D efforts and investment in new

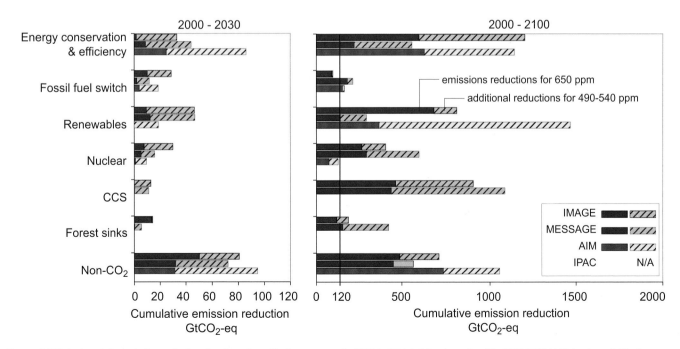

Figure SPM.9: *Cumulative emissions reductions for alternative mitigation measures for 2000 to 2030 (left-hand panel) and for 2000-2100 (right-hand panel). The figure shows illustrative scenarios from four models (AIM, IMAGE, IPAC and MESSAGE) aiming at the stabilization at 490-540 ppm CO$_2$-eq and levels of 650 ppm CO$_2$-eq, respectively. Dark bars denote reductions for a target of 650 ppm CO$_2$-eq and light bars the additional reductions to achieve 490-540 ppm CO$_2$-eq. Note that some models do not consider mitigation through forest sink enhancement (AIM and IPAC) or CCS (AIM) and that the share of low-carbon energy options in total energy supply is also determined by inclusion of these options in the baseline. CCS includes carbon capture and storage from biomass. Forest sinks include reducing emissions from deforestation. [Figure 3.23]*

technologies during the next few decades. This requires that barriers to development, acquisition, deployment and diffusion of technologies are effectively addressed.
- Appropriate incentives could address these barriers and help realize the goals across a wide portfolio of technologies. [2.7, 3.3, 3.4, 3.6, 4.3, 4.4, 4.6].

20. In 2050[30] global average macro-economic costs for multi-gas mitigation towards stabilization between 710 and 445 ppm CO_2-eq, are between a 1% gain to a 5.5% decrease of global GDP (see Table SPM.6). For specific countries and sectors, costs vary considerably from the global average. (See Box SPM.3 and SPM.4 for the methodologies and assumptions and paragraph 5 for explanation of negative costs) *(high agreement, medium evidence).*

21. Decision-making about the appropriate level of global mitigation over time involves an iterative risk management process that includes mitigation and adaptation, taking into account actual and avoided climate change damages, co-benefits, sustainability, equity, and attitudes to risk. Choices about the scale and timing of GHG mitigation involve balancing the economic costs of more rapid emission reductions now against the corresponding medium-term and long-term climate risks of delay *[high agreement, much evidence].*
- Limited and early analytical results from integrated analyses of the costs and benefits of mitigation indicate that these are broadly comparable in magnitude, but do not as yet permit an unambiguous determination of an emissions pathway or stabilization level where benefits exceed costs [3.5].

- Integrated assessment of the economic costs and benefits of different mitigation pathways shows that the economically optimal timing and level of mitigation depends upon the uncertain shape and character of the assumed climate change damage cost curve. To illustrate this dependency:

  o if the climate change damage cost curve grows slowly and regularly, and there is good foresight (which increases the potential for timely adaptation), later and less stringent mitigation is economically justified;
  o alternatively if the damage cost curve increases steeply, or contains non-linearities (e.g. vulnerability thresholds or even small probabilities of catastrophic events), earlier and more stringent mitigation is economically justified [3.6].
- Climate sensitivity is a key uncertainty for mitigation scenarios that aim to meet a specific temperature level. Studies show that if climate sensitivity is high then the timing and level of mitigation is earlier and more stringent than when it is low [3.5, 3.6].
- Delayed emission reductions lead to investments that lock in more emission-intensive infrastructure and development pathways. This significantly constrains the opportunities to achieve lower stabilization levels (as shown in Table SPM.5) and increases the risk of more severe climate change impacts [3.4, 3.1, 3.5, 3.6]

Table SPM.6: *Estimated global macro-economic costs in 2050 relative to the baseline for least-cost trajectories towards different long-term stabilization targets[a] [3.3, 13.3]*

Stabilization levels (ppm CO_2-eq)	Median GDP reduction[b] (%)	Range of GDP reduction[b], [c] (%)	Reduction of average annual GDP growth rates[b], [d] (percentage points)
590-710	0.5	-1 - 2	<0.05
535-590	1.3	slightly negative - 4	<0.1
445-535[e]	not available	<5.5	<0.12

Notes:
[a] This corresponds to the full literature across all baselines and mitigation scenarios that provide GDP numbers.
[b] This is global GDP based market exchange rates.
[c] The median and the 10th and 90th percentile range of the analyzed data are given.
[d] The calculation of the reduction of the annual growth rate is based on the average reduction during the period until 2050 that would result in the indicated GDP decrease in 2050.
[e] The number of studies is relatively small and they generally use low baselines. High emissions baselines generally lead to higher costs.

30 Cost estimates for 2030 are presented in paragraph 5.

E. Policies, measures and instruments to mitigate climate change

22. **A wide variety of national policies and instruments are available to governments to create the incentives for mitigation action. Their applicability depends on national circumstances and an understanding of their interactions, but experience from implementation in various countries and sectors shows there are advantages and disadvantages for any given instrument** *(high agreement, much evidence).*
    - Four main criteria are used to evaluate policies and instruments: environmental effectiveness, cost effectiveness, distributional effects, including equity, and institutional feasibility [13.2].
    - All instruments can be designed well or poorly, and be stringent or lax. In addition, monitoring to improve implementation is an important issue for all instruments. General findings about the performance of policies are: [7.9, 12.2, 13.2]
      - *Integrating climate policies in broader development policies* makes implementation and overcoming barriers easier.
      - *Regulations and standards* generally provide some certainty about emission levels. They may be preferable to other instruments when information or other barriers prevent producers and consumers from responding to price signals. However, they may not induce innovations and more advanced technologies.
      - *Taxes and charges* can set a price for carbon, but cannot guarantee a particular level of emissions. Literature identifies taxes as an efficient way of internalizing costs of GHG emissions.
      - *Tradable permits* will establish a carbon price. The volume of allowed emissions determines their environmental effectiveness, while the allocation of permits has distributional consequences. Fluctuation in the price of carbon makes it difficult to estimate the total cost of complying with emission permits.
      - *Financial incentives* (subsidies and tax credits) are frequently used by governments to stimulate the development and diffusion of new technologies. While economic costs are generally higher than for the instruments listed above, they are often critical to overcome barriers.
      - *Voluntary agreements* between industry and governments are politically attractive, raise awareness among stakeholders, and have played a role in the evolution of many national policies. The majority of agreements has not achieved significant emissions reductions beyond business as usual. However, some recent agreements, in a few countries, have accelerated the application of best available technology and led to measurable emission reductions.
      - *Information instruments* (e.g. awareness campaigns) may positively affect environmental quality by promoting informed choices and possibly contributing to behavioural change, however, their impact on emissions has not been measured yet.
      - *RD&D* can stimulate technological advances, reduce costs, and enable progress toward stabilization.
    - Some corporations, local and regional authorities, NGOs and civil groups are adopting a wide variety of voluntary actions. These voluntary actions may limit GHG emissions, stimulate innovative policies, and encourage the deployment of new technologies. On their own, they generally have limited impact on the national or regional level emissions [13.4].
    - Lessons learned from specific sector application of national policies and instruments are shown in Table SPM.7.

23. **Policies that provide a real or implicit price of carbon could create incentives for producers and consumers to significantly invest in low-GHG products, technologies and processes. Such policies could include economic instruments, government funding and regulation** *(high agreement, much evidence).*
    - An effective carbon-price signal could realize significant mitigation potential in all sectors [11.3, 13.2].
    - Modelling studies, consistent with stabilization at around 550 ppm CO_2-eq by 2100 (see Box SPM.3), show carbon prices rising to 20 to 80 US\$/tCO_2-eq by 2030 and 30 to 155 US\$/tCO_2-eq by 2050. For the same stabilization level, studies since TAR that take into account induced technological change lower these price ranges to 5 to 65 US\$/tCO_2-eq in 2030 and 15 to 130 US\$/tCO_2-eq in 2050 [3.3, 11.4, 11.5].
    - Most top-down, as well as some 2050 bottom-up assessments, suggest that real or implicit carbon prices of 20 to 50 US\$/tCO_2-eq, sustained or increased over decades, could lead to a power generation sector with low-GHG emissions by 2050 and make many mitigation options in the end-use sectors economically attractive. [4.4, 11.6]
    - Barriers to the implementation of mitigation options are manifold and vary by country and sector. They can be related to financial, technological, institutional, informational and behavioural aspects [4.5, 5.5, 6.7, 7.6, 8.6, 9.6, 10.5].

Table SPM.7: *Selected sectoral policies, measures and instruments that have shown to be environmentally effective in the respective sector in at least a number of national cases.*

Sector	Policies[a], measures and instruments shown to be environmentally effective	Key constraints or opportunities
Energy supply [4.5]	Reduction of fossil fuel subsidies Taxes or carbon charges on fossil fuels	Resistance by vested interests may make them difficult to implement
	Feed-in tariffs for renewable energy technologies Renewable energy obligations Producer subsidies	May be appropriate to create markets for low emissions technologies
Transport [5.5]	Mandatory fuel economy, biofuel blending and CO_2 standards for road transport	Partial coverage of vehicle fleet may limit effectiveness
	Taxes on vehicle purchase, registration, use and motor fuels, road and parking pricing	Effectiveness may drop with higher incomes
	Influence mobility needs through land use regulations, and infrastructure planning Investment in attractive public transport facilities and non-motorised forms of transport	Particularly appropriate for countries that are building up their transportation systems
Buildings [6.8]	Appliance standards and labelling	Periodic revision of standards needed
	Building codes and certification	Attractive for new buildings. Enforcement can be difficult
	Demand-side management programmes	Need for regulations so that utilities may profit
	Public sector leadership programmes, including procurement	Government purchasing can expand demand for energy-efficient products
	Incentives for energy service companies (ESCOs)	Success factor: Access to third party financing
Industry [7.9]	Provision of benchmark information Performance standards Subsidies, tax credits	May be appropriate to stimulate technology uptake. Stability of national policy important in view of international competitiveness
	Tradable permits	Predictable allocation mechanisms and stable price signals important for investments
	Voluntary agreements	Success factors include: clear targets, a baseline scenario, third party involvement in design and review and formal provisions of monitoring, close cooperation between government and industry
Agriculture [8.6, 8.7, 8.8]	Financial incentives and regulations for improved land management, maintaining soil carbon content, efficient use of fertilizers and irrigation	May encourage synergy with sustainable development and with reducing vulnerability to climate change, thereby overcoming barriers to implementation
Forestry/ forests [9.6]	Financial incentives (national and international) to increase forest area, to reduce deforestation, and to maintain and manage forests	Constraints include lack of investment capital and land tenure issues. Can help poverty alleviation
	Land use regulation and enforcement	
Waste management [10.5]	Financial incentives for improved waste and wastewater management	May stimulate technology diffusion
	Renewable energy incentives or obligations	Local availability of low-cost fuel
	Waste management regulations	Most effectively applied at national level with enforcement strategies

Note:
a) Public RD & D investment in low emissions technologies have proven to be effective in all sectors

24. **Government support through financial contributions, tax credits, standard setting and market creation is important for effective technology development, innovation and deployment. Transfer of technology to developing countries depends on enabling conditions and financing** *(high agreement, much evidence).*
   - Public benefits of RD&D investments are bigger than the benefits captured by the private sector, justifying government support of RD&D.
   - Government funding in real absolute terms for most energy research programmes has been flat or declining for nearly two decades (even after the UNFCCC came into force) and is now about half of the 1980 level [2.7, 3.4, 4.5, 11.5, 13.2].

- Governments have a crucial supportive role in providing appropriate enabling environment, such as, institutional, policy, legal and regulatory frameworks[31], to sustain investment flows and for effective technology transfer – without which it may be difficult to achieve emission reductions at a significant scale. Mobilizing financing of incremental costs of low-carbon technologies is important. International technology agreements could strengthen the knowledge infrastructure [13.3].
- The potential beneficial effect of technology transfer to developing countries brought about by Annex I countries action may be substantial, but no reliable estimates are available [11.7].
- Financial flows to developing countries through Clean Development Mechanism (CDM) projects have the potential to reach levels of the order of several billions US$ per year[32], which is higher than the flows through the Global Environment Facility (GEF), comparable to the energy oriented development assistance flows, but at least an order of magnitude lower than total foreign direct investment flows. The financial flows through CDM, GEF and development assistance for technology transfer have so far been limited and geographically unequally distributed [12.3, 13.3].

25. **Notable achievements of the UNFCCC and its Kyoto Protocol are the establishment of a global response to the climate problem, stimulation of an array of national policies, the creation of an international carbon market and the establishment of new institutional mechanisms that may provide the foundation for future mitigation efforts** (high agreement, much evidence).
    - The impact of the Protocol's first commitment period relative to global emissions is projected to be limited. Its economic impacts on participating Annex-B countries are projected to be smaller than presented in TAR, that showed 0.2-2% lower GDP in 2012 without emissions trading, and 0.1-1.1% lower GDP with emissions trading among Annex-B countries [1.4, 11.4, 13.3].
26. **The literature identifies many options for achieving reductions of global GHG emissions at the international level through cooperation. It also suggests that successful agreements are environmentally effective, cost-effective, incorporate distributional considerations and equity, and are institutionally feasible** (high agreement, much evidence).
    - Greater cooperative efforts to reduce emissions will help to reduce global costs for achieving a given level of mitigation, or will improve environmental effectiveness [13.3].
    - Improving, and expanding the scope of, market mechanisms (such as emission trading, Joint

Implementation and CDM) could reduce overall mitigation costs [13.3].
- Efforts to address climate change can include diverse elements such as emissions targets; sectoral, local, sub-national and regional actions; RD&D programmes; adopting common policies; implementing development oriented actions; or expanding financing instruments. These elements can be implemented in an integrated fashion, but comparing the efforts made by different countries quantitatively would be complex and resource intensive [13.3].
- Actions that could be taken by participating countries can be differentiated both in terms of when such action is undertaken, who participates and what the action will be. Actions can be binding or non-binding, include fixed or dynamic targets, and participation can be static or vary over time [13.3].

F. Sustainable development and climate change mitigation

27. **Making development more sustainable by changing development paths can make a major contribution to climate change mitigation, but implementation may require resources to overcome multiple barriers. There is a growing understanding of the possibilities to choose and implement mitigation options in several sectors to realize synergies and avoid conflicts with other dimensions of sustainable development** (high agreement, much evidence).
    - Irrespective of the scale of mitigation measures, adaptation measures are necessary [1.2].
    - Addressing climate change can be considered an integral element of sustainable development policies. National circumstances and the strengths of institutions determine how development policies impact GHG emissions. Changes in development paths emerge from the interactions of public and private decision processes involving government, business and civil society, many of which are not traditionally considered as climate policy. This process is most effective when actors participate equitably and decentralized decision making processes are coordinated [2.2, 3.3, 12.2].
    - Climate change and other sustainable development policies are often but not always synergistic. There is growing evidence that decisions about macroeconomic policy, agricultural policy, multilateral development bank lending, insurance practices, electricity market reform, energy security and forest conservation, for example, which are often treated as being apart from

---

31 See the IPCC Special Report on Methodological and Technological Issues in Technology Transfer.
32 Depends strongly on the market price that has fluctuated between 4 and 26 US$/tCO2-eq and based on approximately 1000 CDM proposed plus registered projects likely to generate more than 1.3 billion emission reduction credits before 2012.

climate policy, can significantly reduce emissions. On the other hand, decisions about improving rural access to modern energy sources for example may not have much influence on global GHG emissions [12.2].

- Climate change policies related to energy efficiency and renewable energy are often economically beneficial, improve energy security and reduce local pollutant emissions. Other energy supply mitigation options can be designed to also achieve sustainable development benefits such as avoided displacement of local populations, job creation, and health benefits [4.5,12.3].

- Reducing both loss of natural habitat and deforestation can have significant biodiversity, soil and water conservation benefits, and can be implemented in a socially and economically sustainable manner. Forestation and bioenergy plantations can lead to restoration of degraded land, manage water runoff, retain soil carbon and benefit rural economies, but could compete with land for food production and may be negative for biodiversity, if not properly designed [9.7, 12.3].

- There are also good possibilities for reinforcing sustainable development through mitigation actions in the waste management, transportation and buildings sectors [5.4, 6.6, 10.5, 12.3].

- Making development more sustainable can enhance both mitigative and adaptive capacity, and reduce emissions and vulnerability to climate change. Synergies between mitigation and adaptation can exist, for example properly designed biomass production, formation of protected areas, land management, energy use in buildings and forestry. In other situations, there may be trade-offs, such as increased GHG emissions due to increased consumption of energy related to adaptive responses [2.5, 3.5, 4.5, 6.9, 7.8, 8.5, 9.5, 11.9, 12.1].

G.   Gaps in knowledge

28. **There are still relevant gaps in currently available knowledge regarding some aspects of mitigation of climate change, especially in developing countries. Additional research addressing those gaps would further reduce uncertainties and thus facilitate decision-making related to mitigation of climate change [TS.14].**

Summary for Policymakers

Endbox 1: Uncertainty representation

Uncertainty is an inherent feature of any assessment. The fourth assessment report clarifies the uncertainties associated with essential statements.

Fundamental differences between the underlying disciplinary sciences of the three Working Group reports make a common approach impractical. The "likelihood" approach applied in "Climate change 2007, the physical science basis" and the "confidence" and "likelihood" approaches used in "Climate change 2007, impacts, adaptation, and vulnerability" were judged to be inadequate to deal with the specific uncertainties involved in this mitigation report, as here human choices are considered.

In this report a two-dimensional scale is used for the treatment of uncertainty. The scale is based on the expert judgment of the authors of WGIII on the level of concurrence in the literature on a particular finding (level of agreement), and the number and quality of independent sources qualifying under the IPCC rules upon which the finding is based (amount of evidence[1]) (see Table SPM.E.1). This is not a quantitative approach, from which probabilities relating to uncertainty can be derived.

Table SPM.E.1: *Qualitative definition of uncertainty*

High agreement, limited evidence	High agreement, medium evidence	High agreement, much evidence
Medium agreement, limited evidence	Medium agreement, medium evidence	Medium agreement, much evidence
Low agreement, limited evidence	Low agreement, medium evidence	Low agreement, much evidence

Level of agreement (on a particular finding) ↑

Amount of evidence[33] (number and quality of independent sources)

Because the future is inherently uncertain, scenarios i.e. internally consistent images of different futures - not predictions of the future - have been used extensively in this report.

---

33 "Evidence" in this report is defined as: Information or signs indicating whether a belief or proposition is true or valid. See Glossary.

Contribution of Working Group III to the
Fourth Assessment Report of the
Intergovernmental Panel on Climate Change

Technical Summary

Authors:

Terry Barker (UK), Igor Bashmakov (Russia), Lenny Bernstein (USA), Jean E. Bogner (USA), Peter Bosch (The Netherlands), Rutu Dave (The Netherlands), Ogunlade Davidson (Sierra Leone), Brian S. Fisher (Australia), Sujata Gupta (India), Kirsten Halsnæs (Denmark), BertJan Heij (The Netherlands), Suzana Kahn Ribeiro (Brazil), Shigeki Kobayashi (Japan), Mark D. Levine (USA), Daniel L. Martino (Uruguay), Omar Masera (Mexico), Bert Metz (The Netherlands), Leo Meyer (The Netherlands), Gert-Jan Nabuurs (The Netherlands), Adil Najam (Pakistan), Nebojsa Nakicenovic (Austria/Montenegro), Hans-Holger Rogner (Germany), Joyashree Roy (India), Jayant Sathaye (USA), Robert Schock (USA), Priyadarshi Shukla (India), Ralph E. H. Sims (New Zealand), Pete Smith (UK), Dennis A. Tirpak (USA), Diana Urge-Vorsatz (Hungary), Dadi Zhou (PR China)

Review Editor:

Mukiri wa Githendu (Kenya)

This Technical Summary should be cited as:

Barker T., I. Bashmakov, L. Bernstein, J. E. Bogner, P. R. Bosch, R. Dave, O. R. Davidson, B. S. Fisher, S. Gupta, K. Halsnæs, G.J. Heij, S. Kahn Ribeiro, S. Kobayashi, M. D. Levine, D. L. Martino, O. Masera, B. Metz, L. A. Meyer, G.-J. Nabuurs, A. Najam, N. Nakicenovic, H. -H. Rogner, J. Roy, J. Sathaye, R. Schock, P. Shukla, R. E. H. Sims, P. Smith, D. A. Tirpak, D. Urge-Vorsatz, D. Zhou, 2007: Technical Summary. In: *Climate Change 2007: Mitigation. Contribution of Working Group III to the Fourth Assessment Report of the Intergovernmental Panel on Climate Change* [B. Metz, O. R. Davidson, P. R. Bosch, R. Dave, L. A. Meyer (eds)], Cambridge University Press, Cambridge, United Kingdom and New York, NY, USA.

Table of contents

1 Introduction

Structure of the report, the rationale behind it, the role of cross-cutting themes and framing issues

The main aim of this report is to assess options for mitigating climate change. Several aspects link climate change with development issues. This report explores these links in detail, and illustrates where climate change and sustainable development are mutually reinforcing.

Economic development needs, resource endowments and mitigative and adaptive capacities differ across regions. There is no one-size-fits-all approach to the climate change problem, and solutions need to be regionally differentiated to reflect different socio-economic conditions and, to a lesser extent, geographical differences. Although this report has a global focus, an attempt is made to differentiate the assessment of scientific and technical findings for the various regions.

Given that mitigation options vary significantly between economic sectors, it was decided to use the economic sectors to organize the material on short- to medium-term mitigation options. Contrary to what was done in the Third Assessment Report, all relevant aspects of sectoral mitigation options, such as technology, cost, policies etc., are discussed together, to provide the user with a comprehensive discussion of the sectoral mitigation options.

Consequently, the report has four parts. Part A (Chapters 1 and 2) includes the introduction and sets out the frameworks to describe mitigation of climate change in the context of other policies and decision-making. It introduces important concepts (e.g., risk and uncertainty, mitigation and adaptation relationships, distributional and equity aspects and regional integration) and defines important terms used throughout the report. Part B (Chapter 3) assesses long-term stabilization targets, how to get there and what the associated costs are, by examining mitigation scenarios for ranges of stability targets. The relation between adaptation, mitigation and climate change damage avoided is also discussed, in the light of decision-making regarding stabilization (Art. 2 UNFCCC). Part C (Chapters 4–10) focuses on the detailed description of the various sectors responsible for greenhouse gas (GHG) emissions, the short- to medium-term mitigation options and costs in these sectors, the policies for achieving mitigation, the barriers to getting there and the relationship with adaptation and other policies that affect GHG emissions. Part D (Chapters 11–13) assesses cross-sectoral issues, sustainable development and national and international aspects. Chapter 11 covers the aggregated mitigation potential, macro-economic impacts, technology development and transfer, synergies, and trade-offs with other policies and cross-border influences (or spill-over effects). Chapter 12 links climate mitigation with sustainable development. Chapter 13 assesses domestic climate policies and various forms of international cooperation. This Technical Summary has an additional Chapter 14, which deals with gaps in knowledge.

Past, present and future: emission trends

Emissions of the GHGs covered by the Kyoto Protocol increased by about 70% (from 28.7 to. 49.0 $GtCO_2$-eq) from 1970–2004 (by 24% from 1990–2004), with carbon dioxide (CO_2) being the largest source, having grown by about 80% (see Figure TS.1). The largest growth in CO_2 emissions has come from power generation and road transport. Methane (CH_4) emissions rose by about 40% from 1970, with an 85% increase from the combustion and use of fossil fuels. Agriculture, however, is the largest source of CH_4 emissions. Nitrous oxide (N_2O) emissions grew by about 50%, due mainly to increased use of fertilizer and the growth of agriculture. Industrial emission of N_2O fell during this period *(high agreement, much evidence)* [1.3].

Emissions of ozone-depleting substances (ODS) controlled under the Montreal Protocol (which includes GHGs chlorofluorocarbons (CFCs), hydrochlorofluorocarbons (HCFCs)), increased from a low level in 1970 to about 7.5 $GtCO_2$-eq in 1990 (about 20% of total GHG emissions, not shown in the Figure TS.1), but then decreased to about 1.5 $GtCO_2$-eq in 2004, and are projected to decrease further due to the phase-out of CFCs in developing countries. Emissions of the fluorinated gases (F-gases) (hydrofluorocarbons (HFCs), perfluorocarbons (PFCs) and SF_6) controlled under the Kyoto Protocol grew rapidly (primarily HFCs) during the 1990s as they replaced ODS to a substantial extent and were estimated at about 0.5 $GtCO_2$eq in 2004 (about 1.1% of total emissions on a 100-year global warming potential (GWP) basis) *(high agreement, much evidence)* [1.3].

Atmospheric CO_2 concentrations have increased by almost 100 ppm since their pre-industrial level, reaching 379 ppm in 2005, with mean annual growth rates in the 2000-2005 period higher than in the 1990s. The total CO_2-equivalent (CO_2-eq) concentration of all long-lived GHGs is now about 455 ppm CO_2-eq. Incorporating the cooling effect of aerosols, other air pollutants and gases released from land-use change into the equivalent concentration, leads to an effective 311-435 ppm CO_2-eq concentration *(high agreement, much evidence)*.

Considerable uncertainties still surround the estimates of anthropogenic aerosol emissions. As regards global sulphur emissions, these appear to have declined from 75 ± 10 MtS in 1990 to 55-62 MtS in 2000. Data on non-sulphur aerosols are sparse and highly speculative. *(medium agreement, medium evidence)*.

In 2004, energy supply accounted for about 26% of GHG emissions, industry 19%, gases released from land-use change and forestry 17%, agriculture 14%, transport 13%, residential, commercial and service sectors 8% and waste 3% (see Figure TS.2). These figures should be seen as indicative, as some uncertainty remains, particularly with regards to CH_4 and N_2O emissions (error margin estimated to be in the order of 30-50%) and CO_2 emissions from agriculture and forestry with an even higher error margin *(high agreement, medium evidence)* [1.3].

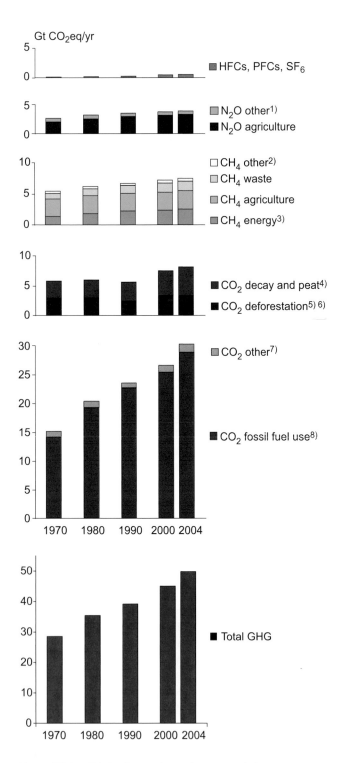

Gt CO₂eq/yr

■ HFCs, PFCs, SF₆

■ N₂O other[1)]
■ N₂O agriculture

□ CH₄ other[2)]
▨ CH₄ waste
▨ CH₄ agriculture
▨ CH₄ energy[3)]

■ CO₂ decay and peat[4)]
■ CO₂ deforestation[5) 6)]

▨ CO₂ other[7)]

■ CO₂ fossil fuel use[8)]

1970 1980 1990 2000 2004

■ Total GHG

1970 1980 1990 2000 2004

Figure TS.1a: *Global anthropogenic greenhouse gas emissions, 1970–2004. One hundred year global warming potentials (GWPs) from IPCC 1996 (SAR) were used to convert emissions to CO₂-eq. (see the UNFCCC reporting guidelines).*
Gases are those reported under UNFCCC reporting guidelines. The uncertainty in the graph is quite large for CH₄ and N₂O (in the order of 30-50%) and even larger for CO₂ from agriculture and forestry. [Figure 1.1a].

Notes:
[1)] Other N₂O includes industrial processes, deforestation/ savannah burning, waste water and waste incineration.
[2)] Other is CH₄ from industrial processes and savannah burning.
[3)] Including emissions from bioenergy production and use
[4)] CO₂ emissions from decay (decomposition) of above ground biomass that remains after logging and deforestation and CO₂ from peat fires and decay of drained peat soils.
[5)] As well as traditional biomass use at 10% of total, assuming 90% is from sustainable biomass production. Corrected for the 10% of carbon in biomass that is assumed to remain as charcoal after combustion.
[6)] For large-scale forest and scrubland biomass burning averaged data for 1997-2002 based on Global Fire Emissions Data base satellite data.
[7)] Cement production and natural gas flaring.
[8)] Fossil fuel use includes emissions from feedstocks.

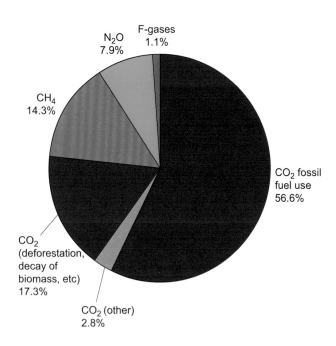

N₂O
7.9%

F-gases
1.1%

CH₄
14.3%

CO₂ fossil
fuel use
56.6%

CO₂
(deforestation, decay of biomass, etc)
17.3%

CO₂ (other)
2.8%

Figure TS.1b: *Global anthropogenic greenhousegas emissions in 2004 [Figure 1.1b].*

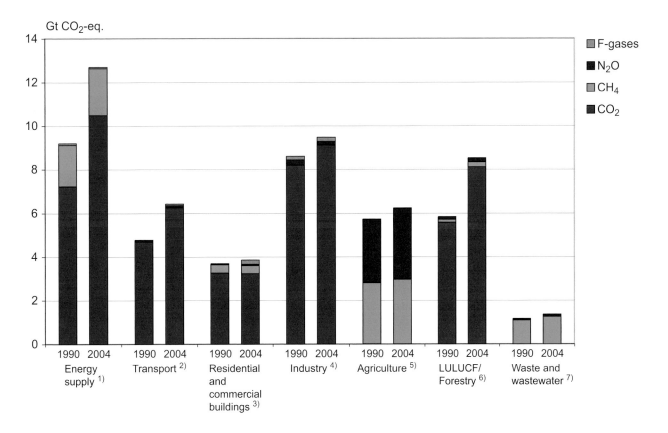

Gt CO$_2$-eq.

Legend:
- F-gases
- N$_2$O
- CH$_4$
- CO$_2$

X-axis labels:
- 1990 2004 Energy supply [1]
- 1990 2004 Transport [2]
- 1990 2004 Residential and commercial buildings [3]
- 1990 2004 Industry [4]
- 1990 2004 Agriculture [5]
- 1990 2004 LULUCF/Forestry [6]
- 1990 2004 Waste and wastewater [7]

Figure TS.2a: *GHG emissions by sector in 1990 and 2004 100-year GWPs from IPCC 1996 (Second Assessment Report (SAR)) were used to convert emissions to CO$_2$-eq. The uncertainty in the graph is quite large for CH$_4$ and N$_2$O (in the order of 30–50%) and even larger for CO$_2$ from agriculture and forestry. For large-scale biomass burning, averaged activity data for 1997–2002 were used from Global Fire Emissions Database based on satellite data. Peat (fire and decay) emissions are based on recent data from WL/Delft Hydraulics. [Figure 1.3a]*

Notes to Figure TS.2a and 2b:
1) Excluding refineries, coke ovens etc., which are included in industry.
2) Including international transport (bunkers), excluding fisheries. Excluding off-road agricultural and forestry vehicles and machinery.
3) Including traditional biomass use. Emissions in Chapter 6 are also reported on the basis of end-use allocation (including the sector's share in emissions caused by centralized electricity generation) so that any mitigation achievements in the sector resulting from lower electricity use are credited to the sector.
4) Including refineries, coke ovens etc. Emissions reported in Chapter 7 are also reported on the basis of end-use allocation (including the sector's share in emissions caused by centralized electricity generation) so that any mitigation achievements in the sector resulting from lower electricity use are credited to the sector.
5) Including agricultural waste burning and savannah burning (non-CO$_2$). CO$_2$ emissions and/or removals from agricultural soils are not estimated in this database.
6) Data include CO$_2$ emissions from deforestation, CO$_2$ emissions from decay (decomposition) of above-ground biomass that remains after logging and deforestation, and CO$_2$ from peat fires and decay of drained peat soils. Chapter 9 reports emissions from deforestation only.
7) Includes landfill CH$_4$, wastewater CH$_4$ and N$_2$O, and CO$_2$ from waste incineration (fossil carbon only).

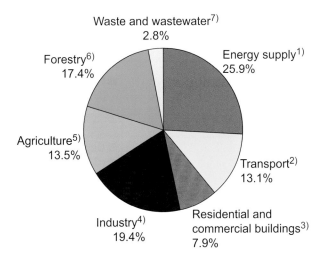

Pie chart labels:
- Waste and wastewater [7] 2.8%
- Energy supply [1] 25.9%
- Forestry [6] 17.4%
- Agriculture [5] 13.5%
- Industry [4] 19.4%
- Residential and commercial buildings [3] 7.9%
- Transport [2] 13.1%

Figure TS.2b: *GHG emissions by sector in 2004 [Figure 1.3b].*

Figure TS.3 identifies the individual contributions to energy-related CO_2 emissions from changes in population, income per capita (gross domestic product (GDP) expressed in terms of purchasing-power parity per person - GDP_{ppp}/cap[1]), energy intensity (Total Primary Energy Supply (TPES)/GDP_{ppp}), and carbon intensity (CO_2/TPES). Some of these factors boost CO_2 emissions (bars above the zero line), while others lower them (bar below the zero line). The actual change in emissions per decade is shown by the dashed black lines. According to Figure TS.3, the increase in population and GDP_{ppp}/cap (and therefore energy use per capita) have outweighed and are projected to continue to outweigh the decrease in energy intensities (TPES/GDP_{ppp}) and conceal the fact that CO_2 emissions per unit of GDP_{ppp} are 40% lower today than during the early 1970s and have declined faster than primary energy per unit of GDP_{ppp} or CO_2 per unit of primary energy. The carbon intensity of energy supply (CO_2/TPES) had an offsetting effect on CO_2 emissions between the mid 1980s and 2000, but has since been increasing and is projected to have no such effect after 2010 *(high agreement, much evidence)* [1.3].

In 2004, Annex I countries had 20% of the world's population, but accounted for 46% of global GHG emissions, and the 80% in Non-Annex I countries for only 54%. The contrast between the region with the highest per capita GHG emissions (North America) and the lowest (Non-Annex I South Asia) is even more pronounced (see Figure TS.4a): 5% of the world's population (North America) emits 19.4%,

while 30.3% (Non-Annex I South Asia) emits 13.1%. A different picture emerges if the metric GHG emissions per unit of GDP_{ppp} is used (see Figure TS.4b). In these terms, Annex I countries generated 57% of gross world product with a GHG intensity of production of 0.68 kg CO_2-eq/US\$ GDP_{ppp} (non-Annex I countries 1.06 kg CO_2-eq/US\$ GDP_{ppp}) *(high agreement, much evidence)* [1.3].

Global energy use and supply – the main drivers of GHG emissions – is projected to continue to grow, especially as developing countries pursue industrialization. Should there be no change in energy policies, the energy mix supplied to run the global economy in the 2025–30 timeframe will essentially remain unchanged, with more than 80% of energy supply based on fossil fuels with consequent implications for GHG emissions. On this basis, the projected emissions of energy-related CO_2 in 2030 are 40–110% higher than in 2000, with two thirds to three quarters of this increase originating in non-Annex I countries, though per capita emissions in developed countries will remain substantially higher, that is 9.6 tCO_2/cap to 15.1 tCO_2/cap in Annex I regions versus 2.8 tCO_2/cap to 5.1 tCO_2/cap in non-Annex I regions *(high agreement, much evidence)* [1.3].

For 2030, projections of total GHG emissions (Kyoto gases) consistently show an increase of 25–90% compared with 2000, with more recent projections higher than earlier ones *(high agreement, much evidence)*.

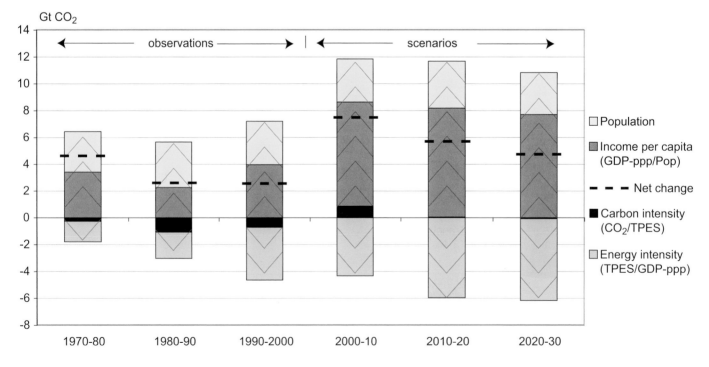

Figure TS.3: *Decomposition of global energy-related CO_2 emission changes at the global scale for three past and three future decades [Figure 1.6].*

1   The GDP_{ppp} metric is used for illustrative purposes only for this report.

For 2100, the SRES[2] range (a 40% decline to 250% increase compared with 2000) is still valid. More recent projections tend to be higher: increase of 90% to 250% compared with 2000 (see Figure TS.5). Scenarios that account for climate policies, whose implementation is currently under discussion, also show global emissions rising for many decades.

Developing countries (e.g., Brazil, China, India and Mexico) that have undertaken efforts for reasons other than climate change have reduced their emissions growth over the past three decades by approximately 500 million tonnes CO_2 per year; that is, more than the reductions required from Annex I countries by the Kyoto Protocol. Many of these efforts are motivated by economic development and poverty alleviation, energy security and local environmental protection. The most promising policy approaches, therefore, seem to be those that capitalize on natural synergies between climate protection and development priorities to advance both simultaneously (*high agreement, medium evidence*) [1.3].

International response

The United Nations Framework Convention on Climate Change (UNFCCC) is the main vehicle for promoting international responses to climate change. It entered into force in March 1994 and has achieved near universal ratification – 189 of the 194 UN member states (December 2006). A Dialogue on Long-Term Cooperation Action to Address Climate Change by Enhancing Implementation of the Convention was set up at CMP1[3] in 2005, taking the form of an open and non-binding exchange of views and information in support of enhanced implementation of the Convention.

The first addition to the treaty, the Kyoto Protocol, was adopted in 1997 and entered into force in February 2005. As of February 2007, 168 states and the European Economic Community have ratified the Protocol. Under Article 3.1 of the Kyoto Protocol, Annex I Parties in aggregate agreed to reduce

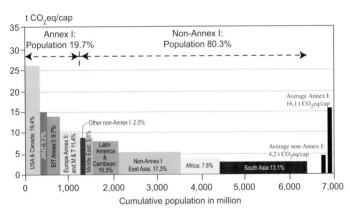

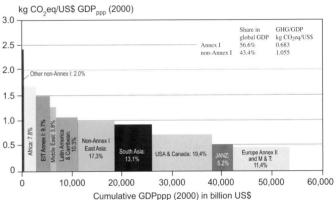

Figure TS.4a: *Distribution of regional per capita GHG emissions (all Kyoto gases including those from land-use) over the population of different country groupings in 2004. The percentages in the bars indicate a region's share in global GHG emissions [Figure 1.4a].*

Figure TS.4b: *Distribution of regional GHG emissions (all Kyoto gases including those from land-use) per US$ of GDPppp over the GDP of different country groupings in 2004. The percentages in the bars indicate a region's share in global GHG emissions [Figure 1.4b].*

Note: Countries are grouped according to the classification of the UNFCCC and its Kyoto Protocol; this means that countries that have joined the European Union since then are still listed under EIT Annex I. A full set of data for all countries for 2004 was not available. The countries in each of the regional groupings include:
- **EIT Annex I:** Belarus, Bulgaria, Croatia, Czech Republic, Estonia, Hungary, Latvia, Lithuania, Poland, Romania, Russian Federation, Slovakia, Slovenia, Ukraine
- **Europe Annex II & M&T:** Austria, Belgium, Denmark, Finland, France, Germany, Greece, Iceland, Ireland, Italy, Liechtenstein, Luxembourg, Netherlands, Norway, Portugal, Spain, Sweden, Switzerland, United Kingdom; Monaco and Turkey
- **JANZ:** Japan, Australia, New Zealand.
- **Middle East:** Bahrain, Islamic Republic of Iran, Israel, Jordan, Kuwait, Lebanon, Oman, Qatar, Saudi Arabia, Syria, United Arab Emirates, Yemen
- **Latin America & the Caribbean:** Antigua & Barbuda, Argentina, Bahamas, Barbados, Belize, Bolivia, Brazil, Chile, Colombia, Costa Rica, Cuba, Dominica, Dominican Republic, Ecuador, El Salvador, Grenada, Guatemala, Guyana, Haiti, Honduras, Jamaica, Mexico, Nicaragua, Panama, Paraguay, Peru, Saint Lucia, St. Kitts-Nevis-Anguilla, St. Vincent-Grenadines, Suriname, Trinidad and Tobago, Uruguay, Venezuela
- **Non-Annex I East Asia:** Cambodia, China, Korea (DPR), Laos (PDR), Mongolia, Republic of Korea, Viet Nam.
- **South Asia:** Afghanistan, Bangladesh, Bhutan, Comoros, Cook Islands, Fiji, India, Indonesia, Kiribati, Malaysia, Maldives, Marshall Islands, Micronesia, (Federated States of), Myanmar, Nauru, Niue, Nepal, Pakistan, Palau, Papua New Guinea, Philippine, Samoa, Singapore, Solomon Islands, Sri Lanka, Thailand, Timor-Leste, Tonga, Tuvalu, Vanuatu
- **North America:** Canada, United States of America.
- **Other non-Annex I:** Albania, Armenia, Azerbaijan, Bosnia Herzegovina, Cyprus, Georgia, Kazakhstan, Kyrgyzstan, Malta, Moldova, San Marino, Serbia, Tajikistan, Turkmenistan, Uzbekistan, Republic of Macedonia
- **Africa:** Algeria, Angola, Benin, Botswana, Burkina Faso, Burundi, Cameroon, Cape Verde, Central African Republic, Chad, Congo, Democratic Republic of Congo, Côte d'Ivoire, Djibouti, Egypt, Equatorial Guinea, Eritrea, Ethiopia, Gabon, Gambia, Ghana, Guinea, Guinea-Bissau, Kenya, Lesotho, Liberia, Libya, Madagascar, Malawi, Mali, Mauritania, Mauritius, Morocco, Mozambique, Namibia, Niger, Nigeria, Rwanda, Sao Tome and Principe, Senegal, Seychelles, Sierra Leone, South Africa, Sudan, Swaziland, Togo, Tunisia, Uganda, United Republic of Tanzania, Zambia, Zimbabwe.

---

2   SRES refers to scenarios described in the IPCC Special Report on Emission Scenarios (IPCC, 2000b). The A1 family of scenarios describes a future with very rapid economic growth, low population growth and rapid introduction of new and more efficient technologies. B1 describes a convergent world, with the same global population that peaks in mid century and declines thereafter, with rapid changes in economic structures. B2 describes a world 'in which emphasis is on local solutions to economic, social, and environmental sustainability'. It features moderate population growth, intermediate levels of economic development, and less rapid and more diverse technological change than the A1B scenario.
3   The Conference of the Parties (COP) is the supreme body of the Convention also serves as the Meeting of the Parties (MOP) for the Protocol. CMP1 is the first meeting of the Conference of the Parties acting as the Meeting of the Parties of the Kyoto Protocol.

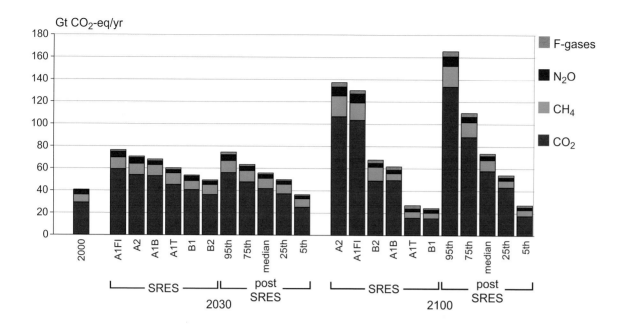

Figure TS.5: *Global GHG emissions for 2000 and projected baseline emissions for 2030 and 2100 from IPCC SRES and the post-SRES literature. The figure provides the emissions from the six illustrative SRES scenarios. It also provides the frequency distribution of the emissions in the post-SRES scenarios (5th, 25th, median, 75th, 95th percentile), as covered in Chapter 3. F-gases cover HFCs, PFCs and SF_6 [Figure 1.7].*

their overall GHG emissions to at least 5% below 1990 levels. The entry into force of the Kyoto Protocol marks a first, though modest, step towards achieving the ultimate objective of the UFCCC to avoid dangerous anthropogenic interference with the climate system. Its full implementation by all the Protocol signatories, however, would still be far from reversing overall global GHG-emission trends. The strengths of the Kyoto Protocol are its provision for market mechanisms such as GHG-emission trading and its institutional architecture. One weakness of the Protocol, however, is its non-ratification by some significant GHG emitters. A new Ad Hoc Working Group (AWG) on the Commitments of Annex I Countries under the Kyoto Protocol beyond 2012 was set up at CMP1, and agreed at CMP2 that the second review of Article 9 of the Kyoto Protocol will take place in 2008.

There are also voluntary international initiatives to develop and implement new technologies to reduce GHG emissions. These include: the Carbon Sequestration Leadership Forum (promoting CO_2 capture and storage); the Hydrogen partnership; the Methane to Markets Partnership, and the Asia-Pacific Partnership for Clean Development and Climate (2005), which includes Australia, USA, Japan, China, India and South-Korea. Climate change has also become an important growing concern of the G8 since its meeting in Gleneagles, Scotland in 2005. At that meeting, a plan of action was developed which tasked the International Energy Agency, the World Bank and the Renewable Energy and Energy Efficiency Partnership with supporting their efforts. Additionally, Gleneagles created a Clean Energy, Climate Change and Sustainable Development Dialogue process for the largest emitters. The International Energy Agency (IEA) and the World Bank were charged with advising that dialogue process [1.4].

Article 2 of the Convention and mitigation

Article 2 of the UNFCCC requires that dangerous interference with the climate system be prevented and hence the stabilization of atmospheric GHG concentrations at levels and within a time frame that would achieve this objective. The criteria in Article 2 that specify (risks of) dangerous anthropogenic climate change include: food security, protection of ecosystems and sustainable economic development. Implementing Article 2 implies dealing with a number of complex issues:

What level of climate change is dangerous?

Decisions made in relation to Article 2 would determine the level of climate change that is set as the goal for policy, and have fundamental implications for emission-reduction pathways as well as the scale of adaptation required. Choosing a stabilization level implies balancing the risks of climate change (from gradual change and extreme events, and irreversible change of the climate, including those to food security, ecosystems and sustainable development) against the risks of response measures that may threaten economic sustainability. Although any judgment on 'dangerous interference' is necessarily a social and political one, depending on the level of risk deemed acceptable, large emission reductions are unavoidable if stabilization is to be achieved. The lower the stabilization level, the earlier these large reductions have to be realized *(high agreement, much evidence)* [1.2].

Sustainable development:

Projected anthropogenic climate change appears likely to adversely affect sustainable development, with the effects tending to increase with higher GHG concentrations (WGII AR4, Chapter 19). Properly designed climate change responses can be an integral part of sustainable development and the two can be mutually reinforcing. Mitigation of climate change can conserve or enhance natural capital (ecosystems, the environment as sources and sinks for economic activities) and prevent or avoid damage to human systems and, thereby contribute to the overall productivity of capital needed for socio-economic development, including mitigative and adaptive capacity. In turn, sustainable development paths can reduce vulnerability to climate change and reduce GHG emissions *(medium agreement, much evidence)* [1.2].

Distributional issues:

Climate change is subject to a very asymmetric distribution of present emissions and future impacts and vulnerabilities. Equity can be elaborated in terms of distributing the costs of mitigation or adaptation, distributing future emission rights and ensuring institutional and procedural fairness. Because the industrialized nations are the source of most past and current GHG emissions and have the technical and financial capability to act, the Convention places the heaviest burden for the first steps in mitigating climate change on them. This is enshrined in the principle of 'common but differentiated responsibilities' *(high agreement, much evidence)* [1.2].

Timing:

Due to the inertia of both climate and socio-economic systems, the benefits of mitigation actions initiated now may result in significant avoided climate change only after several decades. This means that mitigation actions need to start in the short term in order to have medium- and longer-term benefits and to avoid lock-in of carbon-intensive technologies *(high agreement, much evidence)* [1.2].

Mitigation and adaptation:

Adaptation and mitigation are two types of policy response to climate change, which can be complementary, substitutable or independent of each other. Irrespective of the scale of mitigation measures, adaptation measures will be required anyway, due to the inertia in the climate system. Over the next 20 years or so, even the most aggressive climate policy can do little to avoid warming already 'loaded' into the climate system. The benefits of avoided climate change will only accrue beyond that time. Over longer time frames, beyond the next few decades, mitigation investments have a greater potential to avoid climate change damage and this potential is larger than the adaptation options that can currently be envisaged *(medium agreement, medium evidence)* [1.2].

Risk and uncertainty:

An important aspect in the implementation of Article 2 is the uncertainty involved in assessing the risk and severity of climate change impacts and evaluating the level of mitigation action (and its costs) needed to reduce the risk. Given this uncertainty, decision-making on the implementation of Article 2 would benefit from the incorporation of risk-management principles. A precautionary and anticipatory risk-management approach would incorporate adaptation and preventive mitigation measures based on the costs and benefits of avoided climate change damage, taking into account the (small) chance of worst-case outcomes *(medium agreement, medium evidence)* [1.2].

2 Framing issues

Climate change mitigation and sustainable development

There is a two-way relationship between climate change and development. On the one hand vulnerability to climate change is framed and strongly influenced by development patterns and income levels. Decisions about technology, investment, trade, poverty, community rights, social policies or governance, which may seem unrelated to climate policy, may have profound impacts on emissions, the extent of mitigation required, and the cost and benefits that result [2.2.3].

On the other hand, climate change itself, and adaptation and mitigation policies could have significant positive impacts on development in the sense that development can be made more sustainable. This leads to the notion that climate change policies can be considered 1) in their own right ('climate first'); or 2) as an integral element of sustainable-development policies ('development first'). Framing the debate as a sustainable development problem rather than a solely environmental one may better address the needs of countries, while acknowledging that the driving forces for emissions are linked to the underlying development path [2.2.3].

Development paths evolve as a result of economic and social transactions, which are influenced by government policies, private sector initiatives and by the preferences and choices of consumers. These include a broad number of policies related to nature conservation, legal frameworks, property rights, rule of law, taxes and regulation, production, security and safety of food, consumption patterns, human and institutional capacity building efforts, R&D, financial schemes, technology transfer, energy efficiency and energy options. These policies do not usually emerge and become implemented as part of a general development-policy package, but are normally targeted towards more specific policy goals like air-pollution standards, food security and health issues, GHG-emission reduction, income generation by specific groups, or development of industries for green technologies. However, significant impacts can arise from such policies on sustainability and greenhouse mitigation and the outcomes of adaptation. The strong relationship between mitigation of climate change and development applies in both developed and developing countries. Chapter 12 and to some extent Chapters 4–11 address these issues in more detail [2.2.5; 2.2.7].

Emerging literature has identified methodological approaches to identify, characterize and analyze the interactions between sustainable development and climate change responses. Several authors have suggested that sustainable development can be addressed as a framework for jointly assessing social, human, environmental and economic dimensions. One way to address these dimensions is to use a number of economic, environmental, human and social indicators to assess the impacts of policies on sustainable development, including both quantitative and qualitative measurement standards *(high agreement, limited evidence)* [2.2.4].

Decision-making, risk and uncertainty

Mitigation policies are developed in response to concerns about the risk of climate change impacts. However, deciding on a proper reaction to these concerns means dealing with uncertainties. Risk refers to cases for which the probability of outcomes and its consequences can be ascertained through well-established theories with reliable, complete data, while uncertainty refers to situations in which the appropriate data may be fragmentary or unavailable. Causes of uncertainty include insufficient or contradictory evidence as well as human behaviour. The human dimensions of uncertainty, especially coordination and strategic behaviour issues, constitute a major part of the uncertainties related to climate change mitigation *(high agreement, much evidence)* [2.3.3; 2.3.4].

Decision-support analysis can assist decision makers, especially if there is no optimum policy that everybody can agree on. For this, a number of analytical approaches are available, each with their own strengths and weaknesses, which help to keep the information content of the climate change problem within the cognitive limits of the large number of decision makers and support a more informed and effective dialogue among the many parties involved. There are, however, significant problems in identifying, measuring and quantifying the many variables that are important inputs to any decision-support analysis framework – particularly impacts on natural systems and human health that do not have a market value, and for which all approaches are simplifications of the reality *(high agreement, much evidence)* [2.3.7].

When many decision makers with different value systems are involved in a decision, it is helpful to be as clear as possible about the value judgments underpinning any analytic outcomes they are expected to draw on. This can be particularly difficult and subtle where analysis aims to illuminate choices associated with high levels of uncertainty and risk *(medium agreement, medium evidence)* [2.3.2; 2.3.7].

Integrated assessments can inform decision makers of the relationship between geophysical climate change, climate-impact predictions, adaptation potentials and the costs of emission reductions and the benefits of avoided climate change damage. These assessments have frameworks to deal with incomplete or imprecise data.

To communicate the uncertainties involved, this report uses the terms in Table TS.1 to describe the relative levels of expert agreement on the respective statements in the light of the underlying literature (in rows) and the number and quality of independent sources qualifying under IPCC rules[4] upon which a finding is based (in columns). The other approaches of 'likelihood' and 'confidence' are not used in this report as human choices are concerned, and none of the other approaches used provides sufficient characterization of the uncertainties involved in mitigation *(high agreement, much evidence)* [2.4].

Table TS.1: *Qualitative definition of uncertainty [Table 2.2].*

High agreement, limited evidence	High agreement, medium evidence	High agreement, much evidence
Medium agreement, limited evidence	Medium agreement, medium evidence	Medium agreement, much evidence
Low agreement, limited evidence	Low agreement, medium evidence	Low agreement, much evidence

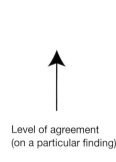

Level of agreement (on a particular finding)

Amount of evidence (number and quality of independent sources)

Note: This table is based on two dimensions of uncertainty: the amount of evidence[5] and the level of agreement. The amount of evidence available about a given technology is assessed by examining the number and quality of independent sources of information. The level of agreement expresses the subjective probability of the results being in a certain realm.

---

4   IPCC rules permit the use of both peer-reviewed literature and non-peer-reviewed literature that the authors deem to be of equivalent quality.
5   'Evidence' in this report is defined as: Information or signs indicating whether a belief or proposition is true or valid. See Glossary.

Costs, benefits, concepts including private and social cost perspectives and relationships with other decision-making frameworks

There are different ways of defining the potential for mitigation and it is therefore important to specify what potential is meant. 'Potential' is used to express the degree of GHG reduction that can be achieved by a mitigation option with a given cost per tonne of carbon avoided over a given period, compared with a baseline or reference case. The measure is usually expressed as million tonnes carbon- or CO_2-equivalent emissions avoided compared with baseline emissions [2.4.3].

Market potential is the mitigation potential based on private costs and private discount rates[6], which might be expected to occur under forecast market conditions, including policies and measures currently in place, noting that barriers limit actual uptake.

Economic potential is the amount of GHG mitigation, which takes into account social costs and benefits and social discount rates[7] assuming that market efficiency is improved by policies and measures and barriers are removed. However, current bottom-up and top-down studies of economic potential have limitations in considering life-style choices and in including all externalities such as local air pollution.

Technical potential is the amount by which it is possible to reduce GHG emissions by implementing a technology or practice that has already been demonstrated. There is no specific reference to costs here, only to 'practical constraints', although implicit economic considerations are taken into account in some cases. *(high agreement, much evidence)* [2.4.3].

Studies of market potential can be used to inform policy makers about mitigation potential with existing policies and barriers, while studies of economic potentials show what might be achieved if appropriate new and additional policies were put into place to remove barriers and include social costs and benefits. The economic potential is therefore generally greater than the market potential.

Mitigation potential is estimated using different types of approaches. There are two broad classes – "bottom-up" and "top-down" approaches, which primarily have been used to assess the economic potential:
- **Bottom-up studies** are based on assessment of mitigation options, emphasizing specific technologies and regulations. They are typically sectoral studies taking the macro-economy as unchanged. Sector estimates have been aggregated, as in the TAR, to provide an estimate of global mitigation potential for this assessment.
- **Top-down studies** assess the economy-wide potential of mitigation options. They use globally consistent frameworks and aggregated information about mitigation options and capture macro-economic and market feedbacks.

Bottom-up studies in particular are useful for the assessment of specific policy options at sectoral level, e.g. options for improving energy efficiency, while top-down studies are useful for assessing cross-sectoral and economy-wide climate change policies, such as carbon taxes and stabilization policies. Bottom-up and top-down models have become more similar since the TAR as top-down models have incorporated more technological mitigation options (see Chapter 11) and bottom-up models have incorporated more macroeconomic and market feedbacks as well as adopting barrier analysis into their model structures.

Mitigation and adaptation relationships; capacities and policies

Climate change mitigation and adaptation have some common elements, they may be complementary, substitutable, independent or competitive in dealing with climate change, and also have very different characteristics and timescales [2.5].

Both adaptation and mitigation make demands on the capacity of societies, which are intimately connected to social and economic development. The responses to climate change depend on exposure to climate risk, society's natural and man-made capital assets, human capital and institutions as well as income. Together these will define a society's adaptive and mitigative capacities. Policies that support development and those that enhance its adaptive and mitigative capacities may, but need not, have much in common. Policies may be chosen to have synergetic impacts on the natural system and the socio-economic system but difficult trade-offs may sometimes have to be made. Key factors that determine the capacity of individual stakeholders and societies to implement climate change mitigation and adaptation include: access to resources; markets; finance; information, and a number of governance issues *(medium agreement, limited evidence)* [2.5.2].

Distributional and equity aspects

Decisions on climate change have large implications for local, national, inter-regional and intergenerational equity, and the application of different equity approaches has major implications for policy recommendations as well as for the distribution of the costs and benefits of climate policies [2.6].

Different approaches to social justice can be applied to the evaluation of the equity consequences of climate change policies. As the IPCC Third Assessment Report (TAR) suggested, given strong subjective preferences for certain equity principles among different stakeholders, it is more effective to look for practical approaches that combine equity principles. Equity approaches vary from traditional economic approaches to rights-

---

6 Private costs and discount rates reflect the perspective of private consumers and companies; see Glossary for a fuller description.
7 Social costs and discount rates reflect the perspective of society. Social discount rates are lower than those used by private investors; see Glossary for a fuller description.

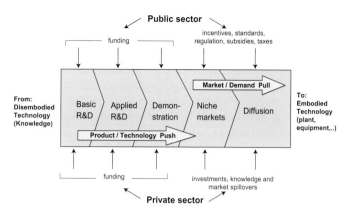

Figure TS.6: *The technology development cycle and its main driving forces [Figure 2.3].*

Note: important overlaps and feedbacks exist between the stylized technology life-cycle phases illustrated here. The figure therefore does not suggest a 'linear' model of innovation. It is important to recognize the need for finer terminological distinction of 'technology', particularly when discussing different mitigation and adaptation options.

based approaches. An economic approach would be to assess welfare losses and gains to different groups and the society at large, while a rights-based approach would focus on rights, for example, in terms of emissions per capita or GDP allowed for all countries, irrespective of the costs of mitigation or the mitigative capacity. The literature also includes a capability approach that puts the emphasis on opportunities and freedom, which in terms of climate policy can be interpreted as the capacity to mitigate or to adapt or to avoid being vulnerable to climate change *(medium agreement, medium evidence)* [2.6.3].

Technology research, development, deployment, diffusion and transfer

The pace and cost of any response to climate change concerns will also depend critically on the cost, performance, and availability of technologies that can lower emissions in the future, although other factors such as growth in wealth and population are also highly important [2.7].

Technology simultaneously influences the size of the climate change problem and the cost of its solution. Technology is the broad set of competences and tools covering know-how, experience and equipment, used by humans to produce services and transform resources. The principal role of technology in mitigating GHG emissions is in controlling the social cost of limiting the emissions. Many studies show the significant economic value of the improvements in emission-mitigating technologies that are currently in use and the development and deployment of advanced emission-mitigation technologies *(high agreement, much evidence)* [2.7.1].

A broad portfolio of technologies can be expected to play a role in meeting the goal of the UNFCCC and managing the risk of climate change, because of the need for large emission reductions, the large variation in national circumstances and

the uncertainty about the performance of individual options. Climate policies are not the only determinant of technological change. However, a review of future scenarios (see Chapter 3) indicates that the overall rate of change of technologies in the absence of climate policies might be as large as, if not larger than, the influence of the climate policies themselves *(high agreement, much evidence)* [2.7.1].

Technological change is particularly important over the long-term time scales characteristic of climate change. Decade- or century-long time scales are typical for the lags involved between technological innovation and widespread diffusion and of the capital turnover rates characteristic of long-lived energy capital stock and infrastructures.

Many approaches are used to split up the process of technological change into distinct phases. One is to consider technological change as roughly a two-part process: 1) conceiving, creating and developing new technologies or enhancing existing technologies – advancing the 'technological frontier'; 2) the diffusion or deployment of these technologies. Our understanding of technology and its role in addressing climate change is improving continuously. The processes by which technologies are created, developed, deployed and eventually replaced, however, are complex (see Figure TS.6) and no simple descriptions of these processes exist. Technology development and deployment is characterized by two public goods problems. First, the level of R&D is sub-optimal because private decision-makers cannot capture the full value of private investments. Second, there is a classical environmental externality problem, in that private markets do not reflect the full costs of climate change *(high agreement, much evidence)* [2.7.2].

Three important sources of technological change are R&D, learning and spill-overs.
- R&D encompasses a broad set of activities in which firms, governments or other entities expend resources specifically to gain new knowledge that can be embodied in new or improved technology.
- Learning is the aggregate outcome of complex underlying sources of technology advance that frequently include important contributions from R&D, spill-overs and economies of scale.
- Spill-overs refer to the transfer of the knowledge or the economic benefits of innovation from one individual, firm, industry or other entity, or from one technology to another.

On the whole, empirical and theoretical evidence strongly suggest that all three of these play important roles in technological advance, and there is no compelling reason to believe that one is broadly more important than the others. As spill-overs from other sectors have had an enormous effect on innovation in the energy sector, a robust and broad technological base may be as important for the development of technologies pertinent to climate change as explicit climate change or energy research. A broad portfolio of research is needed, because it is not possible to identify winners and losers ex-ante. The sources of technological change are frequently subsumed under the general drivers 'supply push' (e.g., via R&D) or 'demand pull' (e.g., via learning). These are, however, not simply substitutes, but may have highly complementary interactions (*high agreement, much evidence*) [2.7.2].

On technology transfer, the main findings of the IPCC Special Report on Methodological and Technological Issues of Technology Transfer (2000) remain valid: that a suitable enabling environment needs to be created in host and recipient countries (*high agreement, much evidence*) [2.7.3].

Regional Dimensions

Climate change studies have used various different regional definitions, depending on the character of the problem considered and differences in methodological approaches. The multitude of possible regional representations hinders the comparability and transfer of information between the various types of studies done for specific regions and scales. This report largely has chosen a pragmatic ways of analysing regional information and presenting findings [2.8].

3 Issues related to mitigation in the long-term context

Baseline scenario drivers

Population projections are now generally lower than in the IPCC Special Report on Emission Scenarios (SRES), based on new data indicating that birth rates in many parts of the world have fallen sharply. So far, these new population projections have not been implemented in many of the new emissions scenarios in the literature. The studies that have incorporated them result in more or less the same overall emissions levels, due to changes in other driving factors such as economic growth (*high agreement, much evidence*) [3.2.1].

Economic growth perspectives have not changed much. There is a considerable overlap in the GDP numbers published, with a slight downwards shift of the median of the new scenarios by about 7% compared with the median in the pre-SRES scenario literature. The data suggest no appreciable change in the distribution of GDP projections. Economic growth projections for Africa, Latin America and the Middle East are lower than in the SRES scenarios (*high agreement, much evidence*) [3.2.1].

Baseline scenario emissions (all gases and sectors)
The resulting span of energy-related and industrial CO_2 emissions in 2100 across baseline scenarios in the post-SRES

literature is very large, ranging from 17 to around 135 $GtCO_2$-eq (4.6-36.8 GtC)[8], about the same as the SRES range (Figure TS.7). Different reasons may contribute to the fact that emissions have not declined despite somewhat lower projections for population and GDP. All other factors being equal, lower population projections would result in lower emissions. In the scenarios that use lower projections, however, changes in other drivers of emissions have partly offset the consequences of lower populations. Few studies incorporated lower population projections, but where they did, they showed that lower population is offset by higher rates of economic growth, and/or a shift toward a more carbon-intensive energy system, such as a shift to coal because of increasing oil and gas prices. The majority of scenarios indicate an increase in emissions during most of the century. However, there are some baseline (reference) scenarios both in the new and older literature where emissions peak and then decline (*high agreement, much evidence*) [3.2.2].

Baseline land-related GHG emissions are projected to increase with growing cropland requirements, but at a slower rate than energy-related emissions. As far as CO_2 emissions from land-use change (mostly deforestation) are concerned, post-SRES scenarios show a similar trend to SRES scenarios: a slow decline, possibly leading to zero net emissions by the end of the century.

Emissions of non-CO_2 GHGs as a group (mostly from agriculture) are projected to increase, but somewhat less rapidly than CO_2 emissions, because the most important sources of CH_4 and N_2O are agricultural activities, and agriculture is growing less than energy use. Emission projections from the recent literature are similar to SRES. Recent non-CO_2 GHG emission baseline scenarios suggest that agricultural CH_4 and N_2O emissions will increase until the end of this century, potentially doubling in some baselines. While the emissions of some fluorinated compounds are projected to decrease, many are expected to grow substantially because of the rapid growth rate of some emitting industries and the replacement of ODS with HFCs (*high agreement, medium evidence*) [3.2.2].

Noticeable changes have occurred in projections of the emissions of the aerosol precursors SO_2 and NO_x since SRES. Recent literature shows a slower short-term growth of these emissions than SRES. As a consequence also the long-term ranges of both emissions sources are lower in the recent literature. Recent scenarios project sulphur emissions to peak earlier and at lower levels than in SRES. A small number of new scenarios have begun to explore emission pathways for black and organic carbon (*high agreement, medium evidence*) [3.2.2].

In general, the comparison of SRES and new scenarios in the literature shows that the ranges of the main driving forces and emissions have not changed very much.

---

8   This is the 5th to 95th percentile of the full distribution

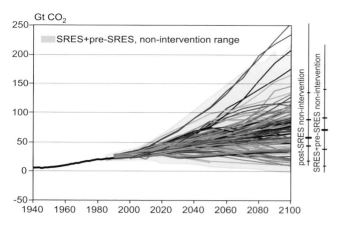

Figure TS.7: *Comparison of the SRES and pre-SRES energy-related and industrial CO_2 emission scenarios in the literature with the post-SRES scenarios [Figure 3.8].*

Note: Two vertical bars on the right extend from the minimum to maximum of the distribution of scenarios and indicate the 5th, 25th, 50th, 75th and the 95th percentiles of the distributions by 2100.

GDP metrics

For long-term scenarios, economic growth is usually reported in the form of growth in GDP or gross national product (GNP). To get a meaningful comparison of the real size of economic activities over time and between countries, GDP is reported in constant prices taken from a base year.

The choice of the conversion factor, Market Exchange Rate (MER) or Purchasing Power Parity (PPP), depends on the type of analysis being undertaken. However, when it comes to calculating emissions (or other physical measures like energy), the choice between MER and PPP-based representations of GDP should not matter, since emission intensity will change (in a compensating manner) when the GDP numbers change. Thus, if a consistent set of metrics is employed, the choice of metric should not appreciably affect the final emission level. A number of new studies in the literature concur that the actual choice of exchange rates does not itself have an appreciable effect on long-term emission projections. In the case of SRES, the emissions trajectories are the same whether economic activities in the four scenario families are measured in MER or PPP.

There are studies that find some differences in emission levels between PPP and MER-based estimates. These results depend critically on convergence assumptions, among other things. In some of the short-term scenarios (with a horizon to 2030) a bottom-up approach is taken where assumptions about productivity growth and investment/saving decisions are the main drivers of growth in the models. In long-term scenarios, a top-down approach is more commonly used where actual growth rates are more directly prescribed on the basis of convergence or other assumptions about long-term growth potentials. Different results can also be due to inconsistencies in adjusting the metrics of energy efficiency improvement when moving from MER to PPP-based calculations.

Evidence from the limited number of new PPP-based studies indicates that the choice of metric for GDP (MER or PPP) does not appreciably affect the projected emissions, when the metrics are used consistently. The differences, if any, are small compared with the uncertainties caused by assumptions on other parameters, for example, technological change. The debate clearly shows, however, the need for modellers to be more transparent in explaining conversion factors as well as taking care in making assumptions on exogenous factors (*high agreement, much evidence*) [3.2.1].

Stabilization scenarios

A commonly used target in the literature is stabilization of CO_2 concentrations in the atmosphere. If more than one GHG is studied, a useful alternative is to formulate a GHG-concentration target in terms of CO_2-equivalent concentration or radiative forcing, thereby weighting the concentrations of the different gases by their radiative properties. Another option is to stabilize or target global mean temperature. The advantage of radiative-forcing targets over temperature targets is that the calculation of radiative forcing does not depend on climate sensitivity. The disadvantage is that a wide range of temperature impacts is possible for each radiative-forcing level. Temperature targets, on the other hand, have the important advantage of being more directly linked to climate change impacts. Another approach is to calculate the risks or the probability of exceeding particular values of global annual mean temperature rise since pre-industrial times for specific stabilization or radiative-forcing targets.

There is a clear and strong correlation between the CO_2-equivalent concentrations (or radiative forcing) and the CO_2-only concentrations by 2100 in the published studies, because CO_2 is the most important contributor to radiative forcing. Based on this relationship, to facilitate scenario comparison and assessment, stabilization scenarios (both multi-gas and CO_2-only studies) have been grouped into different categories that vary in the stringency of the targets (Table TS.2).

Essentially, any specific concentration or radiative-forcing target requires emissions to fall to very low levels as the removal processes of the ocean and terrestrial systems saturate. Higher stabilization targets do push back the timing of this ultimate result beyond 2100. However, to reach a given stabilization target, emissions must ultimately be reduced well below current levels. For achievement of the stabilization categories I and II, negative net emissions are required towards the end of the century in many scenarios considered (Figure TS. 8) (*high agreement, much evidence*) [3.3.5].

The timing of emission reductions depends on the stringency of the stabilization target. Stringent targets require an earlier peak in CO_2 emissions (see Figure TS.8). In the majority of the scenarios in the most stringent stabilization category (I), emissions are required to decline before 2015 and be further reduced to less

Table TS.2: *Classification of recent (Post-Third Assessment Report) stabilization scenarios according to different stabilization targets and alternative stabilization metrics [Table 3.5].*

Category	Additional radiative forcing (W/m²)	CO₂ concentration (ppm)	CO₂-eq concentration (ppm)	Global mean temperature increase above pre-industrial at equilibrium, using "best estimate" climate sensitivity[a], [b] (°C)	Peaking year for CO₂ emissions[c]	Change in global CO₂ emissions in 2050 (% of 2000 emissions)[c]	No. of assessed scenarios
I	2.5-3.0	350-400	445-490	2.0-2.4	2000 - 2015	-85 to -50	6
II	3.0-3.5	400-440	490-535	2.4-2.8	2000 - 2020	-60 to -30	18
III	3.5-4.0	440-485	535-590	2.8-3.2	2010 - 2030	-30 to +5	21
IV	4.0-5.0	485-570	590-710	3.2-4.0	2020 - 2060	+10 to +60	118
V	5.0-6.0	570-660	710-855	4.0-4.9	2050 - 2080	+25 to +85	9
VI	6.0-7.5	660-790	855-1130	4.9-6.1	2060 - 2090	+90 to +140	5
						Total	177

Notes:

[a] Note that global mean temperature at equilibrium is different from expected global mean temperatures in 2100 due to the inertia of the climate system.

[b] The simple relationships $T_{eq} = T_{2 \times CO2} \times \ln([CO_2]/278)/\ln(2)$ and $\Delta Q = 5.35 \times \ln ([CO_2]/278)$ are used. Non-linearities in the feedbacks (including e.g., ice cover and carbon cycle) may cause time dependence of the effective climate sensitivity, as well as leading to larger uncertainties for greater warming levels. The best-estimate climate sensitivity (3 °C) refers to the most likely value, that is, the mode of the climate sensitivity PDF consistent with the WGI assessment of climate sensitivity and drawn from additional consideration of Box 10.2, Figure 2, in the WGI AR4.

[c] Ranges correspond to the 15th to 85th percentile of the Post-Third Assessment Report (TAR) scenario distribution. CO₂ emissions are shown, so multi-gas scenarios can be compared with CO₂-only scenarios.

Note that the classification needs to be used with care. Each category includes a range of studies going from the upper to the lower boundary. The classification of studies was done on the basis of the reported targets (thus including modelling uncertainties). In addition, the relationship that was used to relate different stabilization metrics is also subject to uncertainty (see Figure 3.16).

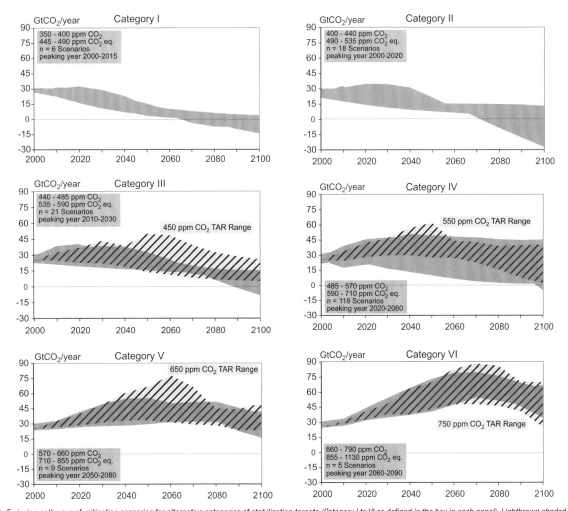

Figure TS.8: *Emission pathways of mitigation scenarios for alternative categories of stabilization targets (Category I to VI as defined in the box in each panel). Lightbrown shaded areas give the CO₂ emissions for the recent mitigation scenarios developed post-TAR. Green shaded and hatched areas depict the range of more than 80 TAR stabilization scenarios (Morita et al., 2001). Category I and II scenarios explore stabilization targets below the lowest of TAR. Base year emissions may differ between models due to differences in sector and industry coverage. To reach the lower stabilization levels some scenarios deploy removal of CO₂ from the atmosphere (negative emissions) using technologies such as biomass energy production utilizing carbon capture and storage [Figure 3.17].*

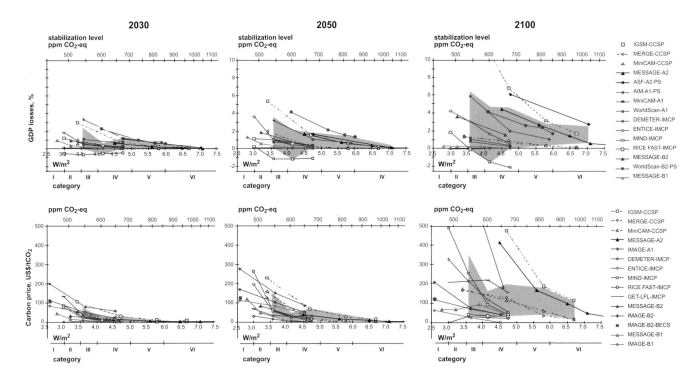

Figure TS.9: *Relationship between the cost of mitigation and long-term stabilization targets (radiative forcing compared with pre-industrial level, W/m² and CO₂-eq concentrations) [Figure 3.25].*

Notes: Panels give costs measured as percentage loss of GDP (top), and carbon price (bottom). Left-hand panels for 2030, middle panels for 2050 and right-hand panels for 2100. Individual coloured lines denote selected studies with representative cost dynamics from very high to very low cost estimates. Scenarios from models sharing similar baseline assumptions are shown in the same colour. The grey shaded range represents the 80th percentile of TAR and post-TAR scenarios. Solid lines show representative scenarios considering all radiatively active gases. Dashed lines represent multi-gas scenarios where the target is defined by the six Kyoto gases (other multi-gas scenarios consider all radiatively active gases). CO_2 stabilization scenarios are added based on the relationship between CO_2 concentration and the radiative-forcing targets given in Figure 3.16.

than 50% of today's emissions by 2050. For category III, global emissions in the scenarios generally peak around 2010–2030, followed by a return to 2000 levels on average around 2040. For category IV, the median emissions peak around 2040 (Figure TS.9) *(high agreement, much evidence)*.

The costs of stabilization depend on the stabilization target and level, the baseline and the portfolio of technologies considered, as well as the rate of technological change. Global mitigation costs[9] rise with lower stabilization levels and with higher baseline emissions. Costs in 2050 for multi-gas stabilization at 650 ppm CO_2-eq (cat IV) are between a 2% loss or a one procent increase[10] of GDP in 2050. For 550 ppm CO_2-eq (cat III) these costs are a range of a very small increase to 4% loss of GDP[11]. For stabilization levels between 445 and 535 ppm CO_2-eq. costs are lower than 5.5% loss of GDP, but the number of studies is limited and they generally use low baselines.

A multi-gas approach and inclusion of carbon sinks generally reduces costs substantially compared with CO_2 emission abatement only. Global average costs of stabilization are uncertain, because assumptions on baselines and mitigation options in models vary a lot and have a major impact. For some countries, sectors or shorter time periods, costs could vary considerably from the global and long-term average *(high agreement, much evidence)* [3.3.5].

Recent stabilization studies have found that land-use mitigation options (both non-CO_2 and CO_2) provide cost-effective abatement flexibility in achieving 2100 stabilization targets. In some scenarios, increased commercial biomass energy (solid and liquid fuel) is significant in stabilization, providing 5–30% of cumulative abatement and potentially 10–25% of total primary energy over the century, especially as a net negative emissions strategy that combines biomass energy with CO_2 capture and storage.

---

9   Studies on mitigation portfolios and macro-economic costs assessed in this report are based on a global least-cost approach, with optimal mitigation portfolios and without allocation of emission allowances to regions. If regions are excluded or non-optimal portfolios are chosen, global costs will go up. The variation in mitigation portfolios and their costs for a given stabilization level is caused by different assumptions, such as on baselines (lower baselines give lower costs), GHGs and mitigation options considered (more gases and mitigation options give lower costs), cost curves for mitigation options and rate of technological change.
10 The median and the 10th–90th percentile range of the analysed data are given.
11 Loss of GDP of 4% in 2050 is equivalent to a reduction of the annual GDP growth rate of about 0.1 percentage points.

The baseline choice is crucial in determining the nature and cost of stabilization. This influence is due mainly to different assumptions about technological change in the baseline scenarios.

The role of technologies

Virtually all scenarios assume that technological and structural changes occur during this century, leading to relative reduction of emissions compared with the hypothetical case of attempting to 'keep' the emission intensities of GDP and economic structures the same as today (see Chapter 2, Section 2.9.1.3].

Baseline scenarios usually assume significant technological change and diffusion of new and advanced technologies. In mitigation scenarios there is additional technological change 'induced' through various policies and measures. Long-term stabilization scenarios highlight the importance of technology improvements, advanced technologies, learning by doing and endogenous technology change both for achieving the stabilization targets and for cost reduction. While the technology improvement and use of advanced technologies have been introduced in scenarios largely exogenously in most of the literature, new literature covers learning-by-doing and endogenous technological change. These newer scenarios show higher benefits of early action, as models assume that early

deployment of technologies leads to benefits of learning and cost reductions *(high agreement, much evidence)* [3.4].

The different scenario categories also reflect different contributions of mitigation measures. However, all stabilization scenarios concur that 60–80% of all reductions would come from the energy and industry sectors. Non-CO_2 gases and land-use would contribute the remaining 30–40% (see for illustrative examples Figure TS. 10). New studies exploring more stringent stabilization levels indicate that a wider portfolio of technologies is needed. Those could include nuclear, carbon capture and storage (CCS) and bioenergy with carbon capture and geologic storage (BECS) *(high agreement, much evidence)* [3.3.5].

Mitigation and adaptation in the light of climate change impacts and decision-making under uncertainties

Concern about key vulnerabilities and notions of what is dangerous climate change will affect decisions about long-term climate change objectives and hence mitigation pathways. Key vulnerabilities traverse different human and natural systems and exist at different levels of temperature change. More stringent stabilization scenarios achieve more stringent climate targets and lower the risk of triggering key vulnerabilities related to climate change. Using the 'best estimate' of climate sensitivity[12],

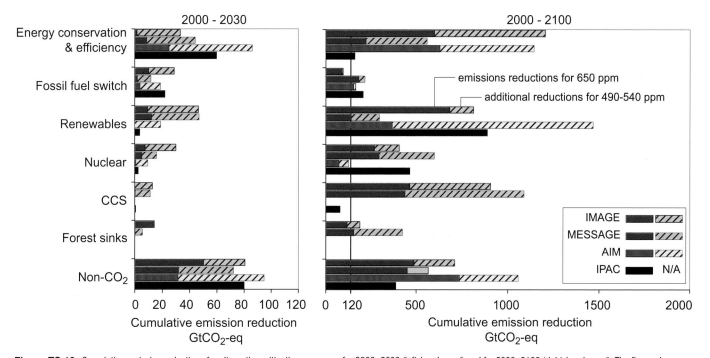

Figure TS.10: *Cumulative emission reductions for alternative mitigation measures for 2000–2030 (left-hand panel) and for 2000–2100 (right-hand panel). The figure shows illustrative scenarios from four models (AIM, IMAGE, IPAC and MESSAGE) aiming at the stabilization at low (490–540 ppm CO_2-eq) and intermediate levels (650 ppm CO_2-eq) respectively. Dark bars denote reductions for a target of 650 ppm CO_2-eq and light bars the additional reductions to achieve 490–540 ppm CO_2-eq. Note that some models do not consider mitigation through forest sink enhancement (AIM and IPAC) or CCS (AIM) and that the share of low-carbon energy options in total energy supply is also determined by inclusion of these options in the baseline. CCS includes carbon capture and storage from biomass. Forest sinks include reducing emissions from deforestation [Figure 3.23].*

---

12   The equilibrium climate sensitivity is a measure of the climate system response to sustained radiative forcing. It is not a projection but is defined as the global average surface warming following a doubling of carbon dioxide concentrations [AR4 WGI SPM].

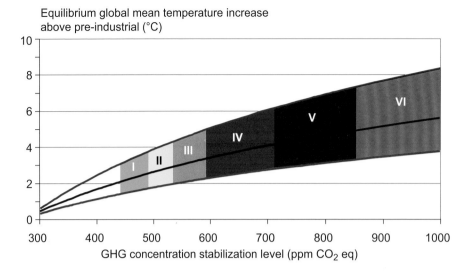

Figure TS.11: *Stabilization scenario categories as reported in Figure TS.8 (coloured bands) and their relationship to equilibrium global mean temperature change above pre-industrial temperatures [Figure 3.38].*

Notes: Middle (black) line – 'best estimate' climate sensitivity of 3°C; upper (red) line – upper bound of likely range of climate sensitivity of 4.5°C; lower (blue) line – lower bound of likely range of climate sensitivity of 2°C. Coloured shading shows the concentration bands for stabilization of GHGs in the atmosphere corresponding to the stabilization scenario categories I to VI as indicated in Table TS.2.

the most stringent scenarios (stabilizing at 445–490 ppm CO_2-eq) could limit global mean temperature increases to 2-2.4°C above pre-industrial, at equilibrium, requiring emissions to peak within 10 years and to be around 50% of current levels by 2050. Scenarios stabilizing at 535-590 ppm CO_2-eq could limit the increase to 2.8-3.2°C above pre-industrial and those at 590-710 CO_2-eq to 3.2-4°C, requiring emissions to peak within the next 25 and 55 years respectively (see Figure TS.11) [3.3, 3.5].

The risk of higher climate sensitivities increases the probability of exceeding any threshold for specific key vulnerabilities. Emission scenarios that lead to temporary overshooting of concentration ceilings can lead to higher rates of climate change over the century and increase the probability of exceeding key vulnerability thresholds. Results from studies exploring the effect of carbon cycle and climate feedbacks indicate that the above-mentioned concentration levels and the associated warming of a given emissions scenario might be an underestimate. With higher climate sensitivity, earlier and more stringent mitigation measures are necessary to reach the same concentration level.

Decision-making about the appropriate level of mitigation is an iterative risk-management process considering investment in mitigation and adaptation, co-benefits of undertaking climate change decisions and the damages due to climate change. It is intertwined with decisions on sustainability, equity and development pathways. Cost-benefit analysis, as one of the available tools, tries to quantify climate change damage in monetary terms (as social cost of carbon (SCC) or time-discounted damage). Due to large uncertainties and difficulties in quantifying non-market damage, it is still difficult to estimate SCC with confidence. Results depend on a large number of

normative and empirical assumptions that are not known with any certainty. Limited and early analytical results from integrated analyses of the costs and benefits of mitigation indicate that these are broadly comparable in magnitude, but do not as yet permit an unambiguous determination of an emissions pathway or stabilization level where benefits exceed costs. Integrated assessment of the economic costs and benefits of different mitigation pathways shows that the economically optimal timing and level of mitigation depends upon the uncertain shape and character of the assumed climate change damage cost curve.

To illustrate this dependency:
- if the climate change damage cost curve grows slowly and regularly, and there is good foresight (which increases the potential for timely adaptation), later and less stringent mitigation is economically justified;
- alternatively if the damage cost curve increases steeply, or contains non-linearities (e.g. vulnerability thresholds or even small probabilities of catastrophic events), earlier and more stringent mitigation is economically justified (*high agreement, much evidence*) [3.6.1].

Linkages between short term and long term

For any chosen GHG-stabilization target, near-term decisions can be made regarding mitigation opportunities to help maintain a consistent emissions trajectory within a range of long-term stabilization targets. Economy-wide modelling of long-term global stabilization targets can help inform near-term mitigation choices. A compilation of results from short-and long-term models using scenarios with stabilization targets in the 3–5 W/m² range (category II to III), reveals that in 2030, for carbon prices of less than 20 US$/tCO$_2$-eq, emission reductions of in the range of 9-18 $GtCO_2$-eq/yr across all GHGs can be expected. For carbon prices less than 50 US\$/tCO_2-eq this range is 14–23 $GtCO_2$-eq/yr and for carbon prices less than US\$100/tCO_2-eq it is 17-26 $GtCO_2$-eq/yr. (*high agreement, much evidence*).

Three important considerations need to be remembered with regard to the reported marginal costs. First, these mitigation scenarios assume complete 'what' and 'where' flexibility; that is, there is full substitution among GHGs, and reductions take place anywhere in the world as soon as the models begin their analyses. Second, the marginal costs of realizing these levels of mitigation increase in the time horizon beyond 2030. Third, at the economic-sector level, emission-reduction potential for all GHGs varies significantly across the different model scenarios (*high agreement, much evidence*) [3.6.2].

A risk management or 'hedging' approach can assist policy-makers to advance mitigation decisions in the absence of a long-term target and in the face of large uncertainties related to the cost of mitigation, the efficacy of adaptation and the negative impacts of climate change. The extent and the timing of the desirable hedging strategy will depend on the stakes, the odds and societies' attitudes to risks, for example, with respect to risks of abrupt change in geophysical systems and other key vulnerabilities. A variety of integrated assessment approaches exist to assess mitigation benefits in the context of policy decisions related to such long-term climate goals. There will be ample opportunity for learning and mid-course corrections as new information becomes available. However, actions in the short term will largely determine long-term global mean temperatures and thus what corresponding climate change impacts can be avoided. Delayed emission reductions lead to investments that lock in more emission-intensive infrastructure and development pathways. This significantly constrains the opportunities to achieve lower stabilization levels and increases the risk of more severe climate change impacts. Hence, analysis of near-term decisions should not be decoupled from analysis that considers long-term climate change outcomes (*high agreement, much evidence*) [3.6; 3.5.2].

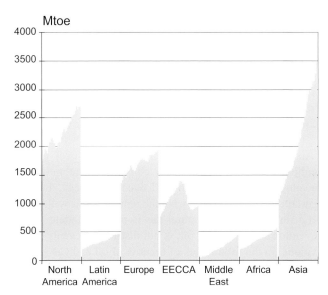

Figure TS.12: *Annual primary energy consumption, including traditional biomass, 1971 to 2003 [Figure 4.2].*

Note: EECCA = countries of Eastern Europe, the Caucasus and Central Asia. 1000 Mtoe = 42 EJ.

compromise energy access, equity and sustainable development of the poorest countries and interfere with reaching poverty-reduction targets that, in turn, imply improved access to electricity, modern cooking and heating fuels and transportation (*high agreement, much evidence*) [4.2.4].

Total fossil fuel consumption has increased steadily during the past three decades. Consumption of nuclear energy has continued to grow, though at a slower rate than in the 1980s. Large hydro and geothermal energy are relatively static. Between 1970 and 2004, the share of fossil fuels dropped from 86% to 81%. Wind and solar are growing most rapidly, but from a very low base (Figure TS.13) (*high agreement, much evidence*) [4.2].

4 Energy supply

Status of the sector and development until 2030

Global energy demand continues to grow, but with regional differences. The annual average growth of global primary energy consumption was 1.4 % per year in the 1990–2004 period. This was lower than in the previous two decades due to the economic transition in Eastern Europe, the Caucasus and Central Asia, but energy consumption in that region is now moving upwards again (Figure TS.12) (*high agreement, much evidence*) [4.2.1].

Rapid growth in energy consumption per capita is occurring in many developing countries. Africa is the region with the lowest per capita consumption. Increasing prices of oil and gas

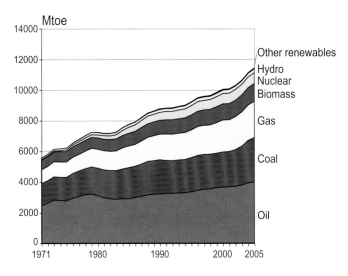

Figure TS.13: *World primary energy consumption by fuel type. [Figure 4.5]*

Most business-as-usual (BAU) scenarios point to continued growth of world population (although at lower rates than predicted decades ago) and GDP, leading to a significant growth in energy demand. High energy-demand growth rates in Asia (3.2% per year 1990–2004) are projected to continue and to be met mainly by fossil fuels *(high agreement, much evidence)* [4.2].

Absolute fossil fuel scarcity at the global level is not a significant factor in considering climate change mitigation. Conventional oil production will eventually peak, but it is uncertain exactly when and what the repercussions will be. The energy in conventional natural gas is more abundant than in conventional oil but, like oil, is not distributed evenly around the globe. In the future, lack of security of oil and gas supplies for consuming nations may drive a shift to coal, nuclear power and/or renewable energy. There is also a trend towards more efficient and convenient energy carriers (electricity, and liquid and gaseous fuels) instead of solids *(high agreement, much evidence)* [4.3.1].

In all regions of the world, emphasis on security of supply has grown since the Third Assessment Report (TAR). This is coupled with reduced investments in infrastructure, increased global demand, political instability in key areas and the threats of conflict, terrorism and extreme weather events. New energy infrastructure investments in developing countries and upgrades of capacity in developed countries opens a window of opportunity for exploiting the co-benefits of choices in the energy mix in order to lower GHG emissions from what they otherwise would be *(high agreement, much evidence)* [4.2.4; 4.1].

The conundrum for many governments has become how best to meet the ever growing demand for reliable energy services while limiting the economic costs to their constituents, ensuring energy security, reducing dependence on imported energy sources and minimizing emissions of the associated GHGs and other pollutants. Selection of energy-supply systems for each region of the world will depend on their development, existing infrastructure and the local comparative costs of the available energy resources *(high agreement, much evidence)* [4.1].

If fossil fuel prices remain high, demand may decrease temporarily until other hydrocarbon reserves in the form of oil sands, oil shales, coal-to-liquids, gas-to-liquids etc. become commercially viable. Should this happen, emissions will increase further as the carbon intensity increases, unless carbon dioxide capture and storage (CCS) is applied. Due to increased energy security concerns and recent increases in gas prices, there is growing interest in new, more efficient, coal-based power plants. A critical issue for future GHG emissions is how quickly new coal plants are going to be equipped with CCS technology, which will increase the costs of electricity. Whether building 'capture ready' plants is more cost-effective than retrofitting plants or building a new plant integrated with CCS

depends on economic and technical assumptions. Continuing high fossil fuel prices may also trigger more nuclear and/or renewable energy, although price volatility will be a disincentive for investors. Concerns about safety, weapons proliferation and waste remain as constraints for nuclear power. Hydrogen may also eventually contribute as an energy carrier with low carbon emissions, dependent on the source of the hydrogen and the successful uptake of CCS for hydrogen production from coal or gas. Renewable energy must either be used in a distributed manner or will need to be concentrated to meet the intensive energy demands of cities and industries, because, unlike fossil fuel sources, the sources of renewable energy are widely distributed with low energy returns per exploited area *(medium agreement, medium evidence)* [4.3].

If energy demand continues to grow along the current trajectory, an improved infrastructure and conversion system will, by 2030, require a total cumulative investment of over US$$_{2005}$ 20 trillion (20×10^{12}). For comparison, the total capital investment by the global energy industry is currently around 300 billion US$ per year (300×10^9) *(medium agreement, medium evidence)* [4.1].

Global and regional emission trends

With the exception of the countries in Eastern Europe, the Caucasus and Central Asia (where emissions declined post-1990 but are now rising again) and Europe (currently stable), carbon emissions have continued to rise. Business-as-usual emissions to 2030 will increase significantly. Without effective policy actions, global CO_2 emissions from fossil fuel combustion are predicted to rise at a minimum of more than 40%, from around 25 GtCO$_2$-eq/yr (6.6 GtC-eq) in 2000 to 37-53 GtCO$_2$-eq/yr (10-14 GtC-eq) by 2030 [4.2.3].

In 2004, emissions from power generation and heat supply alone were 12.7 GtCO$_2$-eq (26% of total emissions) including 2.2 GtCO$_2$eq from CH$_4$. In 2030, according to the World Energy Outlook 2006 baseline, these will have increased to 17.7 GtCO$_2$-eq. *(high agreement, much evidence)* [4.2.2].

Description and assessment of mitigation technologies and practices, options, potentials and costs in the electricity generation sector

The electricity sector has a significant mitigation potential using a range of technologies (Table TS.3). The economic potential for mitigation of each individual technology is based on what might be a realistic deployment expectation of the various technologies using all efforts, but given practical constraints on rate of uptake, public acceptance, capacity building and commercialization. Competition between options and the influence of end-use energy conservation and efficiency improvement is not included [4.4].

A wide range of energy-supply mitigation options are available and cost effective at carbon prices of <20US$/tCO$_2$

Table TS.3: *Potential GHG emissions avoided by 2030 for selected electricity generation mitigation technologies (in excess of the IEA World Energy Outlook (2004) Reference baseline) employed in isolation with estimated mitigation potential shares spread across each cost range (2006 US$/tCO2-eq) [Table 4.19].*

	Regional groupings	Mitigation potential; total emissions saved in 2030 (GtCO$_2$-eq)	Mitigation potential (%) for specific carbon price ranges (US$/tCO$_2$-eq avoided)				
			<0	0-20	20-50	50-100	>100
Fuel switch and plant efficiency	OECD[a]	0.39		100			
	EIT[b]	0.04		100			
	Non-OECD	0.64		100			
	World	1.07					
Nuclear	OECD	0.93	50	50			
	EIT	0.23	50	50			
	Non-OECD	0.72	50	50			
	World	1.88					
Hydro	OECD	0.39	85	15			
	EIT	0.00					
	Non-OECD	0.48	25	35	40		
	World	0.87					
Wind	OECD	0.45	35	40	25		
	EIT	0.06	35	45	20		
	Non-OECD	0.42	35	50	15		
	World	0.93					
Bio-energy	OECD	0.20	20	25	40	15	
	EIT	0.07	20	25	40	15	
	Non-OECD	0.95	20	30	45	5	
	World	1.22					
Geothermal	OECD	0.09	35	40	25		
	EIT	0.03	35	45	20		
	Non-OECD	0.31	35	50	15		
	World	0.43					
Solar PV and concentrated solar power	OECD	0.03				20	80
	EIT	0.01				20	80
	Non-OECD	0.21				25	75
	World	0.25					
CCS + coal	OECD	0.28			100		
	EIT	0.01			100		
	Non-OECD	0.20			100		
	World	0.49					
CCS + gas	OECD	0.09				100	
	EIT	0.04			30	70	
	Non-OECD	0.09				100	
	World	0.22					

Notes:
[a]   Organization for Economic Cooperation and Development
[b]   Economies in Transition

including fuel switching and power-plant efficiency improvements, nuclear power and renewable energy systems. CCS will become cost effective at higher carbon prices. Other options still under development include advanced nuclear power, advanced renewables, second-generation biofuels and, in the longer term, the possible use of hydrogen as an energy carrier *(high agreement, much evidence)* [4.3, 4.4].

Since the estimates in Table TS.3 are for the mitigation potentials of individual options without considering the actual supply mix, they cannot be added. An additional analysis of the supply mix to avoid double counting was therefore carried out.

For this analysis, it was assumed that the capacity of thermal electricity generation capacity would be substituted gradually and new power plants would be built to comply with demand, under the following conditions:

1) Switching from coal to gas was assumed for 20% of the coal plants, as this is the cheapest option.

2) The replacement of existing fossil fuel plants and the building of new plants up to 2030 to meet increasing power demand was shared between efficient fossil fuel plants, renewables, nuclear and coal and gas-fired plants with CCS. No early retirement of plants or stranded assets was assumed.

3) Low- or zero-carbon technologies are employed proportional

Table TS.4: *Projected power demand increase from 2010 to 2030 as met by new, more efficient additional and replacement plants and the resulting mitigation potential above the World Energy Outlook 2004 baseline [Table 4.20].*

	Power plant efficiencies by 2030 (based on IEA 2004a)[a] (%)	Existing mix of power generation in 2010 (TWh)	Generation from additional new plant by 2030 (TWh)	Generation from new plant replacing old, existing 2010 plant by 2030 (TWh)	Share of mix of generation of total new and replacement plant built by 2030 including CCS at various carbon prices (US$/tCO$_2$-eq)[b]			Total GtCO$_2$-eq avoided by fuel switching, CCS and displacing some fossil fuel generation with low-carbon options of wind, solar, geothermal, hydro, nuclear and biomass		
					<20 US$/TWh	<50 US$/TWh	<100 US$/TWh	<20 US$/t	<50 US$/t	<100 US$/t
OECD		11,302	2942	4521		7463		1.58	2.58	2.66
Coal	41	4079	657	1632	899	121	0			
Oil	40	472	−163C	189	13	2	0			
Gas	48	2374	1771	950	1793	637	458			
Nuclear	33	2462	−325	985	2084	2084	1777			
Hydro	100	1402	127	561	1295	1295	1111			
Biomass	28	237	168	95	263	499	509			
Other renew	63	276	707	110	1116	1544	1526			
CCS					0	1282	2082			
Economies In Transition (EIT)		1746	722	698		1420		0.32	0.42	0.49
Coal	32	381	13	152	72	46	29			
Oil	29	69	−8	28	11	7	4			
Gas	39	652	672	261	537	357	240			
Nuclear	33	292	−20	117	442	442	442			
Hydro	100	338	35	135	170	170	170			
Biomass	48	4	7	2	47	109	121			
Other renew	36	10	23	4	142	167	191			
CCS					0	123	222			
Non-OECD/ EIT		7137	7807	2855		10662		2.06	3.44	4.08
Coal	38	3232	3729	1293	2807	1697	1133			
Oil	38	646	166	258	297	179	120			
Gas	46	1401	2459	560	3114	2279	1856			
Nuclear	33	231	289	92	1356	1356	1356			
Hydro	100	1472	874	589	1463	2106	2106			
Biomass	19	85	126	34	621	1294	1443			
Other renew	28	70	164	28	1004	1154	1303			
CCS					0	598	1345			
Total		20185	11471	8074		19545		3.95	6.44	7.22

Notes:
a) Implied efficiencies calculated from WEO 2004 (IEA, 2004b) = Power output (EJ)/Estimated power input (EJ). See Appendix 1, Chapter 11.
b) At higher carbon prices, more coal, oil and gas power generation is displaced by low- and zero-carbon options. Since nuclear and hydro are cost competitive at <20US$/tCO$_2$-eq in most regions (Chapter 4, Table 4.4.4), their share remains constant.
c) Negative data depicts a decline in generation, which was included in the analysis.

to their estimated maximum shares in electricity generation in 2030. These shares are based on the literature, taking into account resource availability, relative costs and variability of supply related to intermittency issues in the power grid, and were differentiated according to carbon cost levels.

The resulting economic mitigation potential for the energy-supply sector by 2030 from improved thermal power-plant efficiency, fuel switching and the implementation of more nuclear, renewables, fuel switching and CCS to meet growing demand is around 7.2 GtCO$_2$-eq at carbon prices <100 US$/tCO$_2$-eq. At costs <20 US$/tCO$_2$-eq the reduction potential is estimated at 3.9 GtCO$_2$-eq (Table TS.4). At this carbon price level, the share of renewable energy in electricity

generation would increase from 20% in 2010 to about 30% in 2030. At carbon prices <50 US$/tCO$_2$-eq, the share would increase to 35% of total electricity generation. The share of nuclear energy would be about 18% in 2030 at carbon prices <50 US$/tCO$_2$-eq, and would not change much at higher prices as other technologies would be competitive.

For assessment of the economic potential, maximum technical shares for the employment of low- or zero-carbon technologies were assumed and the estimate is therefore at the high end of the wide range found in the literature. If, for instance, only 70% of the assumed shares is reached, the mitigation potential at carbon prices <100 US$/tCO$_2$-eq would be almost halved. Potential savings in electricity demand in end-use sectors reduce the need for mitigation measures in the power sector. When the impact of mitigation measures in the building and industry sectors on electricity demand (outlined in Chapter 11) is taken into account, a lower mitigation potential for the energy-supply sector results than the stand-alone figure reported here *(medium agreement, limited evidence)* [4.4].

Interactions of mitigation options with vulnerability and adaptation

Many energy systems are themselves vulnerable to climate change. Fossil fuel based offshore and coastal oil and gas extraction systems are vulnerable to extreme weather events. Cooling of conventional and nuclear power plants may become problematic if river waters are warmer. Renewable energy resources can also be affected adversely by climate change (such as solar systems impacted by changes in cloud cover; hydropower generation influenced by changes in river discharge, glaciers and snow melt; windpower influenced by changing wind velocity; and energy crop yields reduced by drought and higher temperatures). Some adaptation measures to climate change, like air-conditioning and water pumps use energy and may contribute to even higher CO_2 emissions, and thus necessitate even more mitigation *(high agreement, limited evidence)* [4.5.5].

Effectiveness of and experience with climate policies, potentials, barriers, opportunities and implementation issues

The need for immediate short-term action in order to make any significant impact in the longer term has become apparent, as has the need to apply the whole spectrum of policy instruments, since no single instrument will enable a large-scale transition in energy-supply systems on a global basis. Large-scale energy conversion technologies have a life of several decades and hence a turnover of only 1–3% per year. That means that policy decisions taken today will affect the rate of deployment of carbon-emitting technologies for several decades. They will have profound consequences on development paths, especially in a rapidly developing world [4.1].

Economic and regulatory instruments have been employed. Approaches to encourage the greater uptake of low-carbon energy-supply systems include reducing fossil fuel subsidies and stimulating front-runners in specific technologies through active government involvement in market creation (such as in Denmark for wind energy and Japan with solar photovoltaic (PV)). Reducing fossil fuel subsidies has been difficult, as it meets resistance by vested interests. In terms of support for renewable-electricity projects, feed-in-tariffs have been more effective than green certificate trading systems based on quotas. However, with increasing shares of renewables in the power mix, the adjustment of such tariffs becomes an issue. Tradable permit systems and the use of the Kyoto flexible mechanisms are expected to contribute substantially to emission reductions *(medium agreement, medium evidence)* [4.5].

Integrated and non-climate policies and co-benefits of mitigation policies

Co-benefits of GHG mitigation in the energy supply sector can be substantial. When applying cost-effective energy-efficiency measures, there is an immediate economic benefit to consumers from lower energy costs. Other co-benefits in terms of energy supply security, technological innovation, air-pollution abatement and employment also typically result at the local scale. This is especially true for renewables which can reduce import dependency and in many cases minimize transmission losses and costs. Electricity, transport fuels and heat supplied by renewable energy are less prone to price fluctuations, but in many cases have higher costs. As renewable energy technologies can be more labour-intensive than conventional technologies per unit of energy output, more employment will result. High investment costs of new energy system infrastructures can, however, be a major barrier to their implementation.

Developing countries that continue to experience high economic growth will require significant increases in energy services that are currently being met mainly by fossil fuels. Increasing access to modern energy services can have multiple benefits. Their use can help improve air quality, particularly in large urban areas, and lead to a decrease in GHG emissions. An estimated 2400 GW of new power plants plus the related infrastructure will need to be built in developing countries by 2030 to meet increased consumer demand, requiring an investment of around 5 trillion US$ (5×10^{12}). If well directed, such large investments provide opportunities for sustainable development. The integration of development policies with GHG mitigation objectives can deliver the advantages mentioned above and contribute to development goals pertaining to employment, poverty and equity. Analysis of possible policies should take into account these co-benefits. However, it should be noted again that, in specific circumstances, pursuing air-pollution abatement or energy security aims can lead to more energy use and related GHG emissions.

Liberalization and privatization policies to develop free energy markets aim to provide greater competition and lower consumer prices but have not always been successful in this regard, often resulting in a lack of capital investment and scant regard for environmental impacts *(high agreement, much evidence)* [4.2.4; 4.5.2; 4.5.3; 4.5.4].

Technology research, development, diffusion and transfer

Investment in energy technology R&D has declined overall since the levels achieved in the late 1970s that resulted from the oil crisis. Between 1980 and 2002, public energy-related R&D investment declined by 50% in real terms. Current levels have risen, but may still be inadequate to develop the technologies needed to reduce GHG emissions and meet growing energy demand. Greater public and private investment will be required for rapid deployment of low-carbon energy technologies. Improved energy conversion technologies, energy transport and storage methods, load management, co-generation and community-based services will have to be developed *(high agreement, limited evidence)* [4.5.6].

Long-term outlook

Outlooks from both the IEA and World Energy Council project increases in primary energy demand of between 40 and 150% by 2050 over today's demand, depending on the scenarios for population and economic growth and the rate of technology development. Electricity use is expected to grow by between 110 and 260%. Both organizations realize that business-as-usual scenarios are not sustainable. It is well accepted that even with good decision-making and co-operation between the public and private sectors, the necessary transition will take time and the sooner it is begun the lower the costs will be *(high agreement, much evidence)* [4.2.3].

5   Transport and its infrastructure

Status and development of the sector

Transport activity is increasing around the world as economies grow. This is especially true in many areas of the developing world where globalization is expanding trade flows, and rising personal incomes are amplifying demand for motorized mobility. Current transportation activity is mainly driven by internal combustion engines powered by petroleum fuels (95% of the 83 EJ of world transport energy use in 2004). As a consequence, petroleum use closely follows the growth in transportation activity. In 2004, transport energy amounted to 26% of total world energy use. In the developed world, transport energy use continues to increase at slightly more than 1% per year; passenger transport currently consumes 60–75% of total transport energy there. In developing countries, transport energy use is rising faster (3 to 5% per year) and is projected to grow from 31% in 2002 to 43% of world transport energy use by 2025 [5.2.1, 5.2.2].

Transport activity is expected to grow robustly over the next several decades. Unless there is a major shift away from current patterns of energy use, projections foresee a continued growth in world transportation energy use of 2% per year, with energy use and carbon emissions about 80% above 2002 levels by 2030 [5.2.2]. In developed economies, motor vehicle ownership approaches five to eight cars for every ten inhabitants (Figure TS.14). In the developing world, levels of vehicle ownership are much lower; non-motorized transport plays a significant role, and there is a greater reliance on two- and three-wheeled motorized vehicles and public transport. The motorization of transport in the developing world is, however, expected to grow rapidly in

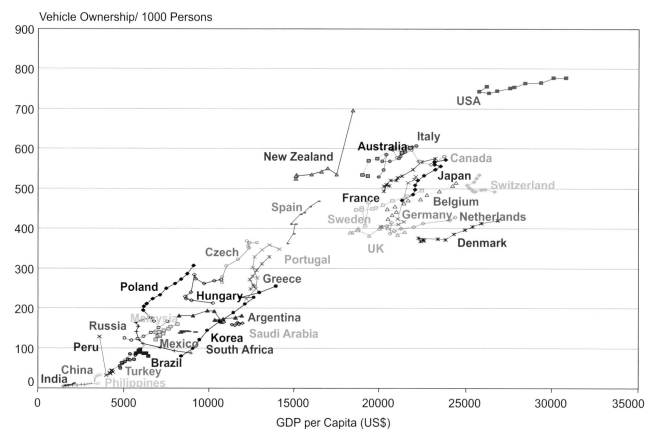

Figure TS.14: *Vehicle ownership and income per capita as a time line per country [Figure 5.2].*

Note: data are for 1900–2002, but the years plotted vary by country, depending on data availability.

the coming decades. As incomes grow and the value of travellers' time increases, travellers are expected to choose faster modes of transport, shifting from non-motorized to automotive, to air and high-speed rail. Increasing speed has generally led to greater energy intensity and higher GHG emissions.

In addition to GHG emissions, the motorization of transport has created congestion and air-pollution problems in large cities all around the world (*high agreement, much evidence*) [5.2.1; 5.2.2; 5.5.4].

Emission trends

In 2004, the contribution of transport to total energy-related GHG emissions was about 23%, with emissions of CO_2 and N_2O amounting to about 6.3-6.4 GtCO$_2$-eq. Transport sector CO_2 emissions (6.2 GtCO$_2$-eq. in 2004) have increased by around 27% since 1990 and its growth rate is the highest among the end-user sectors. Road transport currently accounts for 74% of total transport CO_2 emissions. The share of non-OECD countries is 36% now and will increase rapidly to 46% by 2030 if current trends continue (*high agreement, medium evidence*) [5.2.2].

The transport sector also contributes small amounts of CH_4 and N_2O emissions from fuel combustion and F-gases from vehicle air-conditioning. CH_4 emissions are between 0.1–0.3% of total transport GHG emissions, N_2O between 2.0 and 2.8% (all figures based on US, Japan and EU data only). Emissions of F gases (CFC-12 + HFC-134a + HCFC-22) worldwide in 2003 were 4.9% of total transport CO_2 emissions (*medium agreement, limited evidence*) [5.2.1].

Estimates of CO_2 emissions from global aviation increased by a factor of about 1.5, from 330 MtCO$_2$/yr in 1990 to 480 MtCO$_2$/yr in 2000, and accounted for about 2% of total anthropogenic CO_2 emissions. Aviation CO_2 emissions are projected to continue to grow strongly. In the absence of additional measures, projected annual improvements in aircraft fuel efficiency of the order of 1–2% will be largely surpassed by traffic growth of around 5% each year, leading to a projected increase in emissions of 3–4% per year (*high agreement, medium evidence*). Moreover, the overall climate impact of aviation is much greater than the impact of CO_2 alone. As well as emitting CO_2, aircraft contribute to climate change through the emission of nitrogen oxides (NO_x), which are particularly effective in forming the GHG ozone when emitted at cruise altitudes. Aircraft also trigger the formation of condensation trails, or contrails, which are suspected of enhancing the formation of cirrus clouds, which add to the overall global warming effect. These effects are estimated to be about two to four times greater than those of aviation's CO_2 alone, even without considering the potential impact of cirrus cloud enhancement. The environmental effectiveness of future mitigation policies for aviation will depend on the extent to which these non-CO_2 effects are also addressed (*high agreement, medium evidence*) [5.2.1; 5.2.2].

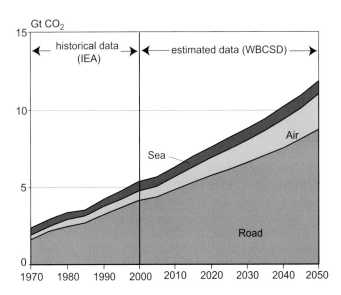

Figure TS.15: *Historical and projected CO$_2$ emissions from transport [Figure 5.4].*

All of the projections discussed above assume that world oil supplies will be more than adequate to support the expected growth in transport activity. There is ongoing debate, however, about whether the world is nearing a peak in conventional oil production that would require a significant and rapid transition to alternative energy sources. There is no shortage of alternative energy sources, including oil sands and oil shales, coal-to-liquids, biofuels, electricity and hydrogen. Among these alternatives, unconventional fossil carbon resources would produce the least expensive fuels most compatible with the existing transportation infrastructure. Unfortunately, tapping into these fossil resources to power transportation would increase upstream carbon emissions and greatly increase the input of carbon into the atmosphere [5.2.2; 5.3].

Description and assessment of mitigation technologies and practices, options, potentials and costs

Transport is distinguished from other energy-using sectors by its predominant reliance on a single fossil resource and by the infeasibility of capturing carbon emissions from transport vehicles with any known technologies. It is also important to view GHG-emission reductions in conjunction with air pollution, congestion and energy security (oil import) problems. Solutions therefore have to try to optimize improvement of transportation problems as a whole, not just GHG emissions [5.5.4].

There have been significant developments in mitigation technologies since the Third Assessment Report (TAR), and significant research, development and demonstration programmes on hydrogen-powered fuel-cell vehicles have been launched around the globe. In addition, there are still many opportunities for improvement of conventional technologies. Biofuels continue to be important in certain markets and have much greater potential for the future. With regard to non-CO_2 emissions, vehicle air-conditioning systems based on low GWP refrigerants have been developed [5.3].

Road traffic: efficient technologies and alternative fuels

Since the TAR, the energy efficiency of road vehicles has improved by the market success of cleaner directed-injection turbocharged (TDI) diesels and the continued market penetration of many incremental efficiency technologies; hybrid vehicles have also played a role, though their market penetration is currently small. Further technological advances are expected for hybrid vehicles and TDI diesel engines. A combination of these with other technologies, including materials substitution, reduced aerodynamic drag, reduced rolling resistance, reduced engine friction and pumping losses, has the potential to approximately double the fuel economy of 'new' light-duty vehicles by 2030, thereby roughly halving carbon emissions per vehicle mile travelled (note that this is only for a new car and not the fleet average) *(medium agreement, medium evidence)* [5.3.1].

Biofuels have the potential to replace a substantial part, but not all, petroleum use by transport. A recent IEA report estimated that the share of biofuels could increase to about 10% by 2030 at costs of 25 US$/tCO$_2$-eq, which includes a small contribution from biofuels from cellulosic biomass. The potential strongly depends on production efficiency, the development of advanced techniques such as conversion of cellulose by enzymatic processes or by gasification and synthesis, costs, and competition with other uses of land. Currently the cost and performance of ethanol in terms of CO_2 emissions avoided is unfavourable, except for production from sugarcane in low-wage countries (Figure TS.16) *(medium agreement, medium evidence)* [5.3.1].

The economic and market potential of hydrogen vehicles remains uncertain. Electric vehicles with high efficiency (more than 90%), but low driving range and short battery life have a limited market penetration. For both options, the emissions are determined by the production of hydrogen and electricity. If hydrogen is produced from coal or gas with CCS (currently the cheapest way) or from biomass, solar, nuclear or wind energy, well-to-wheel carbon emissions could be nearly eliminated. Further technological advances and/or cost reductions would be required in fuel-cells, hydrogen storage, hydrogen or electricity production with low- or zero-carbon emissions, and batteries *(high agreement, medium evidence)* [5.3.1].

The total mitigation potential in 2030 of the energy-efficiency options applied to light duty vehicles would be around 0.7–0.8 GtCO$_2$-eq in 2030 at costs lower than 100 US$/tCO$_2$. Data are not sufficient to provide a similar estimate for heavy-duty vehicles. The use of current and advanced biofuels, as mentioned above, would give an additional reduction potential of another 600–1500 MtCO$_2$-eq

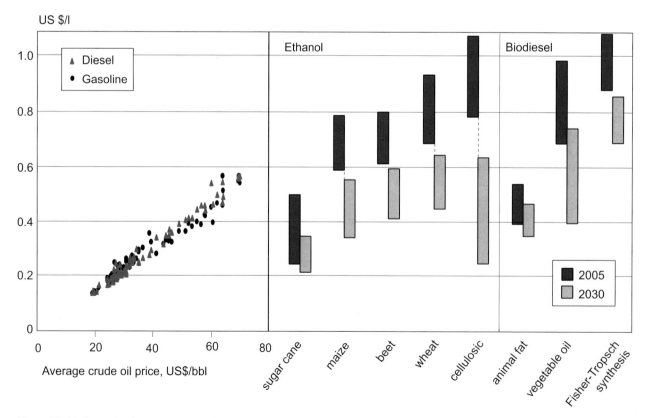

Figure TS.16: *Comparison between current and future biofuel production costs versus gasoline and diesel ex-refinery (FOB) prices for a range of crude oil prices [Figure 5.9].*
Note: prices exclude taxes.

in 2030 at costs lower than 25 US$/tCO$_2$ *(low agreement, limited evidence)* [5.4.2].

A critical threat to the potential for future reduction of CO$_2$ emissions from use of fuel economy technologies is that they can be used to increase vehicle power and size rather than to improve the overall fuel economy and reduce carbon emissions. The preference of the market for power and size has consumed much of the potential for GHG mitigation reduction achieved over the past two decades. If this trend continues, it will significantly diminish the GHG mitigation potential of the advanced technologies described above *(high agreement, much evidence)* [5.2; 5.3].

Air traffic

The fuel efficiency of civil aviation can be improved by a variety of means including technology, operation and management of air traffic. Technology developments might offer a 20% improvement in fuel efficiency over 1997 levels by 2015, with a 40–50% improvement likely by 2050. As civil aviation continues to grow at around 5% each year, such improvements are unlikely to keep carbon emissions from global air travel from increasing. The introduction of biofuels could mitigate some of aviation's carbon emissions, if biofuels can be developed to meet the demanding specifications of the aviation industry, although both the costs of such fuels and the emissions from their production process are uncertain at this time *(medium agreement, medium evidence)* [5.3.3].

Aircraft operations can be optimized for energy use (with minimum CO$_2$ emissions) by minimizing taxiing time, flying at optimal cruise altitudes, flying minimum-distance great-circle routes, and minimizing holding and stacking around airports. The GHG-reduction potential of such strategies has been estimated at 6–12%. More recently, researchers have begun to address the potential for minimizing the total climate impact of aircraft operations, including ozone impacts, contrails and nitrogen oxides emissions. The mitigation potential in 2030 for aviation is 280 MtCO$_2$/yr at costs <100 US$/tCO$_2$ *(medium agreement, medium evidence)* [5.4.2].

Marine transport

Since the TAR, an International Maritime Organization (IMO) assessment found that a combination of technical measures could reduce carbon emissions by 4–20% in older ships and 5–30% in new ships by applying state-of-the-art knowledge, such as hull and propeller design and maintenance. However, due to the long lifetime of engines, it will take decades before measures on existing ships are implemented on a significant scale. The short-term potential for operational measures, including route-planning and speed reduction, ranged from 1–40%. The study estimated a maximum reduction of emissions of the world fleet of about 18% by 2010 and 28% by 2020, when all measures were to be implemented. The data do not allow an estimate of an absolute mitigation potential figure and the mitigation potential is not expected to be sufficient to offset the growth in shipping activity over the same period *(medium agreement, medium evidence)* [5.3.4].

Rail transport

The main opportunities for mitigating GHG emissions associated with rail transport are improving aerodynamics, reduction of train weight, introducing regenerative braking and on-board energy storage and, of course, mitigating the GHG emissions from electricity generation. There are no estimates available of total mitigation potential and costs [5.3.2].

Modal shifts and public transport

Providing public transports systems and their related infrastructure and promoting non-motorized transport can contribute to GHG mitigation. However, local conditions determine how much transport can be shifted to less energy-intensive modes. Occupancy rates and the primary energy sources of the transport modes further determine the mitigation potential [5.3.1].

The energy requirements of urban transport are strongly influenced by the density and spatial structure of the built environment, as well as by the location, extent and nature of the transport infrastructure. Large-capacity buses, light-rail transit and metro or suburban rail are increasingly being used for the expansion of public transport. Bus Rapid Transit systems have relatively low capital and operational costs, but it is uncertain if they can be implemented in developing countries with the same success as in South America. If the share of buses in passenger transport were to increase by 5–10%, then CO$_2$ emissions would fall by 4-9% at costs in the order of US$ 60-70/tCO$_2$ [5.3.1].

More than 30% of the trips made by cars in Europe are for less than 3 km and 50% for less than 5 km. Although the figures may differ for other continents, there is potential for mitigation by shifting from cars to non-motorized transport (walking and cycling), or preventing a growth of car transport at the expense of non-motorized transport. Mitigation potentials are highly dependent on local conditions, but there are substantial co-benefits in terms of air quality, congestion and road safety *(high agreement, much evidence)* [5.3.1].

Overall mitigation potential in the transport sector

The overall potential and cost for CO$_2$ mitigation can only be partially estimated due to lack of data for heavy-duty vehicles, rail transport, shipping and modal split change/ public transport promotion. The total economic potential for improved efficiency of light-duty vehicles and aeroplanes and substituting biofuels for conventional fossil fuels, for a carbon price up to 100 US$/tCO$_2$-eq, is estimated to be about 1600–2550 MtCO$_2$. This is an underestimate of potential for mitigation in the transport sector *(high agreement, medium evidence)* [5.4.2].

Effectiveness of and experience with climate policies, potentials, barriers and opportunities/ implementation issues

Policies and measures for surface transport

Given the positive effects of higher population densities on public transport use, walking, cycling and CO_2 emissions, better integrated spatial planning is an important policy element in the transportation sector. There are some good examples for large cities in several countries. Transportation Demand Management (TDM) can be effective in reducing private vehicle travel if rigorously implemented and supported. Soft measures, such as the provision of information and the use of communication strategies and educational techniques have encouraged a change in personal behaviour leading to a reduction in the use of the car by 14% in an Australian city, 12% in a German city and 13% in a Swedish city *(medium agreement, medium evidence)* [5.5.1].

Fuel-economy standards or CO_2 standards have been effective in reducing GHG emissions, but so far, transport growth has overwhelmed their impact. Most industrialized and some developing countries have set fuel-economy standards for new light-duty vehicles. The forms and stringency of standards vary widely, from uniform, mandatory corporate average standards, through graduated standards by vehicle weight class or size, to voluntary industry-wide standards. Fuel economy standards have been universally effective, depending on their stringency, in improving vehicle fuel economy, increasing on-road fleet-average fuel economy and reducing fuel use and carbon emissions. In some countries, fuel-economy standards have been strongly opposed by segments of the automotive industry on a variety of grounds, ranging from economic efficiency to safety. The overall effectiveness of standards can be significantly enhanced if combined with fiscal incentives and consumer information *(high agreement, much evidence)* [5.5.1].

Taxes on vehicle purchase, registration, use and motor fuels, as well as road and parking pricing policies are important determinants of vehicle-energy use and GHG emissions. They are employed by different countries to raise general revenue, to partially internalize the external costs of vehicle use or to control congestion of public roads. An important reason for fuel or CO_2 tax having limited effects is that price elasticities tend to be substantially smaller than the income elasticities of demand. In the long run, the income elasticity of demand is a factor 1.5–3 higher than the price elasticity of total transport demand, meaning that price signals become less effective with increasing incomes. Rebates on vehicle purchase and registration taxes for fuel-efficient vehicles have been shown to be effective. Road and parking pricing policies are applied in several cities, with marked effects on passenger car traffic *(high agreement, much evidence)* [5.5.1].

Many governments have introduced or are intending to implement policies to promote biofuels in national emission abatement strategies. Since the benefit of biofuels for CO_2 mitigation comes mainly from the well-to-tank part, incentives for biofuels are more effective climate policies if they are tied to entire well-to-wheels CO_2 efficiencies. Thus preferential tax rates, subsidies and quotas for fuel blending should be calibrated to the benefits in terms of net CO_2 savings over the entire well-to-wheel cycle associated with each fuel. In order to avoid the negative effects of biofuel production on sustainable development (e.g., biodiversity impacts), additional conditions could be tied to incentives for biofuels.

Policies and measures for aviation and marine transport

In order to reduce emissions from air and marine transport resulting from the combustion of bunker fuels, new policy frameworks need to be developed. Both the International Civil Aviation Organization (ICAO) and IMO have studied options for limiting GHG emissions. However, neither has yet been able to devise a suitable framework for implementing policies. ICAO, however, has endorsed the concept of an open, international emission-trading system implemented through a voluntary scheme, or the incorporation of international aviation into existing emission-trading systems.

For aviation, both fuel or emission charges and trading would have the potential to reduce emissions considerably. The geographical scope (routes and operators covered), the amount of allowances to be allocated to the aviation sector and the coverage of non-CO_2 climate impacts will be key design elements in determining the effectiveness of emissions trading for reducing the impacts of aviation on climate. Emission charges or trading would lead to an increase in fuel costs that will have a positive impact on engine efficiency [5.5.2].

Current policy initiatives in the shipping sector are mostly based on voluntary schemes, using indexes for the fuel efficiency of ships. Environmentally differentiated port dues are being used in a few places. Other policies to limit shipping emissions would be the inclusion of international shipping in international emissions-trading schemes, fuel taxes and regulatory instruments *(high agreement, medium evidence)* [5.5.2].

Integrated and non-climate policies affecting emissions of GHGs and co-benefits of GHG mitigation policies

Transport planning and policy have recently placed more weight on sustainable development aspects. This includes reducing oil imports, improved air quality, reducing noise pollution, increasing safety, reducing congestion and improving access to transport facilities. Such policies can have important synergies with reducing GHG emissions *(high agreement, medium evidence)* [5.5.4; 5.5.5].

6   Residential and commercial buildings

Status of the sector and emission trends

In 2004, direct GHG emissions from the buildings sector (excluding emissions from electricity use) were about 5 $GtCO_2$-eq/yr (3 $GtCO_2$-eq/yr CO_2; 0.1 $GtCO_2$-eq/yr N_2O; 0.4 $GtCO_2$-eq/yr CH_4 and 1.5 $GtCO_2$-eq/yr halocarbons). The last figure includes F-gases covered by the Montreal protocol and about 0.1–0.2 $GtCO_2$-eq/yr of HFCs. As mitigation in this sector includes many measures aimed at saving electricity, the mitigation potential is generally calculated including electricity saving measures. For comparison, emission figures of the building sector are often presented including emissions from electricity use in the sector . When including the emissions from electricity use, energy-related CO_2 emissions from the buildings sector were 8.6 Gt/yr, or 33% of the global total in 2004. Total GHG emissions, including the emissions from electricity use, are then estimated at 10.6 Gt CO_2eq/yr (*high agreement, medium evidence*) [6.2].

Future carbon emissions from energy use in buildings

The literature for the buildings sector uses a mixture of baselines. Therefore, for this chapter, a building sector baseline was defined, somewhere between SRES B2 and A1B[2], with 14.3 $GtCO_2$-eq GHG emissions (including emissions from electricity use) in 2030. The corresponding emissions in the SRES B2 and A1B scenarios are 11.4 and 15.6 $GtCO_2$. In the SRES B2 scenario (Figure TS.17), which is based on relatively lower economic growth, North America and Non-Annex I East Asia account for the largest portion of the increase in emissions. In the SRES A1B scenario, which shows rapid economic growth, all the CO_2 emissions increase is in the developing world: Asia, Middle East and North Africa, Latin America, and Sub-Saharan Africa, in that order. Overall, average annual CO_2 emission growth between 2004 and 2030 is 1.5% in Scenario B2 and 2.4% in Scenario A1B (*high agreement, medium evidence*) [6.2, 6.3].

Mitigation technologies and practices

Measures to reduce GHG emissions from buildings fall into one of three categories: 1) reducing energy consumption[13] and embodied energy in buildings; 2) switching to low-carbon fuels, including a higher share of renewable energy; 3) controlling emissions of non-CO_2 GHG gases. Many current technologies allow building energy consumption to be reduced through better thermal envelopes[14], improved design methods and building operations, more efficient equipment,and reductions in demand for energy services. The relative importance of heating and cooling depends on climate and thus varies regionally, while the effectiveness of passive design techniques also depends on climate, with important distinctions between hot-humid and hot-arid regions. Occupant behaviour, including avoiding unnecessary operation of equipment and adaptive rather than invariant temperature standards for heating and cooling, is also a significant factor in limiting building energy use (*high agreement, much evidence*) [6.4].

Mitigation potential of the building sector

Substantial CO_2 emission reduction from energy use in buildings can be achieved over the coming years compared with projected emissions. The considerable experience in a wide variety of technologies, practices and systems for energy efficiency and an equally rich experience with policies and programmes that promote energy efficiency in buildings lend considerable confidence to this view. A significant portion of these savings can be achieved in ways that reduce life-cycle costs, thus providing reductions in CO_2 emissions that have a net negative cost (generally higher investment cost but lower operating cost) (*high agreement, much evidence*) [6.4; 6.5].

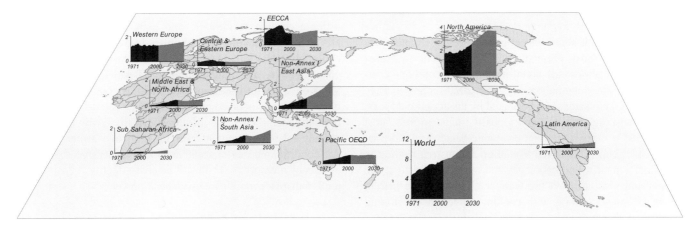

Figure TS.17: *CO_2 emissions (GtCO_2) from buildings including emissions from the use of electricity, 1971–2030 [Figure 6.2].*

Note: Dark red – historic emissions; light red – projection according to SRES B2 scenario. EECCA=Countries of Eastern Europe, the Caucasus and Central Asia.

---

13   This counts all forms of energy use in buildings, including electricity.
14   The term 'thermal envelope' refers to the shell of a building as a barrier to unwanted heat or mass transfer between the interior of the building and outside.

Table TS.5: *GHG emissions reduction potential for the buildings stock in 2020[a] [Table 6.2].*

Economic region	Countries/country groups reviewed for region	Potential as % of national baseline for buildings[b]	Measures covering the largest potential	Measures providing the cheapest mitigation options
Developed countries	USA, EU-15, Canada, Greece, Australia, Republic of Korea, United Kingdom, Germany, Japan	Technical: 21%-54%[c] Economic (<US$ 0/tCO$_2$-eq): 12%-25%[d] Market: 15%-37%	1. Shell retrofit, inc. insulation, esp. windows and walls; 2. Space heating systems; 3. Efficient lights, especially shift to compact fluorescent lamps (CFL) and efficient ballasts.	1. Appliances such as efficient TVs and peripherals (both on-mode and standby), refrigerators and freezers, ventilators and air-conditioners; 2. Water heating equipment; 3. Lighting best practices.
Economies in Transition	Hungary, Russia, Poland, Croatia, as a group: Latvia, Lithuania, Estonia, Slovakia, Slovenia, Hungary, Malta, Cyprus, Poland, the Czech Republic	Technical: 26%-47%[e] Economic (<US$ 0/tCO$_2$eq): 13%-37%[f] Market: 14%	1. Pre- and post- insulation and replacement of building components, esp. windows; 2. Efficient lighting, esp. shift to CFLs; 3. Efficient appliances such as refrigerators and water heaters.	1. Efficient lighting and its controls; 2. Water and space heating control systems; 3. Retrofit and replacement of building components, esp. windows.
Developing countries	Myanmar, India, Indonesia, Argentine, Brazil, China, Ecuador, Thailand, Pakistan, South Africa	Technical: 18%-41% Economic (<US$ 0/tCO$_2$eq): 13%-52%[g] Market: 23%	1. Efficient lights, esp. shift to CFLs, light retrofit, and kerosene lamps; 2. Various types of improved cooking stoves, esp. biomass stoves, followed by LPG and kerosene stoves; 3. Efficient appliances such as air-conditioners and refrigerators.	1. Improved lights, esp. shift to CFLs light retrofit, and efficient kerosene lamps; 2. Various types of improved cooking stoves, esp. biomass based, followed by kerosene stoves; 3. Efficient electric appliances such as refrigerators and air-conditioners.

Notes:
[a] Except for EU-15, Greece, Canada, India, and Russia, for which the target year was 2010, and Hungary, Ecuador and South Africa, for which the target was 2030.
[b] The fact that the market potential is higher than the economic potential for developed countries is explained by limitation of studies considering only one type of potential, so information for some studies likely having higher economic potential is missing.
[c] Both for 2010, if the approximate formula of Potential $_{2020}$ = 1 – (1 – Potential $_{2010}$)$^{20/10}$ is used to extrapolate the potential as percentage of the baseline into the future (the year 2000 is assumed as a start year), this interval would be 38%–79%.
[d] Both for 2010, if suggested extrapolation formula is used, this interval would be 22%–44%.
[e] The last figure is for 2010, corresponds to 72% in 2020 if the extrapolation formula is used.
[f] The first figure is for 2010, corresponds to 24% in 2020 if the extrapolation formula is used.
[g] The last figure is for 2030, corresponds to 38% in 2020 if the suggested extrapolation formula is applied to derive the intermediate potential.

These conclusions are supported by a survey of 80 studies (Table TS.5), which show that efficient lighting technologies are among the most promising GHG-abatement measures in buildings in almost all countries, in terms of both cost-effectiveness and potential savings. By 2020, approximately 760 Mt of CO_2 emissions can be abated by the adoption of least life-cycle cost lighting systems globally, at an average cost of -160 US$/tCO$_2$ (i.e., at a net economic benefit). In terms of the size of savings, improved insulation and district heating in the colder climates and efficiency measures related to space cooling and ventilation in the warmer climates come first in almost all studies, along with cooking stoves in developing countries. Other measures that rank high in terms of savings potential are solar water heating, efficient appliances and energy-management systems.

As far as cost effectiveness is concerned, efficient cooking stoves rank second after lighting in developing countries, while the measures in second place in the industrialized countries differ according to climatic and geographic region. Almost all the studies examining economies in transition (typically in cooler climates) found heating-related measures to be the most cost effective, including insulation of walls, roofs, windows and floors, as well as improved heating controls for district heating. In developed countries, appliance-related measures are typically identified as the most cost-effective, with upgrades of cooling-related equipment ranking high in warmer climates. Air-conditioning savings can be more expensive than other efficiency measures but can still be cost-effective, because they tend to displace more expensive peak power.

In individual new buildings, it is possible to achieve 75% or more energy savings compared with recent current practice, generally at little or no extra cost. Realizing these savings requires an integrated design process involving architects, engineers, contractors and clients, with full consideration of opportunities for passively reducing the energy demands of buildings [6.4.1].

Table TS.6: *Global CO$_2$ mitigation potential projections for 2020, as a function of costs [Table 6.3].*

World regions	Baseline emissions in 2020	CO$_2$ mitigation potentials as share of the baseline CO$_2$ emission projections in cost categories in 2020 (costs in US$/tCO$_2$-eq)				CO$_2$ mitigation potentials in absolute values in cost categories in 2020, GtCO$_2$-eq (costs in US$/tCO$_2$-eq)			
	GtCO$_2$-eq	<0	0-20	20-100	<100	<0	0-20	20-100	<100
Globe	11.1	29%	3%	4%	36%	3.2	0.35	0.45	4.0
OECD (-EIT)	4.8	27%	3%	2%	32%	1.3	0.10	0.10	1.6
EIT	1.3	29%	12%	23%	64%	0.4	0.15	0.30	0.85
Non-OECD	5.0	30%	2%	1%	32%	1.5	0.10	0.05	1.6

Note: The aggregated global potential as a function of cost and region is based on 17 studies that reported potentials in detail as a function of costs.

Addressing GHG mitigation in buildings in developing countries is of particular importance. Cooking stoves can be made to burn more efficiently and combust particles more completely, thus benefiting village dwellers through improved indoor-air quality, while reducing GHG emissions. Local sources of improved, low GHG materials can be identified. In urban areas, and increasingly in rural ones, there is a need for all the modern technologies used in industrialized countries to reduce GHG emissions [6.4.3].

Emerging areas for energy savings in commercial buildings include the application of controls and information technology to continuously monitor, diagnose and communicate faults in commercial buildings ('intelligent control'); and systems approaches to reduce the need for ventilation, cooling, and dehumidification. Advanced windows, passive solar design, techniques for eliminating leaks in buildings and ducts, energy-efficient appliances, and controlling standby and idle power consumption as well as solid-state lighting are also important in both residential and commercial sectors (*high agreement, much evidence)* [6.5].

Occupant behaviour, culture and consumer choice and use of technologies are major determinants of energy use in buildings and play a fundamental role in determining CO$_2$ emissions. However, the potential reduction through non-technological options is rarely assessed and the potential leverage of policies over these is poorly understood (*high agreement, medium evidence)*.

There are opportunities to reduce direct emissions of fluorinated gases in the buildings sector significantly through the global application of best practices and recovery methods, with mitigation potential for all F-gases of 0.7 GtCO$_2$-eq in 2015. Mitigation of halocarbon refrigerants mainly involves avoiding leakage from air conditioners and refrigeration equipment (e.g., during normal use, maintenance and at end of life) and reducing the use of halocarbons in new equipment. A key factor determining whether this potential will be realized is the costs associated with implementation of the measures to achieve the emission reduction. These vary considerably, from a net benefit to 300 US$/tCO$_2$-eq. (*high agreement, much evidence*) [6.5].

Mitigation potential of the building sector

There is a global potential to reduce approximately 30% of the projected baseline emissions from the residential and commercial sectors cost effectively by 2020 (Table TS.6). At least a further 3% of baseline emissions can be avoided at costs up to 20 US$/tCO$_2$-eq and 4% more if costs up to 100 US$/tCO$_2$-eq are considered. However, due to the large opportunities at low costs, the high-cost potential has only been assessed to a limited extent, and thus this figure is an underestimate. Using the global baseline emission projections for buildings[15], these estimates represent a reduction of about 3.2, 3.6, and 4.0 Gtons of CO$_2$-eq in 2020, at zero, 20 US$/tCO$_2$-eq, and 100 US$/tCO$_2$-eq, respectively (*high agreement, much evidence*) [6.5].

The real potential is likely to be higher, because not all end-use efficiency options were considered by the studies; non-technological options and their often significant co-benefits were omitted as were advanced integrated high-efficiency buildings. However, the market potential is much smaller than the economic potential.

Given limited information for 2030, the 2020 findings for the economic potential to 2030 have been extrapolated to enable comparisons with other sectors. The estimates are given in Table TS.7. Extrapolation of the potentials to 2030 suggests that, globally, about 4.5, 5.0 and 5.6 GtCO$_2$-eq/yr could be reduced at costs of <0, <20 and <100 US$/tCO$_2$-eq respectively. This is equivalent to 30, 35, and 40% of the projected baseline emissions. These figures are associated with significantly lower levels of certainty than the 2020 ones due to very limited research available for 2030 (*medium agreement, low evidence*).

The outlook for the long-term future, assuming options in the building sector with a cost up to US$ 25/tCO$_2$-eq, identifies a potential of about 7.7 GtCO$_2$eq reductions in 2050.

15 The baseline CO$_2$ emission projections were calculated on the basis of the 17 studies used for deriving the global potential (if a study did not contain a baseline, projections from another national mitigation report were used).

Table TS.7: *Global CO$_2$ mitigation potential projections for 2030, as a function of cost, based on extrapolation from the 2020 numbers, in GtCO$_2$ [Table 6.4].*

Mitigation option	Region	Baseline projections in 2030	Potential costs at below 100 US$/tCO$_2$-eq		Potential in different cost categories		
			Low	High	<0 US$/tCO$_2$ <0 US$/tC	0-20 US$/tCO$_2$ 0-73 US$/tC	20-100 US$/tCO$_2$ 73-367 US$/tC
Electricity savings[a]	OECD	3.4	0.75	0.95	0.85	0.0	0.0
	EIT	0.40	0.15	0.20	0.20	0.0	0.0
	Non-OECD/EIT	4.5	1.7	2.4	1.9	0.1	0.1
Fuel savings	OECD	2.0	1.0	1.2	0.85	0.2	0.1
	EIT	1.0	0.55	0.85	0.20	0.2	0.3
	Non-OECD/EIT	3.0	0.70	0.80	0.65	0.1	0.0
Total	OECD	5.4	1.8	2.2	1.7	0.2	0.1
	EIT	1.4	0.70	1.1	0.40	0.2	0.3
	Non-OECD/EIT	7.5	2.4	3.2	2.5	0.1	0.0
	Global	14.3	4.8	6.4	4.5	0.5	0.7

Note:
[a] The absolute values of the potentials resulting from electricity savings in Table TS.8 and Chapter 11, Table 11.3 do not coincide due to application of different baselines; however, the potential estimates as percentage of the baseline are the same in both cases. Also Table 11.3 excludes the share of emission reductions which is already taken into account by the energy supply sector, while Table TS.7 does not separate this potential.

Interactions of mitigation options with vulnerability and adaptation

If the world experiences warming, energy use for heating in temperate climates will decline (e.g., Europe, parts of Asia and North America), and for cooling will increase in most world regions. Several studies indicate that, in countries with moderate climates, the increase in electricity for additional cooling will outweigh the decrease for heating, and in Southern Europe a significant increase in summer peak demand is expected. Depending on the generation mix in particular countries, the net effect of warming on CO$_2$ emissions may be an increase even where overall demand for final energy declines. This causes a positive feedback loop: more mechanical cooling emits more GHGs, thereby exacerbating warming (*medium agreement, medium evidence*).

Investments in the buildings sector may reduce the overall cost of climate change by simultaneously addressing mitigation and adaptation. The most important of these synergies includes reduced cooling needs or energy use through measures such as application of integrated building design, passive solar construction, heat pumps with high efficiency for heating and cooling, adaptive window glazing, high-efficiency appliances emitting less waste heat, and retrofits including increased insulation, optimized for specific climates, and storm-proofing. Appropriate urban planning, including increasing green areas as well as cool roofs in cities, has proved to be an efficient way of limiting the 'heat island' effect, thereby reducing cooling needs and the likelihood of urban fires. Adaptive comfort, where occupants accept higher indoor (comfort) temperatures when the outside temperature is high, is now often incorporated in design considerations (*high agreement, medium evidence*) [6.9].

Effectiveness of and experience with policies for reducing CO$_2$ emissions from energy use in buildings

Realizing such emissions reductions up to 2020 requires the rapid design, implementation and enforcement of strong policies promoting energy efficiency for buildings and equipment, renewable energy (where cost-effective), and advanced design techniques for new buildings (*high agreement, much evidence*) [6.5].

There are, however, substantial barriers that need to be overcome to achieve the high indicated negative and low cost mitigation potential. These include hidden costs, mismatches between incentives and benefits (e.g., between landlords and tenants), limitations in access to financing, subsidies on energy prices, as well as fragmentation of the industry and the design process. These barriers are especially strong and diverse in the residential and commercial sectors; overcoming them is therefore only possible through a diverse portfolio of policy instruments combined with good enforcement (*high agreement, medium evidence*).

A wide range of policies has been shown in many countries to be successful in cutting GHG emissions from buildings. Table TS.8 summarizes the key policy tools applied and compares them according to the effectiveness of the policy instrument, based on selected best practices. Most instruments reviewed can achieve significant energy and CO$_2$ savings. In an evaluation of 60 policy evaluations from about 30 countries, the highest CO$_2$ emission reductions were achieved through building codes, appliance standards and tax-exemption policies. Appliance standards, energy-efficiency obligations and quotas, demand-side management programmes and mandatory labelling were found to be among the most cost-effective policy tools. Subsidies and energy or carbon taxes were the least cost-effective instrument. Information programmes are also cost

Table TS.8: *The impact and effectiveness of selected policy instruments aimed at mitigating GHG emissions in the buildings sector using best practices [Table 6.6].*

Policy instrument	Emission reduction effectiveness[a]	Cost-effectiveness[b]	Special conditions for success, major strengths and limitations, co-benefits
Appliance standards	High	High	Factors for success: periodic update of standards, independent control, information, communication and education.
Building codes	High	Medium	No incentive to improve beyond target. Only effective if enforced.
Public leadership programmes, inc. procurement regulations	High	High/Medium	Can be used effectively to demonstrate new technologies and practices. Mandatory programmes have higher potential than voluntary ones. Factor for success: ambitious energy efficiency labelling and testing.
Energy efficiency obligations and quotas	High	High	Continuous improvements necessary: new EE measures, short term incentives to transform markets, etc.
Demand-side management programmes	High	High	Tend to be more cost-effective for commercial sector than for residences.
Energy performance contracting/ESCO support[c]	High	Medium	Strength: no need for public spending or market intervention, co-benefit of improved competitiveness.
Energy efficiency certificate schemes	Medium	Medium	No long-term experience. Transaction costs can be high. Institutional structures needed. Profound interactions with existing policies. Benefits for employment.
Kyoto Protocol flexible mechanisms[d]	Low	Low	So far limited number of CDM &JI projects in buildings.
Taxation (on CO_2 or fuels)	Low	Low	Effect depends on price elasticity. Revenues can be earmarked for further efficiency. More effective when combined with other tools.
Tax exemptions/ reductions	High	High	If properly structured, stimulate introduction of highly efficient equipment and new buildings.
Capital subsidies, grants, subsidised loans	High	Low	Positive for low-income households, risk of free-riders, may induce pioneering investments.
Labelling and certification programmes	Medium/High	High	Mandatory programmes more effective than voluntary ones. Effectiveness can be boosted by combination with other instruments and regular updates.
Voluntary and negotiated agreements	Medium/High	Medium	Can be effective when regulations are difficult to enforce. Effective if combined with financial incentives, and threat of regulation.
Education and information programmes	Low/Medium	High	More applicable in residential sector than commercial. Success condition: best applied in combination with other measures.
Mandatory audit and energy management requirement	High, but variable	Medium	Most effective if combined with other measures such as financial incentives.
Detailed billing and disclosure programmes	Medium	Medium	Success conditions: combination with other measures and periodic evaluation.

Notes:
a) includes ease of implementation; feasibility and simplicity of enforcement; applicability in many locations; and other factors contributing to overall magnitude of realized savings.
b) Cost-effectiveness is related to specific societal cost per carbon emissions avoided.
c) Energy service companies.
d) Joint Implementation, Clean Development Mechanism, International Emissions Trading (includes the Green Investment Scheme).

effective, particularly when they accompany most other policy measures (*medium agreement, medium evidence*) [6.8].

Policies and measures that aim at reducing leakage or discourage the use of refrigerants containing fluorine may reduce emissions of F-gases substantially in future years *(high agreement, medium evidence)* [6.8.4].

The limited overall impact of policies so far is due to several factors: 1) slow implementation processes; 2) the lack of regular updating of building codes (requirements of many policies are often close to common practices, despite the fact that CO_2-neutral construction without major financial sacrifices is already possible) and appliance standards and labelling; 3) inadequate funding; 4) insufficient enforcement. In developing countries and economies in transition, implementation of energy-efficiency policies is compromised by a lack of concrete implementation combined with poor or non-existent enforcement mechanisms. Another challenge is to promote GHG-abatement measures for the building shell of existing buildings due to the long time periods between regular building retrofits and the slow turnover of buildings in developed countries *(high agreement, much evidence)* [6.8].

Co-benefits and links to sustainable development

Energy efficiency and utilization of renewable energy in buildings offer synergies between sustainable development and GHG abatement. The most relevant of these for the least developed countries are safe and efficient cooking stoves that, while cutting GHG emissions, significantly reduce mortality and morbidity by reducing indoor air pollution. Safe and efficient cooking stoves also reduce the workload for women and children who typically gather the fuel for traditional stoves and decrease the demands on scarce natural resources. Reduction in outdoor air pollution is another significant co-benefit.

In general, in developed and developing countries, improved energy efficiency in buildings and the clean and efficient use of locally available renewable energy resources results in:
- substantial savings in energy-related investment, since efficiency is less costly than new supply;
- funds freed up for other purposes, such as infrastructure investments;
- improved system reliability and energy security;
- increased access to energy services;
- reduced fuel poverty;
- improvement of local environmental quality;
- positive effects on employment, by creating new business opportunities and through the multiplier effects of spending money saved on energy costs in another way.

There is increasing evidence that well-designed energy-efficient buildings often promote occupant productivity and health *(high agreement, medium evidence)* [6.9].

Support from industrialized countries for the development and implementation of policies to increase energy efficiency of buildings and equipment in developing countries and economies in transition could contribute substantially to reductions in the growth of CO_2 emissions and improve the welfare of the population. Devoting international aid or other public and private funds aimed at sustainable development to energy efficiency and renewable energy initiatives in buildings can achieve a multitude of development objectives and result in long-lasting impacts. The transfer of knowledge, expertise and know-how from developed to developing countries can facilitate the adoption of photovoltaics (PV), including PV-powered light emitting diode-based (LED) lighting, high-insulation building materials, efficient appliances and lighting, integrated design, building energy-management systems, and solar cooling. However, capital financing will also be needed [6.8.3].

Technology research, development, deployment, diffusion and transfer

Although many practical and cost-effective technologies and practices are available today, research and development is needed in such areas as: high-performance control systems[16]; advanced window glazing; new materials for insulated panels; various systems to utilize passive and other renewable energy sources; phase-change materials to increase thermal storage; high-performance ground-source reversible heat pumps; integrated appliances and other equipment to use waste heat; novel cooling technologies, and the use of community-wide networks to supply heating, cooling and electricity to buildings. Demonstrations of these technologies and systems, and training of professionals, are necessary steps toward bringing those new technologies to market [6.8.3].

Long-term-outlook

Long-term GHG reduction in buildings needs to start soon because of the slow turnover of the building stock. To achieve large-scale savings in new buildings in the longer term, new approaches to integrated design and operation of buildings need to be taught, spread, and put into large-scale practice as soon as possible. Such training is currently not available for the majority of professionals in the building industry. Because of the important role of non-technological opportunities in buildings, ambitious GHG reductions may require a cultural shift towards a society that embraces climate protection and sustainable development among its fundamental values, leading to social pressure for building construction and use with much reduced environmental footprints *(high agreement, medium evidence)* [6.4.1; 6.8.1].

7   Industry

Status of the sector, development trends and implications

Energy-intensive industries, iron and steel, non-ferrous metals, chemicals and fertilizer, petroleum-refining, cement, and pulp and paper, account for about 85% of the industry sector's energy consumption in most countries. Since energy use in other sectors grew faster, the sector's share in global primary energy use declined from 40% in 1971 to 37% in 2004 [7.1.3].

Much of this energy-intensive industry is now located in developing countries. Overall, in 2003, developing countries accounted for 42% of global steel production, 57% of global nitrogen fertilizer production, 78% of global cement manufacture, and about 50% of global aluminium production. In 2004, developing countries accounted for 46% of final energy

---

16 Advanced control systems need to be created that permit the integration of all energy service functions in the design and subsequent operation of commercial buildings ('intelligent control').

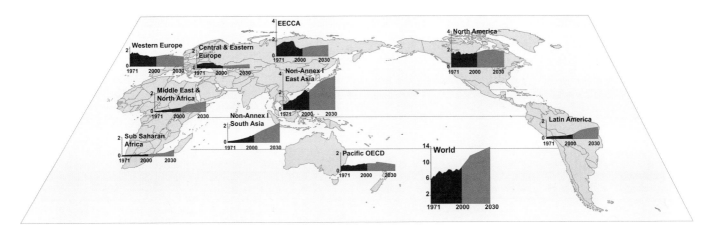

Figure TS.18: *Industrial sector energy-related CO₂ emissions (GtCO₂; including electricity use), 1971–2030. [Table 7.1, 7.2].*

Note: Dark red – historic emissions; light red – projections according to SRES B2 scenario. Data extracted from Price et al. (2006). EECCA = Countries of Eastern Europe, the Caucasus and Central Asia.

use by industry, developed country for 43% and economies in transition for 11%. Many facilities (for aluminium, cement and fertilizer industries) in developing nations are new and include the latest technology with lowest specific energy use. However, as in industrialized countries, many older, inefficient facilities remain. This creates a huge demand for investment in developing countries to improve energy efficiency and achieve emission reductions. The strong growth of energy-intensive industries during the 20th century is expected to continue as population and GDP increase [7.1.2; 7.1.3].

Though large-scale production dominates these energy-intensive industries globally, small- and medium-sized enterprises (SMEs) have significant shares in many developing countries. While regulations and international competition are moving large industrial enterprises towards the use of environmentally sound technology, SMEs may not have the economic or technical capacity to install the necessary control equipment or are slower to innovate. These SME limitations create special challenges for efforts to mitigate GHG emissions (*high agreement, much evidence*) [7.1.1].

Emission trends (global and regional)

Direct GHG emissions from industry are currently about 7.2 $GtCO_2$-eq. As the mitigation options discussed in this chapter include measures aimed at reducing the industrial use of electricity, emissions including those from electricity use are important for comparison. Total industrial sector GHG emissions were about 12 $GtCO_2$-eq in 2004, about 25% of the global total. CO_2 emissions (including electricity use) from the industrial sector grew from 6.0 $GtCO_2$ in 1971 to 9.9 $GtCO_2$ in 2004. In 2004, developed nations accounted for 35% of total energy-related CO_2 emissions, economies in transition for 11% and developing nations for 53% (see Figure TS.18). Industry also emits CO_2 from non-energy uses of fossil fuels and from non-fossil fuel sources. In 2000,

these were estimated to total 1.7 $GtCO_2$ (*high agreement, much evidence*) [7.1.3].

Industrial processes also emit other GHGs, including HFC-23 from the manufacture of HCFC-22; PFCs from aluminium smelting and semiconductor processing; SF_6 from use in flat panel screens (liquid crystal display) and semi-conductors, magnesium die casting, electrical equipment, aluminium melting, and others, and CH_4 and N_2O from chemical industry sources and food-industry waste streams. Total emission from these sources was about 0.4 $GtCO_2$-eq in 2000 (*medium agreement, medium evidence*) [7.1.3].

The projections for industrial CO_2 emissions for 2030 under the SRES-B2[2] scenarios are around 14 $GtCO_2$ (including electricity use) (see Figure TS.18). The highest average growth

Table TS.9: *Projected industrial sector emissions of non-CO₂ GHGs, MtCO₂-eq/yr [Table 7.3].*

Region	1990	2000	2010	2030
Pacific OECD	38	53	47	49
North America	147	117	96	147
Western Europe	159	96	92	109
Central and Eastern Europe	31	21	22	27
EECCA	37	20	21	26
Developing Asia	34	91	118	230
Latin America	17	18	21	38
Sub Saharan Africa	6	10	11	21
Middle East and North Africa	2	3	10	20
World	470	428	438	668

Note:
Emissions from refrigeration equipment used in industrial processes included; emissions from all other refrigeration and air-conditioning applications excluded. EECCA = the countries of Eastern Europe, the Caucasus and Central Asia.

Table **TS.10**: *Examples of industrial technology for reducing GHG emissions (not comprehensive). Technologies in italics are under demonstration or development [Table 7.5].*

Sector	Energy efficiency	Fuel switching	Power recovery	Renewables	Feedstock change	Product change	Material efficiency	Non-CO$_2$ GHG	CO$_2$ capture and storage
Sector wide	Benchmarking; Energy management systems; Efficient motor systems, boilers, furnaces, lighting and heating/ventilation/air conditioning; Process integration	Coal to natural gas and oil	Cogeneration	Biomass, Biogas, PV, Wind turbines, Hydropower	Recycled inputs				*Oxy-fuel combustion, CO$_2$ separation from flue gas*
Iron & steel	Smelt reduction, Near net shape casting, Scrap preheating, Dry coke quenching	Natural gas, oil or plastic injection into the BF	Top-gas pressure recovery, By-product gas combined cycle	Charcoal	Scrap	High strength steel	Recycling, High strength steel, Reduction process losses	n/a	*Hydrogen reduction, oxygen use in blast furnaces*
Non-ferrous metals	*Inert anodes,* Efficient cell designs				Scrap		Recycling, thinner film and coating	PFC/SF$_6$ controls	
Chemicals	Membrane separations, Reactive distillation	Natural gas	Pre-coupled gas turbine, Pressure recovery turbine, H$_2$ recovery		Recycled plastics, bio-feedstock	Linear low density polyethylene, high-perf. plastics	Recycling, Thinner film and coating, Reduced process losses	N$_2$O, PFCs, CFCs and HFCs control	*CO$_2$ storage from ammonia, ethylene oxide processes*
Petroleum refining	Membrane separation Refinery gas	Natural gas	Pressure recovery turbine, hydrogen recovery	Biofuels	Bio-feedstock		(reduction in transport not included here)	Control technology for N$_2$O/CH$_4$	*From hydrogen production*
Cement	Precalciner kiln, Roller mill, *fluidized bed kiln*	Waste fuels, Biogas, Biomass	Drying with gas turbine, power recovery	Biomass fuels, Biogas	Slags, pozzolanes	Blended cement *Geo-polymers*		n/a	*Oxyfuel combustion in kiln*
Glass	Cullet preheating Oxyfuel furnace	Natural gas	*Air bottoming cycle*	n/a	Increased cullet use	High-strength thin containers	Recycling	n/a	*Oxyfuel$_L$ combustion*
Pulp and paper	Efficient pulping, Efficient drying, Shoe press, Condebelt drying	Biomass, Landfill gas	*Black liquor gasification combined cycle*	Biomass fuels (bark, black liquor)	Recycling, Non-wood fibres	Fibre orientation, Thinner paper	Reduction cutting and process losses	n/a	*Oxyfuel combustion in lime kiln*
Food	Efficient drying, Membranes	Biogas, Natural gas	Anaerobic digestion, Gasification	Biomass, By-products, Solar drying			Reduction process losses, Closed water use		

rates in industrial-sector CO_2 emissions are projected for developing countries. Growth in the regions of Central and Eastern Europe, the Caucasus and Central Asia, and Developing Asia is projected to slow in both scenarios for 2000–2030. CO_2 emissions are expected to decline in the Pacific OECD, North America and Western Europe regions for B2 after 2010. For non-CO_2 GHG emissions from the industrial sector, emissions by 2030 are projected to increase globally by a factor of 1.4, from 470 $MtCO_2$-eq. (130 MtC-eq) in 1990 to 670 $MtCO_2$-eq (180 MtC-eq.) in 2030 assuming no further action is taken to control these emissions. Mitigation efforts led to a decrease in non-CO_2 GHG emissions between 1990 and 2000, and many programmes for additional control are underway (see Table TS.9) (*high agreement, medium evidence*) [7.1.3].

Description and assessment of mitigation technologies and practices, options and potentials, costs and sustainability

Historically, the industrial sector has achieved reductions in energy intensity and emission intensity through adoption of energy efficiency and specific mitigation technologies, particularly in energy-intensive industries. The aluminium industry reported >70% reduction in PFC-emission intensity over the period 1990–2004 and the ammonia industry reported that plants designed in 2004 have a 50% reduction in energy intensity compared with those designed in 1960. Continuing to modernize ammonia-production facilities around the world will result in further energy-efficiency improvements. Reductions in refining energy intensity have also been reported [7.4.2, 7.4.3, 7.4.4].

The low technical and economic capacity of SMEs pose challenges for the diffusion of sound environmental technology, though some innovative R&D is taking place in SMEs.

A wide range of measures and technologies have the potential to reduce industrial GHG emissions. These technologies can be grouped into the categories of energy efficiency, fuel switching, power recovery, renewables, feedstock change, product change and material efficiency (Table TS.10). Within each category, some technologies, such as the use of more efficient electric motors, are broadly applicable across all industries, while others, such as top-gas pressure recovery in blast furnaces, are process-specific.

Later in the period to 2030, there will be a substantial additional potential from further energy- efficiency improvements and application of Carbon Capture and Storage (CCS)[17] and non-GHG process technologies. Examples of such new technologies that are currently in the R&D phase include inert electrodes for aluminium manufacture and hydrogen for metal production (*high agreement, much evidence*) [7.2, 7.3, 7.4].

Mitigation potentials and costs in 2030 have been estimated in an industry-by-industry assessment of energy-intensive industries and an overall assessment of other industries. The approach yielded mitigation potentials of about 1.1 $GtCO_2$-eq at a cost of <20 US$/$tCO_2$ (74 US$/tC-eq); about 3.5 $GtCO_2$-eq at costs below <50 US$/$tCO_2$ (180 US$/tC-eq); and about 4 $GtCO_2$-eq/yr (0.60–1.4 GtC-eq/yr) at costs <US$100/$tCO_2$-eq (<US$370/tC-eq) under the B2 scenario. The largest mitigation potentials are in the steel, cement and pulp and paper industries, and in the control of non-CO_2 gases, and much of the potential is available at <50 US$/$tCO_2$-eq (<US$ 180/tC-eq). Application of CCS technology offers a large additional potential, albeit at higher cost.

A recently completed global study for nine groups of technologies indicates a mitigation potential for the industrial sector of 2.5-3.0 $GtCO_2$-eq/yr (0.68-0.82 GtC-eq/yr) in 2030 at costs of <25 US$/$tCO_2$ (< 92US$/tC) (2004$). While the estimate of mitigation potential is in the range found in this assessment, the estimate of mitigation cost is significantly lower (*medium agreement, medium evidence*) [7.5].

Interaction of mitigation options with vulnerability and adaptation

Linkages between adaptation and mitigation in the industrial sector are limited. Many mitigation options (e.g., energy efficiency, heat and power recovery, recycling) are not vulnerable to climate change and therefore create no adaptation link. Others, such as fuels or feedstock switching (e.g. to biomass or other renewable energy sources) may be vulnerable to climate change [7.8].

Effectiveness of and experience with climate policies, potentials, barriers and opportunities/implementation issues

Full use of available mitigation options is not being made in either industrialized or developing nations. In many areas of the world, GHG mitigation is not demanded by either the market or government regulation. In these areas, companies will invest in GHG mitigation to the extent that other factors provide a return for their investments. This return can be economic; for example, energy-efficiency projects that provide an economic pay-out, or can be in terms of achieving larger corporate goals, for example, a commitment to sustainable development. The economic potential as outlined above will only be realized if policies and regulations are in place. Relevant in this respect is that, as noted above, most energy-intensive industries are located in developing countries. Slow rate of capital stock turnover is also a barrier in many industries, as is the lack of the financial and technical resources needed to implement mitigation options, and limitations in the ability of industrial firms, particularly small and medium-sized enterprises, to

---

17 See IPCC Special Report on CO_2 Capture and Storage

access and absorb information about available options (*high agreement, much evidence*) [7.9.1].

Voluntary agreements between industry and government to reduce energy use and GHG emissions have been used since the early 1990s. Well-designed agreements, which set realistic targets and have sufficient government support, often as part of a larger environmental policy package, and a real threat of increased government regulation or energy/GHG taxes if targets are not achieved, can provide more than business-as-usual energy savings or emission reductions. Some have accelerated the application of best available technology and led to reductions in emissions compared with the baseline, particularly in countries with traditions of close cooperation between government and industry. However, the majority of voluntary agreements have not achieved significant emission reductions beyond business-as-usual. Corporations, sub-national governments, non-government organizations (NGOs) and civil groups are adopting a wide variety of voluntary actions, independent of government authorities, which may limit GHG emissions, stimulate innovative policies, and encourage the deployment of new technologies. By themselves, however, they generally have limited impact.

Policies that reduce the barriers to adoption of cost-effective, low-GHG emission technologies (e.g., lack of information, absence of standards and unavailability of affordable financing for first purchases of modern technology) can be effective. Many countries, both developed and developing, have financial schemes available to promote energy saving in industry. According to a World Energy Council survey, 28 countries provide some sort of grant or subsidy for industrial energy-efficiency projects. Fiscal measures are also frequently used to stimulate energy savings in industry. However, a drawback to financial incentives is that they are often also used by investors who would have made the investment without the incentive. Possible solutions to improve cost-effectiveness are to restrict schemes to specific target groups and/or techniques (selected lists of equipment, only innovative technologies), or use a direct criterion of cost-effectiveness [7.9.3].

Several national, regional or sectoral CO_2 emissions trading systems either exist or are being developed. The further refinement of these trading systems could be informed by evidence that suggests that in some important aspects, participants from industrial sectors face a significantly different situation to those from the electricity sector. For instance, responses to carbon emission price in industry tend to be slower because of the more limited technology portfolio and absence of short-term fuel-switching possibilities, making predictable allocation mechanisms and stable price signals a more important issue for industry [7.9.4].

As noted in the TAR, industrial enterprises of all sizes are vulnerable to changes in government policy and consumer preferences. That is why a stable policy regime is so important for industry (*high agreement, much evidence*) [7.9].

Integrated and non-climate policies affecting emissions of greenhouse gases

Policies aimed at balancing energy security, environmental protection and economic development can have a positive or negative impact on mitigation. Sustainable development policies focusing on energy efficiency, dematerialization, and use of renewables support GHG mitigation objectives. Waste-management policies reduce industrial sector GHG emissions by reducing energy use through the re-use of products. Air-pollutant reduction measures can have synergy with GHG-emissions reduction when reduction is achieved by shifting to low-carbon fuels, but do not always reduce GHG emissions as many require the use of additional energy.

In addition to implementing the mitigation options discussed above, achieving sustainable development will require industrial development pathways that minimize the need for future mitigation (*high agreement, medium evidence*). Large companies have greater resources, and usually more incentives, to factor environmental and social considerations into their operations than small and medium enterprises (SMEs), but SMEs provide the bulk of employment and manufacturing capacity in many countries. Integrating SME development strategy into broader national strategies for development is consistent with sustainable development objectives. Energy-intensive industries are now committing to a number of measures towards human capital development, health and safety, community development etc., which are consistent with the goal of corporate social responsibility (*high agreement, much evidence*) [7.7; 7.8].

Co-benefits of greenhouse gas mitigation policies

The co-benefits of industrial GHG mitigation include: reduced emissions of air pollutants, and waste (which in turn reduce environmental compliance and waste disposal costs), increased production and product quality, lower maintenance and operating costs, an improved working environment, and other benefits such as decreased liability, improved public image and worker morale, and delaying or reducing capital expenditures. The reduction of energy use can indirectly contribute to reduced health impacts of air pollutants particularly where no air-pollution regulation exists (*high agreement, much evidence*) [7.10].

Technology research, development, deployment, diffusion and transfer

Commercially available industrial technology provides a very large potential to reduce GHG emissions. However, even with the application of this technology, many industrial processes would still require much more energy than the thermodynamic ideal, suggesting a large additional potential for energy-efficiency improvement and GHG mitigation potential. In addition, some industrial processes emit GHGs that are independent of heat and power use. Commercial technology to eliminate these emissions does not currently exist for some of these processes, for example, development of an inert electrode to eliminate process emissions from aluminium manufacture and the use of hydrogen to reduce iron and non-ferrous metal ores. These new technologies must also meet a host of other criteria, including cost competitiveness, safety and regulatory requirements, as well as winning customer acceptance. Industrial technology research, development, deployment and diffusion are carried out both by governments and companies, ideally in complementary roles. Because of the large economic risks inherent in technologies with GHG emission mitigation as the main purpose, government programmes are likely to be needed in order to facilitate a sufficient level of research and development. It is appropriate for governments to identify fundamental barriers to technology and find solutions to overcome these barriers, but companies should bear the risks and capture the rewards of commercialization.

In addition, government information, energy audits, reporting, and benchmarking programmes promote technology transfer and diffusion. The key factors determining private-sector technology deployment and diffusion are competitive advantage, consumer acceptance, country-specific characteristics, protection of intellectual property rights, and regulatory frameworks (*medium agreement, medium evidence*) [7.11].

Long-term outlook

Many technologies offer long-term potential for mitigating industrial GHG emissions, but interest has focused on three areas: biological processing, use of hydrogen and nanotechnology.

Given the complexity of the industrial sector, achieving low GHG emissions is the sum of many cross-cutting and individual sector transitions. Because of the speed of capital stock turnover in at least some branches of industry, inertia by 'technology lock-in' may occur. Retrofitting provides opportunities in the meantime, but basic changes in technology occur only when the capital stock is installed or replaced (*high agreement, much evidence*) [7.12].

8　Agriculture

Status of the sector, future trends in production and consumption, and implications

Technological developments have allowed remarkable progress in agricultural output per unit of land, increasing per capita food availability despite a consistent decline in per capita agricultural land area (*high agreement, much evidence*). However, progress has been uneven across the world, with rural poverty and malnutrition remaining in some countries. The share of animal products in the diet has increased progressively in developing countries, while remaining constant in the developed world (*high agreement, much evidence*).

Production of food and fibre has more than kept pace with the sharp increase in demand in a more populated world, so that the global average daily availability of calories per capita has increased, though with regional exceptions. However, this growth has been at the expense of increasing pressure on the environment and dwindling natural resources, and has not solved problems of food security and widespread child malnutrition in poor countries (*high agreement, much evidence*).

The absolute area of global arable land has increased to about 1400 Mha, an overall increase of 8% since the 1960s (5% decrease in developed countries and 22% increase in developing countries). This trend is expected to continue into the future, with a projected additional 500 Mha converted to agriculture from 1997–2020, mostly in Latin America and Sub-Saharan Africa (*medium agreement, limited evidence*).

Economic growth and changing lifestyles in some developing countries are causing a growing demand for meat and dairy products. From 1967–1997, meat demand in developing countries rose from 11 to 24 kg per capita per year, achieving an annual growth rate of more than 5% by the end of that period. Further increases in global meat demand (about 60% by 2020) are projected, mostly in developing regions such as South and Southeast Asia, and Sub-Saharan Africa (*medium agreement, much evidence*) [8.2].

Emission trends

For 2005, agriculture accounted for an estimated emission of 5.1 to 6.1 $GtCO_2$-eq (10–12% of total global anthropogenic emissions of GHGs). CH_4 contributed 3.3 $GtCO_2$-eq and N_2O 2.8 $GtCO_2$-eq. Of global anthropogenic emissions in 2005, agriculture accounted for about 60% of N_2O and about 50% of CH_4 (*medium agreement, medium evidence*). Despite large annual exchanges of CO_2 between the atmosphere and agricultural lands, the net flux is estimated to be approximately balanced, with net CO_2 emissions of only around 0.04 $GtCO_2$/yr (emissions from electricity and fuel use in agriculture are covered in the buildings and transport sector respectively) (*low agreement, limited evidence*) [8.3].

Trends in GHG emissions in agriculture are responsive to global changes: increases are expected as diets change and population growth increases food demand. Future climate change may eventually release more soil carbon (though the effect is uncertain as climate change may also increase soil carbon inputs through high production). Emerging technologies may permit reductions of emissions per unit of food produced, but absolute emissions are likely to grow (*medium agreement, medium evidence*).

Without additional policies, agricultural N_2O and CH_4 emissions are projected to increase by 35–60% and ~60%, respectively, to 2030, thus increasing more rapidly than the 14% increase of non-CO_2 GHG observed from 1990 to 2005 (*medium agreement, limited evidence*) [8.3.2].

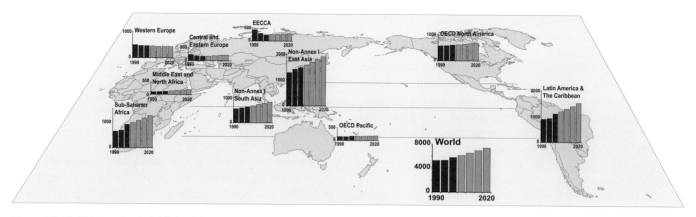

Figure TS.19: *Historic and projected N₂O and CH₄ emissions (MtCO₂-eq.) in the agricultural sector of ten world regions, 1990–2020 [Figure 8.2].*

Note: EECCA=Countries of Eastern Europe, the Caucasus and Central Asia.

Both the magnitude of the emissions and the relative importance of the different sources vary widely among world regions (Figure TS.19). In 2005, the group of five regions consisting mostly of non-Annex I countries were responsible for 74% of total agricultural emissions [8.3].

Mitigation technologies, practices, options, potentials and costs

Considering all gases, the economic potentials for agricultural mitigation by 2030 are estimated to be about 1600, 2700 and 4300 MtCO₂-eq/yr at carbon prices of up to 20, 50 and 100 US$/tCO₂-eq, respectively for a SRES B2 baseline (see Table TS.11) (*medium agreement, limited evidence*) [8.4.3].

Improved agricultural management can reduce net GHG emissions, often affecting more than one GHG. The effectiveness of these practices depends on factors such as climate, soil type and farming system (*high agreement, much evidence*).

About 90% of the total mitigation arises from sink enhancement (soil C sequestration) and about 10% from emission reduction (*medium agreement, medium evidence*). The most prominent mitigation options in agriculture (with potentials shown in Mt

Table TS.11: *Estimates of global agricultural economic GHG mitigation potential (MtCO₂-eq/yr) by 2030 under different assumed carbon prices for a SRES B2 baseline [Table 8.7].*

	Carbon price (US$/tCO₂-eq)		
	Up to 20	**Up to 50**	**Up to 100**
OECD	330 (60-470)	540 (300-780)	870 (460-1280)
EIT	160 (30-240)	270 (150-390)	440 (230-640)
Non-OECD/ EIT	1140 (210-1660)	1880 (1040-2740)	3050 (1610-4480)

Note:
figures in brackets show standard deviation around the mean estimate, potential excluding energy-efficiency measures and fossil fuel offsets from bioenergy.

CO₂eq/yr for carbon prices up to 100 US$/tCO₂-eq by 2030) are (see also Figure TS.20):
- restoration of cultivated organic soils (1260)
- improved cropland management (including agronomy, nutrient management, tillage/residue management and water management (including irrigation and drainage) and set-aside / agro-forestry (1110)
- improved grazing land management (including grazing intensity, increased productivity, nutrient management, fire management and species introduction (810)
- restoration of degraded lands (using erosion control, organic amendments and nutrient amendments (690).

Lower, but still substantial mitigation potential is provided by:
- rice management (210)
- livestock management (including improved feeding practices, dietary additives, breeding and other structural changes, and improved manure management (improved storage and handling and anaerobic digestion) (260) (*medium agreement, limited evidence*).

In addition, 770 MtCO₂-eq/yr could be provided by 2030 by improved energy efficiency in agriculture. This amount is, however, for a large part included in the mitigation potential of buildings and transport [8.1; 8.4].

At lower carbon prices, low cost measures most similar to current practice are favoured (e.g., cropland management options), but at higher carbon prices, more expensive measures with higher mitigation potentials per unit area are favoured (e.g., restoration of cultivated organic / peaty soils; Figure TS.20) (*medium agreement, limited evidence*) [8.4.3].

GHG emissions could also be reduced by substitution of fossil fuels by energy production from agricultural feedstocks (e.g., crop residues, dung, energy crops), which are counted in energy end-use sectors (particularly energy supply and transport). There are no accurate estimates of future agricultural biomass supply, with figures ranging from 22 EJ/yr in 2025

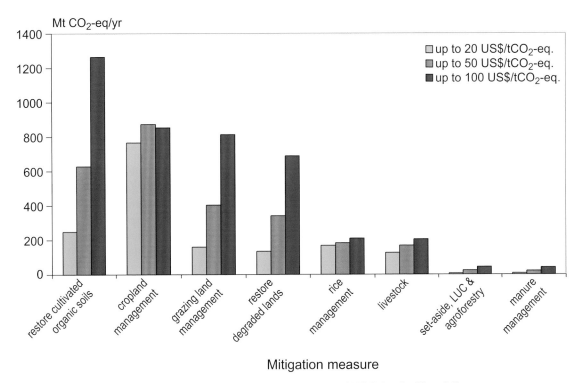

Figure TS.20: *Potential for GHG agricultural mitigation in 2030 at a range of carbon prices for a SRES B2 baseline [Figure 8.9].*
Note: B2 scenario shown, though the pattern is similar for all SRES scenarios. Energy-efficiency measures (770 $MtCO_2$-eq) are included in the mitigation potential of the buildings and energy sector.

to more than 400 EJ/yr in 2050. The actual contribution of agriculture to the mitigation potential by using bioenergy depends, however, on the relative prices of fuels and the balance of demand and supply. Top-down assessments that include assumptions on such a balance estimate the economic mitigation potential of biomass energy supplied from agriculture to be 70–1260 $MtCO_2$-eq/yr at up to 20 US$/$tCO_2$-eq, and 560–2320 $MtCO_2$-eq/yr at up to 50 US$/$tCO_2$-eq. There are no estimates for the additional potential from top-down models at carbon prices up to 100 US$/$tCO_2$-eq, but the estimate for prices above 100 US$/$tCO_2$-eq is 2720 $MtCO_2$-eq/yr. These potentials represent mitigation of 5–80%, and 20–90% of all other agricultural mitigation measures combined, at carbon prices of up to 20, and up to 50 US$/$tCO_2$-eq, respectively. Above the level where agricultural products and residues form the sole feedstock, bioenergy competes with other land-uses for available land, water and other resources The mitigation potentials of bioenergy and improved energy efficiency are not included in Table TS.11 or Figure TS.20, as the potential is counted in the user sectors, mainly transport and buildings, respectively (*medium agreement, medium evidence*) [8.4.4].

The estimates of mitigation potential in the agricultural sector are towards the lower end of the ranges indicated in the Second Assessment Report (SAR) and TAR. This is due mainly to the different time scales considered (2030 here versus 2050 in TAR). In the medium term, much of the mitigation potential is derived from removal of CO_2 from the atmosphere and its

conversion to soil carbon, but the magnitude of this process will diminish as soil carbon approaches maximum levels, and long-term mitigation will rely increasingly on reducing emissions of N_2O, CH_4, and CO_2 from energy use, the benefits of which persist indefinitely (*high agreement, much evidence*) [8.4.3].

Interactions of mitigation options with vulnerability and adaptation

Agricultural actions to mitigate GHGs could: a) reduce vulnerability (e.g. if soil carbon sequestration reduces the impacts of drought) or b) increase vulnerability (e.g., if heavy dependence on biomass energy makes energy supply more sensitive to climatic extremes). Policies to encourage mitigation and/or adaptation in agriculture may need to consider these interactions (*medium agreement, limited evidence*). Similarly, adaptation-driven actions may either a) favour mitigation (e.g., return of residues to fields to improve water-holding capacity will also sequester carbon) or b) hamper mitigation (e.g., use of more nitrogen fertilizer to overcome falling yields, leading to increased N_2O emissions). Strategies that simultaneously increase adaptive capacity, reduce vulnerability and mitigate climate change are likely to present fewer adoption barriers than those with conflicting impacts. For example increasing soil organic matter content can both improve fertility and reduce the impact of drought, improving adaptive capacity, making agriculture less vulnerable to climate change, while also sequestering carbon (*medium agreement, medium evidence*) [8.5].

Effectiveness of climate policies: opportunities, barriers and implementation issues

Actual levels of GHG mitigation practices in the agricultural sector are below the economic potential for the measures reported above (*medium agreement, limited evidence*). Little progress in implementation has been made because of the costs of implementation and other barriers, including: pressure on agricultural land, demand for agricultural products, competing demands for water as well as various social, institutional and educational barriers (*medium agreement, limited evidence*). Soil carbon sequestration in European croplands, for instance, is likely to be negligible by 2010, despite significant economic potential. Many of these barriers will not be overcome without policy/economic incentives (*medium agreement, limited evidence*) [8.6].

Integrated and non-climate policies affecting emissions of greenhouse gases

The adoption of mitigation practices will often be driven largely by goals not directly related to climate change. This leads to varying mitigation responses among regions, and contributes to uncertainty in estimates of future global mitigation potential. Policies most effective at reducing emissions may be those that also achieve other societal goals. Some rural development policies undertaken to fight poverty, such as water management and agro-forestry, are synergistic with mitigation (*medium agreement, limited evidence*). For example, agro-forestry undertaken to produce fuel wood or to buffer farm incomes against climate variation may also increase carbon sequestration. In many regions, agricultural mitigation options are influenced most by non-climate policies, including macro-economic, agricultural and environmental policies. Such policies may be based on UN conventions (e.g., Biodiversity and Desertification), but are often driven by national or regional issues. Among the most beneficial non-climate policies are those that promote sustainable use of soils, water and other resources in agriculture since these help to increase soil carbon stocks and minimize resource (energy, fertilizer) waste (*high agreement, medium evidence*) [8.7].

Co-benefits of greenhouse gas mitigation policies

Some agricultural practices yield purely 'win-win' outcomes, but most involve trade-offs. Agro-ecosystems are inherently complex. The co-benefits and trade-offs of an agricultural practice may vary from place to place because of differences in climate, soil or the way the practice is adopted (*high agreement, medium evidence*).

In producing bioenergy, for example, if the feedstock is crop residues, soil organic matter may be depleted as less carbon is returned, thus reducing soil quality; conversely, if the feedstock is a densely-rooted perennial crop, soil organic matter may be replenished, thereby improving soil quality.

Many agricultural mitigation activities show synergy with the goals of sustainability. Mitigation policies that encourage efficient use of fertilizers, maintain soil carbon and sustain agricultural production are likely to have the greatest synergy with sustainable development (*high agreement, medium evidence*).

For example, increasing soil carbon can also improve food security and economic returns. Other mitigation options have less certain impacts on sustainable development. For example, the use of some organic amendments may improve carbon sequestration, but impacts on water quality may vary depending on the amendment. Co-benefits often arise from improved efficiency, reduced cost and environmental co-benefits. Trade-offs relate to competition for land, reduced agricultural productivity and environmental stresses (*medium agreement, limited evidence*) [8.4.5].

Technology research, development, deployment, diffusion and transfer

Many of the mitigation strategies outlined for the agriculture sector employ existing technology. For example, reduction in emissions per unit of production will be achieved by increases in crop yields and animal productivity. Such increases in productivity can occur through a wide range of practices – better management, genetically modified crops, improved cultivars, fertilizer-recommendation systems, precision agriculture, improved animal breeds, improved animal nutrition, dietary additives and growth promoters, improved animal fertility, bioenergy feed stocks, anaerobic slurry digestion and CH_4 capture systems – all of which reflect existing technology (*high agreement, much evidence*). Some strategies involve new uses of existing technologies. For example, oils have been used in animal diets for many years to increase dietary energy content, but their role and feasibility as a CH_4 suppressant is still new and not fully defined. For some technologies, more research and development will be needed [8.9].

Long-term outlook

Global food demand may double by 2050, leading to intensified production practices (e.g., increasing use of nitrogen fertilizer). In addition, projected increases in the consumption of livestock products will increase CH_4 and N_2O emissions if livestock numbers increase, leading to growing emissions in the baseline after 2030. (*high agreement, medium evidence*). Agricultural mitigation measures will help to reduce GHG emissions per unit of product, relative to the baseline. However, until 2030 only about 10% of the mitigation potential is related to CH_4 and N_2O. Deployment of new mitigation practices for livestock systems and fertilizer applications will be essential to prevent an increase in emissions from agriculture after 2030.

Projecting long-term mitigation potentials is also hampered by other uncertainties. For example, the effects of climate change are unclear: future climate change may reduce soil

Table TS.12: *Estimates of forest area, net changes in forest area (negative numbers indicating decrease), carbon stock in living biomass and growing stock in 1990, 2000 and 2005 [Table 9.1].*

Region	Forest area (mill. ha) 2005	Annual change (mill. ha/yr) 1990-2000	Annual change (mill. ha/yr) 2000-2005	Carbon stock in living biomass (MtCO$_2$) 1990	2000	2005	Growing stock in 2005 (million m³)
Africa	635.412	-4.4	-4.0	241267	228067	222933	64957
Asia	571.577	-0.8	1.0	150700	130533	119533	47111
Europe a)	1001.394	0.9	0.7	154000	158033	160967	107264
North and Central America	705.849	-0.3	-0.3	150333	153633	155467	78582
Oceania	206.254	-0.4	-0.4	42533	41800	41800	7361
South America	831.540	-3.8	-4.3	358233	345400	335500	128944
World	3952.026	-8.9	-7.3	1097067	1057467	1036200	434219

Note:
a) including whole Russian Federation.

carbon-sequestration rates, or could even release soil carbon, though the effect is uncertain as climate change may also increase soil carbon inputs through higher plant production. Some studies have suggested that technological improvements could potentially counteract the negative impacts of climate change on cropland and grassland soil carbon stocks, making technological improvement a key factor in future GHG mitigation. Such technologies could, for example, act through increasing production, thereby increasing carbon returns to the soil and reducing the demand for fresh cropland. (*high agreement, medium evidence*) [8.10].

9 Forestry

Since the TAR, new mitigation estimates have become available from the local scale to the global scale. Major economic reviews and global assessments have become available. There is early research into the integration of mitigation and adaptation options and the linkages to sustainable development. There is increased attention on reducing emissions from deforestation as a low cost mitigation option, one that will have significant positive side effects. There is some evidence that climate change impacts can also constrain the mitigation potential of forests.

Status of the sector, development trends including production and consumption, and implications

Global forest cover is 3952 million ha (Table TS.12), about 30% of the world's land area. Most relevant for the carbon cycle is that between 2000 and 2005 gross deforestation continued at a rate of 12.9 million ha/yr, mainly as a result of converting forests to agricultural land, but also due to expansion of settlements and infrastructure, often for logging. In the 1990s, gross deforestation was slightly higher, 13.1 million ha/yr. Due to afforestation, landscape restoration and natural expansion of forests, the net loss of forest between 2000 and 2005 was 7.3 million ha/yr, with the largest losses in South America, Africa and Southeast Asia. This net rate of loss was lower than the 8.9 million ha/yr loss in the 1990s (*medium agreement, medium evidence*) [9.2.1].

Emission sources and sinks; trends

On the global scale, during the last decade of the 20th century, deforestation in the tropics and forest regrowth in the temperate zone and parts of the boreal zone remained the major factors responsible for CO$_2$ emissions and removals, respectively (Table TS.12, Figure TS.21). Emissions from deforestation in the 1990s are estimated at 5.8 GtCO$_2$/yr.

However, the extent to which the loss of carbon due to tropical deforestation is offset by expanding forest areas and accumulating woody biomass in the boreal and temperate zone is an area of disagreement between actual land observations and estimates using top-down models. The top-down methods based on inversion of atmospheric transport models estimate the net terrestrial carbon sink for the 1990s, the balance of sinks in northern latitudes and sources in the tropics, to be about 9.5 GtCO$_2$. The new estimates are consistent with the increase previously found in the terrestrial carbon sink in the 1990s over the 1980s, but the new sink estimates and the rate of increase may be smaller than previously reported. The residual sink estimate resulting from inversion of atmospheric transport models is significantly higher than any global sink estimate based on land observations.

The growing understanding of the complexity of the effects of land-surface change on the climate system shows the importance of considering the role of surface albedo, the fluxes of sensible and latent heat, evaporation and other factors in formulating policy for climate change mitigation in the forest

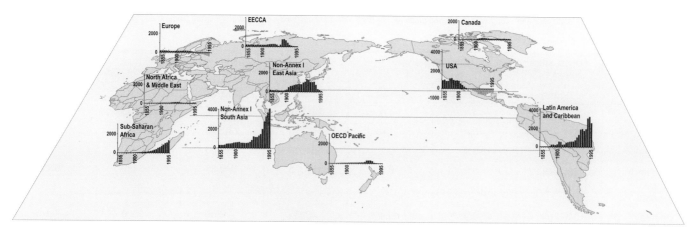

Figure TS.21: *Historical forest carbon balance (MtCO$_2$) per region, 1855–2000 [Figure 9.2].*

Notes: green = sink. EECCA =Countries of Eastern Europe, the Caucasus and Central Asia. Data averaged per 5-year period; year marks starting year of period.

sector. Complex modelling tools are needed to fully consider the climatic effect of changing land surface and to manage carbon stocks in the biosphere, but are not yet available. The potential effect of projected climate change on the net carbon balance in forests remains uncertain [9.3; 9.4].

As even the current functioning of the biosphere is uncertain, projecting the carbon balance of the global forestry sector remains very difficult. Generally, there is a lack of widely accepted studies and thus a lack of baselines. Trends for development in non-OECD countries, and thus of the deforestation rate, are unclear. In OECD countries and in economies in transition, development of management trends, the wood market, and impacts of climate change remain unclear. Long-term models as reported in Chapter 3, show baseline CO$_2$ emissions from land-use change and forestry in 2030 that are the same or slightly lower than in 2000 (medium agreement, medium evidence) [9.3; 9.4].

Description and assessment of mitigation technologies and practices, options and potentials, costs and sustainability

Terrestrial carbon dynamics are characterized by long periods of small rates of carbon uptake per hectare, interrupted by short periods of rapid and large releases of carbon during disturbances or harvest. While individual stands in a forest may be sources or sinks, the carbon balance of the forest is determined by the sum of the net balance of all stands.

Options available to reduce emissions by sources and/or increase removals by sinks in the forest sector are grouped into four general categories:
• maintaining or increasing the forest area;
• maintaining or increasing the site-level carbon density;
• maintaining or increasing the landscape-level carbon density and

• increasing off-site carbon stocks in wood products and enhancing product and fuel substitution.

Each mitigation activity has a characteristic time sequence of actions, carbon benefits and costs (Figure TS.22). Relative to a baseline, the largest short-term gains are always achieved through mitigation activities aimed at avoiding emissions (reduced deforestation or degradation, fire protection, slash burning, etc.).

	Mitigation Activities	Type of Impact	Timing of Impact	Timing of Cost
1A	Increase forest area (e.g. new forests)	⇑		
1B	Maintain forest area (e.g. prevent deforestation, LUC)	⬇		
2A	Increase site-level C density (e.g. intensive management, fertilize)	⇑		
2B	Maintain site-level C density (e.g. avoid degradation)	⬇		
3A	Increase landscape-scale C stocks (e.g. SFM, agriculture, etc.)	⇑		
3B	Maintain landscape-scale C stocks (e.g. suppress disturbances)	⬇		
4A	Increase off-site C in products (but must also meet 1B, 2B and 3B)	⇑		
4B	Increase bioenergy and substitution (but must also meet 1B, 2B and 3B)	⬇		

Legend

Type of Impact		Timing (change in Carbon over time)		Timing of cost (dollars ($) over time)	
Enhance sink	⇑	Delayed		Delayed	
Reduce source	⬇	Immediate		Up-front	
		Sustained or repeatable		On-going	

Figure TS.22: *Generalized summary of the options available in the forest sector and their type and timing of effects on carbon stocks and the timing of costs [Figure 9.4].*

All forest-management activities aimed at increasing site-level and landscape-level carbon density are common practices that are technically feasible, but the extent and area over which they can be implemented could be increased considerably. Economic considerations are typically the main constraint, because retaining additional carbon on site delays revenues from harvest.

In the long term, a sustainable forest-management strategy aimed at maintaining or increasing forest carbon stocks, while producing an annual yield of timber, fibre or energy from the forest, will generate the largest sustained mitigation benefit.

Regional modelling assessments

Bottom-up regional studies show that forestry mitigation options have the economic potential (at costs up to 100 US$/tCO_2-eq) to contribute 1.3-4.2 MtCO_2/yr (average 2.7 GtCO_2/yr) in 2030 excluding bioenergy. About 50% can be achieved at a cost under 20 US$/tCO_2 (1.6 GtCO_2/yr) with large differences between regions. The combined effects of reduced deforestation and degradation, afforestation, forest management, agro-forestry and bioenergy have the potential to increase from the present to 2030 and beyond. This analysis assumes gradual implementation of mitigation activities starting now (*medium agreement, medium evidence*) [9.4.4].

Global top-down models predict mitigation potentials of 13.8 GtCO_2-eq/yr in 2030 at carbon prices less than or equal to 100 US$/tCO_2. The sum of regional predictions is 22% of this value for the same year. Regional studies tend to use more detailed data and consider a wider range of mitigation options, and thus may more accurately reflect regional circumstances and constraints than simpler, more aggregated global models. However, regional studies vary in model structure, coverage, analytical approach and assumptions (including baseline

assumptions). Further research is required to narrow the gap in the estimates of mitigation potential from global and regional assessments (*medium agreement, medium evidence*) [9.4.3].

The best estimate of the economic mitigation potential for the forestry sector at this stage therefore cannot be more certain than a range between 2.7 and 13.8 GtCO_2/yr in 2030, for costs <100 US$/tCO_2; for costs <20 US$/tCO_2 the range is 1.6 to 5 GtCO_2/yr. About 65% of the total mitigation potential (up to 100 US$/tCO_2-eq) is located in the tropics and about 50% of the total could be achieved by reducing emissions from deforestation (*low agreement, medium evidence*).

Forestry can also contribute to the provision of bioenergy from forest residues. The potential of bioenergy, however, is counted in the power supply, transportation (biofuels), industry and building sectors (see Chapter 11 for an overview). Based on bottom-up studies of potential biomass supply from forestry, and assuming that all of that will be used (which depends entirely on the cost of forestry biomass compared with other sources) a contribution in the order of 0.4 GtCO_2/yr could come from forestry.

Global top-down models are starting to provide insight on where and which of the carbon mitigation options can best be allocated on the globe (Figure TS.24).

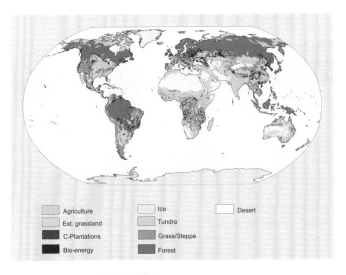

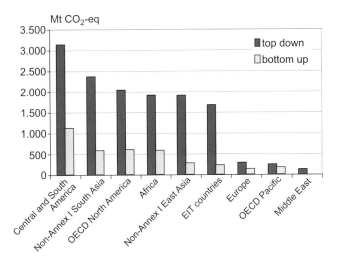

Figure TS.23: *Comparison of outcomes of economic mitigation potential at <100 US$/tCO_2-eq in 2030 in the forestry sector, as based on top-down global models versus the regional modelling results [Figure 9.13].*

Figure TS.24: *Allocation of global afforestation activities as given by two global top-down models. Top: location of bioenergy and carbon plantations in the world in 2100; bottom: percentage of a grid cell afforested in 2100 [Figure 9.11].*

Interactions of mitigation options with vulnerability and adaptation

Mitigation activities for forestry can be designed to be compatible with adapting to climate change, maintaining biodiversity and promoting sustainable development. Comparing environmental and social co-benefits and costs with the carbon benefit will highlight trade-offs and synergies and help promote sustainable development.

The literature on the interaction between forestry mitigation and climate change is in its infancy. Forests are likely to be impacted by climate change, which could reduce their mitigation potential. A primary management adaptation option is to reduce as many ancillary stresses on the forest as possible. Maintaining widely dispersed and viable populations of individual species minimizes the probability of localized catastrophic events causing species extinction. Formation of protected areas or nature reserves is an example of mitigation as well as adaptation. Protecting areas (with corridors) also leads to conservation of biodiversity, in turn reducing vulnerability to climate change.

Forestry-mitigation projects provide adaptation co-benefits for other sectors. Examples include agro-forestry reducing the vulnerability to drought of rain-fed crop income, mangroves reducing the vulnerability of coastal settlements, and shelter belts slowing desertification *(medium agreement, medium evidence)* [9.5].

Effectiveness of and experience with climate policies, potentials, barriers and opportunities/ implementation issues

Forestry can make a very significant contribution to a low cost global mitigation portfolio that provides synergies with adaptation and sustainable development. Chapter 9 of this report identifies a whole set of options and policies to achieve this mitigation potential. However, this opportunity has so far not been taken because of the current institutional context, lack of incentives for forest managers and lack of enforcement of existing regulations. Without better policy instruments, only a small portion of this potential is likely to be realized.

Realization of the mitigation potential requires institutional capacity, investment capital, technology, R&D and transfer, as well as appropriate (international) policies and incentives. In many regions, their absence has been a barrier to implementation of forestry-mitigation activities. Notable exceptions exist, however, such as regional successes in reducing deforestation rates and implementing afforestation programmes *(high agreement, much evidence)*.

Multiple and location-specific strategies are required to guide mitigation policies in the sector. The optimum choices depend on the current state of the forests, the dominant drivers of forest change, and the anticipated future dynamics of the forests within each region. Participation of all stakeholders and policy-makers

is necessary to promote mitigation projects and design an optimal mix of measures. Integration of mitigation in the forestry sector into land-use planning could be important in this respect.

Most existing policies to slow tropical deforestation have had minimal impact due to lack of regulatory and institutional capacity or countervailing profitability incentives. In addition to more dedicated enforcement of regulations, well-constructed carbon markets or other environmental service payment schemes may help overcome barriers to reducing deforestation by providing positive financial incentives for retaining forest cover.

There have been several proposals to operationalize activities post 2012, including market-based as well as non-market based approaches; for example, through a dedicated fund to voluntarily reduce emissions from deforestation. Policy measures such as subsidies and tax exemptions have been used successfully to encourage afforestation and reforestation both in developed and developing countries. Care must be taken, however, to avoid possible negative environmental and social impacts of large-scale plantation establishment.

Despite relative low costs and many potential positive side effects of afforestation and reforestation under the Clean Development Mechanism (CDM), not many project activities are yet being implemented due to a number of barriers, including the late agreement on and complexity of the rules governing afforestation and reforestation CDM project activities. The requirements for forestry mitigation projects to become viable on a larger scale include certainty over future commitments, streamlined and simplified rules, and reductions in transaction costs. Standardization of project assessment can play an important role in overcoming uncertainties among potential buyers, investors and project participants *(high agreement, medium evidence)* [9.6].

Forests and Sustainable Development

While the assessment in the forestry chapter identifies remaining uncertainties about the magnitude of the mitigation benefits and costs, the technologies and knowledge required to implement mitigation activities exist today. Forestry can make a significant and sustained contribution to a global mitigation portfolio, while also meeting a wide range of social, economic and ecological objectives. Important co-benefits can be gained by considering forestry mitigation options as an element of broader land-management plans.

Plantations can contribute positively, for example, to employment, economic growth, exports, renewable energy supply and poverty alleviation. In some instances, plantations may also lead to negative social impacts such as loss of grazing land and source of traditional livelihoods. Agro-forestry can produce a wide range of economic, social and environmental benefits; probably wider than large-scale afforestation. Since ancillary benefits tend to be local rather than global, identifying and accounting for them can reduce or partially compensate the costs of the mitigation measures *(high agreement, medium evidence)* [9.7].

Technology research, development, deployment, diffusion and transfer

The deployment, diffusion and transfer of technologies such as improved forest-management systems, forest practices and processing technologies including bioenergy, are key to improving the economic and social viability of the different mitigation options. Governments could play a critical role in providing targeted financial and technical support, promoting the participation of communities, institutions and NGOs *(high agreement, much evidence)* [9.8].

Long-term outlook

Uncertainties in the carbon cycle, the uncertain impacts of climate change on forests and its many dynamic feedbacks, time-lags in the emission-sequestration processes, as well as uncertainties in future socio-economic paths (e.g., to what extent deforestation can be substantially reduced in the coming decades) cause large variations in future carbon balance projections for forests.

Overall, it is expected that in the long-term, mitigation activities will help increase the carbon sink, with the net balance depending on the region. Boreal primary forests will either be small sources or sinks depending on the net effect of enhancement of growth versus a loss of soil organic matter and emissions from increased fires. Temperate forests will probably continue to be net carbon sinks, favoured also by enhanced forest growth due to climate change. In the tropical regions, human-induced land-use changes are expected to continue to drive the dynamics for decades. Beyond 2040, depending very particularly on the effectiveness of policies aimed at reducing forest degradation and deforestation, tropical forests may become net sinks, depending on the influence of climate change. Also, in the medium to long term, commercial bioenergy is expected to become increasingly important.

Developing optimum regional strategies for climate change mitigation involving forests will require complex analyses of the trade-offs (synergies and competition) in land-use between forestry and other land-uses, trade-offs between forest conservation for carbon storage and other environmental services such as biodiversity and watershed conservation and sustainable forest harvesting to provide society with carbon-containing fibre, timber and bioenergy resources, and trade-offs among utilization strategies of harvested wood products aimed at maximizing storage in long-lived products, recycling, and use for bioenergy [9.9].

10 Waste management

Status of the sector, development trends and implications

Waste generation is related to population, affluence and urbanization. Current global rates of post-consumer waste generation are estimated to be 900-1300 Mt/yr. Rates have been increasing in recent years, especially in developing countries with rapid population growth, economic growth and urbanization. In highly developed countries, a current goal is to decouple waste generation from economic driving forces such as GDP — recent trends suggest that per capita rates of post-consumer waste generation may be peaking as a result of recycling, re-use, waste minimization, and other initiatives *(medium agreement, medium evidence)* [10.1, 10.2].

Post-consumer waste is a small contributor to global GHG emissions (<5%), with landfill CH_4 accounting for >50% of current emissions. Secondary sources of emissions are wastewater CH_4 and N_2O; in addition, minor emissions of CO_2 result from incineration of waste containing fossil carbon. In general, there are large uncertainties with respect to quantification of direct emissions, indirect emissions and mitigation potentials for the waste sector, which could be reduced by consistent and coordinated data collection and analysis at the national level. There are currently no inventory methods for annual quantification of GHG emissions from waste transport, nor for annual emissions of fluorinated gases from post-consumer waste *(high agreement, much evidence)* [10.3].

It is important to emphasize that post-consumer waste constitutes a significant renewable energy resource that can be exploited through thermal processes (incineration and industrial co-combustion), landfill gas utilization and use of anaerobic digester biogas. Waste has an economic advantage in comparison to many biomass resources because it is regularly collected at public expense. The energy content of waste can be most efficiently exploited using thermal processes: during combustion, energy is obtained directly from biomass (paper products, wood, natural textiles, food) and from fossil carbon sources (plastics, synthetic textiles). Assuming an average heating value of 9 GJ/t, global waste contains >8 EJ of available energy, which could increase to 13 EJ (nearly 2% of primary energy demand) in 2030 *(medium agreement, medium evidence)* [10.1]. Currently, more than 130 million tonnes/yr of waste are combusted worldwide, which is equivalent to >1 EJ/yr. The recovery of landfill CH_4 as a source of renewable energy was commercialized more than 30 years ago with a current energy value of >0.2 EJ/yr. Along with thermal processes, landfill gas and anaerobic digester gas can provide important local sources of supplemental energy *(high agreement, much evidence)* [10.1, 10.3].

Because of landfill gas recovery and complementary measures (increased recycling and decreased landfilling through the implementation of alternative technologies), emissions of CH_4 from landfills in developed countries have been largely stabilized. Choices for mature, large-scale waste management technologies to avoid or reduce GHG emissions compared with landfilling include incineration for waste-to-energy and biological processes such as composting or mechanical-biological treatment (MBT). However, in developing countries, landfill CH_4 emissions are increasing as more controlled (anaerobic) landfilling practices are being implemented. This is especially true for rapidly urbanizing areas where engineered landfills provide a more environmentally acceptable waste-disposal strategy than open dumpsites by reducing disease vectors, toxic odours, uncontrolled combustion and pollutant emissions to air, water and soil. Paradoxically, higher GHG emissions occur as the aerobic production of CO_2 (by burning and aerobic decomposition) is shifted to anaerobic production of CH_4. To a large extent, this is the same transition to sanitary landfilling that occurred in many developed countries during 1950–1970. The increased CH_4 emissions can be mitigated by accelerating the introduction of engineered gas recovery, aided by Kyoto mechanisms such as CDM and Joint Implementation (JI). As of late October 2006, landfill gas recovery projects accounted for 12% of the average annual Certified Emission Reductions (CERs) under CDM. In addition, alternative waste management strategies such as recycling and composting can be implemented in developing countries. Composting can provide an affordable, sustainable alternative to engineered landfills, especially where more labour-intensive, lower-technology strategies are applied to selected biodegradable waste streams (*high agreement, medium evidence*) [10.3].

Recycling, re-use and waste minimization initiatives, both public and private, are indirectly reducing GHG emissions by decreasing the mass of waste requiring disposal. Depending on regulations, policies, markets, economic priorities and local constraints, developed countries are implementing increasingly higher recycling rates to conserve resources, offset fossil fuel use, and avoid GHG generation. Quantification of global recycling rates is not currently possible because of varying baselines and definitions; however, local reductions of >50% have been achieved. Recycling could be expanded practically in many countries to achieve additional reductions. In developing countries, waste scavenging and informal recycling are common practices. Through various diversion and small-scale recycling activities, those who make their living from decentralized waste management can significantly reduce the mass of waste that requires more centralized solutions. Studies indicate that low-technology recycling activities can also generate significant employment through creative microfinance and other small-scale investments. The challenge is to provide safer, healthier working conditions than currently experienced by waste scavengers at uncontrolled dumpsites (*medium agreement, medium evidence*) [10.3].

For wastewater, only about 60% of the global population has sanitation coverage (sewerage). For wastewater treatment, almost 90% of the population in developed countries but less than 30% in developing countries has improved sanitation (including sewerage and waste water treatment, septic tanks, or latrines). In addition to GHG mitigation, improved sanitation and wastewater management provide a wide range of health and environmental co-benefits (*high agreement, much evidence*) [10.2, 10.3].

With respect to both waste and wastewater management in developing countries, two key constraints to sustainable development are the lack of financial resources and the selection of appropriate and truly sustainable technologies for a particular setting. It is a significant and costly challenge to implementing waste and wastewater collection, transport, recycling, treatment and residuals management in many developing countries. However, the implementation of sustainable waste and wastewater infrastructure yields multiple co-benefits to assist with the implementation of Millennium Development Goals (MDGs) via improved public health, conservation of water resources, and reduction of untreated discharges to air, surface water, groundwater, soils and coastal zones (*high agreement, much evidence*) [10.4].

Emission trends

With total 2005 emissions of approximately 1300 $MtCO_2$-eq/yr, the waste sector contributes about 2–3% of total GHG emissions from Annex I and EIT countries and 4–5% from non-Annex I countries (see Table TS.13). For 2005–2020, business-as-usual (BAU) projections indicate that landfill CH_4 will remain the largest source at 55–60% of the total. Landfill CH_4 emissions are stabilizing and decreasing in many developed countries as a result of increased landfill gas recovery combined with waste diversion from landfills through recycling, waste minimization and alternative thermal and biological waste management strategies. However, landfill CH_4 emissions are increasing in developing countries because of larger quantities of municipal solid waste from rising urban populations, increasing economic development and, to some extent, the replacement of open burning and dumping by engineered landfills. Without additional measures, a 50% increase in landfill CH_4 emissions from 2005 to 2020 is projected, mainly from the Non-Annex I countries. Wastewater emissions of CH_4 and N_2O from developing countries are also rising rapidly with increasing urbanization and population. Moreover, because the wastewater emissions in Table TS.13 are based on human sewage only and are not available for all developing countries, these emissions are underestimated (*high agreement, medium evidence*) [10.1, 10.2, 10.3, 10.4].

Table TS.13: *Trends for GHG emissions from waste using 1996 and 2006 UNFCCC inventory guidelines, extrapolations and BAU projections (MtCO$_2$-eq, rounded) [Table 10.3].*

Source	1990	1995	2000	2005	2010	2015	2020	Notes
Landfill CH$_4$	550	585	590	635	700	795	910	Averaged using 1996/2006 guidelines
Wastewater[a] CH$_4$	450	490	520	590	600	630	670	1996 guidelines
Wastewater[a] N$_2$O	80	90	90	100	100	100	100	1996 guidelines
Incineration CO$_2$	40	40	50	50	50	60	60	2006 guidelines
Total	1120	1205	1250	1375	1450	1585	1740	

Note:
[a] wastewater emissions are underestimated - see text.

Description and assessment of mitigation technologies and practices, options and potentials, costs and sustainability

Existing waste management technologies can effectively mitigate GHG emissions from this sector – a wide range of mature, low- to high-technology, environmentally-effective strategies are commercially available to mitigate emissions and provide co-benefits for improved public health and safety, soil protection, pollution prevention and local energy supply. Collectively, these technologies can directly reduce GHG emissions (through landfill CH$_4$ recovery and utilization, improved landfill practices, engineered wastewater management, utilization of anaerobic digester biogas) or avoid significant GHG generation (through controlled composting of organic waste, state-of-the-art incineration, expanded sanitation coverage). In addition, waste minimization, recycling and re-use represent an important and increasing potential for indirect reduction of GHG emissions through the conservation of raw materials, improved energy and resource efficiency and fossil fuel avoidance. For developing countries, environmentally responsible waste management at an appropriate level of technology promotes sustainable development and improves public health *(high agreement, much evidence)* [10.4].

Because waste management decisions are often made locally without concurrent quantification of GHG mitigation, the importance of the waste sector for reducing global GHG emissions has been underestimated *(high agreement, medium evidence)* [10.1; 10.4]. Flexible strategies and financial incentives can expand waste management options to achieve GHG mitigation goals – in the context of integrated waste management, local technology decisions are a function of many competing variables, including waste quantity and characteristics, cost and financing issues, regulatory constraints and infrastructure requirements, including available land area and collection/transportation considerations. Life-cycle assessment (LCA) can provide decision-support tools *(high agreement, much evidence)* [10.4].

Landfill CH$_4$ emissions are directly reduced through engineered gas extraction and recovery systems consisting of vertical wells and/or horizontal collectors. In addition, landfill gas offsets the use of fossil fuels for industrial or commercial process heating, onsite generation of electricity or as a feedstock for synthetic natural gas fuels. Commercial recovery of landfill CH$_4$ has occurred at full scale since 1975 with documented utilization in 2003 at 1150 plants recovering 105 MtCO$_2$–eq/yr. Because there are also many projects that flare gas without utilization, the total recovery is likely to be at least double this figure *(high agreement, medium evidence)* [10.1; 10.4]. A linear regression using historical data from the early 1980s to 2003 indicates a growth rate for landfill CH$_4$ utilization of approximately 5% per year. In addition to landfill gas recovery, the further development and implementation of landfill 'biocovers' can provide an additional low cost, biological strategy to mitigate emissions since landfill CH$_4$ (and non-methane volatile organic compounds (NMVOCs)) emissions are also reduced by aerobic microbial oxidation in landfill-cover soils *(high agreement, much evidence)* [10.4].

Incineration and industrial co-combustion for waste-to-energy provide significant renewable energy benefits and fossil fuel offsets at >600 plants worldwide, while producing very minor GHG emissions compared with landfilling. Thermal processes with advanced emission controls are a proven technology but more costly than controlled landfilling with landfill gas recovery *(high agreement, medium evidence)* [10.4].

Controlled biological processes can also provide important GHG mitigation strategies, preferably using source-separated waste streams. Aerobic composting of waste avoids GHG generation and is an appropriate strategy for many developed and developing countries, either as a stand-alone process or as part of mechanical-biological treatment. In many developing countries, notably China and India, small-scale low-technology anaerobic digestion has also been practised for decades. Since higher-technology incineration and composting plants have proved unsustainable in a number of developing countries, lower-technology composting or anaerobic digestion can be implemented to provide sustainable waste management solutions *(high agreement, medium evidence)* [10.4].

For 2030, the total economic reduction potential for CH$_4$ emissions from landfilled waste at costs of <20 US$/tCO$_2$-eq

Table TS.14: *Ranges for economic mitigation potential for regional landfill CH_4 emissions at various cost categories in 2030, see notes [Table 10.5].*

Region	Projected emissions in 2030 (MtCO$_2$-eq)	Total economic mitigation potential at <100 US$/tCO$_2$-eq (MtCO$_2$-eq)	Economic mitigation potential (MtCO$_2$-eq) at various cost categories (US$/tCO$_2$-eq)			
			<0	0-20	20-50	50-100
OECD	360	100-200	100-120	20-100	0-7	1
EIT	180	100	30-60	20-80	5	1-10
Non-OECD	960	200-700	200-300	30-100	0-200	0-70
Global	1500	400-1000	300-500	70-300	5-200	10-70

Notes:
[1] Costs and potentials for wastewater mitigation are not available.
[2] Regional numbers are rounded to reflect the uncertainty in the estimates and may not equal global totals.
[3] Landfill carbon sequestration not considered.
[4] The timing of measures limiting landfill disposal affects the annual mitigation potential in 2030. The upper limits assume that landfill disposal is limited in the coming years to 15% of the waste generated globally. The lower limits reflect a more realistic timing for implementation of measures reducing landfill disposal.

ranges between 400 and 800 MtCO$_2$-eq. Of this total, 300–500 MtCO$_2$-eq/yr has negative cost (Table TS.14). For the long term, if energy prices continue to increase, there will be more profound changes in waste management strategies related to energy and materials recovery in both developed and developing countries. Thermal processes, which have higher unit costs than landfilling, become more viable as energy prices increase. Because landfills continue to produce CH$_4$ for many decades, both thermal and biological processes are complementary to increased landfill gas recovery over shorter time frames *(high agreement, limited evidence)* [10.4].

For wastewater, increased levels of improved sanitation in developing countries can provide multiple benefits for GHG mitigation, improved public health, conservation of water resources and reduction of untreated discharges to water and soils. Historically, urban sanitation in developed countries has focused on centralized sewerage and wastewater treatment plants, which are too expensive for rural areas with low population density and may not be practical to implement in rapidly growing, peri-urban areas with high population density. It has been demonstrated that a combination of low cost technology with concentrated efforts for community acceptance, participation and management can successfully expand sanitation coverage. Wastewater is also a secondary water resource in countries with water shortages where water re-use and recyling could assist many developing and developed countries with irregular water supplies. These measures also encourage smaller wastewater treatment plants with reduced nutrient loads and proportionally lower GHG emissions. Estimates of global or regional mitigation costs and potentials for wastewater are not currently available *(high agreement, limited evidence)* [10.4].

Effectiveness of and experience with climate policies, potentials, barriers and opportunities/ implementation issues

Because landfill CH$_4$ is the dominant GHG from this sector, a major strategy is the implementation of standards that encourage or mandate landfill CH$_4$ recovery. In developed countries, landfill CH$_4$ recovery has increased as a result of direct regulations requiring landfill gas capture, voluntary measures including GHG-emissions credits trading and financial incentives (including tax credits) for renewable energy or green power. In developing countries, it is anticipated that landfill CH$_4$ recovery will increase during the next two decades as controlled landfilling is phased in as a major waste disposal strategy. JI and the CDM have already proved to be useful mechanisms for external investment from industrialized countries, especially for landfill gas recovery projects where the lack of financing is a major impediment. The benefits are twofold: reduced GHG emissions with energy benefits from landfill CH$_4$ plus upgraded landfill design and operations. Currently (late October 2006), under the CDM, the annual average CERs for the 33 landfill gas recovery projects constitute about 12% of the total. Most of these projects (Figure TS.25) are located in Latin-American countries (72% of landfill gas CERs), dominated by Brazil (9 projects; 48% of CERs) *(high agreement, medium evidence)* [10.4].

In the EU, landfill gas recovery is mandated at existing sites, while the landfilling of organic waste is being phased out via the landfill directive (1999/31/EC). This directive requires, by 2016, a 65% reduction relative to 1995 in the mass of biodegradable organic waste that is landfilled annually. As a result, post-consumer waste is being diverted to incineration and to mechanical and biological treatment (MBT) before landfilling to recover recyclables and reduce the organic carbon content. In 2002, EU waste-to-energy plants generated about 40 million GJ of electrical and 110 million GJ of thermal energy, while between 1990 and 2002, landfill CH$_4$ emissions in the EU decreased by

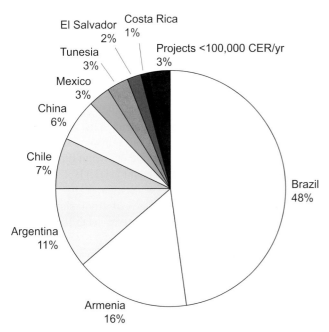

Figure TS.25: *Distribution of landfill gas CDM projects based on average annual CERs for registered projects late October, 2006 [Figure 10.9].*

Note: Includes 11 MtCO$_2$-eq/yr CERs for landfill CH$_4$ out of 91 MtCO$_2$-eq/yr total. Projects <100,000 CERs/yr are located in Israel, Bolivia, Bangladesh and Malaysia.

almost 30% due to the landfill directive and related national legislation *(high agreement, much evidence)* [10.4, 10.5].

Integrated and non-climate policies affecting emissions of greenhouse gases: GHG mitigation as the co-benefit of waste policies and regulations; role of sustainable development

GHG mitigation is often not the primary driver, but is itself a co-benefit of policies and measures in the waste sector that address broad environmental objectives, encourage energy recovery from waste, reduce use of virgin materials, restrict choices for ultimate waste disposal, promote waste recycling and re-use and encourage waste minimization. Policies and measures to promote waste minimization, re-use and recycling indirectly reduce GHG emissions from waste. These measures include Extended Producer Responsibility (EPR), unit pricing (or PAYT/'Pay As You Throw') and landfill taxes. Other measures include separate and efficient collection of recyclables together with both unit pricing and landfill tax systems. Some Asian countries are encouraging 'circular economy' or 'sound material-cycle society' as a new development strategy whose core concept is the circular (closed) flow of materials and the use of raw materials and energy through multiple phases. Because of limited data, differing baselines and other regional conditions, it is not currently possible to quantify the global effectiveness of these strategies in reducing GHG emissions *(medium agreement, medium evidence)* [10.5].

In many countries, waste and wastewater management policies are closely integrated with environmental policies

and regulations pertaining to air, water and soil quality as well as to renewable energy initiatives. Renewable-energy programmes include requirements for electricity generation from renewable sources, mandates for utilities to purchase power from small renewable providers, renewable energy tax credits, and green power initiatives, which allow consumers to choose renewable providers. In general, the decentralization of electricity generation capacity via renewables can provide strong incentives for electrical generation from landfill CH$_4$ and thermal processes for waste-to-energy *(high agreement, much evidence)* [10.5].

Although policy instruments in the waste sector consist mainly of regulations, there are also economic measures in a number of countries to encourage particular waste management technologies, recycling and waste minimization. These include incinerator subsidies or tax exemptions for waste-to-energy. Thermal processes can most efficiently exploit the energy value of post-consumer waste, but must include emission controls to limit emissions of secondary air pollutants. Subsidies for the construction of incinerators have been implemented in several countries, usually combined with standards for energy efficiency. Tax exemptions for electricity generated by waste incinerators and for waste disposal with energy recovery have also been adopted *(high agreement, much evidence)* [10.5].

The co-benefits of effective and sustainable waste and wastewater collection, transport, recycling, treatment and disposal include GHG mitigation, improved public health, conservation of water resources and reductions in the discharge of untreated pollutants to air, soil, surface water and groundwater. Because there are many examples of abandoned waste and wastewater plants in developing countries, it must be stressed that a key aspect of sustainable development is the selection of appropriate technologies that can be sustained within the specific local infrastructure *(high agreement, medium evidence)* [10.5].

Technology research, development and diffusion

In general, the waste sector is characterized by mature technologies that require further diffusion in developing countries. Advances under development include:

- Landfilling: Implementation of optimized gas collection systems at an early stage of landfill development to increase long-term gas collection efficiency. Optimization of landfill biodegradation (bioreactors) to provide greater process control and shorter waste degradation lifetimes. Construction of landfill 'biocovers' that optimize microbial oxidation of CH$_4$ and NMVOCs to minimize emissions.
- Biological processes: For developing countries, lower-technology, affordable sustainable composting and anaerobic digestion strategies for source-separated biodegradable waste.
- Thermal processes: Advanced waste-to-energy technologies that can provide higher thermal and electrical efficiencies

than current incinerators (10–20% net electrical efficiency). Increased implementation of industrial co-combustion using feedstocks from various waste fractions to offset fossil fuels. Gasification and pyrolysis of source-separated waste fractions in combination with improved, lower-cost separation technologies for production of fuels and feedstocks.

- Recycling, re-use, waste minimization, pre-treatment (improved mechanical-biological treatment processes) Innovations in recycling technology and process improvements resulting in decreased use of virgin materials, energy conservation, and fossil fuel offsets. Development of innovative but low-technology recycling solutions for developing countries.

- Wastewater: New low-technology ecological designs for improved sanitation at the household and small community level, which can be implemented sustainably for efficient small-scale wastewater treatment and water conservation in both developed and developing countries *(high agreement, limited evidence)* [10.5; 10.6].

Long-term outlook, systems transitions

To minimize future GHG emissions from the waste sector, it is important to preserve local options for a wide range of integrated and sustainable management strategies. Furthermore, primary reductions in waste generation through recycling, re-use, and waste minimization can provide substantial benefits for the conservation of raw materials and energy. Over the long term, because landfills continue to produce CH_4 for decades, landfill gas recovery will be required at existing landfills even as many countries change to non-landfilling technologies such as incineration, industrial co-combustion, mechanical-biological treatment, large-scale composting and anaerobic digestion. In addition, the 'back-up' landfill will continue to be a critical component of municipal solid waste planning. In developing countries, investment in improved waste and wastewater management confers significant co-benefits for public health and safety, environmental protection and infrastructure development.

11 Mitigation from a cross-sectoral perspective

Mitigation options across sectors

While many of the technological, behavioural and policy options mentioned in Chapters 4–10 concern specific sectors, some technologies and policies reach across many sectors; for example, the use of biomass and the switch from high-carbon fuels to gas affect energy supply, transport, industry and buildings. Apart from potentials for common technologies, these examples also highlight possible competition for resources, such as finance and R&D support [11.2.1].

The bottom-up compilation of mitigation potentials by sector is complicated by interactions and spill-overs between sectors, over time and over regions and markets. A series of formal procedures has been used to remove potential double counting, such as reduction of the capacity needed in the power sector due to electricity saving in industry and the buildings sector. An integration of sector potentials in this way is required to summarize the sectoral assessments of Chapters 4–10. The uncertainty of the outcome is influenced by issues of comparability of sector calculations, difference in coverage between the sectors (e.g., the transport sector) and the aggregation itself, in which only the main and direct sector interactions have been taken into account [11.3.1].

The top-down estimates were derived from stabilization scenarios, i.e., runs towards long-term stabilization of atmospheric GHG concentration [3.6].

Figure TS.26A and Table TS.15 show that the bottom-up assessments emphasize the opportunities for no-regrets options in many sectors, with a bottom-up estimate for all sectors by 2030 of about 6 GtCO$_2$-eq at negative costs; that is, net benefits. A large share of the no-regrets options is in the building sector. The total for bottom-up low cost options (no-regrets and other options costing less than 20 US$/tCO$_2$-eq) is around 13 GtCO$_2$-eq (ranges are discussed below). There are additional bottom-up potentials of around 6 and 4 GtCO$_2$-eq at additional costs of <50 and 100 US$/tCO$_2$-eq respectively *(medium agreement, medium evidence)* [11.3.1].

There are several qualifications to these estimates in addition to those mentioned above. First, in the bottom-up estimates a set of emission-reduction options, mainly for co-generation, parts of the transport sector and non-technical options such as behavioural changes, are excluded because the available literature did not allow a reliable assessment. It is estimated that the bottom-up potentials are therefore underestimated by 10–15%. Second, the chapters identify a number of key sensitivities that have not been quantified, relating to energy prices, discount rates and the scaling-up of regional results for the agricultural and forestry options. Third, there is a lack of estimates for many EIT countries and substantial parts of the non-OECD/EIT region [11.3.1].

The estimates of potentials at carbon prices <20 US$/tCO$_2$-eq are lower than the TAR bottom-up estimates that were evaluated for carbon prices <27 US$/tCO$_2$-eq, due to better information in recent literature *(high agreement, much evidence)*.

Figure TS.15 and Table TS.16 show that the overall bottom-up potentials are comparable with those of the 2030 results from top-down models, as reported in Chapter 3.

At the sectoral level, there are larger differences between bottom-up and top-down, mainly because the sector definitions in top-down models often differ from those in bottom-up assessments (table TS.17). Although there are slight differences

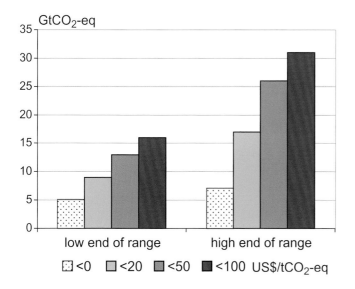

Figure TS.26A: *Global economic mitigation potential in 2030 estimated from bottom-up studies. Data from Table TS.15. [Figure 11.3].*

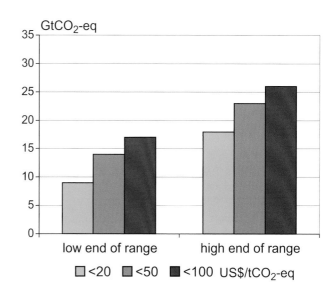

Figure TS.26B: *Global economic mitigation potential in 2030 estimated from top-down studies. Data from Table TS.16. [Figure 11.3].*

between the baselines assumed for top-down and bottom-up assessments, the results are close enough to provide a robust estimate of the overall economic mitigation potential by 2030. The mitigation potential at carbon prices of <100 US$/tCO$_2$-eq is about 25–50% of 2030 baseline emissions *(high agreement, much evidence)*.

Table TS.17 shows that for point-of-emission analysis[18] a large part of the long-term mitigation potential is in the energy-

supply sector. However, for an end-use sector analysis as used for the results in Figure TS.27, the highest potential lies in the building and agriculture sectors. For agriculture and forestry, top-down estimates are lower than those from bottom-up studies. This is because these sectors are generally not well covered in top-down models. The energy supply and industry estimates from top-down models are generally higher than those from bottom-up assessments *(high agreement, medium evidence)* [11.3.1].

Table TS.15: *Global economic mitigation potential in 2030 estimated from bottom-up studies [11.3].*

Carbon price (US$/tCO$_2$-eq)	Economic potential (GtCO$_2$-eq/yr)	Reduction relative to SRES A1 B (68 GtCO$_2$-eq/yr) (%)	Reduction relative to SRES B2 (49 GtCO$_2$-eq/yr) (%)
0	5-7	7-10	10-14
20	9-17	14-25	19-35
50	13-26	20-38	27-52
100	16-31	23-46	32-63

Table TS.16: *Global economic mitigation potential in 2030 estimated from top-down studies [11.3].*

Carbon price (US$/tCO$_2$-eq)	Economic potential (GtCO$_2$-eq/yr)	Reduction relative to SRES A1 B (68 GtCO$_2$-eq/yr) (%)	Reduction relative to SRES B2 (49 GtCO$_2$-eq/yr) (%)
20	9-18	13-27	18-37
50	14-23	21-34	29-47
100	17-26	25-38	35-53

18   In a point-of-emission analysis, emissions from electricity use are allocated to the energy-supply sector. In an end-use sector analysis, emissions from electricity are allocated to the respective end-use sector (particularly relevant for industry and buildings).

Table TS.17: *Economic potential for sectoral mitigation by 2030: comparison of bottom-up (from Table 11.3) and top-down estimates (from Section 3.6) [Table 11.5].*

Chapter of report	Sectors	Sector-based ('bottom-up') potential by 2030 (GtCO$_2$-eq/yr)				Economy-wide model ('top-down') snapshot of mitigation by 2030 (GtCO$_2$-eq/yr)	
		End-use sector allocation (allocation of electricity savings to end-use sectors)		Point-of-emissions allocation (emission reductions from end-use electricity savings allocated to energy supply sector)			
		Carbon price <20 US$/tCO$_2$-eq					
		Low	High	Low	High	Low	High
4	Energy supply & conversion	1.2	2.4	4.4	6.4	3.9	9.7
5	Transport	1.3	2.1	1.3	2.1	0.1	1.6
6	Buildings	4.9	6.1	1.9	2.3	0.3	1.1
7	Industry	0.7	1.5	0.5	1.3	1.2	3.2
8	Agriculture	0.3	2.4	0.3	2.4	0..6	1.2
9	Forestry	0.6	1.9	0.6	1.9	0.2	0.8
10	Waste	0.3	0.8	0.3	0.8	0.7	0.9
11	Total	9.3	17.1	9.1	17.9	8.7	17.9
		Carbon price <50 US$/tCO$_2$-eq					
4	Energy supply & conversion	2.2	4.2	5.6	8.4	6.7	12.4
5	Transport	1.5	2.3	1.5	2.3	0.5	1.9
6	Buildings	4.9	6.1	1.9	2.3	0.4	1.3
7	Industry	2.2	4.7	1.6	4.5	2.2	4.3
8	Agriculture	1.4	3.9	1.4	3.9	0.8	1.4
9	Forestry	1.0	3.2	1.0	3.2	0.2	0.8
10	Waste	0.4	1.0	0.4	1.0	0.8	1.0
11	Total	13.3	25.7	13.2	25.8	13.7	22.6
		Carbon price <100 US$/tCO$_2$-eq					
4	Energy supply & conversion	2.4	4.7	6.3	9.3	8.7	14.5
5	Transport	1.6	2.5	1.6	2.5	0.8	2.5
6	Buildings	5.4	6.7	2.3	2.9	0.6	1.5
7	Industry	2.5	5.5	1.7	4.7	3.0	5.0
8	Agriculture	2.3	6.4	2.3	6.4	0.9	1.5
9	Forestry	1.3	4.2	1.3	4.2	0.2	0.8
10	Waste	0.4	1.0	0.4	1.0	0.9	1.1
11	Total	15.8	31.1	15.8	31.1	16.8	26.2

Sources: Tables 3.16, 3.17 and 11.3
See notes to Tables 3.16, 3.17 and 11.3, and Annex 11.1.

Bioenergy options are important for many sectors by 2030, with substantial growth potential beyond, although no complete integrated studies are available for supply-demand balances. Key preconditions for such contributions are the development of biomass capacity (energy crops) in balance with investments in agricultural practices, logistic capacity and markets, together with commercialization of second-generation biofuel production. Sustainable biomass production and use could ensure that issues in relation to competition for land and food, water resources, biodiversity and socio-economic impacts are not creating obstacles *(high agreement, limited evidence)* [11.3.1.4].

Apart from the mitigation options mentioned in the sectoral Chapters 4–10, geo-engineering solutions to the enhanced greenhouse effect have been proposed. However, options

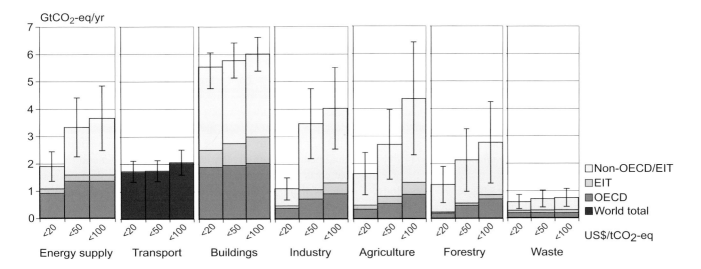

Figure TS.27: *Estimated sectoral economic potential for global mitigation for different regions as a function of carbon price in 2030 from bottom-up studies, compared to the respective baselines assumed in the sector assessments. A full explanation of the derivation of this figure is found in Section 11.3.*

Notes:
1. The ranges for global economic potentials as assessed in each sector are shown by vertical lines. The ranges are based on end-use allocations of emissions, meaning that emissions of electricity use are counted towards the end-use sectors and not to the energy supply sector.
2. The estimated potentials have been constrained by the availability of studies particularly at high carbon price levels.
3. Sectors used different baselines. For industry the SRES B2 baseline was taken, for energy supply and transport the WEO 2004 baseline was used; the building sector is based on a baseline in between SRES B2 and A1B; for waste, SRES A1B driving forces were used to construct a waste specific baseline, agriculture and forestry used baselines that mostly used B2 driving forces.
4. Only global totals for transport are shown because international aviation is included [5.4].
5. Categories excluded are: non-CO_2 emissions in buildings and transport, part of material efficiency options, heat production and cogeneration in energy supply, heavy duty vehicles, shipping and high-occupancy passenger transport, most high-cost options for buildings, wastewater treatment, emission reduction from coal mines and gas pipelines, fluorinated gases from energy supply and transport. The underestimation of the total economic potential from these emissions is of the order of 10-15%.

to remove CO_2 directly from the air, for example, by iron fertilization of the oceans, or to block sunlight, remain largely speculative and may have a risk of unknown side effects. Blocking sunlight does not affect the expected escalation in atmospheric CO_2 levels, but could reduce or eliminate the associated warming. This disconnection of the link between CO_2 concentration and global temperature could have beneficial consequences, for example, in increasing the productivity of agriculture and forestry (in as far as CO_2 fertilization is effective), but they do not mitigate or address other impacts such as further acidification of the oceans. Detailed cost estimates for these options have not been published and they are without a clear institutional framework for implementation *(medium agreement, limited evidence)* [11.2.2].

Mitigation costs across sectors and macro-economic costs

The costs of implementing the Kyoto Protocol are estimated to be much lower than the TAR estimates due to US rejection of the Protocol. With full use of the Kyoto flexible mechanisms, costs are estimated at less than 0.05% of Annex B (without US) GDP (TAR Annex B: 0.1–1.1%). Without flexible mechanisms,

costs are now estimated at less than 0.1% (TAR 0.2–2%) *(high agreement, much evidence)* [11.4].

Modelling studies of post-2012 mitigation have been assessed in relation to their global effects on CO_2 abatement by 2030, the carbon prices required and their effects on GDP or GNP (for the long-term effects of stabilization after 2030 see Chapter 3). For Category IV[19] pathways (stabilization around 650 ppm CO_2-eq) with CO_2 abatement less than 20% below baseline and up to 25 US$/t$CO_2$ carbon prices, studies suggest that gross world product would be, at worst, some 0.7% below baseline by 2030, consistent with the median of 0.2% and the 10–90 percentile range of –0.6 to 1.2% for the full set of scenarios given in Chapter 3.

Effects are more uncertain for the more stringent Category III pathways (stabilization around 550 ppm CO_2-eq) with CO_2 abatement less than 40% and up to 50 US$/t$CO_2$ carbon prices, with most studies suggesting costs less than 1% of global gross world product, consistent with the median of 0.6% and the 10–90 percentile range of 0 to 2.5% for the full set in Chapter 3. Again, the estimates are heavily dependent on approaches and assumptions. The few studies with baselines that require

---

19 See Chapter 3 for the definition of Category III and IV pathways.

higher CO_2 reductions to achieve the targets require higher carbon prices and most report higher GDP costs. For category I and II studies (stabilization between 445 and 535 ppm CO_2-eq) costs are less than 3% GDP loss, but the number of studies is relatively small and they generally use low baselines. The lower estimates of the studies assessed here, compared with the full set of studies reported in Chapter 3, are caused mainly by a larger share of studies that allow for enhanced technological innovation triggered by policies, particularly for more stringent mitigation scenarios *(high agreement, medium evidence)* [11.4].

All approaches indicate that no single sector or technology will be able to address the mitigation challenge successfully on its own, suggesting the need for a diversified portfolio based on a variety of criteria. Top-down assessments agree with the bottom-up results in suggesting that carbon prices around 20-50 US$/t$CO_2$-eq (73-183 US$/tC-eq) are sufficient to drive large-scale fuel-switching and make both CCS and low-carbon power sources economic as technologies mature. Incentives of this order might also play an important role in avoiding deforestation. The various short- and long-term models come up with differing estimates, the variation of which can be explained mainly by approaches and assumptions regarding the use of revenues from carbon taxes or permits, treatment of technological change, degree of substitutability between internationally traded products, and the disaggregation of product and regional markets *(high agreement, much evidence)* [11.4, 11.5, 11.6].

The development of the carbon price and the corresponding emission reductions will determine the level at which atmospheric GHG concentrations can be stabilized. Models suggest that a predictable and ongoing gradual increase in the carbon price that would reach 20–50 US/tCO_2$-eq by 2020–2030 corresponds with Category III stabilization (550 ppm CO_2-eq). For Category IV (650 ppm CO_2-eq), such a price level could be reached after 2030. For stabilization at levels between 450 and 550 ppm CO_2-eq, carbon prices of up to 100 US$/t$CO_2$-eq need to be reached by around 2030 *(medium agreement, medium evidence)* [11.4, 11.5, 11.6].

In all cases, short-term pathways towards lower stabilization levels, particularly for Category III and below, would require many additional measures around energy efficiency, low-carbon energy supply, other mitigation actions and avoidance of investment in very long-lived carbon-intensive capital stock. Studies of decision-making under uncertainty emphasize the need for stronger early action, particularly on long-lived infrastructure and other capital stock. Energy sector infrastructure (including power stations) alone is projected to require at least US$ 20 trillion investment to 2030 and the options for stabilization will be heavily constrained by the nature and carbon intensity of this investment. Initial estimates for lower carbon scenarios show a large redirection of investment, with net additional investments ranging from negligible to less than 5% *(high agreement, much evidence)* [11.6].

As regards portfolio analysis of government actions, a general finding is that a portfolio of options that attempts to balance emission reductions across sectors in a manner that appears equitable (e.g., by equal percentage reduction), is likely to be more costly than an approach primarily guided by cost-effectiveness. Portfolios of energy options across sectors that include low-carbon technologies will reduce risks and costs, because fossil fuel prices are expected to be more volatile relative to the costs of alternatives, in addition to the usual benefits from diversification. A second general finding is that costs will be reduced if options that correct the two market failures of climate change damages and technological innovation benefits are combined, for example, by recycling revenues from permit auctions to support energy-efficiency and low-carbon innovations *(high agreement, medium evidence)* [11.4].

Technological change across sectors

A major development since the TAR has been the inclusion in many top-down models of endogenous technological change. Using different approaches, modelling studies suggest that allowing for endogenous technological change may lead to substantial reductions in carbon prices as well as GDP costs, compared with most of the models in use at the time of the TAR (when technological change was assumed to be included in the baseline and largely independent of mitigation policies and action). Studies without induced technological change show that carbon prices rising to 20 to 80 US$/t$CO_2$-eq by 2030 and 30 to 155 US$/t$CO_2$-eq by 2050 are consistent with stabilization at around 550 ppm CO_2-eq by 2100. For the same stabilization level, studies since TAR that take into account induced technological change lower these price ranges to 5 to 65 US$/t$CO_2$eq in 2030 and 15 to 130 US$/t$CO_2$-eq in 2050. The degree to which costs are reduced hinges critically on the assumptions about the returns from climate change mitigation R&D expenditures, spill-overs between sectors and regions, crowding-out of other R&D, and, in models including learning-by-doing, learning rates *(high agreement, much evidence)* [11.5].

Major technological shifts like carbon capture and storage, advanced renewables, advanced nuclear and hydrogen require a long transition as learning-by-doing accumulates and markets expand. Improvement of end-use efficiency therefore offers more important opportunities in the short term. This is illustrated by the relatively high share of the buildings and industry sector in the 2030 potentials (Table TS.17). Other options and sectors may play a more significant role in the second half of the century (see Chapter 3) *(high agreement, much evidence)* [11.6].

Spill-over effects from mitigation in Annex I countries on Non-Annex I countries

Spill-over effects of mitigation from a cross-sectoral perspective are the effects of mitigation policies and measures in one country or group of countries on sectors in other countries. One aspect of spill-over is so-called 'carbon leakage':

the increase in CO_2 emissions outside the countries taking domestic measures divided by the emission reductions within these countries. The simple indicator of carbon leakage does not cover the complexity and range of effects, which include changes in the pattern and magnitude of global emissions. Modelling studies provide wide-ranging outcomes on carbon leakages depending on their assumptions regarding returns to scale, behaviour in the energy-intensive industry, trade elasticities and other factors. As in the TAR, the estimates of carbon leakage from implementation of the Kyoto Protocol are generally in the range of 5–20% by 2010. Empirical studies on the energy-intensive industries with exemptions under the EU Emission Trading Scheme (ETS) highlight that transport costs, local market conditions, product variety and incomplete information favour local production, and conclude that carbon leakage is unlikely to be substantial *(medium agreement, medium evidence)* [11.7].

Effects of existing mitigation actions on competitiveness have been studied. The empirical evidence seems to indicate that losses of competitiveness in countries implementing Kyoto are not significant, confirming a finding in the TAR. The potential beneficial effect of technology transfer to developing countries arising from technological development brought about by Annex I action may be substantial for energy-intensive industries, but has not so far been quantified in a reliable manner *(medium agreement, low evidence)* [11.7].

Perhaps one of the most important ways in which spill-overs from mitigation actions in one region affect others is through the effect on world fossil fuel prices. When a region reduces its fossil fuel demand because of mitigation policy, it will reduce the world demand for that commodity and so put downward pressure on the prices. Depending on the response of the fossil fuel producers, oil, gas or coal prices may fall, leading to loss of revenues by the producers, and lower costs of imports for the consumers. As in the TAR, nearly all modelling studies that have been reviewed show more pronounced adverse effects on oil-producing countries than on most Annex I countries that are taking the abatement measures. Oil-price protection strategies may limit income losses in the oil-producing countries *(high agreement, limited evidence)* [11.7].

Co-benefits of mitigation

Many recent studies have demonstrated significant benefits of carbon-mitigation strategies on human health, mainly because they also reduce other airborne emissions, for example, SO_2, NO_x and particulate matter. This is projected to result in the prevention of tens of thousands of premature deaths in Asian and Latin American countries annually, and several thousands in Europe. However, monetization of mortality risks remains controversial, and hence a large range of benefit estimates can be found in the literature. However, all studies agree that the monetized health benefits may offset a substantial fraction of the mitigation costs *(high agreement, much evidence)* [11.8].

In addition, the benefits of avoided emissions of air pollutants have been estimated for agricultural production and the impact of acid precipitation on natural ecosystems. Such near-term benefits provide the basis for a no-regrets GHG-reduction policy, in which substantial advantages accrue even if the impact of human-induced climate change turns out to be less than current projections show. Including co-benefits other than those for human health and agricultural productivity (e.g., increased energy security and employment) would further enhance the cost savings *(high agreement, limited evidence)* [11.8].

A wealth of new literature has pointed out that addressing climate change and air pollution simultaneously through a single set of measures and policies offers potentially large reductions in the costs of air-pollution control. An integrated approach is needed to address those pollutants and processes for which trade-offs exist. This is, for instance, the case for NO_x controls for vehicles and nitric acid plants, which may increase N_2O emissions, or the increased use of energy-efficient diesel vehicles, which emit relatively more fine particulate matter than their gasoline equivalents *(high agreement, much evidence)* [11.8].

Adaptation and mitigation

There can be synergies or trade-offs between policy options that can support adaptation and mitigation. The synergy potential is high for biomass energy options, land-use management and other land-management approaches. Synergies between mitigation and adaptation could provide a unique contribution to rural development, particularly in least-developed countries: many actions focusing on sustainable natural resource management could provide both significant adaptation benefits and mitigation benefits, mostly in the form of carbon sequestration. However, in other cases there may be trade-offs, such as the growth of energy crops that may affect food supply and forestry cover, thereby increasing vulnerability to the impacts of climate change *(medium agreement, limited evidence)* [11.9].

12 Sustainable development and mitigation

Relationship between sustainable development and climate change mitigation

The concept of sustainable development was adopted by the World Commission on Environment and Development and there is agreement that sustainable development involves a comprehensive and integrated approach to economic, social and environmental processes. Discussions on sustainable development, however, have focused primarily on the environmental and economic dimensions. The importance of social, political and cultural factors is only now getting more recognition. Integration is essential in order to articulate development trajectories that are sustainable, including addressing the climate change problem [12.1].

Although still in the early stages, there is growing use of indicators to measure and manage the sustainability of development at the macro and sectoral levels, which is driven in part by the increasing emphasis on accountability in the context of governance and strategy initiatives. At the sectoral level, progress towards sustainable development is beginning to be measured and reported by industry and governments using, *inter alia*, green certification, monitoring tools or emissions registries. Review of the indicators shows, however, that few macro-indicators include measures of progress with respect to climate change (*high agreement, much evidence*) [12.1.3].

Climate change is influenced not only by the climate-specific policies that are put in place (the 'climate first approach'), but also by the mix of development choices that are made and the development trajectories that these policies lead to (the 'develop- ment first approach') - a point reinforced by global scenario analysis published since the TAR. Making development more sustainable by changing development paths can thus make a significant contribution to climate goals. It is important to note, however, that changing development pathways is not about choosing a mapped- out path, but rather about navigating through an uncharted and evolving landscape (*high agreement, much evidence*) [12.1.1].

It has further been argued that sustainable development might decrease the vulnerability of all countries, and particularly of developing countries, to climate change impacts. Framing the debate as a development problem rather than an environmental one may better address the immediate goals of all countries, particularly developing countries and their special vulnerability to climate change, while at the same time addressing the driving forces for emissions that are linked to the underlying development path [12.1.2].

Making development more sustainable

Decision-making on sustainable development and climate change mitigation is no longer solely the purview of governments. The literature recognizes the shift to a more inclusive concept of governance, which includes the contributions of various levels of government, the private sector, non-governmental actors and civil society. The more that climate change issues are mainstreamed as part of the planning perspective at the appropriate level of implementation, and the more all these relevant parties are involved in the decision-making process in a meaningful way, the more likely are they to achieve the desired goals (*high agreement, medium evidence*) [12.2.1].

Regarding governments, a substantial body of political theory identifies and explains the existence of national policy styles or political cultures. The underlying assumption of this work is that individual countries tend to process problems in a specific manner, regardless of the distinctiveness or specific features of any specific problem; a national 'way of doing things'. Furthermore, the choice of policy instruments is affected by the institutional capacity of governments to implement the instrument. This implies that the preferred mix of policy decisions and their effectiveness in terms of sustainable development and climate change mitigation depend strongly on national characteristics (*high agreement, much evidence*). However, our understanding of which types of policies will work best in countries with particular national characteristics remains sketchy [12.2.3].

The private sector is a central player in ecological and sustainability stewardship. Over the past 25 years, there has been a progressive increase in the number of companies that are taking steps to address sustainability issues at either the firm or industry level. Although there has been progress, the private sector has the capacity to play a much greater role in making development more sustainable if awareness that this will probably benefit its performance grows (*medium agreement, medium evidence*) [12.2.3].

Citizen groups play a significant role in stimulating sustainable development and are critical actors in implementing sustainable development policy. Apart from implementing sustainable development projects themselves, they can push for policy reform by awareness-raising, advocacy and agitation. They can also pull policy action by filling the gaps and providing policy services, including in the areas of policy innovation, monitoring and research. Interactions can take the form of partnerships or be through stakeholder dialogues that can provide citizens' groups with a lever for increasing pressure on both governments and industry (*high agreement, medium evidence*) [12.2.3].

Deliberative public-private partnerships work most effectively when investors, local governments and citizen groups are willing to work together to implement new technologies, and provide arenas to discuss such technologies that are locally inclusive (*high agreement, medium evidence*) [12.2.3].

Implications of development choices for climate change mitigation

In a heterogeneous world, an understanding of different regional conditions and priorities is essential for mainstreaming climate change policies into sustainable-development strategies. Region- and country-specific case studies demonstrate that different development paths and policies can achieve notable emissions reductions, depending on the capacity to realize sustainability and climate change objectives [12.3].

In industrialized countries, climate change continues to be regarded mainly as a separate, environmental problem to be addressed through specific climate change policies. A fundamental and broad discussion in society on the implications of development pathways for climate change in general and climate change mitigation in particular in the industrialized countries has not been seriously initiated. Priority mitigation areas for countries in this group may be in energy efficiency, renewable energy, CCS, etc. However, low-emission pathways apply not only to energy choices. In some regions, land-use development, particularly infrastructure expansion, is identified as a key variable determining future GHG emissions [12.2.1; 12.3.1].

Economies in transition as a single group no longer exist. Nevertheless, Central and Eastern Europe and the countries of Eastern Europe, the Caucasus and Central Asia (EECCA) do share some common features in socio-economic development and in climate change mitigation and sustainable development. Measures to decouple economic and emission growth would be especially important for this group [12.2.1; 12.3.1].

Some large developing countries are projected to increase their emissions at a faster rate than the industrialized world and the rest of developing nations as they are in the stage of rapid industrialization. For these countries, climate change mitigation and sustainable-development policies can complement one another; however, additional financial and technological resources would enhance their capacity to pursue a low-carbon path of development [12.2.1; 12.3.1].

For most other developing countries, adaptive and mitigative capacities are low and development aid can help to reduce their vulnerability to climate change. It can also help to reduce their emissions growth while addressing energy-security and energy-access problems. CDM can provide financial resources for such developments. Members of the Organization of the Petroleum-Exporting Countries (OPEC) are unique in the sense that they may be adversely affected by development paths that reduce the demand for fossil fuels. Diversification of their economies is high on their agenda [12.2.1; 12.3.1].

Some general conclusions emerge from the case studies reviewed in this chapter on how changes in development pathways at the sectoral level have (or could) lower emissions (*high agreement, medium evidence*) [12.2.4]:

- GHG emissions are influenced by, but not rigidly linked to, economic growth: policy choices can make a difference.
- Sectors where effective production is far below the maximum feasible production with the same amount of inputs – that is, sectors that are far from their production frontier – have opportunities to adopt 'win-win-win' policies, that is, policies that free up resources and bolster growth, meet other sustainable-development goals and also reduce GHG emissions relative to baseline.
- Sectors where production is close to the optimal given available inputs – i.e., sectors that are closer to the production frontier – also have opportunities to reduce emissions by meeting other sustainable development goals. However, the closer one gets to the production frontier, the more trade-offs are likely to appear.

- What matters is not only that a 'good' choice is made at a certain point in time, but also that the initial policy is sustained for a long time – sometimes several decades – to really have effects.
- It is often not one policy decision, but an array of decisions that are needed to influence emissions. This raises the issue of coordination between policies in several sectors and at various scales.

Mainstreaming requires that non-climate policies, programmes and/or individual actions take climate change mitigation into consideration, in both developing and developed countries. However, merely piggybacking climate change on to an existing political agenda is unlikely to succeed. The ease or difficulty with which mainstreaming is accomplished will depend on both mitigation technologies or practices, and the underlying development path. Weighing other development benefits against climate benefits will be a key basis for choosing development sectors for mainstreaming. Decisions about macro-economic policy, agricultural policy, multilateral development bank lending, insurance practices, electricity market reform, energy security, and forest conservation, for example, which are often treated as being apart from climate policy, can have profound impacts on emissions, the extent of mitigation required, and the costs and benefits that result. However, in some cases, such as shifting from biomass cooking to liquid petroleum gas (LPG) in rural areas in developing countries, it may be rational to disregard climate change considerations because of the small increase in emissions when compared with its development benefits (see Table TS.18) (*high agreement, medium evidence*) [12.2.4].

In general terms, there is a high level of agreement on the qualitative findings in this chapter about the linkages between mitigation and sustainable development: the two are linked, and synergies and trade-offs can be identified. However, the literature about the links and more particularly, about how these links can be put into action in order to capture synergies and avoid trade-offs, is as yet sparse. The same applies to good practice guidance for integrating climate change considerations into relevant non-climate policies, including analysis of the roles of different actors. Elaborating possible development paths that nations and regions can pursue – beyond more narrowly conceived GHG emissions scenarios or scenarios that ignore climate change – can provide the context for new analysis of the links, but may require new methodological tools (*high agreement, limited evidence*) [12.2.4].

Implications of mitigation choices for sustainable development trajectories

There is a growing understanding of the opportunities to choose mitigation options and their implementation in such a way that there will be no conflict with or even benefits for other dimensions of sustainable development; or, where trade-offs are inevitable, to allow rational choices to be made. A summary of

Table TS.18: *Mainstreaming climate change into development choices – selected examples [Table 12.3].*

Selected sectors	Non-climate policy instruments and actions that are candidates for mainstreaming	Primary decision-makers and actors	Global GHG emissions by sector that could be addressed by non-climate policies (% of global GHG emissions)[a, d]		Comments
Macro economy	Implement non-climate taxes/subsidies and/or other fiscal and regulatory policies that promote SD	State (governments at all levels)	100	Total global GHG emissions	Combination of economic, regulatory, and infrastructure non-climate policies could be used to address total global emissions.
Forestry	Adoption of forest conservation and sustainable management practices	State (governments at all levels) and civil society (NGOs)	7	GHG emissions from deforestation	Legislation/regulations to halt deforestation, improve forest management, and provide alternative livelihoods can reduce GHG emissions and provide other environmental benefits.
Electricity	Adoption of cost-effective renewables, demand-side management programmes, and reduction of transmission and distribution losses	State (regulatory commissions), market (utility companies) and, civil society (NGOs, consumer groups)	20[b]	Electricity sector CO_2 emissions (excluding auto producers)	Rising share of GHG-intensive electricity generation is a global concern that can be addressed through non-climate policies.
Petroleum imports	Diversifying imported and domestic fuel mix and reducing economy's energy intensity to improve energy security	State and market (fossil fuel industry)	20[b]	CO_2 emissions associated with global crude oil and product imports	Diversification of energy sources to address oil security concerns could be achieved such that GHG emissions are not increased.
Rural energy in developing countries	Policies to promote rural LPG, kerosene and electricity for cooking	State and market (utilities and petroleum companies), civil society (NGOs)	<2[c]	GHG emissions from biomass fuel use, not including aerosols	Biomass used for rural cooking causes health impacts due to indoor air pollution, and releases aerosols that add to global warming. Displacing all biomass used for rural cooking in developing countries with LPG would emit 0.70 GtCO$_2$-eq., a relatively modest amount compared with 2004 total global GHG emissions.
Insurance for building and transport sectors	Differentiated premiums, liability insurance exclusions, improved terms for green products	State and market (insurance companies)	20	Transport and building sector GHG emissions	Escalating damages due to climate change are a source of concern to insurance industry. Insurance industry could address these through the types of policies noted here.
International finance	Country and sector strategies and project lending that reduces emissions	State (international financial institutions) and market (commercial banks)	25[b]	CO_2 emissions from developing countries (non-Annex I)	International financial institutions can adopt practices so that loans for GHG-intensive projects in developing countries that lock-in future emissions are avoided.

Notes:
a) Data from Chapter 1 unless noted otherwise.
b) CO_2 emissions from fossil fuel combustion only; IEA (2006).
c) CO_2 emissions only. Authors estimate, see text.
d) Emissions indicate the relative importance of sectors in 2004. Sectoral emissions are not mutually exclusive, may overlap, and hence sum up to more than total global emissions, which are shown in the Macro economy row.

Table TS.19: *Sectoral mitigation options and sustainable development (economic, local environmental and social) considerations: synergies and trade-offs [Table 12.4].*

Sector and mitigation options	Potential SD synergies and conditions for implementation	Potential SD trade-offs
Energy supply and use: Chapters 4-7		
Energy efficiency improvement in all sectors (buildings, transportation, industry, and energy supply) (Chapters 4-7)	- Almost always cost-effective, reduces or eliminates local pollutant emissions and consequent health impacts, improves indoor comfort and reduces indoor noise levels, creates business opportunities and jobs and improves energy security - Government and industry programmes can help overcome lack of information and principal agent problems - Programmes can be implemented at all levels of government and industry - Important to ensure that low-income household energy needs are given due consideration, and that the process and consequences of implementing mitigation options are, or the result is, gender-neutral	- Indoor air pollution and health impacts of improving the thermal efficiency of biomass cooking stoves in developing country rural areas are uncertain
Fuel switching and other options in the transportation and buildings sectors (Chapters 5 and 6)	- CO_2 reduction costs may be offset by increased health benefits - Promotion of public transport and non-motorized transport has large and consistent social benefits - Switching from solid fuels to modern fuels for cooking and heating indoors can reduce indoor air pollution and increase free time for women in developing countries - Institutionalizing planning systems for CO_2 reduction through coordination between national and local governments is important for drawing up common strategies for sustainable transportation systems	- Diesel engines are generally more fuel-efficient than gasoline engines and thus have lower CO_2 emissions, but increase particle emissions. - Other measures (CNG buses, hybrid diesel-electric buses and taxi renovation) may provide little climate benefit.
Replacing imported fossil fuels with domestic alternative energy sources (DAES) (Chapter 4)	- Important to ensure that DAES is cost-effective - Reduces local air pollutant emissions. - Can create new indigenous industries (e.g., Brazil ethanol programme) and hence generate employment	- Balance of trade improvement is traded off against increased capital required for investment - Fossil fuel-exporting countries may face reduced exports - Hydropower plants may displace local populations and cause environmental damage to water bodies and biodiversity
Replacing domestic fossil fuel with imported alternative energy sources (IAES) (Chapter 4)	- Almost always reduces local pollutant emissions - Implementation may be more rapid than DAES - Important to ensure that IAES is cost-effective - Economies and societies of energy-exporting countries would benefit	- Could reduce energy security - Balance of trade may worsen but capital needs may decline
Forestry sector: Chapter 9		
Afforestation	- Can reduce wasteland, arrest soil degradation, and manage water runoff - Can retain soil carbon stocks if soil disturbance at planting and harvesting is minimized - Can be implemented as agroforestry plantations that enhance food production - Can generate rural employment and create rural industry - Clear delineation of property rights would expedite implementation of forestation programmes	- Use of scarce land could compete with agricultural land and diminish food security while increasing food costs - Monoculture plantations can reduce biodiversity and are more vulnerable to disease - Conversion of floodplain and wetland could hamper ecological functions
Avoided deforestation	- Can retain biodiversity, water and soil management benefits, and local rainfall patterns - Reduce local haze and air pollution from forest fires - If suitably managed, it can bring revenue from ecotourism and from sustainably harvested timber sales - Successful implementation requires involving local dwellers in land management and/or providing them alternative livelihoods, enforcing laws to prevent migrants from encroaching on forest land.	- Can result in loss of economic welfare for certain stakeholders in forest exploitation (land owners, migrant workers) - Reduced timber supply may lead to reduced timber exports and increased use of GHG-intensive construction materials - Can result in deforestation with consequent SD implications elsewhere
Forest Management	- See afforestation	- Fertilizer application can increase N_2O production and nitrate runoff degrading local (ground)water quality - Prevention of fires and pests has short term benefits but can increase fuel stock for later fires unless managed properly

Table TS.19. Continued.

Sector and mitigation options	Potential SD synergies and conditions for implementation	Potential SD trade-offs
Bio-energy (chapter 8 en 9)		
Bio-energy production	- Mostly positive when practised with crop residues (shells, husks, bagasse and/or tree trimmings). - Creates rural employment. - Planting crops/trees exclusively for bio-energy requires that adequate agricultural land and labour is available to avoid competition with food production	- Can have negative environmental consequences if practised unsustainably - biodiversity loss, water resource competition, increased use of fertilizer and pesticides. - Potential problem with food security (location-specific) and increased food costs.
Agriculture: Chapter 8		
Cropland management (management of nutrients, tillage, residues, and agroforestry; water, rice, and set-aside)	- Improved nutrient management can improve groundwater quality and environmental health of the cultivated ecosystem	- Changes in water policies could lead to clash of interests and threaten social cohesiveness - Could lead to water overuse
Grazing land management	- Improves livestock productivity, reduces desertification, and provide social security for the poor - Requires laws and enforcement to ban free grazing	
Livestock management	- Mix of traditional rice cultivation and livestock management would enhance incomes even in semi-arid and arid regions	
Waste management: Chapter 10		
Engineered sanitary landfilling with landfill gas recovery to capture methane gas	- Can eliminate uncontrolled dumping and open burning of waste, improving health and safety for workers and residents. - Sites can provide local energy benefits and public spaces for recreation and other social purposes within the urban infrastructure.	- When done unsustainably can cause leaching that leads to soil and groundwater contamination with potentially negative health impacts
Biological processes for waste and wastewater (composting, anaerobic digestion, aerobic and anaerobic wastewater processes)	- Can destroy pathogens and provide useful soil amendments if properly implemented using source-separated organic waste or collected wastewater. - Can generate employment - Anaerobic processes can provide energy benefits from CH_4 recovery and use.	- A source of odours and water pollution if not properly controlled and monitored.
Incineration and other thermal processes	- Obtain the most energy benefit from waste.	- Expensive relative to controlled landfilling and composting. - Unsustainable in developing countries if technical infrastructure not present. - Additional investment for air pollution controls and source separation needed to prevent emissions of heavy metals and other air toxics.
Recycling, re-use, and waste minimization	- Provide local employment as well as reductions in energy and raw materials for recycled products. - Can be aided by NGO efforts, private capital for recycling industries, enforcement of environmental regulations, and urban planning to segregate waste treatment and disposal activities from community life.	- Uncontrolled waste scavenging results in severe health and safety problems for those who make their living from waste - Development of local recycling industries requires capital.

Note: Material in this table is drawn from the Chapters 4–11. Where new material is introduced, it is referenced in the accompanying text below, which describes the SD implications of mitigation options in each sector.

the sustainable development implications of the main climate change mitigation options is given in Table TS.19 [12.3].

The sustainable development benefits of mitigation options vary within a sector and between regions *(high agreement, much evidence)*:

- Generally, mitigation options that improve the productivity of resource use, whether energy, water, or land, yield positive benefits across all three dimensions of sustainable development. Other categories of mitigation options have a more uncertain impact and depend on the wider socio-economic context within which the option is being implemented.
- Climate-related policies such as energy efficiency and renewable energy are often economically beneficial, improve energy security and reduce local pollutant emissions. Many energy-supply mitigation options can be designed to also achieve sustainable development benefits such as avoided displacement of local populations, job creation and health benefits.
- Reducing deforestation can have significant biodiversity, soil and water conservation benefits, but may result in a loss of economic welfare for some stakeholders. Appropriately designed forestation and bioenergy plantations can lead to restoration of degraded land, manage water runoff, retain soil carbon and benefit rural economies, but may compete with land for food production and be negative for biodiversity.
- There are good possibilities for reinforcing sustainable development through mitigation actions in most sectors, but particularly in the waste management, transportation and buildings sectors, notably through decreased energy use and reduced pollution [12.3].

13 Policies, instruments and co-operative agreements

Introduction

This chapter discusses national policy instruments and their implementation, initiatives of the private sector, local governments and non-governmental organizations, and cooperative international agreements. Wherever feasible, national policies and international agreements are discussed in the context of four principle criteria by which they can be evaluated; that is, environmental effectiveness, cost-effectiveness, distributional considerations and institutional feasibility. There are a number of additional criteria that could also be explicitly considered, such as effects on competitiveness and administrative costs. Criteria may be applied by governments in making ex-ante choices among instruments and in ex-post evaluation of the performance of instruments [13.1].

National policy instruments, their implementation and interactions

The literature continues to reflect that a wide variety of national policies and measures are available to governments to limit or reduce GHG emissions. These include: regulations and standards, taxes and charges, tradable permits, voluntary agreements, phasing out subsidies and providing financial incentives, research and development and information instruments. Other policies, such as those affecting trade, foreign direct investments and social development goals can also affect GHG emissions. In general, climate change policies, if integrated with other government polices, can contribute to sustainable development in both developed and developing countries (see Chapter 12) [13.1].

Reducing emissions across all sectors and gases requires a portfolio of policies tailored to fit specific national circumstances. While the literature identifies advantages and disadvantages for any given instrument, the above-mentioned criteria are widely used by policy makers to select and evaluate policies.

All instruments can be designed well or poorly, stringent or lax. Instruments need to be adjusted over time and supplemented with a workable system of monitoring and enforcement. Furthermore, instruments may interact with existing institutions and regulations in other sectors of society *(high agreement, much evidence)* [13.1].

The literature provides a good deal of information to assess how well different instruments meet the above-mentioned criteria (see Table TS.20) [13.2]. Most notably, it suggests that:

- **Regulatory measures and standards** generally provide environmental certainty. They may be preferable when lack of information or other barriers prevent firms and consumers from responding to price signals. Regulatory standards do not generally give polluters incentives to develop new technologies to reduce pollution, but there are a few examples whereby technology innovation has been spurred by regulatory standards. Standards are common practice in the building sector and there is strong innovation. Although relatively few regulatory standards have been adopted solely to reduce GHG emissions, standards have reduced these gases as a co-benefit *(high agreement, much evidence)* [13.2].
- **Taxes and charges** (which can be applied to carbon or all GHGs) are given high marks for cost effectiveness since they provide some assurance regarding the marginal cost of pollution control. They cannot guarantee a particular level of emissions, but conceptually taxes can be designed to be environmentally effective. Taxes can be politically difficult to implement and adjust. As with regulations, their environmental effectiveness depends on their stringency. As with nearly all other policy instruments, care is needed to prevent perverse effects *(high agreement, much evidence)* [13.2].

Table TS.20: *National environmental policy instruments and evaluative criteria [Table 13.1].*

Instrument	Criteria			
	Environmental effectiveness	Cost-effectiveness	Meets distributional considerations	Institutional feasibility
Regulations and standards	Emission levels set directly, though subject to exceptions Depends on deferrals and compliance	Depends on design; uniform application often leads to higher overall compliance costs	Depends on level playing field; small/new actors may be disadvantaged	Depends on technical capacity; popular with regulators, in countries with weak functioning markets
Taxes and charges	Depends on ability to set tax at a level that induces behavioural change	Better with broad application; higher administrative costs where institutions are weak	Regressive; can be improved with revenue recycling	Often politically unpopular; may be difficult to enforce with underdeveloped institutions
Tradable permits	Depends on emissions cap, participation and compliance	Decreases with limited participation and fewer sectors	Depends on initial permit allocation, may pose difficulties for small emitters	Requires well-functioning markets and complementary institutions
Voluntary agreements	Depends on programme design, including clear targets, a baseline scenario, third-party involvement in design and review, and monitoring provisions	Depends on flexibility and extent of government incentives, rewards and penalties	Benefits accrue only to participants	Often politically popular; requires significant number of administrative staff
Subsidies and other incentives	Depends on programme design; less certain than regulations/ standards.	Depends on level and programme design; can be market-distorting	Benefits selected participants; possibly some that do not need it	Popular with recipients; potential resistance from vested interests. Can be difficult to phase out
Research and development	Depends on consistent funding, when technologies are developed, and polices for diffusion. May have high benefits in long-term	Depends on programme design and the degree of risk	Initially benefits selected participants, Potentially easy for funds to be misallocated	Requires many separate decisions; Depends on research capacity and long-term funding

Note: Evaluations are predicated on assumptions that instruments are representative of best practice rather than theoretically perfect. This assessment is based primarily on experiences and literature from developed countries, since peer-reviewed articles on the effectiveness of instruments in other countries were limited. Applicability in specific countries, sectors and circumstances – particularly developing countries and economies in transition – may differ greatly. Environmental and cost effectiveness may be enhanced when instruments are strategically combined and adapted to local circumstances.

- **Tradable permits** are an increasingly popular economic instrument to control conventional pollutants and GHGs at the sectoral, national and international level. The volume of emissions allowed determines the carbon price and the environmental effectiveness of this instrument, while the distribution of allowances has implications for competitiveness. Experience has shown that banking provisions can provide significant temporal flexibility and that compliance provisions must be carefully designed, if a permit system is to be effective (*high agreement, much evidence*). Uncertainty in the price of emission reductions under a trading system makes it difficult, a priori, to estimate the total cost of meeting reduction targets [13.2].

- **Voluntary agreements between industry and governments** and information campaigns are politically attractive, raise awareness among stakeholders and have played a role in the evolution of many national policies. The majority of voluntary agreements has not achieved significant emission reductions beyond business-as-usual. However, some

recent agreements in a few countries have accelerated the application of best available technology and led to measurable reductions of emissions compared with the baseline (*high agreement, much evidence*). Success factors include clear targets, a baseline scenario, third-party involvement in design and review, and formal provisions for monitoring [13.2].

- **Voluntary actions:** Corporations, sub-national governments, NGOs and civil groups are adopting a wide variety of voluntary actions, independent of government authorities, which may limit GHG emissions, stimulate innovative policies and encourage the deployment of new technologies. By themselves, they generally have limited impact at the national or regional level [13.2].

- **Financial incentives** (subsidies and tax credits) are frequently used by governments to stimulate the diffusion of new, less GHG-emitting technologies. While the economic costs of such programmes are often higher than for the instruments listed above, they are often critical to overcome barriers to the penetration of new technologies (*high agreement, much* *evidence*). As with other policies, incentive programmes must be carefully designed to avoid perverse market effects. Direct and indirect subsidies for fossil fuel use and agriculture remain common practice in many countries, although those for coal have declined over the past decade in many OECD countries and in some developing countries (See also Chapter 2, 7 and 11) [13.2].

- **Government support for research and development** is a special type of incentive, which can be an important instrument to ensure that low GHG-emitting technologies will be available in the long-term. However, government funding for many energy-research programmes dropped after the oil crisis in the 1970s and stayed constant, even after the UNFCCC was ratified. Substantial additional investments in, and policies for, R&D are needed to ensure that technologies are ready for commercialization in order to arrive at stabilization of GHGs in the atmosphere (see Chapter 3), along with economic and regulatory instruments to promote their deployment and diffusion (*high agreement, much evidence*) [13.2.1].

- **Information instruments** – sometimes called public disclosure requirements – may positively affect environmental quality by allowing consumers to make better-informed choices. There is only limited evidence that the provision of information can achieve emissions reductions, but it can improve the effectiveness of other policies (*high agreement, much evidence*) [13.2].

Applying an environmentally effective and economically efficient instrument mix requires a good understanding of the environmental issue to be addressed, of the links with other policy areas and the interactions between the different instruments in the mix. In practice, climate-related policies are seldom applied in complete isolation, as they overlap with other national polices relating to the environment, forestry, agriculture, waste management, transport and energy, and in many cases require more than one instrument (*high agreement, much evidence*) [13.2].

Initiatives of sub-national governments, corporations and non-governmental organizations

The preponderance of the literature reviews nationally based governmental instruments, but corporations, local- and regional authorities, NGOs and civil groups can also play a key role and are adopting a wide variety of actions, independent of government authorities, to reduce emissions of GHGs. Corporate actions range from voluntary initiatives to emissions targets and, in a few cases, internal trading systems. The reasons corporations undertake independent actions include the desire to influence or pre-empt government action, to create financial value, and to differentiate a company and its products. Actions by regional, state, provincial and local governments include renewable portfolio standards, energy-efficiency programmes, emission registries and sectoral cap-and-trade mechanisms. These actions are undertaken to influence national policies, address stakeholder concerns, create incentives for new industries, or

create environmental co-benefits. NGOs promote programmes to reduce emissions through public advocacy, litigation and stakeholder dialogue. Many of the above actions may limit GHG emissions, stimulate innovative policies, encourage the deployment of new technologies and spur experimentation with new institutions, but by themselves generally have limited impact. To achieve significant emission reductions, these actions must lead to changes in national policies (*high agreement, much evidence*) [13.4].

International agreements (climate change agreements and other arrangements)

The UNFCCC and its Kyoto Protocol have set a significant precedent as a means of solving a long-term international environmental problem, but are only the first steps towards implementation of an international response strategy to combat climate change. The Kyoto Protocol's most notable achievements are the stimulation of an array of national policies, the creation of an international carbon market and the establishment of new institutional mechanisms. Its economic impacts on the participating countries are yet to be demonstrated. The CDM, in particular, has created a large project pipeline and mobilized substantial financial resources, but it has faced methodological challenges regarding the determination of baselines and additionality. The protocol has also stimulated the development of emissions trading systems, but a fully global system has not been implemented. The Kyoto Protocol is currently constrained by the modest emission limits and will have a limited effect on atmospheric concentrations. It would be more effective if the first commitment period were to be followed up by measures to achieve deeper reductions and the implementation of policy instruments covering a higher share of global emissions (*high agreement, much evidence*) [13.3].

Many options are identified in the literature for achieving emission reductions both under and outside the Convention and its Kyoto Protocol, for example: revising the form and stringency of emission targets; expanding the scope of sectoral and sub-national agreements; developing and adopting common policies; enhancing international RD&D technology programmes; implementing development-oriented actions, and expanding financing instruments (*high agreement, much evidence*). Integrating diverse elements such as international R&D cooperation and cap-and-trade programmes within an agreement is possible, but comparing the efforts made by different countries would be complex and resource-intensive (*medium agreement, medium evidence*) [13.3].

There is a broad consensus in the literature that a successful agreement will have to be environmentally effective, cost-effective, incorporate distributional considerations and equity, and be institutionally feasible (*high agreement, much evidence*) [13.3].

A great deal of new literature is available on potential structures for and the substance of future international agreements. As has been noted in previous IPCC reports, because climate change is a globally common problem, any approach that does not include a larger share of global emissions will be more costly or less environmentally effective. *(high agreement, much evidence)* (See Chapter 3) [13.3].

Most proposals for future agreements in the literature include a discussion of goals, specific actions, timetables, participation, institutional arrangements, reporting and compliance provisions. Other elements address incentives, non-participation and non-compliance penalties *(high agreement, much evidence)* [13.3].

Goals

The specification of clear goals is an important element of any climate agreement. They can both provide a common vision about the near-term direction and offer longer-term certainty, which is called for by business. Goal-setting also helps structure commitments and institutions, provides an incentive to stimulate action and helps establish criteria against which to measure the success in implementing measures *(high agreement, much evidence)* [13.3].

The choice of the long-term ambition significantly influences the necessary short-term action and therefore the design of the international regime. Abatement costs depend on the goal, vary with region and depend on the allocation of emission allowances among regions and the level of participation (high agreement, much evidence) [13.3].

Options for the design of international regimes can incorporate goals for the short, medium and long term. One option is to set a goal for long-term GHG concentrations or a temperature stabilization goal. Such a goal might be based on physical impacts to be avoided or conceptually on the basis of the monetary and non-monetary damages to be avoided. An alternative to agreeing on specific CO_2 concentration or temperature levels is an agreement on specific long-term actions such as a technology R&D and diffusion target – for example, 'eliminating carbon emissions from the energy sector by 2060'. An advantage of such a goal is that it might be linked to specific actions *(high agreement, much evidence)* [13.3].

Another option would be to adopt a 'hedging strategy', defined as a shorter-term goal on global emissions, from which it is still possible to reach a range of desirable long-term goals. Once the short-term goal is reached, decisions on next steps can be made in light of new knowledge and decreased levels of uncertainty *(medium agreement, medium evidence)* [13.3].

Participation

Participation of states in international agreements can vary from very modest to extensive. Actions to be taken by participating countries can be differentiated both in terms of when such action is undertaken, who takes the action and what the action will be. States participating in the same 'tier' would have the same (or broadly similar) types of commitments. Decisions on how to allocate states to tiers can be based on formalized quantitative or qualitative criteria, or be 'ad hoc'. Under the principle of sovereignty, states may choose the tier into which they are grouped *(high agreement, much evidence)* [13.3].

An agreement can have static participation or may change over time. In the latter case, states can 'graduate' from one tier of commitments to another. Graduation can be linked to passing of quantitative thresholds for certain parameters (or combinations of parameters) that have been predefined in the agreement, such as emissions, cumulative emissions, GDP per capita, relative contribution to temperature increase or other measures of development, such as the human development index (HDI) *(high agreement, much evidence)* [13.3].

Some argue that an international agreement needs to include only the major emitters to be effective, since the largest 15 countries (including the EU-25 as one) make up 80% of global GHG emissions. Others assert that those with historical responsibility must act first. Still another view holds that technology development is the critical factor for a global solution to climate change, and thus agreements must specifically target technology development in Annex I countries – which in turn could offset some or all emissions leakage in Non-Annex I countries. Others suggest that a climate regime is not exclusively about mitigation, but also encompasses adaptation – and that a far wider array of countries is vulnerable to climate change and must be included in any agreement *(high agreement, much evidence)* [13.3].

Regime stringency: linking goals, participation and timing

Under most equity interpretations, developed countries as a group would need to reduce their emissions significantly by 2020 (10–40% below 1990 levels) and to still lower levels by 2050 (40–95% below 1990 levels) for low to medium stabilization levels (450–550ppm CO_2-eq) (see also Chapter 3). Under most of the regime designs considered for such stabilization levels, developing-country emissions need to deviate below their projected baseline emissions within the next few decades *(high agreement, much evidence)*. For most countries, the choice of the long-term ambition level will be more important than the design of the emission-reduction regime [13.3].

The total global costs are highly dependent on the baseline scenario, marginal abatement cost estimates, the assumed concentration stabilization level (see also Chapters 3 and 11) and the level (size of the coalition) and degree of participation (how and when allowances are allocated). If, for example some major emitting regions do not participate in the reductions immediately, the global costs of the participating regions will be higher if the goal is maintained (see also Chapter 3). Regional abatement costs are dependent on the allocation of emission allowances to regions, particularly the timing. However, the assumed stabilization level and baseline scenario are more important in determining regional costs [11.4; 13.3].

Commitments, timetables and actions

There is a significant body of new literature that identifies and evaluates a diverse set of options for commitments that could be taken by different groups. The most frequently evaluated type of commitment is the binding absolute emission reduction cap as included in the Kyoto Protocol for Annex I countries. The broad conclusion from the literature is that such regimes provide certainty about future emission levels of the participating countries (assuming caps are met). Many authors propose that caps be reached using a variety of 'flexibility' approaches, incorporating multiple GHGs and sectors as well as multiple countries through emission trading and/or project-based mechanisms (*high agreement, much evidence*) [13.3].

While a variety of authors propose that absolute caps be applied to all countries in the future, many have raised concerns that the rigidity of such an approach may unreasonably restrict economic growth. While no consensus approach has emerged, the literature provides multiple alternatives to address this problem, including 'dynamic targets' (where the obligation evolves over time), and limits on prices (capping the costs of compliance at a given level – which while limiting costs, would also lead to exceeding the environmental target). These options aim at maintaining the advantages of international emissions trading while providing more flexibility in compliance (*high agreement, much evidence*). However, there is a trade-off between costs and certainty in achieving an emissions level. [13.3]

Market mechanisms

International market-based approaches can offer a cost-effective means of addressing climate change if they incorporate a broad coverage of countries and sectors. So far, only a few domestic emissions-trading systems are in place, the EU ETS being by far the largest effort to establish such a scheme, with over 11,500 plants allocated and authorized to buy and sell allowances (*high agreement, high evidence*) [13.2].

Although the Clean Development Mechanism is developing rapidly, the total financial flows for technology transfer have so far been limited. Governments, multilateral organizations and private firms have established nearly 6 billion US$ in carbon funds for carbon-reduction projects, mainly through the CDM. Financial flows to developing countries through CDM projects are reaching levels in the order of several billion US$/yr. This is higher than the flows through the Global Environment Facility (GEF), comparable to the energy-oriented development assistance flows, but at least an order of magnitude lower than all foreign direct investment (FDI) flows (*high agreement, much evidence*) [13.3].

Many have asserted that a key element of a successful climate change agreement will be its ability to stimulate the development and transfer of technology – without which it may be difficult to achieve emission reductions on a significant scale. Transfer of technology to developing countries depends mainly on investments. Creating enabling conditions for investments and technology uptake and international technology agreements are important. One mechanism for technology transfer is to establish innovative ways of mobilizing investments to cover the incremental cost of mitigating and adapting to climate change. International technology agreements could strengthen the knowledge infrastructure (*high agreement, much evidence*) [13.3].

A number of researchers have suggested that sectoral approaches may provide an appropriate framework for post-Kyoto agreements. Under such a system, specified targets could be set, starting with particular sectors or industries that are particularly important, politically easier to address, globally homogeneous or relatively insulated from competition with other sectors. Sectoral agreement may provide an additional degree of policy flexibility and make comparing efforts within a sector between countries easier, but may be less cost-effective, since trading within a single sector will be inherently more costly than trading across all sectors (*high agreement, much evidence*) [13.3].

Coordination/harmonization of policies

Coordinated policies and measures could be an alternative to or complement internationally agreed targets for emission reductions. A number of policies have been discussed in the literature that would achieve this goal, including taxes (such as carbon or energy taxes); trade coordination/liberalization; R&D; sectoral policies and policies that modify foreign direct investment. Under one proposal, all participating nations – industrialized and developing alike – would tax their domestic carbon usage at a common rate, thereby achieving cost-effectiveness. Others note that while an equal carbon price across countries is economically efficient, it may not be politically feasible in the context of existing tax distortions (*high agreement, much evidence*) [13.3].

Non-climate policies and links to sustainable development

There is considerable interaction between policies and measures taken at the national and sub-national level with actions taken by the private sector and between climate change mitigation and adaptation policies and policies in other areas. There are a number of non-climate national policies that can have an important influence on GHG emissions (see Chapter 12) (*high agreement, much evidence*). New research on future international agreements could focus on understanding the inter-linkages between climate policies, non-climate policies and sustainable development, and how to accelerate the adoption of existing technology and policy tools [13.3].

An overview of how various approaches to international climate change agreements, as discussed above, perform against the criteria, given in the introduction, is presented in

Table TS.21: *Assessment of international agreements on climate change[a] [Table 13.3].*

Approach	Environmental effectiveness	Cost effectiveness	Meets distributional considerations	Institutional feasibility
National emission targets and international emission trading (including offsets)	Depends on participation, and compliance	Decreases with limited participation and reduced gas and sector coverage	Depends on initial allocation	Depends on capacity to prepare inventories and compliance. Defections weaken regime stability
Sectoral agreements	Not all sectors amenable to such agreements, limiting overall effectiveness. Effectiveness depends on whether agreement is binding or non-binding	Lack of trading across sectors increases overall costs, although may be cost-effective within individual sectors. Competitive concerns reduced within each sector	Depends on participation. Within-sector competitiveness concerns alleviated if treated equally at global level	Requires many separate decisions and technical capacity. Each sector may require cross-country institutions to manage agreements
Coordinated policies and measures	Individual measures can be effective; emission levels may be uncertain; success will be a function of compliance	Depends on policy design	Extent of coordination could limit national flexibility; but may increase equity	Depends on number of countries; (easier among smaller groups of countries than at the global level)
Cooperation on Technology RD & D[b]	Depends on funding, when technologies are developed and policies for diffusion	Varies with degree of R&D risk Cooperation reduces individual national risk	Intellectual property concerns may negate the benefits of cooperation	Requires many separate decisions. Depends on research capacity and long-term funding
Development-oriented actions	Depends on national policies and design to create synergies	Depends on the extent of synergies with other development objectives	Depends on distributional effects of development policies	Depends on priority given to sustainable development in national policies and goals of national institutions
Financial mechanisms	Depends on funding	Depends on country and project type	Depends on project and country selection criteria	Depends on national institutions
Capacity building	Varies over time and depends on critical mass	Depends on programme design	Depends on selection of recipient group	Depends on country and institutional frameworks

a)   The table examines each approach based on its capacity to meet its internal goals – not in relation to achieving a global environmental goal. If such targets are to be achieved, a combination of instruments needs to be adopted. Not all approaches have equivalent evaluation in the literature; evidence for individual elements of the matrix varies.

b)   Research, Development and Demonstration

Table TS.21. Future international agreements would have stronger support if they meet these criteria *(high agreement, much evidence)* [13.3].

14   Gaps in knowledge

Gaps in knowledge refer to two aspects of climate change mitigation:

- Where additional data collection, modelling and analysis could narrow knowledge gaps, and the resulting improved knowledge and empirical experience could assist decision-making on climate change mitigation measures and policies; to some extent, these gaps are reflected in the uncertainty statements in this report.
- Where research and development could improve mitigation technologies and/or reduce their costs. This important aspect is not treated in this section, but is addressed in the chapters where relevant.

Emission data sets and projections

Despite a wide variety of data sources and databases underlying this report, there are still gaps in accurate and reliable emission data by sector and specific processes, especially with regard to non-CO_2 GHGs, organic or black carbon, and CO_2 from various sources, such as deforestation, decay of biomass and peat fires. Consistent treatment of non-CO_2 GHGs in the methodologies underlying scenarios for future GHG emissions is often lacking [Chapters 1 and 3].

Links between climate change and other policies

A key innovation of this report is the integrated approach between the assessment of climate change mitigation and wider development choices, such as the impacts of (sustainable) development policies on GHG-emission levels and vice versa.

However, there is still a lack of empirical evidence on the magnitude and direction of the interdependence and interaction of sustainable development and climate change, of mitigation and adaptation relationships in relation to development aspects, and the equity implications of both. The literature on the linkages between mitigation and sustainable development and, more particularly, on how to capture synergies and minimize trade-offs, taking into account state, market and civil society's role, is still sparse. New research is required into the linkages between climate change and national and local policies (including but not limited to energy security, water, health, air pollution, forestry, agriculture) that might lead to politically feasible, economically attractive and environmentally beneficial outcomes. It would also be helpful to elaborate potential development paths that nations and regions can pursue, which would provide links between climate protection and development issues. Inclusion of macro-indicators for sustainable development that can track progress could support such analysis [Chapters 2, 12 and 13].

Studies of costs and potentials

The available studies of mitigation potentials and costs differ in their methodological treatment and do not cover all sectors, GHGs or countries. Because of different assumptions, for example, with respect to the baseline and definitions of potentials and costs, their comparability is often limited. Also, the number of studies on mitigation costs, potentials and instruments for countries belonging to Economies in Transition and most developing regions is smaller than for developed and selected (major) developing countries.

This report compares costs and mitigation potentials based on bottom-up data from sectoral analyses with top-down costs and potential data from integrated models. The match at the sectoral level is still limited, partly because of lack of or incomplete data from bottom-up studies and differences in sector definitions and baseline assumptions. There is a need for integrated studies that combine top-down and bottom-up elements [Chapters 3, 4, 5, 6, 7, 8, 9 and 10].

Another important gap is the knowledge on spill-over effects (the effects of domestic or sectoral mitigation measures on other countries or sectors). Studies indicate a large range (leakage effects[20] from implementation of the Kyoto Protocol of between 5 and 20% by 2010), but are lacking an empirical basis. More empirical studies would be helpful [Chapter 11].

The understanding of future mitigation potentials and costs depends not only on the expected impact of RD&D on technology performance characteristics but also on 'technology learning', technology diffusion and transfer which are often not taken into account in mitigation studies. The studies on the influence of technological change on mitigation costs mostly have a weak empirical basis and are often conflicting.

Implementation of a mitigation potential may compete with other activities. For instance, the biomass potentials are large, but there may be trade-offs with food production, forestry or nature conservation. The extent to which the biomass potential can be deployed over time is still poorly understood.

In general, there is a continued need for a better understanding of how rates of adoption of climate-mitigation technologies are related to national and regional climate and non-climate policies, market mechanisms (investments, changing consumer preferences), human behaviour and technology evolution, change in production systems, trade and finance and institutional arrangements.

---

20   Carbon leakage is an aspect of spill-overs and is the increase in CO2 emissions outside countries taking domestic measures divided by the emission reductions in these countries.

1

Introduction

Coordinating Lead Authors:
H-Holger Rogner (Germany), Dadi Zhou (China)

Lead Authors:
Rick Bradley (USA), Philippe Crabbé (Canada), Ottmar Edenhofer (Germany), Bill Hare (Australia), Lambert Kuijpers (The Netherlands), Mitsutsune Yamaguchi (Japan)

Contributing Authors:
Nicolas Lefevre (France/USA), Jos Olivier (The Netherlands), Hongwei Yang (China)

Review Editors:
Hoesung Lee (Republic of Korea) and Richard Odingo (Kenya)

This chapter should be cited as:
Rogner, H.-H., D. Zhou, R. Bradley. P. Crabbé, O. Edenhofer, B.Hare (Australia), L. Kuijpers, M. Yamaguchi, 2007: Introduction. In Climate Change 2007: Mitigation. Contribution of Working Group III to the Fourth Assessment Report of the Intergovernmental Panel on Climate Change [B. Metz, O.R. Davidson, P.R. Bosch, R. Dave, L.A. Meyer (eds)], Cambridge University Press, Cambridge, United Kingdom and New York, NY, USA.

Table of Contents

EXECUTIVE SUMMARY

The ultimate objective of the United Nations Framework Convention on Climate Change (UNFCCC) is to achieve the stabilization of greenhouse gas (GHG) concentrations in the atmosphere at a level that would prevent dangerous anthropogenic interference with the climate system. Such a level should be achieved within a time frame sufficient to allow ecosystems to adapt naturally to climate change, to ensure that food production is not threatened and to enable economic development to proceed in a sustainable manner (Article 2).

This Chapter discusses Article 2 of the Convention within the framework of the main options and conditions under which it is to be implemented, reflects on past and future GHG emission trends, highlights the institutional mechanisms currently in place for the implementation of climate change and sustainable development objectives, summarizes changes from previous assessments and provides a brief roadmap for the 'Climate Change 2007: Mitigation of Climate Change' assessment.

Defining what is dangerous anthropogenic interference with the climate system and, consequently, the limits to be set for policy purposes are complex tasks that can only be partially based on science, as such definitions inherently involve normative judgments. Decisions made in relation to Article 2 will determine the level of GHG concentrations in the atmosphere (or the corresponding climate change) that is set as the goal for policy and have fundamental implications for emission reduction pathways as well as the scale of adaptation required. The choice of a stabilization level implies the balancing of the risks of climate change (risks of gradual change and of extreme events, risk of irreversible change of the climate, including risks for food security, ecosystems and sustainable development) against the risk of response measures that may threaten economic sustainability. There is little consensus as to what constitutes anthropogenic interference with the climate system and, thereby, on how to operationalize Article 2 (*high agreement, much evidence*).

Although any definition of 'dangerous interference' is by necessity based on its social and political ramifications and, as such, depends on the level of risk deemed acceptable, deep emission reductions are unavoidable in order to achieve stabilization. The lower the stabilization level, the earlier these deep reductions have to be realized (*high agreement, much evidence*).

At the present time total annual emissions of GHGs are rising. Over the last three decades, GHG emissions have increased by an average of 1.6% per year[1] with carbon dioxide (CO_2) emissions from the use of fossil fuels growing at a rate of 1.9% per year. In the absence of additional policy actions,

these emission trends are expected to continue. It is projected that – with current policy settings – global energy demand and associated supply patterns based on fossil fuels – the main drivers of GHG emissions – will continue to grow. Atmospheric CO_2 concentrations have increased by almost 100 ppm in comparison to its preindustrial level, reaching 379 ppm in 2005, with mean annual growth rates in the 2000–2005 period that were higher than those in the 1990s. The total CO_2 equivalent (CO_2-eq) concentration of all long-lived GHGs is currently estimated to be about 455 ppm CO_2-eq, although the effect of aerosols, other air pollutants and land-use change reduces the net effect to levels ranging from 311 to 435 ppm CO_2-eq (*high agreement, much evidence*).

Despite continuous improvements in energy intensities, global energy use and supply are projected to continue to grow, especially as developing countries pursue industrialization. Should there be no substantial change in energy policies, the energy mix supplied to run the global economy in the 2025–2030 time frame will essentially remain unchanged – more than 80% of the energy supply will be based on fossil fuels, with consequent implications for GHG emissions. On this basis, the projected emissions of energy-related CO_2 in 2030 are 40–110 % higher than in 2000 (with two thirds to three quarters of this increase originating in non-Annex I countries), although per capita emissions in developed countries will remain substantially higher. For 2030, GHG emission projections (Kyoto gases) consistently show a 25–90% increase compared to 2000, with more recent projections being higher than earlier ones (*high agreement, much evidence*).

The numerous mitigation measures that have been undertaken by many Parties to the UNFCCC and the entry into force of the Kyoto Protocol in February 2005 (all of which are steps towards the implementation of Article 2) are inadequate for reversing overall GHG emission trends. The experience within the European Union (EU) has demonstrated that while climate policies can be – and are being – effective, they are often difficult to fully implement and coordinate, and require continual improvement in order to achieve objectives. In overall terms, however, the impacts of population growth, economic development, patterns of technological investment and consumption continue to eclipse the improvement in energy intensities and decarbonization. Regional differentiation is important when addressing climate change mitigation – economic development needs, resource endowments and mitigative and adaptive capacities – are too diverse across regions for a 'one-size fits all' approach (*high agreement, much evidence*).

Properly designed climate change policies can be part and parcel of sustainable development, and the two can be mutually reinforcing. Sustainable development paths can

---

1 Total GHG (Kyoto gases) emissions in 2004 amounted to 49.0 GtCO$_2$-eq, which is up from 28.7 GtCO$_2$-eq in 1970 – a 70% increase between 1970 and 2004. In 1990 global GHG emissions were 39.4 GtCO$_2$-eq.

reduce GHG emissions and reduce vulnerability to climate change. Projected climate changes can exacerbate poverty and undermine sustainable development, especially in least-developed countries. Hence, global mitigation efforts can enhance sustainable development prospects in part by reducing the risk of adverse impacts of climate change. Mitigation can also provide co-benefits, such as improved health outcomes. Mainstreaming climate change mitigation is thus an integral part of sustainable development *(medium agreement, much evidence)*.

This chapter concludes with a road map of this report. Although the structure of this report (Fourth Assessment Report (AR4)) resembles the Third Assessment Report (TAR), there are distinct differences. The AR4 assigns greater weight to (1) a more detailed resolution of sectoral mitigation options and costs, (2) regional differentiation, (3) emphasizing cross-cutting issues (e.g. risks and uncertainties, decision and policy making, costs and potentials, biomass, the relationships between mitigation, adaptation and sustainable development, air pollution and climate, regional aspects and issues related to the implementation of UNFCCC Article 2), and (4) the integration of all these aspects.

1.1 Introduction

The assessment 'Climate Change 2007: Mitigation of Climate Change' is designed to provide authoritative, timely information on all aspects of technologies and socio-economic policies, including cost-effective measures to control greenhouse gas (GHG) emissions. A thorough understanding of future GHG emissions and their drivers, available mitigation options, mitigation potentials and associated costs and ancillary benefits is especially important to support negotiations on future reductions in global emissions.

This chapter starts with a discussion of the key issues involved in Article 2 of the United Nations Framework Convention on Climate Change (UNFCCC) and of the relationship of these to emission pathways and broad mitigation options. The sections that follow reflect on past and future GHG emission trends, highlight the institutional mechanisms in place for the implementation of climate change and sustainable development objectives, summarize changes from previous assessments and provides a concise roadmap to the 'Climate Change 2007: Mitigation of Climate Change' assessment.

1.2 Ultimate objective of the UNFCCC

The UNFCCC was adopted in May 1992 in New York and opened for signature at the 'Rio Earth Summit' in Rio de Janeiro a month later. It entered into force in March 1994 and has achieved near universal ratification with ratification by 189 countries of the 194 UN member states (December 2006)[2].

1.2.1 Article 2 of the Convention

Article 2 of the UNFCCC specifies the ultimate objective of the Convention and states:

'The ultimate objective of this Convention and any related legal instruments that the Conference of the Parties may adopt is to achieve, in accordance with the relevant provisions of the Convention, stabilization of greenhouse gas concentrations in the atmosphere at a level that would prevent dangerous anthropogenic interference with the climate system. Such a level should be achieved within a time frame sufficient to allow ecosystems to adapt naturally to climate change, to ensure that food production is not threatened and to enable economic development to proceed in a sustainable manner' (UN, 1992).

The criterion that relates to enabling economic development to proceed in a sustainable manner is a double-edged sword. Projected anthropogenic climate change appears likely to adversely affect sustainable development, with adverse effects tending to increase with higher levels of climate change and

GHG concentrations (IPCC, 2007b, SPM and Chapter 19). Conversely, costly mitigation measures could have adverse effects on economic development. This dilemma facing policymakers results in (a varying degree of) tension that is manifested in the debate over the scale of the interventions and the balance to be adopted between climate policy (mitigation and adaptation) and economic development.

The assessment of impacts, vulnerability and adaptation potentials is likely to be important for determining the levels and rates of climate change which would result in ecosystems, food production or economic development being threatened to a level sufficient to be defined as dangerous. Vulnerabilities to anthropogenic climate change are strongly regionally differentiated, with often those in the weakest economic and political position being the most susceptible to damages (IPCC, 2007b, Chapter 19, Tables 19.1 and 19.3.3).

Limits to climate change or other changes to the climate system that are deemed necessary to prevent dangerous anthropogenic interference with the climate system can be defined in terms of various – and often quite different – criteria, such as concentration stabilization at a certain level, global mean temperature or sea level rise or levels of ocean acidification. Whichever limit is chosen, its implementation would require the development of consistent emission pathways and levels of mitigation (Chapter 3).

1.2.2 What is dangerous interference with the climate system?

Defining what is dangerous interference with the climate system is a complex task that can only be partially supported by science, as it inherently involves normative judgements. There are different approaches to defining danger, and an interpretation of Article 2 is likely to rely on scientific, ethical, cultural, political and/or legal judgements. As such, the agreement(s) reached among the Parties in terms of what may constitute unacceptable impacts on the climate system, food production, ecosystems or sustainable economic development will represent a synthesis of these different perspectives.

Over the past two decades several expert groups have sought to define levels of climate change that could be tolerable or intolerable, or which could be characterized by different levels of risk. In the late 1980s, the World Meteorological Organization (WMO)/International Council of Scientific Unions (ICSU)/ UN Environment Programme (UNEP) Advisory Group on Greenhouse Gases (AGGG) identified two main temperature indicators or thresholds with different levels of risk (Rijsberman and Swart, 1990). Based on the available knowledge at the time a 2°C increase was determined to be 'an upper limit beyond which the risks of grave damage to ecosystems, and of non-linear responses, are expected to increase rapidly'. This early

---

2   http://unfccc.int/essential_background/convention/items/2627.php. 190 ratifications - one from the European Union.

work also identified the rate of change to be of importance to determining the level of risk, a conclusion that has subsequently been confirmed qualitatively (IPCC, 2007b, Chapters 4 and 19). More recently, others in the scientific community have reached conclusions that point in a similar direction 'that global warming of more than 1°C, relative to 2000, will constitute "dangerous" climate change as judged from likely effects on sea level and extermination of species' (Hansen *et al.*, 2006). Probabilistic assessments have also been made that demonstrate how scientific uncertainties, different normative judgments on acceptable risks to different systems (Mastrandrea and Schneider, 2004) and/or interference with the climate system (Harvey, 2007) affect the levels of change or interference set as goals for policy (IPCC, 2007b, Chapter 19). From an economic perspective, the Stern Review (Stern, 2006) found that in order to minimise the most harmful consequences of climate change, concentrations would need to be stabilized below 550 ppm CO_2-eq. The Review further argues that any delay in reducing emissions would be 'would be costly and dangerous'. This latter conclusion is at variance with the conclusions drawn from earlier economic analyses which support a slow 'ramp up' of climate policy action (Nordhaus, 2006) and, it has been argued, is a consequence of the approach taken by the Stern Review to intergenerational equity (Dasgupta, 2006).

The IPCC Third Assessment Report (TAR) identified five broad categories of reasons for concern that are relevant to Article 2: (1) Risks to unique and threatened systems, (2) risks from extreme climatic events, (3) regional distribution of impacts, (4) aggregate impacts and (5) risks from large-scale discontinuities. The Fourth Assessment Report (AR4) focuses on Key Vulnerabilities relevant to Article 2, which are broadly categorized into biological systems, social systems, geophysical systems, extreme events and regional systems (IPCC, 2007b, Chapter 19). The implications of different interpretations of dangerous anthropogenic interference for future emission pathways are reviewed in IPCC (2007b), Chapter 9 and also in Chapter 3 of this report. The literature confirms that climate policy can substantially reduce the risk of crossing thresholds deemed dangerous (IPCC, 2007b, SPM and Chapter 19; Chapter 3, Section 3.5.2 of this report).

While the works cited above are principally scientific (expert-led) assessments, there is also an example of governments seeking to define acceptable levels of climate change based on interpretations of scientific findings. In 2005, the EU Council (25 Heads of Government of the European Union) agreed that – with a view to achieving the ultimate objective of the Convention – the global annual mean surface temperature increase should not exceed 2°C above pre-industrial levels (CEU, 2005).

1.2.3 Issues related to the implementation of Article 2

Decisions made in relation to Article 2 will determine the level of climate change that is set as the goal for policy and have fundamental implications for emission reduction pathways, the feasibility, timing and scale of adaptation required and the magnitude of unavoidable losses. The emission pathways which correspond to different GHG or radiative forcing stabilization levels and consequential global warming are reviewed in Chapter 3 (see Tables 3.9 and 3.10). The potential consequences of two hypothetical limits can provide an indication of the differing scales of mitigation action that depend on Article 2 decisions: A 2°C above pre-industrial limit on global warming would implies that emissions peak within the next decade and be reduced to less than 50% of the current level by 2050[3]; a 4°C limit would imply that emissions may not have to peak until well after the middle of the century and could still be well above 2000 levels in 2100. In relation to the first hypothetical limt, the latter would have higher levels of adaptation costs and unavoidable losses, but carry lower mitigation costs.

Issues related to the mitigation, adaptation and sustainable development aspects of the implementation of Article 2 thus include, among others, the linkages between sustainable development and the adverse effects climate change, the need for equity and cooperation and the recognition of common but differentiated responsibilities and respective capabilities as well as the precautionary principle (see Section 1.4.1 for more detail on relevant UNFCCC Articles that frame these issues). In this context, risk management issues which take into account several key aspects of the climate change problem, such as inertia, irreversibility, the risk of abrupt or catastrophic changes and uncertainty, are introduced in this section and discussed in more detail in Chapters 2, 3 and 11.

1.2.3.1 Sustainable development

Sustainable development has environmental, economic and social dimensions (see Chapter 2, Section 2.1). Properly designed climate change responses can be part and parcel of sustainable development, and the two can be mutually reinforcing (Section 2.1). Mitigation, by limiting climate change, can conserve or enhance natural capital (ecosystems, the environment as sources and sinks for economic activities) and prevent or avoid damage to human systems and, thereby, contribute to the overall productivity of capital needed for socio-economic development, including mitigative and adaptive capacity. In turn, sustainable development paths can reduce vulnerability to climate change and reduce GHG emissions. The projected climate changes can exacerbate poverty and thereby undermine sustainable development (see, for example, IPCC, 2007b, Chapters 6, Section 9.7 and 20.8.3), especially in

---

3   For the best-guess climate sensitivity and the lowest range of multigas stabilization scenarios found in the literature which show a warming of about 2-2.4°C above preindustrial temperatures (Chapter 3, section 3.5.2 and Table 3.10).

developing countries, which are the most dependent on natural capital and lack financial resources (see Chapter 2 and Stern (2006)). Hence global mitigation efforts can enhance sustainable development prospects in part by reducing the risk of adverse impacts of climate change (see also Chapter 12).

1.2.3.2   Adaptation and mitigation

Adaptation and mitigation can be complementary, substitutable or independent of each other (see IPCC, 2007b, Chapter 18). If complementary, adaptation reduces the costs of climate change impacts and thus reduces the benefits of mitigation. Although adaptation and mitigation may be substitutable up to a certain point, they are never perfect substitutes for each other since mitigation will always be required to avoid 'dangerous' and irreversible changes to the climate system. Irrespective of the scale of the mitigation measures that are implemented in the next 10–20 years, adaptation measures will still be required due to the inertia in the climate system. As reported in IPCC, 2007b, Chapter 19 (and also noted in Stern (2006)), changes in the climate are already causing setbacks to economic and social development in some developing countries with temperature increases of less than 1°C. Unabated climate change would increase the risks and costs very substantially (IPCC, 2007b, Chapter 19). Both adaptation and mitigation depend on capital assets, including social capital, and both affect capital vulnerability and GHG emissions (see Chapter 2, Section 2.5.2). Through this mutual dependence, both are tied to sustainable development (see Sections 2.5, 11.8 and 11.9, 12.2 and 12.3).

The stabilization of GHG concentrations and, in particular, of the main greenhouse gas, CO_2, requires substantial emission reductions, well beyond those built into existing agreements such as the Kyoto Protocol. The timing and rate of these reductions depend on the level of the climate goal chosen (see Chapter 3.3.5.1).

1.2.3.3   Inertia

Inertia in both the climate and socio-economic systems would need to be taken into account when mitigation actions are being considered. Mitigation actions aimed at specific climate goals would need to factor in the response times of the climate system, including those of the carbon cycle, atmosphere and oceans. A large part of the atmospheric response to radiative forcing occurs on decadal time scales, but a substantial component is linked to the century time scales of the oceanic response to the same forcing changes (Meehl et al., 2007). Once GHG concentrations are stabilized global mean temperature would very likely stabilize within a few decades, although a further slight increase may still occur over several centuries (Meehl et al., 2007). The rise in sea level, however, would continue for many centuries after GHG stabilization due to both ongoing heat uptake by the oceans and the long time scale of ice sheet response to warming (Meehl et al., 2007). The time scales

for mitigation are linked to technological, social, economic, demographic and political factors. Inertia is a characteristic of the energy system with its long-life infrastructures, and this inertia is highly relevant to how fast GHG concentrations can be stabilized (Chapter 11.6.5). Adaptation measures similarly exhibit a range of time scales, and there can be substantial lead times required before measures can be implemented and subsequently take effect, particularly when it involves infrastructure (IPCC, 2007b, Chapter 17).

The consequence of inertia in both the climate and socio-economic systems is that benefits from mitigation actions initiated now – in the short term – would lead to significant changes in the climate being avoided several decades further on. This means that mitigation actions need to be implemented in the short term in order to have medium- and long-term benefits and to avoid the lock in of carbon intensive technologies (Chapter 11.6.5).

1.2.3.4   Uncertainty and risk

Uncertainty in knowledge is an important aspect in the implementation of Article 2, whether it is assessing future GHG emissions or the severity of climate change impacts and regional changes, evaluating these impacts over many generations, estimating mitigation costs or evaluating the level of mitigation action needed to reduce risk. Notwithstanding these uncertainties, mitigation will reduce the risk of both global mean and regional changes and the risk of abrupt changes in the climate system (see Chapter 2, Section 2.3).

There may be risks associated with rapid and/or abrupt changes in the climate and the climate system as a result of human interference (Solomon et al., 2007; IPCC, 2007b, Chapter 19 Tables 19.1 and 19.3.5-7). These include changes in climate variability (El Nino Southern Oscillation, monsoons); a high likelihood that warming will lead to an increase in the risk of many extreme events, including floods, droughts, heat waves and fires, with increasing levels of adverse impacts; a risk that a 1–2°C sustained global warming (versus the temperature at present) would lead to a commitment to a large sea-level rise due to at least the partial deglaciation of both ice sheets; an uncertain risk of a shutdown of the North Atlantic Meridional Overturning Circulation; a large increase in the intensity of tropical cyclones with increasing levels of adverse impacts as temperatures increase; the risk that positive feedbacks from warming may cause the release of CO_2 or methane (CH_4) from the terrestrial biosphere and soils (IPCC, 2007b, Chapter 19 Tables 19.1 and 19.3.5-7). In the latter case, a positive climate–carbon cycle feedback would reduce the land and ocean uptake of CO_2, implying a reduction of the allowable emissions required to achieve a given atmospheric CO_2 stabilization level (Meehl et al., 2007, Executive Summary).

1.2.3.5   *Irreversibility*

Irreversibility is an important aspect of the climate change issue, with implications for mitigation and adaptation responses. The response of the climate system to anthropogenic forcing is likely to be irreversible over human time scales, and much of the damage is likely to be irreversible even over longer time scales. Mitigation and adaptation will often require investments involving sunk (irreversible) costs in new technologies and practices (Sections 2.2.3, 11.6.5; IPCC, 2007b, Chapter 17). Decision-makers will need to take into account these environmental, socio-economic and technological irreversibilities in deciding on the timing and scale of mitigation action.

1.2.3.6   *Public good*

The climate system tends to be overused (excessive GHG concentrations) because of its natural availability as a resource whose access is open to all free of charge. In contrast, climate protection tends to be underprovided. In general, the benefits of avoided climate change are spatially indivisible, freely available to all (non-excludability), irrespective of whether one is contributing to the regime costs or not. As regime benefits by one individual (nation) do not diminish their availability to others (non-rivalry), it is difficult to enforce binding commitments on the use of the climate system[4] (Kaul *et al.*, 1999; 2003). This may result in 'free riding', a situation in which mitigation costs are borne by some individuals (nations) while others (the 'free riders') succeed in evading them but still enjoy the benefits of the mitigation commitments of the former.

The incentive to evade mitigation costs increases with the degree of substitutability among individual mitigation efforts (mitigation is largely additive) and with the inequality of the distribution of net benefits among regime participants. However, individual mitigation costs decrease with efficient mitigation actions undertaken by others. Because mitigation efforts are additive, the larger the number of participants, the smaller the individual cost of providing the public good – in this case, climate system stabilization. Cooperation requires the sharing of both information on climate change and technologies through technology transfers as well as the coordination of national actions lest the efforts required by the climate regime be underprovided.

1.3.3.7   *Equity*

Equity is an ethical construct that demands the articulation and implementation of choices with respect to the distribution of rights to benefits and the responsibilities for bearing the costs resulting from particular circumstances – for example,

climate change – within and among communities, including future generations. Climate change is subject to a very asymmetric distribution of present emissions and future impacts and vulnerabilities. Equity can be elaborated in terms of distributing the costs of mitigation or adaptation, distributing future emission rights and ensuring institutional and procedural fairness (Chapter 13, Section 13.3.4.3). Equity also exhibits preventative (avoidance of damage inflicted on others), retributive (sanctions), and corrective elements (e.g. 'common but differentiated responsibilities') (Chapter 2, Section 2.6), each of which has an important place in the international response to the climate change problem (Chapter 13).

1.3   Energy, emissions and trends in Research and Development – are we on track?

1.3.1   Review of the last three decades

Since pre-industrial times, increasing emissions of GHGs due to human activities have led to a marked increase in atmospheric concentrations of the long-lived GHG gases carbon dioxide (CO_2), CH_4, and nitrous oxide (N_2O), perfluorocarbons PFCs, hydrofluorocarbons (HFCs) and sulphur hexafluoride (SF_6) and ozone-depleting substances (ODS; chlorofluorocarbons (CFCs), hydrochlorofluorocarbons (HCFCs), halons) and the human-induced radiative forcing of the Earth's climate is largely due to the increases in these concentrations. The predominant sources of the increase in GHGs are from the combustion of fossil fuels. Atmospheric CO_2 concentrations have increased by almost 100 ppm in comparison to its preindustrial levels, reaching 379 ppm in 2005, with mean annual growth rates in the 2000–2005 period that were higher than those in the 1990s.

The direct effect of all the long-lived GHGs is substantial, with the total CO_2 equivalent concentration of these gases currently being estimated to be around 455 ppm CO_2-eq[5] (range: 433–477 ppm CO_2-eq). The effects of aerosol and land-use changes reduce radiative forcing so that the net forcing of human activities is in the range of 311 to 435 ppm CO_2-eq, with a central estimate of about 375 ppm CO_2-eq.

A variety of sources exist for determining global and regional GHG and other climate forcing agent trends. Each source has its strengths and weaknesses and uncertainties. The EDGAR database (Olivier *et al.*, 2005, 2006) contains global GHG emission trends categorized by broad sectors for the period 1970–2004, and Marland *et al.* (2006) report CO_2 emissions on a global basis. Both databases show a similar temporal evolution of emissions. Since 1970, the global warming potential (GWP)-

---

4   Resulting in a prisoners' dilemma situation because of insufficient incentives to cooperate.
5   Radiative forcing (Forster *et al.*, 2007) is converted to CO_2 equivalents using the inversion of the expression Q (W/m^2) = 5.35 × ln (CO_2/278) (see Solomon *et al.*, 2007, Table TS-2 footnote b).

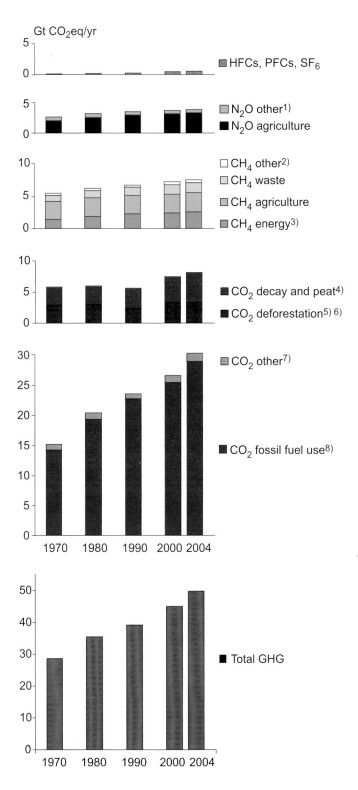

Gt CO_2eq/yr

■ HFCs, PFCs, SF_6

■ N_2O other[1]
■ N_2O agriculture

□ CH_4 other[2]
□ CH_4 waste
■ CH_4 agriculture
■ CH_4 energy[3]

■ CO_2 decay and peat[4]
■ CO_2 deforestation[5) 6]

■ CO_2 other[7]

■ CO_2 fossil fuel use[8]

1970 1980 1990 2000 2004

■ Total GHG

1970   1980   1990   2000 2004

Figure 1.1a *Global anthropogenic greenhouse gas trends, 1970–2004.*

One-hundred year global warming potentials (GWPs) from the Intergovernmental Panel on Climate Change (IPCC) 1996 (SAR) were used to convert emissions to CO_2 equivalents (see the UNFCCC reporting guidelines). Gases are those reported under UNFCCC reporting guidelines. The uncertainty in the graph is quite large for CH_4 and N_2O (of the order of 30–50%) and even larger for CO_2 from agriculture and forestry.

Notes:

1. Other N_2O includes industrial processes, deforestation/savannah burning, waste water and waste incineration.
2. Other is CH_4 from industrial processes and savannah burning.
3. Including emissions from bio energy production and use.
4. CO_2 emissions from decay (decomposition) of above ground biomass that remains after logging and deforestation and CO_2 from peat fires and decay of drained peat soils.
5. As well as traditional biomass use at 10% of total, assuming 90% is from sustainable biomass production. Corrected for the 10% of carbon in biomass that is assumed to remain as charcoal after combustion.
6. For large-scale forest and scrubland biomass burning averaged data for 1997-2002 based on Global Fire Emissions Data base satellite data.
7. Cement production and natural gas flaring.
8. Fossil fuel use includes emissions from feedstocks.

Source: Adapted from Olivier et al., 2005; 2006; Hooijer et al., 2006

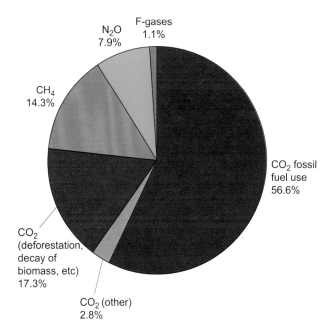

Figure 1.1b *Global anthropogenic greenhouse gas emissions in 2004.*

Source: Adapted from Olivier et al., 2005, 2006

weighted emissions of GHGs (not including ODS which are controlled under the Montreal Protocol), have increased by approximately 70%, (24% since 1990), with CO_2 being the largest source, having grown by approximately 80% (28% since 1990) to represent 77% of total anthropogenic emissions in 2004 (74% in 1990) (Figure 1.1). Radiative forcing as a result of increases in atmospheric CO_2 concentrations caused

by human activities since the preindustrial era predominates over all other radiative forcing agents (IPCC, 2007a, SPM). Total CH_4 emissions have risen by about 40% from 1970 (11% from 1990), and on a sectoral basis there has been an 84% (12% from 1990) increase from combustion and the use of fossil fuels, while agricultural emissions have remained roughly stable due to compensating falls and increases in rice and livestock

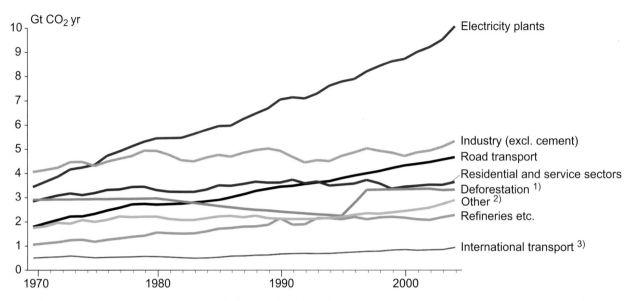

Figure 1.2: *Sources of global CO_2 emissions, 1970–2004 (only direct emissions by sector).*
[1] Including fuelwood at 10% net contribution. For large-scale biomass burning, averaged data for 1997–2002 are based on the Global Fire Emissions Database satellite data (van der Werf *et al.*, 2003). Including decomposition and peat fires (Hooijer et al., 2006). Excluding fossil fuel fires.
[2] Other domestic surface transport, non-energetic use of fuels, cement production and venting/flaring of gas from oil production.
[3] Including aviation and marine transport.

Source: Adapted from Olivier et al., 2005; 2006).

production, respectively. N_2O emissions have grown by 50% since 1970 (11% since 1990), mainly due to the increased use of fertilizer and the aggregate growth of agriculture. Industrial process emissions of N_2O have fallen during this period.

The use and emissions of all fluorinated gases (including those controlled under the Montreal Protocol) decreased substantially during 1990–2004. The emissions, concentrations and radiative forcing of one type of fluorinated gas, the HFCs, grew rapidly during this period as these replaced ODS; in 2004, CFCs were estimated to constitute about 1.1% of the total GHG emissions (100-year GWP) basis. Current annual emissions of all fluorinated gases are estimated at 2.5 $GtCO_2$-eq, with HFCs at 0.4 $GtCO_2$-eq. The stocks of these gases are much larger and currently represent about 21 $GtCO_2$-eq.

The largest growth in CO_2 emissions has come from the power generation and road transport sectors, with the industry, households and the service sector[6] remaining at approximately the same levels between 1970 and 2004 (Figure 1.2). By 2004, CO_2 emissions from power generation represented over 27% of the total anthropogenic CO_2 emissions and the power sector was by far its most important source. Following the sectoral breakdown adopted in this report (Chapters 4–10), in 2004 about 26% of GHG emissions were derived from energy supply (electricity and heat generation), about 19% from industry, 14%

from agriculture[7], 17% from land use and land-use change[8], 13% from transport, 8% from the residential, commercial and service sectors and 3% from waste (see Figure 1.3). These values should be regarded as indicative only as some uncertainty remains, particularly with regards to CH_4 and N_2O emissions, for which the error margin is estimated to be in the order of 30–50%, and CO_2 emissions from agriculture, which have an even larger error margin.

Since 1970, GHG emissions from the energy supply sector have grown by over 145%, while those from the transport sector have grown by over 120%; as such, these two sectors show the largest growth in GHG emissions. The industry sector's emissions have grown by close to 65%, LULUCF (land use, land-use change and forestry) by 40% while the agriculture sector (27%) and residential/commercial sector (26%) have experienced the slowest growth between 1970 and 2004.

The land-use change and forestry sector plays a significant role in the overall carbon balance of the atmosphere. However, data in this area are more uncertain than those for other sectors. The Edgar database indicates that, in 2004, the share of CO_2 emissions from deforestation and the loss of carbon from soil decay after logging constituted approximately 7–16% of the total GHG emissions (not including ODS) and between 11 and 28% of fossil CO_2 emissions. Estimates vary considerably.

---

6   Direct emissions by sector; i.e., data do not include indirect emissions.
7   N_2O and CH_4 emissions (CO_2 emissions are small; compare with Chapter 8) and not counting land clearance. The proportion of emissions of N_2O and CH_4 are higher – around 85 and 45% (±5%), respectively. Emissions from agricultural soils not related to land clearance are quite small – of an order of 40 $MtCO_2$ per year in 2005 (Chapter 8).
8   Deforestation, including biofuel combustion, assuming 90% sustainable production, biomass burning, CO_2 emissions from the decay of aboveground biomass after logging and deforestation and from peat fires and decay of peat soils.

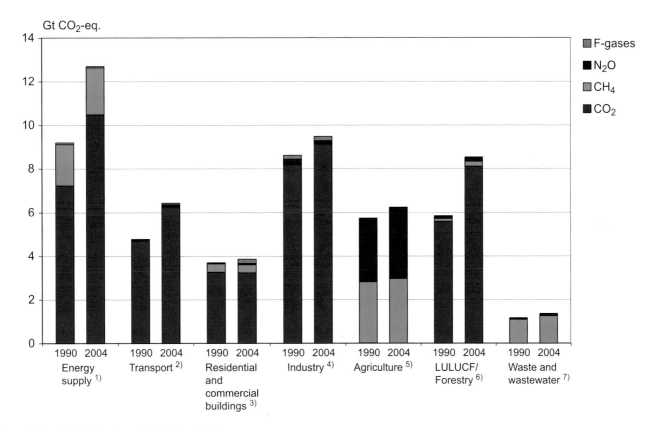

Figure 1.3a: *GHG emissions by sector in 1990 and 2004.*

Source: Adapted from Olivier et al., 2005, 2006.

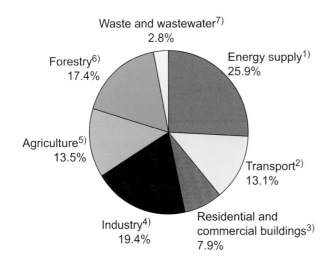

Figure 1.3b: *GHG emissions by sector in 2004.*

Source: Adapted from Olivier et al., 2005; 2006.

One-hundred year GWPs from IPCC, 1996 (Second Assessment Report) were used to convert emissions to CO_2 equivalents. The uncertainty in the graph is quite large for CH_4 and N_2O (of the order of 30–50%) and even larger for CO_2 from agriculture and forestry. For large-scale biomass burning, averaged activity data for 1997–2002 were used from the Global Fire Emissions Database based on satellite data. Peat (fire and decay) emissions are based on recent data from WL/Delft Hydraulics.

[1] Excluding refineries, coke ovens which are included in industry.
[2] Including international transport (bunkers), excluding fisheries; excluding off-road agricultural and forestry vehicles and machinery.
[3] Including traditional biomass use. Emissions reported in Chapter 6 include the sector's share in emissions caused by centralized electricity generation so that any mitigation achievements in the sector resulting from lower electricity use are credited to the sector.
[4] Including refineries and coke ovens. Emissions reported in Chapter 7 include the sector's share in emissions caused by centralized electricity generation so that any mitigation achievements in the sector resulting from lower electricity use are credited to the sector.
[5] Including agricultural waste burning and savannah burning (non-CO_2). CO_2 emissions and/or removals from agricultural soils are not estimated in this database.
[6] Data include CO_2 emissions from deforestation, CO_2 emissions from decay (decomposition) of aboveground biomass that remains after logging and deforestation and CO_2 from peat fires and decay of drained peat soils. Chapter 9 reports emissions from deforestation only.
[7] Includes landfill CH_4, wastewater CH_4 and N_2O, and CO_2 from waste incineration (fossil carbon only).

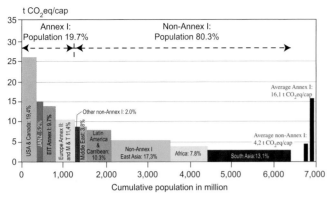

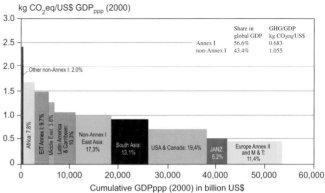

Figure 1.4a: *Distribution of regional per capita GHG emissions (all Kyoto gases including those from land-use) over the population of different country groupings in 2004. The percentages in the bars indicate a region's share in global GHG emissions.*

Source: Adapted from Bolin and Khesgi, 2001) using IEA and EDGAR 3.2 database information (Olivier et al., 2005, 2006).

Figure 1.4b: *Distribution of regional GHG emissions (all Kyoto gases including those from land-use) per USD of GDP$_{ppp}$ over the GDP of different country groupings in 2004. The percentages in the bars indicate a region's share in global GHG emissions.*

Source: IEA and EDGAR 3.2 database information (Olivier et al., 2005, 2006).

Note: Countries are grouped according to the classification of the UNFCCC and its Kyoto Protocol; this means that countries that have joined the European Union since then are still listed under EIT Annex I. A full set of data for all countries for 2004 was not available. The countries in each of the regional groupings include:

- **EIT Annex I**: Belarus, Bulgaria, Croatia, Czech Republic, Estonia, Hungary, Latvia, Lithuania, Poland, Romania, Russian Federation, Slovakia, Slovenia, Ukraine
- **Europe Annex II & M&T**: Austria, Belgium, Denmark, Finland, France, Germany, Greece, Iceland, Ireland, Italy, Liechtenstein, Luxembourg, Netherlands, Norway, Portugal, Spain, Sweden, Switzerland, United Kingdom; Monaco and Turkey
- **JANZ**: Japan, Australia, New Zealand.
- **Middle East**: Bahrain, Islamic Republic of Iran, Israel, Jordan, Kuwait, Lebanon, Oman, Qatar, Saudi Arabia, Syria, United Arab Emirates, Yemen
- **Latin America & the Caribbean**: Antigua & Barbuda, Argentina, Bahamas, Barbados, Belize, Bolivia, Brazil, Chile, Colombia, Costa Rica, Cuba, Dominica, Dominican Republic, Ecuador, El Salvador, Grenada, Guatemala, Guyana, Haiti, Honduras, Jamaica, Mexico, Nicaragua, Panama, Paraguay, Peru, Saint Lucia, St. Kitts-Nevis-Anguilla, St. Vincent-Grenadines, Suriname, Trinidad and Tobago, Uruguay, Venezuela
- **Non-Annex I East Asia**: Cambodia, China, Korea (DPR), Laos (PDR), Mongolia, Republic of Korea, Viet Nam.
- **South Asia**: Afghanistan, Bangladesh, Bhutan, Comoros, Cook Islands, Fiji, India, Indonesia, Kiribati, Malaysia, Maldives, Marshall Islands, Micronesia, (Federated States of), Myanmar, Nauru, Niue, Nepal, Pakistan, Palau, Papua New Guinea, Philippine, Samoa, Singapore, Solomon Islands, Sri Lanka, Thailand, Timor-Leste, Tonga, Tuvalu, Vanuatu
- **North America**: Canada, United States of America.
- **Other non-Annex I**: Albania, Armenia, Azerbaijan, Bosnia Herzegovina, Cyprus, Georgia, Kazakhstan, Kyrgyzstan, Malta, Moldova, San Marino, Serbia, Tajikistan, Turkmenistan, Uzbekistan, Republic of Macedonia
- **Africa**: Algeria, Angola, Benin, Botswana, Burkina Faso, Burundi, Cameroon, Cape Verde, Central African Republic, Chad, Congo, Democratic Republic of Congo, Côte d'Ivoire, Djibouti, Egypt, Equatorial Guinea, Eritrea, Ethiopia, Gabon, Gambia, Ghana, Guinea, Guinea-Bissau, Kenya, Lesotho, Liberia, Libya, Madagascar, Malawi, Mali, Mauritania, Mauritius, Morocco, Mozambique, Namibia, Niger, Nigeria, Rwanda, Sao Tome and Principe, Senegal, Seychelles, Sierra Leone, South Africa, Sudan, Swaziland, Togo, Tunisia, Uganda, United Republic of Tanzania, Zambia, Zimbabwe

There are large emissions from deforestation and other land-use change activities in the tropics; these have been estimated in IPCC (2007a) for the 1990s to have been 5.9 GtCO$_2$-eq, with a large uncertainty range of 1.8–9.9 GtCO$_2$-eq (Denman *et al.*, 2007). This is about 25% (range: 8–42%) of all fossil fuel and cement emissions during the 1990s. The underlying factors accounting for the large range in the estimates of tropical deforestation and land-use changes emissions are complex and not fully resolved at this time (Ramankutty *et al.*, 2006). For the Annex I Parties that have reported LULUCF sector data to the UNFCCC (including agricultural soils and forests) since 1990, the aggregate net sink reported for emissions and removals over the period up to 2004 average out to approximately 1.3 GtCO$_2$-eq (range: –1.5 to –0.9 GtCO$_2$-eq)[9].

On a geographic basis, there are important differences between regions. North America, Asia and the Middle East have

driven the rise in emissions since 1972. The former countries of the Soviet Union have shown significant reductions in CO$_2$ emissions since 1990, reaching a level slightly lower than that in 1972. Developed countries (UNFCCC Annex I countries) hold a 20% share in the world population but account for 46.4% of global GHG emissions. In contrast, the 80% of the world population living in developing countries (non-Annex I countries) account for 53.6% of GHG emissions (see Figure 1.4a). Based on the metric of GHG emission per unit of economic output (GHG/GDP$_{ppp}$)[10], Annex I countries generally display lower GHG intensities per unit of economic production process than non-Annex I countries (see Figure 1.4b).

The promotion of energy efficiency improvements and fuel switching are among the most frequently applied policy measures that result in mitigation of GHG emissions. Although they may not necessarily be targeted at GHG emission

---

9   Data for the Russian Federation is not included in the UNFCCC data set. Chapter 7 estimates the Russian sink for 1990–2000 to be 370–740 MtCO$_2$/year, which would add up to approximately 28–57% of the average sink reported here.
10 The GDP$_{ppp}$ metric is used for illustrative purposes only for this report.

mitigation, such policy measures do have a strong impact in lowering the emission level from where it would be otherwise.

According to an analysis of GHG mitigation activities in selected developing countries by Chandler *et al.* (2002), the substitution of gasoline-fuelled cars with ethanol-fuelled cars and that of conventional CHP (combined heat and power; also cogeneration) plants with sugar-cane bagasse CHP plants in Brazil resulted in an estimated carbon emission abatement of 23.5 $MtCO_2$ in 2000 (actual emissions in 2000: 334 $MtCO_2$). According to the same study, economic and energy reforms in China curbed the use of low-grade coal, resulting in avoided emissions of some 366 $MtCO_2$ (actual emissions: 3,100 $MtCO_2$). In India, energy policy initiatives including demand-side efficiency improvements are estimated to have reduced emissions by 66 $MtCO_2$ (compared with the actual emission level of 1,060 $MtCO_2$). In Mexico, the switch to natural gas, the promotion of efficiency improvements and lower deforestation are estimated to have resulted in 37 $MtCO_2$ of emission reductions, compared with actual emissions of 685 $MtCO_2$.

For the EU-25 countries, the European Environment Agency (EEA, 2006) provides a rough estimate of the avoided CO_2 emissions from public electricity and heat generation due to efficiency improvements and fuel switching. If the efficiency and fuel mix had remained at their 1990 values, emissions in 2003 would have been some 34% above actual emissions, however linking these reductions to specific policies was found to be difficult. For the UK and Germany about 60% of the reductions from 1990 to 2000 were found to be due to factors other than the effects of climate-related policies (Eichhammer *et al.*, 2001, 2002).

Since 2000, however, many more policies have been put into place, including those falling under the European Climate Change Programme (ECCP), and significant progress has been made, including the establishment of the EU Emissions Trading Scheme (EU ETS) (CEC, 2006). A review of the effectiveness of the first stage of the ECCP reported that about one third of the potential reductions had been fully implemented by mid 2006[11]. Overall EU-25 emissions in 2004 were 0.9% lower than in the base year, and the European Commission (EC) assessed the EC Kyoto target (8% reduction relative to the base year) to be within reach under the conditions that (1) all additional measures currently under discussion are put into force in time, (2) Kyoto mechanisms are used to the full extent planned and (3) removals from Articles 3.3 and 3.4 activities (carbon sinks) contribute to the extent projected (CEC, 2006). Overall this shows that climate policies can be effective, but that they are difficult to fully implement and require continual improvement in order to achieve the desired objectives.

1.3.1.1   *Energy supply*

Global primary energy use almost doubled from 5,363 Mtoe (225 EJ) in 1970 to 11,223 Mtoe (470 EJ) in 2004, with an average annual growth of 2.2% over this period. Fossil fuels accounted for 81% of total energy use in 2004 – slightly down from the 86% more than 30 years ago, mainly due to the increase in the use of nuclear energy. Despite the substantial growth of non-traditional renewable forms of energy, especially wind power, over the last decade, the share of renewables (including traditional biomass) in the primary energy mix has not changed compared with 1970 (see Chapter 4, Section 4.2).

1.3.1.2   *Intensities*

The Kaya identity (Kaya, 1990) is a decomposition that expresses the level of energy related CO_2 emissions as the product of four indicators: (1) carbon intensity (CO_2 emissions per unit of total primary energy supply (TPES)), (2) energy intensity (TPES per unit of GDP), (3) gross domestic product per capita (GDP/cap) and (4) population. The global average growth rate of CO_2 emissions between 1970 and 2004 of 1.9% per year is the result of the following annual growth rates: population 1.6%, GDP/cap[12] 1.8%, energy-intensity of –1.2% and carbon-intensity –0.2% (Figure 1.5).

A decomposition analysis according to the refined Laspeyeres index method (Sun, 1998; Sun and Ang, 2000) is shown in Figure 1.6. Each of the three stacked bars refers to 10-year periods and indicates how the net change in CO_2 emissions of that decade can be attributed to the four indicators of the Kaya identity. These contributions – to tonnes of CO_2 emissions – can be positive or negative, and their sum equals the net emission change (shown for each decade by the black line).

GDP/capita and population growth were the main drivers of the increase in global emissions during the last three decades of the 20th century. However, consistently declining energy intensities indicate structural changes in the global energy system. The role of carbon intensity in offsetting emission growth has been declining over the last two decades. The reduction in carbon intensity of energy supply was the strongest between 1980 and 1990 due to the delayed effect of the oil price shocks of the 1970s, and it approached zero towards the year 2000 and reversed after 2000 At the global scale, declining carbon and energy intensities have been unable to offset income effects and population growth and, consequently, carbon emissions have risen. Under the reference scenario of the International Energy Agency (IEA, 2006a) these trends are expected to remain valid until 2030; in particular, energy is not expected to be further decarbonized under this baseline scenario.

---

11 See Table 1 of CEC (2006). Second stage ECCP (ECCP2) policies are being finalized.
12 Purchasing power parity (PPP) at 2000 prices and exchange rates.

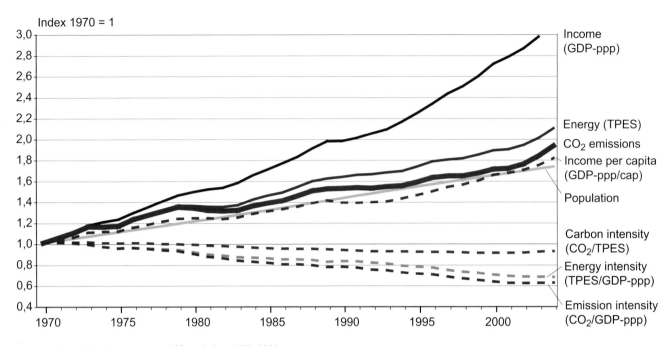

Figure 1.5: *Intensities of energy use and CO_2 emissions, 1970–2004.*

Data Source: IEA data

Of the major countries and groups of countries – North America, Western Europe, Japan, China, India, Brazil, Transition Economies – only the Transition Economies (refers to 1993–2003 only) and, to a lesser extent, the group of the EU15 have reduced their CO_2 emissions in absolute terms.

The decline of the carbon content of energy (CO_2/TPES) was the highest in Western Europe, but the effect led only to a slight reduction of CO_2 in absolute terms. Together with Western Europe and the Transition Countries, USA/Canada, Japan and – to a much lesser extent – Brazil have also reduced their carbon intensity.

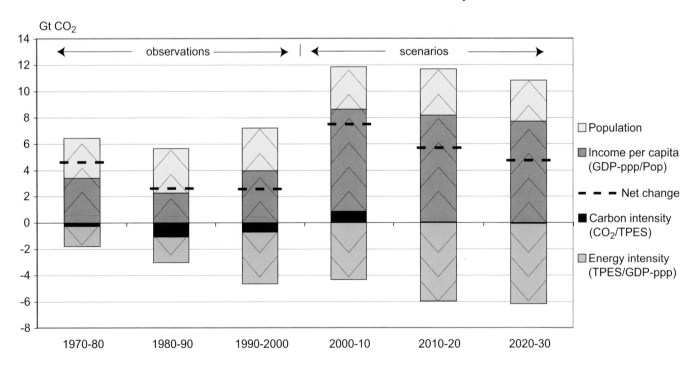

Figure 1.6: *Decomposition of global energy-related CO_2 emission changes at the global scale for three historical and three future decades.*

Sources: IEA data World Energy Outlook 2006 (IEA, 2006a)

Declining energy intensities observed in China and India have been partially offset by increasing carbon intensities (CO_2/TPES) in these countries. It appears that rising carbon intensities accompany the early stages of the industrialization process, which is closely linked to accelerated electricity generation mainly based on fossil fuels (primarily coal). In addition, the emerging but rapidly growing transport sector is fuelled by oil, which further contributes to increasing carbon intensities. Stepped-up fossil fuel use, GDP/capita growth and, to a lesser extent, population growth have resulted in the dramatic increase in carbon emissions in India and China.

The Transition Economies of Eastern Europe and the former Soviet Union suffered declining per capita incomes during the 1990s as a result of their contracting economies and, concurrently, total GHG emissions were greatly reduced. However, the continued low level of energy efficiency in using coal, oil and gas has allowed only moderate improvements in carbon and energy intensities. Despite the economic decline during the 1990s, this group of countries accounted for 12% of global CO_2 emissions in 2003 (Marland *et al., 2006*).

The challenge – an absolute reduction of global GHG emissions – is daunting. It presupposes a reduction of energy and carbon intensities at a faster rate than income and population growth taken together. Admittedly, there are many possible combinations of the four Kaya identity components, but with the scope and legitimacy of population control subject to ongoing debate, the remaining two technology-oriented factors, energy and carbon intensities, have to bear the main burden.

1.3.1.3   Energy security

With international oil prices fluctuating around 70 USD per barrel (Brent Crude in the first half of 2006; EIA, 2006a) and with prices of internationally traded natural gas, coal and uranium following suit, concerns of energy supply security are back on the agenda of many public and private sector institutions. Consequently, there is renewed public interest in alternatives to fossil fuels, especially to oil, resulting in new technology initiatives to promote hydrogen, biofuels, nuclear power and renewables (Section 1.3.1.3). Higher oil prices also tend to open up larger markets for more carbon-intensive liquid fuel production systems, such as shale oil or tar sands. However, first and foremost, energy security concerns tend to invigorate a higher reliance on indigenous energy supplies and resources. Regions where coal is the dominant domestic energy resource tend to use more coal, especially for electricity generation, which increases GHG emissions. In recent years, intensified coal use has been observed for a variety of reasons in developing Asian countries, the USA and some European countries. In a number of countries, the changing relative prices of coal to natural gas have changed the dispatch order in power generation in favour of coal.

Energy security also means access to affordable energy services by those people – largely in developing countries – who currently lack such access. It is part and parcel of sustainable development and plays a non-negligible role in mitigating climate change. Striving for enhanced energy security can impact GHG emissions in opposite ways. On the one hand, GHG emissions may be reduced as the result of a further stimulation of rational energy use, efficiency improvements, innovation and the development of alternative energy technologies with inherent climate benefits. On the other hand, measures supporting energy security may lead to higher GHG emissions due to stepped-up use of indigenous coal or the development of lower quality and unconventional oil resources.

1.3.2   Future outlook

1.3.2.1   *Energy supply*

A variety of projections of the energy picture have been made for the coming decades. These differ in terms of their modelling structure and input assumptions and, in particular, on the evolution of policy in the coming decades. For example, the IEA's World Energy Outlook 2006 reference case (IEA, 2006a) and the the International Energy Outlook of the Energy Information Agency in the USA reference case (EIA, 2006b) have both developed sets of scenarios; however, all of these scenarios project a continued dependence on fossil fuels (see Chapter 4 for past global energy mixes and future energy demand and supply projections). Should there be no change in energy policies, the energy mix supplied to run the global economy in the 2025–2030 time frame will essentially remain unchanged with about 80% (IEA, 2006a) of the energy supply based on fossil fuels. In other words, the energy economy may evolve, but not radically change unless policies change.

According to the IEA and EIA projections, coal (1.8–2.5% per year), oil (1.3–1.4% per year) and natural gas (2.0–2.4% per year) all continue to grow in the period up to 2030. Among the non-fossil fuels, nuclear (0.7–1.0% per year), hydro (2.0% per year), biomass and waste, including non-commercial biomass (1.3% per year), and other renewables (6.6% per year)[13] also continue to grow over the projection period. The growth of new renewables, while robust, starts from a relatively small base. Sectoral growth in energy demand is principally in the electricity generation and transport sectors, and together these will account for 67% of the increase in global energy demand up to 2030 (IEA 2006a).

1.3.2.2   *CO_2 emissions*

Global growth in fossil fuel demand has a significant effect on the growth of energy-related CO_2 emissions: both the IEA and the U.S. EIA project growth of more than 55% in their respective forecast periods. The IEA projects a 1.7% per year

---

13 EIA reports only an aggregate annual growth rate for all renewables of about 2.4% per year.

growth rate to 2030, while the U.S. EIA projects a 2.0% per year rate in the absence of additional policies. According to IEA projections, emissions will reach 40.4 $GtCO_2$ in 2030, an increase of 14.3 $GtCO_2$ over the 2004 level. SRES[14] (IPCC, 2000a) CO_2 emissions from energy use for 2030 are in the range 37.2–53.6 $GtCO_2$, which is similar to the levels projected in the EMF-21[15] (EMF, 2004) scenarios reviewed in Chapter 3, Section 3.2.2 (35.9–52.1 $GtCO_2$). Relative to the approximately 25.5 $GtCO_2$ emissions in 2000 (see Fig 1.1), fossil fuel-sourced CO_2 emissions are projected to increase by 40–110% by 2030 in the absence of climate policies in these scenarios (see Figure 1.7).

As the bulk of the growing energy demand occurs in developing countries, the CO_2 emission growth accordingly is dominated by developing countries. The latter would contribute two thirds to three quarters of the IEA-projected increase in global energy-related emissions. Developing countries, which accounted for 40% of total fossil fuel-related CO_2 emissions in 2004, are projected to overtake the Organization for Economic Co-operation and Development (OECD) as the leading contributor to global CO_2 fossil fuel emissions in the early part of the next decade.

The CO_2 emission projections account for both growth in energy demand and changes in the fuel mix. The IEA projects the share of total energy-related emissions accounted for by gas to increase from 20% in 2004 to 22% in 2030, while the share of coal increases from 41% to 43% and oil drops by approximately 4%, from 39% to 35%, respectively, of the total. On the basis of sectoral shares at the global level, power generation grows from a 41% to a 44% share, while the 20% share of transport is unchanged. The fastest emissions growth rate is in power generation – at 2.0% per year – followed by transport at 1.7% per year. The industry sector grows at 1.6% per year, the residential/commercial sector at 1% per year and international marine and aviation emissions at 0.7% per year.

The SRES range of energy-related CO_2 emissions for 2100 is much larger, 15.8–111.2 $GtCO_2$, while the EMF-21 scenario range for 2100 is 53.6–101.4 $GtCO_2$.

1.3.2.3   Non- CO_2 gases

Methane. Atmospheric CH_4 concentrations have increased throughout most of the 20th century, but growth rates have been close to zero over the 1999–2005 period (Solomon et al., 2007; 2.1.1) due to relatively constant emissions during this period equaling atmospheric removal rates (Solomon et al., 2007; 2.1.1). Human emissions continue to dominate the total CH_4 emissions budget (Solomon et al., 2007; 7.4.1). Agriculture and forestry developments are assessed in Chapters 8 and 9,

respectively, in terms of their impact on the CH_4 sink/source balance and mitigation strategies; waste handling is likewise assessed in Chapter 10.

The future increase in CH_4 concentrations up to 2030 according to the SRES scenarios ranges from 8.1 $GtCO_2$-eq to 10.3 $GtCO_2$-eq (increase of 19–51% compared to 2000), and the increase under the Energy Modeling Forum (EMF)-21 baseline scenarios is quite similar (7.5 $GtCO_2$-eq to 11.3 $GtCO_2$-eq/yr). By 2100, the projected SRES increase in CH_4 concentrations ranges from 5 $GtCO_2$-eq to 18.7 $GtCO_2$-eq (a change of –27% to +175% compared to 2000) and that of the EMF-21 ranges from 5.9 to 29.2 $GtCO_2$-eq (a change of –2% to +390%).

Montreal gases. Emissions of ODS gases (also GHGs) controlled under the Montreal Protocol (CFCs, HCFCs) increased from a very low amount during the 1950–1960s to a substantial percentage – approximately 20% – of total GHG emissions by 1975. This percentage fluctuated slightly during the period between 1975 and 1989, but once the phase-out of CFCs was implemented, the ODS share in total GHG emissions fell rapidly, first to 8% (1995) and then to 4% (2000). Radiative forcing from these gases peaked in 2003 and is beginning to decline (Forster et al., 2007).

After 2000, ODS contributed 3–4% to total GHG emissions (Olivier et al., 2005, 2006). The ODS share is projected to decrease yet further due to the CFC phase-out in developing countries. Emissions of ODS are estimated at 0.5–1.15 Gt CO_2-eq for the year 2015, dependent on the scenario chosen (IPCC, 2005); this would be about 1–2% of total GHG emissions for the year 2015, if emissions of all other GHGs are estimated at about 55 Gt CO_2-eq (for the year 2015). The percentage of HCFC emissions in the total of CFC and HCFC emissions for the year 2015 is projected to be about 70%, independent of the scenario chosen.

Nitrous oxide. Atmospheric concentrations of N_2O have been continuously increasing at an approximately constant growth rate since 1980 (IPCC, 2007a, SPM). Industrial sources, agriculture, forestry and waste developments are assessed in this report in terms of their impact on the N_2O sink/source balance and mitigation strategies. The SRES emissions for 2030 range from 3 $GtCO_2$-eq to 5.3 $GtCO_2$-eq (a change of –13% to 55% compared to 2000). For comparison, the recent EMF-21 baseline range for 2030 is quite close to this (2.8 $GtCO_2$-eq to 5.4 $GtCO_2$-eq, an increase of –17% to 58% compared to 2000). By 2100, the range projected by the SRES scenarios is 2.6 $GtCO_2$-eq to 8.1 $GtCO_2$-eq (an increase of –23% to 140% compared to 2000), whereas the EMF-21 range is a little higher (3.2 $GtCO_2$-eq to 11.5 $GtCO_2$-eq, or an increase of –5% to 240% compared to 2000).

---

14 SRES is the IPCC Special Report on Emissions Scenarios (IPCC, 2000a). The ranges reported here are for the five SRES Marker scenarios.
15 EMF-21 Energy Modeling Forum Study 21: Multi-gas Mitigation scenarios (EMF, 2004)

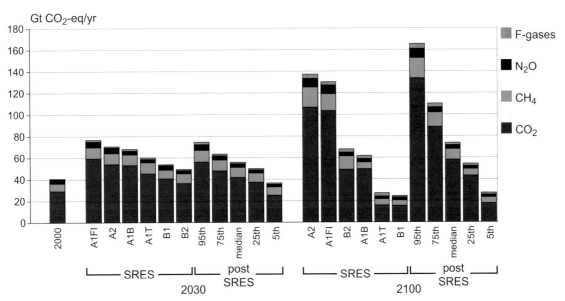

Figure 1.7 *Global GHG emissions for 2000 and projected baseline emissions for 2030 and 2100 from IPCC SRES and the post-SRES literature. The figure provides the emissions from the six illustrative SRES scenarios. It also provides the frequency distribution of the emissions in the post-SRES scenarios (5th, 25th, median, 75th, 95th percentile), as covered in Chapter 3. F-gases include HFCs, PFCs and SF$_6$*

Fluorinated gases. Concentrations of many of these gases have increased by large factors (i.e., 1.3 and 4.3) between 1998 and 2005, and their radiative forcing is rapidly increasing (from low levels) by roughly 10% per year (Forster *et al.*, 2007). Any projection of overall environmental impacts and emissions is complicated by the fact that several major applications retain the bulk of their fluorinated gases during their respective life cycles, resulting in the accumulation of significant stocks that need to be responsibly managed when these applications are eventually decommissioned. A comprehensive review of such assessments was published in an earlier IPCC Special Report (IPCC, 2005). This review reported growth in HFC emissions from about 0.4 GtCO$_2$-eq in 2002 to 1.2 GtCO$_2$-eq per year in 2015. Chapter 3 also describes in some detail the results of long-term GHG emissions scenarios. The range projected by SRES scenarios for 2030 is 1.0–1.6 GtCO$_2$-eq (increase of 190–360% compared to 2000) and the EMF-21 baseline scenarios are quite close to this (1.2–1.7 GtCO$_2$-eq per year, an increase of 115–240% compared to 2000). By 2100, the SRES range is 1.4–4 GtCO$_2$-eq per year (an increase of 300% to more than 1000 % compared to 2000), whereas the new EMF-21 baseline scenarios are higher still (1.9–6.3 GtCO$_2$-eq).

Air pollutants and other radiative substances. As noted above, some air pollutants, such as sulphur aerosol, have a significant effect on the climate system, although considerable uncertainties still surround the estimates of anthropogenic aerosol emissions. Data on non-sulphur aerosols are sparse and highly speculative, but in terms of global sulphur emissions, these appear to have declined from a range of 75 ± 10 MtS in 1990 to 55–62 MtS in 2000. Sulphur emissions from fossil fuel combustion lead to the formation of aerosols that affect regional climate and precipitation patterns and also reduce

radiative forcing. There has been a slowing in the growth of sulphur emissions in recent decades, and more recent emission scenarios show lower emissions than earlier ones (Chapter 3, Section 3.2.2). Other air pollutants, such as NO$_x$ and black and organic carbon, are also important climatologically and adversely affect human health. The likely future development of these emissions is described in Section 3.2.2.

1.3.2.4 Total GHG emissions

Without additional policies global GHG emissions (including those from deforestation) are projected to increase between 25% and 90% by 2030 relative to 2000 (see Figure 1.7). Fossil fuel dominance is expected to continue up to 2030 and beyond; consequently, CO$_2$ emissions from energy use tend to grow faster than total GHGs, increasing by 1.2–2.5% over that period. Two thirds to three quarters of the increase in CO$_2$ emissions are projected to come from developing countries, although the average per capita CO$_2$ emissions in developing country regions will remain substantially lower (2.8– 5.1 tCO$_2$ per capita) than those in developed country regions (9.6–15.1 tCO$_2$ per capita).

By 2100, the range in the GHG emission projections is much wider from a 40% reduction to an increase of 250% compared to 2000. Scenarios that account for climate policies currently under discussion for implementation also show global emissions rising for many decades. With the atmospheric concentrations of GHGs thus unlikely to stabilize in this century (even for the low SRES scenario) without major policy changes, from an emissions perspective, we are not on track for meeting the objectives of UNFCCC Article 2.

1.3.3 Technology research, development and deployment: needs and trends

1.3.3.1 Research and development

Technology research and development (R&D) are important for altering the emission trends shown in the previous sections. In the absence of measures fostering the development of climate-friendly technologies and/or a lack of incentives for their deployment, however, it is not a priori obvious in which direction R&D will influence emissions. Because of the longevity of energy infrastructures (lock-in effect), it is the near-term investment decisions in the development, deployment and diffusion of technologies that will determine the long-term development of the energy system and its emissions (Gritsevskyi and Nakicenovic, 2002).

Generally speaking, it would be economically impossible without technology research, development, demonstration, deployment and diffusion (RDDD&D) and induced technology change (ITC), to stabilize GHG concentrations at a level that would prevent dangerous anthropogenic interference with the climate system. Government support is crucial at the development stage, but private investment will gradually replace the former for deployment (creating necessary market transformation) and for diffusion (successful market penetration).

However, RDDD&D alone is insufficient and effective climate policies are also required (Baker *et al.*, 2006). A recent international modelling comparison exercise (Edenhofer *et al.*, 2006) has shown that ITC not only has the potential to reduce mitigation costs substantially but that it is also essential to the stabilization of concentration levels of CO_2, avoiding dangerous anthropogenic interference.

There are various types of technologies that can play significant roles in mitigating climate change, including energy efficiency improvements throughout the energy system (especially at the end use side); solar, wind, nuclear fission and fusion and geothermal, biomass and clean fossil technologies, including carbon capture and storage; energy from waste; hydrogen production from non-fossil energy sources and fuel cells (Pacala and Socolow, 2004; IEA, 2006b). Some are in their infancy and require public RDDD&D support, while others are more mature and need only market incentives for their deployment and diffusion. Some also need persevering efforts for public acceptance (Tokushige *et al.*, 2006) as well as the resolution of legal and liability issues.

1.3.3.2 Research and development expenditures

The most rapid growth in public-sector energy related technology R&D[16] occurred in the aftermath of the oil price shocks of the 1970s. There is no evidence yet of a similar response from the latest price surges. A technology R&D response to the challenge of climate mitigation has not occurred. Energy technology R&D has remained roughly constant over the last 15 years despite the fact that climate change has become a focus of international policy development. Energy technology R&D is one policy lever that governments have for encouraging a more climate friendly capital, a strengthened publicly funded commitment to technology development could play an important role in altering the trends in GHG emissions.

International cooperation in the field of technology R&D may provide the leverage to otherwise insufficient national R&D budgets. Several international partnerships on the development of cleaner technologies have been created (see Section 1.4.2).

1.4 Institutional architecture

The institutional architecture for climate change, energy and sustainable development in principal covers a wide range of different entities and processes. At the international level, these include the Millennium Development Goals, the World Summit on Sustainable Development in 2002 and its Johannesburg Plan for Implementation (JPOI) and the UN Commission on Sustainable Development (CSD), all of which have broad and important connections to climate change in the context of sustainable development, energy and poverty eradication. Other international fora that are important to advancing the agenda for sustainable development and climate change include – but are not limited to – the UN General Assembly, the G8 Dialogue on Climate Change, Clean Energy and Sustainable Development, OECD, the World Trade Organization (WTO; which pursues trade liberalization, important for technology transfers), IEA and the World Bank. More regional fora include regional banks, the EU and the Asia-Pacific Partnership on Clean Development and Climate for transferring and deploying clean technologies and building up human and institutional capacity. Chapter 2.1 discusses these issues in detail, and they are further evaluated in Chapter 12. This chapter focuses specifically on the UNFCCC and its Kyoto Protocol and with technology cooperation and transfer.

1.4.1 UNFCCC and its Kyoto Protocol

The UNFCCC pursues its ultimate objective, Article 2 (Section 1.2.1), on the basis of several guiding principles laid down in Article 3 of the Convention:
- Equity, which is expressed as "common but differentiated responsibilities" that assigns the lead in mitigation to developed countries (Article 3.1) and that takes the needs and special circumstances of developing countries into account (Article 3.2).
- A precautionary principle, which says that "where there are

---

16 Data for IEA member countries only.

threats of serious or irreversible damage, lack of full scientific certainty should not be used as a reason for postponing such measures, taking into account that policies and measures to deal with climate change should be cost-effective so as to ensure global benefits at the lowest possible cost" (Article 3.3).

- A right to and an obligation to promote sustainable development (Article 3.4).
- An obligation to cooperate in sharing information about climate change, technologies through technology transfers, and the coordination of national actions (Article 3.7)

Based on the principle of common but differentiated responsibilities, Annex I countries are committed to adopt policies and measures aimed at returning – individually or jointly – their GHG emissions to earlier levels by the year 2000 (Article 4.2). Following the decision of the first Conference of the Parties[17] (COP1) in Berlin in 1995 that these commitments were inadequate, the Kyoto Protocol was negotiated and adopted by consensus at COP3, in Kyoto in 1997, and entered into force on 16 February 2005. This was preceded by the detailed negotiation of the implementing rules and agreements for the Protocol – the Marrakech Accords – that were concluded at COP7 in Marrakech and adopted in Montreal at CMP1[18]. As of December 2006, the Protocol has been ratified by 165 countries. While Australia and the United States, both parties to UNFCCC, signed the protocol, both have stated an intention not to ratify.

Several key features of the Protocol are relevant to the issues raised later in this report:
- Each Party listed in Annex B of the Protocol is assigned a legally binding quantified GHG emission limitation and/or reduction measured in CO_2 equivalents for the first commitment period 2008–2012. In aggregate, these Parties are expected to reduce their overall GHG emissions by "at least 5 per cent below 1990 levels in the commitment period 2008 to 2012" (Article 3.1). Some flexibility is shown towards economies in transition who may nominate a base year or period other than 1990 (Article 3.5, 3.7).
- Six classes of gases are listed in Annex A of the Protocol: CO_2, CH_4, N_2O, HFCs, PFCs and SF_6. Emissions from international aviation and maritime transport are not included.
- The so-called Kyoto flexibility mechanisms allow Annex B Parties to obtain emission allowances achieved outside their national borders but supplemental to domestic action, which is expected to be a "significant element of the effort" (Article 6.1 (d), Article17, CMP1[19]). These mechanisms are: an international emission trading system, Joint Implementation (JI) projects in Economies in Transition, projects undertaken as of year 2000 in developing (non-Annex I) countries

under the Clean Development Mechanism (CDM) and carbon sink projects in Annex B countries.
- A set of procedures for emission monitoring, reporting, verification and compliance has been adopted at CMP1 under Articles 5, 7, 8 and 18.

In accordance with Article 3.9, the Parties to the Protocol at CMP1 began the process of negotiating commitments for the Annex B Parties for the second commitment period, creating – the 'Ad Hoc Working Group on Further Commitments for Annex I Parties under the Kyoto Protocol' (AWG), with the requirement that negotiations be completed so that that the first and second commitment periods are contiguous. Work continued at CMP2 in Nairobi and in 2007 the AWG will work on, amongst other thing, ranges of emission reduction objectives of Annex I Parties with due attention to the conditions mentioned in Article 2 of the Convention (see 1.2.1). The task is to consider that "according to the scenarios of the TAR, global emissions of carbon dioxide have to be reduced to very low levels, well below half of levels in 2000, in order to stabilize their concentrations in the atmosphere" (see Chapters 3 and 13).

In addition, CMP2 started preparations for the second review of the Protocol under Article 9, which in principle covers all aspects of the Protocol, and set 2008 as the date for this review.

Under the UNFCCC, a Dialogue on Long-Term Cooperation Action to Address Climate Change by Enhancing Implementation of the Convention (the Dialogue) was established at COP11 in 2005, met during 2006 and is to conclude at COP13 in 2007. The Dialogue is "without prejudice to any future negotiations, commitments, process, framework or mandate under the Convention, to exchange experiences and analyse strategic approaches for long-term cooperative action to address climate change".

1.4.2 Technology cooperation and transfer

Effective and efficient mitigation of climate change depends on the rate of global diffusion and transfer of new as well as existing technologies. To share information and development costs, international cooperation initiatives for RDDD&D, such as the Carbon Sequestration Leadership Forum (CSLF), the International Partnership for Hydrogen Economy (IPHE), the Generation IV International Forum (GIF), the Methane to Markets Partnership and the Renewable Energy & Energy Efficiency Partnership (REEEP), the Global Bioenergy Partnership and the ITER fusion project, were undertaken. Their mandates range from basic R&D and market demonstration to barrier removals for commercialization/diffusion. In addition,

---

17 The Conference of the Parties (COP), which is the supreme body of the Convention, also serves as the Meeting of the Parties (MOP) for the Protocol. Parties to the Convention that are not Parties to the Protocol will be able to participate in Protocol-related meetings as observers (Article 13).
18 CMP1: First meeting of the Conference of the Parties acting as the Meeting of the Parties of the Kyoto Protocol.
19 Decisions can be found at http://unfccc.int/documentation/decisions/items/3597.php?dec=j&such=j&volltext=/CMP.1#beg

there are 40 'implementing agreements' facilitating international cooperation on RDDD&D under IEA auspices, covering all of the key new technologies of energy supply and end use with the exception of nuclear fission (IEA, 2005).

Regional cooperation may be effective as well. Asia-Pacific Partnership of Clean Development and Climate (APPCDC), which was established by Australia, China, India, Japan, Korea and the USA in January 2006, aims to address increased energy needs and associated challenges, including air pollution, energy security, and climate change, by enhancing the development, deployment and transfer of cleaner, more efficient technologies. In September 2005, the EU concluded agreements with India and China, respectively, with the aim of promoting the development of cleaner technologies (India) and low carbon technologies (China).

Bilateral sector-based cooperation agreements also exist. One example is the Japan/China agreement on energy efficiency in the steel industry, concluded in July 2005 (JISF, 2005). These sector-based initiatives may be an effective tool for technology transfer and mitigating GHG emissions.

It is expected that CDM and JI under the Kyoto Protocol will play important role for technology transfer as well.

1.5 Changes from previous assessments and roadmap

1.5.1 Previous assessments

The IPCC was set up in 1988 by UNEP and WMO with three working groups: to assess available scientific information on climate change (WGI), to assess environmental and socio-economic impacts (WGII) and to formulate response strategies (WGIII).

The First Assessment Report (FAR) (IPCC, 1991) dealt with the anthropogenic alteration of the climate system through CO_2 emissions, potential impacts and available cost-effective response measures in terms of mitigation, mainly in the form of carbon taxes without much concern for equity issues (IPCC, 2001, Chapter 1).

For the Second Assessment Report (SAR), in 1996, Working Groups II and III were reorganized (IPCC, 1996). WGII dealt with adaptation and mitigation, and WGIII dealt with the socio-economic cross-cutting issues related to costing climate change's impacts and providing cost-benefit analysis (CBA) for use in decision-making. The socio-institutional context was emphasized as well as the issues of equity, development, and sustainability (IPCC, 2001, Chapter 1).

For the Third Assessment Report (TAR) (IPCC, 2001), Working Groups II and III were again reorganized to deal with adaptation and mitigation, respectively. The concept of mitigative capacity was introduced, and the focus attention was shifted to sustainability concerns (IPCC, 2001, Chapter 1.1). Four cross-cutting issues were identified: costing methods, uncertainties, decision analysis frameworks and development, equity and sustainability (IPCC, 2000b).

The Fourth Assessment Report (AR4) summarizes the information contained in previous IPCC reports - including the IPCC special reports on Carbon Dioxide Capture and Storage, on Safeguarding the Ozone Layer and on the Global Climate System published since TAR - and assesses the scientific literature published since 2000.

Although the structure of AR4 resembles the macro-outline of the TAR, there are distinct differences between them. The AR4 assigns greater weight to (1) a more detailed resolution of sectoral mitigation options and costs; (2) regional differentiation; (3) emphasizing previous and new cross-cutting issues, such as risks and uncertainties, decision- and policy-making, costs and potentials and the relationships between mitigation, adaptation and sustainable development, air pollution and climate, regional aspects and the issues related to the implementation of UNFCCC Article 2; and (4) the integration of all these aspects.

1.5.2 Roadmap

This report assesses options for mitigating climate change. It has four major parts, A–D.

Part A comprises Chapter 1, an Introduction and Chapter 2, which is on 'framing issues'. Chapter 2 introduces the report's cross-cutting themes, which are listed above, and outlines how these themes are treated in subsequent chapters. It also introduces important concepts (e.g. cost-benefit analysis and regional integration) and defines important terms used throughout the report.

Part B consists of one chapter, Chapter 3. This chapter reviews and analyzes baseline (non-mitigation) and stabilization scenarios in the literature that have appeared since the publications of the IPCC SRES and the TAR. It pays particular attention to the literature that criticizes the IPCC SRES scenarios and concludes that uncertainties and baseline emissions have not changed very much. It discusses the driving forces for GHG emissions and mitigation in the short and medium terms and emphasizes the role of technology relative to social, economic and institutional inertia. It also examines the relation between adaptation, mitigation and avoided climate change damage in the light of decision-making on atmospheric GHG concentrations (Article 2 UNFCCC).

Part C consists of seven chapters, each of which assesses sequence mitigation options in different sectors. Chapter 4 addresses the energy supply sector, including carbon capture and storage; Chapter 5 transport and associated infrastructures; Chapter 6 the residential, commercial and service sectors; Chapter 7 the industrial sector, including internal recycling and the reuse of industrial wastes; Chapters 8 and 9 the agricultural and forestry sectors, respectively, including land use and biological carbon sequestration; Chapter 10 waste management, post-consumer recycling and reuse.

These seven chapters use a common template and cover all relevant aspects of GHG mitigation, including costs, mitigation potentials, policies, technology development, technology transfer, mitigation aspects of the three dimensions of sustainable development, system changes and long-term options. They provide the integrated picture that was absent in the TAR. Where supporting literature is available, they address important differences across regions.

Part D comprises three chapters (11–13) that focus on major cross-sectoral considerations. Chapter 11 assesses the aggregated short-/medium-term mitigation potential, macro-economic impacts, economic instruments, technology development and transfer and cross-border influences (or spill-over effects). Chapter 12 links climate mitigation with sustainable development and assesses the GHG emission impacts of implementing the Millennium Development Goals and other sustainable development policies and targets. Chapter 13 assesses domestic climate policy instruments and the interaction between domestic climate policies and various forms of international cooperation and reviews climate change as a global common issue in the context of sustainable development objectives and policies. It summarizes relevant treaties, cooperative development agreements, private–public partnerships and private sector initiatives and their relationship to climate objectives.

REFERENCES

Baker, T., H. Pan, J. Kohler, R. Warren, and S. Winne, 2006: Avoiding Dangerous Climate Change by Inducing Technological Progress: Scenarios Using a Large-Scale Econometric Model. In *Avoiding Dangerous Climate Change,* H.J. Schellnhuber (editor in chief), Cambridge University Press, pp. 361-371.

Bolin, B. and H.S. Kheshgi, 2001: On strategies for reducing greenhouse gas emissions. Proceedings of the National Academy of Sciences of the United States of America, 24 April, 2001, **98**(9), pp. 4850-4854.

CEC, 2006: Report from the Commission: Progress Towards Achieving the Kyoto Objectives. SEC(2006) 1412, Commission of the European Communities, Brussels, Belgium.

CEU, 2005: Presidency Conclusions - Brussels, 22 and 23 March 2005 - IV. *Climate Change,* European Commission, Council of the European Union, Brussels, Belgium. <http://www.eu2005.lu/en/actualites/conseil/2005/03/23conseileuropen/ceconcl.pdf> accessed 25 April 2006.

Chandler, W., R. Schaeffer, Z. Dadi, P.R. Shukla, F. Tudela, O. Davidson and S. Alpan-Atamer, 2002: Climate change mitigation in developing countries: Brazil, China, India, Mexico, South Africa, and Turkey. Pew Center on Global Climate Change.

Dasgupta, P., 2006: Comments on the Stern Review's Economics of Climate Change. <http://www.econ.cam.ac.uk/faculty/dasgupta/STERN.pdf> accessed 15. December 2006.

Denman, K.L., G. Brasseur, A. Chidthaisong, P. Ciais, P.M. Cox, R.E. Dickinson, D. Hauglustaine, C. Heinze, E. Holland, D. Jacob, U. Lohmann, S Ramachandran, P.L. da Silva Dias, S.C. Wofsy and X. Zhang, 2007: Couplings Between Changes in the Climate System and Biogeochemistry. In: *Climate Change 2007: The Physical Science Basis. Contribution of Working Group I to the Fourth Assessment Report of the Intergovernmental Panel on Climate Change* [Solomon, S., D. Qin, M. Manning, Z. Chen, M. Marquis, K.B. Averyt, M.Tignor and H.L. Miller (eds.)]. Cambridge University Press, Cambridge, United Kingdom and New York, NY, USA.

Edenhofer, O., Lessmann, K., Kemfert, C., Grubb, M. and Koehler, J., 2006: Induced Technological Change: Exploring its Implications for the Economics of Atmospheric Stabilisation. Synthesis Report from the Innovation Modelling Comparison Exercise. In *Endogenous Technological Change and the Economics of Atmospheric Stabilisation,* O. Edenhofer, C. Carraro, J. Köhler, M. Grubb. *The Energy Journal,* Special Issue #1.

EEA, 2006: EN09_EU-25_Policy_effectiveness (Underpinning Energy and Environment indicator fact-sheets for Energy and environment in the European Union - Tracking progress towards integration, EEA Report No 8/2006). European Environment Agency.

EIA, 2006a: U.S. Department of Energy, Energy Information Administration, Washington, D.C., 20585, <http://www.eia.doe.gov/emeu/international/crude1.html>, accessed 15. December 2006.

EIA, 2006b: International Energy Outlook 2006. DOE/EIA-0484(2006), U.S. Department of Energy, Energy Information Administration, Washington, D.C., 20585.

Eichhammer, W., F. Gagelmann, J. Schleich and J. Chesshire, 2002: Reasons and Perspectives for Emission Reductions in Germany and the UK. Fraunhofer Institute, <http://www.isi.fhg.de/publ/downloads/isi02a08/emission-reduction.pdf>, accessed 24 January 2007.

Eichhammer, W., U. Boede, F. Gagelmann, E. Jochem, J. Schleich, B. Schlomann, J. Chesshire and H-J. Ziesing, 2001: Greenhouse Gas Reductions in Germany and the UK - Coincidence or Policy Induced. An analysis for international climate policy. Fraunhofer Institute, Science Policy and Technology Research (SPRU), Deutsches Institut für Wirtschaftsforschung (DIW). <http://www.isi.fhg.de/publ/downloads/isi01b20/greenhouse-gas.pdf>, accessed 20. April 2006.

EMF, 2004: Energy Modeling Forum Study 21: Multi-gas Mitigation Scenarios. Energy Modeling Forum, <http://www.stanford.edu/group/EMF>, accessed 10. December 2006.

Forster, P., V. Ramaswamy, P. Artaxo, T. Berntsen, R. Betts, D.W. Fahey, J. Haywood, J. Lean, D.C. Lowe, G. Myhre, J. Nganga, R. Prinn, G. Raga, M. Schulz and R. Van Dorland, 2007: Changes in Atmospheric Constituents and in Radiative Forcing. In *Climate Change 2007: The Physical Science Basis. Contribution of Working Group I to the Fourth Assessment Report of the Intergovernmental Panel on Climate Change* [Solomon, S., D. Qin, M. Manning, Z. Chen, M. Marquis, K.B. Averyt, M.Tignor and H.L. Miller (eds.)]. Cambridge University Press, Cambridge, United Kingdom and New York, NY, USA.

Gritsevsky, A., and N. Nakicenovic, 2002: Modelling uncertainty of induced technological change. In: *Technological change and the environment,* A. Grubler, N. Nakicenovic, W.D. Nordhaus, (eds.). Resources for the Future, pp. 251-279.

Hansen, J., M. Sato, R. Ruedy, K. Lo, D.W. Lea, and M. Medina-Elizade, 2006: Global temperature change. PNAS: 0606291103.

Harvey, L.D.D., 2007: Dangerous Anthropogenic Interference, Dangerous Climatic Change, and Harmful Climate Change: Non Trivial Distinctions with Significant Policy Implications. *Climate Change,* **82**(1-2), pp.1-25. <http://www.springerlink.com/content/k761w23w3506/> accessed 07/06/07.

Hooijer, A., M. Silvius, H. Wösten, and S. Page, 2006: PEAT- CO_2, Assessment of CO_2 emissions from drained peatlands in SE Asia. Delft Hydraulics report Q3943.

IEA, 2005: Energy Technologies at the cutting edge. International Energy Agency, Paris, France.

IEA, 2006a: World Energy Outlook 2006. Paris, France.

IEA, 2006b: Energy Technology Perspectives: Scenarios and Strategies to 2050. Paris, France.

IPCC, 1991: The First Assessment Report of the Intergovernmental Panel on Climate Change (IPCC). Cambridge University Press, Cambridge.

IPCC, 1996: The Second Assessment Report of the Intergovernmental Panel on Climate Change (IPCC). Cambridge University Press, Cambridge.

IPCC, 2000a: Emissions Scenarios. [Nakicenovic, N. and R. Swart (eds.)]. Special Report of the Intergovernmental Panel on Climate Change (IPCC). Cambridge University Press, Cambridge, 570 pp.

IPCC, 2000b: Guidance papers on the Cross Cutting Issues of the Third Assessment Report of the Intergovernmental Panel on Climate Change (IPCC) [R. Pachauri, T. Taniguchi and K. Tanaka (eds.)]. Geneva, Switzerland.

IPCC, 2001: Climate Change 2001: Mitigation - Contribution of Working Group III to the Third Assessment Report of the Intergovernmental Panel on Climate Change (IPCC) [Metz, B., O. Davidson, R. Swart, and J. Pan (eds.)]. Cambridge University Press, Cambridge, 700 pp.

IPCC, 2005: Safeguarding the Ozone Layer and the Global Climate System: issues related to Hydrofluorocarbons and Perfluorocarbons [Metz, B., L. Kuijpers, S. Solomon, S.O. Andersen, O. Davidson, J. Pons, D.de Jager, T. Kestin, M. Manning, and L. Meyer (eds.)]. Special Report of the Intergovernmental Panel on Climate Change (IPCC), Cambridge University Press, Cambridge.

IPCC, 2007a: Climate Change 2007: The Physical Science Basis. Contribution of Working Group I to the Fourth Assessment Report of the Intergovernmental Panel on Climate Change [Solomon, S., D. Qin, M. Manning, Z. Chen, M. Marquis, K.B.M.Tignor and H.L. Miller (eds.)]. Cambridge University Press, Cambridge, United Kingdom and New York, NY, USA, 996 pp.

IPCC, 2007b: Climate Change 2007: Impacts, Adaptation and Vulnerability. Contribution of Working Group II to the Fourth Assessment Report of the Intergovernmental Panel on Climate Change [Parry, M.L., O.F. Canziani, J.P. Palutikof, P.J. van der Linden, C.E. Hanson (eds.)]. Cambridge University Press, Cambridge, United Kingdom and New York, NY, USA.

JISF, 2005; Japan Iron and Steel federation <http://www.jisf.or.jp/en/activity/050715.html>, accessed 15. July 2006.

Kaul, I., I. Grunberg, and M.A. Stern, 1999: Global Public Goods, Oxford University Press 1999.

Kaul, I., P. Conceiçao, K. Le Gouven, and R.U. Mendoz, 2003: Providing Global Public Goods, Oxford University Press.

Kaya, Y, 1990: Impact of Carbon Dioxide Emission Control on GNP Growth: Interpretation of Proposed Scenarios. Paper presented to the IPCC Energy and Industry Subgroup, Response Strategies Working Group, Paris.

Marland, G., T.A. Boden, and R.J. Andres, 2006: Global, Regional, and National Fossil Fuel CO_2 Emissions. In *Trends: A Compendium of Data on Global Change.* Carbon Dioxide Information Analysis Center, Oak Ridge National Laboratory, U.S. Department of Energy, Oak Ridge, Tenn., USA.

Mastrandrea, M.D., and S.H. Schneider, 2004: Probabilistic Integrated Assessment of 'Dangerous' Climate Change. *Science,* **304**(5670), pp. 571-575.

Meehl, G.A., T.F. Stocker, W.D. Collins, P. Friedlingstein, A.T. Gaye, J.M. Gregory, A. Kitoh, R. Knutti, J.M. Murphy, A. Noda, S.C.B. Raper, I.G. Watterson, A.J. Weaver and Z.-C. Zhao, 2007: Global Climate Projections. In: *Climate Change 2007: The Physical Science Basis. Contribution of Working Group I to the Fourth Assessment Report of the Intergovernmental Panel on Climate Change* [Solomon, S., D. Qin, M. Manning, Z. Chen, M. Marquis, K.B. Averyt, M. Tignor and H.L. Miller (eds.)]. Cambridge University Press, Cambridge, United Kingdom and New York, NY, USA.

Nordhaus, W.D., 2006: The *Stern Review* on the Economics of Climate Change. National Bureau of Economic Research, Working Paper 12741. Cambridge, Massachusetts, USA.

Olivier, J.G.J., J.A. Van Aardenne, F. Dentener, V. Pagliari, L.N. Ganzeveld, and J.A.H.W. Peters, 2005: Recent trends in global greenhouse gas emissions: regional trends 1970-2000 and spatial distribution of key sources in 2000. *Environmental Science,* **2**(2-3), pp. 81-99. DOI: 10.1080/15693430500400345. <http://www.mnp.nl/edgar/global_overview/>, accessed 5. December 2006.

Olivier, J.G.J., T. Pulles and J.A. van Aardenne, 2006: Part III: Greenhouse gas emissions: 1. Shares and trends in greenhouse gas emissions; 2. Sources and Methods; Greenhouse gas emissions for 1990, 1995 and 2000. In *CO_2 emissions from fuel combustion 1971-2004*, 2006 Edition, pp. III.1-III.41. International Energy Agency (IEA), Paris. ISBN 92-64-10891-2 (paper) 92-64-02766-1 (CD ROM) (2006).

Pacala, S. and R. Socolow, 2004: Stabilization Wedges: Solving the Climate Problem for the Next 50 Years with Current Technologies. *Science,* **305**, pp. 968-972.

Ramankutty, N., *et al.*, 2006: Challenges to estimating carbon emissions from tropical deforestation. *Global Change Biology* (published article online: 28-Nov-2006 doi: 10.1111/j.1365-2486.2006.01272.x).

Rijsberman, F.J., and R.J. Swart (eds.), 1990: Targets and Indicators of Climate Change. Stockholm Environment Institute, 1666 pp.

Solomon, S., D. Qin, M. Manning, R.B. Alley, T. Berntsen, N.L. Bindoff, Z. Chen, A. Chidthaisong, J.M. Gregory, G.C. Hegerl, M. Heimann, B. Hewitson, B.J. Hoskins, F. Joos, J. Jouzel, V. Kattsov, U. Lohmann, T. Matsuno, M. Molina, N. Nicholls, J.Overpeck, G. Raga, V. Ramaswamy, J. Ren, M. Rusticucci, R. Somerville, T.F. Stocker, P. Whetton, R.A. Wood and D. Wratt, 2007: Technical Summary. In: *Climate Change 2007: The Physical Science Basis. Contribution of Working Group I to the Fourth Assessment Report of the Intergovernmental Panel on Climate Change* [Solomon, S., D. Qin, M. Manning, Z. Chen, M. Marquis, K.B. Averyt, M. Tignor and H.L. Miller (eds.)]. Cambridge University Press, Cambridge, United Kingdom and New York, NY, USA.

Stern, N., 2006: The Stern Review: The Economics of Climate Change. <http://www.hm-treasury.gov.uk/independent_reviews/stern_review_economics_climate_change/stern_review_report.cf>, accessed 28. November 2006.

Sun, J.W., 1998: Changes in energy consumption and energy intensity: A complete decomposition model. *Energy Economics* **20**(1), pp. 85-100.

Sun, J.W. and B.W. Ang, 2000: Some properties of an exact energy decomposition model. *Energy* **25**(12), pp. 1177-1188.

Tokushige, K., K. Akimoto, and T. Tomoda, 2006: Public Acceptance and Risk-benefit Perception of CO_2 Geological Storage for Global Warming Mitigation in Japan, *Mitigation and Adaptation Strategies for Global Change.* <http://springerlink.metapress.com/content/fj451110258r3259/?p=7fa647f50f534>871b6234fdde58fac6a&pi=0> accessed January 10, 2007.

UN, 1992: United Nations Framework Convention on Climate Change, United Nations, New York.

Van der Werf, G.R., J.T. Randerson, G.J. Collatz, and L. Giglio, 2003: Carbon emissions from fires in tropical and subtropical ecosystems. *Global Change Biology,* **9**, pp. 547-562.

2

Framing Issues

Coordinating Lead Authors:
Kirsten Halsnæs (Denmark), Priyadarshi Shukla (India)

Lead Authors:
Dilip Ahuja (India), Grace Akumu (Kenya), Roger Beale (Australia), Jae Edmonds (USA), Christian Gollier (Belgium), Arnulf Grübler (Austria), Minh Ha Duong (France), Anil Markandya (UK), Mack McFarland (USA), Elena Nikitina (Russia), Taishi Sugiyama (Japan), Arturo Villavicencio (Equador), Ji Zou (PR China)

Contributing Authors:
Terry Barker (UK), Leon Clarke (USA), Amit Garg (India)

Review Editors:
Ismail Elgizouli (Sudan), Elizabeth Malone (USA)

This chapter should be cited as:
Halsnæs, K., P. Shukla, D. Ahuja, G. Akumu, R. Beale, J. Edmonds, C. Gollier, A. Grübler, M. Ha Duong, A. Markandya, M. McFarland, E. Nikitina, T. Sugiyama, A. Villavicencio, J. Zou, 2007: Framing issues. In Climate Change 2007: Mitigation. Contribution of Working Group III to the Fourth Assessment Report of the Intergovernmental Panel on Climate Change [B. Metz, O. R. Davidson, P. R. Bosch, R. Dave, L. A. Meyer (eds)], Cambridge University Press, Cambridge, United Kingdom and New York, NY, USA

Table of Contents

EXECUTIVE SUMMARY

This chapter frames climate change mitigation policies in the context of general development issues and recognizes that there is a two-way relationship between climate change and sustainable development. These relationships create a wide potential for linking climate change and sustainable development policies, and an emerging literature has identified methodological approaches and specific policies that can be used to explore synergies and tradeoffs between climate change and economic, social, and environmental sustainability dimensions.

Decision-making about climate change policies is a very complex and demanding task since there is no single decision-maker and different stakeholders assign different values to climate change impacts and to the costs and benefits of policy actions. However, many new initiatives emerge from governmental cooperation efforts, the business sector and NGOs (non-governmental organizations), so various coalitions presently play an increasing role. A large number of analytical approaches can be used to support decision-making, and progress has been made both in integrated assessment models, policy dialogues and other decision support tools.

Like most policy-making, climate policy involves trading off risks and uncertainties. Risks and uncertainties have not only natural but also human and social dimensions. They arise from missing, incomplete and imperfect evidence, from voluntary or involuntary limits to information management, from difficulties in incorporating some variables into formal analysis, as well as from the inherently unpredictable elements of complex systems. An increasing international literature considers how the limits of the evidence basis and other sources of uncertainties can be estimated.

Costs and benefits of climate change mitigation policies can be assessed (subject to the uncertainties noted above) at project, firm, technology, sectoral, community, regional, national or multinational levels. Inputs can include financial, economic, ecological and social factors. In formal cost-benefit analyses, the discount rate is one major determinant of the present value of costs and benefits, since climate change, and mitigation/adaptation measures all involve impacts spread over very long time periods. Much of the literature uses constant discount rates at a level estimated to reflect time preference rates as used when assessing typical large investments. Some recent literature also includes recommendations about using time-decreasing discount rates, which reflect uncertainty about future economic growth, fairness and intra-generational distribution, and observed individual choices. Based on this, some countries officially recommend using time-decreasing discount rates for long time horizons.

The potential linkages between climate change mitigation and adaptation policies have been explored in an emerging literature. It is concluded that there is a number of factors that condition societies' or individual stakeholders' capacity to implement climate change mitigation and adaptation policies including social, economic, and environmental costs, access to resources, credit, and the decision-making capacity in itself.

Climate change has considerable implications for intra-generational and inter-generational equity, and the application of different equity approaches has major implications for policy recommendations, as well as for the implied distribution of costs and benefits of climate policies. Different approaches to social justice can be applied when evaluating equity consequences of climate change policies. They span traditional economic approaches where equity appears in terms of the aggregated welfare consequences of adaptation and mitigation policies, and rights-based approaches that argue that social actions are to be judged in relation to the defined rights of individuals.

The cost and pace of any response to climate change concerns will critically depend on the social context, as well as the cost, performance, and availability of technologies. Technological change is particularly important over the long-term time scales that are characteristic of climate change. Decade (or longer) time scales are typical for the gaps involved between technological innovation and widespread diffusion, and of the capital turnover rates characteristic for long-term energy capital stock and infrastructures. The development and deployment of technology is a dynamic process that arises through the actions of human beings, and different social and economic systems have different proclivities to induce technological change, involving a different set of actors and institutions in each step. The state of technology and technology change, as well as human capital and other resources, can differ significantly from country to country and sector to sector, depending on the starting point of infrastructure, technical capacity, the readiness of markets to provide commercial opportunities and policy frameworks.

The climate change mitigation framing issues in general are characterized by high agreement/much evidence relating to the range of theoretical and methodological issues that are relevant in assessing mitigation options. Sustainable development and climate change, mitigation and adaptation relationships, and equity consequences of mitigation policies are areas where there is conceptual agreement on the range of possible approaches, but relatively few lessons can be learned from studies, since these are still limited *(high agreement, limited evidence)*. Other issues, such as mitigation cost concepts and technological change are very mature in the mitigation policy literature, and there is high agreement/much evidence relating to theory, modelling, and other applications. In the same way, decision-making approaches and various tools and approaches are characterized by high agreement on the range of conceptual issues *(high agreement, much evidence)*, but there is significant divergence in the applications, primarily since some approaches have been applied widely and others have only been applied to a more limited extent *(high agreement, limited evidence)*. There is some debate about which of these framing methodologies and issues relating to mitigation options are most important, reflecting (amongst other things) different ethical choices – to this extent at least there is an irreducible level of uncertainty *(high agreement, limited evidence)*.

2.1 Climate change and Sustainable Development

2.1.1 Introduction

This section introduces the relationship between sustainable development (SD) and climate change and presents a number of key concepts that can be used to frame studies of these relationships. Climate change and sustainable development are considered in several places throughout this report. Chapter 12 provides a general overview of the issues, while more specific issues relating to short- and long-term mitigation issues are addressed in Chapters 3 (Section 3.1) and 11 (Section 11.6). Sectoral issues are covered in Chapters 4-10 (Sections 4.5.4, 5.5.5, 6.9.2, 7.7, 8.4.5, 9.7, and in 10.6). Furthermore, the IPCC (2007b) addresses SD and climate change in Chapters 18 and 20.

2.1.2 Background

The IPCC's Third Assessment Report (TAR; IPCC, 2001) included considerations concerning SD and climate change. These issues were addressed particularly by Working Group II and III, as well as the Synthesis report. The TAR included a rather broad treatment of SD (Metz *et al.*, 2002). The report noted three broad classes of analyses or perspectives: efficiency and cost-effectiveness, equity and sustainable development, and global sustainability and societal learning.

Since the TAR, literature on sustainable development and climate change has attempted to further develop approaches that can be used to assess specific development and climate policy options and choices in this context (Beg *et al.*, 2002; Cohen *et al.*, 1998; Munasinghe and Swart, 2000; Schneider, 2001; Banuri *et al.*, 2001; Halsnæs and Verhagen, 2007; Halsnæs, 2002; Halsnæs and Shukla, 2007, Markandya and Halsnæs, 2002a; Metz *et al.* 2002; Munasinghe and Swart, 2005; Najam and Rahman, 2003; Smit *et al.*, 2001; Swart *et al.*,. 2003; Wilbanks, 2003). These have included discussions about how distinctions can be made between natural processes and feedbacks, and human and social interactions that influence the natural systems and that can be influenced by policy choices (Barker, 2003). These choices include immediate and very specific climate policy responses as well as more general policies on development pathways and the capacity for climate change adaptation and mitigation. See also Chapter 12 of this report and Chapter 18 of IPCC (2007b) for a more extensive discussion of these issues.

Policies and institutions that focus on development also affect greenhouse gas (GHG) emissions and vulnerability. Moreover, these same policies and institutions constrain or facilitate mitigation and adaptation. These indirect effects can be positive or negative, and several studies have therefore suggested the integration of climate change adaptation and mitigation perspectives into development policies, since sustainable development requires coping with climate change

and thereby will make development more sustainable (Davidson *et al.*, 2003; Munasinghe and Swart, 2005; Halsnæs and Shukla, 2007).

Climate change adaptation and mitigation can also be the focus of policy interventions and SD can be considered as an issue that is indirectly influenced. Such climate policies can tend to focus on sectoral policies, projects and policy instruments, which meet the adaptation and mitigation goals, but are not necessarily strongly linked to all the economic, social, and environmental dimensions of sustainable development. In this case climate change policy implementation in practice can encounter some conflicts between general development goals and the goal of protecting the global environment. Furthermore, climate policies that do not take economic and social considerations into account might not be sustainable in the long run.

In conclusion, one might then distinguish between climate change policies that emerge as an integrated element of general sustainable development policies, and more specific adaptation and mitigation policies that are selected and assessed primarily in their capacity to address climate change. Examples of the first category of policies can be energy efficiency measures, energy access and affordability, water management systems, and food security options, while examples of more specific adaptation and mitigation policies can be flood control, climate information systems, and the introduction of carbon taxes. It is worth noticing that the impacts on sustainable development and climate change adaptation and mitigation of all these policy examples are very context specific, so it cannot in general be concluded whether a policy supports sustainable development and climate change jointly or if there are serious tradeoffs between economic and social perspectives and climate change (see also Chapter 12 of this report and Chapter 18 of IPCC (2007b) for a more extensive discussion).

2.1.3 The dual relationship between climate change and Sustainable Development

There is a dual relationship between sustainable development and climate change. On the one hand, climate change influences key natural and human living conditions and thereby also the basis for social and economic development, while on the other hand, society's priorities on sustainable development influence both the GHG emissions that are causing climate change and the vulnerability.

Climate policies can be more effective when consistently embedded within broader strategies designed to make national and regional development paths more sustainable. This occurs because the impact of climate variability and change, climate policy responses, and associated socio-economic development will affect the ability of countries to achieve sustainable development goals. Conversely, the pursuit of those goals will in turn affect the opportunities for, and success of, climate policies.

Climate change impacts on development prospects have also been described in an interagency project on poverty and climate change as 'Climate Change will compound existing poverty. Its adverse impacts will be most striking in the developing nations because of their dependence on natural resources, and their limited capacity to adapt to a changing climate. Within these countries, the poorest, who have the least resources and the least capacity to adapt, are the most vulnerable' (African Development Bank *et al.*, 2003).

Recognizing the dual relationship between SD and climate change points to a need for the exploration of policies that jointly address SD and climate change. A number of international study programmes, including the Development and Climate project (Halsnæs and Verhagen, 2007), and an OECD development and environment directorate programme (Beg *et al.,* 2002) explore the potential of SD-based climate change policies. Other activities include projects by the World Resources Institute (Baumert *et al.,* 2002), and the PEW Centre (Heller and Shukla, 2003). Furthermore, the international literature also includes work by Cohen *et al.*, 1998; Banuri and Weyant, 2001; Munasinghe and Swart 2000; Metz *et al.*, 2002; Munasinghe and Swart, 2005; Schneider *et al.*, 2000; Najam and Rahman, 2003; Smit *et al.*, 2001; Swart *et al.*, 2003; and Wilbanks, 2003).

2.1.4   The Sustainable Development concept

Sustainable development (SD) has been discussed extensively in the theoretical literature since the concept was adopted as an overarching goal of economic and social development by UN agencies, by the Agenda 21 nations, and by many local governments and private-sector actors. The SD literature largely emerged as a reaction to a growing interest in considering the interactions and potential conflicts between economic development and the environment. SD was defined by the World Commission on Environment and Development in the report *Our Common Future* as 'development that meets the needs of the present without compromising the ability of future generations to meet their own needs' (WCED, 1987).

The literature includes many alternative theoretical and applied definitions of sustainable development. The theoretical work spans hundreds of studies that are based on economic theory, complex systems approaches, ecological science and other approaches that derive conditions for how development paths can meet SD criteria. Furthermore, the SD literature emphasizes a number of key social justice issues including inter- and intra-generational equity. These issues are dealt with in Section 2.6.

Since a comprehensive discussion of the theoretical literature on sustainable development is beyond the scope of this report, a pragmatic approach limits us to consider how development can

be made more sustainable.

The debate on sustainability has generated a great deal of research and policy discussion on the meaning, measurability and feasibility of sustainable development. Despite the intrinsic ambiguity in the concept of sustainability, it is now perceived as an irreducible holistic concept where economic, social, and environmental issues are interdependent dimensions that must be approached within a unified framework (Hardi and Barg, 1997; Dresner, 2002; Meadows, 1998). However, the interpretation and valuation of these dimensions have given rise to a diversity of approaches.

A growing body of concepts and models, which explores reality from different angles and in a variety of contexts, has emerged in recent years in response to the inability of normal disciplinary science to deal with complexity and systems – the challenges of sustainability. The outlines of this new framework, known under the loose term of 'Systems Thinking', are, by their very nature, transdisciplinary and synthetic (Kay and Foster, 1999). An international group of ecologists, economists, social scientists and mathematicians has laid the principles and basis of an integrative theory of systems change (Holling 2001). This new theory is based on the idea that systems of nature and human systems, as well as combined human and nature systems and social-ecological systems, are interlinked in never-ending adaptive cycles of growth, accumulation, restructuring, and renewal within hierarchical structures (Holling *et al.*, 2002).

A core element in the economic literature on SD is the focus on growth and the use of man-made, natural, and social capital. The fact that there are three different types of capital that can contribute to economic growth has led to a distinction between weak and strong sustainability, as discussed by Pearce and Turner (1990), and Rennings and Wiggering (1997). Weak sustainability describes a situation where it is assumed that the total capital is maintained and that the three different elements of the capital stock can, to some extent, be used to substitute each other in a sustainable solution. On the other hand, strong sustainability requires each of the three types of capital to be maintained in its own right, at least at some minimum level. An example of an application of the strong sustainability concept is Herman Daly's criteria, which state that renewable resources must be harvested at (or below) some predetermined stock level, and renewable substitutes must be developed to offset the use of exhaustible resources (Daly, 1990). Furthermore, pollution emissions should be limited to the assimilative capacity of the environment.

Arrow *et al.*, 2004, in a joint authorship between leading economists and ecologists, present an approach for evaluating alternative criteria for consumption[1], seen over time in a sustainable development perspective. Inter-temporal consumption and utility are introduced here as measurement

---

1   Consumption should here be understood in a broad sense as including all sorts of goods that are elements in a social welfare function.

points for sustainable development. One of the determinants of consumption and utility is the productive base of society, which consists of capital assets such as manufactured capital, human capital, and natural capital. The productive base also includes the knowledge base of society and institutions.

Although institutions are often understood as part of the capital assets, Arrow *et al.* (2004) only consider institutions in their capacity as guiding the allocation of resources, including capital assets. Institutions in this context include the legal structure, formal and informal markets, various government agencies, inter-personal networks, and the rules and norms that guide their behaviour. Seen from an SD perspective, the issue is then: how, and to what extent, can policies and institutional frameworks for these influence the productive basis of society and thereby make development patterns more sustainable.

The literature includes other views of capital assets that will consider institutions and sustainable development policies as being part of the social capital element in society's productive base. Lehtonen (2004) provides an overview of the discussion on social capital and other assets. He concludes that despite capabilities and social capital concepts not yet being at the practical application stage, the concepts can be used as useful metaphors, which can help to structure thoughts across different disciplines. Lehtonen refers to analysis of social-environmental dimensions by the OECD (1998) that addresses aspects such as demography, health, employment, equity, information, training, and a number of governance issues, as an example of a pragmatic approach to including social elements in sustainability studies.

Arrow *et al.,* (2004) summarize the controversy between economists and ecologists by saying that ecologists have deemed current consumption patterns to be excessive or deficient in relation to sustainable development, while economists have focused more on the ability of the economy to maintain living standards. It is concluded here that the sustainability criterion implies that inter-temporal welfare should be optimized in order to ensure that current consumption is not excessive.[2] However, the optimal level of current consumption cannot be determined (i.e. due to various uncertainties). Theoretical considerations therefore focus instead on factors that make current consumption more or less sustainable. These factors include the relationship between market rates of return on investments and social discount rates, and the relationship between market prices of consumption goods (including capital goods) and the social costs of these commodities.

Some basic principles are therefore emerging from the international sustainability literature, which helps to establish commonly held principles of sustainable development. These include, for instance, the welfare of future generations, the

maintenance of essential biophysical life support systems, more universal participation in development processes and decision-making, and the achievement of an acceptable standard of human well-being (Swart *et al.*, 2003; Meadowcroft, 1997; WCED, 1987).

In the more specific context of climate change policies, the controversy between different sustainability approaches has shown up in relation to discussions on key vulnerabilities; see Section 2.5.2 for more details.

2.1.5   Development paradigms

Assessment of SD and climate change in the context of this report considers how current development can be made more sustainable. The focus is on how development goals, such as health, education, and energy, food, and water access can be achieved without compromising the global climate.

When applying such a pragmatic approach to the concept of SD it is important to recognize that major conceptual understandings and assumptions rely on the underlying development paradigms and analytical approaches that are used in studies. The understanding of development goals and the tradeoffs between different policy objectives depends on the development paradigm applied, and the following section will provide a number of examples on how policy recommendations about SD and climate change depend on alternative understandings of development as such.

A large number of the models that have been used for mitigation studies are applications of economic paradigms. Studies that are based on economic theory typically include a specification of a number of goals that are considered as important elements in welfare or human wellbeing. Some economic paradigms focus on the welfare function of the economy, assuming efficient resource allocation (such as in neoclassical economics), and do not consider deviations from this state and ways to overcome these. In terms of analyzing development and climate linkages, this approach will see climate change mitigation as an effort that adds a cost to the optimal economic state.[3] However, there is a very rich climate mitigation cost literature that concludes that market imperfections in practice often create a potential for mitigation policies that can help to increase the efficiency of energy markets and thereby generate indirect cost savings that can make mitigation policies economically attractive (IPCC, 1996, Chapters 8 and 9; IPCC, 2001, Chapters 7 and 8). The character of such market imperfections is discussed further in Section 2.4.

Other development paradigms based on institutional economics focus more on how markets and other information-

---

2 Arrow *et al.* (2004) state that 'actual consumption today is excessive if lowering it and increasing investment (or reducing disinvestment) in capital assets could raise future utility enough to more than compensate (even after discounting) for the loss in current utility'.
3 Take the benefits of avoided climate change into consideration.

sharing mechanisms establish a framework for economic interactions. Recent development research has included studies on the role of institutions as a critical component in an economy's capacity to use resources optimally. Institutions are understood here in a broad sense, as being a core allocation mechanism and as the structure of society that organizes markets and other information sharing (Peet and Hartwick, 1999).

In this context, climate policy issues can include considerations about how climate change mitigation can be integrated into the institutional structure of an economy. More specifically, such studies can examine various market and non-market incentives for different actors to undertake mitigation policies and how institutional capacities for these policies can be strengthened. Furthermore, institutional policies in support of climate change mitigation can also be related to governance and political systems – see a more elaborate discussion in Chapter 12, Section 12.2.3.

Weak institutions have a lot of implications for the capacity to adapt or mitigate to climate change, as well as in relation to the implementation of development policies. A review of the social capital literature related to economic aspects and the implications for climate change mitigation policies concludes that, in most cases, successful implementation of GHG emission-reduction options will depend on additional measures to increase the potential market and the number of exchanges. This can involve strengthening the incentives for exchange (prices, capital markets, information efforts etc.), introducing new actors (institutional and human capacity efforts), and reducing the risks of participation (legal framework, information, general policy context of market regulation). All these measures depend on the nature of the formal institutions, the social groups of society, and the interactions between them (Olhoff, 2002). See also Chapter 12 of this report for a more extensive discussion of the political science and sociological literature in this area.

Key theoretical contributions to the economic growth and development debate also include work by A. Sen (1999) and P. Dasgupta (1993) concerning capabilities and human well-being. Dasgupta, in his inquiry into well-being and destitution, concludes that 'our citizens' achievements are the wrong things to look at. We should be looking at the extent to which they enjoy the freedom to achieve their ends, no matter what their ends turn out to be. The problem is that the extent of such freedoms depends upon the degree to which citizens make use of income and basic needs'. (Dasgupta, 1993, pp. 54). Following this, Dasgupta recommends studying the distribution of resources, as opposed to outcomes (which, for example, can be measured in terms of welfare). The access to income and basic needs are seen as a fundamental basis for human well-being and these needs include education, food, energy, medical care etc. that individuals can use as inputs to meeting their individual desires.

See also Section 2.6, where the equity dimensions of basic needs and well-being approaches are discussed in more detail.

In the context of capabilities and human well-being, climate change policies can then include considerations regarding the extent to which these policies can support the access of individuals to specific resources as well as freedoms.

The capability approaches taken by Sen and Dasgupta have been extended by some authors from focusing on individuals to also covering societies (Ballet et al., 2003; Lehtonen, 2004). It is argued here that, when designing policies, one needs to look at the effects of economic and environmental policies on the social dimension, including individualistic as well as social capabilities, and that these two elements are not always in harmony.

2.1.6   International frameworks for evaluating Sustainable Development and climate change links

Studies that assess the sustainable development impacts of climate change (and vice versa) when they are considering short to medium-term perspectives will be dealing with a number of key current development challenges. This section provides a short introduction to international policy initiatives and decisions that currently offer a framework for addressing development goals.

A key framework that can be used to organize the evaluation of SD and climate change linkages is the WEHAB[4] framework that was introduced by the World Summit on Sustainable Development in 2002 (WSSD, 2002). The WEHAB sectors reflect the areas selected by the parties at the WSSD meeting to emphasize that particular actions were needed in order to implement Agenda 21. Seen from a climate change policy evaluation perspective it would be relevant to add a few more sectors to the WEHAB group in order to facilitate a comprehensive coverage of major SD and climate change linkages. These sectors include human settlements tourism, industry, and transportation. It would also be relevant to consider demography, institutions and various cultural issues and values as cross-cutting sectoral issues.

Climate change policy aspects can also be linked to the Millennium Development Goals (MDG) that were adopted as major policy targets by the WSSD. The MDGs include nine general goals to eradicate poverty and hunger, health, education, natural resource utilization and preservation, and global partnerships that are formulated for the timeframe up to 2015 (UNDP, 2003a).

---

4   WEHAB stands for Water, Energy, Health, Agriculture, and Biodiversity.

A recent report by the CSD (Commission on Sustainable Development) includes a practical plan for how to achieve the Millennium Development Goals (CSD, 2005). Climate change is explicitly mentioned in the CSD report as a factor that could worsen the situation of the poor and make it more difficult to meet the MDGs. Furthermore, CSD (2005) suggests adding a number of energy goals to the MDGs (i.e. to reflect energy security and the role that energy access can play in poverty alleviation). Adding energy as a separate component in the MDG framework will establish a stronger link between MDGs and climate change mitigation.

Several international studies and agency initiatives have assessed how the MDGs can be linked to goals for energy-, food-, and water access and to climate change impacts, vulnerability, and adaptation (African Development Bank *et al.*, 2003), and an example of how the link between climate change and MDGs can be further developed to include both adaptation and mitigation is shown in Table 2.1. A linkage between MDGs and development goals is also described very specifically by Shukla (2003) and Shukla *et al.* (2003) in relation to the official Indian 10th plan for 2002–2007. In the same way, the Millenium Ecosystem Assessment (MEA) presents a global picture of the relationship between the net gains in human well-being and economic development based on a growing cost through degradation of ecosystem services, and demonstrates how this can pose a barrier to achieving the MDGs (MEA, 2005).

Measuring progress towards SD requires the development and systematic use of a robust set of indicators and measures. Agenda 21 (1992) explicitly recognizes in Chapter 40 that a pre-requisite for action is the collection of data at various levels (local, provincial, national and international), indicating the status and trends of the planet's ecosystems, natural resources, pollution and socio-economy.

The OECD Ministerial Council decided in 2001 that the regular Economic Surveys of OECD countries should include an evaluation of SD dimensions, and a process for agreeing on SD indicators. These will be used in regular OECD peer reviews of government policies and performance. From the OECD menu of SD issues, the approach is to select a few areas that will be examined in depth, based on specific country relevance (OECD, 2003).

The first OECD evaluation of this kind was structured around three topics that member countries could select from the following list of seven policy areas (OECD, 2004):
- Improving environmental areas:
  - Reducing GHG emissions
  - Reducing air pollutants
  - Reducing water pollution
  - Moving towards sustainable use of renewable and non-renewable natural resources
  - Reducing and improving waste management

- Improving living standards in developing countries.
- Ensuring sustainable retirement income policies.

Most of the attention in the country choice was given to the environmental areas, while evaluation of improving living standards in developing countries was given relatively little attention in this first attempt.

The use of SD indicators for policy evaluations has been applied in technical studies of SD and climate change (Munasinghe, 2002; Atkinson *et al.*, 1997; Markandya *et al.*, 2002). These studies address SD dimensions based on a number of economic, environmental, human and social indicators, including both quantitative and qualitative measurement standards. A practical tool applied in several countries, called the Action Impact Matrix (AIM), has been used to identify, prioritize, and address climate and development synergies and tradeoffs (Munasinghe and Swart, 2005).

All together, it can be concluded that many international institutions and methodological frameworks offer approaches for measuring various SD dimensions, and that these have been related to broader development and economic policies by CSD, the WSSD, and the OECD. Many indexes and measurement approaches exist but, until now, relatively few studies have measured climate change in the context of these indexes. In this way, there is still a relatively weak link between actual measurements of and climate change links.

2.1.7 Implementation of Sustainable Development and climate change policies

SD and climate change are influenced by a number of key policy decisions related to economic, social and environmental issues, as well as by business-sector initiatives, private households and many other stakeholders, and these decisions are again framed by government policies, markets, information sharing, culture, and a number of other factors. Some of the decisions that are critically important in this context are investments, use of natural resources, energy consumption, land use, technology choice, and consumption and lifestyle, all of which can lead to both increasing and decreasing GHG emission intensities, which again will have implications for the scope of the mitigation challenge. Seen in a longer-term perspective these decisions are critical determinants for development pathways.

There has been an evolution in our understanding of how SD and climate change mitigation decisions are taken by societies. In particular, this includes a shift from governments that are defined by the nation/state to a more inclusive concept of governance, which recognizes various levels of government (global, transnational/regional, and local), as well as the roles of the private sector, non-governmental actors and civil society. Chapter 12, Section 12.2.3, includes a comprehensive

Table 2.1: *Relationship between MDGs, energy-, food-, and water access, and climate change*

MDG goals	Sectoral themes	Climate change links
To halve (between 1990 and 2015), the proportion of the world's population whose income is below 1US$ a day	Energy: Energy for local enterprises Lighting to facilitate income generation Energy for machinery Employment related to energy provision Food/water: Increased food production Improved water supply Employment	Energy: GHG emissions. Adaptive and mitigative capacity increase due to higher income levels and decreased dependence on natural resources, production costs etc. Food/water: GHG emissions Increased productivity of agriculture can reduce climate change vulnerability. Improved water management and effective use can help adaptation and mitigation. Increased water needs for energy production
To reduce by two-thirds (between 1990 and 2015), the death rate for children under the age of five years	Energy: Energy supply can support health clinics Reduced air pollution from traditional fuels Reduced time spent on fuel collection can increase the time spent on children's health care Food/water: Improved health due to increased supply of high-quality food and clean water Reduced time spent on food and water provision can increase the time spent on children's health care Improved waste and wastewater treatment	Energy: GHG emissions Food/water: Health improvements will decrease vulnerability to climate change and the adaptive capacity Decreased methane and nitrous oxide emissions
To reduce by three-quarters (between 1990 and 2015) the rate of maternal mortality	Energy: Energy provision for health clinics Reduced air pollution from traditional fuels and other health improvements. Food/water: Improved health due to increased supply of high-quality food and clean water Time savings on food and water provision can increase the time spent on children's health care	Energy: GHG emissions Food/water: Health improvements will decrease vulnerability to climate change and the adaptive capacity
Combat HIV/AIDS, malaria and other major diseases	Energy: Energy for health clinics Cooling of vaccines and medicine Food/water: Health improvements from cleaner water supply Food production practices that reduce malaria potential	Energy: GHG emissions from increased health clinic services, but health improvements can also reduce the health service demand Food/water: Health improvements will decrease vulnerability to climate change and the adaptive capacity
To stop the unsustainable exploitation of natural resources	Energy: Deforestation caused by woodfuel collection Use of exhaustible resources Food/water: Land degradation	Energy: GHG emissions Carbon sequestration Food/water: Carbon sequestration Improved production conditions for land-use activities will increase the adaptive and mitigative capacity
To halve (between 1990 and 2015), the proportion of people who are unable to reach and afford safe drinking water	Energy: Energy for pumping and distribution systems, and for desalination and water treatment Water: Improved water systems	Energy: GHG emissions Water: Reduced vulnerability and enhanced adaptive capacity

Source: based on Davidson et al., (2003).

assessment of how state, market, civil society and partnerships play a role in sustainable development and climate change policies.

2.2 Decision-making

2.2.1 The 'public good' character of climate change

Mitigation costs are exclusive to the extent that they may be borne by some individuals (nations) while others might evade them (free-riding) or might actually gain a trade/investment benefit from not acting (carbon leakage). The incentive to evade taking mitigation action increases with the substitutability of individual mitigation efforts and with the inequality of the distribution of net benefits. However, individual mitigation efforts (costs) decrease with efficient mitigation actions undertaken by others.

The unequal distribution of climate benefits from mitigation action, of the marginal costs of mitigation action and of the ability to pay emission reduction costs raises equity issues and increases the difficulty of securing agreement. In a strategic environment, leadership from a significant GHG emitter may provide an incentive for others to follow suit by lowering their costs (Grasso, 2004; ODS, 2002).

Additional understandings come from political science, which emphasizes the importance of analyzing the full range of factors that have a bearing on decisions by nation states, including domestic pressures from the public and affected interest groups, the role of norms and the contribution of NGOs to the negotiation processes. *Case studies* of many MEAs (Multilateral Environmental Agreements) have provided insights, particularly on the institutional, cultural, political and historical dimensions that influence outcomes (Cairncross, 2004). A weakness of this approach is that the conclusions can differ depending on the choice of cases and the way in which the analysis is implemented. However, such ex-post analysis of the relevant policies often provides deep insights that are more accessible to policymakers, rather than theoretical thinking or numeric models.

2.2.2 Long time horizons

Climate policy raises questions of inter-generational equity and changing preferences, which inevitably affect the social weighting of environmental and economic outcomes, due to the long-term character of the impacts (for a survey see Bromley and Paavola, 2002).

However, studies traditionally assume that preferences will be stable over the long time frames involved in the assessment of climate policy options. To the extent that no value is attached to the retention of future options, the preferences of the present generation are implicitly given priority in much of this analysis. As time passes, preferences will be influenced by information, education, social and organizational affiliation, income distribution and a number of cultural values (Palacios-Huerta and Santos, 2002). Institutional frameworks are likely to develop to assist groups, companies and individuals to form preferences in relation to climate change policy options. The institutions can include provision of information and general education programmes, research and assessments, and various frameworks that can facilitate collective decision-making that recognizes the common 'global good' character of climate change.

At an analytic level, the choice of discount rates can have a profound affect on valuation outcomes – this is an important issue in its own right and is discussed in Section 2.4.1.

2.2.3 Irreversibility and the implications for decision-making

Human impacts on the climate system through greenhouse gas emissions may change the climate so much that it is impossible (or extremely difficult and costly) to return it to its original state – in this sense the changes are irreversible (Scheffer *et al.*, 2001; Schneider, 2004). Some irreversibility will almost certainly occur. For example, there is a quasi-certain irreversibility of a millennia time scale in the presence, in the atmosphere, of 22% of the emitted CO_2 (Solomon *et al.*, 2007). However, the speed and nature of these changes, the tipping point at which change may accelerate and when environmentally, socially and economically significant effects become irreversible, and the cost and effectiveness of mitigation and adaptation responses are all uncertain, to a greater or lesser extent.

The combination of environmental irreversibility, together with these uncertainties (Baker, 2005; Narain *et al.*, 2004; Webster, 2002; Epstein, 1980) means that decision-makers have to think carefully about:
a) The timing and sequencing of decisions to preserve options.
b) The opportunity to sequence decisions to allow for learning about climate science, technology development and social factors (Baker, 2005; Kansuntisukmongko, 2004).
c) Whether the damage caused by increases in greenhouse concentrations in the atmosphere will increase proportionally and gradually or whether there is a risk of sudden, non-linear changes, and similarly whether the costs of reducing emissions change uniformly with time and the depth of reduction required, or are they possibly subject to thresholds or other non-linear effects.
d) Whether the irreversible damages are clustered in particular parts of the world or have a general effect, and
e) whether there is a potential that these irreversible damages will be catastrophically severe for some, many or even all communities (Cline, 2005).

Just as there are risks of irreversible climate changes, decisions to reduce GHG emissions can require actions that are essentially irreversible. For example, once made, these long-lived, large-scale investments in low-emission technologies are irreversible. If the assumptions about future policies and the directions of climate science on which these investments are made prove to be wrong, they would become 'stranded' assets. The risks (perceived by investors) associated with irreversibility of this nature further complicate decision-making on abatement action (Keller *et al.*, 2004; Pindyck, 2002; Kolstad, 1996; Sullivan *et al.*, 2006; Hamilton and Kenber, 2006).

Without special actions by governments to overcome their natural inertia, economic and social systems might delay too long in reacting to climate risks, thus leading to irreversible climate changes. Ambitious climate-protection goals would require new investments (physical and intellectual) in climate-friendly technologies (efficiency improvements, renewables, nuclear power, carbon capture and storage), which are higher in cost than current technologies or otherwise divert scarce resources. From an economic point of view these investments are essentially irreversible. As the scale of the investment and the proportion of research and development costs increase, so the private economic risks associated with irreversibility also increase. Therefore, in the presence of uncertainty concerning future policy towards GHG emission reduction, future carbon prices or stabilization targets, investors are reluctant to undertake large-scale irreversible investments (sunk costs) without some form of upfront government support.

2.2.4   Risk of catastrophic or abrupt change

The possibility of abrupt climate change and/or abrupt changes in the earth system triggered by climate change, with potentially catastrophic consequences, cannot be ruled out (Meehl *et al.*, 2007). Disintegration of the West Antarctic Ice Sheet (See Meehl *et al.*, 2007), if it occurred, could raise sea level by 4-6 metres over several centuries. A shutdown of the North Atlantic Thermohaline Circulation (See Meehl *et al.*, 2007) could have far-reaching, adverse ecological and agricultural consequences (See IPCC, 2007b, Chapter 17), although some studies raise the possibility that the isolated, economic costs of this event might not be as high as assumed (See Meehl *et al.*, 2007). Increases in the frequency of droughts (Salinger, 2005) or a higher intensity of tropical cyclones (See Meehl *et al.*, 2007) could occur. Positive feedback from warming may cause the release of carbon or methane from the terrestrial biosphere and oceans (See Meehl *et al.*, 2007), which would add to the mitigation required.

Much conventional decision-making analysis is based on the assumption that it is possible to model and compare all the outcomes from the full range of alternative climate policies. It also assumes there is a smooth trade-off between the different dimensions of each policy outcome; that a probability distribution provides an expected value for each outcome, and that there is a unique best solution – the one with the highest expected value. Consequently, it could suggest that a policy which risked a catastrophically bad outcome with a very low probability might be valued higher than one which completely avoided the possibility of catastrophe and produced merely a bad outcome, but with a very high probability of occurrence.

Assumptions that it is always possible to 'trade off' more of one dimension (e.g. economic growth) for less of another (e.g. species protection) – that there is always a price at which we are comfortable to 'dispense with' a species in the wild (e.g. polar bears), an ecological community or indigenous cultures are problematic for many people. This also applies to assumptions that decision-makers value economic (and other) gains and losses symmetrically – that a dollar gained should always assumed to be valued equally to one that is lost, and that it is possible and appropriate to assume that the current generation's preferences will remain stable over time.

Recent literature drawing on experimental economics and behavioural sciences suggests that these assumptions are an incomplete description of the way in which humans really make decisions. This literature suggests that preferences may be lexicographical (i.e. it is not possible to 'trade off' between different dimensions of alternative possible outcomes – there may be an aversion at any 'price' to losing particular species, ecosystems or communities), that attitudes to gains and losses might not be symmetrical (losses valued more highly than gains of an equivalent magnitude), and that low-probability extreme outcomes are overweighted when making choices (Tversky and Kahneman, 1992; Quiggin, 1982). This literature suggests that under these circumstances the conventional decision axiom of choosing the policy set that maximizes the expected (monetary) value of the outcomes might not be appropriate. Non-conventional decision criteria (e.g. avoiding policy sets which imply the possibility, even if at a very low probability, of specific unacceptable outcomes) might be required to make robust decisions (Chichilnisky, 2000; Lempert and Schlesinger, 2000; Kriegler *et al.*, 2006).

No one analytic approach is optimal. Decision-making inevitably involves applying normative rules. Some normative rules are described in Section 2.2.7 and in Section 2.6.

2.2.5   Sequential decision-making

Uncertainty is a steadfast companion when analyzing the climate system, assessing future GHG emissions or the severity of climate change impacts, evaluating these impacts over many generations or estimating mitigation costs. The typology of uncertainties is explored fully in Section 2.3 below. Uncertainties of differing types exist in key socio-economic factors and scientific phenomena.

The climate issue is a long-term problem requiring long-term solutions. Policymakers need to find ways to explore appropriate long-term objectives and to make judgments about how compatible short-term abatement options are with long-term objectives. There is an increased focus on non-conventional (robust) decision rules (see Section 2.2.7 below), which preserve future options by avoiding unacceptable risks.

Climate change decision-making is not a once-and-for-all event. Rather it is a process that will take place over decades and in many different geographic, institutional and political settings. Furthermore, it does not occur at discrete intervals but is driven by the pace of the scientific and political process. Some uncertainties will decrease with time – for example in relation to the effectiveness of mitigation actions and the availability of low-emission technologies, as well as with respect to the science itself. The likelihood that better information might improve the quality of decisions (the value of information) can support increased investment in knowledge accumulation and its application, as well as a more refined ordering of decisions through time. Learning is an integral part of the decision-making process. This is also referred to as 'act then learn, then act again' (Manne and Richels, 1992; Valverde *et al.*, 1999).

Uncertainties about climate policies at a decadal scale are a source of concern for many climate-relevant investments in the private sector (for example power generation), which have long expected economic lives.

It is important to recognize, however, that some level of uncertainty is unavoidable and that at times the acquisition of knowledge can increase, not decrease, uncertainty. Decisions will nevertheless have to be made.

2.2.6   Dealing with risks and uncertainty in decision-making

Given the multi-dimensionality of risk and uncertainty discussed in Section 2.3, the governance of these deep uncertainties as suggested by Godard *et al.* (2002, p. 21) rests on three pillars: precaution, risk hedging, and crisis prevention and management.

The 1992 UNFCCC Article 3 (Principles) states that the Parties should take precautionary measures to anticipate, prevent or minimize the causes of climate change and mitigate its adverse effects. Where there are threats of serious or irreversible damage, lack of full scientific certainty should not be used as a reason for postponing such measures, taking into account that policies and measures to deal with climate change should be cost-effective in order to ensure global benefits at the lowest possible cost.[5]

While the precautionary principle appears in many other international treaties, from a scientific perspective the concept of precaution is subject to a plurality of interpretations. To frame the discussions on precaution, three key points should be considered first.

First, 'precaution' relates to decision-making in situations of deep uncertainty. It applies in the absence of sufficient data or conclusive or precise probabilistic descriptions of the risks (Cheve and Congar, 2000; Henry and Henry, 2002) or in circumstances where the possibility of unforeseen contingencies or the possibility of irreversibility (Gollier *et al.*, 2000) is suspected.

Second, in addition to that uncertainty/risk dimension, there is also a time dimension of precaution: the precautionary principle recognizes that policy action should not always wait for scientific certainty (see also the costs and decision-making sections of this chapter).

Third, the precautionary principle cuts both ways because in many cases, as Graham and Wiener (1995) noted, environmental choices are trade-offs between one risk and another risk. For example, mitigating climate change may involve more extensive use of nuclear power. Goklany (2002) has suggested a framework for decision-making under the precautionary principle that considers trade-offs between competing risks.

There is no single agreed definition of precautionary decision-making in the scientific literature.

The risk of catastrophes is commercially important, particularly for reinsurers that are large companies whose business is to sell insurance to other insurance companies (see IPCC, 2007b, Chapter 7, Box 7.2). In the context of globalization and consolidation, many reinsurers are actively developing new instruments to trade some of their risk on the deeper financial markets. These instruments include options, swaps and catastrophe bonds.

At the same time, governments are also developing new kinds of public-private partnership to cope with market failures, uncertainties and really big cataclysms. On a global scale, it can be argued that the best form of insurance is to increase the systemic resilience of the human society through scientific research, technical, economic and social development. This requires the broad participation of society in order to succeed.

Mills (2005) concludes that the future role of insurance in helping society to cope with climate change is uncertain. Insurers may rise to the occasion and become more proactive players in improving the science and crafting responses, or they may retreat from oncoming risks, thereby shifting a greater burden to governments and individuals.

---

5   Section 2.6 discusses the ethical questions concerning burden and quantity of proof, as well as procedural issues.

2.2.7   Decision support tools

Decisions concerning the appropriate responses to climate risks require insights into a variety of possible futures over short to very long time frames and into linkages between biophysical and human systems, as well as ethical alternatives. Structured analysis – both numerical and case-based – can 'aid understanding by managing and analyzing information and alternatives' (Arrow *et al.*, 1996a, referenced in Bell *et al.*, 2001). Integrated Assessment Models (IAMs) in particular have improved greatly in terms of the richness with which they represent the biophysical, social and economic systems and the feedbacks between them. They have increasingly explored a variety of decision rules or other means of testing alternative policies. Without structured analysis it is extremely difficult, if not impossible, to understand the possible effects of alternative policy choices that face decision-makers. Structured analysis can assist choices of preferred policies within interests (for example at the national level) as well as negotiating outcomes between interests (by making regional costs and benefits clearer).

The use of projections and scenarios is one way to develop understanding about choices in the context of unpredictability. These are discussed in detail in Chapter 3.

A large number of analytical approaches can be used as a support to decision-making. IPCC (2001) Chapter 10, provides an extensive overview of decision-making approaches and reviews their applicability at geopolitical levels and in climate policy domains. The review includes decision analysis, cost-benefit analysis, cost-effectiveness analysis, tolerable windows/ safe-landing/guard-rail approaches, game theory, portfolio theory, public finance theory, ethical and cultural prescriptive rules, and various policy dialogue exercises. Integrated assessment, multi-attribute analysis and green accounting approaches are also commonly used decision support tools in climate change debates.

A major distinction between cost benefit-analysis, cost-effectiveness analysis, and multi-attribute analysis and different applications of these relates to the extent in which monetary values are used to represent the impacts considered. Cost-benefit analysis aims to assign monetary values to the full range of costs and benefits. This involves at least two important assumptions – that it is possible to 'trade off' or compensate between impacts on different values in a way that can be expressed in monetary values, and that it is possible to ascertain estimates of these 'compensation' values for non-market impacts, such as air pollution, health and biodiversity. By definition, the benefits and costs of climate change policies involve many of such issues, so climate change economic analysis embodies a lot of complicated valuation issues. Section 2.4 goes more into depth about approaches that can be used to value non-markets impacts and the question of discounting.

In multi-attribute analysis, instead of using values derived from markets or from non-market valuation techniques, different dimensions (impacts) are assigned weights – through a stakeholder consultation process, by engaging a panel of experts or by the analyst making explicit decisions. This approach can use quantitative data, qualitative information or a mixture of both. Developing an overall score or ranking for each option allows alternative policies to be assessed, even under conditions of weak comparability. Different functional forms can be used for the aggregation process.

Policy optimization models aim to support the selection of policy/decision strategies and can be divided into a number of types:
- Cost-benefit approaches, which try to balance the costs and benefits of climate policies (including making allowances for uncertainties).
- Target-based approaches, which optimize policy responses, given targets for emission or climate change impacts (again in some instances explicitly acknowledging uncertainties).
- Approaches, which incorporate decision strategies (such as sequential act-learn-act decision-making, hedging strategies etc.) for dealing with uncertainty (often embedded in cost-benefit frameworks).

Another approach is to start with a policy or policies and evaluate the implications of their application. Policy evaluation approaches include:
- Deterministic projection approaches, in which each input and output takes on a single value.
- A stochastic projection approach, in which at least some inputs and outputs take on a range of value.
- Exploratory modelling.
- Public participation processes, such as citizens juries, consultation, and polling.

IAMs aim to combine key elements of biophysical and economic systems into a decision-making framework with various levels of detail on the different sub-components and systems. These models include all different variations on the extent to use monetary values, the integration of uncertainty, and on the formulation of the policy problem with regard to optimization, policy evaluation and stochastic projections. Current integrated assessment research uses one or more of the following methods (Rotmans and Dowlatabadi, 1998):
- Computer-aided IAMs to analyze the behavior of complex systems
- Simulation gaming in which complex systems are represented by simpler ones with relevant behavioral similarity.
- Scenarios as tools to explore a variety of possible images of the future.
- Qualitative integrated assessments based on a limited, heterogeneous data set, without using any model.

A difficulty with large, global models or frameworks is that it is not easy to reflect regional impacts, or equity considerations between regions or stakeholder groups. This is particularly true of 'global' cost-benefit approaches, where it is particularly difficult to estimate a marginal benefit curve, as regional differences are likely to be considerable. Such approaches have difficulty in assisting decision-making where there are many decision-makers and multiple interests and values to be taken into account.

Variants of the safe landing/tolerable windows/guard rails approach emphasize the role of regional/national decision-makers by providing them the opportunity to nominate perceived unacceptable impacts of climate change (for their region or globally), and the limit to tolerable socio-economic costs of mitigation measures they would be prepared to accept to avoid that damage (e.g. Toth 2004). Modelling efforts (in an integrated assessment model linking climate and economic variables, and with explicit assumptions about burden sharing through emissions allocations and trading) are then directed at identifying the sets of feasible mitigation paths – known as 'emissions corridors' – consistent with these constraints. To the extent that there is some overlap between the acceptable 'emissions corridors', the conditions for agreement on mitigation action do exist.

Green accounting attempts to integrate a broader set of social welfare measures into macro-economic studies. These measures can be related to a broad set of social, environmental, and development-oriented policy aspects. The approach has most commonly been used in order to integrate environmental impacts, such as local air pollution, GHG emissions, waste generation, and other polluting substances, into macro-economic studies. Green accounting approaches include both monetary valuation approaches that attempt to calculate a 'green national product' (where the economic values of pollutants are subtracted from the national product), and accounting systems that include quantitative non-monetary pollution data.

Halsnæs and Markandya (2002) recognize that decision analysis methods exhibit a number of commonalities in assumptions. The standard approach goes through the selection of GHG emission-reduction options, selection of impact areas that are influenced by policies as for example costs, local air pollution, employment, GHG emissions, and health, definition of baseline case, assessment of the impacts of implementing the GHG emission-reduction policies under consideration, and application of a valuation framework that can be used to compare different policy impacts.

Sociological analysis includes the understanding of how society operates in terms of beliefs, values, attitudes, behaviour, social norms, social structure, regarding climate change. This analysis includes both quantitative and qualitative approaches, such as general surveys, statistics analysis, focus groups, public participation processes, media content analysis, Delphi etc.

All analytical approaches (explicitly or implicitly) have to consider the described elements, whether this is done in order to collect quantitative information that is used in formalized approaches or to provide qualitative information and focus for policy dialogues. Different decision-making approaches will often involve very similar technical analysis in relation to several elements. For example, multi-criteria-analysis, as well as cost-benefit analysis (as, for example, applied in integrated assessment optimization modelling frameworks) and green accounting may use similar inputs and analysis for many model components, but critically diverge when it comes to determining the valuation approach applied to the assessment of multiple policy impacts.

2.3   Risk and uncertainty

2.3.1   How are risk and uncertainty communicated in this report?

Communicating about risk and uncertainty is difficult because uncertainty is multi-dimensional and there are different practical and philosophical approaches to it. In this report, 'risk' is understood to mean the 'combination of the probability of an event and its consequences', as defined in the risk management standard ISO/IEC Guide 73 (2002). This definition allows a variety of ways of combining probabilities and consequences, one of which is expected loss, defined as the 'product of probability and loss'. The fundamental distinction between 'risk' and 'uncertainty' is as introduced by economist Frank Knight (1921), that risk refers to cases for which the probability of outcomes can be ascertained through well-established theories with reliable complete data, while uncertainty refers to situations in which the appropriate data might be fragmentary or unavailable.

Dealing effectively with the communication of risk and uncertainty is an important goal for the scientific assessment of long-term environmental policies. In IPCC assessment reports, an explicit effort is made to enhance consistency in the treatment of uncertainties through a report-wide coordination effort to harmonize the concepts and vocabulary used. The Third Assessment Report common guidelines to describe levels of confidence were elaborated by Moss and Schneider (2000). The actual application of this framework differed across the three IPCC working groups and across chapters within the groups. It led to consistent treatment of uncertainties within Working Group I (focusing on uncertainties and probabilities, see Sommerville et al., 2007, Section 1.6) and Working Group II (focusing on risks and confidence levels, see IPCC, 2007b, Section 1.1), although consistency across these groups was not achieved. The authors of Working Group III did not systematically apply the guidelines.

Box 2.1 Risk and uncertainty vocabulary used in this report

Uncertainty cannot always be quantified, and thus the vocabulary displayed in Table 2.2 is used to qualitatively describe the degree of scientific understanding behind a finding or about an issue. See text for discussion of Table 2.2's dimensions, the amount of evidence and the level of agreement.

Tabel 2.2: *Qualitative definition of uncertainty*

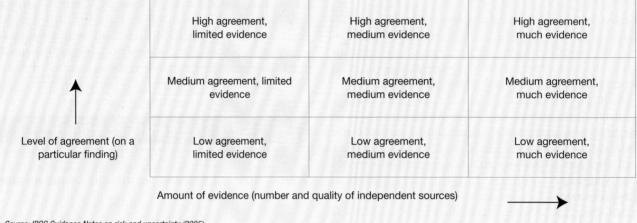

Source: IPCC Guidance Notes on risk and uncertainty (2005).

The most important insight arising from an interdisciplinary assessment of uncertainty is its conceptual diversity. There is no linear scale going from 'perfect knowledge' to 'total uncertainty'. The literature suggests a 'pedigree' approach for characterizing the quality of information (for example the NUSAP approach by Van der Sluijs *et al.*, 2003). This involves examining at least the amount and reliability of *evidence*[6] supporting the information and the level of agreement of the information sources.

The degree of consensus among the available studies is a critical parameter for the quality of information. The *level of agreement* regarding the benefits and drawbacks of a certain technology describes the extent to which the sources of information point in the same direction. Table 2.2's vocabulary is used to qualify IPCC findings along these two dimensions. Because mitigation mostly involves the future of technical and social systems, Table 2.2 is used here to qualify the robustness of findings, and more precise expressions regarding quantified likelihood or levels of confidence are used only when there is high agreement and much evidence, such as converging results from a number of controlled field experiments.

Where findings depend on the future of a dynamic system, it is important to consider the possibility of extreme or/and irreversible outcomes, the potential for resolution (or persistence) of uncertainties in time, and the human dimensions. Rare events with extreme and/or irreversible outcomes are difficult or impossible to assess with ordinary statistics, but receive special attention in the literature.

2.3.2 Typologies of risk and uncertainty

The literature on risk and uncertainty offers many typologies, often comprising the following classes:

Randomness: risk often refers to situations where there is a well-founded probability distribution in typologies of uncertainty. For example, assuming an unchanged climate, the potential annual supply of wind, sun or hydropower in a given area is only known statistically. In situations of randomness, expected utility maximization is a standard decision-making framework.

Possibility: the degree of 'not-implausibility' of a future can be defined rigorously using the notion of acceptable odds, see De Finetti (1937) and Shackle (1949). While it is scientifically controversial to assign a precise probability distribution to a variable in the far distant future determined by social choices such as the global temperature in 2100, some outcomes are not as plausible as others (see the controversy on scenarios in Box 2.2). There are few possibility models related to environmental or energy economics.

Knightian or Deep Uncertainty: the seminal work by Knight (1921) describes a class of situations where the list of outcomes is known, but the probabilities are imprecise. Under deep uncertainty, reporting a range of plausible values allows decision-makers to apply their own views on precaution. Two families of criteria have been proposed for decision-making in this situation. One family associates a real-valued generalized expected utility to each choice (see Ellesberg, 2001), while

---

6   "Evidence" in this report is defined as: Information or signs indicating whether a belief or proposition is true or valid. See Glossary.

the other discards the completeness axiom on the grounds that under deep uncertainty alternative choices may sometimes be incomparable (see Bewley, 2002; Walley, 1991). Results of climate policy analysis under deep uncertainty with imprecise probabilities (Kriegler, 2005; Kriegler *et al.* 2006) are consistent with the previous findings using classical models.

Structural uncertainty: is characterized by « unknown unknowns ». No model (or discourse) can include all variables and relationships. In energy-economics models, for example, there can easily be structural uncertainty regarding the treatment of the informal sector, market efficiency, or the choice between a Keynesian or a neoclassical view of macro-economic dynamics. Structural uncertainty is attenuated when convergent results are obtained from a variety of different models using different methods, and also when results rely more on direct observations (data) rather than on calculations.

Fuzzyness or vagueness: describes the nature of things that do not fall sharply into one category or another, such as the meaning of 'sustainable development' or 'mitigation costs'. One way to communicate the fuzzyness of the variables determining the 'Reasons for concern' about climate change is to use smooth gradients of colours, varying continuously from green to red (see IPCC, 2001a, Figure SPM 2, also known as the 'burning embers' diagram). Fuzzy modelling has rarely been used in the climate change mitigation literature so far.

Uncertainty is not only caused by missing information about the state of the world, but also by human volition: global environmental protection is the outcome of social interactions. Not mentioning taboos, psychological and social aspects, these include:

Surprise: which means a discrepancy between a stimulus and pre-established knowledge (Kagan, 2002). Complex systems, both natural and human, exhibit behaviour that was not imagined by observers until it actually happened. By allowing decision-makers to become familiar (in advance) with a number of diverse but plausible futures, scenarios are one way of reducing surprises.

Metaphysical: describes things that are not assigned a truth level because it is generally agreed that they cannot be verified, such as the mysteries of faith, personal tastes or belief systems. Such issues are represented in models by critical parameters, such as discount rates or risk-aversion coefficients. While these parameters cannot be judged to be true or false they can have

---

Box 2.2 The controversy on quantifying the beliefs in IPCC SRES scenarios

Between its Second and Third Assessment Reports, the Intergovernmental Panel on Climate Change elaborated long-term greenhouse gas emissions scenarios, in part to drive global ocean-atmosphere general circulation models, and ultimately to assess the urgency of action to prevent the risk of climatic change. Using these scenarios led the IPCC to report a range of global warming over the next century from 1.4–5.8°C, without being able to report any likelihood considerations. This range turned out to be controversial, as it dramatically revised the top-range value, which was previously 3.5°C. Yet some combinations of values that lead to high emissions, such as high per-capita income growth and high population growth, appear less likely than other combinations. The debate then fell into the ongoing controversy between the makers and the users of scenarios.

Schneider (2001) and Reilly et al. (2001) argued that the absence of any probability assignment would lead to confusion, as users select arbitrary scenarios or assume equi-probability. As a remedy, Reilly et al. estimated that the 90% confidence limits were 1.1–4.5°C. Using different methods, Wigley and Raper (2001) found 1.7–4.9°C for this 1990 to 2100 warming.

Grübler et al. (2002) and Allen et al. (2001) argued that good scientific arguments preclude determining objective probabilities or the likelihood that future events will occur. They explained why it was the unanimous view of the IPCC report's lead authors that no method of assigning probabilities to a 100-year climate forecast was sufficiently widely accepted and documented to pass the review process. They underlined the difficulty of assigning reliable probabilities to social and economic trends in the latter half of the 21st century, the difficulty of obtaining consensus range for quintiles such as climate sensitivity, and the possibility of a non-linear geophysical response.

Dessai and Hulme (2004) argued that scenarios could not be meaningfully assigned a probability, except relative to other specific scenarios. While a specific scenario has an infinitesimal probability given the infinity of possible futures, taken as a representative of a cluster of very similar scenarios, it can subjectively be judged more or less likely than another. Nonetheless, a set of scenarios cannot be effectively used to objectively generate a probability distribution for a parameter that is specified in each scenario.

In spite of the difficulty, there is an increasing tendency to estimate probability distribution functions for climate sensitivity, discussed extensively in IPCC (2007a), see Chapter 9, Sections 9.6.2 and 9.6.3 and Chapter 10, Sections 10.5.2 and 10.5.4.

a bearing on both behaviour and environmental policy-making. Thompson and Raynor (1998) argue that, rather than being obstacles to be overcome, the uneasy coexistence of different conceptions of natural vulnerability and societal fairness is a source of resilience and the key to the institutional plurality that actually enables us to apprehend and adapt to our ever-changing circumstances.

Strategic uncertainty: involves the fact that information is a strategic tool for rational agents. The response to climate change requires coordination at international and national level. Strategic uncertainty is usually formalized with game theory, assuming that one party in a transaction has more (or better) information than the other. The informed party may thus be able to extract a rent from this advantage. Information asymmetry is an important issue for the regulation of firms by governments and for international agreements. Both adverse selection and moral hazards are key factors in designing efficient mechanisms to mitigate climate change.

2.3.3   Costs, benefits and uncertainties

In spite of scientific progress, there is still much uncertainty about future climate change and its mitigation costs. Given observed risk attitudes, the desirability of preventive efforts should be measured not only by the reduction in the expected (average) damages, but also by the value of the reduced risks and uncertainties that such efforts yield. The difficulty is how to value the societal benefits included in these risk reductions. Uncertainty concerning mitigation costs adds an additional level of difficulty in determining the optimal risk-prevention strategies, since the difference between two independent uncertain quantities is relatively more uncertain than related to the individual.

How can we decide whether a risk is acceptable to society? Cost-benefit analysis alone cannot represent all aspects of climate change policy evaluation, and Section 2.2 on Decision-making discusses a variety of tools. In the private sector, another practical way to deal with these risks has been to pay attention to the Value-At-Risk (VAR): in addition to using the mean and the variance of the outcome, a norm is set on the most unfavourable percentile (usually 0.05) of the distribution of outcomes at a given future date.

However, in the language of cost-benefit analysis, an acceptable risk means that its benefits to society exceed its costs. The standard rule used by public and private decision-makers in a wide variety of fields (from road safety to long-term investments in the energy sector) is that a risk will be acceptable if the expected net present value is positive. Arrow and Lind (1970) justify this criterion when the policy's benefits and costs have known probabilities, and when agents can diversify their own risk through insurance and other markets. For most of the economic analysis of climate change, these assumptions are disputable, and have been discussed in the economic literature.

First, risks associated with climate change cannot easily be diversified using insurance and financial instruments. Atmospheric events are faced by everyone at the same time in the same region. This reduces the potential benefit of any mutual risk-sharing agreement. A solution would be to share risks internationally, but this is difficult to implement, and its efficiency depends upon the correlation of the regional damages. Inability to diversify risks, combined with the risk aversion observed in most public and private decision-makers, implies that there is an additional benefit to preventive efforts coming from the reduced variability of future damages. If these monetized damages are expressed as a percentage of GDP, the marginal benefit of prevention can be estimated as the marginal expected increase in GDP, with some adjustments for the marginal reduction in the variance of damages.

Second, in most instances, objective probabilities are difficult to estimate. Furthermore, a number of climate change impacts involve health, biodiversity, and future generations, and the value of changes in these assets is difficult to capture fully in estimates of economic costs and benefits (see Section 2.4 on costs). Where we cannot measure risks and consequences precisely, we cannot simply maximize net benefits mechanically. This does not mean that we should abandon the usefulness of cost-benefit analysis, but it should be used as an input, among others in climate change policy decisions. The literature on how to account for ambiguity in the total economic value is growing, even if there is no agreed standard.

Finally, Gollier (2001) suggests that a sophisticated interpretation of the Precautionary Principle is compatible with economic principles in general, and with cost-benefit analyses in particular. The timing of the decision process and the resolution of the uncertainty should be taken into account, in particular when waiting before implementing a preventive action as an option. Waiting, and thereby late reactions, yield a cost when risks happen to be worse than initially expected, but yield an option value and cost savings in cases where risks happen to be smaller than expected. Standard dynamic programming methods can be used to estimate these option values.

2.4   Cost and benefit concepts, including private and social cost perspectives and relationships to other decision-making frameworks

2.4.1   Definitions

Mitigation costs can be measured at project, technology, sector, and macro-economic levels, and various geographical boundaries can be applied to the costing studies (see a definition of geographical boundaries in Section 2.8).

The project, technology, sector, and macro-economic levels can be defined as follows:

- **Project:** A project-level analysis considers a 'stand-alone' activity that is assumed not to have significant indirect economic impacts on markets and prices (both demand and supply) beyond the activity itself. The activity can be the implementation of specific technical facilities, infrastructure, demand-side regulations, information efforts, technical standards, etc. Methodological frameworks to assess the project-level impacts include cost-benefit analysis, cost-effectiveness analysis, and lifecycle analysis.

- **Technology:** A technology-level analysis considers a specific GHG mitigation technology, usually with several applications in different projects and sectors. The literature on technologies covers their technical characteristics, especially evidence on learning curves as the technology diffuses and matures. The technology analysis can use analytical approaches that are similar to project-level analysis.

- **Sector:** Sector-level analysis considers sectoral policies in a 'partial-equilibrium' context, for which other sectors and macro-economic variables are assumed to be given. The policies can include economic instruments related to prices, taxes, trade, and financing, specific large-scale investment projects, and demand-side regulation efforts. Methodological frameworks for sectoral assessments include various partial equilibrium models and technical simulation models for the energy sector, agriculture, forestry, and the transportation sector.

- **Macro-economic:** A macro-economic analysis considers the impacts of policies across all sectors and markets. The policies include all sorts of economic policies, such as taxes, subsidies, monetary policies, specific investment programmes, and technology and innovation policies. Methodological frameworks include various macro-economic models, such as general equilibrium models, Keynesian econometric models, and Integrated Assessment Models (IAMs), among others.

In comparing project, technology, sector, and macro-economic cost estimates it is important to bear in mind that cost estimates based on applying taxes in a macro-economic model are not comparable with abatement costs calculated at other assessment levels. This, for example, is because a carbon tax will apply to all GHG emissions, while abatement costs at project, technology or sector level will only reflect the costs of emission reductions.

Private and social costs: Costs can be measured from a private as well as from a social perspective. Individual decision-makers (including both private companies and households) are influenced by various cost elements, such as the costs of input

to a production process, labour and land costs, financial interest rates, equipment costs, fuel costs, consumer prices etc., which are key private cost components. However, the activities of individuals may also cause externalities, for example emissions that influence the utility of other individuals, but which are not taken into consideration by the individuals causing them. A social cost perspective includes the value of these externalities.

External costs: These typically arise when markets fail to provide a link between the person who creates the 'externality' and the person who is affected by it, or more generally when property rights for the relevant resources are not well defined.[7] In the case of GHG emissions, those who will eventually suffer from the impacts of climate change do not have a well-defined 'property right' in terms of a given climate or an atmosphere with given GHG concentrations, so market forces and/or bargaining arrangements cannot work directly as a means to balance the costs and benefits of GHG emissions and climate change. However, the failure to take into account external costs, in cases like climate change, may be due not only to the lack of property rights, but also the lack of full information and non-zero transaction costs related to policy implementation.

Private, financial, and social costs are estimated on the basis of different prices. The private cost component is generally based on market prices that face individuals. Thus, if a project involves an investment of US$ 5 million, as estimated by the inputs of land, materials, labour and equipment, that figure is used as the private cost. That may not be the full cost, however, as far as the estimation of social cost is concerned, because markets can be distorted by regulations and other policies as well as by limited competition that prevent prices from reflecting real resource scarcities. If, for example, the labour input is being paid more than its value in alternative employment, the private cost is higher than the social cost. Conversely, if market prices of polluting fuels do not include values that reflect the environmental costs, these prices will be lower than the social cost. Social costs should be based on market prices, but with eventual adjustments of these with shadow prices, to bring them into line with opportunity costs.

In conclusion, the key cost concepts are defined as follows:
- Private costs are the costs facing individual decision-makers based on actual market prices.
- Social costs are the private costs plus the costs of externalities. The prices are derived from market prices, where opportunity costs are taken into account.

Other cost concepts that are commonly used in the literature are 'financial costs' and 'economic costs'. Financial costs, in line with private costs, are derived on the basis of market prices that face individuals. Financial costs are typically used to assess

---

7   Coase, 1960, page 2 in his essay on The Problem of Social Cost, noted that externality problems would be solved in a 'completely satisfactory manner: when the damaging business has to pay for all damage caused and the pricing system works smoothly' (strictly speaking, this means that the operation of a pricing system is without cost).

the costs of financing specific investment projects. Economic costs, like social costs, assess the costs based on market prices adjusted with opportunity costs. Different from social costs, by definition they do not take all externalities into account.

2.4.2   Major cost determinants

A number of factors are critically important when determining costs, and it is important to understand their character and role when comparing mitigation costs across different studies, as occurs in Chapters 3-11 of this report, which compares costs across different models and which are based on different approaches.

The critical cost factors are based on different theoretical and methodological paradigms, as well as on specific applications of approaches. This section considers a number of factors including discounting, market efficiency assumptions, the treatment of externalities, valuation issues and techniques related to climate change damages[8] and other policy impacts, as well as implementation and transactions costs, and gives guidance on how to understand and assess these aspects within the context of climate change mitigation costing studies. For a more in-depth review of these issues see IPCC, 2001, Chapters 7 and 8.

2.4.2.1   Discount rates

Climate change impacts and mitigation policies have long-term characters, and cost analysis of climate change policies therefore involve a comparison of economic flows that occur at different points in time. The choice of discount rate has a very big influence on the result of any climate change cost analysis.

The debate on discount rates is a long-standing one. As the SAR (Second Assessment Report) notes (IPCC, 1996, Chapter 4), there are two approaches to discounting: a prescriptive approach[9] based on what rates of discount should be applied, and a descriptive approach based on what rates of discount people (savers as well as investors) actually apply in their day-to-day decisions. Investing in a project where the return is less than the standard interest rate makes the investor poorer. This descriptive approach based on a simple arbitrage argument justifies using the after-tax interest rate as the discount rate. The SAR notes that the former leads to relatively low rates of discount (around 2-3% in real terms) and the latter to relatively higher rates (at least 4% after tax and, in some cases, very much higher rates). The importance of choosing different levels of discount rates can be seen, for example when considering the value of US$ 1 million in 100 years from now. The present value of this amount is around US$ 52,000 if a 3% discount rate is used, but only around US$ 3,000 if a discount rate of 6% is used.

The prescriptive approach applies to the so-called social discount rate, which is the sum of the rate of pure time-preference and the rate of increased welfare derived from higher per-capita incomes in the future. The social discount rate can thus be described by two parameters: a rate of pure preference for the present (or rate of impatience, see Loewenstein and Prelec (1992)) δ, and a factor γ that reflects the elasticity of marginal utility to changes in consumption. The socially efficient discount rate r is linked to the rate of growth of GDP per capita, g in the following formula:[10]

$$r = \delta + \gamma g$$

Intuitively, as suggested by this formula, a larger growth in the economy should induce us to make less effort for the future. This is achieved by raising the discount rate. In an inter-generational framework, the parameter δ characterizes our ethical attitude towards future generations. Using this formula, the SAR recommended using a discount rate of 2-4%. It is fair to consider $\delta = 0$ and a growth rate of GDP per capita of 1-2% per year for developed countries and a higher rate for developing countries that anticipate larger growth rates.

Portney and Weyant (1999) provide a good overview of the literature on the issue of inter-generational equity and discounting.

The descriptive approach takes into consideration the market rate of return to safe investments, whereby funds can be conceptually invested in risk-free projects that earn such returns, with the proceeds being used to increase the consumption for future generations. A simple arbitrage argument to recommend the use of a real risk-free rate, such as the discount rate, is proposed.

The descriptive approach relies on the assumption that credit markets are efficient, so that the equilibrium interest rate reflects both the rate of return of capital and the householders' willingness to improve their future. The international literature includes several studies that recommend different discount rates in accordance with this principle. One of them is Dimson *et al.*, 2000, that assesses the average real risk-free rate in developed countries to have been below 2% per year over the 20th century, and on this basis, suggests the use of a low discount rate. This rate is not incompatible with the much larger rates of return requested by shareholders on financial markets (which can be

---

8   Despite the fact that this report focuses on mitigation policies, many economic studies are structured as an integrated assessment of the costs of climate change mitigation and the benefits of avoided damages, and some of the issues related to valuation of climate change damages are therefore an integral part of mitigation studies and are briefly discussed as such in this chapter.
9   The prescriptive approach has often been termed the 'ethical approach' in the literature.
10   This formula is commonly known as the Ramsey rule.

as high as 10–15%), because these rates include a premium to compensate for risk. However, the descriptive approach has several drawbacks. First, it relies on the assumption of efficient financial markets, which is not a credible assumption, both as a result of market frictions and the inability of future generations to participate in financial markets over these time horizons. Second, financial markets do not offer liquid riskless assets for time horizons exceeding 30 years, which implies that the interest rates for most maturities relevant for the climate change problem cannot be observed.

Lowering the discount rate, as in the precriptive approach, increases the weight of future generations in cost-benefit analyses. However, it is not clear that it is necessarily more ethical to use a low (or lower) discount rate on the notion that it protects future generations, because that could also deprive current generations from fixing urgent problems in order to benefit future generations who are more likely to have more resources available.

For discounting over very long time horizons (e.g. periods beyond 30 years), an emerging literature suggests that the discount rate should decrease over time. Different theoretical positions advocate for such an approach based on arguments concerning the uncertainty of future discount rates and economic growth, future fairness and intra-generational distribution, and on observed individual choices of discount rates (Oxera, 2002). The different theoretical arguments lead to different recommendations about the level of discount rates.

Weitzman (2001) showed that if there is some uncertainty on the future return to capital, and if society is risk-neutral, the year-to-year discount rate should fall progressively to its smallest possible value. Newell and Pizer (2004) arrived at a similar conclusion. It is important to observe that this declining rate comes on top of the variable short-term discount rate, which should be frequently adapted to the conditions of the market interest rate.

It is also important to link the long-term macro-economic uncertainty with the uncertainty concerning the future benefits of our current preventive investments. Obviously, it is efficient to bias our efforts towards investments that perform particularly well in the worse states (i.e., states in which the economy collapses). The standard approach to tackle this is to add a risk premium to the benefits of these investments rather than to modify the discount rate, which should remain a universal exchange rate between current and future sure consumption, for the sake of comparability and transparency of the cost-benefit analysis. Using standard financial price modelling, this risk premium is proportional to the covariance between the future benefit and the future GDP.

Whereas it seems reasonable in the above formula to use a rate of growth of GDP per capita of g=1-2% for the next decade, there is much more uncertainty about which growth rate to use for longer time horizons. It is intuitive that, in the long run, the existence of an uncertain growth should reduce the discount rates for these distant time horizons. Calibrating a normative model on this idea, Gollier (2002a, 2002b, 2004) recommended using a decreasing term structure of discount rate, from 5% in the short term to 2% in the long term. In an equivalent model, but with different assumptions on the growth process, Weitzman (1998, 2004) proposed using a zero discount rate for time horizons around 50 years, with the discount rate being negative for longer time horizons. These models are in line with the important literature on the term structure of interest rates, as initiated by Vasicek (1977) and Cox, Ingersoll and Ross (1985). The main difference is the time horizon under scrutiny, with a longer horizon allowing considerable more general specifications for the stochastic process that drives the shape of the yield curve.

Despite theoretical disputes about the use of time-declining discount rates, the UK government has officially recommended such rates for official approval of projects with long-term impacts. The recommendation here is to use a 3.5% rate for 1-30 years, a 3% rate for 31-75 years, a 2.5% rate for 76-125 years, a 2% rate for 125-200 years, 1.5% for 201-300 years, and 1% for longer periods (Oxera, 2002). Similarly, France decided in 2004 to replace its constant discount rate of 8% with a 4% discount rate for maturities below 30 years, and a discount rate that decreases to 2% for longer maturities.[11] Finally, the US government's Office of Management and Budget recognizes the possibility of declining rates (see appendix D of US, 2003).

It is important to remember that these rates discount certainty-equivalent cash flows. This discussion does not solve the question of how to compute certainty equivalents when the project's cash flows are uncertain. For climate change impacts, the assumed long-term nature of the problem is the key issue here. The benefits of reduced GHG emissions vary according to the time of emissions reduction, with the atmospheric GHG concentration at the reduction time, and with the total GHG concentrations more than 100 years after the emissions reduction. Because these benefits are only probabilistic, the standard cost-benefit analysis can be adjusted with a transformation of the random benefit into its certainty equivalent for each maturity. In a second step, the flow of certainty-equivalent cash flows is discounted at the rates recommended above.

For mitigation effects with a shorter time horizon, a country must base its decisions (at least partly) on discount rates that reflect the opportunity cost of capital. In developed countries, rates of around 4–6% are probably justified. Rates of this level

---

11 This should be interpreted as using a discount factor equaling $(1.04)^{-t}$ if the time horizon t is less than 30 years, and a discount rate equaling $(1.04)^{-30}(1.02)^{-(t-30)}$ if t is more than 30 years.

are in fact used for the appraisal of public sector projects in the European Union (EU) (Watts, 1999). In developing countries, the rate could be as high as 10–12%. The international banks use these rates, for example, in appraising investment projects in developing countries. It is more of a challenge, therefore, to argue that climate change mitigation projects should face different rates, unless the mitigation project is of very long duration. These rates do not reflect private rates of return and the discount rates that are used by many private companies, which typically need to be considerably higher to justify investments, and are potentially between 10% and 25%.

2.4.2.2 Market efficiency

The costs of climate change mitigation policies depend on the efficiency of markets, and market assumptions are important in relation to baseline cases, to policy cases, as well as in relation to the actual cost of implementing policy options. For example, the electricity market (and thereby the price of electricity that private consumers and industry face) has direct implications on the efficiency (and thereby GHG emissions) related to appliances and equipment in use.

In practice, markets and public-sector activities will always exhibit a number of distortions and imperfections, such as lack of information, distorted price signals, lack of competition, and/or institutional failures related to regulation, inadequate delineation of property rights, distortion-inducing fiscal systems, and limited financial markets. Proper mitigation cost analysis should take these imperfections into consideration and assess implementation costs that include these imperfections (see Section 2.4.2.3 for a definition of implementation costs).

Many project level and sectoral mitigation costing studies have identified a potential for GHG reduction options with a negative cost, implying that the benefits, including co-benefits, of implementing these options are greater than the costs. Such negative cost options are commonly referred to as 'no-regret options'.[12]

The costs and benefits included in the assessment of no-regret options, in principle, are all impacts of the options including externalities. External impacts can relate to environmental side-impacts, and distortions in markets for labour, land, energy resources, and various other areas. A presumption for the existence of no-regret options is that there are:
- **Market imperfections** that generate efficiency losses. Reducing the existing market or institutional failures and other barriers that impede adoption of cost-effective emission reduction measures, can lower private costs compared to current practice (Larson *et al.*, 2003; Harris

et al., 2000; Vine *et al.*, 2003). This can also reduce private costs overall.
- **Co-benefits**: Climate change mitigation measures will have effects on other societal issues. For example, reducing carbon emissions will often result in the simultaneous reduction in local and regional air pollution (Dessues and O'Connor, 2003; Dudek *et al.*, 2003; Markandya and Rubbelke, 2004; Gielen and Chen, 2001; O'Connor *et al.*, 2003). It is likely that mitigation strategies will also affect transportation, agriculture, land-use practices and waste management and will have an impact on other issues of social concern, such as employment, and energy security. However, not all of these effects will be positive; careful policy selection and design can better ensure positive effects and minimize negative impacts. In some cases, the magnitude of co-benefits of mitigation may be comparable to the costs of the mitigating measures, adding to the no-regrets potential, although estimates are difficult to make and vary widely.[13]
- **Double dividend**: Instruments (such as taxes or auctioned permits) provide revenues to the government. If used to finance reductions in existing distortionary taxes ('revenue recycling'), these revenues reduce the economic cost of achieving greenhouse gas reductions. The magnitude of this offset depends on the existing tax structure, type of tax cuts, labour market conditions, and method of recycling (Bay and Upmann, 2004; Chiroleu-Assouline and Fodha, 2005; Murray, *et al.*, 2005). Under some circumstances, it is possible that the economic benefits may exceed the costs of mitigation. Contrary, it has also been argued that eventual tax distortions should be eliminated anyway, and that the benefits of reducing these therefore cannot be assigned as a benefit of GHG emission reduction policies.

The existence of market imperfections, or co-benefits, and double dividends that are not integrated into markets are also key factors explaining why no-regret actions are not taken. The no-regret concept has, in practice, been used differently in costing studies, and has usually not included all the external costs and implementation costs associated with a given policy strategy.[14]

2.4.2.3 Transaction and implementation costs

In practice, the implementation of climate change mitigation policies requires some transaction and implementation costs. The implementation costs relate to the efforts needed to change existing rules and regulations, capacity-building efforts, information, training and education, and other institutional efforts needed to put a policy into place. Assuming that these implementation requirements are in place, there might still be costs involved in carrying through a given transaction,

---

12 By convention, when assessing the costs of GHG emission reductions, the benefits do not include the impacts associated with avoided climate change damages.
13 It should be recognised that, under a variety of circumstances, it may be more efficient to obtain air pollution reductions through controls targeted at such pollutants rather than coupling them with efforts to reduce GHG emissions, even if the latter results in some air pollution reductions.
14 This is due to difficulties in assessing all external costs and implementation costs, and reflects the incompleteness of the elements that have been addressed in the studies.

for example related to legal requirements of verifying and certifying emission reduction, as in the case of CDM projects. These costs are termed 'transaction costs'. The transaction costs can therefore be defined as the costs of undertaking a business activity or implementing a climate mitigation policy, given that appropriate implementation efforts have been (or are being) created to establish a benign market environment for this activity.

Implementation policies and related costs include various elements related to market creation and broader institutional policies. In principle, mitigation studies (where possible) should include a full assessment of the cost of implementation requirements such as market reforms, information, establishment of legal systems, tax and subsidy reforms, and institutional and human capacity efforts.

In practice, few studies have included a full representation of implementation costs. This is because the analytical approaches applied cannot address all relevant implementation aspects, and because the actual costs of implementing a policy can be difficult to assess *ex ante*. However, as part of the implementation of the emission reduction requirements of the Kyoto Protocol, many countries have gained new experiences in the effectiveness of implementation efforts, which can provide a basis for further improvements of implementation costs analysis.

2.4.2.4   *Issues related to the valuation of non-market aspects*

A basic problem in climate change studies is that a number of social impacts are involved that go beyond the scope of what is reflected in current market prices. These include impacts on human health, nature conservation, biodiversity, natural and historical heritage, as well as potential abrupt changes to ecosystems. Furthermore, complicated valuation issues arise in relation to both market- and non-market areas, since climate change policies involve impacts over very long time horizons, where future generations are affected, as well as intra-generational issues, where relatively wealthy and relatively poor countries face different costs and benefits of climate change impacts, adaptation and mitigation policies. Valuation of climate change policy outcomes therefore also involves assigning values to the welfare of different generations and to individuals and societies living at very different welfare levels today.

The valuation of inter-generational climate change policy impacts involves issues related to comparing impacts occurring at different points in time as discussed in Section 2.4.2.1 on discount rates, as well as issues in relation to uncertainty about the preferences of future generations. Since these preferences are unknown today many studies assume, in a simplified way, that consumer preferences will stay unchanged over time. An overview of some of the literature on the preferences of future generations is given by Dasgupta *et al.,* (1999).

Other limitations in the valuation of climate change policy impacts are related to specific practical and ethical aspects of valuing human lives and injuries. A number of techniques can be used to value impacts on human health – the costs of mortality, for example, can be measured in relation to the statistical values of life, the avoided costs of health care, or in relation to the value of human capital on the labour market. Applications of valuation techniques that involve estimating the statistical values of life will face difficulties in determining values that reflect people in a fair and meaningful way, even with very different income levels around the world. There are obviously a lot of ethical controversies involved in valuing human health impacts. In the Third Assessment Report the IPCC recognized these difficulties and recommended that studies that include monetary values of statistical values of life should use uniform average global per-capita income weights in order to treat all human beings as equal (IPCC, 2001, Chapter 7).

2.4.3   Mitigation potentials and related costs

Chapters 3-11 report the costs of climate change mitigation at global, regional, sectoral, and technology level and, in order to ensure consistency and transparency across the cost estimates reported in these chapters, it has been agreed to use a number of key concepts and definitions that are outlined in this section. Furthermore, the following paragraphs also outline how the concepts relate to mitigation cost concepts that have been used in previous IPCC reports, in order to allow different cost estimates to be compared and eventual differences to be understood.

A commonly used output format for climate change mitigation cost studies means reporting the GHG emission reduction in quantitative terms that can be achieved at a given cost. The potential terminology is often used in a very 'loose' way, which makes it difficult to compare numbers across studies. The aim of the following is to overcome such lack of transparency in cost results based on a definition of major cost and GHG emission reduction variables to be used when estimating potentials.

The term 'potential' is used to report the quantity of GHG mitigation compared with a baseline or reference case that can be achieved by a mitigation option with a given cost (per tonne) of carbon avoided over a given period. The measure is usually expressed as million tonnes carbon- or CO_2-equivalent of avoided emissions, compared with baseline emissions. The given cost per tonne (or 'unit cost') is usually within a range of monetary values at a particular location (e.g. for wind-generated electricity), such as costs less than x US$ per tonne of CO_2- or carbon-equivalent reduction (US$/tC-eq). The monetary values can be defined as private or social unit costs: private unit costs are based on market prices, while social unit costs reflect market prices, but also take externalities associated with the mitigation into consideration. The prices are real prices adjusted for inflation rates.

2.4.3.1 Definitions of barriers, opportunities and potentials

The terms used in this assessment are those used in the Third Assessment Report (TAR). However, the precise definitions are revised and explanations for the revisions are given in the footnotes.

A 'barrier' to mitigation potential is any obstacle to reaching a potential that can be overcome by policies and measures. (From this point onwards, 'policies' will be assumed to include policies, measures, programmes and portfolios of policies.) An 'opportunity' is the application of technologies or policies[15] to reduce costs and barriers, find new potentials and increase existing ones. Potentials, barriers and opportunities all tend to be context-specific and vary across localities and over time.

'Market potential' indicates the amount of GHG mitigation that might be expected to occur under forecast market conditions, including policies and measures in place at the time.[16] It is based on private unit costs and discount rates, as they appear in the base year and as they are expected to change in the absence of any additional policies and measures. In other words, as in the TAR, market potential is the conventional assessment of the mitigation potential at current market price, with all barriers, hidden costs, etc. in place. The baseline is usually historical emissions or model projections, assuming zero social cost of carbon and no additional mitigation policies. However, if action is taken to improve the functioning of the markets, to reduce barriers and create opportunities (e.g. policies of market transformation to raise standards of energy efficiency via labelling), then mitigation potentials will become higher.

In order to bring in social costs, and to show clearly that this potential includes both market and non-market costs, 'economic potential' is defined as the potential for cost-effective GHG mitigation when non-market social costs and benefits are included with market costs and benefits in assessing the options[17] for particular levels of carbon prices in US$/tCO$_2$ and US$/tC-eq. (as affected by mitigation policies) and when using social discount rates instead of private ones. This includes externalities (i.e. non-market costs and benefits such as environmental co-benefits). Note that estimates of economic potential do not normally assume that the underlying structure of consumer preferences has changed. This is the proper theoretical definition of the economic potential, however, as used in most studies, it is the amount of GHG mitigation that is cost-effective for a given carbon price, based on social cost pricing and discount rates (including energy savings but without most externalities), and this is also the case for the studies that were reported in the TAR (IPCC, 2001, Chapters 3, 8 and 9).

There is also a technical potential and a physical potential that, by definition, are not dependent on policies.

The 'technical potential' is the amount by which it is possible to reduce greenhouse gas emissions or improve energy efficiency by implementing a technology or practice that has already been demonstrated. There is no specific reference to costs here, only to 'practical constraints', although in some cases implicit economic considerations are taken into account. Finally the 'physical potential' is the theoretical (thermodynamic) and sometimes, in practice, rather uncertain upper limit to mitigation, which also relies on the development of new technologies.

A number of key assumptions are used to calculate potentials. Some of the major ones are related to:
- Transformation of economic flows to net present values (NVP) or levelised costs. It is consistent here to use the financial rate of return in the discounting of private costs, and a social discount rate in social cost calculations
- Treatment of GHG emission reductions that occur at different points in time. Some studies add quantitative units of GHG reductions over the lifetime of the policy, and others apply discount rates to arrive at net present values of carbon reductions.

The implementation of climate change mitigation policies will involve the use of various economic instruments, information efforts, technical standards, and other policies and measures. Such policy efforts will all have impacts on consumer preferences and taste as well as on technological innovations. The policy efforts (in the short term) can be considered as an implementation cost, and can also be considered as such in the longer term, if transactions costs of policies are successfully reduced, implying that market and social- and economic potentials are increased at a given unit cost.

---

15 Including behaviour and lifestyle changes.
16 The TAR (IPCC, 2001), p. 352 defines market potential as 'the amount of GHG mitigation that might be expected to occur under forecast market conditions, with no changes in policy or implementation of measures whose primary purpose is the mitigation of GHGs'. This definition might be interpreted to imply that market potential includes no implementation of GHG policies. However many European countries have already implemented mitigation policies. It is a substantial research exercise in counterfactual analysis to untangle the effects of past mitigation policies in the current levels of prices and costs and hence mitigation potential. The proposed definition simply clarifies this point.
17 IPCC (2001), Chapter 5 defines 'economic potential' as 'the level of GHG mitigation that could be achieved if all technologies that are cost-effective from the consumers' point of view were implemented' (p. 352). This definition therefore introduces the concept of the consumer as distinct from the market. This is deeply confusing because it loses the connection with market valuations without explanation. Who is to decide how the consumers' point of view is different from the market valuation of costs? On what basis are they to choose these costs? The definition also does not explicitly introduce the social cost of carbon and other non-market valuations necessary to account for externalities and missing markets and it is not readily comparable with the IPCC (2001), Chapter 3 definition of economic potentials. The proposed definition for this report applies to the large body of relevant literature that assesses mitigation potential at different values of the social cost of carbon, and clearly introduces non-market valuations for externalities and time preferences. The proposed definition also matches that actually used in IPCC (2001) Chapter 3, where such potentials are discussed 'at zero social cost' (e.g. p. 203).

2.5 Mitigation, vulnerability and adaptation relationships

2.5.1 Integrating mitigation and adaptation in a development context – adaptive and mitigative capacities

The TAR (IPCC, 2001) introduced a new set of discussions about the institutional and developmental context of climate change mitigation and adaptation policies. One of the conclusions from that discussion was that the capacity for implementing specific mitigation and adaptation policies depends on man-made and natural capital and on institutions. Broadly speaking, institutions should be understood here as including markets and other information-sharing mechanisms, legal frameworks, as well as formal and informal networks.

Subsequent work by Adger (2001a) further emphasizes the role of social capital in adaptation. Adger refers to a definition by Woolcock and Narayan (2000, p. 226), which states that social capital is made up of 'the norms and networks that enable people to act collectively'. According to Adger there are two different views within the main areas of the international literature that are important to climate change issues namely: 1) whether social capital only exists outside the state, and 2) whether social capital is a cause, or simply a symptom, of a progressive and perhaps flexible and adaptive society. The first issue relates to how important planned adaptation and government initiatives can be, and the second considers the macro-level functioning of society and the implications for adaptive capacity.

Adger observes that the role that social capital, networks and state-civil society linkages play in adaptive capacity can be observed in historical and present-day contexts by examining the institutions of resource management and collective action in climate-sensitive sectors and social groups, highlighting a number of such experiences in adaptation to climate change. The examples include an assessment of the importance of social contacts and socio-economic status in relation to excess mortality due to extreme heating, coastal defence in the UK, and coastal protection in Vietnam, where the adaptive capacity in different areas is assessed within the context of resource availability and the entitlements of individuals and groups (Kelly and Adger, 1999). A literature assessment (IPCC, 2007b, Chapter 20) includes a wider range of examples of historical studies of development patterns, thus confirming that social capital has played a key role in economic growth and stability.

IPCC (2001), Chapter 1 initiated a very preliminary discussion about the concept of *mitigative capacity*. Mitigative capacity (in this context) is seen as a critical component of a country's ability to respond to the mitigation challenge, and the capacity, as in the case of adaptation, largely reflects man-made and natural capital and institutions. It is concluded that development, equity and sustainability objectives, as well as

past and future development trajectories, play critical roles in determining the capacity for specific mitigation options. Following that, it can be expected that policies designed to pursue development, equity and/or sustainability objectives might be very benign framework conditions for implementing cost-effective climate change mitigation policies. The final conclusion is that, due to the inherent uncertainties involved in climate change policies, enhancing mitigative capacity can be a policy objective in itself.

It is important to recognize here that the institutional aspects of the adaptive and mitigative capacities refer to a number of elements that have a 'public-good character' as well as general social resources. These elements will be common framework conditions for implementing a broad range of policies, including climate change and more general development issues. This means that the basis for a nation's policy-implementing capacity exhibits many similarities across different sectors, and that capacity-enhancing efforts in this area will have many joint benefits.

There may be major differences in the character of the adaptive and mitigative capacity in relation to sectoral focus and to the range of technical options and policy instruments that apply to adaptation and mitigation respectively. Furthermore, assessing the efficiency and implementability of specific policy options depends on local institutions, including markets and human and social capital, where it can be expected that some main strengths and weaknesses will be similar for different sectors of an economy.

As previously mentioned, the responses to climate change depend on the adaptive and mitigative capacities and on the specific mitigation and adaptation policies adopted. Policies that enhance adaptive and mitigative capacities can include a wide range of general development policies, such as market reforms, education and training, improving governance, health services, infrastructure investments etc.

The actual outcome of implementing specific mitigation and adaptation policies is influenced by the adaptive and mitigative capacity, and the outcome of adaptation and mitigation policies also depends on a number of key characteristics of the socio-economic system, such as economic growth patterns, technology, population, governance, and environmental policies.

It is expected that there may be numerous synergies and tradeoffs between the adaptive and mitigative capacity elements of the socio-economic and natural systems, as well as between specific adaptation and mitigation policies. Building more motorways, for example, can generate more traffic and more GHG emissions. However, the motorways can also improve market access, make agriculture less vulnerable to climate change, help in evacuation prior to big storms, and can support general economic growth (and thereby investments in new efficient production technologies). Similarly, increased fertiliser use in agriculture can increase productivity and reduce climate change vulnerability, but it can also influence the potential for carbon sequestration and can increase GHG emissions.

2.5.2 Mitigation, adaptation and climate change impacts

The discussion on mitigation and adaptation policy portfolios has a global as well as a national/regional dimension. It should be recognized that mitigation and adaptation are very different regarding time frame and distribution of benefits. Dang et al. (2003, Table 1) highlights a number of important commonalities and differences between mitigation and adaptation policies. Both policy areas can be related to sustainable development goals, but differ according to the direct benefits that are global and long term for mitigation, while being local and shorter term for adaptation. Furthermore adaptation can be both reactive (to experienced climate change) and proactive, while mitigation can only be proactive in relation to benefits from avoided climate change occurring over centuries. Dang et al. (2003, Table 4) also points out that there can be conflicts between adaptation and mitigation in relation to the implementation of specific national policy options. For example, installing air-conditioning systems in buildings is an adaptation option, but energy requirements can increase GHG emissions, and thus climate change.

In relation to the trade-off between mitigation and adaptation, Schneider (2004) points out that when long-term integrated assessment studies are used to assess the net benefits of avoided climate change (including adaptation options) versus the costs of GHG emission reduction measures, the full range of possible climate outcomes, including impacts that remain highly uncertain such as surprises and other climate irreversibility, should be included. Without taking these uncertain events into consideration, decision-makers will tend to be more willing to accept prospective future risks rather than attempt to avoid them through abatement. It is worth noting here that,when faced with the risk of a major damage, human beings may make their judgment based on the consequences of the damage rather than on probabilities of events. Schneider concludes that it is not clear that climate surprises have a low probability, they are just very uncertain at present, and he suggests taking these uncertainties into consideration in integrated assessment models, by adjusting the climate change damage estimates. The adjustments suggested include using historical data for estimating the losses of extreme events, valuing ecosystem services, subjective probability assessments of monetary damage estimates, and the use of a discount rate that decreases over time in order to give high values to future generations.

In this way the issues of jointly targeting mitigation and adaptation has an element of decision-making under uncertainty, due to the complexity of the environmental and human systems and their interactions. Kuntz-Duriseti (2004) suggests dealing with this uncertainty by combining economic analysis and precautionary principles, including an insurance premium system, hedging strategies, and inclusion of low-probability events in risk assessments.

A common approach of many regional and national developing country studies on mitigation and adaptation policies has been to focus on the assessment of context-specific vulnerabilities to climate change. Given this, a number of studies and national capacity-building efforts have considered how adaptation and mitigation policies can be integrated into national development and environmental policies, and how they can be supported by financial transfers, domestic funds, and linked to foreign direct investments (IINC, 2004; CINC, 2004). The Danish Climate and Development Action Program aims at a two-leg strategy, where climate impacts, vulnerabilities, and adaptation are assessed as an integral part of development plans and actions in Danish partner countries, and where GHG emission impacts and mitigation options are considered as part of policy implementation (Danida, 2005).

Burton et al. (2002) suggest that research on adaptation should focus on assessing the social and economic determinants of vulnerability in a development context. The focus of the vulnerability assessment according to this framework should be on short-term impacts, i.e. should try to assess recent and future climate variability and extremes, economic and non-economic damages and the distribution of these. Based on this, adaptation policies should be addressed as a coping strategy against vulnerability and potential barriers, obstacles, and the role of various stakeholders and the public sector should be considered. Kelly and Adger (2000) developed an approach for assessing vulnerabilities and concluded that the vulnerability and security of any group is determined by resource availability and entitlements. The approach is applied to impacts from tropical storms in coastal areas in Vietnam.

On a global scale, there is a growing recognition of the significant role that developing countries play in determining the success of global climate change policies, including mitigation and adaptation policy options (Müller, 2002). Many governments of developing countries have started to realize that they should no longer discuss whether to implement any measures against climate change, but how drastic these measures should be, and how climate policies can be an integral part of national sustainable development paths (SAINC, 2003; IINC, 2004; BINC, 2004; CINC, 2004; MOST, 2004).

2.6 Distributional and equity aspects

This section discusses how different equity concepts can be applied to the evaluation of climate change policies and provides examples on how the climate literature has addressed equity issues. See also Chapter 20 in IPCC (2007b), and Chapters 12 and 13 of this report for additional discussions on the equity

Table 2.3: *Measures of Inter-country Equity*

	GNI *Per Capita* US$		Life Expectancy (LE) Years		Literacy (ILL) %	
	Average	C.Var	Average	C.Var	Average	C.Var
1980/90	3,764	4,915	61.2	0.18	72.5	25.3
2001	7,350	10,217	65.1	0.21	79.2	21.4
% Change Average		95%		6%		9%
% Change Co. Var.		6%		14%		-22%

Notes: Literacy rates are for 1990 and 2001. GNI and LE data are for 1980, 1990, and 2001. Ninety-nine countries are included in the sample. Coefficient of variation is the standard deviation of a series divided by the mean. The standard deviation is given by the formula:

$$s = \sqrt{\sum_{i=1}^{i=n}(x_i - \bar{x})^2 \Big/ (n-1)}$$
Where '*x*' refers to the value of a particular observation, '$\bar{x}$' is the mean of the sample and '*n*' is the number of observations.

Source: WB, 2005 (World Development Indicators)

dimensions of sustainable development and climate change policies.

2.6.1 Development opportunities and equity

Traditionally, success in development has been measured in economic terms – increase in Gross National Income (GNI) *per capita* remains the most common measure[18]. Likewise, income distribution has been one of the key components in equity, both within and between countries, and has been measured in terms of inequalities of income, through measures such as the 'GINI' coefficient.[19] [20] Although a great deal has been written in recent years on the components of well-being, the development literature has been slow to adopt a broader set of indicators of this concept, especially as far as equity in well-being is concerned, despite the fact that some authors have argued that absolute changes in income and other indicators of human well-being (e.g. education, mortality rates, water, sanitation etc.) are just as important as the distribution within these indicators (Maddison, 2003; Goklany, 2001).

Probably the most important and forceful critic of the traditional indicators has been Sen (1992, 1999). Sen's vision of development encompasses not only economic goods and services but also individuals' health and life expectancy, their education and access to public goods, the economic and social security that they enjoy, and their freedom to participate freely in economic interchange and social decision-making. While his criticism is widely acknowledged as addressing important shortcomings in the traditional literature, the ideas still have not been made fully operational. Sen speaks of 'substantive freedoms' and 'capabilities' rather than goods and services as the key goals of development and provides compelling examples of how his concepts can paint a different picture of progress in development compared to that of changes in GNI. It

remains the case, however, that actual indicators of equity still do not cover the breadth of components identified by Sen.

The UNDP Human Development Index (HDI) is an important attempt to widen the indicators of development, and initially included *per capita* national income, life expectancy at birth and the literacy rate. However, it is important to recognize that no single all-encompassing indicator can be constructed, will be understandable or useful to either policymakers or the public, so different indexes have to be used that reflect different issues and purposes.

Rather than synthesizing these three components into a single index, as the HDI has done, we can also look at changes in the inter-country equity of the individual components. Table 2.3[21] provides data for the period 1980–2001 for per capita national income (GNI) and life expectancy at birth (LE) and from 1990 to 2001 for the literacy rate (ILL). The increase in average GNI has been much faster over this period than those of life expectancy and literacy rates. The increase in coefficient of variations for GNI per capita (by 6%) and life expectancy (by 14%) therefore show an increase in dispersion over this period, indicating a wider disparity of these parameters across countries. However, literacy rates have become more equal, with a decline in the coefficient of variation by 22% (see Table 2.3). However, a study by Goklany (2002) concluded that inequality between countries does not necessarily translate into inequality between individuals.

As Sen notes, the problem of inequality becomes magnified when attention is shifted from income inequality to inequality of 'substantive freedoms and capabilities', as a result of a 'coupling' of the different dimensions – individuals who are likely to suffer from higher mortality and who are illiterate are also likely to have lower incomes and a lower ability to convert

---

18 The Gross National Income measures the income of all citizens, including income from abroad. GDP is different to GNI as it excludes income from abroad.

19 The GINI coefficient is a measurement standard for the total income that needs to be redistributed if all income was equally distributed. A 0 value means that all are equal, while a 1 value implies considerable inequality.

20 When income distribution is used in equity assessments it is important to recognize that such measures do not include all aspects of justice and equity.

21 Ideally one should use purchasing power (PPP) adjusted GNI, but data on GNI_{ppp} is much more limited for the earlier period. For LE and ILL we also looked at a larger dataset of 142 countries, and found very similar results.

incomes into capabilities and good standards of living. While this is certainly true at the individual level, at the country level the correlation appears to be declining.

This wider analysis of equity has important implications for sharing the costs of mitigation and for assessing the impacts of climate change (see Chapter 1 for a more detailed discussion of climate change impacts and the reference to the UNFCCC Article 2). As generally known, the impacts of climate change are distributed very unequally across the planet, hurting the vulnerable and poor countries of the tropics much more than the richer countries in the temperate regions. Moreover, these impacts do not work exclusively, or even mainly, through changes in real incomes. The well-being of future generations will be affected through the effects of climate change on health, economic insecurity and other factors. As far as the costs of actions to reduce GHGs are concerned, measures that may be the least costly in overall terms are often not the ones that are the most equitable – see Sections 2.6.4 and 2.6.5 for a further discussion of the links between mitigation policy and equity.

2.6.2 Uncertainty as a frame for distributional and equity aspects

Gollier, 2001 outlines a framework for assessing the equity implications of climate change uncertainty, where he considers risk aversion for different income groups. The proposition (generally supported by empirical evidence) is that the relative risk aversion of individuals decreases with increasing wealth (Gollier, 2001), implying that the compensation that an individual asks for in order to accept a risk decreases relative to his income with increasing income. However, the absolute risk aversion – or the total compensation required in order to accept a risk – increases with wealth. It means that a given absolute risk level is considered to be more important to poorer people than to richer, and the comparatively higher risk aversion of poorer people suggests that larger investments in climate change mitigation and adaptation policies are preferred if these risks are borne by the poor rather than the rich.

A similar argument can be applied in relation to the equity consequences of increased climate variability and extreme events. Climate change may increase the possibility of large, abrupt and unwelcome regional or global climatic events. A coping strategy against variability and extreme events can be income-smoothing measures, where individuals even out their income over time through savings and investments. Poorer people with a lower propensity to save, and with less access to credit makers, have smaller possibilities to cope with climate variability and extreme events through such income-smoothing measures, and they will therefore be more vulnerable.

2.6.3 Alternative approaches to social justice

Widening our understanding of equity does not provide us with a rule for ranking different outcomes, except to say that, other things being equal, a less inequitable outcome is preferable to a more inequitable one. But how should one measure outcomes in terms of equity and what do we do when other things are not equal?

The traditional economic approach to resource allocation has been based on utilitarianism, in which a policy is considered to be desirable if no other policy or action is feasible that yields a higher aggregate utility for society. This requires three underlying assumptions:
(a) All choices are judged in terms of their consequences, and not in terms of the actions they entail.
(b) These choices are valued in terms of the utility they generate to individuals and no attention is paid to the implications of the choices for aspects such as rights, duties etc.
(c) The individual utilities are added up to give the sum of utility for society as a whole.
In this way the social welfare evaluation relies on the assumption that there is a net social surplus if the winners can compensate the losers and still be better off themselves. It should be recognized here that philosophers dispute that efficiency is a form of equity.

This approach has been the backbone of welfare economics, including the use of cost-benefit analysis (CBA) as a tool for selecting between options. Under CBA all benefits are added up, as are the costs, and the net benefit – the difference between the benefits and costs – is calculated. The option with the highest net benefit is considered the most desirable.[22] If utilities were proportional to money benefits and 'disutilities' were proportional to money costs, this method would amount to choosing to maximize utilities. Since most economists accept that this proportionality does not hold, they extend the CBA by either (a) asking the decision-maker to take account of the distributional implications of the option as a separate factor, in addition to the calculated net benefit; or (b) weighting costs or benefits by a factor that reflects the relationship between utility and the income of the person receiving that cost or benefit. For details of these methods in the context of climate change, see Markandya and Halsnaes (2002b).[23]

An alternative approach to allocating resources, which is derived from an ethical perspective and has existed for at least as long as the utilitarian approach described above (which has its modern origins in the late 18th century by Jeremy Bentham), is based on the view that social actions are to be judged by whether or not they conform to a 'social contract' that defines the rights and duties of individuals in society. The view was

---

22 This is considerably simplified; ignoring the time dimension and market imperfections in valuing costs and benefits but the principle remains valid.
23 The ability of CBA to combine equity and utility through these means has been challenged by philosophers who argue that there could be serious ethical problems with combining the two when benefits and costs are as hugely disaggregated, as is the case with climate change. See Brown, 2002.

inspired by the work of Kant and Hegel and finds its greater articulation in the writing of Rousseau and the French 19th century philosophers.[24] In this position, for example, a society may predetermine that an individual has the right to be protected from serious negative health damage as a result of social actions. Hence no action, even if it increased utility, could be tolerated if it violated the rights and duties of individuals.

Modern philosophers who have developed the 'rights' view include Rawls, who argued that it is not utilities that matter but the distribution of 'primary goods, which include, in addition to income, "rights, liberties and opportunities and… the social basis of self respect"' (Rawls, 1971). Rawls argued further that social justice demanded that society be judged in terms of the level of well-being of its worst-off member. At the other end of the political spectrum, Nozick and the modern libertarians contend that personal liberties and property rights have (with very few exceptions) absolute precedence over objectives such as the reduction of poverty and deprivation (Nozick, 1974).

More recently, however, some ethical philosophers have found fault with both the 'modified' utilitarian view and the rights-based approach, on a number of grounds. Sen, for example, has argued that options should be judged not only in terms of their consequences, but also in terms of procedures. He advocates a focus on the capabilities of individuals to choose a life that one has reason to value. A person's capability refers to the alternative combinations of 'functionings', where functionings can be more popularly described as 'lifestyles' (Sen, 1999, pp. 74-75). What matters are not only the realized functionings, but also the capability set of alternatives, differently from a utilitarian-based approach that focuses only on the outcomes. In particular, the freedom to make the choices and engage in social and market transactions is worth something in its own right.

Sen criticizes the 'rights-based' equity approaches for not taking into consideration the fact that individuals are different and the actual consequences of giving them specific rights will vary between individuals, so rights should be seen in the context of capabilities. Both apply, because individuals have different preferences and thereby value primary inputs, for example, differently, and because their capability to use different rights also differ. Along these lines, Sen further argues that his capability-based approach can facilitate easier inter-personal comparisons than utilitarianism, since it does not suggest aggregating all individuals, but rather presenting information both on the capability sets available to individuals and their actual achievements.

What implications does this debate have in the context of climate change? One is that rights and capabilities need to be viewed in an international context. An example of an approach based on global equity would be to entitle every individual alive at a given date an equal per capita share in the intrinsic capacity of the earth to absorb GHGs. Countries whose total emissions exceeded this aggregate value would then compensate those below the value. In accordance with a utilitarian approach this compensation would be based on an estimate of the aggregate economic welfare lost by countries due to climate change, seen in relation to their own emissions. In contrast, the capability-based approach would argue for reduced capabilities associated with climate change.

As suggested above, societies do not (in practice) follow a strict utilitarian view of social justice and they do indeed recognize that citizens have certain basic rights in terms of housing, medical care etc. Equally, they do not subscribe to a clear 'rights' view of social justice either. Social choices are then a compromise between a utilitarian solution that focuses on consequences and one that recognizes basic rights in a more fundamental way. Much of the political and philosophical debate is about which rights are valid in this context – a debate that shows little sign of resolution. For climate change there are many options that need to be evaluated, in terms of their consequences for the lives of individuals who will be impacted by them. It is perfectly reasonable for the policymakers to exclude those that would result in major social disruptions, or large number of deaths, without recourse to a CBA. Equally, choices that avoid such negative consequences can be regarded as essential, even if the case for them cannot be made on CBA grounds. Details of where such rules should apply and where choices can be left to the more conventional CBA have yet to be worked out, and this remains an urgent part of the agenda for climate change studies.

As an alternative to social-justice-based equity methods, eco-centric approaches assign intrinsic value to nature as such (Botzler and Armstrong, 1998). This value can be specified in terms of diversity, avoided damages, harmony, stability, and beauty, and these values should be respected by human beings in their interaction with nature. In relation to climate change policies the issue here becomes one of specifying the value of nature such that it can be addressed as specific constraints that are to be respected beyond what is reflected in estimates of costs and benefits and other social impacts.

2.6.4 Equity consequences of different policy instruments

All sorts of climate change policies related to vulnerabilities, adaptation, and mitigation will have impacts on intra- and inter-generational equity. These equity impacts apply at the global, international, regional, national and sub-national levels.

Article 3 of the UNFCCC (1992, sometimes referred to as 'the equity article') states that Parties should protect the

---

24 For a discussion of this debate in an economic context, see Phelps, 1973.

climate system on the basis of equity and in accordance with their common but differentiated responsibilities and respective capabilities. Accordingly, the developed country Parties should take the lead in combating climate change and the adverse effects thereof. Numerous approaches exist in the climate change discourse on how these principles can be implemented. Some of these have been presented to policymakers (both formally and informally) and have been subject to rigorous analysis by academics, civil society and policymakers over long periods of time.

The equity debate has major implications for how different stakeholders judge different instruments for reducing greenhouse gases (GHG) and for adapting to the inevitable impacts of climate change.

With respect to the measures for reducing GHGs, the central equity question has focused on how the burden should be shared across countries (Markandya and Halsnaes, 2002b; Agarwal and Narain, 1991; Baer and Templet, 2001; Shukla, 2005). On a *utilitarian basis*, assuming declining marginal utility, the case for the richer countries undertaking more of the burden is strong – they are the ones to whom the opportunity cost of such actions would have less welfare implications. However, assuming constant marginal utility, one could come to the conclusion that the costs of climate change mitigation that richer countries will face are very large compared with the benefits of the avoided climate change damages in poorer countries. In this way, utilitarian-based approaches can lead to different conclusions, depending on how welfare losses experienced by poorer people are represented in the social welfare function.

Using a *'rights' basis* it would be difficult to make the case for the poorer countries to bear a significant share of the burden of climate change mitigation costs. Formal property rights for GHG emissions allowances are not defined, but based on justice arguments equal allocation to all human beings has been proposed. This would give more emissions rights to developing countries – more than the level of GHGs they currently emit. Hence such a rights-based allocation would impose more significant costs on the industrialized countries, although now, as emissions in the developing world increased, they too, at some point in time, would have to undertake some emissions reductions.

The literature includes a number of comparative studies on equity outcomes of different international climate change agreements. Some of these studies consider equity in terms of the consequences of different climate change policies, while others address equity in relation to rights that nations or individuals should enjoy in relation to GHG emission and the global atmosphere.

Equity concerns have also been addressed in a more pragmatic way as a necessary element in international agreements in order to facilitate consensus. Müller (2001) discusses fairness of emission allocations and that of the burden distribution that takes all climate impacts and reduction costs into consideration and concludes that there is no solution that can be considered as the right and fair one far out in the future. The issue is rather to agree on an acceptable 'fairness harmonization procedure', where an emission allocation is initially chosen and compensation payments are negotiated once the costs and benefits actually occur.

Rose *et al.* (1998) provide reasons why equity considerations are particularly important in relation to climate change agreements. First, country contributions will depend on voluntary compliance and it must therefore be expected that countries will react according to what they consider to be fair,[25] which will be influenced by their understanding of equity. Second, appealing to global economic efficiency is not enough to get countries together, due to the large disparities in current welfare and in welfare changes implied by efficient climate policies.

Studies that focus on the net costs of climate change mitigation versus the benefits of avoided climate change give a major emphasis to the economic consequences of the policies, while libertarian-oriented equity studies focus on emission rights, rights of the global atmosphere, basic human living conditions etc. (Wesley and Peterson, 1999). Studies that focus on the net policy costs will tend to address equity in terms of a total outcome of policies, while the libertarian studies focus more on initial equity conditions that should be applied to *ex ante* emission allocation rules, without explicitly taken equity consequences into consideration.

Given the uncertainties inherent in climate change impacts and their economic and social implications, it is difficult to conduct comprehensive and reliable consequence studies that can be used for an *ex ante* determination of equity principles for climate change agreements. Furthermore, social welfare functions and other value functions, when applied to the assessment of the costs and benefits of global climate change policies, run into a number of crucial equity questions. These include issues that are related to the asymmetry between the concentration of major GHG emission sources in industrialized countries and the relatively large expected damages in developing countries, the treatment of individuals with different income levels in the social welfare function, and a number of inter-generational issues.

Rights-based approaches have been extensively used as a basis for suggestions on structuring international climate change

---

25   What countries consider as 'fair' may be in conflict with their narrow self-interest. Hence there is a problem with resolving the influence of these two determinants of national contributions to reducing GHGs. One pragmatic element in the resolution could be that the difference between the long-term self interest and what is fair is much smaller than that between narrow self-interest and fairness.

agreements around emission allocation rules or compensation mechanisms. Various allocation rules have been examined, including emissions per capita principles, emissions per GDP, grandfathering, liability-based compensation for climate change damages etc. These different allocation rules have been supported with different arguments and with reference to equity principles. An overview and assessment of the various rights-based equity principles and their consequences on emission allocations and costs are included in Rose *et al.* (1998), Valliancourt and Waaub (2004), Leimbach (2003), Tol and Verheyen (2004) and Panayotou *et al.* (2002).

While there is consensus in the literature about how rules should be assessed in relation to specific moral criteria, there is much less agreement on what criteria should apply (e.g. should they be based on libertarian or egalitarian rights-based approaches, or on utilitarian approaches).

A particular difficulty in establishing international agreements on emission allocation rules is that the application of equity in this *ex ante* way can imply the very large transfer of wealth across nations or other legal entities that are assigned emission quotas, at a time where abatement costs, as well as climate change impacts, are relatively uncertain (Halsnæs and Olhoff, 2005). These uncertainties make it difficult for different parties to assess the consequences of accepting given emission allocation rules and to balance emission allocations against climate damages suffered in different parts of the world (Panayotou *et al.*, 2002).

Practical discussions about equity questions in international climate change negotiations have reflected, to a large extent, specific interests of various stakeholders, more than principal moral questions or considerations about the vulnerability of poorer countries. Arguments concerning property rights, for example, have been used by energy-intensive industries to advocate emission allocations based on grandfathering principles that will give high permits to their own stakeholders (that are large past emitters), and population-rich countries have, in some cases, advocated that fair emission allocation rules imply equal per capita emissions, which will give them high emission quotas.

Vaillancourt and Waaub (2004) suggest designing emission allocation criteria on the basis of the involvement of different decision-makers in selecting and weighing equity principles for emission allocations, and using these as inputs to a multi-criteria approach. The criteria include population basis, basic needs, polluter pays, GDP intensity, efficiency and geographical issues, without a specified structure on inter-relationships between the different areas. In this way, the approach primarily facilitates the involvement of stakeholders in discussions about equity.

2.6.5 Economic efficiency and eventual trade-offs with equity

For more than a decade the literature has covered studies that review the economic efficiency of climate change mitigation policies and, to some extent, also discuss different emission allocation rules and the derived equity consequences (IPCC, 1996, Chapter 11; IPCC, 2001, Chapters 6 and 8). Given that markets for GHG emission permits work well in terms of competition, transparency and low transaction costs, trade-offs between economic efficiency and equity (resulting from the distribution of emission rights) do not need to occur. In this ideal case, equity and economic efficiency can be addressed separately, where equity is taken care of in the design of emission allocation rules, and economic efficiency is promoted by the market system.

In practice, however, emission markets do not live up to these ideal conditions and the allocation of emission permits, both in international and domestic settings, will have an influence on the structure and functioning of emission markets, so trade-offs between what seems to be equitable emission allocations and economic efficiency can often occur (Shukla, 2005). Some of the issues that have been raised in relation to the facilitation of equity concerns through initial emission permit allocations include the large differences in emission permits and related market power that different countries would have (Halsnæs and Olhoff, 2005).

2.7 Technology

The cost and pace of any response to climate change concerns will also depend critically on the cost, performance, and availability of technologies that can lower emissions in the future. These technologies include both end-use (demand) as well as production (supply) technologies. Technological change is particularly important over the long time scales characteristic of climate change. Decade or century-long time scales are typical for the lags involved between technological innovation and widespread diffusion and of the capital turnover rates characteristic for long-lived energy capital stock and infrastructures (IPCC, 2001, 2002).

The development and deployment of technology is a dynamic process involving feedbacks. Each phase of this process may involve a different set of actors and institutions. The state of technology and technology change can differ significantly from country to country and sector to sector, depending on the starting point of infrastructure, technical capacity, the readiness of markets to provide commercial opportunities and policy frameworks. This section considers foundational issues related to the creation and deployment of new technology.

'Technology' refers to more than simply devices. Technology includes hardware (machines, devices, infrastructure networks etc.), software (i.e. knowledge/routines required for the production and use of technological hardware), as well as organizational/institutional settings that frame incentives and deployment structures (such as standards) for the generation and use of technology (for a review, compare Grubler, 1998).[26] Both the development of hybrid car engines and the development of Internet retailing mechanisms represent technological changes.

Many frameworks have been developed to simplify the process of technological change into a set of discrete phases. A common definitional framework frequently includes the following phases:
(1) Invention (novel concept or idea, as a result of research, development, and demonstration efforts).
(2) Innovation (first market introduction of these ideas).
(3) Niche markets (initial, small-scale applications that are economically feasible under specific conditions).
(4) Diffusion (widespread adoption and the evolution into mature markets, ending eventually in decline) (see Figure 2.3 below).

While the importance of technology to climate change is widely understood, there are differing viewpoints on the feasibility of current technology to address climate change and the role of new technology. On the one hand, Hoffert et al. (2002) and others have called for a major increase in research funding now to develop innovative technological options because, in this view, existing technologies cannot achieve the deep emission cuts that could be needed to mitigate future change. On the other hand, Pacala and Socolow (2004) advance the view that a range of known current technologies could be deployed, starting now and over the next 50 years, to place society on track to stabilize CO_2 concentrations at 500 ± 50 parts per million. In their view, research for innovative technology is needed but only to develop technologies that might be used in the second half of the century and beyond. Still a third viewpoint is that the matter is better cast in terms of cost, in addition to technical feasibility (e.g. Edmonds et al., 1997; Edmonds, 2004; Nakicenovic and Riahi, 2002) From this viewpoint, today's technology is, indeed, sufficient to bring about the requisite emissions reductions, but the underlying question is not technical feasibility but the degree to which resources would need to be reallocated from other societal goals (e.g. health care, education) to accommodate emissions mitigation. The role of new technology, in this view, is to lower the costs to achieve societal goals.

From the perspective of (commercial) availability and costs it is important to differentiate between the short-term and the long-term, and between technical and economic feasibility. A technology, currently at a pilot plant development stage and thus not available commercially, has no short-term potential to reduce emissions, but might have considerable potential once commercialized. Conversely, a technology, currently available commercially, but only at high cost, might have a short-term emission reduction potential in the (unlikely) case of extremely strong short-term policy signals (e.g. high carbon prices), but might have considerable potential in the long-term if the costs of the technology can be reduced. Corresponding mitigation technology assessments are therefore most useful when they differentiate between short/medium-term and long-term technology options, (commercial) availability status, costs, and the resulting (different) mitigation potentials of individual technology options. Frequently, the resulting ranking of individual technological options with respect to emissions reduction potentials and costs/yields emission abatement 'supply curves' illustrate how much emission reductions can be achieved, at what costs, over the short- to medium-term as well as in the longer-term.

2.7.1 Technology and climate change

Recognizing the importance of technology over the long-term introduces an important element of uncertainty into the climate change debate, as direction and pace of future technological change cannot be predicted. Technological innovation and deployment are responsive to climate policy signals, for example in form of carbon taxes, although the extent and rate of this response can be as uncertain as the timing and magnitude of the policy signal. Reducing such uncertainties, for instance through long-term, predictable policy frameworks and signals, are therefore important. The usual approach consists of formulating alternative scenarios of plausible future developments. These, however, are constrained by inherent biases in technology assessment and uncertainties concerning the response of technological change to climate policy. There is also widespread recognition in the literature that it is highly unlikely that a single 'silver bullet' technology exists that can solve the climate problem, so the issue is not one of identifying singular technologies, but rather ensembles, or portfolios of technologies. This applies to both mitigation and adaptation technologies. These technologies have inter-dependencies and cross-enhancement ('spillover') potentials, which adds another important element of uncertainty into the analysis. Despite these problems of uncertainty and ignorance, insights are available from multiple fields.

Extensive literature surveys on the importance of technological change on the extent of possible climate change and on feasibility and costs of climate policies are provided by Clarke and Weyant (2002), Grubb et al. (2002), Grübler et al. (1999), Jaffe et al. (2003) and Löschel (2002) among others.

---

26 It is also important to note that important linkages exist between technological and behavioural change. A frequently discussed phenomenon is so-called 'take-back' or 'rebound' effects, e.g. a change in consumption behaviour after the adoption of energy efficiency improvement measures (e.g. driving longer distances after purchasing a more energy-efficient car). Compare the review by Schipper and Grubb, 2000.

Quantitative illustrations have been published in a number of important scenario studies including the IPCC SAR (IPCC, 1996) and SRES (IPCC, 2000), the scenarios of the World Energy Council (WEC, Nakicenovic *et al.*, 1998a) as well as from climate policy model inter-comparison projects such as EMF-19 (Energy Modelling Forum) (Weyant, 2004b), the EU-based Innovation Modeling Comparison Project (IMCP) (Edenhofer *et al.*, 2006) and the multi-model calculations of climate 'stabilization' scenarios summarized in the TAR (IPCC, 2001). In a new development since the TAR, technology has also moved to the forefront of a number of international and national climate policy initiatives, including the Global Energy Technology Strategy (GTSP, 2001), the Japanese 'New Earth 21' Project (RITE, 2003), the US 21 Technology Roadmap (NETL, 2004), or the European Union's World Energy Technology Outlook (WETO, 2003).

The subsequent review first discusses the importance of technological change in 'no-climate policy' (or so-called 'reference' or 'baseline') scenarios, and hence the magnitude of possible climate change. The review then considers the role of alternative technology assumptions in climate policy ('stabilization') scenarios. The review continues by presenting a discussion of the multitude of mechanisms underlying technological change that need to be considered when discussing policy options to further the availability and economics of mitigation and adaptation technologies.

2.7.1.1   *Technological change in no-climate policy (reference) scenarios*

The importance of technological change for future GHG emission levels and hence the magnitude of possible climate change has been recognized ever since the earliest literature reviews (Ausubel and Nordhaus, 1983). Subsequent important literature assessments (e.g. Alcamo *et al.,* 1995; Nakicenovic *et al.*, 1998b; Edmonds *et al.*, 1997; SRES, 2000) have examined the impact of alternative technology assumptions on future levels of GHG emissions. For instance, the SRES (2000) report concluded technology to be of similar importance for future GHG emissions as population and economic growth combined. A conceptual simple illustration of the importance of technology is provided by comparing individual GHG emission scenarios that share comparable assumptions on population and economic growth, such as in the Low Emitting Energy Supply Systems (LESS) scenarios developed for the IPCC SAR (1996) or within the IPCC SRES (2000) A1 scenario family, where for a comparable level of energy service demand, the (no-climate-policy) scenarios span a range of between 1038 (A1T) and 2128 (A1FI) GtC cumulative (1990-2100) emissions, reflecting different assumptions on availability and development of low- versus high-emission technologies. Yet another way of illustrating the importance of technology assumptions in baseline scenarios is to compare given scenarios with a hypothetical baseline in which no technological change is assumed to occur at all. For instance, GTSP (2001) and Edmonds *et al.* (1997, see

also Figure 3.32 in Chapter 3) illustrate the effect of changing reference case technology assumptions on CO_2 emissions and concentrations based on the IPCC IS92a scenario by holding technology at 1990 levels to reveal the degree to which advances in technology are already embedded in the non-climate-policy reference case, a conclusion also confirmed by Gerlagh and Zwaan, 2004. As in the other scenario studies reviewed, the degree to which technological change assumptions are reflected in the scenario baseline by far dominates future projected emission levels. The importance of technology is further magnified when climate policies are considered. See for example, the stabilization scenarios reviewed in IPCC TAR (2001) and also Figure 2.1 below.

Perhaps the most exhaustive examination of the influence of technological uncertainty to date is the modelling study reported by Gritsevskyi and Nakicenovic (2000). Their model simulations, consisting of 130,000 scenarios that span a carbon emission range of 6 to 33 GtC by 2100 (Figure 2.1), provided a systematic exploration of contingent uncertainties of long-term technological change spanning a comparable range of future emissions as almost the entirety of the no-climate policy emissions scenario literature (see Chapter 3 for an update of the scenario literature). The study also identified some 13,000 scenarios (out of an entire scenario ensemble of 130,000) regrouped into a set of 53 technology dynamics that are all 'optimal' in the sense that they satisfy the same cost minimum in the objective function, but with a bimodal distribution in terms of emissions outcomes. In other words, considering full endogenous technological uncertainty produces a pattern of 'technological lock-in' into alternatively low or high emissions futures that are equal in terms of their energy systems costs.

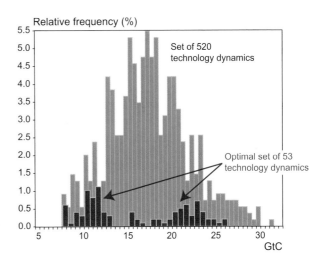

Figure 2.1: *Emission impacts of exploring the full spectrum of technological uncertainty in a given scenario without climate policies. Relative frequency (percent) of 130,000 scenarios of full technological uncertainty regrouped into 520 sets of technology dynamics with their corresponding carbon emissions by 2100. Also shown is a subset of 13,000 scenarios grouped into 53 sets of technology dynamics that are all 'optimal' in the sense of satisfying a cost minimization criterion in the objective function. See text for further discussion. 1 Gt C = 3.7 Gt CO_2*

Source: Adapted from Gritsevskyi and Nakicenovic, 2000.

This finding is consistent with the extensive literature on technological 'path dependency' and 'lock-in phenomena' (e.g. Arthur, 1989) as also increasingly reflected in the scenario literature (e.g. Nakicenovic *et al.*, 1998b and the literature review in Chapter 3). This casts doubts on the plausibility of central tendency technology and emissions scenarios. It also shows that the variation in baseline cases could generate a distribution of minimum costs of the global energy system where low-emission baseline scenarios could be as cheap as their high-emission counterparts.

The results also illustrate the value of technology policy as a hedging strategy aiming at lowering future carbon emissions, even in the absence of directed climate policies, as the costs of reducing emissions even further from a given baseline are *ceteris paribus* proportionally lower with lower baseline emissions.

2.7.1.2   *Technological change in climate policy scenarios*

In addition to the technology assumptions that enter typical 'no-climate policy' baselines, technology availability and the response of technology development and adoption rates to a variety of climate policies also play a critical role. The assessment of which alternative technologies are deployed in meeting given GHG emission limitations or as a function of *ex ante* assumed climate policy variables, such as carbon taxes, again entails calculations that span many decades into the future and typically rely on (no-climate policy) baseline scenarios (discussed above).

Previous IPCC assessments have discussed in detail the differences that have arisen with respect to feasibility and costs of emission reductions between two broad categories of modelling approaches: 'bottom-up' engineering-type models versus 'top-down' macro-economic models. Bottom-up models usually tend to suggest that mitigation can yield financial and economic benefits, depending on the adoption of best-available technologies and the development of new technologies. Conversely, top-down studies have tended to suggest that mitigation policies have economic costs because markets are assumed to have adopted all efficient options already. The TAR offered an extensive analysis of the relationship between technological, socio-economic, economic and market potential of emission reductions, with some discussion of the various barriers that help to explain the differences between the different modeling approaches. A new finding in the underlying literature (see, for example, the review in Weyant, 2004a) is that the traditional distinction between 'bottom-up' (engineering) and 'top down' (macro-economic) models is becoming increasingly blurred as 'top down' models incorporate increasing technology detail, while 'bottom up' models increasingly

incorporate price effects and macro-economic feedbacks, as well as adoption barrier analysis, into their model structures. The knowledge gained through successive rounds of model inter-comparisons, such as implemented within the Energy Modeling Forum (EMF) and similar exercises, has shown that the traditional dichotomy between 'optimistic' (i.e. bottom-up) and 'pessimistic' (i.e. top-down) views on feasibility and costs of meeting alternative stabilization targets is therefore less an issue of methodology, but rather the consequence of alternative assumptions on availability and costs of low- and zero-GHG-emitting technologies. However, in their meta-analysis of post-SRES model results, Barker *et al.* (2002) have also shown that model structure continues to be of importance.

Given the infancy of empirical studies and resulting models that capture in detail the various inter-related inducement mechanisms of technological change in policy models, salient uncertainties continue to be best described through explorative model exercises under a range of (exogenous) technology development scenarios. Which mitigative technologies are deployed, how much, when and where depend on three sets of model and scenario assumptions. First, assumptions on which technologies are used in the reference ('no policy') case, in itself a complex result of scenario assumptions concerning future demand growth, resource availability, and exogenous technology-specific scenario assumptions. Second, technology deployment portfolios depend on the magnitude of the emission constraint, increasing with lower stabilization targets. Finally, results depend critically on assumptions concerning future availability and relative costs of mitigative technologies that determine the optimal technology mix for any given combination of baseline scenarios with alternative stabilization levels or climate policy variables considered.

2.7.1.3   *Technological change and the costs of achieving climate targets*

Rates of technological change are also critical determinants of the costs of achieving particular environmental targets. It is widely acknowledged that technological change has been a critical factor in both cost reductions and quality improvements of a wide variety of processes and products.[27] Assuming that technologies in the future improve similarly to that observed in the past enables experts to quantify the cost impacts of technology improvements in controlled modeling experiments. For instance, Edmonds *et al.* (1997, compare Figure 3.36 in Chapter 3) analyzed the carbon implications of technological progress consistent with historical rates of energy technology change. Other studies have confirmed Edmonds' (1997) conclusion on the paramount importance of future availability and costs of low-emission technologies and the significant economic benefits of improved technology

---

27 Perhaps one of the most dramatic historical empirical studies is provided by Nordhaus (1997) who has analyzed the case of illumination since antiquity, illustrating that the costs per lumen-hour have decreased by approximately a factor of 1,000 over the last 200 years. Empirical studies into computers and semiconductors indicate cost declines of up to a factor of 100,000 (Victor and Ausubel, 2002; Irwin and Klenov, 1994). Comparable studies for environmental technologies are scarce.

that, when compounded over many decades, can add up to trillions of dollars. (For a discussion of corresponding 'value of technological innovation' studies see Edmonds and Smith (2006) and Section 3.4, particularly Figure 3.36 in Chapter 3). However, to date, model calculations offer no guidance on the likelihood or uncertainty of realizing 'advanced technology' scenarios. However, there is an increasing number of studies (e.g. Gerlagh and Van der Zwaan, 2006) that explore the mechanisms and policy instruments that would need to be set in place in order to *induce* such drastic technological changes.

The treatment of technological change in an emissions and climate policy modeling framework can have a huge effect on estimates of the cost of meeting any environmental target. Models in which technological change is dominated by experience (learning) curve effects, show that the cost of stabilizing GHG concentrations could be in the range of a few tenths of a percent of GDP, or even lower (in some models even becoming negative) – a finding also confirmed by other modelling studies (e.g. Rao *et al.*, 2005) and consistent with the results of the study by Gritsevskyi and Nakicenovic (2000) reviewed above, which also showed identical costs of 'high' versus 'low' long-term emission futures. This contrasts with the traditional view that the long-term costs[28] of climate stabilization could be very high, amounting to several percentage points of economic output (see also the review in IPCC, 2001).

Given the persistent uncertainty of what constitutes 'dangerous interference with the climate system' and the resulting uncertainty on ultimate climate stabilization targets, another important finding related to technology economics emerges from the available literature. Differences in the cost of meeting a prescribed CO_2 concentration target across alternative technology development pathways that could unfold in the absence of climate policies are more important than cost differences between alternative stabilization levels *within* a given technology-reference scenario. In other words, the overall 'reference' technology pathway can be equally, if not more, important in determining the costs of a given scenario as the stringency of the ultimate climate stabilization target chosen (confer Figure 2.2).

In a series of alternative stabilization runs imposed on the SRES A1 scenarios, chosen for ease of comparability as sharing similar energy demands, Roehrl and Riahi (2000) confer also IPCC (2001) have explored the cost differences between four alternative baselines and their corresponding stabilization targets, ranging from 750 ppmv all the way down to 450 ppmv. In their calculations, the cost differences between alternative baselines are also linked to differences in baseline emissions:

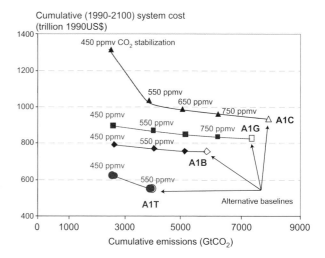

Figure 2.2: *The impacts of different technology assumptions on energy systems costs and emissions (cumulative 1990–2100, systems costs (undiscounted) in trillion US$) in no-climate policy baseline (reference) scenarios (based on the SRES A1 scenario family that share identical population and GDP growth assumptions) and in illustrative stabilization scenarios (750, 650, 550 and 450 ppm respectively). For comparison: the total cumulative (undiscounted) GDP of the scenarios is around 30,000 trillion US$ over the 1990–2100 time period.*

Source: Roehrl and Riahi (2000).

advanced post-fossil fuel technologies yield both lower overall systems costs as well as lower baseline emissions and hence lower costs of meeting a specified climate target (confer the differences between the A1C and A1T scenarios in Figure 2.2). Their findings are consistent with the pattern identified by Edmonds *et al.* (1997) and Gerlagh and Van der Zwaan (2003). Cost differences are generally much larger between alternative technology baselines, characterized by differing assumptions concerning availability and costs of technologies, rather than between alternative stabilization levels. The IEA (2004) World Energy Outlook also confirms this conclusion, and highlights the differential investment patterns entailed by alternative technological pathways.[29] The results from the available literature thus confirm the value of advances in technology importance in lowering future 'baseline' emissions in order to enhance feasibility, flexibility, and economics of meeting alternative stabilization targets, in lowering overall systems costs, as well as in lowering the costs of meeting alternative stabilization targets.

A robust analytical finding arising from detailed technology-specific studies is that the economic benefits of technology improvements (i.e. from cost reductions) are highly non-linear, arising from the cumulative nature of technological change, from interdependence and spillover effects, and from potential

---

28 Note here that this statement only refers to the (very) long term, i.e. a time horizon in which existing capital stock and technologies will have been turned over and replaced by newer vintages. In the short term (and using currently or near-term available technologies) the costs of climate policy scenarios are invariably higher than their unconstrained counterparts.

29 The IEA (2004) 'alternative scenario', while having comparable total systems costs, would entail an important shift in investments away from fossil-fuel-intensive energy supply options towards energy efficiency improvements, a pattern also identified in the scenario study of Nakicenovic *et al.* (1998b).

increasing returns to adoption (i.e. costs decline with increasing market deployment of a given technology).[30] (A detailed review covering the multitude of sources of technological change, including the aforementioned effects, is provided in Chapter 11, Section 11.5, discussing so-called 'induced technological change' models).

2.7.2 Technological change

Changes in technology do not arise autonomously – they arise through the actions of human beings, and different social and economic systems have different proclivities to induce technological change. The range of actors participating in the process of technological change spans the full range of those that use technology, design and manufacture technology, and create new knowledge.

The process of technological change has several defining characteristics. First, the process is highly uncertain and unpredictable. Firms planning research toward a well-defined technical goal must plan without full knowledge regarding the potential cost, time frame, and even the ultimate success. Further, the history of technological development is rife with small and large examples of serendipitous discoveries, (e.g. Teflon) whose application is far beyond, or different, than their intended use.

A second defining characteristic of technological change is the transferable, public-good nature of knowledge. Once created, the value of technological knowledge is difficult to fully appropriate; some or all eventually spills over to others, and in doing so the knowledge is not depleted. This characteristic of knowledge has both benefits and drawbacks. On the one hand, an important discovery by a single individual, such as penicillin, can be utilized worldwide. Knowledge of penicillin is a public good and therefore one person's use of this knowledge does not preclude another person from using this same knowledge – unlike for capital or labour, where use in one task precludes use in an alternative task. On the other hand, the understanding by potential innovators that any new knowledge might eventually spill over to others limits expected profits and therefore dampens private-sector innovative activity. Thus intellectual property rights can serve both as a barrier and an aid in technology change. A final, third feature of technological change is its cumulativeness, which is also frequently related to spillover effects.

There are numerous paradigms used to separate the process of technological change into distinct phases. One approach is to consider technological change as roughly a two-part process,

which includes:

(1) The process of conceiving, creating, and developing new technologies or enhancing existing technologies – the process of advancing the 'technological frontier'.
(2) The process of diffusing or deploying these technologies.

These two processes are inextricably tied. The set of available technology defines what might be deployed, and the use of technology affords learning that can guide R&D programmes or directly improve technology through learning-by-doing. The two processes are also linked temporally. The set of technologies that find their way into use necessarily lags the technological frontier. The useful life of technologies – their natural turnover rate – helps to drive the time relationship. Car lifespans can be in the order of 15 years, but the associated infrastructure – roads, filling stations, vehicle manufacturing facilities – have significantly longer lifespans, and electric power plants may be used for a half-century or more; hence, the average car is substantially younger than the average coal-fired power plant and much of its associated infrastructure. The nature of the capital stock (e.g. flexifuel cars that can use both conventional petrol and ethanol) is also important in determining diffusion speed.

2.7.2.1 *The sources of technological change*

New technology arises from a range of interacting drivers. The literature (for a review see, for example, Freeman, 1994, and Grubler, 1998) divides these drivers into three broad, overlapping categories: R&D, learning-by-doing, and spillovers. These drivers are distinctly different[31] from other mechanisms that influence the costs of a given technology, such as. through economies of scale effects (see Box 2.3 below). Each of these entails different agents, investment needs, financial institutions and is affected by the policy environment. These are briefly discussed below, followed by a discussion of the empirical evidence supporting the importance of these sources and the linkages between them.

Research and Development (R&D): R&D encompasses a broad set of activities in which firms, governments, or other entities expend resources specifically to improve technology or gain new knowledge. While R&D covers a broad continuum, it is often parsed into two categories: applied R&D and fundamental research, and entails both science and engineering (and requires science and engineering education). Applied R&D focuses on improving specific, well-defined technologies (e.g. fuel cells). Fundamental research focuses on broader and more fundamental areas of understanding. Fundamental research may be mission-oriented (e.g. fundamental biological research

---

30 This is frequently referred to as a 'learning-by-doing' phenomenon. However, the linkages between technology costs and market deployment are complex, covering a whole host of influencing factors including (traditional) economics of larger market size, economies of scale in manufacturing, innovation-driven technology improvements, geographical and inter-industry spillover effects, as well as learning-by-doing (experience curve) phenomena proper. For (one of the few available) empirical studies analyzing the relative contribution of their various effects on cost improvements see Nemet (2005). A more detailed discussion is provided in Chapter 11.

31 However, there are important relations between economies of scale and technological change in terms that scaling up usually also requires changes in manufacturing technologies, even if the technology manufactured remains unchanged.

intended to provide a long-term knowledge base to fight cancer or create fuels) or focus on new knowledge creation without explicit consideration of use (see Stokes (1997) regarding this distinction). Both applied R&D and fundamental research are interactive: fundamental research in a range of disciplines or research areas, from materials to high-speed computing, can create a pool of knowledge and ideas that might then be further developed through applied R&D. Obstacles in applied R&D can also feed research priorities back to fundamental research. As a rule of thumb, the private sector takes an increasingly prominent role in the R&D enterprise the further along the process toward commercial application. Similar terms found in the literature include: Research, Development, and Demonstration (RD&D), and Research, Development, Demonstration, and Deployment (RDD&D or RD³). These concepts highlight the importance of linking basic and applied research to initial applications of new technologies that are an important feedback and learning mechanism for R&D proper.

R&D from across the economic spectrum is important to climate change. Energy-focused R&D, basic or applied, as well as R&D in other climate-relevant sectors (e.g. agriculture) can directly influence the greenhouse gas emissions associated with these sectors (CO_2, CH_4). At the same time, R&D in seemingly unrelated sectors may also provide spillover benefits to climate-relevant sectors. For example, advances in computers over the last several decades have enhanced the performance of the majority of energy production and use technologies.

Learning-by-doing: Learning-by-doing refers to the technology-advancing benefits that arise through the use or production of technology, i.e. *market deployment*. The more that an individual or an organization repeats a task, the more adept or efficient that organization or individual becomes at that task. In early descriptions (for example, Wright, 1936), learning-by-doing referred to improvements in manufacturing labour productivity for a single product and production line. Over time, the application of learning-by-doing has been expanded to the level of larger-scale organizations, such as an entire firm producing a particular product. Improvements in coordination, scheduling, design, material inputs, and manufacturing technologies can increase labour productivity, and this broader definition of learning-by-doing therefore reflects experience gained at all levels in the organization, including engineering, management, and even sales and marketing (see, Hirsh, 1956; Baloff, 1966; Yelle, 1979; Montgomery and Day, 1985; Argote and Epple, 1990).

There are clearly important interactions between learning-by-doing and R&D. The production and use of technologies provides important feedbacks to the R&D process, identifying key areas for improvement or important roadblocks. In addition, the distinction between learning-by-doing and R&D is blurred at the edges: for example, everyday technology design improvements lie at the boundary of these two processes.

Spillovers: Spillovers refer to the transfer of knowledge or the economic benefits of innovation from one individual, firm, industry, or other entity to another. The gas turbine in electricity production, 3-D seismic imaging in oil exploration, oil platform technologies and wave energy, and computers are all spillovers in a range of energy technologies. For each of these obvious cases of spillovers there are also innumerable, more subtle instances. The ability to identify and exploit advances in unrelated fields is one of the prime drivers of innovation and improvement. Such advances draw from an enabling environment that supports education, research and industrial capacity.

There are several dimensions to spillovers. Spillovers can occur between:
(1) Firms within an industry in and within countries (intra-industry spillovers).
(2) Industries (inter-industry spillovers).
(3) Countries (international spillovers).
The latter have received considerable attention in the climate literature (e.g. Grubb *et al.*, 2002). Spillovers create a positive externality for the recipient industry, sector or country, but also limit (but not eliminate) the ability of those that create new knowledge to appropriate the economic returns from their efforts, which can reduce private incentives to invest in technological advance (see Arrow, 1962), and is cited as a primary justification for government intervention in markets for innovation.

Spillovers are not necessarily free. The benefits of spillovers may require effort on the part of the receiving firms, industries, or countries. Explicit effort is often required to exploit knowledge that spills over, whether that knowledge is an explicit industrial process or new knowledge from the foundations of science (see Cohen and Levinthal, 1989). The opportunities created by spillovers are one of the primary sources of knowledge that underlies innovation (see Klevorick, *et al.*, 1995). There are different channels by which innovativions may spillover. For instance, the productivity achieved by a firm or an industry depends not only on its own R&D effort, but also on the pool of general knowledge to which it has access. There are also so-called 'rent spillovers', such as R&D leading to quality changes embodied in new and improved outputs which not necessarily yield higher prices. Finally, spillovers are frequent for products with high market rivalry effects (e.g. through reverse engineering or industrial espionage). However it is inherently difficult to distinguish clearly between these various channels of spillovers.

Over the last half century, a substantial empirical literature has developed, outside the climate or energy contexts, which explores the sources of technological advance. Because of the complexity of technological advance and the sizable range of forces and actors involved, this literature has proceeded largely through partial views, considering one or a small number of sources, or one or a small number of technologies. On the whole, the evidence strongly suggests that all three of the

Box 2.3 Economies of scale

Economies of scale refer to the decreases in the average cost of production that come with an increase in production levels, assuming a constant level of technology. Economies of scale may arise, for example, because of fixed production costs that can be spread over larger and larger quantities as production increases, thereby decreasing average costs. Economies of scale are not a source of technological advance, but rather a characteristic of production. However, the two concepts are often intertwined, as increased production levels can bring down costs both through learning-by-doing and economies of scale. It is for this reason that economies of scale have often been used as a justification for using experience curves or learning curves in integrated assessment models.

sources highlighted above – R&D, learning-by-doing, and spillovers – play important roles in technological advance and there is no compelling reason to believe that one is broadly more important than the others. The evidence also suggests that these sources are not simply substitutes, but may have highly complementary interactions. For example, learning from producing and using technologies provides important market and technical information that can guide both public and private R&D efforts.

Beginning with Griliches's study of hybrid corn (see Griliches, 1992), economists have conducted econometric studies linking R&D to productivity (see Griliches, 1992, Nadiri, 1993, and the Australian Industry Commission, 1995 for reviews of this literature). These studies have used a wide range of methodologies and have explored both public and private R&D in several countries. As a body of work, the literature strongly suggests substantial returns from R&D, social rates well above private rates in the case of private R&D (implying that firms are unable to fully appropriate the benefits of their R&D), and large spillover benefits. Griliches (1992) writes that '… there have been a significant number of reasonably well done studies all pointing in the same direction: R&D spillovers are present, their magnitude may be quite large, and social rates of return remain significantly above private rates'.

Since at least the mid-1930s (see Wright, 1936), researchers have also conducted statistical analyses on 'learning curves' correlating increasing cumulative production volumes and technological advance. Early studies focused heavily on military applications, notably wartime ship and airframe manufacture (see Alchian, 1963 and Rapping, 1965). From 1970 through to the mid-1980s, use of experience curves was widely recommended for corporate strategy development. More recently, statistical analyses have been applied to emerging energy technologies such as wind and solar power. (Good summaries of the experience curve literature can be found in Yelle, 1979; Dutton and Thomas, 1984. Energy technology experience curves may be found in Zimmerman, 1982; Joskow and Rose, 1982; Christiansson, 1995; McDonald and Schrattenholzer, 2001).

Based on the strength of these correlations, large-scale energy and environmental models are increasingly using 'experience curves' or 'learning curves' to capture the response of technologies to increasing use (e.g. Messner, 1997; IEA,

2000; Rao et al., 2005; and the review by Clarke and Weyant, 2002). These curves correlate cumulative production volume to per-unit costs or other measures of technological advance.

An important methodological issue arising in the use of these curves is that the statistical correlations on which they are based do not address the causal relationships underlying the correlations between cumulative production and declining costs, and few studies address the uncertainties inherent in any learning phenomenon (including negative learning). Because these curves often consider technologies over long time frames and many stages of technology evolution, they must incorporate the full range of sources that might affect technological advance or costs and performance more generally, including economies of scale, changes in industry structure, own-industry R&D, and spillovers from other industries and from government R&D. Together, these sources of advance reduce costs, open up larger markets, and result in increasing cumulative volume (see Ghemawat, 1985; Day and Montgomery, 1983; Alberts, 1989; Soderholm and Sundqvist, 2003). Hence, the causal relationships necessarily operate both from cumulative volume to technological advance and from technological advance to cumulative volume.

A number of studies have attempted to probe more deeply into the sources of advance underlying these correlations (see, for example, Rapping, 1965; Lieberman, 1984; Hirsh, 1956; Zimmerman, 1982; Joskow and Rose, 1985; Soderholm and Sundqvist, 2003, and Nemet, 2005). On the whole, these studies continue to support the presence of learning-by-doing effects, but also make clear that other sources can also be important and can influence the learning rate. This conclusion is also confirmed by recent studies following a so-called 'two-factor-learning-curve' hypothesis that incorporates both R&D and cumulative production volume as drivers of technological advance within a production function framework (see, for example, Kouvaritakis et al., 2000). However, Soderholm and Sundqvist (2003) conclude that 'the problem of omitted variable bias needs to be taken seriously' in this type of approach, in addition to empirical difficulties that arise, because of the absence of public and private sector technology-specific R&D statistics and due to significant co-linearity and auto-correlation of parameters (e.g. Miketa and Schrattenholzer, 2004).

More broadly, these studies, along with related theoretical work, suggest the need for further exploration of the drivers behind technological advance and the need to develop more explicit models of the interactions between sources. For example, while the two-factor-learning-curves include both R&D and cumulative volume as drivers, they often assume a substitutability of the two forms of knowledge generation that is at odds with the (by now widely accepted) importance of feedback effects between 'supply push' and 'demand pull' drivers of technological change (compare Freeman, 1994). Hence, while modelling paradigms such as two-factor-learning-curves might be valuable methodological steps on the modelling front, they remain largely exploratory. For a (critical) discussion and suggestion for an alternative approach see, for example, Otto *et al.*, 2005.

A range of additional lines of research has explored the sources of technological advance. Authors have pursued the impacts of 'general-purpose technologies', such as rotary motion (Bresnahan and Trajtenberg, 1992), electricity and electric motors (Rosenberg, 1982), chemical engineering (Rosenberg, 1998), and binary logic and computers (Bresnahan and Trajtenberg, 1992). Klevorick *et al.* (1995) explored the sources of technological opportunity that firms exploit in advancing technology, finding important roles for a range of knowledge sources, depending on the industry and the application. A number of authors (see, for example, Jaffe and Palmer 1996; Lanjouw and Mody 1996; Taylor *et al.,* 2003; Brunnermier and Cohen, 2003; Newell *et al.* 1998) have explored the empirical link between environmental regulation and technological advance in environmental technologies. This body of literature indicates an important relationship between environmental regulation and innovative activity on environmental technologies. On the other hand, this literature also indicates that not all technological advance can be attributed to the response to environmental regulation. Finally, there has been a long line of empirical research exploring whether technological advance is induced primarily through the appearance of new technological opportunities (technology-push) or through the response to perceived market demand (market pull). (See, for example, Schmookler, 1962; Langrish *et al.,* 1972; Myers and Marquis, 1969; Mowery and Rosenberg, 1979; Rosenberg 1982; Mowery and Rosenberg, 1989; Utterback, 1996; Rycroft and Kash 1999). Over time, a consensus has emerged that 'the old debate about the relative relevance of "technology push" versus "market pull" in delivering new products and processes has become an anachronism. In many cases one cannot say with confidence that either breakthroughs in research "cause" commercial success or that the generation of successful products or processes was a predictable "effect" of having the capability to read user demands or other market signals accurately' (Rycroft and Kash, 1999).

2.7.2.2 *Development and commercialization: drivers, barriers and opportunities*

Development and diffusion or commercialization of new technology is largely a private-sector endeavour driven by market incentives. The public sector can play an important role in coordination and co-funding of these activities and (through policies) in structuring market incentives. Firms choose to develop and deploy new technologies to gain market advantages that lead to greater profits. Technological change comprises a whole host of activities that include R&D, innovations, demonstration projects, commercial deployment and widespread use, and involves a wide range of actors ranging from academic scientists and engineers, to industrial research labs, consultants, firms, regulators, suppliers and customers. When creating and disseminating revolutionary (currently non-existent) technologies, the path to development may proceed sequentially through the various phases, but for existing technology, interactions can occur between all phases, for example, studies of limitations in currently deployed technologies may spark innovation in fundamental academic research. The ability to identify and exploit advances in unrelated fields (advanced diagnostics and probes, computer monitoring and modelling, control systems, materials and fabrication) is one of the prime drivers of innovation and improvement. Such advances draw from an enabling environment that supports education, research and industrial capacity.

The behaviour of competing firms plays a key role in the innovation process. Especially in their efforts to develop and introduce new non-commercial technology into a sustainable commercial operations, firms require not only the ability to innovate and to finance costly hardware, but also the managerial and technical skills to operate them and successfully market the products, particularly in the early stages of deployment and diffusion. The development of proprietary intellectual property and managerial know-how are key ingredients in establishing competitive advantage with new technology, but they can be costly and difficult to sustain.

Several factors must therefore be considered prominently with respect to the process of technology development and commercialization. A detailed review of these factors is included in the IPCC Special Report on Technology Transfer (SRTT) and the discussion below provides a summary and update, which draws on Flannery and Khesghi (2005) and OECD (2006). Factors to consider in development and commercialization of new technologies include:
* First, the lengthy timescale for deployment of advanced energy technologies.
* Second, the range of barriers that innovative technologies must successfully overcome if they are to enter into widespread commercial use.

- Third, the role of governments in creating an enabling framework to enhance the dissemination of innovative commercial technology created by private companies.
- Fourth, absorptive capacity and technological capabilities are also important determinants of innovation and diffusion.

New technologies must overcome a range of technical and market hurdles to enter into widespread commercial use. Important factors include:
- Performance.
- Cost.
- Consumer acceptance.
- Safety.
- Financial risks, available financing instruments.
- Enabling infrastructure.
- Incentive structures for firms (e.g. licensing fees, royalties, policy environment, etc.).
- Regulatory compliance.
- Environmental impacts.

The diffusion potential for a new technology depends on all above factors. If a technology fails even in one of these dimensions it will not achieve significant global penetration. While reducing greenhouse gas emissions should be an important objective in technological research, it is not the only factor.

Another factor is that the lengthy timescale for deployment of advanced energy technologies has a substantive impact on private-sector behaviour. Even with successful innovation in energy technology, the time necessary for new technology to make a widespread global impact on emissions will be lengthy. Timescales are long, both due to the long lifespan of existing productive capital stock, and the major investment in hardware and infrastructure that is required for significant market penetration. During the time that advanced technology is being deployed, both incremental and revolutionary changes may occur in the technologies under consideration, and in those that compete with them.

One consequence of the long time scales involved with energy technology is that, at any point in time, there will inevitably be a significant spread in the efficiency and performance of the existing equipment deployed. While this presents an opportunity for advanced technology to reduce emissions, the overall investment required to prematurely replace a significant fraction of sunk capital can be prohibitive. Another consequence of the long time scale and high cost of equipment is that it is difficult to discern long-term technological winners and losers in evolving markets.

A third factor is enabling infrastructure. Infrastructure can be interpreted broadly. Key features have been described in

numerous studies and assessments (e.g. IPIECA, 1995), and include: rule of law, safety, secure living environment for workers and communities, open markets, realization of mutual benefits, protection of intellectual property, movement of goods, capital and people, and respect for the needs of host governments and communities. These conditions are not unique for private companies. Many of them also are essential for successful public investment in technology and infrastructure.[32]

2.7.2.3   *The public-sector role in technological change*

Given the importance of technology in determining both the magnitude of future GHG emission levels as well as feasibility and costs of emission reduction efforts, technology policy considerations are increasingly considered in climate policy analyses. Ongoing debate centers on the relative importance of two differing policy approaches: technology-push (through efforts to stimulate research and development) and demand-pull (through measures that demand reduced emissions or enhanced efficiency). Technology-push emphasizes the role of policies that stimulate research and development, especially those aimed at lowering the costs of meeting long-term objectives with technology that today is very far from economic in existing markets. This might include such measures as public-funded R&D or R&D tax credits. Demand-pull emphasizes the use of instruments to enhance the demand for lower-emission technologies, thereby increasing private incentives to improve these technologies and inducing any learning-by-doing effects. Demand-pull instruments might include emissions taxes or more direct approaches, such as renewable portfolio standards, adoption subsidies, or direct public-sector investments (see Figure 2.3).

Two market failures are at issue when developing policies to stimulate technology development. The first is the failure to internalize the environmental costs of climate change, reducing the demand for climate-friendly technologies and thereby reducing private-sector innovation incentives and learning-by-doing. The second is a broad suite of private-sector innovation market failures that hold back and otherwise distort private-sector investment in technological advance, irrespective of environmental concerns (confer Jaffe *et al.*, 2005). Chief among these is the inability to appropriate the benefits of knowledge creation. From an economic standpoint, two market failures require two policy instruments: addressing two market failures with a single instrument will only lead to second-best solutions (see, for example, Goulder and Schneider, 1999). Hence, it is well understood that the optimal policy approach should include both technology-push and demand-pull instruments. While patents and various intellectual property protection (e.g. proprietary know-how) seeks to reward innovators, such protection is inherently imperfect, especially in global markets where such protections are not uniformly enforced by all governments.

---

32   These and other issues required for successful dissemination of technology were the subject of an entire IPCC Special Report (IPCC, 2000)

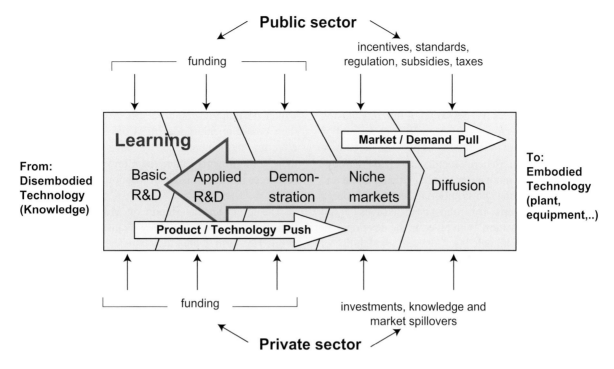

Figure 2.3 : *Technology development cycle and its main driving forces. Note that important overlaps and feedbacks exist between the stylized technology life-cycle phases illustrated here and therefore the illustration does not suggest a 'linear' model of innovation.*

Source: Adapted from Foxon (2003) and Grubb (2005).

Similarly, in the early adoption of technology learning-by-doing (by producers) or learning-by-using (by consumers) may lower the cost to all future users, but in a way that may not fully reward the frontrunners.[33] Similarly, lack of information by investors and potential consumers of innovative technologies may slow the diffusion of technologies into markets. The 'huge uncertainties surrounding the future impacts of climate change, the magnitude of the policy response, and thus the likely returns to R&D investment' exacerbate these technological spillover problems (Jaffe *et al.*, 2005).

The outstanding questions revolve around the relative combinations of instruments and around how effective single-policy approaches might be. Within this context, a number of authors (e.g. Montgomery and Smith, 2005) have argued that fundamental long-term shifts in technology to mitigate greenhouse gas emissions cannot be achieved through emissions-constraining policies alone, and short-term cap and trade emission-reduction policies provide insufficient incentives for R&D into long-term technology options. Conversely, Popp (2002) demonstrated how energy R&D is responsive to price signals, suggesting that without emissions constraints R&D into new low-emission technologies may face a serious lack of incentives and credible policy signals. The argument that emissions-based policies will induce long-term technology

innovation relies primarily on two lines of reasoning (Goulder 2004; Grubb, 2005). The first is that the anticipation of future targets, based on a so-called announcement effect, will stimulate firms to invest in research and development and ultimately to invest in advanced, currently non-commercial technology (the credibility and effectiveness of this effect, however, being challenged by Montgomery and Smith, 2005). The second is that early investment, perhaps through incentives, mandates, or government procurement programmes, will initiate a cycle of learning-by-doing that will ultimately promote innovation in the form of continuous improvement, which will drive down the cost of future investments in these technologies. This issue is especially critical in the scaling up of niche-market applications of new technologies (e.g. renewables) where mobilizing finance and lowering investment risks are important (see, for example, IEA, 2003, or Hamilton, 2005). In their comparative analysis of alternative policy instruments Goulder and Schneider (1999) found that when comparing a policy with only R&D subsidies to an emissions tax, the emissions-based policies performed substantially better.

Irrespective of the mix between demand-pull and technology-push instruments, a number of strong conclusions have emerged with respect to the appropriate policies to stimulate technological advance. First, it is widely understood that flexible, incentive-

---

33 However, there are many other factors, in addition to appropriating returns from innovation, that influence the incentive structure of firms, including 'first mover' advantages, market power, use of complementary assets, etc. (for a review see Levin *et al.*, 1987).

oriented policies are more likely to foster low-cost compliance pathways than those that impose prescriptive regulatory approaches (Jaffe *et al.*, 2005). A second robust conclusion is the need for public policy to promote a broad portfolio of research, because results cannot be guaranteed since it is impossible to *ex ante* identify

technical winners or losers (GTSP, 2001). A third conclusion is that more than explicit climate change or energy research is critical for the development of technologies pertinent to climate change. Spillovers from non-energy sectors have had enormous impacts on energy-sector innovation, implying that a broad and robust technological base may be as important as applied energy sector or similar R&D efforts. This robust base involves the full 'national systems of innovation'[34] involved in the development and use of technological knowledge. Cost and availability of enabling infrastructure can be especially important factors that limit technology uptake in developing countries.[35] Here enabling infrastructure would include management and regulatory capacity, as well as associated hardware and public infrastructure.

2.7.3   The international dimension in technology development and deployment: technology transfer

Article 4.5 of the Convention states that developed country Parties 'shall take all practicable steps to promote, facilitate, and finance, as appropriate, the transfer of, or access to, environmentally sound technologies and know-how to other Parties, particularly developing country Parties, to enable them to implement the provisions of the Convention', and to 'support the development and enhancement of endogenous capacities and technologies of developing country Parties'.

Similarly Article 10(c) of the Kyoto Protocol reiterated that all Parties shall: 'cooperate in the promotion of effective modalities for the development, application, and diffusion of, and take all practicable steps to promote, facilitate and finance, as appropriate, the transfer of, or access to, environmentally sound technologies, know-how, practices and processes pertinent to climate change, in particular to developing countries, including the formulation of policies and programmes for the effective transfer of environmentally sound technologies that are publicly owned or in the public domain and the creation of an enabling environment for the private sector, to promote and enhance the transfer of, and access to, environmentally sound technologies'.

Technology transfer is particularly relevant because of the great interest by developing countries in this issue. This interest arises from the fact that many developing countries are in a

phase of massive infrastructure build up. Delays in technology transfer could therefore lead to a lock-in in high-emissions systems for decades to come (e.g. Zou and Xuyan, 2005). Progress on this matter has usually been linked to progress on other matters of specific interest to developed countries. Thus Article 4.7 of the Convention is categorical that 'the extent to which developing country Parties will effectively implement their commitments under the Convention will depend on the effective implementation by developed country Parties of their commitments under the Convention related to financial resources and the transfer of technology'.

The IPCC Special Report on Methodological and Techno-logical Issues on Technology Transfer (SRTT) (IPCC, 2000) defined the term 'technology transfer' as a broad set of processes covering the flows of know-how, experience and equipment for mitigating and adapting to climate change amongst different stakeholders. A recent survey of the literature is provided in Keller (2004) and reviews with special reference to developing countries are included in Philibert (2005) and Lefevre (2005). The definition of technology transfer in the SRTT and the relevant literature is wider than implied by any particular article of the Convention or the Protocol. The term 'transfer' was defined to 'encompass diffusion of technologies and technology cooperation across and within countries'. It also 'comprises the process of learning to understand, utilize and replicate the technology, including the capacity to choose and adapt to local conditions and integrate it with indigenous technologies'.

This IPCC report acknowledged that the 'theme of techno-logy transfer is highly interdisciplinary and has been approached from a variety of perspectives, including business, law, finance, micro-economics, international trade, international political economy, environment, geography, anthropology, education, communication, and labour studies'.

Having defined technology transfer so broadly, the report (IPCC, 2000, p. 17) concluded that 'although there are numerous frameworks and models put forth to cover different aspects of technology transfer, *there are no corresponding overarching theories*' (emphasis added). Consequently there is no framework that encompasses such a broad definition of technology transfer.

The aforementioned report identified different stages of technology transfer and different pathways through which it is accomplished. These stages of technology transfer are: identification of needs, choice of technology, and assessment of conditions of transfer, agreement and implementation. Evaluation and adjustment or adaptation to local conditions, and replication are other important stages. Pathways for technology

---

34   The literature on national innovation systems highlights in particular the institutional dimensions governing the feedback between supply-push and demand-pull, and the inter-action between the public and private sectors that are distinctly different across countries. A detailed review of this literature is beyond the scope of this assessment. For an overview see, for example Lundvall, 2002, and Nelson and Nelson, 2002.

35   In this context, the concept of technological 'leapfrogging' (Goldemberg, 1991), and the resulting requirements for an enabling environment for radical technological change, is frequently discussed in the literature. For a critical review see, for example, Gallagher (2006).

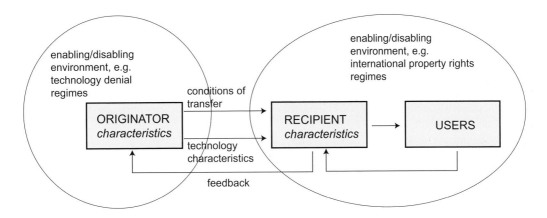

Figure 2.4: *A general framework for factors affecting technology transfer and subsequent innovation.*

transfer vary depending on the sector, technology type and maturity and country circumstances. Given this variety and complexity, the report concluded that there is no pre-set answer to enhancing technology transfer.

There is no international database tracking the flow of ESTs (environmentally sound technologies). Little is known about how much climate-relevant equipment is transferred, and even less about the transfer of know-how, practices and processes, and most international analyses rely on proxy variables. It is well known that the nature of financial flows from OECD countries to developing countries has changed over the last 15 years. Overseas development assistance (ODA) has declined and been overtaken by private sources of foreign direct investments (WDI, 2005). International financial statistics only reflect the quantity and not the quality of FDI. They also say nothing about what fraction is a transfer of ESTs. Despite its decline, ODA is still critical for the poorest countries, particularly when it is aimed at developing basic capacities to acquire, adapt, and use foreign technologies.

IPCC (2000, p. 22) summarized the historical experience as a 'failure of top-down, technology-focused development'. Some developing country policymakers believe that payments for technology are beyond their means and that international technology transfer contributes little to technological development in the recipient country (UNDP, 2000). Many failures of technology transfer have resulted from an absence of human and institutional capacity (IPCC, 2000, p. 118).

There are several modes to encourage technology transfer to developing countries, from technical assistance and technology grants, to capacity building and policy development cooperation. The priorities for these modes shift as host countries develop economically. Technology demonstration projects can play an important role early in the industrialization process. As the economy grows, policy development cooperation, such as assistance to develop energy-efficiency standards or to create an enabling environment for technology diffusion, becomes

more important. Ohshita and Ortolano (2003) studied past experiences of demonstration projects using cleaner energy technologies in developing countries through assistance by international organizations as well as developed countries. They found that demonstration projects raised awareness of cleaner energy technologies in the technology transfer process, but were not very successful in diffusing the technologies more widely in the target developing countries. For China in particular, demonstration projects played an important role in the past, when the economy began shifting from a centrally planned system to a more open, market-based system. There is increasing recognition that other modes of technology diffusion may now be more suitable for China. Given the continued high growth of the Chinese economy, donors have been shifting their assistance programmes from technology demonstration to policy development assistance (Ohsita, 2006).

Figure 2.4 shows one attempt to create a framework for all forms of technology transfer. In all forms technology transfer, especially across countries, at least seven characteristics are important. These are:
1. The characteristics of the technology.
2. The characteristics of the originator of the transfer.
3. The enabling (or disabling) environment in the country of origin.
4. The conditions of the transfer.
5. The characteristics of the recipient.
6. The enabling (or disabling) environment in the host country.
7. The ultimately valuable post-transfer steps, i.e. assimilation, replication and innovation.

Each of these characteristics are discussed below.

Characteristics of the originator of the transfer. Initially, there was a widespread tendency to think of technology transfer in supply-side terms – the initial choice and acquisition of technology (Brooks, 1995) and a lack of corresponding focus on the other factors that influence the successful outcome of technology transfer, such as enabling environment, institutions and finance.

The environment in the country of origin can be conducive or disabling for technology transfer. The public sector continues to be an important driver in the development of ESTs. Of the 22 barriers listed in the technical summary of the IPCC Report (2000) as barriers to technology transfer, 21 relate to the enabling environment of recipient countries. Many governments transfer or license the patents arising out of publicly funded efforts to the private sector as a part of their industrial policy, and then the transferred patents follow the rules of privately owned technologies (IPCC, 2000, p. 25).

One should also consider the 'imperfect' nature of technology markets:
(1) While some of the components of technology are of a public-good nature, others have an important tacit nature.
(2) Technology markets are normally very concentrated on the supply side, and bargaining power is unevenly distributed.
(3) The strategic nature of technologies normally includes limiting clauses and other restrictions in transfer contracts (for a discussion see Arora *et al.*, 2001; Kumar, 1998).

Technology Denial Regimes[36] in the country of origin also sometimes constitute a barrier to technology transfer, especially for multiple-use technologies. Thus supercomputers can be used for climate modelling and global circulation models and also to design missiles.

The conditions of the transfer. Most technologies are transferred in such a way that the originators also benefit from the transfer and this helps to establish strong incentives for proper management and maintenance of the technologies. The conditions of the transfer will primarily depend on the transfer pathway used, as mentioned above. Common pathways include government assistance programmes, direct purchases, trade, licensing, foreign direct investment, joint ventures, cooperative research agreements, co-production agreements, education and training and government direct investment. Developing countries have argued for the transfer of ESTs and corresponding know-how, on favourable, concessional and preferential terms (Agenda 21, 1992, Chapter 34). There have been instances in the pharmaceutical industry when certain drugs benefiting developing countries have been licensed either free or on concessional terms.

The characteristics of the recipient. The recipient must understand local needs and demands; and must possess the ability to assess, select, import, adapt, and adopt or utilize appropriate technologies.

The enabling (or disabling) environment in the host country. Many of the barriers to technology transfer that are listed in the IPCC Report (IPCC, 2000, p. 19) relate to the lack of an enabling (or a disabling) environment in the recipient country for the transfer of ESTs. A shift in focus, from technology transfer per se to the framework represented in Figure 2.4, leads to an equal emphasis on the human and institutional capacity in the receiving country. A crucial dimension of the enabling environment is an adequate science and educational infrastructure. It must be recognized that capacity building to develop this infrastructure is a slow and complex process, to which long-term commitments are essential.

A recipient's ability to absorb and use new technology effectively also improves its ability to develop innovations. Unfortunately, the capacity to innovate and replicate is poorly developed in developing countries (STAP, 1996). However, the engineering and management skills required in acquiring the capacity to optimize and innovate are non-trivial. The technology-importing firm needs to display what has been called 'active technological behaviour'. Firms that do not do this are left in a vicious circle of technological dependence and stagnation (UNDP, 2000).

2.8   Regional dimensions

Climate change studies have used various different regional definitions depending on the character of the problem considered and differences in methodological approaches. Regional studies can be organized according to geographical criteria, political organizational structures, trade relations, climatic conditions, stage of industrialization or other socio-economic criteria relevant to adaptive and mitigative capacity (Duque and Ramos, 2004; Ott *et al.*, 2004; Pan, 2004a).

Some classifications are based on so-called 'normative criteria' such as membership of countries in UN fora and agreements. Differentiation into Annex-1 and non-Annex-1 countries is specified in the UNFCCC, although the classification of certain countries has been a matter of some dispute. Annex-1 countries are further sub-divided into those that are undergoing a transition to market economies. Figure 13.2 in Chapter 13 shows the current country groupings under the Climate Convention, OECD and the European Union. Some Economies in Transition (Rabinovitch and Leitman, 1993) and developing countries are members of the OECD, and some developing countries have income levels that are higher than developed nations (Baumert *et al.*, 2004; Ott *et al.*, 2004). Given the complexities of the criteria used in country groupings, in this report the terms 'developed countries', 'economies in

---

36 Regulatory criteria denying access to certain technologies to individual countries or groups of countries.

transition' (together forming the industrialized countries) and 'developing countries' are commonly used; categories that are primarily of a socio-economic nature.

In climate mitigation studies, there are often two types of regional breakdowns used – physio-geographic or socio-economic. Data on insolation (relevant to solar power), rainfall (relevant to hydrower), temperature, precipitation and soil type (relevant to the potential for carbon sequestration) are examples of physio-geographic classifications useful in climate change mitigation studies.

The multitude of possible regional representations hinders the comparability and transfer of information between the various types of studies implemented for specific regions and scales. Data availability also determines what kinds of aggregation are possible. Proxies are used when data is not available. This report has generally chosen a pragmatic way of analyzing regional information and presenting findings. Readers should bear in mind that any regional classification masks sub-regional differences.

REFERENCES

Adger, W.N., 2001a: Social Capital and Climate Change. Working Paper, **8**.

African, D.B., A.D. Bank, U.K. Department for International Development, E.C. Directorate-General for Development, G. Federal Ministry for Economic Cooperation and Development, T.N. Ministry of Foreign Affairs - Development Cooperation, O.f.E.C.a. Development, U.N.D. Programme, U.N.E. Programme, and T.W. Bank, 2003: Poverty and Climate Change. Reducing the vulnerability of the poor.

Agarwal, A. and S. Narain, 1991: Global warming in an unequal world: A case of ecocolonialism. Centre for Science and Environment, New Delhi.

Agenda 21, 1992: The Rio Declaration on Environment and Development, and the Statement of principles for the Sustainable Management of Forests. Rio de Janeiro, Brazil.

Alberts, W., 1989: The experience curve doctrine reconsidered. *Journal of Marketing*, **53**, pp. 36-49.

Alcamo, J., A. Bouwman, J. Edmonds, A. Grübler, T. Morita, and A. Sugandhy, 1995: An evaluation of the IPCC IS92 emission scenarios. In *Climate Change 1994: Radiative forcing of Climate Change and an evaluation of the IPCC IS92 emission scenarios.* Cambridge University Press, Cambridge, pp. 247-304.

Alchian, A., 1963: Reliability of progress curves in airframe production. *Econometrica*, **31**(4), pp. 679-693.

Allen, M., S. Raper, and J. Mitchell, 2001: Uncertainty in the IPCC's third assessment report. Science, **293**(July 20), pp. 430-433.

Argote, L. and D. Epple, 1990: Learning curves in manufacturing. *Science*, **247**(4945), pp. 920-924.

Arora, A., A. Fosfuri, and A. Gambardella, 2001: Markets for technology. Cambridge, Mass., The MIT Press.

Arrow, K. and Lind, 1970: Uncertainty and the evaluation of public investment decision. *American Economic Review*, **60**, pp. 364-378.

Arrow, K., 1962: The economic implications of learning by doing. *The Review of Economic Studies*, **29**(3), pp. 155-173.

Arrow, K., P. Dasgupta, L. Goulder, G. Daily, P. Ehrlich, G. Heal, S. Levin, K.-G. Mäler, S. Schneider, D. Starret, and B. Walker, 2004: Are we consuming too much? *Journal of Economic Perspectives*, **18**(3), pp. 147-172.

Arthur, W.B., 1989: Competing technologies, increasing returns, and lock-in by historical events. *The Economic Journal*, **99**, pp. 116-131.

Atkinson, G., R. Dubourg, K. Hamilton, M. Munasinghe, D. Pearce, and C. Young, 1997: Measuring sustainable development. Macroeconomics and the Environment.

Australian Industry Commission, 1995: Research and development. Australian Government Publishing Service.

Ausubel, J.H. and W.D. Nordhaus, 1983: A review of estimates of future carbon dioxide emissions. In *Changing Climate.* National Research Council, National Academy of Sciences, National Academy Press, pp. 153-185.

Baer, P. and P. Templet, 2001: GLEAM: A simple model for the analysis of equity in policies to regulate greenhouse gas emissions through tradable permits. In *The Sustainability of Long Term Growth: Socioeconomic and Ecological Perspectives* (M. Munasinghe and O. Sunkel), Cheltenham, UK, Edward Elgar Publishing Company.

Baker, E., 2005: Uncertainty and learning in a strategic environment: Global Climate Change. *Resource and Energy Economics*, **27**, pp. 19-40.

Ballet, J., J.-L. Dubois, and F.-R. Mahieu, 2003: Le développement socialement durable: un moyen d'intégrer capacités et durabilité? In *Third Conference on the Capability Approach,* 6-9 September 2003, University of Pavia.

Baloff, N., 1966: The learning curve - some controversial issues. *The Journal of Industrial Economics*, **14**(3), pp. 275-282.

Banuri, T. and J. Weyant, 2001: Setting the stage: Climate Change and sustainable development. In *Mitigation.* IPCC Working Group III, Third Assessment Report. Cambridge University Press, Cambridge.

Barker, T., 2003: Representing global, climate change, adaptation and mitigation. *Global Environmental Change*, **13**, pp. 1-6.

Barker, T., J. Köhler, and M. Villena, 2002: The costs of greenhouse gas abatement: A meta-analysis of post-SRES mitigation scenarios, *Environmental Economics and Policy Studies*, **5**(2), pp.135-166.

Baumert, K., J. Pershing, *et al.*, 2004: Climate data: insights and observations. Pew Centre on Climate Change, Washington, D.C.

Baumert, K.A., O. Blanchard, S. Llosa, and J. Perkaus, 2002: Building on the Kyoto Protocol: Options for protecting the climate. World Resources Institute, Washington D.C.

Bay, T. and A. Upmann, 2004: On the double dividend under imperfect competition. *Environmental and Resource Economics*, **28**(2), pp. 169-194.

Beg, N., J.C. Morlot, O. Davidson, Y. Afrane-Okesse, L. Tyani, F. Denton, Y. Sokona, J.P. Thomas, E.L.L. Rovere, J.K. Parikh, K. Parikh, and A. Rahman, 2002: Linkages between climate change and sustainable development. *Climate Policy*, **2**(5), pp. 129-144.

Bell, M.L., B.F. Hobbs, E.M. Elliot, T. Hugh Ellis, and Z. Robinson, 2001: An evaluation of multi-criteria methods. Department of Geography and Environmental Engineering, The Johns Hopkins University, DOI: 10.1002/mcda.305. In Integrated Assessment of Climate Policy. *USA Journal of Multi-Criteria Decision Analysis,* **10**, pp. 229-256.

Bewley, T.F., 2002: Knightian decision theory - part I. *Decisions in Economics and Finance*, **25**(2), pp. 79-110.

BINC, 2004: Part III, Chapter 1. In *Brazil's Initial National Communication to UNFCCC.*

Botzler, R.G. and S.J. Armstrong, 1998: Environmental ethics: Divergence and convergence. Boston.

Bresnahan, T. and M. Trajtenberg, 1992: General purpose technologies: engines of growth? National Bureau of Economic Research.

Bromley, D.W. and J. Paavola, 2002: Economics, ethics, and environmental policy: Contested choices. Blackwell, Oxford.

Brooks, H., 1995: Review of C. Freeman: The economics of hope- essays on technical change, economic growth and the environment. *Minerva*, **33**(2), pp. 199-203.

Brown, D., 2002: American heat: Ethical problems with the United States response to global warming. Lanham, MD: Rowan & Littlefield.

Brunnermier, S. and M. Cohen, 2003: Determinants of environmental innovation in us manufacturing industries. *Journal of Environmental Economics and Management*, **45**, pp. 278-293.

Burton, I., S. Huq, B. Lim, O. Pilifosova, and E.L. Shipper, 2002: From impacts assessment to adaptation priorities: the shaping of adaptation policy. *Climate Policy*, **2**(2-3), pp. 145-159.

Cairncross, F., 2004: What makes environmental treaties work? *Conservation Spring 2004*, **5**(2).

Chevé, M. and R. Congar, 2000: Optimal pollution control under imprecise risk and irreversibility.

Chichilnisky, G., 2000: An axiomatic approach to choice under uncertainty with catastrophic risks. *Resource and Energy Economics,* **22**, pp. 221-231.

Chiroleu-Assouline, M. and M. Fodha, 2005: Double dividend with involuntary unemployment: Efficiency and intergenerational equity. *Environmental and Resource Economics*, **31**(4), pp. 389-403.

Christiansson, L., 1995: Diffusion and learning curves of renewable energy technologies. International Institute for Applied Systems Analysis.

CINC, 2004: China's Initial National Communication to UNFCCC.

Clarke, L.E. and J.P. Weyant, 2002: Modeling induced technological change: An overview. In *Technological Change and the Environment.* RFF Press, Washington D.C, pp. 320-363.

Cline, W.R. 2005: Climate Change. In *Global Crises, Global Solutions*, B. Lomborg (ed.), Cambridge University Press, c.1.

Coase, R.H., 1960: The problem of social cost. *The Journal of Law and Economics*, **III**, October 1960.

Cohen, S., D. Demeritt, J. Robinson, and D.S. Rothman, 1998: Climate change and sustainable development: Towards dialogue. *Global Environmental Change*, **8**(4), pp 341-371.

Cohen, W. and D. Levinthal, 1989: Innovation and learning: The two faces of R&D. *The Economic Journal*, **99**, pp. 569-596.

Cox, J., J. Ingersoll, and S.S. Ross, 1985: A theory of the term structure of interest rates. *Econometrica*, **53**, pp. 385-403.

CSD, 2005: Investing in development: A practical plan to achieve the millenium development goals. Earthscan London.

Daly, H.E., 1990: Toward some operational principles of sustainable development. *Ecological Economics*, **2**, pp. 1-6.

Dang, H.H., A. Michaelowa, and D.D. Tuan, 2003: Synergy of adaptation and mitigation strategies in the context of sustainable development: The case of Vietnam. *Climate Policy*, **3**, pp. S81-S96.

Danida, 2005: Danish climate and development action programme. A tool kit for climate proofing Danish development assistance. <http://www.um.dk/Publikationer/Danida/English/Enviroment/ DanishClimateAndDevelopmentActionProgramme/samlet.pdf>, accessed 31/05/07.

Dasgupta, P., 1993: An inquiry into well-being and destitution. Oxford University Press, New York.

Dasgupta, P., K.-G. Mäler, and S. Barrett, 1999: Intergenerational equity, social discount rates, and global warming: in Portney. In *Discounting and Intergenerational Equity. Resources for the Future*, Washington D.C.

Davidson, O., K. Halsnaes, S. Huq, M. Kok, B. Metz, Y. Sokona, and J. Verhagen, 2003: Development and climate: The case of Sub-Saharan Africa. *Climate Policy*, **3**(no Suppl.1), pp. 97-113.

Day, G. and D. Montgomery, 1983: Diagnosing the experience curve. *Journal of Marketing*, **47**, pp. 44-58.

De Finetti, B., 1937: La prévision: Ses lois logiques, ses sources subjectives. Annales de l'Institut Henri Poincaré 7, Paris, 1-68, translated into English by Henry E. Kyburg Jr., Foresight: Its logical laws, its subjective sources. In *Studies in Subjective Probability*, H.E. Kyburg Jr. and H.E. Smokler (eds.), pp. 53-118, Wiley, New York. <http://www.numdam.org/item?id=AIHP_1937__7_1_1_0>, accessed 31/05/07.

Dessai, S. and M. Hulme, 2004: Does climate adaptation policy need probabilities? *Climate Policy*, **4**, pp. 107-128.

Dessues, S. and D. O' Connor, 2003: Climate policy without tears: CGE-based ancillary benefits estimates for Chile. *Environmental and Resource Economics*, **25**, pp. 287-317.

Dimson, E., P. Marsh, and M.M. Staunton, 2000: The Millenium book: A century of investment returns. ABN-AMRO, Londen.

Dresner, S., 2002: The principles of sustainability. Earthscan Publication, London.

Dudek, D., A. Golub, and E. Strukova, 2003: Ancillary benefits of reducing greenhouse gas emissions in transitional economies. *World Development*, **31**(10), pp. 1759-1769.

Duque, J.C. and R. Ramos, 2004: Design of homogeneous territorial units: A methodological proposal. 44th Congress of the European Regional Science Association. Porto.

Dutton, J. and A. Thomas, 1984: Treating progress functions as a managerial opportunity. *The Academy of Management Review*, **9**(2), pp. 235-247.

Edenhofer, O., K. Lessmann, C. Kemfert, M. Grubb, and J. Koehler, 2006: Induced technological change: Exploring its implication for the economics of atmospheric stabilisation. Insights from the Innovation Modelling Comparison Project. *The Energy Journal* (Special Issue, in press).

Edmonds, J., M. Wise, and J. Dooley, 1997: Atmospheric stabilization and the role of energy technology. In *Climate Change Policy, Risk Prioritization and U. S. Economic Growth.* American Council for Capital Formation, Washington D.C, pp. 71-94.

Edmonds, J.A. and S.J. Smith 2006: The technology of two degrees. In *Avoiding Dangerous Climate Change,* H.J. Schellnhuber, W. Cramer, N. Nakicenovic, T. Wigley and G. Yohe (eds). Cambridge University Press, Cambridge, pp. 385-392.

Edmonds, J.A., 2004: Technology options for a long-term mitigation response to Climate Change. In *2004 Earth Technologies Forum*, April 13, Pacific Northwest National Laboratory, Richland, WA.

Ellesberg, D., 2001: Risk, ambiguity and decision. Routledge.

Epstein, L. 1980: Decision making and the temporal resolution of uncertainty. *International Economic Review*, **21**(2), pp. 269-82.

Flannery, B.P. and H.S. Khesghi, 2005: An industry perspective on successful development and global commercialization of innovative technologies for GHG mitigation. In *Intergovernmental Panel on Climate Change Workshop on Industry Technology Development, Transfer and Diffusion*, September 2004, Tokyo.

Foxon, T.J., 2003: Inducing innovation for a low-carbon future: Drivers, barriers and policies. The Carbon Trust, London.

Freeman, C., 1994: The economics of technological change. *Cambridge Journal of Economics*, **18**, pp. 463-514.

Gallagher, K.S., 2006: Limits to leapfrogging in energy technologies? Evidence from the Chinese automobile industry. *Energy Policy*, **34**(4), pp. 383-394.

Gerlagh, R. and B.C.C. van der Zwaan, 2003: Gross World Product and consumption in a global warming model with endogenous technological change. *Resource and Energy Economics*, **25**, pp. 35-57.

Gerlagh, R. and B.C.C. van der Zwaan, 2004: A sensitivity analysis on timing and costs of greenhouse gas abatement, calculations with DEMETER. *Climatic Change*, **65**, pp. 39-71.

Gerlagh, R. and B.C.C. van der Zwaan, 2006: Options and instruments for a deep cut in CO_2 emissions: Carbon dioxide capture or renewables, taxes or subsidies? *The Energy Journal.*

Ghemawat, P., 1985: Building strategy on the experience curve. Harvard Business Review, pp. 143-149.

Gielen, D. and C.H. Chen, 2001: The CO_2 emission reduction benefits of Chinese energy policies and environmental policies: A case study for Shanghai, period 1995-2020. *Ecological Economics*, **39**(2), pp. 257-270.

Godard, O., C. Henry, P. Lagadec, and E. Michel-Kerjan, 2002: Traité des nouveaux risques. Précaution, crise, assurance. *Folio Actuel*, **100**.

Goklany, I.M., 2001: Economic growth and the state of humanity: Political economy research center, Policy Study 21.

Goklany, I.M., 2002: The globalization of human well-being. Policy Analysis, No. 447. Cato Institute, Washington D.C.

Goldemberg, J., 1991: 'Leap-frogging': A new energy policy for developing countries. *WEC Journal,* December 1991, pp. 27-30.

Gollier, C., 2001: Should we beware of the precautionary principle? *Economic Policy,* **16,** pp. 301-328.

Gollier, C., 2002a: Discounting an uncertain future. *Journal of Public Economics,* **85,** pp. 149-166.

Gollier, C., 2002b: Time horizon and the discount rate. *Journal of Economic Theory,* **107,** pp. 463-473.

Gollier, C., 2004: The consumption-based determinants of the term structure of discount rates, mimeo. University of Toulouse.

Gollier, C., B. Jullien, and N. Treich, 2000: Scientific progress and irreversibility: an economic interpretation of the 'Precautionary Principle'. *Journal of Public Economics,* **75,** pp. 229-253.

Goulder, L. and S. Schneider, 1999: Induced technological change and the attractiveness of CO_2 abatement policies. *Resource and Energy Economics,* **21,** pp. 211-253.

Goulder, L.H., 2004: Induced technological change and climate policy. The Pew Center.

Graham, J.D. and J.B. Wiener (eds.), 1995: Risk versus risk: Tradeoffs in protecting health and the environment. Vol. I 'Fundamentals'. Harvard University Press.

Grasso, M., 2004: Climate Change, the global public good. Dipartimento di Sociologia e Ricerca Sociale, Università degli Studi di Milano, Bicocca, WP. <129.3.20.41/eps/othr/papers/0405/0405010.pdf>, accessed 31/05/07.

Griliches, Z., 1992: The search for R&D spillovers. *Scandinavian Journal of Economics,* **94,** pp. 29-47 (Supplement).

Gritsevskyi, A. and N. Nakicenovic, 2000: Modeling uncertainty of induced technological change. *Energy Policy,* **28**(13), pp. 907-921.

Grubb, M., 2004: Technology innovation and Climate Change policy: An overview of issues and options. *Keio economic studies,* Vol. **41,** 2, pp. 103-132. <http://www.econ.cam.ac.uk/faculty/grubb/publications/J38.pdf>, accessed 31/05/07.

Grubb, M., J. Köhler, and D. Anderson, 2002: Induced technical change. In Energy And Environmental Modeling: Analytic Approaches and Policy Implications. *Annual Review of Energy and Environment,* **27,** pp. 271-308.

Grubler, A., 1998: Technology and global change. Cambridge University Press, Cambridge.

Grübler, A., N. Nakicenovic, and D.G. Victor, 1999: Dynamics of energy technologies and global change. *Energy Policy,* **27,** pp. 247-280.

Grübler, A., N. Nakicenovic, and W.D. Nordhaus (eds.), 2002: Technological change and the environment. Washington, D.C.

GTSP, 2001: Addressing Climate Change: An initial report of an international public-private collaboration. Battelle Memorial Institute, Washington, D.C.

Halsnæs, K. and A. Markandya, 2002: Analytical approaches for decision-making, sustainable development and greenhouse gas emission-reduction policies. In *Climate Change and sustainable development. Prospects for developing countries.* Earthscan Publications Ltd., London, pp. 129-162.

Halsnæs, K. and A. Olhoff., 2005: International markets for greenhouse gas emission reduction policies - Possibilities for integrating developing countries. *Energy Policy,* **33**(18), pp. 2313-2525.

Halsnæs, K. and J. Verhagen, 2007: Development based climate change adaptation and mitigation - Conceptual issues and lessons learned in studies in developing countries. In *Mitigation and Adaptation Strategies for Global Change,* first published online <http://www.springerlink.com/content/102962> accessed 05/06/07, doi 10.1007/s11027-007-9093-6.

Halsnæs, K. and P.S. Shukla, 2007: Sustainable development as a framework for developing country participation in international climate policies. In *Mitigation and Adaptation Strategies for Global Change,*

first published online <http://www.springerlink.com/content/102962> accessed 05/06/07, doi 10.1007/s11027-006-9079-9.

Halsnæs, K., 2002: A review of the literature on Climate Change and sustainable development. In *Climate Change and Sustainable Development.* Earthscan. 10, 9 ed.

Hamilton, K. and M. Kenber, April 2006: Business views on international climate and energy policy. Report commissioned by UK Government.

Hamilton, K., 2005: The finance-policy gap: Policy conditions for attracting long-term investment. In *The Finance of Climate Change,* K. Tang (ed.), Risk Books, London.

Hardi, P. and S. Barg, 1997: Measuring sustainable development: Review of current practice. Industry Canada, Occasional Paper Number 17. Ottawa, Ontario.

Harris, J., J. Anderson, and W. Shafron, 2000: Investment in energy efficiency: a survey of Australian firms. *Energy Policy,* **28** (2000), pp. 867-876.

Heller, T. and P.R. Shukla, 2003: Development and climate: Engaging developing countries. Pew Centre on Global Climate Change, Arlington, USA.

Henry, C. and M. Henry, 2002: Formalization and applications of the Precautionary Principle Cahier du Laboratoire d'Econometrie de l'Ecole Polytechnique, n° 2002-008.

Hirsch, W., 1956: Firm progress ratios. *Econometrica,* **24**(2), pp. 136-143.

Hoffert, M.I., K. Caldeira, G. Benford, D.R. Criswell, C. Green, H. Herzog, A.K. Jain, H.S. Kheshgi, K.S. Lackner, J.S. Lewis, H.D. Lightfoot, W. Manheimer, J.C. Mankins, M.E. Mauel, L.J. Perkins, M.E. Schlesinger, T. Volk, and T.M.L. Wigley, 2002: Advanced technology paths to global climate stability: Energy for a greenhouse planet. *Science,* **298,** pp. 981-987.

Holling, C.S., 2001: Understanding the complexity of economic, ecological, and social systems. *Ecosystems,* **4,** pp. 390-405.

Holling, C.S., L. Gunderson, and G. Peterson, 2002: Sustainability and panarchies. In *Panarchy: Understanding Transformations in Human and Natural Systems,* L. Gunderson and C.S. Holling (ed.), 2002: Island Press, Washington D.C.

IEA, 2000: Experience curves for energy technology policy. International Energy Agency, Paris.

IEA, 2003: World energy investment outlook. International Energy Agency, Paris.

IEA, 2004: World energy outlook. International Energy Agency, Paris.

IINC, 2004: India's initial national communication to UNFCCC, Chapter 6.

IPCC, 1996: Climate Change 1995: Economic and social dimensions of Climate Change. Cambridge University Press, Cambridge.

IPCC, 2000: *Methodological and technological issues on technology transfer.* Special Report of the Intergovernmental Panel on Climate Change (IPCC), Cambridge University Press, Cambridge.

IPCC, 2001: *Climate Change 2001: Mitigation.* Contribution of Working Group III to the Third Assessment Report of the Intergovernmental Panel on Climate Change (IPCC). Cambridge University Press, Cambridge.

IPCC, 2001a: *Climate Change 2001: Impacts, adaptation and vulnerability.* Contribution of Working Group II to the Third Assessment Report of Intergovernmental Panel on Climate Change. Cambridge University Press, Cambridge.

IPCC, 2002: *Synthesis Report.* Contributions of Working Group I, II and III to the Third Assessment Report of the Intergovernmental Panel on Climate Change (IPCC). Cambridge University Press, Cambridge. IPIECA and UNEP, London.

IPCC, 2007a: Climate Change 2007: The Physical Science Basis. Contribution of Working Group I to the Fourth Assessment Report of the Intergovernmental Panel on Climate Change [Solomon, S., D. Qin, M. Manning, Z. Chen, M. Marquis, K.B.M.Tignor and H.L. Miller (eds.)]. Cambridge University Press, Cambridge, United Kingdom and New York, NY, USA, 996 pp.

IPCC, 2007b: Climate Change 2007: Impacts, Adaptation and Vulnerability. Contribution of Working Group II to the Fourth Assessment Report of the Intergovernmental Panel on Climate Change [Parry, M.L., O.F. Canziani, J.T. Palutikof, P.J. van der Linden, C.E. Hanson (eds.)]. Cambridge University Press, Cambridge, United Kingdom and New York, NY, USA.

IPIECA, 1995: Technology cooperation and capacity building: The oil industry experience.

Irwin, D.A. and P.J. Klenow, 1994: Learning-by-doing spillovers in the semiconductor industry. *Journal of Political Economy*, **102**(6), pp. 1200-1227.

Jaffe, A. and K. Palmer, 1996: Environmental regulation and innovation: A panel data study. National Bureau of Economic Research.

Jaffe, A.B., R.G. Newell and R.N. Stavins, 2003: Technological change and the environment. In *Handbook of Environmental Economics*, K.-G. Mäler and J. R. Vincent (eds), Elsevier Science B.V., pp. 461-516.

Jaffe, A.B., R.G. Newell, and R.N. Stavins, 2005: A tale of two market failures: Technology and environmental policy. *Ecological Economics*, **54**(2-3), pp. 164-174.

Joskow, P. and N. Rose, 1985: The effects of technological change, experience, and environmental regulation on the construction cost of coal-burning generation units. RAND *Journal of Economics*, **16**(1), pp. 1-27.

Kagan, J., 2002: Surprise, uncertainty, and mental structures. Harvard University Press.

Kansuntisukmongko, C., 2004: Irreversibility and learning in global climate change problem: Linear and single-quadratic models. HKK Conference, University of Waterloo. June 1999, <are.berkeley. edu/courses/envres_seminar/s2004/PaperChalotorn.pdf>, accessed 31/05/07.

Kay, J. and J. Foster, 1999: About teaching systems thinking. In proceedings of the HKK Conference, G. Savage and P. Roe (eds.), University of Waterloo, June 1999.

Keller, K., B.M. Bolker, and D. Bradford, 2004: Uncertain climate thresholds and optimal economic growth. *Journal of Environmental Economics and Management*, **48**, pp. 733-741.

Keller, W. 2004: International technology diffusion. *Journal of Economic Literature*, **42**(3), pp. 752-782.

Kelly, P.M. and W.N. Adger, 1999: Assessing vulnerability to climate change and facilitating adaptation. UK.

Kelly, P.M. and W.N. Adger, 2000: Theory and practice in assessing vulnerability to climate change and facilitating adaptation. *Climatic Change*, **47**, pp. 325-352.

Klevorick, A., R. Levin, R. Nelson, and S. Winter, 1995: On the sources and significance of interindustry differences in technological opportunities. *Research Policy*, **24**, pp. 185-205.

Knight, F.H., 1921: Risk, uncertainty and profit. Houghton Mifflin, Boston.

Kolstad, C.D., 1996: Learning and stock effects in environmental regulation: the case of greenhouse gas emissions. *Journal of Environmental Economics and Management*, **31**(2), pp.1-18.

Kouvaritakis, N., A. Soria, and S. Isoard, 2000: Endogenous learning in world post-Kyoto scenarios: Application of the POLES model under adaptive expectations. *International Journal of Global Energy Issues*, **14**(1-4), pp. 222-248.

Kriegler, E., 2005: *Imprecise probability analysis for integrated assessment of climate change*. PhD Thesis University of Potsdam, Germany.

Kriegler, E., H. Held, and T. Bruckner, 2006: Climate protection strategies under ambiguity about catastrophic consequences. In *Decision Making and Risk Management in Sustainability Science*. J. Kropp and J. Scheffran (eds.), Nova Scientific Publications New York.

Kumar, N., 1998: Globalization, foreign direct investment and technology transfer. Routledge, London.

Kuntz-Duriseti, K., 2004: Evaluating the economic value of the precautionary principle: using cost benefit analysis to place a value on precaution. *Environmental Science and Policy*, **7**, pp. 291-301.

Langrish, J., M. Gibbons, W. Evans, and F. Jevons, 1972: Wealth from knowledge: Studies of innovation in industry. MacMillion.

Lanjouw, J. and A. Mody, 1996: Innovation and the international diffusion of environmentally responsive technology. *Research Policy*, **25**, pp. 549-571.

Larson, E.D., W. Zongxin, P. DeLaquil, C. Wenying, and G. Pengfei, 2003: Future implications of China∕s energy-technology choices. *Energy Policy*, **31** (2003), pp. 1189-1204.

Lefèvre, N., 2005: Deploying climate-friendly technologies through collaboration with developing countries. IEA Information Paper, IEA, Paris.

Lehtonen, M., 2004: The environmental-social interface of sustainable development: capabilities, social capital, institutions. *Ecological Economics*, **49**, pp. 199-214.

Leimbach, M., 2003: Equity and carbon emission trading: A model analysis. *Energy Policy*, **31**, pp. 1033-1044.

Lempert, R.J. and M.E. Schlesinger, 2000: The impacts of climate variability on near-term policy choices and the value of information, *Climatic Change*, **45**(1), pp. 129-161.

Levin, R.C., A.K. Klevorick, R.R. Nelson, and S. Winter, 1987: Appropriating the returns from industrial research and development. *Brookings Papers on Economic Activity*, **3**, pp. 783-820.

Lieberman, M., 1984: The learning curve and pricing in the chemical processing industries. Rand. *Journal of Economics*, **15**(2), pp. 213-228.

Loewenstein, G. and D. Prelec, 1992: Choices over time. New York, Russell Sage Foundation.

Löschel, A., 2002: Technological change in economic models of environmental policy: A survey. *Ecological Economics*, **43**, pp. 105-126.

Lundvall, B.-Å. 2002: *Innovation, growth and social cohesion: The Danish Model*, Edward Elgar Publishing Limited, Cheltenham:

Maddison, A., 2003: The world economy: Historical statistics. Development Centre of the Organisation for Economic Cooperation and Development. Paris, France.

Manne, A. and R. Richels, 1992: Buying greenhouse insurance - the economic costs of carbon dioxide emission limits. MIT Press, Cambridge, MA.

Markandya, A. and D.T.G. Rubbelke, 2004: Ancillary benefits of climate policy. *Jahrbucher fur Nationalokonomie un Statistik*, **224 (4)**, pp. 488-503.

Markandya, A. and K. Halsnaes, 2002a: Climate Change and Sustainable Development. Earthscan, London.

Markandya, A. and K. Halsnaes, 2002b: Developing countries and climate change. In *The Economics of Climate Change*, A. Owen and N. Hanley (eds.), Routledge (2004).

Markandya, A., K. Halsnæs, P. Mason, and A. Olhoff, 2002: A conceptual framework for analysing in the context of sustainable development. In *Climate Change and sustainable development - Prospects for developing countries*. Earthscan Publications Ltd., London, pp. 15-48.

McDonald, A. and L. Schrattenholzer, 2001: Learning rates for energy technologies. *Energy Policy*, **29**, pp. 255-261.

MEA, 2005: Millenium Ecosystem Assessment, 2005. Island Press, Washington, D.C. World Resources Institute.

Meadowcroft, J., 1997: Planning for sustainable development: Insights from the literatures of political science. *European Journal of Political Research*, **31**, pp. 427-454.

Meadows, D., 1998: Indicators and information systems for sustainable development. A Report to Balaton Group. September 1998.

Meehl, G.A., T.F. Stocker, W.D. Collins, P. Friedlingstein, A.T. Gaye, J.M. Gregory, A. Kitoh, R. Knutti, J.M. Murphy, A. Noda, S.C.B. Raper, I.G. Watterson, A.J. Weaver and Z.-C. Zhao, 2007: Global Climate Projections. In: *Climate Change 2007: The Physical Science Basis. Contribution of Working Group I to the Fourth Assessment Report of the Intergovernmental Panel on Climate Change* [Solomon, S., D. Qin, M. Manning, Z. Chen, M. Marquis, K.B. Averyt, M. Tignor and H.L. Miller (eds.)]. Cambridge University Press, Cambridge, United Kingdom and New York, NY, USA.

Messner, S., 1997: Endogenized technological learning in an energy systems model. *Journal of Evolutionary Economics*, **7**(3), pp. 291-313.

Metz, B., *et al.*, 2002: Towards an equitable climate change regime: compatibility with Article 2 of the climate change convention and the link with sustainable development. *Climate Policy*, **2**, pp. 211-230.

Miketa, A. and L. Schrattenholzer, 2004: Experiments with a methodology to model the role of R&D expenditures in energy technology learning processes: First results. *Energy Policy*, **32**, pp. 1679-1692.

Mills, E., 2005: Insurance in a climate of change. *Science*, **309**(5737)(August 12), pp. 1040-1044.

Montgomery, D. and G. Day, 1985: Experience curves: Evidence, empirical issues, and applications. In *Strategic Marketing and Management*. John Wiley & Sons.

Montgomery, W.D. and A.E. Smith, 2005: Price quantity and technology strategies for climate change policy. CRA International, <http://www.crai.com/.%5Cpubs%5Cpub_4141.pdf>, accessed 31/05/07.

Moss, R.H. and S.H. Schneider, 2000: Uncertainties in the IPCC TAR: Recommendations to lead authors for more consistent assessment and reporting. In *Guidance Papers on the Cross Cutting Issues of the Third Assessment Report of the IPCC*. World Meteorological Organization, Geneva, pp. 33-51.

MOST, 2004: Clean Development Mechanism in China: Taking a proactive and sustainable approach. Chinese Ministry of Science and Technology, The Deutsche Gesellschaft für Technische Zusammenarbeit, German Technical Cooperation unit (GTZ) of the Federal Ministry for Economic Cooperation and Development, Swiss State Secretariat for Economic Affairs (SECO), World Bank, and others.

Mowery, C. and N. Rosenberg, 1989: Technology and the pursuit of economic growth. Cambridge University Press, Cambridge.

Mowery, D. and N. Rosenberg, 1979: The influence of market demand upon innovation: A critical review of some recent studies. *Research Policy*, **8** (2), pp. 102-153.

Müller, B., 2001: Varieties of distributive justice in climate change, an editorial comment. *Climatic Change*, **48** (2001), pp. 273-288.

Müller, B., 2002: Equity in climate change: The great divide. Clime Asia (COP8 Special Issue).

Munasinghe, M. and R. Swart (eds.), 2000: Climate change and its linkages with development, equity and sustainability. Intergovernmental Panel on Climate Change (IPCC), Geneva.

Munasinghe, M. and R. Swart, 2005: Primer on Climate Change and sustainable development. Cambridge University Press, UK.

Munasinghe, M., 2002: The sustainomics transdisciplinary meta-framework for making development more sustainable: applications to energy issues. *International Journal of Sustainable Development*, **4**(2), pp. 6-54.

Murray, B.C., A. Keeler, and W.N. Thurman, 2005: Tax interaction effects, environmental regulation and rule of thumb. Adjustments to Social Costs. *Environmental and Resource Economics*, **30**(1), pp. 73-92.

Myers, S. and D. Marquis, 1969: Successful industrial innovation. National Science Foundation.

Nadiri, M., 1993: Innovations and technological spillovers. National Bureau of Economic Research.

Najam, A. and A. Rahman, 2003: Integrating sustainable development into the fourth IPCC assessment. *Climate Policy* (special issue on Climate Change and sustainable development), **3** (supplement 1), pp. S9-S17.

Nakicenovic, N. and K. Riahi, 2002: An assessment of technological change across selected energy scenarios. World Energy Council and International institute for Applied Systems Analysis. <http://www.iiasa.ac.at/Research/TNT/WEB/Publications/Technological_Change_in_Energy/technological_change_in_energy.html>, accessed 31/05/07

Nakicenovic, N., A. Grubler, and A. McDonald (eds), 1998a: Global Energy Perspectives. Cambridge University Press.

Nakicenovic, N., N. Victor, and T. Morita, 1998b: Emissions scenario database and review of scenarios. *Mitigation and Adaptation Strategies for Global Change*, **3**, pp. 95-120.

Narain, U., M. Hanemann and A. Fisher, 2004: The temporal resolution of uncertainty and the irreversibility effect. CUDARE Working Paper 935, Department of Agricultural and Resource Economics, University of California, Berkeley.

Nelson, R.R., and K. Nelson, 2002: Technology, institutions and innovation systems. *Research Policy*, **31**(20), pp. 265-272.

Nemet, G., 2005: Beyond the learning curve: factors influencing cost reductions in photovoltaics. *Energy Policy* (in press), corrected proof available online <http://www.sciencedirect.com>, accessed 1 August 2005.

NETL, 2004: Carbon sequestration technology roadmap and program plan. National Energy Technology Laboratory, Pittsburgh, PN.

Newell, R. and W. Pizer, 2004: Uncertain discount rates in climate policy analysis. *Energy Policy* (32), pp. 519-529.

Newell, R., A. Jae, and R. Stavins, 1998: The induced innovation hypothesis and energy-saving technological change. National Bureau of Economic Research.

Nordhaus, W.D., 1997: Do Real-Output and Real-Wage Measures Capture Reality? The History of Lighting Suggests Not, in Timothy F. Breshnahan and Robert J. Gordon (eds), The Economics of New Goods, pp 26-69, The University of Chicago Press, Chicago.

Nozick, R., 1974: Anarchy state and utopia. Basic Books, New York.

O'Connor, D., F. Zhai, K. Aunan, T. Berntsen and H. Vennemo, 2003: Agricultural and human health impacts of climate policy in China: A general equilibrium analysis with special reference to Guangdong. OECD Development Centre Technical Papers No. 206.

ODS, Office of Development Studies, 2002: Profiling the provision status of global public goods. New York: United Nations Development Programme.

OECD, 1998: OECD environmental performance reviews: Second cycle work plan OECD Environment Monograph, OCDE/GD, **35**(97), 67 pp.

OECD, 2003: OECD environmental indicators: Development, measurement and use. Reference paper, Paris.

OECD, 2004: Sustainable development in OECD countries: Getting the policies right. Paris.

OECD, 2006: Innovation in energy technology: Comparing national innovation systems at the sectoral level. Paris.

Ohsita, S.B., 2006: Cooperation mechanisms: A shift towards policy development cooperation. In *Cooperative Climate: Energy Efficiency Action in East Asia*. T. Sugiyama and S.B. Ohshita (ed.), International Institute for Sustainable Development (IISD): Winnepeg.

Ohshita, S.B. and Ortolano, L., 2003: From demonstration to diffusion: the gap in Japan's environmental technology cooperation with China, *Int. J. Technology Transfer and Commercialisation* **2**(4): pp. 351-368.

Olhoff, A, 2002: Assessing social capital aspects of Climate Change projects. In *Climate Change and Sustainable Development. Prospects for Developing Countries*, A. Markandya and K. Halsnæs (ed.), Earthscan.

O'Neill, B., *et al.*, 2003: Planning for future energy resources. *Science*, **300**, pp. 581-582.

Ott, H.E., H. Winkler, *et al.*, 2004: South-North dialogue on equity in the greenhouse. A proposal for an adequate and equitable global climate agreement. Eschborn, GTZ.

Otto, V., A. Löschel, and R. Dellink, 2005: Energy biased technological change: A CGE analysis. FEEM Working paper 90.2005. <http://www.feem.it/NR/rdonlyres/CF5B910E-8433-4828-A220-420CFC2D7F60/1638/9006.pdf>, accessed 31/05/07.

Oxera, 2002: A social time preference rate for use in long term discounting the office of the Deputy Prime Minister, Department for Transport, and Department of the Environment, and Food and Rural Affairs. UK.

Pacala, S. and R. Socolow, 2004: Stabilization wedges: Solving the climate problem for the next 50 years with current technologies. *Science*, **305**, pp. 968-972.

Palacios-Huerta, I. and T. Santos, 2002: A theory of markets, institutions and endogenous preferences. *Journal of Public Economics*, Accepted 8 July.

Pan, J., 2004a: China's industrialization and reduction of greenhouse emissions. *China and the World Economy* **12**, no.3. <old.iwep.org.cn/wec/2004_5-6/china's%20industrialization.pdf> accessed 31/05/07.

Panayotou, T., J.F. Sachs, and A.P. Zwane, 2002: Compensation for 'meaningful participation'. In Climate Change Control: A Modest Proposal and Empirical Analysis. *Journal of Environmental Economics and Management*, **43**(2002), pp. 437-454.

Pearce, D.W. and R.K. Turner, 1990: Economics of natural resources and the environment. Baltimore, MD, John Hopkins University Press.

Peet, R. and E. Hartwick, 1999: Theories of development. The Guildford Press, New York.

Phelps, E.S. (ed.), 1973: Economic justice - Selected readings. Penguin Education, New York.

Philibert, C., 2005: Energy demand, energy technologies and climate stabilisation. Proccedings of the IPCC Expert Meeting on Industrial Technology Development, Transfer and Diffusion, September 21-23, 2004, Tokyo.

Pindyck, R., 2002: Optimal timing problems in environmental economics. *Journal of Economic Dynamics and Control*, **26**, pp. 1677-1697.

Popp, D., 2002: Induced innovation and energy prices. *American Economic Review*, **92**(1), pp. 160-180.

Portney, P.R. and J. Weyant (eds.), 1999: Discounting and intergenerational equity. John Hopkins University Press, Baltimore, MD.

Quiggin, J., 1982: A theory of anticipated utility. *Journal of Economic Behavior and Organization*, **3**(4), pp. 323-343.

Rabinovitch, J. and J. Leitman, 1993: *Environmental innovation and management in Curitiba, Brasil.* UNDP/UNCHS/World Bank, Washington, D.C, 62 pp.

Rao, S., I. Keppo, and K. Riahi, 2005: Importance of technological change and spillovers in effective long-term climate policy. *Energy* (Special Issue, forthcoming).

Rapping, L., 1965: Learning and World War II production functions. *The Review of Economics and Statistics*, **47**(1), pp. 81-86.

Rawls, J., 1971: *A theory of justice.* Ma. Harvard University Press, Cambridge.

Reilly, J., P.H. Stone, C.E. Forest, M.D. Webster, H.D. Jacoby, and R.G. Prinn, 2001: Uncertainty and climate change assessments. *Science*, **293**(July 20), pp. 430-433.

Rennings, K. and H. Wiggering, 1997: Steps towards indicators of sustainable development: Linking economic and ecological concepts. *Ecological Economics, ***20**, pp. 25-36.
   Risk, Decision and Policy, **5**(2), pp. 151-164.

RITE, 2003: New Earth 21 Project. Research Institute of Innovative Technologies for the Earth. Kyoto, Japan, Available online at <http://www.rite.or.jp/English/lab/syslab/research/new-earth/new-earth.html>, accessed 31/05/07.

Roehrl, A.R. and K. Riahi, 2000: Technology dynamics and greenhouse gas emissions mitigation - A cost assessment. *Technological Forecasting and Social Change*, **63**(2-3), pp. 231-261.

Rose, A., B. Stevens, J. Edmonds, and M. Wise, 1998: International equity and differentiation in global warming policy. An application to Tradeable Emission Permits. *Environment and Resource Economics*, **12** (1998), pp. 25-51.

Rosenberg, N., 1982: Inside the black box: Technology and economics. Cambridge University Press, Cambridge.

Rosenberg, N., 1998: Chemical engineering as a general purpose technology. In *General Purpose Technologies and Economic Growth.* MIT Press.

Rotmans, J. and H. Dowlatabadi, 1998: Integrated assessment modeling. Columbus, OH, Battelle, pp. 292-377.

Rycroft, R. and D. Kash, 1999: The complexity challenge: Technological innovation for the 21st Century. Cassell Academic.

SAINC, 2003: South Africa's initial national communication to UNFCCC.

Salinger, M., 2005: Climate variability and change: Past, present and future - An overview. , **70**(1-2), pp. 9-29.

Scheffer, M.S., S. Carpenter, J.A. Foley, C. Folke, and B. Walker, 2001: Catastrophic shifts in ecosystems. *Nature*, **413**, October, pp. 591-596.

Schipper, L. and M. Grubb, 2000: On the rebound? Feedback between energy intensities and energy uses in IEA countries. *Energy Policy*, **28**(6), pp. 367-388.

Schmookler, J., 1962: Economic sources of inventive activity. *Journal of Economic History*, **22**(1).

Schneider, S., 2004: Abrupt non-linear Climate Change, irreversibility and surprise. *Global Environmental Change*, **14**, pp. 245-258.

Schneider, S.H., 2001: What is 'dangerous' climate change? *Nature*, **411**(May 3), pp. 17-19.

Schneider, S.H., W.E. Easterling, and L.O. Mearns, 2000: Adaptation: Sensitivity to natural variability, agent assumptions and dynamic climate changes. *Climatic Change,* **45**(1), pp. 203-221.

Sen, A., 1992: Inequality reexamined. Clarendon Press, Oxford.

Sen, A., 1999: Development as freedom. Oxford University Press.

Shackle, G.L.S., 1949: A non-additive measure of uncertainty. *Review of Economic Studies*, **17**(1) pp. 70-74.

Shukla, P.R., 2003: Development and climate. Workshop on Development and Climate. <http://developmentfirst.org/india/background.pdf>, accessed 31/05/07. New Delhi, India.

Shukla, P.R, 2005: Aligning justice and efficiency in the global climate change regime: A developing country perspective. In *Perspectives on Climate Change: Science, Economics, Politics and Ethics,* W. Sinnott-Armstrong and R. Howarth (eds.), *Advances in The Economics of Environmental Resources,* **5**, Elsevier Ltd., pp. 121-144.

Shukla, P.R., S.K. Sharma, N.H. Ravindranath, A. Garg, and S. Bhattacharya (eds.), 2003: Climate change and India: Vulnerability assessment and adaptation. Universities Press (India) Private Limited, Hyderabad, India.

Smit, B. and O. Pilifosova, 2001: Adaptation to climate change in context of sustainable development and equity. In *Climate Change 2001: Impacts, Adaptation and Vulnerability.* J.J. McCarthy, O.F. Canziani, N.A. Leary, D.J. Dokken, and K. White (ed.), Cambridge Unversity Press, Cambridge.

Soderholm, P. and T. Sundqvist, 2003: Learning curve analysis for energy technologies: Theoretical and econometric issues. Paper presented at the Annual Meeting of the International Energy Workshop (IEW). Laxenburg, Austria.

Solomon, S., D. Qin, M. Manning, R.B. Alley, T. Berntsen, N.L. Bindoff, Z. Chen, A. Chidthaisong, J.M. Gregory, G.C. Hegerl, M. Heimann, B. Hewitson, B.J. Hoskins, F. Joos, J. Jouzel, V. Kattsov, U. Lohmann, T. Matsuno, M. Molina, N. Nicholls, J. Overpeck, G. Raga, V. Ramaswamy, J. Ren, M. Rusticucci, R. Somerville, T.F. Stocker, P. Whetton, R.A. Wood and D. Wratt, 2007: Technical Summary. In: *Climate Change 2007: The Physical Science Basis. Contribution of Working Group I to the Fourth Assessment Report of the Intergovernmental Panel on Climate Change* [Solomon, S., D. Qin, M. Manning, Z. Chen, M. Marquis, K.B. Averyt, M. Tignor and H.L. Miller (eds.)]. Cambridge University Press, Cambridge, United Kingdom and New York, NY, USA, pp 19-92.

Somerville, R., H. Le Treut, U. Cubasch, Y. Ding, C. Mauritzen, A. Mokssit, T. Peterson and M. Prather, 2007: Historical Overview of Climate Change. In: *Climate Change 2007: The Physical Science Basis. Contribution of Working Group I to the Fourth Assessment Report of the Intergovernmental Panel on Climate Change* [Solomon, S., D. Qin, M. Manning, Z. Chen, M. Marquis, K.B. Averyt, M. Tignor and H.L. Miller (eds.)]. Cambridge University Press, Cambridge, United Kingdom and New York, NY, USA.

SRES, 2000: *Emission scenarios.* Special report of the Intergovernmental Panel on Climate Change (IPCC).

STAP, 1996: Scientific and technical advisory panel. Global Environmental Facility.

Stokes, D., 1997: Pasteur's quadrant: Basic science and technological innovation. Brookings Institution Press.

Sullivan, R., and W. Blyth, August 2006: Climate change policy uncertainty and the electricity industry: Implications and unintended consequences. Briefing Paper, Chatham House, <www.chathamhouse. org.uk>, accessed 31/05/07.

Swart, R., *et al.*, 2003: Climate change and sustainable development: Expanding the options. *Climate Policy*, Special Issue on Climate Change and Sustainable Development, **3**(S1), pp. S19-S40.

Taylor, M., E. Rubin, and D. Hounshell, 2003: *Environmental Science & Technology*. Effect of government actions on technological innovation for SO_2 control, **37**(20), pp. 4527-4534.

Thompson, M. and S. Rayner, 1998: Cultural discourses. Choice and Climate Change [4 Volume Set] (S. Rayner and E.L. Malone (eds), Battelle Press, Columbus, Ohio, USA, ISBN 1-57477-044-6, Chapter 4, pp. 265-344.

Tol, R.S.J. and R. Verheyen, 2004: State responsibility and compensation for climate change damages - a legal and economic assessment. *Energy Policy*, **32**, pp. 1109 -1130.

Toth, F.L., 2004: Coupling climate and economic dynamics: recent achievements and unresolved problems. In *The Coupling of Climate and Economic Dynamics: Essays on integrated assessment.* Kluwer Academic Publishers, Boston.

Tversky and Kahneman, 1992: Cumulative representation of uncertainty. *Journal of Risk and Uncertainty*, **5**, 297-323.

UNDP, 2000: World Energy assessment: energy and the challenge of sustainable development. UNDP. UNDESA and the World Energy Council. New York.

UNDP, 2003a: Global indicators (Millennium Development Goals).

UNFCCC, 1992: United Nations Framework Convention on Climate Change. Climate Change Secretariat, Bonn.

US, 2003: Informing Regulatory Decisions: 2003 Report to Congress on the Costs and Benefits of Federal Regulations and Unfunded Mandates on State, Local, and Tribal Entities, Office of Management and Budget, Office of Information and Regulatory Affairs, <http://www. whitehouse.gov/omb/inforeg/2003_cost-ben_final_rpt.pdf>, accessed 31/05/07.

Utterback, K., 1996: Mastering the dynamics of Innovation. Harvard Business School Press.

Vaillancourt, K. and J.-P. Waaub, 2004: Equity in international greenhouse gas abatement scenarios: A multicriteria approach. *European Journal of Operational Research*, **153**, pp. 489-505.

Valverde, L.J.A., Jr., H.D. Jacoby, and G.M. Kaufman, 1999: Sequential climate decisions under uncertainty: An integrated framework. *USA Environmental Modeling and Assessment,* **4**, pp. 87-101.

Van der Sluijs, J., P. Kloprogge, J. Risbey, and J. Ravetz, 2003: Towards a synthesis of qualitative and quantitative uncertainty assessment: Applications of the Numeral, Unit, Spread, Assessment, Pedigree (NUSAP). System Communication to the International Workshop on Uncertainty, Sensitivity, and Parameter Estimation for Multimedia Environmental Modeling, August 19-21, 2003, Rockville, Maryland, USA, <http://www.nusap.net/>, accessed 31/05/07.

Vasicek, O., 1977: An equilibrium characterization of the term structure. *Journal of Financial Economics*, **5**, pp. 177-188.

Victor, N. and J. Ausubel, 2002: DRAMs as model prganisms for study of technological evolution. *Technological Forecasting and Social Change*, **69**(3), pp. 243-262.

Vine, E., J. Hamerin, N. Eyre, D. Crossley, M. Maloney, and G. Watt, 2003: Public policy analysis of energy efficiency and load management in changing electricity business. *Energy Policy*, **31 (2003)**, pp. 405-430.

Walley, P., 1991: Statistical reasoning with imprecise probabilities. Chapman and Hall, London.

Watts, W., 1999: Discounting and sustainability. The European Commission, DGII Brussels.

WB, 2005: World development indicators databank. World Bank.

WCED, 1987: Our common future. Oxford University Press, World Commission on Environment and Development, London.

WDI, 2005: World development indicators. The World Bank, Washington D.C.

Webster, M.D., 2002: The curious role of 'learning' in climate policy: Should we wait for more Data? *The Energy Journal*, **23**(2), pp. 97-119.

Weitzman, M.L., 1998: Why the far-distant future should be discounted at its lowest possible rate. *Journal of Environmental Economics and Management*, **36**, pp. 201-208.

Weitzman, M.L., 2001: Gamma discounting. *American Economic Review*, **91**, pp. 260-271.

Weitzman, M.L., 2004: Statistical discounting of an uncertain distant future, mimeo. Harvard University.

Wesley, E. and F. Peterson, 1999: The ethics of burden sharing in the global greenhouse. *Journal of Agricultural and Environmental Ethics*, **11**, pp. 167-196.

WETO, 2003: World energy, technology and climate policy outlook 2030. European Commission, Directorate-General for Research Energy, Brussels.

Weyant, J.P., 2004a: Introduction and overview (to EMF-19). *Energy Economics*, **26**, pp. 501-515.

Weyant, J.P., 2004b: *Energy Economics* (Special issue), **26**(4), pp. 501-755.

Wigley, T.M.L. and S.C.B. Raper, 2001: Interpretation of high projections for global-mean warming. *Science*, **293**, pp. 451-454.

Wilbanks, T., 2003: Integrating climate change and sustainable development in a place-based context. *Climate Policy*, **3** (S1), pp. S147-S154.

Woolcock, M. and D. Narayan, 2000: Social capital: Implications for development theory, research, and policy. *Oxford Journals*, Vol. **15**, No 2, pp. 225-249.

Wright, T., 1936: Factors affecting the cost of airplanes. *Journal of Aeronautical Sciences*, **3**(2), pp. 122-128.

WSSD, 2002: Water, energy, health, agriculture and biodiversity. Synthesis of the framework paper of the Working Group on WEHAB. UN WSSD A/conf.199/L.4.

Yelle, L., 1979: The learning curve: Historical review and comprehensive survey. *Decision Sciences*, **10**, pp. 302-328.

Zimmerman, M., 1982: Learning effects and the commercialization of new energy technologies: the case of nuclear power. *Bell Journal of Economics*, **13**(2), pp. 290-31.

Zou, J. and Y. Xu, 2005: Transfer and development of technologies: an important measure to response to Climate Change, environmental protection (published in Chinese), No. 1.

3

Issues related to mitigation in the long-term context

Coordinating Lead Authors:

Brian Fisher (Australia), Nebojsa Nakicenovic (Austria/Montenegro)

Lead Authors:

Knut Alfsen (Norway), Jan Corfee Morlot (France/USA), Francisco de la Chesnaye (USA), Jean-Charles Hourcade (France),
Kejun Jiang (China), Mikiko Kainuma (Japan), Emilio La Rovere (Brazil), Anna Matysek (Australia), Ashish Rana (India),
Keywan Riahi (Austria), Richard Richels (USA), Steven Rose (USA), Detlef Van Vuuren (The Netherlands), Rachel Warren (UK)

Contributing Authors:

Phillipe Ambrosi (France), Fatih Birol (Turkey), Daniel Bouille (Argentina), Christa Clapp (USA), Bas Eickhout (The Netherlands),
Tatsuya Hanaoka (Japan), Michael D. Mastrandrea (USA), Yuzuru Matsuoko (Japan), Brian O'Neill (USA), Hugh Pitcher (USA), Shilpa Rao
(India), Ferenc Toth (Hungary)

Review Editors:

John Weyant (USA), Mustafa Babiker (Kuwait)

This chapter should be cited as:

Fisher, B.S., N. Nakicenovic, K. Alfsen, J. Corfee Morlot, F. de la Chesnaye, J.-Ch. Hourcade, K. Jiang, M. Kainuma, E. La Rovere,
A. Matysek, A. Rana, K. Riahi, R. Richels, S. Rose, D. van Vuuren, R. Warren, 2007: Issues related to mitigation in the long term context,
In Climate Change 2007: Mitigation. Contribution of Working Group III to the Fourth Assessment Report of the Inter-governmental
Panel on Climate Change [B. Metz, O.R. Davidson, P.R. Bosch, R. Dave, L.A. Meyer (eds)], Cambridge University Press, Cambridge,
United Kingdom and New York, NY, USA.

Table of Contents

EXECUTIVE SUMMARY

This chapter documents baseline and stabilization scenarios in the literature since the publication of the IPCC Special Report on Emissions Scenarios (SRES) (Nakicenovic *et al.*, 2000) and Third Assessment Report (TAR, Morita *et al.*, 2001). It reviews the use of the SRES reference and TAR stabilization scenarios and compares them with new scenarios that have been developed during the past five years. Of special relevance is how ranges published for driving forces and emissions in the newer literature compare with those used in the TAR, SRES and pre-SRES scenarios. This chapter focuses particularly on the scenarios that stabilize atmospheric concentrations of greenhouse gases (GHGs). The multi-gas stabilization scenarios represent a significant change in the new literature compared to the TAR, which focused mostly on carbon dioxide (CO_2) emissions. They also explore lower levels and a wider range of stabilization than in the TAR.

The foremost finding from the comparison of the SRES and new scenarios in the literature is that the ranges of main driving forces and emissions have not changed very much *(high agreement, much evidence)*. Overall, the emission ranges from scenarios without climate policy reported before and after the SRES have not changed appreciably. Some changes are noted for population and economic growth assumptions. Population scenarios from major demographic institutions are lower than they were at the time of the SRES, but so far they have not been fully implemented in the emissions scenarios in the literature. All other factors being equal, lower population projections are likely to result in lower emissions. However, in the scenarios that used lower projections, changes in other drivers of emissions have offset their impact. Regional medium-term (2030) economic projections for some developing country regions are currently lower than the highest scenarios used in the SRES. Otherwise, economic growth perspectives have not changed much, even though they are among the most intensely debated aspects of the SRES scenarios. In terms of emissions, the most noticeable changes occurred for projections of SO_x and NO_x emissions. As short-term trends have moved down, the range of projections for both is currently lower than the range published before the SRES. A small number of new scenarios have begun to explore emission pathways for black and organic carbon.

Baseline land-related CO_2 and non-CO_2 GHG emissions remain significant, with continued but slowing land conversion and increased use of high-emitting agricultural intensification practices due to rising global food demand and shifts in dietary preferences towards meat consumption. The post-SRES scenarios suggest a degree of agreement that the decline in annual land-use change carbon emissions will, over time, be less dramatic (slower) than those suggested by many of the SRES scenarios. Global long-term land-use scenarios are scarce in numbers but growing, with the majority of the new literature since the SRES contributing new forestry and biomass scenarios. However, the explicit modelling of land-use in long-term global scenarios is still relatively immature, with significant opportunities for improvement.

In the debate on the use of exchange rates, market exchange rates (MER) or purchasing power parities (PPP), evidence from the limited number of new PPP-based studies indicates that the choice of metric for gross domestic product (GDP), MER or PPP, does not appreciably affect the projected emissions, when metrics are used consistently. The differences, if any, are small compared to the uncertainties caused by assumptions on other parameters, e.g. technological change *(high agreement, much evidence)*.

The numerical expression of GDP clearly depends on conversion measures; thus GDP expressed in PPP will deviate from GDP expressed in MER, more so for developing countries. The choice of conversion factor (MER or PPP) depends on the type of analysis or comparison being undertaken. However, when it comes to calculating emissions (or other physical measures, such as energy), the choice between MER-based or PPP-based representations of GDP should not matter, since emission intensities will change (in a compensating manner) when the GDP numbers change. Thus, if a consistent set of metrics is employed, the choice of MER or PPP should not appreciably affect the final emission levels *(high agreement, medium evidence)*. This supports the SRES in the sense that the use of MER or PPP does not, in itself, lead to significantly different emission projections outside the range of the literature *(high agreement, much evidence)*. In the case of the SRES, the emissions trajectories were the same whether economic activities in the four scenario families were measured in MER or PPP.

Some studies find differences in emission levels between using PPP-based and MER-based estimates. These results critically depend on, among other things, convergence assumptions *(high agreement, medium evidence)*. In some of the short-term scenarios (with a horizon to 2030) a 'bottom-up' approach is taken, where assumptions about productivity growth and investment and saving decisions are the main drivers of growth in the models. In long-term scenario models, a 'top-down' approach is more commonly used, where actual growth rates are more directly prescribed based on convergence or other assumptions about long-term growth potentials. Different results can also be due to inconsistencies in adjusting the metrics of energy efficiency improvement when moving from MER-based to PPP-based calculations.

There is a clear and strong correlation between the CO_2-equivalent concentrations (or radiative forcing) of the published studies and the CO_2-only concentrations by 2100, because CO_2 is the most important contributor to radiative forcing. Based on this relationship, to facilitate scenario comparison and assessment, stabilization scenarios (both multi-gas and CO_2-only studies) have been grouped in this chapter into different categories that vary in the stringency of the targets, from low to high radiative forcing, CO_2-equivalent concentrations and CO_2-only concentrations by 2100, respectively.

Essentially, any specific concentration or radiative forcing target, from the lowest to the highest, requires emissions to eventually fall to very low levels as the removal processes of the ocean and terrestrial systems saturate. For low to medium targets, this would need to occur during this century, but higher stabilization targets can push back the timing of such reductions to beyond 2100. However, to reach a given stabilization target, emissions must ultimately be reduced well below current levels. For achievement of the very low stabilization targets from many high baseline scenarios, negative net emissions are required towards the end of the century. Mitigation efforts over the next two or three decades will have a large impact on opportunities to achieve lower stabilization levels *(high agreement, much evidence)*.

The timing of emission reductions depends on the stringency of the stabilization target. Lowest stabilization targets require an earlier peak of CO_2 and CO_2-equivalent emissions. In the majority of the scenarios in the most stringent stabilization category (a stabilization level below 490 ppmv CO_2-equivalent), emissions are required to decline before 2015 and are further reduced to less than 50% of today's emissions by 2050. For somewhat higher stabilization levels (e.g. below 590 ppmv CO_2-equivalent) global emissions in the scenarios generally peak around 2010–2030, followed by a return to 2000 levels, on average around 2040. For high stabilization levels (e.g. below 710 ppmv CO_2-equivalent) the median emissions peak around 2040 *(high agreement, much evidence)*.

Long-term stabilization scenarios highlight the importance of technology improvements, advanced technologies, learning-by-doing, and induced technological change, both for achieving the stabilization targets and cost reduction *(high agreement, much evidence)*. While the technology improvement and use of advanced technologies have been employed in scenarios largely exogenously in most of the literature, new literature covers learning-by-doing and endogenous technological change. The latter scenarios show different technology dynamics and ways in which technologies are deployed, while maintaining the key role of technology in achieving stabilization and cost reduction.

Decarbonization trends are persistent in the majority of intervention and non-intervention scenarios *(high agreement, much evidence)*. The medians of scenario sets indicate decarbonization rates of around 0.9 (pre-TAR) and 0.6 (post-TAR) compared to historical rates of about 0.3% per year. Improvements of carbon intensity of energy supply and the whole economic need to be much faster than in the past for the low stabilization levels. On the upper end of the range, decarbonization rates of up to 2.5% per year are observed in more stringent stabilization scenarios, where complete transition away from carbon-intensive fuels is considered.

The scenarios that report quantitative results with drastic CO_2 reduction targets of 60–80% in 2050 (compared to today's emission levels) require increased rates of energy intensity and carbon intensity improvement by 2–3 times their historical levels. This is found to require different sets of mitigation options across regions, with varying shares of nuclear energy, carbon capture and storage (CCS), hydrogen, and biomass.

The costs of stabilization crucially depend on the choice of the baseline, related technological change and resulting baseline emissions; stabilization target and level; and the portfolio of technologies considered *(high agreement, much evidence)*. Additional factors include assumptions with regard to the use of flexible instruments and with respect to revenue recycling. Some literature identifies low-cost technology clusters that allow for endogenous technological learning with uncertainty. This suggests that a decarbonized economy may not cost any more than a carbon-intensive one, if technological learning is taken into account.

There are different metrics for reporting costs of emission reductions, although most models report them in macro-economic indicators, particularly GDP losses. For stabilization at 4–5 W/m^2 (or ~ 590–710 ppmv CO_2-equivalent) macro-economic costs range from -1 to 2% of GDP below baseline in 2050. For a more stringent target of 3.5–4.0 W/m^2 (~ 535–590 ppmv CO_2-equivalent) the costs range from slightly negative to 4% GDP loss *(high agreement, much evidence)*. GDP losses in the lowest stabilization scenarios in the literature (445-535 ppmv CO_2-equivalent) are generally below 5.5% by 2050, however the number of studies are relatively limited and are developed from predominantly low baselines *(high agreement, medium evidence)*.

Multi-gas emission-reduction scenarios are able to meet climate targets at substantially lower costs compared to CO_2-only strategies (for the same targets, *high agreement, much evidence*). Inclusion of non-CO_2 gases provides a more diversified approach that offers greater flexibility in the timing of the reduction programme.

Including land-use mitigation options as abatement strategies provides greater flexibility and cost-effectiveness for achieving stabilization *(high agreement, medium evidence)*. Even if land activities are not considered as mitigation alternatives by policy, consideration of land (land-use and land cover) is crucial in climate stabilization for its significant atmospheric inputs and withdrawals (emissions, sequestration, and albedo). Recent stabilization studies indicate that land-use mitigation options could provide 15–40% of total cumulative abatement over the century. Agriculture and forestry mitigation options are projected to be cost-effective abatement strategies across the entire century. In some scenarios, increased commercial biomass energy (solid and liquid fuel) is a significant abatement strategy, providing 5–30% of cumulative abatement and potentially 1–15% of total primary energy over the century.

Decision-making concerning the appropriate level of mitigation in a cost-benefit context is an iterative risk-management process that considers investment in mitigation and adaptation, co-benefits of undertaking climate change decisions and the damages due to climate change. It is intertwined with development decisions and pathways. Cost-benefit analysis tries to quantify climate change damages in monetary terms as the social cost of carbon (SCC) or time-discounted damages. Due to considerable uncertainties and difficulties in quantifying non-market damages, it is difficult to estimate SCC with confidence. Results depend on a large number of normative and empirical assumptions that are not known with any certainty. SCC estimates in the literature vary by three orders of magnitude. Often they are likely to be understated and will increase a few percent per year (i.e. 2.4% for carbon-only and 2–4% for the social costs of other greenhouse gases (IPCC, 2007b, Chapter 20). SCC estimates for 2030 range between 8 and 189 US$/$tCO_2$-equivalent (IPCC, 2007b, Chapter 20), which compares to carbon prices between 1 to 24 US$/$tCO_2$-equivalent for mitigations scenarios stabilizing between 485-570 ppmv CO_2-equivalent) and 31 to 121 US$/$tCO_2$-equivalent for scenarios stabilizing between 440-485 ppmv CO_2-equivalent, respectively *(high agreement, limited evidence)*.

For any given stabilization pathway, a higher climate sensitivity raises the probability of exceeding temperature thresholds for key vulnerabilities *(high agreement, much evidence)*. For example, policymakers may want to use the highest values of climate sensitivity (i.e. 4.5°C) within the 'likely' range of 2–4.5°C set out by IPCC (2007a, Chapter 10) to guide decisions, which would mean that achieving a target of 2°C (above the pre-industrial level), at equilibrium, is already outside the range of scenarios considered in this chapter, whilst a target of 3°C (above the pre-industrial level) would imply stringent mitigation scenarios, with emissions peaking within 10 years. Using the 'best estimate' assumption of climate sensitivity, the most stringent scenarios (stabilizing at 445–490 ppmv CO_2-equivalent) could limit global mean temperature increases to 2–2.4°C above the pre-industrial level, at equilibrium, requiring emissions to peak before 2015 and to be around 50% of current levels by 2050. Scenarios stabilizing

at 535–590 ppmv CO_2-equivalent could limit the increase to 2.8–3.2°C above the pre-industrial level and those at 590–710 CO_2-equivalent to 3.2–4°C, requiring emissions to peak within the next 25 and 55 years, respectively *(high agreement, medium evidence)*.

Decisions to delay emission reductions seriously constrain opportunities to achieve low stabilization targets (e.g. stabilizing concentrations from 445–535 ppmv CO_2-equivalent), and raise the risk of progressively more severe climate change impacts and key vulnerabilities occurring.

The risk of climate feedbacks is generally not included in the above analysis. Feedbacks between the carbon cycle and climate change affect the required mitigation for a particular stabilization level of atmospheric CO_2 concentration. These feedbacks are expected to increase the fraction of anthropogenic emissions that remains in the atmosphere as the climate system warms. Therefore, the emission reductions to meet a particular stabilization level reported in the mitigation studies assessed here might be underestimated.

Short-term mitigation and adaptation decisions are related to long-term climate goals *(high agreement, much evidence)*. A risk management or 'hedging' approach can assist policymakers to advance mitigation decisions in the absence of a long-term target and in the face of considerable uncertainties relating to the cost of mitigation, the efficacy of adaptation and the negative impacts of climate change. The extent and the timing of the desirable hedging strategy will depend on the stakes, the odds and societies' attitudes to risks, for example with respect to risks of abrupt change in geo-physical systems and other key vulnerabilities. A variety of integrated assessment approaches exist to assess mitigation benefits in the context of policy decisions relating to such long-term climate goals. There will be ample opportunity for learning and mid-course corrections as new information becomes available. However, actions in the short term will largely determine what future climate change impacts can be avoided. Hence, analysis of short-term decisions should not be decoupled from analysis that considers long-term climate change outcomes *(high agreement, much evidence)*.

3.1 Emissions scenarios

The evolution of future greenhouse gas emissions and their underlying driving forces is highly uncertain, as reflected in the wide range of future emissions pathways across (more than 750) emission scenarios in the literature. This chapter assesses this literature, focusing especially on new multi-gas baseline scenarios produced since the publication of the IPCC Special Report on Emissions Scenarios (SRES) (Nakicenovic *et al.*, 2000) and on new multi-gas mitigation scenarios in the literature since the publication of the IPCC Third Assessment Report (TAR, Working Group III, Chapter 2, Morita *et al.*, 2001). This literature is referred to as 'post-SRES' scenarios.

The SRES scenarios were representative of some 500 emissions scenarios in the literature, grouped as A1, A2, B1 and B2, at the time of their publication in 2000. Of special relevance in this review is the question of how representative the SRES ranges of driving forces and emission levels are of the newer scenarios in the literature, and how representative the TAR stabilization levels and mitigation options are compared with the new multi-gas stabilization scenarios. Other important aspects of this review include methodological, data and other advances since the time the SRES scenarios were developed.

This chapter uses the results of the Energy Modeling Forum (EMF-21) scenarios and the new Innovation Modelling Comparison Project (IMCP) network scenarios. In contrast to SRES and post-SRES scenarios, these new modelling-comparison activities are not based on fully harmonized baseline scenario assumptions, but rather on 'modeller's choice' scenarios. Thus, further uncertainties have been introduced due to different assumptions and different modelling approaches. Another emerging complication is that even baseline (also called reference) scenarios include some explicit policies directed at emissions reduction, notably due to the Kyoto Protocol entering into force, and other climate-related policies that are being implemented in many parts of the world.

Another difficulty in straightforward comparisons is that the information and documentation of the scenarios in the literature varies considerably.

3.1.1 The definition and purpose of scenarios

Scenarios describe possible future developments. They can be used in an exploratory manner or for a scientific assessment in order to understand the functioning of an investigated system (Carpenter *et al.*, 2005).

In the context of the IPCC assessments, scenarios are directed at exploring possible future emissions pathways, their main underlying driving forces and how these might be affected by policy interventions. The IPCC evaluation of emissions scenarios in 1994 identified four main purposes of emissions scenarios (Alcamo *et al.*, 1995):

- To provide input for evaluating climatic and environmental consequences of alternative future GHG emissions in the absence of specific measures to reduce such emissions or enhance GHG sinks.
- To provide similar input for cases with specific alternative policy interventions to reduce GHG emissions and enhance sinks.
- To provide input for assessing mitigation and adaptation possibilities, and their costs, in different regions and economic sectors.
- To provide input to negotiations of possible agreements to reduce GHG emissions.

Scenario definitions in the literature differ depending on the purpose of the scenarios and how they were developed. The SRES report (Nakicenovic *et al.*, 2000) defines a scenario as a plausible description of how the future might develop, based on a coherent and internally consistent set of assumptions ('scenario logic') about the key relationships and driving forces (e.g. rate of technology change or prices). Some studies in the literature apply the term 'scenario' to 'best-guess' or forecast types of projections. Such studies do not aim primarily at exploring alternative futures, but rather at identifying 'most likely' outcomes. Probabilistic studies represent a different approach, in which the range of outcomes is based on a consistent estimate of the probability density function (PDF) for crucial input parameters. In these cases, outcomes are associated with an explicit estimate of likelihood, albeit one with a substantial subjective component. Examples include probabilistic projections for population (Lutz and Sanderson, 2001) and CO_2 emissions (Webster *et al.*, 2002, 2003; O'Neill, 2004).

3.1.1.1 Types of scenarios

The scenario literature can be split into two largely non-overlapping streams – quantitative modelling and qualitative narratives (Morita *et al.*, 2001). This dualism mirrors the twin challenges of providing systematic and replicable quantitative representation, on the one hand, and contrasting social visions and non-quantifiable descriptors, on the other (Raskin *et al.*, 2005). It is particularly noteworthy that recent developments in scenario analysis are beginning to bridge this difficult gap (Nakicenovic *et al.*, 2000; Morita *et al.*, 2001; and Carpenter *et al.*, 2005).

3.1.1.2 Narrative storylines and modelling

The literature based on narrative storylines that describe futures is rich going back to the first global studies of the 1970s (e.g. Kahn *et al.*, 1976; Kahn and Weiner, 1967) and is also well represented in more recent literature (e.g. Peterson and Peterson, 1994; Gallopin *et al.*, 1997; Raskin *et al.*, 1998; Glenn and Gordon, 1997). Well known are the Shell scenarios that are principally based on narrative stories with illustrative quantification of salient driving forces and scenario outcomes (Wack, 1985a, 1985b; Schwartz, 1991; Shell, 2005).

Catastrophic futures feature prominently in the narrative scenarios literature. They typically involve large-scale environmental or economic collapse, extrapolating current unfavourable conditions and trends in many regions.[1] Many of these scenarios suggest that catastrophic developments may draw the world into a state of chaos within one or two decades. Greenhouse-gas emissions might be low in such scenarios because of low or negative economic growth, but seem unlikely to receive much attention in any case, in the light of more immediate problems. This report does not analyze such futures, except where cases provide emissions pathways.

3.1.1.3   Global futures scenarios

Global futures scenarios are deeply rooted in the long history of narrative scenarios (Carpenter et al., 2005; UNEP, 2002). The direct antecedents of contemporary scenarios lie with the future studies of the 1970s (Raskin et al., 2005). These responded to emerging concerns about the long-term sufficiency of natural resources to support expanding global populations and economies. This first wave of global scenarios included ambitious mathematical simulation models (Meadows et al., 1972; Mesarovic and Pestel, 1974) as well as speculative narrative (Kahn et al., 1976). At this time, scenario analysis was first used at Royal Dutch/Shell as a strategic management technique (Wack, 1985a, 1985b; Schwartz, 1991).

A second round of integrated global analysis began in the late 1980s and 1990s, prompted by concerns with climate change and sustainable development. These included narratives of alternative futures ranging from 'optimistic' and 'pessimistic' worlds to consideration of 'surprising' futures (Burrows et al., 1991; the Central Planning Bureau of the Netherlands, 1992; Kaplan, 1994; Svedin and Aniansson, 1987; Toth et al., 1989). The long-term nature of the climate change issue introduced a new dimension and has resulted in a rich new literature of global emissions scenarios, starting from the IPCC IS92 scenarios (Pepper et al., 1992; Leggett et al., 1992) and most recent scenario comparisons projects (e.g. EMF and IMCP). The first decades of scenario assessment paved the way by showing the power – and limits – of both deterministic modelling and descriptive future analyses. A central challenge of global scenario exercises today is to unify these two aspects by blending the objectivity and clarity of quantification with the richness of narrative (Raskin et al., 2005).

3.1.2   Introduction to mitigation and stabilization scenarios

Climate change intervention, control, or mitigation scenarios capture measures and policies for reducing GHG emissions with respect to some baseline (or reference) scenario. They contain emission profiles, as well as costs associated with the emissions

reduction, but often do not quantify the benefits of reduced impacts from climate change. Stabilization scenarios are mitigation scenarios that aim at a pre-specified GHG reduction pathway, leading to stabilization of GHG concentrations in the atmosphere.

For the purposes of this chapter, a scenario is identified as a mitigation or intervention scenario if it meets one of the following two conditions:
- It incorporates specific climate change targets, which may include absolute or relative GHG limits, GHG concentration levels (e.g. CO_2 or CO_2-equivalent (CO_2-eq) stabilization scenarios), or maximum allowable changes in temperature or sea level.
- It includes explicit or implicit policies and/or measures of which the primary goal is to reduce CO_2 or a broader range of GHG emissions (e.g. a carbon tax, carbon cap or a policy encouraging the use of renewable energy).

Some scenarios in the literature are difficult to classify as mitigation (intervention) or baseline (reference or non-intervention), such as those developed to assess sustainable development (SD) paths. These studies consider futures that require radical policy and behavioural changes to achieve a transition to a postulated sustainable development pathway. Greenpeace formulated one of the first such scenarios (Lazarus et al., 1993). Many sustainable development scenarios are also included in this assessment. Where they do not include explicit policies, as in the case of SRES scenarios, they can be classified as baseline or non-intervention scenarios. For example, the SRES B1 family of reference scenarios can be characterized as having many elements of a sustainability transition that lead to generally low GHG emissions, even though the scenarios do not include policies or measures explicitly directed at emissions mitigation.

Another type of mitigation (intervention or climate policy) scenario approach specifies future 'worlds' that are internally consistent with some specified climate target (e.g. a global temperature increase of no more than 1°C by 2100), and then works backwards to develop feasible emission trajectories and emission driver combinations leading to these targets. Such scenarios, also referred to as 'safe landing' or 'tolerable window' scenarios, imply the necessary development and implementation of climate policies intended to achieve these targets in the most efficient way (Morita et al., 2001). A number of such new multi-gas stabilization scenarios are assessed in this chapter.

Confusion can arise when the inclusion of 'non-climate-related' policies in a reference (non-intervention) scenario has the effect of significantly reducing GHG emissions. For example, energy efficiency or land-use policies that reduce

---

1   Prominent examples of such scenarios include the 'Retrenchment' (Kinsman, 1990), the 'Dark Side of the Market World' or 'Change without Progress' (Schwartz, 1991), the 'Barbarization' (Gallopin et al., 1997) and 'A Passive Mean World' (Glenn and Gordon, 1997).

GHG emissions may be adopted for reasons that are not related to climate policies and may therefore be included in a non-intervention scenario. Such a scenario may include GHG emissions that are lower than some intervention scenarios. The root cause of this potential confusion is that, in practice, many policies can both reduce GHG emissions and achieve other goals (so-called multiple benefits). Whether such policies are assumed to be adopted for climate or non-climate policy-related reasons is determined by the scenario developer, based on the underlying scenario narrative. While this is a problem in terms of making a clear distinction between intervention and non-intervention scenarios, it is at the same time an opportunity. Because many decisions are not made for reasons of climate change alone, measures implemented for reasons other than climate change can have a significant impact on GHG emissions, opening up many new possibilities for mitigation (Morita *et al.*, 2001).

3.1.3 Development trends and the lock-in effect of infrastructure choices

An important consideration in scenario generation is the nature of the economic development process and whether (and to what extent) developing countries will follow the development pathways of industrialized countries with respect to energy use and GHG emissions. The 'lock-in' effects of infrastructure, technology and product design choices made by industrialized countries in the post-World War II period of low energy prices are responsible for the major recent increase in world GHG emissions. A simple mimicking by developing countries of the development paradigm established by industrialized countries could lead to a very large increase in global GHG emissions (see Chapter 2). It may be noted, however, that energy/GDP elasticities in industrialized countries have first increased in successive stages of industrialization, with acceleration during the 1950s and 1960s, but have fallen sharply since then, due to factors such as relative growth of services in GDP share, technical progress induced by higher oil prices and energy conservation efforts.

In developing countries, where a major part of the infrastructure necessary to meet development needs is still to be built, the spectrum of future options is considerably wider than in industrialized countries (e.g. on energy, see IEA, 2004). The spatial distribution of the population and economic activities is still not settled, opening the possibility of adopting industrial policies directed towards rural development and integrated urban, regional, and transportation planning, thereby avoiding urban sprawl and facilitating more efficient transportation and energy systems. The main issue is the magnitude and viability to tap the potential for technological 'leapfrogging', whereby developing countries can bypass emissions-intensive intermediate technology and jump straight to cleaner technologies. There are technical possibilities for less energy-intensive development patterns in the long run, leading to low carbon futures in southern countries that are compatible with

national objectives (see e.g. La Rovere and Americano, 2002). Section 12.2 of Chapter 12 develops this argument further.

On the other hand, the barriers to such development pathways should not be underestimated, going from financial constraints to cultural behaviours in industrialized and developing countries, including the lack of appropriate institution building. One of the key findings of the reviewed literature is the long-term implications for GHG emissions of short- and medium-term decisions concerning the building of new infrastructure, particularly in developing countries (see e.g. La Rovere and Americano, 2002; IEA, 2004).

3.1.4 Economic growth and convergence

Determinants of long-term GDP per person include labour force and its productivity projections. Labour force utilization depends on factors such as the number of working-age people, the level of structural unemployment and hours worked per worker. Demographic change is still the major determinant of the baseline labour supply (Martins and Nicoletti, 2005). Long-term projections of labour productivity primarily depend on improvements in labour quality (capacity building) and the pace of technical change associated with building up the capital-output ratio and the quality of capital.

The literature examining production functions shows increasing returns because of an expanding stock of human capital and, as a result of specialization and investment in 'knowledge' capital (Meier, 2001; Aghion and Howitt, 1998), suggests that economic 'catch-up' and convergence strongly depend on the forces of 'technological congruence' and 'social capability' between the productivity leader and the followers (see the subsequent sub-section on institutional frameworks and Section 3.4 on the role of technological change).

The economic convergence literature (Abramovitz, 1986; Baumol, 1986), using a standard neoclassical economic growth setup following Solow (1956), found evidence of convergence only between the richest countries. Other research efforts documented 'conditional convergence' – meaning that countries appeared to reach their own steady states at a fairly uniform rate of 2% per year (Barro, 1991; Mankiw *et al.*, 1992). Jones (1997) found that the future steady-state distribution of per person income will be broadly similar to the 1990 distribution. Important differences would continue to arise among the bottom two-thirds of the income distribution, thus confirming past trends. Total factor productivity (TFP) levels and convergence for the evolution of income distribution are also important. Expected catch-up, and even overtaking per-person incomes, as well as changes in leaders in the world distribution of income, are among some of the findings in this literature. Quah (1993, 1996) found that the world is moving towards a bimodal income distribution. Some recent assessments demonstrate divergence, not convergence (World Bank, 2002; Halloy and Lockwood, 2005; UNSD, 2005).

Convergence is limited for a number of reasons, such as imperfect mobility of factors (notably labour); different endowments (notably human capital); market segmentation (notably services); and limited technology diffusion. Social inertia (as referred to in Chapter 2, see Section 2.2.3) also contributes to delay convergence. Therefore only limited catch-up can be factored in baseline scenarios: while capital quality is likely to push up productivity growth in most countries, especially in those lagging behind, labour quality is likely to drag down productivity growth in a number of countries, unless there are massive investments in education. However, appropriate policies may accelerate the adoption of new technologies and create incentives for human capital formation and thus accelerate convergence (Martins and Nicoletti, 2005). Nelson and Fagerberg, arguing within an evolutionary paradigm, have different perspectives on the convergence issue (Fagerberg, 1995; Fagerberg and Godinho, 2005; UNIDO, 2005). It should be acknowledged that the old theoretical controversy about steady-state economics and limits to growth still continues (Georgescu-Roegen, 1971).

The above discussion provides the economic background for the range of assumptions on the long-term convergence of income between developing and developed countries (measured by GDP per person) found in the scenario literature. The annual rate of income convergence between 11 world regions in the SRES scenarios falls within the range of less than 0.5% in the A2 scenario family to less than 2% in A1 (both in PPP and MER metrics). The highest rate of income convergence in the SRES is similar to the observed convergence, during the period 1950–1990, of 90 regions in Europe (Barro and Sala-i-Martin 1997). However, Grübler et al. (2006) note that extending convergence analysis to national or sub-national level would suggest that income disparities are larger than suggested by simple inter-regional comparisons and that scenarios of (relative) income convergence are highly sensitive to the spatial level of aggregation used in the analysis. An important finding from the sensitivity analysis performed is that less convergence generally yields higher emissions (Grübler et al., 2004). In B2, an income ratio (between 11 world regions, in market exchange rates) of seven corresponds to CO_2 emissions of 14.2 GtC in 2100, while shifting this income ratio to 16 would lead to CO_2 emissions of 15.5 GtC in 2100. Results pointing in the same direction were also obtained for A2. This can be explained by slower TFP growth, slower capital turnover, and less 'technological congruence', leading to slower adoption of low-emission technologies in developing countries. On the other hand, as climate stabilization scenarios require global application of climate policies and convergence in the adoption of low-emission technologies, they are less compatible with low economic convergence scenarios.

3.1.5   Development pathways and GHG emissions

In the long run, the links between economic development and GHG emissions depend not only on the growth rate (measured in aggregate terms), but also on the nature and structure of this growth. Comparative studies aiming to explain these differences help to determine the main factors that will ultimately influence the amount of GHG emissions, given an assumed overall rate of economic growth (Jung et al., 2000; see also examples discussed in Section 12.2 of Chapter 12).

- Structural changes in the production system, namely the role of high or low energy-intensive industries and services.
- Technological patterns in sectors such as energy, transportation, building, waste, agriculture and forestry – the treatment of technology in economic models has received considerable attention and triggered the most difficult debates within the scientific community working in this field (Edmonds and Clarke, 2005; Grubb et al., 2005; Shukla, 2005; Worrell, 2005; Köhler et al., 2006).
- Geographical distribution of activities encompassing both human settlements and urban structures in a given territory, and its twofold impact on the evolution of land use, and on mobility needs and transportation requirements.
- Consumption patterns – existing differences between countries are mainly due to inequalities in income distribution, but for a given income per person, parameters such as housing patterns, leisure styles, or the durability and rate of obsolescence of consumption goods will have a critical influence on long-run emission profiles.
- Trade patterns – the degree of protectionism and the creation of regional blocks can influence access to the best available technologies, inter alia, and constraints on financial flows can limit the capacity of developing countries to build their infrastructure.

These different relationships between development pathways and GHG emissions may (or may not) be captured in models used for long-term world scenarios, by changes in aggregated variables (e.g. per person income) or through more disaggregated economic parameters, such as the structure of expenses devoted to a given need (e.g. heating, transport or food, or the share of energy and transportation in the production function of industrial sectors). This means that alternative configurations of these underlying factors can be combined to give internally consistent socio-economic scenarios with identical rates of economic growth. It would be false to say that current economic models ignore these factors. They are to some extent captured by changes in economic parameters, such as the structure of household expenses devoted to heating, transportation or food; the share of each activity in the total household budget; and the share of energy and transportation costs in total costs in the industrial sector.

These parameters remain important, but the outcome in terms of GHG emissions will also depend on dynamic links between technology, consumption patterns, transportation and urban infrastructure, urban planning, and rural-urban distribution of the population (see also Chapters 2 and 11 for more extensive discussions of some of these issues).

3.1.6 Institutional frameworks

Recent research has included studies on the role of institutions as a critical component in an economy's capacity to use resources optimally (Ostrom, 1990; Ostrom *et al.*, 2002) and interventions that alter institutional structure are among the most accepted solutions in recent times for shaping economic structure and its associated energy use and emissions. Three important aspects of institutional structure are:

1. The extent of centralization and participation in decisions.
2. The extent (spanning from local to global) and nature of decision mechanisms.
3. Processes for effective interventions (e.g. the mix of market and regulatory processes).

Institutional structures vary considerably across nations, even those with similar levels of economic development. Although no consensus exists on the desirability of a specific type of institutional framework, experience suggests that more participative processes help to build trust and social capital to better manage the environmental 'commons' (World Bank, 1992; Beierle and Cayford, 2002; Ostrom *et al.*, 2002; Rydin, 2003). Other relevant developments may include greater use of market mechanisms and institutions to enhance global cooperation and more effectively manage global environmental issues (see also Chapter 12).

A weak institutional structure basically explains why an economy can be in a position that is significantly below the theoretically efficient production frontier, with several economists terming it as a 'missing link' in the production function (Meier, 2001). Furthermore, weak institutions also cause frictions in economic exchange processes, resulting in high transaction costs.

The existence of weak institutions in developing countries has implications for the capacity to adapt to or mitigate climate change. A review of the social capital literature and the implications for climate change mitigation policies concludes that successful implementation of GHG emission-reduction options will generally depend on additional measures to increase the potential market and the number of exchanges. This can involve strengthening the incentives for exchange (prices, capital markets, information efforts, etc.), introducing new actors (institutional and human capacity efforts), and reducing the risks of participating (legal framework, information, general policy context of market regulation). The measures all depend on the nature of the formal institutions, the social groups in society, and the interaction between them (see Chapter 2 and Halsnaes, 2002).

Some of the climate change policy recommendations that are inspired by institutional economics include general capacity-building programmes, and local enterprise and finance development, for example in the form of soft loans, in addition to educational and training programmes (Halsnaes, 2002, see also Chapters 2 and 12).

In today's less industrialized regions, there is a large and relatively unskilled part of the population that is not yet involved in the formal economy. In many regions industrialization leads to wage differentials that draw these people into the more productive, formal economy, causing accelerated urbanization in the process. This is why labour force growth in these regions contributes significantly to GDP growth. The concerns relating to the informal economy are twofold:

1. Whether historical development patterns and relationships among key underlying variables will hold constant in the projections period.
2. Whether there are important feedbacks between the evolution of a particular sector and the overall development pattern that would affect GHG emissions (Shukla, 2005).

Social and cultural processes shape institutions and the way in which they function. Social norms of ownership and distribution have a vital influence on the structure of production and consumption, as well as the quality and extent of the social 'infrastructure' sectors, such as education, which are paramount to capacity building and technological progress. Unlike institutions, social and culture processes are often more inflexible and difficult to influence. However, specific sectors, such as education, are amenable to interventions. Barring some negative features, such as segregation, there is no consensus as to the interventions that are necessary or desirable to alter social and cultural processes. On the other hand, understanding their role is crucial for assessing the evolution of the social infrastructure that underpins technological progress and human welfare (Jung *et al.*, 2000) as well as evolving perceptions and social understanding of climate change risk (see Rayner and Malone, 1998; Douglas and Wildavsky, 1982; Slovic, 2000).

While institutional arrangements are sometimes described as part of storylines, scenario specifications generally do not include explicit assumptions about them. The role of institutions in the implementation of development choices and its implications to climate change mitigation are discussed further in Section 12.2 of Chapter 12.

3.2 Baseline scenarios

3.2.1 Drivers of emissions

Trajectories of future emissions are determined by complex dynamic processes that are influenced by factors such as demographic and socio-economic development, as well as technological and institutional change. An often-used identity to describe changes in some of these factors is based on the IPAT identity (Impact = Population x Affluence x Technology – see Holdren, 2000; Ehrlich and Holdren, 1971) and in emissions modelling is often called the 'Kaya identity' (see Section 3.2.1.4 and Yamaji *et al.*, 1991). These two relationships state that energy-related emissions are a function of population growth,

GDP per person, changes in energy intensity, and carbon intensity of energy consumption. These factors are discussed in Section 3.2.1 to describe new information published on baseline scenarios since the TAR. There are more than 800 emission scenarios in the literature, including almost 400 baseline (non-intervention) scenarios. Many of these scenarios were collected during the IPCC SRES and TAR processes (Morita and Lee, 1998) and made available through the Internet. Systematic reviews of the baseline and mitigation scenarios were reported in the SRES (Nakicenovic *et al.*, 2000) and the TAR (Morita *et al.*, 2001), respectively. The corresponding databases have been updated and extended recently (Nakicenovic *et al.*, 2006; Hanaoka *et al.*, 2006).[2] The recent scenario literature is discussed and compared with the earlier scenarios in this section.

3.2.1.1 Population projections

Current population projections reflect less global population growth than was expected at the time the TAR was published. Since the early 1990s demographers have revised their outlook on future population downward, based mainly on new data indicating that birth rates in many parts of the world have fallen sharply.

Recent projections indicate a small downward revision to the medium (or 'best guess') outlook and to the high end of the uncertainty range, and a larger downward revision to the low end of the uncertainty range (Van Vuuren and O'Neill, 2006). This global result is driven primarily by changes in outlook for the Asia and the Africa-Latin America-Middle East (ALM) region. On a more detailed level, trends are driven by changes in the outlook for Sub-Saharan Africa, the Middle East and North Africa region, and the East Asia region, where recent data show lower than expected fertility rates, as well as a much more pessimistic view on the extent and duration of the HIV/AIDS crisis in Sub-Saharan Africa. In contrast, in the OECD region, updated projections are somewhat higher than previous estimates. This comes from changes in assumptions regarding migration (in the case of the UN projections), or to a more optimistic projection of future life expectancy (in the case of International Institute for Applied Systems Analysis (IIASA) projections). In the Eastern Europe and Central Asia (Reforming Economic, REF) region, projections have been revised downward, especially by the UN, driven mainly by recent data showing very low fertility levels and mortality rates that are quite high relative to other industrialized countries.

Lutz *et al.* (2004), UN (2004) and Fisher *et al.* (2006) have produced updated projections for the world that extend to 2100. The most recent central projections for global population are

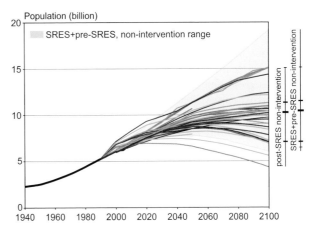

Figure 3.1: *Comparison of population assumptions in post-SRES emissions scenarios with those used in previous scenarios. Blue shaded areas span the range of 84 population scenarios used in SRES or pre-SRES emissions scenarios; individual curves show population assumptions in 117 emissions scenarios in the literature since 2000. The two vertical bars on the right extend from the minimum to maximum of the distribution of scenarios by 2100. The horizontal bars indicate the 5th, 25th, 50th, 75th and the 95th percentiles of the distributions.*

Data source: After Nakicenovic et al., 2006.

1.4–2.0 billion (13–19%) lower than the medium population scenario of 10.4 billion used in the SRES B2 scenarios. As was the case with the outlook for 2050, the long-term changes at the global level are driven by the developing-country regions (Asia and ALM), with the changes particularly large in China, the Middle East and North Africa, and Sub-Saharan Africa.

Most of the SRES scenarios still fall within the plausible range of population outcomes, according to more recent literature (see Figure 3.1). However, the high end of the SRES population range now falls above the range of recent projections from IIASA and the UN. This is a particular problem for population projections in East Asia, the Middle East, North Africa and the Former Soviet Union, where the differences are large enough to strain credibility (Van Vuuren and O'Neill, 2006). In addition, the population assumptions in SRES and the vast majority of more recent emissions scenarios do not cover the low end of the current range of population projections well. New scenario exercises will need to take the lower population projections into account. All other factors being equal, lower population projections are likely to result in lower emissions. However, a small number of recent studies that have used updated and lower population projections (Carpenter *et al.*, 2005; Van Vuuren *et al.*, 2007; Riahi *et al.*, 2006) indicate that changes in other drivers of emissions might partly offset the impact of lower population assumptions, thus leading to no significant changes in emissions.

---

2 It should be noted that the sources of scenario data vary. For some scenarios the data comes directly from the modelling teams. In other cases it has been assembled from the literature or from other scenario comparison exercises such as EMF-19, EMF-21, and IMCP. For this assessment the scenario databases from Nakicenovic et al. (2006) and Hanaoka et al. (2006) were updated with the most recent information. The scenarios published before the year 2000 were retrieved from the database during SRES and TAR. The databases from Nakicenovic et al. (2006) and Hanaoka et al. (2006) can be accessed on the following websites: http://iiasa.ac.at/Research/TNT/WEB/scenario_database.html and www-cger.nies.go.jp/scenario.

3.2.1.2 Economic development

Economic activity is a dominant driver of energy demand and thus of greenhouse gas emissions. This activity is usually reported as gross domestic product (GDP), often measured in per-person (per-capita) terms. To derive meaningful comparisons over time, changes in price levels must be taken into account and corrected by reporting activities as constant prices taken from a base year. One way of reducing the effects of different base years employed across various studies is to report real growth rates for changes in economic output. Therefore, the focus below is on real growth rates rather than on absolute numbers.

Given that countries and regions use particular currencies, another difficulty arises in aggregating and comparing economic output across countries and world regions. There are two main approaches: using an observed market exchange rate (MER) in a fixed year or using a purchasing power parity rate (PPP) (see Box 3.1). GDP trajectories in the large majority of long-term scenarios in the literature are calibrated in MER. A few dozen scenarios exist that use PPP exchange rates, but most of them are shorter-term, generally running until the year 2030.

3.2.1.3 GDP growth rates in the new literature

Many of the long-term economic projections in the literature have been specifically developed for climate-related scenario work. Figure 3.2 compares the global GDP range of 153 baseline scenarios from the pre-SRES and SRES literature with 130 new scenarios developed since SRES (post-SRES). There is a considerable overlap in the GDP numbers published, with a slight downward shift of the median in the new scenarios (by about 7%) compared to the median in the pre-SRES scenario literature. The data suggests no appreciable change in the distribution of GDP projections.

A comparison of some recent shorter-term global GDP projections using the SRES scenarios is illustrated in Figure 3.3. The SRES scenarios project a very wide range of global economic per-person growth rates from 1% (A2) to 3.1% (A1) to 2030, both based on MER. This range is somewhat wider than that covered by the USDOE (2004) high and low scenarios (1.2–2.5%). The central projections of USDOE, IEA and the World Bank all contain growth rates of around 1.5–1.9%, thus occurring in the middle of the range of the SRES scenarios. Other medium-term energy scenarios are also reported to have growth rates in this range (IEA, 2004).

Regionally, for the OECD, Eastern Europe and Central Asia (REF) regions, the correspondence between SRES outcomes and recent scenarios is relatively good, although the SRES GDP growth rates are somewhat conservative. In the ASIA region, the SRES range and its median value are just above that of recent studies. The differences between the SRES outcomes and more recent projections are largest in the ALM region (covering Africa, Latin America and the Middle East). Here,

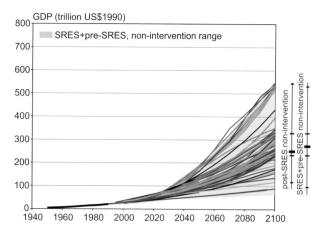

Figure 3.2: *Comparison of GDP projections in post-SRES emissions scenarios with those used in previous scenarios. The median of the new scenarios is about 7% below the median of the pre-SRES and SRES scenario literature. The two vertical bars on the right extend from the minimum to maximum of the distribution of scenarios by 2100. The horizontal bars indicate the 5th, 25th, 50th, 75th and the 95th percentiles of the distributions.*

the A1 and B1 scenarios clearly lie above the upper end of the range of current projections (4%–5%), while A2 and B2 fall near the centre of the range (1.4–1.7%). The recent short-term projections reported here contain an assumption that current barriers to economic growth in these regions will slow growth, at least until 2015.

3.2.1.4 The use of MER in economic and emissions scenarios modelling

The uses of MER-based economic projections in SRES have recently been criticized (Castles and Henderson, 2003a, 2003b; Henderson, 2005). The vast majority of scenarios published in the literature use MER-based economic projections. Some exceptions exist, for example, MESSAGE in SRES, and more recent scenarios using the MERGE model (Manne and Richels, 2003), along with shorter term scenarios to 2030, including the G-Cubed model (McKibbin *et al.*, 2004a, 2004b), the International Energy Outlook (USDOE, 2004), the IEA World Energy Outlook (IEA, 2004) and the POLES model used by the European Commission (2003). The main criticism of the MER-based models is that GDP data for world regions are not corrected with respect to purchasing power parities (PPP) in most of the model runs. The implied consequence is that the economic activity levels in non-OECD countries generally appear to be lower than they actually are when measured in PPP units. In addition, the high growth SRES scenarios (A1 and B1 families) assume that regions tend to conditionally converge in terms of relative per-person income across regions (see Section 3.1.4). According to the critics, the use of MER, together with the assumption of conditional convergence, lead to overstated economic growth in the poorer regions and excessive growth in energy demand and emission levels.

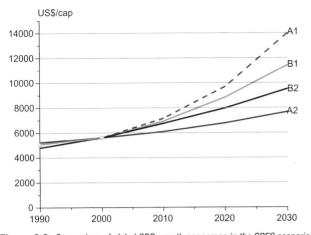
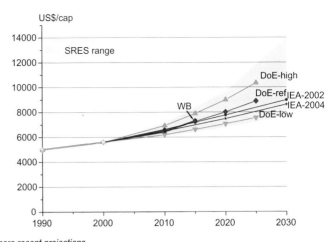

Figure 3.3: *Comparison of global GDP growth per person in the SRES scenarios and more recent projections.*

Notes: SRES = (Nakicenovic *et al.*, 2000), WB = World Bank (World Bank, 2004), DoE = assumptions used by US Department of Energy (USDOE, 2004), IEA assumptions used by IEA (IEA, 2002 and 2004); (Van Vuuren and O'Neill, 2006).

A team of SRES researchers responded to this criticism, indicating that the use of MER or PPP data does not in itself lead to different emission projections outside the range of the literature. In addition, they stated that the use of PPP data in most scenarios models was (and still is) infeasible, due to lack of required data in PPP terms, for example price elasticities and social accounting matrices (Nakicenovic *et al.*, 2003; Grübler *et al.*, 2004). A growing number of other researchers have also indicated different opinions on this issue or explored it in a more quantitative sense (e.g. Dixon and Rimmer, 2005; Nordhaus, 2006b; Manne and Richels, 2003; McKibbin *et al.*, 2004a, 2004b; Holtsmark and Alfsen, 2004a, 2004b; Van Vuuren and Alfsen, 2006).

There are at least three strands to this debate. The first is whether *economic projections* based on MER are appropriate, and thus whether the economic growth rates reported in the SRES and other MER-based scenarios are reasonable and robust. The second is whether the choice of the exchange rate matters when it comes to *emission scenarios*. The third is whether it is possible, or practical, to develop robust scenarios given the sparseness of relevant and required PPP data. While the GDP data are available in PPP, other economic scenario characteristics, such as capital and operational cost of energy facilities, are usually available either in domestic currencies or MER. Full model calibration in PPP for regional and global models is still difficult due to the lack of underlying data. This could be one of the reasons why a vast majority of long-term emissions scenarios continues to be calibrated in MER.

On the question of whether PPP or MER should be employed in economic scenarios, the general recommendations are to use PPP where practical.[3] This is certainly necessary when

comparisons of income levels across regions are of concern. On the other hand, models that analyse international trade and include trade as part of their economic projections, are better served by MER data given that trade takes place between countries in actual market prices. Thus, the choice of conversion factor depends on the type of analysis or comparison being undertaken.

For principle and practical reasons, Nordhaus (2005) recommends that economic growth scenarios should be constructed by using regional or national accounting figures (including growth rates) for each region, but using PPP exchange rates for aggregating regions and updating over time by use of a superlative price index. In contrast, Timmer (2005) actually prefers the use of MER data in long-term modelling, as such data are more readily available, and many international relations within the model are based on MER. Others (e.g. Van Vuuren and Alfsen, 2006) also argue that the use of MER data in long-term modelling is often preferable, given that model parameters are usually estimated on MER data and international trade within the models is based on MER. The real economic consequences of the choice of conversion rates will obviously depend on how the scenarios are constructed, as well as on the type of model used for quantifying the scenarios. In some of the short-term scenarios (with a horizon to 2030) a bottom-up approach is taken where assumptions about productivity growth and investment/saving decisions are the main drivers of growth in the models (e.g. McKibbin *et al.*, 2004a, 2004b). In long-term scenario models, a top-down approach is more commonly used where the actual growth rates are prescribed more directly, based on convergence or other assumptions about long-term growth potentials.

---

3   See, for example, UN (1993), (para 1.38): 'When the objective is to compare the volumes of goods or services produced or consumed per head, data in national currencies must be converted into a common currency by means of purchasing power parities and not exchange rates. It is well known that, in general, neither market nor fixed exchange rates reflect the relative internal purchasing powers of different currencies. When exchange rates are used to convert GDP, or other statistics, into a common currency the prices at which goods and services in high-income countries are valued tend to be higher than in low-income countries, thus exaggerating the differences in real incomes between them. Exchange rate converted data must not, therefore, be interpreted as measures of the relative volumes of goods and services concerned.'

Box 3.1 Market Exchange Rates and Purchasing Power Parity

To aggregate or compare economic output from various countries, GDP data must be converted into a common unit. This conversion can be based on observed market exchange (MER) rates or purchasing power parity (PPP) rates where, in the latter, a correction is made for differences in price levels between countries. The PPP approach is considered to be the better alternative if data is used for welfare or income comparisons across countries or regions. Market exchange rates usually undervalue the purchasing power of currencies in developing countries, see Figure 3.4.

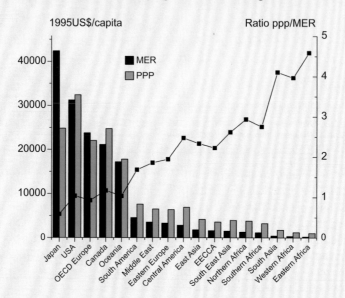

Figure 3.4: *Regional GDP per person, expressed in MER and PPP on the basis of World Bank data aggregated to 17 global regions.*

Note: The left y-axis and columns compare absolute data, while the right y-axis and line graph compare the ratio between PPP and MER data. EECCA = countries of Eastern Europe, the Caucasus and Central Asia.

Source: Van Vuuren and Alfsen, 2006.

Clearly, deriving PPP exchange rates requires analysis of a relatively large amount of data. Hence, methods have been devised to derive PPP rates for new years on the basis of price indices. Unfortunately, there is currently no single method or price index favoured for doing this, resulting in different sets of PPP rates (e.g. from the OECD, Eurostat, World Bank and Penn World Tables) although the differences tend to be small.

When it comes to emission projections, it is important to note that in a fully disaggregated (by country) multi-sector economic model of the global economy, aggregate index numbers play no role and the choice between PPP and MER conversion of income levels does not arise. However, in an aggregated model with consistent specifications (i.e. where model parameter estimation and model calibrations are all carried out based on consistent use of conversion factors), the effects of the choice of conversion measure on emissions should approximately cancel out. The reason can be illustrated by using the Kaya identity, which decomposes the emissions as follows:

GHG = Population x GDP per person x Emissions per GDP

or:

$$GHG = POP \times \left(\frac{GDP}{POP} \right) \times \left(\frac{GHG}{GDP} \right)$$

where GHG stands for greenhouse gas emissions, GDP stands for economic output, and POP stands for population size.[4]

Given this relationship, emission scenarios can be represented, explicitly based on estimates of population development, economic growth, and development of emission intensity.

Population is often projected to grow along a pre-described (exogenous) path, while economic activity and emission intensities are projected based on differing assumptions from scenario to scenario. The economic growth path can be based on historical growth rates, convergence assumptions, or on fundamental growth factors, such as saving and investment behaviour, productivity changes, etc. Similarly, future emission intensities can be projected based on historical experience,

---

4   Other components could be introduced in the identity, such as energy use, without changing the argument.

economic factors, such as labour productivity or other key factors determining structural changes in an economy, or technological development. The numerical expression of GDP clearly depends on conversion measures; thus GDP expressed in PPP will deviate from GDP expressed in MER, particularly for developing countries. However, when it comes to calculating emissions (or other physical measures such as energy), the Kaya identity shows that the choice between MER-based or PPP-based representations of GDP will not matter, since emission intensity will change (in a compensating manner) when the GDP numbers change. While using PPP values necessitates using lower economic growth rates for developing countries under the convergence assumption, it is also necessary to adjust the relationship between income and demand for energy with lower economic growth, leading to slower improvements in energy intensities. Thus, if a consistent set of metrics is employed, the choice of metric should not appreciably affect the final emission level.

In their modelling work, Manne and Richels (2003) and McKibbin *et al.* (2004a, 2004b) find some differences in emission levels between using PPP-based and MER-based estimates. Analysis of their work indicates that these results critically depend on, among other things, the combination of convergence assumptions and the mathematical approximation used between MER-GDP and PPP-GDP. In the Manne and Richels work for instance, autonomous efficiency improvement (AEI) is determined as a percentage of economic growth and estimated on the basis of MER data. In going from MER to PPP, the economic growth rate declines as expected, leading to a decline in the autonomous efficiency improvement. However, it is not clear whether it is realistic not to change the AEI rate when changing conversion measure. On the other hand, Holtsmark and Alfsen (2004a, 2004b), showed that in their simple model consistent replacement of the metric (PPP for MER) – for income levels as well as for underlying technology relationships – leads to a full cancellation of the impact of choice of metric on projected emission levels.

To summarize: available evidence indicates that the differences between projected emissions using MER exchange rates and PPP exchange rates are small in comparison to the uncertainties represented by the range of scenarios and the likely impacts of other parameters and assumptions made in developing scenarios, for example, technological change. However, the debate clearly shows the need for modellers to be more transparent in explaining conversion factors, as well as taking care in determining exogenous factors used for their economic and emission scenarios.

3.2.1.5   *Energy use*

Future evolution of energy systems is a fundamental determinant of GHG emissions. In most models, energy demand growth is a function of key driving forces such as demographic change and the level and nature of human activities such as

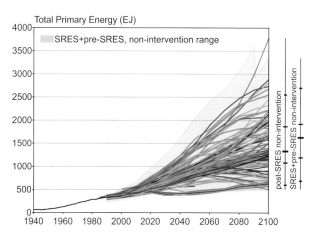

Figure 3.5: *Comparison of 153 SRES and pre-SRES baseline energy scenarios in the literature compared with the 133 more recent, post-SRES scenarios. The ranges are comparable, with small changes on the lower and upper boundaries.*

Note: The two vertical bars on the right extend from the minimum to maximum of the distribution of scenarios by 2100. The horizontal bars indicate the 5th, 25th, 50th, 75th and the 95th percentiles of the distributions.

mobility, information processing, and industry. The type of energy consumed is also important. While Chapters 4 through 11 report on medium-term projections for different parts of the energy system, long-term energy projections are reported here. Figure 3.5 compares the range of the 153 SRES and pre-SRES scenarios with 133 new, post-SRES, long-term energy scenarios in the literature. The ranges are comparable, with small changes on the lower and upper boundaries, and a shift downwards with respect to the median development. In general, the energy growth observed in the newer scenarios does not deviate significantly from the previous ranges as reported in the SRES report. However, most of the scenarios reported here have not adapted the lower population levels discussed in Section 3.2.1.1.

In general, this situation also exists for underlying trends as represented by changes in energy intensity, expressed as gigajoule (GJ)/GDP, and change in the carbon intensity of the energy system (CO_2/GJ) as shown in Figure 3.6. In all scenarios, energy intensity improves significantly across the century – with a mean annual intensity improvement of 1%. The 90% range of the annual average intensity improvement is between 0.5% and 1.9% (which is fairly consistent with historic variation in this factor). Actually, this range implies a difference in total energy consumption in 2100 of more than 300% – indicating the importance of the uncertainty associated with this ratio. The carbon intensity is more constant in scenarios without climate policy. The mean annual long-term improvement rate over the course of the 21st century is 0.4%, while the uncertainty range is again relatively large (from -0.2 to 1.5%). At the high end of this range, some scenarios assume that energy technologies without CO_2 emissions become competitive without climate policy as a result of increasing fossil fuel prices and rapid technology progress for carbon-free technologies. Scenarios with a low carbon-intensity improvement coincide with scenarios with a

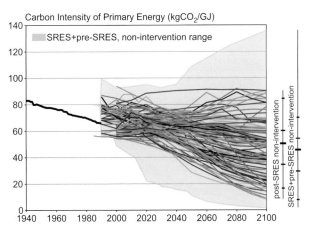
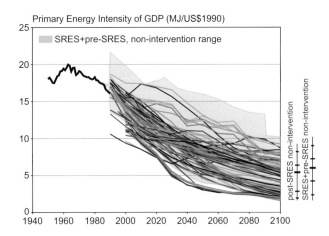

Figure 3.6: *Development of carbon intensity of energy (left) and primary energy intensity of GDP (right). Historical development and projections from SRES and pre-SRES scenarios compared to post-SRES scenarios.*

Note: The blue coloured range illustrates the range of 142 carbon intensity and 114 energy intensity – SRES and pre-SRES non-intervention scenarios.

Source: After Nakicenovic et al., 2006.

large fossil fuel base, less resistance to coal consumption or lower technology development rates for fossil-free energy technologies. The long-term historical trend is one of declining carbon intensities. However, since 2000, carbon intensities are increasing slightly, primarily due to the increasing use of coal. Only a few scenarios assume the continuation of the present trend of increasing carbon intensities. One of the reasons for this may be that just a few of the recent scenarios include the effects of high oil prices.

3.2.1.6   *Land-use change and land-use management*

Understanding land-use and land-cover changes is crucial to understanding climate change. Even if land activities are not considered as subject to mitigation policy, the impact of land-use change on emissions, sequestration, and albedo plays an important role in radiative forcing and the carbon cycle.

Over the past several centuries, human intervention has markedly changed land surface characteristics, in particular through large-scale land conversion for cultivation (Vitousek *et al.*, 1997). Land-cover changes have an impact on atmospheric composition and climate via two mechanisms: biogeophysical and biogeochemical. Biogeophysical mechanisms include the effects of changes in surface roughness, transpiration, and albedo that, over the past millennium, are thought to have had a global cooling effect (Brovkin *et al.*, 1999). Biogeochemical effects result from direct emissions of CO_2 into the atmosphere from deforestation. Cumulative emissions from historical land-cover conversion for the period 1920–1992 have been estimated to be between 206 and 333 Pg CO_2 (McGuire *et al.*, 2001), and as much as 572 Pg CO_2 for the entire industrial period 1850–2000, roughly one-third of total anthropogenic carbon emissions over this period (Houghton, 2003). In addition, land management activities (e.g. cropland fertilization and water management, manure management and forest rotation lengths) also affect land-based emissions of CO_2 and non-CO_2 GHGs,

where agricultural land management activities are estimated to be responsible for the majority of global anthropogenic methane (CH_4) and nitrous oxide (N_2O) emissions. For example, USEPA (2006a) estimated that agricultural activities were responsible for approximately 52% and 84% of global anthropogenic CH_4 or N_2O emissions respectively in the year 2000, with a net contribution from non-CO_2 GHGs of 14% of all anthropogenic greenhouse gas emissions in that year.

Projected changes in land use were not explicitly represented in carbon cycle studies until recently. Previous studies into the effects of future land-use changes on the global carbon cycle employed trend extrapolations (Cramer *et al.*, 2004), extreme assumptions about future land-use changes (House *et al.*, 2002), or derived trends of land-use change from the SRES storylines (Levy *et al.*, 2004). However, recent studies (e.g. Brovkin *et al.*, 2006; Matthews *et al.*, 2003; Gitz and Ciais, 2004) have shown that land use, as well as feedbacks in the society-biosphere-atmosphere system (e.g. Strengers *et al.*, 2004), must be considered in order to achieve realistic estimates of the future development of the carbon cycle; thereby providing further motivation for ongoing development to explicitly model land and land-use drivers in global integrated assessment and climate economic frameworks. For example, in a model comparison study of six climate models of intermediate complexity, Brovkin *et al.* (2006) concluded that land-use changes contributed to a decrease in global mean annual temperature in the range of 0.13–0.25°C, mainly during the 19th century and the first half of the 20th century, which is in line with conclusions from other studies, such as Matthews *et al.* (2003).

In general, land-use drivers influence either the demand for land-based products and services (e.g. food, timber, bio-energy crops, and ecosystem services) or land-use production possibilities and opportunity costs (e.g. yield-improving technologies, temperature and precipitation changes, and CO_2 fertilization). Non-market values – both use and non-use such as environmental services and species existence values respectively – will also shape land-use outcomes.

Food demand is a dominant land-use driver, and population and economic growth are the most significant food demand drivers through per person consumption. Total world food consumption is expected to increase by over 50% by 2030 (Bruinsma, 2003). Moreover, economic growth is expected to generate significant structural change in consumption patterns, with diets shifting to include more livestock products and fewer staples such as roots and tubers. As a result, per person meat consumption is expected to show a strong global increase, in the order of 25% by 2030, with faster growth in developing and transitional countries of more than 40% and 30%, respectively (Bruinsma, 2003; Cassman *et al.*, 2003). The Millennium Ecosystem Assessment (MEA) scenarios projected that global average meat consumption would increase from 36 kg/person in 1997 to 41–70 kg/person by 2050, with corresponding increases in overall food and livestock feed demands (Carpenter *et al.*, 2005). Additional cropland is expected to be required to support these projected increases in demand. Beyond 2050, food demand is expected to level off with slow-down of population growth.

Technological change is also a critical driver of land use, and a critical assumption in land-use projections. For example, Sands and Leimbach (2003) suggest that, globally, 800 million hectares of cropland expansion could be avoided with a 1% annual growth in crop yields. Similarly, Kurosawa (2006) estimates decreased cropland requirements of 18% by 2050, relative to 2000, with 2% annual growth in global average crop yields. Alternatively, the MEA scenarios implement a more complex representation of yield growth projections that, in addition to autonomous technological change, reflect the changes in production practices, investments, technology transfer, environmental degradation, and climate change. The net effect is positive, but shows declining productivity growth over time for some commodities, due in large part to diminishing marginal technical productivity gains and environmental degradation. In all these studies, increasing (decreasing) net productivity per hectare results in reduced (increased) cropland demand.

Also important to land-use projections are potential changes in climate. For instance, rising temperatures and CO_2 fertilization may improve regional crop yields in the short term, thereby reducing pressure for additional cropland and resulting in increased afforestation. However, modelling the beneficial impacts of CO_2 fertilization is not as straightforward as once thought. Recent results suggest: lower crop productivity improvements in the field than shown previously with laboratory results (e.g. Ainsworth and Long, 2005); likely increases in tropospheric ozone and smog associated with higher temperatures that will depress plant growth and partially offset CO_2 fertilization; expected increases in the variability of annual yields; CO_2 effects favouring C3 plants (e.g. wheat, barley, potatoes, rice) over C4 plants (e.g. maize, sugar cane, sorghum, millet) while temperature increases favour C4 over

C3 plants; potential decreased nutritional content in plants subjected to CO_2 fertilization and increased frequency of temperature extremes; and increases in forest disturbance frequency and intensity. See IPCC (2007b, Chapter 5) for an overall discussion of these issues and this literature. Long-term projections need to consider these issues, as well as examining the potential limitations or saturation points of plant responses. However, to date, long-term scenarios from integrated assessment models are only just beginning to represent climate feedbacks on terrestrial ecosystems, much less fully account for the many effects. Current integrated assessment representations only consider CO_2 fertilization and changes in yearly average temperature, if they consider climate change effects at all (e.g. USCCSP, 2006; Van Vuuren *et al.*, 2007).

Only a few global studies have focused on long-term (century) land-use projections. The most comprehensive studies, in terms of sector and land-type coverage, are the SRES (Nakicenovic *et al.*, 2000), the SRES implementation with the IMAGE model (Strengers *et al.*, 2004), the scenarios from the Global Scenarios Group (Raskin *et al.*, 2002), UNEP's Global Environment Outlook (UNEP, 2002), the Millennium Ecosystem Assessment (Carpenter *et al.*, 2005), and some of the EMF-21 Study models (Kurosawa, 2006; Van Vuuren *et al.*, 2006a; Rao and Riahi, 2006; Jakeman and Fisher, 2006; Riahi *et al.*, 2006; Van Vuuren *et al.*, 2007). Recent sector-specific economic studies have also contributed global land-use projections for climate analysis, especially for forestry (Sands and Leimbach, 2003; Sohngen and Mendelsohn, 2003, 2007; Sathaye *et al.*, 2006; Sohngen and Sedjo, 2006). In general, the post-SRES scenarios, though scarce in number for agricultural land use, have projected increasing global cropland areas, smaller forest-land areas, and mixed results for changes in global grassland (Figure 3.7). Unlike the SRES land-use scenarios that span a broader range while representing diverse storylines, the post-SRES scenarios, for forestry in particular, illustrate greater convergence across models on projected land-use change.

Most post-SRES global scenarios project significant changes in agricultural land caused primarily by regional changes in food demand and production technology. Scenarios with larger amounts of land used for agriculture result from assumptions about higher population growth rates, higher food demands, and lower rates of technological improvement that generate negligible increases in crop yields. Combined, these effects are projected to lead to a sizeable expansion (up to 40%) of agricultural land between 1995 and 2100 (Figure 3.7). Conversely, lower population growth and food demand, and more rapid technological change, are projected to result in lower demand for agricultural land (as much as 20% less global agricultural acreage by the end of the century). In the short-term, almost all scenarios suggest an increase in cropland acreage and decline in forest land to meet projected increases in food, feed, and livestock grazing demands over the next few decades. Cropland changes range from -18% to +69% by 2050 relative to 2000 (from -123 to +1158 million hectares)

(A)

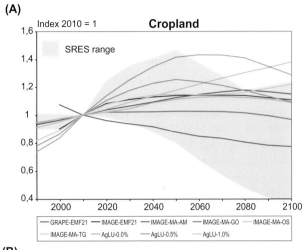

(B)

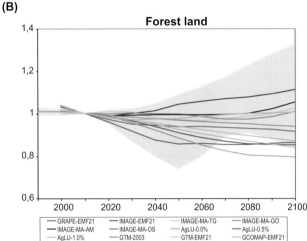

(C)

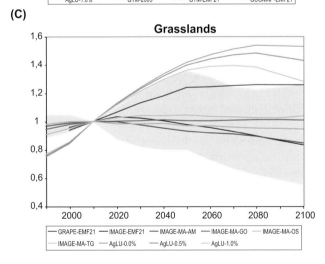

Figure 3.7: *Global cropland (a), forest land (b) and grassland (c) projections.*

Notes: shaded areas indicate SRES scenario ranges, post-SRES scenarios denoted with solid lines. IMAGE-EMF21 = Van Vuuren *et al.* (2006a) scenario from EMF-21 Study; IMAGE-MA-xx = Millennium Ecosystem Assessment (Carpenter *et al.*, 2005) scenarios from the IMAGE model for four storylines (GO = Global Orchestration, OS = Order from Strength, AM = Adapting Mosaic, TG = TechnoGarden); AgLU-x.x% = Sands and Leimbach (2003) scenarios with x.x% annual growth in crop yield; GTM-2003 = Sohngen and Mendelsohn (2003) global forest scenario; GTM-EMF21 = Sohngen and Sedjo (2006) global forest scenario from EMF-21 Study; GCOMAP-EMF21 = Sathaye *et al.* (2006) global forest scenario from EMF-21 Study; GRAPE-EMF21 = Kurosawa (2006) scenario from EMF-21 Study.

and forest-land changes range from -18% to +3% (from -680 to +94 million hectares) by 2050. The changes in global forest generally mirror the agricultural scenarios; thereby, illustrating both the positive and negative aspects of some existing global land modelling. Most of the long-term scenarios assume that forest trends are driven almost exclusively by cropland expansion or contraction, and only deal superficially with driving forces, such as global trade in agricultural and forest products and conservation demands.

Without incentives or technological innovation, biomass crops are currently not projected to assume a large share of global business as usual land cover – no more than about 4% by 2100. Until long-run energy price expectations rise (due to a carbon price, economic scarcity, or other force), biomass and other less economical energy supply technologies (some with higher greenhouse gas emission characteristics than biomass), are not expected to assume more significant baseline roles.

3.2.2 Emissions

There is still a large span of CO_2 emissions across baseline scenarios in the literature, with emissions in 2100 ranging from 10 $GtCO_2$ to around 250 $GtCO_2$. The wide range of future emissions is a result of the uncertainties in the main driving forces, such as population growth, economic development, and energy production, conversion, and end use, as described in the previous section.

3.2.2.1 *CO_2 emissions from energy and industry*

This category of emissions encompasses CO_2 emissions from burning fossil fuels, and industrial emissions from cement production and sometimes feedstocks.[5] Figure 3.8 compares the range of the pre-SRES and SRES baseline scenarios with the post-SRES baseline scenarios. The figure shows that the scenario range has remained almost the same since the SRES. There seems to have been an upwards shift on the high and low end, but careful consideration of the data shows that this is caused by only very few scenarios and the change is therefore not significant. The median of the recent scenario distribution has shifted downwards slightly, from 75 $GtCO_2$ by 2100 (pre-SRES and SRES) to about 60 $GtCO_2$ (post SRES). The median of the recent literature therefore corresponds roughly to emissions levels of the intermediate SRES-B2 scenarios. The majority of scenarios, both pre-SRES and post-SRES, indicate an increase in emissions across most of the century, resulting in a range of 2100 emissions of 17–135 $GtCO_2$ emissions from energy and industry (90th percentile of the full scenario distribution). Also the range of emissions depicted by the SRES scenarios is consistent with the range of other emission

---

5   It should be noted, however, that there are sometimes considerable ambiguities on what is actually included in emissions scenarios reported in the literature. Some of the CO_2 emissions paths included in the ranges may therefore also include non-energy emissions such as those from land-use changes.

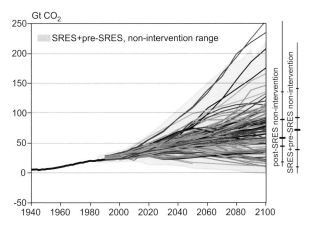

Figure 3.8: *Comparison of the SRES and pre-SRES energy-related and industrial CO₂ emissions scenarios in the literature with the post-SRES scenarios.*

Note: The two vertical bars on the right extend from the minimum to maximum of the distribution of scenarios and indicate the 5th, 25th, 50th, 75th and the 95th percentiles of the distributions by 2100.

Source: After Nakicenovic et al., 2006

scenarios reported in the literature; both in the short and long term (see Van Vuuren and O'Neill, 2006).

Several reasons may contribute to the fact that emissions have not declined in spite of somewhat lower projections for population and GDP. An important reason is that the lower demographic projections are only recently being integrated into emission scenario literature. Second, indirect impacts in the models are likely to offset part of the direct impacts. For instance, lower energy demand leads to lower fossil fuel depletion, thus allowing for a higher share of fossil fuels in the total energy mix over a longer period of time. Finally, in recent years there has been increasing attention to the interpretation of fossil fuel reserves reported in the literature. Some models may have decreased oil and gas use in this context, leading to higher coal use (and thus higher emissions).

Analysis of scenario literature using the Kaya identity shows that pre-SRES and post-SRES baseline scenarios indicate a continuous decline of the primary energy intensity (EJ/GDP), while the change in carbon intensity (CO₂/E) is much slower – or even stable (see Figure 3.6 and Section 3.2.1.5) in the post-SRES scenarios. In other words, in the absence of climate policy, structural change and energy efficiency improvement do contribute to lower emissions, but changes in the energy mix have a much smaller (or even zero) contribution. This conclusion is true for both the pre-SRES, SRES, as well as the post-SRES scenario literature.

Baseline or reference emissions projections generally come from three types of studies:
1. Studies meant to represent a 'best-guess' of what might happen if present-day trends and behaviour continue.
2. Studies with multiple baseline scenarios under comprehensively different assumptions (storylines).
3. Studies based on a probabilistic approach.

In literature, since the TAR, there has been some discussion of the purpose of these approaches (see Schneider, 2001; Grübler *et al.*, 2002; Webster *et al.*, 2002). Figure 3.9 (left panel) shows a comparison of the outcomes of some prominent examples of these approaches by comparing the outcome of baselines scenarios reported in the set of EMF-21 scenarios, representing the 'best-guess' approach, to the outcomes of the SRES scenarios, representing the storyline approach. In the right panel the SRES range is compared to the probabilistic approach (see Webster *et al.*, 2002; Richels *et al.*, 2004, for the probability studies).

The figure shows that the range of different models participating in the EMF-21 study is somewhat smaller than those from SRES and the probabilistic approach. The range of EMF-21 scenarios result from different modelling approaches

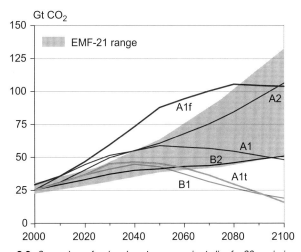

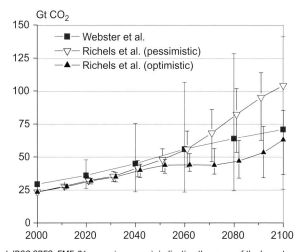

Figure 3.9: *Comparison of various long-term scenario studies for CO₂ emissions. Left panel: IPCC SRES, EMF-21 range (grey area), indicating the range of the lowest and highest reported values in the EMF-21 study (Weyant et al., 2006). Right panel: Webster et al. (2002) and Richels et al. (2004), indicating the mean (markers) and 95% intervals of the reported ranges of these studies (for the latter, showing the 95% interval of the combined range for optimistic and pessimistic technology).*

and from modeller's insights into 'the mostly likely values' for driving forces. The two probabilistic studies and SRES explicitly assume more radical developments, but the number of studies involved is smaller. This leads to the low end of scenarios for the second category having very specific assumptions on development that may lead to low greenhouse gas emissions. The range of scenarios in the probabilistic studies tends to be between these extremes. Overall, the three different approaches seem to lead to consistent results, confirming the range of emissions reported in Figure 3.8 and confirming the emission range of scenarios used for the TAR.

3.2.2.2 Anthropogenic land emissions and sequestration

Some of the first global integrated assessment scenario analyses to account for land-use-related emissions were the IS92 scenario set (Leggett *et al.*, 1992) and the SRES scenarios (Nakicenovic *et al.*, 2000). However, out of the six SRES models, only four dealt with land use specifically (MiniCAM, MARIA, IMAGE 2.1, AIM), of which MiniCAM and MARIA used more simplified land-use modules. ASF and MESSAGE also simulated land-use emissions, however ASF did not have a specific land-use module and MESSAGE incorporated land-use results from the AIM model (Nakicenovic *et al.*, 2000). Although SRES was a seminal contribution to scenario development, the treatment of land-use emissions was not the focus of this assessment; and, therefore, neither was the modelling of land-use drivers, land management alternatives, and the many emissions sources, sinks, and GHGs associated with land.

While some recent assessments, such as UNEP's Third Global Environment Outlook (UNEP, 2002) and the Millennium Ecosystem Assessment (Carpenter *et al.*, 2005), have evaluated land-based environmental outcomes (global environment and ecosystem goods and services respectively), the Energy Modelling Forum's 21st Study (EMF-21) was the first large-scale exercise with a special focus on land as a climate issue. In EMF-21, the integrated assessment models incorporated non-CO_2 greenhouse gases, such as those from agriculture, and carbon sequestration in managed terrestrial ecosystems (Kurosawa, 2006; Van Vuuren *et al.*, 2006a; Rao and Riahi, 2006; Jakeman and Fisher, 2006). A few additional papers have subsequently improved upon their EMF-21 work (Riahi *et al.*, 2006; Van Vuuren *et al.*, 2007). In general, the land-use change carbon emissions scenarios since SRES project high global annual net releases of carbon in the near future that decline over time, leading to net sequestration by the end of the century in some scenarios (see Figure 3.10). The clustering of the non-harmonized post-SRES scenarios in Figure 3.10 suggests a degree of expert agreement that the decline in annual land-use change carbon emissions over time will be less dramatic (slower) than suggested by many of the SRES scenarios. Many of the post-SRES scenarios project a decrease in net deforestation pressure over time, as population growth slows and crop and livestock productivity increase; and, despite

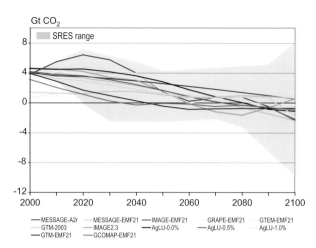

Figure 3.10: *Baseline land-use change and forestry carbon net emissions.*
Notes: MESSAGE-EMF21 = Rao and Riahi (2006) scenario from EMF-21 Study; GTEM-EMF21 = Jakeman and Fisher (2006) scenario from EMF-21 Study; MESSAGE-A2r = Riahi *et al.* (2006) scenario with revised SRES-A2 baseline; IMAGE 2.3 = Van Vuuren *et al.* (2007) scenario; see Figure 3.7 notes for additional scenario references. The IMAGE 2.3 LUCF baseline scenario also emits non-CO_2 emissions (CH_4 and N_2O) of 0.26, 0.30, 0.16 $GtCO_2$-eq in 2030, 2050, and 2100, respectively.

continued projected loss of forest area in some scenarios (Figure 3.7), carbon uptake from afforestation and reforestation result in net sequestration.

There also seems to be a consensus in recent non-CO_2 GHG emission baseline scenarios that agricultural CH_4 and N_2O emissions will increase until the end of this century, potentially doubling in some baselines (see Table 3.1; Kurosawa, 2006; Van Vuuren *et al.*, 2006a; Rao and Riahi, 2006; Jakeman and Fisher, 2006; Riahi *et al.*, 2006; Van Vuuren *et al.*, 2007). The modelling of agricultural emission sources varies across scenarios, with livestock and rice paddy methane and crop soil nitrous oxide emissions consistently represented. However, the handling of emissions from biomass burning and fossil fuel combustion are inconsistent across models; and cropland soil carbon fluxes are generally not reported, probably due to the fact that soil carbon sequestration mitigation options are not currently represented in these models.

As noted in Section 3.2.1.6 climate change feedbacks could have a significant influence on long-term land use and, to date, are only partially represented in long-term modelling of land scenarios. Similarly, climate feedbacks can also affect land-based emissions. For instance, rising temperatures and CO_2 fertilization can influence the amount of carbon that can be sequestered by land and may also lead to increased afforestation due to higher crop yields. Climate feedbacks in the carbon cycle could be extremely important. For instance, Leemans *et al.* (2002) showed that CO_2 fertilization and soil respiration could be as important as the socio-economic drivers in determining the land-use emissions range.

In addition, potentially important additional climate feed-backs in the carbon-climate system are currently not accounted

Table 3.1: *Baseline global agricultural non-CO_2 greenhouse gas emissions from various long-term stabilization scenarios (GtCO$_2$-eq).*

Scenario	Non-CO_2 GHG agricultural emissions sources represented*	GtCO$_2$-eq									
		CH$_4$					N$_2$O				
		2000	2020	2050	2070	2100	2000	2020	2050	2070	2100
GTEM-EMF21	Enteric, manure, paddy rice, soil (N$_2$O)	2.09	2.88	4.28	nm	nm	1.95	2.60	3.64	nm	nm
MESSAGE-EMF21	Enteric, manure, paddy rice, soil (N$_2$O)	2.58	3.42	6.05	6.00	5.06	2.57	3.48	4.65	3.79	2.32
IMAGE-EMF21	Enteric, manure, paddy rice, soil (N$_2$O and CO$_2$), biomass & agriculture waste burning, land clearing	3.07	4.15	4.34	4.37	4.55	2.02	2.75	3.11	3.23	3.27
GRAPE-EMF21	Enteric, manure, paddy rice, soil (N$_2$O), biomass & agricultural waste burning	2.59	2.65	2.85	2.82	2.76	2.79	3.31	3.84	3.93	4.06
MESSAGE-A2r	Enteric, manure, paddy rice, soil (N$_2$O)	2.58	3.43	4.78	5.52	6.57	2.57	3.48	4.37	4.77	5.22
IMAGE 2.3	Enteric, manure, paddy rice, soil (N$_2$O and CO$_2$), biomass & agricultural waste burning, land clearing	3.36	3.95	4.41	4.52	4.46	2.05	2.48	2.93	3.07	3.06

* CO_2 emissions from fossil fuel combustion are tracked as well, but frequently reported (and mitigated) under other sector headings (e.g. energy, transportation).

Notes: SAR GWPs used to compute carbon equivalent emissions. nm = not modelled. The GTEM-EMF21 scenario ran through 2050. See Figure 3.7 and 3.10 notes for the scenario references.

for in integrated assessment scenarios. Specifically, new insights suggest that soil drying and forest dieback may naturally reduce terrestrial carbon sequestration (Cox *et al.*, 2000). However, these studies, as well as studies that try to capture changes in climate due to land-use change (Sitch *et al.*, 2005) have thus far not been able to provide definitive guidance. A modelling system that fully couples land use change scenarios with a dynamic climate-carbon system is required in the future for such an assessment.

3.2.2.3   Non-CO_2 greenhouse gas emissions

The emissions scenario chapter in the TAR (Morita *et al.*, 2001) recommended that future research should include GHGs other than CO_2 in new scenarios work. The reason was that, at that time, certainly regarding mitigation, most of the scenarios literature was still primarily focused on CO_2 emissions from energy. Nevertheless, some multi-gas scenario work existed, including the SRES baseline scenarios, but also some other modelling efforts (Manne and Richels, 2001; Babiker *et al.*, 2001; Tol, 1999). The most important non-CO_2 gases include: methane (CH_4), nitrous oxide (N_2O), and a group of fluorinated compounds (F-gases, i.e., HFCs, PFCs, and SF$_6$). Since the TAR, the number of modelling groups producing long-term emission scenarios for non-CO_2 gases has dramatically increased. As a result, the quantity and quality of non-CO_2 emissions scenarios has improved appreciably.

Unlike CO_2 where the main emissions-related sectors are

few (i.e. energy, industry, and land use), non-CO_2 emissions originate from a larger and more diverse set of economic sectors. Table 3.2 provides a list of the major GHG emitting sectors and their corresponding emissions, estimated for 2000. Note that there is significant uncertainty concerning emissions from some sources of the non-CO_2 gases, and the table summarizes the central values from Weyant *et al.* (2006) which has been used in long-term multi-gas scenario studies of the EMF-21. To make the non-CO_2 emissions comparable to those of CO_2, the common practice is to compare and aggregate emissions by using global warming potentials (GWPs).

The most important work on non-CO_2 GHG emissions scenarios has been done in the context of EMF-21 (De la Chesnaye and Weyant, 2006). The EMF-21 study updated the capability of long-term integrated assessment models for modelling non-CO_2 GHG emissions. The results of the study are illustrated in Figure 3.11.

Evaluating the long-term projections of anthropogenic methane emissions from the EMF-21 data shows a significant range in the estimates, but this range is consistent with that found in the SRES. The methane emission differences in the SRES are due to the different storylines. The differences in the EMF-21 reference cases are mainly due to changes in the economic activity level projected in key sectors by each of the models[6]. This could include, for example, increased agriculture production or increased supply of natural gas and below-ground coal in the energy sector. In addition, different modelling groups

---

6   In the EMF-21 study, reference case scenarios were considered to be 'modeller's choice', where harmonization of input parameters and exogenous assumptions was not sought.

Table 3.2: *Global Anthropogenic GHG Emissions for 2000 at sector level, as used in EMF-21 studies (MtCO$_2$-eq/yr).*

Sector sub-total & percent of total	Sub-sectors	CO$_2$	CH$_4$	N$_2$O	F-gases
ENERGY	Coal	8,133	451		
	Natural gas	4,800	895		
25,098	Petroleum syst.	10,476	62		
67%	Stationary/Mobile sources		59	224	
LUCF^a and AGRICULTURE	LUCF and agriculture (net)	3,435			
	Soils			2,607	
	Biomass		491	187	
	Enteric fermentation		1,745	-	
9,543	Manure management		224	205	
25%	Rice		649	-	
INDUSTRY	Cement	829			
	Adipic & nitric acid production			158	
	HFC-23				95
	PFCs				106
1,434	SF6				55
4%	Substitution of ODS^b				191
WASTE	Landfills		781		
1,448	Wastewater		565	81	
4%	Other		11	11	
Total all GHG	37,524	27,671	5,933	3,472	447
	Gas as percent of total	74%	16%	9%	1%

Notes: ^{a} LUCF is Land-use change and forestry.
   ^{b} HFCs are used as substitutes for ODSs in a range of applications
Sources: Weyant et al, 2006.

employed various methods of representing methane emissions in their models and also made different assumptions about how specific methane emission factors for each economic sector change over time. Finally, the degree to which agricultural activities are represented in the models differs substantially. For example, some models represent all agricultural output as one large commodity, 'agriculture', while others have considerable disaggregation. Interestingly, the latter group of models tend to find slower emissions growth rates (see Van Vuuren *et al.*, 2006b).

The range of long-term projections of anthropogenic nitrous oxide emissions is wider than for methane in the EMF-21 data. Note that for N$_2$O, base year emissions of the different models differ substantially. Two factors may contribute to this. First, different definitions exist as to what should be regarded as human-induced and natural emissions in the case of N$_2$O emissions from soils. Second, some models do not include all emission sources.

The last group of non-CO$_2$ gases are fluorinated compounds, which include hydrofluorocarbons (HFCs), perfluorocarbons (PFCs), and sulphur hexafluoride (SF$_6$). The total global emissions of these gases are almost 450 MtCO$_2$-eq, or slightly over 1% of all GHG for 2000. While the emissions of some fluorinated compounds are projected to decrease, many are expected to grow substantially because of the rapid growth rate of some emitting industries (e.g. semiconductor manufacture

and magnesium production and processing), and the replacement of ozone-depleting substances (ODSs) with HFCs. Long-term projections of these fluorinated GHGs are generated by a fewer number of models, but still show a wide range in the results over the century. Total emissions of non-CO$_2$ GHGs are projected to increase, but somewhat less rapidly than CO$_2$ emissions, due to agricultural activities growing less than energy use.

3.2.2.4   *Scenarios for air pollutants and other radiative substances*

Sulphur dioxide emission scenarios

Sulphur emissions are relevant for climate change modelling as they contribute to the formation of aerosols, which affect precipitation patterns and, taken together, reduce radiative forcing. Sulphur emissions also contribute to regional and local air pollution. Global sulphur dioxide emissions have grown approximately in parallel with the increase in fossil fuel use (Smith *et al.*, 2001, 2004; Stern, 2005). However, since around the late 1970s, the growth in emissions has slowed considerably (Grübler, 2002). Implementation of emissions controls, a shift to lower sulphur fuels in most industrialized countries, and the economic transition process in Eastern Europe and the Former Soviet Union have contributed to the lowering of global sulphur emissions (Smith *et al.*, 2001). Conversely, with accelerated economic development, the growth of sulphur emissions in many parts of Asia has been high in recent decades, although growth rates have moderated recently (Streets *et al.*, 2000;

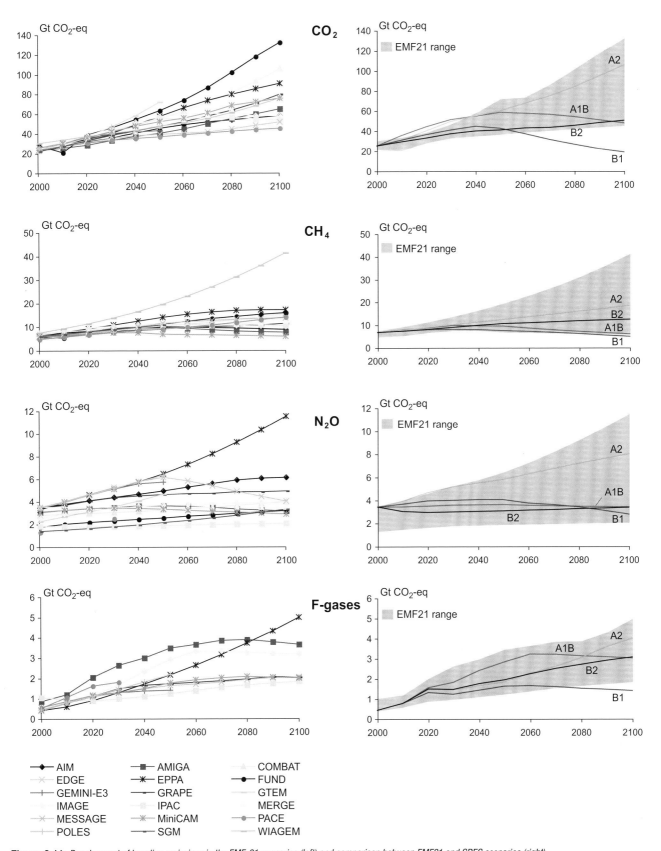

Figure 3.11: *Development of baseline emissions in the EMF-21 scenarios (left) and comparison between EMF21 and SRES scenarios (right).*

Source: De la Chesnaye and Weyant , 2006; see also Van Vuuren et al., 2006b

Stern, 2005; Cofala *et al.*, 2006; Smith *et al.*, 2004). A review of the recent literature indicates that there is some uncertainty concerning present global anthropogenic sulphur emissions, with estimates for the year 2000 ranging between 55.2 MtS (Stern, 2005), 57.5 MtS (Cofala *et al.*, 2006) and 62 MtS (Smith *et al.*, 2004).[7]

Many empirical studies have explored the relationship between sulphur emissions and related drivers, such as economic development (see for example, Smith *et al.*, 2004). The main driving factors that have been identified are increasing income, changes in the energy mix, and a greater focus on air pollution abatement (as a consequence of increasing affluence). Together, these factors may result in an inverted U-shaped pattern of SO_2 emissions, where emissions increase during early stages of industrialization, peak and then fall at higher levels of income, following a Kuznets curve (World Bank, 1992). This general trend is also apparent in most of the recent emissions scenarios in the literature.

Over time, new scenarios have generally produced lower SO_2 emissions projections. A comprehensive comparison of the SRES and more recent sulphur-emission scenarios is given in Van Vuuren and O'Neill (2006). Figure 3.12 illustrates that the resulting spread of sulphur emissions over the medium term (up to the year 2050) is predominantly due to the varying assumptions about the timing of future emissions control, particularly in developing countries[8]. Scenarios at the lower boundary assume the rapid introduction of sulphur-control technologies on a global scale, and hence, a reversal of historical trends and declining emissions in the initial years of the scenario. Conversely, the upper boundaries of emissions are characterized by a rapid increase over coming decades, primarily driven by the increasing use of coal and oil at relatively low levels of sulphur control (SRES A1 and A2).

The comparison shows that overall the SRES scenarios are fairly consistent with recent projections concerning the long-term uncertainty range (Smith *et al.*, 2004; see Figure 3.12). However, the emissions peak over the short-term of some high emissions scenarios in SRES, which lie above the upper boundary estimates of the recent scenarios. There are two main reasons for this difference. First, recent sulphur inventories for the year 2000 have shifted downward. Second, and perhaps more importantly, new information on present and planned sulphur legislation in some developing countries, such as India (Carmichael *et al.*, 2002) and China (Streets *et al.*, 2001) has become available. Anticipating this change in legislation, recent scenarios project sulphur emissions to peak earlier and at lower levels compared to the SRES. Also the lower boundary projections of the recent literature have shifted downward slightly compared to the SRES scenario.

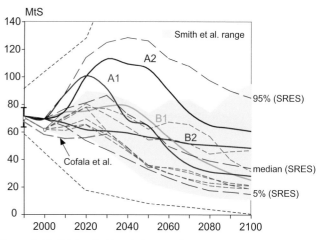

Figure 3.12: *Sulphur dioxide emission scenarios.*

Notes: Thick coloured lines depict the four SRES marker scenarios and the black dashed lines show the median, 5th and 95th percentile of the frequency distribution for the full ensemble of all 40 SRES scenarios. The blue area (and the thin dashed lines in blue) illustrates individual scenarios and the range of Smith *et al.* (2004). Dotted lines indicate the minimum and maximum of sulphur emissions scenarios developed pre-SRES.

NO_x emission scenarios

The most important sources of NO_x emissions are fossil fuel combustion and industrial processes, which combined with other sources such as natural and anthropogenic soil release, biomass burning, lightning, and atmospheric processes, amount to around 25 MtN per year. Considerable uncertainties exist, particularly around the natural sources (Prather *et al.*, 1995; Olivier *et al.*, 1998; Olivier and Berdowski, 2001; Cofala *et al.* (2006). Fossil fuel combustion in the electric power and transport sectors is the largest source of NO_x, with emissions largely being related to the combustion practice. In recent years, emissions from fossil fuel use in North America and Europe are either constant or declining. Emissions have been increasing in most parts of Asia and other developing parts of the world, mainly due to the growing transport sector (Cofala *et al.*, 2006; Smith, 2005; WBCSD, 2004). However in the longer term, most studies project that NO_x emissions in developing countries will saturate and eventually decline, following the trend in the developed world. However, the pace of this trend is uncertain. Emissions are projected to peak in the developing world as early as 2015 (WBCSD, 2004, focusing on the transport sector) and, in worst cases, around the end of this century (see the high emissions projection of Smith, 2005).

There have been very few global scenarios for NO_x emissions since the earlier IS92 scenarios and the SRES. An important characteristic of these (baseline) scenarios is that they consider air pollution legislation (in the absence of any climate policy). Some scenarios, such as those by Bouwman and van Vuuren (1999) and Collins *et al.* (2000) often use IS92a as a 'loose' baseline, with new abatement policies added. Many scenarios

---

7   Note that the Cofala *et al.* (2006) inventory does not include emissions from biomass burning, international shipping and aircraft. In order to enhance comparability between the inventories, emissions from these sources (6 MtS globally) have been added to the original Cofala *et al.* (2006) values.

8   The Amann (2002) projections were replaced by the recently updated IIASA-RAINS projection from Cofala *et al.* (2006).

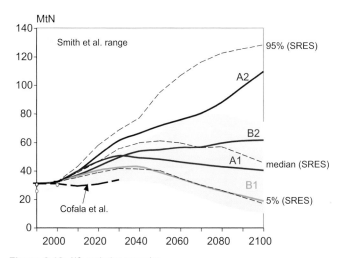

Figure 3.13: *NOx emission scenarios.*

Notes: Thick coloured lines depict the four SRES marker scenarios and the black dashed lines show the median, 5th and 95th percentile of the frequency distribution for the full ensemble of all 40 SRES scenarios. The blue area illustrates the range of the recent Smith (2005) projections.

report rising NO_x emissions up to the 2020s (Figure 3.13), with the lower boundary given by the short-term Cofala *et al.* (2006) reference scenario, projecting emissions to stay at about present levels for the next two to three decades. In the most recent longer-term scenarios (Smith, 2005), NO_x emissions range between 32 MtN and 47 MtN by 2020, which corresponds to an increase in emissions of around 6–50% compared to 2000. The long-term spread is considerably larger, ranging from 9 MtN to 74 MtN by 2100 (see Figure 3.13). The majority of the SRES scenarios (70%) lie within the range of the new Smith (2005) scenarios. However, the upper and lower boundaries of the range of the recent projections have shifted downward compared to the SRES.

Emission scenarios for black and organic carbon

Black and organic carbon emissions (BC and OC) are mainly formed by incomplete combustion, as well as from gaseous precursors (Penner *et al.*, 1993; Gray and Cass, 1998). The main sources of BC and OC emissions include fossil fuel combustion in industry, power generation, traffic and residential sectors, as well as biomass and agriculture waste burning. Natural sources, such as forest fires and savannah burning, are other major contributors. There has recently been some research suggesting that carbonaceous aerosols may contribute to global

warming (Hansen *et al.*, 2000; Andrae, 2001; Jacobson, 2001; Ramaswamy *et al.*, 2001). However, the uncertainty concerning the effects of BC and OC on the change in radiative forcing and hence global warming is still high (see Jacobson, 2001; and Penner *et al.*, 2004).

In the past, BC and OC emissions have been poorly represented in economic and systems engineering models due to unavailability of data. For example, in the IPCC's Third Assessment Report, BC and OC estimates were developed by using CO emissions (IPCC, 2001b). One of the main reasons for this has been the lack of adequate global inventories for different emission sources. However, some detailed global and regional emission inventories of BC and OC have recently become available (Table 3.3). In addition, some detailed regional inventories are also available including Streets *et al.* (2003) and Kupiainen and Klimont (2004). While many of these are comprehensive with regard to detail, considerable uncertainty still exists in the inventories, mainly due to the variety in combustion techniques for different fuels as well as measurement techniques. In order to represent these uncertainties, some studies, such as Bond *et al.* (2004), provide high, low and 'best-guess' values.

The development in the inventories has resulted in the possibility of estimating future BC and OC emissions. Streets *et al.* (2004) use the fuel-use information and technological change in the SRES scenarios to develop estimates of BC and OC emissions from both contained combustion as well as natural sources for all the SRES scenarios until 2050. Rao *et al.* (2005) and Smith and Wigley (2006) estimate BC and OC emissions until 2100 for two IPCC SRES scenarios, with an assumption of increasing affluence leading to an additional premium on local air quality. Liousse *et al.* (2005) use the fuel-mix and other detail in various energy scenarios and obtain corresponding BC and OC emissions.

The inclusion of technological development is an important factor in estimating future BC and OC emissions because, even though absolute fossil fuel use may increase, a combination of economic growth, increased environmental consciousness, technology development and legislation could imply decreased pollutant emissions (Figure 3.14). Liousse *et al.* (2005) neglect the effects of technological change leading to much higher

Table 3.3: *Emission inventories for black and organic carbon (Tg/yr).*

Source	Estimate year	Black carbon	Organic carbon
Penner *et al.*, 1993	1980	13	-
Cooke and Wilson, 1996	1984	14[a]	-
Cooke *et al.*, 1999	1984	5-6.6[a]	7-10[a]
Bond *et al.*, 2004	1996	4.7 (3-10)	8.9 (5-17)
Liousse *et al.*, 1996		12.3	81
Junker and Liousse, 2006	1997	5.7	9.5

Note: [a] Emissions from fossil-fuel use

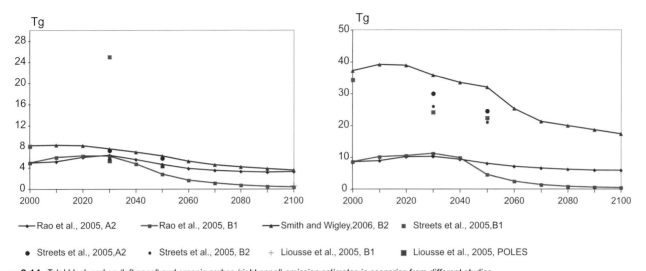

Figure 3.14: *Total black carbon (left panel) and organic carbon (right panel) emission estimates in scenarios from different studies.*
Notes: Rao *et al.*, (2005) include emissions from contained combustion only. Liousse *et al.* (2005), A2 not included as 2100 value of 100 Tg lies way above range.

emission estimates for BC emissions in the long-term in some cases, as compared to other studies such as Streets *et al.* (2004), Rao *et al.* (2005) and Smith and Wigley (2006), all of which show declining emissions in the long-term. Another important factor that Rao *et al.* (2005) also account for is current and proposed environmental legislation. This suggests the necessity for technology-rich frameworks that capture structural and technological change, as well as policy dynamics in the energy system in order to estimate future BC and OC emissions.

Both Streets *et al.* (2004) and Rao *et al.* (2005) show a general decline in BC and OC emissions in developed countries, as well as in regions such as East Asia (including China). In other developing regions, such as Africa and South Asia, slower technology penetration rates lead to much lower emission reductions. There is a large decline in emissions from the residential sector in the developing countries, due to the gradual replacement of traditional fuels and technologies with more efficient ones. Transport-related emissions in both industrialized and developing countries decline in the long-term due to stringent regulations, technology improvements and fuel switching.

To summarize, an important feature of the recent scenario literature is the long-term decline in BC/OC emission intensities per unit of energy use (or economic activity). The majority of the above studies thus indicate that the long-term BC and OC emissions might be decoupled from the trajectory of CO_2 emissions.

3.3   Mitigation scenarios

3.3.1   Introduction

This section contains a discussion of methodological issues (Sections 3.3.2–3.3.4), followed by a focus on the main

characteristics of different groups of mitigation scenarios, with specific attention paid to new literature on non-CO_2 gases and land use (Sections 3.3.5.5 and 3.3.5.6). Finally, short-term scenarios with a regional or national focus are discussed in Section 3.3.6.

3.3.2   Definition of a stabilization target

Mitigation scenarios explore the feasibility and costs of achieving specified climate change or emissions targets, often in comparison to a corresponding baseline scenario. The specified target itself is an important modelling and policy issue. Because Article 2 of United Nations Framework Convention on Climate Change (UNFCCC) states as its objective the 'stabilization of greenhouse gas concentrations in the atmosphere at a level that would prevent dangerous anthropogenic interference with the climate system', most long-term mitigation studies have focused their efforts on GHG concentration stabilization scenarios. However, several other climate change targets may be chosen, for example the rate of temperature change, radiative forcing, or climate change impacts (see e.g. Richels *et al.*, 2004; Van Vuuren *et al.*, 2006b; Corfee-Morlot *et al.*, 2005). In general, selecting a climate policy target early in the cause-effect chain of human activities to climate change impacts, such as emissions stabilization, increases the certainty of achieving required reduction measures, while increasing the uncertainty on climate change impacts (see Table 3.4). Selecting a climate target further down the cause-effect chain (e.g. temperature change, or even avoided climate impacts) provides for greater specification of a desired climate target, but decreases certainty of the emission reductions required to reach that target.

A commonly used target has been the stabilization of the atmospheric CO_2 concentration. If more than one GHG is included, most studies use the corresponding target of stabilizing radiative forcing, thereby weighting the concentrations of the different gases by their radiative properties. The advantage of radiative forcing targets over temperature targets is that

Table 3.4: *Advantages and disadvantages of using different stabilization targets.*

Target	Advantages	Disadvantages
Mitigation costs	Lowest uncertainty on costs.	Very large uncertainty on global mean temperature increase and impacts. Very large uncertainty on global mean temperature increase and impacts. Either needs a different metric to allow for aggregating different gases (e.g. GWPs) or forfeits opportunity of substitution.
Emissions mitigation	Lower uncertainty on costs.	Does not allow for substitution among gases, thus losing the opportunity for multi-gas cost reductions. Indirect link to the objective of climate policy (e.g. impacts).
Concentrations of different greenhouse gases	Can be translated relatively easily into emission profiles (reducing uncertainty on costs).	Allows a wide range of CO_2-only stabilization targets due to substitutability between CO_2 and non-CO_2 emissions.
Radiative forcing	Easy translation to emission targets, thus not including climate sensitivity in costs calculations. Does allow for full flexibility in substitution among gases. Connects well to earlier work on CO_2 stabilization. Can be expressed in terms of CO_2-eq concentration target, if preferred for communication with policymakers.	Indirect link to the objective of climate policy (e.g. impacts).
Global mean temperature	Metric is also used to organize impact literature; and as has shown to be a reasonable proxy for impacts	Large uncertainty on required emissions reduction as result of the uncertainty in climate sensitivity and thus costs.
Impacts	Direct link to objective of climate polices.	Very large uncertainties in required emission reductions and costs.

Based on: Van Vuuren *et al.*, 2006b.

the consequences for emission trajectories do not depend on climate sensitivity, which adds an important uncertainty. The disadvantage is that a wide range of temperature impacts is possible for each radiative forcing level. By contrast, temperature targets provide a more direct first-order indicator of potential climate change impacts, but are less practical to implement in the real world, because of the uncertainty about the required emissions reductions.

Another approach is to calculate risks or the probability of exceeding particular values of global annual mean temperature rise (see also Table 3.9). For example, Den Elzen and Meinshausen (2006) and Hare and Meinshausen (2006) used different probability density functions of climate sensitivity in the MAGICC simple climate model to estimate relationships between the probability of achieving climate targets and required emission reductions. Studies by Richels *et al.* (2004), Yohe *et al.* (2004), Den Elzen *et al.* (2006), Keppo *et al.* (2006), and Kypreos (2006) have used a similar probabilistic concept in an economic context. The studies analyze the relationship between potential mitigation costs and the increase in probability of meeting specific temperature targets.

The choice of different targets is not only relevant because it leads to different uncertainty ranges, but also because it leads to different strategies. Stabilization of one type of target, such as temperature, does not imply stabilization of other possible targets, such as rising sea levels, radiative forcing, concentrations or emissions. For instance, a cost-effective way to stabilize

temperature is not radiative forcing stabilization, but rather to allow radiative forcing to peak at a certain concentration, and then decrease with additional emissions reductions so as to avoid (delayed) further warming and stabilize global mean temperature (see Meinshausen, 2006; Kheshgi *et al.*, 2005; Den Elzen *et al.*, 2006). Finally, targets can also be defined to limit a rate of change, such as the rate of temperature change. While such targets have the advantage of providing a link to impacts related to the rate of climate change, strategies to achieve them may be more sensitive to uncertainties and thus, require careful planning. The rate of temperature change targets, for instance, may be difficult to achieve in the short-term even, using multi-gas approaches (Manne and Richels, 2006; Van Vuuren *et al.*, 2006a).

3.3.3   How to define substitution among gases

In multi-gas studies, a method is needed to compare different greenhouse gases with different atmospheric lifetimes and radiative properties. Ideally, the method would allow for substitution between gases in order to achieve mitigation cost reductions, although it may not be suitable to ensure equivalence in measuring climate impact. Fuglestvedt *et al.* (2003) provide a comprehensive overview of the different methods that have been proposed, along with their advantages and disadvantages. One of these methods, CO_2-eq emissions based on Global Warming Potentials (GWP), has been adopted by current climate policies, such as the Kyoto Protocol and the US climate policy (White House, 2002). Despite the continuing scientific and economic

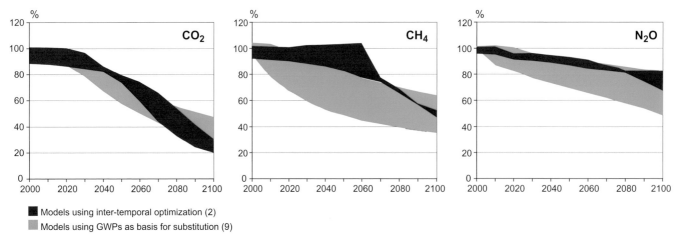

Figure 3.15: *Reduction of emissions in the stabilization strategies aiming for stabilization at 4.5 W/m² (multi-gas strategies) in EMF-21.*

Notes: Range for models using GWPs (blue; standard deviation) versus those not using them (purple; full range). For the first group, all nine reporting long-term models were used. For the second category, results of two of the three reporting models were used (the other model shows the same pattern with respect to the distribution among gases, but has a far higher overall reduction rate and, as such, an outlier).

Data source: De la Chesnaye and Weyant, 2006.

debate on the use of GWPs (i.e. they are not based on economic considerations and use an arbitrary time horizon) the concept is in use under the UNFCCC, the Kyoto Protocol, and the US climate policy. In addition, no alternative measure has attained comparable status to date.

Useful overviews of the mitigation and economic implication of substitution metrics are provided by Bradford (2001) and Godal (2003). Models that use inter-temporal optimization can avoid the use of substitution metrics (such as GWPs) by optimizing the reductions of all gases simultaneously under a chosen climate target. Inter-temporal optimization or perfect foresight models assume that economic agents know future prices and make decisions to minimize costs. Manne and Richels (2001) show, using their model, that using GWPs as the basis of substitution did not lead to the cost-optimal path (minimizing welfare losses) for the long-term targets analyzed. In particular, reducing methane early had no benefit for reaching the long-term target, given its short lifespan in the atmosphere. In the recent EMF-21 study some models validated this result (see De la Chesnaye and Weyant, 2006). Figure 3.15 shows the projected EMF-21 CO_2, CH_4, N_2O, and F-gas reductions across models stabilizing radiative forcing at 4.5 W/m². Most of the EMF-21 models based substitution between gases on GWPs. However, three models substituted gases on the basis of inter-temporal optimization. While (for most of the gases) there are no systematic differences between the results from the two groups, for methane and some F-gases (not shown), there are clear differences related to the very different lifespans of these gases. The models that do not use GWPs, do not substantially reduce CH_4 until the end of the time horizon. However, for models using GWPs, the reduction of CH_4 emissions in the first three decades is substantial: here, CH_4 reductions become a cost-effective short-term abatement strategy, despite the short lifespan (Van Vuuren *et al.*, 2006b). It should be noted that if

a short-term climate target is selected (e.g. rate of temperature change) then inter-temporal optimization models would also favour early methane reductions.

While GWPs do not necessarily lead to the most cost-effective stabilization solution (given a long-term target), they can still be a practical choice: in real-life policies an exchange metric is needed to facilitate emissions trading between gases within a specified time period. Allowing such exchanges creates the opportunity for cost savings through 'what and where flexibility'. It is appropriate to ask what are the costs of using GWPs versus not using them and whether other 'real world' metrics exist that could perform better. O'Neill (2003) and Johansson *et al.* (2006) have argued that the disadvantages of GWPs are likely to be outweighed by the advantages, by showing that the cost difference between a multi-gas strategy and a CO_2-only strategy is much larger than the difference between a GWP-based multi-gas strategy and a cost-optimal strategy. Aaheim *et al.* (2006) found that the cost of using GWPs compared to optimal weights, depends on the ambition of climate policies. Postponing the early CH_4 reductions of the GWP-based strategy, as is suggested by inter-temporal optimization, generally leads to larger temperature increases during the 2000–2020 period. This is because the increased reduction of CO_2 from the energy sector also leads to reduction of sulphur emissions (hence the cooling associated with sulphur-based aerosols) but allows the potential to be used later in the century.

3.3.4   Emission pathways

Emission pathway studies often focus on specific questions with respect to the consequences of timing (in terms of environmental impacts) or overall reduction rates needed for specific long-term targets, (e.g. the emission pathways developed by Wigley *et al.*, 1996). A specific issue raised in the literature on emission pathways since the TAR has concerned a temporary overshoot of the target (concentration, forcing, or temperature). Meinshausen (2006) used a simple carbon-cycle model to illustrate that for low-concentration targets (i.e. below 3 W/m²/ 450 ppmv CO_2-eq) overshoot is inevitable, given the feasible maximum rate of reduction. Wigley (2003) argued that overshoot profiles may give important economic benefits. In response, O'Neill and Oppenheimer (2004) showed that the associated incremental warming of large overshoots may significantly increase the risks of exceeding critical climate thresholds to which ecosystems are known to be able to adapt. Other emission pathways that lead to less extreme concentration overshoots may provide a sensible compromise between these two results. For instance, the 'peaking strategies' chosen by Den Elzen *et al.* (2006) show that it is possible to increase the likelihood of meeting the long-term temperature target or to reach targets with a similar likelihood at lower costs. Similar arguments for analyzing overshoot strategies are made by Harvey (2004), and Kheshgi *et al.* (2005).

3.3.5   Long-term stabilization scenarios

A large number of studies on climate stabilization have been published since the TAR. Several model comparison projects contributed to the new literature, including the Energy Modelling Forum's EMF-19 (Weyant, 2004) and EMF-21 studies (De la Chesnaye and Weyant, 2006), that focused on technology change and multi-gas studies, respectively, the IMCP (International Model Comparison Project), which focused on technological change (Edenhofer *et al.*, 2006), and the US Climate Change Science Programme (USCCSP, 2006). The updated emission scenario database (Hanaoka *et al.*, 2006; Nakicenovic *et al.*, 2006) includes a total of 151 new mitigation scenarios published since the SRES.

Comparison of mitigation scenarios is more complicated now than at the time of the TAR because:
- Parts of the modelling community have expanded their analysis to include non-CO_2 gases, while others have continued to focus solely on CO_2. As discussed in the previous section, multi-gas mitigation scenarios use different targets, thus making comparison more complicated.
- Some recent studies have developed scenarios that do not stabilize radiative forcing (or temperature) – but show a peak before the end of the modelling time horizon (in most cases 2100).
- At the time of the TAR, many studies used the SRES scenarios as baselines for their mitigation analyses, providing a comparable set of assumptions. Now, there is a broader range of underlying assumptions.

This section introduces some metrics to group the CO_2-only and multi-gas scenarios so that they are reasonably comparable. In Figure 3.16 the reported CO_2 concentrations in 2100 are plotted against the 2100 total radiative forcing (relative to pre-industrial times). Figure 3.16 shows that a

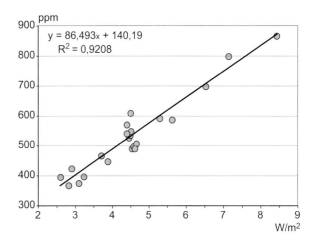

Figure 3.16: *Relationship of total radiative forcing vis-à-vis CO_2 concentration for the year 2100 (25 multi-gas stabilization scenarios for alternative stabilization targets).*

relationship exists between the two indicators. This can be explained by the fact that CO_2 forms by far the most important contributor to radiative forcing – and subsequently, a reduction in radiative forcing needs to coincide with a reduction in CO_2 concentration. The existing spread across the studies is caused by several factors, including differences in the abatement rate among alternative gases, differences in specific forcing values for GHGs and other radiative gases (particularly aerosols), and differences in the atmospheric chemistry and carbon cycle models that are used. Here, the relationship is used to classify the available mitigation literature into six categories that vary in the stringency of the climate targets. The most stringent group includes those scenarios that aim to stabilize radiative forcing below 3 W/m². This group also includes all CO_2-only scenarios that stabilize CO_2 concentrations below 400 ppmv. In contrast, the least stringent group of mitigation scenarios have a radiative forcing in 2100 above 6 W/m² – associated with CO_2 concentrations above 660 ppmv. By far the most studied group of scenarios are those that aim to stabilize radiative forcing at 4–5 W/m² or 485–570 ppmv CO_2 (see Table 3.5).

The classification of scenarios, as given in Table 3.5, permits the comparison of multi-gas and CO_2-only stabilization scenarios according to groups of scenarios with comparable level of mitigation stringency. The studies have been classified on the basis of the reported targets, using the relationship from Figure 3.16 to permit comparability of studies using different stabilization metrics. The following section uses these categories (I to VI) to analyze the underlying dynamics of stabilization scenarios as a function of the stabilization target. However, it should be noted that the classification is subject to uncertainty and should thus to be used with care.

3.3.5.1   *Emission reductions and timing*

Figure 3.17 shows the projected CO_2 emissions associated with the new mitigation scenarios. In addition, the figure depicts the range of the TAR stabilization scenarios (more than 80

Table 3.5: *Classification of recent (post-TAR) stabilization scenarios according to different stabilization targets and alternative stabilization metrics. Groups of stabilization targets were defined using the relationship in Figure 3.16.*

Category	Additional radiative forcing W/m^2	CO$_2$ concentration ppm	CO$_2$-eq concentration ppm	Peaking year for CO$_2$ emissions[a] year	Change in global emissions in 2050 (% of 2000 emissions)[1] %	No. of scenarios
I	2.5-3.0	350-400	445-490	2000-2015	-85 to -50	6
II	3.0-3.5	400-440	490-535	2000-2020	-60 to -30	18
III	3.5-4.0	440-485	535-590	2010-2030	-30 to +5	21
IV	4.0-5.0	485-570	590-710	2020-2060	+10 to +60	118
V	5.0-6.0	570-660	710-855	2050-2080	+25 to +85	9
VI	6.0-7.5	660-790	855-1130	2060-2090	+90 to +140	5
Total						177

Note: [a] Ranges correspond to the 15th to 85th percentile of the Post-TAR scenario distribution.
Note that the classification needs to be used with care. Each category includes a range of studies going from the upper to the lower boundary. The classification of studies was done on the basis of the reported targets (thus including modeling uncertainties). In addition, also the relationship, which was used to relate different stabilization metrics, is subject to uncertainty (see Figure 3.16).

scenarios) (Morita *et al.*, 2001). Independent of the stabilization level, scenarios show that the scale of the emissions reductions, relative to the reference scenario, increases over time. Higher stabilization targets do push back the timing of most reductions, even beyond 2100.

An increasing body of literature assesses the attainability of very low targets of below 450 ppmv CO$_2$ (e.g. Van Vuuren *et al.*, 2007; Riahi *et al.*, 2006). These scenarios from class I and II extend the lower boundary beyond the range of the TAR stabilization scenarios of 450 ppmv CO$_2$ (see upper panels of Figure 3.17). The attainability of such low targets is shown to depend on: 1) using a wide range of different reduction options; and 2) the technology 'readiness' of advanced technologies, in particular the combination of bio-energy, carbon capture and geologic storage (BECCS). If biomass is grown sustainably, this combination may lead to negative emissions (Williams, 1998; Herzog *et al.*, 2005), Rao and Riahi (2006), Azar *et al.* (2006) and Van Vuuren *et al.* (2007) all find that such negative emissions technologies might be essential for achieving very stringent targets.

The emission range for the scenarios with low and intermediate targets between 3.5 and 5 W/m^2 (scenarios in categories III and IV) are consistent with the range of the 450 and 550 ppmv CO$_2$ scenarios in the TAR. Emissions in this category tend to show peak emissions around 2040 – with emissions in 2100 similar to, or slightly below, emissions today. Although for these categories less rapid and forceful reductions are required than for the more stringent targets, studies focusing on these stabilization categories find that a wide portfolio of reduction measures would be needed to achieve such emission pathways in a cost-effective way.

The two highest categories of stabilization scenarios (V and VI) overlap with low-medium category baseline scenarios (see

Section 3.2). This partly explains the relatively small number of new studies on these categories. The emission profiles of these scenarios are found to be consistent with the emissions ranges as published in the TAR.

There is a relatively strong relationship between the cumulative CO$_2$ emissions in the 2000–2100 period and the stringency of climate targets (see Figure 3.18). The uncertainties associated with individual stabilization levels (shown by the different percentiles[9]) are primarily due to the ranges associated with individual stabilization categories, substitutability of CO$_2$ and non-CO$_2$-emissions, different model parameterizations of the carbon cycle, but they are also partly due to differences in emissions pathways (delayed reduction pathways can allow for somewhat higher cumulative emissions). In general, scenarios aiming for targets below 3 W/m^2 require cumulative CO$_2$ emissions of around 1100 GtCO$_2$ (range of 800–1500 GtCO$_2$). The cumulative emissions increase for subsequently less stringent targets. The middle category (4–5 W/m^2) requires emissions to be in the order of 3000 GtCO$_2$ (range of 2270–3920 GtCO$_2$). The highest category (>6 W/m^2) exhibits emissions, on average, around 5020 GtCO$_2$ (range of 4400–6600 GtCO$_2$).

The timing of emission reductions also depends on the stringency of the stabilization target. Timing of climate policy has always been an important topic in the scenario literature. While some studies argue for early action for smooth transitions and stimulating technology development (e.g. Azar and Dowlatabadi, 1999, Van Vuuren and De Vries, 2001), others emphasize delayed response to benefit from better technology and higher CO$_2$ fertilization rates from natural systems at later points in time (e.g. Wigley *et al.*, 1996; Tol, 2000; for a more elaborate discussion on timing see also Section 3.6). This implies that a given stabilization target can be consistent with a range of interim targets. Nevertheless, stringent targets require an earlier peak of CO$_2$ emissions (see Figure 3.19 and

---

9   Note that the percentiles are used to illustrate the statistical properties of the scenario distributions, and should not be interpreted as likelihoods in any probabilistic context.

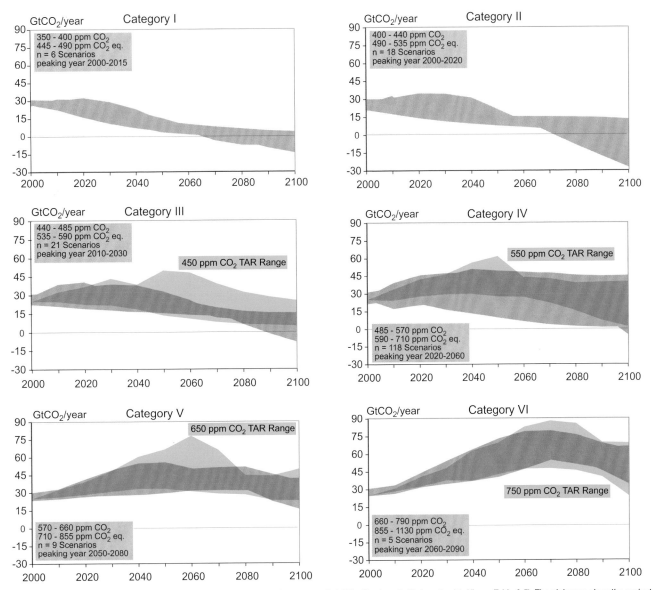

Figure 3.17: *Emissions pathways of mitigation scenarios for alternative groups of stabilization targets (Categories I to VI, see Table 3.5). The pink area gives the projected CO_2 emissions for the recent mitigation scenarios developed post-TAR. Green shaded areas depict the range of more than 80 TAR stabilization scenarios (Morita et al., 2001). Category I and II scenarios explore stabilization targets below the lowest target of the TAR.*

Source: After Nakicenovic et al., 2006, and Hanaoka et al., 2006

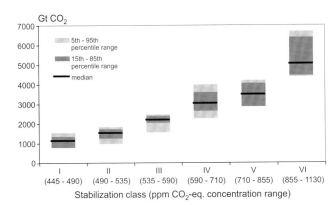

Figure 3.18: *Relationship between the scenario's cumulative carbon dioxide emissions (2000–2100) and the stabilization target (stabilization categories I to VI, of Table 3.5).*

Data source: After Nakicenovic et al., 2006, and Hanaoka et al., 2006

Table 3.5). In the majority of the scenarios concerning the most stringent group (< 3 W/m²), emissions start to decline before 2015, and are further reduced to less than 50% of today's emissions by 2050 (Table 3.5). The emissions profiles of these scenarios indicate the need for short-term infrastructure investments for a comparatively early decarbonization of the energy system. Achieving these low-emission trajectories requires a comprehensive global mitigation effort, including a further tightening of existing climate policies in Annex I countries, and simultaneous emission mitigation in developing countries, where most of the increase in emissions is expected in the coming decades. For the medium stringency group (4-5 W/m²) the peak of global emissions generally occurs around 2010 to 2030; followed by a return to 2000 levels, on average, around 2040 (with the majority of these scenarios returning to 2000 emissions levels between 2020 and 2060). For targets between 5–6 W/m², the median emissions peak around 2070. The figure also indicates that the uncertainty range is relatively small for the more stringent targets, illustrating the reduced flexibility of the emissions path and the requirement for early mitigation. The less stringent categories allow more flexibility in timing. Most of the stringent stabilization scenarios of category I (and some II scenarios) assume a temporal overshoot of the stabilization target (GHG concentration, radiative forcing, or temperature change) before the eventual date of stabilization between 2100 and 2150. Recent studies indicate that while such 'overshoot' strategies might be inevitable for very low targets (given the climate system and socio-economic inertia), they might also provide important economic benefits. At the same time, however, studies note that the associated rate of warming from large overshoots might significantly increase the risk of exceeding critical climate thresholds. (For further discussion, see Section 3.3.4.)

The right-hand panel of Figure 3.19 illustrates the time at which CO_2 emissions will have to return to present levels. For stringent stabilization targets (below 4 W/m²; category I, II and III) emissions return to present levels, on average, before the middle of this century, that is about one to two decades after the year in which emissions peak. In most of the scenarios for the highest stabilization category (above 6 W/m²; category VI) emissions could stay above present levels throughout the century.

The absolute level of the required emissions reduction does not only depend on the stabilization target, but also on the baseline emissions (see Hourcade and Shukla, 2001). This is clearly shown in the right-hand panel of Figure 3.20, which illustrates the relationship between the cumulative baseline emissions and the cumulative emissions reductions for the stabilization scenarios (by 2100). In general, scenarios with high baseline emissions require a higher reduction rate to reach the

same reduction target: this implies that the different reduction categories need to show up as diagonals in figure 3.20. This is indeed the case for the range of studies and the 'category averages' (large triangles). As indicated in the figure, a scenario with high baseline emissions requires much deeper emission reduction in order to reach a medium stabilization target (sometimes more than 3600 $GtCO_2$) than a scenario with low baseline emissions to reach the most stringent targets (in some cases less than 1800 $GtCO_2$). For the same target (e.g. category IV) reduction may differ from 370 to 5500 $GtCO_2$. This comes from the large spread of emissions in the baseline scenarios. While scenarios for both stringent and less-stringent targets have been developed from low and high baseline scenarios, the data suggests that, on average, mitigation scenarios aimed at the most stringent targets start from the lowest baseline scenarios.

In the short-term (2030), the relationship between emission reduction and baseline is less clear, given the flexibility in the timing of emission reductions (left-hand panel in Figure 3.20). While the averages of the various stabilization categories are aligned in a similar way to those discussed for 2100 (with exception of category I, for which the scenario sample is smaller than for the other categories); the uncertainty ranges here are very large.

3.3.5.2 GHG abatement measures

The abatement of GHG emissions can be achieved through a wide portfolio of measures in the energy, industry, agricultural and forest sectors (see also Edmonds et al., 2004b; Pacala and Socolow, 2004; Metz and Van Vuuren, 2006). Measures for reducing CO_2 emissions range from structural changes in the energy system and replacement of carbon-intensive fossil fuels by cleaner alternatives (such as a switch from coal to natural gas, or the enhanced use of nuclear and renewable energy), to demand-side measures geared towards energy conservation and efficiency improvements. In addition, capturing carbon

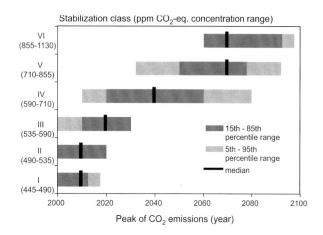

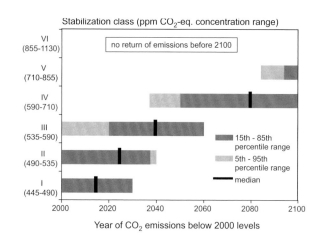

Figure 3.19: *Relationship between the stringency of the stabilization target (category I to VI) and 1) the time at which CO_2 emissions have to peak (left-hand panel), and 2) the year when emissions return to present (2000) levels.*

Data source: After Nakicenovic et al., 2006, and Hanaoka et al., 2006.

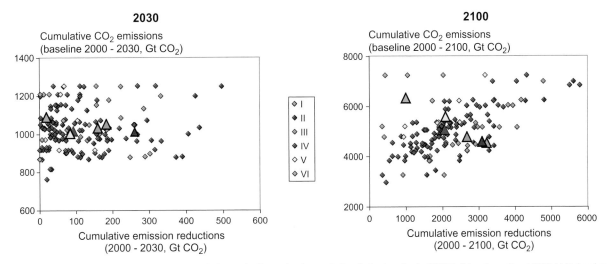

Figure 3.20: *Relationship between required cumulative emissions reduction and carbon emissions in the baseline by 2030 (left-hand panel) and 2100 (right-hand panel).*
Notes: Coloured rectangles denote individual scenarios for alternative stabilization targets (categories I to VI). The large triangles indicate the averages for each category.
Data source: After Nakicenovic et al., 2006, and Hanaoka et al., 2006

during energy conversion processes with subsequent storage in geological formations (CCS) provides an approach for reducing emissions. Another important option for CO_2 emission reduction encompasses the enhancement of forest sinks through afforestation, reforestation activities and avoided deforestation.

In the energy sector the aforementioned options can be grouped into two principal measures for achieving CO_2 reductions:
1. Improving the efficiency of energy use (or measures geared towards energy conservation).
2. Reducing the emissions per unit of energy consumption.

The latter comprises the aggregated effect of structural changes in the energy systems and the application of CCS. A response index has been calculated (based on the full set of stabilization scenarios from the database) in order to explore the importance of these two strategies. This index is equal to the ratio of the reductions achieved through energy efficiency over those achieved by carbon-intensity improvements (Figure 3.21). Similar to Morita *et al.* (2001), it was discovered that the mitigation response to reduce CO_2 emissions would shift over time, from initially focusing on energy efficiency reductions in the beginning of the 21st century to more carbon-intensity reduction in the latter half of the century (Figure 3.21). The amount of reductions coming from carbon-intensity improvement is more important for the most stringent scenarios. The main reason is that, in the second half of the century, increasing costs of further energy efficiency improvements and decreasing costs of low-carbon or carbon-free energy sources make the latter category relatively more attractive. This trend is

also visible in the scenario results of model comparison studies (Weyant, 2004; Edenhofer *et al.*, 2006).

In addition to measures for reducing CO_2 emissions from energy and industry, emission reductions can also be achieved from other gases and sources. Figure 3.22 illustrates the relative contribution of measures towards achieving climate stabilization from three main sources:
1. CO_2 from energy and industry.
2. CO_2 from land-use change.
3. The full basket of non-CO_2 emissions from all relevant sources.

The figure compares the contribution of these measures towards achieving stabilization for a wide range of targets (between 2.6 and 5.3 W/m² by 2100) and baseline scenarios. An important conclusion across all stabilization levels and baseline scenarios is the central role of emissions reductions in the energy and industry sectors. All stabilization studies are consistent in that (independent of the baseline or target uncertainty) more than 65% of total emissions reduction would occur in this sector. The non-CO_2 gases and land-use-related CO_2 emissions (including forests) are seen to contribute together up to 35% of total emissions reductions.[10] However, as noted further above, the majority of recent studies indicate the relative importance of the latter two sectors for the cost-effectiveness of integrated multi-gas GHG abatement strategies (see also Section 3.3.5.4 on CO_2-only versus multi-gas mitigation and 3.3.5.5 on land-use).

The strongest divergence across the scenarios concerns the contribution of land-use-related mitigation. The results range

---

10 Most of the models include an aggregated representation of the forest sector comprising the joint effects of deforestation, afforestation and avoided deforestation.

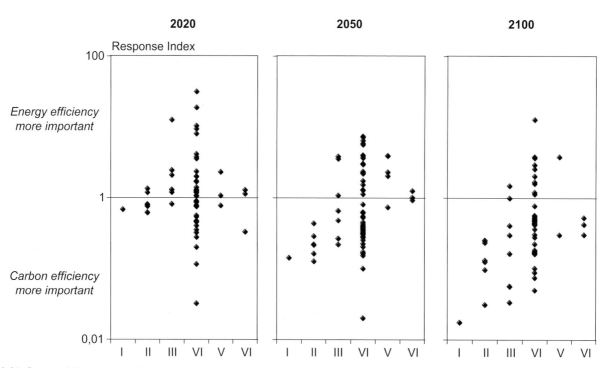

Figure 3.21: *Response index to assess priority setting in energy-intensity reduction (more than 1) or in carbon-intensity reduction (less than 1) for post-TAR stabilization scenarios.*

Note: The panels show the development of the index for the years 2020, 2050, and 2100 (66, 77, and 59 scenarios, respectively, for which data on energy, GDP and carbon emissions were available.

Data source: After Nakicenovic et al., 2006, and Hanaoka et al., 2006)

from negative contributions of land-use change to potential emissions savings of more than 1100 GtCO$_2$ over the course of the century (Figure 3.22). The primary reason for this is the considerable uncertainty with respect to future competition for land between dedicated bio-energy plantations and potential gains from carbon savings in terrestrial sinks. Some scenarios, for example, project massive expansion of dedicated bio-energy plantations, leading to an increase in emissions due to net deforestation (compared to the baseline).

An illustrative example for the further breakdown of mitigation options is shown in Figure 3.23. The figure shows stabilization scenarios for a range of targets (about 3–4.5 W/m^2) based on four illustrative models (IMAGE, MESSAGE, AIM and IPAC) for which sufficient data were available. The scenarios share similar stabilization targets, but differ with respect to salient assumptions for technological change, long-term abatement potentials, as well as model methodology and structure. The scenarios are also based on different baseline scenarios. For example, cumulative baseline emissions over the course of the century range between 6000 GtCO$_2$-eq in MESSAGE and IPAC scenarios to more than 7000 GtCO$_2$-eq in the IMAGE and AIM scenarios. Figure 3.24 shows the primary energy mix of the baseline and the mitigation scenarios.

It should be noted that the figure shows reduction on top of the baseline (e.g. other renewables may already make a large baseline contribution). Above all, Figure 3.23 illustrates the

importance of using a wide portfolio of reduction measures, with many categories of measures, showing contributions of more than a few hundred GtCO$_2$ over the course of the century. In terms of the contribution of different options, there is agreement for some options, while there is disagreement for others. The category types that have a large potential over the long term (2000–2100) in at least one model include energy conservation, carbon capture and storage, renewables, nuclear and non-CO$_2$ gases. These options could thus constitute an important part of the mitigation portfolio. However, the differences between the models also emphasize the impact of different assumptions and the associated uncertainty (e.g. for renewables, results can vary strongly depending on whether they are already used in the baseline, and how this category competes against other zero or low-emission options in the power sector, such as nuclear and CCS). The figure also illustrates that the limitations of the mitigation portfolio with respect to CCS or forest sinks (AIM and IPAC) would lead to relatively higher contributions of other options, in particular nuclear (IPAC) and renewables (AIM).

Figure 3.23 also illustrates the increase in emissions reductions necessary to strengthen the target from 4.5 to about 3–3.6 W/m^2. Most of the mitigation options increase their contribution significantly by up to a factor of more than two. This effect is particularly strong over the short term (2000–2030), indicating the need for early abatement in meeting stringent stabilization targets. Another important conclusion from the figure is that CCS and forest sink options are playing a relatively modest

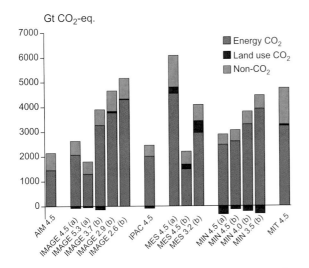

Figure 3.22: *Cumulative contribution of alternative mitigation measures by source (2000–2100)*

Note: Contributions for a wide range of stabilization targets (2.6–5.3 W/m²), indicated after the model name) and alternative baseline scenarios. Mitigation scenarios using the same baseline are indicated for each model as (a) and (b) reprectively.

Data source: EMF-21, Smith and Wigley., 2006; Van Vuuren et al., 2007; Riahi et al., 2006

role in the short-term mitigation portfolio, particularly for the intermediate stabilization target (4.5 W/m²). The results thus indicate that the widespread deployment of these options might require relatively more time compared to the other options and

also relatively higher carbon prices (see also Figure 3.25 on increasing carbon prices over time).

As noted above, assumptions with regards to the baseline can have significant implications for the contribution of individual mitigation options in achieving stabilization. Figure 3.24 clearly shows that the baseline assumptions of the four models differ, and that these differences play a role in explaining some of the results. For instance, the MESSAGE model already includes a large amount of renewables in its baseline and further expansion is relatively costly. Nevertheless, some common trends among the models may also be observed. First of all, almost all cases show a clear reduction in primary energy use. Second, in all models coal use is significantly reduced under the climate policy scenarios, compared to the baseline. It should be noted that in those models that consider CCS, the remaining fossil fuel use is mostly in combination with carbon capture and storage. In 2030, oil use is only modestly reduced by climate policies – this also applies to natural gas use. In 2100, both oil and gas are reduced compared to the baseline in most models. Finally, renewable energy and nuclear power increase in all models – although the distribution across these two options differs.

3.3.5.3 Stabilization costs

Models use different metrics to report the costs of emission reductions. Top-down general equilibrium models tend to report GDP losses, while system-engineering partial equilibrium

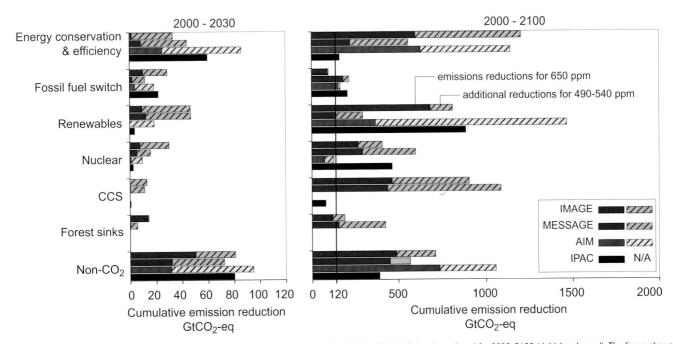

Figure 3.23: *Cumulative emissions reductions for alternative mitigation measures for 2000 to 2030 (left-hand panel) and for 2000-2100 (right-hand panel). The figure shows illustrative scenarios from four models (AIM, IMAGE, IPAC and MESSAGE) for stabilization levels of 490-540 ppmv CO₂-eq and levels of 650 ppmv CO₂-eq, respectively. Dark bars denote reductions for a target of 650 ppmv CO₂-eq and light bars the additional reductions to achieve 490-540 ppmv CO₂-eq. Note that some models do not consider mitigation through forest sink enhancement (AIM and IPAC) or CCS (AIM) and that the share of low-carbon energy options in total energy supply is also determined by inclusion of these options in the baseline. CCS includes carbon capture and storage from biomass. Forest sinks include reducing emissions from deforestation.*

Data source: Van Vuuren et al. (2007); Riahi et al. (2006); Hijioka, et al. (2006); Masui et al. (2006); Jiang et al. (2006).

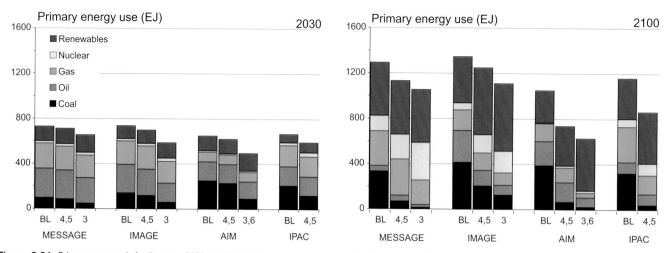

Figure 3.24: *Primary energy mix for the years 2030 and 2100. Illustrative scenarios aim at stabilizing radiative forcing at low (3–3.6 W/m²) and intermediate levels (4.5 W/m2) respectively.*

Note: BL= Baseline. For the corresponding contribution of individual mitigation measures in (in GtCO₂) see also Figure 3.23.

models usually report the increase in energy system costs or the net present value (NPV) of the abatement costs. A common cost indicator is also the marginal cost/price of emissions reduction (US$/tC or US$/tCO₂).

Figure 3.25 shows the relationship between stabilization targets and alternative measures of mitigation costs, comprising GDP losses, net present value of abatement, and carbon price in terms of US$ /tCO₂-eq.

It is important to note that for the following reported cost estimates, the vast majority of the models assume transparent markets, no transaction costs, and thus perfect implementation of policy measures throughout the 21ˢᵗ century, leading to the universal adoption of cost-effective mitigation measures, such as carbon taxes or universal cap and trade programmes. These assumptions generally result in equal carbon prices across all regions and countries equivalent to global, least-cost estimates. Relaxation of these modelling assumptions, alone or in combination (e.g. mitigation-only in Annex I countries, no emissions trading, or CO₂-only mitigation), will lead to an appreciable increase in all cost categories.

The grey shaded area in Figure 3.25 illustrates the 10th–90th percentile of the mitigation cost ranges of recent studies, including the TAR. The area includes only those recent scenarios in the literature that report cost estimates based on a comprehensive mitigation analysis, defined as those that have a sufficiently wide portfolio of mitigation measures.[11] The selection was made on a case-by-case basis for each scenario considered in this assessment. The Figure also shows results from selected illustrative studies (coloured lines). These studies report costs for a range of stabilization targets and are

representative of the overall cost dynamics of the full set of scenarios. They show cases with high-, intermediate- and low-cost estimates (sometimes exceeding the 80th (i.e. 10th–90th) percentile range on the upper and lower boundaries of the grey-shaded area). The colour coding is used to distinguish between individual mitigation studies that are based on similar baseline assumptions. Generally, mitigation costs (for comparable stabilization targets) are higher from baseline scenarios with relatively high baseline emissions (brown and red lines). By the same token, intermediate or low baseline assumptions result in relatively lower cost estimates (blue and green lines).

Figure 3.25a shows that the majority of studies find that GDP losses increase with the stringency of the target, even though there is considerable uncertainty with respect to the range of losses. Barker *et al.* (2006) found that, after allowing for baseline emissions, the differences can be explained by:

• The spread of assumptions in modelling-induced technical change.
• The use of revenues from taxes and permit auctions.
• The use of flexibility mechanisms (i.e. emissions trading, multi-gas mitigation, and banking).
• The use of backstop technologies.
• Allowing for climate policy related co-benefits.
• Other specific modelling assumptions.

Weyant (2000) lists similar factors but also includes the number and type of technologies covered, and the possible substitution between cost factors (elasticities). A limited set of studies finds negative GDP losses (economic gains) that arise from the assumption that a model's baseline is assumed to be a non-optimal pathway and incorporates market imperfections. In these models, climate policies steer economies in the direction

---

11 The assessment of mitigation costs excludes stabilization scenarios that assume major limitation of the mitigation portfolio. For example, our assessment of costs does not include stabilization scenarios that exclude non-CO₂ mitigation options for achieving multi-gas targets (for cost implications of CO2-only mitigation see also Section 3.3.5.4). The assessment nevertheless includes CO₂ stabilization scenarios that focus on single-gas stabilization of CO₂ concentrations. The relationship between the stabilization metrics given in Figure 3.16 is used to achieve comparability of multi-gas and CO₂ stabilization scenarios.

a) Selected studies reporting GDP losses

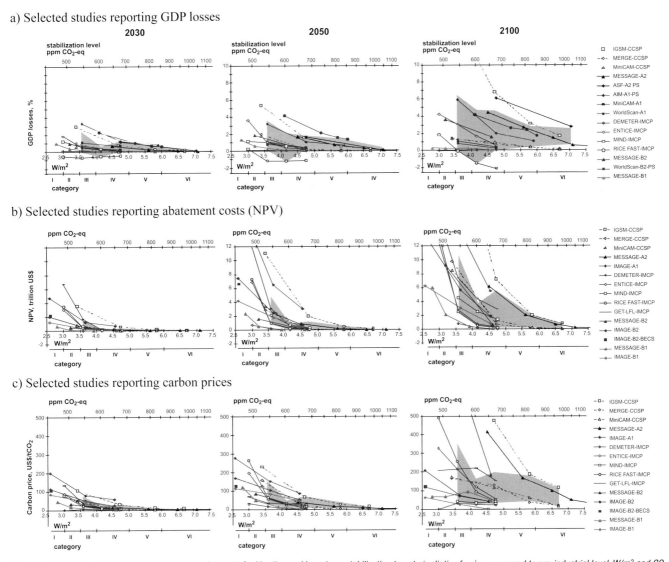

Figure 3.25: *Figure 3.25 Relationship between the cost of mitigation and long-term stabilization targets (radiative forcing compared to pre-industrial level, W/m² and CO₂-eq concentrations).*

Notes: These panels show costs measured as a % loss of GDP (top), net present value of cumulative abatement costs (middle), and carbon price (bottom). The left-hand panels give costs for 2030, the middle panel for 2050, and the right-hand panel for 2100 repectively. Individual coloured lines denote selected studies with representative cost dynamics from very high to very low cost estimates. Scenarios from models sharing similar baseline assumptions are shown in the same colour. The grey-shaded range represents the 80th percentile of the TAR and post-TAR scenarios. NPV calculations are based on a discount rate of 5%. Solid lines show representative scenarios considering all radiatively active gases. CO₂ stabilization scenarios are added based on the relationship between CO₂ concentration and the radiative forcing targets shown in Figure 3.16. Dashed lines represent multi-gas scenarios, where the target is defined by the six Kyoto gases (other multi-gas scenarios consider all radiatively active gases).

Data sources: CCSP scenarios (USCCSP, 2006); IMCP scenarios (Edenhofer et al., 2006); Post-SRES (PS) scenarios (Morita et al., 2001); Azar et al., 2006; Riahi et al., 2006; Van Vuuren et al., 2007.

of reducing these imperfections, for example by promoting more investment into research and development and thus achieving higher productivity, promoting higher employment rates, or removing distortionary taxes.

The left-hand side panel of Figure 3.25a shows that for 2030, GDP losses in the vast majority of the studies (more than 90% of the scenarios) are generally below 1% for the target categories V and VI. Also in the majority of the category III and IV scenarios (70% of the scenarios) GDP losses are below 1%. However, it is important to note that for categories III and IV costs are higher, on average, and show a wider range than those for categories V

and VI. For instance, for category IV the interval lying between the 10th and 90th percentile varies from about 0.6% gain to about 1.2% loss. For category III, this range is shifted upwards (0.2–2.5%). This is also indicated by the median GDP losses by 2030, which increases from below 0.2% for categories V and VI, to about 0.2% for the category IV scenarios, and to about 0.6% for category III scenarios. GDP losses of the lowest stabilization categories (I & II) are generally below 3% by 2030, however the number of studies are relatively limited and in these scenarios stabilization is achieved predominantly from low baselines. The absolute GDP losses by 2030 correspond on average to a reduction of the *annual* GDP growth rate of less than 0.06 percentage points for the scenarios of category IV, and less than 0.1 and 0.12 percentage points for the categories III and I&II, respectively.

GDP losses by 2050 (middle panel of Figure 3.25a) are comparatively higher than the estimates for 2030. For example, for category IV scenarios the range is between -1% and 2% GDP loss compared to baseline (median 0.5%), and for category III scenarios the range is from slightly negative to 4% (median 1.3%). The Stern review (2006), looking at the costs of stabilization in 2050 for a comparable category (500–550 CO_2-eq) found a similar range of between -2% and +5%. For the studies that also explore different baselines (in addition to multiple stabilization levels), Figure 3.25a also shows that high emission baselines (e.g. high SRES-A1 or A2 baselines) tend to lead to higher costs. However, the uncertainty range across the models is at least of a similar magnitude. Generally, models that combine assumptions of very slow or incremental technological change with high baseline emissions (e.g. IGSM-CCSP) tend to show the relatively highest costs (Figure 3.25a). GDP losses of the lowest stabilization categories (I & II) are generally below 5.5% by 2050, however the number of studies are relatively limited and in these scenarios stabilization is achieved predominantly from low baselines. The absolute GDP losses numbers for 2050 reported above correspond on average to a reduction of the *annual* GDP growth rate of less than 0.05 percentage points for the scenarios of category IV, and less than 0.1 and 0.12 percentage points for the categories III and I&II, respectively.

Finally, the most right-hand side panel of Figure 3.25a shows that GDP losses show a bigger spread and tend to be somewhat higher by 2100. GDP losses are between 0.3% and 3% for category V scenarios and -1.6% to about 5% for category IV scenarios. Highest costs are given by category III (from slightly negative costs up to 6.5%). The sample size for category I is not large enough for a statistical analysis. Similarly, for category II scenarios, the range is not shown as the stabilization scenarios of category II are predominantly based on low or intermediate baselines, and thus the resulting range would not be comparable to those from the other stabilization categories. However, individual studies indicate that costs become higher for more stringent targets (see, for example, studies highlighted in green and blue for the lowest stabilization categories in Figure 3.25a).[12]

The results for the net present value of cumulative abatement costs show a similar picture (Figure 3.25b). However, given the fact that abatement costs only capture direct costs, this cost estimate is by definition more certain.[13] The interval from the 10th to the 90th percentile in 2100 ranges from nearly zero to about 11 trillion US$. The highest level corresponds to

around 2–3% of the NPV of global GDP over the same period. Again, on the basis of comparison across models, it is clear that costs depend both on the stabilization level and baseline emissions. In general, the spread of costs for each stabilization category seems to be of a similar order to the differences across stabilization scenarios from different baselines. In 2030, the interval covering 80% of the NPV estimates runs from around 0–0.3 trillion for category IV scenarios. The majority of the more stringent (category III) scenarios range between 0.2 to about 1.6 trillion US$. In 2050, typical numbers for category IV are around 0.1–1.2 trillion US$ and, for category III, this is 1–5 trillion US$ (or below about 1% of the NPV of GDP). By 2100 the NPV estimates increase further, with the range up to 5 trillion for category IV scenarios and up to 11 trillion for category III scenarios, respectively. The results of these studies, published since the TAR, are consistent with the numbers presented in the TAR, although the new studies extend results to substantially lower stabilization levels.

Finally, a similar trend is found for carbon price estimates. In 2030, typical carbon prices across the range of models and baselines for a 4.5 W/m^2 stabilization target (category IV) range from around 1–24 US$/$tCO_2$ (80% of estimates), with the median of about 11 US$/$tCO_2$. For category III, the corresponding prices are somewhat higher and range from 18–79 US$/$tCO_2$ (with the median of the scenarios around 45 US$/$tCO_2$). Most individual studies for the most stringent category cluster around prices of about 100 US$/$tCO_2$.[14] Carbon prices by 2050 are comparatively higher than those in 2030. For example, costs of category IV scenarios by 2050 range between 5 and 65 US$/$tCO_2$, and those for category III range between 30 and 155 US$/$tCO_2$. Carbon prices in 2100 vary over a much wider range – mostly reflecting uncertainty in baseline emissions and technology development. For the medium target of 4.5 W/m^2, typical carbon prices in 2100 range from 25–200 US$/$tCO_2$ (80% of estimates). This is primarily a consequence of the nature of this metric, which often represents costs at the margin. Costs tend to slowly increase for more stringent targets – with a range between the 10th and 90th percentile of more than 35 to about 350 US$/$tCO_2$ for category III.

3.3.5.4   The role of non-CO_2 GHGs

As also illustrated by the scenario assessment in the previous sections, more and more attention has been paid since the TAR to incorporating non-CO_2 gases into climate mitigation and stabilization analyses. As a result, there is now a body of literature (e.g. Van Vuuren *et al.*, 2006b; De la Chesnaye and Weyant, 2006; De la Chesnaye *et al.*, 2007) showing that mitigation costs for these sectors can be lower than for energy-

---

12 If not otherwise mentioned, the discussion of the cost ranges (Figure 3.25) refers to the 80th percentile of the TAR and post-TAR scenario distribution (see the grey area in Figure 3.25).

13 NPV calculations are based on carbon tax projections of the scenarios, using a discount rate of 5%, and assuming that the average cost of abatement would be half the marginal price of carbon. Some studies report abatement costs themselves, but for consistency this data was not used. The assumption of using half the marginal price of carbon results in a slight overestimation.

14 Note that the scenarios of the lowest stabilization categories (I and II) are mainly based on intermediate and low baseline scenarios.

related CO_2 sectors. As a result, when all these options are employed in a multi-gas mitigation policy, there is a significant potential for reduced costs, for a given climate policy objective, versus the same policy when CO_2 is the only GHG directly mitigated. These cost savings can be especially important where carbon dioxide is not the dominant gas, on a percentage basis, for a particular economic sector and even for a particular region. While the previous sections have focused on the joint assessment of CO_2 and multi-gas mitigation scenarios, this section explores the specific role of non-CO_2 emitting sectors.[15]

A number of parallel numerical experiments have been carried out by the Energy Modelling Forum (EMF-21; De la Chesnaye and Weyant, 2006). The overall conclusion is that economic benefits of multi-gas strategies are robust across all models. This is even true, despite the fact that different methods were used in the study to compare the relative contribution of these gases in climate forcing (see Section 3.3.3). The EMF-21 study specifically focused on comparing stabilization scenarios aiming for 4.5 W/m^2 compared to pre-industrial levels. There were two cases employed to achieve the mitigation target:
1. Directly mitigate CO_2 emissions from the energy sector (with some indirect reduction in non-CO_2 gases).
2. Mitigate all available GHG in costs-effective approaches using full 'what' flexibility.

In the CO_2-only mitigation scenario, all models significantly reduced CO_2 emissions, on average by about 75% in 2100 compared to baseline scenarios. Models still indicated some emission reductions for CH_4 and N_2O as a result of systemic changes in the energy system. Emissions of CH_4 were reduced by about 20% and N_2O by about 10% (Figure 3.26).

In the multi-gas mitigation scenario, all models found that an appreciable percentage of the emission reductions occur through reductions of non-CO_2 gases, which then results in smaller required reductions of CO_2. The emission reduction for CO_2 in 2100 therefore drops (on average) from 75% to 67%. This percentage is still rather high, caused by the large share of CO_2 in total emissions (on average, 60% in 2100) and partly due to the exhaustion of reduction options for non-CO_2 gases. The reductions of CH_4 across the different models averages around 50%, with remaining emissions coming from sources for which no reduction options were identified, such as CH_4 emissions from enteric fermentation. For N_2O, the increased reduction in the multi-gas strategy is not as large as for CH_4 (almost 40%). The main reason is that the identified potential for emission reductions for the main sources of N_2O emissions, fertilizer use and animal manure, is still limited. Finally, for the fluorinated gases, high reduction rates (about 75%) are found across the different models.

Although the contributions of different gases change sharply over time, there is a considerable spread among the different models. Many models project relatively early reductions of both CH_4 and the fluorinated gases under the multi-gas case. However, the subset of models that does not use GWPs as the substitution metric for the relative contributions of the different gases to the overall target, but does assume inter-temporal optimization in minimizing abatement costs, do not start to reduce CH_4 emissions substantially until the end of the period. The increased flexibility of a multi-gas mitigation strategy is seen to have significant implications for the costs of stabilization across all models participating in the EMF-21. These scenarios concur that multi-gas mitigation is significantly cheaper than CO_2-only. The potential reductions of the GHG price ranges in the majority of the studies between 30% and 85% (See Figure 3.27).

Finally, the EMF-21 research also showed that, for some sources of non-CO_2 gases, the identified reduction potential is still very limited (e.g. most agricultural sources for N_2O emissions). For long-term scenarios (and more stringent targets) in particular, identifying how this potential may develop in time is a crucial research question. Attempts to estimate the maximum feasible reductions (and the development of potential over time) have been made in Van Vuuren et al. (2007).

3.3.5.5   Land use

Changes in land-use practices are regarded as an important component of long-term strategies to mitigate climate change. Modifications to land-use activities can reduce emissions of both CO_2 and non-CO_2 gases (CH_4 and N_2O), increase sequestration of atmospheric CO_2 into plant biomass and soils, and produce biomass fuel substitutes for fossil fuels (see Chapters 4, 8, and 9 of this report for discussions of detailed land-related mitigation alternatives). Available information before the TAR suggested that land has the technical potential to sequester up to an additional 319 billion tonnes of CO_2 ($GtCO_2$) by 2050 in global forests alone (IPCC, 1996a; IPCC, 2000; IPCC, 2001a). In addition, current technologies are capable of substantially reducing CH_4 and N_2O emissions from agriculture (see Chapter 8). A number of global biomass energy potential assessments have also been conducted (see Berndes et al. 2003 for an overview).[16]

The explicit modelling of land-based climate change mitigation in long-term global scenarios is relatively new and rapidly developing. As a result, assessment of the long-term role of global land-based mitigation was not formally addressed by the Special Report on Land use, Land-use Change, and Forestry (IPCC, 2000) or the TAR. This section assesses the modelling of land in long-term climate stabilization and the relationship to detailed global forestry mitigation estimates from partial equilibrium sectoral models that model 100-year carbon price trajectories.

---

15 Note that the multi-gas stabilization scenarios, which consider only CO_2 abatement options (discussed in this section), are not considered in the overall mitigation cost assessment of Section 3.3.5.3.
16 Most of the assessments are conducted with large regional spatial resolutions; exceptions are Fischer and Schrattenholzer (2001), Sørensen (1999), and Hoogwijk et al. (2005).

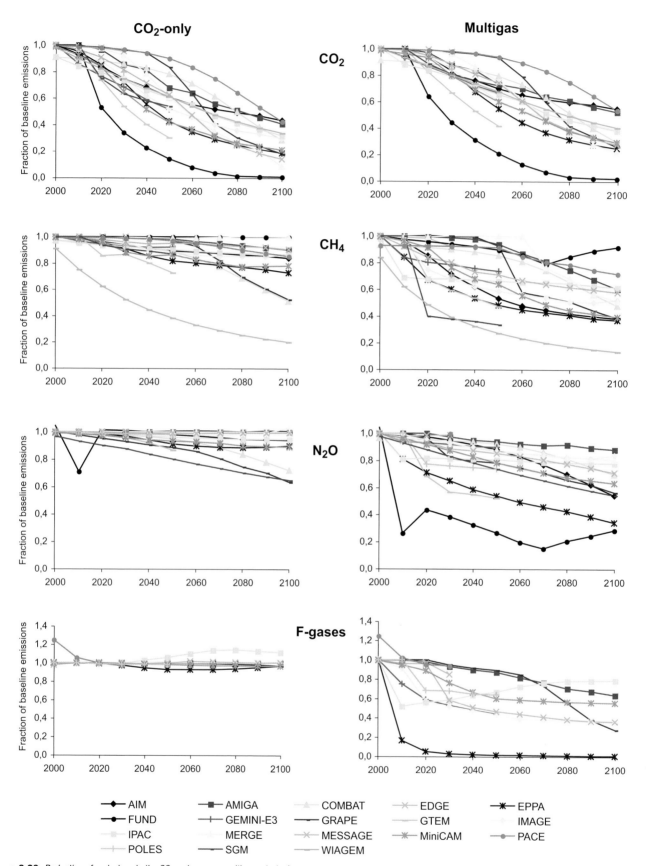

Figure 3.26: *Reduction of emissions in the CO₂-only versus multi-gas strategies.*

Source: De la Chesnaye and Weyant, 2006 (see also Van Vuuren et al., 2006b)

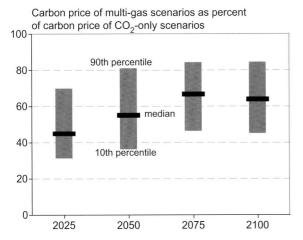

Figure 3.27: *Reduction in GHG abatement price (%) in multi-gas stabilization scenarios compared to CO_2-only cases. Ranges correspond to alternative scenarios for a stabilization target of 4.5 W/m².*

Data source: De la Chesnaye and Weyant, 2006

Development of, among other things, global sectoral land mitigation models (e.g. Sohngen and Sedjo, 2006), bottom-up agricultural mitigation costs for specific technologies (e.g. USEPA, 2006b), and biomass technical potential studies (e.g. Hoogwijk *et al.*, 2005) has facilitated the formal incorporation of land mitigation in long-term integrated assessment of climate change stabilization strategies. Hoogwijk *et al.* (2005), for example, estimated the potential of abandoned agricultural lands for providing biomass for primary energy demand and identified the technical biomass supply limits of this land type (e.g. under the SRES A2 scenario, abandoned agricultural lands could provide for 20% of 2001 total energy demand). Sands and Leimbach (2003) conducted one of the first studies to explicitly explore land-based mitigation in stabilization, suggesting that the total cost of stabilization could be reduced by including land strategies in the set of eligible mitigation options (energy crops in this case). The Energy Modelling Forum Study-21 (EMF-21; De la Chesnaye and Weyant, 2006) was the first coordinated stabilization modelling effort to include an explicit evaluation of the relative role of land in stabilization; however, only a few models participated. Building on their EMF-21 efforts, some modelling teams have also generated even more recent stabilization scenarios with revised land modelling. These studies are conspicuously different in the specifics of their modelling of land and land-based mitigation (Rose *et al.*, 2007). Differences in the types of land considered, emissions sources, and mitigation alternatives and implementation imply different opportunities and opportunity costs for land-related mitigation; and, therefore, different outcomes.

Four of the modelling teams in the EMF-21 study directly explored the question of the cost-effectiveness of including land-based mitigation in stabilization solutions and found that including these options (both non-CO_2 and CO_2) provided greater flexibility and was cost-effective for stabilizing radiative forcing at 4.5 W/m² (Kurosawa, 2006; Van Vuuren *et al.*, 2006a; Rao and Riahi, 2006; Jakeman and Fisher, 2006). Jakeman and Fisher (2006), for example, found that including land-use change and forestry mitigation options reduced the emissions reduction burden on all other emissions sources such that the projected decline in global real GDP associated with achieving stabilization was reduced to 2.3% at 2050 (3.4 trillion US$), versus losses of around 7.1% (10.6 trillion US$) and 3.3% (4.9 trillion US$) for the CO_2-only and multi-gas scenarios, respectively.[17] Unfortunately, none of the EMF-21 papers isolated the GDP effects associated with biomass fuel substitution or agricultural non-CO_2 abatement. However, given agriculture's small estimated share of total abatement (discussed below), the GDP savings associated with agricultural non-CO_2 abatement could be expected to be modest overall, though potentially strategically significant to the dynamics of mitigation portfolios. Biomass, on the other hand, may have a substantial abatement role and therefore a large effect on the economic cost of stabilization. Notably, strategies for increasing cropland soil carbon have not been incorporated to date into this class of models (see Chapter 8 for an estimate of the short-term potential for enhancing agricultural soil carbon).

Figure 3.28 presents the projected mitigation from forestry, agriculture, and biomass for the EMF-21 4.5 W/m² stabilization scenarios, as well as additional scenarios produced by the MESSAGE and IMAGE models – an approximate 3 W/m² scenario from Rao and Riahi (2006), a 4.5 W/m² scenario from Riahi *et al.* (2006), and approximately 4.5, 3.7, and 2.9 W/m² scenarios from Van Vuuren *et al.* (2007) (see Rose *et al.*, 2007, for a synthesis). While there are clearly different land-based mitigation pathways being taken across models for the same stabilization target, and across targets with the same model and assumptions, some general observations can be made. First, forestry, agriculture, and biomass are called upon to provide significant cost-effective mitigation contributions (Rose *et al.*, 2007). In the short-term (2000–2030), forest, agriculture, and biomass together could account for cumulative abatement of 10–65 GtCO$_2$-eq, with 15–60% of the total abatement considered by the available studies, and forest/agricultural non-CO_2 abatement providing at least three quarters of total land abatement.[18] Over the entire century (2000–2100), cumulative land-based abatement of approximately 345–1260 GtCO$_2$-eq is estimated to be cost-effective, accounting for 15–40% of total cumulative abatement. Forestry, agriculture, and biomass abatement levels are each projected to grow annually with relatively stable annual increases in agricultural mitigation and gradual deployment of biomass mitigation, which accelerates dramatically in the last half of the century to become the dominant land-mitigation strategy.

---

17 All values here are given in constant US dollars at 2000 prices.
18 The high percentage arises because some scenarios project that the required overall abatement from 2000–2030 is modest, and forestry and agricultural abatement options cost-effectively provide the majority of abatement.

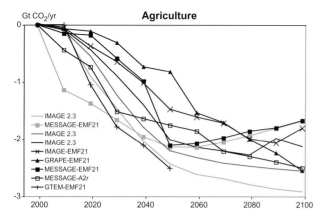

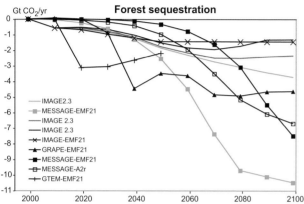

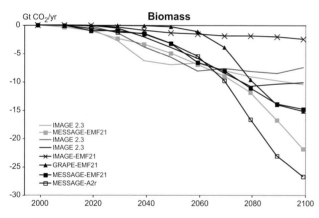

Figure 3.28: *Cost-effective agriculture, forest, and commercial biomass annual greenhouse gas emissions abatement from baselines from various 2100 stabilization scenarios (note y-axes have different ranges).*

Notes: The colour of the line indicates the 2100 stabilization target modelled: green < 3.25 W/m^2 (< 420 CO$_2$ concentration, < 510 CO$_2$-eq concentration), pink 3.25–4 (42–490, 510–590), and dark blue 4–5 (490–570, 59–710). The IMAGE-EMF21 and IMAGE 2.3 forest results are net of deforestation carbon losses induced by bio-energy crop extensification. These carbon losses are accounted for under forestry by the other scenarios. The MESSAGE-EMF21 results are taken from the sensitivity analysis of Rao and Riahi (2006). The GTEM-EMF21 scenarios ran through 2050 and the GTEM agriculture mitigation results include fossil fuel emissions reductions in agriculture (5-7% of the annual agricultural abatement). Scenario references: IMAGE-EMF21 (Van Vuuren *et al.*, 2006a); MESSAGE-EMF21 (Rao and Riahi, 2006); MESSAGE-A2r (Riahi *et al.*, 2006); GRAPE-EMF21 (Kurosawa, 2006); GTEM-EMF21 (Jakeman and Fisher, 2006); and IMAGE 2.3 (Van Vuuren *et al.*, 2007).

Source: Rose et al. (2007)

Figures 3.28 and 3.29 show that additional land-based abatement is expected to be cost-effective with tighter stabilization targets and/or higher baseline emissions (e.g. see the IMAGE 2.3 results for various stabilization targets and the MESSAGE 4.5 W/m^2 stabilization results with B2 (EMF-21) and A2r baselines). Biomass is largely responsible for the additional abatement; however, agricultural and forestry abatement are also expected to increase. How they might increase is model and time dependent. In general, the overall mitigation role of agricultural abatement of rice methane, livestock methane, nitrous oxide (enteric and manure) and soil nitrous oxide is projected to be modest throughout the time horizon, with some suggestion of increased importance in early decades.

However, there are substantial uncertainties. There is little agreement about the magnitudes of abatement (Figures 3.28 and 3.29). The scenarios disagree about the role of agricultural strategies targeting CH$_4$ versus N$_2$O, as well as the timing and annual growth of forestry abatement, with some scenarios suggesting substantial early deployment of forest abatement, while others suggest gradual annual growth or increasing annual growth.

A number of the recent scenarios suggest that biomass energy alternatives could be essential for stabilization, especially as a mitigation strategy that combines the terrestrial sequestration mitigation benefits associated with bio-energy CO$_2$ capture and storage (BECCS), where CO$_2$ emissions are captured during biomass energy combustion for storage in geologic formations (e.g. Rao and Riahi, 2006; Riahi *et al.*, 2006; Kurosawa, 2006; Van Vuuren *et al.*, 2007; USCCSP, 2006). BECCS has also been suggested as a potential rapid-response prevention strategy for abrupt climate change. Across stabilization scenarios, absolute emissions reductions from biomass are projected to grow slowly in the first half of the century, and then rapidly in the second half, as new biomass processing and mitigation technologies become available. Figure 3.28 suggests biomass mitigation of up to 7 GtCO$_2$/yr in 2050 and 27 GtCO$_2$/yr in 2100, for cumulative abatement over the century of 115–749 GtCO$_2$ (Figure 3.29). Figure 3.30 shows the amount of commercial biomass primary energy utilized in various stabilization scenarios. For example, in 2050, the additional biomass energy provides approximately 5–55 EJ for a 2100 stabilization target of 4–5 W/m^2 and approximately 40–115 EJ for 3.25–4 W/m^2, accounting for about 0–10 and 5–20% of 2050 total primary energy respectively (USCCSP, 2006; Rose *et al.*, 2007). Over the century, the additional bio-energy accounts for 500–6,700 EJ for targets of 4–5 W/m^2 and 6100–8000 EJ for targets of 3.25–4 W/m^2 (1–9% and 9–13% of total primary energy, respectively).

More biomass energy is supplied with tighter stabilization targets, but how much is required for any particular target depends on the confluence of the many different modelling assumptions. Modelled demands for biomass include electric power and end-use sectors (transportation, buildings, industry, and non-energy uses). Current scenarios suggest that electric power is

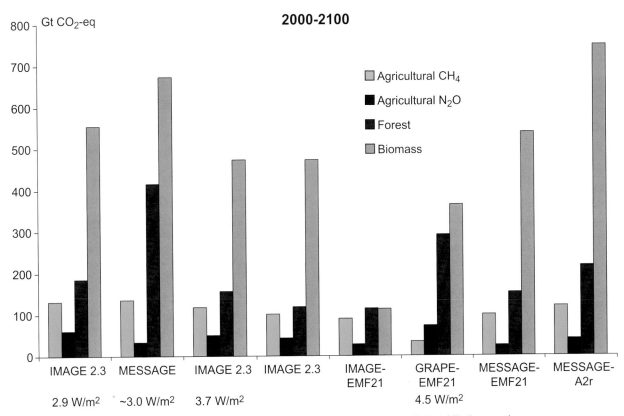

Figure 3.29: *Cumulative cost-effective agricultural, forestry, and biomass abatement 2000–2100 from various 2100 stabilization scenarios.*

Source: Rose et al. (2007)

projected to dominate biomass demand in the initial decades and, in general, with less stringent stabilization targets. Later in the century (and for more stringent targets) transportation is projected to dominate biomass use. When biomass is combined with BECCS, biomass mitigation shifts to the power sector late in the century, to take advantage of the net negative emissions from the combined abatement option, such that BECCS could represent a signifant share of cumulative biomass abatement over the century (e.g. 30–50% of total biomass abatement from MESSAGE in Figure 3.29).

To date, detailed analyses of large-scale biomass conversion with CO_2 capture and storage is scarce. As a result, current integrated assessment BECCS scenarios are based on a limited and uncertain understanding of the technology. In general, further research is necessary to characterize biomass' long-term mitigation potential, especially in terms of land area and water requirements, constraints, and opportunity costs, infrastructure possibilities, cost estimates (collection, transportation, and processing), conversion and end-use technologies, and ecosystem externalities. In particular, present studies are relatively poor in representing land competition with food supply and timber production, which has a significant influence on the economic potential of bio-energy crops (an exception is Sands and Leimbach, 2003).

Terrestrial mitigation projections are expected to be regionally unique, while still linked across time and space by

changes in global physical and economic forces. For example, Rao and Riahi (2006) offer intuitive results on the potential role of agricultural methane and nitrous oxide mitigation across industrialized and developing country groups, finding that agriculture is expected to form a larger share of the developing countries' total mitigation portfolio; and, developing countries

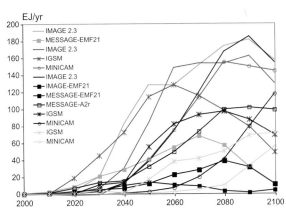

Figure 3.30: *Commercial biomass primary energy scenarios above baseline from various 2100 stabilization scenarios.*

Notes: The colour of the line indicates the 2100 stabilization target modelled: green < 3.25 W/m² (< 420 CO_2 concentration, < 510 CO_2-eq concentration), pink 3.25–4 (420–490, 510–590), dark blue 4–5 (490–570, 590–710), and light blue 5–6 (570–660, 710–860). Scenario references: IMAGE-EMF21 (Van Vuuren et al., 2006a); MESSAGE-EMF21 (Rao and Riahi, 2006); MESSAGE-A2r (Riahi et al., 2006); IMAGE 2.3 (Van Vuuren et al., 2007); IGSM and MiniCAM (USCCSP, 2006).

Source: Rose et al. (2007) ; USCCSP (2006)

are likely to provide the vast majority of global agricultural mitigation. Some aggregate regional forest mitigation results also are discussed below. However, given the paucity of published regional results from integrated assessment models, it is currently not possible to assess the regional land-use abatement potential in stabilization. Future research should direct attention to this issue in order to more fully characterize mitigation potential.

In addition to the stabilization scenarios discussed thus far from integrated assessment and climate economic models, the literature includes long-term mitigation scenarios from global land sector economic models (e.g. Sohngen and Sedjo, 2006; Sathaye *et al.*, 2006; Sands and Leimbach, 2003). Therefore, a comparison is prudent. The sectoral models use exogenous carbon price paths to simulate different climate policies and assumptions. It is possible to compare the stabilization and sectoral scenarios using these carbon price paths. Stabilization (e.g. EMF-21, discussed above) and 'optimal' (e.g. Sohngen and Mendelsohn, 2003) climate abatement policies suggest that carbon prices will rise over time.[19] Table 3.6 compares the forest mitigation outcomes from stabilization and sectoral scenarios that have similar carbon price trajectories (Rose *et al.*, 2007).[20] Rising carbon prices will provide incentives for additional forest area, longer rotations, and more intensive management to increase carbon storage. Higher effective energy prices might also encourage shorter rotations for joint production of forest bioenergy feedstocks.

Table 3.6 shows that the vast majority of forest mitigation is projected to occur in the second half of the century, with tropical regions in all but one scenario in Table 3.6 assuming a larger share of global forest sequestration/mitigation than temperate regions. The IMAGE results from EMF-21 are discussed separately below. Lower initial carbon prices shift early period mitigation to the temperate regions since, at that time, carbon incentives are inadequate for arresting deforestation. The sectoral models project that tropical forest mitigation activities are expected to be heavily dominated by land-use change activities (reduced deforestation and afforestation), while land management activities (increasing inputs, changing rotation length, adjusting age or species composition) are expected to be the slightly dominant strategies in temperate regions. The current stabilization scenarios model more limited and aggregated forestry GHG abatement technologies that do not distinguish the detailed responses seen in the sectoral models.

The sectoral models, in particular, Sohngen and Sedjo (2006), suggest substantially more mitigation in the second half of the century compared to the stabilization scenarios. A number of factors are likely to be contributing to this deviation from the integrated assessment model results. First and foremost, is that Sohngen and Sedjo explicitly model future markets, which none of the integrated assessment models are currently

capable of doing. Therefore, a low carbon price that is expected to increase rapidly results in a postponement of additional sequestration actions in Sohngen and Sedjo until the price (benefit) of sequestration is greater. Endogenously modelling forest biophysical and economic dynamics will be a significant future challenge for integrated assessment models. Conversely, the integrated assessment models may be producing a somewhat more muted forest sequestration response given:

(i) Their explicit consideration of competing mitigation alternatives across all sectors and regions, and, in some cases, land-use alternatives.

(ii) Their more limited set of forest-related abatement options, with all integrated assessment models modelling afforestation strategies, but only some considering avoided deforestation, and none modelling forest management options at this point.

(iii) Some integrated assessment models (including those in Table 3.6) sequentially allocate land, satisfying population food and feed-demand growth requirements first.

(iv) Climate feedbacks in integrated assessment models can lead to terrestrial carbon losses relative to the baseline.

The IMAGE results in Table 3.6 provide a dramatic illustration of the potential implications and importance of some of these counterbalancing effects. Despite the planting of additional forest plantations in the IMAGE scenario, net tropical forest carbon stocks decline (relative to the baseline) due to deforestation induced by bioenergy crop extensification, as well as reduced CO_2 fertilization that affects forest carbon uptake, especially in tropical forests, and decreases crop productivity, where the latter effect induces greater expansion of food crops onto fallow lands, thereby displacing stored carbon.

In addition to reducing uncertainty about the magnitude and timing of land-based mitigation, biomass potential, and regional potential, there are a number of other important outcomes from changes in land that should be tracked and reported in order to properly evaluate long-term land mitigation. Of particular importance to climate stabilization are the albedo implications of land-use change, which can offset emissions reducing land-use change (Betts, 2000; Schaeffer *et al.*, 2006), as well as the potential climate-driven changes in forest disturbance frequency and intensity that could affect the effectiveness of forest mitigation strategies. Non-climate implications should also be considered. As shown in the Millennium Ecosystem Assessment (Carpenter *et al.*, 2005), land use has implications for social welfare (e.g. food security, clean water access), environmental services (water quality, soil retention), and economic welfare (output prices and production).

A number of relevant key baseline land modelling challenges have already been discussed in Sections 3.2.1.6 and 3.2.2.2. Central to future long-term land mitigation modelling are

---

19 Optimal is defined in economic terms as the equating of the marginal benefits and costs of abatement.
20 Rose *et al.* (2007) report the carbon price paths from numerous stabilization and sectoral mitigation scenarios.

Table 3.6: *Cumulative forest carbon stock gains above baseline by 2020, 2050 and 2100, from long-term global forestry and stabilization scenarios (GtCO$_2$).*

US$2.73/tCO$_2$ (in 2010) + 5% per year				
		2020	2050	2100
Sathaye et al., (2006)	World	na	91.3	353.8
	Temperate	na	25.3	118.8
	Tropics	na	55.1	242.0
Sohngen and Sedjo (2006) original baseline	World	0.0	22.7	537.5
	Temperate	3.3	8.1	207.9
	Tropics	-3.3	14.7	329.6
Sohngen and Sedjo (2006) accelerated deforestation baseline	World	1.5	15.0	487.3
	Temperate	1.1	12.1	212.7
	Tropics	0.7	2.9	275.0
Stabilization at 4.5 W/m^2 (~650 CO$_2$-eq ppmv) by 2100				
		2020	2050	2100
GRAPE-EMF21	World	-0.6	70.3	291.9
	Temperate	-0.2	10.0	45.2
	Tropics	-0.5	60.3	246.7
IMAGE-EMF21	World	-22.5	-13.4	10.4
	Temperate	14.1	31.9	78.3
	Tropics	-36.6	-45.3	-67.9
MESSAGE-EMF21[*]	World	0.0	3.5	152.5
	Temperate	0.0	0.1	23.4
	Tropics	0.0	3.4	129.1

Notes: * Results based on the 4.5 W/m^2 MESSAGE scenario from the sensitivity analysis of Rao and Riahi (2006).
Tropics: Central America, South America, Sub-Saharian Africa, South Asia, Southeast Asia. Temperate: North America, Western and Central Europe, Former Soviet Union, East Asia, Oceania, Japan. Na = data not available.
Source: Stabilization data assembled from Rose et al. (2007)

improvements in the dynamic modelling of regional land use and land-use competition and mitigation cost estimates, as well as modelling of the implications of climate change for land-use and land mitigation opportunities. The total cost of any land-based mitigation strategy should include the opportunity costs of land, which are dynamic and regionally unique functions of changing regional biophysical and economic circumstances. In addition, the results presented in this section do not consider climate shifts that could dramatically alter land-use conditions, such as a permanent El-Nino-like state in tropical regions (Cox *et al.*, 1999).

To summarize, recent stabilization studies have found that land-use mitigation options (both non-CO$_2$ and CO$_2$) provide cost-effective abatement flexibility in achieving 2100 stabilization targets, in the order of 345–1260 GtCO$_2$-eq (15–40%) of cumulative abatement over the century. In some scenarios, increased commercial biomass energy (solid and liquid fuel) is significant in stabilization, providing 115–749 GtCO$_2$-eq (5–30%) of cumulative abatement and 500–9500 EJ of additional bio-energy above the baseline over the century (potentially 1–15% of total primary energy), especially as a net negative emissions strategy that combines biomass energy with CO$_2$ capture and storage. Agriculture and forestry mitigation options are projected to be cost effective short-term and long-term abatement strategies. Global forestry models project greater additional forest sequestration than found in stabilization scenarios, a result attributable in part to differences in the modelling of forest dynamics and general economic feedbacks. Overall, the explicit modelling of land-based climate change

mitigation in long-term global scenarios is relatively immature, with significant opportunities for improving baseline and mitigation land-use scenarios.

3.3.5.6　*Air pollutants, including co-benefits*

Quantitative analysis on a global scale on the implications of climate mitigation for air pollutants such as SO$_2$, NO$_x$, CO, VOC (volatile organic compounds), BC (black carbon) and OC (organic carbon), are relatively scarce. Air pollutants and greenhouse gases are often emitted by the same sources, and changes in the activity of these sources affect both types of emissions. Previous studies have focused on purely ancillary benefits to air pollution that accrue from a climate mitigation objective, but recently there is a focus on integrating air quality and climate concerns, thus analyzing the co-benefits of such policies. Several recent reviews have summarized the issues related to such benefits (OECD 2000, 2003). They cover absolute air pollutant emission reductions, monetary value of reduced pollution, the climatic impacts of such reductions and the improved health effects due to reduced pollution.

The magnitude of such benefits largely depends on the assumptions of future policies and technological change in the baseline against which they are measured, as discussed in Morgenstern (2000). For example, Smith *et al.* (2005) and Rao *et al.* (2005) assume an overall growth in environmental awareness and formulation of new environmental policies with increased affluence in the baseline scenario, and thus reduced air pollution, even in the absence of any climate policies. The pace of this trend differs significantly across pollutants and baseline scenarios, and may or may not have an obvious effect on greenhouse gases. An added aspect of ancillary benefit measurement is the representation of technological options. Some emission-control technologies reduce both air pollutants and greenhouse gases, such as selective catalytic reduction (SCR) on gas boilers, which reduces not only NO_x, but also N_2O, CO and CH_4 (IPCC, 1997). But there are also examples where, at least in principle, emission-control technologies aimed at a certain pollutant could increase emissions of other pollutants. For example, substituting more fuel-efficient diesel engines for petrol engines might lead to higher PM/black carbon emissions (Kupiainen and Klimont 2004). Thus estimating co-benefits of climate mitigation should include adequate sectoral representation of emission sources, a wide range of substitution possibilities, assumptions on technological change and a clear representation of current environmental legislation.

Only a few studies have explored the longer-term ancillary benefits of climate policies. Alcamo (2002) and Mayerhofer et al. (2002) assess in detail the linkages between regional air pollution and climate change in Europe. They emphasize important co-benefits between climate policy and air pollution control but also indicate that, depending on assumptions, air pollution policies in Europe will play a greater role in air pollutant reductions than climate policy. Smith and Wigley (2006) suggest that there will be a slight reduction in global sulphur aerosols as a result of long-term multi-gas climate stabilization. Rao et al. (2005) and Smith and Wigley (2006) find that climate policies can reduce cumulative BC and OC emissions by providing the impetus for adoption of cleaner fuels and advanced technologies. In addition, the inclusion of co-benefits for air pollution can have significant impacts on the cost effectiveness of both the climate policy and air pollution policy under consideration. Van Harmelen et al. (2002) find that to comply with agreed upon or future policies to reduce regional air pollution in Europe, mitigation costs are implied, but these are reduced by 50–70% for SO_2 and around 50% for NO_x when combined with GHG policies. Similarly, in the shorter-term, Van Vuuren et al. (2006c) find that for the Kyoto Protocol, about half the costs of climate policy might be recovered from reduced air pollution control costs. The exact benefits, however, critically depend on how the Kyoto Protocol is implemented.

The different spatial and temporal scale of greenhouse gases and air pollutants is a major difficulty in evaluating ancillary benefits. Swart et al. (2004) stress the need for new analytical bridges between these different spatial and temporal scales. Rypdal et al. (2005) suggest the possibility of including some local pollutants, such as CO and VOCs, in global climate agreement with others (e.g. NO_x) and aerosols being regulated by regional agreements. It should be noted that some air pollutants, such as sulphate and carbonaceous aerosols, exert radiative forcing and thus global warming, but their contribution is uncertain. Smith and Wigley (2006) find that the attendant reduced aerosol

cooling from sulphates can more than offset the reduction in warming that accrues from reduced GHGs. On the other hand, air pollutants such as NO_x, CO and VOCs act as indirect greenhouse gases having an influence for example via their impact on OH (hydroxil) radicals and therefore the lifetime of direct greenhouse gases (e.g. methane and HFCs). Further, the climatic effects of some pollutants, such as BC and OC aerosols, remain unclear.

While there has been a lot of recent research in estimating co-benefits of joint GHG and air pollution policies, most current studies do not have a comprehensive treatment of co-benefits in terms of reduction costs and the related health and climate impacts in the long-term, thus indicating the need for more research in this area.

3.3.6 Characteristics of regional and national mitigation scenarios

Table 3.7 summarizes selected national mitigation scenarios. There are broadly two types of national scenarios that focus on climate mitigation. First, there are the scenarios that study mitigation options and related costs under a given national emissions cap and trade regime. The second are the national scenarios that focus on evaluation of climate mitigation measures and policies in the absence of specific emissions targets. The former type of analysis has been mainly undertaken in the studies in the European Union and Japan. The latter type has been explored in the USA, Canada and Japan. There is also an increasing body of literature, mainly in developing countries, which analyses national GHG emissions in the context of domestic concerns, such as energy security and environmental co-benefits. Many of these developing country analyses do not explicitly address emissions mitigation. In contrast to global studies, regional scenario analyses have focused on shorter time horizons, typically up to between 2030 and 2050.

A number of scenario studies have been conducted for various countries within Europe. These studies explore a wide range of emission caps, taking into account local circumstances and potentials for technology implementation. Many of these studies have used specific burden-sharing allocation schemes, such as the contraction and convergence (C&C) approach (GCI, 2005) for calculating the allocation of worldwide emissions to estimate national emissions ceilings. The UK's Energy White Paper (DTI, 2003) examined measures to achieve a 60% reduction in CO_2 emissions by 2050 as compared to the current level. Several studies have explored renewable energy options, for example, the possibility of expanding the share of renewable energy and the resulting prospects for clean hydrogen production from renewable energy sources in Germany (Deutscher Bundestag, 2002; Fischedick and Nitsch, 2002; Fischedick et al., 2005). A European study, the COOL project (Tuinstra et al., 2002; Treffers et al., 2005), has explored the possibilities of reducing emissions in the Netherlands by 80% in 2050 compared to 1990 levels. In France, the Inter Ministerial Task Force on Climate Change (MIES, 2004) has examined mitigation options that

Table 3.7: *National scenarios with quantification up to 2050 and beyond.*

Country	Author/Agency	Model	Time horizon	Target variables	Base year	Target of reduction to the value of the base year
USA	Hanson *et al.* (2004)	AMIGA[1]	2000-2050	-	2000	(about 44% in 2050)
Canada	Natural Resource Canada (NRCan) (2000)	N.A.	2000-2050	GHG emissions	2000	(53% in 2050)
India	Nair *et al.* (2003)	Integrated modelling framework[1,3]	1995-2100	Cumulative CO_2 emissions		550 ppmv, 650 pmv
	Shukla *et al.* (2006)	ERB[2]	1990-2095	CO_2 emissions		550 ppmv
China Netherlands	Chen (2005)	MARKAL-MACRO[2,3]	2000-2050	CO_2 emissions	Reference	5-45% in 2050
	Van Vuuren *et al.* (2003)	IMAGE/TIMER[2,4]	1995-2050	GHG emissions	1995	-
	Jiang and Xiulian (2003)	IPAC-emission[2,3]	1990-2100	GHG emissions	1990	-
	Tuinstra et al. (2002) (COOL)		1990-2050	GHG emissions	1990	80% in 2050
Germany	Deutscher Bundestag (2002)	WI[4], IER	2000-2050	CO_2 emissions	1990	80% in 2050
UK	Department of Trade and Industry (DTI) (2003)	MARKAL[3]	2000-2050	CO_2 emissions	2000	45%, 60%, 70% in 2050
France	Interministerial Task Force on Climate Change (MIES) (2004)	N.A.	2000-2050	CO_2 emissions	2000	0.5 tC/cap (70% in 2050)
Australia	Ahammad et al. (2006)	GTEM[1]	2000-2050	GHG emissions	1990	50% in 2050
Japan	Japan LCS Project (2005)	AIM/Material[1] MENOCO[4]	2000-2050	CO_2 emissions	1990	60-80% in 2050
	Ministry of Economy, Trade and Industry (2005)	GRAPE[3]	2000-2100	CO_2/GDP	2000	1/3 in 2050, 1/10 in 2100
	Masui *et al.* (2006)	AIM/Material[1]	2000-2050	CO_2 emissions	1990	74% in 2050
	Akimoto *et al.* (2004)	Optimization model[3]	2000-2050	CO_2 emissions	2000	0.5% /yr (21% in 2050)
	Japan Atomic Industrial Forum (JAIF) (2004)	MARKAL[3]	2000-2050	CO_2 emissions	2010 (1990)	40% in 2050

Notes: model types: 1: CGE-type top-down model, 2: other type of top-down model, 3: bottom-up technology model with optimization, 4: bottom-up technology model without optimization.

could lead to significant reductions in per capita emissions intensity. Savolainen *et al.* (2003) and Lehtila *et al.* (2005) have conducted a series of scenario analyses in order to assess technological potentials in Finland for a number of options that include wind power, electricity-saving possibilities in households and office appliances, and emission abatement of fluorinated GHGs.

Scenario studies in the USA have explored the implications of climate mitigation for energy security (Hanson *et al.*, 2004). For example, Mintzer *et al.* (2003) developed a set of scenarios describing three divergent paths for US energy supply and use from 2000 through 2035. These scenarios were used to identify key technologies, important energy policy decisions, and strategic investment choices that may enhance energy security, environmental protection, and economic development.

A wide range of scenario studies have also been conducted to estimate potential emissions reductions and associated costs for Japan. For example, Masui *et al.* (2006) developed a set of scenarios that explore the implications of severe emissions cutbacks of between 60 and 80% CO_2 by 2050 (compared to 1990). Another important study by Akimoto *et al.* (2004) evaluates the possibilities of introducing the CCS option and its economic implications for Japan.

National scenarios pertaining to developing countries such as China and India mainly analyze future emission trajectories under various scenarios that include considerations such as economic growth, technology development, structure changes, globalization of world markets, and impacts of mitigation options. Unlike the scenarios developed for the European countries, most of the developing-country scenarios do not

Table 3.8: *Developed countries scenarios with more than 40% reduction (compared to 2000 emissions), and some Chinese scenarios: CO_2 emission changes from 2000 to 2050; Energy intensity and carbon intensity in 2000, and their changes from 2000 up to 2050.*

(A) CO_2 emission changes, energy intensity, and carbon intensity in 2000.

Country	CO_2 emission change [%] (2000-2050) ◆: BaU scenarios ○: Intervention scenarios (less than 40%) +: Drastic reduction scenarios (equal and more than 40%)	Initial value (2000) Energy intensity [toe/1000 95$(MER)]	Carbon intensity [ton CO_2/ktoe]
China	59.5 ... 377.8	0.97	2.61
Japan	-74.2 ... 33.9	0.09	2.26
Germany	-76.0 ... -15.9	0.13	2.43
France	-69.8 ... 38.8	0.15	1.46
UK	-69.9 ... -11.3	0.18	2.26
USA	-46.2 ... 65.7	0.26	2.47

(B) Changes in energy intensity and carbon intensity.

Country	CO_2 emission reduction factors — Energy intensity — Annual change in energy intensity (2000-2050) (%/year)	Carbon intensity — Annual change in carbon intensity (including CCS) (2000-2050) (%/year)	Share of CCS in carbon intensity reduction (2000-2050) (%)
China	-4.02 ... -2.47	-1.04 ... -0.33	0 +
Japan	-2.82 ... -0.82	-1.87 ... -0.73	0 + ... 36.9 ... 58.14
Germany	-2.83 ... -1.41	-2.73 ... -1.28	0 + ... 78.3 ... 84.2
France	-2.26 ... -1.33	-2.65 ... -1.75	0 + ... 61.4 ... 100
UK	-3.05 ... -2.52	-2.39 ... -0.61	4.1 ... 79.9
USA	-2.70 ... -2.20	-1.29 ... -0.54	38.8 ... 65.5

Notes: Data sources: China: Jiang and Xulian (2003), Van Vuuren et al. (2003), Japan: Masui et al. (2006), Akimoto et al. (2004), JAIF (2004), Germany: Deutscher Bundestag (2002), France: MIES (2004), UK: DTI (2003), USA: Hanson et al. (2004). The coloured areas show the range of the global model results of EMF-21 with the target of 4.5 W/m². The range of EU-15 is shown for European countries

specify limits on emissions (Van Vuuren *et al.*, 2003; Jiang and Xiulian, 2003). Chen (2005) shows that structural change can be a more important contributor to CO_2 reduction than technology efficiency improvement. The scenario construction for India pays specific attention to developing-country dynamics, underlying the multiple socio-economic transitions during the century, including demographic transitions (Shukla *et al.*, 2006). Nair *et al.* (2003) studied potential shifts away from coal-intensive baselines to the use of natural gas and renewables.

There are several country scenarios that consider drastic reduction of CO_2 emissions. In these studies, which consider 60–80% reductions of CO_2 in 2050, rates of improvement in

energy intensity and carbon intensity increase by about two to three times their historical levels (Kawase *et al.*, 2006).

Table 3.8 summarizes scenarios with more than 40% CO_2 reductions (2000–2050) in several developed countries. The table also includes some Chinese scenarios with deep cuts of CO_2 emissions compared to the reference cases. Physical indicators of the Chinese economy show that current efficiency is below the OECD average in most sectors, thus indicating a greater scope for improvement (Jiang and Xiulian, 2003). It should be noted that comparing the energy intensity of the Chinese economy on the basis of market exchanges rates to OECD averages suggests even larger differences, but this is

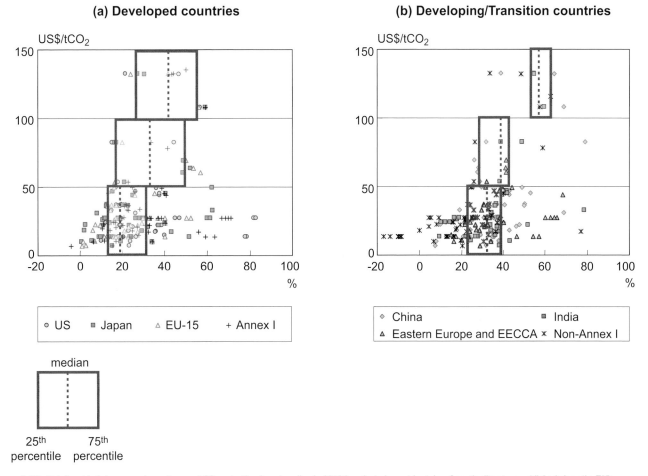

Figure 3.31: *Relationship between carbon prices and CO₂ reduction from baseline in 2050 in selected countries taken from the literature published since the TAR.*
Note: The red box shows the range between the 25th and 75th percentile of the scenarios for each price range. EECCA= Countries of Eastern Europe, the Caucasus and Central Asia.

misleading given the differences in purchasing power (PPP-corrected energy intensity data gives a somewhat better basis for comparison, but still suffers from uncertainty about the data and different economic structures).

In the countries with low energy intensity levels in 2000 (such as Japan, Germany and France), the scenarios specify solutions for meeting long-term drastic reduction goals by carbon intensity improvement measures, such as shifting to natural gas in the UK, renewable energy in the Netherlands, and CCS in certain scenarios in France, Germany, the UK and the USA. France has a scenario where CCS accounts for 100% of carbon intensity improvement. Most of the scenarios with drastic CO₂ reductions for the USA and the UK assume the introduction of CCS.

The light yellow coloured area in Table 3.8 shows the range of the global model results of EMF-21 with the stabilization target of 4.5 W/m². Most country results show the need for greater improvement in carbon intensity during 2000 to 2050 compared to the global results. The results of scenario analysis since the TAR show that energy intensity improvement is superior to

carbon intensity reduction in the first half of the 21st century, but that carbon intensity reduction becomes more dominant in the latter half of the century (Hanaoka *et al.*, 2006).

3.3.6.1 Costs of mitigation in regional and country scenarios

Figure 3.31 shows the relationship between carbon prices and the CO₂ mitigation rates from the baseline in 2050 in some major countries and regions such as the USA, Japan, EU-15, India, China, Former Soviet Union (FSU) and Eastern Europe, taken from the literature since the TAR (Hanaoka *et al.*, 2006). In the developing countries there are many scenarios where relatively high CO₂ reductions are projected even with low carbon prices. With high prices in the range of 100-150 US$/tCO₂ (in 2000 US dollars) more CO₂ reductions are expected in China and India than in developed countries when the same level of carbon price is applied.

3.4 The role of technologies in long-term mitigation and stabilization: research, development, deployment, diffusion and transfer

Technology is among the central driving forces of GHG emissions. It is one of the main determinants of economic development, consumption patterns and thus human well-being. At the same time, technology and technological change offer the main possibilities for reducing future emissions and achieving the eventual stabilization of atmospheric concentrations of GHGs (see Chapter 2, Section 2.7.1.2, which assesses the role of technology in climate change mitigation, including long-term emissions and stabilization scenarios).

The ways in which technology reduces future GHG emissions in long-term emission scenarios include:
- Improving technology efficiencies and thereby reducing emissions per unit service (output). These measures are enhanced when complemented by energy conservation and rational use of energy.
- Replacing carbon-intensive sources of energy by less intensive ones, such as switching from coal to natural gas. These measures can also be complemented by efficiency improvements (e.g. combined cycle natural gas power plants are more efficient than modern coal power plants) thereby further reducing emissions.
- Introducing carbon capture and storage to abate uncontrolled emissions. This option could be applied at some time in the future, in conjunction with essentially all electricity generation technologies, many other energy conversion technologies and energy-intensive processes using fossil energy sources as well as biomass (in which case it corresponds to net carbon removal from the atmosphere).
- Introducing carbon-free renewable energy sources ranging from a larger role for hydro and wind power, photovoltaics and solar thermal power plants, modern biomass (that can be carbon-neutral, resulting in zero net carbon emissions) and other advanced renewable technologies.
- Enhancing the role of nuclear power as another carbon-free source of energy. This would require a further increase in the nuclear share of global energy, depending on the development of 'inherently' safe reactors and fuel cycles, resolution of the technical issues associated with long-term storage of fissile materials and improvement of national and international non-proliferation agreements.
- New technology configurations and systems, e.g. hydrogen as a carbon-free carrier to complement electricity, fuel cells and new storage technologies.
- Reducing GHG and CO_2 emissions from agriculture and land use in general critically depends on the diffusion of new technologies and practices that could include less fertilizer-intensive production and improvement of tillage and livestock management.

Virtually all scenarios assume that technological and structural changes occur during this century, leading to relative reductions in emissions compared to the hypothetical case of attempting to 'keep' emissions intensities of GDP and structure the same as today (see Chapter 2, Section 2.7.1.1, which discusses the role of technology in baseline scenarios). Figure 3.32 shows such a hypothetical range of cumulative emissions under the assumption of 'freezing' technology and structural change in all scenarios at current levels, but letting populations change and economies develop as assumed in the original scenarios (Nakicenovic *et al.*, 2006). To show this, the energy intensity of GDP and the carbon intensity of energy are kept constant. The bars in the figure indicate the central tendencies of the scenarios in the literature by giving the cumulative emissions ranges between the 25th and the 75th percentile of the scenarios in the scenario database.[21] The hypothetical cumulative emissions (without technology and structural change) range from about 9000 (25th percentile) to 12000 (75th percentile), with a median of about 10400 $GtCO_2$ by 2100.

The next bar in Figure 3.32 shows cumulative emissions by keeping carbon intensity of energy constant while allowing energy intensity of GDP to evolve as originally specified in the underlying scenarios. This in itself reduces the cumulative emissions substantively, by more than 40% to almost 50% (75th and 25th percentiles, respectively). Thus, structural economic changes and more efficient use of energy lead to significant reductions of energy requirements across the scenarios as incorporated in the baselines, indicating that the baseline already includes vigorous carbon saving. In other words, this means that many new technologies and changes that lead to lower relative emissions are assumed in the baseline. Any mitigation measures and policies need to go beyond these baseline assumptions.

The next bar in Figure 3.32 also allows carbon intensities of energy to change as originally assumed in the underlying scenarios. Again, the baseline assumptions lead to further and substantial reductions of cumulative emissions, by some 13% to more than 20% (25th and 75th percentile, respectively), *or less than half the emissions*, as compared to the case of no improvement in energy or carbon intensities. This results in the original cumulative emissions as specified by reference scenarios in the literature, from 4050 (25th percentile) to 5400 (75th percentile), with a median of 4730 $GtCO_2$ by 2100. It should be noted that this range is for the 25th to the 75th percentile only. In contrast, the full range of cumulative emissions across 56 scenarios in the database is from 2075 to 7240 $GtCO_2$.[22]

---

21 The outliers, above the 75th and below the 25th percentile are discussed in more detail in the subsequent sections.
22 The cumulative emissions range represents a huge increase compared to the historical experience. Cumulative global emissions were about 1100 $GtCO_2$ from the 1860s to today, a very small fraction indeed of future expected emissions across the scenarios.

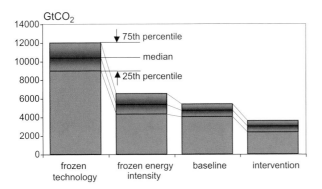

Figure 3.32: *Median, 25th and 75th percentile of global cumulative carbon emissions by 2100 in the scenarios developed since 2001.*

Note: The range labelled 'frozen technology' refers to hypothetical futures without improvement in energy and carbon intensities in the scenarios; the range labelled 'frozen energy intensity' refers to hypothetical futures where only carbon intensity of energy is kept constant, while energy intensity of GDP is left the same as originally assumed in scenarios; the range labelled CO_2 baseline refer to the 83 baseline scenarios in the database, while the region labelled CO_2 intervention includes 211 mitigation and/or stabilization scenarios.

Source: After Nakicenovic et al. (2006)

The next and final step is to compare the cumulative emissions across baseline scenarios with those in the mitigation and stabilization variants of the same scenarios. Figure 3.32 shows (in the last bar) yet another significant reduction of future cumulative emissions from 2370 to 3610 (corresponding to the 25th to the 75th percentile of the full scenario range), with a median of 3010 $GtCO_2$ by 2100. This corresponds to about 70% emissions reduction across mitigation scenarios, compared to the hypothetical case of no changes in energy and carbon intensities and still a large, or about a 30%, reduction compared to the respective baseline scenarios.[23]

This illustrates the importance of technology and structural changes, both in reference and mitigation scenarios. However, this is an aggregated illustration across all scenarios and different mitigation levels for cumulative emissions. Thus, it is useful to also give a more specific illustrative example. Figure 3.33 gives such an illustration by showing the importance of technological change assumptions in both reference and mitigation scenarios for a 550 ppmv concentration target based on four SRES scenarios. Such analyses are increasingly becoming available. For instance, Placet *et al.* (2004) provide a detailed study of possible technology development pathways under climate stabilization for the US government Climate Change Technology Program. To illustrate the importance of technological change, actual projected scenario values in the original SRES no-climate policy scenarios are compared with a hypothetical case with frozen 1990 structures and technologies for both energy supply and end-use. The difference (denoted by a grey shaded area in Figure 3.33) illustrates the impact of technological change, which leads to improved efficiency and 'decarbonization' in energy systems already incorporated into the baseline emission scenario.

The impacts of technological options leading to emission reductions are illustrated by the colour-shaded areas in Figure 3.33, regrouped into three categories: demand reductions (e.g. through deployment of more efficient end-use technologies, such as lighting or vehicles), fuel switching (substituting high-GHG-emitting technologies for low- or zero-emitting technologies such as renewables or nuclear), and finally, CO_2 capture and storage technologies. The mix in the mitigative technology portfolio required to reduce emissions from the reference scenario level to that consistent with the illustrative 550 ppmv stabilization target varies as a function of the baseline scenario underlying the model calculations (shown in Figure 3.33), as well as with the degree of stringency of the stabilization target adopted (not shown in Figure 3.33). An interesting finding from a large number of modelling studies is that scenarios with higher degrees of technology diversification (e.g. scenario A1B in Figure 3.33) also lead to a higher degree of flexibility with respect of meeting alternative climate (e.g. stabilization) targets and generally also to lower overall costs compared with less diversified technology scenarios. This illustrative example also confirms the conclusion reached in Section 3.3 that was based on a broader range of scenario literature.

This brief assessment of the role of technology across scenarios indicates that there is a significant technological change and diffusion of new and advanced technologies already assumed in the baselines and additional technological change 'induced' through various policies and measures in the mitigation scenarios. The newer literature on induced technological change assessed in the previous sections, along with other scenarios (e.g. Grübler *et al.*, 2002; and Köhler *et al.*, 2006, see also Chapter 11), also affirms this conclusion.

3.4.1 Carbon-free energy and decarbonization

3.4.1.1 Decarbonization trends

Decarbonization denotes the declining average carbon intensity of primary energy over time. Although decarbonization of the world's energy system is comparatively slow (0.3% per year), the trend has persisted throughout the past two centuries (Nakicenovic, 1996). The overall tendency towards lower carbon intensities is due to the continuous replacement of fuels with high carbon content by those with low carbon content; however, intensities are currently increasing in some developing regions. In short- to medium-term scenarios such a declining tendency for carbon intensity may not be as discernable as across the longer-term literature, e.g. in the World Energy Outlook 2004 (IEA, 2004), the reference scenario to 2030 shows the replacement of gas for other fossil fuels as well as cleaner fuels due to limited growth of nuclear and bioenergy.

Another effect contributing towards reduced carbon intensity of the economy is the declining energy requirements per unit

---

23 In comparison, the full range of cumulative emissions from mitigation and stabilization scenarios in the database runs from 785 to 6794 $GtCO_2$.

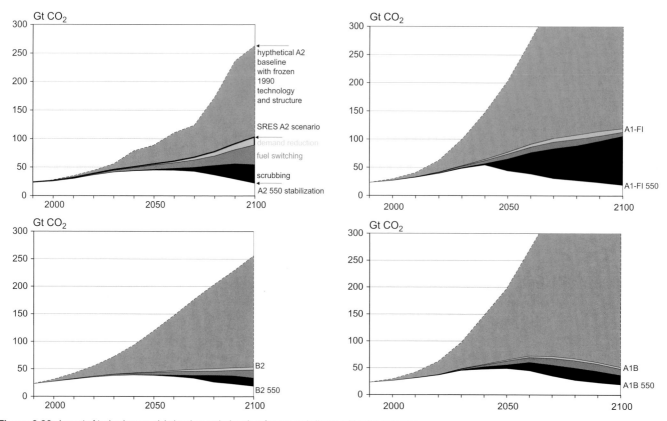

Figure 3.33: *Impact of technology on global carbon emissions in reference and climate mitigation scenarios.*

Note: Global carbon emissions (GtC) in four scenarios developed within the IPCC SRES and TAR (A2, B2 top and bottom of left panel; A1FI and A1B top and bottom of right panel). The grey-shaded area indicates the difference in emissions between the original no-climate policy reference scenario compared with a hypothetical scenario assuming frozen 1990 energy efficiency and technology, illustrating the impact of technological change incorporated already into the reference scenario. Colour-shaded areas show the impact of various additional technology options deployed in imposing a 550 ppmv CO_2 stabilization constraint on the respective reference scenario, including energy conservation (blue), substitution of high-carbon by low- or zero-carbon technologies (orange), as well as carbon capture and sequestration (black). Of particular interest are the two A1 scenarios shown on the right-hand side of the panel that share identical (low) population and (high) economic growth assumptions, thus making differences in technology assumptions more directly comparable.

Source: Adapted from Nakicenovic et al. (2000), IPCC (2001a), Riahi and Roehrl (2001), and Edmonds (2004).

of GDP, or energy intensity of GDP. Globally, energy intensity has been declining more rapidly than carbon intensity of energy (0.9% per year) during the past two centuries (Nakicenovic, 1996). Consequently, carbon intensity of GDP declined globally at about 1.2% per year.

The carbon intensity of energy and energy intensities of GDP were shown in Section 3.2 of this chapter, Figure 3.6, for the full scenario sample in the scenario database compared to the newer (developed after 2001) non-intervention scenarios. As in Sections 3.2 and 3.3, the range of the scenarios in the literature until 2001 is compared with recent projections from scenarios developed after 2001 (Nakicenovic *et al.*, 2005).

The majority of the scenarios in the literature portray a similar and persistent decarbonization trend as observed in the past. In particular, the medians of the scenario sets indicate energy decarbonization rates of about 0.9% (pre-2001 literature median) and 0.6% (post-2001 median) per year, which is a significantly more rapid decrease compared to the historical rates of about 0.3% per year. Decarbonization of GDP is also more rapid (about 2.5% per year for both pre- and post-2001

literature medians) compared with the historical rates of about 1.2% per year. As expected, the intervention and stabilization scenarios have significantly higher decarbonization rates and the post-2001 scenarios include a few with significantly more rapid decarbonization of energy, even extending into the negative range. This means that towards the end of the century these more extreme decarbonization scenarios foresee net carbon removal from the atmosphere, e.g. through carbon capture and storage in conjunction with large amounts of biomass energy. Such developments represent a radical paradigm shift compared to the current and more short-term energy systems, implying significant and radical technological changes.

In contrast, the scenarios that are most intensive in the use of fossil fuels lead to practically no reduction in carbon intensity of energy, while all scenarios portray decarbonization of GDP. For example, the upper boundary of the recent scenarios developed after 2001 depict slightly increasing (about 0.3% per year) carbon intensities of energy (A2 reference scenario, Mori (2003), see Figure 3.8, comparing carbon emissions across scenarios in the literature presented in Section 3.2). Most notably, a few scenarios developed before 2001 follow an opposite path compared to other scenarios: decarbonization of primary energy with decreasing energy efficiency until 2040, followed by rapidly increasing ratios of CO_2 per unit of primary energy after 2040 – in other words, recarbonization. In the long term, these scenarios lie well above the range spanned by the new scenarios, indicating a shift towards more rapid CO_2 intensity improvements in the recent literature (Nakicenovic *et al.*, 2006). In contrast, there are just a very few scenarios in the post-2100 literature that envisage increases in carbon intensity of energy.

The highest rates of decarbonization of energy (up to 2.5% per year for the recent scenarios) are from scenarios that include a complete transition in the energy system away from carbon-intensive fossil fuels. Clearly, the majority of these scenarios are intervention scenarios, although some non-intervention scenarios show drastic reductions in CO_2 intensities due to reasons other than climate policies (e.g. the combination of sustainable development policies and technology push measures to promote renewable hydrogen systems). The relatively fast decarbonization rate of intervention scenarios is also illustrated by the median of the post-2001 intervention scenarios, which depict an average rate of improvement of 1.1% per year over the course of the century, compared to just 0.3% for the non-intervention scenarios. Note, nevertheless, that the modest increase in carbon intensity of energy improvements in the intervention scenarios above the 75th percentile of the distribution of the recent scenarios. The vast majority of these scenarios represent sensitivity analysis; have climate policies for mitigation of non-CO_2 greenhouse gas emissions (methane emissions policies: Reilly *et al.*, 2006); or have comparatively modest CO_2 reductions measures, such as the implementation of a relatively minor carbon tax of 10 US\$/tC (about 2.7 US\$/tCO_2) over the course of the century (e.g. Kurosawa, 2004). Although these scenarios are categorized according to our definition as intervention scenarios, they do not necessarily lead to the stabilization of atmospheric CO_2 concentrations.

3.4.1.2   Key factors for carbon-free energy and decarbonization development

All of the technological options assumed to contribute towards further decarbonization and reduction of future GHG emissions require further research and development (R&D) to improve their technical performance, reduce costs and achieve social acceptability. In addition, deployment of carbon-saving technologies needs to be applied at ever-larger scales in order to benefit from potentials of technological learning that can result in further improved costs and economic characteristics of new technologies. Most importantly, appropriate institutional and policy inducements are required to enhance widespread diffusion and transfer of these technologies.

Full replacement of dominant technologies in the energy systems is generally a long process. In the past, the major energy technology transitions have lasted more than half a century, such as the transition from coal as the dominant energy source

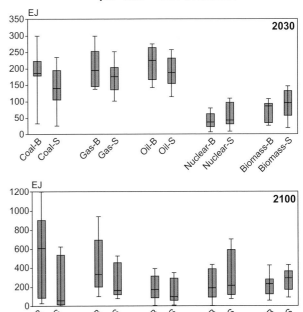

pre-TAR + TAR Scenarios

Figure 3.34: *Deployment of primary energy technologies across pre-2001 scenarios by 2030 and 2100: Left-side 'error' bars show baseline (non-intervention) scenarios and right-side ones show intervention and stabilization scenarios. The full ranges of the distributions (full vertical line with two extreme tic marks), the 25th and 75th percentiles (blue area) and the median (middle tic mark) are also shown.*

in the world some 80 years ago, to the dominance of crude oil during the 1970s. Achieving such a transition in the future towards lower GHG intensities is one of the major technological challenges addressed in mitigation and stabilization scenarios.

Figures 3.34 and 3.35 show the ranges of energy technology deployment across scenarios by 2030 and 2100 for baseline (non-intervention) and intervention (including stabilization) scenarios, respectively. The deployment of energy technologies in general, and of new technologies in particular, is significant indeed, even through the 2030 period, but especially by 2100. The deployment ranges should be compared with the current total global primary energy requirements of some 440 EJ in 2000. Coal, oil and gas reach median deployment levels ranging from some 150 to 250 EJ by 2030. The variation is significantly higher by 2100, but even medians reach levels of close to 600 EJ for coal in reference scenarios, thereby exceeding by 50% the current deployment of all primary energy technologies in the world. Deployment of nuclear and biomass is comparatively lower, in the range of about 50–100 EJ by 2030 and up to ten times as much by 2100. This all indicates that radical technological changes occur across the range of scenarios.

The deployment ranges are large for each of the technologies but do not differ much when comparing the pre-2001 with post-2001 scenarios over both time periods, up to 2030 and 2100. Thus, while technology deployments are large in the mean and variance, the patterns have changed little in the new (compared

221

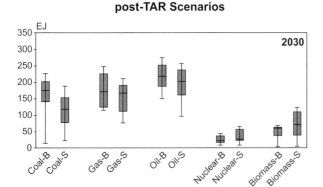

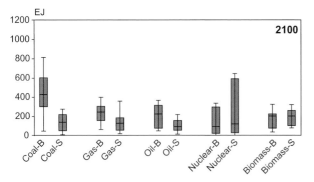

Figure 3.35: *Deployment of primary energy technologies across post-2001 scenarios by 2030 and 2100: Left-side 'error' bars show baseline (non-intervention) scenarios and right-side ones show intervention and stabilization scenarios. The full ranges of the distributions (full vertical line with two extreme tic marks), the 25th and 75th percentiles (blue area) and the median (middle tic mark) are also shown.*

with the older) scenarios. What is significant in both sets of literature is the radically different structure and portfolio of technologies between baseline and stabilization scenarios. Mitigation generally means significantly less coal, somewhat less natural gas and consistently more nuclear and biomass. What cannot be seen from this comparison, due to the lack of data and information about the scenarios, is the extent to which carbon capture and storage is deployed in mitigation scenarios. However, it is very likely that most of the coal and much of the natural gas deployment across stabilization scenarios occurs in conjunction with carbon capture and storage. The overall conclusion is that mitigation and stabilization in emissions scenarios have a significant inducement on diffusion rates of carbon-saving and zero-carbon energy technologies.

3.4.2 RD&D and investment patterns

As mentioned in Chapter 2, the private sector is leading global research and development of technologies that are close to market deployment, while public funding is essential for the longer term and basic research. R&D efforts in the energy area are especially important for GHG emissions reduction.

Accelerating the availability of advanced and new technologies will be central to greatly reducing CO_2 emissions

from energy and other sources. Innovation in energy technology will be integral to meeting the objective of emission reduction. Investment and incentives will be needed for all components of the innovation system – research and development (R&D), demonstration, market introduction and its feedback to development, flows of information and knowledge, and the scientific research that could lead to new technological advances.

Thus, sufficient investment will be required to ensure that the best technologies are brought to market in a timely manner. These investments, and the resulting deployment of new technologies, provide an economic value. Model calculations enable economists to quantify the value of improved technologies as illustrated for two technologies in Figure 3.36.

Generally, economic benefits from improved technology increase non-linearly with:
1. The distance to *current* economic characteristics (or the ones assumed to be characteristic of the scenario baseline).
2. The stringency of environmental targets.
3. The comprehensiveness and diversity of a particular technology portfolio considered in the analysis.

Thus, the larger the improvement of future technology characteristics compared to current ones, the lower the stabilization target, and the more comprehensive the suite of available technologies, the greater will be the economic value of improvements in technology.

These results lend further credence to technology R&D and deployment incentives policies (for example prices[24]) as 'hedging' strategies addressing climate change. However, given the current insufficient understanding of the complexity of driving forces underlying technological innovation and cost improvements, cost-benefit or economic 'return on investment', calculations have (to date) not been attempted in the literature, due at least in part to a paucity of empirical technology-specific data on R&D and niche-market deployment expenditures and the considerable uncertainties involved in linking 'inputs' (R&D and market stimulation costs) to 'outputs' (technology improvements and cost reductions).

3.4.3 Dynamics and drivers of technological change, barriers (timing of technology deployment, learning)

3.4.3.1 Summary from the TAR

The IPCC-TAR concluded that reduction of greenhouse gas emissions is highly dependent on both technological innovation and implementation of technologies (a conclusion broadly confirmed in Chapter 2, Section 2.7.1). However, the rate of introduction of new technologies, and the drivers for adoption are different across different parts of the world, particularly

---

24 See Newell et al., 1999.

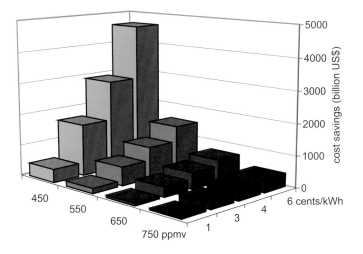

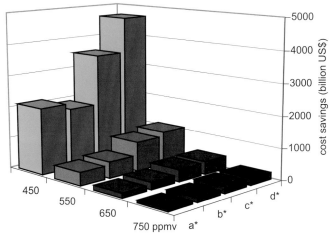

a* soil carbon sequestration only
b* central power plant with CCS
c* central power plant and hydrogen production with CCS
d* all types of carbon capture and sequestration

Figure 3.36: *The value of improved technology.*

Note: Modelling studies enable experts to calculate the economic value of technology improvements that increase particularly drastically with increasing stringency of stabilization targets (750, 650, 500, and 450 ppmv, respectively) imposed on a reference scenario (modelling after the IS92a scenario in this particular modelling study). Detailed model representation of technological interdependencies and competition and substitution is needed for a comprehensive assessment of the economic value of technology improvements. Left panel: cost savings (billions of 1996 US$) compared to the reference scenario when lowering the costs of solar photovoltaics (PV) from a reference value of 9 US cents per kWh (top) by 1, 3, 4, and 6 cents/kWh, respectively. For instance, the value of reducing PV costs from 9 to 3 cents per kWh could amount to up to 1.5 trillion US$ in an illustrative 550 ppmv stabilization scenario compared to the reference scenario in which costs remain at 9 cents/kWh). Right panel: cost savings resulting from availability of an ever larger and diversified portfolio of carbon capture and sequestration technologies. For instance, adding soil carbon sequestration to the portfolio of carbon capture and sequestration technology options (forest-sector measures were not included in the study) reduces costs by 1.1 trillion US$ in an illustrative 450 ppmv stabilization scenario. Removing all carbon capture sequestration technologies would triple the costs of stabilization for all concentration levels analyzed.

Source: GTSP, 2001.

in industrial market economies, economies in transition and developing countries. To some extent this is reflected in global emissions scenarios as they often involve technological change at a level that includes a dozen or so world regions. This usually involves making more region-specific assumptions about future performance, costs and investment needs for new and low-carbon technologies.

There are multiple policy approaches to encourage technological innovation and change. Through regulation of energy markets, environmental regulations, energy efficiency standards, financial and other market-based incentives, such as energy and emission taxes, governments can induce technology changes and influence the level of innovations. In emissions scenarios, this is reflected in assumptions about policy instruments such as taxes, emissions permits, technology standards, costs, and lower and upper boundaries of technology diffusion.

3.4.3.2   Dynamics of technology

R&D, technological learning, and spillovers are the three broad categories of drivers behind technological change. These are discussed in Chapter 2, Section 2.7, and Chapter 11, Section 11.5. The main conclusion is that, on the whole, all

three of the sources of induced technological change (ITC) play important roles in technological advance. Here, we focus on the dynamics of technology and ITC in emissions and stabilization scenarios.

Emissions scenarios generally treat technological change as an exogenous assumption about costs, market penetration and other technology characteristics, with some notable exceptions such as in Gritsevskyi and Nakicenovic (2000). Hourcade and Shukla (2001), in their review of scenarios from top-down models, indicate that technology assumptions are a critical factor that affects the timing and cost of emission abatement in the models. They identify widely differing costs of stabilization at 550 ppmv by 2050, of between 0.2 and 1.75% of GDP, mainly influenced by the size of the emissions in the baseline.

The International Modelling Comparison Project (IMCP) (Edenhofer *et al.*, 2006) compared the treatment relating to technological change in many models covering a wide range of approaches. The economies for technological change were simulated in three groups: effects through R&D expenditures, learning-by-doing (LBD) or specialization and scale. IMCP finds that ITC reduces costs of stabilization, but in a wide range, depending on the flexibility of the investment decisions and the range of mitigation options in the models. It should be noted, however, that induced technological change is not a 'free lunch', as it requires higher upfront investment and deployment of new technologies in order to achieve cost-reductions thereafter. This can lead to lower overall mitigation costs.

All models indicate that real carbon prices for stabilization targets rise with time in the early years, with some models showing a decline in the optimal price after 2050 due to the accumulated effects of LBD and positive spillovers on economic growth. Another robust result is that ITC can reduce costs when models include low carbon energy sources (such as renewables, nuclear, and carbon capture and sequestration), as well as energy efficiency and energy savings. Finally, policy uncertainty is seen as an issue. Long-term and credible abatement targets and policies will reduce some of the uncertainties around the investment decisions and are crucial to the transformation of the energy system.

ITC broadens the scope of technology-related policies and usually increases the benefits of early action, which accelerates deployment and cost-reductions of low-carbon technologies (Barker et al., 2006; Sijm, 2004; Gritsevskyi and Nakicenovic, 2000). This is due to the cumulative nature of ITC as treated in the new modelling approaches. Early deployment of costly technologies leads to learning benefits and lower costs as diffusion progresses. In contrast, scenarios with exogenous technology assumptions imply waiting for better technologies to arrive in the future, though this too may result in reduced costs of emission reduction (European Commission, 2003).

Other recent work also confirms these findings. For example, Manne and Richels (2004) and Goulder (2004) also found that ITC lowers mitigation costs and that more extensive reductions in GHGs are justified than with exogenous technical change. Nakicenovic and Riahi (2003) noted how the assumption about the availability of future technologies was a strong driver of stabilization costs. Edmonds et al. (2004a) studied stabilization at 550 ppmv CO_2 in the SRES B2 world using the MiniCAM model and showed a reduction in costs of a factor of 2.5 in 2100 using a baseline incorporating technical change. Edmonds et al. consider advanced technology development to be far more important as a driver of emission reductions than carbon taxes. Weyant (2004) concluded that stabilization will require the large-scale development of new energy technologies, and that costs would be reduced if many technologies are developed in parallel and there is early adoption of policies to encourage technology development.

The results from the bottom-up and more technology-specific modelling approaches give a different perspective. Following the work of the IIASA in particular, models investigating induced technical change emerged during the mid- and late-1990s. These models show that ITC can alter results in many ways. In the previous sections of this chapter the authors have also illustrated that the baseline choice is crucial in determining the nature (and by implication also the cost) of stabilization.

However, this influence is itself largely due to the different assumptions made about technological change in the baseline scenarios. Gritsevskyi and Nakicenovic (2000) identified some 53 clusters of least-cost technologies, allowing for endogenous technological learning with uncertainty. This suggests that a decarbonized economy may not cost any more than a carbon-intensive one, if technology learning curves are taken into account. Other key findings are that there is a large diversity across alternative energy technology strategies, a finding that was confirmed in IMCP (Edenhofer et al., 2006). These results suggest that it is not possible to choose an 'optimal' direction for energy system development. Some modelling reported in the TAR suggests that a reduction (up to 5 GtC a year) by 2020 (some 50% of baseline projections) might be achieved by current technologies, half of the reduction at no direct cost, the other half at direct costs of less than 100 US$/tC-equivalent (27 US$/tCO_2-eq).

3.4.3.3 Barriers to technology transfer, diffusion and deployment for long-term mitigation

Chapter 2, Section 2.7.2 includes a discussion of the barriers to development and commercialization of technologies. Barriers to technology transfer vary according to the specific context from sector to sector and can manifest themselves differently in developed and developing countries, and in economies-in-transition (EITs). These barriers range from a lack of information; insufficient human capabilities; political and economic barriers (such as the lack of capital, high transaction costs, lack of full cost pricing, and trade and policy barriers); institutional and structural barriers; lack of understanding of local needs; business limitations (such as risk aversion in financial institutions); institutional limitations (such as insufficient legal protection); and inadequate environmental codes and standards.

3.4.3.4 Dynamics in developing countries and timing of technology deployment

National policies in developing countries necessarily focus on more fundamental priorities of development, such as poverty alleviation and providing basic living conditions for their populations, and it is unlikely that short-term national policies would be driven by environmental concerns. National policies driven by energy security concerns can, however, have strong alignment with climate goals. The success of policies that address short-term development concerns will determine the pace at which the quality of life in the developing and the developed world converges over the long term.

In the long term, the key drivers of technological change in developing countries will depend on three 'changes' that are simultaneous and inseparable within the context of development: exogenous behavioural changes or changes in social infrastructure; endogenous policies driven by 'development goals'; and any induced change from climate policies (Shukla et al., 2006).

3.5 Interaction between mitigation and adaptation, in the light of climate change impacts and decision-making under long-term uncertainty

3.5.1 The interaction between mitigation and adaptation, an iterative climate policy process

Responses to climate change include a portfolio of measures:
a. Mitigation – actions that reduce net carbon emissions and limit long-term climate change.
b. Adaptation – actions that help human and natural systems to adjust to climate change.
c. Research on new technologies, on institutional designs and on climate and impacts science, which should reduce uncertainties and facilitate future decisions (Richels *et al.*, 2004; Caldeira *et al.*, 2003; Yohe *et al.*, 2004).

A key question for policy is what combination of short-term and long-term actions will minimize the total costs of climate change, in whatever form these costs are expressed, across mitigation, adaptation and the residual climate impacts that society is either prepared or forced to tolerate. Although there are different views on the form and dynamics of such trade-offs in climate policies, there is a consensus that they should be aligned with (sustainable) development policies, since the latter determine the capacity to mitigate and to adapt in the future (TAR, Hourcade and Shukla, 2001). In all cases, policy decisions will have to be made with incomplete understanding of the magnitude and timing of climate change, of its likely consequences, and of the cost and effectiveness of response measures.

3.5.1.1 An iterative risk-management framework to articulate options

Previous IPCC reports conclude that climate change decision-making is not a once-and-for-all event, but an iterative risk-management process that is likely to take place over decades, where there will be opportunities for learning and mid-course corrections in the light of new information (Lempert *et al.*, 1994; Keller *et al.*, 2006).

This iterative process can be described using a decision tree (Figure 3.37), where the square nodes represent decisions, the circles represent the reduction of uncertainty and the arrows indicate the range of decisions and outcomes. Some nodes summarize today's options – how much should be invested in mitigation, in adaptation, in expanding mitigative and adaptive capacity, or in research to reduce uncertainty? Other nodes represent opportunities to learn and make mid-course corrections. This picture is a caricature of real decision processes, which are continuous, overlapping and iterative.

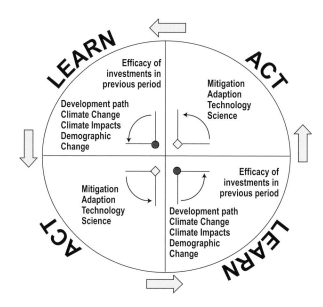

Figure 3.37: *The Iterative Nature of the Climate Policy Process.*

However, it is useful to conceptually put the many determinants of any short-term strategy in a context of progressive resolution of uncertainty.

3.5.1.2 Qualitative insights into interactions between mitigaton, adaptation and development

Until recently, a main focus in the policy and integrated assessment literature has been on comparing mitigation costs and avoided damages. Since the TAR, attention has shifted towards the interaction between mitigation and adaptation in reducing damages in a risk-management framework. This has accompanied a growing realization that some climate change in the coming decades is inevitable.

Limited treatment of adaptation in climate policy assessments is still a problem and a number of reasons explain this. First, the focus of the international climate change negotiations has largely been on mitigation (perhaps because attention to adaptation could be viewed as 'giving up' on mitigation) even though the importance of adaptation is underlined in Article 4 of the UNFCCC and Article 10 of the Kyoto Protocol. Second, adaptation is largely undertaken at the local scale, by individual households, farmers, companies or local governments; it is thus difficult to target through coordinated international incentives, and is more complicated to handle quantitatively by models in global scenarios. Third, it is difficult to generalize the ways that individuals or communities are likely to adapt to specific impacts. However, the literature is evolving quickly and recent work is available in a number of regions; for example, in Finland (Carter *et al.*, 2005), the UK (West and Gawith, 2005), Canada (Cohen *et al.*, 2004) and the USA (e.g. California, Hayhoe *et al.*, 2004).

Despite the scarcity of global systematic assessments (Tol, 2005a), some interesting insights into the interaction between adaptation and mitigation emerge from recent regional-scale studies. Some adaptation measures are 'no-regret' measures and should be undertaken anyway (Agrawala, 2005), such as preservation of mangroves in coastal zones, which provide a buffer for increased coastal flood risk due to climate change and help to maintain healthy marine ecosystems (Nicholls *et al.*, 2006). A few may be synergistic with mitigation (Bosello, 2005) such as investing in more efficient buildings that will limit human vulnerability to increasingly frequent heatwaves and also reduce energy use, hence emissions. But many adaptation options involve net costs with a risk of committing to irreversible and misplaced investment given the considerable uncertainty about climate change at a local scale. Given this uncertainty, and the fact that learning about adaptation to climate change imposes some costs and takes time (Kelly *et al.*, 2005), mis-allocation of investments may occur, or the rate of long-term investment in adaptation strategies may slow (Kokic *et al.*, 2005; Kelly *et al.*, 2005).

Finally, the interactions between adaptation and mitigation are intertwined with development pathways. A key issue is to understand at what point (over)investment in mitigation or adaptation might limit funds available for development, and thus reduce future adaptive capacity (Sachs, 2004; Tol, 2005a; Tol and Yohe, 2006). Another issue concerns the point at which climate change damages, and the associated investment in adaptation, could crowd out more productive investments later and harm development (Kemfert, 2002; Bosello and Zhang, 2005; Kemfert and Schumacher, 2005). The answer to these questions depends upon modelling assumptions that drive repercussions in other sectors of the economy and other regions and the potential impacts on economic growth. These are 'higher-order' social costs of climate change from a series of climate-change-induced shocks; they include the relative influence of: a) the cross-sectoral interactions across all major sectors and regions; b) a crowding out effect that slows down capital accumulation and technical progress, especially if technical change is endogenous. These indirect impacts reduce development and adaptive capacity and may be in the same order of magnitude, or greater than, the direct impact of climate change (Fankhauser and Tol, 2005; Roson and Tol, 2006; Kemfert, 2006).

Both the magnitude and the sign of the indirect macro-economic impacts of climate change are conditional upon the growth dynamics of the countries concerned. When confronted by the same mitigation policies and the same climate change impacts, economies experiencing strong disequilibrium (including 'poverty traps') and large market and institutional imperfections will not react in the same way as countries that are on a steady and high economic growth pathway. The latter are near what economists call their 'production frontier' (the maximum of production attainable at a given point in time); the former are more vulnerable to any climatic shock or badly

calibrated mitigation policies, but symmetrically offer more opportunities for synergies between mitigation, adaptation and development policies (Shukla *et al.*, 2006). On the adaptation side for example, Tol and Dowlatabadi (2001) demonstrate that there is significant potential to reduce vulnerability to the spread of malaria in Africa. In some circumstances, mitigation measures can be aligned with development policies and alleviate important sources of vulnerability in these countries, such as dependency on oil imports or local pollution. But this involves transition costs over the coming 10–20 years (higher domestic energy prices, higher investments in the energy sector), which in turn suggests opportunities for international cooperative mechanisms to minimize these costs.

Bosello (2005) shows complementarity between adaptation, mitigation and investment in R&D, whilst others consider these as substitutes (Tol, 2005a). Schneider and Lane (2006) consider that mitigation and adaptation only trade off for small temperature increments where adaptation might be cheaper, whereas for larger temperature increases mitigation is always the cheaper option. Goklany (2003) promotes the view that the contribution by climate change to hunger, malaria, coastal flooding, and water stress (as measured by populations at risk) is small compared to that of non-climate-change-related factors, and that through the 2080s, efforts to reduce vulnerability would be more cost-effective in reducing these problems than mitigation. This analysis neglects critical thresholds at the regional level (such as the temperature ceiling on feasibility of regional crop growth) and at the global level (such as the onset of ice sheet melting or release of methane from permafrost), and, like many others studies, it neglects the impacts of extreme weather events. It also promotes a very optimistic view of adaptive capacity, which is increasingly challenged in the literature (Tompkins and Adger, 2005). An adaptation-only policy scenario in the coming decades leads to an even greater challenge for adaptation in decades to follow, owing to the inertia of the climate system. In the absence of mitigation, temperature rises will be much greater than would otherwise occur with pusuant impacts on economic development (IPCC, 2007b, Chapter 19.3.7; Stern 2006). Hence adaptation alone is insufficient to avoid the serious risks due to climate change (see Table 3.11; also IPCC, 2007b, Chapter 19, Table 19.1).

To summarize, adaptation and mitigation are thus increasingly viewed as complementary (on the global scale), whilst locally there are examples of both synergies and conflicts between the two (IPCC, 2007b, Chapter 18). Less action on mitigation raises the risk of greater climate-change-induced damages to economic development and natural systems and implies a greater need for adaptation. Some authors maintain that adaptation and mitigation are substitutes, because of competition for funds, whilst others claim that such tradeoffs occur only at the margin when considering incremental temperature change and incremental policy action, because for large temperature changes mitigation is always cheaper than adaptation.

Table 3.9: *Global mean temperature increase at equilibrium, greenhouse gas concentration and radiative forcing. Equilibrium temperatures here are calculated using estimates of climate sensitivity and do not take into account the full range of bio-geophysical feedbacks that may occur.*

Equilibrium temperature increase in °C above pre-industrial temperature	CO_2-eq concentration and radiative forcing corresponding to best estimate of climate sensitivity for warming level in column 1[1,2]		CO_2-eq concentration that would be expected to limit warming below level in column 1 with an estimated likelihood of about 80% [3]
	CO_2-equivalent (ppm)	Radiative forcing (W/m²)	
0.6	319	0.7	305
1.6	402	2.0	356
2.0	441	2.5	378
2.6	507	3.2	415
3.0	556	3.7	441
3.6	639	4.5	484
4.0	701	4.9	515
4.6	805	5.7	565
5.0	883	6.2	601
5.6	1014	6.9	659
6.0	1112	7.4	701
6.6	1277	8.2	768

Note: see Figure 3.38 on page 228 for footnotes.

3.5.2 Linking emission scenarios to changes in global mean temperature, impacts and key vulnerabilities

In a risk-management framework, a first step to understanding the environmental consequences of mitigation strategies is to look at links between various stabilization levels for concentrations of greenhouse gases in the atmosphere, and the global mean temperature change relative to a particular baseline. A second step is to link levels of temperature change and key vulnerabilities. Climate models indicate significant uncertainty at both levels. Figure 3.38 shows CO_2-eq concentrations that would limit warming at equilibrium below the temperatures indicated above pre-industrial levels, for 'best estimate' climate sensitivity, and for the likely range of climate sensitivity (see Meehl *et al.*, 2007, Section 10.7, and Table 10.8; and the notes to Figure 3.38). It also shows the corresponding radiative forcing levels and their relationship to equilibrium temperature and CO_2-eq concentrations. The table and the figure illustrate how lower temperature constraints require lower stabilization levels, and also that, if the potential for climate sensitivities is higher than the 'best estimate' and is taken into account, the constraint becomes more stringent. These more stringent constraints lower the risks of exceeding the threshold.

Figure 3.38 and Table 3.10 provide an overview of how emission scenarios (Section 3.3) relate to different stabilization targets and to the likelihood of staying below certain equilibrium warming levels. For example, respecting constraints of 2°C above pre-industrial levels, at equilibrium, is already outside the range of scenarios considered in this chapter, if the higher values of likely climate sensitivity are taken into account (red curve in Figure 3.38), whilst a constraint of respecting 3°C above pre-industrial levels implies the most stringent of the category I scenarios, with emissions peaking in no more than the next 10 years, again if the higher likely values of climate sensitivity are taken into account. Using the 'best estimate' of climate sensitivity (i.e. the estimated mode) as a guide for establishing targets, implies the need for less stringent emission constraints. This 'best estimate' assumption shows that the most stringent (category I) scenarios could limit global mean temperature increases to 2°C–2.4°C above pre-industrial levels, at equilibrium, requiring emissions to peak within 10 years. Similarly, limiting temperature increases to 2°C above pre-industrial levels can only be reached at the lowest end of the concentration interval found in the scenarios of category I (i.e. about 450 ppmv CO_2-eq using 'best estimate' assumptions). By comparison, using the same 'best estimate' assumptions, category II scenarios could limit the increase to 2.8°C–3.2°C above pre-industrial levels at equilibrium, requiring emissions to peak within the next 25 years, whilst category IV scenarios could limit the increase to 3.2°C–4°C above pre-industrial at equilibrium requiring emissions to peak within the next 55 years. Note that Table 3.10 category IV scenarios could result in temperature increases as high as 6.1°C above pre-industrial levels, when the likely range for the value of climate sensitivity is taken into account. Hence, setting policy on the basis of a 'best estimate' climate sensitivity accepts a significant risk of exceeding the temperature thresholds, since the climate sensitivity could be higher than the best estimate.

Table 3.11 highlights a number of climate change impacts and key vulnerabilities organized as a function of global mean temperature rise (IPCC, 2007b, Chapter 19). The table highlights a selection of key vulnerabilities representative of categories covered in Chapter 19 (Table 19.1) in IPCC (2007b).

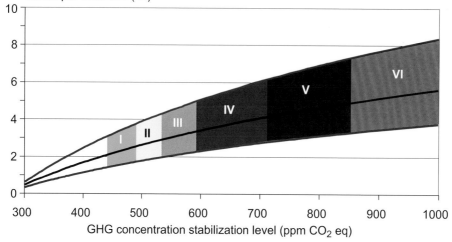

Figure 3.38: *Relationship between global mean equilibrium temperature change and stabilization concentration of greenhouse gases using: (i) 'best estimate' climate sensitivity of 3°C (black), (ii) upper boundary of likely range of climate sensitivity of 4.5°C (red), (iii) lower boundary of likely range of climate sensitivity of 2°C (blue) (see also Table 3.9).*

Notes:

1. IPCC (2007a) finds that the climate sensitivity is likely to be in the range 2°C –4.5°C, with a 'best estimate' of about 3°C, very unlikely to be less than 1.5°C and values substantially higher than 4.5°C 'cannot be excluded' (IPCC (2007a, SPM).

2. The simple relationship $T_{eq} = T_{2xCO_2} \times \ln([CO_2]/280)/\ln(2)$ is used (see Meehl *et al.* (2007), Section 10.7, and Table 10.8), with upper and lower values of T_{2xCO_2} of 2 and 4.5°C.

3. Non-linearities in the feedbacks (including e.g. ice cover and carbon cycle) may cause time dependence of the effective climate sensitivity, as well as leading to larger uncertainties for greater warming levels. This likelihood level is consistent with the IPCC Working Group I assessment of climate sensitivity, see Note 1, and drawn from additional consideration of Box 10.2, Figure 2, in IPCC (2007a).

The *italic text* in Table 3.11 highlights *examples* of avoided impacts derived from ensuring that temperatures are constrained to any particular temperature range compared to a higher one. For example, significant benefits result from constraining temperature change to not more than 1.6°C–2.6°C above pre-industrial levels. These benefits would include lowering (with different levels of confidence) the risk of: widespread deglaciation of the Greenland Ice Sheet; avoiding large-scale transformation of ecosystems and degradation of coral reefs; preventing terrestrial vegetation becoming a carbon source; constraining species extinction to between 10–40%; preserving many unique habitats (see IPCC, 2007b, Chapter 4, Table 4.1 and Figure 4.5) including much of the Arctic; reducing increases in flooding, drought, and fire; reducing water quality declines, and preventing global net declines in food production. Other benefits of this constraint, not shown in the Table 3.11, include reducing the risks of extreme weather events, and of at least partial deglaciation of the West Antarctic Ice Sheet (WAIS), see also IPCC, 2007b, Section 19.3.7. By comparison, for 'best guess' climate sensitivity, attaining these benefits becomes unlikely if emission reductions are postponed beyond the next 15 years to a time period between the next 15–55 years. Such postponement also results in increasing risks of a breakdown of the Meridional Overturning Circulation (IPCC, 2007b, Table 19.1).

Even for a 2.6°C –3.6°C temperature rise above pre-industrial levels there is also medium confidence in net negative impacts in many developed countries (IPCC, 2007b, Section 19.3.7). For emission-reduction scenarios resulting in likely temperature increases in excess of 3.6°C above pre-industrial levels, successively more severe impacts result. Low temperature constraints are necessary to avoid significant increases in the impacts in less developed regions of the world and in polar regions, since many market sectors in developing countries are already affected below 2.6°C above pre-industrial levels (IPCC, 2007b, Section 19.3.7), and indigenous populations in high latitude areas already face significant adverse impacts.

It is possible to use stablization metrics (i.e. global mean temperature increase, concentrations in ppmv CO_2-eq or radiative forcing in W/m2) in combination with the mitigation scenarios literature to assess the cost of alternative mitigation pathways that respect a given equilibrium temperature, key vulnerability (KV) or impact threshold. Whatever the target, both early and delayed-action mitigation pathways are possible, including 'overshoot' pathways that temporarily exceed this level. A delayed mitigation response leads to lower discounted costs of mitigation, but accelerates the rate of change and the risk of transiently overshooting pre-determined targets (IPCC, 2007b, Section 19.4.2).

A strict comparison between mitigation scenarios and KVs is not feasible as the KVs in Table 3.11 refer to realized transient temperatures in the 21st century rather than equilibrium temperatures, but a less rigorous comparison is still useful. Avoidance of many KVs requires temperature change in 2100 to be below 2°C above 1990 levels (or 2.6°C above pre-industrial levels). Using equilibrium temperature as a guide, impacts or KV could be less than expected, for example if impacts

Table 3.10: *Properties of emissions pathways for alternative ranges of CO_2 and CO_2-eq stabilization targets. Post-TAR stabilization scenarios in the scenario database (see also Sections 3.2 and 3.3); data source: after Nakicenovic et al., 2006 and Hanaoka et al., 2006)*

Class	Anthropogenic addition to radiative forcing at stabilization (W/m²)	Multi-gas concentration level (ppmv CO_2-eq)	Stabilization level for CO_2 only, consistent with multi-gas level (ppmv CO_2)	Number of scenario studies	Global mean temperature C increase above pre-industrial at equilibrium, using best estimate of climate sensitivity[c]	Likely range of global mean temperature C increase above pre-industrial at equilibrium[a]	Peaking year for CO_2 emissions[b]	Change in global emissions in 2050 (% of 2000 emissions)[b]
I	2.5-3.0	445-490	350-400	6	2.0-2.4	1.4-3.6	2000-2015	-85 to -50
II	3.0-3.5	490-535	400-440	18	2.4-2.8	1.6-4.2	2000-2020	-60 to -30
III	3.5-4.0	535-590	440-485	21	2.8-3.2	1.9-4.9	2010-2030	-30 to +5
IV	4.0-5.0	590-710	485-570	118	3.2-4.0	2.2-6.1	2020-2060	+10 to +60
V	5.0-6.0	710-855	570-660	9	4.0-4.9	2.7-7.3	2050-2080	+25 to +85
VI	6.0-7.5	855-1130	660-790	5	4.9-6.1	3.2-8.5	2060-2090	+90 to +140

Notes:
a. Warming for each stabilization class is calculated based on the variation of climate sensitivity between 2°C –4.5°C, which corresponds to the likely range of climate sensitivity as defined by Meehl *et al.* (2007,Chapter 10).
b. Ranges correspond to the 70% percentile of the post-TAR scenario distribution.
c. 'Best estimate' refers to the most likely value of climate sensitivity, i.e. the mode (see Meehl *et al.* (2007,Chapter 10) and Table 3.9

do not occur until the 22nd century, because there is more time for adaptation. Or they might be greater than expected, as temperatures in the 21st century may transiently overshoot the equilibrium, or stocks at risk (such as human populations) might be larger. Some studies explore the link between transient and equilibrium temperature change for alternative emission pathways (O'Neill and Oppenheimer, 2004; Schneider and Mastrandrea, 2005; Meinshausen, 2006).

It is transient climate change, rather than equilibrium change, that will drive impacts. More research is required to address the question of emission pathways and transient climate changes and their links to impacts.[25] In the meantime, equilibrium temperature change may be interpreted as a gross indicator of change, and given the caveats above, as a rough guide for policymakers' consideration of KV and mitigation options to avoid KV.

3.5.3   Information for integrated assessment of response strategies

Based upon a better understanding of the links between concentration levels, magnitude and rate of warming and key vulnerabilities, the next step in integrated assessment is to make informed decisions by combining information on climate science, impact analysis and economic analysis within a consistent analytical framework. These exercises can be grouped into three main categories depending on the way uncertainty is dealt with, the degree of complexity and multi-disciplinary nature of models and on the degree of ambition in terms of normative insights:
1. Assessment and sensitivity analysis of climate targets.
2. Inverse analyses to determine emission-reduction corridors (trajectories) to avoid certain levels of climate change or of climate impacts.

3. Monetary assessment of climate change damages.

Section 3.6 discusses how this information is used in economic analyses to determine optimal emission pathways.

3.5.3.1   *Scenario and sensitivity analysis of climate targets*

Probabilistic scenario analysis can be used to assess the risk of overshooting some climate target or to produce probabilistic projections that quantify the likelihood of a particular outcome. Targets for such analysis can be expressed in several different ways: absolute global mean temperature rise by 2100, rate of climate change, other thresholds beyond which dangerous anthropogenic interference (DAI) may occur, or additional numbers of people at risk to various stresses. For example, Arnell *et al.* (2002) show that such stresses (conversion of forests to grasslands, coastal flood risk, water stress) are far less at 550 ppmv than at 750 ppmv.

Recent Integrated Assessment Models (IAM) literature reflects a renewed attention to climate sensitivity as a key driver of climate dynamics (Den Elzen and Meinshausen, 2006; Hare and Meinshausen, 2006; Harvey, 2006; Keller *et al.*, 2006; Mastrandrea and Schneider, 2004; Meehl *et al.*, 2005; Meinshausen *et al.*, 2006, Meinshausen, 2006; O'Neill and Oppeinheimer, 2002, 2004; Schneider and Lane, 2004; Wigley, 2005). The consideration of a full range of possible climate sensitivity increases the probability of exceeding thresholds for specific DAI. It also magnifies the consequence of delaying mitigation efforts. Hare and Meinshausen (2006) estimate that each 10-year delay in mitigation implies an additional 0.2°C–0.3°C warming over a 100–400 year time horizon. For a climate sensitivity of 3°C, Harvey (2006) shows that immediate mitigation is required to constrain temperature rise to roughly

---

25 See IPCC (2007b, Section 19.4, Figure 19.2) and Meehl *et al.* (2007, Section 10.7) for further discussion of equilibrium and transient temperature increases in relation to stabilization pathways

Table 3.11: *Examples of key vulnerabilities (taken from IPCC, 2007b, Table 19.1).*

GMT range relative to 1990 (pre-industrial)	Geophysical systems Example: Greenland ice sheet[a] (IPCC, 2007b: 6.3; 19.3.5.2; IPCC, 2007a: 4.7.4; 6.4.3.3; 10.7.4.3; 10.7.4.4)	Global biological systems Example: terrestrial ecosystems[b] (IPCC, 2007b: 4.4.11; 1.3.4; 1.3.5)	Global social systems Example: water[c] (IPCC, 2007b: 3 ES; 3.4.3; 13.4.3)	Global social systems Example: food supply[c] (IPCC, 2007b: 5.6.1; 5.6.4)	Regional systems Example: Polar Regions[d] (IPCC, 2007b: 15.4.1; 15.4.2; 15.4.6; 15.4.7)	Extreme events Example: fire risk[e] (IPCC, 2007a: 7.3; IPCC, 2007b: 1.3.6)
>4 (>4-6)	Near-total deglaciation**	Large-scale transformation of ecosystems and ecosystem services** At least 35% of species committed to extinction (3°C)**	Severity of floods, droughts, erosion, water quality deterioration will increase with increasing climate change***	Further declines in global food production o/*	Continued warming likely to lead to further loss of ice cover and permafrost**. Arctic ecosystems further threatened**, although net ecosystem productivity estimated to increase (o)	Frequency and intensity likely to be greater, especially in boreal forests and dry peat lands after melting of permafrost*
3-4 (3.6-4.6)	Commitment to widespread* to near-total deglaciation* 2-7 m sea level rise over centuries to millennia	Global vegetation becomes net source of C above 2-3°C */**	Sea level rise will extend areas of salinization of ground water, decreasing freshwater availability in coastal areas****		While some economic opportunities will open up (e.g. shipping), traditional ways of life will be disrupted**	
2-3 (2.6-3.6)	*Lowers risk of near-total deglaciation*	Widespread disturbance, sensitive to rate of climate change and land use*** 20 to 50% species committed to extinction* *Avoids widespread disturbance to ecosystems and their services***, and constrains species losses*	Hundreds of millions people would face reduced water supplies (**)	Global food production peaks and begins to decrease o/* (1-3°C) *Lowers risk of further declines in global food production associated with higher temperatures*		
1-2 (1.6-2.6)	Localized deglaciation (already observed due to local warming), extent would increase with temperature**	10-40% of species committed to extinction* *Reduces extinctions to below 20-50%; prevents vegetation becoming carbon source*/** Many ecosystems already affected***	Increased flooding and drought severity** *Lowers risk of floods, droughts, deteriorating water quality*** and reduced water supplied for hundreds of millions of people**	Reduced low latitude production*. Increased high latitude production* (1-3°C)	Climate change is already having substantial impacts on societal and ecological systems***	Increased fire frequency and intensity in many areas, particularly where drought increases**
0-1 (0.6-1.6)	*Lowers risk of widespread** to near-total deglaciation*	*Reduces extinctions to below 10-30%; reduces disturbance levels****		*Increased global production o/* Lowers risk of decrease in global food production and reduces regional losses (or gains) o/*	*Reduced loss of ice cover and permafrost; limits risk to Arctic ecosystems and limits disruption of traditional ways of life**	*Lowers risk of more frequent and more intense fires in many areas**

Notes:
Plain text shows predicted vulnerabilities in various temperature ranges for global annual mean temperature rise relative to 1990. Italic text shows benefits (or damage avoided) upon constraining temperature increase to lower compared to higher temperature ranges.
Confidence symbol legend: o low confidence; * medium confidence; ** high confidence; *** very high confidence
Excerpts from IPCC, 2007b, Table 19.1;
a. Refer to IPCC (2007b, Table 19.1) for further information, also concerning bio-geochemical cycles, West Antarctic Ice Sheet and Meridional Overturning Circulation
b. Refer to IPCC (2007b, Table 19.1) for further information, also concerning marine and freshwater ecosystems
c. Refer to IPCC (2007b, Table 19.1) for further information, also concerning infrastructure, health, migration and conflict, and aggregate market impacts
d. Refer to IPCC (2007b, Table 19.1) for further information and also for other regions
e. Refer to IPCC (2007b, Table 19.1) for further information, also concerning tropical cyclones, flooding, extreme heat, and drought

2°C above pre-industrial levels. Only in the unlikely situation where climate sensitivity is 1°C or lower would immediate mitigation not be necessary.[26] Harvey also points out that, even in the case of a 2°C threshold (above pre-industrial levels), acidification of the ocean would still occur and that this might not be considered safe.

Another focus of sensitivity analysis is on mitigation scenarios that overshoot and eventually return to a given stabilization or temperature target (Kheshgi, 2004; Wigley, 2005; Harvey, 2004; Izrael and Semenov, 2005; Kheshgi et al., 2005; Meinshausen et al., 2006). Schneider and Mastrandrea (2005) find that this risk of exceeding a threshold of 2°C above pre-industrial levels is increased by 70% for an overshoot scenario stabilizing at 500 ppmv CO_2-eq (as compared to a scenario stabilizing at 500 ppmv CO_2-eq). Such overshoot scenarios are likely to be necessary if there is a decision to achieve stablization of GHG concentrations close to (or at) today's levels. They are indeed likely to lower the costs of mitigation but, in turn, raise the risk of exceeding such thresholds (Keller et al., 2006; Schneider and Lane, 2004) and may limit the ability to adapt by increasing the rate of climate change, at least temporarily (Hare and Meinshausen, 2006). O'Neill and Oppenheimer (2004) find that the transient temperature up to 2100 is equally, or more, controlled by the pathway to stabilization than by the stabilization target, and that overshooting can lead to a peak temperature increase that is higher than in the long-term (equilibrium) warming.

The last and important contribution of this approach is to test the sensitivity of results to carbon cycle and climate change feedbacks (Cox et al., 2000; Friedlingstein et al., 2001; Matthews, 2005) and other factors that may affect carbon cycle dynamics, such as deforestation (Gitz and Ciais, 2003). For example, carbon cycle feedbacks amplify warming (Meehl et al., 2007) and are omitted from most other studies that thus underestimate the risks of exceeding (or overshooting) temperature targets for a given effort of mitigation in the energy sector only. This could increase warming by up to 1°C in 2100, according to a simple model (Meehl et al., 2007). The amplification, together with further potential amplification due to feedbacks of uncertain magnitude, such as the potential release of methane from permafrost, peat bogs and seafloor clathrates (Meehl et al., 2007) are also not included in the analysis presented in Figure 3.38 and Table 3.10. This analysis reflects only known feedbacks for which the magnitude can be estimated and are included in General Circulation Models (GCMs). Hence, scenario and sensitivity analysis shows that the risks of exceeding a given temperature threshold for a given temperature target may be higher than that shown in Table 3.10 and Figure 3.38.

3.5.3.2   Inverse modelling and guardrail analysis

Inverse modelling approaches such as Safe Landing Analysis (Swart et al., 1998) and Tolerable Windows Approach (Toth,

2003), aim to define a guardrail of allowable emissions for sets of unacceptable impacts or intolerable mitigation costs. They explore how the set of viable emissions pathways is constrained by parameters such as the starting date, the rate of emission reductions, or the environmental constraints. They provide insights into the influence of short-term decisions on long-term targets by delineating allowable emissions corridor, but they do not prescribe unique emissions pathways, as per cost-effectiveness or costs-benefit analysis.

For example, Toth et al. (2002) draw on climate impact response functions (CIRFs) by Füssel and van Minnen (2001) that use detailed biophysical models to estimate regionally specific, non-monetized impacts for different sectors (i.e. agricultural production, forestry, water runoff and biome changes). They show that the business-as-usual scenario of GHG emissions (which resembles the SRES A2 scenario) to 2040 precludes the possibility of limiting the worldwide transformation of ecosystems to 30% or less, even with very high willingness to pay for the mitigation of GHG emissions afterwards. Some applications of guardrail analyses assess the relationship between emission pathways and abrupt change such as thermohaline circulation (THC) collapse (Rahmstorf and Zickfeld, 2005). The latter study concludes that stringent mitigation policy reduces the probability of THC collapse but cannot entirely avoid the risk of shutdown.

Corfee-Morlot and Höhne (2003) conclude that only low stabilization targets (e.g. 450 ppmv CO_2 or 550 ppmv CO_2-eq) significantly reduce the likelihood of climate change impacts. They use an inverse analysis to conclude that more than half of the SRES (baseline) emission scenarios leave this objective virtually out of reach as of 2020.

More generally, referring to Table 3.10, if the peaking of global emissions is postponed beyond the next 15 years to a time period somewhere between the next 15–55 years, then constraining global temperature rise to below 2°C above 1990 (2.6°C above pre-industrial levels) becomes unlikely (using 'best estimate' assumptions of climate sensitivity), resulting in increased risks of the impacts listed in Table 3.11 and discussed in Section 3.5.2.

3.5.3.3   Cost-benefit analysis, damage cost estimates and social costs of carbon

The above analysis provides a means of eliminating those emissions scenarios that are outside sets of pre-determined guardrails for climate protection and provides the raw material for cost-effectiveness analysis of optimal pathways for GHG emissions. If one wants to determine these pathways through a cost-benefit analysis it is necessary to assess the trade-off between mitigation, adaptation and damages, and consequently, to measure damages in the same monetary metric as mitigation

---

26 This is below the range accepted by IPCC Working Group I.

and adaptation expenditures. Such assessment can be carried out directly in the form of 'willingness to pay for' avoiding certain physical consequences.

Some argue that it is necessary to specify more precisely why certain impacts are undesirable and to comprehensively itemize the economic consequences of climate change in monetary terms. The credibility of such efforts has often been questioned, given the uncertainty surrounding climate impacts and the efficacy of societal responses to them, plus the controversial meaning of a monetary metric across different regions and generations (Jacoby, 2004). This explains why few economists have taken the step of monetizing global climate impacts. At the time of the TAR, only three such comprehensive studies had been published (Mendelsohn *et al.*, 2000; Nordhaus and Boyer, 2000; and Tol, 2002a, 2002b). Their estimates ranged from negligible to 1.5% of the GDP for a global mean temperature rise of +2.5°C and Nordhaus and Boyer carefully warned: 'Along the economically efficient emission path, the long-run global average temperature after 500 years is projected to increase 6.2°C over the 1900 global climate. While we have only the foggiest idea of what this would imply in terms of ecological, economic, and social outcomes, it would make the most thoughtful people, even economists, nervous to induce such a large environmental change. Given the potential for unintended and potentially disastrous consequences....'

Progress has been made since the TAR in assessing the impacts of climate change. Nonetheless, as noted in Watkiss *et al.* (2005), estimates of the social costs of carbon (SCC) in the recent literature still reflect an incomplete subset of relevant impacts; many significant impacts have not yet been monetized (see also IPCC, 2007b; for SCC see IPCC (2007b, Section 20.6) and others are calibrated in numeraires that may defy monetization for some time to come. Existing reviews of available SCC estimates show that they span several orders of magnitude – ranges that reflect uncertainties in climate sensitivity, response lags, discount rates, the treatment of equity, the valuation of economic and non-economic impacts, and the treatment of possible catastrophic losses (IPCC, 2007b, Chapter 20). The majority of available estimates in the literature also capture only impacts driven by lower levels of climate change (e.g. 3°C above 1990 levels). IPCC (2007b) highlights available estimates of SCC that run from -3 to 95 US$ /tCO$_2$ from one survey, but also note that another survey includes a few estimates as high as 400 US$/tCO$_2$ (IPCC, 2007b, Chapter 20, ES and Section 20.6.1). However the lower boundary of this range includes studies where climate change is presumed to be low and aggregate benefits accrue. Moreover, none of the aggregate estimates reflect the significant differences in impacts that will be felt across different regions; nor do they capture any of the social costs of other greenhouse gases. A more recent estimate by Stern (2006) is at the high end of these estimates (at 85 US$/tCO$_2$) because an extremely low discount rate (of 1.4%) is used in calculating damages that include additional costs attributed to abrupt change and increases in global mean temperature for

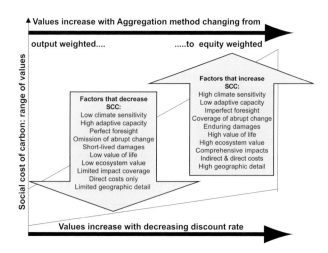

Figure 3.39: *Factors influencing the social costs of carbon.*
Source: Downing et al., 2005

some scenarios in excess of 7°C (Nordhaus, 2006a; Yohe, 2006; Tol and Yohe, 2006). The long-term high-temperature scenarios are due to inclusion of feedback processes. IPCC (2007b) also highlights the fact that the social costs of carbon and other greenhouse gases could increase over time by 2–4% per year (IPCC, 2007b; Chapter 20, ES and Section 20.6.1).

For a given level of climate change, the discrepancies in estimates of the social costs of carbon can be explained by a number of parameters highlighted in Figure 3.39. These stem from two different types of questions: normative and empirical. Key normative parameters include the inter-temporal aggregation of damages through discount rates and aggregation methods for impacts across diverse populations within the same time period (Azar and Lindgren, 2003; Howarth, 2003; Mastrandrea and Schneider, 2004) and are responsible for much of the variation.

The other parameters relate to the empirical validity of their assessment, given the poor quality of data and the difficulty of predicting how society will react to climate impacts in a given sector, at a given scale in future decades. Pearce (2003) suggests that climate damages and SCC may be over-estimated due to the omission of possible amenity benefits in warmer climates or high-latitude regions (Maddison 2001) and possible agricultural benefits. However, overall, it is likely that current SCC estimates are understated due to the omission of significant impacts that have not yet been monetized (IPCC, 2007b, Chapters 19 and 20; Watkiss *et al.*, 2005).

Key empirical parameters that increase the social value of damages include:
- *Climate sensitivity and response lag*. Equilibrium temperature rise for a doubling of CO_2, and the modelled response time of climate to such a change in forcing. Hope (2006) in his PAGE 2002 model found that, as climate sensitivity was varied from 1.5°–5°C, the model identified a strong correlation with SCC.

- *Coverage of abrupt or catastrophic changes*, such as the crossing of the THC threshold (Keller *et al.*, 2000 and 2004; Mastrandrea and Schneider, 2001; Hall and Behl, 2006) or the release of methane from permafrost and the weakening of carbon sinks. The Stern Review (2006) finds that such abrupt changes may more than double the market damages (e.g. from 2.1% to 5% of global GDP) if temperatures were to rise by 7.4°C in 2200.

- *Inclusion and social value of non-market impacts*: what value will future generations place on impacts, such as the quality of landscape or biodiversity?

- *Valuation methods for market impacts such as the value of life.*

- *Adaptative capacity*: social costs will be magnified if climate change impacts fall on fragile economies.

- *Predictive capacity*: studies finding efficient adaptation assume that actors decide using perfect foresight (after a learning process; see Mendelsohn and Williams, 2004). Higher costs are found if one considers the volatility of climate signals and transaction costs. For agriculture, Parry *et al.* (2004) shows the costs of a mismatch between expectations and real climate change (sunk costs, value of real estates, and of capital stock).

- *Geographic downscaling*: using a geographic-economic cross-sectional (1990) database, Nordhaus (2006a) concludes that this downscaling leads to increased damage costs, from previous 0.7% estimates to 3% of world output for a 3°C increase in global mean temperature.

- *The propagation* of local economic and social shocks: this blurs the distinction between winners and losers. The magnitude of this type of indirect impact depends on the existence of compensation mechanisms, including direct assistance and insurance as well as on how the cross-sectoral interdependences and transition costs are captured by models (see Section 3.5.1).

The influence of this set of parameters, which is set differently in various studies, explains the wide range of estimates for the SCC.

In an economically-efficient mitigation response, the marginal costs of mitigation should be equated to the marginal benefits of emission reduction. The marginal benefits are the avoided damages for an additional tonne of carbon abated within a given emission pathway, also known as the SCC. As discussed in Section 3.6, both sides of this equation are uncertain, which is why a sequential or iterative decision-making framework, with progressive resolution of information, is needed. Despite a paucity of analytical results in this area, it is possible to draw on today's literature to make a first comparison between the range of SCC estimates and the range of marginal costs of mitigation across different scenarios. IPCC (2007b, Chapter 20) reviews ranges of SCC from available literature. Allowing for a range of SCC between 4–95 US$/tCO$_2$ (14–350 US$/tC from Tol (2005b) median and 95[th] percentile estimates) and assuming a 2.4% per year increase (IPCC, 2007b, Chapter 20), produces a range of estimates for 2030 of 8–189 US$/tCO$_2$. The mitigation studies in this chapter suggest carbon prices in 2030 of 1–24 US$/tCO$_2$-eq for category IV scenarios, 18–79 US$/tCO$_2$-eq for category III scenarios, and 31–121 US$/tCO$_2$-eq for category I and II scenarios (see Sections 3.3 and 3.6).

3.6 Links between short-term emissions trends, envisaged policies and long-term climate policy targets

In selecting the most appropriate portfolio of policies to deal with climate change, it is important to distinguish between the case of 'certainty', where the ultimate target is known from the outset, and a 'probabilistic' case, where there is uncertainty about the level of a 'dangerous interference' and about the costs of greenhouse gas abatement.

In the case of certainty, the choice of emissions pathway can be seen as a pure GHG budget problem, depending on a host of parameters (discounting, technical change, socio-economic inertia, carbon cycle and climate dynamics, to name the most critical) that shape its allocation across time. The IPCC Second and Third Assessment Reports demonstrated why this approach is an oversimplification and therefore misleading. Policymakers are not required to make once-and-for-all decisions, binding their successors over very long time horizons, and there will be ample opportunities for mid-course adjustments in the light of new information. The choice of short-term abatement rate (and adaptation strategies) involves balancing the economic risks of rapid abatement now and the reshaping of the capital stock that could later be proven unnecessary, against the corresponding risks of delay. Delay may entail more drastic adaptation measures and more rapid emissions reductions later to avoid serious damages, thus necessitating premature retirement of future capital stock or taking the risk of losing the option of reaching a certain target altogether (IPCC, 1996b, SPM).

The calculation of such short-term 'optimal' decisions in a cost-benefit framework assumes the existence of a metaphorical 'benevolent planner' mandated by cooperative stakeholders. The planner maximizes total welfare under given economic, technical and climate conditions, given subjective visions of climate risks and attitudes towards risks. A risk-taking society might choose to delay action and take the (small) risk of triggering significant and possibly irreversible abrupt change impacts over the long-term. If society is averse to risk – that is, interested in avoiding worst-case outcomes – it would prefer hedging behaviour, implying more intense and earlier mitigation efforts.

A significant amount of material has been produced since the SAR and the TAR to upgrade our understanding of the parameters influencing the decisions about the appropriate timing of climate action in a hedging perspective. We review these recent developments, starting with insights from a body of literature drawing on analytical models or compact IAMs. We then assess the findings from the literature for short-term sectoral emission and mitigation estimates from top-down economy-wide models.

3.6.1 Insights into the choice of a short-term hedging strategy in the context of long-term uncertainty

There are two main ways of framing the decision-making approaches for addressing the climate change mitigation and adaptation strategies. They depend on different metrics used to assess the benefits of climate policies:

a. A cost-effectiveness analysis that minimizes the discounted costs of meeting various climate constraints (concentration ceiling, temperature targets, rate of global warming).

b. A cost-benefit analysis that employs monetary estimates of the damages caused by climate change and finds the optimal emissions pathway by minimizing the discounted present value of adaptation and mitigation costs, co-benefits and residual damages.

The choice between indicators of the mitigation benefits reflects a judgment on the quality of the available information and its ability to serve as a common basis in the decision-making process. Actually the necessary time to obtain comprehensive, non-controversial estimates of climate policy benefits imposes a trade-off between the measurement **accuracy** of indicators describing the benefits of climate policies (which diminishes as one moves down the causal chain from global warming to impacts and as one downscales simulation results) and their **relevance**, that is their capacity to translate information that policymakers may desire, ideally prior to a fully-informed decision. Using a set of environmental constraints is simply a way of considering that, beyond such constraints, the threat of climate change might become unacceptable; in a monetary-metric, or valuation approach, the same expectation can be translated through using damage curves with *dangerous* thresholds. The only serious source of divergence between the two approaches is the discount rate. Within a cost-effectiveness framework, environmental constraints are not influenced by discounting. Conversely, in a cost-benefit framework, some benefits occur later than costs and thus have a lower weighting when discounted.

3.6.1.1 *Influence of passing from concentration targets to temperature targets in a cost-effectiveness framework*

New studies such as Den Elzen *et al.* (2006) confirm previous results. They establish that reaching a concentration target as low as 450 ppmv CO_2-eq, under even optimistic assumptions of full participation, poses significant challenges in the 2030–2040 timeframe, with rapidly increasing emission reduction rates and rising costs. In a stochastic cost-effectiveness framework,

reaching such targets requires a significant and early emissions reduction with respect to respective baselines.

But concentration ceilings are a poor surrogate for climate change risks: they bypass many links from atmospheric chemistry to ultimate damages and they only refer to long-term implications of global warming. A better proxy of climate change impacts can be found in global mean temperature: every regional assessment of climate change impacts refers to this parameter, making it easier for stakeholders to grasp the stakes of global warming for their region; one can also take into account the rate of climate change, a major determinant of impacts and damages.

Therefore, with a noticeable acceleration in the last few years, the scientific community has concentrated on assessing climate policies in the context of climate stabilization around various temperature targets. These contributions have mainly examined the influence of the uncertainty about climate sensitivity on the allowable (short-term) GHGs emissions budget and on the corresponding stringency of the climatic constraints, either through sensitivity analyses (Böhringer *et al.*, 2006; Caldeira *et al.*, 2003; Den Elzen and Meinshausen, 2006; Richels *et al.*, 2004) or within an optimal control frame-work (Ambrosi *et al.*, 2003; Yohe *et al.*, 2004).

On the whole, these studies reach similar conclusions, outlining the significance of uncertainty about climate sensitivity. Ambrosi *et al.* (2003) demonstrates the information value of climate sensitivity before 2030, given the significant economic regrets from a precautionary climate policy in the presence of uncertainty about this parameter. Such information might not be available soon (i.e. at least 50 years could be necessary – Kelly *et al.*, 2000). Yohe *et al.* (2004) thus conclude: 'uncertainty (about climate sensitivity) is the reason for acting in the near term and uncertainty cannot be used as a justification for doing nothing'.

A few authors analyze the trade-off between a costly acceleration of mitigation costs and a (temporary) overshoot of targets, and the climate impacts of this overshoot. Ambrosi *et al.* (2003) did so through a willingness to pay for not interfering with the climate system. They show that allowing for overshoot of an *ex-ante* target significantly decreases the required acceleration of decarbonization and the peak of abatement costs, but does not drastically change the level of abatement in the first period. However, the overshoot may significantly increase climate change damages as discussed above (see Section 3.5). Another result is that higher climate sensitivity magnifies the rate of warming, which in turn exacerbates adaptation difficulties, and leads to stringent abatement policy recommendations for the coming decades (Ambrosi, 2007). This result is robust for the choice of discount rate; uncertainty about the rate constraint is proven to be more important for short-term decisions than uncertainty about the magnitude of warming. Therefore, research should be aimed at better characterizing early climate change risks with a view to helping decision-makers in agreeing on a safe guardrail to limit the rate of global warming.

3.6.1.2   Implications of assumptions concerning damage functions in cost-benefit analysis

What is remarkable in cost-benefit studies of the optimal timing of mitigation is that the shape (or curvature) of the damage function matters even more than the ultimate level of damages – a fact long established by Peck and Teisberg (1995). With damage functions exhibiting smooth and regular damages (such as power functions with integer exponents or polynomial functions), GHG abatement is postponed. This is because, for several decades, the temporal rate of increase in marginal climate change damage remains low enough to conclude that investments to accelerate the rate of economic growth are more socially profitable that investing in abatement.

This result changes if singularities in the damage curve represent non-linear events. Including even small probabilities of catastrophic 'nasty surprises' may substantially alter optimal short-term carbon taxes (Mastrandea and Schneider, 2004; Azar and Lindgren, 2003). Many other authors report similar findings (Azar and Schneider, 2001; Howarth, 2003; Dumas and Ha-Duong, 2005; Baranzini et al., 2003), whilst Hall and Behl (2006) suggest a damage function reflecting climate instability needs to include discontinuities in capital stock and the rate of return on capital, and hysteresis with respect to heating and cooling – resulting in a non-convex optimization function such that economic optimization models can provide no solution. But these surprises may be caused by forces other than large catastrophic events. They may also be triggered by smooth climate changes that exceed a vulnerability threshold (e.g. shocks to agricultural systems in developing countries leading to starvation) or by policies that lead to maladaptations to climate change.

In the case of an irreversible THC collapse, Keller et al. (2004) point out another seemingly paradoxical result: if a climate catastrophe seems very likely within a short-term time horizon, it might be economically sound to accept its consequences instead of investing in expensive mitigation to avoid the inevitable. This shows that temporary overshoot of a pre-determined target may be preferable to bearing the social costs of an exaggerated reduction in emission, as well as the need to be attentive to 'windows of opportunity' for abatement action. The converse argument is that timely abatement measures, especially in the case of ITC, can reduce long-term mitigation costs and avoid some of the catastrophic events. In this respect, limited differences in GMT curves for different emissions pathways within coming decades are often misinterpreted. It does not imply that early mitigation activities would make no material difference to long-term warming. On the contrary, if the social value of the damages is high enough to justify deep emission cuts decades from now, then early action is necessary due to inertia in socio-economic systems. For example, one challenge is to avoid further build-up of carbon-intensive capital stock.

3.6.2   Evaluation of short-term mitigation opportunities in long-term stabilization scenarios

3.6.2.1   Studies reporting short-term sectoral reduction levels

While there are many potential emissions pathways to a particular stabilization target from a specific year, it is possible to define emissions trajectories based on short-term mitigation opportunities that are consistent with a given stabilization target. This section assesses scenario results (by sector) from top-down models for the year 2030, to evaluate the range of short-term mitigation opportunities in long-term stabilization scenarios. To put these identified mitigation opportunities in context, Chapter 11, Section 11.3 compares the short-term mitigation estimates across all of the economic sectors.

Many of the modelling scenarios represented in this section were an outcome from the Energy Modelling Forum Study 21 (EMF-21), which focused specifically on multi-gas strategies to address climate change stabilization (see De la Chesnaye and Weyant, 2006). Models that were evaluated in this assessment are listed in Table 3.12.

For each model, the resulting emissions in the mitigation case for each economic sector in 2030 were compared to projected emissions in a reference case. Results were compared across a range of stabilization targets. For more detail on the relationship between stabilization targets defined in concentrations, radiative forcing and temperature, see Section 3.3.2.

Key assumptions and attributes vary across the models evaluated, thus having an impact on the results. Most of the top-down models evaluated have a time horizon beyond 2050 such as AIM, IPAC, IMAGE, GRAPE, MiniCAM, MERGE, MESSAGE, and WIAGEM. Top-down models with a time horizon up to 2050, such as POLES and SGM, were also evaluated. The models also vary in their solution concept. Some models provide a solution based on inter-temporal optimization, allowing mitigation options to be adopted with perfect foresight as to what the future carbon price will be. Other models are based on a recursive dynamic, allowing mitigation options to be adopted based only on today's carbon price. Recursive dynamic models tend to show higher carbon prices to achieve the same emission reductions as in inter-temporal optimization models, because emitters do not have the foresight to take early mitigation actions that may have been cheaper (for more discussion on modelling approaches, refer to Section 3.3.3).

Three important considerations need to be remembered with regard to the reported carbon prices. First, these mitigation scenarios assume complete 'what' and 'where' flexibility (i.e.

Table 3.12: *Top-down models assessed for mitigation opportunities in 2030*

Model	Model type	Solution concept	Time horizon	Modelling team and reference
AIM (Asian-Pacific Integrated Model)	Multi-Sector General Equilibrium	Recursive Dynamic	Beyond 2050	NIES/Kyoto Univ., Japan Fujino et al., 2006.
GRAPE (Global Relationship Assessment to Protect the Environment)	Aggregate General Equilibrium	Inter-temporal Optimization	Inter-temporal Optimization	Institute for Applied Energy, Japan Kurosawa, 2006.
IMAGE (Integrated Model to Assess The Global Environment)	Market Equilibrium	Recursive Dynamic	Beyond 2050	Netherlands Env. Assessment Agency Van Vuuren et al., Energy Journal, 2006a. (IMAGE 2.2) Van Vuuren et al., Climatic Change, 2007. (IMAGE 2.3)
IPAC (Integrated Projection Assessments for China)	Multi-Sector General Equilibrium	Recursive Dynamic	Beyond 2050	Energy Research Institute, China Jiang et al., 2006.
MERGE (Model for Evaluating Regional and Global Effects of GHG Reduction Policies)	Aggregate General Equilibrium	Inter-temporal Optimization	Beyond 2050	EPRI & PNNL/Univ. Maryland, U.S. USCCSP, 2006.
MESSAGE-MACRO (Model for Energy Supply Strategy Alternatives and Their General Environmental Impact)	Hybrid: Systems Engineering & Market Equilibrium	Inter-temporal Optimization	Beyond 2050	International Institute for Applied Systems Analysis, Austria Rao and Riahi, 2006.
MiniCam (Mini-Climate Assessment Model)	Market Equilibrium	Recursive Dynamic	Beyond 2050	PNNL/Univ. Maryland, U.S. Smith and Wigley, 2006.
SGM (Second Generation Model)	Multi-Sector General Equilibrium	Recursive Dynamic	Up to 2050	PNNL/Univ. Maryland and EPA, U.S. Fawcett and Sands, 2006.
POLES (Prospective Outlook on Long-Term Energy Systems)	Market Equilibrium	Recursive Dynamic	Up to 2050	LEPII-EPE & ENERDATA, France Criqui et al., 2006.
WIAGEM (World Integrated Applied General Equilibrium Model)	Multi-Sector General Equilibrium	Inter-temporal Optimization	Beyond 2050	Humboldt University and DIW Berlin, Germany Kemfert et al., 2006.

Source: Weyant et al., 2006.

there is full substitution among GHGs and reductions take place anywhere in the world, according to the principle of least cost). Limiting the degree of flexibility in these mitigation scenarios, such as limiting mitigation only to CO_2, removing major countries or regions from undertaking mitigation, or both, will increase carbon prices, all else being equal. Second, the carbon prices of realizing these levels of mitigation increase in the time horizon beyond 2030. See Figure 3.25 for an illustration of carbon prices across longer time horizons from top-down scenarios. Third, at the economic sector level, estimated emission reduction for all greenhouse gases varies significantly across the different model scenarios, in part because each model uses sector definitions specific to that type of model.

Across all the models, the long-term target in the stabilization scenarios could be met through the mitigation of multiple greenhouse gases (CO_2, CH_4, N_2O and high-GWP gases). However the specific mitigation options and the treatment of technological progress vary across the models. For example, only some of the models include carbon capture and storage as a mitigation option (GRAPE, IMAGE, IPAC, MiniCAM,

and MESSAGE). Some models also include forest sinks as a mitigation option. The model results shown in Table 3.13 do not include forest sinks as a mitigation option, while the results shown in Table 3.14 do include forest sinks, as described in further detail below.

Table 3.13 illustrates the amount of global GHG mitigation reported by sector for the year 2030 across a range of multi-gas stabilization targets. Across the higher *Category IV* stabilization target scenarios, emission reductions of 3–31% from the reference case emissions across all greenhouse gases can be achieved for a carbon price of 2–57 US$/$tCO_2$-eq. The results from the POLES models fall into the higher end of the price range, in part due to the recursive dynamic nature of the model, and also due to its shorter time horizon over which to plan. The results from the GRAPE model fall into the lower end of the price range, which is the only inter-temporally optimizing model shown in the higher stabilization scenarios. In the GRAPE results, only 3% of the emissions are reduced by 2030, implying that the majority of the mitigation necessary to meet the target is undertaken beyond 2030. In scenarios

Table 3.13: *Global emission reductions from top-down models in 2030 by sector for multi-gas scenarios.*

Model		POLES	IPAC	AIM	GRAPE	MiniCAM	SGM	MERGE	WIAGEM
Stabilization category		Category VI						Category II	Category I
Stabilization target		550 ppmv	550 ppmv	4.5 W/m² from pre-Industrial	4.5 W/m² from pre-Industrial	4.5 W/m² from pre-Industrial	From MiniCAM trajectory	3.4 W/m² from pre-Industrial	2% from pre-Industrial
Carbon price in 2030 (2000 US$/tCO$_2$-eq)		57	14	29	2	12	21	192	9
Reference emissions 2030 Total all gases (GtCO$_2$-eq)		53.0	55.3	49.4	57.0	54.2	53.5	47.2	43.1
Sector Mitigation estimates in 2030 (total all gases GtCO$_2$-eq)	Energy supply: electric	9.5	6.4	5.2	0.5	7.3	3.1	9.5	7.0
	Energy supply: non-electric	3.0	0.6	1.1	0.0	1.5	1.6[a]	3.2	1.7
	Transportation demand	0.5	0.8	0.5	0.1	0.2	0.4[a]	Included in Energy supply	Included in Energy supply
	Buildings demand	1.0	0.6	0.5	0.4	0.3	Included in Energy supply	Included in Energy supply	Included in Energy supply
	Industry demand	1.9	1.2	0.5	Included in Buildings demand	1.7	Included in Energy supply	Included in Energy supply	Included in Energy supply
	Industry production	0.8	0.0	0.8	0.3[h]	0.2[d]	1.7[a]	3.6[b]	3.6
	Agriculture	(0.2)	(1.0)[e]	2.0	0.6	0.3	1.7	Included in industry production	1.1
	Forestry	No mitigation options modelled						No mitigation options modelled	
	Waste management	Included in another sector	0.0[g]	Included in Buildings demand	0.0[f]	0.3	0.5	Included in Industry production	No mitigation options modelled
Global total		16.4	8.7	10.6	1.9	11.9	11.2[a]	16.3	15.5[c]
Mitigation as % of reference emissions		31%	16%	21%	3%	22%	21%	35%	35%

Notes:
[a] SGM sector mitigation estimates for Transportation Demand and Industry Production are not complete global representation due to varying levels of regional aggregation.
[b] MERGE sector mitigation estimates for Industry Production, Agriculture, and Waste Management are aggregated. No Forestry mitigation options were modelled.
[c] WIAGEM sector mitigation estimates do not sum to global total due to the breakout of the household and chemical sectors.
[d] MiniCAM CO$_2$ mitigation from Industrial Production is accounted for in the Industry Demand.
[e] Higher IPAC Agriculture emissions in the stabilization scenario than in the reference case reflects the loss of permanent forest due to growing bioenergy crops.
[f] GRAPE Waste sector mitigation reflects only GDP activity factor changes in 2030, and reflects emission factor reductions in later years.
[g] IPAC Waste sector cost-effective mitigation options are included in the baseline.
[h] GRAPE CO$_2$ from cement production is included in Buildings Demand.

with lower *Category I and II* stabilization targets, higher levels of short-term mitigation are required to achieve the target in the long run, resulting in a higher range of prices. Emission reductions of approximately 35% can be achieved at a price of 9–92 US$/tCO$_2$-eq.

Several of the models included in the EMF-21 study also ran multi-gas scenarios that included forest sinks as a mitigation option. Table 3.14 shows the 2030 mitigation estimates for these scenarios that model net land-use change (including forest carbon sinks) as a mitigation option. When terrestrial sinks are modelled as a mitigation option, it can lessen the pressure to mitigate in other sectors. Further discussion of forest

sequestration as a mitigation option is presented in Section 3.3.5.5. Across the higher *Category IV* stabilization target scenarios, emission reductions of 4–24% from the reference case emissions across all greenhouse gases can be achieved at a price of 2–21 US$/tCO$_2$-eq. In scenarios with lower *Category I and II* stabilization targets, emission reductions of 26–40% can be achieved at a price of 31–121 US$/tCO$_2$-eq.

3.6.2.2 Assessment of reduction levels at different marginal prices

To put these identified mitigation opportunities into context they will be compared with mitigation estimates from bottom-up

Table 3.14: *Global emission reductions from top-down models in 2030 (by sector) for multi-gas plus sinks scenarios.*

Model		GRAPE	IMAGE 2.2	IMAGE 2.3	MESSAGE	MESSAGE	IMAGE 2.3	IMAGE 2.3	MESSAGE
Stabilization categories		*Category VI*					Category III	Category I/II	
Stabilization target		4.5 Wm^2 from pre-Industrial	4.5 Wm^2 from pre-Industrial	4.5 Wm^2 from pre-Industrial	B2 scenario, 4.5 Wm^2 from pre-Industrial	A2 scenario, 4.5 Wm^2 from pre-Industrial	3.7 Wm^2 from pre-Industrial	3.0 Wm^2 from pre-Industrial	B2 scenario, 3.0 Wm^2 from pre-Industrial
Carbon price in 2030 (2000 US$/tCO$_2$-eq)		2	18	21	6	15	50	121	31
Reference emissions 2030 Total all gases (GtCO$_2$-eq)		57.0	65.5	59.7	57.8	70.9	59.7	59.7	57.8
Sector mitigaiton estimates in 2030 (total all gases GtCO$_2$-eq)	Energy supply: electric	0.5	2.4	1.7	1.1	7.3	3.9	8.7	4.3
	Energy supply: non-electric	0.0	2.2	1.6	0.5	3.5	2.3	3.7	2.2
	Transportation demand	0.0	1.3	0.7	0.3	1.0	1.5	2.8	2.2
	Buildings demand	0.3	0.8	0.3	0.5	1.2	0.5	1.0	1.4
	Industry demand	Included in Buildings demand	0.8	0.5	0.1	0.4	1.6	3.2	0.8
	Industry production	0.1[b]	1.1	0.8	0.3	0.6	1.1	2.0	0.8
	Agriculture	0.3	0.7	0.6	0.6	1.5	1.0	1.2	1.7
	Forestry	0.9	1.4	0.3	0.0	0.2	0.2	0.2	0.6
	Waste management	0.0[a]	0.7	1.0	0.9	1.1	1.0	1.1	0.9
Global total		2.1	11.5	7.6	4.4	16.8	13.0	24.0	15.0
Mitigation as % reference emissions		4%	18%	13%	8%	24%	40%	40%	26%

Notes:
[a] GRAPE Waste sector mitigation reflects only GDP activity factor changes in 2030, and reflects emission factor reductions in later years.
[b] GRAPE CO$_2$ from cement production is included in Buildings Demand.

models. Chapters 4 through 10 describe mitigation technologies available in specific economic sectors. Chapter 11, Section 11.3 compares the short-term mitigation estimates across all of the economic sectors for selected marginal costs levels (20, 50 and 100 US$/tCO$_2$-eq). For that purpose, we have plotted the permit price and (sectoral) reduction levels of the different studies. These plots have been used to explore whether the combination of the studies suggests certain likely reduction levels at the three target levels of 20, 50 and 100 US$/tCO$_2$-eq. As far more studies were available that reported economy-wide reduction levels than the ones that provided sectoral information, we were able to use a formal statistical method for the former. For the latter, a statistical method was also applied, but outcomes have been used with more care.

Economy-wide reduction levels

Figure 3.40 shows the available data from studies that report economy-wide reduction levels (multi-gas) and permit prices. The data has been taken from the emission scenario database (Hanaoka *et al.*, 2006; Nakicenovic *et al.*, 2006) – and information directly reported in the context of EMF-21 (De la Chesnaye and Weyant, 2006) and IMCP (Edenhofer *et al.*, 2006). The total sets suggest some form of a relationship with

studies reporting higher permit prices: also, in general, reporting higher reduction levels.

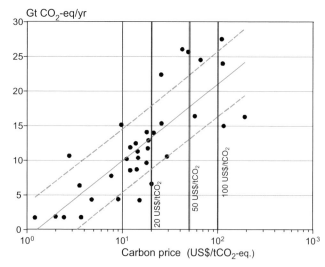

Figure 3.40: *Permit price versus level of emission reduction – total economy in 2030 (the natural logarithm of the permit price is used for the x-axis). The uncertainty range indicated is the 68% interval.*

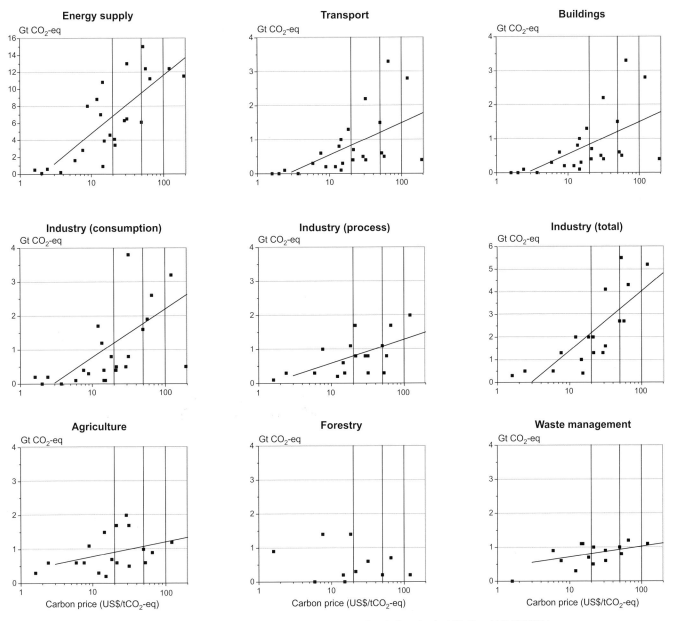

Figure 3.41: *Permit price versus emission reduction level – several sectors in 2030 (vertical lines indicate levels at 20, 50 and 100 US$/tCO₂).*

Obviously, a considerable range of results is also found – this is a function of factors such as:
- Model uncertainties, including technology assumptions and inertia.
- Assumed baseline developments.
- The trajectory of the permit price prior to 2030.

The suggested relationship across the total is linear if permit prices are plotted on a logarithmic scale as shown in Figure 3.40. In other words, the relationship between the two variables is logarithmic, which is a form that is consistent with the general form of marginal abatement curves reported in literature: increasing reduction levels for higher prices, but diminishing returns at higher prices as the reduction tends to reach a theoretical maximum. The figure not only shows the best-guess regression

line, but also 68% confidence interval. The latter can be used to derive the 68 percentile interval of the reduction potential for the 20 and 100 US$/tCO$_2$-eq price levels, which are 13.3 ± 4.6 GtCO$_2$-eq/yr and 21.5 ± 4.7 GtCO$_2$-eq/yr, respectively.

Sectoral estimates

A more limited set of studies reported sectoral reduction levels. The same plot as Figure 3.40 has been made for the sectoral data (see Figure 3.41), again plotting the logarithm of the permit price against emission reduction levels. The data here are directly taken from Table 3.13 and Table 3.14. As less data are available, the statistical analysis becomes less robust. Nevertheless, for most sectors, a similarly formed relationship was found across the set of studies as for the economy-wide potential (logarithmic relationship showing increasing reduction

Table 3.15: *Reduction potential at various marginal prices, averages across different models (low and high indicate one standard deviation variation).*

	20 US$/tCO$_2$-eq		50 US$/tCO$_2$-eq		100 US$/tCO$_2$-eq	
	Low	High	Low	High	Low	High
Energy supply	3.9	9.7	6.7	12.4	8.7	14.5
Transport	0.1	1.6	0.5	1.9	0.8	2.5
Buildings	0.2	1.1	0.4	1.3	0.6	1.5
Industry	1.2	3.2	2.2	4.3	3.0	5.0
Agriculture	0.6	1.2	0.8	1.4	0.9	1.5
Forestry	0.2	0.8	0.2	0.8	0.2	0.8
Waste	0.7	0.9	0.8	1.0	0.9	1.1
Overall[1]	8.7	17.9	13.7	22.6	16.8	26.2

Note: 1) The overall potential has been estimated separately from the sectoral totals.

levels at relatively low prices, and a much slower increase at higher prices). As expected, in several sectors, the spread across models in the 2030 set is larger than in the economy-wide estimates.

In general, a relatively strong relationship is found in the sectors for energy supply, transport, and industrial energy consumption. The relationship between the price and emission reduction level is less clear in other sectors – and more-or-less absent for the limited reported data on the forestry sector. It should be noted here that definitions across studies may be less well-defined – and also, forest sector emissions may actually increase in mitigation scenarios as a result of net deforestation due to bio-energy production.

It should be noted that emission data (and thus also reduction levels) are reported on a 'point of emission basis' (emissions are reported for the sectors in which the emissions occur). For example, the efficiency improvements in end-use sectors for electricity lead to reductions in the energy supply sector. Likewise, using bio-energy leads to emission reductions in the end-use sectors, but at the same time (in some models) may lead to increases in emissions for forestry, due to associated land-use changes. The latter may explain differences in the way that data from top-down models are represented elsewhere in this report, as here (in most cases) only the emission changes from mitigation measures in the forestry sector itself are reported. It also explains why the potential in some of the end-use sectors is relatively small, as emission reductions from electricity savings are reported elsewhere.

Reported estimates

On the basis of the available data, the following ranges have been estimated for the reduction potential at a 20, 50 and 100 US$/tCO$_2$-eq price (Table 3.15). As estimates have been made independently, the total of the different sectors does not add up to the overall range (as expected, the sum of the sectors gives a slightly wider range).

The largest potential is found in energy supply – covering both the electricity sector and energy supply – with a relatively high capability of responding to permit prices. Relatively high reduction levels are also found for the industry sector. Relatively small reduction levels are reported for the forestry sector and the waste management sector.

REFERENCES

Aaheim, H.A., H. Asbjørn, J.S. Fuglestvedt, and O. Godal, 2006: Costs savings of a flexible multi-gas climate policy, *The Energy Journal*, Vol. Multi-Greenhouse gas Mitigation and Climate Policy, Special Issue No.3, pp. 485-502.

Abramovitz, M., 1986: Catching-up, forging ahead, and falling behind. *Journal of Economic History*, **46**(2), pp. 385-406.

Aghion, P.P. and P. Howitt, 1998: *Endogenous growth theory*. MIT Press, Cambridge, 694 pp.

Agrawala, S. (ed.), 2005: Bridge over troubled waters: Linking climate change and development. OECD Publishing, Paris, 154 pp.

Ahammad, H., A. Matysek, B.S. Fisher, R. Curtotti, A. Gurney, G. Jakeman, E. Heyhoe, and D. Gunasekera, 2006: Economic impact of climate change policy: the role of technology and economic instruments. ABARE Research report 06.7, Canberra, 66 pp. <http://www.abareconomics.com/publications_html/climate/climate_06/cc_policy_nu.pdf>, accessed 1 June 2007.

Ainsworth, E.A., S.P. Long, 2005: What have we learned from 15 years of free-air CO$_2$ enrichment (FACE)? A meta-analytic review of the responses of photosynthesis, canopy properties and plant production to rising CO$_2$. *New Phytologist*, **165**, pp. 351-372.

Akimoto, K., H. Kotsubo, T. Asami, X. Li, M. Uno, T. Tomoda, and T. Ohsumi, 2004: Evaluation of carbon dioxide sequestration in Japan with a mathematical model. *Energy*, **29**(9-10), pp. 1537-1549.

Alcamo, J., A. Bouwman, J. Edmonds, A. Grübler, T. Morita, and A. Sugandhy, 1995: An evaluation of the IPCC IS92 emission scenarios. In *Climate change 1994, radiative forcing of climate change and an evaluation of the IPCC IS92 emission scenarios*. Cambridge University Press, Cambridge, pp. 233-304.

Alcamo, J. (ed.), 2002: Linkages between regional air pollution and climate change in Europe. *Environmental Science & Policy*, **5**(4), pp. 255-365.

Amann, M., 2002: Emission trends of anthropogenic air pollutants in the Northern Hemisphere. Air pollution as a climate forcing. Goddard Institute for Space Studies, Honolulu, Hawaii, 29 April – 3 May 2002. <http://www.giss.nasa.gov/meetings/pollution2002/d1_amann.html>, accessed 1 June 2007.

Ambrosi, P., J.C. Hourcade, S. Hallegatte, F. Lecocq, P. Dumas, and M.H. Duong, 2003: Optimal control models and elicitation of attitudes towards climate damages. *Environmental Modeling and Assessment*, **8**(3), pp. 133-147.

Ambrosi, P., 2007: Mind the rate! Why the rate of global climate change matters, and how much. In *Conference Volume of the 6th CESIfo Venice Summer Institute - David Bradford Memorial Conference on the Design of Climate Policy*, R. Guesnerie and H. Tulkens (eds.), MIT Press, Cambridge MA.

Andrae, M.O., 2001: The dark side of aerosols. *Nature*, **409**(6821), pp. 671-672.

Arnell, N.W., M.G.R. Cannell, M. Hulme, R.S. Kovats, J.F.B. Mitchell, R.J. Nicholls, M.L. Parry, M.T.J. Livermore and A. White, 2002: The Consequences of CO_2 stabilisation for the impacts of climate change. *Climatic Change*, **53**(4), pp. 413-446.

Azar, C. and H. Dowlatabadi, 1999: A review of technical change in assessment of climate policy. *Annual Review of Energy and the Environment*, **24**, pp. 513-544.

Azar, C. and S.H. Schneider, 2001: Are uncertainties in climate and energy systems a justification for stronger near-term mitigation policies? Pew Center on Global Climate Change, 52 pp, <http://stephenschneider.stanford.edu/Publications/PDF_Papers/UncertaintiesInClimate.pdf>, accessed 1 June 2007.

Azar, C. and K. Lindgren, 2003: Catastrophic events and stochastic cost-benefit analysis of climate change. An editorial comment. *Climatic Change*, **56**(3), pp. 245-255.

Azar, C., K. Lindgren, E. Larson, and K. Möllersten, 2006: Carbon capture and storage from fossil fuels and biomass - Costs and potential role in stabilizing the atmosphere. *Climatic Change*, **74**(1-3), pp. 47-79.

Babiker, M.H., J.M. Reilly, M. Mayer, R.S. Eckaus, I. Sue Wing, and R.C. Hyman, 2001: The emissions prediction and policy analysis (EPPA) model: revisions, sensitivities and comparisons of results. MIT, Cambridge MA. <http://web.mit.edu/globalchange/www/MITJPSPGC_Rpt71.pdf>, accessed 1 June 2007.

Baranzini, A., M. Chesney, and J. Morisset, 2003: The impact of possible climate catastrophes on global warming policy. *Energy Policy*, **31**(8), pp. 691-701.

Barker, T., H. Pan, J. Köhler, R. Warren, S. Winne, 2006: Avoiding dangerous climate change by inducing technological progress: Scenarios using a large-scale econometric model. In *Avoiding Dangerous Climate Change*, H.J. Schellnhuber, W. Cramer, N. Nakicenovic, T. Wigley and G. Yohe (eds.), Cambridge University Press, Cambridge, pp. 361-373.

Barro, R., 1991: Economic growth in a cross section of countries. *Quarterly Journal of Economics*, **106**, pp. 407-443.

Barro, R.J., and X. Sala-i-Martin, 1997: Technological diffusion, convergence, and growth. *Journal of Economic Growth*, **2**(1), pp. 1-26.

Baumol, W.J., 1986: Productivity growth, convergence, and welfare: what the long-run data show. *American Economic Review*, **76**(5), pp. 1072-1085.

Beierle, T.C. and J. Cayford, 2002: *Democracy in practice: public participation in environmental decisions*. Resources for the Future, Washington DC, 158 pp.

Berndes, G., M. Hoogwijk, and R. van den Broek, 2003: The contribution of biomass in the future global energy system: a review of 17 studies. *Biomass and Bioenergy*, **25**(1), pp. 1-28.

Betts, R.A., 2000: Offset of the potential carbon sink from boreal forestation by decreases in surface albedo. *Nature*, **408**(6809), pp. 187-190.

Böhringer, C., A. Löschel, and T.F. Rutherford, 2006: Efficiency gains from "what"-flexibility in climate policy - An integrated CGE assessment, *The Energy Journal*, Vol. Multi-Greenhouse Gas Mitigation and Climate Policy, Special Issue, No.3, pp. 405-424.

Bond, T.C., D.G. Streets, K.F. Yarber, S.M. Nelson, J.-H. Woo, and Z. Klimont, 2004: A technology-based global inventory of black and organic carbon emissions from combustion. *Journal of Geophysical Research*, **109**(14), pp. D14203.

Bosello, F. 2005: Adaptation and mitigation to global climate change: Conflicting strategies? Insights from an empirical integrated assessment exercise. University of Venice, 41 pp. <http://www.feem-web.it/ess05/files/Bosello.pdf>, accessed 1 June 2007.

Bosello, F. and J. Zhang, 2005: Assessing climate change impacts: Agriculture. Fondazione Eni Enrico Mattei, Nota di lavoro 94.2005, Venice, 39 pp. <http://www.feem.it/NR/rdonlyres/8D8F51C4-EC35-4BC5-AA8F-67739BBC793C/1650/9405.pdf>, accessed 1 June 2007.

Bouwman, A.F. and D.P. van Vuuren, 1999: *Global assessment of acidification and eutrophication of natural ecosystems*. Rijksinstituut voor Volksgezondheid en Milieu (RIVM), Report 402001012, Bilthoven, 64 pp. <http://www.mnp.nl/bibliotheek/rapporten/402001012.pdf>, accessed 1 June 2007.

Bradford, D.F., 2001: Time, money and tradeoffs. *Nature*, **410**(6829), pp. 649-650.

Brovkin, V., A. Ganopolski, M. Claussen, C. Kubatzki, and V. Petoukhov, 1999: Modelling climate response to historical land cover change. *Global Ecology and Biogeography*, **8**(6), pp. 509-517.

Brovkin, V., M. Claussen, E. Driesschaert, T. Fichefet, D. Kicklighter, M.F. Loutre, H. D. Matthews, N. Ramankutty, M. Schaeffer, and A. Sokolov, 2006: Biogeophysical effects of historical land cover changes simulated by six Earth system models of intermediate complexity. *Climate Dynamics*, **26**(6), pp. 587-600.

Bruinsma, J.E. (ed.), 2003: *World agriculture: towards 2015/2030. An FAO perspective. Earthscan*, London, 432 pp.

Burrows, B., A. Mayne, and P. Newbury, 1991: *Into the 21st century: a handbook for a sustainable future*. Adamantine Press, Twickenham, 442 pp.

Caldeira, K., A.K. Jain, and M.I. Hoffert, 2003: Climate sensitivity uncertainty and the need for energy without CO_2 emission. *Science*, **299**(5615), pp. 2052-2054.

Carmichael, G.R., D.G. Streets, G. Calori, M. Amann, M.Z. Jacobson, J. Hansen, and H. Ueda, 2002: Changing trends in sulphur emissions in Asia: implications for acid deposition, air pollution, and climate. *Environmental Science and Technology*, **36**(22), pp. 4707- 4713.

Carpenter, S.R., P.L. Pingali, E.M. Bennet, M.B. Zurek (eds.), 2005: Ecosystems and Human Well-being: Scenarios, Vol.II, *Millennium Ecosystem Assessment (MA)*, Island Press, Chicago, 561 pp.

Carter, T.R., K. Jylhä, A. Perrels, S. Fronzek, and S. Kankaanpää, 2005: Alternative futures for considering adaptation to climate change in Finland. Finadapt Working Paper 2, Finnish Environment Institute, Helsinki, 42 pp. <http://www.ymparisto.fi/download.asp?contentid=44018&lan=en>, accessed 1 June 2007.

Cassman, K.G., A. Dobermann, D.T. Walters, and H. Yang, 2003: Meeting cereal demand while protecting natural resources and improving environmental quality. *Annual Review of Environmental Resources*, **28**, pp. 315-358.

Castles, I. and D. Henderson, 2003a: Economics, emissions scenarios and the work of the IPCC. Energy and Environment, 14(4), pp. 415- 435.

Castles, I. and D. Henderson, 2003b: The IPCC emission scenarios: an economic-statistical critique. *Energy and Environment*, **14**(2-3), pp. 159-185.

Central Planning Bureau, 1992: *Scanning the future: a long-term scenario study of the world economy 1990-2015*. SDU Publishers, The Hague, 246 pp.

Chen, W., 2005: The costs of mitigating carbon emissions in China: Findings from China MARKAL-MACRO modeling. *Energy Policy*, **33**(7), pp. 885-896.

Cofala, J., M. Amann, Z. Klimont, and R. Mechler, 2006: Scenarios of world anthroprogenic emissions of air pollutants and methane up to 2030. IIASA Interim Report IR-06-023, International Institute of Applied Systems Analysis, Laxenburg. <http://www.iiasa.ac.at/rains/global_emiss/IR-06-023.pdf>, accessed 1 June 2007.

Cohen, S.J., D. Neilsen, and R. Welbourn (eds.), 2004: Expanding the dialogue on climate change and water management on the Okanagan Basin, British Columbia. Natural Resources Canada, Ottawa, 241 pp. <http://www.ires.ubc.ca/downloads/publications/layout_Okanagan_final.pdf>, accessed 8 June 2007.

Collins, W.J., D.E. Stevenson, C.E. Johnson, and R.G. Derwent, 2000: The European regional ozone distribution and its links with the global scale for the years 1992 and 2015. *Atmospheric Environment*, **34**(2), pp. 255-267.

Cooke, W.F. and J.J.N. Wilson, 1996: A global black carbon aerosol model. *Journal of Geophysical Research*, **101**(14), pp. 19395-19409.

Cooke, W.F., C. Liousse, H. Cachier, and J. Feichter, 1999: Construction of a 1° x 1° fossil fuel emission data set for carbonaceous aerosol and implementation and radiative impact in the ECHAM4 model. *Journal of Geophysical Research*, **104**(D18), pp. 22137-22162.

Corfee-Morlot, J. and N. Höhne, 2003: Climate change: long-term targets and short-term commitments. *Global Environmental Change*, **13**(4), pp. 277-293.

Corfee-Morlot, J., J. Smith, S. Agrawala, T. Franck, 2005: Long-term goals and post-2012 commitments: Where do we go from here with climate policy? *Climate Policy*, **5**(3), pp. 251-272.

Cox, P.M., R.A. Betts, C.B. Bunton, R.L.H. Essery, P.R. Rowntree, and J. Smith, 1999: The impact of new land surface physics on the GCM simulation of climate and climate sensitivity. *Climate Dynamics*, **15**(3), pp. 183-203.

Cox, P.M., R.A. Betts, C.D. Jones, S.A. Spall, and I.J. Totterdell, 2000: Acceleration of global warming due to carbon-cycle feedbacks in a coupled climate model. *Nature*, **408**(6809), pp. 184-187.

Cramer, W., A. Bondeau, S. Schaphoff, W. Lucht, B.E. Smith, and S. Sitch, 2004: Tropical forests and the global carbon cycle: impacts of atmospheric carbon dioxide climate change and rate of deforestations. *Philosophical Transactions of the Royal Society of London Series B*, **359**(1443), pp. 331-343.

Criqui, P., P. Russ, and D. Debye, 2006: Impacts of multi-gas strategies for greenhouse gas emission abatement: Insights from a partial equilibrium model. *The Energy Journal*, Vol. Multi-gas Mitigation and Climate Policy, Special Issue No.3, pp. 251-274.

De la Chesnaye, F.C. and J.P. Weyant (eds.), 2006: Multi-gas mitigation and climate policy. *The Energy Journal*, Special Issue No.3, 520 pp.

De la Chesnaye, F.C., C. Delhotal, B. DeAngelo, D. Ottinger-Schaefer, and D. Godwin, 2007: Past, present, and future of non-CO_2 gas mitigation analysis. In *Human-Induced Climate Change: An Interdisciplinary Assessment*. M. Schlesinger, H. Kheshgi, J. Smith, F. de la Chesnaye, J.M. Reilly, T. Wilson, C. Kolstad (eds.), Cambridge University Press, Cambridge.

Den Elzen, M.G.J. and M. Meinshausen, 2006: Multi-gas emission pathways for meeting the EU 2 C climate target. In *Avoiding Dangerous Climate Change*. H.J. Schellnhuber, W. Cramer, N. Nakicenovic, T. Wigley, and G. Yohe (eds.), Cambridge University Press, Cambridge.

Den Elzen, M.G.J., M. Meinshausen, and D.P. Van Vuuren, 2006: Multi-gas emission envelopes to meet greenhouse gas concentration targets: Costs versus certainty of limiting temperature increase. In *Stabilizing greenhouse gas concentrations at low levels*, D.P. van Vuuren, M.G.J. Den Elzen, P.L. Lucas, B. Eickhout, B.J. Strengers, B. Van Ruijven, M.M. Berk, H.J.M. De Vries, S.J. Wonink, R. Van den Houdt, R. Oostenrijk, M. Hoogwijk, M. Meinshausen, Netherlands Environmental Assessment Agency (MNP), Bilthoven, 273 pp.

Deutscher Bundestag, 2002: Enquete commission on sustainable energy supply against the background of globalisation and liberalisation: summary of the final report. Berlin. 80 pp. <http://www.bundestag.de/parlament/gremien/kommissionen/archiv14/ener/schlussbericht/engl.pdf>, accessed 1 June 2007.

Dixon, P.B. and M.T. Rimmer, 2005: Analysing convergence with a multi-country computable general equilibrium model: PPP versus MER. Proceedings of 8th Annual Conference on Global Economic Analysis, Lübeck. <https://www.gtap.agecon.purdue.edu/resources/download/2254.pdf>, accessed 1 June 2007.

Downing, T., D. Anthoff, R. Butterfield, M. Ceronsky, M. Grubb, J. Guo, C. Hepburn, C. Hope, A. Hunt, A. Li, A. Markandya, S. Moss, A. Nyong, R.S.J. Tol, and P. Watkiss. 2005. Social cost of carbon: a closer look at uncertainty. Department for Environment, Food and Rural Affairs, London, 95 pp. <http://www.defra.gov.uk/environment/climatechange/research/carboncost/pdf/sei-scc-report.pdf>, accessed 1 June 2007.

Douglas, M. and A. Wildavsky, 1982: *Risk and culture: an essay on the selection of technological and environmental damages*. University of California Press, Berkeley.

DTI, 2003: Energy White Paper: Our energy future - creating a low carbon economy. Department of Trade and Industry, London, 142 pp. <http://www.dti.gov.uk/files/file10719.pdf>, accessed 1 June 2007.

Dumas, P. and M. Ha-Duong, 2005: An abrupt stochastic damage function to analyse climate policy benefits: essays on integrated assessment. In *The Coupling of climate and economic dynamics: essays on integrated assessment*. A. Haurie and L. Viguier (eds.), Springer Netherlands, pp. 97-111.

Edenhofer, O., K. Lessmann, C. Kemfert, M. Grubb, J. Köhler, 2006: Induced technological change: Exploring its implications for the economics of atmospheric stabilization; Synthesis report from the innovation modeling comparison project. *The Energy Journal*, Vol. Endogenous Technological Change and the Economics of Atmospheric Stabilisation, Special Issue No.1, pp. 57-108.

Edmonds, J.A., 2004: Technology options for a long-term mitigation response to climate change. 2004 Earth Technologies Forum, 13 April 2004, Pacific Northwest National Laboratory, Washington DC.

Edmonds, J., J. Clarke, J. Dooley, S.H. Kim, and S.J. Smith, 2004a: Stabilization of CO_2 in a B2 world: insights on the roles of carbon capture and disposal, hydrogen, and transportation technologies. *Energy Economics*, **26**(4), pp. 517-537.

Edmonds, J.A., J. Clarke, J. Dooley, S.H. Kim, S.J. Smith, 2004b: Modeling greenhouse gas energy technology responses to climate change. *Energy*, **29**(9-10), pp. 1529-1536.

Edmonds, J. and L. Clarke, 2005: Endogeneous technological change in long-term emissions and stabilization scenarios. IPCC Expert Meeting on Emissions Scenarios, 12 - 14 January 2005, Washington DC.

Ehrlich, P.R. and J.P. Holdren, 1971: Impact of population growth. *Science*, **171**(3977), pp. 1212-1217.

European Commission, 2003: World energy, technology and climate policy outlook 2030 (WETO). EUR 20366, Office of Official Publications of European Communities, Luxembourg. <http://ec.europa.eu/research/energy/pdf/weto_final_report.pdf>, accessed 1 June 2007.

Fagerberg, J., 1995: User-producer interaction, learning and competitive advantage. *Cambridge Journal of Economics*, **19**(1), pp. 243-256.

Fagerberg, J. and M. Godinho, 2005: Innovation and catching-up. In *The Oxford Handbook of Innovation*, J. Fagerberg, D.C. Mowery, R.R. Nelson (eds.), Oxford University Press, Oxford, pp. 514-542.

Fawcett, A.A. and R.D. Sands, 2006: Non-CO_2 greenhouse gases in the second generation model. *The Energy Journal*, Vol. Multi-Greenhouse Gas Mitigation and Climate Policy. Special Issue No.3, pp. 305-322.

Fischedick, M. and J. Nitsch, 2002 (eds.): Long-term scenarios for a sustainable energy future in Germany. Wuppertal Institute for Climate, Environment, Energy; Wissenschaftszentrum Nordrhein-Westfalen, Wuppertal. <http://archive.greenpeace.org.au/climate/pdfs/Wuppertal_report.pdf>, accessed 1 June 2007.

Fischedick, M., J. Nitsch, and S. Ramesohl, 2005: The role of hydrogen for the long-term development of sustainable energy systems - a case study for Germany. *Solar Energy*, **78**(5), pp. 678-686.

Fischer, G. and L. Schrattenholzer, 2001: Global bioenergy potentials through 2050. *Biomass and Bioenergy*, **20**(3), pp. 151-159.

Fisher, B.S., G. Jakeman, H.M. Pant, M. Schwoon, R.S.J. Tol, 2006: CHIMP: A simple population model for use in integrated assessment of global environmental change. *The Integrated Assessment Journal*, **6**(3), pp. 1-33.

Fankhauser, S. and R.S.J. Tol, 2005: On climate change and economic growth. *Resource and Energy Economics*, **27**(1), pp. 1-17.

Friedlingstein, P., P. Bopp, P. Ciais, J.-L. Dufresne, L. Fairhead, H. LeTreut, P. Monfray, and J. Orr, 2001: Positive feedback between future climate change and the carbon cycle. *Geophysical Research Letters*, **28**(8), pp. 1543-1546.

Füssel, H.-M. and J.G. van Minnen, 2001: Climate impact response functions for terrestrial ecosystems. *The Integrated Assessment Journal*, **2**, pp. 183-197.

Fuglestvedt, J.S., T.K. Berntsen, O. Godal, R. Sausen, K.P. Shine, T. Skodvin, 2003: Metrics of climate change: Assessing radiative forcing and emission indices. *Climatic Change*, **58**(3), pp. 267-331.

Fujino, J., R. Nair, M. Kainuma, T. Masui, and Y. Matsuoka, 2006: Multi-gas mitigation analysis on stabilization scenarios using AIM global model. *The Energy Journal*, Vol. Multi-Greenhouse Gas Mitigation and Climate Policy, Special Issue No.3, pp. 343-354.

GCI, 2005: GCI Briefing: Contraction & Convergence. <http://www.gci.org.uk/briefings/ICE.pdf> accessed April 2006.

Gallopin, G., A. Hammond, P. Raskin, and R. Swart, 1997: Branch points: global scenarios and human choice: a resource paper of the Global Scenario Group. Stockholm Environment Institute, Stockholm, 55 pp. <http://www.tellus.org/seib/publications/branchpt.pdf>, accessed 1 June 2007.

Georgescu-Roegen, 1971: *The Entropy Law and the Economic Process*. Harvard University Press, Cambridge MA, 476 pp.

Gitz, V. and P. Ciais, 2003: Amplifying effects of land-use change on future atmospheric CO_2 levels. *Global Biogeochemical cycles*, **17**(1), pp. 24-39.

Gitz, V. and P. Ciais, 2004: Future expansion of agriculture and pasture acts to amplify atmospheric CO_2 in response to fossil fuel and land-use change emissions. *Climate Change*, **67**(2-3), pp. 161-184.

Glenn, J.C. and T.J. Gordon, 1997: *1997 State of the future: implications for actions today*. American Council for the United Nations University, Washington DC, 202 pp.

Godal, O., 2003: The IPCC's assessment of multidisciplinary issues: The case of greenhouse gas indices. *Climate Change*, **58**(3), pp. 243-249.

Goklany, I.M., 2003: Relative contributions of global warming to various climate sensitive risks, and their implications for adaptation and mitigation. *Energy and Environment*, **14**(6), pp. 797-822.

Goulder, L.H., 2004: Induced technological change and climate policy. Pew Center on Global Climate Change, Washington DC, 47 pp. <http://www.pewclimate.org/docUploads/ITC%5FReport%5FF2%2E pdf>, accessed 1 June 2007.

Gray, H.A. and G.R. Cass, 1998: Source contributions to atmospheric fine carbon particle concentrations. *Atmospheric Environment*, **32**(22), pp. 3805-3825.

Gritsevskyi, A. and N. Nakicenovic, 2000: Modeling uncertainty of induced technological change. *Energy Policy*, **28**(13), pp. 907-921.

Grubb, M., O. Edenhofer, and C. Kemfert, 2005: Technological change and the innovation modeling comparison project. IPCC Expert Meeting on Emissions Scenarios, 12 - 14 January 2005, Washington DC.

Grübler, A., 2002: Trends in global emissions: Carbon, sulphur and nitrogen, *Encyclopedia of Global Environmental Change*, 3, pp. 35-53.

Grübler, A., N. Nakicenovic, and W.D. Nordhaus (eds.), 2002: *Technological change and the environment*. Resources for the Future Press, Washington DC, 407 pp.

Grübler, A., N. Nakicenovic, J. Alcamo, G. Davis, J. Fenhann, B. Hare, S. Mori, B. Pepper, H. Pitcher, K. Riahi, H.H. Rogner, E.L. La Rovere, A. Sankovski, M. Schlesinger, R.P. Shukla, R. Swart, N. Victor, and T.Y. Jung, 2004: Emissions scenarios: a final response. *Energy and Environment*, **15**(1), pp. 11-24.

Grübler, A., B. O'Neill, K. Riahi, V. Chirkov, A. Goujon, P. Kolp, I. Prommer, S. Scherbov, and E. Slentoe, 2006: Regional, national, and spatially explicit scenarios of demographic and economic change based on SRES. *Technological Forecasting and Social Change*, Special Issue, 74(8–9). doi:10.1016/j.techfore.2006.05.023.

GTSP, 2001: Global Energy Technology Strategy: Addressing Climate Change, 60 pp. <http://www.pnl.gov/gtsp/docs/infind/cover.pdf>, accessed 1 June 2007.

Hall, D.C. and R.J. Behl, 2006: Integrating economic analysis and the science of climate instability. *Ecological Economics*, **57**(3), pp. 442-465.

Halloy, S.R.P. and J.A. Lockwood, 2005: Ethical implications of the laws of pattern abundance distribution. *Emergence: Complexity and Organization*, **7**(2), pp. 41-53.

Halsnaes, K., 2002: A review of the literature on climate change and sustainable development. In *Climate change and sustainable development: prospects for developing countries*. A. Markandya and Kirsten Halsnaes (eds.), Earthscan Publications, London, pp. 272.

Hanaoka, T., R. Kawase, M. Kainuma, Y. Matsuoka, H. Ishii, and K. Oka, 2006: Greenhouse gas emissions scenarios database and regional mitigation analysis. CGER-D038-2006, National Institute for Environmental Studies, Tsukuba, 106 pp. <http://www-cger.nies.go.jp/publication/D038/all_D038.pdf>, accessed 1 June 2007.

Hansen, J., M. Sato, R. Ruedy, A. Lacis, and V. Oinas, 2000: Global warming in the twenty-first century: an alternative scenario. *Proceedings of the National Academy of Sciences of the United States of America*, **97**(18), pp. 9875-9880.

Hanson, D.A., I. Mintzer, J.A. Laitner, and J.A. Leonard, 2004: Engines of growth: energy challenges, opportunities and uncertainties in the 21st century. Argonne National Laboratory, Argonne, IL, 59 pp. <http://www.ari.vt.edu/hydrogen/Resources/BackDoc/Doc/engines_growth.pdf>, accessed 1 June 2007.

Hare, B. and M. Meinshausen, 2006: How much warming are we committed to and how much can be avoided? *Climatic Change*, 75(1-2), pp. 111-149.

Harvey, L.D.D., 2004: Declining temporal effectiveness of carbon sequestration: implications for compliance with the United Nations Framework Convention on Climate Change. *Climatic Change*, **63**(3), pp. 259-290.

Harvey, L.D.D., 2006: Uncertainties in global warming science and near-term emission policies, *Climate Policy*, **6**(5), pp. 573–584.

Hayhoe, K., D. Cayan, C.B. Field, P.C. Frumhoff, E.P. Maurer, N.L. Miller, S.C. Moser, S.H. Schneider, K.N. Cahill, E.E. Cleland, L. Dale, R. Drapek, R.M. Hanemann, L.S. Kalkstein, J. Lenihan, C.K. Lunch, R.P. Neilson, S.C. Sheridan, and J.H. Verville, 2004: Emissions pathways, climate change, and impacts on California. *Proceedings of the National Academy of Sciences of the United States of America*, **101**(34), pp. 12422-12427.

Henderson, D., 2005: The treatment of economic issues by the Intergovernmental Panel on Climate Change. *Energy & Environment*, **16**(2), pp. 321-326.

Herzog, H., K. Smekens, P. Dadhich, J. Dooley, Y. Fujii, O. Hohmayer, K. Riahi, 2005: Cost and economic potential. In: *IPCC Special Report on Carbon Dioxide Capture and Storage*. Intergovernmental Panel on Climate Change, Cambridge University Press, Cambridge, pp. 339-362.

Hijioka, Y, T. Masui, K. Takahashi, Y. Matsuoka, and H. Harasawa, 2006: Development of a support tool for greenhouse gas emissions control policy to help mitigate the impact of global warming. *Environmental Economics and Policy Studies*, **7**(3), pp. 331-345.

Holdren, J.P., 2000: Environmental degradation: population, affluence, technology, and sociopolitical factors. *Environment*, **42**(6), pp. 4-5.

Holtsmark, B.J. and K.H. Alfsen, 2004a: PPP correction of the IPCC emission scenarios - does it matter? *Climatic Change*, **68**(1-2), pp. 11-19.

Holtsmark, B.J. and K.H. Alfsen, 2004b: The use of PPP or MER in the construction of emission scenarios is more than a question of 'metrics'. *Climate Policy*, **4**(2), pp. 205-216.

Hoogwijk, M., A. Faaij, B. Eickhout, B. de Vries and W. Turkenburg, 2005: Potential of biomass energy out to 2100, for four IPCC SRES land-use scenarios. *Biomass and Bioenergy*, **29**(4), pp. 225-257.

Hope, C., 2006: The Marginal Impact of CO_2 from PAGE 2002: an Integrated Assessment Model Incorporating the IPCC's Five Reasons for Concern. *The Integrated Assessment Journal*, **6**(1), pp. 19-56.

Houghton, R.A., 2003: Revised estimates of the annual net flux of carbon to the atmosphere from changes in land use and land management 1850-2000. *Tellus, Serie B: Chemical and Physical Meteorology*, **55b**(2), pp. 378-390.

Hourcade, J.-C. and P. Shukla, 2001: Global, regional and national costs and ancillary benefits of mitigation. In *Climate change 2001: Mitigation. Contribution of Working Group III to the Third Assessment Report of the Intergovernmental Panel on Climate Change*. Cambridge University Press, Cambridge MA, pp. 702.

House, J.I., I.C. Prentice, and C. Le Quere, 2002: Maximum impacts of future reforestation or deforestation on atmospheric CO_2. *Global Change Biology*, **8**(11), pp. 1047-1052.

Howarth, R.B., 2003: Catastrophic outcomes in the economics of climate change. An editorial comment. *Climatic Change*, **56**(3), pp. 257-263.

IEA, 2002: *World Energy Outlook 2002*. International Energy Agency, Paris, 530 pp.

IEA, 2004: *World Energy Outlook 2004*. International Energy Agency, Paris, 570 pp.

IPCC, 1996a: Climate Change 1995: impacts, adaptations and mitigation of climate change: scientific-technical analysis. Contribution of Working Group II to the Second Assessment Report of the IPCC. R.T. Watson, M.C. Zinyowera, R.H. Moss [eds], Intergovernmental Panel on Climate Change, Cambridge University Press, Cambridge.

IPCC, 1996b: Climate Change 1995: economic and social dimensions of climate change. Contribution of Working Group III to the Second Assessment Report of the IPCC. J.B. Bruce, H. Lee, E.F. Haites [eds], Intergovernmental Panel on Climate Change, Cambridge University Press, Cambridge.

IPCC, 1997: Revised 1996 IPCC Guidelines for National Greenhouse Inventories. Intergovernmental Panel on Climate Change, IPCC/OECD/IEA, Paris, France.

IPCC, 2000: Land use, land-use change and forestry. A special report of the IPCC. Intergovernmental Panel on Climate Change, Cambridge University Press, Cambridge UK, 377 pp.

IPCC, 2001a: Climate change 2001: Mitigation. Contribution of Working Group III to the Third Assessment Report of the IPCC. Intergovernmental Panel on Climate Change, Cambridge University Press, Cambridge, UK, 702 pp.

IPCC, 2001b: Climate Change 2001: The Scientific Basis. Contribution of Working Group I to the Third Assessment Report of the IPCC. Intergovernmental Panel on Climate Change, Intergovernmental Panel on Climate Change, Cambridge University Press, UK, 944 pp.

IPCC, 2007a: Climate Change 2007: The Physical Science Basis. Contribution of Working Group I to the Fourth Assessment Report of the Intergovernmental Panel on Climate Change [Solomon, S., D. Qin, M. Manning, Z. Chen, M. Marquis, K.B.M.Tignor and H.L. Miller (eds.)]. Cambridge University Press, Cambridge, United Kingdom and New York, NY, USA, 996 pp.

IPCC, 2007b: Climate Change 2007: Impacts, Adaptation and Vulnerability. Contribution of Working Group II to the Fourth Assessment Report of the Intergovernmental Panel on Climate Change [Parry, M.L., O.F. Canziani, J.P. Palutikof, P.J. van der Linden, C.E. Hanson (eds.)]. Cambridge University Press, Cambridge, United Kingdom and New York, NY, USA.

Izrael, Y.A. and S.M. Semenov, 2005: Calculations of a change in CO_2 concentration in the atmosphere for some stabilization scenarios of global emissions using a model of minimal complexity. *Russian Meteorology and Hydrology*, **1**, pp. 1-8.

Jacobson, M.Z., 2001: Strong radiative heating due to the mixing state of black carbon in atmospheric aerosols. *Nature*, **409**(6821), pp. 695-697.

Jacoby, H.D., 2004: Informing climate policy given incommensurable benefits estimates. *Global Environment Change*, **14**(3), pp. 287-297.

Jakeman, G. and B.S. Fisher, 2006: Benefits of multi-gas mitigation: An application of the global trade and environment model (GTEM). *The Energy Journal*, Multi-Greenhouse Gas Mitigation and Climate Policy, Special Issue No.3, pp. 323-342.

JAIF, 2004: 2050 Nuclear Vision and Roadmap. Japan Atomic Industrial Forum, Inc, Tokyo. <http://www.jaif.or.jp/english/news/2005/0317vision.html>, accessed 1 June 2007.

Japan LCS Project, 2005: Backcasting from 2050, Japan Low Carbon Society Scenarios Toward 2050. LCS Research Booklet No.1, National Institute for Environmental Studies, Tsukuba, 11 pp, <http://2050.nies.go.jp/material/lcs_booklet/lcs_booklet.html>, accessed 1 June 2007.

Jiang, K. and H. Xiulian, 2003: Long-term GHG emissions scenarios for China. Second Conference on Climate Change, Beijing.

Jiang, K., H. Xiulian, and S. Zhu, 2006: Multi-gas mitigation analysis by IPAC. *The Energy Journal*, Vol. Multi-Greenhouse Gas Mitigation and Climate Policy, Special Issue No.3, pp. 425-440.

Johansson, D.J.A., U.M. Persson and C. Azar, 2006: The cost of using global warming potentials: Analysing the trade off between CO_2, CH_4 and N_2O. *Climatic Change*, **77**(3-4), pp. 291-309.

Jones, C.I., 1997: Convergence revisited. *Journal of Economic Growth*, **2**(2), pp. 131-153.

Jung, T.Y., E.L. La Rovere, H. Gaj, P.R. Shukla, and D. Zhou, 2000: Structural changes in developing countries and their implication to energy-related CO_2 emissions. *Technological Forecasting and Social Change*, **63**(2-3), pp. 111-136.

Junker, C. and C. Liousse, 2006: A Global Emission Inventory of Carbonaceous Aerosol from historic records of Fossil Fuel and Biofuel consumption for the Period 1860 - 1997, *Atmospheric Chemistry and Physics Discussions*, **6**, pp. 4897–4927

Kahn, H. and A. Weiner, 1967: The year 2000: a framework for speculation on the next thirty-three years. Macmillan, New York, 431 pp.

Kahn, H., W. Brown, and L. Martel, 1976: *The next 200 years: a scenario for America and the World*. Morrow, New York, 241 pp.

Kaplan, R.D., 1994: The coming anarchy. *The Atlantic Monthly*, **272**(2), pp. 44-76.

Kawase, R., Y. Matsuoka, and J. Fujino, 2006: Decomposition analysis of CO_2 emission in long-term climate stabilization scenarios. *Energy Policy*, **34**(15), pp. 2113-2122.

Keller, K., K. Tan, F.M.M. Morel, and D.F. Bradford, 2000: Preserving the ocean circulation: implications for climate policy. *Climatic Change*, **47**(1-2), pp. 17-43.

Keller, K., B.M. Bolker, and D.F. Bradford, 2004: Uncertain climate thresholds and economic optimal growth. *Journal of Environmental Economics and Management*, **48**, pp. 723-741.

Keller, K., G. Yohe, and M. Schlesinger, 2006: Managing the risks of climate thresholds: Uncertainties and information needs. *Climatic Change*, doi:10.1007/s10584-006-9114-6. <http://www.geosc.psu.edu/~kkeller/publications.html>, accessed 1 June 2007.

Kelly, D.L., C.D. Kolstad, M.E. Schlesinger, and N.G. Andronova, 2000: Learning about climate sensitivity from the instrumental temperature record. <http://www.econ.ucsb.edu/papers/wp32-98.pdf>, accessed 1 June 2007.

Kelly, D.L., C.D. Kolstad, and G.T. Mitchell, 2005: Adjustment costs from environmental change. *Journal of Environmental Economics and Management*, **50**(3), pp. 468-495.

Kemfert, C., 2002: Global economic implications of alternative climate policy strategies. *Environmental Science and Policy*, **5**(5), pp. 367-384.

Kemfert, C. and K. Schumacher, 2005: Costs of inaction and costs of action in climate protection: assessment of costs of inaction or delayed action of climate protection and climate change. Final Report: Project FKZ 904 41 362 for the Federal Ministry for the Environment, Berlin. <http://www.diw.de/deutsch/produkte/publikationen/diwkompakt/docs/diwkompakt_2005-013.pdf>, accessed 1 June 2007.

Kemfert, C., 2006: An integrated assessment of economy, energy and climate. The model WIAGEM - A reply to comment by Roson and Tol. *The Integrated Assessment Journal*, **6**(3), pp. 45-49.

Kemfert, C., T.P. Truong, and T. Bruckner, 2006: Economic impact assessment of climate change - A multi-gas investigation with WIAGEM-GTAPEL-ICM. *The Energy Journal*, Multi-Greenhouse Gas Mitigation and Climate Policy, Special Issue No.3, pp. 441-460.

Keppo, I., B. O'Neill, and K. Riahi, 2006: Probabilistic temperature change projections and energy system implications of greenhouse gas emission scenarios. *Technological Forecasting and Social Change*, Special Issue, 74(8–9), doi:10.1016/j.techfore.2006.05.024.

Kheshgi, H.S. 2004: Evasion of CO_2 injected into the ocean in the context of CO_2 stabilization. *Energy*, **29**(9-10), pp. 1479-1486.

Kheshgi, H.S., S.J. Smith, and J.A. Edmonds, 2005: Emissions and atmospheric CO_2 stabilization: Long-term limits and paths. *Mitigation and Adaptation Strategies for Global Change*, **10**(2), pp. 213-220.

Kinsman, F., 1990: *Millennium: towards tomorrow's society*. W.H. Allen, London.

Kokic, P., A. Heaney, L. Pechey, S. Crimp, and B.S. Fisher, 2005: Climate change: predicting the impacts on agriculture: a case study. *Australian Commodities*, **12**(1), pp. 161-170.

Köhler, J., M. Grubb, D. Popp and O. Edenhofer, 2006: The transition to endogenous technical change in climate-economy models: A technical overview to the innovation modeling comparison project. *The Energy Journal*, Endogenous Technological Change and the Economics of Atmospheric Stabilisation, Special Issue No. 1.

Kupiainen, K. and Z. Klimont, 2004: Primary emissions of submicron and carbonaceous particles in Europe and the potential for their control. IR-04-079, International Institute for Applied Systems Analysis Interim Report, Laxenburg. <http://www.iiasa.ac.at/rains/reports/ir-04-079.pdf>, accessed 1 June 2007.

Kurosawa, A., 2004: Carbon concentration target and technological choice. *Energy Economics*, **26**(4), pp. 675-684.

Kurosawa, A., 2006: Multi-gas mitigation: An economic analysis using GRAPE model. *The Energy Journal*, Vol. Multi-Greenhouse Gas Mitigation and Climate Policy, Special Issue No.3, pp. 275-288.

Kypreos, S., 2006: Stabilizing global temperature change below thresholds; A Monte Carlo analyses with MERGE. Paul Scherrer Institut, Villigen, 28 pp. <http://eem.web.psi.ch/Publications/Books_Journals/2006_Kypreos_stabilizing_with_merge.pdf>, accessed 1 June 2007.

La Rovere, E.L. and B.B. Americano, 2002: Domestic actions contributing to the mitigation of GHG emissions from power generation in Brazil. *Climate Policy*, **2**(2-3), pp. 247-254.

Lazarus, M., L. Greber, J. Hall, C. Bartels, S. Bernow, E. Hansen, P. Raskin, and D. Von Hippel, 1993: Towards a fossil free energy future: the next energy transition. Stockholm Environment Institute, Boston Center, Boston. Greenpeace International, Amsterdam.

Leemans, R., B. Eickhout, B. Strengers, L. Bouwman, and M. Schaeffer, 2002: The consequences of uncertainties in land use, climate and vegetation responses on the terrestrial carbon. *Science in China*, **45**(Supplement), pp. 126-141.

Leggett, J., W.J. Pepper, and R.J. Swart , J. Edmonds, L.G. Meira Filho, I. Mintzer, M.X. Wang, and J. Watson, 1992: Emissions scenarios for IPCC: an update. In Climate change 1992: the supplementary report to the IPCC Scientific Assessment. Cambridge University Press, Cambridge.

Lehtila, A., I. Savolainen, and S. Syri, 2005: The role of technology development in greenhouse gas emissions reduction: the case of Finland. *Energy*, **30**(14), pp. 2738-2758.

Lempert, R.J., M.E. Schlesinger, and J.K. Hammitt, 1994: The impact of potential abrupt climate changes on near-term policy changes. *Climatic Change*, **26**(4), pp. 351-376.

Levy, P.E., A.D. Friend, A. White, and M.G.R. Cannell, 2004: The influence of land use change on global-scale fluxes of carbon from terrestrial ecosystems. *Climatic Change*, **67**(2-3), pp. 185-209.

Liousse, C., J.E. Penner, C. Chuang, J.J. Walton, H. Eddleman, and H. Cachier, 1996: A global three-dimensional model study of carbonaceous aerosols. *Journal of Geophysical Research*, **101**(14), pp. 19411-19432.

Liousse, C., B. Guillaume, C. Junker, C. Michel, H. Cachier, B. Guinot, P. Criqui, S. Mima, and J.M. Gregoire, 2005: Management and impact on climate change (a french initiative). GICC report, Carbonaceous Aerosol Emission Inventories from 1860 to 2100, March. <http://medias.obs-mip.fr/gicc/interface/projet.php?4/01>, accessed 1 June 2007.

Lutz, W. and W. Sanderson, 2001: The end of world population growth. *Nature*, **412**(6846), pp. 543-545.

Lutz, W., W.C. Sanderson, and S. Scherbov, 2004: *The end of world population growth. In: The end of world population growth in the 21st century: New challenges for human capital formation and sustainable development*. W. Lutz and W. Sandersen (eds.), Earthscan Publications, London, pp. 17-83.

Maddison, D., 2001: *The amenity value of global climate*. Earthscan, London, 240 pp.

Mankiw, N.G., D. Romer, and D.N. Weil, 1992: A contribution to the empirics of economic growth. *Quarterly Journal of Economics*, **107**(2), pp. 407-437.

Manne, A.S. and R.G. Richels, 2001: An alternative approach to establishing trade-offs among greenhouse gases. *Nature*, **410**(6829), pp. 675-677.

Manne, A. and R. Richels, 2003: Market exchange rates or purchasing power parity: does the choice make a difference in the climate debate? AEI-Brookings Joint Center for Regulatory Studies, Working Paper No. 03-11, 14 pp. <http://ssrn.com/abstract=449265>, accessed 1 June 2007.

Manne, A. and R. Richels, 2004: The impact of learning-by-doing on the timing and costs of CO_2 abatement. *Energy Economics*, **26**(4), pp. 603-619.

Manne, A.S. and R.G. Richels, 2006: The role of non-CO_2 greenhouse gases and carbon sinks in meeting climate objectives. *The Energy Journal*, Vol. Multi-Greenhouse Gas Mitigation and Climate Policy, Special Issue No.3, pp. 393-404.

Martins, J.O. and G. Nicoletti, 2005: Long-term economic growth projections: could they be made less arbitrary? IPCC Expert Meeting on Emissions Scenarios, 12 - 14 January 2005, Washington DC.

Mastrandrea, M.D., and S.H. Schneider, 2001: Integrated assessment of abrupt climatic changes. *Climate Policy*, **1**(4), pp. 433-449.

Mastrandrea, M.D. and S.H. Schneider, 2004: Probabilistic integrated assessment of 'dangerous' climate change. *Science*, **304**(5670), pp. 571-575.

Masui, T., Y. Matsuoka, and M. Kainuma, 2006: Long-term CO_2 emission reduction scenarios in Japan. *Environmental Economics and Policy Studies*, **7**(3), pp. 347-366.

Matthews, H.D., A.J. Weaver, M. Eby, and K.J. Meissner, 2003: Radiative forcing of climate by historical land cover change. *Geophysical Research Letters*, **30**(2), pp. 1055-1059.

Matthews, H.D., 2005: Decrease of emissions required to stabilize atmospheric CO_2 due to positive carbon cycle-climate feedbacks. *Geophysical Research Letters*, **32**, L21707.

Mayerhofer, P., B. de Vries, M.G.J. den Elzen, D.P. van Vuuren, J. Onigkeit, M. Posch, and R. Guardans, 2002: Long-term, consistent scenarios of emissions, deposition and climate change in Europe. *Environmental Science and Policy*, **5**(4), pp. 273-305.

McGuire, A.D., S. Sitch, J.S. Clein, R. Dargaville, G. Esser, J. Foler, M. Heimann, F. Joos, J. Kaplan, D.W. Kicklighter, R.A. Meier, J.M. Melillo, B. Moore, I.C. Prentice, N. Ramankutty, T. Reichenau, A. Schloss, H. Tian, L.J. Williams, and U. Wittenberg, 2001: Carbon balance of the terrestrial biosphere in the twentieth century: analyses of CO_2, climate and land use effects with four process-based ecosystem models. *Global Biogeochemical Cycles*, **15**(1), pp. 183-206.

McKibbin, W.J., D. Pearce, and A. Stegman, 2004a: Can the IPCC SRES be improved? *Energy and Environment*, **15**(3), pp. 351-362.

McKibbin, W.J., D. Pearce, and A. Stegman, 2004b: Long-run projections for climate change scenarios. Centre for Applied Macroeconomics Analysis Working Paper Series, Australian National University, Canberra, 71 pp. <http://cama.anu.edu.au/Working%20Papers/Papers/2004/McKibbinPaper12004.pdf>, accessed 1 June 2007.

Meadows, D.H., D.L. Meadows, J. Randers, and W.W. Behrens, 1972: The limits to growth: a report for the club of Rome's project on the predicament of mankind. Universe Books, New York.

Meehl, G.A., W.M. Washington, W.D. Collins, J.M. Arblaster, A. Hu, L.E. Buja, W.G. Strand, and H. Teng, 2005: How much more global warming and sea level rise? *Science*, **307**(5716), pp. 1769-1772.

Meehl, G. A., T. F. Stocker, W. D. Collins, P. Friedlingstein, A. T. Gaye, J. M. Gregory, A. Kitoh, R. Knutti, J. M. Murphy, A. Noda, S. C. B. Raper, I. G. Watterson, A. J. Weaver, Z.-C. Zhao, 2007: Global Climate Projections. In *Climate Change 2007: The Physical Science Basis. Contribution of Working Group I to the Fourth Assessment Report of the Intergovernmental Panel on Climate Change*, S. Solomon, D. Qin, M. Manning, Z. Chen, M. Marquis, K. B. Averyt, M. Tignor, and H. L. Miller (eds.), Cambridge University Press, Cambridge, UK and New York, NY, USA.

Meier, G.M., 2001: The old generation of development economists and the new. In *Frontiers of development economics. The Future in perspective*. G.M. Meier and J.M. Stiglitz (eds.), Oxford University Press, New York.

Meinshausen, M., 2006: What does a 2 °C Target mean for greenhouse gas concentrations? A brief analysis based on multi-gas emission pathways and several climate sensitivity uncertainty estimates. In *Avoiding Dangerous Climate Change*. H.J. Schellnhuber, W. Cramer, N. Nakicenovic, T. Wigley, and G. Yohe (eds.), Cambridge University Press, Cambridge.

Meinshausen, M., B. Hare, T.M.M. Wigley, D. van Vuuren, M.G.J. den Elzen, and R. Swart, 2006: Multi-gas emissions pathways to meet climate targets. *Climatic Change*, **75**(1-2), pp. 151-194.

Mendelsohn, R., W. Morrison, M.E. Schlesinger, and N.G. Andronova, 2000: Country-specific market impacts of climate change. *Climatic Change*, **45**(3-4), pp. 553.

Mendelsohn, R. and L. Williams, 2004: Comparing forecasts of the global impacts of climate change. *Mitigation and Adaptation Strategies for Global Change*, **9**(4), pp. 315-333.

Mesarovic, M.D. and E. Pestel, 1974: *Mankind at the turning point: the second report to the Club of Rome*. Dutton, New York, 210 pp.

Metz, B. and D. P. Van Vuuren, 2006: How, and at what costs, can low-level stabilisation be achieved? - An overview. In *Avoiding Dangerous Climate Change*. H.J. Schellnhuber, W. Cramer, N. Nakicenovic, T. Wigley, and G. Yohe (eds.), Cambridge University Press, Cambridge, UK.

METI, 2005: Strategic Technology Roadmap (Energy Sector) - Energy Technology Vision 2100-. Ministry of Economy, Trade and Industry, The Institute of Applied Energy, Tokyo, 96 pp. <http://www.iae.or.jp/2100/main.pdf>, accessed 1 June 2007.

MIES, 2004: Reducing CO_2 emissions fourfold in France by 2050: introduction to the debate. La Mission Interministérielle de l'Effet de Serre, Paris, 40 pp. <http://www.climnet.org/pubs/1004_France_Factor%204.pdf>, accessed 1 June 2007.

Mintzer, I., J.A. Leonard, and P. Schwartz, 2003: U.S. energy scenarios for the 21st century. Pew Center on Global Climate Change, Arlington, 88 pp. <http://www.pewclimate.org/docUploads/EnergyScenarios.pdf>, accessed 1 June 2007.

Morgenstern, 2000: Baseline issues in the estimation of ancillary benefits of greenhouse gas mitigation policies, Ancillary benefits and costs of greenhouse gas mitigation. OECD Proceedings of an IPCC Co-sponsored Workshop, 27-29 March 2000, in Washington DC, OECD Publishing, Paris, pp. 95-122.

Mori, S., 2003: Issues on global warming, energy and food from the long-term point of view. *Journal of the Japan Institute of Energy*, **82**(1), pp. 25-30.

Morita, T. and H.-C. Lee, 1998: Appendix: IPCC Emissions Scenarios Database, *Mitigation and Adaptation Strategies for Global Change*, **3**, pp. 121-131.

Morita, T., J. Robinson, A. Adegbulugbe, J. Alcamo, D. Herbert, E. Lebre La Rovere, N. Nakicenovic, H. Pitcher, P. Raskin, K. Riahi, A. Sankovski, V. Sololov, B. de Vries and D. Dadi, 2001: Greenhouse gas emission mitigation scenarios and implications. In *Climate change 2001: Mitigation, Report of Working Group III of the IPCC*. Cambridge University Press, Cambridge, pp. 115-166.

Nair, R., P.R. Shukla, M. Kapshe, A. Garg, and A. Rana, 2003: Analysis of long-term energy and carbon emission scenarios for India. *Mitigation and Adaptation Strategies for Global Change*, **8**(1), pp. 53-69.

Nakicenovic, N., 1996: Freeing energy from carbon. *Daedalus*, **125**(3), pp. 95-112.

Nakicenovic, N., J. Alcamo, G. Davis, B. de Vries, J. Fenham, S. Gaffin, K. Gregory, A. Grübler, T.-Y. Jung, T. Kram, E.L. La Rovere, L. Michaelis, S. Mori, T. Morita, W. Pepper, H. Pitcher, L. Price, K. Riahi, A. Reohrl, H.H. Rogner, A. Sankovski, M. Schlesinger, P. Shukla, S. Smith, R. Swart, S. van Rooijen, N. Victor, and Z. Dadi, 2000: *Special report on emissions scenarios. Working Group III, Intergovernmental Panel on Climate Change (IPCC)*. Cambridge University Press, Cambridge, 595 pp.

Nakicenovic, N. and K. Riahi, 2003: Model runs with MESSAGE in the context of the further development of the Kyoto-Protocol. Wissenschaftlicher Beirat der Bundesregierung Globale Umweltveränderungen, Berlin, 54 pp. <http://www.wbgu.de/wbgu_sn2003_ex03.pdf>, accessed 1 June 2007.

Nakicenovic, N., A. Grübler, S. Gaffin, T.T. Jung, T. Kram, T. Morita, H. Pitcher, K. Riahi, M. Schlesinger, P.R. Shukla, D. van Vuuren, G. Davis, L. Michaelis, R. Swart, and N. Victor, 2003: IPCC SRES revisited: a response. *Energy and Environment*, **14**(2-3), pp. 187-214.

Nakicenovic, N., J. McGalde, S. Ma, J. Alcamo, E. Bennett, W. Cramer, J. Robinson, F.L. Toth, M. Zurek, 2005: Lessons learned for scenario analysis. In Ecosystems and Human Well-being: Scenarios. Carpenter, S.R., P.L. Pingali, E.M. Bennet, M.B. Zurek (eds.), Vol.II, Millennium Ecosystem Assessment (MA), Island Press, Chicago, pp. 449-468.

Nakicenovic, N., P. Kolp, K. Riahi, M. Kainuma, T. Hanaoka, 2006: Assessment of emissions scenarios revisited. *Environmental Economics and Policy Studies*, **7**(3), pp. 137-173.

Newell, R.G., A.B. Jaffe, and R.N. Stavins, 1999: The induced innovation hypothesis and energy-saving technological change. *Quarterly Journal of Economics*, **114**(3), pp. 941-975.

Nicholls, R.J., S.E. Hansen, J.A. Lowe, D.A.Vaughan, T. Lenton, A. Ganopolski, R.S.J. Tol, and A.T. Vafeidis, 2006: Metrics for assessing the economic benefits of climate change policies: sea level rise, Organisation for Economic Co-operation and Development, ENV/EPOC/GSP(2006)3/FINAL, OECD Publishing, Paris, pp. 125. <http://www.oecd.org/dataoecd/19/63/37320819.pdf>, accessed 1 June 2007.

Nordhaus, W.D. and J. Boyer, 2000: Warming the world: economic models of global warming. MIT Press, Cambridge MA, 244 pp.

Nordhaus, W.D., 2005: Should modelers use purchasing power parity or market exchange rates in global modelling systems. IPCC Seminar on Emission Scenarios, Intergovernmental Panel on Climate Change (IPCC), 12 - 14 January 2005, Washington DC.

Nordhaus, W.D., 2006a: Geography and macroeconomics: New data and new findings. *Proceedings of the National Academy of Sciences of the United States of America*, **103**(10), pp. 3510-3517.

Nordhaus, W.D., 2006b: The "Stern" review on the economics of climate change. National Bureau of Economic Research, Working Paper No. 12741, pp. 10. <http://ssrn.com/abstract=948654>, accessed 1 June 2007.

NRCan, 2000: Energy Technology Futures. 2050 Study: Greening the pump. National Resources Canada, Ottowa.

OECD, 2000: Ancillary benefits and costs of greenhouse gas mitigation. Organisation for Economic Co-operation and Development Proceedings of an IPCC co-sponsored workshop in Washington DC, 27-29 March 2000, OECD Publishing, Paris.

OECD, 2003: Organisation for Economic Co-operation and Development Workshop on the benefits of climate policy: improving information for policy makers. 12-13 December 2002, OECD Publishing, Paris.

Olivier, J.G.J., A.F. Bouwman, K.W. Van der Hoek, and J.J.M. Berdowski, 1998: Global air emission inventories for anthropogenic sources of NO$_x$, NH$_3$ and N$_2$O in 1990. Environmental Pollution, 102(Supplement 1), pp. 135-148.

Olivier, J.G.J. and J.J.M. Berdowski, 2001: Global emissions sources and sinks. In *The Climate System*. J. Berdowski, R. Guicherit and B.J. Heij (eds.), A.A. Balkema/Swets & Zeitlinger, Lisse, pp. 33-78.

O'Neill, B.C. and M. Oppenheimer, 2002: Dangerous climate impacts and the Kyoto Protocol, *Science*, **296**(5575), pp. 1971-1972.

O'Neill, B.C., 2003; Economics, natural science, and the costs of Global Warming Potentials. *Climatic Change*, **58**(3), pp. 251-260.

O'Neill, B.C., 2004: Conditional probabilistic population projections: an application to climate change. *International statistical review*, **72**(2), pp. 167-184.

O'Neill, B.C. and M. Oppenheimer, 2004: Climate change impacts sensitive to the concentration stabilization path. *Proceedings of the National Academy of Science of the United States of Amercia*, **101**(47), pp. 16411-16416.

Ostrom, E., 1990: *Governing the commons: The evolution of institutions for collective action*. Cambridge University Press, Cambridge, 298 pp.

Ostrom, E., T. Dietz, N. Dolsak, P.C. Stern, S. Stonich, and E. Weber (eds.), 2002: *The drama of the commons*. National Academy Press, Washington DC, 521 pp.

Pacala, S. and R. Socolow, 2004: Stabilization wedges: Solving the climate problem for the next 50 years with current technologies. *Science*, **305**(5686), pp. 968 - 972.

Parry, M.L., C. Rosenzweig, A. Iglesias, M. Livermore, and G. Fischer, 2004: Effects of climate change on global food production under SRES emissions and socio economic scenarios. *Global Environmental Change*, **14**(1), pp. 53-67.

Pearce, D., 2003: The social cost of carbon and its policy implications. *Oxford Review of Economic Policy*, **19**(3), pp. 362-384.

Peck, S.C., and T.J. Teisberg, 1995: International CO$_2$ emissions control: an analysis using CETA. *Energy Policy*, **23**(4-5), pp. 297-308.

Penner, J.E., H. Eddleman, and T. Novakov, 1993: Towards the development of a global inventory for black carbon emissions. *Atmospheric Environment*, **27A**(8), pp. 1277-1295.

Penner, J.E., X. Dong, and Y. Chen, 2004: Observational evidence of a change in radiative forcing due to the indirect aerosol effect. *Nature*, **427**(6971), pp. 231-234.

Pepper, W., J. Leggett, R. Swart, J. Wasson, J. Edmonds, and I. Mintzer, 1992: Emission scenarios for the IPCC. An update: assumptions, methodology and results. In: *Climate change 1992: supplementary report to the IPCC scientific assessment*. Cambridge University Press, Cambridge.

Peterson, J.A. and L.F. Peterson, 1994: Ice retreat from the neoglacial maxima in the Puncak Jayakesuma area, Republic of Indonesia. *Zeitschrift fur Gletscherkunde und Glazialgeologie*, **30**, pp. 1-9.

Placet, M., K.K. Humphreys, and N.M. Mahasean, 2004: Climate change technology scenarios: Energy, emissions and economic implications. PNNL-14800, Pacific Northwest National Laboratory, US, 104 pp. <http://www.pnl.gov/energy/climate/climate_change-technology_scenarios.pdf>, accessed 1 June 2007.

Prather, M., R. Derwent, P. Erhalt, P. Fraser, E. Sanhueza, and X. Zhou, 1995: Other trace gases and atmospheric chemistry. In Climate change 1994 - radiative forcing of climate change. Intergovernmental Panel on Climate Change, Cambridge University Press, Cambridge, pp. 73-126.

Quah, D., 1993: Galton's fallacy and tests of the convergence hypothesis. The Scandinavian Journal of Economics, **95**(4), pp. 427-443.

Quah, D.T., 1996: Convergence empirics across economies with (some) capital mobility. *Journal of Economic Growth*, **1**(1), pp. 95-124.

Rahmstorf, S. and K. Zickfeld, 2005: Thermohaline Circulation Changes: A question of risk assessment. *Climatic Change*, **68**(1-2), pp. 241-247.

Ramaswamy, V., O. Boucher, J. Haigh, D. Hauglustaine, J. Haywood, G. Myhre, T. Nakajima, G.Y. Shi, and S. Solomon, 2001: Radiative forcing of climate change. In Climate change 2001: the scientific basis. Contribution of Working Group I to the Third Assessment Report to the Intergovernmental Panel on Climate Change. Cambridge University Press, New York, 881 pp.

Rao, S., K. Riahi, K. Kupiainen, and Z. Klimont, 2005: Long-term scenarios for black and organic carbon emissions. *Environmental Sciences*, **2**(2-3), pp. 205-216.

Rao, S. and K. Riahi, 2006: The role of non-CO$_2$ greenhouse gases in climate change mitigation: long-term scenarios for the 21st century. *The Energy Journal*, Vol. Multi-Greenhouse Gas Mitigation and Climate Policy. Special Issue No.3, pp. 177-200.

Raskin, P., G. Gallopin, P. Gutman, A. Hammond, and R. Swart, 1998: Bending the curve: toward global sustainability. A report of the Global Scenario Group. Stockholm Environment Institute, Stockholm, 144 pp. <http://www.tellus.org/seib/publications/bendingthecurve.pdf>, accessed 1 June 2007.

Raskin, P., T. Banuri, G. Gallopin, P. Gutman, A. Hammond, R. Kates, and R. Swart, 2002: *Great transition: the promise and lure of the times ahead*. A report of the Global Scenario Group. Stockholm Environment Institute, Stockholm, 111 pp. <http://www.tellus.org/seib/publications/Great_Transitions.pdf>, accessed 1 June 2007.

Raskin, P., F. Monks, T. Ribeiro, van D. Vuuren, and M. Zurek, 2005: Global scenarios in historical perspective. In Ecosystems and Human Well-being: Scenarios. Carpenter, S.R., P.L. Pingali, E.M. Bennet, M.B. Zurek (eds.), Vol. II, Millennium Ecosystem Assessment Report, Island Press, Washington DC.

Rayner, S., and E.L. Malone (eds.), 1998: *Human choice and climate change: the societal framework*. Battelle Press, Columbus, 536 pp.

Reilly, J., M. Sarofim, S. Paltsev, and R. Prinn, 2006: The role of non-CO$_2$ GHGs in climate policy: analysis using the MIT IGSM. *The Energy Journal* Vol. Multi-Greenhouse Gas Mitigation and Climate Policy, Special Issue No.3, p. 503-520.

Riahi, K. and R.A. Roehrl, 2001: Energy technology strategies for carbon dioxide mitigation and sustainable development. *Environmental Economics and Policy Studies,* **3**(2), pp. 89-123.

Riahi, K., A. Grübler, and N. Nakicenovic, 2006: Scenarios of long-term socio-economic and environmental development under climate stabilization. *Technological Forecasting and Social Change*, Special Issue, 74(8–9). doi:10.1016/j.techfore.2006.05.026.

Richels, R.G., A.S. Manne, and T.M.L. Wigley, 2004: Moving beyond concentrations: the challenge of limiting temperature change. AEI-Brookings Joint Center for Regulatory Studies, Working Paper No. 04-11, 32 pp. <http://ssrn.com/abstract=545742>, accessed 1 June 2007.

Rypdal, K., T. Berntsen, J. Fuglestvedt, K. Aunan, A. Torvanger, F. Stordal, J. Pacyna and L. Nygaard, 2005: Tropospheric ozone and aerosols in climate agreements: scientific and political challenges. *Environmental Science and Policy*, **8**(1), pp. 29-43.

Rose, S., H. Ahammad, B. Eickhout, B. Fisher, A. Kurosawa, S. Rao, K. Riahi, and D. van Vuuren, 2007: Land in climate stabilization modeling: Initial observations. Energy Modeling Forum Report, Stanford University, <http://www.stanford.edu/group/EMF/projects/group21/EMF21sinkspagenew.htm>, accessed 1 June 2007.

Roson, R. and R.S.J. Tol, 2006: An integrated assessment of economy-energy-climate -- the model WIAGEM: a comment. *The Integrated Assessment Journal*, **6**(1), pp. 75-82.

Rydin, Y., 2003: *Conflict, consensus and rationality in environmental planning: an institutional discourse approach*. Oxford University Press, Oxford, 216 pp.

Sachs, J.D., 2004: Seeking a global solution: the Copenhagen Consensus neglects the need to tackle climate change. *Nature*, **430**(7001), pp. 725-726.

Sands, R.D. and M. Leimbach, 2003: Modeling agriculture and land use in and integrated assessment framework. *Climatic Change*, **56**(1-2), pp. 185-210.

Sathaye, J., W. Makundi, L. Dale, P. Chan, and K. Andrasko, 2006: GHG mitigation potential, costs and benefits in global forests: A dynamic partial equilibrium approach. *The Energy Journal*, Vol. Multi-Greenhouse Gas Mitigation and Climate Policy, Special Issue No.3, pp. 127-162.

Savolainen, I., M. Ohlstrom, and S. Soimakallio, 2003: *Climate challenge for technology: views and results from the CLIMTECH program*. TEKES, The National Technology Agency of Finland, Helsinki.

Schaeffer, M., B. Eickhout, M. Hoogwijk, B. Strengers, D. van Vuuren, R. Leemans, and T. Opsteegh, 2006: CO_2 and albedo climate impacts of extratropical carbon and biomass plantations. *Global Biogeochemical Cycles*, **20**(2), GB2020.

Schneider, S.H., 2001: What is 'Dangerous' Climate Change? *Nature* **411**(6833), pp. 17-19.

Schneider, S. and J. Lane, 2004: Abrupt non-linear climate change and climate policy. In *The Benefits of Climate Change Policy: Analytical and Framework Issues*. J. Corfee-Morlot and S. Agrawala (eds.), OECD Publishing, Paris.

Schneider, S.H. and M.D. Mastrandrea, 2005: Probabilistic assessment of 'dangerous' climate change and emissions pathways. *Proceedings of the National Academy of Sciences of the United States of America*, **102**(44), pp. 15728-15735.

Schneider, S.H. and J. Lane, 2006: Dangers and thresholds in climate change and the implications for justice in *Fairness in Adaptation to Climate Change*, W.N. Adger, J. Paavola, S. Huq, and M.J. Mace (eds.), MIT Press, Cambridge MA.

Schwartz, P., 1991: *The art of the long view*. Doubleday, New York, 258 pp.

Shell, 2005: *The Shell Global Scenarios to 2025. The future business environment: trends, trade-offs and choices*. Institute for International Economics, 220 pp.

Shukla, P.R., 2005: The role of endogenous technology development in long-term mitigation and stabilization scenarios: a developing country perspective. IPCC Expert Meeting on Emissions Scenarios. Washington DC.

Shukla, P.R., A. Rana, A. Garg, M. Kapshe, and R. Nair, 2006: Global climate change stabilization regimes and Indian emission scenarios: lessons for modeling of developing country transitions. *Environmental Economics and Policy Studies*, **7**(3), pp. 205-231.

Sijm, J.B.M., 2004: Induced technological change and spillovers in climate policy modeling. An assessment. Energy Research Centre of the Netherlands, ZG Petten, 80 pp. <http://www.ecn.nl/docs/library/report/2004/c04073.pdf>, accessed 1 June 2007.

Sitch, S., V. Brovkin, W. von Bloh, D. Van Vuuren, B. Eickhout, A. Ganopolski, 2005: Impacts of future land cover on atmospheric CO_2 and climate. *Global Biogeochemical Cycles*, **19**(2), pp. 1-15.

Slovic, P., 2000: *The perception of risk*. Earthscan, London, 518 pp.

Smith, S.J., H. Pitcher, and T.M.L. Wigley, 2001: Global and regional anthropogenic sulfur dioxide emissions. *Global and Planetary Change*, **29**(1-2), pp. 99-119.

Smith, S.J., R. Andres, E. Conception, and J. Lurz, 2004: Historical sulfur dioxide emissions, 1850-2000: methods and results. Joint Global Research Institute, College Park, 14 pp. <http://www.pnl.gov/main/publications/external/technical_reports/PNNL-14537.pdf>, accessed 1 June 2007.

Smith, S.J., 2005: Income and pollutant emissions in the ObjECTS MiniCAM Model. *Journal of Environment and Development*, **14**(1), pp.175-196.

Smith, S.J., H. Pitcher, and T.M.L. Wigley, 2005: Future sulfur dioxide emissions. *Climatic Change*, **73**(3), pp. 267-318.

Smith, S.J. and T.M.L. Wigley, 2006: Multi-gas forcing stabilization with the MiniCAM. *The Energy Journal*, Vol. Multi-Greenhouse Gas Mitigation and Climate Policy, Special Issue No.3, pp. 373-392.

Sohngen, B. and R. Mendelsohn, 2003: An optimal control model of forest carbon sequestration. *American Journal of Agricultural Economics*, **85**(2), pp. 448-457.

Sohngen, B. and R. Sedjo, 2006: Carbon sequestration in global forests under different carbon price regimes. *The Energy Journal*, Vol. Multi-Greenhouse Gas Mitigation and Climate Policy, Special Issue, No.3, pp. 108-126.

Sohngen, B. and R. Mendelsohn, 2007: A sensitivity analysis of carbon sequestration. In *Human-induced climate change: An Interdisciplinary assessment*. M. Schlesinger, H. Kheshgi, J. Smith, F. de la Chesnaye, J.M. Reilly, T. Wilson, C. Kolstad (Eds.), Cambridge University Press, Cambridge.

Solow, R.M., 1956: A contribution to the theory of economic growth. *Quarterly Journal of Economics*, **70**(1), pp. 65-94.

Sørensen, B., 1999: *Long-term scenarios for global energy demand and supply - four global greenhouse mitigation scenarios*. Roskilde University.

Stern, D.I., 2005: Global sulfur emissions from 1850 to 2000. *Chemosphere*, **58**(2), pp. 163-175.

Stern, N., 2006: Stern review on the economics of climate change. HM Treasury, London. <http://www.hm-treasury.gov.uk/independent_reviews/ stern_review_economics_climate_change/sternreview_index.cfm>, accessed 1 June 2007.

Streets, D.G., N.Y. Tsai, H. Akimoto, and K. Oka, 2000: Sulfur dioxide emissions in Asia in the period 1985-1997. *Atmospheric Environment*, . **34**(26), pp. 4413-4424.

Streets, D.G., K. Jiang, X. Hu, J.E. Sinton, X.-Q. Zhang, D. Xu, M.Z. Jacobson, and J.E. Hansen, 2001: Recent reductions in China's greenhouse gas emissions. *Science*, **294**(5548), pp. 1835-1837.

Streets, D.G., T.C. Bond, G.R. Carmichael, S.D. Fernandes, Q.Fu, D.He, Z.Klimont, S.M.Nelson, N.Y. Tsai, M.Q. Wang, J.-H. Hu, and K.F. Yarber. 2003: An inventory of gaseous and primary aerosol emissions in Asia in the year 2000. *Journal of Geophysical Research*, **108**, (D21, 8809).

Streets, D.G., T.C. Bond, T. Lee, and C. Jang, 2004: On the future of carbonaceous aerosol emissions. *Journal of Geophysical Research*, **109**(24), pp. 1-19.

Strengers, B., R. Leemans, B. Eickhout, B. de Vries, and L. Bouwman, 2004: The land-use projections and resulting emissions in the IPCC SRES scenarios as simulated by the IMAGE 2.2 model. *GeoJournal*, **61**(4), pp. 381-393.

Svedin, U. and B. Aniansson (eds.), 1987: Surprising futures: notes from an international workshop on long-term development. Friiberg Manor, Sweden, January 1986. Swedish Council for Planning and Coordination of Research, Stockholm, 128 pp.

Swart, R.S., M. Berk, M. Janssen, and G.J.J. Kreileman, 1998: The safe landing approach: Risks and trade-offs in climate change, In *Global Change Scenarios of the 21st Century: Results from the IMAGE 2 Model*. J. Alcamo, G.J.J. Kreileman, and R. Leemans (eds.), Elsevier, London, pp. 193-218.

Swart, R., M. Amann, F. Raes, and W. Tuinstra, 2004: A good climate for clean air: linkages between climate change and air pollution. *Climatic Change*, **66**(3), pp. 263-269.

Timmer, H., 2005: PPP vs. MER: A view from the World Bank. IPCC Expert Meeting on Emission Scenarios, 12 - 14 January 2005, Washington DC.

Tol, R.S.J., 1999: The marginal costs of greenhouse gas emissions. *Energy Journal*, **20**(1), pp. 61-81.

Tol, R.S.J., 2000: Timing of greenhouse gas emission reduction. *Pacific and Asian Journal of Energy*, **10**(1), pp. 63-68.

Tol, R.S.J. and H. Dowlatabadi, 2001: Vector-borne diseases, development & climate change, *The Integrated Assessment Journal*, **2**, pp. 173-181.

Tol, R.S.J., 2002a: Estimates of the damage costs of climate change. Part 1: Benchmark estimates. *Environmental and Resource Economics*, **21**(1), pp. 47-73.

Tol, R.S.J., 2002b: Estimates of the damage costs of climate change. Part 2: Dynamic estimates. *Environmental and Resource Economics*, **21**(1), pp. 135-160.

Tol, R.S.J., 2005a: Adaptation and mitigation: trade-offs in substance and methods. *Environmental Science & Policy*, **8**, pp. 572-578.

Tol, R.S.J., 2005b: The marginal damage costs of climate change: an assessment of the uncertainties. *Energy Policy*, **33**(16), pp. 2064-2074.

Tol, R.S.J. and G. Yohe, 2006: On dangerous climate change and dangerous emission reduction. In *Avoiding Dangerous Climate Change*, H.J. Schellnhuber, W. Cramer, N. Nakicenovic, T. Wigley, and G. Yohe (eds.), Cambridge University Press, Cambridge, pp. 291-298.

Tompkins, E.L. and W.N. Adger, 2005: Defining response capacity to enhance climate change policy. *Environmental Science and Policy*, **8**(6), pp. 562-571.

Toth, F.L., E. Hizsnyik, and W. Clark (eds.), 1989: Scenarios of socioeconomic development for studies of global environmental change: a critical review. RR-89-4, International Institute for Applied Systems, Laxenburg.

Toth, F.L., T. Bruckner, H.-M. Füssel, M. Leimbach, G. Petschel-Held and H.J. Schellnhuber 2002: Exploring options for global climate policy: a new analytical framework. *Environment*, **44**(5), pp. 22-34.

Toth, F.L. (ed.), 2003: Integrated Assessment of Climate Protection Strategies. *Climatic Change*, Special Issue, **56**(1-2).

Treffers, D.J., A.P.C. Faaij, J. Spakman and A.Seebregts, 2005: Exploring the possibilities for setting up sustainable energy systems for the long-term: Two visions for the Dutch energy system in 2050. *Energy Policy*, **33**(13), pp. 1723-1743.

Tuinstra, W., M. Berk, M. Hisschemöller, L. Hordijk, B. Metz and A.P.J. Mol (eds.), 2002: Climate OptiOns for the Long-term (COOL) - Synthesis Report. National Reference Point Report 954281, DA Zoetermeer, 118 pp. <http://www2.wau.nl/cool/reports/COOLVolumeAdef.pdf>, accessed 1 June 2007.

UN, 1993: System of National Accounts. United Nations, New York.

UN, 2004: World population to 2300. Department of Economic and Social Affairs, Population Division, United Nations, New York, 254 pp. <http://www.un.org/esa/population/publications/longrange2/WorldPop2300final.pdf>, accessed 1 June 2007.

UNEP, 2002: Global Environment Outlook 3. Past, present and future perspectives. United Nations Environment Programme, Earthscan, London. <http://www.unep.org/geo/geo3/english/pdf.htm>, accessed 1 June 2007.

UNIDO, 2005: Capability building for catching-up. United Nations Industrial Development Organization, Industrial Development Report, Vienna, 204 pp. <http://www.unido.org/doc/5156>, accessed 1 June 2007.

UNSD, 2005: Progress towards the Millennium Development Goals, 1990-2005. United Nations Statistics Division, New York, <http://unstats.un.org/unsd/mi/mi_coverfinal.htm>, accessed 1 June 2007.

USCCSP, 2006: Scenarios of greenhouse gas emissions and atmospheric concentrations. Report by the United States Climate Change Science Program and approved by the Climate Change Science Program Product Development Advisory Committee, 212 pp. <http://www.climatescience.gov/Library/sap/sap2-1/sap2-1a-draft3-all.pdf>, accessed 1 June 2007.

USDOE, 2004: International Energy Outlook. United States Department of Energy - Energy Information Administration, Washington DC, 248 pp. <http://tonto.eia.doe.gov/FTPROOT/forecasting/0484(2004).pdf>, accessed 1 June 2007.

USEPA, 2006a: Global anthropogenic emissions of non-CO_2 greenhouse gases 1990-2020. United States Environmental Protection Agency Report 430-R-06-003, Washington DC. <http://www.epa.gov/nonco2/econ-inv/international.html>, accessed 1 June 2007.

USEPA, 2006b: Global mitigation of non-CO_2 greenhouse gases. United States Environmental Protection Agency Report 430-R-06-005, Washington DC. <http://www.epa.gov/nonco2/econ-inv/international.html>, accessed 1 June 2007.

Van Harmelen, T., J. Bakker, B. de Vries, D. van Vuuren, M. den Elzen, and P. Mayerhofer, 2002: Long-term reductions in costs of controlling regional air pollution in Europe due to climate policy. *Environmental Science & Policy*, **5**(4), pp. 349-365.

Van Vuuren, D.P. and H.J.M. de Vries, 2001: Mitigation scenarios in a world oriented at sustainable development: the role of technology, efficiency and timing. *Climate Policy*, **1**(2), pp. 189-210.

Van Vuuren, D., Z. Fengqi, B. de Vries, J. Kejun, C. Graveland, and L. Yun, 2003: Energy and emission scenarios for China in the 21st century - exploration of baseline development and mitigation options. *Energy Policy*, **31**(4), pp. 369-387.

Van Vuuren, D.P. and K.H. Alfsen, 2006: PPP Versus MER: searching for answers in a multi-dimensional debate. *Climatic Change*, **75**(1-2), pp. 47-57.

Van Vuuren, D. and B.C. O'Neill, 2006: The consistency of IPCC's SRES scenarios to 1990-2000 trends and recent projections. *Climatic Change*, 75(1-2), pp. 9-46.

Van Vuuren, D.P., B. Eickhout, P.L. Lucas, and M.G.J. den Elzen, 2006a: Long-term multi-gas scenarios to stabilise radiative forcing - Exploring costs and benefits within an integrated assessment framework, *The Energy Journal*. Vol. Multi-Greenhouse Gas Mitigation and Climate Policy, Special Issue No.3, pp. 201-234.

Van Vuuren, D., J. Weyant, and F. de la Chesnaye, 2006b: Multi-gas scenarios to stabilise radiative forcing. *Energy Economics*, **28**(1), pp. 102-120.

Van Vuuren, D.P., J. Cofala, H.E. Eerens, R. Oostenrijk, C. Heyes, Z. Klimont, M.G.J. den Elzen, and M. Amann, 2006c: Exploring the ancillary benefits of the Kyoto Protocol for air pollution in Europe. *Energy Policy*, **34**(4), pp. 444-460.

Van Vuuren, D.P., M.G.J. den Elzen, P.L. Lucas, B. Eickhout, B.J. Strengers, B. van Ruijven, S. Wonink, R. van Houdt, 2007: Stabilizing greenhouse gas concentrations at low levels: an assessment of reduction strategies and costs. *Climatic Change*, **81**(2), pp. 119-159.

Vitousek, P.M., H.A. Mooney, J. Lubchenco, and J.M. Melillo, 1997: Human domination of Earth's ecosystems. *Science*, **277**(5325), pp. 494-499.

Wack, P., 1985a: Scenarios: shooting the rapids. *Harvard Business Review*, 63(6), pp. 139-150.

Wack, P., 1985b: Scenarios: uncharted waters ahead. *Harvard Business Review*, **63**(5), pp. 72-89.

Watkiss, P., D. Anthoff, T. Downing, C. Hepburn, C. Hope, A. Hunt, and R.S.J. Tol, 2005: The social cost of carbon - methodological approaches for using SCC estimates in policy assessment. AEA Technology Environment, Didcot, 125 pp. <http://socialcostofcarbon.aeat.com/html/report.htm>, accessed 1 June 2007.

WBCSD, 2004: Mobility 2030: Meeting the challenges to sustainability. World Business Council for Sustainable Development, Conches-Geneva. <http://www.wbcsd.org/web/publications/mobility/mobility-full.pdf>, accessed 1 June 2007.

Webster, M.D., M. Babiker, M. Mayer, J.M. Reilly, J. Harnisch, R. Hyman, M.C. Sarofim and C. Wang, 2002: Uncertainty in emissions projections for climate models. Atmospheric Environment, **36**(22), pp. 3659-3670.

Webster, M.D., C. Forest, J. Reilly, M. Babiker, D. Kicklighter, M. Mayer, R. Prinn, M. Sarofim, A. Sokolov, P. Stone and C. Wang, 2003: Uncertainty analysis of climate change and policy response. *Climatic Change*, **61**(3), pp. 295-320.

West, C. and M. Gawith, 2005: Measuring progress: preparing for climate change through the UK Climate Impacts Programme. UK Climate Impacts Programme, Oxford, 72 pp.

Weyant, J.P., 2000: An introduction to the economics of climate change policy. Pew Center on Global Climate Change. Washington DC, 56 pp.

Weyant, J.P. (ed.), 2004: EMF 19 alternative technology strategies for climate change policy. *Energy Economics*, **26**(4), pp. 501-755.

Weyant, J.P., F. de la Chesnaye, and G. Blanford, 2006: Overview of EMF – 21: Multigas mitigation and climate policy. *The Energy Journal*, Vol. Multi-Greenhouse Gas Mitigation and Climate Policy, Special Issue No.3, pp. 1-32.

White House, 2002: Executive summary of Bush Climate Change Initiative. Washington DC. <http://www.whitehouse.gov/news/releases/2002/02/climatechange.html>, accessed 1 June 2007.

Wigley, T.M.L., R. Richels, and J.A. Edmonds, 1996: Economic and environmental choices in the stablization of atmospheric CO_2 concentrations. *Nature*, **379**(6562), pp. 240-243.

Wigley, T.M.L., 2003: Modeling climate change under no-policy and policy emissions pathways. Benefits of climate policy: improving information for policy makers. Organization for Economic Co-operation and Development, OECD Publishing, Paris, 32 pp.

Wigley, T.M.L. 2005: The Climate Change Commitment. *Science*, **307**(5716), pp. 1766-1769.

Williams, R.H., 1998: Fuel decarbonization for fuel cell applications and sequestration of the separated CO_2. In *Eco-restructuring: Implications for Sustainable Development*. R.U. Ayres and P.M. Weaver (eds.), United Nations University Press, Tokyo, pp. 180-222.

World Bank, 1992: *World development report 1992: development and the environment*. Oxford University Press, Oxford, 308 pp.

World Bank, 2002: Income poverty - trends in inequality. <http://www.worldbank.org/poverty/data/trends/inequal.htm>, accessed 1 June 2007.

World Bank, 2004: *World Economic Prospects 2004*. World Bank, Washington DC.

Worrell, E., 2005: The use and development of bottom-up technological scenarios. IPCC Expert Meeting on Emissions Scenarios. Meeting Report, 12-14 January 2005, Washington DC.

Yamaji, K., R. Matsuhashi, Y. Nagata, Y. Kaya, 1991: An integrated systems for CO_2 /Energy/GNP Analysis: Case studies on economic measures for CO_2 reduction in Japan. Workshop on CO_2 Reduction and Removal: Measures for the Next Century, 19-21 March 1991, International Institute for Applied Systems Analysis, Laxenburg.

Yohe, G., N. Andronova, and M. Schlesinger, 2004: To hedge or not against an uncertain climate future? *Science*, **306**(5695), pp. 416-417.

Yohe, G., 2006: Some thoughts on the damage estimates presented in the Stern Review - An Editorial. *The Integrated Assessment Journal*, **6**(3), pp. 65-72.

4

Energy Supply

Coordinating Lead Authors:

Ralph E.H. Sims (New Zealand), Robert N. Schock (USA)

Lead Authors:

Anthony Adegbululgbe (Nigeria), Jørgen Fenhann (Denmark), Inga Konstantinaviciute (Lithuania), William Moomaw (USA),
Hassan B. Nimir (Sudan), Bernhard Schlamadinger (Austria), Julio Torres-Martínez (Cuba), Clive Turner (South Africa),
Yohji Uchiyama (Japan), Seppo J.V. Vuori (Finland), Njeri Wamukonya (Kenya), Xiliang Zhang (China)

Contributing Authors:

Arne Asmussen (Germany), Stephen Gehl (USA), Michael Golay (USA), Eric Martinot (USA)

Review Editors:

Hans Larsen (Denmark), José Roberto Moreira (Brazil)

This chapter should be cited as:

R.E.H. Sims, R.N. Schock, A. Adegbululgbe, J. Fenhann, I. Konstantinaviciute, W. Moomaw, H.B. Nimir, B. Schlamadinger,
J. Torres-Martínez, C. Turner, Y. Uchiyama, S.J.V. Vuori, N. Wamukonya, X. Zhang, 2007: Energy supply. In Climate Change 2007:
Mitigation. Contribution of Working Group III to the Fourth Assessment Report of the Intergovernmental Panel on Climate Change
[B. Metz, O.R. Davidson, P.R. Bosch, R. Dave, L.A. Meyer (eds)], Cambridge University Press, Cambridge, United Kingdom and New York,
NY, USA.

Table of Contents

EXECUTIVE SUMMARY

Annual total greenhouse gas (GHG) emissions arising from the global energy supply sector continue to increase. Combustion of fossil fuels continues to dominate a global energy market that is striving to meet the ever-increasing demand for heat, electricity and transport fuels. GHG emissions from fossil fuels have increased each year since the IPCC 2001 Third Assessment Report (TAR) (IPCC,2001), despite greater deployment of low- and zero-carbon technologies, (particularly those utilizing renewable energy); the implementation of various policy support mechanisms by many states and countries; the advent of carbon trading in some regions, and a substantial increase in world energy commodity prices. Without the near-term introduction of supportive and effective policy actions by governments, energy-related GHG emissions, mainly from fossil fuel combustion, are projected to rise by over 50% from 26.1 $GtCO_2eq$ (7.1 GtC) in 2004 to 37–40 $GtCO_2$ (10.1–10.9 GtC) by 2030. Mitigation has therefore become even more challenging.

Global dependence on fossil fuels has led to the release of over 1100 $GtCO_2$ into the atmosphere since the mid-19th century. Currently, energy-related GHG emissions, mainly from fossil fuel combustion for heat supply, electricity generation and transport, account for around 70% of total emissions including carbon dioxide, methane and some traces of nitrous oxide (Chapter 1). To continue to extract and combust the world's rich endowment of oil, coal, peat, and natural gas at current or increasing rates, and so release more of the stored carbon into the atmosphere, is no longer environmentally sustainable, unless carbon dioxide capture and storage (CCS) technologies currently being developed can be widely deployed (*high agreement, much evidence*).

There are regional and societal variations in the demand for energy services. The highest per-capita demand is by those living in Organisation for Economic Co-operation and Development (OECD) economies, but currently, the most rapid growth is in many developing countries. Energy access, equity and sustainable development are compromised by higher and rapidly fluctuating prices for oil and gas. These factors may increase incentives to deploy carbon-free and low-carbon energy technologies, but conversely, could also encourage the market uptake of coal and cheaper unconventional hydrocarbons and technologies with consequent increases in carbon dioxide (CO_2) emissions.

Energy access for all will require making available basic and affordable energy services using a range of energy resources and innovative conversion technologies while minimizing GHG emissions, adverse effects on human health, and other local and regional environmental impacts. To accomplish this would require governments, the global energy industry and society as a whole to collaborate on an unprecedented scale. The method used to achieve optimum integration of heating, cooling, electricity and transport fuel provision with more efficient energy systems will vary with the region, local growth rate of energy demand, existing infrastructure and by identifying all the co-benefits (*high agreement, much evidence*).

The wide range of energy sources and carriers that provide energy services need to offer long-term security of supply, be affordable and have minimal impact on the environment. However, these three government goals often compete. There are sufficient reserves of most types of energy resources to last at least several decades at current rates of use when using technologies with high energy-conversion efficient designs. How best to use these resources in an environmentally acceptable manner while providing for the needs of growing populations and developing economies is a great challenge.

- Conventional oil reserves will eventually peak as will natural gas reserves, but it is uncertain exactly when and what will be the nature of the transition to alternative liquid fuels such as coal-to-liquids, gas-to-liquids, oil shales, tar sands, heavy oils, and biofuels. It is still uncertain how and to what extent these alternatives will reach the market and what the resultant changes in global GHG emissions will be as a result.

- Conventional natural gas reserves are more abundant in energy terms than conventional oil, but they are also distributed less evenly across regions. Unconventional gas resources are also abundant, but future economic development of these resources is uncertain.

- Coal is unevenly distributed, but remains abundant. It can be converted to liquids, gases, heat and power, although more intense utilization will demand viable CCS technologies if GHG emissions from its use are to be limited.

- There is a trend towards using energy carriers with increased efficiency and convenience, particularly away from solid fuels to liquid and gaseous fuels and electricity.

- Nuclear energy, already at about 7% of total primary energy, could make an increasing contribution to carbon-free electricity and heat in the future. The major barriers are: long-term fuel resource constraints without recycling; economics; safety; waste management; security; proliferation, and adverse public opinion.

- Renewable energy sources (with the exception of large hydro) are widely dispersed compared with fossil fuels, which are concentrated at individual locations and require distribution. Hence, renewable energy must either be used in a distributed manner or concentrated to meet the higher energy demands of cities and industries.

- Non-hydro renewable energy-supply technologies, particularly solar, wind, geothermal and biomass, are currently small overall contributors to global heat and electricity supply, but are the most rapidly increasing. Costs, as well as social and environmental barriers, are restricting this growth. Therefore, increased rates of deployment may need supportive government policies and measures.

- Traditional biomass for domestic heating and cooking still accounts for more than 10% of global energy supplies but could eventually be replaced, mainly by modern biomass and other renewable energy systems as well as by fossil-based domestic fuels such as kerosene and liquefied petroleum gas (LPG) *(high agreement, much evidence – except traditional biomass)*.

Security of energy supply issues and perceived future benefits from strategic investments may not necessarily encourage the greater uptake of lower carbon-emitting technologies. The various concerns about the future security of conventional oil, gas and electricity supplies could aid the transition to more low-carbon technologies such as nuclear, renewables and CCS. However, these same concerns could also encourage the greater uptake of unconventional oil and gaseous fuels as well as increase demand for coal and lignite in countries with abundant national supplies and seeking national energy-supply security.

Addressing environmental impacts usually depends on the introduction of regulations and tax incentives rather than relying on market mechanisms. Large-scale energy-conversion plants with a life of 30–100 years give a slow rate of turnover of around 1–3% per year. Thus, decisions taken today that support the deployment of carbon-emitting technologies, especially in countries seeking supply security to provide sustainable development paths, could have profound effects on GHG emissions for the next several decades. Smaller-scale, distributed energy plants using local energy resources and low- or zero-carbon emitting technologies, can give added reliability, be built more quickly and be efficient by utilizing both heat and power outputs locally (including for cooling).

Distributed electricity systems can help reduce transmission losses and offset the high investment costs of upgrading distribution networks that are close to full capacity.

More energy-efficient technologies can also improve supply security by reducing future energy-supply demands and any associated GHG emissions. However, the present adoption path for these, together with low- and zero-carbon supply technologies, as shown by business-as-usual baseline scenarios, will not reduce emissions significantly.

The transition from surplus fossil fuel resources to constrained gas and oil carriers, and subsequently to new energy supply and conversion technologies, has begun. However it faces regulatory and acceptance barriers to rapid implementation and market competition alone may not lead to reduced GHG emissions. The energy systems of many nations are evolving from their historic dependence on fossil fuels in response to the climate change threat, market failure of the supply chain, and increasing reliance on global energy markets, thereby necessitating the wiser use of energy in all sectors. A rapid transition toward new energy supply systems with reduced carbon intensity needs to be managed to minimize economic, social and technological risks and to co-opt those stakeholders who retain strong interests in maintaining the status quo. The electricity, building and industry sectors are beginning

to become more proactive and help governments make the transition happen. Sustainable energy systems emerging as a result of government, business and private interactions should not be selected on cost and GHG mitigation potential alone but also on their other co-benefits.

Innovative supply-side technologies, on becoming fully commercial, may enhance access to clean energy, improve energy security and promote environmental protection at local, regional and global levels. They include thermal power plant designs based on gasification; combined cycle and super-critical boilers using natural gas as a bridging fuel; the further development and uptake of CCS; second-generation renewable energy systems; and advanced nuclear technologies. More efficient energy supply technologies such as these are best combined with improved end-use efficiency technologies to give a closer matching of energy supply with demand in order to reduce both losses and GHG emissions.

Energy services are fundamental to achieving sustainable development. In many developing countries, provision of adequate, affordable and reliable energy services has been insufficient to reduce poverty and improve standards of living. To provide such energy services for everyone in an environmentally sound way will require major investments in the energy-supply chain, conversion technologies and infrastructure (particularly in rural areas) *(high agreement, much evidence)*.

There is no single economic technical solution to reduce GHG emissions from the energy sector. There is however good mitigation potential available based on several zero-or low-carbon commercial options ready for increased deployment at costs below 20 US$/tCO$_2$ avoided or under research development. The future choice of supply technologies will depend on the timing of successful developments for advanced nuclear, advanced coal and gas, and second-generation renewable energy technologies. Other technologies, such as CCS, second-generation biofuels, concentrated solar power, ocean energy and biomass gasification, may make additional contributions in due course. The necessary transition will involve more sustained public and private investment in research, development, demonstration and deployment (RD[3]) to better understand our energy resources, to further develop cost-effective and -efficient low- or zero-carbon emitting technologies, and to encourage their rapid deployment and diffusion. Research investment in energy has varied greatly from country to country, but in most cases has declined significantly in recent years since the levels achieved soon after the oil shocks during the 1970s.

Using the wide range of available low- and zero-carbon technologies (including large hydro, bioenergy, other renewables, nuclear and CCS together with improved power-plant efficiency and fuel switching from coal to gas), the total mitigation potential by 2030 for the electricity sector alone, at carbon prices below 20 US$/tCO$_2$-eq, ranges between 2.0 and 4.2 GtCO$_2$-eq/yr. At the high end of this range, the over 70% share of fossil fuel-based power generation in the baseline drops to 55% of the total. Developing countries could provide around half of this potential. This range corresponds well with the TAR analysis potential of 1.3–2.5 $GtCO_2$-eq/yr at 27 US$/$tCO_2$-eq avoided, given that the TAR was only up to 2020 and that, since it was published in 2001, there has been an increase in development and deployment of renewable energy technologies, a better understanding of CCS techniques and a greater acceptance of improved designs of nuclear power plants.

For investment costs up to 50 US$/$tCO_2$-eq, the total mitigation potential by 2030 rises to between 3.0 and 6.4 $GtCO_2$-eq/yr avoided. Up to 100 US$/$tCO_2$-eq avoided, the total potential is between 4.0 and 7.2 $GtCO_2$-eq/yr, mainly coming from non-OECD/EIT countries (*medium agreement, limited evidence*).

There is high agreement in the projections that global energy supply will continue to grow and in the types of energy likely to be used by 2030. However, there is only medium confidence in the regional energy demand assumptions and the future mix of conversion technologies to be used. Overall, the future costs and technical potentials identified should provide a reasonable basis for considering strategies and decisions over the next several decades.

No single policy instrument will ensure the desired transition to a future secure and decarbonized world. Policies will need to be regionally specific and both energy and non-energy co-benefits should be taken into account. Internalizing environmental costs requires development of policy initiatives, long-term vision and leadership based on sound science and economic analysis. Effective policies supporting energy-supply technology development and deployment are crucial to the uptake of low-carbon emission systems and should be regionally specific. A range of policies is already in place to encourage the development and deployment of low-carbon-emitting technologies in OECD countries as well as in non-OECD countries including Brazil, Mexico, China and India. Policies in several countries have resulted in the successful implementation of renewable energy systems to give proven benefits linked with energy access, distributed energy, health, equity and sustainable development. Nuclear energy policies are also receiving renewed attention. However, the consumption of fossil fuels, at times heavily subsidized by governments, will remain dominant in all regions to meet ever-increasing energy demands unless future policies take into account the full costs of environmental, climate change and health issues resulting from their use.

Energy sector reform is critical to sustainable energy development and includes reviewing and reforming subsidies, establishing credible regulatory frameworks, developing policy environments through regulatory interventions, and creating market-based approaches such as emissions trading. Energy security has recently become an important policy driver. Privatization of the electricity sector has secured energy supply and provided cheaper energy services in some countries in the short term, but has led to contrary effects elsewhere due to increasing competition, which, in turn, leads to deferred investments in plant and infrastructure due to longer-term uncertainties. In developed countries, reliance on only a few suppliers, and threats of natural disasters, terrorist attacks and future uncertainty about imported energy supplies add to the concerns. For developing countries lack of security and higher world-energy prices constrain endeavours to accelerate access to modern energy services that would help to decrease poverty, improve health, increase productivity, enhance competition and thus improve their economies (*high agreement, much evidence*).

In short, the world is not on course to achieve a sustainable energy future. The global energy supply will continue to be dominated by fossil fuels for several decades. To reduce the resultant GHG emissions will require a transition to zero- and low-carbon technologies. This can happen over time as business opportunities and co-benefits are identified. However, more rapid deployment of zero- and low-carbon technologies will require policy intervention with respect to the complex and interrelated issues of: security of energy supply; removal of structural advantages for fossil fuels; minimizing related environmental impacts, and achieving the goals for sustainable development.

4.1   Introduction

This chapter addresses the energy-supply sector and analyses the cost and potential of greenhouse gas (GHG) mitigation from the uptake of low- and zero-carbon-emitting technologies (including carbon capture and storage) over the course of the next two to three decades. Business-as-usual fossil-fuel use to meet future growth in energy demand will produce significant increases in GHG emissions. To make a transition by 2030 will be challenging. Detailed descriptions of the various technologies have been kept to a minimum, especially for those that have changed little since the Third Assessment Report (TAR) as they are well covered elsewhere (e.g., IEA, 2006a).

The main goal of all energy transformations is to provide energy services that improve quality of life (e.g. health, life expectancy and comfort) and productivity (Hall *et al.,* 2004). A supply of secure, equitable, affordable and sustainable energy is vital to future prosperity. Approximately 45% of final consumer energy is used for low-temperature heat (cooking, water and space heating, drying), 10% for high-temperature industrial process heat, 15% for electric motors, lighting and electronics and 30% for transport. The CO_2 emissions from meeting this energy demand using mainly fossil fuels account for around 80% of total global emissions (IEA, 2006b). Demands for all forms of energy continue to rise to meet expanding economies and increases in world population. Rising prices and concerns about insecure energy supplies will compromise growth in fossil fuel consumption.

Energy supply is intimately tied in with development in the broad sense. At present, the one billion people living in developed (OECD) countries consume around half of the 470 EJ current annual global primary energy use (IEA, 2006b), whereas the one billion *poorest* people in developing countries consume only around 4%, mainly in the form of traditional biomass used inefficiently for cooking and heating. The United Nations has set Millennium Development Goals to eradicate poverty, raise living standards and encourage sustainable economic and social development (UN, 2000). Economic policies aimed at sustainable development can bring a variety of co-benefits including utilizing new energy technologies and improved access to adequate and affordable modern energy services. This will determine how many humans can expect to achieve a decent standard of living in the future (Section 4.5.4; Chapter 3).

There are risks to being unprepared for future energy-supply constraints and disruptions. Currently, fossil fuels provide almost 80% of world energy supply; a transition away from their traditional use to zero- and low-carbon-emitting modern energy systems (including carbon dioxide capture and storage (CCS) (IPCC, 2005), as well as improved energy efficiency, would be part solutions to GHG-emission reduction. It is yet to be determined which technologies will facilitate this transition and which policies will provide appropriate impetus, although security of energy supply, aligned with GHG-reduction goals, are co-policy drivers for many governments wishing to ensure that future generations will be able to provide for their own well-being without their need for energy services being compromised.

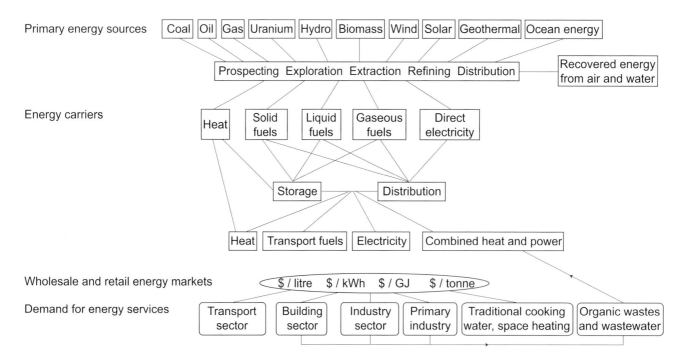

Figure 4.1: *Complex interactions between primary energy sources and energy carriers to meet societal needs for energy services as used by the transport (Chapter 5), buildings (6), industry (7) and primary industry (8 and 9) sectors.*

A mix of options to lower the energy per unit of GDP and carbon intensity of energy systems (as well as lowering the energy intensity of end uses) will be needed to achieve a truly sustainable energy future in a decarbonized world. Energy-related GHG emissions are a by-product of the conversion and delivery sector (which includes extraction/refining, electricity generation and direct transport of energy carriers in pipelines, wires, ships, etc.), as well as the energy end-use sectors (transport, buildings, industry, agriculture, forestry and waste), as outlined in Chapters 5 to 10 (Figure 4.1).

In all regions of the world energy demand has grown in recent years (Figure 4.2). A 65% global increase above the 2004 primary energy demand (464 EJ, 11,204 Mtoe) is anticipated by 2030 under business as usual (IEA, 2006b). Major investment will be needed, mostly in developing countries. As a result, without effective mitigation, total energy-related carbon dioxide emissions (including transformations, own use and losses) will rise from 26.1 $GtCO_2$ (7.2 GtC) in 2004 to around 37–40 $GtCO_2$ (11.1 GtC) in 2030 (IEA, 2006b; Price and de la Rue du Can, 2006), possibly even higher (Fisher, 2006), assuming modest energy-efficiency improvements are made to technologies currently in use. This means that all cost-effective means of reducing carbon emissions would need to be deployed in order to slow down the rate of increase of atmospheric concentrations (WBCSD, 2004; Stern, 2006).

Implementing any major energy transition will take time. The penetration rates of emerging energy technologies depend on the expected lifetime of capital stock, equipment and the relative cost. Some large-scale energy-conversion plants can have an operational life of up to 100 years giving a slow rate of turnover, but around 2–3% per year replacement rate is more usual (Section 4.4.3). There is, therefore, some resistance to

change, and breakthroughs in technology to increase penetration rate are rare.

Technology only diffuses rapidly once it can compete economically with existing alternatives or offers added value (e.g. greater convenience), often made possible by the introduction of new regulatory frameworks. It took decades to provide the large-scale electricity and natural-gas infrastructures now common in many countries. Power stations, gas and electricity distribution networks and buildings are usually replaced only at the end of their useful life, so early action to stabilize atmospheric GHGs to have minimal impact on future GDP, it is important to avoid building 'more of the same' (Stern, 2006).

Total annual capital investment by the global energy industry is currently around 300 billion US$. Even allowing for improved energy efficiency, if global energy demand continues to grow along the anticipated trajectory, by 2030 the investment over this period in energy-carrier and -conversion systems will be over 20 trillion (10^{12}) US$, being around 10% of world total investment or 1% of cumulative global GDP (IEA, 2006b). This will require investment in energy-supply systems of around 830 billion US$/yr, mainly to provide an additional 3.5 TW of electricity-generation plant and transmission networks, particularly in developing countries, and provide opportunities for a shift towards more sustainable energy systems. Future investment in state-of-the-art technologies in countries without embedded infrastructure may be possible by 'leapfrogging' rather than following a similar historic course of development to that of OECD nations. New financing facilities are being considered because of the G8 Gleneagles Communiqué on *Climate Change, Clean Energy and Sustainable Development* of July 2005 (World Bank, 2006).

It is uncertain how future investments will best meet future energy demand while achieving atmospheric GHG stabilization goals. There are many possible scenarios somewhere between the following extremes (WEC, 2004a).
- *High demand growth*, giving very large productivity increases and wealth. Being technology- and resource-intensive, investment in technological changes would yield rapid stock turnover with consequent improvements in energy intensity and efficiency.
- *Reduced energy demand*, with an investment goal to reduce CO_2 emissions by one per cent per year by 2100. This would be technologically challenging and assumes unprecedented progressive international cooperation focused explicitly on developing a low-carbon economy that is both equitable and sustainable, requiring improvements in end-use efficiency and aggressive changes in lifestyle to emphasize resource conservation and dematerialization.

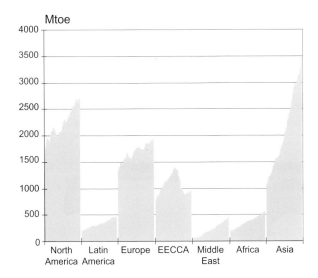

Figure 4.2: *Global annual primary energy demand (including traditional biomass), 1971 – 2003 by region.*

Note: EECCA = countries of Eastern Europe, the Caucasus and Central Asia.
1000 Mtoe = 42 EJ.
Source: IEA, 2004a.

The last century has seen a decline in the use of solids relative to liquids and gases. In the future, the use of gases is expected to increase (Section 4.3.1). The share of liquids will probably remain constant but with a gradual transition from conventional oil (Section 4.3.1.3) toward coal-to-liquids, unconventional oils (Section 4.3.1.4) and modern biomass (Section 4.3.3.3).

A robust mix of energy sources (fossil, renewable and nuclear), combined with improved end-use efficiency, will almost certainly be required to meet the growing demand for energy services, particularly in many developing countries. Technological development, decentralized non-grid networks, diversity of energy-supply systems and affordable energy services are imperative to meeting future demand. In many OECD countries, historical records show a decrease in energy per capita. Energy reduction per unit of GDP is also becoming apparent with respect to energy supplies in developing countries such as China (Larson *et al.*, 2003).

4.1.1 Summary of Third Assessment Report (TAR)

Energy-supply and end-use-efficiency technology options (Table 3.36, TAR) showed special promise for reducing CO_2 emissions from the industrial and energy sectors. Opportunities included more efficient electrical power generation from fossil fuels, greater use of renewable technologies and nuclear power, utilization of transport biofuels, biological carbon sequestration and CCS. It was estimated that potential reductions of 350–700 MtC/yr (1.28–2.57 GtCO$_2$-eq/yr) were possible in the energy supply and conversion sector by 2020 for <100 US$/C (27.3 US$/tCO$_2$) (Table 3.37, TAR) divided equally between developed and developing countries. Improved end-use efficiency held greater potential for reductions.

There are still obstacles to implementing the low-carbon technologies and measures identified in the TAR. These include a lack of human and institutional capacity; regulatory impediments and imperfect capital markets that discourage investment, including for decentralized systems; uncertain rates of return on investment; high trade tariffs on emission-lowering technologies; lack of market information, and intellectual property rights issues. Adoption of renewable energy is constrained by high investment costs, lack of capital, government support for fossil fuels and lack of government support mechanisms.

The problem of 'lock-in' by existing technologies and the economic, political, regulatory, and social systems that support them were seen as major barriers to the introduction of low-emission technologies in all types of economies. This has not changed. Several technological innovations such as ground-source heat pumps, solar photovoltaic (PV) roofing, and offshore wind turbines have been recently introduced into the market as a result of multiple drivers including economic profit or productivity gains, non-energy-related benefits, tax incentives, environmental benefits, performance efficiency and other regulations. Lower GHG emissions were not always a major driver in their adoption. Policy changes in development assistance (Renewables, 2004) and direct foreign investment provide opportunities to introduce low-emission technologies to developing countries more rapidly.

4.2 Status of the sector

Providing energy services from a range of sources to meet society's demands should offer security of supply, be affordable and have minimal impact on the environment. However, these three goals often conflict. Recent liberalization of energy markets in many countries has led to cheaper energy services in the short term, but in the longer term, investments with longer write-off periods and often lower returns (including nuclear power plants and oil refineries) are not always being made due to the need to maximize value for short-term shareholders. Energy-supply security has improved in some countries but deteriorated elsewhere due to increasing competition, which, because of insecurity, leads to deferred investments in grid and plants. Addressing environmental impacts, including climate

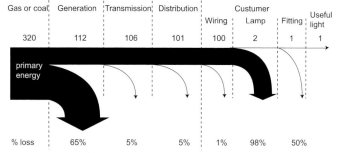

a) Thermal-power energy and losses in the production of one unit of useful light energy.

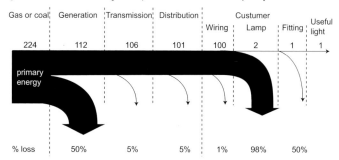

b) Investment in more efficient gas-fired power stations reduces fuel inputs by around 30%.

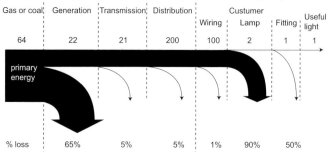

c) Investment in energy-saving compact fluorescent lightbulbs reduces fuel inputs by around 80%.

Figure 4.3: *The conversion from primary energy to carriers and end-uses is an essential driver of efficiency, exemplified here by the case of lighting. Primary fuel inputs can be reduced using more efficient generation plants, but also to a greater degree by more energy-efficient technologies (as described in Chapters 5, 6 and 7)*

Source: Cleland, 2005.

change, usually depends on laws and tax incentives rather than market mechanisms (Section 13.2.1.1).

Primary energy sources are: fossil carbon fuels; geothermal heat; fissionable, fertile and fusionable nuclides; gravitational (tides) and rotational forces (ocean currents), and the solar flux. These must be extracted, collected, concentrated, transformed, transported, distributed and stored (if necessary) using technologies that consume some energy at every step of the supply chain (Figure 4.3). The solar flux provides both

intermittent energy forms including wind, waves and sunlight, and stored energy in biomass, ocean thermal gradients and hydrologic supplies. Energy carriers such as heat, electricity and solid, liquid and gaseous fuels deliver useful energy services. The conversion of primary energy-to-energy carriers and eventually to energy services creates losses, which, together with distribution losses, represent inefficiencies and cost of delivery (Figure 4.4).

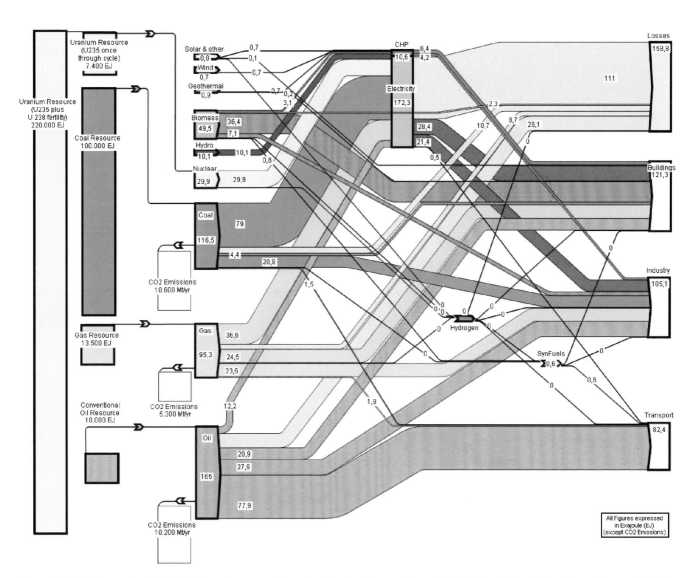

Figure 4.4: *Global energy flows (EJ in 2004) from primary energy through carriers to end-uses and losses. Related carbon dioxide emissions from coal, gas and oil combustion are also shown, as well as resources (vertical bars to the left).*

Notes: See also Table 4.2. Note that the IEA (2006b) data on known reserves and estimated resources, as used here, differ from the data in Table 4.2 that uses a breakdown in conventional and unconventional. The latter category may include some quantities shown as resources in Figure 4.26.
1) The current capacity of energy carriers is shown by the width of the lines.
2) Further energy conversion steps may take place in the end-use sectors, such as the conversion of natural gas into heat and/or electricity on site at the individual consumer level.
3) 'Buildings' include residential, commercial, public service and agricultural.
4) Peat is included with coal. Organic waste is included with biomass.
5) The resource efficiency ratio by which fast-neutron technology increases the power-generation capability per tonne of natural uranium varies greatly from the OECD assessment of 30:1 (OECD, 2006b). In this diagram the ratio used is up to 240:1 (OECD,2006c).
6) Comparisons can be made with SRES B2 scenario projections for 2030 energy supply, as shown in Figure 4.26.
Source: IEA, 2006b.

Analysis of energy supply should be integrated with energy carriers and end use since all these aspects are inextricably and reciprocally dependent. Energy-efficiency improvements in the conversion of primary energy resources into energy carriers during mining, refining, generation etc. continue to occur but are relatively modest. Reducing energy demand by the consumer using more efficient industrial practices, buildings, vehicles and appliances also reduces energy losses (and hence CO_2 emissions) along the supply chain and is usually cheaper and more efficient than increasing the supply capacity (Chapters 5, 6 and 7 and Figure 4.3).

Since 1971, oil and coal remain the most important primary energy sources with coal increasing its share significantly since 2000 (Figure 4.5). Growth slowed in 2005 and the total share of fossil fuels dropped from 86% in 1971 to 81% in 2004, (IEA, 2006b) excluding wind, solar, geothermal, bioenergy and biofuels, as well as non-traded traditional biomass. Combustible biomass and wastes contributed approximately 10% of primary energy consumption (IEA, 2006b) with more than 80% used for traditional fuels for cooking and heating in developing countries.

Around 40% of global primary energy was used as fuel to generate 17,408 TWh of electricity in 2004 (Figure 4.4). Electricity generation has had an average growth rate of 2.8%/yr since 1995 and is expected to continue growing at a rate of 2.5–3.1%/yr until 2030 (IEA, 2006b; Enerdata, 2004). In 2005, hard coal and lignite fuels were used to generate 40% of world electricity production with natural gas providing 20%, nuclear 16%, hydro 16%, oil 7% and other renewables 2.1% (IEA 2006b). Non-hydro renewable energy power plants have

expanded substantially in the past decade with wind turbine and solar PV installations growing by over 30% annually. However, they still supply only a small portion of electricity generation (Enerdata, 2004).

Many consumers of petroleum and, to a lesser degree, natural gas depend to varying but significant amounts of fuels imported from distant, often politically unstable regions of the world and transported through a number of locations equally vulnerable to disruptions. For example, in 2004 16.5–17 Mbbl/d of oil was shipped through the Straits of Hormuz in the Persian Gulf and 11.7 Mbbl/d through the Straits of Malacca in Asia (EIA/DOE, 2005). A disruption in supply at either of these points could have a severe impact on global oil markets. Political unrest in some oil and gas producing regions of Middle East, Africa and Latin America has also highlighted the vulnerability of supply. When international trade in oil and gas expands in the near future, the risks of supply disruption may increase leading to more serious impacts (IEA, 2004b; CIEP, 2004). This is a current driver for shifting to less vulnerable renewable energy resources.

Whereas fossil fuel sources of around 100,000 W/m² land area have been discovered at individual locations, extracted and then distributed, renewable energy is usually widely dispersed at densities of 1–5 W/m² and hence must either be used in a distributed manner or concentrated to meet the high energy demands of cities and industries. For renewable energy systems, variations in climate may produce future uncertainties result from dry years for hydro, poor crop yields for biomass, increased cloud cover and materials costs for solar, and variability in annual wind speeds. However, over their lifetime they are relatively price-stable sources and in a mixed portfolio of technologies can avoid losses from fluctuating oil, gas and power prices (Awerbuch and Sauter, 2005) unless their owner also has to sell based on volatile short-term prices (Roques *et al.*, 2006). World oil and gas prices in 2005 and 2006 were significantly higher than most pre-2005 scenario models predicted. This might lead to a reduction in transportation use and GHG emissions (Chapter 5), but conversely could also encourage a shift to coal-fired power plants. Hence, high energy prices do not necessarily mean increased investments in low carbon technologies or lower GHG emissions.

For nuclear power, investment uncertainties exist due to financial markets commanding a higher interest rate to cover perceived risks, thus increasing the cost of capital and thereby generation costs. Increasing environmental concerns will also raise the costs of obtaining permits. Conversely, surplus uranium supplies may possibly lower fuel prices, but this represents a relatively low fraction of generation costs compared with fossil-fuel power stations (Hagen *et al.*, 2005).

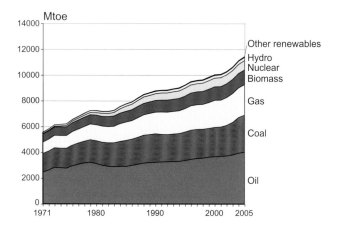

Figure 4.5: *World primary energy consumption by fuel type.*

Note: The IEA convention is to assume a 33% conversion efficiency when calculating the primary energy equivalent of nuclear energy from gross generation. The conversion efficiencies of a fossil fuel or nuclear power plant are typically about 33% due to heat losses whereas the energy in stored water (and other non-thermal means) is converted in turbines at efficiencies approaching 100%. Thus, for a much lower energy equivalent, hydro can produce the same amount of electricity as a thermal plant without a system to utilize the waste heat. 1000 Mtoe = 42 EJ.

Source: IEA, 2006b.

4.2.1 Global development trends in the energy sector (production and consumption)

From 1900 to 2000, world primary energy increased more than ten-fold, while world population rose only four-fold from 1.6 billion to 6.1 billion. Most energy forecasts predict considerable growth in demand in the coming decades due to increasing economic growth rates throughout the world but especially in developing countries. Global primary-energy consumption rose from 238 EJ in 1972 to 464 EJ in 2004 (Chapter 1). During the period 1972 to 1990, the average annual growth was 2.4%/yr, dropping to 1.4%/yr from 1990 to 2004 due to the dramatic decrease in energy consumption in the former Soviet Union (FSU) (Figure 4.2) and to energy intensity improvements in OECD countries. The highest growth rate in the last 14 years was in Asia (3.2%/yr).

Low electrification rates correlate with slow socio-economic development. The average rates in the Middle East, North Africa, East Asia/China and Latin America have resulted in grid connection for over 85% of their populations, whereas sub-Saharan Africa is only 23% (but only 8% in rural regions) and South Asia is 41% (30% in rural regions) (IEA, 2005c).

There is a large discrepancy between primary energy consumption per capita of 336 GJ/yr for the average North American to around 26 GJ/yr for the average African (Enerdata, 2004). The region with the lowest per-capita consumption has changed from Asian developing countries in 1972 to African countries today.

4.2.2 Emission trends of all GHGs

Growing global dependence on coal, oil and natural gas since the mid-19th century has led to the release of over 1100 $GtCO_2$ into the atmosphere (IPCC, 2001). Global CO_2 emissions from fuel combustion (around 70% of total GHG emissions and 80% of total CO_2) temporarily stabilized after the two oil crises in 1973 and 1979 before growth continued (Figure 4.6). (Emission data can be found at UNFCCC, 2006 and EEA, 2005). Analyses of potential CO_2 reductions for energy-supply options (for example IPCC, 2001; Sims *et al.,* 2003a; IEA/NEA, 2005; IEA, 2006b) showed that emissions from the energy-supply sector have grown at over 1.5% per year from around 20 $GtCO_2$ (5.5 GtC) in 1990 to over 26 $GtCO_2$ (7 GtC) by 2005.

The European Union's CO_2 emissions almost stabilized in this period mainly due to reductions by Germany, Sweden, and UK, but offset by increases by other EU-15 members (BP, 2004) such that total CO_2 emissions had risen 6.5% by 2004. Other OECD country emissions increased by 20% during the same period, Brazil by 68%, and Asia by 104%. From 1990 to 2005, China's CO_2 emissions increased from 676 to 1,491 $MtCO_2$/yr to become 18.7% of global emissions (IEEJ, 2005; BP, 2006) second only to the US. Carbon emissions from non OECD Europe and the FSU dropped by 38% between 1989

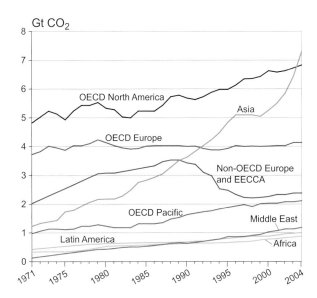

Gt CO_2

Figure 4.6: *Global trends in carbon dioxide emissions from fuel combustion by region from 1971 to 2004.*

Note: EECCA = countries of Eastern Europe, the Caucasus and Central Asia.
Source: IEA, 2006b.

and 1999 but have since started to increase as their economies rebound.

Natural gas and nuclear gained an increased market share after the oil crises in the 1970s and continue to play a role in lowering GHG emissions, along with renewable energy. Continuous technical progress towards non-carbon energy technologies and energy-efficiency improvements leads to an annual decline in carbon intensity. The carbon intensity of global primary energy use declined from 78 gCO_2/MJ in 1973 to 61 gCO_2/MJ in 2000 (BP, 2005) mainly due to diversification of energy supply away from oil. China's carbon-intensity reduction was around 5%/yr during the period 1980 to 2000 with 3%/yr expected out to 2050 (Chen, 2005), although recent revision of China's GDP growth for 2004 by government officials may affect this prediction. The US has decreased its GHG intensity (GHG/unit GDP) by 2% in 2003 and 2.5% in 2004 (Snow, 2006) although actual emissions rose.

For the power generation and heat supply sector, emissions were 12.7.$GtCO_2$-eq in 2004 (26% of total) including 2.2 $GtCO_2$-eq from methane (31% of total) and traces of N_2O (Chapter 1). In 2030, according to the World Energy Outlook 2006 baseline (IEA, 2006b), these will have increased to 17.7 $GtCO_2$-eq. During combustion of fossil fuels and biomass, nitrous oxide, as well as methane, is produced. Methane emissions from natural gas production, transmission and distribution are uncertain (UNFCCC, 2004). The losses to the atmosphere reported to the UNFCCC in 2002 were in the range 0.3–1.6% of the natural gas consumed. For more than a decade, emissions from flaring and venting of the gas associated with oil extraction have remained stable at about 0.3 $GtCO_2$-eq/yr. Developing countries accounted for more than 85% of this emission source (GGFR, 2004).

Coal bed methane (CBM, Section 4.3.1.2) is naturally contained in coal seams and adjacent rock strata. Unless it is intentionally drained and captured from the coal and rock the process of coal extraction will continue to liberate methane into the atmosphere. Around 10% of total anthropogenic methane emissions in the USA are from this source (US EPA, 2003). The 13 major coal-producing countries together produce 85% of worldwide CBM estimated to be 0.24 $GtCO_2$-eq in 2000. China was the largest emitter (0.1 $GtCO_2$-eq) followed by the USA (0.04 $GtCO_2$-eq), and Ukraine (0.03 $GtCO_2$-eq). Total CBM emissions are expected to exceed 0.3 $GtCO_2$-eq in 2020 (US EPA, 2003) unless mitigation projects are implemented.

Other GHGs are produced by the energy sector but in relatively low volumes. SF_6 is widely used in high-voltage gas-insulated substations, switches and circuit breakers because of its high di-electric constant and electrical insulating properties (Section 7.4.8). Its 100-year global warming potential (GWP) is 23,900 times that of CO_2 and it has a natural lifetime in the atmosphere of 3200 years, making it among the most potent of heat-trapping gases. Approximately 80% of SF_6 sales go to power utilities and electric power equipment manufacturers. The US government formed a partnership with 62 electric power generators and utilities (being about 35% of the USA

power grid) to voluntarily reduce leakage of SF_6 from electrical equipment and the release rate dropped from 17% of stocks to 9% between 1999 and 2002. This represented a 10% reduction from the 1999 baseline to 0.014 $GtCO_2$-eq (EPA, 2003). Australia and the Netherlands also have programmes to reduce SF_6 emissions and a voluntary agreement in Norway should lead to 13% reductions by 2005 and 30% by 2010 below their 2000 release rates. CFC-114 is used as a coolant in gaseous diffusion enrichment for nuclear power, but its GHG contribution is small compared to CO_2 emissions (Dones et al., 2005).

4.2.3 Regional development trends

World primary energy demand is projected to reach 650–890EJ by 2030 based on A1 and B2 SRES scenarios and the Reference scenario of the IEA's World Energy Outlook 2004 (Price and de la Rue du Can, 2006). All three scenarios show Asia could surpass North American energy demand by around 2010 and be close to doubling it by 2030. Africa, the Middle East and Latin America could double their energy demand by 2030; sub-Saharan Africa and the Former Soviet Union may both reach 60–70 EJ, and Pacific OECD and Central and Eastern Europe will be less than 40 EJ each. Demand is more evenly distributed among regions in the B2 scenario, with Central and Eastern Europe and the Pacific OECD region reducing future demand. A similar pattern is evident for final consumer energy (Table 4.1).

Table 4.1: Final energy consumption and carbon dioxide emissions for all sectors by region to 2030 based on assumptions from three baseline scenarios.

Region	WEO 2004 Reference				SRES A1 Marker				SRES B2 Marker			
	2002	2010	2020	2030	2000	2010	2020	2030	2000	2010	2020	2030
Final energy (EJ)												
Pacific OECD	23.6	26.6	29.5	30.9	21.5	24.6	29.8	36.6	23.5	26.5	30.0	32.3
Canada/US	70.2	78.3	87.4	94.6	71.3	79.3	89.8	99.2	71.0	82.4	93.3	104.1
Europe	51.5	56.7	62.3	66.5	52.0	58.9	67.6	74.6	46.9	51.3	54.4	57.9
EIT	27.0	31.0	35.9	40.5	38.4	42.6	50.1	58.8	32.0	37.5	44.8	52.7
Latin America	18.6	23.0	29.7	37.6	23.5	42.1	63.2	81.7	20.9	27.8	33.1	39.6
Africa/Middle East	28.4	35.4	44.8	54.3	36.4	57.2	87.6	123.7	25.6	32.6	40.2	53.1
Asia	66.8	83.1	105.3	128.3	71.5	100.6	143.9	194.6	69.4	92.5	122.0	157.5
World	**286.2**	**334.0**	**395.0**	**452.8**	**314.6**	**405.3**	**532.0**	**669.1**	**289.2**	**350.6**	**417.6**	**497.2**
Emissions ($GtCO_2$)												
Pacific OECD	2.12	2.32	2.52	2.53	2.42	2.62	2.89	3.12	2.10	2.33	2.28	2.10
Canada/US	6.47	7.24	7.88	8.32	5.84	6.08	6.13	5.97	6.61	7.63	8.36	8.43
Europe	4.12	4.45	4.81	4.90	4.21	4.53	4.74	4.73	3.95	4.04	4.07	4.13
EIT	2.39	2.79	3.21	3.54	2.97	3.45	3.71	3.85	3.23	3.26	3.66	4.08
Latin America	1.34	1.678	2.21	2.89	1.67	3.38	4.99	6.16	1.41	1.99	2.29	2.69
Africa/Middle East	2.01	2.51	3.40	4.21	2.50	4.89	7.55	10.29	1.98	2.39	2.85	3.90
Asia	5.52	7.33	9.91	12.66	5.82	9.85	14.32	18.53	5.58	7.47	9.65	12.12
Int. marine bunkers	0.46	0.47	0.48	0.51								
World	**23.98**	**28.33**	**33.93**	**39.03**	**25.42**	**34.81**	**44.33**	**52.65**	**24.86**	**29.10**	**33.15**	**37.46**

Source: Price and De la Rue du Can, 2006

The World Energy Council projected 2000 data out to 2050 for three selected scenarios with varying population estimates (WEC, 2004d). The IEA (2003c) and IPCC SRES scenarios (Chapter 3) did likewise. Implications of sustainable development were that primary energy demands are likely to experience a 40 to 150% increase, with emissions rising to between 48 and 55 GtCO$_2$/yr. This presents difficulties for the energy-supply side to meet energy demand. It requires technical progress and capital provision, and provides challenges for minimizing the environmental consequences and sustainability of the dynamic system. Electricity is expected to grow even more rapidly than primary energy by between 110 and 260% up to 2050, presenting even more challenges in needing to build power production and transmission facilities, mostly in developing countries.

The Asia-Pacific region has almost 30% of proven coal resources but otherwise is highly dependent on imported energy, particularly oil, which is now the largest source of primary commercial energy consumed in the region. In 2003, 82% of imported oil came from the Middle East and the region will continue to depend on OPEC countries. A continuation of China's rapid annual economic growth of 9.67% from 1990 to 2003 (CSY, 2005) will result in continued new energy demand, primary energy consumption having increased steadily since the 1980s. Energy consumption in 2003 reached 49 EJ. High air pollution in China is directly related to energy consumption, particularly from coal combustion that produces 70% of national particulate emissions, 90% of SO$_2$, 67% of N$_2$O and 70% of CO$_2$ (BP, 2004).

Increased use of natural gas has recently occurred throughout the Asian region, although its share of 12% of primary energy remains lower than the 23% and 17% shares in the United States and the European Union, respectively (BP, 2006). A liquefied natural gas (LNG) market has recently emerged in the region, dominated by Japan, South Korea and Spain, who together provide about 68% of worldwide trade flows.

Primary energy consumption in the Asia-Pacific region due to continued overall economic growth and increasing transport fuel demand is estimated to increase by 1.0% annually over the period 2002–2030 in OECD Asia, 2.6% in China, 2.1% in India, and 2.7% in Indonesia (IEA, 2004a). This will then account for 42% of the increase in world primary-energy demand. The region could be faced with overall energy resource shortages in the coming decades (Komiyama *et al.*, 2005). Energy security risks are likely to increase and stricter environmental restrictions on fossil fuel consumption could be imposed. Nuclear power (Section 4.3.2), hydropower (Section 4.3.3.1) and other renewables (Section 4.3.3) may play a greater role in electricity generation to meet the ever-rising demand.

For economies in transition (EIT, mainly from the former Soviet Union), the total primary energy consumption in 2000 (Figure 4.6) was only 70% of the 1990 level (Enerdata, 2004)

and a sharp downturn in GHG emissions resulted. Although increasing more recently, emissions remain some 30% below 1990 levels (IEA, 2003a; Figure 4.2). Despite the economic and political transformations, energy systems in EIT countries are still characterized by overcapacity in electricity production, high dependency on fossil-fuel imports and inefficient use (IEA, 2003b). Market reforms have been accompanied with the opening of these economies, leading to their integration into the European and global economies. Growth is likely to accelerate faster in those countries that have achieved EU membership (IEA, 2003b). The total primary-energy consumption of EIT has increased by 2% per year since 2000 and is expected to increase steadily over the next couple of decades as income levels and economic outputs expand, unless energy efficiency manages to stabilize demand.

Latin America, Africa and the Middle East are expected to double their energy demand over the next two to three decades and to retain their shares of global energy demand (IEA, 2005a; Price and de la Rue du Can, 2006). Policies in developing countries aimed at energy-supply security, reducing environmental impacts and encouraging a free market economy (Section 4.5.1.1) may help encourage market efficiency, energy conservation, common oil-reserve storage, investment in resource exploration, implementation of the Clean Development Mechanism (CDM) and international carbon emission trading. International cooperation will continue to play a role in the development of energy resources and improvement of industrial productivity.

4.2.4 Implications of sustainable development and energy access

Analysis from 125 countries indicated that well-being and level of development correlate with the degree of modern energy services consumed per capita in each country (Bailis *et al.*, 2005) (Figure 4.7).

Lack of energy access frustrates the aspirations of many developing countries (OECD, 2004a). Without improvement, the United Nations' Millennium Development Goals (MDGs)

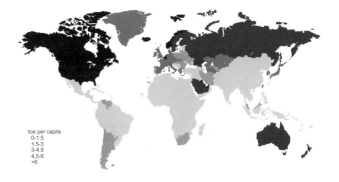

toe per capita
0-1.5
1.5-3
3-4.5
4.5-6
<6

Figure 4.7: *Global annual energy consumption per capita by region (toe/capita).*
Source: BP, 2004.

of halving the proportion of people living on less than a dollar a day by 2015 (UN, 2000) will be difficult to meet. Achieving this target implies a need for increased access to electricity and expansion of modern cooking and heating fuels for millions of people in developing countries mainly in South Asia and sub-Saharan Africa (IEA, 2005a). Historical electricity access rates of 40 million people per annum in the 1980s and 30 million per annum in the 1990s suggest that current efforts to achieve the MDGs will need to be greatly exceeded. By 2030, around 2400 GW of new power plant capacity will be needed in developing countries (100 GW/yr), which, together with the necessary infrastructure, will require around 5 trillion US$ investment (IEA, 2006b).

Ecological implications of energy supply result from coal and uranium mining, oil extraction, oil and gas transport, deforestation, erosion and river-flow disturbance. Certain synergetic effects can be achieved between renewable energy generation and ecological values such as reforestation and landscape structural improvements, but these are relatively minor.

4.3 Primary energy resource potentials, supply chain and conversion technologies

This section discusses primary-supply and secondary-energy (carrier) technologies. Technologies that have developed little since the TAR are covered in detail elsewhere (e.g., IEA, 2006a). Energy flows proceed from primary sources through carriers to provide services for end-users (Figure 4.3). The status of energy sources and carriers is reviewed here along with their available resource potential and usage, conversion technologies, costs and environmental impacts. An analysis is made of the potential contributions due to further technological development for each resource to meet the world's growing energy needs, but also to reduce atmospheric GHG emissions. Assessments of global energy reserves, resources and fluxes, together with cost ranges and sustainability issues, are summarized in Table 4.2.

Table 4.2: *Generalized data for global energy resources (including potential reserves), annual rate of use (490 EJ in 2005), share of primary energy supply and comments on associated environmental impacts.*

Energy class	Specific energy source[a]	Estimated available energy resource[b] (EJ)	Rate of use in 2005 (EJ/yr)[c]	2005 share of total supply (%)	Comments on environmental impacts
Fossil energy	Coal (conventional)	>100,000	120	25	Average 92.0 gCO$_2$/MJ
	Coal (unconventional)	32,000	0		
	Peat[d]	large	0.2	<0.1	
	Gas (conventional)	13,500	100	21	Average 52.4 gCO$_2$/MJ
	Gas (unconventional)	18,000	Small		Unknown, likely higher
	Coalbed methane	>8,000?	1.5	0.3	
	Tight sands	8,000	3.3	0.7	
	Hydrates	>60,000	0		
	Oil (conventional)	10,000	160	33	Average 76.3 gCO$_2$/MJ
	Oil (unconventional)	35,000	3	0.6	Unknown, likely higher
Nuclear	Uranium[e]	7,400	26	5.3	Spent fuel disposition
	Uranium recycle[f]	220,000	Very small		Waste disposal
	Fusion	5 x 109 estimated	0		Tritium handling
Renewable[g]	Hydro (>10 MW)	60 /yr	25	5.1	Land-use impacts
	Hydro (< 10 MW)	2 /yr	0.8	0.2	
	Wind	600 /yr	0.95	0.2	
	Biomass (modern)	250 /yr	9	1.8	Likely land-use for crops
	Biomass (traditional)		37	7.6	Air pollution
	Geothermal	5,000 /yr	2	0.4	Waterway contamination
	Solar PV	1,600 /yr	0.2	<0.1	Toxics in manufacturing
	Concentrating solar	50 /yr[h]	0.03	0.1	Small
	Ocean (all sources)	7/yr (exploitable)	<1	0	Land and coastal issues.

Notes:
[a] See Glossary for definitions of conventional and unconventional.
[b] Various sources contain ranges, some wider than others (e.g., those for conventional oil cluster much more closely than those for biomass). For the purposes of this assessment of mitigation potentials these values, generalized to a first approximation with some very uncertain, are more than adequate.
[c] Hydro and wind are treated as equivalent energy to fossil and biomass since the conversion losses are much less (www.iea.org/textbase/stats/questionaire/faq.asp)
[d] Peat land area under active production is approximately 230,000 ha. This is about 0.05% of the global peat land area of 400 million hectares (WEC, 2004c).
[e] Once-through thermal reactors.
[f] Light-water and fast-spectrum reactors with plutonium recycle
[g] Data from 2005 is at www.ren21.net/globalstatusreport/issuesGroup.asp
[h] Very uncertain. The potential of the Mediterranean area alone has been estimated by one source to be 8000 EJ/yr (http://www.dlr.de/tt/med-csp)
Sources: Data from BP, 2006; WEC, 2004c; IEA, 2006b; IAEA, 2005c; USGS, 2000; Martinot, 2005; Johansson, 2004; Hall, 2003; Encyclopaedia of Energy, 2004.

4.3.1 Fossil fuels

Fossil energy resources remain abundant but contain significant amounts of carbon that are normally released during combustion. The proven and probable reserves of oil and gas are enough to last for decades and in the case of coal, centuries (Table. 4.2). Possible undiscovered resources extend these projections even further.

Fossil fuels supplied 80% of world primary energy demand in 2004 (IEA, 2006b) and their use is expected to grow in absolute terms over the next 20–30 years in the absence of policies to promote low-carbon emission sources. Excluding traditional biomass, the largest constituent was oil (35%), then coal (25%) and gas (21%) (BP, 2005). In 2003 alone, world oil consumption increased by 3.4%, gas by 3.3% and coal by 6.3% (WEC, 2004a). Oil accounted for 95% of the land-, water- and air-transport sector demand (IEA, 2005d) and, since there is no evidence of saturation in the market for transportation services (WEC, 2004a), this percentage is projected to rise (IEA, 2003c). IEA (2005b) projected that oil demand will grow between 2002 and 2030 (by 44% in absolute terms), gas demand will almost double, and CO_2 emissions will increase by 62% (which lies between the SRES A1 and B2 scenario estimates of +101% and +55%, respectively; Table 4.1).

Fossil energy use is responsible for about 85% of the anthropogenic CO_2 emissions produced annually (IEA, 2003d). Natural gas is the fossil fuel that produces the lowest amount of GHG per unit of energy consumed and is therefore favoured in mitigation strategies. Fossil fuels have enjoyed economic advantages that other technologies may not be able to overcome, although there has been a recent trend for fossil fuel prices to increase and renewable energy prices to decrease because of continued productivity improvements and economies of scale. All fossil fuel options will continue to be used if matters are left solely to the market place to determine choice of energy conversion technologies. If GHGs are to be reduced significantly, either current uses of fossil energy will have to shift toward low- and zero-carbon sources, and/or technologies will have to be adopted that capture and store the CO_2 emissions. The development and implementation of low-carbon technologies and deployment on a larger scale requires considerable investment, which, however, should be compared with overall high investments in future energy infrastructure (see Section 4.1).

4.3.1.1 Coal and peat

Coal is the world's most abundant fossil fuel and continues to be a vital resource in many countries (IEA, 2003e). In 2005, coal accounted for around 25% of total world energy consumption primarily in the electricity and industrial sectors (BP, 2005; US EIA, 2005; Enerdata, 2004). Global proven recoverable reserves of coal are about 22,000 EJ (BP, 2004; WEC, 2004b) with another 11,000 EJ of probable reserves and

an estimated additional possible resource of 100,000 EJ for all types. Although coal deposits are widely distributed, over half of the world's recoverable reserves are located in the US (27%), Russia (17%) and China (13%). India, Australia, South Africa, Ukraine, Kazakhstan and the former Yugoslavia account for an additional 33% (US DOE, 2005). Two thirds of the proven reserves are hard coal (anthracite and bituminous) and the remainder are sub-bituminous and lignite. Together these resources represent stores of over 12,800 $GtCO_2$. Consumption was around 120 EJ/yr in 2005, which introduced approximately 9.2 $GtCO_2$/yr into the atmosphere.

Peat (partially decayed plant matter together with minerals) has been used as a fuel for thousands of years, particularly in Northern Europe. In Finland, it provides 7% of electricity and 19% of district heating.

Technologies

The demand for coal is expected to more than double by 2030 and the IEA has estimated that more than 4500 GW of new power plants (half in developing countries) will be required in this period (IEA, 2004a). The implementation of modern high-efficiency and clean utilization coal technologies is key to the development of economies if effects on society and environment are to be minimized (Section 4.5.4).

Most installed coal-fired electricity-generating plants are of a conventional subcritical pulverized fuel design, with typical efficiencies of about 35% for the more modern units. Supercritical steam plants are in commercial use in many developed countries and are being installed in greater numbers in developing countries such as China (Philibert and Podkanski, 2005). Current supercritical technologies employ steam temperatures of up to 600°C and pressures of 280 bar delivering fuel to electricity-cycle efficiencies of about 42% (Moore, 2005). Conversion efficiencies of almost 50% are possible in the best supercritical plants, but are more costly (Equitech, 2005; IPCC, 2001; Danish Energy Authority, 2005). Improved efficiencies have reduced the amount of waste heat and CO_2 that would otherwise have been emitted per unit of electricity generation.

Technologies have changed little since the TAR. Supercritical plants are now built to an international standard, however, and a CSIRO (2005) project is under way to investigate the production of ultra-clean coal that reduces ash below 0.25%, sulphur to low levels and, with combined-cycle direct-fired turbines, can reduce GHG emissions by 24% per kWh, compared with conventional coal power stations.

Gasifying coal prior to conversion to heat reduces the emissions of sulphur, nitrogen oxides, and mercury, resulting in a much cleaner fuel while reducing the cost of capturing CO_2 emissions from the flue gas where that is conducted. Continued development of conventional combustion integrated gasification combined cycle (IGCC) systems is expected to further reduce emissions.

Coal-to-liquids (CTL) is well understood and regaining interest, but will increase GHG emissions significantly without CCS (Section 4.3.6). Liquefaction can be performed by direct solvent extraction and hydrogenation of the resulting liquid at up to 67% efficiency (DTI, 1999) or indirectly by gasification then producing liquids by Fischer-Tropsch catalytic synthesis as in the three SASOL plants in South Africa. These produce 0.15 Mbbl/day of synthetic diesel fuel (80%) plus naphtha (20%) at 37–50% thermal efficiency. Lower-quality coals would reduce the thermal efficiency whereas co-production with electricity and heat (at a 1:8 ratio) could increase it and reduce the liquid fuel costs by around 10%.

Production costs of CTL appear competitive when crude oil is around 35–45 US$/bbl, assuming a coal price of 1 US$/GJ. Converting lignite at 0.50 US$/GJ close to the mine could compete with production costs of about 30 US$/bbl. The CTL process is less sensitive to feedstock prices than the gas-to-liquids (GTL) process, but the capital costs are much higher (IEA, 2005e). An 80,000 barrel per day CTL installation would cost about 5 billion US$ and would need at least 2–4 Gt of coal reserves available to be viable.

4.3.1.2 Gaseous fuels

Conventional natural gas

Natural gas production has been increasing in the Middle East and Asia–Oceania regions since the 1980s. Globally, from 1994–2004, it showed an annual growth rate of 2.3%. During 2005, 11% of natural gas was produced in the Middle East, while Europe and Eurasia produced 38%, and North America 27% (BP, 2006). Natural gas presently accounts for 21% of global consumption of modern energy at around 100 EJ/yr, contributing around 5.5 $GtCO_2$ annually to the atmosphere.

Proven global reserves of natural gas are estimated to be 6500 EJ (BP, 2006; WEC, 2004c; USGS, 2004b). Almost three quarters are located in the Middle East, and the transitional economies of the FSU and Eastern Europe. Russia, Iran and Qatar together account for about 56% of gas reserves, whereas the remaining reserves are more evenly distributed on a regional basis including North Africa (BP, 2006). Probable reserves and possible undiscovered resources that expect to be added over the next 25 years account for 2500 EJ and 4500 EJ respectively (USGS, 2004a), although other estimates are less optimistic.

Natural gas-fired power generation has grown rapidly since the 1980s because it is relatively superior to other fossil-fuel technologies in terms of investment costs, fuel efficiency, operating flexibility, rapid deployment and environmental benefits, especially when fuel costs were relatively low. Combined cycle, gas turbine (CCGT) plants produce less CO_2 per unit energy output than coal or oil technologies because of the higher hydrogen-carbon ratio of methane and the relatively high thermal efficiency of the technology. A large number of CCGT plants currently being planned, built, or operating are in the 100–500 MW_e size range. Advanced gas turbines

currently under development, such as so-called 'H' designs, may have efficiencies approaching 60% using high combustion temperatures, steam-cooled turbine blades and more complex steam cycles.

Despite rising prices, natural gas is forecast to continue to be the fastest-growing primary fossil fuel energy source worldwide (IEA, 2006b), maintaining average growth of 2.0% annually and rising to 161 EJ consumption in 2025. The industrial sector is projected to account for nearly 23% of global natural gas demand in 2030, with a similar amount used to supply new and replacement electric power generation. The share of natural gas used to generate electricity worldwide is projected to increase from 25% of primary energy in 2004 to 31% in 2030 (IEA, 2006b).

LNG

Meeting future increases in global natural gas demand for direct use by the industrial and commercial sectors as well as for power generation will require development and scale-up of liquefied natural gas (LNG) as an energy carrier. LNG transportation already accounts for 26% of total international natural gas trade in 2002, or about 6% of world natural gas consumption and is expected to increase substantially.

The Pacific Basin is the largest LNG-producing region in the world, supplying around 50% of all global exports in 2002 (US EIA, 2005). The share of total US natural gas consumption met by net imports of LNG is expected to grow from about 1% in 2002 to 15% (4.5 EJ) in 2015 and to over 20% (6.8 EJ) in 2025. Losses during the LNG liquefaction process are estimated to be 7 to 13% of the energy content of the withdrawn natural gas being larger than the typical loss of pipeline transportation over 2000 km.

LPG

Liquefied petroleum gas (LPG) is a mixture of propane, butane, and other hydrocarbons produced as a by-product of natural gas processing and crude oil refining. Total global consumption of LPG amounted to over 10 EJ in 2004 (MCH/WLPGA, 2005), equivalent to 10% of global natural gas consumption (Venn, 2005). Growth is likely to be modest with current share maintained.

Unconventional natural gas

Methane stored in a variety of geologically complex, unconventional reservoirs, such as tight gas sands, fractured shales, coal beds and hydrates, is more abundant than conventional gas (Table 4.2). Development and distribution of these unconventional gas resources remain limited worldwide, but there is growing interest in selected tight gas sands and coal-bed methane (CBM). Probable CBM resources in the US alone are estimated to be almost 800 EJ but less than 110 EJ is believed to be economically recoverable (USGS, 2004b) unless gas prices rise significantly. Worldwide resources may be larger than 8000 EJ, but a scarcity of basic information on the gas content of coal resources makes this number highly speculative.

Large quantities of tight gas are known to exist in geologically complex formations with low permeability, particularly in the US, where most exploration and production has been undertaken. However, only a small percentage is economically viable with existing technology and current US annual production has stabilized between 2.7 and 3.8 EJ.

Methane gas hydrates occur naturally in abundance worldwide and are stable as deep marine sediments on the ocean floor at depths greater than 300m and in polar permafrost regions at shallower depths. The amount of carbon bound in hydrates is not well understood, but is estimated to be twice as large as in all other known fossil fuels (USGS, 2004a). Hydrates may provide an enormous resource with estimates varying from 60,000 EJ (USGS, 2004a) to 800,000 EJ (Encyclopedia of Energy, 2004). Recovering the methane is difficult, however, and represents a significant environmental problem if unintentionally released to the atmosphere during extraction. Safe and economic extraction technologies are yet to be developed (USGS, 2004a). Hydrates also contain high levels of CO_2 that may have to be captured to produce pipeline-quality gas (Encyclopedia of Energy, 2004).

The GTL process is gaining renewed interest due to higher oil prices, particularly for developing uneconomic natural gas reserves such as those associated with oil extraction at isolated gas fields which lie far from markets. As for CTL, the natural gas is turned into synthesis gas, which is converted by the Fischer-Tropsch process to synthetic fuels. At present, at least nine commercial GTL projects are progressing through various development stages in gas-rich countries such as Qatar, Iran, Russia, Nigeria, Australia, Malaysia and Algeria with worldwide production estimated at 0.58 Mbbl/day (FACTS, 2005). GTL conversion technologies are around 55% efficient and can help bring some of the estimated 6000 EJ of stranded gas resources to market. Production costs vary depending on gas prices, but where stranded gas is available at 0.5 US$/GJ production costs are around 30 US$ a barrel (IEA 2006a). Higher CO_2 emissions per unit consumed compared with conventional oil products.

4.3.1.3   Petroleum fuels

Conventional oil products extracted from crude oil-well bores and processed by primary, secondary or tertiary methods represent about 37% of total world energy consumption (Figure 4.4 and Table 4.2) with major resources concentrated in relatively few countries. Two thirds of proven crude oil reserves are located in the Middle East and North Africa (IEA, 2005a).

Known or proven reserves are those extractable at today's prices and technologies. Additional probable and possible resources are based on historical experience in geological basins. While new discoveries have lagged behind production for more than 20 years, reserve additions from all sources including discoveries, extensions, revisions and improvements in oil recovery continue to outpace production (IEA, 2005b).

Various studies and models have been used to forecast future oil production (US EIA, 2004; Bentley, 2005). Geological models take into consideration the volume and quality of hydrocarbons but do not include economic effects on price, which in turn has a direct effect on supply and the overall rate of recovery. Mathematical models generally use the historical as well as the observed patterns of production to estimate a peak (or several peaks) reached when half the reserves are consumed.

Assessments of the amount of oil consumed, the amount remaining for extraction, and whether the peak oil tipping point is close or not, have been very controversial (Hirsch et al., 2005). Estimates of the ultimate extractable resource (proven + probable + possible reserves) with which the world was endowed have varied from less than 5730 EJ to 34,000 EJ (1000 to 6000 Gbbl), though the more recent predictions have all ranged between 11,500–17,000 EJ (2000–3000 Gbbl) (Figure 4.8). Over time, the prediction trend showed increasing resource estimates in the 1940s and 1950s as more fields were discovered. However, the very optimistic estimates of the 1970s were later discredited and a relatively constant estimate has since been observed.

Specific analyses include Bentley (2002b), who concluded that 4870 EJ had been consumed by 1998 and that 6300 EJ will have been extracted by 2008. The US Geological Survey (USGS, 2000) the World Petroleum Congress and the IFP agreed that approximately 4580 EJ (800 Gbbl) have been consumed in the past 150 years and 5730 EJ (1000 Gbbl) of proven reserves remain. Other detailed analyses (e.g. USGS, 2000) also estimated there are 4150 EJ of probable and possible resources still available for extraction. Thus, the total available potential proven reserves plus resources of around 10,000 EJ (BP, 2004; WEC, 2004b) should be sufficient for about 70 years'

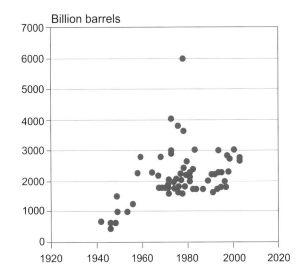

Figure 4.8: *Estimates of the global ultimate extractable conventional oil resource by year of publications.*

Source: Based on Bentley, 2002a; Andrews and Udal, 2003.

supply at present rates of consumption. Since consumption rates will continue to rise, however, 30 to 40 years' supply is a more reasonable estimate (Hallock *et al.,* 2004). Burning this amount of petroleum resources would release approximately 700 $GtCO_2$ (200 GtC) into the atmosphere, about two thirds the amount released to date from all fossil-fuel consumption. Opportunities for energy-efficiency improvements in oil refineries and associated chemical plants are covered in Chapter 7.

4.3.1.4 *Unconventional oil*

As conventional oil supplies become scarce and extraction costs increase, unconventional liquid fuels, in addition to CTL and GTL, will become more economically attractive, but offset by greater environmental costs (Williams *et al.,* 2006). Oil that requires extra processing such as from shales, heavy oils and oil (tar) sands is classified as unconventional. Resource estimates are uncertain, but together contributed around 3% of world oil production in 2005 (2.8 EJ) and could reach 4.6 EJ by 2020 (USGS, 2000) and up to 6 EJ by 2030 (IEA, 2005a). The oil industry has the potential to diversify the product mix, thereby adding to fuel-supply security, but higher environmental impacts may result and investment in new infrastructure would be needed.

Heavy oil reserves are greater than 6870 EJ (1200 Gbbl) of oil equivalent with around 1550 EJ technically recoverable. The Orinoco Delta, Venezuela has a total resource of 1500 EJ with current production of 1.2 EJ/yr (WEC, 2004c). Plans for 2009 are to apply deep-conversion, delayed coking technology to produce 0.6 Mbbl/day of high-value transport fuels.

Oil shales (kerogen that has not completed the full geological conversion to oil due to insufficient heat and pressure) represent a potential resource of 20,000 EJ with a current production of just 0.024 EJ/yr, mostly in the US, Brazil, China and Estonia. Around 80% of the total resource lies in the western US with 500 Gbbl of medium-quality reserves from rocks yielding 95 L of oil per tonne but with 1000 Gbbl potential if utilizing lower-quality rock. Mining and upgrading of oil shale to syncrude fuel costs around 11 US$/bbl. As with oil sands (below), the availability of abundant water is an issue.

Around 80% of the known global tar sand resource of 15,000 EJ is in Alberta, Canada, which has a current production of 1.6 EJ/yr, representing around 15% of national oil demand. Around 310 Gbbl is recoverable (CAPP, 2006). Production of around 2 Mbbl/day by 2010 could provide more than half of Canada's projected total oil production with 4 Mbbl/day possible by 2020. Total resources represent at least 400 Gt of stored carbon and will probably be added to as more are discovered, assuming that natural gas and water (steam) to extract the hydrocarbons are available at a reasonable cost.

Technologies for recovering tar sands include open cast (surface) mining where the deposits are shallow enough (which

accounts for 10% of the resource but 80% of current extraction), or injection of steam into wells *in situ* to reduce the viscosity of the oil prior to extraction. Mining requires over 100m³ of natural gas per barrel of bitumen extracted and in situ around 25m³. In both cases cleaning and upgrading to a level suitable for refining consumes a further 25–50m³ per barrel of oil feedstock. The mining process uses about four litres of water to produce one litre of oil but produces a refinable product. The *in situ* process uses about two litres of water to one of oil, but the very heavy product needs cleaning and diluting (usually with naptha) at the refinery or sent to an upgrader to yield syncrude at an energy efficiency of around 75% (NEB, 2006). The energy efficiency of oil sand upgrading is around 75%. Mining, producing and upgrading oil sands presently costs about 15 US$/bbl (IEA, 2006a) but new greenfield projects would cost around 30–35 US$/bbl due to project-cost inflation in recent years (NEB, 2006). If CCS is integrated, then an additional 5 US$ per barrel at least should be added. Comparable costs for conventional oil are 4–6 US$/bbl for exploration and production and 1–2 US$/bbl for refining.

Mining of oil sands leaves behind large quantities of pollutants and areas of disturbed land.

The total CO_2 emitted per unit of energy during production of liquid unconventional oils is greater than for a unit of conventional oil products due to higher energy inputs for extraction and processing. Net emissions amount to 15–34 $kgCO_2$ (4–9 kgC) per GJ of transport fuel compared with around 5-10 $kgCO_2$ (1.3-2.7 kgC) per GJ for conventional oil (IEA, 2005d, Woyllinowicz *et al.,* 2005). Oil sands currently produce around 3–4 times the pre-combustion emissions ($CO_2/$ GJ liquid fuel) compared with conventional oil extraction and refining, whereas large-scale production of oil-shale processing would be about 5 times, GTL 3–4 times, and CTL around 7–8 times when using sub-bituminous coal. The Athabascan oil-sands project has refining energy expenditures of 1 GJ energy input per 6 GJ bitumen processed, producing emissions of 11 $kgCO_2$ (3 kgC) per GJ from refining alone, but with a voluntary reduction goal of 50% by 2010 (Shell, 2006).

4.3.2 Nuclear energy

In 2005, 2626 TWh of electricity (16% of the world total) was generated by nuclear power, requiring about 65,500 t of natural uranium (WNA, 2006a). As of December 2006, 442 nuclear power plants were in operation with a total installed capacity of about 370 GW_e (WNA, 2006a). Six plants were in long-term shutdown and since 2000, the construction of 21 new reactors has begun (IAEA, 2006). The US has the largest number of reactors and France the highest percentage hare of total electricity generation. Many more reactors are either planned or proposed, mostly in China, India, Japan, Korea, Russia, South Africa and the US (WNA, 2006a). Nuclear power capacity forecasts out to 2030 (IAEA, 2005c; WNA, 2005a; Maeda, 2005; Nuclear News, 2005) vary between 279 and

740 GW$_e$ when proposed new plants and the decommissioning of old plants are both considered. In Japan 55 nuclear reactors currently provide nearly a third of total national electricity with one to be shut down in 2010. Immediate plans for construction of new reactors have been scaled down due to anticipated reduced power demand due to greater efficiency and population decline (METI, 2005). The Japanese target is now to expand the current installed 50 GW$_e$ to 61 GW$_e$ by adding 13 new reactors with nine operating by 2015 to provide around 40% of total electricity (JAEC, 2005). In China there are nine reactors in operation, two under construction and proposals for between 28 and 40 new ones by 2020 (WNA, 2006b; IAEA, 2006) giving a total capacity of 41–46 GW$_e$ (Dellero & Chessé, 2006). To meet future fuel demand, China has ratified a safeguards agreement (ANSTO, 2006) enabling the future purchase of thousands of tonnes of uranium from Australia, which has 40% of the world's reserves. In India seven reactors are under construction, with plans for 16 more to give 20 GW$_e$ of nuclear capacity installed by 2020 (Mago, 2004).

Improved safety and economics are objectives of new designs of reactors. The worldwide operational performance has improved and the 2003–2005 average unit capacity factor was 83.3% (IAEA, 2006). The average capacity factors in the US increased from less than 60% to 90.9% between 1980 and 2005, while average marginal electricity-production costs (operation, maintenance and fuel costs) declined from 33 US$/MWh in 1988 to 17 US$/MWh in 2005 (NEI, 2006).

The economic competitiveness of nuclear power depends on plant-specific features, number of plants previously built, annual hours of operation and local circumstances. Full life-cycle cost analyses have been used to compare nuclear-generation costs with coal, gas or renewable systems (Section 4.4.2; Figure 4.27) (IEA/NEA, 2005) including:

- investment (around 45–70% of total generation costs for design, construction, refurbishing, decommissioning and expense schedule during the construction period);
- operation and maintenance (around 15–40% for operating and support staff, training, security, and periodic maintenance); and
- fuel cycle (around 10–20% for purchasing, converting and enriching uranium, fuel fabrication, spent fuel conditioning, reprocessing, transport and disposal of the spent fuel).

Decommissioning costs are below 500 US$/kW (undiscounted) for water reactors (OECD, 2003) but around 2500 US$/kW for gas-cooled (e.g. Magnox) reactors due to radioactive waste volumes normalized by power output being about ten times higher. The decommissioning and clean-up of the entire UK Sellafield site, including facilities not related to commercial nuclear power production, has been estimated to cost £31.8 billion or approximately 60 billion US$ (NDA, 2006).

Total life-cycle GHG emissions per unit of electricity produced from nuclear power are below 40 gCO$_2$-eq/kWh (10 gC-eq/kWh), similar to those for renewable energy sources (Figure 4.18). (WEC, 2004a; Vattenfall, 2005). Nuclear power is therefore an effective GHG mitigation option, especially through license extensions of existing plants enabling investments in retro-fitting and upgrading. Nuclear power currently avoids approximately 2.2–2.6 GtCO$_2$/yr if that power were instead produced from coal (WNA, 2003; Rogner, 2003) or 1.5 GtCO$_2$/yr if using the world average CO$_2$ emissions for electricity production in 2000 of 540 gCO$_2$/kWh (WEC, 2001). However, Storm van Leeuwen and Smith (2005) give much higher figures for the GHG emissions from ore processing and construction and decommissioning of nuclear power plants.

4.3.2.1   Risks and environmental impacts

Regulations demand that public and occupational radiation doses from the operation of nuclear facilities be kept as low as reasonably achievable and below statutory limits. Mining, milling, power-plant operation and reprocessing of spent fuel dominate the collective radiation doses (OECD, 2000). Protective actions for mill-tailing piles and ponds have been demonstrated to be effective when applied to prevent or reduce long-term impacts from radon emanation. In the framework of the IAEA's Nuclear Safety Convention (IAEA, 1994), the IAEA member countries have agreed to maintain high safety culture to continuously improve the safety of nuclear facilities. However, risks of radiation leakage resulting from accidents at a power plant or during the transport of spent fuel remain controversial.

Operators of nuclear power plants are usually liable for any damage to third parties caused by an incident at their installation regardless of fault (UIC, 2005), as defined by both international conventions and national legislation. In 2004, the contracting parties to the OECD Paris and Brussels Conventions signed Amending Protocols setting the minimum liability limit at 700 million € with additional compensation up to 800 € through public funds. Many non-OECD countries have similar arrangements through the IAEA's Vienna Convention. In the US, the national Price-Anderson Act provides compensation up to 300 million US$ covered by an insurance paid by each reactor and also by a reactor-operator pool from the 104 reactors, which provides 10.4 billion US$.

4.3.2.2   Nuclear-waste management, disposal and proliferation aspects

The main safety objective of nuclear waste management (IAEA, 1997; IAEA, 2005b) is that human health and the environment need to be protected now and in the future without imposing undue burdens on future generations. Repositories are in operation for the disposal of low- and medium-level radioactive wastes in several countries but none yet exist for high-level waste (HLW) such as spent light-water reactor (LWR) fuel. Deep geological repositories are the most extensively studied option but resolution of both technical and political/societal issues is still needed.

In 2001, the Finnish Parliament agreed to site a spent fuel repository near the Olkiluoto nuclear power plant. After detailed rock-characterization studies, construction is scheduled to start soon after 2010 with commissioning planned for around 2020. In Sweden, a repository-siting process is concentrating on the comparison of several site alternatives close to the Oskarshamn and Forsmark nuclear power plants. In the US, the Yucca Mountain area has been chosen, amidst much controversy, as the preferred site for a HLW repository and extensive site-characterization and design studies are underway, although not without significant opposition. It is not expected to begin accepting HLW before 2015. France is also progressing on deep geological disposal as the reference solution for long-lived radioactive HLW and sets 2015 as the target date for licensing a repository and 2025 for opening it (DGEMP, 2006). Spent-fuel reprocessing and recycling of separate actinides would significantly reduce the volume and radionuclide inventory of HLW.

The enrichment of uranium (U-235), reprocessing of spent fuel and plutonium separation are critical steps for nuclear-weapons proliferation. The Treaty on Non-Proliferation of Nuclear Weapons (NPT) has been ratified by nearly 190 countries. Compliance with the terms of the NPT is verified and monitored by the IAEA. Improving proliferation resistance is a key objective in the development of next-generation nuclear reactors and associated advanced fuel-cycle technologies. For once-through uranium systems, stocks of plutonium are continuously built up in the spent fuel, but only become accessible if reprocessed. Recycling through fast-spectrum reactors on the other hand allows most of this material to be burned up in the reactor to generate more power, although there are vulnerabilities in the reprocessing step and hence still the need for careful safeguards. Advanced reprocessing and partitioning and transmutation technologies could minimize the volumes and toxicity of wastes for geological disposal, yet uncertainties about proliferation-risk and cost remain.

4.3.2.3   Development of future nuclear-power systems

Present designs of reactors are classed as Generations I through III (Figure 4.9). Generation III+ advanced reactors are now being planned and could first become operational during the period 2010–2020 (GIF, 2002) and state-of-the-art thereafter to meet anticipated growth in demand. These evolutionary reactor designs claim to have improved economics, simpler safety systems with the impacts of severe accidents limited to the close vicinity of the reactor site. Examples include the European design of a pressurized water reactor (EPR) scheduled to be operating in Finland around 2010 and the Flamanville 3 reactor planned in France.

Generation IV nuclear-energy technologies that may become operational after about 2030 employ advanced closed-fuel cycle systems with more efficient use of uranium and thorium resources. Advanced designs are being pursued mainly by the Generation-IV International Forum (GIF, a group of ten nations plus the EU and coordinated by the US Department of Energy)

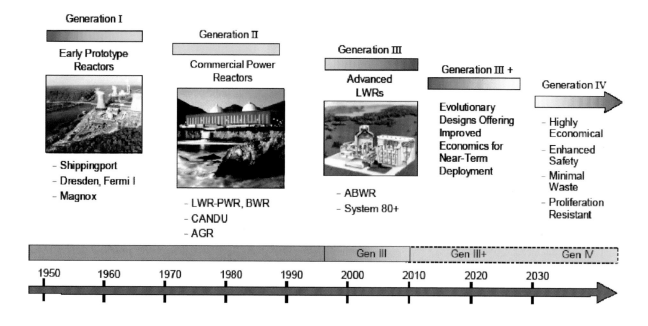

Figure 4.9: *Evolution of nuclear power systems from Generation I commercial reactors in the 1950s up to the future Generation IV systems which could be operational after about 2030.*

Notes: LWR = light-water reactor; PWR = pressurized water reactor; BWR = boiling-water reactor; ABWR = advanced boiling-water reactor; CANDU = Canada Deuterium Uranium.
Source: GIF, 2002.

as well as the International Project on Innovative Nuclear Reactors and Fuel Cycles (INPRO) coordinated by the IAEA. The Global Nuclear Energy Partnership (US DOE, 2006), proposed by the US, has similar objectives. These initiatives focus on the development of reactors and fuel cycles that provide economically competitive, safe and environmentally sound energy services based on technology designs that exclude severe accidents, involve proliferation-resistant fuel cycles decoupled from any fuel-resource constraints, and minimize HLW. Much additional technology development would be needed to meet these long-term goals so strategic public RD&D funding is required, since there is limited industrial/commercial interest at this early stage.

GIF has developed a framework to plan and conduct international cooperative research on advanced (breeder or burner) nuclear-energy systems (GIF, 2002) including three designs of fast-neutron reactor, (sodium-cooled, gas-cooled and lead-cooled) as well as high-temperature reactors. Reactor concepts capable of producing high-temperature nuclear heat are intended to be employed also for hydrogen generation, either by electrolysis or directly by special thermo-chemical water-splitting processes or steam reforming. There is also an ongoing development project by the South African utility ESKOM for an innovative high-temperature, pebble-bed modular reactor. Specific features include its smaller unit size, modularity, improved safety by use of passive features, lower power production costs and the direct gas-cycle design utilizing the Brayton cycle (Koster *et al.*, 2003; NER, 2004). The supercritical light-water reactor is also one of the GIF concepts intended to be operated under supercritical water pressure and temperature conditions. Conceivably, some of these concepts may come into practical use and offer better prospects for future use of nuclear power.

Experience of the past three decades has shown that nuclear power can be beneficial if employed carefully, but can cause great problems if not. It has the potential for an expanded role as a cost-effective mitigation option, but the problems of potential reactor accidents, nuclear waste management and disposal and nuclear weapon proliferation will still be constraining factors.

4.3.2.4   Uranium exploration, extraction and refining

In the long term, the potential of nuclear power is dependent upon the uranium resources available. Reserve estimates of the uranium resource vary with assumptions for its use (Figure 4.10). Used in typical light-water reactors (LWR) the identified resources of 4.7 Mt uranium, at prices up to 130 US$/kg, correspond to about 2400 EJ of primary energy and should be sufficient for about 100 years' supply (OECD, 2006b) at the 2004 level of consumption. The total conventional proven (identified) and probable (yet undiscovered) uranium resources are about 14.8 Mt (7400 EJ). There are also unconventional uranium resources such as those contained in phosphate minerals, which are recoverable for between 60 and 100 US$/kg (OECD, 2004a).

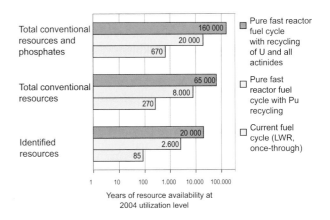

Figure 4.10: *Estimated years of uranium-resource availability for various nuclear technologies at 2004 nuclear-power utilization levels.*
Source: OECD, 2006b; OECD, 2006c.

If used in present reactor designs with a 'once-through' fuel cycle, only a small percentage of the energy content is utilized from the fissile isotope U-235 (0.7% in natural uranium). Uranium reserves would last only a few hundred years at current rate of consumption (Figure 4.10). With fast-spectrum reactors operated in a 'closed' fuel cycle by reprocessing the spent fuel and extracting the unused uranium and plutonium produced, the reserves of natural uranium may be extended to several thousand years at current consumption levels. In the recycle option, fast-spectrum reactors utilize depleted uranium and only plutonium is recycled so that the uranium-resource efficiency is increased by a factor of 30 (Figure 4.10; OECD, 2001). Thereby the estimated enhanced resource availability of total conventional uranium resources corresponds to about 220,000 EJ primary energy (Table 4.2). Even if the nuclear industry expands significantly, sufficient fuel is available for centuries. If advanced breeder reactors could be designed in the future to efficiently utilize recycled or depleted uranium and all actinides, then the resource utilization efficiency would be further improved by an additional factor of eight (OECD, 2006c).

Nuclear fuels could also be based on thorium with proven and probable resources being about 4.5 Mt (OECD, 2004a). Thorium-based fast-spectrum reactors appear capable of at least doubling the effective resource base, but the technology remains to be developed to ascertain its commercial feasibility (IAEA, 2005a). There are not yet sufficient commercial incentives for thorium-based reactors except perhaps in India. The thorium fuel cycle is claimed to be more proliferation-resistant than other fuel cycles since it produces fissionable U-233 instead of fissionable plutonium, and, as a by-product, U-232 that has a daughter nuclide emitting high-energy photons.

4.3.2.5   Nuclear fusion

Energy from the fusion of heavy hydrogen fuel (deuterium, tritium) is actively being pursued as a long-term almost inexhaustible supply of energy with helium as the by-product. The scientific feasibility of fusion energy has been proven, but technical feasibility remains to be demonstrated in experimental facilities. A major international effort, the proposed international thermonuclear experimental reactor (ITER, 2006), aims to demonstrate magnetic containment of sustained, self-heated plasma under fusion temperatures. This 10 billion US$ pilot plant to be built in France is planned to operate for 20 years and will resolve many scientific and engineering challenges. Commercialization of fusion-power production is thought to become viable by about 2050, assuming initial demonstration is successful (Smith et al., 2006a; Cook et al., 2005).

4.3.3   Renewable energy

Renewable energy accounted for over 15% of world primary energy supply in 2004, including traditional biomass (7–8%), large hydro-electricity (5.3%, being 16% of electricity generated[1]), and other 'new' renewables (2.5%) (Table 4.2). Under the business-as-usual case of continued growing energy demand, renewables are not expected to greatly increase their market share over the next few decades without continued and sustained policy intervention. For example, IEA (2006b) projected in the Reference scenario that renewables will have dropped to a 13.7 % share of global primary energy (20.8 % of electricity) in 2030, or under the Alternative Policy scenario will have risen to 16.2 % (25.3 % of electricity).

Renewable-energy systems can contribute to the security of energy supply and protection of the environment. These and other benefits of renewable energy systems were defined in a declaration by 154 nations at the Renewables 2004 conference held in Bonn (Renewables, 2004). Renewable-energy technologies can be broadly classified into four categories:

1) *technologically mature with established markets in at least several countries*:– large and small hydro, woody biomass combustion, geothermal, landfill gas, crystalline silicon PV solar water heating, onshore wind, bioethanol from sugars and starch (mainly Brazil and US);

2) *technologically mature but with relatively new and immature markets in a small number of countries*:– municipal solid waste-to-energy, anaerobic digestion, biodiesel, co-firing of biomass, concentrating solar dishes and troughs, solar-assisted air conditioning, mini- and micro-hydro and offshore wind;

3) *under technological development with demonstrations or small-scale commercial application, but approaching wider market introduction*:– thin-film PV, concentrating PV, tidal range and currents, wave power, biomass gasification and pyrolysis, bioethanol from ligno-cellulose and solar thermal towers; and

4) *still in technology research stages*:– organic and inorganic nanotechnology solar cells, artificial photosynthesis,

biological hydrogen production involving biomass, algae and bacteria, biorefineries, ocean thermal and saline gradients, and ocean currents.

The most mature renewable technologies (large hydro, biomass combustion, and geothermal) have, for the most part, been able to compete in today's energy markets without policy support. Solar water heating, solar PV in remote areas, wind farms on exceptional sites, bioethanol from sugar cane, and forest residues for combined heat and power (CHP) are also competitive today in the best locations. In countries with the most mature markets, several forms of 'new' renewable energy can compete with conventional energy sources on an average-cost basis, especially where environmental externalities and fossil fuel price risks are taken into account. In countries where market deployment is slow due to less than optimal resources, higher costs (relative to conventional fuels) and/or a variety of market and social barriers, these technologies still require government support (IEA, 2006e). Typical construction costs for new renewable energy power plants are high, between 1000 and 2500 US$/kW, but on the best sites they can generate power for around 30–40 US$/MWh thanks to low operation, maintenance and fuel costs (Martinot, 2005; NREL, 2005). Costs are very variable, however, due to the diversity of resources on specific sites (Table 4.7). In areas where the industry is growing, many sites with good wind, geothermal, biomass and hydro resources have already been utilized. The less mature technologies are not yet competitive but costs continue to decline due to increased learning experience as exemplified by wind, solar and bioethanol (Figure 4.11).

Many renewable energy sources are variable over hourly, daily and/or seasonal time frames. Energy-storage technologies

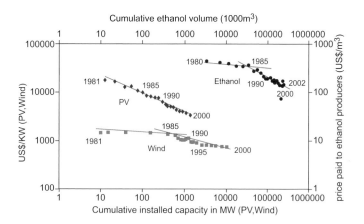

Figure 4.11: *Investment costs and penetration rates for PV, wind and bioethanol systems showing cost reductions of 20% due to technological development and learning experience for every doubling of capacity once the technology has matured.*

Source: Johansson et al., 2004.

---

1   Proportions of electricity production were calculated using the energy content of the electricity.

272

may be needed, particularly for wind, wave and solar, though stored hydro reserves, geothermal and bioenergy systems can all be used as dispatchable back-up sources as can thermal power plants. Studies on intermittency and interconnection issues with the grid are ongoing (e.g., Gul and Stenzel, 2005; UKERC, 2006; Outhred and MacGill, 2006).

A wide range of policies and measures exist to enhance the deployment of renewable energy (IEA, 2004c; Martinot *et al.*, 2005; Section 4.5). Over 49 nations, including all EU countries along with a number of developing countries such as Brazil, China, Colombia, Egypt, India, Malaysia, Mali, Mexico, Philippines, South Africa and Thailand, and many individual states/provinces of the USA, Canada and Australia have set renewable energy targets. Some targets focus on electricity, while others include renewable heating and cooling and/or biofuels. By 2004, at least 30 states/provinces and two countries had mandates in place for blending bioethanol or biodiesel with petroleum fuels.

Since the TAR, several large international companies such as General Electric, Siemens, Shell and BP have invested further in renewable energy along with a wide range of public and private sources. Commercial banks such as Fortis, ANZ Bank and Royal Bank of Canada are financing a growing number of projects; commodity traders and financial investment firms such as Fimat, Goldman Sachs and Morgan Stanley are acquiring renewable energy companies; traditional utilities are developing their own renewable energy projects; commercial reinsurance companies such as Swiss Re and Munich Re are offering insurance products targeting renewable energy, and venture capital investors are observing market projections for wind and PV. New CDM-supported and carbon-finance projects for renewables are emerging and the OECD has improved the terms for Export Credit Arrangements for renewable energy by extending repayment terms (Martinot *et al*., 2005).

There has also been increasing support for renewable energy deployment in developing countries, not only from international development and aid agencies, but also from large and small local financiers with support from donor governments and market facilitators to reduce their risks. As one example, total donor funding pledges or requirements in the Bonn Renewables 2004 Action Programme amounted to around 50 billion US$ (Renewables, 2004). Total investment in new renewable energy capacity in 2005 was 38 billion US$, excluding large hydropower, which itself was another 15–20 billion US$ (Martinot *et al.,* 2006).

Numerous detailed and comprehensive reports, websites, and conference proceedings on renewable energy resources, conversion technologies, industry trends and government support policies have been produced since the TAR (e.g., Renewables, 2004; BIREC, 2005; Martinot *et al*, 2005; IEA, 2004d; IEA, 2005d; IEA 2006a; IEA 2006c; WEC, 2004c; ISES, 2005; WREC, 2006; WREA, 2005). The following

sections address only the key points relating to progress in each major renewable energy source.

4.3.3.1 *Hydroelectricity*

Large (>10 MW) hydroelectricity systems accounted for over 2800 TWh of consumer energy in 2004 (BP, 2006) and provided 16% of global electricity (90% of renewable electricity). Hydro projects under construction could increase the share of electricity by about 4.5% on completion (WEC, 2004d) and new projects could be deployed to provide a further 6000 TWh/yr or more of electricity economically (BP, 2004; IEA, 2006a), mainly in developing countries. Repowering existing plants with more powerful and efficient turbine designs can be cost effective whatever the plant scale. Where hydro expansion is occurring, particularly in China and India, major social disruptions, ecological impacts on existing river ecosystems and fisheries and related evaporative water losses are stimulating public opposition. These and environmental concerns may mean that obtaining resource permits is a constraint.

Small (<10 MW) and micro (<1 MW) hydropower systems, usually run-of-river schemes, have provided electricity to many rural communities in developing countries such as Nepal. Their present generation output is uncertain with predictions ranging from 4 TWh/yr (WEC, 2004d) to 9% of total hydropower output at 250 TWh/yr (Martinot *et al.,* 2006). The global technical potential of small and micro hydro is around 150–200 GW with many unexploited resource sites available. About 75% of water reservoirs in the world were built for irrigation, flood control and urban water-supply schemes and many could have small hydropower generation retrofits added. Generating costs range from 20 to 90 US$/MWh but with additional costs needed for power connection and distribution. These costs can be prohibitive in remote areas, even for mini-grids, and some form of financial assistance from aid programmes or governments is often necessary.

The high level of flexibility of hydro plants enables peak loads in electricity demand to be followed. Some schemes, such as the 12.6 GW Itaipu plant in Brazil/Paraguay, are run as baseload generators with an average capacity factor of >80%, whereas others (as in the 24 GW of pumped storage plant in Japan) are used mainly as fast-response peaking plants, giving a factor closer to 40% capacity. Evaluations of hybrid hydro/wind systems, hydro/hydrogen systems and low-head run-of-river systems are under review (IEA, 2006d).

GHG emissions vary with reservoir location, power density (W capacity per m^2 flooded), flow rate, and whether dam or run-or-river plant. Recently, the GHG footprint of hydropower reservoirs has been questioned (Fearnside, 2004; UNESCO, 2006). Some reservoirs have been shown to absorb CO_2 at their surface, but most emit small amounts as water conveys carbon in the natural carbon cycle (Tremblay, 2005). High emissions of CH_4 have been recorded at shallow, plateau-type

273

tropical reservoirs where the natural carbon cycle is most productive (Delmas, 2005). Deep water reservoirs at similar low latitudes tend to exhibit lower emissions. Methane from natural floodplains and wetlands may be suppressed if they are inundated by a new reservoir since the methane is oxidized as it rises through the covering water column (Huttunen, 2005; dos Santos, 2005). Methane formation in freshwater produces by-product carbon compounds (phenolic and humic acids) that effectively sequester the carbon involved (Sikar, 2005). For shallow tropical reservoirs, further research is needed to establish the extent to which these may increase methane emissions.

Several Brazilian hydro-reservoirs were compared using life-cycle analyses with combined-cycle natural gas turbine (CCGT) plants of 50% efficiency (dos Santos et al., 2004). Emissions from flooded reservoirs tended to be less per kWh generated than those produced from the CCGT power plants. Large hydropower complexes with greater power density had the best environmental performance, whereas those with lower power density produced similar GHG emissions to the CCGT plants. For most hydro projects, life-cycle assessments have shown low overall net GHG emissions (WEC, 2004a; UNESCO, 2006). Since measuring the incremental anthropogenic-related emissions from freshwater reservoirs remains uncertain, the Executive Board of the UN Framework Convention on Climate Change (UNFCCC) has excluded large hydro projects with significant water storage from the CDM. The IPCC Guidelines for National GHG Inventories (2006) recommended using estimates for induced changes in the carbon stocks.

Whether or not large hydro systems bring benefits to the poorest has also been questioned (Collier, 2006; though this argument is not exclusive to hydro). The multiple benefits of hydro-electricity, including irrigation and water-supply resource creation, rapid response to grid-demand fluctuations due to peaks or intermittent renewables, recreational lakes and flood control, need to be taken into account for any given development. Several sustainability guidelines and an assessment protocol have been produced by the industry (IHA, 2006; Hydro Tasmania, 2005; WCD, 2000).

4.3.3.2 Wind

Wind provided around 0.5% of the total 17,408 TWh global electricity production in 2004 (IEA, 2006b) but its technical potential greatly exceeds this (WEC, 2004d; GWEC, 2006). Installed capacity increased from 2.3 GW in 1991 to 59.3 GW at the end of 2005 when it generated 119 TWh at an average capacity factor of around 23%. New wind installation capacity has grown at an average of 28% per year since 2000, with a record 40% increase in 2005 (BTM, 2006) due to lower costs, greater government support through feed-in tariff and renewable energy certificate policies (Section 4.5), and improved technology development. Total offshore wind capacity reached 679 MW at the end of 2005 (BTM, 2006), with the expectation

that it will grow rapidly due to higher mean annual wind-speed conditions offsetting the higher costs and public resistance being less. Various best-practices guidelines have been produced and issues such as noise, electromagnetic (EMF) interference, airline flight paths, land-use, protection of areas with high landscape value, and bird and bat strike, are better understood but remain constraints. Most bird species exhibit an avoidance reaction to wind turbines, which reduces the probability of collision (NERI, 2004).

The average size of wind turbines has increased in the last 25 years from less than 50 kW in the early 1980s to the largest commercially available in 2006 at around 5MW and having a rotor diameter of over 120 m. The average turbine size being sold in 2006 was around 1.6–2 MW but there is also a market for smaller turbines <100 kW. In Denmark, wind energy accounted for 18.5% of electricity generation in 2004, and 25% in West Denmark where 2.4 GW is installed, giving the highest generation per capita in the world.

Capital costs for land-based wind turbines can be below 900 US$/kW with 25% for the tower and 75% for the rotor and nacelle, although price increases have occurred due to supply shortages and increases in steel prices. Total costs of an onshore wind farm range from 1000–1400 US$/kW, depending on location, road access, proximity to load, etc. Operation and maintenance costs vary from 1% of investment costs in year one, rising to 4.5% after 15 years. This means that on good sites with low surface roughness and capacity factors exceeding 35%, power can be generated for around 30–50 US$/MWh (IEA, 2006c; Morthorst, 2004; Figure 4.12).

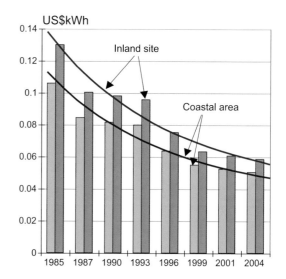

Figure 4.12: *Development of wind-generation costs based on Danish experience since 1985 with variations shown due to land surface and terrain variations (as indicated by roughness indicator classes which equal 0 for open water and up to 3 for rugged terrain).*

Source: Morthorst, 2004.

A global study of 7500 surface stations showed mean annual wind speeds at 80 m above ground exceeded 6.9 m/s with most potential found in Northern Europe along the North Sea, the southern tip of South America, Tasmania, the Great Lakes region, and the northeastern and western coasts of Canada and the US. A technical potential of 72 TW installed global capacity at 20% average capacity factor would generate 126,000 TWh/yr (Archer and Jacobsen, 2005). This is five times the assumed global production of electricity in 2030 (IEA, 2006b) and double the 600 EJ potential capacity estimated by Johansson et al. (2004) (Table 4.2).

The main wind-energy investments have been in Europe, Japan, China, USA and India (Wind Force 12, 2005). The Global Wind Energy Council assumed this will change and has estimated more widespread installed capacity of 1250 GW by 2020 to supply 12% of the world's electricity. The European Wind Energy Association set a target of 75 GW (168 TWh) for EU-15 countries in 2010 and 180 GW (425 TWh) in 2020 (EWEA, 2004). Several Australian and USA states have similar ambitious targets, mainly to meet the increasing demand for power rather than to displace nuclear or fossil-fuel plants. Rapid growth in several developing countries including China, Mexico, Brazil and India is expected since private investment interest is increasing (Martinot et al., 2005).

The fluctuating nature of the wind constrains the contribution to total electricity demand in order to maintain system reliability. To supply over 20% would require more accurate forecasting (Giebel, 2005), regulations that ensure wind has priority access to the grid, demand-side response measures, increases in the use of operational reserves in the power system (Gul and Stenzel, 2005) or development of energy storage systems (EWEA, 2005; Mazza and Hammerschlag, 2003). The additional cost burden in Denmark to provide reliability was claimed to be between 1–1.5 billion € (Bendtsen, 2003) and 2–2.5 billion € per annum (Krogsgaard, 2001). However, the costs for back-up power decrease drastically with larger grid area, larger area containing distributed wind turbines and greater share of flexible hydro and natural-gas-fired power plants (Morthorst, 2004).

A trend to replace older and smaller wind turbines with larger, more efficient, quieter and more reliable designs gives higher power outputs from the same site often at a lower density of turbines per hectare. Costs vary widely with location (Table 4.7). Sites with wind speeds of less than 7–8 m/s are not currently economically viable without some form of government support if conventional power-generation costs are above 50 US$/Wh (Oxera, 2005). A number of technologies are under development in order to maximize energy capture for lower wind-speed sites. These include: optimized turbine designs; larger turbines; taller towers; the use of carbon-fibre technology to replace glass-reinforced polymer in longer wind-turbine blades; maintenance strategies for offshore turbines to overcome difficulties with access during bad weather/rough seas; more accurate aero-elastic models and more advanced control strategies to keep the wind loads within the turbine design limits.

4.3.3.3 Biomass and bioenergy

Biomass continues to be the world's major source of food, stock fodder and fibre as well as a renewable resource of hydrocarbons for use as a source of heat, electricity, liquid fuels and chemicals. Woody biomass and straw can be used as materials, which can be recycled for energy at the end of their life. Biomass sources include forest, agricultural and livestock residues, short-rotation forest plantations, dedicated herbaceous energy crops, the organic component of municipal solid waste (MSW), and other organic waste streams. These are used as feedstocks to produce energy carriers in the form of solid fuels (chips, pellets, briquettes, logs), liquid fuels (methanol, ethanol, butanol, biodiesel), gaseous fuels (synthesis gas, biogas, hydrogen), electricity and heat. Biomass resources and bioenergy use are discussed in several other chapters (Fig. 4.13) as outlined

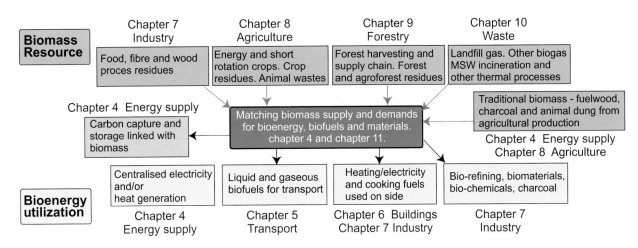

Figure 4.13: *Biomass supplies originate from a wide range of sources and, after conversion in many designs of plants from domestic to industrial scales, are converted to useful forms of bioenergy.*

in Chapter 11. This chapter 4 concentrates on the conversion technologies of biomass resources to provide bioenergy in the form of heat and electricity to the energy market.

Bioenergy carriers range from a simple firewood log to a highly refined gaseous fuel or liquid biofuel. Different biomass products suit different situations and specific objectives for using biomass are affected by the quantity, quality and cost of feedstock available, location of the consumers, type and value of energy services required, and the specific co-products or benefits (IEA Bioenergy, 2005). Prior to conversion, biomass feedstocks tend to have lower energy density per volume or mass compared with equivalent fossil fuels. This makes collection, transport, storage and handling more costly per unit of energy (Sims, 2002). These costs can be minimized if the biomass can be sourced from a location where it is already concentrated, such as wood-processing residues or sugar plant.

Globally, biomass currently provides around 46 EJ of bioenergy in the form of combustible biomass and wastes, liquid biofuels, renewable MSW, solid biomass/charcoal, and gaseous fuels. This share is estimated to be over 10% of global primary energy, but with over two thirds consumed in developing countries as traditional biomass for household use (IEA, 2006b). Around 8.6 EJ/yr of modern biomass is used for heat and power generation (Figure 4.14). Conversion is based on inefficient combustion, often combined with significant local and indoor

air pollution and unsustainable use of biomass resources such as native vegetation (Venkataraman et al., 2004).

Residues from industrialized farming, plantation forests and food- and fibre-processing operations that are currently collected worldwide and used in modern bioenergy conversion plants are difficult to quantify but probably supply approximately 6 EJ/yr. They can be classified as primary, secondary and tertiary (Figure 4.15). Current combustion of over 130 Mt of MSW provides more than 1 EJ/yr though this includes plastics, etc. (Chapter 10). Landfill gas also contributes to biomass supply at over 0.2 EJ/yr (Chapter 10).

A wide range of conversion technologies is under continuous development to produce bioenergy carriers for both small- and large-scale applications. Organic residues and wastes are often cost-effective feedstocks for bioenergy conversion plants, resulting in niche markets for forest, food processing and other industries. Industrial use of biomass in OECD countries was 5.6 EJ in 2002 (IEA, 2004a), mainly in the form of black liquor in pulp mills, biogas in food processing plants, and bark, sawdust, rice husks etc. in process heat boilers.

The use of biomass, particularly sugarcane bagasse, for cogeneration (CHP) and industrial, domestic and district heating continues to expand (Martinot et al., 2005). Combustion for heat and steam generation remains state of the art, but

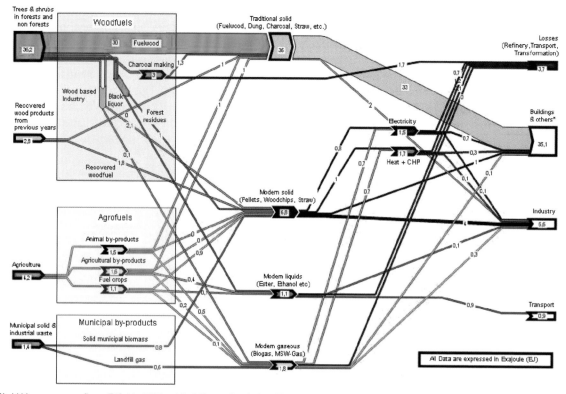

Figure 4.14: *World biomass energy flows (EJ/yr) in 2004 and their thermochemical and biochemical conversion routes to produce heat, electricity and biofuels for use by the major sectors.*

Note: much of the data is very uncertain, although a useful indication of biomass resource flows and bioenergy outputs still results.

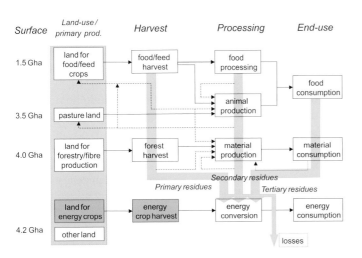

Figure 4.15: *Biomass sources from land used for primary production can be processed for energy with residues available from primary, secondary and tertiary activities.*

Source: van den Broek, 2000.

advancing technologies include second-generation biofuels (Chapter 5), biomass integrated-gasification combined-cycle (BIGCC), co-firing (with coal or gas), and pyrolysis. Many are close to commercial maturity but awaiting further technical breakthroughs and demonstrations to increase efficiency and further bring down costs.

Biochemical conversion using enzymes to convert ligno-cellulose to sugars that, in turn can be converted to bioethanol, biodiesel, di-methyl ester, hydrogen and chemical intermediates in biorefineries is not yet commercial. Biochemical- and Fischer-Tropsch-based thermochemical synthesis processes can be integrated in a single biorefinery such that the biomass carbohydrate fraction is converted to ethanol and the lignin-rich residue gasified and used to produce heat for process energy, electricity and/or fuels, thus greatly increasing the overall system efficiency to 70–80% (OECD, 2004b; Sims, 2004).

Combustion and co-firing
Biomass can be combined with fossil-fuel technologies by co-firing solid biomass particles with coal; mixing synthesis gas, landfill gas or biogas with natural gas prior to combustion. There has been rapid progress since the TAR in the development of the co-utilisation of biomass materials in coal-fired boiler plants. Worldwide more than 150 coal-fired power plants in the 50–700 MW$_e$ range have operational experience of co-firing with woody biomass or wastes, at least on a trial basis (IEA, 2004c). Commercially significant lignites, bituminous and sub-bituminous coals, anthracites and petroleum coke have all been co-fired up to 15% by energy content with a very wide range of biomass material, including herbaceous and woody materials, wet and dry agricultural residues and energy crops. This experience has shown how the technical risks associated with co-firing in different types of coal-fired power plants can be reduced to an acceptable level through proper selection of

biomass type and co-firing technology. It is a relatively low-cost, low-risk method of adding biomass capacity, particularly in countries where coal-fired plants are prevalent.

Gaseous fuels
Gasification of biomass (or coal, Section 4.3.1.1) to synthesis (producer) gas, mainly CO and H$_2$, has a relatively high conversion efficiency (40–45%) when used to generate electricity through a gas engine or gas turbine. The gas produced can also be used as feedstock for a range of liquid biofuels. Development of efficient BIGCC systems is nearing commercial realization, but the challenges of gas clean-up remain. Several pilot and demonstration projects have been evaluated with varying degrees of success (IEA, 2006d).

Recovery of methane from anaerobic digestion plants has increased since the TAR. More than 4500 installations (including landfill-gas recovery plants) in Europe, corresponding to 3.3 Mt methane or 92 PJ/yr, were operating in 2002 with a total market potential estimated to be 770 PJ (assuming 28 Mt methane will be produced) in 2020 (Jönsson, 2004). Biogas can be used to produce electricity and/or heat. It can also be fed into natural gas grids or distributed to filling stations for use in dedicated or dual gas-fuelled vehicles, although this requires biogas upgrading (Section 10.4).

Costs and reduction opportunities
Costs vary widely for biomass fuel sources giving electricity costs commonly between 0.05 and 0.12 US$/kWh (Martinot *et al.*, 2005) or even lower where the disposal cost of the biomass is avoided. Cost reductions can occur due to technical learning and capital/labour substitution. For example, capital investment costs for a high-pressure, direct-gasification combined-cycle plant up to 50 MW are estimated to fall from over 2000 US$/kW to around 1100 US$/kW by 2030, with operating costs, including delivered fuel supply, also declining to give possible generation costs down to 0.03 US$/kWh (Martinot *et al.*, 2005; Specker, 2006; EIA/DOE, 2006). Commercial small-scale options using steam turbines, Stirling engines, organic Rankin-cycle systems etc. can generate power for up to 0.12 US$/kWh, but with the opportunity to further reduce the capital costs by mass production and experience.

4.3.3.4   *Geothermal*

Geothermal resources from low-enthalpy fields located in sedimentary basins of geologically stable platforms have long been used for direct heat extraction for building and district heating, industrial processing, domestic water and space heating, leisure and balneotherapy applications. High-quality high-enthalpy fields (located in geodynamically active regions with high-temperature natural steam reached by drilling at depths less than 2 km) where temperatures are above 250°C allow for direct electricity production using binary power plants (with low boiling-point transfer fluids and heat exchangers), organic Rankin-cycle systems or steam turbines. Plant capacity factors range from 40 to 95%, with some therefore suitable for base load (WEC, 2004b). Useful heat and power produced globally is around 2 EJ/yr (Table 4.2).

Fields of natural steam are rare. Most are a mixture of steam and hot water requiring single- or double-flash systems to separate out the hot water, which can then be used in binary plants or for direct use of the heat (Martinot *et al.*, 2005). Binary systems have become state-of-the-art technologies but often with additional cost. Re-injection of the fluids maintains a constant pressure in the reservoir and hence increases the life of the field, as well as overcoming any concerns at environmental impacts. Sustainability concerns relating to land subsidence, heat-extraction rates exceeding natural replenishment (Bromley and Currie, 2003), chemical pollution of waterways (e.g. with arsenic), and associated CO_2 emissions have resulted in some geothermal power-plant permits being declined. This could be partly overcome by re-injection techniques. Deeper drilling up to 8 km to reach molten rock magma resources may become cost effective in future. Deeper drilling technology could also help to develop widely abundant hot dry rocks where water is injected into artificially fractured rocks and heat extracted as steam. Pilot schemes exist but tend not to be cost effective at this stage. In addition, the growth of ground-to-air heat pumps for heating buildings (Chapter 6) is expected to increase.

Capital costs have declined by around 50% from the 3000–5000 US$/kW in the 1980s for all plant types (with binary cycle plants being the more costly). Power-generation costs vary with high- and low-enthalpy fields, shallow or deep resource, size of field, resource-permit conditions, temperature of resource and the applications for any excess heat (IEA, 2006d; Table 4.7). Operating costs increase if CO_2 emissions released either entail a carbon charge or require CCS.

Several advanced energy-conversion technologies are becoming available to enhance the use of geothermal heat, including combined-cycle for steam resources, trilateral cycles for binary total-flow resources, remote detection of hot zones during exploration, absorption/regeneration cycles (e.g., heat pumps) and improved power-generation technologies (WEC, 2004c). Improvements in characterizing underground reservoirs, low-cost drilling techniques, more efficient conversion systems and utilization of deeper reservoirs are expected to improve the uptake of geothermal resources as will a decline in the market value for extractable co-products such as silica, zinc, manganese and lithium (IEA, 2006d).

4.3.3.5   Solar thermal electric

The proportion of solar radiation that reaches the Earth's surface is more than 10,000 times the current annual global energy consumption. Annual surface insolation varies with latitude, ranging between averages of 1000 W/m² in temperate regions and 1200 W/m² in low-latitude dry desert areas.

Concentrating solar power (CSP) plants are categorized according to whether the solar flux is concentrated by parabolic trough-shaped mirror reflectors (30–100 suns concentration), central tower receivers requiring numerous heliostats (500–1000 suns), or parabolic dish-shaped reflectors (1000–10,000 suns). The receivers transfer the solar heat to a working fluid, which, in turn, transfers it to a thermal power-conversion system based on Rankine, Brayton, combined or Stirling cycles. To give a secure and reliable supply with capacity factors at around 50% rising to 70% by 2020 (US DOE, 2005), solar intermittency problems can be overcome by using supplementary energy from associated natural gas, coal or bioenergy systems (IEA, 2006g) as well as by storing surplus heat.

Solar thermal power-generating plants are best sited at lower latitudes in areas receiving high levels of direct insolation. In these areas, 1 km² of land is enough to generate around 125 GWh/yr from a 50 MW plant at 10% conversion of solar energy to electricity (Philibert, 2004). Thus about 1% of the world's desert areas (240,000 km²), if linked to demand centres by high-voltage DC cables, could, in theory, be sufficient to meet total global electricity demand as forecast out to 2030 (Philibert, 2006; IEA, 2006b). CSP could also be linked with desalination in these regions or used to produce hydrogen fuel or metals.

The most mature CSP technology is solar troughs with a maximum peak efficiency of 21% in terms of conversion of direct solar radiation into grid electricity. Tower technology has been successfully demonstrated by two 10 MW systems in the USA with commercial development giving long-term levelized energy costs similar to trough technology. Advanced technologies include troughs with direct steam generation, Fresnel collectors, which can reduce costs by 20%, energy storage including molten salt, integrated combined-cycle systems and advanced Stirling dishes. The latter are arousing renewed interest and could provide opportunities for further cost reductions (WEC, 2004d; IEA 2004b).

Technical potential estimates for global CSP vary widely from 630 GW$_e$ installed by 2040 (Aringhoff *et al.*, 2003) to 4700 GW$_e$ by 2030 (IEA, 2003h; Table 4.2). Installed capacity is 354 MW$_e$ from nine plants in California ranging from 14 to 80 MW$_e$ with over 2 million m² of parabolic troughs. Connected to the grid during 1984–1991, these generate around 400 GWh/yr at 100–126 US$/MWh (WEC, 2004d). New projects totalling over 1400 MW are being constructed or planned in 11 countries including Spain (500 MW supported by a new feed-in tariff) (ESTIA, 2004; Martinot *et al.*, 2005) and Israel for the first of several 100 MW plants (Sagie, 2005). The African Development Bank has financed a 50 MW combined-cycle plant in Morocco that will generate 55 GWh/yr, and two new Stirling dish projects totalling 800 MW$_e$ planned for the Mojave Desert, USA (ISES, 2005) are estimated to generate at below 90 US$/MWh (Stirling, 2005). Installed capacity of 21.5 GW$_e$, if reached by 2020, would produce 54.6 TWh/yr with a further possible increase leading towards 5% coverage of world electricity demand by 2040.

4.3.3.6   Solar photovoltaic (PV)

Electricity generated directly by utilizing solar photons to create free electrons in a PV cell is estimated to have a technical potential of at least 450,000 TWh/yr (Renewables, 2004; WEC, 2004d). However, realizing this potential will be severely limited by land, energy-storage and investment constraints. Estimates of current global installed peak capacity vary widely, including 2400 MW (Greenpeace, 2004); 3100 MW (Maycock, 2003); >4000MW generating more than 21 TWh (Martinot et al., 2005) and 5000 MW (Greenpeace, 2006). Half the potential may be grid-connected, primarily in Germany, Japan and California, and grow at annual rates of 50–60% in contrast to more modest rates of 15–20% for off-grid PV. Expansion is taking place at around 30% per year in developing countries where around 20% of all new global PV capacity was installed in 2004, mainly in rural areas where grid electricity is either not available or unreliable (WEC, 2004c). Decentralized generation by solar PV is already economically feasible for villages with long distances to a distribution grid and where providing basic lighting and radio is socially desirable. Annual PV module production grew from 740 MW in 2003 to 1700 MW in 2005, with new manufacturing plant capacity built to meet growing demand (Martinot et al., 2005). Japan is the world market leader, producing over half the present annual production (IEA, 2003f). However, solar generation remains at only 0.004% of total world power.

Most commercially available solar PV modules are based on crystalline silicon cells with monocrystalline at up to 18% efficiency, having 33.2% of the market share. Polycrystalline cells at up to 15% efficiency are cheaper per W_p (peak Watt) and have 56.3% market share. Modules costing 3–4 US$/$W_p$ can be installed for around 6–7 US$/$W_p$ from which electricity can be generated for around 250 US$/MWh in high sunshine regions (US Climate Change Technology Program, 2005). Cost reductions are expected to continue (UNDP, 2000; Figure 4.11), partly depending on the future world price for silicon; solar-cell efficiency improvements as a result of R&D investment; mass production of solar panels and learning through project experience. Costs in new buildings can be reduced where PV systems are designed to be an integral part of the roof, walls or even windows.

Thinner cell materials have prospects for cost reduction, including thin-film silicon cells (8.8% of market share in 2003), thin-film copper indium diselenide cells (0.7% of market share), photochemical cells and polymer cells. Commercial thin-film cells have efficiencies up to 8%, but 10–12% should be feasible within the next few years. Experimental multilayer cells have reached higher efficiencies but their cost remains high. Work to reduce the cost of manufacturing, using low-cost polymer materials, and developing new materials such as quantum dots and nano-structures, could allow the solar resource to be more fully exploited. Combining solar thermal and PV power-generation systems into one unit has good potential as using the heat produced from cooling the PV cells would make it more efficient (Bakker et al., 2005).

4.3.3.7   Solar heating and cooling

Solar heating and cooling of buildings can reduce conventional fuel consumption and reduce peak electricity loads. Buildings can be designed to use efficient solar collection for passive space heating and cooling (Chapter 6), active heating of water and space using glazed and circulating fluid collectors, and active cooling using absorption chillers or desiccant regeneration (US Climate Change Technology Program, 2003). There is a risk of lower performance due to shading of windows or solar collectors by new building construction or nearby trees. Local 'shading' regulations can prevent such conflicts by identifying a protected 'solar envelope' (Duncan, 2005). A wide range of design measures, technologies and opportunities are covered by the IEA Solar Heating and Cooling implementing agreement (www.iea-shc.org).

Active systems of capturing solar energy for direct heat are used mainly in small-scale, low-temperature, domestic hot water installations; heating of building space; swimming pools; crop drying; cook stoves; industrial processes; desalination plants and solar-assisted district heating. The estimated annual global solar thermal-collector yield of domestic hot water systems alone is around 80 TWh (0.3 EJ) with the installations growing by 20% per year. Annual solar thermal energy use depends on the area of collectors in operation, the solar radiation levels available and the technologies used including both unglazed and glazed systems. Unglazed collectors, mainly used to heat swimming pools in the USA and Europe, represented about 28 million m^2 in 2003.

More than 130 million m^2 of glazed collector area was installed worldwide by the end of 2003 to provide around 0.5 EJ of heat from around 91 GW$_{th}$ capacity (Weiss et al., 2005). In 2005, around 125 million m^2 (88 GW$_{th}$) of active solar hot-water collectors existed, excluding swimming pool heating (Martinot et al, 2005). China is the world's largest market for glazed domestic solar hot-water systems with 80% of annual global installations and existing capacity of 79 million m^2 (55 GW$_{th}$) at the end of 2005. Most new installations in China are now evacuated-tube in contrast with Europe (the second-largest market), where most collectors are flat-plate (Zhang et al., 2005). Domestic solar hot-water systems are also expanding rapidly in other developing countries. Estimated annual energy yields for glazed flat-plate collectors range between 400 kWh/m^2 in Germany and 1000 kWh/m^2 in Israel (IEA, 2004d). In Austria, annual solar yields were estimated to be 300 kWh/m^2 for unglazed, 350 kWh/m^2 for flat-plate, and 550 kWh/m^2 for evacuated tube collectors (Weiss et al., 2005). The retail price for a solar water heater unit for a family home differs with location and any government support schemes. Installed costs range from around 700 US$ in Greece for a thermo-siphon system with a 2.4 m² collector and 150 L tank, to 2300 US$ in Germany for a pumped system with antifreeze device. Systems manufactured in China are typically 200–300 US$ each.

Nearly 100 commercial solar cooling technologies exist in Europe, representing 24,000 m² with a cooling power of 9 MW_{th}. High potential energy savings compared with conventional electric vapour-pressure air-conditioning systems do not offset the higher costs (Philibert and Podkanski, 2005).

4.3.3.8 Ocean energy

The potential marine-energy resource of wind-driven waves, gravitational tidal ranges, thermal gradients between warm surface water and colder water at depths of >1000 m, salinity gradients, and marine currents is huge (Renewables, 2004), but what is exploitable as the economic potential is low. All the related technologies (with the exception of three tidal-range barrages amounting to 260 MW, including La Rance that has generated 600 GWh/yr since 1967) are at an early stage of development with the only two commercial wave-power projects totalling 750 kW. To combat the harsh environment, installed costs are usually high. The marine-energy industry is now in a similar stage of development to the wind industry in the 1980s (Carbon Trust, 2005). Since oceans are used by a range of stakeholders, siting devices will involve considerable consultation.

The best *wave-energy* climates (Figure 4.16) have deep-water power densities of 60–70 kW/m but fall to about 20 kW/m at the foreshore. Around 2% of the world's 800,000 km of

coastline exceeds 30 kW/m, giving a technical potential of around 500 GW assuming offshore wave-energy devices have 40% efficiency. The total economic potential is estimated to be well below this (WEC, 2004d) with generating cost estimates around 80–110 US$/MWh highly uncertain, since no truly commercial scale plant exists (IEA, 2006d).

Extracting electrical energy from *marine currents* could yield in excess of 10 TWh/yr (0.4 EJ/yr) if major estuaries with large tidal fluctuations could be tapped, but cost estimates range from 450–1350 US$/MWh (IEA, 2006a). A 1 km-stretch of permanent turbines built in the Agulhas current off the coast of South Africa, for example, could give 100 MW of power (Nel, 2003). However, environmental effects on tidal mud flats, wading birds, invertebrates etc. would need careful analysis. In order for these new technologies to enter the market, sustained government and public support is needed.

Ocean thermal and *saline gradient* energy-conversion systems remain in the research stage and it is still too early to estimate their technical potential. Initial applications have been for building air conditioning (www.makai.com/p-pipelines/) for desalination in open- and hybrid-cycle plants using surface condensers and in future could benefit tropical island nations where power is presently provided by expensive diesel generators.

4.3.4 Energy carriers

Energy carriers include electricity and heat as well as solid, liquid and gaseous fuels. They occupy intermediate steps in the energy-supply chain between primary sources and end-use applications. An energy carrier is thus a transmitter of energy. For reasons of both convenience and economy, energy carriers

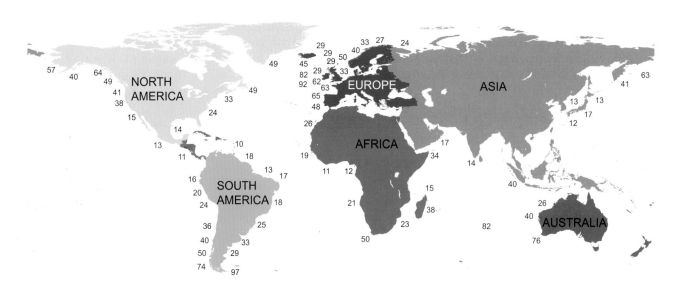

Figure 4.16: *Annual average wave-power density flux (kW/m at deep water)*
Source: Wavegen, 2004.

have shown a continual shift from solids to liquids and more recently from liquids to gases (WEC, 2004b), a trend that is expected to continue. At present, about one third of final energy carriers reach consumers in solid form (as coal and biomass, which are the primary cause of many local, regional and indoor air-pollution problems associated with traditional domestic uses); one third in liquid form (consisting primarily of oil products used in transportation); and one third through distribution grids in the form of electricity and gas. The share of all grid-oriented energy carriers could increase to about one half of all consumer energy by 2100.

New energy carriers such as hydrogen (Section 4.3.4.3) will only begin to make an impact around 2050, whereas the development of smaller scale decentralized energy systems and micro-grids (Section 4.3.8) could occur much sooner (Datta *et al.*, 2002; IEA, 2004d). Technology issues surrounding energy carriers involve the conversion of primary to secondary energy, transporting the secondary energy, in some cases storing it prior to use, and converting it to useful end-use applications (Figure 4.17).

Where a conversion process transforms primary energy near the source of production (e.g. passive solar heating) a carrier is not involved. In other cases, such as natural gas or woody biomass, the primary-energy source also becomes the carrier and also stores the energy. Over long distances, the primary transportation technologies for gaseous and liquid materials are pipelines, shipping tankers and road tankers; for solids they are rail wagons, boats and trucks, and for electricity wire conductors. Heat can also be stored but is normally transmitted over only short distances of 1–2 km.

Each energy-conversion step in the supply chain invokes additional costs for capital investment in equipment, energy losses and carbon emissions. These directly affect the ability of an energy path to compete in the marketplace. The final benefit/cost calculus ultimately determines market penetration of an energy carrier and hence the associated energy source and end-use technology.

Hydrocarbon substances produced from fossil fuels and biomass are utilized widely as energy carriers in solid, slurry, liquid or gaseous forms (Table 4.3). Coal, oil, natural gas and biomass can be used to produce a variety of synthetic liquids and gases for transport fuels, industrial processes and domestic heating and cooking, including petroleum products refined from crude oil. Liquid hydrocarbons have relatively high energy densities that are superior for transport and storage properties.

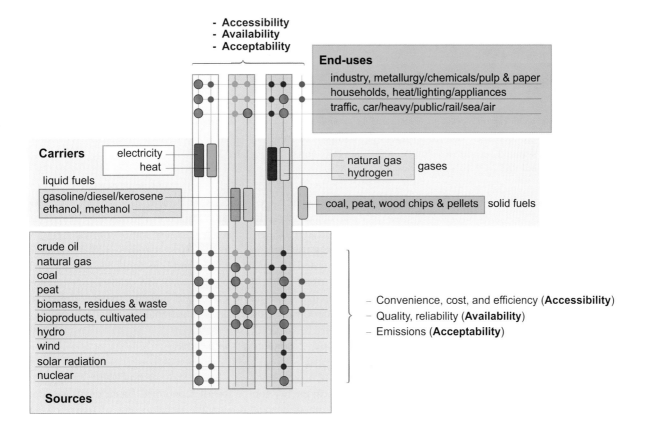

Figure 4.17: *Dynamic interplay between energy sources, energy carriers and energy end-uses.*
Energy sources are shown at the lower left; carriers in the middle; and end-uses at the upper right. Important intersections are noted with circles, small blue for transformations to solid energy carriers and small pink to liquid or gaseous carriers. Large green circles are critical transformations for future energy systems.
Source: WEC, 2004d.

Table 4.3: *Energy carriers of hydrocarbon substances.*

Primary energy	Energy carriers of secondary energy			
	Solid	**Slurry**	**Liquid**	**Gas**
Coal	Pulverized coal Coke	Coal/water mix Coal/oil mix	Coal to liquid (CTL) Synthetic fuel	Coal gas Producer gas Blast furnace gas Water gas Gasified fuel Hydrogen
Oil			Oil refinery products	Oil gas Synthetic gas Hydrogen
Natural gas			LNG, LPG Gas to liquid (GTL) GTL alcoholics Di-methyl ethers	Methane Hydrogen
Biomass	Wood residues Energy crops Refuse derived fuel (RDF)		Methanol Ethanol Biodiesel esters Di-methyl ethers	Methane Producer gas Hydrogen

4.3.4.1   Electricity

Electricity is the highest-value energy carrier because it is clean at the point of use and has so many end-use applications to enhance personal and economic productivity. It is effective as a source of motive power (motors), lighting, heating and cooling and as the prerequisite for electronics and computer systems. Electricity is growing faster as a share of energy end-uses (Figure 4.18) than other direct-combustion uses of fuels with the result that electricity intensity (Electricity/GDP) has remained relatively constant even though the overall global-energy intensity (Energy/GDP) continues to decrease. If electricity intensity continues to decrease due to efficiency increases, future electricity demand could be lower than otherwise forecast (Sections 4.4.4 and 11.3.1).

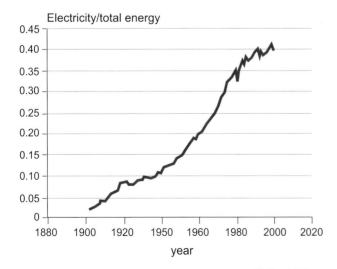

Figure 4.18: *Ratio of electricity to total primary energy in the US since 1900.*
Source: EPRI, 2003.

Life-cycle GHG-emission analyses of power-generation plants (WEC 2004a; Vattenfall, 2005; Dones *et al.,* 2005; van de Vate, 2002; Spadaro, 2000; Uchiyama and Yamamoto, 1995; Hondo, 2005) show the relatively high CO_2 emissions from fossil-fuel combustion are 10–20 times higher than the indirect emissions associated with the total energy requirements for plant construction and operation during the plant's life (Figure 4.19). Substitution by nuclear or renewable energy decreases carbon emissions per kWh by the difference between the full-energy-chain emission coefficients and allowing for varying plant-capacity factors (WEC 2004a; Sims *et al.*, 2003a). The average thermal efficiency for electricity-generation plants has improved from 30% in 1990 to 36% in 2002, thereby reducing GHG emissions.

Electricity generated from traditional coal-fired, steam-power plants is expected to be displaced over time with more advanced technologies such as CCGT or advanced coal to reduce the production of GHG and increase the overall efficiency of energy use. Previous IPCC (2001) and WEC (2001) scenarios suggested that nuclear, CCGT and CCS could become dominant electricity-sector technologies early this century (Section 4.4). Although CCS can play a role, its potential may be limited and hence some consider it as a transitional bridging technology.

4.3.4.2   Heat and heat pumps

Heat, whether from fossil fuels or renewable energy, is a critical energy source for all economies. Its efficient use could play an important role in the development of transition and developing economies (UN, 2004; IEA, 2004e). It is used in industrial processes for food processing, petroleum refining, timber drying, pulp production, etc. (Chapter 7), as well as in commercial and residential buildings for space heating, hot

tonnes CO$_2$-eq/GWh$_{el}$

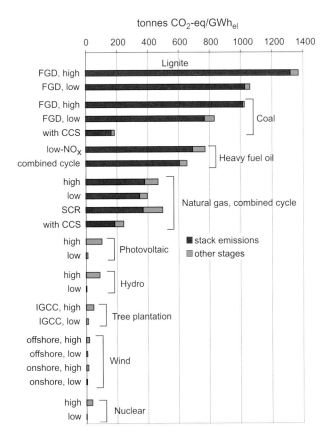

Figure 4.19: *GHG emissions for alternative electricity-generation systems.*

Notes: 1 tCO$_2$ –eq/GWh = 0.27 tC –eq/GWh. The high estimate for hydro includes possible GHG emissions from reservoirs (Section 4.3.3.1)

Source: WEC, 2004a

water and cooking (Chapter 6). Many industries cogenerate both heat and electricity as an integral part of their production process (Section 4.3.5; Chapter 7), in most cases being used on-site, but at times sold for other uses off-site such as district heating schemes.

Heating and cooling using renewable energy (Section 4.3.3.7) can compete with fossil fuels (IEA, 2006f). In some instances, the best use of modern biomass will be co-firing with coal at blends up to 5–10% biomass or with natural gas.

Heat pumps can be used for simple air-to-air space heating, air-to-water heating, and for utilizing waste heat in domestic, commercial and industrial applications (Chapter 6). Thermo-dynamically reverse Carnot-cycle heat pumps are more demand-side technologies but also linked with sustainable energy supply by concentrating low-grade solar heat in air and water. Their efficiency is evaluated by the coefficient of performance (COP), with COPs of 3 to 4 available commercially and over 6 using advanced turbo-refrigeration (www.mhi.co.jp/aircon/). A combination of CCGT with advanced heat-pump technology could reduce carbon emissions from supplying heat more than using a conventional gas-fired CHP plant of similar capacity.

4.3.4.3 Liquid and gaseous fuels

Coal, natural gas, petroleum and biomass can all be used to produce a variety of liquid fuels for transport, industrial processes, power generation and, in some regions of the world, domestic heating. These include petroleum products from crude oil or coal; methanol from coal or natural gas; ethanol and fatty acid esters (biodiesel) from biomass; liquefied natural gas; and synthetic diesel fuel and di-methyl ether from coal or biomass. Of these, crude oil is the most energy-efficient fuel to transport over long distances from source to refinery and then to distribute to product demand points. After petrol, diesel oil and other light and medium distillates are extracted at the refinery, the residues are used to produce bitumen and heavy fuel oil used as an energy source for industrial processes, oil-fired power plants and shipping.

Gaseous fuels provide a great deal of the heating requirements in the developed world and increased use can lead to lower GHG and air-pollution emissions.

Hydrogen

Realizing hydrogen as an energy carrier depends on low-cost, high-efficiency methods for production, transport and storage. Most commercial hydrogen production today is based on steam reforming of methane, but electrolysis of water (especially using carbon-free electricity from renewable or nuclear energy) or splitting water thermo-chemically may be viable approaches in the future. Electrolysis may be favoured by development of fuel cells that require a low level of impurities. Current costs of electrolysers are high but declining. Producing hydrogen from fossil fuels on a large scale will need integration of CCS if GHG emissions are to be avoided. A number of routes to produce hydrogen from solar energy are also technically feasible (Figure 4.20).

Hydrogen has potential as an energy-storage medium for electricity production or transport fuel when needed. The prospects for a future hydrogen economy will depend

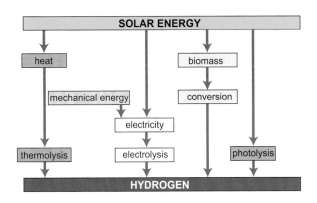

Figure 4.20: *Routes to hydrogen-energy carriers from solar-energy sources.*
Source: EPRI, 2003

on developing competitively priced fuel cells for stationary applications or vehicles, but fuel cells are unlikely to become fully commercial for one or two decades. International cooperative programmes, such as the IEA Hydrogen Implementing Agreement (IEA, 2005f), and more recently the International Partnership for the Hydrogen Economy (www. iphe.net) aim to advance RD&D on hydrogen and fuel cells across the application spectrum (IEA, 2003g; EERE, 2005).

Hydrogen fuel cells may eventually become commercially viable electricity generators, but because of current costs, complexity and state of development, they may only begin to penetrate the market later this century (IEA, 2005g). Ultimately, hydrogen fuel could be produced in association with CCS leading to low-emission transport fuels. Multi-fuel integrated-energy systems or 'energyplexes' (Yamashita and Barreto, 2005) could co-produce electricity, hydrogen and liquid fuels with overall high-conversion efficiencies, low emissions and also facilitating CCS. FutureGen is a US initiative to build the world's first integrated CCS and hydrogen-production research power plant (US DOE, 2004).

4.3.5   Combined heat and power (CHP)

Up to two thirds of the primary energy used to generate electricity in conventional thermal power plants is lost in the form of heat. Switching from condensing steam turbines to CHP (cogeneration) plants produces electricity but captures the excess heat for use by municipalities for district heating, commercial buildings (Chapter 6) or industrial processes (Chapter 7). CHP is usually implemented as a distributed energy resource (Jimison, 2004), the heat energy usually coming from steam turbines and internal combustion engines. Current CHP designs can boost overall conversion efficiencies to over 80%, leading to cost savings (Table 4.4) and hence to significant carbon-emissions reductions per kWh generated. About 75% of district heat in Finland, for example, is provided from CHP plants with typical overall annual efficiencies of 85–90% (Helynen, 2005).

CHP plants can range from less than 5 kW$_e$ from micro-gas-turbines, fuel cells, gasifiers and Stirling engines (Whispergen, 2005) to 500 MW$_e$. A wide variety of fuels is possible including

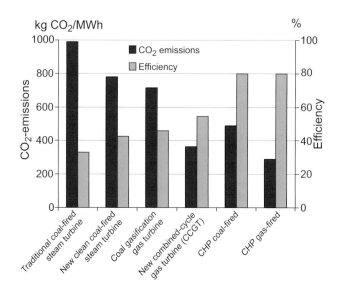

Figure 4.21: *Carbon dioxide emissions and conversion efficiencies of selected coal and gas-fired power generation and CHP plants.*

Note: CHP coal- fired and CHP gas-fired assume more of the available heat is utilized from coal than from gas to both give 80%.

biomass (Kirjavainen *et al.*, 2004), with individual installations accepting more than one fuel. A well-designed and operated CHP scheme will provide better energy efficiency than a conventional plant, leading to both energy and cost savings (UNEP, 2004; EDUCOGEN, 2001). Besides the advantage of cost reductions because of higher efficiency, CHP has the environmental benefit of reducing 160–500 gCO$_2$/kWh, given a fossil-fuel baseline for the heat and electricity generation.

4.3.6   Carbon dioxide capture and storage (CCS)

The potential to separate CO$_2$ from point sources, transport it and store it in isolation from the atmosphere was covered in an IPCC Special Report (IPCC, 2005). Uncertainties relate to proving the technologies, anticipating environmental impacts and how governments should incentivise uptake, possibly by regulation (OECD/IEA, 2005) or by carbon charges, setting a price on carbon emissions. Capture of CO$_2$ can best be applied

Table 4.4: *Characteristics of CHP (cogeneration) plants*

Technology	Fuel	Capacity MW	Electrical efficiency (%)	Overall efficiency (%)
Steam turbine	Any combustible	0.5-500	17-35	60-80
Gas turbine	Gasous & liquid	0.25-50+	25-42	65-87
Combined cycle	Gasous & liquid	3-300+	35-55	73-90
Diesel and Otto engines	Gasous & liquid	0.003-20	25-45	65-92
Micro-turbines	Gasous & liquid	0.05-0.5	15-30	60-85
Fuel cells	Gasous & liquid	0.003-3+	37-50	85-90
Stirling engines	Gasous & liquid	0.003-1.5	30-40	65-85

to large carbon point sources including coal-, gas- or biomass-fired electric power-generation or cogeneration (CHP) facilities, major energy-using industries, synthetic fuel plants, natural gas fields and chemical facilities for producing hydrogen, ammonia, cement and coke. Potential storage methods include injection into underground geological formations, in the deep ocean or industrial fixation as inorganic carbonates (Figure 4.22). Application of CCS for biomass sources (such as when co-fired with coal) could result in the net removal of CO_2 from the atmosphere.

Injection of CO_2 in suitable geological reservoirs could lead to permanent storage of CO_2. Geological storage is the most mature of the storage methods, with a number of commercial projects in operation. Ocean storage, however, is in the research phase and will not retain CO_2 permanently as the CO_2 will re-equilibrate with the atmosphere over the course of several centuries. Industrial fixation through the formation of mineral carbonates requires a large amount of energy and costs are high. Significant technological breakthroughs will be needed before deployment can be considered.

Estimates of the role CCS will play over the course of the century to reduce GHG emissions vary. It has been seen as a 'transitional technology', with deployment anticipated from 2015 onwards, peaking after 2050 as existing heat and power-plant stock is turned over, and declining thereafter as the decarbonization of energy sources progresses (IEA, 2006a).

Other studies show a more rapid deployment starting around the same time, but with continuous expansion even towards the end of the century (IPCC, 2005). Yet other studies show no significant use of CCS until 2050, relying more on energy efficiency and renewable energy (IPCC, 2005). Long-term analyses by use of integrated assessment models, although using a simplified carbon cycle (Read and Lermit, 2005; Smith, 2006b), indicated that a combination of bioenergy technologies together with CCS could decrease costs and increase attainability of low stabilization levels (below 450 ppmv).

New power plants built today could be designed and located to be CCS-ready if rapid deployment is desired (Gibbins *et al.*, 2006). All types of power plants can be made CCS-ready, although the costs and technical measures vary between different types of power plants. However, beyond space reservation for the capture, installation and siting of the plant to enable access to storage reservoirs, significant capital pre-investments at build time do not appear to be justified by the cost reductions that can be achieved (Bohm, 2006; Sekar, 2005). Although generic outline engineering studies for retro-fitting capture technologies to natural-gas GTCC plants have been undertaken, detailed reports on CCS-ready plant-design studies are not yet in the public domain.

Storage of CO_2 can be achieved in deep saline formations, oil and gas reservoirs and deep unminable coal seams using injection and monitoring techniques similar to those utilized by

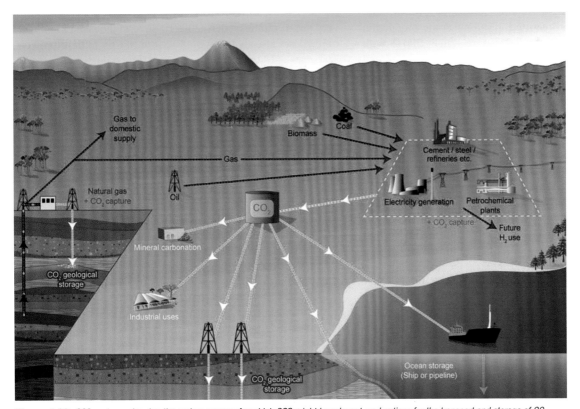

Figure 4.22: *CCS systems showing the carbon sources for which CCS might be relevant, and options for the transport and storage of CO_2.*
Source: IPCC, 2005.

the oil and gas industry. Of the different types of potential storage formations, storage in coal formations is the least developed. If injected into suitable saline formations or into oil and gas fields at depths below 800 m, various physical and geochemical trapping mechanisms prevent the CO_2 from migrating to the surface. Projects in all kinds of reservoirs are planned.

Storage capacity in oil and gas fields, saline formations and coal beds is uncertain. The IPCC (IPCC, 2005) reported 675 to 900 $GtCO_2$ for the relatively well-characterized gas and oil fields, more than 1000 $GtCO_2$ (possibly up to an order of magnitude higher) for saline formations, and up to 200 $GtCO_2$ for coal beds. Bradshaw *et al.* (2006) highlighted the incomparability of localized storage-capacity data that use different assumptions and methodologies. They also criticized any top-down estimate of storage capacity not based on a detailed site characterization and a clear methodology, and emphasized the value of conservative estimates. In the literature, however, specific estimates were based on top-down data and varied beyond the range cited in the IPCC (2005). For instance, a potential of >4000 $GtCO_2$ was reported for saline formations in North America alone (Dooley *et al.*, 2005) and between 560 and 1170 $GtCO_2$ for injection in oil and gas fields (Plouchart *et al.*, 2006). Agreement on a common methodology for storage capacity estimates on the country- and region-level is needed to give a more reliable estimate of storage capacities.

Biological removal of CO_2 from an exhaust stream is possible by passing the stack emissions through an algae or bacterial solution in sunlight. Removal rates of 80% for CO_2 and 86% for NO_X have been reported, resulting in the production of 130,000 litres/ha/yr of biodiesel (Greenfuels 2004) with residues utilized as animal feed. Other unconventional biological approaches to CCS or fuel production have been reported (Greenshift, 2005; Patrinos, 2006). Another possibility is the capture of CO_2 from air. Studies claim costs less than 75 US$/$tCO_2$ and energy requirements of a minimum of 30% using a recovery cycle with $Ca(OH)_2$ as a sorbent. However, no experimental data on the complete process are yet available to demonstrate the concept, its energy use and engineering costs.

Before the option of ocean injection can be deployed, significant research is needed into its potential biological impacts to clarify the nature and scope of environmental consequences, especially in the longer term (IPCC, 2005). Concerns surrounding geological storage include the risk of seismic activity causing a rapid release of CO_2 and the impact of old and poorly sealed well bores on the storage integrity of depleted oil and gas fields. Risks in CO_2 transportation include rupture or leaking of pipelines, possibly leading to the accumulation of a dangerous level of CO_2 in the air. Dry CO_2 is not corrosive to pipelines even if it contains contaminants, but it becomes corrosive when moisture is present. Any moisture therefore needs to be removed to prevent corrosion and avoid the high cost of constructing pipes made from corrosion-resistant material. Transport of CO_2 by ship is feasible under

specific conditions, but is currently carried out only on a small scale due to limited demand (IPCC, 2005).

Clarification of the nature and scope of long-term environmental consequences of ocean storage requires further research (IPCC, 2005). Concerns around geological storage include rapid release of CO_2 as a consequence of seismic activity and the impact of old and poorly sealed well bores on the storage integrity of depleted oil and gas fields Risks are estimated to be comparable to those of similar operations (IPCC, 2005). For CO_2 pipelines, accident numbers reported are very low, although there are risks of rupture or leaking leading to local accumulation of CO_2 in the air to dangerous levels (IPCC, 2005).

4.3.6.1 Costs

Cost estimates of the components of a CCS system vary widely depending on the base case and the wide range of source, transport and storage options (Table 4.5). In most systems, the cost of capture (including compression) is the largest component, but this could be reduced by 20–30% over the next few decades using technologies still in the research phase as well as by upscaling and learning from experience (IPCC, 2005). The extra energy required is a further cost consideration. CO_2 storage is economically feasible under conditions specific to enhanced oil recovery (EOR), and in saline formations, avoiding carbon tax charges for offshore gas fields in Norway. Pipeline transport of CO_2 operates as a mature market technology (IPCC, 2005), costing 1–5 US$/$tCO_2$ per 100 km (high end for very large volumes) (IEA, 2006a). Several thousand kilometres of pipelines already transport 40 Mt/yr of CO_2 to EOR projects. The costs of transport and storage of CO_2 could decrease slowly as technology matures further and the plant scale increases.

4.3.7 Transmission, distribution, and storage

A critical requirement for providing energy at locations where it is converted into useful services is a system to move the converted energy (e.g. refined products, electricity, heat) and store it ready for meeting a demand. Any leakage or losses (Figure 4.23) result in increased GHG emissions per unit of useful consumer energy delivered as well as lost revenue.

Electricity transmission networks cover hundreds of kilometres and have successfully provided the vital supply chain link between generators and consumers for decades. The fundamental architecture of these networks has been developed to meet the needs of large, predominantly fossil fuel-based generation technologies, often located remotely from demand centres and hence requiring transmission over long distances to provide consumers with energy services.

Transmission and distribution networks account for 54% of the global capital assets of electric power (IEA, 2004d). Aging equipment, network congestion and extreme peak load

Table 4.5: *Current cost ranges for the components of a CCS system applied to a given type of power plant or industrial source*

CCS system components	Cost range	Remarks
Capture from a coal- or gas-fired power plant	15-75 US$/tCO$_2$ net captured	Net costs of captured CO$_2$ compared to the same plant without capture
Capture from hydrogen and ammonia production or gas processing	5-55 US$/tCO$_2$ net captured	Applies to high-purity sources requiring simple drying and compression
Capture from other industrial sources	25-115 US$/tCO$_2$ net captured	Range reflects use of a number of different technologies and fuels
Transport	1-8 US$/tCO$_2$ transported	Per 250 km pipeline or shipping for mass flow rates of 5 (high end) to 40 (low end) MtCO$_2$/yr.
Geological storage[a]	0.5-8 US$/tCO$_2$ net injected	Excluding potential revenues from EOR or ECBM.
Geological storage: monitoring and verification	0.1-0.3 US$/tCO$_2$ injected	This covers pre-injection, injection, and post-injection monitoring, and depends on the regulatory requirements
Ocean storage	5-30 US$/tCO$_2$ net injected	Including offshore transportation of 100–500 km, excluding monitoring and verification
Mineral carbonation	50-100 US$/tCO$_2$ net mineralized	Range for the best case studied. Includes additional energy use for carbonation

[a] Over the long term, there may be additional costs for remediation and liabilities

Source: IPCC, 2005.

demands contribute to losses and low reliability, especially in developing countries, such that substantial upgrading is often required. Existing infrastructure will need to be modernized to improve security, information and controls, and to incorporate low-emission energy systems. Future infrastructure and control systems will need to become more complex in order to handle higher, more variable loads; to recognize and dispatch small-scale generators; and to enable the integration of intermittent and decentralized sources without reduced system performance as it relates to higher load flow, frequency oscillations, and

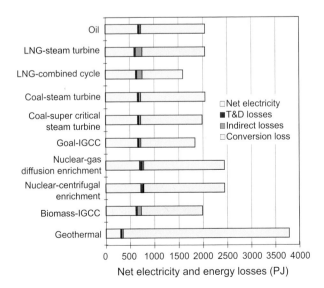

Figure 4.23: *Comparison of net electricity production per 1000MW$_e$ of installed capacity for a range of power-generation technology systems in Japan.*

Note: Analysed over a 30-year plant life, and showing primary fuel-use efficiency losses and transmission losses assuming greater distances for larger scale plants. Transport and distribution losses were taken as 4% for fossil fuel and bioenergy, 7% for nuclear.

Source: Data updated from Uchiyama, 1996.

voltage quality (IEA, 2006a). New networks being built should have these features incorporated, though due to private investors seeking to minimize investment costs, this is rarely the case. The demands of future systems may be significantly less than might be otherwise anticipated through increased use of distributed energy (IEA, 2003c).

Superconducting cables, sensors and rapid response controls that could help to reduce electricity costs and line losses are all under development. Superconductors may incorporate hydrogen as both cryogenic coolant and energy carrier. System management will be improved by providing advanced information on grid behaviour; incorporating devices to route current flows on the grid; introduce real-time pricing and other demand-side technologies including smart meters and better system planning. The energy security challenges that many OECD countries currently face from technical failures, theft, physical threats to infrastructure and geopolitical actions are concerns that can be overcome in part by greater deployment of distributed energy systems to change the electricity-generation landscape (IEA, 2006g).

4.3.7.1   Energy storage

Energy storage allows the energy-supply system to operate more or less independently from the energy-demand system. It addresses four major needs: utilizing energy supplies when short-term demand does not exist; responding to short-term fluctuations in demand (stationary or mobile); recovering wasted energy (e.g. braking in mobile applications), and meeting stationary transmission expansion requirements (Testor *et al.*, 2005). Storage is of critical importance if variable low-carbon energy options such as wind and solar are to be better utilized, and if existing thermal or nuclear systems are to be optimized for peak performance in terms of efficiencies and thus emissions. Advanced energy-storage systems include mechanical (flywheels, pneumatic), electrochemical (advanced batteries, reversible fuel cells, hydrogen), purely electric or magnetic (super- and ultra-capacitors, superconducting magnetic storage), pumped-water (hydro) storage, thermal (heat) and compressed air. Adding any of these storage systems necessarily decreases the energy efficiency of the entire system (WEC, 2004d). Overall, cycle efficiencies today range from 60% for pumped hydro to over 90% for flywheels and super-capacitors (Testor *et al.*, 2005). Electric charge carriers such as vanadium redox batteries and capacitors are under evaluation but have low energy density and high cost. Cost and durability (cycle life) of the high-technology systems remains the big challenge, possibly to be met by more advanced materials and fabrication. Energy storage has a key role for small local systems where reliability is an important feature.

4.3.8   Decentralized energy

Decentralized (or distributed) energy systems (DES) located close to customer loads often employ small- to medium-scale facilities to provide multiple-energy services referred to as 'polygeneration'. Grid-connected DES are already commercial in both densely populated urban markets requiring supply reliability and peak shedding as well as in the form of mini-grids in rural markets with high grid connection costs and abundant renewable energy resources. Diesel-generating sets are an option, but will generally emit more CO_2 per kWh than a power grid system. Renewable-energy systems connected to the grid or used instead of diesel gensets will reduce GHG emissions. The merits of DES include:

- reduced need for costly transmission systems and shorter times to bring on-stream;
- substantially reduced grid power losses over long transmission distances resulting in deferred costs for upgrading transmission and distribution infrastructure capacity to meet a growing load;
- improved reliability of industrial parks, information technology and data management systems including stock markets, banks and credit card providers where outages would prove to be very costly (IEA, 2006g);
- proximity to demand for heating and cooling systems which, for fossil fuels, can increase the total energy recovered from 40–50% up to 70–85% with corresponding reductions in CO_2 emissions of 50% or more;
- zero-carbon, renewable energy sources such as solar, wind and biomass are widely distributed and useful resources for DES. However, developing decentralized mini-power grids is usual practice if these sources are to make significant local contributions to electricity supply and emission reductions.

There are added expenses, power limitations and reliability issues with DES. The World Alliance for Decentralized Energy (WADE, 2005) reported that at the end of 2004, just 7.2% of global electric power generation was supplied by decentralized systems, having a total capacity of 281.9 GW$_e$. Capacity of DES expanded by 11.4% between 2002 and 2004, much of it as combined heat and power (CHP) using natural gas or biomass to combine electric power generation with the capture and use of waste heat for space heating, industrial and residential hot water, or for cooling. Growth in the USA, where capacity stands at 80 GW$_e$, has been relatively slow because of regulatory barriers and the rising price of natural gas. The European market is expected to expand following the 2003 Cogeneration Directive from the European Commission, while India has added decentralized generation to enhance system reliability. Brazil, Australia and elsewhere are adding CHP facilities that use bagasse from their sugar and ethanol processing. Brazil has the potential to generate 11% of its electricity from this source. China is also adding small amounts of decentralized electric power in some of its major cities (50 GW$_e$ in 2004), but central power still dominates. Japan is promoting the use of natural gas-fuelled CHP with a target of almost 5000 MW by 2010 to save over 11 MtCO$_2$ (Kantei, 2006). In 2005, 24% of global electricity markets from all newly installed power plants were claimed to be from DES (WADE, 2006).

The trend towards DES is growing, especially for distributed electricity generation (DG), in which local energy sources (often renewable) are utilized or energy is carried as a fuel to a point at or near the location of consumption where it is then converted to electricity and distributed locally. As well as wind, geothermal and biomass-fuelled technologies, DG systems can use a wide range of fuels to run diesel generators, gas engines, small and micro-turbines, and Stirling engines with power outputs down to <1 kW$_e$ and widely varying power-heat output ratios between 1:3 and 1:36 (IEA, 2006a). The motive power of a vehicle to supply electricity could be used. Hydrogen (Section 4.3.4.3), could fuel modified internal combustion engines to provide a near-term option, or fuel cells in the longer term (Gehl, 2004). A critical objective, however, will be to first increase the power density of fuel cells, reduce the installed costs and store the hydrogen safely.

Small-to-medium CHP systems at a scale of 1–40 MW$_e$ are in common use as the heat can be usefully employed on site or locally. CCS systems will probably not be economic at such a small scale. Mass production of technologies as demand increases will help reduce the current high costs of around 5000 US$/kW$_e$ for many small systems. Reciprocating engine generator sets are commercially available; micro CHP Stirling engine systems are close to market (Whispergen, 2005) and fuel cells with the highest power-heat ratio need significant capital cost reductions.

The recent growth in DG technologies, mainly diesel-generation based, to provide reliable back-up systems, is apparent in North America (Figure 4.24). Technology advances

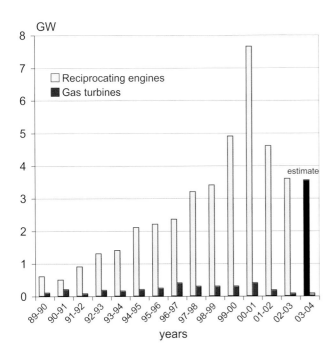

Figure 4.24: *Recent growth in distributed electricity generation using fossil-fuel resources in North America.*
Source: EPRI, 2003.

may encourage the emergence of a new generation of higher-value energy services, including power quality and information-related services based on fuel cells with good reliability.

Flexible alternating-current transmission systems (FACTS) are now being employed as components using information technology (IT) and solid-state electronics to control power flow. Numerous generators can then be controlled by the utility or line company to match the ever-changing load demand. Improved grid stability can result from appliances such as cool stores shedding load and generation plants starting up in response to system frequency variations. In addition, price sensitivities and real-time metering could be used to stimulate selected appliances to be used off-peak. IT could provide a better quality product and services for customers, but in itself may not reduce emissions if say peak load is switched to base load and the utility uses gas for peaking plants and coal for base-load plants. It could, however, enable the greater integration of more low-carbon-emitting technologies into the grid. The intermittent nature of many forms of renewable energy may require some form of energy storage or the use of a mix of energy sources and load responses to provide system reliability. To optimize the integration of intermittent renewable energy systems, IT could be used to determine generator preference and priority through a predetermined merit order based on both availability and market price.

4.3.9   Recovered energy

Surplus heat generated during the manufacturing process by some industries such as fertilizer manufacturing, can be used on site to provide process heat and power. This is covered in Chapter 7.

4.4 Mitigation costs and potentials of energy supply

Assessing future costs and potentials across the range of energy-supply options is challenging. It is linked to the uncertainties of political support initiatives, technological development, future energy and carbon prices, the level of private and public investment, the rate of technology transfer and public acceptance, experience learning and capacity building and future levels of subsidies and support mechanisms. Just one such example of the complexity of determining the cost, potential and period before commercial delivery of a technology is the hydrogen economy. It encompasses all these uncertainties leading to considerable debate on its future technical and economic potential, and indeed whether a hydrogen economy will ever become feasible at all, and if so, when (USCCTP, 2005; IEA, 2003b).

Bioenergy also exemplifies the difficulties when analysing current costs and potentials for a technology as it is based on a broad range of energy sources, geographic locations, technologies, markets and biomass-production systems. In addition, future projections are largely dependent upon RD&D success and economies of plant scale. Bioethanol from ligno-cellulose, for example, has been researched for over three decades with little commercial success to date. So there can be little certainty over the timing of future successes despite the recent advances of several novel biotechnology applications. Energy technological learning is nevertheless an established fact (WEC, 2001; Johansson, 2004; Section 2.7) and gives some confidence in projections for future market penetration.

4.4.1.   Carbon dioxide emissions from energy supply by 2030

A few selected baseline (IEA 2006b, WEO Reference; SRES A1; SRES B2 (Table 4.1); ABARE Reference) and policy mitigation scenarios (IEA 2006b, WEO Alternative policy; ABARE Global Technology and ABARE Global Technology +CCS) out to 2030 illustrate the wide range of possible future energy-sector mixes (Figure 4.25). They give widely differing views of future energy-supply systems, the primary-energy mix and the related GHG emissions. Higher energy prices (as experienced in 2005/06), projections that they will remain high (Section 4.3.1) or current assessments of CCS deployment rates (Section 4.3.6) are not always included in the scenarios. Hence, more recent studies (for example IEA 2006b, IEA 2006d; Fisher, 2006) are perhaps more useful for evaluating future energy supply potentials, though they still vary markedly.

The ABARE global model, based on an original version produced for the Asia Pacific Partnership (US, Australia, Japan, China, India, Korea) (Fisher, 2006), is useful for mitigation analysis as it accounts for both higher energy prices and CCS opportunities. However, it does not separate 'modern biomass' from 'other renewables', and the modellers had also assumed that CCS would play a more significant mitigation role after 2050, rather than by the 2030 timeframe discussed here. The reference case ('Ref' in Figure 4.25) is a projection of key economic, energy and technology variables assuming the continuation of current or already announced future government policies and no significant shifts in climate policy. The Global Technology scenario (ABARE 'Tech') assumed that development and transfer of advanced energy-efficient technologies will occur at an accelerated rate compared with the reference case. Collaborative action from 2006 was assumed to affect technology development and transfer between several leading developed countries and hence lead to more rapid uptake of advanced technologies in electricity, transport and key industry sectors. The 'Tech+CCS' scenario assumed similar technology developments and transfer rates for electricity, transport and key industry sectors, but in addition CCS was utilized in all new coal- and gas-fired electricity generation plant from 2015 in US, Australia and Annex I countries and from 2020 in China, India and Korea.

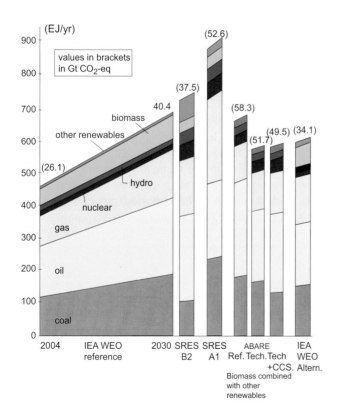

Figure 4.25: *Indicative comparison of selected primary energy-supply baseline (reference) and policy scenarios from 2004 to 2030 and related total energy-related emissions in 2004 and 2030 (GtCO2-eq)*

Note: The IEA (2006b) Beyond Alternative Policy scenario (not shown) depicts that energy-related emissions could be reduced to 2004 levels.

Source: Based on IEA, 2006b; IPCC, 2001; Price and de la Rue du Can, 2006; Fisher, 2006.

Table 4.6: *Estimated carbon dioxide emissions from fossil-fuel use in the energy sector for 2002 and 2030 (MtCO2 /yr).*

		2002	2030
Transport (includes marine bunkers)		5999	10631
Industry, of which:		9013	13400
Electricity		4088	6667
Heat:	- coal	2086	2413
	- oil	1436	2098
	- gas	1403	2222
Buildings, of which:		8967	14994
Electricity		5012	9607
Heat:	- coal	495	356
	- oil	1841	2693
	- gas	1618	2338
Total		**23979[a]**	**39025**

[a] WEO, 2006 (IEA 2006b, unavailable at the time of the analysis) gives total CO_2 emissions as 26,079 $MtCO_2$ for 2004

Source: Price and de la Rue du Can, 2006.

4.4.2 Cost analyses

This section places emphasis on the costs and mitigation potentials of the electricity-supply sector. Heat and CHP potentials are more difficult to determine due to lack of available data, and transport potentials are analysed in Chapter 5.

Cost estimates are sensitive to assumptions used and inherent data inconsistencies. They vary over time and with location and chosen technology. There is a tendency for some countries, particularly where regulations are lax, to select the cheapest technology option (at times using second-hand plant) regardless of total emission or environmental impact (Royal Academy of Engineering, 2004; Sims *et al*, 2003a). Here, based upon full life-cycle analyses in the literature, only broad cost comparisons are possible due to the wide variations in specific site costs and variations in labour charges, currency exchange rates, discount rates used, and plant capacity factors. Cost uncertainties in the electricity sector also exist due to the rate of market liberalization and the debate over the maximum level of intermittent renewable energy sources acceptable to the grid without leading to reliability issues and needing costly back-up.

One analysis compared the levelized investment, operations and maintenance (O&M), fuel and total generation costs from 27 coal-fired, 23 gas-fired, 13 nuclear, 19 wind- and 8 hydro-power plants, either operational or planned in several countries (IEA/NEA, 2005). The technologies and plant types included several units under construction or due to be commissioned before 2015, but for which cost estimates had been developed through paper studies or project bids (Figure 4.27). The economic competitiveness of selected electricity-generation systems depends upon plant-specific features. The projected total levelized generation cost ranges tend to overlap (Figure

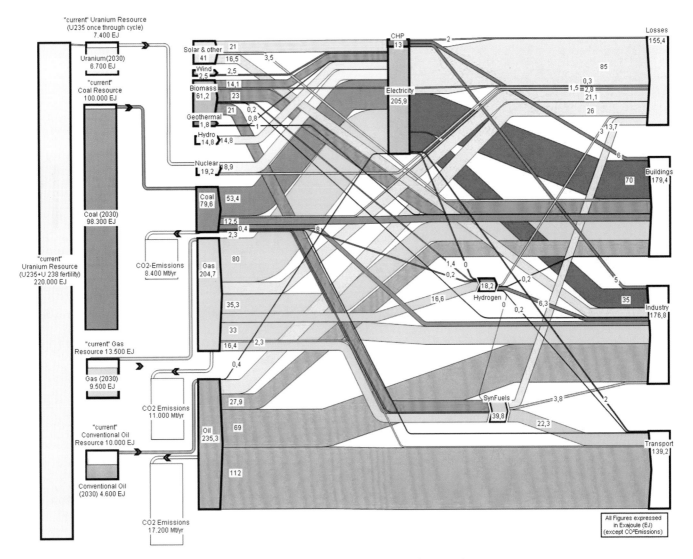

Figure 4.26: *Predicted world energy sources to meet growing demand by 2030 based on updated SRES B2 scenario.*

Note: Related CO$_2$ emissions from coal, gas and oil are also shown, as well as resources in 2004 (see Figure 4.4) and their depletion between 2004 and 2030 (vertical bars to the left). The resource efficiency ratio by which fast-neutron technology increases the power-generation capability per tonne of natural uranium varies greatly from the OECD assessment of 30:1 (OECD, 2006b). In this diagram the ratio used is up to 240:1 (OECD, 2006c).

Source: IPCC, 2001; IIASA 1998

4.27) showing that under favourable circumstances, and given possible future carbon charge additions, all technologies can be economically justified as a component in a diversified energy technology portfolio.

Construction cost assumptions ranged between 1000 and 1500 US$/kW$_e$ for coal plants; 400 and 800 US$/kW for CCGT; 1000 and 2000 US$/kW for wind; 1000 and 2000 US$/kW for nuclear and 1400 and 7000 US$/kW for hydro. Capacity factors of 85% were adopted for coal, gas and nuclear as baseload; 50% for hydro; 17 to 38% for onshore wind-power plants, and 40 to 45% for offshore wind. The costs of nuclear waste management and disposal, refurbishing and decommissioning were accounted for in all the studies reviewed, but remain uncertain. For example, decommissioning costs of a German pressurized water reactor were 155 €/kW, being 10% of the capital investment costs (IEA/ NEA, 2005). A further study, however, calculated life-cycle costs

of nuclear power to be far higher at between 47 and 70 US$/MWh by 2030 (MIT, 2003). Another cost comparison between coal, gas and nuclear options based upon five studies (WNA, 2005b) showed that nuclear was up to 40% more costly than coal or gas in two studies, but cheaper in the other three. Such projected costs depend on country- and project-specific conditions and variations in assumptions made, such as the economic lifetime of the plants and capacity factors. For example, nuclear and renewable energy plants could become more competitive if gas and coal prices rise and if the externality costs associated with CO$_2$ emissions are included.

In this regard, a European study (EU, 2005) evaluated external costs for a number of power-generation options (Figure 4.28) emphasizing the zero- or low-carbon-emitting benefits of nuclear and renewables and reinforcing the benefits

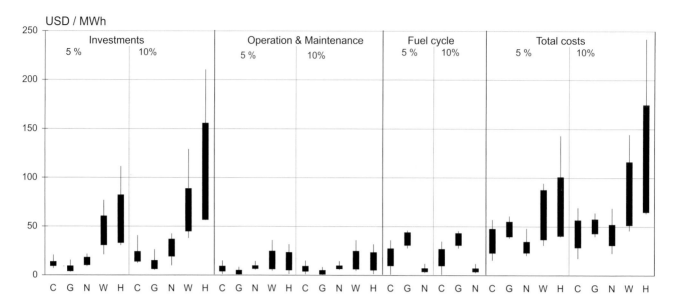

Figure 4.27: *Projected power-generating levelized costs for actual and planned coal (C), gas (G) nuclear (N), wind (W) and hydro (H) power plants with assumed capital interest rates of 5 or 10%.*

Notes: Bars depict 10 and 90 percentiles and lines extend to show minimum and maximum estimates. Other analyses provide different cost ranges (Table 4.7), exemplifying the uncertainties resulting from the discount rates and other underlying assumptions used.

Source: IEA/NEA, 2005.

of CHP systems (Section 4.3.5) (even though only less efficient, small-scale CHP plants were included in the analysis). This comparison highlights the value from conducting full life-cycle analyses when comparing energy-supply systems and costs (Section 4.5.3).

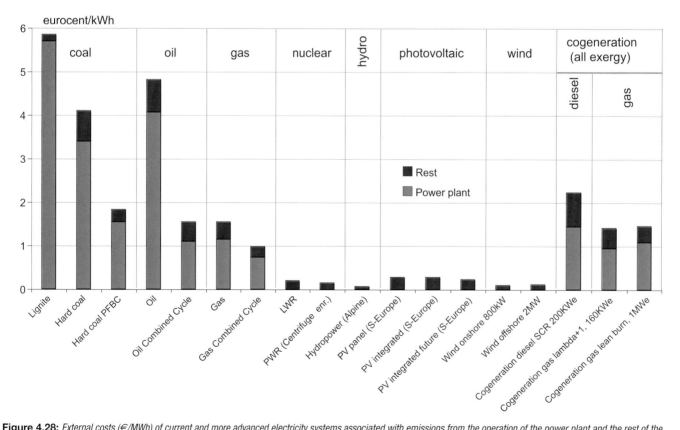

Figure 4.28: *External costs (€/MWh) of current and more advanced electricity systems associated with emissions from the operation of the power plant and the rest of the fuel-supply chain (EU, 2005). 'Rest' is the external cost related to the fuel cycle (1 € = 1.3 US$ approximately).*

Table 4.7: *The technical potential energy resource and fluxes available, potential associated carbon and projected costs (US$ 2006) in 2030 for a range of energy resources and carriers.*

Energy resources and carriers	Technical potential EJ[a]	Approximate inherent carbon (GtC)	Present energy costs[c] US$ (2005)	Projected costs in 2030		Additional references
				Investment US$/W$_e$[d]	Generation US$/MWh	
Oil	10,000-35,000[e]	200-1300	~9/GJ ~50/bbl ~48/MWh	n/a	50-100	Wall Street Journal, daily commodity prices
Natural gas	18,000-60,000	170-860	~5-7/GJ ~37/MWh	0.2-0.8	40-60 +CCS 60-90	EIA/DOE, 2006 IPCC, 2005
Coal	130,000	3500	~3-4.5/GJ ~20/MWh	0.4-1.4	40-55 +CCS 60-85	EIA/DOE, 2006 IPCC, 2005
Nuclear power	7400 (220,000)[f]	*[b]	10-120	1.5-3.0	25-75	IAEA, 2006 Figures 4.27, 4.28
Hydro > 10MW	1250	*	20-100/MWh	1.0-3.0	30-70	
Solar PV	40,000	*	250-1600/MWh	0.6-1.2	60-250	
Solar CSP	50	*	120-450/MWh	2.0-4.0	50-180	
Wind	15,000	*	40-90 MWh	0.4-1.2	30-80	
Geothermal	50	*	40-100/MWh	1.0-2.0	30-80	
Ocean	large	*	80-400/MWh	?	70-200	
Biomass - heat and power	Modern 9	6000	30-120/MWh 8-12/GJ	0.4-1.2	30-100	
Biofuels	1.2	*	8-30/GJ	?	23-75 c/l	Chapter 5, Figure 5.9
Hydrogen carrier	0.1	?	50/GJ	?	?	US NAE, 2004

Notes:
[a] From Table 4.2. Generalized potential for extractable energy: for fossil fuels the remaining extractable resources; for renewable energy likely cumulative by 2030
[b] * = small amount
[c] Prices volatile. Include old and new plants operating in 2006. Electricity costs for conversion efficiencies of 35% for fossil, nuclear and biomass
[d] Excluding carbon dioxide capture and storage
[e] Includes probable and unconventional oil and gas reserves
[f] At 130 US$/kg and assuming all remaining uranium, either used in once-through thermal reactors or recycled through light-water reactors and in fast reactors utilizing depleted uranium and the plutonium produced (in parentheses)

Source: Data from IEA, 2005a; IEA, 2006b; Johansson et al., 2004; IEA, 2004a; Fisher, 2006; IIASA/WEC, 1998; MIT, 2003.

A summary of cost-estimate ranges for the specific technologies as discussed in Section 4.3 is presented in Table 4.7. Costs and technical potentials out to 2030 show that abundant supplies of primary-energy resources will remain available. Despite uncertainty due to the wide range of assumptions, renewable energy fluxes and uranium resources are in sufficient supply to meet global primary-energy demands well past 2030 (Table 4.7). Proven and probable fossil-fuel reserves are also large, but concern over environmental impacts from combusting them could drive a transition to non-carbon energy sources. The speed of such a transition occurring depends, *inter alia*, on a number of things: how quickly investment costs can be driven down; confirmation that future life-cycle cost assessments for nuclear power, CCS and renewables are realistic; true valuation of external costs and their inclusion in energy prices; and what policies are established to improve energy security and reduce GHG emissions (University of Chicago, 2004).

4.4.3   Evaluation of costs and potentials for low-carbon, energy-supply technologies

As there are several interactions between the mitigation options that have been described in Section 4.3, the following sections assess the aggregated mitigation potential of the energy sector in three steps based on the literature and using the World Energy Outlook 2004 'Reference' scenario as the baseline (IEA, 2004a):

- The mitigation potentials in excess of the baseline are quantified for a number of technologies individually (Sections 4.4.3.1–4.4.3.6).
- A mix of technologies to meet the projected electricity demand by 2030 is compiled for OECD, EIT and non-OECD/ EIT country regions (Section 4.4.4) assuming competition between technologies, improved efficiency of conversion over time and that real-world constraints exist when building new (additional and replacement) plants and infrastructure.
- The interaction of the energy supply sector with end-use power demands from the building and industry sectors is

Table 4.8: *Baseline data from the World Energy Outlook 2004 Reference scenario.*

	Primary-energy fuel consumed for heat and electricity production in 2030 (EJ/yr)	Primary-energy fuel consumed for electricity in 2030[a] (EJ /yr)	Final electricity demand in 2030 (TWh/yr)	Increase in new power demand 2002 to 2030 (TWh)	Total emissions from electricity in 2030 (GtCO$_2$-eq/yr)
OECD	118.6	115.4	14,244	4,488	5.98
EIT	29.3	22.1	2,468	983	1.17
Non-OECD	128.5	125.3	14,944	10,111	8.62
World	276.4	262.8	31,656	15,582	15.77

[a] Final electricity generation was based on the electrical efficiencies calculated from 2002 data (IEA, 2004a Appendix 1) including a correction for the share of final heat in the total final energy consumption (see Chapter 11).

Source: IEA, 2004a.

then analysed (Section 11.3). Any savings of electricity and heat resulting from the uptake of energy-efficiency measures will result in some reduction in total demand for energy, and hence lower the mitigation potential of the energy supply sector.

Mitigation in the electricity supply sector can be achieved by optimization of generation plant-conversion efficiencies, fossil-fuel switching, substitution by nuclear power (Section 4.3.2) and/or renewable energy (4.3.3) and by CCS (4.3.6). These low-carbon energy technologies and systems are unlikely to be widely deployed unless they become cheaper than traditional generation or if policies to support their uptake (such as carbon pricing or government subsidies and incentives) are adopted.

The costs (Table 4.7) and mitigation potentials for the major energy-supply technologies are compared and quantified out to 2030 based on assumptions taken from the literature, particularly the recent IEA Energy Technology Perspectives (ETP) report (IEA, 2006a). The assessment of the electricity-supply sector potentials are partly based on the TAR assessment[2] but use more recent data and revised assumptions. Heat and CHP potentials (Section 4.3.5) were difficult to assess as reliable data are unavailable. For this reason the IEA aggregates commercial heat with power (IEA, 2004a, 2005a, 2006b). An estimate of the potential mitigation from increased CHP uptake by industry by 2050 was 0.2–0.4 GtCO$_2$ (IEA, 2006a), but is uncertain so heat is not included here.

The 2030 electricity sector baseline (Table 4.8; IEA, 2004a) was chosen because the SRES B2 scenario (Figure 4.26) provided insufficient detail and the latest WEO (IEA, 2006b) had not been published at the time. Estimates of the 2030 global demand for power are disaggregated for OECD, EIT, and non-OECD/EIT regions. The WEO 2004 baseline assumed that the 44% of coal in the power-generation primary fuel mix in 2002 would change to 42% by 2030; oil from 8% to 4%; gas 21% to

29%; nuclear 18% to 12%; hydro would remain the same at 6% (using the direct equivalent method); biomass 2% to 4%, and other renewables 1% to 3%.

This analysis quantifies the mitigation potential at the high end of the range for each technology by 2030 above the baseline. It assumes each technology will be implemented as much as economically and technically possible, but is limited by the practical constraints of stock turnover, rate of increase of manufacturing capacity, training of specialist expertise, etc. The assumptions used are compared with other analyses reported in the literature. Since, in reality, each technology will be constrained by what will be happening elsewhere in the energy-supply sector, they could never reach this total 'maximum' potential collectively, so these individual potentials cannot be directly added together to obtain a projected 'real' potential. Further analysis based on a possible future mix of generation technologies is therefore provided in Section 4.4.4 and further in Chapter 11, accounting for energy savings reducing the total demand. Emission factors per GJ primary fuel for CO$_2$, N$_2$O and CH$_4$ (IPCC, 1997) were used in the analysis but the non-CO$_2$ gases accounted for less than 1% of emissions.

4.4.3.1 *Plant efficiency and fuel switching*

Reductions in CO$_2$ emissions can be gained by improving the efficiency of existing power generation plants by employing more advanced technologies using the same amount of fuel. For example, a 27% reduction in emissions (gCO$_2$/kWh) is possible by replacing a 35% efficient coal-fired steam turbine with a 48% efficient plant using advanced steam, pulverized-coal technology (Table 4.9). Replacing a natural gas single-cycle turbine with a combined cycle (CCGT) of similar output capacity would help reduce CO$_2$ emissions per unit of output by around 36%.

Switching from coal to gas increases the efficiency of the power plant because of higher operating temperatures, and

---

2   The TAR (IPCC, 2001) estimated potential emission reductions of 1.3–2.5 GtCO$_2$ (0.35–0.7 GtC) by 2020 for less than 27 US$/tCO$_2$ (100 US$/tC) based on fuel switching from coal to gas; deployment of nuclear, hydro, geothermal, wind, biomass and solar thermal; the early uptake of CCS; and co-firing of biomass with coal.

Table 4.9: *Reduction in CO_2 emission coefficient by fuel substitution and energy conversion efficiency in electricity generation.*

Existing generation technology			Mitigation substitution option			Emission reduction per unit of output
Energy source	Efficiency (%)	Emission coefficient (gCO_2/kWh)	Switching option	Efficiency (%)	Emission coefficient (gCO_2/kWh)	(gCO_2/kWh)
Coal, steam turbine	35	973	Pulverised coal, advanced steam	48	710	-263
Coal, steam turbine	35	973	Natural gas, combined cycle	50	404	-569
Fuel oil, steam turbine	35	796	Natural gas, combined cycle	50	404	-392
Diesel oil, generator set	33	808	Natural gas, combined cycle	50	404	-404
Natural gas, single cycle	32	631	Natural gas, combined cycle	50	404	-227

Source: Danish Energy Authority, 2005.

when used together with the more efficient combined-cycle results in even higher efficiencies (IEA, 2006a). Emission savings (gCO_2-eq/kWh) were calculated before and after each substitution option (based on IPCC 1996 emission factors). The baseline scenario (IEA, 2004a) assumed a 5% CO_2 reduction from fossil-fuel mix changes (coal to gas, oil to gas etc.) and a further 7% reduction in the Alternative Policy scenario from fuel switching in end uses (see Chapters 6 and 7). By 2030, natural gas CCGT plants displacing coal, new advanced steam coal plants displacing less-efficient designs, and the introduction of new coal IGCC plants to replace traditional steam plants could provide a potential between 0.5 and 1.4 $GtCO_2$ depending on the timing and sequence of economics and policy measures (IEA, 2006a). IEA analysis also showed that up to 50 GW of stationary gas-fired fuel cells could be operating by 2030, growing to around 3% of all power generation capacity by 2050 and giving about 0.5 Gt CO_2 emissions reduction (IEA, 2006j). This potential is uncertain, however, as it relies on appropriate fuel-cell development and is not included here.

By 2030, a proportion of old heat and power plants will have been replaced with more modern plants having higher energy efficiencies. New plants will also have been built to meet the growing world demand. It is assumed that after 2010

only the most efficient plant designs available will be built, though this is unlikely and will therefore increase future CO_2 emissions above the potential reductions. The coal that could be displaced by gas and the additional gas power generation required is assessed by region (Table 4.10). A plant life time of 50 years; a 2%-per-year replacement rate in all regions starting in 2010; 20% of existing coal plants replaced by 2030 and 50% of all new-build thermal plants fuelled by gas, are among the most relevant assumptions. The cost of fuel switching partly depends on the difference between coal and gas prices. For example if mitigation costs below 20 US$/t$CO_2$-eq avoided, this would imply a relatively small price gap between coal and gas, although since fuel switching to a significant degree would affect natural gas prices, actual future costs are difficult to estimate with accuracy. Generation costs are assumed to be 40–55 US$/MWh for coal-fired and 40–60 US$/MWh for gas-fired power plants.

4.4.3.2   Nuclear

Proposed and existing fossil fuel power plants could be partly replaced by nuclear power plants to provide electricity

Table 4.10: *Potential GHG emission reductions by 2030 from coal-to-gas fuel switching and improved efficiency of existing plant.*

	Coal displaced by gas and improved efficiency (EJ/yr)	Additional gas power required (TWh/yr)	Emissions avoided ($GtCO_2$-eq/yr)	Cost ranges (US$/t$CO_2$-eq)	
				Lowest	Highest
OECD	7.18	947	0.39	0	12
EIT	0.73	79	0.04	0	10
Non-OECD	10.92	1392	0.64	0	11
World	18.83	2418	1.07		

Table 4.11: *Potential GHG emission reduction and cost ranges in 2030 from nuclear-fission displacing fossil-fuel power plants.*

	Potential contribution to electricity mix (%)	Additional generation above baseline (TWh/yr)	Emissions avoided (GtCO$_2$-eq/yr)	Cost ranges (US$/tCO$_2$-eq)	
				Lowest	Highest
OECD	25	1424	0.93	-24	25
EIT	25	345	0.23	-23	22
Non-OECD	10	974	0.72	-21	21
World	18	2743	1.88		

Table 4.12: *Potential GHG emission reduction and cost ranges in 2030 as a result of hydro power displacing fossil-fuel thermal power plants.*

	Potential contribution to electricity mix (%)	Additional generation above baseline (TWh/yr)	Net emissions avoided (GtCO$_2$-eq/yr)	Cost ranges (US$/tCO$_2$-eq)	
				Lowest	Highest
OECD	15	608	0.39	-16	3
EIT	15	0	0.0	0	0
Non-OECD	20	643	0.48	-14	41
World	17	1251	0.87		

and heat. Since the nuclear plant and fuel system consumes only small quantities of fossil fuels in the fuel cycle, net CO$_2$ emissions could be lowered significantly. Assessments of future potential for nuclear power are uncertain and controversial. The 2006 WEO Alternative scenario (IEA, 2006b) anticipated a 50% increase in nuclear energy (to 4106 TWh/yr) by 2030. The ETP report (IEA, 2006a) assumed a mitigation potential of 0.4–1.3 GtCO$_2$ by 2030 from the construction of Generation II, III, III+ and IV nuclear plants (Section 4.3.2). From a review of the literature and the various scenario projections described above (for example Figure 4.25), it is assumed that by 2030 18% of total global power-generation capacity could come from existing nuclear power plants as well as new plants displacing proposed new coal, gas and oil plants in proportion to their current share of the baseline (Table 4.11). The rate of build required is possible (given the nuclear industry's track record for building reactors in the 1970s) and generating costs of 25–75 US$/MWh are assumed (Section 4.4.2). However, there is still some controversy regarding the relatively low costs shown by comparative life-cycle analysis assessments reported in the literature (Section 4.4.2) and used here.

4.4.3.3   Renewable energy

Fossil fuels can be partly replaced by renewable energy sources to provide heat (from biomass, geothermal or solar) or electricity (from wind, solar, hydro, geothermal and bioenergy generation) or by CHP plants. Ocean energy is immature and assumed unlikely to make a significant contribution to overall power needs by 2030. Net GHG emissions avoided are used in

the analysis since most renewable energy systems emit small amounts of GHG from the fossil fuels used for manufacturing, transport, installation and from any cement or steel used in their construction. Overall, net GHG emissions are generally low for renewable energy systems (Figure 4.19) with the possible exception of some biofuels for transport, where fossil fuels are used to grow the crop and process the biofuel.

Hydro

The ETP (IEA, 2006a) stated the technical potential of hydropower to be 14,000 TWh/yr, of which around 6000 TWh/yr (56 EJ) could be realistic to develop (IHA, 2006). The WEO Alternative scenario (IEA, 2006b) assumed an increased share for hydro generation above baseline, reaching 4903 TWh/yr by 2030[3]. IEA (2006a) suggested hydropower (both small and large) could offset fossil-fuel power plants to give a mitigation potential between 0.3–1.0 GtCO$_2$/yr by 2030. Here it is assumed that enough existing and new sites will be available to contribute around 5500 TWh/yr (17% of total electricity generation) by 2030 as a result of displacing coal, gas and oil plants based on their current share of the base load (Table 4.12). Future costs range from 30–70 US$/MWh for good sites with high hydrostatic heads, close proximity to load demand, and with good all-year-round flow rates. Smaller plants and those installed in less-favourable terrain at a distance from load could cost more. GHG emissions from construction of hydro dams and possible release of methane from resulting reservoirs (Section 4.3.3.1) are uncertain and not included here.

---

3   Although nuclear (Table 4.11) and hydro both offer a similar contribution to the global electricity mix today and by 2030, their emission reduction potentials differ due to variations in assumptions of regional shares and baseline. Estimates in the baseline were 4248 TWh yr[-1] from hydro by 2030 compared with 2929 TWh yr[-1] from nuclear (IEA, 2004a).

Table 4.13: *Potential GHG emission reduction and costs in 2030 from wind power displacing fossil-fuel thermal power plants.*

	Potential contribution to electricity mix (%)	Additional generation above baseline (TWh/yr)	Net extra emissions reductions (GtCO$_2$-eq/yr)	Cost ranges (US$/tCO$_2$-eq)	
				Lowest	Highest
OECD	10	687	0.45	-16	33
EIT	5	99	0.06	-16	30
Non-OECD	5	572	0.42	-14	27
World	7	1358	0.93		

Table 4.14: *Potential emissions reduction and cost ranges in 2030 from bioenergy displacing fossil-fuel thermal power plants.*

	Potential contribution to electricity mix (%)	Additional generation above baseline (TWh/yr)	Net emissions reductions (GtCO$_2$-eq/yr)	Cost ranges (US$/tCO$_2$-eq)	
				Lowest	Highest
OECD	5	307	0.20	-16	63
EIT	5	112	0.07	-16	60
Non-OECD	10	1283	0.95	-14	54
World	7	1702	1.22		

Wind

The 2006 WEO Reference scenario baseline (IEA, 2006b) assumed 1132 TWh/yr (3.3% of total global electricity) of wind generation in 2030 rising to a 4.8% share in the Alternative Policy scenario. However, wind industry 'advanced' scenarios are more optimistic, forecasting up to a 29.1% share for wind by 2030 with a mitigation potential of 3.1 GtCO$_2$/yr (GWEC, 2006). The ETP mitigation potential assessment (IEA, 2006a) for on- and offshore wind power by 2030 ranged between 0.3 and 1.0 GtCO$_2$/yr. In this analysis on- and offshore wind power is assumed to reach a 7% share by 2030, mainly in OECD countries, and to displace new and existing fossil-fuel power plants according to the relevant shares of coal, oil and gas in the baseline for each region (Table 4.13). Intermittency issues on most grids would not be limiting at these low levels given suitable control and back-up systems in place. The costs are very site specific and range from 30 US$/MWh on good sites to 80 US$/MWh on poorer sites that would also need to be developed if this 7% share of the total mix is to be met.

Bioenergy (excluding biofuels for transport)

Large global resources of biomass could exist by 2030 (Chapters 8, 9 and 10), but confidence in estimating the bioenergy heat and power potential is low since there will be competition for these feedstocks for biomaterials, chemicals and biofuels. Bioenergy in its various forms (landfill gas, combined heat and power, biogas, direct combustion for heat etc.) presently contributes 2.6% to the OECD power mix, 0.4% to EIT and 1.5% to non-OECD. The WEO 2006 (IEA, 2006b) assumed 805 TWh of biomass power generation in 2030 rising 22% to 983 TWh under the Alternative scenario to then give 3% of total electricity generation. The ETP gave a bioenergy potential ranging between 0.1 and 0.3 GtCO$_2$ /yr by 2030. The

baseline (IEA, 2004a) assumed biomass and waste for heat and power generation will rise from 2% of primary fuel use (3.2 EJ) in 2002 to 4% (10.8 EJ) by 2030.

Heat and CHP estimates are wide ranging so cannot be included in this analysis, even though the bioenergy potential could be significant. However, any heat previously utilized from displaced coal and gas CHP plants could easily be supplied from biomass, with more biomass available for use in stand-alone heat plants (Chapter 11). In this analysis, a 5% share in OECD regions is assumed feasible, relying on co-firing in existing and new coal plants and with 7–8% of the total replacement capacity built being bioenergy plants. In EIT regions, the available forest biomass could be utilized to gain 5% share and in non-OECD regions, where there are less stock turnover issues than in the OECD, 10% of power could come from new bioenergy plants (Table 4.14). A total potential by 2030 of 5% is assumed based on costs of 30–100 US$/MWh. The biomass feedstock required to meet these potentials, assuming thermal-conversion efficiencies of 20–30%, would be around 9–13 EJ in OECD, 1–3 EJ in EIT, and 18–27 EJ in non-OECD regions. Little additional bioenergy capacity above that already assumed in the baseline is anticipated in EIT regions where only a small contribution is expected compared with developing countries. Small inputs of fossil fuels are often used to produce, transport and convert the biomass (IEA, 2006h), but the same is true when using the fossil fuels it replaces. Since both are of a similar order of magnitude, and these emissions are already accounted for in the overall total for fossil fuels, bioenergy is credited with zero emissions (in compliance with IPCC guidelines).

Table 4.15: *Potential emissions reduction and cost ranges in 2030 from geothermal displacing fossil-fuel thermal power plants.*

	Potential contribution in electricity mix (%)	Additional geothermal (TWh/yr)	Net emissions avoided (GtCO$_2$-eq/yr)	Cost ranges (US$/tCO$_2$-eq)	
				Lowest	Highest
OECD	2	137	0.09	-16	33
EIT	2	44	0.03	-16	30
Non-OECD	3	413	0.31	-14	27
World	2	594	0.43		

Table 4.16: *Potential emission reduction and cost ranges in 2030 from solar PV and CSP displacing fossil-fuel thermal power plants.*

	Potential contribution to electricity mix (%)	Additional generation above baseline (TWh/yr)	Emissions reductions (GtCO$_2$-eq/yr)	Cost ranges (US$/tCO$_2$-eq)	
				Lowest	Highest
OECD	1	44	0.03	61	294
EIT	1	21	0.01	60	288
Non-OECD	2	275	0.21	53	257
World	2	340	0.25		

Geothermal

The installed geothermal-generation capacity of over 8.9 GW$_e$ in 24 countries produced 56.8 TWh (0.3%) of global electricity in 2004 and is growing at around 20%/yr (Bertani, 2005) with the baseline giving 0.05% of total generation by 2030. IEA WEO 2006 (IEA, 2006b) assumed 174 TWh/yr by 2030 rising 6% to 185 TWh under the Alternative scenario. The ETP (IEA, 2006a) gave a potential of 0.1–0.3 GtCO$_2$-eq/yr by 2030.

In this analysis, generation costs of 30–80 US$/MWh are assumed to provide a 2% share of the total 2030 energy mix. Direct heat applications are not included. Although CO$_2$ emissions are assumed to be zero, as for other renewables, this may not always be the case depending on underground CO$_2$ released during the heat extraction.

Solar

Concentrating solar power (CSP) and photovoltaics (PV) can theoretically gain a maximum 1–2% share of the global electricity mix by 2030 even at high costs. The 2006 WEO Reference scenario (IEA, 2006b) estimated 142 TWh/yr of PV generation in 2030 rising to 237 TWh in the Alternative scenario but still at <1% of total generation. EPRI (2003) assessed total PV capacity to be 205 GW by 2020 generating 282 TWh/yr or about 1% of global electricity demand. Other analyses range from over 20% of global electricity generation by 2040 (Jäger-Waldau, 2003) to 0.008% by 2030 with mitigation potential for both PV and CSP likely to be <0.1 GtCO$_2$ in 2030 (IEA, 2006a) The calculated minimum costs for even the best sites resulted in relatively high costs per tonne CO$_2$ avoided (Table 4.16). The baseline (IEA, 2004a) gave the total solar potential as 466 TWh or 1.4% of total generation in 2030.

In this analysis, generating costs from CSP plants could fall sufficiently to compete at around 50–180 US$/MWh by 2030 (Trieb, 2005; IEA, 2006a). PV installed costs could decline to around 60–250 US$/MWh, the wide range being due to the various technologies being installed on buildings at numerous sites, some with lower solar irradiation levels. Penetration into OECD and EIT markets is assumed to remain small with more support for developing country electrification.

4.4.3.4   Carbon dioxide capture and storage

In the absence of explicit policies, CCS is unlikely to be deployed on a large scale by 2030 (IPCC, 2005). The total CO$_2$ storage potential for each region (Hendriks *et al.* 2004; Table 4.17) appears to be sufficient for storage over the next few decades, although capacity assessments are still under debate (IPCC, 2005). The proximity of a CCS plant to a storage site affects the cost, but this level of analysis was not considered here. CCS does not appear in the baseline (IEA, 2004a). Penetration by 2030 is uncertain as it depends both on the carbon price and the rate of technological advances in costs and performance.

Coal CCS

ABARE (Fisher, 2006) suggested that worldwide by 2030, 1811 TWh/yr would be generated from coal with CCS (17 EJ); 7871 TWh (73 EJ) from coal without; 1492 TWh (14 EJ) from gas with; and 6315 TWh (59 EJ) from gas without. CCS would thus result in around 4.4 GtCO$_2$ of GHG emissions avoided in 2030 giving a 17% reduction from the reference base case level (Figure 4.25). In contrast, the ETP mitigation assessments for CCS with coal plants ranged between only 0.3 and 1.0 GtCO$_2$ in 2030 (IEA, 2006a), given that commercial-scale CCS demonstration will be needed before widespread deployment.

Table 4.17: *Potential emissions reduction and cost ranges in 2030 from CCS used with coal-fired power plants.*

	Share of plants with CCS (%)	Coal-fired power generation with CCS	Annual emissions avoided (GtCO$_2$-eq/yr)	Total potential storage volume[a] (GtCO$_2$)	Cost ranges (US$/tCO$_2$-eq)	
					Lowest	Highest
OECD	9	388	0.28	71-1025	28	42
EIT	4	14	0.01	114-1250	22	33
Non-OECD	4	253	0.20	291-3600	26	39
World	6	655	0.49	476-5875		

[a] Hendriks et al, 2004

Table 4.18: *Potential emissions reduction and cost ranges in 2030 from CCS used with gas-fired power plants.*

	Share of plants with CCS (%)	Gas-fired power generation with CCS (TWh/yr)	Annual emission avoided (GtCO$_2$-eq/yr)	Total capture from both coal + gas, 2015-2030 (GtCO$_2$)	Cost ranges (US$/tCO$_2$-eq)	
					Lowest	Highest
OECD	7	243	0.09	8.37	52	79
EIT	5	78	0.04	2.03	43	64
Non-OECD	5	276	0.09	12.56	51	76
World	6	597	0.22	22.96		

In this analysis, CCS is assumed to begin only after 2015 in OECD countries and after 2020 elsewhere, linked mainly with advanced steam coal plants installed with flue gas separation, although these IGCC plants and oxyfuel systems are only just entering the market (Dow Jones, 2006). Assuming a 50-year life of coal plants (IEA, 2006a) and that 30% of new coal plants built in OECD and 20% elsewhere will be equipped with CCS, then the replacement rate of old plants by new designs with CCS incorporated is 0.6% per year in OECD and 0.4% elsewhere. Then 9% of total new and existing coal-fired plants will have CCS by 2030 in the OECD region and 4% elsewhere. Assuming 90% of the CO$_2$ can be captured and a reduced fuel-to-electricity conversion efficiency of 30% (leading to less power available for sale – IPCC, 2005), then the additional overall costs range between 20 and 30 US$/MWh depending on the ease of CO$_2$ transport and storage specific to each plant (Table 4.17).

Gas CCS
The assumed life of a CCGT plant is 40 years, and with 20% of new gas-fired plants utilizing CCS starting in 2015 in OECD countries and 2020 elsewhere, then the replacement rate of old plants by new designs integrating CCS is 0.5% per year. By 2030 7% of all OECD gas plants will have CCS and 5% elsewhere. Assuming 90% of the CO$_2$ is captured, a reduction of gas-fired power plant conversion efficiency of 15% (IPCC, 2005), and an additional overall cost ranging between 20 and 30 US$/MWh generated, then the costs and potentials by 2030 (compared with the IEA (2004a) baseline of no CCS) are assessed (Table 4.18). The costs for both coal and gas CCS compare well with the IPCC (2005) range of 15–75 US$/tCO$_2$ (Table 4.5).

4.4.3.5   *Summary*

The cost ranges (US$/tCO$_2$-eq avoided) for each of the technologies analysed in Section 4.4.3 are compared (Table 4.19). The percentage share of the total potential is shown spread across the defined cost class ranges for each region and technology. This assumes that a linear relationship exists between the lowest and highest costs as presented in Section 4.4.3 for each technology and region.

Since each technology is assumed to be promoted individually and crowding-out by other technologies under real-world constraints is ignored, the potentials in Table 4.19 are independent and cannot be added together.

4.4.4   Electricity-supply sector mitigation potential and cost of GHG emission avoidance

To provide a more realistic indication of the total mitigation potential for the global electricity sector, further analysis is conducted based on the literature, and assuming that no additional energy-efficiency measures in the building and industry sectors will occur beyond those already in the baseline. (Section 11.3.1 accounts for the impacts of energy efficiency on the heat and power-supply sector). The WEO 2004 baseline (IEA, 2004a) is used, based on data from Price and de la Rue du Can, (2006). The fuel-to-electricity conversion efficiencies were derived from the correction of the heat share in the WEO 2004 data, by assuming the share of heat in the total primary energy supply was constant from 2002 onwards.

Table 4.19: *Potential GHG emissions avoided by 2030 for selected, electricity generation mitigation technologies (in excess of the World Energy Outlook 2004 Reference baseline, IEA, 2004a) if developed in isolation and with the estimated mitigation potential shares spread across each cost range (2006 US$/tCO$_2$-eq) for each region.*

	Regional groupings	Mitigation potential; total emissions saved in 2030 (GtCO$_2$-eq)	Mitigation potential (%) spread over cost ranges (US$/tCO$_2$-eq avoided)				
			<0	0-20	20-50	50-100	>100
Fuelswitch and plant efficiency	OECD	0.39		100			
	EIT	0.04		100			
	Non-OECD	0.64		100			
	World	**1.07**					
Nuclear	OECD	0.93	50	50			
	EIT	0.23	50	50			
	Non-OECD	0.72	50	50			
	World	**1.88**					
Hydro	OECD	0.39	85	15			
	EIT	0.00					
	Non-OECD	0.48	25	35	40		
	World	**0.87**					
Wind	OECD	0.45	35	40	25		
	EIT	0.06	35	45	20		
	Non-OECD	0.42	35	50	15		
	World	**0.93**					
Bioenergy	OECD	0.20	20	25	40	15	
	EIT	0.07	20	25	40	15	
	Non-OECD	0.95	20	30	45	5	
	World	**1.22**					
Geothermal	OECD	0.09	35	40	25		
	EIT	0.03	35	45	20		
	Non-OECD	0.31	35	50	15		
	World	**0.43**					
Solar PV and CSP	OECD	0.03				20	80
	EIT	0.01				20	80
	Non-OECD	0.21				25	75
	World	**0.25**					
CCS + coal	OECD	0.28			100		
	EIT	0.01			100		
	Non-OECD	0.20			100		
	World	**0.49**					
CCS + gas	OECD	0.09				100	
	EIT	0.04			30	70	
	Non-OECD	0.09				100	
	World	**0.22**					

The baseline

By 2010 total power demand is 20,185 TWh with 13,306 TWh generation coming from fossil fuels (65.9% share of the total generation mix), 3894 TWh from all renewables (19.3%), and 2985 TWh from nuclear (14.8%). Resulting emissions are 11.4 GtCO$_2$-eq. By 2030 the increased electricity demand of 31,656 TWh is met by 22,602 TWh generated from fossil fuels, 6,126 TWh from renewables, and 2,929 TWh from nuclear power. The fossil-fuel primary energy consumed for electricity generation in 2030 produces 15.77 GtCO$_2$-eq of emissions (IEA, 2004a; Table 4.8).

New electricity generation plants to be built between 2010 and 2030 are to provide additional generating capacity to meet the projected increase in power demand, and to replace capacity of old, existing plants being retired during the same period. Additional capacity built after 2010, consumes an additional 82.5 EJ/yr of primary energy in order to generate 11,471 TWh/yr more electricity by 2030. Replacement capacity built during the period consumes 72 EJ/yr in 2030 and generates 8074 TWh/yr. Therefore, the total generation from new plants in the baseline is 19,545 TWh/yr by 2030, of which 14,618 TWh/yr comes from fossil-fuel plants (75%), 3787 TWh/yr from other renewables (19%), and 1140 TWh/yr from nuclear power (6%) (IEA, 2004a).

Sector analysis from 2010 to 2030

The potential for the global electricity sector to reduce baseline GHG emissions as a result of the greater uptake of low- and zero-carbon-emitting technologies is assessed. The method employed is outlined below. Fossil-fuel switching from coal to gas; substitution of coal, gas and oil plants with nuclear, hydro, bioenergy and other renewables (wind, geothermal, solar PV and solar CSP), and the uptake of CCS are all included.

- For each major world country-grouping (OECD Pacific, US and Canada, OECD Europe, EIT, East Asia, South Asia, China, Latin America, Mexico, Middle East and Africa), WEO 2004 baseline data (Price and de la Rue du Can, 2006; IEA, 2004a) are used to show the capacity of fossil-fuel thermal electricity generation per year that could be substituted after 2010, assuming a linear replacement rate and a 50-year life for existing coal, gas and oil plants. The results are then aggregated into OECD, EIT and non-OECD/EIT regional groupings.
- New generation plants built by 2030 to meet the increasing power demand are shared between fossil fuel, renewables, nuclear and, after 2015, coal and gas-fired plants with CCS. The shares of total power generation assumed for each of these technologies by 2030 are based on the literature (Section 4.4.3), but also depend partly on their relative costs (Table 4.19). The relatively high shares assumed for nuclear and renewable energy, particularly in OECD countries, are debatable, but supported to some extent by European Commission projections (EC, 2007).
- No early retirements of plant or stranded assets are contemplated (although in reality a faster replacement rate of existing fossil-fuel capacity could be possible given more stringent policies in future to reduce GHG emissions). The assumed replacement rates of old fossil-fuel plant capacity by nuclear, and renewable electricity, and the uptake of CCS technologies, are each based on the regional power mix shares of coal, gas and oil plants operating in the baseline.
- In reality, the future value of carbon will likely affect the actual generation shares for each technology, as will any mitigation policies in place before 2030 that encourage reductions of GHG emissions from specific components of the energy-supply sector.

- It is assumed that after 2010 only power plants with higher conversion efficiencies (Table 4.20) are built.
- As fuel switching from coal to natural gas supply is assessed to be an option with relatively low costs, this is implemented first with 20% of new proposed coal-fired power plants substituted by gas-fired technologies in all regions (based on Section 4.4.3.1).
- It is assumed that, where cost-effective, some of the new fossil-fuel plants required according to the baseline (after adjustments for the previous step) are displaced by low- and zero-carbon-intensive technologies (wind, geothermal, hydro, bioenergy, solar, nuclear and CCS) in proportion to their relative costs and potential deployment rates. The resulting GHG emissions avoided are assessed.
- It is assumed that by 2030, wind, solar CSP and solar PV plants that displace new and replacement fossil-fuel generation are partly constrained by related environmental impact issues, the relatively high costs for some renewable plants compared to coal, gas and nuclear, and intermittency issues in power grids. However, developments in energy-storage technologies, supportive policy trends and recognition of co-benefits are assumed to partly offset these constraints. Priority grid access for renewables is also assumed. Thus, reasonably high shares in the mix become feasible (Table 4.20).
- The share of electricity generation from each technology assumes that the maximum resource available is not exceeded. The available energy resources are evaluated on a regional basis to ensure all assumptions can be met in principle.
  - Any volumes of biomass needed above those available from agricultural and forest residues (Chapters 8 and 9) will need to be purpose-grown, so could be constrained by land and water availability. While there is some uncertainty in this respect, there should be sufficient production possible in all regions to meet the generation from bioenergy as projected in this analysis.
  - Uranium fuel supplies for nuclear plants should meet the assumed growth in demand, especially given the anticipation of 'Gen III' plant designs with fuel recycling coming on stream before 2030.
  - There is sufficient storage capacity for sequestering the estimated capture of CO_2 volumes in all regions given the anticipated rate of growth of CCS over the next few decades (Hendricks *et al.*, 2004).
- CCS projects for both coal- and gas-fired power plants are deployed only after 2015, assuming commercial developments are unavailable until then.

Table 4.20: Projected power demand increase from 2010 to 2030 as met by new, more efficient additional and replacement plants that will displace 60% of existing plants at the end of their life. The potential mitigation above the baseline of GHG avoided for <20 US$/t, <50 US$/t and <100 US$/tCO$_2$eq results from fuel switching from coal to gas, a portion of fossil-fuel generation being displaced by nuclear, renewable energy and bioenergy in each region and CCS.

	Power plant efficiencies by 2030 (based on IEA 2004a)[a] (%)	Existing mix of power generation in 2010 (TWh)	Generation from additional new plant by 2030 (TWh)	Generation from new plant replacing old, existing 2010 plant by 2030 (TWh)	Share of mix of generation of total new and replacement plant built by 2030 including CCS at various costs of US$/tCO$_2$-eq avoided[b] <20 US$/t TWh	<50 US$/t TWh	<100 US$/t TWh	Total GtCO$_2$-eq avoided by fuel switching, CCS and displacing some fossil fuel generation with low carbon option of wind, solar, geothermal, hydro, nuclear and biomass <20 US$/t	<50 US$/t	<100 US$/t
OECD		**11302**	**2942**	**4521**	**7463**			**1.58**	**2.58**	**2.66**
Coal	41	4079	657	1632	899	121	0			
Oil	40	472	-163[c]	189	13	2	0			
Gas	48	2374	1771	950	1793	637	458			
Nuclear	33	2462	-325	985	2084	2084	1777			
Hydro	100	1402	127	561	1295	1295	1111			
Biomass	28	237	168	95	263	499	509			
Other renewables	63	276	707	110	1116	1544	1526			
CCS					0	1282	2082			
EIT		**1746**	**722**	**698**	**1420**			**0.32**	**0.42**	**0.49**
Coal	32	381	13	152	72	46	29			
Oil	29	69	-8	28	11	7	4			
Gas	39	652	672	261	537	357	240			
Nuclear	33	292	-20	117	442	442	442			
Hydro	100	338	35	135	170	170	170			
Biomass	48	4	7	2	47	109	121			
Other renewables	36	10	23	4	142	167	191			
CCS					0	123	222			
Non-OECD/EIT		**7137**	**7807**	**2855**	**10662**			**2.06**	**3.44**	**4.08**
Coal	38	3232	3729	1293	2807	1697	1133			
Oil	38	646	166	258	297	179	120			
Gas	46	1401	2459	560	3114	2279	1856			
Nuclear	33	231	289	92	1356	1356	1356			
Hydro	100	1472	874	589	1463	2106	2106			
Biomass	19	85	126	34	621	1294	1443			
Other renewables	28	70	164	28	1004	1154	1303			
CCS					0	598	1345			
TOTAL		**20185**	**11471**	**8074**	**19545**			**3.95**	**6.44**	**7.22**

[a] Implied efficiencies calculated from WEO 2004 (IEA, 2004b) = Power output (EJ) / Estimated power input (EJ). See Appendix 1, Chapter 11.
[b] At higher costs of US$/tCO$_2$ avoided, more coal, oil and gas power generation is displaced by low- and zero-carbon options. Since nuclear and hydro are cost competitive at <20 US$/tCO$_2$-eq avoided in most regions (Table 4.9), and the rate of building new plants is constrained, their share remains constant.
[c] Negative data depicts a decline in generation, which was included in the analysis.

Source: Based on IEA, 2004a.

4.4.4.1 Mitigation potentials of the electricity supply sector

Based on the method described above and the results from the analysis (Table 4.20), the following conclusions can be drawn.

With reference to the baseline:
a) power plants existing in 2010 that remain in operation by 2030 (Table 4.20), including coal, oil and gas-fired, continue to generate 12,111 TWh/yr in 2030 (38% of the total power demand) and produce 5.77 $GtCO_2$-eq/yr of emissions;
b) new additional plants to be built over the 20-year period from 2010 generate 11,471 TWh/yr by 2030 and new plants built to replace old plants generate 8074 TWh/yr;
c) the share of all new build plants burning coal, oil and gas produce around 10 $GtCO_2$-eq/yr by 2030, thereby giving total baseline emissions of 15.77 $GtCO_2$-eq/yr (Table 4.8).

For costs < 20 US\$/tCO$_2$-eq avoided:
a) The baseline generation from fossil fuel-fired plants in 2030 of 22,602 TWh (including 14,618 TWh from new generation) reduces by 22.5% to 17,525 TWh (including 9541 TWh of new build generation) due to the increased uptake of low- and zero-carbon technologies. This is a reduction from the 71% of total generation in the baseline to 55%.
b) Of this total, fuel switching from coal to gas results in additional gas-fired plants generating 1,495 TWh/yr by 2030, mainly in non-OECD/EIT countries, and thereby avoiding 0.67 $GtCO_2$-eq/yr of emissions.
c) Renewable energy generation increases from the 2030 baseline of 6126 TWh/yr to 7904 TWh/yr (6122 TWh/yr from new generation plus 2336 TWh/yr remaining in operation from 2010). The share of generation increases from 19.4% in 2010 to 26.7% by 2030.
d) The nuclear power baseline of 2929 TWh/yr by 2030 (9.3% of total generation) increases to 5673 TWh/yr (17.9% of generation), of which 3882 TWh/yr is from newly built plants.
e) Overall, GHG emissions are reduced by 3.95 $GtCO_2$-eq giving 25.0% lower emissions than in the baseline. Around half of this potential occurs in non-OECD/EIT countries with OECD countries providing most of the remainder.
f) Should just 70% of the individual power-generation shares assumed above for all the mitigation technologies be achieved by 2030, the mitigation potential would reduce to 1.69 $GtCO_2$-eq.
g) This range is in reasonable agreement with the TAR analysis potential of 1.3 to 2.5 $GtCO_2$-eq/yr for less than 27 US\$/tCO$_2$-eq avoided (IPCC, 2001), because this potential was only out to 2020, the baseline has since been adjusted, and since the TAR was published there has been increased acceptance for improved designs of nuclear power plants, an increase in development and deployment of renewable energy technologies and a better understanding of CCS technologies.

For costs <50 US\$/tCO$_2$-eq avoided:
a) Fossil-fuel generation reduces further to 13,308 TWh/yr (of which 5324 TWh/yr is from new build plants) and accounts for 42% of total generation.
b) Renewable-energy generation increases to 10,673 TWh/yr by 2030 giving a 33.7% share of total generation. Solar PV and CSP are more costly (Table 4.19) so they can only offer substitution for fossil fuels above 50 US\$/tCO$_2$-eq avoided.
c) Nuclear power share of total generation remains similar since other technologies now compete.
d) CCS now becomes competitive and 2003 TWh/yr is generated by coal and gas-fired plants with CCS systems installed.
e) Overall GHG emissions in 2030 are now reduced by 6.44 $GtCO_2$-eq/yr below the baseline, although if only 70% of the assumed shares of total power generation for all the mitigation technologies are reached by 2030, the potential declines to 3.05 $GtCO_2$-eq. Non-OECD/EIT countries continue to provide half of the mitigation potential.

For costs <100 US\$/tCO$_2$-eq avoided:
a) As more low- and zero-carbon technologies become competitive, fossil-fuel generation without CCS further reduces to 11,824 TWh in 2030 and is now only 37% of total generation.
b) New renewable energy generation increases to 8481 TWh/yr by 2030, which together with the plants remaining in operation from 2010, gives a 34% share of total generation.
c) New nuclear power provides 3574 TWh to give 17% of total generation.
d) Coal- and gas-fired plants with CCS account for 3650 TWh/yr by 2030 or 12% of total generation.
e) The overall mitigation potential of the electricity sector is 7.22 $GtCO_2$-eq/yr which is a reduction of around 45% of GHGs below the baseline. If only 70% of the assumed shares of power generation by all low- and zero-emission technologies are achieved, then the potential would be around 45% lower at 3.97 $GtCO_2$-eq. Non-OECD/EIT countries contribute over half the total potential.

No single technological option has sufficient mitigation potential to meet the economic potential of the electricity-generation sector. To achieve these potentials by 2030, the relatively high investment costs, the difficulties in rapidly building sufficient capacity and expertise, and the threats resulting from introducing new low-carbon technologies as perceived by the incumbents in the existing markets, will all need to be addressed.

This analysis concentrates on the individual mitigation potentials for each technology at the high end of the wide range found in the literature (Figure 4.29b; IEA, 2006a; IEA, 2006b). This serves to illustrate that significant reductions in emissions from the energy-supply sector are technically and economically feasible using both the range of technology solutions currently

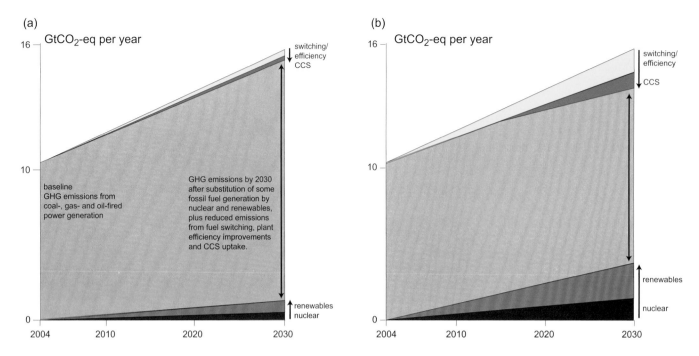

Figure 4.29: *Indicative low(a) and high(b) range estimates of the mitigation potential in the electricity sector based on substitution of existing fossil-fuel thermal power stations with nuclear and renewable energy power generation, coupled with energy-efficiency improvements in power-generation plants and transmission, including switching from coal to gas and the uptake of CCS. CHP and heat are not included, nor electricity savings from energy-efficiency measures in the building and industry sectors.*
Source: Based on IEA, 2006a; IEA, 2006b.

available and those close to market. Reducing the individual assumed shares of the technologies in the 2030 power generation mix by 30% gives less ambitious potentials that are closer to the lower end of the ranges found in the literature (Figure 4.29a). Energy-efficiency savings of electricity use in the buildings (Chapter 6) and industry (Chapter 7) sectors will reduce these total emissions potentials (Section 11.3.1).

4.4.4.2   *Uncertainties*

The wide range of energy supply-related potentials in the literature is due to the many uncertainties and assumptions involved. This analysis of the costs and mitigation potential for energy-supply technologies through to 2030 involved the following degrees of confidence.

- There is high agreement on the energy types and amounts of current global and regional energy sources used in the baseline (with the exception of traditional biomass, for which data are uncertain) because the several sources of those estimates are in close agreement.
- There is high agreement that energy supply will grow between now and 2030 with medium confidence in projections of the total energy demand by 2030. Most assumptions about population and energy use in various scenarios do not diverge greatly until after 2030, although past experience suggests that projections, even over a 25-year period, can be erroneous.
- Estimates of specific potentials out to 2030 for electricity-supply technologies based on specific studies have only low

agreement that a single value can be estimated accurately. However, there is medium confidence that the true potential of a mixture of supply technologies lies somewhere within the range estimated.
- The actual distribution of new technologies in 2030 can be estimated with medium confidence by using trend analyses, technology assessments, economic models and other techniques, but cannot take into account changing national policies and preferences, future carbon-price factors, and the unanticipated evolution of technologies or their cost. Current rates of adoption for particular technologies have been identified but there is low to medium agreement that these rates may continue until 2030.
- Despite the methodological limitations, the future costs and technical potentials identified provide a medium confidence for considering strategies and decisions over the next several decades. The analysis falls within the range of other projections for specific technologies.

4.4.4.3   *Transport biofuels*

Assessments for the uptake of biofuels range between 20 and 25% of global transport road fuels by 2050 and beyond (Chapter 5). The 2006 WEO (IEA, 2006b) Reference scenario predicted biofuels will supply 4% of road fuels by 2030 with greater potential up to 7% under the Alternative Policy scenario. To achieve double this penetration, as envisaged under the Beyond Alternative Policy scenario, would avoid around 0.5$GtCO_2$/ yr, but is likely to require large-scale introduction of

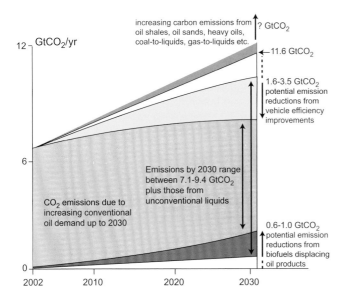

Figure 4.30: *Potential increased emissions from the greater uptake of unconventional oils by 2030 could offset potential reductions from both biofuels and vehicle-efficiency improvements, but will be subject to the future availability and price of conventional oil.*

Source: Based on IEA, 2006b.

second-generation biofuels from ligno-cellulosic conversions. Based on ETP assumptions (IEA, 2006a), the mitigation potential of biofuels by 2030 is likely to be less than from vehicle efficiency improvements (Chapter 5; Figure 4.30).

Transport emissions of 6.7 $GtCO_2$ in 2002 will increase under business as usual to 11.6 $GtCO_2$ by 2030, but could be reduced by efficiency improvements together with the increased uptake of biofuels to emit between 7.1 and 9.4 $GtCO_2$ (IEA, 2006a). This mitigation potential of between 2.2 and 4.5 $GtCO_2$, however, could be partially offset by the increased uptake of unconventional liquid fuels (Section 4.3.1.4). Their potential is uncertain as, being more costly per litre to produce, they will be dependent partly on the future oil price and level of reserves. Overall then, the emissions from transport fuels up to 2030 will probably continue to rise (Chapter 5).

4.4.4.4   *Heating and cooling*

The wide range of fuels and applications used for temperature modifications and the poor data base of existing heat and refrigeration plants makes the mitigation potential for heating and cooling difficult to assess. IEA (2006a) calculated the mitigation potential by 2030 for buildings (Chapter 6) of up to 2.6 $GtCO_2$/yr, including 0.1-0.3 $GtCO_2$/yr for solar systems, and up to 0.6 $GtCO_2$/yr for industry (Chapter 7). The mitigation potentials of CHP and trigeneration (heating, cooling and power generation) have not been assessed here.

4.5   Policies and instruments

4.5.1   Emission reduction policies

The reduction of GHG emissions from energy-supply systems is being actively pursued through a variety of government policies and private sector research. There are many technologies, behavioural changes and infrastructural developments that could be adopted to reduce the environmental impacts of current energy-supply systems (see Chapter 13). Whereas planning policies provide background for climate-change mitigation programmes, most climate policies relating to energy supply tend to come from three policy 'families' (OECD, 2002a):

- economic instruments (e.g. subsidies, taxes, tax exemption and tax credit);
- regulatory instruments (e.g. mandated targets, minimum performance standards, vehicle-exhaust emission controls); and
- policy processes (e.g. voluntary agreements and consultation, dissemination of information, strategic planning).

In addition, governments support RD&D programmes with financial incentives or direct investment to stimulate the development and deployment of new innovative energy-conversion technologies and create markets for them (Section 4.5.6).

Many GHG emission-reduction policies undertaken to date aim to achieve multiple objectives. These include market and subsidy reform, particularly in the energy sector (Table 4.21). In addition, governments are using a variety of approaches to overcome market barriers to energy-efficiency improvements and other 'win-win' actions.

Selecting policies and measures is not an easy task. It depends on many factors, including costs, potential capacity, the extent to which emissions must be reduced, environmental and economic impacts, rates at which the technology can be introduced, government resources available and social factors such as public acceptance. When implementing policies and measures, governments could consider the impacts of measures on other economies such as the specific needs and concerns of least developed countries arising from the adverse effects of climate change, on those nations that rely heavily on income generated from fossil-fuel exports, and on oil-importing developing countries.

4.5.1.1   *Emission-reduction policies for energy supply*

Subsidies, incentives and market mechanisms presently used to promote fossil fuels, nuclear power and renewables may need some redirection to achieve more rapid decarbonization of the energy supply.

Table 4.21: *Examples of policy measures given general policy objectives and options to reduce GHG emissions from the energy-supply sector.*

Policy objectives \ Policy options	Economic instruments	Regulatory instruments	Policy processes		
			Voluntary agreements	Dissemination of information and strategic planning	Technological RD&D and deployment
Energy efficiency	• Higher energy taxes • Lower energy subsidies • Power plant GHG taxes • Fiscal incentives • Tradable emissions permits	• Power plant minimum efficient standards • Best available technologies prescriptions	• Voluntary commitments to improve power plant efficiency	• Information and education campaigns.	• Cleaner power generation from fossil fuels
Energy source switching	• GHG taxes • Tradable emissions permits • Fiscal incentives	• Power plant fuel portfolio standards	• Voluntary commitments to fuel portfolio changes	• Information and education campaigns.	• Increased power generation from renewable, nuclear, and hydrogen as an energy carrier
Renewable energy	• Capital grants • Feed-in tariffs • Quota obligation and permit trading • GHG taxes • Tradable emissions permits	• Targets • Supportive transmission tariffs and transmission access	• Voluntary agreements to install renewable energy capacity	• Information and education campaigns • Green electricity validation	• Increased power generation from renewable energy sources
Carbon capture and storage	• GHG taxes • Tradable emissions permits	• Emissions restrictions for major point source emitters	• Voluntary agreements to develop and deploy CCS	• Information campaigns	• Chemical and biological sequestration • Sequestration in underground geological formations

Subsidies and other incentives

The effects of various policies and subsidies that support fossil-fuel use have been reviewed (IEA, 2001; OECD, 2002b; Saunders and Schneider, 2000). Government subsidies in the global energy sector are in the order of 250–300 billion US$/yr, of which around 2–3% supports renewable energy (de Moor, 2001; UNDP 2004a). An OECD study showed that global CO_2 emissions could be reduced by more than 6% and real income increased by 0.1% by 2010 if support mechanisms on fossil fuels used by industry and the power-generation sector were removed (OECD, 2002b). However, subsidies are difficult to remove and reforms would need to be conducted in a gradual and programmed fashion to soften any financial hardship.

For both environmental and energy-security reasons, many industrialized countries have introduced, and later increased, grant support schemes for producing electricity, heat and transport fuels based on nuclear or renewable energy resources and on installing more energy-efficient power-generation plant. For example, the US has recently introduced federal loan guarantees that could cover up to 80% of the project costs, production tax credits worth 6 billion US$, and 2 billion US$ of risk coverage for investments in new nuclear plants (Energy Policy Act, 2005). To comply with the 2003 renewable energy directive, all European countries have installed feed-in tariffs or

tradable permit schemes for renewable electricity (EEA, 2004; EU, 2003). Several developing countries including China, Brazil, India and a number of others have adopted similar policies.

Quantitative targets

Setting goals and quantitative targets for low-carbon energy at both national and regional levels increases the size of the markets and provides greater policy stability for project developers. For example, EU-15 members agreed on targets to increase their share of renewable primary energy to 12% of total energy by 2010 including electricity to 22% and biofuels to 5.75% (EU, 2001; EU 2003). The Latin American and Caribbean Initiative, signed in May 2002 included a target of 10% renewable energy by 2010 (Goldemberg, 2004). The South African Government mandated an additional 10 TWh renewable energy contribution by 2013 (being 4% of final energy consumption) to the existing contribution of 115 TWh/yr mainly from fuel wood and waste (DME, 2003). Many other countries outlined similar targets at the major renewable energy conference in Bonn (Renewables, 2004) attended by 154 governments, but not to the extent that emissions will be reduced below business as usual.

Feed-in tariffs/Quota obligations

Quota obligations with tradable permits for renewable

energy and feed-in tariffs have been used in many countries to accelerate the transition to renewable energy systems (Martinot, 2005). Both policies essentially serve different purposes, but they both help promote renewable energy (Lauber, 2004). Price-based, feed-in tariffs (providing long price certainty for renewable energy producers) have been compared with quantity-based instruments, including quotas, green certificates and competitive bidding (Sawin, 2003a; Menanteau *et al.*, 2003; Lauber, 2004). The total level of support provided for preferential power tariffs in EU-15, in particular Germany, Italy and Spain, exceeded 1 billion € in 2001 (EEA, 2004).

Experience confirms that incentives to support 'green power' by rewarding performance are preferable to a capital investment grant, because they encourage market deployment while also promoting increases in production efficiency (Neuhoff, 2004). In terms of installed renewable energy capacity, better results have been obtained with price-based than with quantity-based approaches (EC, 2005; Ragwitz *et al.*, 2005; Fouquet *et al.*, 2005). In theory, this difference should not exist as bidding prices that are set at the same level as feed-in tariffs should logically give rise to comparable capacities being installed. The discrepancy can be explained by the higher certainty of current feed-in tariff schemes and the stronger incentive effect of guaranteed prices.

The potential advantages offered by green certificate trading systems based on fixed quotas are encouraging a number of countries and states to introduce such schemes to meet renewable energy goals in an economically efficient way. Such systems can encourage more precise control over quotas, create competition among producers and provide incentives to lower costs (Menanteau *et al.*, 2003). Quota-obligation systems are only beginning to have an effect on capacity additions, in part because they are still new. However, about 75% of the wind capacity installed in the US between 1998 and 2004 occurred in states with renewable energy standards. Experience shows that if certificates are delivered under long-term agreements, effectiveness and compliance can be high (Linden *et al.*, 2005; UCS, 2005).

Tradable permit systems and CDM

In recent years, domestic and international tradable emission permit systems have received recognition as a means of lowering the costs of meeting climate-change targets. Creating carbon markets can help economies identify and realize economic ways to reduce GHG emissions and other energy-related pollutants, or to improve efficiency of energy use. The cost of achieving the Kyoto Protocol targets in OECD regions could fall from 0.2% of GDP without trading to 0.1% (Newman *et al.*, 2002) as a result of introducing emission trading in an international regime. Emission trading, such as the European and CDM schemes, is designed to result in immediate GHG reductions, but CDM also has long-term aspects, since the projects must assist developing countries in achieving sustainable development (see Chapter 13). The CDM successfully registered 450 projects by

the end of 2006 under the UNFCCC by the Executive Board with many more in the pipeline. Since the first project entered the pipeline in December 2003, 76% of projects belong to the energy sector. If all the 1300 projects in the pipeline at the end of 2006 are successfully registered with the UNFCCC and perform as expected, an accumulated emission reduction of more than 1400 $MtCO_2$-eq by end of 2012 can be expected (UNEP, 2006).

Information instruments

Education, technical training and public awareness are essential complements to GHG mitigation policies. They provide direct and continuous incentives to think, act and buy 'green' energy and to use energy wisely. Green power schemes, where consumers may choose to pay more for electricity generated primarily from renewable energy sources, are an example of combining information with real choice for the consumer (Newman *et al.*, 2002). Voluntary energy and emissions savings programmes, such as Energy Star (EPA, 2005a), Gas Star (EPA, 2005b) and Coalbed Methane Outreach (EPA, 2005c) serve to effectively disseminate relevant information and reduce knowledge barriers to the efficient and clean use of energy. These programmes include public education aspects, but are also built on industry/government partnerships. However, uncertainties on the effectiveness of information instruments for climate-change mitigation remain. More sociological research would improve the knowledge on adequacy of information instruments (Chapter 13).

Technology development and deployment

The need for further investments in R&D of all low-carbon-emission technologies, tied with the efficient marketing of these products, is vital to climate policy. Programmes supporting 'clean technology' development and diffusion are a traditional focus of energy and environmental policies because energy innovations face barriers all along the energy-supply chain (from R&D, to demonstration projects, to widespread deployment). Direct government support is often necessary to hasten deployment of radically new technologies due to a lack of industry investment. This suggests that there is a role for the public sector in increasing investment directly and in correcting market and regulatory obstacles that inhibit investment in new technology through a variety of fiscal instruments such as tax deduction incentives (Energy Policy Act, 2005; Jaffe *et al*, 2005).

Following the two oil crises in the 1970s, public expenditure for energy RD&D rose steeply, but then fell steadily in industrial countries from 15 billion US$ in 1980 to about 7 billion US$ in 2000 (2002 prices and exchange rates). Shares of IEA member-country support for energy R&D over the period 1974–2002 were about 8% for renewable energy, 6% for fossil fuel, 18% for energy efficiency, 47% for nuclear energy and 20% on other items (IEA, 2004b). During this period, a number of national governments (e.g. US, Germany, United Kingdom, France, Spain and Italy) made major cuts in their support for energy

R&D. Public spending on energy RD&D increased in Japan, Switzerland, Denmark and Finland and remained stable in other OECD countries (Goldemberg and Johannson, 2004).

Technology deployment is a critical activity and learning from market experience is fundamental to the complicated process of advancing a technology toward economic efficiency while encouraging the development of large-scale, private-sector infrastructure (IEA, 2003h). This justifies new technology deployment support by governments (Section 4.5.6).

4.5.1.2 Policy implementation experiences—successes and failures

Experiences of early policy implementation in the 1990s to reduce GHG emissions exist all over the world. This section lists and evaluates some examples. The fast penetration of wind power in Denmark was due to a regulated, favourable feed-in tariff. However, a new energy act in 1999 changed the policy to one based on the trading of green certificates. This created considerable uncertainty for investors and led to a significant reduction in annual investments in wind power plants during recent years (Johansson and Turkenburg, 2004).

In Germany, a comprehensive renewable energy promotion approach launched at the beginning of the 1990s led to it becoming the world leader in terms of installed wind capacity, and second in terms of installed PV capacity. The basic elements of the German approach are a combination of policy instruments, favourable feed-in tariffs and security of support to reduce investment risks (Johansson and Turkenburg, 2004).

When Spain passed a feed-in law in 1994, relatively few wind turbines were in operation. By the end of 2002, the country ranked second in the world, but had less success with solar PV in spite of having high solar radiation levels and setting PV tariffs similar to those in Germany. Little PV capacity was installed initially because regulations to enable legal grid connection were not established until 2001 when national technical standards for safe grid connection were implemented. PV producers who sold electricity into the grid, including individual households, had to register as businesses in order not to pay income tax on their sales (Sawin, 2003a). Significant growth in Spanish PV manufacturing in recent years is more attributable to the neighbouring German market (Ristau, 2003).

In 1990, the UK government launched the first of several rounds of competitive bidding for renewable energy contracts, known as the Non-Fossil Fuel Obligation (NFFO). The successive tendering procedures resulted in regular decreases in the prices for awarded contract value for wind and other renewable electricity projects. The average price for project proposals, irrespective of the technology involved, decreased from 0.067 €/kWh in 1994 to 0.042 €/kWh by 1998, being only 0.015 €/kWh above the wholesale electricity pool reference purchase price for the corresponding period

(Menanteau et al., 2001). Due to only relatively small volumes of renewable electricity being realized through the tender process, the government changed to a support mechanism by placing an obligation on electricity suppliers to sell a minimum percentage of power from new renewable energy sources. The annual growth rate of electricity generation by eligible renewable energy plants has significantly increased since the introduction of the obligation in April 2002 (OFGEM, 2005).

Swedish renewable energy policy during the 1970s and 1980s focused on strong efforts in technology research and demonstration. Subsequently market development took off during the 1990s when taxes and subsidies created favourable economic conditions for new investments and fuel switching. The use of biomass increased substantially during the 1990s (for example forest residues for district heating increased from 13 PJ in 1990 to 65 PJ in 2001). Increased carbon taxes created strong incentives for fuel switching from cheaper electric and oil-fired boiler for district heating to biomass cogeneration. The increase of biomass utilization led to development of the technology for biomass extraction from forests, production of short-rotation coppice *Salix* and implementation of more efficient district heating conversion technologies (Johansson, 2004).

Japan launched a 'Solar Roofs' programme in 1994 to promote PV through low-interest loans, a comprehensive education and awareness programme and rebates for grid-connected residential systems. In 1997, the rebates were opened to owners and developers of housing complexes and Japan become the world's largest installer of PV modules (Haas, 2002). Government promotion included publicity on television and in newspapers (IEA, 2003f). Total capacity increased at an average of more than 42% annually between 1994 and 2002 with more than 420 MW installed leading to a 75% cost reduction per Watt (Maycock, 2003; IEA, 2003f). The rebates declined gradually from 50% of installed cost in 1994 to 12% in 2002 when the programme ended. Japan is now the world's leading manufacturer, having surpassed the US in the late 1990s.

China's State Development and Planning Commission launched a renewable energy Township Electrification Program in 2001 to provide electricity to remote rural areas by means of stand-alone renewable energy power systems. During 2002–2004, almost 700 townships received village-scale solar PV stations of approximately 30–150 kW (about 20 MW total), of which few were hybrid systems with wind power (about 800 kW total of wind). Overall, the government provided 240 million US$ to subsidize the capital costs of equipment and around one million rural dwellers were provided with electricity from PV, wind-PV hybrid, and small hydropower systems (Martinot, 2005). Given the difficulties of other rural electrification projects using PV (ERC, 2004), it is too early to assess the effectiveness of this programme.

The California expansion plan to aid the installation of a million roofs of solar power in the residential sector in the next ten years was signed into law in August 2006 (Environment California, 2006). The law increased the cap on net metering from 0.5% of a utility's load to 2.5%. A solar rebate programme will be created and it will be mandatory that solar panels become a standard option for new homebuyers.

4.5.2 Air quality and pollution

The Johannesburg Plan of Implementation (UNDESA, 2002) called on all countries to develop more sustainable consumption and production patterns. Policies and measures to promote such pathways will automatically result in a reduction of GHG emissions and be useful to control air pollution (Section 11.8). Non-toxic CO_2 emissions from combustion processes have no detrimental effects on a local or regional scale, whereas toxic emissions such as SO_2 and particulates can have local health impacts as well as potentially wider detrimental environmental impacts.

The need for uncontaminated food and clean water to maintain general health have been recognized and addressed for a long time. However, only in recent decades has the importance of clean air to health been seriously noted (WHO, 2003). Major health problems suffered by women and children in the developing world (acute respiratory infection, chronic obstructive lung disease, cancer and pulmonary diseases) have been attributed to a lack of access to high-quality modern energy for cooking (Smith, 2002; Smith et al., 2000a; Lang et al., 2002; Bruce et al., 2000). The World Health Organisation (WHO, 2002) ranked indoor air pollution from burning solid fuels as the fourth most important health-risk factor in least developed countries where 40% of the world's population live, and is estimated to be responsible for 2.7% of the global burden of disease (Figure 4.31). It has been estimated that half

a million children and women die each year in India alone from indoor air pollution (Smith et al., 2000a). A study of indoor smoke levels conducted in Kenya revealed 24-hour average respirable particulate concentrations as high as 5526 $\mu g/m^3$ compared with the EPA standards for acceptable annual levels of 50 $\mu g/m^3$ (ITDG, 2003) and the EU standard for PM_{10} of 40 $\mu g/m^3$ (European Council Directive 99/30/EC). Another comprehensive study in Zimbabwe showed that those who came from households using wood, dung or straw for cooking were more than twice as likely to have suffered from acute respiratory disease than those coming from households using LPG, natural gas or electricity (Mishra, 2003).

Feasible and cost-effective solutions to poor air quality in both urban and rural areas need to be urgently identified and implemented (World Bank, 1998). Increasing access to modern energy services can help alleviate air-quality problems as well as realize a decrease in GHG emissions as greater overall efficiency is often achieved over the entire domestic energy cycle, starting from the provision of primary energy up to the eventual end-use. For instance, a shift from burning crop residues to LPG, kerosene, ethanol gel or biogas could decrease indoor air pollution by approximately 95% (Smith et al., 2000b; Sims et al, 2003b; Goldemberg et al., 2004; Larson and Yang, 2004).

Policies and measures aimed at increasing sustainability through reduction of energy use, energy-efficiency improvements, switching from the use of fossil fuels, and reducing the production of process wastes, will result in a simultaneous lowering of GHG emissions and reduced air pollution. Conversely, there are cases where measures taken to improve air quality can result in a simultaneous increase in the quantity of GHGs emitted. This is most likely to occur in those developing countries experiencing a phase of strong economic growth, but where it may not be economically feasible or desirable to move rapidly away from the use of an indigenous primary energy source such as oil or coal (Brendow, 2004).

Most regulations for air quality rely on limiting emissions of pollutants, often incorporating ambient air-quality guidelines or standards (Sloss et al., 2003). Although regulations to limit CO_2 emissions could be incorporated as command and control clauses in most of the existing legislative schemes, no country has so far attempted to do so. Rather, emissions trading has emerged as the preferred method of effecting global GHG mitigation, both within and outside the auspices of the Kyoto Protocol (Sloss et al., 2003).

Ambient air-quality standards or guidelines are usually set in terms of protecting health or ecosystems. They are thus applicable only at or near ground level where acceptable concentrations of gaseous emissions such as SO_2 can often be achieved through atmospheric dispersion using a tall stack as opposed to physical removal by scrubbers. Tall stacks avoid excessive ground-level concentrations of gaseous pollutants and

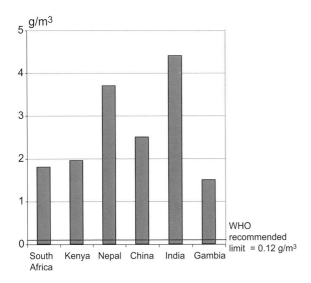

Figure 4.31: *Indoor levels of particulate concentrations emitted from wood fuel combustion in selected developing countries*

Source: Karekezi and Kithyoma, 2003.

are still in use at the majority of existing industrial installations and power plants around the world. If the use of tall stacks is precluded due to stringent limits being set for ambient SO_2 concentrations mandate, then the alternative of SO_2 scrubbers or other end-of-pipe removal equipment will require energy for its operation and thus divert it away from the production process. This leads to an overall decrease in cycle efficiency with a concomitant increase in CO_2 emissions. Sorbent extraction or other processes necessary to support scrubber operations also have GHG emissions associated with them. This effectively amounts to trading off a potential local or regional acid rain problem against a larger global climate problem. The overall costs of damage due to unmitigated CO_2 emissions have been estimated to greatly exceed those from regional acidification impacts arising from insufficient control of SO_2 emissions (Chae and Hope, 2003).

Air-quality legislation needs to be approached using the principles of integrated pollution prevention and control if unexpected and unwanted climate impacts on a global scale are to be avoided (Nalbandian, 2002). Adopting a multi-parameter approach could be useful. A US proposal calls for a cap and trade scheme for the power sector, simultaneously covering SO_2, NO_x, mercury and CO_2, which would specifically avoid conflicts with conventional regulations. Facilities would be required to optimize control strategies across all four pollutants (Burtraw and Toman, 2000). An approach developed for Mexico City showed that linear programming, applied to a database comprising emission-reduction information derived separately for air pollutants and GHGs, could provide a useful decision support tool to analyse least-cost strategies for meeting co-control targets for multiple pollutants (West et al., 2004).

4.5.3   Co-benefits of mitigation policies

Mitigation policies relating to energy efficiency of plants, fuel switching, renewable energy uptake and nuclear power, may have several objectives that imply a diverse range of co-benefits. These include the mitigation of air-pollution impacts, energy-supply security (by increased energy diversity), technological innovation, reduced fuel cost, employment and reducing urban migration. Reducing GHG emissions in the energy sector yields a global impact, but the co-benefits are typically experienced on a local or regional level. The variety of co-benefits stemming from GHG mitigation policies and the utilization of new energy technologies can be an integral part of economic policies that strive to facilitate sustainable development. These include improved health, employment and industrial development, and are explored in Chapter 11. This section therefore only covers aspects specifically related to energy supply. Quantitative information remains primarily limited to health effects with many co-effects not quantified due to a lack of information.

Fuel switching and the growth of energy-efficiency programmes (Swart et al., 2003) can lead to air-quality improvements and economic benefits as well as reduced GHG emissions (Beg, 2002). The relatively high capital costs for many renewable energy technologies are offset by the fuel input having minimal or zero cost and not prone to price fluctuations, as is the case with fossil fuels (Janssen, 2002). Nuclear energy shares many of the same market co-benefits as renewables (Hagen et al., 2005). Benefits of GHG mitigation may only be expected by future generations, but co-benefits are often detectable to the current generation.

Co-benefits of mitigation can be important decision criteria in analyses by policymakers, but often neglected (Jochem and Madlener, 2002). There are many cases where the net co-benefits are not monetised, quantified or even identified by decision-makers and businesses. Due consideration of co-benefits can significantly influence policy decisions concerning the level and timing of GHG mitigation action. There may be significant economic advantages to the national stimulation of technical innovation and possible spillover effects, with developing countries benefiting from innovation stimulated by GHG mitigation in industrialized countries. Most aspects of co-benefits have short-term effects, but they support long-term mitigation policies by creating a central link to sustainable development objectives (Kessels and Bakker, 2005). To date, most analyses have calculated GHG mitigation costs by dividing the incremental costs of 'mitigation technologies' by the amount of GHG avoided. This implicitly attributes all the costs to GHG-emission reduction and the co-benefits are seen as ancillary. Ideally, one would attribute the incremental costs to the various co-benefits by attempting to weight them. This could lead to significantly lower costs of GHG reductions since the other co-benefits would carry a share of the costs together with a change in the cost ranking of mitigation options (Schlamadinger et al. 2006).

The reduced costs of new technologies due to experience, and the incentives for further improvement due to competition, can be co-benefits of climate-change policies (Jochem and Madlener, 2002). New energy technologies are typically more expensive during their market-introduction phase but substantial learning experience can usually be achieved to reduce costs and enhance skill levels (Barreto, 2001; Herzog et al., 2001; IEA, 2000; McDonald and Schrattenholzer, 2001; NCOE, 2004). Increased net employment and trade of technologies and services are useful co-benefits given high unemployment in many countries. Employment is created at different levels, from research and manufacturing to distribution, installation and maintenance. Renewable-energy technologies are more labour-intensive than conventional technologies for the same energy output (Kamman et al., 2004). For example, solar PV generates 5.65 person-years of employment per 1 million US$ investment (over ten years) and the wind-energy industry 5.7 person-years. In contrast, every million dollars invested in the coal industry generates only 3.96 person-years of employment over the same time period (Singh and Fehrs, 2001). In South Africa, the development of renewable energy technologies could lead to the creation of over 36,000 direct jobs by 2020 (Austin et al., 2003) while more than 900,000 new jobs could be created across Europe by 2020 as a result of the increased use of renewable energy (EUFORES, 2004).

4.5.4 Implications of energy supply on sustainable development

The connection between climate-change mitigation and sustainable development is covered extensively in Chapter 12. The impact of the mitigation efforts from the energy-supply sector can be illustrated using the taxonomy of sustainability criteria and the indicators behind it. An analysis of the sustainability indicators mentioned in 750 project design documents submitted for validation under the CDM up to the end of 2005 (Olsen and Fenhann, 2006) indicated renewable energy projects provide the most sustainable impacts. Examples include biomass energy to create employment; geothermal and hydro to give a positive balance of payment; fossil-fuel switching to reduce emissions of SO_2 and NO_x; coal bed methane capture to reduce the number of explosions/accidents; and solar PV to create improved and increased access to electricity, employment, welfare and better learning possibilities.

4.5.4.1 Health and environment

Energy interlinks with health in two contradictory ways. It is essential to support the provision of health services, but energy conversion and consumption can have negative health impacts (Section 11.8). For example, in the UK, a lack of insufficient home heating has been identified as a principal cause of high levels of winter deaths (London Health Commission, 2003), but emissions from oil, gas, wood and coal combustion can add to reduced air quality and respiratory diseases.

The historical dilemma between energy supply and health can be demonstrated for various sectors, although it should be noted that recent times have seen major improvements. For instance, whereas epidemiological studies have shown that oil production in developed countries is not accompanied by significant health risks due to application of effective abatement technology, a Kazakhstan study compared the health costs between the city of Atyrau (with a high rate of pollution from oil extraction) and Astana (without). Health costs per household in Atyrau were twice as high as in Astana. The study also showed that the annual benefits of investments in abatement technologies were at least five times higher than the virtual annual abatement costs. A key barrier to investment in abatement technologies was the differentiated responsibility, as household health costs are borne by individuals, while the earnings from oil extraction accrue to the local authorities (Netalieva et al., 2005).

Accidental spills during oil-product transportation are damaging to the environment and health. There have been many spills at sea resulting in the destruction of fauna and flora, but the frequency of such incidents has declined sharply in recent times (Huijer, 2005). There are also spills originating from cracks in pipelines due to failure or sabotage. For example, it was estimated that the trans-Ecuadorian pipeline alone has spilt 400,000 litres of crude oil since it opened in 1972. Spills at oil refineries are also not uncommon. Verweij (2003) reported that in South Africa more than one million litres of petrol leaked from the refinery pipeline systems into the soil in 2001, thus contaminating ground water. One of the most recent oil spills occurred in Nanchital, Mexico in December 2004, where it was estimated that 5000 barrels of crude oil spilled from the pipeline with much of it going into the Coatzacoalcos River. Pemex, the company owning the pipeline, indicated a willingness to compensate the more than 250 local fishermen and the owners of the 200 hardest-hit homes. Coal mining is also hazardous with many thousands of fatalities each year. Exposure to coal dust has also been associated with accelerated loss of lung function (Beeckman and Wang, 2001).

4.5.4.2 Equity and shared responsibility

Economies with a high dependence on oil exports tend to have a poorer economic performance (Leite and Weidmann, 1999). The local energy needs of the host countries may be overlooked by their governments in the quest for foreign earnings from energy exports. Inadequate returns to the energy resource-rich communities have resulted in organized resistance against oil-extraction companies. Insecurities associated with oil supplies also result in high military expenditure as shown by OPEC countries (Karl and Gary, 2004).

The advent of reform in the energy sector increases inequalities. Notably electricity tariffs have generally shifted upwards after commencement of reforms (Wamukonya, 2003; Dubash, 2003) making electricity even more inaccessible to the lower-income earners. There are many genuine efforts to address such issues (World Bank, 2005), although much still needs to be done (Lort-Phillips and Herringshaw, 2006). Companies whose origin countries have stringent mandatory disclosure requirements are reported to perform best on transparency. Public private partnerships in developing countries are starting to make inroads into the issue of inequity and to harmonize practices between the developed and developing world. One such example is the Global Gas Flaring Reduction Partnership (World Bank, 2004a) aimed at reducing wasteful flaring and conserving the hydrocarbon resources for utilization by the host country.

4.5.4.3 Barriers to providing energy sources for sustainable development

The high investment cost required to build energy-system infrastructure is a major barrier to sustainable development. The IEA (2004a) estimated that 5 trillion US$ will be needed to meet electricity demand in developing countries by 2030. To meet all the eight Millennium Development Goals will require an annual average investment of 20 billion US$ to develop energy infrastructure and deliver energy services (UNDP, 2004b). Access to finance for investment in energy systems, especially in developing countries, has, nonetheless, been declining.

Available infrastructure also dictates energy types and use patterns. For instance, in a study on Peruvian household demand for clean fuels, Jack (2004) found that urban dwellers were more likely to use clean fuels than rural householders, due to the availability of the necessary infrastructure. Investment costs necessary to capture natural gas and divert it into energy systems and curb flaring and venting are a barrier, even though efforts are being made to overcome this problem (World Bank, 2004a). It is estimated that over 110 billion m^3 of natural gas are flared and vented worldwide annually, equivalent to the total annual gas consumption of France and Germany (ESMAP, 2004).

Levels of investment vary across regions, with the most needy receiving the least resources. Between 1990 and 2001, private investments to developing and transition countries for power projects were about 207 billion US$. Nearly 43% went to Latin America and the Caribbean, 33% to East Asia and the Pacific and approximately 1.5% to sub-Saharan Africa (Kessides, 2004). Accessibility and affordability of clean fuels remains a major barrier in many developing countries, exacerbated when complex supply systems are required that lead to high transaction costs.

Corruption, bureaucracy and mismanagement of energy resources have often prevented the use of proceeds emanating from extraction of energy resources from being used to provide local energy systems to meet sustainable development needs. Forms of corruption have encompassed such schemes as:

- the granting of lucrative power purchase agreements (PPAs) by politicians, who then benefit from receiving a share of guaranteed prices considerably higher than the international market price (Shorrock, 2002; Vallete and Wysham, 2002);
- suspending plant operations, thereby compromising access to electricity and persuading government agencies to pay high premiums for political risk insurance (Hall and Lobina, 2004); and
- granting of lucrative sole-supplier trading rights for gas supplies (Lovei and McKechnie, 2000).

Oil-backed loans have contributed to high foreign debts in many oil-producing countries at the expense of the poor majority (IMF, 2001; Global Witness, 2004). Despite heavy debts, such countries continue to sign for such loans (AEI, 2003) and potential revenues are used as collateral to finance government external debt rather than to reduce poverty or promote sustainable development. These loans are typically provided at higher interest rates than conventional concessionary loans (World Bank, 2004b) and so the majority of the local population fail to benefit from high oil prices (IRIN, 2004). The problem could be overcome by legal frameworks that enable the channelling of revenue into investments that provide energy systems and promote sustainable development in communities affected by energy-resource extraction. In the meantime, the problem remains a key barrier to sustainable development and, although several countries including Peru, Nigeria and Gabon have mandated enabling mechanisms for such transfers, progress in implementing these measures has been slow (Gary and Karl, 2003).

Poor policies in the international financing sector hinder the establishment of energy systems for sustainable development. A review of the extractive industries (World Bank, 2004b), for example, revealed that the World Bank group and the International Finance Corporation (IFC) have been investing in oil- and gas-extractive activities that have negative impacts on poverty alleviation and sustainable development. The review, somewhat controversially, recommended that the banks should pull out of oil, gas and coal projects by 2008.

Population growth and higher per-capita energy demand are forcing the transition of supply patterns from potentially sustainable systems to unsustainable ones. Efficient use of biomass can reduce CO_2 emissions, but can only be sustained if supplies are adequate to satisfy demand without depleting carbon stocks by deforestation (Section 4.3.3.3). If supplies are inadequate, it may be necessary to shift demand to fossil fuels to prevent overharvesting. In Niger, for example, despite the concerted efforts through a long-term World Bank funded project, it is not possible to provide sufficient woody biomass on a sustainable basis. As a result, the government has launched a campaign to encourage consumers, particularly industry, to shift from wood to coal and has re-launched a 3000 t/yr production unit, distributed 300 t of coal to Niamey, and produced 3800 coal-burning stoves (ISNA, 2004). Further, in the electricity sector, PPAs that are not favourable to the establishment of generation plants that promote sustainable development are increasingly common. These include long-term PPAs with payments made in foreign currency denominations, leaving the power sector extremely vulnerable to macro-economic shocks as demonstrated by the 1998 Asian crisis (Wamukonya, 2003).

4.5.4.4 Strategies for providing energy for sustainable development

Although the provision of improved energy services is not mentioned specifically in the formal Millennium Development Goals (MDGs) framework, it is a vital factor. Electrification and other energy-supply strategies should target income generation if they are to be economically sustainable. It is important to focus on improving productive uses of energy as a way of contributing to income generation by providing services and not as an end in themselves. It has been argued that the traditional top-down approaches to reform the power sector – motivated by macroeconomic factors and not aimed at improving access for the poor – should be replaced by bottom-up ones with communities at the centre of the decision process (GNESD, 2006).

4.5.5   Vulnerability and adaptation

It is essential to look at how the various components of the energy-supply chain might be affected by climate change. At the same time, it is desirable to assess current adaptation measures and their adequacy to handle potential vulnerability. A robust predictive skill is required to ensure that any mitigation programmes adopted now will still function adequately if altered climatic conditions prevail in the future.

Official aid investments in developing countries are often more focused on recovery from disaster than on the creation of adaptive capacity. Lending agencies and donors will need to reform their investment policies accordingly to mitigate this problem (Monirul, 2004). Many developing countries are particularly vulnerable to extremes of normal climatic variability that are expected to be exacerbated by climate change. Assessing the vulnerability of energy supply to climatic events and longer-term climate change needs to be country- or region-specific. The magnitude and frequency of extreme weather events such as ice storms, tornadoes and cyclones is predicted to change, as may annual rainfall, cloud cover and sunshine hours. This is likely to increase the vulnerability of the various components of the energy-supply infrastructure such as transmission lines and control systems.

Sea-level rise, tropical cyclones and large ocean waves may hamper offshore oil and gas exploration and extraction of these fossil fuels. Higher ambient temperatures may affect the efficiency and capacity ratings of fossil-fuel-powered combustion turbines. In addition, electricity transmission losses may increase due to higher ambient temperatures. Renewable-energy systems may be adversely affected (Sims, 2003), for example if solar power generation and water heating are impacted by increased cloud cover. Lower precipitation and higher evaporation due to higher ambient temperatures may cause lower water levels in storage lakes or rivers that will affect the outputs of hydro-electric power stations. Energy crop yields could be reduced due to new pests and weather changes and more extreme storm events could damage wind turbines and ocean energy devices. The need to take measures to lessen the impacts on energy systems resulting from their intrinsic vulnerability to climate change will remain a challenge for the foreseeable future.

4.5.6   Technology Research, Development, Demonstration, plus Deployment (RD³)

Future investments in RD³ will, in part, determine:
- future security of energy supplies;
- accessibility, availability and affordability of desired energy services;
- attainment of sustainable development;
- free-market distribution of energy supplies to all countries;
- deployment of low-carbon energy carriers and conversion technologies;

- the quantities of GHGs emitted for the rest of this century; and
- achievement, or otherwise, of GHG stabilization concentration levels.

Technology can play an important role in reducing the energy intensity of an economy (He and Zhang, 2006; He et al., 2006). In addition to new and improved energy-conversion technologies, such concepts as novel supply structures, distributed energy systems, grid optimization techniques, energy transport and storage methods, load management, co-generation and community-based services will have to be developed and improved (Luther, 2004). The knowledge base required to transform the energy supply and utilization system will then need to be created and expanded.

Major innovations that will shape society will require a foundation of strong basic research (Friedman, 2003). Areas of generic scientific research in material-, chemical-, bio-, and geo-sciences that could be particularly important to energy supply need to be reviewed. Progress in basic research could lead to new materials and technologies that can radically reduce costs or reveal new approaches to providing energy services. For example, the development of fibre optics from generic research investment resulted in their current use to extract greater volumes of oil or gas from a reservoir than had been previously possible.

Cross-disciplinary collaborations between many scientific areas, including applied research and social science, are needed for successful introduction of new energy supply and end-use technologies necessary to combat the unprecedented challenge of supporting human growth and progress while protecting global and local environments. Integrating scientific progress into energy and environmental policies is difficult and has not always received the attention it deserves (IEA, 2003a). Successful introduction of new technologies into the market requires careful coordination with governments to encourage, or at least not to hinder, their introduction. There is no single area of research that will secure a reliable future supply of energy. A diverse range of energy sources will be utilized and hence a broad range of fundamental research will be needed. Setting global priorities for technology development should be based on quantitative assessments of possible emissions and their abatement paths, but guidelines would first need to be developed (OECD, 2006a).

4.5.6.1   Public and private funding

Almost all (98%) of total OECD energy R&D investment has been by only ten IEA member countries (Margolis and Kammen, 1999; WEC, 2001). The amount declined by 50% between the peak of 1980 (following the oil price shocks) and 2002 in real terms (Figure 4.32). Expenditure on nuclear technologies, integrated over time, has been many times higher than investment in renewable energies. The end of the cold

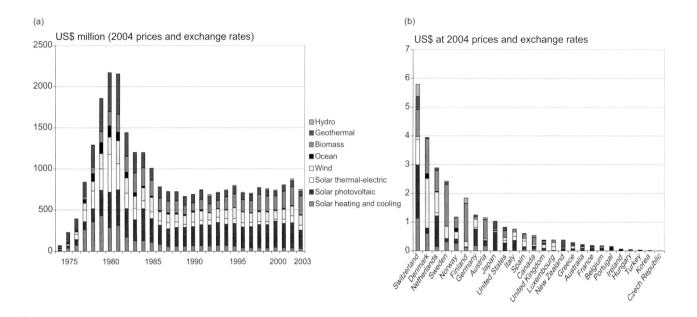

Figure 4.32: *IEA member government budgets for total renewable energy R&D annual investments for 1974–2003 (left, a) and investment per capita, averaged between 1990 and 2003 (right, b).*
Source: IEA, 2006d.

war and lower fossil-fuel prices decreased the level of public attention on energy planning in the 1980s, and global energy R&D investment has yet to return to these levels despite growing concerns about energy security and climate change (Chapter 13).

Ultimately, it is only by creating a demand-pull market (rather than supply-push) that technological development, learning from experience, economies of scale in production and related cost reductions can result. As markets expand and new industries grow (the wind industry for example), more private investment in R&D results, which is often more successful than public research (Sawin, 2003b).

The private sector invests a significant amount in energy RD[3] to seek competitive advantage through improved technology and risk avoidance in relation to commercialization. Firms tend to focus on incremental technology improvements to gain profits in the short term. R&D spending by firms in the energy industry is particularly low with utilities investing only 1% of total sales in US, UK and the Netherlands compared with the 3% R&D-to-sales ratio for manufacturing, and up to 8% for pharmaceutical, computer and communication industries.

If government policies relating to strategic research can ensure long-term markets for new technologies, then industries can see their potential, perform their own R&D and complement public research institutions (Luther, 2004). Fixed pricing laws to encourage the uptake of new energy-supply technologies have been successful but do not usually result in novel concepts. Further innovation is encouraged once

manufacturers and utilities begin to generate profits from a new technology. They then invest more in R&D to lower costs and further increase profit margins (Menanteau *et al.*, 2003). Under government mandatory quota systems (as used to stimulate renewable energy projects in several countries – Section 4.5.1), consumers tend to benefit the most and hence producers receive insufficient profit to invest in R&D.

Recent trends in both public and private energy RD[3] funding indicate that the role of 'technology push' in reducing GHG emissions is often overvalued and may not be fully understood. Subsidies and externalities (both social and environmental) affect energy markets and tend to support conventional sources of energy. Intervention to encourage R&D and adoption of renewable energy technologies, together with private investment and the more intelligent use of natural and social sciences is warranted (Hall and Lobina, 2004). Obtaining a useful balance between public and private research investment can be achieved by using partnerships between government, research institutions and firms.

Current levels of public and private energy-supply R&D investment are unlikely to be adequate to reduce global GHG emissions while providing the world with the energy needs of the developing nations (Edmonds and Smith, 2006). Success in long-term energy-supply R&D is associated with near-term investments to ensure that future energy services are delivered cost-effectively and barriers to implementation are identified and removed. Sustainable development and providing access to modern energy services for the poor have added challenges to R&D investment (IEA, 2004a; IEA 2006a; Chapter 13).

REFERENCES

AEI, 2003: $1.15 billion oil backed fresh cash, Africa Energy Intelligence. American Enterprises Institute for public policy research, Washington D.C.

Andrews, S. and R. Udal, 2003: Oil prophets: looking at world oil studies over time. Association for Study of Peak Oil Conference, Paris.

ANSTO, 2006: Australia's possible role in the nuclear fuel cycle, ANSTO's submission to the uranium mining, processing and nuclear energy review, <http://www.pmc.gov.au/umpner/submissions/211_sub_umpner.pdf> accessed 05/06/07.

Archer, C.L. and M.Z. Jacobsen, 2005: Evaluation of global wind power. *Journal of Geophysical Research*, **110**(D12110), doi:10.1029/2004JD005462.

Aringhoff, R., C. Aubrey, G. Brakmann, and S. Teske, 2003: Solar thermal power 2020, Greenpeace International/European Solar Thermal Power Industry Association, Netherlands.

Austin, G., A. Williams, G. Morris, R. Spalding-Feche, and R. Worthington, 2003: Employment potential of renewable energy in South Africa. Earthlife Africa, Johannesburg and World Wildlife Fund (WWF), Denmark, November, 104 pp.

Awerbuch, S. and R. Sauter, 2005: Exploiting the oil-GDP effect to support renewables deployment. Paper No. 129, SPRU Electronic Working Paper Series, <http://www.sussex.ac.uk/spru/> accessed 05/06/07.

Bakker, M., H.A. Zondag, M.J. Elswijk, K.J. Strootman, and M.J.M. Jong, 2005: Performance and costs of a roof-sized PV/thermal array combined with a ground coupled heat pump. *Solar Energy*, **78**, pp. 331-339.

Bailis, R., M. Ezzati, and D.M. Kammen, 2005: Mortality and greenhouse gas impacts of biomass and petroleum energy futures in Africa. *Science*, **308**(5718), pp. 98-103.

Barreto, L., 2001: *Technological learning in energy optimisation models and deployment of emerging technologies.* PhD thesis, Swiss Federal Institute of Technology, Zurich, Switzerland.

Beeckman, L.A. and M.L. Wang, 2001: Rapid declines in FEVI and subsequent respiratory symptoms, illness and mortality in coal miners in the United States. *American Journal of Respiratory and Critical Care Medicine*, **163**(3 Pt 1), pp. 633-639.

Beg, N., 2002: Ancillary benefits and costs of GHG mitigation: policy conclusion. Publication Service, OECD, Paris, France, 43 pp.

Bendtsen, B., 2003: Parliamentary answer to question S4640, Danish Parliament, September 2, <http://www.ft.dk/Samling/20021/spor_sv/S4640.htm> accessed 05/06/07.

Bentley, R.W., 2002a: Oil forecasts past and present. *Energy Exploration and Exploitation*, **20**(6), pp. 481-492.

Bentley, R.W., 2002b: Global oil and gas depletion: an overview. *Energy Policy*, **30**, pp. 189-205.

Bentley, R.W., 2005: Global oil depletion - methodologies and results. 4th International Workshop on oil depletion. Lisbon, May 2005.

Bertani, R., 2005: Worldwide geothermal generation 2001-2005 state of the art. Proceedings of the World Geothermal Congress 2005: April, Antalya, Turkey.

BIREC, 2005: Beijing International Renewable Energy Conference, Beijing, November, <www.birec.ch> accessed 05/06/07.

Bohm, M.C., 2006: *Capture-ready power plants - Options, technologies and economics*, MSc Thesis, MIT. <http://sequestration.mit.edu/pdf/Mark_Bohm_Thesis.pdf>, accessed 05/06/07.

BP, 2004: Statistical review of world energy, BP Oil Company Ltd., London. <www.bp.com/subsection.do?categoryId=95&contentId=2006480> accessed 05/06/07.

BP, 2005: BP Statistical Review of world energy. BP Oil Company Ltd., London.

BP, 2006: BP Statistical review of world energy. BP Oil Company Ltd., London. <www.bp.com> accessed 05/06/07.

Bradshaw, J., S. Bachu, D. Bonijoly, R. Burruss, S. Holloway, N.P. Christenssen, and O.M. Mathiassen, 2006: CO_2 storage capacity estimation: issues and development of standards. Paper at the 8th Greenhouse gas technologies conference, June 19-22, 2006: Trondheim, Norway.

Brendow, K., 2004: World coal perspectives to 2030. World Energy Council, Geneva and London.

Broek, R. van den, 2000: *Sustainability of biomass electricity systems - An assessment of costs, macro-economic and environmental impacts in Nicaragua, Ireland and the Netherlands*. PhD Thesis, Utrecht University, 215 pp.

Bromley, C.J. and S. Currie, 2003: Analysis of subsidence at Crown Road, Taupo: a consequence of declining groundwater. Proceedings of the 25th New Zealand Geothermal Workshop, Auckland University, pp. 113-120.

Bruce, N., R. Perez-Padilla, and R. Albalak, 2000: Indoor air pollution in developing countries: a major environmental and public health challenge. *Bulletin of the World Health Organization*, **78**, Geneva, Switzerland, pp. 1078-1092.

BTM, 2006: World Market Update 2005: International wind energy development status by end of 2005 and Forecast 2006-2010, BTM Consult, Aps, Denmark, <www.btm.dk> accessed 05/06/07.

Burtraw, D. and M. Toman, 2000: Ancillary benefits of greenhouse gas mitigation policies. *Climate Change Issues Brief*, **7**, Resources for the Future, Washington DC

CAPP, 2006: Canadian Association of Petroleum Producers, <www.capp.ca/raw.asp?x=1&dt=NTV&e=PDF&dn=103586> accessed 05/06/07.

Carbon Trust, 2005: Marine energy challenge. <www.carbontrust.co.uk> accessed 05/06/07.

Chae, Y. and C. Hope, 2003: Integrated assessment of CO_2 and SO_2 policies in North East Asia. *Climate Policy*, **3**, Supplement 1, pp. S57-S79.

Chen, W., 2005: The costs of mitigating carbon emissions in China: findings from China MARKAL-MACRO modeling. *Energy Policy*, **33**, pp. 885-896.

CIEP, 2004: Study on energy supply security and geopolitics. Final report, Clingendael International Energy Programme, Institute for International Relations 'Clingendael', The Hague, The Netherlands.

Cleland, D., 2005: Sustainable energy use and management, Proceedings of the conference; People and Energy - how do we use it? Christchurch, Royal Society of New Zealand, Miscellaneous series, **66**, pp. 96-102.

CSY, 2005: China Statistical Yearbook, National Bureau of Statistics of China.

Collier, U., 2006: Meeting Africa's energy needs - the costs and benefits of hydro power, Worldwide Fund for Nature, Oxfam and Wateraid joint report <http://allafrica.com/stories/200603060496.html> accessed 05/06/07.

Cook, I., N. Taylor, D. Ward, L. Baker, and T. Hender, 2005: Accelerated development of fusion power. Report UKAEA FUS521, EURATOM/UKAEA.

CSIRO, 2005: Coal preparation, ultra clean coal. <http://www.det.csiro.au/science/lee_cc/ultra_clean_coal.htm> assessed 08/08/07.

Danish Energy Authority, 2005: Technology data for electricity and heat generation plants, <http://www.ens.dk/graphics/Publikationer/Forsyning_UK/Technology_Data_for_Electricity_and_Heat_Generating_Plants/index.htm> accessed 05/06/07.

Datta, K.E., T. Feiler, K.R. Rábago, N.J. Swisher, A. Lehmann, and K. Wicker, 2002: Small is profitable. Rocky Mountain Institute, Snowmass, Colorado, ISBN 1-881071-07-3. <www.rmi.org>, accessed 05/06/07.

de Moor, A., 2001: Towards a grand deal on subsidies and climate change. *Journal Natural Resources Forum*, **25**(2), May.

Dellero, N. and F. Chessé, 2006: New nuclear plant economics. IYNC 2006 conference paper 234, Stockholm, Sweden - Olkiluoto, Finland, 18-23 June, AREVA, World Nuclear Association, <http://www.world-nuclear.org/info/inf63.htm> accessed 05/06/07.

Delmas, R., 2005: Long term greenhouse gas emissions from the hydroelectric reservoir of Petit Saut (French Guiana) and potential impacts. Global Warming and Hydroelectric Reservoirs, CDD 363.73874, pp. 117-124.

DGEMP, 2006: Radioactive Materials and Waste, Planning Act of 28 June 2006, National Policy for the Sustainable Management of Radioactive Materials and Waste. Direction generale de l'énergie et des matieres primaires (DGEMP), France.

DME, 2003: White paper on renewable energy. Department of Minerals and Energy, Pretoria, South Africa.

Dones, R, T. Heck, M. Faist Emmenegger, and N. Jungbluth, 2005: Life cycle inventories for the nuclear and natural gas energy systems, and examples of uncertainty analysis. *International Journal LCA,* **10**(1), pp. 10-23.

dos Santos, A.M, L.P. Rosaa, B. Sikard, E. Sikarb, and E.O. dos Santos, 2004: Gross greenhouse gas fluxes from hydro-power reservoirs compared to thermal power plants. aIVIG/COPPE/UFRJ and Energy Planning Program/COPPE/UFRJ, University of Grande Rio and University of Sao Paulo, Sao Carlos, Brazil.

dos Santos, M.A. (ed.), 2005: Carbon budget in tropical reservoirs. *Global Warming and Hydroelectric Reservoirs*, CDD 363.73874, pp 95-100.

Dooley, J.J., R.T. Dahowski, C.L. Davidson, S. Bachu, N. Gupta and J. Gale, 2005: IEA greenhouse gas programme. Proceedings of 7th International Conference on Greenhouse Gas Control Technologies, *Volume 1: Peer-Reviewed Papers and Plenary Presentations* (E.S. Rubin, D.W. Keith and C.F. Gilboy, eds.), Cheltenham, UK, <http://www.ghgt7.ca> accessed 02/07/07

Dow Jones, 2006: Kratwerk mit pilotanlage fur CO_2 abscheidung in Beitrib. *Dow Jones Trade News Emissions*, 19 May, **10**, 11 pp.

DTI, 1999: Coal liquefaction - technology status report. Department of Trade and Industry, London, UK.

Dubash, N.K., 2003: Revisiting electricity reform: the case for a sustainable development approach, Utilities Policy, **11**, pp. 143-154

Duncan, A., 2005: *Solar building developments.* Master Applied Science thesis, Massey University Library, Palmerston North, New Zealand.

EC, 2005: The support of electricity from renewable energy sources. European Commission COM(2005) 627 final, Brussels, 50 pp.

EC, 2007: World energy technology outlook 2050 (WETO-H2) Report. Directorate-General for Research, European Commission, January, <http://ec.europa.eu/research/energy/gp/gp_pu/article_1100_en.htm> accessed 05/06/07.

Edmonds, J. and S. Smith, 2006: The technology of two degrees, 'Avoiding Dangerous Climate Change'. H.J. Schellenhuber, W. Cramer, N. Nakicenovic, T. Wrigley, and G. Yohe (eds), Cambridge University Press, pp. 385-391.

EDUCOGEN, 2001: A guide to cogeneration, produced under the auspices of Contract N° XVII/4.1031/P/99-159. The European Association for the Promotion of Cogeneration, 51 pp. <www.cogen.org> accessed 05/06/07.

EEA, 2004: Energy subsidies in the European Union - a brief overview. European Environment Agency, Technical report, Office for Official Publications of the European Communities, Luxembourg, 20 pp.

EERE, 2005: A national vision of America's transition to a hydrogen economy - to 2030 and beyond. <http:/www.eere.energy.gov/hydrogenandfuelcells/annual_report.html> accessed 05/06/07.

EIA/DOE, 2005: Energy Information Administration, U.S. Department of Energy <www.eia.doe.gov/cabs/World_Oil_Transit_Chokepoints/Hormuz.html> and <www.eia.doe.gov/cabs/World_Oil_Transit_Chokepoints/Malacca.html> accessed 05/06/07.

EIA/DOE, 2006: International Energy Outlook 2006. <www.eia.doe.gov/oiaf/ieo/index.html> accessed 05/06/07.

Encyclopaedia of Energy, 2004: C.J. Cleveland (ed.), Elsevier Academic Press, Oxford. **4**, pp. 257-272.

Enerdata, 2004: A global energy data base including OECD data. Grenoble, France.

Energy Policy Act, 2005: http://energycommerce.house.gov/ accessed 05/06/07.

Environment California, 2006: <www.environmentcalifornia.org/newsroom/energy/energy-program-news/million-solar-roofs-bill-sb-1-signed-into-law#1H202IC_O9f0derEDnExjw> accessed 05/06/07.

EPA, 2003: SF6 emissions reduction partnership for electric power systems. 2002 Partnership Report, <http://www.epa.gov/highgwp/electricpower-sf6/pdf/2002-partnership-report.pdf> accessed 05/06/07.

EPA, 2005a: Energy Star. Environmental Protection Agency, <http://www.energystar.gov/> accessed 05/06/07.

EPA, 2005b: Natural gas star program. Environmental Protection Agency, <http://www.epa.gov/gasstar/> accessed 05/06/07.

EPA, 2005c: Coalbed methane outreach program. Environmental Protection Agency, <http://www.epa.gov/cmop/> accessed 05/06/07.

EPRI, 2003: Electricity technology roadmap, 2003 summary and synthesis. Report No.1010929, Electric Power Research Institute, Palo Alto, California, USA.

Equitech, 2005: Equitech International. LLC, <http://www.equitechllc.com/papers/carbonemissions.html> accessed 05/06/07.

ERC, 2004: Solar electrification by the concession approach in the rural Eastern Cape; Phase 2 monitoring survey November 2003: <www.erc.uct.ac.za> accessed 05/06/07.

ESMAP, 2004: Regulation of associated gas flaring and venting: a global overview and lessons from international experience. Energy Sector Management Assistance Programme (ESMAP) report 285/04, World Bank, Washington D.C.

ESTIA, 2004: Exploiting the heat from the sun to combat climate change. European Solar Thermal Industry Association and Greenpeace, Solar Thermal Power 2020, UK.

EU, 2001: Promotion of electricity from renewable energy sources in the internal electricity market. Directive 2001/77/EC of the European Parliament and of the Council, *Official Journal of the European Communities*, 27 October, pp. 33-40.

EU, 2003: Directive 2003/30/EC of the European Parliament and of the Council on the promotion of the use of biofuels or other renewable fuels for transport. *Official Journal of the European Communities*, 17.5.2003: pp. 42-46.

EU, 2005: Externalities of energy: extension of accounting framework and policy applications (ExternE-Pol). Final Report of project, <http://www.externe.info/expoltec.pdf> accessed 05/06/07.

EUFORES, 2004: The impact of renewables on employment and economic growth. European Commission ALTENER program, <http://www.eufores.org/> accessed 05/06/07.

EWEA, 2004: Wind energy - the facts - an analysis of wind energy in the EU-25. European Wind Energy Association, Brussels.

EWEA, 2005: Large scale integration of wind energy in the European power supply system: analysis, issues and recommendations. <http://www.ewea.org/fileadmin/ewea_documents/documents/publications/grid/051215_Grid_report_summary.pdf> accessed 05/06/07.

EEA, 2005: Greenhouse gas emission trends and projections in Europe 2005: European Environment Agency. <http://reports.eea.eu.int> accessed 05/06/07.

FACTS Inc., 2005: Gas data book: Asia-Pacific natural gas & LNG. Honolulu, Hawaii.

Fearnside, P.M., 2004: Greenhouse gas emissions from hydroelectric dams: controversies provide a springboard for rethinking a supposedly clean energy source: Editorial Comment. *Climatic Change*, **66**, pp. 1-2.

Fisher, B., 2006: Industry related technology research. Proceedings of the Expert Review Meeting, Intergovernmetnal Panel on Climate Change (IPCC) WG III, Capetown, 17-19 January.

Fouquet, D., C. Grots, J. Sawin, N. Vassilakos, 2005: Reflections on a possible unified EU financial support scheme for renewable energy systems (RES): a comparison of minimum-price and quota systems and an analysis of market conditions. Brussels and Washington D.C., January 2005, 26 pp.

Friedman, 2003: Will innovation flourish in the future? *The Industrial Physicist*, January, pp. 22-25.

Gary, I. and T.L. Karl, 2003: Bottom of the barrel: Africa oil boom and the poor. Report Catholic Relief Services, Baltimore MD, 110 pp.

Gehl, S., 2004: Generation technology choices: near and long term. US DOE/EPRI Annual Energy Outlook Conference, Washington D.C., www.eia.doe.gov/oiaf/archive/aeo04/conf/pdf/gehl.pdf accessed 05/06/07.

Gibbins, J., S. Haszeldine, S. Holloway, J. Pearce, J. Oakey, S. Shackley, and C. Turley, 2006: Scope for future CO_2 emission reductions from electricity generation through the deployment of carbon capture and storage technologies. In *Avoiding Dangerous Climate Change*, H.J. Schellnhuber (ed.), Chapter 40, Cambridge University Press, ISBN: 13 978-0-521-86471-8 hardback, ISBN: 10 0-521-86471-2 paperback, 379 pp. <http://www.defra.gov.uk/environment/climatechange/research/dangerous-cc/index.htm> accessed 05/06/07.

Giebel, G., 2005: Wind power prediction using ensembles. Risø National Laboratory, ISBN 87-550-3464-0, September, 43 pp.

GIF, 2002: Technology roadmap for generation IV nuclear energy systems. US DOE Nuclear Energy Research Advisory Committee and the Generation IV International Forum (GIF), December. <http://gif.inel.gov/roadmap/pdfs/gen_iv_roadmap.pdf> accessed 05/06/07.

GGFR, 2004: Global gas flaring reductions - a public/private partnership. Report number 6, September.

Global Witness, 2004: Global policy forum (GPF), New York USA, <http://www.globalpolicy.org/> accessed 05/06/07.

GNESD, 2006: Electricity and development in Africa, Asia and Latin America. Consolidated Report of regional workshop on Global Network on Sustainable Development (GNESD), January, <www.gnesd.org/Downloadables/GnesdRegionalWorkshop.pdf> accessed 05/06/07.

Goldemberg, J., 2004: The case for renewable energies. Thematic background paper, International Conference for Renewable Energies, Bonn, June, <www.renewables2004.de> accessed 05/06/07.

Goldemberg, J. and T.B. Johansson, 2004: World Energy Assessment; Overview 2004 update. United Nations Development Programme, USA, 88 pp.

Goldemberg, J., T.B. Johansso, A.K.N. Reddy, and R.H. Williams, 2004: A global clean cooking fuel initative. *Energy for Sustainable Development*, **8**(3), pp. 5-12.

Greenpeace, 2006: Solar generation. K. McDonald (ed.), Greenpeace International, Amsterdam.

Greenfuels, 2004: Greenfuels Technology Corp. announces first field results from MIT Cogeneration plant beta test site. CK Test report, <http://www.greenfuelonline.com/gf_files/CK_Test_Report.pdf> accessed 05/06/07.

Greenshift, 2005: Greenshift acquires rights to patented carbon dioxide reduction technology. <http://www.greenshift.com/news.php?id=97> accessed 05/06/07.

Greenpeace, 2004: http://www.greenpeace.org.ar/cop10ing/SolarGeneration.pdf accessed 05/06/07.

Gul, T., and T. Stenzel, 2005: Variability of wind power and other renewables - management options and strategies. International Energy Agency working paper, OECD/IEA, Paris.

GWEC, 2006: Global wind energy outlook. Global Wind Energy Council, Bruxelles and Greenpeace, Amsterdam, September, 56 pp., <www.gwec.net> accessed 05/06/07.

Haas, R., 2002: Building PV markets: the impact of financial incentives. *Journal Renewable Energy World*, **5**(4), pp. 184-201.

Hagen, R.E, J.R. Moens, and Z.D. Nikodem, 2005: Impact of U.S. nuclear generation on greenhouse gas emissions. Energy Information Administration, U.S. Department of Energy, <http://tonto.eia.doe.gov/FTPROOT/nuclear/ghg.pdf> accessed 05/06/07.

Hall, C., P. Tharakan, J. Hallock, C. Cleveland, and J. Jefferson, 2004: Hydrocarbons and the evolution of human culture. November, *Nature*, **426**, pp. 318-322, <www.nature.com/nature> accessed 05/06/07.

Hall, D.O., 2003: Electricity in Latin America. Public Services International Research Unit (PSIRU), London UK, 16 pp.

Hall, D.O. and E. Lobina, 2004: Private and public interests in water and energy. *Natural Resources Forum* **28**, pp. 286-277.

Hallock, J.L, Tharakan P.J., C.A.S. Hall, M. Jefferson, and W. Wu, 2004: Forecasting the limits to the availability and diversity of global conventional oil supply. *Energy*, **29**, pp. 1673-1696, Elsevier.

He, J.K., A.L. Zhang, and C.S. Shang, 2006: Assessment of GHG mitigation using INET energy systems model. *Journal of Tsinghua University*, **36**(10), pp. 68-73.

He, J.K. and X.L. Zhang, 2006: Analysis of China's energy consumption intensity reduction tendency during the 11th Five-Year-Plan period. *China Soft Science*, **184**, pp. 33-38, April.

Helynen, S., 2005: Distributed generation for sustainable development. Proceedings on Technologies for Sustainable Energy Development in the Long Term, L. Peterson and H. Larsen (eds.), Risoe-R-1517(EN), 459 pp., <www.risoe.dk/rispubl/SYS/ris-r-1517.htm> accessed 05/06/07.

Hendriks, C., W. Graus and F. van Bergen, 2004: Global carbon dioxide storage potential. Ecofys, TNO, Utrecht, Netherlands.

Herzog, A.V., T.E. Lipman, J.L. Edwards, and D.M. Kammen, 2001: Renewable energy: a viable choice. *Environment*, **43**(10).

Hirsch, R.L., R. Bezdek, and R. Wendling, 2005: Peaking of world oil production: impacts, mitigation, and risk management. Report to the American Administration, Washington D.C.

Hondo, H., 2005: Life cycle GHG emission analysis of power generation system: Japanese case. *Energy*, **30**, pp. 2042-2056.

Huijer, K., 2005: Trends in oil spills from tanker ships 1995-2004: International Tanker Owners Pollution Federation (ITOPF). London.

Huttunen, J., 2005: Long-term net methane release from Finnish hydro reservoirs. *Global Warming and Hydroelectric Reservoirs*, CDD 363.73874, pp. 125-127.

Hydro Tasmania, 2005: Sustainable guidelines for hydro power developments. Produced by Hydro Tasmania for the International Hydro Power Association.

IAEA, 1997: Joint convention on the safety of spent fuel management and on the safety of radioactive waste management. INFCIRC/546, 24 December 1997 (entered into force on 18 June 2001); <http://www.iaea.org/Publications/Documents/Infcircs/1997/infcirc546.pdf> accessed 05/06/07.

IAEA, 1994: Convention on nuclear safety. International Atomic Energy Agency <http://www.iaea.org/Publications/Documents/Infcircs/Others/inf449.shtml> accessed 05/06/07.

IAEA, 2005a: Thorium fuel cycle - potential benefits and challenges. International Atomic Energy Agency, Vienna 2005: <www-pub.iaea.org/MTCD/publications/PDF/TE_1450_web.pdf> accessed 05/06/07.

IAEA, 2005b: Draft safety requirements: geological disposal of radioactive waste. Report no. WS-R-4, International Atomic Energy Agency, Vienna 2005, <http://www-pub.iaea.org/MTCD/publications/PDF/Pub1231_web.pdf> accessed 05/06/07.

IAEA, 2005c: Energy, electricity and nuclear power estimates for the period up to 2030. Reference Data Series No.1, International Atomic Energy Agency, ISBN 92-0-1011-4289, July, Vienna, Austria.

IAEA, 2006: Power reactor information system (PRIS). <http://www.iaea.org/programmes/a2/> accessed 05/06/07.

IEA, 2000: Experience curves for energy policy assessment. International Energy Agency, Paris, pp. 127.

IEA, 2001: Impact of sustained high oil prices on developing countries. International Energy Agency Economic Analysis Division Working Party, Publication Service, OECD, Paris, France.

IEA, 2003a: From oil crisis to climate challenge - understanding CO_2 emission trends in IEA countries. International Energy Agency, OECD Publication Service, Paris, 22 pp.

IEA, 2003b: Economies in transition, the IEA and renewable energy. Public Information Forum, International Energy Agency, Renewable Energy Working Party, Budapest, Hungary, OECD, Paris. <http://www.iea.org/dbtw-wpd/textbase/work/2003/budapest/background.pdf> accessed 05/06/07.

IEA, 2003c: Energy to 2050: scenarios for a sustainable future. International Energy Agency, OECD, Paris.

317

IEA, 2003d: Building the cost curves for the industrial sources of non-CO_2 greenhouse gases. Greenhouse Gas R&D Programme, Report No. PH4/25, OECD, Paris, <www.iea.org> accessed 05/06/07.

IEA, 2003e: Developments in the international coal market in 2001-2. Coal Industry Advisory Board, International Energy Agency, OECD, Paris, <http://www.iea.org/textbase/papers/2002/ciabmark.pdf> accessed 05/06/07.

IEA, 2003f: National Survey Report of PV Power Applications in Japan 2002. Prepared by Kiyoshi Shino, Osamu Ikki, Tokyo, Japan, 27, International Energy Agency, Paris.

IEA, 2003g: Proceedings of the seminar *Towards Hydrogen - R & D priorities to create a hydrogen infrastructure.* OECD, Paris.

IEA, 2003h: Technology innovation, development and diffusion. Information paper, International Energy Agency, OECD, Paris.

IEA, 2004a: World Energy Outlook 2004: International Energy Agency, OECD Paris.

IEA, 2004b: Renewable energy - market and policy trends in IEA countries. International Energy Agency, OECD, Paris. <www.iea.org/textbase/nppdf/free/2004/renewable1.pdf> accessed 05/06/07.

IEA, 2004c: Policies of IEA countries - 2004 review. International Energy Agency, Publication Service, OECD, Paris, 539 pp, <http://www.oecd.org/dataoecd/25/9/34008620.pdf> accessed 02/07/07.

IEA, 2004d: Renewable Energy Working Party workshop *Distributed generation- key issues, challenges, roles for its integration into mainstream energy systems,* 1 March 2004: International Energy Agency, OECD, Paris. <www.iea.org> accessed 02/07/07.

IEA, 2004e: Coming in from the cold - improving district heating policy in transition economies. International Energy Agency, OECD, Paris.

IEA Bioenergy, 2005: Benefits of bioenergy. International Energy Agency, <http://www.ieabioenergy.com/library/179_BenefitsofBioenergy.pdf> accessed 02/07/07.

IEA, 2005a: World Energy Outlook 2005: Middle East and North Africa Insights. International Energy Agency, OECD, Paris.

IEA, 2005b: International Energy Agency, OECD, Paris, <http://www.iea.org/Textbase/stats/> accessed 02/07/07.

IEA, 2005c: Renewables information 2005. International Energy Agency, OECD Paris.

IEA, 2005d: Energy to 2050: scenarios for a sustainable future. International Energy Agency, OECD Paris.

IEA, 2005e: Resources to reserves: oil and gas technologies for the energy markets of the future. International Energy Agency, Paris (books@iea.org).

IEA, 2005f: Hydrogen implementing agreement. International Energy Agency, OECD, Paris.

IEA, 2005g: Prospects for hydrogen and fuel cells. International Energy Agency, OECD, Paris.

IEA, 2006a : Global energy technology perspectives, International Energy Agency, OECD, Paris. <www.iea.org> accessed 02/07/07.

IEA, 2006b: World energy outlook 2006. International Energy Agency, OECD Publication Service, OECD, Paris. <www.iea.org> accessed 02/07/07.

IEA, 2006c: Key world energy statistics. <http://www.iea.org/textbase/nppdf/free/2006/key2006.pdf> accessed 02/07/07.

IEA, 2006d: Renewable energy RD & D priorities report. International Energy Agency, OECD, Paris. <www.iea.org> accessed 02/07/07.

IEA, 2006e: Policies and measures database. International Energy Agency, OECD, Paris, <www.iea.org> accessed 02/07/07.

IEA, 2006f: Renewable energy heating and cooling. Seminar proceedings, April, International Energy Agency, OECD, Paris. <www.iea.org> accessed 02/07/07.

IEA, 2006g : Contribution of renewables to energy security. Ölz S, Sims R E H and Kirchner N, Information report, International Energy Agency, OECD, Paris. <www.iea.org>

IEA, 2006h: Bioenergy Implementing Agreement. International Energy Agency, OECD Paris, <http://www.joanneum.at/iea-bioenergy-task38/publications/faq/> accessed 02/07/07.

IEA, 2006j: Prospects for hydrogen and fuel cells. International Energy Agency, OECD, Paris.

IEA/NEA, 2005: Projected costs of generating electricity - 2005 update. Joint Report, Nuclear Energy Agency and International Energy Agency, IEA/OECD, Paris.

IEEJ, 2005: Handbook of energy & economic statistics in Japan. The Energy Data and Modelling Centre, Institute of Energy Economics, Japan.

IIASA, 1998: Energy in a finite world, paths to a sustainable future. International Institute for Applied Systems Analysis, Vienna, Ballinger Publishing Company, ISBN 0-88410-641-1, 225 pp.

IIASA/WEC, 1998: Global Energy Perspectives. N. Nakicenovic, A. Grübler, and A. McDonald, (eds.), International Institute for Applied Systems Analysis, Vienna, and World Energy Council, Cambridge University Press, 1998.

IHA, 2006: Sustainability assessment protocol. International Hydropower Association, <www.hydropower.org> accessed 02/07/07.

IMF, 2001: International Monetary Fund, the Democratic Republic of Congo. 2001 article iv consultation and discussions on staff-monitored program, Country Report No. 01/114, Washington D.C.

IPCC, 1997: *Revised 1996 IPCC guidelines for national greenhouse gas inventories.* Intergovernmental Panel on Climate Change (IPCC), Volume 3.

IPCC, 2001: *Climate Change 2001: Mitigation* - Contribution of Working Group III to the Third Assessment Report of the Intergovernmental Panel on Climate Change (IPCC) [Metz, B., O. Davidson, R. Swart, and J. Pan (eds.)]. Cambridge University Press, Cambridge, 700 pp.

IPCC, 2005: *IPCC special report on Carbon dioxide Capture and Storage.* Prepared by Working Group III of the Intergovernmental Panel on Climate Change [Metz, B., O. Davidson, H.C. de Coninck, M. Loos and L.A. Meyer (eds.)], Cambridge University Press, Cambridge, United kingdom and New York, NY, USa, 442 pp.

IPCC, 2006: *IPCC Guidelines for national greenhouse gas inventories.* Report of the Intergovernmental Panel on Climate Change, Cambridge University Press, Cambridge. <http://www.ipcc-nggip.iges.or.jp/public/2006gl/index.htm> accessed 02/07/07.

IRIN, 2004: Angola: frustration as oil windfall spending neglects the poor. United Nations Integrated Regional Information Networks, <http://www.irinnews.org/report.asp?ReportID=43013&SelectRegion=Southern_Africa> accessed 02/07/07.

ISES, 2005: Proceedings on International Solar Energy Society (ISES). 2005 Solar World Congress, D.Y. Goswami, S. Vijayaraghaven, and R. Campbell-Howe R (eds.), Orlando, Florida, August.

ISNA, 2004: From wood to coal in an effort to stop deforestation. Inter Services news agency (IPS), Rome, <http://www.ipsnews.net/interna.asp?idnews=25947> accessed 02/07/07.

ITDG, 2003: Sustainable energy for poverty reduction: an action plan. ISES, 2005: Proceedings of the International Solar Energy Society (ISES), 2005 Solar World Congress, D. Goswami, D.Y. Vijayaraghaven, and R. Campbell-Howe (eds.), Orlando, Florida, August.

ITER, 2006: International Thermonuclear Experimental Reactor (ITER). <http://www.iter.org/> accessed 02/07/07.

Jack, D., 2004: Income, household energy and health. Kennedy School of Government, Harvard MA, 16 pp.

JAEC, 2005: Framework for nuclear energy policy. Japan Atomic Energy Commission <http://www.aec.go.jp/jicst/NC/tyoki/tyoki_e.htm> accessed 02/07/07.

Jaffe, A.B., R.G. Newell, and R.N. Stavins, 2005: A tale of two market failures - technology and environmental policy. *Ecological Economics*, **54**, pp. 164-174.

Jäger-Waldau, A., 2003: PV status report 2003 - research, solar cell production and market implementation in Japan, USA and European Union. Joint Research Centre, European Commission.

Janssen, R., 2002: Renewable energy into the mainstream. International Energy Agency, Publication Service, OECD, Paris, France, 54 pp.

Jimison, J., 2004: The advantages of combined heat and power. Workshop on Distributed Generation and Combined Heat and Power, April, EPRI (Electric Power Research Institute), Palo Alto, California.

Jochem, E. and R. Madlener: 2002: The forgotten benefits of climate change mitigation: innovation, technological leapfrogging, employment and sustainable development. OECD Workshop on the Benefits of Climate policy: Improving Information for Policy Makers, 12-13 December, Paris, France.

Johansson, T.B., 2004: The potentials of renewable energy: thematic background paper, *Renewables, 2004*. International Conference for Renewable Energies, Bonn, http://www.renewables2004.de/pdf/tbp/TBP10-potentials.pdf > accessed 02/07/07.

Johansson, T.B. and W. Turkenburg, 2004: Policies for renewable energy in the European Union and its member states - an overview. *Energy for Sustainable Development*, **VIII**(1), pp. 5-24.

Johansson, T.B., K. McCormick, L. Neij, and W. Turkenburg, 2004: The potentials of renewable energy, Thematic background paper for International conference "Renewables2004", Bonn, June, <www.renewables2004.de> accessed 02/07/07.

Jönsson, 2004: [website] <http://www.novaenergie.ch/iea-bioenergy-task37/Dokumente/06%20biogasupgrading.pdf> accessed 02/07/07.

Kamman, D.M., K. Kapadia, and M. Fripp, 2004: Putting renewables to work: how many jobs can the clean energy industry generate? Renewable and Appropriate Energy Laboratory (RAEL) report, University of California, Berkeley, 24 pp.

Kantei, 2006: <http://www.kantei.go.jp/foreign/policy/kyoto/050428plan_e.pdf> accessed 02/07/07.

Karekezi, S. and W. Kithyoma, 2003: Renewable and rural energy overview in sub-Saharan Africa. African Energy Policy Research Network (AFREPREN)/FWD Secretariat, Nairobi, Kenya.

Karl, T.L. and I. Gary, 2004: The global record. Petropolitics special report, January 2004: Foreign Policy in Focus (FPIF), Silver City, New Mexico and Washington D.C., 8 pp.

Kessels, J.R. and S.J.A. Bakker, 2005: ESCAPE: energy security & climate policy evaluation. ECN-C--05-032, ECN, Petten, Netherlands, <http://www.ecn.nl/library/reports/2005/c05032.html> accessed 02/07/07.

Kessides, I.N., 2004: Reforming infrastructure privatisation regulation and competition. World Bank policy research report, a co-publication of World Bank and Oxford University Press, Oxford, UK.

Kirjavainen, M., T. Savola, and M. Salomon, 2004: Small scale biomass CHP technologies: situation in Finland, Sweden and Denmark. OPET report 12, contract to European Commission, Directorate General for Energy and Transport, <http://www.opet-chp.net/download/wp2/small_scale_biomass_chp_technologies.pdf > accessed 02/07/07.

Komiyama, R., Z. Li and K. Ito, 2005: World energy outlook in 2020 focusing on China's energy impacts on the world and Northeast Asia. *International Journal on Global Energy Issues*, **24**(3/4), pp. 183-210.

Koster, A., H.D. Matzner, and D.R. Nichols, 2003: PBMR design for the future. *Nuclear Engineering and Design*, **222**(2-3) pp. 231-245, June.

Krogsgaard, O.T., 2001: Energiepolitiken som vinden blaeser, (Energy policy as the wind blows). *Politiken*, 14 January.

Lang, Q., R.S. Chapman, D.M. Schreinemachers, L. Tian and X. He, 2002: Household stove improvement and risk of lung cancer in Xuanwei China. *Journal of the National Cancer Institute*, **94**(11), pp. 826-834.

Larson, E.D. and H. Yang, 2004: DME from coal as a household cooking fuel in China. Energy for Sustainable Development, **8**(3), pp. 115-126.

Larson, E.D., Z.X. Wu, P. Delaquil, W.Y. Chen, and P.F. Gao, 2003: Future implications of China-technology choices. *Energy Policy*, **31**(12), pp. 1189-1204.

Lauber, V., 2004: REFIT and RPS: options for a harmonized community framework. *Energy Policy*, **32**(12), pp. 1405-1414.

Leite, C. and J. Weidmann, 1999: Does mother nature corrupt? Natural resources, corruption and economic growth. IMF working paper WP/99/85, International Monetary Fund, Washington D.C.

Linden, N.H., M.A. Uyterlinde, C. Vrolijk, L.J. Nilsson, J. Khan, K. Astrand, K. Ericsson, and R. Wiser, 2005: Review of international experience with renewable energy obligation support mechanisms. ECN, Petten, Netherlands, May 2005.

London Health Commission, 2003: Energy and health: making the link. City Hall, London.

Lort-Phillips, E. and V. Herrigshaw, 2006: Beyond the rhetoric, measuring transparency in the oil and gas industry. In *Global Corruption Report 2006: Transparency International*, Pluto Press, London.

Lovei, L. and A. McKechnie, 2000: The costs of corruption for the poor - the energy sector, public policy for the private sector. Note no. 207, The World Bank, Washington D.C.

Luther, J., 2004: Research and development - the basis for wide-spread employment of renewable energies. International Conference for Renewable Energies, thematic background paper, Bonn, June, <www.renewables2004.de> accessed 02/07/07.

Maeda, H., 2005: The global nuclear fuel market - supply and demand 2005-2030. WNA Market Report, World Nuclear Association Annual Symposium 2005, <http://world-nuclear.org/sym/2005/maeda.htm> accessed 02/07/07.

Mago, R., 2004: Nuclear power - an option to meet the long term electricity needs of the country. Nuclear Power Corporation of India Ltd, Mumbai, India.

Margolis, R.M. and D.M. Kammen, 1999: Underinvestment: the energy technology and R&D policy challenge. *Science*, **285**, pp. 690-692.

Martinot, E. (lead author) *et al.*, 2005: Renewables 2005: Global status report. REN21 Renewable Energy Policy Network, <http://www.ren21.net/globalstatusreport/g2005.asp> accessed 02/07/07.

Martinot, E. *et al.*, 2006: Renewables 2006 global status report. REN21 Renewable Energy Policy Network, www.ren21.net/globalstatusreport/issuesGroup.asp accessed 02/07/07.

Maycock, P., 2003: PV market update. *Journal Renewable Energy World*, **6**(4), July/August 2003, pp. 84-101.

Mazza, P. and R. Hammerschlag, 2003: Carrying the energy future: comparing hydrogen and electricity for transmission, storage and transportation. Institute for Lifecycle Environmental Assessment, Seattle, Washington, 98122-0437, <www.ilea.org> accessed 02/07/07.

McDonald, A. and D. Schrattenholzer, 2001: Learning rates for energy technologies. *Journal of Energy Policy*, **29**, pp. 255-261.

MCH/WLPGA, 2005: Statistical Review of global LP gas. MCH Oil & Gas Consultancy/World LP Gas Association.

Menanteau, P., F.D. Land, and M-L. Lamy, 2001: Prices versus quantities: Environmental policies for promoting the development of renewable energy. Institut d'Economie et de Politique de l'Energie (IEPE) report, May 2001: Grenoble, France, 23 pp.

Menanteau, P., D. Finon, and M-L. Lamy, 2003: Prices versus quantities: choosing policies for promoting the development of renewable energy. *Energy Policy*, **31**, pp. 799-812.

METI, 2005: Japan's Energy Outlook 2030. Ministry of Economy, Trade and Industry, Japan.

MIT, 2003: The future of nuclear power. An interdisciplinary MIT study, Massachusetts Institute of Technology, E.J. Moniz and J. Deutch, Co-Chairs, <http://web.mit.edu/nuclearpower/> accessed 02/07/07.

Mishra, V., 2003: Indoor air pollution from biomass combustion and acute respiratory illness in preschool age children in Zimbabwe. *International Journal of Epidemiology,* **2003**(32), pp. 847-853.

Monirul, Q.M.M., 2004: Climate change and the Canadian energy sector: report on vulnerability impact and adaptation. Environment Canada.

Moore, T., 2005: Coal based generation at the crossroads. *EPRI Journal*, Summer, Palo Alto, CA.

Morthorst, P.E., 2004: Wind power - status and perspectives. In *Future Technologies for a Sustainable Electricity System*, Cambridge University Press, Cambridge.

Nalbandian, H., 2002: Prospects for integrated pollution control in pulverised coal-fired power plant. International Energy Agency, Clean Coal Centre, London.

NCOE, 2004: Ending the Energy Stalemate. National Commission on Energy Policy, Department of Energy, Washington D.C., 148 pp.

NDA, 2006: Strategy. Nuclear Decommissioning Authority, April, <http://www.nda.gov.uk/documents/loader.cfm?url=/commonspot/security/getfile.cfm&pageid=4957> accessed 02/07/07.

NEB, 2006: Oil sands - opportunities and challenges to 2015. Canadian National Energy Board <www.neb.gc.ca> accessed 02/07/07.

319

NEI, 2006: Nuclear facts. Nuclear Energy Institute, <http://www.nei.org/doc.asp?catnum=2&catid=106> accessed 02/07/07.

Nel, W., 2003: Feasibility of energy plant in Agulhas ocean current. Eskom Report No RES/SC/03/22509, Eskom Holdings Ltd., Johannesburg.

NER, 2004: National integrated resource plan 2003/4: reference case. ISEP * Eskom (Resources and Strategy), Energy Research Institute, University of Cape Town and the National Electricity Regulator, Pretoria, <www.ner.org.za> accessed 02/07/07.

NERI, 2004: Investigations of migratory birds during operation of Horns Rev offshore wind farm. National Environmental Research Institute, Ministry of the Environment, Denmark.

Netalieva, I., J. Wesseler, and W. Heijman, 2005: Health costs caused by oil extraction air emissions and the benefits from abatement: the case of Kazakhstan. *Energy Policy,* **33**, pp. 1169-1177.

Neuhoff, K., 2004: Large scale deployment of renewables for electricity generation. Publication Service, OECD, Paris, France, 40 pp.

Newman, J., N. Beg, J. Corfee-Morlot, G. McGlynn, and J. Ellis, 2002: Climate change and energy: trends, drivers, outlook and policy options. OECD Publication Service, Paris, France, 102 pp.

NREL, 2005: *Renewable Energy Cost Trends.* Golden Colorado, <www.nrel.gov/analysis/docs/cost curves 2005.ppt> accessed 02/07/07.

Nuclear News, 2005: WNA report forecasts three scenarios for nuclear's growth. *Nuclear News,* November 2005: pp. 60-62, 69.

OECD, 2000: The radiological impacts of spent nuclear fuel management options. Comparative study, OECD, Paris, <http://www.nea.fr/html/rp/reports/2000/nea2328-PARCOM-ENG.pdf> accessed 02/07/07.

OECD, 2001: Trend in the nuclear fuel cycle, economic, environmental and social aspects. OECD, Paris. <http://www1.oecd.org/publications/e-book/6602011e.pdf> accessed 02/07/07.

OECD, 2002a: Dealing with climate change - policies and measure in IEA member countries, Organisation for Economic Cooperation and Development Publication Service, OECD, Paris, France, 151 pp.

OECD, 2002b: Reforming energy subsidies. UN Environmental Programme and Organisation for Economic Cooperation and Development, OECD/IEA, Oxford, UK, 31 pp.

OECD, 2003: Decommissioning nuclear power plants: policies, strategies and costs. OECD Nuclear Energy Agency, 105 pp. ISBN 92-64-10431-3. <http://www1.oecd.org/publications/e-book/6603221E.pdf> accessed 02/07/07.

OECD, 2004a: Uranium 2003: resources, production and demand. Joint Report by the OECD Nuclear Energy Agency and the International Atomic Energy Agency, OECD, Paris.

OECD, 2004b: Biomass and Agriculture - sustainability markets and policies. Organisation of Economic and Co-operative Development, Paris, ISBN 92-64-10555-7.

OECD, 2006a: Global science forum on scientific challenges for energy research. Paris, May, <www.oecd.org/sti/gsf/energy> accessed 02/07/07.

OECD, 2006b: Uranium 2005: resources, production and demand. Joint Report by the OECD Nuclear Energy Agency and the International Atomic Energy Agency, OECD/NEA, ISBN: 92-64-02425-5388 pp.

OECD, 2006c: Advanced nuclear fuel cycles and radioactive waste management. ISBN: 92-64-02485-9, OECD/NEA, 248 pp.

OECD/IEA, 2005: Legal aspects of storing CO_2. Joint International Energy Agency / Carbon Sequestration Leadership Forum workshop, Paris, July 2004.

OFGEM, 2005: Renewables Obligation - an overview of the second year. Office for Gas and Electricity Markets, http://www.ofgem.gov.uk/Media/FactSheets/Documents1/9693-renewablesfsupdatefeb05.pdf> accessed 02/07/07.

Olsen, K.H. and J. Fenhann, 2006: Sustainable development benefits of clean development projects. CD4CDM Working Paper Series, Working Paper No. 2, Risoe, October 2006

Outhred, H. and I. MacGill, 2006: Integrating wind energy in the Australian National Electricity market. Proceedings of the World Renewable Energy Congress IX, August, ISBN 008 44671 X.

Oxera, 2005: What is the impact of limiting ROC eligibility for low-cost renewable generation technologies? Final report to the Department of Trade and Industry, London, 44 pp. Oxera Consulting Ltd, Oxford.

Patrinos, A., 2006: The promises of biotechnology. Synthetic Genomics Inc, USA, paper, OECD conference *Scientific Challenges for Energy Research,* Paris, 17-18 May.

Philibert, C., 2004: Case study 1: concentrating solar power technologies. International Energy Technology Collaboration and Climate Change Mitigation, OECD Environmental Directorate, IEA, Paris.

Philibert, C. and J. Podkanski, 2005: International energy technology collaboration and climate change mitigation case study 4: clean coal technologies. International Energy Agency, Paris.

Philibert, C., 2006: Barriers to the diffusion of solar thermal technologies. OECD and IEA Information Paper, International Energy Agency, Paris.

Plouchart, G., A. Fradet, and A.Prieur, 2006: The potential of CO_2 Capture and Storage of worldwide electricity related greenhouse gases emissions in oil and gas fields. Paper at the 8th Greenhouse gas technologies conference, June 19 - 22, 2006: Trondheim, Norway.

Price, L. and S. de la Rue du Can, 2006: Sectoral trends in global energy use and greenhouse gas emissions. Lawrence Berkeley National Laboratory, LBNL-56144.

Ragwitz, M., J. Schleich, C. Huber, G. Resch, T. Faber, M. Voogt, R. Coenraads, H. Cleijne, and P. Bodo,.2005: Analyses of the EU renewable energy sources' evolution up to 2020 (FORRES 2020, Fraunhofer Institute, Karlsruhe, Germany..

Read, P. and J.R. Lermit, 2005: Bioenergy with carbon storage(BECS): a sequential decision approach to the threat of abrupt climate change. *Energy,* **30**(14), pp. 2654-2671.

Renewables, 2004: International Conference for Renewable Energies. Proceedings, Conference Report, Outcomes and Documentation, Political Declaration/International Action Programme/Policy Recommendations for Renewable Energy, 1-4 June 2004, Bonn, Germany, 54 pp., <www.renewables2004.de> accessed 02/07/07.

Ristau, O., 2003: Sunny prospects on the Costa del Sol. *Journal New Energy,* **3**(2003), pp. 48-50.

Roques, F.A., W. J. Nuttall, D.M. Newbery, R. de Neufville, and S. Connors, 2006: Nuclear power: a hedge against uncertain gas and carbon prices? *The Energy Journal,* **27**(4), pp. 1-23.

Rogner, H.H., 2003: Nuclear power and climate change. World Climate Change Conference (WCCC), Moscow (20 September-3 October), <http://www.world-nuclear.org/wgs/cop9/cop-9_holger.pdf> accessed 02/07/07.

Royal Academy of Engineering, 2004: The cost of generating electricity. ISBN 1-903496-11-X, London, <http://www.nowap.co.uk/docs/generation_costs_report.pdf> accessed 02/07/07.

Sagie, D., 2005: Feasibility study of 100MWe solar power plant. Proceedings of the International Solar Energy Society (ISES), 2005 Solar World Congress, D.Y. Goswami, S. Vijayaraghaven, and R. Campbell-Howe (eds.), Orlando, Florida, August.

Saunders, A. and K. Schneider, 2000: Removing energy subsidies in developing and transition economies, energy markets and the new millennium: Economic Environment, Security of Supply. 23rd Annual IAEE International Conference, Sydney, Australia, 21 pp.

Sawin, J.L, 2003a: National policy instruments: policy lessons for the advancement & diffusion of renewable energy technologies around the world. Thematic background paper, pp. 56, International Conference for Renewable Energy, 1-4 June 2004: Bonn, Germany, <www.renewables2004.de> accessed 02/07/07.

Sawin, J.L., 2003b: Charting a new energy future, in *State of the World 2003.* L. Starke, W.W. Norton and Company (eds.), New York, pp 85-109.

Schlamadinger, B., R. Edwards, K.A. Byrne, A. Cowie, A. Faaij, C. Green, S. Fijan-Parlov, L. Gustavsson, T. Hatton, N. Heding, K. Kwant, K. Pingoud, M. Ringer, K. Robertson, B. Solberg, S. Soimakallio, and S. Woess-Gallasch, 2006: Optimizing the greenhouse gas benefits of bioenergy systems. Proceedings of the 14th European Biomass Conference, 17-21 October 2005: Paris, France, pp. 2029-2032.

Sekar, R.S., 2005: *Carbon dioxide capture from coal-fired power plants: a real options analysis.* MSc Thesis, MIT. <http://sequestration.mit.edu/pdf/LFEE_2005-002_RP.pdf> accessed 02/07/07.

Shell, 2006: Sustainable choices, stakeholder voices: 2005. Sustainable Development Report, pp. 36.

Shorrock, T., 2002: Enron's Asia misadventure. Asia Times 29 January, <http://www.atimes.com/global-econ/DA29Dj01.html> accessed 02/07/07.

Sikar, E., 2005: Greenhouse gases and initial findings on the carbon circulation in two reservoirs and their watersheds. *Verhandlung Internationale Vereinigung für Limnologie,* **27**.

Sims, R.E.H., 2002: The Brilliance of Bioenergy - the business and the practice. James & James, (Science Publishers) London, 316 pp. ISBN 1-902916-28-X.

Sims, R.E.H., 2003: Renewable energy - a response to climate change. Proceedings of the International Solar Energy Society World Congress 2001: Adelaide, W.Y. Saman and W.W.S. Charters (eds.), *Australian and New Zealand Solar Energy Society,* **1**, pp. 69-79.

Sims, R.E.H., H-H. Rogner, and K. Gregory, 2003a: Carbon emission and mitigation cost comparisons between fossil fuel, nuclear and renewable energy resources for electricity generation. *Energy Policy,* **31**, pp. 1315-1326.

Sims, R.E.H., N. El Bassam, R.P. Overend, K.O. Lim, K. Lwin and P. Chaturvedi, 2003b: Bioenergy options for a cleaner environment - in developed and developing countries. Elsevier, Oxford, 184 pp., ISBN 0 08 0443516.

Sims, R.E.H., 2004: Biomass, Bioenergy and Biomaterials - future prospects, 'Biomass and Agriculture - sustainability markets and policies'. Pp. 37-61 ISBN 92-64-10555-7. OECD, Paris.

Singh, V. and J. Fehrs, 2001: The work that goes into renewable energy. Research Report, Renewable Energy Policy Project (REPP), November 2001: No.13, 28, Washington D.C.

Sloss, L., C. Henderson, and J. Topper, 2003: Environmental standards and controls and their influence on development of clean coal technologies. International Energy Agency Clean Coal Centre, London.

Smith, K.R., 2002: Indoor air pollution in developing countries: recommendations for research. *Indoor Air,* **12**, pp. 198-207.

Smith, K.R., J. Zhang, R. Uma, W.N. Kishore, V. Joshi, and M.A.K. Khalil, 2000a: Greenhouse implications of household fuels: an analysis for India. *Annual Review of Energy and Environment* **25**, pp. 741-763.

Smith, K.R., J.M. Samet, I. Romieu, and N. Bruce, 2000b: Indoor air pollution in developing countries and acute lower respiratory infections in children. *Thorax,* **55**, pp. 518-532.

Smith, S., A. Brenket, and J. Edmonds, 2006b: Biomass with carbon dioxide capture and storage in a carbon-constrained world. GHGT8 proceedings, Trondheim, Norway.

Snow, T., White House Press Briefing, 2006: <http://www.whitehouse.gov/news/releases/2006/10/20061031-8.html>, <http://www.whitehouse.gov/news/releases/2006/10/20061031-8.html#> accessed 31 October 2006.

Spadaro K, 2000: Greenhouse gas emissions of electricity generation chains - Assessing the difference. International Atomic Energy Agency (IAEA) Bulletin 42/2/2000.

Specker, S., 2006: Generation technologies in a carbon-constrained world. Electric Power Research Institute, Palo Alto, California, March.

Stern, N., 2006: The economics of climate change. <http://www.hm-treasury.gov.uk/independent_reviews/stern_review_economics_climate_change/sternreview_index.cfm> accessed 02/07/07.

Stirling, 2005: Concentrated solar power development company. <http://www.stirlingenergy.com/news.asp?Type=stirling> accessed 02/07/07.

Storm van Leeuwen, J.W. and P. Smith, 2005: Nuclear power: the energy balance, <http://www.stormsmith.nl/> accessed 02/07/07.

Swart, R., J. Robinson, and S. Cohen, 2003: Climate change and sustainable development: expanding the options. *Climate Policy,* **3S1**, pp. 19-40.

Testor, J.W., E.M. Drake, M.W. Golay, M.J. Driscoll, and W.A. Peters, 2005: Sustainable energy - choosing among options. MIT Press, Cambridge, Massachusetts.

Tremblay, A. (ed.), 2005: Greenhouse gas emissions - fluxes and processes. Hydroelectric Reservoirs and Natural Environments. Springer, Berlin.

Trieb, F., 2005: Concentrating solar power for the Mediterranean region (MED-CSP). Study prepared for the German Federal Ministry for the Environment, Nature Conservation and Nuclear Safety (BMU), March, <http://www.dlr.de/tt/> accessed 02/07/07.

Uchiyama, Y., 1996: Life cycle analysis of electricity generation and supply systems, Electricity, Health and the Environment: Comparative Assessment in Support of Decision Making. IAEA, Vienna.

Uchiyama, Y. and H. Yamamoto, 1995: Life cycle analysis of power generation plants, CRIEPI report Y94009.

UCS, 2005: Renewable electricity standards at work in the States. <http://www.ucsusa.org/clean_energy/clean_energy_policies/res-at-work-in-the-states.html> accessed 02/07/07.

UKERC, 2006: The costs and impacts of intermittency. UK Energy Research Centre, Imperial College, London, ISBN 1 90314 404 3, <www.ukerc.ac.uk/content/view/259/953> accessed 02/07/07.

UN, 2000: General Assembly of the United Nations Millennium Declaration. 8th Plenary meeting, September, New York.

UN, 2004: Integration of the economies in transition into the world economy. A report on the activities of the Economic Commission for Europe, Department Policy and Planning Office, <http://www.un.org/esa/policy/reports/e_i_t/ece.pdf> accessed 02/07/07.

UNDESA, 2002: The Johannesburg plan of implementation. United Nations Department of Economic and Social Affairs, New York.

UNDP, 2000: Energy and the challenge of sustainability. World Energy Assessment, United Nations Development Programme, UNDESA and World Energy Council, ISBN 9211261260, New York, <http://www.undp.org/energy/activities/wea/drafts-frame.html> accessed 02/07/07.

UNDP, 2004a: World energy assessment, overview 2004: an update of 'Energy and the challenge of sustainability, world energy assessment'. Published in 2000: UNDESA/WEC, United Nations Development Program, New York.

UNDP, 2004b: Access to modern energy services can have a decisive impact on reducing poverty. http://www2.undp.org.yu/files/news/20041119_energy_poverty.pdf> accessed 02/07/07.

UNEP, 2004: Energy technology fact sheet, Cogeneration. UNEP Division of Technology, Industry and Economics - Energy and Ozone Action Unit, <http://www.cogen.org/Downloadables/Publications/Fact_Sheet_CHP.pdf> accessed 02/07/07.

UNEP, 2006: The UNEP Risoe CDM/JI pipeline. October, <www.cd4cdm.org/Publications/CDMPipeline.xls> accessed 02/07/07.

UNESCO, 2006: Workshop on GHG from freshwater reservoirs. 5-6 December, United Nations Educational, Scientific and Cultural Organization Headquarters, Paris, France, UNESCO IHP-VI website, <www.unesco.org/water/ihp/> accessed 02/07/07.

University of Chicago, 2004: The economic future of nuclear power. <http://www.anl.gov/Special_Reports/NuclEconAug04.pdf> accessed 02/07/07.

UIC, 2005: Civil liability for nuclear damage. Uranium Information Centre, Nuclear Issues Briefing Paper # 70, December, <http://www.uic.com.au/nip70.htm> accessed 02/07/07.

UNFCCC, 2004: Synthesis and assessment report of the greenhouse gas inventories submitted in 2004 from Annex-I countries. Pp. 66, 75, 93, and 94.

UNFCCC, 2006: Definition of renewable biomass. Minutes of CDM EB23, Annex 18, <http://cdm.unfccc.int/EB/Meetings/023/eb23_repan18.pdf> accessed 02/07/07.

USCCTP, 2005: United States Climate Change Technology Program, <http://www.climatetechnology.gov/library/2003/tech-options/> accessed 02/07/07.

US DOE, 2004: FutureGen-integrated hydrogen. Electric Power Production and Carbon Sequestration Research Initiative, <http://www.fossil.energy.gov/programs/powersystems/futuregen> accessed 02/07/07.

US DOE, 2005: Projected benefits of federal energy efficiency and renewable energy programs (FY2006- FY2050), Long-term benefits analysis of EERE's Programs (Chapter 5). Department of Energy, Washington D.C.

US DOE, 2006: Global nuclear energy partnership (GNEP). <http://www.gnep.energy.gov/> accessed 02/07/07.

US EIA, 2004: Long term world oil supply. Washington D.C. <http://www.eia.doe.gov/pub/oil_gas/petroleum/presentations/2004/worldoilsupply/oilsupply04> gevonden <http://tonto.eia.doe.gov/FTPROOT/features/longterm.pdf> accessed 02/07/07.

US EIA, 2005: International energy outlook 2005. U.S. Energy Information Administration, Office of Integrated Analysis and Forecasting, U.S. Department of Energy, Washington, D.C., 186 pp, <http://www.eia.doe.gov/oiaf/ieo/index.html> accessed 03/07/07.

US EPA, 2003: Assessment of the worldwide market potential for oxidizing coal mine ventilation air methane. United States Environmental Protection Agency, Washington D.C., July.

USGS, 2000: United States Geological Survey, World Assessment Summary.

USGS, 2004a: United States Geological Survey fact sheet, December 2. <http://marine.usgs.gov/fact-sheets/gas-hydrates/title.html> accessed 03/07/07.

USGS, 2004b: United States Geological Survey fact sheet, December 1. < http://energy.usgs.gov/factsheets/Coalbed/coalmeth.html> accessed 03/07/07.

US NAE 2004, The Hydrogen Economy: Opportunities, Costs, Barriers and R&D Needs, U.S. National Academy of Engineering, Washington, D.C., 2004.

Vallette, J. and D. Wysham, 2002: Enron's pawns - how public institutions bankrolled Enron's globalization game. Sustainable Energy and Economy Network, Institute of Policy Studies, Washington D.C.

Vate, J.F. van de, 2002: Full-energy-chain greenhouse-gas emissions: a comparison between nuclear power, hydropower, solar power and wind power. *International Journal of Risk Assessment and Management 2002,* **3**(1), pp. 59-74.

Vattenfall, 2005: Certified environmental product declaration of electricity from Forsmark NPP (updated 2005) and Ringhals NPP. S-P-00021& S-P-00026, June 2004. <http://www.environdec.com/reg/021/Chapters/Dokument/EPD-FKA-2005.pdf>, <http://www.environdec.com/reg/026/Chapters/Dokument/EPD-Ringhals.pdf> accessed 03/07/07.

Venkataraman, C., G. Habib, A. Eiguren-Fernandez, A.H. Miguel, and S.K. Friedlander, 2004: Residential biofuels in South Asia: carbonaceous aerosol emissions and climate impacts, *Science,* **307**(5714), pp.1454-1456.

Venn, J., 2005: Rapid access to modern energy services using LP gas. *Energy & Environment,* **16**(5), Multi-Science Publishing Co. Ltd., Brentwood.

Verweij, M., 2003: Leaking pipelines - Shell in South Africa. The Globalization and Energy Project, Milieudefensie, Amsterdam, Netherlands, 17 pp.

WADE, 2005: World survey of decentralized energy 2005. Edinburgh, UK,

WADE, 2006: World survey of decentralized energy 2006. <www.localpower.org> accessed 03/07/07.

Wamukonya, N., 2003: Power sector reform in developing countries: mismatched agendas. In *Electricity Reform: Social and Environmental Challenges,* N. Wamukonya (ed.), United Nations Environmental Programme, Roskilde, pp. 7-47.

Wavegen, 2004: <www.wavegen.com> accessed 03/07/07.

WBCSD, 2004: Energy and climate change - facts and trends to 2050. World Business Council for Sustainable Development, Geneva, Switzerland, <www.wbcsd.org> accessed 03/07/07.

WCD, 2000: Dams and development - a new framework for decision making. World Commission on Dams, <www.dams.org> accessed 03/07/07.

WEC, 2001: Performance of generating plant, World Energy Council, London.

WEC, 2004a: Comparison of energy systems using life cycle assessment - special report. World Energy Council, July, London.

WEC, 2004b: Oil and gas exploration & production: reserves, costs, contracts. Technip, 102 pp.

WEC, 2004c: 2004 Survey of energy resources. World Energy Council, London.

WEC, 2004d: Energy end-use technologies for the 21st Century. World Energy Council, London, UK, 128 pp., <www.worldenergy.org/wec-geis> accessed 03/07/07.

Weiss, W., I. Bergmann, and G. Faninger, 2005: Solar heating worldwide: markets and contribution to energy supply. Solar Heating and Cooling Programme, International Energy Agency, Paris, <www.iea-shc.org> accessed 03/07/07.

West, J.J., P. Osnaya, I. Laguna, J. Martínez, and A. Fernández, 2004: Co-control of urban air pollutants and greenhouse gases in Mexico City. *Environmental Science and Technology,* **38**(13), pp. 3474 -3481.

Whispergen, 2005: Small Stirling engine combined heat and power plants. Whispertech Company Ltd., Christchurch, New Zealand, <www.whispertech.co.nz> accessed 03/07/07.

WHO, 2002: The world health report 2002 - reducing risks, promoting healthy life. World Health Organisation, Geneva, Switzerland, 248 pp.

WHO, 2003: Exposure assessment in studies on the chronic effects of long-term exposure to air pollution. World Health Organisation report on a WHO/HEI Workshop, Bonn, Germany, 4-5 February 2002: Geneva, Switzerland.

Williams, R.H., E.D. Larson, and H. Jin, 2006: Synthetic fuels in a world of high oil and carbon prices. Proceedings of the 8th International Conference on Greenhouse Gas Control Technologies, Trondheim, Norway, June.

Wind Force 12, 2005: Global Wind Energy Council and Greenpeace, <http://www.gwec.net/index.php?id=8> accessed 03/07/07.

World Bank, 1998: Urban air quality management, pollution prevention and abatement handbook. Washington D.C.

World Bank, 2004a: A voluntary standard for global gas flaring and venting reduction. The International Bank for Reconstruction and Development/ The World Bank,Washington D.C.

World Bank, 2004b: Striking a better balance - the World Bank group and extractive industries: the final report of the extractive industries review. Washington D.C., 52 pp.

World Bank, 2005: An open letter to the Catholic Relief Services and bank information centre in response to the report 'Chad's Oil: Miracle or Mirage for the poor?'. News release no: 2005/366/AFR, Washington D.C.

World Bank, 2006: Clean energy and development: towards an investment framework. Environmental and Socially Sustainable Development Vice Presidency, World Bank, Washington D.C., March.

WNA, 2003: Global warming. Issue Brief, January 2003, World Nuclear Association, <www.world-nuclear.org/info/inf59.htm> accessed 03/07/07.

WNA, 2005a: The global nuclear fuel market: supply and demand 2005-2030. World Nuclear Association report, <www.world-nuclear.org/info/> accessed 03/07/07.

WNA, 2005b: The new economics of nuclear power. World Nuclear Association report, <http://www.world-nuclear.org/reference/pdf/economics.pdf > accessed 03/07/07.

WNA, 2006a: World nuclear power reactors 2005-06 and uranium requirements. World Nuclear Association, May 2006, <http://www.world-nuclear.org/info/reactors.htm> accessed 03/07/07.

WNA, 2006b: Nuclear power in China. Issue Brief, World Nuclear Association, February 2006: <www.world-nuclear.org/info/inf63.htm> accessed 03/07/07.

Woyllinowicz, D., C. Severson-Baker, and M. Raynolds, 2005: Oil sand fever. The Pembina Institute for Appropriate Development, Canada, <www.pembina.org> accessed 03/07/07.

WREC, 2006: Proceedings of the 8th World Renewable Energy Congress, August, Firenze.

WREA, 2005: Proceedings of World Renewable Energy Assembly, World Council of Renewable Energy, Bonn, November, <www.wrea2005.org> accessed 03/07/07.

Yamashita, K. and L. Barreto, 2005: Energyplexes for the 21st Century. *Energy,* **30**, pp. 2453- .

Zhang, X.L., R. Chen, and J.K. He, 2005: Renewable energy development and utilisation in China: opportunities and challenges. *International Journal on Global Energy Issues,* **24**(3/4), pp. 249-258.

5

Transport and its infrastructure

Coordinating Lead Authors:
Suzana Kahn Ribeiro (Brazil), Shigeki Kobayashi (Japan)

Lead Authors:
Michel Beuthe (Belgium), Jorge Gasca (Mexico), David Greene (USA), David S. Lee (UK), Yasunori Muromachi (Japan), Peter J. Newton (UK), Steven Plotkin (USA), Daniel Sperling (USA), Ron Wit (The Netherlands), Peter J. Zhou (Zimbabwe)

Contributing Authors:
Hiroshi Hata (Japan), Ralph Sims (New Zealand), Kjell Olav Skjolsvik (Norway)

Review Editors:
Ranjan Bose (India), Haroon Kheshgi (USA)

This chapter should be cited as:
Kahn Ribeiro, S., S. Kobayashi, M. Beuthe, J. Gasca, D. Greene, D. S. Lee, Y. Muromachi, P. J. Newton, S. Plotkin, D. Sperling, R. Wit, P. J. Zhou, 2007: Transport and its infrastructure. In Climate Change 2007: Mitigation. Contribution of Working Group III to the Fourth Assessment Report of the Intergovernmental Panel on Climate Change [B. Metz, O.R. Davidson, P.R. Bosch, R. Dave, L.A. Meyer (eds)], Cambridge University Press, Cambridge, United Kingdom and New York, NY, USA.

Table of Contents

EXECUTIVE SUMMARY

Transport activity, a key component of economic development and human welfare, is increasing around the world as economies grow. For most policymakers, the most pressing problems associated with this increasing transport activity are traffic fatalities and injuries, congestion, air pollution and petroleum dependence. These problems are especially acute in the most rapidly growing economies of the developing world. Mitigating greenhouse gas (GHG) emissions can take its place among these other transport priorities by emphasizing synergies and co-benefits *(high agreement, much evidence)*.

Transport predominantly relies on a single fossil resource, petroleum that supplies 95% of the total energy used by world transport. In 2004, transport was responsible for 23% of world energy-related GHG emissions with about three quarters coming from road vehicles. Over the past decade, transport's GHG emissions have increased at a faster rate than any other energy using sector *(high agreement, much evidence)*.

Transport activity will continue to increase in the future as economic growth fuels transport demand and the availability of transport drives development, by facilitating specialization and trade. The majority of the world's population still does not have access to personal vehicles and many do not have access to any form of motorized transport. However, this situation is rapidly changing.

Freight transport has been growing even more rapidly than passenger transport and is expected to continue to do so in the future. Urban freight movements are predominantly by truck, while international freight is dominated by ocean shipping. The modal distribution of intercity freight varies greatly across regions. For example, in the United States, all modes participate substantially, while in Europe, trucking has a higher market share (in tkm[1]), compared to rail *(high agreement, much evidence)*.

Transport activity is expected to grow robustly over the next several decades. Unless there is a major shift away from current patterns of energy use, world transport energy use is projected to increase at the rate of about 2% per year, with the highest rates of growth in the emerging economies, and total transport energy use and carbon emissions is projected to be about 80% higher than current levels by 2030 *(medium agreement, medium evidence)*.

There is an ongoing debate about whether the world is nearing a peak in conventional oil production that will require a significant and rapid transition to alternative energy resources. There is no shortage of alternative energy sources, including oil sands, shale oil, coal-to-liquids, biofuels, electricity and hydrogen. Among these alternatives, unconventional fossil carbon resources would produce less expensive fuels most compatible with the existing transport infrastructure, but lead to increased carbon emissions *(medium agreement, medium evidence)*.

In 2004, the transport sector produced 6.3 GtCO$_2$ emissions (23% of world energy-related CO$_2$ emissions) and its growth rate is highest among the end-user sectors. Road transport currently accounts for 74% of total transport CO$_2$ emissions. The share of non-OECD countries is 36% now and will increase rapidly to 46% by 2030 if current trends continue *(high agreement, much evidence)*. The transport sector also contributes small amounts of CH$_4$ and N$_2$O emissions from fuel combustion and F-gases (fluorinated gases) from vehicle air conditioning. CH$_4$ emissions are between 0.1–0.3% of total transport GHG emissions, N$_2$O between 2.0 and 2.8% (based on US, Japan and EU data only). Worldwide emissions of F-gases (CFC-12+HFC-134a+HCFC-22) in 2003 were 0.3–0.6 GtCO$_2$-eq, about 5–10% of total transport CO$_2$ emissions *(medium agreement, limited evidence)*.

When assessing mitigation options it is important to consider their lifecycle GHG impacts. This is especially true for choices among alternative fuels but also applies to a lesser degree to the manufacturing processes and materials composition of advanced technologies. Electricity and hydrogen can offer the opportunity to 'de-carbonise' the transport energy system although the actual full cycle carbon reduction depends upon the way electricity and hydrogen are produced. Assessment of mitigation potential in the transport sector through the year 2030 is uncertain because the potential depends on:

- World oil supply and its impact on fuel prices and the economic viability of alternative transport fuels;
- R&D outcomes in several areas, especially biomass fuel production technology and its sustainability in massive scale, as well as battery longevity, cost and specific energy.

Another problem for a credible assessment is the limited number and scope of available studies of mitigation potential and cost.

Improving energy efficiency offers an excellent opportunity for transport GHG mitigation through 2030. Carbon emissions from 'new' light-duty road vehicles could be reduced by up to 50% by 2030 compared to currently produced models, assuming continued technological advances and strong policies to ensure that technologies are applied to increasing fuel economy rather than spent on increased horsepower and vehicle mass. Material substitution and advanced design could reduce the weight of light-duty vehicles by 20–30%. Since the TAR (Third Assessment Report), energy efficiency of road vehicles has improved by the market success of cleaner direct-injection turbocharged (TDI) diesels and the continued market penetration of numerous incremental efficiency technologies.

---

1 ton-km, "ton" refers to metric ton, unless otherwise stated.

Hybrid vehicles have also played a role, though their market penetration is currently small. Reductions in drag coefficients of 20–50% seem achievable for heavy intercity trucks, with consequent reductions in fuel use of 10–20%. Hybrid technology is applicable to trucks and buses that operate in urban environments, and the diesel engine's efficiency may be improved by 10% or more. Prospects for mitigation are strongly dependent on the advancement of transport technologies.

There are also important opportunities to increase the operating efficiencies of transport vehicles. Road vehicle efficiency might be improved by 5–20% through strategies such as eco-driving styles, increased load factors, improved maintenance, in-vehicle technological aids, more efficient replacement tyres, reduced idling and better traffic management and route choice *(medium agreement, medium evidence)*.

The total mitigation potential in 2030 of the energy efficiency options applied to light duty vehicles would be around 0.7–0.8 $GtCO_2$-eq in 2030 at costs <100 US$/$tCO_2$. Data is not sufficient to provide a similar estimate for heavy-duty vehicles. The use of current and advanced biofuels would give an additional reduction potential of another 600–1500 $MtCO_2$-eq in 2030 at costs <25 US$/$tCO_2$ *(low agreement, limited evidence)*.

Although rail transport is one of the most energy efficient modes today, substantial opportunities for further efficiency improvements remain. Reduced aerodynamic drag, lower train weight, regenerative breaking and higher efficiency propulsion systems can make significant reductions in rail energy use. Shipping, also one of the least energy intensive modes, still has some potential for increased energy efficiency. Studies assessing both technical and operational approaches have concluded that energy efficiency opportunities of a few percent to up to 40% are possible *(medium agreement, medium evidence)*.

Passenger jet aircraft produced today are 70% more fuel efficient than the equivalent aircraft produced 40 years ago and continued improvement is expected. A 20% improvement over 1997 aircraft efficiency is likely by 2015 and possibly 40 to 50% improvement is anticipated by 2050. Still greater efficiency gains will depend on the potential of novel designs such as the blended wing body, or propulsion systems such as the unducted turbofan. For 2030 the estimated mitigation potential is 150 $MtCO_2$ at carbon prices less than 50 US$/$tCO_2$ and 280 $MtCO_2$ at carbon prices less than 100 US$/$tCO_2$ *(medium agreement, medium evidence)*. However, without policy intervention, projected annual improvements in aircraft fuel efficiency of the order of 1–2%, will be surpassed by annual traffic growth of around 5% each year, leading to an annual increase of CO_2 emissions of 3–4% per year *(high agreement, much evidence)*.

Biofuels have the potential to replace a substantial part but not all petroleum use by transport. A recent IEA analysis estimates that biofuels' share of transport fuel could increase to about 10% in 2030. The economic potential in 2030 from biofuel application is estimated at 600–1500 $MtCO_2$-eq/yr at a cost of <25 US$/$tCO_2$-eq. The introduction of flexfuel vehicles able to use any mixture of gasoline[2] and ethanol rejuvenated the market for ethanol as a motor fuel in Brazil by protecting motorists from wide swings in the price of either fuel. The global potential for biofuels will depend on the success of technologies to utilise cellulose biomass *(medium agreement, medium evidence)*.

Providing public transports systems and their related infrastructure and promoting non-motorised transport can contribute to GHG mitigation. However, local conditions determine how much transport can be shifted to less energy intensive modes. Occupancy rates and primary energy sources of the transport mode further determine the mitigation impact. The energy requirements for urban transport are strongly influenced by the density and spatial structure of the built environment, as well as by location, extent and nature of transport infrastructure. If the share of buses in passenger transport in typical Latin American cities would increase by 5–10%, then CO_2 emissions could go down by 4–9% at costs of the order of 60–70 US$/$tCO_2$ *(low agreement, limited evidence)*.

The few worldwide assessments of transport's GHG mitigation potential completed since the TAR indicate that significant reductions in the expected 80% increase in transport GHG emission by 2030 will require both major advances in technology and implementation via strong, comprehensive policies *(medium agreement, limited evidence)*.

The mitigation potential by 2030 for the transport sector is estimated to be about 1600–2550 $MtCO_2$ for a carbon price less than 100 US$/$tCO_2$. This is only a partial assessment, based on biofuel use throughout the transport sector and efficiency improvements in light-duty vehicles and aircraft and does not cover the potential for heavy-duty vehicles, rail transport, shipping, and modal split change and public transport promotion and is therefore an underestimation. Much of this potential appears to be located in OECD North America and Europe. This potential is measured as the further reduction in CO_2 emissions from a Reference scenario, which already assumes a substantial use of biofuels and significant improvements in fuel efficiency based on a continuation of current trends. This estimate of mitigation costs and potentials is highly uncertain. There remains a critical need for comprehensive and consistent assessments of the worldwide potential to mitigate transport's GHG emissions *(low agreement, limited evidence)*.

While transport demand certainly responds to price signals, the demand for vehicles, vehicle travel and fuel use are significantly price inelastic. As a result, large increases in prices or taxes are required to make major changes in GHG emissions.

---

2   US term for petrol.

Many countries do heavily tax motor fuels and have lower rates of fuel consumption and vehicle use than countries with low fuel taxes *(high agreement, much evidence)*.

Fuel economy regulations have been effective in slowing the growth of GHG emissions, but so far growth of transport activity has overwhelmed their impact. They have been adopted by most developed economies as well as key developing economies, though in widely varying form, from uniform, mandatory corporate average standards, to graduated standards by vehicle weight class or size, to voluntary industry-wide standards. The overall effectiveness of standards can be significantly enhanced if combined with fiscal incentives and consumer information *(medium agreement, medium evidence)*.

A wide array of transport demand management (TDM) strategies have been employed in different circumstances around the world, primarily to manage traffic congestion and reduce air pollution. TDMs can be effective in reducing private vehicle travel if rigorously implemented and supported *(high agreement, low evidence)*.

In order to reduce emissions from air and marine transport resulting from the combustion of bunker fuels, new policy frameworks need to be developed. However ICAO endorsed the concept of an open, international emission trading system for the air transport sector, implemented through a voluntary scheme, or incorporation of international aviation into existing emission trading systems. Environmentally differentiated port dues are being used in a few places. Other policies to affect shipping emissions would be the inclusion of international shipping in international emissions trading schemes, fuel taxes and regulatory instruments *(high agreement, much evidence)*.

Since currently available mitigation options will probably not be enough to prevent growth in transport's emissions, technology research and development is essential in order to create the potential for future, significant reductions in transport GHG emissions. This holds, amongst others, for hydrogen fuel cell, advanced biofuel conversion and improved batteries for electric and hybrid vehicles *(high agreement, medium evidence)*.

The best choice of policy options will vary across regions. Not only levels of economic development, but the nature of economic activity, geography, population density and culture all influence the effectiveness and desirability of policies affecting modal choices, infrastructure investments and transport demand management measures *(high agreement, much evidence)*.

5.1 Introduction

Mobility is an essential human need. Human survival and societal interaction depend on the ability to move people and goods. Efficient mobility systems are essential facilitators of economic development. Cities could not exist and global trade could not occur without systems to transport people and goods cheaply and efficiently (WBCSD, 2002).

Since motorized transport relies on oil for virtually all its fuel and accounts for almost half of world oil consumption, the transport sector faces a challenging future, given its dependence on oil. In this chapter, existing and future options and potentials to reduce greenhouse gases (GHG) are assessed.

GHG emission reduction will be only one of several key issues in transport during the coming decades and will not be the foremost issue in many areas. In developing countries especially, increasing demand for private vehicles is outpacing the supply of transport infrastructure – including both road networks and public transit networks. The result is growing congestion and air pollution,[3] and a rise in traffic fatalities. Further, the predominant reliance on private vehicles for passenger travel is creating substantial societal strains as economically disadvantaged populations are left out of the rapid growth in mobility. In many countries, concerns about transport will likely focus on the local traffic, pollution, safety and equity effects. The global warming issue in transport will have to be addressed in the context of the broader goal of sustainable development.

5.2 Current status[4] and future trends

5.2.1 Transport today

The transport sector plays a crucial and growing role in world energy use and emissions of GHGs. In 2004, transport energy use amounted to 26% of total world energy use and the transport sector was responsible for about 23% of world energy-related GHG emissions (IEA, 2006b). The 1990–2002 growth rate of energy consumption in the transport sector was highest among all the end-use sectors. Of a total of 77 EJ[5] of total transport energy use, road vehicles account for more than three-quarters, with light-duty vehicles and freight trucks having the lion's share (see Table 5.1). Virtually all (95%) of transport energy comes from oil-based fuels, largely diesel (23.6 EJ, or about 31% of total energy) and gasoline (36.4 EJ, 47%). One consequence of this dependence, coupled with the only moderate differences in carbon content of the various

Table 5.1: *World transport energy use in 2000, by mode*

Mode	Energy use (EJ)	Share (%)
Light-duty vehicles (LDVs)	34.2	44.5
2-wheelers	1.2	1.6
Heavy freight trucks	12.48	16.2
Medium freight trucks	6.77	8.8
Buses	4.76	6.2
Rail	1.19	1.5
Air	8.95	11.6
Shipping	7.32	9.5
Total	76.87	100

Source: WBCSD, 2004b.

oil-based fuels, is that the CO_2 emissions from the different transport sub-sectors are approximately proportional to their energy use (Figure 5.1).

Economic development and transport are inextricably linked. Development increases transport demand, while availability of transport stimulates even more development by allowing trade and economic specialization. Industrialization and growing specialization have created the need for large shipments of goods and materials over substantial distances; accelerating globalization has greatly increased these flows.

Urbanization has been extremely rapid in the past century. About 75% of people in the industrialized world and 40% in the developing world now live in urban areas. Also, cities have grown larger, with 19 cities now having a population over 10 million. A parallel trend has been the decentralization of cities – they have spread out faster than they have grown in population, with rapid growth in suburban areas and the rise of 'edge cities' in the outer suburbs. This decentralization has created both a growing demand for travel and an urban pattern that is not easily served by public transport. The result has been a rapid increase in personal vehicles – not only cars but also 2-wheelers – and a declining share of transit. Further, the lower-density development and the greater distances needed to access jobs and services have seen the decline of walking and bicycling as a share of total travel (WBCSD, 2002).

Another crucial aspect of our transport system is that much of the world is not yet motorized because of low incomes. The majority of the world's population does not have access to personal vehicles, and many do not even have access to motorized public transport services of any sort. Thirty-three percent of China's population and 75% of Ethiopia's still did not have access to all-weather transport (e.g., with roads passable

---

3 Although congestion and air pollution are also found in developed countries, they are exacerbated by developing country conditions.

4 The primary source for the 'current status' part of this discussion is WBCSD (World Business Council for Sustainable Development) Mobility 2001 (2002), prepared by Massachusetts Institute of Technology and Charles River Associates Incorporated.

5 83 EJ in 2004 (IEA, 2006b).

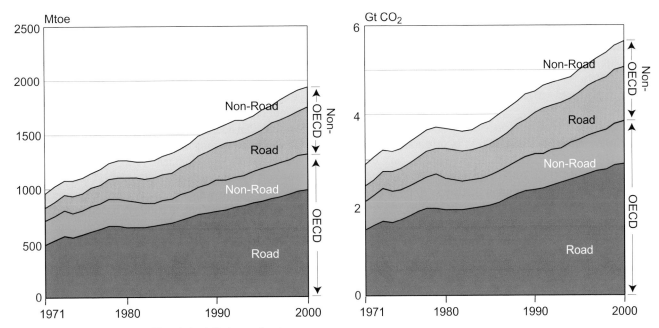

Figure 5.1: *Energy consumption and CO_2 emission in the transport sector*
Source: IEA, 2006c,d

most of the year). Walking more than 10 km/day[6] each way to farms, schools and clinics is not unusual in rural areas of the developing world, particularly sub-Saharan Africa, but also in parts of Asia and Latin America. Commuting by public transport is very costly for the urban poor, taking, for example, 14% of the income of the poor in Manila compared with 7% of the income of the non-poor (World Bank, 1996). If and when these areas develop and their population's incomes rise, the prospects for a vast expansion of motorization and increase in fossil fuel use and GHG emissions is very real. And these prospects are exacerbated by the evidence that the most attractive form of transport for most people as their incomes rise is the motorized personal vehicle, which is seen as a status symbol as well as being faster, flexible, convenient and more comfortable than public transport. Further aggravating the energy and environmental concerns of the expansion of motorization is the large-scale importation of used vehicles into the developing world. Although increased access to activities and services will contribute greatly to living standards, a critical goal will be to improve access while reducing the adverse consequences of motorization, including GHG emissions.

Another factor that has accelerated the increase in transport energy use and carbon emissions is the gradual growth in the size, weight and power of passenger vehicles, especially in the industrialized world. Although the efficiency of vehicle technology has improved steadily over time, much of the benefit of these improvements have gone towards increased power and size at the expense of improved fuel efficiency. For example, the US Environmental Protection Agency has concluded that

the US new Light-duty Vehicle (LDV) fleet fuel economy in 2005 would have been 24% higher had the fleet remained at the weight and performance distribution it had in 1987. Instead, over that time period, it became 27% heavier and 30% faster in 0–60 mph (0–97 km/h) time, and achieved 5% poorer fuel economy (Heavenrich, 2005). In other words, if power and size had been held constant during this period, the fuel consumption rates of light-duty vehicles would have dropped more than 1% per year.

Worldwide travel studies have shown that the average time budget for travel is roughly constant worldwide, with the relative speed of travel determining distances travelled yearly (Schafer, 2000). As incomes have risen, travellers have shifted to faster – and more energy-intensive – modes, from walking and bicycling to public transport to automobiles and, for longer trips, to aircraft. And as income and travel have risen, the percentage of trips made by automobiles has risen. Automobile travel now accounts for 15–30% of total trips in the developing world, but 50% in Western Europe and 90% in the United States. The world auto fleet has grown with exceptional rapidity – between 1950 and 1997, the fleet increased from about 50 million vehicles to 580 million vehicles, five times faster than the growth in population. In China, for example, vehicle sales (not including scooters, motorcycles and locally manufactured rural vehicles) have increased from 2.4 million in 2001 to 5.6 million in 2005[7] and further to 7.2 million in 2006.[8] 2-wheeled scooters and motorcycles have also played an important role in the developing world and in warmer parts of Europe, with a current world fleet of a few hundred million

6   6.21 miles/day.
7   Automotive News Data Center: http://www.autonews.com/apps/pbcs.dll/search?Category=DATACENTER01archive.
8   China Association of Automobile Manufacturers 2007.1.17: http://60.195.249.78/caam/caam.web/Detail.asp?id=359#

vehicles (WBSCD, 2002). Non-motorized transport continues to dominate the developing world. Even in Latin America and Europe, walking accounts for 20–40% of all trips in many cities (WBCSD, 2002). Bicycles continue to play a major role in much of Asia and scattered cities elsewhere, including Amsterdam and Copenhagen.

Public transport plays a crucial role in urban areas. Buses, though declining in importance against private cars in the industrialized world (EC, 2005; Japanese Statistical Bureau, 2006; US Bureau of Transportation Statistics, 2005) and some emerging economies, are increasing their role elsewhere, serving up to 45% of trips in some areas. Paratransit – primarily minibus jitneys run by private operators – has been rapidly taking market share from the formal public-sector bus systems in many areas, now accounting for 35% of trips in South Africa, 40% in Caracas and Bogota and up to 65% in Manila and other southeast Asian cities (WBCSD, 2002). Heavy rail transit systems are generally found only in the largest, densest cities of the industrialized world and a few of the upper-tier developing world cities.

Intercity and international travel is growing rapidly, driven by growing international investments and reduced trade restrictions, increases in international migration and rising incomes that fuel a desire for increased recreational travel. In the United States, intercity travel already accounts for about one-fifth of total travel and is dominated by auto and air. European and Japanese intercity travel combines auto and air travel with fast rail travel. In the developing world, on the other hand, intercity travel is dominated by bus and conventional rail travel, though air travel is growing rapidly in some areas – 12% per year in China, for example. Worldwide passenger air travel is growing 5% annually – a faster rate of growth than any other travel mode (WBCSD, 2002).

Industrialization and globalization have also stimulated freight transport, which now consumes 35% of all transport energy, or 27 exajoules (out of 77 total) (WBCSD, 2004b). Freight transport is considerably more conscious of energy efficiency considerations than passenger travel because of pressure on shippers to cut costs, however this can be offset by pressure to increase speeds and reliability and provide smaller 'just-in-time' shipments. The result has been that, although the energy-efficiency of specific modes has been increasing, there has been an ongoing movement to the faster and more energy-intensive modes. Consequently, rail and domestic waterways' shares of total freight movement have been declining, while highway's share has been increasing and air freight, though it remains a small share, has been growing rapidly. Some breakdowns:

- Urban freight is dominated by trucks of all sizes.
- Regional freight is dominated by large trucks, with bulk commodities carried by rail and pipelines and some water transport.

- National or continental freight is carried by a combination of large trucks on higher speed roads, rail and ship.
- International freight is dominated by ocean shipping. The bulk of international freight is carried aboard extremely large ships carrying bulk dry cargo (e.g., iron ore), container freight or fuel and chemicals (tankers).
- There is considerable variation in freight transport around the world, depending on geography, available infrastructure and economic development. The United States' freight transport system, which has the highest total traffic in the world, is one in which all modes participate substantially. Russia's freight system, in contrast, is dominated by rail and pipelines, whereas Europe's freight systems are dominated by trucking with a market share of 72% (tkm) in EU-25 countries, while rail's market share is just 16.4% despite its extensive network.[9] China's freight system uses rail as its largest carrier, with substantial contributions from trucks and shipping (EC, 2005).

Global estimates of direct GHG emissions of the transport sector are based on fuel use. The contribution of transport to total GHG emissions was about 23%, with emissions of CO_2 and N_2O amounting to about 6300–6400 $MtCO_2$-eq in 2004. Transport sector CO_2 emissions have increased by around 27% since 1990 (IEA, 2006d). For sub-sectors such as aviation and marine transport, estimates based on more detailed information are available. Estimates of global aviation CO_2 emissions using a consistent inventory methodology have recently been made by Lee et al. (2005). These showed an increase by approximately a factor of 1.5 from 331 $MtCO_2$/yr in 1990 to 480 $MtCO_2$/yr in 2000. For seagoing shipping, fuel usage has previously been derived from energy statistics (e.g., Olivier et al., 1996; Corbett et al., 1999; Endresen et al., 2003). More recently, efforts have been committed to constructing inventories using activity-based statistics on shipping movements (Corbett and Köhler, 2003; Eyring et al., 2005a). This has resulted in a substantial discrepancy. Estimated CO_2 emissions vary accordingly. This has prompted debate over inventory methodologies in the literature (Endresen et al., 2004; Corbett and Köhler, 2004). It is noteworthy that the NO_x emissions estimates also vary strongly between the different studies (Eyring et al., 2005a).

5.2.2 Transport in the future

There seems little doubt that, short of worldwide economic collapse, transport activity will continue to grow at a rapid pace for the foreseeable future. However, the shape of that demand and the means by which it will be satisfied depend on several factors.

First, it is not clear whether oil can continue to be the dominant feedstock of transport. There is an on-going debate about the date when conventional oil production will peak, with many arguing that this will occur within the next few decades

---

9   This rather small share is the result of priority given to passenger transport and market fragmentation between rival national rail systems.

Box 5.1: Non-CO$_2$ climate impacts

When considering the mitigation potential for the transport sector, it is important to understand the effects that it has on climate change. Whilst the principal GHG emitted is CO$_2$, other pollutants and effects may be important and control/ mitigation of these may have either technological or operational trade-offs.

Individual sectors have not been studied in great detail, with the exception of aviation. Whilst surface vehicular transport has a large fraction of global emissions of CO$_2$, its radiative forcing (RF) impact is little studied. Vehicle emissions of NO$_x$, VOCs and CO contribute to the formation of tropospheric O$_3$, a powerful GHG; moreover, black carbon and organic carbon may affect RF from this sector. Shipping has a variety of associated emissions, similar in many respects to surface vehicular transport. One of shipping's particular features is the observed formation of low-level clouds ('ship-tracks'), which has a negative RF effect. The potential coverage of these clouds and its associated RF is poorly studied, but one study estimates a negative forcing of 0.110 W/m^2 (Capaldo *et al.*, 1999), which is potentially much larger than its positive forcing from CO$_2$ and it is possible that the overall forcing from shipping may be negative, although this requires more study. However, a distinction should be drawn between RF and an actual climate effect in terms of global temperature change or sea-level rise; the latter being much more complicated to estimate.

Non-CO$_2$ emissions (CH$_4$ and N$_2$O) from road transport in major Annex I parties are listed in UNFCC GHG inventory data. The refrigerant banks and emission trend of F-gases (CFC-12 + HFC-134a) from air-conditioning are reported in the recent IPCC special report on Safeguarding the Ozone Layer and the Global Climate System (IPCC, 2005). Since a rapid switch from CFC-12 to HFC-134a, which has a much lower GWP index, is taking place, the total amount of F-gases is increasing due to the increase in vehicles with air-conditioning, but total emission in CO$_2$-eq is decreasing and forecasted to continue to decrease. Using the recent ADEME data (2006) on F-gas emissions, the shares of emissions from transport sectors for CO$_2$, CH$_4$, N$_2$O and F-gases (CFC-12 + HFC-134a+HCFC-22) are:

	CO$_2$ (%)	CH$_4$ (%)	N$_2$O (%)	F-gas (%)
USA	88.4	0.2	2.0	8.9
Japan	96.0	0.1	2.5	1.4
EU	95.3	0.3	2.8	1.7

Worldwide F-gas emissions in 2003 were reported to be 610 MtCO$_2$-eq in IPCC (2005), but more recent ADEME data (ADEME, 2006) was about 310 Mt CO$_2$-eq (CFC-12 207, HFC-134a 89, HCFC-22 10 MtCO$_2$-eq), which is about 5% of total transport CO$_2$ emission. It can be seen that non-CO$_2$ emissions from the transport sector are considerably smaller than the CO$_2$ emissions. Also, air-conditioning uses significant quantities of energy, with consequent CO$_2$ emissions from the fuel used to supply this energy. Although this depends strongly on the climate conditions, it is reported to be 2.5–7.5% of vehicle energy consumption (IPCC, 2005).

Aviation has a larger impact on radiative forcing than that from its CO$_2$ forcing alone. This was estimated for 1992 and a range of 2050 scenarios by IPCC (1999) and updated for 2000 by Sausen *et al.* (2005) using more recent scientific knowledge and data. Aviation emissions impact radiative forcing in positive (warming) and negative (cooling) ways as follows: CO$_2$ (+25.3 mW/m^2); O$_3$ production from NO$_x$ emissions (+21.9 mW/m^2); ambient CH$_4$ reduction as a result of NO$_x$ emissions (–10.4 mW/m^2); H$_2$O (+2.0 mW/m^2); sulphate particles (–3.5 mW/m^2); soot particles (+2.5 mW/m^2); contrails (+10.0 mW/m^2); cirrus cloud enhancement (10–80 mW/m^2). These effects result in a total aviation radiative forcing for 2000 of 47.8 mW/m^2, excluding cirrus cloud enhancement, for which no best estimate could be made, as was the case for IPCC (1999). Forster et al. (2007) assumed that aviation radiative forcing (0.048 W/m^2 in 2000, which excludes cirrus) to have grown by no more that 10% between 2000 and 2005. Forster et al. (2007) estimate a total net anthropogenic radiative forcing in 2005 of 1.6 W/m^2 (range 0.6–2.4 W/m^2). Aviation therefore accounts for around 3% of the anthropogenic radiative forcing in 2005 (range 2–8%). This 90% confidence range is skewed towards lower percentages and does not account for uncertainty in the aviation forcings.

(though others, including some of the major multinational oil companies, strongly oppose this view). Transport can be fuelled by multiple alternative sources, beginning with liquid fuels from unconventional oil (very heavy oil, oil sands and oil shale), natural gas or coal, or biomass. Other alternatives include gaseous fuels such as natural gas or hydrogen and electricity, with both hydrogen and electricity capable of being produced from a variety of feedstocks. However, all of these alternatives are costly, and several – especially liquids from fossil resources – can increase GHG emissions significantly without carbon sequestration.

Vehicle Ownership/ 1000 Persons

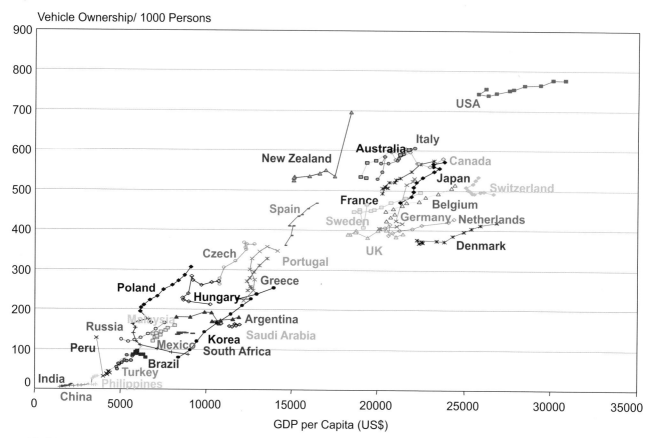

Figure 5.2: *Vehicle ownership as a function of per capita income*

Note: plotted years vary by country depending on data availability.

Data source: World Bank, 2004.

Second, the growth rate and shape of economic development, the primary driver of transport demand, is uncertain. If China and India as well as other Asian countries continue to rapidly industrialize, and if Latin America and Africa fulfil much of their economic potential, transport demand will grow with extreme rapidity over the next several decades. Even in the most conservative economic scenarios though, considerable growth in travel is likely.

Third, transport technology has been evolving rapidly. The energy efficiency of the different modes, vehicle technologies, and fuels, as well as their cost and desirability, will be strongly affected by technology developments in the future. For example, although hybrid electric drive trains have made a strong early showing in the Japanese and US markets, their ultimate degree of market penetration will depend strongly on further cost reductions. Other near-term options include the migration of light-duty diesel from Europe to other regions. Longer term opportunities requiring more advanced technology include new biomass fuels beyond those made from sugar cane in Brazil and corn in the USA, fuel cells running on hydrogen and battery-powered electric vehicles.

Fourth, as incomes in the developing nations grow, transport infrastructure will grow rapidly. Current trends point towards

growing dependence on private cars, but other alternatives exist (as demonstrated by cities such as Curitiba and Bogota with their rapid bus transit systems). Also, as seen in Figure 5.2, the intensity of car ownership varies widely around the world even when differences in income are accounted for, so different countries have made very different choices as they have developed. The future choices made by both governments and travellers will have huge implications for future transport energy demand and CO_2 emissions in these countries.

Most projections of transport energy consumption and GHG emissions have developed Reference Cases that try to imagine what the future would look like if governments essentially continued their existing policies without adapting to new conditions. These Reference Cases establish a baseline against which changes caused by new policies and measures can be measured, and illustrate the types of problems and issues that will face governments in the future.

Two widely cited projections of world transport energy use are the Reference Cases in the ongoing world energy forecasts of the United States Energy Information Administration, 'International Energy Outlook 2005' (EIA, 2005) and the International Energy Agency, World Energy Outlook 2004 (IEA, 2004a). A recent study by the World Business Council on

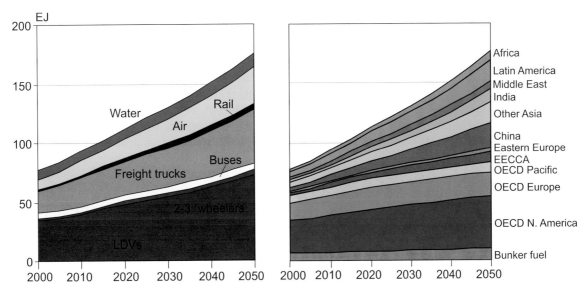

Figure 5.3: *Projection of transport energy consumption by region and mode*
Source: WBCSD, 2004a.

Sustainable Development, 'Mobility 2030', also developed a projection of world transport energy use. Because the WBCSD forecast was undertaken by IEA personnel (WBCSD, 2004b), the WEO 2004 and Mobility 2030 forecasts are quite similar. The WEO 2006 (IEA, 2006b) includes higher oil price assumptions than previously. Its projections therefore tend to be somewhat lower than the two other studies.

The three forecasts all assume that world oil supplies will be sufficient to accommodate the large projected increases in oil demand, and that world economies continue to grow without significant disruptions. With this caveat, all three forecast robust growth in world transport energy use over the next few decades, at a rate of around 2% per year. This means that transport energy use in 2030 will be about 80% higher than in 2002 (see Figure 5.3). Almost all of this new consumption is expected to be in petroleum fuels, which the forecasts project will remain between 93% and slightly over 95% of transport fuel use over the period. As a result, CO_2 emissions will essentially grow in lockstep with energy consumption (see Figure 5.4).

Another important conclusion is that there will be a significant regional shift in transport energy consumption, with the emerging economies gaining significantly in share (Figure 5.3). EIA's International Energy Outlook 2005, as well as the IEA, projects a robust 3.6% per year growth rate for these economies, while the IEA's more recent WEO 2006 projects transport demand growth of 3.2%. In China, the number of cars has been growing at a rate of 20% per year, and personal travel has increased by a factor of five over the past 20 years. At its projected 6% rate of growth, China's transport energy use would nearly quadruple between 2002 and 2025, from 4.3 EJ in 2002 to 16.4 EJ in 2025. China's neighbour India's transport energy is projected to grow at 4.7% per year during this period and countries such as Thailand, Indonesia, Malaysia and Singapore

will see growth rates above 3% per year. Similarly, the Middle East, Africa and Central and South America will see transport energy growth rates at or near 3% per year. The net effect is that the emerging economies' share of world transport energy use would grow in the EIA forecasts from 31% in 2002 to 43% in 2025. In 2004, the transport sector produced 6.2 $GtCO_2$ emissions (23% of world energy-related CO_2 emissions). The share of Non-OECD countries is 36% now and will increase rapidly to 46% by 2030 if current trends continue.

In contrast, transport energy use in the mature market economies is projected to grow more slowly. EIA forecasts 1.2% per year and IEA forecasts 1.3% per year for the OECD nations. EIA projects transport energy in the United States to grow at 1.7% per year, with moderate population and travel growth and only modest improvement in efficiency. Western Europe's transport energy is projected to grow at a much slower 0.4% per year, because of slower population growth, high fuel taxes and significant improvements in efficiency. IEA projects a considerably higher 1.4% per year for OECD Europe. Japan, with an aging population, high taxes and low birth rates, is projected to grow at only 0.2% per year. These rates would lead to 2002–2025 increases of 46%, 10% and 5%, for the USA, Western Europe and Japan, respectively. These economies' share of world transport energy would decline from 62% in 2002 to 51% in 2025.

The sectors propelling worldwide transport energy growth are primarily light-duty vehicles, freight trucks and air travel. The Mobility 2030 study projects that these three sectors will be responsible for 38, 27 and 23%, respectively, of the total 100 EJ growth in transport energy that it foresees in the 2000–2050 period. The WBCSD/SMP reference case projection indicates that the number of LDVs will grow to about 1.3 billion by 2030 and to just over 2 billion by 2050, which is almost three times

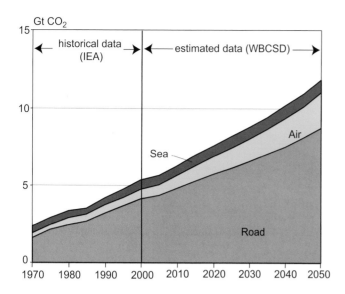

Figure 5.4: *Historical and projected CO₂ emission from transport by modes, 1970–2050*

Source: IEA, 2005; WBCSD, 2004b.

higher than the present level (Figure 5.5). Nearly all of this increase will be in the developing world.

Aviation

Civil aviation is one of the world's fastest growing transport means. ICAO (2006) analysis shows that aviation scheduled traffic (revenue passenger-km, RPK) has grown at an average annual rate of 3.8% between 2001 and 2005 despite the downturn from the terrorist attacks and SARS *(Severe Acute Respiratory*

Syndrome) during this period, and is currently growing at 5.9% per year. These figures disguise regional differences in growth rate: for example, Europe-Asia/Pacific traffic grew at 12.2% and North American domestic traffic grew at 2.6% per year in 2005. ICAO's outlook for the future forecasts a passenger traffic demand growth of 4.3% per year to 2020. Industry forecasts offer similar prospects for growth: the Airbus Global Market Forecast (Airbus, 2004) and Boeing Current Market Outlook (Boeing, 2006) suggest passenger traffic growth trends of 5.3% and 4.9% respectively, and freight trends at 5.9% and 6.1% respectively over the next 20 or 25 years. In summary, these forecasts and others predict a global average annual passenger traffic growth of around 5% – passenger traffic doubling in 15 years – with freight traffic growing at a faster rate that passenger traffic, although from a smaller base.

The primary energy source for civil aviation is kerosene. Trends in energy use from aviation growth have been modelled using the Aero2K model, using unconstrained demand growth forecasts from Airbus and UK Department of Trade and Industry. The model results suggest that by 2025 traffic will increase by a factor of 2.6 from 2002, resulting in global aviation fuel consumption increasing by a factor of 2.1 (QinetiQ, 2004). Aero2k model results suggest that aviation emissions were approximately 492 $MtCO_2$ and 2.06 $MtNO_x$ in 2002 and will increase to 1029 and 3.31 Mt respectively by 2025.

Several organizations have constructed scenarios of aviation emissions to 2050 (Figure 5.6), including:
- IPCC (1999) under various technology and GDP assumptions (IS92a, e and c). Emissions were most strongly affected by

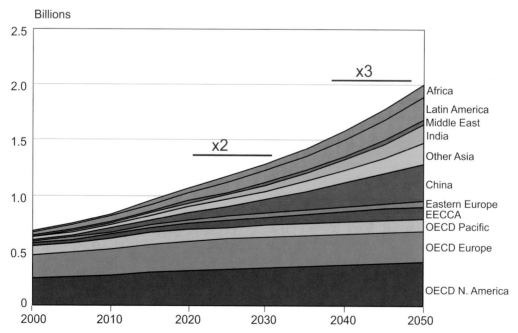

Figure 5.5: *Total stock of light-duty vehicles by region*

Source: WBCSD, 2004a.

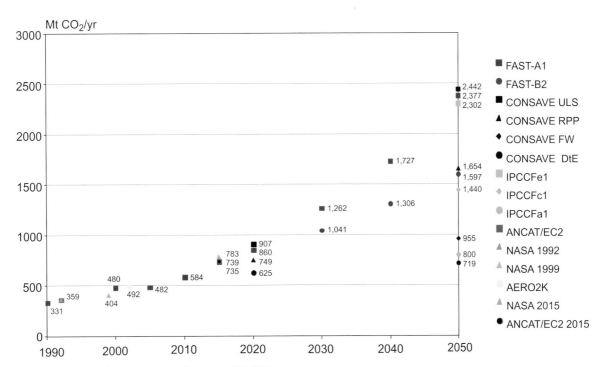

Figure 5.6: *Comparison of global CO$_2$ emissions of civil aviation, 1990–2050*

the GDP assumptions, with technology assumptions having only a second order effect;

- CONSAVE 2050, a European project has produced further 2050 scenarios (Berghof *et al.*, 2005). Three of the four CONSAVE scenarios are claimed to be broadly consistent with IPCC SRES scenarios A1, A2 and B1. The results were not greatly different from those of IPCC (1999);

- Owen and Lee (2005) projected aviation emissions for years 2005 through to 2020 by using ICAO-FESG forecast statistics of RPK (FESG, 2003) and a scenario methodology applied thereafter according to A1 and B2 GDP assumptions similarly to IPCC (1999).

The three estimates of civil aviation CO$_2$ emissions in 2050 from IPCC (1999) show an increase by factors of 2.3, 4.0 and 6.4 over 1992; CONSAVE (Berghof *et al.*, 2005) four scenarios indicate increases of factors of 1.5, 1.9, 3.4 and 5.0 over 2002 emissions (QinetiQ, 2004); and FAST A1 and B2 results (Owen and Lee, 2006) indicate increases by factors of 3.3 and 5.0 over 2000 emissions.

Shipping

Around 90% of global merchandise is transported by sea. For many countries sea transport represents the most important mode of transport for trade. For example, for Brazil, Chile and Peru over 95% of exports in volume terms (nearly 75% in value terms) are seaborne. Economic growth and the increased integration in the world economy of countries from far-east and southeast Asia is contributing to the increase of international marine transport. Developments in China are now considered to be one of the most important stimulus to growth for the tanker, chemical, bulk and container trades (OECD, 2004b).

World seaborne trade in ton-miles recorded another consecutive annual increase in 2005, after growing by 5.1%. Crude oil and oil products dominate the demand for shipping services in terms of ton-miles (40% in 2005) (UN, 2006), indicating that demand growth will continue in the future. During 2005, the world merchant fleet expanded by 7.2%. The fleets of oil tankers and dry bulk carriers, which together make up 72.9% of the total world fleet, increased by 5.4%. There was a 13.3% increase in the container ship fleet, whose share of total fleet is 12%.

Eyring *et al.* (2005a) provided a set of carbon emission projections out to 2050 (Eyring *et al.*, 2005b) based upon four traffic demand scenarios corresponding to SRES A1, A2, B1, B2 (GDP) and four technology scenarios which are summarized below in Table 5.2.

The resultant range of potential emissions is shown in Figure 5.7.

5.3 Mitigation technologies and strategies

Many technologies and strategies are at hand to reduce the growth or even, eventually, reverse transport GHG emissions. Most of the technology options discussed here were mentioned in the TAR. The most promising strategy for the near term is incremental improvements in current vehicle technologies. Advanced technologies that provide great promise include greater use of electric-drive technologies, including hybrid-

Table 5.2: *Summary of shipping technology scenarios*

Technology scenario 1 (TS1) – 'Clean scenario'	Technology scenario 2 (TS2) – 'Medium scenario'	Technology scenario 3 (TS3) – 'IMO compliant scenario'	Technology scenario 4 (TS4) – 'BAU'
Low S content fuel (1%/0.5%), aggressive NO$_x$ reductions	Relatively low S content fuel (1.8%/1.2%), moderate NO$_x$ reduction	High S content fuel (2%/2%), NO$_x$ reductions according to IMO stringency only	High S content fuel (2%/2%), NO$_x$ reductions according to IMO stringency only
Fleet = 75% diesel, 25% alternative plant	Fleet = 75% diesel, 25% alternative plant	Fleet = 75% diesel, 25% alternative plant	Fleet = 100% diesel

Note: The fuel S percentages refer to values assumed in (2020/2050).

Source: Eyring et al. 2005b.

electric power trains, fuel cells and battery electric vehicles. The use of alternative fuels such as natural gas, biofuels, electricity and hydrogen, in combination with improved conventional and advanced technologies; provide the potential for even larger reductions.

Even with all these improved technologies and fuels, it is expected that petroleum will retain its dominant share of transport energy use and that transport GHG emissions will continue to increase into the foreseeable future. Only with sharp changes in economic growth, major behavioural shifts, and/or major policy intervention would transport GHG emissions decrease substantially.

5.3.1 Road transport

GHG emissions associated with vehicles can be reduced by four types of measures:
1. Reducing the loads (weight, rolling and air resistance and accessory loads) on the vehicle, thus reducing the work needed to operate it;
2. Increasing the efficiency of converting the fuel energy to work, by improving drive train efficiency and recapturing energy losses;

3. Changing to a less carbon-intensive fuel; and
4. Reducing emissions of non-CO$_2$ GHGs from vehicle exhaust and climate controls.

The loads on the vehicle consist of the force needed to accelerate the vehicle, to overcome inertia; vehicle weight when climbing slopes; the rolling resistance of the tyres; aerodynamic forces; and accessory loads. In urban stop-and-go driving, aerodynamic forces play little role, but rolling resistance and especially inertial forces are critical. In steady highway driving, aerodynamic forces dominate, because these forces increase with the square of velocity; aerodynamic forces at 90 km/h[10] are four times the forces at 45 km/h. Reducing inertial loads is accomplished by reducing vehicle weight, with improved design and greater use of lightweight materials. Reducing tyre losses is accomplished by improving tyre design and materials, to reduce the tyres' rolling resistance coefficient, as well as by maintaining proper tyre pressure; weight reduction also contributes, because tyre losses are a linear function of vehicle weight. And reducing aerodynamic forces is accomplished by changing the shape of the vehicle, smoothing vehicle surfaces, reducing the vehicle's cross-section, controlling airflow under the vehicle and other measures. Measures to reduce the heating and cooling needs of the passengers, for example by changing window glass to reflect incoming solar radiation, are included in the group of measures.

Increasing the efficiency with which the chemical energy in the fuel is transformed into work, to move the vehicle and provide comfort and other services to passengers, will also reduce GHG emissions. This includes measures to improve engine efficiency and the efficiency of the rest of the drive train and accessories, including air conditioning and heating. The range of measures here is quite great; for example, engine efficiency can be improved by three different kinds of measures, increasing thermodynamic efficiency, reducing frictional losses and reducing pumping losses (these losses are the energy needed to pump air and fuel into the cylinders and push out the exhaust) and each kind of measure can be addressed by a great number of design, material and technology changes. Improvements in transmissions can reduce losses in the transmission itself and help engines to operate in their most

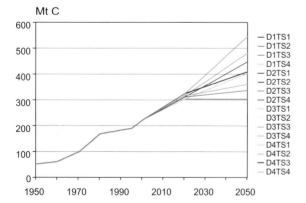

Mt C

Figure 5.7: *Historical and projected CO$_2$ emissions of seagoing shipping, 1990-2050*

Note: See Table 5.2 for the explanation of the scenarios.

Source: adapted from Eyring et al., 2005a,b.

10 1 km/h = 0.621 mph

efficient modes. Also, some of the energy used to overcome inertia and accelerate the vehicle – normally lost when the vehicle is slowed, to aerodynamic forces and rolling resistance as well to the mechanical brakes (as heat) – may be recaptured as electrical energy if regenerative braking is available (see the discussion of hybrid electric drive trains).

The use of different liquid fuels, in blends with gasoline and diesel or as 'neat fuels' require minimal or no changes to the vehicle, while a variety of gaseous fuels and electricity would require major changes. Alternative liquid fuels include ethanol, biodiesel and methanol, and synthetic gasoline and diesel made from natural gas, coal, or other feedstocks. Gaseous fuels include natural gas, propane, dimethyl ether (a diesel substitute) and hydrogen. Each fuel can be made from multiple sources, with a wide range of GHG emission consequences. In evaluating the effects of different fuels on GHG emissions, it is crucial to consider GHG emissions associated with fuel production and distribution in addition to vehicle tailpipe emissions (see the section on well-to-wheels analysis). For example, the consumption of hydrogen produces no emissions aside from water directly from the vehicle, but GHG emissions from hydrogen production can be quite high if the hydrogen is produced from fossil fuels (unless the carbon dioxide from the hydrogen production is sequestered).

The sections that follow discuss a number of technology, design and fuel measures to reduce GHG emissions from vehicles.

5.3.1.1   *Reducing vehicle loads*

Lightweight materials

A 10% weight reduction from a total vehicle weight can improve fuel economy by 4–8%, depending on changes in vehicle size and whether or not the engine is downsized. There are several ways to reduce vehicle weight; including switching to high strength steels (HSS), replacing steel by lighter materials such as Al, Mg and plastics, evolution of lighter design concepts and forming technologies. The amount of lighter materials in vehicles has been progressively increasing over time, although not always resulting in weight reductions and better fuel economy if they are used to increase the size or performance of the vehicle. In fact, the average weight of a vehicle in the USA and Japan has increased by 10–20% in the last 10 years (JAMA, 2002; Haight, 2003), partly due to increased concern for safety and customers' desire for greater comfort.

Steel is still the main material used in vehicles, currently averaging 70% of kerb weight. Aluminium usage has grown to roughly 100 kg per average passenger car, mainly in the engine, drive train and chassis in the form of castings and forgings. Aluminium is twice as strong as an equal weight of steel, allowing the designer to provide strong, yet lightweight structures. Aluminium use in body structures is limited, but there are a few commercial vehicles with all Al bodies (e.g., Audi's A2 and A8). Where more than 200 kg of Al is used and secondary weight reductions are gained by down-sizing the engine and suspension – more than 11–13% weight reduction can be achieved. Ford's P2000 concept car[11] has demonstrated that up to 300 kg of Al can be used in a 900 kg vehicle.

Magnesium has a density of 1.7–1.8 g/cc[12], about 1/4 that of steel, while attaining a similar (volumetric) strength. Major hurdles for automobile application of magnesium are its high cost and performances issues such as low creep strength and contact corrosion susceptibility. At present, the use of magnesium in vehicle is limited to only 0.1–0.3% of the whole weight. However, its usage in North American-built family vehicles has been expanding by 10 to 14% annually in recent years. Aluminium has grown at 4–6%; plastics by 1–1.8%; and high strength steels by 3.5–4%. Since the amount of energy required to produce Mg and also Al is large compared with steel, LCA analysis is important in evaluating these materials' potential for CO_2 emission reduction (Helms and Lambrecht, 2006). Also, the extent of recycling is an important issue for these metals.

The use of plastics in vehicles has increased to about 8% of total vehicle weight, which corresponds to 100-120 kg per vehicle. The growth rate of plastics content has been decreasing in recent years however, probably due to concerns about recycling, given that most of the plastic goes to the automobile shredder residue (ASR) at the end of vehicle life. Fibre-reinforced plastic (FRP) is now widely used in aviation, but its application to automobiles is limited due to its high cost and long processing time. However, its weight reduction potential is very high, maybe as much as 60%. Examples of FRP structures manufactured using RTM (resin transfer method) technology are wheel housings or entire floor assemblies. For a compact-size car, this would make it possible to reduce the weight; of a floor assembly (including wheel housings) by 60%, or 22 kg per car compared to a steel floor assembly. Research examples of plastics use in the chassis are leaf or coil springs manufactured from fibre composite plastic. Weight reduction potentials of up to 63% have been achieved in demonstrators using glass and/or carbon fibre structures (Friedricht, 2002).

Aside from the effect of the growing use of non-steel materials, the reduction in the average weight of steel in a car is driven by the growing shift from conventional steels to high strength steels (HSS). There are various types of HSS, from relatively low strength grade (around 400 MPa) such as solution-hardened and precipitation-hardened HSS to very high strength grade (980–1400 MPa) such as TRIP steel and tempered martensitic HSS. At present, the average usage per vehicle of HSS is 160 kg (11% of whole weight) in the USA

---

11 SAE International (Society of Automotive Engineers): The aluminum angle, automotive engineering on-line, http://www.sae.org/automag/metals/10.htm.
12 Specific gravity 1738

and 75 kg (7%) in Japan. In the latest Mercedes A-class vehicle, HSS comprises 67% of body structure weight. The international ULSAB-AVC project (Ultra Light Steel Auto Body – Advanced Vehicle Concept) investigated intensive use of HSS, including advanced HSS, and demonstrated that using HSS as much as possible can reduce vehicle weight by 214 kg (–19%) and 472 kg (–32%) for small and medium passenger cars respectively. In this concept, the total usage of HSS in body and closures structures is 280–330 kg, of which over 80% is advanced HSS (Nippon Steel, 2002).

Since heavy-duty vehicles such as articulated trucks are much heavier than passenger vehicles, their weight reduction potential is much larger. It is possible to reduce the weight of tractor and trailer combination by more than 3000 kg by replacing steel with aluminium (EAA, 2001).

Aerodynamics improvement

Improvements have been made in the aerodynamic performance of vehicles over the past decade, but substantial additional improvements are possible. Improvement in aerodynamic performance offers important gains for vehicles operating at higher speeds, e.g., long-distance trucks and light-duty vehicles operating outside congested urban areas. For example a 10% reduction in the coefficient of drag (C_D) of a medium sized passenger car would yield only about a 1% reduction in average vehicle forces on the US city cycle (with 31.4 km/h average speed), whereas the same drag reduction on the US highway cycle, with average speed of 77.2 km/h, would yield about a 4% reduction in average forces.[13] These reductions in vehicle forces translate reasonably well into similar reductions in fuel consumption for most vehicles, but variations in engine efficiency with vehicle force may negate some of the benefit from drag reduction unless engine power and gearing are adjusted to take full advantage of the reduction.

For light-duty vehicles, styling and functional requirements (especially for light-duty trucks) may limit the scope of improvement. However, some vehicles introduced within the past five years demonstrate that improvement potential still remains for the fleet. The Lexus 430, a conservatively styled sedan, attains a C_D (coefficient of aerodynamic drag) of 0.26 versus a fleet average of over 0.3 for the US passenger car fleet. Other fleet-leading examples are:
- Toyota Prius, Mercedes E-class sedans, 0.26
- Volkswagen Passat, Mercedes C240, BMW 320i, 0.27

For light trucks, General Motors' 2005 truck fleet has reduced average C_D by 5–7% by sealing unnecessary holes in the front of the vehicles, lowering their air dams, smoothing their undersides and so forth (SAE International, 2004).

The current generation of heavy-duty trucks in the United States has average C_Ds ranging from 0.55 for tractor-trailers to 0.65 for tractor-tandem trailers. These trucks generally have spoilers at the top of their cabs to reduce air drag, but substantial further improvements are available. C_D reductions of about 0.15, or 25% or so (worth about 12% reduced fuel consumption at a steady 65 mph[14]), can be obtained with a package of base flaps (simple flat plates mounted on the edges of the back end of a trailer) and side skirts (McCallen et al., 2004). The US Department of Energy's 2012 research goals for heavy-duty trucks (USDOE, 2000)[15] include a 20% reduction (from a 2002 baseline, with C_D of 0.625) in aerodynamic drag for a 'class 8' tractor-trailer combination.[16] C_D reductions of 50% and higher, coupled with potential benefits in safety (from better braking and roll and stability control), may be possible with pneumatic (air blowing) devices (Englar, 2001). A complete package of aerodynamic improvements for a heavy-duty truck, including pneumatic blowing, might save about 15–20% of fuel for trucks operating primarily on uncongested highways, at a cost of about 5000 US$ in the near-term, with substantial cost reductions possible over time (Vyas et al., 2002).

The importance of aerodynamic forces at higher speeds implies that reduction of vehicle highway cruising speeds can save fuel and some nations have used speed limits as fuel conservation measures, e.g., the US during the period following the 1973 oil embargo. US tests on nine vehicles with model years from 1988 to 1997 demonstrated an average 17.1% fuel economy loss in driving at 70 mph compared to 55 mph (ORNL, 2006). Recent tests on six contemporary vehicles, including two hybrids, showed similar results – the average fuel economy loss was 26.5% in driving at 80 mph compared to 60 mph, and 27.2% in driving at 70 mph compared to 50 mph (Duoba et al., 2005).

Mobil Air Conditioning (MAC) systems

MAC systems contribute to GHG emissions in two ways by direct emissions from leakage of refrigerant and indirect emissions from fuel consumption. Since 1990 significant progress has been made in limiting refrigerant emissions due to the implementation of the Montreal Protocol. The rapid switch from CFC-12 (GWP 8100) to HFC-134a (GWP 1300) has led to the decrease in the CO_2-eq emissions from about 850 $MtCO_2$-eq in 1990 to 609 $MtCO_2$-eq in 2003, despite the continued growth of the MAC system fleet (IPCC, 2005).

Refrigerant emissions can be decreased by using new refrigerants with a much lower GWP, such as HFC-152a or CO_2, restricting refrigerant sales to certified service professionals and better servicing and disposal practices. Although the feasibility of CO_2 refrigerant has been demonstrated, a number of technical hurdles have still to be overcome.

---

13 The precise value would depend on the value of the initial C_D as well as other aspects of the car's design.
14 1 mph = 1.6 km/h
15 Http://www.eere.energy.gov/vehiclesandfuels/about/partnerships/21centurytruck/21ct_goals.shtml.
16 These are heavy-duty highway trucks with separate trailers, but less than 5 axles – the standard long-haul truck in the U.S.

Since the energy consumption for MAC is estimated to be 2.5–7.5% of total vehicle energy consumption, a number of solutions have to be developed in order to limit the energy consumption of MAC, such as improvements of the design of MAC systems, including the control system and airflow management.

5.3.1.2 *Improving drive train efficiency*

Advanced Direct Injection Gasoline / Diesel Engines and transmissions.

New engine and transmission technologies have entered the light-duty vehicle fleets of Europe, the USA and Japan, and could yield substantial reductions in carbon emissions if more widely used.

Direct injection diesel engines yielding about 35% greater fuel economy than conventional gasoline engines are being used in about half the light-duty vehicles being sold in European markets, but are little used in Japan and the USA (European taxes on diesel fuel generally are substantially lower than on gasoline, which boosts diesel share). Euro 4 emission standards were enforced in 2005, with Euro 5 (still undefined) to follow around 2009–2010. These standards, plus Tier 2 standards in the USA, will challenge diesel NO_x controls, adding cost and possibly reducing fuel efficiency somewhat. Euro 4/Tier 2 compliant diesels for light-duty vehicles, obtaining 30% better fuel efficiency than conventional gasoline engines, may cost about 2000–3000 US$ more than gasoline engines (EEA, 2003).

Improvements to gasoline engines include direct injection. Mercedes' M271 turbocharged direct injection engine is estimated to attain 18% reduced fuel consumption, part of which is due to intake valve control and other engine technologies (SAE International, 2003a); cylinder shutoff during low load conditions (Honda Odyssey V6, Chrysler Hemi, GM V8s) (SAE International, 2003a) and improved valve timing and lift controls.

Transmissions are also being substantially improved. Mercedes, GM, Ford, Chrysler, Volkswagen and Audi are introducing advanced 6 and 7 speed automatics in their luxury vehicles, with strong estimated fuel economy improvements ranging from 4–8% over a 4-speed automatic for the Ford/GM 6-speed to a claimed 13% over a manual, plus faster acceleration, for the VW/Audi BorgWarner 6-speed (SAE International, 2003b). If they follow the traditional path for such technology, these transmissions will eventually be rolled into the fleet. Also, continuously variable transmissions (CVTs), which previously had been limited to low power drive trains, are gradually rising in their power-handling capabilities and are moving into large vehicles.

The best diesel engines currently used in heavy-duty trucks are very efficient, achieving peak efficiencies in the 45–46% range (USDOE, 2000). Although recent advances in engine and drive train technology for heavy-duty trucks have focused on emissions reductions, current research programmes in the US Department of Energy are aiming at 10–20% improvements in engine efficiency within ten years (USDOE, 2000), with further improvements of up to 25% foreseen if significant departures from the traditional diesel engine platform can be achieved.

Engines and drive trains can also be made more efficient by turning off the engine while idling and drawing energy from other sources. The potential for reducing idling emissions in heavy-duty trucks is significant. In the USA, a nationwide survey found that, on average, a long-haul truck consumed about 1,600 gallons, or 6,100 litres, per year from idling during driver rest periods. A variety of behavioural and technological practices could be pursued to save fuel. A technological fix is to switch to grid connections or use onboard auxiliary power units during idling (Lutsey *et al.*, 2004).

Despite the continued tightening of emissions standards for both light-duty vehicles and freight trucks, there are remaining concerns about the gap between tested emissions and on-road emissions, particularly for diesel engines. Current EU emissions testing uses test cycles that are considerably gentler than seen in actual driving, allowing manufacturers to design drive trains so that they pass emissions tests but 'achieve better fuel efficiency or other performance enhancement at the cost of higher emissions during operation on the road (ECMT, 2006).' Other concerns involve excessive threshold limits demanded of onboard diagnostics systems, aftermarket mechanical changes (replacement of computer chips, disconnection of exhaust gas recirculation systems) and failure to maintain required fluid levels in Selective Catalytic Reduction systems (ECMT, 2006). Similar concerns in the USA led to the phase-in between 2000 and 2004 of a more aggressive driving cycle (the US06 cycle) to emission tests for LDVs; however, the emission limits tied to this cycle were not updated when new Tier 2 emission standards were promulgated, so concerns about onroad emissions, especially for diesels, will apply to the USA as well.

Hybrid drive trains

Hybrid-electric drive trains combine a fuel-driven power source, such as a conventional internal combustion engine (ICE) with an electric drive train – electric motor/generator and battery (or ultracapacitor) - in various combinations.[17] In current hybrids, the battery is recharged only by regenerative braking and engine charging, without external charging from the grid. 'Plug-in hybrids,' which would obtain part of their energy from the electric grid, can be an option but require a larger battery and perhaps a larger motor. Hybrids save energy by:

---

17 A hybrid drive train could use an alternative to an electric drive train, for example a hydraulic storage and power delivery system. The U.S. Environmental Protection Agency has designed such a system.

- Shutting the engine down when the vehicle is stopped (and possibly during braking or coasting);
- Recovering braking losses by using the electric motor to brake and using the electricity generated to recharge the battery;
- Using the motor to boost power during acceleration, allowing engine downsizing and improving average engine efficiency;
- Using the motor instead of the engine at low load (in some configurations), eliminating engine operation during its lowest efficiency mode;
- Allowing the use of a more efficient cycle than the standard Otto cycle (in some hybrids);
- Shifting power steering and other accessories to (more efficient) electric operation.

Since the 1998 introduction of the Toyota Prius hybrid in the Japanese market, hybrid electric drive train technology has advanced substantially, expanding its markets, developing in alternative forms that offer different combinations of costs and benefits and improving component technologies and system designs. Hybrids now range from simple belt-drive alternator-starter systems offering perhaps 7 or 8% fuel economy benefit under US driving conditions to 'full hybrids' such as the Prius offering perhaps 40–50% fuel economy benefits[18] (the Prius itself more than doubles the fuel economy average – on the US test – of the combined 2004 US model year compact and medium size classes, although some portion of this gain is due to additional efficiency measures). Also, hybrids may improve fuel efficiency by substantially more than this in congested urban driving conditions, so might be particularly useful for urban taxis and other vehicles making frequent stops. Hybrid sales have expanded rapidly: in the United States, sales were about 7,800 in 2000 and have risen rapidly, to 207,000 in 2005[19]; worldwide hybrid sales were about 541,000 in 2005 (IEA Hybrid Website, 2006).

Improvements made to the Prius since its introduction demonstrate how hybrid technology is developing. For example, the power density of Prius's nickel metal hydride batteries has improved from 600 W/kg[1] in 1998 to 1250 W/kg[1] in 2004 - a 108% improvement. Similarly, the batteries' specific energy has increased 37% during the same period (EEA, 2003). Higher voltage in the 2004 Prius allows higher motor power with reduced electrical losses and a new braking-by-wire system maximizes recapture of braking energy. The 1998 Prius compact sedan attained 42 mpg on the US CAFE cycle, with 0–60 mph acceleration time of 14.5 seconds; the 2004 version is larger (medium size) but attains 55 mpg and a 0–60 of 10.5 seconds. Prius-type hybrid systems will add about 4,000 US$ to the price of a medium sized sedan (EEA, 2003), but continued cost reduction and development efforts should gradually reduce costs.

Hybridization can yield benefits in addition to directly improving fuel efficiency, including (depending on the design) enhanced performance (with reduced fuel efficiency benefits in some designs), less expensive 4-wheel drive systems, provision of electric power for off-vehicle use (e.g., GM Silverado hybrid), and ease of introducing more efficient transmissions such as automated manuals (using the motor to reduce shift shock).

Hybrid drive trains' strong benefits in congested stop-and-go travel mesh well with some heavier-duty applications, including urban buses and urban delivery vehicles. An initial generation of hybrid buses in New York City obtained about a 10% improvement in fuel economy as well as improved acceleration capacity and substantially reduced emissions (Foyt, 2005). More recently, a different design achieved a 45% fuel economy increase in NYC operation (not including summer, where the increase should be lower) (Chandler et al., 2006). Fedex has claimed a 57% fuel economy improvement for its E700 diesel hybrid delivery vehicles (Green Car Congress, 2004).

Hybrid applications extend to two and three-wheelers, as well, because these often operate in crowded urban areas in stop-and-go operation. Honda has developed a 50 cc hybrid scooter prototype that offers about a one-third reduction in fuel use and GHG emissions compared to similar 50 cc scooters (Honda, 2004). However, sales of two and three-wheeled vehicles in most markets are extremely price sensitive, so the extent of any potential market for hybrid technology may be quite limited.

Plug-in hybrids, or PHEVs, are a merging of hybrid electric and battery electric. PHEVs get some of their energy from the electricity grid. Plug-in hybrid technology could be useful for both light-duty vehicles and for a variety of medium duty vehicles, including urban buses and delivery vehicles. Substantial market success of PHEV technology is, however, likely to depend strongly on further battery development, in particular on reducing battery cost and specific energy and increasing battery lifetimes.

PHEVs' potential to reduce oil use is clear – they can use electricity to 'fuel' a substantial portion of miles driven. The US Electric Power Research Institute (EPRI, 2001) estimates that 30 km hybrids (those that have the capability to operate up to 30 km solely on electricity from the battery) can substitute electricity for gasoline for approximately 30–40% of miles driven in the USA. With larger batteries and motors, the vehicles could replace even more mileage. However, their potential to reduce GHG emissions more than that achieved by current hybrids depends on their sources of electricity. For regions that rely on relatively low-carbon electricity for off-peak power, e.g., natural gas combined cycle power, GHG reductions over the PHEV's lifecycle will be substantial; in contrast, PHEVs in areas that rely on coal-fired power could have increased lifecycle

---

18 Precise values are somewhat controversial because of disagreements about the fuel economy impact of other fuel-saving measures on the vehicles.
19 Based on sales data from http://electricdrive.org/index.php?tg=articles&topics=7 and J.D. Power.

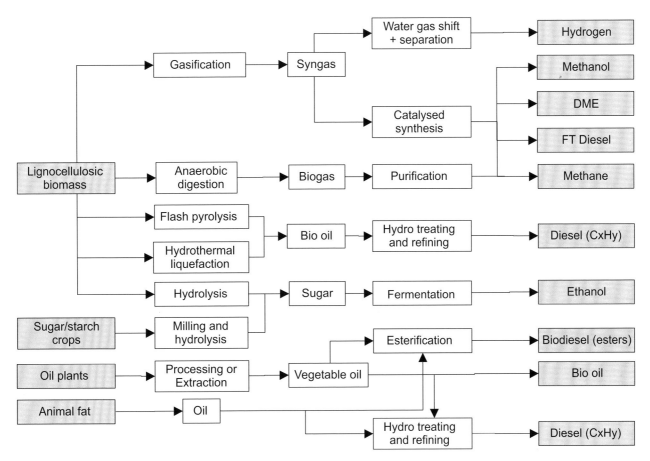

Figure 5.8: *Overview of conversion routes from crops to biofuels*
Source: Adapted from Hamelinck and Faaij, 2006.

carbon emissions. In the long-term, movement to a low-carbon electricity sector could allow PHEVs to play a major role in reducing transport sector GHG emissions.

5.3.1.3   Alternative fuels

Biofuels

The term biofuels describes fuel produced from biomass. A variety of techniques can be used to convert a variety of CO_2 neutral biomass feedstocks into a variety of fuels. These fuels include carbon-containing liquids such as ethanol, methanol, biodiesel, di-methyl esters (DME) and Fischer-Tropsch liquids, as well as carbon-free hydrogen. Figure 5.8 shows some main routes to produce biofuels: extraction of vegetable oils, fermentation of sugars to alcohol, gasification and chemical synthetic diesel, biodiesel and bio oil. In addition, there are more experimental processes, such as photobiological processes that produce hydrogen directly.

Biofuels can be used either 'pure' or as a blend with other automotive fuels. There is a large interest in developing biofuel technologies, not only to reduce GHG emission but more so to decrease the enormous transport sector dependence on imported oil. There are two biofuels currently used in the world for transport purposes – ethanol and biodiesel.

Ethanol is currently made primarily by the fermentation of sugars produced by plants such as sugar cane, sugar beet and corn. Ethanol is used in large quantities in Brazil where it is made from sugar cane, in the USA where it is made from corn, but only in very small quantities elsewhere.

Ethanol is blended with gasoline at concentrations of 5–10% on a volume basis in North America and Europe. In Brazil ethanol is used either in its pure form replacing gasoline, or as a blend with gasoline at a concentration of 20–25%. The production of ethanol fuelled cars in Brazil achieved 96% market share in 1985, but sharply declining shortly thereafter to near zero. Ethanol vehicle sales declined because ethanol producers shifted to sugar production and consumers lost confidence in reliable ethanol supply. A 25% blend of ethanol has continued to be used. With the subsequent introduction of flexfuel cars (see Box 5.2), ethanol fuel sales have increased. However, the sugar cane experience in Brazil will be difficult to replicate elsewhere. Land is plentiful, the sugar industry is highly efficient, the crop residues (bagasse) are abundant and easily used for process energy, and a strong integrated R&D capability has been developed in cane growing and processing.

In various parts of Asia and Africa, biofuels are receiving increasing attention and there is some experience with ethanol-

Box 5.2 Flexfuel vehicle (FFV)[20]

Particularly in Brazil where there is large ethanol availability as an automotive fuel there has been a substantial increase in sales of flexfuel vehicles (FFV). Flexfuel vehicle sales in Brazil represent about 81% (Nov. 2006) of the market share of light-duty vehicles. The use of FFVs facilitates the introduction of new fuels. The incremental vehicle cost is small, about 100 US$.

The FFVs were developed with systems that allow the use of one or more liquid fuels, stored in the same tank. This system is applied to OTTO cycle engines and enables the vehicles to run on gasoline, ethanol or both in a mixture, according to the fuel availability. The combustion control is done through an electronic device, which identifies the fuel being used and then the engine control system makes the suitable adjustments allowing the running of the engine in the most adequate condition.

One of the greatest advantages of FFVs is their flexibility to choose their fuel depending mainly on price. The disadvantage is that the engine cannot be optimized for the attributes of a single fuel, resulting in foregone efficiency and higher pollutant emissions (though the latter problem can be largely addressed with sophisticated sensors and computer controls, as it is in the USA).

In the USA[21], the number of FFVs is close to 6 million and some US manufacturers are planning to expand their sales. However, unlike in the Brazilian experience, ethanol has not been widely available at fuel stations (other than as a 10% blend) and thus the vehicles rarely fuel with ethanol. Their popularity in the USA is due to special fuel economy credits available to the manufacturer.

gasoline blending of up to 20%. Ethanol is being produced from sugar cane in Africa and from corn in small amounts in Asia. Biodiesel production is being considered from Jatropha (a drought resistant crop) that can be produced in most parts of Africa (Yamba and Matsika, 2004). It is estimated that with 10% ethanol-gasoline blending and 20% biodiesel-diesel blending in southern Africa, a reduction of 2.5 $MtCO_2$ and 9.4 $MtCO_2$ respectively per annum can be realized. Malaysian palm oil and US soybean oils are currently being used as biodiesel transport fuel in limited quantities and other oilseed crops are being considered elsewhere.

For the future, the conversion of ligno-cellulosic sources into biofuels is the most attractive biomass option. Ligno-cellulosic sources are grasses and woody material. These include crop residues, such as wheat and rice straw, and corn stalks and leaves, as well as dedicated energy crops. Cellulosic crops are attractive because they have much higher yields per hectare than sugar and starch crops, they may be grown in areas unsuitable for grains and other food/feed crops and thus do not compete with food, and the energy use is far less, resulting in much greater GHG reductions than with corn and most food crops (IEA, 2006a).

A few small experimental cellulosic conversion plants were being built in the USA in 2006 to convert crop residues (e.g., wheat straw) into ethanol, but considerably more R&D investment is needed to make these processes commercial.

These investments are beginning to be made. In 2006 BP announced it was committing 1 billion US$ to develop new biofuels, with special emphasis on bio-butanol, a liquid that can be easily blended with gasoline. Other large energy companies were also starting to invest substantial sums in biofuels R&D in 2006, along with the US Department of Energy, to increase plant yields, develop plants that are better matched with process conversion technologies and to improve the conversion processes. The energy companies in particular are seeking biofuels other than ethanol that would be more compatible with the existing petroleum distribution system.[22]

Biodiesel is less promising in terms of cost and production potential than cellulosic fuels but is receiving increasing attention. Bioesters are produced by a chemical reaction between vegetable or animal oil and alcohol, such as ethanol or methanol. Their properties are similar to those of diesel oil, allowing blending of bioesters with diesel or the use of 100% bioesters in diesel engines, and they are all called biodiesel. Blends of 20% biodiesel with 80% petroleum diesel (B20) can generally be used in unmodified diesel engines.[23]

Diesel fuel can also be produced through thermochemical hydrocracking of vegetable oil and animal fats. This technology has reached the demonstration stage. In Finland and Brazil[24] a commercial production project is under way. The advantage of the hydrocracked biodiesel is its stability and compatibility with conventional diesel (Koyama *et al.*, 2006).

20 http://www.eere.energy.gov/afdc/afv/eth_vehicles.html, http://en.wikipedia.org/wiki/Flexible-fuel_vehicle.
21 http://www.epa.gov/smartway/growandgo/documents/factsheet-e85.htm.
22 http://www.greencarcongress.com/2006/06/bp_and_dupont_t.html.
23 http://www.eere.energy.gov/afdc/altfuel/biodiesel.html.

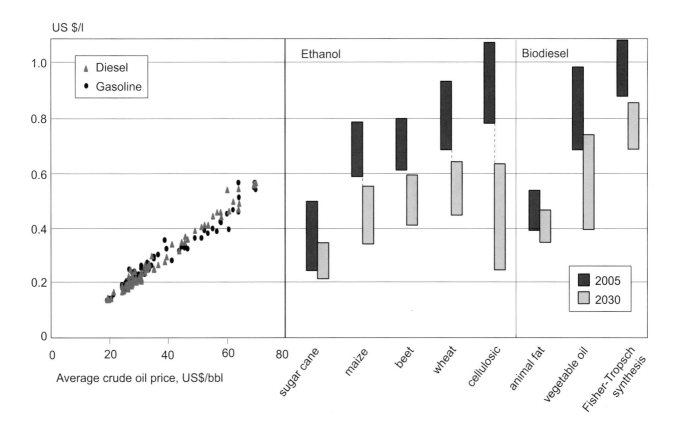

Figure 5.9: *Comparison of cost for various biofuels with those for gasoline and diesel*
Source: IEA, 2006b.

A large drawback of biodiesel fuels is the very high cost of feedstocks. If waste oils are used the cost can be competitive, but the quantity of waste oils is miniscule compared to transport energy consumption. If crops are used, the feedstock costs are generally far higher than for sugar, starch or cellulosic materials. These costs are unlikely to drop since they are the same highly developed crops used for foods and food processing. Indeed, if diverted to energy use, the oil feedstock costs are likely to increase still further, creating a direct conflict with food production. The least expensive oil feedstock at present is palm oil. Research is ongoing into new ways of producing oils. The promising feedstock seems to be algae, but cost and scale issues are still uncertain.

For 2030 IEA (2006a) reports mitigation potentials for bioethanol between 500–1200 MtCO$_2$, with possibly up to 100–300 MtCO$_2$ of that for ligno-cellulosic ethanol (or some other bio-liquid). The long-term potential for ligno-cellulosic fuels beyond 2030 is even greater. For biodiesel, it reports mitigation potential between 100–300 MtCO$_2$.

The GHG reduction potential of biofuels, especially with cellulosic materials, is very large but uncertain. IEA estimated the total mitigation potential of biofuels in the transport sector in 2050 to range from 1800 to 2300 MtCO$_2$ at 25 US$/tCO$_2$-

eq. based on scenarios with a respective replacement of 13 and 25% of transport energy demand by biofuels (IEA, 2006a). The reduction uncertainty is huge because of uncertainties related to costs and GHG impacts.

Only in Brazil is biofuel competitive with oil at 50 US$ per barrel or less. All others cost more. As indicated in Figure 5.9, biofuel production costs are expected to drop considerably, especially with cellulosic feedstocks. But even if the processing costs are reduced, the scale issue is problematic. These facilities have large economies of scale. However, there are large diseconomies of scale in feedstock production (Sperling, 1985). The cost of transporting bulky feedstock materials to a central point increases exponentially, and it is difficult assembling large amount of contiguous land to serve single large processing facilities.

Another uncertainty is the well-to-wheel reduction in GHGs by these various biofuels. The calculations are very complex because of uncertainties in how to allocate GHG emissions across the various products likely to be produced in the bio-refinery facilities, how to handle the effects of alternative uses of land, and so on, and the large variations in how the crops are grown and harvested, as well as the uncertain efficiencies and design configurations of future process technologies and

24 Brasil Energy, No.397-July/August (2006), 40:"H-Bio, The Clean Diesel".

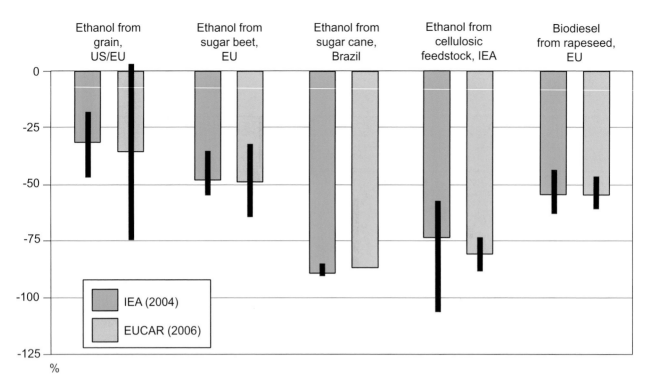

Figure 5.10: *Reduction of well-to-wheels GHG emissions compared to conventionally fuelled vehicles*
Note: bars indicate range of estimates.
Source: IEA, 2004c; EUCAR/CONCAWE/JRC, 2006.

bio-engineering plant materials. Typical examples are shown in Figure 5.10.

Ethanol from sugar cane, as produced in Brazil, provides significant reductions in GHG emissions compared to gasoline and diesel fuel on a 'well-to-wheels' basis. These large reductions result from the relatively energy efficient nature of sugar cane production, the use of bagasse (the cellulosic stalks and leaves) as process energy and the highly advanced state of Brazilian sugar farming and processing. Ongoing research over the years has improved crop yields, farming practices and process technologies. In some facilities the bagasse is being used to cogenerate electricity which is sold back to the electricity grid.

In contrast, the GHG benefits of ethanol made from corn are minor (Ribeiro & Yones-Ibrahim, 2001). Lifecycle estimates range from a net loss to gains of about 30%, relative to gasoline made from conventional oil. Farrell *et al.* (2006) evaluates the many studies and concludes that on average the reductions are probably about 13% compared to gasoline from conventional oil. The corn-ethanol benefits are minimal because corn farming and processing are energy intensive.

Biofuels might play an important role in addressing GHG emissions in the transport sector, depending on their production pathway (Figure 5.10). In the years to come, some biofuels may become economically competitive, as the result of increased biomass yields, developments of plants that are better suited

to energy production, improved cellulosic conversion processes and even entirely new energy crops and conversion processes. In most cases, it will require entirely new businesses and industries. The example of ethanol in Brazil is a model. The question is the extent to which this model can be replicated elsewhere with other energy crops and production processes.

The biofuel potential is limited by:
- The amount of available agricultural land (and in case of competing uses for that land) for traditional and dedicated energy crops;
- The quantity of economically recoverable agricultural and silvicultural waste streams;
- The availability of proven and cost-effective conversion technology.

Another barrier to increasing the potential is that the production of biofuels on a massive scale may require deforestation and the release of soil carbon as mentioned in Chapter 8.4. Another important point on biofuels is a view from the cost-effectiveness among the sectors. When comparing the use of biofuel in the transport sector with its use in power stations, the latter is more favourable from a cost-effectiveness point of view (ECMT, 2007).

Natural Gas (CNG / LNG / GTL)
Natural gas, which is mainly methane (CH_4), can be used directly in vehicles or converted into more compact fuels. It may be stored in compressed (CNG) or liquefied (LNG) form

on the vehicle. Also, natural gas may be converted in large petrochemical plants into petroleum-like fuels (the process is known as GTL, or gas-to-liquid). The use of natural gas as a feedstock for hydrogen is described in the hydrogen section.

CNG and LNG combustion characteristics are appropriate for spark ignition engines. Their high octane rating, about 120, allows a higher compression ratio than is possible using gasoline, which can increase engine efficiency. This requires that the vehicle be dedicated to CNG or LNG, however. Many current vehicles using CNG are converted from gasoline vehicles or manufactured as bifuel vehicles, with two fuel tanks. Bifuel vehicles cannot take full advantage of CNG's high octane ratio.

CNG has been popular in polluted cities because of its good emission characteristics. However, in modern vehicles with exhaust gas after-treatment devices, the non-CO_2 emissions from gasoline engines are similar to CNG, and consequently CNG loses its emission advantages in term of local pollutants; however it produces less CO_2. Important constraints on its use are the need for a separate refuelling infrastructure system and higher vehicle costs – because CNG is stored under high pressure in larger and heavier fuel tanks.

Gas-to-liquids (GTL) processes can produce a range of liquid transport fuels using Fischer-Tropsch or other conversion technologies. The main GTL fuel produced will be synthetic sulphur-free diesel fuel, although other fuels can also be produced. GTL processes may be a major source of liquid fuels if conventional oil production cannot keep up with growing demand, but the current processes are relatively inefficient: 61–65% (EUCAR/CONCAWE/JRC, 2006) and would lead to increased GHG emissions unless the CO_2 generated is sequestered.

DME can be made from natural gas, but it can also be produced by gasifying biomass, coal or even waste. It can be stored in liquid form at 5–10 bar pressure at normal temperature. This pressure is considerably lower than that required to store natural gas on board vehicles (200 bar). A major advantage of DME is its high cetane rating, which means that self-ignition will be easier. The high cetane rating makes DME suitable for use in efficient diesel engines.

DME is still at the experimental stage and it is still too early to say whether it will be commercially viable. During experiments, DME has been shown to produce lower emissions of hydrocarbons, nitric oxides and carbon monoxide than diesel and zero emissions of soot (Kajitani et al., 2005). There is no current developed distribution network for DME, although it has similarities to LPG and can use a similar distribution system. DME has a potential to reduce GHG emissions since it

has a lower carbon intensity (15 tC/TJ) than petroleum products (18.9–20.2 tC/TJ) (IPCC, 1996).

Hydrogen / Fuel Cells

During the last decade, fuel cell vehicles (FCVs) have attracted growing attention and have made significant technological progress. Drivers for development of FCVs are global warming (FCVs fuelled by hydrogen have zero CO_2 emission and high efficiency), air quality (zero tailpipe emissions), and energy security (hydrogen will be produced from a wide range of sources), and the potential to provide new desirable customer attributes (low noise, new designs).

There are several types of FCVs; direct-drive and hybrid power train architectures fuelled by pure hydrogen, methanol and hydrocarbons (gasoline, naphtha). FCVs with liquid fuels have advantages in terms of fuel storage and infrastructure, but they need on-board fuel reformers (fuel processors), which leads to lower vehicle efficiency (30–50% loss), longer start-up time, slower response and higher cost. Because of these disadvantages and rapid progress on direct hydrogen systems, nearly all auto manufacturers are now focused on the pure hydrogen FCV. Significant technological progress has been made since TAR including: improved fuel cell durability, cold start (sub-freezing) operation, increased range of operation, and dramatically reduced costs (although FCV drive train costs remain at least an order of magnitude greater than internal combustion engine (ICE) drive train costs) (Murakami and Uchibori, 2006).

In addition, many demonstration projects have been initiated since TAR[25]. Since 2000, members of the California Fuel Cell Partnership have placed 87 light-duty FCVs and 5 FC buses in California, which have travelled over 590,000 km on California's roads and highways. In 2002–2003, Japanese automakers began leasing FCVs in Japan and the USA, now totalling 17 vehicles. In 2004, US DOE started government/industry partnership 'learning demonstrations' for testing, demonstrating and validating hydrogen fuel cell vehicles and infrastructure and vehicle/infrastructure interfaces for complete system solutions. In Europe, there are several partnerships for FCV demonstration such as CUTE (Clean Urban Transport for Europe), CEP (Clean Energy Partnership) and ECTOS (Ecological City Transport System), using more than 30 buses and 20 passenger cars.

The recent US (NRC/NAE, 2004) and EU (JRC/IPTS, 2004) analyses conclude:
Although the potential of FCVs for reducing GHG emissions is very high there are currently many barriers to be overcome before that potential can be realized in a commercial market. These are:
- To develop durable, safe, and environmentally desirable fuel cell systems and hydrogen storage systems and reduce the

---

25 See the report of JHFC, Current status of overseas FCV demonstration, http://www.jhfc.jp/j/data/data/h17/11_h17seminar_e.pdf.

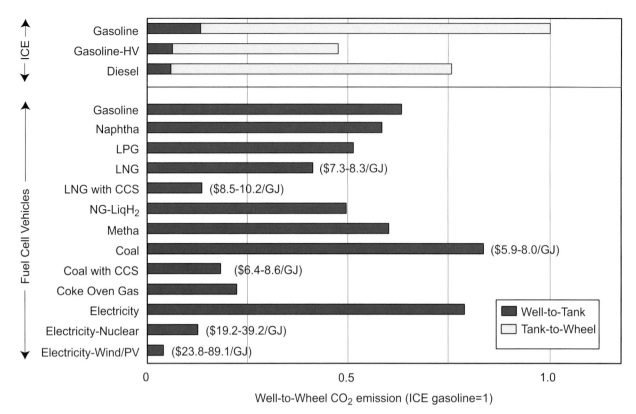

Figure 5.11: *Well-to-wheel CO₂ emission for major pathways of hydrogen with some estimates of hydrogen production cost (numbers in parentheses)*
Source: Toyota/Mizuho, 2004; NRC/NAE, 2004.

cost of fuel cell and storage components to be competitive with today's ICEs;

- To develop the infrastructure to provide hydrogen for the light-duty vehicle user;
- To sharply reduce the costs of hydrogen production from renewable energy sources over a time frame of decades. Or to capture and store ('sequester') the carbon dioxide byproduct of hydrogen production from fossil fuels.

Public acceptance must also be secured in order to create demand for this technology. The IEA echoes these points while also noting that deployment of large-scale hydrogen infrastructure at this point would be premature, as some of the key technical issues that are still being worked on, such as fuel cell operating conditions and hydrogen on-board storage options, may have a considerable impact on the choice of hydrogen production, distribution and refuelling (IEA, 2005).

The GHG impact of FCVs depends on the hydrogen production path and the technical efficiency achieved by vehicles and H_2 production technology. At the present technology level with FCV tank-to-wheel efficiency of about 50% and where hydrogen can be produced from natural gas at 60% efficiency, well-to-wheel (WTW) CO_2 emissions can be reduced by 50–60% compared to current conventional gasoline vehicles. In the future, those efficiencies will increase and the potential of WTW CO_2 reduction can be increased to nearly 70%. If hydrogen is derived from water by electrolysis using electricity produced

using renewable energy such as solar and wind, or nuclear energy, the entire system from fuel production to end-use in the vehicle has the potential to be a truly 'zero emissions'. The same is almost true for hydrogen derived from fossil sources where as much as 90% of the CO_2 produced during hydrogen manufacture is captured and stored (see Figure 5.11).

FCV costs are expected to be much higher than conventional ICE vehicles, at least in the years immediately following their introduction and H_2 costs may exceed gasoline costs. Costs for both the vehicles and fuel will almost certainly fall over time with larger-scale production and the effects of learning, but the long-term costs are highly uncertain. Figure 5.11 shows both well-to-wheels emissions estimates for several FCV pathways and their competing conventional pathways, as well as cost estimates for some of the hydrogen pathways.

Although fuel cells have been the primary focus of research on potential hydrogen use in the transport sector, some automakers envision hydrogen ICEs as a useful bridge technology for introducing hydrogen into the sector and have built prototype vehicles using hydrogen. Mazda has started to lease bi-fuel (hydrogen or gasoline) vehicles using rotary engines and BMW has also converted a 7-series sedan to bi-fuel operation using liquefied hydrogen (Kiesgen *et al.*, 2006) and is going to lease them in 2007. Available research implies that a direct injected turbocharged hydrogen engine could potentially achieve efficiency greater than a DI diesel (Wimmer *et al.*,

2005), although research and development challenges remain, including advanced sealing technology to insure against leakage with high pressure injection.

Electric vehicles

Fuel cell and hybrid vehicles gain their energy from chemical fuels, converting them into electricity onboard. Pure electric vehicles operating today are either powered from off-board electricity delivered through a conductive contact – usually buses with overhead wires or trains with electrified 'third rails' – or by electricity acquired from the grid and stored on-board in batteries. Future all-electric vehicles might use inductive charging to acquire electricity, or use ultracapacitors or flywheels in combination with batteries to store electricity on board.

The electric vehicles are driven by electric motors with high efficiencies of more than 90%, but their short driving range and short battery life have limited the market penetration. Even a limited driving range of 300 km requires a large volume of batteries weighing more than 400 kg (JHFC, 2006). Although the potential of CO_2 reduction strongly depends on the power mix, well-to-wheels CO_2 emission can be reduced by more than 50% compared to conventional gasoline-ICE (JHFC, 2006).

Vehicle electrification requires a more powerful, sophisticated and reliable energy-storage component than lead-acid batteries. These storage components will be used to start the car and also operate powerful by-wire control systems, store regenerative braking energy and to operate the powerful motor drives needed for hybrid or electric vehicles. Nickel metal hydride (NiMH) batteries currently dominate the power-assist hybrid market and Li ion batteries dominate the portable battery business. Both are being aggressively developed for broader automotive applications. The energy density has been increased to 170 Wh kg^{-1} and 500 Wh L^{-1} for small-size commercial Li ion batteries (Sanyo, 2005) and 130 Wh kg^{-1} and 310 Wh L^{-1} for large-size EV batteries (Yuasa, 2000). While NiMH has been able to maintain hybrid vehicle high-volume business, Li ion batteries are starting to capture niche market applications (e.g., the idle-stop model of Toyota's Vitz). The major hurdle left for Li ion batteries is their high cost.

Ultracapacitors offer long life and high power but low energy density and high current cost. Prospects for cost reduction and energy enhancement and the possibility of coupling the capacitor with the battery are attracting the attention of energy storage developers and automotive power technologists alike. The energy density of ultracapacitors has increased to 15–20 Wh kg^{-1} (Power System, 2005), compared with 40–60 Wh kg^{-1} for Ni-MH batteries. The cost of these advanced capacitors is in the range of several 10s of dollars/Wh, about one order of magnitude higher than Li batteries.

5.3.1.4 Well-to-wheels analysis of technical mitigation options

Life cycle analysis (LCA) is the most systematic and comprehensive method for the assessment of the environmental impacts of transport technologies. However, non-availability, uncertainty or variability of data limit its application. One key difficulty is deciding where to draw the boundary for the analysis; another is treating the byproducts of fuel production systems and their GHG emission credits. Also in some cases, LCA data varies strongly across regions

For automobiles, the life-cycle chain can be divided into the fuel cycle (extraction of crude oil, fuel processing, fuel transport and fuel use during operation of vehicle) and vehicle cycle (material production, vehicle manufacturing and disposal at the end-of-life). For a typical internal combustion engine (ICE) vehicle, 70–90% of energy consumption and GHG emissions take place during the fuel cycle, depending on vehicle efficiency, driving mode and lifetime driving distance (Toyota, 2004).

Recent studies of the Well-to-wheels CO_2 emissions of conventional and alternative fuels and vehicle propulsion concepts include a GM/ANL (2005) analysis for North America, EU-CAR/CONCAWE/JRC (2006) for Europe and Toyota/Mizuho (2004) for Japan. Some results are shown in Fig. 5.12. Some of the differences, as apparent from Figure 5.12 for ICE-gasoline and ICE-D (diesel) reflect difference in the oil producing regions and regional differences in gasoline and diesel fuel requirements and processing equipment in refineries.

The Well-to-wheel CO_2 emissions shown in Fig. 5.12 are for three groups of vehicle/fuel combinations – ICE/fossil fuel, ICE/biofuel and FCV. The full well-to-wheels CO_2 emissions depend on not only the drive train efficiency (TTW: tank-to-wheel) but also the emissions during the fuel processing (WTT: well-to-tank). ICE-CNG (compressed natural gas) has 15–25% lower emissions than ICE-G (Gasoline) because natural gas is a lower-carbon fuel and ICE-D (Diesel) has 16–24% lower emissions due to the high efficiency of the diesel engine. The results for hybrids vary among the analyses due to different assumptions of vehicle efficiency and different driving cycles. Although Toyota's analysis is based on Prius, and using Japanese 10–15 driving cycle, the potential for CO_2 reduction is 20–30% in general.

Table 5.3 summarizes the results and provides an overview of implementation barriers. The lifecycle emissions of ICE vehicles using biofuels and fuel cell vehicles are extremely dependent on the fuel pathways. For ICE-Biofuel, the CO_2 reduction potential is very large (30–90%), though world potential is limited by high production costs for several biomass pathways and land availability. The GHG reduction potential for the natural gas-sourced hydrogen FCV is moderate, but lifecycle emissions can be dramatically reduced by using CCS

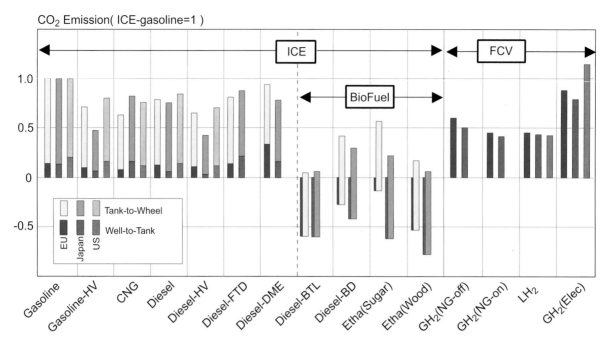

CO$_2$ Emission(ICE-gasoline=1)

Figure 5.12: *Comparison of three studies on Well-to-wheels CO$_2$ emission analyses*
Note: See text for an explanation of the legend. All the results are normalized by the value of ICE-G (gasoline).
Source: EUCAR/CONCAWE/JRC, 2006; GM/ANL, 2005; Toyota/Mizuho, 2004.

(carbon capture and storage) technology during H$_2$ production (FCEV-H2ccs in Table 5.3). Using renewable energy such as C-neutral biomass as a feedstock or clean electricity as an energy source (FCEV-RE-H$_2$) also will yield very low emissions.

5.3.1.5   Road transport: mode shifts

Personal motor vehicles consume much more energy and emit far more GHGs per passenger-km than other surface passenger modes. And the number of cars (and light trucks) continues to increase virtually everywhere in the world. Growth in GHG emissions can be reduced by restraining the growth in personal vehicle ownership. Such a strategy can, however, only be successful if high levels of mobility and accessibility can be provided by alternative means.

In general, collective modes of transport use less energy and generate less GHGs than private cars. Walking and biking emit even less. There is important worldwide mitigation potential if public and non-motorised transport trip share loss is reversed. The challenge is to improve public transport systems in order to preserve or augment the market share of low-emitting modes. If public transport gets more passengers, it is possible to increase the frequency of departures, which in turn may attract new passengers (Akerman and Hojer, 2006).

The USA is somewhat of an anomaly, though. In the USA, passenger travel by cars generates about the same GHG

emissions as bus and air travel on a passenger-km basis (ORNL Transportation Energy Databook; ORNL, 2006). That is mostly because buses have low load factors in the USA. Thus, in the USA, a bus-based strategy or policy will not necessarily lower GHG emissions. Shifting passengers to bus is not simply a matter of filling empty seats. To attract more passengers, it is necessary to enhance transit service. That means more buses operating more frequently – which means more GHG emissions. It is even worse than that, because transit service is already offered where ridership[26] demand is greatest. Adding more service means targeting less dense corridors or adding more service on an existing route. There are good reasons to promote transit use in the USA, but energy use and GHGs are not among them.

Virtually everywhere else in the world, though, transit is used more intensively and therefore has a GHG advantage relative to cars. Table 5.4 shows the broad average GHG emissions from different vehicles and transport modes in a developing country context. GHG emissions per passenger-km are lowest for transit vehicles and two-wheelers. It also highlights the fact that combining alternative fuels with public transport modes can reduce emissions even further.

It is difficult to generalize, though, because of substantial differences across nations and regions. The types of buses, occupancy factors, and even topography and weather can affect emissions. For example, buses in India and China tend

---

26 The number of passengers using a specific form of public transport.

Table 5.3: *Reduction of Well-to-wheels GHG emissions for various drive train/fuel combinations*

		Drive train/Fuel	GHG reduction (%)	Barriers
ICE	Fossil fuel	Gasoline (2010)	12-16	
		Diesel	16-24	Emissions (NO_x, PM)
		CNG	15-25	Infrastructure, storage
		G-HEV	20-52	Cost, battery
		D-HEV	29-57	Cost
	Biofuel	Ethanol-Cereal	30-65	Cost, availability (biomass, land), competition with food
		Ethanol-Sugar	79-87	
		Biodiesel	47-78	
		Advanced biofuel (cellulosic ethanol)	70-95	Technology, cost, environmental impact, competition with usage of other sectors
	H_2	H_2-ICE	6-16	H2 storage, cost
				Cost, infrastructure, deregulation
FCV		FCEV	43-59	
		FCEV+H_2ccs	78-86	Technology (stack, storage), cost, durability
		FCEV+RE-H_2	89-99	

Source: EUCAR/CONCAWE/JRC, 2006, GM/ANL, 2005 and Toyota/Mizuho (2004).

to be more fuel-efficient than those in the industrialized world, primarily because they have considerably smaller engines and lack air conditioning (Sperling and Salon, 2002).

Public transport

In addition to reducing transport emissions, public transport is considered favourably from a socially sustainable point of view because it gives higher mobility to people who do not have access to car. It is also attractive from an economically sustainable perspective since public transport provides more capacity at less marginal cost. It is less expensive to provide additional capacity by expanding bus service than building new roads or bridges. The expansion of public transport in the form of large capacity buses, light rail transit and metro or suburban rail can be feasible mitigation options for the transport sector.

The development of new rail services can be an effective measure for diverting car users to carbon-efficient mode while providing existing public transport users with upgraded service. However, a major hurdle is higher capital and possibly operating cost of the project. Rail is attractive and effective at generating high ridership in very dense cities. During the 1990s, less capital-intensive public transport projects such as light rail transit (LRT) were planned and constructed in Europe, North America and Japan. The LRT systems were successful in some regions, including a number of French cities where land use and transport planning is often well integrated (Hylen and Pharoah, 2002), but less so in other cities especially in the USA (Richmond, 2001; Mackett and Edwards, 1998), where more attention has been paid to this recently.

Around the world, the concept of bus rapid transit (BRT) is gaining much attention as a substitute for LRT and as an enhancement of conventional bus service. BRT is not new. Plans and studies for various BRT type alternatives have been prepared since the 1930s and a major BRT system was installed in Curitiba, Brazil in the 1970s (Levinson *et al.*, 2002). But only since about 2000 has the successful Brazilian experience gained serious attention from cities elsewhere.

BRT is 'a mass transit system using exclusive right of way lanes that mimic the rapidity and performance of metro systems, but utilizes bus technology rather than rail vehicle technology' (Wright, 2004). BRT systems can be seen as enhanced bus service and an intermediate mode between conventional bus service and heavy rail systems. BRT includes features such as exclusive right of way lanes, rapid boarding and alighting, free transfers between routes and preboard fare collection and fare verification, as well as enclosed stations that are safe and comfortable, clear route maps, signage and real-time information displays, modal integration at stations and terminals, clean vehicle technologies and excellence in marketing and customer service. To be most effective, BRT systems (like other transport initiatives) should be part of a comprehensive strategy that includes increasing vehicle and fuel taxes, strict land-use controls, limits and higher fees on parking, and integrating transit systems into a broader package of mobility for all types of travellers (IEA, 2002b).

Most BRT systems today are being delivered in the range of 1–15 million US$/km, depending upon the capacity requirements and complexity of the project. By contrast, elevated rail systems and underground metro systems can cost from 50 million US$

Table 5.4: *GHG Emissions from vehicles and transport modes in developing countries*

	Load factor (average occupancy)	CO$_2$-eq emissions per passenger-km (full energy cycle)
Car (gasoline)	2.5	130-170
Car (diesel)	2.5	85-120
Car (natural gas)	2.5	100-135
Car (electric)[a]	2.0	30-100
Scooter (two-stroke)	1.5	60-90
Scooter (four-stroke)	1.5	40-60
Minibus (gasoline)	12.0	50-70
Minibus (diesel)	12.0	40-60
Bus (diesel)	40.0	20-30
Bus (natural gas)	40.0	25-35
Bus (hydrogen fuel cell)[b]	40.0	15-25
Rail Transit[c]	75% full	20-50

Note: All numbers in this table are estimates and approximations and are best treated as illustrative.

a) Ranges are due largely to varying mixes of carbon and non-carbon energy sources (ranging from about 20–80% coal), and also the assumption that the battery electric vehicle will tend to be somewhat smaller than conventional cars.
b) Hydrogen is assumed to be made from natural gas.
c) Assumes heavy urban rail technology ('Metro') powered by electricity generated from a mix of coal, natural gas and hydropower, with high passenger use (75% of seats filled on average).

Source: Sperling and Salon, 2002.

Table 5.5: *Cost and potential estimated for BRT in Bogota*

O & M[a] (%)	Fuel price per barrel (US$)	Cost (US$/tCO$_2$)
20	40	11.22
20	60	7.60
50	40	12.20
50	60	15.84

Note: Assuming 20% of the urban population uses the BRT each day.

a) Operation and maintenance (O & M) costs are expected to be high as the measure involves high demand for management and implementation beyond putting up the infrastructure.

Source: estimate based on Bogata CDM Project (footnote 27)

to over 200 million US$/km (Wright, 2004). BRT systems now operate in several cities throughout North America, Europe, Latin America, Australia, New Zealand and Asia. The largest and most successful systems to date are in Latin America in Bogotá, Curitiba and Mexico City (Karekezi *et al.*, 2003).

Analysing the Bogotá Clean Development Mechanism project gives an insight into the cost and potential of implementing BRT in large cities. The CDM project shows the potential of moving about 20% of the city population per day on the BRT that mainly constitutes putting up dedicated bus lanes (130 km), articulated buses (1200) and 500 other large buses operating on feeder routes. The project is supported by an integrated fare system, centralized coordinated fleet control and improved bus management[27]. Using the investment costs, an assumed operation and maintenance of 20–50%[28] of investment costs per year, fuel costs of 40 to 60 US$ per barrel in 2030 and a discount rate of 4%, a BRT lifespan of 30 years, the cost of implementing BRT in the city of Bogotá was estimated to range from 7.6 US$/tCO$_2$ to 15.84 US$/tCO$_2$ depending on the price of fuel and operation and maintenance (Table 5.5). Comparing with results of Winkelman (2006), BRT cost estimates ranged from 14-66 US$/tCO$_2$ depending on the BRT package involved

(Table 5.6). The potential for CO$_2$ reduction for the city of Bogotá was determined to average 247,000 tCO$_2$ per annum or 7.4 million tCO$_2$ over a 30 year lifespan of the project.

Non-motorized transport (NMT)

The prospect for the reduction in CO$_2$ emissions by switching from cars to non-motorized transport (NMT) such as walking and cycling is dependent on local conditions. In the Netherlands, where 47% of trips are made by NMT, the NMT plays a substantial role up to distances of 7.5 km and walking up to 2.5 km (Rietveld, 2001). As more than 30% of trips made in cars in Europe cover distances of less than 3 km and 50% are less than 5 km (EC, 1999), NMT can possibly reduce car use in terms of trips and, to a lesser extent, in terms of kilometres. While the trend has been away from NMT, there is considerable potential to revive interest in NMT. In the Netherlands, with strong policies and cultural commitment, the modal share of bicycle and walking for accessing trains from home is about 35 to 40% and 25% respectively (Rietveld, 2001).

Walking and cycling are highly sensitive to the local built environment (ECMT, 2004a; Lee and Mouden, 2006). In Denmark, where the modal share of cycling is 18%, urban planners seek to enhance walking and cycling by shortening journey distances and providing better cycling infrastructure (Dill and Carr 2003, Page, 2005). In the UK where over 60% of people live within a 15 minute bicycle ride of a station, NMT could be increased by offering convenient, secure bicycle parking at stations and improved bicycle carriage on trains (ECMT, 2004a).

Safety is an important concern. NMT users have a much higher risk per trip of being involved in an accident than those using cars, especially in developing countries where most NMT users cannot afford to own a car (Mohan and Tiwari, 1999). Safety can be improved through traffic engineering and campaigns to educate drivers. An important co-benefit of NMT,

---

27 http://cdm.unfccc.int/Projects/DB/DNV-CUK1159192623.07/view.html.
28 O & M costs are expected to be high as the measure involves high demand for management and implementation beyond putting up the infrastructure.

gaining increasing attention in many countries, is public health (National Academies studies in the USA; Pucher, 2004).

In Bogotá, in 1998, 70% of the private car trips were under 3 km. This percentage is lower today thanks to the bike and pedestrian facilities. The design of streets was so hostile to bicycle travel that by 1998 bicycle trips accounted for less than 1% of total trips. After some 250 km of new bicycle facilities were constructed by 2001 ridership had increased to 4% of total trips. In most of Africa and in much of southern Asia, bicyclists and other non-motorised and animal traction vehicles are generally tolerated on the roadways by authorities. Non-motorised goods transport is often important for intermodal goods transport. A special form of rickshaw is used in Bangladesh, the bicycle van, which has basically the same design as a rickshaw (Hook, 2003).

Mitigation potential of modal shifts for passenger transport

Rapid motorization in the developing world is beginning to have a large effect on global GHG emissions. But motorization can evolve in quite different ways at very different rates. The amount of GHG emissions can be considerably reduced by offering strong public transport, integrating transit with efficient land use, enhancing walking and cycling, encouraging minicars and electric two-wheelers and providing incentives for efficient vehicles and low-GHG fuels. Few studies have analyzed the potential effect of multiple strategies in developing nations, partly because of a severe lack of reliable data and the very large differences in vehicle mix and travel patterns among varying areas.

Wright and Fulton (2005) estimated that a 5% increase in BRT mode share against a 1% mode share decrease of private automobiles, taxis and walking, plus a 2% share decrease of mini-buses can reduce CO_2 emissions by 4% at an estimated cost of 66 US$/tCO_2 in typical Latin American cities. A 5% or 4% increase in walking or cycling mode share in the same scenario analysis can also reduce CO_2 emissions by 7% or 4% at an estimated cost of 17 or 15 US$/tCO_2, respectively (Table 5.6). Although the assumptions of a single infrastructure unit cost and its constant impact on modal share in the analysis might be too simple, even shifting relatively small percentages of mode share to public transport or NMT can be worthwhile, because of a 1% reduction in mode share of private automobiles represents over 1 $MtCO_2$ through the 20-year project period.

Figure 5.13 shows the GHG transport emission results, normalized to year 2000 emissions, of four scenario analyses of developing nations and cities (Sperling and Salon, 2002). For three of the four cases, the 'high' scenarios are 'business-as-usual' scenarios assuming extrapolation of observable and emerging trends with an essentially passive government presence in transport policy. The exception is Shanghai, which is growing and changing so rapidly that 'business-as-usual' has little meaning. In this case the high scenario assumes both rapid

Table 5.6: *CO_2 reduction potential and cost per tCO_2 reduced using public transit policies in typical Latin American cities*

Transport measure	GHG reduc-tion potential (%)	Cost per tCO₂ (US$)
BRT mode share increases from 0-5%	3.9	66
BRT mode share increases from 0-10%	8.6	59
Walking share increases from 20-25%	6.9	17
Bike share increases from 0-5%	3.9	15
Bike mode share increases from 1-10%	8.4	14
Package (BRT, pedestrian up-grades, cycleways)	25.1	30

Source: Wright and Fulton, 2005.

motorization and rapid population increases, with the execution of planned investments in highway infrastructure while at the same time efforts to shift to public transport falter (Zhou and Sperling, 2001).

5.3.1.6   *Improving driving practices (eco-driving)*

Fuel consumption of vehicles can be reduced through changes in driving practices. Fuel-efficient driving practices, with conventional combustion vehicles, include smoother deceleration and acceleration, keeping engine revolutions low, shutting off the engine when idling, reducing maximum speeds and maintaining proper tyre pressure (IEA, 2001). Results from studies conducted in Europe and the USA suggested possible improvement of 5–20% in fuel economy from eco-driving training. The mitigation costs of CO_2 by eco-driving training were mostly estimated to be negative (ECMT/IEA, 2005).

Eco-driving training can be attained with formal training programmes or on-board technology aids. It applies to drivers of all types of vehicles, from minicars to heavy-duty trucks. The major challenge is how to motivate drivers to participate in the programme, and how to make drivers maintain an efficient driving style long after participating (IEA, 2001). In the Netherlands, eco-driving training is provided as part of driving school curricula (ECMT/IEA, 2005).

5.3.2   Rail

Railway transport is widely used in many countries. In Europe and Japan, electricity is a major energy source for rail, while diesel is a major source in North America. Coal is also still used in some developing countries. Rail's main roles are high speed passenger transport between large (remote) cities, high density commuter transport in the city and freight transport over long distances. Railway transport competes with other transport modes, such as air, ship, trucks and private vehicles. Major

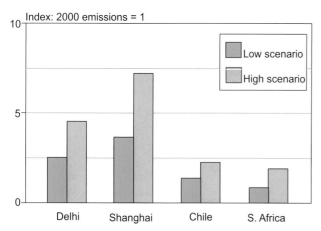

Figure 5.13: *Projections for transport GHG emissions in 2020 for some cities of developing countries*

Notes: Components of the Low 2020 scenario:

Delhi (Bose and Sperling, 2001): Completion of planned busways and rail transit, land-use planning for high density development around railway stations, network of dedicated bus lanes, promotion of bicycle use, including purchase subsidies and special lanes, promotion of car sharing, major push for more natural gas use in vehicles, economic re-straints on personal vehicles.

Shanghai (Zhou and Sperling, 2001): Emphasis on rapid rail system growth, high density development at railway sta-tions, bicycle promotion with new bike lanes and parking at transit stations, auto industry focus on minicars and farm cars rather than larger vehicles, incentives for use of high tech in minicars – electric, hybrid, fuel cell drive trains, promotion of car sharing.

Chile (O'Ryan et al., 2002): Overall focus on stronger use of market-based policy to insure that vehicle users pay the full costs of driving, internalizing costs of pollution and congestion, parking surcharges and restrictions, vehicle fees, and road usage fees, improvements in bus and rail systems, encourage-ment of minicars, with lenient usage and parking rules and strong commitment to alternative fuels, especially natural gas. By 2020, all taxis and 10% of other light and medium vehicles will use natural gas; all new buses will use hydrogen, improvements in bus and rail sys-tems.

South Africa (Prozzi et al., 2002): Land-use policies towards more efficient growth patterns, strong push to improve public transport, including use of bus-ways in dense corridors, provision of new and better buses, strong government oversight of the minibus jitney industry, incentives to moderate private car use, coal-based synfuels shifts to imported natural gas as a feedstock

Source: Sperling and Salon, 2002.

R&D goals for railway transport are higher speeds, improved comfort, cost reductions, better safety and better punctuality. Many energy efficiency technologies for railways are discussed in the web site of the International Union of Railways.[29] R&D programmes aimed at CO_2 reduction include:

Reducing aerodynamic resistance

For high speed trains such as the Japanese Shinkansen, French TGV and German ICE, aerodynamic resistance dominates vehicle loads. It is important to reduce this resistance to reduce energy consumption and CO_2 emissions. Aerodynamic resistance is determined by the shape of the train. Therefore, research has been carried out to find the optimum shape by using computer simulation and wind tunnel testing. The latest series 700 Shinkansen train has reduced aerodynamic resistance by 31% compared with the first generation Shinkansen.

Reducing train weight

Reduction of train weight is an effective way to reduce energy consumption and CO_2 emission. Aluminium car bodies, lightweight bogies and lighter propulsion equipments are proven weight reduction measures.

Regenerative braking

Regenerative brakes have been used in railways for three decades, but with limited applications. For current systems, the electric energy generated by braking is used through a catenary for powering other trains, reducing energy consumption and CO_2 emissions. However, regenerative braking energy cannot be effectively used when there is no train running near a braking train. Recently research in energy storage device onboard or trackside is progressing in several countries. Lithium ion batteries, ultracapacitors and flywheels are candidates for such energy storage devices.

Higher efficiency propulsion system

Recent research on rail propulsion has focused on superconducting on-board transformers and permanent magnet synchronous traction motors.

Apart from the above technologies mainly for electric trains, there are several promising technologies for diesel swichers, including common rail injection system and hybridization/on-board use of braking energy in diesel-electric vehicles (see the web site of the International Union of Railways),

5.3.3 Aviation

Fuel efficiency is a major consideration for aircraft operators as fuel currently represents around 20% of total operating costs for modern aircraft (2005 data, according to ICAO estimates[30] for the scheduled airlines of Contracting States). Both aircraft and engine manufacturers pursue technological developments to reduce fuel consumption to a practical minimum. There are no fuel efficiency certification standards for civil aviation. ICAO[31] has discussed the question of whether such a standard would be desirable, but has been unable to develop any form of parameter from the information available that correlates sufficiently well with the aircraft/engine performance and is therefore unable to define a fuel efficiency parameter that might be used for a standard at this time. 'Point' certification could drive manufacturers to comply with the regulatory requirement, possibly at the expense of fuel consumption for other operational conditions and missions. Market pressures therefore determine fuel efficiency and CO_2 emissions.

---

29 Energy Efficiency Technologies for Railways: http://www.railway-energy.org/tfee/index.php.
30 ICAO Estimates for the scheduled airlines of Contracting States, 2005.
31 Doc. 9836, CAEP/6, 2004.

Box 5.3 Constraints on aviation technology development

Technology developments in civil aviation are brought to the marketplace only after rigorous airworthiness and safety testing. The engineering and safety standards that apply, along with exacting weight minimisation, reliability and maintainability requirements, impose constraints to technology development and diffusion that do not necessarily apply to the same degree for other transport modes. Some of these certification requirements for engines are as follows:

- Altitude relight to 30000ft – the engine must be capable of relighting under severe adverse conditions
- Engine starting capability between –50°C–+50°C
- Ice, hail and water ingestion
- Fan blade off test – blade to be contained and engine to run down to idle
- ETOPS (extended range operations) clearance – demonstrable engine reliability to allow single engine flight for up to 240 minutes for twin-engine aircraft

In addition, the need to comply with stringent engine emissions and aircraft noise standards, to offer products that allow aircraft to remain commercially viable for three decades or more and to meet the most stringent safety requirements impose significant costs for developments. Moreover, a level of engineering excellence beyond that demanded for other vehicles is the norm. It is under these exacting conditions that improvements are delivered thus affect the rate at which improvements can be offered.

Technology developments

Aviation's dependence on fossil fuels, likely to continue for the foreseeable future, drives a continuing trend of fuel efficiency improvement through aerodynamic improvements, weight reductions and engine fuel efficient developments. New technology is developed not only to be introduced into new engines, but also, where possible, to be incorporated into engines in current production. Fuel efficiency improvements also confer greater range capability and extend the operability of aircraft. Evolutionary developments of engine and airframe technology have resulted in a positive trend of fuel efficiency improvements since the passenger jet aircraft entered service, but more radical technologies are now being explored to continue this trend.

Engine developments

Engine developments require a balancing of the emissions produced to both satisfy operational need (fuel efficiency) and regulatory need (NO_x, CO, smoke and HC). This emissions performance balance must also reflect the need to deliver safety, reliability, cost and noise performance for the industry. Developments that reduce weight, reduce aerodynamic drag or improve the operation of the aircraft can offer all-round benefits. Emissions – and noise – regulatory compliance hinders the quest for improved fuel efficiency, and is often most difficult for those engines having the highest pressure ratios (PR). Higher PRs increase the temperature of the air used for combustion in the engine, exacerbating the NO_x emissions challenge. Increasing an engine's pressure ratio is one of the options engine manufacturers use to improve engine efficiency. Higher pressure ratios are likely to be a continuing trend in engine development, possibly requiring revolutionary NO_x control techniques to maintain compliance with NO_x certification standards.

A further consideration is the need to balance not only emissions trade-offs, but the inevitable trade-off between emissions and noise performance from the engine and aircraft. For example, the engine may be optimised for minimum NO_x emissions, at which design point the engine will burn more fuel than it might otherwise have done. A similar design compromise may reduce noise and such performance optimisation must be conducted against engine operability requirements described in Box 5.3.

Aircraft developments

Fuel efficiency improvements are available through improvements to the airframe, as well as the engine. Most modern civil jet aircraft have low-mounted swept wings and are powered by two or four turbofan engines mounted beneath the wings. Such subsonic aircraft are about 70% more fuel efficient per passenger-km than 40 years ago. The majority of this gain has been achieved through engine improvements and the remainder from airframe design improvements. A 20% improvement in fuel efficiency of individual aircraft types is projected by 2015 and a 40–50% improvement by 2050 relative to equivalent aircraft produced today (IPCC, 1999). The current aircraft configuration is highly evolved, but has scope for further improvement. Technological developments have to be demonstrated to offer proven benefits before they will be adopted in the aviation industry, and this coupled with the overriding safety requirements and a product lifetime that has 60% of aircraft in service at 30 years age (ICAO, 2003) results in slower change than might be seen in other transport forms.

For the near term, lightweight composite materials for the majority of the aircraft structure are beginning to appear and promise significant weight reductions and fuel burn benefits. The use of composites, for example in the Boeing 787 aircraft

Table 5.7: *Weight breakdown for four kerosene-fuelled configurations with the same payload and range*

Configuration	Empty weight (t)	Payload (t)	Fuel (t)	Max TOW (t)
Baseline	236	86	178 (100%)	500
BWB	207	86	137 (77%)	430
Laminar Flying Wing (LFW)	226	86	83 (47%)	395
LFW with UDF	219	86	72 (40%)	377

Source: GbD, 2001.

(that has yet to enter service), could reduce fuel consumption by 20% below that of the aircraft the B787 will replace[32]. Other developments, such as the use of winglets, the use of fuselage airflow control devices and weight reductions have been studied by aircraft manufacturers and can reduce fuel consumption by around 7%[33]. But these can have limited practical applicability – for example, the additional fuel burn imposed by the weight of winglets can negate any fuel efficiency advantage for short haul operations.

Longer term, some studies suggest that a new aircraft configuration might be necessary to realise a step change in aircraft fuel efficiency. Alternative aircraft concepts such as blended wing bodies or high aspect ratio/low sweep configuration aircraft designs might accomplish major fuel savings for some operations. The blended wing body (flying wing) is not a new concept and in theory holds the prospect of significant fuel burn reductions: estimates suggest 20–30% compared with an equivalent sized conventional aircraft carrying the same payload (GbD, 2001; Leifsson and Mason, 2005). The benefits of this tailless design result from the minimised skin friction drag, as the tail surfaces and some engine/fuselage integration can be eliminated. Its development for the future will depend on a viable market case and will incur significant design, development and production costs.

Laminar flow technology (reduced airframe drag through control of the boundary layer) is likely to provide additional aerodynamic efficiency potential for the airframe, especially for long-range aircraft. This technology extends the smooth boundary layer of undisturbed airflow over more of the aerodynamic structure, in some cases requiring artificial means to promote laminar flow beyond its natural extent by suction of the disturbed flow through the aerodynamic surface. Such systems have been the subject of research work in recent times, but are still far from a flightworthy application. Long-term technical and economic viability have yet to be proven, despite studies suggesting that fuel burn could be reduced by between 10 and 20% for suitable missions (Braslow, 1999).

In 2001 the Greener by Design (GbD) technology subgroup of the Royal Aeronautical Society considered a range of possible future technologies for the long-term development of the aviation industry and their possible environmental benefits (GbD, 2001). It offered a view of the fuel burn reduction benefits that some advanced concepts might offer. Concepts considered included alternative aircraft configurations such as the blended wing body and the laminar flying wing, and the use of an unducted fan (open rotor) power plant. The study concluded that these two aircraft concepts could offer significant fuel burn reduction potential compared with a conventional aircraft design carrying an equivalent payload. Other studies (Leifsson and Mason, 2005) have suggested similar results. Table 5.7 summarises, from the GbD results, the theoretical fuel savings of these future designs relative to a baseline conventional swept wing aircraft for a 12,500 km design range, with the percentage fuel burn requirements for the mission.

Further reduction in both NO_x and CO_2 emission could be achieved by advances in airframe and propulsion systems which reduce fuel burn. In propulsion, the open rotor offers significant reductions in fuel burn over the turbofan engines used typically on current passenger jets. However, aircraft speed is reduced below typical jet aircraft speeds as a consequence of propeller tip speed limits and therefore this technology may be more suitable for short- and medium-haul operations where speed may be less important. The global average flight length in 2005 was 1239 km (ICAO, 2006) and many flights are over shorter distances than this average. However, rotor noise from such devices would need to be controlled within acceptable (regulatory) limits.

In summary, airframe and engine technology developments, weight reduction through increased used of advanced structural composites, and drag reduction, particularly through the application of laminar flow control, hold the promise of further aviation fuel burn reductions over the long term. Such developments will only be accepted by the aviation industry should they offer an advantage over existing products and meet demanding safety and reliability criteria.

Alternative fuels for aviation

Kerosene is the primary fuel for civil aviation, but alternative fuels have been examined. These are summarised in Box 5.4.

---

32 http://www.boeing.bom/commercial/787family/specs.
33 NASA, www.nasa.gov/centers/dryden/about/Organisations/Technology/Facts?

Box 5.4 Alternative fuels for aviation

The applicability of alternative and renewable fuels for civil aviation has been examined by many countries, for both the environmental benefit that might be produced and to address energy security issues. One study, The Potential for Renewable Energy Sources in Aviation (PRESAV, 2003) concluded that biodiesel, Fischer-Tropsch synthetic kerosene liquefied hydrogen (H_2) could be suitable for aviation application. Fuel cost would be an issue as in comparative terms, in 2003, conventional aviation kerosene cost 4.6 US$/GJ whereas the cost of biodiesel, FT kerosene and H_2 would be in the respective ranges of 33.5–52.6 US$, 8–31.7 US$, 21.5–53.8 US$/GJ. In the and elsewhere, synthetic kerosene production is being studied the engine company Pratt and Whitney noted in a presentation (Biddle, 2006) that synthetic kerosene could be 'economically viable when crude prices reach (up to) 59 US$/barrel'. However, any alternative fuel for commercial aircraft will need to be compatible with aviation kerosene (to obviate the need for tank and system flushing on re-fuelling) and meet a comprehensive performance and safety specification.

A potential non-carbon fuel is hydrogen and there have been several studies on its use in aviation. An EC study (Airbus, 2004) developed a conceptual basis for applicability, safety, and the full environmental compatibility for a transition from kerosene to hydrogen for aviation. The study concluded that conventional aircraft designs could be modified to accommodate the larger tank sizes necessary for hydrogen fuels. However, the increased drag due to the increased fuselage volume would increase the energy consumption of the aircraft by between 9% and 14%. The weight of the aircraft structure might increase by around 23% as a result, and the maximum take-off weight would vary between +4.4% to –14.8% dependent on aircraft size, configuration and mission. The hydrogen production process would produce CO_2 unless renewable energy was used and the lack of hydrogen production and delivery infrastructure would be a major obstacle. The primary environmental benefit from the use of hydrogen fuel would be the prevention of CO_2 emissions during aircraft operation. But hydrogen fuelled aircraft would produce around 2.6 times more water vapour than the use of kerosene and water vapour is a GHG. The earliest implementation of this technology was suggested as between 15–20 years, provided that research work was pursued at an appropriate level. The operating cost of hydrogen-powered aircraft remains unattractive under today's economic conditions.

The introduction of biofuels could mitigate some of aviation's carbon emissions, if biofuels can be developed to meet the demanding specifications of the aviation industry, although both the costs of such fuels and the emissions from their production process are uncertain at this time.

Aviation potential practices

The operational system for aviation is principally governed by air traffic management constraints. If aircraft were to operate for minimum fuel use (and CO_2 emissions), the following constraints would be modified: taxi-time would be minimized; aircraft would fly at their optimum cruising altitude (for load and mission distance); aircraft would fly minimum distance between departure and destination (i.e., great circle distances) but modified to take account of prevailing winds; no holding/stacking would be applied.

Another type of operational system/mitigation potential is to consider the total climate impact of aviation. Such studies are in their infancy but were the subject of a major European project 'TRADE-OFF'. In this project different methods were devised to minimize the total radiative forcing impact of aviation; in practice this implies varying the cruise altitudes as O_3 formation, contrails (and presumably cirrus cloud enhancement) are all sensitive to this parameter. For example, Fichter et al. (2005) found in a parametric study that contrail coverage could be reduced by approximately 45% by flying the global fleet 6,000 feet lower, but at a fuel penalty of 6% compared with a base case. Williams et al. (2003) also found that regional contrail coverage was reduced by flying lower with a penalty on fuel usage. By flying lower, NO_x emissions tend to increase also, but the removal rate of NO_x is more efficient at lower altitudes: this, compounded with a lower radiative efficiency of O_3 at lower altitudes meant that flying lower could also imply lower O_3 forcing (Grewe et al., 2002). Impacts on cirrus cloud enhancement cannot currently be modelled in the same way, since current estimates of aviation effects on cirrus are rudimentary and based upon statistical analyses of air traffic and satellite data of cloud coverage (Stordal et al., 2005) rather than modelling. However, as Fichter et al. (2005) note, to a first order, one might expect aviation-induced cirrus cloud to scale with contrails. The overall 'trade-offs' are complex to analyse since CO_2 forcing is long lasting, being an integral over time. Moreover, the uncertainties on some aviation forcings (notably contrail and cirrus) are still high, such that the overall radiative forcing consequences of changing cruise altitudes need to be considered as a time-integrated scenario, which has not yet been done. However, if contrails prove to be worth avoiding, then such drastic action of reducing all aircraft cruising altitudes need not be done, as pointed out by Mannstein et al. (2005), since contrails can be easily avoided – in principle – by relatively small changes in flight level, due to the shallowness of ice supersaturation layers. However, this more finely tuned operational change would not necessarily apply to O_3 formation as the magnitude is a continuous process rather than the case of contrails that are either short-lived or persistent. Further intensive research of the impacts is required to determine whether such operational measures can be environmentally beneficial.

ATM (Air Traffic Management) environmental benefits

The goal of RVSM (Reduced Vertical Separation Minimum) is to reduce the vertical separation above flight level (FL) 290 from the current 610 m (2000 ft) minimum to 305 m (1000 ft) minimum. This will allow aircraft to safely fly more optimum profiles, gain fuel savings and increase airspace capacity. The process of safely changing this separation standard requires a study to assess the actual performance of airspace users under the current separation (610 m) and potential performance under the new standard (305 m). In 1988, the ICAO Review of General Concept of Separation Panel (RGCSP) completed this study and concluded that safe implementation of the 305 m separation standard was technically feasible.

A Eurocontrol study (Jelinek *et al.*, 2002) tested the hypothesis that the implementation of RVSM would lead to reduced aviation emissions and fuel burn, since the use of RVSM offers the possibility to optimise flight profiles more readily than in the pre-existing ATC (Air Traffic Control) regime. RVSM introduces six additional flight levels between FL290 and FL410 for all States involved in the EUR RVSM programme. The study analysed the effect from three days of actual traffic just before implementation of RVSM in the European ATC region, with three traffic days immediately after implementation of RVSM. It concluded that a clear trend of increasing environmental benefit was shown. Total fuel burn, equating to CO_2 and H_2O emissions, was reduced by between 1.6–2.3% per year for airlines operating in the European RVSM area. This annual saving in fuel burn translates to around 310,000 tonnes annually, for the year 2003.

Lower flight speeds

Speed comes at a cost in terms of fuel burn, although modern jet aircraft are designed to fly at optimum speeds and altitudes to maximise the efficiencies of their design. Flying slower would be a possibility, but a different engine would be required in order to maximise the efficiencies from such operation. The propfan – this being a conventional gas turbine powering a highly efficient rotating propeller system, as an open rotor or unducted fan – is already an established technology and was developed during the late 1980s in response to a significant increase in fuel cost at the time. The scimitar shaped blades are designed to minimise aerodynamic problems associated with high blade speeds, although one problem created is the noise generated by such devices. The fuel efficiency gains from unducted fans, which essentially function as ultra high bypass ratio turbofans, are significant and require the adoption of lower aircraft speeds in order to minimise the helical mach number at the rotating blade tip. Typically the maximum cruise speed would be less than 400 miles per hour, compared with 550 mph[34] for conventional jet aircraft. In the event the aero acoustic problem associated with propfans could be overcome, such aircraft might be suitable for short-haul operations where

speed has less importance. But there would be the need to influence passenger choice: propeller driven aircraft are often perceived as old fashioned and dangerous and many passengers are reluctant to use such aircraft.

5.3.4 Shipping

In the past few years, the International Maritime Organization (IMO) has started research and discussions on the mitigation of GHG emissions by the shipping industry. The potential of technical measures to reduce CO_2 emissions was estimated at 5–30% in new ships and 4–20% in old ships. These reductions could be achieved by applying current energy-saving technologies vis-à-vis hydrodynamics (hull and propeller) and machinery on new and existing ships (Marintek, 2000).

The vast majority of marine propulsion and auxiliary plants onboard ocean-going ships are diesel engines. In terms of the maximum installed engine output of all civilian ships above 100 gross tonnes (GT), 96% of this energy is produced by diesel power. These engines typically have service lives of 30 years or more. It will therefore be a long time before technical measures can be implemented in the fleet on any significant scale. This implies that operational emission abatement measures on *existing* ships, such as speed reduction, load optimization, maintenance, fleet planning, etc., should play an important role if policy is to be effective before 2020.

Marintek (2000) estimates the short-term potential of operational measures at 1–40%. These CO_2 reductions could in particular be achieved by fleet optimization and routing and speed reduction. A general quantification of the potential is uncertain, because ship utilization varies across different segments of shipping and the operational aspects of shipping are not well defined.

The long-term reduction potential, assuming implementation of technical or operational measures, was estimated for the major fuel consuming segments[35] of the world fleet as specific case studies. The result of this analysis was that the estimated CO_2 emission reduction potential of the world fleet would be 17.6% in 2010 and 28.2% in 2020. Even though this potential is significant, it was noted that this would not be sufficient to compensate for the effects of projected fleet growth (Marintek, 2000). Speed reduction was found to offer the greatest potential for reduction, followed by implementation of new and improved technology. Speed reduction is probably only economically feasible if policy incentives, such as CO_2 trading or emissions charges are introduced.

A significant shift from a primarily diesel-only fleet to a fleet that uses alternative fuels and energy sources cannot be expected until 2020, as most of the promising alternative

---

34   1 mph = 1.6 km/h
35   In fact four segments covering 80% of the fuel consumption were assessed: tank, bulk, container and general cargo ships.

techniques are not yet tested to an extent that they can compete with diesel engines (Eyring *et al.*, 2005b). Furthermore, the availability of alternative fuels is currently limited and time is needed to establish the infrastructure for alternative fuels. For these reasons, in the short term switching to alternative fuels provides a limited potential in general, but a significant potential for segments where a switch from diesel to natural gas is possible (Skjølsvik, 2005). Switching from diesel to natural gas has a 20% CO_2 reduction potential and is being pursued as a measure in Norway for inland ferries and offshore supply vessels operating on the Norwegian Continental Shelf. The main obstacle to the increased utilization of natural gas is the access to LNG (Liquefied Natural Gas) and the technology's level of costs compared to traditional ship solutions based on traditional fuel (Skjølsvik, 2005). A co-benefit of a switch from diesel to natural gas is that it also reduces emissions of SO_x and NO_x that contribute to local air pollution in the vicinity of ports.

For the long-term (2050), the economical CO_2 reduction potential might be large. One potential option is a combination of solar panels and sails. The use of large sails for super tankers is currently being tested in Germany and looks promising and may even be a cost-effective measure in the short term in case oil prices continue to soar. The use of large sails does not require fleet turnover but can be added to existing vessels (retrofit). The introduction of hydrogen-propelled ships and the use of fuel cell power at least for the auxiliary engines seem to be a possibility as well. For larger vessels capable and reliable fuel-cell-based ship propulsion systems are still a long way into the future, but might be possible in 2050 (Eyring *et al.*, 2005b). Altmann *et al.* (2004) concluded that fuel cells offer the potential for significant environmental improvements both in air quality and climate protection. Local pollutant emissions and GHG emissions can be eliminated almost entirely over the full life cycle using renewable primary energies. The direct use of natural gas in high temperature fuel cells employed in large ships and the use of natural gas derived hydrogen in fuel cells installed in small ships allows for a GHG emission reduction of 20–40%.

5.4 Mitigation potential

As discussed earlier, under 'business-as-usual' conditions with assumed adequate supplies of petroleum, GHG emissions from transport are expected to grow steadily during the next few decades, yielding about an 80% increase from 2002–2030 or 2.1% per year. This growth will not be evenly distributed; IEA projections of annual CO_2 growth rates for 2002–2030 range from 1.3% for the OECD nations to 3.6% for the developing countries. The potential for reducing this growth will vary widely across countries and regions, as will the appropriate policies and measures that can accomplish such reduction.

Analyses of the potential for reducing GHG emissions in the transport sector are largely limited to national or sub-national studies or to examinations of technologies at the vehicle level, for example well-to-wheel analyses of alternative fuels and drive trains for light-duty vehicles. The TAR presented the results of several studies for the years 2010 and 2020 (Table 3.16 of the TAR), with virtually all limited to single countries or to the EU or OECD. Many of these studies indicated that substantial reductions in transport GHG emissions could be achieved at negative or minimal costs, although these results generally used optimistic assumptions about future technology costs and/or did not consider trade-offs between vehicle efficiency and other (valued) vehicle characteristics. Studies undertaken since the TAR have tended to reach conclusions generally in agreement with these earlier studies, though recent studies have focused more on transitions to hydrogen used in fuel cell vehicles.

This section will discuss some available studies and provide estimates of GHG emissions reduction potential and costs/tonne of carbon emissions reduced for a limited set of mitigation measures. These estimates do not properly reflect the wide range of measures available, many of which would likely be undertaken primarily to achieve goals other than GHG reduction (or saving energy), for example to provide mobility to the poor, reduce air pollution and traffic reduce congestion. The estimates do not include:

- Measures to reduce shipping emissions;
- Changes in urban structure that would reduce travel demand and enhance the use of mass transit, walking and bicycling;
- Transport demand management measures, including parking 'cash out', road pricing, inner city entry charges, etc.

5.4.1 Available worldwide studies

Two recent studies – the International Energy Agency's *World Energy Outlook* (IEA, 2004a) and the World Business Council on Sustainable Development's *Mobility 2030* (WBCSD, 2004a) – examined worldwide mitigation potential but were limited in scope. The IEA study focused on a few relatively modest measures and the WBCSD examined the impact of specified technology penetrations on the road vehicle sector (the study sponsors are primarily oil companies and automobile manufacturers) without regard to either cost or the policies needed to achieve such results. In addition, IEA has developed a simple worldwide scenario for light-duty vehicles that also explores radical reductions in GHG emissions.

World Energy Outlook postulates an 'Alternative scenario' to their Reference scenario projection described earlier, in which vehicle fuel efficiency is improved, there are increased sales of alternative-fuel vehicles and the fuels themselves and demand side measures reduce transport demand and encourage a switch to alternative and less energy intensive transport modes. Some specific examples of technology changes and policy measures are:

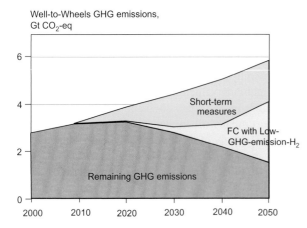

Well-to-Wheels GHG emissions, Gt CO$_2$-eq

Figure 5.14: *Two possible scenarios for GHG reductions in Light-duty vehicles*
Source: IEA, 2004b.

- In the United States and Canada, vehicle fuel efficiency is nearly 20% better in 2030 than in the Reference scenario and hybrid and fuel-cell powered vehicles make up 15% of the stock of light-duty vehicles in 2030;
- Average fuel efficiency in the developing countries and transition economies are 10–15% higher than in the Reference scenarios;
- Measures to slow traffic growth and move to more efficient modes reduce road traffic by 5% in the European Union and 6% in Japan. Similarly, road freight is reduced by 8% in the EU and 10% in Japan.

The net reductions in transport energy consumption and CO$_2$ emissions in 2030 are 315 Mtoe, or 9.6% and 997 MtC, or 11.4%, respectively compared to the Reference scenario. This represents a 2002–2030 reduction in the annual growth rate of energy consumption from 2.1-1.3% per year, a significant accomplishment but one which still allows transport energy to grow by 57% during the period. CO$_2$ emissions grow a bit less because of the shift to fuels with less carbon intensity, primarily natural gas and biofuels.

IEA has also produced a technology brief that examines a simple scenario for reducing world GHG emissions from the transport sector (IEA, 2004b). The scenario includes a range of short-term actions, coupled with the development and deployment of fuel-cell vehicles and a low-carbon hydrogen fuel infrastructure. For the long-term actions, deployment of fuel-cell vehicles would aim for a 10% share of light-duty vehicle sales by 2030 and 100% by 2050, with a 75% per-vehicle reduction in GHG emissions by 2050 compared to gasoline vehicles. The short-term measures for light-duty vehicles are:
- Improvements in fuel economy of gasoline and diesel vehicles, ranging from 15% (in comparison to the IEA reference case) by 2020 to 35% by 2050;
- Growing penetration of hybrid vehicles, to 50% of sales by 2040;
- Widespread introduction of biofuels, with 50% lower well-

to-wheels GHG emissions per km than gasoline, with a 25% penetration by 2050;
- Reduced travel demand, compared to the reference case, of 20% by 2050.

Figure 5.14 shows the light-duty vehicle GHG emissions results of the scenario. The penetration of fuel cell vehicles by itself brings emissions back to their 2000-levels by 2050. Coupled with the nearer-term measures, GHG emissions peak in 2020 and retreat to half of their 2000-level by 2050.

The Mobility 2030 study examined a scenario postulating very large increases in the penetration of fuel efficient technologies into road vehicles, coupled with improvements in vehicle use, assuming different time frames for industrialized and developing nations.

The technologies and their fuel consumption and carbon emissions savings referenced to current gasoline ICEs were:

Technology	Carbon reduced/vehicle (%)
1. Diesels	18
2. Hybridization	30 (36 for diesel hybrids)
3. Biofuels	20-80
4. Fuel cells with fossil hydrogen	45
5. Carbon-neutral hydrogen	100

Figure 5.15 shows the effect of a scenario postulating the market penetration of all of the technologies as well as an assumed change in consumer preferences for larger vehicles and improved traffic flows. The scenario assumes that diesels make up 45% of light-duty vehicles and medium trucks by 2030; that half of all sales in these vehicle classes are hybrids, also by 2030; that one-third of all motor vehicle liquid fuels are biofuels (mostly advanced) by 2050; that half of LDV and medium truck vehicle sales are fuel cells by 2050, with the hydrogen beginning as fossil-based but gradually moving to 80% carbon neutral by 2050; that better traffic flow and other efficiency measures reduce GHG emissions by 10%; and that the underlying efficiency of light-duty vehicles improves by 0.6% per year due to steady improvements (e.g., better aerodynamics and tyres) and to reduced consumer preference for size and power. In this scenario, GHG emissions return to their 2000-level by 2050.

Mobility 2030's authors make it quite clear that for this 'mixed' scenario to be even remotely possible will require overcoming many major obstacles. The introduction and widespread use of hydrogen fuel cell vehicles for example requires huge reductions in the costs of fuel cells; breakthroughs in on-board hydrogen storage; major advances in hydrogen production; overcoming the built-in advantages of the current gasoline and diesel fuel infrastructure; demonstration and commercialization of carbon

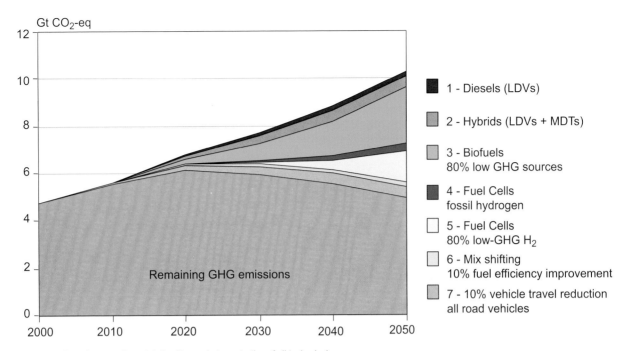

Figure 5.15: *The effect of a scenario postulating the market penetration of all technologies*
Source: WBCSD, 2004a.

sequestration technologies for fossil fuel hydrogen production (at least if GHG emission goals are to be reached); and a host of other R&D, engineering and policy successes.

Table 5.8 summarizes technical potentials for various mitigation options for the transport sector. As mentioned above, there are few studies dealing with worldwide analysis. In most of these studies, potentials are evaluated based on top-down scenario analysis. For combinations of specific power train technologies and fuels, well-to-wheels analyses are used to examine the various supply pathways. Technical potentials for operating practices, policies and behaviours are more difficult to isolate from economic and market potential and are usually derived from case studies or modelling analyses. Uncertainty is a key factor at all stages of assessment, from technology performance and cost to market acceptance.

5.4.2 Estimate of world mitigation costs and potentials in 2030

By extrapolating from recent analyses from the IEA and others an estimate can be given of the cost and potential for reducing transport CO_2 emissions. This section covers improving the efficiency of light-duty vehicles and aircraft, and the substitution of conventional fossil fuels by biofuels throughout the transport sector (though primarily in road vehicles). As noted above, these estimates do not represent the full range of options available to reduce GHG emissions in the transport sector.

5.4.2.1 Light-duty Vehicles

The following estimate of the overall GHG emissions reduction potential and costs for improving the efficiency of the world's light-duty vehicle fleet (thus reducing carbon emissions), is based on the IEA Reference Case, as documented in a spreadsheet model developed by the IEA for the Mobility 2030 project (WBSCD, 2004b). The cost estimates for total mitigation potential are provided in terms of 'societal' costs of reductions in GHG emissions, measured in US$/tonne of carbon (tC) or carbon dioxide (CO_2); the costs are the net of higher vehicle costs minus discounted lifetime fuel savings. Fuel savings benefits are measured in terms of the untaxed cost of the fuels at the retail level, and future savings are discounted at a low societal rate of 4% per year. These costs are not the same as those that would be faced by consumers, who would face the full taxed costs of fuel, would almost certainly use a higher discount rate, and might value only a few years of fuel savings. Also, they do not include the consumer costs of forgoing further increases in vehicle performance and weight. Over the past few decades, increasing acceleration performance and vehicle weight have stifled increases in fuel economy for light-duty vehicles and these trends must be stopped if substantial progress is to be made in fleet efficiency. Because consumers value factors such as vehicle performance, stopping these trends will have a perceived cost – but there is little information about its magnitude.

The potential improvements in light-duty fuel economy assumed in the analysis, and the costs of these improvements, are based on the scenarios in the MIT study summarised in Box 5.5. The efficiency improvements as mentioned in this study are

Table 5.8: *Summary table of CO2 mitigation potentials in transport sector taken from several studies*

Study	Mitigation measure/policy	Region	CO$_2$ reduction (%)				CO$_2$ reduction (Mt)			
			2010	2020	2030	2050	2010	2020	2030	2050
IEA 2004a	Alternative scenario	World	2.2	6.8	11.4		133	505	997	
		OECD	2	6.9	11.5		77	308	557	
		Developing countries	2.8	6.8	11.4		49	170	381	
		Transition economies	2.3	6.2	11.2		8	27	59	
IEA 2001	Improving Tech for Fuel Economy	OECD		30	40					
	Diesel				5-15					
IEA 2002a	All scenarios included	NA	6.6	14.4			148	358		
	All scenarios included	Western Europe	6.6	15.6			76	209		
	All scenarios included	Japan	8.3	16.1			28	61		
IEA 2004d	Improving fuel economy	World			18					
	Biofuels				12					
	FCV with hydrogen refuelling				7					
	COMBINING THESE THREE				30					
IEA 2004b	Reduction in fuel use per km	World		15	25	35				
	Blend of biofuels			5	8	13				
	Reduction in growth of LDV travel			5	10	20				
	using hydrogen in vehicle			0	3	75				
ACEEE 2001	A-scenario	USA	9.9	26.3			132	418		
	B-scenario		11.8	30.6			158	488		
	C-scenario		13.2	33.4			176	532		
MIT 2004		USA			(2035)					
	Baseline			3.4	16.8					
	Medium HEV			5.2	29.9					
	Composite			14.9	44.4					
	Combined policies		3-6	14-24	32-50					
Greene and Schafer 2003	Efficiency standards	USA		(2015)						
	Light-duty vehicles			6	18					
	Heavy trucks			2	3					
	Commercial aircraft			1	2					
	Replacement & alternative fuels									
	Low-carbon replacement fuels			2	7					
	Hydrogen fuel (all LDV fuel)			1	4					
	Pricing policies									
	Low-carbon fuel subsidy			2	6					
	Carbon pricing			3	6					
	Variabilization			6	9					
	Behavioural									
	Land use & infrastructure			3	5					
	System efficiency			0	1					
	Climate change education			1	2					
	Fuel economy information			1	1					
	Total			22	48					
WEC 2004	New technologies	World		30		46				
WBCSD 2004b	Road transport	World								
	Diesels (LDVs)			0.9	2.1	1.8		61	160	181
	Hybrids (LDVs and MDTs)			2.4	6.1	6.1		161	474	623
	Biofuels-80% low GHG sources			5.7	15.6	29.5		386	1207	3030
	Fuel Cells-fossil hydrogen			5.9	16.7	32.7		400	1293	3364
	Fuel Cells-80% low-GHG hydrogen			5.9	17.2	45.3		400	1333	4650
	Mix shifting 10% FE improvement			6.7	18.8	47.3		451	1455	4864
	10% Vehicle travel reduction-all vehicles			9.4	22.8	51.9		639	1765	5335

Box 5.5 Fuel economy benefits of multiple efficiency technologies

Several studies have examined the fuel economy benefits of simultaneously applying multiple efficiency technologies to light-duty vehicles. However, most of these are difficult to compare because they examine various types of vehicles, on different driving cycles, using different technology assumptions, for different time frames. The Massachusetts Institute of Technology has developed such an assessment for 2020 (MIT, 2000) with documentation of basic assumptions though with few details about the specific technologies that achieve these values, for a medium size passenger car driving over the official US Environmental Protection Agency driving cycle (Heywood et al., 2003). There are two levels of technology improvement – 'baseline' and 'advanced,' with the latter level of improvement further subdivided into conventional and hybrid drive trains.

Some of the key features of the 2020 vehicles are:
- Vehicle mass is reduced by 15% (baseline) and 22% (advanced) by a combination of greater use of high strength steel, aluminium and plastics coupled with advanced design;
- Tyre rolling resistance coefficient is reduced from the current .009 to .008 (baseline) and .006 (advanced);
- Drag coefficient is reduced to 0.27 (baseline) and 0.22 (advanced). The baseline level is at the level of the best current vehicles, while the advanced level should be readily obtainable for the best vehicles in 2020, but seems quite ambitious for a fleet average;
- Indicated engine efficiency increases to 41% in both baseline and advanced versions. This level of efficiency would likely require direct injection, full valve control (and possibly camless valves) and advanced engine combustion strategies.

The combined effects of applying this full range of technologies are quite dramatic (Table 5.9). From current test values of 30.6 mpg (7.69 litres/100 km) as a 2001 reference, baseline 2020 gasoline vehicles obtain 43.2 mpg (5.44 L/100 km), advanced gasoline vehicles 49.2 mpg (4.78 L/100 km) and gasoline hybrids 70.7 mpg (3.33 L/100 km); advanced diesels obtain 58.1 mpg (4.05 L/100 km) and diesel hybrids 82.5 mpg (2.85 L/100 km) (note that on-road values will be *at least* 15% lower). In comparison, Ricardo Consulting Engineers (Owen and Gordon, 2002) estimate the potential for achieving 92 g/km CO_2 emissions, equivalent to 68.6 mpg (3.43 L/100 km), for an advanced diesel hybrid medium size car 'without' substantive non-drive train improvements. This is probably a bit more optimistic than the MIT analysis when accounting for the additional effects of reduced vehicle mass, tyre rolling resistance and aerodynamic drag coefficient.

These values should be placed in context. First, the advanced vehicles represent 'leading edge' vehicles which must then be introduced more widely into the new vehicle fleet over a number of years and may take several years (if ever) to represent an 'average' vehicle. Second, the estimated fuel economy values are attainable only if trends towards ever-increasing vehicle performance are stifled; this may be difficult to achieve.

discounted somewhat to take into account the period in which the full benefits can be achieved. Further, fleet penetration of the technology advances are assumed to be delayed by 5 years in developing nations; however, because developing nation fleets are growing rapidly, higher efficiency vehicles, once introduced, may become a large fraction of the total fleet in these nations within a relatively short time. The technology assumptions for two 'efficiency scenarios' are as follows (Table 5.9a).

The high efficiency and medium efficiency scenarios achieve the following improvements in efficiency for the new light-duty vehicle fleet (Table 5.9b):

Table 5.10 shows the light-duty vehicle fuel consumption and (vehicle only) CO_2 emissions for the Reference scenario and the High and medium efficiency scenarios. In the Reference case, LDV fuel consumption increases by nearly 60% by 2030; the High Efficiency Case cuts this increase to 26% and the Medium efficiency scenario cuts it to 42%. For the OECD nations, the Reference Case projects only a 22% increase by 2030, primarily

because of moderate growth in travel demand, with the High efficiency scenario actually reducing fuel consumption in this group of nations by 9% and the Medium efficiency scenario reducing growth to only 6%. This regional decrease (or modest increase) in fuel use is overwhelmed by the rapid growth in the world's total fleet size and overall travel demand and the slower uptake of efficiency technologies in the developing nations. Because no change in the use of biofuels was assumed in this analysis, the CO_2 emissions in the scenarios essentially track the energy consumption paths discussed above. Figure 5.16 shows the GHG emissions path for the three scenarios, resulting in a mitigation potential of about 800 (High) and 400 (Medium) $MtCO_2$ in 2030.

Table 5.11 shows the cost of the reductions in GHG emissions in US$/t$CO_2$ for those reductions obtained by the 2030 new vehicle fleet over its lifetime, assuming oil prices of 30 US$, 40 US$, 50 US$ and 60 US$/bbl over the vehicles' lifetime.[36] Note that the costs in Table 5.11 do not apply to the carbon reductions achieved in that year by the entire LDV fleet (from Table

Table 5.9a: *Fuel economy and cost assumptions for cost and potentials analysis*

Medium size car	MPG (L/100 km)	Incr from Ref (%)	Cost (%)	ΔCost (US$)[a]
2001 reference	30.6 (7.69)	0	100	0
2030 baseline	43.2 (5.55)	41	105	1,000
2030 advanced	49.2 (4.78)	61	113	2,600
2030 hybrid	70.7 (3.33)	131	123	4,600
2030 diesel	58.1 (4.05)	90	119	3,800
2030 diesel hybrid	82.5 (2.85)	170	128	5,600

a) Cost differential based on a reference 20,000 US$ vehicle. See Box 5.5 for the definitions of the vehicle types.

Source: adapted from MIT (2000), as explained in the text.

Table 5.9b: *Efficiency improvements new light-duty vehicle fleet*

Region	% improvement from 2001 levels, high/medium			
	2015	2020	2025	2030
North America	**30/15**	**45/25**	**70/32**	**80/40**
Europe	30/25	40/30	55/35	70/40
Emerging Asia/Pacific	30/25	40/30	65/35	75/40
Rest of world	0/12+	30/20+	45/25+	60/30+

5.10), because those reductions are associated with successive waves of high efficiency vehicles entering the fleet during the approximately 15 year period before (and including) 2030.

The Table 5.11 results show that the 'social cost of carbon reduction' for light-duty vehicles varies dramatically across regions and with fuel prices (since the cost is the *net* of technology costs minus the value of fuel savings). The results are also quite different for the High and Medium efficiency scenarios, primarily because the estimated technology costs begin to rise more steeply at higher efficiency levels, raising the average cost/tonne of CO_2 in the High efficiency scenario. For the High efficiency scenario, CO_2 reduction costs are very high for the OECD countries aside from North America, even at 60 US$/bbl oil prices, reflecting the ambitious (and expensive) increases in that scenario, the relatively high efficiencies of those regions' fleets in the Reference Case, and the relatively low km/vehicle/year driven outside North America; on the other hand, the costs of the moderate increases in the Medium efficiency scenario are low to negative for all regions, reflecting the availability of moderate cost technologies capable of raising average vehicle efficiencies up to 30–40% or so.

The values in Table 5.11 are sensitive to several important assumptions:

- Technology costs: the costs assumed here appear to be considerably higher than those assumed in WEO 2006 (IEA, 2006a).

- Discount rates: the analysis assumes a low social discount rate of 4% in keeping with the purpose of the analysis. As noted, vehicle purchasers would undoubtedly use higher rates and would value fuel savings at retail fuel prices rather than the untaxed values used here; they might also only value a few years of fuel savings rather than the lifetime savings assumed here. WEO 2006 on the other hand, used a zero discount rate, substantially reducing the net cost of carbon reduction.

- Vehicle km travelled (vkt): this analysis used the IEA/WBCSD spreadsheet's assumption of constant vkt over time and applied these values to new cars. Actual driving patterns will depend on the balance of increasing road infrastructure and rapidly increasing fleet size in developing nations. Unless infrastructure keeps pace with growing fleet size, which will be difficult, the assumption of constant vkt/vehicle may prove accurate or even optimistic.

- Efficiency gains assumed in the Reference scenario: the Reference scenario assumed significant gains in most areas (aside from North America), which makes the Efficiency scenarios more expensive.

Table 5.12 shows the economic potential for reducing CO_2 emissions in the 2030 fleet of new LDVs as a function of world oil price.[37] The values show that much of the economic potential is available at a net savings, 'if consumer preference for power and other efficiency-robbing vehicle attributes is ignored'. Even at 30 US$/bbl oil prices, over half of the total

---

36 Note, however, that these results do not take into account changes in travel demand that would occur with changing fuel price and changes in Reference case vehicle efficiency levels. At higher oil prices, the Reference case would likely have less travel and higher vehicle efficiency; this would, in turn, reduce the oil savings and GHG reductions obtained by the Efficiency case and would likely raise the costs/tonne C from the values shown here.

Table 5.10: *Regional and worldwide Light-duty vehicle CO_2 emissions (vehicle only) and fuel consumption, efficiency and reference cases*

	CO₂ emissions (Mt)				Energy use (EJ)			
	2000	2030			2000	2030		
		Reference	High	Medium		Reference	High	Medium
OECD North America	1226	1623	1178	1392	17.7	23.4	17.0	20.0
OECD Europe	488	535	431	479	7.0	7.5	6.0	6.7
OECD Pacific	220	219	176	197	3.2	3.2	2.6	2.9
EECCA	84	229	188	209	1.2	3.3	2.7	3.0
Eastern Europe	49	82	68	74	0.7	1.2	1.0	1.0
China	46	303	267	287	0.7	4.4	3.8	4.1
Other Asia	54	174	148	160	0.8	2.5	2.1	2.3
India	22	103	87	95	0.3	1.5	1.2	1.4
Middle East	27	67	57	62	0.4	1.0	0.8	0.9
Latin America	110	294	251	273	1.6	4.2	3.6	3.9
Africa	53	167	152	162	0.8	2.4	2.2	2.3
Total	2379	3797	3004	3390	34.2	54.4	43.1	48.6

Note: EECCA = countries of EasternEurope, the Caucasus and Central Asia.

Table 5.11: *Cost of CO_2 reduction in new 2030 LDVs*

	CO₂ reduction cost (US$/tCO₂)							
	High efficiency case				Medium efficiency case			
	30 US$/bbl 0.39 US$/L	40 US$/bbl 0.45 US$/L	50 US$/bbl 0.51 US$/L	60 US$/bbl 0.60 US$/L	30 US$/bbl 0.39 US$/L	40 US$/bbl 0.45 US$/L	50 US$/bbl 0.51 US$/L	60 US$/bbl 0.60 US$/L
OECD North America	5	-16	-37	-68	-72	-93	-114	-146
OECD Europe	131	110	89	58	14	-7	-28	-60
OECD Pacific	231	210	189	157	-14	-36	-57	-88
EECCA	81	60	39	8	-54	-76	-97	-128
Eastern Europe	181	160	139	107	-18	-39	-60	-92
China	23	2	-19	-51	-23	-44	-65	-97
Other Asia	19	-2	-23	-55	-23	-44	-65	-96
India	62	41	20	-12	9	-12	-33	-65
Middle East	-15	-36	-57	-89	-49	-70	-91	-122
Latin America	-6	-27	-48	-79	-42	-63	-84	-116
Africa	10	-12	-33	-64	-33	-54	-75	-106

Note: EECCA = countries of EasternEurope, the Caucasus and Central Asia.

(<100 US$/tCO₂) potential is available at a net savings over the vehicle lifetime; at 40 US$/bbl, over 90% of the 718 Mt total potential is available at a net savings.

The regional detail, not shown in Table 5.12, is illuminating. In the High Efficiency scenario, of 793 Mt of total potential,

445 Mt are in OECD North America and are available at a net savings at 40 US$/bbl oil (and at less than 20 US$/tCO₂ at 30 US$/bbl oil). The next highest regional potential is in OECD Europe at 104 Mt, but this potential is more expensive: at 30 US$/bbl oil. Only 56 Mt is available below 100 US$/tCO₂, and becomes available at below 100 US$/tCO₂ only at 60 US$/bbl

---

37 These results do not take account of the effect higher oil prices would have on LDV efficiency in the Reference Scenario. This efficiency level would be expected to be a strong function of oil price, that is, it would be higher for higher prices. Consequently, the technology cost of improving vehicle efficiency further would also be higher – reducing the economic potential.

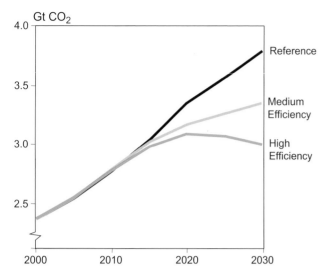

Figure 5.16: *Light-duty vehicle CO_2 emissions for three scenarios*

oil. China has the next highest total emissions (2030 Reference case emissions of 303 Mt) but only a moderate potential of 36 Mt. This potential is fully available at a net savings only if oil is 50 US$/bbl or higher – perhaps not surprising because China has ambitious fuel economy standards embedded in the Reference Case and has relatively low driving rates, which make further improvements more difficult and expensive.

5.4.2.2   Aircraft

QinetiQ (UK)[38] analysed the fuel consumption and CO_2 trends for a simple global aviation growth scenario to provide an indicative view on the extent that technology and other developments might mitigate aviation emissions. The ICAO traffic forecast (ICAO/FESG, 2003) defined traffic growth to 2030 from which a future fleet composition was developed, using a range of current and future aircraft types where their introduction could be assumed, as well as representative aircraft types based on seat capacity. Fuel burn and emissions were calculated using known emissions performance and projections for future aircraft where necessary.

The analysis assumed a range of technology options as follows:
- Case 1 assumed no technology change from 2002 to 2030; using the extrapolated traffic forecast from ICAO FESG – this case shows only the effects of traffic growth on emissions.
- Case 2 – as Case 1, but assumes all new aircraft deliveries after 2005 would be 'best available technology at a 2005 (BAT)' performance standard, and with specific new aircraft (A380, A350, B787) delivered from 2008.
- Case 3 – as Case 1, but with assumed annual fleet fuel efficiency improvements as per 'Greene' and DTI (IPCC

1999, Chapter 9, Table 9.15). This assumes a fleet efficiency improvement trend of 1.3% per year to 2010, assumed then to decline to 1.0% per year to 2020 and 0.5% per year thereafter. This is the reference case.
- Case 4 – as Case 3, plus the assumption that a 50 US$/tCO$_2$ cost will produce a further 0.5% fuel efficiency improvement per annum from 2005, as suggested by the cost-potential estimates of Wit *et al.*, (2002), that assume technologies such as winglets, fuselage skin treatments (riblets) and further weight reductions and engine developments will be introduced by airlines.
- Case 5 – as Case 3, plus the assumption of 100 US$/tCO$_2$ cost, producing a 1.0% fuel efficiency improvement per annum from 2005 (Wit *et al.*, 2002), again influencing the introduction of additional technologies as above.

The results of this analysis are summarised in Table 5.13.

Case 2 is a simple representation of planned industry developments and shows their effect to 2030, ignoring further technology developments. This is an artificial case, as on-going efficiency improvements would occur as a matter of course, but it shows that these planned fleet developments alone might save 14% of the CO_2 that the 'no technology change' of Case 1 would have produced. Case 3 should be regarded as the 'base case' from which benefits are measured, as this case reflects an agreed fuel efficiency trend assumed for some of the calculations produced in the IPCC Special Report (1999). This results in a further 11% reduction in CO_2 by 2030 compared with Case 2. Cases 4 and 5 assume that a carbon cost will drive additional technology developments from 2005 – no additional demand effect has been assumed. These show further CO_2 reduction of 11.8% and 22.2% compared with 'base case' 3 over the same period from technologies that are assumed to be more attractive than hitherto. However, even the most ambitious scenario suggests that CO_2 production will increase by almost 100% from the base year. The cost potentials for Cases 4 and 5 are based on one study and further studies may refine these results. There is limited literature in the public domain on costs of mitigation technologies. The effects of more advanced technology developments, such as the blended wing body, are not modelled here, as these developments are assumed to take place after 2030.

The analysis suggests that aviation emissions will continue to grow as a result of continued demand for civil aviation. Assuming the historical fuel efficiency trend produced by industry developments will continue (albeit at a declining level), carbon emissions will also grow, but at a lower rate than traffic. Carbon pricing could effect further emissions reductions if the aviation industry introduces further technology measures in response.

---

38 http://www.dti.gov.uk/files/file35675.pdf

Table 5.12: *Economic potential of LDV mitigation technologies as a function of world oil price, for new vehicles in 2030*

World oil price (US$/bbl)	Region	Economic potential (MtCO₂)				
		Cost ranges (US$/tCO₂)				
		<100	<0	0-20	20-50	50-100
30	OECD	523	253	270	0	0
	EIT	49	28	0	0	21
	Other	146	88	30	20	8
	World	718	369	300	20	29
40	OECD	523	523	0	0	0
	EIT	49	28	0	0	21
	Other	146	118	20	8	0
	World	718	669	20	8	21
50	OECD	571	523	0	0	48
	EIT	49	28	0	21	0
	Other	146	138	8	0	0
	World	766	689	8	21	48
60	OECD	571	523	0	0	48
	EIT	49	28	21	0	0
	Other	146	146	0	0	0
	World	766	697	21	0	48

5.4.2.3   Biofuels

IEA has projected the potential worldwide increased use of biofuels in the transport sector assuming successful technology development and policy measures reducing barriers to biomass deployment and providing economic incentives.

IEA's *World Energy Outlook 2006* (IEA, 2006b) develops an Alternative policy scenario that adds 55 Mtoe biofuels above baseline levels of 92 Mtoe by 2030, which increases the biofuels share of total transport fuel demand from 3 to 5%. In this scenario, all of the biofuels are produced by conventional technology, that is ethanol from starch and sugar crops and biodiesel from oil crops. Assuming an average CO_2 reduction from gasoline use of 25%,[39] this would reduce transport CO_2 emissions by 36 Mt.

Furthermore, according to the Beyond the Alternative policy scenario (BAPS), which assumed more energy savings and emission reductions through a set of technological

breakthroughs, biofuels use in road transport would double compared to the Alternative policy scenario.

A second IEA report, Energy Technology Perspectives 2006 (IEA, 2006a), evaluates a series of more ambitious scenarios that yield biomass displacement of 13–25% of transport energy demand by 2050, compared to Baseline levels of 3% displacement. Two scenarios, called Accelerated Technology (ACT) Map and TECH Plus, assume economic incentives equivalent to 25 US$/tCO₂, increased support for research and development, demonstration, and deployment programmes, and policy instruments to overcome commercialization barriers. Both scenarios have optimistic assumptions about the success of efforts to reduce fuel production costs, increase crop yields, and so forth. In the ACT Map scenario, transport biofuels production reaches 480 Mtoe in 2050, accounting for 13% of total transport demand; in TECH Plus, biofuels represents 25% of transport energy demand by 2050. These displacements yield CO_2 reductions (below the Baseline levels) of 1800 MtCO₂ in Map and 2300 MtCO₂ in TECH Plus, with the major

Table 5.13: *Summaries of CO₂ mitigation potential analysis in aviation*

Aviation technology	2002 CO₂ (Mt)	2030 CO₂ (Mt)	Ratio (2030/2002)
Case 1 (no technological change)	489.29	1,609.74	3.29
Case 2 (BAT new aircrafts)	489.29	1,395.06	2.85
Case 3 (base)	489.29	1,247.02 (100%)	2.55
Case 4 (50 US$/tCO₂-eq)	489.29	1,100.15 (88%)	2.25
Case 5 (100 US$/tCO₂-eq)	489.29	969.96 (78%)	1.98

---

39 IEA cites the following estimates for biofuels CO_2 reduction when used as a replacement fuel: Corn in the U.S., –13%; ethanol in Europe, –30%; ethanol in Brazil, –90%; sugar beets to ethanol in Europe, –40 to –60%; rapeseed-derived biodiesel in Europe, –40 to –60%.

contributors being biodiesel from Fischer Tropsch conversion and ethanol from both sugar crops and cellulosic feedstocks; biodiesel from vegetable oil and ethanol from grains represent somewhat lower shares.

Although the report does not provide quantitative estimates of CO_2 reduction in 2030, it presents qualitative information (Table 3.5 of the IEA report) that implies that 2030-levels of biodiesel from vegetable oil and ethanol from grain and sugar crops are similar to 2050-levels, but biodiesel from Fischer Tropsch conversion, a major source in 2050, plays little role in 2030 and cellulosic ethanol is also significantly lower in 2030 than in 2050. The implied 2030 potential from the two scenarios appears to be about 600–1500 $MtCO_2$.

5.4.2.4   Totals

The estimates discussed above can be summarized as follows:
Light-duty vehicles:
    718–766 $MtCO_2$ at carbon prices less than 100 US$/$tCO_2$
    689–718 $MtCO_2$ at carbon prices less than 50 US$/$tCO_2$
    669–718 $MtCO_2$ at carbon prices less than 20 US$/$tCO_2$
    369–697 $MtCO_2$ at carbon prices less than 0 US$/$tCO_2$
Aircraft:
    150 MtCO2 at carbon prices less than 50 US$/$tCO_2$
    280 $MtCO_2$ at carbon prices less than 100 US$/$tCO_2$
Biofuels:
    600–1500 $MtCO_2$ at carbon prices less than 25 US$/$tCO_2$

Although presumably the potential for biofuels penetration would be higher above the cited 25 US$/$tCO_2$ carbon price, the total potential for a carbon price of 100 US$/$tCO_2$ for the three mitigation sources is about 1600–2550 $MtCO_2$. Much of this potential appears to be located in OECD North America and Europe. Note, however, that the potential is measured as the 'further' reduction in CO_2 emissions from the Reference scenario, which assumes that substantial amounts of biofuels will be produced in Brazil and elsewhere and significant improvements in fuel efficiency will occur in China and in other industrializing nations without further policy measures.

5.5   Policies and measures

This section provides policies and measures for the transport sector, considering experiences of countries and regions in achieving both energy savings (and hence GHG reduction) and sustainable transport systems. An overall policy consideration at the national level and international levels is presented in Chapter 13.3

The policies and measures that have been considered in this section that are commonly applied for the sector and can be effective are:

- Land use and transport planning;
- Taxation and pricing;
- Regulatory and operational instruments (e.g., traffic management, control and information);
- Fuel economy standards – road transport;
- Transport demand management;
- Non-climate policies influencing GHG emissions;
- Co-benefits and ancillary benefits.

This section discusses climate policies related to GHG from international aviation and shipping separately, reflecting the international coordination that is required for effective reduction strategies in these sectors. Both sectors are subject to a global legal framework and mitigation policies applied on a unilateral basis may reduce its environmental effectiveness due to evasion (Wit et al., 2004).

5.5.1   Surface transport

A wide array of policies and strategies has been employed in different circumstances around the world to restrain vehicle usage, manage traffic congestion and reduce energy use, GHGs, and air pollution. There tends to be considerable overlap among these policies and strategies, often with synergistic effects. The recent history almost everywhere in the world has been increasing travel, bigger vehicles, decreasing land-use densities and sprawling cities. But some cities are far less dependent on motor vehicles and far denser than others, even at the same incomes. The potential exists to greatly reduce transport energy use and GHG emissions by shaping the design of cities, restraining motorization and altering the attributes of vehicles and fuels. Indeed, slowing the growth in vehicle use through land-use planning and through policies that restrain increases in vehicle use would be an important accomplishment. Planning and policy to restrain vehicles and densify land use not only lead to reduced GHG emissions, but also reduced pollution, traffic congestion, oil use, and infrastructure expenditures and are generally consistent with social equity goals as well.

5.5.1.1   Land use and transport planning

Energy use for urban transport is determined by a number of factors, including the location of employment and residential locations. In recent decades, most cities have been increasing their dependence on the automobile and decreasing dependence on public transport. In some cases increasing motorization is the result of deliberate planning – what became known as 'predict and provide' (The Royal Commission on Transport and the Environment, 1994; Goodwin, 1999). This planning and programming process played a central role in developed countries during the second half of the 20th century. In many developing countries, the process of motorization and road building is less organized, but is generally following the same motorization path, often at an accelerated rate.

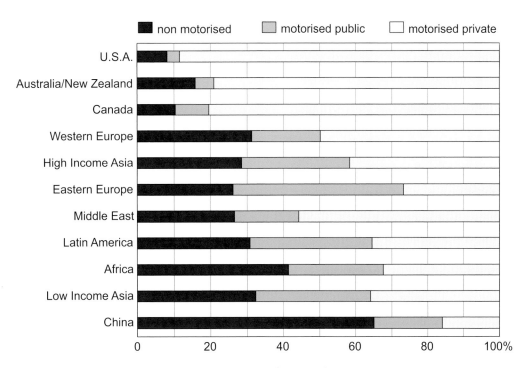

Figure 5.17: *Modal split for the cities represented in the Millennium Cities Database for Sustainable Transport by region*
Source: Kenworthy & Laube, 2002.

Income plays a central role in explaining motorization. But cities of similar wealth often have very different rates of motorizsation. Mode shares vary dramatically across cities, even within single countries. The share of trips by walking, cycling and public transport is 50% or higher in most Asian, African and Latin American cities, and even in Japan and Western Europe (Figure 5.17). Coordination of land use and transport planning is key to maintaining these high mode shares.

Kenworthy and Laube (1999) pointed out that high urban densities are associated with lower levels of car ownership and car use and higher levels of transit use. These densities are decreasing almost everywhere. Perhaps the most important strategy and highest priority to slow motorization is to strengthen local institutions, particularly in urban areas (Sperling and Salon, 2002).

Some Asian cities with strong governments, especially Hong Kong, Singapore and Shanghai are actively and effectively pursuing strategies to slow motorization by providing high quality public transport and coordinating land use and transport planning (Cullinane, 2002; Willoughby, 2001; Cameron *et al.*, 2004; Sperling and Salon, 2002).

There are many other examples of successfully integrated land use and transport planning, including Stockholm and Portland, Oregon (USA) (Abbott, 2002; Lundqvist, 2003). They mostly couple mixed-use and compact land use development with better public transport access to minimize auto dependence. The effectiveness of these initiatives in reducing sprawl is the subject of debate, especially in the USA (Song and Knaap, 2004; Gordon and Richardson, 1997; Ewing, 1997). There are several

arguments that the settlement pattern is largely determined, so changes in land use are marginal, or that travel behaviour may be more susceptible to policy interventions than land-use preferences (Richardson and Bae, 2004). Ewing and Cervero (2001) found that typical elasticity of vehicle-km travelled with respect to local density is –0.05, while Pickrell (1999) noted that reduction in auto use become significant only at densities of 4000 people or more per square kilometre – densities rarely observed in US suburbs, but often reached elsewhere (Newman and Kenworthy, 1999). Coordinated transport and land-use methods might have greater benefits in the developing world where dense mixed land use prevails and car ownership rate is low. Curitiba is a prime example of coordinated citywide transport and land-use planning (Gilat and Sussman, 2003; Cervero, 1998).

The effectiveness of policies in shifting passengers from cars to buses and rails is uncertain. The literature on elasticity with respect to other prices (cross price elasticity) is not abundant and likely to vary according to the context (Hensher, 2001). The Transport Research Laboratory guide showed several cross price elasticity estimates with considerable variance in preceding studies (TRL, 2004). Goodwin (1992) gave an average cross elasticity of public transport demand with respect to petrol prices of +0.34. Jong and Gunn (2001) also gave an average cross elasticity of public transport trips with respect to fuel price and car time of +0.33 and +0.27 in the short term and +0.07 and +0.15 in the long term.

The literature on mode shifts from cars to new rail services is also limited. A monitoring study of Manchester indicated that about 11% of the passengers on the new light rail would have

Table 5.14: *Taxes and pricing in the transport sector in developing and developed countries*

Instrument	Developing countries/EIT	Developed countries
Tax incentives to promote use of natural gas	Pakistan, Argentina, Colombia, Russia	Italy, Germany, Australia, Ireland, Canada, UK, Belgium
Incentives to promote natural gas vehicles	Malaysia, Egypt	Belgium, UK, USA, Australia, Ireland
Annual road tax differentiated by vintage	Singapore and India (fixed span and scrapping)	Germany
Emission trading	Chile	
Congestion pricing including Area Licensing Scheme; vehicle registration fees; annual circulation tax	Chile, Singapore	Norway, Belgium
Vehicle taxes based on emissions-tax deductions on cleaner cars e.g., battery operated or alternative fuel vehicles	South Korea	Austria, Britain, Belgium, Germany, Japan, The Netherlands, Sweden
Carbon tax by size of engine	Zimbabwe	
Cross subsidization of cleaner fuels (ethanol blending by gasoline tax - through imposition of lower surcharge or excise duty exemption)	India	

Source: Adapted from Pandey and Bhardwaj, 2000; Gupta, 1999 and European Natural Gas Vehicle Association, 2002.

otherwise used their cars for their trips (Mackett and Edwards, 1998), while a Japanese study of four domestic rails and monorails showed that 10–30% of passengers on these modes were diverted from car mode. The majority of the passengers were transferred from alternative bus and rail routes (Japanese Ministry of Land, Infrastructure and Transport and Institute of Highway Economics, 2004). The Transport Research Laboratory guide (2004) contained international evidence of diversion rates from car to new urban rail ranging from 5–30%. These diversion rates are partly related to car mode share, in the sense that car share is so high in the USA and Australia that ridership on new rail systems is more likely to come from cars in those countries (Booz Allen & Hamilton 1999, cited in Transport Research Laboratory, 2004). It is also known that patronage of metros for cities in the developing world has been drawn almost exclusively from existing public transport users or through generation effects (Fouracre *et al.*, 2003).

The literature suggests that in general, single policies or initiatives tend to have a rather modest effect on the motorization process. The key to restraining motorization is to cluster a number of initiatives and policies, including improved transit service, improved facilities for NMT (Non-motorized transport) and market and regulatory instruments to restrain car ownership and use (Sperling and Salon, 2002). Various pricing and regulatory instruments are addressed below.

Investment appraisal is an important issue in transport planning and policy. The most widely applied appraisal technique in transport is cost benefit analysis (CBA) (Nijkamp *et al.*, 2003). In CBA, the cost of CO_2 emissions can be indirectly included in the vehicle operating cost or directly counted at an estimated price, but some form of robustness testing is useful in the latter case. Alternatively, the amount of CO_2 emissions is

listed on an appraisal summary table of Multi-Criteria Analysis (MCA) as a part of non-monetized benefits and costs (Mackie and Nellthorp, 2001; Grant-Muller *et al.*, 2001; Forkenbrock and Weisbrod, 2001; Japanese Study Group on Road Investment Evaluation, 2000). To the extent that the cost of CO_2 emissions has a relatively important weight in these assessments, investments in unnecessarily carbon-intensive projects might be avoided. Strategic CBA can further make transport planning and policy carbon-efficient by extending CBA to cover multi-modal investment alternatives, while Strategic Environmental Assessment (SEA) can accomplish it by including multi-sector elements. (ECMT, 2000; ECMT, 2004b).

5.5.1.2 *Taxation and pricing*

Transport pricing refers to the collection of measures used to alter market prices by influencing the purchase or use of a vehicle. Typically measures applied to road transport are fuel pricing and taxation, vehicle license/registration fees, annual circulation taxes, tolls and road charges and parking charges. Table 5.14 presents an overview of examples of taxes and pricing measures that have been applied in some developing and developed countries.

Pricing, taxes and charges, apart from raising revenue for governments, are expected to influence travel demand and hence fuel demand and it is on this basis that GHG reduction can be realized.

Transport pricing can offer important gains in social welfare. For the UK, France and Germany together, (OECD, 2003) estimates net welfare gains to society of optimal charges (set at the marginal social cost level) at over 20 billion €/yr (22.6 US$/yr).

Box 5.6 Examples of pricing policies for heavy-duty vehicles

Switzerland: In January 2001, trucks of maximum 35 tonnes weight were allowed on Swiss territory (previously 28 tonnes) and a tax of 1.00 cent/tkm (for the vehicle middle emission category) was imposed on trucks above 3.5 tonnes on all roads. It replaced a previous fixed tax on heavy-duty vehicles. The tax is raised electronically. Since 2005, the tax is higher at 1.60 cent/tkm, but 40 tonnes trucks are allowed. Over the period 2001–2003, it was estimated that it contributed to an 11.9% decrease in vehicle-km and a 3.5% decrease in tonnes-km of domestic traffic. The tax led to an improved carriers' productivity and it is anticipated that, for that reason, emissions of CO_2 and NO_x would decrease over the period 2001–2007 by 6–8%. On the other hand transit traffic, which amounts to 10% of total traffic, was also affected in a similar way by the new tax regime, so that the number of HDL has been decreasing at a rate of about 2–3% per year, while, at the same time, increasing in terms of tonnes-km (ARE, 2004b; 2006). A part of the revenues are used to finance improvements to the rail network.

Germany: A new toll system was introduced in January 2005 for all trucks with a maximum weight of 12 tonnes and above. This so-called LKW-MAUT tax is levied on superhighways on the base of the distance driven; its cost varies between 9 and 14 Eurocents according to the number of axles and the emission category of the truck. Payments are made via a GPS system, at manual payment terminals or by Internet. The receipts will be used to improve the transport networks of Germany. The system introduction appears successful, but it is too early to assess its impacts.

Although the focus here is on transport pricing options to limit CO_2 emissions, it should be recognized that many projects and policies with that effect are not focused on GHG emissions but rather on other objectives. A pricing policy may well aim simultaneously at reducing local pollution and GHG emissions, accidents, noise and congestion, as well as generating State revenue for enlarging of social welfare and/or infrastructure construction and maintenance. Every benefit with respect to these objectives may then be assessed simultaneously through CBA or MCA; they may be called co-benefits. Governments can take these co-benefits into account when considering the introduction of transport pricing such as for fuel. This is all the more important since a project could be not worth realising if only one particular benefit is considered, whereas it could very well be proved beneficial when adding all the co-benefits.

Taxes

Empirically, throughout the last 30 years, regions with relatively low fuel prices have low fuel economy (USA, Canada, Australia) and regions where relatively high fuel prices apply (due to fuel taxes) have better car fuel economy (Japan and European countries). For example, fuel taxes are about 8 times higher in the UK than in the USA, resulting in fuel prices that are about three times higher. UK vehicles are about twice as fuel-efficient; mileage travelled is about 20% lower and vehicle ownership is lower as well. This also results in lower average per capita fuel expenditures. Clearly, automobile use is sensitive to cost differences in the long run (VTPI, 2005). In theory, long run impact of increases in fuel prices on fuel consumption are likely to be about 2 to 3 times greater than short run impact (VTPI, 2005). Based on the price elasticities (Goodwin et al., 2004) judged to be the best defined results for developed countries, if the real price of fuel rises by 10% and stays at that level, the volume of fuel consumed by road vehicles will fall by about 2.5% within a year, building up to a

reduction of over 6% in the longer run (about 5 years or so), as shown in Table 5.15.

An important reason why a fuel or CO_2 tax would have limited effects is that price elasticities tend to be substantially smaller than the income elasticities of demand. In the long run the income elasticity of demand is a factor 1.5–3 higher than the price elasticity of total transport demand (Goodwin et al., 2004). In developing countries, where incomes are lower, the demand response to price changes may be significantly more elastic.

Recent evidence suggests that the effect of CO_2 taxes and high fuel prices may be having a shrinking effect in the more car-dependent societies. While the evidence is solid that price elasticities indicated in Table 5.15 and used by Goodwin were indeed around –0.25 (i.e., 2.5% reduction in fuel for every 10% increase in price), in earlier years, new evidence indicates a quite different story. Small and Van Dender (2007) found that price elasticities in the USA dropped to about –0.11 in the late 1990s, and Hughes et al. (2006) found that they dropped even further in 2001–2006, to about –0.04. The explanation seems to be that people in the USA have become so dependent on their vehicles that they have little choice but to adapt to higher prices. One might argue that these are short term elasticities, but the erratic nature of gasoline prices in the USA (and the world) result in drivers never exhibiting long-term behavior. Prices drop before they seriously consider changing work or home locations or even buying more efficient vehicles. If oil prices continue to cycle up and down, as many expect, drivers may continue to cling to their current behaviors. If so, CO_2 taxes would have small and shrinking effects in the USA and other countries where cars are most common.

Table 5.15: *Impact of a permanent increase in real fuel prices by 10%*

	Short run/within 1 year (%)	Long run/5 years (%)
Traffic volume	-1	-3
Fuel consumption	-2.5	-6
Vehicle fuel efficiency	-1.5	-4
Vehicle ownership	Less than -1	-2.5

Source: Goodwin et al. 2004.

As an alternative to fuel taxes, registration and circulation taxes can be used to incentivise the purchase (directly) and manufacturing (indirectly) of fuel-efficient cars. This could be done through a revenue neutral fee system, where fuel-efficient cars receive a rebate and guzzler cars are faced with an extra fee. There is evidence that incentives given through registration taxes are more effective than incentives given through annual circulation taxes (Annema *et al.*, 2001). Buyers of new cars do not expect to be able to pass on increased registration taxes when selling the vehicle. Due to refunds on registration taxes for cars that were relatively fuel efficient compared to similar sized cars, the percentage of cars sold in the two most fuel efficient classes increased from 0.3%–3.2% (cars over 20% more fuel efficient than average) and from 9.5%–16.1% (for cars between 10 and 20% more fuel efficient than average) in the Netherlands (ADAC, 2005). After the abolishment of the refunds, shares decreased again. COWI (2002) modelled the impact on fuel efficiency of reforming current registration and circulation taxes so they would depend fully on the CO_2 emissions of new cars. Calculated reduction percentages varied from 3.3–8.5% for 9 European countries, depending on their current tax bases.

Niederberger (2005) outlines a voluntary agreement with the Swiss government under which the oil industry took responsibility for GHG emissions from the road transport sector, which they supply with fuel. As of 1 October 2005, Swiss oil importers voluntarily contribute the equivalent of about 5 cents per gallon (approx. 80 million US$ annually) into a climate protection fund that is invested via a non-profit (non-governmental) foundation into climate mitigation projects domestically and abroad (via the emerging carbon market mechanisms of the Kyoto Protocol). Cost savings (compared with an incentive tax) are huge and the private sector is in charge of investing the funds effectively. A similar system in the USA could generate 9 billion US$ in funds annually to incentivize clean alternative fuels and energy efficient vehicles, which could lower US dependency on foreign fuel sources. This policy is also credible from a sustainable development perspective than the alternative CO_2 tax, since the high CO_2 tax would have led to large-scale shifts in tank tourism – and bookkeeping GHG reductions for Switzerland – although the real reductions would have been less than half of the total effect and neighbouring countries would have been left with the excess emissions.

Licensing and parking charges

The most renowned area licensing and parking charges scheme has been applied in Singapore with effective reduction in total vehicular traffic and hence energy (petroleum) demand (Fwa, 2002). The area licensing scheme in Singapore resulted in 1.043 GJ per day energy savings with private vehicular traffic reducing by 75% (Fwa, 2002).

Unfortunately there is currently a lack of data on potential GHG savings associated with policy, institutional and fiscal reforms/measures with respect to transport particularly in other developing countries. General estimates of reduction in use of private vehicle operators resulting from fuel pricing and taxing are 15–20% (World Bank, 2002; Martin *et al.*, 1995).

Table 5.16: *Potential energy and GHG savings from pricing, taxes and charges for road transport*

Tax/pricing measure	Potential energy/GHG savings or transport improvements	Reference
Optimal road pricing based on congestion charging (London, UK)	20% reduction in CO_2 emissions as a result of 18% reduction in traffic	Transport for London (2005)
Congestion pricing of the Namsan Tunnels (Seoul, South Korea)	34% reduction of peak passenger traffic volume. Traffic flow from 20 to 30 km/hr.	World Bank (2002)
Fuel pricing and taxation	15-20% for vehicle operators.	Martin et al. (1995)
Area Licensing Scheme (Singapore)	1.043 GJ/day energy savings. Vehicular traffic reduced by 50%. Private traffic reduced by 75%. Travel speed increased 20 to 33 km/hr.	Fwa (2002)
Urban gasoline tax (Canada)	1.4 Mton by 2010 2.6 Mton by 2020	Transportation in Canada; www.tc.gc.ca/pol/en/Report/anre1999/tc9905be.htm
Congestion charge trial in Stockholm (2005-2006)	13% reduction of CO_2	http://www.stockholmsforsoket.se/templates/page.aspx?id=2453

Box 5.7 Policies to promote biofuels

Policies to promote biofuels are prominent in national emissions abatement strategies. Since benefits of biofuels for CO_2 mitigation mainly come from the well-to-tank part, incentives for biofuels are more effective climate policies if they are tied to the whole well-to-wheels CO_2 efficiencies. Thus preferential tax rates, subsi-dies and quotas for fuel blending should be calibrated to the benefits in terms of net CO_2 savings over the whole well-to-wheel cycle associated with each fuel. Development of an index of CO_2 savings by fuel type would be useful and if agreed internationally could help to liberalise markets for new fuels. Indexing incentives would also help to avoid discrimination between feedstocks. Subsidies that support production of specific crops risk being counterproductive to emission policies in the long run (ECMT, 2007). In order to avoid negative effects of biofuel production on sustainable development (e.g. biodiversity impacts), additional conditions could be tied to incentives for biofuels.

The following incentives for biofuels are implemented or in the policy pipeline (Hamelinck, *et al*. 2005):
Brazil was one of the first countries to implement policies to stimulate biofuel consumption. Currently, flexible fuel vehicles are eligible for federal value-added tax reductions ranging from 15–28%. In addition, all gasoline should meet a legal alcohol content requirement of 20–24%.

Motivated by the biofuels directive in the European Union, the EU member states have implemented a variety of policies. Most of the member states have implemented an excise duty relief. Austria, Spain, Sweden, the Netherlands and the UK have implemented an obligation or intend to implement an obligation in the coming years. Sweden and Austria also implemented a CO_2 tax.

The American Jobs creation act of 2004 provides tax incentives for alcohol and biodiesel fuels. The credits have been set at 0.5–1 US$/gallon (about 0.11–0.21 €/litre). Some 39 states have developed additional policy programmes or mechanisms to support the increase use of biofuel. The types of measures range from tax exemptions on resources required to manufacturing or distributing biofuels (e.g. labour, buildings); have obligatory targets for governmental fleets and provide tax exemptions or subsidies when purchasing more flexible vehicles. One estimate is that total subsidies in the US for biofuels were 5.1–6.8 billion US$ in 2006, about half in the form of fuel excise tax reductions, and another substantial amount for growing corn used for ethanol.

New blending mandates have also appeared in China, Canada, Colombia, Malaysia and Thailand. Four provinces in China added dates for blending in major cities, bringing to nine the number of provinces with blending mandates (REN21, 2006).

5.5.1.3   Regulatory and operational measures

Although pricing and fiscal instruments are obvious tools for government policy, they are often not very effective, as reflected by the potential reduction in fuel savings (IEA, 2003). Potential effective (and cost-effective) non-fiscal measures that can be effective in an oil crisis are regulatory measures such as:

- Lower speed limits on motorways;
- High occupancy vehicle requirements for certain roads and networks;
- Vehicle maintenance requirements;
- Odd/even number plate and other driving restrictions;
- Providing information on CO_2 emission performances of vehicles (labelling);
- Establishing carbon standards for fuels;
- Direct traffic restrictions (e.g., no entry into business district);
- Free/expanded urban public transport;
- Encouraging alternatives to travel (e.g., greater telecommuting);
- Emergency switching from road to rail freight;
- Reducing congestion through removal of night-time/ weekend driving bans for freight.

IEA (2003) indicates that such measures could contribute to significant oil savings. This is a typical case where a portfolio of measures is applied together and they would work well with adequate systems of monitoring and enforcement.

For the measures to be implemented effectively considerable preparatory work is necessary and Table 5.17 shows examples of what could be done to ensure the measures proposed above can be effective in oil savings.

The combined effect of these regulatory measures used to target light-duty vehicles (in addition to blending non-petroleum fuels with gasoline and diesel) is estimated to be a reduction of 15% of daily fuel consumption.

In OECD countries vehicles consume 10–20% more fuel per km than indicated by their rated efficiency. It is estimated that 5–10% reduction in fuel consumption can be achieved by stronger inspection and vehicle maintenance programmes, adoption of on board technologies, more widespread driver training and better enforcement and control of vehicle speeds.

Table 5.17: *Preparations required to implement some regulatory measures*

Measure to be implemented	Preparatory work
Speed limits[a]	• Install electronic speed limit system • Change the law
Carpool days	• System of finding rides • Car parks • High occupancy car lanes
Energy efficient car and driving choice from home	• On board efficient indicator systems • Driver training • Information on efficient car purchases
Telecommuting days	• Telecommuting programmes and protocols • Practice
Clean car choice	• Public awareness of car consumption • Labeling based on CO_2 performance
Car free days	• Biking/walking/transit facilities • Home/job commuting reduced

a) The Swedish road administration has calculated the effect of regulatory measures on speed. Exceeding speed limits on the Swedish road network gives an extra CO_2 emission of 0.7Mt on an annual basis (compared to total emissions of 20 Mt). A large part of this can be tackled using speed cameras and in the future intelligent speed adaptation in vehicles. Besides this, reduction of speed limits (by 10 km/h except for the least densely populated areas where there is no alternative to the private car) could result in a similar amount of CO_2 reduction.

Source: Adapted from IEA, 2003.

Vehicle travel demand can be reduced by 10–15% by aggressively combining infrastructure improvements, intelligent transport technologies and systems (e.g., better routing systems and congestion reduction), information systems and better transit systems in addition to road pricing.

Another regulatory approach, under consideration in California as part of its 2006 Global Warming Solutions Act, is carbon-based fuel standards. Fuel suppliers would be required to reduce the carbon content of their fuels according to a tightening schedule. For instance, gasoline from conventional oil would be rated at 1.0, ethanol from corn and natural gas at 0.8, electricity for vehicles at 0.6 and so on. The fuel suppliers would be allowed to trade and bank credits and car makers would be required to produce vehicles at an amount that corresponds to the planned sales of alternative fuels. Reductions of 5% or more in transport fuel GHGs by 2020 are envisioned, with much greater reductions in later years.

5.5.1.4 Fuel economy standards – road transport

Most industrialized nations now impose fuel economy requirements (or their equivalent in CO_2 emissions requirements)

on new light-duty vehicles (Plotkin, 2004; An and Sauer, 2004). The first standards were imposed by the United States in 1975, requiring 27.5 mpg (8.55 L/100 km) corporate fleet averages for new passenger cars and 20.7 mpg (11.36 L/100 km) for light trucks (based on tests instituted by the US Environmental Protection Agency, using the 'CAFE' driving cycle) by 1985. The passenger car standard remains unchanged, whereas the light truck standard has recently been increased to 22.2 mpg (10.6 L/100 km) for the 2007 model year and to 23.5 mpg (10.0 L/100 km) in model year 2010.[40] Additional standards (some voluntary) include:

- European Union: a 2008 fleet wide requirement[41] of 140 gCO_2/km, about 41 mpg (5.74 L/100 km) of gasoline equivalent, using the New European Driving Cycle (NEDC), based on a Voluntary Agreement between the EU and the European manufacturers, with the Korean and Japanese manufacturers following in 2009. Recent slowing of the rate of efficiency improvement has raised doubts that the manufacturers will achieve the 2008 and 2009 targets (Kageson, 2005).
- Japan: a 2010 target of about 35.5 mpg (6.6 L/100 km) for new gasoline passenger vehicles, using the Japan 10/15 driving cycle based on weight-class standards.
- China: weight-class standards that are applied to each new vehicle using the NEDC driving cycle, with target years of 2005 and 2008. At the historical mix of vehicles, the standards are equivalent to fleet targets of about 30.4 mpg (7.7 L/100 km) by 2005 and 32.5 mpg (7.2 L/100 km) by 2008 (An and Sauer, 2004).
- Australia: a 2010 target for new vehicles of 18% reduction in average fuel consumption relative to the 2002 passenger car fleet, corresponding to 6.8 L/100 km, or 34.6 mpg. (DfT, 2003), based on a voluntary agreement between industry and government.
- The State of California has established GHG emission standards for new light-duty vehicles designed to reduce per-vehicle emissions by 22% in 2012 and 30% by 2016. Several US states have decided to adopt these standards, as well. At the time of writing, US industry and the federal government were fighting these standards in the courts.

The NEDC and Japan 10/15 driving cycles are slower than the US CAFE cycle and, for most vehicles (though probably not for hybrids), will yield lower measured fuel economy levels than the CAFE cycle for the same vehicles. Consequently, if they reach their targets, the EU, Japanese and Chinese fleets are likely to achieve fuel economies higher than implied by the values above if measured on the US test. A suggested correction factor (for the undiscounted test results) is 1.13 for the EU and China and 1.35 for Japan (An and Sauer, 2004), though these are likely to be at the high end of the possible range of values

---

40 In 2011, manufacturers must comply with a reformed system where required CAFE levels depend on the manufacturer's fleet mix based on vehicle "footprint," or track width wheelbase (NHTSA CAFE website, 2006).

41 There are no specific corporate requirements for the entire new light-duty vehicle fleet.

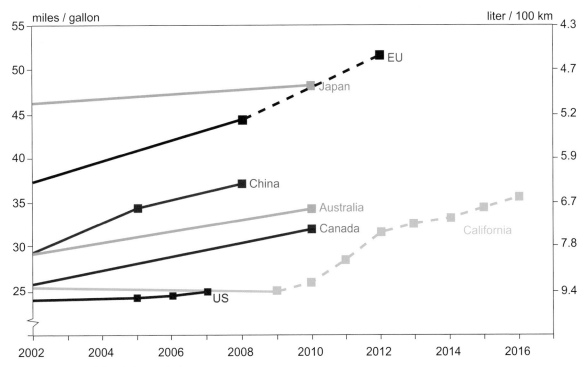

Figure 5.18: *Fuel economy and GHG emission standards*

Note: all the fuel economy targets represent test values based on artificial driving cycles. The standards in the EU and Australia are based on voluntary agreements. In most cases, actual on-road fuel economy values will be lower; for example, the US publishes fuel economy estimates for individual LDVs that are about 15% lower than the test val-ues and even these values appear to be optimistic. Miles/gallon is per US gallon.

for such factors.[42] Figure 5.18 shows the 'corrected' comparison of standards.

Recent studies of the costs and fuel savings potential of technology improvements indicate considerable opportunity to achieve further fleet fuel economy gains from more stringent standards. For example, the US National Research Council (NRC, 2002) estimates that US light-duty vehicle fuel economy can be increased by 25–33% within 15 years with existing technologies that cost less than the value of fuel saved. A study by Ricardo Consulting Engineers for the UK Department for Transport (Owen and Gordon, 2002) develops a step-wise series of improvements in a baseline diesel passenger car that yields a 38% reduction in CO_2 emissions (a 61% increase in fuel economy), to 92 g/km, by 2013 using parallel hybrid technology at an incremental cost of 2300–3,100 £ (4200–5700 US$) with a 15,300 £ (28,000 US$) baseline vehicle. Even where fuel savings will outweigh the cost of new technologies, however, the market will not necessarily adopt these technologies by itself (or achieve the maximum fuel economy benefits from the technologies even if they are adopted). Two crucial deterrents are, first, that the buyers of new vehicles tend to consider only the first three years or so of fuel savings (NRC, 2002; Annema *et al.*, 2001), and second, that vehicle buyers will take some

of the benefits of the technologies in higher power and greater size rather than in improved fuel economy. Further, potential benefits for consumers over the vehicle's lifetime are generally small, while risks for producers are high (Greene, 2005). Also, neither the purchasers of new vehicles nor their manufacturers will take into account the climate effects of the vehicles.

Strong criticisms have been raised about fuel economy standards, particularly concerning claimed adverse safety implications of weight reductions supposedly demanded by higher standards and increased driving caused by the lower fuel costs (per mile or km) associated with higher fuel economy.

The safety debate is complex and not easily summarized. Although there is no doubt that adding weight to a vehicle improves its safety in some types of crashes, it does so at the expense of other vehicles; further, heavy light trucks have been shown to be no safer, and in some cases less safe than lighter passenger cars, primarily because of their high rollover risk (Ross *et al.*, 2006). The US National Highway Traffic Safety Administration (NHTSA) has claimed that fleet wide weight reductions 'reduce' fleet safety (Kahane, 2003), but this conclusion is strongly disputed (DRI, 2004; NRC, 2002). An important concern with the NHTSA analysis is that it does not

42 These values are derived by simulating US vehicles running on the CAFE, NEDC, and Japan 10.15 cycles and comparing their estimated fuel economies. Because car manu-facturers design their vehicles to do well on the cycles on which they will be tested, the US vehicles are likely to do a bit worse on the NEDC and Japan 10.15 cycles than they would have had they been designed for those cycles. This will somewhat exaggerate the estimated differences between the cycles in their effects on fuel economy.

separate the effects of vehicle weight and size. In any case, other factors, e.g., overall vehicle design and safety equipment, driver characteristics, road design, speed limits and alcohol regulation and enforcement play a more significant role in vehicle safety than does average weight.

Some have argued that increases in driving associated with reduced fuel cost per mile will nullify the benefits of fuel economy regulations. Increased driving 'is' likely, but it will be modest and decline with higher income and increased motorization. Recent data implies that a driving 'rebound' would reduce the GHG reduction (and reduce oil consumption) benefits from higher standards by about 10% in the United States (Small and Van Dender, 2007) but more than this in less wealthy and less motorized countries.

In deciding to institute a new fuel economy standard, governments should consider the following:

- Basing stringency decisions on existing standards elsewhere requires careful consideration of differences between the home market and compared markets in fuel quality and availability; fuel economy testing methods; types and sizes of vehicles sold; road conditions that may affect the robustness of key technologies; and conditions that may affect the availability of technologies, for example, availability of sophisticated repair facilities.
- There are a number of different approaches to selecting stringency levels for new standards. Japan selected its weight class standards by examining 'top runners' – exemplary vehicles in each weight class that could serve as viable targets for future fleet wide improvements. Another approach is to examine the costs and fuel saving effects of packages of available technologies on several typical vehicles, applying the results to the new vehicle fleet (NRC, 2002). Other analyses have derived cost curves (percent increase in fuel economy compared with technology cost) for available technology and applied these to corporate or national fleets (Plotkin *et al.*, 2002). These approaches are not technology-forcing, since they focus on technologies that have already entered the fleet in mass-market form. More ambitious standards could demand the introduction of emerging technologies. Selection of the appropriate level of stringency depends, of course, on national goals and concerns. Further, the selection of enforcement deadlines should account for limitations on the speed with which vehicle manufacturers can redesign multiple models and introduce the new models on a schedule that avoids severe economic disruption.
- The structure of the standard is as important as its level of stringency. Basing target fuel economy on vehicle weight (Japan, China) or engine size (Taiwan, South Korea) will tend to even out the degree of difficulty the standards impose on competing automakers, but will reduce the potential fuel economy gains that can be expected (because weight-based standards eliminate weight reduction and engine-size-based standards eliminate engine downsizing as viable means of

achieving the standards). Basing the standard on vehicle wheelbase times track width may provide safety benefits by providing a positive incentive to maintain or increase these attributes. Using a uniform standard for all vehicles or for large classes of vehicles (as in the US) is simple and easy to explain, but creates quite different challenges on different manufacturers depending on the market segments they focus on.

- Allowing trading of fuel economy 'credits' among different vehicles or vehicle categories in an automaker's fleet, or even among competing automakers, will reduce the overall cost of standards without reducing the total societal benefits, but may incur political costs from accusations of allowing companies or individuals to 'buy their way out' of efficiency requirements.
- Alternatives (or additions) to standards are worth investigating. For example, 'feebates', which award cash rebates to new vehicles whose fuel economy is above a designated level (often the fleet average) and charge a fee to vehicles with lower fuel economy, may be an effective market-based measure to increase fleet fuel economy. An important advantage of feebates is that they provide a 'continuous' incentive to improve fuel economy, because an automaker can always gain a market advantage by introducing vehicles that are more efficient than the current average.

5.5.1.5 *Transport Demand Management*

Transport Demand Management (TDM) is a formal designation for programmes in many countries that improve performance of roads by reducing traffic volumes (Litman, 2003). There are many potential TDM strategies in these programmes with a variety of impacts. Some improve transport diversity (the travel options available to users). Others provide incentives for users to reduce driving, changing the frequency, mode, destination, route or timing of their travel. Some reduce the need for physical travel through mobility substitutes or more efficient land use. Some involve policy reforms to correct current distortions in transport planning practices. TDM is particularly appropriate in developing country cities, because of its low costs, multiple benefits and potential to redirect the motorization process. In many cases, effective TDM during early stages of development can avoid problems that would result if communities become too automobile dependent. This can help support a developing country's economic, social and environmental objectives (Gwilliam *et al.*, 2004).

The set of strategies to be implemented will vary depending on each country's demographic, geographic and political conditions. TDM strategies can have cumulative and synergetic impacts, so it is important to evaluate a set of TDM programmes as a package, rather than as an individual programme. Effective strategies usually include a combination of positive incentives to use alternative modes ('carrots' or 'sweeteners') and negative incentives to discourage driving ('sticks' or 'levellers'). Recent literature gives a comprehensive overview of these programmes with several case studies (May *et al.*, 2003; Litman, 2003; WCTRS and IPTS, 2004). Some major strategies such as pricing and land-use planning are addressed above. Below is a selective review of additional TDM strategies with significant potential to reduce vehicle travel and GHGs.

Employer travel reduction strategies gained prominence from a late 1980s regulation in southern California that required employers with 100 or more employees to adopt incentives and rules to reduce the number of car trips by employees commuting to work (Giuliano *et al.*, 1993). The State of Washington in the USA kept a state law requiring travel plans in its most urban areas for employers with 100 or more staff. The law reduced the percentage of employees in the targeted organizations who drove to work from 72–68% and affected about 12% of all trips made in the area. In the Netherlands, the reduction in single occupant commute trips from a travel plan averaged 5–15%. In the UK, in very broad terms, the average effectiveness of UK travel plans might be 6% in trips by drive alone to work and 0.74% in the total vehicle-km travelled to work by car. The overall effectiveness was critically dependent on both individual effectiveness and levels of plan take-up (Rye, 2002).

Parking supply for employees is so expensive that employers naturally have an incentive to reduce parking demand. The literature found the price elasticity of parking demand for commuting at –0.31 to –0.58 (Deuker *et al.*, 1998) and –0.3 (Veca and Kuzmyak, 2005) based on a non-zero initial parking price. The State of California enacted legislation that required employers with 50 or more persons who provided parking subsidies to offer employees the option to choose cash in lieu of a leased parking space, in a so-called parking cash-out programme. In eight case studies of employers who complied with the cash-out programme, the solo driver share fell from 76% before cashing out to 63% after cashing out, leading to the reduction in vehicle-km for commuting by 12%. If all the commuters who park free in easily cashed-out parking spaces were offered the cash option in the USA, it would reduce vehicle-km travelled per year by 6.3 billion (Shoup, 1997).

Reducing car travel or CO_2 emissions by substituting telecommuting for actual commuting has often been cited in the literature, but the empirical results are limited. In the USA, a micro-scale study estimated that 1.5% of the total workforce telecommuted on any day, eliminating at most 1% of total household vehicle-km travelled (Mokhtarian, 1998), while a macro-scale study suggested that telecommuting reduced annual vehicle-km by 0–2% (Choo *et al.*, 2005).

Reduction of CO_2 emissions by hard measures, such as car restraint, often faces public opposition even when the proposed measures prove effective. Soft measures, such as a provision of information and use of communication strategies and educational techniques (OECD, 2004a) can be used for supporting the promotion of hard measures. Soft measures can also be directly

helpful in encouraging a change in personal behaviour leading to an efficient driving style and reduction in the use of the car (Jones, 2004). Well organized soft measures were found to be effective for reducing car travel while maintaining a low cost. Following travel awareness campaigns in the UK, the concept of Individualized marketing, a programme based on a targeted, personalized, customized marketing approach, was developed and applied in several cities for reducing the use of the car. The programme reduced car trips by 14% in an Australian city, 12% in a German city and 13% in a Swedish city. The Travel Blending technique was a similar programme based on four special kits for giving travel-feedback to the participants. This programme reduced vehicle-km travelled by 11% in an Australian city. The monitoring study after the programme implementation in Australian cities also showed that the reduction in car travel was maintained (Brog *et al.*, 2004; Taylor and Ampt, 2003). Japanese cases of travel-feedback programmes supported the effectiveness of soft measures for reducing car travel. The summary of the travel-feedback programmes in residential areas, workplaces and schools indicated that car use was reduced by 12% and CO_2 emissions by 19%. It also implied that the travel-feedback programmes with a behavioural plan requiring a participant to make a plan for a change showed better results than programmes without one (Fujii and Taniguchi, 2005).

5.5.2   Aviation and shipping

In order to reduce emissions from air and marine transport resulting from the combustion of bunker fuels, new policy frameworks need to be developed. Both the ICAO and IMO have studied options for limiting GHG emissions. However, neither has as yet been able to devise a suitable framework for implementing effective mitigation policies.

5.5.2.1   *Aviation*

IPCC (1999), ICAO/FESG (2004a,b), Wit *et al.* (2002 and 2005), Cames and Deuber (2004), Arthur Andersen (2001) and others have examined potential economic instruments for mitigating climate effects from aviation.

At the global level no support exists for the introduction of kerosene taxes. The ICAO policy on exemption of aviation fuel from taxation has been called into question mainly in European states that impose taxes on fuel used by other transport modes and other sources of GHGs. A study by Resource Analysis (1999) shows that introducing a charge or tax on aviation fuel at a 'regional' level for international flights would give rise to considerable distortions in competition and may need amendment of bilateral air service agreements. In addition, the effectiveness of a kerosene tax imposed on a regional scale would be reduced as airlines could take 'untaxed' fuel onboard into the taxed area (the so-called tankering effect).

Wit and Dings (2002) analyzed the economic and environmental impacts of en-route emission charges for all flights in European airspace. Using a scenario-based approach and an assumed charge level of 50 US$/tCO$_2$, the study found a cut in forecast aviation CO$_2$ emissions in EU airspace of about 11 Mt (9%) in 2010. This result would accrue partly (50%) from technical and operational measures by airlines and partly from reduced air transport demand. The study found also that an en-route emission charge in European airspace designed in a non-discriminative manner would have no significant impact on competition between European and non-European carriers.

In a study prepared for CAEP/6, the Forecasting and Economic Analysis Support Group (ICAO/FESG, 2004a) considered the potential economic and environmental impacts of various charges and emission trading schemes. For the period 1998–2010, the effects of a global CO$_2$ charge with a levy equivalent to 0.02 US$/kg to 0.50 US$/kg jet fuel show a reduction in global CO$_2$ emissions of 1–18%. This effect is mainly caused by demand effects (75%). The AERO modelling system was used to conduct the analyses (Pulles, 2002).

As part of the analysis of open emission trading systems for CAEP/6, an impact assessment was made of different emission trading systems identified in ICF *et al.* (2004). The ICAO/FESG report (2004b) showed that under a Cap-and-Trade system for aviation, total air transport demand will be reduced by about 1% compared to a base case scenario (FESG2010). In this calculation, a 2010 target of 95% of the 1990-level was assumed for aviation on routes from and to Annex-I countries and the more developed non-Annex-I countries such as China, Hong Kong, Thailand, Singapore, Korea and Brazil. Furthermore a permit price of 20 US$/tCO$_2$ was assumed. Given the relative high abatement costs in the aviation sector, this scenario would imply that the aviation sector would buy permits from other sectors for about 3.3 billion US$.

In view of the difficulty of reaching global consensus on mitigation policies to reduce GHG emissions from international aviation, the European Commission decided to prepare climate policies for aviation. On 20 December 2006 the European Commission presented a legislative proposal that brings aviation emissions into the existing EU Emissions Trading Scheme (EU ETS). The proposed directive will cover emissions from flights within the EU from 2011 and all flights to and from EU airports from 2012. Both EU and foreign aircraft operators would be covered. The environmental impact of the proposal may be significant because aviation emissions, which are currently growing rapidly, will be capped at their average level in 2004–2006. By 2020 it is estimated by model analysis that a total of 183 MtCO$_2$ will be reduced per year on the flights covered, a 46% reduction compared with business-as-usual. However, aviation reduces the bulk of this amount through purchasing allowances from other sectors and through additional supply of Joint Implementation and Clean Development Mechanism credits. In 2020 aviation reduces its own emissions by 3% below business-as-usual (EC, 2006).

If emission trading or emission charges were applied to the aviation sector in isolation, the two instruments would in principle be equivalent in terms of cost-effectiveness. However, combining the reduction target for aviation with the emission trading scheme of other sectors increases overall economic efficiency by allowing the same amount of reductions to be made at a lower overall cost to society. Therefore, if aviation were to achieve the same environmental goal under emission trading and emission charges, the economic costs for the sector and for the economy as a whole would be lower if this was done under an emission trading scheme including other sectors rather than under a charging system for aviation only.

Alternative policy instruments that may be considered are voluntary measures or fuel taxation for domestic flights. Fuel for domestic flights, which are less vulnerable to economic distortions, is already taxed in countries such as the USA, Japan, India and the Netherlands. In parallel to the introduction of economic instruments such as emission trading, governments could improve air traffic management.

Policies to address the full climate impact of aviation

A major difficulty in developing a mitigation policy for the climate impacts of aviation is how to cover non-CO$_2$ climate impacts, such as the emission of nitrogen oxides (NO$_x$) and the formation of condensation trails and cirrus clouds (see also Box 5.1 in section 5.2). IPCC (1999) estimated these effects to be about 2 to 4 times greater than those of CO$_2$ alone, even without considering the potential impact of cirrus cloud enhancement. This means that the perceived environmental effectiveness of any mitigation policy will depend on the extent to which these non-CO$_2$ climate effects are also taken into account.

Different approaches may be considered to account for non-CO$_2$ climate impacts from aviation (Wit et al., 2005). A first possible approach is where initially only CO$_2$ from aviation is included in for example an emission trading system, but flanking instruments are implemented in parallel such as differentiation of airport charges according to NO$_x$ emissions.

Another possible approach is, in case of emission trading for aviation, a requirement to surrender a number of emission permits corresponding to its CO$_2$ emissions multiplied by a precautionary average factor reflecting the climate impacts of non-CO$_2$ impacts. It should be emphasised that the metric that is a suitable candidate for incorporating the non-CO$_2$ climate impacts of aviation in a single metric that can be used as a multiplier requires further development, being fairly theoretical at present. The feasibility of arriving at operational methodologies for addressing the full climate impact of aviation depends not only on improving scientific understanding of non-CO$_2$ impacts, but also on the potential for measuring or calculating these impacts on individual flights.

5.5.2.2   *Shipping*

CO₂ emission indexing scheme

The International Maritime Organisation (IMO), a specialized UN agency, has adopted a strategy with regard to policies and measures, focusing mainly on further development of a CO_2 emission indexing scheme for ships and further evaluation of technical, operational and market-based solutions.

The basic idea behind a CO_2 emission index is that it describes the CO_2 efficiency (i.e., the fuel efficiency) of a ship, i.e., the CO_2 emission per tonne cargo per nautical mile. This index could, in the future, assess both the technical features (e.g., hull design) and operational features of the ship (e.g., speed).

In June 2005, at the 53rd session of the Marine Environment Protection Committee of IMO (IMO, 2005), interim guidelines for voluntary ship CO_2 emission indexing for use in trials were approved. The Interim Guidelines should be used to establish a common approach for trials on voluntary CO_2 emission indexing, which enable shipowners to evaluate the performance of their fleet with regard to CO_2 emissions. The indexing scheme will also provide useful information on a ship's performance with regard to fuel efficiency and may thus be used for benchmarking purposes. The interim guidelines will later be updated, taking into account experience from new trials as reported by industry, organisations and administrations.

A number of hurdles have to be overcome before such a system could become operational. The main bottleneck appears to be that there is major variation in the fuel efficiency of similar ships, which is not yet well understood (Wit *et al.*, 2004). This is illustrated by research by the German delegation of IMO's Working Group on GHG emission reduction (IMO, 2004), in which the specific energy efficiency (i.e., a CO_2 emission index) was calculated for a range of container ships, taking into account engine design factors rather than operational data. The results of this study show that there is considerable scatter in the specific engine efficiency of the ships investigated, which could not be properly explained by the deadweight of the ships, year of build, ship speed and several other ship design characteristics. The paper therefore concludes that the design of any CO_2 indexing scheme and its differentiation according to ship type and characteristics, requires in-depth investigation. Before such a system can be used in an incentive scheme, the reasons for the data scatter need to be understood. This is a prerequisite for reliable prediction of the economic, competitive and environmental effects of any incentive based on this method.

Voluntary use and reporting results of CO_2 emission indexing may not directly result in GHG emission reductions, although it may well raise awareness and trigger certain initial moves towards 'self regulation'. It might also be a first step in the process of designing and implementing some of the other policy options. Reporting of the results of CO_2 emission indexing could thus generate a significant impetus to the further development and implementation of this index, since it would lead to widespread experience with the CO_2 indexing methodology, including reporting procedure and monitoring, for shipping companies as well as for administrations of states.

In the longer term, in order to be more effective, governments may consider using CO_2 indexing via the following paths:
1. The indexing of ship operational performance is introduced as a voluntary measure and over time developed and adopted as a standard;
2. Based on the experience with the standard, it will act as a new functional requirement when new buildings are ordered, hence over time the operational index will affect the requirements from ship owners related to the energy efficiency of new ships;
3. Differentiation of en route emission charges or existing port dues on the basis of a CO_2 index performance;
4. To use the CO_2 index of specific ship categories as a baseline in a (voluntary) baseline-and-credit programme.

Economic instruments for international shipping

There are currently only a few cases of countries or ports introducing economic instruments to create incentives to reduce shipping emissions. Examples include environmentally differentiated fairway dues in Sweden, the Green Award scheme[43] in place in 35 ports around the world, the Green Shipping bonus in Hamburg and environmental differentiation of tonnage tax in Norway. None of these incentives are based on GHG emissions, but generally relate to fuel sulphur content, engine emissions (mainly NO_x), ship safety features and management quality.

Harrison *et al.* (2004) explored the feasibility of a broad range of market-based approaches to regulate atmospheric emissions from seagoing ship in EU sea areas. The study focused primarily on policies to reduce the air pollutants SO_2 and NO_x, but the approaches adopted may to a certain extent also be applicable to other emissions, including CO_2. According to a follow-up study by Harrison *et al.* (2005) the main obstacles to a programme of voluntary port dues differentiation are to provide an adequate level of incentive, alleviating ports' competitive concerns and reconciling differentiation with specially negotiated charges. Swedish experience suggests that when combined with a centrally determined mandatory charging programme, these problems may be surmountable. However, in many cases a voluntary system would not likely be viable and other approaches to emissions reductions may therefore be required.

An alternative economic instrument, such as a fuel tax is vulnerable to evasion; that is ships may avoid the tax by taking

---

43 www.greenaward.org

fuel on board outside the taxed area. Offshore bunker supply is already common practice to avoid paying port fees or being constrained by loading limits in ports. Thus even a global fuel tax could be hard to implement to avoid evasion, as an authority at the port state level would have to collect the tax (ECON, 2003). A CO_2-based route charge or a (global) sectoral emission trading scheme would overcome this problem if monitoring is based on the carbon content of actual fuel consumption on a single journey. As yet there is no international literature that analyzes the latter two policy options. Governments may therefore consider investigating the feasibility and effectiveness of emission charges and emission trading as policy instruments to reduce GHG emissions from international shipping.

5.5.3   Non-climate policies

Climate change is a minor factor in decision making and policy in the transport sector in most countries. Policies and measures are often primarily intended to achieve energy security and/or sustainable development benefits that include improvements in air pollution, congestion, access to transport facilities and recovery of expenditure on infrastructure development. Achieving GHG reduction is therefore often seen as a co-benefit of policies and measures intended for sustainable transport in the countries. On the other hand, there are many transport policies that lead to an increase in GHG emissions. Depending on their orientation, transport subsidies can do both.

The impact of transport subsidies
Globally, transport subsidies are significant in economic terms. Van Beers and Van den Bergh (2001) estimated that in the mid-1990s transport subsidies amounted to 225 billion US$, or approximately 0.85% of the world GDP. They estimated that transport subsidies affect over 40% of world trade. In a competitive environment (not necessarily under full competition), subsidies decrease the price of transport. This results in the use of transport above its equilibrium value and most of the time also results in higher emissions, although this depends on the type of subsidy. Secondly, they decrease the incentive to economise on fuel, either by driving efficiently or by buying a fuel-efficient vehicle.

A quantitative appraisal of the effect of subsidies on GHG emissions is very complicated (Nash et al., 2002). Not only have shifts between fuels and transport modes to be taken into account, but the relation between transport and the production structure also needs to be analysed. As a result, reliable quantitative assessments are almost non-existent (OECD, 2004a). Qualitative appraisals are less problematic. Transport subsidies that definitely raise the level of GHG emissions include subsidies on fossil transport fuels, subsidies on commuting and subsidies on infrastructure investments.

Many, mostly oil producing, countries provide their inhabitants with transport fuels below the world price. Some countries spend more than 4% of their GDP on transport fuel subsidies (Esfahani, 2001). Many European countries and Japan have special fiscal arrangements for commuting expenses. In most of these countries, taxpayers can deduct real expenses or a fixed sum from their income (Bach, 2003). By reducing the incentive to move closer to work, these tax schemes enhance transport use and emissions.

Not all transport subsidies result in higher emissions of GHGs. Some subsidies stimulate the use of climate-friendly fuels. In many countries, excise duty exemptions on compressed natural or petroleum gas and on biofuels exist (e.g., Riedy, 2003). If these subsidies result in a change in the fuel mix, without resulting in more transport movements, they may actually decrease emissions of GHGs.

The most heavily subsidised form of transport is probably public transport, notably suburban and regional passenger rail services. In the USA, fares only cover 25% of the costs, in Europe 50% (Brueckner, 2004). Although public transport generally emits fewer GHGs per passenger-km, the net effect of these subsidies has not been quantified. It depends on the balance between increased GHG emissions due to higher demand (due to lower 'subsidised' fares) and substitution of relatively less efficient transport modes.

5.5.4   Co-benefits and ancillary benefits

The literature uses the term ancillary benefits when focusing primarily on one policy area, and recognizing there may be benefits with regard to other policy objectives. One speaks of co-benefits when looking from an integrated perspective. This section focuses on co-benefits and ancillary benefits of transport policies. Chapter 11.6 provides a general discussion of the benefits and linkages related to air pollution policies.

As mentioned above, several different benefits can result from one particular policy. In the field of transport, local air pollutants and GHGs have a common source in motorized traffic, which may also induce congestion, noise and accidents. Addressing these problems simultaneously, if possible, offers the potential of large cost reductions, as well as reductions of health and ecosystems risks. A recent review of costs of road transport emissions, and particularly of particulates PM2.5, for European countries strongly supports that view (HEATCO, 2006). Tackling these problems would also contribute to more effective planning of transport, land use and environmental policy (UN, 2002; Stead et al., 2004). This suggests that it would be worthwhile to direct some research towards the linkages between these effects.

Model studies indicate a potential saving of up to 40% of European air pollution control costs if the changes in the energy systems that are necessary for compliance with the Kyoto protocol were simultaneously implemented (Syri et al., 2001). For China, the costs of a 5–10% CO2 reduction

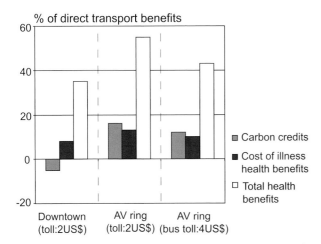

Figure 5.19: *Co-benefits from different mitigation measures in Santiago de Chile*
Note: toll is applied for cars/busses to enter downtown area or inside the Americo Vespucio ring around the city.

Source: Cifuentes and Jorquera, 2002.

would be compensated by increased health benefits from the accompanying reduction in particulate matter (Aunan *et al.*, 1998). McKinley *et al.* (2003) analyzed several integrated environmental strategies for Mexico City. They conclude that measures to improve the efficiency of transport are the key to joint local/global air pollution control in Mexico City. The three measures in this category that were analyzed, taxi fleet renovation, metro expansion and hybrid buses, all have monetized public health benefits that are larger than their costs when the appropriate time horizon is considered.

A simulation of freight traffic over the Belgian network indicated that a policy of internalizing the marginal social costs caused by freight transport types would induce a change in the modal shares of trucking, rail and inland waterways transport. Trucking would decrease by 26% and the congestion cost it created by 44%. It was estimated that the total cost of pollution and GHG emissions (together) would decrease by 15.4%, the losses from accidents diminish by 24%, the cost of noise by 20% and wear and tear by 27%. At the same time, the total energy consumption by the three modes would decrease by 21% (Beuthe *et al.*, 2002).

Other examples of worthwhile policies can be given. The policy of increasing trucks' weight and best practices awareness in Sweden, UK and the Netherlands lead to a consolidation of loads that resulted in economic benefits as well as environmental benefits, including a decrease in CO_2 emissions (MacKinnon, 2005; Leonardi and Baumgartner, 2004). Likewise, the Swiss heavy vehicle fee policy also leads to better loaded vehicles and a decrease of 7% in CO_2 emissions (ARE, 2004a).

Obviously, promotion of non-motorized transport (NMT) has the large and consistent co-benefits of GHG reduction, air quality and people health improvement (Mohan and Tiwari, 1999).

In the City of London a congestion charge was introduced in February 2003, to reduce congestion. Simultaneous with the introduction of the charge, investment in public transport increased to provide a good alternative. The charge is a fee for motorists driving into the central London area. It was introduced in February 2003. Initially set at 5 £/day (Monday to Friday, between 7 am and 6.30 pm), it was raised to 8 £ in July 2005. The charge will be extended to a larger area in 2007. On a cost-benefit rating, the results of the charge are not altogether clear (Prud'homme and Bocarejo, 2005, Mackie, 2005). However, it contributed to a 30% decrease of the traffic by the chargeable vehicles in the area and less congestion, to higher speed of private vehicles (+20%) and buses (+7%), and to an increased use of public transport, plus more walking and bicycling. The charge has had substantial ancillary benefits with respect to air quality and climate policy. All the volume and substitution effects in the charging zone has led to an estimated reductions in CO_2 emissions of 20%. Primary emissions of NO_x and PM10 fell by 16% after one year of introduction (Transport for London, 2006). A variant of that scheme has been in operation since 1975 in Singapore with similar results; Stockholm is presently experimenting with such a system, Trondheim, Oslo and Durham are other examples.

Under the Integrated Environmental Strategies Program of the US EPA, analysis of public health and environmental benefits of integrated strategies for GHG mitigation and local environmental improvement is supported and promoted in developing countries. A mix of measures for Chile has been proposed, aimed primarily at local air pollution abatement and energy saving. Measures in the transport sector (CNG buses, hybrid diesel-electric buses and taxi renovation) proved to provide little ancillary benefits in the field of climate policy, see Figure 5.19. Only congestion charges were expected to have substantial ancillary benefits for GHG reduction (Cifuentes *et al.*, 2001, Cifuentes & Jorquera, 2002).

While there are many synergies in emission controls for air pollution and climate change, there are also trade-offs. Diesel engines are generally more fuel-efficient than gasoline engines and thus have lower CO_2 emissions, but increase particle emissions. Air quality driven measures, like obligatory particle matter (PM) and NO_x filters and in-engine measures, do not result in higher fuel use if appropriate technologies are used, like Selective Catalytic Reduction (SCR)- NO_x catalyst.

5.5.5 Sustainable Development impacts of mitigation options and considerations on the link of adaptation with mitigation.

Within the transport sector there are five mitigation options with a clear link between sustainable development, adaptation and mitigation. These areas are biofuels, energy efficient, public transport, non-motorised transport and urban planning. Implementing these options would generally have positive social, environmental and economic side effects. The economic effects of using bio-energy and encouraging public transport systems, however, need to be evaluated on a case-by-case basis. For transport there are no obvious links between mitigation and adaptation policies and the impact on GHG emissions due to adaptation is expected to be negligible.

Mitigation and sustainable development is discussed from a much wider perspective, including the other sectors, in Chapter 12, Section 12.2.4.

5.6 Key uncertainties and gaps in knowledge

Key uncertainties in assessment of mitigation potential in the transport sector through the year 2030 are:
- World oil supply and its impact on prices and alternative transport fuels;
- R&D outcomes in several areas, especially biomass fuel production technology and its sustainability if used on a massive scale, and batteries. These outcomes will strongly influence the future costs and performance of a wide range of transport technologies.

The degree to which the potential can be realized will crucially depend on the priority that developed and developing countries give to GHG emissions mitigation.

A key gap in knowledge is the lack of comprehensive and consistent assessments of the worldwide potential and cost to mitigate transport's GHG emissions. There are also important gaps in basic statistics and information on transport energy consumption and GHG mitigation, especially in developing countries.

REFERENCES

Abbot, C., 2002: Planning a Sustainable City: The Promise and Performance of Portland's Urban Growth Boundary. In *Urban Sprawl*, G.D. Squires (ed.), The Urban Land Institute, pp. 207-235.

ACEEE, 2001: Technical Options for Improving the Fuel Economy of U.S. Cars and Light Trucks by 2010-2015. DeCicco, J., F. An, and M. Ross, American Council for an Energy-Efficient Economy, <www.aceee.org/pubs/t012.htm> accessed 30/05/07.

ADAC, 2005: Study on the effectiveness of Directive 1999/94/EC relating to the availability of consumer information on fuel economy and CO_2 emissions in respect of the marketing of new passenger cars. München, March 2003.

ADEME, 2006: Inventories of the worldwide fleets of refrigerating and air-conditioning equipment in order to determine refrigerant emissions: the 1990 to 2003 updating, ADEME, Paris.

Airbus, 2004: Global Market Forecast 2004-2023. <http://www.airbus.com/store/mm_repository/pdf/att00003033/media_object_file_GMF2004_full_issue.pdf> accessed 30/05/07.

Akerman, J. and M. Hojer, 2006: How Much Transport Can the Climate Stand? - Sweden on a Sustainable Path in 2050. *Energy Policy*, **34**, pp. 1944-1957.

Altmann, M., M. Weinberger, and W. Weindorf, 2004: Life Cycle Analysis results of fuel cell ships; Recommendations for improving cost effectiveness and reducing environmental impacts. Contract no. G3RD-CT-2002-00823 in the framework of the FCSHIP project of the European Commission, Final report July 2004, 58 pp.

An, F. and A. Sauer, 2004: Comparison of Passenger Vehicle Fuel Economy and Greenhouse Gas Emission Standards around the World. Pew Center on Global Climate Change, Arlington, 33 pp.

Annema, J.A., E. Bakker, R. Haaijer, J. Perdok, and J. Rouwendal, 2001: Stimuleren van verkoop van zuinige auto's; De effecten van drie prijsmaatregelen op de CO_2-uitstoot van personenauto's (summary in English). RIVM, Bilthoven, The Netherlands, 49 pp.

ARE, 2004a: Die Schwerverkehrsabgabe der Schweiz. Bundesamt für Raumentwicklung, Bern.

ARE, 2004b: Entwicklung des Strassengüterverkehrs nach Einführung von LSVA und 34t-limite. Bundesamt für Raumentwicklung, Bern.

ARE, 2006: Equitable et efficiente, la redevance sur le trafic des poids lourds liée aux prestations RPLP en Suisse. Office Fédéral du Développement Territorial, Bern.

Arthur Andersen Consulting, 2001: Emission trading for aviation - Workstream 3, key findings and conclusions. Report prepared for the International Air Transport Association (IATA).

Aunan, K., G. Patzay, H.A. Aaheim, and H.M. Seip, 1998: Health and environmental benefits from air pollution reduction in Hungary. *Science of the Total Environment*, **212**, pp. 245-268.

Bach, S., 2003: Entfernungspauschale: Kürzung gerechtfertigt. *DIW Wochenbericht* **70**(40), pp. 602-608.

Beers, C. van, and J.C.J.M. van den Bergh, 2001: Perseverance of perverse subsidies and their impact on trade and environment. *Ecological Economics*, **36**, pp. 457-486.

Berghof, R., A. Schmitt, C. Eyers, K. Haag, J. Middel, M. Hepting, A. Grübler, and R. Hancox, 2005: CONSAVE 2050 Constrained Scenarios on Aviation and Emissions.

Beuthe, M., F. Degransart, J-F Geerts, and B. Jourquin, 2002: External costs of the Belgian interurban freight traffic: a network analysis of their internalisation. *Transportation Research*, **D7**, 4, pp. 285-301.

Biddle, T., 2006: Approval of a Fully Synthetic Fuel for Gas Turbine Engines. TRB 23 Jan. 2006, <http://www.trbav030.org/pdf2006/265_Biddle.pdf> accessed 30/05/07.

Boeing, 2006: Current Market Outlook 2006, <http://www.boeing.com/commercial/cmo/pdf/CMO_06.pdf> accessed 30/05/07

Booz Allen & Hamilton, 1999: Effects of Public Transport System Changes on Mode Switching and Road Traffic Levels. Transfund New Zealand.

Bose, R., and D. Sperling, 2001: Transportation in Developing Countries: Greenhouse Gas Scenarios for Delhi, India. Pew Center on Global Climate Change, Arlington, 43 pp.

Braslow, A.L., 1999: A History of Suction-Type Laminar-Flow Control with Emphasis on Flight Research, http://www.nasa.gov/centers/dryden/pdf/88792main_Laminar.pdf

Brog, W., E. Erl, and N. Mense, 2004: Individualized Marketing: Changing Travel Behavior for a Better Environment. In *Communicating Environmentally Sustainable Transport: The Role of Soft Measures*. OECD (ed.), pp. 83-97.

Brueckner, J.K., 2004: Transport Subsidies, System Choice, and Urban Sprawl. Unpublished paper, Department of Economics, University of Illinois at Urbana-Champaign, 26 pp.

Cameron, I., T.J. Lyons, and J.R. Kenworthy, 2004: Trends in Vehicle Kilometres of Travel in World Cities, 1960-1990: Underlying Drives and Policy Responses. *Transport Policy*, **11**, pp. 287-298.

Cames, M. and O. Deuber, 2004: Emission trading in international aviation. Oeko Institute V, Berlin, ISBN 3-934490-0.

Capaldo, K., J.J. Corbett, P. Kasibhatla, P.S. Fishbeck, and S.N. Pandis, 1999: Effects of ship emission on sulphur cycling and radiative climate forcing over the ocean. *Nature* **400**, pp. 743-746.

Cervero, R., 1998: The Transit Metropolis. Island Press, 464 pp.

Chandler, K., E. Eberts, and L. Eudy, 2006: New York City Transit Hybrid and CNG Buses: Interim Evaluation Results. National Renewable Energy Laboratory Technical Report NREL/TP-540-38843, January.

Choo, S., P.L. Mokhtarian, and I. Salomon, 2005: Does Telecommuting Reduce Vehicle-Miles Traveled? An Aggregate Time Series for the U.S. Transportation. **32**, pp. 37-64.

Cifuentes, L. and H. Jorquera, 2002: IES Developments in Chile. October 2002, <http://www.epa.gov/ies/documents/chile/cifuentespres.pdf> accessed 30/05/07.

Cifuentes, L., H. Jorquera, E. Sauma, and F. Soto, 2001: International CO-Controls Analysis Program. Final Report, December 2001, the Catholic University of Chile, 62pp.

Corbett, J.J. and H.W. Köhler, 2003: Updated emissions from ocean shipping. *Journal of Geophysical Research*, **108**(D20), 4650 pp., doi:10.1029/2003JD003751.

Corbett, J.J. and H.W. Köhler, 2004: Considering alternative input parameters in an activity-based ship fuel consumption and emissions model: Reply to comment by Øyvind Endresen et al. on 'Updated emissions from ocean shipping.' *Journal of Geophysical Research*, **109**, D23303, doi:10.1029/2004JD005030.

Corbett, J.J., P.S. Fischbeck, and S.N. Pandis, 1999: Global nitrogen and sulfur inventories for oceangoing ships. *Journal of Geophysical Research*, **104**(3), pp. 3457-3470.

COWI, 2002: Fiscal measures to reduce CO_2 emissions from new passenger cars. Main report, EC, Brussel, 191 pp.

Cullinane, C., 2002: The Relationship between Car Ownership and Public Transport Provision: A Case Study of Hong Kong. *Transport Policy*, **9**, pp. 29-39.

Deuker, K.J., J.G. Strathman, and M.J. Bianco, 1998: Strategies to Attract Auto Users to Public Transportation. TCRP Report 40, Transportation Research Board, 105 pp.

DfT, 2003: Carbon to Hydrogen -Roadmap for Passenger Cars: Update of the Study. Department for Transport, UK.

Dill, J. and T. Carr, 2003: Bicycle Commuting and Facilities in Major U.S. Cities: If You Build Them, Commuters Will Use Them. *Transportation Research*, **1828**, pp. 116-123.

DRI, 2004: A Review of the Results in the 1997 Kahane, 2002 DRI, 2003 DRI, and 2003 Kahane. Reports on the Effects of Passenger Car and Light Truck Weight and Size on Fatality Risk. R.M. Van Auken, and J.W. Zellner, Dynamic Research Inc. DRI-TR-04-02, Torrance, CA.

Duoba, M., H. Lohse-Busch, and T. Bohn, 2005: Investigating Vehicle Fuel Economy Robustness of Conventional and Hybrid Electric Vehicles. EVS-21, January.

EAA, 2001: Moving up to aluminium -The future of road transport. European Aluminium Association, Brussels, 19 pp.

EC, 1999: Cycling: The Way Ahead for Towns and Cities. European Commission, Brussels, 61 pp.

EC, 2005: Energy and Transport in Figures 2005: Part 3: Transport. European Commission, Directorate-General for Energy and Transport, Brussels, Belgium.

EC, 2006: Impact assessment. Commission Staff Working Document COM (2006) 818 final, accompanying document to the Proposal for the European Parlement and the Council Amending Directive 2003/87/EC, Brussels.

ECMT, 2000: Strategic Environmental Assessment. OECD, European Conference of Ministers of Transport, 91 pp.

ECMT, 2004a: National Policies to Promote Cycling. OECD, 91 pp.

ECMT, 2004b: Transport and Spatial Policies: The Role of Regulatory and Fiscal Incentives. OECD, 180 pp.

ECMT, 2006: Reducing NO_x Emissions on the Road; Ensuring Future Exhausts Emission Limits Deliver Air Quality Standards, OECD, 50 pp.

ECMT, 2007: Cutting Transport CO_2 Emissions - What Progress?, OECD, 264 pp.

ECMT/IEA, 2005: Making cars more fuel efficient; Technology for real improvements on the road, Paris, 2005.

ECON, 2003: GHG Emissions from International Shipping and Aviation. ECON Report no 38400, Oslo, Norway, ISBN 82-7645-577-8, 40 pp.

EEA, 2003: Europe's Environment: The Third Assessment. Environmental assessment report No 10. European Environment Agency, Copenhagen.

EIA DOE, 2005: International Energy Outlook 2005, DOE/EIA-0484. Energy Information Administration.

Endresen, Ø., E. Søgård, J.K. Sundet, S.B. Dalsøren, I.S.A. Isaksen, T.F. Berglen, and G. Gravir, 2003: Emission from international sea transportation and environmental impact. *Journal of Geophysical Research*, **108**, 4560 pp, doi:10.1029/2002JD002898.

Endresen, Ø., E. Sørgård, J. Bakke, and I.S.A. Isaksen, 2004: Substantiation of a lower estimate for the bunker inventory: Comment on 'Updated emissions from ocean shipping' by J.J. Corbett and H.W. Koehler. *Journal of Geophysical Research*, **109**, D23302, doi.

Englar, R.J., 2001: Advanced Aerodynamic Devices to Improve the Performance, Economics, Handling and Safety of Heavy Vehicles. SAE paper 2001-01-2072.

EPRI, 2001: Comparing the Benefits and Impacts of Hybrid Electric Vehicle Options. Report 1000349, Electric Power Research Institute, Palo Alto, CA, July 2001.

Esfahani, H.S., 2001: A Political Economy Model of Resource Pricing with Evidence from the Fuel Market. ERS Working paper 200134, Economic Research Forum, Cairo, Egypt, 24 pp.

EUCAR/CONCAWE/JRC, 2006: Well-to-Wheels Analysis of Future Automotive Fuels and Powertrains in the European Context. <http://ies.jrc.ec.europa.eu/WTW> accessed 30/05/07.

European Natural Gas Vehicle Association, 2002: <www.engva.org> accessed 30/05/07.

Ewing, R., 1997: Is Los Angeles-Style Sprawl Desirable? *Journal of American Planning Association*, **63**(1), pp. 96-126.

Ewing, R. and R. Cervero, 2001: Travel and the Built Environment: A Synthesis. *Transportation Research*, **1780**, pp. 87-113.

Eyring, V., H.W. Köhler, J. van Aardenne, and A. Lauer, 2005a: Emissions from international shipping: 1. The last 50 years. *Journal of Geophysical Research*, **110**, D17305, doi: 10.1029/2004JD005619.

Eyring, V., H.W. Köhler, A. Lauer, and B. Lemper, 2005b: Emissions from international shipping: 2. Impact of future technologies on scenarios until 2050. *Journal of Geophysical Research*, **110**, D17306, doi:10.1029/2004JD005620.

Farrell, A.E., R.J. Plevin, B.T. Turner, A.D. Jones, M. O'Hare, and D.M. Kammen, 2006: Ethanol can contribute to energy and environmental goals. *Science*, **311**, pp. 506-508.

FESG, 2003: Report of the FESG/CAEP-6 traffic and fleet forecast (forecasting sub-group of FESG). ICAO-CAEP FESG, Montreal.

Fichter, C., S. Marquart, R. Sausen and D.S. Lee, 2005: The impact of cruise altitude on contrails and related radiative forcing. *Meteorologische Zeitschrift*, **14**(4), pp. 563-572.

Forkenbrock, D.J. and G.E. Weisbrod, 2001: Guidebook for Assessing the Social and Economic Effects of Transportation Projects. NCHRP Report 456, Transportation Research Board, 242 pp.

Forster, P., V. Ramaswamy, P. Artaxo, T. Berntsen, R. Betts, D.W. Fahey, J. Haywood, J. Lean, D.C. Lowe, G. Myhre, J. Nganga, R. Prinn, G. Raga, M. Schulz and R. Van Dorland, 2007: Changes in Atmospheric Constituents and in Radiative Forcing. In: Climate Change 2007: The Physical Science Basis. Contribution of Working Group I to the Fourth Assessment Report of the Intergovernmental Panel on Climate Change [Solomon, S., D. Qin, M. Manning, Z. Chen, M. Marquis, K.B. Averyt, M.Tignor and H.L. Miller (eds.)]. Cambridge University Press, Cambridge, United Kingdom and New York, NY, USA.

Fouracre, P., C. Dunkerley, and G. Gardner, 2003: Mass Rapid Transit System for Cities in the Developing World. *Transport Reviews*, **23**(3), pp. 299-310.

Foyt, G., 2005: Demonstration and Evaluation of Hybrid Diesel-Electric Transit Buses - Final Report. Connecticut Academy of Science and Engineering report CT-170-1884-F-05-10, Hartford, CT, 33 pp.

Friedricht, H.E., 2002: Leightbau und Werkstoffinnovationen im Fahrzeugbau. *ATZ*, **104**(3), pp. 258-266.

Fujii, S. and A. Taniguchi, 2005: Travel Feedback Programs: Communicative Mobility Management Measures for Changing Travel Behavior. The Eastern Asia Society for Transportation Studies, 6, CD.

Fwa, T.F. 2002: Transportation Planning and Management for Sustainable Development - Singapore's Experience. Paper presented at a Brainstorming Session on Non-Technology Options for Stimulating Modal Shifts in City Transport Systems held in Nairobi, Kenya. STAP/GEF, Washington D.C.

GbD, 2001: The Technology Challenge. Greener by Design, <http://www.greenerbydesign.org.uk/> accessed 30/05/07.

Gilat, M. and J.M. Sussman, 2003: Coordinated Transportation and Land Use Planning in the Developing World: Case of Mexico City. *Transportation Research*, **1859**, pp. 102-109.

Giuliano, G., K. Hwang, and M. Wachs, 1993: Employee Trip Reduction in Southern California: First Year Results. *Transportation Research*, **27A**(2), pp. 125-137.

GM/ANL, 2005: Well-to-Wheels Analysis of Advanced Fuel/Vehicle Systems-A North American Study of Energy Use, Greenhouse Gas Emissions, and Criteria Pollutant Emissions. <http://www.transportation.anl.gov/pdfs/TA/339.pdf> accessed 30/05/07.

Goodwin, P., 1992: A Review of New Demand Elasticities with Special Reference to Short and Long Run Effects of Price Changes. *Journal of Transport Economics and Policy*, **26**, pp. 139-154.

Goodwin, P., 1999: Transformation of Transport Policy in Great Britain. *Transportation Research*, **33**(7/8), pp. 655-669.

Goodwin, P., J. Dargay, and M. Hanly, 2004: Elasticities of Road Traffic and Fuel Consumption With Respect to Price and Income: A Review. *Transport Reviews*, **24**(3), 275-292.

Gordon, P. and H.W. Richardson, 1997: Are Compact Cities a Desirable Planning Goal? *Journal of American Planning Association*, **63**(1), pp. 95-106.

Grant-Muller, S.M., P.J. Mackie, H. Nellthorp, and A.D. Pearman, 2001: Economic Appraisal of European Transport Projects-the State of the Art Revisited. *Transport Reviews*, **21**(2), pp. 237-262.

Green Car Congress, 2004: [website] <http://www.greencarcongress.com/2004/10/fedex_hybrid_up.html> accessed 30/05/07.

Greene, D.L. and A. Schafer, 2003: Reducing Greenhouse Gas Emissions from U.S. Transportation. PEW Center, Arlington, 68 pp.

Greene, D.L., 2005: Improving the Nation's Energy Security: Can Cars and Trucks be Made More Fuel Efficient? Testimony to the U.S. House of Representatives, Committee on Science, February.

Grewe, V., M. Dameris, C. Fichter, and D.S. Lee, 2002: Impact of aircraft NO_x emissions. Part 2: effects of lowering flight altitudes. *Meteorologische Zeitschrift*, **11**, pp. 197-205.

Gupta, S., 1999: Country environment review - India, policy measures for sustainable development Manila: Asian Development Bank [Discussion paper].

Gwilliam, K., K. Kojima, and T. Johnson, 2004: Reducing Air Pollution from Urban Transport. The World Bank, 16 pp.

Haight, B., 2003: Advanced Technologies are opening doors for new materials. Automotive Industries, April 2003.

Hamelinck, C.N., R. Janzic, A. Blake, R. van den Broek, 2005: Policy Incentive Options for Liquid Biofuels Development in Ireland. SEI, 56 pp.

Hamelinck, C.N. and A.P.C. Faaij, 2006: Outlook for advanced biofuels. *Energy Policy*, **34**(17), 3268-3283.

Harrison, D., D. Radov and J. Patchett, 2004: Evaluation of the feasibility of Alternative Market-Based Mechanisms to Promote Low-Emission Shipping in European Sea Areas. NERA Economic Consulting, London, UK, 106 pp.

Harrison, D., D. Radov, J. Patchett, P. Klevnas, A. Lenkoski, P. Reschke and A. Foss, 2005: Economic instruments for reducing ship emissions in the European Union. NERA Economic Consulting, London, UK, 117 pp.

HEATCO, 2006: Proposal for harmonised guidelines, Deliverable 5 to the EU Commission. IER, Developing Harmonised European Approaches for Transport Costing and Project assessment, Germany.

Heavenrich, R.M., 2005: Light-Duty Automotive Technology and Fuel Economy Trends, 1975 Through 2005, U.S. Environmental Protection Agency Report EPA-420-R-05-001, July.

Helms, H. and U. Lambrecht, 2006: The potential contribution of light-weighting to reduce transport energy consumption. *The International Journal of Life Cycle Assessment*. <http://dx.doi.org/10.1065/lca2006.07.258>, 7 pp.

Hensher, D.A., 2001: Chapter 8 Modal Diversion. In *Handbook of Transport Systems and Traffic Control*. Hensher, D.A. and K.J. Button, (ed.), Pergamon, pp. 107-123.

Heywood, J. B., M. A. Weiss, A. Schafer, S. A. Bassene, and V. K. Natarajan, 2003: The Performance of Future ICE and Fuel Cell Powered Vehicles and Their Potential Fleet Impact. Massachusetts Institute of Technology, Laboratory for Energy and the Environment, MIT LFEE 2003-004 RP, December.

Honda, 2004: Honda Develops Hybrid Scooter Prototype. <http://world.honda.com/news/2004/2040824_02.html> accessed 30/05/07.

Hook, W., 2003: Preserving and Expanding the Role of Non-motorised Transport. GTZ, Eschborn, Germany, 35 pp.

Hughes, J.E., C.R. Knittel, and D. Sperling, 2006: Evidence of a Shift in the Short-Run Price Elasticity of Gasoline Demand. Institute of Transportation Studies, University of California, Davis UCD-ITS-RR-06-16.

Hylen, B. and T. Pharoah, 2002: Making Tracks-Light Rail in England and France. Swedish National Road and Transport Research Institute, 91 pp.

ICAO, 2003: CAEP FESG Report to CAEP/6 2003.

ICAO, 2006: Form A - ICAO Reporting form for Air Carrier Traffic.

ICAO/FESG, 2003: FESG CAEP-SG20031-IP/8 10/6/03. Steering Group Meeting Report of the FESG/CAEP6 Traffic and Fleet Forecast (Forecasting sub-group of FESG), Orlando SG meeting, June 2003.

ICAO/FESG, 2004a: Analysis of voluntary agreements and open emission trading for the limitation of CO_2 emissions from aviation with the AERO modeling system Part I. Montreal, Canada, 57 pp.

ICAO/FESG, 2004b: Analysis of open emission trading systems for the limitation of CO_2 emissions from aviation with the AERO modeling system. Montreal, Canada, 77 pp.

ICF, 2004: Designing a greenhouse gas emissions trading system for international aviation, Study carried out for ICAO, London, UK

IEA Hybrid, 2006: International Energy Agency Implementing Agreement on Hybrid and Electric Vehicles. <http://www.ieahev.org/evs_hevs_count.html> accessed 30/05/07.

IEA, 2001: Saving Oil and Reducing CO_2 Emissions in Transport. International Energy Agency, OECD, 194 pp.

IEA, 2002a: Transportation Projections in OECD Regions - Detailed report. International Energy Agency, 164 pp.

IEA, 2002b: Bus Systems for the Future: Achieving Sustainable Transport Worldwide. International Energy Agency, 188 pp.

IEA, 2003: Transport Technologies and policies for energy security and CO_2 Reductions. Energy technology policy and collaboration papers, International Energy Agency, ETPC paper no 02/2003.

IEA, 2004a: World Energy Outlook 2004. International Energy Agency, 570 pp.

IEA, 2004b: Energy Technologies for a Sustainable Future: Transport. International Energy Agency, Technology Brief, 40 pp.

IEA, 2004c: Biofuels for Transport: An International Perspective. International Energy Agency, Paris, 210 pp.

IEA, 2004d: Reducing Oil Consumption in Transport - Combining Three Approaches. IEA/EET working paper by L. Fulton, International Energy Agency, Paris, 24 pp.

IEA, 2005: Prospects for Hydrogen and Fuel Cells. International Energy Agency, Paris, 253 pp.

IEA, 2006a: Energy Technology Perspectives 2006; Scenarios & Strategies to 2050. International Energy Agency, Paris, 479 pp.

IEA, 2006b: World Energy Outlook 2006. International Energy Agency, Paris, 596 pp.

IEA, 2006c: Energy Balances of Non-OECD countries, 2003-2004. International Energy Agency, Paris, 468pp.

IEA, 2006d: CO_2 Emissions from Fuel Combustion 1971-2004. International Energy Agency, Paris, 548pp.

IMO, 2004: MEPC 51/INF.2, Statistical investigation of containership design with regard to emission indexing. Submitted by Germany to the 51ste of the Marine Environment Protection Committee of IMO.

IMO, 2005: MEPC 53/WP.11, Prevention of air pollution from ships. Report of the Working Group, 21 July 2005.

IPCC, 1996: IPCC Guidelines for National Greenhouse Gas Inventories - workbook. Intergovernmental Panel on Climate Change.

IPCC, 1999: *Aviation and the Global Atmosphere* [Penner, J.E., D.H. Lister, D.J. Griggs, D.J. Dokken and M. McFarland (eds)]. Special report of the Intergovernmental Panel on Climate Change (IPCC) Working Groups I and III, Cambridge University Press, Cambridge.

IPCC, 2005: *Safeguarding the Ozone Layer and the Global Climate System*. Special Report of the Intergovernmental Panel on Climate Change (IPCC), Cambridge University Press, Cambridge, 478 pp.

JAMA, 2002: The Motor Industry of Japan 2002. Japan Automobile Manufacturers Association.

Japanese Statistical Bureau, 2006: Historical Statistics of Japan. Chapter 12: Transport, Statistics Bureau, Tokyo.

Jelinek, F., S. Carlier, J. Smith, and A. Quesne, 2002: The EUR RVSM implementation Project-Environmental Benefit Analysis. Eurocontrol, Brussels, Belgium, 77 pp.

JHFC, 2006: JHFC Well-to-Wheel Efficiency Analysis Results. JARI, Tsukuba, 114 pp.

JMLIT and IHE, 2004: The Impact Study of New Public Transport on Road Traffic. IHE, Japanese Ministry of Land, Infrastructure and Transport and Institute of Highway Economics, Japan, 169 pp.

Jones, P., 2004: Comments on the Roles of Soft measures in Achieving EST. In *Communicating Environmentally Sustainable Transport: The Role of Soft Measures*. OECD (ed.), pp. 63-66.

Jong, G.D. and H. Gunn, 2001: Recent Evidence on Car Cost and Time Elasticities of Travel Demand in Europe. *Journal of Transport Economics and Policy*, **35**, pp. 137-160.

JRC/IPTS, 2004: Potential for Hydrogen as a Fuel for Transport in the Long Term (2020-2030) -Full Background Report.

JSGRIE, 2000: Guidelines for the Evaluation of Road Investment Projects. Japan Research Institute, Japanese Study Group on Road Investment Evaluation, 188 pp.

Kageson, J., 2005: Reducing CO_2 Emissions from New Cars, European Federation for Transport and Environment.

Kahane, C.J., 2003: Relationships between Vehicle Size and Fatality Risk in Model Year 1985-93 Passenger Cars and Light Trucks. Technical Report No. DOT HS 808 570, National Highway Traffic Safety Administration, Washington, D.C.

Kajitani, S., M. Takeda, A. Hoshimiya, S. Kobori, and M. Kato, 2005: The Concept and Experimental Result of DME Engine Operated at Stoichiometric Mixture. Presented at International Symposia on Alcohol Fuels (September 26-28, 2005), see <http://www.eri.ucr.edu/ISAFXVCD/ISAFXVPP/CnERDEOS.pdf> accessed 30/05/07.

Karekezi, S., L. Majoro, and T. Johnson, 2003: Climate Change Mitigation in the Urban Transport Sector: Priorities for the World Bank. World Bank, 53 pp.

Kenworthy, J.R. and F.B. Laube, 1999: Patterns of Automobile Dependence in Cities: An International Overview of Key Physical and Economic Dimensions with Some Implications for Urban Policy. *Transportation Research*, 33A(7/8), pp. 691-723.

Kenworthy, J. and F. Laube, 2002: Urban transport patterns in a global sample of cities & their linkages to transport infrastructure, land use, economics & environment. *World Transport Policy & Practice*, **8**(3), pp. 5-19.

Kiesgen, G., M. Kluting, C. Bock, and H. Fischer, 2006: The New 12-Cylinder Hydrogen Engine in the 7 Series: The H_2 ICE Age Has Begun. SAE Technical Paper 2006-01-0431.

Koyama, A., H. Iki, Y. Ikawa, M. Hirose, K. Tsurutani, and H. Hayashi, 2006: Proceedings of the JSAE Annual Congress, #20065913.

Lee, D.S., B. Owen, A. Graham, C. Fichter, L.L. Lim, and D. Dimitriu, 2005: International aviation emissions allocations - present day and historical. Manchester Metropolitan University, Centre for Air Transport and the Environment, Report CATE 2005-3(C)-2.

Lee, C. and A.V. Mouden, 2006: The 3Ds+R: Quantifying Land Use and Urban Form Correlates of Walking. *Transportation Research*, **11D**(3), pp. 204-215.

Leifsson, L.T. and W.H. Mason, 2005: The Blended Wing Body Aircraft, Virginia Polytechnic Institute and State University Blacksburg, VA, <http://www.aoe.vt.edu/research/groups/bwb/papers/TheBWBAircraft.pdf> accessed 30/05/07.

Leonardi, J. and M. Baumgartner, 2004: CO_2 efficiency in road freight transportation: Status quo, measures and potential. *Transportation Research*, **D9**(6), pp. 451-464.

Levinson, H.S., S. Zimmerman, J. Clinger, and S.C. Rutherford, 2002: Bus Rapid Transit: An overview. *Journal of Public Transportation*, **5**(2), pp. 1-30.

Litman, T., 2003: The online TDM Encyclopedia: Mobility Management Information Gateway. *Transport Policy*, **10**, pp. 245-249.

Lundqvist, L., 2003: Multifunctional land use and mobility. In *Multifunctional Land Use*, P. Nijkamp, C.A. Rodenburg and R. Vreeker (ed.), Vrije Universiteit Amsterdam and Habiforum, pp. 37-40.

Lutsey, N., C.J. Brodrick, D. Sperling, and C. Oglesby, 2004: Heavy-duty truck idling characteristics: Results from a national truck survey. *Transportation Research Record*, **1880**, pp. 29-38.

Mackett, R.L. and M. Edwards, 1998: The impact of new urban public transport systems: Will the expectations be met? *Transportation Research*, **32A**(4), pp. 231-245.

Mackie, P. and J. Nellthorp, 2001: Chapter 10 Cost-benefit analysis in transport. In *Handbook of Transport Systems and Traffic Control*, D.A. Hensher and K.J. Button, (eds.), Pergamon, pp. 143-174.

Mackie, P., 2005: The London congestion charge: a tentative economic appraisal. A comment on the paper by Prud'homme and Bocajero, *Transport Policy*, **12**, pp. 288-290.

MacKinnon, A.C., 2005: The economic and environmental benefits of increasing maximum truck weight: the British experience. *Transportation Research*, **D10**(1), pp. 79-95.

Mannstein, H., P. Spichtinger, and K. Gierens, 2005: A note on how to avoid contrail cirrus. *Transportation Research*, **10D**(5), 421-426.

Marintek, 2000: Study of greenhouse gas emissions from ships. Report to the IMO, by Marintek, Det Norske Veritas, Econ, Carnegie Mellon University.

Martin, D., D. Moon, S. Collings, and A. Lewis, 1995: Mechanisms for improved energy efficiency in transport. Overseas Development Administration, London, UK.

May, A.D., A.F. Jopson, and B. Matthews, 2003: Research challenges in urban transport policy. *Transport Policy*, **10**, pp. 157-164.

McCallen, R.C., K. Salari, J. Ortega, L. DeChant, B. Hassan, C. Roy, W.D. Pointer, F. Browand, M. Hammache, T.Y. Hsu, A. Leonard, M. Rubel, P. Chatalain, R. Englar, J. Ross, D. Satran, J.T. Heineck, S. Walker, D. Yaste, and B. Storms, 2004: DOE's effort to reduce truck aerodynamic drag-joint experiments and computations lead to smart design. Lawrence Livermore National Laboratory paper UCRL-CONF-204819, 34th AIAA Fluid Dynamics Conference and Exhibit, Portland, OR, United States, June 22. <http://www.llnl.gov/tid/lof/documents/pdf/308799.pdf> accessed 30/05/07

McKinley, G., Zuk, M., Hojer, M., Ávalos, M., González, I., Hernández, M., Iniestra, R., Laguna, I., Martínez, M.A., Osnaya, P., Reynales, L.M., Valdés, R., and J. Martínez, 2003: The local benefits of global air pollution control in Mexico City. Instituto Nacional de Ecologia, Instituto Nacional de Salud Publica, Mexico.

MIT, 2000: On the road in 2020: A life-cycle analysis of new automobile technologies. Massachusetts Institute of Technology, Energy Laboratory Report #MIT EL 00-003, 153 pp.

MIT, 2004: Coordinated policy measures for reducing the fuel consumption of the U.S. light-duty vehicle fleet. Bandivadekar, A.P. and J.B. Heywood, Massachusetts Institute of Technology, Laboratory for Energy and the Environment Report LFEE 2004-001, 76 pp.

Mohan, D. and D. Tiwari, 1999: Sustainable transport systems, linkages between environmental issues. Public transport, non-motorised transport and safety. *Economic and Political Weekly*, **XXXIV**(25), June 19, 1580-1596.

Mokhtarian, P.L., 1998: A synthetic approach to estimating the impacts of telecommuting on travel. *Urban Studies*, **35**(2), pp. 215-241.

Murakami, Y. and K. Uchibori, 2006: Development of fuel cell vehicle with next-generation fuel cell stack. SAE Paper 2006-01-0034.

Nash, C., P. Bickel, R. Friedrich, H. Link and L. Stewart, 2002: The environmental impact of transport subsidies. Paper prepared for the OECD Workshop on Environmentally Harmful Subsidies, Paris, 35 pp.

Newman, P. and J. Kenworthy, 1999: Sustainablity and cities: Overcoming automobile dependence. Island Press, Washington, D. C. 442 pp.

NHTSA CAFE, 2006: Light truck fuel economy standards. <http://www.nhtsa.dot.gov/portal/site/nhtsa/menuitem.43ac99aefa80569cdba046a0/> accessed 30/05/07.

Niederberger, A., 2005: The Swiss climate penny: An innovative approach to transport sector emissions. *Transport Policy*, **12**(4), pp. 303-313.

Nijkamp, P., B. Ubbels, and E. Verhoef, 2003: Chapter 18. Transport Investment Appraisal and the Environment. In *Handbook of Transport and the Environment*, D.A. Hensher and K.J. Button, (eds.). Pergamon, Amsterdam,pp. 333-355.

Nippon Steel, 2002: Advanced technology of Nippon Steel contributes to ULSAB-AVC Program. *Nippon Steel News*, **295**, September 2002.

NRC, 2002: Effectiveness and impact of Corporate Average Fuel Economy (CAFE) Standards. National Research Council, National Academy Press, Washington, D.C., 184 pp.

NRC/NAE, 2004: The Hydrogen Economy. National Academy Press, Washington D.C. 240 pp.

O'Ryan, R., D. Sperling, M. Delucchi, and T. Turrentine, 2002: Transportation in Developing Countries: Greenhouse Gas Scenarios for Chile. Pew Center on Global Climate Change, Arlington, 56 pp.

OECD, 2003: Reforming transport taxes. Paris, 199 pp

OECD, 2004a: Synthesis report on environmentally harmful subsidies. OECD publication SG/SD(2004)3/FINAL, Paris, France Priorities for the World Bank. African Energy.

OECD, 2004b: Current international shipping market trends - community maritime policy and legislative initiatives. OECD Workshop on Maritime Transport, Paris.

Olivier, J.G.J., A.F. Bouwman, C.W.M. van der Maas, J.J.M. Berdowski, C. Veldt, J.P.J. Bloos, A.J.H. Visschedijk, P.Y.J. Zandveld, and J.L. Haverlag, 1996: Description of EDGAR Version 2.0: A set of global emission inventories of greenhouse gases and ozone-depleting substances for all anthropogenic and most natural sources on a per country basis and on 1°x1° grid. RIVM, Bilthoven, The Netherlands, 171 pp.

ORNL (Oak Ridge National Laboratory), 2006: Transportation Energy Data Book: Edition 25. Oak Ridge National Laboratory report ORNL-6974, 332pp.

Owen, N. and R. Gordon, 2002: Carbon to hydrogen - roadmaps for passenger cars: A study for the transport and the Department of Trade and Industry. Ricardo Consulting Engineers, Ltd, November 2002, RCEF.0124.31.9901, 174 pp.

Owen, B. and D.S. Lee, 2005: International aviation emissions - future cases. Study on the allocation of emissions from international aviation to the UK inventory - CPEG7. CATE-2005-3(C)-3, Centre for Air Transport and the Environment, Manchester Metropolitan University.

Owen, B. and D.S. Lee, 2006: Allocation of international aviation emissions from scheduled air traffic-Future cases, 2005 to 2050 (Report 3 of 3). Manchester Metropolitan University, Centre for Air Transport and the Environment, CATE-2006-3(C)-3A, Manchester, UK, 37 pp.

Page, M., 2005: Chapter 34. Non-motorized transport policy. In *Handbook of Transport Strategy, Policy and Institutions*, D.A. Hensher and K.J. Button, (eds.), Pergamon, pp. 581-596.

Pandey, R and G. Bhardwaj, 2000: Economic policy instruments for controlling vehicular air pollution. New Delhi: National Institute of Public Finance and Policy.

Pickrell, D., 1999: Transportation and land use. In *Essays in Transportation Economics and Policy*, J. Gomez-Ibanez, W.B. Tye, and C. Winston (eds.), Brookings, pp. 403-435.

Plotkin, S., D. Greene, and K.G. Duleep, 2002: Examining the potential for voluntary fuel economy standards in the United States and Canada. Argonne National Laboratory report ANL/ESD/02-5, October 2002.

Plotkin, S., 2004: Fuel economy initiatives: International comparisons. In *Encyclopedia of Energy*, **2**, 2004 Elsevier.

Power System, 2005: Press release 2005.6.27. Development of High Power and High Energy Density Capacitor (in Japanese). <http://www.powersystems.co.jp/newsrelease/20050627nscreleaser1-1.pdf> accessed 30/05/07.

PRESAV, 2003: The potential for Renewable Energy Sources in aviation. Imperial College, London, 78 pp.

Prozzi, J.P., C. Naude, and D. Sperling, 2002: Transportation in Developing Countries: Greenhouse Gas Scenarios for South Africa. Pew Center on Global Climate Change, Arlington, 52 pp.

Prud'homme, R. and J.P. Bocarejo, 2005: The London congestion charge: a tentative economic appraisal. *Transport Policy*, **12**, pp. 279-287.

Pucher, J., 2004: Chapter 8. Public Transportation. In *The Geography of Urban Transportation*, S. Hanson and G. Giuliano (eds.), Guilford, pp. 199-236.

Pulles, J.W., 2002: Aviation emissions and evaluation of reduction options/ AERO. Ministry of Transport, Public Works and Watermanagement, Directorate-General of Civil Aviation, The Hague, The Netherlands.

QinetiQ, 2004: AERO2k Global Aviation Emissions Inventories for 2002 and 2025. QinetiQ Ltd, Hampshire, UK.

REN21. 2006: Renewables -Global Status Report- 2006 Update, REN21, Paris, 35 pp.

Resource Analysis, MVA Limited, Dutch National Aerospace Laboratory and International Institute of Air and Space Law, 1999: Analysis of the taxation of aircraft fuel. Produced for European Commission, Delft, The Netherlands.

Ribeiro, S.K. and P.S. Yones-Ibrahim, 2001: Global warming and transport in Brazil - Ethanol alternative. *International Journal of Vehicle Design 2001*, **27**(1/2/3/4), pp. 118-128.

Richardson, H.W. and C.H. Bae, 2004: Chapter 15. Transportation and urban compactness. In *Handbook of Transport Geography and Spatial Systems*, D.A. Hensher, K.J. Button, K.E. Haynes, and P.R. Stopher (eds.), Pergamon, pp. 333-355.

Richmond, J., 2001: A whole-system approach to evaluating urban transit investments. *Transport Reviews*, **21**(2), pp. 141-179.

Riedy, C., 2003: Subsidies that encourage fossil fuel use in Australia. Working paper CR2003/01, Institute for sustainable futures, Sydney, Australia, 39 pp.

Rietveld, P., 2001: Chapter 19 Biking and Walking: the Position of Non-motorized Transport Modes in Transport Systems. In *Handbook of Transport Systems and Traffic Control*. D.A. Hensher and K.J. Button, (eds.), Pergamon, pp. 299-319.

Ross, M., D. Patel, and T. Wenzel, 2006: Vehicle design and the physics of traffic safety. *Physics Today*, January, pp. 49-54.

Rye, T., 2002: Travel Plans: Do They Work? *Transport Policy*, **9**(4), pp. 287-298.

SAE International, 2003a: A powerful mix. *Automotive Engineering International*, **111**(5), pp. 50-56.

SAE International, 2003b: A different automatic. *Automotive Engineering International*, **111**(7), pp. 32-36.

SAE International, 2004: Trucks get aerodynamic touch. *Automotive Engineering International*, **112**(7), pp. 67-69.

Sanyo, 2005: On-line catalog, <http://www.sanyo.co.jp/energy/english/product/lithiumion_2.html> accessed 30/05/07.

Sausen, R., I. Isaksen, V. Grewe, D. Hauglustaine, D.S. Lee, G. Myhre, M.O. Köhler, G. Pitari, U. Schumann, F. Stordal and C. Zerefos, 2005: Aviation radiative forcing in 2000: an update on IPCC (1999). *Meteorologische Zeitschrift*, **114**, pp. 555-561.

Schafer, A., 2000: Regularities in Travel Demand: An International Perspective. *Journal of Transportation and Statistics*, **3**(3), pp. 1-31.

Shoup, D., 1997: Evaluating the effects of California's parking cash-out law: eight case studies. *Transport Policy*, **4**(4), pp. 201-216.

Skjølsvik, K.O., 2005: Natural gas used as ship fuel; Norway heads the development. In *Navigare*, February 2005, Oslo, Norway.

Small, K.A. and K. Van Dender, 2007: Fuel Efficiency and Motor Vehicle Travel: The Declining Rebound Effect. *The Energy Journal*, **28**(1), pp. 25-51.

Song, Y. and G.J. Knaap, 2004: Measuring urban form: is Portland winning the war on sprawl? *Journal of American Planning Association*, **70**(2), pp. 210-225.

Sperling, D., 1985: An analytical framework for siting and sizing biomass fuel plants. *Energy*, **9**, pp. 1033-1040.

Sperling, D. and D. Salon, 2002: Transportation in developing countries: An overview of greenhouse gas reduction strategies. Pew Center on Global Climate Change, Arlington, 40 pp.

Stead, D., H. Geerlings and E. Meijers, 2004: Policy integration in practice: the integration of land use planning, transport and environmental policy-making in Denmark, England and Germany. Delft University Press, Delft.

Stordal, F., G. Myhre, E.J.G. Stordal, W.B. Rossow, D.S. Lee, D.W. Arlander and T. Svenby, 2005: Is there a trend in cirrus cloud cover due to aircraft traffic? *Atmospheric Chemistry and Physics*, **5**, pp. 2155-2162.

Syri, S., M. Amann, P. Capros, L. Mantzos, J. Cofala, and Z. Klimont, 2001: Low-CO$_2$ energy pathways and regional air pollution in Europe. *Energy Policy*, **29**, pp. 871-884.

Taylor, M.A.P. and E.S. Ampt, 2003: Travelling Smarter Down Under: Policies for Voluntary Travel Behavior Change in Australia. *Transport Policy*, **10**, pp. 165-177.

The Royal Commission on Transport and the Environment, 1994: Transport and the Environment. Oxford, 325 pp.

Toyota, 2004: Environmental & Social Report 2004. <http://www.toyota.co.jp/en/environmental_rep/04/download/index.html> accessed 30/05/07.

Toyota/Mizuho, 2004: Well-to-Wheel Analysis of Greenhouse Gas Emissions of Automotive Fuels in the Japanese Context. Mizuho, Tokyo, 122 pp.

Transport for London, 2005: Central London congestion charging, Impacts monitoring; Third annual report. 162pp.

Transport for London, 2006: Central London congestion charging. Fourth annual report, 215pp.

TRL, 2004: The Demand for Public Transport: A Practical Guide. Transport Research Laboratory, *TRL Report*, **593**, 237 pp.

U.S. Bureau of Transportation Statistics, 2005: Transportation Statistics Annual Report. Bureau of Transportation Statistics, Washington, US, 351pp.

U.S. DOE, 2000: Technology Roadmap for the 21st Century Truck Program. Department of Energy, 21CT-001, 196pp.

UN, 2002: Plan of implementation of the World Summit on Sustainable Development. United Nations, New York.

UN, 2006: Review of Maritime Transport. United Nations, New York/Geneva, 160pp.

Veca, E. and J.R. Kuzmyak, 2005: Traveler Response to Transport System Changes: Chapter 13- Parking Pricing and Fees. TCRP Report 95, Transportation Research Board, 62 pp.

VTPI (Victoria Transport Policy Institute), 2005: TDM Encyclopedia, Transportation Elasticities, <http://www.vtpi.org/tdm/tdm11.htm> accessed 30/05/07.

Vyas, A., C. Saricks, and F. Stodolsky, 2002: The Potential Effect of Future Energy-Efficiency and Emissions-Improving Technologies on Fuel Consumption of Heavy Trucks. Argonne National Laboratory report ANL/ESD/02-4, August.

WBCSD, 2002: Mobility 2001: World Mobility at the End of the Twentieth Century, and its Sustainability. World Business Council for Sustainable Development, <http://www.wbcsd.ch/> accessed 30/05/07.

WBCSD, 2004a: Mobility 2030: Meeting the Challenges to Sustainability. <http://www.wbcsd.ch/> accessed 30/05/07.

WBSCD, 2004b: IEA/SMP Model Documentation and Reference Projection. Fulton, L. and G. Eads, <http://www.wbcsd.org/web/publications/mobility/smp-model-document.pdf> accessed 30/05/07

WCTRS and ITPS, 2004: Urban Transport and the Environment: An International Perspective, World Conference on Transport Research Society and Institute for Transport Policy Studies, Elsevier, 515 pp.

WEC, 2004: Energy End-Use Technologies for the 21st Century. WEC, London, UK 128 pp.

Weiss, M.A., J.B. Heywood, E.M. Drake, A. Schafer, and F.F. AuYeung, 2000: On the Road in 2020: A Life-cycle Analysis of New Automobile Technologies, MIT Energy Laboratory, <http://web.mit.edu/energylab/www/> accessed 30/05/07.

Williams, V., R.B. Noland, and R. Toumi, 2003: Reducing the climate change impacts of aviation by restricting cruise altitudes. *Transportation Research*, **D7**, pp. 451-464.

Willoughby, C., 2001: Singapore's Motorisation Policies: 1960-2000. *Transport Policy*, **8**, pp. 125-139.

Wimmer, A., T. Wallner, J. Ringler, and F. Gerbig, 2005: H$_2$-Direct Injection - A Highly Promising Combustion Concept. SAE Technical Paper 2005-01-0108.

Winkelman, S., 2006: Transportation, the Clean Development Mechanism and International Climate Policy. Center for Clean Air Policy, 8th meeting of the Transportation Research Board. Washington D.C., 24 January 2006.

Wit, R.C.N. and J.M.W. Dings, 2002: Economic incentives to mitigate greenhouse gas emissions from international aviation. CE Delft, Study commissioned by the European Commission, Delft, The Netherlands, 200 pp.

Wit, R.C.N., B. Boon, A. van Velzen, M. Cames, O. Deuber, and D.S. Lee, 2005: Giving Wings to Emission Trading; Inclusion of Aviation under the European Emission Trading System (ETS); Design and Impacts. CE Delft, 245 pp.

Wit, R.C.N., B. Kampman, B. Boon, P.E. Meijer, J. Olivier, and D.S. Lee, 2004: Climate impacts from international aviation and shipping; State-of-the-art on climatic impacts, allocation and mitigation policies. CE Delft, The Netherlands, 104 pp.

World Bank, 1996: Sustainable Transport: Priorities for Policy Reform, A World Bank Publication.

World Bank, 2002: Cities on the Move - A World Bank Urban Transport Strategy Review. Private Sector Development and Infrastructure Department, Washington, D.C, 206 pp.

World Bank, 2004: World Development Indicators CD-ROM, World Bank, Washington, D.C.

Wright, L., 2004: Bus Rapid Transit Planning Guide. GTZ, Eschborn, Germany, 225 pp.

Wright, L., L. Fulton, 2005: Climate Change Mitigation and Transport in Developing Nations. *Transport Reviews*, **25**(6), 691-717.

Yamba, F.D. and E. Matsika, 2004: Proceedings of the IPCC Expert Meeting on Industrial Technology Development, Transfer and Diffusion. Tokyo, Japan, 21-23 September 2004, pp. 278-288.

Yuasa, 2000: Press release 2000.4.20 - Development of high capacity Li batteries with Mn type cathode (in Japanese). <http://www.gs-yuasa.com/jp/news/ycj/topick/top20000420.html> ac-cessed 30/05/07.

Zhou, H. and D. Sperling, 2001: Transportation in Developing Countries: Greenhouse Gas Scenarios for Shanghai. China, Pew Center on Global Climate Change, Arlington, 43 pp.

6

Residential and commercial[1] buildings

Coordinating Lead Authors:
Mark Levine (USA), Diana Ürge-Vorsatz (Hungary)

Lead Authors:
Kornelis Blok (The Netherlands), Luis Geng (Peru), Danny Harvey (Canada), Siwei Lang (China), Geoffrey Levermore (UK),
Anthony Mongameli Mehlwana (South Africa), Sevastian Mirasgedis (Greece), Aleksandra Novikova (Russia), Jacques Rilling (France),
Hiroshi Yoshino (Japan)

Contributing Authors:
Paolo Bertoldi (Italy), Brenda Boardman (UK), Marilyn Brown (USA), Suzanne Joosen (The Netherlands), Phillipe Haves (USA),
Jeff Harris (USA), Mithra Moezzi (USA)

Review Editors:
Eberhard Jochem (Germany), Huaqing Xu (PR China)

This chapter should be cited as:
Levine, M., D. Ürge-Vorsatz, K. Blok, L. Geng, D. Harvey, S. Lang, G. Levermore, A. Mongameli Mehlwana, S. Mirasgedis, A. Novikova,
J. Rilling, H. Yoshino, 2007: Residential and commercial buildings. In Climate Change 2007: Mitigation. Contribution of Working Group III
to the Fourth Assessment Report of the Intergovernmental Panel on Climate Change [B. Metz, O.R. Davidson, P.R. Bosch, R. Dave, L.A.
Meyer (eds)], Cambridge University Press, Cambridge, United Kingdom and New York, NY, USA.

---

1 The category of non-residential buildings is referred to by different names in the literature, including commercial, tertiary, public, office, and municipal. In this chapter we consider all non-domestic residential buildings under the "commercial" sector.

Table of Contents

EXECUTIVE SUMMARY

In 2004, emissions from the buildings sector including through electricity use were about 8.6 $GtCO_2$, 0.1 $GtCO_2$-eq N_2O, 0.4 $GtCO_2$-eq CH_4 and 1.5 $GtCO_2$-eq halocarbons (including CFCs and HCFCs). Using an accounting system that attributes CO_2 emissions to electricity supply rather than buildings end-uses, the direct energy-related carbon dioxide emissions of the building sector are about 3 Gt/yr.

For the buildings sector the literature uses a variety of baselines. Therefore a baseline was derived for this sector based on the literature, resulting in emissions between the B2 and A1B SRES scenarios, with 11.1 Gt of emissions of CO_2 in 2020 and 14.3 $GtCO_2$ in 2030 (including electricity emissions but omitting halocarbons, which could conceivably be substantially phased out by 2030).

Measures to reduce greenhouse gas (GHG) emissions from buildings fall into one of three categories: reducing energy consumption and embodied energy in buildings, switching to low-carbon fuels including a higher share of renewable energy, or controlling the emissions of non-CO_2 GHG gases.[2] This chapter devotes most attention to improving energy efficiency in new and existing buildings, which encompasses the most diverse, largest and most cost-effective mitigation opportunities in buildings.

The key conclusion of the chapter is that substantial reductions in CO_2 emissions from energy use in buildings can be achieved over the coming years using mature technologies for energy efficiency that already exist widely and that have been successfully used *(high agreement, much evidence)*. A significant portion of these savings can be achieved in ways that reduce life-cycle costs, thus providing reductions in CO_2 emissions that have a net benefit rather than cost. However, due to the long lifetime of buildings and their equipment, as well as the strong and numerous market barriers prevailing in this sector, many buildings do not apply these basic technologies to the level life-cycle cost minimisation would warrant *(high agreement, much evidence)*.

Our survey of the literature (80 studies) indicates that there is a global potential to reduce approximately 29% of the projected baseline emissions by 2020 cost-effectively in the residential and commercial sectors, the highest among all sectors studied in this report *(high agreement, much evidence)*. Additionally at least 3% of baseline emissions can be avoided at costs up to 20 US$/$tCO_2$ and 4% more if costs up to 100 US$/$tCO_2$ are considered. However, due to the large opportunities at low-costs, the high-cost potential has been assessed to a limited extent, and thus this figure is an underestimate *(high agreement, much evidence)*.

Using the global baseline CO_2 emission projections for buildings, these estimates represent a reduction of approximately 3.2, 3.6 and 4.0 $GtCO_2$/yr in 2020, at zero, 20 US$/$tCO_2$ and 100 US$/$tCO_2$ respectively. Our extrapolation of the potentials to the year 2030 suggests that, globally, about 4.5, 5.0 and 5.6 $GtCO_2$ at negative cost, <20 US$ and <100 US$/$tCO_2$-eq respectively, can be reduced (approximately 30, 35 and 40% of the projected baseline emissions) *(medium agreement, limited evidence)*. These numbers are associated with significantly lower levels of certainty than the 2020 ones due to very limited research available for 2030.

While occupant behaviour, culture and consumer choice and use of technologies are also major determinants of energy use in buildings and play a fundamental role in determining CO_2 emissions *(high agreement, limited evidence)*, the potential reduction through non-technological options is rarely assessed and the potential leverage of policies over these is poorly understood. Due to the limited number of demand-side end-use efficiency options considered by the studies, the omission of non-technological options and the often significant co-benefits, as well as the exclusion of advanced integrated highly efficiency buildings, the real potential is likely to be higher *(high agreement, limited evidence)*.

There is a broad array of accessible and cost-effective technologies and know-how that have not as yet been widely adopted, which can abate GHG emissions in buildings to a significant extent. These include passive solar design, high-efficiency lighting and appliances[3], highly efficient ventilation and cooling systems, solar water heaters, insulation materials and techniques, high-reflectivity building materials and multiple glazing. The largest savings in energy use (75% or higher) occur for new buildings, through designing and operating buildings as complete systems. Realizing these savings requires an integrated design process involving architects, engineers, contractors and clients, with full consideration of opportunities for passively reducing building energy demands. Over the whole building stock the largest portion of carbon savings by 2030 is in retrofitting existing buildings and replacing energy-using equipment due to the slow turnover of the stock *(high agreement, much evidence)*.

Implementing carbon mitigation options in buildings is associated with a wide range of co-benefits. While financial assessment has been limited, it is estimated that their overall value may be higher than those of the energy savings benefits *(medium agreement, limited evidence)*. Economic co-benefits include the creation of jobs and business opportunities, increased economic competitiveness and energy security. Other co-benefits include social welfare benefits for low-income households, increased access to energy services, improved indoor and outdoor air quality, as well as increased comfort,

---

2   Fuel switching is largely the province of Chapter 4, energy supply.
3   By appliances, we mean all electricity-using devices, with the exception of equipment used for heating, cooling and lighting.

health and quality of life. In developing countries, safe and high-efficiency cooking devices and high-efficiency electric lighting would not only abate substantial GHG emissions, but would reduce mortality and morbidity due to indoor air pollution by millions of cases worldwide annually *(high agreement, medium evidence)*.

There are, however, substantial market barriers that need to be overcome and a faster pace of well-enforced policies and programmes pursued for energy efficiency and de-carbonisation to achieve the indicated high negative and low-cost mitigation potential. These barriers include high costs of gathering reliable information on energy efficiency measures, lack of proper incentives (e.g., between landlords who would pay for efficiency and tenants who realize the benefits), limitations in access to financing, subsidies on energy prices, as well as the fragmentation of the building industry and the design process into many professions, trades, work stages and industries. These barriers are especially strong and diverse in the residential and commercial sectors; therefore, overcoming them is only possible through a diverse portfolio of policy instruments *(high agreement, medium evidence)*.

Energy efficiency and utilisation of renewable energy in buildings offer a large portfolio of options where synergies between sustainable development and GHG abatement exist. The most relevant of these for the least developed countries are safe and efficient cooking stoves that, while cutting GHG emissions, significantly, reduce mortality and morbidity by reducing indoor air pollution. Such devices also reduce the workload for women and children and decrease the demands placed on scarce natural resources. Reduced energy payments resulting from energy-efficiency and utilisation of building-level renewable energy resources improve social welfare and enhance access to energy services.

A variety of government policies have been demonstrated to be successful in many countries in reducing energy-related CO_2 emissions in buildings *(high agreement, much evidence)*. Among these are continuously updated appliance standards and building energy codes and labelling, energy pricing measures and financial incentives, utility demand-side management programmes, public sector energy leadership programmes including procurement policies, education and training initiatives and the promotion of energy service companies. The greatest challenge is the development of effective strategies for retrofitting existing buildings due to their slow turnover. Since climate change literacy, awareness of technological, cultural and behavioural choices are important preconditions to fully operating policies, applying these policy approaches needs to go hand in hand with programmes that increase consumer access to information and awareness and knowledge through education.

To sum up, while buildings offer the largest share of cost-effective opportunities for GHG mitigation among the sectors examined in this report, achieving a lower carbon future will require very significant efforts to enhance programmes and policies for energy efficiency in buildings and low-carbon energy sources well beyond what is happening today.

6.1   Introduction

Measures to reduce greenhouse gas (GHG) emissions from buildings fall into one of three categories: reducing energy consumption[4] and embodied energy in buildings, switching to low-carbon fuels including a higher share of renewable energy, or controlling the emissions of non-CO_2 GHG gases. Renewable and low-carbon energy can be supplied to buildings or generated on-site by distributed generation technologies. Steps to de-carbonise electricity generation can eliminate a substantial share of present emissions in buildings. Chapter 4 describes the options for centralized renewable energy generation, while this chapter covers building-level options for low-carbon electricity generation on-site. This chapter devotes most attention to energy efficiency in new and existing buildings, as fuel switching is largely covered elsewhere in this report (Chapter 4). Non-CO_2 GHGs are treated in depth in the IPCC special report on safeguarding the ozone layer and the climate system (IPCC/TEAP, 2005), but some of the most significant issues related to buildings are discussed in this chapter as well.

A very large number of technologies that are commercially available and tested in practice can substantially reduce energy use while providing the same services and often substantial co-benefits. After a review of recent trends in building energy use followed by a description of scenarios of energy use and associated GHG emissions, this chapter provides an overview of the various possibilities in buildings to reduce GHG emissions. Next, a selection of these technologies and practices is illustrated by a few examples, demonstrating the plethora of opportunities to achieve GHG emission reductions as significant as 70–80%. This is followed by a discussion of co-benefits from reducing GHG emissions from buildings, and a review of studies that have estimated the magnitude and costs of potential GHG reductions worldwide.

In spite of the availability of these high-efficiency technologies and practices, energy use in buildings continues to be much higher than necessary. There are many reasons for this energy waste in buildings. The chapter continues with identifying the key barriers that prevent rational decision-making in energy-related choices affecting energy use in buildings. Countries throughout the world have applied a variety of policies in order to deal with these market imperfections. The following sections offer an insight into the experiences with the various policy instruments applied in buildings to cut GHG emissions worldwide. The past five years have shown increasing application of these policies in many countries in Europe and growing interest in several key developing and transition economies. In spite of this fact, global CO_2 emissions resulting from energy use in buildings have increased at an average of 2.7% per year in the past five years for which data is available (1999–2004). The substantial barriers that need to be overcome and the relatively slow pace

of policies and programmes for energy efficiency will provide major challenges to rapid achievement of low-emission buildings.

6.2   Trends in buildings sector emissions

In 2004, direct emissions from the buildings sector (excluding the emissions from electricity use) were about 3 $GtCO_2$, 0.4 $GtCO_2$-eq CH_4, 0.1 $GtCO_2$-eq N_2O and 1.5 $GtCO_2$-eq halocarbons (including CFCs and HCFCs). As mitigation in this sector includes a lot of measures aimed at electricity saving it is useful to compare the mitigation potential with carbon dioxide emissions, including those through the use of electricity. When including the emissions from electricity use, energy-related carbon dioxide emissions were 8.6 Gt/yr (Price *et al.*, 2006), or almost a quarter of the global total carbon dioxide emissions as reported in Chapter 1. IEA estimates a somewhat higher fraction of carbon dioxide emissions due to buildings.

Figure 6.1 shows the estimated emissions of CO_2 from energy use in buildings from two different perspectives. The bar at the left represents emissions of CO_2 from all energy end-uses in buildings. The bar at the right represents only those emissions from direct combustion of fossil fuels. Because the electricity can be derived from fuels with lower carbon content than current fuels, CO_2 emissions from electricity use in buildings can also be altered on the supply side.

Carbon dioxide emissions, including through the use of electricity in buildings, grew from 1971 to 2004 at an annual rate of 2%, – about equal to the overall growth rate

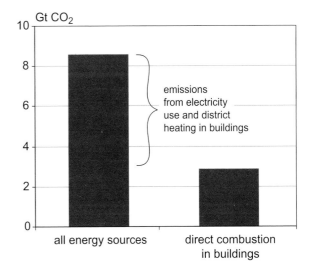

Figure 6.1: *Carbon dioxide emissions from energy, 2004*
Sources: IEA, 2006e and Price et al. 2006.

---

4   This counts all forms of energy use in buildings, including electricity.

of CO_2 emissions from all uses of energy. CO_2 emissions for commercial buildings grew at 2.5% per year and at 1.7% per year for residential buildings during this period. The largest regional increases in CO_2 emissions (including through the use of electricity) for commercial buildings were from developing Asia (30%), North America (29%) and OECD Pacific (18%). The largest regional increase in CO_2 emissions for residential buildings was from Developing Asia accounting for 42% and Middle East/North Africa with 19%.

During the past seven years since the IPCC Third Assessment Report (TAR, IPCC, 2001), CO_2 emissions (including through the use of electricity) in residential buildings have increased at a much slower rate than the 30-year trend (annual rate of 0.1% versus trend of 1.4%) and emissions associated with commercial buildings have grown at a faster rate (3.0% per year in last five years) than the 30-year trend (2.2%) (Price *et al.*, 2006).

Non-CO_2 emissions (largely halocarbons, CFCs, and HCFCs, covered under the Montreal Protocol and HFCs) from cooling and refrigeration contribute more than 15% of the 8.6 $GtCO_2$ emissions associated with buildings. About 1.5 $GtCO_2$-eq of

halocarbon (HFCs, CFCs and HCFCs) emissions, or 60% of the total halocarbon emissions was due to refrigerants and blowing agents for use in buildings (refrigerators, air conditioners and insulation) in 2002. Emissions due to these uses are projected to remain about constant until 2015 and decline if effective policies are pursued (IPCC/TEAP, 2005).

6.3 Scenarios of carbon emissions resulting from energy use in buildings

Figure 6.2 shows the results for the buildings sector of disaggregating two of the emissions scenarios produced for the IPCC Special Report on Emissions Scenarios (SRES) (IPCC, 2000), Scenarios A1B and B2, into ten world regions (Price *et al.*, 2006). These scenarios show a range of projected buildings related CO_2 emissions (including through the use of electricity): from 8.6 $GtCO_2$ emissions in 2004 to 11.4 and 15.6 $GtCO_2$ emissions in 2030 (B2 and A1B respectively), representing an approximately 30% share of total CO_2 emissions in both scenarios. In Scenario B2, which has lower economic growth, especially in

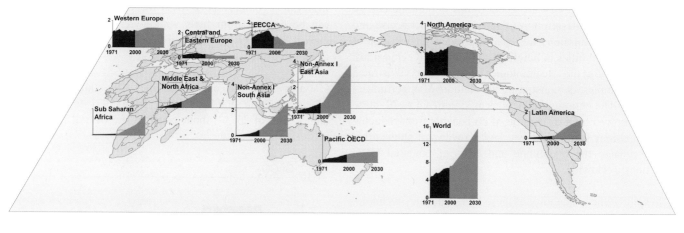

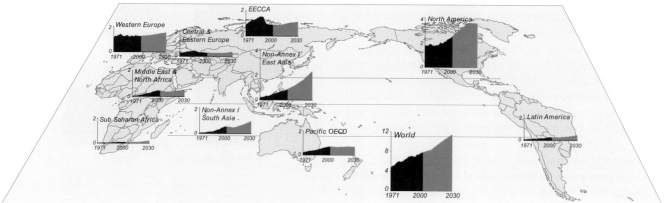

Figure 6.2: *CO_2 emissions including through the use of electricity: A1B (top) and B2 (bottom) IPCC (SRES) scenarios*

Note: Dark red – historic emissions 1971–2000 based on Price *et al.* (2006) modifications of IEA data. Light red – projections 2001–2030 data based on Price *et al.* (2006) disaggregation of SRES data; 2000–2010 data adjusted to actual 2000 carbon dioxide emissions. EECCA = Countries of Eastern Europe, the Caucasus and Central Asia.

the developing world (except China), two regions account for the largest portion of increased CO_2 emissions from 2004 to 2030: North America and Developing Asia. In Scenario A1B (which shows rapid economic growth, especially in developing nations), all of the increase in CO_2 emissions occurs in the developing world: Developing Asia, Middle East/North Africa, Latin America and sub-Saharan Africa, in that order. Overall, average annual CO_2 emissions growth is 1.5% in Scenario B2 and 2.4% in Scenario A1B over the 26-year period.

For the purpose of estimating the CO_2 mitigation potential in buildings, a baseline was derived based on the review of several studies. This baseline represents an aggregation of national and regional baselines reported in the studies (see Box 6.1). The building sector baseline derived and used in this chapter shows emissions between the B2 and A1B (SRES) scenarios, with 11.1 Gt of CO_2-eq emissions in 2020 and 14.3 Gt in 2030 (including electricity emissions).

6.4 GHG mitigation options in buildings and equipment

There is an extensive array of technologies that can be used to abate GHG emissions in new and existing residential and commercial buildings. Prior to discussing options for reducing specific end-uses of energy in buildings it is useful to present an overview of energy end-uses in the residential and commercial sectors, where such information is available and to review some principles of energy-efficient design and operation that are broadly applicable. Figure 6.3 presents a breakdown of energy end-use in the residential and commercial sector for the United States and China. The single largest user of energy in residential buildings in both regions is space heating, followed by water heating (China) and other uses – primarily electric appliances (USA). The order of the next largest uses are reversed in China and the United States, suggesting that electric appliances will increase in use over time in China. The end-uses in commercial buildings are much less similar between China and the United States. For China, heating is by far the largest end-use. For the United States, the largest end-use is other (plug loads involving office equipment and small appliances).

Water heating is the second end-use in China; it is not significant in commercial buildings in the United States. Lighting and cooling are similarly important as the third and fourth largest user in both countries.

The single largest use of energy in residential buildings in both regions is for space heating, followed by water heating. Space heating is also the single largest use of energy in commercial

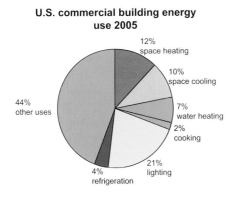

U.S. commercial building energy use 2005

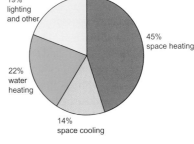

China commercial building energy use 2000

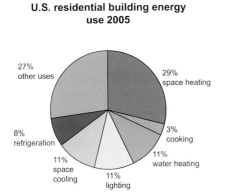

U.S. residential building energy use 2005

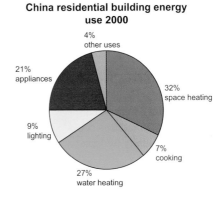

China residential building energy use 2000

Figure 6.3: *Breakdown of residential and commercial sector energy use in United States (2005) and China (2000).*

Sources: EIA, 2006 and Zhou, 2007.

buildings in the EU, accounting for up to 2/3 of total energy use and is undoubtedly dominant in the cold regions of China and in the Former Soviet Union. Lighting is sometimes the largest single use of electricity in commercial buildings, although in hot climates, air conditioning tends to be the single largest use of electricity.

6.4.1 Overview of energy efficiency principles

Design strategies for energy-efficient buildings include reducing loads, selecting systems that make the most effective use of ambient energy sources and heat sinks and using efficient equipment and effective control strategies. An integrated design approach is required to ensure that the architectural elements and the engineering systems work effectively together.

6.4.1.1 Reduce heating, cooling and lighting loads

A simple strategy for reducing heating and cooling loads is to isolate the building from the environment by using high levels of insulation, optimizing the glazing area and minimizing the infiltration of outside air. This approach is most appropriate for cold, overcast climates. A more effective strategy in most other climates is to use the building envelope as a filter, selectively accepting or rejecting solar radiation and outside air, depending on the need for heating, cooling, ventilation and lighting at that time and using the heat capacity of the building structure to shift thermal loads on a time scale of hours to days.

6.4.1.2 Utilize active solar energy and other environmental heat sources and sinks

Active solar energy systems can provide electricity generation, hot water and space conditioning. The ground, ground water, aquifers and open bodies of water, and less so air, can be used selectively as heat sources or sinks, either directly or by using heat pumps. Space cooling methods that dissipate heat directly to natural heat sinks without the use of refrigeration cycles (evaporative cooling, radiative cooling to the night sky, earth-pipe cooling) can be used.

6.4.1.3 Increase efficiency of appliances, heating and cooling equipment and ventilation

The efficiency of equipment in buildings continues to increase in most industrialized and many developing countries, as it has over the past quarter-century. Increasing the efficiency – and where possible reducing the number and size – of appliances, lighting and other equipment within conditioned spaces reduces energy consumption directly and also reduces cooling loads but increases heating loads, although usually by lesser amounts and possibly for different fuel types.

6.4.1.4 Implement commissioning and improve operations and maintenance

The actual performance of a building depends as much on the quality of construction as on the quality of the design itself. Building commissioning is a quality control process that includes design review, functional testing of energy-consuming systems and components, and clear documentation for the owner and operators. Actual building energy performance also depends critically on how well the building is operated and maintained. Continuous performance monitoring, automated diagnostics and improved operator training are complementary approaches to improving the operation of commercial buildings in particular.

6.4.1.5 Change behaviour

The energy use of a building also depends on the behaviour and decisions of occupants and owners. Classic studies at Princeton University showed energy use variations of more than a factor of two between houses that were identical but had different occupants (Socolow, 1978). Levermore (1985) found a variation of 40% gas consumption and 54% electricity consumption in nine identical children's homes in a small area of London. When those in charge of the homes knew that their consumption was being monitored, the electricity consumption fell. Behaviour of the occupants of non-residential buildings also has a substantial impact on energy use, especially when the lighting, heating and ventilation are controlled manually (Ueno et al., 2006).

6.4.1.6 Utilize system approaches to building design

Evaluation of the opportunities to reduce energy use in buildings can be done at the level of individual energy-using *devices* or at the level of building 'systems' (including building energy management systems and human behaviour). Energy efficiency strategies focused on individual energy-using devices or design features are often limited to incremental improvements. Examining the building as an entire system can lead to entirely different design solutions. This can result in new buildings that use much less energy but are no more expensive than conventional buildings.

The systems approach in turn requires an integrated design process (IDP), in which the building performance is optimized through an iterative process that involves all members of the design team from the beginning. The steps in the most basic IDP for a commercial building include (i) selecting a high-performance envelope and highly efficient equipment, properly sized; (ii) incorporating a building energy management system that optimises the equipment operation and human behaviour, and (iii) fully commissioning and maintaining the equipment (Todesco, 2004). These steps alone can usually achieve energy savings in the order of 35–50% for a new commercial building, compared to standard practice, while utilization of more advanced or less conventional approaches has often achieved savings in the order of 50–80% (Harvey, 2006).

6.4.1.7   Consider building form, orientation and related attributes

At the early design stages, key decisions – usually made by the architect – can greatly influence the subsequent opportunities to reduce building energy use. These include building form, orientation, self-shading, height-to-floor-area ratio and decisions affecting the opportunities for and effectiveness of passive ventilation and cooling. Many elements of traditional building designs in both developed and developing countries have been effective in reducing heating and cooling loads. Urban design, including the clustering of buildings and mixing of different building types within a given area greatly affect the opportunities for and cost of district heating and cooling systems (Section 6.4.7) as well as transport energy demand and the shares of different transport modes (Chapter 5, Section 5.5.1).

6.4.1.8   Minimize halocarbon emissions

Many building components – notably air conditioning and refrigeration systems, foam products used for insulation and other purposes and fire protection systems – may emit greenhouse gases with relatively high global-warming potentials. These chemicals include chlorofluorocarbons, hydrochlorofluorocarbons, halons (bromine-containing fluorocarbons) and hydrofluorocarbons (HFCs). While the consumption of the first three is being eliminated through the Montreal Protocol and various national and regional regulations, their on-going emission is still the subject of strategies discussed in the IPCC special report (IPCC/TEAP, 2005). Meanwhile, the use and emissions of HFCs, mostly as replacements for the three ozone-depleting substances, are increasing worldwide.

For many air conditioning and refrigeration applications, the CO_2 emitted during the generation of electricity to power the equipment will typically vastly outweigh the equivalent emissions of the HFC refrigerant. Some exceptions to this general rule exist and two building-related emission sources – CFC chillers and HFC supermarket refrigeration systems – are discussed further. In addition to these applications, some emission mitigation from air conditioning and refrigeration systems is achievable through easy, low-cost options including education and training, proper design and installation, refrigerant leakage monitoring and responsible use and handling of refrigerants throughout the equipment lifecycle.

Like air conditioning and refrigeration systems, most foams and fire protection systems are designed to exhibit low leak rates, and therefore often only emit small portions of the total fluorocarbon under normal use conditions. Upon decommissioning of the building and removal and/or destruction of foam products and fire protection systems, however, large

portions of the remaining fluorocarbon content may be released, particularly if no specific measures are adopted to prevent such release. This raises the need to ensure that proper end-of-life management protocols are followed to avoid these unnecessary emissions.

6.4.2   Thermal envelope

The term 'thermal envelope' refers to the shell of the building as a barrier to unwanted heat or mass transfer between the interior of the building and the outside conditions. The effectiveness of the thermal envelope depends on (i) the insulation levels in the walls, ceiling and ground or basement floor, including factors such as moisture condensation and thermal bridges that affect insulation performance; (ii) the thermal properties of windows and doors; and (iii) the rate of exchange of inside and outside air, which in turn depends on the air-tightness of the envelope and driving forces such as wind, inside-outside temperature differences and air pressure differences due to mechanical ventilation systems or warm/cool air distribution.

Improvements in the thermal envelope can reduce heating requirements by a factor of two to four compared to standard practice, at a few percent of the total cost of residential buildings, and at little to no net incremental cost in commercial buildings when downsizing of heating and cooling systems is accounted for (Demirbilek et al., 2000; Hamada et al., 2003; Hastings, 2004). A number of advanced houses have been built in various cold-climate countries around the world that use as little as 10% of the heating energy of houses built according to the local national building code (Badescu and Sicre, 2003; Hamada et al., 2003; Hastings, 2004). Reducing the envelope and air exchange heat loss by a factor of two reduces the heating requirement by more than a factor of two because of solar gains and internal heat gains from equipment, occupants and lighting. In countries with mild winters but still requiring heating (including many developing countries), modest (and therefore less costly) amounts of insulation can readily reduce heating requirements by a factor of two or more, as well as substantially reducing indoor summer temperatures, thereby improving comfort (in the absence of air conditioning) or reducing summer cooling energy use (Taylor et al., 2000; Florides et al., 2002; Safarzadeh and Bahadori, 2005).

6.4.2.1   Insulation

The choice of insulation material needs to maximize long-term thermal performance of the building element overall. As mentioned previously, this involves consideration of remaining thermal bridges and any water ingress, or other factor, which could result in deterioration of performance over time. For existing buildings, space may be at a premium and the most efficient insulation materials may be needed to minimize thicknesses required. Where upgrading of existing elements is essentially voluntary, minimization of cost and disturbance is equally important and a range of post-applied technologies can be considered, including cavity wall insulation, spray foams and rolled loft insulation. Only a few specific applications with effective control of end-of life emissions have been identified in which foams containing high GWP blowing agents will lead to lower overall climate impacts than hydrocarbon or CO_2 solutions. However, where this is the case, care should still be taken to optimize life-cycle management techniques in order to minimize blowing agent emissions (see 6.4.15).

6.4.2.2   Windows

The thermal performance of windows has improved greatly through the use of multiple glazing layers, low-conductivity gases (argon in particular) between glazing layers, low-emissivity coatings on one or more glazing surfaces and use of framing materials (such as extruded fibreglass) with very low conductivity. Operable (openable) windows are available with heat flows that have only 25–35% of the heat loss of standard non-coated double-glazed (15 to 20% of single-glazed) windows. Glazing that reflects or absorbs a large fraction of the incident solar radiation reduces solar heat gain by up to 75%, thus reducing cooling loads. In spite of these technical improvements, the costs of glazing and windows has remained constant or even dropped in real terms (Jakob and Madlener, 2004). A major U.S. Department of Energy program is developing electrochromic and gasochromic windows which can dynamically respond to heating and cooling in different seasons.

6.4.2.3   Air leakage

In cold climates, uncontrolled exchange of air between the inside and outside of a building can be responsible for up to half of the total heat loss. In hot-humid climates, air leakage can be a significant source of indoor humidity. In residential construction, installation in walls of a continuous impermeable barrier, combined with other measures such as weather-stripping, can reduce rates of air leakage by a factor of five to ten compared to standard practice in most jurisdictions in North America, Europe and the cold-climate regions of Asia (Harvey, 2006).

In addition to leakage through the building envelope, recent research in the United States has demonstrated that leaks in ducts for distributing air for heating and cooling can increase heating and cooling energy requirements by 20–40% (Sherman and Jump, 1997; O'Neal et al., 2002; Francisco et al., 2004). A technology in early commercial use in the United States seals leaks by spraying fine particles into ducts. The sticky particles collect at leakage sites and seal them permanently. This technology is cost-effective for many residential and commercial buildings; it achieves lower costs by avoiding the labour needed to replace or manually repair leaky ducts.

6.4.3   Heating systems

6.4.3.1   Passive solar heating

Passive solar heating can involve extensive sun-facing glazing, various wall- or roof-mounted solar air collectors, double-façade wall construction, airflow windows, thermally massive walls behind glazing and preheating or pre-cooling of ventilation air through buried pipes. Technical details concerning conventional and more advanced passive solar heating techniques, real-world examples and data on energy savings are provided in books by Hastings (1994), Hestnes et al. (2003) and Hastings (2004). Aggressive envelope measures combined with optimisation of passive solar heating opportunities, as exemplified by the European Passive House Standard, have achieved reductions in purchased heating energy by factors of five to thirty (i.e., achieving heating levels less than 15 kWh/m^2/yr even in moderately cold climates, compared to 220 and 250–400 kWh/m^2/yr for the average of existing buildings in Germany and Central/Eastern Europe, respectively (Krapmeier and Drössler, 2001; Gauzin-Müller, 2002; Kostengünstige Passivhäuser als europäische Standards, 2005).

6.4.3.2   Space heating systems

In the industrialized nations and in urban areas in developing countries (in cold winter climates), heating is generally provided by a district heating system or by an on-site furnace or boiler. In rural areas of developing countries, heating (when provided at all) is generally from direct burning of biomass. The following sections discuss opportunities to increase energy efficiency in these systems.

Heating systems used primarily in industrialized countries
Multi-unit residences and many single-family residences (especially in Europe) use boilers, which produce steam or hot water that is circulated, generally through radiators. Annual Fuel Utilization Efficiencies (AFUE) values range from 80% to 99% for the boiler, not including distribution losses. Modern residential forced-air furnaces, which are used primarily in North America, have AFUE values ranging from 78% to 97% (again, not including distribution system losses). Old equipment tends to have an efficiency in the range of 60–70%, so new equipment can provide substantial savings (GAMA (Gas Appliance Manufacturers Association), 2005). In both boilers and furnaces, efficiencies greater than about 88% require condensing operation, in which some of the water vapour in the exhaust is condensed in a separate heat exchanger. Condensing boilers are increasingly used in Western Europe due to regulation of new buildings, which require higher-efficiency systems.

Hydronic systems (in which water rather than air is circulated), especially floor radiant heating systems, are capable of greater energy efficiency than forced air systems because of the low energy required to distribute a given amount of heat, low distribution heat losses and absence of induced infiltration of outside air into the house due to poorly balanced air distribution systems (low-temperature systems also make it possible to use low-grade solar thermal energy).

Heat pumps use an energy input (almost always electricity) to transfer heat from a cold medium (the outside air or ground in the winter) to a warmer medium (the warm air or hot water used to distribute heat in a building). During hot weather, the heat pump can operate in reverse, thereby cooling the indoor space. In winter, drawing heat from a relatively warm source (such as the ground rather than the outside air) and distributing the heat at the lowest possible temperature can dramatically improve the heat pump efficiency. Use of the ground rather than the outside air as a heat source reduced measured energy use for heating by 50 to 60% in two US studies (Shonder *et al.*, 2000; Johnson, 2002). Due to the large energy losses (typically 60–65%) in generating electricity from fossil fuels, heat pumps are particularly advantageous for heating when they replace electric-resistance heating, but may not be preferable to direct use of fuels for heating. The ground can also serve as a low-temperature heat sink in summer, increasing the efficiency of air conditioning.

Coal and biomass burning stoves in rural areas of developing countries

Worldwide, about three billion people use solid fuels – biomass and, mainly in China, coal – in household stoves to meet their cooking, water heating and space heating needs. Most of these people live in rural areas with little or no access to commercial sources of fuel or electricity (WEC (World Energy Council and Food and Agriculture Organization), 1999). Statistical information on fuel use in cooking stoves is sketchy, so any estimates of energy use and associated GHG emissions are uncertain.[5] The global total for traditional biofuel use – a good proxy for energy use in household stoves – was about 32 EJ in 2002, compared to commercial energy use worldwide of 401 EJ (IEA, 2004c).

Worldwide, most household stoves use simple designs and local materials that are inefficient, highly polluting and contribute to the overuse of local resources. Studies of China and India have found that if only the Kyoto Protocol basket of GHGs is considered, biomass stoves appear to have lower emission factors than fossil-fuel alternatives (Smith *et al.*, 2000; Edwards *et al.*, 2004). If products of incomplete combustion (PICs) other than methane and N_2O are considered, however, then biomass stove-fuel combinations exhibit GHG emissions three to ten times higher than fossil-fuel alternatives, and in many cases even higher emissions than from stoves burning coal briquettes (Goldemberg *et al.*, 2000). Additional heating effects arise from black carbon emissions associated with wood-burning stoves. Programmes to develop and disseminate more-efficient biomass stoves have been very effective in China, less

so in India and other countries (Barnes *et al.*, 1994; Goldemberg *et al.*, 2000; Sinton *et al.*, 2004). In the long term, stoves that use biogas or biomass-derived liquid fuels offer the greatest potential for significantly reducing the GHG (and black carbon) emissions associated with household use of biomass fuels.

6.4.4 Cooling and cooling loads

Cooling energy can be reduced by: 1) reducing the cooling load on a building, 2) using passive techniques to meet some or all of the load, and 3) improving the efficiency of cooling equipment and thermal distribution systems.

6.4.4.1 *Reducing the cooling load*

Reducing the cooling load depends on the building shape and orientation, the choice of building materials and a whole host of other decisions that are made in the early design stage by the architect and are highly sensitive to climate. In general, recently constructed buildings are no longer adapted to prevailing climate; the same building forms and designs are now seen in Stockholm, New York, Houston, Hong Kong, Singapore and Kuwait. However, the principles of design to reduce cooling load for any climate are well known. In most climates, they include: (i) orienting a building to minimize the wall area facing east or west; (ii) clustering buildings to provide some degree of self shading (as in many traditional communities in hot climates); (iii) using high-reflectivity building materials; (iv) increasing insulation; (v) providing fixed or adjustable shading; (vi) using selective glazing on windows with a low solar heat gain and a high daylight transmission factor and avoiding excessive window area (particularly on east- and west-facing walls); and (vii) utilizing thermal mass to minimize daytime interior temperature peaks. As well, internal heat loads from appliances and lighting can be reduced through the use of efficient equipment and controls.

Increasing the solar reflectivity of roofs and horizontal or near-horizontal surfaces around buildings and planting shade trees can yield dramatic energy savings. The benefits of trees arise both from direct shading and from cooling the ambient air. Rosenfeld *et al.* (1998) computed that a very large-scale, city-wide program of increasing roof and road albedo and planting trees in Los Angeles could yield a total savings in residential cooling energy of 50–60%, with a 24–33% reduction in peak air conditioning loads.

6.4.4.2 *Passive and low-energy cooling techniques*

Purely passive cooling techniques require no mechanical energy input, but can often be greatly enhanced through small amounts of energy to power fans or pumps. A detailed discussion of passive and low-energy cooling techniques can be found in

---

5   Estimates are available for China and India, collectively home to about one third of the world's population. Residential use of solid fuels in China, nearly all used in stoves, was about 9 EJ in 2002, or 18% of all energy use in the country (National Bureau of Statistics, 2004). The corresponding figures for India were 8 EJ and 36% (IEA, 2004c). In both cases, nearly all of this energy is in the form of biomass.

Harvey (2006) and Levermore (2000). Highlights are presented below:

Natural and night-time ventilation

Natural ventilation reduces the need for mechanical cooling by: directly removing warm air when the incoming air is cooler than the outgoing air, reducing the perceived temperature due to the cooling effect of air motion, providing night-time cooling of exposed thermal mass and increasing the acceptable temperature through psychological adaptation when the occupants have control of operable windows. When the outdoor temperature is 30°C, the average preferred temperature in naturally ventilated buildings is 27°C, compared to 25°C in mechanically ventilated buildings (de Dear and Brager, 2002).

Natural ventilation requires a driving force and an adequate number of openings, to produce airflow. Natural ventilation can be induced through pressure differences arising from inside-outside temperature differences or from wind. Design features, both traditional and modern, that create thermal driving forces and/or utilize wind effects include courtyards, atria, wind towers, solar chimneys and operable windows (Holford and Hunt, 2003). In addition to being increasingly employed in commercial buildings in Europe, natural ventilation is starting to be used in multi-story commercial buildings in more temperate climates in North America (McConahey *et al.*, 2002). Natural ventilation can be supplemented with mechanical ventilation as needed.

In climates with a minimum diurnal temperature variation of 5°C to 7°C, natural or mechanically assisted night-time ventilation, in combination with exposed thermal mass, can be very effective in reducing daily temperature peaks and, in some cases, eliminating the need for cooling altogether. Simulations carried out in California indicate that night-time ventilation is sufficient to prevent peak indoor temperatures from exceeding 26°C over 43% of California in houses with an improved envelope and modestly greater thermal mass compared to standard practice (Springer *et al.*, 2000). For Beijing, da Graça et al. (2002) found that thermally and wind-driven night-time ventilation could eliminate the need for air conditioning of a six-unit apartment building during most of the summer if the high risk of condensation during the day due to moist outdoor air coming into contact with the night-cooled indoor surfaces could be reduced.

Evaporative cooling

There are two methods of evaporatively cooling the air supplied to buildings. In a '*direct*' evaporative cooler, water evaporates directly into the air stream to be cooled. In an 'indirect' evaporative cooler, water evaporates into and cools a secondary air stream, which cools the supply air through a heat exchanger without adding moisture. By appropriately combining direct and indirect systems, evaporative cooling can provide comfortable conditions most of the time in most parts of the world.

Subject to availability of water, direct evaporative cooling can be used in arid areas; indirect evaporative cooling extends the region of applicability to somewhat more humid climates. A new indirect-direct evaporative cooler in the development phase indicated savings in annual cooling energy use of 92 to 95% for residences and 89 to 91% for a modular school classroom in simulations for a variety of California climate zones (DEG, 2004).

Other passive cooling techniques

Underground earth-pipe cooling consists of cooling ventilation air by drawing outside air through a buried air duct. Good performance depends on the climate having a substantial annual temperature range. Desiccant dehumidification and cooling involves using a material (desiccant) that removes moisture from air and is regenerated using heat. Solid desiccants are a commercially available technology. The energy used for dehumidification can be reduced by 30 to 50% compared to a conventional overcooling/reheat scheme (50 to 75% savings of conventional sources if solar energy is used to regenerate the desiccant) (Fischer *et al.*, 2002; Niu *et al.*, 2002). In hot-humid climates, desiccant systems can be combined with indirect evaporative cooling, providing an alternative to refrigeration-based air conditioning systems (Belding and Delmas, 1997).

6.4.4.3   *Air conditioners and vapour-compression chillers*

Air conditioners used for houses, apartments and small commercial buildings have a nominal COP (cooling power divided by fan and compressor power, a direct measure of efficiency) ranging from 2.2 to 3.8 in North America and Europe, depending on operating conditions. More efficient mini-split systems are available in Japan, ranging from 4.5 to 6.2 COP for a 2.8 kW cooling capacity unit. Chillers are larger cooling devices that produce chilled water (rather than cooled air) for use in larger commercial buildings. COP generally increases with size, with the largest and most efficient centrifugal chillers having a COP of up to 7.9 under full-load operation and even higher under part-load operation. Although additional energy is used in chiller-based systems for ventilation, circulating chilled water and operating a cooling tower, significant energy savings are possible through the choice of the most efficient cooling equipment in combination with efficient auxiliary systems (see Section 6.4.5.1 for principles).

Air conditioners – from small room-sized units to large building chillers – generally employ a halocarbon refrigerant in a vapour-compression cycle. Although the units are designed to exhibit low refrigerant emission rates, leaks do occur and additional emissions associated with the installation, service and disposal of this equipment can be significant. The emissions will vary widely from one installation to the next and depend greatly on the practices employed at the site. In some cases, the GWP-weighted lifetime emissions of the refrigerant will outweigh the CO_2 emissions associated with the electricity, highlighting the need to consider refrigerant type and handling as well as energy efficiency when making decisions on the purchase, operation, maintenance and replacement of these systems.

Until recently, the penetration of air conditioning in developing countries has been relatively low, typically only used in large office buildings, hotels and high-income homes. That is quickly changing however, with individual apartment and home air conditioning becoming more common in developing countries, reaching even greater levels in developed countries. This is evident in the production trends of typical room-to-house sized units, which increased 26% (35.8 to 45.4 million units) from 1998 to 2001 (IPCC/TEAP, 2005).

6.4.5   Heating, ventilation and air conditioning (HVAC) systems

The term HVAC is generally used in reference to commercial buildings. HVAC systems include filtration and, where required by the climate, humidification and dehumidification as well as heating and cooling. However, energy-efficient houses in climates with seasonal heating are almost airtight, so mechanical ventilation has to be provided (during seasons when windows will be closed), often in combination with the heating and/or cooling system, as in commercial buildings.

6.4.5.1   *Principles of energy-efficient HVAC design*

In the simplest HVAC systems, heating or cooling is provided by circulating a fixed amount of air at a sufficiently warm or cold temperature to maintain the desired room temperature. The rate at which air is circulated in this case is normally much greater than that needed for ventilation to remove contaminants. During the cooling season, the air is supplied at the coldest temperature needed in any zone and reheated as necessary just before entering other zones. There are a number of changes in the design of HVAC systems that can achieve dramatic savings in the energy use for heating, cooling and ventilation. These include (i) using variable-air volume systems so as to minimize simultaneous heating and cooling of air; (ii) using heat exchangers to recover heat or coldness from ventilation exhaust air; (iii) minimizing fan and pump energy consumption by controlling rotation speed; (iv) separating the ventilation from the heating and cooling functions by using chilled or hot water for temperature control and circulating only the volume of air needed for ventilation; (v) separating cooling from dehumidification functions through the use of desiccant dehumidification; (vi) implementing a demand-controlled ventilation system in which ventilation airflow changes with changing building occupancy which alone can save 20 to 30% of total HVAC energy use (Brandemuehl and Braun, 1999); (vii) correctly sizing all components; and (viii) allowing the temperature maintained by the HVAC system to vary seasonally with outdoor conditions (a large body of evidence indicates that the temperature and humidity set-points commonly encountered in air-conditioned buildings are significantly lower than necessary (de Dear and Brager, 1998; Fountain *et al.*, 1999),

while computer simulations by Jaboyedoff *et al.* (2004) and by Jakob *et al.* (2006) indicate that increasing the thermostat by 2°C to 4°C will reduce annual cooling energy use by more than a factor of three for a typical office building in Zurich, and by a factor of two to three if the thermostat setting is increased from 23°C to 27°C for night-time air conditioning of bedrooms in apartments in Hong Kong (Lin and Deng, 2004).

Additional savings can be obtained in 'mixed-mode' buildings, in which natural ventilation is used whenever possible, making use of the extended comfort range associated with operable windows, and mechanical cooling is used only when necessary during periods of very warm weather or high building occupancy.

6.4.5.2   *Alternative HVAC systems in commercial buildings*

The following paragraphs describe two alternatives to conventional HVAC systems in commercial buildings that together can reduce the HVAC system energy use by 30 to 75%. These savings are in addition to the savings arising from reducing heating and cooling loads.

Radiant chilled-ceiling cooling
A room may be cooled by chilling a large fraction of the ceiling by circulating water through pipes or lightweight panels. Chilled ceiling (CC) cooling has been used in Europe since at least the mid-1970s. In Germany during the 1990s, 10% of retrofitted buildings used CC cooling (Behne, 1999). Significant energy savings arise because of the greater effectiveness of water than air in transporting heat and because the chilled water is supplied at 16°C to 20°C rather than at 5°C to 7°C. This allows a higher chiller COP when the chiller operates, but also allows more frequent use of 'water-side free cooling,' in which the chiller is bypassed altogether and water from the cooling tower is used directly for space cooling. For example, a cooling tower alone could directly meet the cooling requirements 97% of the time in Dublin, Ireland and 67% of the time in Milan, Italy if the chilled water is supplied at 18°C (Costelloe and Finn, 2003).

Displacement ventilation
Conventional ventilation relies on turbulent mixing to dilute room air with ventilation air. A superior system is 'displacement ventilation' (DV) in which air is introduced at low speed through many diffusers in the floor or along the sides of a room and is warmed by internal heat sources (occupants, lights, plug-in equipment) as it rises to the top of the room, displacing the air already present. The thermodynamic advantage of displacement ventilation is that the supply air temperature is significantly higher for the same comfort conditions (about 18°C compared with about 13°C in a conventional mixing ventilation system). It also permits significantly smaller airflow.

DV was first applied in northern Europe; by 1989 it had captured 50% of the Scandinavian market for new industrial buildings and 25% for new office buildings (Zhivov and

Rymkevich, 1998). The building industry in North America has been much slower to adopt DV; by the end of the 1990s fewer than 5% of new buildings used under-floor air distribution systems (Lehrer and Bauman, 2003). Overall, DV can reduce energy use for cooling and ventilation by 30 to 60%, depending on the climate (Bourassa et al., 2002; Howe et al., 2003).

6.4.6 Building energy management systems (BEMS)

BEMSs are control systems for individual buildings or groups of buildings that use computers and distributed microprocessors for monitoring, data storage and communication (Levermore, 2000). The BEMS can be centrally located and communicate over telephone or Internet links with remote buildings having 'outstations' so that one energy manager can manage many buildings remotely. With energy meters and temperature, occupancy and lighting sensors connected to a BEMS, faults can be detected manually or using automated fault detection software (Katipamula et al., 1999), which helps avoid energy waste (Burch et al., 1990). With the advent of inexpensive, wireless sensors and advances in information technology, extensive monitoring via the Internet is possible.

Estimates of BEMS energy savings vary considerably: up to 27% (Birtles and John, 1984); between 5% and 40% (Hyvarinen, 1991; Brandemuehl and Bradford, 1999; Brandemuehl and Braun, 1999; Levermore, 2000); up to 20% in space heating energy consumption and 10% for lighting and ventilation; and 5% to 20% overall (Roth et al., 2005).

6.4.6.1 Commissioning

Proper commissioning of the energy systems in a commercial building is a key to efficient operation (Koran, 1994; Kjellman et al., 1996; IEA, 2005; Roth et al., 2005). Building commissioning is a quality control process that begins with the early stages of design. Commissioning helps ensure that the design intent is clear and readily tested, that installation is subjected to on-site inspection and that all systems are tested and functioning properly before the building is accepted. A systems manual is prepared to document the owner's requirements, the design intent (including as-built drawings), equipment performance specifications and control sequences.

Recent results of building commissioning in the USA showed energy savings of up to 38% in cooling and/or 62% in heating and an average higher than 30% (Claridge et al., 2003). A study by Mills et al. (2005) reviewed data from 224 US buildings that had been commissioned or retro-commissioned. The study found that the costs of commissioning new buildings were typically outweighed by construction cost savings due to fewer change orders and that retro-commissioning produced median energy savings of 15% with a median payback period of 8.5 months. It is very difficult to assess the energy benefits of commissioning new buildings due to the lack of a baseline.

6.4.6.2 Operation, maintenance and performance benchmarking

Once a building has been commissioned, there is a need to maintain its operating efficiency. A variety of methods to monitor and evaluate performance and diagnose problems are currently under development (Brambley et al., 2005). Post-occupancy evaluation (POE) is a useful complement to ongoing monitoring of equipment and is also useful for ensuring that the building operates efficiently. A UK study of recently constructed buildings found that the use of POE identified widespread energy wastage (Bordass et al., 2001a; Bordass et al., 2001b).

Cogeneration and District Heating/Cooling
Buildings are usually part of a larger community. If the heating, cooling and electricity needs of a larger collection of buildings can be linked together in an integrated system without major distribution losses, then significant savings in primary energy use are possible – beyond what can be achieved by optimising the design of a single building. Community-scale energy systems also offer significant new opportunities for the use of renewable energy. Key elements of an integrated system can include: 1) district heating networks for the collection of waste or surplus heat and solar thermal energy from dispersed sources and its delivery to where it is needed; 2) district cooling networks for the delivery of chilled water for cooling individual buildings; 3) central production of steam and/or hot water in combination with the generation of electricity (cogeneration) and central production of cold water; 4) production of electricity through photovoltaic panels mounted on or integrated into the building fabric; 5) diurnal storage of heat and coldness produced during off-peak hours or using excess wind-generated electricity; and 6) seasonal underground storage of summer heat and winter coldness.

District heating (DH) is widely used in regions with large fractions of multi-family buildings, providing as much as 60% of heating and hot water energy needs for 70% of the families in Eastern European countries and Russia (OECD/IEA, 2004). While district heating can have major environmental benefits over other sources of heat, including lower specific GHG emissions, systems in these countries suffer from the legacies of past mismanagement and are often obsolete, inefficient and expensive to operate (Lampietti and Meyer, 2003, Ürge-Vorsatz et al., 2006). Making DH more efficient could save 350 million tonnes of CO_2 emissions in these countries annually, accompanied by significant social, economic and political benefits (OECD/IEA, 2004).

The greatest potential improvement in the efficiency of district heating systems is to convert them to cogeneration systems that involve the simultaneous production of electricity and useful heat. For cogeneration to provide an improvement in efficiency, a use has to be found for the waste heat. Centralized production of heat in a district heat system can be more efficient than on-site boilers or furnaces even in the absence of cogeneration and in spite of distribution losses, if a district-heating network is used with heat pumps to upgrade and distribute heat from scattered sources. Examples include waste heat from sewage in Tokyo (Yoshikawa, 1997) and Gothenberg, Sweden (Balmér, 1997) and low-grade geothermal heat in Tianjin, China, that is left over after higher-temperature heat has been used for heating and hot water purposes (Zhao *et al.*, 2003). Waste heat from incineration has been used, particularly in northern Europe.

Chilled water supplied to a district-cooling network can be produced through trigeneration (the simultaneous production of electricity, heat and chilled water), or it can be produced through a centralized chilling plant independent of power generation. District cooling provides an alternative to separate chillers and cooling towers in multi-unit residential buildings that would otherwise use inefficient small air conditioners. In spite of the added costs of pipes and heat exchangers in district heating and cooling networks, the total capital cost can be less than the total cost of heating and cooling units in individual buildings, (Harvey, 2006, Chapter 15). Adequate control systems are critical to the energy-efficient operation of both district cooling and central (building-level) cooling systems.

District heating and cooling systems, especially when combined with some form of thermal energy storage, make it more economically and technically feasible to use renewable sources of energy for heating and cooling. Solar-assisted district heating systems with storage can be designed such that solar energy provides 30 to 95% of total annual heating and hot water requirements under German conditions (Lindenberger *et al.*, 2000). Sweden has been able to switch a large fraction of its building heating energy requirements to biomass energy (plantation forestry) for its district heating systems (Swedish Energy Agency, 2004).

6.4.7 Active collection and transformation of solar energy

Buildings can serve as collectors and transformers of solar energy, meeting a large fraction of their energy needs on a sustainable basis with minimal reliance on connection to energy grids, although for some climates this may only apply during the summer. As previously discussed, solar energy can be used for daylighting, for passive heating and as one of the driving forces for natural ventilation, which can often provide much or all of the required cooling. By combining a high-performance thermal envelope with efficient systems and devices, 50–75% of the heating and cooling energy needs of buildings as constructed under normal practice can either be eliminated or satisfied through passive solar design. Electricity loads, especially in commercial buildings, can be drastically reduced to a level that allows building-integrated photovoltaic panels (BiPV) to meet much of the remaining electrical demand during daytime hours. Photovoltaic panels can be supplemented by other forms of active solar energy, such as solar thermal collectors for hot water, space heating, absorption space cooling and dehumidification.

6.4.7.1 Building-integrated PV (BiPV)

The principles governing photovoltaic (PV) power generation and the prospects for centralized PV production of electricity are discussed in Chapter 4, Section 4.3.3.6. Building-integrated PV (BiPV) consists of PV modules that function as part of the building envelope (curtain walls, roof panels or shingles, shading devices, skylights). BiPV systems are sometimes installed in new 'showcase' buildings even before the systems are generally cost-effective. These early applications will increase the rate at which the cost of BiPVs comes down and the technical performance improves. A recent report presents data on the cost of PV modules and the installed-cost of PV systems in IEA countries (IEA, 2003b). Electricity costs from BiPV at present are in the range of 0.30–0.40 US$/kWh in good locations, but can drop considerably with mass production of PV modules (Payne *et al.*, 2001).

Gutschner *et al.* (2001) have estimated the potential for power production from BiPV in IEA member countries. Estimates of the percentage of present total national electricity demand that could be provided by BiPV range from about 15% (Japan) to almost 60% (USA).

6.4.7.2 Solar thermal energy for heating and hot water

Most solar thermal collectors used in buildings are either flat-plate or evacuated-tube collectors.[6] Integrated PV/thermal collectors (in which the PV panel serves as the outer part of a thermal solar collector) are also commercially available (Bazilian *et al.*, 2001; IEA, 2002). 'Combisystems' are solar systems that provide both space and water heating. Depending on the size of panels and storage tanks, and the building thermal envelope performance, 10 to 60% of the combined hot water and heating demand can be met by solar thermal systems at central and northern European locations. Costs of solar heat have been 0.09–0.13 €/kWh for large domestic hot water systems and 0.40–0.50 €/kWh for combisystems with diurnal storage (Peuser *et al.*, 2002).

Worldwide, over 132 million m^2 of solar collector surface for space heating and hot water were in place by the end of 2003. China accounts for almost 40% of the total (51.4 million m^2), followed by Japan (12.7 million m^2) and Turkey (9.5 million m^2) (Weiss *et al.*, 2005).

---

6   See Peuser *et al.* (2002) and Andén (2003) for technical information.

6.4.8 Domestic hot water

Options to reduce fossil or electrical energy used to produce hot water include (i) use of water saving fixtures, more water-efficient washing machines, cold-water washing and (if used at all) more water-efficient dishwashers (50% typical savings); (ii) use of more efficient and better insulated water heaters or integrated space and hot-water heaters (10–20% savings); (iii) use of tankless (condensing or non-condensing) water heaters, located close to the points of use, to eliminate standby and greatly reduce distribution heat losses (up to 30% savings, depending on the magnitude of standby and distribution losses with centralized tanks); (v) recovery of heat from warm waste water; (vi) use of air-source or exhaust-air heat pumps; and (vii) use of solar thermal water heaters (providing 50–90% of annual hot-water needs, depending on climate). The integrated effect of all of these measures can frequently reach a 90% savings. Heat pumps using CO_2 as a working fluid are an attractive alternative to electric-resistance hot water heaters, with a COP of up to 4.2–4.9 (Saikawa et al., 2001; Yanagihara, 2006).

6.4.9 Lighting systems

Lighting energy use can be reduced by 75 to 90% compared to conventional practice through (i) use of daylighting with occupancy and daylight sensors to dim and switch off electric lighting; (ii) use of the most efficient lighting devices available; and (iii) use of such measures as ambient/task lighting.

6.4.9.1 High efficiency electric lighting

Presently 1.9 $GtCO_2$ are emitted by electric lighting worldwide, equivalent to 70% of the emissions from light passenger vehicles (IEA, 2006b). Continuous improvements in the efficacy[7] of electric lighting devices have occurred during the past decades and can be expected to continue. Advances in lamps have been accompanied by improvements in occupancy sensors and reductions in cost (Garg and Bansal, 2000; McCowan et al., 2002). A reduction in residential lighting energy use of a factor of four to five can be achieved compared to incandescent/halogen lighting.

For lighting systems providing ambient (general space) lighting in commercial buildings, the energy required can be reduced by 50% or more compared to old fluorescent systems through use of efficient lamps (ballasts and reflectors, occupancy sensors, individual or zone switches on lights and lighter colour finishes and furnishings. A further 40 to 80% of the remaining energy use can be saved in perimeter zones through daylighting (Rubinstein and Johnson, 1998; Bodart and Herde, 2002). A simple strategy to further reduce energy use is to provide a relatively low background lighting level, with local levels of greater illumination at individual workstations. This strategy is referred to as 'task/ambient lighting' and is popular in Europe. Not only can this alone cut lighting energy use in half, but it provides a greater degree of individual control over personal lighting levels and can reduce uncomfortable levels of glare and high contrast.

About one third of the world's population depends on fuel-based lighting (such as kerosene, paraffin or diesel), contributing to the major health burden from indoor air pollution in developing countries. While these devices provide only 1% of global lighting, they are responsible for 20% of the lighting-related CO_2 emissions and consume 3% of the world's oil supply. A CFL or LED is about 1000 times more efficient than a kerosene lamp (Mills, 2005). Efforts are underway to promote replacement of kerosene lamps with LEDs in India. Recent advances in light-emitting diode (LED) technology have significantly improved the cost-effectiveness, longevity and overall viability of stand-alone PV-powered task lighting (IEA, 2006b).

6.4.10 Daylighting

Daylighting systems involve the use of natural lighting for the perimeter areas of a building. Such systems have light sensors and actuators to control artificial lighting. Opportunities for daylighting are strongly influenced by architectural decisions early in the design process, such as building form; the provision of inner atria, skylights and clerestories (glazed vertical steps in the roof); and the size, shape and position of windows. IEA (2000) provides a comprehensive sourcebook of conventional and less conventional techniques and technologies for daylighting.

A number of recent studies indicate savings in lighting energy use of 40 to 80% in the daylighted perimeter zones of office buildings (Rubinstein and Johnson, 1998; Jennings et al., 2000; Bodart and Herde, 2002; Reinhart, 2002; Atif and Galasiu, 2003; Li and Lam, 2003). The management of solar heat gain along with daylighting to reduce electric lighting also leads to a reduction in cooling loads. Lee et al. (1998) measured savings for an automated Venetian blind system integrated with office lighting controls, finding that lighting energy savings averaged 35% in winter and ranged from 40 to 75% in summer. Monitored reductions in summer cooling loads were 5 to 25% for a southeast-facing office in Oakland, California building, with even larger reductions in peak cooling loads. Ullah and Lefebvre (2000) reported measured savings of 13 to 32% for cooling plus ventilation energy using automatic blinds in a building in Singapore.

An impediment to more widespread use of daylighting is the linear, sequential nature of the design process. Based on a survey of 18 lighting professionals in the USA, Turnbull

---

7   This is the ratio of light output in lumens to input power in watts.

and Loisos (2000) found that, rather than involving lighting consultants from the very beginning, architects typically make a number of irreversible decisions at an early stage of the design that adversely impact daylighting, 'then' pass on their work to the lighting consultants and electrical engineers to do the lighting design. As a result, the lighting system becomes, *de facto*, strictly an electrical design.

6.4.11 Household appliances, consumer electronics and office equipment

Energy use by household appliances, office equipment and consumer electronics, from now on referred to as 'appliances', is an important fraction of total electricity use in both households and workplaces (Kawamoto *et al.*, 2001; Roth *et al.*, 2002). This equipment is more than 40% of total residential primary energy demand in 11 large OECD nations[8] (IEA, 2004f). The largest growth in electricity demand has been in miscellaneous equipment (home electronics, entertainment, communications, office equipment and small kitchen equipment), which has been evident in all industrialized countries since the early 1980s. Such miscellaneous equipment now accounts for 70% of all residential appliance electricity use in the 11 large OECD nations (IEA, 2004f). Appliances in some developing countries constitute a smaller fraction of residential energy demand. However, the rapid increase in their saturation in many dynamically developing countries such as China, especially in urban areas, demonstrates the expected rise in importance of appliances in the developing world as economies grow (Lawrence Berkeley National Laboratory, 2004).

On a primary energy basis appliances undoubtedly represent a larger portion of total energy use for residential than for commercial buildings. In the United States, for example, they account for almost 55% of total energy consumption in commercial buildings. Miscellaneous equipment and lighting combined account for more than half of total energy consumption in commercial buildings in the United States and Japan (Koomey *et al.*, 2001; Murakami *et al.*, 2006).

The most efficient appliances require a factor of two to five less energy than the least efficient appliances available today. For example, in the USA, the best horizontal-axis clothes-washing machines use less than half the energy of the best vertical-axis machines (FEMP, 2002), while refrigerator/freezer units meeting the current US standard (478 kWh/yr) require about 25% of the energy used by refrigerator/freezers sold in the USA in the late 1970s (about 1800 kWh/yr) and about 50% of energy used in the late 1980s. Available refrigerator/freezers of standard US size use less than 400 kWh/yr (Brown *et al.*, 1998). However, this is still in excess of the average energy use by (generally smaller) refrigerators in Sweden, the Netherlands, Germany and Italy in the late 1990s (IEA, 2004f).

Standby and low power mode use by consumer electronics (i.e., energy used when the machine is turned off) in a typical household in many countries often exceeds the energy used by a refrigerator/freezer unit that meets the latest US standards, that is often more than 500 kWh/yr, (Bertoldi *et al.*, 2002). The growing proliferation of electronic equipment such as set-top boxes for televisions, a wide variety of office equipment (in homes as well as offices) and sundry portable devices with attendant battery chargers – combined with inefficient power supplies (Calwell and Reeder, 2002) and highly inefficient circuit designs that draw unnecessary power in the resting or standby modes – have caused this equipment to be responsible for a large fraction of the electricity demand growth in both residential and commercial buildings in many nations. Efforts are underway especially at the International Energy Agency and several countries (e.g., Korea, Australia, Japan and China) to reduce standby energy use by a factor of two to three (Ross and Meier, 2002; Fung *et al.*, 2003). Electricity use by office equipment may not yet be large compared to electricity use by the HVAC system, but (as noted) it is growing rapidly and is already an important source of internal heat gain in offices and some other commercial buildings. The biggest savings opportunities are: 1) improved power supply efficiency in both active and low-power modes, 2) redesigned computer chips that reduce electricity use in low-power mode, and 3) repeated reminders to users to turn equipment off during non-working hours.

The cooking stove, already referred to in Section 6.4.3.2 for heating, is a major energy-using appliance in developing countries. However, there is particular concern about emissions of products of incomplete combustion described in that section. Two-and-a-half billion people in developing countries depend on biomass, such as wood, dung, charcoal and agricultural residues, to meet their cooking energy needs (IEA, 2006e). Options available to reduce domestic cooking energy needs include: 1) improved efficiency of biomass stoves; 2) improved access to clean cooking fuels, both liquid and gaseous; 3) access to electricity and low-wattage and low-cost appliances for low income households; 4) non-electric options such as solar cookers; 5) efficient gas stoves; and 6) small electric cooking equipment such as microwaves, electric kettles or electric frying pans. Improved biomass stoves can save from 10 to 50% of biomass consumption for the same cooking service (REN21 (Renewable Energy Policy Network), 2005) at the same time reducing indoor air pollution by up to one-half. Although the overall impact on emissions from fuel switching can be either positive or negative, improved modern fuels and greater conversion efficiency would result in emission reductions from all fuels (IEA, 2006e).

---

8   Australia, Denmark, Finland, France, Germany, Italy, Japan, Norway, Sweden, the United Kingdom, and the United States.

6.4.12 Supermarket refrigeration systems

Mitigation options for food-sales and service buildings, especially supermarkets and hypermarkets extend beyond the energy savings mitigation options reviewed so far (e.g., high efficiency electric lighting, daylighting, etc.). Because these buildings often employ large quantities of HFC refrigerants in extensive and often leaky systems, a significant share of total GHG emissions are due to the release of the refrigerant. In all, emissions of the refrigerant can be greater than the emissions due to the system energy use (IPCC/TEAP, 2005).

Two basic mitigation options are reviewed in IPCC/TEAP report: leak reduction and alternative system design. Refrigerant leakage rates are estimated to be around 30% of banked system charge. Leakage rates can be reduced by system design for tightness, maintenance procedures for early detection and repairs of leakage, personnel training, system leakage record keeping and end-of-life recovery of refrigerant. Alternative system design involves for example, applying direct systems using alternative refrigerants, better containment, distributed systems, indirect systems or cascade systems. It was found that up to 60% lower LCCP values can be obtained by alternative system design (IPCC/TEAP, 2005).

6.4.13 Energy savings through retrofits

There is a large stock of existing and inefficient buildings, most of which will still be here in 2025 and even 2050. Our long-term ability to reduce energy use depends critically on the extent to which energy use in these buildings can be reduced when they are renovated. The equipment inside a building, such as the furnace or boiler, water heater, appliances, air conditioner (where present) and lighting is completely replaced over time periods ranging from every few years to every 20–30 years. The building shell – walls, roof, windows and doors – lasts much longer. There are two opportunities to reduce heating and cooling energy use by improving the building envelope: (i) at any time prior to a major renovation, based on simple measures that pay for themselves through reduced energy costs and potential financial support or incentives; and (ii) when renovations are going to be made for other (non-energy) reasons, including replacement of windows and roofs.

6.4.13.1 Conventional retrofits of residential buildings

Cost-effective measures that can be undertaken without a major renovation of residential buildings include: sealing points of air leakage around baseboards, electrical outlets and fixtures, plumbing, the clothes dryer vent, door joists and window joists; weather stripping of windows and doors; and adding insulation in attics, to walls or wall cavities. A Canadian study found that the cost-effective energy savings potential ranges from 25–30% for houses built before the 1940s, to about 12% for houses built in the 1990s (Parker et al., 2000). In a carefully documented retrofit of four representative houses in the York

region of the UK, installation of new window and wooden door frames, sealing of suspended timber ground floors and repair of cracks in plaster reduced the rate of air leakage by a factor of 2.5–3.0 (Bell and Lowe, 2000). This, combined with improved insulation, doors and windows, reduced the heating energy required by an average of 35%. Bell and Lowe (2000) believe that a reduction of 50% could be achieved at modest cost using well-proven (early 1980s) technologies, with a further 30–40% reduction through additional measures.

Studies summarized by Francisco et al. (1998) indicate that air-sealing retrofits alone can save an average of 15–20% of annual heating and air conditioning energy use in US houses. Additional energy savings would arise by insulating pipework and ductwork, particularly in unconditioned spaces. Rosenfeld (1999) refers to an 'AeroSeal' technique (see Sec. 6.4.2.2) that he estimates is already saving three billion US$/yr in energy costs in the USA. Without proper sealing, homes in the USA lose, on average, about one-quarter of the heating and cooling energy through duct leaks in unconditioned spaces – attics, crawl spaces, basements.

In a retrofit of 4003 homes in Louisiana, the heating, cooling and water heating systems were replaced with a ground-source heat pump system. Other measures were installation of attic insulation and use of compact fluorescent lighting and water saving showerheads. Space and hot water heating previously provided by natural gas was supplied instead by electricity (through the heat pump), but total electricity use still decreased by one third (Hughes and Shonder, 1998).

External Insulation and Finishing Systems (EIFSs) provide an excellent opportunity for upgrading the insulation and improving the air-tightness of single- and multi-unit residential buildings, as well as institutional and commercial buildings. This is because of the wide range of external finishes that can be applied, ranging from stone-like to a finish resembling aged plaster. A German company manufacturing some of the components used in EIFSs undertook a major renovation of some of its own 1930s multi-unit residential buildings. The EIFSs in combination with other measures achieved a factor of eight measured reduction in heating energy use (see www.3lh. de). An envelope upgrade of an apartment block in Switzerland reduced the heating requirement by a factor of two, while replacing an oil-fired boiler at 85% seasonal average efficiency with an electric heat pump having a seasonal average COP of 3.2 led to a further large decrease in energy use. The total primary energy requirement decreased by about 75% (Humm, 2000).

6.4.13.2 Conventional retrofits of institutional and commercial buildings

There are numerous published studies showing that energy savings of 50 to 75% can be achieved in commercial buildings through aggressive implementation of integrated sets of measures. These savings can often be justified in terms of the energy-cost savings alone, although in other cases full justification requires consideration of a variety of less tangible benefits. In the early 1990s, a utility in California sponsored a 10 million US$ demonstration of advanced retrofits. In six of seven retrofit projects, an energy savings of 50% was obtained; in the seventh project, a 45% energy savings was achieved. For Rosenfeld (1999), the most interesting result was not that an alert, motivated team could achieve savings of 50% with conventional technology, but that it was very hard to find a team competent enough to achieve these results.

Other, recent examples that are documented in the published literature include:

- A realized savings of 40% in heating, plus cooling, plus ventilation energy use in a Texas office building through conversion of the ventilation system from one with constant to one with variable air flow (Liu and Claridge, 1999);
- A realized savings of 40% of heating energy use through the retrofit of an 1865 two-story office building in Athens, where low-energy was achieved through some passive technologies that required the cooperation of the occupants (Balaras, 2001);
- A realized savings of 74% in cooling energy use in a one-story commercial building in Florida through duct sealing, chiller upgrade and fan controls (Withers and Cummings, 1998);
- Realized savings of 50–70% in heating energy use through retrofits of schools in Europe and Australia (CADDET, 1997);
- Realized fan, cooling and heating energy savings of 59, 63 and 90% respectively in buildings at a university in Texas; roughly half due to standard retrofit and half due to adjustment of the control-system settings (which were typical for North America) to optimal settings (Claridge et al., 2001).

6.4.13.3 Solar retrofits of residential, institutional and commercial buildings

Solar retrofit performed in Europe under the IEA Solar and Cooling Program achieved savings in space heating of 25–80% (Harvey, 2006, Chapter 14). The retrofit examples described above, while achieving dramatic (35–75%) energy savings, rely on making incremental improvements to the existing building components and systems. More radical measures involve re-configuring the building so that it can make direct use of solar energy for heating, cooling and ventilation. The now-completed Task 20 of the IEA's *Solar Heating and Cooling (SHC)* implementing agreement was devoted to solar retrofitting techniques.

Solar renovation measures that have been used are installation of roof- or façade-integrated solar air collectors; roof-mounted or integrated solar DHW heating; transpired solar air collectors; advanced glazing of balconies; external transparent insulation; and construction of a second-skin façade over the original façade. Case studies are presented in Boonstra and Thijssen (1997), Haller et al. (1997) and Voss (2000a), Voss (2000b) and are summarized in Harvey (2006), Chapter 14.

6.4.14 Trade-offs between embodied energy and operating energy

The embodied energy in building materials needs to be considered along with operating energy in order to reduce total lifecycle energy use by buildings. The replacement of materials that require significant amounts of energy to produce (such as concrete and steel) with materials requiring small amounts of energy to produce (such as wood products) will reduce the amount of energy embodied in buildings. Whether this reduces energy use on a lifecycle basis, however, depends on the effect of materials choice on the energy requirements for heating and cooling over the lifetime of the building and whether the materials are recycled at the end of their life (Börjesson and Gustavsson, 2000; Lenzen and Treloar, 2002). For typical standards of building construction, the embodied energy is equivalent to only a few years of operating energy, although there are cases in which the embodied energy can be much higher (Lippke et al., 2004). Thus, over a 50-year time span, reducing the operating energy is normally more important than reducing the embodied energy. However, for traditional buildings in developing countries, the embodied energy can be large compared to the operating energy, as the latter is quite low.

In most circumstances, the choice that minimizes operating energy use also minimizes total lifecycle energy use. In some cases, the high embodied energy in high-performance building envelope elements (such as krypton-filled double- or triple-glazed windows) can be largely offset from savings in the embodied energy of heating and/or cooling equipment (Harvey, 2006, Chapter 3), so a truly holistic approach is needed in analysing the lifecycle energy use of buildings.

6.4.15 Trade-offs involving energy-related emissions and halocarbon emissions

Emissions of halocarbons from building cooling and refrigeration equipment, heat pumps and foam insulation amount to 1.5 GtCO$_2$-eq at present, compared to 8.6 GtCO$_2$ from buildings (including through the use of electricity) (IPCC/TEAP, 2005). Emissions due to these uses are projected only to 2015 and are constant or decline in this period. Halocarbon emissions are thus an important consideration. Issues pertaining to stratospheric ozone and climate are comprehensively reviewed in the recent IPCC/TEAP report (IPCC/TEAP, 2005).

Halocarbons (CFCs, HCFCs and HFCs) are involved as a working fluid in refrigeration equipment (refrigerators, freezers and cold storage facilities for food), heating and cooling of buildings (heat pumps, air conditioners and chillers) and as an blowing agent used in foam insulation for refrigerators, pipes and buildings. All three groups are greenhouse gases. The GWP of HCFCs is generally lower than CFCs. The GWP of HFCs is also generally lower than that of the CFCs, but generally slightly higher than that of the HCFCs. The consumption (production plus imports, minus exports, minus destruction) of CFCs except for critical uses (e.g., medical devices) stopped in 1996 in developed countries, while developing countries have been given to 2010 to eliminate consumption. HCFCs are being phased out, also for reasons of ozone depletion, but will not be completely phased out of production until 2030 in developed countries and 2040 in developing countries. Nevertheless, projected emissions of HCFCs and HFCs (and ongoing emissions from CFC banks) are sufficiently high that scenarios of halocarbon emissions related to buildings in 2015 show almost the same emissions as in 2002 (about 1.5 GtCO$_2$-eq. emissions). For the coming decade or longer, the bank of CFCs in the stock of cooling equipment and foams is so large that particular attention needs to be given to recovering these CFCs.

Lifetime emissions of refrigerants from cooling equipment, expressed as CO$_2$-eq per unit of cooling, have fallen significantly during the past 30 years. Leakage rates are generally in the order of 3%, but rates as high as 10–15% occur. By 2010, it is expected that HFCs will be the only halocarbon refrigerant to be used in air conditioners and heat pumps manufactured in developed countries. Non-halocarbon refrigerants can entail similar efficiency benefits if the heat pump is fully optimised. Thus, both the performance of the heat pump and the impact of halocarbon emissions need to be considered in evaluating the climatic impact of alternative choices for refrigerants.

The climatic impact of air conditioners and most chillers is generally dominated by the energy used to power them. For leakage of HFC refrigerants at rates of 1 to 6%/yr (IPCC/TEAP, 2005) (best practice is about 0.5%/yr) and recovery of 85% of the refrigerant (compared to 70–100% in typical practice) at the end of a 15-year life, refrigerant leakage accounts for only 1 to 5% of the total impact on climate of the cooling equipment to up to 20%, without end-of-life recovery, of the total impact (derived from (IPCC/TEAP, 2005). This demonstrates the importance of end-of-life recovery, which is highly uncertain for HFCs at present. However, for CFC chillers, the high GWP of the refrigerant and the typical high leakage of older CFC-based designs cause the refrigerant to be a significant factor in overall emissions. This demonstrates that emphasis needs to be put on the replacement of CFC chillers in both developed and developing countries for which hydrocarbons are now widely used in EU countries.

The energy/HFC relationship for air conditioners does not hold for most large built-up refrigeration systems, such as those found in supermarkets and hypermarkets. Roughly half of the total equivalent emissions from these systems result from the refrigerant, in case an HFC blend is used. Various designs explored in IPCC/TEAP report (IPCC/TEAP, 2005) indicate

that direct refrigerant emissions can drop from 40–60% of the total emissions in a typical system to 15% for improved systems. The value is 0% for systems using hydrocarbon or ammonia refrigerants. Although some designs may incur a slight increase in energy use, total (energy + refrigerant) emissions are nonetheless significantly reduced.

For foam insulation blown with halocarbons, the benefit of reduced heating energy use can outweigh the effect of leakage of blowing agent when insulating buildings that were previously either poorly insulated or uninsulated (Ashford et al., 2005). However, for high levels of insulation, the opposite becomes true (Harvey, 2007) without end-of-life recovery of the blowing agent. In general terms, the use of methods such as Life Cycle Climate Performance (LCCP) is essential in evaluating the most appropriate course of action in each situation.

6.4.16 Summary of mitigation options in buildings

The key conclusion of section 6.4 is that substantial reductions in CO$_2$ emissions from energy use in buildings can be achieved over the coming years using existing, mature technologies for energy efficiency that already exist widely and that have been successfully used (*high agreement, much evidence*). There is also a broad array of widely accessible and cost-effective technologies and know-how that can abate GHG emissions in buildings to a significant extent that has not as yet been widely adopted.

Table 6.1 summarizes selected key technological opportunities in buildings for GHG abatement in five world regions based on three criteria. Twenty-one typical technologies were selected from those described in section 6.4. As economic and climatic conditions in regions largely determine the applicability and importance of technologies, countries were divided into three economic classes and two climatic types. The three criteria include the maturity of the technology, cost/effectiveness and appropriateness. Appropriateness includes climatic, technological and cultural applicability. For example, direct evaporative cooling is ranked as highly appropriate in dry and warm climates but it is not appropriate in humid and warm climates. The assessment of some technologies depends on other factors, too. For instance, the heat pump system depends on the energy source and whether it is applied to heating or cooling. In these cases, variable evaluation is indicated in the table.

Table 6.1: *Applicability of energy efficiency technologies in different regions. Selected are illustrative technologies, with an emphasis on advanced systems, the rating of which is different between countries*

Energy efficiency or emission reduction technology	Developing countries — Cold climate			Developing countries — Warm climate			OECD — Cold climate			OECD — Warm climate			Economies in transition, Continental			Reference
	Technology stage	Cost/ effectiveness	Appropriateness	Technology stage	Cost/ effectiveness	Appropriateness	Technology stage	Cost/ effectiveness	Appropriateness	Technology stage	Cost/ effectiveness	Appropriateness	Technology stage	Cost/ effectiveness	Appropriateness	
Structural insulation panels																6.4.2.1
Multiple glazing layers						[1] [2]										6.4.2.2
Passive solar heating																6.4.3.1
Heat pumps	[3]			[4]	[5] [6]	[7] [8]	[9]			[10] [11]	[12] [13]	[14] [15]	[16]			6.4.3.3 / 6.4.8
Biomass derived liquid fuel stove																6.4.3.2
High-reflectivity bldg. materials																6.4.4.1
Thermal mass to minimize daytime interior temperature peaks						[17]						[19]				6.4.2
Direct evaporative cooler						[18] [21] [22]						[20] [23] [24]				6.4.4.2
Solar thermal water heater																6.4.8
Cogeneration																6.4.6
District heating & cooling system																6.4.6
PV																6.4.7.1
Air to air heat exchanger																6.4.5.1
High efficiency lightning (FL)										μ						6.4.10
High efficiency lightning (LED)																6.4.10
HC-based domestic refrigerator												[25] [26]				6.4.11
HC or CO_2 air conditioners			μ			μ						μ [27] [28]				6.4.4.3
Advance supermarket technologies																6.4.12

Table 6.1: *Applicability of energy efficiency technologies in different regions. Selected are illustrative technologies, with an emphasis on advanced systems, the rating of which is different between countries*

Energy efficiency or emission reduction technology	Developing countries						OECD						Economies in transition, Continental			Reference
	Cold climate			Warm climate			Cold climate			Warm climate						
	Technology stage	Cost/ effectiveness	Appropri-ateness	Technology stage	Cost/ effectiveness	Appropri-ateness	Technology stage	Cost/ effectiveness	Appropri-ateness	Technology stage	Cost/ effectiveness	Appropri-ateness	Technology stage	Cost/ effectiveness	Appropri-ateness	
Variable speed drives for pumps and fans	~	○	○	~	○	○	~	○	○	~	○	○	~	○	○	6.4.4.2
Advanced control system based on BEMS	●	○	○	●	○	○	●	○	○	●	○	○	●	○	○	6.4.6

Notes:

1 For heat block type; 2 For Low-E; 3 Limited to ground heat source etc.; 4 For air conditioning; 5 For hot water; 6 For cooling; 7 For hot water; 8 For hot water; 9 Limited to ground heat source, etc.; 10 For cooling; 11 For hot water; 12 For hot water; 13 For cooling; 14 For hot water; 15 For cooling; 16 Limited to ground heat source, etc.; 17 In high humidity region; 18 In arid region; 19 In high humidity region; 20 In arid region; 21 In high humidity region; 22 In arid region; 23 In high humidity region; 24 In arid region; 25 United States; 26 South European Union; 27 United States; 28 South European Union.

Evaluation ranks:

Visual representation	Stage of technology	Cost/Effectiveness	Appropriateness
•	Research phase (including laboratory and development) [R]	Expensive/Not effective [$$/-]	Not appropriate {-}
○ (grey)	Demonstration phase [D]	Expensive/Effective [$$/+]	Appropriate {+}
● (large)	Economically feasible under specific conditions [E]	Cheap/Effective [$/+]	Highly appropriate {++}
~	Mature Market (widespread commercially available without specific governmental support) {M}	'~' Not available	'~' Not available
μ	No Mature Market (not necessarily available/not necessarily mature market)		

6.5 Potential for and costs of greenhouse gas mitigation in buildings

The previous sections have demonstrated that there is already a plethora of technological, systemic and management options available in buildings to substantially reduce GHG emissions. This section aims at quantifying the reduction potential these options represent, as well as the costs associated with their implementation.

6.5.1 Recent advances in potential estimations from around the world

Chapter 3 of the TAR (IPCC, 2001) provided an overview of the global GHG emission reduction potential for the residential and commercial sectors, based on the work of IPCC (1996) and Brown *et al.* (1998). An update of this assessment has been conducted for this report, based on a review of 80 recent studies from 36 countries and 11 country groups, spanning all inhabited continents. While the current appraisal concentrates on new results since the TAR, a few older studies were also revisited if no recent study was located to represent a geopolitical region in order to provide more complete global coverage. Table 6.2 reviews the findings of a selection of major studies on CO_2 mitigation potential from various countries around the world that could be characterized in a common framework. Since the studies apply a variety of assumptions and analytical methods, these results should be compared with caution (see the notes for each row, for methodological aspects of such a comparison exercise).

According to Table 6.2, estimates of technical potential range from 18% of baseline CO_2 emissions in Pakistan (Asian Development Bank, 1998) where only a limited number of options were considered, to 54% in 2010[9] in a Greek study (Mirasgedis *et al.*, 2004) that covered a very comprehensive range of measures in the residential sector. The estimates of economic potential[10] vary from 12% in EU-15 in 2010[11] (Joosen and Blok, 2001) to 52% in Ecuador in 2030[12] (FEDEMA, 1999). Estimates of market potential[13] range from 14% in Croatia, focusing on four options only (UNFCCC NC1 of Croatia, 2001), to 37% in the USA, where a wide range of policies were appraised (Koomey *et al.*, 2001).

Our calculations based on the results of the reviewed studies (see Box 6.1) suggest that, globally, approximately 29% of the projected baseline emissions by 2020[14] can be avoided cost-effectively through mitigation measures in the residential

and commercial sectors *(high agreement, much evidence)*. Additionally at least 3% of baseline emissions can be avoided at costs up to 20 US$/tCO_2 and 4% more if costs up to 100 US$/tCO_2 are considered. Although due to the large opportunities at low-costs, the high-cost potential has been assessed to a limited extent and thus this figure is an underestimate *(high agreement, much evidence)*. These estimates represent a reduction of approximately 3.2, 3.6 and 4.0 billion tonnes of CO_2-eq in 2020, at zero, 20 US$/tCO_2 and 100 US$/tCO_2, respectively. Due to the limited number of demand-side end-use efficiency options considered by the studies, the omission of non-technological options, the often significant co-benefits, as well as the exclusion of advanced integrated highly efficiency buildings, the real potential is likely to be higher *(high agreement, low evidence)*. While occupant behaviour, culture and consumer choice as well as use of technologies are also major determinants of energy consumption in buildings and play a fundamental role in determining CO_2 emissions, the potential reduction through non-technological options is not assessed. These figures are very similar to those reported in the TAR for 2020, indicating the dynamics of GHG reduction opportunities. As previous estimates of additional energy efficiency and GHG reduction potential begin to be captured in a new baseline, they tend to be replaced by the identification of new energy-efficiency and GHG-mitigation options. For comparison with other sectors these potentials have been extrapolated to 2030. The robustness of these figures is significantly lower than those for 2020 due to the lack of research for this year. The extrapolation of the potentials to the year 2030 suggests that, globally, at least 31% of the projected baseline emissions can be mitigated cost-effectively by 2030 in the buildings sector. Additionally at least 4% of baseline emissions can be avoided at costs up to 20 US$/tCO_2 and 5% more at costs up to 100 US$/tCO_2 *(medium agreement, low evidence)*[15]. This mitigation potential would result in a reduction of approximately 4.5, 5.0 and 5.6 billion tonnes of CO_2-eq at zero, 20 US$/tCO_2 and 100 US$/tCO_2, respectively, in 2030. Both for 2020 and 2030, low-cost potentials are highest in the building sector from all sectors assessed in this report (see Table 11.3). The outlook to the long-term future assuming options in the building sector with a cost up to 25 US$/tCO_2 identifies the potential of approximately 7.7 billion tonnes of CO_2 in 2050 (IEA, 2006d).

The literature on future non-CO_2 emissions and potentials for their mitigation have been recently reviewed in the IPCC/TEAP report (2005). The report identifies that there are opportunities to reduce direct emissions significantly through the global application of best practices and recovery methods, with a reduction potential of about 665 million tonnes of CO_2-eq of

---

9. If the approx. formula of Potential $_{2020}$ = 1 - (1 - Potential $_{2010}$)$^{20/10}$ is used to extrapolate the potential as percentage of the baseline into the future (2000 is assumed as a start year), this corresponds to approx. 78% CO_2 savings in 2020.
10. In this chapter we refer to 'cost-effective' or 'economic' potential, to remain consistent with the energy-efficiency literature, considering a zero-carbon price.
11. Corresponds to an approx. 22% potential in 2020 if the extrapolation formula is used.
12. Corresponds to an approx. 38% potential in 2020 if the formula is applied to derive the intermediate potential.
13. For definitions of technical, economic, market and enhanced market potential, see Chapter 2 Section 2.4.3.1.
14. The baseline CO_2 emission projections were calculated on the basis of the reviewed studies, and are a composite of business-as-usual and frozen efficiency baseline.
15. These are the average figures of the low and high scenario of the potential developed for 2030.

Table 6.2: *Carbon dioxide emissions reduction potential for residential commercial sectors*

Case studies providing information for demand-side measures

Country/ region	Reference	Type of potential	Description of mitigation scenarios	Potential Million tCO$_2$	Baseline (%)	Measures with lowest costs	Measures with highest potential	Notes
EU-15	Joosen and Blok, 2001	Technical	25 options: retrofit (insulation); heating systems; new zero & low energy buildings, lights, office equipment & appliances; solar and geo-thermal heat production; BEMS for electricity, space heating and cooling.	310	21%	1. Efficient TV and peripheries; 2. Efficient refrigerators & freezers; 3. Lighting Best Practice.	1.Retrofit: insulated windows; 2.Retrofit: wall insulation; 3.BEMS for space heating and cooling.	[1].4%; [4].Fr-ef.; [5].TY 2010.
		Economic		175	12%			
Canada	Jaccard and Associates, 2002	Market	Mainly fuel switch in water and space heating, hot water efficiency and the multi-residential retrofit program in households; landfill gas, building shell efficiency actions and fuel switch in commerce.	22	24%	n.a. (not listed in the study)	1.Electricity demand reductions; 2.Commercial landfill gas; 3.Furnaces & shell improvements.	[1].10%; [5].TY 2010.
Greece	Mirasgedis et al., 2004	Technical	14 technological options: fuel switch, controls, insulation, lights, air conditioning and others.	13	54%	1. Replacement of central boilers; 2. Use of roof ventilators; 3. Replacement of AC.	1.Shell, esp. insulation; 2.Lighting & water heating; 3.Space heating systems.	[1].6%; [4].Fr-ef; [5].TY 2010; [7].R only.
		Economic		6	25%			
UK	DEFRA, 2006	Technical	41 options: insulation; low-e double glazing windows; various appliances; heating controls; better IT equipment, more efficient motors, shift to CFLs, BEMS, etc.	46	24% (res. only)	1. Efficient fridge/ freezers; 2. Efficient chest freezers; 3. Efficient dishwashers.	1.Efficient gas boilers; 2.Cavity insulation; 3.Loft insulation.	[1].7-5%: R/C; [4].BL: Johnston et al., 2005; [5].BY 2005.
Australia	Australian Greenhouse Office, 2005	Market	Fridges and other appliances, air conditioners, water heating, swimming pool equipment, chillers, ballasts, standards, greenlight Australia plan, refrigerated cabinets, water dispensers, standby.	18	15%	1.Standby programs; 2.MEPS for appliances 1999; 3.TVs on-mode.	1.Packaged air-conditioners; 2.Ballast program in 2003; 3.Fluorescent bulbs.	[1].5%; [4]. BL: scenario without measures; [5].BY 2005.
Estonia	Kallaste et al., 1999	Market	4 insulation measures: 3d window glass, new insulation into houses, renovation of roofs, additional attic insulation.	0.4	2.5% of nation. emis.	1. New insulation; 2. Attic insulation; 3. 3d window glass.	1.New insulation; 2.3d window glass; 3.Attic insulation.	[1].6%; [5]. BY 1995; TY 2025.
China	ERI, 2004	Enhanced market	Key policies: energy conservation standards, heat price reform, standards & labelling for appliances, energy efficiency projects, etc.	422	23%	n.a. (not listed in the study)	n.a. (not listed in the study)	[1].N.a.
New EU Member States[a]	Petersdorff et al., 2005	Technical	Building envelope esp. insulation of walls, roofs, cellar/ground floor, windows with lower U-value; and renewal of energy supply.	62	-	1. Roof insulation; 2. Wall insulation; 3. Floor Insulation.	1. Window replacement; 2. Wall insulation; 3. Roof insulation.	[1] 6%; [4] Fr-ef; [5] BY 2006; TY 2015.
Hungary	Szlavik et al., 1999	Technical	25 technological options and measures: building envelope, space heating, hot water supply, ventilation, awareness, lighting, appliances.	22	45%	1. Individual metering of hot water; 2. Water flow controllers; 3. Retrofitted windows.	1. Post insulation; 2. Retrofit of windows; 3. Replacement of windows.	[1] 3%; [5] TY 2030.
		Economic		15	31%			
Myanmar	Asian Development Bank, 1998	Economic	5 options: shift to CFLs, switch to efficient biomass and LPG cooking stoves, improved kerosene lamps, efficient air conditioners.	3	N.a.	1. Biomass cooking stoves, 2. Kerosene lamps, 3. CFLs.	1. Biomass cooking stoves, 2. LPG cooking stoves, 3. CFLs.	[1] 10%.

Table 6.2. *Continued.*

Country/ region	Reference	Type of potential	Description of mitigation scenarios	Potential — Million tCO$_2$	Potential — Baseline (%)	Measures with lowest costs	Measures with highest potential	Notes
India	Reddy and Balachandra, 2006	Market	Lighting: mixture of incandescents, fluorescent tubes and CFLs, exchange of traditional kerosene and wood stoves and water heaters for efficient equipment.	17	33%	1. Efficient packages of lighting; 2. Kerosene stoves; 3. Wood stoves.	1. Wood stoves, 2. Efficient packages of lighting; 3. Kerosene stoves.	[1] n.a.; [5] TY 2010; [7] R only.
Republic of Korea	Asian Development Bank, 1998	Economic	7 options: heating - condensing gas boilers, solar hot water systems, insulation standards; cooling - air conditioners; improved lights - shift to fluorescents and CFLs; efficient motors and inverters.	20	17%	1. Heating: gas boilers, solar hot water, insulation standards; 2. Air conditioners; 3. Inverters & motors.	1. Improved lights; 2. Motors & inverters; 3. Gas boiler RES, solar hot water system, insulation standards.	[1] 8.5%; [5] BY1998.
Ecuador	FEDEMA, 1999	Economic	6 main options: improvements of appliances, lighting systems, electricity end-uses esp. in rural areas and in the services, solar water heating, public lighting.	7	52%	1. Lights; 2. Electric appliances (esp. rural areas); 3. Electricity end-use in services.	1. Improved electricity end-uses in rural areas; 2. Electric appliances (esp. in rural areas); 3. Light systems.	[1] 10%; [5] TY 2030.
Thailand	Asian Development Bank, 1998	Technical	3 technological programs: lighting (shift to fluorescents), refrigerator (insulation and compressors) and air-conditioning.	15	31%	1. Lighting, 2. Efficient refrigerators, 3. Air conditioning.	1. Efficient air-conditioning, 2. Efficient refrigerators, 3 Lighting.	[1] 10%; [5] BY 1997; [7] R only.
Thailand	Asian Development Bank, 1998	Economic		6.1	13%			
Pakistan	Asian Development Bank, 1998	Technical	Energy efficiency improvements of electric appliances and other end-use devices such as lights, fans, refrigerators, water heaters and improvement of building design.	7	18%	1. Improved lights, 2. Efficient ceiling fans, 3. More efficient refrigerators.	1. Efficient ceiling fans, 2. Improved lights, 3. Improved building design.	[1] 8%; [5] BY 1998.
Pakistan	Asian Development Bank, 1998	Economic		6	16%			
Indonesia	AIM, 2004	Technical	9 Main technological options: energy efficient appliances such as refrigerator and air conditioners, lights (shift from incandescents to fluorescents), kerosene, electricity and gas water heater, kerosene and gas heater, wall and window insulation and others.	13.5	25%	Efficient refrigerator, electricity and gas water heater, kerosene and gas heater, lights, air-conditioners (not ranked).	1. Efficient refrigerators, 2. Fluorescent lamps, 3. Efficient electric water heater (ranking at negative marginal cost).	[1] 5%; [2] Integrated Assessment; [4] Fr-ef.; [7] R only.
Indonesia	AIM, 2004	Economic		10.2	19%			
Argentina	AIM, 2004	Technical		3.8	22%			
Argentina	AIM, 2004	Economic		3.3	19%			
Brazil	AIM, 2004	Technical		5.4	41%			
Brazil	AIM, 2004	Economic		4.8	36%			
Poland	Gaj and Sadowski, 1997	Technical	13 options: lights in streets, commerce & households, gas boiler controls, appliances, heat meters, thermal insulation for walls and roofs, window tightening & replacement, fuel switch from coal to gas, solar or biomass, DH boilers.	43	26%	1. Efficient street lighting; 2. Improved controls of small gas boilers; 3. Efficient lighting in commerce.	1. Insulation of walls, 2. Improvement of home appliances, 3. Fuel switching from coal to gas, solar and biomass.	[1] n.a.; [4] BL: UNFCCC NC3 of Poland, 2001.
Poland	Gaj and Sadowski, 1997	Economic		30	18%			
Russia	Izrael et al., 1999	Technical	Downsized thermal generators (boilers and heaters), thermal insulation (improved panels, doors, balconies, windows), heat and hot water meters and controls, hot water distribution devices, electric appliances.	182	47%	n.a. (not listed in the study)	n.a. (not listed in the study)	[1] n.a.; [4] BL: UNFCCC NC3 of Russia, 2002; PNNL & CENEf, 2004; [5] TY 2010.
Russia	Izrael et al., 1999	Economic		52	13%			

Table 6.2. *Continued.*

Country/ region	Reference	Type of potential	Description of mitigation scenarios	Potential		Measures with lowest costs	Measures with highest potential	Notes
				Million tCO$_2$	Baseline (%)			
South Africa	De Villiers and Matibe, 2000 De Villiers, 2000	Technical	21 options: light practices; new & retrofits HVAC; stoves, thermal envelope; fuel switch in heaters; standards & labelling; for hot water: improved insulation, heat pumps, efficient use; solar heating.	41	23%	1. Energy star equipment; 2. Lighting retrofit; 3. New lighting systems.	1. Hybrid solar water heaters; 2. New building thermal design; 3. New HVAC systems.	[1] 6%; [4] Fr-ef.; [5] BY 2001; TY 2030.
		Economic		37	20%			
Croatia	UNFCCC NC1 of Croatia, 2001	Market	Electricity savings for not heating purposes (low energy bulbs, more efficient appliances, improved motors), solar energy use increase, thermal insulation improvement.	2	14%	1. Bulbs & appliances; 2. Solar energy use increase; 3. Insulation improvement	1. Insulation improvement; 2. Solar energy use increase; 3. Bulbs & appliances.	[1] n.a.
Studies providing information about both supply and demand-side options not separating them								
New EU Member States[a]	Lechtenboh- mer et al., 2005	Economic	Improvement in space and water heating, appliances and lighting, cooling/freezing, air-conditioning, cooking, motors, process heat, renewable energies, reduced emissions from electricity generation.	81	37%	n.a. (not listed in the study)	R: 1.Insulation; 2.Heating systems, fuel switch, DH&CHP; C: 1. Energy efficiency, 2. Renewables.	[1] 3-5%; [5] BY 2005; [7] C includes agriculture.
USA	Koomey et al., 2001	Market	Voluntary labelling, deployment programmes, building codes, new efficiency standards, government procurement, implementation of tax credits, expansion of cost-shared federal R&D expenditures.	898	37%	n.a. (The study did not examine a GHG potential supply cost curve).	1. Lighting; 2. Space cooling; 3.Space heating.	[1] 7%; [5] BY 1997.
Japan	Murakami et al., 2006	Technical	15 options: new and retrofit insulation, double glazing window, home appliances (water & space heating/cooling, lighting, cooking), PVs, solar heating, shift to energy efficient living style, low-carbon electricity generation.	46	28%	n.a. (not listed in the study)	1. Water heater; 2. Space heater; 3. Home appliances.	[1] n.a.; [7] R only.
Germany	Martinsen et al., 2002	Technical	Two options: fuel switch from coal and oil to natural gas and biomass and heat insulation.	31	26%	n.a. (not listed in the study)	1.Heat insulation; 2.Fuel switch from coal & oil to gas & biomass.	[1] n.a.; [5] BY 2002; [7] R only.

Notes specify those parameters which are different from those identified below (the number of a note is the number of the model parameter):
a) Hungary, Slovakia, Slovenia, Estonia, Latvia, Lithuania, Poland and the Czech Republic
1. Discount Rate (DR) belongs to the interval [3%; 10%]. 2. Most models are Bottom-up (BU). 3. All models consider CO$_2$. If a study considered GHGs, CO$_2$ only was analysed, if the study assessed C, potential was converted into CO$_2$: 4. Baseline (BL) is Business-as-Usual Scenario (BAU) or similar (Frozen efficiency scenario is abbreviated as fr-ef). 5. Base year (BY) is 2000; Target year (TY) is 2020. 6. Costs covered: cost of incremental reduction, abatement costs, costs of avoided or saved or mitigated CO$_2$, marginal costs. 7. Estimations are made for Residential (R) and commercial (C) sectors in sum.

Box 6.1: Methodology for the global assessment of potentials and costs of CO_2 mitigation in buildings

This chapter evaluated the potential for GHG mitigation in buildings and associated costs based on the review of existing national and regional potential estimates. For this purpose, over 80 studies containing bottom-up mitigation potential estimates for buildings were identified from 36 countries and 11 country groups covering all inhabited continents. One study (AIM, 2004) covered the entire planet, but it was not suitable for the purposes of this report, as it assessed a very limited number of mitigation options.

To allow the comparison of studies in a common framework, their main results and related assumptions were processed and inserted into a database containing the key characteristics of the methods used and results. To eliminate the major effects of different methodological assumptions, only those studies were selected for further analysis whose assumptions fell into a range of common criteria. For instance, studies were only used for further assessment if their discount rates fell in the interval of 3–10%. For studies which did not report their baseline projections, these were taken from the latest available National Communications to the UNFCCC, or other recent related reports.

Table 6.2 presents the results of a selection of major mitigation studies meeting such criteria for different parts of the world. For definitions of various mitigation potentials see Chapter 2, Section 2.4.3.1.

The next step was to aggregate the results into global and regional potential estimates, as a function of CO_2 costs. Only three studies covered a 2030 target year and they were for countries with insignificant global emissions, thus this was only possible for 2020 in the first iteration. Since few studies reported potentials as a function of cost (typically only technical/economic or market potentials were reported), only 17 studies from the remaining subset meeting our other selection criteria could be used. IPCC SRES or WEO scenarios could not be used as a baseline because little information is available for these on the technology assumptions in buildings. In order to make sure the potentials are entirely consistent with the baseline, an average baseline was created from the studies used for the global potential estimates. For the global potential estimates and the baseline construction, the world was split into seven regions[16]. For each such region, two to four studies were located, thus dividing each region into two to four sub-regions represented by these marker countries in terms of emission growth rates and potential as a percentage of baseline. CO_2 baseline emissions in the seven regions were estimated starting with 2000 IPCC A1B and B2 (SRES) data and applying the CO_2 growth rates calculated for each region as the population weighted average CO_2 baseline growth rates of two to four sub-regions. The baseline projections were estimated for 2000–2020 based on mainly 2020 data from the studies; these trends were prolonged for the period 2020–2030. Since three of the seventeen studies used a frozen efficiency baseline, the baseline used in this chapter can be considered a business-as-usual one with some frozen efficiency elements. The resulting baseline is higher than the B2 (SRES) scenario but lower than A1B (SRES) and WEO scenarios.

Analogously, CO_2 potentials as a percentage of the baseline in cost categories (US\$/tCO_2: (<0); (0;20); (20;100)) were calculated based on population weighted average potentials in the sub-regions for each cost category .While the three studies using a frozen efficiency baseline result in a relatively higher potential than in studies using a BAU baseline, this does not compromise the validity of the global potential, since for the regions applying a frozen efficiency baseline, the latter baseline was used in calculating the global total. The results of these estimates are presented in Table 6.3.

As mentioned above, only three studies covered the baseline or mitigation potential for 2030. Therefore these figures were derived by extrapolating the 2020 figures to 2030. Since the simple exponential formula used for such extrapolations by other sectors was found to yield disputably high or low results in some cases, a modified exponential function was used which allows regulating the maximum potential considered theoretically achievable for different regions[17]. The results of the projections are presented in Table 6.4

---

16 OECD North America, OECD Pacific, Western Europe, Transition Economies, Latin America, Africa and Middle East, and Asia

17 $X(t) = X_{saturation} - C\,e^{-kt}$ (reached from the differential equation: $dx/dt = k\,(X_{saturation} - x)$), saturation illustrates that the closer potential is to this upper limit, the lower potential growth rate is experienced, and the potential does not exceed the maximum judged reasonable. C can be found from the starting conditions (in year 2000); thus if we know the potential in 2020, then:

$$X_{2030} = X_{saturation}(1 - EXP\ \frac{30}{20}\ LN\ (1 - \frac{X_{2020}}{X_{saturation}}))$$

direct emissions in 2015, as compared to the BAU scenario. About 40% of this potential is attributed to HFC emission reduction covered by the Kyoto Protocol to the UNFCCC, while HCFCs and CFCs regulated by the Montreal Protocol contribute about 60% of the potential. A key factor determining whether this potential will be realized is the costs associated with the implementation of the measures to achieve the emission reduction. These vary considerably from a net benefit to 300 US$/tCO$_2$-eq. Refrigeration applications and stationary and mobile air conditioning contribute most to global direct GHG emissions. Action in these sub-sectors could therefore have a substantial influence on future emissions of HCFCs and HFCs. The available literature does not contain reliable estimates for non-CO$_2$ mitigation potentials in the long-term future, including the year 2030. Therefore, the 2015 figures can serve as low estimates of the potentials in 2030, taking into account that upcoming progressive policies in many countries have already led to new products with very low non-CO$_2$ emissions as compared to their previous analogues.

6.5.2 Recent advances in estimating the costs of GHG mitigation in buildings

Table 6.3 below and Table 6.4 provide information on the GHG abatement potentials in buildings as a function of costs and world regions for 2020 (Table 6.3) and for 2030 (Table 6.4). These demonstrate that the majority of measures for CO$_2$ abatement in buildings are cost-effective. The table also demonstrates that measures to save electricity in buildings typically offer larger and cheaper options to abate CO$_2$ emissions than measures related to fuel savings. This is especially true for developing countries located in warmer regions, which have less need for space and water heating.

6.5.3 Supply curves of conserved carbon dioxide

CO$_2$ conservation supply curves relate the quantity of CO$_2$ emissions that can be reduced by certain technological or other measures to the cost per unit CO$_2$ savings (Sathaye and Meyers,

Table 6.3: *CO$_2$ mitigation potential projections in 2020 as a function of CO$_2$ cost*

World regions	Baseline emissions in 2020 (GtCO$_2$-eq)	CO$_2$ mitigation potentials as share of the baseline CO$_2$ emission projections in cost categories in 2020 (costs in US$/tCO$_2$-eq)				CO$_2$ mitigation potentials in absolute values in cost categories in 2020, GtCO$_2$ (costs in US$/tCO$_2$-eq)			
		<0	0-20	20-100	<100	<0	0-20	20-100	<100
Globe	11.1	29%	3%	4%	36%	3.2	0.35	0.45	4.0
OECD (-EIT)	4.8	27%	3%	2%	32%	1.3	0.10	0.10	1.6
EIT	1.3	29%	12%	23%	64%	0.4	0.15	0.30	0.85
Non-OECD	5.0	30%	2%	1%	32%	1.5	0.10	0.05	1.6

Table 6.4: *Extrapolated CO$_2$ mitigation potential in 2030 as a function of CO$_2$ cost, GtCO$_2$*

Mitigation option	Region	Baseline projections in 2030	Potential at costs at below 100 US$/tCO$_2$-eq		Potential in different cost categories		
					<0 US$/tCO$_2$	0-20 US$/tCO$_2$	20-100 US$/tCO$_2$
			Low	High	<0 US$/tC	0-73 US$/tC	73-367 US$/tC
Electricity savings[a]	OECD	3.4	0.75	0.95	0.85	0.0	0.0
	EIT	0.4	0.15	0.20	0.20	0.0	0.0
	Non-OECD/EIT	4.5	1.7	2.4	1.9	0.1	0.1
Fuel savings	OECD	2.0	1.0	1.2	0.85	0.2	0.1
	EIT	1.0	0.55	0.85	0.20	0.2	0.3
	Non-OECD/EIT	3.0	0.70	0.80	0.65	0.1	0.0
Total	OECD	5.4	1.8	2.2	1.7	0.2	0.1
	EIT	1.4	0.70	1.1	0.40	0.2	0.3
	Non-OECD/EIT	7.5	2.4	3.2	2.5	0.1	0.0
	Global	14.3	4.8	6.4	4.5	0.5	0.7

Note: a) The absolute values of the potentials resulting from electricity savings in Table 6.4 and Table 11.3 do not coincide due to application of different baselines. Table 6.4 uses the baseline constructed on the basis of the reviewed studies while Table 11.3 applies WEO 2004 baseline (IEA, 2004e) to calculate CO$_2$ emission reductions from electricity savings. The potential estimates as a percentage of the baseline are the same in both cases. Also Table 11.3 excludes the share of emission reductions which is already taken into account by the energy supply sector, while Table 6.4 does not separate this potential.

1995). The measures, or packages of measures, are considered in order of growing marginal CO_2 abatement cost, therefore forming a 'supply curve' for the commodity of CO_2 reduction.

Figure 6.4 depicts the potentials for CO_2 abatement as a function of costs for eight selected recent detailed studies from different world regions. The steepness of the curves, that is the rate at which the costs of the measures increase as more of the potential is captured, varies substantially by country and by study. While the shape of each supply curve is profoundly influenced by the underlying assumptions and methods used in the study, the figure attests that opportunities for cost-effective and low-cost CO_2 mitigation in buildings are abundant in each world region. All eight studies covered here identified measures at negative costs. The supply curves of developing countries and economies in transition are characterized by a flat slope and lie, in general, lower than the curves of developed countries. The flat slope justifies the general perception (for instance, which provided the main rationale for the Kyoto Flexibility Mechanisms) that there is a higher abundance of 'low-hanging fruit' in these countries. More concretely, the net costs of GHG mitigation in buildings in these countries do not grow rapidly even over 30–50% of emissions reductions. For developed countries, the baseline scenario assumes that many of the low-cost opportunities are already captured due to progressive policies in place or in the pipeline.

6.5.4 Most attractive measures in buildings

From a policy-design perspective, it is important to understand which technologies/end-uses entail the lowest unit abatement costs for society, as well as which ones offer the largest abatement potential. This section reviews the most attractive mitigation options in terms of overall potential. Both Table 6.4 and Table 11.3 in Chapter 11 demonstrate that CO_2-saving options are largest from fuel use in developed countries and countries in transition due to their more northern locations and, thus, larger potential for heat-saving measures. Conversely, electricity savings constitute the largest potential in developing countries located in the south, where the majority of emissions in the buildings sector are associated with appliances and cooling. This distribution of the potential also explains the difference in mitigation costs between developing and developed countries. The shift to more efficient appliances quickly pays back, while building shell retrofits and fuel switching, together providing approximately half of the potential in developed countries, are more expensive.

While it is impossible to draw universal conclusions regarding individual measures and end-uses, Table 6.2 attests that efficient lighting technologies are among the most promising measures in buildings, in terms of both cost-effectiveness and size of potential savings in almost all countries. The IEA

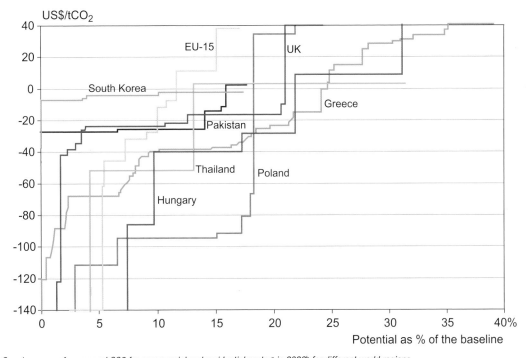

Figure 6.4: *Supply curves of conserved CO2 for commercial and residential sector[a] in 2020[b] for different world regions*

Notes:
a) Except for the UK, Thailand and Greece, for which the supply curves are for the residential sector only.
b) Except for EU-15 and Greece, for which the target year is 2010 and Hungary, for which the target year is 2030. Each step on the curve represents a type of measure, such as improved lighting or added insulation. The length of a step on the 'X' axis shows the abatement potential represented by the measure, while the cost of the measure is indicated by the value of the step on the 'Y' axis.

Sources for data: Joosen and Blok, 2001; Asian Development Bank, 1998; De Villiers and Matibe, 2000; De Villiers, 2000; Szlavik et al., 1999; DEFRA, 2006; Mirasgedis et al., 2004; Gaj and Sadowski, 1997.

(2006b) estimates that by 2020, approximately 760 Mt of CO_2 emissions can be abated by the adoption of least life-cycle cost lighting systems globally, at an average cost of US$–161/t$CO_2$. In developing countries, efficient cooking stoves rank second, while the second-place measures differ in the industrialized countries by climatic and geographic region. Almost all studies examining economies in transition (typically in cooler climates) have found heating-related measures to be most cost-effective, including insulation of walls, roofs, windows and floors, as well as improved heating controls for district heat. In developed countries, appliance-related measures are typically identified as the most cost-effective, with cooling-related equipment upgrades ranking high in the warmer climates.

In terms of the size of savings, improved insulation and district heating in the colder climates and efficiency measures related to space conditioning in the warmer climates come first in almost all studies,18 along with cooking stoves in developing countries. Other measures that rank high in terms of savings potential are solar water heating, efficient lighting and efficient appliances, as well as building energy management systems.

6.5.5 Energy and cost savings through use of the Integrated Design Process (IDP)

Despite the usefulness of supply curves for policy-making, the methods used to create them rarely consider buildings as integrated systems; instead, they focus on the energy savings potential of incremental improvements to individual energy-using devices. As demonstrated in the first part of this chapter, integrated building design not only can generate savings that are greater than achievable through individual measures, but can also improve cost-effectiveness. This suggests that studies relying solely on component estimates may underestimate the abatement potential or overestimate the costs, compared with a systems approach to building energy efficiency. Recent published analyses show that, with an integrated approach, (i) the cost of saving energy can go down as the amount of energy saved goes up, and (ii) highly energy-efficient buildings can cost less than buildings built according to standard practice (Harvey, 2006; Chapter 13).

6.6 Co-benefits of GHG mitigation in the residential and commercial sectors

Co-benefits of mitigation policies should be an important decision element for decision-makers in both the residential and commercial sectors. Although these co-benefits are often not quantified, monetized, or perhaps even identified by the decision-makers or economic modellers (Jochem and Madlener, 2003), they can still play a crucial role in making GHG emissions mitigation a higher priority. This is especially true in less economically advanced countries, where environmentalism – and climate change specifically – may not have a strong tradition or a priority role in either the policy agenda or the daily concerns of citizens. In these circumstances, every opportunity for policy integration can be of value in order to reach climate change mitigation goals.

6.6.1 Reduction in local/regional air pollution

Climate mitigation through energy efficiency in the residential and commercial sectors will improve local and regional air quality, particularly in large cities, contributing to improved public health (e.g., increased life expectancy, reduced emergency room visits, reduced asthma attacks, fewer lost working days) and avoidance of structural damage to buildings and public works. As an example in China, replacement of residential coal burning by large boiler houses providing district heating is among the abatement options providing the largest net benefit per tonne of CO_2 reduction, when the health benefits from improved ambient air conditions are accounted for (Mestl et al., 2005). A study in Greece (Mirasgedis et al., 2004) found that the economic GHG emissions abatement potential in the residential sector could be increased by almost 80% if the co-benefits from improved air quality are taken into account. Beyond the general synergies between improved air quality and climate change mitigation described in Chapter 11 (see Section 11.8.1), some of the most important co-benefits in the households of developing countries are due to reduced indoor air pollution through certain mitigation measures, discussed in sections 6.6.2 and 6.1.1.

6.6.2 Improved health, quality of life and comfort

In the least developed countries, one of the most important opportunities for achieving GHG mitigation as well as sustainable development in buildings is to focus on the health-related benefits of clean domestic energy services, including safe cooking. Indoor air pollution is a key environmental and public health peril for countless of the world's poorest, most vulnerable people. Approximately three billion people worldwide rely on biomass (wood, charcoal, crop residues and dung) and coal to meet their household cooking and heating energy needs (ITDG, 2002). Smoke from burning these fuels contributes to acute respiratory infections in young children and chronic obstructive pulmonary disease in adults. These health problems are responsible for nearly all of the 2.2 million deaths attributable to indoor air pollution each year, over 98% of which are in developing countries (Gopalan and Saksena, 1999; Smith et al., 2004), (See Box 6.2). In addition, women and children also bear the brunt of the work of collecting biomass fuel. Clean-burning cooking stoves not only save substantial amounts of GHG emissions, but also prevent many of these

---

18 Note that several studies covered only electricity-related measures, and thus excluded some heating options.

Box 6.2: Traditional biomass-based cooking has severe health effects

In South Africa, children living in homes with wood stoves are almost five times more likely than others to develop respiratory infections severe enough to require hospitalization. In Tanzania, children younger than five years who die of acute respiratory infection are three times more likely than healthy children to have been sleeping in a room with an open cooking stove. In the Gambia, children carried on their mothers' backs as the mothers cook over smoky stoves contract pneumonia at a rate 2.5 times higher than unexposed children. In Colombia, women exposed to smoke during cooking are over three times more likely than others to suffer from chronic lung disease. In Mexico, urban women who use coal for cooking and heating over many years are subject to a risk of lung cancer two to six times higher than women who use gas. Rural coal smoke exposure can increase lung cancer risks by a factor of nine or more. In India, smoke exposure has been associated with a 50% increase in stillbirths.

Cleaner-burning improved cooking stoves (ICS), outlined in the previous sections of this chapter, help address many of the problems associated with traditional cooking methods. The benefits derived from ICS are: 1) reduced health risks for women and children due to improved indoor air quality; 2) reduced risks associated with fuel collection; 3) cost-effective and efficient energy use, which eases the pressure on the natural biomass resource; 4) a reduction in the amount of money spent on fuel in urban areas; and 5) a reduction in fuel collection and cooking time, which translates into an increase in time available for other economic and developmental activities.

Source: UN, 2002

health problems and provide many other benefits identified in Box 6.2.

In developed countries, the diffusion of new technologies for energy use and/or savings in residential and commercial buildings contributes to an improved quality of life and increases the value of buildings. Jakob (2006) lists examples of this type of co-benefit, such as improved thermal comfort (fewer cold surfaces such as windows) and the substantially reduced level of outdoor noise infiltration in residential or commercial buildings due to triple-glazed windows or high-performance wall and roof insulation. At noisy locations, an improvement of 10–15 dB could result in gross economic benefits up to the amount of 3–7% of the rental income from a building (Jakob, 2006). Lastly, better-insulated buildings eliminate moisture problems associated with, for example, thermal bridges and damp basements and thus reduce the risk of mould build-up and associated health risks.

6.6.3 Improved productivity

There is increasing evidence that well-designed, energy efficient buildings often have the co-benefits of improving occupant productivity and health (Leaman and Bordass, 1999; Fisk, 2000; Fisk, 2002). Assessing these productivity gains is difficult (CIBSE (The Chartered Institution of Building Services Engineers), 1999) but in a study of 16 buildings in the UK, occupants estimated that their productivity was influenced by the environment by between –10% and +11% (Leaman and Bordass, 1999).

The implementation of new technologies for GHG emissions mitigation achieves substantial learning and economies of scale, resulting in cost reductions. Jacob and Madlener (2004) analyzed the technological progress and marginal cost developments for energy efficiency measures related to the building envelope using data for the time period 1975 to 2001 in Switzerland. The analysis yields technical progress factors of around 3% per annum for wall insulation and 3.3% per annum for double glazing windows, while real prices decreases of 0.6% since 1985 for facades and 25% over the last 30 years for double glazing windows (Jacob and Madlener, 2004).

6.6.4 Employment creation and new business opportunities

Most studies agree that energy-efficiency investments will have positive effects on employment, directly by creating new business opportunities and indirectly through the economic multiplier effects of spending the money saved on energy costs in other ways (Laitner et al., 1998; Jochem and Madlener, 2003). Providing energy-efficiency services has proven to be a lucrative business opportunity. Experts estimate a market opportunity of € 5–10 billion in energy service markets in Europe (Butson, 1998). The data on energy service company (ESCO) industry revenues in Section 6.8.3.5 demonstrates that the energy services business appears to be both a very promising and a quickly growing business sector worldwide. The European Commission (2005) estimates that a 20% reduction in EU energy consumption by 2020 can potentially create (directly or indirectly) as many as one million new jobs in Europe, especially in the area of semi-skilled labour in the buildings trades (Jeeninga et al., 1999; European Commission, 2003).

6.6.5 Improved social welfare and poverty alleviation

Improving residential energy efficiency helps households cope with the burden of paying utility bills and helps them afford adequate energy services. One study estimated that an average EU household could save € 200–1000 (US$ 248–1240) in utility costs through cost-effective improvements in energy efficiency (European Commission, 2005). Reducing the economic burden of utility bills is an important co-benefit of energy efficiency for less affluent households. This is especially true in former communist countries and others (e.g., in Asia and Latin America) where energy subsidies have been removed and energy expenditures are a major burden for much of the population (Ürge-Vorsatz et al., 2006). In economies in transition, this situation provides an opportunity to redirect those social programmes aimed at compensating for increasing energy costs towards energy-efficiency efforts. In this way resources can be invested in long-term bill reduction through energy efficiency instead of one-time subsidies to help pay current utility bills (Ürge-Vorsatz and Miladinova, 2005).

Fuel poverty, or the inability to afford basic energy services to meet minimal needs or comfort standards, is also found in even the wealthiest countries. In the UK in 1996, about 20% of all households were estimated to live in fuel poverty. The number of annual excess winter deaths, estimated by the UK Department of Health at around 30 thousand annually between 1997 and 2005, can largely be attributed to inadequate heating (Boardman, 1991; DoH (UK Department of Health), 2000). Improving energy efficiency in these homes is a major component of strategies to eradicate fuel poverty.

In developing countries, energy-efficient household equipment and low-energy building design can contribute to poverty alleviation through minimizing energy expenditures, therefore making more energy services affordable for low-income households (Goldemberg, 2000). Clean and efficient utilization of locally available renewable energy sources reduces or replaces the need for energy and fuel purchases, increasing the access to energy services. Therefore, sustainable development strategies aimed at improving social welfare go hand-in-hand with energy efficiency and renewable energy development.

6.6.6 Energy security

Additional co-benefits of building-level GHG mitigation include improved energy security and system reliability (IEA, 2004f), discussed in more detail in Chapter 4. Improving end-use energy efficiency is among the top priorities on the European Commission's agenda to increase energy security, with the recognition that energy efficiency is likely to generate additional macro-economic benefits because reduced energy imports will improve the trade balances of importing countries (European Commission, 2003).

6.6.7 Summary of co-benefits

In summary, investments in residential and commercial building energy efficiency and renewable energy technologies can yield a wide spectrum of benefits well beyond the value of saved energy and reduced GHG emissions. Several climate mitigation studies focusing on the buildings sector maintain that, if co-benefits of the various mitigation options are included in the economic analysis, their economic attractiveness may increase considerably – along with their priority levels in the view of decision-makers (Jakob et al., 2002; Mirasgedis et al., 2004; Banfi et al., 2006). Strategic alliances with other policy fields, such as employment, competitiveness, health, environment, social welfare, poverty alleviation and energy security, can provide broader societal support for climate change mitigation goals and may improve the economics of climate mitigation efforts substantially through sharing the costs or enhancing the dividends (European Commission, 2005). In developing countries, residential and commercial-sector energy efficiency and modern technologies to utilize locally available renewable energy forms, can form essential components of sustainable development strategies.

6.7 Barriers to adopting building technologies and practices that reduce GHG emissions

The previous sections have shown the significant cost-effective potential for CO_2 mitigation through energy efficiency in buildings. The question often arises: If these represent profitable investment opportunities, or energy cost savings foregone by households and businesses, why are these opportunities not pursued? If there are profits to be made, why do markets not capture these potentials?

Certain characteristics of markets, technologies and end-users can inhibit rational, energy-saving choices in building design, construction and operation, as well as in the purchase and use of appliances. The Carbon Trust (2005) suggests a classification of these barriers into four main categories: financial costs/benefits; hidden costs/benefits; real market failures; and behavioural/organizational non-optimalities. Table 6.5 gives characteristic examples of barriers that fall into these four main categories. The most important among them that pertain to buildings are discussed below in further detail.

6.7.1 Limitations of the traditional building design process and fragmented market structure

One of the most significant barriers to energy-efficient building design is that buildings are complex systems. While the typical design process is linear and sequential, minimizing energy use requires optimizing the system as a whole by systematically addressing building form, orientation, envelope,

Table 6.5: *Taxonomy of barriers that hinder the penetration of energy efficient technologies/practices in the buildings sector*

Barrier categories	Definition	Examples
Financial costs/benefits	Ratio of investment cost to value of energy savings	Higher up-front costs for more efficient equipment Lack of access to financing Energy subsidies Lack of internalization of environmental, health and other external costs
Hidden costs/benefits	Cost or risks (real or perceived) that are not captured directly in financial flows	Costs and risks due to potential incompatibilities, performance risks, transaction costs etc. Poor power quality, particularly in some developing countries
Market failures	Market structures and constraints that prevent the consistent trade-off between specific energy-efficient investment and the energy saving benefits	Limitations of the typical building design process Fragmented market structure Landlord/tenant split and misplaced incentives Administrative and regulatory barriers (e.g., in the incorporation of distributed generation technologies) Imperfect information
Behavioural and organizational non-optimalities	Behavioural characteristics of individuals and organizational characteristics of companies that hinder energy efficiency technologies and practices	Tendency to ignore small opportunities for energy conservation Organizational failures (e.g., internal split incentives) Non-payment and electricity theft Tradition, behaviour, lack of awareness and lifestyle Corruption

Source: Carbon Trust, 2005.

glazing area and a host of interaction and control issues involving the building's mechanical and electrical systems.

Compounding the flaws in the typical design process is fragmentation in the building industry as a whole. Assuring the long-term energy performance and sustainability of buildings is all the more difficult when decisions at each stage of design, construction and operation involve multiple stakeholders. This division of responsibilities often contributes to suboptimal results (e.g., under-investment in energy-efficient approaches to envelope design because of a failure to capitalize on opportunities to down-size HVAC equipment). In Switzerland, this barrier is being addressed by the integration of architects into the selection and installation of energy-using devices in buildings (Jefferson, 2000); while the European Directive on the Energy Performance of Buildings in the EU (see Box 6.3) aims to bring engineers in at early stages of the design process through its whole-building, performance-based approach.

6.7.2 Misplaced incentives

Misplaced incentives, or the agent-principal barrier takes place when intermediaries are involved in decisions to purchase energy-saving technologies, or agents responsible for investment decisions are different from those benefiting from the energy savings, for instance due to fragmented institutional organizational structures. This limits the consumer's role and often leads to an under-emphasis on investments in energy efficiency. For example, in residential buildings, landlords often provide the AC equipment and major appliances and decide on building renovation, while the tenant pays the energy bill. As a result, the landlord is not likely to invest in energy efficiency, since he or she is not the one rewarded for the investment (Scott,

1997; Schleich and Gruber, 2007). Decisions about the energy features of a building (e.g., whether to install high-efficiency windows or lighting) are often made by agents not responsible for the energy bills or not using the equipment, divorcing the interests of the builder/investor and the occupant. For example, in many countries the energy bills of hospitals are paid from central public funds while investment expenditures must come either from the institution itself or from the local government (Rezessy *et al.*, 2006). Finally, the prevailing selection criteria and fee structures for building designers may emphasize initial costs over life-cycle costs, hindering energy-efficiency considerations (Lovins, 1992; Jones *et al.*, 2002).

6.7.3 Energy subsidies, non-payment and theft

In many countries, electricity historically has been subsidized to residential customers (and sometimes to commercial or government customers as well), creating a disincentive for energy efficiency. This is particularly the case in many developing countries and historically in Eastern Europe and the former Soviet Union – for example widespread fuel poverty in Russia has driven the government to subsidize energy costs (Gritsevich, 2000). Energy pricing that does not reflect the long-term marginal costs of energy, including direct subsidies to some customers, hinders the penetration of efficient technologies (Alam *et al.*, 1998).

However, the abrupt lifting of historically prevailing subsidies may also have adverse effects. After major tariff increases, non-payment has been reported to be a serious issue in some countries. In the late 1990s, energy bill collection rates in Albania, Armenia and Georgia were around 60% of billings. Besides non-payment, electricity theft has been occurring on a large scale in many countries – estimates show that distribution losses due to theft are as high as 50% in some states in India (New Delhi, Orissa and Jammu-Kashmir) (EIA (Energy Information Administration), 2004). Even in the United States, it has been estimated to cost utilities billions of dollars each year (Suriyamongkol, 2002). The failure of recipients to pay in full for energy services tends to induce waste and discourage energy efficiency.

6.7.4 Regulatory barriers

A range of regulatory barriers has been shown to stand in the way of building-level distributed generation technologies such as PV, reciprocating engines, gas turbines and fuel cells (Alderfer et al., 2000). In many countries, these barriers include variations in environmental permitting requirements, which impose significant burdens on project developers. Similar variations in metering policies cause confusion in the marketplace and represent barriers to distributed generation. Public procurement regulations often inhibit the involvement of ESCOs or the implementation of energy performance contracts. Finally, in some countries the rental market is regulated in a way that discourages investments in general and energy-efficient investments in particular.

6.7.5 Small project size, transaction costs and perceived risk

Many energy-efficiency projects and ventures in buildings are too small to attract the attention of investors and financial institutions. Small project size, coupled with disproportionately high transaction costs – these are costs related to verifying technical information, preparing viable projects and negotiating and executing contracts – prevent some energy-efficiency investments. Furthermore, the small share of energy expenditures in the disposable incomes of affluent population groups, and the opportunity costs involved with spending the often limited free time of these groups on finding and implementing the efficient solutions, severely limits the incentives for improved efficiency in the residential sector. Similarly, small enterprises often receive higher returns on their investments into marketing or other business-related activities than investing their resources, including human resources, into energy-related activities. Conservative, asset-based lending practices of financial institutions, a limited understanding of energy-efficiency technologies on the part of both lenders and their consumers, lack of traditions in energy performance contracting, volatile prices for fuel (and in some markets, electricity), and small, non-diversified portfolios of energy projects all increase the perception of market and technology risk (Ostertag, 2003; Westling, 2003; Vine, 2005). As discussed in Section 6.8 below, policies can be adopted that can help reduce these transaction costs, thus improving the economics and financing options for energy-efficiency investments.

6.7.6 Imperfect information

Information about energy-efficiency options is often incomplete, unavailable, expensive and difficult to obtain or trust. In addition, few small enterprises in the building industry have access to sufficient training in new technologies, new standards, new regulations and best practices. This insufficient knowledge is compounded by uncertainties associated with energy price fluctuations (Hassett and Metcalf, 1993). It is particularly difficult to learn about the performance and costs of energy-efficient technologies and practices, because their benefits are often not directly observable. For example, households typically receive an energy bill that provides no breakdown of individual end-uses and no information on GHG emissions, while infrequent meter readings (e.g., once a year, as is typical in many EU countries) provide insufficient feedback to consumers on their energy use and on the potential impact of their efficiency investments. Trading off energy savings against higher purchase prices for many energy-efficient products involves comparing the time-discounted value of the energy savings with the present cost of the equipment – a calculation that can be difficult for purchasers to understand and compute.

6.7.7 Culture, behaviour, lifestyle and the rebound effect

Another broad category of barriers stems from the cultural and behavioural characteristics of individuals. The potential impact of lifestyle and tradition on energy use is most easily seen by cross-country comparisons. For example, dishwasher usage was 21% of residential energy use in UK residences in 1998 but 51% in Sweden (European Commission, 2001). Cold water is traditionally used for clothes washing in China (Biermayer and Lin, 2004) whereas hot water washing is common in Europe. Similarly, there are substantial differences among countries in how lighting is used at night, room temperatures considered comfortable, preferred temperatures of food or drink, the operating hours of commercial buildings, the size and composition of households, etc. (IEA, 1997; Chappells and Shove, 2004). Variation across countries in quantity of energy used per capita, which is large both at economy and household levels (IEA, 1997), can be explained only partly by weather and wealth; this is also appropriately attributed to different lifestyles. Even in identical houses with the same number of residents, energy consumption has been shown to differ by a factor of two or more (Socolow, 1978). Studies aimed at understanding these issues suggest that while lifestyle, traditions and culture can act as barriers, retaining and supporting lower-consuming lifestyles may also be effective in constraining GHG emissions (e.g., EEA, 2001).

The 'rebound effect' has often been cited as a barrier to the implementation of energy-efficiency policies. This takes place when increased energy efficiency is accompanied by increased demand for energy services (Moezzi and Diamond, 2005). The literature is divided about the magnitude of this effect (Herring, 2006).

6.7.8 Other barriers

Due to space limitations, not all barriers to energy efficiency identified in Table 6.5 can be detailed here. Other important barriers in the buildings sector include the limited availability of capital and limited access to capital markets of low-income households and small businesses, especially in developing countries (Reddy, 1991); limited availability of energy-efficient equipment along the retail chain (Brown et al., 1991); the case of poor power quality in some developing countries interfering with the operation of the electronics needed for energy efficient end-use devices (EAP UNDP, 2000); and the inadequate levels of energy services (e.g., insufficient illumination levels in schools, or unsafe wiring) in many public buildings in developing countries and economies in transition. This latter problem can severely limit the cost-effectiveness of efficiency investments, since a proposed efficiency upgrade must also address these issues, offsetting most or all of the energy and cost savings associated with improved efficiency and in turn make it difficult to secure financing or pay back a loan from energy cost savings.

6.8 Policies to promote GHG mitigation in buildings

Preceding sections have demonstrated the high potential for reducing GHG emissions in buildings through cost-effective energy-efficiency measures and distributed (renewable) energy generation technologies. The previous section has demonstrated that even the cost-effective part of the potential is unlikely to be captured by markets alone, due to the high number of barriers. Although there is no quantitative or qualitative evidence in the literature, it is possible that barriers to the implementation of economically attractive GHG reduction measures are the most numerous and strongest in the building sector, especially in households. Since policies can reduce or eliminate barriers and associated transaction costs (Brown, 2001), special efforts targeted at removing the barriers in the buildings sector may be especially warranted for GHG mitigation efforts.

Sections 6.8.1–6.8.5 describe a selection of the major instruments summarized in Table 6.6 that complement the more general discussion of Chapter 13, with a focus on policy tools specific to or specially applied to buildings. The rest of Table 6.6 is discussed in Section 6.8.5.

6.8.1 Policies and programmes aimed at building construction, retrofits, and installed equipment and systems

6.8.1.1 Building codes

Building regulations originally addressed questions related to safety and the protection of occupants. Oil price shocks in the 1970s led most OECD countries to extend their regulations to include energy efficiency. Nineteen out of twenty OECD countries surveyed have such energy standards and regulations, although coverage varies among countries (OECD, 2003).

Building energy codes may be classified as follows: 1) Overall performance-based codes that require compliance with an annual energy consumption level or energy cost budget, calculated using a standard method. This type of code provides flexibility but requires well-trained professionals for implementation; 2) Prescriptive codes that set separate performance levels for major envelope and equipment components, such as minimum thermal resistance of walls, maximum window heat loss/gain and minimum boiler efficiency. There are also examples of codes addressing electricity demand. Several cantons in Switzerland specify maximum installed electric loads for lighting ventilation and cooling in new commercial buildings (SIA, 2006); and 3) A combination of an overall performance requirement plus some component performance requirements, such as wall insulation and maximum window area.

Energy codes are often considered to be an important driver for improved energy efficiency in new buildings. However, the implementation of these codes in practice needs to be well prepared and to be monitored and verified. Compliance can be difficult to enforce and varies among countries and localities (XENERGY, 2001; City of Fort Collins, 2002; OECD, 2003; Ürge-Vorsatz et al., 2003).

Prescriptive codes are often easier to enforce than performance-based codes (Australian Greenhouse Office, 2000; City of Fort Collins, 2002; Smith and McCullough, 2001). However, there is a clear trend in many countries towards performance-based codes that address the overall energy consumption of the buildings. This trend reflects the fact that performance-based policies allow optimization of integrated design and leave room for the creativity of designers and innovative technologies. However, successful implementation of performance-based codes requires education and training – of both building officials and inspectors – and demonstration projects showing that the building code can be achieved without much additional cost and without technical problems (Joosen, 2006). New software-based design and education tools, including continuous e-learning tools, are examples of tools that can provide good design techniques, continuous learning by professionals, easier inspection methods and virtual testing of new technologies for construction and building systems.

Public policies in many countries are also increasingly addressing energy efficiency in existing buildings. For instance, the EU Commission introduced the Directive on the Energy Performance of Buildings in 2003 (see Box 6.3), which standardized and strengthened building energy-efficiency requirements for all EU Member States. To date, most codes for existing buildings include requirements for minimum levels of performance of the components used to retrofit building elements or installations. In some countries, the codes may even prohibit the use of certain technologies – for example Sweden's prohibition of direct electric resistance heating systems, which has led to the rapid introduction of heat pumps in the last five years. Finally, the EU Directive also mandated regular inspection and maintenance of boilers and space conditioning installations in existing buildings (see Box 6.3).

According to the OECD (2003), there is still much room for further upgrading building energy-efficiency codes throughout the OECD member countries. To remain effective, these codes have to be regularly upgraded as technologies improve and costs of energy-efficient features and equipment decline. Setting flexible (e.g., performance-based) codes can help keep compliance costs low and may provide more incentives for innovation.

6.8.1.2 Building certification and labelling systems

The purpose of building labelling and certification is to overcome barriers relating to the lack of information, the high transaction costs, the long lifetime of buildings and the problem of displaced incentives between the builder and buyer, or between the owner and tenant. Certification and labelling schemes can be either mandatory or voluntary.

With the introduction of the EU Directive on the Energy Performance of Buildings (see Box 6.3), building certification is to be instituted throughout Europe. Voluntary certification and/or labelling systems have also been developed for building products such as windows, insulation materials and HVAC components in North America, the EU and a few other countries (McMahon and Wiel, 2001; Menanteau, 2001). The voluntary Energy Star Buildings rating and Energy Star Homes label in the USA and the NF-MI voluntary certificate for houses in France have proven to be effective in ensuring compliance with energy code requirements and sometimes in achieving higher performance levels (Hicks and Von Neida, 1999). Switzerland has developed the 'Minergie' label for new buildings that have a 50% lower energy demand than buildings fulfilling the mandatory requirements; such buildings typically require roughly 6% additional investment costs (OPET Network, 2004). Several local governments in Japan apply the Comprehensive Assessment System for Building Environmental Efficiency (CASBEE) (IBEC, 2006). The Australian city of Canberra (ACT) has a requirement for all houses to be energy-efficiency rated on sale. The impact on the market has been to place a financial value on energy efficiency through a well-informed marketplace (ACT, 2006).

6.8.1.3 Education, training and energy audit programmes

Lack of awareness of energy-savings opportunities among practicing architects, engineers, interior designers and

---

Box 6.3: The European Directive on the Energy Performance of Buildings

One of the most advanced and comprehensive pieces of regulation targeted at the improvement of energy efficiency in buildings is the new European Union Directive on the Energy Performance of Buildings (European Commission, 2002). The Directive introduces four major actions. The *first action* is the establishment of 'common methodology for calculating the integrated energy performance of buildings', which may be differentiated at the regional level. The *second action* is to require member states to 'apply the new methods to minimum energy performance standards' for new buildings. The Directive also requires that a non-residential building, when it is renovated, be brought to the level of efficiency of new buildings. This latter requirement is a very important action due to the slow turnover and renovation cycle of buildings, and considering that major renovations to inefficient older buildings may occur several times before they are finally removed from the stock. This represents a pioneer effort in energy-efficiency policy; it is one of the few policies worldwide to target existing buildings. The *third action* is to set up 'certification schemes for new and existing buildings' (both residential and non-residential), and in the case of public buildings to require the public display of energy performance certificates. These certificates are intended to address the landlord/tenant barrier, by facilitating the transfer of information on the relative energy performance of buildings and apartments. Information from the certification process must be made available for new and existing commercial buildings and for dwellings when they are constructed, sold, or rented. *The last action* mandates Member States to establish 'regular inspection and assessment of boilers and heating/cooling installations'.

The European Climate Change Programme (ECCP, 2001) estimated that CO_2 emissions to be tapped by implementation of this directive by 2010 are 35–45 million tCO_2-eq at costs below 20 EUR/tCO_2-eq, which is 16–20% of the total cost-effective potential associated with buildings at these costs in 2010.

professionals in the building industry, including plumbers and electricians, is a major impediment to the construction of low-energy buildings. In part, this reflects inadequate training at universities and technical schools, where the curricula often mirror the fragmentation seen in the building design profession. There is a significant need in most countries to create comprehensive, integrated programmes at universities and other educational establishments to train the future building professional in the design and construction of low-energy buildings. The value of such programmes is significantly enhanced if they have an outreach component to upgrade the skills and knowledge of practicing professionals – for example, by assisting in the use of computer simulation tools as part of the integrated design process.

The education of end-users and raising their awareness about energy-efficiency opportunities is also important. Good explanation (e.g., user-friendly manuals) is often a condition for proper installation and functioning of energy-efficient buildings and components. Since optimal operation and regular maintenance are often as important as the technological efficiency in determining overall energy consumption of equipment, accessible information and awareness raising about these issues during and after purchase are necessary. This need for widespread education is beginning to be reflected in the curricula of some countries: Japan's and Germany's schools increasingly teach the importance of energy savings (ECCJ, 2006; Hamburger-bildungsserver, 2006). Better education is also relevant for professionals such as plumbers and electricians. Incentives for consumers are generally needed along with the information programs to have significant effect (Shipworth, 2000).

Energy audit programmes assist consumers in identifying opportunities for upgrading the energy efficiency of buildings.

Occasionally with financial support from government or utility companies, these programmes may provide trained energy auditors to conduct on-site inspections of buildings, perform most of the calculations for the building owner and offer recommendations for energy-efficiency investments or operational measures, as well as other cost-saving actions (e.g., reducing peak electrical demand, fuel-switching). The implementation of the audit recommendations can be voluntary for the owner, or mandated-such as in the Czech Republic and Bulgaria, which require that installations with energy consumption above a certain limit conduct an energy-efficiency audit and implement the low-cost measures (Ürge-Vorsatz *et al.*, 2003). In India, all large commercial buildings have to conduct an energy audit at specified intervals of time (The Energy Conservation Act, 2001). The EU EPB Directive mandates audits and the display of the resulting certificate in an increasing number of situations (see Box 6.3).

6.8.2 Policies and programmes aimed at appliances, lighting and office/consumer plug loads

Appliances, equipment (including information and communication technology) and lighting systems in buildings typically have very different characteristics from those of the building shell and installed equipment, including lower investment costs, shorter lifetimes, different ownership characteristics and simpler installation and maintenance. Thus, the barriers to energy-efficient alternatives are also different to some extent, warranting a different policy approach. This section provides an overview of policies specific to appliances, lighting and plug-in equipment.

Box 6.4: Global efforts to combat unneeded standby and low-power mode consumption in appliances

Standby and low-power-mode (LoPoMo) electricity consumption of appliances is growing dramatically worldwide, while technologies exist that can eliminate or reduce a significant share of related emissions. The IEA (2001) estimated that standby power and LoPoMo waste may account for as much as 1% of global CO_2 emissions and 2.2% of OECD electricity consumption. Lebot *et al.* (2000) estimated that the total standby power consumption in an average household could be reduced by 72%, which would result in emission reductions of 49 million tCO_2 in the OECD. Various instruments – including minimum energy efficiency performance standards (MEPS), labelling, voluntary agreements, quality marks, incentives, tax rebates and energy-efficient procurement policies – are applied globally to reduce the standby consumption in buildings (Commission of the European Communities, 1999), but most of them capture only a small share of this potential. The international expert community has been urging a 1-Watt target (IEA, 2001). In 2002, the Australian government introduced a 'one-watt' plan aimed at reducing the standby power consumption of individual products to less than one watt. To reach this, the National Appliance and Equipment Energy Efficiency Committee has introduced a range of voluntary and mandatory measures to reduce standby – including voluntary labelling, product surveys, MEPS, industry agreements and mandatory labelling (Commonwealth of Australia, 2002). As of mid-2006, the only mandatory standard regarding standby losses in the world has been introduced in California (California Energy Commission, 2006), although in the USA the Energy Policy Act of 2005 directed the USDOE to evaluate and adopt low standby power standards for battery chargers.

6.8.2.1   Standards and labelling

Energy-efficiency performance standards and labels (S&L) for appliances and lighting are increasingly proving to be effective vehicles for transforming markets and stimulating adoption of new, more efficient technologies and products. Since the 1990s, 57 countries have legislated efficiency standards and/or labels, applied to a total of 46 products as of 2004 (Wiel and McMahon, 2005). Today, S&L programmes are among the most cost-effective instruments across the economy to reduce GHG emissions, with typically large negative costs (see Table 6.6). Products subject to standards or labels cover all end-uses and fuel types, with a focus on appliances, information and communications devices, lighting, heating and cooling equipment and other energy-consuming products.

Endorsement and comparison labels[19] induce manufacturers to improve energy efficiency and provide the means to inform consumers of the product's relative or absolute performance and (sometimes) energy operating costs. According to studies evaluating the effectiveness of labels (Thorne and Egan, 2002), those that show the annual energy cost savings appear to be more effective than labels that present life-cycle cost savings. An advantage of a 'categorical' labelling scheme, showing a number of stars or an A-B-C rating, is that it is often easiest for consumers to understand and to transfer their understanding of the categories from one product purchase to others. The categories also provide a useful framework for implementing rebates, tax incentives, or preferential public procurement programmes, while categorical labels on HVAC and other installed equipment make it easy for the building inspector to check for code compliance. A downside of a categorical labelling system can be that if standards are not revised from time to time, there is no stimulus to the manufacturers to develop more efficient appliances and the whole market will be able to deliver appliances fitting the highest efficiency class.

Despite widely divergent approaches, national S&L programmes have resulted in significant cost-effective GHG savings. The US programme of national, mandatory energy-efficiency standards began in 1978. By 2004, the programme had developed (and, in 17 cases, updated) standards for 39 residential and commercial products. The total federal expenditure for implementing the US appliance standards adopted so far (US$ 2 per household) is estimated to have induced US$ 1270 per household of net-present-value savings during the lifetimes of the products affected. Projected annual residential carbon reductions in 2020 due to these appliance standards amount roughly to 9% of projected US residential carbon emissions in the 2020 (base case) (Meyers et al., 2002). In addition, the US Energy Star endorsement label programme estimates savings of 13.2 million tCO_2-eq and US$ 4.2 billion

in 2004 (US EPA, 2005), and projects that the programme will save 0.7 billion tonnes of CO_2 over the period 2003 to 2010, growing to 1.8 billion tonnes of CO_2 over the period 2003 to 2020, if the market target penetration is reached (Webber et al., 2003). According to the IEA (2003a), GHG abatement through appliance standards and labelling in Europe by 2020 will be achieved at a cost of –65 US$/$tCO_2$ in North America and –169 €/tCO_2 (–191 US$/$tCO_2$) (i.e., both at substantial 'net benefit'). An evaluation of the impact of the EU appliance-labelling scheme showed a dramatic shift in the efficiency of refrigerators sold in the EU in the first decade of its S&L programme, as displayed in Figure 6.5 (Bertoldi, 2000). Japan imposes stringent energy efficiency standards on equipment through its 'Top Runner Programme' by distinctly setting the target values based on the most energy-efficient model on the market at the time of the value-setting process. Energy-efficiency values and a rating mark are voluntarily displayed in promotional materials so that consumers can consider energy-efficiency when purchasing (Murakoshi and Nakagami, 2005).

A recent IEA report (2003a) concludes that, without existing policy measures such as energy labelling, voluntary agreements, and MEPS, electricity consumption in OECD countries in 2020 would be about 12% (393 TWh) higher than is now predicted. The report further concludes that the current policies are on course to produce cumulative net cost savings of € 137 billion (US$ 155 billion) in OECD-Europe from 1990 to 2020. As large as these benefits are, the report found that much greater benefits could be attained if existing policies were strengthened.

A study of China's energy-efficiency standards (Fridley and Lin, 2004) estimated savings from eight new MEPS and nine energy-efficiency endorsement labels. The study concluded that, during the first 10 years of implementation, these measures will save 200 TWh (equivalent to all of China's residential electricity consumption in 2002) and 250 $MtCO_2$. Among other countries, Korea shows similar evidence of the impact of labelling, as does the EU (CLASP, 2006). Recently, Australia transformed its S&L programme in order to aggressively improve energy efficiency (NAEEEC, 2006).

In the past few years, strong regional and global S&L efforts have also emerged, offering a more coordinated pathway to promote S&L and improve the cost-effectiveness and market impact of the programmes. One of these pathways is regional harmonization. The IEA (2003b) identifies several forms of multilateral cooperation, including: 'collaboration' in the design of tests, labels and standards; 'harmonization' of the test procedures and the energy-efficiency thresholds used in labels and standards; and 'coordination' of programme implementation and monitoring efforts. However, while easing certain trade restrictions, harmonization of standards and testing methods

---

19 Endorsement labels (or "quality marks") define a group of products as "efficient" when they meet pre-specified criteria, while comparison labels allow buyers to compare the efficiency of products based on factual information about their absolute or relative performance.

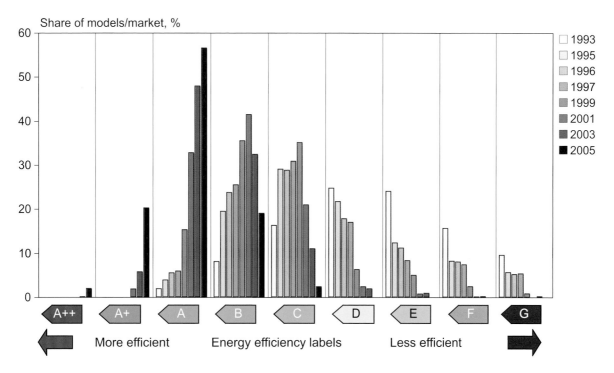

Figure 6.5: *The Impact of the EU Appliance Label (A++ to G, with G being the least efficient) on the Market of Cold Appliances in EU-25.*
Source: CECED, 2005.

can have the unintended consequence of overcoming cultural and other differences that affect consumer preferences, possibly leading to increased levels of energy consumption (Moezzi and Maithili, 2002; Biermayer and Lin, 2004).

6.8.2.2   Voluntary agreements

Voluntary agreements, in which the government and manufacturers agree to a mutually acceptable level of energy use per product, are being used in place of, or in conjunction with, mandatory MEPS to improve the energy efficiency of appliances and equipment. In the European context, this includes a wide range of industry actions such as industry covenants, negotiated agreements, long-term agreements, self-regulation, codes of conduct, benchmarking and monitoring schemes (Rezessy and Bertoldi, 2005). Voluntary measures can cover equipment, building design and operation and public, and private sector energy management policies and practices. Examples include Green Lights in the EU and the Energy Star programmes in the USA, as well as successful EU actions for the reduction of standby losses and efficiency improvement of washing machines and cold appliances. Industry often favours voluntary agreements to avoid the introduction of mandatory standards (Bertoldi, 1999). For the public authorities, voluntary agreements offer a faster approach than mandatory regulation and are often acceptable if they include the following three elements: (i) commitments by those manufacturers accounting for most of the equipment sold, (ii) quantified commitments to significant improvements in the energy efficiencies of the

equipment over a reasonable time scale, and (iii) an effective monitoring scheme (Commission of the European Communities, 1999). Voluntary agreements are considered especially useful in conjunction with other instruments and if mandatory measures are available as a backup or to encourage industry to deliver the targeted savings, such as for the case of cold appliances in the EU (Commission of the European Communities, 1999; Jäger-Waldau, 2004).

6.8.3   Cross-cutting policies and programmes that support energy efficiency and/or CO$_2$ mitigation in buildings

This section reviews a range of policies and programmes that do not focus specifically on either buildings and installed equipment, or on appliances and smaller plug-in devices in buildings, but may support energy efficiency and emissions reductions – including effects across other end-use sectors.

6.8.3.1   Utility demand-side management programmes

One of the most successful approaches to achieving energy efficiency in buildings in the USA has been utility-run demand-side management (DSM) programmes. However, there are important disincentives that need to be removed or lowered for utilities to be motivated in pursuing DSM programmes. The most important of these difficulties, (i.e., that utilities make profits from selling electricity, not from reducing sales) can be overcome by regulatory changes in which the utility will avoid revenue losses from reduced sales, and in some cases also receive profits from successful execution of DSM programmes.

The major large-scale experience with utility DSM has been in the United States primarily in the west coast and New England, but now spreading to other parts of the country. Spending on DSM was US$ 1.35 billion in 2003 (York and Kushler, 2005), and since California is more than doubling its expenditure to US$ 700 million/yr for the next three years, DSM spending in the United States will increase substantially.

These programmes have had a major impact. For the United States as a whole, where DSM investments have been 0.5% of revenues, savings are estimated to be 1.9% of revenues. For California, cumulative annual savings are estimated to be 7.5% of sales, while DSM investment has been less than 2% (1.2% in 2003). Overall, for each of the years 1996 through 2003, DSM has produced average annual savings of about 33.5 MtCO$_2$-eq annually for the USA, an annual net savings of more than US$ 3.7 billion (York and Kushler, 2005).

There are numerous opportunities to expand utility DSM programmes: in the United States, by having other states catch up with the leaders (especially California at present), much more so in Europe and other OECD countries, which have little experience with such programmes offered by utilities, and over time in developing countries, as well.

6.8.3.2 Energy prices, pricing schemes, energy price subsidies and taxes

Market-based energy pricing and energy taxes represent a broad measure for saving energy in buildings. The effect of energy taxes depends on energy price elasticity, that is the percent change in energy demand associated with each 1% change in price. In general, residential energy price elasticities are low in the richest countries. In the UK, long-run price elasticity for the household sector is only –0.19 (Eyre, 1998), in the Netherlands –0.25 (Jeeninga and Boots, 2001) and in Texas only –0.08 (Bernstein and Griffin, 2005). However, if energy expenditures reach a significant proportion of disposable incomes, as in many developing countries and economies in transition, elasticities – and therefore the expected impact of taxes and subsidy removal – may be higher, although literature is sparse on the subject. In Indonesia, price elasticity was –0.57 in the period from 1973 to 1990 and in Pakistan –0.33 (De Vita et al., 2006). Low elasticity means that taxes on their own have little impact; it is behavioural and structural barriers that need to be addressed (Carbon Trust, 2005). To have a significant impact on CO$_2$ emission reduction, excise taxes have to be substantial. This is only the case in a few countries (Figure 6.6): the share of excise tax compared to total fuel price differs considerably by country.

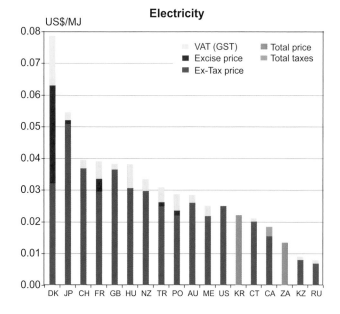
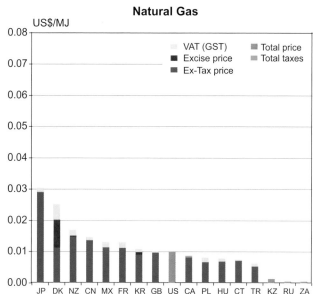

Figure 6.6: *Electricity and gas prices and taxes for households in 2004*

Notes: Total price is listed when no breakdown available to show taxes; total taxes are provided when no breakdown on excise and VAT (GST). Country name abbreviations (according to the ISO codes except Chinese Taipei): DK – Denmark, JP – Japan, CH – Switzerland, FR – France, GB – United Kingdom, HU – Hungary, TR – Turkey, PO – Poland, NZ – New Zealand, AU - Australia, MX – Mexico, US – United States of America, KR – South Korea, CT – Chinese Taipei, CA – Canada, ZA – South Africa*, KZ – Kazakhstan, RU – Russia. * South Africa data is for 2003.

Sources: IEA, 2006a; RAO, 2006.

In stark contrast to imposing energy taxes, energy prices are *subsidized* in many countries. This results in under-pricing of energy, which reduces the incentive to use it more efficiently. Energy subsidies are also typically much larger, per GJ, in developing and transition countries than in most industrial economies (Markandya, 2000). The total value of energy subsidies of eight of the largest non-OECD countries (China, Russia, India, Indonesia, Iran, South Africa, Venezuela and Kazakhstan), covering almost 60% of total non-OECD energy demand, was around US$ 95 billion in 1998 (UNEP OECD/IEA, 2002). In 1999, the IEA estimated that removing the energy subsidies in those eight countries would reduce primary energy use by 13%, lower CO_2 emissions by 16% and raise GDP by almost 1%.

While it may be economically and environmentally desirable, it is a socially sensitive task to remove end-user subsidies, especially in the residential sector. Since the bulk of these subsidies are found in countries with low incomes and high fuel-poverty rates, their removal can cause a substantial financial burden for families and even institutions. This, in turn, can lead to bankruptcy, increased payment arrears, energy theft and generally increased social tensions (ERRA/LGI, 2002; Ürge-Vorsatz et al., 2003), ultimately leading to disincentives to improve efficiency. Therefore, a drastic subsidy removal is often accompanied by social compensation programmes. One potentially important form of alternative compensation – although not frequently used to date – is assistance to low-income households to invest in energy-saving measures that reduce fuel costs and GHG emissions in the long term as opposed to direct cash assistance providing short-term relief (ERRA/LGI, 2002). For a number of years, the US government has provided 1.5–2.0 billion US$/yr to help low-income households pay their energy bills (LIHEAP, 2005), and smaller amounts budgeted for grants to 'weatherize' many of these same households with efficiency measures that help to permanently reduce monthly fuel and electricity bills (Schweitzer and Berry, 1999).

Some forms of energy subsidies can have positive energy and environmental effects. For example, subsidies on oil products and electricity in developing countries reduce deforestation and also reduce indoor pollution as poor, rural households switch away from traditional energy sources, such as wood, straw, crop residues and dung. These positive effects, however, can be better achieved through other means – e.g., the introduction of safe and efficient cookers and heaters utilizing these renewable sources. The challenge is to design and reform energy subsidies so they favour the efficient and environmentally sound use of energy systems (UNEP OECD/IEA, 2002)

6.8.3.3   Investment subsidies, financial incentives and other fiscal measures

As noted in Section 6.5.5, applying an integrated design process (IDP) can result in buildings that use 35–70% less energy than conventional designs, at little or no additional capital cost, but with a potential increase in the design cost. Providing financial incentives for the design process rather than financial incentives for the capital cost of the building is an approach used in several regions, such as by Canada in its Commercial Building Incentive Program (Larsson, 2001), by California in its Savings By Design programme and in Germany under the *SolarBau* programme (Reinhart et al., 2000).

Going beyond IDP, other measures – particularly those that include renewable energy options – entail significant added capital costs. Many developed countries offer incentives for such measures (IEA, 2004f). Types of financial support include subsidies, tax reduction (or tax credit) schemes and preferential loans or funds, with investment subsidies being the most frequently used (IEA, 2004f). Capital subsidy programmes and tax exemption schemes for both new construction and existing buildings have been introduced in nine OECD countries out of 20 surveyed (OECD, 2003). Several countries (USA, France, Belgium, UK and the Netherlands) combine their financial incentive policy for the existing building stock with social policy to assist low-income households (IEA, 2004a; VROM, 2006; USDOE, 2006). Increasingly, eligibility requirements for financial support are tied to CO_2 emission reduction (IEA, 2004a; KfW Group, 2006). Within the Energy Star Homes programme in the USA, houses that meet the energy-efficiency standard are eligible for a special mortgage (Nevin and Watson, 1998; Energystar, 2006). Financial incentives for the purchase of energy-efficient appliances are in place in some countries, including Mexico, the USA, Belgium, Japan and Greece (Boardman, 2004; IEA, 2004f). Incentives also encourage connection to district heating in Austria, Denmark and Italy.

There has been limited assessment of the efficiency of these schemes. The cost-effectiveness of subsidy-type schemes can vary widely, depending on programme design. Joosen et al. (2004) have estimated that subsidy programmes for residential buildings cost Dutch society 32–105 US$/tCO$_2$, whereas this range for the commercial sector was between 64 and 123 US$/tCO$_2$. A variety of financial incentives available simultaneously may make the decision process difficult; simplicity of the schemes might be an asset (Barnerjee and Solomon, 2003). A combination of government financial incentives and private bank loans may be more effective than a government-subsidized loan, as may combining building rating or labelling with a loan-especially when the labelling scheme has public approval.

6.8.3.4   Public sector leadership programmes and public procurement policies

Government agencies, and ultimately taxpayers, are responsible for a wide range of energy-consuming facilities and services such as government office buildings, schools and health care facilities. The government itself is often a country's largest consumer of energy and largest buyer of energy-using equipment. The US federal government spends over US$ 10 billion/yr for energy-using equipment (Harris and Johnson, 2000). Government policies and actions can thus contribute, both directly and indirectly, to energy savings and associated GHG reductions (Van Wie McGrory *et al.*, 2002). A recent study for several EU countries (Borg *et al.*, 2003) found a potential for direct energy savings of 20% or more in EU government facilities and operations. According to the USDOE's Federal Energy Management Program (FEMP), average energy intensity (site energy per square meter) in federal buildings has been reduced by about 25% since 1985, while average energy intensity in US commercial buildings has stayed roughly constant (USDOE/EERE, 2005; USDOE/FEMP, 2005).

Indirect beneficial impacts occur when Governments act effectively as market leaders. First, government buying power can create or expand demand for energy-efficient products and services. Second, visible government energy-saving actions can serve as an example for others. Public sector energy efficiency programmes fall into five categories (Harris *et al.*, 2005): (i) Policies and targets (energy/cost savings; CO_2 reductions); (ii) Public buildings (energy-saving retrofit and operation of existing facilities, as well as sustainability in new construction), (iii) Energy-efficient government procurement; (iv) Efficiency and renewable energy use in public infrastructure (transit, roads, water and other public services); and (v) Information, training, incentives and recognition of leadership by agencies and individuals. The following paragraphs provide selected examples.

The EU Directive on Energy Performance of Buildings discussed above and in Box 6.3, includes special requirements for public building certification. UK policy requires all new and refurbished government buildings to be rated under the British Research Establishment Environmental Assessment Method (BREEAM), which includes credits for energy efficiency and reduced CO_2 emissions. New government buildings must achieve a BREEAM rating of 'Excellent,' while major refurbishments require a 'Good' rating (UK/DEFRA, 2004). In the USA, a recent law requires new federal buildings to be designed 30% better for energy performance than that required by current commercial and residential building codes (U.S. Congress, 2005).

Energy-efficient government purchasing and public procurement can be powerful market tools. (Borg *et al.*, 2003; Harris *et al.*, 2004). Energy-efficient government procurement policies are in place in several EU countries, as well as in Japan, Korea, Mexico, China and the USA (Harris *et al.*, 2005). In the USA, in 2005, Congress passed a law mandating that all federal agencies specify and buy efficient products that qualify for the Energy Star label, or (in cases where that label does not apply) products designated by USDOE/FEMP as being among the top 25th percentile of efficient products (US Congress, 2005). Federal purchasing policies are expected to save 1.1 million tonnes CO_2-eq and US$ 224 million/yr in 2010 (Harris and Johnson, 2000).

Public procurement policies can have their greatest impact on the market when they are based on widely harmonized energy-efficiency specifications that can send a strong market signal to manufacturers and suppliers (Borg *et al.*, 2003). If US agencies at all levels of government adopt the federal efficiency criteria for their own purchases, estimated annual electricity savings in the USA would be 10.8 million tonnes CO_2-eq, allowing for at least one billion US$/yr savings on public energy bills (Harris and Johnson, 2000).

6.8.3.5 Promotion of energy service companies (ESCOs) and energy performance contracting (EPC)

While not a 'policy instrument', ESCOs have become favoured vehicles to deliver energy-efficiency improvements and are promoted by a number of policies. An ESCO is a company that offers energy services, such as energy analysis and audits, energy management, project design and implementation, maintenance and operation, monitoring and evaluation of savings, property/facility management, energy and/or equipment supply and provision of energy services (e.g., space heating, lighting). ESCOs guarantee the energy savings and/or the provision of a specified level of energy service at lower cost by taking responsibility for energy-efficiency investments or/ and improved maintenance and operation of the facility. This is typically executed legally through an arrangement called 'energy performance contracting' (EPC). In many cases, the ESCO's compensation is directly tied to the energy savings achieved. ESCOs can also directly provide or arrange for project financing, or assist with financing by providing an energy (cost) savings guarantee for their projects. Finally, ESCOs often retain an ongoing operational role, provide training to on-site personnel, and take responsibility for measuring and verifying the savings over the term of the project loan.

In 2006, the US ESCO market is considered the most advanced in the world (Goldman *et al.*, 2005), with revenues reaching about US$ 2 billion in 2002 (Lin and Deng, 2004). Most US ESCO activity (approximately 75%) is in the public sector. The market for energy-efficiency services in Western Europe was estimated to be € 150 million/yr in 2000, while the market potential was estimated at € 5–10 billion/yr (Butson, 1998; Bertoldi and Starter, 2003). Germany and Austria are the ESCO leaders in Europe, with street-lighting projects among the most common demand-side EPC projects, and public buildings the most targeted sector (Bertoldi *et al.*, 2005; Rezessy *et al.*, 2005). Between 1998 and 2003, 600–700 public buildings were renovated in Austria using energy performance contracting by ESCOs. Austria is now using EPCs to renovate 50% of the total floor area of federal buildings (Leutgöb, 2003). In Germany, more than 200 EPCs have been signed since the mid-1990s, primarily for public buildings (Seefeldt, 2003). In Japan, the ESCO market is growing quickly, with a focus on the commercial and public sectors (office buildings and hospitals) (Murakoshi and Nakagami, 2003). In India and Mexico, ESCOs also have targeted at least 50% of their activity in the public and commercial sectors (Vine, 2005). Most ESCOs do not target the residential sector, although exceptions exist (e.g., in Nepal and South Africa).

ESCOs greatly facilitate the access of building owners and operators to technical expertise and innovative project financing. They can play a central role in improving energy efficiency without burdening public budgets and regulatory intervention to markets. However, the ESCO industry does not always develop on its own and policies and initiatives may be necessary to kick-start the market. The commitment of federal and municipal authorities to use ESCOs for their energy-efficiency projects, along with supportive policies and public-private partnerships has been crucial in countries such as Germany and Austria (Brand and Geissler, 2003). In some cases, obligations imposed on electricity companies have fostered the development of ESCO activities, as in the case of Brazil, where power utilities are required to invest 1% of their net operating revenues in energy efficiency.

6.8.3.6   Energy-efficiency obligations and tradable energy-efficiency certificates

Recognising that traditional energy policy tools have not achieved the magnitude of carbon savings needed to meet climate stabilization targets, a few new innovative instruments are being introduced or planned in a number of countries. Among them are the so-called 'white certificates', a cap-and-trade scheme (or, in some cases, an obligation without the trading element) applied to achieve energy efficiency improvements. The basic principle is an obligation for some category of economic actors (e.g., utility companies, product manufacturers or distributors and large consumers) to meet specified energy savings or programme-delivery goals, potentially coupled with a trading system based on verified and certified savings achieved (or expected) for energy-efficiency measures (the 'white' certificate) (ECEEE, 2004; Oikonomou et al., 2004). Energy efficiency obligation programmes without certificate trading have been operating in the UK since 1994 and in Flanders (Belgium) since 2003; white certificate schemes with a trading element were in place in 2006 in Italy, France and New South Wales. Other European countries have announced their intention to introduce similar schemes.

Capturing the desired benefit of certificate trading schemes – that is minimising the costs of meeting energy savings goals – depends on the liquidity of the market. There is a trade-off between liquidity, crucial to minimizing the costs, and manageability and transaction costs. Where transaction costs turn out to be very high, a simple energy savings obligation for electricity and gas distributors, without the complication of trading, may be a better way to deliver the desired outcome (Bertoldi and Rezessy, 2006). Since the first white certificate schemes are just starting, it remains to be seen whether this policy instrument will deliver the expected level of savings and at what cost.

In the UK, the Energy Efficiency Commitment (EEC) requires that all large gas and electricity suppliers deliver a certain quantity of energy savings by assisting customers to take energy-efficiency actions in their homes. The delivered overall savings of the first phase, 87 TWh, largely exceeded the target of 65 TWh and the target has since been increased to 130.2 TWh (Lees, 2006).

6.8.3.7   The Kyoto Protocol's Flexibility Mechanisms

The flexibility mechanisms of the Kyoto Protocol (KP), especially the clean development mechanism (CDM) and joint implementation (JI), could offer major benefits for buildings in developing countries and economies in transition, in terms of financing, transfer of advanced technologies and know-how, building of local capacity and demonstration effects (Woerdman, 2000; Grubb et al., 2002). Buildings should be prime targets for project-based mechanisms due to the variety and magnitude of cost-effective potentials (see section 6.5). For instance, Trexler and Associates (Margaree Consultants, 2003) estimated that building and appliance efficiency accounts for 32% of total potential in CDM in 2010 under 0 US$/tCO$_2$ and 20% under 20 US$/tCO$_2$. However, evidence until 2006 shows that little of this potential is expected to be unlocked during the first commitment period (Novikova et al., 2006). After initial enthusiasm in the activities implemented jointly (AIJ) phase, where 18 out of 156 registered projects were targeted to buildings, JI and CDM experience to date suggests that this pilot phase brought disappointment in building-related projects. As of February 2006, only four CDM projects out of 149 projects registered or seeking validation were for buildings, and none of the 152 approved and submitted JI projects was due to invest in buildings (Novikova et al., 2006).

While it is too early to conclude that the Kyoto Protocols's project-based mechanisms do not work well for buildings, there are no indications that this trend will reverse. A number of barriers prevent these mechanisms from fully mobilizing their benefits for buildings (Tangen and Heggelund, 2003; ECON Analysis, 2005). Chief among these is the proportionately high transaction costs due to the relatively small size of building-related projects: although these costs are around 100 €/tCO$_2$ (124 US$/tCO$_2$) for building-related projects, they amount only to 0.1 €/tCO$_2$ (0.12 US$/tCO$_2$) for very large-scale projects (Michaelowa and Jotzo, 2005). While a few hypothetical solutions have been suggested to overcome the barriers (Novikova et al., 2006), their implementation is uncertain. Another major chance opens for buildings in former communist countries with large emission surpluses through Green Investment Schemes, or the 'greening' of these surplus emission units, if they are constructed to accommodate small-scale energy-efficiency investments better than CDM or JI, potentially delivering over a billion tonnes of real CO$_2$ reductions.

In summary, if the KP is here to stay, the architecture of the flexible mechanisms could be revisited to address these shortcomings, so that the major opportunities from buildings in developing countries and EITs do not stay unutilised. A potential criterion for appraising climate regimes – in terms of their success in leveraging lowest costs mitigation options, as well as in meeting sustainable development goals – could be their success in promoting buildings-level investments in developing countries and economies in transition, reflecting their recognized importance in minimized-cost global emission mitigation efforts.

6.8.3.8 *Technology research, development, demonstration and deployment (RD&D)*

Section 6.4 attested that there is already a broad array of accessible and cost-effective technologies and know-how that can abate GHG emissions in existing and new buildings to a significant extent that have not been widely adopted yet. At the same time, several recently developed technologies, including high performance windows, active glazing, vacuum insulated panels, phase change materials to increase building thermal mass, high performance reversible heat pumps and many other technologies may be combined with integrated passive solar design and result in up to 80% reduction of building energy consumption and GHG emissions. Large-scale GHG reduction in buildings requires fast and large-scale dissemination and transfer in many countries, including efficient and continuous training of professionals in the integrated approach to design and optimized use of combinations of technologies. Integrated intelligent building control systems, building- or community-level renewable energy generation, heat and coldness networks, coupled to building renewable energy capture components and intelligent management of the local energy market need more research, development and demonstration, and could develop significantly in the next two decades.

Between 1996 and 2003 the annual worldwide RD&D budget for energy efficiency in buildings has been approximately US$ 225–280 million/yr (IEA, 2004d). The USA has been the leading country in energy research and development for buildings for over a decade. Despite the decline in US funds by 2/3rd between 1993 and 2003, down from a peak of US$ 180 million, the USA is still responsible for half of the total global expenditures (IEA, 2004d). Substantial buildings-related energy-efficiency RD&D is also sponsored in Japan (15% of global expenditure).

The overall share of energy-efficiency in total energy RD&D expenditure is low, especially compared to its envisioned role in global GHG mitigation needs. In the period from 2001 to 2005 on average only 14% of all energy RD&D expenditure in IEA countries has been designated for energy-efficiency improvement (IEA, 2006c), whereas its contribution to CO_2 emission reduction needs by 2050 is 45% according to the most commonly used 'Map' scenario of the IEA (2006d). The share dedicated to energy efficiency improvements in buildings was only 3%, in stark contrast with their 18% projected role in

the envisioned necessary 32 Gt global CO_2 reduction by 2050 (IEA, 2006d).

6.8.4 Policies affecting non-CO_2 gases

In the buildings sector, non-CO_2 greenhouse gases (halocarbons) are used as the working fluid in most vapour-compression cooling equipment, and as a blowing agent in some insulation foams including polyurethane spray foam. Background in this report is in Section 6.4.15, which is in turn a brief summary of IPCC/TEAP (2005).

6.8.4.1 *Stationary refrigeration, air conditioning and heat pump applications*

A number of countries have established legislative and voluntary regimes to control emissions and use of fluorinated gases. In Europe, a number of countries have existing policies that aim at reducing leakage or discouraging the use of refrigerants containing fluorine. Regulations in the Netherlands minimize leakage rates through improved maintenance and regular inspection. Substantial taxes for refrigerants containing fluorine are levied in Scandinavian countries, and legislation in Luxembourg requires all new large cooling systems to use natural refrigerants (Harmelink *et al.*, 2005). Some countries such as Denmark and Austria have banned the use of HFCs in selected air conditioning and refrigeration applications. In 2006 the EU Regulation 842/2006 entered into force, which requires that all medium and large stationary air conditioning applications in the EU will use certified and trained service personnel, and assures recovery of refrigerants at the end-of-life (Harmelink *et al.*, 2005).

In the USA, it has been illegal under the Clean Air Act since 1995, to vent substitutes for CFC and HCFC refrigerants during maintenance, repair and disposal of air conditioning and refrigeration equipment (US EPA, 2006). Japan, has established a target to limit HFC, PFC and SF6 emissions. Measures to meet this target include voluntary action plans by industries, mandatory recovery systems for HFCs used as refrigerants (since April 2002) and the research and development of alternatives (UNFCCC, 2006). Australia has developed an Ozone Protection and Synthetic Greenhouse Gas Management Act. Measures include supply controls though the licensing of importers, exporters and manufacturers of fluorinated gases and pre-charged refrigeration and air conditioning equipment; end-use regulations on handling, use, recovery, sale and reporting are in place (Australian Government, 2006). Canada has established a National Action Plan for the Environmental Control of ODS and their Halocarbon Alternatives (NAP). This ensures that HFCs are only used in applications where they replace ODS and requires recovery, recycling and reclamation for CFCs, HCFCs and HFCs (Canadian Council of Ministers of the Environment, 2001).

6.8.4.2 *Insulating foams and SF₆ in sound-insulating glazing*

Within the European Union, Denmark and Austria have introduced legislation to ban the use of HFC for the production of several foam types (Cheminfo, 2004). Since 2006 the EU Regulation 842/2006 on certain Fluorinated Gases limits emissions and certain uses of fluorinated gases (European Commission, 2006), banning the use of HFCs in One-Component Foam from 2008, except where required to meet national safety standards. Japan has established a target to limit HFC, PFC and SF6 emissions. Measures to meet this target include voluntary action plans by industries, improved containment during the production process, less blowing agent per product, improved productivity per product and the use of non-fluorocarbon low GWP alternatives. Australia has developed an act for industries covered by the Montreal Protocol and extended voluntary arrangements for non-Montreal Protocol industries. Measures include supply controls though the licensing of importers, exporters and manufacturers of HFCs.

Although there are no international proposals to phase out the use of HFCs in foams, the high costs of HFCs have naturally contributed to the minimization of their use in formulations (often by use with co-blowing agents) and by early replacement by alternative technologies based primarily on CO_2, water or hydrocarbons (e.g. pentane). There is more regulatory uncertainty at regional level and in Japan some pressure exists to stop HFC-use in the foam sector. In Europe, the recently published F-Gas regulation (European Commission, 2006) only impacts the use of HFCs in one component foam (OCF) which is used primarily for gap filling in the construction sector. However, there is a requirement to put in place provisions for recovery of blowing agent at end-of-life where such provisions are technically feasible and do not entail disproportionate cost.

6.8.5 Policy options for GHG abatement in buildings: summary and conclusion

Section 6.8 demonstrates that there is a variety of government policies, programmes, and market mechanisms in many countries for successfully reducing energy-related CO_2 emissions in buildings (*high agreement, medium evidence*). Table 6.6 (below) reviews 20 of the most important policy tools used in buildings according to two criteria from the list of criteria suggested in Chapter 13 (of the ones for which literature was available in policy evaluations): emission reduction effectiveness and cost-effectiveness. Sixty-six ex post (with a few exceptions) policy evaluation studies were identified from over 30 countries and country groups that served as a basis for the assessment.

The first column in Table 6.6 identifies the key policy instruments grouped by four major categories using a typology synthesized from several sources including Grubb (1991); Crossley *et al.* (2000) and Verbruggen and Bongaerts (2003): (i) control and regulatory mechanisms, (ii) economic and market-based instruments, (iii) financial instruments and incentives, and (iv) support and information programmes and voluntary action. The second column identifies a selection of countries where the policy instrument is applied[20]. Then, the effectiveness in achieving CO_2 reduction and cost-effectiveness were rated qualitatively based on available literature as well as quantitatively based on one or more selected case studies. Since any instrument can perform poorly if not designed carefully, or if its implementation and enforcement are compromised, the qualitative and quantitative comparisons are based on identified best practices, in order to demonstrate what impact an instrument can achieve if applied well. Finally, the table lists special conditions for success, major strengths and limitations, and co-benefits.

While the 66 studies represent the majority of such evaluations available in the public domain in 2006, this sample still leaves few studies in certain categories. Therefore, the comparative findings of this assessment should be viewed as indicative rather than conclusive. Although a general caveat of comparative policy assessments is that policies act as parts of portfolios and therefore the impact of an individual instrument is difficult to delineate from those of other tools, this concern affects the assessment to a limited extent since the literature used already completed this disaggregation before evaluating individual instruments.

All the instruments reviewed can achieve significant energy and CO_2 savings; however the costs per tonne of CO_2 saved diverge greatly. In our sample, appliance standard, building code, labelling and tax exemption policies achieved the highest CO_2 emission reductions. Appliance standards, energy efficiency obligations, demand-side management programmes, public benefit charges and mandatory labelling were among the most cost-effective policy tools in the sample, all achieving significant energy savings at negative costs. Investment subsidies (as opposed to rebates for purchases of energy efficient appliances) were revealed as the least cost-effective instrument. Tax reductions for investments in energy efficiency appeared more effective than taxation. Labelling and voluntary programmes can lead to large savings at low-costs if they are combined with other policy instruments. Finally, information programmes can also achieve significant savings and effectively accompany most other policy measures.

---

20 Since we made a strong effort to highlight best practices from developing countries where possible, major front-running developed countries where the instrument is applied may not be listed in each applicable row of the table.

Table 6.6: *The impact and effectiveness of various policy instruments aimed to mitigate GHG emission in the buildings sector*

Policy instrument[a]	Examples of countries	Effectiveness[b]	Energy or emission reductions for selected best practices[b]	Cost-effectiveness	Cost of GHG emission reduction for selected best practices[c]	Special conditions for success, major strengths and limitations, co-benefits	References
Control and regulatory mechanisms							
Appliance standards	EU, US, JP, AU, BR, CN	High	JP: 31 M tCO$_2$ in 2010; CN: 240 MtCO$_2$ in 10 yrs; US: 2.5% of electricity use in 2000 = 65 MtCO$_2$, 6.5% = 223.87 MtCO$_2$ in 2010.	High	AU: −15 $/tCO$_2$ in 2012; US: −65 $/tCO$_2$ in 2020; EU: −194 $/tCO$_2$ in 2020.	Factors for success: periodical update of standards, independent control, information, communication and education.	IEA, 2005; Schlomann et al. 2001; Gillingham et al., 2004; ECS, 2002; World Energy Council, 2004; Australian Greenhouse Office, 2005; IEA 2003a; Fridley and Lin, 2004.
Building codes	SG, PH, DZ, EG, US, GB, CN, EU	High	HK: 1% of total electricity saved; US: 79.6 MtCO$_2$ in 2000; EU: 35–45 MtCO$_2$, max 60% energy savings in new buildings.	Medium	NL: from −189 $/tCO$_2$ to −5 $/tCO$_2$ for end-users, 46–109 $/tCO$_2$ for society.	No incentive to improve beyond target. Only effective if enforced.	World Energy Council, 2001; Lee & Yik, 2004; Schaefer et al., 2000; Joosen et al., 2004; Geller et al., 2006; ECCP, 2001.
Procurement regulations	US, EU, CN, MX, KR, JP	High	MX: 4 cities saved 3.3 ktCO$_2$-eq in one year; CN: 3.6 MtCO$_2$ expected; EU: 20–44 MtCO$_2$ potential.	Medium	MX: $1Million in purchases saves $726,000/yr; EU: <21 $/tCO$_2$.	Success factors: enabling legislation, energy efficiency labelling & testing, ambitious energy efficiency specifications.	Borg et al., 2003; Harris et al., 2005; Van Wie McGrory et al., 2006.
Mandatory labelling and certification programmes	US, CA, AU, JP, MX, CN, CR, EU	High	AU: 5 M tCO$_2$ savings 1992–2000; DK: 3.568 MtCO$_2$.	High	AU: −30 $/tCO$_2$ abated.	Effectiveness can be boosted by combination with other instrument and regular updates.	World Energy Council, 2001; OPET network, 2004; Holt & Harrington, 2003.
Energy efficiency obligations and quotas	GB, BE, FR, IT, DK, IE	High	GB: 1.4 MtCO$_2$/yr.	High	Flanders: −216 $/tCO$_2$ for households, −60 $/tCO2 for other sector in 2003; GB: −139 $/tCO$_2$.	Continuous improvements necessary: new energy efficiency measures, short-term incentives to transform markets etc.	UK government, 2006; Sorell, 2003; Lees, 2006; Collys, 2005; Bertoldi & Rezessy, 2006; Defra, 2006.
Utility demand-side management programmes	US, CH, DK, NL, DE, AT	High	US: 36.7 MtCO$_2$ in 2000.	High	US: Average costs approx. −35 $/tCO$_2$.	DSM programmes for commercial sector tend to be more cost-effective than those for residences.	IEA, 2005; Kushler et al., 2004.
Economic and market-based instruments							
Energy performance contracting	DE, AT, FR, SE, FI, US, JP, HU	High	FR, SE, US, FI: 20–40% of buildings energy saved; EU:40–55MtCO$_2$ by 2010; US: 3.2 MtCO$_2$/yr.	Medium	EU: mostly at no cost, rest at <22 $/tCO2; US: Public sector: B/C ratio 1.6, Priv. sector: 2.1	Strength: no need for public spending or market intervention, co-benefit of improved competitiveness.	ECCP, 2003; OPET network, 2004; Singer, 2002; IEA, 2003a; World Energy Council, 2004; Goldman et al., 2005.

Table 6.6. Continued.

Policy instrument[a]	Examples of countries	Effectiveness[b]	Energy or emission reductions for selected best practices	Cost-effectiveness	Cost of GHG emission reduction for selected best practices[c]	Special conditions for success, major strengths and limitations, co-benefits	References
Co-operative procurement	DE, IT, GB, SE, AT, IE, JP, PO, SK, CH	High	Varies, German telecom company: up to 60% energy savings for specific units.	High	0: Energy-efficient purchasing relies on funds that would have been spent anyway.	Success condition: energy efficiency needs to be prioritized in purchasing decisions.	Oak Ridge National Laboratory, 2001; Le Fur 2002; Borg et al., 2003.
Energy efficiency certificate schemes	IT, FR	Medium	IT: 3.64 Mt CO_2 eq by 2009 expected.	Medium	n.a.	No long-term experience yet. Transaction costs can be high. Monitoring and verification crucial. Benefits for employment.	OPET network, 2004; Bertoldi & Rezessy, 2006; Lees, 2006; Defra, 2006.
Kyoto Protocol flexible mechanisms[d]	CN, TH, CEE (JI & AIJ)	Low	CEE: 220 K tCO_2 in 2000.	Low	63 $/$tCO_2$.	So far limited number of CDM & JI projects in buildings.	ECS, 2005; Novikova. et al., 2006.
Financial instruments and incentives							
Taxation (on CO_2 or household fuels)	NO, DE, GB, NL, DK, CH	Low	DE: household consumption reduced by 0.9%.	Low		Effect depends on price elasticity. Revenues can be earmarked for further efficiency. More effective when combined with other tools.	World Energy Council, 2001; Kohlhaas, 2005.
Tax exemptions / reductions	US, FR, NL, KO	High	US: 88 $MtCO_2$ in 2006.	High	Overall B/C ratio – Commercial buildings: 5.4 – New homes: 1.6.	If properly structured, stimulate introduction of highly efficient equipment and new buildings.	Quinlan et al., 2001; Geller & Attali, 2005.
Public benefit charges	BE, DK, FR, NL, US states	Medium/ low	US: 0.1–0.8% of total electricity sales saved /yr, average of 0.4%.	high in reported cases	From –53 US$/ tCO_2 to –17 $/$tCO_2$.		Western Regional Air Partnership, 2000; Kushler et al., 2004.
Capital subsidies, grants, subsidized loans	JP, SI, NL, DE, CH, US, HK, GB	High	SI: up to 24% energy savings for buildings, GB: 3.3 $MtCO_2$; US:29.1 Mio BTU/yr gas savings.	Low	NL: 41–105 US$/ tCO_2 for soc; GB:29 US$/$tCO_2$ for soc, –66 $/$tCO_2$ for end-user.	Positive for low-income households, risk of free-riders, may induce pioneering investments.	ECS, 2001; Martin et al., 1998; Schaefer et al., 2000; Geller et al., 2006; Berry & Schweitzer, 2003; Joosen et al., 2004; Shorrock, 2001.
Support, information and voluntary action							
Voluntary certification and labelling	DE, CH, US, TH, BR, FR	Medium/ high	BR: 169.6 $ktCO_2$ in 1998, US: 13.2 $MtCO_2$ in 2004, 2.1 bio tCO_2-eq in total by 2010; TH: 192 tCO_2.	High	BR: US$ 20 million saved.	Effective with financial incentives, voluntary agreements and regulations.	OPET network, 2004; Word Energy Council, 2001; Geller et al., 2006; Egan et al., 2000; Webber et al., 2003.

Table 6.6. Continued.

Policy instrument[a]	Applicability	Effectiveness[b]	Energy or emission reductions for selected best practices	Cost-effectiveness	Cost of GHG emission reduction for selected best practices[c]	Special conditions for success, major strengths and limitations, co-benefits	References
Voluntary and negotiated agreements	Mainly Western Europe, JP, US	Medium/High	US: 88 $MtCO_2$-eq/yr UK: 15.8 $MtCO_2$	Medium	GB: 54.5–104 US$/$tCO_2$ (Climate Change Agreements).	Can be effective when regulations are difficult to enforce. Effective if combined with financial incentives and threat of regulation.	Geller et al., 2006; Cottrell, 2004.
Public leadership programmes	NZ, MX, PH, AR, BR, EC	High	De: 25% public sector CO_2 reduction over 15 years.	High	US DOE/FEMP estimates 4 US$ savings for every 1 US$ of public funds invested.	Can be used to demonstrate new technologies and practices. Mandatory programmes have higher potential than voluntary ones.	Borg et al., 2003; Harris et al., 2005; Van Wie McGrory et al., 2006; OPET, 2004.
Awareness raising, education / information campaigns	DK, US, GB, CA, BR, JP	Low/ Medium	GB: Energy Efficiency Advice Centres: 10.4 K tCO_2 annually.	High	BR: –66 US$/ tCO_2; GB: 8 US$/ tCO_2 (for all programmes of Energy Trust).	More applicable in residential sector than commercial.	Bender et al., 2004; Dias et al., 2004; Darby, 2006; IEA, 2005; Lutzenhiser, 1993; Ueno et al. 2006; Energy Saving Trust, 2005.
Mandatory audit & energy management requirement	US, FR, NZ, EG, AU, CZ	High, but variable	US: Weatherization Program: 22% saved in weatherized households.	Medium	US Weatherization Program: BC-ratio: 2.4.	Most effective if combined with other measures such as financial incentives	World Energy Council, 2001
Detailed billing and disclosure programmes	ON, IT, SE, FI, JP, NO, CL	Medium	Up to 20% energy savings.	Medium	n.a.	Success conditions: combination with other measures and periodic evaluation. Comparability with other households is positive.	Crossley et al., 2000; Darby 2000; Roberts & Baker, 2003; Energywatch, 2005.

Notes:
Country name abbreviations (according to the ISO codes except California, Ontario, Central and Eastern Europe and European Union): DZ – Algeria, AR – Argentina, AU – Australia, AT – Austria, BE – Belgium, BR – Brazil, CL – California, CA – Canada, CEE – Central and Eastern Europe, CN – China, CR – Costa Rica, CZ – Czech Republic, DE – Germany, Denmark – DK, EC – Ecuador, EG – Egypt, EU – European Union, FI – Finland, FR – France, GB – United Kingdom, HK – Hong Kong, HU – Hungary, IN – India, IE – Ireland, IT – Italy, JP – Japan, KR – Korea (South), MX – Mexico, NL – Netherlands, NO – Norway, ON – Ontario, NZ – New Zealand, NG – Nigeria, PH – Philippines, PO – Poland, SG – Singapore, SK – Slovakia, SI – Slovenia, CH – Switzerland, SE – Sweden, TH – Thailand, US – United States.
a) For definitions of the instruments see: Crossley et al. (1999), Crossley et al. (2000), Vine et al. (2003) and Wuppertal Institute (2002).
b) Effectiveness of CO2 emission reduction: includes ease of implementation; feasibility and simplicity of enforcement; applicability in many locations; and other factors contributing to overall magnitude of realized savings.
c) Cost-effectiveness is related to specific societal cost per unit of carbon emissions avoided. Energy savings were recalculated into emission savings using the following references for the emission factors: Davis (2003), UNEP (2000), Center for Clean Air Policy (2001). The country-specific energy price was subtracted from the cost of saved energy in order to account for the financial benefits of energy savings (Koomey and Krause, 1989), if they were not considered originally.
d) Kyoto flexible mechanisms: Joint Implementation (JI), Clean Development Mechanism (CDM), International Emissions Trading (includes the Green Investment Schemes).

The effectiveness of economic instruments, information programmes and regulation can be substantially enhanced if these are appropriately combined into policy packages that take advantage of synergistic effects (Ott *et al.*, 2005). A typical example is the co-ordination of energy audit programmes with economic instruments, such as energy taxes and capital subsidy schemes. In addition, ESCOs can flourish when public procurement legislation accommodates EPCs and includes ambitious energy-efficiency or renewable energy provisions, or in the presence of an energy-saving obligation.

Section 6.8 demonstrates that, during the last decades, many new policies have been initiated. However, so far only incremental progress has been achieved by these policies. In most developed countries, the energy consumption in buildings is still increasing (IEA, 2004f). Although some of this growth is offset by increased efficiency of major energy-consuming appliances, overall consumption continues to increase due to the growing demand for amenities, such as new electric appliances and increased comfort. The limited overall impact of policies so far is due to several factors: (i) slow implementation processes (e.g., as of 2006, not all European countries are on time with the implementation of the EU Buildings Directive); (ii) the lack of regular updating of building codes (requirements of many policies are often close to common practices, despite the fact that CO_2-neutral construction without major financial sacrifices is already possible) and appliance standards and labelling; and (iii) insufficient enforcement. In addition, Section 6.7 demonstrated that barriers in the building sector are numerous; diverse by region, sector and end-user group, and are especially strong.

There is no single policy instrument that can capture the entire potential for GHG mitigation. Due to the especially strong and diverse barriers in the residential and commercial sectors, overcoming these is only possible through a diverse portfolio of policy instruments for effective and far-reaching GHG abatement and for taking advantage of synergistic effects. Since climate change literacy, awareness of technological, cultural and behavioural choices and their impacts on emissions are important preconditions to fully operating policies, these policy approaches need to go hand in hand with programmes that increase consumer access to information, awareness and knowledge *(high agreement, medium evidence)*.

In summary, significant CO_2 and other GHG savings can be achieved in buildings, often at net benefit to society (in addition to avoided climate change) and also meeting many other sustainable development and economic objectives, but this requires a stronger political commitment and more ambitious policy-making than today, including careful design of policies as well as enforcement and regular monitoring.

6.9 Interactions of mitigation options with vulnerability, adaptation and sustainable development

6.9.1 Interactions of mitigation options with vulnerability and adaptation

In formulating climate change strategies, mitigation efforts need to be balanced with those aimed at adaptation. There are interactions between vulnerability, adaptation and mitigation in buildings through climatic conditions and energy systems. As a result of a warming climate, heating energy consumption will decline, but energy demand for cooling will increase while at the same time passive cooling techniques will become less effective. The net impact of these changes on GHG emissions is related to the available choice of primary energy used and the efficiency of technologies that are used for heating and cooling needs. Mansur *et al.* (2005) find that the combination of climate warming and fuel switching in US buildings from fuels to electricity results in increases in the overall energy demand, especially electricity. Other studies indicate that in European countries with moderate climate the increase in electricity for additional cooling is higher than the decrease for heating demand in winter (Levermore *et al.*, 2004; Aebischer *et al.*, 2006; Mirasgedis *et al.*, 2006). Aebischer *et al.* (2006) finds that in Europe there is likely to be a net increase in power demand in all but the most northerly countries, and in the south a significant increase in summer peak demand is expected. Depending on the generation mix in particular countries, the net effect on carbon dioxide emissions may be an increase even where overall demand for final energy declines. Since in many countries electricity generation is largely based on fossil fuels, the resulting net difference between heating reduction and cooling increases may significantly increase the total amount of GHG emissions. This causes a positive feedback loop: more mechanical cooling emits more GHGs, thereby exacerbating warming, although the effect maybe moderate.

Vulnerability of energy demand to climate is country- and region-specific. For instance, a temperature increase of 2°C is associated with an 11.6% increase in residential per capita electricity use in Florida, but with a 7.2% decrease in Washington DC (Sailor, 2001). Increased net energy demand translates into increased welfare losses. Mansur *et al.* (2005) found that, for a 5°C increase in temperature by 2100, the annual welfare loss in increased energy expenditures is predicted to reach US$ 40 billion for US households.

Fortunately, there are many potential synergies where investments in the buildings sector may reduce the overall cost of climate change-in terms of both mitigation and adaptation. For instance, if new buildings are constructed, the design can address both mitigation and adaptation aspects. Among the most important of these are reduced cooling loads. For instance, using advanced insulation techniques and passive solar design to reduce the expected increase in air conditioning load. In addition, if high-efficiency electric appliances are used, the savings are increased due to reduced electricity demand for air conditioning, especially in commercial buildings. Roof retrofits can incorporate increased insulation and storm security in one investment. In addition, the integrated design of well-insulated, air-tight buildings, with efficient air management and energy systems, leads not only to lower GHG emissions, but also to reduced thermal stress to occupants, reducing extreme weather-related mortality and other health effects. Furthermore, adaptive comfort, where occupants accept higher indoor (comfort) temperatures when the outside temperature is high, is now incorporated in design considerations, especially for predominantly naturally ventilated buildings (see Box 6.5).

Policies that actively promote integrated building solutions for both mitigating and adapting to climate change are especially important for the buildings sector. It has been observed that building users responding to a warmer climate generally choose options that increase cooling energy consumption rather than other means, such as insulation, shading, or ventilation, which consume less energy. A prime example of this is the tendency of occupants of existing, poorly performing buildings (mainly in developing countries) to buy portable air conditioning units. These trends – which clearly will accelerate in warmer summers to come – may result in a significant increase of GHG emissions from the sector, enhancing the positive feedback process. However, well-designed policies supporting less energy-intensive cooling alternatives can help combat these trends (see Box 6.5 and Section 6.4.4.1). Good urban planning, including increasing green areas as well as cool roofs in cities, has proven to be an efficient way to limit the heat island effect, which also aggravates the increased cooling needs (Sailor, 2002).

6.9.2 Synergies with sustainability in developing countries

The failure of numerous development strategies in the least developed countries, most of them in Africa, to yield the expected results has been attributed to the fact that the strategies failed to address the core needs of such countries – these are economic growth, poverty alleviation and employment creation (OECD, 2001). Often a tension exists between the main agenda of most of these countries (poverty alleviation through increased access to energy) and climate change concerns. Increased access to modern energy for the mostly rural population has been a priority in recent years. Most countries, therefore, place more policy emphasis on increasing the supply of petroleum and electricity than on renewables or energy efficiency (Karakezi and Ranja, 2002). The success of climate change mitigation policies depends largely on the positive management of these tensions. GHG reduction strategies in developing countries have a higher chance of success if they are 'embedded' in poverty eradication efforts, rather than executed independently.

Fortunately, buildings offer perhaps the largest portfolio of options where such synergies can be identified. Matrices in Chapter 12 demonstrate that the impact of mitigation options in the building sector on sustainable development, for both industrialized countries and developing countries, is reported to be positive for all of the criteria used. Both Sections 6.6 above and Box 6.1 discuss many of the opportunities for positive synergies in detail; the next paragraph revisits a few of them.

The dual challenges of climate change and sustainable development were strongly emphasised in the 2002 Millennium Development Goals (MDGs). GHG mitigation strategies are more realizable if they work mutually with MDGs towards the realization of these set objectives. For example, MDG goal seven is to ensure sustainable development, in part by reducing the proportion of people using solid fuels which will lead to the reduction of indoor air pollution (see sections 6.6.1). GHG mitigation and public health are co-benefactors in the achievement of this goal. Similarly, increased energy efficiency in buildings, or considering energy efficiency as the guiding principle during the construction of new homes, will result in both reduced energy bills – enhancing the affordability of increased energy services – and GHG abatement. If technologies that utilise locally available renewable resources in an efficient and clean way are used broadly, this provides access to 'free'

Box 6.5: Mitigation and adaptation case study: Japanese dress codes

In 2005, the Ministry of the Environment (MOE) in Japan widely encouraged businesses and the public to set air conditioning thermostats in offices to around 28°C during summer. As a part of the campaign, MOE has been promoting summer business styles ('Cool Biz') to encourage business people to wear cool and comfortable clothes, allowing them to work efficiently in these warmer offices.

In 2005, a survey of 562 respondents by the MOE (Murakami et al., 2006) showed that 96% of the respondents were aware of 'Cool Biz' and 33% answered that their offices set the thermostat higher than in previous years. Based on this result, CO_2 emissions were reduced by approximately 460,000 tonnes in 2005, which is equivalent to the amount of CO_2 emitted from about one million Japanese households for one month. MOE will continue to encourage offices to set air conditioning in offices at 28°C and will continue to promote 'Cool Biz.'

energy to impoverished communities for many years and contributes to meeting other MDGs.

However, for the poorest people in both developing countries and industrialised countries, the main barrier to energy-efficiency and renewable energy investments is the availability of financing for the investments. Devoting international aid or other public and private funds aimed at sustainable development to energy efficiency and renewable energy initiatives in buildings can achieve a multitude of development objectives and result in a long-lasting impact. These investments need not necessarily be executed through public subsidies, but may increasingly be achieved through innovative financing schemes, such as ESCOs or public-private partnerships. These schemes offer win-win opportunities, and leverage and strengthen markets (Blair *et al.*, 2005).

With a few exceptions, energy policies and practices in residential and commercial buildings in sub-Saharan Africa (SSA) do not take efficiency into consideration. However, energy efficiency in buildings has recently been recognised as one of the ways of increasing energy security and benefiting the environment, through energy savings (Winkler *et al.*, 2002). South Africa, for example, has drafted an energy-efficiency strategy to promote efficiency in buildings (DME, 2004). Such policies can be promoted in other SSA countries by linking energy efficiency in buildings directly to the countries' development agendas, by demonstrating how energy efficiency practises contribute to energy security. The positive impacts of these practices, including GHG mitigation, could then be considered as co-benefits.

6.10   Critical gaps in knowledge

During the review of the global literature, a few important areas have been identified which are not adequately researched or documented. First, there is a critical lack of literature and data about GHG emissions and mitigation options in developing countries. Whereas the situation is somewhat better in developed regions, in the vast majority of countries detailed end-use data is poorly collected or reported publicly, making analyses and policy recommendations insufficiently robust. Furthermore, there is a severe lack of robust, comprehensive, detailed and up-to-date bottom-up assessments of GHG reduction opportunities and associated costs in buildings worldwide, preferably using a harmonized methodology for analysis. In existing assessments of mitigation options, co-benefits are typically not included, and in general, there is an important need to quantify and monetize these so that they can be integrated into policy decision frameworks. Moreover, there is a critical lack of understanding, characterisation and taxonomization of non-technological options to reduce GHG emissions. These are rarely included in global GHG mitigation assessment models, potentially largely underestimating overall potentials. However, our policy leverage to realise these options is also poorly understood.

Finally, literature on energy price elasticities in the residential and commercial sectors in the different regions is very limited, while essential for the design of any policies influencing energy tariffs, including GHG taxes and subsidy removal.

REFERENCES

ACT, 2006: <http://www.actpla.act.gov.au/topics/property_purchases/sales/energy_efficiency>, accessed on July 3, 2007.

Aebischer, B., G. Henderson, and D. Catenazzi, 2006: Impact of climate change on energy demand in the Swiss service sector - and application to Europe. In *Proceedings of the International Conference on Improving Energy Efficiency in Commercial Buildings (IEECB '06)*, April 2006, Germany.

AIM, 2004: GHG Emissions and Climate Change. National Institute for Environmental Studies, Tsukuba, <http://www-iam.nies.go.jp/aim/aimpamph/pdf/All1115.pdf>, accessed on July 3, 2007.

Alam, M., J. Sathaye, and D. Barnes, 1998: Urban household energy use in India: Efficiency and policy implications. *Energy Policy*, **26**(11), pp. 885-892.

Alderfer, R.B., M.M. Eldridge, T.J. Starrs, 2000: Making connections: Case studies of interconnection barriers and their impact on distributed power projects. National Renewable Energy Laboratory, Golden, CO.

Andén, L., 2003: Solar Installations. Practical Applications for the Built Environment. James and James, London.

Ashford, P., J. Wu, M. Jeffs, S. Kocchi, P. Vodianitskaia, S. Lee, D. Nott, B. Johnson, A. Ambrose, J. Mutton, T. Maine, and B. Veenendaal, 2005: Chapter 7. Foams. In *IPCC/TEAP Special Report on Safeguarding the Ozone Layer and the Global Climate System: Issues related to Hydrofluorocarbons and Perfluorocarbons*.

Asian Development Bank, 1998: Asia Least-cost Greenhouse Gas Abatement Strategy. ALGAS Series: Thailand, Republic of Korea, Myanmar, Pakistan. <http://www.adb.org/Documents/Reports/ALGAS/default.asp>, accessed on July 5, 2007.

Atif, M.R. and A.D. Galasiu, 2003: Energy performance of daylight-linked automatic lighting control systems in large atrium spaces: report on two field-monitored case studies. *Energy and Buildings*, **35**, pp. 441-461.

Australian Government, 2006: Operation of the Ozone Protection and Synthetic Greenhouse Gas Management Act 1989.

Australian Greenhouse Office, 2000: Impact of minimum performance requirements for class 1 buildings in Victoria.

Australian Greenhouse Office, 2005: National appliances and equipment programs: When you keep measuring it, you know even more about it. Projected impacts 2005-2020. <www.energyrating.gov.au/library/pubs/200505-projectimpacts-agosummary.pdf>, accessed on July 3, 2007.

Badescu, V. and B. Sicre, 2003: Renewable energy for passive house heating Part 1. Building description. *Energy and Buildings*, **35**, pp. 1077-1084.

Balaras, C.A., 2001: Energy retrofit of a neoclassic office building - Social aspects and lessons learned. *ASHRAE Transactions*, **107**(1), pp. 191-197.

Balmér, P., 1997: Energy conscious waste water treatment plant. *CADDET Energy Efficiency Newsletter*, June 1997, pp. 11-12.

Banfi, S., M. Farsi, M. Filippini, and M. Jakob, 2006: Willingness to pay for energy-saving measures in residential buildings. *Energy Economic*, <http://dx.doi.org/10.1016/j.eneco.2006.06.001>, accessed on July 5, 2007.

Barnerjee, A. and B.D. Solomon, 2003: Eco-labeling for energy efficiency and sustainability: a meta-evaluation of US programs. *Energy Policy*, **31**, pp. 109-123.

Barnes, D.F., K. Openshaw, K.R. Smith, and R.v.d. Plas, 1994: What makes people cook with improved biomass stoves? Technical Paper No. 242. World Bank, Washington, D.C.

Bazilian, M.D., F. Leenders, B.G.C. van Der Ree, and D. Prasad, 2001: Photovoltaic cogeneration in the built environment. *Solar Energy*, **71**, pp. 57-69.

Behne, M., 1999: Indoor air quality in rooms with cooled ceilings. Mixing ventilation or rather displacement ventilation? *Energy and Buildings*, **30**, pp. 155-166.

Belding, W.A. and M.P.F. Delmas, 1997: Novel desiccant cooling system using indirect evaporative cooler. *ASHRAE Transactions*, **103**(1), pp. 841-847.

Bell, M. and R. Lowe, 2000: Energy efficient modernisation of housing: a UK case study. *Energy and Buildings*, **32**, pp. 267-280.

Bender, S., M. Moezzi, M.H. Gossard, L. Lutzenhiser, 2004: Using mass media to influence energy consumption behavior: California's 2001 Flex Your Power campaign as a case study. In *2004 ACEEE Summer Study on Energy Efficiency in Buildings*, ACEEE Press.

Bernstein, M.A. and J. Griffin, 2005: Regional differences in the price-elasticity of demand for energy. National Renewable Energy Laboratory.

Bertoldi, P. and O. Starter, 2003: Combining long term agreements with emissions trading: An overview of the current EU energy efficiency policies for the industrial sector and a proposal for a new industrial efficiency policy. In *ACEEE 2003 Summer Study on Energy Efficiency in Industry*, Washington, D.C.

Bertoldi, P., 1999: The use of long term agreements to improve energy efficiency in the industrial sector: overview of European experiences and proposal for a common framework. In *ACEEE 1999 Summer study on energy efficiency in industry*, Washington, D.C.

Bertoldi, P., 2000: The European strategy for reducing standby losses in consumer electronics: status and results. In *Proceeding of the ACEEE 2000 Summer Study on Energy Efficiency in Buildings*, Washington D.C.

Bertoldi, P., B. Aebischer, C. Edlington, C. Hershberg, B. Lebot, J. Lin, T. Marker, A. Meier, H. Nakagami, Y. Shibata, H.-P. Siderius, C. Webber, 2002: Standby power use: How big is the problem? What policies and technical solutions can address it? In *Proceedings of the ACEEE 2002 Summer Study on Energy Efficiency in Buildings*, **7**, American Council for an Energy Efficient Economy, Washington.

Bertoldi, P., S. Rezessy, and D. Ürge-Vorsatz, 2005: Tradable certificates for energy savings: opportunities, challenges and prospects for integration with other market instruments in the energy sector. *Energy and Environment*, **16**(6), pp. 959-992.

Bertoldi, P., and S. Rezessy, 2006: Tradable certificates for energy savings (White Certificates). Joint Research Center of the European Commission. <http://re.jrc.ec.europa.eu/energyefficiency/index.htm>, accessed on April 12, 2007.

Berry, L., M. Schweitzer, 2003: Metaevaluation of National Weatherization Assistance Program Based on State Studies 1993-2002. ORNL/CON-488.

Biermayer, P. and J. Lin, 2004: Clothes washer standards in China - the problem of water and energy trade-offs in establishing efficiency standards. In *2005 ACEEE Summer Study on Energy Efficiency in Buildings*, ACEEE Press.

Birtles, A.B. and R.W. John, 1984: Study of the performance of an energy management system. BSERT, London.

Blair, H.L., M. Dworkin, and B. Sachs, 2005: The efficiency utility: a model for replication. In *ECEEE Summer Studies 2005*, pp. 237-242.

Boardman, B., 1991: Fuel Poverty. Bellhaven Press.

Boardman, B., 2004: New directions for household energy efficiency: evidence from the UK. *Energy Policy*, **32**, pp. 1921-1933.

Bodart, M. and A.D. Herde, 2002: Global energy savings in office buildings by the use of daylighting. *Energy and Buildings*, **34**, pp. 421-429.

Boonstra, C. and I. Thijssen, 1997: Solar energy in building renovation. James & James, London, 64 pp.

Bordass, B., R. Cohen, M. Standeven, and A. Leaman, 2001a: Assessing building performance in use 3: energy performance of the Probe buildings. *Building Research and Information*, **29**(2), pp. 114-128.

Bordass, B., A. Leaman, and P. Ruyssevelt, 2001b: Assessing building performance in use 5: conclusions and implications. *Building Research & Information*, **29**(2).

Borg, N., Y. Blume, S. Thomas, W. Irrek, H. Ritter, A. Gula, A. Figórski, S. Attali, A. Waldmann, A. Loozen, L. Pagliano, A. Pindar, H.-F. Lund, P. Lund, L. Bångens, J. Harris, E. Sandberg, and G. Westring, 2003: Harnessing the power of the public purse: Final report from the European PROST study on energy efficiency in the public sector. EU-SAVE, Stockholm.

Börjesson, P. and L. Gustavsson, 2000: Greenhouse gas balances in building construction: wood versus concrete from life-cycle and forest land-use perspectives. *Energy Policy*, **28**, pp. 575-588.

Bourassa, N., P. Haves, and J. Huang, 2002: A computer simulation appraisal of non-residential low energy cooling systems in California. In *2002 ACEEE Summer Study on Energy Efficiency in Buildings*, American Council for an Energy Efficient Economy, Washington, pp. 41-53.

Brambley, M.R., D. Hansen, P. Haves, D.R. Holmberg, S.C. McDonald, and K.W.R.P. Torcellini, 2005: Advanced sensors and controls for building applications: Market assessment and potential R&D pathways. Prepared for US DOE Office of Energy Efficiency and Renewable Energy, Building Technologies Program, Pacific Northwest National Laboratory.

Brand, M. and M. Geissler, 2003: Innovations in CHP and lighting: best practice in the public & building sector.

Brandemuehl, M.J. and J.E.. Braun, 1999: The impact of demand-controlled and economizer ventilation strategies on energy use in buildings. *ASHRAE Transactions*, **105**(2), pp. 39-50.

Brandemuehl, M.J. and M.J. Bradford, 1999: Optimal supervisory control of cooling plants without storage. Final Report on ASHRAE Research Project 823.

Brown, M.A., 2001: Market barriers to energy efficiency. *Energy Policy*, **29**(14), pp. 1197-1208.

Brown, M.A., L.G. Berry, and R. Goel, 1991: Guidelines for successfully transferring government-sponsored innovations. *Research Policy*, **20**(2), pp. 121-143.

Brown, M.A., M.D. Levine, J.P. Romm, A.H. Rosenfeld, and J.G. Koomey, 1998: Engineering-economic studies of energy technologies to reduce greenhouse emissions: Opportunities and challenges. *Annual Review of Energy and the Environment*, **23**, pp. 287-385.

Burch, J., K. Subbarao, and A. Lekov, 1990: Short-term energy monitoring in a large commercial building. *ASHRAE Transactions*, **96**(1), pp. 1459-1477.

Butson, J., 1998: The potential for energy service companies in the European Union. Amsterdam.

CADDET, 1997: CADDET Energy Efficiency Newsletter (March 1997).

California Energy Commission, 2006: 2006 Appliance Efficiency Regulations (California Code of Regulations, Title 20, Sections 1601 through 1608), <http://www.energy.ca.gov/appliances/index.html>, accessed on July 3, 2007.

Calwell, C. and T. Reeder, 2002: Power supplies: A hidden opportunity for energy savings. A NRDC report prepared for Natural Resources Defense Council, San Francisco, CA.

Canadian Council of Ministers of the Environment, 2001: National action plan for the environmental control of ozone-depleting substances (ODS) and their halocarbon alternatives.

Carbon Trust, 2005: The UK Climate Change Programme: potential evolution for business and the public sector. Technical Report available online: <http://www.carbontrust.co.uk/Publications/publicationdetail. htm?productid=CTC518>, accessed on July 3, 2007.

CECED (European Committee on Household Appliance Manufacturers), 2005: CECED Unilateral Commitment on reducing energy consumption on household refrigerators and freezers. 2nd Annual report for 2004 to the Commission of the European Communities.

Center for Clean Air Policy, 2001: Identifying investment opportunities for the Clean Development Mechanism (CDM) in Brazil's Industrial Sector, Washington D.C.

Chappells, H. and E. Shove, 2004: Comfort: A review of philosophies and paradigms.

Cheminfo, 2004: International management instruments regarding HFCs, PFCs and SF6. Cheminfo for Environment Canada.

CIBSE, 1999: Environmental factors affecting office worker performance: review of evidence, TM24. The Chartered Institution of Building Services Engineers, London, UK.

City of Fort Collins, 2002: Evaluation of new home energy efficiency: An assessment of the 1996 Fort Collins residential energy code and benchmark study of design, construction and performance for homes built between 1994 and 1999. Fort Collins, Colorado, <http://www.fcgov.com/utilities/pdf/newhome-eval.pdf>, accessed on July 3, 2007.

CLASP (Collaborative Labeling and Appliance Standards Program), 2006: General information about standards and labelling. Effectiveness of energy-efficiency labels and standards <http://www.claspoline.org/>, accessed on July 5, 2007.

Claridge, D.E., M. Liu, and W.D. Turner, 2003: Commissioning of existing buildings - state of the technology and its implementation. Proceedings of the International Short Symposium on HVAC Commissioning. Kyoto, Japan.

Claridge, D.E., M. Liu, S. Deng, W.D. Turner, J.S. Haberl, S.U. Lee, M. Abbas, H. Bruner, and B.V.S.U. Lee, 2001: Cutting heating and cooling use almost in half without capital expenditure in a previously retrofit building. European Council for an energy efficient economy, 2001 Summer Proceedings. *ECEEE*, **4**, pp. 74-85.

Collys, A., 2005: The Flanders (BE) regional utility obligations. Presentation at the European Parliament. 3 May 2005. Presentation, Ministry of Flanders, Department of Natural Resources and Energy <http://195.178.164.205/library_links/downloads/ESD/Bottom-up.3March05.Collys.ppt>, accessed on July 5, 2007.

Commission of the European Communities, 1999: Communication from the Commission to the Council and European Parliament on policy instruments to reduce stand by losses of consumer electronic equipment. Brussels. September 15, 1999.

Commonwealth of Australia, 2002: National Appliance and Equipment Energy Efficiency Program, "Money isn't all you're saving: Australia's Standby Power Strategy 2002-2012".

Costelloe, B. and D. Finn, 2003: Indirect evaporative cooling potential in air-water systems in temperate climates. *Energy and Buildings*, **35**, pp. 573-591.

Cottrell, J., 2004: Ecotaxes in Germany and the United Kingdom- a business view. Green Budget Germany Conference Report. Heinrich-Böll Foundation, Berlin.

Crossley, D., J. Hamrin, E. Vine, N. Eyre, 1999: Public policy implications of mechanisms for promoting energy efficiency and load management in changing electricity businesses. Hornsby Heights, Task VI of the International Energy Agency Demand-Side Management Program.

Crossley, D., M. Maloney, and G. Watt, 2000: Developing mechanisms for promoting demand-side management and energy efficiency in changing electricity businesses. Research Report No 3. Task VI of the International Energy Agency Demand-Side Management Programme. Task VI of the International Energy Agency Demand-Side Management Program, Hornsby Heights.

Da Graça, G.C., Q. Chen, L.R. Glicksman, and L.K. Norfold, 2002: Simulation of wind-driven ventilative cooling systems for an apartment building in Beijing and Shanghai. *Energy and Buildings*, **34**, pp. 1-11.

Darby, S., 2000: Making it obvious: designing feedback into energy consumption, Appendix 4 of Boardman and Darby 2000. In *Effective Advice. Energy efficiency and the disadvantaged by B. Boardman, and S. Darby, ed.* Environmental Change Institute, University of Oxford, Oxford.

Darby, S., 2006: Social learning and public policy: Lessons from an energy-conscious village. *Energy Policy,* 34(17), pp. 2929-2940.

Davis, K., 2003: Greenhouse Gas Emission Factor Review. Final Technical Memorandum. Edison Mission Energy. Irvine, CA, Austin, TX.

De Dear, R.J. and G.S. Brager, 1998: Developing an adaptive model of thermal comfort and preference. *ASHRAE Transactions*, **104**(Part 1A), pp. 145-167.

De Dear, R.J. and G.S. Brager, 2002: Thermal comfort in naturally ventilated buildings: revisions to ASHRAE Standard 55. *Energy and Buildings*, **34**, pp. 549-561.

De Vita, G., K. Andresen, and L.C. Hunt, 2006: An empirical analysis of energy demand in Namibia. *Energy Policy*, **34**, pp. 3447-3463.

De Villiers, M. and K. Matibe, 2000: Greenhouse gas baseline and mitigation options for the residential sector. Energy and Development Research Centre, University of Cape Town.

De Villiers, M., 2000: Greenhouse gas baseline and mitigation options for the commercial sector. South African Country study, Department of Environmental Affairs and Tourism, Pretoria.

DEFRA, 2006: Review and development of carbon abatement curves for available technologies as part of the Energy Efficiency Innovation Review. Final Report by ENVIROS Consulting Ltd.

DEG, 2004: Development of an improved two-stage evaporative cooling system. Davis Energy Group, California Energy Commission Report P500-04-016, 62 pp. <http://www.energy.ca.gov/pier/final_project_reports/500-04-016.html>, accessed on July 3, 2007.

Demirbilek, F.N., U.G. Yalçiner, M.N. Inanici, A. Ecevit, and O.S. Demirbilek, 2000: Energy conscious dwelling design for Ankara. *Building and Environment*, **35**, pp. 33-40.

Dias, R., C. Mattos, J.A.P. Balestieri, 2004: Energy education: breaking up the rational energy use barriers. *Energy Policy*, 32(11), pp. 1339-1347.

DME, 2004: Energy efficiency strategy for the Republic of South Africa. Department of Minerals and Energy.

DoH, 2000: Fuel poverty and health. Circular from the UK Department of Health 2000, <www.dh.gov.uk/assetRoot/04/01/38/47/04013847.pdf>, accessed on July 3, 2007.

EAP UNDP, 2000: Sustainable energy strategies: materials for decision-makers. Energy and Atmosphere Programme of United Nations Development Programme, Chapter 5, 22 pp.

ECCJ, 2006: The Energy Conservation Center Japan <http://www.eccj.or.jp/top_runner/index.html>, accessed on July 3, 2007.

ECCP, 2001: Long Report. European Climate Change Programme.

ECCP, 2003: Second ECCP Progress Report: Can We Meet Our Kyoto targets, Technical Report. European Climate Change Programme.

ECEEE, 2004: Presentations in frame of Energy Services and Energy Efficiency Directive.

ECON Analysis, 2005: BASREC Regional handbook on procedures for Joint Implementation in the Baltic Sea region. Energy Unit, CBSS Secretariat, Stockholm.

ECS, 2002: Fiscal policies for improving energy efficiency. Taxation, grants and subsidies. Energy Charter Secretariat, Brussels, Belgium.

ECS, 2005: Carbon trading and energy efficiency. Integrating energy efficiency and environmental policies. Energy Charter Secretariat, Brussels, Belgium.

EFA, 2002: Mechanisms for promoting societal demand management. Independent Pricing and Regulatory Tribunal (IPART) of New South Wales.

Edwards, R.D., K.R. Smith, J. Zhang, and Y. Ma, 2004: Implications of changes in household stoves and fuel use in China. *Energy Policy*, **32**(3), pp. 395-411.

EEA, 2001: Indicator fact sheet signals - chapter Households. European Environment Agency.

EIA, 2004: International Energy Outlook. Energy Information Administration (EIA), Washington.

EIA, 2006: *Annual Energy Outlook.* Energy Information Administration (EIA), Washington.

Egan, C., C. Payne, J. Thorne, 2000: Interim Findings of an Evaluation of the U.S. EnergyGuide Label. In *Proceedings: 2000 ACEEE Summer Study.* LBL-46061.

Energy Saving Trust, 2005: Annual Report 2004/2005. <http://www.est.org.uk/uploads/documents/aboutest/CO111_Annual_report_04-05_final.pdf>, accessed on July 3, 2007.

Energystar, 2006: Energystar is a government-backed program helping businesses and individuals protect the environment through superior energy efficiency. <http://www.energystar.gov/>, accessed 12/07/07.

Energywatch, 2005: Get smart: bring meters into the 21st century <http://www.energywatch.org.uk/uploads/Smart_meters.pdf>, accessed on July 5, 2007.

ERI (National Development and Reform Commission), 2004: China National Energy Strategy and Policy to 2020: Scenario Analysis on Energy Demand.

ERRA/LGI, L.G.a.P.S.R.I., 2002: Meeting low-income consumers' needs. Energy Regulators, Regional Association/LGI Training material No 5.

European Commission, 2001: Green Paper - Towards a European strategy for the security of energy supply, Technical Document. <http://www.ec.europa.eu>, accessed on July 3, 2007.

European Commission, 2002: Directive 2002/91/EC of the European Parliament and of the Council of 16 December 2002 on the energy performance of buildings. *Official Journal of the European Communities,* L 1/65 of 4.1.2003.

European Commission, 2003: Proposal for a Directive of the European Parliament and of the Council on energy end-use efficiency and energy services.

European Commission, 2005: Green paper on energy efficiency. Doing more with less.

European Commission, 2006: Regulation (EC) No 842/2006 of the European Parliament and of the Council of 17 May 2006 on certain fluorinated greenhouse gases. *Official Journal of European Union,* L 161/1 of 14.06.2006.

Eyre, N., 1998: A golden age or a false dawn? Energy efficiency in UK competitive energy markets. *Energy Policy,* **26**(12), pp. 963-972.

FEDEMA, 1999: Economics of GHG limitations. Country study series, Ecuador. UNEP Collaborating Centre for Energy and Environment, Riso National Laboratory, Denmark.

FEMP, 2002: Federal Energy Management Program (FEMP, USA) Technology Profiles: Front loading clothes washers.

Fischer, J.C., J.R. Sand, B. Elkin, and K. Mescher, 2002: Active desiccant, total energy recovery hybrid system optimizes humidity control, IAQ, and energy efficiency in an existing dormitory facility. *ASHRAE Transactions,* **108** (Part 2), pp. 537-545.

Fisk, W.J., 2000: Health and productivity gains from better indoor environments and their implications for the U.S. Department of Energy. Washington, D.C. E-Vision 2000 Conference, <http://eetd.lbl.gov/IED/viaq/pubs/LBNL-47458.pdf>, accessed November 16, 2005.

Fisk, W.J., 2002: How IEQ affects health, productivity. *ASHRAE Journal,* **44**(5).

Florides, G.A., S.A. Tassou, S.A. Kalogirou, and L.C. Wrobel, 2002: Measures used to lower building energy consumption and their cost effectiveness. *Applied Energy,* **73**, pp. 299-328.

Fountain, M.E., E. Arens, T. Xu, F.S. Bauman, and M. Oguru, 1999: An investigation of thermal comfort at high humidities'. *ASHRAE Transactions,* **105**, pp. 94-103.

Francisco, P.W., B. Davis, D. Baylon, and L. Palmiter, 2004: Heat pumps system performance in northern climates. *ASHARE Translations,* **110**(Part 1), pp. 442-451.

Francisco, P.W., L. Palmiter, and B. Davis, 1998: Modeling the thermal distribution efficiency of ducts: comparisons to measured results. *Energy and Buildings,* **28**, pp. 287-297.

Fridley, D. and J. Lin, 2004: Private communications at Lawrence Berkeley National Laboratory (LBNL) regarding the contents of a study conducted for the U.S. Energy Foundation by the China Center for the Certification of Energy Conservation Products (CECP), the China National Institute of Standardization (CNIS), and LBNL.

Fung, A.S., A. Aulenback, A. Ferguson, and Ugursal, 2003: Standby power requirements of household appliances in Canada. *Energy and Buildings,* **35**, pp. 217-228.

Gaj, H. and M. Sadowski, 1997. Climate Change Mitigation: Case Studies from Poland. Advanced International Studies Unit, Pacific Northwest National Laboratory. Washington, D.C., 2004.

GAMA, 2005: March 2005 Consumers Directory of certified efficiency ratings for heating and water heating equipment. Gas Appliance Manufacturers Association.

Garg, V. and N.K. Bansal, 2000: Smart occupancy sensors to reduce energy consumption. *Energy and Buildings,* **32**, pp. 81-87.

Gauzin-Müller, D., 2002: Sustainable architecture and urbanism. Birkhäuser, Basel, 255 pp.

Geller, H. and S. Attali, 2005: The experience with energy efficiency policies and programmes in IEA countries: learning from the critics. IEA Information Paper.

Geller, H., P. Harrington, A.H. Rosenfeld, S. Tanishimad and F. Unander, 2006: Polices for increasing energy efficiency: Thirty years of experience in OECD countries. *Energy Policy,* 34(5), pp. 556-573.

Gillingham, K., R. Newell, K. Palmer, 2004: The effectiveness and cost of energy efficiency programmes. Resources. Resources for the Future, Technical paper.

Goldemberg, J., 2000: World Energy Assessment. Energy and the challenge of sustainability. United Nations Development (UNDP), New York.

Goldemberg, J., A.K.N. Reddy, K.R. Smith, and R.H. Williams., 2000: Rural energy in developing countries. Chapter 10. In *World energy assessment: energy and the challenge of sustainability,* United Nations Development Program.

Goldman, C., N. Hopper, and J. Osborn, 2005: Review of US ESCO industry market trends: an empirical analysis of project data. *Energy Policy,* **33**, pp. 387-405.

Gopalan, H.N.B. and S. Saksena (eds.), 1999: Domestic environment and health of women and children. Replika Press, Delhi.

Gritsevich, I., 2000: Motivation and decision criteria in private households, companies and administration on energy efficiency in Russia. In *Proceedings of the IPCC Expert Meeting on Conceptual Frameworks for Mitigation Assessment from the Perspective of Social Science,* 21-22 March 2000, Karlsruhe, Germany.

Grubb, M., 1991: Energy policies and the greenhouse effect. Vol. 1: Policy appraisal. Dartmouth, Aldershot.

Grubb, M., C. Vrolijk, and D. Brack, 2002: The Kyoto Protocol. A Guide and Assessment. 2nd ed. Harvest-Print, Moscow, 2 ed.

Gutschner, M. and Task-7 Members, 2001: Potential for building integrated photovoltaics. International Energy Agency, Photovoltaic Power Systems Programme, Task 7, Paris, Summary available from <www.iea-pvps.org>, accessed on July 3, 2007.

Haller, A., E. Schweizer, P.O. Braun, and K. Voss, 1997: Transparent insulation in building renovation. James & James, London, 16 pp.

Hamada, Y., M. Nakamura, K. Ochifuji, S. Yokoyama, and K. Nagano, 2003: Development of a database of low energy homes around the world and analysis of their trends. Renewable Energy, 28, pp. 321-328.

Hamburger-bildungsserver, 2006: <http://www.hamburger-bildungsserver.de/index.phtml?site=themen.klima>, accessed on July 3, 2007.

Harmelink, M., S. Joosen, and J. Harnisch, 2005: Cost-effectiveness of non-CO_2 greenhouse gas emission reduction measures implemented in the period 1990-2003. Ecofys, on behalf of SenterNovem.

Harris, J. and F. Johnson, 2000: Potential energy, cost, and CO_2 savings from energy-efficient government purchasing. In *Proceedings of the ACEEE Summer Study on Energy-efficient Buildings,* Asilomar, CA.

Harris, J., M. Brown, J. Deakin, S. Jurovics, A. Kahn and E. Wisniewski, J. Mapp and B. Smith, M. Podeszwa, A. Thomas, 2004: Energy-efficient purchasing by state and local government: triggering a landslide down the slippery slope to market transformation. In *Proceedings of the 2004 ACEEE Summer Study on Energy Efficiency in Buildings,* Asilomar, CA.

Harris, J., B. Aebischer, J. Glickman, G. Magnin, A. Meier and J. Vliegand, 2005: Public sector leadership: Transforming the market for efficient products and services. In *Proceedings of the 2005 ECEEE Summer Study: Energy savings: What works & who delivers?* Mandelieu, France.

Harvey, L.D.D., 2006: A Handbook on low-energy buildings and district energy systems: fundamentals, techniques, and examples. James and James, London.

Harvey, L.D.D., 2007: Net climatic impact of solid foam insulation produced with halocarbon and non-halocarbon blowing agents. *Buildings and Environment,* **42**(8), pp. 2860-2879.

Hassett, K. and G. Metcalf, 1993: Energy conservation investments: do consumers discount the future correctly? *Energy Policy,* **21**(6), pp. 710-716.

Hastings, S.R., 1994: Passive solar commercial and institutional buildings: a sourcebook of examples and design insights. John Wiley, Chichester, 454 pp.

Hastings, S.R., 2004: Breaking the 'heating barrier'. Learning from the first houses without conventional heating. *Energy and Buildings,* **36**, pp. 373-380.

Herring, H., 2006: Energy efficiency - a critical view. *Energy,* **31**(1), pp. 10-20.

Hestnes, A.G., R. Hastings, and B. Saxhof (eds.), 2003: Solar energy houses: strategies, technologies, examples, Second Edition. James & James, London, 202 pp.

Hicks, T.W. and B. von Neida, 1999: An evaluation of America's first ENERGY STAR Buildings: the class of 1999. In *ACEEE Summer Study on Energy Efficiency in Buildings,* pp. 4177-4185.

Holford, J.M. and G.R. Hunt, 2003: Fundamental atrium design for natural ventilation. *Building and Environment,* 38, pp. 409-426.

Holt, S., L. Harrington, 2003: Lessons learnt from Australia's standards and labelling programme. In *ECEEE summer study 2003,* ECEEE.

Howe, M., D. Holland, and A. Livchak, 2003: Displacement ventilation - Smart way to deal with increased heat gains in the telecommunication equipment room. *ASHRAE Transactions,* **109**(Part 1), pp. 323-327.

Hughes, P.J. and J.A. Shonder, 1998: The evaluation of a 4000-home geothermal heat pump retrofit at Fort Polk, Louisiana: Final Report. Oak Ridge National Laboratory.

Humm, O., 2000: Ecology and economy when retrofitting apartment buildings. *IEA Heat Pump Centre Newsletter,* **15**(4), pp. 17-18.

Hyvarinen, J., 1991: Cost benefit assessment methods for BEMS. International Energy Agency Annex 16, Building and Energy Management Systems: User Guidance. AIVC, Coventry, U.K.

IBEC, 2006: Institute for Building Environment and Energy Conservation [website] <http://www.ibec.or.jp/CASBEE/english/index.htm>, accessed on July 3, 2007.

IEA, 1997: The link between energy and human activity. International Energy Agency.

IEA, 2000: Daylighting in buildings: A sourcebook on daylighting systems and components. International Energy Agency Available from Lawrence Berkeley National Laboratory (eetd.lbl.gov) as LBNL-47493.

IEA, 2001: Things that go blip in the night. Standby power and how to limit it. International Energy Agency, Paris.

IEA, 2002: Energy conservation in buildings and community systems programme strategic plan 2002-2007. International Energy Agency.

IEA, 2003a: Cool appliances: policy strategies for energy efficient homes. International Energy Agency, Paris.

IEA, 2003b: Trends in photovoltaic applications: survey report of selected IEA countries between 1992 and 2002. International Energy Agency, Photovoltaic Power Systems Programme, Task 1, Paris, <www.iea-pvps.org>, accessed on July 3, 2007.

IEA, 2004a: *Dealing with Climate Change, policies and measures database.* IEA, Paris.

IEA, 2004b: *Energy Balances of Non-OECD Countries.* IEA, Paris.

IEA, 2004c: *Energy Conservation in Buildings & Community Systems.* Annual report 2002-2003, Birmingham, UK.

IEA, 2004d: *IEA RD&D database.* Edition 2004 with 2003 data. IEA, Paris.

IEA, 2004e: World Energy Outlook 2004. IEA, Paris.

IEA, 2004f: *30 Years of energy use in IEA countries.* IEA, Paris.

IEA, 2005: Saving electricity in a hurry - dealing with temporary shortfalls in electricity supplies. International Energy Agency, Organization for Economic Cooperation and Development, Paris.

IEA, 2006a: *Energy prices and taxes.* Quarterly Statistics. Second quarter, 2006, OECD/IEA, Paris.

IEA, 2006b: *Light's Labour's Lost.* Policies for energy-efficient lighting. IEA, Paris.

IEA, 2006c: *IEA RD&D database.* Edition 2006 with 2005 data. IEA, Paris.

IEA, 2006d: *Energy Technology Perspectives.* IEA, Paris.

IEA, 2006e: *World Energy Outlook 2006.* IEA, Paris.

IPCC, 1996: Climate Change 1995 - Economic and social dimensions of climate change. Contribution of Working Group III to the Second Assessment Report of the Intergovernmental Panel on Climate Change. J.P. Bruce, H. Lee, and E.F. Haites [eds]. Cambridge University Press, Cambridge, United Kingdom and New York, NY, USA, 448 pp.

IPCC, 2000: *Emissions Scenarios: Special Report of the Intergovernmental Panel on Climate Change (IPCC).*

IPCC, 2001: *Climate Change 2001: Mitigation. Contribution of Working Group III to the Third Assessment Report of the Intergovernmental Panel on Climate Change (IPCC).* B. Metz, O. Davidson, R. Swart, and J. Pan [eds.]. Cambridge University Press, Cambridge, United Kingdom and New York, NY, USA, 752 pp.

IPCC/TEAP, 2005: *Safeguarding the ozone layer and the global climate system: Issues related to hydrofluorocarbons and perfluorocarbons.* Special Report, Cambridge University Press, Cambridge, 478 pp.

ITDG, 2002: Smoke, health and household energy. Issues Paper compiled for DFID - EngKaR, (project no. R8021 Liz Bates - September 2002).

Izrael, Yu.A., S.I. Avdjushin, I.M. Nazarov, Yu.A. Anokhin, A.O. Kokorin, A.I. Nakhutin, and A.F. Yakovlev, 1999: Russian Federation Climate Change Country Study. Climate Change, Action Plan. Final Report. Russian Federal Service for Hydrometeorology and Environmental Monitoring: Moscow.

Jaboyedoff, P., C.A. Roulet, V. Dorer, A. Weber, and A. Pfeiffer, 2004: Energy in air-handling units - results of the AIRLESS European project. *Energy and Buildings,* **36**, pp. 391-399.

Jaccard, M.K. and Associates, 2002: Construction and analysis of sectoral, regional and national cost curves of GHG abatement of Canada. Part IV: Final Analysis Report. Submitted to Michel Francoeur of Natural Resources Canada, Vancouver. Contract No: NRCan-01-0332.

Jäger-Waldau, A. (ed), 2004: Status Report 2004 - Energy end use efficiency and electricity from biomass, wind and photovoltaics and in the European Union.

Jakob, M. and R. Madlener, 2004: Riding down the experience curve for energy efficient building envelopes: the Swiss case for 1970-2020. *International Journal of Energy Technology and Policy,* **2**(1-2), pp. 153-178.

Jakob, M., 2006: Marginal costs and co-benefits of energy efficiency investments. The case of the Swiss residential sector. *Energy Policy,* **34**(2), pp. 172-187.

Jakob, M., E. Jochem, and K. Christen, 2002: Marginal costs and benefits of heat loss protection in residential buildings. Swiss Federal Office of Energy (BFE).

Jakob, M., Jochem, E., Honegger, A., Baumgartner, Menti, U., Plüss, I., 2006: Grenzkosten bei forcierten. Energie-Effizienzmassnahmen und optimierter Gebäudetechnik bei Wirtschaftsbauten (Marginal costs of Energy-efficiency measures and improved building technology for buildings of the commercial sectors). CEPE ETH Zurich, Amstein+Walthert, HTA Luzern on behalf of Swiss Federal Office of Energy (SFOE), Zürich/Bern.

Jeeninga, H. and M.G. Boots, 2001: Development of the domestic energy consumption in the liberalised energy market, effects on purchase and use behavior. ECN Beleidsstudies, Petten, the Netherlands.

Jeeninga, H., C. Weber, I. Mäenpää, F.R. García, V. Wiltshire, and J. Wade, 1999: Employment impacts of energy conservation schemes in the residential sector. Calculation of direct and indirect employment effects using a dedicated input/output simulation approach.

Jefferson, M., 2000: Energy policies for sustainable development. In *World Energy Assessment: Energy and the challenge of sustainability.* UNDP, New York, 425 pp.

Jennings, J.D., F.M. Rubinstein, D. DiBartolomeo, and S.L. Blanc, 2000: Comparison of control options in private offices in an advanced lighting controls testbed. Lawrence Berkeley National Laboratory, 26 pp.

Jochem, E. and R. Madlener, 2003: The forgotten benefits of climate change mitigation: Innovation, technological leapfrogging, employment, and sustainable development. In *Workshop on the Benefits of Climate Policy: Improving Information for Policy Makers.*

Johnson, W.S., 2002: Field tests of two residential direct exchange geothermal heat pumps. *ASHRAE Transactions,* **108**(Part 2), pp. 99-106.

Jones, D.W., D.J. Bjornstad, and L.A. Greer, 2002: Energy efficiency, building productivity, and the commercial buildings market. Oak Ridge National Laboratory, Oak Ridge, TN.

Johnston, D., R. Lowe, M. Bell, 2005: An exploration of the technical feasibility of achieving CO_2 emission reductions in excess of 60% within the UK housing stock by the year 2050. *Energy Policy* (33), pp. 1643-1659.

Joosen, S., K. Blok, 2001: Economic evaluation of sectoral emission reduction objectives for climate change. Economic evaluation of carbon dioxide emission reduction in the household and services sectors in the EU. Bottom-up analysis, European Commission by Ecofys, 44 pp. <http://europa.eu.int/comm/environment/enveco>, accessed on July 5, 2007.

Joosen, S., 2006: Evaluation of the Dutch energy performance standard in the residential and services sector. Ecofys report prepared within Energy Intelligence for Europe program, contract number EIE-2003-114, Utrecht, the Netherlands.

Joosen, S., M. Harmelink, and K. Blok, 2004: Evaluation climate policy in the built environment 1995-2002. Ecofys report prepared for Dutch ministry of housing, spatial planning and the environment, Utrecht, the Netherlands, 235 pp.

Kallaste, T., T. Pallo, M.-P. Esop, O. Liik, M. Valdma, M. Landsberg, A. Ots, V. Selg, A. Purin, A. Martins, I. Roos, A. Kull, and R. Rost, 1999: *Economics of GHG Limitations. Country Case Study. Estonia.* UNEP Collaborating Centre for Energy and Environment, Riso National Laboratory, Denmark.

Karakezi, S. and T. Ranja, 2002: Renewable energy technologies in Africa. Zed Books Ltd, London and New Jersey.

Katipamula, S., R.G. Pratt, D.P. Chassin, Z.T. Taylor, K. Gowri, and M.R. Brambley, 1999: Automated fault detection and diagnostics for outdoor-air ventilation systems and economizers: Methodology and results from field testing. ASHRAE transactions. **105**(1), pp. 555-567.

Kawamoto, K., J. Koomey, B. Nordman, R.E. Brown, M.A. Piette, M. Ting, and A.K. Meier, 2001: Electricity used by office equipment and network equipment in the U.S., Lawrence Berkeley National Laboratory, Berkeley, CA.

KfW Group, 2006: The CO_2 Building Rehabilitation Programme <http://www.kfw-foerderbank.de/EN_Home/Housing_Construction/KfWCO2Buil.jsp>, accessed on July 3, 2007.

Kjellman, C.R., R.G. Haasl, and C.A. Chappell, 1996: Evaluating commissioning as an energy-saving measure. *ASHRAE Transactions,* **102**(1).

Kohlhaas, M., 2005: Gesamtwirtschaftliche Effekte der ökologischen Steuerreform. (Macroeconomic effects of the ecological tax reform) DIW, Berlin, 2006.

Koomey, J.G., C.A. Webber, C.S. Atkinson, and A. Nicholls, 2001: Addressing energy-related challenges for the U.S. buildings sector: results from the clean energy futures study. *Energy Policy* (also LBNL-47356), **29**(14), pp. 1209-1222.

Koomey, J., and F. Krause, 1989: Unit costs of carbon savings from urban trees, Rural trees, and electricity conservation: a utility cost perspective, Lawrence Berkeley Laboratory LBL-27872, Presented at the Heat Island Workshop LBL-27311.

Koran, W.E., 1994: Expanding the scope of commissioning: monitoring shows the benefits. *ASHRAE Transactions,* **100**(1), pp. 1393-1399.

Kostengünstige Passivhäuser als europäische Standards, 2005: <www.cepheus.de>, accessed July 3, 2007.

Krapmeier, H. and E. Drössler (eds.), 2001: CEPHEUS: Living comfort without heating. Springer-Verlag, Vienna, 139 pp.

Kushler, M., D. York, P. Witte, 2004: Five years. In *An examination of the first half decade of Public Benefits Energy Efficiency Policies.* American Council for an Energy Efficient Economy Main Report and appendices, Report: U042, April 2004, <http://www.aceee.org/pubs/u041.pdf>, accessed on July 5, 2007.

Laitner, S., S. Bernow, and J. DeCicco, 1998: Employment and other macroeconomic benefits of an innovation-led climate strategy for the United States. *Energy Policy,* **26**(5), pp. 425-432.

Lampietti, J.A., A.S. Meyer, 2003: Coping with the cold: Heating strategies for Eastern Europe and Central Asia's urban poor, World Bank Report.

Larsson, N., 2001: Canadian green building strategies. The 18th International Conference on Passive and Low Energy Architecture. Florianopolis, Brazil.

Lawrence Berkeley National Laboratory, 2004: China Energy Databook. Version 6, 2006.

Leaman, A. and B. Bordass, 1999: Productivity in buildings: the 'killer' variables. *Building Research & Information,* **27**(1), pp. 4-19.

Lebot, B., A. Meier, and A. Anglade, 2000: Global implications of standby power use. In *ACEEE Summer study on energy efficiency in buildings,* ACEEE, Asilomar, CA.

Lechtenbohmer, S., V. Grimm, D. Mitze, S. Thomas, M. Wissner, 2005: *Target 2020: Policies and measures to reduce greenhouse gas emissions in the EU.* WWF European Policy Office, Wuppertal.

Lee, E.S., D.L. DiBartolomeo, and S.E. Selkowitz, 1998: Thermal and daylighting performance of an automated Venetian blind and lighting system in a full-scale private office. *Energy and Buildings,* **29**, pp. 47-63.

Lee, W.L. and F.W. Yik, 2004: Regulatory and voluntary approaches for enhancing building energy efficiency. *Progress in Energy and Combustion Science,* **30**, 477-499.

Lees, E., 2006: Evaluation of the energy efficiency commitment 2002-05.

Le Fur, B., 2002 : *Panorama des dispositifs d'économie d'énergie en place dans les pays de L'Union Européenne.* Club d'Amélioration de l'Habitat, Paris.

Lehrer, D. and F. Bauman, 2003: Hype vs. reality: new research findings on underfloor air distribution systems. In *Greenbuild 2003,* November 2003, Pittsburgh, PA, 12 pp.

Lenzen, M. and G. Treloar, 2002: Embodied energy in buildings: wood versus concrete - reply to Börjesson and Gustavsson. *Energy Policy,* **30**, pp. 249-255.

Leutgöb, K., 2003: The role of energy agencies in developing the 'classical' EPC-market in Austria. In *Proceedings of First Pan-European Conference on Energy Services Companies: Creating the Market for Energy Services Companies (ESCOs) Industry in Europe.* Milan (Italy), 22-23 May 2003.

Levermore, G., D. Chow, P. Jones, and D. Lister, 2004: Accuracy of modelled extremes of temperature and climate change and its implications for the built environment in the UK. Technical Report 14. Tyndall Centre for Climate Change Research.

Levermore, G.J., 1985: Monitoring and targeting; motivation and training. In *Energy Management Experience Conference,* A.F.C. Sherratt (ed.), CICC, Cambridge, UK, pp. 21-30.

Levermore, G.J., 2000: Building energy management systems; application to low-energy HVAC and natural ventilation control. Second edition. E&FN Spon, Taylor & Francis Group, London.

Li, D.H.W. and J.C. Lam, 2003: An investigation of daylighting performance and energy saving in a daylight corridor. *Energy and Buildings,* **35**, pp. 365-373.

LIHEAP, 2005: Low Income Home Energy Assistance Program <http://www.liheap.ncat.org/Funding/lhemhist.htm>, accessed April 12, 2007, US Department of Health and Human Services, Administration for Families and Children.

Lin, Z. and S. Deng, 2004: A study on the characteristics of night time bedroom cooling load in tropics and subtropics. *Building and Environment*, **39**, pp. 1101-1114.

Lindenberger, D., T. Bruckner, H.M. Groscurth, and R. Kümmel, 2000: Optimization of solar district heating systems: seasonal storage, heat pumps, and cogeneration. *Energy*, **25**, pp. 591-608.

Lippke, B., J. Wilson, J. Perez-Garcia, J. Bowyer, and J. Meil, 2004: CORRIM: life-cycle environmental performance of renewable building materials. *Forest Products Journal*, **54**(6), pp. 8-19.

Liu, M. and D.E. Claridge, 1999: Converting dual-duct constant-volume systems to variable-volume systems without retrofitting the terminal boxes. *ASHRAE Transactions*, **105**(1), pp. 66-70.

Lovins, A., 1992: Energy-efficient buildings: Institutional barriers and opportunities. Strategic Issues Paper, E-Source, Inc., Boulder, CO.

Lutzenhiser, L., 1993: Social and behavioral aspects of energy use. *Annual Review of Energy and Environment*, **18**.

Mansur, E., R. Mendelsohn, and W. Morrison, 2005: A discrete-continuous choice model of climate change impacts on energy. Technical paper.

Margaree Consultants (ed.), 2003: Estimating the market potential for the clean development mechanism: Review of models and lessons learned. Washington D.C.

Martinsen, D., P. Markewitz, and S. Vögele, 2002: Roads to carbon reduction in Germany. International Workshop by Energy Modelling Forum, IEA and IIASA, 24-26 June 2003, Laxenburg, Austria <http://www.iiasa.ac.at/Research/ECS/IEW2003/Papers/2003P_martinsen.pdf>, accessed on July 5, 2007.

Markandya, 2000: Energy subsidy reform: an integrated framework to assess the direct and indirect environmental, economics and social costs and benefits of energy subsidy reform. Working paper prepared on behalf of UNEP, Division of Technology, Industry and Economics, December 2000.

Martin, Y. and Y. Carsalade, 1998: La maîtrise de l'énergie. La documentation Française, January 1998.

McConahey, E., P. Haves, and T. Christ, 2002: The integration of engineering and architecture: a perspective on natural ventilation for the new San Francisco federal building. In *Proceedings of the 2002 ACEEE Summer study on energy efficiency in buildings*, August, Asilomar, CA.

McCowan, B., T. Coughlin, P. Bergeron, and G. Epstein., 2002: High performance lighting options for school facilities. In *2002 ACEEE Summer study on energy efficiency in buildings*, American Council for an Energy Efficient Economy, Washington, pp. 253-268.

McMahon, J.F. and S. Wiel, 2001: Energy - efficiency labels and standards: A guidebook for appliances, equipment and lighting. LBNL, Environmental Energy Technologies Division, 237 pp.

Menanteau, P., 2001: Are voluntary agreements an alternative policy to efficiency standards for transforming the electrical appliances market. In *European Council for Energy Efficient Economy (ECEEE) summer study*, ECEEE, Stockholm.

Mestl, S.H., K. Aunan, F. Jinghua, H.M. Seip, J.M. Skjelvik, and H. Vennemo, 2005: Cleaner production as climate investment - integrated assessment in Taiyuan City, China. *Journal of Cleaner Production*, **13**, pp. 57-70.

Meyers, S., J.E. McMahon, M. McNeil, and X. Liu, 2002: Impacts of US federal energy efficiency standards for residential appliances. Lawrence Berkeley National Laboratory.

Michaelowa, A. and F. Jotzo, 2005: Transaction costs, institutional rigidities and the size of the clean development mechanism. *Energy Policy*, **33**, pp. 511-523.

Mills, E., 2005: The specter of fuel-based lighting. *Science*, **308**(27 May), pp. 1263-1264.

Mills, E., H. Friedman, T. Powell, N. Bourassa, D. Claridge, T. Haasl, and M.A. Piette, 2005: The cost-effectiveness of commercial-buildings commissioning: a meta-analysis of 224 case studies. *Journal of Architectural Engineering* (October 2005), pp. 20-24.

Mirasgedis, S., E. Georgopoulou, Y. Sarafidis, C. Balaras, A. Gaglia, and D.P. Lalas, 2004: CO_2 emission reduction policies in the Greek residential sector: a methodological framework for their economic evaluation. *Energy Conversion and Management*, **45**, pp. 537-557.

Mirasgedis, S., Y. Sarafidis, E. Georgopoulou, V. Kotroni, K. Lagouvardos, and D.P. Lalas, 2006: Modeling Framework for estimating the impacts of climate change on electricity demand at regional level: case of Greece, energy conversion and management.

Moezzi, M. and I. Maithili, 2002: What else is transferred along with energy efficiency? In *2004 ACEEE Summer Study on Energy Efficiency in Buildings*, ACEEE Press, Washington, D.C.

Moezzi, M. and R. Diamond, 2005: Is efficiency enough? Towards a new framework for carbon savings in the California residential sector. California Energy Commission.

Murakami, S., M.D. Levine, H. Yoshino, T. Inoue, T. Ikaga, Y. Shimoda, S. Miura, T. Sera, M. Nishio, Y. Sakamoto and W. Fujisaki, 2006: Energy consumption, efficiency, conservation and greenhouse gas mitigation in Japan's building sector. Lawrence Berkeley National Laboratory, LBNL-60424.

Murakoshi, C. and H. Nakagami, 2003: Present condition of ESCO business for carrying out climate change countermeasures in Japan. In *Proceedings of the ECEEE 2003 summer study*, European Council for Energy Efficient Economy, Stockholm, pp. 885-892.

Murakoshi, C. and H. Nakagami, 2005: New challenges of Japanese energy efficiency program by Top Runner approach.

NAEEEC, 2006: National appliance & equipment energy efficiency program.

National Bureau of Statistics, 2004: China Energy Statistical Yearbook (2000-2002). China Statistics Press, Beijing.

Nevin, R. and G. Watson, 1998: Evidence of rational market valuations for home energy efficiency. *Appraisal Journal,* October, pp. 401-409.

Niu, J.L., L.Z. Zhang, and H.G. Zuo, 2002: Energy savings potential of chilled-ceiling combined with desiccant cooling in hot and humid climates. *Energy and Buildings*, **34**, pp. 487-405.

Novikova, A., D. Ürge-Vorsatz, and C. Liang, 2006: The 'Magic' of the Kyoto Mechanisms: Will it work for buildings? American Council for an Energy Efficient Economy Summer Study 2006. California, USA.

Oak Ridge National Laboratory, 2001: Improving the methods used to evaluate voluntary energy efficiency. Programs Report for DOE and EPA, <http://www.ornl.gov/~webworks/cppr/y(2001/pres/111104.pdf>, accessed on July 5, 2007.

OECD, 2001: Working Party on development cooperation and environment policy guidance for strategies for sustainable development. Draft Document. DCD/DAC/ENV(2001)9, March 2001.

OECD, 2003: Environmentally sustainable buildings - Challenges and policies.

OECD/IEA, 2004: Coming from the cold. Improving district heating policy in transition economies, OECD/IEA, Paris.

Oikonomou, V., M. Patel, L. Muncada, T. Johnssons, and U. Farinellie, 2004: A qualitative analysis of White, Green Certificates and EU CO_2 allowances, Phase II of the White and Green project. Utrecht University Copernicus Institute and contribution of Lund University Sweden, Utrecht, the Netherlands.

O'Neal, D.L., A. Rodriguez, M. Davis, and S. Kondepudi, 2002: Return air leakage impact on air conditioner performance in humid climates. *Journal of Solar Energy Engineering*, **124**, pp. 63-69.

OPET Network, 2004: OPET building European research project: final reports <http://www.managenergy.net/indexes/I315.htm>, accessed on July 5, 2007.

Ostertag, K., 2003: No-regret potentials in energy conservation - An analysis of their relevance, size and determinants. Physica-Verlag, Heidelberg, New York.

Ott, W., M. Jakob, M. Baur, Y. Kaufmann, A. Ott, and A. Binz, 2005: Mobilisierung der energetischen Erneuerungspotenziale im Wohnbaubestand. On behalf of the research program Energiewirtschaftliche Grundlagen (EWG) of the Swiss Federal Office of Energy (SFOE), Bern.

PNNL and CENEf, 2004: *National inventory of energy-related emissions of greenhouse gases in Russia.* Moscow.

Parker, P., I.H. Rowlands, and D. Scott, 2000: Assessing the potential to reduce greenhouse gas emissions in Waterloo region houses: Is the Kyoto target possible? *Environments,* **28**(3), pp. 29-56.

Payne, A., R. Duke, and R.H. Williams., 2001: Accelerating residential PV expansion: supply analysis for competitive electricity markets. *Energy Policy,* **29**, pp. 787-800.

Petersdorff, C., T. Boermans, J. Harnisch, S. Joosen, F. Wouters, 2005: Cost effective climate protection in the building stock of the New EU Member States. Beyond the EU Energy perfomance of Buildings Directive. ECOFYS, Germany.

Peuser, F.A., K.H. Remmers, and M. Schnauss, 2002: Solar thermal systems: Successful planning and construction. James & James & Solarprazis, Berlin, 364 pp.

Price, L., S. De la Rue du Can, S. Sinton, E. Worrell, N. Zhou, J. Sathaye, and M. Levine, 2006: Sectoral trends in global energy use and greenhouse gas emissions. Lawrence Berkeley National Laboratory, Berkeley, CA.

Quinlan, P., H. Geller, S. Nadel, 2001: Tax incentives for innovative energy-efficient technologies (updated). American Council for an Energy-Efficient Economy, Washington D.C. Report Number E013.

RAO, 2006: Electricity prices for households. United Energy Systems of Russia, <http://www.rao-ees.ru/ru/info/about/main_gacts/show. cgi?tarif1204.htm>, accessed on December 5, 2006.

Reddy, A., 1991: Barriers to improvements in energy efficiency. Paper presented at the International Workshop on Reducing Carbon Emissions from the Developing World.

Reddy, B.S., and P. Balachandra, 2006: Dynamics of technology shift in the households sector - implications for clean development mechanism. *Energy Policy,* 34(2006), pp. 2586-2599.

Reinhart, C.F., 2002: Effects of interior design on the daylight availability in open plan offices. In *2002 ACEEE Summer Study on Energy Efficiency in Buildings,* American Council for an Energy Efficient Economy, Washington, pp. 309-322.

Reinhart, C.F., K. Voss, A. Wagner, and G. Löhnert, 2000: Lean buildings: Energy efficient com-mercial buildings in Germany. In *Proceedings of the 2000 ACEEE Summer Study on Energy Efficiency in Buildings,* American Council for an Energy Efficient Economy, pp. 257-298.

REN21, 2005: Renewables 2005 Global Status Report. Renewable Energy Policy Network, Worldwatch Institute, Washington D.C.

Rezessy, S. and P. Bertoldi, 2005: Are voluntary agreements an effective energy policy instrument? Insights and experiences from Europe. In *American Council for Energy Efficient Economy (ACEEE) Summer Study on energy efficiency in industry,* ACEEE.

Rezessy, S., K. Dimitrov, D. Ürge-Vorsatz, and S. Baruch, 2006: Municipalities and energy efficiency in countries in transition: Review of factors that determine municipal involvement in the markets for energy services and energy efficient equipment, or how to augment the role of municipalities as market players. *Energy Policy,* **34**(2), pp. 223-237.

Rezessy, S., P. Bertoldi, J. Adnot, and M. Dupont, 2005: Energy service companies in Europe: assembling the puzzle. In *Preliminary analysis of the results to date from the first European ESCO database,* ECEEE summer study.

Roberts, S., and W. Baker, 2003: Towards effective energy information. Improving consumer feedback on energy consumption. Center for sustainable energy. Con/Spec/2003/16, <http://www.cse.org.uk/pdf/pub1014.pdf>, accessed on July, 2007.

Rosenfeld, A.H., 1999: The art of energy efficiency: Protecting the environment with better technology. *Annual Review of Energy and the Environment,* 24, pp. 33-82.

Rosenfeld, A.H., H. Akbari, J.J. Romm, and M. Pomerantz, 1998: Cool communities: strategies for heat island mitigation and smog reduction. *Energy and Buildings,* **28**, pp. 51-62.

Ross, J.P. and A. Meier., 2002: Measurements of whole-house standby power consumption in California homes. *Energy,* **27**, pp. 861-868.

Roth, K.W., F. Goldstein, and J. Kleinman, 2002: Energy consumption by office and telecommunications equipment in commercial buildings. Volume 1: Energy Consumption Baseline, Arthur D. Little Inc., Cambridge (MA), 201 pp. <http://www.eere.energy.gov/buildings/info/documents/pdfs/office_telecom-vol1_final.pdf>, accessed on July 3, 2007.

Roth, K., P. Llana, W. Detlef, and J. Brodrick, 2005: Automated whole building diagnostics. *ASHRAE Journal,* **47**(5).

Rubinstein, F. and S. Johnson, 1998: Advanced lighting program development (BG9702800) Final Report. Lawrence Berkeley National Laboratory, Berkeley, California, 28 pp.

Safarzadeh, H. and M.N. Bahadori, 2005: Passive cooling effects of courtyards. *Building and Environment,* **40**, pp. 89-104.

Saikawa, M., K. Hashimoto, M. Itoh, H. Sakakibara, and T. Kobayakawa, 2001: Development of CO_2 heat pump water heater for residential use. *Transactions of Japan Society of Refrigeration and Air Conditioning Engineers,* **18**(3), pp. 225-232.

Sailor, D., 2001: Relating residential and commercial sector electricity loads to climate-evaluating state level sensitivities and vulnerabilities. *Energy 2001,* **26**(7), pp. 645-657.

Sailor, D.J., 2002: Urban Heat Island: Opportunities and challenges for mitigation and adaptation. Toronto-North American Heat Island Summit, Portland State University.

Sathaye, J. and S. Meyers, 1995: Greenhouse gas mitigation assessment: a guidebook. Kluwer Academic Publishers, Dordrecht/Boston/London.

Schaefer, C., C. Weber, H. Voss-Uhlenbrock, A. Schuler, F. Oosterhuis, E. Nieuwlaar, R. Angioletti, E. Kjellsson, S. Leth-Petersen, M. Togeby, and J. Munksgaard, 2000: *Effective policy instruments for energy efficiency in residential space heating - an International Empirical analysis.* University of Stuttgart, Institute for Energy.

Schleich, J. and E. Gruber, 2007: Beyond case studies: Barriers to energy efficiency in commerce and the services sectors. *Energy Economics* (forthcoming).

Schlomann, B., W. Eichhammer, and E. Gruber, 2001: Labelling of electrical appliances - An evaluation of the Energy Labelling Ordinance in Germany and resulting recommendations for energy efficiency policy. In *Proceedings of the European Council for an Energy Efficient Economy (ECEEE),* Summer Study 2001.

Schweitzer, M. and L. Berry, 1999: Metaevaluation of National Weatherization Assistance Program based on State Studies, 1996-1998. Oak Ridge National Laboratory, Oak Ridge, TN.

Scott, S., 1997: Household energy efficiency in Ireland: A replication study of ownership of energy saving items. *Energy Economics,* **19**, pp. 187-208.

Seefeldt, F., 2003: Energy Performance Contracting - success in Austria and Germany - dead end for Europe? In *Proceedings of the European Council for Energy Efficient Economy 2003 Summer Study.* European Council for an Energy-Efficient Economy, Stockholm.

Sherman, M.H. and D.A. Jump., 1997: Thermal energy conservation in buildings. In *CRC Handbook of Energy Efficiency.* CRC Press, Boca Raton, Florida, pp. 269-303.

Shipworth, M., 2000: *Motivating home energy action - A handbook of what works.* The Australian Greenhouse Office.

Shonder, J.A., M.A. Martin, H.A. McLain, and P.J. Hughes, 2000: Comparative analysis of life-cycle costs of geothermal heat pumps and three conventional HVAC systems. *ASHRAE Transactions,* **106**(Part 2), pp. 51-560.

Shorrock, L., 2001: Assessing the effect of grants for home energy efficiency improvements. In *Proceedings of the European Council for an Energy Efficient Economy (ECEEE) Summer Study 2001.*

SIA, 2006: A Swiss building standard no.380/4, information available at <www.sia.ch/ and www.bfe.admin.ch/>, accessed on July 3, 2007.

Singer, T., 2002: *IEA DSM Task X- Performance Contracting- Country Report: United States.* International Energy Agency, Paris.

Sinton, J.E., K.R. Smith, J.W. Peabody, L. Yaping, Z. Xiliang, R. Edwards, and G. Quan, 2004: An assessment of programs to promote improved household stoves in China. *Energy for Sustainable Development,* **8**(3), pp. 33-52.

Smith, D.L. and J.J. McCullough, 2001: Alternative code implementation strategies for states <http://www.energycodes.gov/implement/pdfs/strategies.pdf>, accessed on July 3, 2007.

Smith, K., S. Mehta, and M. Maeusezahl-Feuz, 2004: Indoor smoke from solid fuels. In *Comparative quantification of health risks: Global and regional burden of disease due to selected major risk factors.* World Health Organization, Geneva, pp. 1437-1495.

Smith, K.R., R. Uma, V.V.N. Kishore, J. Zhang, V. Joshi, and M.A.K. Khalil, 2000: Greenhouse implications of household fuels: An analysis for India. *Annual Review of Energy and Environment,* **25**, pp. 741-63.

Socolow, R.H., 1978: Saving energy in the home: Princeton's experiments at Twin Rivers. Ballinger Press, Cambridge, MA.

Sorrell, S., 2003: Who owns the carbon? Interactions between the EU Emissions Trading Scheme and the UK Renewables Obligation and Energy Efficiency Commitment. *Energy and Environment,* **14**(5), pp. 677-703.

Springer, D., G. Loisos, and L. Rainer, 2000: Non-compressor cooling alternatives for reducing residential peak load. In *2000 ACEEE Summer Study on Energy Efficiency in Buildings,* American Council for an Energy Efficient Economy, Washington, pp. 319-330.

Suriyamongkol, D., 2002: Non-technical losses in electrical power systems.

Swedish Energy Agency, 2004: Energy in Sweden: Facts and figures 2004. Swedish Energy Agency, Eskilstuna, 40 pp.

Szlavik, M., T. Palvolgyi, D. Urge-Vorsatz, M. Fule, B. Bakoss, G. Bartus, M. Bulla, M. Csutora, K. Czimber, M. Fule, P. Kaderjak, K. Kosi, Z. Lontay, Z. Somogyi, K. Staub, L. Szendrodi, J. Szlavik, T. Tajthy, L. Valko, and G. Zilahy, 1999: *Economics of GHG Limitations. Country Case Study. Hungary.* UNEP Collaborating Centre for Energy and Environment, Riso National Laboratory: Denmark.

Tangen, K. and G. Heggelund, 2003: Will the Clean Development Mechanism be effectively implemented in China? *Climate Policy,* **3**(3), pp. 303-307.

Taylor, P.B., E.H. Mathews, M. Kleingeld, and G.W. Taljaard, 2000: The effect of ceiling insulation on indoor comfort. *Building and Environment,* **35**, pp. 339-346.

The Energy Conservation Act, 2001: Extraordinary. Part II. Section 1(October-1): Chapter I, IV and The schedule. The Gazette of India.

Thorne, J. and C. Egan, 2002: An evaluation of the federal trade commission's energy guide appliance label: Final Report and Recommendations. ACEEE, pp. 39, <http://aceee.org/Pubs/a021full.pdf/>, accessed on July 3, 2007.

Todesco, G., 2004: Integrated design and HVAC equipment sizing. *ASHRAE Journal,* **46**(9), pp. S42-S47.

Turnbull, P.W. and G.A. Loisos, 2000: Baselines and barriers: Current design practices in daylighting. In *ACEEE Summer Study on Energy Efficiency in Buildings,* American Council for an Energy Efficient Economy, Washington, pp. 329-336.

Ueno, T., F. Sano, O. Saeki, and K. Tsuji, 2006: Effectiveness of an energy-consumption information system on energy savings in residential houses based on monitored data. *Applied Energy,* **83**, pp. 166-183.

UK government, 2006: Climate Change. The UK programme 2006, Crown.

UK/DEFRA, 2004: Framework for sustainable development on the Government estate, Part E - Energy. UK Sustainable Development Unit Department for Environment Food and Rural Affairs.

Ullah, M.B. and G. Lefebvre, 2000: Estimation of annual energy-saving contribution of an automated blind system. *ASHRAE Transactions,* **106**(Part 2), pp. 408-418.

UN, 2002: Global challenge global opportunity. Trends in sustainable development. Published by The United Nations Department of Economic and Social Affairs for the World Summit on Sustainable Development Johannesburg, 26 August - 4 September, 2002, <http://www.un.org/esa/sustdev/publications/critical_trends_report_2002.pdf>, accessed on July 5, 2007.

UNEP, 2000: *The GHG Indicator: UNEP guidelines for calculating greenhouse gas emissions for businesses and non-commercial organisations.* Geneva/Paris.

UNEP OECD/IEA, 2002: Reforming energy subsidies, An explanatory summary of the issues and challenges in removing or modifying subsidies on energy that undermine the pursuit of sustainable development. UNEP Division of Technology, Industry and Economics.

UNFCCC NC1 of Croatia, 2001: The first national communication of the Republic of Croatia to the United Nations Framework Convention on Climate Change (UNFCCC). Ministry of Environmental Protection and Physical Planning of Croatia, Zagreb.

UNFCCC 3NC of Poland, 2001: Republic of Poland. Third National Communication to the Conference of the Parties to the United Nations Framework Convention on Climate Change (UNFCCC). Ministry of Environment and National Fund for Environmental Protection and Water Management of Poland, Institute of Environmental Protection, Warsaw.

UNFCCC 3NC of Russia, 2002: Third National Communication to the Conference of the Parties to the United Nations Framework Convention on Climate Change (UNFCCC) of the Russian Federation. Inter-Agency Climate Change Commission of the Russian Federation, Moscow.

UNFCCC, 2006: Promotion of measures to limit HFC, PFC and SF6 emissions in Japan.

Ürge-Vorsatz, D. and G. Miladinova, 2005: Energy efficiency policy in an enlarged European Union: the Eastern perspective. In *European Council for an energy efficient economy summer study,* pp. 253-265.

Ürge-Vorsatz, D., G. Miladinova, and L. Paizs, 2006: Energy in transition: From the iron curtain to the European Union. *Energy Policy,* **34**, pp. 2279-2297.

Ürge-Vorsatz, D., L. Mez, G. Miladinova, A. Antipas, M. Bursik, A. Baniak, J. Jánossy, D. Nezamoutinova, J. Beranek, and G. Drucker, 2003: The impact of structural changes in the energy sector of CEE countries on the creation of a sustainable energy path: Special focus on investments in environmentally friendly energies and impact of such a sustainable energy path on employment and access conditions for low income consumers. European Parliament, Luxembourg.

US Congress, 2005: Energy Policy Act of 2005 (Public Law 109-58). Washington D.C.

USDOE/EERE, 2005: Indicators of energy intensity in the United States. U.S. Department of Energy Office of Energy Efficiency and Renewable Energy.

USDOE, 2006: Weatherization Assistance Program. U.S. Department of Energy Energy Efficiency and Renewable Energy.

USDOE/FEMP, 2005: FY 2004 Year in Review. U.S. Department of Energy, F.E.M.P.

US EPA, 2005: Investing in our future. Energy Star and other voluntary programs, 2004 Annual report, United States Environmental Protection Agency, Washington, D.C.

US EPA, 2006: Clean Air Act. Washington, United States <http://www.epa.gov/oar/caa/>, accessed on July 3, 2007.

Van Wie McGrory, L., J. Harris, M. Breceda Lapeyre, S. Campbell, M. della Cava, J. González Martínez, S. Meyer, and A.M. Romo, 2002: Market leadership by example: Government sector energy efficiency in developing countries. In *Proceedings: 2002 ACEEE Summer Study on Energy Efficiency in Buildings.*

Van Wie McGrory, L., D.Fridley, .J. Harris, P. Coleman, R. Williams, E. Villasenor, and R. Rodriguez, 2006: Two paths to transforming markets through public sector energy efficiency bottom up versus top down. In *Proceedings: ACEEE 2005 Summer Study.*

Verbruggen, A. and V. Bongaerts, 2003: Workshop documentation. In *SAVE 2001 project*, "Bringing energy services to the liberalized market (BEST)".

Vine, E., J. Hanrin, N. Eyre, D. Crossley, M. Maloney, and G. Watt, 2003: Public policy analysis of energy efficiency and load management in changing electricity businesses. Energy Policy, 31, pp. 405-430.

Vine, E., 2005: An international survey of the energy service company (ESCO) industry. *Energy Policy*, **33**(5), pp. 691-704.

VROM, 2006: Subsidy energy savings for low income households (Teli). VROM, (Ministry of Housing Spatial Planning and the Environment) the Netherlands.

Voss, K., 2000a: Solar energy in building renovation - results and experience of international demonstration buildings. *Energy and Buildings*, **32**, pp. 291-302.

Voss, K., 2000b: Solar renovation demonstration projects, results and experience. James & James, London, 24 pp.

Webber, C., R. Brown, M. McWhinney, and J. Koomey, 2003: 2002 Status report: savings estimates for the energy star voluntary labeling program. Lawrence Berkeley National Laboratory.

World Energy Council, 1999: The challenge of rural energy poverty in developing countries. World Energy Council and Food and Agriculture Organization, <http://www.worldenergy.org/wec-geis/publications/reports/rural/exec_summary/exec_summary.asp>, accessed on July 3, 2007, London.

World Energy Council, 2001: *Energy Efficiency Policies and Indicators.* London.

World Energy Council, 2004: *Energy Efficiency: A Worldwide review. Indicators, Policies, Evaluation.*

Weiss, W., I. Bergmann, and G. Faninger, 2005: Solar heating worldwide, markets and contribution to the energy supply 2003. IEA Solar Heating and Cooling Programme, Graz, Austria.

Western Regional Air Partnership, 2000: Air Pollution Prevention Meeting Summary <http://wrapair.org/forums/ap2/meetings/000531/000531sum.html>, accessed on July 5, 2007.

Westling, H., 2003: Performance contracting. Summary Report from the IEA DSM Task X within the IEA DSM Implementing Agreement. International Energy Agency, Paris.

Wiel, S. and J.E. McMahon, 2005: Energy-efficiency labels and standards: a guidebook for appliances, equipment, and lighting. 2nd edition. Collaborative Labeling and Appliance Standards Program (CLASP). Washington D.C., CLASP.

Winkler, H., R. Spalding-Fecher, and L. Tyani, 2002: Multi-project baseline for potential CDM projects in the electricity sector in South Africa. In *Developing Energy Solutions for Climate Change.* Energy & Development Research Centre: University of Cape Town, Cape Town.

Withers, C.R. and J.B. Cummings, 1998: Ventilation, humidity, and energy impacts of uncontrolled airflow in a light commercial building. *ASHRAE Transactions*, **104**(2), pp. 733-742.

Woerdman, E., 2000: Implementing the Kyoto protocol: Why JI and CDM show more promise than international emission trading. *Energy Policy*, **28**(1), pp. 29-38.

Wuppertal Institute, 2002: Bringing energy efficiency to the liberalised electricity and gas markets how energy companies and others can assist end-users in improving energy efficiency, and how policy can reward such action. Wuppertal: Wuppertal Inst. für Klima, Umwelt, Energie.

XENERGY, 2001: Impact analysis of the Massachusetts 1998 residential energy code revisions. Prepared for the Massachusetts Board of Building Regulations and Standards, Boston.

Yanagihara, T., 2006: Heat pumps the trump card in fight against global warming. *The Japan Journal* (October 2006), pp. 22-24.

York, D. and M. Kushler, 2005: ACEEE's 3rd National scorecard on utility and public benefits energy efficiency programs: A national review and update of state-level activities. ACEEE, Washington, D.C., October 2005.

Yoshikawa, S., 1997: Japanese DHC system uses untreated sewage as a heat source. CADDET Energy Efficiency Newsletter, June 1997, pp. 8-10.

Zhao, P.C., L. Zhao, G.L. Ding, and C.L. Zhang, 2003: Temperature matching method of selecting working fluids for geothermal heat pumps. *Applied Thermal Engineering*, **23**, pp. 179-195.

Zhivov, A.M. and A.A. Rymkevich, 1998: Comparison of heating and cooling energy consumption by HVAC system with mixing and displacement air distribution for a restaurant dining area in different climates. *ASHRAE Transactions*, **104**(Part 2), pp. 473-484.

Zhou, N., 2007: Energy use in China: Sectoral trends and future outlook. LBNL, Energy Analysis Division. LBNL-61904.

7

Industry

Coordinating Lead Authors:

Lenny Bernstein (USA), Joyashree Roy (India)

Lead Authors:

K. Casey Delhotal (USA), Jochen Harnisch (Germany), Ryuji Matsuhashi (Japan), Lynn Price (USA), Kanako Tanaka (Japan),
Ernst Worrell (The Netherlands), Francis Yamba (Zambia), Zhou Fengqi (China)

Contributing Authors:

Stephane de la Rue du Can (France), Dolf Gielen (The Netherlands), Suzanne Joosen (The Netherlands), Manaswita Konar (India),
Anna Matysek (Australia), Reid Miner (USA), Teruo Okazaki (Japan), Johan Sanders (The Netherlands),
Claudia Sheinbaum Parado (Mexico)

Review Editors:

Olav Hohmeyer (Germany), Shigetaka Seki (Japan)

This chapter should be cited as:

Bernstein, L., J. Roy, K. C. Delhotal, J. Harnisch, R. Matsuhashi, L. Price, K. Tanaka, E. Worrell, F. Yamba, Z. Fengqi, 2007: Industry.
In Climate Change 2007: Mitigation. Contribution of Working Group III to the Fourth Assessment Report of the Intergovernmental Panel
on Climate Change [B. Metz, O.R. Davidson, P.R. Bosch, R. Dave, L.A. Meyer (eds)], Cambridge University Press, Cambridge, United
Kingdom and New York, NY, USA.

Table of Contents

EXECUTIVE SUMMARY

Industrial sector emissions of greenhouse gases (GHGs) include carbon dioxide (CO_2) from energy use, from non-energy uses of fossil fuels and from non-fossil fuel sources (e.g., cement manufacture); as well as non-CO_2 gases.

- Energy-related CO_2 emissions (including emissions from electricity use) from the industrial sector grew from 6.0 $GtCO_2$ (1.6 GtC) in 1971 to 9.9 $GtCO_2$ (2.7 GtC) in 2004. Direct CO_2 emissions totalled 5.1 Gt (1.4 GtC), the balance being indirect emissions associated with the generation of electricity and other energy carriers. However, since energy use in other sectors grew faster, the industrial sector's share of global primary energy use declined from 40% in 1971 to 37% in 2004. In 2004, developed nations accounted for 35%; transition economies 11%; and developing nations 53% of industrial sector energy-related CO_2 emissions.
- CO_2 emissions from non-energy uses of fossil fuels and from non-fossil fuel sources were estimated at 1.7 Gt (0.46 GtC) in 2000.
- Non-CO_2 GHGs include: HFC-23 from HCFC-22 manufacture, PFCs from aluminium smelting and semiconductor processing, SF_6 from use in electrical switchgear and magnesium processing and CH_4 and N_2O from the chemical and food industries. Total emissions from these sources (excluding the food industry, due to lack of data) decreased from 470 $MtCO_2$-eq (130 MtC-eq) in 1990 to 430 $MtCO_2$-eq (120 MtC-eq) in 2000.

Direct GHG emissions from the industrial sector are currently about 7.2 $GtCO_2$-eq (2.0 GtC-eq), and total emissions, including indirect emissions, are about 12 $GtCO_2$-eq (3.3 GtC-eq) *(high agreement, much evidence)*.

Approximately 85% of the industrial sector's energy use in 2004 was in the energy-intensive industries: iron and steel, non-ferrous metals, chemicals and fertilizers, petroleum refining, minerals (cement, lime, glass and ceramics) and pulp and paper. In 2003, developing countries accounted for 42% of iron and steel production, 57% of nitrogen fertilizer production, 78% of cement manufacture and about 50% of primary aluminium production. Many industrial facilities in developing nations are new and include the latest technology with the lowest specific energy use. However, many older, inefficient facilities remain in both industrialized and developing countries. In developing countries, there continues to be a huge demand for technology transfer to upgrade industrial facilities to improve energy efficiency and reduce emissions *(high agreement, much evidence)*.

Many options exist for mitigating GHG emissions from the industrial sector *(high agreement, much evidence)*. These options can be divided into three categories:

- Sector-wide options, for example more efficient electric motors and motor-driven systems; high efficiency boilers and process heaters; fuel switching, including the use of waste materials; and recycling.
- Process-specific options, for example the use of the bio-energy contained in food and pulp and paper industry wastes, turbines to recover the energy contained in pressurized blast furnace gas, and control strategies to minimize PFC emissions from aluminium manufacture.
- Operating procedures, for example control of steam and compressed air leaks, reduction of air leaks into furnaces, optimum use of insulation, and optimization of equipment size to ensure high capacity utilization.

Mitigation potential and cost in 2030 have been estimated through an industry-by-industry assessment for energy-intensive industries and an overall assessment for other industries. The approach yielded mitigation potentials at a cost of <100 US$/ tCO_2-eq (<370 US$/tC-eq) of 2.0 to 5.1 $GtCO_2$-eq/yr (0.6 to 1.4 GtC-eq/yr) under the B2 scenario[1]. The largest mitigation potentials are located in the steel, cement, and pulp and paper industries and in the control of non-CO_2 gases. Much of the potential is available at <50 US$/$tCO_2$-eq (<180 US$/tC-eq). Application of carbon capture and storage (CCS) technology offers a large additional potential, albeit at higher cost *(medium agreement, medium evidence)*.

Key uncertainties in the projection of mitigation potential and cost in 2030 are the rate of technology development and diffusion, the cost of future technology, future energy and carbon prices, the level of industry activity in 2030, and climate and non-climate policy drivers. Key gaps in knowledge are the base case energy intensity for specific industries, especially in economies-in-transition, and consumer preferences.

Full use of available mitigation options is not being made in either industrialized or developing nations. In many areas of the world, GHG mitigation is not demanded by either the market or government regulations. In these areas, companies will invest in GHG mitigation if other factors provide a return on their investment. This return can be economic, for example energy efficiency projects that provide an economic payout, or it can be in terms of achieving larger corporate goals, for example a commitment to sustainable development. The slow rate of capital stock turnover is also a barrier in many industries, as is the lack of the financial and technical resources needed to implement mitigation options, and limitations in the ability of

---

1 A1B and B2 refer to scenarios described in the IPCC Special Report on Emission Scenarios (IPCC, 2000b). The A1 family of scenarios describe a future with very rapid econoic growth, low population growth, and rapid introduction of new and more efficient technologies. B2 describes a world 'in which emphasis is on local solutions to economic, social, and environmental sustainability'. It features moderate population growth, intermediate levels of economic development, and less rapid and more diverse technological change than the A1B scenario.

industrial firms to access and absorb technological information about available options *(high agreement, much evidence)*.

Industry GHG investment decisions, many of which have long-term consequences, will continue to be driven by consumer preferences, costs, competitiveness and government regulation. A policy environment that encourages the implementation of existing and new mitigation technologies could lead to lower GHG emissions. Policy portfolios that reduce the barriers to the adoption of cost-effective, low-GHG-emission technology can be effective *(medium agreement, medium evidence)*.

Achieving sustainable development will require the implementation of cleaner production processes without compromising employment potential. Large companies have greater resources, and usually more incentives, to factor environmental and social considerations into their operations than small and medium enterprises (SMEs), but SMEs provide the bulk of employment and manufacturing capacity in many developing countries. Integrating SME development strategy into the broader national strategies for development is consistent with sustainable development objectives *(high agreement, much evidence)*.

Industry is vulnerable to the impacts of climate change, particularly to the impacts of extreme weather. Companies can adapt to these potential impacts by designing facilities that are resistant to projected changes in weather and climate, relocating plants to less vulnerable locations, and diversifying raw material sources, especially agricultural or forestry inputs. Industry is also vulnerable to the impacts of changes in consumer preference and government regulation in response to the threat of climate change. Companies can respond to these by mitigating their own emissions and developing lower-emission products *(high agreement, much evidence)*.

While existing technologies can significantly reduce industrial GHG emissions, new and lower-cost technologies will be needed to meet long-term mitigation objectives. Examples of new technologies include: development of an inert electrode to eliminate process emissions from aluminium manufacture; use of carbon capture and storage in the ammonia, cement and steel industries; and use of hydrogen to reduce iron and non-ferrous metal ores *(medium agreement, medium evidence)*.

Both the public and the private sectors have important roles in the development of low-GHG-emission technologies that will be needed to meet long-term mitigation objectives. Governments are often more willing than companies to fund the higher risk, earlier stages of the R&D process, while companies should assume the risks associated with actual commercialisation. The Kyoto Protocol's Clean Development Mechanism (CDM) and Joint Implementation (JI), and a variety of bilateral and multilateral programmes, have the deployment, transfer and diffusion of mitigation technology as one of their goals *(high agreement, much evidence)*.

Voluntary agreements between industry and government to reduce energy use and GHG emissions have been used since the early 1990s. Well-designed agreements, which set realistic targets, include sufficient government support, often as part of a larger environmental policy package, and include a real threat of increased government regulation or energy/GHG taxes if targets are not achieved, can provide more than business-as-usual energy savings or emission reductions. Some voluntary actions by industry, which involve commitments by individual companies or groups of companies, have achieved substantial emission reductions. Both voluntary agreements and actions also serve to change attitudes, increase awareness, lower barriers to innovation and technology adoption, and facilitate co-operation with stakeholders *(medium agreement, much evidence)*.

7.1 Introduction

This chapter addresses past, ongoing, and short (to 2010) and medium-term (to 2030) future actions that can be taken to mitigate GHG emissions from the manufacturing and process industries.[2]

Globally, and in most countries, CO_2 accounts for more than 90% of CO_2-eq GHG emissions from the industrial sector (Price et al., 2006; US EPA, 2006b). These CO_2 emissions arise from three sources: (1) the use of fossil fuels for energy, either directly by industry for heat and power generation or indirectly in the generation of purchased electricity and steam; (2) non-energy uses of fossil fuels in chemical processing and metal smelting; and (3) non-fossil fuel sources, for example cement and lime manufacture. Industrial processes also emit other GHGs, e.g.:

- Nitrous oxide (N_2O) is emitted as a byproduct of adipic acid, nitric acid and caprolactam production;
- HFC-23 is emitted as a byproduct of HCFC-22 production, a refrigerant, and also used in fluoroplastics manufacture;
- Perfluorocarbons (PFCs) are emitted as byproducts of aluminium smelting and in semiconductor manufacture;
- Sulphur hexafluoride (SF_6) is emitted in the manufacture, use and, decommissioning of gas insulated electrical switchgear, during the production of flat screen panels and semiconductors, from magnesium die casting and other industrial applications;
- Methane (CH_4) is emitted as a byproduct of some chemical processes; and
- CH_4 and N_2O can be emitted by food industry waste streams.

Many GHG emission mitigation options have been developed for the industrial sector. They fall into three categories: operating procedures, sector-wide technologies and process-specific technologies. A sampling of these options is discussed in Sections 7.2–7.4. The short- and medium-term potential for and cost of all classes of options are discussed in Section 7.5, barriers to the application of these options are addressed in Section 7.6 and the implication of industrial mitigation for sustainable development is discussed in Section 7.7.

Section 7.8 discusses the sector's vulnerability to climate change and options for adaptation. A number of policies have been designed either to encourage voluntary GHG emission reductions from the industrial sector or to mandate such reductions. Section 7.9 describes these policies and the experience gained to date. Co-benefits of reducing GHG emissions from the industrial sector are discussed in Section 7.10. Development of new technology is key to the cost-

effective control of industrial GHG emissions. Section 7.11 discusses research, development, deployment and diffusion in the industrial sector and Section 7.12, the long-term (post-2030) technologies for GHG emissions reduction from the industrial sector. Section 7.13 summarizes gaps in knowledge.

7.1.1 Status of the sector

This chapter focuses on the mitigation of GHGs from energy-intensive industries: iron and steel, non-ferrous metals, chemicals (including fertilisers), petroleum refining, minerals (cement, lime, glass and ceramics) and pulp and paper, which account for most of the sector's energy consumption in most countries (Dasgupta and Roy, 2000; IEA, 2003a,b; Sinton and Fridley, 2000). The food processing industry is also important because it represents a large share of industrial energy consumption in many non-industrialized countries. Each of these industries is discussed in detail in Section 7.4.

Globally, large enterprises dominate these industries. However, small- and medium-sized enterprises (SMEs) are important in developing nations. For example, in India, SMEs have significant shares in the metals, chemicals, food and pulp and paper industries (GOI, 2005). There are 39.8 million SMEs in China, accounting for 99% of the country's enterprises, 50% of asset value, 60% of turnover, 60% of exports and 75% of employment (APEC, 2002). While regulations are moving large industrial enterprises towards the use of environmentally sound technology, SMEs may not have the economic or technical capacity to install the necessary control equipment (Chaudhuri and Gupta, 2003; Gupta, 2002) or are slower to innovate (Swamidass, 2003). These SME limitations create special challenges for efforts to mitigate GHG emissions. However, innovative R&D for SMEs is also taking place for this sector (See Section 7.7).

7.1.2 Development trends

The production of energy-intensive industrial goods has grown dramatically and is expected to continue growing as population and per capita income increase. Since 1970, global annual production of cement increased 271%; aluminium, 223%; steel, 84% (USGS, 2005), ammonia, 200% (IFA, 2005) and paper, 180% (FAO, 2006).

Much of the world's energy-intensive industry is now located in developing nations. China is the world's largest producer of steel (IISI, 2005), aluminium and cement (USGS, 2005). In 2003, developing countries accounted for 42% of global steel production (IISI, 2005), 57% of global nitrogen fertilizer production (IFA, 2004), 78% of global cement manufacture and about 50% of global primary aluminium production (USGS,

---

2 For the purposes of this chapter, industry includes the food processing and paper and pulp industries, but the growing of food crops and trees is covered in Chapters 8 and 9 respectively. The production of biofuels is covered in Chapter 4. This chapter also discusses energy conversions, such as combined heat and power and coke ovens, and waste management that take place within industrial plants. These activities also take place in dedicated facilities, which are discussed in Chapters 4 and 10 respectively.

Table 7.1: *Industrial sector final energy, primary energy and energy-related carbon dioxide emissions, nine world regions, 1971–2004*

	Final energy (EJ)			Primary energy (EJ)			Energy-related carbon dioxide, including indirect emissions from electricity use (MtCO$_2$)		
	1971	1990	2004	1971	1990	2004	1971	1990	2004
Pacific OECD	6.02	8.04	10.31	8.29	11.47	14.63	524	710	853
North America	20.21	19.15	22.66	25.88	26.04	28.87	1,512	1,472	1512
Western Europe	14.78	14.88	16.60	19.57	20.06	21.52	1,380	1,187	1126
Central and Eastern Europe	3.75	4.52	2.81	5.46	7.04	3.89	424	529	263
EECCA	11.23	18.59	9.87	15.67	24.63	13.89	1,095	1,631	856
Developing Asia	7.34	19.88	34.51	9.38	26.61	54.22	714	2,012	4098
Latin America	2.79	5.94	8.22	3.58	7.53	10.87	178	327	469
Sub-Saharan Africa	1.24	2.11	2.49	1.70	2.98	3.60	98	178	209
Middle East/North Africa	0.83	4.01	6.78	1.08	4.89	8.63	65	277	470
World	68.18	97.13	114.25	90.61	131.25	160.13	5,990	8,324	9855

Notes: EECCA = countries of Eastern Europe, the Caucasus and Central Asia. Biomass energy included. Industrial sector 'final energy' use excludes energy consumed in refineries and other energy conversion operations, power plants, coal transformation plants, etc. However, this energy is included in 'primary energy'. Upstream energy consumption was reallocated by weighting electricity, petroleum and coal products consumption with primary factors reflecting energy use and loses in energy industries. Final energy includes feedstock energy consumed, for example in the chemical industry. 'CO$_2$ emissions' in this table are higher than in IEA's Manufacturing Industries and Construction category because they include upstream CO$_2$ emissions allocated to the consumption of secondary energy products, such as electricity and petroleum fuels. To reallocate upstream CO$_2$ emissions to final energy consumption, we calculate CO$_2$ emission factors, which are multiplied by the sector's use of secondary energy.

Source: Price et al., 2006.

2005). Since many facilities in developing nations are new, they sometimes incorporate the latest technology and have the lowest specific emission rates (BEE, 2006; IEA, 2006c). This has been demonstrated in the aluminium (Navarro *et al.*, 2003), cement (BEE, 2003), fertilizer (Swaminathan and Sukalac, 2004) and steel industries (Tata Steel, Ltd., 2005). However, due to the continuing need to upgrade existing facilities, there is a huge demand for technology transfer (hardware, software and know-how) to developing nations to achieve energy efficiency and emissions reduction in their industrial sectors *(high agreement, much evidence).*

New rules introduced both domestically and through the multilateral trade system, foreign buyers, insurance companies, and banks require SMEs to comply with higher technical (e.g., technical barriers to trade), environmental (ISO, 1996), and labour standards (ENDS-Directory, 2006). These efforts can be in conflict with pressures for economic growth and increased employment, for example in China, where the government's efforts to ban the use of small-scale coke-producing facilities for energy efficiency and environmental reasons have been unsuccessful due to the high demand for this product (IEA, 2006a).

Competition within the developing world for export markets, foreign investment, and resources is intensifying. Multinational enterprises seeking out new markets and investments offer both large enterprises (Rock, 2005) and capable SMEs the opportunity to insert themselves into global value chains through subcontracting linkages, while at the same time increasing competitive pressure on other enterprises, which could lose their existing markets. Against this backdrop, SMEs, SME associations, support institutions, and governments in transition and developing countries face the challenge of adopting new approaches and fostering SME competitiveness. Integration of SME development strategy in the broader national strategies for technology development, sustainable development and/ or poverty reduction and growth is under consideration in transition and developing countries (GOI, 2004).

7.1.3 Emission trends

Total industrial sector GHG emissions are currently estimated to be about 12 GtCO$_2$-eq/yr (3.3 GtC-eq/yr) *(high agreement, much evidence).* Global and sectoral data on final energy use, primary energy use[3], and energy-related CO$_2$ emissions including indirect emissions related to electricity use, for 1971 to 2004 (Price *et al.*, 2006), are shown in Table 7.1. In

---

3   Primary energy associated with electricity and heat consumption was calculated by multiplying the amount of elec-tricity and heat consumed by each end-use sector by eletricity and heat primary factors. Primary factors were derived as the ratio of fuel inputs at power plants to electricity or heat delivered. Fuel inputs for electricity production were separated from inputs to heat production, with fuel inputs in combined heat and power plants being separated into fuel inputs for electricity and heat production according to the shares of electricity and heat produced in these plants. In order to calculate primary energy for non-fossil fuel (hydro, nuclear, renewables), we followed the direct equivalent method (SRES method): the primary energy of the non-fossil fuel energy is accounted for at the level of secondary energy, that is, the first usable energy form or "currency" available to the energy system (IPCC, 2000b).

Table 7.2: *Projected industrial sector final energy, primary energy and energy-related CO$_2$ emissions, based on SRES Scenarios, 2010–2030.*

A1B Scenario

	Final energy (EJ)			Primary energy (EJ)			Energy-related carbon dioxide, including indirect emissions from electricity use (MtCO$_2$)		
	2010	2020	2030	2010	2020	2030	2010	2020	2030
Pacific OECD	10.04	10.68	11.63	14.19	14.25	14.52	1,170	1,169	1,137
North America	24.95	26.81	28.34	32.32	32.84	32.94	1,875	1,782	1,650
Western Europe	16.84	18.68	20.10	24.76	25.45	25.47	1,273	1,226	1,158
Central and Eastern Europe	6.86	7.74	8.57	9.28	10.28	10.99	589	608	594
EECCA	20.82	24.12	27.74	28.83	32.20	35.43	1,764	1,848	1,853
Developing Asia	39.49	54.00	72.50	62.09	84.64	109.33	4,827	6,231	7,340
Latin America	18.20	26.58	33.13	29.14	38.72	51.09	1,492	2,045	2,417
Sub-Saharan Africa	7.01	10.45	13.70	13.27	19.04	27.40	833	1,286	1,534
Middle East/North Africa	14.54	22.21	29.17	20.34	29.20	39.32	1,342	1,888	2,224
World	158.75	201.27	244.89	234.32	286.63	346.48	15,165	18,081	19,908

B2 Scenario

	Final energy (EJ)			Primary energy (EJ)			Energy-related carbon dioxide including indirect emissions from electricity use (MtCO$_2$)		
	2010	2020	2030	2010	2020	2030	2010	2020	2030
Pacific OECD	10.83	11.64	11.38	14.27	14.17	12.83	980	836	688
North America	20.23	20.82	21.81	28.64	29.28	29.18	1,916	1,899	1,725
Western Europe	14.98	14.66	14.35	19.72	18.56	17.69	1,270	1,154	1,063
Central and Eastern Europe	3.42	4.30	5.03	4.44	5.28	6.06	327	380	424
EECCA	12.65	14.74	16.96	16.06	19.06	22.33	1,093	1,146	1,208
Developing Asia	40.68	53.62	67.63	55.29	72.42	90.54	4,115	4,960	5,785
Latin America	11.46	15.08	18.24	15.78	20.10	24.84	950	1,146	1,254
Sub-Saharan Africa	2.75	4.96	10.02	4.33	7.53	14.51	260	345	665
Middle East/North Africa	8.12	9.67	12.48	13.90	15.51	19.22	791	888	1,080
World	125.13	149.49	177.90	172.44	201.92	237.19	11,703	12,755	13,892

Note: Biomass energy included, EECCA = countries of Eastern Europe, the Caucasus and Central Asia.

Source: Price et al. (2006).

1971, the industrial sector used 91 EJ of primary energy, 40% of the global total of 227 EJ. By 2004, industry's share of global primary energy use declined to 37%.

The developing nations' share of industrial CO$_2$ emissions from energy use grew from 18% in 1971 to 53% in 2004. In 2004, energy use by the industrial sector resulted in emissions of 9.9 GtCO$_2$ (2.7 GtC), 37% of global CO$_2$ emissions from energy use. Direct CO$_2$ emissions totalled 5.1 Gt (1.4 GtC), the balance being indirect emissions associated with the generation of electricity and other energy carriers. In 2000, CO$_2$ emissions from non-energy uses of fossil fuels (e.g., production of petro-chemicals) and from non-fossil fuel sources (e.g., cement

manufacture) were estimated to be 1.7 GtCO$_2$ (0.46 GtC) (Olivier and Peters, 2005). As shown in Table 7.3, industrial emissions of non-CO$_2$ gases totalled about 0.4 GtCO$_2$-eq (0.1 GtC-eq) in 2000 and are projected to be at about the same level in 2010. Direct GHG emissions from the industrial sector are currently about 7.2 GtCO$_2$-eq (2.0 GtC-eq), and total emissions, including indirect emissions, are about 12 GtCO$_2$-eq (3.3 GtC-eq).

Table 7.2 shows the results for the industrial sector of the disaggregation of two of the emission scenarios (see footnote 1), A1B and B2, produced for the IPCC Special Report on Emissions Scenarios (SRES) (IPCC, 2000b) into four subsectors

Table 7.3: *Projected industrial sector emissions of non-CO₂ GHGs, MtCO₂-eq/yr*

Region	1990	2000	2010	2030
Pacific OECD	38	53	47	49
North America	147	117	96	147
Western Europe	159	96	92	109
Central and Eastern Europe	31	21	22	27
EECCA	37	20	21	26
Developing Asia	34	91	118	230
Latin America	17	18	21	38
Sub-Saharan Africa	6	10	11	21
Middle East/North Africa	2	3	10	20
World	470	428	438	668

Notes: Emissions from refrigeration equipment used in industrial processes included; emissions from all other refrigeration and air conditioning applications excluded. EECCA = countries of Eastern Europe, the Caucasus and Central Asia.

Source: US EPA, 2006b.

Table 7.4: *Projected baseline industrial sector emissions of non-CO₂ GHGs*

Industrial sector	Emissions (MtCO₂-eq/yr)			
	1990	2000	2010	2030
N_2O emissions from adipic/nitric acid production	223	154	164	190
HFC/PFC emissions from substitutes for ozone-depleting substances[a]	0	52	93	198
HFC-23 emissions from HCFC-22 production	77	96	45	106
SF_6 emission from use of electrical equipment (excluding manufacture)	42	27	46	74
PFC emission from aluminium production	98	58	39	51
PFC and SF_6 emissions from semiconductor manufacture	9	23	35	20
SF_6 emissions from magnesium production	12	9	4	9
N_2O emission from caprolactam manufacture	8	10	13	20
Total	470	428	438	668

[a] Emissions from refrigeration equipment used in industrial processes included; emissions from all other refrigeration and air conditioning applications excluded.

Source: US EPA, 2006a,b.

and nine world regions (Price *et al.*, 2006). These projections show energy-related industrial CO_2 emissions of 14 and 20 $GtCO_2$ in 2030 for the B2 and A1B scenarios, respectively. In both scenarios, CO_2 emissions from industrial energy use are expected to grow significantly in the developing countries, while remaining essentially constant in the A1 scenario and declining in the B2 scenario for the industrialized countries and countries with economies-in-transition.

Table 7.3 shows projections of non-CO_2 GHG emissions from the industrial sector to 2030 extrapolated from data to 2020 (US EPA 2006a,b). US EPA provides the only comprehensive data set with baselines and mitigation costs over this time frame for all gases and all sectors. However, baselines differ substantially for sectors covered by other studies, for example IPCC/TEAP (2005). As a result of mitigation actions, non-CO_2

GHG emissions decreased from 1990 to 2000, and there are many programmes underway to further reduce these emissions (See Sections 7.4.2 and 7.4.8.). Therefore Table 7.3 shows the US EPA's 'technology adoption' scenario, which assumes continued compliance with voluntary industrial targets. Table 7.4 shows these emissions by industrial process.[4]

7.2 Industrial mitigation matrix

A wide range of technologies have the potential for reducing industrial GHG emissions *(high agreement, much evidence)*. They can be grouped into categories, for example energy efficiency, fuel switching and power recovery. Within each category, some technologies, such as the use of more efficient

---

4   Tables 7.3 and 7.4 include HFC emissions from refrigeration equipment used in industrial processes and food storage, but not HFC emissions from other refrigeration and air conditioning applications. The tables also do not include HFCs from foams or non-CO_2 emissions from the food industry. Foams should be considered in the buildings sector. Global emissions from the food industry are not available, but are believed to be small compared with the totals presented in these tables.

Table 7.5: *Selected examples of industrial technology for reducing greenhouse-gas emissions (not comprehensive). Technologies in italics are under demonstration or development*

Sector	Energy efficiency	Fuel switching	Power recovery	Renewables	Feedstock change	Product change	Material efficiency	Non-CO$_2$ GHG	CO$_2$ sequestration
Sector wide	Benchmarking; Energy management systems; Efficient motor systems, boilers, furnaces, lighting and HVAC; Process integration	Coal to natural gas and oil	Cogeneration	Biomass, Biogas, PV, Wind turbines, Hydropower	Recycled inputs				*Oxy-fuel combustion, CO$_2$ separation from flue gas*
Iron & Steel	Smelt reduction, Near net shape casting, Scrap preheating, Dry coke quenching	Natural gas, oil or plastic injection into the BF	Top-gas pressure recovery, Byproduct gas combined cycle	Charcoal	Scrap	High strength steel	Recycling, High strength steel, Reduction process losses	n.a.	*Hydrogen reduction, Oxygen use in blast furnaces*
Non-Ferrous Metals	*Inert anodes*, Efficient cell designs				Scrap		Recycling, thinner film and coating	PFC/SF$_6$ controls	
Chemicals	Membrane separations, Reactive distillation	Natural gas	Pre-coupled gas turbine, Pressure recovery turbine, H$_2$ recovery		Recycled plastics, biofeedstock	Linear low density polyethylene, high-performance Plastics	Recycling, Thinner film and coating, Reduced process losses	N$_2$O, PFCs, CFCs and HFCs control	*Application to ammonia, ethylene oxide processes*
Petroleum Refining	Membrane separation Refinery gas	Natural gas	Pressure recovery turbine, hydrogen recovery	Biofuels	Bio-feedstock		Increased efficiency transport sector	Control technology for N$_2$O/CH$_4$	*From hydrogen production*
Cement	Precalciner kiln, Roller mill, *fluidized bed kiln*	Waste fuels, Biogas, Biomass	Drying with gas turbine, power recovery	Biomass fuels, Biogas	Slags, pozzolanes	Blended cement *Geo-polymers*		n.a.	*O$_2$ combustion in kiln*
Glass	Cullet preheating Oxyfuel furnace	Natural gas	*Air bottoming cycle*	n.a.	Increased cullet use	High-strength thin containers	Re-usable containers	n.a.	*O$_2$ combustion*
Pulp and Paper	Efficient pulping, Efficient drying, Shoe press, Condebelt drying	Biomass, Landfill gas	*Black liquor gasification combined cycle*	Biomass fuels (bark, black liquor)	Recycling, Non-wood fibres	Fibre orientation, Thinner paper	Reduction cutting and process losses	n.a.	*O$_2$ combustion in lime kiln*
Food	Efficient drying, Membranes	Biogas, Natural gas	Anaerobic digestion, Gasification	Biomass, Biogas, Solar drying			Reduction process losses, Closed water use		

electric motors and motor systems, are broadly applicable across all industries; while others, such as top-gas pressure recovery in blast furnaces, are process-specific. Table 7.5 presents selected examples of both classes of technologies for a number of industries. The table is not comprehensive and does not cover all industries or GHG mitigation technologies.

7.3 Industrial sector-wide operating procedures and technologies

This section discusses sector-wide mitigation options. Barriers to the implementation of these options are discussed in Section 7.6.

7.3.1 Management practices, including benchmarking

Management tools are available to reduce GHG emissions, often without capital investment or increased operating costs. Staff training in both skills and the company's general approach to energy efficiency for use in their day-to-day practices has been shown to be beneficial (Caffal, 1995). Programmes, for example reward systems that provide regular feedback on staff behaviour, have had good results.

Even when energy is a significant cost for an industry, opportunities for improvement may be missed because of organizational barriers. Energy audit and management programmes create a foundation for improvement and provide guidance for managing energy throughout an organization. Several countries have instituted voluntary corporate energy management standards, for example Canada (Natural Resources Canada, n.d.), Denmark (Gudbjerg, 2005) and the USA (ANSI, 2005). Others, for example India, through the Bureau of Energy Efficiency (GOI 2004, 2005), promote energy audits. Integration of energy management systems into broader industrial management systems, allowing energy use to be managed for continuous improvement in the same manner as labour, waste and other inputs are managed, is highly beneficial (McKane et al., 2005). Documentation of existing practices and planned improvements is essential to achieving a transition from energy efficiency programmes and projects dependent on individuals to processes and practices that are part of the corporate culture. Software tools are available to help identify energy saving opportunities (US DOE, n.d.-a; US EPA, n.d.).

Energy Audits and Management Systems. Companies of all sizes use energy audits to identify opportunities for reducing energy use, which in turn reduces GHG emissions. For example, in 2000, Exxon Mobil implemented its Global Energy Management System with the goal of achieving a 15% reduction in energy use in its refineries and chemical plants (Eidt, 2004). Okazaki et al. (2004) estimate that approximately 10% of total energy consumption in steel making could be saved through improved energy and materials management. Mozorov and Nikiforov (2002) reported an even larger 21.6% efficiency improvement in a Russian iron and steel facility. For SMEs in Germany, Schleich (2004) reported that energy audits help overcome several barriers to energy efficiency, including missing information about energy consumption patterns and energy saving measures. Schleich also found that energy audits conducted by engineering firms were more effective than those conducted by utilities or trade associations.

GHG Inventory and Reporting Systems. Understanding the sources and magnitudes of its GHG emissions gives industry the capability to develop business strategies to adapt to changing government and consumer requirements. Protocols for inventory development and reporting have been developed; the Greenhouse Gas Protocol developed by the World Resources Institute and World Business Council for Sustainable Development (WRI/WBCSD, 2004) is the most broadly used. The Protocol defines an accounting and reporting standard that companies can use to ensure that their measurements are accurate and complete. Several industries (e.g., aluminium, cement, chemical and pulp and paper) have developed specific calculation tools to implement the Protocol. Other calculation tools have been developed to estimate GHG emissions from office-based business operations and to quantify the uncertainty in GHG measurement and estimation (WRI/WBCSD, 2005). Within the European Union, GHG reporting guidelines have been developed for companies participating in the EU Emission Trading System.

GHG Management Systems. Environmental quality management systems such as ISO 14001 (ISO, 1996), are being used by many companies to build capacity for GHG emission reduction. For example, the US petroleum industry developed their own standard based on systems developed by various companies (API, 2005). The GHG emissions reduction opportunities identified by these management systems are evaluated using normal business criteria, and those meeting the current business or regulatory requirements are adopted. Those not adopted represent additional capacity that could be used if business, government, or consumer requirements change.

Benchmarking. Companies can use benchmarking to compare their operations with those of others, to industry average, or to best practice, to determine whether they have opportunities to improve energy efficiency or reduce GHG emissions. Benchmarking is widely used in industry, but benchmarking programmes must be carefully designed to comply with laws ensuring fair competition, and companies must develop their own procedures for using the information generated through these programmes. The petroleum industry has the longest experience with energy efficiency benchmarking through the use of an industry-accepted index developed by a private company (Barats, 2005). Many benchmarking programmes are developed through trade associations or ad hoc consortia of companies, and their details are often proprietary. However, ten Canadian potash operations published the details of their benchmarking exercise (CFI, 2003), which showed that increased employee awareness and training was the most frequently identified opportunity for improved energy performance. The success of the aluminium industry's programmes is discussed in Section 7.4.2.

Several governments have supported the development of benchmarking programmes in various forms, for example Canada, Flanders (Belgium), the Netherlands, Norway and the USA. As part of its energy and climate policy the Dutch government has reached an agreement with its energy-intensive industry that is explicitly based on industry's energy efficiency performance relative to that of comparable industries worldwide. Industry is required to achieve world best practice in terms of energy efficiency. In return, the government refrains from implementing additional climate policies. By 2002 this programme involved companies using 94% of the energy consumed by industry in the Netherlands. Phylipsen et al. (2002) critiqued the agreement, and conclude that it would avoid emissions of 4 to 9 $MtCO_2$ (1.1 to 2.5 MtC) in 2012 compared to a business-as-usual scenario, but that these emission reductions were smaller than those that would be achieved by a continuation of the Long-Term Agreements with industry (which ended in 2000) that called for a 2%/yr improvement in energy efficiency. The Flemish covenant, agreed in 2002, uses a similar approach. As of 1 January 2005, 177 companies had joined the covenant, which projects cumulative emissions saving of 2.45 $MtCO_2$ (0.67 MtC) in 2012 (Government of Flanders, 2005).

In the USA, EPA's Energy STAR for Industry programme has developed a benchmarking system for selected industries, for example automotive assembly plants, cement and wet corn milling (Boyd, 2005). The system is used by programme participants to evaluate the performance of their individual plants against a distribution of the energy performance of US peers. Other benchmarking programmes compare individual facilities to world best practice (Galitsky et al., 2004).

7.3.2 Energy efficiency

IEA (2006a) reports 'The energy intensity of most industrial processes is at least 50% higher than the theoretical minimum determined by the laws of thermodynamics. Many processes have very low energy efficiency and average energy use is much higher than the best available technology would permit.' This provides a significant opportunity for reducing energy use and its associated CO_2 emissions.

The major factors affecting energy efficiency of industrial plants are: choice and optimization of technology, operating procedures and maintenance, and capacity utilization, that is the fraction of maximum capacity at which the process is operating. Many studies (US DOE, 2004; IGEN/BEE; n.d.) have shown that large amounts of energy can be saved and CO_2 emissions avoided by strict adherence to carefully designed operating and maintenance procedures. Steam and compressed air leaks, poorly maintained insulation, air leaks into boilers and furnaces and similar problems all contribute to excess energy use. Quantification of the amount of CO_2 emission that could be avoided is difficult, because, while it is well known that these problems exist, the information on their extent is case-specific. Low capacity utilization is associated with more frequent shutdowns and poorer thermal integration, both of which lower energy efficiency and raise CO_2 emissions.

In view of the low energy efficiency of industries in many developing counties, in particular Africa (UNIDO, 2001), application of industry-wide technologies and measures can yield technical and economic benefits, while at the same time enhance environmental integrity. Application of housekeeping and general maintenance on older, less-efficient plants can yield energy savings of 10–20%. Low-cost/minor capital measures (combustion efficiency optimisation, recovery and use of exhaust gases, use of correctly sized, high efficiency electric motors and insulation, etc.) show energy savings of 20–30%. Higher capital expenditure measures (automatic combustion control, improved design features for optimisation of piping sizing, and air intake sizing, and use of variable speed drive motors, automatic load control systems and process residuals) can result in energy savings of 40–50% (UNIDO, 2001, Bakaya-Kyahurwa, 2004).

Electric motor driven systems provide a large potential for improvement of industry-wide energy efficiency. De Keulenaer et al., (2004) report that motor-driven systems account for approximately 65% of the electricity consumed by EU-25 industry. Xenergy (1998) gave similar figures for the USA, where motor-driven systems account for 63% of industrial electricity use. The efficiency of motor-driven systems can be increased by improving the efficiency of the electric motor through reducing losses in the motor windings, using better magnetic steel, improving the aerodynamics of the motor and improving manufacturing tolerances. However, the motor is only one part of the system, and maximizing efficiency requires properly sizing of all components, improving the efficiency of the end-use devices (pumps, fans, etc.), reducing electrical and mechanical transmission losses, and the use of proper operation and maintenance procedures. Implementing high-efficiency motor driven systems, or improving existing ones, in the EU-25 could save about 30% of the energy consumption, up to 202 TWh/yr, and avoid emissions of up to 100 $MtCO_2$/yr (27.2 MtC/yr) (De Keulenaer et al., 2004). In the USA, use of more efficient electric motor systems could save over 100 TWh/yr by 2010, and avoid emissions of 90 $MtCO_2$/yr (24.5 MtC/yr) (Xenergy, 1998). A study of the use of variable speed drives in selected African food processing plants, petroleum refineries, and municipal utility companies with a total motor capacity of 70,000 kW resulted in a potential saving of 100 $ktCO_2$-eq/yr (27 ktC/yr), or between 30–40%, at an internal rate of return of 40% (CEEEZ, 2003). IEA (2006b) estimates the global potential to be >20–25%, but a number of barriers have limited the optimization of motor systems (See Section 7.6).

Typical estimates indicate that about 20% of compressed air is lost through leakage. US DOE has developed best practices to identify and eliminate sources of leakage (US DOE, n.d.-a). IEA (2006a) estimates that steam generation consumes about 15% of global final industrial energy use. The efficiency of current steam boilers can be as high as 85%, while research in the USA aims to develop boilers with an efficiency of 94%. However, in practice, average efficiencies are often much lower. Efficiency measures exist for both boilers and distribution systems. Besides general maintenance, these include improved insulation, combustion controls and leak repair in the boiler, improved steam traps and condensate recovery. Studies in the USA identified energy-efficiency opportunities with economically attractive potentials up to 18–20% (Einstein *et al.*, 2001; US DOE, 2002). Boiler systems can also be upgraded to cogeneration systems.

Efficient high-pressure boilers using process residuals like bagasse are now available (Cornland *et al.*, 2001) and can be used to replace traditional boilers (15–25 bar) in the sugar industry. The high-pressure steam is used to generate electricity for own use with a surplus available for export to the grid (see also 7.3.4). For example, a boiler with a 60 MW steam turbine system in a 400 t/hour sugar factory could provide a potential surplus of 40 MW of zero-carbon electricity, saving 400 ktCO$_2$/yr (Yamba and Matsika, 2003). Similar technology installed at an Indian sugar mill increased the crushing period from 150 to 180 days, and exported an average of 10 MW of zero carbon electricity to the grid (Sobhanbabu, 2003).

Furnaces and process heaters, many of which are tailored for specific applications, can be further optimized to reduce energy use and emissions. Efficiency improvements are found in most new furnaces (Berntsson *et al.*, 1997). Research is underway to further optimize combustion processes by improving furnace and burner designs, preheating combustion air, optimizing combustion controls (Martin *et al.*, 2000); and using oxygen enrichment or oxy-fuel burners (See Section 7.3.7). These techniques are already being applied in specific applications.

7.3.3   Fuel switching, including the use of waste materials

While some industrial processes require specific fuels (e.g., metallurgical coke for iron ore reduction)[5], many industries use fuel for steam generation and/or process heat, with the choice of fuel being determined by cost, fuel availability and environmental regulations. The TAR (IPCC, 2001a) limited its consideration of industrial fuel switching to switches within fossil fuels (replacing coal with oil or natural gas), and concluded, based on a comparison of average and lowest carbon intensities for eight industries, that such switches could reduce

CO$_2$ emissions by 10–20%. These values are still applicable. A variety of industries are using methane from landfills as a boiler fuel (US EPA, 2005).

Waste materials (tyres, plastics, used oils and solvents and sewerage sludge) are being used by a number of industries. Even though many of these materials are derived from fossil fuels, they can reduce CO$_2$ emissions compared to an alternative in which they were landfilled or burned without energy recovery. The steel industry has developed technology to use wastes such as plastics (Ziebek and Stanek, 2001) as alternative fuel and feedstock's. Pretreated plastic wastes have been recycled in coke ovens and blast furnaces (Okuwaki, 2004), reducing CO$_2$ emissions by reducing both emissions from incineration and the demand for fossil fuels. In Japan, use of plastics wastes in steel has resulted in a net emissions reduction of 0.6 MtCO$_2$-eq/yr (Okazaki *et al.*, 2004). Incineration of wastes (e.g., tyres, municipal and hazardous waste) in cement kilns is one of the most efficient methods of disposing of these materials (Cordi and Lombardi, 2004; Houillon and Jolliet, 2005). Heidelberg Cement (2006) reported using 78% waste materials (tyres, animal meal and grease, and sewerage sludge) as fuel for one of its cement kilns. The cement industry, particularly in Japan, is investing to allow the use of municipal waste as fuel (Morimoto *et al.*, 2006). Cement companies in India are using non-fossil fuels, including agricultural wastes, sewage, domestic refuse and used tyres, as well as wide range of waste solvents and other organic liquids; coupled with improved burners and burning systems (Jain, 2005).

Humphreys and Mahasenan (2002) estimated that global CO$_2$ emissions could be reduced by 12% through increased use of waste fuels. However, IEA (2006a) notes that use of waste materials is limited by their availability, Also, use of these materials for fuel must address their variable composition, and comply with all applicable environmental regulations, including control of airborne toxic materials.

7.3.4   Heat and power recovery

Energy recovery provides major energy efficiency and mitigation opportunities in virtual all industries. Energy recovery techniques are old, but large potentials still exist (Bergmeier, 2003). Energy recovery can take different forms: heat, power and fuel recovery. Fuel recovery options are discussed in the specific industry sectors in Section 7.4. While water (steam) is the most used energy recovery medium, the use of chemical heat sinks in heat pumps, organic Rankine cycles and chemical recuperative gas turbines, allow heat recovery at lower temperatures. Energy-efficient process designs are often based on increased internal energy recovery, making it hard to define the technology or determine the mitigation potential.

---

5   Options for fuel switching in those processes are discussed in Section 7.4.

Heat is used and generated at specific temperatures and pressures and discarded afterwards. The discarded heat can be re-used in other processes onsite, or used to preheat incoming water and combustion air. New, more efficient heat exchangers or more robust (e.g., low-corrosion) heat exchangers are being developed continuously, improving the profitability of enhanced heat recovery. In industrial sites the use of low-temperature waste heat is often limited, except for preheating boiler feed water. Using heat pumps allows recovery of the low-temperature heat for the production of higher temperature steam.

While there is a significant potential for heat recovery in most industrial facilities, it is important to design heat recovery systems that are energy-efficient and cost-effective (i.e., process integration). Even in new designs, process integration can identify additional opportunities for energy efficiency improvement. Typically, cost-effective energy savings of 5 to 40% are found in process integration analyses in almost all industries (Martin et al., 2000; IEA-IETS, n.d.). The wide variation makes it hard to estimate the overall potential for energy-efficiency improvement and GHG mitigation. However, Martin et al. (2000) estimated the potential fuel savings from process integration in US industry to be 10% above the gain for conventional heat recovery systems. Einstein et al. (2001) and the US DOE (2002) estimated an energy savings potential of 5 to 10% above conventional heat recovery techniques.

Power can be recovered from processes operating at elevated pressures using even small pressure differences to produce electricity through pressure recovery turbines. Examples of pressure recovery opportunities are blast furnaces, fluid catalytic crackers and natural gas grids (at sites where pressure is reduced before distribution and use). Power recovery may also include the use of pressure recovery turbines instead of pressure relief valves in steam networks and organic Rankine cycles from low-temperature waste streams. Bailey and Worrell (2005) found a potential savings of 1 to 2% of all power produced in the USA, which would mitigate 21 $MtCO_2$ (5.7 MtC).

Cogeneration (also called Combined Heat and Power, CHP) involves using energy losses in power production to generate heat for industrial processes and district heating, providing significantly higher system efficiencies. Cogeneration technology is discussed in Section 4.3.5. Industrial cogeneration is an important part of power generation in Germany and the Netherlands, and is the majority of installed cogeneration capacity in many countries. Laurin et al. (2004) estimated that currently installed cogeneration capacity in Canada provided a net emission reduction of almost 30 $MtCO_2$/yr (8.18 MtC/yr). Cogeneration is also well established in the paper, sugar and chemical industries in India, but not in the cement industry due to lack of indigenously proven technology suitable for high dust loads. The Indian government is recommending adoption of technology already in use in China, Japan and Southeast Asian countries (Raina, 2002).

There is still a large potential for cogeneration. Mitigation potential for industrial cogeneration is estimated at almost 150 $MtCO_2$ (40 MtC) for the USA (Lemar, 2001), and 334 $MtCO_2$ (91.1 MtC) for Europe (De Beer et al., 2001). Studies also have been performed for specific countries, for example Brazil (Szklo et al., 2004), although the CO_2 emissions mitigation impact is not always specified.

7.3.5 Renewable energy

The use of biomass is well established in some industries. The pulp and paper industry uses biomass for much of its energy needs (See Section 7.4.6.). In many developing countries the sugar industry uses bagasse and the edible oils industry uses byproduct wastes to generate steam and/or electricity (See Section 7.4.7.). The use of bagasse for energy is likely to grow as more becomes available as a byproduct of sugar-based ethanol production (Kaltner et al., 2005). When economically attractive, other industries use biomass fuels, for example charcoal in blast furnaces in Brazil (Kim and Worrell, 2002a). These applications will reduce CO_2 emissions, but will only achieve zero net CO_2 emissions if the biomass is grown sustainably.

Industry also can use solar or wind generated electricity, if it is available. The potential for this technology is discussed in Section 4.3.3. The food and jute industries make use of solar energy for drying in appropriate climates (Das and Roy, 1994). The African Rural Energy Enterprise Development initiative is promoting the use of solar food driers in Mali and Tanzania to preserve fresh produce for local use and for the commercial market (AREED, 2000). Concentrating solar power could be used to provide process heat for industrial purposes, though there are currently no commercial applications (IEA-SolarPACES, n.d.).

7.3.6 Materials efficiency and recycling

Materials efficiency refers to the reduction of energy use by the appropriate choice of materials and recycling. Many of these options are applicable to the transport and building sectors and are discussed in Chapter 5, section 5.3.1 and Chapter 6, section 6.4. Recycling is the best-documented material efficiency option for the industrial sector. Recycling of steel in electric arc furnaces accounts about a third of world production and typically uses 60–70% less energy (De Beer et al., 1998). This technology, and options for further energy savings, are discussed in Section 7.4.1. Recycling aluminium requires only 5% of the energy of primary aluminium production. Recycled aluminium from used products and sources outside the aluminium industry now constitutes 33% of world supply and is forecast to rise to 40% by 2025 (IAI, 2006b, Martcheck, 2006). Recycling is also an important energy saving factor in other non-ferrous metal industries, as well as the glass and plastics industries (GOI, various issues). Recycling occurs both internally within plants and externally in the waste management sector (See Section 10.4.5).

Materials substitution, for example the addition of wastes (blast furnace slag, fly ash) and geo-polymers to clinker to reduce CO_2 emissions from cement manufacture (See Section 7.4.5.1), is also applicable to the industrial sector. Some materials substitution options, for example the production of lightweight materials for vehicles, can increase GHG emissions from the industrial sector, which will be more than offset by the reduction of emissions from other sectors (See Section 7.4.9). Use of bio-materials is a special case of materials substitution. No projections of the GHG mitigation potential of this option were found in the literature.

7.3.7 Carbon dioxide Capture and Storage (CCS), including oxy-fuel combustion

CCS involves generating a stream with a high concentration of CO_2, then either storing it geologically, in the ocean, or in mineral carbonates, or using it for industrial purposes. The IPCC Special Report on CCS (IPCC, 2005b) provides a full description of this technology, including its potential application in industry. It also discusses industrial uses of CO_2, including its temporary retention in beverages, which are small compared to total industrial emissions of CO_2.

Large quantities of hydrogen are produced as feedstock for petroleum refining, and the production of ammonia and other chemicals. Hydrogen manufacture produces a CO_2-rich by-product stream, which is a potential candidate for CCS technology. IPCC (2005b) estimated the representative cost of CO_2 storage from hydrogen manufacture at 15 US$/t$CO_2$ (55 US$/tC). Transport (250 km pipeline) injection and monitoring would add another 2 to 16 US$/t$CO_2$ (7 to 60 US$/tC) to costs.

CO_2 emissions from steel making are also a candidate for CCS technology. IEA (2006a) estimates that CCS could reduce CO_2 emissions from blast furnaces and DRI (direct reduction iron) plants by about 0.1 GtCO_2 (0.03 GtC) in 2030 at a cost of 20 to 30 US$/t$CO_2$ (73 to 110 US$/tC). Smelt reduction also allow the integration of CCS into the production of iron. CCS has also been investigated for the cement industry. Anderson and Newell (2004) estimate that it is possible to reduce CO_2 emissions by 65 to 70%, at costs of 50 to 250 US$/t$CO_2$ (183–917 US$/tC). IEA (2006a) estimates the potential for this application at up to 0.25 GtCO_2 (0.07 GtC) in 2030.

Oxy-fuel combustion can be used to produce a CO_2-rich flue gas, suitable for CCS, from any combustion process. In the past, oxy-fuel combustion has been considered impractical because of its high flame temperature. However, Gross et al. (2003), report on the development of technology that allows oxy-fuel combustion to be used in industrial furnaces with conventional materials. Tests in an aluminium remelting furnace showed up to 73% reduction in natural gas use compared to a conventional air-natural gas furnace. When the energy required to produce oxygen is taken into account, overall energy saving is reduced to 50 to 60% (Jupiter Oxygen Corp., 2006). Lower but still

impressive energy efficiency improvements have been obtained in other applications, up to 50% in steel remelting furnaces, up to 45% in small glass-making furnaces, and up to 15% in large glass-making furnaces (NRC, 2001). The technology has also been demonstrated using coal and waste oils as fuel. Since much less nitrogen is present in the combustion chamber, NO_x emissions are very low, even without external control, and the system is compatible with integrated pollution removal technology for the control of mercury, sulphur and particulate emissions as well as CO_2 (Ochs et al., 2005).

Industry does not currently use CCS as a mitigation option, because of its high cost. However, assuming that the R&D currently underway on lowering CCS cost is successful, application of this technology to industrial CO_2 sources should begin before 2030 and be wide-spread after that date.

7.4 Process-specific technologies and measures

This section discusses process specific mitigation options. Barriers to the implementation of these options are discussed in Section 7.6. The section focuses on energy intensive industries: iron and steel, non-ferrous metals, chemicals, petroleum refining, minerals (cement, lime and glass) and pulp and paper. IEA (2006a) reported that these industries (ex-petroleum refining) accounted for 72% of industrial final energy use in 2003. With petroleum refining, the total is about 85%. A subsection covers the food industry, which is not a major contributor to global industrial GHG emissions, but is a large contributor to these emissions in many developing countries. Subsections also cover other industries and inter-industry options, where the use of one industry's waste as a feedstock or energy source by another industry can reduce overall emissions (See Section 7.4.9). All the industries discussed in this section can benefit from application of the sector-wide technologies (process optimization, energy efficiency, etc.) discussed in Section 7.3. The application of these technologies will not be discussed again.

7.4.1 Iron and steel

Steel is by far the world's most important metal, with a global production of 1129 Mt in 2005. In 2004, the most important steel producers were China (26%), EU-25 (19%), Japan (11%), USA (10%) and Russia (6%) (IISI, 2005). Three routes are used to make steel. In the primary route (about 60%), used in almost 50 countries, iron ore is reduced to iron in blast furnaces using mostly coke or coal, then processed into steel. In the second route (about 35%), scrap steel is melted in electric-arc furnaces to produce crude steel that is further processed. This process uses only 30 to 40% of the energy of the primary route, with CO_2 emissions reduction being a function of the source of electricity (De Beer et al., 1998). The remaining steel production (about 5%), uses natural gas to produce direct reduced iron (DRI). DRI cannot be used in primary steel plants, and is mainly used as an alternative iron input in electric arc furnaces, which can result in a reduction of up to 50% in CO_2 emissions compared with primary steel making (IEA, 2006a). Use of DRI is expected to increase in the future (Hidalgo *et al.*, 2005).

Global steel industry CO_2 emissions are estimated to be 1500 to 1600 $MtCO_2$ (410 to 440 MtC), including emissions from coke manufacture and indirect emissions due to power consumption, or about 6 to 7% of global anthropogenic emissions (Kim and Worrell, 2002a). The total is higher for some countries, for example steel production accounts for over 10% of China's energy use and about 10% of its anthropogenic CO_2 emissions (Price *et al.*, 2002). Emissions per tonne of steel vary widely between countries: 1.25 tCO_2 (0.35 tC) in Brazil, 1.6 tCO_2 (0.44 tC) in Korea and Mexico, 2.0 tCO_2 (0.54 tC) in the USA, and 3.1 to 3.8 tCO_2 (0.84 to 1.04 tC) in China and India (Kim and Worrell, 2002a). The differences are based on the production routes used, product mix, production energy efficiency, fuel mix, carbon intensity of the fuel mix, and electricity carbon intensity.

Energy Efficiency. Iron and steel production is a combination of batch processes. Steel industry efforts to improve energy efficiency include enhancing continuous production processes to reduce heat loss, increasing recovery of waste energy and process gases, and efficient design of electric arc furnaces, for example scrap preheating, high-capacity furnaces, foamy slagging and fuel and oxygen injection. Continuous casting, introduced in the 1970s and 1980s, saves both energy and material, and now accounts for 88% of global steel production (IISI, 2005). Figure 7.1 shows the technical potential[6] for CO_2 emission reductions by region in 2030 for full diffusion of eight cost-effective and/or well developed energy savings technologies under the SRES B2 scenario, using a methodology developed by Tanaka *et al.* (2005, 2006).

The potential for energy efficiency improvement varies based on the production route used, product mix, energy and carbon intensities of fuel and electricity, and the boundaries chosen for the evaluation. Tanaka *et al.* (2006) also used a Monte Carlo approach to estimate the uncertainty in their projections of technical potential for three steel making technologies. Kim and Worrell (2002a) estimated economic potential by taking industry structure into account. They benchmarked the energy efficiency of steel production to the best practice performance in five countries with over 50% of world steel production, finding potential CO_2 emission reductions due to energy efficiency improvement varying from 15% (Japan) to 40% (China, India and the USA). While China has made significant improvements in energy efficiency, reducing energy consumption per tonne

of steel from 29.3 GJ in 1990 to 23.0 GJ in 2000[7] (Price *et al.*, 2002), there is still considerable potential for energy efficiency improvement and CO_2 emission mitigation (Kim and Worrell, 2002a). Planned improvements include greater use of continuous casting and near-net shape casting, injection of pulverized coal, increased heat and energy recovery and improved furnace technology (Zhou *et al.*, 2003). A study in 2000 estimated the 2010 global technical potential for energy efficiency improvement with existing technologies at 24% (De Beer *et al.*, 2000a) and that an additional 5% could be achieved by 2020 using advanced technologies such as smelt reduction and near net shape casting.

ULCOS (Ultra-Low CO_2 Steel making), a consortium of 48 European companies and organizations, has as its goal the development of steel making technology that reduces CO_2 emission by at least 50%. The technologies being evaluated, including CCS, biomass and hydrogen reduction, show a potential for controlling emissions to 0.5 to 1.5 tCO_2/t (0.14 to 0.41 tC/t) steel (Birat, 2005). Economics may limit the achievable emission reduction potential. A study of the US steel industry found a 2010 technical potential for energy-efficiency improvement of 24% (Worrell *et al.*, 2001a), but economic potential, using a 30% hurdle rate, was only 18%, even accounting for the full benefits of the energy efficiency measures (Worrell *et al.*, 2003). A similar study of the European steel industry found an economic potential of less than 13% (De Beer *et al.*, 2001). These studies focused mainly on retrofit options. However, potential savings could be realized by a combination of capital stock turnover and retrofit of existing equipment. A recent analysis of the efficiency improvement of electric arc furnaces in the US steel industry found that the average efficiency improvement between 1990 and 2002 was 1.3%/yr, of which 0.7% was due to capital stock turnover and 0.5% due to retrofit of existing furnaces (Worrell and Biermans, 2005). Future efficiency developments will aim at further process Data is pluralintegration. The most important are near net shape casting (Martin *et al.*, 2000), with current applications at numerous plants around the world; and smelt reduction, which integrates ore agglomeration, coke making and iron production in a single process, offering an energy-efficient alternative at small to medium scales (De Beer *et al.*, 1998).

Fuel Switching. Coal (in the form of coke) is the main fuel in the iron and steel industry because it provides both the reducing agent and the flow characteristics required by blast furnaces in the production of iron. Steel-making processes produce large volumes of byproducts (e.g., coke oven and blast furnace gas) that are used as fuel. Hence, a change in coke use will affect the energy balance of an integrated iron and steel plant.

---

6   See Section 2.4.3.1 for definitions of mitigation potential.
7   China uses various indicators to present energy intensity, including "comprehensive" and "comparable" energy intensity. The indicators are not always easily comparable to energy intensities from other countries or regions. The above figures use the comparable energy intensity, which is a constructed indicator, making it impossible to compare to those of other studies. Only a detailed assessment of the energy data can result in an internationally comparable indicator (Price *et al.*, 2002).

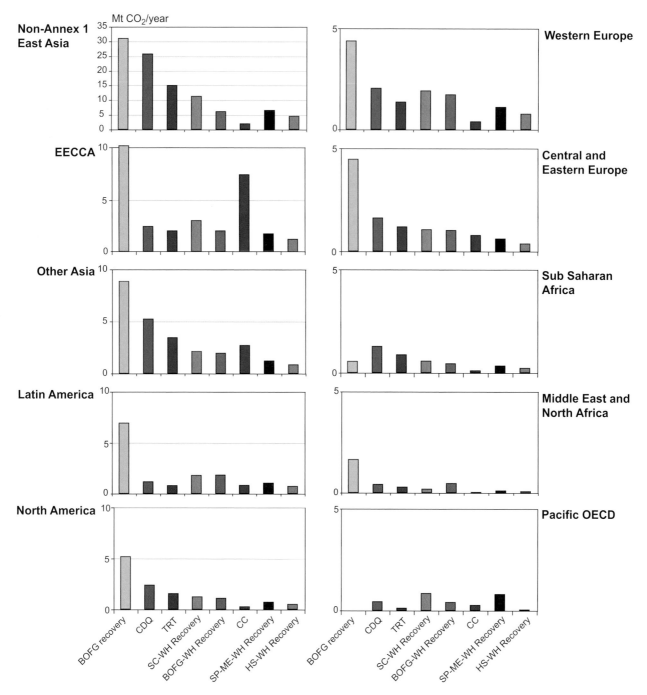

Figure 7.1: *CO₂ reduction potential of eight energy saving technologies in 2030*

CDQ = Coke Dry Quenching, HS = Hot Stove, TRT = Top Pressure Recovery Turbine, SC = Sinter Cooling, CC = Continuous Casting, SP = Sinter Plant, BOFG = Basic Oxygen Furnace Gas, ME = Main Exhaust, WH = Waste Heat

Note: B2 Scenario, CO₂ emission reduction based on energy saving assuming 100% diffusion in 2030 less current diffusion rates.

Source: Tanaka, 2006.

Technology enabling the use of oil, natural gas and pulverized coal to replace coke in iron-making has long been available. Use of this technology has been dictated by the relative costs of the fuels and the process limitations in iron-making furnaces. Use of oil and natural gas could reduce CO₂ emissions. More recently, the steel industry has developed technologies that use wastes, such as plastics, as alternative fuel and raw materials

(Ziebek and Stanek, 2001). Pretreated plastic wastes have been recycled in coke ovens and blast furnaces (Okuwaki, 2004), reducing CO₂ emissions by reducing emissions from incineration and the demand for fossil fuels. In Brazil, charcoal is used as an alternative to coke in blast furnaces. While recent data are not available, use of charcoal declined in the late 1990s, as merchant coke became cheaper (Kim and Worrell,

2002a). The use of hydrogen to reduce iron ore is a longer-term technology discussed in Section 7.12. CCS is another longer-term technology that might be applicable to steel making (see section 7.3.7).

7.4.2   Non-ferrous metals

The commercially relevant non-ferrous metals and specific and total CO_2 emissions from electrode and reductant use are shown in Table 7.6. Annual production of these metals ranges from approximately 30 Mt for aluminium to a few hundred kilotonnes for metals and alloys of less commercial importance. Production volumes are fairly low compared to some of the world's key industrial materials like cement, steel, or paper. However, primary production of some of these metals from ore can be far more energy intensive. In addition, the production of these metals can result in the emission of high-GWP GHGs, for example PFCs in aluminium or SF_6 in magnesium, which can add significantly to CO_2-eq emissions.

Generally, the following production steps need to be considered: mining, ore refining and enrichment, primary smelting, secondary smelting, metal refining, rolling and casting. For most non-ferrous metals, primary smelting is the most energy-intensive step, but significant levels of emissions of fluorinated GHGs have been reported from the refining and casting steps.

7.4.2.1   *Aluminium*

Global primary aluminium production was 29.9 Mt in 2004 (IAI, 2006b) and has grown an average of 5% per year over the last ten years. Production is expected to grow by 3% per year for the next ten years. Recycled aluminium production was approximately 14 Mt in 2004 and is also expected to double by 2020 (Marchek, 2006).

Primary aluminium metal (Al) is produced by the electrolytic reduction of alumina (Al_2O_3) in a highly energy-intensive process. In addition to the CO_2 emissions associated with electricity generation, the process itself is GHG-intensive. It involves a reaction between Al_2O_3 and a carbon anode: $2 Al_2O_3 + 3 C = 4 Al + 3 CO_2$. In the electrolysis cell, Al_2O_3 is dissolved in molten cryolite (Na_3AlF_6). If the flow of Al_2O_3 to the anode is lower than required, cryolite will react with the anode to form PFCs, CF_4 and C_2F_6 (IAI, 2001). CF_4 has a GWP[8] of 6500 and C_2F_6, which accounts for about 10% of the mix, has a GWP of 9200 (IPCC, 1995). These emissions can be significantly reduced by careful attention to operating procedures and more use of computer-control. Even larger reductions in emissions can be achieved by upgrading older cell technology (for example., Vertical Stud Södeberg or Side Worked Prebake) by addition of point feeders to better control alumina feeding. The

Table 7.6: *Emission factors and estimated global emissions from electrode use and reductant use for various non-ferrous metals*

	CO_2 emissions (tCO_2/t product)	Global CO_2 emissions (ktCO_2)
Primary aluminium	1.55	44,700
Ferrosilicon	2.92	10,500
Ferrochromium	1.63	9,500
Silicomanganese	1.66	5,800
Calcium carbide	1.10	4,475
Magnesium	0.05	4,000
Silicon metal	4.85	3,500
Lead	0.64	3,270
Zinc	0.43	3,175
Others		6,000
Total		91,000

Note: Indirect emissions and non-CO_2 greenhouse-gas emissions are not included.

Source: Sjardin, 2003.

cost of such a retrofit can be recovered through the improved productivity. Use of the newer technologies, which require a major retrofit, can cost up to 27 US$/t$CO_2$-eq (99 US$/tC-eq) (US EPA, 2006a).

Members of the International Aluminium Institute (IAI), responsible for more than 70% of the world's primary aluminium production, have committed to an 80% reduction in PFC emissions intensity for the industry as a whole, and to a 10% reduction in smelting energy intensity by 2010 compared to 1990 for IAI member companies. IAI data (IAI, 2006a) shows a reduction in CF_4 emissions intensity from 0.60 to 0.16 kg/t Al, and a reduction in C_2F_6 emissions intensity from 0.058 to 0.016 kg/t Al between 1990 and 2004, with best available technology having a median emission rate of only 0.05 kg CF_4/t in 2004. Overall, PFC emissions from the electrolysis process dropped from 4.4 to 1.2 tCO_2-eq/t (1.2 to 0.3 tC-eq/t) Al metal produced. IAI data (IAI, 2006b) show a 6% reduction in smelting energy use between 1990 and 2004.

Benchmarking has been used to identify opportunities for emission reductions. The steps taken to control these emissions have been mainly low or no-cost, and have commonly been connected to smelter retrofit, conversion, or replacements (Harnisch *et al.*, 1998; IEA GHG 2000). However, much of the 30% of production from non-IAI members still uses older technology (EDGAR, 2005).

SF_6 (GWP = 23,900 (IPCC, 1995)) has been used for stirring and degassing of molten aluminium in secondary smelters and foundries (Linde, 2005). The process is not very common

---

8   The Global Warming Potentials used in this chapter are those used for national inventory reporting under the UNFCCC. They are the 100-year values reported in the IPCC Second Assessment Report (IPCC, 1995).

Table 7.7: *Greenhouse-gas emission from production of various non-ferrous metals*

Metal	Global emissions (MtCO$_2$-eq/yr)	Source and year
Aluminium		
CO$_2$ - Mining and refining	109	IEA GHG, 2000 for 1995
CO$_2$ - Electrodes	48	IAI, 2006b for 2004
PFC - Emissions	69	EDGAR, 2005 for 2000
CO$_2$ - Electricity	300	IEA GHG, 2001 for 1995
Magnesium		
CO$_2$ - Electrode and cell-feed	4	Sjardin, 2003 for 1995
SF$_6$ - Production and casting	9	US EPA, 2006b for 2000
CO$_2$ - Electricity	Unknown	
CO$_2$ - Other steps in the production process	Unknown	
All other non-ferrous metals		
CO$_2$ - Process	40	Sjardin, 2003
CO$_2$ - Electricity	Unknown	
CO$_2$ - Other steps	Unknown	
All non-ferrous metals	**Approximately 500 (lower bound)**	

because of cost and technical problems (UBA, 2004). Current level of use is unknown, but is believed to be much smaller than SF$_6$ used in magnesium production.

The main potentials for additional CO$_2$-eq emission reductions are a further penetration of state-of-the-art, point feed, prebake smelter technology and process control plus an increase of recycling rates for old-scrap (IEA GHG, 2001). Research is proceeding on development of an inert anode that would eliminate anode-related CO$_2$ and PFC emissions from Al smelting. A commercially viable design is expected by 2020 (The Aluminium Association, 2003). However, IEA (2006a) notes that the ultimate technical feasibility of inert anodes has yet to be proven, despite 25 years of research.

7.4.2.2   Magnesium

Magnesium, produced in low volumes, is very energy intensive. Its growth rate has been high due to increasing use of this lightweight metal in the transport industry. SF$_6$ is quite commonly used as cover gas for casting the primary metal into ingots and for die casting magnesium. Estimates of global SF$_6$ emissions from these sources in 2000 range from about 9 MtCO$_2$-eq (2.4 MtC-eq) (US EPA, 2006a), to about 20 MtCO$_2$-eq (5.5 MtC-eq) (EDGAR, 2005). The later value is about equal to energy related emissions from the production of magnesium. Harnisch and Schwarz (2003) found that the majority of these emissions can be abated for <1.2 US$/tCO$_2$-eq (<4.4 US$/tC-eq) by using SO$_2$, the traditional cover gas, which is toxic and corrosive, or using more advanced fluorinated cover gases with low GWPs. US EPA (2006a) report similar results. Significant parts of the global magnesium industry located in Russia and

China still use SO$_2$ as a cover gas. The International Magnesium Association, which represented about half of global magnesium production in 2002, has committed its member companies to phasing out SF$_6$ use by 2011 (US EPA, 2006a).

7.4.2.3   Total emissions and reduction potentials

Table 7.7 gives the lower bounds for key emission sources in the non-ferrous metal industry. Total annual GHG gas emissions from the non-ferrous metal industry were at least 500 MtCO$_2$-eq (140 MtC-eq) in 2000. The GHG abatement options for the production of non-ferrous metals other than aluminium are still fairly uncertain. In the past, these industries have been considered too small or too complex regarding raw materials, production technologies and product qualities, to be systematically assessed for reduction options.

7.4.3   Chemicals and fertilizers

The chemical industry is highly diverse, with thousands of companies producing tens of thousands of products in quantities varying from a few kilograms to thousand of tonnes. Because of this complexity, reliable data on GHG emissions is not available (Worrell *et al.*, 2000a). The majority of the CO$_2$-eq direct emissions from the chemical industry are in the form of CO$_2$, the largest sources being the production of ethylene and other petrochemicals, ammonia for nitrogen-based fertilizers, and chlorine. These emissions are from both energy use and from venting and incineration of byproducts. In addition, some chemical processes create other GHGs as byproducts, for example N$_2$O from adipic acid, nitric acid and caprolactam manufacture; HFC-23 from HCFC-22 manufacture; and very small amounts of CH_4 from the manufacture of silicon carbide and some petrochemicals. Pharmaceutical manufacture uses relatively little energy, most of which is used in the buildings that house industrial facilities (Galitsky and Worrell, 2004).

The chemical industry makes use of many of the sector-wide technologies described in Section 7.3. Much of the petro-chemical industry is co-located with petroleum refining, creating many opportunities for process integration and cogeneration of heat and electricity. Both industries make use of the energy in byproducts that would otherwise be vented or flared, contributing to GHG emissions. Galitsky and Worrell (2004) identify separations, chemical synthesis and process heating as the major energy consumers in the chemical industry, and list examples of technology advances that could reduce energy consumption in each area, for example improved membranes for separations, more selective catalysts for synthesis and greater process integration to reduce process heating requirements. Longer-term, biological processing offers the potential of lower energy routes to chemical products (See Section 7.12.1).

7.4.3.1   Ethylene

Ethylene, which is used in the production of plastics and many other products, is produced by steam cracking hydrocarbon feedstocks, from ethane to gas oil. Hydrogen, methane, propylene and heavier hydrocarbons are produced as byproducts. The heavier the feedstock, the more and heavier the byproducts, and the more energy consumed per tonne of ethylene produced (Worrell *et al.*, 2000a). Ren *et al.* (2006) report that steam cracking for olefin production is the most energy consuming process in the chemicals industry, accounting for emissions of about 180 $MtCO_2$/yr (49MtC/yr), but that significant reductions are possible. Cracking consumes about 65% of the total energy used in ethylene production, but use of state-of-the-art technologies (e.g., improved furnace and cracking tube materials and cogeneration using furnace exhaust) could save up to about 20% of total energy. The remainder of the energy is used for separation of the ethylene product, typically by low-temperature distillation and compression. Up to 15% total energy can be saved by improved separation and compression techniques (e.g., absorption technologies for separation). Catalytic cracking also offers the potential for reduced energy use, with a savings of up to 20% of total energy. This savings is not additional to the energy savings for improved steam cracking (Ren *et al.*, 2006). Processes have been developed for converting methane in natural gas to olefins as an alternative to steam cracking. However, Ren *et al.* (2005) conclude that the most efficient of these processes uses more than twice as much primary energy as state-of-the-art steam cracking of naphtha.

7.4.3.2   Fertilizer manufacture

Swaminathan and Sukalac (2004) report that the fertilizer industry uses about 1.2% of world energy consumption and is

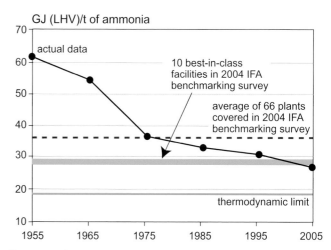

Figure 7.2: *Design energy consumption trends in world ammonia plants*
Sources: Chaudhary, 2001; PSI, 2004.

responsible for about the same share of global GHG emissions. More than 90% of this energy is used in the production of ammonia (NH_3). However, as the result of energy efficiency improvements, modern ammonia plants are designed to use about half the energy per tonne of product than those designed in 1960s, (see Figure 7.2), with design energy consumption dropping from over 60 GJ/t NH_3 in the 1960s to 28 GJ/t NH_3 in the latest design plants, approaching the thermodynamic limit of about 19 GJ/t NH_3, and limiting scope for further efficiency increases. Benchmarking data indicate that the best-in-class performance of operating plants ranges from 28.0 to 29.3 GJ/t NH_3 (Chaudhary, 2001; PSI, 2004).

The newest plants tend to have the best energy performance, and many of them are located in developing countries, which now account for 57% of nitrogen fertilizer production (IFA, 2004). Individual differences in energy performance are mostly determined by feedstock (natural gas compared with heavier hydrocarbons) and the age and size of the ammonia plant (PSI, 2004, Phylipsen *et al.*, 2002). National and regional averages are strongly influenced by whether the sector has undergone restructuring, which tends to drive less efficient producers out of the market (Sukalac, 2005). Ammonia plants that use natural gas as a feed-stock have an energy efficiency advantage over plants that use heavier feedstock's and a high percentage of global ammonia capacity already is based on natural gas. China is an exception in that 67% of its ammonia production is based on coal (CESP, 2004) and small-scale plants account for 90% of the coal-based production. The average energy intensity of Chinese coal-based production is about 53 GJ/t, compared with a global average of 41.4 GJ/t (Giehlen, 2006).

Retrofit of old plants is feasible and offers a potential for improved efficiency. Verduijn and de Wit (2001) concluded that the energy efficiency of large single train ammonia plants, the bulk of existing capacity, could be improved at reasonable cost to levels approaching newly designed plants, provided that the upgrading is accompanied by an increase in capacity.

Significant reductions of CO_2 emissions, below those achieved by state-of-the-art ammonia plants, could be achieved by using low-carbon or carbon-free hydrogen, which could be obtained through the application of CCS technology (see Section 7.3.7), biomass gasification, or electrolysis of water using electricity from nuclear or renewables. About half the ammonia produced for fertilizer is reacted with CO_2 to form urea (UNIDO and IFDC, 1998), but the CO_2 is released when the fertilizer is applied. However, this use of CO_2 reduces the potential for applying CCS technology.

7.4.3.3   Chlorine manufacture

The TAR (IPCC, 2001a) reported on the growing use of more energy-efficient membrane electrolysis cells for chlorine production. There have been no significant developments affecting GHG emissions from chlorine production since the TAR.

7.4.3.4   N₂O emissions from adipic acid, nitric acid and caprolactam manufacture

N_2O emissions from nitric and adipic acid plants account for about 5% of anthropogenic N_2O emissions. Due to significant investment in control technologies by industry in North America, Japan and the EU, worldwide emissions of N_2O (GWP = 310 (IPCC,1995)) from adipic and nitric acid production decreased by 30%, from 223 $MtCO_2$-eq (61 MtC-eq) in 1990 to 154 $MtCO_2$-eq (42 MtC-eq) in 2000 (US EPA 2006b). Some of the reduction was due to the installation of NO control technology to meet regulatory requirements. By 2020, global emission from the manufacture of adipic acid and from the manufacture of nitric acid are projected to grow to 177 $MtCO_2$-eq (48 MtC-eq). Developed nations account for approximately 55% of emissions in both 2000 and 2020 (US EPA, 2006b). Experience in the USA, Japan and the EU shows that thermal destruction can eliminate 96% of the N_2O emitted from an adipic acid plant. Catalytic reduction can eliminate 89% of the N_2O emitted from a typical nitric plant in a developed country (US EPA, 2006a). Mitigation potential at nitric acid plants can range from 70% to almost 100% depending on the catalyst and plant operating conditions (US EPA, 2001, Continental Engineering BV, 2001). Costs range from 2.0 to 5.8 US$/tCO_2-eq (7.3 to 21.2 US$/tC-eq) (2000 US$) using a 20% discount rate and a 40% corporate tax rate, and a maximum mitigation potential of 174 $MtCO_2$-eq (44 MtC-eq) is projected in 2030.

Global N_2O emissions from caprolactam production in 2000 were estimated at 10 to 15 $MtCO_2$-eq (2.7 to 4.1 MtC) (EDGAR, 2005). IPCC (2006) indicates that these emissions can be controlled to a high degree by non-specific catalytic reduction.

7.4.3.5   HFC-23 emissions from HCFC-22 manufacture

On average, 2.3% HFC-23 (GWP = 11,700 (IPCC, 1995)) is produced as a byproduct of HCFC-22 manufacture. The EDGAR database estimated 2000 emissions at 78 $MtCO_2$-eq (21 MtC-eq) (EDGAR, 2005), while the US EPA estimated 96 $MtCO_2$-eq (26 MtC-eq) (US EPA, 2006a). HCFC-22 has been used as a refrigerant, but under the Montreal Protocol its consumption is scheduled to end by 2020 in developed countries and over a longer period in developing countries. However, production of HCFC-22 for use as a feedstock in the manufacture of fluoropolymers, plastics and HFCs is expected to grow, leading to increasing emissions through 2015 in the business-as-usual case. Data on production rates and control technologies are contained in the IPCC Special Report on Safeguarding the Ozone Layer and the Global Climate System (IPCC/TEAP, 2005). Capture and destruction by thermal oxidation is a highly effective option for reducing HFC-23 emissions at a cost of less than 0.20 to 0.35 US$/tCO_2-eq (0.75 to 1.20 US$/tC-eq) (IPCC/TEAP, 2005, US EPA, 2006a).

7.4.4   Petroleum refining

As of the beginning of 2004, there were 735 refineries in 128 countries with a total crude oil distillation capacity of 82.3 million barrels per day. The U.S (20.5%), EU-25 (16.4%), Russia (6.6%), Japan (5.7%) and China (5.5%) had the largest shares of this capacity (EIA, 2005). Petroleum industry operations consume up to 15 to 20% of the energy in crude oil, or 5 to 7% of world primary energy, with refineries consuming most of that energy (Eidt, 2004). Comparison of energy or CO_2 intensities among countries is not practical because refining energy use is a complex function of crude and product slates and processing equipment. Simple metrics (e.g., energy consumed/barrel refined) do not account for that complexity. The shifts towards heavier crude and lower sulphur products will increase refinery energy use and CO_2 emissions. One study indicated that the combination of heavier crude and a 10 ppm maximum gasoline and diesel sulphur content would increase European refinery CO_2 emissions by about 6% (CONCAWE, 2005).

Worrell and Galitsky (2005), based on a survey of US refinery operations, found that most petroleum refineries can economically improve energy efficiency by 10–20%, and provided a list of over 100 potential energy saving steps. Key items included: use of cogeneration, improved heat integration, combustion optimization, control of compressed air and steam leaks and use of efficient electrical devices. The petroleum industry has had long-standing energy efficiency programmes for refineries and the chemical plants with which they are often integrated. These efforts have yielded significant results. Exxon Mobil reported over 35% reduction in energy use in its refineries and chemical plants from 1974 to 1999, and in 2000 instituted a programme whose goal was a further 15% reduction, which would reduce emissions by an additional 12 $MtCO_2$/yr. (Eidt, 2004). Chevron (2005) reported a 24% reduction in its index of energy use between 1992 and 2004. Shell (2005) reported energy efficiency improvements of 3 to 7% at its refineries and chemical plants. Efficiency improvements are expected to continue as technology improves and energy prices rise.

Refineries typically use a wide variety of gaseous and liquid byproducts as fuel. Byproducts that are not used as fuel are flared. Reducing the amount of material flared will increase refinery energy efficiency and decrease CO_2 emissions, and has become an objective for refinery management worldwide, though flare reduction projects are often undertaken to reduce local environmental impacts Munn (2004). No estimate of the incremental reduction in CO_2 emissions is available.

Refineries use hydrogen to remove sulphur and other impurities from products, and to process heavy hydrocarbons into lighter components for use in gasoline and distillate fuels. The hydrogen is supplied from reformer gas, a hydrogen-rich byproduct of catalytic reforming, and a process for upgrading gasoline components. If this source is insufficient for the refinery's needs, hydrogen is manufactured by gasification of fossil fuels. US refineries use about 4% of their energy input to manufacture hydrogen (Worrell and Galitsky, 2005). Hydrogen production produces a CO_2-rich stream, which is a candidate for CCS (see Section 7.3.7).

7.4.5   Minerals

7.4.5.1   Cement

Cement is produced in nearly all countries. Cement consumption is closely related to construction activity and to general economic activity. Global cement production grew from 594 Mt in 1970 to 2200 Mt in 2005, with the vast majority of the growth occurring in developing countries. In 2004 developed countries produced 570 Mt (27% of world production) and developing countries 1560 Mt (73%) (USGS, 2005). China has almost half the world's cement capacity, manufacturing an estimated 1000 Mt in 2005 (47% of global production), followed by India with a production of 130 Mt in 2005 (USGS, 2006). Global cement consumption is growing at about 2.5%/yr.

The production of clinker, the principal component of cement, emits CO_2 from the calcination of limestone. Cement production is also highly energy-intensive. The major energy uses are fuel for the production of clinker and electricity for grinding raw materials and the finished cement. Coal dominates in clinker making. Based on average emission intensities, total emissions in 2003 are estimated at 1587 $MtCO_2$ (432 MtC) to 1697 $MtCO_2$ (462 MtC), or about 5% of global CO_2 emissions, half from process emissions and half from direct energy use. Global average CO_2 emission per tonne cement production is estimated by Worrell *et al.* (2001b) at 814 kg (222 kg C), while Humphreys and Mahasenan (2002) estimated 870 kg (264 kg C). CO_2 emission/t cement vary by region from a low of 700 kg (190 kg C) in Western Europe and 730 kg (200 kg C) in Japan and South Korea, to a high of 900, 930, and 935 kg (245, 253 and 255 kg C) in China, India and the United States (Humphreys and Mahasenan, 2002; Worrell *et al.*, 2001b). The differences in

emission intensity are due (in order of contribution) to differences in the clinker content of the cement produced, energy efficiency, carbon intensity of the clinker fuel and carbon intensity of power generation (Kim and Worrell, 2002b).

Emission intensities have decreased by approximately 0.9%/yr since 1990 in Canada, 0.3%/yr (1970–1999) in the USA, and 1%/yr in Mexico (Nyboer and Tu, 2003; Worrell and Galitsky, 2004; Sheinbaum and Ozawa, 1998). A reduction in energy intensity in India since 1995–1996 has led to a reduction in emissions from the industry despite the increase in output (Dasgupta and Roy, 2001). Analysis of CO_2 emission trends in four major cement-producing countries showed that energy efficiency improvement and reduction of clinker content in cement were the main factors contributing to emission reduction, while the carbon intensity of fuel mix in all countries increased slightly.

Both energy-related and process CO_2 emissions can be reduced. The combined technical potential of these opportunities is estimated at 30% globally, varying between 20 and 50% for different regions (Humphreys and Mahasenan, 2002; Kim and Worrell, 2002b). Energy efficiency improvement has historically been the main contributor to emission reduction. Benchmarking and other studies have demonstrated a technical potential for up to 40% improvement in energy efficiency (Kim and Worrell, 2002b; Worrell *et al.*, 1995). Countries with a high potential still use outdated technologies, like the wet process clinker kiln. Studies for the USA identified 30 opportunities in every production step in the cement-making process and estimated the economic potential for energy efficiency improvement in the US cement industry at 11%, reducing emissions by 5% (Worrell *et al.*, 2000b; Worrell and Galitsky, 2004). The cement industry is capital intensive and equipment has a long lifetime, limiting the economic potential in the short term. The clinker kiln is an ideal candidate for the use of a wide variety of fuels, including waste-derived fuels, such as tyres, plastics, biomass, municipal solid wastes and sewage sludge (see Section 7.3.2). Section 7.3.7 discusses the potential for applying CCS in the cement industry.

Standard Portland cement contains 95% clinker. Clinker production is responsible for the process emissions and most of the energy-related emissions. The use of blended cement, in which clinker is replaced by alternative cementitious materials, for example blast furnace slag, fly ash from coal-fired power stations, and natural pozzolanes, results in lower CO_2 emissions (Josa *et al.*, 2004). Humphreys and Mahasenan (2002) and Worrell *et al.* (1995) estimate the potential for reduction of CO_2 emissions at more than 7%. Current use of blended cement is relatively high in continental Europe and low in the USA and UK. Alternatives for limestone-based cement are also being investigated (Gartner, 2004; Humphreys and Mahasenan, 2002). Geopolymers have been applied in niche markets, but have yet to be proven economical for large-scale application.

7.4.5.2   Lime

Generally lime refers both to high-calcium and dolomitic forms containing magnesium. Lime is produced by burning limestone or dolomite in small-scale vertical or large-scale rotary kilns. While in most industrialized countries the industry is concentrated in a small number of larger corporations, in most developing countries lime kilns are small operations using local technology. Even in industrialized countries like Greece there are independent small-scale vertical kilns in operation. Pulp and sugar mills may have captive lime production to internally regenerate lime. Lime is mainly used in a small number of industries (especially steel, but also chemicals, paper and sugar), mining, as well as for flue gas desulphurization. There are no detailed statistics on global lime production, however Miller (2003) estimated global production at 120 Mt, excluding regenerated lime. The largest producers are China, the USA, Russia, Germany, Mexico and Brazil.

Process CO_2 emissions from the calcination of limestone and dolomite are a function of the amounts of calcium carbonate, magnesium carbonate and impurity in the feedstock, and the degree of calcination. Theoretical process emissions are 785 kg CO_2/t (214 kgC/t) calcium oxide and 1092 kg CO_2/t (298 kgC/t) magnesium oxide produced. Energy use emissions are a function of the efficiency of the process, the fuel used, and indirect emissions from the electric power consumed in the process. In efficient lime kilns about 60% of the emissions are due to de-carbonisation of the raw materials. No estimates of global CO_2 emissions due to lime production are available. In Europe process emissions are estimated at 750 kg CO_2/t (205 kgC/t) lime (IPPC, 2001). For some applications, lime is re-carbonated, mitigating part of the emissions generated in the lime industry. Regeneration of lime in pulp and sugar mills does not necessarily lead to additional CO_2 emissions, as the CO_2 is from biomass sources (Miner and Upton, 2002). Emissions from fuel use vary with the kiln type, energy efficiency and fuel mix. Energy use is 3.6 to 7.5 GJ/t lime in the EU (IPPC, 2001), 7.2 GJ/t in Canada (CIEEDAC, 2004) and for lime kilns in US pulp mills (Miner and Upton, 2002), and up to 13.2 GJ/t for small vertical kilns in Thailand (Dankers, 1995). In Europe, fuel-related emissions are estimated at 0.2 to 0.45 tCO$_2$/t (0.05 to 0.12 tC/t) lime (IPPC, 2001). Electricity use for lime production is 40 to 140 kWh/t lime, depending on the type of kiln and the required fineness of the lime (IPPC, 2001).

Emission reductions are possible by use of more efficient kilns (Dankers, 1995; IPPC, 2001) and through improved management of existing kilns, using similar techniques to the cement industry (see Section 7.4.5.1). Switching to low-fossil carbon fuels can further reduce CO_2 emissions. The use of solar energy has been investigated for small-scale installations (Meier et al., 2004). It may also be possible to reduce lime consumption in some processes, for example the sugar industry (Vaccari et al., 2005).

7.4.5.3   Glass

Glass is produced by melting raw materials (mainly silica, soda ash and limestone), and often cullet (recycled glass), in glass furnaces of different sizes and technologies. Typical furnace designs include: cross-fired or end-fired with regenerative air preheat, recuperative heat recovery and fuel-oxygen firing (EU-BREF Glass, 2001). The industry is capital intensive, furnaces have a lifetime of up to 12 years and there are a limited number of technology providers. Natural gas and fuel oil are the main fuels used by the glass industry. Reliable international statistics on glass production are not available. The global glass industry is dominated by the production of container glass and flat glass. According to industry estimates the global production of container glass was 57 Mt in 2001 (ISO, 2004); production of flat glass was 38 Mt in 2004 (Pilkington, 2005). The production volumes of special glass, domestic glass, mineral wool and glass fibres are each smaller by roughly an order of magnitude.

Beerkens and van Limpt (2001) report the energy intensity of continuous glass furnaces in Europe and the USA as 4 to 10 GJ/t of container glass and 5 to 8.5 GJ/t of flat glass, depending on the size and technology of the furnace and the share of cullet used. The energy consumption for batch production is higher, typically 12.5 to 30 GJ/t of product (Römpp, 1995). Assuming an average energy use of 7 GJ/t of product, half from natural gas and half from fuel oil, yields an emission factor of 450 kg energy related CO_2/t of product. Globally, energy used in the production of container and flat glass results in emissions of approximately 40 to 50 MtCO$_2$ (11 to 14 MtC) per year. Emissions from the decarbonisation of soda ash and limestone can contribute up to 200 kg CO_2/t (55 kgC/t) of product depending on the composition of the glass and the amount of cullet used (EU-BREF Glass, 2001).

The mid-term emission potential for energy efficiency improvements is less than half of what corresponds to the range of efficiencies reported by Beerkens and van Limpt (2001), which also reflect differences in product quality and furnace age. The global potential for emissions reduction from fuel switching is unknown. The main mitigation options in the industry include: improved process control, increased use (up to 100%) of cullet (Kirk-Othmer, 2005), increased furnace size, use of regenerative heating, oxy-fuel technology, batch and cullet pre-heating, reduction of reject rates (Beerkens and van Limpt, 2001), use of natural gas instead of fuel oil, and CO_2 capture for large oxy-fuel furnaces. High caloric value biogas could be used to reduce net CO_2 emissions, but potential new break-through technologies are not in sight.

7.4.5.4   Ceramics

The range of commercial ceramics products is large and includes bricks, roof, wall and floor tiles, refractory ceramics, sanitary ware, tableware and cookware and other products. In terms of volume, the production of bricks and tiles dominate.

The main raw materials used in the brick industry include clay and kaolin. Production technologies and respective energy efficiencies vary tremendously from large industrial operations to cottage and artisan production, which are still very common in many developing countries. The main fuels used in modern industrial kilns are natural gas and fuel oil. Specific energy consumption varies considerably for different products and kiln designs. The EU-BREF Ceramics (2005) reported specific energy consumptions for modern industrial brick production of 1.4 to 2.4 GJ/t of product.

Small-scale kilns – used mainly for brick production – are often used in developing countries. Wood, agricultural residues and coal (FAO, 1993) are the main fuels used, with specific energy consumptions of 0.8 to 2.8 GJ/t of brick for the small- to medium sized kilns, and 2 to 8 GJ/t of brick for the very small-scale kilns used by cottage industries and artisans (FAO, 1993). Producers also utilize the energy contained in the organic fraction of clay and shale as well as in pore forming agents (e.g., sawdust) added to the clay in the production process. CO_2 emissions from the calcination of carbonates contained in clay and shale typically contribute 20 to 50% of total emissions. The current choices of building materials and kiln technologies are closely related to local traditions, climate, and the costs of labour, capital, energy and transport, as well as the availability of alternative fuels, raw materials and construction materials.

Reliable international statistics on the production of ceramics products are not available. Consumption of bricks, tiles and other ceramic products in tonnes per capita per year is estimated at 1.2 in China (Naiwei, 2004); 0.4 in the EU (EU-BREF Ceramics, 2005), 0.1 in the USA (USGS, 2005), and 0.25, 0.12, and 0.05 for Pakistan, India and Bangladesh (FAO, 1993). This suggests that the global production of ceramic products exceeds 2 Gt/yr, leading to the emission of more than 400 MtCO$_2$ (110 MtC) per year from energy use and calcination of carbonates. Additional research to better understand the emission profile and mitigation options for the industry is needed.

GHG mitigation options include the use of more efficient kiln design and operating practices, fuel switching from coal to fuel oil, natural gas and biomass, and partial substitution of clay and shale by alternative raw materials such as fly ash. Mitigation options could also include the use of alternative building materials such as wood or bricks made from lime and sand. However, emissions over the whole life cycle of the products including their impact on the energy performance of the building need to be considered.

7.4.6   Pulp and paper

The pulp and paper industry is a highly diverse and increasing global industry. In 2003, developing countries produced 26% of paper and paperboard and 29% of global wood products; 31% of paper and paperboard output was traded internationally (FAOSTAT, 2006). Direct emissions from the pulp, paper,

paperboard and wood products industries are estimated to be 264 MtCO$_2$/yr (72 MtC/yr) (Miner and Lucier, 2004). The industry's indirect emissions from purchased electricity are less certain, but are estimated to be 130 to 180 MtCO$_2$/yr (35 to 50 MtC/yr) (WBCSD, 2005).

7.4.6.1   *Mitigation options*

Use of biomass fuels: The pulp and paper industry is more reliant on biomass fuels than any other industry. In developed countries biomass provides 64% of the fuels used by wood products facilities and 49% of the fuel used by pulp, paper and paperboard mills (WBCSD, 2005). Most of the biomass fuel used in the pulp and paper industry is spent pulping liquor, which contains dissolved lignin and other materials from the wood that are not used in paper production. The primary biomass fuel in the wood-products sector is manufacturing residuals that are not suitable for use as byproducts.

Use of combined heat and power: In 2002, the pulp and paper industry used cogeneration to produce 40% of its electricity requirements in the USA (US DOE, 2002) and over 30% in the EU (CEPI, 2001), and that use continues to grow.

Black liquor gasification: Black liquor is the residue from chemical processing to produce wood pulp for papermaking. It contains a significant amount of biomass and is currently being burned as a biomass fuel. R&D is underway on gasification of this material to increase the efficiency of energy recovery. Gasification could also create the potential to produce synfuels and apply CCS technology. IEA (2006a) estimates a 10 to 30 MtCO$_2$ (2.7 to 8.1 MtC) mitigation potential for this technology in 2030. While gasification would increase the energy efficiency of pulp and paper plants, the industry as a whole would not become a net exporter of biomass energy (Farahani *et al.*, 2004).

Recycling: Recovery rates for waste paper (defined as the percentage of domestic consumption that is collected for reuse) in developed countries are typically at least 50% and are over 65% in Japan and parts of Europe (WBCSD, 2005). Globally, the utilization rate (defined as the fraction of fibre feedstock supplied by recovered fibre) was about 44% in 2004 (IEA, 2006a). The impact of this recycling is complex, affecting the emissions profile of paper plants, forests and landfills. A number of studies examine the impacts of recycling on life-cycle GHG emissions (Pickens *et al.*, 2002, Bystrom and Lonnstedt, 1997). These and other studies vary in terms of boundary conditions and assumptions about end-of-life management, and none attempt to examine potential indirect impacts of recycling on market-based decisions to leave land in forest rather than convert it to other uses. Although most (but not all) of these studies find that paper recycling reduces life-cycle emissions of GHG compared to other means of managing used paper, the analyses are dependent on study boundary conditions and site-specific factors and it is not yet possible to develop reliable estimates of the global mitigation potential related to recycling. However, both the USA (US EPA, 2002) and EU (EC, 2004) identify paper recycling as a GHG emissions reduction option.

7.4.6.2   Emission reduction potential

Because of increased use of biomass and energy efficiency improvements, the GHG emissions from the pulp and paper industry have been reduced over time. Since 1990, CO_2 emission intensity of the European paper industry has decreased by approximately 25% (WBCSD, 2005), the Australian pulp and paper industry about 20% (A3P, 2006), and the Canadian pulp and paper industry over 40% (FPAC, n.d.). Fossil fuel use by the US pulp and paper industry declined by more than 50% between 1972 and 2002 (AF&PA, 2004). However, despite these improvements, Martin *et al.* (2000) found a technical potential for GHG reduction of 25% and a cost-effective potential of 14% through widespread adoption of 45 energy-saving technologies and measures in the US pulp and paper industry. Möllersten *et al.* (2003) found that CO_2 emissions from the Swedish pulp and paper industry could be reduced by 0.5 to 5.0 $MtCO_2$/yr (0.14 to 1.4 MtC/yr) at negative cost using commercially available technologies, primarily by generating more biomass-based electricity to displace carbon-intensive electricity from the grid. The large variation in the results reflected varying assumptions about the carbon intensity of displaced electricity and the impacts of 'industrial valuation' compared with 'societal valuation' of capital. Inter-country comparisons of energy-intensity in the mid-1990s suggest that fuel consumption by the pulp and paper industry could be reduced by 20% or more in a number of countries by adopting best practices (Farla *et al.*, 1997).

7.4.7   Food

Most food industry products are major commercial commodities, particularly for developing countries, and are quite energy-intensive. The most important products from a climate perspective are sugar, palm oil, starch and corn refining, since these can be a source of fuel products. The sugar cane industry produces 1.2 Gt sugar/yr. (Banda, 2002) from about 1670 mills, mostly located in tropical developing countries (Sims, 2002). Edible oils are another significant product, the exports of which support many developing country economies. Malaysia, the world's largest producer and exporter of palm oil, has 3.5 Mha under palm oil production (UNDP, 2002), whilst Sri Lanka, the world's fourth largest producer of coconut oils, has 0.4 Mha under cultivation (Kumar *et al.*, 2003).

Corn refining, including wet corn milling, has been the fastest growing market for US agriculture over the past twenty years (CRA, 2002). Further growth is projected as a result of the demand for ethanol as an automotive fuel. Corn wet milling is the most energy-intensive food industry, using 15% of total US food industry energy (EIA, 2002). Over 100 technologies and

measures for improving energy efficiency of corn wet milling have been identified (Galitsky *et al.*, 2003).

7.4.7.1   Production processes, emissions and emission intensities

The main production processes for the food industry are almost identical, involving preparatory stages including crushing, processing/refining, drying and packaging. Most produce process residuals, which typically go to waste. Food production requires electricity, process steam and thermal energy, which in most cases are produced from fossil fuels. The major GHG emissions from the food industry are CO_2 from fossil fuel combustion in boilers and furnaces, CH_4 (GWP=21 (IPCC, 1995)) and N_2O (GWP = 310 (IPCC, 1995)) from waste water systems.

The largest source of food industry emissions is CH_4 from waste water treatment, which could be recovered for energy generation. For example, the Malaysian palm oil industry emits an estimated 5.17 $MtCO_2$-eq (1.4 MtC-eq) from open-ponding systems that could generate 2.25 GWh of electricity while significantly reducing GHG emissions (Yeoh, 2004). Emissions from the Thai starch industry (Cohen, 2001) are estimated at 370 $ktCO_2$-eq/yr (101 ktC-eq/yr), 88% were from waste water treatment, 8% from combustion of fuel oil and 4% from grid electricity. Although individual food industry factory emissions are low, their cumulative effect is significant in view of the large numbers of factories in both developed and developing countries. Typical energy intensities estimated at about 11 GJ/t for edible oils, 5 GJ/t for sugar and 10 GJ/t for canning operations (UNIDO, 2002).

7.4.7.2   Mitigation opportunities

The most important mitigation opportunities to reduce food industry GHG emissions in the near- and medium-term include technology and processes related to good housekeeping and improved management, improvements in both cross-cutting systems (e.g., boilers, steam and hot water distribution, pumps, compressors and fans) and process-specific technologies, improved process controls, more efficient process designs and process integration (Galitsky *et al.*, 2001), cogeneration to produce electricity for own use and export (Cornland, 2001), and anaerobic digestion of residues to produce biogas for electricity generation and/or process steam (Yeoh, 2004). These technologies were discussed in Section 7.3, but some specific food industry applications are presented below.

In Brazil, electricity sales to the grid from bagasse cogeneration reached 1.6 TWh in 2005 from an installed capacity of 400 MW. This capacity is expected to increase to 1000 MW with implementation of a government-induced voluntary industry programme (Moreira, 2006). In India, the sugar industry has diversified into cogeneration of power and production of fuel ethanol. Cogeneration began in 1993–1994, and as of 2004 reached 680 MW. Full industry potential is estimated at 3500

MW. In 2001, India instituted a mixed fuel programme requiring use of a 5% ethanol blend, which will create an annual demand for 500 M litres of ethanol (Balasubramaniam, 2005).

Application of traditional boilers with improved combustion and CEST (Condensing Extraction Steam Turbines) in the southern African sugar industry could produce surpluses of 135 MW for irrigation purposes and 1620 MW for export to the national grid (Yamba and Matsika, 2003) in 2010. Sims (2002) found that if all 31 of Australia's existing sugar mills were converted to CEST technology, they could generate 20 TWh/yr of electricity and reduce emissions by 16 $MtCO_2$/yr (4.4 MtC/ yr), assuming they replaced coal-fired electricity generation. Gasifying the biomass and using it in combined cycle gas turbine could double the CO_2 savings (Cornland, 2001). Proposed CDM projects in the Malaysian palm oil industry (UNDP, 2002), and the Thai starch industry (Cohen, 2001) demonstrated that use of advanced anaerobic methane reactors to produce electricity would yield a GHG emission reduction of 56 to 325 $ktCO_2$-eq/yr (15 to 90 MtC-eq/yr). Application of improved energy management practices in the coconut industry (Kumar *et al.*, 2003) and bakery industry (Kannan and Boy, 2003) showed significant saving of 40 to 60 % in energy consumption for the former and a modest saving of 6.5% for the latter. In the long term, use of residue biomass generated from the food industry in state-of-the-art Biomass Integrated Gasifier Combined Cycle (BIG/CC) technologies, could double electricity generation and GHG savings compared to CEST technology (Yamba and Matsika, 2003; Cornland *et al.*, 2001).

Virtually all countries have environmental regulations of varied stringency, which require installations including the food industry to limit final effluent BOD (Biochemical Oxygen Demand) in the waste water before discharge into waterways. Such measures are compelling industries to use more efficient waste water treatment systems. The recently introduced EU-directive requiring Best Available Techniques (BAT) as a condition for environmental permits in the fruit and vegetable processing industry (Dersden *et al.*, 2002) will compel EU industry in this sector to introduce improved waste water purification processes thereby reducing fugitive emissions due to anaerobic reactions.

7.4.8   Other industries

This section covers a selection of other industries with significant emissions of high GWP gases. While some analyses include all emissions of these gases in the industrial sector, this chapter will consider only those which actually occur in the industrial sector. Thus, HFC and PFC emissions from use of automotive and residential air conditioning are covered in Chapter 5, section 5.2.1 and Chapter 6, section 6.8.4 respectively.

The manufacture of semiconductors, liquid crystal display and photovoltaic cells can result in the emissions of PFCs, SF_6, NF_3 and HFC-23 (IPCC, 2006). The technology available to reduce these emissions from semiconductor manufacturing, and the World Semiconductor Council (WSC) commitment to reduce PFC emissions by at least 10% by 2010 from 1995-levels are discussed in the TAR (IPCC, 2001a). US EPA (2006a) reports that emission levels from semiconductor manufacture were about 30 $MtCO_2$-eq (7 MtC-eq) in 2000, and that significant growth in emissions will occur unless the WSC commitment is implemented globally and strengthened after 2010. US EPA (2006a) estimates that this 10% reduction could occur cost-effectively through replacement of C_2F_6 by C_3F_8 (which has a lower GWP), NF_3 remote cleaning of the chemical vapour deposition chamber, or capturing and recycling of SF_6. Emissions from the production of liquid crystal displays and photovoltaic cells, mainly located in Asia, Europe and the USA, are growing rapidly and mitigation options need further research.

SF_6 emissions in 2000 from the production of medium and high voltage electrical transmission and distribution equipment were estimated at about 10 $MtCO_2$-eq (2.8 MtC-eq) (IEA GHG, 2001). These emissions, mainly located in Europe and Japan, are estimated to have declined, despite a 60% growth in production between 1995 and 2003, mainly due to targeted training of staff and improved gas handling and test procedures at production sites. Emissions of SF_6 at the end-of-life of electrical equipment are growing in relevance, and US EPA (2006b) estimates total SF_6 emissions from production, use and disposal of electrical equipment at 27 $MtCO_2$ in 2000 growing to 66 $MtCO_2$ in 2020, if no mitigation actions are taken. Emissions from disposal of electrical equipment could be reduced by implementation of a comprehensive recovery system, addressing all entities involved in handling and dismantling this equipment (Wartmann and Harnisch, 2005).

A third group of industries that emits hydrofluorocarbons (HFCs) includes those manufacturing rigid foams, refrigeration and air conditioning equipment and aerosol cans, as well as industries using fluorinated compounds as solvents or for cleaning purposes. This group of industries previously used ozone-depleting substances (ODS), which are subject to declining production and use quotas defined under the Montreal Protocol. As part of the phase out of ODS, many of them have switched to HFCs as replacements, or intend to do so in the future. Mitigation options include improved containment, training of staff, improved recycling at the end-of-life, the use of very low GWP alternatives, and the application of not-in-kind technologies. A detailed discussion of use patterns, emission projections and mitigation options for these applications can be found in IEA GHG (2001), IPCC/TEAP (2005) and more recent US EPA reports (2006a,b).

IEA GHG (2001) estimated that global fugitive emissions from the production of HFCs will rise from 2 $MtCO_2$-eq (0.6 MtC-eq) in 1996 to 8 $MtCO_2$-eq (2.2 MtC-eq) by 2010. Solvent and cleaning uses of HFCs and PFCs are commonly emissive despite containment and recycling measures. IEA GHG (2001) forecast that these emissions would increase to up to 20 $MtCO_2$-eq/yr (5.5 MtC-eq/yr) by 2020. However other analyses suggest a more moderate growth in emissions from solvent applications to about 5 $MtCO_2$-eq/yr (1.4 MtC-eq/yr) by 2020 (IPCC/TEAP, 2005).

7.4.9   Inter-industry options

Some options for reducing GHG emissions involve more than one industry, and may increase energy use in one industry to achieve a greater reduction in energy use in another industry or for the end-use consumer. For example, the use of granulated slag in Portland cement may increase energy use in the steel industry, but can reduce both energy consumption and CO_2 emissions during cement production by about 40%. Depending on the concrete application, slag content can be as high as 60% of the cement, replacing an equivalent amount of clinker (Cornish and Kerkhoff, 2004). Lightweight materials (high-tensile steel, aluminium, magnesium, plastics and composites) often require more energy to produce than the heavier materials they replace, but their use in vehicles will reduce transport sector energy use, leading to an overall reduction in global energy consumption. Life-cycle calculations (IAI, 2000) indicate that the CO_2 emission reductions in vehicles resulting from the weight reduction achieved by using aluminium more than offsets the GHG emissions from producing the aluminium.

Co-siting of industries can achieve GHG mitigation by allowing the use of byproducts as useful input and by integrating energy systems. In Kalundborg (Denmark) various industries (e.g., cement and pharmaceuticals production and a CHP plant) form an eco-industrial park that serves as an example of the integration of energy and material flows (Heeres et al., 2004). Heat-cascading systems, where waste heat from one industry is used by another, are a promising cross-industry option for saving energy. Based on the Second Law of Thermodynamics, Grothcurth et al. (1989) estimated up to 60% theoretical energy saving potential from heat cascading systems. However, Matsuhashi et al. (2000) found the practical potential of these systems was limited to approximately 5% energy saving. Actual potential will depend on site-specific conditions.

7.5   Short- and medium-term mitigation potential and cost

Limited information is available on mitigation potential and cost[9] in industry, but it is sufficient to develop a global estimate for the industrial sector. Available studies vary widely with respect to system boundaries, baseline, time period, subsectors included, completeness of mitigation measures included, and

economic factors (e.g., costs and discount rates). In many cases study assumptions are not specified, making it impossible to adjust the studies to a common basis, or to quantify overall uncertainty. A full discussion of the basis for evaluating costs in this report appears in Chapter 2.5.

Table 7.8 presents an assessment of the industry-specific literature. Mitigation potential and cost for industrial CO_2 emissions were estimated as follows:
(1) Price et al. (2006)'s estimates for 2030 production rate by industry and geographic area for the SRES A1 and B2 scenarios (IPCC, 2000b) were used.
(2) Literature estimates of mitigation potential were used, where available. In other cases, mitigation potential was estimated by assuming that current best practice could be achieved by all plants in 2030.
(3) Literature estimates of mitigation cost were used, where available. When literature values were not available, expert judgment (informed by the available literature and data) was used to assign costs to mitigation technology.

Cost estimates are reported as 2030 mitigation potential below a given cost level. In most cases it was not possible to develop a marginal abatement cost curve that would allow estimation of mitigation potential as a function of cost. Estimates have not been made for some smaller industries (e.g., glass) and for the food industry. One or more of the critical inputs needed for these estimates were missing.

Table 7.8 should be interpreted with care. It is based on a limited number of studies – sometimes only one study per industry – and implicitly assumes that current trends will continue until 2030. Key uncertainties in the projections include: the rate of technology development and diffusion, the cost of future technology, future energy and carbon prices, the level of industrial activity in 2030, and policy driver, both climate and non-climate. The use of two scenarios, A1B and B2, is an attempt to bracket the range of these uncertainties.

Table 7.8 projects 2030 mitigation potential for the industrial sector at a cost of <100 US$/tCO$_2$-eq (<370 US$/tC-eq) of 3.0 to 6.3 GtCO$_2$-eq/yr (0.8 to 1.7 GtC-eq/yr) under the A1B scenario, and 2.0 to 5.1 GtCO$_2$-eq/yr (0.6 to 1.4 GtC-eq/yr) under the B2 scenario. The largest mitigation potentials are found in the steel, cement, and pulp and paper industries and in the control of non-CO_2 gases. Much of that potential is available at <50 US$/tCO$_2$-eq (<180 US$/tC-eq). Application of CCS technology offers a large additional potential, albeit at higher cost (low agreement, little evidence).

Some data are available on industrial sector mitigation potential and cost by country or region. However, an attempt

---

9   Mitigation potential is the 'economic potential', which is defined as the amount of GHG mitigation that is cost-effective for a given carbon price, with energy savings included, when using social discount rates (3-10%).

to build-up a global estimate from this data was unsuccessful. Information was lacking for the former Soviet Union, Africa, Latin America and parts of Asia.

7.5.1 Electricity savings

Electricity savings are of particular interest, since they feedback into the mitigation potential calculation for the energy sector and because of the potential for double counting of the emissions reductions. Section 7.3.2 indicates that in the EU and USA electric motor driven systems account for about 65% of industrial energy use, and that efficient systems could reduce this use by 30%. About one-third of the savings potential was assumed to be realized in the baseline, resulting in a net mitigation potential of 13% of industrial electricity use. This mitigation potential was included in the estimates of mitigation potential for energy-intensive industries presented in Table 7.8. However, it is also necessary to consider the potential for electricity savings from non-energy-intensive industries, which are large consumers of electricity.

The estimation procedure used to develop these numbers was as follows: Because data could not be found on other countries/regions, US data (EIA, 2002) on electricity use as a fraction of total energy use by industry and on the fraction of electricity use consumed by motor driven systems was taken as representative of global patterns. Based on De Keulenaer *et al.* (2004) and Xenergy (1998), a 30% mitigation potential was assumed. Emission factors to convert electricity savings into CO_2 reductions were derived from IEA data (IEA, 2004). The emission reduction potential from non-energy-intensive industries were calculated by subtracting the savings from energy-intensive industries from total industrial emissions reduction potential. Using the B2 baseline, 49% of total electricity savings are found in industries other than those identified in Table 7.8.

7.5.2 Non-CO$_2$ gases

Table 7.9 shows mitigation potential for non-CO_2 gases in 2030 based on a global study conducted by the US EPA (2006a,b), which projected emission and mitigation costs to 2020. Emissions in 2030 were projected by linear extrapolation by region using 2010 and 2020 data. Mitigation costs were assumed to be constant between 2020 and 2030, and interpolated from US EPA data, which used different cost categories. The analysis uses US EPA's technical adoption scenario, which assumes that industry will continue meeting its voluntary commitments. The SRES A1B and B2 scenarios used as the base case for the rest of this chapter do not include sufficient detail on non-CO_2 gases to allow a comparison of the two approaches. IPCC/TEAP (2005) contains significantly different estimates of 2015 baseline emissions for HFCs and PFCs in some sectors compared to Table 7.9. We note that these emissions are reported by end-use, not by the sectoral approach

used in this report, and that insufficient information is provided to extrapolate to 2030. Caprolactam projections were not found in the literature. They were estimated based on historical data from a variety of industry sources. Mitigation costs and potentials were estimated by applying costs and potential from nitric acid production.

7.5.3 Summary and comparison with other studies

Using the SRES B2 as a baseline (see Section 11.3.1), Table 7.10 summarizes the mitigation potential for the different cost categories. To avoid double counting, the total mitigation potential as given in Table 7.8 has been corrected for changes in emission factors of the transformation sectors to arrive at the figures included in Table 7.10 (see also Chapter 11, table 11.3).

Two recent studies provide bottom-up, global estimates of GHG mitigation potential in the industrial sector in 2030. IEA (2006a) used its Energy Technology Perspectives Model (ETP), which belongs to the MARKAL family of bottom-up modelling tools, to estimate mitigation potential for CO_2 from energy use in the industrial sector to be 5.4 Gt/yr (1.5 GtC/yr) in 2050. IEA's base case was an extrapolation of its World Energy Outlook 2005 Reference Scenario, which projected energy use to 2030. IEA provides ranges for mitigation potential in 2030 for nine groups of technologies totalling about 2.5 to 3.0 GtCO$_2$/yr (0.68 to 0.82 GtC/yr). Mitigation cost is estimated at <25 US$/tCO$_2$ (<92 US$/tC) (2004 US$). While IEA's estimate of mitigation potential is in the range found in this assessment, their estimate of mitigation cost is significantly lower.

ABARE (Matysek *et al.*, 2006) used its general equilibrium model of the world economy (GTEM) to estimate the emission reduction potential associated with widespread adoption of advanced technologies in five key industries: iron and steel, cement, aluminium, pulp and paper, and mining. In the most optimistic ABARE scenario, industrial sector emissions across all gases are reduced by an average of about 1.54 GtCO$_2$-eq/yr (0.42GtC-eq/yr) over the 2001 to 2050 time frame and 2.8 GtCO$_2$-eq/yr (0.77 GtC-eq/yr) over the 2030-2050 time frame, relative to the GTEM reference case, which assumes energy efficiency improvements and continuation of current or announced future government policy. The ABARE carbon dioxide only industry mitigation potential for the period 2030–2050 of approximately 1.94 GtCO$_2$-eq/yr (0.53GtC/yr) falls below the range developed in this assessment. This outcome is the likely result of differences in the modelling approaches used – ABARE's GTEM model is a top down model whereas the mitigation potentials in this assessment are developed using detailed bottom-up methodologies. ABARE did not estimate the cost of these reductions.

The TAR (IPCC, 2001a) developed a bottom up estimate of mitigation potential in 2020 for the industrial sector of 1.4 to 1.6 GtC (5.1 to 5.9 GtCO$_2$) based an SRES B2 scenario baseline

Table 7.8: *Mitigation potential and cost in 2030*

Product	Area[b]	2030 production (Mt)[a]		GHG intensity (tCO$_2$-eq/t prod.)	Mitigation potential (%)	Cost range (US$)	Mitigation potential (MtCO$_2$-eq/yr)	
		A1	B2				A1	B2
CO$_2$ emissions from processes and energy use								
Steel[c,d]	Global	1,163	1,121	1.6-3.8	15-40	20-50	430-1,500	420-1,500
	OECD	370	326	1.6-2.0	15-40	20-50	90-300	80-260
	EIT	162	173	20.-3.8	25-40	20-50	80-240	85-260
	Dev. Nat.	639	623	1.6-3.8	25-40	20-50	260-970	250-940
Primary aluminium[e,f]	Global	39	37	8.4	15-25	<100	53-82	49-75
	OECD	12	11	8.5	15-25	<100	16-25	15-22
	EIT	9	6	8.6	15-25	<100	12-19	8-13
	Dev. Nat.	19	20	8.3	15-25	<100	25-38	26-40
Cement[g,h,i]	Global	6,517	5,251	0.73-0.99	11-40	<50	720-2,100	480-1,700
	OECD	600	555	0.73-0.99	11-40	<50	65-180	50-160
	EIT	362	181	0.81-0.89	11-40	<50	40-120	20-60
	Dev. Nat.	5,555	4,515	0.82-0.93	11-40	<50	610-1,800	410-1,500
Ethylene[j]	Global	329	218	1.33	20	<20	85	58
	OECD	139	148	1.33	20	<20	35	40
	EIT	19	11	1.33	20	<20	5	3
	Dev. Nat.	170	59	1.33	20	<20	45	15
Ammonia[k,l]	Global	218	202	1.6-2.7	25	<20	110	100
	OECD	23	20	1.6-2.7	25	<20	11	10
	EIT	21	23	1.6-2.7	25	<20	10	12
	Dev. Nat.	175	159	1.6-2.7	25	<20	87	80
Petroleum refining[m]	Global	4,691	4,508	0.32-0.64	10-20	Half <20	150-300	140-280
	OECD	2,198	2,095	0.32-0.64	10-20	Half <50	70-140	67-130
	EIT	384	381	0.32-0.64	10-20	"	12-24	12-24
	Dev. Nat.	2,108	2,031	0.32-0.64	10-20	"	68-140	65-130
Pulp and paper[n]	Global	1,321	920	0.22-1.40	5-40	<20	49-420	37-300
	OECD	695	551	0.22-1.40	5-40	<20	28-220	22-180
	EIT	65	39	0.22-1.40	5-40	<20	3-21	2-13
	Dev. Nat.	561	330	0.22-1.40	5-40	<20	18-180	13-110

Notes and sources:

[a] Price *et al.*, 2006.

[b] Global total may not equal sum of regions due to independent rounding.

[c] Kim and Worrell, 2002a.

[d] Expert judgement.

[e] Emission intensity based on IAI Life-Cycle Analysis (IAI, 2003), excluding alumina production and aluminium shaping and rolling. Emissions include anode manufacture, anode oxidation and power and fuel used in the primary smelter. PFC emission included under non-CO$_2$ gases.

[f] Assumes upgrade to current state-of-the art smelter electricity use and 50% penetration of zero emission inert electrode technology by 2030.

[g] Humphreys and Mahasenan, 2002.

[h] Hendriks *et al.*, 1999.

[i] Worrell *et al.*, 1995.

[j] Ren *et al.*, 2005.

[k] Basis for estimate: 10 GJ/t NH$_3$ difference between the average plant and the best available technology (Figure 7.2) and operation on natural gas (Section 7.4.3.2).

[l] Rafiqul *et al.*, 2005.

[m] Worrell and Galitsky, 2005.

[n] Farahani *et al.*, 2004.

[o] The process emissions from ammonia manufacturing (based on natural gas) are about 1.35 tCO$_2$/t NH$_3$ (De Beer, 1998). However, as noted in Section

7.4.3.2, the fertilizer industry uses nearly half of the CO$_2$ it generates for the production of urea and nitrophosphates. The remaining CO$_2$ is suitable for storage. IPCC (2005a) indicates that it should be possible to store essentially all of this remaining CO$_2$ at a cost of <20 US$/t.

[p] IPCC, 2005a.

[q] US refineries use about 4% of their energy input to manufacture hydrogen (Worrell and Galitsky, 2005). Refinery hydrogen production is expected to increase as crude slates become heavier and the demand for clean products increases. We assume that in 2030, 5% of refinery energy use worldwide will be used for hydrogen production, and that the byproduct CO$_2$ will be suitable for carbon storage.

[r] Total potential and application potential derived from IEA, 2006a. Subdivision into regions based in production volumes and carbon intensities. IEA, 2006a does not provide a regional breakdown.

[s] Extrapolated from US EPA, 2006b. This publication does not use the SRES scenarios as baselines.

[t] See Section 7.5.1 for details of the estimation procedure.

[u] Due to gaps in quantitative information (see the text) the column sums in this table do not represent total industry emissions or mitigation potential. Global total may not equal sum of regions due to independent rounding.

[v] The mitigation potential of the main industries include electricity savings. To prevent double counting with the energy supply sector, these are shown separately in Chapter 11.

[w] Mitigation potential for other industries includes only reductions for reduced electricity use for motors. Limited data in the literature did not allow estimation of the potential for other mitigation options in these industries.

Table 7.8: *Continued*

Product	Area[b]	2030 production (Mt)[a]		CCS Potential (tCO$_2$/t)	Mitigation potential (%)	Cost range (US$)	Mitigation potential (MtCO$_2$-eq)	
		A1	B2				A1	B2
Carbon Capture and Storage								
Ammonia[o,p]	Global	218	202	0.5	about 100	<50	150	140
	OECD	23	20	0.5	about 100	<50	15	13
	EIT	21	23	0.5	about 100	<50	14	16
	Dev. Nat.	175	159	0.5	about 100	<50	120	110
Petroleum Refining[m,p,q]	Global	4,691	4,508	0.032-0.064	about 50	<50	75-150	72-150
	OECD	2,198	2,095	0.032-0.064	about 50	<50	35-70	34-70
	EIT	384	381	0.032-0.064	about 50	<50	6-12	6-12
	Dev. Nat.	2,108	2,031	0.032-0.064	about 50	<50	34-70	32-65
Cement[r]	Global	6,517	5,251	0.65-0.89	about 6	<100	250-350	200-280
	OECD	600	555	0.65-0.80	about 6	<100	23-32	22-27
	EIT	362	181	0.73-0.80	about 6	<100	16-17	8-9
	Dev. Nat.	5,555	4,515	0.74-0.84	about 6	<100	210-300	170-240
Iron and Steel	Global	1,163	1,121	0.32-0.76	about 20	<50	70-180	70-170
	OECD	370	326	0.32-0.40	about 20	<50	24-30	21-26
	EIT	162	173	0.40-0.76	about 20	<50	13-25	14-26
	Dev. Nat.	639	623	0.32-0.76	about 20	<50	33-120	35-120
Non-CO$_2$ gases[r]								
	Global	668				37% <0US$	380	
	OECD	305				53% <20US$	160	
	EIT	53				55% <50US$	29	
	Dev. Nat.	310				57%<100US$	190	
Other industries, electricity conservation[s]								
	Global					25% <20	1,100-1,300	410-540
	OECD					25% <50	140-210	65-140
	EIT					50% <100	340-350	71-85
	Dev. Nat.					d	640-700	280-320
Sum[t,u,v,w]	Global						3,000-6,300	2,000-5,100
	OECD						580-1,300	470-1,100
	EIT						540-830	250-510
	Dev. Nat.						2,000-4,300	1,300-3,400

and on the evaluation of specific technologies. Extrapolating the TAR estimate to 2030 would give values above the upper end of the range developed in this assessment. The newer studies used in this assessment take industry-specific conditions into account, which reduces the risk of double counting.

7.6 Barriers to industrial GHG mitigation

Full use of mitigation options is not being made in either industrialized or developing nations *(high agreement, much*

evidence). In many areas of the world, GHG mitigation is neither demanded nor rewarded by the market or government. In these areas, companies will invest in GHG mitigation to the extent that other factors provide a return for their investments. This return can be economic, for example energy-efficiency improvements that show an economic payout. Nicholson (2004) reported that the projects BP undertook to lower its CO$_2$ emissions by 10% increased shareholder value by US$ 650 million. Alternatively, the return can be in terms of achievement of a larger corporate goal, for example DuPont's commitment to cut its GHG emission by two-thirds as part of a larger commitment to sustainable growth (Holliday, 2001).

Table 7.9: *Global mitigation potential in 2030 for non-CO₂ gases*

Source	2030 Baseline emissions (MtCO$_2$-eq)	Mitigation potential by cost category (US$)			
		<0	<20	<50	<100
N$_2$O from adipic and nitric acid production	190	158	158	158	174
N$_2$O from caprolactam production	20	16	16	16	16
PFC from aluminium production	51	1.6	7.6	8.2	8.2
PFC and SF$_6$ from semiconductor manufacture	20	9.6	9.6	10	10
SF$_6$ from use of electrical equipment (excluding manufacture)	74	32	39	39	39
SF$_6$ from magnesium production	9.3	9.2	9.2	9.2	9.2
HFC-23 from HCFC-22 production	106	0	86	86	86
ODS[a] substitutes: aerosols	88	27	27	27	27
ODS substitutes: industrial refrigeration and cooling	80	3.5	3.5	3.5	3.5
ODS substitutes: fire extinguishing	27	0	0	6.3	6.7
ODS substitutes: solvents	4.0	1.2	2.0	2.0	2.0
Total: Global	668	249	357	364	380
OECD[b]	305	135	154	157	158
Economies in Transition	53	27	28	29	29
Developing Nations	309	87	182	187	187

[a] ODS = Ozone-Depleting Substances
[b] Regional information given in references.

Source: Extrapolated from US EPA 2006a,b.

Even though a broad range of cost-effective GHG mitigation technologies exist, a variety of economic barriers prevent their full realisation in both developed and developing countries. Policies and measures must overcome the effective costs of capital (Toman, 2003). Industry needs a stable, transparent policy regime addressing both economic and environmental concerns to reduce the costs of capital.

The slow rate of capital stock turnover in many of the industries covered in this chapter is a barrier to mitigation (Worrell and Biermans, 2005). Excess capacity, as exists in some industries, can further slow capital stock turnover. Policies that encourage capital stock turnover, such as Japan's programme to subsidize the installation of new high performance furnaces (WEC, 2001), will increase GHG mitigation. Companies must also take into consideration the risks involved with adopting a new technology, the payback period of a technology, the appropriate discount rate and transaction costs. Newer, relatively expensive technologies often have longer payback periods and represent a greater risk. Reliability is a key concern of industry, making new technologies less attractive (Rosenberg, 1999). Discount rates vary substantially across industries and little information exists on transaction costs of mitigation options (US EPA, 2003).

Resource constraints are also a significant barrier to mitigation. Unless legally mandated, GHG mitigation will have to compete for financial and technical resources against projects to achieve other company goals. Financial constraints can hinder diffusion of technologies within firms (Canepa and Stoneman, 2004). Projects to increase capacity or bring new products to the market typically have priority, especially in developing countries, where markets are growing rapidly and where a large portion of industrial capacity is in SMEs. Energy efficiency and other GHG mitigation technologies can provide attractive rates of return, but they tend to increase initial capital costs, which can be a barrier in locations where capital is limited. If the technology involved is new to the market in question, even if it is well-demonstrated elsewhere, the problem of raising capital may be further exacerbated (Shashank, 2004). Provision of funding for demonstration of the technology can overcome this barrier (CPCB, 2005).

The rate of technology transfer is another factor limiting the adoption of mitigation technologies. As documented in the IPCC Special Report Methodological and Technological Issues in Technology Transfer (IPCC, 2000c), lack of an enabling environment is a barrier to technology transfer in some countries. Even when an enabling environment is present, the ability of industrial organizations to access and absorb information on technologies is limited. Access to information tends to be more of a problem in developing nations, but all companies, even the largest, have limited technical resources to interpret and translate the available information. The success of programmes such as US DOE's Industrial Technologies Programs (ITP) and of the voluntary information sharing programmes discussed in Section 7.9.2 is evidence of the pervasiveness of this barrier.

Table 7.10: *Estimated economic potentials for GHG mitigation in industry in 2030 for different cost categories using the SRES B2 baseline*

Mitigation option	Region	Economic potential <100 US$/tCO$_2$-eq		Economic potential in different cost categories			
		Cost category (US$/tCO$_2$-eq)		<0	0-20	20-50	50-100
		Cost category (US$/tC-eq)		<0	0-73	73-183	183-367
		Low	High				
		(MtCO$_2$-eq)					
Electricity savings	OECD	300		70	70		150
	EIT	80		20	20		40
	Non-OECD/EIT	450		100	100		250
Other savings, including non-CO$_2$ GHG	OECD	350	900	300	250		50
	EIT	200	450	80	250		20
	Non-OECD/EIT	1,200	3,300	500	1,700		80
Total	OECD	600	1,200	350	350		200
	EIT	250	550	100	250		60
	Non-OECD/EIT	1,600	3,800	600	1,800		300
	Global	2,500	5,500	1,100	2,400		550

McKane *et al.* (2005) provide a case study of the interaction of some of these elements in their analysis of the barriers to the adoption of energy-efficient electric motors and motor-systems. These include: (1) industrial markets that focus on components, not systems; (2) energy efficiency not being a core mission for most industries, which results in a lack of internal support systems for mitigation goals; and (3) lack of technical skills to optimize the systems to the specific application – one size does not fit all. They found industrial energy efficiency standards a useful tool in overcoming these barriers.

7.7 Sustainable Development (SD) implications of industrial GHG mitigation

Although there is no universally accepted, practical definition of SD, the concept has evolved as the integration of economic, social and environmental aims (IPCC, 2000a; Munasinghe, 2002). Companies worldwide adopted Triple Bottom Line (financial, environmental and social responsibility) reporting in the late 1990's. The Global Reporting Initiative (GRI, n.d.), a multi-stakeholder process, has enabled business organizations to account for and better explain their contributions to sustainable development. Companies are also reporting under Sigma Guidelines (The Sigma Project, 2003a), and AA1000 (The Sigma Project, 2003b) and SA 8000 (SAI, 2001) procedures. Many companies are trying to demonstrate that their operations minimize water use and carbon emissions and produce zero solid waste (ITC, 2006). SD consequences can be observed or monitored through various indicators grouped under the three major categories. (See Section 12.1.1 and 12.1.3 for more detail).

However, the SD consequences of mitigation options are not automatic. GHG mitigation, *per se*, has little impact on four of the SD indicators: poverty reduction, empowerment/gender, water pollution and solid waste. The literature indicates that supplementing mitigation options with appropriate national macroeconomic policies, and with social and local waste reduction strategies at the company level (Tata Steel, Ltd., 2005; BEE, 2006), has achieved some sustainability goals. Economy-wide impact studies (Sathaye *et al.*, 2005; Phadke *et al.*, 2005) show that in developing countries, like India, adoption of efficient electricity technology can lead to higher employment and income generation. However, the lack of empirical studies leads to much uncertainty about the SD implications of many mitigation strategies, including use of renewables, fuel switching, feedstock and product changes, control of non-CO$_2$ gases, and CCS. For example, fuel switching can have a positive effect on local air pollution and company profitability, but its impacts on employment are uncertain and will depend on inter-input substitution opportunities.

GHG emissions mitigation policies induce increased innovation that can reduce the energy and capital intensity of industry. However, this could come at the expense of other, even more valuable, productivity-enhancing investments or learning-by-doing efforts (Goulder and Schneider, 1999). If policies are successful in stimulating economic activity, they are also likely to stimulate increased energy use. GHG emissions would increase unless policies decreased the carbon-intensity of economic activity by more than the increase in activity.

Due to energy efficiency improvements and fuel switching in OECD countries (Schipper *et al.*, 2000; Liskas *et al.*, 2000), as well as in developing countries like India (Dasgupta and Roy, 2001), China (Zhang, 2003), Korea (Choi and Ang, 2001; Chang, 2003), Bangladesh (Bain, 2005), and Mexico (Aguayo and Gallagher, 2005), energy and carbon intensity have decreased, for the industry sector in general and for energy-intensive industries in particular. In Mexico, deindustrialization also played a role. For OECD countries, structural change has also played an important role in emissions reduction. However, overall economic activity has increased more rapidly, resulting in higher total carbon emissions.

SMEs have played a part in advancing the SD agenda, for example as part of coordinated supply chain or industrial park initiatives, or by participating in research and innovation in sustainable goods and services (Dutta *et al.*, 2004). US DOE's Industrial Assessment Centers (IACs) are an example of how SMEs can be provided with financial and technical support to assess and identify energy and cost-saving opportunities and training to improve human capital (US DOE, 2003).

7.8 Interaction of mitigation technologies with vulnerability and adaptation

Industry's vulnerability to extreme weather events arises from site characteristics, for example coastal areas or flood-prone river basins *(high agreement, much evidence)*. Because of their financial and technical resources, large industrial organizations typically have a significant adaptive capacity for addressing vulnerability to weather extremes. SMEs typically have fewer financial and technical resources and therefore less adaptive capacity. The food processing industry, which relies on agricultural resources that are vulnerable to extreme weather conditions like floods or droughts, is engaging in dialogue with its supply chain to reduce GHGs emissions. Companies are also attempting to reduce vulnerability through product diversification (Kolk and Pinkse, 2005).

Linkages between adaptation and mitigation in the industrial sector are limited. In areas dependent on hydropower, mitigation options that reduce industrial electricity demand will help in adapting to climate variability or change that affects water supply (Subak *et al.*, 2000). Many mitigation options (e.g., energy efficiency, heat and power recovery, recycling) are not vulnerable to climate change and therefore create no additional adaptation link. Others, such as fuel switching can be vulnerable to climate change under certain circumstances. As the 2005 Atlantic hurricane season demonstrated, the oil and gas infrastructure is vulnerable to weather extremes. Use of solar or biomass energy will be vulnerable to both weather extremes and climate change. Adaptation, the construction of more weather resistant facilities and provision of back-up energy supplies could reduce this vulnerability.

7.9 Effectiveness of and experience with policies

As noted in the TAR (IPCC, 2001b), industrial enterprises of all sizes are vulnerable to changes in government policy and consumer preferences. While the specifics of government climate policies will vary greatly, all will have one of two fundamental objectives: constraining GHG emissions or adapting to existing or projected climate change. And while consumers may become more sensitive to the GHG impacts of the products and services they use, it is almost certain that they will continue to seek the traditional qualities of low-cost, reliability, etc. The challenge to industry will be to continue to provide the goods and services on which society depends in a GHG-constrained world. Industry can respond to the potential for increased government regulation or changes in consumer preferences in two ways: by mitigating its own GHG emissions or by developing new, lower GHG emission products and services. To the extent that industry does this before required by either regulation or the market, it is demonstrating the type of anticipatory, or planned, adaptation advocated in the TAR (IPCC, 2001b).

7.9.1 Kyoto mechanisms (CDM and JI)

The Clean Development Mechanism (CDM) was created under the Kyoto Protocol to allow Annex I countries to obtain GHG emission reduction credits for projects that reduced GHG emission in non-Annex I countries, provided that those projects contributed to the sustainable development of the host country (UNFCCC, 1997). As of November 2006, over 400 projects had been registered, with another 900 in some phase of the approval process. Total emission reduction potential of both approved and proposed projects is nearly 1.5 $GtCO_2$ (410 MtC). The majority of these projects are in the energy sector; as of November 2006, only about 6% of approved CDM projects were in the industrial sector (UNFCCC, CDM, n.d.). The concept of Joint Implementation (JI), GHG-emissions reduction projects carried out jointly by Annex I countries or business from Annex I countries, is mentioned in the UNFCCC, but amplified in the Kyoto Protocol. However, since the Kyoto Protocol does not allow JI credits to be transferred before 2008, progress on JI implementation has been slow. Both CDM and JI build on experience gained in the pilot-phase Activities Implemented Jointly (AIJ) programme created by the UNFCCC in 1995 (UNFCCC, 1995). A fuller discussion of CDM, JI and AIJ appears in Section 13.3.3.

7.9.1.1 *Regional differences*

Project-based mechanisms are still in their early stages of implementation, but significant differences have emerged in the ability of developing countries to take advantage of them. This is particularly true of Africa, which, as of November 2006, lagged behind other regions in their implementation. Only two of fifty AIJ projects were in Africa. None of the twenty projects recently approved under The Netherlands carbon purchase programme, CERUPT, were in Africa (CDM for Sustainable Africa, 2004), and only 3% of the registered CDM projects were in Africa (UNFCCC, CDM, n.d.).

Yamba and Matsika (2004) identified financial, policy, technical and legal barriers inhibiting participation in the CDM in sub-Saharan Africa. Financial barriers pose the greatest challenges: low market value of carbon credits, high CDM transaction costs and lack of financial resources discourage industry participation. Policy barriers include limited awareness of the benefits of CDM and the project approval process in government and the private sector, non-ratification of the Kyoto Protocol, and failure to establish the Designated National Authorities required by CDM. Technical barriers include limited awareness of the availability of energy-saving and other appropriate technologies for potential CDM projects. Legal barriers include limited awareness in government and the private sector of the Kyoto Protocol, and the legal requirements for development of CDM projects. Limited human resources for the development of CDM projects, and CDM's requirements on additionality are additional constraints. Other countries, for example Brazil, China and India (Silayan, 2005), have more capacity for the development of CDM projects. The Government of India (GOI, 2004) has identified energy efficiency in the steel industry as one of the priorities for Indian CDM projects.

7.9.2 Voluntary GHG programmes and agreements

7.9.2.1 *Government-initiated GHG programmes and voluntary agreements*

Government-initiated GHG programmes and agreements that focus on energy-efficiency improvement, reduction of energy-related GHG emissions and reduction of non-CO_2 GHG emissions are found in many countries. Voluntary Agreements are defined as formal agreements that are essentially contracts between government and industry that include negotiated targets with time schedules and commitments on the part of all participating parties (IEA, 1997). Voluntary agreements for energy efficiency improvement and reduction of energy-related GHG emissions by industry have been implemented in industrialized countries since the early 1990s. These agreements fall into three categories: completely voluntary; voluntary with the threat of future taxes or regulation if shown to be ineffective; and voluntary, but associated with an energy or carbon tax (Price, 2005). Agreements that include explicit targets, and exert pressure on industry to meet those targets, are the most effective (UNFCCC, 2002). An essential part of voluntary agreements is government support, including the programme elements such as information-sharing, energy and GHG emissions management, financial assistance, awards and recognition, standards and target-setting (APERC, 2003; CLASP, 2005; Galitsky *et al.*, 2004; WEC, 2004). Voluntary agreements typically cover a period of five to ten years, so that

strategic energy-efficiency investments can be planned and implemented. There are also voluntary agreements covering process emissions in Australia, Bahrain, Brazil, Canada, France, Germany, Japan, the Netherlands, New Zealand, Norway, the UK and the USA (Bartos, 2001; EFCTC, 2000; US EPA, 1999).

Independent assessments find that experience with voluntary agreements has been mixed, with some of the earlier programmes, such as the French Voluntary Agreements on CO_2 Reductions and Finland's Action Programme for Industrial Energy Conservation, appearing to have been poorly designed, failing to meet targets, or only achieving business-as-usual savings (Bossoken, 1999; Chidiak, 2000; Chidiak, 2002; Hansen and Larsen, 1999; OECD, 2002; Starzer, 2000). Recently, a number of voluntary agreement programmes have been modified and strengthened, while additional countries, including some newly industrialized and developing countries, are adopting such agreements in efforts to increase the efficiency of their industrial sectors (Price, 2005). Such strengthened programmes include the French Association des Enterprises por la Réduction de l'Effet de Serre (AERES) agreements, Finland's Agreement on the Promotion of Energy Conservation in Industry, and the German Agreement on Climate Protection (AERES, 2005; IEA, 2004; RWI, 2004). The more successful programmes are typically those that have either an implicit threat of future taxes or regulations, or those that work in conjunction with an energy or carbon tax, such as the Dutch Long-Term Agreements, the Danish Agreement on Industrial Energy Efficiency and the UK Climate Change Agreements (see Box 13.2). Such programmes can provide energy savings beyond business-as-usual (Bjørner and Jensen, 2002; Future Energy Solutions, 2004; Future Energy Solutions, 2005) and are cost-effective (Phylipsen and Blok, 2002). The Long-Term Agreements, for example, stimulated between 27% and 44% (17 to 28 PJ/yr) of the observed energy savings, which was a 50% increase over historical autonomous energy efficiency rates in the Netherlands prior to the agreements (Kerssemeeckers, 2002; Rietbergen *et al.*, 2002). The UK Climate Change Agreements saved 3.5 to 9.8 $MtCO_2$ (1.0 to 2.7 MtC) over the baseline during the first target period (2000–2002) and 5.1 to 8.9 $MtCO_2$ (1.4 to 2.4 MtC) during the second target period (2002–2004) depending upon whether the adjusted steel sector target is accounted for (Future Energy Solutions, 2005).

In addition to the energy and carbon savings, these agreements have important longer-term impacts (Delmas and Terlaak, 2000; Dowd *et al.*, 2001) including:
- Changing attitudes towards and awareness of energy efficiency;
- Reducing barriers to innovation and technology adoption;
- Creating market transformations to establish greater potential for sustainable energy-efficiency investments;
- Promoting positive dynamic interactions between different actors involved in technology research and development, deployment, and market development, and

- Facilitating cooperative arrangements that provide learning mechanisms within an industry.

The most effective agreements are those that set realistic targets, include sufficient government support, often as part of a larger environmental policy package, and include a real threat of increased government regulation or energy/GHG taxes if targets are not achieved (Bjørner and Jensen, 2002; Price, 2005) *(medium agreement, much evidence)*.

7.9.2.2   *Company or industry-initiated voluntary actions*

Many companies participate in GHG emissions reporting programmes as well as take voluntary actions to reduce energy use or GHG emissions through individual corporate programmes, non-governmental organization (NGO) programmes and industry association initiatives. Some of these companies report their GHG emission in annual environmental or sustainable development reports, or in their Corporate Annual Report. Beginning in the late 1990s, a number of individual companies initiated in-house energy or GHG emissions management programmes and made GHG emissions reduction commitments (Margolick and Russell, 2001; PCA, 2002).

Questions have been raised as to whether such initiatives, which operate outside regulatory or legal frameworks, often without standardized monitoring and reporting procedures, just delay the implementation of government-initiated programmes without delivering real emissions reductions (OECD, 2002). Early programmes appear to have produced little benefit. For example, an evaluation of the Germany industry's self-defined global-warming declaration found that achievements in the first reporting period appeared to be equivalent to business-as-usual trends (Jochem and Eichhammer, 1999; Ramesohl and Kristof, 2001). However, more recent efforts appear to have yielded positive results (RWI, 2004). Examples of targets and the actual reductions achieved include:

- DuPont's reduction of GHG emissions by over 72% while holding energy use constant, surpassing its pledge to reduce GHG emissions by 65% by 2010 and hold energy use constant compared to a 1990 baseline (DuPont, 2002; McFarland, 2005);
- BP's target to reduce GHG emissions by 10% in 2010 compared to a 1990 baseline which was reached in 2001 (BP, 2003; BP, 2005), and
- United Technologies Corporation's goal to reduce energy and water consumption by 25% as a percentage of sales by the year 2007 using a 1997 baseline that was exceeded by achieving a 27% energy reduction and 34% water use reduction through 2002 (Rainey and Patilis, 2000; UTC, 2003).

Often these corporate commitments are formalized through GHG reporting programmes or registries such as the World Economic Forum Greenhouse Gas Register where 13 multinational companies disclose the amount of GHGs their worldwide operations produce (WEF, 2005) and through NGO programmes such as the Pew Center on Global Climate Change's Business Environmental Leadership Council (Pew Center on Global Climate Change, 2005), the World Wildlife Fund's Climate Savers Program (WWF, n.d.), as well as programmes of the Chicago Climate Exchange (CCX, 2005).

Industrial trade associations provide another platform for organizing and implementing GHG mitigation programmes:
- The International Aluminium Institute initiated the Aluminium for Future Generations sustainability programme in 2003, which established nine sustainable development voluntary objectives (increased to 12 in 2006), 22 performance indicators, and a programme to provide technical services to member companies (IAI, 2004). Performance to date against GHG mitigation objectives was discussed in Section 7.4.2.1.
- The World Semiconductor Council (WSC), comprised of semiconductor industry associations of the United States, Japan, Europe, Republic of Korea and Chinese Taipei, established a target of reducing PFC emissions by at least 10% below the 1995 baseline level by 2010 (Bartos, 2001).
- The World Business Council for Sustainable Development (WBCSD) started the Cement Sustainability Initiative in 1999 with ten large cement companies and it has now grown to 16 (WBCSD, 2005). The Initiative conducts research related to actions that can be undertaken by cement companies to reduce GHG emissions (Battelle Institute/WBCSD, 2002) and outlines specific member company actions (WBCSD, 2002). As of 2004, 94% of the 619 kilns of CSI member companies had developed CO_2 inventories and three had established emissions reduction targets (WBCSD, 2005).
- By 2003, the Japanese chemical industry had reduced its CO_2 emissions intensity by 9% compared with 1990-levels (Nippon Keidanren, 2004), but due to increased production, overall CO_2 emissions were up by 10.5%.
- The European Chemical Industry Council established a Voluntary Energy Efficiency Programme (VEEP) with a commitment to improve energy efficiency by 20% between 1990 and 2005, provided that no additional energy taxes are introduced (CEFIC, 2002).

In 2003, the members of the International Iron and Steel Institute, representing 38% of global steel production, committed to voluntary reductions in energy and GHG emission intensities. In most countries this programme is too new to provide meaningful results (IISI, 2006). However, as part of a larger voluntary programme in Japan, Japanese steelmakers committed to a voluntary action programme to mitigate climate change with the goal of a 10% reduction in energy consumption in 2010 against 1990. In fiscal year 2003, this programme resulted in a 6.4% reduction in CO_2 intensity emissions against 1990, through improvement of blast furnaces, upgrade of oxygen production plants, installation of regenerative burners and other steps (Nippon Keidanren, 2004).

7.9.3 Financial instruments: taxes, subsidies and access to capital

To date there is limited experience with taxing industrial GHG emissions. France instituted an eco-tax on a range of activities, including N_2O emission from the production of nitric, adipic and glyoxalic acids. The tax rate is modest (37 US$ (2000) per tonne N_2O, or 1.5 US$/$tCO_2$-eq (5.5 US$/tC-eq), but it provides a supplementary incentive for emissions reductions. The UK Climate Change Levy applies to industry only and is levied on all non-household use of coal (0.15 UK pence/kWh or 0.003 US$/kWh), gas (0.15 UK pence/kWh), electricity (0.43 UK pence/kWh or 0.0085 US$/kWh) and non-transport LPG (0.07 UK pence/kWh or 0.0014 US$/kWh). Industry includes agriculture and the public sector. Fuels used for electricity generation or non-energy uses, waste-derived fuels, renewable energy, including quality CHP, which uses specified fuels and meets minimum efficiency standards, are exempt from the tax. The UK Government also provided an 80% discount from the levy for those energy-intensive sectors that agreed to challenging targets for improving their energy efficiency. Climate change agreements have now been concluded with almost all eligible sectors (UK DEFRA, 2006).

In 1999, Germany introduced an eco-tax on the consumption of electricity, gasoline, fuel oil and natural gas. Revenues are recycled to subsidize the public pension system. The tax rate for electricity consumed by industrial consumers is € 0.012/kWh. Very large consumers are exempt to maintain their competitiveness. The impact of this eco-tax on CO_2 emissions is still under discussion (Green Budget Germany, 2004).

Tax reductions are frequently used to stimulate energy savings in industry. Some examples include:
- In the Netherlands, the Energy Investment Deduction (Energie Investeringsaftrek, EIA) stimulates investments in low-energy capital equipment and renewable energy by means of tax deductions (deduction of the fiscal profit of 55% of the investment) (IEA, 2005).
- In France, investments in energy efficiency are stimulated through lease credits. In addition to financing equipment, these credits can also finance associated costs such as construction, land and transport (IEA, 2005).
- The UK's Enhanced Capital Allowance Scheme allows businesses to write off the entire cost of energy-savings technologies specified in the 'Energy Technology List' during the year they make the investment (HM Revenue & Customs, n.d.).
- Australia requires companies receiving more than AU$ 3 million (US$ 2.5 million) of fuel credits to be members of its Greenhouse Challenge Plus programme (Australian Greenhouse Office, n.d.).
- Under Singapore's Income Tax Act, companies that invest in qualifying energy-efficient equipment can write-off the capital expenditure in one year instead of three. (NEEC, 2005).

- In the Republic of Korea, a 5% income tax credit is available for energy-efficiency investments (UNESCAP, 2000).
- Romania has a programme where imported energy-efficient technologies are exempt from customs taxes and the share of company income directed for energy efficiency investments is exempt from income tax (CEEBICNet Market Research, 2004).
- In Mexico, the Ministry of Energy has linked its energy efficiency programmes with Energy Service Companies (ESCOs). These are engineering and financing specialised enterprises that provide integrated energy services with a wide range and flexibility of technologies to the industrial and service sectors (NREL, 2006).

Subsidies are used to stimulate investment in energy-saving measures by reducing investment cost. Subsidies to the industrial sector include: grants, favourable loans and fiscal incentives, such as reduced taxes on energy-efficient equipments, accelerated depreciation, tax credits and tax deductions. Many developed and developing countries have financial schemes to promote industrial energy savings. A WEC survey (WEC, 2004) showed that 28 countries, most in Europe, provide grants or subsidies for industrial energy efficiency projects. Subsides can be fixed amounts, a percentage of the investment (with a ceiling), or be proportional to the amount of energy saved. In Japan, the New Energy and Technology Development Organization (NEDO) pays up to one-third of the cost of each new high performance furnace. NEDO estimates that the project will save 5% of Japan's final energy consumption by 2010 (WEC, 2001). The Korean Energy Management Corporation (KEMCO) provides, long-term, low interest loans to certified companies (IEA, 2005).

Evaluations show that subsidies for industry may lead to energy savings and corresponding GHG emission reductions and can create a larger market for energy efficient technologies (De Beer et al., 2000b; WEC, 2001). Whether the benefits to society outweigh the cost of these programmes, or whether other instruments would have been more cost-effective, has to be evaluated on a case-by-case basis. A drawback to subsidies is that they are often used by investors who would have made the investment without the incentive. Possible approaches for improving their cost-effectiveness include restricting schemes to specific target groups and/or techniques (selected list of equipment, only innovative technologies, etc.), or using a direct criterion of cost-effectiveness.

Investors in developing countries tend to have a weak capital basis. Development and finance institutions therefore often play a critical role in implementing energy efficiency and emission mitigation policies. Their role often goes beyond the provision of project finance and may directly influence technology choice and the direction of innovation (George and Prabhu, 2003). The retreat of national development banks in some developing countries (as a result of both financial liberalisation and financial crises in national governments) may hinder the widespread adoption of mitigation technologies because of lack of financial mechanisms to handle the associated risk.

7.9.4 Regional and national GHG emissions trading programmes

Several established or evolving national, regional or sectoral CO_2 emissions trading systems exist, for example in the EU, the UK, Norway, Denmark, New South Wales (Australia), Canada and several US States. The International Emissions Trading Association (IETA, 2005) provides an overview of systems. This section focuses on issues relevant to the industrial sector. A more in-depth discussion of emission trading can be found in Section 13.2.1.

The results of an assessment of the first two years of the UK scheme (NERA, 2004) show that reduction of non-CO_2 GHG emissions from industrial sources provided the least cost options. It also found that the heterogeneity of industrial emitters may require a tiered approach for the participation of small, medium-sized and large emitters, that is in respect to monitoring and verification, and described the impacts of individual industrial emitters gaining dominating market power on allowance prices.

In January 2005, the European Union Greenhouse Gas Emission Trading Scheme (EU ETS) was launched as the world's largest multi-country, multi-sector GHG emission trading scheme (EC, 2005). A number of assessments have analysed current and projected likely future impacts of the EU-ETS on the industrial sector in the EU (IEA, 2005; Egenhofer *et al.*, 2005). Recurring themes with specific relevance to industry include: allocation approaches based on benchmarking, grandfathering and auctioning; electricity price increases leading to so-called 'windfall profits' in the utility sector; competitiveness of energy-intensive industries; specific provisions for new entrants, closures, capacity expansions, and organic growth; and compliance costs for small emitters. The further refinement of these trading systems could be informed by evidence which suggests that in some important aspects participants from industrial sectors face a significantly different situation from those in the electricity sector (Carbon Trust, 2006):

- The range of products from industry sectors is generally more diverse (e.g., in the paper, glass or ceramics industry) making it difficult to define sector specific best practice values to be used for the allocation of allowances (see discussion in DTI (2005)).
- While grid connections limit electricity to regional or national markets, many industrial products are globally traded commodities, constrained only by transport costs. This increasingly applies as value per mass or volume goes up, that is from bulk ceramics products and cement,

to petrochemicals, to base metals, making the impacts of trading schemes on international competitiveness a matter of varying concern for the different subsectors.

- Only a few industrial sectors (e.g., steel and refineries) are prepared to actively participate in the early phase of trading schemes, leading to reduced liquidity and higher allowance prices, suggesting that specific instruments are needed to increase industrial involvement in trading.
- Responses to carbon emission price in industry tend to be slower because of the more limited technology portfolio and absence of short term fuel switching possibilities, making predictable allocation mechanisms and stable price signals a more important issue for industry.

The EU Commission recently published its findings and recommendations based on the first year of trading under the EU-ETS (EC, 2006a). An EU High Level Group on Competitiveness, Energy and the Environment has been formed to review the impacts of the EU-ETS on industry (EU-HLG, 2006). Issues highlighted in these EU processes include the need for the allocation of credits to be more harmonized across the EU, the need to increase certainty for investors, that is through long-term clarity on allocations, extension of the scheme to other sectors and alleviation of high participation costs for small installations. Industrial sectors sources considered for inclusion in the EU-ETS include CO_2 emissions from ammonia production, N_2O emissions from nitric and adipic acid production and PFC emission from aluminium production (EC, 2006b).

7.9.5 Regulation of non-CO_2 gases

The first regulations on non-CO_2 GHGs are emerging in Europe. A new EU regulation (EC 842/2006) on fluorinated gases includes prohibition of the use of SF_6 in magnesium die casting. The regulation contains a review clause that could lead to further use restrictions. National legislation is in place in Austria, Denmark, Luxembourg, Sweden and Switzerland that limits the use of HFCs in refrigeration equipment, foams and solvents. During the review of permits for large emitters under the EU's Integrated Pollution Prevention and Control (IPPC) Directive (EC, 96/61) a number of facilities have been required to implement best available control technologies for N_2O and fluorinated gases (EC, 2006c).

7.9.6 Energy and technology policies

The IEA's World Energy Outlook 2006 (IEA, 2006c) provides an up-to-date estimate of the impacts of energy policies on the industrial sector[10]. The IEA compares two scenarios, a Reference Scenario, which assumes continuation of policies currently in place, and an Alternate Policy Scenario, which projects the cumulative impact of the more than 1400 energy

---

10   IEA's definition of the industrial sector does not include petroleum refining.

policies being considered by governments worldwide, many of which affect the industrial sector. The Alternate Policy Scenario assumes faster deployment of commercially demonstrated technology, but not technologies that are still to be commercially demonstrated, including CCS and advanced biofuels.

Global industrial energy demand in 2030 in the IEA's Alternate Policy Scenario is 9% (14 EJ) lower than in the Reference Scenario. Industrial sector CO_2 emissions are 12% (0.9 $GtCO_2$) lower. Estimated investment to achieve these savings is US$ 362 billion (2005 US$), US$ 195 billion of which is in electrical equipment. The savings in electricity costs are about three times the investment in electrical equipment. The IEA (2006c) does not provide information on the value of the fuel savings in industry, but clearly it is larger than the investment.

Government is expected to lower financial risk and promote the investment through technology policy, which includes diverse options: budget allocations for R&D on innovative technologies, subsidy or legislation to stimulate specific environmental technologies, or regulation to suppress unsustainable technologies. See for example the US DOE's solicitation for industrial R&D projects (US DOE, n.d.-a) and the Government of India's Central Pollution Control Board Programmes on development and deployment of energy efficient technologies (CPCB, 2005).

7.9.7   Sustainable Development policies

Appropriate sustainable development policies focusing on energy efficiency, dematerialization and use of renewables can support GHG mitigation objectives. For example, the policy options selected by the Commission on Sustainable Development 13th session to provide a supportive environment for new business formation and the development of small enterprises, included:

- Reduce information barriers for energy efficiency technology for industries;
- Build capacity for industry associations, and
- Stimulate technological innovation and change to reduce dependency on imported fuels, to improve local air pollution and to generate local employment (CSD, 2005).

Individual countries are also trying to achieve these objectives. Most policies are stated in general terms, but their implementation would have to include the industrial sector.

The EU's strategy for sustainable development highlights addressing climate change through the reduction of energy use in all sectors and the control of non-CO_2 GHGs (EC, 2001). The UK's sustainable development policy incorporates the UK's emissions trading and climate levy policies for the control of CO_2 emissions from industry (UK DEFRA, 2005). As part of its sustainable development policy, Sweden is emphasizing energy efficiency and a long-term goal of obtaining all energy from

renewable sources (OECD, 2002). China faces a significant challenge in achieving its sustainable development goals, because from 2002 to 2004 its primary energy use grew faster than its GDP, with over two-thirds of that increase coming from coal. In 2005 the Chinese government emphasized that rapid growth must be sustainable and announced the goal of reducing energy consumption per unit of GDP by 20% between 2005 and 2010 (Naughton, 2005). India has launched a series of reforms aimed at achieving industrial sector sustainable development. The 2001 Energy Conservation Act mandated a Bureau of Energy Efficiency charged with ensuring efficient use of energy and use of renewables (GOI, 2004). The Indian Industry Programme for Energy Conservation includes both mandatory and voluntary efforts, with greater emphasis on voluntary approaches (BEE, 2006).

These countries are trying to improve resources use efficiency, waste management, water and air pollution reduction, and enhance use of renewables, while providing health benefits and improved services to communities. Many developed (Sutton, 1998) and developing countries (Jindal Steel and Power, Ltd., 2006; ITC, 2006) encourage companies to help achieve these goals thought dematerialization, habitat restoration, recycling, and commitment to corporate social responsibility.

7.9.8   Air quality policies

Section 4.5.2 contains a more general discussion of the relationships between air quality policies and GHG mitigation. In general air quality and climate change are treated as separate issues in national and international policies, even though most practices and technologies that will reduce GHG emissions will also cause a net reduction of emissions of air pollutants. However, air pollutant reduction measures do not always reduce GHG emissions, as many require the use of additional energy (STAPPA/ALAPCO, 1999). Examples of policies dealing with air pollution and GHG emissions in an integrated fashion include: (1) the EU IPPC Directive (96/61/EC), which lays down a framework requiring Member States to issue operating permits for certain industrial installations, and (2) the Dutch plan for a NO_x emission trading system, which will be implemented through the same legal and administrational infrastructure as the European CO_2 emission trading system (Dekkers, 2003).

7.9.9   Waste management policies

Waste management policies can reduce industrial sector GHG emissions by reducing energy use through the re-use of products (e.g., of refillable bottles) and the use of recycled materials in industrial production processes. Recycled materials significantly reduce the specific energy consumption of the production of paper, glass, steel, aluminium and magnesium. The amount, quality and price of recycled materials are largely determined by waste management policies. These policies can also influence the design of products – including the choice of materials, with its implications for production levels and

Table 7.11: *Co-benefits of greenhouse-gas mitigation or energy-efficiency programmes of selected countries*

Category of Co-benefit	Examples
Health	Reduced medical/hospital visits, reduced lost working days, reduced acute and chronic respiratory symptoms, reduced asthma attacks, increased life expectancy.
Emissions	Reduction of dust, CO, CO_2, NO_x and SO_x; reduced environmental compliance costs.
Waste	Reduced use of primary materials; reduction of waste water, hazardous waste, waste materials; reduced waste disposal costs; use of waste fuels, heat and gas.
Production	Increased yield; improved product quality or purity; improved equipment performance and capacity utilization; reduced process cycle times; increased production reliability; increased customer satisfaction.
Operation and maintenance	Reduced wear on equipment; increased facility reliability; reduced need for engineering controls; lower cooling requirements; lower labour requirements.
Working environment	Improved lighting, temperature control and air quality; reduced noise levels; reduced need for personal protective equipment; increased worker safety.
Other	Decreased liability; improved public image; delayed or reduced capital expenditures; creation of additional space; improved worker morale.

Sources: Aunan et al., 2004; Pye and McKane, 2000; Worrell et al., 2003.

emissions. Prominent examples can be found in the packaging sector, for example the use of cardboard rather than plastic for outer sales packages, or PET instead of conventional materials in the beverage industry. Vertical and horizontal integration of business provides synergies in the use of raw materials and reuse of wastes. The paper and paper boards wastes generated in cigarette packaging and printing are used as raw materials in paper and paper board units (ITC, 2006).

Another important influence of waste policies on industrial GHG emissions is their influence on the availability of secondary 'waste' fuels and raw materials for industrial use. For example, the 'EU Landfill Directive' (EU-OJ, 1999), which limits the maximum organic content of wastes acceptable for landfills, resulted in the restructuring of the European waste sector currently taking place. It makes available substantial amounts of waste containing significant biomass fractions. Typically there is competition between the different uses for these wastes: dedicated incineration in the waste sector, co-combustion in power plants, or combustion in industrial processes, for example cement kilns. In order to provide additional inexpensive disposal routes, several countries have set incentives to promote the use of various wastes in industrial processes in direct competition with dedicated incineration. Emissions trading systems or project-based mechanisms like CDM/JI can provide additional economic incentives to expand the use of secondary fuels or biomass as substitutes for fossil fuels. The impact of switching from a fossil fuel to a secondary fuel on the energy efficiency of the process itself is frequently negative, but is often compensated by energy savings in other parts of the economy.

Mineral wastes, such as fly-ash or blast-furnace slag can have several competing alternative uses in the waste, construction and industrial sectors. The production of cement, brick and stone-wool provides energy saving uses for these materials in industry. For secondary fuels and raw materials, life-cycle

assessment can help to quantify the net effects of these policies on emission across the affected parts of the economy (Smith *et al.*, 2001). The interactions between climate policies and waste policies can be complex, sometimes leading to unexpected results because of major changes of industry practices and material flows induced by minor price differences.

7.10 Co-benefits of industrial GHG mitigation

The TAR explained that 'co-benefits are the benefits from policy options implemented for various reasons at the same time, acknowledging that most policies resulting in GHG mitigation also have other, often at least equally important, rationales' (IPCC, 2001a). Significant co-benefits arise from reduction of emissions, especially local air pollutants. These are discussed in Section 11.8.1. Here we focus on co-benefits of industrial GHG mitigation options that arise due to reduced emissions and waste (which in turn reduce environmental compliance and waste disposal costs), increased production and product quality, improved maintenance and operating costs, an improved working environment, and other benefits such as decreased liability, improved public image and worker morale, and delaying or reducing capital expenditures (see Table 7.11) (Pye and McKane, 2000; Worrell *et al.*, 2003).

A review of forty-one industrial motor system optimization projects implemented between 1995 and 2001 found that twenty-two resulted in reduced maintenance requirements on the motor systems, fourteen showed improvements in productivity in the form of production increases or better product quality, eight reported lower emissions or reduction in purchases of products such as treatment chemicals, six projects forestalled equipment purchases, and others reported increases in production or decreases in product reject rates (Lung *et al.*, 2003). Motor system optimization projects in China are seen as an activity that can reduce operating costs, increase system reliability and contribute to the economic viability of Chinese industrial enterprises faced with increased competition (McKane *et al.*, 2003).

A review of 54 emerging energy-efficient technologies, produced or implemented in the USA, EU, Japan and other industrialized countries for the industrial sector, found that 20 of the technologies had environmental benefits in the areas of 'reduction of wastes' and 'emissions of criteria air pollutants'. The use of such environmentally friendly technologies is often most compelling when it enables the expansion of incremental production capacity without requiring additional environmental permits. In addition, 35 of the technologies had productivity or product quality benefits (Martin *et al.*, 2000).

Quantification of the co-benefits of industrial technologies is often done on a case-by-case basis. One evaluation identified 52 case studies from projects in the USA, the Netherlands, UK, New Zealand, Canada, Norway and Nigeria that monetized non-energy savings. These case studies had an average simple payback time of 4.2 years based on energy savings alone. Addition of the quantified co-benefits reduced the simple payback time to 1.9 years (Worrell *et al.*, 2003). Inclusion of quantified co-benefits in an energy-conservation supply curve for the US iron and steel industry doubled the potential for cost-effective savings (Worrell *et al.*, 2001a; 2003).

Not all co-benefits are easily quantifiable in financial terms (e.g., increased safety or employee satisfaction), there are variations in regulatory regimes vis-à-vis specific emissions and the value of their reduction and there is a lack of time series and plant-level data on co-benefits. Also, there is a need to assess net co-benefits, as negative impacts that may be associated with some technologies, such as increased risk, increased training requirements and production losses during technology installation (Worrell *et al.*, 2003).

7.11 Technology Research, Development, Deployment and Diffusion (RDD&D)

Most industrial processes use at least 50% more than the theoretical minimum energy requirement determined by the laws of thermodynamics, suggesting a large potential for energy-efficiency improvement and GHG emission mitigation (IEA, 2006a). However, RDD&D is required to capture these potential efficiency gains and achieve significant GHG emission reductions. Studies have demonstrated that new technologies are being developed and entering the market continuously, and that new technologies offer further potential for efficiency improvement and cost reduction (Worrell *et al.*, 2002).

While this chapter has tended to discuss technologies only in terms of their GHG emission mitigation potential and cost,

it is important to realize that successful technologies must also meet a host of other performance criteria, including cost competitiveness, safety, and regulatory requirements; as well as winning consumer acceptance. (These topics are discussed in more detail in Section 7.11.2.) While some technology is marketed as energy-efficient, other benefits may drive the development and diffusion of the technology, as evidenced by a case study of impulse drying in the paper industry, in which the driver was productivity (Luiten and Blok, 2004). This is understandable given that energy cost is just one of the drivers for technology development. Innovation and the technology transfer process are discussed in Section 2.8.2.

Technology RDD&D is carried out by both governments (public sector) and companies (private sector). Ideally, the roles of the public and private sectors will be complementary. Flannery (2001) argued that it is appropriate for governments to identify the fundamental barriers to technology and find solutions that improve performance, including environmental, cost and safety performance, and perhaps customer acceptability; but that the private sector should bear the risk and capture the rewards of commercializing technology. Case studies of specific successful energy-efficient technologies, including shoe press in papermaking (Luiten and Blok, 2003a) and strip casting in the steel industry (Luiten and Blok, 2003b), have shown that a better understanding of the technology and the development process is essential in the design of effective government support of technology development. Government can also play an important role in cultivating 'champions' for technology development, and by 'anchoring' energy and climate as important continuous drivers for technology development (Luiten and Blok, 2003a).

While GHG mitigation is not the only objective of energy R&D, IEA studies show a mismatch between R&D spending and the contribution of technologies to reduction of CO_2 emissions. In its analysis of its Accelerated Technology scenarios, IEA (2006a) found that end-use energy efficiency, much of it in the industrial sector, contributed most to mitigation of CO_2 emissions from energy use. It accounted for 39–53% of the projected reduction, except in the scenario that deemphasized these technologies. However, IEA countries spent only 17% of their public energy R&D budgets on energy-efficiency (IEA, 2005).

Many studies have indicated that the technology required to reduce GHG emissions and eventually stabilize their atmospheric concentrations is not currently available (Jacoby, 1998; Hoffert *et al.*, 2002; Edmonds *et al.*, 2003) *(medium agreement, medium evidence)*. While these studies concentrated on energy supply options, they also indicate that significant improvements in end-use energy efficiency will be necessary. Much of the necessary research and development is being carried out in public-private partnerships, for example the US Department of Energy's Industrial Technologies Program (US DOE, n.d.-b).

7.11.1 Public sector

A more complete discussion of public sector policies is presented in Section 7.9 and in Chapter 13. While government use many policies to spur RDD&D in general, this section focuses specifically on programmes aimed at improving energy efficiency and reducing GHG emissions.

7.11.1.1 Domestic policies

Governments are often more willing than companies to fund higher-risk technology research and development. This willingness is articulated in the US Department of Energy's Industrial Technologies Program role statement: 'The programme's primary role is to invest in high-risk, high-value research and development that will reduce industrial energy requirements while stimulating economic productivity and growth' (US DOE, n.d.-a). The Institute for Environment and Sustainability of the EU's Joint Research Centre has a similar mission, albeit focusing on renewable energy (Joint Research Centre, n.d.a), as does the programme of the Japanese government's New Energy and Industrial Technology Development Organization (NEDO, n.d.).

Selection of technology is a crucial step in any technology adoption. Governments can play an important role in technology diffusion by disseminating information about new technologies and by providing an environment that encourages the implementation of energy-efficient technologies. For example, energy audit programmes, provide more targeted information than simple advertising. Audits by the US Department of Energy's Industrial Assessment Center program in SMEs resulted in implementation of about 42% of the suggested measures (Muller and Barnish, 1998). Programmes or policies that promote or require reporting and benchmarking of energy consumption can have a similar function. These programmes have been implemented in many countries, including Canada, Denmark, Germany, the Netherlands, Norway, the UK and the USA (Sun and Williamson, 1999), and in specific industrial sectors such as the petroleum refining, ethylene and aluminium industries. (See Section 7.3.1).

Many of the voluntary programmes discussed in Section 7.9.2 include information exchange activities to promote technology diffusion at the national level and across sectors. For 2004, the US Industrial Technologies Program claimed cumulative energy savings of approximately 5 EJ as the result of diffusion of more than 90 technologies across the US industrial sector (US DOE, 2006). EU programmes, for example Lights of the Future and the Motor Challenge Programme (Joint Research Centre, n.d.b), have similar objectives, as do programmes in other regions.

A wide array of policies has been used and tested in the industrial sector in industrialized countries, with varying success rates (Galitsky et al., 2004; WEC, 2004). No single instrument will reduce all the barriers to technology diffusion; an integrated policy accounting for the characteristics of technologies, stakeholders and regions addressed is needed.

Evenson (2002) suggests that the presence of a domestic research and development programme in a developing country increase the county's ability to adapt and adopt new technologies. Preliminary analysis seems to suggest that newly industrialized countries are becoming more active in the generation of scientific and technical knowledge, although there is no accurate information on the role of technology development and investments in scientific knowledge in developing countries (Amsden and Mourshed, 1997).

7.11.1.2 Foreign or international policies

Industrial RDD&D programmes assume that technologies are easily adapted across regions with little innovation. This is not always the case. While many industrial facilities in developing nations are new and include the latest technology, as in industrialized countries, many older, inefficient facilities remain. The problem is exacerbated by the presence of large numbers of small-scale, much less energy-efficient plants in some developing nations; for example the iron and steel, cement and pulp and paper industries in China, and in the iron and steel industry in India (IEA, 2006a). This creates a huge demand for technology transfer to developing countries to achieve energy efficiency and emissions reductions.

Internationally, there are a growing number of bilateral technology RDD&D programmes to address the slow and potentially sporadic diffusion of technology across borders. A December, 2004 US Department of State Fact Sheet lists 20 bilateral agreements with both developed and developing nations (US Dept. of State, 2004), many of which include RDD&D.

Multilaterally, the UNFCCC has resulted in the creation of two technology diffusion efforts, the Climate Technology Initiative (CTI) and the UNFCCC Secretariat's TT:CLEAR technology transfer database. CTI was established in 1995 by 23 IEA/OECD member countries and the European Commission, and as of 2003 has been recognized as an IEA Implementing Agreement. Its focus is the identification of climate technology needs in developing countries and countries with economies-in-transition, and filling those needs with training, information dissemination and other support activities (CTI, 2005). TT: CLEAR is a more passive technology diffusion mechanism that depends on users accessing the database and finding the information they need (UNFCCC, 2004). Additionally, in 2001, the UNFCCC established an Expert Group on Technology Transfer (EGTT) (UNFCCC, 2001). EGTT has promoted a number of activities including workshops on enabling environments and innovative financing for technology transfer. Ultimately, the Kyoto Protocol's CDM and JI should act as powerful tools for the diffusion of GHG mitigation technology.

IEA implementing agreements, for example the Industrial Energy Related Technology and Systems Agreement (IEA-IETS, n.d.), also provide a multilateral basis for technology transfer. While still in the planning stage, it is hoped that the newly established Asia-Pacific Partnership on Clean Development and Climate will play a key role in technology transfer to China, India and Korea (APP, n.d.)

7.11.2 Private sector

In September, 2004, the IPCC convened an expert meeting on industrial technology development, transfer and diffusion. One of the objectives of the meeting was to identify the key drivers of these processes in the private sector (IPCC, 2005a). Among the key drivers for private sector involvement in the technology process discussed at the meeting were:
- Maintaining competitive advantage in open markets;
- Consumer acceptance in response to environmental stewardship;
- Country-specific characteristics: economic and political as well as its natural resource endowment;
- Scale of facilities, which affects the type of technology that can be deployed;
- Intellectual property rights (IPR): protection of IPR is critical to achieving competitive advantage through technology.
- Regulatory framework, including: government incentives; government policies on GHG emissions reduction, energy security and economic development; rule of law; and investment certainty.

The meeting concluded that each of these drivers could either be stimulants or barriers to the technology process, depending on their level, for example a high level of protection for IPR would stimulate the deployment of innovative technology in a specific country while a low level would be a barrier. However, it was also recognized that these drivers were only indicators and that actual decisions had to consider interactions between the drivers, as well as non-technology factors.

7.12 Long-term outlook, system transitions, decision-making and inertia

7.12.1 Longer-term mitigation options

Many technologies offer long-term potential for mitigating industrial GHG emissions, but interest has focused in three areas:
- Advanced biological processing, in which chemicals are produced by biological reactions that require lower energy input;
- Use of hydrogen for metal smelting, in fuel cells for electricity production, and as a fuel – provided the hydrogen is produced via a low or zero-carbon process – and;

- Nanotechnology, which could provided the basis for more efficient catalysts for chemical processing and for more effective conversion of low-temperature heat into electricity (Hillhouse and Touminen, 2001).

While some applications of these technologies could enter the marketplace by 2030, their widespread application, and impact on GHG emissions, is not expected until post-2030.

7.12.2 System transitions, inertia and decision-making

Given the complexity of the industrial sector, the changes required to achieve low GHG emissions cannot be characterized in terms of a single system transition. For example, development of an inert electrode for aluminium smelting would significantly lower GHG emissions from this process, but would have no impact on emissions from other industries.

Inertia in the industrial sector is characterized by capital stock turnover rate. As discussed in Section 7.6, the capital stock in many industries has lifetimes measured in decades. While opportunities exist for retrofitting some capital stock, basic changes in technology occur only when the capital stock is installed or replaced. This inertia is often referred to as 'technology lock-in', a concept first proposed by Arthur (1988). IEA (2006a) discusses the potential effects of technology lock-in in electric power generation, where much of the capital stock in developed nations will be replaced, and much of the capital stock in developing nations will be installed, in the next few decades. Installation of lower-cost, but less efficient technology will then impact GHG emission for decades thereafter. The same concerns and impacts apply in the industrial sector.

Industrial companies are hierarchical organizations and have well-established decision-making processes. In large companies, these processes have formal methods for incorporating technical and economic information, as well as regulatory requirements, consumer preferences and stakeholder inputs. Procedures in SMEs are often informal, but all successful enterprises have to address the same set of inputs.

7.13 Key uncertainties and gaps in knowledge

Gaps in knowledge are defined as missing information that could be developed by research. Uncertainties are missing information that cannot be developed through research. Key uncertainties in the projection of mitigation potential and cost in 2030 are:
- The rate of technology development and diffusion;
- The cost of future technology;
- Future energy and carbon prices;

- The level of industry activity in 2030; and
- Policy drivers, both climate and non-climate.

Key gaps in knowledge are: base case energy intensity for specific industries, especially in transition economies; co-benefits, SD implications of mitigation options and consumer preferences.

REFERENCES

AERES, 2005: Rapport 2005. Association des Entreprises pour la Réduction de l'Effet de Serre Nanterre, France, <http://www.aeres-asso.org/doc/28-06-06%20AERES_RA%20FR.pdf>, accessed 31/05/07.

AF&PA, 2004: AF&PA Environmental, Health and Safety Verification Program: Year 2002 Report. American Forest and Paper Association, Washington, D.C.

Aguayo, F. and K. Gallagher, 2005: Economic reform, energy and development: The case of Mexican manufacturing. *Energy Policy*, **33** (7), pp. 829-837.

Amsden, A.H. and M. Mourshed, 1997: Scientific publications, patents and technological capabilities in Late-Industrializing Countries. *Technology Analysis & Strategic Management*, **3**, pp. 343-359.

Anderson, S. and R. Newell, 2004: Prospects for carbon capture and storage technologies. *Annual Review of Environmental Resources*, **29**, pp. 109-42.

ANSI, 2005: A management system for energy. ANSI/MSE 2000:2005, American National Standards Institute Washington, D.C., American National Standards Institute.

APEC, 2002: Profiles in SMEs and SME Issues, 1990-2000. Asia-Pacific Economic Cooperation, Singapore, World Scientific Publishing.

APERC, 2003: Energy efficiency programmes in developing and transitional APEC economies. Asia Pacific Energy Research Centre, **Tokyo.**

API, 2005: 3rd API/DOE Conference on voluntary actions by the oil and gas industry to address climate change. September 29-30, 2004, American Petroleum Institute Arlington, Virginia. <www.api.org/ehs/climate/response/upload/3rd_ConferenceSynopsis.pdf>, accessed 31/05/07.

APP, n.d.: Asia pacific partnership on clean development and climate. Asia Pacific Partnership, <www.asiapacificpartnership.org>, accessed 01/06/07.

A3P, 2006: Performance, people and prosperity: A3P Sustainability Action Plan. Australian Plantation Products and Paper Industry Council, <http://www.a3p.asn.au/assets/pdf/PerformancePeopleProsperity_actionplan.pdf>, accessed 01/06/07.

AREED, 2000: Open for business. African Rural Energy Enterprise Development, <http://www.areed.org/status/Open_for_Business.pdf>, accessed 01/06/07.

Arthur, B.: 1988: Self-reinforcing mechanisms in economics. In *The Economy as an Evolving Complex System*. P.W. Anderson and K.J. Arrow (eds.), New York: Addison-Wesley.

Aunan, K., J. Fang, H. Vennemo, K. Oye, and H.M. Seop, 2004: Co-benefits of climate policy: Lessons learned from a study in Shanxi, China. *Energy Policy*, **32**, pp. 567-581.

Australian Greenhouse Office, n.d.: Fuel tax credits and greenhouse challenge plus membership. <www.greenhouse.gov.au/challenge/members/fueltaxcredits.html>, accessed 31/05/07.

Bailey, O. and E. Worrell, 2005: Clean energy technologies, a preliminary inventory of the potential for electricity generation. Lawrence Berkeley National Laboratory, Berkeley, California (LBNL-57451).

Bain, 2005: An analysis of energy consumption in Bangladesh. PhD. Thesis. Jadavpur University, Kolkata, India.

Bakaya-Kyahurwa, E., 2004: Energy efficiency and energy awareness in Botswana; ESI, *Power Journal of Africa*, Issue **2**, 2004, pp. 36-40.

Balasubramaniam, S.V., 2005: Sugar: New dimensions of growth. In *Survey of Indian Industry, 2005*. The Hindu, Chennai, India.

Banda, A., 2002: Electricity production from sugar industries in Africa: A case study of South Africa. M.Sc thesis, University of Cape Town, South Africa.

Barats, C., 2005: Managing greenhouse gas risk: A roadmap for moving forward. Presented at Risk and Insurance Management Society Annual Meeting. Philadelphia, PA, April 20, 2005.

Bartos, S.C., 2001: Semiconductor industry leadership in global climate protection. Future Fab International 10, <www.future-fab.com/documents.asp?d_ID=1144>, accessed 31/05/07.

Battelle Institute/WBCSD, 2002: Toward a sustainable cement industry. World Business Council on Sustainable Development, <www.wbcsd.org/web/publications/batelle-full.pdf>, accessed 31/05/07.

BEE, 2003: Energy management policy - Guidelines for energy-intensive industry of India. Bureau of Energy Efficiency Delhi, Government of India. Ministry of Power.

BEE, 2006: Implementation: Case studies from industries. Bureau of Energy Efficiency, <www.bee-india.nic.in/Implementation/Designated%20Consumers.htm>, accessed 31/05/07.

Beerkens, R.G.C. and J. van Limpt, 2001: Energy efficient benchmarking of glass furnaces; 62nd conference on glass problems. University of Illinois, October 16-17, 2001.

Bergmeier, M., 2003: The history of waste energy recovery in Germany since 1920. *Energy*, **28**, pp. 1359-1374.

Berntsson, T., P-A Franke, and A. Åsblad, 1997. Process heating in the low and medium temperature range. IEA/Caddet, Sittard, The Netherlands.

Birat, J.P., 2005: The UCLOS Program (Ultra Low CO_2 Steelmaking). <http://cordis.europa.eu/estep/events-infostp_en.html>, accessed 31/05/07.

Bjørner, T.B. and H.H. Jensen, 2002: Energy taxes, voluntary agreements and investment subsidies - A micro panel analysis of the effect on Danish industrial companies' energy demand. *Resource and Energy Economics*, **24**, pp. 229-249.

Bossoken, E., 1999: Case study: A comparison between France and The Netherlands of the Voluntary Agreements Policy. Study carried out in the frame of the MURE project, financed by the SAVE programme, European Commission, Directorate General for Energy (DG 17).

Boyd, G.A., 2005. A method for measuring the efficiency gap between average and best practice energy use: The ENERGY STAR Industrial Performance Indicator. *Journal of Industrial Ecology*, **3**, pp. 51-66.

BP, 2003: Defining our path: Sustainability Report 2003. London: BP. <www.bp.com/liveassets/bp_internet/globalbp/STAGING/global_assets/downloads/B/BP_Sustainability_Report_2003.pdf>, accessed 31/05/07.

BP, 2005: Making energy more: Sustainability Report 2005. London: BP. <www.bp.com/liveassets/bp_internet/globalbp/STAGING/global_assets/downloads/S/bp_sustainability_report_2.pdf>, accessed 31/05/07.

Bystrom, S. and L. Lonnstedt, 1997: Paper recycling, environmental and economic impacts. *Resource Conservation and Recycling*, **21**, pp. 109-127.

Caffal, C., 1995. Energy management in industry. Centre for the Analysis and Dissemination of Demonstrated Energy Technologies (CADDET). Analysis Series 17. Sittard, The Netherlands.

Canepa, A. and P. Stoneman, 2004: Financing constraints in the inter firm diffusion of new process technologies. *Journal of Technology Transfer*, **30**, pp. 159-169.

Carbon Trust, 2006: Allocation and competitiveness in the EU emissions trading scheme: Options for Phase II and beyond. London, June, 2006. <www.carbontrust.co.uk/climatechange/policy/allocation_competitiveness.htm>, accessed 31/05/07.

CCX, 2005: Members of the Chicago Climate Exchange ® (CCX®). <http://www.chicagoclimatex.com/about/members.html>, accessed 31/05/07.

CDM for Sustainable Africa, 2004: Synergy: CDM for sustainable Africa - capacity building for Clean Development Mechanisms in Sub-Saharan Africa - final project report. <http://www.rgesd-sustcomm.org/CDM_AFRICA/PDF/Final_Report-CDM_for_Sustainable_Africa_Total.pdf >, accessed 31/05/07.

CEEBICNet Market Research, 2004: Energy Efficiency in Romania. <www.mac.doc.gov/ceebic/country/Romania/EnergyEfficiency.htm>, accessed 31/05/07.

CEEEZ, 2003: Multi-energy efficiency through use of automatic load control for selected industries. Centre for Energy, Environment and Engineering, Zambia Ltd.Project Idea Note (PIN) submitted to 500 ppm, Frankfurt, Germany.

CEFIC, 2002: Managing energy: Voluntary energy efficiency program (VEEP). <http://www.cefic.org/Templates/shwStory.asp?NID=471&HID=149>, accessed 31/05/07.

CEPI, 2001: Potential energy efficiency improvements for the European Pulp and Paper Industry, Confederation of European Paper Industries, Brussels, March 2001.

CESP, 2004: China's sustainable energy scenarios in 2020. China Environmental Science Press. Beijing.

CFI, 2003: Energy benchmarking: Canadian potash production facilities. Canadian Fertilizer Institute, Ottawa, Canada, Canadian Industry Program for Energy Conservation.

Chang, H.J., 2003: New horizons for Korean energy industry - shifting paradigms and challenges ahead. *Energy Policy*, **31**, pp. 1073-1084.

Chaudhary, T.R., 2001: Technological measures of improving productivity: Opportunities and constraints. Presented at the Fertilizer Association of India Seminar "Fertiliser and Agriculture Future Directions," New Delhi, India, 6-8 December 2001.

Chaudhuri, S. and K. Gupta, 2003: Inflow of foreign capital, environmental pollution, and welfare in the presence of an informal sector. Dept. of Economics, University of Calcutta, Kolkata, India.

Chevron, 2005: Corporate Responsibility Report 2005. <www.chevron.com/cr_report/2005/downloads.asp>, accessed 31/05/07.

Chidiak, M., 2000: Voluntary agreements - Implementation and efficiency. The French Country Study: Case Studies in the Sectors of Packaging Glass and Aluminium. Paris: CERNA, Centre d'économie industrielle, Ecole Nationale Supérieure des Mines de Paris.

Chidiak, M., 2002: Lessons from the French experience with voluntary agreements for greenhouse gas reduction. *Journal of Cleaner Production*, **10**, pp. 121-128.

Choi, K.H. and B.W. Ang, 2001: A time-series analysis of energy-related carbon emissions in Korea. *Energy Policy*, **29**, pp. 1155-1161.

CIEEDAC, 2004: Development of energy intensity indicators for Canadian industry 1990-2002. Canadian Industrial Energy End-Use Data and Analysis Center Burnaby, BC, Canada.

CLASP, 2005: S&L Worldwide Summary. Collaborative Labelling and Appliance Standards Program, <www.clasponline.org/main.php>, accessed 31/05/07.

Cohen, A.T., 2001: Prospects for biogas harvesting at Sungunn Wongse Industries (Thailand). Report of a feasibility and technical evaluation.

CONCAWE, 2005: The impact of reducing sulphur to 10 ppm max in European automotive fuels: An Update. Report 8/05. Conservation of Clean Air and Water Europe (CON-CAWE), Brussels, 53 pp.

Continental Engineering BV, 2001: High temperature catalytic reduction of nitrous oxide emissions from nitric acid production plants. Report funded by Dutch Ministry of Housing Spatial Planning and Environment, SenterNovem Project No. 375001/0080, CE Ref. J-4586-02.

Cornish, A.T. and B. Kerkhoff, 2004: European standards and specifications in cements. In *Innovations in Portland Cement Manufacture*. J.I. Bhatty, F.M. Miller and S.H. Kosmatka (eds.), Portland Cement Association, Skokie, IL, USA, pp. 1209-1218.

Cornland, D., F. Johnson, F.D. Yamba, E.W. Chidumayo, M.N. Morales, O. Kalumiana, and S.B. Mtonga-Chidumayo, 2001: Sugar cane resources for sustainable development, Stockholm Environment Institute: ISBN: 91 88714713.

Cordi, A. and L. Lombardi, 2004: End life tyres: Alternative final disposal process compared by LCA. *Energy*, **29**, pp. 2089-2108.

CPCB, 2005: Annual Action Plan 2005-2006. Central Pollution Control Board of India <www.cpcb.nic.in/actionplan/plan2005-ch3.htm>, accessed 31/05/07.

CRA, 2002: Information on corn wet milling. Corn Refiners Association <www.corn.org/whatiscornrefining.htm>, accessed 31/05/07.

CSD, 2005: Report on the thirteenth session (30 April 2004 and 11-22 April 2005). Commission on Sustainable Development UN ECOSOC, Official Records, 2005, Supplement No. 9.

CTI, 2005: Climate Technology Initiative. <www.climatetech.net>, accessed 31/05/07.

Dankers, A.P.H., 1995: The energy situation in Thai lime industries. *Energy for Sustainable Development*, **1**, pp. 36-40.

Das, T.K. and J. Roy, 1994: Cost benefit analysis of the solar water heating system in Khardah Jute Mill and solar cookers under operation in various parts of Calcutta. Report submitted to West Bengal Renewable Energy Development Agency.

Dasgupta, M and J. Roy, 2000: Manufacturing energy use in India: A decomposition analysis. *Asian Journal of Energy and Environment*, **1**, pp. 223-247.

Dasgupta, M. and J. Roy, 2001: Estimation and analysis of carbon dioxide emissions from energy intensive manufacturing industries in India. *International Journal of Energy, Environment and Economics*, **11**, pp. 165-179.

De Beer, J.G., 1998: Potential for industrial energy-efficiency improvement in the long term. PhD Dissertation, Utrecht University, Utrecht, The Netherlands.

De Beer, J.G., E. Worrell, and K. Blok, 1998: Future technologies for energy efficient iron and steelmaking. *Annual Review of Energy and Environment*, **23**, pp. 123-205.

De Beer, J.G., J. Harnisch and M. Kerssemeeckers, 2000a: Greenhouse gas emissions from iron and steel production. IEA Greenhouse Gas R&D Programme, Cheltenham, UK.

De Beer, J.G., M.M.M. Kerssemeeckers, R.F.T. Aalbers, H.R.J. Vollebergh, J. Ossokina, H.L.F de Groot, P. Mulder, and K. Blok, 2000b: Effectiveness of energy subsidies: Research into the effectiveness of energy subsidies and fiscal facilities for the period 1988-1999. Ecofys, Utrecht, the Netherlands.

De Beer, J.G., D. Phylipsen, and J. Bates, 2001: Economic evaluation of carbon dioxide and nitrous oxide emission reductions in industry in the EU. European Commission - DG Environment, Brussels, Belgium.

De Keulenaer, H., R. Belmans, E. Blaustein, D. Chapman, A. De Almeid, and B. De Wachter, 2004: Energy efficient motor driven systems, <http://re.jrc.ec.europa.eu/energyefficiency/pdf/HEM_lo_all%20final.pdf>, accessed 01/06/07.

Dekkers, C., 2003: Netherlands to launch legislation on NO_x and CO_2 emissions trading. Emissions Trader - Emissions Marketing Association, <www.emissions.org/publications/emissions_trader/0303>, accessed 31/05/07.

Delmas, M. and A. Terlaak, 2000: Voluntary agreements for the environment: Innovation and transaction costs. CAVA Working Paper 00/02/13 February.

Dersden, A., P. Vercaenst and P. Pijkmans 2002: Best available techniques (BAT) for the fruit and vegetable processing industry. *Resources, Conservation and Recycling*, **34**, pp. 261-271.

Dowd, J., K. Friedman, and G. Boyd, 2001: How well do voluntary agreements and programs perform at improving industrial energy efficiency. Proceedings of the 2001 ACEEE Summer Study on Energy Efficiency in Industry. Washington, D.C.: American Council for an Energy-Efficient Economy.

DTI, 2005: EU Emissions trading scheme: Benchmark Research for Phase 2. Department of Trade and Industry (UK), Prepared by Entec UK Limited, and NERA Economic Consulting on behalf of the UK Department of Transport and Industry, July 2005.

DuPont, 2002: Sustainable growth: 2002 Progress Report. Wilmington, DC: DuPont. <www1.dupont.com/dupontglobal/corp/documents/US/en_US/news/publications/dupprogress/2002USprogrept.pdf>, accessed 31/05/07.

Dutta, S., P.K. Datta, S. Ghosal, A. Gosh and S.K. Sanyal, 2004: Low-cost pollution control devices for continuous cupolas. *Indian Foundry Journal*, **50**, pp. 29-33.

EC, 2001: A sustainable europe for a better world: A European Union strategy for sustainable development. European Commission, <http://europa.eu.int/eur-lex/en/com/cnc/2001/com2001_0264en01.pdf>, accessed 01/06/07.

EC, 2004: Comprehensive Report 2002-2003 regarding the role of forest products for climate change mitigation. Enterprise DG Unit E.4, Forest-based industries. Brussels: European Commission.

EC, 2005: The European Union Greenhouse Gas Emission Trading Scheme. European Commission, <http://europa.eu.int/comm/environment/climat/pdf/emission_trading2_en.pdf>, accessed 31/05/07.

EC, 2006a: A report to the European Parliament and the Council considering the functioning of the EU Emissions Trading Scheme (ETS). European Commission, Brussels, 2006, <http://ec.europa.eu/environment/climat/emission/review_en.htm>, accessed 31/05/07.

EC, 2006b: Inclusion of additional activities and gases into the EU - Emissions Trading Scheme, European Commission DG Environment, <http://ec.europa.eu/environment/climat/emission/review_en.htm>, accessed 01/06/07.

EC, 2006c: Installations with reported emissions of more than 10% of EU wide total emissions according to EPER, European Commission, DG Environment, January 2006. <http://ec.europa.eu/environment/ippc/pdf/table_largest_emitters_jan_06.pdf>, accessed 31/05/07.

EDGAR, 2005: EDGAR 2005 Fast Track Data Set. MNP - Netherlands Environmental Assessment Agency (MNP), <www.mnp.nl/edgar/model/v32ft2000edgar/>, accessed 31/05/07.

Edmonds, J., J.J. Dooley, S.H. Kim, R. Cesar Izaurralde, N.J. Rosenberg, and G.M. Stokes, 2003: The potential role of biotechnology in addressing the long-term problem of climate change in the context of global energy and economic systems. In *Greenhouse Gas Control Technologies: Proceedings of the Sixth International Conference on Greenhouse Gas Control Technologies*, Kyoto, Japan, Elsevier Science, Oxford, UK, J. Gale and Y. Kaya (eds.), pp. 1427-1433, ISBN 0080442765.

EFCTC, 2000: The United Nations Framework Convention on Climate Change - Fluorochemicals and the Kyoto Protocol. European Fluorocarbon Technical Committee, <www.cefic.be/sector/efctc/kyoto2000-02.htm>, accessed 31/05/07.

Egenhofer, C., N. Fujiwara, and K. Giaoglou, 2005: Business consequences of the EU Emissions Trading Scheme. Brussels, Centre for European Policy Studies, February, 2005.

EIA, 2002: Manufacturing consumption of energy 1998. Energy Information Agency Office of Energy Markets and End Use, Washington, D.C.

EIA, 2005: Annual Energy Outlook 2005: With projections to 2025. Energy Information Agency, Washington, D.C., <www.eia.doe.gov/oiaf/aeo/pdf/0383(2005).pdf>, accessed 31/05/07.

Eidt, B., 2004: Cogeneration opportunities - Global Energy Management System. Presented at the 3rd API/DOE Conference on Voluntary Actions by the Oil and Gas Industry to Address Climate Change, September 29-30, 2004, Arlington, Virginia, <www.api.org/ehs/climate/response/upload/3rd_ConferenceSynopsis.pdf>, accessed 31/05/07.

Einstein, D., E. Worrell, and M. Khrushch, 2001: Steam systems in industry: Energy use and energy efficiency improvement potentials. Proceedings of the 2001 ACEEE Summer Study on Energy Efficiency in Industry - Volume 1, Tarrytown, NY, July 24-27th, 2001, pp. 535-548.

ENDS-Directory, 2006: Getting help through environmental business support schemes. Judith Cowan, Business in the Environment, <http://www.endsdirectory.com/index.cfm?action=articles.view&articleID=bie>, accessed 31/05/07.

EU-HLG, 2006: High Level group (HLG) on competitiveness, energy and the environment. <http://ec.europa.eu/enterprise/environment/hlg/hlg_en.htm>, accessed 31/05/07.

EU-BREF Ceramics, 2005: Integrated pollution prevention and control (IPPC) reference document on best available technology in the ceramic manufacturing industry. <ftp://ftp.jrc.es/pub/eippcb/doc/cer_bref_1206.pdf>, accessed 31/05/07.

EU-BREF Glass, 2001: Integrated Pollution Prevention and Control (IPPC) Reference Document on Best Available Technology in the Glass Manufacturing Industry. <ftp://ftp.jrc.es/pub/eippcb/doc/gls_bref_1201.pdf>, accessed 31/05/07.

EU-OJ, 1999: Council Directive 1999/31/EC of 26 April 1999 on the landfill of waste. Official Journal L 182, 16/07/1999, Pages 0001-0019.

Evenson, R.E., 2002: Induced adaptive invention/innovation convergence in developing countries. In Technological Change and the Environment, A. Grubler, N. Nackicenovic and W.D. Nordhaus (eds.), Resources for the Future, Washington, D.C.

FAO, 1993: Status and development issues of the brick industry in Asia. Regional Wood Energy Development Programme in Asia, FAO: GCP/RAS/154/NET, Bangkok.

FAO, 2006: Statistical database. <http://faostat.fao.org/default.aspx>, accessed 31/05/07.

FAOSTAT, 2006: Forestry data. <http://faostat.fao.org/site/381/default.aspx>, accessed 31/05/07.

Farahani, S., E. Worrell, and G. Bryntse, 2004: CO_2-free paper? *Resources and Conservation Recycling*, **42**, pp. 317-336.

Farla, J., K. Blok, and L. Schipper, 1997: Energy efficiency developments in the paper and pulp industry. *Energy Policy*, **25**, pp. 745-758.

Flannery, B.P., 2001: An industry perspective on carbon management. Presented at *Carbon Management: Implications for R&D in the Chemical Sciences and Technology*, A Workshop Report to the Chemical Sciences Roundtable, (US) National Research Council.

FPAC, n.d.: Environmental stewardship. Forest Products Association of Canada, <www.fpac.ca/en/sustainability/stewardship>, accessed 31/05/07.

Future Energy Solutions, 2004: Climate Change Agreements: Results of the Second Target Period Assessment, Version 1.0., Didcot, UK: AEA Technology. www.defra.gov.uk/environment/ccl/pdf/cca_jul05.pdf

Future Energy Solutions, 2005: Climate Change Agreements: Results of the second target period assessment, Version 1.2. Didcot, UK: AEA Technology. <www.defra.gov.uk/environment/ccl/pdf/cca_aug04.pdf>, accessed 31/05/07.

Galitsky, C., N. Martin, and E. Worrell, 2001: Energy efficiency opportunities for United States breweries. *MBAA Technical Quarterly*, **4**, pp. 189-198.

Galitsky, C., E. Worrell, and M. Ruth, 2003: Energy efficiency improvement and cost saving opportunities for the corn wet milling industry: An ENERGY STAR Guide for Energy and Plant Manager. Berkeley, CA, Lawrence Berkeley National Laboratory (LBNL-52307).

Galitsky, C., L. Price, and E. Worrell, 2004: Energy efficiency programs and policies in the industrial sector in industrialized countries. Berkeley, CA, Lawrence Berkeley National Laboratories (LBNL-54068).

Galitsky, C. and E. Worrell, 2004: Profile of the chemical industry in California. Berkeley, CA, Lawrence Berkeley National Laboratory (LBNL-55668).

Gartner, E., 2004: Industrially interesting approaches to "low-CO_2" cements. *Cement and Concrete Research*, **34**, pp. 1489-98.

Giehlen, D.J., 2006: Energy efficiency and CO_2 emissions reduction: Opportunities for the fertiliser industry. Paper presented at the 2006 IFA Technical Symposium. Vilnius, Lithuania, 25-28 April 2006.

George, G. and G.N. Prabhu, 2003: Developmental financial institutions as technology policy instruments: Implications for innovation and entrepreneurship in emerging economies. *Research Policy*, **32**, pp. 89-108.

GOI, 2004: India's initial national communication to the UN Framework Convention on Climate Change. Government of India Ministry of the Environment and Forest, New Delhi.

GOI, 2005: Annual Report, 2004-2005 of the Ministry of Environment and Forest. Government of India, New Delhi.

GOI, Various issues: Annual survey of industries. Central Statistical Organization, Government of India, Kolkata, India.

Goulder, L.H. and S.H. Schneider, 1999: Induced technological change and the attractiveness of CO_2 abatement policies. *Resource and Energy Economics*, **21**, pp. 211-253.

Government of Flanders, 2005: Energiebenchmarking in Vlaanderen. <www.benchmarking.be/en/home.html>, accessed 31/05/07.

Green Budget Germany, 2004: Ecotax - Memorandum 2004. Förderverein Ökologische Steuerreform e.V., <www.czp.cuni.cz/ekoreforma/EDR/ GBG_Memorandum_2004.pdf>, accessed 01/06/07.

GRI, n.d.: A common framework for sustainability reporting. <www.globalreporting.org>, accessed 31/05/07.

Gross, D., A. Gross, M. Schoenfeld, W. Simmons and B. Patrick, 2003: Jupiter oxygen combustion technology of coal and other fossil fuels in boilers and furnaces. Presented at the 28th International Technical Conference on Coal Utilization and Fuel Systems, Clearwater, FL, USA.

Grothcurth, H.M., R. Kummel, and W.V. Goole, 1989: thermodynamic limits to energy optimization. *Energy*, **14**, pp. 241-258.

Gudbjerg, E., 2005: Energy Management: The Danish standard. Presented at the First European Energy Management Conference, Milan, Italy, 23-24 November 2005.

Gupta, K., 2002: The urban informal sector and environmental pollution: A theoretical analysis. Working Paper No. 02-006. Center for Environment and Development Economics, University of York, York, UK.

Hansen, K. and A. Larsen, 1999. Energy efficiency in industry through voluntary agreements - the implementation of five voluntary agreement schemes and an assessment. Copenhagen: AKF, Amternes og Kommunernes Forskningsinstitut.

Harnisch, J., I. Sue Wing, H.D. Jacoby, and R.G. Prinn, 1998: Primary aluminium production: Climate policy, emissions and costs. MIT Joint Program on the Science and Policy of Global Change Report Series, Report No. 44, Massachusetts Institute of Technology, Cambridge, MA.

Harnisch, J. and W. Schwarz, 2003: Costs and the impact on emissions of potential regulatory framework for reducing emissions of hydrofluorocarbons, perfluorocarbons and sulphur hexafluoride. On behalf of DG Environment of the European Commission, Brussels. <http://europa.eu.int/comm/environment/climat/pdf/ecofys_ oekorecherchestudy.pdf>, accessed 31/05/07.

Heeres, R.R.,*et al.*, 2004: Eco-industrial park initiatives in the USA and the Netherlands: first lessons. *Journal of Cleaner Production*, **12**(8-10), pp. 985-995.

Heidelberg Cement, 2006: Building on sustainability, group sustainability report, 2004/2005. <www.heidelbergcement.com/global/en/company/ sustainability/_channelposting.htm>, accessed 31/05/07.

Hendriks, C.A., E. Worrell, L. Price, N. Martin, and L. Ozawa Meida, 1999: The reduction of greenhouse gas emissions from the cement industry. Cheltenham, UK, IEA Greenhouse Gas R&D Programme, Report PH3/7.

Hidalgo, I., L. Szabo, J.C. Ciscar, and A. Soria, 2005: Technological prospects and CO_2 emissions trading analysis in the iron and steel industry: a global model. *Energy*, **30**, pp. 583-610.

Hillhouse, H.W. and M.T. Touminen, 2001: Modelling the thermoelectric transport properties of nanowires embedded in oriented microporous and mesoporous films. *Microporous and Mesoporous Materials*, **47**, pp. 39-50.

HM Revenue & Customs, n.d.: ECA - 100% Enhanced capital allowances for energy-saving investments. <http://www.hmrc.gov.uk/capital_ allowances/eca-guidance.htm>, accessed 01/06/07.

Hoffert, M.I., K. Caldeira, G. Benford, C. Green, H. Herzog, A.K. Jain, H.S. Kheshgi, K.S. Lackner, J.S. Lewis, H.D. Lightfoot, W. Manheimer, J.C. Mankins, M.E. Mauel, L.J. Perkins, M.E. Schlesinger, T. Volk, and T.M.L. Wigley, 2002: Advanced Technology Paths to Global Climate Stability. *Science*, **298**, pp. 981-987.

Holliday, C., 2001: Sustainable growth, the DuPont Way. Harvard Business Review, September, 2001. Electronic Download R0108J.

Houillon, G. and O. Jolliet, 2005: Life cycle assessment of processes for the treatment of wastewater urban sludge: Energy and global warming analysis. *Journal of Cleaner Production*. **13**, pp. 287-99.

Humphreys, K. and M. Mahasenan, 2002: Towards a sustainable cement industry - Substudy 8: Climate Change. World Business Council for Sustainable Development (WBCSD), Geneva, Switzerland.

IAI, 2000: Life cycle results. International Aluminium Institute, <www. world-aluminium.org/environment/climate/lifecycle2.html>, accessed 31/05/07.

IAI, 2001: Perfluorocarbon emissions reduction programme 1990-2000. International Aluminium Institute, <www.world-aluminium.org/iai/ publications/documents/pfc2000.pdf>, accessed 31/05/07.

IAI, 2003. Life cycle assessment of aluminium: Inventory data for the worldwide primary aluminium industry. International Aluminium Institute, London, 52 pp.

IAI, 2004: The global sustainable development initiative. International Aluminium Institute London, 6 pp.

IAI, 2006a: The International Aluminium Institute's report on the Aluminium Industry's global perfluorocarbon gas emissions reduction programme - Results of the 2004 anode effect survey. <www.world- aluminium.org/iai/publications/documents/pfc2004.pdf>, accessed 31/05/07.

IAI, 2006b: Aluminium for future generations - sustainability update 2005. International Aluminium Institute, <www.world-aluminium.org/iai/ publications/documents/update_2005.pdf>, accessed 31/05/07.

IEA, 1997: Voluntary actions for energy-related CO_2 abatement. OECD/ International Energy Agency, Paris.

IEA, 2003a: Energy balances of Non-OECD Countries: 2000-2001. Paris: IEA.

IEA, 2003b: Energy Balances of OECD Countries: 2000-2001. International Energy Agency, Paris.

IEA, 2004: Energy Policies of IEA Countries: Finland 2003 review. International Energy Agency, Paris

IEA, 2005: R&D Database. International Energy Agency, <www.iea.org/ dbtw-wpd/textbase/stats/rd.asp>, accessed 31/05/07.

IEA, 2006a: Energy Technology Perspectives 2006: Scenarios and strategies to 2050. International Energy Agency, Paris, 484 pp.

IEA, 2006b: Industrial motor systems energy efficiency: Towards a plan of action. International Energy Agency, Paris, 14 pp.

IEA, 2006c: World Energy Outlook 2006. International Energy Agency, Paris, 596 pp.

IEA GHG, 2000: Greenhouse gases from major industrial sources - IV. The aluminium industry; International Energy Agency Greenhouse Gas R&D Programme, Report Number PH3/23, ICF Kaiser, Cheltenham, UK, 88 pp.

IEA GHG, 2001: Abatement of emissions of other greenhouse gases: Engineered Chemicals. IEA Greenhouse Gas R&D Programme, J. Harnisch, O. Stobbe, D. de Jager, Cheltenham, 83 pp.

IEA-IETS, n.d.: Implementing Agreement. International Energy Agency - Industrial Energy-Related Technology and Systems, <www.iea-iets. org>, accessed 31/05/07.

IEA-SolarPACES, n.d.: Task IV: SHIP - Solar heat for industrial processes. International Energy Agency - Concentrating Solar Power, <www. solarpaces.org/task_iv.htm>, accessed 31/05/07.

IETA, 2005: Greenhouse Gas Market 2005: The rubber hits the road. International Emissions Trading Association, <www.ieta.org/ieta/ www/pages/getfile.php?docID=1270>, accessed 31/05/07.

491

IFA, 2004: IFADATA 2004. International Fertilizer Industry Association, <www.fertilizer.org/ifa/statistics/indicators/pocket_production.asp>, accessed 31/05/07.

IFA, 2005: Summary report - World agriculture and fertilizer demand, global fertilizer supply and trade. International Fertilizer Industry Association 2005-2006. <www.fertilizer.org/ifa/publicat/PDF/2005_ council_sevilla_ifa_summary.pdf>, accessed 31/05/07.

IGEN/BEE, n.d.: Success stories. Indo-German Energy Program in cooperation with Bureau of Energy Efficiency under Ministry of Power, <www.energymanagertraining.com/new_success.php>, accessed 31/05/07.

IISI, 2005: World steel in figures, 2005: International Iron and Steel Institute (IISI), Brussels.

IISI, 2006: Steel: The foundation of a sustainable future - Sustainability report of the world steel industry 2005. International Iron and Steel Institute. <www.worldsteel.org/pictures/publicationfiles/SR2005. pdf>, accessed 31/05/07.

IPCC, 1995: *Climate Change 1995: The Science of Climate Change. Intergovernmental Panel on Climate Change (IPCC)*, Cambridge University Press, Cambridge, UK, 572 pp.

IPCC, 2000a: *Development, Equity and Sustainability. Cross Cutting Issues Guidance Papers. Intergovernmental Panel on Climate Change (IPCC)*, Geneva, pp. 69-113.

IPCC, 2000b: *Emission Scenarios. Special Report of the Intergovernmental Panel on Climate Change (IPCC)* [N. Nakicenovic, *et al.*] Cambridge University Press, Cambridge, UK, 598 pp.

IPCC, 2000c: *Methodological and Technological Issues in Technology Transfer. Special Report of the Intergovernmental Panel on Climate Change (IPCC)* [B. Metz, *et al.*] Cambridge University Press, Cambridge, UK, 466 pp.

IPCC, 2001a: *Climate Change 2001: Mitigation. Intergovernmental Panel on Climate Change (IPCC)* [B. Metz, *et al.* (eds)]. Cambridge University Press, Cambridge, UK.

IPCC, 2001b: *Climate Change 2001: Impacts, Adaptation, and Vulnerability. Intergovernmental Panel on Climate Change (IPCC)* [J.J. McCarthy, *et al.* (eds.), Cambridge University Press, Cambridge, UK.

IPCC, 2005a: *Meeting Report: IPCC Expert Meeting on Industrial Technology Development, Transfer and Diffusion. Intergovernmental Panel on Climate Change (IPCC)*, 21-23 Sept. 2004, Tokyo, 35 pp.

IPCC, 2005b: Carbon Capture and Storage. Special Report of the Intergovernmental Panel on Climate Change (IPCC), Cambridge University Press, Cambridge, UK.

IPCC, 2006: 2006 *Inventory Guidelines. Intergovernmental Panel on Climate Change (IPCC)*, Cambridge University Press, Cambridge, UK.

IPCC/TEAP, 2005: *Safeguarding the Ozone Layer and the Global Climate System: Issues Related to Hydrofluorocarbons and Perfluorocarbons. Special Report of the Intergovernmental Panel on Climate Change/ Technical and Economic Assessment Panel*, Cambridge University Press, Cambridge, UK.

IPPC, 2001: Reference document on best available techniques in the cement and lime manufacturing industries. European Commission, Brussels, Belgium.

ISO, 1996: ISO 14001, Environmental Management Systems - Specification with guidance for use. International Standards Organization ISO, Geneva, 14 pp.

ISO, 2004: Business Plan ISO/TC 63 Glass Containers, Geneva: International Standard Organization, 9 pp.

ITC, 2006: For all our tomorrows. Sustainability Report 2005, <www. itcportal.com/sustainability_report_2005/default.html>, accessed 31/05/07.

Jacoby, H.D., 1998: The uses and misuses of technology development as a component of climate policy. MIT Joint Program on the Science and Politics of Global Change, Report No, 43., <http://web.mit.edu/ globalchange/www/rpt43.html>, accessed 31/05/07.

Jain, A.K., 2005: Cement: strong growth performance, survey of Indian industry 2005. The Hindu, Chennai, India.

Jindal Steel and Power, Ltd., 2006: Building a new India, 26th Annual Report 2004-2005. New Delhi, India.

Jochem, E. and W. Eichhammer, 1999: Voluntary agreements as an instrument to substitute regulating and economic instruments. Lessons from the German Voluntary Agreements on CO_2 Reduction. In *Voluntary Approaches in Environmental Policy*, C. Carraro and F. Lévêque (eds.), Kluwer Academic Publisher, Dordrecht, Boston, London.

Joint Research Centre, n.d.a: Mission of the Joint Research Centre. <www.jrc.cec.eu.int>, accessed 31/05/07.

Joint Research Centre, n.d.b.: End-use energy efficiency activities of the European Commission Joint Research Centre. <http://re.jrc.ec.europa. eu/energyefficiency/index.htm>, accessed 01/06/07.

Josa, A., A. Aguado, A. Heino, E. Byars, and A. Cardin, 2004: Comparative analysis of available life-cycle inventories of cement in the EU. *Cement and Concrete Research*, **34**, pp. 1313-1320.

Jupiter Oxygen Corp., 2006: Supply-side energy efficiency and fossil fuel switch. Presented at Carbon Expo, May 11, 2006, Cologne, Germany, <www.jupiteroxygen.com/spotlight/uploads/050206_efficiency.pdf>, accessed 31/05/07.

Kaltner, F.J., G.F.P Zepeda, I.A. Campos, and A.O.F. Mondi, 2005: Liquid biofuels for transportation in Brazil. <www.fbds.org.br/IMG/pdf/doc-116.pdf>, accessed 31/05/07.

Kannan, R. and W. Boy, 2003: Energy management practices in SME- Case study of bakery in Germany. *Energy Conversion and Management*, **44**, pp. 945-959.

Kerssemeeckers, M., 2002: The Dutch long term voluntary agreements on energy efficiency improvement in industry. Utrecht, the Netherlands: Ecofys.

Kim, Y. and E. Worrell, 2002a: International comparison of CO_2 emissions trends in the iron and steel industry. *Energy Policy*, **30**, pp. 827-838.

Kim, Y. and E. Worrell, 2002b: CO_2 emission trends in the cement industry: An international comparison. *Mitigation and Adaptation Strategies for Global Change*, **7**, pp. 115-33.

Kirk-Othmer, 2005: Recycling, Glass. In : C. Philip Ross, Kirk-Othmer, *Encyclopaedia of Chemical Technology*, John Wiley & Sons; Article Online Posting Date: May 13, 2005.

Kolk, A. and J. Pinkse, 2005: Business response to climate change: Identifying emergent strategies. *California Management Review*, **47**(3), pp. 6-20.

Kumar, S., G. Senanakaye, C. Visvanathan, and B. Basu, 2003: Desiccated coconut industry of Sri-Lanka's opportunities for energy efficiency and environmental protection. *Energy Conversion and Management*, **44**(13), pp. 2205-2215.

Laurin, A., J. Nyboer, C. Strickland, N. Rivers. M. Bennett, M. Jaccard, R. Murphy, and B. Sandownik, 2004: Strategic options for combined heat and power in Canada. Office of Energy Efficiency, Natural Resources Canada, Ottawa, ON, Canada.

Lemar, P.L., 2001: The potential impact of policies to promote combined heat and power in US industry. *Energy Policy*, **29**, pp. 1243-1254.

Linde, 2005: Stirring and degassing. Linde Mix 14, <www.us.lindegas. com/international/web/lg/us/likelgus30.nsf/DocByAlias/nav_ industry_alum_stir>, accessed 31/05/07.

Liskas, K., G. Mavrotas, M. Mandaraka, and D. Diakoulaki, 2000: Decomposition of industrial CO_2 emissions: The case of the European Union. *Energy Economics*, **22**, pp. 383-394.

Luiten, E. and K. Blok, 2003a: The success of a simple network in developing an innovative energy-efficient technology. *Energy*, **28**, pp. 361-391.

Luiten, E. and K. Blok, 2003b: Stimulating R&D of industrial energy-efficient technology; the effect of government intervention on the development of strip casting technology. *Energy Policy*, **31**, pp. 1339-1356.

Luiten, E. and K. Blok, 2004: Stimulating R&D of industrial energy-efficient technology. Policy Lessons - Impulse Technology. *Energy Policy*, **32**, pp. 1087-1108.

Lung, R.B., A. McKane, and M. Olzewski, 2003: Industrial Motor system optimization projects in the US: An impact study. Proceedings of the 2003 American Council for an Energy-Efficient Economy Summer Study on Energy Efficiency in Industry. Washington, D.C., ACEEE.

Margolick, M. and D. Russell, 2001: Corporate greenhouse gas reduction targets. Washington, D.C.: Pew Center for Global Climate Change.

Martcheck, K., 2006: Modelling more sustainable aluminium: Case study. International Journal of LCA, 11, pp. 34-37.

Martin, N., E. Worrell, M. Ruth, M., L. Price, R.N. Elliott, A. Shipley and J. Thorne, 2000: Emerging energy-efficient industrial technologies. Washington D.C., American Council for an Energy-efficient economy and Berkeley, CA, Lawrence Berkeley National Laboratory (LBNL-46990).

Matsuhashi, R., M. Shigyo, Y. Yoshida, and H. Ishitani, 2000: Life cycle analysis of systems utilizing waste heat to reduce CO_2 emissions. Proceedings of International Conference on Electrical Engineering, pp. 613-616.

Matysek, A., M. Ford, G. Jakeman, A. Gurney, and B.S. Fisher, 2006: Technology: Its role in economic development and climate change. ABARE Research Report 06.6, Canberra, Australia. <www.abare. gov.au/publications_html/climate/climate_06/cc_technology.pdf>, accessed 31/05/07.

McFarland, M., 2005: Statement of Mack McFarland, Ph.D., Global Environmental Manager, DuPont Fluoroproducts, E.I. DuPont de Nemours and Company, Inc. before the Committee on Science, US House of Representatives, June 8, 2005, <www.house.gov/science/hearings/full05/june8/dupont.pdf>, accessed 31/05/07.

McKane, A., G. Zhou, R. Williams, S. Nadel, and V. Tutterow, 2003: The China Motor Systems Energy Conservation Program: Establishing the foundation for systems energy efficiency. Berkeley, CA:, Lawrence Berkeley National Laboratory (LBNL-52772).

McKane, A., W. Perry, A. Li, T. Li, and R. Williams, 2005: Creating a standard framework for sustainable industrial energy efficiency. Presented at Energy Efficiency in Motor Driven Systems (EEMODS 2005) Conference, Heidelberg, Germany, 5-8 September 2005.

Meier, A., E. Bonaldi, G.M. Cella, W. Lipinski, D. Wullemin, and R. Palumbo, 2004: Design and experimental investigation of a horizontal rotary reactor for the solar thermal production of lime. Energy, 29, pp. 811-821.

Miller, M.M., 2003: Lime. In *2003 Minerals Yearbook*, United States Geological Survey. Reston, VA.

Miner, R. and A. Lucier, 2004: A value chain assessment of the climate change and energy issues affecting the global forest products industry. <http://www.wbcsd.ch/web/projects/forestry/ncasi.pdf>, accessed 01/06/07.

Miner, R. and B. Upton, 2002: Methods for estimating greenhouse gas emissions from lime kilns at Kraft pulp mills. *Energy*, **27**, pp. 729-738.

Möllersten, K., J. Yan, and M. Westermark, 2003: Potential and cost effectiveness of CO_2 reductions through energy measures in Swedish pulp and paper mills. *Energy*, **28**, pp. 691-710.

Moreira, J., 2006: Global biomass energy potential. *Journal of Mitigation and Adaptation Strategies for Global Change*, **11**, pp. 313-333.

Morimoto, K., H.X. Nguyen, M. Chihara, T. Honda and R. Yamamoto, 2006: Proposals for classification and an environmental impact evaluation method for eco-services: Case study of municipal waste treatment in cement production. *Journal of Life Cycle Assessment*, Japan, **2**, pp. 347-354.

Mozorov, A. and G. Nikiforov, 2002: Energy development concept of OAO Magnitogorsk Iron and Steel Works. *Steel Journal*, **1**, Moscow, Russia (original in Russian).

Muller, M.R. and T.J. Barnish, 1998: Evaluation of the former EADC-Program. In *Industrial Energy Efficiency Policies: Understanding Success and Failure*, Martin *et al..* (eds.), Proceedings of a workshop organized by the International Network for Energy Demand Analysis in the Industrial Sector. Utrecht, The Netherlands, June 11-12, 1998, (LBNL-42368).

Munasinghe, M., 2002: Sustainomics transdisciplinary framework for making development more sustainable - application to energy issues. *International Journal of Sustainable Development*, **5**, pp.125-82.

Munn, A., 2004: Emission reduction at Engen refinery in South Durban. Paper presented at the 8th World Congress on Environmental Health, Durban, South Africa, 22-27 February 2004.

Naiwei, S., 2004: The Chinese brick and tile industry - Steady development in 2004. Ziegelindustrie International, Pp. 46-49, 4/2004.

Natural Resources Canada, n.d: Industrial Energy Innovators. <http://oee.nrcan.gc.ca/industrial/opportunities/innovator/index. cfm?attr=24>, accessed 31/05/07.

Naughton, B. 2005: The new common economic program: China's 11th five year plan and what it means. China Leadership Monitor No.16, <www.hoover.org/publications/clm/issues/2898936.html>, accessed 31/05/07.

Navarro, P., G. Gregoric, O. Cobo, and A. Calandra, 2003: A new anode effect quenching procedure. *Light Metals*, 2003, pp. 479-486.

NEDO, n.d.: Industrial technology research and development activities. New Energy and Industrial Technology Development Organization, <www.nedo.go.jp/english/activities/index.html>, accessed 31/05/07.

NEEC, 2005: One-year accelerated depreciation allowance for energy efficient equipment and technology. National Energy Efficiency Committee (Singapore), < http://www.nccc.gov.sg/incentive/home. shtm >, accessed 01/06/07.

NERA, 2004: Review of the first and second years of the UK emissions trading scheme. Prepared for UK Department for Environment, Food and Rural Affairs, D. Radov and P. Klevnäs, NERA, <http:// www.defra.gov.uk/environment/climatechange/trading/uk/pdf/nera-commissionreport.pdf>, accessed 31/05/07.

Nicholson, C., 2004: Protecting the environment - A business perspective. 4th Specialty Conference on Environmental Progress, Bahrain, <www. bp.com/genericarticle.do?categoryId=98&contentId=2017173>, accessed 31/05/07.

Nippon Keidanren, 2004: Results of the fiscal 2004 follow-up to the Keidanren Voluntary Action Plan on the Environment - Section on Global Warming Measures, <http://www.keidanren.or.jp/english/policy/2004/091/index.html>, accessed 31/05/07.

NRC, 2001: Energy Research at DOE: Was it worth it? Energy efficiency and fossil energy research 1978 to 100. National Research Council, Washington D.C, National Academy Press, < http://www.nap.edu/openbook/0309074487/html/>, accessed 01/06/07.

NREL, 2006: Advancing Clean Energy Use in Mexico. National (US) Renewable Energy Laboratory, <www.nrel.gov/docs/fy05osti/38628. pdf>, accessed 31/05/07.

Nyboer, J. and J.J. Tu, 2003: A review of energy consumption & related data: Canadian cement manufacturing industry, 1990 to 2001: Canadian Industry Energy End-use Data and Analysis centre, Simon Fraser University, Vancouver, BC.

Ochs, T., P. Turner, D. Oryshchyn, B. Patrick, A. Gross, C. Summers, W. Simmons, and M. Schoenfeld, 2005: Proof of concept for integrating oxy-fuel combustion and the removal of all pollutants from a coal-fired flame. Presented at the 30th International Technical Conference on Coal Utilization and Fuel Systems, Clearwater, FL, USA.

OECD, 2002: Voluntary approaches for environmental policy: Effectiveness, efficiency and use in policy mixes. Organization for Economic Cooperation and Development (OECD), Paris.

Okazaki, T., M. Nakamura, and K. Kotani, 2004: Voluntary initiatives of Japan's steel industry against global warming. Paper presented at IPCC Industrial Expert Meeting (ITDT) in Tokyo.

Okuwaki, A., 2004: Feedstock recycling of plastics in Japan. *Polymer Degradation and Stabilization*, **85**, pp. 981-988.

Olivier, J.G.J. and J.A.H.W. Peters, 2005: CO_2 from non-energy use of fuels: A global, regional and national perspective based on the IPCC Tier 1 approach. *Resources, Conservation and Recycling*, **45**(3), pp. 210-225.

PCA, 2002: Common elements among advanced greenhouse gas management programs: A discussion paper. New York, Partnership for Climate Action, <www.environmentaldefense.org/documents/1885_PCAbooklet.pdf>, accessed 31/05/07.

Pew Center on Global Climate Change, 2005: Business Environmental Leadership Council (BELC). <www.pewclimate.org/companies_leading_the_way_belc>, accessed 31/05/07.

Phadke, A., J. Sathaye, and S. Padmanabhan, 2005: Economic benefits of reducing Maharashtra's electricity shortage through end-use efficiency improvement. LBNL-57053. Lawrence Berkeley National Laboratory, Berkeley, CA.

Phylipsen, D., K. Blok, E. Worrell, and J. De Beer, 2002: Benchmarking the energy efficiency of Dutch industry: an assessment of the expected effect on energy consumption and CO_2 emissions. *Energy Policy*, **30**, pp. 663-679.

Phylipsen, G.J.M. and K. Blok, 2002: The effectiveness of policies to reduce industrial greenhouse gas emissions. Paper for the AIXG Workshop on Policies to Reduce Greenhouse Gas Emissions in Industry - Successful Approaches and Lessons Learned, 2-3 December 2002, Berlin.

Pickens, J.G., S.T.S. Yuen, and H. Hennings, 2002: Waste management options to reduce greenhouse gas emissions from paper in Australia. *Atmospheric Environment*, **36**, pp. 741-752.

Pilkington, 2005: Flat glass industry - Summary, Pilkington, plc. <www.pilkington.com/about+pilkington/flat+glass+industry/default.htm>, accessed 31/05/07.

Price, L., 2005: Voluntary agreements for energy efficiency or ghg emissions reduction in industry: An assessment of programs around the world. Proceedings of the 2005 American Council for an Energy Efficient Economy Summer Study on Energy Efficiency in Industry, West Point, NY, USA, July, 2005.

Price, L., J. Sinton, E. Worrell, D. Phylipsen, H. Xiulian, and L. Ji, 2002: Energy use and carbon dioxide emissions from steel production in China. *Energy*, **5**, pp. 429-446.

Price, L., S. de la Rue du Can, J. Sinton, E. Worrell, N. Zhou, J. Sathaye, and M. Levine, 2006: Sectoral trends in global energy use and greenhouse gas emissions. LBNL-56144, Lawrence Berkeley National Laboratory, Berkeley, CA.

PSI, 2004: Energy efficiency and CO_2 emissions benchmarking of IFA Ammonia Plants (2002-2003 Operating Period). Plant Surveys International, Inc General Edition. Commissioned by the International Fertilizer Industry Association (IFA).

Pye, M. and A. McKane, 2000: Making a stronger case for industrial energy efficiency by quantifying non-energy benefits. *Resources, Conservation and Recycling*, **28**, pp. 171-183.

Rafiqul, I., C. Weber, B. Lehmann, and A. Voss, 2005: Energy efficiency improvements in ammonia plants: Perspectives and uncertainties. *Energy*, **30**, pp. 2487-2504.

Raina, S.J., 2002: Energy efficiency improvements in Indian Cement Industry. <www.energymanagertraining.com/cement/pdf/IIPEC_NCCBM-2.pdf>, accessed 31/05/07.

Rainey, D. and P. Patilis, 2000: Developing an energy conservation program at United Technologies Corporation. *Corporate Environmental Strategy*, **7**, pp. 333-342.

Ramesohl, S. and K. Kristof, 2001: The declaration of German industry on global warming potential - A dynamic analysis of current performance and future prospects for development. *Journal of Cleaner Production*, **9**, pp. 437-446.

Ren, T., M. Patel, and K. Blok, 2005: Natural gas and steam cracking: Energy use and production cost. Published in the conference proceedings of the American Institute of Chemical Engineers Spring National Meeting, Atlanta, GA, USA, 10-14 April, 2005.

Ren, T., M. Patel, and K. Blok, 2006: Olefins from conventional and heavy feedstocks: Energy use in steam cracking and alternative processes. Accepted for publication in Energy. Available online at <www.ScienceDirect.com>, accessed 31/05/07.

Rietbergen, M.J., J.C.M. Farla, and K. Blok, 2002: Do agreements enhance energy efficiency improvement? Analyzing the Actual Outcome of the Long-Term Agreements on Industrial Energy Efficiency Improvement in The Netherlands. *Journal of Cleaner Production*, **10**, pp. 153-163.

Rock, M.T. and D.T. Angel, 2005: Industrial transformation in the developing world. Chapter 6: Win-Win environmental intensity or technique effects and technological learning: Evidence from Siam City Cement. New York, Oxford University Press, pp. 127-149.

Römpp, 1995: Römpp Chemie Lexikon. J. Falbe and M. Regnitz (eds.), Stuttgart, Germany.

Rosenberg, N., 1999: Inside the black box: Technology and economics. Cambridge University Press.

RWI, 2004: Die Klimaschutzerklärung der Deutschen Wirtschaft, Rheinisch-Westfälisches Institut für Wirtschaftsforschung Essen, Germany: RWI <http://www.rwi-essen.de/pls/portal30/docs/FOLDER/PROJEKTE/CO2MONITORING/MONITORING_DATEIEN/MONITORINGBERICHT_2000-2002.PDF>, accessed 01/06/07.

SAI, 2001: Social Accountability 8000. Social Accountability International, New York, SAI, 8 pp.

Sathaye, J., J. Roy, R. Khaddaria, and S. Das, 2005: Reducing electricity deficit through energy efficiency in India: An evaluation of macroeconomic benefits. LBLN-59092. Berkeley, CA, Lawrence Berkeley National Laboratory.

Schipper, L., F. Unander, and C. Marie-Lilliu, 2000: The IEA Energy Indicators: Analysing emissions on the road to Kyoto. Paris, France, IEA.

Schleich, J., 2004: Do energy audits help reduce barriers to energy efficiency? An empirical analysis for Germany. *International Journal of Energy Technology and Policy*, **2**, pp. 226-239.

Shashank, J., 2004: Energy conservation in the industrial sector: A special report on energy conservation day. New Delhi, Economic Times.

Sheinbaum, C. and L. Ozawa, 1998: Energy use and CO_2 emissions for Mexico's cement industry. *Energy*, **23**, pp. 725-32.

Shell, 2005: The Shell Sustainability Report, 2005. < www.shell.com/shellreport/>, accessed 31/05/07.

Sims, R.E.H, 2002: Waste co-gen plant-rock point sugar. Renewable Energy World, (March-April, 2002), pp. 105-109.

Sinton, J.E. and D.G. Fridley, 2000: What goes up: Recent trends in China's Energy Consumption. *Energy Policy*, **28**, pp. 671-687.

Silayan, A., 2005: Equitable distribution of CDM projects among developing countries. HWWA-Report, Hamburg, Germany, Hamburg Institute of International Economics.

Sjardin, M., 2003: CO_2 emission factors for non-energy use in the non-ferrous metal, ferroalloys and inorganics industry. Copernicus Institute, Utrecht, pp. 63.

Smith, A, K. Brown, S. Olilvie, K. Rushton, and J. Bates, 2001: Waste management options and climate change. AEA Technology for the European Commission, DG ENV. <http://europa.eu.int/comm/environment/waste/studies/pdf/climate_change.pdf>, accessed 31/05/07.

Sobhanbabu, P.R.K., 2003: Sugar cogeneration for power challenges and opportunities. Cane-Cogen, India, Vol. 16. a quarterly newsletter in sugar and cogeneration.

STAPPA/ALAPCO, 1999: Reducing greenhouse gases and air pollution, A menu of harmonized options. Final Report, State and Territorial Air Pollution Program Administrators/Association of Local Air Pollution Control Officials, Washington, pp. 119-163.

Starzer, O., 2000: Negotiated agreements in industry: Successful ways of implementation. Vienna, EVA, Austrian Energy Agency.

Subak, S., J.P. Palutikof, M.D. Agnew, S.J. Watson, C.G. Bentham, M.G.R. Cannell, M. Hulme, S. McNally, J.E. Thornes, D. Waughray, and J.C. Woods, 2000: The impact of the anomalous weather of 1995 on the UK Economy. *Climatic Change*, **44**(1-2), pp. 1-26.

Sukalac, K.E., 2005: Technology transfer to reduce climate change impacts from the fertilizer industry. Presented at UNFCCC COP-11 Side Event "Knowledge Transfer to Reduce Greenhouse Gas Emission: Lessons from the Fertilizer Industry. Montreal, Canada, 6 December 2005.

Sun, C. and M. Williamson, 1999: Industrial energy use benchmarking. In *Proceedings 1999 ACEEE Summer Study on Energy Efficiency in Industry*, American Council for an Energy-Efficient Economy: Washington, D.C.

Sutton, P., 1998: Sustainability: Which industries first? Green Innovations, Inc. <www.green-innovations.asn.au/wot-ind.htm>, accessed 31/05/07.

Swamidass, P.M., 2003: Modelling the adoption rates of manufacturing technology innovations by small US manufacturers: a longitudinal investigation. *Research Policy*, **32**(3), pp. 351-366.

Swaminathan, B. and K.E. Sukalac, 2004: Technology transfer and mitigation of climate change: The fertilizer industry perspective. Presented at the IPCC Expert Meeting on Industrial Technology Development, Transfer and Diffusion, Tokyo, Japan, 21-23 Sept. 2004.

Szklo, A.M., J.B. Soares, and M.T. Tolmasquim, 2004: Economic potential of natural gas-fired co-generation-analysis of Brazil's chemical industry. *Energy Policy*, **32**, pp. 1415-1428.

Tanaka, K., K. Sasaki, and H. Kudoh, 2005: Estimation of CO_2 reduction potentials through energy efficient technologies. *Journal of Institute for Energy Economics in Japan*, **31**, pp. 61-81.

Tanaka, K., R. Matsuhashi, N. Masahiro, and H. Kudo, 2006: CO_2 reduction potential by energy efficient technology in energy intensive industry. <http://eneken.ieej.or.jp/en/data/pdf/324.pdf>, accessed 31/05/07.

Tata Steel, Ltd, 2005: 3rd Social Audit. Mumbai, India.

The Aluminum Association, 2003: Aluminum Industry Technology Roadmap. Washington, DC, 60 pp.

The Sigma Project, 2003a: The Sigma guidelines: Putting sustainable development into practice - A guide for organizations. <www. projectsigma.co.uk/Guidelines/SigmaGuidelines.pdf>, accessed 31/05/07.

The Sigma Project, 2003b: The Sigma Guidelines - Toolkit - The Sigma Guide to the AA1000 Assurance Standard.

Toman, M., 2003: 'Greening' economic development activities for greenhouse gas mitigation. Resources for the Future, Issue Brief, RFF IB 03-02, <www.rff.org>, accessed 31/05/07.

UBA, 2004: Fluorinated greenhouse gases in products and processes: Technical climate protection measures. Umweltbundesamt, <www. umweltbundesamt.de/produkte-e/fckw/index.htm>, accessed 31/05/07. Berlin, Germany, 249 pp.

UK DEFRA, 2005: Securing the future: UK Government Sustainable Development Strategy. United Kingdom Department of the Environment, Food and Rural Affairs, http://www.sustainable-development.gov.uk/publications/uk-strategy/index.htm, accessed 31/05/07.

UK DEFRA, 2006: Climate change agreements - The climate change levy; United Kingdom Department of the Environment, Food and Rural Affairs. <www.defra.gov.uk/environment/ccl/intro.htm>, accessed 31/05/07.

UNDP, 2002: Oil Palm R&D: Malaysia. UN Development Programme.

UNESCAP, 2000: Promotion of energy efficiency in industry and financing of investments. United Nations Economic and Social Commission for Asia and the Pacific, <www.unescap.org/esd/energy/publications/finance/index.html>, accessed 31/05/07.

UNFCCC, 1995: Report of COP-1, Actions Taken. UN Framework Convention on Climate Change, <http://unfccc.int/resource/docs/cop1/07a01.pdf>, accessed 31/05/07.

UNFCCC, 1997: Kyoto Protocol. UN Framework Convention on Climate Change, <http://unfccc.int/resource/docs/convkp/kpeng.pdf>, accessed 31/05/07.

UNFCCC, 2001: The Marrakech Accords. UN Framework Convention on Climate Change, <http://unfccc.int/resource/docs/cop7/13a02.pdf>, accessed 31/05/07.

UNFCCC, 2002: Good practices. Policies and measures among parties included in Annex I to the Convention in their Third National Communications - Report by the Secretariat. UN Framework Convention on Climate Change.

UNFCCC, 2004: Results of the survey on effectiveness of the use of the UNFCCC technology information. UN Framework Convention on Climate Change clearinghouse (TT:CLEAR).

UNFCCC, CDM, n.d.: UN Framework Convention on Climate Change, Clean Development Mechanism. <http://cdm.unfccc.int/Statistics>, accessed 31/05/07.

UNIDO, 2001: Africa industry and climate change project proceedings. UN Industrial Development Organization, UNIDO publication, Vienna.

UNIDO, 2002: Developing national capacity to implement Clean Development Mechanism projects in a selected number of countries in Africa-A case study for Zambia and Zimbabwe UNIDO publication. UN Industrial Development Organization.

UNIDO and IFDC, 1998: Fertilizer manual. UN Industrial Development Organization and International Fertilizer Development Center, Dordrecht, The Netherlands, Kluwer Academic Press, 515 pp.

US DOE, 2002: Steam system opportunity assessment for the pulp and paper, chemical manufacturing, and petroleum refining industries - Main Report. US Department of Energy, Washington, D.C.

US DOE, 2003: Energy matters, the best practices quarterly newsletter for the US Department of Energy's Industrial Technologies Program. US Department of Energy, Summer 2003 issue, pp. 6-7.

US DOE, 2004: 20 ways to save energy now. US Department of Energy, <www.eere.energy.gov/industry/bestpractices/energymatters/pdfs/em_volume29.pdf>, accessed 31/05/07.

US DOE, 2006: Industrial Technologies Program: Summary of Program results for CY 2004. US Department of Energy, <www1.eere.energy.gov/industry/about/pdfs/impacts2006.pdf>, accessed 31/05/07.

US DOE, n.d.-a: Industrial Technologies Program. US Department of Energy, <http://www1.eere.energy.gov/industry>, accessed 31/05/07.

US DOE, n.d.-b.: Solicitations. US Department of Energy, <www1.eere.energy.gov/industry/financial/solicitations.html>, accessed 31/05/07.

US Dept. of State, 2004: Bilateral and regional partnerships. <http://www.state.gov/g/oes/ris/fs/2004/39438.htm>, accessed 31/05/07, <www.state.gov/g/oes/rls/fs/2004/39438.htm>, accessed 31/05/07.

US EPA, 1999: International efforts to reduce perfluorocarbon (PFC) emissions from primary aluminium production. United States Environmental Protection Agency Washington, DC.

US EPA, 2001: Non-CO_2 greenhouse gas emissions from developed countries: 1990-2010. United States Environmental Protection Agency, Washington, DC.

US EPA, 2002: Solid waste management and greenhouse gases: A life-cycle assessment of emissions and sinks. United States Environmental Protection Agency, EPA # 530-R-02-006.

US EPA, 2003: International analysis of methane and nitrous oxide abatement opportunities: report to energy modelling forum. Working Group 21, Appendix C. United States Environmental Protection Agency, <www.epa.gov/nonco2/econ-inv/pdfs/methodologych4.pdf>, accessed 31/05/07.

US EPA, 2005: Landfill methane outreach program. United States Environmental Protection Agency, <www.epa.gov/lmop>, accessed 31/05/07.

US EPA, 2006a: Global mitigation of non-CO_2 greenhouse gas emissions (EPA Report 430-R-06-005). United States Environmental Protection Agency Office of Air and Radiation. Washington, D.C., USA.

US EPA, 2006b: Global anthropogenic emissions of non-CO_2 greenhouse gas 1990-2020 (EPA Report 430-R-06-003). United States Environmental Protection Agency Office of Air and Radiation. Washington, D.C., USA.

US EPA, n.d.: Energy STAR for Industry. United States Environmental Protection Agency, <www.eere.energystar.gov/index.cfm?c=industry.bus_industry>, accessed 31/05/07.

USGS, 2005: Minerals Yearbook 2004. US Geological Survey Reston, VA, USA. <http://minerals.usgs.gov/minerals/pubs/myb.html>, accessed 31/05/07.

USGS, 2006: Minerals commodity summary: cement. Reston, VA, USA. US Geological Survey, <http://minerals.usgs.gov/minerals/pubs/commodity/cement/cemenmcs06.pdf>, accessed 31/05/07.

UTC, 2003: United technologies steps up global conservation efforts. United Technologies Corporation <http://www.utc.com/press/releases/2003-04-22.htm>, accessed 31/05/07.

Vaccari, G., E. Tamburini, G. Sualdino, K. Urbaniec. and J. Klemes. 2005: Overview of the environmental problems in beet sugar processing: Possible solutions. *Journal of Cleaner Production*, **13**, pp. 499-507.

Verduijn, W.D. and J.J. de Wit, 2001: Energy conservation: Key to survival for fertiliser producers? Proceedings No. 479, International Fertiliser Society, York, UK, 24 pp.

Wartmann, S., and J. Harnisch, 2005: Reduction of SF_6 emissions from high and medium voltage equipment in Europe. For the Coordinating Committee for the Associations of Manufacturers of Industrial Electrical Switchgear and Control Gear in the EU (CAPIEL) <www.capiel-elec-tric.com/publicats/Ecofys%20SF6%20Study%20Final%20Report%2022%20Nov%202005.pdf>, accessed 31/05/07.

WBCSD, 2002: The cement sustainability initiative: Our agenda for action. World Business Council for Sustainable Development, <www.wbcsd.org/web/publications/cement-action-plan.pdf>, accessed 31/05/07.

WBCSD, 2005: The Cement Sustainability Initiative: Progress Report. World Business Council for Sustainable Development, <www.wbcsdcement.org/pdf/csi_progress_report_2005.pdf>, accessed 31/05/07.

WEC, 2001: Energy efficiency policies and indicators policies, World Energy Council London.

WEC, 2004: Energy efficiency: A worldwide review - Indicators, policies, evaluation. World Energy Council, London.

WEF, 2005: Global Greenhouse Gas Register. World Economic Forum. <www.weforum.org/en/iniatives/ghg/GreenhouseGasRegister/index.htm>, accessed 31/05/07.

WWF, n.d.: Climate change featured projects: Climate savers. World Wildlife Fund, <www.worldwildlife.org/climate/projects/climateSavers.cfm>, accessed 31/05/07.

Worrell, E., D. Smit, K. Phylipsen, K. Blok, F. van der Vleuten, and J. Jansen, 1995: International comparison of energy efficiency improvement in the cement industry. Proceedings ACEEE 1995 Summer Study on Energy Efficiency in Industry, Volume II, pp. 123-134.

Worrell, E., D. Phylipsen, D. Einstein, and N. Martin, 2000a: Energy use and energy intensity of the US chemical industry. Berkeley, CA, Lawrence Berkeley National Laboratory (LBNL-44314).

Worrell, E., N. Martin, and L.K. Price, 2000b: Potentials for energy efficiency improvement in the us cement industry. *Energy*, **25**, pp. 1189-1214.

Worrell, E., N. Martin, and L.K. Price, 2001a: Energy efficiency and carbon dioxide emissions reduction opportunities in the US iron and steel sector. *Energy*, **26**, pp. 513-36.

Worrell, E., L.K. Price, N. Martin, C. Hendriks, and L. Ozawa Meida, 2001b: Carbon dioxide emissions from the global cement industry. *Annual Review of Energy and Environment*, **26**, pp. 303-29.

Worrell, E., N. Martin, L. Price, M. Ruth, N. Elliott, A. Shipley, and J. Thorne, 2002: Emerging energy efficient technologies for industry. *Energy Engineering Journal*, **99**, pp. 36-55.

Worrell, E., J.A. Laitner, M., Ruth, and H. Finman, 2003: Productivity Benefits of Industrial Energy Efficiency Measures. *Energy*, **28**, pp. 1081-1098.

Worrell, E. and C. Galitsky, 2004: Energy efficiency improvement opportunities for cement making - An ENERGY STAR Guide for Energy and Plant Managers. Berkeley, CA: Lawrence Berkeley National Laboratory, (LBNL 54036).

Worrell, E. and G. Biermans, 2005: Move over! Stock Turnover, Retrofit and Industrial Energy Efficiency. *Energy Policy*, **33**, pp. 949-962.

Worrell, E. and C. Galitsky, 2005: Energy efficiency improvement opportunities for petroleum refineries - An ENERGY STAR Guide for Energy and Plant Managers. Berkeley, CA: Lawrence Berkeley National Laboratory (LBNL 56183).

WRI/WBCSD, 2004: The Greenhouse Gas Protocol: A corporate accounting and reporting standard (revised edition). World Resources Institute/World Business Council for Sustainable Development, 112 pp.

WRI/WBCSD, 2005: GHG Protocol initiative/GHG calculation tools. World Resources Institute/World Business Council for Sustainable Development, <www.ghgprotocol.org>, accessed 31/05/07.

Xenergy, Inc., 1998: Evaluation of the US Department of Energy Motor Challenge Program. <www.eere.energy.gov/industry/bestpractices/pdfs/mceval1_2.pdf>, accessed 31/05/07.

Yamba, F. D. and E.M. Matsika, 2003: Assessment of the potential of the state of the art biomass technologies in contributing to a sustainable SADC removal mitigation energy scenario. Energy technologies for post Kyoto targets in the medium term. Proceedings of RISO International Energy Conference (19-23 May 2003).

Yamba, F.D. and E.M. Matsika, 2004: Factors and barriers influences the transfer of technologies with particular reference to Southern Africa. Presented at the IPCC Expert Meeting on Industrial Technology Development, Transfer and Diffusion, Tokyo (21-23 September 2004).

Yeoh, B.G., 2004: A technical and economic analysis of heat and power generation of palm oil mill effluent (POME) <www.cogen3.net/doc/countryinfo/malaysia/TechnicalEconomicAnalysisCHPPalmEffluent_BG.pdf >, accessed 31/05/07.

Zhang, Z. X., 2003: Why did energy intensity fall in China's industrial sector in the 1990s? The relative importance of structural change and intensity change. East-West Center Working Papers. Environmental Change, Vulnerability and Governance Series.

Zhou, D., Y. Dai, C. Yi, Y. Guo, and Y. Zhu, 2003: China's Sustainable Energy Scenarios in 2020. China Environmental Science Press, Beijing.

Ziebek, A. and W. Stanek, 2001: Forecasting of the energy effects of injecting plastic wastes into the blast furnace in comparison with other auxiliary fuels. *Energy*, **26**, pp. 1159-1173.

8

Agriculture

Coordinating Lead Authors:

Pete Smith (UK), Daniel Martino (Uruguay)

Lead Authors:

Zucong Cai (PR China), Daniel Gwary (Nigeria), Henry Janzen (Canada), Pushpam Kumar (India), Bruce McCarl (USA), Stephen Ogle (USA), Frank O'Mara (Ireland), Charles Rice (USA), Bob Scholes (South Africa), Oleg Sirotenko (Russia)

Contributing Authors:

Mark Howden (Australia), Tim McAllister (Canada), Genxing Pan (China), Vladimir Romanenkov (Russia), Steven Rose (USA), Uwe Schneider (Germany), Sirintornthep Towprayoon (Thailand), Martin Wattenbach (UK)

Review Editors:

Kristin Rypdal (Norway), Mukiri wa Githendu (Kenya)

This chapter should be cited as:

Smith, P., D. Martino, Z. Cai, D. Gwary, H. Janzen, P. Kumar, B. McCarl, S. Ogle, F. O'Mara, C. Rice, B. Scholes, O. Sirotenko, 2007: Agriculture. In Climate Change 2007: Mitigation. Contribution of Working Group III to the Fourth Assessment Report of the Intergovernmental Panel on Climate Change [B. Metz, O.R. Davidson, P.R. Bosch, R. Dave, L.A. Meyer (eds)], Cambridge University Press, Cambridge, United Kingdom and New York, NY, USA.

Table of Contents

EXECUTIVE SUMMARY

Agricultural lands (lands used for agricultural production, consisting of cropland, managed grassland and permanent crops including agro-forestry and bio-energy crops) occupy about 40-50% of the Earth's land surface.

Agriculture accounted for an estimated emission of 5.1 to 6.1 $GtCO_2$-eq/yr in 2005 (10-12% of total global anthropogenic emissions of greenhouse gases (GHGs)). CH_4 contributes 3.3 $GtCO_2$-eq/yr and N_2O 2.8 $GtCO_2$-eq/yr. Of global anthropogenic emissions in 2005, agriculture accounts for about 60% of N_2O and about 50% of CH_4 (*medium agreement, medium evidence*). Despite large annual exchanges of CO_2 between the atmosphere and agricultural lands, the net flux is estimated to be approximately balanced, with CO_2 emissions around 0.04 $GtCO_2$/yr only (emissions from electricity and fuel use are covered in the buildings and transport sector, respectively) (*low agreement, limited evidence*).

Globally, agricultural CH_4 and N_2O emissions have increased by nearly 17% from 1990 to 2005, an average annual emission increase of about 60 $MtCO_2$-eq/yr. During that period, the five regions composed of Non-Annex I countries showed a 32% increase, and were, by 2005, responsible for about three-quarters of total agricultural emissions. The other five regions, mostly Annex I countries, collectively showed a decrease of 12% in the emissions of these gases (*high agreement, much evidence*).

A variety of options exists for mitigation of GHG emissions in agriculture. The most prominent options are improved crop and grazing land management (e.g., improved agronomic practices, nutrient use, tillage, and residue management), restoration of organic soils that are drained for crop production and restoration of degraded lands. Lower but still significant mitigation is possible with improved water and rice management; set-asides, land use change (e.g., conversion of cropland to grassland) and agro-forestry; as well as improved livestock and manure management. Many mitigation opportunities use current technologies and can be implemented immediately, but technological development will be a key driver ensuring the efficacy of additional mitigation measures in the future (*high agreement, much evidence*).

Agricultural GHG mitigation options are found to be cost competitive with non-agricultural options (e.g., energy, transportation, forestry) in achieving long-term (i.e., 2100) climate objectives. Global long-term modelling suggests that non-CO_2 crop and livestock abatement options could cost-effectively contribute 270–1520 $MtCO_2$-eq/yr globally in 2030 with carbon prices up to 20 US$/$tCO_2$-eq and 640–1870 $MtCO_2$-eq/yr with C prices up to 50 US$/$tCO_2$-eq Soil carbon management options are not currently considered in long-term modelling (*medium agreement, limited evidence*).

Considering all gases, the global technical mitigation potential from agriculture (excluding fossil fuel offsets from biomass) by 2030 is estimated to be ~5500-6,000 $MtCO_2$-eq/yr (*medium agreement, medium evidence*). Economic potentials are estimated to be 1500-1600, 2500-2700, and 4000-4300 $MtCO_2$-eq/yr at carbon prices of up to 20, 50 and 100 US$/$tCO_2$-eq, respectively About 70% of the potential lies in non-OECD/EIT countries, 20% in OECD countries and 10% for EIT countries (*medium agreement, limited evidence*).

Soil carbon sequestration (enhanced sinks) is the mechanism responsible for most of the mitigation potential (*high agreement, much evidence*), with an estimated 89% contribution to the technical potantial. Mitigation of CH_4 emissions and N_2O emissions from soils account for 9% and 2%, respectively, of the total mitigation potential (*medium agreement, medium evidence*). The upper and lower limits about the estimates are largely determined by uncertainty in the per-area estimate for each mitigation measure. Overall, principal sources of uncertainties inherent in these mitigation potentials include: a) future level of adoption of mitigation measures (as influenced by barriers to adoption); b) effectiveness of adopted measures in enhancing carbon sinks or reducing N_2O and CH_4 emissions (particularly in tropical areas; reflected in the upper and lower bounds given above); and c) persistence of mitigation, as influenced by future climatic trends, economic conditions, and social behaviour (*medium agreement, limited evidence*).

The role of alternative strategies changes across the range of prices for carbon. At low prices, dominant strategies are those consistent with existing production such as changes in tillage, fertilizer application, livestock diet formulation, and manure management. Higher prices elicit land-use changes that displace existing production, such as biofuels, and allow for use of costly animal feed-based mitigation options. A practice effective in reducing emissions at one site may be less effective or even counterproductive elsewhere. Consequently, there is no universally applicable list of mitigation practices; practices need to be evaluated for individual agricultural systems based on climate, edaphic, social setting, and historical patterns of land use and management (*high agreement, much evidence*).

GHG emissions could also be reduced by substituting fossil fuels with energy produced from agricultural feed stocks (e.g., crop residues, dung, energy crops), which would be counted in sectors using the energy. The contribution of agriculture to the mitigation potential by using bioenergy depends on relative prices of the fuels and the balance of supply and demand. Using top-down models that include assumptions on such a balance the economic mitigation potential for agriculture in 2030 is estimated to be 70-1260 $MtCO_2$-eq/yr at up to 20 US$/$tCO_2$-eq, and 560-2320 $MtCO_2$-eq/yr at up to 50 US$/$tCO_2$-eq There are no estimates for the additional potential from top down models at carbon prices up to 100 US$/$tCO_2$-eq, but the estimate for prices above 100 US$/$tCO_2$-eq is 2720 $MtCO_2$-eq/yr. These potentials represent mitigation of 5-80%, and 20-90% of all other agricultural mitigation measures combined, at carbon prices of up to 20, and up to 50 US$/tCO$_2$-eq, respectively. An additional mitigation of 770 MtCO$_2$-eq/yr could be achieved by 2030 by improved energy efficiency in agriculture, though the mitigation potential is counted mainly in the buildings and transport sectors (*medium agreement, medium evidence*).

Agricultural mitigation measures often have synergy with sustainable development policies, and many explicitly influence social, economic, and environmental aspects of sustainability. Many options also have co-benefits (improved efficiency, reduced cost, environmental co-benefits) as well as trade-offs (e.g., increasing other forms of pollution), and balancing these effects will be necessary for successful implementation (*high agreement, much evidence*).

There are interactions between mitigation and adaptation in the agricultural sector, which may occur simultaneously, but differ in their spatial and geographic characteristics. The main climate change benefits of mitigation actions will emerge over decades, but there may also be short-term benefits if the drivers achieve other policy objectives. Conversely, actions to enhance adaptation to climate change impacts will have consequences in the short and long term. Most mitigation measures are likely robust to future climate change (e.g., nutrient management), but a subset will likely be vulnerable (e.g., irrigation in regions becoming more arid). It may be possible for a vulnerable practice to be modified as the climate changes and to maintain the efficacy of a mitigation measure (*low agreement, limited evidence*).

In many regions, non-climate policies related to macro-economics, agriculture and the environment, have a larger impact on agricultural mitigation than climate policies (*high agreement, much evidence*). Despite significant technical potential for mitigation in agriculture, there is evidence that little progress has been made in the implementation of mitigation measures at the global scale. Barriers to implementation are not likely to be overcome without policy/economic incentives and other programmes, such as those promoting global sharing of innovative technologies.

Current GHG emission rates may escalate in the future due to population growth and changing diets (*high agreement, medium evidence*). Greater demand for food could result in higher emissions of CH$_4$ and N$_2$O if there are more livestock and greater use of nitrogen fertilizers (*high agreement, much evidence*). Deployment of new mitigation practices for livestock systems and fertilizer applications will be essential to prevent an increase in emissions from agriculture after 2030. In addition, soil carbon may be more vulnerable to loss with climate change and other pressures, though increases in production will offset some or all of this carbon loss (*low agreement, limited evidence*).

Overall, the outlook for GHG mitigation in agriculture suggests that there is significant potential (*high agreement, medium evidence*). Current initiatives suggest that synergy between climate change policies, sustainable development and improvement of environmental quality will likely lead the way forward to realize the mitigation potential in this sector.

8.1   Introduction

Agriculture releases to the atmosphere significant amounts of CO_2, CH_4, and N_2O (Cole *et al.*, 1997; IPCC, 2001a; Paustian *et al.*, 2004). CO_2 is released largely from microbial decay or burning of plant litter and soil organic matter (Smith, 2004b; Janzen, 2004). CH_4 is produced when organic materials decompose in oxygen-deprived conditions, notably from fermentative digestion by ruminant livestock, from stored manures, and from rice grown under flooded conditions (Mosier *et al.* 1998). N_2O is generated by the microbial transformation of nitrogen in soils and manures, and is often enhanced where available nitrogen (N) exceeds plant requirements, especially under wet conditions (Oenema *et al.*, 2005; Smith and Conen, 2004). Agricultural greenhouse gas (GHG) fluxes are complex and heterogeneous, but the active management of agricultural systems offers possibilities for mitigation. Many of these mitigation opportunities use current technologies and can be implemented immediately.

This chapter describes the development of GHG emissions from the agricultural sector (Section 8.2), and details agricultural practices that may mitigate GHGs (Section 8.4.1), with many practices affecting more than one GHG by more than one mechanism. These practices include: cropland management; grazing land management/pasture improvement; management of agricultural organic soils; restoration of degraded lands; livestock management; manure/bio-solid management; and bio-energy production.

It is theoretically possible to increase carbon storage in long-lived agricultural products (e.g., strawboards, wool, leather, bio-plastics) but the carbon held in these products has only increased from 37 to 83 MtC per year over the past 40 years. Assuming a first order decay rate of 10 to 20 % per year, this is estimated to be a global net annual removal of 3 to 7 $MtCO_2$ from the atmosphere, which is negligible compared to other mitigation measures. The option is not considered further here.

Smith *et al.* (2007a) recently estimated a global potential mitigation of 770 $MtCO_2$-eq/yr by 2030 from improved energy efficiency in agriculture (e.g., through reduced fossil fuel use), However, this is usually counted in the relevant user sector rather than in agriculture and so is not considered further here. Any savings from improved energy efficiency are discussed in the relevant sections elsewhere in this volume, according to where fossil fuel savings are made, for example, from transport fuels (Chapter 5), or through improved building design (Chapter 6).

8.2 Status of sector, development trends including production and consumption, and implications

Population pressure, technological change, public policies, and economic growth and the cost/price squeeze have been the main drivers of change in the agricultural sector during the last four decades. Production of food and fibre has more than kept pace with the sharp increase in demand in a more populated world. The global average daily availability of calories per capita has increased (Gilland, 2002), with some notable regional exceptions. This growth, however, has been at the expense of increased pressure on the environment, and depletion of natural resources (Tilman *et al.*, 2001; Rees, 2003), while it has not resolved the problems of food security and child malnutrition suffered in poor countries (Conway and Toenniessen, 1999).

Agricultural land occupied 5023 Mha in 2002 (FAOSTAT, 2006). Most of this area was under pasture (3488 Mha, or 69%)

Table 8.1. *Agricultural land use in the last four decades.*

	Area (Mha)					Change 2000s/1960s	
	1961-70	1971-80	1981-90	1991-00	2001-02	%	Mha
1. World							
Agricultural land	4,562	4,684	4,832	4,985	5,023	+10	461
Arable land	1,297	1,331	1,376	1,393	1,405	+8	107
Permanent crops	82	92	104	123	130	+59	49
Permanent pasture	3,182	3,261	3,353	3,469	3,488	+10	306
2. Developed countries							
Agricultural land	1,879	1,883	1,877	1,866	1,838	-2	-41
Arable land	648	649	652	633	613	-5	-35
Permanent crops	23	24	24	24	24	+4	1
Permanent pasture	1,209	1,210	1,201	1,209	1,202	-1	-7
3. Developing countries							
Agricultural land	2,682	2,801	2,955	3,119	3,184	+19	502
Arable land	650	682	724	760	792	+22	142
Permanent crops	59	68	80	99	106	+81	48
Permanent pasture	1,973	2,051	2,152	2,260	2,286	+16	313

Source: FAOSTAT, 2006.

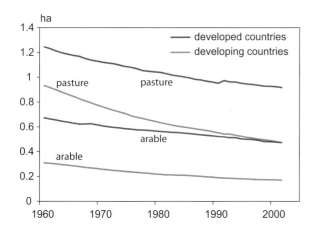

Figure 8.1. *Per-capita area of arable land and pasture, in developing and developing countries.*
Source: FAOSTAT, 2006.

and cropland occupied 1405 Mha (28%). During the last four decades, agricultural land gained almost 500 Mha from other land uses, a change driven largely by increasing demands for food from a growing population. Every year during this period, an average 6 Mha of forestland and 7 Mha of other land were converted to agriculture, a change occurring largely in the developing world (Table 8.1). This trend is projected to continue into the future (Huang *et al.*, 2002; Trewavas, 2002; Fedoroff and Cohen, 1999; Green *et al.*, 2005), and Rosegrant *et al.*, (2001) project that an additional 500 Mha will be converted to agriculture during 1997-2020, mostly in Latin America and Sub-Saharan Africa.

Technological progress has made it possible to achieve remarkable improvements in land productivity, increasing per-capita food availability (Table 8.2), despite a consistent decline in per-capita agricultural land (Figure 8.1). The share of animal products in the diet has increased consistently in the developing countries, while remaining constant in developed countries (Table 8.2). Economic growth and changing lifestyles in some developing countries are causing a growing demand for meat and dairy products, notably in China where current demands

are low. Meat demand in developing countries rose from 11 to 24 kg/capita/yr during the period 1967-1997, achieving an annual growth rate of more than 5% by the end of that period. Rosegrant *et al.* (2001) forecast a further increase of 57% in global meat demand by 2020, mostly in South and Southeast Asia, and Sub-Saharan Africa. The greatest increases in demand are expected for poultry (83 % by 2020; Roy *et al.*, 2002).

Annual GHG emissions from agriculture are expected to increase in coming decades (included in the baseline) due to escalating demands for food and shifts in diet. However, improved management practices and emerging technologies may permit a reduction in emissions per unit of food (or of protein) produced. The main trends in the agricultural sector with the implications for GHG emissions or removals are summarized as follows:

- Growth in land productivity is expected to continue, although at a declining rate, due to decreasing returns from further technological progress, and greater use of marginal land with lower productivity. Use of these marginal lands increases the risk of soil erosion and degradation, with highly uncertain consequences for CO_2 emissions (Lal, 2004a; Van Oost *et al.*, 2004).

- Conservation tillage and zero-tillage are increasingly being adopted, thus reducing the use of energy and often increasing carbon storage in soils. According to FAO (2001), the worldwide area under zero-tillage in 1999 was approximately 50 Mha, representing 3.5% of total arable land. However, such practices are frequently combined with periodical tillage, thus making the assessment of the GHG balance highly uncertain.

- Further improvements in productivity will require higher use of irrigation and fertilizer, increasing the energy demand (for moving water and manufacturing fertilizer; Schlesinger, 1999). Also, irrigation and N fertilization can increase GHG emissions (Mosier, 2001).

- Growing demand for meat may induce further changes in land use (e.g., from forestland to grassland), often increasing CO_2 emissions, and increased demand for animal

Table 8.2: *Per capita food supply in developed and developing countries*

	1961-70	1971-80	1981-90	1991-00	2001-02	Change 2000s/1960s %	cal/d or g/d
1. Developed countries							
Energy, all sources (cal/day)	3049	3181	3269	3223	3309	+9	261
% from animal sources	27	28	28	27	26	-2	--
Protein, all sources (g/day)	92	97	101	99	100	+9	8
% from animal sources	50	55	57	56	56	+12	--
2. Developing countries							
Energy, all sources (cal/day)	2032	2183	2443	2600	2657	+31	625
% from animal sources	8	8	9	12	13	+77	--
Protein, all sources (g/day)	9	11	13	18	21	+123	48
% from animal sources	18	20	22	28	30	+67	--

Source: FAOSTAT, 2006.

feeds (e.g., cereals). Larger herds of beef cattle will cause increased emissions of CH_4 and N_2O, although use of intensive systems (with lower emissions per unit product) is expected to increase faster than growth in grazing-based systems. This may attenuate the expected rise in GHG emissions.

- Intensive production of beef, poultry, and pork is increasingly common, leading to increases in manure with consequent increases in GHG emissions. This is particularly true in the developing regions of South and East Asia, and Latin America, as well as in North America.

- Changes in policies (e.g., subsidies), and regional patterns of production and demand are causing an increase in international trade of agricultural products. This is expected to increase CO_2 emissions, due to greater use of energy for transportation.

- There is an emerging trend for greater use of agricultural products (e.g., bio-plastics bio-fuels and biomass for energy) as substitutes for fossil fuel-based products. This has the potential to reduce GHG emissions in the future.

8.3   Emission trends (global and regional)

With an estimated global emission of non-CO_2 GHGs from agriculture of between 5120 $MtCO_2$-eq/yr (Denman *et al.*, 2007) and 6116 $MtCO_2$-eq/yr (US-EPA, 2006a) in 2005, agriculture accounts for 10-12 % of total global anthropogenic emissions of GHGs. Agriculture contributes about 47% and 58% of total anthropogenic emissions of CH_4 and N_2O, respectively, with a wide range of uncertainty in the estimates of both the agricultural contribution and the anthropogenic total. N_2O emissions from soils and CH_4 from enteric fermentation constitute the largest sources, 38% and 32% of total non-CO_2 emissions from agriculture in 2005, respectively (US-EPA, 2006a). Biomass burning (12%), rice production (11%), and manure management (7%) account for the rest. CO_2 emissions from agricultural soils are not normally estimated separately, but are included in the land use, land use change and forestry sector (e.g., in national GHG inventories). So there are few comparable estimates of emissions of this gas in agriculture. Agricultural lands generate very large CO_2 fluxes both to and from the atmosphere (IPCC, 2001a), but the *net* flux is small. US-EPA, 2006b) estimated a net CO_2 emission of 40 $MtCO_2$-eq from agricultural soils in 2000, less than 1% of global anthropogenic CO_2 emissions.

Both the magnitude of the emissions and the relative importance of the different sources vary widely among world regions (Figure 8.2). In 2005, the group of five regions mostly consisting of non-Annex I countries was responsible for 74% of total agricultural emissions.

In seven of the ten regions, N_2O from soils was the main source of GHGs in the agricultural sector in 2005, mainly associated with N fertilizers and manure applied to soils. In

the other three regions - Latin America and The Caribbean, the countries of Eastern Europe, the Caucasus and Central Asia, and OECD Pacific - CH_4 from enteric fermentation was the dominant source (US-EPA, 2006a). This is due to the large livestock population in these three regions which, in 2004, had a combined stock of cattle and sheep equivalent to 36% and 24% of world totals, respectively (FAO, 2003).

Emissions from rice production and burning of biomass were heavily concentrated in the group of developing countries, with 97% and 92% of world totals, respectively. While CH_4 emissions from rice occurred mostly in South and East Asia, where it is a dominant food source (82% of total emissions), those from biomass burning originated in Sub-Saharan Africa and Latin America and the Caribbean (74% of total). Manure management was the only source for which emissions where higher in the group of developed regions (52%) than in developing regions (48%; US-EPA, 2006a).

The balance between the large fluxes of CO_2 emissions and removals in agricultural land is uncertain. A study by US-EPA (2006b) showed that some countries and regions have net emissions, while others have net removals of CO_2. Except for the countries of Eastern Europe, the Caucasus and Central Asia, which had an annual emission of 26 $MtCO_2$/yr in 2000, all other countries showed very low emissions or removals.

8.3.1   Trends since 1990

Globally, agricultural CH_4 and N_2O emissions increased by 17% from 1990 to 2005, an average annual emission increase of 58 $MtCO_2$-eq/yr (US-EPA, 2006a). Both gases had about the same share of this increase. Three sources together explained 88% of the increase: biomass burning (N_2O and CH_4), enteric fermentation (CH_4) and soil N_2O emissions (US-EPA, 2006a).

During that period, according to US-EPA (2006a; Figure 8.2), the five regions composed of Non-Annex I countries showed a 32% increase in non-CO_2 emissions (equivalent to 73 $MtCO_2$-eq/yr).The other five regions, with mostly Annex I countries, collectively showed a decrease of 12% (equivalent to 15 $MtCO_2$-eq/yr). This was mostly due to non-climate macroeconomic policies in the Central and Eastern European and the countries of Eastern Europe, the Caucasus and Central Asia (see Section 8.7.1 and 8.7.2).

8.3.2   Future global trends

Agricultural N_2O emissions are projected to increase by 35-60% up to 2030 due to increased nitrogen fertilizer use and increased animal manure production (FAO, 2003). Similarly, Mosier and Kroeze (2000) and US-EPA (2006a; Figure 8.2) estimated that N_2O emissions will increase by about 50% by 2020 (relative to 1990). If demands for food increase, and diets shift as projected, then annual emissions of GHGs from agriculture may escalate further. But improved management

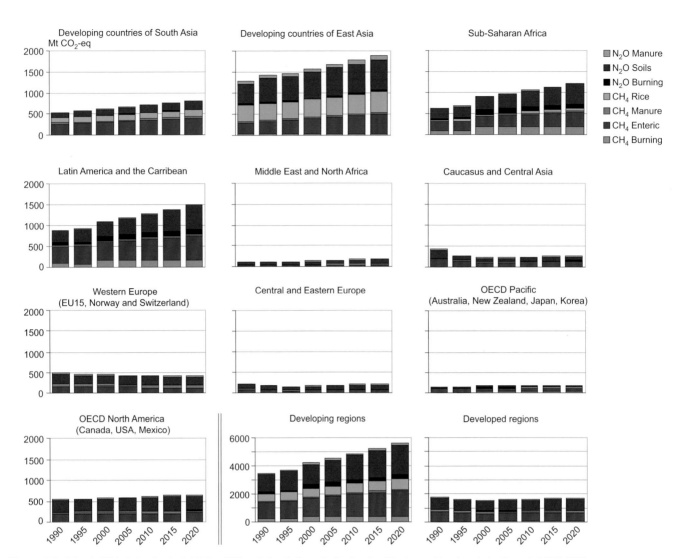

Figure 8.2: *Estimated historical and projected N$_2$O and CH$_4$ emissions in the agricultural sector of the ten world regions during the period 1990-2020.*
Source: Adapted from US-EPA, 2006a.

practices and emerging technologies may permit a reduction in emissions per unit of food (or protein) produced, and perhaps also a reduction in emissions per capita food consumption.

If CH$_4$ emissions grow in direct proportion to increases in livestock numbers, then global livestock-related methane production is expected to increase by 60% up to 2030 (FAO, 2003). However, changes in feeding practices and manure management could ameliorate this increase. US-EPA (2006a) forecast that combined methane emissions from enteric fermentation and manure management will increase by 21% between 2005 and 2020.

The area of rice grown globally is forecast to increase by 4.5% to 2030 (FAO, 2003), so methane emissions from rice production would not be expected to increase substantially. There may even be reductions if less rice is grown under continuous flooding (causing anaerobic soil conditions) as a result of scarcity of water, or if new rice cultivars that emit

less methane are developed and adopted (Wang *et al.*, 1997). However, US-EPA (2006a) projects a 16% increase in CH$_4$ emissions from rice crops between 2005 and 2020, mostly due to a sustained increase in the area of irrigated rice.

No baseline agricultural non-CO$_2$ GHG emission estimates for the year 2030 have been published, but according to US-EPA (2006a), aggregate emissions are projected to increase by ~13% during the decades 2000-2010 and 2010-2020. Assuming similar rates of increase (10-15%) for 2020-2030, agricultural emissions might be expected to rise to 8000–8400, with a mean of 8300 MtCO$_2$-eq by 2030. The future evolution of CO$_2$ emissions from agriculture is uncertain. Due to stable or declining deforestation rates (FAO, 2003), and increased adoption of conservation tillage practices (FAO, 2001), these emissions are likely to decrease or remain at low levels.

8.3.3 Regional trends

The Middle East and North Africa, and Sub-Saharan Africa have the highest projected growth in emissions, with a combined 95% increase in the period 1990 to 2020 (US-EPA, 2006a). Sub-Saharan Africa is the one world region where per-capita food production is either in decline, or roughly constant at a level that is less than adequate (Scholes and Biggs, 2004). This trend is linked to low and declining soil fertility (Sanchez, 2002), and inadequate fertilizer inputs. Although slow, the rising wealth of urban populations is likely to increase demand for livestock products. This would result in intensification of agriculture and expansion to still largely unexploited areas, particularly in South and Central Africa (including Angola, Zambia, DRC, Mozambique and Tanzania), with a consequent increase in GHG emissions.

East Asia is projected to show large increases in GHG emissions from animal sources. According to FAO (FAOSTAT, 2006), total production of meat and milk in Asian developing countries increased more than 12 times and 4 times, respectively, from 2004 to 1961. Since the per-capita consumption of meat and milk is still much lower in these countries than in developed countries, increasing trends are expected to continue for a relatively long time. Accordingly, US-EPA (2006a) forecast increases of 153% and 86% in emissions from enteric fermentation and manure management, respectively, from 1990 to 2020. In South Asia, emissions are increasing mostly because of expanding use of N fertilizers and manure to meet demands for food, resulting from rapid population growth.

In Latin America and the Caribbean, agricultural products are the main source of exports. Significant changes in land use and management have occurred, with forest conversion to cropland and grassland being the most significant, resulting in increased GHG emissions from soils (CO_2 and N_2O). The cattle population has increased linearly from 176 to 379 Mhead between 1961 and 2004, partly offset by a decrease in the sheep population from 125 to 80 Mhead. All other livestock categories have increased in the order of 30 to 600% since 1961. Cropland areas, including rice and soybean, and the use of N fertilizers have also shown dramatic increases (FAOSTAT, 2006). Another major trend in the region is the increased adoption of no-till agriculture, particularly in the Mercosur area (Brazil, Argentina, Paraguay, and Uruguay). This technology is used on ~30 Mha every year in the region, although it is unknown how much of this area is under permanent no-till.

In the countries of Central and Eastern Europe, the Caucasus and Central Asia, agricultural production is, at present, about 60-80% of that in 1990, but is expected to grow by 15-40% above 2001 levels by 2010, driven by the increasing wealth of these countries. A 10-14% increase in arable land area is forecast for the whole of Russia due to agricultural expansion. The

widespread application of intensive management technologies could result in a 2 to 2.5-fold rise in grain and fodder yields, with a consequent reduction of arable land, but may increase N fertilizer use. Decreases in fertilizer N use since 1990 have led to a significant reduction in N_2O emissions. But, under favourable economic conditions, the amount of N fertilizer applied will again increase, although unlikely to reach pre-1990 levels in the near future. US-EPA (2006a) projected a 32% increase in N_2O emissions from soils in these two regions between 2005 and 2020, equivalent to an average rate of increase of 3.5 $MtCO_2$-eq/yr.

OECD North America and OECD Pacific are the only developed regions showing a consistent increase in GHG emissions in the agricultural sector (18% and 21%, respectively between 1990 and 2020; Figure 8.2). In both cases, the trend is largely driven by non-CO_2 emissions from manure management and N_2O emissions from soils. In Oceania, nitrogen fertilizer use has increased exponentially over the past 45 years with a 5 and 2.5 fold increase since 1990 in New Zealand and Australia, respectively. In North America, in contrast, nitrogen fertilizer use has remained stable; the main driver for increasing emissions is management of manure from cattle, poultry and swine production, and manure application to soils. In both regions, conservation policies have resulted in reduced CO_2 emissions from land conversion. Land clearing in Australia has declined by 60% since 1990 with vegetation management policies restricting further clearing, while in North America, some marginal croplands have been returned to woodland or grassland.

Western Europe is the only region where, according to US-EPA (2006a), GHG emissions from agriculture are projected to decrease to 2020 (Figure 8.2). This is associated with the adoption of a number of climate-specific and other environmental policies in the European Union, as well as economic constraints on agriculture, as discussed in Sections 8.7.1 and 8.7.2.

8.4 Description and assessment of mitigation technologies and practices, options and potentials, costs and sustainability

8.4.1 Mitigation technologies and practices

Opportunities for mitigating GHGs in agriculture fall into three broad categories[1], based on the underlying mechanism:

a. Reducing emissions: Agriculture releases to the atmosphere significant amounts of CO_2, CH_4, or N_2O (Cole et al., 1997; IPCC, 2001a; Paustian et al., 2004). The fluxes

---

[1] Smith et al. (2007a) have recently reviewed mechanisms for agricultural GHG mitigation. This section draws largely from that study.

of these gases can be reduced by more efficient management of carbon and nitrogen flows in agricultural ecosystems. For example, practices that deliver added N more efficiently to crops often reduce N_2O emissions (Bouwman, 2001), and managing livestock to make most efficient use of feeds often reduces amounts of CH_4 produced (Clemens and Ahlgrimm, 2001). The approaches that best reduce emissions depend on local conditions, and therefore, vary from region to region.

b. Enhancing removals: Agricultural ecosystems hold large carbon reserves (IPCC, 2001a), mostly in soil organic matter. Historically, these systems have lost more than 50 Pg C (Paustian *et al.*, 1998; Lal, 1999, 2004a), but some of this carbon lost can be recovered through improved management, thereby withdrawing atmospheric CO_2. Any practice that increases the photosynthetic input of carbon and/or slows the return of stored carbon to CO_2 via respiration, fire or erosion will increase carbon reserves, thereby 'sequestering' carbon or building carbon 'sinks'. Many studies, worldwide, have now shown that significant amounts of soil carbon can be stored in this way, through a range of practices, suited to local conditions (Lal, 2004a). Significant amounts of vegetative carbon can also be stored in agro-forestry systems or other perennial plantings on agricultural lands (Albrecht and Kandji, 2003). Agricultural lands also remove CH_4 from the atmosphere by oxidation (but less than forests; Tate *et al.*, 2006), but this effect is small compared to other GHG fluxes (Smith and Conen, 2004).

c. Avoiding (or displacing) emissions: Crops and residues from agricultural lands can be used as a source of fuel, either directly or after conversion to fuels such as ethanol or diesel (Schneider and McCarl, 2003; Cannell, 2003). These bio-energy feedstocks still release CO_2 upon combustion, but now the carbon is of recent atmospheric origin (via photosynthesis), rather than from fossil carbon. The net benefit of these bio-energy sources to the atmosphere is equal to the fossil-derived emissions displaced, less any emissions from producing, transporting, and processing. GHG emissions, notably CO_2, can also be avoided by agricultural management practices that forestall the cultivation of new lands now under forest, grassland, or other non-agricultural vegetation (Foley *et al.*, 2005).

Many practices have been advocated to mitigate emissions through the mechanisms cited above. Often, a practice will affect more than one gas, by more than one mechanism, sometimes in opposite ways, so the net benefit depends on the combined effects on all gases (Robertson and Grace, 2004; Schils *et al.*, 2005; Koga *et al.*, 2006). In addition, the temporal pattern of influence may vary among practices or among gases for a given practice; some emissions are reduced indefinitely, other reductions are temporary (Six *et al.*, 2004; Marland *et al.*, 2003a). Where a practice affects radiative forcing through other

mechanisms such as aerosols or albedo, those impacts also need to be considered (Marland *et al.*, 2003b; Andreae *et al.*, 2005).

The impacts of the mitigation options considered are summarized qualitatively in Table 8.3. Although comprehensive life-cycle analyses are not always possible, given the complexity of many farming systems, the table also includes estimates of the confidence based on expert opinion that the practice can reduce overall net emissions at the site of adoption. Some of these practices also have indirect effects on ecosystems elsewhere. For example, increased productivity in existing croplands could avoid deforestation and its attendant emissions (see also Section 8.8). The most important options are discussed in Section 8.4.1.

8.4.1.1 *Cropland management*

Because often intensively managed, croplands offer many opportunities to impose practices that reduce net GHG emissions (Table 8.3). Mitigation practices in cropland management include the following partly-overlapping categories:

a. Agronomy: Improved agronomic practices that increase yields and generate higher inputs of carbon residue can lead to increased soil carbon storage (Follett, 2001). Examples of such practices include: using improved crop varieties; extending crop rotations, notably those with perennial crops that allocate more carbon below ground; and avoiding or reducing use of bare (unplanted) fallow (West and Post, 2002; Smith, 2004a, b; Lal, 2003, 2004a; Freibauer *et al.*, 2004). Adding more nutrients, when deficient, can also promote soil carbon gains (Alvarez, 2005), but the benefits from N fertilizer can be offset by higher N_2O emissions from soils and CO_2 from fertilizer manufacture (Schlesinger, 1999; Pérez-Ramírez *et al.*, 2003; Robertson, 2004; Gregorich *et al.*, 2005). Emissions per hectare can also be reduced by adopting cropping systems with reduced reliance on fertilizers, pesticides and other inputs (and therefore, the GHG cost of their production: Paustian *et al.*, 2004). An important example is the use of rotations with legume crops (West and Post, 2002; Izaurralde *et al.*, 2001), which reduce reliance on external N inputs although legume-derived N can also be a source of N_2O (Rochette and Janzen, 2005). Another group of agronomic practices are those that provide temporary vegetative cover between successive agricultural crops, or between rows of tree or vine crops. These 'catch' or 'cover' crops add carbon to soils (Barthès *et al.*, 2004; Freibauer *et al.*, 2004) and may also extract plant-available N unused by the preceding crop, thereby reducing N_2O emissions.

b. Nutrient management: Nitrogen applied in fertilizers, manures, biosolids, and other N sources is not always used efficiently by crops (Galloway *et al.*, 2003; Cassman *et al.*, 2003). The surplus N is particularly susceptible to emission

Table 8.3: *Proposed measures for mitigating greenhouse gas emissions from agricultural ecosystems, their apparent effects on reducing emissions of individual gases where adopted (mitigative effect), and an estimate of scientific confidence that the proposed practice can reduce overall net emissions at the site of adoption.*

Measure	Examples	Mitigative effects[a]			Net mitigation[b] (confidence)	
		CO_2	CH_4	N_2O	Agreement	Evidence
Cropland management	Agronomy	+		+/-	***	**
	Nutrient management	+		+	***	**
	Tillage/residue management	+		+/-	**	**
	Water management (irrigation, drainage)	+/-		+	*	*
	Rice management	+/-	+	+/-	**	**
	Agro-forestry	+		+/-	***	*
	Set-aside, land-use change	+	+	+	***	***
Grazing land management/ pasture improvement	Grazing intensity	+/-	+/-	+/-	*	*
	Increased productivity (e.g., fertilization)	+		+/-	**	*
	Nutrient management	+		+/-	**	**
	Fire management	+	+	+/-	*	*
	Species introduction (including legumes)	+		+/-	*	**
Management of organic soils	Avoid drainage of wetlands	+	-	+/-	**	**
Restoration of degraded lands	Erosion control, organic amendments, nutrient amendments	+		+/-	***	**
Livestock management	Improved feeding practices		+	+	***	***
	Specific agents and dietary additives		+		**	***
	Longer term structural and management changes and animal breeding		+	+	**	*
Manure/biosolid management	Improved storage and handling		+	+/-	***	**
	Anaerobic digestion		+	+/-	***	*
	More efficient use as nutrient source	+		+	***	**
Bio-energy	Energy crops, solid, liquid, biogas, residues	+	+/-	+/-	***	**

Notes:

[a]   + denotes reduced emissions or enhanced removal (positive mitigative effect);

- denotes increased emissions or suppressed removal (negative mitigative effect);

+/- denotes uncertain or variable response.

[b]   A qualitative estimate of the confidence in describing the proposed practice as a measure for reducing net emissions of greenhouse gases, expressed as CO_2-eq: Agreement refers to the relative degree of consensus in the literature (the more asterisks, the higher the agreement); Evidence refers to the relative amount of data in support of the proposed effect (the more asterisks, the more evidence).

Source: adapted from Smith et al., 2007a.

of N_2O (McSwiney and Robertson, 2005). Consequently, improving N use efficiency can reduce N_2O emissions and indirectly reduce GHG emissions from N fertilizer manufacture (Schlesinger, 1999). By reducing leaching and volatile losses, improved efficiency of N use can also reduce off-site N_2O emissions. Practices that improve N use efficiency include: adjusting application rates based on precise estimation of crop needs (e.g., precision farming); using slow- or controlled-release fertilizer forms or nitrification inhibitors (which slow the microbial processes leading to N_2O formation); applying N when least susceptible to loss, often just prior to plant uptake (improved timing); placing the N more precisely into the soil to make it more accessible to crops roots; or avoiding N applications in excess of immediate plant requirements (Robertson, 2004; Dalal *et al.*, 2003; Paustian *et al.*, 2004; Cole *et al.*, 1997; Monteny *et al.*, 2006).

c.   Tillage/residue management: Advances in weed control methods and farm machinery now allow many crops to be grown with minimal tillage (reduced tillage) or without tillage (no-till). These practices are now increasingly used throughout the world (e.g., Cerri *et al.*, 2004). Since soil disturbance tends to stimulate soil carbon losses through enhanced decomposition and erosion (Madari *et al.*, 2005), reduced- or no-till agriculture often results in soil carbon gain, but not always (West and Post, 2002; Ogle *et al.*, 2005; Gregorich *et al.*, 2005; Alvarez 2005). Adopting reduced- or no-till may also affect N_2O, emissions but the net effects are inconsistent and not well-quantified globally (Smith and Conen, 2004; Helgason *et al.*, 2005; Li *et al.*, 2005; Cassman *et al.*, 2003). The effect of reduced tillage on N_2O emissions may depend on soil and climatic conditions. In some areas, reduced tillage promotes N_2O emissions, while elsewhere it may reduce emissions or have no measurable influence (Marland *et al.*, 2001). Further, no-tillage systems can reduce CO_2 emissions from energy use (Marland *et al.*, 2003b; Koga *et al.*, 2006). Systems that retain crop residues also tend to increase soil carbon because these residues are the precursors for soil organic matter, the main carbon store in soil. Avoiding the burning of residues (e.g., mechanising sugarcane harvesting, eliminating the need for pre-harvest burning (Cerri *et al.*, 2004)) also avoids emissions of aerosols and GHGs generated from fire, although CO_2 emissions from fuel use may increase.

d. Water management: About 18% of the world's croplands now receive supplementary water through irrigation (Millennium Ecosystem Assessment, 2005). Expanding this area (where water reserves allow) or using more effective irrigation measures can enhance carbon storage in soils through enhanced yields and residue returns (Follett, 2001; Lal, 2004a). But some of these gains may be offset by CO_2 from energy used to deliver the water (Schlesinger 1999; Mosier *et al.*, 2005) or from N_2O emissions from higher moisture and fertilizer N inputs (Liebig *et al.* 2005), The latter effect has not been widely measured. Drainage of croplands lands in humid regions can promote productivity (and hence soil carbon) and perhaps also suppress N_2O emissions by improving aeration (Monteny *et al.*, 2006). Any nitrogen lost through drainage, however, may be susceptible to loss as N_2O.(Reay *et al.* 2003).

e. Rice management: Cultivated wetland rice soils emit significant quantities of methane (Yan *et al.*, 2003). Emissions during the growing season can be reduced by various practices (Yagi *et al.*, 1997; Wassmann *et al.*, 2000; Aulakh *et al.*, 2001). For example, draining wetland rice once or several times during the growing season reduces CH_4 emissions (Smith and Conen, 2004; Yan *et al.*, 2003; Khalil and Shearer, 2006). This benefit, however, may be partly offset by increased N_2O emissions (Akiyama *et al.* 2005), and the practice may be constrained by water supply. Rice cultivars with low exudation rates could offer an important methane mitigation option (Aulakh *et al.*, 2001). In the off-rice season, methane emissions can be reduced by improved water management, especially by keeping the soil as dry as possible and avoiding water logging (Cai *et al.*, 2000 2003; Kang *et al.*, 2002; Xu *et al.*, 2003). Increasing rice production can also enhance soil organic carbon stocks (Pan *et al.*, 2006). Methane emissions can be reduced by adjusting the timing of organic residue additions (e.g., incorporating organic materials in the dry period rather than in flooded periods; Xu *et al.*, 2000; Cai and Xu, 2004), by composting the residues before incorporation, or by producing biogas for use as fuel for energy production (Wang and Shangguan, 1996; Wassmann *et al.*, 2000).

f. Agro-forestry: Agro-forestry is the production of livestock or food crops on land that also grows trees for timber, firewood, or other tree products. It includes shelter belts and riparian zones/buffer strips with woody species. The standing stock of carbon above ground is usually higher than the equivalent land use without trees, and planting trees may also increase soil carbon sequestration (Oelbermann *et al.*, 2004; Guo and Gifford, 2002; Mutuo *et al.*, 2005; Paul *et al.*, 2003). But the effects on N_2O and CH_4 emissions are not well known (Albrecht and Kandji, 2003).

g. Land cover (use) change: One of the most effective methods of reducing emissions is often to allow or encourage the reversion of cropland to another land cover, typically one similar to the native vegetation. The conversion can occur over the entire land area ('set-asides'), or in localized spots, such as grassed waterways, field margins, or shelterbelts (Follett, 2001; Freibauer *et al.*, 2004; Lal, 2004b; Falloon *et al.*, 2004; Ogle *et al.*, 2003). Such land cover change often increases carbon storage. For example, converting arable cropland to grassland typically results in the accrual of soil carbon because of lower soil disturbance and reduced carbon removal in harvested products. Compared to cultivated lands, grasslands may also have reduced N_2O emissions from lower N inputs, and higher rates of CH_4 oxidation, but recovery of oxidation may be slow (Paustian *et al.*, 2004). Similarly, converting drained croplands back to wetlands can result in rapid accumulation of soil carbon (removal of atmospheric CO_2). This conversion may stimulate CH_4 emissions because water logging creates anaerobic conditions (Paustian *et al.*, 2004). Planting trees can also reduce emissions. These practices are considered under agro-forestry (Section 8.4.1.1f); afforestation (Chapter 9), and reafforestation (Chapter 9). Because land cover (or use) conversion comes at the expense of lost agricultural productivity, it is usually an option only on surplus agricultural land or on croplands of marginal productivity.

8.4.1.2 Grazing land management and pasture improvement

Grazing lands occupy much larger areas than croplands (FAOSTAT, 2006) and are usually managed less intensively. The following are examples of practices to reduce GHG emissions and to enhance removals:

a. Grazing intensity: The intensity and timing of grazing can influence the removal, growth, carbon allocation, and flora of grasslands, thereby affecting the amount of carbon accrual in soils (Conant *et al.*, 2001; 2005; Freibauer *et al.*, 2004; Conant and Paustian, 2002; Reeder *et al.*, 2004). Carbon accrual on optimally grazed lands is often greater than on ungrazed or overgrazed lands (Liebig *et al.*, 2005; Rice and Owensby, 2001). The effects are inconsistent, however, owing to the many types of grazing practices employed and the diversity of plant species, soils, and climates involved (Schuman *et al.*, 2001; Derner *et al.*, 2006). The influence of grazing intensity on emission of non-CO_2 gases is not well-established, apart from the direct effects on emissions from adjustments in livestock numbers.

b. Increased productivity: (including fertilization): As for croplands, carbon storage in grazing lands can be improved by a variety of measures that promote productivity. For instance, alleviating nutrient deficiencies by fertilizer or organic amendments increases plant litter returns and, hence, soil carbon storage (Schnabel *et al.*, 2001; Conant *et al.*, 2001). Adding nitrogen, however, often stimulates N_2O emissions (Conant *et al.*, 2005) thereby offsetting some of the benefits. Irrigating grasslands, similarly, can promote soil carbon gains (Conant *et al.*, 2001). The net effect of this practice, however, depends also on emissions from energy use and other activities on the irrigated land (Schlesinger, 1999).

c. Nutrient management: Practices that tailor nutrient additions to plant uptake, such as those described for croplands, can reduce N_2O emissions (Dalal *et al.*, 2003; Follett *et al.*, 2001). Management of nutrients on grazing lands, however, may be complicated by deposition of faeces and urine from livestock, which are not as easily controlled nor as uniformly applied as nutritive amendments in croplands (Oenema *et al.*, 2005).

d. Fire management: On-site biomass burning (not to be confused with bio-energy, where biomass is combusted off-site for energy) contributes to climate change in several ways. Firstly, it releases GHGs, notably CH_4 and, and to a lesser extent, N_2O (the CO_2 released is of recent origin, is absorbed by vegetative regrowth, and is usually not included in GHG inventories). Secondly, it generates hydrocarbon and reactive nitrogen emissions, which react to form tropospheric ozone, a powerful GHG. Thirdly, fires produce a range of smoke aerosols which can have either warming or cooling effects on the atmosphere; the *net* effect is thought to be positive radiative forcing (Andreae *et al.*, 2005; Jones *et al.*, 2003; Venkataraman *et al.*, 2005; Andreae, 2001; Andreae and Merlet, 2001; Anderson *et al.*, 2003; Menon *et al.*, 2002). Fourth, fire reduces the albedo of the land surface for several weeks, causing warming (Beringer *et al.*, 2003). Finally, burning can affect the proportion of woody versus grass cover, notably in savannahs, which occupy about an eighth of the global land surface. Reducing the frequency or intensity of fires typically leads to increased tree and shrub cover, resulting in a CO_2 sink in soil and biomass (Scholes and van der Merwe, 1996). This woody-plant encroachment mechanism saturates over 20-50 years, whereas avoided CH_4 and N_2O emissions continue as long as fires are suppressed.

Mitigation actions involve reducing the frequency or extent of fires through more effective fire suppression; re-

ducing the fuel load by vegetation management; and burning at a time of year when less CH_4 and N_2O are emitted (Korontzi *et al.*, 2003). Although most agricultural-zone fires are ignited by humans, there is evidence that the area burned is ultimately under climatic control (Van Wilgen *et al.*, 2004). In the absence of human ignition, the fire-prone ecosystems would still burn as a result of climatic factors.

e. Species introduction: Introducing grass species with higher productivity, or carbon allocation to deeper roots, has been shown to increase soil carbon. For example, establishing deep-rooted grasses in savannahs has been reported to yield very high rates of carbon accrual (Fisher *et al.*, 1994), although the applicability of these results has not been widely confirmed (Conant *et al.*, 2001; Davidson *et al.*, 1995). In the Brazilian Savannah (Cerrado Biome), integrated crop-livestock systems using Brachiaria grasses and zero tillage are being adopted (Machado and Freitas, 2004). Introducing legumes into grazing lands can promote soil carbon storage (Soussana *et al.*, 2004), through enhanced productivity from the associated N inputs, and perhaps also reduced emissions from fertilizer manufacture if biological N_2 fixation displaces applied N fertilizer N (Sisti *et al.*, 2004; Diekow *et al.*, 2005). Ecological impacts of species introduction need to be considered.

Grazing lands also emit GHGs from livestock, notably CH_4 from ruminants and their manures. Practices for reducing these emissions are considered under Section 8.4.1.5: Livestock management.

8.4.1.3 Management of organic/peaty soils

Organic or peaty soils contain high densities of carbon accumulated over many centuries because decomposition is suppressed by absence of oxygen under flooded conditions. To be used for agriculture, these soils are drained, which aerates the soil, favouring decomposition and therefore, high CO_2 and N_2O fluxes. Methane emissions are usually suppressed after draining, but this effect is far outweighed by pronounced increases in N_2O and CO_2 (Kasimir-Klemedtsson *et al.*, 1997). Emissions from drained organic soils can be reduced to some extent by practices such as avoiding row crops and tubers, avoiding deep ploughing, and maintaining a shallower water table. But the most important mitigation practice is avoiding the drainage of these soils in the first place or re-establishing a high water table (Freibauer *et al.*, 2004).

8.4.1.4 Restoration of degraded lands

A large proportion of agricultural lands has been degraded by excessive disturbance, erosion, organic matter loss, salinization, acidification, or other processes that curtail productivity (Batjes, 1999; Foley *et al.*, 2005; Lal, 2001a, 2003, 2004b). Often, carbon storage in these soils can be partly restored by practices that reclaim productivity including: re-vegetation (e.g., planting grasses); improving fertility by nutrient amendments; applying organic substrates such as manures, biosolids, and composts; reducing tillage and retaining crop residues; and conserving water (Lal, 2001b; 2004b; Bruce *et al.*, 1999; Olsson and Ardö, 2002; Paustian *et al.*, 2004). Where these practices involve higher nitrogen amendments, the benefits of carbon sequestration may be partly offset by higher N_2O emissions.

8.4.1.5   *Livestock management*

Livestock, predominantly ruminants such as cattle and sheep, are important sources of CH_4, accounting for about one-third of global anthropogenic emissions of this gas (US-EPA, 2006a). The methane is produced primarily by enteric fermentation and voided by eructation (Crutzen, 1995; Murray *et al.*, 1976; Kennedy and Milligan, 1978). All livestock generate N_2O emissions from manure as a result of excretion of N in urine and faeces. Practices for reducing CH_4 and N_2O emissions from this source fall into three general categories: improved feeding practices, use of specific agents or dietary additives; and longer-term management changes and animal breeding (Soliva *et al.*, 2006; Monteny *et al.*, 2006).

a.   Improved feeding practices: Methane emissions can be reduced by feeding more concentrates, normally replacing forages (Blaxter and Claperton, 1965; Johnson and Johnson, 1995; Lovett *et al.*, 2003; Beauchemin and McGinn, 2005). Although concentrates may increase daily methane emissions per animal, emissions per kg-feed intake and per kg-product are almost invariably reduced. The magnitude of this reduction per kg-product decreases as production increases. The net benefit of concentrates, however, depends on reduced animal numbers or younger age at slaughter for beef animals, and on how the practice affects land use, the N content of manure and emissions from producing and transporting the concentrates (Phetteplace *et al.*, 2001; Lovett *et al.*, 2006). Other practices that can reduce CH_4 emissions include: adding certain oils or oilseeds to the diet (e.g., Machmüller *et al.*, 2000; Jordan *et al.*, 2006c); improving pasture quality, especially in less developed regions, because this improves animal productivity, and reduces the proportion of energy lost as CH_4 (Leng, 1991; McCrabb *et al.*, 1998; Alcock and Hegarty, 2006); and optimizing protein intake to reduce N excretion and N_2O emissions (Clark *et al.*, 2005).

b.   Specific agents and dietary additives: A wide range of specific agents, mostly aimed at suppressing methanogenesis, has been proposed as dietary additives to reduce CH_4 emissions:

- Ionophores are antibiotics that can reduce methane emissions (Benz and Johnson, 1982; Van Nevel and Demeyer, 1996; McGinn *et al.*, 2004), but their effect may be transitory (Rumpler *et al.*, 1986); and they have been banned in the EU.

- Halogenated compounds inhibit methanogenic bacteria (Wolin *et al.*, 1964; Van Nevel and Demeyer, 1995) but their effects, too, are often transitory and they can have side-effects such as reduced intake.

- Novel plant compounds such as condensed tannins (Pinares-Patiño *et al.*, 2003; Hess *et al.*, 2006), saponins (Lila *et al.*, 2003) or essential oils (Patra *et al.*, 2006; Kamra *et al.*, 2006) may have merit in reducing methane emissions, but these responses may often be obtained through reduced digestibility of the diet.

- Probiotics, such as yeast culture, have shown only small, insignificant effects (McGinn *et al.*, 2004), but selecting strains specifically for methane-reducing ability could improve results (Newbold and Rode, 2006).

- Propionate precursors such as fumarate or malate reduce methane formation by acting as alternative hydrogen acceptors (Newbold *et al.*, 2002). But as response is elicited only at high doses, propionate precursors are, therefore, expensive (Newbold *et al.*, 2005).

- Vaccines against methanogenic bacteria are being developed but are not yet available commercially (Wright *et al.*, 2004).

- Bovine somatotropin (bST) and hormonal growth implants do not specifically suppress CH_4 formation, but by improving animal performance (Bauman, 1992; Schmidely, 1993), they can reduce emissions per-kg of animal product (Johnson *et al.*, 1991; McCrabb, 2001).

c.   Longer-term management changes and animal breeding: Increasing productivity through breeding and better management practices, such as a reduction in the number of replacement heifers, often reduces methane output per unit of animal product (Boadi *et al.*, 2004). Although selecting cattle directly for reduced methane production has been proposed (Kebreab *et al.*, 2006), it is still impractical due to difficulties in accurately measuring methane emissions at a magnitude suitable for breeding programmes. With improved efficiency, meat-producing animals reach slaughter weight at a younger age, with reduced lifetime emissions (Lovett and O'Mara, 2002). However, the whole-system effects of such practices may not always lead to reduced emissions. For example in dairy cattle, intensive selection for higher yield may reduce fertility, requiring more replacement heifers in the herd (Lovett *et al.*, 2006).

8.4.1.6 *Manure management*

Animal manures can release significant amounts of N_2O and CH_4 during storage, but the magnitude of these emissions varies. Methane emissions from manure stored in lagoons or tanks can be reduced by cooling, use of solid covers, mechanically separating solids from slurry, or by capturing the CH_4 emitted (Amon *et al.* 2006; Clemens and Ahlgrimm, 2001; Monteny *et al.* 2001, 2006; Paustian *et al.*, 2004). The manures can also be digested anaerobically to maximize CH_4 retrieval as a renewable energy source (Clemens and Ahlgrimm, 2001; Clemens *et al.*, 2006). Handling manures in solid form (e.g., composting) rather than liquid form can suppress CH_4 emissions, but may increase N_2O formation (Paustian *et al.*, 2004). Preliminary evidence suggests that covering manure heaps can reduce N_2O emissions, but the effect of this practice on CH_4 emissions is variable (Chadwick, 2005). For most animals, worldwide there is limited opportunity for manure management, treatment, or storage; excretion happens in the field and handling for fuel or fertility amendment occurs when it is dry and methane emissions are negligible (Gonzalez-Avalos and Ruiz-Suarez, 2001). To some extent, emissions from manure might be curtailed by altering feeding practices (Külling *et al.*, 2003; Hindrichsen *et al.*, 2006; Kreuzer and Hindrichsen, 2006), or by composting the manure (Pattey *et al.*, 2005; Amon *et al.*, 2001), but if aeration is inadequate CH_4 emissions during composting can still be substantial (*Xu et al.*, 2007). All of these practices require further study from the perspective of their impact on whole life-cycle GHG emissions.

Manures also release GHGs, notably N_2O, after application to cropland or deposition on grazing lands. Practices for reducing these emissions are considered in Subsection 8.4.1.1: Cropland management and Subsection 8.4.1.2: Grazing land management.

8.4.1.7 *Bioenergy*

Increasingly, agricultural crops and residues are seen as sources of feedstocks for energy to displace fossil fuels. A wide range of materials have been proposed for use, including grain, crop residue, cellulosic crops (e.g., switchgrass, sugarcane), and various tree species (Edmonds, 2004; Cerri *et al.*, 2004; Paustian *et al.*, 2004; Sheehan *et al.*, 2004; Dias de Oliveira *et al.*, 2005; Eidman, 2005). These products can be burned directly, but can also be processed further to generate liquid fuels such as ethanol or diesel fuel (Richter, 2004). Such fuels release CO_2 when burned, but this CO_2 is of recent atmospheric origin (via photosynthetic carbon uptake) and displaces CO_2 which otherwise would have come from fossil carbon. The net benefit to atmospheric CO_2, however, depends on energy used in growing and processing the bioenergy feedstock (Spatari *et al.*, 2005).

The competition for other land uses and the environmental impacts need to be considered when planning to use energy crops (e.g., European Environment Agency, 2006). The interactions of an expanding bioenergy sector with other land uses, and impacts on agro-ecosystem services such as food production, biodiversity, soil and nature conservation, and carbon sequestration have not yet been adequately studied, but bottom-up approaches (Smeets *et al.*, 2007) and integrated assessment modelling (Hoogwijk *et al.*, 2005; Hoogwijk, 2004)

offer opportunities to improve understanding. Latin America, Sub-Saharan Africa, and Eastern Europe are promising regions for bio-energy, with additional long-term contributions from Oceania and East and Northeast Asia. The technical potential for biomass production may be developed at low production costs in the range of 2 US$/GJ (Hoogwijk, 2004; Rogner *et al.*, 2000).

Major transitions are required to exploit the large potential for bioenergy. Improving agricultural efficiency in developing countries is a key factor. It is still uncertain to what extent, and how fast, such transitions could be realized in different regions. Under less favourable conditions, the regional bio-energy potential(s) could be quite low. Also, technological developments in converting biomass to energy, as well as long distance biomass supply chains (e.g., those involving intercontinental transport of biomass derived energy carriers) can dramatically improve competitiveness and efficiency of bio-energy (Faaij, 2006; Hamelinck *et al.*, 2004).

8.4.2 Mitigation technologies and practices: per-area estimates of potential

As mitigation practices can affect more than one GHG[2], it is important to consider the impact of mitigation options on all GHGs (Robertson *et al.*, 2000; Smith *et al.*, 2001; Gregorich *et al.*, 2005). For non-livestock mitigation options, ranges for per-area mitigation potentials of each GHG are provided in Table 8.4 (tCO_2-eq/ha/yr).

Mitigation potentials for CO_2 represent the *net* change in soil carbon pools, reflecting the accumulated difference between carbon inputs to the soil after CO_2 uptake by plants, and release of CO_2 by decomposition in soil. Mitigation potentials for N_2O and CH_4 depend solely on emission reductions. Soil carbon stock changes were derived from about 200 studies, and the emission ranges for CH_4 and N_2O were derived using the DAYCENT and DNDC simulation models (IPCC, 2006; US-EPA, 2006b; Smith *et al.*, 2007b; Ogle *et al.*, 2004, 2005).

Table 8.5 presents the mitigation potentials in livestock (dairy cows, beef cattle, sheep, dairy buffalo and other buffalo) for reducing enteric methane emissions via improved feeding practices, specific agents and dietary additives, and longer term structural and management changes/animal breeding. These estimates were derived by Smith *et al.* (2007a) using a model similar to that described in US-EPA (2006b).

Some mitigation measures operate predominantly on one GHG (e.g., dietary management of ruminants to reduce CH_4 emissions) while others have impacts on more than one GHG (e.g., rice management). Moreover, practices may benefit more

---

2   Smith *et al.* (2007a) have recently collated per-area estimates of agricultural GHG mitigation options. This section draws largely from that study.

Table 8.4: *Annual mitigation potentials in each climate region for non-livestock mitigation options*

Climate zone	Activity	Practice	CO$_2$ (tCO$_2$/ha/yr)			CH$_4$ (tCO$_2$-eq/ha/yr)			N$_2$O (tCO$_2$-eq/ha/yr)			All GHG (tCO$_2$-eq/ha/yr)		
			Mean estimate	Low	High	Mean estimate	Low	High	Mean estimate	Low	High	Mean estimate	Low	High
Cool-dry	Croplands	Agronomy	0.29	0.07	0.51	0.00	0.00	0.00	0.10	0.00	0.20	0.39	0.07	0.71
	Croplands	Nutrient management	0.26	-0.22	0.73	0.00	0.00	0.00	0.07	0.01	0.32	0.33	-0.21	1.05
	Croplands	Tillage and residue management	0.15	-0.48	0.77	0.00	0.00	0.00	0.02	-0.04	0.09	0.17	-0.52	0.86
	Croplands	Water management	1.14	-0.55	2.82	0.00	0.00	0.00	0.00	0.00	0.00	1.14	-0.55	2.82
	Croplands	Set-aside and LUC	1.61	-0.07	3.30	0.02	0.00	0.00	2.30	0.00	4.60	3.93	-0.07	7.90
	Croplands	Agro-forestry	0.15	-0.48	0.77	0.00	0.00	0.00	0.02	-0.04	0.09	0.17	-0.52	0.86
	Grasslands	Grazing, fertilization, fire	0.11	-0.55	0.77	0.02	0.01	0.02	0.00	0.00	0.00	0.13	-0.54	0.79
	Organic soils	Restoration	36.67	3.67	69.67	-3.32	-0.05	-15.30	0.16	0.05	0.28	33.51	3.67	54.65
	Degraded lands	Restoration	3.45	-0.37	7.26	0.08	0.04	0.14	0.00	0.00	0.00	3.53	-0.33	7.40
	Manure/ biosolids	Application	1.54	-3.19	6.27	0.00	0.00	0.00	0.00	-0.17	1.30	1.54	-3.36	7.57
	Bioenergy	Soils only	0.15	-0.48	0.77	0.00	0.00	0.00	0.02	-0.04	0.09	0.17	-0.52	0.86
Cool-moist	Croplands	Agronomy	0.88	0.51	1.25	0.00	0.00	0.00	0.10	0.00	0.20	0.98	0.51	1.45
	Croplands	Nutrient management	0.55	0.01	1.10	0.00	0.00	0.00	0.07	0.01	0.32	0.62	0.02	1.42
	Croplands	tillage and residue management	0.51	0.00	1.03	0.00	0.00	0.00	0.02	-0.04	0.09	0.53	-0.04	1.12
	Croplands	Water management	1.14	-0.55	2.82	0.00	0.00	0.00	0.00	0.00	0.00	1.14	-0.55	2.82
	Croplands	Set-aside and LUC	3.04	1.17	4.91	0.02	0.00	0.00	2.30	0.00	4.60	5.36	1.17	9.51
	Croplands	Agro-forestry	0.51	0.00	1.03	0.00	0.00	0.00	0.02	-0.04	0.09	0.53	-0.04	1.12
	Grasslands	Grazing, fertilization, fire	0.81	0.11	1.50	0.00	0.00	0.00	0.00	0.00	0.00	0.80	0.11	1.50
	Organic soils	Restoration	36.67	3.67	69.67	-3.32	-0.05	-15.30	0.16	0.05	0.28	33.51	3.67	54.65
	Degraded lands	Restoration	3.45	-0.37	7.26	1.00	0.69	1.25	0.00	0.00	0.00	4.45	0.32	8.51
	Manure/ biosolids	Application	2.79	-0.62	6.20	0.00	0.00	0.00	0.00	-0.17	1.30	2.79	-0.79	7.50
	Bioenergy	Soils only	0.51	0.00	1.03	0.00	0.00	0.00	0.02	-0.04	0.09	0.53	-0.04	1.12
Warm-dry	Croplands	Agronomy	0.29	0.07	0.51	0.00	0.00	0.00	0.10	0.00	0.20	0.39	0.07	0.71
	Croplands	Nutrient management	0.26	-0.22	0.73	0.00	0.00	0.00	0.07	0.01	0.32	0.33	-0.21	1.05
	Croplands	Tillage and residue management	0.33	-0.73	1.39	0.00	0.00	0.00	0.02	-0.04	0.09	0.35	-0.77	1.48
	Croplands	Water management	1.14	-0.55	2.82	0.00	0.00	0.00	0.00	0.00	0.00	1.14	-0.55	2.82
	Croplands	Set-aside and LUC	1.61	-0.07	3.30	0.02	0.00	0.00	2.30	0.00	4.60	3.93	-0.07	7.90
	Croplands	Agro-forestry	0.33	-0.73	1.39	0.00	0.00	0.00	0.02	-0.04	0.09	0.35	-0.77	1.48
	Grasslands	Grazing, fertilization, fire	0.11	-0.55	0.77	0.00	0.00	0.00	0.00	0.00	0.00	0.11	-0.55	0.77
	Organic soils	Restoration	73.33	7.33	139.33	-3.32	-0.05	-15.30	0.16	0.05	0.28	70.18	7.33	124.31
	Degraded lands	Restoration	3.45	-0.37	7.26	0.00	0.00	0.00	0.00	0.00	0.00	3.45	-0.37	7.26
	Manure/ biosolids	Application	1.54	-3.19	6.27	0.00	0.00	0.00	0.00	-0.17	1.30	1.54	-3.36	7.57
	Bioenergy	Soils only	0.33	-0.73	1.39	0.00	0.00	0.00	0.02	-0.04	0.09	0.35	-0.77	1.48
Warm-moist	Croplands	Agronomy	0.88	0.51	1.25	0.00	0.00	0.00	0.10	0.00	0.20	0.98	0.51	1.45
	Croplands	Nutrient management	0.55	0.01	1.10	0.00	0.00	0.00	0.07	0.01	0.32	0.62	0.02	1.42
	Croplands	Tillage and residue management	0.70	-0.40	1.80	0.00	0.00	0.00	0.02	-0.04	0.09	0.72	-0.44	1.89
	Croplands	Water management	1.14	-0.55	2.82	0.00	0.00	0.00	0.00	0.00	0.00	1.14	-0.55	2.82
	Croplands	Set-aside and LUC	3.04	1.17	4.91	0.02	0.00	0.00	2.30	0.00	4.60	5.36	1.17	9.51
	Croplands	Agro-forestry	0.70	-0.40	1.80	0.00	0.00	0.00	0.02	-0.04	0.09	0.72	-0.44	1.89
	Grasslands	Grazing, fertilization, fire	0.81	0.11	1.50	0.00	0.00	0.00	0.00	0.00	0.00	0.81	0.11	1.50
	Organic soils	Restoration	73.33	7.33	139.33	-3.32	-0.05	-15.30	0.16	0.05	0.28	70.18	7.33	124.31
	Degraded lands	Restoration	3.45	-0.37	7.26	0.00	0.00	0.00	0.00	0.00	0.00	3.45	-0.37	7.26
	Manure/ biosolids	Application	2.79	-0.62	6.20	0.00	0.00	0.00	0.00	-0.17	1.30	2.79	-0.79	7.50
	Bioenergy	Soils only	0.70	-0.40	1.80	0.00	0.00	0.00	0.02	-0.04	0.09	0.72	-0.44	1.89

Notes:

The estimates represent average change in soil carbon stocks (CO$_2$) or emissions of N$_2$O and CH$_4$ on a per hectare basis. Positive values represent CO$_2$ uptake which increases the soil carbon stock, or a reduction in emissions of N$_2$O and CH$_4$.

Estimates of soil carbon storage (CO$_2$ mitigation) for all practices except management of organic soils were derived from about 200 studies (see IPCC, 2006, Grassland and Cropland Chapters of Volume IV, Annexes 5A and 6A) using a linear mixed-effect modelling approach, which is a standard linear regression technique with the inclusion of random effects due to dependencies in data from the same country, site and time series (Ogle et al., 2004, 2005; IPCC, 2006; Smith et al., 2007b). The studies were conducted in regions throughout the world, but temperate studies were more prevalent leading to smaller uncertainties than for estimates for warm tropical climates. Estimates represent annual soil carbon change rate for a 20-year time horizon in the top 30 cm of the soil. Soils under bio-energy crops and agro-forestry were assumed to derive their mitigation potential mainly from cessation of soil disturbance, and given the same estimates as no-till. Management of organic soils was based on emissions under drained conditions from IPCC guidelines (IPCC, 1997). Soil CH$_4$ and N$_2$O emission reduction potentials were derived as follows:

a) for organic soils, N$_2$O emissions were based on the median, low and high nutrient status organic soil N$_2$O emission factors from the IPCC GPG LULUCF (IPCC, 2003) and CH$_4$ emissions were based on low, high and median values from Le Mer and Roger (2001);

b) N$_2$O figures for nutrient management were derived using the DAYCENT simulation model, and include both direct emissions from nitrification/denitrification at the site, as well as indirect N$_2$O emissions associated with volatilization and leaching/runoff of N that is converted into N$_2$O following atmospheric deposition or in waterways, respectively (US-EPA, 2006b; assuming a N reduction to 80% of current application);

c) N$_2$O figures for tillage and residue management were derived using DAYCENT (US-EPA, 2006b; figures for no till);

d) Rice figures were taken directly from US-EPA (2006b) so are not shown here. Low and high values represent the range of a 95% confidence interval. Table 8.4 has mean and uncertainty for change in soil C, N$_2$O and CH$_4$ emissions at the climate region scale, and are not intended for use in assessments at finer scales such as individual farms.

Table 8.5: *Technical reduction potential (proportion of an animal's enteric methane production) for enteric methane emissions due to (i) improved feeding practices, (ii) specific agents and dietary additives and (iii) longer term structural/management change and animal breeding[a]*

AEZ regions	Improved feeding practices[b]					Specific agents and dietary additives[c]					Longer term structural/management change and animal breeding[d]				
	Dairy cows	Beef cattle	Sheep	Dairy buffalo	Non-dairy buffalo	Dairy cows	Beef cattle	Sheep	Dairy buffalo	Non-dairy buffalo	Dairy cows	Beef cattle	Sheep	Dairy buffalo	Non-dairy buffalo
Northern Europe	0.18	0.12	0.04			0.08	0.04	0.004			0.04	0.03	0.003		
Southern. Europe	0.18	0.12	0.04			0.08	0.04	0.004			0.04	0.03	0.003		
Western Europe	0.18	0.12	0.04			0.08	0.04	0.004			0.04	0.03	0.003		
Eastern. Europe	0.11	0.06	0.03			0.04	0.01	0.002			0.03	0.07	0.003		
Russian Federation	0.10	0.05	0.03			0.03	0.04	0.002			0.03	0.06	0.003		
Japan	0.17	0.11	0.04			0.08	0.09	0.004			0.03	0.03	0.003		
South Asia	0.04	0.02	0.02	0.04	0.02	0.01	0.01	0.0005	0.01	0.002	0.01	0.01	0.001	0.01	0.02
East Asia	0.10	0.05	0.03	0.10	0.05	0.03	0.05	0.002	0.03	0.012	0.03	0.06	0.003	0.03	0.07
West Asia	0.06	0.03	0.02	0.06	0.03	0.01	0.02	0.001	0.01	0.004	0.01	0.02	0.001	0.02	0.03
Southeast Asia	0.06	0.03	0.02	0.06	0.03	0.01	0.02	0.001	0.01	0.004	0.01	0.02	0.001	0.02	0.03
Central Asia	0.06	0.03	0.02	0.06	0.03	0.01	0.02	0.001	0.01	0.004	0.01	0.02	0.001	0.02	0.03
Oceania	0.22	0.14	0.06			0.08	0.08	0.004			0.05	0.03	0.004		
North America	0.16	0.11	0.04			0.11	0.09	0.004			0.03	0.03	0.003		
South America	0.06	0.03	0.02			0.03	0.02	0.001			0.02	0.03	0.002		
Central America	0.03	0.02	0.02			0.02	0.01	0.001			0.01	0.02	0.002		
East Africa	0.01	0.01	0.01			0.003	0.004	0.0002			0.004	0.006	0.0004		
West Africa	0.01	0.01	0.01			0.003	0.004	0.0002			0.004	0.006	0.0004		
North Africa	0.01	0.01	0.01			0.003	0.004	0.0002			0.004	0.006	0.0004		
South Africa	0.01	0.01	0.01			0.003	0.004	0.0002			0.004	0.006	0.0004		
Middle Africa	0.01	0.01	0.01			0.003	0.004	0.0002			0.004	0.006	0.0004		

Notes:

[a] The proportional reduction due to application of each practice was estimated from reports in the scientific literature (see footnotes below). These estimates were adjusted for:

(i) proportion of the animal's life where the practice was applicable;

(ii) technical adoption feasibility in a region, such as whether farmers have the necessary knowledge, equipment, extension services, etc. to apply the practice (average dairy cow milk production in each region over the period 2000-2004 was used as an index of the level of technical efficiency in the region, and was used to score a region's technical adoption feasibility);

(iii) proportion of animals in a region that the measure can be applied (i.e. if the measure is already being applied to some animals as in the case of bST use in North America, it is considered to be only applicable to the proportion of animals not currently receiving the product;

(iv) Non-additivity of simultaneous application of multiple measures.

There is evidence in the literature that some measures are not additive when applied simultaneously, such as the use of dietary oils and ionophores, but this is probably not the case with most measures. However, the model used (as described in Smith et al., 2007a) did account for the fact that once one measure is applied, the emissions base for the second measure is reduced, and so on, and a further 20% reduction in mitigation potential was incorporated to account for unknown non-additivity effects. Only measures considered feasible for a region were applied in that region (e.g., bST was not considered for European regions due to the ban on its use in the EU). It was assumed that total production of milk or meat was not affected by application of the practices, so that if a measure increased animal productivity, animal numbers were reduced in order to keep production constant.

[b] Includes replacing roughage with concentrate (Blaxter & Claperton, 1965; Moe & Tyrrell, 1979; Johnson & Johnson, 1995; Yan et al., 2000; Mills et al., 2003; Beauchemin & McGinn, 2005; Lovett et al., 2006), improving forages/inclusion of legumes (Leng, 1991; McCrabb et al., 1998; Woodward et al., 2001; Waghorn et al., 2002; Pinares-Patiño et al., 2003; Alcock & Hegarty, 2006) and feeding extra dietary oil (Machmüller et al., 2000; Dohme et al., 2001; Machmüller et al., 2003, Lovett et al., 2003; McGinn et al., 2004; Beauchemin & McGinn, 2005; Jordan et al., 2006a; Jordan et al., 2006b; Jordan et al., 2006c).

[c] Includes bST (Johnson et al., 1991; Bauman, 1992), growth hormones (McCrabb, 2001), ionophores (Benz & Johnson, 1982; Rumpler et al., 1986; Van Nevel & Demeyer, 1996; McGinn et al., 2004), propionate precursors (McGinn et al., 2004; Beauchemin & McGinn, 2005; Newbold et al., 2005; Wallace et al., 2006).

[d] Includes lifetime management of beef cattle (Johnson et al., 2002; Lovett & O'Mara, 2002) and improved productivity through animal breeding (Ferris et al., 1999; Hansen, 2000; Robertson and Waghorn, 2002; Miglior et al., 2005).

Source: adapted from Smith et al., 2007a.

than one gas (e.g., set-aside/headland management) while others involve a trade-off between gases (e.g., restoration of organic soils). The effectiveness of non-livestock mitigation options are variable across and within climate regions (see Table 8.4). Consequently, a practice that is highly effective in reducing emissions at one site may be less effective or even counterproductive elsewhere. Similarly, effectiveness of livestock options also varies regionally (Table 8.5). This means that there is no universally applicable list of mitigation practices, but that proposed practices will need to be evaluated for individual agricultural systems according to the specific climatic, edaphic, social settings, and historical land use and management.

Assessments can be conducted to evaluate the effectiveness of practices in specific areas, building on findings from the global scale assessment reported here. In addition, such assessments could address GHG emissions associated with energy use and other inputs (e.g., fuel, fertilizers, and pesticides) in a full life cycle analysis for the production system.

The effectiveness of mitigation strategies also changes with time. Some practices, like those which elicit soil carbon gain, have diminishing effectiveness after several decades; others such as methods that reduce energy use may reduce emissions indefinitely. For example, Six et al. (2004) found a strong time dependency of emissions from no-till agriculture, in part because of changing influence of tillage on N_2O emissions.

8.4.3 Global and regional estimates of agricultural GHG mitigation potential

8.4.3.1 Technical potential for GHG mitigation in agriculture

There have been numerous attempts to assess the technical potential for GHG mitigation in agriculture. Most of these have focused on soil carbon sequestration. Estimates in the IPCC Second Assessment Report (SAR; IPCC, 1996) suggested that 400-800 MtC/yr (equivalent to about 1400-2900 MtCO$_2$-

eq/yr) could be sequestered in global agricultural soils with a finite capacity saturating after 50 to100 years. In addition, SAR concluded that 300-1300 MtC (equivalent to about 1100-4800 MtCO$_2$-eq/yr) from fossil fuels could be offset by using 10 to15% of agricultural land to grow energy crops; with crop residues potentially contributing 100-200 MtC (equivalent to about 400-700 MtCO$_2$-eq/yr) to fossil fuel offsets if recovered and burned. Burning residues for bio-energy might increase N_2O emissions but this effect was not quantified.

SAR (IPCC, 1996) estimated that CH_4 emissions from agriculture could be reduced by 15 to 56%, mainly through improved nutrition of ruminants and better management of paddy rice, and that improved management could reduce N_2O

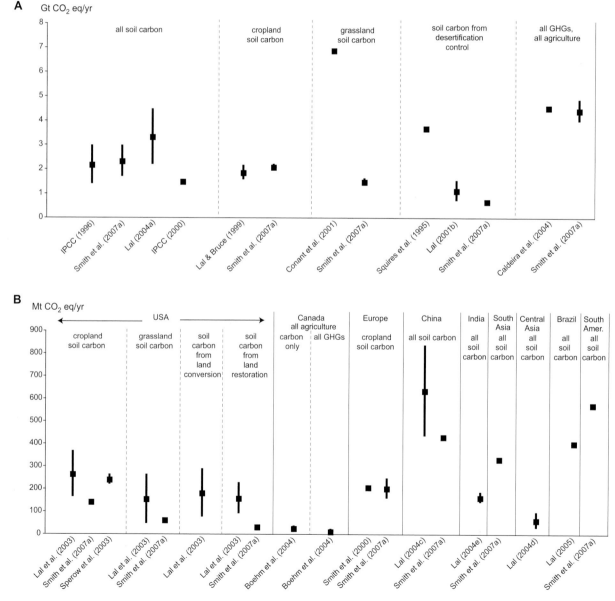

Figure 8.3: *Global (A) and regional (B) estimates of technical mitigation potential by 2030*
Note: Equivalent values for Smith et al. (2007a) are taken from Table 7 of Smith et al., 2007a.

emissions by 9-26%. The document also stated that GHG mitigation techniques will not be adopted by land managers unless they improve profitability but some measures are adopted for reasons other than climate mitigation. Options that both reduce GHG emissions and increase productivity are more likely to be adopted than those which only reduce emissions.

Of published estimates of technical potential, only Caldeira *et al.* (2004) and Smith *et al.* (2007a) provide global estimates considering all GHGs together, and Boehm *et al.* (2004) consider all GHGs for Canada only for 2008. Smith *et al.* (2007a) used per-area or per-animal estimates of mitigation potential for each GHG and multiplied this by the area available for that practice in each region. It was not necessary to use baseline emissions in calculating mitigation potential. US-EPA (2006b) estimated baseline emissions for 2020 for non-CO_2 GHGs as 7250 MtCO_2-eq in 2020 (see Chapter 11; Table 11.4). Non-CO_2 GHG emissions in agriculture are projected to increase by about 13% from 2000 to 2010 and by 13% from 2010 to 2020 (US-EPA, 2006b). Assuming a similar rate of increase as in the period from 2000 to 2020, global agricultural non-CO_2 GHG emissions would be around 8200 MtCO_2-eq in 2030.

The global technical potential for mitigation options in agriculture by 2030, considering all gases, was estimated to be ~4500 by Caldeira *et al.* (2004) and ~5500-6000 MtCO_2-eq/yr by Smith *et al.* (2007a) if considering no economic or other barriers. Economic potentials are considerably lower (see Section 8.4.3.2). Figure 8.3 presents global and regional estimates of agricultural mitigation potential. Of the technical potentials estimated by Smith *et al.* (2007a), about 89% is from soil carbon sequestration, about 9% from mitigation of methane and about 2% from mitigation of soil N_2O emissions (Figure 8.4). The total mitigation potential per region is presented in Figure 8.5.

The uncertainty in the estimates of the technical potential is given in Figure 8.6, which shows one standard deviation either side of the mean estimate (box), and the 95% confidence interval about the mean (line). The range of the standard deviation, and the 95% confidence interval about the mean of 5800 MtCO_2-eq/yr, are 3000-8700, and 300-11400 MtCO_2-eq/yr, respectively, and are largely determined by uncertainty in the per-area estimate for the mitigation measure. For soil carbon sequestration (89% of the total potential), this arises from the mixed linear effects model used to derive the mitigation potentials. The most appropriate mitigation response will vary among regions, and different portfolios of strategies will be developed in different regions, and in countries within a region.

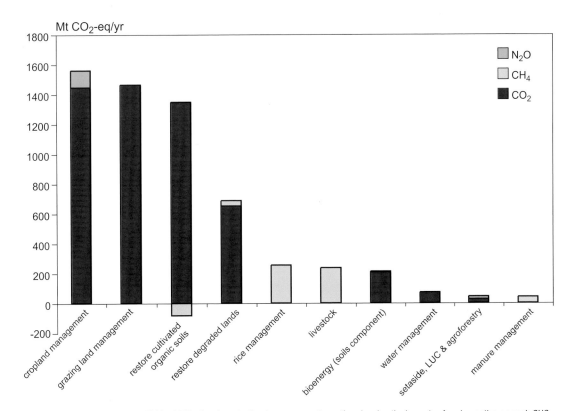

Figure 8.4: *Global technical mitigation potential by 2030 of each agricultural management practice showing the impacts of each practice on each GHG.*
Note: based on the B2 scenario though the pattern is similar for all SRES scenarios.
Source: Drawn from data in Smith et al., 2007a.

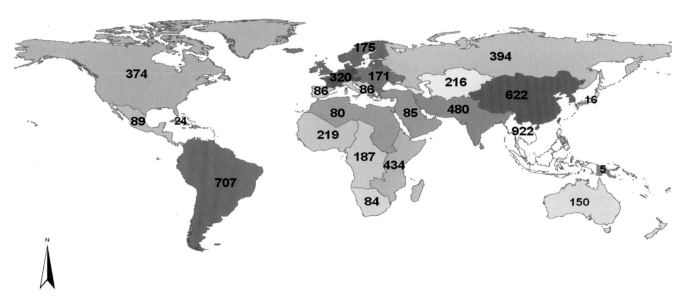

Figure 8.5: *Total technical mitigation potentials (all practices, all GHGs: MtCO2-eq/yr) for each region by 2030, showing mean estimates.*
Note: based on the B2 scenario though the pattern is similar for all SRES scenarios.
Source: Drawn from data in Smith et al., 2007a.

8.4.3.2 Economic potential for GHG mitigation in agriculture

US-EPA (2006b) provided estimates of the agricultural mitigation potential (global and regional) at various assumed carbon prices, for N_2O and CH_4, but not for soil carbon sequestration. Manne & Richels (2004) estimated the economic mitigation potential (at 27 US$/tCO_2$-eq) for soil carbon sequestration only.

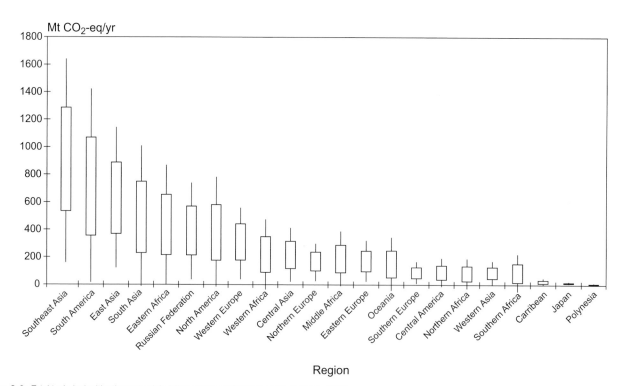

Figure 8.6: *Total technical mitigation potentials (all practices, all GHGs) for each region by 2030*
Note: Boxes show one standard deviation above and below the mean estimate for per-area mitigation potential, and the bars show the 95% confidence interval about the mean. Based on the B2 scenario, although the pattern is similar for all SRES scenarios.
Source: Drawn from data in Smith et al., 2007a.

Table 8.6: *Global agricultural mitigation potential in 2030 from top-down models*

Carbon price	Mitigation (MtCO$_2$-eq/yr)			Number of scenarios
US$/tCO$_2$-eq	CH$_4$	N$_2$O	CH$_4$+N$_2$O	
0-20	0-1116	89-402	267-1518	6
20-50	348-1750	116-1169	643-1866	6
50-100	388	217	604	1
>100	733	475	1208	1

Note: From Chapter 3, Sections 3.3.5 and 3.6.2.

Source: Data assembled from USCCSP, 2006; Rose et al., 2007; Fawcett and Sands, 2006; Smith and Wigley, 2006; Fujino et al., 2006; and Kemfert et al., 2006.

In the IPCC Third Assessment Report (TAR; IPCC, 2001b), estimates of agricultural mitigation potential by 2020 were 350-750 MtC/yr (~1300-2750 MtCO$_2$/yr). The range was mainly caused by large uncertainties about CH$_4$, N$_2$O, and soil-related CO$_2$ emissions. Most reductions will cost between 0 and 100 US$/tC-eq (~0-27 US$/tCO$_2$-eq) with limited opportunities for negative net direct cost options. The analysis of agriculture included only conservation tillage, soil carbon sequestration, nitrogen fertilizer management, enteric methane reduction and rice paddy irrigation and fertilizers. The estimate for global mitigation potential was not broken down by region or practice.

Smith *et al.* (2007a) estimated the GHG mitigation potential in agriculture for all GHGs, for four IPCC SRES scenarios, at a range of carbon prices, globally and for all world regions. Using methods similar to McCarl and Schneider (2001), Smith *et al.* (2007a) used marginal abatement cost (MAC) curves given in US-EPA (2006b) for either region-specific MACs where available for a given practice and region, or global MACs where these were unavailable from US-EPA (2006b).

Recent bottom-up estimates of agricultural mitigation potential of CH$_4$ and N$_2$O from US-EPA (2006b) and DeAngelo *et al.* (2006) have allowed inclusion of agricultural abatement into top-down global modelling of long-term climate stabilization scenario pathways. In the top-down framework, a dynamic cost-effective portfolio of abatement strategies is identified. The portfolio includes the least-cost combination of mitigation strategies from across all sectors of the economy, including agriculture. Initial implementations of agricultural abatement into top-down models have employed a variety of alternative approaches resulting in different decision modelling of agricultural abatement (Rose *et al.*, 2007). Currently, only non-CO$_2$ GHG crop (soil and paddy rice) and livestock (enteric and manure) abatement options are considered by top-down models. In addition, some models also consider emissions from burning of agricultural residues and waste, and fossil fuel combustion CO$_2$ emissions. Top-down estimates of global CH$_4$ and N$_2$O mitigation potential, expressed in CO$_2$ equivalents, are given in Table 8.6 and Figure 8.7.

Comparing mitigation estimates from top-down and bottom-up modelling is not straightforward. Bottom-up mitigation responses are typically constrained to input management (e.g.,

fertilizer quantity, livestock feed type) and cost estimates are partial equilibrium in that input and output market prices are fixed as can be key input quantities such as acreage or production. Top-down mitigation responses include more generic input management responses and changes in output (e.g., shifts from cropland to forest) as well as changes in market prices (e.g., decreases in land prices with increasing production costs due to a carbon tax). Global estimates of economic mitigation potential from different studies at different assumed carbon prices are presented in Figure 8.8.

The top-down 2030 carbon prices, as well as the agricultural mitigation response, reflect the confluence of multiple forces, including differences in implementation of agricultural emissions and mitigation, as well as the stabilization target used, the magnitude of baseline emissions, baseline energy technology options, the eligible set of mitigation options, and the solution algorithm. As a result, the opportunity cost of agricultural mitigation in 2030 is very different across scenarios (i.e., model/baseline/mitigation option combinations). As illustrated by the connecting lines in Figure 8.7, agricultural abatement

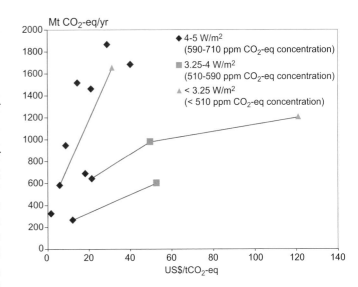

Figure 8.7: *Global agricultural mitigation potential in 2030 from top-down models by carbon price and stabilisation target*

Note: Dashed lines connect results from scenarios where tighter stabilization targets were modelled with the same model and identical baseline characterization and mitigation technologies. From Chapter 3, Sections 3.3.5 and 3.6.2.

Source: Data assembled from USCCSP, 2006; Rose et al., 2007; Fawcett and Sands, 2006; Smith and Wigley, 2006; Fujino et al., 2006; Kemfert et al., 2006.

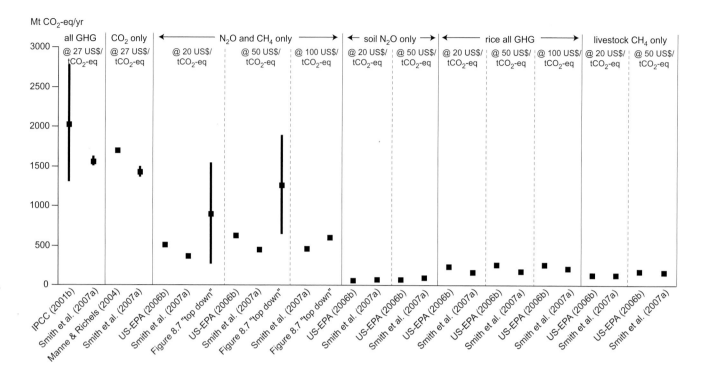

Figure 8.8: *Global economic potentials for agricultural mitigation arising from various practices shown for comparable carbon prices at 2030.*
Notes: US-EPA (2006b) figures are for 2020 rather than 2030. Values for top-down models are taken from ranges given in Figure 8.7.

is projected to increase with the tightness of the stabilization target. On-going model development in top-down land-use modelling is expected to yield more refined characterizations of agricultural alternatives and mitigation potential in the future.

Smith *et al.* (2007a) estimated global economic mitigation potentials for 2030 of 1500-1600, 2500-2700, and 4000-4300 MtCO$_2$-eq/yr at carbon prices of up to 20, 50 and 100 US$/

tCO$_2$-eq., respectively shown for OECD versus EIT versus non-OECD/EIT (Table 8.7). The change in global mitigation potential with increasing carbon price for each practice is shown in Figure 8.9.

Table 8.7: *Estimates of the global agricultural economic GHG mitigation potential (MtCO$_2$-eq/yr) by 2030 under different assumed prices of CO$_2$-equivalents*

SRES Scenario		Price of CO$_2$-eq (US$/tCO$_2$-eq)		
		Up to 20	Up to 50	Up to 100
B1	OECD	310 (60-450)	510 (290-740)	810 (440-1180)
	EIT	150 (30-220)	250 (140-370)	410 (220-590)
	Non-OECD/EIT	1080 (210-1560)	1780 (1000-2580)	2830 (1540-4120)
A1b	OECD	320 (60-460)	520 (290-760)	840 (450-1230)
	EIT	160 (30-230)	260 (150-380)	410 (220-610)
	Non-OECD/EIT	1110 (210-1610)	1820 (1020-2660)	2930 (1570-4290)
B2	OECD	330 (60-470)	540 (300-780)	870 (460-1280)
	EIT	160 (30-240)	270 (150-390)	440 (230-640)
	Non-OECD/EIT	1140 (210-1660)	1880 (1040-2740)	3050 (1610-4480)
A2	OECD	330 (60-480)	540 (300-790)	870 (460-1280)
	EIT	165 (30-240)	270 (150-400)	440 (230-640)
	Non-OECD/EIT	1150 (210-1670)	1890 (1050-2760)	3050 (1620-4480)

Note: Figures in brackets show one standard deviation about the mean estimate.

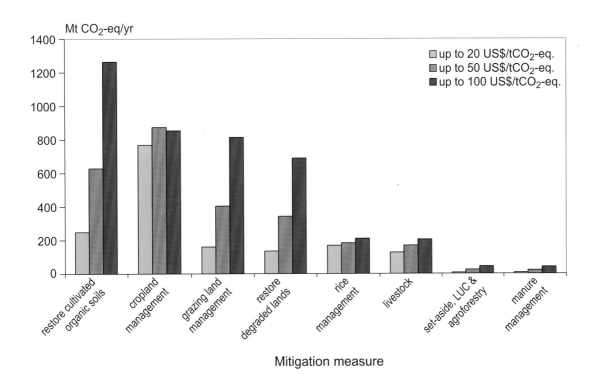

Figure 8.9: *Economic potential for GHG agricultural mitigation by 2030 at a range of prices of CO$_2$-eq*
Note: Based on B2 scenario, although the pattern is similar for all SRES scenarios.
Source: Drawn from data in Smith et al., 2007a.

8.4.4   Bioenergy feed stocks from agriculture

Bioenergy to replace fossil fuels can be generated from agricultural feedstocks, including by-products of agricultural production, and dedicated energy crops.

8.4.4.1   Residues from agriculture

The energy production and GHG mitigation potentials depend on yield/product ratios, and the total agricultural land area as well as type of production system. Less intensive management systems require re-use of residues for maintaining soil fertility. Intensively managed systems allow for higher utilization rates of residues, but also usually deploy crops with higher crop-to-residue ratios.

Estimates of energy production potential from agricultural residues vary between 15 and 70 EJ/yr. The latter figure is based on the regional production of food (in 2003) multiplied by harvesting or processing factors, and assumed recoverability factors. These figures do not subtract the potential competing uses of agricultural residues which, as indicated by (Junginger et al., 2001), can reduce significantly the net availability of agricultural residues for energy or materials. In addition, the expected future availability of residues from agriculture varies widely among studies. Dried dung can also be used as an energy feedstock. The total estimated contribution could be 5 to 55 EJ/yr worldwide, with the range defined by current global use at

the low end, and technical potential at the high end. Utilization in the longer term is uncertain because dung is considered to be a "poor man's fuel".

Organic wastes and residues together could supply 20-125 EJ/yr by 2050, with organic wastes making a significant contribution.

8.4.4.2   Dedicated energy crops

The energy production and GHG mitigation potentials of dedicated energy crops depends on availability of land, which must also meet demands for food as well as for nature protection, sustainable management of soils and water reserves, and other sustainability criteria. Because future biomass resource availability for energy and materials depends on these and other factors, an accurate estimate is difficult to obtain. Berndes *et al.* (2003) in reviewing 17 studies of future biomass availability found no complete integrated assessment and scenario studies. Various studies have arrived at differing figures for the potential contribution of biomass to future global energy supplies, ranging from below 100 EJ/yr to above 400 EJ/yr in 2050. Smeets *et al.* (2007) indicate that ultimate technical potential for energy cropping on current agricultural land, with projected technological progress in agriculture and livestock, could deliver over 800 EJ/yr without jeopardizing the world's food supply. In Hoogwijk *et al.* (2005) and Hoogwijk (2004), the IMAGE 2.2 model was used to analyse biomass production potentials for different SRES scenarios. Biomass production on abandoned agricultural land is calculated at 129 EJ (A2) up to 411 EJ (A1) for 2050 and possibly increasing after that timeframe. 273 EJ (for A1) – 156 EJ (for A2) may be available below US$ 2/GJ production costs. A recent study (Sims *et al.*, 2006) which used lower per-area yield assumptions and bio-energy crop areas projected by the IMAGE 2.2 model suggested more modest potentials (22 EJ/yr) by 2025.

Based on assessment of other studies, Hoogwijk *et al.* (2003), indicated that marginal and degraded lands (including a land surface of 1.7 Gha worldwide) could, be it with lower productivities and higher production costs, contribute another 60-150 EJ. Differences among studies are largely attributable to uncertainty in land availability, energy crop yields, and assumptions on changes in agricultural efficiency. Those with the largest projected potential assume that not only degraded/surplus land are used, but also land currently used for food production (including pasture land, as did Smeets *et al.*, 2007).

Converting the potential biomass production into a mitigation potential is not straightforward. First, the mitigation potential is determined by the lowest supply and demand potentials, so without the full picture (see Chapter 11) no estimate can be made. Second, any potential from bioenergy use will be counted towards the potential of the sectors where bioenergy is used (mainly energy supply and transport). Third, the proportion of the agricultural biomass supply compared to that from the waste or forestry sector cannot be specified due to lack of information on cost curves.

Top-down integrated assessment models can give an estimate of the cost competitiveness of bioenergy mitigation options relative to one another and to other mitigation options in achieving specific climate goals. By taking into account the various bioenergy supplies and demands, these models can give estimates of the combined contribution of the agriculture, waste, and forestry sectors to bioenergy mitigation potential. For achieving long-term climate stabilization targets, the competitive cost-effective mitigation potential of biomass energy (primarily from agriculture) in 2030 is estimated to be 70 to 1260 $MtCO_2$-eq/yr (0-13 EJ/yr) at up to 20 US$/t CO_2-eq, and 560-2320 $MtCO_2$-eq/yr (0-21 EJ/yr) at up to 50 US$/t$CO_2$-eq (Rose et al., 2007, USCCSP, 2006). There are no estimates for the additional potential from top down models at carbon prices up to 100 US$/t$CO_2$-eq, but the estimate for prices above 100 US$/t$CO_2$-eq is 2720 $MtCO_2$-eq/yr (20-45 EJ/yr). This is of the same order of magnitude as the estimate from a synthesis of supply and demand presented in Chapter 11, Section 11.3.1.4. The mitigation potentials estimated by top-down models represent mitigation of 5-80%, and 20-90% of all other agricultural mitigation measures combined, at carbon prices of up to 20, and up to 50 US$/t$CO_2$-eq, respectively.

8.4.5 Potential implications of mitigation options for sustainable development

There are various potential impacts of agricultural GHG mitigation on sustainable development. The impacts of mitigation activities in agriculture, on the constituents and determinants of sustainable development are set out in Table 8.8. Broadly, three constituents of sustainable development have been envisioned as the critical minimum: social, economic, and environmental factors. Table 8.8 presents the degree and direction of the likely impact of the mitigation options. The exact magnitude of the effect, however, depends on the scale and intensity of the mitigation measures, and the sectors and policy arena in which they are undertaken.

Agriculture contributes 4% of global GDP (World Bank, 2003) and provides employment to 1.3 billion people (Dean, 2000). It is a critical sector of the world economy, but uses more water than any other sector. In low-income countries, agriculture uses 87% of total extracted water, while this figure is 74% in middle-income countries and 30% in high-income countries (World Bank, 2003). There are currently 276 Mha of irrigated croplands (FAOSTAT, 2006), a five-fold increase since the beginning of the 20th century. With irrigation increasing, water management is a serious issue. Through proper institutions and effective functioning of markets, water management can be implemented with favourable outcomes for both environmental and economic goals. There is a greater need for policy coherence and innovative responses creating a situation where users are asked to pay the full economic costs of the water. This has special relevance for developing countries. Removal of subsidies in the electricity and water sectors might lead to effective water use in agriculture, through adaptation of appropriate irrigation technology, such as drip irrigation in place of tube well irrigation.

Agriculture contributes nearly half of the CH_4 and N_2O emissions (Bhatia *et al.*, 2004) and rice, nutrient, water and tillage management can help to mitigate these GHGs. By careful drainage and effective institutional support, irrigation costs for farmers can also be reduced, thereby improving economic aspects of sustainable development (Rao, 1994). An appropriate mix of rice cultivation with livestock, known as integrated annual crop-animal systems and traditionally found in West Africa, India and Indonesia and Vietnam, can enhance net income, improve cultivated agro-ecosystems, and enhance human well-being (Millennium Ecosystem Assessment, 2005). Such combinations of livestock and cropping, especially for rice, can improve income generation, even in semi-arid and arid areas of the world.

Groundwater quality may be enhanced and the loss of biodiversity can be influenced by the choice of fertilizer used and use of more targeted pesticides. Further, greater demand for farmyard manure would create income for the animal husbandry sector where usually the poor are engaged. Various country

Table 8.8: *Potential sustainable development consequences of mitigation options*

Activity category	Sustainable development			Notes
	Social	Economic	Environmental	
Croplands – agronomy	?	+	+	1
Croplands – nutrient management	?	+	+	2
Croplands – tillage/residues	?	?	+	3
Croplands – water management	+	+	+	4
Croplands – rice management	+	+	+	5
Croplands – set-aside & LUC	?	-	+	6
Croplands – agro-forestry	+	?	+	7
Grasslands – grazing, nutrients, fire	+	+	+	8
Organic soils – restoration	?	?	+	9
Degraded soils – restoration	+	+	+	10
Biosolid applications	+	-	+/-	11
Bioenergy	+	?	+/-	12
Livestock – feeding	-/?	+	?	13
Livestock – additives	-/?	n/d	n/d	14
Livestock – breeding	-/?	n/d	n/d	14
Manure management	?	n/d	n/d	15

Notes:
+ denotes beneficial impact on component of SD
- denotes negative impact
? denotes uncertain impact
n/d denotes no data

1 Improved yields would mean better economic returns and less land required for new cropland. Societal impact uncertain - impact could be positive but could negatively affect traditional practices.
2 Improved yields would mean better economic returns and less land required for new cropland. Societal impact uncertain - impact could be positive but could negatively affect traditional practices.
3 Improves soil fertility may not increase yield so societal and economic impacts uncertain.
4 All efficiency improvements are positive for sustainability goals and should yield economic benefits even if costs of irrigation are borne by the farmer.
5 Improved yields would mean better economic returns and less land required for new cropland. Societal impacts likely to benign or positive as no large-scale change to traditional practices.
6 Improve soil fertility but less land available for production; potential negative impact on economic returns.
7 Likely environmental benefits, less travel required for fuelwood; positive societal benefits; economic impact uncertain.
8 Improved production would mean better economic returns and less land required for grazing; lower degradation. Societal effects likely to be positive.
9 Organic soil restoration has a host of biodiversity/environmental co-benefits but opportunity cost of crop production lost from this land; economic impact depends upon whether farmers receive payment for the GHG emission reduction.
10 Restoration of degraded lands will provide higher yields and economic returns, less new cropland and provide societal benefits via production stability.
11 Likely environmental benefits though some negative impacts possible (e.g., water pollution) but, depending on the bio-solid system implemented, could increase costs.
12 Bio-energy crops could yield environmental co-benefits or could lead to loss of bio-diversity (depending on the land use they replace). Economic impact uncertain. Social benefits could arise from diversified income stream.
13 Negative/uncertain societal impacts as these practices may not be acceptable due to prevailing cultural practices especially in developing countries. Could improve production and economic returns.
14 Negative/uncertain societal impacts as these practices may not be acceptable due to prevailing cultural practices especially in developing countries. No data (n/d) on economic or environmental impacts.
15 Uncertain societal impacts. No data (n/d) on economic or environmental impacts.

strategy papers on The Millennium Development Goal (MDG) clearly recommend encouragement to animal husbandry (e.g., World Bank, 2005). This is intended to enhance livelihoods and create greater employment. Better nutrient management can also improve environmental sustainability.

Controlling overgrazing through pasture improvement has a favourable impact on livestock productivity (greater income from the same number of livestock) and slows or halts desertification (environmental aspect). It also provides social security to the poorest people during extreme events such as drought (especially in Sub-Saharan Africa). One effective strategy to control overgrazing is the prohibition of free

grazing, as was done in China (Rao, 1994) but approaches in other regions need to take into account cultural and institutional contexts. Dryland and desert areas have the highest number of poor people (Millennium Ecosystem Assessment, 2005) and measures to halt overgrazing, coupled with improved livelihood options (e.g., fisheries in Syria , Israel and other central Asian countries), can help reduce poverty and achieve sustainability goals.

Land cover and tillage management could encourage favourable impacts on environmental goals. A mix of horticulture with optimal crop rotations would promote carbon sequestration and could also improve agro-ecosystem function.

Societal well-being would also be enhanced by providing water and enhanced productivity. While the environmental benefits of tillage/residue management are clear, other impacts are less certain. Land restoration will have positive environmental impacts, but conversion of floodplains and wetlands to agriculture could hamper ecological function (reduced water recharge, bioremediation, nutrient cycling, etc.) and therefore, could have an adverse impact on sustainable development goals (Kumar, 2001).

The other mitigation measures listed in Table 8.8 are context- and location-specific in their influence on sustainable development constituents. Appropriate adoption of mitigation measures is likely in many cases to help achieve environmental goals, but farmers may incur additional costs, reducing their returns and income. This trade-off would be most visible in the short term, but in the long term, synergy amongst the constituents of sustainable development would emerge through improved natural capital. Trade-offs between economic and environmental aspects of sustainable development might become less important if the environmental gains were better acknowledged, quantified, and incorporated in the decision-making framework.

Large-scale production of modern bioenergy crops, partly for export, could generate income and employment for rural regions of world. Nevertheless, these benefits will not necessarily flow to the rural populations that need them most. The net impacts for a region as a whole, including possible changes and improvements in agricultural production methods should be considered when developing biomass and bioenergy production capacity. Although experience around the globe (e.g., Brazil, India biofuels) shows that major socioeconomic benefits can be achieved, new bioenergy production schemes could benefit from the involvement of the regional stakeholders, particularly the farmers. Experience with such schemes needs to be built around the globe.

8.5 Interactions of mitigation options with adaptation and vulnerability

As discussed in Chapters 3, 11 and 12, mitigation, climate change impacts, and adaptation will occur simultaneously and interactively. Mitigation-driven actions in agriculture could have (a) positive adaptation consequences (e.g., carbon sequestration projects with positive drought preparedness aspects) or (b) negative adaptation consequences (e.g., if heavy dependence on biomass energy increases the sensitivity of energy supply to climatic extremes; see Chapter 12, Subsection 12.1.4). Adaptation-driven actions also may have both (a) positive consequences for mitigation (e.g., residue return to fields to improve water holding capacity will also sequester carbon); and (b) negative consequences for mitigation (e.g., increasing use of nitrogen fertilizer to overcome falling yield leading to

increased nitrous oxide emissions). In many cases, actions taken for reasons unrelated to either mitigation or adaptation (see Sections 8.6 and 8.7) may have considerable consequences for either or both(e.g., deforestation for agriculture or other purposes results in carbon loss as well as loss of ecosystems and resilience of local populations). Adaptation to climate change in the agricultural sector is detailed in (IPCC, 2007; Chapter 5).

For mitigation, variables such as growth rates for bioenergy feedstocks, the size of livestock herds, and rates of carbon sequestration in agricultural lands are affected by climate change (Paustian et al., 2004). The extent depends on the sign and magnitude of changes in temperature, soil moisture, and atmospheric CO_2 concentration, which vary regionally (Christensen et al., 2007). All of these factors will alter the mitigation potential; some positively and some negatively. For example: (a) lower growth rates in bioenergy feedstocks will lead to larger emissions from hauling and increased cost; (b) lower livestock growth rates would possibly increase herd size and consequent emissions from manure and enteric fermentation; and (c) increased microbial decomposition under higher temperatures will lower soil carbon sequestration potential. Interactions also occur with adaptation. Butt et al. (2006) and Reilly et al. (2001) found that modified crop mix, land use, and irrigation are all potential adaptations to warmer climates. All would alter the mitigation potential. Some of the key vulnerabilities of agricultural mitigation strategies to climate change, and the implications of adaptation on GHG emissions from agriculture are summarized in Table 8.9.

8.6 Effectiveness of, and experience with, climate policies; potentials, barriers and opportunities/implementation issues

8.6.1 Impact of climate policies

Many recent studies have shown that actual levels of GHG mitigation are far below the technical potential for these measures. The gap between technical potential and realized GHG mitigation occurs due to costs and other barriers to implementation (Smith, 2004b).

Globally and for Europe, Cannell (2003) suggested that, for carbon sequestration and bioenergy-derived fossil fuel offsets, the realistically achievable potential (potential estimated to take account of all barriers) was ~20% of the technical potential. Similar figures were derived by Freibauer et al. (2004) and the European Climate Change Programme (2001) for agricultural carbon sequestration in Europe. Smith et al. (2005a) showed recently that carbon sequestration in Europe is likely to be negligible by the first Commitment Period of the Kyoto Protocol (2008-2012), despite the significant technical potential (e.g.,

Table 8.9: *Some of the key vulnerabilities of agricultural mitigation strategies to climate change and the implications of adaptation on GHG emissions from agriculture*

Agricultural mitigation strategies	Vulnerability of the mitigation option to climate change	Implication for GHG emissions of adaptation actions
Cropland management – agronomy	Vulnerable to decreased rainfall, and in cases near the limit of their climate niche, to higher temperatures.	NO_2 emissions would increase if fertilizer use increased, or if more legumes were planted in response to climate-induced production declines.
Cropland management – nutrient management	Only weakly sensitive to climate change, except in cases where the entire cropping enterprise becomes unviable.	No significant adaptation to effects of climate change possible beyond tailoring of practices to ambient conditions. Therefore, additional GHGs not expected.
Cropland management – tillage/residue management	Sensitive to climate change. Higher temperatures could lower soil carbon sequestration potential. Warmer, wetter climates can increase risk of crop pests and diseases associated with reduced till practices.	Adaptation not anticipated to have a significant GHG effect.
Cropland management – water management	Irrigation is susceptible to climate changes that reduce the availability of water for irrigation or increases crop water demand.	Possible increase in energy-related GHG emissions if greater pumping distances or volumes are required. Adoption of more water-use efficient practices will generally lower GHG emissions.
Cropland management – rice management	Vulnerable to climate-change-induced changes in water availability. Low CH_4 emitting cultivars may be susceptible to changes in temperature beyond their tolerance limits.	Adaptation strategies are limited and not expected to have large GHG consequences.
Cropland management – set-aside and land-use change	Set-asides may be needed to offset loss of productivity on other lands.	Adaptation is either to try to keep production high on non-set-aside land, which could increase GHG emissions, or return some set-asides to production. Increases in GHGs are in both cases fairly small, and less than the case of not having set-asides in the first place. They could be further mitigated by applying low GHG emitting practices in all cases.
Cropland management – agro-forestry	Large changes in climate could make certain forms of agro-forestry unviable in particular situations.	Adaptation of practices and species used to less favourable climates could lead to some loss of CO_2 uptake potential.
Grazing land management/pasture improvement	Fire management can be impacted negatively or positively by climate change depending on ecosystem and sign of climate change. Extreme drying or warming could make marginal grazing lands unviable. Wetter conditions will promote conversion of grazing lands to crops.	Increased fire protection activities can increase GHGs emissions by a small amount, thus reducing the net benefit obtained from reducing fire extent and frequency.
Management of agricultural organic soils	The mitigation measure is sensitive to increases in temperature or decreases in moisture, both of which would decrease the carbon sequestration potential.	Some trade-offs between CO_2 uptake and CH_4 emissions can be expected if the soils become wetter as a result of the adaptation management.
Restoration of degraded lands	The sustainability of restored lands could be vulnerable to increased temperature and/or decreased soil moisture.	Energy used to replant, or fertilizer used to increase establishment, success could lead to small additional GHG emissions.
Livestock – improved feeding practices, specific agents and dietary additives, longer term structural and management changes and animal breeding	Weakly vulnerable to climate change except if it leads to the loss of viability of livestock enterprises in marginal areas or increased cost (or decreased availability) of feed inputs.	No general adaptation strategies. Specific strategies may have minor impacts on GHG emissions, for example, transport of feed supplements from distant locations could lead to increased net GHG emissions.
Manure/biosolid management	Controlled waste digestion generally positively affected by moderately rising temperatures. Where GHGs are not trapped, higher temperatures could hamper management.	If used as a nutrient source on pasture can increase CO_2 uptake and carbon storage.
Bioenergy – energy crops, solid, liquid, biogas, residues	Particular bioenergy crops potentially sensitive to climate change, either positively or negatively. Areas devoted to bioenergy could be under increasing competition with the needs for food agriculture or biodiversity conservation under changing climate.	Generally, results in net CO_2 uptake on land (apart from the fossil-fuel substitution). N_2O emissions would increase if N turnover rates were greater than under previous land uses. Possible positive and negative impacts on net GHG emissions at various stages of the energy chain (cultivation, harvesting, transport, conversion) must be managed.

Smith *et al.*, 2000; Freibauer *et al.*, 2004; Smith, 2004a). The estimates of global economic mitigation potential in 2030 at different costs reported in Smith *et al.* (2007a) were 28, 45 and 73% of technical potential at up to 20, 50 and 100 US$/tCO$_2$-eq, respectively.

In Europe, there is little evidence that climate policy is affecting GHG emissions from agriculture (see Smith *et al.*, 2005a), with most emission reduction occurring through non-climate policy (see Section 8.7; Freibauer *et al.*, 2004). Some countries have agricultural policies designed to reduce GHG emissions (e.g., Belgium), but most do not (Smith *et al.*, 2005a). The European Climate Change Programme (2001) recommended improvement of fertilizer application, set-aside, and reduction of livestock methane emissions (mainly through biogas production) as being the most cost-effective GHG mitigation options for European agriculture.

In North America, the US Global Climate Change Initiative aims to reduce GHG intensity by 18% by 2012. Agricultural sector activities include manure management, reduced tillage, grass plantings, and afforestation of agricultural land. In Canada, agriculture contributes about 10% to national emissions, so mitigation (removals and emission reductions) is considered to be an important contribution to reducing emissions (and at the same time to reduce risk to air, water and soil quality). Various programmes (e.g., AAFC GHG Mitigation programme) encourage voluntary adoption of mitigation practices on farms.

In Oceania, vegetation management policies in Australia have assisted in progressively restricting emissions from land-use change (mainly land clearing for agriculture) to about 60% of 1990 levels. Complementary policies that aim to foster establishment of both commercial and non-commercial forestry and agro-forestry are resulting in significant afforestation of agricultural land in both Australia and New Zealand. Research is being supported to develop cost-effective GHG abatement technologies for livestock (including dietary manipulation and other methods of reducing enteric methane emissions, as well as manure management), agricultural soils (including nutrient and soil management strategies), savannas, and planted forests. The Greenhouse Challenge Plus programme and other partnership initiatives between the Government and industry are facilitating the integration of GHG abatement measures into agricultural management systems.

In Latin America and the Caribbean, climate change mitigation is still not considered in mainstream policy. Most countries have devoted efforts to capacity building for complying with obligations under the UNFCCC, and a few have prepared National Strategy Studies for Kyoto Protocol's Clean Development Mechanism (CDM). Carbon sequestration in agricultural soils has the highest mitigation potential in the region, and its exclusion from the CDM has hindered wider adoption of pertinent practices (e.g., zero tillage).

In Asia, China has policies that reduce GHG emissions, but these were implemented for reasons other than climate policy. These are discussed further in Section 8.7. Currently, there are no policies specifically aimed at reducing GHG emissions. Japan has a number of policies such as Biomass Nippon Strategy, which promotes the utilization of biomass as an alternative energy source, and Environment-Conserving Agriculture, which promotes energy-efficient agricultural machinery, reduction in use of fertilizer, and appropriate management of livestock waste, etc.

In Africa, the impacts of climate policy on agricultural emissions are small. There are no approved CDM projects in Africa related to the reduction of agricultural GHG emissions *per se*. Several projects are under investigation in relation to the restoration of agriculturally-degraded lands, carbon sequestration potential of agro-forestry, and reduction in sugarcane burning. Many countries in Africa have prepared National Strategy Studies for the CDM in complying with obligations under UNFCCC. The main obstacles to implementation of CDM projects in Africa, however, are lack of financial resources, qualified personnel, and the complexity of the CDM.

Agricultural GHG offsets can be encouraged by market-based trading schemes. Offset trading, or trading of credits, allows farmers to obtain credits for reducing their GHG emission reductions. The primary agricultural project types include CH$_4$ capture and destruction, and soil carbon sequestration. Although not included in current projects, measures to reduce N$_2$O emissions could be included in the future. The vast majority of agricultural projects have focused on CH$_4$ reduction from livestock wastes in North America (Canada, Mexico and the United States), South America (Brazil), China, and Eastern Europe. Most of these projects have resulted in the production of Certified Emission Reductions (CERs) from the CDM. Credits are bought and sold through the use of offset aggregators, brokers, and traders. Although the CDM does not currently support soil carbon sequestration projects, emerging markets in Canada and the United States are supporting offset trading from soil carbon sequestration. In Canada, farm groups such as the Saskatchewan Soil Conservation Association (SSCA) encourage farmers to adopt no-till practices in return for carbon offset credits. In the USA, the Pacific Northwest Direct Seed Association offers soil carbon credits generated from no-till management to an energy company The Chicago Climate Exchange (CCX) (www.chicagoclimatex.com/) allows GHG offsets from no-tillage and conversion of cropland to grasslands to be traded by voluntary action through a market trading mechanism. These approaches to agriculturally derived GHG offset will likely expand geographically and in scope. Policy instruments are detailed in Chapter 13 (Section 13.2).

8.6.2 Barriers and opportunities/implementation issues

The commonly mentioned barriers to adoption of carbon sequestration activities on agricultural lands include the following:

Maximum Storage: Carbon sequestration in soils or terrestrial biomass has a maximum capacity for the ecosystem, which may be reached after 15 to 60 years, depending on management practice, management history, and the system (West and Post, 2002). However, sequestration is a rapidly and cheaply deployable mitigation option, until more capital-intensive developments, and longer-lasting actions become available (Caldeira *et al.*, 2004; Sands and McCarl, 2005).

Reversibility: A subsequent change in management can reverse the gains in carbon sequestration over a similar period of time. Not all agricultural mitigation options are reversible; reduction in N_2O and CH_4 emissions, avoided emissions as a result of agricultural energy efficiency gains or substitution of fossil fuels by bio-energy are non-reversible.

Baseline: The GHG net emission reductions need to be assessed relative to a baseline. Selection of an appropriate baseline to measure management-induced soil carbon changes is still an obstacle in some mitigation projects. The extent of practices already in place in project regions will need to be determined for the baseline.

Uncertainty: This has two components: mechanism uncertainty and measurement uncertainty. Uncertainty about the complex biological and ecological processes involved in GHG emissions and carbon storage in agricultural systems makes investors more wary of these options than of more clear-cut industrial mitigation activities. This barrier can be reduced by investment in research. Secondly, agricultural systems exhibit substantial variability between seasons and locations, creating high variability in offset quantities at the farm level. This variability can be reduced by increasing the geographical extent and duration of the accounting unit (e.g., multi-region, multi-year contracts; Kim and McCarl, 2005).

Displacement of Emissions: Adopting certain agricultural mitigation practices may reduce production within implementing regions, which, in turn, may be offset by increased production outside the project region unconstrained by GHG mitigation objectives, reducing the net emission reductions. 'Wall-to-wall' accounting can detect this, and crediting correction factors may need to be employed (Murray *et al.*, 2004; US-EPA, 2005).

Transaction costs: Under an incentive-based system such as a carbon market, the amount of money farmers receive is not the market price, but the market price less brokerage cost. This may be substantial, and is an increasing fraction as the amount of carbon involved diminishes, creating a serious entry barrier for

smallholders. For example, a 50 kt contract needs 25 kha under soil carbon management (uptake ~ 2 tCO_2 ha/yr). In developing countries, this could involve many thousands of farmers.

Measurement and monitoring costs: Mooney *et al.* (2004) argue that such costs are likely to be small (under 2% of the contract), but other studies disagree (Smith, 2004c). In general, measurement costs per carbon-credit sold decrease as the quantity of carbon sequestered and area sampled increase. Methodological advances in measuring soil carbon may reduce costs and increase the sensitivity of change detection. However, improved methods to account for changes in soil bulk density remain a hindrance to quantification of changes in soil carbon stocks (Izaurralde and Rice, 2006). Development of remote sensing, new spectral techniques to measure soil carbon, and modelling offer opportunities to reduce costs but will require evaluation (Izaurralde and Rice, 2006, Brown *et al.*, 2006; Ogle and Paustian, 2005; Gehl and Rice, 2007).

Property rights: Property rights, landholdings, and the lack of a clear single-party land ownership in certain areas may inhibit implementation of management changes.

Other barriers: Other possible barriers to implementation include the availability of capital, the rate of capital stock turnover, the rate of technological development, risk attitudes, need for research and outreach, consistency with traditional practices, pressure for competing uses of agricultural land and water, demand for agricultural products, high costs for certain enabling technologies (e.g., soil tests before fertilization), and ease of compliance (e.g., straw burning is quicker than residue removal and can also control some weeds and diseases, so farmers favour straw burning).

8.7 Integrated and non-climate policies affecting emissions of GHGs

Many policies other than climate policies affect GHG emissions from agriculture. These include other UN conventions such as Biodiversity, Desertification and actions on Sustainable Development (see Section 8.4.5), macroeconomic policy such as EU Common Agricultural Policy (CAP)/CAP reform, international free trade agreements, trading blocks, trade barriers, region-specific programmes, energy policy and price adjustment, and other environmental policies including various environmental/agro-environmental schemes. These are described further below.

8.7.1 Other UN conventions

In Asia, China has introduced laws to convert croplands to forest and grassland in Vulnerable Ecological Zones under the UN Convention on Desertification. This will increase carbon storage and reduce N_2O emissions. Under the UN Convention on Biodiversity, China has initiated a programme that restores croplands close to lakes, the sea, or other natural lands as conservation zones for wildlife. This may increase soil carbon sequestration but, if restored to wetland, could increase CH_4 emissions. In support of UN Sustainable Development guidelines, China has introduced a Land Reclamation Regulation (1988) in which land degraded by, for example, construction or mining is restored for use in agriculture, thereby increasing soil carbon storage. In Europe (including Eastern Europe, the Caucasus and Central Asia) and North America, the UN conventions have had few significant impacts on agricultural GHG emissions. In Europe, the UN Convention on Long Range Trans-boundary Air Pollutants also leads to regulations to control air pollutants (e.g., by regulating N emissions) that could have substantial impacts on emission reductions in the agricultural sector.

8.7.2 Macroeconomic and sectoral policy

Some macro-economic changes, such as the burden of a high external debt in Latin America, triggered the adoption in the 1970s of policies designed for improving the trade balance, mainly by promoting agricultural exports (Tejo, 2004). This resulted in the changes in land use and management (see Section 8.3.3), which are still causing increases in annual GHG emissions today. In other regions, such as the countries of Eastern Europe, the Caucasus and Central Asia and many Central and East European countries, political changes since 1990 have meant agricultural de-intensification with less inputs, and land abandonment, leading to a decrease in agricultural GHG emissions. In Africa, the cultivated area in Southern Africa has increased by 30% since 1960, while agricultural production has doubled (Scholes and Biggs, 2004). The macroeconomic development framework for Africa (NEPAD, 2005) emphasises agriculture-led development. It is, therefore, anticipated that the cropped area will continue to increase, especially in Central, East, and Southern Africa, perhaps at an accelerating rate. In Western Europe, North America, Asia (China) and Oceania, macroeconomic policy has tended to reduce GHG emissions. The declining emission trend in Western Europe is likely a consequence of successive reforms of the Common Agricultural Policy (CAP) since 1992. The 2003 EU CAP reform is expected to lead to further reductions, mainly through reduction of animal numbers (Binfield *et al.*, 2006). The reduced GHG emissions could be offset by activity elsewhere. Various macro-economic policies that potentially affect agricultural GHG emissions in each major world region are presented in Table 8.10.

WTO negotiations, to the extent they move toward free trade, would permit countries to better adjust to climate change and the dislocations in production caused by mitigation activities, by adjusting their import/export mix. International trade agreements such as WTO may also have impacts on the amount and geographical distribution of GHG emissions. If agricultural subsidies are reduced and markets become more open, a shift in production from developed to developing countries would be expected, with the consequent displacement of GHG emissions

to the latter. Since agricultural practices and GHG emissions per unit product differ between countries, such displacement may also cause changes in total emissions from agriculture. In addition, the increase in international flow of agricultural products which may result from trade liberalization could cause higher GHG emissions from the use of transport fuels.

8.7.3 Other environmental policies

In most world regions, environmental policies have been put in place to improve fertility, to reduce erosion and soil loss, and to improve agricultural efficiency. The majority of these environmental policies also reduce GHG emissions. Various environmental policies not implemented specifically to address GHG emissions but potentially affect agricultural GHG emissions in each major world region are presented in Table 8.11.

In all regions, policies to improve other aspects of the environment have been more effective in reducing GHG emissions from agriculture than policies aimed specifically at reducing agricultural GHG emissions (see Section 8.6.1). The importance of identifying these co-benefits when formulating climate and other environmental policy is addressed in Section 8.8.

8.8 Co-benefits and trade-offs of mitigation options

Many of the measures aimed at reducing GHG emissions have other impacts on the productivity and environmental integrity of agricultural ecosystems, mostly positive (Table 8.12). These measures are often adopted mainly for reasons other than GHG mitigation (see Section 8.7.3). Agro-ecosystems are inherently complex and very few practices yield purely win-win outcomes; most involve some trade-offs (DeFries *et al.*, 2004; Viner *et al.*, 2006) above certain levels or intensities of implementation. Specific examples of co-benefits and trade-offs among agricultural GHG mitigation measures include:

- Practices that maintain or increase crop productivity can improve global or regional food security (Lal, 2004a, b; Follett *et al.*, 2005). This co-benefit may become more important as global food demands increase in coming decades (Sanchez and Swaminathan, 2005; Rosegrant and Cline, 2003; FAO, 2003; Millennium Ecosystem Assessment, 2005). Building reserves of soil carbon often also increases the potential productivity of these soils. Furthermore, many of the measures that promote carbon sequestration also prevent degradation by avoiding erosion and improving soil structure. Consequently, many carbon conserving practices sustain or enhance future fertility, productivity and resilience of soil resources (Lal, 2004a; Cerri *et al.*, 2004; Freibauer *et al.*, 2004; Paustian *et al.*, 2004; Kurkalova

Table 8.10: *Summary of various macro-economic policies that potentially affect agricultural GHG emissions, listing policies for each major world region and the potential impact on the emissions of each GHG*

Region	Macro-economic policies potentially affecting agricultural GHG emissions	Impact on CO_2 emissions	Impact on N_2O emissions	Impact on CH_4 emissions
North America	• Energy conservation and energy security policies – promote bio-energy – increase fossil fuel offsets and possibly SOC (USA)	+		
	• Energy price adjustments - encourage agricultural mitigation - more reduced tillage – increase SOC (USA)	+	?	
	• Removal of the Grain Transportation Subsidy shifted production from annual to perennial crops and livestock (Canada)	+	+	+
Latin America	• Policies since 1970s to promote agricultural products exports (Tejo, 2004) resulting in land management change –increasing GHG emissions (Latin America)	-	-	-
	• Promotion of biofuels (e.g., PROALCOOL (Brazil)	+		
	• Brazil and Argentina implemented policies to make compulsory 5% biodiesel in all diesel fuels consumed (Brazil & Argentina)	+		
Europe, the Caucasus and Central Asia	• Common Agricultural Policy (CAP) 2003 - Single Farm Payment decoupled from production - replaces most of the previous area-based payments. Income support conditional to statutory environmental management requirements (e.g., legislation on nitrates) and the obligation to maintain land under permanent pasture (cross-compliance).	+	+	
	• Political changes in Eastern Europe - closure of many intensive pig units – reduced GHG emissions (EU and wider Europe)	+	+	+
	• Macro-economic changes in the countries of Eastern Europe, the Caucasus and Central Asia:			
	a) Abandonment of croplands since 1990 (1.5 Mha); grasslands and regenerating forests sequestering carbon in soils and woody biomass (all countries of Eastern Europe, the Caucasus and Central Asia:)	+	+	
	b) Use of agricultural machinery declined and fossil fuel use per ha of cropland (Romanenkov *et al.*, 2004) - decreased CO_2 (fossil fuel) increased CO_2 (straw burning – all countries of Eastern Europe, the Caucasus and Central Asia	+		
	c) Fertilizer consumption has dropped; 1999 N_2O emissions from agriculture 19.5% of 1990 level (Russia & Belarus).	+	+	
	d) CO_2 emissions from liming have dropped to 8% of the 1990 levels (Russia)	+		
	e) Livestock CH_4 emissions in 1999 were less than 48% of the 1990 level (Russia)	+		+
	f) The use of bare fallowing has declined (88% of the area in bare fallow in 1999 compared to 1990; Agriculture of Russia, 2004) (Russia)	+		
	g) Changes in rotational structure (more perennial grasses) (Russia)			
Africa	• The cultivated area in Southern Africa has increased 30% since 1960, while agricultural production has doubled - agriculture-led development (Scholes and Biggs 2004; NEPAD 2005).Cropped area will continue to increase, especially in Central, East and Southern Africa, perhaps at an accelerating rate.	-	-	-
Asia	• In some areas, croplands are currently in set-aside for economic reasons (China)	+	+	
Oceania	• Australia and New Zealand continue to provide little direct subsidy to agriculture - highly efficient industries that minimize unnecessary inputs and reduce waste - potential for high losses (such as N_2O) is reduced. Continuing tightening of terms of trade for farm enterprises, as well as ongoing relaxation of requirements for agricultural imports, is likely to maintain this focus (Australia and New Zealand)		+	
	• The establishment of comprehensive water markets are expected, over time, to result in reductions in the size of industries such as rice and irrigated dairy with consequent reductions in the emissions from these sectors (Australia)		+	+

Note: + denotes a positive effect (benefit); - denotes a negative effect

Table 8.11: *A non-exhaustive summary of environmental policies that were not implemented specifically to address GHG emissions, but that can affect agricultural GHG emissions. Examples of policies are listed for major world region and the potential impact on the emissions of each GHG is indicated.*

Region	Other environmental policies potentially affecting agricultural GHG emissions	Impact on CO_2 emissions	Impact on N_2O emissions	Impact on CH_4 emissions
North America	• Environmental Quality Incentives Program (EQIP) – cost-sharing and incentive payments for conservation practices on working farms (USA)	+	+	
	• Conservation Reserve Program (CRP) - environmentally sensitive land converted to native grasses, trees, etc. (USA)	+	+	
	• Conservation Security Program (CSP) – assistance promoting conservation on cropland, pasture and range land (and farm woodland) (USA)	+	+	
	• Green cover in Canada and provincial initiatives – encourages shift from annual to perennial crop production on poor quality soils (Canada)	+	+	+
	• Agriculture Policy Framework (APF) programmes to reduce agriculture risks to the environment, including GHG emissions (Canada)	+	+	+
	• Nutrient Management programmes – introduced to improve water quality, may indirectly reduce N_2O emissions (Canada)			
Latin America	• Increasing adoption of environmental policies driven by globalization, consolidation of democratic regimes (Latin America & Caribbean)	+/-	+/-	+/-
	• 14 countries have introduced environmental regulations over the last 20 years – most have implemented measures to protect the environment	+	?	
	• Promotion of no-till agriculture in the Mercosur area (Brazil, Argentina, Uruguay and Paraguay)	+	?	
	• "Program Crop-Livestock Integration" promotes soil carbon, reduced erosion, reduced pathogens, fertility for pastures, no till cropping (Brazil)	+		
Europe, the Caucasus and Central Asia	• EU set aside programme - encouraged carbon sequestering practices, but now replaced by the single farm payment under the new CAP (EU)	+		
	• EU/number of member states - soil action plans to promote soil quality/health/ sustainability, encourages soil carbon sequestration (EU)	+		
	• Encouragement of composting in some EU member states (e.g., Belgium; Sleutel 2005) (EU)	+	+	+
	• EU Water Framework Directive (WFD) promotes careful use of N fertilizer. Impact of WFD on agricultural GHG emissions as yet unclear (EU)	?	+	
	• The ban of burning of field residues in the 1980s (for air quality purposes) enhance soil carbon, reduce N_2O and CH_4 (Smith et al., 1997; 2000) (EU)	+		
	• The dumping ban at sea of sewage sludge in Europe in 1998 - more sludge reached agricultural land (Smith et al., 2000; 2001) (EU)	+		
	• "Vandmiljoplaner" (water environmental plans) for the agricultural sector with clear effect (decrease) of GHGs (Denmark)	+		
	• Land Codes of the Russian Federation, Belarus and the Ukraine - land conservation for promoting soil quality restoration and protection	+	+	
	• "Land Reform Development in Russian Federation" & "Fertility 2006–2010" – plans to promote soil conservation/fertility/sustainability (Russia)	+		
	• Ukrainian law "Land protection" - action plans to promote soil conservation/increase commercial yields/fertility/sustainability (Ukraine)	+	?	+
	• Laws in Belarus such as "State Control of Land Use and Land Protection" encourages carbon sequestration (Belarus)	+	+	
	• Laws in the Ukraine to promote the conversion of degraded lands to set-aside (Ukraine)			
	• Water quality initiatives, for example, Water Codes encourage reforestation and grassland riparian zones (Russia, Ukraine and Belarus)			
	• The ban of fertilizer application in some areas - reduce N_2O emissions (Russia, Belarus, Ukraine) & regional programmes for example, Revival of the Volga			
Africa	• The reduction of the area of rangelands burned - objective of both colonial and post-colonial administrations; renewed efforts (South Africa, 1998)	+	+	+
Asia	• Soil sustainability programmes - N fertilizer added to soils only after soil N testing (China)		+	
	• Regional agricultural development programmes - enhance soil carbon storage (China)	+		
	• Water quality programmes that control non-point source pollution (China)			
	• Air quality legislation - bans straw burning, thus reducing CO_2 (and CH_4 and N_2O) emissions (China)	+	+	+
	• "Township Enterprises" & "Ecological Municipality" - reduce waste disposal, chemical fertilizer and pesticides, and bans straw burning (China)	+	+	+
Oceania	• Wide range of policies to maintain function/conservation of agricultural landscapes, river systems and other ecosystems (Australia and New Zealand)	+	-	
	• Industry changes leading to rapid increase in N fertilizer use over the past decade (250% and 500% increases in Australia and New Zealand, respectively)	+	+	-
	• Increases in intensive livestock production; raised concerns about water quality and the health of riverine/offshore ecosystems (Australia and New Zealand)	+	+	+
	• Policy responses are being developed that include monitoring, regulatory, research and extension components (Australia and New Zealand)			

Note: + denotes a positive effect (benefit); - denotes a negative effect

Table 8.12: *Summary of possible co-benefits and trade-offs of mitigation options in agriculture.*

Measure	Examples	Food security (productivity)	Water quality	Water conservation	Soil quality	Air quality	Bio-diversity, wildlife habitat	Energy conservation	Conservation of other biomes	Aesthetic/ amenity value
Cropland management	Agronomy	+	+/-	+/-	+	+/-	+/-	-	+	+/-
	Nutrient management	-/+	+	+	+	+	+	+	+	
	Tillage/residue management	+	+/-	+	+		+	+	+	
	Water management (irrigation, drainage)	+	+/-	+/-	+/-					
	Rice management	+	+	+/-	+/-	+/-		+	+	
	Agro-forestry	+/-	+/-	-	+	+	+	+	-	+
	Set-aside, land-use change	-	+	+	+	+	+	+	+	+
Grazing land management/ pasture improvement	Grazing intensity	+/-	+/-		+		+			+
	Increased productivity (e.g., fertilization)	+		+	+		+	-	+	+/-
	Nutrient management	+	+/-	+	+	+	+/-		+	+/-
	Fire management	+	+			+				
	Species introduction (including legumes)	+			+		+	+		+
Management of organic soils	Avoid drainage of/restore wetlands	-			+		+	+	-	+
Restoration of degraded lands	Erosion control, organic amendments, nutrient amendments	+	+	+	+		+	+	+	+
Livestock management	Improved feeding practices	+			+/-			+	+	
	Specific agents and dietary additives	+								
	Longer term structural and management changes and animal breeding	+								
Manure/biosolid management	Improved storage and handling	+	+/-		+	+/-		+	-	+
	Anaerobic digestion					+		+	+	
	More efficient use as nutrient source	+	+		+	+	-	+	+	+/-
Bioenergy	Energy crops, solid, liquid, biogas, residues	-					-	+	-	
	References (see footnotes)	a	b	c	d	e	f	g	h	i

Note:+ denotes a positive effect (benefit); – denotes a negative effect (trade-off). The co-benefits and trade-offs vary among regions. Economic costs and benefits, often key driving variables, are considered in Section 8.4.3

Sources:

a *Foley et al., 2005; Lal, 2001a, 2004a;*

b *Mosier, 2002; Freibauer et al., 2004; Paustian et al., 2004; Cerri et al., 2004*

c *Lal, 2002, 2004b; Dias de Oliveira et al., 2005; Rockström, 2003.*

d *Lal, 2001b, Janzen, 2005; Cassman et al., 2003; Cerri et al., 2004; Wander and Nissen, 2004*

e *Mosier, 2001; 2002; Paustian et al., 2004*

f *Foley et al., 2005; Dias de Oliveira et al. 2005; Freibauer et al., 2004; Falloon et al., 2004; Huston and Marland, 2003; Totten et al., 2003*

g *Lal et al., 2003; West and Marland, 2003*

h *Balmford et al., 2005; Trewavas, 2002; Green et al., 2005; West and Marland, 2003*

i *Freibauer et al., 2004*

et al., 2004; Díaz-Zorita *et al.*, 2002). In some instances, where productivity is enhanced through increased inputs, there may be risks of soil depletion through mechanisms such as acidification or salinization (Barak *et al.*, 1997; Díez *et al.*, 2004; Connor, 2004).

- A key potential trade-off is between the production of bio-energy crops and food security. To the extent that bio-energy production uses crop residues, excess agricultural products or surplus land and water, there will be little resultant loss of food production. But above this point, proportional losses of food production will be strongly negative. Food insecurity is determined more by inequity of access to food (at all scales) than by absolute food production insufficiencies, so the impact of this trade-off depends among other things on the economic distributional effects of bio-energy production.

- Fresh water is a dwindling resource in many parts of the world (Rosegrant and Cline, 2003; Rockström, 2003). Agricultural practices for mitigation of GHGs can have both negative and positive effects on water conservation, and on water quality. Where measures promote water use efficiency (e.g., reduced tillage), they provide potential benefits. But in some cases, the practices could intensify water use, thereby reducing stream flow or groundwater reserves (Unkovich, 2003; Dias de Oliveira *et al.*, 2005). For instance, high-productivity, evergreen, deep-rooted bio-energy plantations generally have a higher water use than the land cover they replace (Berndes, 2002, Jackson *et al.*, 2005). Some practices may affect water quality through enhanced leaching of pesticides and nutrients (Freibauer *et al.*, 2004; Machado and Silva, 2001).

- If bio-energy plantations are appropriately located, designed, and managed, they may reduce nutrient leaching and soil erosion and generate additional environmental services such as soil carbon accumulation, improved soil fertility; removal of cadmium and other heavy metals from soils or wastes. They may also increase nutrient recirculation, aid in the treatment of nutrient-rich wastewater and sludge; and provide habitats for biodiversity in the agricultural landscape (Berndes and Börjesson, 2002; Berndes *et al.* 2004; Börjesson and Berndes, 2006).

- Changes to land use and agricultural management can affect biodiversity, both positively and negatively (e.g., Xiang *et al.*, 2006; Feng *et al.*, 2006). For example, intensification of agriculture and large-scale production of biomass energy crops will lead to loss of biodiversity where they occur in biodiversity-rich landscapes (European Environment Agency, 2006). But perennial crops often used for energy production can favour biodiversity, if they displace annual crops or degraded areas (Berndes and Börjesson, 2002).

- Agricultural mitigation practices may influence non-agricultural ecosystems. For example, practices that diminish productivity in existing cropland (e.g., set-aside lands) or divert products to alternate uses (e.g., bio-energy crops) may induce conversion of forests to cropland elsewhere.

Conversely, increasing productivity on existing croplands may 'spare' some forest or grasslands (West and Marland, 2003; Balmford *et al.*, 2005; Mooney *et al.*, 2005). The net effect of such trade-offs on biodiversity and other ecosystem services has not yet been fully quantified (Huston and Marland, 2003; Green *et al.*, 2005).

- Agro-ecosystems have become increasingly dependent on input of reactive nitrogen, much of it added as manufactured fertilizers (Galloway *et al.*, 2003; Galloway, 2004). Practices that reduce N_2O emission often improve the efficiency of N use from these and other sources (e.g., manures), thereby also reducing GHG emissions from fertilizer manufacture and avoiding deleterious effects on water and air quality from N pollutants (Oenema *et al.*, 2005; Dalal *et al.*, 2003; Olesen *et al.*, 2006; Paustian *et al.*, 2004). Suppressing losses of N as N_2O might in some cases increase the risk of losing that N via leaching. Curtailing supplemental N use without a corresponding increase in N-use efficiency will restrict yields, thereby hampering food security.

- Implementation of agricultural GHG mitigation measures may allow expanded use of fossil fuels, and may have some negative effects through emissions of sulphur, mercury and other pollutants (Elbakidze and McCarl, 2007).

The co-benefits and trade-offs of a practice may vary from place to place because of differences in climate, soil, or the way the practice is adopted. In producing bio-energy, for example, if the feedstock is crop residue, that may reduce soil quality by depleting soil organic matter. Conversely, if the feedstock is a densely rooted perennial crop that may replenish organic matter and thereby improve soil quality (Paustian *et al.*, 2004).These few examples, and the general trends described in Table 8.12, demonstrate that GHG mitigation practices on farm lands exert complex, interactive effects on the environment, sometimes far from the site at which they are imposed. The merits of a given practice, therefore, cannot be judged solely on effectiveness of GHG mitigation.

8.9 Technology research, development, deployment, diffusion and transfer

There is much scope for technological developments to reduce GHG emissions in the agricultural sector. For example, increases in crop yields and animal productivity will reduce emissions per unit of production. Such increases in crop and animal productivity will be implemented through improved management and husbandry techniques, such as better management, genetically modified crops, improved cultivars, fertilizer recommendation systems, precision agriculture, improved animal breeds, improved animal nutrition, dietary additives and growth promoters, improved animal fertility, bio-energy crops, anaerobic slurry digestion and methane capture systems. All of these depend to some extent on technological developments. Although technological improvement may have very significant effects, transfer of these technologies is a key requirement for these mitigations to be realized. For example, the efficiency of N use has improved over the last two decades in developed countries, but continues to decline in many developing countries due to barriers to technology transfer (International Fertilizer Industry Association, 2007). Based on technology change scenarios developed by Ewert *et al.* (2005), and derived from extrapolation of current trends in FAO data, Smith *et al.* (2005b) showed that technological improvements could potentially counteract the negative impacts of climate change on cropland and grassland soil carbon stocks in Europe. This and other work (Rounsevell *et al.*, 2006) suggest that technological improvement will be a key factor in GHG mitigation in the future.

In most instances, the cost of employing mitigation strategies will not alter radically in the medium term. There will be some shifts in costs due to changes in prices of agricultural products and inputs, but these are unlikely to be of significant magnitude. Likewise, the potential of most options for CO_2 reduction is unlikely to change greatly. There are some exceptions which fall into two categories: (i) options where the practice or technology is not new, but where the emission reduction potential has not been adequately quantified, such as improved nutrient utilization; and (ii) options where technologies are still being refined such as probiotics in animal diets, or nitrification inhibitors.

Many of the mitigation strategies outlined for agriculture employ existing technology (e.g., crop management, livestock management). With such strategies, the main issue is technology transfer, diffusion, and deployment. Other strategies involve new use of existing technologies. For example, oils have been used in animal diets for many years to increase dietary energy content, but their role as a methane suppressant is relatively new, and the parameters of the technology in terms of scope for methane reduction are only now being defined. Other strategies still require further research to allow viable systems to operate (e.g., bio-energy crops). Finally, many novel mitigation strategies are presently being refined, such as the use of probiotics, novel plant extracts, and the development of vaccines. Thus, there is still a major role for research and development in this area.

Differences between regions can arise due to the state of development of the agricultural industry, the resources available and legislation. For example, the scope to use specific agents and dietary additives in ruminants is much greater in developed than in the developing regions because of cost, opportunity (e.g., it is easier to administer products to animals in confined systems than in free ranging or nomadic systems), and availability of the technology (US-EPA, 2006a). Furthermore, certain technologies are not allowed in some regions, for example, ionophores are banned from use in animal feeding in the EU, and genetically modified crops are not approved for use in some countries.

8.10 Long-term outlook

Trends in GHG emissions in the agricultural sector depend mainly on the level and rate of socio-economic development, human population growth, and diet, application of adequate technologies, climate and non-climate policies, and future climate change. Consequently, mitigation potentials in the agricultural sector are uncertain, making a consensus difficult to achieve and hindering policy making. However, agriculture is a significant contributor to GHG emissions (Section 8.2). Mitigation is unlikely to occur without action, and higher emissions are projected in the future if current trends are left unconstrained. According to current projections, the global population will reach 9 billion by 2050, an increase of about 50% over current levels (Lutz *et al.*, 2001; Cohen, 2003). Because of these increases and changing consumption patterns, some analyses estimate that the production of cereals will need to roughly double in coming decades (Tilman *et al.*, 2001; Roy *et al.*, 2002; Green *et al.*, 2005). Achieving these increases in food production may require more use of N fertilizer, leading to possible increases in N_2O emissions, unless more efficient fertilization techniques and products can be found (Galloway, 2003; Mosier, 2002). Greater demands for food could also increase CH_4 emissions from enteric fermentation if livestock numbers increase in response to demands for meat and other livestock products. As projected by the IMAGE 2.2 model, CO_2, CH_4, and N_2O emissions associated with land use vary greatly between scenarios (Strengers *et al.*, 2004), depending on trends towards globalization or regionalization, and on the emphasis placed on material wealth relative to sustainability and equity.

Some countries are moving forward with climate and non-climate policies, particularly those linked with sustainable development and improving environmental quality as described in Sections 8.6 and 8.7. These policies will likely have direct or synergistic effects on GHG emissions and provide a way forward for mitigation in the agricultural sector. Moreover, global sharing of innovative technologies for efficient use of land resources and agricultural inputs, in an effort to eliminate poverty and malnutrition, will also enhance the likelihood of significant mitigation from the agricultural sector.

Mitigation of GHG emissions associated with various agricultural activities and soil carbon sequestration could be achieved through best management practices, many of which are currently available for implementation. Best management practices are not only essential for mitigating GHG emissions, but also for other facets of environmental protection such as air and water quality management. Uncertainties do exist, but they can be reduced through finer scale assessments of best management practices within countries, evaluating not only the GHG mitigation potential but also the influences of mitigation options on socio-economic conditions and other environmental impacts.

The long-term outlook for development of mitigation practices for livestock systems is encouraging. Continuous improvements in animal breeds are likely, and these will improve the GHG emissions per kg of animal product. Enhanced production efficiency due to structural change or better application of existing technologies is also generally associated with reduced emissions, and there is a trend towards increased efficiency in both developed and developing countries. New technologies may emerge to reduce emissions from livestock such as probiotics, a methane vaccine or methane inhibitors. However, increased world demand for animal products may mean that while emissions per kg of product decline, total emissions may increase.

Recycling of agricultural by-products, such as crop residues and animal manures, and production of energy crops provides opportunities for direct mitigation of GHG emissions from fossil fuel offsets. However, there are barriers in technologies and economics to using agricultural wastes, and in converting energy crops into commercial fuels. The development of innovative technologies is a critical factor in realizing the potential for biofuel production from agricultural wastes and energy crops. This mitigation option could be moved forward with government investment for the development of these technologies, and subsidies for using these forms of energy.

A number of agricultural mitigation options which have limited potential now will likely have increased potential in the long-term. Examples include better use of fertilizer through precision farming, wider use of slow and controlled release fertilizers and of nitrification inhibitors, and other practices that reduce N application (and thus N_2O emissions). Similarly, enhanced N-use efficiency is achievable as technologies such as field diagnostics, fertilizer recommendations from expert/decision support systems and fertilizer placement technologies are developed and more widely used. New fertilizers and water management systems in paddy rice are also likely in the longer term.

Possible changes to climate and atmosphere in coming decades may influence GHG emissions from agriculture, and the effectiveness of practices adopted to minimize them. For example, atmospheric CO_2 concentrations, likely to double within the next century, may affect agro-ecosystems through changes in plant growth rates, plant litter composition, drought tolerance, and nitrogen demands (e.g., Long *et al.*, 2006; Henry *et al.*, 2005; Van Groenigen *et al.*, 2005; Jensen and Christensen, 2004; Torbert *et al.*, 2000; Norby *et al.*, 2001). Similarly, atmospheric nitrogen deposition also affects crop production systems as well as changing temperature regimes, although the effect will depend on the magnitude of change and response of the crop, forage, or livestock species. For example, increasing temperatures are likely to have a positive effect on crop production in colder regions due to a longer growing season (Smith *et al.*, 2005b). In contrast, increasing temperatures could accelerate decomposition of soil organic matter, releasing

stored soil carbon into the atmosphere (Knorr *et al.*, 2005; Fang *et al.*, 2005; Smith *et al.* 2005b). Furthermore, changes in precipitation patterns could change the adaptability of crops or cropping systems selected to reduce GHG emissions. Many of these effects have high levels of uncertainty; but demonstrate that practices chosen to reduce GHG emissions may not have the same effectiveness in coming decades. Consequently, programmes to reduce emissions in the agricultural sector will need to be designed with flexibility for adaptation in response to climate change.

Overall, the outlook for GHG mitigation in agriculture suggests significant potential. Current initiatives suggest that identifying synergies between climate change policies, sustainable development, and improvement of environmental quality will likely lead the way forward to realization of mitigation potential in this sector.

REFERENCES

Akiyama, H., K. Yagi, and X. Yan 2005: Direct N_2O emissions from rice paddy fields: summary of available data. *Global Biogeochemical Cycles*, **19**, GB1005, doi:10.1029/2004GB002378.

Albrecht, A. and S.T. Kandji, 2003: Carbon sequestration in tropical agroforestry systems. *Agriculture, Ecosystems and Environment*, **99**, pp. 15-27.

Alcock, D. and R.S. Hegarty, 2006: Effects of pasture improvement on productivity, gross margin and methane emissions of a grazing sheep enterprise. In *Greenhouse Gases and Animal Agriculture: An Update.* C.R. Soliva, J. Takahashi, and M. Kreuzer (eds.), International Congress Series No. 1293, Elsevier, The Netherlands, pp. 103-106.

Alvarez, R. 2005: A review of nitrogen fertilizer and conservative tillage effects on soil organic storage. *Soil Use and Management*, **21**, pp. 38-52.

Amon, B., T. Amon, J. Boxberger, and C. Wagner-Alt, 2001: Emissions of NH_3, N_2O and CH_4 from dairy cows housed in a farmyard manure tying stall (housing, manure storage, manure spreading). *Nutrient Cycling in Agro-Ecosystems*, **60**, pp. 103-113.

Amon, B., V. Kryvoruchko, T. Amon, and S. Zechmeister-Boltenstern, 2006: Methane, nitrous oxide and ammonia emissions during storage and after application of dairy cattle slurry and influence of slurry treatment. *Agriculture, Ecosystems & Environment*, **112**, pp. 153-162.

Anderson, T.L., R.J. Charlson, S.E. Schwartz, R. Knutti, O. Boucher, H. Rodhe, and J. Heintzenberg, 2003: Climate forcing by aerosols - a hazy picture. *Science*, **300**, pp. 1103-1104.

Andreae, M.O. 2001: The dark side of aerosols. *Nature*, **409**, pp. 671-672.

Andreae, M.O. and P. Merlet, 2001: Emission to trace gases and aerosols from biomass burning. *Global Biogeochemical Cycles*, **15**, pp. 955-966.

Andreae, M.O., C.D. Jones, and P.M. Cox, 2005: Strong present-day aerosol cooling implies a hot future. *Nature*, **435**, 1187 pp.

Aulakh, M.S., R., Wassmann, C. Bueno, and H. Rennenberg, 2001: Impact of root exudates of different cultivars and plant development stages of rice (*Oryza sativa* L.) on methane production in a paddy soil. *Plant and Soil*, **230**, pp. 77-86.

Balmford, A., R.E. Green, and J.P.W. Scharlemann, 2005: Sparing land for nature: exploring the potential impact of changes in agricultural yield on the area needed for crop production. *Global Change Biology*, **11**, pp. 1594-1605.

Barak, P., B.O. Jobe, A.R. Krueger, L.A. Peterson, and D.A. Laird, 1997: Effects of long-term soil acidification due to nitrogen fertilizer inputs in Wisconscin. *Plant and Soil*, **197**, pp. 61-69.

Barthès, B., A. Azontonde, E. Blanchart, C. Girardin, C. Villenave, S. Lesaint, R. Oliver, and C. Feller, 2004: Effect of a legume cover crop (*Mucuna pruriens* var. *utilis*) on soil carbon in an Ultisol under maize cultivation in southern Benin. *Soil Use and Management*, **20**, pp. 231-239.

Batjes, N.H., 1999: Management options for reducing CO_2-concentrations in the atmosphere by increasing carbon sequestration in the soil. Dutch National Research Programme on Global Air Pollution and Climate Change, Project executed by the International Soil Reference and Information Centre, Wageningen, The Netherlands, 114 pp.

Bauman, D.E., 1992: Bovine somatotropin: review of an emerging animal technology. *Journal of Dairy Science*, **75**, pp. 3432-3451.

Beauchemin, K. and S. McGinn, 2005: Methane emissions from feedlot cattle fed barley or corn diets. *Journal of Animal Science*, **83**, pp. 653-661.

Benz, D.A. and D.E. Johnson, 1982: The effect of monensin on energy partitioning by forage fed steers. *Proceedings of the West Section of the American Society of Animal Science*, **33,** 60 pp.

Beringer, J., L.B. Hutley, N.J. Tapper, A. Coutts, A. Kerley, and A.P. O'Grady, 2003: Fire impacts on surface heat, moisture and carbon fluxes from a tropical savanna in northern Australia. *International Journal of Wildland Fire*, **12**, pp. 333-340.

Berndes, G. and P. Börjesson, 2002: Multi-functional biomass production systems. Available at: <http://www.elkraft.ntnu.no/eno/konf_pub/ISES2003/full_paper/6%20MISCELLANEOUS/O6%204.pdf> accessed 26 March 2007.

Berndes, G., 2002: Bioenergy and water: the implications of large-scale bioenergy production for water use and supply. *Global Environmental Change*, **12**, pp. 253-271.

Berndes, G., F. Fredrikson, and P. Borjesson, 2004: Cadmium accumulation and Salix-based phytoextraction on arable land in Sweden. *Agriculture, Ecosystems & Environment*, **103**, pp. 207-223.

Berndes, G., M. Hoogwijk, and R. van den Broek, 2003: The contribution of biomass in the future global energy supply: a review of 17 studies. *Biomass & Bioenergy*, **25**, pp. 1-28.

Bhatia, A., H. Pathak, and P.K. Aggarwal, 2004: Inventory of methane and nitrous oxide emissions from agricultural soils of India and their global warming potential. *Current Science*, **87**, pp. 317-324.

Binfield, J., T. Donnellan, K. Hanrahan, and P. Westhoff, 2006: World Agricultural Trade Reform and the WTO Doha Development Round: Analysis of the Impact on EU and Irish Agriculture. Teagasc, Athenry, Galway, Ireland, 79 pp.

Blaxter, K.L. and J.L. Claperton, 1965: Prediction of the amount of methane produced by ruminants. *British Journal of Nutrition*, **19**, pp. 511-522.

Boadi, D., C. Benchaar, J. Chiquette, and D. Massé, 2004: Mitigation strategies to reduce enteric methane emissions from dairy cows: update review. *Canadian Journal of Animal Science*, **84**, pp. 319-335.

Boehm, M., B. Junkins, R. Desjardins, S. Kulshreshtha, and W. Lindwall, 2004: Sink potential of Canadian agricultural soils. *Climatic Change*, **65**, pp. 297-314.

Börjesson, P. and G. Berndes, 2006: The prospects for willow plantations for wastewater treatment in Sweden. *Biomass & Bioenergy*, **30**, pp. 428-438.

Bouwman, A., 2001: Global Estimates of Gaseous Emissions from Agricultural Land. FAO, Rome, 106 pp.

Brown, D.J., K.D. Shepherd, M.G. Walsh, M.D. Mays, and T.G. Reinsch 2006: Global soil characterization with VNIR diffuse reflectance spectroscopy. *Geoderma*, **132**, pp. 273-290.

Bruce, J.P., M. Frome, E. Haites, H. Janzen, R. Lal, and K. Paustian, 1999: Carbon sequestration in soils. *Journal of Soil and Water Conservation*, **54**, pp. 382-389.

Butt, T.A., B.A. McCarl, and A.O. Kergna, 2006: Policies for reducing agricultural sector vulnerability to climate change in Mali. *Climate Policy*, **5**, pp. 583-598.

Cai, Z.C. and H. Xu, 2004: Options for mitigating CH_4 emissions from rice fields in China. In *Material Circulation through Agro-Ecosystems in East Asia and Assessment of Its Environmental Impact,* Hayashi, Y. (ed.), NIAES Series 5, Tsukuba, pp. 45-55.

Cai, Z.C., H. Tsuruta, and K. Minami, 2000: Methane emissions from rice fields in China: measurements and influencing factors. *Journal of Geophysical Research*, **105 D13**, pp. 17231-17242.

Cai, Z.C., H. Tsuruta, M. Gao, H. Xu, and C.F. Wei, 2003: Options for mitigating methane emission from a permanently flooded rice field. *Global Change Biology,* **9**, pp. 37-45.

Caldeira, K., M.G. Morgan, D. Baldocchi, P.G. Brewer, C.T.A. Chen, G.J. Nabuurs, N. Nakicenovic, and G.P. Robertson, 2004: A portfolio of carbon management options. In *The Global Carbon Cycle. Integrating Humans, Climate, and the Natural World*, C.B. Field, and M.R. Raupach (eds.). SCOPE 62, Island Press, Washington DC, pp.103-129.

Cannell, M.G.R., 2003: Carbon sequestration and biomass energy offset: theoretical, potential and achievable capacities globally, in Europe and the UK. *Biomass & Bioenergy*, **24**, pp. 97-116.

Cassman, K.G., A. Dobermann, D.T. Walters, and H. Yang, 2003: Meeting cereal demand while protecting natural resources and improving environmental quality. *Annual Review of Environment and Resources*, **28**, pp. 315-358.

Cerri, C.C., M. Bernoux, C.E.P. Cerri, and C. Feller, 2004: Carbon cycling and sequestration opportunities in South America: the case of Brazil. *Soil Use and Management*, **20**, pp. 248-254.

Chadwick, D.R., 2005: Emissions of ammonia, nitrous oxide and methane from cattle manure heaps: effect of compaction and covering. *Atmospheric Environment*, **39**, pp. 787-799.

Christensen, J.H., B. Hewitson, A. Busuioc, A. Chen, X. Gao, I. Held, R. Jones, R.K. Kolli, W.-T. Kwon, R. Laprise, V. Magaña Rueda, L. Mearns, C.G. Menéndez, J. Räisänen, A. Rinke, A. Sarr and P. Whetton, 2007: Regional Climate Projections. In: *Climate Change 2007: The Physical Science Basis. Contribution of Working Group I to the Fourth Assessment Report of the Intergovernmental Panel on Climate Change* [Solomon, S., D. Qin, M. Manning, Z. Chen, M. Marquis, K.B. Averyt, M. Tignor and H.L. Miller (eds.)]. Cambridge University Press, Cambridge, United Kingdom and New York, NY, USA.

Clark, H., C. Pinares, and C. de Klein, 2005: Methane and nitrous oxide emissions from grazed grasslands. In *Grassland. A Global Resource,* D. McGilloway (ed.), Wageningen Academic Publishers, Wageningen, The Netherlands, pp. 279-293.

Clemens, J. and H.J. Ahlgrimm, 2001: Greenhouse gases from animal husbandry: mitigation options. *Nutrient Cycling in Agroecosystems*, **60**, pp. 287-300.

Clemens, J., M. Trimborn, P. Weiland, and B. Amon, 2006: Mitigation of greenhouse gas emissions by anaerobic digestion of cattle slurry. *Agriculture, Ecosystems and Environment*, **112**, pp. 171-177.

Cohen, J.E., 2003: Human population: the next half century. *Science*, **302**, pp. 1172-1175.

Cole, C.V., J. Duxbury, J. Freney, O. Heinemeyer, K. Minami, A. Mosier, K. Paustian, N. Rosenberg, N. Sampson, D. Sauerbeck, and Q. Zhao, 1997: Global estimates of potential mitigation of greenhouse gas emissions by agriculture. *Nutrient Cycling in Agroecosystems*, **49**, pp. 221-228.

Conant, R.T. and K. Paustian, 2002: Potential soil carbon sequestration in overgrazed grassland ecosystems. *Global Biogeochemical Cycles*, **16** (4), 1143 pp., doi:10.1029/2001GB001661.

Conant, R.T., K. Paustian, and E.T. Elliott, 2001: Grassland management and conversion into grassland: Effects on soil carbon. *Ecological Applications*, **11**, pp. 343-355.

Conant, R.T., K. Paustian, S.J. Del Grosso, and W.J. Parton, 2005: Nitrogen pools and fluxes in grassland soils sequestering carbon. *Nutrient Cycling in Agroecosystems*, **71**, pp. 239-248.

533

Connor, D.J., 2004: Designing cropping systems for efficient use of limited water in southern Australia. *European Journal of Agronomy,* **21**, pp. 419-431.

Conway, G. and G. Toenniessen, 1999: Feeding the world in the twenty-first century. *Nature,* **402**, pp. C55-C58.

Crutzen, P.J., 1995: The role of methane in atmospheric chemistry and climate. Proceedings of the Eighth International Symposium on Ruminant Physiology. Ruminant Physiology: Digestion, Metabolism, Growth and Reproduction, Von Engelhardt, W., S. Leonhard-Marek, G. Breves, and D. Giesecke (eds.), Ferdinand Enke Verlag, Stuttgart, pp. 291-316.

Dalal, R.C., W. Wang, G.P. Robertson, and W.J. Parton, 2003: Nitrous oxide emission from Australian agricultural lands and mitigation options: a review. *Australian Journal of Soil Research,* **41**, pp. 165-195.

Davidson, E.A., D.C. Nepstad, C. Klink, and S.E. Trumbore, 1995: Pasture soils as carbon sink. *Nature,* **376**, pp. 472-473.

Dean, T., 2000: Development: agriculture workers too poor to buy food. UN IPS, New York, 36 pp.

DeAngelo, B.J., F.C. de la Chesnaye, R.H. Beach, A. Sommer, and B.C. Murray, 2006: Methane and nitrous oxide mitigation in agriculture. Multi-Greenhouse Gas Mitigation and Climate Policy, *Energy Journal,* Special Issue **#3**.Available at: <http://www.iaee.org/en/publications/journal.aspx> accessed 26 March 2007.

DeFries, R.S., J.A. Foley, and G.P. Asner, 2004: Land-use choices: balancing human needs and ecosystem function. *Frontiers in Ecology and Environment,* **2**, pp. 249-257.

Denman, K.L., G. Brasseur, A. Chidthaisong, P. Ciais, P.M. Cox, R.E. Dickinson, D. Hauglustaine, C. Heinze, E. Holland, D. Jacob, U. Lohmann, S Ramachandran, P.L. da Silva Dias, S.C. Wofsy and X. Zhang, 2007: Couplings Between Changes in the Climate System and Biogeochemistry. In: *Climate Change 2007: The Physical Science Basis. Contribution of Working Group I to the Fourth Assessment Report of the Intergovernmental Panel on Climate Change* [Solomon, S., D. Qin, M. Manning, Z. Chen, M. Marquis, K.B. Averyt, M.Tignor and H.L. Miller (eds.)]. Cambridge University Press, Cambridge, United Kingdom and New York, NY, USA.

Derner, J.D., T.W. Boutton, and D.D. Briske, 2006: Grazing and ecosystem carbon storage in the North American Great Plains. *Plant and Soil,* **280**, pp. 77-90.

Dias de Oliveira, M.E., B.E. Vaughan, and E.J. Rykiel, Jr., 2005: Ethanol as fuel: energy, carbon dioxide balances, and ecological footprint. *BioScience,* **55**, pp. 593-602.

Díaz -Zorita, M., G.A. Duarte, and J.H. Grove, 2002: A review of no-till systems and soil management for sustainable crop production in the subhumid and semiarid Pampas of Argentina. *Soil and Tillage Research,* **65**, pp. 1-18.

Diekow, J., J. Mielniczuk, H. Knicker, C. Bayer, D.P. Dick, and I. Kögel-Knabner, I. 2005: Soil C and N stocks as affected by cropping systems and nitrogen fertilization in a southern Brazil Acrisol managed under no-tillage for 17 years. *Soil and Tillage Research,* **81**, pp. 87-95.

Díez, J.A., P. Hernaiz, M.J. Muñoz, A. de la Torre, A. Vallejo, 2004: Impact of pig slurry on soil properties, water salinization, nitrate leaching and crop yield in a four-year experiment in Central Spain. *Soil Use and Management,* **20**, pp. 444-450.

Dohme, F.A., A. Machmuller, A.Wasserfallen, and M. Kreuzer, 2001: Comparative efficiency of various fats rich in medium-chain fatty acids to suppress ruminal methanogenesis as measured with Rusitec. *Canadian Journal of Animal Science,* **80**, pp. 473-482.

Edmonds, J.A., 2004: Climate change and energy technologies. *Mitigation and Adaptation Strategies for Global Change,* **9**, pp. 391-416.

Eidman, V.R., 2005: Agriculture as a producer of energy. In *Agriculture as a Producer and Consumer of Energy,* J.L. Outlaw, K.J. Collins, and J.A. Duffield (eds.), CABI Publishing, Cambridge, MA, pp.30-67.

Elbakidze, L. and B.A. McCarl, 2007: Sequestration offsets versus direct emission reductions: consideration of environmental co-effects. *Ecological Economics,* **60**(3), pp. 564-571.

European Climate Change Programme, 2001: Agriculture. Mitigation potential of Greenhouse Gases in the Agricultural Sector. Working Group 7, Final report, COMM(2000)88. European Commission, Brussels, 17 pp.

European Environment Agency, 2006: How much biomass can Europe use without harming the environment? EEA Briefing 2/2006. Available at: <http://reports.eea.europa.eu/briefing_2005_2/en> (accessed 26 March 2007).

Ewert, F., M.D.A. Rounsevell, I. Reginster, M. Metzger, and R. Leemans, 2005: Future scenarios of European agricultural land use. I: estimating changes in crop productivity. *Agriculture, Ecosystems and Environment,* **107**, pp. 101-116.

Faaij, A., 2006: Modern biomass conversion technologies. *Mitigation and Adaptation Strategies for Global Change,* **11**, pp. 335-367.

Falloon, P., P. Smith, and D.S. Powlson, 2004: Carbon sequestration in arable land - the case for field margins. *Soil Use and Management,* **20**, pp. 240-247.

Fang, C., P. Smith, J.B. Moncrieff, and J.U. Smith, 2005: Similar response of labile and resistant soil organic matter pools to changes in temperature. *Nature,* **433**, pp. 57-59.

FAO, 2001: Soil carbon sequestration for improved land management. World Soil Resources Reports No. 96. FAO, Rome, 58 pp.

FAO, 2003: *World Agriculture: Towards 2015/2030.* An FAO Perspective. FAO, Rome, 97 pp.

FAOSTAT, 2006: FAOSTAT Agricultural Data. Available at: <http://faostat.fao.org/> accessed 26 March 2007.

Fawcett, A.A. and R.D. Sands. 2006: Non-CO_2 Greenhouse Gases in the Second Generation Model. Multi-Greenhouse Gas Mitigation and Climate Policy, *Energy Journal,* Special Issue **#3**, pp 305-322. Available at: <http://www.iaee.org/en/publications/journal.aspx> accessed 26 March 2007.

Fedoroff, N.V. and J.E. Cohen, 1999: Plants and population: is there time? *Proceedings of the National Academy of Sciences USA,* **96**, pp. 5903-5907.

Feng, W., G.X. Pan, S. Qiang, R.H. Li, and J.G. Wei, 2006: Influence of long-term fertilization on soil seed bank diversity of a paddy soil under rice/rape rotation. *Biodiversity Science,* **14** (6), pp. 461-469.

Ferris, C.P., F.J. Gordon, D.C. Patterson, M.G. Porter, and T. Yan, 1999: The effect of genetic merit and concentrate proportion in the diet on nutrient utilization by lactating dairy cows. *Journal of Agricultural Science,* Cambridge **132**, pp. 483-490.

Fisher, M.J., I.M. Rao, M.A. Ayarza, C.E. Lascano, J.I. Sanz, R.J. Thomas, and R.R. Vera,1994: Carbon storage by introduced deep-rooted grasses in the South American savannas. *Nature,* **371**, pp. 236-238.

Foley, J.A., R. DeFries, G. Asner, C. Barford, G. Bonan, S.R. Carpenter, F.S. Chapin, M.T. Coe, G.C. Dailey, H.K. Gibbs, J.H. Helkowski, T. Holloway, E.A. Howard, C.J. Kucharik, C. Monfreda, J.A. Patz, I.C. Prentice, N. Ramankutty, and P.K. Snyder, 2005: Global consequences of land use. *Science,* **309**, pp. 570-574.

Follett, R.F., 2001: Organic carbon pools in grazing land soils. In *The Potential of U.S. Grazing Lands to Sequester Carbon and Mitigate the Greenhouse Effect.* R.F. Follett, J.M. Kimble, and R. Lal (eds.), Lewis Publishers, Boca Raton, Florida, pp. 65-86.

Follett, R.F., J.M. Kimble, and R. Lal, 2001: The potential of U.S. grazing lands to sequester soil carbon. In *The Potential of U.S. Grazing Lands to Sequester Carbon and Mitigate the Greenhouse Effect,* R.F. Follett, J.M. Kimble, and R. Lal (eds.), Lewis Publishers, Boca Raton, Florida, pp. 401-430.

Follett, R.F., S.R. Shafer, M.D. Jawson, and A.J. Franzluebbers, 2005: Research and implementation needs to mitigate greenhouse gas emissions from agriculture in the USA. *Soil and Tillage Research,* **83**, pp. 159-166.

Freibauer, A., M. Rounsevell, P. Smith, and A. Verhagen, 2004: Carbon sequestration in the agricultural soils of Europe. *Geoderma,* **122**, pp. 1-23.

Fujino, J., R. Nair, M. Kainuma, T. Masui, and Y. Matsuoka, 2006: Multi-gas mitigation analysis on stabilization scenarios using AIM global model. Multi-Greenhouse Gas Mitigation and Climate Policy, *Energy Journal,* Special Issue #3, pp 343-354. Available at: <http://www.iaee.org/en/publications/journal.aspx> (accessed 26 March 2007).

Galloway, J.N., 2003: The global nitrogen cycle. *Treatise on Geochemistry,* **8,** pp. 557-583.

Galloway, J.N., F.J. Dentener, D.G. Capone, E.W. Boyer, R.W. Howarth, S.P. Seitzinger, G.P. Asner, C.C. Cleveland, P.A. Green, E.A. Holland, D.M. Karl, A.F. Michaels, J.H. Porter, A.R. Townsend, and C.J. Vörösmarty, 2004: Nitrogen cycles: past, present, and future. *Biogeochemistry,* **70,** pp. 153-226.

Galloway, J.N., J.D. Aber, J.W. Erisman, S.P. Seitzinger, R.W. Howarth, E.B. Cowling, and B.J. Cosby, 2003: The nitrogen cascade. *Bioscience,* **53,** pp. 341-356.

Gehl, R.J., C.W. Rice, 2007: Emerging technologies for in situ measurement of soil carbon. *Climatic Change* **80,** pp. 43-54.

Gilland, B., 2002: World population and food supply. Can food production keep pace with population growth in the next half-century? *Food Policy,* **27,** pp. 47-63.

Gonzalez-Avalos, E. and L.G. Ruiz-Suarez, 2001: Methane emission factors from cattle in Mexico. *Bioresource Technology,* **80,** pp. 63-71.

Green, R.E., S.J. Cornell, J.P.W. Scharlemann, and A. Balmford, 2005: Farming and the fate of wild nature. *Science,* **307,** pp. 550-555.

Gregorich, E.G., P. Rochette, A.J. van den Bygaart, and D.A. Angers, 2005: Greenhouse gas contributions of agricultural soils and potential mitigation practices in Eastern Canada. *Soil and Tillage Research,* **83,** pp. 53-72.

Guo, L.B. and R.M. Gifford, 2002: Soil carbon stocks and land use change: a meta analysis. *Global Change Biology,* **8,** pp. 345-360.

Hamelinck, C.N., R.A.A. Suurs, and A.P.C. Faaij, 2004: Techno-economic analysis of international bio-energy trade chains. *Biomass & Bioenergy,* **29,** pp. 114-134.

Hansen, L.B., 2000: Consequences of selection for milk yield from a geneticist's viewpoint. *Journal of Dairy Science,* **83,** pp. 1145-1150.

Helgason, B.L., H.H. Janzen, M.H. Chantigny, C.F. Drury, B.H. Ellert, E.G. Gregorich, Lemke, E. Pattey, P. Rochette, and C. Wagner-Riddle, 2005: Toward improved coefficients for predicting direct N₂O emissions from soil in Canadian agroecosystems. *Nutrient Cycling in Agroecosystems,* **71,** pp. 87-99.

Henry, H.A.L., E.E. Cleland, C.B. Field, and P.M. Vitousek, 2005: Interactive effects of elevated CO₂, N deposition and climate change on plant litter quality in a California annual grassland. *Oecologia,* **142,** pp. 465-473.

Hess, H.D., T.T. Tiemann, F. Noto, J.E. Carulla, and M. Kruezer, 2006: Strategic use of tannins as means to limit methane emission from ruminant livestock. In *Greenhouse Gases and Animal Agriculture: An Update,* C.R. Soliva, J. Takahashi, and M. Kreuzer (eds.). International Congress Series No. 1293, Elsevier, The Netherlands, pp. 164-167.

Hindrichsen, I.K., H.R. Wettstein, A. Machmüller, and M. Kreuzer, 2006: Methane emission, nutrient degradation and nitrogen turnover in dairy cows and their slurry at different production scenarios with and without concentrate supplementation. *Agriculture, Ecosystems and Environment,* **113,** pp. 150-161.

Hoogwijk, M., 2004: *On the Global and Regional Potential of Renewable Energy Sources.* PhD Thesis, Copernicus Institute, Utrecht University, March 12, 2004. 256 pp.

Hoogwijk, M., A. Faaij, R. van den Broek, G. Berndes, D. Gielen, and W. Turkenburg, 2003: Exploration of the ranges of the global potential of biomass for energy. *Biomass and Bioenergy,* **25,** pp. 119-133.

Hoogwijk, M., A. Faaij, B. Eickhout, B. de Vries, and W. Turkenburg, 2005: Potential of biomass energy out to 2100, for four IPCC SRES land-use scenarios. *Biomass & Bioenergy,* **29,** pp. 225-257.

Huang, J., C. Pray, and S. Rozelle, 2002: Enhancing the crops to feed the poor. *Nature,* **418,** pp. 678-684.

Huston, M.A. and G. Marland, 2003: Carbon management and biodiversity. *Journal of Environmental Management,* **67,** pp. 77-86.

International Fertilizer Industry Association, 2007: Fertilizer consumption statistics. Available at: <http://www.fertilizer.org/ifa/statistics.asp> (accessed 26 March 2007).

IPCC, 1996: *Climate change 1995: The Science of Climate Change. Contribution of Working Group I to the Second Assessment Report of the Intergovernmental Panel on Climate Change (IPCC).* Cambridge University Press, Cambridge.

IPCC, 1997: *Revised 1996 IPCC Guidelines for National Greenhouse Gas Inventories Workbook. Volume 2.* Cambridge University Press, Cambridge.

IPCC, 2000: *Land Use, Land-Use Change and Forestry. Special Report of the Intergovernmental Panel on Climate Change.* Cambridge University Press, Cambridge.

IPCC, 2001a: *Climate Change 2001: The Scientific Basis. Contribution of Working Group I to the Third Assessment Report of the Intergovernmental Panel on Climate Change* [Houghton, J.T., Y. Ding, D.J. Griggs, M. Noguer, P.J. van der Linden, X. Dai, K. Maskell, and C.A. Johnson, (eds.)], Cambridge University Press, 881 pp.

IPCC, 2001b: *Climate Change 2001: Mitigation: Contribution of Working Group III to the Third Assessment Report of the Intergovernmental Panel on Climate Change* [Metz, B., O. Davidson, R. Swart, and J. Pan, (eds.)], Cambridge University Press, 752 pp.

IPCC, 2003: *Good Practice Guidance for Greenhouse Gas Inventories for Land-Use, Land-Use Change & Forestry.* Institute of Global Environmental Strategies (IGES), Kanagawa, Japan.

IPCC, 2006: 2006 National Greenhouse Gas Inventory Guidelines. Institute of Global Environmental Strategies (IGES), Kanagawa, Japan.

IPCC, 2007: *Climate Change 2007: Impacts, Adaptation and Vulnerability. Contribution of Working Group II to the Fourth Assessment Report of the Intergovernmental Panel on Climate Change* [Parry, M.L., O.F. Canziani, J.P Palutikof, P.J. van der Linden, C.E. Hanson (eds.)]. Cambridge University Press, Cambridge, United Kingdom and New York, NY, USA.

Izaurralde, R.C. and C.W. Rice, 2006: Methods and tools for designing pilot soil carbon sequestration projects. In *Carbon Sequestration in Soils of Latin America,* Lal, R., C.C. Cerri, M. Bernoux, J. Etchvers, and C.E. Cerri (eds.), CRC Press, Boca Raton, pp. 457-476.

Izaurralde, R.C., W.B. McGill, J.A. Robertson, N.G. Juma, and J.J. Thurston, 2001: Carbon balance of the Breton classical plots over half a century. *Soil Science Society of America Journal,* **65,** pp. 431-441.

Jackson, R.B., E.G. Jobbágy, R. Avissar, S. Baidya Roy, D. Barrett, C.W. Cook, K.A. Farley, D.C. le Maitre, B.A. McCarl, and B.C. Murray 2005: Trading water for carbon with biological carbon sequestration. *Science,* **310,** pp. 1944-1947.

Janzen, H.H., 2004: Carbon cycling in earth systems - a soil science perspective. *Agriculture, Ecosystems and Environment,* **104,** pp. 399-417.

Janzen, H.H., 2005: Soil carbon: A measure of ecosystem response in a changing world? *Canadian Journal of Soil Science,* **85,** pp. 467-480.

Jensen, B., B.T. Christensen, 2004: Interactions between elevated CO₂ and added N: effects on water use, biomass, and soil ¹⁵N uptake in wheat. *Acta Agriculturae Scandinavica,* Section B 54, pp. 175-184.

Johnson, D.E., G.M. Ward, and J. Torrent, 1991: The environmental impact of bovine somatotropin (bST) use in dairy cattle. *Journal of Dairy Science,* **74S,** 209 pp.

Johnson, D.E., H.W. Phetteplace, and A.F. Seidl, 2002: Methane, nitrous oxide and carbon dioxide emissions from ruminant livestock production systems. In *Greenhouse Gases and Animal Agriculture,* J. Takahashi, and B.A. Young (eds.), Elsevier, Amsterdam, The Netherlands, pp. 77-85.

Johnson, K.A. and D.E. Johnson, 1995: Methane emissions from cattle. *Journal of Animal Science,* **73,** pp. 2483-2492.

535

Jones, C.D., P.M. Cox, R.L.H. Essery, D.L. Roberts, and M.J. Woodage, 2003: Strong carbon cycle feedbacks in a climate model with interactive CO_2 and sulphate aerosols. *Geophysical Research Letters,* **30,** pp. 32.1-32.4.

Jordan, E., D. Kenny, M. Hawkins, R. Malone, D.K. Lovett, and F.P. O'Mara, 2006b: Effect of refined soy oil or whole soybeans on methane output, intake and performance of young bulls. *Journal of Animal Science,* **84,** pp. 2418-2425.

Jordan, E., D.K. Lovett, F.J. Monahan, and F.P. O'Mara, 2006a: Effect of refined coconut oil or copra meal on methane output, intake and performance of beef heifers. *Journal of Animal Science,* **84,** pp. 162-170.

Jordan, E., D.K. Lovett, M. Hawkins, J. Callan, and F.P. O'Mara, 2006c: The effect of varying levels of coconut oil on intake, digestibility and methane output from continental cross beef heifers. *Animal Science,* **82,** pp. 859-865.

Junginger, M., A. Faaij, A. Koopmans, R. van den Broek, and W. Hulscher, 2001: Setting up fuel supply strategies for large scale bio-energy projects - a methodology for developing countries. *Biomass & Bioenergy,* **21,** pp. 259-275.

Kamra, D.N., N. Agarwal, and L.C. Chaudhary, 2006: Inhibition of ruminal methanogenesis by tropical plants containing secondary compounds. In *Greenhouse Gases and Animal Agriculture: An Update,* C.R. Soliva, J. Takahashi, and M. Kreuzer (eds.). International Congress Series No. 1293, Elsevier, The Netherlands, pp. 156-163.

Kang, G.D., Z.C. Cai, and X.Z. Feng, 2002: Importance of water regime during the non-rice growing period in winter in regional variation of CH_4 emissions from rice fields during following rice growing period in China. *Nutrient Cycling in Agroecosystems,* **64,** pp. 95-100.

Kasimir-Klemedtsson, A., L. Klemedtsson, K. Berglund, P. Martikainen, Silvola, and O. Oenema, 1997: Greenhouse gas emissions from farmed organic soils: a review. *Soil Use and Management,* **13,** pp. 245-250.

Kebreab, E., K. Clark, C. Wagner-Riddle, and J. France, 2006: Methane and nitrous oxide emissions from Canadian animal agriculture: A review. *Canadian Journal of Animal Science,* **86,** pp. 135-158.

Kemfert, C., T.P. Truong, and T. Bruckner. 2006: Economic impact assessment of climate change-A multi-gas investigation with WIAGEM-GTAPEL-ICM. Multi-Greenhouse Gas Mitigation and Climate Policy, *Energy Journal,* Special Issue #3. Available at: <http://www.iaee.org/en/publications/journal.aspx> accessed 26 March 2007.

Kennedy, P.M. and L.P. Milligan, 1978: Effects of cold exposure on digestion, microbial synthesis and nitrogen transformation in sheep. *British Journal of Nutrition,* **39,** pp. 105-117.

Khalil, M.A.K. and M.J. Shearer, 2006. Decreasing emissions of methane from rice agriculture. In *Greenhouse Gases and Animal Agriculture: An Update.* Soliva, C.R., J. Takahashi, and M. Kreuzer (eds.), International Congress Series No. 1293, Elsevier, The Netherlands, pp. 33-41.

Kim, M-K. and B.A. McCarl, 2005: Uncertainty Discounting for Land-Based Carbon Sequestration. Presented at International Policy Forum on Greenhouse Gas Management April 2005 Victoria, British Columbia.

Knorr, W., I.C. Prentice, J.I. House, E.A. Holland, 2005: Long-term sensitivity of soil carbon turnover to warming. *Nature,* **433,** pp. 298-301.

Koga, N., T. Sawamoto, and H. Tsuruta 2006: Life cycle inventory-based analysis of greenhouse gas emissions from arable land farming systems in Hokkaido, northern Japan. *Soil Science and Plant Nutrition,* **52,** pp. 564-574.

Korontzi, S., C.O. Justice, and R.J. Scholes, 2003: Influence of timing and spatial extent of savannah fires in southern Africa on atmospheric emissions. *Journal of Arid Environments,* **54,** pp. 395-404.

Kreuzer, M. and I.K. Hindrichsen, 2006: Methane mitigation in ruminants by dietary means: the role of their methane emission from manure. In *Greenhouse Gases and Animal Agriculture: An Update.* C.R. Soliva, J. Takahashi, and M. Kreuzer (eds.). International Congress Series No. 1293, Elsevier, The Netherlands, pp. 199-208.

Külling, D.R., H. Menzi, F. Sutter, P. Lischer, and M. Kreuzer, 2003: Ammonia, nitrous oxide and methane emissions from differently stored dairy manure derived from grass- and hay-based rations. *Nutrient Cycling in Agroecosystems,* **65,** pp. 13-22.

Kumar, P. 2001: *Valuation of Ecological services of Wetland Ecosystems: A Case Study of Yamuna Floodplains in the Corridors of Delhi.* Mimeograph, Institute of Economic Growth, Delhi, India, 137 pp.

Kurkalova, L., C.L. Kling, and J. Zhao, 2004: Multiple benefits of carbon-friendly agricultural practices: Empirical assessment of conservation tillage. *Environmental Management,* **33,** pp. 519-527.

Lal, R. and J.P. Bruce, 1999: The potential of world cropland soils to sequester C and mitigate the greenhouse effect. *Environmental Science and Policy,* **2,** pp. 177-185.

Lal, R., 1999: Soil management and restoration for C sequestration to mitigate the accelerated greenhouse effect. *Progress in Environmental Science,* **1,** pp. 307-326.

Lal, R., 2001a: World cropland soils as a source or sink for atmospheric carbon. *Advances in Agronomy,* **71,** pp. 145-191.

Lal, R., 2001b: Potential of desertification control to sequester carbon and mitigate the greenhouse effect. *Climate Change,* **15,** pp. 35-72.

Lal, R., 2002: Carbon sequestration in dry ecosystems of West Asia and North Africa. *Land Degradation and Management,* **13,** pp. 45-59.

Lal, R., 2003: Global potential of soil carbon sequestration to mitigate the greenhouse effect. *Critical Reviews in Plant Sciences,* **22,** pp. 151-184.

Lal, R., 2004a: Soil carbon sequestration impacts on global climate change and food security. *Science,* **304,** pp. 1623-1627.

Lal, R., 2004b: Soil carbon sequestration to mitigate climate change. *Geoderma,* **123,** pp. 1-22.

Lal, R., 2004c: Offsetting China's CO_2 emissions by soil carbon sequestration. *Climatic Change,* **65,** pp. 263-275.

Lal, R., 2004d: Carbon sequestration in soils of central Asia. *Land Degradation and Development,* **15,** pp. 563-572.

Lal, R., 2004e: Soil carbon sequestration in India. *Climatic Change,* **65,** pp. 277-296.

Lal, R., 2005: Soil carbon sequestration for sustaining agricultural production and improving the environment with particular reference to Brazil. *Journal of Sustainable Agriculture,* **26,** pp. 23-42.

Lal, R., R.F. Follett, and J.M. Kimble, 2003: Achieving soil carbon sequestration in the United States: a challenge to the policy makers. *Soil Science,* **168,** pp. 827-845.

Le Mer, J. and P. Roger, 2001: Production, oxidation, emission and consumption of methane by soils: a review. *European Journal of Soil Biology,* **37,** pp. 25-50.

Leng, R.A., 1991: Improving Ruminant Production and Reducing Methane Emissions from Ruminants by Strategic Supplementation. EPA Report no. 400/1-91/004, Environmental Protection Agency, Washington, D.C.

Li, C., S. Frolking, and K. Butterbach-Bahl, 2005: Carbon sequestration in arable soils is likely to increase nitrous oxide emissions, offsetting reductions in climate radiative forcing. *Climatic Change,* **72,** pp. 321-338.

Liebig, M.A., J.A. Morgan, J.D. Reeder, B.H. Ellert, H.T. Gollany, and G.E. Schuman, 2005: Greenhouse gas contributions and mitigation potential of agricultural practices in northwestern USA and western Canada. *Soil & Tillage Research,* **83,** pp. 25-52.

Lila, Z.A., N. Mohammed, S. Kanda, T. Kamada, and H. Itabashi, 2003: Effect of sarsaponin on ruminal fermentation with particular reference to methane production in vitro. *Journal of Dairy Science,* **86,** pp. 330-336.

Long, S.P., E.A. Ainsworth, A.D.B. Leakey, J. Nosberger, and D.R. Ort, 2006: Food for thought: lower-than-expected crop yield stimulation with rising CO_2 concentrations. *Science,* **312,** pp. 1918-1921.

Lovett, D., S. Lovell, L. Stack, J. Callan, M. Finlay, J. Connolly, and F.P. O'Mara, 2003: Effect of forage/concentrate ratio and dietary coconut oil level on methane output and performance of finishing beef heifers. *Livestock Production Science,* **84,** pp. 135-146.

Lovett, D.K. and F.P. O'Mara, 2002: Estimation of enteric methane emissions originating from the national livestock beef herd: a review of the IPCC default emission factors. *Tearmann,* **2**, pp. 77-83.

Lovett, D.K., L. Shalloo, P. Dillon, and F.P. O'Mara, 2006: A systems approach to quantify greenhouse gas fluxes from pastoral dairy production as affected by management regime. *Agricultural Systems,* **88,** pp. 156-179.

Lutz, W., W. Sanderson, S. Scherbov, 2001: The end of world population growth. *Nature,* **412,** pp. 543-545.

Machado, P.L.O.A. and C.A. Silva, 2001: Soil management under no-tillage systems in the tropics with special reference to Brazil. *Nutrient Cycling in Agroecosystems,* **61,** pp. 119-130.

Machado, P.L.O.A. and P.L. Freitas 2004: No-till farming in Brazil and its impact on food security and environmental quality. In *Sustainable Agriculture and the International Rice-Wheat System,* R. Lal, P.R. Hobbs, N. Uphoff, D.O. Hansen (eds.), Marcel Dekker, New York, pp. 291-310.

Machmüller, A., C.R. Soliva, and M. Kreuzer, 2003: Methane-suppressing effect of myristic acid in sheep as affected by dietary calcium and forage proportion. *British Journal of Nutrition,* **90,** pp. 529-540.

Machmülller, A., D.A. Ossowski, and M. Kreuzer, 2000: Comparative evaluation of the effects of coconut oil, oilseeds and crystalline fat on methane release, digestion and energy balance in lambs. *Animal Feed Science and Technology,* **85,** pp. 41-60.

Madari, B., P.L.O.A. Machado, E. Torres, A.G. Andrade, and L.I.O. Valencia, 2005: No tillage and crop rotation effects on soil aggregation and organic carbon in a Fhodic Ferralsol from southern Brazil. *Soil and Tillage Research,* **80,** pp. 185-200.

Manne, A.S. and R.G. Richels, 2004: A multi-gas approach to climate policy. In *The Global Carbon Cycle. Integrating Humans, Climate, and the Natural World,* C.B. Field and M.R. Raupach (eds.). SCOPE 62, Island Press, Washington DC, pp. 439-452.

Marland, G., B.A. McCarl, and U.A. Schneider, 2001: Soil carbon: policy and economics. *Climatic Change,* **51,** pp. 101-117.

Marland, G., R.A. Pielke Jr., M. Apps, R. Avissar, R.A. Betts, K.J. Davis, P.C. Frumhoff, S.T. Jackson, L.A. Joyce, P. Kauppi, J. Katzenberger, K.G. MacDicken, R.P. Neilson, J.O. Niles, D.S. Niyogi, R.J. Norby, N. Pena, N. Sampson, and Y. Xue, 2003a: The climatic impacts of land surface change and carbon management, and the implications for climate-change mitigation policy. *Climate Policy,* **3,** pp. 149-157.

Marland, G., T.O. West, B. Schlamadinger, and L. Canella, 2003b: Managing soil organic carbon in agriculture: the net effect on greenhouse gas emissions. *Tellus* **55B,** pp. 613-621.

McCarl, B.A. and U.A. Schneider, 2001: Greenhouse gas mitigation in U.S. agriculture and forestry. *Science,* **294,** pp. 2481-2482.

McCrabb, G.C., 2001: Nutritional options for abatement of methane emissions from beef and dairy systems in Australia. In *Greenhouse Gases and Animal Agriculture,* J. Takahashi and B.A. Young (eds.), Elsevier, Amsterdam, pp 115-124.

McCrabb, G.J., M. Kurihara, and R.A. Hunter, 1998: The effect of finishing strategy of lifetime methane production for beef cattle in northern Australia. *Proceedings of the Nutrition Society of Australia,* **22,** 55 pp.

McGinn, S.M., K.A. Beauchemin, T. Coates, and D. Colombatto, 2004: Methane emissions from beef cattle: effects of monensin, sunflower oil, enzymes, yeast, and fumaric acid. *Journal of Animal Science,* **82,** pp. 3346-3356.

McSwiney, C.P. and G.P. Robertson, 2005: Nonlinear response of N_2O flux to incremental fertilizer addition in a continuous maize (*Zea mays* L.) cropping system. *Global Change Biology,* **11,** pp. 1712-1719.

Menon, S., J. Hansen, L. Nazarenko, and Y. Luo, 2002: Climate effects of black carbon aerosols in China and India. *Science,* **297,** pp. 2250-2253.

Miglior, F., B.L. Muir, and B.J. Van Doormaal, 2005: Selection indices in Holstein cattle of various countries. *Journal of Dairy Science,* **88,** pp. 1255-1263.

Millennium Ecosystem Assessment, 2005: Ecosystems and Human Well-Being: Current State and Trends. Findings of the Condition and Trends Working Group. Millennium Ecosystem Assessment Series, Island press, Washington D.C., 815 pp.

Mills, J.A.N., E. Kebreab, C.M. Yates, L.A. Crompton, S.B. Cammell, M.S. Dhanoa, R.E. Agnew, and J. France, 2003: Alternative approaches to predicting methane emissions from dairy cows. *Journal of Animal Science,* **81,** pp. 3141-3150.

Moe, P.W. and H.F. Tyrrell, 1979: Methane production in dairy cows. *Journal of Dairy Science,* **62,** pp. 1583-1586.

Monteny, G.-J., A. Bannink, and D. Chadwick, 2006: Greenhouse gas abatement strategies for animal husbandry. *Agriculture, Ecosystems and Environment,* **112,** pp. 163-170.

Monteny, G.J., C.M. Groenestein, and M.A. Hilhorst, 2001: Interactions and coupling between emissions of methane and nitrous oxide from animal husbandry. *Nutrient Cycling in Agroecosystems,* **60,** pp. 123-132.

Mooney, H., A. Cropper, and W. Reid, 2005: Confronting the human dilemma. *Nature,* **434,** pp. 561-562.

Mooney, S., J.M. Antle, S.M. Capalbo, and K. Paustian, 2004: Influence of project scale on the costs of measuring soil C sequestration. *Environmental Management,* **33(S1),** pp. S252-S263.

Mosier, A. and C. Kroeze, 2000: Potential impact on the global atmospheric N_2O budget of the increased nitrogen input required to meet future global food demands. *Chemosphere-Global Change Science,* **2,** pp. 465-473.

Mosier, A.R., 2001: Exchange of gaseous nitrogen compounds between agricultural systems and the atmosphere. *Plant and Soil,* **228,** pp. 17-27.

Mosier, A.R., 2002: Environmental challenges associated with needed increases in global nitrogen fixation. *Nutrient Cycling in Agroecosystems,* **63,** pp. 101-116.

Mosier, A.R., A.D. Halvorson, G.A. Peterson, G.P. Robertson, and L. Sherrod, 2005: Measurement of net global warming potential in three agroecosystems. *Nutrient Cycling in Agroecosystems,* **72,** pp. 67-76.

Mosier, A.R., J.M. Duxbury, J.R. Freney, O. Heinemeyer, K. Minami, and D.E. Johnson, 1998: Mitigating agricultural emissions of methane. *Climatic Change,* **40,** pp. 39-80.

Murray, B.C., B.A. McCarl, and H-C. Lee, 2004: Estimating leakage from forest carbon sequestration programs. *Land Economics,* **80,** pp. 109-124.

Murray, R.M., A.M. Bryant, and R.A. Leng, 1976: Rate of production of methane in the rumen and the large intestine of sheep. *British Journal of Nutrition,* **36,** pp. 1-14.

Mutuo, P.K., G. Cadisch, A. Albrecht, C.A. Palm, and L. Verchot, 2005: Potential of agroforestry for carbon sequestration and mitigation of greenhouse gas emissions from soils in the tropics. *Nutrient Cycling in Agroecosystems,* **71,** pp. 43-54.

NEPAD, 2005: The NEPAD Framework Document. New Partnership for Africa's Development, <http://www.uneca.org/nepad/> (accessed 26 March 2007).

Newbold, C.J. and L.M. Rode, 2006: Dietary additives to control methanogenesis in the rumen. In *Greenhouse Gases and Animal Agriculture: An Update.* C.R. Soliva, J. Takahashi, and M. Kreuzer (eds.), International Congress Series No. 1293, Elsevier, The Netherlands, pp. 138-147.

Newbold, C.J., J.O. Ouda, S. López, N. Nelson, H. Omed, R.J. Wallace, and A.R. Moss, 2002: Propionate precursors as possible alternative electron acceptors to methane in ruminal fermentation. In *Greenhouse Gases and Animal Agriculture.* J. Takahashi and B.A. Young (eds.), Elsevier, Amsterdam, pp. 151-154.

Newbold, C.J., S. López, N. Nelson, J.O. Ouda, R.J. Wallace, and A.R. Moss, 2005: Proprinate precursors and other metabolic intermediates as possible alternative electron acceptors to methanogenesis in ruminal fermentation in vitro. *British Journal of Nutrition,* **94,** pp. 27-35.

Norby, R.J., M.F. Cotrufo, P. Ineson, E.G. O'Neill, and J.G. Canadell, 2001: Elevated CO_2, litter chemistry, and decomposition: a synthesis. *Oecologia,* **127,** pp. 153-165.

Oelbermann, M., R.P. Voroney, and A.M. Gordon, 2004: Carbon sequestration in tropical and temperate agroforestry systems: a review with examples from Costa Rica and southern Canada. *Agriculture Ecosystems and Environment,* **104,** pp. 359-377.

Oenema, O., N. Wrage, G.L. Velthof, J.W. van Groenigen, J. Dolfing, and P.J. Kuikman, 2005: Trends in global nitrous oxide emissions from animal production systems. *Nutrient Cycling in Agroecosystems,* **72,** pp. 51-65.

Ogle, S.M. and K. Paustian, 2005: Soil organic carbon as an indicator of environmental quality at the national scale: monitoring methods and policy relevance. *Canadian Journal of Soil Science,* **8,** pp. 531-540.

Ogle, S.M., F.J. Breidt, and K. Paustian, 2005: Agricultural management impacts on soil organic carbon storage under moist and dry climatic conditions of temperate and tropical regions. *Biogeochemistry,* **72,** pp. 87-121.

Ogle, S.M., F.J. Breidt, M.D. Eve, and K. Paustian, 2003: Uncertainty in estimating land use and management impacts on soil organic storage for US agricultural lands between 1982 and 1997. *Global Change Biology,* **9,** pp. 1521-1542.

Ogle, S.M., R.T. Conant, and K. Paustian, 2004: Deriving grassland management factors for a carbon accounting approach developed by the Intergovernmental Panel on Climate Change. *Environmental Management,* **33,** pp. 474-484.

Olesen, J.E., K. Schelde, A. Weiske, M.R. Weisbjerg, W.A.H. Asman, and J. Djurhuus, 2006: Modelling greenhouse gas emissions from European conventional and organic dairy farms. *Agriculture, Ecosystems and Environment,* **112,** pp. 207-220.

Olsson, L. and J. Ardö, 2002: Soil carbon sequestration in degraded semiarid agro-ecosystems - perils and potentials. *Ambio,* **31,** pp. 471-477.

Pan, G.X., P. Zhou, X.H. Zhang, L.Q. Li, J.F. Zheng, D.S. Qiu, and Q.H. Chu, 2006: Effect of different fertilization practices on crop C assimilation and soil C sequestration: a case of a paddy under a long-term fertilization trial from the Tai Lake region, China. *Acta Ecologica Sinica,* **26**(11), pp. 3704-3710.

Patra, A.K., D.N. Kamra, and N. Agarwal, 2006: Effect of spices on rumen fermentation, methanogenesis and protozoa counts in in vitro gas production test. In *Greenhouse Gases and Animal Agriculture: An Update,* C.R. Soliva, J. Takahashi, and M. Kreuzer (eds.), International Congress Series No. 1293, Elsevier, The Netherlands, pp. 176-179.

Pattey, E., M.K. Trzcinski, and R.L. Desjardins, 2005: Quantifying the reduction of greenhouse gas emissions as a result of composting dairy and beef cattle manure. *Nutrient Cycling in Agroecosystems,* **72,** pp. 173-187.

Paul, E.A., S.J. Morris, J. Six, K. Paustian, and E.G. Gregorich, 2003: Interpretation of soil carbon and nitrogen dynamics in agricultural and afforested soils. *Soil Science Society of America Journal,* **67,** pp. 1620-1628.

Paustian, K., B.A. Babcock, J. Hatfield, R. Lal, B.A. McCarl, S. McLaughlin, A. Mosier, C. Rice, G.P. Robertson, N.J. Rosenberg, C. Rosenzweig, W.H. Schlesinger, and D. Zilberman, 2004: Agricultural Mitigation of Greenhouse Gases: Science and Policy Options. CAST (Council on Agricultural Science and Technology) Report, R141 2004, ISBN 1-887383-26-3, 120 pp.

Paustian, K., C.V. Cole, D. Sauerbeck, and N. Sampson, 1998: CO_2 mitigation by agriculture: An overview. *Climatic Change,* **40,** pp. 135-162.

Pérez-Ramírez, J., F. Kapteijn, K. Schöffel, and J.A. Moulijn, 2003: Formation and control of N_2O in nitric acid production: Where do we stand *today? Applied Catalysis B: Environmental,* **44,** pp. 117-151.

Phetteplace, H.W., D.E. Johnson, and A.F. Seidl, 2001: Greenhouse gas emissions from simulated beef and dairy livestock systems in the United States. *Nutrient Cycling in Agroecosystems,* **60,** 9-102.

Pinares-Patiño, C.S., M.J. Ulyatt, G.C. Waghorn, C.W. Holmes, T.N. Barry, K.R. Lassey, and D.E. Johnson, 2003: Methane emission by alpaca and sheep fed on lucerne hay or grazed on pastures of perennial ryegrass/white clover or birdsfoot trefoil. *Journal of Agricultural Science,* **140,** pp. 215-226.

Rao, C.H., 1994: *Agricultural Growth, Rural Poverty and Environmental Degradation in India.* Oxford University Press, Delhi.

Reay, D.S., K.A. Smith, and A.C. Edwards, 2003: Nitrous oxide emission from agricultural drainage waters. *Global Change Biology,* **9,** pp. 195-203.

Reeder, J.D., G.E. Schuman, J.A. Morgan, and D.R. Lecain, 2004: Response of organic and inorganic carbon and nitrogen to long-term grazing of the shortgrass steppe. *Environmental Management,* **33,** pp. 485-495.

Rees, W.E., 2003: A blot on the land. *Nature,* **421,** 898 pp.

Reilly, J.M., F. Tubiello, B.A. McCarl, and J. Melillo, 2001: Climate change and agriculture in the United States. In *Climate Change Impacts on the United States: US National Assessment of the Potential Consequences of Climate Variability and Change: Foundation,* Cambridge University Press, Cambridge, Chapter 13, pp 379-403.

Rice, C.W. and C.E. Owensby, 2001: Effects of fire and grazing on soil carbon in rangelands. In *The Potential of U.S. Grazing Lands to Sequester Carbon and Mitigate the Greenhouse Effect.* R. Follet, J.M. Kimble, and R. Lal (eds.), Lewis Publishers, Boca Raton, Florida, pp. 323-342.

Richter, B., 2004: Using ethanol as an energy source. *Science,* **305,** 340 pp.

Robertson, G.P. and P.R. Grace, 2004: Greenhouse gas fluxes in tropical and temperate agriculture: The need for a full-cost accounting of global warming potentials. *Environment, Development and Sustainability,* **6,** pp. 51-63.

Robertson, G.P., 2004: Abatement of nitrous oxide, methane and other non-CO_2 greenhouse gases: the need for a systems approach. In *The global carbon cycle. Integrating Humans, Climate, and the Natural World,* C.B. Field, and M.R. Raupach (eds.). SCOPE 62, Island Press, Washington D.C., pp. 493-506.

Robertson, G.P., E.A. Paul, and R.R. Harwood, 2000: Greenhouse gases in intensive agriculture: Contributions of individual gases to the radiative forcing of the atmosphere. *Science,* **289,** pp. 1922-1925.

Robertson, L.J. and G.C. Waghorn, 2002: Dairy industry perspectives on methane emissions and production from cattle fed pasture or total mixed rations in New Zealand. *Proceedings of the New Zealand Society of Animal Production,* **62,** pp. 213-218.

Rochette, P. and H.H. Janzen, 2005: Towards a revised coefficient for estimating N_2O emissions from legumes. *Nutrient Cycling in Agroecosystems,* **73,** pp. 171-179.

Rockström, J., 2003: Water for food and nature in drought-prone tropics: vapour shift in rain-fed agriculture. *Philosophical Transactions of the Royal Society of London B,* **358,** pp. 1997-2009.

Rogner, H., M. Cabrera, A. Faaij, M. Giroux, D. Hall, V. Kagramanian, Serguei, T.Lefevre, R. Moreira, R. Notstaller, P. Odell, M. Taylor, 2000: *Energy Resources.* In *World Energy Assessment of the United Nations,* J. Goldemberg *et al.* (eds.), UNDP, UNDESA/WEC. UNDP, New York, Chapter 5, pp. 135-171.

Romanenkov, V., I. Romanenko, O. Sirotenko, and L. Shevtsova, 2004: Soil carbon sequestration strategy as a component of integrated agricultural sustainability policy under climate change. In *Proceedings of Russian National Workshop on Research Related to the IHDP on Global Environmental Change,* November 10-12, 2004, IHDP, Moscow, Russia, pp. 180-189.

Rose, S., H. Ahammad, B. Eickhout, B. Fisher, A. Kurosawa, S. Rao, K. Riahi, and D. van Vuuren, 2007. *Land in Climate Stabilization Modeling.* Energy Modeling Forum Report, Stanford University < http://www.stanford.edu/group/EMF/projects/group21/Landuse.pdf > accessed 26 March 2007.

Rosegrant, M., M.S. Paisner, and S. Meijer, 2001: *Long-Term Prospects for Agriculture and the Resource Base*. The World Bank Rural Development Family. Rural Development Strategy Background Paper #1. The World Bank, Washington.

Rosegrant, M.W. and S.A. Cline, 2003: Global food security: challenges and policies. *Science*, **302**, pp. 1917-1919.

Rounsevell, M.D.A, I. Reginster, M.B. Araújo, T.R. Carter, N. Dendoncker, F. Ewert, J.I. House, S. Kankaanpää, R. Leemans, M.J. Metzger, C. Schmit, P. Smith, and G. Tuck, 2006: A coherent set of future land use change scenarios for Europe. *Agriculture Ecosystems and Environment*, **114**, pp. 57-68.

Roy, R.N., R.V. Misra, and A. Montanez, 2002: Decreasing reliance on mineral nitrogen - yet more food. *Ambio*, **31**, pp. 177-183.

Rumpler, W.V., D.E. Johnson, and D.B. Bates, 1986: The effect of high dietary cation concentrations on methanogenesis by steers fed with or without ionophores. *Journal of Animal Science,* **62**, pp. 1737-1741.

Sanchez, P.A. and M.S. Swaminathan, 2005: Cutting world hunger in half. *Science*, **307**, pp. 357-359.

Sanchez, P.A., 2002: Soil fertility and hunger in Africa. *Science*, **295**, pp. 2019-2020.

Sands, R.D. and B.A. McCarl, 2005: Competitiveness of terrestrial greenhouse gas offsets: are they a bridge to the future? In *Abstracts of USDA Symposium on Greenhouse Gases and Carbon Sequestration in Agriculture and Forestry*, March 22-24, USDA, Baltimore, Maryland.

Schils, R.L.M., A. Verhagen, H.F.M. Aarts, and L.B.J. Sebek, 2005: A farm level approach to define successful mitigation strategies for GHG emissions from ruminant livestock systems. *Nutrient Cycling in Agroecosystems,* **71**, pp. 163-175.

Schlesinger, W.H., 1999: Carbon sequestration in soils. *Science,* **284**, pp. 2095

Schmidely, P., 1993: Quantitative review on the use of anabolic hormones in ruminants for meat production. I. Animal performance. *Annales de Zootechie,* **42**, pp. 333-359.

Schnabel, R.R., A.J. Franzluebbers, W.L. Stout, M.A. Sanderson, and J.A. Stuedemann, 2001: The effects of pasture management practices. In *The Potential of U.S. Grazing Lands to Sequester Carbon and Mitigate the Greenhouse Effect* R.F. Follett, J.M. Kimble, and R. Lal (eds.), Lewis Publishers, Boca Raton, Florida, pp. 291-322.

Schneider, U.A. and B.A. McCarl, 2003: Economic potential of biomass based fuels for greenhouse gas emission mitigation. *Environmental and Resource Economics,* **24**, pp. 291-312.

Scholes, R.J. and M.R. van der Merwe, 1996: Sequestration of carbon in savannas and woodlands. *The Environmental Professional,* **18,** pp. 96-103.

Scholes, R.J. and R. Biggs, 2004: Ecosystem services in southern Africa: a regional assessment. CSIR, Pretoria.

Schuman, G.E., J.E. Herrick, and H.H. Janzen, 2001: The dynamics of soil carbon in rangelands. In *The Potential of U.S. Grazing Lands to Sequester Carbon and Mitigate the Greenhouse Effect,* R.F. Follett, J.M. Kimble, and R. Lal (eds.), Lewis Publishers, Boca Raton, Florida, pp. 267-290.

Sheehan, J., A. Aden, K. Paustian, K. Killian, J. Brenner, M. Walsh, and R. Nelson, 2004: Energy and environmental aspects of using corn stover for fuel ethanol. *Journal of Industrial Ecology,* **7**, pp. 117-146.

Sims, R.E.H., A. Hastings, B. Schlamadinger, G. Taylor, and P. Smith, 2006: Energy crops: current status and future prospects. *Global Change Biology,* **12**, pp. 1-23.

Sisti, C.P.J., H.P. Santos, R. Kohhann, B.J.R. Alves, S. Urquiaga, and R.M. Boddey, 2004: Change in carbon and nitrogen stocks in soil under 13 years of conventional or zero tillage in southern Brazil. *Soil and Tillage Research,* **76**, pp. 39-58.

Six, J., S.M. Ogle, F.J. Breidt, R.T. Conant, A.R. Mosier, and K. Paustian, 2004: The potential to mitigate global warming with no-tillage management is only realized when practised in the long term. *Global Change Biology,* **10**, pp.155-160.

Sleutel, S., 2005: *Carbon Sequestration in Cropland Soils: Recent Evolution and Potential of Alternative Management Options*. PhD. thesis, Ghent University, Ghent, Belgium.

Smeets, E.M.W., A.P.C. Faaij, I.M. Lewandowski, and W.C. Turkenburg, 2007: A bottom up quickscan and review of global bio-energy potentials to 2050. *Progress in Energy and Combustion Science*, **33**, pp. 56-106.

Smith, J.U., P. Smith, M. Wattenbach, S. Zaehle, R. Hiederer, R.J.A. Jones, L. Montanarella, M.D.A. Rounsevell, I. Reginster, and F. Ewert, 2005b: Projected changes in mineral soil carbon of European croplands and grasslands, 1990-2080. *Global Change Biology,* **11,** pp. 2141-2152.

Smith, K.A. and F. Conen, 2004: Impacts of land management on fluxes of trace greenhouse gases. *Soil Use and Management,* **20**, pp. 255-263.

Smith, P., 2004a: Carbon sequestration in croplands: the potential in Europe and the global context. *European Journal of Agronomy,* **20**, pp. 229-236.

Smith, P., 2004b: Engineered biological sinks on land. In *The Global Carbon Cycle. Integrating humans, climate, and the natural world*, C.B. Field and M.R. Raupach (eds.). SCOPE 62, Island Press, Washington D.C., pp. 479-491.

Smith, P., 2004c: Monitoring and verification of soil carbon changes under Article 3.4 of the Kyoto Protocol. *Soil Use and Management,* **20**, pp. 264-270.

Smith, P., D. Martino, Z. Cai, D. Gwary, H.H. Janzen, P. Kumar, B. McCarl, S. Ogle, F. O'Mara, C. Rice, R.J. Scholes, O. Sirotenko, M. Howden, T. McAllister, G. Pan, V. Romanenkov, U. Schneider, S. Towprayoon, M. Wattenbach, and J.U. Smith, 2007a: Greenhouse gas mitigation in agriculture. *Philosophical Transactions of the Royal Society, B.,* **363.** doi:10.1098/rstb.2007.2184.

Smith, P., D. Martino, Z. Cai, D. Gwary, H.H. Janzen, P. Kumar, B.A. McCarl, S.M. Ogle, F. O'Mara, C. Rice, R.J. Scholes, O. Sirotenko, M. Howden, T. McAllister, G. Pan, V. Romanenkov, U.A. Schneider, and S. Towprayoon, 2007b: Policy and technological constraints to implementation of greenhouse gas mitigation options in agriculture. *Agriculture, Ecosystems and Environment,* **118,** pp. 6-28.

Smith, P., D.S. Powlson, J.U. Smith, P.D. Falloon, and K. Coleman, 2000: Meeting Europe's climate change commitments: quantitative estimates of the potential for carbon mitigation by agriculture. *Global Change Biology,* **6**, pp. 525-539.

Smith, P., D.S. Powlson, M.J. Glendining, and J.U. Smith, 1997: Potential for carbon sequestration in European soils: preliminary estimates for five scenarios using results from long-term experiments. *Global Change Biology,* **3**, pp. 67-79.

Smith, P., K.W. Goulding, K.A. Smith, D.S. Powlson, J.U. Smith, P.D. Falloon, and K. Coleman, 2001: Enhancing the carbon sink in European agricultural soils: Including trace gas fluxes in estimates of carbon mitigation potential. *Nutrient Cycling in Agroecosystems,* **60**, pp. 237-252.

Smith, P., O. Andrén, T. Karlsson, P. Perälä, K. Regina, M. Rounsevell, and B. Van Wesemael, 2005a: Carbon sequestration potential in European croplands has been overestimated. *Global Change Biology,* **11**, pp. 2153-2163.

Smith, S.J. and T.M.L. Wigley, 2006: Multi-Gas Forcing Stabilization with the MiniCAM. Multi-Greenhouse Gas Mitigation and Climate Policy, *Energy Journal,* Special Issue #**3**. Available at: <http://www.iaee.org/en/publications/journal.aspx> accessed 26 March 2007.

Soliva, C.R., J. Takahashi, and M. Kreuzer (eds.), 2006: *Greenhouse Gases and Animal Agriculture: An Update*. International Congress Series No. 1293, Elsevier, The Netherlands, 377 pp.

Soussana, J.-F., P. Loiseau, N. Viuchard, E. Ceschia, J. Balesdent, T. Chevallier, and D. Arrouays, 2004: Carbon cycling and sequestration opportunities in temperate grasslands. *Soil Use and Management,* **20**, pp. 219-230.

South Africa, 1988: *Fire Act*. Government Printer, Pretoria, South Africa.

Spatari, S., Y. Zhang, and H.L. Maclean, 2005: Life cycle assessment of switchgrass- and corn stover-derived ethanol-fueled automobiles. *Environmental Science and Technology,* **39**, pp. 9750-9758.

Sperow, M., M. Eve, and K. Paustian, 2003: Potential soil C sequestration on U.S. agricultural soils. *Climatic Change,* **57**, pp. 319-339.

Squires, V., E.P. Glenn, and A.T. Ayoub, 1995: *Combating Global Climate Change by Combating Land Degradation*. Proceedings of Workshop in Nairobi, 4-8 Sept. 1995. UNEP, Nairobi, Kenya. 348 pp.

Strengers, B., R. Leemans, B. Eickhout, B. de Vries, and L. Bouwman, 2004: The land-use projections and resulting emissions in the IPCC SRES scenarios as simulated by the IMAGE 2.2 model. *GeoJournal,* **61,** pp. 381-393.

Tate, K.R., D.J. Ross, N.A. Scott, N.J. Rodda, J.A. Townsend, and G.C. Arnold, 2006: Post-harvest patterns of carbon dioxide production, methane uptake and nitrous oxide production in a *Pinus radiata* D. Don plantation. *Forest Ecology and Management,* **228,** pp. 40-50.

Tejo, P., 2004: *Public Policies and Agriculture in Latin America During the 2000's.* CEPAL (Comisión Económica para América Latina), Serie Desarrollo Productivo 152 (in Spanish), Santiago, Chile, 74 pp.

Tilman, D., J. Fargione, B. Wolff, C. D'Antonio, A. Dobson, R. Howarth, D. Schindler, W.H. Schlesinger, D. Simberloff, and D. Swackhamer, 2001: Forecasting agriculturally driven global environmental change. *Science,* **292,** pp. 281-284.

Torbert, H.A., S.A. Prior, H.H. Hogers, and C.W. Wood, 2000: Review of elevated atmospheric CO_2 effects on agro-ecosystems: residue decomposition processes and soil C storage. *Plant and Soil,* **224,** pp. 59-73.

Totten, M., S.I. Pandya, and T. Janson-Smith, 2003: Biodiversity, climate, and the Kyoto Protocol: risks and opportunities. *Frontiers in Ecology and Environment,* **1,** pp. 262-270.

Trewavas, A., 2002: Malthus foiled again and again. *Nature,* **418,** pp. 668-670.

Unkovich, M., 2003: Water use, competition, and crop production in low rainfall, alley farming systems of south-eastern Australia. *Australian Journal of Agricultural Research,* **54,** pp. 751-762.

USCCSP, 2006: *Scenarios of Greenhouse Gas Emissions and Atmospheric Concentrations.* Report by the U.S. Climate Change Science Program and approved by the Climate Change Science Program Product Development Advisory Committee. Clarke, L., J. Edmonds, J. Jacoby, H. Pitcher, J. Reilly, R. Richels (eds.), U.S. Climate Change Science Program.

US-EPA, 2005: *Greenhouse Gas Mitigation Potential in U.S. Forestry and Agriculture.* United States Environmental Protection Agency, EPA 430-R-05-006, November 2005. Washington, D.C., <http://epa.gov/ sequestration/pdf/ghg_part2.pdf> accessed 26 March 2007.

US-EPA, 2006a: *Global Anthropogenic Non-CO_2 Greenhouse Gas Emissions: 19902020.* United States Environmental Protection Agency, EPA 430-R-06-003, June 2006. Washington, D.C., < http://www.epa. gov/nonco2/econ-inv/downloads/GlobalAnthroEmissionsReport.pdf > accessed 26 March 2007.

US-EPA, 2006b: *Global Mitigation of Non-CO_2 Greenhouse Gases.* United States Environmental Protection Agency, EPA 430-R-06-005, Washington, D.C. < http://www.epa.gov/nonco2/econ-inv/downloads/ GlobalMitigationFullReport.pdf > accessed 26 March 2007.

Van Groenigen, K.J., A. Gorissen, J. Six, D. Harris, P.J. Kuikman, J.W. van Groenigen, and C. van Kessel, 2005: Decomposition of ^{14}C-labeled roots in a pasture soil exposed to 10 years of elevated CO_2. *Soil Biology and Biochemistry,* 37, pp. 497-506.

Van Nevel, C.J. and D.I. Demeyer, 1995: Lipolysis and biohydrogenation of soybean oil in the rumen in vitro: Inhibition by antimicrobials. *Journal of Dairy Science,* **78,** pp. 2797-2806.

Van Nevel, C.J. and D.I. Demeyer, 1996: Influence of antibiotics and a deaminase inhibitor on volatile fatty acids and methane production from detergent washed hay and soluble starch by rumen microbes in vitro. *Animal Feed Science and Technology,* 37, pp. 21-31.

Van Oost, K., G. Govers, T.A., Quine, and G. Heckrath, 2004: Comment on 'Managing soil carbon' (I). *Science,* **305,** 1567 pp.

Van Wilgen, B.W., N. Govender, H.C. Biggs, D. Ntsala, and X.N. Funda, 2004: Response of savanna fire regimes to changing fire-management policies in a large African National Park. *Conservation Biology,* **18,** pp. 1533-1540.

Venkataraman, C., G. Habib, A. Eiguren-Fernandez, A.H. Miguel, and S.K. Friedlander, 2005: Residential biofuels in south Asia: carbonaceous aerosol emissions and climate impacts. *Science,* **307,** pp. 1454-1456.

Viner, D., M. Sayer, M. Uyarra, and N. Hodgson, 2006: *Climate Change and the European Countryside: Impacts on Land Management and Response Strategies.* Report prepared for the Country Land and Business Association, UK. Publ., CLA, UK, 180 pp.

Waghorn, G.C., M.H. Tavendale, and D.R. Woodfield, 2002: Methanogenesis from forages fed to sheep. *Proceedings of New Zealand Society of Animal Production,* **64,** pp. 161-171.

Wallace, R.J., T.A. Wood, A. Rowe, J. Price, D.R. Yanez, S.P. Williams, and C.J. Newbold, 2006: Encapsulated fumaric acid as a means of decreasing ruminal methane emissions. In *Greenhouse Gases and Animal Agriculture: An Update.* Soliva, C.R., J. Takahashi and M. Kreuzer (eds.), International Congress Series No. 1293, Elsevier, The Netherlands, pp. 148-151.

Wander, M. and T. Nissen, 2004: Value of soil organic carbon in agricultural lands. *Mitigation and Adaptation Strategies for Global Change,* **9,** pp. 417-431.

Wang, B., H. Neue, and H. Samonte, 1997: Effect of cultivar difference on methane emissions. *Agriculture, Ecosystems and Environment,* **62,** pp. 31-40.

Wang, M.X. and X.J. Shangguan, 1996: CH_4 emission from various rice fields in PR China. *Theoretical and Applied Climatology,* **55,** pp. 129-138.

Wassmann, R., R.S. Lantin, H.U. Neue, L.V. Buendia, T.M. Corton, and Y. Lu, 2000: Characterization of methane emissions from rice fields in Asia. III. Mitigation options and future research needs. *Nutrient Cycling Agroecosystems,* **58,** pp. 23-36.

West, T.O. and G. Marland, 2003: Net carbon flux from agriculture: Carbon emissions, carbon sequestration, crop yield, and land-use change. *Biogeochemistry,* **63,** pp. 73-83.

West, T.O. and W.M. Post, 2002: Soil organic carbon sequestration rates by tillage and crop rotation: A global data analysis. *Soil Science Society of America Journal,* **66,** pp. 1930-1946.

Wolin, E.A., R.S. Wolf, and M.J. Wolin, 1964: Microbial formation of methane. *Journal of Bacteriology,* **87,** pp. 993-998.

Woodward, S.L., G.C. Waghorn, M.J. Ulyatt, and K.R. Lassey, 2001: Early indications that feeding *Lotus* will reduce methane emissions in ruminants. *Proceedings of the New Zealand Society of Animal Production,* **61,** pp. 23-26.

World Bank, 2003: *World Development Indicators* [CD-ROM]. The World Bank, Washington, D.C..

World Bank, 2005: *India Development Strategy Paper 2005-2008.* The World Bank, Washington D.C..

Wright, A.D.G., P. Kennedy, C.J. O'Neill, A.F. Troovey, S. Popovski, S.M. Rea, C.L. Pimm, and L. Klein. 2004: Reducing methane emissions in sheep by immunization against rumen methanogens. *Vaccine,* **22,** pp. 3976-3985.

Xiang, C.G, P.J. Zhang, G.X. Pan, D.S. Qiu, and Q.H. Chu, 2006: Changes in diversity, protein content, and amino acid composition of earthworms from a paddy soil under different long-term fertilizations in the Tai Lake Region, China. *Acta Ecologica Sinica,* **26**(6), pp. 1667-1674.

Xu, H., Z.C. Cai, and H. Tsuruta, 2003: Soil moisture between rice-growing seasons affects methane emission, production, and oxidation. *Soil Science Society of America* Journal, **67,** pp. 1147-1157.

Xu, H., Z.C. Cai, Z.J. Jia, and H. Tsuruta, 2000: Effect of land management in winter crop season on CH_4 emission during the following flooded and rice-growing period. *Nutrient Cycling in Agroecosystems,* **58,** pp. 327-332.

Xu, S., X. Hao, K. Stanford, T. McAllister, F.J. Larney, and J. Wang, 2007: Greenhouse gas emissions during co-composting of cattle mortalities with manure. *Nutrient Cycling in Agroecosystems,* **78,** 177-187.

Yagi, K., H. Tsuruta, and K. Minami, 1997: Possible options for mitigating methane emission from rice cultivation. *Nutrient Cycling in Agroecosystems,* **49,** pp. 213-220.

Yan, T., R.E. Agnew, F.J. Gordon, and M.G. Porter, 2000: Prediction of methane energy output in dairy and beef cattle offered grass silage-based diets. *Livestock Production Science,* **64,** pp. 253-263.

Yan, X., T. Ohara, and H. Akimoto, 2003: Development of region-specific emission factors and estimation of methane emission from rice field in East, Southeast and South Asian countries. *Global Change Biology,* **9,** pp. 237-254.

9

Forestry

Coordinating Lead Authors:
Gert Jan Nabuurs (The Netherlands), Omar Masera (Mexico)

Lead Authors:
Kenneth Andrasko (USA), Pablo Benitez-Ponce (Equador), Rizaldi Boer (Indonesia), Michael Dutschke (Germany), Elnour Elsiddig (Sudan), Justin Ford-Robertson (New Zealand), Peter Frumhoff (USA), Timo Karjalainen (Finland), Olga Krankina (Russia), Werner A. Kurz (Canada), Mitsuo Matsumoto (Japan), Walter Oyhantcabal (Uruguay), Ravindranath N.H. (India), Maria José Sanz Sanchez (Spain), Xiaquan Zhang (China)

Contributing Authors:
Frederic Achard (Italy), Carlos Anaya (Mexico), Sander Brinkman (The Netherlands), Wenjun Chen (Canada), Raymond E. (Ted) Gullison (Canada), Niro Higuchi (Brazil), Monique Hoogwijk (The Netherlands), Esteban Jobbagy (Argentina), G. Cornelis van Kooten (Canada), Franck Lecocq (France), Steven Rose (USA), Bernhard Schlamadinger (Austria), Britaldo Silveira Soares Filho (Brazil), Brent Sohngen (USA), Bart Strengers (The Netherlands), Eveline Trines (The Netherlands)

Review Editors:
Mike Apps (Canada), Eduardo Calvo (Peru)

This chapter should be cited as:
Nabuurs, G.J., O. Masera, K. Andrasko, P. Benitez-Ponce, R. Boer, M. Dutschke, E. Elsiddig, J. Ford-Robertson, P. Frumhoff, T. Karjalainen, O. Krankina, W.A. Kurz, M. Matsumoto, W. Oyhantcabal, N.H. Ravindranath, M.J. Sanz Sanchez, X. Zhang, 2007: Forestry. In Climate Change 2007: Mitigation. Contribution of Working Group III to the Fourth Assessment Report of the Intergovernmental Panel on Climate Change [B. Metz, O.R. Davidson, P.R. Bosch, R. Dave, L.A. Meyer (eds)], Cambridge University Press, Cambridge, United Kingdom and New York, NY, USA.

Table of Contents

EXECUTIVE SUMMARY

During the last decade of the 20th century, deforestation in the tropics and forest regrowth in the temperate zone and parts of the boreal zone remained the major factors responsible for emissions and removals, respectively. However, the extent to which the carbon loss due to tropical deforestation is offset by expanding forest areas and accumulating woody biomass in the boreal and temperate zones is an area of disagreement between land observations and estimates by top-down models. Emissions from deforestation in the 1990s are estimated at 5.8 $GtCO_2$/yr (*medium agreement, medium evidence*).

Bottom-up regional studies show that forestry mitigation options have the economic potential at costs up to 100 US$/$tCO_2$-eq to contribute 1.3-4.2 $GtCO_2$-eq/yr (average 2.7 $GtCO_2$-eq/yr) in 2030. About 50% can be achieved at a cost under 20 US$/$tCO_2$_eq (around 1.6 $GtCO_2$/yr) with large differences between regions. Global top-down models predict far higher mitigation potentials of 13.8 $GtCO_2$-eq/yr in 2030 at carbon prices less than or equal to 100 US$/$tCO_2$-eq. Regional studies tend to use more detailed data and a wider range of mitigation options are reviewed, Thus, these studies may more accurately reflect regional circumstances and constraints than simpler, more aggregate global models. However, regional studies vary in model structure, coverage, analytical approach, and assumptions (including baseline assumptions). In the sectoral comparison in Section 11.3, the more conservative estimate from regional studies is used. Further research is required to narrow the gap in the potential estimates from global and regional assessments (*medium agreement, medium evidence*).

The carbon mitigation potentials from reducing deforestation, forest management, afforestation, and agro-forestry differ greatly by activity, regions, system boundaries and the time horizon over which the options are compared. In the short term, the carbon mitigation benefits of reducing deforestation are greater than the benefits of afforestation. That is because deforestation is the single most important source, with a net loss of forest area between 2000 and 2005 of 7.3 million ha/yr.

Mitigation options by the forestry sector include extending carbon retention in harvested wood products, product substitution, and producing biomass for bioenergy. This carbon is removed from the atmosphere and is available to meet society's needs for timber, fibre, and energy. Biomass from forestry can contribute 12-74 EJ/yr to energy consumption, with a mitigation potential roughly equal to 0.4-4.4 $GtCO_2$/yr depending on the assumption whether biomass replaces coal or gas in power plants (*medium agreement, medium evidence*).

In the long term, a sustainable forest management strategy aimed at maintaining or increasing forest carbon stocks, while producing an annual sustained yield of timber, fibre or energy from the forest, will generate the largest sustained mitigation benefit. Most mitigation activities require up-front investment with benefits and co-benefits typically accruing for many years to decades. The combined effects of reduced deforestation and degradation, afforestation, forest management, agro-forestry and bioenergy have the potential to increase from the present to 2030 and beyond (*medium agreement, medium evidence*).

Global change will impact carbon mitigation in the forest sector but the magnitude and direction of this impact cannot be predicted with confidence as yet. Global change may affect growth and decomposition rates, the area, type, and intensity of natural disturbances, land-use patterns, and other ecological processes (*medium agreement, medium evidence*).

Forestry can make a very significant contribution to a low-cost global mitigation portfolio that provides synergies with adaptation and sustainable development. However, this opportunity is being lost in the current institutional context and lack of political will to implement and has resulted in only a small portion of this potential being realized at present (*high agreement, much evidence*).

Globally, hundreds of millions of households depend on goods and services provided by forests. This underlines the importance of assessing forest sector activities aimed at mitigating climate change in the broader context of sustainable development and community impact. Forestry mitigation activities can be designed to be compatible with adapting to climate change, maintaining biodiversity, and promoting sustainable development. Comparing environmental and social co-benefits and costs with the carbon benefit will highlight trade-offs and synergies, and help promote sustainable development (*low agreement, medium evidence*).

Realization of the mitigation potential requires institutional capacity, investment capital, technology RD and transfer, as well as appropriate policies and incentives, and international cooperation. In many regions, their absence has been a barrier to implementation of forestry mitigation activities. Notable exceptions exist, however, such as regional successes in reducing deforestation rates and implementing large-scale afforestation programmes. Considerable progress has been made in technology development for implementation, monitoring and reporting of carbon benefits but barriers to technology transfer remain (*high agreement, much evidence*).

Forestry mitigation activities implemented under the Kyoto Protocol, including the Clean Development Mechanism (CDM), have to date been limited. Opportunities to increase activities include simplifying procedures, developing certainty over future commitments, reducing transaction costs, and building confidence and capacity among potential buyers, investors and project participants (*high agreement, medium evidence*).

While the assessment in this chapter identifies remaining uncertainties about the magnitude of mitigation benefits and costs, the technologies and knowledge required to implement mitigation activities exist today.

9.1 Introduction

In the context of global change and sustainable development, forest management activities play a key role through mitigation of climate change. However, forests are also affected by climate change and their contribution to mitigation strategies may be influenced by stresses possibly resulting from it. Socio-economically, global forests are important because many citizens depend on the goods, services, and financial values provided by forests. Within this context, mitigation options have to be sought.

The world's forests have a substantial role in the global carbon cycle. IPCC (2007a) reports the latest estimates for the terrestrial sink for the decade 1993-2003 at 3,300 MtCO$_2$/yr, ignoring emissions from land-use change (Denman et al., 2007, Table 7.1). The most likely estimate of these emissions for 1990s is 5,800 MtCO$_2$/yr, which is partly being sequestered on land as well (IPCC, 2007a).

The IPCC Third Assessment Report (TAR) (Kauppi et al., 2001) concluded that the forest sector has a biophysical mitigation potential of 5,380 MtCO$_2$/yr on average up until 2050, whereas the SR LULUCF (IPCC, 2000a) presented a biophysical mitigation potential on all lands of 11670 MtCO$_2$/yr in 2010 (copied in IPCC, 2001, p. 110).

Forest mitigation options include reducing emissions from deforestation and forest degradation, enhancing the sequestration rate in existing and new forests, providing wood fuels as a substitute for fossil fuels, and providing wood products for more energy-intensive materials. Properly designed and implemented, forestry mitigation options will have substantial co-benefits in terms of employment and income generation opportunities, biodiversity and watershed conservation, provision of timber and fibre, as well as aesthetic and recreational services.

Many barriers have been identified that preclude the full use of this mitigation potential. This chapter examines the reasons for the discrepancy between a large theoretical potential and substantial co-benefits versus the rather low implementation rate.

Developments since TAR

Since the IPCC Third Assessment Report (TAR), new mitigation estimates have become available from local to global scale (Sathaye et al., 2007) as well as major economic reviews and global assessments (Stern, 2006). There is early research into the integration of mitigation and adaptation options and the linkages to sustainable development (MEA, 2005a). There is increased attention to reducing emissions from deforestation as a low cost mitigation option, and with significant positive side-effects (Stern, 2006). There is some evidence that climate change impacts can also constrain the forest potential. There are very few multiple land-use studies that examine a wider set of forest functions and economic constraints (Brown et al., 2004). Furthermore, the literature shows a large variation of mitigation estimates, partly due to the natural variability in the system, but partly also due to differences in baseline assumptions and data quality. In addition, Parties to the Convention are improving their estimates through the design of National Systems for Greenhouse Gas (GHG) Inventories.

Basic problems remain. Few major forest-based mitigation analyses have been conducted using new primary data. There is still limited insight regarding impacts on soils, lack of integrated views on the many site-specific studies, hardly any integration with climate impact studies, and limited views in relation to social issues and sustainable development. Little new effort was reported on the development of global baseline scenarios of land-use change and their associated carbon balance, against which mitigation options could be examined. There is limited quantitative information on the cost-benefit ratios of mitigation interventions. Finally, there are still knowledge gaps in how forest mitigation activities may alter, for example, surface hydrology and albedo (IPCC, 2007b: Chapter 4).

This chapter: a) provides an updated estimate of the economic mitigation potential through forests; b) examines the reasons for difference between a large theoretical potential and a low rate of implementation; and c) and integrates the estimates of the economic potential with considerations to both adaptation and mitigation in the context of sustainable development.

9.2 Status of the sector and trends

9.2.1 Forest area

The global forest cover is 3952 million ha (Table 9.1), which is about 30 percent of the world's land area (FAO, 2006a). Most relevant for the carbon cycle is that between 2000 and 2005, gross deforestation continued at a rate of 12.9 million ha/yr. This is mainly as a result of converting forests to agricultural land, but also due to expansion of settlements, infrastructure, and unsustainable logging practices (FAO, 2006a; MEA, 2005b). In the 1990s, gross deforestation was slightly higher, at 13.1 million ha/yr. Due to afforestation, landscape restoration and natural expansion of forests, the most recent estimate of net loss of forest is 7.3 million ha/yr. The loss is still largest in South America, Africa and Southeast Asia (Figure 9.1). This net loss was less than that of 8.9 million ha/yr in the 1990s.

Thus, carbon stocks in forest biomass decreased in Africa, Asia, and South America, but increased in all other regions. According to FAO (2006a), globally net carbon stocks in forest biomass decreased by about 4,000 MtCO$_2$ annually between 1990 and 2005 (Table 9.1).

Table 9.1: *Estimates of forest area, net changes in forest area (negative numbers indicating decrease), carbon stock in living biomass, and growing stock in 1990, 2000, and 2005*

Region	Forest area, (mill. ha)	Annual change (mill. ha/yr)		Carbon stock in living biomass (MtCO$_2$)			Growing stock in 2005
	2005	1990-2000	2000-2005	1990	2000	2005	million m³
Africa	63,5412	-4.4	-4.0	241,267	228,067	222,933	64,957
Asia	571,577	-0.8	1.0	150,700	130,533	119,533	47,111
Europe[a]	1001,394	0.9	0.7	154,000	158,033	160,967	107,264
North and Central America	705,849	-0.3	-0.3	150,333	153,633	155,467	78,582
Oceania	206,254	-0.4	-0.4	42,533	41,800	41,800	7,361
South America	831,540	-3.8	-4.3	358,233	345,400	335,500	128,944
World	3,952,026	-8.9	-7.3	1,097,067	1,057,467	1,036,200	434,219

a) Including all of the Russian Federation
Source: FAO, 2006a

The area of forest plantation was about 140 million ha in 2005 and increased by 2.8 million ha/yr between 2000 and 2005, mostly in Asia (FAO, 2006a). According to the Millennium Ecosystem Assessment (2005b) scenarios, forest area in industrialized regions will increase between 2000 and 2050 by about 60 to 230 million ha. At the same time, the forest area in the developing regions will decrease by about 200 to 490 million ha. In addition to the decreasing forest area globally, forests are severely affected by disturbances such as forest fires, pests (insects and diseases) and climatic events including drought, wind, snow, ice, and floods. All of these factors have also carbon balance implications, as discussed in Sections 9.3 and 9.4. Such disturbances affect roughly 100 million ha of forests annually (FAO, 2006a). Degradation, defined as decrease of density or increase of disturbance in forest classes, affected tropical regions at a rate of 2.4 million ha/yr in the 1990s.

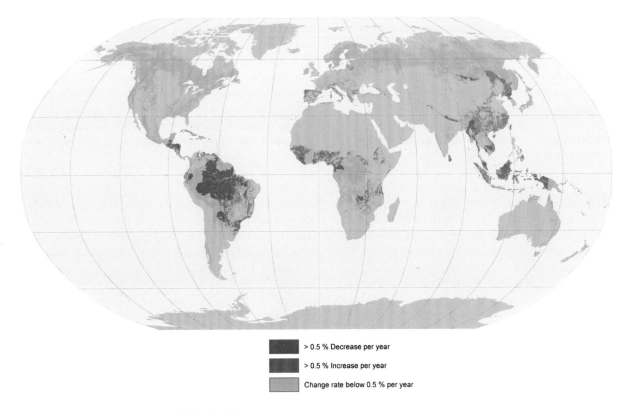

> 0.5 % Decrease per year

> 0.5 % Increase per year

Change rate below 0.5 % per year

Figure 9.1: *Net change in forest area between 2000 and 2005*
Source: FAO, 2006a.

9.2.2   Forest management

Data on progress towards sustainable forest management were collected for the recent global forest resources assessment (FAO, 2006a). These data indicate globally there are many good signs and positive trends (intensive forest plantation and rising conservation efforts), but also negative trends continue (primary forests continue to become degraded or converted to agriculture in some regions). Several tools have been developed in the context of sustainable forest management, including criteria and indicators, national forest programmes, model forests and certification schemes. These tools can also support and provide sound grounds for mitigation of climate change and thus carbon sequestration.

Nearly 90% of forests in industrialized countries are managed "according to a formal or informal management plan" (FAO, 2001). National statistics on forest management plans are not available for many developing countries. However, preliminary estimates show that at least 123 million ha, or about 6% of the total forest area in these countries is covered by a "formal, nationally approved forest management plan covering a period of at least five years." Proper management plans are seen as prerequisites for the development of management strategies that can also include carbon-related objectives.

Market-based development of environmental services from forests, such as biodiversity conservation, carbon sequestration, watershed protection, and nature-based tourism, is receiving attention as a tool for promoting sustainable forest management. Expansion of these markets may remain slow and depends on government intervention (Katila and Puustjärvi, 2004). Nevertheless, development of these markets and behaviour of forest owners may influence roundwood markets and availability of wood for conventional uses, thus potentially limiting substitution possibilities.

9.2.3   Wood supply, production and consumption of forest products

Global wood harvest is about 3 billion m^3 and has been rather stable in the last 15 years (FAO, 2006a). Undoubtedly, the amount of wood removed is higher, as illegally wood removal is not recorded. About 60% of removals are industrial roundwood; the rest is wood fuel (including fuelwood and charcoal). The most wood removal in Africa and substantial proportions in Asia and South America are non-commercial wood fuels. Recently, commercial biomass for bioenergy received a boost because of the high oil prices and the government policies initiated to promote renewable energy sources.

Although accounting for only 5% of global forest cover, forest plantations were estimated in 2000 to supply about 35% of global roundwood harvest and this percentage is expected to increase (FAO, 2006a). Thus, there is a trend towards concentrating the harvest on a smaller forest area. Meeting society's needs for timber through intensive management of a smaller forest area creates opportunities for enhanced forest protection and conservation in other areas, thus contributing to climate change mitigation. With rather stable harvested volumes, the manufacture of forest products has increased as a result of improved processing efficiency. Consumption of forest products is increasing globally, particularly in Asia.

9.3   Regional and global trends in terrestrial greenhouse gas emissions and removals

Mitigation measures will occur against the background of ongoing change in greenhouse gas emissions and removals. Understanding current trends is critical for evaluation of additional effects from mitigation measures. Moreover, the potential for mitigation depends on the legacy of past and present patterns of change in land-use and associated emissions and removals. The contribution of the forest sector to greenhouse gas emissions and removals from the atmosphere remained the subject of active research, which produced an extensive body of literature (Table 9.2 and IPCC, 2007a: Chapter 7 and 10).

Globally during the 1990s, deforestation in the tropics and forest regrowth in the temperate zone and parts of the boreal zone were the major factors responsible for emissions and removals, respectively (Table 9.2; Figure 9.2). However, the extent to which carbon loss due to tropical deforestation is offset by expanding forest areas and accumulating woody biomass in the boreal and temperate zones is the area of disagreement between land observations and estimates by top-down models. The top-down methods based on inversion of atmospheric transport models estimate the net terrestrial carbon sink for the 1990s, which is the balance of sinks in northern latitudes and source in tropics (Gurney et al., 2002). The latest estimates are consistent with the increase found in the terrestrial carbon sink in the 1990s over the 1980s.

Denman et al. (2007) reports the latest estimates for gross residual terrestrial sink for the 1990s at 9,500 MtCO$_2$/yr, while their estimate for emissions from deforestation amounts to 5,800 MtCO$_2$/yr. The residual sink estimate is significantly higher than any land-based global sink estimate and in the upper range of estimates produced by inversion of atmospheric transport models (Table 9.2). It includes the sum of biases in estimates of other global fluxes (fossil fuel burning, cement production, ocean uptake, and land-use change) and the flux in terrestrial ecosystems that are not undergoing change in land use.

Improved spatial resolution allowed separate estimates of the land-atmosphere carbon flux for some continents (Table 9.2). These estimates generally suggest greater sink or smaller source than the bottom-up estimates based on analysis of forest inventories and remote sensing of change in land-cover

(Houghton, 2005). While the estimates of forest expansion and regrowth in the temperate and boreal zones appear relatively well constrained by available data and consistent across published results, the rates of tropical deforestation are uncertain and hotly debated (Table 9.2; Fearnside and Laurance, 2004). Studies based on remote sensing of forest cover report lower rates than UN-ECE/FAO (2000) and lower carbon emissions carbon (Achard *et al.*, 2004).

Recent analyses highlight the important role of other carbon flows. These flows were largely overlooked by earlier research and include carbon export through river systems (Raymond and Cole, 2003), volcanic activity and other geological processes (Richey *et al.*, 2002), transfers of material in and out of products pool (Pacala *et al.*, 2001), and uptake in freshwater ecosystems (Janssens *et al.*, 2003).

Attribution of estimated carbon sink in forests to the short- and long-term effects of the historic land-use change and shifting natural disturbance patterns on one hand, and to the effects of N and CO_2 fertilization and climate change on the other, remains problematic (Houghton, 2003b). For the USA, for example, the fraction of carbon sink attributable to changes in land-use and land management might be as high as 98% (Caspersen *et al.*, 2000), or as low as 40% (Schimel *et al.*, 2001). Forest expansion and regrowth and associated carbon sinks were reported in many regions (Table 9.2; Figure 9.2). The expanding tree cover in South Western USA is attributed to the long-term effects of fire control but the gain in carbon storage was smaller than previously thought. The lack of consensus on factors that control the carbon balance is an obstacle to development of effective mitigations strategies.

Large year-to-year and decade scale variation of regional carbon sinks (Rodenbeck *et al.*, 2003) make it difficult to define distinct trends. The variation reflects the effects of climatic variability, both as a direct impact on vegetation and through the effects of wild fires and other natural disturbances. There are indications that higher temperatures in boreal regions will increase fire frequency; possible drying of the Amazon basin would increase fire frequency there as well (Cox *et al.*, 2004). Global emissions from fires in the 1997/98 El Nino year are estimated at 7,700 $MtCO_2$/yr, 90% from tropics (Werf *et al.*, 2004).

The picture emerging from Table 9.2 is complex because available estimates differ in the land-use types included and in the use of gross fluxes versus net carbon balance, among other variables. This makes it impossible to set a widely accepted baseline for the forestry sector globally. Thus, we had to rely on the baselines used in each regional study separately (Section 9.4.3.1), or used in each global study (Section 9.4.3.3). However, this approach creates large uncertainty in assessing the overall mitigation potential in the forest sector. Baseline CO_2 emissions from land-use change and forestry in 2030 are the same as or slightly lower than in 2000 (see Chapter 3, Figure 3.10).

9.4   Assessment of mitigation options

In this section, a conceptual framework for the assessment of mitigation options is introduced and specific options are briefly described. Literature results are summarized and compared for regional bottom-up approaches, global forest sector models, and global top-down integrated model approaches. The assessment is limited to CO_2 balances and economic costs of the various mitigation options. Broader issues including biodiversity, sustainable development, and interactions with adaptation strategies are discussed in subsequent sections.

9.4.1   Conceptual introduction

Terrestrial carbon dynamics are characterized by long periods of small rates of carbon uptake, interrupted by short periods of

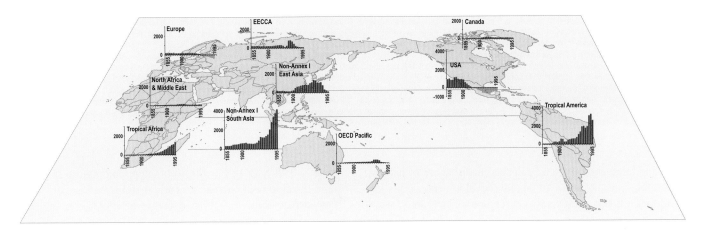

Figure 9.2: *Historical forest carbon balance (MtCO₂) per region, 1855-2000.*

Notes: green = sink. EECCA=Countries of Eastern Europe, the Caucasus and Central Asia. Data averaged per 5-year period, year marks starting year of period.
Source: Houghton, 2003b.

Table 9.2: *Selected estimates of carbon exchange of forests and other terrestrial vegetation with the atmosphere (in MtCO2/yr)*

Regions	Annual carbon flux based on international statistics	Annual carbon flux during 1990s	
	UN-ECE, 2000	Based on inversion of atmospheric transport models	Based on land observations
		MtCO$_2$/yr	
OECD North America		1,833 ± 2,200[9]	0 ÷ 1,100[5]
Separately: Canada	340		293 ± 733[1]
USA	610	2,090 ± 3,337[2]	
OECD Pacific	224		0±733[1]
Europe	316	495 ± 752[6]	0 ± 733[1]
			513[11]
Countries in Transition	1,726	3,777 ± 3,447[2]	1,100 ± 2,933[9]
			1,181 ÷ -1,588[7]
Separately: Russia	1,572	4,767 ± 2,933[9]	1,907± 469[8]
Northern Africa		623 ± 3,593[2]	
Sub-Saharan Africa			-576 ±235[3]
			-440 ± 110[4]
			-1,283 ± 733[1]
Caribbean, Central and South America		-2,310	-1,617 ± 972[3]
			-1,577 ± 733[4]
			-2,750 ± 1,100[1]
Separately: Brazil			± 733[12]
Developing countries of South and East Asia and Middle East		-2,493 ± 2,713[2]	-3,997 ± 1,833[1]
			-1,734 ± 550[3]
			-1,283 ± 550[4]
Separately: China		2,273 ± 2,420[2]	- 110 ± 733[1]
			128 ± 95[13]
			249[14]
Global total		4,767 ± 5,500[9]	-7,993 ± 2,933[1]
		2,567 ± 2,933[10]	-3,300 ÷ 7,700[5]
		4,913[2]	-4,000[15]
		9516[17]	-5,800 [16]
			-8485[18]
Annex I (excluding Russia)			1300[19]

Notes: Positive values represent carbon sink, negative values represent source. Sign ÷ indicates a range of values; sign ± indicates error term.
Because of differences in methods and scope of studies (see footnotes), values from different publications are not directly comparable. They represent a sample of reported results.
1 Houghton 2003a (flux from changes in land use and land management based on land inventories); 2 Gurney et al., 2002 (inversion of atmospheric transport models, estimate for Countries in Transition applies to Europe and boreal Asia; estimate for China applies to temperate Asia); 3 Achard et al., 2004 (estimates based on remote sensing for tropical regions only); 4 DeFries, 2002 (estimates based on remote sensing for tropical regions only); 5 Potter et al., 2003 (NEP estimates based on remote sensing for 1982-1998 and ecosystem modelling, the range reflects inter-annual variability); 6 Janssens et al., 2003 (combined use of inversion and land observations; includes forest, agricultural lands and peatlands between Atlantic Ocean and Ural Mountains, excludes Turkey and Mediterranean isles); 7 Shvidenko and Nilson, 2003 (forests only, range represents difference in calculation methods); 8 Nilsson et al., 2003 (includes all vegetation); 9 Ciais et al., 2000 (inversion of atmospheric transport models, estimate for Russia applies to Siberia only); 10 Plattner et al., 2002 (revised estimate for 1980's is 400±700); 11Nabuurs et al., 2003 (forests only); 12 Houghton et al., 2000 (Brazilian Amazon only, losses from deforestation are offset by regrowth and carbon sink in undisturbed forests); 13 Fang et al., 2005; 14 Pan et al., 2004, 15 FAO, 2006a (global net biomass loss resulting from deforestation and regrowth); 16 Denman *et al.*,2007 (estimate of biomass loss from deforestation), 17 Denman et al.,2007 (Residual terrestrial carbon sink), 18 EDGAR database for agriculture and forestry (see Chapter 1, Figure 1.3a/b (Olivier et al., 2005)). These include emissions from bog fires and delayed emissions from soils after land- use change, 19 (Olivier et al., 2005).

rapid and large carbon releases during disturbances or harvest. Depending on the stage of stand[1] development, individual stands are either carbon sources or carbon sinks (1m^3 of wood stores ~ 0.92 tCO$_2$)^2. For most immature and mature stages of stand development, stands are carbon sinks. At very old ages, ecosystem carbon will either decrease or continue to increase

---

1 In this chapter, 'stand' refers to an area of trees of similar characteristics (e.g., species, age, stand structure or management regime) while 'forest' refers to a larger estate comprising many stands.
2 Assuming a specific wood density of 0.5g dry matter/cm3 and a carbon content of 0.5g C/g dry matter.

slowly with accumulations mostly in dead organic matter and soil carbon pools. In the years following major disturbances, the losses from decay of residual dead organic matter exceed the carbon uptake by regrowth. While individual stands in a forest may be either sources or sinks, the forest carbon balance is determined by the sum of the net balance of all stands. The theoretical maximum carbon storage (saturation) in a forested landscape is attained when all stands are in old-growth state, but this rarely occurs as natural or human disturbances maintain stands of various ages within the forest.

The design of a forest sector mitigation portfolio should consider the trade-offs between increasing forest ecosystem carbon stocks and increasing the sustainable rate of harvest and transfer of carbon to meet human needs (Figure 9.3). The selection of forest sector mitigation strategies should minimize net GHG emissions throughout the forest sector and other sectors affected by these mitigation activities. For example, stopping all forest harvest would increase forest carbon stocks, but would reduce the amount of timber and fibre available to meet societal needs. Other energy-intensive materials, such as concrete, aluminium, steel, and plastics, would be required to replace wood products, resulting in higher GHG emissions (Gustavsson *et al.*, 2006). Afforestation may affect the net GHG balance in other sectors, if for example, forest expansion reduces agricultural land area and leads to farming practices with higher emissions (e.g., more fertilizer use), conversion of land for cropland expansion elsewhere, or increased imports of agricultural products (McCarl and Schneider, 2001). The choice of system boundaries and time horizons affects the ranking of mitigation activities (Figure 9.3).

Forest mitigation strategies should be assessed within the framework of sustainable forest management, and with consideration of the climate impacts of changes to other processes such as albedo and the hydrological cycle (Marland *et al.*, 2003). At present, however, few studies provide such comprehensive assessment.

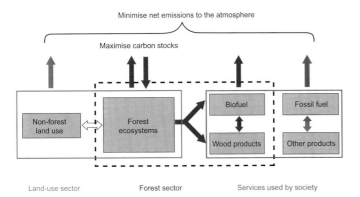

Figure 9.3: *Forest sector mitigation strategies need to be assessed with regard to their impacts on carbon storage in forest ecosystems on sustainable harvest rates and on net GHG emissions across all sectors.*

For the purpose of this discussion, the options available to reduce emissions by sources and/or to increase removals by sinks in the forest sector are grouped into four general categories:

- maintaining or increasing the forest area through reduction of deforestation and degradation and through afforestation/reforestation;
- maintaining or increasing the stand-level carbon density (tonnes of carbon per ha) through the reduction of forest degradation and through planting, site preparation, tree improvement, fertilization, uneven-aged stand management, or other appropriate silviculture techniques;
- maintaining or increasing the landscape-level carbon density using forest conservation, longer forest rotations, fire management, and protection against insects;
- increasing off-site carbon stocks in wood products and enhancing product and fuel substitution using forest-derived biomass to substitute products with high fossil fuel requirements, and increasing the use of biomass-derived energy to substitute fossil fuels.

Each mitigation activity has a characteristic time sequence of actions, carbon benefits and costs (Figure 9.4). Relative to a baseline, the largest short-term gains are always achieved through mitigation activities aimed at emission avoidance (e.g., reduced deforestation or degradation, fire protection, and slash burning). But once an emission has been avoided, carbon stocks on that forest will merely be maintained or increased slightly. In contrast, the benefits from afforestation accumulate over years to decades but require up-front action and expenses. Most forest management activities aimed at enhancing sinks require up-front investments. The duration and magnitude of their carbon benefits differ by region, type of action and initial condition of the forest. In the long term, sustainable forest management strategy aimed at maintaining or increasing forest carbon stocks, while producing an annual yield of timber, fibre, or energy from the forest, will generate the largest sustained mitigation benefit.

Reduction in fossil fuel use in forest management activities, forest nursery operations, transportation and industrial production provides additional opportunities similar to those in other sectors, but are not discussed here (e.g., see Chapter 5, Transportation). The options available in agro-forestry systems are conceptually similar to those in other parts of the forest sector and in the agricultural sector (e.g., non-CO_2 GHG emission management). Mitigation using urban forestry includes increasing the carbon density in settlements, but indirect effects must also be evaluated, such as reducing heating and cooling energy use in houses and office buildings, and changing the albedo of paved parking lots and roads.

9.4.2   Description of mitigation measures

Each of the mitigation activities is briefly described. The development of a portfolio of forest mitigation activities requires

	Mitigation Activities	Type of Impact	Timing of Impact	Timing of Cost
1A	Increase forest area (e.g. new forests)	⇑	∫	⌐
1B	Maintain forest area (e.g. prevent deforestation, LUC)	⇓	⌐	⌐
2A	Increase site-level C density (e.g. intensive management, fertilize)	⇑	∫	⌐
2B	Maintain site-level C density (e.g. avoid degradation)	⇓	⌐	⌐
3A	Increase landscape-scale C stocks (e.g. SFM, agriculture, etc.)	⇑	∫	⌐
3B	Maintain landscape-scale C stocks (e.g. suppress disturbances)	⇓	⌐	⌐
4A	Increase off-site C in products (but must also meet 1B, 2B and 3B)	⇑	⌐	⌐
4B	Increase bioenergy and substitution (but must also meet 1B, 2B and 3B)	⇓	⌐	⌐

Legend

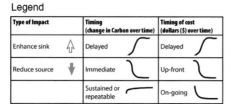

Type of Impact		Timing (change in Carbon over time)		Timing of cost (dollars ($) over time)	
Enhance sink	⇑	Delayed	∫	Delayed	⌐
Reduce source	⇓	Immediate	⌐	Up-front	⌐
		Sustained or repeatable	⌐	On-going	⌐

Figure 9.4: *Generalized summary of forest sector options and type and timing of effects on carbon stocks and the timing of costs* [3]

an understanding of the magnitude and temporal dynamics of the carbon benefits and the associated costs.

9.4.2.1 Maintaining or increasing forest area: reducing deforestation and degradation

Deforestation - human-induced conversion of forest to non-forest land uses - is typically associated with large immediate reductions in forest carbon stock, through land clearing. Forest degradation - reduction in forest biomass through non-sustainable harvest or land-use practices - can also result in substantial reductions of forest carbon stocks from selective logging, fire and other anthropogenic disturbances, and fuelwood collection (Asner *et al.*, 2005).

In some circumstances, deforestation and degradation can be delayed or reduced through complete protection of forests (Soares-Filho *et al.*, 2006), sustainable forest management policies and practices, or by providing economic returns from non-timber forest products and forest uses not involving tree removal (e.g., tourism). Protecting forest from all harvest typically results in maintained or increased forest carbon stocks, but also reduces the wood and land supply to meet other

societal needs.

Reduced deforestation and degradation is the forest mitigation option with the largest and most immediate carbon stock impact in the short term per ha and per year globally (see Section 9.2 and global mitigation assessments below), because large carbon stocks (about 350-900 tCO_2/ha) are not emitted when deforestation is prevented. The mitigation costs of reduced deforestation depend on the cause of deforestation (timber or fuelwood extraction, conversion to agriculture, settlement, or infrastructure), the associated returns from the non-forest land use, the returns from potential alternative forest uses, and on any compensation paid to the individual or institutional landowner to change land-use practices. These costs vary by country and region (Sathaye *et al.*, 2007), as discussed below.

9.4.2.2 Maintaining or increasing forest area: afforestation/reforestation

Afforestation and reforestation are the direct human-induced conversion of non-forest to forest land through planting, seeding, and/or the human-induced promotion of natural seed sources. The two terms are distinguished by how long the non-forest condition has prevailed. For the remainder of this chapter, afforestation is used to imply either afforestation or reforestation. To date, carbon sequestration has rarely been the primary driver of afforestation, but future changes in carbon valuation could result in large increases in the rates of afforestation (US EPA, 2005).

Afforestation typically leads to increases in biomass and dead organic matter carbon pools, and to a lesser extent, in soil carbon pools, whose small, slow increases are often hard to detect within the uncertainty ranges (Paul *et al.*, 2003). Biomass clearing and site preparation prior to afforestation may lead to short-term carbon losses on that site. On sites with low initial soil carbon stocks (e.g., after prolonged cultivation), afforestation can yield considerable soil carbon accumulation rates (e.g., Post and Kwon (2000) report rates of 1 to 1.5 t CO_2/yr). Conversely, on sites with high initial soil carbon stocks, (e.g., some grassland ecosystems) soil carbon stocks can decline following afforestation (e.g., Tate *et al.* (2005) report that in the whole of New Zealand soil carbon losses amount up to 2.2 $MtCO_2$/yr after afforestation). Once harvesting of afforested land commences, forest biomass carbon is transferred into wood products that store carbon for years to many decades. Accumulation of carbon in biomass after afforestation varies greatly by tree species and site, and ranges globally between 1 and 35 t CO_2/ha.yr (Richards and Stokes, 2004).

Afforestation costs vary by land type and region and are affected by the costs of available land, site preparation, and labour. The cost of forest mitigation projects rises significantly

---

3   We thank Mike Apps for a draft of this figure.

when opportunity costs of land are taken into account (VanKooten *et al.*, 2004). A major economic constraint to afforestation is the high initial investment to establish new stands coupled with the several-decade delay until afforested areas generate revenue. The non-carbon benefits of afforestation, such as reduction in erosion or non-consumptive use of forests, however, can more than off-set afforestation cost (Richards and Stokes, 2004).

9.4.2.3   *Forest management to increase stand- and landscape-level carbon density*

Forest management activities to increase stand-level forest carbon stocks include harvest systems that maintain partial forest cover, minimize losses of dead organic matter (including slash) or soil carbon by reducing soil erosion, and by avoiding slash burning and other high-emission activities. Planting after harvest or natural disturbances accelerates tree growth and reduces carbon losses relative to natural regeneration. Economic considerations are typically the main constraint, because retaining additional carbon on site delays revenues from harvest. The potential benefits of carbon sequestration can be diminished where increased use of fertilizer causes greater N_2O emissions. Drainage of forest soils, and specifically of peatlands, may lead to substantial carbon loss due to enhanced respiration (Ikkonen *et al.*, 2001). Moderate drainage, however, can lead to increased peat carbon accumulation (Minkkinen *et al.*, 2002).

Landscape-level carbon stock changes are the sum of stand-level changes, and the impacts of forest management on carbon stocks ultimately need to be evaluated at landscape level. Increasing harvest rotation lengths will increase some carbon pools (e.g., tree boles) and decrease others (e.g., harvested wood products) (Kurz *et al.*, 1998).

9.4.2.4   *Increasing off-site carbon stocks in wood products and enhancing product and fuel substitution*

Wood products derived from sustainably managed forests address the issue of saturation of forest carbon stocks. The annual harvest can be set equal to or below the annual forest increment, thus allowing forest carbon stocks to be maintained or to increase while providing an annual carbon flow to meet society's needs of fibre, timber and energy. The duration of carbon storage in wood products ranges from days (biofuels) to centuries (e.g., houses and furniture). Large accumulations of wood products have occurred in landfills (Micales and Skog, 1997). When used to displace fossil fuels, woodfuels can provide sustained carbon benefits, and constitute a large mitigation option (see Box 9.2).

Wood products can displace more fossil-fuel intensive construction materials such as concrete, steel, aluminium, and plastics, which can result in significant emission reductions (Petersen and Solberg, 2002). Research from Sweden and Finland suggests that constructing apartment buildings with wooden frames instead of concrete frames reduces lifecycle net carbon emissions by 110 to 470 kg CO_2 per square metre of floor area (Gustavsson and Sathre, 2006). The mitigation benefit is greater if wood is first used to replace concrete building material and then after disposal, as biofuel.

9.4.3   Global assessments

For quantification of the economic potential of future mitigation by forests, three approaches are presented in current literature. These are: a) regional bottom-up assessments per country or continent; b) global forest sector models; and c) global multi-sectoral models. An overview of studies for these approaches is presented in Section 9.4.3. The final integrated global conclusion and regional comparison is given in Section 9.4.4. Supply of forest biomass for bioenergy is given in Box 9.2 and incorporated in Section 11.3.1.4, within the energy sector's mitigation potential. For comments on the baselines, see Section 9.3.

9.4.3.1   *Regional bottom-up assessments*

Regional assessments comprise a variety of model results. On the one hand, these assessments are able to take into account the detailed regional specific constraints (in terms of ecological constraints, but also in terms of land owner behaviour and institutional frame). On the other hand, they also vary in assumptions, type of potential addressed, options taken into account, econometrics applied (if any), and the adoption of baselines. Thus, these assessments may have strengths, but when comparing and summing up, they have weaknesses as well. Some of these assessments, by taking into account institutional barriers, are close to a market potential.

Tropics

The available studies about mitigation options differ widely in basic assumptions regarding carbon accounting, costs, land areas, baselines, and other major parameters. The type of mitigation options considered and the time frame of the study affect the total mitigation potential estimated for the tropics. A thorough comparative analysis is, therefore, very difficult. More detailed estimates of economic or market potential for mitigation options by region or country are needed to enable policy makers to make realistic estimates of mitigation potential under various policy, carbon price, and mitigation program eligibility rule scenarios. Examples to build on include Benitez-Ponce *et al. (*2007) and Waterloo *et al.* (2003), highlighting the large potential by avoiding deforestation and enhancing afforestation and reforestation, including bioenergy.

Reducing deforestation

Assumptions of future deforestation rates are key factors in estimates of GHG emissions from forest lands and of mitigation benefits, and vary significantly across studies. In all the studies, however, future deforestation is estimated to remain high in the tropics in the short and medium term. Sathaye *et al.* (2007) estimate that deforestation rates continue in all regions, particularly at high rates in Africa and South America, for a total of just under 600 million ha lost cumulatively by 2050. Using a spatial-explicit model coupled with demographic and economic databases, Soares-Filo *et al.* (2006) predict that, under a business-as-usual scenario, by 2050, projected deforestation trends will eliminate 40% of the current 540 million ha of Amazon forests, releasing approximately 117,000 ± 30,000 MtCO$_2$ of carbon to the atmosphere (Box 9.1).

Reducing deforestation is, thus, a high-priority mitigation option within tropical regions. In addition to the significant carbon gains, substantive environmental and other benefits could be obtained from this option. Successfully implementing mitigation activities to counteract the accelerated loss of tropical forests requires understanding the causes for deforestation, which are multiple and locally based; few generalizations are possible (Chomitz *et al.*, 2006).

Recent studies have been conducted at the national, regional, and global scale to estimate the mitigation potential (areas, carbon benefits and costs) of reducing tropical deforestation. In a short-term context (2008-2012), Jung (2005) estimates that 93% of the total mitigation potential in the tropics corresponds to avoided deforestation. For the Amazon basin, Soares- Filo *et al.* (2006) estimate that by 2050 the cumulative avoided deforestation potential for this region reaches 62,000 MtCO$_2$ under a "governance" scenario (see Box 9.1).

Looking at the long-term, (Sohngen and Sedjo, 2006) estimate that for 27.2 US$/tCO2, deforestation could potentially be virtually eliminated. Over 50 years, this could mean a net cumulative gain of 278,000 MtCO$_2$ relative to the baseline and

422 million additional hectares in forests. For lower prices of 1.36 US$/tCO$_2$, only about 18,000 MtCO$_2$ additional could be sequestered over 50 years. The largest gains in carbon would occur in Southeast Asia, which gains nearly 109,000 MtCO$_2$ for 27.2 US$/tCO$_2$, followed by South America, Africa, and Central America, which would gain 80,000, 70,000, and 22,000 MtCO$_2$ for 27.2 US$/tCO$_2$, respectively (Figure 9.5).

In a study of eight tropical countries covering half of the total forested area, Grieg-Gran (2004) present a best estimate of total costs of avoided deforestation in the form of the net present value of returns from land uses that are prevented, at 5 billion US$ per year. These figures represent costs of 483 US$ to 1050 US$/ha.

Afforestation and reforestation

The assumed land availability for afforestation options depends on the price of carbon and how that competes with existing or other land-use financial returns, barriers to changing land uses, land tenure patterns and legal status, commodity price support, and other social and policy factors.

Cost estimates for carbon sequestration projects for different regions compiled by Cacho *et al.*, (2003) and by Richards and Stokes (2004) show a wide range. The cost is in the range of 0.5 US$ to 7 US$/tCO$_2$ for forestry projects in developing countries, compared to 1.4 US$ to 22 US$/tCO$_2$ for forestry projects in industrialized countries. In the short-term (2008-2012), an estimate of economic potential area available for afforestation/ reforestation under the Clean Development Mechanism (CDM) is estimated to be 5.3 million ha in Africa, Asia and Latin America together, with Asia accounting for 4.4 million ha (Waterloo *et al.*, 2003).

Summing the measures, the cumulative carbon mitigation benefits (Figure 9.6) by 2050 for a scenario of 2.7 US$/ tCO$_2$ + 5% annual carbon price increment for one model are estimated to be 91,400 MtCO$_2$; 59% of it coming from avoided deforestation. These estimates increase for a higher price scenario of 5.4 US$/tCO$_2$ + 3%/yr annual carbon price into 104,800 MtCO$_2$), where 69% of total mitigation comes from avoiding deforestation (Sathaye *et al.*, 2007). During the period 2000-2050, avoided deforestation in South America and Asia dominate by accounting for 49% and 21%, respectively, of the total mitigation potential. When afforestation is considered, Asia dominates. The mitigation potential of the continents Asia, Africa and Latin America dominates the global total mitigation potential for the period up to 2050 and 2100, respectively (Figure 9.6).

In conclusion, the studies report a large variety for mitigation potential in the tropics. All studies indicate that this part of the world has the largest mitigation potential in the forestry sector. For the tropics, the mitigation estimates for lower price ranges (<20 US$/tCO$_2$) are around 1100 MtCO$_2$/yr in 2040, about

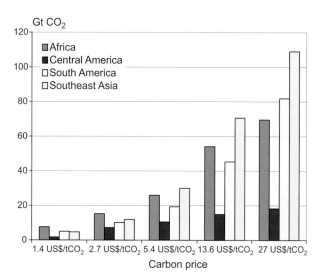

Figure 9.5: *Cumulative carbon gained through avoided deforestation by 2055 over the reference case, by tropical regions under various carbon price scenarios*
Source: Sohngen and Sedjo, 2006.

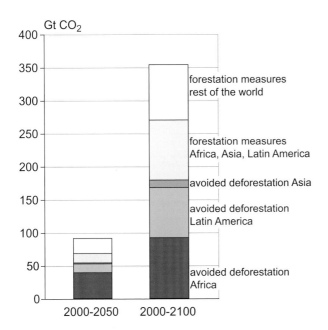

Figure 9.6: *Cumulative mitigation potential (2000-2050 and 2000-2100) according to mitigation options under the 2.7 US$/tCO2 +5%/yr annual carbon price increment*

Source: Sathaye et al., 2007.

half of this potential is located in Central and South America (Sathaye et al., 2007; Soares Filho et al., 2006; Sohngen and Sedjo, 2006). For each of the regions Africa and Southeast Asia, this mitigation potential is estimated at 300 $MtCO_2$/yr in 2040. In the high range of price scenarios (< 100 US$/$tCO_2$), the mitigation estimates are in the range of 3000 to 4000$MtCO_2$/yr in 2040. In the summary overviews in Section 9.4.4, an average estimate of 3500 is used, with the same division over regions: 875, 1750 and 875 for Africa, Latin and South America, and Southeast Asia, respectively. The global economic potential for the tropics ranges from 1100 to 3500 $MtCO_2$/yr in 2040 (Table 9.6).

OECD North America

Figure 9.8 shows the technical potential of management actions aimed at modifying the net carbon balance in Canadian forests (Chen *et al.*, 2000). Of the four scenarios examined, the potential was largest in the scenario aimed at reducing regeneration delays by reforesting after natural disturbances. The second largest estimate was obtained with annual, large-scale (125 million ha) low-intensity (5 kg N/ha/yr) nitrogen fertilization programmes. Neither of these scenarios is realistic,

Box 9.1 Deforestation scenarios for the Amazon Basin

An empirically based, policy-sensitive simulation model of deforestation for the Pan-Amazon basin has been developed (Soares-Filho *et al.*, 2006) (Figure 9.7). Model output for the worst-case scenario (business-as-usual) shows that, by 2050, projected deforestation trends will eliminate 40% of the current 5.4 million km² of Amazon forests, releasing approximately 117,000 $MtCO_2$ cumulatively by 2050. Conversely, under the best-case governance scenario, 4.5 million km² of forest would remain in 2050, which is 83% of the current extent or only 17% deforested, reducing cumulative carbon emissions by 2050 to only 55,000 $MtCO_2$. Current experiments in forest conservation on private properties, markets for ecosystem services, and agro-ecological zoning must be refined and implemented to achieve comprehensive conservation. Part of the financial resources needed for these conservation initiatives could come in the form of carbon credits resulting from the avoidance of 62,000 $MtCO_2$ emissions over 50 years.

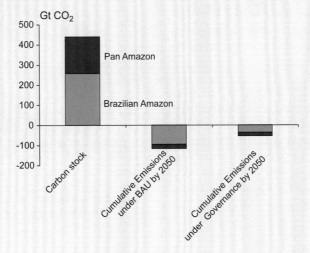

Figure 9.7: *Current carbon stocks for the Pan-Amazon and Brazilian Amazon (left bar) and estimates of cumulative future emission by 2050 from deforestation under BAU (business-as-usual) and governance scenarios.*
Note: The difference between the two scenarios represents an amount equivalent to eight times the carbon emission reduction to be achieved during the first commitment period of the Kyoto Protocol.

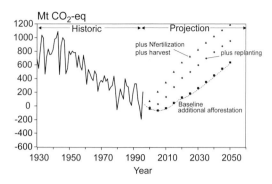
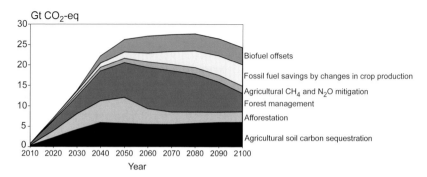

Figure 9.8: *OECD North America: technical potential for the forest sector alone for Canada (left, sink is positive) and the economic potential (at 15 US$/tCO$_2$eq in constant real prices) in the agriculture and forestry sector in the USA (right) Left: Chen et al., 2000; right: US-EPA, 2005*

however, but can be seen as indications of the type of measures and impact on carbon balance (as described by Chen *et al.,* *2000*). Chen's measures sum up to a technical potential of 570 MtCO$_2$/yr. Based on the assumption that the economic potential is about 10% of technical potential (see Section 9.4.3.3. for carbon prices 20 US$/tCO$_2$), the economic potential can be "guesstimated" at around 50-70 MtCO$_2$/yr (Table 9.6).

Other studies have explored the potential of large-scale afforestation in Canada. Mc Kenney *et al.* (2004) project that at a carbon price of 25 US$/tCO$_2$, 7.5 million ha of agricultural land would become economically attractive for poplar plantations. Economic constraints are contributing to the declining trend in afforestation rates in Canada from about 10,000 ha/yr in 1990 to 4,000 ha/yr in 2002 (White and Kurz, 2005).

For the USA, Richards and Stokes (2004) reviewed eight national estimates of forest mitigation and found that carbon prices ranging from 1 to 41 US$/tCO$_2$ generated an economic mitigation potential of 47-2,340 MtCO$_2$/yr from afforestation, 404 MtCO$_2$ from forest management, and 551-2,753 MtCO$_2$/yr from total forest carbon. Sohngen and Mendelsohn (2003) found that a carbon programme with prices rising from 2 US$/tCO$_2$ to 51 US$/tCO$_2$ during this century could induce sequestration of 122 to 306 MtCO$_2$/yr total carbon sequestration, annualized over a 100-year time frame.

US EPA (2005) present that, at 15 US$/tCO$_2$, the mitigation potential of afforestation and forest management (annualized) would amount to 356 MtCO$_2$/yr over a 100-year time frame. At 30 US$/tCO$_2$, this analysis would generate 749 MtCO$_2$ annualized over 100 years. At higher prices and in the long term, the potential was mainly determined by biofuels. With the mitigation potential given above for Canada, the OECD North America sums to a range of 400 to 820 MtCO$_2$/yr in 2040 (Table 9.6).

Europe

Most assessments shown (Figure 9.9) are of the carbon balance of the forest sector of Europe's managed forest as a whole[4]. Additional effects of measures were studied by Cannell (2003), Benitez-Ponce *et al.* (2007), EEA (2005), and Eggers *et al.* (2007). Karjalainen *et al,* (2003) present a projection of the full sector carbon balance (Figure 9.9). Eggers *et al.* (2007) presents the European forest sector carbon sink under two global SRES scenarios, and a maximum difference between scenarios of 197 MtCO$_2$/yr in 2040. Therefore, an additionally achievable sink of 90 to 180 MtCO$_2$/yr was estimated (Table 9.6). Economic analyses were not only done; country studies were done, for example, Hoen and Solberg (1994) for Norway. New European scale economic analyses may be available from the INSEA[5] project, MEACAP project[6], and Carbo Europe [7].

Issues in European forestry where mitigation options can be found include: afforestation of abandoned agricultural

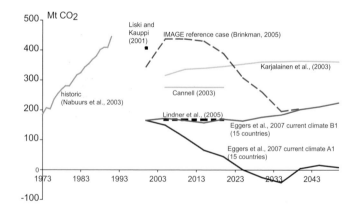

Figure 9.9: *European forest sector carbon sink projections for which various assumptions on implementation rate of measures were made*
Note: positive = sink.

---

4   Europe here excludes the European part of Russia.
5   www.iiasa.ac.at/Research/FOR/INSEA/index.html?sb=19
6   www.ieep.eu/projectMiniSites/meacap/index.php
7   www.carboeurope.org/

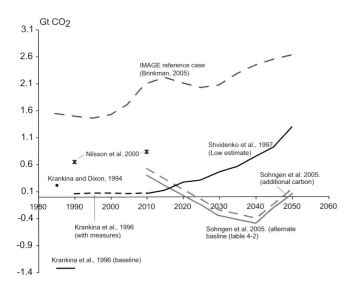

Figure 9.10: *Russian Federation forest sector carbon sink projections, with assumptions regarding implementation rates differing in the various studies*
Note: positive = sink.

lands; bioenergy from complementary fellings; and forest management practices to address carbon saturation in older forests. Furthermore, management of small now under-managed woodlands represent a potential (Viner *et al.,* 2006) and also in combination with adaptation measures in connecting the fragmented nature reserves (Schröter *et al.,* 2005).

Russian Federation

The forests of the Russian Federation include large areas of primary (mostly boreal) forests. Most estimates indicate that the Russian forests are neither a large sink nor a large source. Natural disturbances (fire) play a major role in the carbon balance with emissions up to 1,600 $MtCO_2$/yr (Zhang *et al.,* 2003). Large uncertainty surrounds the estimates for the current carbon balance ((Shvidenko and Nilsson *et al.,* 2003). For the decade 1990-2000, the range of carbon sink values for Russia is 350-750 $MtCO_2$/yr (Nilsson *et al.,* 2003; Izrael *et al.,* 2002). A recent analysis estimated the net sink in Russia at 146-439 $MtCO_2$/yr at present (Sohngen *et al.,* 2005). They projected this baseline to be about 257 $MtCO_2$ per year in 2010, declining to a net source by 2030 as younger forests mature and are harvested. They estimated the economic potential in Russia of afforestation and reforestation at 73-124 $MtCO_2$/yr on average over an 80-year period, for a carbon price of 1.9-3.55 US$/$tCO_2$, and 308-476 $MtCO_2$/yr at prices of more than 27 US$/$tCO_2$ (Figure 9.10). Based on these estimates, the estimated economic mitigation potential would be between 150 and 300 $MtCO_2$/yr in the year 2040 (Table 9.6).

OECD Pacific

Richards and Brack (2004) used estimates of establishment rates for hardwood (short and long rotation) and softwood plantations to model a carbon account for Australia's post-1990 plantation estate. The annual sequestration rate in forests and wood products together is estimated to reach 20 $MtCO_2$/yr in 2020.

New Zealand reached a peak in new planting of around 98,000 ha in 1994 and estimates of stock changes largely depend on afforestation rates (MfE, 2002). If a new planting was maintained at 40,000 ha/yr, the stock increase in forests established since 1990 (117 $MtCO_2$ cumulative since 1990) is estimated to offset *all* increases in emissions in New Zealand since 1990. The total stock increase in all forests would offset *all* emissions increases until 2020.

However, the current new planting rate has declined to 6,000 ha and conversion of 7,000 ha of plantations to pasture has led to net deforestation in the year to March 2005 (MAF, 2006). As a result, the total removal units anticipated to be available during the first commitment period dropped to 56 $MtCO_2$ in 2005 (MfE, 2005). Trotter *et al.* (2005) estimate New Zealand has approximately 1.45 million ha of marginal pastoral land suitable for afforestation. If all of this area was established, total sequestration could range from 10 to 42 $MtCO_2$/yr. This would lead to a removal of approximately 44 to 170 $MtCO_2$ cumulative by 2010 at 13 US$/$tCO_2$.

In Japan, 67% of the land is covered with forests including semi-natural broad-leaved forests and planted coniferous forests mostly. The sequestration potential is estimated in the range of 35 to 70 $MtCO_2$/yr (Matsumoto *et al.*, 2002; Fang *et al.*, 2005), and planted forests account for more than 60% of the carbon sequestration. These assessments show that there is little potential for afforestation and reforestation, while forest management and practices for planted forests including thinning and regeneration are necessary to maintain carbon sequestration and to curb saturation. In addition, there seems to be large potential for bioenergy as a mitigation option.

These three countries for the region lead to an estimate of potential in the range of 85 to 255 $MtCO_2$/yr in 2040 (Table 9.6).

Non-annex I East Asia

East Asia to a large extent formed by China, Korea, and Mongolia has a range of forest covers from a relatively small area of moist tropical forest to large extents of temperate forest and steppe-like shrubland. Country assessments for the forest sector all project a sink ranging from 75 to 400 $MtCO_2$/yr (Zhang and Xu, 2003). Given the large areas and the fast economic development (and thus demand for wood products resulting in increased planting), the additional potential in the region would be in the high range of the country assessments at 150 to 400 $MtCO_2$/yr (Table 9.6). Issues in forestry with which the carbon sequestration goal can be combined sustainably are: reducing degradation of tropical and dry woodlands; halting desertification of the steppes (see Chapter 8); afforestation; and bioenergy from complementary fellings.

9.4.3.2   Global Forest sectoral modelling

Currently, no integrated assessment (Section 9.4.3.3) and climate stabilization economic models (Section 3.3.5) have fully integrated a land use sector with other sectors in the models. Researchers have taken several approaches, however, to account for carbon sequestration in integrated assessment models, either by iterating with the land sector models (e.g., Sohngen and Mendelsohn, 2003), or implementing mitigation response curves generated by the sectoral model (Jakeman and Fisher, 2006). The sectoral model results described here use exogenous carbon price paths to simulate effects of different climate policies and assumptions. The starting point and rate of increase are determined by factors such as the aggressiveness of the abatement policy, abatement option and cost assumptions, and the social discount rate (Sohngen and Sedjo, 2006).

Since TAR, several new global assessments of forest mitigation potential have been produced. These include Benitez-Ponce *et al,* (2004; 2007), Waterloo *et al.* (2003) with a constraints study, Sathaye *et al.* (2007), Strengers *et al.* (2007) Vuuren *et al.* (2007), and Riahi *et al.* (2006). Global estimates are provided that are consistent in methodology across countries and regions, and in terms of measures included. Furthermore, they provide a picture in which the forestry sector is one option that is part of a multi sectoral climate policy and its measures. Thus, these assessments provide insight into whether land-based mitigation is a cost-efficient measure in comparison to other mitigation efforts. Some of these models use a grid-based global land-use model and provide insight into where these models allocate the required afforestation (Figure 9.11).

The IMAGE model (Strengers *et al.*, 2007) allocates bio-energy plantations and carbon plantations mostly in the fringes of the large forest biomes, and in Eastern Europe. The Waterloo study only looked at tropical countries, but found by far the largest potential in China and Brazil. Several models report at the regional level, and project strong avoided deforestation in Africa, the Amazon, and to a lesser extent in Southeast Asia (where land opportunity costs in the timber market are relatively high). Benitez-Ponce (2004) maps geographic distribution of afforestation, adjusted by country risk estimates, under a 50

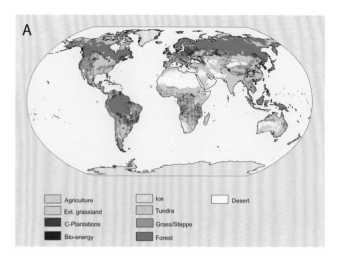

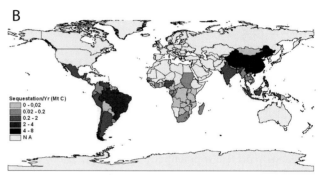

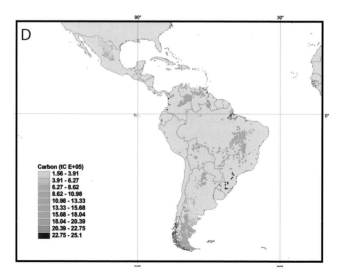

Figure 9.11: *Comparison of allocation of global afforestation in various studies*
(A)   Location of bioenergy and carbon plantations
(B)   Additional sequestration from afforestation per tropical country per year in the period 2008-2012 (MtC/yr),
(C)   Percentage of a grid cell afforested
(D)   Cumulative carbon sequestration through afforestation between 2000 and 2012 in Central and South America).

Source: (A) Strengers et al., 2007; (B) Waterloo et al., 2003; (C) Strengers et al., 2007; (D) Benitez-Ponce et al., 2007.

US$/t carbon price. Afforestation activity is clustered in bands in South-Eastern USA, Southeast Brazil and Northern South America, West Africa, north of Botswana and East Africa, the steppe zone grasslands from Ukraine through European Russia, North- Eastern China, and parts of India, Southeast Asia, and Northern Australia. Hence, forest mitigation is likely to be patchy, but predictable using an overlay of land characteristics, land rental rates, and opportunity costs, risks, and infrastructure capacity.

Several models produced roughly comparable assessments for a set of constant and rising carbon price scenarios in the EMF 21 modelling exercise, from 1.4 US$/tCO$_2$ in 2010 and, rising by 5% per year to 2100, to a 27 US$ constant CO$_2$ price, to 20 US$/tCO$_2$ rising by 1.4 US$/yr though 2050 then capped. This exercise allowed more direct comparison of modelling assumptions than usual. Caveats include: (1) models have varying assumptions about deforestation rates over time, land area in forest in 2000 and beyond, and land available for mitigation; and (2) models have different drivers of land use change (e.g., population and GDP growth for IMAGE, versus land rental rates and timber market demand for GTM).

Global models provide broad trends, but less detail than national or project analyses. Generally global models do not address implementation issues such as transaction costs (likely to vary across activities, regions), barriers, and mitigation programme rules, which tend to drive mitigation potential downward toward true market potential. Political and financial risks in implementing afforestation and reforestation by country were considered by Benitez-Ponce et al. (2007), for example, who found that the sequestration reduced by 59% once the risks were incorporated.

In the last few years, more insight has been gained into carbon supply curves. At a price of 5 US$/tCO$_2$, Sathaye et al. (2007) project a cumulative carbon gain of 10,400 MtCO$_2$ by 2050 (Figure 9.12b). The mitigation results from a combination of avoided deforestation (68%) and afforestation (32%). These results are typical in their very high fraction of mitigation from reduced deforestation. Sohngen and Sedjo (2006) estimate some 80% of carbon benefits in some scenarios from land-use change (e.g., reduced deforestation and afforestation/reforestation) versus some 20% from forest management.

Benitez-Ponce et al. (2007) project that at a price of 13.6 US$/tCO$_2$, the annual sequestration from afforestation and reforestation for the first 20 years amounts to on average 510 MtCO$_2$/yr (Figure 9.12a). For the first 40 years, the average annual sequestration is 805 MtCO$_2$/yr. The single price of 13.6 US$/tCO$_2$ used by Benitez-Ponce et al. (2005) should make afforestation an attractive land-use option in many countries. It covers the range of median values for sequestration costs that Richards and Stokes (2004) give of 1 US$ to 12 US$/tCO$_2$, although VanKooten et al. (2004) present marginal cost results rising far higher. Sathaye et al. (2007) project the economic

potential cumulative carbon gains from afforestation and avoided deforestation together (see also tropics, Section 9.4.3.1.). In the moderate carbon price scenarios, the cumulative carbon gains by 2050 add up to 91,400 to 104,800 MtCO$_2$.

The anticipated carbon price path over time has important implications for forest abatement potential and timing. Rising carbon prices provide an incentive for delaying forest abatement actions to later decades, when it is more profitable (Sohngen and Sedjo, 2006). Carbon price expectations influence forest investment decisions and are, therefore, an important consideration for estimating mitigation potential. Contrary, high constant carbon prices generate significant early mitigation, but the quantity may vary over time. Mitigation strategies need to take into account this temporal dimension if they seek to meet specific mitigation goals at given dates in the future (US EPA, 2005).

Some patterns emerge from the range of estimates reviewed in order to assess the ratio between economic potential and technical potential (Sathaye et al., 2007; Lewandrowski et al., 2004; US EPA, 2005; Richards and Stokes, 2004). The technical potential estimates are generally significantly larger than the economic potential. These studies are difficult to compare, since each estimate uses different assumptions by different analysts. Economic models used for these analyses can generate mitigation potential estimates in competition to other forestry or agricultural sector mitigation options. Generally, they do not specify or account for specific policies and measures and market penetration rates, so few market potential estimates are generated. Many studies do not clearly state which potentials are estimated.

The range of economic potential as a percentage of technical potential is 2% to 100% (the latter against all costs). At carbon prices less than 7 US$/tCO$_2$, the highest estimate of economic potential is 16% of the technical potential. At carbon prices from 27 US$/tCO$_2$ to 50 US$/tCO$_2$, the range of economic potential is estimated to be 58% or higher of the technical potential, a much higher fraction as carbon prices rise. Table 9.3 summarizes mitigation results for four major global forest analyses for a single near-term date of 2030: two forest sector models - GTM (Sohngen and Sedjo, 2006; and GCOMAP (Sathaye et al., 2007), one recent detailed spatially resolved analysis of afforestation (Benitez-Ponce et al., 2007), and one integrated assessment model with detail for the forest sector (IMAGE 2.2, Vuuren et al., 2007). These studies offer roughly comparable results, including global coverage of the forest sector, and land-use competition across at least two forest mitigation options (except Benitez-Ponce et al., 2007). All but the Benitez-Ponce et al. study have been compared by the modelling teams in the EMF 21 modelling exercise (see Sections 3.2.2.3 and 3.3.5) as well.

These global models (Table 9.3) present a large potential for climate mitigation through forestry activities. The global annual

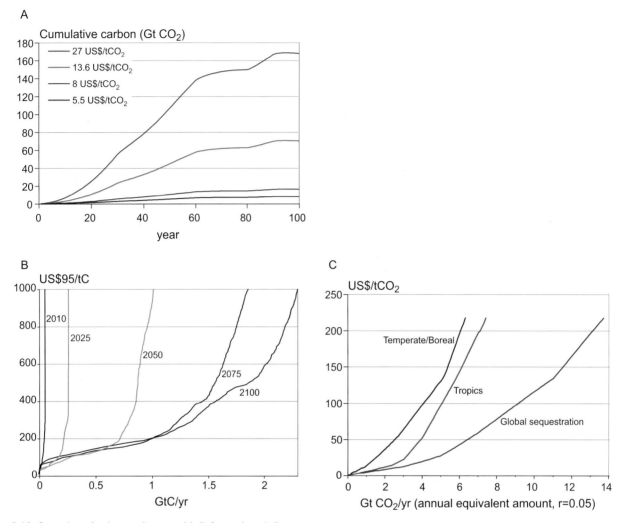

Figure 9.12: *Comparison of carbon supply curves globally from various studies*

(A) Cumulative carbon supply curves: afforestation and reforestation by year and price scenario. At a price of 100 US$/tC after 70 years, some 40 Gt carbon will have been supplied cumulatively from afforestation.

(B) Annual cost-supply curves for abandoned agricultural land in the B2 scenario. For example, at a price of 100 US$/tC, in 2075, some 250 Mt carbon will have been supplied annually from afforestation and reducing deforestation.

(C) Annual marginal cost curves for carbon sequestration in forests: estimates for boreal, temperate, and tropical regions. For example, at a price of 100 US$/tC, some 1400 Mt carbon will have been supplied annually from afforestation and reducing deforestation in 2100.

Sources: (A) Benitez-Ponce et al., 2005; (B) Strengers et al., 2007; (C) Sohngen and Sedjo, 2006.

potential in 2030 is estimated at 13,775 $MtCO_2$/yr (at carbon prices less than or equal to 100 US$/$tCO_2$), 36% (~5000 $MtCO_2$/yr) of which can be achieved under a price of 20 US$/$tCO_2$. Reduced deforestation in Central and South America is the most important measure in a single region with 1,845 $MtCO_2$/yr. The total for the region is the largest for Central and South America with an estimated total potential of 3,100 $MtCO_2$/yr. Regions with a second largest potential, each around 2000 $MtCO_2$, are

Africa, Centrally Planned Asia, other Asia, and USA. These results project significantly higher mitigation than the regional largely bottom-up results. This is somewhat surprising, and likely, the result of the modelling structure, assumptions, and which activities are included. Additional research is required to resolve the various estimates to date using different modelling approaches of the potential magnitude of forestry mitigation of climate change.

Table 9.3: *Potential of mitigation measures of global forestry activities. Global model results indicate annual amount sequestered or emissions avoided, above business as usual, in 2030 for carbon prices 100 US$/tCO$_2$ and less.*

Region	Activity	Potential at costs equal or less than 100 US$/tCO$_2$, in MtCO$_2$/yr in 2030 [1]	Fraction in cost class: 1-20 US$/tCO$_2$	Fraction in cost class: 20-50 US$/tCO$_2$
USA	Afforestation	445	0.3	0.3
	Reduced deforestation	10	0.2	0.3
	Forest management	1,590	0.26	0.32
	TOTAL	2,045	0.26	0.31
Europe	Afforestation	115	0.31	0.24
	Reduced deforestation	10	0.17	0.27
	Forest management	170	0.3	0.19
	TOTAL	295	0.3	0.21
OECD Pacific	Afforestation	115	0.24	0.37
	Reduced deforestation	30	0.48	0.25
	Forest management	110	0.2	0.35
	TOTAL	255	0.25	0.34
Non-annex I East Asia	Afforestation	605	0.26	0.26
	Reduced deforestation	110	0.35	0.29
	Forest management	1,200	0.25	0.28
	TOTAL	1,915	0.26	0.27
Countries in transition	Afforestation	545	0.35	0.3
	Reduced deforestation	85	0.37	0.22
	Forest management	1,055	0.32	0.27
	TOTAL	1,685	0.33	0.28
Central and South America	Afforestation	750	0.39	0.33
	Reduced deforestation	1,845	0.47	0.37
	Forest management	550	0.43	0.35
	TOTAL	3,145	0.44	0.36
Africa	Afforestation	665	0.7	0.16
	Reduced deforestation	1,160	0.7	0.19
	Forest management	100	0.65	0.19
	TOTAL	1,925	0.7	0.18
Other Asia	Afforestation	745	0.39	0.31
	Reduced deforestation	670	0.52	0.23
	Forest management	960	0.54	0.19
	TOTAL	2,375	0.49	0.24
Middle East	Afforestation	60	0.5	0.26
	Reduced deforestation	30	0.78	0.11
	Forest management	45	0.5	0.25
	TOTAL	135	0.57	0.22
TOTAL	Afforestation	4,045	0.4	0.28
	Reduced deforestation	3,950	0.54	0.28
	Forest management	5,780	0.34	0.28
	TOTAL	13,775	0.42	0.28

1) Results average activity estimates reported from three global forest sector models including GTM (Sohngen and Sedjo, 2006), GCOMAP (Sathaye et al., 2007), and IIASA-DIMA (Benitez-Ponce et al., 2007). For each model, output for different price scenarios has been published. The original authors were asked to provide data on carbon supply under various carbon prices. These were summed and resulted in the total carbon supply as given middle column above. Because carbon supply under various price scenarios was requested, fractionation was possible as well.

Two right columns represent the proportion available in the given cost class. None of the models reported mitigation available at negative costs. The column for the carbon supply fraction at costs between 50 and 100 US$/tCO$_2$ can easily be derived as 1- sum of the two right hand columns.

9.4.3.3   Global forest mitigation in climate stabilization analysis

Evaluating the cost-competitiveness of forestry mitigation versus other sector options in achieving climate mitigation goals requires different modelling capabilities. Global integrated assessment and climate economic models are top-down models, generally capable of dynamically representing feedbacks in the economy across sectors and regions and reallocations of inputs, as well as interactions between economic and atmospheric-ocean-terrestrial systems. These models can be used to evaluate long-term climate stabilization scenarios, like achieving a stabilization target of 450 or 650 CO_2-eq by 2100 (see Section 3.3.5). In this framework, the competitive mitigation role of forest abatement options, such as afforestation, can be estimated as part of a dynamic portfolio of the least-cost combination of mitigation options from across all sectors of the economy, including energy, transportation, and agriculture.

To date, researchers have used various approaches to represent terrestrial carbon sequestration in integrated assessment models. These approaches include iterating with the land-sector models (e.g., Sohngen and Mendelsohn, 2003), and implementing mitigation response curves generated by a sectoral model (Jakeman and Fisher, 2006). At present, all integrated assessment models include afforestation strategies, but only some consider avoided deforestation, and none explicitly model forest management mitigation options (e.g., harvest timing: Rose *et al.*, 2007). However, the top-down mitigation estimates account for economic feedbacks, as well as for some biophysical feedbacks such as climate and CO_2 fertilization effects on forest growth.

Table 9.4: *Global forest cost-effective mitigation potential in 2030 from climate stabilization scenarios, or 450-650 CO_2-eq atmospheric concentration targets, produced by top-down global integrated assessment models. Forest options are in competition with other sectoral options to generate least-cost mitigation portfolios for achieving long-run stabilization.*

Carbon price in scenario (US$/t$CO_2$-eq)	Mitigation potential in 2030	
	MtCO_2-eq/yr	Number of scenario results
0 - 20	40 - 970	4
20 - 50	604 - 790	3
50 - 100	nd	0
>100	851	1

Notes: Jakeman and Fisher (2006) estimated 2030 forest mitigation of 3,059 $MtCO_2$, well above other estimates, but not included due to an inconsistency inflating their forest mitigation estimates for the early 21st century.
nd = no data.
Source: Section 3.3.5; data from Rose et al., 2007.

The few estimates of global competitive mitigation potential of forestry in climate stabilization in 2030 are given in Table 9.4. Some estimates represent carbon plantation gains only, while others represent net forest carbon stock changes that include plantations as well as deforestation carbon loses induced by bioenergy crops. On-going top-down land-use modelling developments should produce more refined characterization of forestry abatement alternatives and cost-effective mitigation potential in the near future. The results in Table 9.4 suggest a reasonable central estimate of about 700 million tonne CO_2 in 2030 from forestry in competition with other sectors for achieving stabilization, significantly less than the regional bottom-up or global sector top-down estimates in this chapter summarized in Table 9.7.

Box 9.2: Commercial biomass for bioenergy from forests

Current use of biomass from fuelwood and forest residues reaches 33 EJ (see Section 4.3.3). Three main categories of forest residues may be used for energy purposes: primary residues (available from additional stemwood fellings or as residues (branches) from thinning salvage after natural disturbances or final fellings); secondary residues (available from processing forest products) and tertiary residues (available after end use). Various studies have assessed the future potential supply of forest biomass (Yamamoto et al., 2001; Smeets and Faaij, 2007; Fischer and Schrattenholzer, 2001). Furthermore, some global biomass potential studies include forest residues aggregated with crop residue and waste (Sørensen, 1999). At a regional or national scale, studies are more detailed and often include economic considerations (Koopman, 2005; Bhattacharya et al., 2005; Lindner et al., 2005; Cuiping et al., 2004). Typical values of residue recoverability are between 25 and 50 % of the logging residues and between 33 and 80% of processing residues. Lower values are often assumed for developing regions (Yamamoto et al., 2001; Smeets and Faaij, 2007). At a global level, scenario studies on the future energy mixture (IPCC, 2000c; Sørensen, 1999; OECD, 2006) have included residues from the forestry sector in their energy supply (market potential).

The technical potential of primary biomass sources given by the different global studies is aggregated by region in Table Box 9.2. From this table, it can conclude that biomass from forestry can contribute from about a few percent to about 15% (12 to 74 EJ/yr) of current primary energy consumption. It is outside the scope of this chapter to examine all pros and cons of increased production required for biomass for bioenergy (see Section 11.9).

Box 9.2 continued

Table 9.5. *The technical potential of primary biomass for bioenergy from the forest sector at a regional level (in EJ/yr), for the period 2020-2050. The economic potential under 20 US$/tCO$_2$ is assumed to be in the range of 10-20% of these numbers.*

Regions	EJ/yr	
	LOW	HIGH
OECD		
OECD North America	3	11
OECD Europe	1	4
Japan + Australia + New Zealand	1	3
Economies in Transition		
Central and Eastern Europe, the Caucasus and Central Asia	2	10
Non-OECD		
Latin America	1	21
Africa	1	10
Non-Annex I East Asia	1	5
Non-Annex I Other Asia	1	8
Middle East	1	2
World low and high estimates	*12*	*74*
World (based on global studies) assumed economic potential	**14**	**65**

Notes: Conversion factors used: 0.58 tonne dry matter/m^3, a heating value of 15 GJ/tonne air dry matter, and a percentage of 49% carbon of dry matter. For example, 14 EJ (left column) is roughly comparable to 700 million tonnes of dry matter, which is (if assumed this has to come from additional stemwood fellings) comparable to roughly 1.5 billion m^3 of roundwood, half of current global harvesting of wood.
Sources: Fischer and Schrattenholzer, 2001; Ericsson and Nilsson, 2006; Yoshioka et al., 2006; Yamamoto, 2001; Williams, 1995; Walsh et al., 1999; Smeets and Faaij, 2007.

In general, the delivery or production costs of forestry residues are expected to be at a level of 1.0 to 7.7 US$/GJ. Smeets and Faaij (2007) concluded that at a global level, the economic potential of all types of biomass residues is 14 EJ/yr: at the very lower level of estimates in the table. This and the notion that the summation of the column of lower ranges of dry matter supply equals 700 million tonnes (which is assumed stemwood) is half of current global stemwood harvesting) was the reason to estimate the economic potential at 10-20% of above given numbers.

The CO_2 mitigation potential can only be calculated if the actual use and the amount of use of forestry biomass supply are known. This depends on the balance of supply and demand (see bioenergy in Section 11.3.1.4.). However, to give an indication of the order of magnitude of the figures the CO_2-eq emissions avoided have been calculated from the numbers in Table 9.5 using the assumption that biomass replaces either coal (high range) or gas (low range). Based on these calculations[8], the CO_2-eq emissions avoided range from 420 to 4,400 MtCO$_2$/yr for 2030. This is about 5 to 25% of the total CO_2-eq emissions that originate from electricity production in 2030, as reported in the World Energy Outlook (OECD, 2006).

9.4.4   Global summation and comparison

An overview of estimates derived in the regional bottom-up estimates as given in Section 9.4.3.1 are presented in Table 9.6. Based on indications in literature and carbon supply curves, the fraction of the mitigation potential in the cost class < 20 US$/tCO$_2$ was estimated.

Assuming a linear implementation rate of the measures, the values in Table 9.4 were adjusted to 2030 values (the values required in the cross sector summation in Chapter 11, Table 11.3). The 2030 values are presented in Table 9.7 against the values derived from global forest sector models, and from global integrated models for three world regions. The mitigation effect of biomass for bioenergy (see text, Box 9.2) was excluded.

The range of estimates in the literature and presented in Table 9.7 help in understanding the uncertainty surrounding forestry mitigation potential. Bottom-up estimates of mitigation generally include numerous activities in one or more regions

---

8   Assuming that it is used in a biomass combustion plant of 30% conversion efficiency and replaces a coal combustion plant with an efficiency of 48% (see IEA 2002) and a coal CO_2 content of 95 kgCO$_2$/GJ for the high range or a gas IGCC with an efficiency of 49% and a gas CO_2 content of 57 kgCO$_2$/GJ.

Table 9.6: *Summation of regional results (excluding bioenergy) as presented in Section 9.4.3.1 for 2040. Fraction by cost class is derived from Section 9.4.3.1.*

	Economic potential in 2040 (MtCO$_2$/yr) low	Economic potential in 2040 (MtCO$_2$/yr) high	Fraction of total (technical) potential in cost class <20 US$/tCO$_2$
North America	400	820	0.2
Europe	90	180	0.2
Russian Federation	150	300	0.3
Africa	300	875	0.6
OECD Pacific	85	255	0.35
Caribbean, Central and South America	500	1750	0.6
Non Annex I East Asia	150	400	0.3
Non Annex I South Asia	300	875	0.6
Total	**1,975**	**5,455**	

Note: These figures are surrounded by uncertainty. Differences in studies, assumptions, baselines, and price scenarios make a simple summation difficult.

Table 9.7: *Comparison of estimates of economic mitigation potential by major world region and methodology excluding biomass for bioenergy in MtCO$_2$/yr in 2030, at carbon prices less or equal to 100 US$/tCO$_2$. Fraction by cost class is given in Tables 9.3 and 9.6.*

	Regional bottom-up estimate			Global forest sector models	Global integrated assessment models
	Mean	Low	High		
OECD	700	420	980	2,730	
Economies in transition	150	90	210	3,600	
Non-OECD	1,900	760	3,040	7,445	
Global	**2,750[a]**	**1,270**	**4,230**	**13,775**	**700**

[a] Excluding bioenergy (see Box 9.2). Including the emission reduction effect of the economic potential of biomass for bioenergy would yield a total mean emission reduction potential (based on bottom up) of 3140 MtCO$_2$/yr in 2030.

represented in detail. Top-down global modelling of sectors and of long-term climate stabilization scenario pathways generally includes fewer, simplified forest options, but allows competition across all sectors of the economy to generate a portfolio of least-cost mitigation strategies. Comparison of top-down and bottom-up modelling estimates (Figure 9.13) is difficult at present. This stems from differences in how the two approaches represent mitigation options and costs, market dynamics, and the effects of market prices on model and sectoral inputs and outputs such as labour, capital, and land. One important reason that bottom-up results yield a lower potential consistently for every region (Figure 9.13) is that this type of study takes into account (to some degree) barriers to implementation. The bottom-up estimate has, therefore, characteristics of a market potential study, but the degree is unknown.

The uncertainty and differences behind the studies referred to, and the lack of baselines are reasons to be rather conservative with the final estimate for the forestry mitigation potential. Therefore, mostly the bottom-up estimates are used in the final estimate. This stands apart from any preference for a certain type of study. Thus synthesizing the literature, we estimate that forestry mitigation options have the economic potential (at

carbon prices up to 100 US$/tCO$_2$) to contribute between 1270 and 4230 MtCO$_2$/yr in 2030 (medium confidence, medium agreement). About 50% of the medium estimate can be achieved at a cost under 20 US$/tCO$_2$ (= 1550 MtCO$_2$/yr: see Figure 9.14). The combined effects of reduced deforestation and degradation, afforestation, forest management, agro-forestry and bioenergy have the potential to increase gradually from the present to 2030 and beyond. For comparison with other sectors in Chapter 11, Table 11.2, data on cost categories <0 US$/tCO$_2$ and 20-50 US$100/tCO$_2$ have been derived from Tables 9.3 and 9.6, using cost information derived from regional bottom-up studies and global top-down modelling. The cost classes assessed should be seen as rough cost-class indications, as the information in the literature varies a lot. These analyses assume gradual implementation of mitigation activities starting at present.

This sink enhancement/emission avoidance will be located for 65% in the tropics (high confidence, high agreement; Figure 9.14); be found mainly in above-ground biomass; and for 10% achieved through bioenergy (medium confidence, medium agreement). In the short term, this potential is much smaller, with 1180 MtCO$_2$/yr in 2010 (high confidence, medium agreement). Uncertainty from this estimate arises from the variety of studies

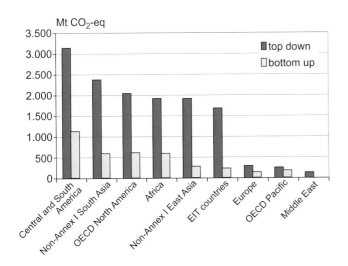

Figure 9.13: *Comparison of estimates of economic mitigation potential in the forestry sector (up to 100 US$/tCO₂ in 2030) as based on global forest sector models (top-down) versus regional modelling results (bottom-up).*

Note: Excluding bioenergy; data from Table 9.3 and Table 9.6.

used, the different assumptions, the different measures taken into account, and not taking into account possible leakage between continents.

These final results allow comparison with earlier IPCC estimates for forestry mitigation potential (Figure 9.15). The estimates for Second Assessment Report (SAR), Third Assessment Report (TAR) and Special Report have to be seen as estimates for a technical potential, and are comparable to our Fourth Assessment Report (AR4) estimates for a carbon dioxide price < 100 US$/tCO₂ (as displayed). As the bars in this figure are lined by the year to which they apply, one would expect an increasing trend towards the right-hand columns. This is not the

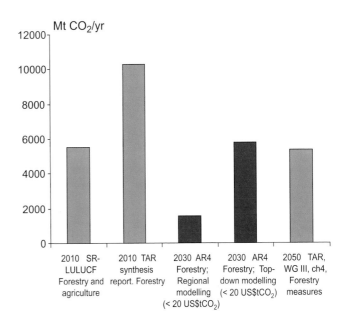

Figure 9.15: *Comparison of estimates of mitigation potential in previous IPCC reports (blue) and the current report (in red).*

Note the difference in years to which the estimate applies, in applied costs, and between forest sector only versus whole LULUCF estimates.

case. Instead a large variety is displayed. There is a trend visible through the consecutive IPCC reports, and not so much through the years to which the estimate applies. When ignoring the TAR synthesis, we start with the highest estimate in SAR (just over 8000 MtCO₂/yr), then follows SR LULUCF with 5500 MtCO₂, and TAR with 5300. Finally, the present report follows with a conservative estimate of 3140 (including bioenergy).

9.5 Interactions with adaptation and vulnerability

Some of the mitigation potential as given in this chapter might be counteracted by adverse effects of climate change on forest ecosystems (Fischlin *et al.*, 2007). Further, mitigation-driven actions in forestry could have positive adaptive consequences (e.g., erosion protection) or negative adaptation consequences (e.g., increase in pest and fires). Similarly, adaptation actions could have positive or negative consequences on mitigation. To avoid trade-offs, it is important to explore options to adapt to new climate circumstances at an early stage through anticipatory adaptation (Robledo *et al.,* 2005). The limits to adaptation stem in part from the way that societies exacerbate rather than ameliorate vulnerability to climate fluctuations (Orlove, 2005) that can also affect mitigation potentials. There are significant opportunities for mitigation and for adapting to climate change, while enhancing the conservation of biodiversity, and achieving other environmental as well as socio-economic benefits. However, mitigation and adaptation have been considered separately in the global negotiations as well as in the literature

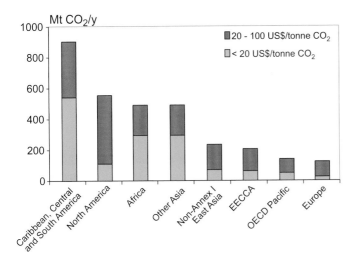

Figure 9.14: *Annual economic mitigation potential in the forestry sector by world region and cost class in 2030.*

Note: EECCA=Countries of Eastern Europe, the Caucasus and Central Asia.

until very recently. Now, the two concepts are seen to be linked, however to achieve synergies may be a challenge (Tol, 2006). In the IPCC Third Assessment Report, potential synergy and trade-off issues were not addressed. This section explores the synergy between mitigation and adaptation in the forest sector (Ravindranath and Sathaye, 2002). The potential and need for incorporating adaptation strategies and practices in mitigation projects is illustrated with a few examples.

9.5.1 Climate impacts on carbon sink and adaptation

In addition to natural factors, forest ecosystems have long been subjected to many human-induced pressures such as land-use change, over-harvesting, overgrazing by livestock, fire, and introduction of new species. Climate change constitutes an additional pressure that could change or endanger these ecosystems. The IPCC Fourth Assessment report (Fischlin *et al.*, 2007 and Easterling *et al.*, 2007) has highlighted the potential impacts of climate change on forest ecosystems. New findings indicate that negative climate change impacts may be stronger than previously projected and positive impacts are being over-estimated as well as the uncertainty on predictions.

Recent literature indicates that the projected potential positive effect of climate change as well as the estimated carbon sink in mature forests may be substantially threatened by enhancing or changing the regime of disturbances in forests such as fire, pests, drought, and heat waves, affecting forestry production including timber (Fuhrer *et al.,* 2007; Sohngen *et al.*, 2005; Ciais *et al.,* 2005).

Most model limitations persist; models do not include key ecological processes, and feedbacks. There are still inconsistencies between the models used by ecologists to estimate the effects of climate change on forest production and composition, and the models used by foresters to predict forest yield (Easterling *et al.*, 2007). Despite the achievements and individual strengths of the selected modelling approaches, core problems of global land-use modelling have not yet been resolved. For a new generation of integrated large-scale land-use models, a transparent structure would be desirable (Heistermann *et al.,* 2006).

Global change, including the impacts of climate change, can affect the mitigation potential of the forestry sector by either increasing (nitrogen deposition and CO_2 fertilization), or decreasing (negative impacts of air pollution,) the carbon sequestration. But, recent studies suggest that the beneficial impacts of climate change are being overestimated by ignoring some of the feedbacks (Körner, 2004) and assumption of linear responses. Also, the negative impacts may be larger than expected (Schroter *et al.,* 2005), with either some effects remaining incompletely understood (Betts *et al.,* 2004) or impossible to separate one from the other.

9.5.2 Mitigation and adaptation synergies

The mitigation and adaptation trade-offs and synergies in the forestry sector are dealt with in Klein *et al.* (2007). Many of the response strategies to address climate change, such as Global Environmental Facility (GEF) and Clean Development Mechanism (CDM), Activities under Article 3.3 and Article 3.4 and the Adaptation Fund aim at implementation of either mitigation or adaptation technologies or policies. It is necessary to promote synergy in planning and implementation of forestry mitigation and adaptation projects to derive maximum benefit to the global environment as well as local communities or economies, for example promoting adaptive forest management (McGinley & Finegan, 2003). However, recent analyses not specifically focused on the Forestry sector point out that it may be difficult to enhance synergies. This is due to the different actors involved in mitigation and adaptation, competitive use of funds, and the fact that in many cases both activities take place at different implementation levels (Tol, 2006). It should also be taken into account that activities to address mitigation and adaptation in the forestry sector are planned and implemented locally.

It is likely that adaptation practices will be easier to implement in forest plantations than in natural forests. Several adaptation strategies or practices can be used in the forest sector, including changes in land use choice (Kabat *et al.,* 2005), management intensity, hardwood/softwood species mix, timber growth and harvesting patterns within and between regions, changes in rotation periods, salvaging dead timber, shifting to species more productive under the new climatic conditions, landscape planning to minimize fire and insect damage, and to provide connectivity, and adjusting to altered wood size and quality (Spittlehouse and Stewart, 2003). A primary aim of adaptive management is to reduce as many ancillary stresses on the forest resource as possible. Maintaining widely dispersed and viable populations of individual species minimizes the probability that localized catastrophic events will cause extinction (Fischlin *et al.,* 2007). While regrowth of trees due to effective protection will lead to carbon sequestration, adaptive management of protected areas also leads to conservation of biodiversity and reduced vulnerability to climate change. For example, ecological corridors create opportunities for migration of flora and fauna, which facilitates adaptation to changing climate.

Adaptation practices could be incorporated synergistically in most mitigation projects in the forest sector. However, in some cases, mitigation strategies could also have adverse implications for watersheds in arid and semi-arid regions (UK FRP, 2005) and biodiversity (Caparros and Jacquemont, 2003). To achieve an optimum link between adaptation and mitigation activities, it is necessary to clearly define who does the activity, where and what are the activities for each case. Several principles can be defined (Murdiyarso *et al.,* 2005): prioritizing mitigation activities that help to reduce pressure on natural resources, including vulnerability to climate change as

Table 9.8: *Adaptation and mitigation matrix*

Mitigation option	Vulnerability of the mitigation option to climate change	Adaptation options	Implications for GHG emissions due to adaptation
A. Increasing or maintaining the forest area			
Reducing deforestation and forest degradation	Vulnerable to changes in rainfall, higher temperatures (native forest dieback, pest attack, fire and, droughts)	Fire and pest management Protected area management Linking corridors of protected areas	No or marginal implications for GHG emissions, positive if the effect of perturbations induced by climate change can be reduced
Afforestation / Reforestation	Vulnerable to changes in rainfall, and higher temperatures (increase of forest fires, pests, dieback due to drought)	Species mix at different scales Fire and pest management Increase biodiversity in plantations by multi-species plantations. Introduction of irrigation and fertilisation Soil conservation	No or marginal implications for GHG emissions, positive if the effect of perturbations induced by climate change can be reduced May lead to increase in emissions from soils or use of machinery and fertilizer
B. Changing forest management: increasing carbon density at plot and landscape level			
Forest management in plantations	Vulnerable to changes in rainfall, and higher temperatures (i.e. managed forest dieback due to pest or droughts)	Pest and forest fire management. Adjust rotation periods Species mix at different scales	Marginal implications on GHGs. May lead to increase in emissions from soils or use of machinery or fertilizer use
Forest management in native forest	Vulnerable to changes in rainfall, and higher temperatures (i.e. managed forest dieback due to pest, or droughts)	Pest and fire management Species mix at different scales	No or marginal
C. Substitution of energy intensive materials			
Increasing substitution of fossil energy intensive products by wood products	Stocks in products not vulnerable to climate change		No implications in GHGs emissions
D. Bioenergy			
Bioenergy production from forestry	An intensively managed plantation from where biomass feedstock comes is vulnerable to pests, drought and fire occurrence, but the activity of substitution is not.	Suitable selection of species to cope with changing climate Pest and fire management	No implications for GHG emissions except from fertilizer or machinery use

a risk to be analysed in mitigation activities; and prioritizing mitigation activities that enhance local adaptive capacity, and promoting sustainable livelihoods of local populations.

Considering adaptation to climate change during the planning and implementation of CDM projects in forestry may also reduce risks, although the cost of monitoring performance may become very complex (Murdiyarso *et al.*, 2005). Adaptation and mitigation linkages and vulnerability of mitigation options to climate change are summarized in Table 9.8, which presents four types of mitigation actions.

Reducing deforestation is the dominant mitigation option for tropical regions (Section 9.4). Adaptive practices may be complex. Forest conservation is a critical strategy to promote sustainable development due to its importance for biodiversity conservation, watershed protection and promotion of livelihoods

of forest-dependent communities in existing natural forest (IPCC, 2002).

Afforestation and reforestation are the dominant mitigation options in specific regions (e.g., Europe). Currently, afforestation and reforestation are included under Article 3.3 and in Articles 6 and 12 (CDM) of the Kyoto Protocol. Plantations consisting of multiple species may be an attractive adaptation option as they are more resilient, or less vulnerable, to climate change. The latter as a result of different tolerance to climate change characteristic of each plantation species, different migration abilities, and differential effectiveness of invading species (IPCC, 2002).

Agro-forestry provides an example of a set of innovative practices designed to enhance overall productivity, to increase carbon sequestration, and that can also strengthen the system's ability to cope with adverse impacts of changing climate conditions. Agro-forestry management systems offer important opportunities creating synergies between actions undertaken for mitigation and for adaptation (Verchot *et al.*, 2006). The area suitable for agro-forestry is estimated to be 585-1215 Mha with a technical mitigation potential of 1.1 to 2.2 PgC in terrestrial ecosystems over the next 50 years (Albrecht and Kandji, 2003). Agro-forestry can also help to decrease pressure on natural forests and promote soil conservation, and provide ecological services to livestock.

Bioenergy. Bioenergy plantations are likely to be intensively managed to produce the maximum biomass per unit area. To ensure sustainable supply of biomass feedstock and to reduce vulnerability to climate change, the practices mentioned above for afforestation and reforestation projects need to be explored such as changes in rotation periods, salvage of dead timber, shift to species more productive under the new climatic conditions, mixed species forestry, mosaics of different species and ages, and fire protection measures.

Adaptation and mitigation synergy and sustainable development

The need for integration of mitigation and adaptation strategies to promote sustainable development is presented in Klein *et al.* (2007). The analysis has shown the complementarity or synergy between many of the adaptation options and mitigation (Dang *et al.*, 2003). Promotion of synergy between mitigation and adaptation will also advance sustainable development, since mitigation activities could contribute to reducing the vulnerability of natural ecosystems and socio-economic systems (Ravindranath, 2007). Currently, there are very few ongoing studies on the interaction between mitigation, adaptation and sustainable development (Wilbanks, 2003; Dang *et al.*, 2003). Quantification of synergy is necessary to convince the investors or policy makers (Dang *et al.*, 2003).

The possibility of incorporating adaptation practices into mitigation projects to reduce vulnerability needs to be explored. Particularly, Kyoto Protocol activities under Article 3.3, 3.4 and 12 provide an opportunity to incorporate adaptation practices. Thus, guidelines may be necessary for promoting synergy in mitigation as well as adaptation programmes and projects of the existing UNFCCC and Kyoto Protocol mechanisms as well as emerging mechanisms. Integrating adaptation practices in such mitigation projects would maximize the utility of the investment flow and contribute to enhancing the institutional capacity to cope with risks associated with climate change (Dang *et. al.*, 2003).

9.6 Effectiveness of and experience with policies

This section examines the barriers, opportunities, and implementation issues associated with policies affecting mitigation in the forestry sector. Non-climate policies, that is forest sector policies that affect net greenhouse gas emissions from forests, but that are not designed primarily to achieve climate objectives, as well as policies primarily designed to reduce net forest emissions are considered. Many factors influence the efficacy of forest policies in achieving intended impacts on forest land-use, including land tenure, institutional and regulatory capacity of governments, the financial competitiveness of forestry as a land use, and a society's cultural relationship to forests. Some of these factors typically differ between industrialized and developing countries. For example, in comparison to developing countries, industrialized countries tend to have relatively small amounts of unallocated public lands, and relatively strong institutional and regulatory capacities. Where appropriate, policy options and their effectiveness are examined separately for industrialized and developing countries. Because integrated and non-climate policies are designed primarily to achieve objectives other than net emissions reductions, evaluations of their effectiveness focus primarily on indicators, such as maintenance of forest cover. This provides only partial insight into their potential to mitigate climate change. Under conditions with high potential for leakage, for example, such indicators may overestimate the potential for carbon benefits (Section 9.6.3).

9.6.1 Policies aimed at reducing deforestation

Deforestation in developing countries, the largest source of emissions from the forestry sector, has remained at high levels since 1990 (FAO, 2005). The causes of tropical deforestation are complex, varying across countries and over time in response to different social, cultural, and macroeconomic conditions (Geist and Lambin, 2002). Broadly, three major barriers to enacting effective policies to reduce forest loss are: (i) profitability incentives often run counter to forest conservation and sustainable forest management (Tacconi *et al.*, 2003); (ii) many direct and indirect drivers of deforestation lie outside of the forest sector, especially in agricultural policies and markets (Wunder, 2004); and (iii) limited regulatory and institutional capacity and insufficient resources constrain the ability of many governments to implement forest and related sectoral policies on the ground (Tacconi *et al.*, 2003).

In the face of these challenges, national forest policies designed to slow deforestation on public lands in developing countries have had mixed success:

- In countries where institutional and regulatory capacities are insufficient, new clearing by commercial and small-scale agriculturalists responding to market signals continues to be a dominant driver of deforestation (Wunder, 2004).

- A number of national initiatives are underway to combat illegal logging (Sizer *et al.,* 2005). While these have increased the number of charges and convictions, it is too early to assess their impact on forest degradation and deforestation.
- Legally protecting forests by designating protected areas, indigenous reserves, non-timber forest reserves and community reserves have proven effective in maintaining forest cover in some countries, while in others, a lack of resources and personnel result in the conversion of legally protected forests to other land uses (Mertens *et al.*, 2004).

China (Cohen *et al.*, 2002), the Philippines and Thailand (Granger, 1997) have significantly reduced deforestation rates in response to experiencing severe environmental and public health consequences of forest loss and degradation. In India, the Joint Forest Management programme has been effective in partnering with communities to reduce forest degradation (Bhat *et al.,* 2001). These examples indicate that strong and motivated government institutions and public support are key factors in implementing effective forest policies.

Options for maintaining forests on private lands in developing countries are generally more limited than on public lands, as governments typically have less regulatory control. An important exception is private landholdings in the Brazilian Amazon, where the government requires that landowners maintain 80% of the property under forest cover. Although this regulation has had limited effectiveness in the past (Alves *et al.,* 1999), recent experience with a licensing and monitoring system in the state of Mato Grosso has shown that commitment to enforcement can significantly reduce deforestation rates.

A recently developed approach is for governments to provide environmental service payments to private forest owners in developing countries, thereby providing a direct financial incentive for the retention of forest cover. Relatively high transaction costs and insecure land and resource tenure have thus far limited applications of this approach in many countries (Grieg-Gran, 2004). However, significant potential may exist for developing payment schemes for restoration and retention of forest cover to provide climate mitigation (see below) and watershed protection services.

In addition to national-level policies, numerous international policy initiatives to support countries in their efforts to reduce deforestation have also been attempted:
- Forest policy processes, such as the UN Forum on Forests, and the International Tropical Timber Organization have provided support to national forest planning efforts but have not yet had demonstrable impacts on reducing deforestation (Speth, 2002).
- The World Bank has modified lending policies to reduce the risk of direct negative impacts to forests, but this does not appear to have measurably slowed deforestation (WBOED, 2000).

- The World Bank and G-8 have recently initiated the Forest Law Enforcement and Governance (FLEG) process among producer and consumer nations to combat illegal logging in Asia and Africa (World Bank, 2005). It is too early to assess the effectiveness of these initiatives on conserving forests stocks.
- The Food and Agricultural Organization (FAO) Forestry Programme has for decades provided a broad range of technical support in sustainable forest management (FAO, 2006b); assessing measurable impacts has been limited by the lack of an effective monitoring programme (Dublin and Volante, 2004).

Taken together, non-climate policies have had minimal impact on slowing tropical deforestation, the single largest contribution of land-use change to global carbon emissions. Nevertheless, there are promising examples where countries with adequate resources and political will have been able to slow deforestation. This raises the possibility that, with sufficient institutional capacity, financial incentives, political will and sustained financial resources, it may possible to scale up these efforts. One potential source of additional financing for reducing deforestation in developing countries is through well-constructed carbon markets or other environmental service payment schemes (Winrock International, 2004; Stern, 2006).

Under the UNFCCC and Kyoto Protocol, no climate policies currently exist to reduce emissions from deforestation or forest degradation in developing countries. The decision to exclude avoided deforestation projects from the CDM in the Kyoto Protocol's first commitment period was in part based on methodological concerns. These concerns are particularly associated with additionality and baseline setting and whether leakage could be sufficiently controlled or quantified to allow for robust carbon crediting (Trines *et al.*, 2006). In December 2005, COP-11 established a two-year process to review relevant scientific, technical, and methodological issues and to consider possible policy approaches and positive incentives for reducing emissions from deforestation in developing countries (UNFCCC, 2006).

Recent studies suggests a broad range of possible architectures by which future climate policies might be designed to effectively reduce emissions from tropical deforestation and forest degradation (Schlamadinger *et al.*, 2005; Trines *et al.*, 2006). For example, Santilli *et al.* (2005), propose that non-Annex I countries might, on a voluntary basis, elect to reduce their national emissions from deforestation. The emission reductions could then be credited and sold to governments or international carbon investors at the end of a commitment period, contingent upon agreement to stabilize, or further reduce deforestation rates in the subsequent commitment periods.

One advantage of a national-sectoral approach over a project-based approach to reduce emissions from deforestation relates to leakage, in that any losses in one area could be balanced against gains in other areas. This does not entirely address the leakage problem since the risk of international leakage remains, as occurs in other sectors.

Other proposals emphasize accommodation to diverse national circumstances, including differing levels of development, and include a suggestion of separate targets for separate sectors (Grassl *et al.*, 2003). This includes a "no-lose" target, whereby emission allowances can be sold if the target is reached. No additional emission allowances would have to be bought if the target was not met. A multi-stage approach such that the level of commitment of an individual country increases gradually over time; capacity building and technology research and development; or quantified sectoral emission limitation and reduction commitments similar to Annex 1 commitments under the Kyoto Protocol (Trines *et al.*, 2006).

Proposed financing mechanisms include both carbon market-based instruments (Stern, 2006) and non-market based channels, for example, through a dedicated fund to voluntarily reduce emissions from deforestation (UNFCCC, 2006). Box 9.3 discusses recent technical advances relevant to the effective design and implementation of climate policies aimed at reducing emissions from deforestation and forest degradation.

9.6.2 Policies aimed to promote afforestation and reforestation

Non-climate forest policies have a long history in successful creation of plantation forests on both public and private lands in developing and developed countries. If governments have strong regulatory and institutional capacities, they may successfully control land use on public lands, and state agencies can reforest these lands directly. In cases where such capacities are more limited, governments may enter into joint management agreements with communities, so that both parties share the costs and benefits of plantation establishment (Williams, 2002). Incentives for plantation establishment may take the form of afforestation grants, investment in transportation and roads, energy subsidies, tax exemptions for forestry investments, and tariffs against competing imports (Cossalter and Pye-Smith, 2003). In contrast to conservation of existing forests, the underlying financial incentives to establish plantations may be positive. However, the creation of virtually all significant plantation estates has relied upon government support, at least in the initial stages. This is due, in part, to the illiquidity of the investment, the high cost of capital establishment and long waiting period for financial return.

9.6.3 Policies to improve forest management

Industrialized countries generally have sufficient resources to implement policy changes in public forests. However, the fact that these forests are already managed to relatively high standards may limit possibilities for increasing sequestration through changed management practices (e.g., by changing species mix, lengthening rotations, reducing harvest damage and or accelerating replanting rates). There may be possibilities to reduce harvest rates to increase carbon storage however, for example, by reducing harvest rates and/or harvest damage.

Governments typically have less authority to regulate land use on private lands, and so have relied upon providing incentives to maintain forest cover, or to improve management. These incentives can take the form of tax credits, subsidies, cost sharing, contracts, technical assistance, and environmental service payments. In the United States, for example, several

---

BOX 9.3: Estimating and monitoring carbon emissions from deforestation and degradation

Recent analyses (DeFries *et al.*, 2006; UNFCCC, 2006) indicate considerable progress since the Third Assessment Report and the IPCC Good Practice Guidance for Land Use, Land-Use Change and Forestry (IPCC, 2003) in data acquisition and development of methods and tools for estimating and monitoring carbon emissions from deforestation and forest degradation in developing countries. Remote sensing approaches to monitoring changes in land cover/land use at multiple scales and coverage are now close to operational on a routine basis. Measuring forest degradation through remote sensing is technically more challenging, but methods are being developed (DeFries *et al.*, 2006).

Various methods can be applied, depending on national capabilities, deforestation patterns, and forest characteristics. Standard protocols need to be developed for using remote sensing data, tools and methods that suit both the variety of national circumstances and meet acceptable levels of accuracy. However, quantifying accuracy and ensuring consistent methods over time are more important than establishing consistent methods across countries.

Several developing countries, including India and Brazil, have systems in place for national-scale monitoring of deforestation (DeFries *et al.*, 2006). While well-established methods and tools are available for estimating forest carbon stocks, dedicated investment would be required to expand carbon stock inventories so that reliable carbon estimates can be applied to areas identified as deforested or degraded through remote sensing. With sound data on both change in forest cover and on change in carbon stocks resulting from deforestation and degradation, emissions can be estimated using methods described by the new IPCC Inventory Guidelines (IPCC, 2006).

government programmes promote the establishment, retention, and improved management of forest cover on private lands, often of marginal agricultural quality (Box 9.4; Gaddis *et al.*, 1995).

The lack of robust institutional and regulatory frameworks, trained personnel, and secure land tenure has constrained the effectiveness of forest management in many developing countries (Tacconi *et al.*, 2003; Box 9.5). Africa, for example, had about 649 million forested hectares as of 2000 (FAO, 2001). Of this, only 5.5 million ha (0.8%) had long-term management plans, and only 0.9 million ha (0.1%) were certified to sound forestry standards. Thus far, efforts to improve logging practices in developing countries have met with limited success. For example, reduced-impact logging (RIL) techniques would increase carbon storage over traditional logging, but have not been widely adopted by logging companies, even when they lead to cost savings (Holmes *et al.*, 2002). Nevertheless, there are several examples where large investments in building technical and institutional capacity have dramatically improved forestry practices (Dourojeanni, 1999).

Policies aimed at liberalizing trade in forest products have mixed impacts on forest management practices. Trade liberalization in forest products can enhance competition and can make improved forest management practices more economically attractive in mature markets (Clarke, 2000). But, in the relatively immature markets of many developing countries, liberalization may act to magnify the effects of policy and market failures (Sizer *et al.*, 1999).

The recent FAO forest assessment conservatively estimates that insects, disease and fire annually impact 3.2% of the forests in reporting countries (FAO, 2005). Policies that successfully increase the forest protection against natural disturbance agents may reduce net emissions from forest lands (Richards *et al.*, 2006). In industrialized countries, a history of fire suppression and a lack of thinning treatments have created high fuel loads in many public forests, such that when fires do occur, they release large quantities of carbon (Schelhaas *et al.*, 2003).

A major technical obstacle is designing careful management interventions to reduce fuel loading and to restore landscape heterogeneity to forest structure (USDA Forest Service, 2000). Scaling up their application to large forested areas, such as in Western USA, Northern Canada or Russia, could lead to large gains in the conservation of existing carbon stocks (Sizer *et al.*, 2005). Forest fire prevention and suppression capacities are rudimentary in many developing countries, but trial projects show that with sufficient resources and training, significant reductions in forest fires can be achieved (ITTO, 1999).

Voluntary certification to sustainable forest management standards aims to improve forest management by providing incentives such as increased market access or price premiums to certified producers who meet these standards. Various certification schemes have collectively certified hundreds of millions of hectares in the last decade and certification can result in measurable improvements in management practices (Gullison, 2003). However, voluntary certification efforts to date continue to be challenged in improving the management of forest managers operating at low standards, where the potential for improvement and net emissions reductions are greatest. One possible approach to overcome current barriers in areas with weak forest management practices is to include stepwise or phased approaches to certification (Atyi and Simula, 2002).

9.6.4 Policies to increase substitution of forest-derived biofuels for fossil fuels and biomass for energy-intensive materials

Countries may promote the use of bioenergy for many non-climate reasons, including increasing energy security and promoting rural development (Parris, 2004). Brazil, for example, has a long history of encouraging plantation establishment for the production of industrial charcoal by offering a combination of tax exemption for plantation lands, tax exemption for income originating from plantation companies, and deductibility of funds used to establish plantations (Couto and Betters, 1995). The United States provides a range of incentives for ethanol production including exclusion from excise taxes, mandating clean air performance requirements that created markets for ethanol, and tax incentives and accelerated depreciation schedules for electricity generating equipment that burn biomass (USDOE, 2005). The Australian Government's Mandatory Renewable Energy Target, which seeks to create a market for renewable energy, provides incentives for the development of renewable energy from plantations and wood waste (Government of Australia, 2006).

Building codes and other government policies that, where appropriate, can promote substitution of use of sustainably harvested forest products wood for more energy-intensive construction materials may have substantial potential to reduce net emissions (Murphy, 2004). Private companies and individuals may also modify procurement to prefer or require certified wood from well-managed forests on environmental grounds. Such efforts might be expanded once the climate mitigation benefits of sustainably harvested wood products are more fully recognized.

9.6.5 Strengthening the role of forest policies in mitigating climate change

Policies have generally been most successful in changing forestry activities where they are consistent with underlying profitability incentives, or where there is sufficient political will, financial resources and regulatory capacity for effective implementation. Available evidence suggests that policies that seek to alter forestry activities where these conditions do not apply have had limited effectiveness. Additional factors that influence the potential for non-climate policies to reduce net

---

Box 9.4: Non-climate forest policies as an element of carbon management in the United States

Many programmes in the United States support the establishment, retention, and improved management of forest cover on private lands. These entail contracts and subsidies to private landowners to improve or change land-use management practices. USDA also provides technical information, research services, cost sharing and other financial incentives to improve land management practices, including foresting marginal agricultural lands, and improving the management of existing of forests. Examples include the Conservation Reserve Program; Forestry Incentives Program, and Partners for Wildlife; (Richards et al., 2006). For example, in the 20-year period between 1974 and 1994, the Forestry Incentives Program spent 200 US$ million to fund 1.34 million hectares of tree planting; 0.58 million hectares of stand improvement; and 11 million hectares of site preparation for natural regeneration (Gaddis *et al.*, 1995).

Richards *et al.* (2006) suggest that substantial gains in carbon sequestration and storage could be achieved by increasing the resources and scope of these programmes and through new results-based programmes, which would reward landowners based on the actual carbon they sequester or store.

Box 9.5: Non-climate forest policies as an element of carbon management in Africa

Forest and land use policies across African countries have historically passed through two types of governance: Under *traditional systems controlled by families, traditional leaders and communities*, decisions regarding land allocation, redistribution and protection were the responsibility of local leaders. Most land and resources were under relatively sustainable management by nomadic or agro-pastoralist communities who developed systems to cope with vulnerable conditions. Agriculture was typically limited to shifting cultivation, with forest and range resources managed for multiple benefits.

Under *central government systems*, land-use policies are sectoral-focused, with strong governance in the agricultural sector. Agriculture expansion policies typically dominate land use at the expense of forestry and rangeland management. This has greatly influenced present day forest and range policies and practices and resulted in vast land degradation (IUCN, 2002; 2004).The adoption of centralized land management policies and legislation system has often brought previously community-oriented land management systems into national frameworks, largely without the consent and involvement of local communities. Central control is reflected in large protected areas, with entry of local communities prevented.

Presently, contradiction and conflicts in land-use practices between sectors and communities is common. Negotiations demanding decentralization and equity in resource distribution may lead to changes in land tenure systems in which communities and official organizations will increasingly agree to collaboration and joint management in which civil societies participate. Parastatal institutions, established in some countries, formulate and implement policies and legislation that coordinate between sectors and to encourage community participation in land and resource management.

Land tenure categories characteristically include *private holdings* (5–25% of national area), *communal land* (usually small percentage) and *state lands* (the majority of the land under government control). Each faces many problems generated by conflicting rights of use and legislation that gives greater government control on types of resource use even under conditions of private ownership. Land control system and land allocation policy adopted by central governments often have negative impacts on land and tree tenure. Local communities are not encouraged to plant, conserve and manage trees on government owned land that farmers use on lease systems. Even large-scale farmers who are allocated large areas for cultivation, abandon the land and leave it as bare when it becomes non-productive. Forest lands reserved and registered under community ownership are communally managed on the basis of stakeholder system and shared benefits.

Evidence from many case studies in Sudan suggests that integrated forest management where communities have access rights to forest lands and are involved in management, is a key factor favouring the restoration of forest carbon stocks (IUCN, 2004). These projects provide examples of a collaborative system for the rehabilitation and use of the forest land property based on defined and acceptable criteria for land cultivation by the local people and for renewal of the forest crop.

---

emissions from the forest sector include their ability to (1) provide relatively large net reductions per unit area; (2) be potentially applicable at a large geographic scale; and, (3) have relatively low leakage (Niesten *et al.*, 2002).

By these criteria, promising approaches across both industrialized and developing countries include policies that combat the loss of public forests to natural disturbance agents, and "Payment for Environmental Services" (PES) systems that provide an incentive for the retention of forest cover. In both cases, there are good examples where they have been successfully implemented at small scales, and the impediments to increasing scale are relatively well understood. There is also a successful history of policies to create new forests, and these have led to large on-site reductions in net emissions. Care must be taken, however, to make sure that at plantation creation, there is no displacement of economic or subsistence activities that will lead to forest clearing elsewhere. Policies to increase the substitution of fossil fuels with bioenergy have also had a large positive impact on net emissions. If feedstock is forestry waste, then there is little potential leakage. If new plantations are created for biofuel, then care must be taken to reduce leakage.

Because forestry policies tend not to have climate mitigation as core objective, leakage and other factors that may limit net reductions are generally not considered. This may change as countries begin to integrate climate change mitigation objectives more fully into national forestry policies. Countries where such integration is taking place include Costa Rica, the Dominican Republic, and Peru (Rosenbaum *et al.*, 2004).

9.6.6 Lessons learned from project-based afforestation and reforestation since 2000

Experience is limited by the fact that Joint Implementation is not operational yet, and the first call for afforestation and reforestation (A/R) methodologies under CDM was only issued in September 2004. In addition, the modalities and procedures for CDM A/R as decided in December 2003 are complex. Nevertheless, the capacities built up through the development of projects and related methodologies should not be underestimated. As of November 2006, 27 methodologies were submitted, 17 from Latin America, four from Asia and Africa respectively, and two from Eastern Europe. The four which were approved by the CDM Executive Board relate to projects located in China, Moldova, Albania and Honduras and all consist of planting forests on degraded agricultural land. In anticipation of Joint Implementation, several projects are under development in several Annex I countries in Eastern Europe, notably in Romania, Ukraine and the Czech Republic.

There are voluntary project-based activities in the USA, with a programme for trading certificates established by the Chicago Climate Exchange (Robins, 2005). The Voluntary Reporting (1605 (b)) Program of the US Department of Energy (USDOE, 2005) provides reporting guidelines for forestry activities. Since the Special Report on LULUCF (IPCC, 2000a), there has been methodological progress in several areas discussed below.

9.6.6.1 Leakage

There is no indication that leakage effects are necessarily higher in forestry than in project activities in other sectors but they can be significant (Chomitz, 2002). Some studies distinguish between primary and secondary effects. A primary effect is defined as resulting from agents that perform land

use activities reflected in the baseline. Populations previously active on the project area may shift their activities to other areas. In land protection projects, logging companies may shift operations or buy timber from outside the project area to compensate for reduced supply of the commodity (activity outsourcing). Secondary leakage is not linked to project participants or previous actors on the area. It is often a market effect, where a project increases (by forest plantation) or decreases (deforestation avoidance) wood supply. Quantitative estimates of leakage (Table 9.9) suggest that leakage varies by mitigation activity and region.

The order of magnitude and even the direction of leakage (negative versus positive), however, depend on the project design (Schwarze *et al.*, 2003). Leakage risk is likely to be low if a whole country or sector is involved in the mitigation activity, or if project activities are for subsistence and do not affect timber or other product markets. There are also well-documented methods to minimize leakage of project-based activities. For example, afforestation projects can be combined with biomass energy plants, or they may promote the use of timber as construction material. Fostering agricultural intensification in parallel can minimize negative leakage from increased local land demand. Where a project reduces deforestation, it can also reduce pressure on forest lands, for example, by intensifying the availability of fuel wood from other sources for local communities. Projects can be designed to engage local people formerly responsible for deforestation in alternative income-generating activities (Sohngen and Brown, 2004).

Leakage appears to have a time dimension as well, due to the dynamics of the forest carbon cycle and management (for example, timing of harvest, planting and regrowth, or protection). Analysis in the USA indicates that national afforestation in response to a carbon price of 15 US$/tCO$_2$ would have 39% leakage in the first two decades, but decline to 24% leakage over five to ten decades, due to forest management dynamics (US EPA, 2005).

9.6.6.2 Potential non-permanence of carbon storage

The reversibility of carbon removal from the atmosphere creates liability issues whenever integrating land use in any kind of accounting system. There needs to be a liability for the case that carbon is released back into the atmosphere because Parties to the UNFCCC agreed, "…that reversal of any removal due to land use, land-use change and forestry activities be accounted for at the appropriate point in time" (UNFCCC, 2001). In 2000, the Colombian delegation first presented a proposal to create expiring Certified Emission Reductions under CDM (UNFCCC, 2001). Its basic idea is that the validity of Certified Emission Reductions (CERs) from afforestation and reforestation project activities under CDM is linked to the time of existence of the relating stocks. The principle of temporary crediting gained support over the subsequent years.

Table 9.9: *Forestry mitigation activity leakage estimates by activity, estimation method and region from the literature*

Activity	Region	Leakage estimation method	Estimated leakage rate (% of carbon mitigation)	Source
Afforestation: tropical region estimates				
Afforestation of degraded lands	Kolar district, Karnataka, India hypothetical project	Household wood demand survey	0.02	Ravindranath, et al., 2007
Plantations, forest conservation, agro-forestry of degraded lands	Magat watershed, Philippines hypothetical project	Historical rates of technology adoption	19 – 41	Authors estimates based on Lasco et al., 2007
Afforestation on small landowner parcels	Scolel Té project, Chiapas, Mexico	Household wood demand survey	0 (some positive leakage)	De Jong et al., 2007
Afforestation degraded uplands	Betalghat hypothetical project, Uttaranchal, India	Household wood demand survey	10 from fuelwood, fodder	Hooda et al., 2007
Afforestation, farm forestry	Bazpur hypothetical project, Uttaranchal, India	Household wood demand survey	20 from fuelwood, poles	Hooda et al., 2007
Afforestation: global and temperate region estimates				
Afforestation (plantation establishment)	Global	PEM	0.4-15.6	Sedjo and Sohngen, 2000
Afforestation	USA-wide	PEM	18-42	Murray et al., 2004
Afforestation only	USA-wide	PEM	24	US EPA, 2005
Afforestation and forest management jointly	USA-wide	PEM	-2.8 [a]	US EPA, 2005
Avoided deforestation: tropical region estimates				
Avoided deforestation	Bolivia, Noel Kempff project and national	PEM	2-38 discounted 5-42 undiscounted	Sohngen and Brown, 2004
Avoided deforestation and biofuels: temperate region estimates				
Avoided deforestation	Northeast USA	PEM	41-43	US EPA, 2005
Avoided deforestation	Rest of USA	PEM	0-92	US EPA, 2005
Avoided deforestation	Pacific Northwest USA	PEM	8-16	US EPA, 2005
Avoided deforestation (reduced timber sales)	Pacific Northwest USA	Econometric model	43 West region 58 Continental US 84 US and Canada	Wear and Murray, 2004
Biofuel production (short rotation woody crops)	USA	PEM	0.2	US EPA, 2005

[a] Negative leakage rate means positive leakage; PEM means partial equilibrium model of forest and/or agriculture sector(s).

Source: Sathaye and Andrasko, 2007

Consequently, the Milan Decision 19/CP.9 (UNFCCC, 2003) created two types of expiring CERs: temporary CERs - tCERs and long-term CERs - lCERs. The validity of both credit types is limited and reflected on the actual certificate. The credit owner is liable to replace them when they expire or when the relating stocks are found to be lost at the end of the commitment period. Afforestation and reforestation projects need to be verified first at a time at the discretion of the project participants, and in intervals of exactly five years thereafter. The value of temporary CERs critically depends on the market participants' mitigation cost expectations for future commitment periods. Assuming constant carbon prices, the price for a temporary CER during the first commitment period is estimated to range between 14 and 35 % of that of a permanent CER from any other mitigation

activity (Dutschke, *et al.*, 2005). This solution is safe from the environmental integrity point of view, yet it has created much uncertainty among project developers (Pedroni, 2005).

9.6.6.3   *Additionality and baselines*

A project that claims carbon credits for mitigation needs to demonstrate its additionality by proving that the same mitigation effect would not have taken place without the project. For CDM, the Executive Board's Consolidated Additionality Tool offers a standardized procedure to project developers. Specific for CDM afforestation and reforestation (A/R), there is an area eligibility test along the forest definitions provided under the relevant Decision 11/CP.7 in order to avoid implementation on areas that prior to the project start were forests in 1990 or after. In the modalities and procedures for CDM, there are three different baseline approaches available for A/R. So far, only one has been successfully applied in the four approved methodologies.

9.6.6.4   Monitoring

For project monitoring, there is now an extended guidance available (IPCC, 2006; USDOE, 2005). Monitoring costs depend on many variables, including the project complexity (including the number of stakeholders involved), heterogeneity of the forest type, the number and type of carbon pools, and GHG to be monitored and the appropriate measurement frequencies. There is a trade-off between the completeness of monitoring data and the carbon price that can be achieved: monitoring costs can sum up an important share of a project's transaction costs. Proper design of the monitoring plan is, therefore, essential for the economic viability of forestry projects. If project developers can demonstrate that omitting particular carbon pools from the project's quantification exercise does not constitute an overestimate of the project's GHG benefits, such pools may be left outside the monitoring plan.

9.6.6.5   Options for scaling up

Despite relative low costs and many possible positive side-effects, the pace with which forest carbon projects are being implemented is slow. This is due to a variety of barriers. Barriers can be categorized as economic, risk-related, political/bureaucratic, logistic, and capacity or political will (the latter barrier also occurring in industrialized countries; Trines *et al.*, 2006). One of the most important climate-related barriers is the complexity of the rules for afforestation and reforestation project activities. This leads to uncertainty among project developers and investors. Temporary accounting of credits is a major obstacle for two reasons: (1) The future value of temporary CERs depends on the buyer's confidence in the underlying project. This may limit investor interest in getting involved in project development. (2) The value of temporary CERs hinges on future allowance price expectations because they will have to be replaced in future commitment periods. Furthermore, EU has deferred its decision to accept forestry credits under its emissions trading scheme. Even if EU decided to integrate these credits, this would come too late to take effect in the first commitment period because trees need time to grow. Given the low value of temporary CERs, transaction costs have a higher share in afforestation and reforestation than in energy mitigation projects. Simplified small-scale rules were introduced in order to reduce transaction costs, but the maximum size of 8 kilotonnes of average annual CO_2 net removal limits their viability.

For forestry mitigation projects to become viable on a larger scale, certainty over future commitments is needed because forestry needs a long planning horizon. Rules need to be streamlined, based on the experience gathered so far. Standardization of project assessment can play important roles to overcome uncertainty among potential buyers and investors, and to prevent negative social and environmental impacts.

9.7   Forests and Sustainable Development

Sustainable forest management of both natural and planted forests is essential to achieving sustainable development. It is a means to reduce poverty, reduce deforestation, halt the loss of forest biodiversity, and reduce land and resource degradation, and contribute to climate change mitigation. Forests play an important role in stabilization of greenhouse gas concentrations in the atmosphere while promoting sustainable development (Article 2; Kyoto Protocol). Thus, forests have to be seen in the framework of the multiple dimensions of sustainable development, if the positive co-benefits from forestry mitigation activities have to be maximized. Important environmental, social, and economic ancillary benefits can be gained by considering forestry mitigation options as an element of the broader land management plans.

9.7.1   Conceptual aspects

Forestry policies and measures undertaken to reduce GHG emissions may have significant positive or negative impacts on environmental and sustainable development objectives that are a central focus of other multilateral environmental agreements (MEAs), including UN Convention on Biological Diversity (CBD), UN Convention to Combat Desertification (CCD), and Ramsar Convention on Wetlands. In Article 2.1(a, b), Kyoto Protocol, Parties agreed various ways to consider potential impacts of mitigation options and whether and how to establish some common approaches to promoting the sustainable development contributions of forestry measures. In addition, a broad range of issues relating to forest conservation and sustainable forest management have been the focus of recent dialogues under the Intergovernmental Forum on Forests.

Recent studies highlighted that strategic thinking about the transition to a sustainable future is particularly important for land (Swanson *et al.*, 2004). In many countries, a variety of separate sets of social, economic and environmental indicators are used, making it difficult to allow for adequate monitoring and analysis of trade-offs between these interlinked dimensions. Still, sustainable development strategies often remain in the periphery of government decision-making processes; and lack coordination between sub-national and local institutions; and economic instruments are often underutilized.

To manage forest ecosystems in a sustainable way implies knowledge of their main functions, and the effects of human practices. In recent years, scientific literature has shown an increasing attempt to understand integrated and long-term effects of current practices of forest management on sustainable development. But often, environmental or socio-economic effects are considered in isolation, or there is no sufficient understanding of the potential long-term impacts of current practices on sustainable development. Payment for Environmental Services (PES) schemes for forest services (recognizing carbon value) may be foreseen as part of forest management implementation, providing new incentives to change to more sustainable decision patterns. Experience, however, is still fairly limited and is concentrated in a few countries, notably in Latin America, and has had mixed results to date (Wunder, 2004).

Important environmental, social, and economic ancillary benefits can be gained by considering forestry mitigation options as an element of the broad land management plans, pursuing sustainable development paths, involving local people and stakeholders and developing adequate policy frameworks.

9.7.2 Ancillary effects of GHG mitigation policies

Climate mitigation policies may have benefits that go beyond global climate protection and actually accrue at the local level (Dudek et al., 2002). Since ancillary benefits tend to be local, rather than global, identifying and accounting for them can reduce or partially compensate the costs of the mitigation measures. However, forests fulfil many important environmental functions and services that can be enhanced or negatively disturbed by human activities and management decisions. Negative effects can be triggered by some mitigation options under certain circumstances. Positive and negative impacts of mitigation options on sustainable development are presented in Table 9.10.

Stopping or slowing deforestation and forest degradation (loss of carbon density) and sustainable forest management may significantly contribute to avoided emissions, conserve water resources and prevent flooding, reduce run-off, control erosion, reduce river siltation, and protect fisheries and investments in hydroelectric power facilities; and at the same time, preserve biodiversity (Parrotta, 2002). Thus, avoided deforestation has large positive implications for sustainable development. Further, natural forests are a significant source of livelihoods to hundreds and millions of forest-dependent communities.

Plantations provide an option to enhance terrestrial sinks and mitigate climate change. Effects of plantations on sustainable development of rural societies have been diverse, depending on socio-economic and environmental conditions and management regime. Plantations may have either significant positive and /or negative effects (environmental and social effects). They can positively contribute, for example, to employment, economic growth, exports, renewable energy supply and poverty alleviation. In some instances, plantation may also lead to negative social impacts such as loss of grazing land and source of traditional livelihoods.

Large investments have been made in commercial plantations on degraded lands in Asia. However, lack of consultation with stakeholders (state of land tenure and use rights) may result in failure to achieve the pursued results. Better integration between social goals and afforestation is necessary (Farley et al., 2004). As demand increases for lands to afforest, more comprehensive, multidimensional environmental assessment and planning will be required to manage land sustainably.

Agro-forestry can produce a wide range of economic, social and environmental benefits, and probably wider than in case of large-scale afforestation. Agro-forestry systems could be an interesting opportunity for conventional livestock production with low financial returns and negative environmental effects (overgrazing and soil degradation). For many livestock farmers, who may face financial barriers to develop this type of combined systems (e.g., silvo-pastoral systems), payment for environmental services could contribute to the feasibility of these initiatives (Gobbi, 2003). Shadow trees and shelter may have also beneficial effects on livestock production and income, as reported by Bentancourt et al., (2003). Little evidence of local extinctions and invasions of species risking biodiversity has been found when practising agro-forestry (Clavijo et al., 2005).

9.7.3 Implications of mitigation options on water, biodiversity and soil

The Millennium Development Goals (MDGs) aim at poverty reduction, and to improve health, education, gender equality, sanitation and environmental sustainability to promote Sustainable Development. Forest sector can significantly contribute to reducing poverty and improving livelihoods (providing access to forest products such as fuelwood, timber, and non timber products). Land degradation, access to water and food and human health remained at the centre of global attention under the debate on the World Summit on Sustainable Development (WSSD). A focus on five key thematic areas was proposed (Water, Energy, Health, Agriculture, and Biodiversity -WEHAB), driving attention to the fact that managing the natural resources like forest in a sustainable and integrated manner is essential for sustainable development. In this regard, to reverse the current trend in forest degradation as soon as possible, strategies need to be implemented that include targets adopted at national and, where appropriate, regional levels to protect ecosystems and to achieve integrated management of land, water and living resources associated to forest areas, while strengthening regional, national and local capacities.

Literature describing in detail the environmental impacts of different forest activities is still scarce and focuses mostly on planted forests. For these reasons, the discussion focuses more on plantations. It is important to underline that while benefits of climate change mitigation are global, co-benefits and costs tend to be local (OECD, 2002) and, in accordance, trade-offs have to be considered at local level.

Table 9.10: *Sustainable development implications of forestry mitigation*

Activity category	Sustainable development implications		
	Social	**Economic**	**Environmental**
A. Increasing or maintaining the forest area			
Reducing deforestation and forest degradation	*Positive* Promotes livelihood.	*Positive or negative* Provides sustained income for poor communities. Forest protection may reduce local incomes.	*Positive* Biodiversity conservation. Watershed protection. Soil protection. Amenity values (Nature reserves, etc.)
Afforestation/ reforestation	*Positive or negative* Promotes livelihood. Slows population migration to other areas (when a less intense land use is replaced). Displacement of people may occur if the former activity is stopped, and alternate activities are not provided. Influx of outside population has impacts on local population.	*Positive or negative* Creation of employment (when less intense land use is replaced). Increase/decrease of the income of local communities. Provision of forest products (fuelwood, fibre, food construction materials) and other services.	*Positive or negative* Impacts on biodiversity at the tree, stand, or landscape level depend on the ecological context in which they are found. Potential negative impacts in case on biodiversity conservation (mono-specific plantations replacing biodiverse grasslands or shrub lands). Watershed protection (except if water-hungry species are used) . Losses in stream flow. Soil protection. Soil properties might be negatively affected.
B. Changing to sustainable forest management			
Forest management in plantations	*Positive* Promotes livelihood.	*Positive* Creation of employment Increase of the income of local communities. Provision of forest products (fuelwood, fibre, food, construction materials) and other services.	*Positive* Enhance positive impacts and minimize negative implications on biodiversity, water and soils.
Sustainable forest management in native forest	*Positive* Promotes livelihood.	*Positive* Creation of employment. Increase of the income of local communities. Provision of forest products (fuelwood, fibre, food, construction materials) and other services.	*Positive* Sustainable management prevents forest degradation, conserves biodiversity and protects watersheds and soils.
C. Substitution of energy intensive materials			
Substitution of fossil intensive products by wood products	*Positive or negative* Forest owners may benefit. Potential for competition with the agricultural sector (food production, etc.).	*Positive* Increased local income and employment in rural and urban areas. Potential diversification of local economies. Reduced imports.	*Negative* Non-sustainable harvest may lead to loss of forests, biodiversity and soil.
D. Bioenergy			
Bioenergy production from forestry	*Positive or negative* Forest owners may benefit. Potential for competition with the agricultural sector (food production, etc.)	*Positive or negative* Increased local income and employment. Potential diversification of local economies. Provision of renewable and independent energy source. Potential competition with the agricultural sector (food production, etc.)	*Positive or negative* Benefits if production of fuelwood is done in a sustainable way. Mono specific short rotation plantations for energy may negatively affect biodiversity, water and soils, depending on site conditions.

Water cycle: Afforestation may result in better balance in the regional water cycle balance by reducing run-off, flooding, and control of groundwater recharge and watersheds protection. However, massive afforestation grasslands may reduce water flow into other ecosystems and rivers, and affect aquifers layer and recharge, and lead to substantial losses in stream flow (Jackson *et al*, 2005). In addition, some possible changes in soil properties are largely driven by changes in hydrology.

Soils: Intensively managed plantations have nutrient demands that may affect soil fertility and soil properties, for example leading to higher erosion of the uncovered mineral soil surface (Perez-Bidegain *et al.*, 2001; Carrasco-Letellier *et al.*, 2004); and biological properties changes (Sicardi *et al.,* 2004) if the choice of species is not properly matched with site conditions. Regarding chemical properties, increased Na concentrations, exchangeable sodium percentage and soil acidity, and decreased base saturation have been detected in many situations. (Jackson, *et al.*, 2005).In general, afforestation of low soil carbon croplands may present considerable opportunities for carbon sequestration in soil, while afforestation of grazing land can result in relatively smaller increases or decreases in soil carbon (Section 9.4.2.2). Most mitigation options other than monoculture plantations conserve and protect soils and watersheds.

Biodiversity: Plantations can negatively affect biodiversity if they replace biologically rich native grassland or wetland habitats (Wagner, et al., 2006). Also, plantations can have either positive or negative impacts on biodiversity depending on management practices (Quine and Humphrey, 2005). Plantations may act as corridors, source, or barriers for different species, and a tool for landscape restoration (Parrota, 2002). Other forestry mitigation options such as reducing deforestation, agro-forestry, multi-species plantations, and sustainable native forest management lead to biodiversity conservation.

Managing plantations to produce goods (such as timber) while also enhancing ecological services (such as biodiversity) involves several trade-offs. Overcoming them involves a clear understanding of the broader ecological context in which plantations are established as well as participation of the different stakeholders. The primary management objective of most industrial plantations traditionally has been to optimize timber production. This is not usually the case in small-scale plantations owned by farmers, where more weight is given to non-timber products and ecological services. A shift from a stand level to a broader forest and non-forest landscape level approach will be required to achieve a balance between biodiversity and productivity/profitability.

The literature seems to suggest that plantations, mainly industrial plantations, require careful assessment of the potential impacts on soils, hydrological cycle and biodiversity, and that negative impacts could be controlled or minimized if adequate landscape planning and basin management and good practices are introduced. Carbon sequestration strategies

with afforestation of non-forest lands should consider their full environmental consequences. The ultimate balance of co-benefits and impacts depends on the specific site conditions and previous and future land and forest management.

9.8 Technology, R&D, deployment, diffusion and transfer

R&D and technology transfer have a potential to promote forest sector mitigation options by increasing sustainable productivity, conserving biodiversity and enhancing profitability. Technologies are available for promoting mitigation options from national level to forest stand level, and from single forest practices to broader socio-economic approaches (IPCC, 2000b).

Traditional and/or existing techniques in forestry including planting, regeneration, thinning and harvesting are fundamental for implementation of mitigation options such as afforestation, reforestation, and forest management. Further, improvement of such sustainable techniques is required and transfer could build capacity in developing countries. Biotechnology may have an important role especially for afforestation and reforestation. As the area of planted forests including plantations of fast-growing species for carbon sequestration increases, sustainable forestry practices will become more important for both productivity and environment conservation.

The development of suitable low-cost technologies will be necessary for promoting thinning and mitigation options. Moreover, technology will have to be developed for making effective use of small wood, including thinned timber, in forest products and markets. Thinning and tree pruning for fuelwood and fodder are regularly conducted in many developing countries as part of local integrated forest management strategies. Although natural dynamics are part of the forest ecosystem, suppression of forest fires and prevention of insect and pest disease are important for mitigation.

Regarding technology for harvesting and procurement, mechanized forest machines such as harvesters, processors and forwarders developed in Northern Europe and North America have been used around the world for the past few decades. Mechanization under sustainable forest management seems to be effective for promoting mitigation options including product and energy substitution (Karjalainen and Asikainen, 1996). However, harvesting and procurement systems vary due to terrain, type of forest, infrastructure and transport regulations, and appropriate systems also vary by regions and countries. Reduced impact logging is considered in some cases such as in tropical forests (Enters *et al.,* 2002).

There is a wide array of technologies for using biomass from plantations for direct combustion, gasification, pyrolysis, and fermentation (see Section 4.3.3.3). To conserve forest resources, recycling of wood waste material needs to be expanded. Technology for manufacturing waste-derived board has almost been established, but further R&D will be necessary to re-use waste sawn timber, or to recycle it as lumber. While these technologies often need large infrastructure and incentives in industrialized countries, practical devices such as new generations of efficient wood-burning cooking stoves (Masera *et al.,* 2005) have proved effective in developing countries. They are effective as a means to reduce the use of wood fuels derived from forests, at the same time providing tangible sustainable development benefits for local people, such as reduction in indoor air pollution levels.

Technological R&D for estimation of carbon stocks and fluxes is fundamental not only for monitoring but also for evaluating policies. Practical methods for estimating carbon stocks and fluxes based on forest inventories and remote sensing have been recommended in the Good Practice Guidance for LULUCF (IPCC, 2003). Over the last three decades, earth observation satellites have increased in number and sophistication (DeFries *et al.,* 2006). High-resolution satellite images have become available, so new research on remote sensing has begun on using satellite radar and LIDAR (light detection and ranging) for estimating forest biomass (Hirata *et al.,* 2003). Remote sensing methods are expected to play an increasing role in future assessments, especially as a tool for mapping land cover and its change over time. However, converting these maps into estimates of carbon sources and sinks remains a challenge and will continue to depend on in-situ measurements and modelling.

Large-scale estimations of the forest sector and its carbon balance have been carried out with models such as the CBM-CFS2 (Kurz and Apps, 2006), CO2FIX V.2 model (Masera *et al.,* 2003), EFISCEN (Nabuurs *et al.,* 2005, 2006), Full CAM (Richards and Evans, 2004), and GORCAM (Schlamadinger and Marland, 1996).

Micrometeorological observation of carbon dioxide exchange between the terrestrial ecosystem and the atmosphere has been carried out in various countries (Ohtani, 2005). Based on the observation, a global network FLUXNET (Baldocchi *et al.,* 2001) and regional networks including AmeriFlux, EUROFLUX, AsiaFlux and OzNet are being enlarged for stronger relationships.

New technologies for monitoring and verification including remote sensing, carbon flux modelling, micrometeorological observation and socio-economic approaches described above will facilitate the implementation of mitigation options. Furthermore, the integration of scientific knowledge, practical techniques, socio-economic and political approaches will become increasingly significant for mitigation technologies in the forest sector.

Few forest-based mitigation analyses have been conducted using primary data. There is still limited insight regarding impacts on soils, lack of integrated views on the many site-specific studies, hardly any integration with climate impact studies, and limited views in relation to social issues and sustainable development. Little new effort was reported on the development of global baseline scenarios of land-use change and their associated carbon balance, against which mitigation options could be examined. There is limited quantitative information on the cost-benefit ratios of mitigation interventions.

Technology deployment, diffusion and transfer in the forestry sector provide a significant opportunity to help mitigate climate change and adapt to potential changes in the climate. Apart from reducing GHG emissions or enhancing the carbon sinks, technology transfer strategies in the forest sector have the potential to provide tangible socio-economic and local and global environmental benefits, contributing to sustainable development (IPCC, 2000b). Especially, technologies for improving productivity, sustainable forest management, monitoring, and verification are required in developing countries. However, existing financial and institutional mechanism, information and technical capacity are inadequate. Thus, new policies, measures and institutions are required to promote technology transfer in the forest sector.

For technology deployment, diffusion and transfer, governments could play a critical role in: a) providing targeted financial and technical support through multilateral agencies (World Bank, FAO, UNDP, UNEP), in developing and enforcing the regulations to implement mitigation options; b) promoting the participation of communities, institutions and NGOs in forestry projects; and c) creating conditions to enable the participation of industry and farmers with adequate guidelines to ensure forest management and practices as mitigation options. In addition, the role of private sector funding of projects needs to be promoted under the new initiatives, including the proposed flexible mechanisms under the Kyoto Protocol. The Global Environmental Facility (GEF) could fund projects that actively promote technology transfer and capacity building in addition to the mitigation aspects (IPCC, 2000b).

9.9 Long-term outlook

Mitigation measures up to 2030 can prevent the biosphere going into a net source globally. The longer-term mitigation prospects (beyond 2030) within the forestry sector will be influenced by the interrelationship of a complex set of environmental, socio-economic and political factors. The history of land-use and forest management processes in the last century, particularly within the temperate and boreal regions, as well as on the recent patterns of land-use will have a critical effect on the mitigation potential.

Several studies have shown that uncertainties in the contemporary carbon cycle, the uncertain future impacts of climatic change and its many dynamic feedbacks can cause large variation in future carbon balance projections (Lewis *et al.*, 2005). Other scenarios suggest that net deforestation pressure will slow over time as population growth slows and crop and livestock productivity increase. Despite continued projected loss of forest area, carbon uptake from afforestation and reforestation could result in net sequestration (Section 3.2.2).

Also, the impacts of climate change on forests will be a major source of uncertainty regarding future projections (Viner *et al.*, 2006). Other issues that will have an effect on the long-term mitigation potential include future sectoral changes within forestry, changes in other economic sectors, as well as political and social change, and the particular development paths within industrialized and developing countries beyond the first half of the 21st century. The actual mitigation potential will depend ultimately on solving structural problems linked to the sustainable management of forests. Such structural problems include securing land tenure and land rights of indigenous people, reducing poverty levels in rural areas and the rural-urban divide, and providing disincentives to short-term behaviour of economic actors and others. Considering that forests store more carbon dioxide than the entire atmosphere (Stern, 2006), the role of forests is critical.

Forestry mitigation projections are expected to be regionally unique, while still linked across time and space by changes in global physical and economic forces. Overall, it is expected that boreal primary forests will either be sources or sinks depending on the net effect of some enhancement of growth due to climate change versus a loss of soil organic matter and emissions from increased fires. The temperate forests in USA, Europe, China and Oceania, will probably continue to be net carbon sinks, favoured also by enhanced forest growth due to climate change. In the tropical regions, the human induced land-use changes are expected to continue to drive the dynamics for decades. In the meantime, the enhanced growth of large areas of primary forests, secondary regrowth, and increasing plantation areas will also increase the sink. Beyond 2040, depending on the extent and effectiveness of forest mitigation activities within tropical areas, and very particularly on the effectiveness of policies aimed at reducing forest degradation and deforestation, tropical forest may become net sinks. In the medium to long term as well, commercial bio-energy is expected to become increasingly important.

In the long-term, carbon will only be one of the goals that drive land-use decisions. Within each region, local solutions have to be found that optimize all goals and aim at integrated and sustainable land use. Developing the optimum regional strategies for climate change mitigation involving forests will require complex analyses of the trade-offs (synergies and competition) in land-use between forestry and other land uses,

the trade-offs between forest conservation for carbon storage and other environmental services such as biodiversity and watershed conservation and sustainable forest harvesting to provide society with carbon-containing fibre, timber and bio-energy resources, and the trade-offs among utilization strategies of harvested wood products aimed at maximizing storage in long-lived products, recycling, and use for bioenergy.

REFERENCES

Achard, F., H.D. Eva, P. Mayaux, H.-J. Stibig, and A. Belward, 2004: Improved estimates of net carbon emissions from land cover change in the Tropics for the 1990's. *Global Biogeochemical Cycles,* **18**, GB2008, doi:10.1029/2003GB002142.

Albrecht, A. and S.T. Kandji, 2003: Carbon sequestration in tropical agroforestry systems. *Agriculture, Ecosystems & Environment*, **99**(1-3), pp. 15-27.

Alves, D.S., J.L.G. Pereira, C.L. de Sousa, J.V. Soares, and F. Yamaguchi, 1999: Characterizing landscape changes in central Rondonia using Landsat TM imagery. *International Journal of Remote Sensing*, **20**, pp. 2877-2882.

Asner, G.P., D.E. Knapp, E.N. Broadbent, P.J.C. Oliveira, M. Keller, and J.N. Silva, 2005: Selective Logging in the Brazilian Amazon. *Science,* **310**(5747), pp. 480-482.

Atyi, R.E. and M. Simula, 2002: Forest certification: pending challenges for tropical timber. *Tropical Forest Update,* **12**(3), pp. 3-5.

Baldocchi, D., E. Falge, L. Gu, R. Olson, D. Hollinger, S. Running, P. Anthoni, Ch. Bernhofer, K. Davis, R. Evans, J. Fuentes, A. Goldstein, G. Katul, B. Law, X. Lee, Y. Malhi, T. Meyers, W. Munger, W. Oechel, K.T. Paw U, K. Pilegaard, H.P. Schmid, R. Valentini, S. Verma, T. Vesala, K. Wilson, and S. Wofsy, 2001: FLUXNET: A new tool to study the temporal and spatial variability of ecosystem-scale carbon dioxide, water vapor and energy flux densities. *Bulletin of the American Meteorological Society*, **82**(11), pp. 2415-2434.

Benitez-Ponce, P.C. 2005: *Essays on the economics of forestry based carbon mitigation.* PhD Thesis, Wageningen Agricultural University.

Benítez-Ponce, P.C., I. Mc Callum, M. Obersteiner, and Y. Yamagata. 2004: *Global supply for carbon sequestration: identifying least-cost afforestation sites under country risk considerations.* IIASA Interim Report IR-04-022, Laxenburg, Austria.

Benítez-Ponce, P.C., I. McCallum, M. Obersteiner, and Y. Yamagata. 2007: Global potential for carbon sequestration: geographical distribution, country risk and policy implications. *Ecological Economics,* **60**, pp. 572-583.

Bentancourt, K., M. Ibrahim, C. Harvey, and B. Vargas, 2003: Effect of tree cover on animal behavior in dual purpose cattle farms in Matiguas, Matagalpa, Nicaragua. *Agroforestería en las Américas,* **10**, pp. 47-51.

Betts, R.A., P.M. Cox, M. Collins, P.P. Harris, C. Huntingford, and C.D. Jones, 2004: The role of ecosystem-atmosphere interactions in simulated Amazonian precipitation decrease and forest dieback under global climate warming. *Theoretical and Applied Climatology*, **78**(1-3), pp. 157-175.

Bhat, D.M, K.S. Murali, and N.H. Ravindranath, 2001: Formation and recovery of secondary forests in india: A particular reference to western ghats in South India. *Journal of Tropical Forest Science,* **13**(4), pp. 601-620.

Bhattacharya, S.C., P. Abdul Salam, H.L. Pham, and N.H. Ravindranath, 2005: Sustainable biomass production for energy in selected Asian Countries. *Biomass and Bioenergy*, **25**, pp. 471-482.

Brinkman, S., 2005: IMAGE 2.2. Carbon cycle analysis. Brinkman climate change consultant. The Hague. The Netherlands. 37 pp.

Brown, S., A. Dushku, T. Pearson, D. Shoch, J. Winsten, S. Sweet, and J. Kadyszewski, 2004: Carbon supply from changes in management of forest, range, and agricultural lands of California. Winrock International for California Energy Commission, 144 pp. + app.

Cacho, O.J., R L. Hean, and R M. Wise, 2003: Carbon-accounting methods and reforestation incentives**.** *The Australian Journal of Agricultural and Resource Economics,* **47**, pp. 153-179.

Cannell, M.G.R., 2003: Carbon sequestration and biomass energy offset: theoretical, potential and achievable capacities globally, in Europe and the UK. *Biomass and Bioenergy*, **24**, pp. 97-116.

Caparros, A. and F. Jacquemont, 2003: Conflicts between biodiversity and carbon sequestration programs: economic and legal implications. *Ecological Economics*, **46**, pp. 143-157.

Carrasco-Letellier, L., G. Eguren, C. Castiñeira, O. Parra, and D. Panario, 2004: Preliminary study of prairies forested with Eucalyptus sp. at the Nortwestern Uruguayan soils. *Environmental Pollution,* **127**, pp. 49-55.

Caspersen, J.P., S.W. Pacala, J.C. Jenkins, G.C. Hurtt, P.R. Moorcroft, and R.A. Birdsey, 2000: Contributions of land-use history to carbon accumulation in U.S. forests. *Science* **290**, pp. 1148-1151.

Chen, W., J.M. Chen, D.T. Price, J. Cihlar, and J. Liu, 2000: Carbon offset potentials of four alternative forest management strategies in Canada: A simulation study. *Mitigation and Adaptation Strategies for Global Change,* **5**, pp. 143-169.

Chomitz, K.M., 2002: Baseline, leakage and measurement issues: how do forestry and energy projects compare? *Climate Policy,* **2**, pp. 35-49.

Chomitz, K.M., P. Buys, G. DeLuca, T.S. Thomas, and S. Wertz-Kanounnikoff, 2006: At Loggerheads? Agricultural expansion, poverty reduction, and the environment in the tropics. The World Bank, Washington, D.C., 284 pp.

Ciais, P, P. Peylin P, and P. Bousquet, 2000: Regional biospheric carbon fluxes as inferred from atmospheric CO_2 measurements. *Ecological Applications,* **10**(6), pp. 1574-1589.

Ciais, P., M. Reichstein, N. Viovy, A. Granier, J. Ogée, V. Allard, M. Aubinet, N. Buchmann, Chr. Bernhofer, A. Carrara, F. Chevallier, N. De Noblet, A.D. Friend, P. Friedlingstein, T. Grünwald, B. Heinesch, P. Keronen, A. Knohl, G. Krinner, D. Loustau, G. Manca, G. Matteucci, F. Miglietta, J.M. Ourcival, D. Papale, K. Pilegaard, S. Rambal, G. Seufert, J.F. Soussana, M.J. Sanz, E.D. Schulze, T. Vesala, and R. Valentini, 2005: Europe-wide reduction in primary productivity caused by the heat and drought in 2003. *Nature* (London), **437**(7058), pp. 529-533.

Clarke, M., 2000: Real barriers to trade in forest products. Working Paper 2000/4, New Zealand Institute of Economic Research, Wellington, 20 pp.

Clavijo, M., M. Nordenstahl, P. Gundel, and E. Jobbágy, 2005: Poplar afforestation effects on grasslands structure and composition in the flooding pampas. *Rangeland Ecology & Management*, **58**, pp. 474-479.

Cohen, D.H., L. Lee, and I. Vertinsky, 2002: China's natural forest protection program (NFPP): impact on trade policies regarding wood. Prepared for CIDA with the Research Center for Ecological and Environmental Economics, Chinese Academy of Social Sciences, 63 pp.

Cossalter, C. and C. Pye-Smith, 2003: Fast-wood Forestry: myths and realities. Center for International Forestry Research, Bogor, Indonesia, 59 pp.

Couto, L. and R. Betters, 1995: Short-rotation eucalypt plantations in Brazil: social and environmental issues. ORNL/TM-12846. Oak Ridge National Laboratory, Oak Ridge, Tennessee. <http://bioenergy.ornl. gov/reports/euc-braz/index.html>. accessed 20 September 2005.

Cox, P.M., R.A. Betts, M. Collins, P.P. Harris, C. Huntingford, and C.D. Jones, 2004: Amazonian forest dieback under climate-carbon cycle projections for the 21st century. *Theoretical and Applied Climatology,* **78**, pp. 137-156.

Cuiping, L., Yanyongjie, W. Chuangzhi, and H. Haitao: Study on the distribution and quantity of biomass residues resource in China. *Biomass and Bioenergy* 2004, **27**, pp. 111-117.

Dang, H.H., A. Michaelowa, and D.D. Tuan, 2003: Synergy of adaptation and mitigation strategies in the context of sustainable development: in the case of Vietnam. *Climate Policy,* **3S1**, pp. S81- S96.

De Jong, B.H.J., E.E. Bazán, and S.Q. Montalvo, 2007 (in print): Application of the Climafor baseline to determine leakage: The case of Scolel Té. *Mitigation and Adaptation Strategies for Global Change.*

DeFries, R., 2002: Past and future sensitivity of primary production to human modification of the landscape. *Geophysical Research Letters*, **29**(7), Art. No. 1132.

DeFries, R., F. Achard, S. Brown, M. Herold, D. Murdiyarso, B. Schlamadinger, and C. DeSouza, 2006: Reducing greenhouse gas emissions from deforestation in developing countries: Considerations for monitoring and measuring. Report of the Global Terrestrial Observing System (GTOS) number **46**, GOFC-GOLD report 26, 23 pp.<www.fao.org/gtos/pubs.html> accessed 11 June 2007.

Denman, K.L., G. Brasseur, A. Chidthaisong, Ph. Ciais, P. Cox, R.E. Dickinson, D. Hauglustaine, C. Heinze, E. Holland, D. Jacob, U. Lohmann, S. Ramachandran, P.L. da Silva Dias, S.C. Wofsy, X. Zhang, 2007: Couplings Between Changes in the Climate System and Biogeochemistry, Chapter 7 in: *Climate Change 2007: The Physical Science Basis*, The IPCC Fourth Assessment Report, Intergovernmental Panel on Climate Change, Cambridge University Press, Cambridge.

Dourojeanni, M.J., 1999: The future of Latin American forests. Environment Division Working Paper, Interamerican Development Bank, Washington D.C.

Dublin, H. and C. Volante, 2004: Biodiversity Program Study 2004. Global Environment Facility Office of Monitoring and Evaluation, Washington, D.C., 141 pp.

Dudek, D., A. Golub, and E. Strukova. 2002: Ancillary benefits of reducing greenhouse gas emissions in transitional economies. Working Paper, Environmental Defence. Washington, D.C.

Dutschke, M., B. Schlamadinger, J.L.P. Wong, and M. Rumberg, 2005: Value and risks of expiring carbon credits from affor-estation and reforestation projects under the CDM. *Climate Policy,* **5**(1).

Easterling, W., P. Aggarwal, P. Batima, K. Brander, L. Erda, M. Howden, A. Kirilenko, J. Morton, J.F. Soussana, J. Schmidhuber, F. Tubiello, Food, Fibre, and Forest Products, Chapter 5 in: *Climate Change 2007: Climate Change Impacts, Adaptation and Vulnerability*, The IPCC Fourth Assessment Report, Cambridge University Press, Cambridge.

EEA, 2005: How much bioenergy can Europe provide without harming the environment? EEA report 07/2006. European Environment Agency, Copenhagen.

Eggers, J., M. Lindner, S. Zudin, S. Zaehle, J. Liski, and G.J. Nabuurs, 2007. Forestry in Europe under changing climate and land use. Proceedings of the OECD Conference 'Forestry: A Sectoral Response to Climate Change', 21-23 November 2006. In: P. Freer-Smith (ed.), Forestry Commission, UK.

Enters, T., P.B. Durst, G.B. Applegate, P.C.S. Kho, G. Man, T. Enters, P.B. Durst, G.B. Applegate, P.C.S. Kho, and G. Man, 2002: Trading forest carbon to promote the adoption of reduced impact logging, Applying reduced impact logging to advance sustainable forest management. International conference proceedings, Kuching, Malaysia, RAP-Publication, pp. 261-274.

Ericsson, K. and L.J. Nilsson, 2006: Assessment of the potential biomass supply in Europe using a resource-focused approach. *Biomass and Bioenergy*, **30**(1), pp. 1-15.

Fang, J., T. Oikawa, T. Kato and W. Mo, 2005: Biomass carbon accumulation by Japan's forests from 1947 to 1995. *Global Biogeochemical Cycles*, **19**, GB2004.

FAO, 2001: Global Forest Resources Assessment 2000. Main report. FAO Forestry Paper 140, 479 pp.

FAO, 2005: State of the world's forests 2005. 153 pp.

FAO, 2006a: Global Forest Resources Assessment 2005. Progress towards sustainable forest management. FAO Forestry Paper 147, 320 pp.

FAO, 2006b: Summaries of FAO's work in forestry. Rome, Italy. <http://www.fao.org/forestry/foris/webview/forestry2/index.jsp?siteId=3741&sitetreeId=11467&langId=1&geoId=0> accessed 27 October 2006.

Farley, K., E.G. Jobbágy and R.B. Jackson, 2004: Effects of afforestation on water yield: a global synthesis with implications for policy. Center on Global Change, Duke University, Durham. Department of Biology and Nicholas School of the Environment and Earth Sciences, Duke University, Durham. Grupo de Estudios Ambientales - IMASL, Universidad Nacional de San Luis & CONICET, Argentina.

Fearnside, P.M. and W.F. Laurance, 2004: Tropical deforestation and greenhouse-gas emissions. *Ecological Applications,* **14**(4), pp. 982-986.

Fischer, G. and L. Schrattenholzer, 2001: Global bioenergy potentials through 2050. *Biomass and Bioenergy,* **20**, pp. 151-159.

Fischlin, A., G.F. Midgley, J. Price, R. Leemans, B. Gopal, C. Turley, M. Rounsevell, P. Dube, J. Tarazona, A. Velichko, 2007: Ecosystems, their Properties, Goods, and Services, Chapter 4 in: *Climate Change 2007: Climate Change Impacts, Adaptation and Vulnerability,* The IPCC Fourth Assessment Report, Cambridge University Press, Cambridge.

Fuhrer, J., M. Benitson, A. Fischlin, C. Frei, S. Goyette, K. Jasper, and C. Pfister, 2007: Climate risks and their impact on agriculture and forests in Switzerland. *Climatic change,* **72** (accepted in print).

Gaddis, D.A., B.D. New, F.W. Cubbage, R.C. Abt, and R.J. Moulton, 1995: Accomplishments and economic evaluations of the Forestry Incentive Program: a review. SCFER Working Paper (78), pp. 1-52. Southeastern Center for Forest Economics Research, Research Triangle Park, NC.

Geist, H.J. and E.F. Lambin, 2002: Proximate causes and underlying driving forces of tropical deforestation. *BioScience,* **52**, pp. 143-150.

Gobbi, J., 2003: Financial behavior of investment in sylvopastoral systems in cattle farms of Esparza, Costa Rica. *Agroforestería en las Américas,* **10**, pp. 52-60.

Government of Australia, 2006: Biofuels Capital Grants (BCG) - Fact Sheet. <http://www.ausindustry.gov.au/content/content.cfm?ObjectID=8B98D9B4-D244-43EF-A4AA7D42B39DFE1B> accessed 27 October 2006.

Granger, A., 1997: Bringing tropical deforestation under control change programme. Global environmental change programme briefings, Number 16. Economic and Social Research Council, Global Environmental Change Programme, Sussex, U.K., 4 pp. <http://www.sussex.ac.uk/Units/gec/pubs/briefing/brief-16.pdf> accessed 1 September 2005.

Grassl, H., J. Kokott, M. Kulessa, J. Luther, F. Nuscheler, R. Sauerborn, H.-J. Schellnhuber, R. Schubert, and E.-D. Schulze: 2003. Climate protection strategies for the 21st century: Kyoto and beyond. Berlin: German Advisory Council on Global Change (WBGU).

Grieg-Gran, M., 2004: Making environmental service payments work for the poor: some experiences from Latin America. International Fund for Agricultural Development Governing Council Side Event. <www.ifad.org/events/gc/27/side/presentation/ieed.ppt> accessed 1 September 2005.

Gullison, R.E., 2003: Does forest certification conserve biodiversity? *Oryx,* **37**(2), pp. 153-165.

Gurney, K.R., R.M. Law, A.S. Denning, P.J. Rayner, D. Baker, P. Bousquet, L. Bruhwiler, Y-H. Chen, P. Ciais, S. Fan, I.Y. Fung, M. Gloor, M. Heimann, K. Higuchi, J. John, T. Maki, S. Maksyutov, K. Masarie, P. Peylin, M. Prather, B.C. Pak, J. Randerson, J. Sarmiento, S. Taguchi, T. Takahashi, and C-W. Yuen, 2002: Towards robust regional estimates of CO_2 sources and sinks using atmospheric transport models. *Nature,* **415**, pp. 626-629.

Gustavsson, L. and R. Sathre, 2006: Variability in energy and carbon dioxide balances of wood and concrete building materials. *Building and Environment,* **41**, pp. 940-951.

Gustavsson, L., K. Pingoud, and R. Sathre, 2006: Carbon dioxide balance of wood substitution: comparing concrete and wood-framed buildings. *Mitigation and Adaptation Strategies for Global Change,* **11**, pp. 667-691.

Heistermann, M., C. Müller, and K. Ronneberger, 2006: Review land in sight? Achievements, deficits and potentials of continental to global scale land-use modelling. *Agriculture, Ecosystems and Environment,* **114**(2006), pp. 141-158.

Hirata, Y., Y. Akiyama, H. Saito, A. Miyamoto, M. Fukuda, and T. Nisizono, 2003: Estimating forest canopy structure using helicopter-borne LIDAR measurement. P. Corona *et al.* (eds), Advances in forest inventory for sustainable forest management and biodiversity monitoring, Kluwer Academic Publishers, pp. 125-134.

Hoen, H.F. and B. Solberg, 1994: Potential and efficiency of carbon sequestration in forest biomass through silvicultural management. *Forest Science,* **40**, pp. 429-451.

Holmes, T.P., G.M. Blate, J.C. Zweede, R. Pereira Jr., P. Barreto, F. Boltz, and R. Bauch, 2002: Financial and ecological indicators of reduced impact logging performance in the eastern Amazon. *Forest Ecology and Management,* **163**, pp. 93-110.

Hooda, N., M. Gera, K. Andrasko, J.A. Sathaye, M.K. Gupta, H.B. Vasistha, M. Chandran, and S.S. Rassaily, 2007: Community and farm forestry climate mitigation projects: case studies from Uttaranchal, India. *Mitigation and Adaptation Strategies for Global Change.* **12**(6) pp. 1099-1130.

Houghton, R.A., 2005: Aboveground forest biomass and the global carbon balance. *Global Change Biology,* **11**(6), pp. 945-958.

Houghton, R.A., 2003a: Why are estimates of the terrestrial carbon balance so different? *Global change biology,* **9**, pp. 500-509.

Houghton, R.A., 2003b: Revised estimates of the annual net flux of carbon to the atmosphere from changes in land use and land management 1850-2000. *Tellus Series B Chemical and Physical Meteorology,* **55**(2), pp. 378-390.

Houghton, R.A., D.L. Skole, C.A. Nobre, J.L. Hackler, K.T. Lawrence, and W.H. Chomentowski, 2000: Annual fluxes or carbon from deforestation and regrowth in the Brazilian Amazon. *Nature,* **403**, pp. 301-304.

IEA, 2002: Sustainable Production of woody biomass for energy, a position paper prepared by IEA Bioenergy. <http://www.ieabioenergy.com/> accessed 11 June 2007.

Ikkonen, E.N., V.K. Kurets, S.I. Grabovik, and S.N. Drozdov, 2001: The rate of carbon dioxide emission into the atmosphere from a southern Karelian mesooligotrophic bog. *Russian Journal of Ecology,* **32**(6), pp. 382-385.

IPCC, 2000a: *Land use, land-use change and forestry.* Special report of the Intergovernmental Panel on Climate Change (IPCC). Cambridge University Press, Cambridge, 377 pp.

IPCC, 2000b: *Methodological and technological issues in the technology transfer.* Intergovernmental Panel on Climate Change (IPCC). Cambridge University Press, Cambridge.

IPCC, 2000c: *Emission Scenarios.* Special Report of the Intergovernmental Panel on Climate Change (IPCC). Cambridge University Press, Cambridge.

IPCC, 2001: *Climate Change 2001: Synthesis Report.* A Contribution of Working Groups I, II, and III to the Third Assessment Report of the Intergovernmental Panel on Climate Change [Watson, R.T. and the Core Writing Team (eds.)]. Cambridge University Press, Cambridge, United Kingdom, and New York, NY, USA, 398 pp.

IPCC, 2002: *Climate and Biodiversity.* IPCC Technical Paper V [Gitay, H., A. Suarez., R.T. Watson, and D.J. Dokken (eds.)]. Intergovernmental Panel on Climate Change (IPCC).

IPCC, 2003: *Good practice guidance for land use, land-use change and forestry.* IPCC-IGES, Japan.

IPCC, 2006: *2006 IPCC guidelines for national greenhouse gas inventories.* Prepared by the National Greenhouse Gas Inventories Programme [Eggleston H.S., L. Buenia, K. Miwa, T. Ngara, and K. Tanabe (eds)]. IPCC-IGES, Japan.

IPCC, 2007a: Climate Change 2007: The Physical Science Basis. Contribution of Working Group I to the Fourth Assessment Report of the Intergovernmental Panel on Climate Change [Solomon, S., D. Qin, M. Manning, Z. Chen, M. Marquis, K.B.M.Tignor and H.L. Miller (eds.)]. Cambridge University Press, Cambridge, United Kingdom and New York, NY, USA, 996 pp.

IPCC, 2007b: Climate Change 2007: Impacts, Adaptation and Vulnerability. Contribution of Working Group II to the Fourth Assessment Report of the Intergovernmental Panel on Climate Change [Parry, M.L., O.F. Canziani, J.P. Palutikof, P.J. van der Linden, C.E. Hanson (eds.)]. Cambridge University Press, Cambridge, United Kingdom and New York, NY, USA.

ITTO, 1999: International cross sectoral forum on forest fire management in South East Asia. International Tropical Timber Organization, Jakarta, Indonesia, 7 And 8 December 1998. Report of the Meeting. <http://www.fire.uni-freiburg.de/programmes/itto/cross.pdf> accessed 12 April 2006.

IUCN, 2002: Analysis of stakeholder power and responsibilities in community involvement in forest management in Eastern and Southern Africa. IUCN, Nairobi.

IUCN, 2004: Community-based natural resource management in the IGAD Region. IUCN, Nairobi.

Izrael, Y.A., M.L. Gytarsky, R.T. Karaban, A.L. Lelyakin, and I.M. Nazarov: 2002: Consequences of climate change for forestry and carbon dioxide sink in Russian forests. Izvestiya, *Atmospheric and Oceanic Physics,* 38(Suppl. 1), pp. S84-S98.

Jackson, R.B., E.G. Jobbágy, R Avissar, S. Baidya Roy, D.J. Barrett, Ch.W. Cook, K.A. Farley, D.C. le Maitre, B.A. McCarl, and B.C. Murray, 2005: Trading water for carbon with biological carbon sequestration. *Science,* 310, pp. 1944-1947.

Jakeman, G. and B.S. Fisher, 2006: Benefits of multi-gas mitigation: an application of the global trade and environment model (GTEM). *Energy Journal,* Special Issue pp. 323-342.

Janssens, I.A., A. Freibauer, P. Ciais, P. Smith, G.-J. Nabuurs, G. Folberth, B. Schlamadinger, R.W.A. Hutjes, R. Ceulemans, E.-D. Schulze, R. Valentini, and A.J. Dolman, 2003: Europe's terrestrial biosphere absorbs 7 to 12% of European anthropogenic CO_2 emissions. *Science,* 300, pp. 1538-1542.

Jung, M., 2005: The role of forestry sinks in the CDM-analysing the effects of policy decisions on the carbon market. HWWA discussion paper 241, Hamburg Institute of International Economics, 32 pp.

Kabat, P., W. Vierssen, J. Veraart, P. Vellinga, and J. Aerts, 2005: Climate proofing the Netherlands. *Nature,* 438, pp. 283-284.

Karjalainen, T. and A. Asikainen, 1996: Greenhouse gas emissions from the use of primary energy in forest operations and long-distance transportation of timber in Finland, *Forestry,* 69, pp. 215-228.

Karjalainen, T., A. Pussinen, J. Liski, G.-J. Nabuurs, T. Eggers, T. Lapvetelainen, and T. Kaipainen, 2003: Scenario analysis of the impacts of forest management and climate change on the European forest sector carbon budget. *Forest Policy and Economics,* 5, pp. 141-155.

Katila, M. and E. Puustjärvi, 2004: Markets for forests environmental services: reality and potential. *Unasylva,* 219(55, 2004/4), pp.53-58.

Kauppi, P., R.J. Sedjo, M. Apps, C. Cerri, T. Fujimori, H. Janzen, O. Krankina, W. Makundi, G. Marland, O. Masera, G.J. Nabuurs, W. Razali, and N.H. Ravindranath, 2001: Technical and economic potential of options to enhance, maintain and manage biological carbon reservoirs and geo-engineering. In *Mitigation 2001. The IPCC Third Assessment Report,* [Metz, B., *et al.,* (eds.)], Cambridge, Cambridge University Press.

Klein, R.J.T., S. Huq, F. Denton, T. Downing, R.G. Richels, J.B. Robinson, F.L. Toth, 2007: Inter-relationships between Adaptation and Mitigation, Chapter 18 in: *Climate Change 2007: Climate Change Impacts, Adaptation and Vulnerability,* The IPCC Fourth Assessment Report, Cambridge University Press, Cambridge.

Koopman, A., 2005: Biomass energy demand and supply for South and South-East Asia - assessing the resource base. *Biomass and Bioenergy,* 28, pp 133-150.

Körner, C., 2004: Through enhanced tree dynamics carbon dioxide enrichment may cause tropical forests to lose carbon. *Philosophical transactions of the Royal Society of London - Series A,* 359(1443), pp. 493-498.

Krankina ON, Harmon ME, Winjum K. 1996. Carbon storage and sequestration in Russian Forest Sector. Ambio 25: 284-288.

Krankina, O.N., R.K. **Dixon** 1994. Forest management options to conserve and sequester terrestrial carbon in the Russian Federation World Resour. Rev. 6: 88-101.

Kurz, W.A. and M.J. Apps, 2006: Developing Canada's national forest carbon monitoring, Accounting and reporting system to meet the reporting requirements of the Kyoto Protocol. *Mitigation and Adaptation Strategies for Global Change,* 11, pp. 33-43.

Kurz, W.A., S.J. Beukema, and M.J. Apps, 1998: Carbon budget implications of the transition from natural to managed disturbance regimes in forest landscapes. *Mitigation and Adaptation Strategies for Global Change,* 2, pp. 405-421.

Lasco, R.D., F.B. Pulhin, and R.F. Sales, 2007 (accepted, in print): An analysis of a forestry carbon sequestration project in the Philippines: the case of upper Magat watershed. *Mitigation and Adaptation Strategies for Global Change.*

Lewandrowski, J., M. Peters, C. Jones, R. House, M. Sperow, M. Eve, and K. Paustian, 2004: Economics of Sequestering Carbon in the U.S. Agricultural Sector. *Technical Bulletin,* TB1909, 69 pp.

Lewis, S., O. Phillips, T. Baker, Y. Malhi, and J. Lloyd, 2005: Tropical forests and atmospheric carbon dioxide: Current knowledge and potential future scenarios. Oxford University Press, 260 pp.

Lindner, M., J. Meyer, Th. Eggers, and A. Moiseyev, 2005: Environmentally enhanced bio-energy potential from European forests. A report commissioned by the European Environment Agency through the European Topic Centre on Biodiversity, Paris. European Forest Institute, Joensuu, Finland.

MAF, 2006: The permanent forest sink initative. Bulletin Issue 3, December 2006. Ministry of Agriculture and Forestry, Wellington, New Zealand. <http://www.maf.govt.nz/forestry/pfsi/bulletin/issue-3/index.htm> accessed 11 June 2007.

Marland, G., R.A. Pielke Sr, M. Apps, R. Avissar, R.A. Betts, K.J. Davis, P.C. Frumhoff, S.T. Jackson, L.A. Joyce, P. Kauppi, J. Katzenberger, K.G. MacDicken, R.P. Neilson, J.O. Niles, D.D.S. Niyogi, R.J. Norby, N. Pena, N. Sampson, and Y. Xue, 2003: The climatic impacts of land surface change and carbon management, and the implications for climate-change mitigation policy. *Climate Policy,* 3, pp. 149-157.

Masera, O.R., J.F. Garza Caligaris, M. Kanninen, T. Karjalainen, J. Liski, G.J. Nabuurs, A. Pussinen, B.H.J.d. Jong, G.M.J. Mohren, and B.H.J. de Jong, 2003, Modelling carbon sequestration in afforestation, agroforestry and forest management projects: the CO2FIX V.2 approach. *Ecological modeling,* 164, pp. 177-199.

Masera, O.R., R. Díaz, and V. Berrueta, 2005: From cookstoves to cooking systems: The integrated program on sustainable household energy use in Mexico. *Energy for Sustainable Development,* 9(5), pp. 25-36.

Matsumoto, M., H. Kanomata, and M. Fukuda, 2002: Distribution of carbon uptakes by Japanese forest. *J Forest Research,* 113, 91 pp. (in Japanese).

McCarl, B.A. and U.A. Schneider, 2001: Greenhouse gas mitigation in U.S. agriculture and forestry. *Science,* 294, pp. 2481-2482.

McGinley, K. and B. Finegan, 2003: The ecological sustainability of tropical forest management: evaluation of the national forest management standards of Costa Rica and Nicaragua, with emphasis on the need for adaptive management. *Forest Policy and Economics,* 5(2003), pp. 421-431.

McKenney, D.W., D. Yemshanov, G. Fox, and E. Ramlal, 2004: Cost estimates for carbon sequestration from fast growing poplar plantations in Canada. *Forest Policy and Economics.* 6: 345–358

MEA, 2005a: Millennium Ecosystem Assessment. Ecosystems and Human Well-being: Synthesis. Island Press, Washington, D.C., 137 pp.

MEA, 2005b: Millennium Ecosystem Assessment. Ecosystems and human well-being: Scenarios. Findings of the Scenarios Working Group. Island Press, Washington D.C.

Mertens, B., D. Kaimowitz, A. Puntodewo, J. Vanclay, and P. Mendez, 2004: Modeling deforestation at distinct geographic scales and time periods in Santa Cruz, Bolivia. *International Regional Science Review*, **27**(3), pp. 271-296.

MfE, 2002: National Communication 2001: New Zealand's Third National Communication under the Framework Convention on Climate Change. Ministry for the Environment, Wellington, NZ. <http://www.climatechange.govt.nz/resources/reports/index.html> accessed 11 June 2007.

MfE, 2005: Projected balance of units during the first commitment period of the Kyoto Protocol. Ministry for the Environment, Wellington, NZ. <http://www.climatechange.govt.nz/resources/reports/index.html> accessed 11 June 2007.

Micales, J.A. and K.E. Skog, 1997: The decomposition of forest products in landfills. *International Biodeterioration & Biodegradation*, **39**(2-3), pp. 145-158.

Minkkinen, K., R. Korhonen, I. Savolainen, and J. Laine, 2002: Carbon balance and radiative forcing of Finnish peatland 1900-2100, the impacts of drainage. *Global Change Biology*, **8**, pp. 785-799.

Murdiyarso, D., 2005: Linkages between mitigation and adaptation in land-use change and forestry activities. In *Tropical forests and adaptation to climate change: In search of synergies.* C. Robledo, C. Kanninen, and L. Pedroni (eds.). Bogor, Indonesia: Center for International Forestry Research (CIFOR), 186 pp.

Murphy, R., 2004: Timber and the circle of life. ITTO. *Tropical Forest Update,* **14**(3), pp. 12-14.

Murray, B.C., B.A. McCarl, and H. Lee, 2004: Estimating leakage from forest carbon sequestration programs. *Land Economics,* **80**(1), pp. 109-124.

Nabuurs, G.J., M.J. Schelhaas, G.M.J. Mohren, and C.B. Field, 2003: Temporal evolution of the European Forest sector carbon sink 1950-1999. *Global Change Biology,* **9**, pp.152-160.

Nabuurs, G.J., I.J.J. van den Wyngaert, W.D. Daamen, A.T.F. Helmink, W.J.M. de Groot, W.C. Knol, H. Kramer, and P.J. Kuikman, 2005: National system of greenhouse gas reporting for forest and nature areas under UNFCCC in the Netherlands Wageningen: Alterra, 2005 (Alterra-rapport 1035.1), 57 pp.

Nabuurs, G.J., J. van Brusselen, A. Pussinen, and M.J. Schelhaas, 2006: Future harvesting pressure on European forests. *European Journal of Forest Research*, **126** (3) pp. 401-412. doi: 10.1007/s10342-006-0158-y.

Niesten, E., P.C. Frumhoff, M. Manion, and J.J. Hardner, 2002: Designing a carbon market that protects forests in developing countries. *Philosophical Transactions of the Royal Society, London A*, pp. 1875-1888.

Nilsson, S., E.A. Vaganov, A.Z. Shvidenko, V. Stolbovoi, V.A. Rozhkov, I. MacCallum, and M. Ionas, 2003: Carbon budget of vegetation ecosystems of Russia. *Doklady Earth Sciences*, **393A**(9), pp. 1281-1283.

Nilsson, S., A. Shvidenko, V. Stolbovoi, M. Gluck, M. Jonas, M. Obersteiner, 2000: Full Carbon Account for Russia. IIASA Interim Report IR-00-021, 190 p.

OECD, 2002: Ancillary benefits of reducing Greenhouse gas emissions in Transitional Economies.

OECD, 2006: World Energy Outlook. IEA, Paris.

Ohtani, Y., N. Saigusa, S. Yamamoto, Y. Mizoguchi, T. Watanabe, Y. Yasuda, and S. Murayama, 2005: Characteristics of CO_2 fluxes in cool-temperate coniferous and deciduous broadleaf forests in Japan. *Phyton,* **45**, pp. 73-80.

Olivier, J.G.J., J.A. Van Aardenne, F. Dentener, V. Pagliari, L.N. Ganzeveld, and J.A.H.W. Peters, 2005: Recent trends in global greenhouse gas emissions: regional trends 1970-2000 and spatial distribution of key sources in 2000. *Environmental Science*, **2**(2-3), pp. 81-99.

Orlove, B., 2005: Human adaptation to climate change: a review of three historical cases and some general perspectives. *Environmental Science & Policy*, **8**(6), pp. 589-600.

Pacala, S.W., G.C. Hurtt, D. Baker, P. Peylin, R.A. Houghton, R.A. Birdsey, L. Heath, E.T. Sundquist, R.F. Stallard, P. Ciais, P. Moorcroft, J.P. Caspersen, E. Shevliakova, B. Moore, G. Kohlmaier, E. Holland, M. Gloor, M.E. Harmon, S.-M. Fan, J.L. Sarmiento, C.L. Goodale, and D. Schimel, 2001: Consistent land- and atmosphere-based U.S. carbon sink estimates. *Science*: **292**, pp. 2316-2320.

Pan, Y., T. Luo, R. Birdsey, J. Hom, and J. Melillo, 2004: New estimates of carbon storage and sequestration in China's forests: effects of age-class and method on inventory-based carbon estimation. *Climatic Change*, **67**(2), pp. 211-236.

Parris, K., 2004: Agriculture, biomass, sustainability and policy: an overview. In *Biomass and agriculture: sustainability, markets and policies.* K. Parris and T. Poincet (eds.), OECD Publication Service, Paris, pp. 27-36.

Parrotta, J.A., 2002: Restoration and management of degraded tropical forest landscapes. In *Modern Trends in Applied Terrestrial Ecology.* R.S. Ambasht and N.K. Ambasht (eds.), Kluwer Academic/Plenum Press, New York, pp. 135-148 (Chapter 7).

Paul, K.I., P.J. Polglase, and G.P. Richards, 2003: Predicted change in soil carbon following afforestation or reforestation, and analysis of controlling factors by linking a C accounting model (CAMFor) to models of forest growth (3PG), litter decomposition (GENDEC) and soil C turnover (RothC). *Forest Ecology and Management*, **177**, 485 pp.

Pedroni, L., 2005: Carbon accounting for sinks in the CDM after CoP-9. *Climate Policy,* **5**, pp. 407-418.

Pérez Bidegain, M., P.F. García, and R. Methol, 2001: Long-term effect of tillage intensity for Eucalyptus grandis planting on some soil physical properties in an Uruguayan Alfisol. 3rd International Conference on Land Degradation and Meeting of IUSS Subcommission C-Soil and Water Conservation. September 17-21 2001- Rio de Janeiro - Brazil.

Petersen, A.K. and B. Solberg, 2002: Greenhouse gas emissions, life-cycle inventory and cost-efficiency of using laminated wood instead of steel construction. - Case: beams at Gardermoen airport. *Environmental Science and Policy*, **5**, pp. 169-182.

Plattner, G.-K., F. Joos, and T.F. Stocker, 2002: Revision of the global carbon budget due to changing air-sea oxygen fluxes. *Global Biogeochemical Cycles*, **16**(4), 1096 pp., doi:10.1029/2001GB001746.

Post, W.M. and K.C. Kwon, 2000: Soil carbon sequestration and land-use change: processes and potential. *Global Change Biology,* **6**, pp. 317-328.

Potter, C., S. Klooster, R. Myneni, V. Genovese, P.N. Tan, and V. Kumar, 2003: Continental-scale comparisons of terrestrial carbon sinks estimated from satellite data and ecosystem modeling 1982-1998. *Global and Planetary Change,* **39**(3-4), pp. 201-213.

Quine, C. and J. Humphrey, 2005: Stand Management and biodiversity. In *International Conference: Biodiversity and Conservation Biology in Plantation Forests.* Bordeaux, France.

Rao, S. and K. Riahi, 2006: The role of non-CO_2 greenhouse gases in climate change mitigation: Long-term scenarios for the 21st century. *Energy Journal*, Special Issue pp.177-200.

Ravindranath, N.H. and J.A. Sathaye, 2002: Climate change and developing countries. Kluwer Academic Publishers.

Ravindranath, N.H., I. K. Murthy, P. Sudha, V. Ramprasad, M.D.V. Nagendra, C.A. Sahana, K.G. Srivathsa, and H. Khan, 2007: Methodological issues in forestry mitigation projects: A case study of Kolar district. *Mitigation and Adaptation Strategies for Global Change,* **12** (6) pp. 1077-1098.

Ravindranath, N.H., 2007: Mitigation and adaptation synergy in forest sector. *Mitigation and Adaptation Strategies for Global Change.* **12** (5) pp. 843-853.

Raymond, P.A. and J.J. Cole, 2003: Increase in the export of alkalinity from North America's largest river. *Science,* **301**, pp. 88-91.

Riahi, K., A. Gruebler, and N. Nakicenovic, 2006: Scenarios of long-term socio-economic and environmental development under climate stabilization. *Technological Forecasting and Social Change,* Special Issue.

Richards, G.P. and C. Brack: A modelled carbon account for Australia's post-1990 plantation estate. *Australian Forestry,* **67**(4), pp. 289-300.

Richards, G.P. and D.M.W. Evans, 2004: Development of a carbon accouning model (FullCAM Vers.1.0) for the Australian continent. *Australian Forestry,* **67**, pp. 277-283.

Richards, K.R., C. Stokes, 2004: A review of forest carbon sequestration cost studies: a dozen years of research. *Climatic Change,* **63**, pp. 1-48.

Richards, K.R., R.N. Sampson, and S. Brown, 2006: Agricultural and forest lands: U.S. carbon policy strategies. Pew Center on Global Climate Change, Arlington, VA.

Richey, J.E., J.M. Melack, A.K. Aufdenkampe, V.M. Ballester, and L.L. Hess, 2002: Outgassing from Amazonian rivers and wetlands as a large tropical source of atmospheric CO_2. *Nature,* **416**(6881), pp. 617-620.

Robins, A., 2005: Ecosystem Services Markets. Seattle, University of Washington.

Robledo, C., C. Kanninen, and L. Pedroni (eds.), 2005: Tropical forests and adaptation to climate change. In search of synergies. Bogor, Indonesia: Center for International Forestry Research (CIFOR), 186 pp.

Rodenbeck, C., S. Houweling, M. Gloor, and M. Heimann, 2003: CO_2 flux history 1982-2001 inferred from atmospheric data using a global inversion of atmospheric transport. *Atmospheric Chemistry and Physics,* **3**, pp. 1919-1964.

Rose, S., H. Ahammad, B. Eickhout, B. Fisher, A. Kurosawa, S. Rao, K. Riahi, and D. van Vuuren, 2007: Land in climate stabilization modeling: Initial observations. Energy Modeling Forum Report, Stanford University, <http://www.stanford.edu/group/EMF/projects/group21/EMF21sinkspagenew.htm> accessed January 2007.

Rosenbaum, K.L., D. Schoene, and A. Mekouar, 2004: Climate change and the forest sector: possible national and subnational legislation. Food and Agriculture Organization of the United Nations, Rome, 60 pp.

Santilli, M.P., P. Moutinho, S. Schwartzman, D. Nepstad, L. Curran, and C. Nobre, 2005: Tropical deforestation and the Kyoto Protocol: an editorial essay. *Climatic Change,* **71**, pp. 267-276.

Sathaye, J.A., W. Makundi, L. Dale, P. Chan and K. Andrasko, 2007: GHG mitigation potential, costs and benefits in global forests: A dynamic partial equilibrium approach. *Energy Journal,* Special Issue 3, pp. 127-172.

Sathaye, J.A. and K. Andrasko, 2007: Special issue on estimation of baselines and leakage in carbon mitigation forestry projects: Editorial. *Mitigation and Adaptation Strategies for Global Change,* **12**(6) pp. 963-970.

Schelhaas, M.J., G.J. Nabuurs, and A. Schuck, 2003: Natural disturbances in the European forests in the 19th and the 20th centuries. *Global Change Biology,* **9**, pp. 1620-1633.

Schimel, D.S., J.I. House, K.A. Hibbard, P. Bousquet, P. Ciais, P. Peylin, B.H. Braswell, M.J. Apps, D. Baker, A. Bondeau, J. Canadell, G. Churkina, W. Cramer, A.S. Denning, C.B. Field, P. Friedlingstein, C. Goodale, M. Heimann, R.A. Houghton, J.M. Melillo, B. III, Moore., D. Murdiyarso, I. Noble, S.W. Pacala, I.C. Prentice, M.R. Raupach, P.J. Rayner, R.J. Scholes, W.L. Steffen, and C. Wirth, 2001: Recent patterns and mechanisms of carbon exchange by terrestrial ecosystems. *Nature,* **414**(6860), pp. 169-172.

Schlamadinger, B. and G. Marland, 1996: The role of forest and bioenergy strategies in the global carbon cycle. *Biomass and Bioenergy,* **10**, pp. 275-300.

Schlamadinger, B., L. Ciccarese, M. Dutschke, P.M. Fearnside, S. Brown, and D. Murdiyarso, 2005: Should we include avoidance of deforestation in the international response to climate change? In *Carbon forestry: who will benefit?* D. Murdiyarso, and H. Herawati (eds.). Proceedings of Workshop on Carbon Sequestration and Sustainable Livelihoods, held in Bogor on 16-17 February 2005. Bogor, Indonesia, CIFOR, pp. 26-41.

Schröter, D., W. Cramer, R. Leemans, I.C. Prentice, M.B. Araujo, N.W. Arnell, A. Bondeau, H. Bugmann, T.R. Carter, C.A. Gracia, A.C. Dela Vega-Leinert, M.J. Metxger, J. Meyer, T.D. Mitchell, I. Reginster, M. Rounsevell, S. Sabate, S. Sitch, B. Smith, J. Smith, P. Smith, M.T. Sykes, K. Thonicke, W. Thuiller, G. Tuck, S. Zaehle, and B. Zierl, 2005: Ecosystem service supply and vulnerability to global change in Europe. *Science,* **310**(5752), pp. 1333-1337.

Schwarze, R., J.O. Niles, and J. Olander, 2003: Understanding and managing leakage in forest-based green-house-gas-mitigation projects. Capturing carbon and conserving biodiversity: The market approach. I.R. Swingland, London, The Royal Society.

Sedjo, R. and B. Sohngen, 2000: Forestry sequestration of CO_2 and markets for timber. Washington, D.C., Resources for the Future, Discussion Paper 00-03, 51 pp.

Shvidenko, A. and S. Nilsson, 2003: A synthesis of the impact of Russian forests on the global carbon budget for 1961-1998. *Tellus Series B,* **55**(2), pp. 391-415.

Shvidenko A., Nilsson S., Rozhkov V.A. 1997. Possibilities for increased carbon sequestration through the implementation of rational forest management in Russia. Water, Air & Soil Pollution, 94:137-162

Sicardi, M., F. García-Prechac, and L. Fromi, 2004: Soil microbial indicators sensitive to land use conversion from pastures to commercial Eucalyptus grandis (Hill ex Maiden) plantations in Uruguay. *Applied Soil Ecology,* **27**, pp. 125-133.

Sizer, N., D. Downes, and D. Kaimowitz, 1999: Tree trade - liberalization of international commerce in forest products: risks and opportunities. World Resources Institute Forest Notes, Washington, D.C., 24 pp.

Sizer, N., S. Bass, and J. Mayers (Coordinating Lead Authors), 2005: Wood, fuelwood and non-wood forest products. In *Millennium Ecosystem Assessment, 2005: Policy Responses: Findings of the Responses Working Group.* Ecosystems and Human Well-being, **3**, pp. 257-293. Island Press, Washington, D.C.

Smeets, E.M.W. and A.P.C. Faaij, 2007: Bioenergy potentials from forestry in 2050, an assessment of drivers that determine the potential. <http://www.bioenergytrade.org/downloads/smeetsandfaaijbioenergyfromforestryclimaticcha.pdf> accessed 11 June 2007.

Soares-Filho, B.S., D. Nepstad, L. Curran, E. Voll, G. Cerqueira, R.A. Garcia, C.A. Ramos, A. Mcdonald, P. Lefebvre, and P. Schlesinger, 2006: Modelling conservation in the Amazon basin. *Nature,* **440**, pp. 520-523.

Sohngen, B. and R. Sedjo, 2006: Carbon sequestration costs in global forests. *Energy Journal,* Special Issue, pp. 109-126.

Sohngen, B. and S. Brown, 2004: Measuring leakage from carbon projects in open economies: a stop timber harvesting project case study. *Canadian Journal of Forest Research,* **34**, pp. 829-39.

Sohngen, B., and R. Mendelsohn, 2003: An optimal control model of forest carbon sequestration. *American Journal of Agricultural Economics,* **85**(2), pp. 448-457.

Sohngen, S., K. Andrasko, M. Gytarsky, G. Korovin, L. Laestadius, B. Murray, A. Utkin, and D. Zamolodchikov, 2005: Stocks and flows: carbon inventory and mitigation potential of the Russian forest and land base. Report of the World Resources Institute, Washington D.C.

Sørensen, 1999: Long-term scenarios for global energy demand and supply: Four global greenhouse mitigation scenarios. Energy & Environment Group, Roskilde University, Roskilde.

Speth, J.G., 2002: Recycling Environmentalism. *Foreign Policy,* July/August, pp. 74-76.

Spittlehouse, D.L. and R.B. Stewart, 2003: Adaptation to climate change in forest management. *Journal of Ecosystems and Management,* **4**, pp. 1-11.

Stern, N., 2006: Stern Review: the Economics of Climate Change. HM Treasury, Cambridge University Press, UK.

Strengers, B., J. Van Minnen and B. Eickhout, 2007 (in print): The role of carbon plantations in mitigating climate change: potentials and costs. *Climatic change.*

Swanson, D., L. Pinter, F. Bregha, A. Volkery, and K. Jacob, 2004: Sustainable development and national government actions. IISD, Commentary.

Tacconi, L., M. Boscolo, and D. Brack, 2003: National and international policies to control illegal forest activities. Center for International Forest Research, Jakarta.

Tate, K.R., R.H. Wilde, D.J. Giltrap, W.T. Baisden, S. Saggar, N.A. Trustrum, N.A. Scott, and J.R. Barton, 2005: Soil organic carbon stocks and flows in New Zealand: System development, measurement and modelling. *Canadian Journal of Soil Science,* **85***(4),* pp. 481-489.

Tol, R.S.J., 2006: Adaptation and mitigation: trade-offs in substance and methods. *Environmental Science & Policy,* **8,** pp. 572-578.

Trines, E., N. Höhne, M. Jung, M. Skutsch, A. Petsonk, G. Silva-Chavez, P. Smith, G.-J. Nabuurs, P. Verweij, B. and Schlamadinger, 2006: Integrating agriculture, forestry and other land use in future climate regimes: Methodological issues and policy options. WAB Report 500101002. Netherlands Environmental Assessment Agency, Bilthoven, the Netherlands.

Trotter, C., K. Tate, N. Scott, J. Townsend, H. Wilde, S. Lambie, M. Marden, and T. Pinkney, 2005: Afforestation/reforestation of New Zealand marginal pasture lands by indigenous shrublands: the potential for Kyoto forest sinks. *Annals of Forest Science,* **62,** pp. 865-871.

UK FRP, 2005: From the mountain to the tap: How land use and water management can work for the rural poor. UK FRP, London., United Kingdom Forestry Research Programme.

UN-ECE/FAO, 2000: Forest Resources of Europe, CIS, North America, Australia, Japan and New Zealand (industrialized temperate/boreal countries), UN-ECE/FAO Contribution to the Global Forest Resources Assessment 2000. United Nations, New York, NY, USA and Geneva, Switzerland. *Geneva Timber and Forest Study Papers,* **17,** 445 pp.

UNFCCC, 2001: Land use, land use change and forestry (LULUCF) projects in the CDM: Ex-piring CERs. FCCC/SB/2000/MISC.4/Add.2, **5,** pp. 22-36.

UNFCCC, 2003: United Nations Framework Convention on Climate Change. Decision 19/CP.9.

UNFCCC, 2006: Background paper for the Workshop on Reducing Emissions from Deforestation in Developing Countries, Part I: Scientific, socio-economic, technical and methodological issues related to deforestation in developing countries. Working paper No. 1 (a), 30 August-1 September, Rome, Italy.

US EPA, 2005: *Greenhouse gas mitigation potential in U.S. forestry and agriculture.* Washington, DC, EPA 430-R-006, November, 150 pp.

USDA Forest Service, 2000: Protecting people and sustaining resources in rire-adapted ecosystems: A cohesive strategy, the Forest Service Management response to the General Accounting Office report GAO/RCED-99-65. 13 October, US Department of Agriculture (USDA) Forestry Service, Washington, D.C., pp. 85.

USDOE, 2005: Biomass Program. United States Department of Energy <http://www.eere.energy.gov/biomass/> accessed 12 October 2005.

VanKooten, G.C., A.J. Eagle, J. Manley, and T. Smolak, 2004: How costly are carbon offsets? A meta-analysis of carbon forest sink. *Environmental science & policy,* **7,** pp. 239-251.

Verchot, L.V., J. Mackensen, S. Kandji, M. van Noordwijk, T. Tomich, C. Ong, A. Albrecht, C. Bantilan, K.V. Anupama, and C. Palm, 2006: Opportunities for linking adaptation and mitigation in agroforestry systems. In *Tropical forests and adaptation to climate change: In search of synergies,* C. Robledo, M. Kanninen, L. Pedroni (eds.), Bogor, Indonesia: Center for International Forestry Research (CIFOR).

Viner, D., M. Sayer, M. Uyarra, and N. Hodgson, 2006: Climate change and the European countryside: Impacts on land management and response strategies. Report prepared for the Country Land and Business Association., UK Publishing CLA, UK, 180 pp.

Vuuren, D. Van, M. den Elzen, P. Lucas, B. Eickhout, B. Strengers, B van Ruijven, S. Wonink, and R. van Houdt, 2007: Stabilizing greenhouse gas concentrations at low levels: an assessment of reduction strategies and costs. *Climatic Change,* **81** (2) pp. 119-159.

Wagner, R.G., K.M. Little, B. Richardson, and K. Mcnabb, 2006: The role of vegetation management for enhancing productivity of the world's forests. *Forestry,* **79**(1), pp. 57-79.

Walsh, M.E., L.R. Perlack, A. Turhollow, D. de la Torre Ugarte, D.A. Becker, R.L. Graham, S.E. Slinsky, and D.E. Ray, 2000: Oak Ridge National Laboratory, Biomass Feedstock Availability in the United States: 1999 State Level Analysis, April 30, 1999, updated January, 2000: Oak Ridge, TN 37831-6205.

Waterloo, M.J., P.H. Spiertz, H. Diemont, I. Emmer, E. Aalders, R. Wichink-Kruit, and P. Kabat, 2003: Criteria potentials and costs of forestry activities to sequester carbon within the framework of the Clean Development Mechanism. Alterra Rapport 777, Wageningen, 136 pp.

WBOED, 2000: Striking the right balance: World Bank forest strategy. World Bank Operations Evaluation Department, *Precis,* 203, 6 pp.

Wear, D. and B.C. Murray, 2004: Federal timber restrictions, interregional spillovers, and the impact on U.S. softwood markets. *Journal of Environmental Economics and Management,* **47**(2), pp. 307-330.

Werf, G.R van der, J.T. Randerson, G.J. Collatz, L. Giglio, P.S. Kasibhatla, A.F. Arellano Jr., S.C. Olsen, E.S. Kasischke, 2004: Continental-scale partitioning of fire emissions during the 1997 to 2001 El Niño/ La Niña period. *Science,* **303,** pp. 73-76.

White, T.M. and W.A. Kurz, 2005: Afforestation on private land in Canada from 1990 to 2002 estimated from historical records. *The Forestry Chronicle,* **81,** pp. 491-497.

Wilbanks, T.J., 2003: Integrating climate change and sustainable development in a place-based context. *Climate Policy,* **3,** pp. 147-154.

Williams, R.H., 1995: Variants of a low CO_2-emitting energy supply system (LESS) for the world - prepared for the IPCC Second Assessment Report Working Group IIa, Energy Supply Mitigation Options, Richland: Pacific Northwest Laboratories, 39 pp.

Williams, J., 2002: Financial and other incentives for plantation establishment. Proceedings of the International Conference on Timber Plantation Development. <http://www.fao.org/documents/show_cdr.asp?url_file=//docrep/005/ac781e/AC781E00.htm>. accesssed 1 September 2005.

Winrock International, 2004: Financial incentives to communities for stewardship of environmental resources: feasibility study. Arlington, VA, 60 pp.

World Bank, 2005: Forest Governance Program. <http://lnweb18.worldbank.org/ESSD/ardext.nsf/14ByDocName/ForestGovernanceProgram> accessed 9 September 2005.

Wunder, S., 2004: Policy options for stabilising the forest frontier: A global perspective. In *Land Use, Nature Conservation and the Stability of Rainforest Margins in Southeast Asia,* M. Gerold, M. Fremerey, and E. Guhardja (eds). Springer-Verlag Berlin, pp. 3-25.

Yamamoto, H., J. Fujino, and K. Yamaji, 2001: Evaluation of bioenergy potential with a multi-regional global-land-use-and-energy model. *Biomass and Bioenergy,* **21,** pp. 185-203.

Yoshioka, T., K. Aruga, T. Nitami, H. Sakai, and H. Kobayashi, 2006: A case study on the costs and the fuel consumption of harvesting, transporting, and chipping chains for logging residues in Japan. *Biomass and Bioenergy,* **30**(4), pp. 342-348.

Zhang, X.Q. and D. Xu, 2003: Potential carbon sequestration in China's forests. *Environmental Science & Policy,* **6,** pp. 421-432.

Zhang, Y.H., M.J. Wooster, O. Tutubalina, and G.LW. Perry, 2003: Monthly burned area and forest fire carbon emission estimates for the Russian Federation from SPOT VGT. *Remote Sensing of Environment,* **87,** pp. 1-15.

10

Waste Management

Coordinating Lead Authors:

Jean Bogner (USA)

Lead Authors:

Mohammed Abdelrafie Ahmed (Sudan), Cristobal Diaz (Cuba), Andre Faaij (The Netherlands), Qingxian Gao (China),

Seiji Hashimoto (Japan), Katarina Mareckova (Slovakia), Riitta Pipatti (Finland), Tianzhu Zhang (China)

Contributing Authors:

Luis Diaz (USA), Peter Kjeldsen (Denmark), Suvi Monni (Finland)

Review Editors:

Robert Gregory (UK), R.T.M. Sutamihardja (Indonesia)

This chapter should be cited as:

Bogner, J., M. Abdelrafie Ahmed, C. Diaz, A. Faaij, Q. Gao, S. Hashimoto, K. Mareckova, R. Pipatti, T. Zhang, Waste Management, In Climate Change 2007: Mitigation. Contribution of Working Group III to the Fourth Assessment Report of the Intergovernmental Panel on Climate Change [B. Metz, O.R. Davidson, P.R. Bosch, R. Dave, L.A. Meyer (eds)], Cambridge University Press, Cambridge, United Kingdom and New York, NY, USA.

Table of Contents

EXECUTIVE SUMMARY

Post-consumer waste is a small contributor to global greenhouse gas (GHG) emissions (<5%) with total emissions of approximately 1300 $MtCO_2$-eq in 2005. The largest source is landfill methane (CH_4), followed by wastewater CH_4 and nitrous oxide (N_2O); in addition, minor emissions of carbon dioxide (CO_2) result from incineration of waste containing fossil carbon (C) (plastics; synthetic textiles) *(high evidence, high agreement)*. There are large uncertainties with respect to direct emissions, indirect emissions and mitigation potentials for the waste sector. These uncertainties could be reduced by consistent national definitions, coordinated local and international data collection, standardized data analysis and field validation of models *(medium evidence, high agreement)*. With respect to annual emissions of fluorinated gases from post-consumer waste, there are no existing national inventory methods for the waste sector, so these emissions are not currently quantified. If quantified in the future, recent data indicating anaerobic biodegradation of chlorofluorocarbons (CFCs) and hydrochlorofluorocarbons (HCFCs) in landfill settings should be considered *(low evidence, high agreement)*.

Existing waste-management practices can provide effective mitigation of GHG emissions from this sector: a wide range of mature, environmentally-effective technologies are available to mitigate emissions and provide public health, environmental protection, and sustainable development co-benefits. Collectively, these technologies can directly reduce GHG emissions (through landfill gas recovery, improved landfill practices, engineered wastewater management) or avoid significant GHG generation (through controlled composting of organic waste, state-of-the-art incineration and expanded sanitation coverage) *(high evidence, high agreement)*. In addition, waste minimization, recycling and re-use represent an important and increasing potential for indirect reduction of GHG emissions through the conservation of raw materials, improved energy and resource efficiency and fossil fuel avoidance *(medium evidence, high agreement)*.

Because waste management decisions are often made locally without concurrent quantification of GHG mitigation, the importance of the waste sector for reducing global GHG emissions has been underestimated *(medium evidence, high agreement)*. Flexible strategies and financial incentives can expand waste management options to achieve GHG mitigation goals – in the context of integrated waste management, local technology decisions are a function of many competing variables, including waste quantity and characteristics, cost and financing issues, infrastructure requirements including available land area, collection and transport considerations, and regulatory constraints. Life cycle assessment (LCA) can provide decision-support tools *(high evidence, high agreement)*.

Commercial recovery of landfill CH_4 as a source of renewable energy has been practised at full scale since 1975 and currently exceeds 105 $MtCO_2$-eq, yr. Because of landfill gas recovery and complementary measures (increased recycling, decreased landfilling, use of alternative waste-management technologies), landfill CH_4 emissions from developed countries have been largely stabilized *(high evidence, high agreement)*. However, landfill CH_4 emissions from developing countries are increasing as more controlled (anaerobic) landfilling practices are implemented; these emissions could be reduced by both accelerating the introduction of engineered gas recovery and encouraging alternative waste management strategies *(medium evidence, medium agreement)*.

Incineration and industrial co-combustion for waste-to-energy provide significant renewable energy benefits and fossil fuel offsets. Currently, >130 million tonnes of waste per year are incinerated at over 600 plants *(high evidence, high agreement)*. Thermal processes with advanced emission controls are proven technology but more costly than controlled landfilling with landfill gas recovery; however, thermal processes may become more viable as energy prices increase. Because landfills produce CH_4 for decades, incineration, composting and other strategies that reduce landfilled waste are complementary mitigation measures to landfill gas recovery in the short- to medium-term *(medium evidence, medium agreement)*.

Aided by Kyoto mechanisms such as the Clean Development Mechanism (CDM) and Joint Implementation (JI), as well as other measures to increase worldwide rates of landfill CH_4 recovery, the total global economic mitigation potential for reducing landfill CH_4 emissions in 2030 is estimated to be >1000 $MtCO_2$-eq (or 70% of estimated emissions) at costs below 100 US$/tCO_2$-eq/yr. Most of this potential is achievable at negative to low costs: 20–30% of projected emissions for 2030 can be reduced at negative cost and 30–50% at costs <20 US$/tCO_2$-eq/yr. At higher costs, more significant emission reductions are achievable, with most of the additional mitigation potential coming from thermal processes for waste-to-energy *(medium evidence, medium agreement)*.

Increased infrastructure for wastewater management in developing countries can provide multiple benefits for GHG mitigation, improved public health, conservation of water resources, and reduction of untreated discharges to surface water, groundwater, soils and coastal zones. There are numerous mature technologies that can be implemented to improve wastewater collection, transport, re-use, recycling, treatment and residuals management *(high evidence, high agreement)*. With respect to both waste and wastewater management for developing countries, key constraints on sustainable development include the local availability of capital as well as the selection of appropriate and truly sustainable technology in a particular setting *(high evidence, high agreement)*.

587

10.1   Introduction

Waste generation is closely linked to population, urbanization and affluence. The archaeologist E.W. Haury wrote: 'Whichever way one views the mounds [of waste], as garbage piles to avoid, or as symbols of a way of life, they…are the features more productive of information than any others.' (1976, p.80). Archaeological excavations have yielded thicker cultural layers from periods of prosperity; correspondingly, modern waste-generation rates can be correlated to various indicators of affluence, including gross domestic product (GDP)/cap, energy consumption/cap, , and private final consumption/cap (Bingemer and Crutzen, 1987; Richards, 1989; Rathje *et al.*, 1992; Mertins *et al.*, 1999; US EPA, 1999; Nakicenovic *et al.*, 2000; Bogner and Matthews, 2003; OECD, 2004). In developed countries seeking to reduce waste generation, a current goal is to decouple waste generation from economic driving forces such as GDP (OECD, 2003; Giegrich and Vogt, 2005; EEA, 2005). In most developed and developing countries with increasing population, prosperity and urbanization, it remains a major challenge for municipalities to collect, recycle, treat and dispose of increasing quantities of solid waste and wastewater. A cornerstone of sustainable development is the establishment of affordable, effective and truly sustainable waste management practices in developing countries. It must be further emphasized that multiple public health, safety and environmental co-benefits accrue from effective waste management practices which concurrently reduce GHG emissions and improve the quality of life, promote public health, prevent water and soil contamination, conserve natural resources and provide renewable energy benefits.

The major GHG emissions from the waste sector are landfill CH_4 and, secondarily, wastewater CH_4 and N_2O. In addition, the incineration of fossil carbon results in minor emissions of CO_2. Chapter 10 focuses on mitigation of GHG emissions from *post-consumer* waste, as well as emissions from municipal wastewater and high biochemical oxygen demand (BOD) industrial wastewaters conveyed to public treatment facilities. Other chapters in this volume address *pre-consumer* GHG emissions from waste within the industrial (Chapter 7) and energy (Chapter 4) sectors which are managed within those respective sectors. Other chapters address agricultural wastes and manures (Chapter 8), forestry residues (Chapter 9) and related energy supply issues including district heating (Chapter 6) and transportation biofuels (Chapter 5). National data are not available to quantify GHG emissions associated with waste transport, including reductions that might be achieved through lower collection frequencies, higher routing efficiencies or substitution of renewable fuels; however, all of these measures can be locally beneficial to reduce emissions.

It should be noted that a separate chapter on post-consumer waste is new for the Fourth Assessment report; in the Third Assessment Report (TAR), GHG mitigation strategies for waste were discussed primarily within the industrial sector (Ackerman,

2000; IPCC, 2001a). It must also be stressed that there are high uncertainties regarding global GHG emissions from waste which result from national and regional differences in definitions, data collection and statistical analysis. Because of space constraints, this chapter does not include detailed discussion of waste management technologies, nor does this chapter prescribe to any one particular technology. Rather, this chapter focuses on the GHG mitigation aspects of the following strategies: landfill CH_4 recovery and utilization; optimizing methanotrophic CH_4 oxidation in landfill cover soils; alternative strategies to landfilling for GHG avoidance (composting; incineration and other thermal processes; mechanical and biological treatment (MBT)); waste reduction through recycling, and expanded wastewater management to minimize GHG generation and emissions. In addition, using available but very limited data, this chapter will discuss emissions of non-methane volatile organic compounds (NMVOCs) from waste and end-of-life issues associated with fluorinated gases.

The mitigation of GHG emissions from waste must be addressed in the context of integrated waste management. Most technologies for waste management are mature and have been successfully implemented for decades in many countries. Nevertheless, there is significant potential for accelerating both the direct reduction of GHG emissions from waste as well as extended implications for indirect reductions within other sectors. LCA is an essential tool for consideration of both the direct and indirect impacts of waste management technologies and policies (Thorneloe *et al.*, 2002; 2005; WRAP, 2006). Because direct emissions represent only a portion of the life cycle impacts of various waste management strategies (Ackerman, 2000), this chapter includes complementary strategies for GHG avoidance, indirect GHG mitigation and use of waste as a source of renewable energy to provide fossil fuel offsets. Using LCA and other decision-support tools, there are many combined mitigation strategies that can be cost-effectively implemented by the public or private sector. Landfill CH_4 recovery and optimized wastewater treatment can directly reduce GHG emissions. GHG generation can be largely avoided through controlled aerobic composting and thermal processes such as incineration for waste-to-energy. Moreover, waste prevention, minimization, material recovery, recycling and re-use represent a growing potential for indirect reduction of GHG emissions through decreased waste generation, lower raw material consumption, reduced energy demand and fossil fuel avoidance. Recent studies (e.g., Smith *et al.*, 2001; WRAP, 2006) have begun to comprehensively quantify the significant benefits of recycling for indirect reductions of GHG emissions from the waste sector.

Post-consumer waste is a significant renewable energy resource whose energy value can be exploited through thermal processes (incineration and industrial co-combustion), landfill gas utilization and the use of anaerobic digester biogas. Waste has an economic advantage in comparison to many biomass resources because it is regularly collected at public expense

(See also Section 11.3.1.4). The energy content of waste can be more efficiently exploited using thermal processes than with the production of biogas: during combustion, energy is directly derived both from biomass (paper products, wood, natural textiles, food) and fossil carbon sources (plastics, synthetic textiles). The heating value of mixed municipal waste ranges from <6 to >14 MJ/kg (Khan and Abu-Ghararath, 1991; EIPPC Bureau, 2006). Thermal processes are most effective at the upper end of this range where high values approach low-grade coals (lignite). Using a conservative value of 900 Mt/yr for total waste generation in 2002 (discussed in Box 10.1 below), the energy potential of waste is approximately 5–13 EJ/yr. Assuming an average heating value of 9 GJ/t for mixed waste (Dornburg and Faaij, 2006) and converting to energy equivalents, global waste in 2002 contained about 8 EJ of available energy, which could increase to 13 EJ in 2030 using waste projections in Monni *et al.* (2006). Currently, more than 130 million tonnes per year of waste are combusted worldwide (Themelis, 2003), which is equivalent to >1 EJ/yr (assuming 9 GJ/t). The biogas fuels from waste – landfill gas and digester gas – typically have a heating value of 16–22 MJ/Nm^3, depending directly on the CH_4 content. Both are used extensively worldwide for process heating and on-site electrical generation; more rarely, landfill gas may be upgraded to a substitute natural gas product. Conservatively, the energy value of landfill gas currently being utilized is >0.2 EJ/yr (using data from Willumsen, 2003).

An overview of carbon flows through waste management systems addresses the issue of carbon storage versus carbon turnover for major waste-management strategies including landfilling, incineration and composting (Figure 10.1). Because landfills function as relatively inefficient anaerobic digesters, significant long-term carbon storage occurs in landfills, which is addressed in the 2006 IPCC Guidelines for National Greenhouse Gas Inventories (IPCC, 2006). Landfill CH_4 is the major gaseous C emission from waste; there are also minor emissions of CO_2 from incinerated fossil carbon (plastics). The CO_2 emissions from biomass sources – including the CO_2 in landfill gas, the CO_2 from composting, and CO_2 from incineration of waste biomass – are not taken into account in GHG inventories as these are covered by changes in biomass stocks in the land-use, land-use change and forestry sectors.

A process-oriented perspective on the major GHG emissions from the waste sector is provided in Figure 10.2. In the context of a landfill CH_4 mass balance (Figure 10.2a), emissions are one of several possible pathways for the CH_4 produced by anaerobic methanogenic microorganisms in landfills; other pathways include recovery, oxidation by aerobic methanotrophic microorganisms in cover soils, and two longer-term pathways: lateral migration and internal storage (Bogner and Spokas, 1993; Spokas *et al.*, 2006). With regard to emissions from wastewater transport and treatment (Figure 10.2b), the CH_4 is microbially produced under strict anaerobic conditions as in landfills, while the N_2O is an intermediate product of microbial nitrogen cycling promoted by conditions of reduced aeration, high moisture and abundant nitrogen. Both GHGs can be produced and emitted at many stages between wastewater sources and final disposal.

It is important to stress that both the CH_4 and N_2O from the waste sector are microbially produced and consumed with rates controlled by temperature, moisture, pH, available substrates, microbial competition and many other factors. As a result, CH_4 and N_2O generation, microbial consumption, and net emission rates routinely exhibit temporal and spatial variability over many orders of magnitude, exacerbating the problem of developing credible national estimates. The N_2O from landfills is considered an insignificant source globally (Bogner *et al.*, 1999; Rinne *et al.*, 2005), but may need to be considered locally where cover soils are amended with sewage sludge (Borjesson and Svensson, 1997a) or aerobic/semi-aerobic landfilling practices are implemented (Tsujimoto *et al.*, 1994). Substantial emissions of CH_4 and N_2O can occur during wastewater transport in closed sewers and in conjunction with anaerobic or aerobic treatment. In many developing countries, in addition to GHG emissions, open sewers and uncontrolled solid waste disposal sites result in serious public health problems resulting from pathogenic microorganisms, toxic odours and disease vectors.

Major issues surrounding the costs and potentials for mitigating GHG emissions from waste include definition of system boundaries and selection of models with correct baseline assumptions and regionalized costs, as discussed in the TAR (IPCC, 2001a). Quantifying mitigation costs and potentials (Section 10.4.7) for the waste sector remains a challenge due to national and regional data uncertainties as well as the variety of mature technologies whose diffusion is limited by local costs, policies, regulations, available land area, public perceptions and other social development factors. Discussion of technologies

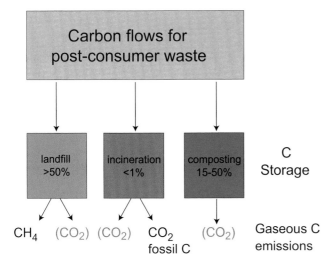

Figure 10.1: *Carbon flows through major waste management systems including C storage and gaseous C emissions. The CO_2 from biomass is not included in GHG inventories for waste.*
References for C storage are: Huber-Humer, 2004; Zinati et al., 2001; Barlaz, 1998; Bramryd, 1997; Bogner, 1992.

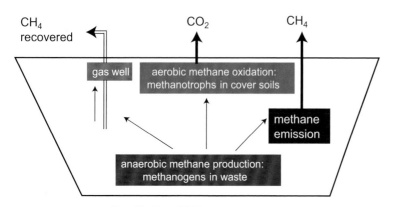

Figure 10.2: *Pathways for GHG emissions from landfills and wastewater systems:*

Figure 10.2a: *Simplified landfill CH4 mass balance: pathways for CH_4 generated in landfilled waste, including CH_4 emitted, recovered and oxidized.*
Note: Not shown are two longer-term CH_4 pathways: lateral CH4 mitigation and internal changes in CH_4 storage (Bogner and Spokas, 1993; Spokas et al., 2006) Methane can be stored in shallow sediments for several thousand years (Coleman, 1979).

<u>Simplified Landfill Methane Mass Balance</u>
Methane (CH_4) produced (mass/time) = $\Sigma(CH_4$ recovered + CH_4 emitted + CH_4 oxidized)

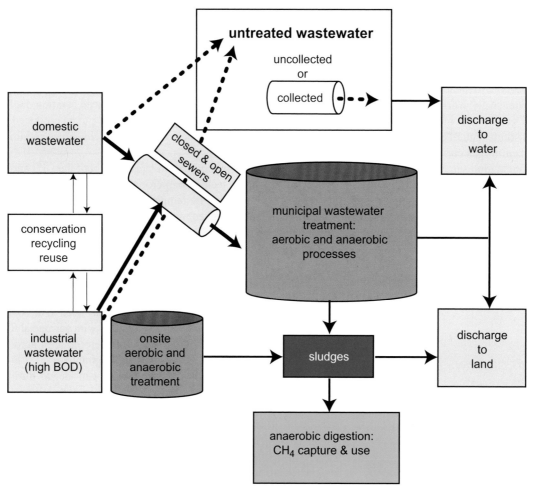

Figure 10.2b: *Overview of wastewater systems.*
Note: The major GHG emissions from wastewater – CH_4 and N_2O – can be emitted during all stages from sources to disposal, but especially when collection and treatment are lacking. N_2O results from microbial N cycling under reduced aeration; CH_4 results from anaerobic microbial decomposition of organic C substrates in soils, surface waters or coastal zones.

and mitigation strategies in this chapter (Section 10.4) includes a range of approaches from low-technology/low-cost to high-technology/high-cost measures. Often there is no single best option; rather, there are multiple measures available to decision-makers at the municipal level where several technologies may be collectively implemented to reduce GHG emissions and achieve public health, environmental protection and sustainable development objectives.

10.2 Status of the waste management sector

10.2.1 Waste generation

The availability and quality of annual data are major problems for the waste sector. Solid waste and wastewater data are lacking for many countries, data quality is variable, definitions are not uniform, and interannual variability is often not well quantified. There are three major approaches that have been used to estimate global waste generation: 1) data from national waste statistics or surveys, including IPCC methodologies (IPCC, 2006); 2) estimates based on population (e.g., SRES waste scenarios), and 3) the use of a proxy variable linked to demographic or economic indicators for which national data are annually collected. The SRES waste scenarios, using population as the major driver, projected continuous increases in waste and wastewater CH_4 emissions to 2030 (A1B-AIM), 2050 (B1-AIM), or 2100 (A2-ASF; B2-MESSAGE), resulting in current and future emissions significantly higher than those derived from IPCC inventory procedures (Nakicenovic et al., 2000) (See also Section 10.3). A major reason is that waste generation rates are related to affluence as well as population – richer societies are characterized by higher rates of waste generation per capita, while less affluent societies generate less waste and practise informal recycling/re-use initiatives that reduce the waste per capita to be collected at the municipal level. The third strategy is to use proxy or surrogate variables based on statistically significant relationships between waste generation

per capita and demographic variables, which encompass both population and affluence, including GDP per capita (Richards, 1989; Mertins et al., 1999) and energy consumption per capita (Bogner and Matthews, 2003). The use of proxy variables, validated using reliable datasets, can provide a cross-check on uncertain national data. Moreover, the use of a surrogate provides a reasonable methodology for a large number of countries where data do not exist, a consistent methodology for both developed and developing countries and a procedure that facilitates annual updates and trend analysis using readily available data (Bogner and Matthews, 2003). The box below illustrates 1971–2002 trends for regional solid-waste generation using the surrogate of energy consumption per capita. Using UNFCCC-reported values for percentage biodegradable organic carbon in waste for each country, this box also shows trends for landfill carbon storage based upon the reported data.

Solid waste generation rates range from <0.1 t/cap/yr in low-income countries to >0.8 t/cap/yr in high-income industrialized countries (Table 10.1). Even though labour costs are lower in developing countries, waste management can constitute a larger percentage of municipal income because of higher equipment and fuel costs (Cointreau-Levine, 1994). By 1990, many developed countries had initiated comprehensive recycling programmes. It is important to recognize that the percentages of waste recycled, composted, incinerated or landfilled differ greatly amongst municipalities due to multiple factors, including local economics, national policies, regulatory restrictions, public perceptions and infrastructure requirements

---

Box 10.1: 1971–2002 Regional trends for solid waste generation and landfill carbon storage using a proxy variable.

Solid-waste generation rates are a function of both population and prosperity, but data are lacking or questionable for many countries. This results in high uncertainties for GHG emissions estimates, especially from developing countries. One strategy is to use a proxy variable for which national statistics are available on an annual basis for all countries. For example, using national solid-waste data from 1975–1995 that were reliably referenced to a given base year, Bogner and Matthews (2003) developed simple linear regression models for waste generation per capita for developed and developing countries. These empirical models were based on energy consumption per capita as an indicator of affluence and a proxy for waste generation per capita; the surrogate relationship was applied to annual national data using either total population (developed countries) or urban population (developing countries). The methodology was validated using post-1995 data which had not been used to develop the original model relationships. The results by region for 1971–2002 (Figure 10.3a) indicate that approximately 900 Mt of waste were generated in 2002. Unlike projections based on population alone, this figure also shows regional waste-generation trends that decrease and increase in tandem with major economic trends. For comparison, recent waste-generation estimates by Monni et al. (2006) using 2006 inventory guidelines, indicated about 1250 Mt of waste generated in 2000. Figure 10.3b showing annual carbon storage in landfills was developed using the same base data as Figure 10.3a with the percentage of landfilled waste for each country (reported to UNFCCC) and a conservative assumption of 50% carbon storage (Bogner, 1992; Barlaz, 1998). This storage is long-term: under the anaerobic conditions in landfills, lignin does not degrade significantly (Chen et al., 2004), while some cellulosic fractions are also non-degraded. The annual totals for the mid-1980s and later (>30 MtC/yr) exceed estimates in the literature for the annual quantity of organic carbon partitioned to long-term geologic storage in marine environments as a precursor to future fossil fuels (Bogner, 1992). It should be noted that the anaerobic burial of waste in landfills (with resulting carbon storage) has been widely implemented in developed countries only since the 1960s and 1970s.

Box 10.1 continued

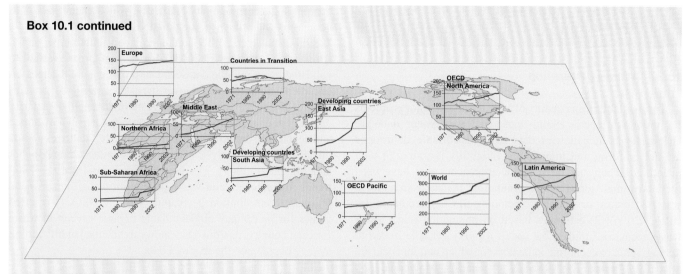

Figure 10.3a: *Annual rates of post-consumer waste generation 1971–2002 (Tg) using energy consumption surrogate.*

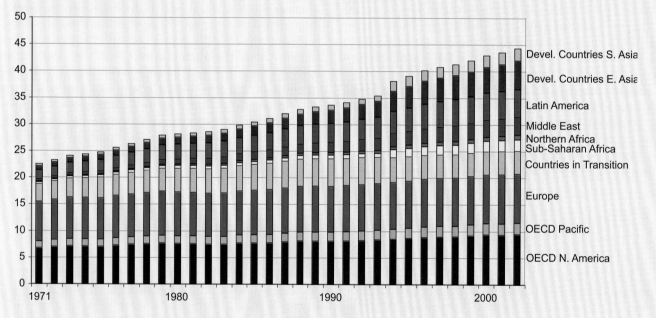

Figure 10.3b: *Minimum annual rates of carbon storage in landfills from 1971–2002 (Tg C).*

Table 10.1: *Municipal solid waste-generation rates and relative income levels*

Country	Low income	Middle income	High income
Annual income (US$/cap/yr)	825-3255	3256-10065	>10066
Municipal solid waste generation rate (t/cap/yr)	0.1-0.6	0.2-0.5	0.3 to >0.8

Note: Income levels as defined by World Bank (www.worldbank.org/data/wdi2005).

Sources: Bernache-Perez et al., 2001; CalRecovery, 2004, 2005; Diaz and Eggerth, 2002; Griffiths and Williams, 2005; Idris et al., 2003; Kaseva et al., 2002; Ojeda-Benitez and Beraud-Lozano, 2003; Huang et al., 2006; US EPA, 2003.

10.2.2 Wastewater generation

Most countries do not compile annual statistics on the total volume of municipal wastewater generated, transported and treated. In general, about 60% of the global population has sanitation coverage (sewerage) with very high levels (>90%) characteristic for the population of North America (including Mexico), Europe and Oceania, although in the last two regions rural areas decrease to approximately 75% and 80%, respectively (DESA, 2005; Jouravlev, 2004; PNUD, 2005; WHO/UNICEF/WSSCC, 2000, WHO-UNICEF, 2005; World Bank, 2005a). In developing countries, rates of sewerage are very low for rural areas of Africa, Latin America and Asia, where septic tanks and latrines predominate. For 'improved sanitation' (including sewerage + wastewater treatment, septic tanks and latrines), almost 90% of the population in developed countries, but only about 30% of the population in developing countries, has access to improved sanitation (Jouravlev, 2004; World Bank, 2005a, b). Many countries in Eastern Europe and Central Asia lack reliable benchmarks for the early 1990s. Regional trends (Figure 10.4) indicate improved sanitation levels of <50% for Eastern and Southern Asia and Sub-Saharan Africa (World Bank and IMF, 2006). In Sub-Saharan Africa, at least 450 million people lack adequate sanitation. In both Southern and Eastern Asia, rapid urbanization is posing a challenge for the development of wastewater infrastructure. The highly urbanized region of Latin America and the Caribbean has also made slow progress in providing wastewater treatment. In the Middle East and North Africa, the countries of Egypt, Tunisia and Morocco have made significant progress in expanding wastewater-treatment infrastructure (World Bank and IMF, 2006). Nevertheless, globally, it has been estimated that 2.6 billion people lack improved sanitation (WHO-UNICEF, 2005).

Estimates for CH_4 and N_2O emissions from wastewater treatment require data on degradable organic matter (BOD; COD[1]) and nitrogen. Nitrogen content can be estimated using Food and Agriculture Organization (FAO) data on protein consumption, and either the application of wastewater treatment, or its absence, determines the emissions. Aerobic treatment plants produce negligible or very small emissions, whereas in anaerobic lagoons or latrines 50–80% of the CH_4 potential can be produced and emitted. In addition, one must take into account the established infrastructure for wastewater treatment in developed countries and the lack of both infrastructure and financial resources in developing countries where open sewers or informally ponded wastewaters often result in uncontrolled discharges to surface water, soils, and coastal zones, as well as the generation of N_2O and CH_4. The majority of urban wastewater treatment facilities are publicly operated and only about 14% of the total private investment in water and sewerage in the late 1990s was applied to the financing of wastewater collection and treatment, mainly to protect drinking water supplies (Silva, 1998; World Bank 1997).

Most wastewaters within the industrial and agricultural sectors are discussed in Chapters 7 and 8, respectively. However, highly organic industrial wastewaters are addressed in this chapter, because they are frequently conveyed to municipal treatment facilities. Table 10.2 summarizes estimates for total and regional 1990 and 2001 generation in terms of kilograms of BOD per day or kilograms of BOD per worker per day, based on measurements of plant-level water quality (World Bank, 2005a). The table indicates that total global generation decreased >10% between 1990 and 2001; however, increases of 15% or more

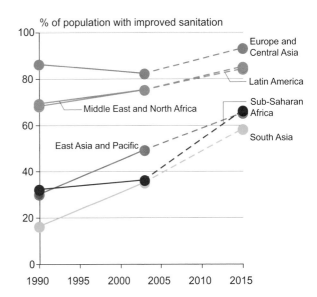

Figure 10.4: *Regional data for 1990 and 2003 with 2015 Millenium Development Goal (MDG) targets for the share of population with access to improved sanitation (sewerage + wastewater treatment, septic system, or latrine).*

Source: World Bank and IMF (2006)

were observed for the Middle East and the developing countries of South Asia.

10.2.3 Development trends for waste and wastewater

Waste and wastewater management are highly regulated within the municipal infrastructure under a wide range of existing regulatory goals to protect human health and the environment; promote waste minimization and recycling; restrict certain types of waste management activities; and reduce impacts to residents, surface water, groundwater and soils. Thus, activities related to waste and wastewater management are, and will continue to be, controlled by national regulations, regional restrictions, and local planning guidelines that address waste and wastewater transport, recycling, treatment, disposal, utilization, and energy use. For developing countries, a wide range of waste management legislation and policies have been implemented with evolving structure and enforcement; it is expected that regulatory frameworks in developing countries will become more stringent in parallel with development trends.

Depending on regulations, policies, economic priorities and practical local limits, developed countries will be characterized by increasingly higher rates of waste recycling and pre-treatment to conserve resources and avoid GHG generation. Recent studies have documented recycling levels of >50%

---

1 BOD (Biological or Biochemical Oxygen Demand) measures the quantity of oxygen consumed by aerobically biodegradable organic C in wastewater. COD (Chemical Oxygen Demand) measures the quantity of oxygen consumed by chemical oxidation of C in wastewater (including both aerobic/anaerobic biodegradable and non-biodegradable C).

Table 10.2: *Regional and global 1990 and 2001 generation of high BOD industrial wastewaters often treated by municipal wastewater systems.*

Regions	Kg BOD/day [Total, Rounded] (1000s)		Kg BOD/worker/ day		Primary metals (%)	Paper and pulp (%)	Chemicals (%)	Food and beverages (%)	Textiles (%)
Year	1990	2001	1990	2001	2001	2001	2001	2001	2001
1. OECD North America	3100	2600	0.20	0.17	9	15	11	44	7
2. OECD Pacific	2200	1700	0.15	0.18	8	20	6	46	7
3. Europe	5200	4800	0.18	0.17	9	22	9	40	7
4. Countries in transition	3400	2400	0.15	0.21	13	8	6	50	14
5. Sub-Saharan Africa	590	510	0.23	0.25	3	12	6	60	13
6. North Africa	410	390	0.20	0.18	10	4	6	50	25
7. Middle East	260	300	0.19	0.19	9	12	10	52	11
8. Caribbean, Central and South America	1500	1300	0.23	0.24	5	11	8	61	11
9. Developing countries, East Asia	8300	7700	0.14	0.16	11	14	10	36	15
10. Developing countries, South Asia	1700	2000	0.18	0.16	5	7	6	42	35
Total for 1-4 (developed)	**13900**	**11500**							
Total for 5-10 (developing)	**12800**	**12200**							

Note: Percentages are included for major industrial sectors (all other sectors <10% of total BOD).
Source: World Bank, 2005a.

for specific waste fractions in some developed countries (i.e., Swedish Environmental Protection Agency, 2005). Recent US data indicate about 25% diversion, including more than 20 states that prohibit landfilling of garden waste (Simmons *et al.*, 2006). In developing countries, a high level of labour-intensive informal recycling often occurs. Via various diversion and small-scale recycling activities, those who make their living from decentralized waste management can significantly reduce the mass of waste that requires more centralized solutions; however, the challenge for the future is to provide safer, healthier working conditions than currently experienced by scavengers on uncontrolled dumpsites. Available studies indicate that recycling activities by this sector can generate significant employment, especially for women, through creative microfinance and other small-scale investments. For example, in Cairo, available studies indicate that 7–8 daily jobs per ton of waste and recycling of >50% of collected waste can be attained (Iskandar, 2001).

Trends for sanitary landfilling and alternative waste-management technologies differ amongst countries. In the EU, the future landfilling of organic waste is being phased out via the landfill directive (Council Directive 1999/31/EC), while engineered gas recovery is required at existing sites (EU, 1999). This directive requires that, by 2016, the mass of biodegradable organic waste annually landfilled must be reduced 65% relative to landfilled waste in 1995. Several countries (Germany, Austria, Denmark, Netherlands, Sweden) have accelerated the EU schedule through more stringent bans on landfilling of organic waste. As a result, increasing

quantities of post-consumer waste are now being diverted to incineration, as well as to MBT before landfilling to 1) recover recyclables and 2) reduce the organic carbon content by a partial aerobic composting or anaerobic digestion (Stegmann, 2005). The MBT residuals are often, but not always, landfilled after achieving organic carbon reductions to comply with the EU landfill directive. Depending on the types and quality control of various separation and treatment processes, a variety of useful recycled streams are also produced. Incineration for waste-to-energy has been widely implemented in many European countries for decades. In 2002, EU WTE plants generated 41 million GJ of electrical energy and 110 million GJ of thermal energy (Themelis, 2003). Rates of incineration are expected to increase in parallel with implemention of the landfill directive, especially in countries such as the UK with historically lower rates of incineration compared to other European countries. In North America, Australia and New Zealand, controlled landfilling is continuing as a dominant method for large-scale waste disposal with mandated compliance to both landfilling and air-quality regulations. In parallel, larger quantities of landfill CH_4 are annually being recovered, both to comply with air-quality regulations and to provide energy, assisted by national tax credits and local renewable-energy/green power initiatives (see Section 10.5). The US, Canada, Australia and other countries are currently studying and considering the widespread implementation of 'bioreactor' landfills to compress the time period during which high rates of CH_4 generation occur (Reinhart and Townsend, 1998; Reinhart *et al.*, 2002; Berge *et al.*, 2005); bioreactors will also require the early implementation of engineered gas extraction. Incineration has not been widely implemented in these countries due to historically low landfill tipping fees in many regions, negative public perceptions and high capital costs. In Japan, where open space is very limited for construction of waste management infrastructure, very high rates of both recycling and incineration are practised and are expected to continue into the future. Historically, there have also been 'semi-aerobic' Japanese landfills with potential for N_2O generation (Tsujimoto et al., 1994). Similar aerobic (with air) landfill practices have also been studied or implemented in Europe and the US for reduced CH_4 generation rates as an alternative to, or in combination with, anaerobic (without air) practices (Ritzkowski and Stegmann, 2005).

In many developing countries, current trends suggest that increases in controlled landfilling resulting in anaerobic decomposition of organic waste will be implemented in parallel with increased urbanization. For rapidly growing 'mega cities', engineered landfills provide a waste disposal solution that is more environmentally acceptable than open dumpsites and uncontrolled burning of waste. There are also persuasive public health reasons for implementing controlled landfilling – urban residents produce more solid waste per capita than rural inhabitants, and large amounts of uncontrolled refuse accumulating in areas of high population density are linked to vermin and disease (Christensen, 1989). The process of converting open dumping and burning to engineered landfills implies control of waste placement, compaction, the use of cover materials, implementation of surface water diversion and drainage, and management of leachate and gas, perhaps applying an intermediate level of technology consistent with limited financial resources (Savage et al., 1998). These practices shift the production of CO_2 (by burning and aerobic decomposition) to anaerobic production of CH_4. This is largely the same transition that occurred in many developed countries in the 1950–1970 time frame. Paradoxically, this results in higher rates of CH_4 generation and emissions than previous open-dumping and burning practices. In addition, many developed and developing countries have historically implemented large-scale aerobic composting of waste. This has often been applied to mixed waste, which, in practice, is similar to implementing an initial aerobic MBT process. However, source-separated biodegradable waste streams are preferable to mixed waste in order to produce higher quality compost products for horticultural and other uses (Diaz et al., 2002; Perla, 1997). In developing countries, composting can provide an affordable, sustainable alternative to controlled landfilling, especially where more labour-intensive lower technology strategies are applied to selected biodegradable wastes (Hoornweg et al., 1999). It remains to be seen if mechanized recycling and more costly alternatives such as incineration and MBT will be widely implemented in developing countries. Where decisions regarding waste management are made at the local level by communities with limited financial resources seeking the least-cost environmentally acceptable solution – often this is landfilling or composting (Hoornweg, 1999; Hoornweg et al., 1999; Johannessen and Boyer, 1999). Accelerating the introduction of landfill gas extraction and utilization can mitigate the effect of increased CH_4 generation at engineered landfills. Although Kyoto mechanisms such as CDM and JI have already proven useful in this regard, the post-2012 situation is unclear.

With regard to wastewater trends, a current priority in developing countries is to increase the historically low rates of wastewater collection and treatment. One of the Millennium Development Goals (MDGs) is to reduce by 50% the number of people without access to safe sanitation by 2015. One strategy may be to encourage more on-site sanitation rather than expensive transport of sewerage to centralized treatment plants: this strategy has been successful in Dakar, Senegal, at the cost of about 400 US$ per household. It has been estimated that, for sanitation, the annual investment must increase from 4 billion US$ to 18 billion US$ to achieve the MDG target, mostly in East Asia, South Asia and Sub-Saharan Africa (World Bank, 2005a).

10.3   Emission trends

10.3.1   Global overview

Quantifying global trends requires annual national data on waste production and management practices. Estimates for many countries are uncertain because data are lacking, inconsistent or incomplete; therefore, the standardization of terminology for national waste statistics would greatly improve data quality for this sector. Most developing countries use default data on waste generation per capita with inter-annual changes assumed to be proportional to total or urban population. Developed countries use more detailed methodologies, activity data and emission factors, as well as national statistics and surveys, and are sharing their methods through bilateral and multilateral initiatives.

For landfill CH_4, the largest GHG emission from the waste sector, emissions continue several decades after waste disposal; thus, the estimation of emission trends requires models that include temporal trends. Methane is also emitted during wastewater transport, sewage treatment processes and leakages from anaerobic digestion of waste or wastewater sludges. The major sources of N_2O are human sewage and wastewater treatment. The CO_2 from the non-biomass portion of incinerated waste is a small source of GHG emissions. The IPCC 2006 Guidelines also provide methodologies for CO_2, CH_4 and N_2O emissions from open burning of waste and for CH_4 and N_2O emissions from composting and anaerobic digestion of biowaste. Open burning of waste in developing countries is a significant local source of air pollution, constituting a health risk for nearby communities. Composting and other biological treatments emit very small quantities of GHGs but were included in 2006 IPCC Guidelines for completeness.

Table 10.3: *Trends for GHG emissions from waste using (a) 1996 and (b) 2006 IPCC inventory guidelines, extrapolations, and projections (MtCO$_2$-eq, rounded)*

Source	1990	1995	2000	2005	2010	2015	2020	2030	2050
Landfill CH$_4$[a]	760	770	730	750	760	790	820		
Landfill CH$_4$[b]	340	400	450	520	640	800	1000	1500	2900
Landfill CH$_4$ (average of [a] and [b])	550	585	590	635	700	795	910		
Wastewater CH$_4$[a]	450	490	520	590	600	630	670		
Wastewater N$_2$O[a]	80	90	90	100	100	100	100		
Incineration CO$_2$[b]	40	40	50	50	60	60	60	70	80
Total GHG emissions	1120	1205	1250	1345	1460	1585	1740		

Notes: Emissions estimates and projections as follows:
[a] Based on reported emissions from national inventories and national communications, and (for non-reporting countries) on 1996 inventory guidelines and extrapolations (US EPA, 2006).
[b] Based on 2006 inventory guidelines and BAU projection (Monni et al., 2006).
Total includes landfill CH$_4$ (average), wastewater CH$_4$, wastewater N$_2$O and incineration CO$_2$.

Overall, the waste sector contributes <5% of global GHG emissions. Table 10.3 compares estimated emissions and trends from two studies: US EPA (2006) and Monni *et al.* (2006). The US EPA (2006) study collected data from national inventories and projections reported to the United Nations Framework Convention on Climate Change (UNFCCC) and supplemented data gaps with estimates and extrapolations based on IPCC default data and simple mass balance calculations using the 1996 IPCC Tier 1 methodology for landfill CH$_4$. Monni *et al.* (2006) calculated a time series for landfill CH$_4$ using the first-order decay (FOD) methodology and default data in the 2006 IPCC Guidelines, taking into account the time lag in landfill emissions compared to year of disposal. The estimates by Monni *et al.* (2006) are lower than US EPA (2006) for the period 1990–2005 because the former reflect slower growth in emissions relative to the growth in waste. However, the future projected growth in emissions by Monni *et al.* (2006) is higher, because recent European decreases in landfilling are reflected more slowly in the future projections. For comparison, the reported 1995 CH$_4$ emissions from landfills and wastewater from national inventories were approximately 1000 MtCO$_2$eq (UNFCCC, 2005). In general, data from Non-Annex I countries are limited and usually available only for 1994 (or 1990). In the TAR, annual global CH$_4$ and N$_2$O emissions from all sources were approximately 600 Tg CH$_4$/yr and 17.7 Tg N/yr as N$_2$O (IPCC, 2001b). The direct comparison of reported emissions in Table 10.3 with the SRES A1 and B2 scenarios (Nakicenovic *et al.*, 2000) for GHG emissions from waste is problematical: the SRES do not include landfill-gas recovery (commercial since 1975) and project continuous increases in CH$_4$ emissions based only on population increases to 2030 (AIB-AIM) or 2100 (B2-MESSAGE), resulting in very high emission estimates of >4000 MtCO$_2$-eq/yr for 2050.

Table 10.3 indicates that total emissions have historically increased and will continue to increase (Monni *et al.*, 2006; US EPA, 2006; *see also* Scheehle and Kruger, 2006). However, between 1990 and 2003, the percentage of total global GHG emissions from the waste sector declined 14–19% for Annex I and EIT countries (UNFCCC, 2005). The waste sector contributed 2–3% of the global GHG total for Annex I and EIT countries for 2003, but a higher percentage (4.3%) for non-Annex I countries (various reporting years from 1990–2000) (UNFCCC, 2005). In developed countries, landfill CH$_4$ emissions are stabilizing due to increased landfill CH$_4$ recovery, decreased landfilling, and decreased waste generation as a result of local waste management decisions including recycling, local economic conditions and policy initiatives. On the other hand, rapid increases in population and urbanization in developing countries are resulting in increases in GHG emissions from waste, especially CH$_4$ from landfills and both CH$_4$ and N$_2$O from wastewater. CH$_4$ emissions from wastewater alone are expected to increase almost 50% between 1990 and 2020, especially in the rapidly developing countries of Eastern and Southern Asia (US EPA, 2006; *Table 10.3*). Estimates of global N$_2$O emissions from wastewater are incomplete and based only on human sewage treatment, but these indicate an increase of 25% between 1990 and 2020 (*Table 10.3*). It is important to emphasize, however, that these are business-as-usual (BAU) scenarios, and actual emissions could be much lower if additional measures are in place. Future reductions in emissions from the waste sector will partially depend on the post-2012 availability of Kyoto mechanisms such the CDM and JI.

Uncertainties for the estimates in Table 10.3 are difficult to assess and vary by source. According to 2006 IPCC Guidelines (IPCC, 2006), uncertainties can range from 10–30% (for countries with good annual waste data) to more than twofold (for countries without annual data). The use of default data and the Tier 1 mass balance method (from 1996 inventory guidelines) for many developing countries would be the major source of uncertainty in both the US EPA (2006) study and reported GHG emissions (IPCC, 2006). Estimates by Monni *et al.* (2006) were sensitive to the relationship between waste generation and GDP, with an estimated range of uncertainty for the baseline for 2030 of –48% to +24%. Additional sources of uncertainty include the use of default data for waste generation, plus the suitability of parameters and chosen methods for individual countries. However, although country-specific uncertainties may be large, the uncertainties by region and over time are estimated to be smaller.

10.3.2 Landfill CH_4: regional trends

Landfill CH_4 has historically been the largest source of GHG emissions from the waste sector. The growth in landfill emissions has diminished during the last 20 years due to increased rates of landfill CH_4 recovery in many countries and decreased rates of landfilling in the EU. The recovery and utilization of landfill CH_4 as a source of renewable energy was first commercialized in 1975 and is now being implemented at >1150 plants worldwide with emission reductions of >105 $MtCO_2$-eq/yr (Willumsen, 2003; Bogner and Matthews, 2003). This number should be considered a minimum because there are also many sites that recover and flare landfill gas without energy recovery. Figure 10.5 compares regional emissions estimates for five-year intervals from 1990–2020 (US EPA, 2006) to annual historical estimates from 1971–2002 (Bogner and Matthews, 2003). The trends converge for Europe and the OECD Pacific, but there are differences for North America and Asia related to differences in methodologies and assumptions.

A comparison of the present rate of landfill CH_4 recovery to estimated global emissions (Table 10.3) indicates that the minimum recovery and utilization rates discussed above (>105 $MtCO_2$-eq yr) currently exceed the average projected increase from 2005 to 2010. Thus, it is reasonable to state that landfill CH_4 recovery is beginning to stabilize emissions from this source. A linear regression using historical data from the early 1980s to 2003 indicates a conservative growth rate for landfill CH_4 utilization of approximately 5% per year (Bogner and Matthews, 2003). For the EU-15, trends indicate that landfill CH_4 emissions are declining substantially. Between 1990 and 2002, landfill CH_4 emissions decreased by almost 30% (Deuber *et al.*, 2005) due to the early implementation of the landfill directive (1999/31/EC) and similar national legislation intended to both reduce the landfilling of biodegradable waste and increase landfill CH_4 recovery at existing sites. By 2010, GHG emissions from waste in the EU are projected to be more than 50% below 1990 levels due to these initiatives (EEA, 2004).

For developing countries, as discussed in the previous section (10.3.1), rates of landfill CH_4 emissions are expected to increase concurrently with increased landfilling. However, incentives such as the CDM can accelerate rates of landfill CH_4 recovery and use in parallel with improved landfilling practices. In addition, since substantial CH_4 can be emitted both before and after the period of active gas recovery, sites should be encouraged, where feasible, to install horizontal gas collection

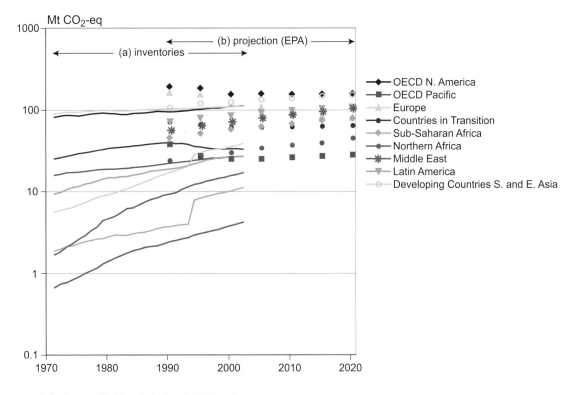

Figure 10.5: *Regional landfill CH_4 emission trends (MtCO$_2$-eq).*

Notes: Includes a) Annual historic emission trends from Bogner and Matthews (2003), extended through 2002; b) Emission estimates for five-year intervals from 1990–2020 using 1996 inventory procedures, extrapolations and projections (US EPA, 2006).

systems concurrent with filling and implement solutions to mitigate residual emissions after closure (such as landfill biocovers to microbially oxidize CH_4—see section 10.4.2).

10.3.3 Wastewater and human sewage CH_4 and N_2O: regional trends

CH_4 and N_2O can be produced and emitted during municipal and industrial wastewater collection and treatment, depending on transport, treatment and operating conditions. The resulting sludges can also microbially generate CH_4 and N_2O, which may be emitted without gas capture. In developed countries, these emissions are typically small and incidental because of extensive infrastructure for wastewater treatment, usually relying on centralized treatment. With anaerobic processes, biogas is produced and CH_4 can be emitted if control measures are lacking; however, the biogas can also be used for process heating or onsite electrical generation.

In developing countries, due to rapid population growth and urbanization without concurrent development of wastewater infrastructure, CH_4 and N_2O emissions from wastewater are generally higher than in developed countries. This can be seen by examining the 1990 estimated CH_4 and N_2O emissions and projected trends to 2020 from wastewater and human sewage (UNFCCC/IPCC, 2004; US EPA, 2006). However, data reliability for many developing countries is uncertain. Decentralized 'natural' treatment processes and septic tanks in developing countries may also result in relatively large emissions of CH_4 and N_2O, particularly in China, India and Indonesia where wastewater volumes are increasing rapidly with economic development (Scheehle and Doorn, 2003).

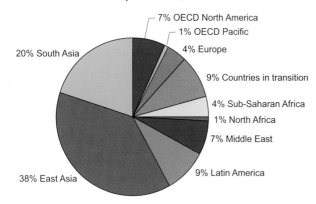

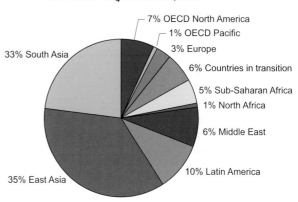

Figure 10.6a: *Regional distribution of CH_4 emissions from wastewater and human sewage in 1990 and 2020. See Table 10.3 for total emissions.*

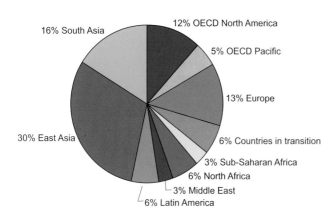

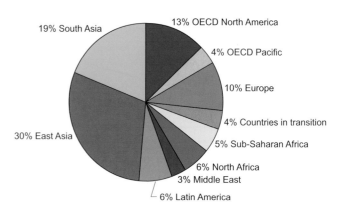

Figure 10.6b: *Regional distribution of N2O emissions from human sewage in 1990 and 2020. See Table 10.3 for total emissions.*

Notes: The US estimates include industrial wastewater and septic tanks, which are not reported by all developed countries.
Source: UNFCCC/IPCC (2004)

The highest regional percentages for CH_4 emissions from wastewater are from Asia (especially China, India). Other countries with high emissions in their respective regions include Turkey, Bulgaria, Iran, Brazil, Nigeria and Egypt. Total global emissions of CH_4 from wastewater handling are expected to rise by more than 45% from 1990 to 2020 (Table 10.3) with much of the increase from the developing countries of East and South Asia, the middle East, the Caribbean, and Central and South America. The EU has projected lower emissions in 2020 relative to 1990 (US EPA, 2006).

The contribution of human sewage to atmospheric N_2O is very low with emissions of 80–100 $MtCO_2$-eq/yr during the period 1990–2020 (Table 10.3) compared to current total global anthropogenic N_2O emissions of about 3500 $MtCO_2$-eq (US EPA, 2006). Emission estimates for N_2O from sewage for Asia, Africa, South America and the Caribbean are significantly underestimated since limited data are available, but it is estimated that these countries accounted for >70% of global emissions in 1990 (UNFCCC/IPCC, 2004). Compared with 1990, it is expected that global emissions will rise by about 20% by 2020 (Table 10.3). The regions with the highest relative N_2O emissions are the developing countries of East Asia, the developing countries of South Asia, Europe and the OECD North America (Figure 10.6b). Regions whose emissions are expected to increase the most by 2020 (with regional increases of 40 to 95%) are Africa, the Middle East, the developing countries of S and E Asia, the Caribbean, and Central and South America (US EPA, 2006). The only regions expected to have lower emissions in 2020 relative to 1990 are Europe and the EIT Countries.

10.3.4 CO_2 from waste incineration

Compared to landfilling, waste incineration and other thermal processes avoid most GHG generation, resulting only in minor emissions of CO_2 from fossil C sources, including plastics and synthetic textiles. Estimated current GHG emissions from waste incineration are small, around 40 $MtCO_2$-eq/yr, or less than one tenth of landfill CH_4 emissions. Recent data for the EU-15 indicate CO_2 emissions from incineration of about 9 $MtCO_2$-eq/yr (EIPPC Bureau, 2006). Future trends will depend on energy price fluctuations, as well as incentives and costs for GHG mitigation. Monni et al. (2006) estimated that incinerator emissions would grow to 80–230 $MtCO_2$-eq/yr by 2050 (not including fossil fuel offsets due to energy recovery).

Major contributors to this minor source would be the developed countries with high rates of incineration, including Japan (>70% of waste incinerated), Denmark and Luxembourg (>50% of waste), as well as France, Sweden, the Netherlands and Switzerland. Incineration rates are increasing in most European countries as a result of the EU Landfill Directive. In 2003, about 17% of municipal solid waste was incinerated with energy recovery in the EU-25 (Eurostat, 2003; Statistics Finland, 2005). More recent data for the EU-15 (EIPCC, 2006)

indicate that 20–25% of the total municipal solid waste is incinerated at over 400 plants with an average capacity of about 500 t/d (range of 170–1400 t/d). In the US, only about 14% of waste is incinerated (US EPA, 2005), primarily in the more densely populated eastern states. Thorneloe et al. (2002), using a life cycle approach, estimated that US plants reduced GHG emissions by 11 $MtCO_2$-eq/yr when fossil-fuel offsets were taken into account.

In developing countries, controlled incineration of waste is infrequently practised because of high capital and operating costs, as well as a history of previous unsustainable projects. The uncontrolled burning of waste for volume reduction in these countries is still a common practice that contributes to urban air pollution (Hoornweg, 1999). Incineration is also not the technology of choice for wet waste, and municipal waste in many developing countries contains a high percentage of food waste with high moisture contents. In some developing countries, however, the rate of waste incineration is increasing. In China, for example, waste incineration has increased rapidly from 1.7% of municipal waste in 2000 to 5% in 2005 (including 67 plants). (Du et al., 2006a, 2006b; National Bureau of Statistics of China, 2006).

10.4 Mitigation of post-consumer emissions from waste

10.4.1 Waste management and GHG-mitigation technologies

A wide range of mature technologies is available to mitigate GHG emissions from waste. These technologies include landfilling with landfill gas recovery (reduces CH_4 emissions), post-consumer recycling (avoids waste generation), composting of selected waste fractions (avoids GHG generation), and processes that reduce GHG generation compared to landfilling (thermal processes including incineration and industrial co-combustion, MBT with landfilling of residuals, and anaerobic digestion). Therefore, the mitigation of GHG emissions from waste relies on multiple technologies whose application depends on local, regional and national drivers for both waste management and GHG mitigation. There are many appropriate low- to high-technology strategies discussed in this section (see Figure 10.7 for a qualitative comparison of technologies). At the 'high technology' end, there are also advanced thermal processes for waste such as pyrolysis and gasification, which are beginning to be applied in the EU, Japan and elsewhere. Because of variable feedstocks and high unit costs, these processes have not been routinely applied to mixed municipal waste at large scale (thousands of tonnes per day). Costs and potentials are addressed in Section 10.4.7.

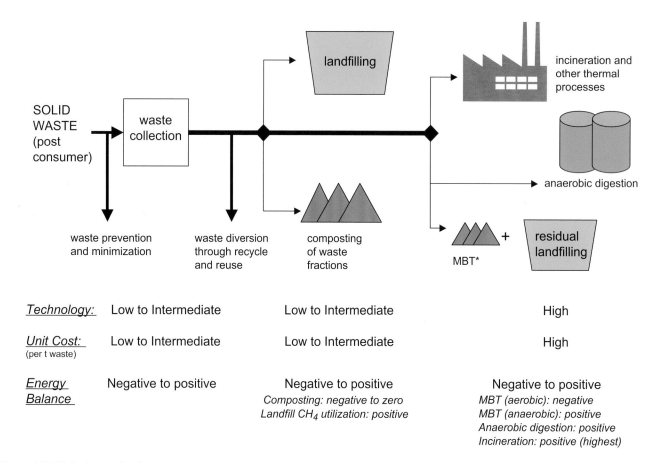

Figure 10.7: *Technology gradient for waste management: major low- to high-technology options applicable to large-scale urban waste management*

Note: MBT=Mechanical Biological Treatment.

10.4.2 CH$_4$ management at landfills

Global CH$_4$ emissions from landfills are estimated to be 500–800 MtCO$_2$-eq/yr(US EPA, 2006; Monni et al. 2006; Bogner and Matthews 2003). However, direct field measurements of landfill CH$_4$ emissions at small scale (<1m^2) can vary over seven orders of magnitude (0.0001–>1000 g CH$_4$/m^2/d) depending on waste composition, cover materials, soil moisture, temperature and other variables (Bogner *et al.*, 1997a). Results from a limited number of whole landfill CH$_4$ emissions measurements in Europe, the US and South Africa are in the range of about 0.1–1.0 tCH$_4$/ha/d (Nozhevnikova *et al.*, 1993; Oonk and Boom, 1995; Borjesson, 1996; Czepiel *et al.*, 1996; Hovde *et al.*, 1995; Mosher *et al.*, 1999; Tregoures *et al.*, 1999; Galle *et al.*, 2001; Morris, 2001; Scharf *et al.*, 2002).

The implementation of an active landfill gas extraction system using vertical wells or horizontal collectors is the single most important mitigation measure to reduce emissions. Intensive field studies of the CH$_4$ mass balance at cells with a variety of design and management practices have shown that >90% recovery can be achieved at cells with final cover and an efficient gas extraction system (Spokas *et al.*, 2006). Some sites may have less efficient or only partial gas extraction systems and

there are fugitive emissions from landfilled waste prior to and after the implementation of active gas extraction; thus estimates of 'lifetime' recovery efficiencies may be as low as 20% (Oonk and Boom, 1995), which argues for early implementation of gas recovery. Some measures that can be implemented to improve overall gas collection are installation of horizontal gas collection systems concurrent with filling, frequent monitoring and remediation of edge and piping leakages, installation of secondary perimeter extraction systems for gas migration and emissions control, and frequent inspection and maintenance of cover materials. Currently, landfill CH$_4$ is being used to fuel industrial boilers; to generate electricity using internal combustion engines, gas turbines or steam turbines; and to produce a substitute natural gas after removal of CO$_2$ and trace components. Although electrical output ranges from small 30 kWe microturbines to 50 MWe steam turbine generators, most plants are in the 1–15 MWe range. Significant barriers to increased diffusion of landfill gas utilization, especially where it has not been previously implemented, can be local reluctance from electrical utilities to include small power producers and from gas utilities/pipeline companies to transport small percentages of upgraded landfill gas in natural gas pipelines.

A secondary control on landfill CH_4 emissions is CH_4 oxidation by indigenous methanotrophic microorganisms in cover soils. Landfill soils attain the highest rates of CH_4 oxidation recorded in the literature, with rates many times higher than in wetland settings. CH_4 oxidation rates at landfills can vary over several orders of magnitude and range from negligible to 100% of the CH_4 flux to the cover. Under circumstances of high oxidation potential and low flux of landfill CH_4 from the landfill, it has been demonstrated that atmospheric CH_4 may be oxidized at the landfill surface (Bogner et al., 1995; 1997b; 1999; 2005; Borjesson and Svensson, 1997b). In such cases, the landfill cover soils function as a sink rather than a source of atmospheric CH_4. The thickness, physical properties moisture content, and temperature of cover soils directly affect oxidation, because rates are limited by the transport of CH_4 upward from anaerobic zones and O_2 downward from the atmosphere. Laboratory studies have shown that oxidation rates in landfill cover soils may be as high as 150–250 g $CH_4/m^2/d$ (Kightley et al., 1995; de Visscher et al., 1999). Recent field studies have demonstrated that oxidation rates can be greater than 200 g/m^2/d in thick, compost-amended 'biocovers' engineered to optimize oxidation (Bogner et al., 2005; Huber-Humer, 2004). The prototype biocover design includes an underlying coarse-grained gas distribution layer to provide more uniform fluxes to the biocover above (Huber-Humer, 2004). Furthermore, engineered biocovers have been shown to effectively oxidize CH_4 over multiple annual cycles in northern temperate climates (Humer-Humer, 2004). In addition to biocovers, it is also possible to design passive or active methanotrophic biofilters to reduce landfill CH_4 emissions (Gebert and Gröngröft, 2006; Streese and Stegmann, 2005). In field settings, stable C isotopic techniques have proven extremely useful to quantify the fraction of CH_4 that is oxidized in landfill cover soils (Chanton and Liptay, 2000; de Visscher et al., 2004; Powelson et al., 2007). A secondary benefit of CH_4 oxidation in cover soils is the co-oxidation of many non-CH_4 organic compounds, especially aromatic and lower chlorinated compounds, thereby reducing their emissions to the atmosphere (Scheutz et al., 2003a).

Other measures to reduce landfill CH_4 emissions include installation of geomembrane composite covers (required in the US as final cover); design and installation of secondary perimeter gas extraction systems for additional gas recovery; and implementation of bioreactor landfill designs so that the period of active gas production is compressed while early gas extraction is implemented.

Landfills are a significant source of CH_4 emissions, but they are also a long-term sink for carbon (Bogner, 1992; Barlaz, 1998. See Figure 10.1 and Box 10.1). Since lignin is recalcitrant and cellulosic fractions decompose slowly, a minimum of 50% of the organic carbon landfilled is not typically converted to biogas carbon but remains in the landfill (See references cited on Figure 10.1). Carbon storage makes landfilling a more competitive alternative from a climate change perspective, especially where landfill gas recovery is combined with energy use (Flugsrud et al. 2001; Micales and Skog, 1997; Pingoud et al. 1996; Pipatti and Savolainen, 1996; Pipatti and Wihersaari, 1998). The fraction of carbon storage in landfills can vary over a wide range, depending on original waste composition and landfill conditions (for example, see Hashimoto and Moriguchi, 2004 for a review addressing harvested wood products).

10.4.3 Incineration and other thermal processes for waste-to-energy

These processes include incineration with and without energy recovery, production of refuse-derived fuel (RDF), and industrial co-combustion (including cement kilns: see Onuma et al., 2004 and Section 7.3.3). Incineration reduces the mass of waste and can offset fossil-fuel use; in addition, GHG emissions are avoided, except for the small contribution from fossil carbon (Consonni et al., 2005). Incineration has been widely applied in many developed countries, especially those with limited space for landfilling such as Japan and many European countries. Globally, about 130 million tonnes of waste are annually combusted in >600 plants in 35 countries (Themelis, 2003).

Waste incinerators have been extensively used for more than 20 years with increasingly stringent emission standards in Japan, the EU, the US and other countries. Mass burning is relatively expensive and, depending on plant scale and flue-gas treatment, currently ranges from about 95–150 €/t waste (87–140 US$/t) (Faaij et al., 1998; EIPPC Bureau, 2006). Waste-to-energy plants can also produce useful heat or electricity, which improves process economics. Japanese incinerators have routinely implemented energy recovery or power generation (Japan Ministry of the Environment, 2006). In northern Europe, urban incinerators have historically supplied fuel for district heating of residential and commercial buildings. Starting in the 1980s, large waste incinerators with stringent emission standards have been widely deployed in Germany, the Netherlands and other European countries. Typically such plants have a capacity of about 1 Mt waste/yr, moving grate boilers (which allow mass burning of waste with diverse properties), low steam pressures and temperatures (to avoid corrosion) and extensive flue gas cleaning to conform with EU Directive 2000/76/EC. In 2002, European incinerators for waste-to-energy generated 41 million GJ electrical energy and 110 million GJ thermal energy (Themelis, 2003). Typical electrical efficiencies are 15% to >20% with more efficient designs becoming available. In recent years, more advanced combustion concepts have penetrated the market, including fluidized bed technology.

10.4.4 Biological treatment including composting, anaerobic digestion, and MBT (Mechanical Biological Treatment)

Many developed and developing countries practise composting and anaerobic digestion of mixed waste or biodegradable waste fractions (kitchen or restaurant wastes, garden waste, sewage sludge). Both processes are best applied to source-separated waste fractions: anaerobic digestion is particularly appropriate for wet wastes, while composting is often appropriate for drier feedstocks. Composting decomposes waste aerobically into CO_2, water and a humic fraction; some carbon storage also occurs in the residual compost (see references on Figure 10.1). Composting can be sustainable at reasonable cost in developing countries; however, choosing more labour-intensive processes over highly mechanized technology at large scale is typically more appropriate and sustainable; Hoornweg et al. (1999) give examples from India and other countries. Depending on compost quality, there are many potential applications for compost in agriculture, horticulture, soil stabilization and soil improvement (increased organic matter, higher water-holding capacity) (Cointreau, 2001). However, CH_4 and N_2O can both be formed during composting by poor management and the initiation of semi-aerobic (N_2O) or anaerobic (CH_4) conditions; recent studies also indicate potential production of CH_4 and N_2O in well-managed systems (Hobson et al., 2005).

Anaerobic digestion produces biogas ($CH_4 + CO_2$) and biosolids. In particular, Denmark, Germany, Belgium and France have implemented anaerobic digestion systems for waste processing, with the resulting biogas used for process heating, onsite electrical generation and other uses. Minor quantities of CH_4 can be vented from digesters during start-ups, shutdowns and malfunctions. However, the GHG emissions from controlled biological treatment are small in comparison to uncontrolled CH_4 emissions from landfills without gas recovery (e.g. Petersen et al. 1998; Hellebrand 1998; Vesterinen 1996; Beck-Friis, 2001; Detzel et al. 2003). The advantages of biological treatment over landfilling are reduced volume and more rapid waste stabilization. Depending on quality, the residual solids can be recycled as fertilizer or soil amendments, used as a CH_4-oxidizing biocovers on landfills (Barlaz et al., 2004; Huber-Humer, 2004), or landfilled at reduced volumes with lower CH_4 emissions.

Mechanical biological treatment (MBT) of waste is now being widely implemented in Germany, Austria, Italy and other EU countries. In 2004, there were 15 facilities in Austria, 60 in Germany and more than 90 in Italy; the total throughput was approximately 13 million tonnes with larger plants having a capacity of 600–1300 tonnes/day (Diaz et al., 2006). Mixed waste is subjected to a series of mechanical and biological operations to reduce volume and achieve partial stabilization of the organic carbon. Typically, mechanical operations (sorting, shredding, crushing) first produce a series of waste fractions for recycling or for subsequent treatment (including combustion or secondary biological processes). The biological steps consist of either aerobic composting or anaerobic digestion. Composting can occur either in open windrows or in closed buildings with gas collection and treatment. In-vessel anaerobic digestion of selected organic fractions produces biogas for energy use. Compost products and digestion residuals can have potential horticultural or agricultural applications; some MBT residuals are landfilled, or soil-like residuals can be used as landfill cover. Under landfill conditions, residual materials retain some potential for CH_4 generation (Bockreis and Steinberg, 2005). Reductions of as much as 40–60% of the original organic carbon are possible with MBT (Kaartinen, 2004). Compared with landfilling, MBT can theoretically reduce CH_4 generation by as much as 90% (Kuehle-Weidemeier and Doedens, 2003). In practice, reductions are smaller and dependent on the specific MBT processes employed (see Binner, 2002).

10.4.5 Waste reduction, re-use and recycling

Quantifying the GHG-reduction benefits of waste minimization, recycling and re-use requires the application of LCA tools (Smith et al., 2001). Recycling reduces GHG emissions through lower energy demand for production (avoided fossil fuel) and by substitution of recycled feedstocks for virgin materials. Efficient use of materials also reduces waste. Material efficiency can be defined as a reduction in primary materials for a particular purpose, such as packaging or construction, with no negative impact on existing human activities. At several stages in the life cycle of a product, material efficiency can be increased by more efficient design, material substitution, product recycling, material recycling and quality cascading (use of recycled material for a secondary product with lower quality demands). Both material recycling and quality cascading occur in many countries at large scale for metals recovery (steel, aluminium) and recycling of paper, plastics and wood. All these measures lead to indirect energy savings, reductions in GHG emissions, and avoidance of GHG generation. This is especially true for products resulting from energy-intensive production processes such as metals, glass, plastic and paper (Tuhkanen et al., 2001).

The magnitude of avoided GHG-emissions benefits from recycling is highly dependent on the specific materials involved, the recovery rates for those materials, the local options for managing materials, and (for energy offsets) the specific fossil fuel avoided (Smith et al., 2001). Therefore, existing studies are often not comparable with respect to the assumptions and calculations employed. Nevertheless, virtually all developed countries have implemented comprehensive national, regional or local recycling programmes. For example, Smith et al. (2001) thoroughly addressed the GHG-emission benefits from recycling across the EU, and Pimenteira et al. (2004) quantified GHG emission reductions from recycling in Brazil.

10.4.6 Wastewater and sludge treatment

There are many available technologies for wastewater management, collection, treatment, re-use and disposal, ranging from natural purification processes to energy-intensive advanced technologies. Although decision-making tools are available that include environmental trade-offs and costs (Ho, 2000), systematic global studies of GHG-reduction potentials and costs for wastewater are still needed. When efficiently applied, wastewater transport and treatment technologies reduce or eliminate GHG generation and emissions; in addition, wastewater management promotes water conservation by preventing pollution, reducing the volume of pollutants, and requiring a smaller volume of water to be treated. Because the size of treatment systems is primarily governed by the volume of water to be treated rather than the mass loading of nitrogen and other pollutants, smaller volumes mean that smaller treatment plants with lower capital costs can be more extensively deployed. Wastewater collection and transport includes conventional (deep) sewerage and simplified (shallow) sewerage. Deep sewerage in developed countries has high capital and operational costs. Simplified (shallow) sewerage in both developing and developed countries uses smaller-diameter piping and shallower excavations, resulting in lower capital costs (30–50%) than deep systems.

Wastewater treatment removes pollutants using a variety of technologies. Small wastewater treatment systems include pit latrines, composting toilets and septic tanks. Septic tanks are inexpensive and widely used in both developed and developing countries. Improved on-site treatment systems used in developing countries include inverted trench systems and aerated treatment units. More advanced treatment systems include activated sludge treatment, trickling filters, anaerobic or facultative lagoons, anaerobic digestion and constructed wetlands. Depending on scale, many of these systems have been used in both developed and developing countries. Activated sludge treatment is considered the conventional method for large-scale treatment of sewage. In addition, separation of black water and grey water can reduce the overall energy requirements for treatment (UNEP/GPA-UNESCO/IHE, 2004). Pretreatment or limitation of industrial wastes is often necessary to limit excessive pollutant loads to municipal systems, especially when wastewaters are contaminated with heavy metals. Sludges (or biosolids) are the product of most wastewater treatment systems. Options for sludge treatment include stabilization, thickening, dewatering, anaerobic digestion, agricultural re-use, drying and incineration. The use of composted sludge as a soil conditioner in agriculture and horticulture recycles carbon, nitrogen and phosphorus (and other elements essential for plant growth). Heavy metals and some toxic chemicals are difficult to remove from sludge; either the limitation of industrial inputs or wastewater pretreatment is needed for agricultural use of sludges. Lower quality uses for sludge may include mine site rehabilitation, highway landscaping, or landfill cover (including biocovers). Some sludges are landfilled, but this practice may result in increased volatile siloxanes and H_2S in the landfill gas. Treated wastewater can either be re-used or discharged, but re-use is the most desirable option for agricultural and horticultural irrigation, fish aquaculture, artificial recharge of aquifers, or industrial applications.

10.4.7 Waste management and mitigation costs and potentials

In the waste sector, it is often not possible to clearly separate costs for GHG mitigation from costs for waste management. In addition, waste management costs can exhibit high variability depending on local conditions. Therefore the baseline and cost assumptions, local availability of technologies, and economic and social development issues for alternative waste management strategies need to be carefully defined. An older study by de Jager and Blok (1996) assumed a 20-year project life to compare the cost-effectiveness of various options for mitigating CH_4 emissions from waste in the Netherlands, with costs ranging from –2 US$/t$CO_2$-eq for landfilling with gas recovery and on-site electrical generation to >370 US$/t$CO_2$-eq for incineration. In general, for landfill CH_4 recovery and utilization, project economics are highly site-specific and dependent on the financial arrangements as well as the distribution of benefits, risks and responsibilities among multiple partners. Some representative unit costs for landfill-gas recovery and utilization (all in 2003 US$/kW installed power) are: 200–400 for gas collection; 200–300 for gas conditioning (blower/compressor, dehydration, flare); 850–1200 for internal combustion engine/generator; and 250–350 for planning and design (Willumsen, 2003).

Smith *et al.* (2001) highlighted major cost differences between EU member states for mitigating GHG emissions from waste. Based on fees (including taxes) for countries with data, this study compared emissions and costs for various waste management practices with respect to direct GHG emissions, carbon sequestration, transport emissions, avoided emissions from recycling due to material and energy savings, and avoided emissions from fossil-fuel substitution via thermal processes and biogas (including landfill gas). Recycling costs are highly dependent on the waste material recycled. Overall, the financial success of any recycling venture is dependent on the current market value of the recycled products. The price obtained for recovered materials is typically lower than separation/reprocessing costs, which can be, in turn, higher than the cost of virgin materials – thus recycling activities usually require subsidies (except for aluminium and paper recycling). Recycling, composting and anaerobic digestion can provide large potential emission reductions, but further implementation is dependent on reducing the cost of separate collection (10–400 €/t waste (9–380 US$/t)) and, for composting, establishing local markets for the compost product. Costs for composting can range from 20–170 €/t waste (18–156 US$/t) and are typically 35 €/t waste (32 US$/t) for open-windrow operations and 50 €/t waste (46 US$/t) for in-vessel processes. When the replaced fossil fuel is coal, both mass incineration and co-combustion offer comparable and less expensive GHG-emission reductions compared to recycling (averaging 64 €/t waste (59 US$/t), with a range of 30–150 €/t waste (28–140 US$/t)). Landfill disposal is the most inexpensive waste management option in the EU (averaging 56 €/t waste (52 US$/t), ranging from 10–160 €/t waste (9–147 US$/t), including taxes), but it is also the largest source of GHG emissions. With improved gas management, landfill emissions can be significantly reduced at low cost. However, landfilling costs in the EU are increasing due to increasingly stringent regulations, taxes and declining capacity. Although there is only sparse information regarding MBT costs, German costs are about 90 €/t waste (83 US$/t, including landfill disposal fees); recent data suggest that, in the future, MBT may become more cost-competitive with landfilling and incineration.

Costs and potentials for reducing GHG emissions from waste are usually based on landfill CH_4 as the baseline (Bates and Haworth, 2001; Delhotal *et al.* 2006; Monni *et al.* 2006; Nakicenovic *et al.*, 2000; Pipatti and Wihersaari 1998). When reporting to the UNFCCC, most developed countries take the dynamics of landfill gas generation into account; however, most developing countries and non-reporting countries do not. Basing their study on reported emissions and projections, Delhotal *et al.* (2006) estimated break-even costs for GHG abatement from landfill gas utilization that ranged from about −20 to +70 US$/$tCO_2$-eq, with the lower value for direct use in industrial boilers and the higher value for on-site electrical generation. From the same study, break-even costs (all in US$/$tCO_2$-eq) were approximately 25 for landfill-gas flaring; 240–270 for composting; 40–430 for anaerobic digestion; 360 for MBT and 270 for incineration. These costs were based on the EMF-21 study (US EPA, 2003), which assumed a 15-year technology lifetime, 10% discount rate and 40% tax rate.

Compared to thermal and biological processes which only affect future emissions, landfill CH_4 is generated from waste landfilled in previous decades, and gas recovery, in turn, reduces emissions from waste landfilled in previous years. Most existing studies for the waste sector do not consider these temporal issues. Monni *et al.* (2006) developed baseline and mitigation scenarios for solid waste management using the first order decay (FOD) methodology in the 2006 IPCC Guidelines, which takes into account the timing of emissions. The baseline scenario by Monni *et al.* (2006) assumed that: 1) waste generation will increase with growing population and GDP (using the same population and GDP data as SRES scenario A1b); 2) waste management strategies will not change significantly, and 3) landfill gas recovery and utilization will continue to increase at the historical rate of 5% per year in developed countries (Bogner and Matthews, 2003; Willumsen, 2003). Mitigation scenarios were developed for 2030 and 2050 which focus on increased landfill gas recovery, increased recycling, and increased incineration. In the increased landfill gas recovery scenario, recovery was estimated to increase 15% per year, with most of the increase in developing countries because of CDM or similar incentives (above baseline of current CDM projects). This growth rate is about triple the current rate and corresponds to a reasonable upper limit, taking into account the fact that recovery in developed countries has already reached high levels, so that increases would come mainly from developing countries, where current lack of funding is a barrier to deployment. Landfill gas

recovery was capped at 75% of estimated annual CH_4 generation for developed countries and 50% for developing countries in both the baseline and increased landfill gas recovery scenarios. In the increased incineration scenario, incineration grew 5% each year in the countries where waste incineration occurred in 2000. For OECD countries where no incineration took place in 2000, 1% of the waste generated was assumed to be incinerated in 2012. In non-OECD countries, 1% waste incineration was assumed to be reached only in 2030. The maximum rate of incineration that could be implemented was 85% of the waste generated. The increased recycling scenario assumed a growth in paper and cardboard recycling in all parts of the world using a technical maximum of 60% recycling (CEPI, 2003). This maximum was assumed to be reached in 2050. In the mitigation scenarios, only direct emission reductions compared to the baseline CH_4 emissions from landfills were estimated – thus avoided emissions from recycled materials, reduced energy use, or fossil fuel offsets were not included. In the baseline scenario (Figure 10.8), emissions increase threefold during the period from 1990 to 2030 and more than fivefold by 2050. These growth rates do not include current or planned legislation relating to either waste minimization or landfilling – thus future emissions may be overestimated. Most of the increase comes from non-OECD countries whose current emissions are smaller because of lower waste generation and a higher percentage of waste degrading aerobically. The mitigation scenarios show that reductions by individual measures in 2030 range from 5–20% of total emissions and increase proportionally with time. In 2050, the corresponding range is approximately 10–30%. As the measures in the scenarios are largely additive, total mitigation potentials of approximately 30% in 2030 and 50% in 2050 are projected relative to the baseline. Nevertheless, the estimated abatement potential is not capable of mitigating the growth in emissions.

The baseline emission estimates in the Delhotal *et al.* (2006) study are based on similar assumptions to the Monni *et al.* (2006) study: population and GDP growth with increasing amounts of landfilled waste in developing countries. Baselines also include documented or expected changes in disposal rates due to composting and recycling, as well as the effects of landfill-gas recovery. In Delhotal *et al.* (2006), emissions increase by about 30% between 2000 and 2020; therefore, the growth in emissions to 2020 is more moderate than in Monni *et al.* (2006). This more moderate growth can be attributed to the inclusion of current and planned policies and measures to reduce emissions, plus the fact that historical emissions from prior landfilled waste were only partially considered.

Scenario development in both studies was complemented with estimates on maximum mitigation potentials at given marginal cost levels using the baseline scenarios as the starting point. Monni *et al.* (2006) derived annual regional waste-generation estimates for the Global Times model by using static aggregate emission coefficients calibrated to regional FOD models. Some modifications to the assumptions used in the

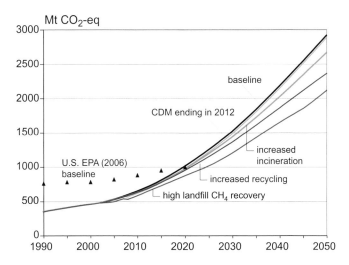

Figure 10.8: *Global CH_4 emissions from landfills in baseline scenario compared to the following mitigation scenarios: increased incineration, CDM ending by 2012 (end of the first Kyoto commitment period), increased recycling, and high landfill CH_4 recovery rates including continuation of CDM after 2012 (Monni et al., 2006). The emission reductions estimated in the mitigation scenarios are largely additional to 2050. This figure also includes the US EPA (2006) baseline scenario for landfill CH_4 emissions from Delhotal et al. (2006).*

Delhotal *et al.* (2006) and Monni *et al.* (2006) both conclude that substantial emission reductions can be achieved at low or negative costs (see Table 10.4). At higher costs, more significant reductions would be possible (more than 80% of baseline emissions) with most of the additional mitigation potential coming from thermal processes for waste-to-energy. Since combustion of waste results in minor fossil CO_2 emissions, these were considered in the calculations, but Table 10.4 only includes emissions reductions from landfill CH_4. In general, direct GHG emission reductions from implementation of thermal processes are much less than indirect reductions due to fossil fuel replacement, where that occurs. The emission reduction potentials for 2030 shown in Table 10.4 are assessed using a steady-state approach that can overestimate near-term annual reductions but gives more realistic values when integrated over time.

The economic mitigation potentials for the year 2030 in Table 10.5 take the dynamics of landfill gas generation into account. These estimates are derived from the static, long-term mitigation potentials previously shown in Table 10.4 (Monni *et al.* 2006). The upper limits of the ranges assume that landfill disposal is limited in the coming years so that only 15% of the waste generated globally is landfilled after 2010. This would mean that by 2030 the maximum economic potential would be almost 70% of the global emissions (see Table 10.5). The lower limits of the table have been scaled down to reflect a more realistic timing of implementation in accordance with emissions in the high landfill gas recovery (HR) and increased incineration (II) scenarios (Monni *et al.*, 2006).

It must be emphasized that there are large uncertainties in costs and potentials for mitigation of GHG emissions from waste due to the uncertainty of waste statistics for many countries and emissions methodologies that are relatively unsophisticated. It is also important to point out that the cost estimates are global

scenario development were also made; for example, recycling was excluded due to its economic complexity, biological treatment was included and the technical efficiency of landfill-gas recovery was assumed the same in all regions (75%). Cost data were taken from various sources (de Feber & Gielen, 2000; OECD, 2004; Hoornweg, 1999).

As in the EMF-21 study (US EPA, 2003), both Delhotal *et al.* (2006) and Monni *et al.* (2006) assumed the same capital costs for all regions, but used regionalized labour costs for operations and maintenance.

Table 10.4: *Economic reduction potential for CH_4 emissions from landfilled waste by level of marginal costs for 2020 and 2030 based on steady state models[a].*

2020 (Delhotal et al., 2006)	US$/tCO$_2$-equivalent				
	0	**15**	**30**	**45**	**60**
OECD	12%	40%	46%	67%	92%
EIT	NA	NA	NA	NA	NA
Non-OECD	NA	NA	NA	NA	NA
Global	12%	41%	50%	57%	88%
2030 (Monni et al., 2006)	**0**	**10**	**20**	**50**	**100**
OECD	48%	86%	89%	94%	95%
EIT	31%	80%	93%	99%	100%
Non-OECD	32%	38%	50%	77%	88%
Global	35%	53%	63%	83%	91%

[a] *The steady-state approach tends to overestimate the near-term annual reduction potential but gives more realistic results when integrated over time.*

Table 10.5: *Economic potential for mitigation of regional landfill CH$_4$ emissions at various cost categories in 2030 (from estimates by Monni et al., 2006). See notes.*

Region	Projected emissions for 2030	Total economic mitigation potential (MtCO$_2$-eq) at <100 US$/tCO$_2$-eq	Economic mitigation potential (MtCO$_2$-eq) at various cost categories (US$/tCO$_2$-eq)			
			<0	0-20	20-50	50-100
OECD	360	100-200	100-120	20-100	0-7	1
EIT	180	100	30-60	20-80	5	1-10
Non-OECD	960	200-700	200-300	30-100	0-200	0-70
Global	1500	400-1000	300-500	70-300	5-200	10-70

Notes:
1. Costs and potentials for wastewater mitigation are not available.
2. Regional numbers are rounded to reflect the uncertainty in the estimates and may not equal global totals.
3. Landfill carbon sequestration is not considered.
4. The timing of measures limiting landfill disposal affect the annual mitigation potential in 2030. The upper limits of the ranges given assume that landfill disposal is limited in the coming years to 15% of the waste generated globally. The lower limits correspond to the sum of the mitigation potential in the high recycling and increased incineration scenarios in the Monni et al. 2006 study.

averages and therefore not necessarily applicable to local conditions.

10.4.8 Fluorinated gases: end-of-life issues, data and trends in the waste sector

The CFCs and HCFCs regulated as ozone-depleting substances (ODS) under the Montreal Protocol can persist for many decades in post-consumer waste and occur as trace components in landfill gas (Scheutz *et al.*, 2003). The HFCs regulated under the Kyoto Protocol are promoted as substitutions for the ODS. High global-warming potential (GWP) fluorinated gases have been used for more than 70 years; the most important are the chlorofluorocarbons (CFCs), hydrochlorofluorocarbons (HCFCs) and the hydrofluorocarbons (HFCs) with the existing bank of CFCs and HCFCs estimated to be >1.5 Mt and 0.75 Mt, respectively (TFFEoL, 2005; IPCC, 2005). These gases have been used as refrigerants, solvents, blowing agents for foams and as chemical intermediates. End-of-life issues in the waste sector are mainly relevant for the foams; for other products, release will occur during use or just after end-of-life. For the rigid foams, releases during use are small (Kjeldsen and Jensen, 2001, Kjeldsen and Scheutz, 2003, Scheutz *et al*, 2003b), so most of the original content is still present at the end of their useful life. The rigid foams include polyurethane and polystyrene used as insulation in appliances and buildings; in these, CFC-11 and CFC-12 were the main blowing agents until the mid-1990s. After the mid-1990s, HCFC-22, HCFC-141b and HCFC-142b with HFC-134a have been used (CALEB, 2000). Considering that home appliances are the foam-containing product with the lowest lifetime (average maximum lifetime 15 years, TFFEoL, 2005), a significant fraction of the CFC-11 in appliances has already entered waste management systems. Building insulation has a much longer lifetime (estimated to 30-80 years, Gamlen *et al.*, 1986) and most of the fluorinated gases in building insulation have not yet reached the end of their useful life (TFFEoL, 2005). Daniel *et al.* (2007) discuss the uncertainties and some possible temporal trends for depletion of CFC-11 and CFC-12 banks.

Consumer products containing fluorinated gases are managed in different ways. After 2001, landfill disposal of appliances was prohibited in the EU (IPCC, 2005), resulting in appliance-recycling facilities. A similar system was established in Japan in 2001 (IPCC, 2005). For other developed countries, appliance foams are often buried in landfills, either directly or following shredding and metals recycling. For rigid foams, shredding results in an instantaneous release with the fraction released related to the final particle size (Kjeldsen and Scheutz, 2003). A recent study estimating CFC-11 releases after shredding at three American facilities showed that 60–90% of the CFC remains and is slowly released following landfill disposal (Scheutz *et al.*, 2005a). In the US and other countries, appliances typically undergo mechanical recovery of ferrous metals with landfill disposal of residuals. A study has shown that 8–40% of the CFC-11 is lost during segregation (Scheutz *et al.*, 2002; Fredenslund *et al.*, 2005). Then, during landfilling, the compactors shred residual foam materials and further enhance instantaneous gaseous releases.

In the anaerobic landfill environment, some fluorinated gases may be biodegraded because CFCs and, to some extent, HCFCs can undergo dechlorination (Scheutz *et al.*, 2003b). Potentially this may result in the production of more toxic intermediate degradation products (e.g., for CFC-11, the degradation products can be HCFC-21 and HCFC-31). However, recent laboratory experiments have indicated rapid CFC-11 degradation with only minor production of toxic intermediates (Scheutz *et al.*, 2005b). HFCs have not been shown to undergo either anaerobic or aerobic degradation. Thus, landfill attenuation processes may decrease emissions of some fluorinated gases, but not of others. However, data are entirely lacking for PFCs, and field studies are needed to verify that CFCs and HCFCs are being attenuated *in situ* in order to guide future policy decisions.

10.4.9 Air quality issues: NMVOCs and combustion emissions

Landfill gas contains trace concentrations of aromatic, chlorinated and fluorinated hydrocarbons, reduced sulphur gases and other species. High hydrocarbon destruction efficiencies are typically achieved in enclosed flares (>99%), which are recommended over lower-efficiency open flares. Hydrogen sulphide is mainly a problem at landfills which co-dispose large quantities of construction and demolition debris containing gypsum board. Emissions of NO_x can sometimes be a problem for permitting landfill gas engines in strict air quality regions.

At landfill sites, recent field studies have indicated that NMVOC fluxes through final cover materials are very small with both positive and negative fluxes ranging from approximately 10^{-8} to 10^{-4} g/m/d for individual species (Scheutz et al., 2003a; Bogner et al., 2003; Barlaz et al., 2004). In general, the emitted compounds consist of species recalcitrant to aerobic degradation (especially higher chlorinated compounds), while low to negative emissions (uptake from the atmosphere) are observed for species which are readily degradable in aerobic cover soils, such as the aromatics and vinyl chloride (Scheutz et al., 2003a).

Uncontrolled emissions resulting from waste incineration are not permitted in developed countries, and incinerators are equipped with advanced emission controls. Modern incinerators must meet stringent emission-control standards in Japan, the EU, the US and other developed countries (EIPPC Bureau, 2006). For reducing incinerator emissions of volatile heavy metals and dioxins/dibenzofurans, the removal of batteries, other electronic waste and polyvinyl chloride (PVC) plastics is recommended prior to combustion (EIPPC Bureau, 2006).

10.5 Policies and measures: waste management and climate

GHG emissions from waste are directly affected by numerous policy and regulatory strategies that encourage energy recovery from waste, restrict choices for ultimate waste disposal, promote waste recycling and re-use, and encourage waste minimization. In many developed countries, especially Japan and the EU, waste-management policies are closely related to and integrated with climate policies. Although policy instruments within the waste sector consist mainly of regulations, there are also economic measures to promote recycling, waste minimization and selected waste management technologies. In industrialized countries, waste minimization and recycling are encouraged through both policy and regulatory drivers. In developing countries, major policies are aimed at restricting the uncontrolled dumping of waste. Table 10.6 provides an overview of policies and measures, some of which are discussed below.

10.5.1 Reducing landfill CH_4 emissions

There are two major strategies to reduce landfill CH_4 emissions: implementation of standards that require or encourage landfill CH_4 recovery and a reduction in the quantity of biodegradable waste that is landfilled. In the US, landfill CH_4 emissions are regulated indirectly under the Clean Air Act (CAA) Amendments/New Source Performance Standards (NSPS) by applying a landfill-gas generation model, either measured or default mixing ratios for total non-methane organic compounds (NMOCs), and restricting the emissions of NMOCs. Larger quantities of landfill CH_4 are also being annually recovered to both comply with air-quality regulations and provide energy, assisted by national tax credits and local renewable-energy/green-power initiatives. As discussed above, the EU landfill directive (1999/31/EC) requires a phased reduction in landfilled biodegradable waste to 50% of 1995 levels by 2009 and 35% by 2016, as well as the collection and flaring of landfill gas at existing sites (Commission of the European Community, 2001). However, increases in the availability of landfill alternatives (recycling, composting, incineration, anaerobic digestion and MBT) are required to achieve these regulatory goals (Price, 2001).

Landfill CH_4 recovery has also been encouraged by economic and regulatory incentives. In the UK, for example, the Non Fossil Fuel Obligation, requiring a portion of electrical generation capacity from non-fossil sources, provided a major incentive for landfill gas-to-electricity projects during the 1980s and 1990s. It has now been replaced by the Renewables Obligation. In the US, as mentioned above, the implementation of CAA regulations in the early 1990s provided a regulatory driver for gas recovery at large landfills; in parallel, the US EPA Landfill Methane Outreach Program provides technical support, tools and resources to facilitate landfill gas utilization projects in the US and abroad. Also, periodic tax credits in the US have provided an economic incentive for landfill gas utilization – for example, almost 50 of the 400+ commercial projects in the US started up in 1998, just before the expiration of federal tax credits. A small US tax credit has again become available for landfill gas and other renewable energy sources; in addition, some states also provide economic incentives through tax structures or renewable energy credits and bonds. Other drivers include state requirements that a portion of electrical energy be derived from renewables, green-power programmes (which allow consumers to select renewable providers), regional programmes to reduce GHG emissions (the RGGI/ Regional GHG Initiative in the northeastern states; a state programme in California) and voluntary markets (such as the Chicago Climate Exchange with binding commitments by members to reduce GHG emissions).

In non-Annex I countries, it is anticipated that landfill CH_4 recovery will increase significantly in the developing countries of Asia, South America and Africa during the next two decades as controlled landfilling is phased in as a major waste-disposal

Table 10.6: *Examples of policies and measures for the waste management sector.*

Policies and measures	Activity affected	GHG affected	Type of instruments
Reducing landfill CH$_4$ emissions			
Standards for landfill performance to reduce landfill CH$_4$ emissions by capture and combustion of landfill gas with or without energy recovery	Management of landfill sites	CH$_4$	Regulation Economic Incentive
Reduction in biodegradable waste that is landfilled.	Disposal of biodegradable waste	CH$_4$	Regulation
Promoting incineration and other thermal processes for waste-to-energy			
Subsidies for construction of incinerator combined with standards for energy efficiency	Performance standards for incinerators	CO$_2$ CH$_4$	Regulation
Tax exemption for electricity generated by waste incineration with energy recovery	Energy recovery from incineration of waste	CO$_2$ CH$_4$	Economic incentive
Promoting waste minimization, re-use and recovery			
Extended Producer Responsibility (EPR)	Manufacture of products Recovery of used products Disposal of waste	CO$_2$ CH$_4$ Fluorinated gases	Regulation Voluntary
Unit pricing / Variable rate pricing / Pay-as-you-throw (PAYT)	Recovery of used products Disposal of waste	CO$_2$ CH$_4$	Economic incentive
Landfill tax	Recovery of used products Disposal of waste	CO$_2$ CH$_4$	Regulation
Separate collection and recovery of specific waste fractions	Recovery of used products Disposal of waste	CO$_2$ CH$_4$	Subsidy
Promotion of the use of recycled products	Manufacturing of products	CO$_2$ CH$_4$	Regulation Voluntary
Wastewater and sludge treatment			
Collection of CH$_4$ from wastewater treatment system	Management of wastewater treatment system	CH$_4$	Regulation Voluntary
Post-consumer management of fluorinated gases			
Substitutes for gases used commercially	Production of fluorinated gases	Fluorinated gases	Regulation Economic incentive Voluntary
Collection of fluorinated gases from end-of-life products	Management of end-of-life products	Fluorinated gases	Regulation Voluntary
JI and CDM in waste management sector			
JI and CDM	Landfill gas and biogas recovery	CO$_2$ CH$_4$	Kyoto mechanism

strategy. Where this occurs in parallel with deregulated electrical markets and more decentralized electrical generation, it can provide a strong driver for increased landfill CH$_4$ recovery with energy use. Significantly, both JI in the EIT countries and the recent availability of the Clean Development Mechanism (CDM) in developing countries are providing strong economic incentives for improved landfilling practices (to permit gas extraction) and landfill CH$_4$ recovery. Box 10.2 summarizes the important role of landfill CH$_4$ recovery within CDM and gives an example of a successful project in Brazil.

10.5.2 Incineration and other thermal processes for waste-to-energy

Thermal processes can efficiently exploit the energy value of post-consumer waste, but the high cost of incineration with

emission controls restricts its sustainable application in many developing countries. Subsidies for construction of incinerators have been implemented in several countries, usually combined with standards for energy efficiency (Austrian Federal Government, 2001; Government of Japan, 1997). Tax exemptions for electricity generated by waste incinerators (Government of the Netherlands, 2001) and for waste disposal with energy recovery (Government of Norway, 2002) have been adopted. In Sweden, it has been illegal to landfill pre-sorted combustible waste since 2002 (Swedish Environmental Protection Agency, 2005). Landfill taxes have also been implemented in a number of EU countries to elevate the cost of landfilling to encourage more costly alternatives (incineration, industrial co-combustion, MBT). In the UK, the landfill tax has also been used as a funding mechanism for environmental and community projects, as discussed by Morris *et al.* (2000) and Grigg and Read (2001).

10.5.3   Waste minimization, re-use and recycling

Widely implemented policies include Extended Producer Responsibility (EPR), unit pricing (or PAYT/Pay As You Throw) and landfill taxes. Waste reduction can also be promoted by recycling programmes, waste minimization and other measures (Miranda *et al.*, 1994; Fullerton and Kinnaman, 1996). The EPR regulations extend producer responsibility to the post-consumer period, thus providing a strong incentive to redesign products using fewer materials as well as those with increased recycling potential (OECD, 2001). Initially, EPR programmes were reported to be expensive (Hanisch, 2000), but the EPR concept is very broad: a number of successful schemes have been implemented in various countries for diverse waste fractions such as packaging waste, old vehicles and electronic equipment. EPR programmes range in complexity and cost, but waste reductions have been reported in many countries and regions. In Germany, the 1994 Closed Substance Cycle and Waste Management Act, other laws and voluntary agreements have restructured waste management over the past 15 years (Giegrich and Vogt, 2005).

Unit pricing has been widely adopted to decrease landfilled waste and increase recycling (Miranda *et al.*, 1996). Some municipalities have reported a secondary increase in waste generation after an initial decrease following implementation of unit pricing, but the ten-year sustainability of these programmes has been demonstrated (Yamakawa and Ueta, 2002).

Separate and efficient collection of recyclable materials is needed with both PAYT and landfill tax systems. For kerbside programmes, the percentage recycled is related to the efficiency of kerbside collection and the duration of the programme (Jenkins *et al.*, 2003). Other policies and measures include local subsidies and educational programmes for collection of recyclables, domestic composting of biodegradable waste and procurement of recycled products (green procurement). In the US, for example, 21 states have requirements for separate collection of garden (green) waste, which is diverted to composting or used as an alternative daily cover on landfills.

10.5.4   Policies and measures on fluorinated gases

The HFCs regulated under the Kyoto Protocol substitute for the ODS. A number of countries have adopted collection systems for products still in use based on voluntary agreements (Austrian Federal Government, 2001) or EPR regulations for appliances (Government of Japan, 2002). Both the EU and Japan have successfully prohibited landfill disposal of appliances containing ODS foams after 2001 (TFFEoL, 2005).

10.5.5   Clean Development Mechanism/Joint Implementation

Because lack of financing is a major impediment to improved waste and wastewater management in EIT and developing countries, the JI and CDM have been useful mechanisms for obtaining external investment from industrialized countries. As described in Section 10.3, open dumping and burning are common waste disposal methods in many developing countries, where GHG emissions occur concurrently with odours, public health and safety problems, and environmental degradation. In addition, developing countries often do not have existing infrastructure for collection and treatment of municipal wastewaters. Thus, the benefits from JI and CDM are twofold: improving waste management practices and reducing GHG emissions. To date, CDM has assisted many landfill gas recovery projects (see Box 10.2) while improving landfill operations, because adequate cover materials are required to minimize air intrusion during gas extraction (to prevent internal landfill fires). The validation of CDM projects requires attention to baselines, additionality and other criteria contained in approved methodologies (Hiramatsu *et al.*, 2003); however, for landfill gas CDM projects, certified emission reductions (CERs, with units of tCO2-eq) are determined directly from quantification of the CH_4 captured and combusted. In many countries, the anaerobic digestion of wastewaters and sludges could produce a useful biogas for heating use or onsite electrical generation (Government of Japan, 1997; Government of Republic of Poland, 2001); such projects could also be suitable for JI and CDM. In the future, waste sector projects involving municipal wastewater treatment, carbon storage in landfills or compost, and avoided GHG emissions due to recycling, composting, or incineration could potentially be implemented pending the development of approved methodologies.

10.5.6   Non-climate policies affecting GHG emissions from waste

The EIT and many developing countries have implemented market-oriented structural reforms that affect GHG emissions. As GDP is a key parameter to predict waste generation (Daskalopoulos *et al.*, 1998), economic growth affects the consumption of materials, the production of waste, and hence GHG emissions from the waste sector. Decoupling waste generation from economic and demographic drivers, or dematerialization, is often discussed in the context of sustainable development. Many developed countries have reported recent decoupling trends (OECD, 2002a), but the literature shows no absolute decline in material consumption in developed countries (Bringezu *et al.*, 2004). In other words, solid waste generation does not support an environmental Kuznets curve (Dinda, 2004), because environmental problems related to waste are not fully internalized. In Asia, Japan and China are both encouraging 'circular economy' or 'sound material-cycle society' as a new development strategy, whose core concept is the circular (closed) flow of materials and the use of raw materials and energy through multiple phases (Japan Ministry of the Environment, 2003; Yuan *et al.*, 2006). This approach is expected to achieve efficient economic growth while discharging fewer pollutants.

609

Box 10.2: Significant role of landfill gas recovery for CDM projects: overview and example

As of late October 2006, 376 CDM projects had achieved registration. These include 33 landfill gas projects, which collectively total 12% of the annual average CERs (12 million of approximately 91 million CERs per year). (http://cdm.unfccc.int/Projects/registered.html). The pie chart shows the distribution of landfill gas CERs by country. Most of these projects are located in Latin America and the Carribean region (72% of landfill gas CERs), dominated by Brazil (nine projects; 48% of CERs). Some projects are flaring gas, while others are using the gas for on-site electrical generation or direct-use projects (including leachate evaporation). Although eventual landfill gas utilization is desirable, an initial flaring project under CDM can simplify the CDM process (fewer participants, lower capital cost) and permit definition of composite gas quantity and quality prior to capital investment in engines or other utilization hardware.

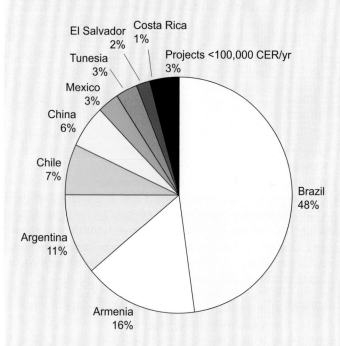

Figure 10.9: *Distribution of landfill gas CDM projects based on average annual CERs for registered projects late October 2006 (unfccc.org). Includes 10.9 Mt CERs for landfill CH$_4$ of 91 Mt total CERs. Projects <100,000 CERs/yr are located in Israel, Bolivia, Bangladesh and Malaysia*

An example of a successful Brazilian project is the ONYX SASA Landfill Gas Recovery Project at the VES landfill, Trémembé, Sao Paulo State (Figure 10.10). The recovered landfill gas is flared and used to evaporate leachate. As of December, 2005, approximately 93,600 CERs had been delivered (Veolia Environmental Services, 2005).

Figure 10.10: *ONYX SASA Landfill Gas Recovery Project .VES landfill, Trémembé, Sao Paulo State*

In 2002, the Johannesburg Summit adopted the Millennium Development Goals to reduce the number of people without access to sanitation services by 50% via the financial, technical and capacity-building expertise of the international community. If achieved, the Johannesburg Summit goals would significantly reduce GHG emissions from wastewater.

10.5.7 Co-benefits of GHG mitigation policies

Most policies and measures in the waste sector address broad environmental objectives, such as preventing pollution, mitigating odours, preserving open space and maintaining air, soil and water quality (Burnley, 2001). Thus, reductions in GHG emissions frequently occur as a co-benefit of regulations and policies not undertaken primarily for the purpose of climate-change mitigation (Austrian Federal Government, 2001). For

example, the EU Landfill Directive is primarily concerned with preventing pollution of water, soil and air (Burnley, 2001).

10.6 Long-term considerations and sustainable development

10.6.1 Municipal solid waste management

GHG emissions from waste can be effectively mitigated by current technologies. Many existing technologies are also cost effective; for example, landfill gas recovery for energy use can be profitable in many developed countries. However, in developing countries, a major barrier to the diffusion of technologies is lack of capital – thus the CDM, which is increasingly being implemented for landfill gas recovery projects, provides a major incentive for both improved waste management and GHG emission reductions. For the long term, more profound changes in waste management strategy are expected in both developed and developing countries, including more emphasis on waste minimization, recycling, re-use and energy recovery. Huhtala (1997) studied optimal recycling rates for municipal solid waste using a model that included recycling costs and consumer preferences; results suggested that a recycling rate of 50% was achievable, economically justified and environmentally preferable. This rate has already been achieved in many countries for the more valuable waste fractions such as metals and paper (OECD, 2002b).

Decisions for alternative waste management strategies are often made locally; however, there are also regional drivers based on national regulatory and policy decisions. Selected waste management options also determine GHG mitigation options. For the many countries which continue to rely on landfilling, increased utilization of landfill CH_4 can provide a cost-effective mitigation strategy. The combination of gas utilization for energy with biocover landfill cover designs to increase CH_4 oxidation can largely mitigate site-specific CH_4 emissions (Huber-Humer, 2004; Barlaz et al., 2004). These technologies are simple ('low technology') and can be readily deployed at any site. Moreover, R&D to improve gas-collection efficiency, design biogas engines and turbines with higher efficiency, and develop more cost-effective gas purification technologies are underway. These improvements will be largely incremental but will increase options, decrease costs, and remove existing barriers for expanded applications of these technologies.

Advances in waste-to-energy have benefited from general advances in biomass combustion; thus the more advanced technologies such as fluidized bed combustion with emissions control can provide significant future mitigation potential for the waste sector. When the fossil fuel offset is also taken into account, the positive impact on GHG reduction can be even greater (e.g., Lohiniva et al. 2002; Pipatti and Savolainen 1996; Consonni et al. 2005). High cost, however, is a major barrier to the increased implementation of waste-to-energy. Incineration has often proven to be unsustainable in developing countries – thus thermal processes are expected to be primarily (but not exclusively) deployed in developed countries. Advanced combustion technologies are expected to become more competitive as energy prices increase and renewable energy sources gain larger market share.

Anaerobic digestion as part of MBT, or as a stand-alone process for either wastewater or selected wastes (high moisture), is expected to continue in the future as part of the mix of mature waste management technologies. In general, anaerobic digestion technologies incur lower capital costs than incineration; however, in terms of national GHG mitigation potential and energy offsets, their potential is more limited than landfill CH_4 recovery and incineration. When compared to composting, anaerobic digestion has advantages with respect to energy benefits (biogas), reduced process times and reduced volume of residuals; however, as applied in developed countries, it typically incurs higher capital costs. Projects where mixed municipal waste was anaerobically digested (e.g., the Valorga project) have been largely discontinued in favour of projects using specific biodegradable fractions such as food waste. In some developing countries such as China and India, small-scale digestion of biowaste streams with CH_4 recovery and use has been successfully deployed for decades as an inexpensive local waste-to-energy strategy – many other countries could also benefit from similar small-scale projects. For both as a primary wastewater treatment process or for secondary treatment of sludges from aerobic processes, anaerobic digestion under higher temperature using thermophilic regimes or two-stage processes can provide shorter retention times with higher rates of biogas production.

Regarding the future of up-front recycling and separation technologies, it is expected that wider implementation of incrementally-improving technologies will provide more rigorous process control for recycled waste streams transported to secondary markets or secondary processes, including paper and aluminium recycling, composting and incineration. If analysed within an LCA perspective, waste can be considered a resource, and these improvements should result in more advantageous material and energy balances for both individual components and urban waste streams as a whole. For developing countries, provided sufficient measures are in place to protect workers and the local environment, more labour-intensive recycling practices can be introduced and sustained to conserve materials, gain energy benefits and reduce GHG emissions. In general, existing studies on the mitigation potential for recycling yield variable results because of the differing assumptions and methodologies applied; however, recent studies (i.e., Myllymaa et al., 2005) are beginning to quantitatively examine the environmental benefits of alternative waste strategies, including recycling.

10.6.2 Wastewater management

Although current GHG emissions from wastewater are lower than emissions from waste, it is recognized that there are substantial emissions which are not quantified by current estimates, especially from septic tanks, latrines and uncontrolled discharges in developing countries. Nevertheless, the quantity of wastewater collected and treated is increasing in many countries in order to maintain and improve potable water quality, as well for other public health and environmental protection benefits. Concurrently, GHG emissions from wastewater will decrease relative to future increases in wastewater collection and treatment.

For developing countries, it is a significant challenge to develop and implement innovative, low-cost but effective and sustainable measures to achieve a basic level of improved sanitation (Moe and Reingans, 2006). Historically, sanitation

Table 10.7: *Summary of adaptation, mitigation and sustainable development issues for the waste sector.*

Technologies and practices	Vulnerability to climate change	Adaptation implications & strategies to minimize emissions	Sustainable development dimensions			Comments
			Social	Economic	Environmental	
Recycling, reuse & waste minimization	Indirect low vulnerability or no vulnerability	Minimal implications	Usually positive Negative for waste scavenging without public health or safety controls	Positive Job creation	Positive Negative for waste scavenging from open dumpsites with air and water pollution	Indirect benefits for reducing GHG emissions from waste Reduces use of energy and raw materials. Requires implementation of health and safety provisions for workers
Controlled landfilling with landfill gas recovery and utilization	Indirect low vulnerability or positive effects: Higher temperatures increase rates of microbial methane oxidation rates in cover materials	Minimal implications May be regulatory mandates or economic incentives Replaces fossil fuels for process heat or electrical generation	Positive Odour reduction (non-CH_4 gases)	Positive Job creation Energy recovery potential	Positive Negative for improperly managed sites with air and water pollution	Primary control on landfill CH_4 emissions with >1200 commercial projects Important local source of renewable energy: replaces fossil fuels Landfill gas projects comprise 12% of annual registered CERs under CDM[a] Oxidation of CH_4 and NMVOCs in cover soils is a smaller secondary control on emissions
Controlled landfilling without landfill gas recovery	Indirect low vulnerability or positive effects: Higher temperatures increase rates of microbial methane oxidation rates in cover materials	Minimal implications Gas monitoring and control still required	Positive Odour reduction (non-CH_4 gases)	Positive Job creation	Positive Negative for improperly managed sites with air and water pollution	Use of cover soils and oxidation in cover soils reduce rate of CH_4 and NMVOC emissions
Optimizing microbial methane oxidation in landfill cover soils ('biocovers')	Indirect low vulnerability or positive effects: Increased rates at higher temperatures	Minimal implications or positive effects	Positive Odour reduction (non-CH_4 gases)	Positive Job creation	Positive Negative for improperly designed or managed biocovers with GHG emissions and NMVOC emissions	Important secondary control on landfill CH_4 emissions and emissions of NMVOCs Utilizes other secondary materials (compost, composted sludges) Low-cost low-technology strategy for developing countries
Uncontrolled disposal (open dumping & burning)	Highly vulnerable Detrimental effects: warmer temp. promote pathogen growth and disease vectors	Exacerbates adaptation problems Recommend implementation of more controlled disposal and recycling practices	Negative	Negative	Negative	Consider alternative lower-cost medium technology solutions (e.g., landfill with controlled waste placement, compaction, and daily cover materials)
Thermal processes including incineration, industrial co-combustion, and more advanced processes for waste-to-energy (e.g., fluidized bed technology with advanced flue gas cleaning)	Low vulnerability	Minimal implications Requires source control and emission controls to prevent emissions of heavy metals, acid gases, dioxins and other air toxics	Positive Odour reduction (non-CH_4 gases)	Positive Job creation Energy recovery potential	Positive Negative for improperly designed or managed facilities without air pollution controls	Reduces GHG emissions relative to landfilling Costly, but can provide significant mitigation potential for the waste sector, especially in the short term Replaces fossil fuels
Aerobic biological treatment (composting) Also a component of mechanical biological treatment (MBT)	Indirect low vulnerability or positive effects: Higher temperatures increase rates of biological processes (Q_{10})	Minimal implications or positive effects Produces CO_2 (biomass) and compost Reduces volume, stabilizes organic C, and destroys pathogens	Positive Odour reduction (non-CH_4 gases)	Positive Job creation Use of compost products	Positive Negative for improperly designed or managed facilities with odours, air and water pollution	Reduces GHG emissions Can produce useful secondary materials (compost) provided there is quality control on material inputs and operations Can emit N_2O and CH_4 under reduced aeration or anaerobic conditions
Anaerobic biological treatment (anaerobic digestion) Also a component of mechanical-biological treatment (MBT)	Indirect low vulnerability or positive effects: Higher temperatures increase rates of biological processes	Minimal implications Produces CH_4, CO_2, and biosolids under highly controlled conditions Biosolids require management	Positive Odour reduction (non-CH_4 gases)	Positive Job creation Energy recovery potential Use of residual biosolids	Positive Negative for improperly designed or managed facilities with, odours, air and water pollution	Reduces GHG emissions CH_4 in biogas can replace fossil fuels for process heat or electrical generation Can emit minor quantities of CH_4 during start-ups, shutdowns and malfunctions
Wastewater control and treatment (aerobic or anaerobic)	Highly vulnerable Detrimental effects in absence of wastewater control and treatment: Warmer temperatures promote pathogen growth and poor public health	Large adaptation implications High potential for reducing uncontrolled GHG emissions Residuals (biosolids) from aerobic treatment may be anaerobically digested	Positive Odour reduction (non-CH_4 gases)	Positive Job creation Energy recovery potential from anaerobic processes Use of sludges and other residual biosolids	Positive Negative for improperly designed or managed facilities with odours, air and water pollution and GHG emissions	Wide range of available technologies to collect, treat, recycle and re-use wastewater Wide range of costs CH_4 from anaerobic processes replaces fossil fuels for process heat or electrical generation Need to design and operate to minimize N_2O and CH_4 emissions during transport and treatment

a *http://cdm.unfccc.int/Projects/registerd.html*, October 2006 in developed countries has included costly centralized sewerage and wastewater treatment plants, which do not offer appropriate sustainable solutions for either rural areas in developing countries with low population density or unplanned, rapidly growing, peri-urban areas with high population density (Montgomery and Elimelech, 2007). It has been demonstrated that a combination of low-cost technology with concentrated efforts for community acceptance, participation and management can successfully expand sanitation coverage; for example, in India more than one million pit latrines have been built and maintained since 1970 (Lenton *et al.*, 2005). The combination of household water treatment and 'point-of-use' low-technology improved sanitation in the form of pit latrines or septic systems has been shown to lower diarrhoeal diseases by >30% (Fewtrell *et al.*, 2005).

Wastewater is also a secondary water resource in countries with water shortages. Future trends in wastewater technology include buildings where black water and grey water are separated, recycling the former for fertilizer and the latter for toilets. In addition, low-water use toilets (3–5 L) and ecological sanitation approaches (including ecological toilets), where nutrients are safely recycled into productive agriculture and the environment, are being used in Mexico, Zimbabwe, China, and Sweden (Esrey *et al.*, 2003). These could also be applied in many developing and developed countries, especially where there are water shortages, irregular water supplies, or where additional measures for conservation of water resources are needed. All of these measures also encourage smaller wastewater treatment plants with reduced nutrient loads and proportionally lower GHG emissions.

10.6.3 Adaptation, mitigation and sustainable development in the waste sector

In addition to providing mitigation of GHG emissions, improved public health, and environmental benefits, solid waste and wastewater technologies confer significant co-benefits for adaptation, mitigation and sustainable development (Table 10.7; see also Section 12.3.4). In developing countries, improved waste and wastewater management using low- or medium-technology strategies are recommended to provide significant GHG mitigation and public health benefits at lower cost. Some of these strategies include small-scale wastewater management such as septic tanks and recycling of grey water, construction of medium-technology landfills with controlled waste placement and use of daily cover (perhaps including a final biocover to optimize CH_4 oxidation), and controlled composting of organic waste.

The major impediment in developing countries is the lack of capital, which jeopardizes improvements in waste and wastewater management. Developing countries may also lack access to advanced technologies. However, technologies must be sustainable in the long term, and there are many examples of advanced, but unsustainable, technologies for

waste management that have been implemented in developing countries. Therefore, the selection of truly sustainable waste and wastewater strategies is very important for both the mitigation of GHG emissions and for improved urban infrastructure.

REFERENCES

Ackerman, F., 2000: Waste Management and Climate Change. *Local Environment,* **5**(2), pp. 223-229.

Austrian Federal Government, 2001: Third National Climate Report of the Austrian Federal Government. Vienna, Austria.

Barlaz, M., 1998: Carbon storage during biodegradation of municipal solid waste components in laboratory-scale landfills. *Global Biogeochemical Cycles,* **12**(2), pp. 373-380.

Barlaz, M., R. Green, J. Chanton, R.D. Goldsmith, and G. Hater, 2004: Evaluation of a biologically-active cover for mitigation of landfill gas emissions. *Environmental Science and Technology,* **38**(18), pp. 4891-4899.

Bates, J. and A. Haworth, 2001: Economic evaluation of emission reductions of methane in the waste sector in the EU: Bottom-up analysis. Final Report to DG Environment, European Commission by Ecofys Energy and Environment, by AEA Technology Environment and National Technical University of Athens as part of Economic Evaluation of Sectoral Emission Reduction Objective for Climate Change, 73 pp.

Beck-Friis, B.G. 2001: *Emissions of ammonia, N_2O, and CH_4 during composting of organic household waste.* PhD Thesis, Swedish University of Agricultural Sciences, Uppsala, 331 pp.

Berge, N., D. Reinhart, and T. Townsend, 2005: A review of the fate of nitrogen in bioreactor landfills. *Critical Reviews in Environmental Science and Technology,* **35**(4), pp. 365-399.

Bernache-Perez, G., S. Sánchez-Colón, A.M. Garmendia, A. Dávila-Villarreal, and M.E. Sánchez-Salazar, 2001: Solid waste characterization study in Guadalajara Metropolitan Zone, Mexico. *Waste Management & Research,* **19**, pp. 413-424.

Bingemer, H.G. and P.J. Crutzen, 1987: The production of CH_4 from solid wastes. *Journal of Geophysical Research,* **92**(D2), pp. 2182-2187.

Binner, E., 2002: The impact of mechanical-biological pretreatment on the landfill behaviour of solid wastes. Proceedings of the workshop on Biowaste, Brussels, April 8-10, 2002. pp. 16.

Bockreis, B. and I. Steinberg, 2005: Influence of mechanical-biological waste pre-treatment methods on gas formation in landfills. *Waste Management,* **25**, pp. 337-343.

Bogner, J., 1992: Anaerobic burial of refuse in landfills: increased atmospheric methane and implications for increased carbon storage. *Ecological Bulletin,* **42**, pp. 98-108.

Bogner, J. and E. Matthews, 2003: Global methane emissions from landfills: New methodology and annual estimates 1980-1996. *Global Biogeochemical Cycles,* **17**, pp. 34-1 to 34-18.

Bogner, J., M. Meadows, and P. Czepiel, 1997a: Fluxes of methane between landfills and the atmosphere: natural and engineered controls. *Soil Use and Management,* **13**, pp. 268-277.

Bogner, J., C. Scheutz, J. Chanton, D. Blake, M. Morcet, C. Aran, and P. Kjeldsen, 2003: Field measurement of non-methane organic compound emissions from landfill cover soils. Proceedings of the Sardinia '03, International Solid and Hazardous Waste Symposium, published by CISA, University of Cagliari, Sardinia.

Bogner, J. and K. Spokas, 1993: Landfill CH_4: rates, fates, and role in global carbon cycle. *Chemosphere,* **26**(1-4), pp. 366-386.

Bogner, J., K. Spokas, and E. Burton, 1997b: Kinetics of methane oxidation in landfill cover materials: major controls, a whole-landfill oxidation experiment, and modeling of net methane emissions. *Environmental Science and Technology,* **31**, pp. 2504-2614.

Bogner, J., K. Spokas, and E. Burton, 1999a: Temporal variations in greenhouse gas emissions at a midlatitude landfill. *Journal of Environmental Quality*, **28**, pp. 278-288.

Bogner, J., K. Spokas, E. Burton, R. Sweeney, and V. Corona, 1995: Landfills as atmospheric methane sources and sinks. *Chemosphere*, **31**(9), pp. 4119-4130.

Bogner, J., K. Spokas, J. Chanton, D. Powelson, and T. Abichou, 2005: Modeling landfill methane emissions from biocovers: a combined theoretical-empirical approach. Proceedings of the Sardinia '05, International Solid and Hazardous Waste Symposium, published by CISA, University of Cagliari, Sardinia.

Borjesson, G., 1996: *Methane oxidation in landfill cover soils.* Doctoral Thesis, Dept. of Microbiology, Swedish University of Agricultural Sciences, Uppsala, Sweden.

Borjesson, G. and B. Svensson, 1997a: Nitrous oxide release from covering soil layers of landfills in Sweden. *Tellus*, **49B**, pp. 357-363.

Borjesson, G. and B. Svensson, 1997b: Seasonal and diurnal methane emissions from a landfill and their regulation by methane oxidation. *Waste Management and Research*, **15**(1), pp. 33-54.

Bramryd, T., 1997: Landfilling in the perspective of the global CO_2 balance. Proceedings of the Sardinia '97, International Landfill Symposium, October 1997, published by CISA, University of Cagliari, Sardinia.

Bringezu, S., H. Schutz, S. Steger, and J. Baudisch, 2004: International comparison of resource use and its relation to economic growth: the development of total material requirement, direct material inputs and hidden flows and the structure of TMR. *Ecological Economics*, **51**, pp. 97-124.

Burnley, S., 2001: The impact of the European landfill directive on waste management in the United Kingdom. *Resources, Conservation and Recycling*, **32**, pp. 349-358.

CALEB, 2000: *Development of a global emission function for blowing agents used in closed cell foam.* Final report prepared for AFEAS. Caleb Management Services, Bristol, United Kingdom.

CalRecovery, Inc., 2004: Waste analysis and characterization study. Asian Development Bank, Report TA - 3848-PHI.

CalRecovery, Inc., 2005: Solid waste management. Report to Division of Technology, Industry, and Economics, International Environmental Technology Centre, UNEP, Japan, Vols. 1 and 2. <www.unep.or.jp/Ietc/Publications/spc/Solid_Waste_Management/index.asp>, accessed 13/08/07.

CEPI, 2003: Summary of the study on non-collectable and non-recyclable paper products. Confederation of European Paper Industries (CEPI), Brussels, Belgium.

Chanton, J. and K. Liptay, 2000: Seasonal variation in methane oxidation in a landfill cover soil as determined by an *in situ* stable isotope technique. *Global Biogeochemical Cycles*, **14**, pp. 51-60.

Chen, L., M. Nanny, D. Knappe, T. Wagner, and N. Ratasuk, 2004: Chemical characterization and sorption capacity measurements of degraded newsprint from a landfill. *Environmental Science and Technology*, **38**, pp. 3542-3550.

Christensen, T., 1989: Environmental aspects of sanitary landfilling. In *Sanitary Landfilling*, T. Christensen, *et al.,* (ed.), Academic Press, San Diego, pp. 19-29.

Cointreau, S., 2001: Declaration of principles for sustainable and integrated solid waste management, World Bank, Washington, D.C., 4 pp.

Cointreau-Levine, S., 1994: Private sector participation in municipal solid waste services in developing countries, Vol.1, The Formal Sector. *Urban Management and the Environment*, **13**, UNDP/UNCHS (United Nations Centre for Human Settlements), World Bank, Washington, D.C., 52 pp.

Coleman, D.D., 1979: The origin of drift-gas deposits as determined by radiocarbon dating of methane. In *Radiocarbon Dating,* R. Berger and H.E. Seuss (eds), University of California Press, Berkeley, pp. 365-387.

Commission of the European Community, 2001: Third Communication from the European Community under the UN Framework Convention on Climate Change, Brussels.

Consonni, S., M. Giugliano, and M. Grosso, 2005: Alternative strategies for energy recovery from municipal solid waste. Part B: emission and cost estimates. *Waste Management*, **25**, pp. 137-148.

Czepiel, P., B. Mosher, R. Harriss, J.H. Shorter, J.B. McManus, C.E. Kolb, E. Allwine, and B. Lamb, 1996: Landfill methane emissions measured by enclosure and atmospheric tracer methods. *Journal of Geophysical Research*, **D101, pp.** 16711-16719.

Daniel, J., G.J.M. Velders, S. Solomon, M. McFarland, and S.A. Montzka, 2007: Present and future sources and emissions of halocarbons: Toward new constraints. *Journal of Geophysical Research*, **112**(D02301), pp. 1-11.

Daskalopoulos, E., O. Badr, and S.D. Probert, 1998: Municipal solid waste: a prediction methodology for the generation rate and composition in the European Union countries and the United States of America, *Resources. Conservation and Recycling*, **24**, pp. 155-166.

De Feber, M. and D. Gielen, 2000: Biomass for greenhouse gas emission reduction. Task 7, Energy Technology Characterisation. Netherlands Energy Research Foundation, Report nr. ECN-C--99-078.

De Jager, D. and K. Blok, 1996: Cost-effectiveness of emission-reduction measures for CH_4 in the Netherlands. *Energy Conservation Management*, **37**, pp. 1181-1186.

De Visscher, A., D. Thomas, P. Boeckx, and O. Van Cleemput, 1999: Methane oxidation in simulated landfill cover soil environments. *Environmental Science and Technology*, **33**(11), pp. 1854-1859.

De Visscher, A., I. De Pourcq, and J. Chanton, 2004: Isotope fractionation effects by diffusion and methane oxidation in landfill cover soils. *Journal of Geophysical Research*, **109** (D18), paper D18111.

Delhotal, C., F. de la Chesnaye, A. Gardiner, J. Bates, and A. Sankovski, 2006: Estimating potential reductions of methane and nitrous oxide emissions from waste, energy and industry. *The Energy Journal*. Special Issue: Multi-Greenhouse Gas Mitigation and Climate Policy, F.C. de la Chesnaye and J. Weyant (eds).

DESA/UN, 2005: Country, regional and global estimates on water and sanitation, statistics division. Dept. of Economic and Social Affairs/UN, MDG Report - WHO/UNICEF WES estimate, <http://www.unicef.org/wes/mdgreport> accessed 29/06/07.

Detzel, A., R. Vogt, H. Fehrenbach, F. Knappe, and U. Gromke, 2003: Anpassung der deutschen Methodik zur rechnerischen Emissionsermittlung und internationale Richtlinien: Teilberich Abfall/Abwasser. IFEU Institut - Öko-Institut e.V. 77 pp.

Deuber, O., M. Cames, S. Poetzsch, and J. Repenning, 2005: Analysis of greenhouse gas emissions of European countries with regard to the impact of policies and measures. Report by Öko-Institut to the German Umweltbundesamt, Berlin, 253 pp.

Diaz, L.F. and L.L. Eggerth, 2002: Waste Characterization Study. Ulaanbaatar, Mongolia, WHO/WPRO, Manila, Philippines.

Diaz, L.F., G. Savage, and C. Goluke, 2002: Solid waste composting. In *Handbook of Solid Waste Management*, G. Tchobanoglous and F. Kreith (eds.), McGraw-Hill, Chapter 12.

Diaz, L.F., A. Chiumenti, G. Savage, and L. Eggerth, 2006: Managing the organic fraction of municipal solid waste. *Biocycle*, **47**, pp. 50-53.

Dinda, S., 2004: Environmental Kuznets curve hypothesis: a survey. *Ecological Economics*, **49**, pp. 431-455.

Dornberg, V. and A. Faaij, 2006: Optimising waste treatment systems. Part B: Analyses and scenarios for The Netherlands. *Resources Conservation & Recycling*, **48**, pp. 227-248.

Du, W., Q. Gao, E. Zhang, 2006a: The emission status and composition analysis of municipal solid waste in China. *Journal of Research of Environmental Sciences*, **19**, pp. 85-90.

Du, W., Q. Gao, E. Zhang, 2006b: The treatment and trend analysis of municipal solid waste in China. *Journal of Research of Environmental Sciences,* **19**, pp. 115-120.

EEA, 2004: Greenhouse gas emission trends and projections in Europe 2004. European Environment Agency (EEA) Report no5/2004, Progress EU and its member states towards achieving their Kyoto targets, Luxembourg, ISSN 1725-9177, 40 pp.

EEA, 2005: European Environment Outlook. EU Report 4/2005, ISSN 1725-9177, Luxembourg, published by the European Environment Agency (EEA), Copenhagen, 92 pp.

EIPPC Bureau, 2006: Reference document on the best available techniques for waste incineration (BREF), integrated pollution prevention and control. European IPPC, Seville, Spain, 602 pp. <http://eippcb.jrc.es>

Esrey, S., I. Andersson, A. Hillers, and R. Sawyer, 2003 (2nd edition): Closing the loop - ecological sanitation for food security. Swedish International Development Agency (SIDA), *Publications on Water Resources, 18*, Mexico City, Mexico.

EU, 1999: Council Directive 1999/31/3C of 26 April 1999 on the landfill of waste. Official Journal of the European Communities 16.7.1999.

Eurostat, 2003: Waste generated and treated in Europe. Data 1990-2001. Office for Official Publications of the European Communities, Luxemburg, 140 pp.

Faaij, A., M. Hekkert., E. Worrell, and A. van Wijk, 1998: Optimization of the final waste treatment system in the Netherlands. *Resources, Conservation, and Recycling, 22*, pp. 47-82.

Fewtrell, L., R. Kaufmann, D. Kay, W. Enanoria, L. Haller, and J. Colford, 2005: Water, sanitation, and hygiene interventions to reduce diarrhoea in less developed countries: a systematic review and meta-analysis. *Lancet, 5*, pp. 42-52.

Flugsrud, K., B. Hoem, E. Kvingedal, and K. Rypdal, 2001: Estimating net emissions of CO_2 from harvested wood products. SFT report 1831/200, Norwegian Pollution Control Authority, Oslo, 47 pp. <http://www.sft.no/publikasjoner/luft/1831/ta1831.pdf>, accessed 29/06/07.

Fredenslund, A.M., C. Scheutz, and P. Kjeldsen, 2005: Disposal of refrigerators-freezers in the U.S.: State of Practice. Report by Environment and Resources. Technical University of Denmark (DTU). Lyngby.

Fullerton, D. and T.C. Kinnaman, 1996: Household responses to pricing garbage by the bag. *American Economic Review, 86* (4), pp. 971-984.

Galle, B., J. Samuelsson, B. Svensson, and G. Borjesson, 2001: Measurements of methane emissions from landfills using a time correlation tracer method based on FTIR absorption spectroscopy. *Environmental Science and Technology, 35*(1), pp. 21-25.

Gamlen, P.H., B.C. Lane, and P.M. Midgley, 1986: The production and release to the atmosphere of CCl_3F and CCl_2F_2 (Chlorofluorocarbons CFC-11 and CFC-12). *Atmospheric Environment, 20*(6), pp. 1077-1085.

Gebert, J. and A. Gröngröft, 2006: Passive landfill gas emission - influence of atmospheric pressure and implications for the operation of methane-oxidising biofilters. *Waste Management, 26*, pp. 245-251.

Giegrich, J. and R. Vogt, 2005: The contribution of waste management to sustainable development in Germany. Umweltbundesamt Report FKZ 203 92 309, Berlin.

Government of Japan, 1997: Japan's Second National Communication to the UNFCCC, Tokyo.

Government of Japan, 2002: Japan's Third National Communication to the UNFCCC, Tokyo.

Government of Norway, 2002: Norway's Third National Communication to the UNFCCC, Oslo.

Government of Republic of Poland, 2001: Third National Communication to the COP of the UNFCCC, Warsaw.

Government of the Netherlands, 2001: Third Netherlands National Communication on Climate Change Policies, The Hague.

Griffiths, A.J. and K.P. Williams, 2005: Thermal treatment options. *Waste Management World*, July-August 2005.

Grigg, S.V.L. and A.D. Read, 2001: A discussion on the various methods of application for landfill tax credit funding for environmental and community projects. *Resources, Conservation and Recycling, 32*, pp. 389-409.

Hanisch, C., 2000: Is extended producer responsibility effective? *Environmental Science and Technology, 34*(7), pp. 170A-175A.

Hashimoto, S. and Y. Moriguchi, 2004: Data book: material and carbon flow of harvested wood in Japan. CGER Report D034, National Institute for Environmental Studies, Japan, Tsukuba. 40 pp. <http://www-cger.nies.go.jp/> accessed 29/06/07.

Haury, E.W., 1976: The Hohokam: Desert Farmers and Craftsmen, University of Arizona Press, Tucson, Arizona.

Hellebrand, H.J., 1998: Emissions of N_2O and other trace gases during composting of grass and green waste. *Journal of Agricultural Engineering Research, 69*, pp. 365-375.

Hiramatsu, A., K. Hanaki, and T. Aramaki, 2003: Baseline options and greenhouse gas emission reduction of clean development mechanism project in urban solid waste management. *Mitigation and Adaptation Strategies for Global Change, 8*, pp. 293-310.

Ho, G., 2000: Proceedings of the Regional Workshop on sustainable wastewater and stormwater management in Latin America and the Caribbean. 27-31 March 2000, International Environmental Technology Centre, Report Series - Issue 10, United Nations, Osaka, Shiga, pp. 115-156.

Hobson, A., J. Frederickson, and N. Dise, 2005: CH_4 and N_2O from mechanically turned windrow and vermicomposting systems following in-vessel pre-treatment. *Waste Management, 25*, pp. 345-352.

Hoornweg, D., 1999: What a waste: solid waste management in Asia. Report of Urban Development Sector Unit, East Asia and Pacific Region, World Bank, Washington, D.C.

Hoornweg, D., L. Thomas, and L. Otten, 1999: Composting and its applicability in developing countries. Urban Waste Management Working Paper 8, Urban Development Division, World Bank, Washington, DC. 46 pp.

Hovde, D.C., A.C. Stanton, T.P. Meyers, and D.R. Matt, 1995: Methane emissions from a landfill measured by eddy correlation using a fast-response diode laser sensor. *Journal of Atmospheric Chemistry, 20*, pp. 141-162.

Huang, Q., Q. Wang, L. Dong, B. Xi, and B. Zhou, 2006: The current situation of solid waste management in China. *The Journal of Material Cycles and Waste Management, 8*, pp. 63-69.

Huber-Humer, M., 2004: *Abatement of landfill methane emissions by microbial oxidation in biocovers made of compost.* PhD Thesis, University of Natural Resources and Applied Life Sciences (BOKU), Vienna, 279 pp.

Huhtala, A., 1997: A post-consumer waste management model for determining optimal levels of recycling and landfilling. *Environmental and Resource Economics, 10*, pp. 301-314.

Idris, A., *et al.*, 2003: Overview of municipal solid waste landfill sites in Malaysia. Proceedings of the 2nd Workshop on Material Cycles and Waste Management in Asia, Tsukuba, Japan.

IPCC, 1996: Greenhouse gas inventory reference manual: Revised 1996 IPCC guidelines for national greenhouse gas inventories, Reference manual Vol. 3, J.T. Houghton, L.G. Meira Filho, B. Lim, K. Treanton, I. Mamaty, Y. Bonduki, D.J. Griggs and B.A. Callender [Eds]. IPCC/OECD/IEA. UK Meteorological Office, Bracknell, pp. 6.15-6.23.

IPCC, 2000: *Emissions Scenarios* [Nakicenovic, N., and R. Swart (eds.)]. Special Report of the Intergovernmental Panel on Climate Change (IPCC). Cambridge University Press, Cambridge, 570 pp.

IPCC, 2001a: *Climate Change 2001: Mitigation. Contribution of Working Group III to the third assessment report of the Intergovernmental Panel on Climate Change (IPCC).* Cambridge University Press, Cambridge.

IPCC, 2001b: *Climate Change 2001: The Scientific Basis. Contribution of Working Group I to the third assessment report of the Intergovernmental Panel on Climate Change (IPCC).* Cambridge University Press, Cambridge.

IPCC, 2005: *Safeguarding the Ozone Layer and the Global Climate System: issues related to Hydrofluorocarbons and Perfluorocarbons.* Special Report of the Intergovernmental Panel on Climate Change (IPCC). Cambridge University Press, Cambridge.

IPCC, 2006: *IPCC Guidelines for National Greenhouse Gas Inventories.* IPCC/IGES, Hayama, Japan. http://www.ipcc-nggip.iges.or.jp/public/2006gl/ppd.htm

Iskandar, L., 2001: The informal solid waste sector in Egypt: prospects for formalization, published by Community and Institutional Development, Cairo, Egypt, 65 pp.

Japan Ministry of the Environment, 2003: Fundamental plan for establishing a sound material-cycle society. <http://www.env.go.jp/en/recycle/smcs/f_plan.pdf> accessed 25/06/07.

Japan Ministry of the Environment, 2006: State of discharge and treatment of municipal solid waste in FY 2004.

Jenkins, R.R., S.A. Martinez, K. Palmer, and M.J. Podolsky, 2003: The determinants of household recycling: a material-specific analysis of recycling program features and unit pricing. *Journal of Environmental Economics and Management,* **45**(2), pp. 294-318.

Johannessen, L.M. and G. Boyer, 1999: Observations of solid waste landfills in developing countries: Africa, Asia, and Latin America. Report by Urban Development Division, Waste Management Anchor Team, World Bank, Washington, D.C.

Jouravlev, A., 2004: Los servicios de agua potable y saneamiento en el umbral del siglo XXI. Santiago de Chile, julio 2004, CEPAL - Serie Recursos Naturales e Infraestructura No 74.

Kaartinen, T., 2004: *Sustainable disposal of residual fractions of MSW to future landfills.* M.S. Thesis, Technical University of Helsinki, Espoo, Finland. In Finnish.

Kaseva, M.E., S.B. Mbuligwe, and G. Kassenga, 2002: Recycling inorganic domestic solid wastes: results from a pilot study in Dar es Salaam City, Tanzania. *Resources Conservation and Recycling,* **35,** pp. 243-257.

Khan, M.Z.A. and A.H. Abu-Ghararath, 1991: New approach for estimating energy content of municipal solid waste. *Journal of Environmental Engineering,* 117, 376-380.

Kightley, D., D. Nedwell, and M. Cooper, 1995: Capacity for methane oxidation in landfill cover soils measured in laboratory-scale microcosms. *Applied and Environmental Microbiology,* **61**, pp. 592-601.

Kjeldsen, P. and M.H. Jensen, 2001: Release of CFC-11 from disposal of polyurethane from waste. *Environmental Science and Technology,* **35**, pp. 3055-3063.

Kjeldsen, P. and C. Scheutz, 2003: Short and long term releases of fluorocarbons from disposal of polyurethane foam waste. *Environmental Science and Technology,* **37**, pp. 5071-5079.

Kuehle-Weidemeier, M. and H. Doedens, 2003: Landfilling and properties of mechanical-biological treated municipal waste. Proceedings of the Sardinia '03, International Solid and Hazardous Waste Symposium, October 2005, published by CISA, University of Cagliari, Sardinia.

Lenton, R., A.M. Wright, and K. Lewis, 2005: Health, dignity, and development: what will it take? UN Millennium Project Task Force on Water and Sanitation, published by Earthscan, London.

Lohiniva, E., K. Sipilä, T. Mäkinen, and L. Hietanen, 2002: Waste-to-energy and greenhouse gas emissions. Espoo, VTT - Tiedotteita - Research Notes 2139, 119 pp. In Finnish (published in English in 2006).

Mertins, L., C. Vinolas, A. Bargallo, G. Sommer, and J. Renau, 1999: Development and application of waste factors - An overview. Technical Report No. 37, European Environment Agency, Copenhagen.

Micales, J.A. and K.E. Skog, 1997: The decomposition of forest products in landfills. *International Biodeterioration and Biodegradation,* **39**(2-3), pp. 145-158.

Miranda, M.L., S.D. Bauer, and J.E. Aldy, 1996: Unit pricing programs for residential municipal solid waste: an assessment of the literatures. Report prepared for U.S. Environmental Protection Agency, Washington D.C.

Miranda, M.L., J.W. Everett, D. Blume, and B.A. Roy, 1994: Market-based incentives and residential municipal solid waste. *Journal of Policy Analysis and Management,* **13**(4), pp. 681-698.

Moe, C.L., and R.D. Reingans, 2006: Global challenges in water, sanitation, and health. *Journal Water Health,* **4**, pp. 41-57.

Monni, S., R. Pipatti, A. Lehtilä, I. Savolainen, and S. Syri, 2006: Global climate change mitigation scenarios for solid waste management. Espoo, Technical Research Centre of Finland. VTT Publications, No. 603, pp 51.

Montgomery, M.A. and M. Elimelech, 2007: Water and sanitation in developing countries: including health in the equation. *Environmental Science and Technology,* **41**, pp. 16-24.

Morris, J.R., P.S. Phillips, and A.D. Read, 2000: The UK landfill tax: financial implications for local authorities. *Public Money and Management,* **20**(3), pp. 51-54.

Morris, J.R., 2001: *Effects of waste composition on landfill processes under semi-arid conditions.* PhD Thesis, Faculty of Engineering, University of the Witwatersrand, Johannesburg, S. Africa. 1052 pp.

Mosher, B., P. Czepiel, R. Harriss, J. Shorter, C. Kolb, J.B. McManus, E. Allwine, and B. Lamb, 1999: Methane emissions at nine landfill sites in the northeastern United States. *Environmental Science and Technology,* **33**(12), pp. 2088-2094.

Myllymaa, T., H. Dahlbo, M. Ollikainen, S. Peltola and M. Melanen, 2005: A method for implementing life cycle surveys of waste management alternatives: environmental and cost effects. Helsinki, Suomen Ympäristö - Finnish Environment 750, 108 pp.

National Bureau of Statistics of China, 2006: China Statistical Yearbook 2005/2006, China Statistics Press, Beijing.

Nozhevnikova, A.N., A.B. Lifshitz, V.S. Lebedev, and G.A. Zavarin, 1993: Emissions of methane into the atmosphere from landfills in the former USSR. *Chemosphere,* **26**(1-4), pp. 401-417.

OECD, 2001: Extended producer responsibility: A guidance manual for governments. OECD publishers, Paris.

OECD, 2002a: Indicators to measure decoupling of environmental pressure from economic growth. OECD Environment Directorate. SG/SD(2002)1/FINAL, 108 pp.

OECD, 2002b: Environmental data waste compendium 2002. Environmental Performance and Information Division, OECD Environment Directorate. Working Group on Environmental Information and Outlooks. 27 pp.

OECD, 2003: OECD Environmental Data Compendium 2002. Paris. <www.oecd.org> accessed 25/06/07.

OECD, 2004: Towards waste prevention performance indicators. OECD Environment Directorate. Working Group on Waste Prevention and Recycling and Working Group on Environmental Information and Outlooks. 197 pp.

Ojeda-Benitez, S., and J.L. Beraud-Lozano, 2003: Characterization and quantification of household solid wastes in a Mexican city. *Resources, Conservation and Recycling,* **39**(3), pp. 211-222.

Onuma, E., Y. Izumi, and H. Muramatsu, 2004: Consideration of CO_2 from alternative fuels in the cement industry. <http://arch.rivm.nl/env/int/ipcc/docs/ITDT/ITDT%20Energy%20Intensive%20Industry%20Session.pdf>, accessed 25/06/07.

Oonk, H. and T. Boom, 1995: Landfill gas formation, recovery and emissions. TNO-report 95-130, TNO, Apeldoorn, the Netherlands.

Perla, M., 1997: Community composting in developing countries. *Biocycle,* June, pp. 48-51.

Petersen, S.O., A.M. Lind, and S.G. Sommer, 1998: Nitrogen and organic matter losses during storage of cattle and pig manure. *Journal of Agricultural Science,* **130**, pp. 69-79.

Pimenteira, C.A.P., A.S. Pereira, L.B. Oliveira, L.P. Rosa, M.M. Reis, and R.M. Henriques, 2004: Energy conservation and CO_2 emission reductions due to recycling in Brazil, *Waste Management,* **24**, pp. 889-897.

Pingoud, K., A-L. Perälä, S. Soimakallio, and A. Pussinen, 1996: Greenhouse impact of the Finnish forest sector including forest products and waste management. *Ambio,* **25**, pp. 318-326.

Pipatti, R. and I. Savolainen, 1996: Role of energy production in the control of greenhouse gas emissions from waste management. *Energy Conservation Management,* **37**(6-8), pp. 1105-1110.

Pipatti, R. and M. Wihersaari, 1998: Cost-effectiveness of alternative strategies in mitigating the greenhouse impact of waste management in three communities of different size. *Mitigation and Adaption Strategies for Global Change,* **2**, pp. 337-358.

PNUD/UNDP, 2005: Informe sobre desarrollo humano 2005: la cooperación internacional ante una encrucijada, Cuadro 7 - Agua, Saneamiento y Nutrición. Publicado para el Programa de las Naciones Unidas para el Desarrollo, Ediciones Mundi - Prensa 2005, Barcelona, Spain. <http://hdr.undp.org/statistics> accessed 29/06/07.

Powelson, D., J. Chanton, and T. Abichou, 2007: Methane oxidation in biofilters measured by mass-balance and stable isotope methods. *Environmental Science and Technology,* **41**, pp. 620-625.

Price, J.L., 2001: The landfill directive and the challenge ahead: demands and pressures on the UK householder. *Resources, Conservation and Recycling*, **32**, pp. 333-348.

Rathje, W.L., W.W. Hughes, D.C. Wilson, M.K. Tani, G.H. Archer, R.G. Hunt, and T.W. Jones, 1992: The archaeology of contemporary landfills. *American Antiquity*, **57**(3), pp. 437-447.

Reinhart, D. and T. Townsend, 1998: Landfill bioreactor design and operation. Lewis/CRC Press, Boca Raton, Florida.

Reinhart, D., T. Townsend, and P. McCreanor, 2002: The status of bioreactor landfills. *Waste Management and Research*, **20**, pp. 67-81.

Richards, K., 1989: Landfill gas: working with Gaia. *Biodeterioration Abstracts*, **3**(4), pp. 317-331.

Rinne, J., M. Pihlatie, A. Lohila, T. Thum, M. Aurela, J-P. Tuovinen, T. Laurila, and R.Vesala, 2005: N_2O emissions from a municipal landfill. *Environmental Science and Technology*, **39**, pp. 7790-7793.

Ritzkowski, M. and R. Stegmann, 2005: Reduction of GHG emissions by landfill in situ aeration. Proceedings of the Sardinia '05, International Solid and Hazardous Waste Symposium, October 2005, published by CISA, University of Cagliari, Sardinia.

Savage, G., L.F. Diaz, C. Golueke, C. Martone, and R. Ham, 1998: Guidance for landfilling waste in economically developing countries. ISWA Working Group on Sanitary Landfilling, published under U.S. EPA Contract 68-C4-0022. ISBN No. 87-90402-07-3, 300 pp.

Scharf, H., H. Oonk, A. Hensen, and D.M.M. van Rijn, 2002: Emission measurements as a tool to improve methane emission estimates. Proceedings of the Intercontinental Landfill Research Symposium, Asheville, NC.

Scheehle, E. and M. Doorn, 2003: Improvements to the U.S. wastewater CH_4 and N_2O emissions estimates. Proceedings 12th International Emissions Inventory Conference, "Emissions Inventories—Applying New Technologies", San Diego, California, April 29-May 3, 2003, published by U.S. EPA, Washington, DC., <www.epa.gov/ttn/chief/conference/ei12>, accessed 12/08/07.

Scheehle, E., and D. Kruger, 2006: Global anthropogenic methane and nitrous oxide emissions, in Yatchew, A., ed., Multi-Greenhouse Gas Mitigation and Climate Policy, Special Issue #3 *Energy Journal*, 520 pp.

Scheutz, C. and P. Kjeldsen, 2002: Determination of the fraction of blowing agent released from refrigerator/freezer foam after decommissioning the product. Report Environment and Resources, Technical University of Denmark (DTU), Lygnby, 72pp.

Scheutz, C., J. Bogner, J. Chanton, D. Blake, M. Morcet, and P. Kjeldsen, 2003a: Comparative oxidation and net emissions of CH_4 and selected non-methane organic compounds in landfill cover soils. *Environmental Science and Technology*, **37**, pp. 5143-5149.

Scheutz, C., A.M. Fredenslund, P. Kjeldsen, 2003b: Attenuation of alternative blowing agents in landfills. Report Environment and Resources, Technical University of Denmark (DTU), Lyngby, 66 pp.

Silva, G., 1998: Private participation in the water and sewerage sector - recent trends. In *Public Policy for the Private Sector (Viewpoint): Water*. World Bank, Washington D.C., pp. 12-20.

Simmons, P., N. Goldstein, S. Kaufman, N. Themelis, and J. Thompson, Jr., 2006: The state of garbage in America. *Biocycle*, **47**, pp. 26-35.

Smith, A., K. Brown, S. Ogilvie, K. Rushton, and J. Bates, 2001: Waste management options and climate change. Final Report ED21158R4.1 to the European Commission, DG Environment, AEA Technology, Oxfordshire, 205 pp.

Spokas, K., J. Bogner, J. Chanton, M. Morcet, C. Aran, C. Graff, Y. Moreau-le-Golvan, N. Bureau, and I. Hebe, 2006: Methane mass balance at three landfill sites: what is the efficiency of capture by gas collection systems? *Waste Management*, **26**, pp. 516-525.

Statistics Finland, 2005: Environment Statistics. *Environment and Natural Resources*, 2, Helsinki, 208 pp.

Stegmann, R., 2005: Mechanical biological pretreatment of municipal solid waste. Proceedings of the Sardinia '05, International Waste Management and Landfill Symposium, October, 2005, CISA, University of Cagliari, Sardinia.

Streese, J., and R. Stegmann, 2005: Potentials and limitations of biofilters for methane oxidation. Proceedings of the Sardinia '05, International Waste Management and Landfill Symposium, October, 2005, CISA, University of Cagliari, Sardinia.

Swedish Environmental Protection Agency, 2005: A strategy for sustainable waste management: Sweden's Waste Plan, published by the Swedish Environmental Protection Agency, 54 pp.

TFFEoL, 2005: Report of the UNEP task force on foam end-of-life issues. Task force on Foam End-of-Life Issues, May 2005, Montreal Protocol on Substances that Deplete the Ozone Layer. UNEP Technology and Economic Assessment Panel, 113 pp.

Themelis, N., 2003: An overview of the global waste-to-energy industry. *Waste Management World*, 2003-2004 Review Issue July-August 2003, pp. 40-47.

Thorneloe, S., K. Weitz, S. Nishtala, S. Yarkosky, and M. Zannes, 2002: The impact of municipal solid waste management on greenhouse gas emissions in the United States. *Journal of the Air & Waste Management Association*, **52**, pp. 1000-1011.

Thorneloe, S., K. Weitz, and J. Jambeck, 2005: Moving from solid waste disposal to materials management in the United States. Proceedings of the Sardinia '05, International Solid and Hazardous Waste Symposium, published by CISA, University of Cagliari, Sardinia.

Tregoures, A., A. Beneito, P. Berne, M.A. Gonze, J.C. Sabroux, D. Savanne, Z. Pokryszka, C. Tauziede, P. Cellier, P.Laville, R. Milward, A. Arnaud, and R. Burkhalter, 1999: Comparison of seven methods for measuring methane flux at a municipal solid waste landfill site. *Waste Management and Research*, **17**, pp. 453-458.

Tsujimoto, Y., J. Masuda, J. Fukuyama, and H. Ito, 1994: N_2O emissions at solid waste disposal sites in Osaka City. *Air Waste*, **44**, pp. 1313-1314.

Tuhkanen, S., R. Pipatti, K. Sipilä, and T. Mäkinen, 2001: The effect of new solid waste treatment systems on greenhouse gas emissions. In *Greenhouse Gas Control Technologies. Proceeding of the Fifth International Conference on Greenhouse Gas Control Technologies (GHGT-5)*. D.J. Williams, R.A. Durie, P. Mcmullan, C.A.J. Paulson, and A.Y. Smith, (eds). Collingwood: CSIRO Publishing, pp. 1236-1241.

UNEP/GPA-UNESCO/IHE, 2004: Improving municipal wastewater management in coastal cities. Training Manual, version 1, February 2004, UNEP/GPA Coordination Office, The Hague, Netherlands, pp. 49-81 and 103-117.

UNFCCC, 2005: Key GHG Data. United Nations Framework Convention on Climate Change, <http://www.unfccc.int/resource/docs/publications/keyghg.pdf> accessed 25/06/07.

UNFCCC/IPCC, 2004, Historical [1990] Greenhouse Gas Emissions Data. <unfccc.int/ghg_emissions_data/items/3800.php>, accessed 12/08/07.

US EPA, 1999: National source reduction characterization report for municipal solid waste in the United States. EPA 530R-99-034, Office of Solid Waste and Emergency Response, Washington, D.C.

US EPA, 2003: International analysis of methane and nitrous oxide abatement opportunities. Report to Energy Modeling Forum, Working Group 21. U.S. Environmental Protection Agency June, 2003. <http://www.epa.gov/methane/intlanalyses.html>.

US EPA, 2005: Municipal solid waste generation, recycling and disposal in the United States: facts and figures for 2003. Washington, D.C., USA, <http://www.epa.gov/garbage/pubs/msw03rpt.pdf> accessed 25/06/07.

US EPA, 2006: Global anthropogenic non-CO_2 greenhouse gas emissions: 1990-2020. Office of Atmospheric Programs, Climate Change Division. <http://www.epa.gov/ngs/econ-inv/downloads/GlobalAnthroEmissionsReport.pdf> accessed 25/06/07.

Veolia Environmental Services, 2005: Monitoring Report, Onyx SASA Landfill Gas Recovery Project, Tremembe, Brazil. Period 1 January 2003 to 31 December 2005.

Vesterinen, R., 1996: Greenhouse gas emissions from composting. SIHTI-Research Programme, Seminar 13-14 March 1996, Hanasaari, Espoo, Finland, 3 pp. In Finnish.

WHO-UNICEF, 2005: Joint Monitoring Programme: Worldwide Sanitation. <http://www.who.int/water_sanitation_health/monitoring/jmp2005/en/index.html> accessed 25/06/07.

WHO/UNICEF/WSSCC, 2000: Global water supply and sanitation assessment: 2000 Report. WHO/UNICEF Joint Monitoring Programme for Water Supply and Sanitation, ISBN924156202 1, 2000. <http://www.who.int>

Willumsen, H.C., 2003: Landfill gas plants: number and type worldwide. Proceedings of the Sardinia '05, International Solid and Hazardous Waste Symposium, October 2005, published by CISA, University of Cagliari, Sardinia.

World Bank, 1997: Toolkits for private sector participation in water and sanitation. World Bank, Washington D.C.

World Bank, 2005a: World Development Indicators 2005: Environment, Table 2.15 Disease Prevention: Coverage and Quality. Table 3.6: Water pollution, Table 3.11: Urban Environment, <http://www.worldbank.org/data/wdi2005/wditext> accessed 25/06/07.

World Bank, 2005b: Global Monitoring Report 2005, Chapter 3: Scaling up service delivery, WHO and UNICEF joint monitoring program, Washington, D.C.

World Bank and IMF, 2006: Global Monitoring Report 2006: Ensuring Environmental Sustainability Target 10. World Bank and International Monetary Fund, Washington, D.C.

WRAP, 2006: Environmental benefits of recycling, an international review of life cycle comparisons for key materials in the UK recycling sector. Waste and Resources Action Program, Peer reviewed report prepared by H. Wenzel *et al.*, Danish Technical University, published by WRAP, Banbury, Oxfordshire, England.

Yamakawa, H. and K. Ueta, 2002: Waste reduction through variable charging programs: its sustainability and contributing factors. *Journal of Material Cycles and Waste Management,* **4**, pp. 77-86.

Yuan, Z., J. Bi, and Y. Moriguchi, 2006: The circular economy-a new development strategy in China. *Journal of Industrial Ecology* **10**, pp. 4-8.

Zinati, G.M., Y.C. Li, and H.H. Bryan, 2001: Utilization of compost increases organic carbon and its humin, humic, and fulvic acid fractions in calcareous soil. *Compost Science & Utilization* **9**, pp. 156-162.

11

Mitigation from a cross-sectoral perspective

Coordinating Lead Authors:
Terry Barker (UK) and Igor Bashmakov (Russia)

Lead Authors:
Awwad Alharthi (Saudi Arabia), Markus Amann (Austria), Luis Cifuentes (Chile), John Drexhage (Canada), Maosheng Duan (China), Ottmar Edenhofer (Germany), Brian Flannery (USA), Michael Grubb (UK), Monique Hoogwijk (Netherlands), Francis I. Ibitoye (Nigeria), Catrinus J. Jepma (Netherlands), William A. Pizer (USA), Kenji Yamaji (Japan)

Contributing Authors:
Shimon Awerbuch † (USA), Lenny Bernstein (USA), Andre Faaij (Netherlands), Hitoshi Hayami (Japan), Tom Heggedal (Norway), Snorre Kverndokk (Norway), John Latham (UK), Axel Michaelowa (Germany), David Popp (USA), Peter Read (New Zealand), Stefan P. Schleicher (Austria), Mike Smith (UK), Ferenc Toth (Hungary)

Review Editors:
David Hawkins (USA), Aviel Verbruggen (Belgium)

This chapter should be cited as:
Barker, T., I. Bashmakov, A. Alharthi, M. Amann, L. Cifuentes, J. Drexhage, M. Duan, O. Edenhofer, B. Flannery, M. Grubb, M. Hoogwijk, F. I. Ibitoye, C. J. Jepma, W.A. Pizer, K. Yamaji, 2007: Mitigation from a cross-sectoral perspective. In Climate Change 2007: Mitigation. Contribution of Working Group III to the Fourth Assessment Report of the Intergovernmental Panel on Climate Change [B. Metz, O.R. Davidson, P.R. Bosch, R. Dave, L.A. Meyer (eds)], Cambridge University Press, Cambridge, United Kingdom and New York, NY, USA.

Table of Contents

EXECUTIVE SUMMARY

Mitigation potentials and costs from sectoral studies

The economic potentials for GHG mitigation at different costs have been reviewed for 2030 on the basis of bottom-up studies. The review confirms the Third Assessment Report (TAR) finding that there are substantial opportunities for mitigation levels of about 6 $GtCO_2$-eq involving net benefits (costs less than 0), with a large share being located in the buildings sector. Additional potentials are 7 $GtCO_2$-eq at a unit cost (carbon price) of less than 20 US$/t$CO_2$-eq, with the total, low-cost, potential being in the range of 9 to 18 $GtCO_2$-eq. The total range is estimated to be 13 to 26 $GtCO_2$-eq, at a cost of less than 50 US$/t$CO_2$-eq and 16 to 31 $GtCO_2$-eq at a cost of less than 100 US$/t$CO_2$-eq (370 US$/tC-eq). As reported in Chapter 3, these ranges are comparable with those suggested by the top-down models for these carbon prices by 2030, although there are differences in sectoral attribution *(medium agreement, medium evidence)*.

No one sector or technology can address the entire mitigation challenge. This suggests that a diversified portfolio is required based on a variety of criteria. All the main sectors contribute to the total. In the lower-cost range, and measured according to end-use attribution,[1] the potentials for electricity savings are largest in buildings and agriculture. When attribution is based on point of emission,[2] energy supply makes the largest contribution *(high agreement, much evidence)*.

These estimated ranges reflect some key sensitivities to baseline fossil fuel prices (most studies use relatively low fossil fuel prices) and discount rates. The estimates are derived from the underlying literature, in which the assumptions adopted are not usually entirely comparable and where the coverage of countries, sectors and gases is limited.

Bioenergy

These estimates assume that bioenergy options will be important for many sectors by 2030, with substantial growth potential beyond, although no complete integrated studies are available for supply-demand balances. The usefulness of these options depends on the development of biomass capacity (energy crops) in balance with investments in agricultural practices, logistic capacity, and markets, together with the commercialization of second-generation biofuel production. Sustainable biomass production and use imply the resolution of issues relating to competition for land and food, water resources, biodiversity and socio-economic impact.

Unconventional options

The aim of geo-engineering options is to remove CO_2 directly from the air, for example through ocean fertilization, or to block sunlight. However, little is known about effectiveness, costs or potential side-effects of the options. Blocking sunlight does not affect the expected escalation in atmospheric CO_2 levels, but could reduce or eliminate the associated warming. Disconnecting CO_2 concentration and global temperature in this way could induce other effects, such as the further acidification of the oceans *(medium agreement, limited evidence)*.

Carbon prices and macro-economic costs of mitigation to 2030

Diverse evidence indicates that carbon prices in the range 20–50 US$/t$CO_2$ (US$75–185/tC), reached globally by 2020–2030 and sustained or increased thereafter, would deliver deep emission reductions by mid-century consistent with stabilization at around 550ppm CO_2-eq (Category III levels, see Table 3.10) if implemented in a stable and predictable fashion. Such prices would deliver these emission savings by creating incentives large enough to switch ongoing investment in the world's electricity systems to low-carbon options, to promote additional energy efficiency, and to halt deforestation and reward afforestation.[3] For purposes of comparison, it can be pointed out that prices in the EU ETS in 2005–2006 varied between 6 and 40 US$/t$CO_2$. The emission reductions will be greater (or the price levels required for a given trajectory lower in the range indicated) to the extent that carbon prices are accompanied by expanding investment in technology RD&D and targeted market-building incentives *(high agreement, much evidence)*.

Pathways towards 650ppm CO_2-eq (Category IV levels; see Table 3.10) could be compatible with such price levels being deferred until after 2030. Studies by the International Energy Agency suggest that a mid-range pathway between Categories III and IV, which returns emissions to present levels by 2050, would require global carbon prices to rise to 25 US$/t$CO_2$ by 2030 and be maintained at this level along with substantial investment in low-carbon energy technologies and supply *(high agreement, much evidence)*.

Effects of the measures on GDP or GNP by 2030 vary accordingly (and depend on many other assumptions). For the 650ppm CO_2-eq pathways requiring reductions of 20% global CO_2 or less below baseline, those modelling studies that allow for induced technological change involve lower costs than the full range of studies reported in Chapter 3, depending on policy mix and incentives for the innovation and deployment of low-carbon technologies. Costs for more stringent targets of 550 ppm CO_2-eq requiring 40% CO_2 abatement or less show an

---

1   In Chapters 4 to 10, the emissions avoided as a result of the electricity saved in various mitigation options are attributed to the end-use sectors using average carbon content for power generation.

2   In 'point-of-emission' attribution, as adopted in Chapter 4, all emissions from power generation are attributed to the energy sector.

3   The forestry chapter also notes that a continuous rise in carbon prices poses a problem: forest sequestration might be deferred to increase profits given higher prices in the future. Seen from this perspective, a more rapid carbon price rise followed by a period of stable carbon prices could encourage more sequestration.

621

even more pronounced reduction in costs compared to the full range (*high agreement, much evidence*).

Mitigation costs depend critically on the baseline, the modelling approaches and the policy assumptions. Costs are lower with low-emission baselines and when the models allow technological change to accelerate as carbon prices rise. Costs are reduced with the implementation of Kyoto flexibility mechanisms over countries, gases and time. If revenues are raised from carbon taxes or emission schemes, costs are lowered if the revenues provide the opportunity to reform the tax system, or are used to encourage low-carbon technologies and remove barriers to mitigation *(high agreement, much evidence)*.

Innovation and costs

All studies make it clear that innovation is needed to deliver currently non-commercial technologies in the long term in order to stabilize greenhouse gas concentrations *(high agreement, much evidence)*.

A major development since the TAR has been the inclusion in many top-down models of endogenous technological change. Using different approaches, modelling studies suggest that allowing for endogenous technological change reduces carbon prices as well as GDP costs, this in comparison with those studies that largely assumed that technological change was independent of mitigation policies and action. These reductions are substantial in some studies *(medium agreement, limited evidence)*.

Attempts to balance emission reductions equally across sectors (without trading) are likely to be more costly than an approach primarily guided by cost efficiency. Another general finding is that costs will be reduced if policies that correct the two relevant market failures are combined by incorporating the damage resulting from climate change in carbon prices, and the benefits of technological innovation in support for low-carbon innovation. An example is the recycling of revenues from tradeable permit auctions to support energy efficiency and low-carbon innovations. Low-carbon technologies can also diversify technology portfolios, thereby reducing risk *(high agreement, much evidence)*.

Incentives and investment

The literature emphasizes the need for a range of cross-sectoral measures in addition to carbon pricing, notably in relation to regulatory and behavioural aspects of energy efficiency, innovation, and infrastructure. Addressing market and regulatory failures surrounding energy efficiency, and providing information and support programmes can increase responsiveness to price instruments and also deliver direct emission savings *(high agreement, much evidence)*.

Innovation may be greatly accelerated by direct measures and one robust conclusion from many reviews is the need for public policy to promote a broad portfolio of research. The

literature also emphasizes the need for a range of incentives that are appropriate to different stages of technology development, with multiple and mutually supporting policies that combine technology push and pull in the various stages of the 'innovation chain' from R&D through the various stages of commercialization and market deployment. In addition, the development of cost-effective technologies will be rewarded by well-designed carbon tax or cap and trade schemes through increased profitability and deployment. Even so, in some cases, the short-term market response to climate policies may lock in existing technologies and inhibit the adoption of more fruitful options in the longer term *(high agreement, much evidence)*.

Mitigation is not a discrete action: investment, in higher or lower carbon options, is occurring all the time. The estimated investment required is around \$20 trillion in the energy sector alone out to 2030. Many energy sector and land use investments cover several decades; buildings, urban and transport infrastructure, and some industrial equipment may influence emission patterns over the century. Emission trajectories and the potential to achieve stabilization levels, particularly in Categories A and B, will be heavily influenced by the nature of these investments. Diverse policies that deter investment in long-lived carbon-intensive infrastructure and reward low-carbon investment could maintain options for these stabilization levels at lower costs *(high agreement, much evidence)*.

However, current measures are too uncertain and short-term to deliver much lower-carbon investment. The perceived risks involved mean that the private sector will only commit the required finance if there are incentives (from carbon pricing and other measures) that are clearer, more predictable, longer-term and more robust than provided for by current policies *(high agreement, much evidence)*.

Spillover effects from Annex I action

Estimates of carbon leakage rates for action under Kyoto range from 5 to 20% as a result of a loss of price competitiveness, but they remain very uncertain. The potential beneficial effect of technology transfer to developing countries arising from technological development brought about by Annex I action may be substantial for energy-intensive industries. However, it has not yet been quantified reliably. As far as existing mitigation actions, such as the EU ETS, are concerned, the empirical evidence seems to indicate that competitive losses are not significant, confirming a finding in the TAR *(medium agreement, limited evidence)*.

Perhaps one of the most important ways in which spillover from mitigation action in one region affects others is through its effect on world fossil fuel prices. When a region reduces its local fossil fuel demand as a result of mitigation policy, it will reduce world demand for that commodity and so put downward pressure on prices. Depending on the response from fossil-fuel producers, oil, gas or coal prices may fall, leading to loss of revenue for the producers, and lower costs of imports for the consumers. Nearly all modelling studies that have been reviewed indicate more pronounced adverse effects on countries with high shares of oil output in GDP than on most of the Annex I countries taking abatement action *(high agreement, much evidence)*.

Co-benefits of mitigation action

Co-benefits of action in the form of reduced air pollution, more energy security or more rural employment offset mitigation costs. While the studies use different methodological approaches, there is general consensus for all world regions analyzed that near-term health and other benefits from GHG reductions can be substantial, both in industrialized and developing countries. However, the benefits are highly dependent on the policies, technologies and sectors chosen. In developing countries, much of the health benefit could result from improvements in the efficiency of, or switching away from, the traditional use of coal and biomass. Such near-term co-benefits of GHG control provide the opportunity for a true no-regrets GHG reduction policy in which substantial advantages accrue even if the impact of human-induced climate change itself turns out to be less than that indicated by current projections *(high agreement, much evidence)*.

Adaptation and mitigation from a sectoral perspective

Mitigation action for bioenergy and land use for sinks are expected to have the most important implications for adaptation. There is a growing awareness of the unique contribution that synergies between mitigation and adaptation could provide for the rural poor, particularly in the least developed countries: many actions focusing on sustainable policies for managing natural resources could provide both significant adaptation benefits and mitigation benefits, mostly in the form of carbon sink enhancement *(high agreement, limited evidence)*.

11.1   Introduction

This chapter takes a cross-sectoral approach to mitigation options and costs, and brings together the information in Chapters 4 to 10 to assess overall mitigation potential. It compares these sectoral estimates with the top-down estimates from Chapter 3, adopting a more short- and medium-term perspective, taking the assessment to 2030. It assesses the cross-sectoral and macro-economic cost literatures since the Third Assessment Report (TAR) (IPCC, 2001), and those covering the transition to a low-carbon economy, spillovers and co-benefits of mitigation.

The chapter starts with an overview of the cross-cutting options for mitigation policy (Section 11.2), including technologies that cut across sectors, such as hydrogen-based systems and options not covered in earlier chapters, examples being ocean fertilization, cloud creation and bio- and geo-engineering. Section 11.3 covers overall mitigation potential by sector, bringing together the various options, presenting

the assessment of the sectoral implications of mitigation, and comparing bottom-up with top-down estimates. Section 11.4 covers the literature on the macro-economic costs of mitigation.

Since the TAR, there is much more literature on the quantitative implications of introducing endogenous technological change into the models. Many studies suggest that higher carbon prices and other climate policies will accelerate the adoption of low-carbon technologies and lower macroeconomic costs, with estimates ranging from a negligible amount to negative costs (net benefits). Section 11.5 describes the effects of introducing endogenous technological change into the models, and particularly the effects of inducing technological change through climate policies.

The remainder of the chapter looks at interactions of various kinds: Section 11.6 links the medium-term to the long-term issues discussed in Chapter 3, linking the shorter-term costs and social prices of carbon to the longer-term stabilization targets; 11.7 covers spillovers from action in one group of countries on the rest of the world; 11.8 covers co-benefits (particularly local air quality benefits) and costs; and 11.9 deals with synergies and trade-offs between mitigation and adaptation.

11.2   Technological options for cross-sectoral mitigation: description and characterization

This section covers technologies that affect many sectors (11.2.1) and other technologies that cannot be attributed to any of the sectors covered in Chapters 4 to 10 (geo-engineering options etc. in 11.2.2). The detailed consolidation and synthesis of the mitigation potentials and costs provided in Chapters 4 to 10 are covered in the next section, 11.3.

11.2.1 Cross-sectoral technological options

Cross-sectoral mitigation technologies can be broken down into three categories in which the implementation of the technology:
1. occurs in parallel in more than one sector;
2. could involve interaction between sectors, or
3. could create competition among sectors for scarce resources.

Some of the technologies implemented in parallel have been discussed earlier in this report. Efficient electric motor-driven systems are used in the industrial sector (Section 7.3.2) and are also a part of many of the technologies for the buildings sector, e.g. efficient heating, ventilation and air conditioning systems (Section 6.4.5). Solar PV can be used in the energy sector for centralized electricity generation (Section 4.3.3.6) and in the buildings sector for distributed electricity generation (Section

6.4.7). Any improvement in these technologies in one sector will benefit the other sectors.

On a broad scale, information technology (IT) is implemented in parallel across sectors as a component of many end-use technologies, but the cumulative impact of its use has not been analyzed. For example, IT is the basis for integrating the control of various building systems, and has the potential to reduce building energy consumption (Section 6.4.6). IT is also the key to the performance of hybrids and other advanced vehicle technologies (Section 5.3.1.2). Smart end-use devices (household appliances, etc) could use IT to program their operation at times when electricity demand is low. This could reduce peak demand for electricity, resulting in a shift to base load generation, which is usually more efficient (Hirst, 2006). The impact of such a switch on CO_2 emissions is unknown, because it is easy to construct cases where shifts from peak load to base load would increase CO_2 emissions (e.g., natural-gas-fired peak load, but coal-fired base load). General improvements in IT, e.g. cheaper computer chips, will benefit all sectors, but applications have to be tailored to the specific end-use. Of course, the net impact of IT on greenhouse gas emissions could result either in net reductions or gains, depending for example on whether or not efficiency gains are offset by increases in production.

An example of a group of technologies that could involve interaction between sectors is gasification/hydrogen/carbon dioxide capture and storage (CCS) technology (IPCC, 2005 and Chapter 4.3.6). While these technologies can be discussed separately, they are interrelated and being applied as a group enhances their CO_2-emission mitigation potential. For example, CCS can be applied as a post-combustion technology, in which case it will increase the amount of resource needed to generate a unit of heat or electricity. Using a pre-combustion approach, i.e. gasifying fossil fuels to produce hydrogen that can be used in fuel cells or directly in combustion engines, may improve overall energy efficiency. However, unless CCS is used to mitigate the CO_2 by-product from this process, the use of that hydrogen will offer only modest benefits. (See Section 5.3.1.4 for a comparison of fuel cell and hybrid vehicles.) Adding CCS would make hydrogen an energy carrier, providing a low CO_2 emission approach for transportation, buildings, or industrial applications. Implementation of fuel cells in stationary applications could provide valuable learning for vehicle application; in addition, fuel cell vehicles could provide electric power to homes and buildings (Romeri, 2004).

In the longer term, hydrogen could be manufactured by gasifying biomass – an approach which has the potential to achieve negative CO_2 emissions (IPCC, 2005) – or through electrolysis using carbon-free sources of electricity, a zero CO_2 option. In the even longer term, it may be possible to produce hydrogen by other processes, e.g. biologically, using genetically-modified organisms (GCEP, 2005). However, none of these longer-term technologies are likely to have a significant impact before 2030, the time frame for this analysis.

Biomass is an example of a cross-sectoral technology which may compete for resources. Any assessment of the use of biomass, e.g., as a source of transportation fuels, must consider competing demands from other sectors for the creation and utilization of biomass resources. Technical breakthroughs could allow biomass to make a larger future contribution to world energy needs. Such breakthroughs could also stimulate the investments required to improve biomass productivity for fuel, food and fibre. See Chapter 4 and Section 11.3.

Another example of resource competition involves natural gas. Natural gas availability could limit the application of some short- to medium-term mitigation technology. Switching to lower carbon fuels, e.g. from coal to natural gas for electricity generation, or from gasoline or diesel to natural gas for vehicles, is a commonly cited short-term option. Because of its higher hydrogen content, natural gas is also the preferred fossil fuel for hydrogen manufacture. Discussion of these options in one sector rarely takes natural gas demand from other sectors into account.

In conclusion, there are several important interactions between technologies across sectors that are seldom taken into account. This is an area of energy system modelling that requires further investigation.

11.2.2 Ocean fertilization and other geo-engineering options

Since the TAR, a body of literature has developed on alternative, geo-engineering techniques for mitigating climate change. This section focuses on apparently promising techniques: ocean fertilization, geo-engineering methods for capturing and safely sequestering CO_2 and reducing the amount of sunlight absorbed by the earth's atmospheric system. These options tend to be speculative and many of their environmental side-effects have yet to be assessed; detailed cost estimates have not been published; and they are without a clear institutional framework for implementation. Conventional carbon capture and storage is covered in Chapter 4, Section 4.3.6 and the IPCC Special Report (2005) on the topic.

11.2.2.1 *Iron and nitrogen fertilization of the oceans*

Iron fertilization of the oceans may be a strategy for removing CO_2 from the atmosphere. The idea is that it stimulates the growth of phytoplankton and therefore sequesters CO_2 in the form of particulate organic carbon (POC). There have been eleven field studies in different ocean regions with the primary aim of examining the impact of iron as a limiting nutrient for phytoplankton by the addition of small quantities (1–10 tonnes) of iron sulphate to the surface ocean. In addition, commercial tests are being pursued with the combined (and conflicting) aims of increasing ocean carbon sequestration and productivity. It should be noted, however, that iron addition will only stimulate phytoplankton growth in ~30% of the oceans (the Southern

Ocean, the equatorial Pacific and the Sub-Arctic Pacific), where iron depletion prevails. Only two experiments to date (Buesseler and Boyd, 2003) have reported on the second phase, the sinking and vertical transport of the increased phytoplankton biomass to depths below the main thermocline (>120m). The efficiency of sequestration of the phytoplankton carbon is low (<10%), with the biomass being largely recycled back to CO_2 in the upper water column (Boyd et al., 2004). This suggests that the field-study estimates of the actual carbon sequestered per unit iron (and per dollar) are over-estimates. The cost of large-scale and long-term fertilization will also be offset by CO_2 release/emission during the acquisition, transportation and release of large volumes of iron in remote oceanic regions. Potential negative effects of iron fertilization include the increased production of methane and nitrous oxide, deoxygenation of intermediate waters and changes in phytoplankton community composition that may cause toxic blooms and/or promote changes further along the food chain. None of these effects have been directly identified in experiments to date, partly due to the time and space constraints.

Nitrogen fertilization is another option (Jones, 2004) with similar problems and consequences.

11.2.2.2 Technologically-varied solar radiative forcing

The basic principle of these technologies is to reduce the amount of sunlight accepted by the earth's system by an amount sufficient to compensate for the heating resulting from enhanced atmospheric CO_2 concentrations. For CO_2 levels projected for 2100, this corresponds to a reduction of about 2%. Three techniques are considered:

A. Deflector System at Earth-Sun L-1[4] point. The principle underlying this idea (e.g. Seifritz (1989), Teller et al. (2004), Angel (2006)) is to install a barrier to sunlight measuring about 106 km[2] at or close to the L-1 point. Teller et al. estimate that its mass would be about 3000 t, consisting of a 30μm metallic screen with 25nm ribs.[5] They envisage it being spun in situ, and emplaced by one shuttle flight a year over 100 years. It should have essentially zero maintenance. The cost has not yet been determined. Computations by Govindasamy et al. (2003) suggest that this scheme could markedly reduce regional and seasonal climate change.

B. Stratospheric Reflecting Aerosols. This technique involves the controlled scattering of incoming sunlight with airborne sub-microscopic particles that would have a stratospheric residence time of about 5 years. Teller et al. (2004) suggest that the particles could be: (a) dielectrics; (b) metals; (c) resonant scatterers. Crutzen (2006) proposes (d) sulphur particles. The implications of these schemes, particularly with regard to stratospheric chemistry, feasibility and costs, require further assessment (Cicerone, 2006).

C. Albedo Enhancement of Atmospheric Clouds. This scheme (Latham, 1990; 2002) involves seeding low-level marine stratocumulus clouds – which cover about a quarter of the Earth's surface – with micrometre-sized aerosol, formed by atomizing seawater. The resulting increases in droplet number concentrations in the clouds raises cloud albedo for incoming sunlight, resulting in cooling which could be controlled (Bower et al., 2006) and be sufficient to compensate for global warming. The required seawater atomization rate is about 10 m3/sec. The costs would be substantially less than for the techniques mentioned under B. An advantage is that the only raw material required is seawater but, while the physics of this process are reasonably well understood, the meteorological ramifications need further study.

These schemes do not affect the expected escalation in atmospheric CO_2 levels, but could reduce or eliminate the associated warming. Disconnecting CO_2 concentration and global temperature in this way could have beneficial consequences such as increases in the productivity of agriculture and forestry. However, there are also risks and this approach will not mitigate or address other effects such as increasing ocean acidification (see IPCC, 2007b, Section 4.4.9).

11.3 Overall mitigation potential and costs, including portfolio analysis and cross-sectoral modelling

This section synthesizes and aggregates the estimates from chapters 4 to 10 and reviews the literature investigating cross-sectoral effects. The aim is to identify current knowledge about the integrated mitigation potential and/or costs covering more than two sectors. There are many specific policies for reducing GHG emissions (see Chapter 13). Non-climate policies may also yield substantial GHG reductions as co-benefits (see Section 11.8 and Chapter 12). All these policies have direct sectoral effects. They also have indirect cross-sectoral effects, which are covered in this section and which diffuse across countries. For example, domestic policies promoting a new technology to reduce the energy use of domestic lighting lead to reductions in emissions of GHG from electricity generation. They may also result in more exports of the new technology and, potentially, additional energy savings in other countries. This section also looks at studies relating to a portfolio analysis of mitigation options.

---

4 This is the L-1 Lagrange point between the sun and the earth.
5 μm stands for micrometre and Nm stands for nanometre (see glossary).

11.3.1 Integrated summary of sectoral emission potentials

Chapters 4 to 10 assessed the economic potential of GHG mitigation at a sectoral scale for the time frame out to 2030 (for a discussion of the different definitions of potential, see Chapter 2). These bottom-up estimates are derived using a variety of literature sources and various methodologies, as discussed in the underlying chapters. This section derives ranges of aggregate economic potentials for GHG mitigation over different costs (i.e. carbon prices) at year-2000 prices.

11.3.1.1 Problems in aggregating emissions

In compiling estimates of this kind, various issues must be considered:

Comparability: There is no common, standardized approach in the underlying literature that is used systematically for assessing the mitigation potential. The comparability of data is therefore far from perfect. The comparability problem was addressed by using a common format to bring together the variety of data found in the literature (as shown below in section 11.3.1.3 and Table 11.3), acknowledging that any aberrations due to a lack of a common methodological base may in part cancel each other out in the aggregation process. Some extrapolations were necessary, for example in the residential sector where the literature mostly refers to 2020. The final result can be considered the best result that is possible and it is accurate within the uncertainty ranges provided.

Coverage: Chapters 4 to 10 together cover virtually all sources of greenhouse gas emissions. However, for parts of some sectors, it was not possible to derive emission reduction potentials from the literature. Furthermore, no quantified emission reduction potentials were available for some options. This leads to a certain under-estimation of the emission reduction potential as discussed in Section 11.3.1.3. The under-estimation of the total mitigation potential is limited, but not negligible.

Baselines: Ideally, emission reduction potentials should adopt a common baseline. Some emission scenarios, such as those developed for the Special Report on Emission Scenarios (IPCC, 2000), are suitable for worldwide, sectoral and multi-gas coverage. However, for a number of sectors, such baselines are not detailed enough to serve as a basis for making bottom-up emission reduction calculations. The baselines used are described and discussed further in Section 11.3.1.2.

Aggregation: The aggregation of mitigation potentials for various sectors is complicated by the fact that mitigation action in one sector may affect mitigation potential in another. There is a risk of double counting of potentials. The problem and the procedures used to overcome this risk are explained in Section 11.3.1.3. In addition the baselines differ to some extent.

11.3.1.2 The baseline

All mitigation potentials have to be estimated against a baseline. The main baseline scenarios used for compiling the assessments in the chapters are the SRES B2 and A1B marker scenarios (IPCC, 2000) and the World Energy Outlook 2004 (WEO2004) (IEA, 2004). The assumed emissions in the three baseline scenarios vary in magnitude and regional distribution. The baseline scenarios B2 and WEO2004 are comparable in the main assumptions for population, GDP and energy use. Figure 11.1 shows that the emissions are also comparable. Scenario A1B, which assumes relatively higher economic growth, shows substantially higher emissions in countries outside the OECD/ EIT region.

The crude oil prices assumed in SRES B2 and WEO2004 are of the same order of magnitude. The oil prices in the SRES scenarios vary across studies. For the MESSAGE model (B2 scenario), the price is about 25 US$/barrel (Riahi *et al.*, 2006). In the case of the WEO2004, for example, the oil price assumed in 2030 is 29 US$/barrel. These prices (and all other energy price assumptions) are substantially lower than those prevailing in 2006 and assumed for later projections (IEA, 2005 and 2006b). The 2002–6 rises in world energy prices are also reflected in the energy futures markets for at least another five to ten years. In fact, the rise in crude oil prices during this period, some 50 US$/barrel, is comparable to the impact of a 100 US$/tCO$_2$-eq increase in the price of carbon. However, it is still uncertain whether these price increases will have a significant impact on the long-term energy price trend.

Higher energy prices and further action on mitigation may reinforce each other in their impact on mitigation potential, although it is still uncertain how and to what extent. On the one hand, for instance, economies of scale may facilitate the introduction of some new technologies if supported by a higher

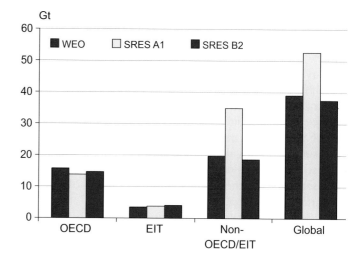

Figure 11.1: *Energy-related CO$_2$-only emissions per world region for the year 2030 in the World Energy Outlook, and in the SRES B2 and A1B scenarios*
Source: Price et al., 2006.

Table 11.1: *Overview of the global emissions for the year 2004 and the baseline emissions for all GHGs adopted for the year 2030 (in GtCO$_2$-eq)*

	Global emissions 2004 (allocated to the end-use sector)[a, c]	Global emissions 2004 (point of emissions)[a, b]	Type of baseline used[d]	Global emissions 2030 (allocated to the end-use sector)	Global emissions 2030 (point of emissions)
Energy supply	– [j]	12.7	WEO	– [j, f]	15.8 [f]
Transport	6.4	6.4	WEO	10.6 [f]	10.6 [f]
Buildings	9.2	3.9	Own	14.3 [f]	5.9 [e) f]
Industry	12.0	9.5	B2/USEPA	14.6	8.5 [g]
Agriculture	6.6	6.6	B2/FAO	8.3	8.3
LULUCF/Forestry[k]	5.8	5.8	Own	5.8 [h]	5.8 [h]
Waste[i]	1.4	1.4	A1B	2.1	2.1

Notes:
a)   The emissions in the year 2004 as reported in the sectoral chapters and Chapter 1, Figure 1.3a/b.
b)   The allocation to point of emission means that the emissions are allocated to the sector where the emission takes place. For example, electricity emissions are allocated to the power sector. There is a difference between the sum when allocating the emissions in different ways. This is explained by the exclusion of electricity emissions from the agricultural and transport sectors due to lack of data and by the exclusion of emissions from conversion of energy as most end-use emissions are based on final energy supply.
c)   'Allocated to the end-use sector' means that the emissions are allocated to the sectors that use the energy. For example, electricity emissions are allocated to the end-use sectors, mainly buildings and industry. Emissions from extraction and distribution are not included here.
d)   See text for further clarification on the type of baselines used.
e)   This figure is based on the assumption that the share of electricity-related emissions in the constructed baseline in Chapter 6 is the same as for the SRES B2 scenario. According to Price et al. (2006), the electricity-related emissions amount to 59%. 59% of the baseline (14.3 GtCO2-eq) is 8.4 GtCO2-eq. The remaining emissions are allocated to the buildings sector.
f)   2030 emissions of the F-gases are not available for the Transport, Buildings, and Energy Supply sectors.
g)   Source: Price et al., 2006.
h)   No baseline emissions for the year 2030 from the forestry sector are reported. See 9.4.3. On the basis of top-down models, it can be expected that the emissions in 2030 will be similar to 2004.
i)   The data for waste include waste disposal, wastewater and incineration. The emissions from wastewater treatment are for the years 2005 and 2020.
j)   The emissions from conversion losses are not included due to lack of data.
k)   Note that the peat fires and other bog fires, as mentioned in Chapter 1, are not included here. Nor are they included in Chapters 8 and 9.

energy price trend. On the other hand, it is also conceivable that, once some cost-effective innovation has already been triggered by higher energy prices, any further mitigation action through policies and measures may become more costly and difficult. Finally, although general energy prices rises will encourage energy efficiency, the mix of the different fuel prices is also important. Oil and gas prices have risen substantially in relation to coal prices 2002–6, and this will encourage greater use of coal, for example in electricity generation, increasing GHG emissions.

As a rule, the SRES B2 and WEO2004 baselines were both used for the synthesis of the emission mitigation potentials by sector. Most chapters have reported the mitigation potential for at least one of these baseline scenarios. There are a few exceptions. Chapter 5 (transportation) uses a different, more suitable, scenario (WBCSD, 2004). However, it is comparable to WEO2004. Chapter 6 (buildings) constructed a baseline scenario with CO$_2$ emissions between those of the SRES B2 and A1B marker scenarios taken from the literature (see Section 6.5). The agriculture and forestry sectors based their mitigation potential on changes in land use as deduced from various scenarios (including marker scenarios, see Sections 8.4.3 and 9.4.3). The SRES scenarios did not include enough detail for

the waste sector, so Chapter 10 used the GDP and population figures from SRES A1B and the methodologies described in IPCC Guidelines 2006 (see Section 10.4.7).

Table 11.1 compares the emissions of the different sectoral baselines for 2004 and 2030 against a background of the end-use and point-of-emission allocation of emissions attributed to electricity use. Since the 2030 data are from studies that differ in terms of coverage and comparability, they should not be directly aggregated across the different sectors and therefore no totals across all sectors are shown in Table 11.1[6]. An important difference between the WEO baseline and SRES B2 is that the WEO emissions do not include all non-CO$_2$ GHG emissions.

11.3.1.3   Synthesizing the potentials from Chapters 4 to 10 involving electricity

When aggregating the sectoral mitigation potentials, the links between sectors need to be considered (Figure 11.2). For example, the options in electricity supply interact with those for electricity demand in the buildings and industry sectors. On the supply side, fossil-fuel electricity can be substituted by low-CO$_2$ or CO$_2$-free technologies such as renewable sources, nuclear energy, bioenergy or fossil fuel in combination with

6   However, since the ranges allow for uncertainties in the baseline, they can be aggregated under specific assumptions and these ranges are shown below.

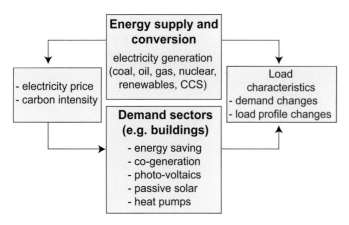

Figure 11.2: *Interaction of CO_2 mitigation measures between electricity supply and demand sectors*

carbon capture and storage. On the demand side, the buildings and the industrial sectors have options for electricity savings. The emission reductions from these two sets of options cannot be aggregated since emission reductions in demand reduce the potential for those in supply and vice-versa.

To overcome this problem, the following approach was adopted: The World Energy Outlook (IEA, 2004) for the year 2030 was used as the baseline. The potentials from electricity savings in the buildings and the industry sectors were estimated first. Electricity savings then reduce demand for electricity. This sequence was followed because electricity savings can be achieved at relatively low cost and their implementation can therefore be expected first. Electricity savings were converted to emission reductions using the average carbon intensity of the electricity supply in the baseline for the year 2030. In reality, it can be expected that electricity savings would result in a relatively larger reduction in fossil-fuel electricity generation than electricity generation involving low marginal costs such as renewables and nuclear. This is because, in the operating system, low-cost generation is normally called on before high-cost generation. However, this response depends on local conditions and it is not appropriate to consider it here. However, it does imply that the emission reductions for electricity savings reported here are an under-estimate. This under-estimate becomes more pronounced with higher carbon prices, and higher marginal costs for fossil fuels.

The detailed sequence is as follows:

1. Electricity savings from the measures in the buildings and industry sector were subtracted from the baseline supply estimates to obtain the corrected electricity supply for 2030.
2. No early withdrawal of plant or stranded assets is assumed. Low-carbon options can therefore only be applied to new electricity supply.
3. The new electricity supply required to 2030 was calculated from 1) additional new capacity between 2010 and 2030 and 2) capacity replaced in the period 2010–2030 after an assumed average plant lifetime of 50 years (see Chapter 4.4.3).

4. The new electricity supply required was divided between available low-carbon supply options. As the cost estimates were lowest for a fuel switch from coal to natural gas supply, it was assumed that this would take place first. In accordance with Chapter 4 it was assumed that 20% of the new coal plants required would be substituted by gas technologies.
5. An assessment was made of the prevented emissions from the other low-carbon substitution options after the fuel switch. The following technologies were taken into account: renewables (wind and geothermal), bioenergy, hydro, nuclear and CCS. It was assumed that the fossil fuel requirement in the baseline (after adjustments for the previous step) was met by these low-carbon intensive technologies. The substitution was made on the basis of relative maximum technical potential. The same breakdown as in Section 4.4.3 was used for the low-carbon options.
6. It was then possible to estimate the resulting mitigation potential for the energy sector, after savings in the end-use sectors buildings and industry.
7. For the buildings and industry sectors, the mitigation potential was broken down into emission savings resulting from less electricity use and the remainder.
8. For sectors other than energy, buildings and industry, the data given in the chapters were used for the overall aggregation.

When evaluating mitigation potential in the energy supply sector, the calculations in Chapter 4 did not subtract the electricity savings from the buildings and industry sectors (see Chapter 4, Table 4.19). Adopting this order (which is not the preferred order, as explained above) implies first taking all the mitigation measures in the energy sector and then applying the electricity savings from buildings and industry sectors. This would result in different mitigation potentials for each of the sectors and mitigation measures, although the total will not change. See Appendix 11.1 for a further discussion of the methodology and details of the calculation.

In the case of the other sectors, the data given in the chapters were used for the overall aggregation. The mitigation potential for the buildings and industry sectors was broken down into emission savings for lower electricity use and the remainder, so that the potential could be re-allocated where necessary to the power sector.

11.3.1.4 Synthesizing the potentials from Chapters 4 to 10 involving biomass

Biomass supplies originate in agriculture (residues and cropping), forestry, waste supplies, and in biomass processing industries (such as the paper & pulp and sugar industries). Key applications for biomass are conversion to heat, power, transportation fuels and biomaterials. Information about biomass supplies and utilization is distributed over the relevant chapters in this report and no complete integrated studies are available for biomass supply-demand balances and biomass potential.

Biomass demand from different sectors

Demand for biomass as transportation fuel involves the production of biofuels from agricultural crops such as sugar cane, rape seed, corn, etc., as well as potentially 'second-generation' biofuels produced from lignocellulosic biomass. The first category dominates in the shorter term. The penetration of second-generation biofuels depends on the speed of technological development and the market penetration of gasification technology for synfuels and hydrolysis technology for the production of ethanol from woody biomass. Demand projections for primary biomass in Chapter 5 are largely based on WEO-IEA (2006) global projections, with a relatively wide range of about 14 to 40 EJ of primary biomass, or 8–25 EJ of fuel. However, there are also higher estimates ranging from 45 to 85 EJ demand for primary biomass in 2030 (or roughly 30–50 EJ of fuel) (see Chapter 5).

Demand for biomass for power and heat is considered in Chapter 4 (energy). Demand for biomass for heat and power will be strongly influenced by the availability and introduction of competing technologies such as CCS, nuclear power, wind energy, solar heating, etc. The projected demand in 2030 for biomass would be around 28–43 EJ according to the data used in Section 4.4.3.3. These estimates focus on electricity generation. Heat is not explicitly modelled or estimated in the WEO, resulting in an under-estimate of total demand for biomass.

Industry is an important user of biomass for energy, most notably the paper & pulp industry and the sugar industry, which both use residues for generating process energy (steam and electricity). Chapter 7 highlights improvements in energy production from such residues, most notably the deployment of efficient gasification/combined cycle technology that could strongly improve efficiencies in, for example, pulp and sugar mills. Mitigation potentials reducing the demand for such commodities or raising the recycling rate for paper will not result in *additional* biomass demand. Biomass can also be used for the production of chemicals and plastics, and as a reducing agent for steel production (charcoal) and for construction purposes (replacing, for example, metals or concrete). Projections for such production routes and subsequent demand for biomass feedstocks are not included in this report, since their deployment is expected to be very limited (see Chapter 7).

In the *built environment*, biomass is used in particular for heating for both non-commercial uses (and also as cooking fuel) and in modern stoves. The use of biomass for domestic heating could represent a significant mitigation potential. No quantitative estimates are available of future biomass demand for the built environment (for example, heating with pellets or cooking fuels) (Chapter 6).

Biomass supplies

Biomass production on agricultural and degraded lands. Table 11.2 summarizes the biomass supply energy potentials

discussed in Chapters 8 (agriculture), 9 (Forestry) and 10 (waste). Those potentials are accompanied by considerable uncertainties. In addition, the estimates are derived from scenarios for the year 2050. The largest contribution could come from energy crops on arable land, assuming that efficiency improvements in agriculture are fast enough to outpace food demand so as to avoid increased pressure on forests and nature areas. Section 8.4.4.2 provides a range from 20–400 EJ. The highest estimate is a technical potential for 2050. Technically, the potentials for such efficiency increases are very large, but the extent to which such potentials can be exploited over time is still poorly studied. Studies assume the successful introduction of biomass production in key regions as Latin America, Sub-Saharan Africa, Eastern Europe and Oceania, combined with gradual improvements in agricultural practice and management (including livestock). However, such development schemes – that could also generate substantial additional income for rural regions that can export biomass – are uncertain, and implementation depends on many factors such as trade policies, agricultural policies, the establishment of sustainability frameworks such as certification, and investments in infrastructure and conventional agriculture (see also Faaij & Domac, 2006).

In addition, the use of degraded lands for biomass production (as in reforestation schemes: 8–110 EJ) could contribute significantly. Although biomass production with such low yields generally results in more expensive biomass supplies, competition with food production is almost absent and various co-benefits, such as the regeneration of soils (and carbon storage), improved water retention, and protection from erosion may also offset some of the establishment costs. An important example of such biomass production schemes at the moment is the establishment of jatropha crops (oil seeds, also spelled jathropa) on marginal lands.

Biomass residues and wastes. Table 11.2 also depicts the energy potentials in residues from forestry (12–74 EJ/yr) and agriculture (15–70 EJ/yr) as well as waste (13 EJ/yr). Those biomass resource categories are largely available before 2030, but also somewhat uncertain. The uncertainty comes from possible competing uses (for example, the increased use of biomaterials such as fibreboard production from forest residues and the use of agro-residues for fodder and fertilizer) and differing assumptions about the sustainability criteria deployed with respect to forest management and agriculture intensity. The current energy potential of waste is approximately 8 EJ/yr, which could increase to 13 EJ in 2030. The biogas fuel potentials from waste, landfill gas and digester gas are much smaller.

Synthesis of biomass supply & demand

A proper comparison of demand and supply is not possible since most of the estimates for supply relate to 2050. Demand has been assessed for 2030. Taking this into account, the lower end of the biomass supply (estimated at about 125 EJ/yr) exceeds the lower estimate of biomass demand (estimated

Table 11.2: *Biomass supply potentials and biomass demand in EJ based on Chapters 4 to 10*

Sector	Supply	Demand			
	Biomass supplies to 2050	Energy supply biomass demand 2030	Transport biomass demand 2030	Built environment	Industry
Agriculture				Relevant, in particular in developing countries as cooking fuel	Sugar industry significant. Food & beverage industry. No quantitative estimate on use for new biomaterials (e.g. bio-plastics) not significant for 2030.
Residues	15-70				
Dung	5-55				
Energy crops on arable land and pastures	20-300				
Crops on degraded lands	60-150				
Forestry	12-74	Key application	Relevant for second-generation biofuels	Relevant	
Waste	13	Power and heat production	Possibly via gasification	Minimal	Cement industry
Industry	Process residues				Relevant; paper & pulp industry
Total supply primary biomass	125-760				
Total demand primary biomass	70-130	28-43 (electricity) Heat excluded	45-85	Relevant (currently several dozens of EJ; additional demand may be limited)	Significant demand; paper & pulp and sugar industry use own process residues; additional demand expected to be limited

at 70 EJ/yr). However, demand does not include estimates of domestic biomass use (such as cooking fuel, although that use may diminish over time depending on development pathways in developing countries), increased biomass for production of heat (although additional demand in this area may be limited) and biomass use in industry (excluding the possible demand of biomass for new biomaterials). It seems that this demand can be met by biomass residues from forestry, agriculture, waste and dung and a limited contribution from energy crops. Such a 'low biomass demand' pathway may develop from the use of agricultural crops with more limited potentials, lower GHG mitigation impact and less attractive economic prospects, in particular in temperate climate regions. The major exception here is sugar-cane-based ethanol production.

The estimated high biomass demand consists of an estimated maximum use of biomass for power production and the constrained growth of production of biofuels when the WEO projections are taken into consideration (25 EJ/yr biofuels and about 40 EJ/yr primary biomass demand). Total combined demand for biomass for power and fuels adds up to about 130 EJ/yr. Clearly, a more substantial contribution from energy crops (perhaps in part from degraded lands, for example producing jatropha oil seeds) is required to cover total demand of this magnitude, but this still seems feasible, even for 2030; the low-end estimate for energy crops for agricultural land is 50 EJ/yr, which is in line with the 40 EJ/yr primary projected demand for biofuels.

However, as was also acknowledged in the WEO, the demand for biomass as biofuels in around 2030 will depend in particular on the commercialization of second-generation biofuel technologies (i.e. the large-scale gasification of biomass for the production of synfuels as Fischer-Tropsch diesel, methanol or DME, and the hydrolysis of lignocellulosic biomass for the production of ethanol). According to Hamelinck and Faaij (2006), such technologies offer competitive biofuel production compared to oil priced at between 40–50 US$/barrel (assuming biomass prices of around 2 US$/GJ). In Chapter 5, Figure 5.9 (IEA, 2006b), however, assumes higher biofuel costs. Another key option is the wider deployment of sugar cane for ethanol production, especially on a larger scale using state-of-the art mills, and possibly in combination with hydrolysis technology and additional ethanol production from bagasse (as argued by Moreira, 2006 and other authors). The availability of such technologies before 2020 may lead to an acceleration of biofuel production and use, even before 2030. Biofuels may therefore become the most important demand factor for biomass, especially in the longer term (i.e. beyond 2030).

A more problematic situation arises when the development of biomass resources (both residues and cultivated biomass) fails to keep up with demand. Although the higher end of biomass supply estimates (2050) is well above the maximum projected biomass demand for 2030, the net availability of biomass in 2030 will be considerably lower than the 2050 estimates. If biomass supplies fall short, this is likely to lead to significant price increases for raw materials. This would have a direct effect on the economic feasibility of various biomass applications. Generally, biomass feedstock costs can cover 30–50% of the production costs of secondary energy carriers, so increasing feedstock prices will quickly reduce the increase in biomass demand (but simultaneously stimulate investments in biomass production). To date, there has been very little research into interactions of this kind, especially at the global scale.

Comparing mitigation estimates for top-down and bottom-up modelling is not straightforward. Bottom-up mitigation responses are typically more detailed and derived from more constrained modelling exercises. Cost estimates are therefore in partial equilibrium in that input and output market prices are fixed, as may be key input quantities such as acreage or capital. Top-down mitigation responses consider more generic mitigation technologies and changes in outputs and inputs (such as shifts from food crops or forests to energy crops) as well as changes in market prices (such as land prices as competition for land increases). In addition, current top-down models optimistically assume the simultaneous global adoption of a coordinated climate policy with an unconstrained, or almost unconstrained, set of mitigation options across sectors. A review of top-down studies (Chapter 3 data assembled from Rose *et al.* (2007) and US CCSP (2006)) results in a range for total projected biomass use over all cost categories of 20 to 79 EJ/yr (defined as solid and liquid, requiring a conversion ratio from primary biomass to fuels). This is, on average, half the range for estimates obtained via bottom-up information from the various chapters.

Given the relatively small number of relevant scenario studies available to date, it is fair to say that the role of biomass in long-term stabilization (beyond 2030) will be very significant but that it is subject to relatively large uncertainties. Further research is required to improve our insight into the potential. A number of key factors influencing biomass mitigation potential are worth noting: the baseline economic growth and energy supply alternatives, assumptions about technological change (such as the rate of development of cellulosic ethanol conversion technology), land use competition, and mitigation alternatives (overall and land-related).

Given the lack of studies of how biomass resources may be distributed over various demand sectors, we do not suggest any allocation of the different biomass supplies to various applications. Furthermore, the net avoidance costs per ton of CO_2 of biomass usage depend on a wide variety of factors, including the biomass resource and supply (logistics) costs, conversion costs (which in turn depend on the availability of improved or advanced technologies) and reference fossil fuel prices, most notably of oil.

11.3.1.5 Estimates of mitigation potentials from Chapters 4 to 10

Table 11.3 uses the procedures outlined above to bring together the estimates for the economic potentials for GHG mitigation from Chapters 4 to 10. It was not possible to break down the potential into the desired cost categories for all sectors. Where appropriate, then, the cells in the table have been merged to account for the fact that the numbers represent the total of two cost categories. Only the potentials in the cost categories up to 100 US$/tCO$_2$-eq are reported here. Some of the chapters also report numbers for the potential in higher cost categories. This is the case for Chapter 5 (transport) and Chapter 8 (agriculture).

Table 11.3 suggests that the economic potential for reducing GHG emissions at costs below 100 US$/tCO$_2$ ranges[7] from 16 to 30 GtCO$_2$-eq. The contributions of each sector to the totals are in the order of magnitude 2 to 6 GtCO$_2$-eq (mid-range numbers), except for the waste sector (0.4 to 1 GtCO$_2$-eq). The mitigation potentials at the lowest cost are estimated for the buildings sector. Based on the literature assessment presented in Chapter 6 it can be concluded that over 80% of the buildings potential can be identified at negative cost. However, significant barriers need to be overcome to achieve these potentials. See Chapter 6 for more information on these barriers.

In all sectors, except for the transport sector, the highest economic potential for emission reduction is thought to be in the non-OECD/EIT region. In relative terms, although it is not possible to be exact because baselines across sectors are different, the emission reduction options at costs below 100 US$/tCO$_2$-eq are in the range of 30 to 50% of the totalled baseline. This is an indicative figure as it is compiled from a range of different baselines.

A number of comments should be made on the overview presented in Table 11.3.

First, a set of emission reduction options have been excluded from the analysis, because the available literature did not allow for a reliable assessment of the potential.[8]

- Emission reduction estimates of fluorinated gases from energy supply, transport and buildings are not included in the sector mitigation potentials from Chapters 4 to 6. For these sectors, the special IPCC report on ozone and climate (IPCC & TEAP, 2005) reported a mitigation potential for HFCs of 0.44 GtCO$_2$-eq for the year 2015 (a mitigation potential of 0.46 GtCO$_2$-eq was reported for CFCs and HCFCs).

---

7   Note that the range is found by aggregating the low or the high potentials per sector. As the errors in the potentials by sector are not correlated, counting up the errors using error propagation rules would lead to a range about half this size. However, given all the uncertainties, and in order to make statements with enough confidence, the full range reported here is used.

8   As indicated in the notes to Table 11.1, bog fires in the forestry sector have also been excluded from the emissions and therefore from the reduction potential as well. The emissions may be significant (in the order of 3 GtCO2-eq), see Chapter 1.

Table 11.3: *Estimated economic potentials for GHG mitigation at a sectoral level in 2030 for different cost categories using the SRES B2 and IEA World Energy Outlook (2004) baselines*

Sector	Mitigation option[a]	Region	Economic potential <100 US$/tCO$_2$-eq[c] Cost cat. US$/tCO$_2$-eq / Cost cat. US$/tC-eq — Low	High	Economic potential in different cost categories[d],[e] — <0 / <0	0-20 / 0-73	20-50 / 73-183	50-100 / 183-367
			Low	*High*	Gt CO$_2$-eq			
Energy supply[e] (see also 4.4)	All options in energy supply excl. electricity savings in other sectors	OECD	0.90	1.7	0.9		0.50	0
		EIT	0.20	0.25	0.15		0.06	0
		Non-OECD/EIT	1.3	2.7	0.80		0.90	0.35
		Global	2.4	4.7	1.9		1.4	0.35
Transport[b],[e],[g] (see also 5.6)	Total	OECD	0.50	0.55	0.25	0.25	0	0
		EIT	0.05	0.05	0.03	0	0	0.02
		Non-OECD/EIT	0.15	0.15	0.10	0.03	0.02	0
		Global[b]	1.6	2.5	0.35	1.4	0.15	0.15
Buildings (see also 6.4)[f],[h]	Electricity savings	OECD	0.8	1.0	0.95	0.00	0	
		EIT	0.2	0.3	0.25	0	0	
		Non-OECD/EIT	2.0	2.5	2.1	0.05	0.05	
	Fuel savings	OECD	1.0	1.3	0.85	0.15	0.15	
		EIT	0.6	0.8	0.2	0.15	0.35	
		Non-OECD/EIT	0.7	0.8	0.65	0.10	0.01	
	Total	OECD	1.8	2.3	1.8	0.15	0.15	
		EIT	0.9	1.1	0.45	0.15	0.35	
		Non-OECD/EIT	2.7	3.3	2.7	0.15	0.10	
		Global	5.4	6.7	5.0	0.50	0.60	
Industry (see also 7.5)	Electricity savings	OECD	0.30		0.07		0.07	0.15
		EIT	0.08		0.02		0.02	0.040
		Non-OECD/EIT	0.45		0.10		0.10	0.25
	Other savings, including non-CO$_2$ GHG	OECD	0.35	0.90	0.30		0.25	0.05
		EIT	0.20	0.45	0.08		0.25	0.02
		Non-OECD/EIT	1.2	3.3	0.50		1.7	0.08
	Total	OECD	0.60	1.2	0.35		0.35	0.20
		EIT	0.25	0.55	0.10		0.25	0.06
		Non-OECD/EIT	1.6	3.8	0.60		1.8	0.30
		Global	2.5	5.5	1.1		2.4	0.55
Agriculture (see also 8.4)	All options	OECD	0.45	1.3	0.30		0.20	0.30
		EIT	0.25	0.65	0.15		0.10	0.15
		Non-OECD/EIT	1.6	4.5	1.1		0.75	1.2
		Global	2.3	6.4	1.6		1.1	1.7
Forestry (see also 9.4)	All options	OECD	0.40	1.0	0.01	0.25	0.30	0.25
		EIT	0.09	0.20	0	0.05	0.05	0.05
		Non-OECD/EIT	0.75	3.0	0.15	0.90	0.55	0.35
		Global	1.3	4.2	0.15	1.1	0.90	0.65
Waste (see also 10.4)	All options	OECD	0.10	0.20	0.10	0.06	0.00	0.00
		EIT	0.10	0.10	0.05	0.05	0.00	0.00
		Non-OECD/EIT	0.20	0.70	0.25	0.07	0.10	0.04
		Global	0.40	1.0	0.40	0.18	0.10	0.04
All sectors[i]	All options	OECD	4.9	7.4	2.2	2.1	1.3	1.1
		EIT	1.8	2.8	0.55	0.65	0.50	1.0
		Non-OECD/EIT	8.3	16.8	3.3	3.6	4.1	2.4
		Global	15.8	31.1	6.1	7.4	6.0	4.5

Notes:

a) Several reduction options are not included due to limited literature sources. This underestimation could be about 10–15%; see below.

b) For transport, the regional data by cost category do not add up to the global potential: regional (cost) distribution is available for LDV only. Due to the lack of international agreement about the regional allocation of aviation emissions, only global cost distributions are available for aviation. A lack of data means that only global figures are presented for biofuels, and not cost distribution.

c) The ranges indicated by the potential are derived differently for each chapter. See underlying chapters for more information.

d) The economic potential figures per cost category are mid-range numbers.

e) The mitigation potential for the use of biomass is allocated to the transport and power sector. See the discussion on biomass energy in 11.3.1.4.

f) For the buildings sector the literature mainly focuses on low-cost mitigation options, and the potential in high-cost categories may be underestimated. The zero may represent an underestimation of the emissions.

g) The '0' means zero, 0.00 means a value below 5 Mton.

h) The electricity savings in the end-use sectors Buildings and Industry are the high estimates. The electricity savings would be significantly lower if the order of measurement were to be reversed; the substitution potential in the energy sector would have been assessed before electricity savings (see Appendix 11.1).

i) The tourism sector is included in the buildings and the transport sector.

- The potential for combined generation of heat and power in the energy supply sector has not added to the other potentials as it is uncertain (see Section 4.4.3). IEA (2006a) quotes a potential here of 0.2 to 0.4 $GtCO_2$-eq.
- The potential emissions reduction for coal mining and gas pipelines has not been included in the reductions from the power sector. De Jager *et al.* (2001) indicated that the CH_4 emissions from coal mining in 2020 might be in the order of 0.65 $GtCO_2$-eq. Reductions of 70 to 90% with a penetration level of 40% might be possible, resulting for 2020 in the order of 0.20 $GtCO_2$-eq. Higher reduction potentials of 0.47 $GtCO_2$-eq for CH_4 from coal mining have also been mentioned (Delhotal *et al.*, 2006).
- Emission reductions in freight transport (heavy duty vehicles), public transport, and marine transport have not been included. In the transport sector, only the mitigation potential for light duty vehicle efficiency improvement (LDV), air planes and biofuels for road transport has been assessed. Because LDV represents roughly two-thirds of transportation by road, and because road transportation represents roughly three-quarters of transport as a whole (air, water, and rail transport represent roughly 11, 9 and 3 percent of overall transport respectively), the estimate for LDV broadly reflects half of the transport activity for which a mitigation potential of over 0.70 $GtCO_2$-eq is reported. In the case of marine transport, the literature studies discussed in Section 5.3.4 indicate that large reductions are possible compared to the current standard but this might not be significant when comparing to a baseline. See also Table 5.8 for indicative potentials for some of the options.
- Non-technical options in the transport sector, like speed limits and changes in modal split or behaviour changes, are not taken into account (an indication of the order of magnitude for Latin American cities is given in Table 5.6).
- For the buildings sector, most literature sources focused on low-cost mitigation options and so high-cost options are less well represented. Behavioural changes in the buildings sector have not been included; some of these raise energy demand, examples being rebound effects from improvements in energy efficiency.[9]
- In the industry sector, the fuel savings have only been estimated for the energy-intensive sectors representing approximately 50% of fuel use in manufacturing industry.
- The TAR stated an emission reduction estimate of 2.20 $GtCO_2$-eq in 2020 for material efficiency. Chapter 7 does not include material efficiency, except for recycling for selected industries, in the estimate of the industrial emission reduction potential. To avoid double counting, the TAR estimate should not be added to the potentials of Chapter 7. However, it is likely that the potential for material efficiency significantly exceeds that for recycling for selected industries only given in Chapter 7.

In conclusion, the options excluded represent significant potentials that justify future analysis. These options represent about 10 to 15% of the potential reported in Table 11.3; this magnitude is not such that the conclusions of the bottom-up analysis would change substantially.

Secondly, the chapters identified a number of key sensitivities that have not been quantified. Note that the key sensitivities are different for the different sectors.

In general, higher energy prices will have some impact on the mitigation potentials presented here (i.e. those with costs below 100 US$/$tCO_2$-eq), but the impact is expected to be generally limited, except for the transport sector (see below). No major options have been identified exceeding 100 US$/$tCO_2$-eq that could move to below 100 US$/$tCO_2$-eq. However, this is only true of the fairly static approach presented here. The costs and potential of technologies in 2030 may be different if energy prices remain high for several decades compared to the situation if they return to the levels of the 1990s. High energy prices may also impact the baseline since the fuel mix will change and lower emissions can be expected. Note that options in some areas, such as agriculture and forestry and non-CO_2 greenhouse gases (about one third of the potential reported), are not affected by energy prices, or much less so.

More specifically, an important sensitivity for the transport sector is the future oil price. The total potential for the LDV in transport increases by 7% as prices rise from 30 to 60 US$/barrel. However, the potential at costs <0 US$/$tCO_2$ increases much more – by almost 90% – because of the fuel saving effect. (See Section 5.4.2).

- Discount rates that formed the basis of the analysis are – as reported in the individual chapters – in the range of 3 to 10%, with the majority of studies using the lower end of this range. Lower discount rates (e.g. 2%) would imply some shift to lower cost ranges, without substantially affecting the total potential. Moving to higher discount rates would have a particular impact on the potential in the highest cost range, which makes up 15 to 20% of the total potential.
- Agriculture and forestry potential estimates are based on long-term experimental results under current climate conditions. Given moderate deviations in the climate expected by 2030, the mitigation estimates are considered quite robust.

Thirdly, potentials with costs below zero US$/$tCO_2$-eq are presented in Table 11.3. The potential at negative cost is considerable. There is evidence from business studies showing the existence of mitigation options at negative cost (for example, The Climate Group, 2005). For a discussion of the reasons for mitigation options at negative costs, see IPCC (2001), Chapters 3 and 7; and Chapters 2 and 6, and Section 11.6 of this report.

---

9   Greening et al. (2000, p.399), in a survey of the rebound effect (in which efficiency improvements lead to more use of energy), remark that 'rebound is not high enough to mitigate the importance of energy efficiency as a way of reducing carbon emissions. However, climate policies that rely only on energy efficiency technologies may need reinforcement by market instruments such as fuel taxes and other incentive mechanisms.'

These remarks do not affect the validity of the overall findings, i.e. that the economic potential at costs below 100 US$/tCO$_2$-eq ranges from 16 to 31 GtCO$_2$-eq. However, they reflect a basic shortcoming of the bottom-up analysis. For individual countries, sectors or gases, the literature includes excellent bottom-up analysis of mitigation potentials. However, they are usually not comparable and their coverage of countries/sectors/gases is limited.

The following gaps in the literature have been identified. Firstly, there is no harmonized integrated standard for bottom-up analysis that compares all future economic potentials. Harmonization is considered important for, *inter alia*, target years, discount rates, price scenarios. Secondly, there is a lack of bottom-up estimates of mitigation potentials, including those for rebound effects of energy-efficiency policies for transport and buildings, for regions such as many EIT countries and substantial parts of the non-OECD/EIT grouping.

11.3.1.6 Comparison with the Third Assessment Report (TAR)

Table 11.4 compares the estimates in this report (AR4) for 2030 with those from the TAR for 2020, which were evaluated at costs less than 27 US$/tCO$_2$-eq (100 US$/tCO$_2$-eq). The last column shows the AR4 estimates for potentials at costs of less than 20 US$/tCO$_2$-eq, which are more comparable with those from the TAR. Overall, the estimated bottom-up economic potential has been revised downwards compared to that in the TAR, even though this report has a longer time horizon than the TAR. Only the buildings sector has been revised upwards in this cost category. For the forestry sector, the economic potential now is significantly lower compared to TAR. However, the TAR numbers for the forestry potential were not specified in terms of cost levels and are more comparable with the < 100

US$/tCO$_2$-eq potential in this report. Even then, they are much higher because they are based on top-down global forest models. These models generally give much higher values then bottom-up studies, as reflected in Chapter 9 of this report. The industry sector is estimated to have a lower potential at costs below 20 US$/tCO$_2$-eq, partly due to a lack of data available for use in the AR4 analyses. Only electricity savings have been included for light industry. In addition, the potential for CHP was allocated to the industry sector in the TAR and was not covered in this report. The most important difference between the TAR and the current analysis is that, in the TAR, material efficiency in a wide sense has been included in the industry sector. In this report, only some aspects of material efficiency have been included, namely in Chapter 7.

The updated estimates might be expected to be higher due to:
- The greater range of economic potentials, extending up to 100 US$/tCO2, compared to less than 27.3 US$/tCO$_2$ (100 US$/tC) in the TAR;
- The different time frame: 2030 compared to 2020 in the TAR.

However, the overall estimated bottom-up economic potential has been revised downwards somewhat, compared to that in the TAR, especially considering that the AR4 estimates allow for about five more years of technological change. Part of the difference is caused by the lower coverage of mitigation options up to 2030 in the AR4 literature.

11.3.1.7 Conclusions of bottom-up potential estimates

When comparing the emission reduction potentials as presented in Table 11.4 with the baseline emissions, it can be concluded that the total economic potential at costs below 20 US$/tCO$_2$eq ranges from 15 to 30% of the total added-up baseline.

Table 11.4: *Comparison of potential global emission reductions for 2030 with the global estimates for 2020 from the Third Assessment Report (TAR) in GtCO$_2$-eq*

Sector	Options	TAR potential emissions reductions by 2020 at costs <27.3 US$/tCO$_2$ [a)]		AR4 potential emissions reductions by 2030 at costs <20 US$/tCO$_2$ [b)]	
	Estimate	Low	High	Low	High
Energy supply and conversion		1.3	2.6	1.2	2.4
Transport	CO$_2$ only	1.1	2.6	1.3	2.1
Buildings	CO$_2$ only	3.7	4.0	4.9	6.1
Industry	CO$_2$ only			0.70	1.5
- energy efficiency		2.6	3.3		
- material efficiency		2.2	2.2		
	non-CO$_2$	0.37	0.37		
Agriculture[c)]	C-sinks and non CO$_2$ [c)]	1.3	2.8	0.30	2.4
Forestry		(11.7)[d)]	(11.7)	0.55	1.9
Waste	CH$_4$ only	0.73	0.73	0.35	0.85
Total		13.2[e)]	18.5[e)]	9.3	17.1

Notes:
a) The TAR range excludes options with costs above 27.3 US$/tCO$_2$ (100 US$/tC) (except for non-CO$_2$ GHGs), and options that will not be adopted through the use of generally accepted policies (p. 264). Differences are due to rounding off.
b) This is the sum of the potential reduction at negative costs and below 20 US$/tCO2. See, however, notes to Table 11.2.
c) Note that TAR estimates are for non-CO$_2$ emissions only. The AR4 estimates also include soil C sequestration (about 90% of the mitigation potential).
d) TAR copied the estimate of Special Report on LULUCF for 2010, which was seen as a technical potential.
e) The 2020 emissions for the SRES B2 baseline was estimated at 49.5 Gton CO2-eq (IPCC, 2000)

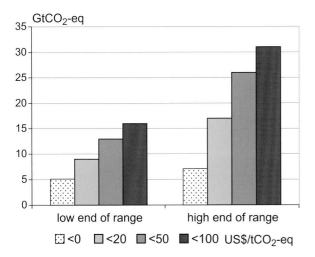
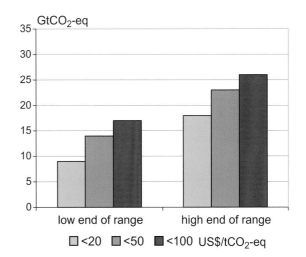

Figure 11.3: *Economic mitigation potential in different cost categories as compared to the baseline*

Notes: The ranges reported in Tables 3.13 and 3.14 were used for comparison for the top-down studies. 'High' and 'low' refer to the high and low ends of the economic potential range reported.

The economic potential up to 100 US$/tCO$_2$eq is about 30 to 50% of emissions in 2030. There is medium evidence for these conclusions because, although a significant amount of literature is available, there are gaps and regional biases, and baselines are different. There is also medium agreement on these conclusions because there is literature for each sector with substantial ranges but the ranges may not capture all the uncertainties that exist. Although there are differences in relative mitigation potentials and specific mitigation costs between sectors (e.g. the buildings sector has a large share of low-cost options), it is clear that the total mitigation potential is spread across the various sectors. Substantial emission reductions can only be achieved if most of the sectors contribute to the emission reduction. In addition, there are barriers that need to be overcome if these potentials are to materialize.

11.3.2 Comparing bottom-up and top-down sectoral potentials for 2030

Table 11.5 and Figure 11.3 bring together the ranges of economic potentials synthesized from Chapters 4 to 10, as discussed in 11.3.1, with the ranges of top-down sectoral estimates for 2030 presented in Chapter 3. The bottom-up estimates are shown with the potentials from end-use electricity savings attributed (1) to the end-use sectors, i.e. to the buildings and industry sectors primarily responsible for the electricity use and (2) upstream, at the point of emission to the energy supply sector. The top-down ranges are provided by an analysis of the data from multi-gas studies for 2030 reported in Section 3.6. A relationship has been estimated between the absolute reductions in total GHGs and the carbon prices required to achieve them (see Appendix 3.1). Ranges for mitigation potential have been calculated for a 68% confidence interval for carbon prices at 20 and 100 US$(2000)/tCO$_2$-eq. The ranges are shown in the last two columns of Table 11.5.

The ranges of bottom-up and top-down aggregate estimates of potentials overlap substantially under all cost ceilings except for the no-regrets bottom-up options. This contrasts with the comparison in the TAR, where top-down costs were higher. It is not the case that bottom-up approaches systematically generate higher abatement potentials. This change comes largely from lower costs in the top-down models, because some have introduced multi-gas abatement and have introduced more bottom-up features, such as induced technological change, which also tend to reduce costs.

Two further points can be made with regard to the comparison of bottom-up and top-down results:
1) Sector definitions differ between top-down and bottom-up approaches. The sectoral data presented here are not fully comparable. The main difference is that the electricity savings are allocated to the power sector in the top-down models compared to the end-use sectors in Table 11.3. Both allocation approaches are presented in Table 11.5.
2) At a sector level however, there are some systematic and striking discrepancies:
   Energy supply. The top-down models indicate a higher emission reduction. This can be explained in part by differences in the mitigation options that are included in the top-down models and not included in the bottom-up approach. Examples are: reductions in extraction and distribution, reductions of other non-CO$_2$ emissions, and reductions through the increased use of CHP. Further, different estimates of the inertia of the substitution are expected to play a role. In bottom-up estimates, fuel substitution is assumed only after end-use savings whereas top-down models adopt a more continuous approach. Finally, the top-down estimates include the effects of energy savings in other sectors and structural changes. For example, a reduction in oil use also implies a reduction in emissions from refineries. These effects are excluded from the bottom-up estimates.

Table 11.5.: *Economic potential for sectoral mitigation by 2030: comparison of bottom-up and top-down estimates*

Chapter of report	Estimate	Sector-based ('bottom-up') potential by 2030 (GtCO$_2$-eq/yr)				Economy-wide model ('top-down') snapshot of mitigation by 2030 (GtCO$_2$-eq/yr)	
		Downstream (indirect) allocation of electricity savings to end-use sectors		Point-of-emissions allocation (emission savings from end-use electricity savings allocated to energy supply sector)			
		Low	High	Low	High	Low	High
\multicolumn{8}{c}{'Low cost' emission reductions: carbon price <20 US$/tCO$_2$-eq}
4	Energy supply	1.2	2.4	4.4	6.4	3.9	9.7
5	Transport	1.3	2.1	1.3	2.1	0.1	1.6
6	Buildings	4.9	6.1	1.9	2.3	0.3	1.1
7	Industry	0.7	1.5	0.5	1.3	1.2	3.2
8	Agriculture	0.3	2.4	0.3	2.4	0.6	1.2
9	Forestry	0.6	1.9	0.6	1.9	0.2	0.8
10	Waste	0.3	0.8	0.3	0.8	0.7	0.9
11	Total	9.3	17.1	9.1	17.9	8.7	17.9
\multicolumn{8}{c}{'Medium cost' emission reductions: carbon price <50 US$/tCO$_2$-eq}
4	Energy supply	2.2	4.2	5.6	8.4	6.7	12.4
5	Transport	1.5	2.3	1.5	2.3	0.5	1.9
6	Buildings	4.9	6.1	1.9	2.3	0.4	1.3
7	Industry	2.2	4.7	1.6	4.5	2.2	4.3
8	Agriculture	1.4	3.9	1.4	3.9	0.8	1.4
9	Forestry	1.0	3.2	1.0	3.2	0.2	0.8
10	Waste	0.4	1.0	0.4	1.0	0.8	1
11	Total	13.3	25.7	13.2	25.8	13.7	22.6
\multicolumn{8}{c}{'High cost' emission reductions: carbon price <100 US$/tCO$_2$-eq}
4	Energy supply	2.4	4.7	6.3	9.3	8.7	14.5
5	Transport	1.6	2.5	1.6	2.5	0.8	2.5
6	Buildings	5.4	6.7	2.3	2.9	0.6	1,5
7	Industry	2.5	5.5	1.7	4.7	3	5
8	Agriculture	2.3	6.4	2.3	6.4	0.9	1.5
9	Forestry	1.3	4.2	1.3	4.2	0.2	0.8
10	Waste	0.4	1.0	0.4	1	0.9	1.1
11	Total	15.8	31.1	15.8	31.1	16.8	26.2

Sources: Tables 3.16, 3.17 and 11.2 and Edenhofer et al., 2006
See notes to Tables 3.16, 3.17 and 11.2 and Appendix 11.1.

Buildings. Top-down models give estimates of reduction potentials from the buildings sector that are lower than those from bottom-up assessments. This is because the top-down models look only at responses to price signals, whereas most of the potential in the buildings sector is thought to be from 'negative cost' measures that would be primarily realized through other kinds of interventions (such as buildings or appliance standards). Top-down models assume that the regulatory environments of 'reference' and 'abatement' cases are similar, so that any negative cost potential is either neglected or assumed to be included in baseline.

Agriculture and forestry. The estimates from bottom-up assessments were higher than those found in top-down studies, particularly at higher cost levels. These sectors are often not covered well by top-down models due to their specific character. An additional explanation is that the data from the top-down estimates include additional deforestation (negative mitigation potential) due to biomass energy plantations. This factor is not included in the bottom-up estimates.

Industry. The top-down models generate higher estimates of reduction potentials in industry than the bottom-up assessments. One of the reasons could be that top-down models allow for product substitution, which is often excluded in bottom-up sector analysis; equally, top-down models may have a greater tendency to allow for innovation over time.

The overall bottom-up potential, both at low and high carbon prices, is consistent with that of 2030 results from top-down models as reported in Chapter 3, Section 3.6.2 for a limited set of models. For carbon prices <20 US$/tCO$_2$-eq, the ranges are 10–17 GtCO$_2$-eq/yr for bottom-up, as opposed to 9–18 GtCO$_2$-eq/yr for top-down studies. For carbon prices <50 US$/tCO$_2$-eq, the ranges are 14–25 GtCO$_2$-eq/yr for bottom-up versus 14–23 GtCO$_2$-eq/yr for top-down studies. For carbon prices <100 US$/tCO$_2$-eq the ranges are 16–30 GtCO$_2$-eq/yr and 17–26 GtCO$_2$-eq/yr for bottom-up and top-down respectively. As explained above, the differences between bottom-up and top-down are larger at the sector level.

11.3.3 Studies of interactions between energy supply and demand

This section looks at literature dealing specifically with the modelling of interactions between energy supply and demand. It first considers the carbon content of electricity, a crucial feature of the cross-sectoral aggregation of potentials discussed above, and then the effect of mitigation on energy prices. The studies emphasize the dependence of mitigation potentials from end-use electricity savings on the generation mix.

11.3.3.1 *The carbon content of electricity*

As discussed above, there are many interactions between CO_2 mitigation measures in the demand and supply of energy. Particularly in the case of electricity, consumers are unaware of the types and volumes of primary energy required for generating electricity. The electricity producer determines the power generation mix, which depends on the load characteristics. The CO_2 mitigation measures not only affect the generation mix (supply side) through the load characteristics. They are also influenced by the price.

Iwafune *et al.* (2001a; 2001b; 2001c), and Kraines *et al.* (2001) discuss the effects of the interactions between electricity supply and demand sectors in the Virtual Tokyo model. Demand-side options and supply-side options are considered simultaneously, with changes in the optimal mix in power generation reflecting changes in the load profile caused by the introductions of demand-side options such as the enhanced insulation of buildings and installation of photovoltaic (PV) modules on rooftops. The economic indicators used for demand-side behaviours are investment pay-back time and marginal CO_2 abatement cost. Typical results of Iwafune *et al.* (2001a) are that the introduction of demand-side measures reduces electricity demand in Tokyo by 3.5%, reducing CO_2 emissions from power supply by 7.6%. The CO_2 emission intensity of the reduced electricity demand is more than two times higher than the average CO_2 intensity of electricity supply because reductions in electricity demands caused by the saving of building energy demand and/or the installation of PV modules occur mainly in daytime when more carbon-intensive fuels are used. A similar 'wedge' – in this case between the average carbon intensity of electricity supply and the carbon value of electricity savings – was found, in the UK system, to depend upon the price of EU ETS allowances, with high ETS prices increasing the carbon value of end-use savings by around 40% as coal is pushed to the margin of power generation (Grubb and Wilde, 2005).

Komiyama *et al.* (2003) evaluate the total system effect in terms of CO_2 emission reduction by introducing co-generation (CHP, combined heat and power) in residential and commercial sectors, using a long-term optimal generation-mix model to allow for the indirect effects on CO_2 emissions from power generation. In a standard scenario, where the first technology to be substituted is oil-fired power, followed later by LNG CC and

IGCC, the installation of CHP reduces CO_2 emission in the total system. However, in a different scenario, the CO_2 reduction effect of CHP introduction may be substantially lower. For example, the effect is negligible when highly efficient CCGTs (combined cycle gas turbines) are dominant at baseline and replaced by CHP. Furthermore, in the albeit unlikely case of nuclear power being competitive at baseline but replaced by CHP, the total CO_2 emission from the energy system increases with CHP installation. These results suggest that the CO_2 reduction potential associated with the introduction of CHP should be evaluated with caution.

11.3.3.2 *The effects of rising energy prices on mitigation*

Price responses to energy demand can be much larger when energy prices are rising than when they are falling, but responses in conventional modelling are symmetric. The mitigation response to policy may therefore be much larger when energy prices are rising. This phenomenon is addressed in literature about asymmetrical price responses and the effects of technological change (Gately and Huntington, 2002; Griffin and Shulman, 2005). Bashmakov (2006) also argues for asymmetrical responses in the analysis of what is called the economics of constants and variables: the existence of very stable energy costs to income proportions, which can be observed over the longer period of statistical observations in many countries. He argues that there are thresholds for total energy costs such as a ratio of GDP or gross output, and energy costs by transportation and residential sector as shares of personal income. If rising energy prices push the ratios towards the given thresholds, then the dynamics of energy-demand price responses are changed. The effect on real income can become sufficient to reduce GDP growth, mobility and the level of indoor comfort. Carbon taxes and permits become more effective the closer the ratio is to the threshold, so the same rates and prices generate different results depending on the relationship of the energy costs to income or of the gross output ratio to the threshold.

11.3.4 Regional cross-sectoral effects of greenhouse gas mitigation policies to 2025

Various estimates of cross-sectoral mitigation potential for specific regions have been published, usually as reports commissioned by governments. Unfortunately, however, the issue of attributing costs to cross-sectoral effects of greenhouse gas mitigation policies has not been reported extensively since the TAR, and literature on this topic is consequently sparse.

In one of the few studies to examine the sectoral effects of mitigation policies across countries, Meyer and Lutz (2002), using the COMPASS model, carried out a simulation of the effects of carbon taxes or the G7 countries, which include some of the biggest energy users. The authors assumed the introduction of a carbon tax of 1US\$ per ton of CO_2 in 2001 in all of these countries, rising linearly to 10 US\$ in 2010, with revenues used to lower social security contributions. Table 11.6

Table 11.6: *Impact on sectoral output of 1 US$/tCO$_2$ tax in 2001 rising to 10 US$/tCO$_2$ by 2010*

	USA	Japan	Germany	France	Italy	UK	Canada
	% difference from business-as-usual gross output in 2010						
Food processing	-2.02	-0.27	-0.32	-0.36	-0.29	-0.69	-1.83
Petroleum and coal products	-2.87	-0.33	-0.82	-0.50	-0.47	-2.42	-3.67
Iron and steel	-1.35	-0.28	-0.33	-0.45	-0.48	-0.82	-1.60
Machinery	-1.06	-0.22	-0.26	-0.29	-0.48	-0.72	-1.11
Motor vehicles	-1.41	-0.42	-0.33	-0.47	-0.40	-0.74	-1.92
Construction	-1.01	-0.02	-0.13	-0.21	-0.39	-0.78	-1.06
All industries	-1.74	-0.18	-0.32	-0.33	-0.35	-0.75	-1.71

Source: Meyer and Lutz (2002)

shows the effects on output: the decline in petroleum and coal products will be highest, with the effects on construction being mild. The scale of the effects differs substantially between countries, depending on the energy intensities of the economies and the carbon content of this energy, with effects on output being much larger in US and Canadian industries.

One major cross-sectoral study (EU DG Environment, 2001) brings together low-cost mitigation options and shows their effects across sectors and regions. It shows how a Kyoto-style target (8% reduction of EU GHGs below 1990/95 by 2010) can be achieved for the EU-15 member states with options costing less than 20 US$/tCO$_2$. The study assesses the direct and indirect outcomes using a top-down model (PRIMES) for energy-related CO$_2$ and a bottom-up model (GENESIS) for all other GHGs. The synthesis of the results is presented in Table 11.7. This multi-gas study considers all GHGs, but assumes that the JI and CDM flexibility instruments are not used. The study shows the wide variations in cost-effective mitigation across sectors. The largest reductions compared to the 1990/95 baselines are in the energy and energy-intensive sectors, whereas there is an increase of 25% in the transport sector compared to 1990/95 emissions. Note also the large reductions in methane and N$_2$O in the achievement of the overall target as shown in the lower panel of the table. The results are, however, dominated by bottom-up energy-engineering assumptions since PRIMES is a partial-equilibrium model. Consequently, the GDP effects of the options are not provided. These potentials can be compared to those at less than 20 US$/tCO$_2$ in Table 11.3 above for the sectoral synthesis for the OECD. The EU potentials are similarly concentrated in the buildings sector, but with a larger share for industry, and a lower one for transport, reflecting the high existing taxes on transport fuels in the EU.

Masui *et al.* (2005) report the effects of a tax and sectoral subsidy regime for Japan to achieve the Kyoto target by 2010, in which carbon tax revenues are used to subsidize additional investments to reduce greenhouse gases. The investment costs are shown in Table 11.8 for each sector. The table shows that about about 9 US$/tCO$_2$ (3,400 Japanese Yen/tC) will be required as carbon tax and most of the investment will be in energy-saving measures in the buildings sector (Residential and

Commercial). The macro-economic effects for this study are reported in Section 11.4.3.4.

Schumacher and Sands (2006) model the response of German GHG emissions to various technology and carbon policy assumptions over the next few decades using the SGM model for Germany. Accounting for advanced technologies such as coal IGCC, NGCC, CCS, and wind power, they show that emission reductions can be achieved at substantially lower marginal abatement costs in the long run with new advanced electricity generating technologies in place. In a scenario assuming a carbon price of 50 US$/tCO$_2$ giving a 15% reduction of CO$_2$ below baseline by 2020, they show that, with the new and advanced technologies, the electricity sector would account for the largest share of emissions reductions (around 50% of total emissions reductions), followed by other (non-energy-intensive) industries and households. The effects on gross output are very uneven across sectors: energy transformation is 9% below base, but other industry, services and agriculture (and GDP) are 0.7% below base by 2050.

The effects of different policy mixes on sectoral outcomes are shown in the US EIA (2005) analysis of the National Commission on Energy Policy's (NCEP) 2004 proposals. These involve reductions in the US emissions in GHGs of about 11% by 2025 below a reference case, including an analysis of the cap-and-trade component, (involving a safety valve limiting the maximum cost of emissions permits to US$ (2003)8.50/tCO$_2$ through to 2025) and a no-safety-valve case (in which the cost rises to US$(2003) 35/tCO$_2$ and the GHG reduction to 15% by 2025). The effects on CO$_2$ emissions by broad sector are shown in Figure 11.4. Note that the NCEP scenario includes the cap-and-trade scheme (with a safety valve) shown separately in the figure and that the no-safety-valve scenario is additional to the NCEP scenario. The NCEP scenario includes substantial energy efficiency policies for transportation and buildings. This explains the relatively large contributions of these sectors in this scenario. The cap-and-trade schemes mainly affect the electricity sector, since the price of coal-fired generation rises relative to other generation technologies. For discussion of macro-economic estimates of mitigation costs for the US from this study and others, see Section 11.4.3.1.

Table 11.7: *Sectoral results from top-down energy modelling (PRIMES for energy-related CO_2) and bottom-up modelling (of non-CO_2 GHGs). The table shows the distribution of direct and total (direct and indirect) emissions of GHGs in 1990/1995, in the 2010 baseline and in the most cost-effective solution for 2010, where emissions are reduced by 8% compared to the 1990/1995 level. The top table gives the breakdown into sectors and the bottom table the breakdown into gases.*

EU-15 Emission breakdown per sector (top-down)	Direct emissions (MtCO₂-eq)					Direct and indirect emissions (MtCO₂-eq)				
	Emissions in 1990/95	Baseline emissions in 2010	Cost-effective objective 2010	Change from 1990/95	Change from 2010 baseline	Emissions in 1990/95	Baseline emissions in 2010	Cost-effective objective 2010	Change from 1990/95	Change from 2010 baseline
Energy supply[a),b)]	**1190**	**1206**	**1054**	**-11%**	**-13%**	**58**	**45**	**42**	**-27%**	**-6%**
CO₂ (energy-related)	1132	1161	1011	-11%	-13%					
auto-producers	124	278	229	85%	-18%					
utilities	836	772	667	-20%	-14%					
other	172	111	115	-33%	4%					
Non-CO₂	58	45	42	-27%	-6%	58	45	42	-27%	-6%
Non-CO₂ fossil fuel[c)]	**95**	**61**	**51**	**-46%**	**-16%**	**95**	**61**	**51**	**-46%**	**-16%**
Industry[b)]	894	759	665	-26%	-12%	1383	1282	1125	-19%	-12%
Iron and steel	196	158	145	-26%	-9%	253	200	183	-28%	-9%
Non-ferrous metals	24	22	13	-47%	-40%	66	42	30	-54%	-28%
Chemicals	243	121	81	-66%	-33%	362	257	201	-44%	-22%
Building Materials	201	212	208	3%	-2%	237	240	232	-2%	-3%
Paper and Pulp	29	22	20	-32%	-9%	69	106	92	34%	-13%
Food, drink, tobacco	46	35	26	-42%	-24%	89	107	91	2%	-15%
Other industries	155	189	172	11%	-9%	308	331	295	-4%	-11%
Transport	**753**	**984**	**946**	**26%**	**-4%**	**778**	**1019**	**975**	**25%**	**-4%**
CO₂ (energy-related)	735	919	887	21%	-4%	760	953	916	21%	-4%
road	624	741	724	16%	-2%	624	741	724	16%	-2%
train	9	2	2	-83%	-8%	34	36	31	-10%	-14%
aviation[d)]	82	150	135	65%	-10%	82	150	135	65%	-10%
inland navigation	21	27	26	26%	-2%	21	27	26	26%	-2%
Non-CO₂ (road)	18	65	59	222%	-10%	18	84	143	681%	70%
Households	**447**	**445**	**420**	**-6%**	**-6%**	**792**	**748**	**684**	**-14%**	**-9%**
Services	**176**	**200**	**170**	**-3%**	**-15%**	**448**	**500**	**428**	**-4%**	**-14%**
Agriculture	**417**	**398**	**382**	**-8%**	**-4%**	**417**	**398**	**382**	**-8%**	**-4%**
Waste	**166**	**137**	**119**	**-28%**	**-13%**	**166**	**137**	**119**	**-28%**	**-13%**
Total	**4138**	**4190**	**3807**	**-8%**	**-9%**	**4138**	**4190**	**3807**	**-8%**	**-9%**

Breakdown per gas	Emissions in 1990/95	Baseline emissions in 2010	Cost-effective objective 2010	Change from 1990/95	Change from 2010 baseline
CO₂ energy-related	3068	3193	2922	-5%	-8%
CO₂ other	164	183	182	11%	-1%
Methane	462	380	345	-25%	-9%
Nitrous oxide	376	317	282	-25%	-11%
HFCs	52	84	54	3%	-36%
PFCs	10	25	19	87%	-27%
SF6	5	7	3	-41%	-53%
Total	**4138**	**4190**	**3807**	**-8%**	**-9%**

Notes:
a) The direct CO_2 emissions of energy supply are allocated to the energy demand sectors in the right-hand part of the table representing direct and indirect emissions. Refineries are included in the energy supply sector.
b) Industrial boilers are allocated to industrial sectors.
c) Non-CO_2 GHG emissions from fossil fuel extraction, transport and distribution.
d) Due to missing data, emission data for aviation include international aviation, which is excluded in the IPCC inventory methodology.

Source: EU DG Environment, 2001.
http://europa.eu.int/comm/environment/enveco/climate_change/summary_report_policy_makers.pdf

Table 11.8: *Carbon tax rate and required additional investments for CO$_2$ abatement in Japan*

Sector	Subsidized measures and devices	Additional money grant (billion US$/yr)
Industrial sector	Boiler conversion control, High-performance motor, High-performance industrial furnace, Waste plastic injection blast furnace, LDF with closed LDG recovery, High-efficiency continuous annealing, Diffuser bleaching device, High-efficiency clinker cooler, Biomass power generation	0.95
Residential sector	High-efficiency air conditioner, High-efficiency gas stove, Solar water heater, High-efficiency gas cooking device, High-efficiency television, High-efficiency VCR, Latent heat recovery type water heater, High-efficiency illuminator, High-efficiency refrigerator, Standby electricity saving, Insulation	3.33
Commercial sector	High-efficiency electric refrigerator, High-efficiency air conditioner, High-efficiency gas absorption heat pump, High-efficiency gas boiler, Latent heat recovery type boiler, Solar water heater, High-efficiency gas cooking device, High-frequency inverter lighting with timer, High-efficiency vending machine, Amorphous transformer, Standby electricity saving, Ventilation with heat exchanger, Insulation	1.83
Transportation sector	High-efficiency gasoline private car, High-efficiency diesel car, Hybrid taxi, High-efficiency diesel bus, High-efficiency small-sized truck, High-efficiency standard-sized track	1.00
Forest management	Plantation, Weeding, Tree thinning, Multilayered thinning, Improvement of natural forests	1.84
Total		8.96
Required carbon tax rate (US$/tCO$_2$)		**8.7**

Source: Masui et al. (2005).

11.3.5 Portfolio analysis of mitigation options

Portfolio analysis in this context is the study of the mix of actions available to reduce emissions or adapt to climate change and to business in diversifying their investments against risk.

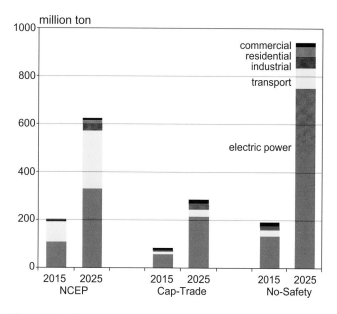

Figure 11.4: *Carbon dioxide reductions by sector in the NCEP, Cap-Trade, and No-Safety Cases, 2015 and 2025*

Notes: National Energy Modeling System, runs BING-ICE-CAP.D021005C BING-CAP.D021005A, and BING-NOCAP.D020805A. Note that NCEP includes technology mandates, and Cap-Trade is without technology mandates.

Source: US Energy Information Administration (EIA)(2005, p.15).

One issue is the allocation of GHG abatement across sectors or regions. Capros and Mantzos (2000) show that, within the EU, equal percentage reductions across sectors cost more than twice as much as a least cost distribution (which can be obtained by, for example, allowing trade between sectors); see Table 11.9. The table also shows the gains through international trading both across the EU and in Annex I, confirming the benefits reported in the TAR from a wide range of previous literature.

The reference case assumes that the Kyoto commitment is implemented separately by domestic action in each EU member state. The alternative reference case assumes that, within a member state, the overall emission reduction target of the burden-sharing agreement applies equally to each individual sector of the economy, with allocation evidently being more expensive than the least-cost approach in the reference case.

A related issue is the allocation of CO$_2$ emission reductions under Kyoto to sources in the EU Emissions Trading Scheme (EU ETS), as compared all non-ETS sources. Klepper and Peterson (2006), using a CGE model, conclude that ETS National Allocation Plans reduce the allowance price in the ETS below the implicit tax necessary for reaching the Kyoto targets in the non-ETS sectors, implying significant distortion. The limited use of CDM and JI to meet the allocations would result in a negative effect on welfare of close to 1% in 2012 relative to 'business as usual'; this assumes that EU Member States do not import more than 50% of their required reductions and that they do not import 'hot air'. Unrestricted trading in CDM and JI credits and allowances would result in an allocation where the Kyoto target can be met with hardly any welfare costs.

Table 11.9: *The effects of EU-wide and Annex B trading on compliance cost, savings and marginal abatement costs in 2010*

	Compliance cost	Savings against Reference Case		Savings against Alternative Reference Case		Marginal abatement cost (US$/tCO$_2$)	
	million US$	million US$	%	million US$	%	for sectors participating in EU-wide trading	for other sectors
No EU-wide trading							
Reference case: burden-sharing target implemented at least cost across sectors within a member state	9026	n.a.	n.a.	11482	56.0	n.a.	54.3
Alternative reference case: burden-sharing target allocated uniformly to all sectors within a member state	20508	-11482	-127.2	n.a.	n.a.	n.a.	125.8
EU-wide trading							
Energy suppliers	7158	1868	20.7	13350	65.1	32.3	45.3
Energy suppliers and energy-intensive industries	6863	2163	24.0	13645	66.5	33.3	43.3
All sectors	5957	3069	34.0	14551	71.0	32.6	32.6
Annex B trading: All sectors	4639	4387	48.6	15869	77.4	17.7	17.7

Notes: A negative sign means a cost increase. A positive sign means a cost saving. It is assumed that the international allowance price would be 17.7 US$/tCO$_2$. Compliance cost and savings are on an annual basis. Original results in € have been converted to US$ at €1 for 1US$.

Source: adapted from Capros and Mantzos (2000, p.8).

Jaccard *et al.* (2002) evaluate the cost of climate policy in Canada. They compare the costs of achieving the Canadian Kyoto target in 2010 (using the CIMS model) for equal sector targets or one national target. According to their estimates, the electricity, residential, and commercial/institutional sectors contribute more, at lower marginal costs, to reductions when there is one national target, while the industry and transportation sectors contribute less. For example, the marginal cost for the electricity sector is about 20 US$/tCO$_2$-eq for the sector target and 80 US$/tCO$_2$-eq for the national target, while those of industrial sector are 200 and 80 US$/tCO$_2$-eq respectively.

Both studies illustrate a general finding that a portfolio of options which attempts to balance emission reductions across sectors with 'equal percentage reductions' is more costly than optimizing the policy mix for cost effectiveness.

Another aspect of mitigation options is the opportunity afforded by portfolio analysis to reduce risks and costs. Because fossil fuel prices are uncertain and variable, there are potential benefits in portfolios of energy supply sources that increase diversity so as to include, in particular, sources such as renewables and nuclear, the costs of which do not depend on fossil fuel prices. Long-standing methods from finance theory can help to quantify a new low-carbon technology's contribution to overall risk, and to quantify costs associated with the development of a set of options for GHG mitigation and energy security. The portfolio approach differs from the traditional stand-alone cost approach in that it introduces

market risk and includes inter-relationships between the costs of different technologies (Awerbuch, 2006, MITI). New technologies that diversify the generating mix and low-carbon options tend to be quantifiably more diverse than business-as-usual reliance on fossil fuels (see Stirling, 1994; 1996; Grubb *et al.*, 2006). Moreover, in contrast to the expected year-to-year variability of fossil fuel prices (which can be estimated from historic patterns), operating costs for wind, solar, nuclear and other capital-intensive non-fossil technologies are largely uncorrelated to fossil fuel prices.

Theory, supported by application, suggests that risk-optimized generating mixes will include larger shares of wind, geothermal and other fixed-cost renewables, even where these technologies cost more than gas and coal generation. Optimal mixes will also enhance energy security while simultaneously minimizing expected generating cost and risk. Awerbuch, Stirling, Jansen and Beurskens (2006) explore the limitations of the mean-variance portfolio (MVP) approach, and compare MVP optimal generating mixes to 'maximum diversity' mixes that also provide protection against uncertainty, ignorance and 'surprise'. They find that the optimal mixes in both cases contain larger shares of wind energy.

These findings suggest that portfolios of cross-sector energy options that include low-carbon technologies and products will reduce risks and costs, simply because fossil fuel prices are more volatile relative to other costs, in addition to the usual benefits from diversification.

11.4   Macro-economic effects

The main conclusions from the TAR on the macro-economic costs of mitigation can be summarized as follows. Mitigation costs can be substantially reduced through a portfolio of policy instruments, including those that help to overcome barriers, with emissions trading in particular expected to reduce the costs. However, mitigation costs may be significant for particular sectors and countries over some periods and the costs tend to rise with more stringent levels of atmospheric stabilization. Unplanned and unexpected policies with sudden short-term effects may cost much more for the same eventual results than planned and expected policies with gradual effects. Near-term anticipatory action in mitigation and adaptation would reduce risks and provide benefits because of the inertia in climate, ecological and socio-economic systems. The effectiveness of adaptation and mitigation is increased and costs are reduced if they are integrated with policies for sustainable development.

Since the TAR, the Kyoto Protocol has come into force and there has been a range of domestic initiatives in different countries. This has led to diverse modelling activities addressing the Kyoto Protocol as implemented (without the United States and Australia), post-Kyoto strategies, and more intricate domestic policies, providing more refined estimates of mitigation costs, through more accurate representation of policy implementation, improved modelling techniques, and improved meta-analysis of existing results.

11.4.1   Measures of economic costs

Chapter 2 discusses cost concepts. Here we report, where available, the prices associated with CO_2 emissions, and negative or positive impacts on GDP, welfare and employment.

The TAR reviewed studies of climate policy interactions with the existing tax system. These interactions change the aggregate impact of a climate policy by changing the costs associated with taxes in other markets. They also point to the opportunity for climate policy – through carbon taxes or auctioned permits – to generate government revenue and, in turn, to reduce other taxes and their associated burden. The TAR pointed to this opportunity as a way to reduce climate policy costs. Since the TAR, additional studies have extended the debate (Bach *et al.*, 2002; Roson, 2003). Meanwhile, such arguments have been the basis of the UK Climate Change Levy and linked reduction in National Insurance Contributions, small auctions under the EU ETS and US NO_x Budget Program, large proposed auctions under the Regional Greenhouse Gas Initiative in the United States, and proposals in the United States, Japan, and New Zealand for carbon taxes.

11.4.2   Policy analysis of the effects of the Kyoto Protocol

Most analyses discussed in the TAR focused on national emission policies under the Kyoto Protocol in the form of an economy-wide tax or tradeable permit system. This has continued to be an active area of policy modelling since the Kyoto Protocol came into force. Global cost studies of the Kyoto Protocol since the TAR have considered more detailed implementation questions and their likely impact on overall cost. Chief among these have been the impact of the Bonn-Marrakesh agreements concerning sink budgets, the non-participation of the United States, and banking and the use of 'hot air' (Manne and Richels, 2001; Böhringer, 2002; den Elzen and de Moor, 2002; Löschel and Zhang, 2002; Böhringer and Löschel, 2003; McKibbin and Wilcoxen, 2004; Klepper and Peterson, 2005).

U.S. non-participation in the Kyoto Protocol, coupled with the increase in sink budgets in Bonn and Marrakech, implies that the target for Annex B countries as a whole will likely be met with virtually no effort. In other words, excess allowances in Russian and Ukraine (referred to as 'hot air') roughly equal the shortfall in Europe, Japan, Canada, and other countries. However, some of these same studies emphasize that strategic behaviour by Russia and Ukraine, acting as a supply cartel and/or choosing to bank allowances until the next commitment period, will lead to a positive emission price (Löschel and Zhang, 2002; Böhringer and Löschel, 2003; Maeda, 2003; Klepper and Peterson, 2005). For example, Böhringer and Löschel (2003) use a large-scale static CGE model of the world economy to analyse the costs of Kyoto in different scenarios with and without Annex B emissions trading and U.S. participation. GDP costs of Kyoto for 2010 without US participation, with Annex B trading, but without use of 'hot air' are estimated at 0.03% for Annex B (without US) for a carbon price of 13 US$/$tCO_2$, with a 6.6% reduction in total Annex-B CO_2. Regional GDP costs are 0.05% for the EU15 and Japan, and 0.1% for Canada, with benefits of 0.2% for the European Economies in Transition and 0.4% for Russia and other countries in Eastern Europe, the Caucasus and Central Asia. Without Annex B trading, the costs are estimated at 0.08% for Annex B (without US).

National and regional studies cited below also suggest similar low or negligible costs for the ratified Kyoto Protocol for Canada, the EU and Japan compared with the estimates in the TAR, depending on the extent of trade in emission permits and CDM/JI certificates. The importance of CDM supply and other assumptions on the carbon price is shown in Figure 11.5 (den Elzen and de Moor, 2002).

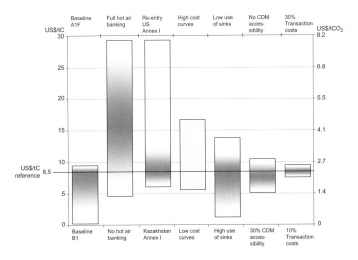

Figure 11.5: *Key sensitivities for the emission permit price from the FAIR model applied to the Kyoto Protocol under the Bonn-Marrakesh Accords*

The following key factors and associated assumptions were chosen for the analysis:

- Baseline emissions: LOW reflects the B1 scenario and HIGH the A1F scenario (IMAGE team, 2001). Our reference is the A1B scenario.
- Hot Air Banking: the LOW case reflects no banking of hot air, with all hot air being banked in the HIGH case; the reference case is one in which hot air banking is optimal for the Annex-I FSU.
- Marginal Abatement Cost (MAC) curves: the MAC curves of WorldScan are used in the reference case while the MAC curves of the POLES model represent the HIGH case.
- Participation Annex I: at the LOW end, we examined the participation of Kazakhstan while the HIGH end reflects US re-entry.
- Sinks: a LOW case has been constructed by assuming that CDM sink credits are capped to 0.5 percent of base year emissions (instead of 1 percent), carbon credits from forest management based on data submitted by the Parties (which are lower than the reported values in Appendix Z, see (Pronk, 2001) and low estimates for carbon credits from agricultural and grassland management using the ALTERRA ACSD model (Nabuurs et al., 2000). The HIGH case reflects sinks credits based on high ACSD estimates for agricultural and grassland management and maximum carbon credits from forest management as reported in Appendix Z. In total, the LOW case implies 70 MtC while the HIGH case implies 195 MtC of carbon credits from sink-related activities. The Marrakesh Accords represent the reference case of 120 MtC.
- CDM Accessibility Factor: this reflects the operational availability of viable CDM projects and is set at 10 percent of the theoretical maximum in the reference case. In the LOW case, we assume no accessibility, while in the HIGH case the factor is set at 30 percent.
- Transaction Costs: the transaction costs associated with the use of the Kyoto Mechanisms are set at 20 percent in the reference case, 10 percent in the LOW case and 30 percent in the HIGH case.

Source: den Elzen and Both (2002, p.43).

11.4.3 National and regional studies of responses to mitigation policies

As individual countries have begun contemplating domestic policy responses (see Chapter 13), an increasing number of studies have focused on more detailed national cost assessments. This increased detail includes both more careful representation of proposed and actual policy responses and more disaggregated results by sector, region, and consumer group. This detail is difficult to achieve in the context of a global model. We briefly summarize the results of studies for various countries/blocks on the basis of the literature available.

11.4.3.1 Policy studies for the United States

Both Fischer and Morgenstern (2006) and Lasky (2003) identify treatment of international trade and the disaggregation of the energy sector as important factors leading to differences in the cost of Kyoto for the US economy. Lasky also identifies energy-demand elasticities and sensitivities to higher inflation as important factors. He concludes that the cost of the US joining Kyoto under Annex I permit trading is between -0.5 to -1.2% of GDP by 2010, with a standardized energy-price sensitivity, and including non-CO_2 gases and sinks, but excluding recycling benefits and any ancillary benefits from improved air quality. The cost falls to 0.2% of GDP with global trading of permits. Barker and Ekins (2004) review the large number of modelling studies dealing with the costs of Kyoto for the US economy that were available when the US administration decided to withdraw from the process. These include the World Resources Institute's meta-analysis (Repetto and Austin, 1997), the EMF-16 studies (Weyant and Hill, 1999) and the US Administration's own study discussed above (EIA, 1998). The review confirms Lasky's range of costs but offsets these with benefits from recycling the revenues from permit auctioning and the environmental benefits of lower air pollution. These co-benefits of mitigation are discussed in Section 11.8 below.

Following U.S. rejection of the Kyoto Protocol, there have been a number of policy proposals in the United States focusing on climate change, most notably two proposed during 2005 Congressional debates over comprehensive energy legislation (the Bingaman and McCain-Lieberman proposals, the Regional Greenhouse Gas Initiative, the Pavley Bill in California, and the earlier proposal by the National Commission on Energy Policy). The costs and other consequences of those proposals are summarized in Table 11.10, as compiled by Morgenstern (2005) from studies by the U.S. Energy Information Administration (1998; 2004; 2005). The sectoral implications of (EIA, 2005) are discussed above in Section 11.3.3.

All estimates derive from EIA's NEMS model, a hybrid top-down, bottom-up model that contains a detailed representation of energy technologies, energy demand, and primary energy supply, coupled with an aggregate model of economic activity (Holte and Kydes, 1997; Kydes, 2000; Gabriel et al., 2001). While the estimates were conducted over a period of seven years, with changes occurring in the baseline, the model produces a remarkably consistent set of estimates, with most physical quantities (including emission reductions) varying more or less linearly with carbon price, and potential absolute GDP impacts varying with the price squared. Real GDP impacts, which include business cycle effects, are less consistent and depend on both policy timing and assumptions about revenue recycling. For example, the real GDP loss of 0.64% shown for 'Kyoto+9%' is reduced to 0.3% by 2020 when recycling benefits are taken into account (EIA, 1998).

Table 11.10: *The EIA's analysis of the Kyoto Protocol, McCain-Lieberman proposal, and Bingman/NCEP proposal: United States in 2020*

	Bingman	McCain-Lieberman	'Kyoto +9%'[a]
GHG emissions (% domestic reduction compared to baseline)	4.5	17.8	23.9
GHG emissions reductions (million metric tons CO_2 reduced per year in 2010)	404	1346	1690
Allowance price (2000 US$ per ton CO_2)	8	33	40
Coal use (% change from baseline)	-5.7	-37.4	-72.1
Coal use (% change from 2003)	14.5	-23.2	-68.9
Natural gas use (% change from baseline)	0.6	4.6	10.3
Electricity price (% change from baseline)	3.4	19.4	44.6
Potential GDP (% change compared to baseline)[b]	-0.02	-0.13	-0.36
Real GDP (% change compared to baseline)[b]	-0.09	-0.22	-0.64

Notes:
a) Kyoto (+9%) refers to a scenario where offsets make up 9% of the U.S. target, thereby allowing domestic emissions to rise 9% above the Kyoto target.
b) GDP in 2020 is estimated to be roughly 20 trillion US$ in 2020, so each 1/100th of a percentage point (0.01%)-equals 2 billion US$. Potential GDP is the level of GDP consistent with long-run growth that fully utilizes available resources.

Source: Morgenstern (2005).

In addition, EIA (2005) analyses the 2004 scenario of the National Commission on Energy Policy. The estimated cost is 0.4% of the reference case GDP by 2025 and the overall growth of the economy is 'not materially altered' (p. 42). However, no costs were included for the implementation of the 'CAFE' transportation sector portion of the NCEP programme that produced most of the emission reductions.

As an independent, government statistical agency, EIA's modelling results tend to be at the centre of most policy debates in the United States. Researchers at MIT (Paltsev *et al.*, 2003) also provided estimates of impacts associated with the McCain-Lieberman proposal that had similar allowance prices but found roughly one-quarter to one-third of the GDP costs reported in the EIA analyses. This is partly explained by the fact that the EIA uses an econometric model to compute GDP costs derived from historic experience in the face of energy price shocks. The MIT and other CGE models assume that, to a large extent, aggregate costs equal the accumulated marginal costs of abatement, typically yielding lower costs than the econometric models (Repetto and Austin, 1997; EIA, 2003).

A threshold question in the McCain-Lieberman discussion has been whether the exclusion of small sources below 10,000 metric tons (e.g. households and agriculture) would alter the efficiency of the program. Pizer *et al.* (2006) use a CGE model to show that exclusion of these sectors has little impact on costs. However, excluding industry roughly doubles costs while implementing alternative CO_2-reducing policies in the power and transport sectors (a renewable energy standard in the power sector and fuel economy standards for cars) results in costs that are ten times higher.

The states in the U.S. have put forward climate policy proposals. An analysis of a package of eight efficiency measures using a CGE model (Roland-Holst, 2006) indicates that it will reduce GHG emissions by some 30% by 2020 – about half of the Californian target of returning to 1990 CO_2 levels by 2020 – with a net benefit of 2.4% for the state's output and a small increase in employment (Hanemann *et al.*, 2006). These results, driven by bottom-up estimates of potential savings in the vehicle and building efficiency, remain controversial, as the debate over vehicle fuel economy standards demonstrates (see NHTSA, 2006 for a discussion of bottom-up estimates and issues).

11.4.3.2   Policy studies for Canada

Jaccard *et al.* (2003) provide estimates of the costs of reaching the Kyoto targets in Canada as part of their wider effort to reconcile top-down and bottom-up modelling results. Using their benchmark run, and assuming compliance without international trading, they find an allowance price of 100 US$/$tCO_2$-eq with an associated GDP loss of nearly 3% in 2010. They note that, while these costs are in line with similar studies of reduction costs in the United States conducted by EIA, they are considerably higher than alternative results for Canada derived from a bottom-up model, and they predict a roughly 33 US$/$tCO_2$-eq allowance price. The authors then show how, by making what they consider longer-run assumptions – lower capital and greater price sensitivity – they can duplicate the lower GDP costs in their model.

11.4.3.3   Policy studies for Europe

Since the TAR, many studies have analysed the macro-economic costs in Europe of committing to Kyoto or other targets, different trade regimes, and multiple greenhouse gases. We report results from some of the key studies below.

An important development within the European Union has been the production of additional detailed results from individual member states. Viguier *et al.* (2003) provide a comparison of four model estimates of the costs of meeting Kyoto targets without trading (except for the EU estimate) based on the 1998 burden sharing agreement replicated in Table

Table 11.11: *A comparison of estimates of domestic carbon prices, welfare, GDP, and Terms of Trade for domestic emissions trading without international allowance trading (except for the EU total) to achieve the 2010 Kyoto target.*

Model	Domestic carbon prices (2000 US$/tCO₂)				Reduction in consumption (%)	Reduction in GDP (%)	Improvement in Terms of Trade (%)
	EPPA	GTEM	POLES	PRIMES	EPPA-EU	EPPA-EU	EPPA-EU
Germany	35.4	52.6	31.8	26.2	0.63	1.17	1.10
France	40.4	-	65.4	42.8	0.67	1.11	1.11
UK	27.1	33.6	39.5	36.6	0.96	1.14	-0.77
Italy	43.7	-	104.6	51.4	1.01	1.47	1.54
Rest of EU	47.6	-	-	65.7	1.23	2.12	1.07
Spain	54.7	-	-	39.8	2.83	4.76	2.06
Finland	64.5	85.9	-	44.6	1.90	2.73	1.67
Netherlands	87.1	-	-	159.3	4.92	7.19	0.55
Sweden	92.2	106.4	-	65.1	3.47	5.11	1.18
Denmark	114.5	118.9	-	56.2	3.97	5.72	-0.74
EU	47.3	46.1	55.9	40.1	Not available	Not available	Not available
USA	68.1	-	52.6	-	0.49	1.01	2.39
Japan	59.8	-	70.8	-	0.22	0.49	2.70

Source: Viguier et al. (2003, p.478)

11.11. EPPA and GTEM are both CGE models, while POLES and PRIMES are partial-equilibrium models with considerable energy sector detail. Viguier *et al.* (2003) explain differences between model results in terms of baseline forecasts and estimates of abatement costs. Germany, for example, has lower baseline emission forecasts in both POLES and PRIMES, but at the same time higher abatement costs. The net effect is that national carbon prices are estimated to be lowest in Germany in POLES and PRIMES while EPPA and GTEM find lower costs in the United Kingdom. Overall, the two general-equilibrium models find similar EU-wide costs located between the POLES and PRIMES estimates.

Viguier *et al.* (2003) go on to discuss the differential consequences across European countries. They find that other measures of cost – welfare and GDP losses – generally follow the pattern of estimated allowances prices, as allowance prices reflect the marginal abatement costs. France, the United Kingdom, and Germany face lower costs and Scandinavian countries generally face higher costs. Terms of trade generally improve for European countries, except for the United Kingdom and Denmark, the former owing to its position as a net exporter of oil and the latter owing to its very low share of fuels and energy-intensive goods in its basket of imports.

There are other studies estimating the equilibrium price in the European market with emissions trading and savings,

as compared to a no-trade case (see also 13.2.1.3). An early study by IPTS (2000) calculates the clearing price in the EU market in 2010 at about 50 US$(2000)/tCO₂ using the POLES model, with a 25% cost reduction arising from emissions trading among countries, and Germany and the UK emerging as net sellers. A more recent study by Criqui and Kitous (2003), which also uses the POLES model, finds even larger gains and lower prices: the equilibrium allowance price is 26 US$(2000)/tCO₂[10], and trading among countries reduces total compliance costs by almost 60%. Without any competition from non-trading European countries and the other Annex B countries on the JI and CDM credits market, they further estimate that the allowance price collapses from 26 US$/tCO₂ to less than 5 US$/tCO₂, and the annual compliance costs are reduced by another 60%. Using the PRIMES model, Svendsen and Vesterdal (2003) find reductions in costs of 13% from trading within the electricity sector in the EU, 32% EU economy-wide trading, and 40% from Annex B trading. Klepper and Peterson (2004; 2006) consider the division between trading and non-trading sectors in the EU, and emphasize the potential inefficiency of generous allocation plans if the non-trading sectors are forced to make up the difference without significant use of the Kyoto mechanisms.

Eyckmans *et al.* (2000) investigate the EU Burden Sharing Agreement on the distribution of the Kyoto emissions reduction target over the EU member states, without the EUETS. Even

---

10 Prices in euros in the citation have been converted at 2000 average rates of $1 to 1 euro.

if only cost efficiency is taken into account, they argue that the burden sharing agreement does not go far enough towards equalizing marginal abatement costs among the member states. For instance, some poorer EU member states have been allowed to increase their emissions considerably, but still their allowances are too low. Introducing a measure of inequality aversion reinforces most of the conclusions.

Other studies have looked at the savings from a multigas approach in Europe. The European Commission (1999) finds that, at a cost below about 50 US$/$CO_2$-eq, 42% of total reduction needed may come from non-CO_2 emissions. Burniaux (2000) finds that a multigas approach reduces the costs of implementing the Kyoto Protocol in the European Union by about one third. For Eastern European countries, the reduction in costs will be even higher when they use a multigas approach. Jensen and Thelle (2001) find similar results using the EDGE model to include non-CO_2 gases, with EU welfare costs falling from about 0.09% to 0.06% in 2010 compared to the baseline.

Babiker *et al.* (2003) use the EPPA-EU model to study the idea that emission permits trade may negatively impact welfare in some cases because of the presence of non-optimal taxation in the pre-trade situation. The selling of permits pushes up a country's carbon price. When a rise in price comes on top of an already distorted fuel price, this results in an additional negative effect on welfare, which might outweigh the gains from sales of permits. It is a negative price effect and a positive income effect. This study finds that some countries, like Scandinavian countries or Spain (mainly importers of carbon permits), would be better off with international trading. Others, like the United Kingdom, Germany or France (mainly exporters of permits) are worse off with trading than without.

In summary, the costs of committing to the Kyoto Protocol may be less than 0.1% of GDP in Europe with flexible trading. U.S. rejection of the Kyoto Protocol reduces the costs of Kyoto in Europe if there are flexible mechanisms in place but, because of the effects of trading terms, pushes costs upwards in the absence of emissions trading or other flexible mechanisms. The permit prices and costs depend on restrictions on trade and the possible exercise of market power in the emission permit market. Multiple greenhouse gas abatement will reduce costs compared to a situation with only CO_2 abatement, a point also emphasized in Section 3.3.5.4.

11.4.3.4 Policy studies for Japan

Masui *et al.* (2005) examine the effects of a carbon tax in Japan to meet the Kyoto target using the AIM (Asia-Pacific Integrated Model). By 2010, a carbon tax with lump-sum recycling leads to an average GDP loss of 0.16% and a tax of 115 US$/$tCO_2$. A tax and subsidy regime with carbon tax revenue used to subsidize CO_2 reduction investments leads to an average GDP loss of 0.03% and a tax of 9 US$/$tCO_2$. By contrast, Hunt and Ninomiya (2005) look at emission trends and argue that as long as growth

is less than 1% per year, and the carbon intensity of energy does not rise, Japan should be able to achieve its target, for example through the Kyoto Target Achievement Plan. If growth is closer to 2% per year, the plan will not suffice.

11.4.3.5 Policy studies for China

The ERI (2003) report on three alternatives for China's development to 2020 presents effects of policies that reduce CO_2 emissions in a 'green growth' scenario. For the same GDP growth of 7% per year, policies of accelerated economic reform, increased energy efficiency standards, higher taxes on vehicle fuels and more use of low-carbon technologies in power generation reduce the growth of CO_2 to 1.7% per year compared to 3.6% per year in an 'ordinary effort' scenario.

Chen (2005) presents a comparison of assumed marginal abatement cost curves and GDP costs associated with various reduction efforts in China in different models (see Figure 11.6 and Table 11.12 below). Chen (p. 891) discusses the reasons for the differences, which are largely due to differences in baselines and assumptions about available technologies and substitution between fossil and non-fossil energy. GDP costs for 2010 vary: 0.2 and 1.5% reduction compared to baseline, associated with a 20% reduction in CO_2 compared to baseline. Garbaccio *et al.* (1999) consider smaller CO_2 reductions – between 5 and 15% – and find not only lower costs, but potentially positive GDP effects after only a few years owing to a double-dividend effect.

11.4.4 Post-Kyoto studies

The macro-economic cost measure adopted in the literature on mitigation costs is generally GDP or gross marketed world output, excluding valuations of environmental costs and benefits.

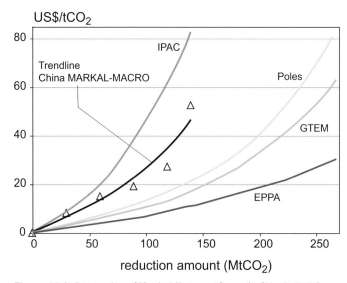

Figure 11.6: *A comparison of Marginal Abatement Curves for China in 2010 from different models*

Source: adapted from Chen (2005, p. 891)

Table 11.12: *A comparison of GDP loss rates for China across models in 2010*

Model	Emission reduction compared to baseline (%)	Marginal carbon abatement cost ((2000)US$/tCO$_2$)	GDP (GNP) loss relative to baseline (%)
GLOBAL 2100	20	25	1.0
	30	50	1.9
GREEN	20	4	0.3
	30	7	0.5
Zhang's CGE model	20	7	1.5
	30	13	2.8
China MARKAL-MACRO	20	18	0.7
	30	22	1.0
	40	35	1.7

Notes: Marginal carbon abatement costs were originally measured at 1990 prices in GLOBAL 2100, at 1985 prices in GREEN, at 1987 prices in Zhang's CGE model, and at 1995 prices in China MARKAL-MACRO. They were converted to 2000 prices for comparison with other carbon prices in the chapter.

Source: adapted from Chen (2005, p.894)

11.4.4.1 A comparison of the macro-economic costs of mitigation to 2030 from modelling studies

Since the TAR, groups of modellers have found a reduction in expected macro-economic costs as a result of the use of multigas options (EMF21, Weyant *et al.*, 2006) (see Section 3.3.5.4) and because carbon prices affect technological change in the models (EMF19, IMCP) (see Section 11.5). Figure 11.7 summarizes the 2030 data brought together in these studies as well in as other post-TAR Category III (stabilization at around 550ppm CO$_2$-eq) studies covered in Chapter 3.[11] The figure is in 3 parts, showing (a) the carbon prices in US$(2000) by 2030 (typically a rising trend) and their effects on CO$_2$ emissions, (b) the effects of CO$_2$ abatement on GDP, and (c) the relationship between carbon prices and gross world output (GDP). All data are differences from the baseline projections for 2030. The studies are grouped around two of the stabilization categories set out in Chapter 3 (Table 3.5), with corresponding insights.

Category IV stabilization trajectories from 25 scenarios: In most models (24 of the 25 scenarios[12]) the 'optimal' trajectory towards stabilization at 4.5W/m^2 (EMF21 studies), or the near-equivalent 550 ppm CO$_2$-only (IMCP and EMF19), requires abatement at less than 20% CO$_2$ compared to baseline by 2030, with correspondingly low-carbon prices (mostly below 20 US$/tCO$_2$-eq, all prices in 2000 US$). Costs are less than 0.7% global GDP, consistent with the median of 0.2% and the 10–90 percentile range –0.6 to 1.2% for the full set of scenarios given in Chapter 3 (see Figure 3.14). Carbon prices in the EMF21 multigas studies for 4.5W/m^2 by 2030 average 18 US$/tCO$_2$-eq, and span 1.2–26 US$/tCO$_2$-eq, except one at 110 US$/tCO$_2$-eq.

Carbon prices in the corresponding 550 ppm CO$_2$-only studies in EMF19 average 14 US$/tCO$_2$ and span 3-19 US$/ tCO$_2$-eq, except one at 50 US$/tCO$_2$. Six of the IMCP 550 ppm CO$_2$-only models have 2030 prices in the range 7–12 US$/tCO$_2$, but four have low to zero prices in 2030, bringing the average to only 6 US$/tCO$_2$.

Category III stabilization trajectories from 12 scenarios: In 11 of the 12 post-TAR scenarios,[13] abatement is less than 40% of CO$_2$ by 2030. Costs are below 1% GDP, consistent with the median of 0.6% and the 10–90 percentile range 0 to 2.5% for the full set in Chapter 3, which also has a range of 18–79 US$/tCO$_2$-eq for carbon prices (see Figure 3.14). The largest comparable dataset available in this category is the IMCP 450ppm CO$_2$-only studies. Most of these produce a carbon price by 2030 in the range 20–45 US$/tCO$_2$, with one higher outlier, and a mean of 31 US$/tCO$_2$ (just over 110 US$/tC). The other Category III models nearly all give higher prices.

The lower estimates of costs and carbon prices for studies assessed here, in comparison with the full set of studies reported in Chapter 3, are mainly caused by a larger share of studies that allow for enhanced technological innovation triggered by climate policies; see 11.5 below. The impact of endogenous technological change is greater for more stringent mitigation scenarios.

Figures 11.7 (a) and (c) show how the carbon prices affect CO$_2$ and global GDP in the models. Note that carbon prices are rising (not shown in Figure 11.7) – sharply for some of the higher numbers – from lower levels in 2020 and also

---

11 These include three scenarios in the U.S. Climate Change Science Program (US CCSP, 2006). Note that the cost assessment presented here is based on a smaller set of scenarios than the assessment in Chapter 3. While Chapter 3 uses the full set of scenarios, including the post-SRES of the TAR, the assessment here relies on post-TAR studies that report information for macro-economic costs. In other words, modelling studies that do not give integrated GDP results are excluded from Figure 11.7 and the associated discussion in this chapter. While Chapter 3 focuses primarily on the assessment of representative cost ranges covering a larger sample, this chapter focuses on the comparative analysis of different post-TAR studies exploring the relationship between the cost indicators and their determinants in the models.

12 The excluded scenario is also an outlier in that FUND is the only EMF21 model to show a declining path for carbon prices, which fall to near zero by 2100 (Weyant *et al*, 2006, p. 25).

13 These scenarios exclude post-SRES results, which did not report carbon prices; see footnote 11. The Category III outlier scenario comes from the CCSP-IGSM model. The price rises to 1651 US$/tCO2 by 2100. This high price is partly due to the assumption of the limited substitution of fossil fuels by electricity as an energy source for transportation: 'In the IGSM scenarios, fuel demand for transportation, where electricity is not an option and for which biofuels supply is insufficient, continues to be a substantial source of emissions.' (US CCSP, 2006, p. 4–21).

(a) Carbon Prices and CO$_2$, 2030

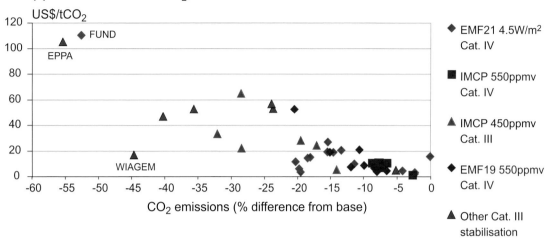

(b) Gross world product and CO$_2$, 2030

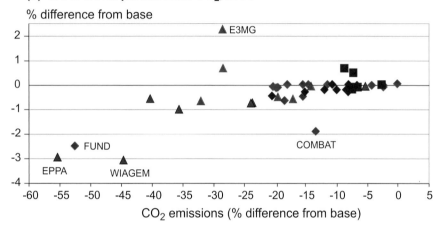

(c) Carbon Prices and Gross World Product, 2030

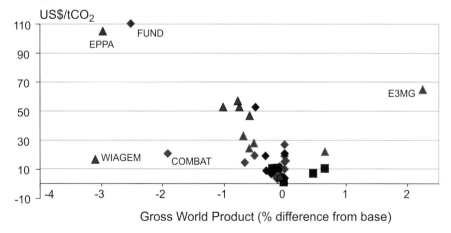

Figure 11.7: *Year 2030 estimated carbon prices and gross-world-product (GDP) costs of various pathways to stabilization targets*

Notes: Figure 11.7 shows, for 2030, the carbon price, CO_2 abatement relative to the baseline, and global GDP differences from baseline for five different sets of stabilization studies: EMF21 radiative forcing at 4.5 W/m^2 (multigas); IMCP at 550 and 450ppm (CO_2-only with induced technological change); EMF19 at 550ppm (CO_2-only with induced technological change) and 6 studies in category III included in Figure 3.24. The results as shown exclude incomplete sets (i.e. data have to be available for all three variables shown). The EMF21 results exclude studies unsuitable for near-term analysis (e.g. substantial effects for a past year). The IMCP results exclude those from two experimental/partial studies. The breakdown into Category III and IV scenarios treats CO_2-only studies as if they also allow for cost-effective non-CO_2 multigas GHG mitigation (see Table 3.14). Note that prices and outputs are based on various definitions, so the figures are indicative only. The price bases in the original studies vary and have been converted to 2000 US$.

Sources: Weyant, 2004; Masui et al., 2005; Edenhofer et al., 2006b, Weyant et al., 2006 and Chapter 3.

after 2030. Most models considered in this analysis therefore suggest that the 20–50 US$/tCO$_2$ cost category of the sector studies is the carbon price level which, if reached globally by 2020–2030, delivers trajectories compatible with subsequent stabilization at mid-category III levels. The corresponding CO$_2$ reduction by 2030 is 5–40% relative to baseline (which varies between studies, with higher baselines giving higher reduction percentages in 2030).

Figure 11.7 (b) shows the CO$_2$ abatement plotted against world GDP. In most studies, higher abatement is associated with higher loss of GDP. The relationships vary, and two models in particular stand out as radically different from others (E3MG and FUND). Three models in the IMCP predict GDP gains under different assumptions.[14] These prices and costs are largely determined by the approaches and assumptions adopted by the modellers, with GDP outcomes being strongly affected by assumptions about technology costs and change processes (see 11.5 below), the use of revenues from permits and taxes (see above), and capital stock and inertia (considered in 11.6) (Barker *et al.*, 2006a; Fischer and Morgenstern, 2006).

11.4.4.2   Other modelling studies

Bollen *et al.* (2004), using Worldscan (a global CGE model), consider the consequences of post-Kyoto policies seeking a 30% reduction for Annex B countries below 1990 levels by 2020. They do not include the CDM, sinks or induced technological change in the modelling. Like most studies, they find dramatically lower costs when global trading occurs. With only Annex I countries participating in emissions trading, the high-growth benchmark case shows an allowance price of about 130 US$(2000)/tCO$_2$, and a 2.2% reduction below baseline for Annex I GDP. With global trading, the allowance price is about 17 US$(2000)/tCO$_2$ and there is a much lower loss of 0.6% in GDP. In a more modest scenario that focuses exclusively on maintaining the current Kyoto targets for all Annex B countries, Russ, Ciscar, and Szabo (2005) estimate a 7 US$(2000)/tCO$_2$ price and a 0.02–0.05% GDP loss in 2025 with global trading (using the POLES and GEM-E3 models).

A number of other studies consider post-Kyoto impact out to 2025 or 2050 based on approaches to stabilization, typically at 550 ppm CO$_2$-eq (category III of Table 3.10) (longer-term strategies are discussed in Chapter 3; discussions of policy mechanisms are covered in Chapter 13). For example, Den Elzen *et al.* (2005), using the IMAGE-FAIR modelling system, show that different assumptions about business-as-usual emission levels and abatement cost curves lead to a range of marginal costs of between 50 US$ and 200 US$/tCO$_2$-eq and of total direct abatement costs of between 0.4 and 1.4% of world GDP in 2050, consistent with a recent EU report (EEA, 2006).

The Stern Review (2006), which was commissioned by the UK Treasury, also considers a range of modelling results. Drawing on estimates from two studies, it reports the costs of an emissions trajectory leading to stabilization at around 500–550ppm CO$_2$-eq. One of the two studies (Anderson, 2006) calculates estimates of annual abatement costs (i.e. not the macro-economic costs) of 0.3% of GDP for 2015, 0.7% for 2025 and 1% for 2050 from an engineering analysis based on several underlying reports of future technology costs. His uncertainty analysis, exploring baseline uncertainties about technology costs and fuel prices, shows a 95% prediction range of costs from –0.5% to +4% of GDP for 2050. The other study is a meta-analysis by Barker et al. (2006a) and looks at the macro-economic costs in terms of GDP effects. The study aims to explain the different estimates of costs for given reductions in global CO$_2$ in terms of the model characteristics and policy assumptions adopted in the studies. With favourable assumptions about international flexibility mechanisms, the responsiveness and cost of low-carbon technological change, and tax reform recycling revenues to reduce burdensome taxes, costs are lowered, and in some cases become negative (i.e. GDP is higher than baseline).

In summary, various post-2012 Kyoto studies have been completed since the TAR. Nearly all those focusing on 550 and 650 ppm CO$_2$-eq stabilization targets (Categories B and C, Table 3.10) with a 5–40% reduction in global CO$_2$ below baseline by 2030, find total costs of about 1% or lower of global GDP by 2030. The critical assumption in these studies is global emissions trading, but there is limited consideration of multi-gas stabilization and endogenous technological change across the studies, and no co-benefits. The few studies with baselines that require higher CO$_2$ reductions to achieve the targets require higher carbon prices and report higher GDP costs. As noted in Sections 11.5 (induced technological change), 3.3.5.4 (multi-gas approaches), and 11.8.1 (co-benefits), these considerations all tend to lower costs or provide non-climate benefits, perhaps substantially.

11.4.5   Differences between models

Research has continued to focus on differences in various cost estimates between models (Weyant, 2000; Weyant, 2001; Lasky, 2003; Weyant, 2003; Barker *et al.*, 2006a; Fischer and Morgenstern, 2006). Weyant (2001) argues that the five major determinants of costs are: projections for base case GHG emissions; climate policy (flexibility, for example); substitution possibilities for producers and consumers; the rate and process of technological change; and the characterization of mitigation benefits. Turning to the base case, he notes the importance of assumptions about population and economic activity, resource availability and prices, and technology availability and costs.

---

14 E3MG (Barker et al., 2006b) takes a Post Keynesian approach, allowing under-used resources in the global economy to be taken up for the extra low-carbon investment induced by climate policies when permit/tax revenues are recycled by reducing indirect taxes. Such a response to revenue recyling is a feature of regional studies reported in the TAR (p. 518). FEEM-RICE (Bosetti et al., 2006) allows international cooperation in climate policies to increase the productivity of R&D investment. ENTICE-BR (Popp, 2006a), in a scenario which assumes a high elasticity of substitution between backstop and fossil fuels, shows increasing global output above baseline with more stringent stabilization targets (p. 173).

The key policy feature is flexibility, in other words whether trading over companies, nations, gases, and time is allowed. Substitution possibilities are governed by assumptions about the malleability of capital, economic foresight, and technology detail. Technology modelling includes assumptions about whether technological change is endogenous or exogenous, and whether technology costs drop with increasing use of technologies. Finally, mitigation benefits may be included in varying degrees in different models.

The factors accounting for differences between cost estimates can be divided into three groups: features inherent in the economies being studied (for example, high substitution possibilities at low cost), assumptions about policy (such as the use of international trading in emission permits, or the recycling of auction revenues), and simplifying assumptions chosen by the model builders to represent the economy (how many sector or regions are included in the model). The first two sets of factors can be controlled by specifying the countries and time-scales of the mitigation action, and the exact details of the policies, as in the EMF-16 studies. However, differences in modellers' approaches and assumptions persist in the treatment of substitution and technology. The various factors can be disentangled by a meta-analysis of published finding (this may include an analysis of analyses). This technique was first used in this context by Repetto and Austin (1997) in a mitigation-cost analysis of GDP costs for the US economy. Fischer and Morgenstern (2006) conduct a similar meta-analysis dealing with the carbon prices (taken to be the marginal abatement costs) of achieving Kyoto targets reported by the EMF-16 studies and discussed in the TAR (Weyant and Hill, 1999).

The crucial finding of these meta-analyses is that most of the differences between models are accounted for by the modellers' assumptions. For example, the strongest factor leading to lower carbon prices is the assumption of high substitutability between internationally-traded products. Other factors leading to lower prices include the greater disaggregation of product and regional markets. This suggests that any particular set of results about costs may well be the outcome of the particular assumptions and characterization of the problem chosen by the model builder, and these results may not be replicated by others choosing different assumptions.

Like earlier studies, the comparison of model results in Barker et al., (2006a) emphasizes that the uncertainty in costs estimates comes from both policy and modelling approaches as well as the baseline adopted. Uncertainty about policy is associated with the design of the abatement policies and measures (flexibility over countries, greenhouse gases and time) and with the use of carbon taxes or auctioned CO_2 permits to provide the opportunity for beneficial reforms of the tax system or incentives for low-carbon innovation. In addition,

targeted reductions in fossil-fuel use resulting from climate policies can yield benefits in terms of non-climate policy e.g. reductions in local air pollution. Uncertainty about the modelling approaches is associated with the extent to which substitution is allowed in terms of backstop technology, whether the economy responds efficiently (in terms of the use of CGE models), and whether technological change is assumed to respond to carbon prices, the topic of the next section. Uncertainty about the baseline is associated with assumptions adopted for rates of technological change and economic growth, and future prices of fossil fuels.

11.5 Technology and the costs of mitigation

11.5.1 Endogenous and exogenous technological development and diffusion

A major development since the TAR has been the treatment of technological change in many models as endogenous – and therefore potentially induced by climate policy – compared to previous assumptions of exogenous technological change that is unaffected by climate policies (see glossary for definitions). This section discusses the effect of the new endogenous approach on emission permit prices, carbon tax rates, GDP and/ or economic welfare, and policy modelling (Chapter 2, Section 2.7.1 discusses the concepts and definitions, and Chapter 13 provides a broader discussion of mitigation and technology policy choices).

The TAR reported that most models make exogenous assumptions about technological change (9.4.2.3) and that there continues to be active debate about whether the rate of aggregate technological change will respond to climate policies (7.3.4.1). The TAR also reported that endogenizing technological change could shift the optimal timing of mitigation forward or backward (8.4.5). The direction depends on whether technological change is driven by R&D investments (suggesting less mitigation now and more mitigation later, when costs decline) or by accumulation of experience induced by the policies (suggesting an acceleration in mitigation to gain that experience, and lower costs, earlier). Overall, the TAR noted that differences in exogenous technology assumptions were a central determinate of differences in estimated mitigation costs and other impacts.

Table 11.13 lists the implications for modelling of exogenous and endogenous technological change[15] and demonstrates the challenges for research. The table shows that, at least in their simplified forms, the two types of innovation processes potentially have very different policy implications in a number of different dimensions.

---

15 See 'technological change' in the Glossary.

Table 11.13: *Implications of modelling exogenous and endogenous technological change*

	Exogenous technological change	Endogenous technological change
Process	Technological change depends on autonomous trends	Technological change develops based on behavioural responses, particularly (a) choices about R&D investments that lower future costs; and (b) levels of current technology use that lower future technology cost via learning-by-doing
Modelling implications		
Modelling term	Exogenous	Endogenous / induced
Typical main parameters	Autonomous Energy Efficiency Index (AEEI)	Spillovers to learning / return to R&D / cost of R&D / Learning rate
Optimization implications (note: not all modelling exercises are dynamically optimized)	Single optimum with standard techniques	Potential for multiple-equilibria; unclear whether identified solutions are local or global optima
Economic/policy implications		
Implications for long-run economics of climate change	Atmospheric stabilization below approximately 550 ppm CO_2 likely to be very costly without explicit assumption of change in autonomous technology trends.	Stringent atmospheric stabilization may or may not be very costly, depending on implicit assumptions about responsiveness of endogenous technological trends.
Policy instruments that can be modelled	Taxes and tradable permits	Taxes and tradable permits as well as R&D and investment incentives / subsidies
Timing implications for mitigation and mitigation costs associated with cost minimization	Arbitrage conditions suggest that the social unit cost of carbon should rise over time roughly at the rate of interest.	Learning-by-doing implies that larger (and more costly) efforts are justified earlier as a way to lower future costs.
'First mover' economics	Costs with few benefits	Potential benefits of technological leadership, depending on assumed appropriability of knowledge
International spillover / leakage implications	Spillovers generally negative (abatement in one region leads to industrial migration that increases emissions elsewhere)	In addition to negative spillovers from emission leakage / industrial migration, there are also positive spillovers (international diffusion of cleaner technologies induced by abatement help to reduce emissions in other regions)

The role of technology assumptions in models continues to be viewed as a critical determinant of GDP and welfare costs, and emission permit prices or carbon tax rates (Barker *et al.*, 2002; Fischer and Morgenstern, 2006). These analyses cover large numbers of modelling studies undertaken before 2000 and regard the treatment of technology as influential in reducing costs and carbon prices, but find that the cross-model results on the issue are conflicting, uncertain and weak. Since the TAR, there has been considerable focus on the role of technology, especially in top-down and hybrid modelling, in estimating the impact of mitigation policies. However, syntheses of this work tend to reveal wide differences in the theoretical approaches, and results that are strongly dependent on a wide range of assumptions adopted (Barker *et al.*, 2006a; Stern, 2006), about which there is little agreement (DeCanio, 2003).

The approaches to modelling technological change (see Section 2.7.2.1), include (1) explicit investment in research and development (R&D) that increases the stock of knowledge, (2) the (typically) cost-free accumulation of applying that knowledge through 'learning-by-doing' (LBD); and (3) spillover effects. These approaches are in addition to simple analyses of sensitivity to cost assumptions, especially when

technological change is treated as exogenous. There have been many reviews (see Clarke and Weyant, 2002; Grubb *et al.*, 2002b; Löschel, 2002; Jaffe *et al.*, 2003; Goulder, 2004; Weyant, 2004; Smulders, 2005; Grübler *et al.* 2002; Vollebergh and Kemfert, 2005; Clarke *et al.*, 2006; Edenhofer *et al.*, 2006b; Köhler *et al.*, 2006; Newell *et al.*, 2006; Popp, 2006b; Sue Wing, 2006; Sue Wing and Popp, 2006). One feature that emerges from the studies is the considerable variety in the treatment of technological change and its relationship to economic growth. Another is the substantial reductions in costs apparent in some studies when endogenous technological change is introduced, comparable to previously estimated cost savings from *ad hoc* increases in the exogenous rate of technological change (Kopp, 2001) or in the modelling of advanced technologies (Placet *et al.*, 2004 p. 5.2 & 8.10).

This section reviews the effect of endogenizing technological change on model estimates of the costs of mitigation. It follows the majority of the literature and takes a cost-effectiveness approach to assess the costs associated with particular emission or cumulative emission goals, such as post-2012 CO_2 reduction below 1990 levels or medium-term pathways to stabilization.

Table 11.14: *Technology policies and modelling approaches*

Policies	Modelling approach	Key points for measuring costs
R&D in low-GHG products and processes from: • Corporate tax incentives for R&D (supply-push R&D) • More government-funded R&D (supply-push R&D)	• Explicit modelling of R&D stock(s) that are choice variables, like capital, and enter the production function for various (low-carbon) goods. • R&D policies can be modelled as explicit increases in R&D supply or subsidies for the R&D price.	The assumed rate of return from R&D, typically based on an assumption that there are substantial spillovers and that the rate of return to R&D is several times higher than conventional investment at the margin due to spillover. Another important point is the assumed cost of R&D input, which may be high if it is taken from other R&D (crowding-out)
Learning-by-doing: • Purchase requirements or subsidies for new, low-GHG products • Corporate tax incentives for investment in low-GHG products and processes	More production from a given technology lowers costs.	Rate at which increases in output lowers costs and long-run potential for costs to fall.

The review shows that endogenizing technological change – via R&D responses and learning-by-doing – lowers costs, perhaps substantially, relative to estimates where the path of technological change is fixed from the baseline. The degree to which costs are reduced hinges critically on assumptions about the returns from climate change R&D, spillovers (across sectors and regions) associated with climate change R&D, crowding-out associated with climate change R&D, and (in models with learning-by-doing) assumed learning rates. Table 11.14 shows the policies that have been modelled to induce technological change, and how they have been introduced into the models.

The policies are in two groups: effects through R&D expenditure, and those through learning-by-doing.Unfortunately, our empirical understanding of these phenomena over long periods of time is no better than our ability to forecast exogenous rates of change. As Popp (2006b) notes, none of the ETC models he reviews use empirical estimates of technological change to calibrate the models because, until recently, there were few empirical studies of innovation and environmental policy. So although we are confident that mitigation costs will be lower than those predicted by models assuming historically-based, exogenous rates of technological change, views continue to differ about how much lower they will be.

11.5.2 Effects of modelling sectoral technologies on estimated mitigation costs

The Energy Modelling Forum conducted a comparative study (EMF19) with the aim of determining how models for global climate change policy analyses represent current and potential future energy technologies, and technological change. The study assesses how assumptions about technology development – whether endogenous or exogenous – might affect estimates of aggregate costs for a 550 ppm CO_2 concentration stabilization target. The modellers emphasize the detailed representations for one or more technologies within integrated frameworks. Weyant (2004) summarizes the results, which indicate low GDP costs and a wide range of estimated carbon tax rates hinging on assumptions about baseline emission growth, as

well as technology developments with regard to carbon capture, nuclear, renewables, and end-use efficiency. Figure 11.8 shows that the carbon tax rates are very low before 2050, with all models indicating values below about 14 US$/tCO$_2$ to 2030. Six of the nine models generate values below 27 US$/tCO$_2$ by 2050. By comparison, the EU ETS price of carbon reached nearly 35 US$/tCO$_2$ in August 2005 and again in April 2006.

Perhaps more revealing in the EMF-19 study are the specific features chosen by various modelling teams in their respective papers. Six teams focused on carbon capture and storage (Edmonds *et al.*, 2004; Kurosawa, 2004; McFarland *et al.*, 2004; Riahi *et al.*, 2004; Sands, 2004; Smekens-Ramirez Morales, 2004), one on nuclear (Mori and Saito, 2004), one on renewables (van Vuuren *et al.*, 2004), two on end-use efficiency (Akimoto *et al.*, 2004; Hanson and Laitner, 2004), and one on an unspecified carbon-free technology (Manne and Richels, 2004).

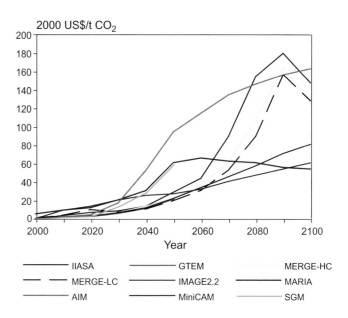

Figure 11.8: *Carbon price projections for the 550 ppm CO₂-only stabilization scenario*
Source: Weyant (2004).

The impacts associated with varying technology assumptions within a given model ranged from a net economic gain, to substantial cuts in the cost of stabilization, to almost no effect on the cost of stabilization.

Despite the wide range of results, they suggest some overarching conclusions (Weyant, 2004). First, technological development, however and under whatever policy it unfolds, is a (if not *the*) critical factor determining the long-term costs and benefits of mitigation. Second, there is no obvious silver bullet: a variety of technologies may be important depending on local circumstances in the future, and a portfolio of investments will be necessary to achieve significant mitigation at lower costs. Third, major technology shifts like carbon capture, advanced nuclear, and hydrogen require a long transition as learning-by-doing accumulates and markets expand so that they tend to play a more significant role in the second half of the century. By contrast, end-use efficiency may provide major opportunities in the shorter term.

11.5.3 The costs of mitigation with and without endogenous technological change

Modellers have pursued two broad approaches to endogenizing technological change, usually independently of each other: explicit modelling of R&D activities that contribute to a knowledge stock and reduce costs, and the accumulation of knowledge through learning-by-doing. Sijm (2004) and Edenhofer *et al.* (2006b) provide detailed comparative assessments of different implementations of both approaches with a focus on mitigation costs when endogenous technology effects are 'switched on'. Their syntheses provide a useful window for understanding the variation in results and how policies might induce technological change.

In his review, Sijm (2004) distinguishes top-down models that mostly focus on explicit R&D effects, and bottom-up models that focus mostly on LBD effects. Among the top-down models, which are described in Table 11.15, he finds considerable variation in the effect of including Endogenous Technological Change (ETC). While some models find a large reduction in mitigation costs (e.g. Popp, 2006a), some find small impacts (e.g. Nordhaus, 2002). These differences can be attributed to:
- the extent of substitution allowed of low-carbon fuels for high-carbon fuels. When this factor is included, the reduction in costs is more pronounced, and the higher it is, the greater the reduction.
- the degree of 'crowding-out' associated with energy R&D expenditures. If new energy R&D is assumed to be in addition to existing R&D, this will generate larger reductions in mitigation than if new energy R&D is assumed to lead to a reduction in R&D elsewhere.
- the approach to spillover. In addition to justifying higher rates of return from R&D, spillover implies that the market outcome with too little investment could be improved by policy intervention.

- the degree of differentiation among R&D activities, the assumed rates of return from those activities, and the capacity of R&D activities to lower costs for low-carbon technologies.
- the rate of learning if LBD is included. Higher rates imply larger reductions in mitigation costs with ETC included.

The first point is that the way low-carbon and high-carbon energy are treated in the models –whether as complements or substitutes – is critical is determining the flexibility of the model to low-carbon innovation and costs of mitigation. Models that do not allow high levels of substitution between low-carbon and high-carbon energy (Goulder and Mathai, 2000; Nordhaus, 2002; Popp, 2006b) indicate that R&D has less impact than those that do, e.g. by introducing a carbon-free backstop technology (Gerlagh and Lise, 2005; Popp, 2006b). Similar results are found more widely for LBD and R&D models: the more substitution possibilities allowed in the models, the lower the costs (Edenhofer *et al.*, 2006a, p.104).

When providing evidence to support the second point – the studies of induced R&D effects via the stock of knowledge – Goulder and Schneider (1999), Goulder and Mathai (2000), Nordhaus (2002), Buonanno *et al.* (2003) and Popp (2004) differ considerably about the extent of crowding-out. In other words, does R&D have an above-average rate of return and does an increase in R&D to support the carbon-saving technologies come from ordinary production activities (no crowding-out), or equally valuable R&D in other areas (crowding-out)? Nordhaus (2002) assumes complete crowding-out in which carbon-saving R&D has a social rate of return that is four times the private rate of return but, because it is assumed that it replaces other equally valuable R&D activities, it costs four times as much as conventional investment. At the other extreme, Buonanno *et al.* (2003) consider spillovers that lead to similarly high social rates of return, but without the higher opportunity costs. Not surprisingly, Nordhaus finds very modest mitigation cost savings and Buonanno *et al.* find enormous savings. In general, induced technological change in a general-equilibrium framework has its own opportunity costs, which may reduce the potential for cost reduction in CGE models substantially.

Popp (2006b), in turn, suggests on the basis of the empirical evidence that half of the R&D spending on energy in the 1970s and 1980s took place at the expense of other R&D. Something between full and partial crowding-out appears more recently in Gerlagh and Lise (2005). Goulder and Matthai (2000) provide an example of the importance for cost reduction of parameters describing returns from R&D and capacity for innovation. They compare both R&D as new knowledge and learning-by-doing (LBD), finding a 29% reduction in the marginal costs with R&D by 2050 and 39% with LBD. As they note, however, this reflects the calibration of their model to a 30% cost saving based on Manne and Richels' assumptions (1992). The model results simply reflect the choice of calibrated parameter values.

Table 11.15: *Treatment of endogenous technological change (ETC) in some global top-down integrated assessment models*

Study	Model	ETC channel	Number of production sectors	Number of regions	Major results (impact of ETC)	Comments	Focus of analysis
Barker et al., 2006b	E3MG, econometric	LBD and R&D	41	20	Cumulative investments and R&D spending determine energy demand via a technology index. Learning curves for energy technologies (electricity generation). Cumulative investments and R&D spending determine exports via a technology index.	Econometric model. Investments beyond baseline levels trigger a Keynesian multiplier effect. Sectoral R&D intensities stay constant overtime	Long-term costs of stabilization Income and production losses
Bollen, 2004	WorldScan CGE	R&D R&D (and occasionally LBD)	12	12	ETC magnifies income losses.	Includes international spillovers. No crowding-out effect	Compliance costs of Kyoto protocol
Buonanno et al., 2003	FEEM-RICE optimal growth	R&D and LBD	1	8	Direct abatement costs are lower, but total costs are higher. ET ceilings have adverse effects on equity and efficiency.	Factor substitution in Cobb-Douglas production.	Impact of emissions trading (+ restrictions)
Bosetti et al., 2006	FEEM-RICE	LBD	1	8	An index of energy technological change increases elasticity of substitution. Learning-by-doing in abatement and R&D investments raise the index. Energy technological change explicitly decreases carbon intensity.		Experimental model exploring high inertia.
Crassous et al., 2006	IMACLIM-R GCE	R&D and LBD	1	5	Cumulative investments drive energy efficiency. Fuel prices drive energy efficiency in transportation and residential sector. Learning curves for energy technologies (electricity generation).	Endogenous labour productivity, capital deepening.	
Edenhofer et al., 2006a	MIND Optimal growth	LBD	1	1	R&D investments improve energy efficiency. Factor substitution in a constant-elasticity-of-substitution (CES) production function. Carbon-free energy from backstop technologies (renewables) and CCS. Learning-by-doing for renewable energy. R&D investments in labour productivity. Learning-by-doing in resource extraction		
Gerlagh and Van der Zwaan, 2003	DEMETER Optimal growth	LBD	1	1	Costs are significantly lower. Transition to carbon-free energy. Lower tax profile. Early abatement	Results are sensitive to elasticity of substitution between technologies as well as to the learning rate for non-carbon energy	Optimal tax profile Optimal abatement profile Abatement costs
Gerlagh, 2006	DEMETER-1 CCS	LBD	1	1	Factor substitution in CES production. Carbon-free energy from renewables and CCS. Learning-by-doing for both and for fossil fuels.		

Note: See sources for details of models.

Sources: The table is derived from Sijm (2004) and Edenhofer et al. (2006b).

Table 11.15: *Continued*

Study	Model	ETC channel	Number of production sectors	Number of regions	Major results (impact of ETC)	Comments	Focus of analysis
Goulder and Mathai, 2000	Partial cost-function model with central planner	R&D LBD	1	1	Lower time profile of optimal carbon taxes. Impact on optimal abatement varies depending on ETC channel. Impact on overall costs and cumulative abatement varies, but may be quite large	Deterministic One instrument High aggregation Weak database	Optimal carbon tax profile Optimal abatement profile
Goulder and Schneider, 1999	CGE multisectoral model	R&D	7	1	Gross costs increase due to R&D crowding-out effect. Net benefits decrease.	Lack of empirical calibration Focus on U.S. Full 'crowding-out' effect	Abatement costs and benefits
Kverndokk et al., 2004	CGE model for a small open economy	LBD	1	1	Innovation subsidy is more important in the short term than a carbon tax. Innovation subsidy may lead to 'picking a winner' and 'lock in'	Numerical illustrative model	Optimal timing and mixture of policy instruments Welfare effects of technology subsidies
Masui et al., 2006	AIM/Dynamic - Global	R&D	9	6	Factor substitution in CES production. Investments in energy conservation capital increase energy efficiency for coal, oil, gas and electricity. Carbon-free energy from backstop technology (nuclear/renewables).		Focus on energy efficiency with limited supply-side substitution.
Nordhaus, 2002	R&DICE optimal growth	R&D	1	8	ETC impact is lower than substitution impact and quite modest in early decades.	Deterministic Full 'crowding-out' of R&D High aggregation	Factor substitution versus ETC Carbon intensity Optimal carbon tax
Popp, 2004	ENTICE, optimal growth	R&D	1	1	Impact on cost is significant. Impact on emissions and global temperature is small	Partial crowding-out effect	Welfare costs Sensitivity analysis of R&D parameters
Popp, 2006a	ENTICE-BR	R&D	1	1	Factor substitution in Cobb-Douglas production. R&D investments in energy efficiency knowledge stock. Carbon-free energy from generic backstop technology	R&D investments lower price of energy from backstop technology.	
Rao et al., 2006	MESSAGE/MACRO CGE	LBD	1	11	Carbon-free energy from backstop technologies (renewables), carbon scrubbing & sequestration). Learning curves for electricity generation and renewable hydrogen production	Factor substitution in CES production in MACRO.	
Rosendahl, 2004	Builds on Goulder and Mathai (2000)	LBD	1	2	Restrictions on emissions trading are cost-effective. Optimal carbon tax in Annex I region is increased with external spillovers	Outcomes are sensitive to learning rate, discount rate and slope of abatement curve	Optimal carbon tax (or permit price) over time in two regions Optimal emissions trading +restrictions

Table 11.16: *Learning rates (%) for electricity generating technologies in bottom-up energy system models*

Learning	(a) One-factor learning curves				(b) Two-factor learning curves			
	ERIS	MARKAL	MERGE-ETL	MESSAGE	ERIS		MERGE-ETL	
					LDR	LSR	LDR	LSR
Advanced coal	5	6	6	7	11	5	6	4
Natural gas combined cycle	10	11	11	15	24	2	11	1
New nuclear	5	4	4	7	4	2	4	2
Fuel cell	18	13	19	–	19	11	19	11
Wind power	8	11	12	15	16	7	12	6
Solar PV	18	19	19	28	25	10	19	10

Notes:
- Learning rates are defined as the percentage reduction in unit cost associated with a doubling of output. The acronym LDR stands for Learning-by-Doing Rates and LSR for Learning-by-Searching Rates in two-factor learning curves. In two-factor learning curves, cumulative capacity and cumulative R&D (or 'knowledge stock') are used to represent market experience (learning-by-doing) and knowledge accumulated through R&D activities, respectively.
- In MERGE-ETL, endogenous technological progress is applied to 8 energy technologies: six power plants (integrated coal gasification with combined cycle, gas, turbine with combined cycle, gas fuel cell, new nuclear designs, wind turbine and solar photovoltaic) and two plants producing hydrogen (from biomass and solar photovoltaic). Furthermore, compared to the original MERGE model, Bahn and Kypreos (2002; 2003a) have introduced two new power plants (using coal and gas) with CO2 capture and disposal in depleted oil and gas reservoirs. Like the MARKAL model, the ERIS model is a bottom-up energy system model. Both studies mentioned above cover six learning technologies. MESSAGE is also a bottom-up system engineering model. Like the other bottom-up energy system models, it determines how much of the available resources and technologies are used to satisfy a particular end-use demand, subject to various constraints, while minimizing total discounted energy system costs.
- For a review of the literature on learning curves, including 42 learning rates of energy technologies, see McDonald and Schrattenholzer, (2001).
- For a discussion and explanation of similar (and even wider) variations in estimated learning rates for wind power, see Söderholm and Sundqvist (2003) and Neij et al. (2003a; 2003b)

Sources: Sijm (2004), Messner (1997), Seebregts et al. (1999), Kypreos and Bahn (2003), and Barreto and Klaassen (2004), Barreto (2001), Barreto and Kypreos (2004), and Bahn and Kypreos (2003b).

By contrast with the results for top-down models, Sijm (2004) finds considerably more consistency among bottom-up models, where the effects of learning-by-doing typically reduce costs by 20% to 40% over the next half century, and by 60% to 80% over the next century. Importantly, however, these numbers are relative to a static technology alternative. To demonstrate the influence of this assumption, van Vuuren et al. (2004) run their model without a carbon constraint, but with learning, to identify a baseline level of technological change. Their approach roughly halves the estimated effect of ETC on mitigation.

The variations in the estimated effects of learning on costs in bottom-up models are driven primarily by variations in the assumed rate of learning (in other words, the extent of the reduction in costs for each doubling of installed capacity). Estimates of these rates vary, depending on whether they are assumed or econometrically estimated, and whether they derive from expert elicitation or historical studies. These learning rates vary between four leading models by as much as a factor of two for a given technology, as shown in Table 11.16.

The modelling of LBD is beset with problems. Model solutions become more complex because costs can fall indefinitely, depending on the extent of the market. Avoidance of multiple solutions typically requires the modeller to constrain the penetration of new technologies, making one element of the cost reduction effectively exogenous. Since many low-carbon technologies are compared with mature energy technologies

early in the learning process, it becomes inevitable that their adoption spreads and that they eventually take over as carbon prices fall. Finally, the approach often assumes that diffusion and accompanying R&D are cost-free, although the investments required for the technologies with high learning rates may be comparable with those that are replaced.

In addition, the measurement of learning rates poses econometric problems. It is difficult to separate the effects of time trends, economies of scale and R&D from those of LBD (Isoard and Soria, 2001) and different functional forms and data periods yield different estimates, so the learning rates may be more uncertain than suggested by their treatment in the models. When controls for the effects of other variables are included, such as crowding-out effects, the influence of LBD in some models may become very small compared to the effect of R&D (Köhler et al., 2006; Popp, 2006c).

A second survey of ETC effects on aggregate mitigation costs comes from the Innovation Modelling Comparison Project (IMCP) (Edenhofer et al., 2006b). Rather than reviewing previous results, the IMCP engaged modelling teams to report results for specific concentration scenarios and, in particular, with and without ETC. Like the van Vuuren et al. (2004) study noted earlier, the IMCP creates a baseline technology path with ETC but without an explicit climate policy. This baseline technology path can then be either fixed, as autonomous technological change, or allowed to change in response to the climate policy.

(a) Averaged effects of including ETC on carbon price

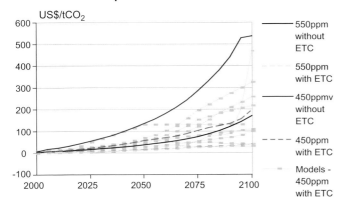

(b) Averaged effects of including ETC on CO$_2$ emissions

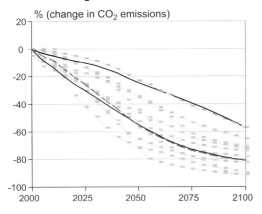

(c) Averaged effects of including ETC on GDP

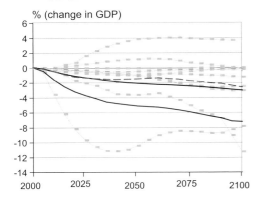

Figure 11.9: *Averaged effects of including ETC on carbon tax rates, CO$_2$ emissions and GDP: 9 global models 2000–2100 for the 450 ppm and 550 ppm CO$_2$-only stabilization scenarios*

Notes: The figures show the simple averages of results from 9 global models 2000–2100 for (a) carbon tax rates and emission permit prices in US$(2000)/tCO$_2$, (b) changes in CO$_2$ (% difference from base) and (c) changes in global GDP (% difference from base). The results are shown with and without endogenous technological change. The grey background lines show the range from the individual models for 450 ppm with ETC. See source for details of models.

Source: adapted from Edenhofer et al. (2006b).

Table 11.15 also summarizes the treatment of technological change in the IMCP models; in principle, the wide range of approaches provides additional confidence in the results when common patterns emerge. Like Sijm (2004), Edenhofer *et al.* (2006b) find that, while ETC reduces mitigation costs, there continues to be a wide range of quantitative results: some are close to zero and others generate substantial reductions in costs.

Figure 11.9 shows the effects of introducing ETC into the models for the 550 and 450 ppm CO$_2$ stabilization scenarios 2000–2100. The reductions in carbon prices and GDP are substantial for many studies in both stabilization cases when ETC is introduced. The effects on CO$_2$ reductions show that including ETC in the models leads to earlier reductions in emissions. It should be noted that the reduction of costs in IMCP models is not mainly driven by LBD. The assumptions about the crowding-out of conventional R&D by low-carbon R&D and the availability of mitigation options (models have different sets of options) are more important factors determining costs and mitigation profiles than LBD (Edenhofer *et al.*, 2006b, p.101–104). One major research challenge is to test the influence of these aspects of ETC on current technologies by econometric and backcasting methods, fitting the models to historical data.

Figure 11.9 emphasizes the range and the uncertainty of the results for induced technological change[16] from climate policies. The potential of ETC to reduce mitigation costs varies remarkably between different model types. For a 450 ppm CO$_2$-only concentration stabilization level at the upper end of the range, including ETC in the model reduces mitigation costs by about 90%, but at the lower end it makes no difference (Edenhofer *et al.* 2006b, p. 74). The averages also somewhat exaggerate the effects of ETC because there are other assumptions that affect the costs, as evident in a meta-analysis of the macro-economic costs of mitigation undertaken for the UK Treasury's Stern Review (Barker *et al.*, 2006b). An example is the use of tax/permit revenues, as discussed in 11.4.4 above. This study combines the IMCP results on costs with earlier data on post-SRES scenarios (Repetto and Austin, 1997; Morita *et al.*, 2000) so that the effects of other assumptions can be identified. The average effects of including ETC in the IMCP models by 2030 for pathways to 550 and 450 ppm CO$_2$-only are reduced from 1.1 and 2.7% of global GDP compared to baseline, as shown in Figure 11.9, to 0.4 and 1.3% respectively using the full equation of the meta-analysis, which allows for individual model outliers, time and scenario effects as well as the approaches and assumptions adopted by the modellers. In other words, allowing for technologies to respond to climate policies reduces the GDP costs of Category III stabilization, as estimated by the IMCP models, by 1.3% by 2030. Costs across models of 2.1% without ETC, but allowing for emissions trading and backstop technologies, are reduced to 0.8% GDP by 2030 with ETC. The ETC effects become more

---

16 When a model includes ETC, further change can generally be induced by economic policies. Hence the term 'induced technological change' (ITC); ITC cannot be studied within a model unless it simulates ETC. See Glossary on 'technological change'.

pronounced in the reduction of costs for later years and as the stabilization targets become more stringent, partly due to the associated extra increases in the required carbon prices.

Edenhofer *et al.* conclude that the results for effects of ETC depend on:

- baseline effects: baseline assumptions about the role of technology that generate relatively low emission scenarios can leave little opportunity for further ETC effects;

- the assumption of the inefficient use of resources in the baseline (distinct from market failure associated with greenhouse gas emissions and climate change): this provides opportunities for policy to improve otherwise inefficient private decisions and may even raise welfare. Spillovers were an example of this in the Sijm (2004) discussion; some simulations also include inefficient energy investment decisions.

- how the investment decision is modelled: recursive savings decisions, as opposed to foresight and intertemporal opti-mization, provide less opportunity for investment and R&D to expand. In the Sijm (2004) context, less responsiveness in aggregate investment and R&D implies more crowding-out.

- the modelling of substitution towards a backstop techno-logy (such as a carbon-free energy source available at constant, albeit initially high, marginal cost): this can substantially affect the results. For example, if investment in the technology is endogenous and involves learning-by-doing, costs can fall dramatically. Popp (2006a, p.168) goes further, and shows that the addition of a backstop technology by itself can have a larger effect on mitigation costs than the addition of LBD. These results are also confirmed by the IMCP study (Edenhofer *et al.*, 2006b, p.214). However, investment in backstop technologies requires time-consistent policies (Montgomery and Smith, 2006). It is therefore debatable to what extent the indicated potential for cost reduction can be realized under real-world conditions where a global, long-term and time-consistent climate policy has yet to be implemented.

11.5.4 Modelling policies that induce technological change

Most of the studies discussed so far consider only how endogenous technological change affects the cost associated with correcting the market failure of damaging GHG emissions through market-based approaches to carbon taxes and/or emissions trading schemes. However, when spillovers from low-carbon innovation are introduced into the modelling of ETC, for example where the social rate of return exceeds the private rate of return from R&D because innovators cannot capture all the benefits of their investment, there is a second market failure. This implies that at least two instruments should be included for policy optimization (Clarke and Weyant, 2002, p.332; Fischer, 2003; Jaffe *et al.*, 2005). Even without the spillover effect, however, the advantage of models with endogenous technological change is their potential to model the

effect of technology policy, distinct from mitigation policy, or in tandem. As discussed in Chapter 13, there has been increasing interest in such policies.

Surprising, few models have explored this question of mitigation versus technology policies, and they have focused instead on the cost assessments reviewed above. Those studies that have looked at this question find that technology policies alone tend to have smaller impacts on emissions than mitigation policies (Nordhaus, 2002; Fischer and Newell, 2004; Popp, 2006b; Yang and Nordhaus, 2006). In other words, it is more important to encourage the use of technologies than to encourage their development. On the other hand, with the existence of spillovers, technology policies alone may lead to larger welfare gains (Otto *et al.*, 2006). However, the same study points out that an even better policy (in terms of improving welfare) is to fix the R&D market failure throughout the economy. Given the difficulty in correcting the economy-wide market failure (e.g., through more effective patent protection or significantly increased government spending on research), it may be unrealistic to expect successful correction within the narrow area of energy R&D. This is true despite our ability to model such results.

However, this does open up the possibility of portfolios of policies utilizing some of the revenues from emission permit auctions to provide incentives for low-carbon technological innovation. An example is the approach of Masui *et al.* (2005) for Japan discussed in 11.3.4. Weber *et al.* (2005), using a long-run calibrated global growth model, conclude that '...increasing the fraction of carbon taxes recycled into subsidizing investments in mitigation technologies not only reduces global warming, but also enhances economic growth by freeing business resources, which are then available for investments in human and physical capital' (p. 321).

Unlike the studies that assess the effects of technology and mitigation policies on emissions and welfare in a simulation model, Popp (2002) examines the empirical effect of both energy prices and government spending on US patent activities in 11 energy technologies in the period 1970–1998. He finds that while energy prices have a swift and significant effect on shifting the mix of patents towards energy-related activities, government-sponsored energy R&D has no significant effect. While not addressing efforts to encourage private-sector R&D, this work casts doubt on the effectiveness of government-sponsored low-GHG research by itself as a mitigation option.

11.6 From medium-term to long-term mitigation costs and potentials

We now consider how the sectoral and macroeconomic analyses to 2030 relate to the stabilization-oriented studies of Chapter 3; this leads to a focus on the transitions in the second quarter of the century.

The section concludes by considering wider dimensions of timing and strategy.

11.6.1 Structural trends in the transition

Most studies suggest that GHG mitigation shifts over time from energy efficiency improvements to the decarbonization of supply. This is the clear trend in the global scenarios survey in Chapter 3 (Figure 3.23), and also in the time-path plots of energy against carbon intensity changes in the models in the IMCP studies (Edenhofer *et al.*, 2006b). It is also true of the national long-term studies surveyed in Chapter 3 (Table 3.7); of the detailed sectoral assessments of Chapters 4–10; and the IEA's ETP study (IEA, 2006a). In the first quarter of this century, the majority of global emission savings are associated with end-use savings in buildings and, to a lesser degree, in industry and transport. Moreover, despite important savings in electricity use in these sectors, economies in mitigation scenarios tend to become more electrified (Edmonds *et al.*, 2006). In the second quarter of this century, the degree of decarbonization of supplies starts to dominate efficiency savings as a result of a mix of strategies including CCS and diverse low-carbon energy sources. In the IEA study, the power sector consists of more than 50% non-fossil generation by 2050, and half of the remainder is made up of coal plants with CCS. The power sector still tends to dominate emission savings by 2030, even at lower carbon prices (see also Table 11.5), but obviously the degree of decarbonization is less.

There are two reasons for these trends. First, there are strong indications in the literature that improvements in energy efficiency with current technologies have greater potential at lower cost (see Chapters 5–7). This is apparent from the sectoral assessments summarized in Table 11.3, where energy efficiency accounts for nearly all the potential available at negative cost (particularly in buildings), and at least as much as the potential available from switching to lower carbon fuels and technologies in energy supply, for costs in the range up to 20 US$/tCO$_2$ -eq. The second reason is that most models assume some inertia in the capital stock and diffusion of supply-side technologies, but not of many demand-side technologies. This slows down the penetration of low-carbon supply sources even when carbon prices rise enough (or when costs fall sufficiently) to make them economic. Some end-use technologies (such as appliances or vehicles) do have a capital lifetime that is much shorter than major supply-side investments; but there are very important caveats to this, as discussed below.

For the analysis of transitions during the first quarter of this century, then, most of the relevant modelling literature emphasizes, for stabilization between 650 and 550 ppm CO$_2$-eq (categories III and IV in table 3.5), energy supply and other sectors such as forestry in which mitigation potentials are dominated by long-lifetime, medium-cost options.

11.6.2 Carbon prices by 2030 and after in global stabilization studies

Many analyses in this report emphasize that efficient mitigation will require a mix of incentives: regulatory measures to overcome barriers to energy efficiency; funding and other support for innovation; and carbon prices to improve the economic attractiveness of energy efficiency and of low-carbon sources, and to provide incentives for low-carbon innovation and CCS. Most of the regulatory and R&D measures are sector-specific and are discussed in the respective sectoral chapters (4–10). Some implications of innovation processes are discussed below. Most global models focus on the additional costs of mitigation in the form of shadow prices or marginal costs, and the resulting changes that would be delivered by carbon prices. The carbon prices reached by 2030 are discussed in Section 11.4.4 above. The levels and trends in these prices are crucial to the transition processes.

The time trend of carbon prices after 2030 is important but specific to each model. Some models maintain a constant rate of price increase that largely reflects the discount rate employed (they establish an emissions time-path to reflect this). Two models in the EMF studies, for example, assume increases in carbon prices of about 5.5% per year and over 6% per year that are constant throughout the century. In this approach, carbon prices roughly treble over the period 2030–2050, and every two decades thereafter. Two models in the IMCP studies also use constant, and much lower, growth rates for prices that vary with the stabilization constraint. Edenhofer *et al.* (2006b) find that real carbon prices for stabilization targets rise with time in the early years for all models, with some models showing a decline in the optimal price after 2050 due to the accumulated effects of LBD and positive spillovers on economic growth. In these cases, a high-price policy in the earlier years may generate innovation that provides benefits in later years. In all these models, the *rates* of change frequently reflect intrinsic model parameters (notably the discount rate) and do not depend much on the stabilization target, which is reached by adjusting the starting carbon price instead. However, most but not all models with endogenous technical change have rates of carbon price increase that decline over time, and two models actually result in carbon price falls as technological systems mature.

A carbon price that rises over time is a natural feature of an efficient trajectory towards stabilization. The macro-economic cost depends on the *average* mitigation cost, which tends to rise more slowly and may decline with technical progress. The Stern review illustrates and explains scenarios in which rising carbon prices accompany declining average costs over time (Stern, 2006).

11.6.3 Price levels required for deep mid-century emission reductions: the wider evidence

Several other lines of evidence shed light on the carbon prices required to deliver transitions to deep mid-century CO$_2$

reductions. By contrast with the rising prices in most 'optimal' stabilization trajectories, some global models have been run with constant prices. Perhaps the most extensive, the IEA-ETP (2006a) study (MAP scenario), returns global CO_2 emissions roughly equal to 2005 levels by 2050 (more than halving emissions from reference). This is consistent with a trajectory towards category III stabilization at around 550 ppm CO_2-eq, with carbon prices rising to 24 US(2000)\$/tCO_2 (\$87/tC) by 2030 and then remaining fixed. The IEA study emphasizes the combination of end-use efficiency in buildings, industry and transport, together with the decarbonization of power generation as indicated. In other global studies that report sectoral results, the power sector dominates emission savings even in the weaker category IV scenarios. Some other models with detailed energy sectors do not force a constantly rising price or display periods of relatively stable prices along with stable or declining emissions.[17]

The carbon price results are consistent with the technology cost analyses in Chapter 4. These suggest that price levels in the 20–50 US\$/tCO_2 range should make both CCS and a diversity of low-carbon power generation technologies economic on a global scale, with correspondingly large reduction potential attributed to the power sector in this cost range (Table 11.8). Newell, Jaffe & Stavins (2006) focus on the economics of CCS at prices within this range, noting that the carbon prices required will depend not only on CCS technology but also upon the reference alternative. Schumacher and Sands (2006) also focus on CCS but, in the context of the German energy system, conclude that a similar range is critical 'for CCS as well as advanced wind technologies to play a major role' (p. 3941). Riahi et al. (2004) project coal-based CCS costs up to 53 US/t$CO2$. Corresponding reductions may accrue, whether a carbon price is considered to be implemented directly or as the incentive from certified emission reduction (CER) credits. Shrestha (2004) projects that 'business-as-usual' shares of coal in power generation by 2025 will be 46%, 78% and 85% in Vietnam, Sri Lanka and Thailand respectively, but an effective CER price of 20 US\$/tCO_2 from 2006 onwards would reduce the share of coal to 18%, 0% and 45% respectively in the three countries by 2025. Natural gas and, to a lesser extent renewables, oil and electricity imports are the main beneficiaries.

The sectoral results from Chapters 4–10 (Table 11.3) suggest that carbon prices in the range 20–50 US\$/tCO_2-eq could deliver substantial emission reductions from most sectors. Of the total potential identified below 100 US\$/tCO_2-eq across all sectors, more than 80% is estimated to be economic at a cost below 50 US\$/tCO_2-eq. Moreover, the lowest proportions are for agriculture (56%) and forestry (76%). Of the main sectors for which carbon cap-and-trade is being applied or considered

at present, costs below 50 US\$/tCO_2-eq account for 90% of the identified potential in energy supply and 86% in industry, whereas the proportion of the total below 20 US\$/tCO_2-eq is about half (52%) and a quarter (27%) respectively for these sectors. This underlines the conclusion that carbon prices in the 20–50 US\$/tCO_2-eq range would be critical to securing major changes in these principal industrial emitting sectors.

The bottom-up estimates of emission reductions available at less than 50 US\$/tCO_2-eq for the total energy sector (supply, buildings, industry and transport) span 11.5–15 GtCO_2-eq/yr (Table 11.3, full range). This is strikingly similar to the range of CO_2 reductions by 2030 that global top-down studies consider to be necessary for trajectories consistent with stabilization in the Category III range (Figure 11.7 (a), in the range 25–40% reduction of CO_2 which, against the central baseline projection of 37–40 GtCO_2-eq (WEO/A2) for energy-related emissions that is used for the bottom-up estimates, equates to 10–16 GtCO_2-eq. Incidentally, this also equates to global emissions in 2030 that are roughly at present levels).

The capital stock lifetime of industrial and forestry systems (discussed further below) means that it takes some decades for the impact of a given carbon price to work its way through in terms of delivered reductions.[18] The assessment of timing is complicated by the fact that most global stabilization studies model a steadily rising price with 'perfect foresight'. However, Figure 11.7(c) confirms that almost all models project prices of at least 20 US\$/tCO_2-eq by 2030, and some breach the 50 US\$/tCO_2-eq level earlier in that decade, as might be expected in order to secure the required reductions by 2030. Applying the same statistical framing as Chapter 3, the analysis of price trends confirms that global carbon prices in more than 80% of the Category III stabilization studies cross within the range \$20–50/tCO_2-eq during the decade 2020–30. These diverse strands of evidence therefore suggest a high level of confidence that carbon prices of 20–50 US\$/tCO_2-eq (75–185 US\$/tC-eq) reached globally in 2020–2030 and sustained or increased thereafter would deliver deep emission reductions by mid-century consistent with stabilization at about 550 ppm CO_2-eq (category III levels). To depict the impact in the models, such prices would have to be implemented in a stable and predictable manner and all investors would need to plan accordingly, at the discount rates embodied in the models.

Carbon prices at these levels would deliver these changes by largely decarbonizing the world's electricity systems, by providing a substantial incentive for additional energy efficiency and, if extended to land use, by providing major incentives to halt deforestation and reward afforestation.[19] By comparison,

---

17 Specifically, the GET-LFL 450 ppmCO_2 run has a peak in carbon prices at 37 US\$/tCO_2 in 2020 followed by 2-3 decades of slight decline; the DNE21+ 450 ppmCO_2 run model rises sharply to about US\$30/tCO_2 in 2020 followed by slow increase for a decade, then rises to 64 US\$/tCO_2 in 2040 followed by slow increases out to 2070. See Hedenus, Azar and Lindgren (2006) and Sano et al. (2006) respectively.

18 The perfect foresight assumed in many of the global models complicates the assessment of timing; see 11.6.6 below.

19 The forestry chapter also notes that continuously rising carbon prices pose a problem in that forest sequestration might be deferred to gain more advantage from future higher prices; seen from this perspective, a more rapid carbon price rise followed by period of stable carbon prices could encourage more sequestration.

prices in the EU ETS in 2005 peaked close to 30 euros (about 40 US$)/tCO$_2$. Transition scenarios for non-energy sectors (in particular agriculture and deforestation) are reported in the respective sectoral chapters and in some of the multi-gas studies in Chapter 3.

Particularly in models that embody some economies of scale/learning-by-doing, prices maintained at such levels largely decarbonize the power sector over a period of decades. Some of the models display a second period with a similar pattern, later and at higher prices, as fuel cell-based transport matures and diffuses. In integrated Category III scenarios, such scenarios can also deliver more potential abatement in the transport sector (at a higher cost), partly because several of the low-carbon transport technologies depend on the prior availability of low-carbon electricity. Assumptions about the availability of petroleum and the costs of carbon-based 'backstop' liquid fuels also tend to be very important considerations in terms of the associated net costs (Edmonds *et al.*, 2004; Edenhofer *et al.*, 2006b; Hedenus *et al.*, 2006).

The price in the 20 to 50 US$/tCO$_2$ range required to deliver such changes – and answers to the questions of whether and by how much further carbon prices might need to rise in the longer term – depend upon developments in three other main areas: the contribution of voluntary and regulatory measures associated with energy efficiency; the extent and impact of complementary policies associated with innovation and infrastructure; and the credibility, stability and conviction that the private sector attributes to the price-based measures. We consider each in turn.

11.6.4 Complementary measures for deep emission reductions

The sectoral and multi-gas studies indicate that substantial emission savings are still available at low cost (< 20 US$/tCO$_2$), particularly from buildings (Chapter 6) and end-use efficiencies in a number of industrial sectors (Table 7.8); many governments are therefore already well embarked upon policies to exploit these low-cost opportunities. The IEA's World Energy Outlook (IEA, 2006b, Part 2) estimates that such measures could contribute a 16% reduction below the reference level by 2030. This would be an important contribution, but clearly insufficient to get close to halting or reversing global emissions growth in the absence of price-based measures.

Innovation will also be crucial for deep reductions by mid-century and in the longer term. Some of the technologies required to deliver ongoing emission reductions out to 2050 are already commercialized, but others (such as CCS) are not (see sector chapters). Deeper emission reductions will get more and more difficult over time without accelerated innovation bringing down costs, and increasing the diversity, of low-carbon options. Achieving the mitigation scenarios indicated therefore requires adequate progress in a range of relevant industries

based on low-carbon technology (Weyant, 2004; IEA, 2006b). Chapter 2 has laid out the basic principles for low-carbon innovation and Chapter 3 the long-term role of technologies in stabilization scenarios. The sectoral chapters discussed the specific technologies and Section 11.5 covered the post-TAR modelling of induced technological change. This section briefly assesses the insights relating to innovation that are relevant to transitions in the second quarter of this century.

The conceptual relationship between such innovation investments and measures relating to carbon pricing is sketched out in Figure 11.10. Most low-carbon technologies (at least for supply) are currently much more expensive than carbon-based fuels. As R&D, investment and associated learning accumulates, their costs will decline, and market scale may grow. Rising carbon prices bring forward the time when they become competitive (indeed, many such technologies might never become competitive without carbon pricing). The faster the rise in carbon prices – particularly if industry can project such increases with confidence in a clear and stable policy environment – the sooner such technologies will become competitive and the greater the overall economic returns from the initial learning investment.

However, the literature also emphasizes that carbon pricing alone is insufficient. Sanden and Azar (2005) argue that carbon cap-and-trade is important for diffusion – 'picking technologies from the shelf' – but insufficient for innovation – 'replenishing the shelf'. Foxon (2003) emphasizes the interaction of environmental and knowledge market failures, arguing that this creates 'systemic' obstacles that require government action beyond simply fixing the two market failures (of climate damages and technology spillovers) independently.

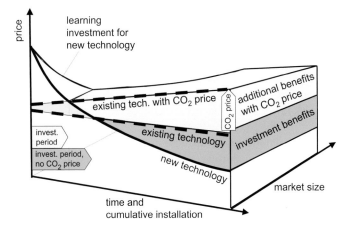

Figure 11.10: *Relationship between learning investments and carbon prices*

Notes: The figure illustrates cost relationships for new low-carbon technology as experience and scale build over time. Initially introduction is characterized by relatively high current costs and a very small market share, and requires a high unit rate for 'learning investment'. With increasing scale and learning, costs move towards existing, higher-carbon sources, the costs of which may also be declining, but more slowly. Rising carbon prices over time bring forward the time when the new technology may be competitive without additional support, and may greatly magnify the economic returns from the initial learning investment.

Source: Adapted with author's permission from Neuhoff (2004).

Table 11.17: *Observed and estimated lifetimes of major GHG-related capital stock*

Typical lifetime of capital stock			Structures with influence > 100 years
Less than 30 years	30-60 years	60-100 years	
Domestic appliances Water heating and HVAC systems Lighting Vehicles	Agriculture Mining Construction Food Paper Bulk chemicals Primary aluminium Other manufacturing	Glass manufacturing Cement manufacturing Steel manufacturing Metals-based durables	Roads Urban infrastructure Some buildings

Source: IEA (2000); industrial process data from Worrel and Biermans (2005).

There is therefore general consensus in the literature that, whilst emission reduction (including pricing) mechanisms are a necessary component for delivering such innovation, they are not sufficient: efficient innovation requires even more government action.

This underlines the complexity of measures required to drive adequate innovation. On the basis of four general lessons from US technology policy, Alic *et al.* (2003) derive various specific conclusions for action.[20] They break them down into direct R&D funding, support for deployment, and support for education and training. However, they also underline that 'technology policies alone cannot adequately respond to global climate change. They must be complemented by regulatory and/or energy pricing policies that create incentives for innovation and adoption of improved or alternative technologies ... the technological response will depend critically on environmental and energy policies as well as technology polices.'

Philibert (2005) places climate technology policy in the context of the wider experience of US, European and IEA technology programmes and present initiatives, and discusses explicitly the international dimensions associated with globalization, export credit, diffusion, standards and explicit technology negotiations. Grubb (2004) outlines at least six different possible forms of international technology-oriented agreements that could, in principle, help to foster global moves towards lower-carbon energy structures (see Chapter 13).

The common theme in all these studies is the need for multiple and mutually supporting policies that combine technology push and pull across the various stages of the 'innovation chain', so as to foster more effective innovation and more rapid diffusion of low-carbon technologies, both nationally (the tax and subsidy regime for Japan discussed in 11.4.3.4, for example) and internationally. Most studies also emphasize the need for feedback enabling policy to learn from experience and experimentation – using 'learning-by-doing' in the process of policy development itself.

11.6.5 Capital stock and inertia determinants of transitions in the second quarter of the century

The scope for change, and the rate of transition, will be constrained by the inertia of the relevant systems. The IPCC SAR Summary for Policymakers noted that 'the choice of abatement paths involves balancing the economic risks of rapid abatement now (that premature capital stock retirement will later be proved unnecessary) against the corresponding risk of delay (that more rapid reduction will then be required, necessitating premature retirement of future capital stock).' Capital stock is therefore a central consideration.

The time scales of stock turnover vary enormously between different economic sectors, but appear to be very long for most greenhouse-gas emitting sectors. Typical investment time scales are several decades for forestry, coal mining and transporting facilities, oil & gas production, refineries, and power generation. On the demand side, observed time scales for typical industrial stock using energy are estimated to range from decades to a century (Worrell and Biermans, 2005; see Table 11.17). The time scales for other end-use infrastructure (e.g. processes, building stock, roads and rail) may be even longer, though components (such as heaters, cars) may have considerably faster turnover.

However, Lempert *et al.* (2002) caution against overly simplistic interpretations of nameplate lifetimes, emphasizing that they 'are not significant drivers [of retirement decisions] in the absence of policy or market incentives' and that 'capital has no fixed cycle'. This can be crucial to rates of decarbonization. A study of the US paper industry found that 'an increase in the rate of capital turnover is the most important factor in permanently changing carbon emission profiles and energy efficiency' (Davidsdottir and Ruth, 2004). Similarly, emission reductions in the UK power sector were largely driven by the retirement of old, inefficient coal plant during the 1990s, through sulphur regulations which meant plant owners were faced with the choice of either retrofitting stock or retiring it (Eyre, 2001).

---

20 Their four general lessons are: (i) Technology innovation is a complex process involving invention, development, adoption, learning and diffusion; (ii) Gains from new technologies are realized only with widespread adoption, a process that takes considerable time and typically depends on a lengthy sequence of incremental improvements that enhance performance and reduce costs; (iii) Technology learning is the essential step that paces adoption and diffusion; (iv) Technology innovation is a highly uncertain process.

Such micro-level 'tipping points' at which investment decisions need to be taken may offer ongoing opportunities for lower cost abatement.

Energy system inertia provides another dimension to the time scales involved. It has taken at least 50 years for each major energy source to move from 1% penetration to a major position in global supplies. Such long time scales – and the even longer periods associated with interactions between systems – imply that, for stabilization, higher inertia brings forward the date at which abatement must begin to start meeting any given constraint, and lowers the subsequent emissions trajectory (Ha-Duong et al., 1997). In the context of stabilization at 550 ppm CO_2, van Vuuren et al. (2004) and Schwoon and Tol (2006) demonstrate that higher inertia in the energy system brings forward mitigation.[21]

However, beyond a certain point, inertia can also dramatically increase the cost of stabilization, particularly when infrastructure constraints are likely to limit the growth of new industries more than established ones. Manne & Richels (2004) illustrate that if global total contributions from new (renewable) power sources are limited to 1% by 2010 and treble each decade thereafter, the world has little choice other than to continue expanding carbon-intensive power systems out to around 2030. This feature appears to drive their finding of high costs for 450 ppm CO_2 stabilization, since much of this stock then has to be retired in subsequent decades to meet the constraint. This pattern contrasts sharply with some other studies, such the MIT study (McFarland et al., 2004) that states an opposing time profile based partly upon the rapid deployment of natural gas plant, including CCS. Crassous et al. (2006) also find high costs by assuming that long-lived infrastructure construction continues without foresight over the century. If low-carbon transport technologies do not become available quickly enough, the economy is squeezed as carbon controls tighten. They also show that the early adoption of appropriate infrastructure avoids this squeeze and allows lower costs for carbon control. Drawing partly on more sociological literature, and the systems innovation literature (Unruh, 2002), tends to support a view that we are now 'locked in' to carbon-intensive systems, with profound implications: 'Carbon lock-in arises through technological, organizational, social and institutional co-evolution ... due to the self-referential nature of [this process], escape conditions are unlikely to be generated internally.'

Lock-in is less of a problem for new investment in rapidly developing countries where the CDM is currently the principal economic incentive to decarbonize new investments. The Shrestha (2004) study cited above illustrates how the structure of power sectors could be radically different depending upon the value of Certified Emission Reduction (CER) units. Their finding that an effective CER price of 20 US$/t$CO_2$ from 2006 onwards could drive a radical switch of investment from new coal plants and primarily to natural gas and renewables in the three Asian countries studied would not only represent a large saving in CO_2 emission, but a totally different capital endowment that would sustain far lower emission trajectories after 2030. Again, this supports the conclusion that carbon prices of this order play a very important role, with their potential to forestall the construction of carbon-intensive stock in developing countries. Diverse policies that deter investment in long-lived carbon-intensive infrastructure and reward low-carbon investment may maintain options for low stabilization levels in Category I and II at lower costs.

At a global scale, van Vuuren et al. (2004) present a systematic set of results showing the effects of different time profiles for carbon prices in studies that combine the representation of inertia and induced innovation. A carbon price that rises linearly to 82 US$/t$CO_2$ by 2030 reduces emissions by 40% by 2030 if the tax is introduced in 2020 and raised sharply, but by 55% if it is introduced in 2000 and increased more slowly. Van Vuuren et al. do not describe the impact on subsequent trajectories, but clearly the capital stock endowment differs substantially. Moreover, Lecocq et al. (1998) demonstrate that, in the face of uncertainty, an efficient approach may include greater effort directed at reducing investment in longer-lived carbon intensive infrastructure, over and above the incentives of any uniform carbon price.

Chapter 3 (Section 3.6) emphasizes the importance of 'hedging' strategies based upon sequential 'act-then-learn' decision-making. Mitigation over the next couple of decades that would be consistent with enabling stabilization at lower levels (Categories I, II or III) does not irrevocably commit the world to such levels. The major numerical addition to the literature in this vein appears to be that of Mori (2006). Using the MARIA model, he analyses optimal strategies to limit the global temperature increase to 2.5 °C given uncertainty about climate sensitivity in the range of 1.5-4.5 °C per doubling of CO_2-equivalent. When there is no uncertainty, only the above-average sensitivities require significant mitigation in the next few decades. In the context of uncertainty, however, the optimal strategy is to keep global emissions relatively constant at present levels until the uncertainty is resolved, after which they may rise or decline depending upon the findings.

11.6.6 Investment and incentive stability

The longevity of capital stock, projections of rapidly growing global emissions under 'business-as-usual', and the importance of industrial scale and learning in low-carbon technology industries all illustrate the central role of investment in relation to the climate change problem. As discussed in Chapter 4, the IEA (2004) estimates that about US$20 trillion will be invested in energy supplies up to 2030, half to two-thirds of which is associated with power generation.

---

21 Specifically, van Vuuren et al. (2004, p. 599) state that including inertia 'results in a 10% reduction of global emissions after 5 years and 35% reduction after 30 years'.

Several major studies shed light upon the investment implications of low-carbon scenarios over the next few decades. The World Bank (2006) estimates that to 'significantly de-carbonize power production' would require incremental investments of 'up to' US $40bn per year globally, of which about US$30bn per year would be in non-OECD countries. However, in a comprehensive scenario, this would be offset by the reduced investment requirements resulting from improved end-use efficiency. The IEA WEO (2006b) 'alternative policy scenario' estimates that an increased investment of US$2.4 trillion in improved efficiency would be more than offset by US$3 trillion savings in supply investments. The more aggressive IEA 'Map' scenario (IEA, 2006a), that returns emissions to 2005 levels by 2050 (and is consistent with trajectories towards stabilization between 550 and 650 ppm CO_2-eq) as discussed above, reflects greater impact as a result of switching investment from more to less carbon-intensive paths. Investments across renewables, nuclear and CCS are projected of US$7.9 trillion, US$4.5 trillion of which is offset directly by the reduced investment required in fossil-fuel power plants. Most of the rest is offset by the reduced need for transmission and distribution investment and fuel savings arising from increased energy efficiency. The net additional cost for the Map scenario out to 2050 is only US$100 billion, about 0.5% of total projected sector investments.

Because the net cost estimates arise from balancing supply and demand, there is considerable uncertainty. The World Bank figure for incremental low-carbon power generation costs, for example, is much higher, at close to 10% of projected total investment costs, but does not fully offset these against end-use savings, or co-benefits. It is clear that low-carbon paths consistent with the IEA Map result of returning global CO_2 emissions to present levels involve a large redirection of investment, but the net additional cost based on this limited set of studies is likely to be less than 5–10% of the total investment requirements, and may be negligible. The studies collectively emphasize that the choice of path over the next few decades will have profound implications for the structure of capital stock, and its carbon intensity, well into the second half of this century and even beyond.

Much of this investment will come from the private sector. However, the associated literature emphasizes that current signals are inadequate and that the effectiveness of carbon pricing depends critically upon its credibility and predictability. For example, the perceived uncertainty with respect to the EU ETS after 2012 deters companies from investing on the basis of price. The credit agency Standard and Poor's (2005) state that 'this uncertainty has and will result in delays to investment decisions'. Sullivan and Blyth (2006) analyse the economics of investment in conditions of uncertainty and concur that the perceived uncertainties make it optimal for companies to defer investment and to keep old power plants running instead. This could even

increase emissions. Consequently, the 'electricity or carbon prices required to stimulate investment in low-carbon technology may be higher than expected...' due to the uncertainties. This underlines the present gap between the modelling abstraction of perfect foresight, and the real-world uncertainties. The costs of mitigation will be reduced only to the extent that governments can make clear and credible commitments about future carbon controls that are sufficient for the private sector to see as 'bankable' in project investment appraisals.

11.6.7 Some generic features of long-term national studies

Finally, the rapidly growing number of national goals and strategies oriented towards securing ambitious CO_2 reduction goals, typically by 60–80% below present levels in industrialized countries, are relevant to the understanding of low-carbon transitions for the first half of this century. Some quantitative findings from some long-term national modelling studies have been summarized in Chapter 3, and some shorter-term studies earlier in this chapter.[22] Additional studies of long-term mitigation in developing countries are beginning to emerge (e.g. Jiang and Hu, 2006; Shukla et al., 2006). The range and number of national analyses, scenarios and strategies devoted to mitigation targets is beyond the scope of this section but, in general, they suggest that there are a number of common 'high-level' features that underpin some main messages of the academic literature in terms of the need for a combination of:
- innovation-related action on all fronts, both R&D and market-based learning-by-doing stimulated by a variety of instruments;
- measures that establish a long-term, stable and predictable price for carbon to encourage lower carbon investment, particularly but not exclusively in power sector investments;
- measures that span the range of non-CO_2 gases so as to capture the 'low-hanging fruit' across the economy;
- measures relating to long-lived capital stock, especially buildings and energy infrastructure;
- institution- and option-building including considerations relating both to system structures, and policy experimentation with review processes to learn which are the most effective and efficient policies in delivering such radical long-term changes as knowledge about climate impacts accumulates.

11.7 International spillover effects

11.7.1 The nature and importance of spillover

Spillover effects of mitigation in a cross-sectoral perspective are the effects that mitigation policies and measures in

22 In addition to some of the specific economy-modelling studies referred to in preceding sections as indicated, strategic national studies written up in the academic literature include the Dutch COOL project (Treffers et al., 2005), and analysis for long-term targets in the UK (Johnston et al., 2005) and Japan (Masui et al, 2006).

one country or group of countries have on sectors in other countries. (Inter-generational consequences, which are the effects of actions taken by the present generation on future generations, are covered in Chapter 2.) Spillover effects are an important element in the evaluation of environmental policies in economies globally linked through trade, foreign direct investment, technology transfer and information. Due to spillover effects, it is difficult to determine precisely the net mitigation potential for sectors and regions, and the effects of policies. An added complication is that the effects may be displaced over time. The measurement of the effects is also complex because effects are often indirect and secondary, although they can also accumulate to make local or regional mitigation action either ineffective or the source of global transformation. Much of the literature recognizes the existence of spillover effects. However, uncertainty and disagreement about time scale, cost, technology development, modelling approaches, policy and investment pathways lead to uncertainty about their extent and therefore the overall mitigation potentials.

The same spillover effect will be seen differently depending on the point of view adopted. Multiple differences between regions and nations imply differing, and perhaps contradictory views, about mitigation policies and their implementation. These differences emanate from the diverse and sometimes distinct natural endowments and social structures of those regions, as well as differences in the financial ability to cope with the costs that may be incurred as a result of the implementation of these policies. Methodologies that are developed for market-based industrialized economies may not be completely relevant for the economies of developing countries.

Some researchers who use general-equilibrium models (e.g. Babiker, 2005) conclude that spillover will, given certain assumptions, render mitigation action ineffective or worse if it is confined to Annex I countries. Other researchers (e.g. Grubb et al., 2002a; Sijm et al., 2004) argue that spillovers from Annex I action, implemented via induced technological change, could have substantial effects on sustainable development, with emission intensities from developing countries being a fraction of what they would be otherwise. 'However, no global models yet exist that could credibly quantify directly the process of global diffusion of induced technological change.' (Grubb et al., 2002b, p.303). It is important to empha-size the uncertainties in estimating spillover effects, as well as uncertainties in estimating potential mitigation costs and benefits. In the modelling of spillovers through international trade, researchers rely on different approaches (bottom-up or top-down, for example), different assumptions (perfect/imperfect or 'Armington' substitution) and estimates of parameters when signs and magnitudes are disputed. Many of the models used to estimate the costs of mitigation focus on substitution effects and set aside information, policy and political spillovers, as well as the induced development and diffusion of technologies.

11.7.2 Carbon leakage

Carbon leakage is defined as the increase in CO_2 emissions outside the countries taking domestic mitigation action divided by the reduction in the emissions of these countries. It has been demonstrated that an increase in local fossil fuel prices resulting, for example, from mitigation policies may lead to the re-allocation of production to regions with less stringent mitigation rules (or with no rules at all), leading to higher emissions in those regions and therefore to carbon leakage. Furthermore, a decrease in global fossil fuel demand and resulting lower fossil fuel prices may lead to increased fossil fuel consumption in non-mitigating countries and therefore to carbon leakage as well. However, the investment climate in many developing countries may be such that they are not ready yet to take advantage of such leakage. Different emission constraints in different regions may also affect the technology choice and emission profiles in regions with fewer or no constraints because of the spillover of learning (this is discussed in Section 11.7.6).

Since the TAR, the literature has extended earlier-equilibrium analysis to include effects of trade liberalization and increasing returns in energy-intensive industries. A new empirical literature has also developed. The literature on carbon leakage since the TAR has introduced a new dimension to the analysis of the subject: the potential carbon leakage from projects intended for developing countries to help them reduce carbon emissions. One example is Gundimeda (2004) in the case of India (discussed in Section 11.7.3 below).

11.7.2.1 Equilibrium modelling of carbon leakage from the Kyoto Protocol

Paltsev (2001) uses a static global-equilibrium model GTAP-EG to analyse the effects of the Kyoto Protocol. He reports a leakage rate of 10.5%, with an uncertainty range of 5–15% covering different assumptions about aggregation, trade elasticities and capital mobility, but his main purpose is to trace back non-Annex B increases in CO_2 to their sources in the regions and sectors of Annex B. The chemicals and iron and steel sectors make the highest contributions (20% and 16% respectively), with the EU being the largest regional source (41% of total leakage). The highest bilateral leakage is from the EU to China (over 10% of the total). Kuik and Gerlagh (2003) use a similar GTAP-E model and conclude that, for Annex I Kyoto-style action, the main reason for leakage is the reduction in world energy prices, rather than substitution within Annex I. They find that the central estimate of 11% leakage is sensitive to assumptions about trade-substitution elasticities and fossil-fuel supply elasticities and to lower import tariffs under the Uruguay Round, and they state a range of 6% to 17% leakage.

In a more recent study, Babiker (2005), using a model with different assumptions about production and competition in the energy-intensive sector, reports a range of global leakage rates between 25% and 130%, depending on the assumptions adopted. The main reasons for the higher estimates are the inclusion of increasing returns to scale, strategic behaviour in the energy-intensive industry and the assumption of homogeneous products. Rates above 100% would imply that mitigation action in one region leads to more global GHG emissions rather than less.

However, other studies point to real world conditions that make these outcomes unlikely. Significant carbon leakage arises when internationally tradeable energy-intensive production moves abroad to non-abating regions. This is frequently referred to as a competitiveness concern. In industrialized countries, these sectors account for 15–20% of CO_2 emissions (IEA, 2004). Results with high leakage therefore reflect conditions in which countries implement policies that lead to most emission savings being obtained by industrial relocation (to areas of lower-cost, and in some cases less efficient, production), rather than in the less mobile sectors (such as power generation, domestic, services etc). In practice, most countries have tended to adjust policies to avoid any such outcome (for example through derogation, exemption or protection for such sectors).

Sijm *et al.* (2004) provide a literature review and an assessment of the potential effects of Annex I mitigation associated with the EU emissions trading scheme (ETS) for carbon leakage, and especially in developing countries. Technological spillovers discussed in this paper are considered in section 11.7.6 below. In the empirical analysis of effects in energy-intensive industries, the modelling studies reporting high leakage rates look at many other factors in addition to price competitiveness. They conclude that, in practice, carbon leakage is unlikely to be substantial because transport costs, local market conditions, product variety and incomplete information all favour local production. They argue that a simple indicator of carbon leakage is insufficient for policymaking. Szabo *et al.* (2006) report production leakage estimates of 29% by 2010 for cement given an EU ETS allowance price of about 50 US$/tCO_2 and a detailed model of the global industry. Leakage rates rise with the allowance price. More generally, Reinaud (2005) surveys estimates of leakage for energy-intensive industries (steel, cement, newsprint and aluminium) assuming the EU ETS. She comes to a similar conclusion to Sijm *et al.* (2004) and finds that, with the free allocation of CO_2 allowances, 'any leakage would be considerably lower than previously projected, at least in the near term.' (p. 10). However, 'the ambiguous results of the empirical studies in both positive and negative spillovers warrant further research in this field.' (p.179).

11.7.3 Spillover impact on sustainable development via the Kyoto mechanisms and compensation

The Kyoto mechanisms may also result in spillover effects that offset their additionality. Gundimeda (2004) considers how the clean development mechanism (CDM) might work in India. (The CDM is considered in detail in Chapter 13.) The paper examines the effects of CDM projects involving land-use change and forestry on the livelihoods of the rural

poor. It concludes that, for CDM to be sustainable and to result in sustainable development for local people, three important criteria must be met: (1) in sequestration projects, local use of forestry (as firewood, for example) should also be an integral part of the project (2) management of the common lands by the rural poor through proper design of the rules for the sustenance of user groups; and (3) ensuring that the maximum revenue from carbon sequestration is channelled to the rural poor. 'Otherwise CDM would just result in either [carbon] leakage [e.g. through unplanned use of forestry for firewood] ... or have negative welfare implications for the poor' (p. 329).

Kemfert (2002) considers the spillover and competitiveness effects of the Kyoto mechanisms used separately (CDM, CDM with sinks, joint implementation (JI) and emissions trading (ET)) using a general-equilibrium model – WIAGEM – with Kyoto-style action (including the USA) continuing until 2050. The study shows the full welfare effect (% difference from business as usual) in 2050, broken down into the effects of domestic action, competitiveness and spillovers. It is notable that the mechanisms have a very small impact on welfare. At most, as an outlier, there is a 0.7% increase for countries in transition (REC) for emissions trading and a 0.1% decrease for the EU-15 for joint implementation. The CDM is found to improve welfare most in developing countries. However, the model does not include induced technological change or environmental co-benefits and it assumes full employment in all countries. If the CDM is assumed to result in more technological development, a more productive use of labour or an improvement in air or water quality, then the environmental and welfare effects in non-Annex I countries will be much larger than those reported.

Böhringer and Rutherford (2004) use a CGE model to assess the implications of UNFCCC articles 4.8 and 4.9 dealing with compensation. They conclude that 'spillover effects are an important consequence of multilateral carbon abatement policies. Emission mitigation by individual developed regions may not only significantly affect development and performance in non-abating developing countries, but may also cause large changes in the economic costs of emission abatement for other industrialized nations.' They estimate that the US should pay OPEC and Mexico estimated compensation of 0.7 billion US$ annually to offset the adverse impacts on these regions and that the EU should pay the same amount to the US to account for the positive spillover.

11.7.4 Impact of mitigation action on competitiveness (trade, investment, labour, sector structure)

The international competitiveness of economies and sectors is affected by mitigation action (see surveys by Boltho (1996), Adams (1997) and Barker and Köhler (1998)). In the long run, exchange rates change to compensate for a persistent loss of national competitiveness, but this is a general effect and particular sectors can become more or less competitive.

In the short run, the higher costs of fossil fuels lead to a loss in sectoral price competitiveness, especially in energy-intensive industries. The effects of domestic mitigation action on a region's international competitiveness are broken down by the literature into the effects on price and non-price competitiveness. This section covers price competitiveness, while technological spillover effects are discussed in Section 11.7.6 below.

In general, energy efficiency policies intended for GHG mitigation will tend to improve competitiveness (see Section 11.6.3 above). Zhang and Baranzini (2004) have reviewed empirical studies on the effects of Annex 1 action on international competitiveness. They conclude that 'empirical studies on existing carbon/energy taxes seem to indicate that competitive losses are not significant'. They therefore support the conclusions of the TAR, namely that 'reported effects on international competitiveness are very small and that at the firm and sector level, given well-designed policies, there will not be a significant loss of competitiveness from tax-based policies to achieve targets similar to those of the Kyoto Protocol.' (p.589). Baron and ECONEnergy (1997) looked at carbon prices similar to those expected to be necessary to implement the Kyoto Protocol (see 11.4.3.3). They report a static analysis of the cost increases from a tax of 27 US$/tCO$_2$ on four energy-intensive sectors in 9 OECD economies (iron and steel, other metals, paper and pulp, and chemicals). Average cost increases are very low – less than about 3% for most country sectors studied – with higher cost increases in Canada (all 4 sectors), Australia (both metal sectors) and Belgium (iron and steel).

However, action by Annex I governments (the EU, Denmark, Norway, Sweden, UK) have generally exempted or provided special treatment for energy-intensive industries. Babiker *et al.* (2003) suggest that this is a potentially expensive way of maintaining competitiveness, and recommend a tax and subsidy scheme instead. One reason for such exemptions being expensive is that, for a given target, non-exempt sectors require a higher tax rate, with mitigation at higher cost.

The impact of mitigation policies on trade within a region and between regions as a result of spillover is linked through capital flows from one country to another (within a region) or from one region to another, as individual investors and firms look for a higher rate of return on their investments which are considered by the receiving countries to be Foreign Direct Investment (FDI). Different market regulations and the flow of goods and services are influenced by mitigation policies, and the resulting spillover make 'measuring the welfare cost of climate change policies a real challenge, raising difficult issues of micro- and macro-economics: cost-benefit analysis on the one hand, foreign trade and international specialization on the second hand' (Bernard and Vielle, 2003). Partly for these reasons, the literature is sparse and the effects of different mitigation policies on FDI, trade, investment and labour market development within and between regions and any spillover effects are important areas for further research.

11.7.5 Effect of mitigation on energy prices

As discussed in 11.7.2, perhaps one of the most important ways in which spillovers from mitigation action in one region affect the others is through their effect on world energy prices. When a region reduces its fossil fuel demand as a result of mitigation policy, it will reduce the world demand for that commodity and so put downward pressure on the prices. Depending on responses from producers of fossil fuels, oil, gas or coal prices may fall, leading to losses of revenue for the producers, and lower import costs for the consumers. Demand for alternative, low-carbon fuels may increase. Three distinct spillover effects have been identified for non-mitigating countries. First, income for producers of fossil fuels will decline as the quantity sold is reduced, causing welfare losses and unemployment along with associated problems. Second, consuming nations will face lower prices for imported energy and may reduce subsidies or allow domestic energy prices to fall so that they tend to consume more, leading to carbon leakage as discussed above. Third, those non-mitigating countries producing low-carbon or alternative fuels will see an increase in demand and prices, with potentially positive effects on the markets for bioenergy.

11.7.5.1 Effects of Annex I action reported in the TAR

The TAR reviewed studies (based on CGE models with no induced technological change) of Annex I action in the form of a carbon tax or emissions trading schemes. The TAR (pp. 541–6) reported that, for abatement in Annex I, 'it was universally found that most non-Annex I economies that suffered welfare losses under uniform independent abatement suffered smaller welfare losses under emission trading' (p. 542). The magnitude of these losses is reduced under the less stringent Kyoto targets compared to assumptions about more stringent targets in pre-Kyoto studies. Some non-Annex I regions that would experience a welfare loss under the more stringent targets experience a mild welfare gain under the less stringent Kyoto targets. Similarities in regions identified as gainers and losers were quite marked. Oil-importing countries relying on exports of energy-intensive goods are gainers. Economies that rely on oil exports experience losses, with no clear-cut results for other countries.

The TAR considered the effect of OPEC acting as a cartel (pp. 543-4) and concludes that any OPEC response will have a modest effect on the loss of wealth to oil producers and the level on emission permit prices in mitigating regions. Analyses pertaining to the group of oil-exporting non-Annex I countries report costs differently, and the costs include, *inter alia*, reductions in projected oil revenues. Emissions trading reallocates mitigation to lower-cost options. The study reporting the lowest costs shows reductions of 0.2% of projected GDP with no emissions trading and less than 0.05% of projected GDP with Annex B emissions trading in 2010. The study reporting the highest costs shows a reduction of 25% in projected oil revenues with no emissions trading, and 13% in projected oil revenues with Annex B emissions trading in 2010. These studies did not consider policies and measures, other than Annex B emissions trading, that could lessen the impact on non-Annex I, oil-exporting countries, and therefore tend to overstate both the costs to these countries and overall costs. The effects on these countries can be further reduced by the removal of subsidies for fossil fuels, energy tax restructuring according to carbon content, the increased use of natural gas, and diversification of the economies of non-Annex I, oil-exporting countries (IPCC, 2001, p. 60).

11.7.5.2   Effect of mitigation on oil prices and oil exporters' revenues

The literature has hardly advanced since the TAR. GHG mitigation is expected to reduce oil prices, but the regional effects on GDP and welfare are mixed. Some studies point to gains by Annex I countries and losses to the developing countries, while others note losses in both of varying magnitudes, depending on different assumptions in the models. Studies that consider welfare gains/losses and international trade in Annex I countries also lead to mixed results, even if subsidies plus incentives and ancillary benefits are taken into account (Bernstein *et al.*, 1999; Pershing, 2000; Barnett *et al.*, 2004).

The highest modelling costs for implementing the Kyoto Protocol quoted by Barnett *et al.* (2004) for action in all Annex I countries are for OPEC: a 13% loss of oil revenues in the GCubed model (IPCC, 2001, p. 572). The scenarios underlying these costs assume Annex B action, including the USA and Australia, with a CO_2 tax but no allowances for non-CO_2 gases, sinks, targeted recycling of revenues or ancillary benefits. The outcome for OPEC is that its share of the world oil market falls compared to baseline projections. The authors argue that these costs will be lower following the Marrakech Accord; they are also lower because the US and Australia are not part of the Kyoto process, so the extent of mitigation action will be less than that modelled. All model estimates reviewed by Barnett *et al.* show that OPEC countries will see an increase in demand for oil but that this increase will be slowed by mitigation efforts following the Kyoto Protocol.

The use of OPEC market power could reduce negative effects, but this is uncertain (Barnett *et al.*, 2004, p. 2085). OPEC's World Energy Model assumes that OPEC production remains at baseline levels in the scenarios. This results in excess market supplies, since oil demand will be reduced. This leads to an estimate of OPEC losses of 63 billion US$ a year or about 10% of GDP, compared with 2% if supply is restricted in line with demand. Another scenario estimates the effect of an oil-price protection strategy, assuming that all major oil-producing countries in non-Annex B and in the former Soviet Union act together with OPEC. The conclusion is that OPEC losses would be substantially reduced. Another interesting feature of these results is that the losses as a percentage of 1999 GDP vary substantially across economies: from between 3.3% for Qatar to 0.07% for Indonesia by 2010.

Awerbuch and Sauter (2006) assess the effect of a 10% increase in the share of renewables in global electricity generation (which would reduce CO_2 by about 3% by 2030, compared with 16% in the IEA scenario). They suggest that the global oil price reduction would be in the range of 3 to 10%, with world GDP gains of 0.2 to 0.6%. Once again, the substantial increase expected in oil exporters' revenues would be reduced, although oil-importing countries would benefit.

Nearly all modelling studies that have been reviewed show more pronounced adverse effects on countries with high shares of oil output in GDP than on most of the Annex I countries taking the abatement measures.

11.7.6 Technological spillover

Mitigation action may lead to more advances in mitigation technologies. Transfer of these technologies, typically from industrialized nations to developing countries, is another avenue for spillover effects. However, as discussed in Chapter 2, effective transfer implies that developing countries have an active role in both the development and the adaptation of the technologies. The transfer also implies changes in flows of capital, production and trade between regions.

Sijm *et al.* (2004) assess the spillover effects of technological change. They divide the literature into two groups, depending on their 'top-down' or 'bottom-up' approach to modelling. (See the discussion on the topic in Section 11.3 above.) Most top-down modelling studies omit the effect or show it playing a minor role. The authors argue that the potential beneficial effect of technology transfer to developing countries arising from technological development brought about by Annex I action may be substantial for energy-intensive industries, but has so far not been quantified in a reliable manner. 'Even in a world of pricing CO_2 emissions, there is a good chance that net spillover effects are positive given the unexploited no-regret potentials and the technology and know-how transfer by foreign trade and educational impulses from Annex I countries to Non-Annex I countries.' (p. 179).

However, results from bottom-up and top-down models are strongly influenced by assumptions and data transformations and that lead to high levels of uncertainty. 'Innovation and technical progress are only portrayed superficially in the predominant environmental economic top-down models, and that the assumption of perfect factor substitution does not correctly mirror actual production conditions in many energy-intensive production sectors. Bottom-up models, on the other hand, neglect macroeconomic interdependencies between the modelled sector and the general economy.' (Lutz *et al.*, 2005). The effects of spillovers combined with learning-by-doing are explored specifically using bottom-up models by Barreto and Kypreos (2002) using MARKAL, and by Barreto and Klaassen (2004) using ERIS. They find that, owing to the presence of spillovers, the imposition of emission constraints in the

Annex I region may induce technological change and, hence, emission reductions in the non-Annex I region even when the latter region does not face emission constraints itself.

The existence of spillover effects also changes the theoretical conclusions in the economics literature. In the pure competition-equilibrium model, the most efficient policy is an equal rate of carbon tax for every sector and region. Rosendahl (2004) shows that, for maximum efficiency with spillovers and learning-by-doing, the carbon tax should be higher in those sectors and regions with the highest potential for technological progress. This is a general argument for stronger mitigation in those sectors and countries where technological progress is most likely to be accelerated by higher taxes on carbon use. In a game-theory context, with the shared benefits of R&D improving energy efficiencies, Kemfert (2004, p. 463) finds that 'full cooperation on climate control and technological improvements benefits all nations in comparison to a unilateral strategy.'

Although the technologies for CO_2 reduction in the electricity sectors are accessible, their dissemination still faces some challenges, especially in economies with low purchasing power and educational levels (Kumar *et al.*, 2003). An additional issue is that technology sharing by the fossil-fuel energy suppliers has been severely limited to date, probably due to the industrial organization of coal, oil and gas production, which is dominated by a few large private and state companies. Unlike, for example, new IT technologies, which quickly become industry standards, newly developed energy-related technology providing a competitive advantage generally becomes available to competitors slowly. However, modelling of the spillovers and the evolution of technologies, as well as structural changes in corporate management, require a better understanding of knowledge production and the knowledge transfer process within and between industries, and of the role and efficiency of transfer institutions such as universities, technology transfer centres and consultancy companies (Haag and Liedl, 2001).

11.8   Synergies and trade-offs with other policy areas

Anthropogenic GHG emissions are intricately linked to the structure of consumption patterns and levels of activity, which themselves are driven by a wide range of non-climate-related policy interests. These include policies on air quality, public health, energy security, poverty reduction, trade, FDI/investment regimes, industrial development, agriculture, population, urban and rural development, taxation and fiscal policies. There are therefore common drivers behind policies addressing economic development and poverty alleviation, employment, energy security, and local environmental protection on the one hand, and GHG mitigation on the other. Put another way, there are multiple drivers for actions that reduce emissions, and they produce multiple benefits.

Potential synergies and trade-offs between measures directed at non-climate objectives and GHG mitigation have been addressed by an increasing number of studies. The literature points out that, in most cases, climate mitigation is not the goal, but rather an outgrowth of efforts driven by economic, security, or local environmental concerns. The most promising policy approaches, then, will be those that capitalize on natural synergies between climate protection and development priorities to advance both simultaneously. Policies directed towards other environmental problems, such as air pollution, can often be adapted at low or no cost to reduce greenhouse gas emissions simultaneously. Such integration/policy coherence is especially relevant for developing countries, where economic and social development – not climate change mitigation – are the top priorities (Chandler *et al.*, 2002). Since the TAR, a wealth of new literature has addressed potential synergies and trade-offs between GHG mitigation and air pollution control, employment and energy security concerns.

11.8.1   Interaction between GHG mitigation and air pollution control

Many of the traditional air pollutants and GHGs have common sources. Their emissions interact in the atmosphere and, separately or jointly, they cause a variety of environmental effects at the local, regional and global scales. Since the TAR, a wealth of new literature has pointed out that capturing synergies and avoiding trade-offs when addressing the two problems simultaneously through a single set of technologies or policy measures offers potentially large cost reductions and additional benefits.

However, there are important differences at the temporal and spatial scales between air pollution control and climate change effects. Benefits from reduced air pollution are more certain; they occur earlier, and closer to the places where measures are taken, while climate impact is long-term and global. These mismatches of scales are mirrored by a separation of the current scientific and policy frameworks that address these problems (Swart *et al.*, 2004; Rypdal *et al.*, 2005).

Since the TAR, numerous studies have identified a variety of co-benefits of greenhouse gas mitigation on air pollution for industrialized and developing countries. In many cases, when measured using standard economic techniques, the health and environmental benefits add up to substantial fractions of the direct mitigation costs. More recent studies have found that decarbonization strategies generate significant direct cost savings because of reduced air pollution costs, highlighting the urgency of an integrated approach for greenhouse gas mitigation and air pollution control strategies.

11.8.1.1   *Co-benefits of greenhouse gas mitigation on air pollution*

A variety of analytical methods have been applied to identify co-benefits of greenhouse gas mitigation and air pollution.

Some assessments are entirely bottom-up and static, and focus on a single sector or sub-sector. Others include multi-sector or economy-wide general-equilibrium effects, taking a combination of bottom-up and top-down approaches. In addition, there are numerous methodological distinctions between studies. There are, for example, different baseline emission projections, air quality modelling techniques, health impact assessments, valuation methods, etc. These methodological differences, together with the scarcity of data, are a major source of uncertainties when estimating co-benefits. While the recent literature provides new insights into individual co-benefits (for example in the areas of health, agriculture, ecosystems, cost savings, etc.), it is still a challenge to derive a complete picture of total co-benefits.

11.8.1.2   Co-benefits for human health

Epidemiological studies have identified consistent associations between human health (mortality and morbidity) and exposure to fine particulate matter and ground-level ozone, both in industrialized and developing countries (WHO, 2003; HEI, 2004). Because the burning of fossil fuels is linked to both climate change and air pollution, lowering the amount of fuel combusted will lead to lower carbon emissions as well as lower health and environmental impacts from reduced emissions of air pollutants and their precursors.

Since the TAR, an increasing number of studies have demonstrated that carbon mitigation strategies result in significant benefits, not only as a result of improved air quality in cities, but also from reduced levels of regional air pollution. These benefits affect a larger share of the population and result from lower levels of secondary air pollutants. Although the literature employs a variety of methodological approaches, a consistent picture emerges from the studies conducted for industrialized regions in Europe and North America, as well as for developing countries in Latin America and Asia (see Table 11.18). Mitigation strategies aiming at moderate reductions of carbon emissions in the next 10 to 20 years (typically involving CO_2 reductions between 10 to 20% compared to the business-as-usual baseline) also reduce SO_2 emissions by 10 to 20%, and NO_x and PM emissions by 5 to 10%. The associated health impacts are substantial. They depend, *inter alia*, on the level at which air pollution emissions are controlled and how strongly the source sector contributes to population exposure. Studies calculate for Asian and Latin American countries several tens of thousands of premature deaths that could be avoided annually as a side-effect of moderate CO_2 mitigation strategies (Wang and Smith, 1999; Aunan *et al.*, 2003; O'Connor *et al.*, 2003; Vennemo *et al.*, 2006 for China; Bussolo and O'Connor, 2001 for India; Cifuentes *et al.*, 2001a; Dessus and O'Connor, 2003; McKinley *et al.*, 2005 for Latin America). Studies for Europe (Bye *et al.*, 2002; van Vuuren *et al.*, 2006), North America (Caton and Constable, 2000; Burtraw *et al.*, 2003) and Korea (Han, 2001; Joh *et al.*, 2003) reveal fewer, but nevertheless substantial, health benefits from moderate CO_2 mitigation

strategies, typically in the order of several thousand premature deaths that could be avoided annually.

Several authors conducted an economic valuation of these health effects in order to arrive at a monetary quantification of the benefits, which can then be directly compared with mitigation costs. While the monetization of health benefits remains controversial, especially with respect to the monetary value attributed to mortality risks in an international context, calculated benefits range from 2 US$/tCO$_2$ (Burtraw *et al.*, 2003; Joh *et al.*, 2003) up to a hundred or more US$/tCO2 (Han, 2001; Aunan *et al.*, 2004; Morgenstern *et al.*, 2004). This wide range is partially explained by differences in methodological approaches. The lower estimates emerge from studies that consider health impacts from only one air pollutant (such as SO_2 or NO_x), while the higher estimates cover multiple pollutants, including fine particulate matter, which has been recently shown to have the greatest impact. Differences in mortality evaluation methods and results also constitute a substantial source of discrepancy in the estimated value of health impact as well.

The benefits also largely depend on the source sector in which the mitigation measure is implemented. Decarbonization strategies that reduce fossil fuel consumption in sectors with a strong impact on population exposure (such as domestic stoves for heating and cooking, especially in developing countries) can typically result in health benefits that are 40 times greater than a reduction in emissions from centralized facilities with high stacks such as power plants (Wang and Smith, 1999). Mestl *et al.*, (2005) show that the local health benefits of reducing emissions from power plants in China are small compared to abating emissions from area sources and small industrial boilers. A third factor is the extent to which air pollution emission controls have already been applied. Health benefits are larger in countries and sectors where pollutants are normally emitted in an uncontrolled way, for instance for small combustion sources in developing countries.

Despite the large range of benefit estimates, all studies agree that monetized health benefits make up a substantial fraction of mitigation costs. Depending on the stringency of the mitigation level, the source sector, the measure and the monetary value attributed to mortality risks, health benefits range from 30 to 50% of estimated mitigation costs (Burtraw *et al.*, 2003; Proost and Regemorter, 2003) up to a factor of three to four (Aunan *et al.*, 2004; McKinley *et al.*, 2005). Particularly in developing countries, several of the studies reviewed indicate that there is scope for measures with benefits that exceed mitigation costs (no-regret measures).

Such potential for no-regret measures in developing countries are consistently confirmed by studies applying a general-equilibrium modelling approach, which takes into account economic feedback within the economy. Bussolo & O'Connor (2001) estimate that the potential for CO_2 mitigation in India for 2010, without a net loss in welfare, is between 13

Table 11.18: *Implications for air-quality co-benefits from GHG mitigation studies*

Authors	Country	Target year	Sector	Delta CO₂ emissions	Carbon price (US$/tCO₂)	Impact on air pollutant emissions	Difference in health impacts	Health benefits (US$/tCO₂)	Difference in air pollution control costs	Total benefits
Burtraw et al., 2003	US	2010	Power sector		7			2	1–2 US$/tCO₂	
Caton and Constable, 2000	Canada	2010	All sectors	-9%		SO₂: -9%; NOₓ: -7%; PM: -1%		11 (12–77)		
Wang & Smith, 1999	China	2020	Power sector	15% below BAU	11		4,400–5,200 premature deaths per year			
	China	2020	Domestic sector	15% below BAU	1.4		120,000–180,000 premature deaths per year			
O'Connor, 2003	China	2010	All sources	15% below BAU						No loss in net welfare
Aunan et al., 2004	Shanxi, China	2000	Cogeneration		-30 (net benefit)			32		
			Modified boiler design		-6			23		
			Boiler replacement		-3			32		
			Improved boiler management		9			32		
			Coal washing		22			86		
			Briquetting		27			118		
Kan et al., 2004	Shanghai, China	2010	All sources		24		608–5144 premature deaths per year			
		2020					1189–10462 premature deaths per year			
Li, 2006	Thailand			80–236 MtCO₂ annually						45% lower welfare losses
Vennemo et al., 2006	China	2008–2012	Power production, industrial boilers, steel making, cement, chemical industry		6 for the 80 Mt potential; unknown for the upper estimate	SO₂: 0.5–3 million tons; TSP: 0.2–1.6 million tons	2700 – 38000 lives saved annually (34–161 lives saved per million tons CO₂)	Avoided deaths: 4.1–20; all health effects: 5–44		

Note: The carbon prices in the table are indicative only. They have been converted to US$/tCO₂, but the implicit price bases in the original studies vary and may not be quoted or available.

Table 11.18: *Continued*

Authors	Country	Target year	Sector	Delta CO$_2$ emissions	Carbon price (US$/tCO$_2$)	Impact on air pollutant emissions	Difference in health impacts	Health benefits (US$/tCO$_2$)	Difference in air pollution control costs	Total benefits
Morgenstern, 2004	Taiyuan, China		Phase-out of small boilers	80%		-95%		38-175 US$/tCO$_2$		
Bussolo & O'Connor, 2001	India		All sources	13-23% below BAU						No welfare loss
Joh *et al.*, 2003	Korea	2020	All sources	5-15%				2 US$/tCO$_2$		
Han, 2001	Korea	2010	All sources	-10%		SO$_2$: -10% NO$_x$: -9.6% PM: -10%		58-76		
Van Vuuren, 2006	Europe	2020	All sources	4-7%		SO$_2$: 5-14%				
Syri *et al.* 2001	EU-15	2010	All sources	-8%		SO$_2$: 13-40% NO$_x$: 10-15%			-10%	
Proost *et al.*, 2003	Belgium	2010-2030	All sources	7-15%						30% of mitigation costs
Syri *et al.*, 2002	Finland	2010	All sources	Kyoto compliance		SO$_2$: -10% NO$_x$: -5% PM: -5%				
Bye *et al*, 2002	Nordic countries		All sources	20-30%					9-22 US$/ tCO$_2$	0.4% to 1.2% of GDP
Cifuentes *et al.*, 2001a, 2001b	Mexico City, Santiago, Sao Paulo, New York	2020					64,000 premature deaths per year			
West *et al.*, 2004	Mexico City	2010	18 GHG measures (mainly transport)	9%		PM10: -1.3% NO$_x$: 1.4% HC: 3.2%				
McKinley *et al.*, 2005	Mexico City	2020	5 mitigation options	0.8 Mt C/yr (1.1%)			100 premature deaths per year			
Dessus *et al.*, 2003	Santiago de Chile	2010		20% below BAU						No welfare loss

and 23% of the emissions for a business-as-usual scenario. For China, this potential has been estimated by O'Connor (2003) for 2010 at 15 to 20%, and Dessus and O'Connor (2003) arrive at a figure of 20% for Chile compared with the business-as-usual emissions in 2010. Li (2002; 2006) finds for Thailand that inclusion of health impacts reduces the negative impacts on GDP of a carbon tax by 45%, improving welfare for households and resulting in cleaner producers.

11.8.1.3   Co-benefits for agricultural production

While a strong body of literature demonstrates that there are important co-benefits from GHG mitigation and health benefits from improved air quality, there has been less research addressing co-benefits from improved agricultural production. The potential positive, long-term, effect of higher CO_2 concentrations on plants can be counteracted by short-term damage from increased air pollution. The effects of tropospheric ozone exposure on plant tissues and crop yields are well established, and the scientific literature has already been reviewed in US EPA (1996) and EC (1999). Chameides *et al.* (1994) estimate that 10–35% of the world's grain production is in locations where ozone exposure may reduce crop yields. Surface ozone levels are sensitive to, *inter alia*, NO_x and VOC emissions from fossil-fuel-burning power plants, industrial boilers, motor vehicle exhaust, gasoline retail outlets, and N-fertilizer-induced soil emissions of NO_x.

Using an atmospheric ozone formation model and an economic general-equilibrium model, O'Connor *et al.* (2003) find, for a CO_2 mitigation strategy in China, that the monetary benefits from increased agricultural productivity due to lower ground-level ozone are comparable to the health benefits. Together, these benefits would allow China a 15–20% CO_2 reduction without suffering a welfare loss. Agricultural benefits have important distributional implications. When agricultural effects are not taken into consideration, poor rural households experience welfare losses from carbon mitigation even at low levels of abatement. Once agricultural effects are considered, rural households in this study enjoy welfare gains up to a ten percent abatement rate. So while a purely health-based measure of ancillary benefits tends to show benefits from a climate commitment to be urban-biased, a broader definition of benefits alters the picture considerably.

11.8.1.4   Co-benefits for natural ecosystems

A few studies have pointed out co-benefits of decarbonization strategies from reduced air pollution on natural ecosystems. VanVuuren *et al.* (2006) estimate that, in Europe, compared to an energy policy without climate targets, the implementation of the Kyoto protocol would bring acid deposition below the critical loads in an additional 0.6 to 1.4 million hectares of forest ecosystems, and that an additional 2.2 to 4.1 million hectares would be protected from excess nitrogen deposition. The exact area will depend on the actual use of flexible

instruments, which allow for spatial flexibility in the implementation of mitigation measures but do not take into consideration the environmental sensitivities of ecosystems that are affected by the associated air pollution emissions. Syri *et al.* obtained similar results (2001).

While sustainability and the protection of natural ecosystem have turned out to be important policy drivers in the past (for example in the case of the emission reduction protocols of the Convention on Long-range Transboundary Air Pollution for Europe), there is no generally accepted method for quantifying the monetary value of the existence and function of natural ecosystems. It therefore continues to be difficult to include co-benefits on natural ecosystems in a comprehensive monetary cost-benefit calculation of mitigation measures.

11.8.1.5   Avoidance of air-pollution control costs

As pointed out above, the co-benefits from CO_2 mitigation on air pollution impacts have been found to be largest in developing countries, where air pollutants are often emitted without stringent emission control legislation. Most industrialized countries, however, enforce comprehensive legal frameworks to safeguard local air quality, and these frameworks include source-specific performance standards, national or sectoral emission caps, and ambient air quality criteria.

An increasing number of studies demonstrate significant savings from GHG mitigation strategies on the compliance costs for such air quality legislation. When there are source-specific performance standards, fewer plants burning fossil fuels also imply fewer air pollution control devices. If overall emissions in a country are capped, for example through national emission ceilings in the European Union, or by the obligations of the Gothenburg Protocol of the Convention on Long-range Transboundary Air Pollution, the lower consumption of carbonaceous fuels also reduces the costs for complying with such emission ceilings. This is particularly important since, in these conditions, countries can avoid implementing more expensive air pollution control measures. A similar situation applies when there are legal systems requiring compliance with ambient air quality standards. Carbon mitigation strategies that reduce the levels of polluting activities alleviate control requirements for the remaining sources.

Several studies consistently demonstrate the significance of such cost savings for different countries. Syri *et al.* (2001) found that low-carbon strategies could reduce air pollution control costs for complying with the EU national emission ceilings in 2010 by 10 to 20%, depending on the extent to which flexible mechanisms of the Kyoto protocol are applied. For the long-term perspective until 2100, van Harmelen *et al.* (2002) found air pollution (SO_2 and NO_x) control costs without climate policy objectives to be comparable or, in some periods, even higher than the total costs of an integrated strategy that also includes CO_2 mitigation.

The impact of flexible mechanisms on cost savings has been further explored by van Vuuren *et al.* (2006) for Western European countries. If the Kyoto obligations were to be implemented through domestic action alone, CO_2 mitigation measures amounting to 17 billion US$ per year would allow savings on air pollution control costs of 9.4 billion US$ per year. By contrast, if these countries reached compliance by buying permits for 4 billion US$ per year from outside and implemented domestic measures amounting to 1.4 billion US$ per year, air pollution control costs would decline by 2.4 billion US$ per year in these countries. At the same time, the other European countries selling permits (for 4.3 billion US$ per year) would save an additional 0.7 billion US$ per year on their own air pollution control costs due to the additional carbon mitigation measures.

A study of the United States by EIA (1998) estimated that, for a 31% reduction in CO_2 emissions, the associated decline in SO_2 emissions would be so large that the prices for SO_2 allowances will be driven to zero. Burtraw *et al.* (2003) calculated, for a 7 US$/t$CO_2$ carbon tax, savings of 1–2 US$/t$CO_2$. Their finding was that these savings would be generated by reduced investments in SO_2 and NO_x abatement in order to comply with emission caps.

These cost savings are immediate, they do not depend on controversial judgments on the monetary value of mortality risks, and they can be directly harvested by the actors who need to invest in mitigation measures. They therefore add an important component to a comprehensive assessment of the co-benefits of mitigation strategies. While these cost savings predominantly emerge at present in industrialized countries with elaborate air quality regulations, they will gain increasing importance in developing countries as the latter also progressively implement action to achieve sustainable levels of local air quality.

11.8.1.6   *The need for an integrated approach*

While the studies above adopt different methodological approaches, there is general consensus for all the world regions analyzed that near-term benefits from GHG reductions on human health, agriculture and natural ecosystems can be substantial, both in industrialized and developing countries. In addition, decarbonization strategies lead to reduced air pollution control costs. However, the benefits are highly dependent on the technologies and sectors chosen. In developing countries, many of the benefits could result from improvements to the efficiency of, or switching away from, traditional uses of coal and biomass. Such near-term secondary benefits of GHG control provide an opportunity for a true no-regrets GHG reduction policy in which substantial advantages accrue even if the impact of human-induced climate change itself turns out to be less than current projections indicate.

Climate mitigation policies, if developed independently from air pollution policies, will either constrain or reinforce air pollution policies, and vice-versa. The efficiency of a framework depends on the choice and design of the policy instruments, in particular on how well they are integrated. From an economic perspective, policies that may not be regarded as cost-effective from a climate change or an air pollution perspective alone may be found to be cost-effective if both aspects are considered. So piecemeal regulatory treatment of individual pollutants, rather than a comprehensive approach, could lead to stranded investments in equipment (for example, if new conventional air pollutant standards are put into place in advance of carbon dioxide controls at power plants) (Lempert *et al.*, 2002).

On the basis of recent insights into atmospheric chemistry and health impacts, the literature has identified several concrete options for harvesting synergies between air pollution control and GHG mitigation, and has identified other options that induce undesired trade-offs.

The co-control of emissions – in other words controlling two or more distinct pollutants (or gases) that tend to emanate from a single source through a single set of technologies or policy measures – is a key element of any integrated approach. Air pollutants and GHGs are often emitted by the same sources and so changes in the activity levels of these sources affect both types of emissions. Technical emission control measures aiming at the reduction of one type of emissions from a particular source may reduce or increase the emissions of other substances.

In the energy sector, efficiency improvements and the increased use of natural gas can address both problems (resulting in synergy effects), while the desulphurization of flue gases reduces sulphur emissions but can – to a limited extent – increase carbon dioxide emissions (trade-offs). There are also trade-offs for NO_x control measures for vehicles and nitric acid plants, where increases in N_2O emissions are possible. Concerns have been expressed that measures that improve the local environmental performance of coal in electricity generation might result in a lock-in of coal technologies that will make it more difficult to mitigate CO_2 emissions (McDonald, 1999; Unruh, 2000).

In agriculture, some specific measures to abate ammonia emissions could enhance nitrous oxide and/or methane emissions, while other types of measures could reduce the latter. For Europe, Brink *et al.* (2001) have estimated that abating agricultural emissions of ammonia (NH_3) may cause releases of N_2O from this sector that are up to 15% higher than they would be without NH_3 control. There may be substantial differences in the observed effects between various countries, depending on the extent and type of NH_3 control options applied.

11.8.1.7   *Methane/ozone*

Analyzing non-CO_2 greenhouse gases broadens the scope of climate protection and expands opportunities for synergies involving local pollutants since the co-emission of local pollutants and greenhouse gases vary depending on the type of greenhouse gas considered. For example, in addition to its role as a potent GHG, methane acts as a precursor to tropospheric ozone, together with emissions of nitrogen oxides (NO_x), volatile organic compounds (VOC) and carbon monoxide (CO). Whereas reductions in NO_x and VOC emissions influence local surface ozone concentrations, reductions in methane emissions lower the global ozone background and improve surface air quality everywhere. So reducing methane emissions addresses simultaneously both the pursuit of improved ozone air quality and climate change mitigation objectives (Fiore *et al.*, 2002; Dentener *et al.*, 2004). For instance, West *et al.* (2006) estimate the decreases in premature human mortality that can be attributed to lower surface ozone concentrations resulting from methane mitigation. Reducing global anthropogenic methane emissions by 20% starting in 2010 would prevent approximately 30,000 premature all-cause mortalities globally in 2030, and approximately 370,000 between 2010 and 2030. If avoided mortalities are valued at $1 million each, the benefit of 12 US$/tCO$_2$-equivalent exceeds the marginal cost of the methane reduction. These benefits of climate-motivated methane emission reductions are comparable to those estimated in other studies for CO_2.

A review of health impact studies conducted by the World Health Organization finds evidence for negative effects of ozone on human health even at very low concentrations (WHO, 2003). This has turned the attention of air quality management away from ozone peak episodes towards long-term concentrations, both in the industrialized and the developing world. Long-term concentration levels are driven by emissions at the hemisphere scale and are strongly influenced by atmospheric processes involving methane.

Tropospheric ozone, in addition to its health and vegetation effects, is also a potent GHG (IPCC, 2007a). So ozone reductions will not only result in benefits for local air quality, but also reduce radiative forcing. Further work will be necessary to identify mitigation portfolios that include hemispheric or global methane mitigation on the one hand and control of the local ozone precursor emissions on the other in order to maximize benefits for the global radiation balance and local air quality.

11.8.1.8   *Biomass*

Particularly relevant trade-offs have been identified for GHG mitigation strategies that enhance the use of biofuels and diesel. Biofuels from sustainably-grown biomass are considered to be carbon-neutral. They have therefore been proposed as an important element in decarbonization strategies. However, their combustion in household devices under uncontrolled conditions releases large amounts of fine particulate matter and volatile organic compounds, which cause significant negative health impacts. For instance, Streets and Aunan (2005) estimate that the combustion of coal and biofuels in Chinese households

has contributed to about 10–15% of the total global emissions of black carbon during the past two decades. Emissions from these sources have been identified as the major source of health effects from air pollution in developing countries, adding the highest burden of disease (Smith *et al.*, 2004). In addition to the negative health impacts of traditional biomass combustion, there are concerns about the effectiveness of the combustion of biomass in stoves as a climate change mitigation measure due to the loss of efficiency compared to stoves using fossil fuels (Edwards *et al.*, 2004).

However, the controlled combustion of biomass with stringent air quality measures would prevent a substantial proportion of any toxic emissions. This would sometimes be accompanied by increases in efficiency. Furthermore, ethanol and biodiesel can be produced from biomass in medium-to-large industrial installations with air quality control measures that prevent negative health impacts.

11.8.1.9   *Diesel*

Similar concerns apply to attempts to reduce CO_2 emissions through the replacement of gasoline vehicles by more energy-efficient diesel vehicles. Without the most advanced particle filters, that require very-low-sulphur fuel which is not available everywhere, diesel vehicles are a major contributor to population exposure to fine particulate matter, especially of $PM_{2.5}$ and finer. Diesel particles have been shown to be more aggressive than other types of particles, and are also associated with cancer (HEI, 1999). Mitigation strategies that increase the use of diesel vehicles without appropriate emission control devices counteract efforts to manage air quality. At the same time, concern has been expressed in the literature about the radiative effects of the emissions of black carbon and organic matter from diesel vehicles, which might offset the gains from lower CO2 emissions (Jacobson, 2002). Although both the US and the EU are moving towards very stringent emission standards for diesel engines, their adoption by the rest of the world may be delayed by years.

11.8.1.10   *Practical examples of integrated strategies*

The realization of co-benefits has moved beyond a notion or an analytical exercise and is actually reflected increasingly in national regulations and international treaties.

US EPA operates a programme called 'Integrated Environmental Strategies' that is designed to build capacity to conceptualize co-control measures, analyze their co-benefit potential, and encourage the implementation of promising measures in developing countries. The programme has been active in eight developing countries, resulted in numerous assessments at the urban and national levels of co-benefits, and has helped influence policies leading to efficient measures that address local pollution and GHGs together. The programme is outlined in detail in US EPA (2005).

The European Commission, in its European Climate Change (ECCP) and Clean Air For Europe (CAFE) programmes, explores the interactions between the European Union's climate change and air pollution strategies and examines harmonized strategies that maximize the synergies between both policy areas (CEC, 2005).

The 1987 Montreal Protocol on Subsances that deplete the Ozone Layer mandates the phase-out of ozone-depleting substances, CFCs, halons, HBFCs, HCFCs, and methyl bromide. Some of the alternatives to these products, which are used primarily in refrigeration and in air conditioning, and for producing insulating foam, have significant GWPs although these are, in many cases, less than those for the CFCs and HCFCs. They also can improve the energy efficiency of some equipment and products in which they are used. In order to investigate the link between ozone depletion and climate change, a Special Report was produced by IPCC and the Technology and Economic Assessment Panel (TEAP) of the Montreal Protocol (IPCC & TEAP, 2005).

11.8.2 Impacts of GHG mitigation on employment

A number of studies point out that investments in greenhouse gas mitigation could have a greater impact on employment than investments in conventional technologies. The net impact on employment in Europe in the manufacturing and construction industries of a 1% annual improvement in energy efficiency has been shown to induce a positive effect on total employment (Jeeninga et al., 1999). The effect has been shown to be substantially positive, even after taking into account all direct and indirect macro-economic factors such as the reduced consumption of energy, impact on energy prices, reduced VAT, etc. (European Commission, 2003) The strongest effects are seen in the area of semi-skilled labour in the building trades, which also accounts for the strongest regional policy effects. Furthermore, the European Commission (2005) estimates that a 20% saving on present energy consumption in the European Union by 2020 has the potential to create, directly or indirectly, up to one million new jobs in Europe.

Meyer and Lutz (2002) use the COMPASS model to study the carbon taxes for the G7 countries. They find that recycling revenues via social security contributions increases employment by nearly 1% by 2010 in France and Germany, but much less in US and Japan. Bach et al. (2002), using the models PANTHA RHEI and LEAN, find that the modest ecological tax reform enacted in Germany in 1999–2003 increased employment by 0.1 to 0.6% by 2010. This is as much as 250,000 additional jobs. There is also a 2–2.5% reduction in CO_2 emissions and a negligible effect on GDP. The labour intensity of renewable energy sources has been estimated to be approximately 10 times higher in Poland than that of traditional coal power (0.1–0.9 jobs/GWh compared to 0.01–0.1 jobs/GWh). Given this assumption, government targets for renewable energy would create 30,000 new jobs by 2010 (Jeeninga et al., 1999).

In a study of climate policies for California, Hanemann et al. (2006) report small increases in employment for a package of measures focusing on the tightening of regulations affecting emissions.

11.8.3 Impacts of GHG mitigation on energy security

Since the TAR, new literature has addressed the question of energy security and climate change, especially following the rapid increases and fluctuations in commodity prices, particularly oil, in the period 2004-2006. The concept of energy security is usually understood to be an issue of the reliability of energy supplies that is illustrated by the exposure of oil im-porters to world market prices (Bauen, 2006) and, as Sullivan and Blyth (2006) point out, the reliability of electricity systems given the growing penetration of intermittant renewables, which may require back-up generation capacity (but see UKERC, 2006).

The possibilities of synergies and trade-offs between mitigation actions and energy security are very specific to national circumstances, particularly the relevant fuel mixes as a result of evolving energy markets, the sectors being targeted and energy consumption trends (Turton and Barreto, 2006). The transportation sector, in particular, is characterized by strong synergies relating to energy supply: measures replacing oil with domestic biofuels reduce both emissions and reliance on oil imports. Mitigation action for the electricity sector may lead to synergies with energy security. For example, a more decentralized system based on new renewable generation may reduce gas imports. Alternatively, there may be trade-offs. For example, security reasons may lead countries to increase their dependence on internal reserves of coal rather than relying on natural gas imports (Kuik, 2003).

Whether in the form of synergies or trade-offs, there is a growing recognition of the critical linkages that exist between climate change and energy security, and the fact that energy prices still have yet to reflect these 'externalities' effectively (Bauen, 2006). The inability to manage either one of these threats could result in significant economic and social costs (Turton and Barreto, 2006). Measures that successfully address both issues therefore have the potential to provide signficant social and economic benefits. In conclusion, it seems likely that climate change and energy security pressures will become more acute as international development proceeds. Public policies to address either of these issues can take many forms and their combination makes the effects uncertain, implying a gap in understanding their synergies and trade-offs (Blyth and Lefevre, 2004).

11.8.4 Summary

The recent literature has produced an increasing understanding of the interactions between greenhouse gas mitigation and other policy areas. Numerous studies have identified a wide range of co-benefits and quantified them for industrialized and developing countries. However, the literature does not (as yet) provide a complete picture that includes all the different types of co-benefits needed for a comprehensive assessment. Nevertheless, even the co-benefits quantified at present can make up substantial fractions of, or under specific conditions even exceed, direct mitigation costs.

Beyond the recognition of co-benefits, the realization of potential synergies and avoidance of trade-offs requires an integrated approach that considers a single set of technologies or policy measures in order to simultaneously address all relevant areas. There are practical examples of targeted programmes for pinpointing co-benefits and identifying those policy measures that offer most potential for capturing possible synergies.

In the case of low-income countries, the consideration of potential synergies between GHG mitigation and other policy objectives could be even more important than in high-income countries. At present, climate change policies are often still relatively marginal issues in these countries compared to issues such as poverty eradication, food supply, the provision of energy services, employment, transportation and local environmental quality. Accelerated and sustainable development could therefore become a common interest for both local and global communities (Criqui *et al.*, 2003).

11.9   Mitigation and adaptation - synergies and trade-offs

This section brings together the effects of climate change on mitigation action and the effects of mitigation action on adaptation as identified in Chapters 4 to 10 above. The topic of adaptation-mitigation linkage is covered in Chapter 2, Section 6, and IPCC (2007b, Chapter 18), which is the main reference for concepts, definitions, and analyses. The issue of adaptation-mitigation linkages, particularly when exploring synergies, is fairly nascent in the published literature: Barker (2003) and Dessai and Hulme (2003) analyze mitigation and adaptation linkages as fairly distinctive responses within the context of integrated assessment models; while Dang *et al.* (2003) and Klein *et al.* (2003) have more explicitly addressed the issue of whether and how mitigation and adaptation measures could be more effectively integrated as an overall response to the threat of climate change. Tol (2005) argues that adaptation and mitigation are policy substitutes and should be analyzed as an integrated response to climate change. However, they are usually addressed in different policy and institutional contexts, and policies are implemented at different spatial and temporal scales. This hampers analysis and weakens the trade-offs between adaptation and mitigation. An exception is facilitative adaptation (enhancing adaptive capacity). Like mitigation, it requires long-term policies at the macro-level, but they also compete for resources.

At the national level, mitigation and adaptation are often cast as competing priorities for policy makers (Cohen *et al.*, 1998; Michaelowa, 2001). In other words, interest groups will fight about the limited funds available in a country for addressing climate change, providing analyses of how countries might then make optimal decisions about the appropriate adaptation-mitigation 'mix'. Using a public choice model, Michaelowa (2001) finds that mitigation will be preferred by societies with a strong climate protection industry and low mitigation costs. Public pressure for adaptation will depend on the occurrence of extreme weather events. As technical adaptation measures will lead to benefits for closely-knit, clearly defined groups who can organize themselves well in the political process, these will benefit from subsidy-financed programmes. Changes in society will become less attractive as benefits are spread more widely.

Nonetheless, at the local level, there is a growing recognition that there are in fact important overlaps, particularly when natural, energy and sequestration systems intersect. Examples include bioenergy, forestry and agriculture (Morlot and Agrawala, 2004). This recognition is thought to be particularly relevant for developing countries, particularly the least developed countries, which rely extensively on natural resources for their energy and development needs. More specifically, there is a growing literature analyzing opportunities for linking adaptation and mitigation in agroforestry systems (Verchot, 2004; Verchot *et al.*, 2005), in forestry and agriculture (Dang *et al.*, 2003), and in coastal systems (Ehler *et al.*, 1997).

11.9.1   Sectoral mitigation action: links to climate change and adaptation

11.9.1.1   *Energy*

Section 4.5.5 covers the impact of climate change on energy supply, such as extreme events (Easterling *et al.*, 2000), the effect of warming on infrastructure (such as damage to gas and oil pipelines caused by permafrost melt) and changes in water levels for hydro projects (Nelson *et al.*, 2002). There is a broad consensus that a decentralized energy system (4.3.8) might be more robust in coping with extreme events. Areas that clearly link mitigation and adaptation include, in particular, hydro, biomass and nuclear. Changes in rainfall patterns/glacier melting will clearly impact hydro power and future hydro as a feasible carbon-neutral alternative. The same could be said for biomass, in which too much land used for energy crops may affect both food supply and forestry cover, thereby reducing the ability of communities to adapt to the impacts of climate change, reducing food supplies and therefore making them more vulnerable. Nuclear power generation has been vulnerable to shortages of cooling water due to heat-waves resulting in high ambient temperatures, like those in the EU in 2003 and 2006. This problem is expected to intensify with the rise in these climate-related events. There are opportunities for synergies between mitigation and adaptation in the area of energy supply, particularly for rural populations. For example, the opportunity to develop perennial biomass, such as switch grass, would meet rural energy needs and also provide adaptation benefits because of its relatively low water supply requirements (Samson *et al.*, 2000).

11.9.1.2 Transportation

Options for mitigation in transportation are not considered to be vulnerable to climate change. For transport there are no obvious links between mitigation and adaptation. Any adaptation of the system to climate change, e.g. more air conditioning in vehicles, is not expected to have a significant long-term impact on mitigation.

11.9.1.3 Commercial and residential buildings

While it is clear that the impact of climate change on commercial and residential buildings could be massive, particularly as a result of extreme events and sea level rises, there is less appreciation of the major synergies that are possible between adaptation and mitigation. Modern architecture rarely takes the prevailing climate into consideration, even though design options could result in a considerable reduction in the energy load of buildings, and improve their adaptation to a changing climate (Larsson, 2003). Nevertheless, there is a relatively small amount of literature exploring adaptation-mitigation linkages for new and existing buildings. One example is cool-roof technology options for adapting to higher temperatures. These options also provide mitigation advantages by reducing electricity use and CO_2 emissions. At the same time, cool roofs contribute to reducing the formation of ground level ozone. An example of a conflict between adaptation and mitigation is the effect of a sizeable increase in heat-waves in urban centres. An increase of this kind could intensify pressure for the penetration of inefficient air conditioners, increasing power demand and CO_2 emissions, as was the case during the heat-wave of 1–14 August 2003 in Europe.

11.9.1.4 Industry

Synergies and conflicts between mitigation and adaptation in the industry sector are highly site-specific (see 7.8). It is assumed that large firms would not be as vulnerable to flood risks or weather extremes since they have access to more financial and technical resources. There appears to be no literature indicating explicitly how industry could design its manufacturing and operating processes in such a way that, by adapting to possible climate change events, it can also help to reduce GHG emissions associated with their operations. It is obvious, however, that reducing energy demand would be a good adaptive and mitigative strategy if power supply (from hydro power, for example) were at risk from climate change (Subak *et al.*, 2000). Reducing dependence on cooling water may also be a good adaptive strategy in some locations, but the impact on emissions is not clear.

11.9.1.5 Agriculture and forestry

Most of the literature relating to mitigation-adaptation linkages concerns the agriculture and forestry sectors. In particular, there is a growing awareness of the unique contribution that such synergies could provide for the rural poor, particularly in the least developed countries: many measures focusing on sustainable natural resource management policies could provide both significant adaptation and mitigation benefits, mostly in the form of sequestration activities (Gundimeda, 2004; Morlot and Agrawala, 2004; Murdiyarso *et al.*, 2004). Agriculture is, of course, extremely vulnerable to the impact of climate change, that affects all aspects related to crop land management, and particularly areas related to water management (see Sections 8.5 and 8.8). Low-tillage practices are an example of a win-win technology that reduces erosion and the use of fossil fuels. As discussed in the energy section, bioenergy can of course play a significant role in mitigating global GHG emissions, although the full lifecycle implications of bioenergy options, including effects on deforestation and agriculture, need to be taken into account.

In the forestry sector, policies and measures often take neither adaptation nor mitigation into account (Huq and Grubb, 2004). There is increasing recognition that forestry mitigation projects can often have significant adaptation benefits, particularly in the areas of forest conservation, afforestation and reforestation, biomass energy plantations, agro-forestry, and urban forestry. These projects provide shading, and reduce water evaporation and vulnerability to heat stress. And many adaptation projects in the forestry sector can involve mitigation benefits, including soil and water conservation, agroforestry and biodiversity conservation.

With regard to the increase of biomass energy plantations as a mitigation measure (see Section 11.3.1.4), there may be increased competition for land in many regions, with two crucial effects. First, increased pressure to cultivate what are currently non-agricultural areas may reduce the area available to natural ecosystems, increase fragmentation and restrain the natural adaptive capacity. Secondly, increasing land rents might make agronomically viable adaptation options unprofitable. An alternative view is that there is no shortage of land (Bot *et al.*, 2000; Moreira, 2006), but of investment in land. In this view, the remedy consists of revenues derived from the energy sector (through the CDM, for example), both to raise land productivity through carbon-sequestering soil improvement and to co-produce food or fibre with biomass residuals for conversion to bioenergy products (Greene *et al.*, 2004; Read, 2005; Faaij, 2006; Lehmann *et al.*, 2006; Verchot *et al.*, 2005). Recent studies suggest that technological progress in agriculture will outstrip population growth under a variety of SRES scenarios, leaving enough land for bioenergy cropping, in the most optimistic scenario, to meet all forecast demands for primary energy (Hoogwijk *et al.*, 2005).

Mitigation may have a positive effect on adaptation in agriculture, depending on the circumstances. Additional employment in rural areas will raise incomes and reduce migration. Well-designed CDM projects can reduce the use of traditional biomass as fuel (Gundimeda, 2004) and replace it with marketable renewable fuels, providing a double benefit. There may be also benefits from some mitigation measures for human health, increasing the overall adaptive capacity of the population and making it less vulnerable to specific climate impacts (Tol and Dowlatabadi, 2001).

REFERENCES

Adams, J., 1997: Globalisation, trade, and environment. In *Globalisation and environment: Preliminary perspectives.* OECD, Paris, 184 pp.

Akimoto, K., T. Tomoda, Y. Fujii, and K. Yamaji, 2004: Assessment of global warming mitigation options with integrated assessment model DNE21. *Energy Economics,* **26**, pp. 635-653.

Alic, J.A., D.C. Mowery, and E.S. Rubin, 2003: US technology and innovation policies - lessons from climate change. Pew Centre on Global Climate Change, Arlington.

Anderson, D., 2006: Costs and finance of abating carbon emissions in the energy sector. Imperial College London, London, 63 pp.

Angel, R., 2006: Feasibility of cooling the earth with a cloud of small spacecraft near the inner Lagrange Point (L1). Proceedings of the National Academy of Sciences, USA, doi:10.1073/pnas.0608163103.

Aunan, K., H.E. Mestl, H.M. Seip, J. Fang, D. O'Connor, H. Vennemo, and F. Zhai, 2003: Co-benefits of CO_2-reducing policies in China - a matter of scale? *International Journal of Global Environmental Issues,* **3**(3), pp. 287-304.

Aunan, K., J. Fang, H. Vennemo, K. Oye, and H.M. Seip, 2004: Co-benefits of climate policy - lessons learned from a study in Shanxi, China. *Energy Policy,* **32**(4), pp. 567-581.

Awerbuch, S., 2006: Portfolio-based electricity generation planning: policy implications for renewables and energy security. *Mitigation and Adaptation Strategies for Global Change,* **11**(3), pp. 693-710.

Awerbuch, S., A.C. Stirling, J. Jansen, and L. Beurskens, 2006: Full-spectrum portfolio and diversity analysis of energy technologies. In *Managing Enterprise Risk: What the Electric Industry Experience Implies for Contemporary Business.* Elsevier, 248 pp.

Awerbuch, S., and R. Sauter, 2006: Exploiting the oil-GDP effect to support renewables deployment. *Energy Policy* **34**(17), pp. 2805-2819.

Babiker, M.H., 2005: Climate change policy, market structure and carbon leakage. *Journal of International Economics,* **65**(2), pp. 421-445.

Babiker, M.H., P. Criqui, A.D. Ellerman, J. Reilly, and L. Viguier, 2003: Assessing the impact of carbon tax differentiation in the European Union. *Environmental Modeling and Assessment,* **8**(3), pp. 187-197.

Bach, S., M. Kohlhaas, B. Meyer, B. Praetorius, and H. Welsch, 2002: The effects of environmental fiscal reform in Germany: a simulation study. *Energy Policy,* **30**(9), pp. 803-811.

Bahn, O. and S. Kypreos, 2002: MERGE-ETL: An optimization-equilibrium model with two different endogenous technological learning formulations. Paul Scherrer Institute, Villigen, Switzerland.

Bahn, O. and S. Kypreos, 2003a: Incorporating different endogenous learning formulations in MERGE. *International Journal of Global Energy Issues,* **19**(4), pp. 333-358.

Bahn, O. and S. Kypreos, 2003b: A MERGE model with endogenous technological progress. *Environmental Modeling and Assessment,* **8**, pp. 249-259.

Barker, T. and J. Köhler (eds.), 1998: International competitiveness and environmental policies. Edward Elgar, Cheltenham.

Barker, T., 2003: Representing global climate change, adaptation and mitigation. *Global Environmental Change,* **13**(1), pp. 1-6.

Barker, T., and P. Ekins, 2004: The costs of Kyoto for the US economy. *The Energy Journal,* **25**(3), pp. 53-71.

Barker, T., J. Köhler, and M. Villena, 2002: The costs of greenhouse gas abatement: A meta-analysis of Post-SRES mitigation scenarios. *Environmental Economics and Policy Studies,* **5**, pp. 135-166.

Barker, T., M. Qureshi, and J. Köhler, 2006a: The costs of greenhouse gas mitigation with induced technological change: A meta-analysis of estimates in the literature. Working Paper 89, Tyndall Centre for Climate Change Research, Norwich, 63 pp.

Barker, T., P. Haoran, J. Köhler, R. Warren, and S. Winne, 2006b: Decarbonising the global economy with induced technological change: Scenarios to 2100 using E3MG. *The Energy Journal,* **27**, pp. 143-160.

Barnett, J., S. Dessai, and M. Webber, 2004: Will OPEC lose from the Kyoto Protocol? *Energy Policy,* **32**(18), pp. 2077-2088.

Baron, R. and ECON-Energy, 1997: Economic/fiscal instruments: competitiveness issues related to carbon/energy taxation. OECD/IEA, Paris.

Barreto, L., 2001: Technical learning in energy optimization models and deployment of emerging technologies. Swiss Federal Institute of Technology, Zurich.

Barreto, L., and G. Klaassen, 2004: Emissions trading and the role of learning-by-doing spillovers in the 'bottom-up' energy-systems ERIS model. *International Journal of Energy Technology and Policy,* **2**(1-2), pp. 70-95.

Barreto, L., and S. Kypreos, 2002: Multi-regional technological learning in the energy-systems MARKAL model. *International Journal of Global Energy Issues* **17**(3), pp. 189-213.

Barreto, L., and S. Kypreos, 2004: Emissions trading and technology deployment in an energy-systems 'bottom-up' model with technology learning. *European Journal of Operational Research,* **158**(1), pp. 243-261.

Bashmakov, I., 2006: Oil prices: limits to growth and the depths of falling. *Voprosy Economiki,* **3**.

Bauen, A., 2006: Future energy sources and systems - Acting on climate change and energy security. *Journal of Power Sources,* **157**(2), pp. 893-901.

Bernard, A.L., and M. Vielle, 2003: Measuring the welfare cost of climate change policies: A comparative assessment based on the computable general-equilibrium Model GEMINI-E3. *Environmental Modeling & Assessment,* **8**(3), pp. 199-217.

Bernstein, P., W.D. Montgomery, and T.F. Rutherford, 1999: Global impacts of the Kyoto agreement: Results from the MS-MRT model. *Resource and Energy Economics,* **21**, pp. 375-413.

Blyth, W., and N. Lefevre, 2004: Energy security and climate change policy interactions: An assessment framework. IEA Information Paper, International Energy Agency, Paris, 88 pp.

Böhringer, C. and A. Löschel, 2003: Market power and hot air in international emissions trading: the impacts of US withdrawal from the Kyoto Protocol. *Applied Economics,* **35**, pp. 651-663.

Böhringer, C. and T. Rutherford, 2004: Who should pay how much? *Computational Economics,* **23**(1), pp. 71-103.

Böhringer, C., 2002: Climate politics from Kyoto to Bonn: From little to nothing? *The Energy Journal,* **23**(2), pp. 51-71.

Bollen, J., T. Manders, and P. Veenendaal, 2004: How much does a 30% emission reduction cost? Macroeconomic effects of post-Kyoto climate policy in 2020. CPB Netherlands Bureau for Economic Policy Analysis, The Hauge.

Boltho, A., 1996: The assessment: international competitiveness. *Oxford Review of Economic Policy,* **12**(3), pp. 1-16.

Bosetti, V., C. Carraro, and M. Galeotti, 2006: The dynamics and energy intensity in a model of endogenous technological change. *The Energy Journal* (Special Issue: Endogenous Technological Change and the Economics of Atmospheric Stabilisation), pp. 73-88.

Bot, A.J., F.O. Nachtergaele, and A. Young, 2000: Land resource potential and constraints at regional and country levels. FAO, Rome, 126 pp.

Bower, K., T.W. Choularton, J. Latham, J. Sahraei, and S. Salter, 2006: Computational assessment of a proposed technique for global warming mitigation via albedo-enhancement of marine stratocumulus clouds. *Atmospheric Research*, **82**(1-2), pp. 328-336.

Boyd, P.W., C.S. Law, C.S. Wong, Y. Nojiri, A. Tsuda, M. Levasseur, S. Takeda, R. Rivkin, P.J. Harrison, R. Strzepek, J. Gower, R.M. McKay, E. Abraham, M. Arychuk, J. Barwell-Clarke, W. Crawford, D. Crawford, M. Hale, K. Harada, K. Johnson, H. Kiyosawa, I. Kudo, A. Marchetti, W. Miller, J. Needoba, J. Nishioka, H. Ogawa, J. Page, M. Robert, H. Saito, A. Sastri, N. Sherry, T. Soutar, N. Sutherland, Y. Taira, F. Whitney, S.E. Wong, and T. Yoshimura, 2004: The decline and fate of an iron-induced subarctic phytoplankton bloom. *Nature*, **428**, pp. 549-553.

Brink, C., C. Kroeze, and Z. Klimont, 2001: Ammonia abatement and its impact on emissions of nitrous oxide and methane - Part 2: Application for Europe. *Atmospheric Environment*, **35**(36), pp. 6313-6325.

Buesseler, K. and P. Boyd, 2003: Will ocean fertilization work? *Science*, **300**, pp. 67-68.

Buonanno, P., C. Carraro, and M. Galeotti, 2003: Endogenous induced technical change and the costs of Kyoto. *Resource and Energy Economics*, **25**, pp. 11-34.

Burniaux, J., 2000: A multi-gas assessment of the Kyoto Protocol. Economics Department Working Papers No. 270, OECD.

Burtraw, D., A. Krupnick, K. Palmer, A. Paul, M. Toman, and C. Bloyd, 2003: Ancillary benefits of reduced air pollution in the US from moderate greenhouse gas mitigation policies in the electricity sector. *Journal of Environmental Economics and Management* **45**, pp. 650-673.

Bussolo, M., and D. O'Connor, 2001: Clearing the air in India: The economics of climate policy with ancillary benefits. OECD Development Centre, Paris.

Bye, B., S. Kverndokk, and K.E. Rosendahl, 2002: Mitigation costs, distributional effects, and ancillary benefits of carbon policies in the Nordic countries, the U.K., and Ireland. *Mitigation and Adaptation Strategies for Global Change*, **7**(4), pp. 339-366.

Capros, P. and L. Mantzos, 2000: The economic effects of EU-wide industry-level emission trading to reduce greenhouse gases - Results from PRIMES energy systems model. European Commission.

Caton, R. and S. Constable, 2000: Clearing the air: A preliminary analysis of air quality co-benefits from reduced greenhouse gas emissions in Canada. The David Suzuki Foundation, Vancouver.

Chameides, W.L., P.S. Kasibhatla, J. Yienger, and H. Levy II, 1994: Growth of continental-scale metro-agro-plexes, regional ozone pollution, and world food production. *Science*, **264**, pp. 74-77.

Chandler, W., R. Schaeffer, Z. Dadi, P.R. Shukla, F. Tudela, O. Davidson, and S. Alpan-Atamer, 2002: Climate change mitigation in developing countries: Brazil, China, India, Mexico, South Africa, and Turkey. Pew Center on Global Climate Change, Arlington.

Chen, W., 2005: The costs of mitigating carbon emissions in China: finds from China MARKAL-MACRO modeling. *Energy Policy*, **33**(7), pp. 885-896.

Cicerone, R.J., 2006: Geoengineering: encouraging research and overseeing implementation. *Climatic Change*, **77**(3-4), pp. 221-226.

Cifuentes, L., V.H. Borja-Aburto, N. Gouveia, G. Thurston, and D.L. Davis, 2001: Assessing the health benefits of urban air pollution reductions associated with climate change mitigation (2000-2020): Santiago, São Paulo, México City, and New York City. *Environmental Health Perspectives*, **109**(3), pp. 419-425.

Cifuentes, L., V.H. Borja-Aburto, N. Gouveia, G. Thurston, and D.L. Davis, 2001b: Hidden health benefits of greenhouse gas mitigation. *Science*, **239**(5533), pp. 1257-1259.

Clarke, L. and J. Weyant, 2002: Modeling induced technological change: An overview. In *Technological change and the environment*. Resources for the Future Press, Washington D.C, pp. 320-363.

Clarke, L., J. Weyant, and A. Birky, 2006: On the sources of technological change: Assessing the evidence. *Energy Economics*, **28**, pp. 579-595.

Cohen, S., D. Demeritt, J. Robinson, and D.S. Rothman, 1998: Climate change and sustainable development: towards dialogue. *Global Environmental Change*, **8**(4), pp. 341-371.

CEC, 2005: Communication from the Commission to the Council and the European Parliament, Thematic strategy on air pollution, Commission of the European Communities, pp. 1-13.

Crassous, R., J.C. Hourcade, and O. Sassi, 2006: Endogenous structural change and climate targets: Modelling experiments with IMACLIM-R. *Energy Journal*, **27**(Special Issue: Endogenous technological change and the economics of atmospheric stabilisation), pp. 161-178.

Criqui, P. and A. Kitous, 2003: KPI Technical Report: Impacts of linking JI and CDM credits to the European emission allowance trading scheme. DG Environment, Brussels, pp. 1-15.

Criqui, P., A. Kitous, M. Berk, M. den Elzen, B. Eickhout, P. Lucas, D. van Vuuren, N. Kouvaritakis, and D. Vanregemorter, 2003: Greenhouse gas reduction pathways in the UNFCCC process up to 2025. DG Environment, Brussels, pp. 1-33.

Crutzen, P.J., 2006: Albedo enhancement by stratospheric sulfur injections: a contribution to resolve a policy dilemma? *Climatic Change*, **77**(3-4), pp. 211-219.

Dang, H.H., A. Michaelowa, and D.D. Tuan, 2003: Synergy of adaptation and mitigation strategies in the context of sustainable development: the case of Vietnam. *Climate policy*, **3**(1), pp. S81-S96.

Davidsdottir, B. and M. Ruth, 2004: Capital vintage and climate change policies: the case of US pulp and paper. *Environmental Science and Policy*, **7**(3), pp. 221-233.

De Jager, D., C.A. Hendriks, C. Byers, M. van Brummelen, C. Petersdorff, A.H.M. Struker, K. Blok, J. Oonk, S. Gerbens, and G. Zeeman, 2001: Emission reduction of non-CO_2 greenhouse gases. Report no.: 410 200 094, Utrecht, 233 pp.

DeCanio, S.J., 2003: Economic models of climate change: A critique. Palgrave MacMillan, New York, 203 pp.

Delhotal, K.C., F. de la Chesnaye, A. Gardiner, J. Bates, and A. Sankovski, 2006: Mitigation of methane and nitrous oxide emissions from waste, energy and industry. *The Energy Journal* (Special Issue #3, Multi-greenhouse gas mitigation and climate policy), pp. 45-62.

Den Elzen, M. and A. de Moor, 2002: Evaluating the Bonn-Marrakesh Agreement. *Climate Policy*, **2**, pp. 111-117.

Den Elzen, M., and S. Both, 2002: Modelling emissions trading and abatement costs in FAIR 1.1 - Case study: the Kyoto Protocol under the Bonn-Marrakesh Agreement. RIVM Bilthoven, 68 pp.

Den Elzen, M., P. Lucas, and D. van Vuuren, 2005: Abatement costs of post-Kyoto climate regimes. *Energy Policy*, **33**, pp. 2138-2151.

Dentener, F., D. Stevenson, J. Cofala, R. Mechler, M. Amann, P. Bergamaschi, F. Raes, and R. Derwent, 2004: The impact of air pollutant and methane emission controls on tropospheric ozone and radiative forcing: CTM calculations for the period 1990-2030. *Atmospheric Chemistry and Physics Discussion Paper*, **4**, pp. 8471-8538.

Dessai, S. and M. Hulme, 2003: Does climate policy need probabilities? Working paper 34, Tyndall Centre for Climate Change Research, Norwich, 43 pp.

Dessus, S. and D. O'Connor, 2003: Climate policy without tears: CGE-based ancillary benefits estimates for Chile. *Environmental and Resource Economics,* **25**(3), pp. 287-317.

Easterling, D.R., G.A. Meehl, C. Parmesan, S.A. Changnon, T.R. Karl, and L.O. Mearns, 2000: Climate extremes: Observations, modeling, and impacts. *Science*, **289**(5487), pp. 2068-2074.

Edenhofer, O., K. Lessman, and N. Bauer, 2006a: Mitigation strategies and costs of climate protection: The effects of ETC in the hybrid model MIND. *The Energy Journal* (Special Issue: Endogenous Technological Change and the Economics of Atmospheric Stabilisation), pp. 89-104.

Edenhofer, O., K. Lessman, C. Kemfert, M. Grubb, and J. Köhler, 2006b: Induced technological change: Exploring its implications for the economics of atmospheric stabilisation. Synthesis Report from the Innovation Modeling Comparison Project. *Energy Journal* (Special Issue: Endogenous Technological Change and the Economics of Atmospheric Stabilisation), pp. 1-51.

Edmonds, J., J. Clarke, J. Dooley, S.H. Kim, and S.J. Smith, 2004: Stabilisation of CO_2 in a B2 world: insights on the roles of carbon capture and disposal, hydrogen, and transportation technologies. *Energy Economics*, **26**(4), pp. 517-537.

Edmonds, J., T. Wilson, M. Wise, and J.P. Weyant, 2006: Electrification of the economy and CO_2 emissions mitigation. *Environmental Economics and Policy Studies*, **7**(3), pp. 175-204.

Edwards, R.D., K.R. Smith, J. Zhang, and Y. Ma, 2004: Implications of changes in household stoves and fuel use in China. *Energy Policy*, **32**(3), pp. 395-411.

EEA, 2006: Climate change and a European low-carbon energy system. European Environment Agency, Copenhagen.

Ehler, C.N., B. Cicin-Sain, R. Knecht, R. South, and R. Weiher, 1997: Guidelines to assist policy makers and managers of coastal areas in the integration of coastal management programs and national climate-change action plans. *Ocean & Coastal Management*, **37**(1), pp. 7-27.

EIA, 1998: Impacts of the Kyoto Protocol on U.S. energy markets and economic activity. Energy Information Administration, Washington, D.C.

EIA, 2003: Analysis of S.139, the Climate Stewardship Act of 2003. Energy Information Administration, Washington, D.C.

EIA, 2004: Analysis of Senate Amendment 2028, the Climate Stewardship Act of 2003. Energy Information Administration, Washington, D.C.

EIA, 2005: Impacts of modelled recommendations of the National Commission on Energy Policy. Energy Information Administration, Washington, D.C.

ERI, 2003: China's sustainable energy future. Scenarios of Energy and Carbon Emissions. Energy Research Institute of the National Development and Reform Commission, Peoples Republic of China.

EU DG Environment, 2001: Economic evaluation of sectoral emission reduction objectives for climate change - Summary report for policy makers EU DG Environment.

European Commission, 1999: Economic evaluation of quantitative objectives for climate change. COHERENCE, Belgium.

European Commission, 2003: Directive of the European Parliament and of the Council on Energy end-use efficiency and energy services. COM (2003) 739 Final, Brussels.

European Commission, 2005: Green paper on energy efficiency - Doing more with less. Luxembourg office for official publications of the European Communities, 45 pp.

Eyckmans, J., J. Cornillie, and D. Van Regemorter, 2000: Efficiency and equity in the EU Burden Sharing Agreement. Katholieke Universiteit, Leuven.

Eyre, N., 2001: Carbon reduction in the real world: how the UK will surpass its Kyoto obligations. *Climate Policy*, **1**(3), pp. 309-326.

Faaij, A., 2006: Modern biomass conversion technologies mitigation and adaptation strategies. *Global Change*, **11**(2), pp. 335-367.

Faaij, A.P.C. and J. Domac, 2006: Emerging international bioenergy markets and opportunities for socio-economic development. *Energy for Sustainable Development* (Special Issue on emerging international bioenergy markets and opportunities for socio-economic development), **X**(1), March 2006, pp. 7-19.

Fichtner, W., A. Fleury, and O. Rentz, 2003: Effects of CO_2 emission reduction strategies on air pollution. *International Journal of Global Environmental Issues*, **3**(3), pp. 245-265.

Fiore, A.M., D.J. Jacob, B.D. Field, D.G. Streets, S.D. Fernandes, and C. Jang, 2002: Linking ozone pollution and climate change: The case for controlling methane. *Geophysical Research Letters*, **29**(19), pp. 2521-2524.

Fischer, C. and R. Morgenstern, 2006: Carbon abatement costs: Why the wide range of estimates? *The Energy Journal*, **27**(2), pp. 73-86.

Fischer, C. and R. Newell, 2004: Environmental and technology policies for climate change and renewable energy. RFF Discussion paper 04-05.

Fischer, C., 2003: Climate change policy choices and technical innovation. *Minerals and Energy*, **18**(2), pp. 7-15.

Foxon, T.J., 2003: Inducing innovation for a low-carbon future: drivers, barriers and policies. The Carbon Trust, London, 64 pp.

Fritsche, U.R., K. Hünecke, A. Hermann, F. Schulze, and K. Wiegmann, 2006: Sustainability standards for bioenergy. Öko-Institut e.V., Darmstadt, WWF Germany, Frankfurt am Main, November.

Gabriel, S., A. Kydes, and P. Whitman, 2001: The national energy modeling system: A large-scale energy-economic-equilibrium model. *Operations Research*, **49**(1), pp. 14-25.

Garbaccio, R., M.S. Ho, and D.W. Jorgenson, 1999: Controlling carbon emissions in China. *Environment and Development Economics*, **4**, pp. 493-518.

Gately, D. and H.G. Huntington, 2002: The asymmetric effects of changes in price and income on energy and oil demand. *Energy Journal*, **23**(1), pp. 19-55.

GCEP, 2005: Global Climate & Energy Project Brochure. Global Climate & Energy Project, Stanford University, California, 40 pp.

Gerlagh, R. and B. van der Zwaan, 2003: Gross World Product and consumption in a global warming model with endogenous technological change. Resource and Energy Economics, **25**(1), pp. 35-57.

Gerlagh, R. and W. Lise, 2005: Carbon taxes: A drop in the ocean, or a drop that erodes the stone? The effect of carbon taxes on technological change. *Ecological Economics*, **54**(2-3), pp. 241-260.

Gerlagh, R., 2006: ITC in a global growth-climate model with CCS: The value of induced technical change for climate stabilisation. *Energy Journal* (Special Issue: Endogenous Technological Change and the Economics of Atmospheric Stabilisation), pp. 55-72.

Goulder, L.H. and K. Mathai, 2000: Optimal CO_2 abatement in the presence of induced technological change. *Journal of Environmental Economics and Management*, **39**(1), pp. 1-38.

Goulder, L.H. and S.H. Schneider, 1999: Induced technological change and the attractiveness of CO_2 abatement policies. *Resource and Energy Economics*, **21**(3-4), pp. 211-253.

Goulder, L.H., 2004: Induced technological change and climate policy. Pew Centre on Global Climate Change, Arlington.

Govindasamy, B., K. Caldeira, and P.B. Duffy, 2003: Geoengineering Earth's radiation balance to mitigate climate change from a quadrupling of CO_2. *Global and Planetary Change*, **37**(1-2), pp. 157-168.

Greene, N., F.E. Celik, B. Dale, M. Jackson, K. Jayawardhana, H. Jin, E. Larson, M. Laser, L. Lynd, D. MacKenzie, M. Jason, J. McBride, S. McLaughlin, and D. Saccardi, 2004: Growing energy: How biofuels can help end America's oil dependence. Natural Resources Defense Council, 86 pp.

Greening, L., D.L. Greene, and C. Difiglio, 2000: Energy efficiency and consumption - The rebound effect - A survey. *Energy Policy*, **28**, pp. 389-401.

Griffin, J.M. and C.T. Shulman, 2005: Price asymmetry in energy demand models: A proxy for energy-saving technical change? *Energy Journal* **26**(2), pp. 1-22.

Grubb, M. and J. Wilde, 2005: The UK climate change programme: potential evolution for the business and public sector. The Carbon Trust, London, 72 pp.

Grubb, M., 2004: Technology innovation and climate change policy: An overview of issues and options. *Keio Economic Studies*, **41**(2), pp. 103-132.

Grubb, M., C. Hope, and R. Fouquet, 2002a: Climatic implications of the Kyoto Protocol: The contribution of international spillover. *Climatic Change*, **54**(1-2), pp. 11-28.

Grubb, M., J. Köhler, and D. Anderson, 2002b: Induced technical change in energy and environmental Modeling: Analytic approaches and policy implications. *Annual Review of Energy Environment*, **27**, pp. 271-308.

Grubb, M., L. Butler, and P. Twomey, 2006: Diversity and security in UK electricity generation: The influence of low carbon objectives. *Energy Policy* **34**(18), pp. 4050-4062.

Grübler, A., B. O'Neill, K. Riahi, V. Chirkov, A. Goujon, P. Kolp, I. Prommer, S. Scherbov, and E. Slentoe, 2006: Regional, national, and spatially explicit scenarios of demographic and economic change based on SRES. *Technological Forecasting & Social Change,* doi:10.1016/j.techfore.2006.05.023.

Grübler, A., N. Naki_enovi_, and W.D. Nordhaus (eds.), 2002: Technological change and the environment. Resources for the future, Washington, D.C.

Gundimeda, H., 2004: How 'sustainable' is the 'sustainable development objective' of CDM in developing countries like India? *Forest Policy and Economics,* **6**(3-4), pp. 329-343.

Haag, G. and P. Liedl, 2001: Modelling of knowledge, capital formation and innovation behaviour within micro-based profit oriented and correlated decision processes. In *Knowledge, Complexity and Innovation Systems.* Springer, pp. 251-274.

Ha-Duong, M., M. Grubb, and J.C. Hourcade, 1997: Influence of socio-economic inertia and uncertainty on optimal CO_2-emission abatement. *Nature,* **390**(6657), pp. 270-273.

Hamelinck, C. and A.Faaij, 2006: Outlook for advanced biofuels. *Energy Policy,* **34** (17), pp. 3268-3283.

Han, H., 2001: Analysis of the environmental benefits of reductions in greenhouse gas emissions. Korea Environmental Institute Seoul, Report 7.

Hanemann, M., A.E. Farrell, and D. Roland-Holst, 2006: Executive Summary. In *Managing Greenhouse Gas Emissions in California.* The California Climate Change Center at UC Berkeley, pp. 1-6.

Hanson, D. and J.A.S. Laitner, 2004: An integrated analysis of policies that increase investments in advanced energy-efficient/low-carbon technologies. *Energy Economics,* **26**(4), pp. 739-755.

Hedenus, F., C. Azar, and K. Lindgren, 2006: Induced technological change in a limited foresight optimization model. *The Energy Journal* (Special Issue: Endogenous Technological Change and the Economics of Atmospheric Stabilisation), pp. 41-54.

HEI, 1999: Diesel emissions and lung cancer: Epidemiology and quantitative risk assessment. Health effects institute, Cambridge, MA, 82 pp.

HEI, 2004: Health effects of outdoor air pollution in developing countries of Asia: A literature review. Health Effects Institute, Boston, MA, 124 pp.

Hirst, D., 2006: Demand side - Teaching an old dog new tricks. IEE Networks for Sustainable Power, London.

Holte, S.H. and A. Kydes, 1997: The national energy modeling system: Policy analysis and forecasting at the U.S. Department of Energy. In *Systems Modeling for Energy Policy,* John Wiley and Sons Ltd, New York.

Hoogwijk, M., A. Faaij, B. Eickhout, B. de Vries, and W. Turkenburg, 2005: Potential of biomass energy out to 2100, for four IPCC SRES land-use scenarios. *Biomass and Bioenergy,* **29**(4), pp. 225-257.

Hoogwijk, M., A. Faaij, R. van den Broek, G. Berndes, D. Gielen, and W. Turkenburg, 2003: Exploration of the ranges of the global potential of biomass for energy. *Biomass and Bioenergy,* Vol. 25 No.2, pp. 119-133.

Hunt, L.C. and Y. Ninomiya, 2005: Primary energy demand in Japan: An empirical analysis of long-term trends and future CO_2 emissions. *Energy Policy,* **33**(11), pp. 1409-1424.

Huq, S. and M. Grubb, 2004: Scientific assessment of the inter-relationships of mitigation and adaptation. IPCC Expert Meeting. Geneva.

IEA, 2000: World Energy Outlook 2000. International Energy Agency, Paris.

IEA, 2004: World Energy Outlook 2004. International Energy Agency, Paris.

IEA, 2005: World Energy Outlook 2005 - Middle East and North Africa Insights. International Energy Agency, Paris.

IEA, 2006a: Energy Technology Perspectives - Scenarios & Strategies to 2050. International Energy Agency - OECD, 484 pp.

IEA, 2006b: World Energy Outlook 2006. International Energy Agency, Paris.

IMAGE-team, 2001: The IMAGE 2.2 implementation of the SRES scenarios. A comprehensive analysis of emissions, climate change and impacts in the 21st century. National Institute for Public Health and the Environment, Bilthoven.

IPCC & TEAP, 2005: *Safeguarding the ozone layer and the global climate system: Issues related to hydrofluorocarbons and perfluorocarbons.* Special Report of the Intergovernmental Panel on Climate Change (IPCC). Cambridge University Press, Cambridge, 478 pp.

IPCC, 2000: *Emissions Scenarios.* Special Report of the Intergovernmental Panel on Climate Change (IPCC). Cambridge University Press, Cambridge.

IPCC, 2001: *Climate Change 2001: Mitigation* - Contribution of Working Group III to the Third Assessment Report of the Intergovernmental Panel on Climate Change (IPCC) [Metz, B., O. Davidson, R. Swart, and J. Pan (eds.)]. Cambridge University Press, Cambridge, 700 pp.

IPCC, 2005: *Carbon Capture and Storage.* Special Report of the Intergovernmental Panel on Climate Change (IPCC). Cambridge University Press, Cambridge and New York, 442 pp.

IPCC, 2007a: *Climate Change 2007: The Physical Science Basis.* Contribution of Working Group I to the Fourth Assessment Report of the Intergovernmental Panel on Climate Change [Solomon, S., D. Qin, M. Manning, Z. Chen, M. Marquis, K.B.M.Tignor and H.L. Miller (eds.)]. Cambridge University Press, Cambridge, United Kingdom and New York, NY, USA, 996 pp.

IPCC, 2007b: *Climate Change 2007: Impacts, Adaptation and Vulnerability.* Contribution of Working Group II to the Fourth Assessment Report of the Intergovernmental Panel on Climate Change [Parry, M.L., O.F. Canziani, J.P. Palutikof, P.J. van der Linden, C.E. Hanson (eds.)]. Cambridge University Press, Cambridge, United Kingdom and New York, NY, USA.

IPTS, 2000: Preliminary analysis of the implementation of an EU-wide permit trading scheme on CO_2 emissions abatement costs. Results from the POLES-model. Institute for Prospective Technological Studies, 4 pp.

Isoard, S. and A. Soria, 2001: Technical change dynamics: evidence from the emerging renewable energy technologies. *Energy Economics,* **23**(6), pp. 619-636.

Iwafune, Y. and K. Yamaji, 2001a: Comprehensive evaluation of the effect of energy saving technologies in residential sector. *The Transactions of The Institute of Electrical Engineers of Japan,* **121-B**(9), pp. 1076-1084.

Iwafune, Y. and K. Yamaji, 2001b: Comprehensive evaluation of the effect of CO_2 reduction options in commercial and residential sectors considering long-term POWER generation mix. *The Transactions of The Institute of Electrical Engineers of Japan,* **121-B**(12), pp. 1716-1724.

Iwafune, Y. and K. Yamaji, 2001c: Comprehensive evaluation of the effect of energy saving technologies in commercial buildings. *The Transactions of The Institute of Electrical Engineers of Japan,* **121-B**(5), pp. 581-589.

Jaccard, M., J. Nyboer, and B. Sadownik, 2002: The cost of climate policy. UBC Press, Canada.

Jaccard, M., R. Loulou, A. Kanudia, J. Nyboer, A. Bailie, and M. Labriet, 2003: Methodological contrasts in costing greenhouse gas abatement policies: Optimization and simulation modeling of micro-economic effects in Canada. *European Journal of Operational Research,* **1**(145), pp. 148-164.

Jacobson, M.Z., 2002: Control of fossil-fuel particulate black carbon and organic matter, Possibly the most effective method of slowing global warming. *Journal of Geophysical Research,* **107**(19).

Jaffe, A., R. Newell, and R. Stavins, 2003: Technological change and the environment. In *Handbook of Environmental Economics.* Elsevier.

Jaffe, A., R. Newell, and R. Stavins, 2005: A tale of two market failures: Technology and environmental policy. *Ecological Economics,* **54**, pp. 164-174.

Jeeninga, H., C. Weber, I. Mäenpää, F. Rivero García, V. Wiltshire, and J. Wade, 1999: Employment impacts of energy conservation schemes in the residential sector: Calculation of direct and indirect employment effects using a dedicated input/output simulation approach. SAVE Contract XVII/4.1031/D/97-032, ECN.

Jensen, J. and M.H. Thelle, 2001: *What are the gains from a multi-gas strategy?* Working Paper No. 84, FEEM, Milano.

Jiang, K. and X. Hu, 2006: Energy demand and emissions in 2030 in China: Scenarios and policy options. *Environmental Economics and Policy Studies*, **7**(3), pp. 233-250.

Joh, S., Y.-M. Nam, S. ShangGyoo, S. Joohon, and S. Youngchul, 2003: Empirical study of environmental ancillary benefits due to greenhouse gas mitigation in Korea. *International Journal of Sustainable Development*, **6**(3), pp. 311-327.

Jones, I.S.F., 2004: The enhancement of marine productivity for climate stabilisation and food security. In *Handbook of Microalgal Culture*. Blackwell, Oxford.

Kan, H., B. Chen, C. Chen, Q. Fu, and M. Chen, 2004: An evaluation of public health impact of ambient air pollution under various energy scenarios in Shanghai, China. *Atmospheric Environment*, **38**, pp. 95-102.

Kemfert, C., 2002: An integrated assessment model of economy-energy-climate - The model Wiagem. *Integrated Assessment*, **3**(4), pp. 281-298.

Kemfert, C., 2004: Climate coalitions and international trade: assessment of cooperation incentives by issue linkage. *Energy Policy*, **32**(4), pp. 455-465.

Klein, R.J.T., E.L. Schipper, and S. Dessai, 2003: Integrating mitigation and adaptation into climate and development policy: three research questions. Working paper 40, Tyndall Centre for Climate Change Research, Norwich, 20 pp.

Klepper, G. and S. Peterson, 2004: The EU emissions trading scheme: Allowance prices, trade flows, competitiveness effects. *European Environment,* **14**(4), pp. 201-218.

Klepper, G. and S. Peterson, 2005: Trading hot-air. The influence of permit allocation rules, market power and the US withdrawal from the Kyoto Protocol. *Environmental and Resource Economics*, **32**(2), pp. 205-228.

Klepper, G. and S. Peterson, 2006: Emission trading, CDM, JI and more: The climate strategy of the EU. *Energy Journal*, **27**(2), pp. 1-26.

Köhler, J., M. Grubb, D. Popp, and O. Edenhofer, 2006: The transition to endogenous technical change in climate-economy models. *The Energy Journal* (Special Issue: Endogenous Technological Change and the Economics of Atmospheric Stabilisation), pp. 17-55.

Komiyama, R., K. Yamaji, and Y. Fujii, 2003: CO_2 emission reduction and primary energy conservation effects of cogeneration system in commercial and residential sectors. Considering Long-Term Power Generation Mix in Japan. *IEE Japan*, **123B**(5), pp. 577-588.

Kopp, R.J., 2001: Climate policy and the economics of technical advance: Drawing on inventive activity. In *Climate Change Economics and Policy: An RFF Anthology*. RFF Press, Washington.

Kraines, S.B., D.R. Wallace, Y. Iwafune, Y. Yoshida, T. Aramaki, K. Kato, K. Hanaki, H. Ishitani, T. Matsuo, H. Takahashi, K. Yamada, K. Yamaji, Y. Yanagisawa, and H. Komiyama, 2001: An integrated computational infrastructure for a virtual Tokyo: concepts and examples. *Journal of Industrial Ecology*, **5**(1), pp. 35-54.

Kuik, O. and R. Gerlagh, 2003: Trade liberalization and carbon leakage. *The Energy Journal*, **24**(3), pp. 97-120.

Kuik, O., 2003: Climate change policies, energy security and carbon dependency: Trade-offs for the European Union in the longer term. *International Environmental Agreements Politics, Law and Economics*, **3**(3), pp. 221-242.

Kumar, A., S.K. Jain, and N.K. Bansal, 2003: Disseminating energy-efficient technologies: a case study of compact fluorescent lamps (CFLs) in India. *Energy Policy*, **31**(3), pp. 259-272.

Kurosawa, A., 2004: Carbon concentration target and technological choice. *Energy Economics* **26**(4), pp. 675-684.

Kurosawa, A., 2006: Multigas mitigation: an economic analysis using GRAPE model. *The Energy Journal* (Special Issue #3, Multi-Greenhouse Gas Mitigation and Climate Policy), pp. 275-288

Kverndokk, S., T. Rutherford, and K. Rosendahl, 2004: Climate policies and induced technological change: Which to choose, the carrot or the stick? *Environmental and Resource Economics*, **27**(1), pp. 21-41.

Kydes, A.S., 2000: Modeling technology learning in the national energy modeling system. Asian *Journal of Energy and Environment*, **1**(2).

Kypreos, S. and O. Bahn, 2003: A MERGE model with endogenous technological progress. *Environmental Modeling and Assessment*, **8**(3), pp. 249-259.

Larsson, N., 2003: Adapting to climate change in Canada. *Building Research and Information*, **31**(3-4), pp. 231-239.

Lasky, M., 2003: *The economic costs of reducing emissions of greenhouse gases: A survey of economic models*. Technical Paper Series, Congressional Budget Office, Washington, D.C.

Latham, J., 1990: Control of global warming? *Nature*, **347**(6291), pp. 339-340.

Latham, J., 2002: Amelioration of global warming by controlled enhancement of the Albedo and longevity of low-level maritime clouds. *Atmospheric Science Letters*, **3**(2-4), pp. 52-58.

Lecocq, F., J.C. Hourcade, and M. Ha-Duong, 1998: Decision making under uncertainty and inertia constraints: sectoral implications of the when flexibility. *Energy Economics*, **20**(5-6), pp. 539-555.

Lehmann, J., J. Gaunt, and M. Rondon, 2006: Bio-char sequestration in terrestrial ecosystems - a review. *Mitigation and Adaptation Strategies for Global Change*, **11**(2), pp. 395-419.

Lempert, R.J., S.W. Popper, S.A. Resetar, and S.L. Hart, 2002: Capital cycles and the timing of climate change policy. Pew Center on Global Climate Change, Arlington.

Li, J., 2002: Including the feedback of local health improvement in assessing costs and benefits of GHG reduction. *Review of Urban and Regional Development Studies*. **14**(3), pp. 282-304.

Li, J., 2006: A multi-period analysis of a carbon tax including local health feedback: an application to Thailand. *Environment and Development Economics*, **11**(3), pp. 317-342.

Löschel, A. and Z. Zhang, 2002: The economic and environmental implications of the US repudiation of the Kyoto Protocol and the subsequent deals in Bonn and Marrakech. *Weltwirtschaftliches Archiv*, **138**(4), pp. 711-746.

Löschel, A., 2002: Technological change in economic models of environmental policy: a survey. *Ecological Economics*, **43**, pp. 105 - 126.

Lutz, C., B. Meyer, C. Nathani, and J. Schleich, 2005: Endogenous technological change and emissions: the case of the German steel industry. *Energy Policy*, **33**(9), pp. 1143-1154.

Maeda, A., 2003: The emergence of market power in emission rights markets: The role of initial permit distribution. *Journal of Regulatory Economics*, **24**(3), pp. 293-314.

Manne, A. and R. Richels, 1992: Buying greenhouse gas insurance: The economic costs of CO_2 emission limits. MIT Press, Cambridge.

Manne, A. and R. Richels, 2001: US Rejection of the Kyoto Protocol: the impact on compliance costs and CO_2 emissions. Stanford University Colorado, pp. 1-19.

Manne, A. and R. Richels, 2004: The impact of learning-by-doing on the timing and costs of CO_2 abatement. *Energy Economics*, **26**(4), pp. 603- 619.

Masui, T., G. Hibino, J. Fujino, Y. Matsuoka, and M. Kainuma, 2005: Carbon dioxide reduction potential and economic impacts in Japan: Application of AIM. *Environmental Economics and Policy Studies*, **7**(3), pp. 271-284.

Masui, T., T. Hanaoka, S. Hikita, and M. Kainuma, 2006: Assessment of CO_2 reductions and economic impacts considering energy-saving investments. *The Energy Journal* (Special Issue: Endogenous Technological Change and the Economics of Atmospheric Stabilisation), pp. 77-92.

McDonald, A. and L. Schrattenholzer, 2001: Learning rates for energy technologies. *Energy Policy*, **29**(4), pp. 255-261.

McDonald, A., 1999: Priorities for Asia: combating acid deposition and climate change. *Environment*, **41**(4), pp. 4-11.

McFarland, J., J. Reilly, and H. Herzog, 2004: Representing energy technologies in top-down economic models using bottom-up information. *Energy Economics*, **26**(4), pp. 685-707.

McKibbin, W.J. and P.J. Wilcoxen, 2004: Estimates of the costs of Kyoto: Marrakesh versus the McKibbin-Wilcoxen blueprint. *Energy Policy*, **32**(4), pp. 467-479.

McKinley, G., M. Zuk, M. Hojer, M. Avalos, I. Gonzalez, R. Iniestra, I. Laguna, M. Martinez, P. Osnaya, L.M. Reynales, R. Valdes, and J. Martinez, 2005: Quantification of local and global benefits from air pollution control in Mexico City. *Environmental Science & Technology,* **39**(7), pp. 1954-1961.

Messner, S., 1997: Endogenized technological learning in an energy systems model. *Journal of Evolutionary Economics,* 7(3), pp. 291-313.

Mestl, H.E.S., K. Aunan, J. Fang, H.M. Seip, J.M. Skjelvik, and H. Vennemo, 2005: Cleaner production as climate investment - integrated assessment in Taiyuan City, China. *Journal of Cleaner Production,* **13**(1), pp. 57-70.

Meyer, B. and C. Lutz, 2002: Carbon tax and labour compensation - a simulation for G7. In *Economy-Energy-Environment Simulation: Beyond the Kyoto Protocol.* Kluwer Academic Publishers, London, pp. 185-190.

Michaelowa, A., 2001: Mitigation versus adaptation: the political economy of competition between climate policy strategies and the consequences for developing countries. Discussion Paper 153, Hamburg Institute of International Economics, Hamburg, 34 pp.

Montgomery, W.D. and A.E. Smith, 2006: Price, quantity and technology strategies for climate policy. In *Human-Induced Climate Change: An Interdisciplinary Assessment.* Cambridge University Press, Cambridge.

Moreira, J., 2006: Global biomass energy potential. *Journal of Mitigation and Adaptation Strategies for Global Change,* **11**(2), pp. 313-333.

Morgenstern, R., 2005: Prepared statement of Richard D. Morgenstern, PhD U.S. Senate. Committee on Energy and Natural Resources, Washington, D.C.

Morgenstern, R., A. Krupnick, and X. Zhang, 2004: The ancillary carbon benefits of SO_2 reductions from a small-boiler policy in Taiyuan, PRC. *The Journal of Environment & Development,* **13**(2), pp. 140-155.

Mori, S. and T. Saito, 2004: Potentials of hydrogen and nuclear towards global warming mitigation-expansion of an integrated assessment model MARIA and simulations. *Energy Economics,* **26**(4), pp. 565-578.

Mori, S., 2006: Role of integrated assessment and scenario development issues beyond SRES. *Environmental Economics and Policy Studies,* **7**(3), pp. 315-330.

Morita, T., N. Naki_enovi_, and J. Robinson, 2000: Overview of mitigation scenarios for global climate stabilisation based on new IPCC emission scenarios (SRES). *Environmental Economics and Policy Studies,* **3,** pp. 65-88.

Morlot, J.C. and S. Agrawala (eds.), 2004: The benefits of climate change policies: Analytical and framework issues. OECD, Paris, 323 pp.

Murdiyarso, D., L. Lebel, A.N. Gintings, S.M.H. Tampubolon, A. Heil, and M. Wasson, 2004: Policy responses to complex environmental problems: Insights from a science-policy activity on transboundary haze from vegetation fires in Southeast Asia. *Agriculture, Ecosystems and Environment,* **104**(1), pp. 47-56.

Nabuurs, G.J., A.J. Dolman, E. Verkaik, P.J. Kuikman, C.A. van Diepen, A.P. Whitmore, W.P. Daamen, O. Oenema, P. Kabat, and G.M.J. Mohren, 2000: Article 3.3 and 3.4 of the Kyoto Protocol: consequences for industrialised countries' commitment, the monitoring needs, and possible side effects. *Environmental Science & Policy,* **3**(2-3), pp. 123-134.

Neij, L., P. Andersen, and M. Durstewitz, 2003b: The use of experience curves for assessing energy policy programmes. EU/IEA workshop Experience Curves: Tool for Energy Policy Analysis and Design. IEA, Paris.

Neij, L., P. Andersen, M. Durstewitz, P. Helby, M. Hoppe-Kilpper, and P. Morthorst, 2003a: Experience curves: A tool for energy policy assessment. Environmental and Energy Systems Studies, Lund University, Sweden.

Nelson, F.E., O.A. Anisimov, and N.I. Shiklomanov, 2002: Climate change and hazard zonation in the Circum-Arctic Permafrost regions. *Natural Hazards,* **26**(3), pp. 203-225.

Neuhoff, K., 2004: Large scale deployment of renewables for electricity generation. *Oxford Review of Economic Policy,* **21**(1), pp. 88-110.

Newell, R., A. Jaffe, and R. Stavins, 2006: The effects of economic and policy incentives on carbon mitigation technologies. *Energy Economics,* **28**(5-6), pp. 563-578.

NHTSA, 2006: Final regulatory impact analysis, corporate average fuel economy and CAFÉ reform for MY 2008-2011 light trucks. National Highway Traffic and Safety Administration, Washington, D.C.

Nordhaus, W.D., 2002: Modeling induced innovation climate-change policy. In *Technological Change and the Environment.* Resources for the Future Press, Washington, pp. 182-209.

O'Connor, D., F. Zhai, K. Aunan, T. Berntsen, and H. Vennemo, 2003: Agricultural and human health impacts of climate policy in China: A general-equilibrium analysis with special reference to Guangdong. Technical Paper 206, OECD Development Centre, Paris, 85 pp.

Otto, V., A. Löschel, and J. Reilly, 2006: Directed technical change and climate policy. Nota di lavoro 81, FEEM, 41 pp.

Paltsev, S., 2001: The Kyoto Protocol: Regional and sectoral contributions to the carbon leakage. *The Energy Journal,* **22**(4), pp. 53-79.

Paltsev, S., J.M. Reilly, H.D. Jacoby, A.D. Ellerman, and K.H. Tay, 2003: Emissions trading to reduce greenhouse gas emissions in the United States: The McCain-Lieberman Proposal. MIT Press, Cambridge, MA.

Pershing, J., 2000: Fossil fuel implications of climate change mitigation responses. Proceedings of an IPCC Expert Meeting, 14-15 February 2000, Technical Support Unit, IPCC Working Group III.

Philibert, C., 2005: The role of technological development and policies in a post-Kyoto climate regime. *Climate Policy,* **5**(3), pp. 291-308.

Pizer, W.A., D. Burtraw, W. Harrington, R. Newell, and J. Sanchirico, 2006: Modeling economy-wide versus sectoral climate policies using combined aggregate-sectoral models. *Energy Journal,* **27**(3), pp. 135-168.

Placet, M., K.K. Humphreys, and N.M. Mahasenan, 2004: Climate change technology scenarios: Energy, emissions and economic implications. Pacific Northwest National laboratory, Washington.

Popp, D., 2002: Induced innovation and energy prices. *American Economic Review,* **92**(1), pp. 160-180.

Popp, D., 2004: ENTICE: endogenous technological change in the DICE model of global warming. *Journal of Environmental Economics and Management,* **48**(1), pp. 742-768.

Popp, D., 2006a: Comparison of climate policies in the ENTICE-BR model. *The Energy Journal* (Special Issue: Endogenous Technological Change and the Economics of Atmospheric Stabilisation), pp. 163-174.

Popp, D., 2006b: ENTICE-BR: The effects of backstop technology R&D on climate policy models. *Energy Economics,* **28**(2), pp. 188-222.

Popp, D., 2006c: Innovation in climate policy models: Implementing lessons from the economics of R&D. *Energy Economics,* **28**(5-6), pp. 596-609.

Price, L., S. de la Rue du Can, J. Sinton, E. Worrell, Z. Nan, J. Sathaye, and M. Levine, 2006: Sectoral trends in global energy use and greenhouse gas emissions. Report number: LBNL-56144, Lawrence Berkeley National Laboratory.

Pronk, J., 2001: New proposals by the President of COP6. The Hague. pp. 1-24.

Proost, S. and D.V. Regemorter, 2003: Interaction between local air pollution and global warming and its policy implications for Belgium. *International Journal of Global Environmental Issues,* **3**(3), pp. 266-286.

Rao, S. and K. Riahi, 2006: The Role of non-CO_2 greenhouse gases in climate change mitigation: long-term scenarios for the 21st century. *The Energy Journal* (Special Issue #3, Multi-Greenhouse Gas Mitigation and Climate Policy), pp.177-200.

Rao, S., I. Keppo, and K. Riahi, 2006: Importance of technological change and spillovers in long-term climate policy. *The Energy Journal* (Special Issue: Endogenous Technological Change and the Economics of Atmospheric Stabilisation), pp. 105-122.

Read, P., 2005: Carbon cycle management with biotic fixation and long-term sinks. In *Stabilisation 2005.* Cambridge University Press, Cambridge.

Reinaud, J., 2005: Industrial competitiveness under the European Union Emissions Trading Scheme. International Energy Agency, 92 pp.

Repetto, R. and D. Austin, 1997: The costs of climate protection: A guide for the perplexed. World Resources Institute, Washington, D.C.

Riahi, K., A. Grübler, and N. Nakicenovic, 2006: Scenarios of long-term socio-economic and environmental development under climate stabilisation. *Technological Forecasting and Social Change,* doi:10.1016/j.techfore.2006.05.026.

Riahi, K., E.S. Rubin, M.R. Taylor, L. Schrattenholzer, and D. Hounshell, 2004: Technological learning for carbon capture and sequestration technologies. *Energy Economics,* **26**(4), pp. 539- 564.

Roland-Holst, D., 2006: Economic assessment of some California greenhouse gas control policies: Applications of the BEAR model. In *Managing Greenhouse Gas Emissions in California.* The California Climate Change Center at UC Berkeley, pp. 1-74.

Romeri, V., 2004: Hydrogen: A new possible bridge between mobility and distributed generation (CHP). In *19th World Energy Congress,* September 5-9, Sydney, Australia, 12 pp.

Rose, S., H. Ahammad, B. Eickhout, B. Fisher, A. Kurosawa, S. Rao, K. Riahi, and D. van Vuuren, 2007. Land in climate stabilisation modeling: Initial observations. Energy Modeling Forum report, Stanford University, <http://www.stanford.edu/group/EMF/projects/group21/EMF21sinkspagenew.htm>, accessed 16 May 2007.

Rosendahl, K.E., 2004: Cost-effective environmental policy: implications of induced technological change. *Journal of Environmental Economics and Management,* **48**(3), pp. 1099-1121.

Roson, R., 2003: Climate change policies and tax recycling schemes: Simulations with a dynamic general-equilibrium model of the Italian economy. *Review of Urban & Regional Development Studies,* **15**(1), pp. 26-44.

Russ, P., J.C. Ciscar, and L. Szabó, 2005: *Analysis of post-2012 climate policy scenarios with limited participation.* European Commission, 31 pp.

Rypdal, K., T. Berntsen, J.S. Fuglestvedt, K. Aunan, A. Torvanger, F. Stordal, J.M. Pacyna, and L.P. Nygaard, 2005: Tropospheric ozone and aerosols in climate agreements: scientific and political challenges. *Environmental Science and Policy,* **8**(1), pp. 29-43.

Samson, R., M. Drisdelle, L. Mulkins, C. Lapointe, and P. Duxbury, 2000: The use of switchgrass biofuel pellets as a greenhouse gas offset strategy. Bioenergy 2000. Buffalo, New York.

Sandén, B.A. and C. Azar, 2005: Near-term technology policies for long-term climate targets - economy wide versus technology specific approaches. *Energy Policy,* **33**(12), pp. 1557-1576.

Sands, R.D., 2004: Dynamics of carbon abatement in the Second Generation Model. *Energy Economics,* **26**(4), pp. 721-738.

Schumacher, K. and R.D. Sands, 2006: Innovative energy technologies and climate policy in Germany. *Energy Policy,* **34**(18), pp. 3929-3941.

Schwoon, M. and R.S.J. Tol, 2006: Optimal CO_2-abatement with Socio-economic Inertia and Induced Technological Change. *Energy Journal,* **27**(4), pp. 25-60.

Seebregts, A., T. Kram, G.J. Schaeffer, A. Stoffer, S. Kypreos, L. Barreto, S. Messner, and L. Schrattenholzer, 1999: Endogenous technological change in energy systems models - Synthesis of experience with ERIS, MARKAL, and MESSAGE. ECN-C--99-025, Energy research Centre of the Netherlands (ECN), Petten.

Seifritz, W., 1989: Mirrors to halt global warming? *Nature,* **340**(6235), pp. 603.

Shrestha, R.M., 2004: Technological Implications of the Clean Development Mechanism for the power sector in three Asian countries. *International Review for Environmental Strategies,* **5**(1), pp. 273-288.

Shukla, P.R., A. Rana, A. Garg, K. Kapshe, and R. Nair, 2006: Global climate change stabilisation regimes and Indian emission scales: Lessons from modeling of developed country transitions. *Environmental Economics and Policy Studies,* **7**(3), pp. 205-232.

Sijm, J.P.M., 2004: Induced technological change and spillovers in climate policy modelling. ECN, the Netherlands.

Sijm, J.P.M., O.J. Kuik, M. Patel, V. Oikonomou, E. Worrell, P. Lako, E. Annevelink, G.J. Nabuurs, and H.W. Elbersen, 2004: Spillovers of climate policy: An assessment of the incidence of carbon leakage and

induced technological change due to CO_2 abatement measures. ECN, Netherlands, 251 pp.

Smeets, M.W., A.P.C. Faaij, I.M. Lewandowski, and W.C. Turkenburg, 2006: A bottom-up assessment and review of global bioenergy potentials to 2050. *Progress in Energy and Combustion Science,* **33**(1), pp. 56-106. <http://www.sciencedirect.com/science/journal/03601285>, accessed 3 April 2007.

Smekens-Ramirez Morales, K.E.L., 2004: Response from a MARKAL technology model to the EMF scenario assumptions. *Energy Economics,* **26**(4), pp. 655-674.

Smith, K.R., S. Mehta, and M. Maeusezahl-Feuz, 2004: Indoor air pollution from household use of solid fuels. In *Comparative Quantification of Health Risks: Global and Regional Burden of Disease due to Selected Major Risk Factors.* World Health Organization, pp. 1435-1493.

Smulders, S., 2005: Endogenous technological change, natural resources and growth. In *Scarcity and Growth in the New Millennium.* RFF Press, Baltimore.

Söderholm, P. and T. Sundqvist, 2003: Pricing environmental externalities in the power sector: Ethical limits and implications for social choice. *Ecological Economics,* **46**(3), pp. 333-350.

Standard and Poor's, 2005: Climate change credit survey: A study of emissions trading, nuclear power, and renewable energy. Standard and Poor's (ed.), London, 38 pp.

Stern, N., 2006: The economics of climate change: The Stern Review HM Treasury. Cambridge University Press.

Stirling, A.C., 1994: Diversity and ignorance in electricity supply investment: Addressing the solution rather than the problem. *Energy Policy,* **22**(3), pp. 195-216

Stirling, A.C., 1996: On the economics and analysis of diversity. Science Policy Research Unit (SPRU), Paper No. 28, University of Sussex, Brighton, 134 pp.

Streets, D.G. and K. Aunan, 2005: The importance of China's household sector for black carbon emissions. *Geophysical Research Letters,* **32**(12), pp. L12708.

Subak, S., J.P. Palutikof, M.D. Agnew, S.J. Watson, C.G. Bentham, M.G.R. Cannell, M. Hulme, S. McNally, J.E. Thornes, D. Waughray, and J.C. Woods, 2000: The impact of the anomalous weather of 1995 on the U.K. *Economy Climatic Change,* **44**(1-2), pp. 1-26.

Sue Wing, I. and D. Popp, 2006: Representing endogenous technological change in models for climate policy analysis: Theoretical and empirical considerations. In *Managing Greenhouse Gas Emissions in California.* The California Climate Change Center UC Berkeley.

Sue Wing, I., 2006: Representing induced technological change in models for climate policy analysis. *Energy Economics,* **28**(5-6), pp. 539-562.

Sullivan, R. and W. Blyth, 2006: Climate change policy uncertainty and the electricity industry: implications and unintended consequences. EEDP BP 06/02, Chatham House, London, 11 pp.

Svendsen, G.T. and M. Vesterdal, 2003: Potential gains from CO_2 trading in the EU. *European Environment,* **13**(6), pp. 303-313.

Swart, R., M. Amann, F. Raes, and W. Tuinstra, 2004: A good climate for clean air: Linkages between climate change and air pollution. An Editorial Essay. *Climatic Change,* **66**(3), pp. 263-269.

Syri, S., M. Amann, P. Capros, L. Mantzos, J. Cofala, and Z. Klimont, 2001: Low-CO_2 energy pathways and regional air pollution in Europe. *Energy Policy,* **29**(11), pp. 871-884.

Syri, S., N. Karvosenoja, A. Lehtila, T. Laurila, V. Lindfors, and J.P. Tuovinen, 2002: Modeling the impacts of the Finnish climate strategy on air pollution. *Atmospheric Environment,* **36**, pp. 3059-3069.

Szabo, L., I. Hidalgo, J.C. Ciscar, and A. Soria, 2006: CO_2 emission trading within the European Union and Annex B countries: the cement industry case. *Energy Policy,* **34**(1), pp. 72-87.

Teller, E., R. Hyde, M. Ishikawa, J. Nuckolls, and L. Wood, 2004: Active climate stabilisation: Presently-feasible albedo-control approaches to prevention of both types of climate change. Tyndall Centre-Cambridge-MIT Symposium on Macro-Engineering Options for Climate Change Management & Mitigation. Cambridge, UK.

The Climate Group, 2005: Carbon down, profits up. Beacon Press, 2nd ed., 38 pp.

Tol, R.S.J. and H. Dowlatabadi, 2001: Vector-borne diseases, development & climate change. *Integrated Assessment*, **2**(4), pp. 173-181.

Tol, R.S.J., 2005: Adaptation and mitigation: trade-offs in substance and methods. *Environmental Science and Policy*, **8**(6), pp. 572-578.

Turton, H. and L. Barreto, 2006: Long-term security of energy supply and climate change. *Energy Policy*, **34**(15), pp. 2232-2250.

UKERC, 2006: The costs and impacts of intermittency: An assessment of the evidence on the costs and impacts of intermittent generation on the British electricity network. Imperial College, London.

UNDP, 2004: Update World Energy Assessment of the United Nations. UNDP, UNDESA/WEC, New York.

Unruh, G.C., 2000: Understanding carbon lock-in. *Energy Policy*, **28**(12), pp. 817-830.

Unruh, G.C., 2002: Escaping carbon lock-in. *Energy Policy*, **30**(4), pp. 317-325.

US CCSP, 2006: Scenarios of greenhouse gas emissions and atmospheric concentrations. Report by the U.S. Climate Change Science Program and approved by the Climate Change Science Program Product Development Advisory Committee. L. Clarke, J. Edmonds, J. Jacoby, H. Pitcher, J. Reilly, and R. Richels (eds).

US EPA, 1996: Air quality criteria for ozone and related photochemical oxidants. Volume II of III. EPA, New York.

US EPA, 2005: *The integrated environmental strategies handbook: A resource guide for air quality planning.* Washington D.C.

US EPA, 2006: Global anthropogenic non-CO_2 greenhouse gas emissions: 1990-2020. Washington D.C., 269 pp.

Van Dam, J., M. Junginger, A. Faaij, I. Jürgens, G. Best, and U. Fritsche, 2006: *Overview of recent developments in sustainable biomass.* Paper written within the frame of IEA Bioenergy Task 40. 22 December 2006 certification.

Van Harmelen, T., J. Bakker, B. de Vries, D. van Vuuren, M. den Elzen, and P. Mayerhofer, 2002: Long-term reductions in costs of controlling regional air pollution in Europe due to climate policy. *Environmental Science & Policy,* **5**(4), pp. 349-365.

Van Vuuren, D., M. den Elzen, P. Lucas, B. Eickhout, B. Strengers, B. van Ruijven, S. Wonink, R. van Houdt, 2006: Stabilizing greenhouse gas concentrations at low levels: an assessment of reduction strategies and costs. *Climatic Change,* **81**(2), pp. 119-159.

Van Vuuren, D.P., B. de Vries, B. Eickhout, and T. Kram, 2004: Responses to technology and taxes in a simulated world. *Energy Economics*, **26**(4), pp. 579-601.

Van Vuuren, D.P., J. Cofala, H. Eerens, R. Oostenrijk, C. Heyes, Z. Klimont, M. den Elzen, and M. Amann, 2006: Exploring the ancillary benefits of the Kyoto Protocol for Air Pollution in Europe. *Energy Policy* **34**(4), pp. 444-460.

Vennemo, V., K. Aunan, F. Jinghua, P. Holtedahl, H. Tao, and H.M. Seip, 2006: Domestic environmental benefits of China's energy-related CDM potential. *Climatic Change,* **75**(1-2), pp. 215-239.

Verchot, L.V., 2004: Agroforestry in the CDM: an opportunity for synergy between mitigation and adaptation. *CDM Investment Newsletter*, **3**, pp. 3-4.

Verchot, L.V., J. Mackensen, S. Kandji, M. van Noordwijk, T.P. Tomich, C.K. Ong, A. Albrecht, J. Bantilan, J. Anupama, and C.A. Palm, 2005: Opportunities for linking adaptation and mitigation in agroforestry systems. In *Tropical forests and adaptation to climate change: In search of synergie,* C. Robledo, M. Kanninen, and L. Pedroni, (eds). Bogor, Indonesia. Center for International Forestry Research (CIFOR), pp.103-121.

Viguier, L., M.H. Babiker, and J. Reilly, 2003: The costs of the Kyoto Protocol in the European Union. *Energy Policy*, **31**(5), pp. 459-481.

Vollebergh, H.R.J. and C. Kemfert, 2005: The role of technological change for a sustainable development. *Ecological Economics*, **54** (2-3), pp. 133-147.

Wang, X. and K.R. Smith, 1999: Secondary benefits of greenhouse gas control: health impacts in China. *Environmental Science and Technology*, **33**(18), pp. 3056-3061.

WBCSD, 2004: *Mobility 2030; Meeting the challenges to sustainability.* 180 pp.

Weber, M., V. Barth, and K. Hasselmann, 2005: A multi-actor dynamic integrated assessment model (MADIAM) of induced technological change and sustainable economic growth. *Ecological Economics*, **54**(2-3), pp. 306-327.

West, J.J., A.M. Fiore, L.W. Horowitz, and D.L. Mauzerall, 2006: Global health benefits of mitigating ozone pollution with methane emission controls. *PNAS*, **103**(11), pp. 3988-3993.

West, J.J., P. Osnaya, I. Laguna, J. Martiänez, and A. Fernäändez, 2004: Co-control of urban air pollutants and greenhouse gases in Mexico City. *Environmental Science and Technology*, **38**(13), pp. 3474-3481.

Weyant, J.P. and J.N. Hill, 1999: Introduction and overview. The costs of the Kyoto Protocol: A multi-model evaluation. *The Energy Journal*, **20**(Special Issue), pp. vii-xliv.

Weyant, J.P., 2000: *An introduction to the economics of Climate Change Policy.* Pew Center on Global Climate Change, Arlington.

Weyant, J.P., 2001: Economic models: how they work and why their results differ. In *Climate Change: Science, Strategies and Solutions.* Brill Academic Publishers, Boston.

Weyant, J.P., 2003: *Integrated Environmental Assessment: An Overview and Assessment.* The OECD Workshop on Environment and Economic Modeling. Stanford University, Ottawa.

Weyant, J.P., 2004: Introduction and overview. *Energy Economics*, **26**(4), pp. 501-515.

Weyant, J.P., F.C. de la Chesnaye, and G.J. Blanford, 2006: Overview of EMF-21: Multigas mitigation and climate policy. *The Energy Journal*, **27** (Special Issue #3, Multi-Greenhouse Gas Mitigation and Climate Policy), pp. 1-32.

WHO, 2003: Health aspects of air pollution with particulate matter, ozone and nitrogen dioxide. World Health Organization, Report on a WHO Working Group, Bonn, Germany.

World Bank and IMF, 2006: An investment framework for clean energy and development: a progress report. DC2006-0012.

Worrell, E. and G. Biermans, 2005: Move over! Stock turnover, retrofit and industrial energy efficiency. *Energy Policy*, **33**(7), pp. 949-962.

Yang, Z. and W.D. Nordhaus, 2006: Magnitude and direction of technological transfers for mitigating GHG emissions. *Energy Economics*, **28**(5-6), pp. 730-741.

Zhang, Z. and A. Baranzini, 2004: What do we know about carbon taxes? An inquiry into their impacts on competitiveness and distribution of income. *Energy Policy*, **32**(4), pp. 507-518.

Appendix to Chapter 11

Technical description of the assessment of aggregate mitigation potentials from the sectoral literature

1. Methodology for adding up sectoral emission reduction potential

Adding up all the emission reduction potentials at the sectoral level reported in the sectoral chapters will result in double counting for part of the potential. To avoid this, two interactions have been taken into account in the assessment of the total mitigation potential in Chapter 11 (Table 11.3):

- The interaction between the reduction potential from electricity savings in buildings and industry on the one hand and measures in the electricity supply sector on the other (substitution by low-carbon electricity supply). This topic is discussed in this appendix.
- The interaction between the estimated supply and demand of biomass for energy purposes. This topic is covered in Section 1.3.1.4.

1.1. The electricity sector

The two main reduction options for electricity use are:
1) electricity savings in the industry and buildings sector, and
2) substitution in the power sector tending towards low-carbon electricity technologies.

The overall CO_2 emission reduction from the electricity savings in industry and buildings therefore depends on the fuel mix of the power supply and the penetration of low-carbon technologies in that supply.

The methodology chosen to prevent double counting is presented in Figure 11.A1 and described below, step by step.

Step 1: Baseline electricity consumption and emissions
In step 1, 2000–2030 projections were compiled for final electricity consumption, primary energy consumption for electricity production and GHG emissions from the fuels used. The final electricity consumption at the regional basis was taken from the World Energy Outlook 2004 (IEA, 2004).

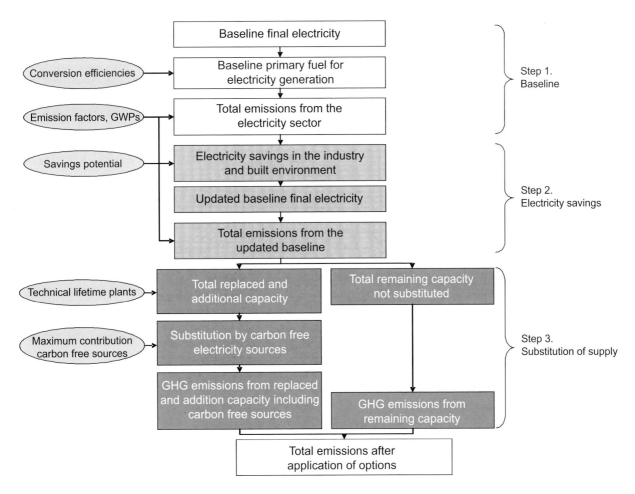

Figure 11.A1: *Methodology for the assessment of the mitigation potential related to electricity consumption; electricity savings and the implementation of low-carbon supply technologies*

To arrive at the primary fuel required for final electricity consumption, an intermediate step is needed. As the World Energy Outlook 2004 provides statistics on primary energy supply for electricity and heat combined, the implicit supplies required for heat were estimated and removed as follows. The primary energy consumption for electricity supply only was calculated on the basis of the efficiencies of combined heat and power, and a correction for the share of heat in total final energy consumption. The share of heat was calculated from the IEA Balances for the year 2002 and assumed to be constant over time. See also Section 4.4.3 for the efficiencies and the baseline in the year 2030.

Finally, using the data on primary fuel required, the GHG emissions were estimated on the basis of the primary fuel supply for power production using the emission factors for primary fuels (IEA, 2005) and the 1996 GWP numbers taken from UNFCCC.

Step 2: The electricity savings

The second step consists of reducing the baseline electricity by the savings from buildings and industry. Electricity savings are found at relatively low costs and they are therefore expected to be implemented first. The maximum electricity savings for the industry and buildings sector were taken from the sectoral chapters. These have been applied using the share of the electricity consumption of the sectors in total electricity consumption (WEO2004). In this step, it was assumed that the savings were equally distributed across the different power sources, including low-carbon sources.

The savings indicated in Table 11.A1 have been used.

In fact, it can be expected that electricity savings will result in higher levels of fossil-fuel electricity generation compared to generation at low marginal cost such as renewables and nuclear. This is because, in the usual operation of electricity systems, low-cost fuels are dispatched before high-cost fuels. But system operation depends on local conditions and it is not

appropriate to consider these here. This consideration implies that the emission reductions for electricity savings reported here are underestimated. Higher carbon prices, and higher marginal costs of fossil fuels, exacerbate this effect.

Finally, the amount of primary fuels needed for power generation has been updated, resulting in lower emissions. The difference between the emissions from the updated baseline and the original baseline gives the avoided emissions; see Table 11.A2 (see also Section 11.3.3).

Step 3: The substitution of generating capacity with low-carbon capacity

The reduction in GHG emissions achieved through substitution towards low-carbon intensive technologies was assessed using the updated electricity demand from step 2.

First, an estimate of the new required generation capacity from 2010 to 2030 was made. It was assumed that low-carbon technologies are only implemented when new capacity is to be installed. The required new capacity to 2030 was calculated from 1) additional capacity between 2010 and 2030 to meet new demand and 2) capacity replaced in the period 2010–2030 after an assumed average plant lifetime of 50 years (see Chapter 4.4.3).

Secondly, the fuel switch from coal to natural gas was considered to be the option involving least cost, so it was assumed that it would be implemented first. Since new gas infrastructure is required, it was assumed in accordance with Chapter 4 that 20% at most of the new required coal plants (in the baseline) could be substituted by gas technologies.

Thirdly, after the fuel switch, emissions avoided from the other low-carbon substitution options were assessed. The following technologies were taken into account: renewables (such as wind, geothermal and solar), bioenergy, hydro, nuclear and CCS. It was assumed that the new fossil-fuel generation required according to the baseline was substituted by low-carbon generation (for each of the cost classes), proportional to the relative maximum technical potential of the technologies. The technologies were assumed to penetrate so as to achieve maximum shares in generation, as described in Table 4.20.

Finally, the new fossil fuel requirement was estimated and the GHG emissions assessed.

The avoided emissions in each of the steps were calculated using the same emission factors as in the baseline indicated above, and they are presented in Table 11.A2.

Table 11.A1: *Main assumptions used in the assessment of the emission reduction potential because of electricity savings in the buildings and industry sector*

	Assumption (%)	Origin
Electricity savings in the industrial sector	13[a]	Section 7.5.1
Electricity savings in the residential sector (mean value)		Section 6.5
OECD	23-26	
EIT	44-55	
Non-OECD	43-48	

Note:
[a] Chapter 7 reports energy savings of 30% compared to frozen efficiency for motor systems. Within the baseline, 10% efficiency improvements can be assumed. In addition, motors take about 65% of the total energy use resulting in electricity savings for 2030 of 13%.

Table 11.A2: *Baseline electricity demand and supply fuel mix with electricity savings (step 2) and mitigation measures in the power sector (step 3) in 2030*

	Primary energy (EJ)	Baseline (1) Secondary energy (TWh)	Baseline (1) Emissions (GtCO₂-eq)	After electricity savings (2) Secondary energy (TWh)	After electricity savings (2) Emissions (GtCO₂-eq)	After electricity savings (2) Emissions avoided compared to 1 (GtCO₂-eq)	After substitution (3) Secondary energy (TWh)	After substitution (3) Emissions (GtCO₂-eq)	After substitution (3) Emissions avoided compared to 2 (GtCO₂-eq)	Total emissions avoided (MtCO₂-eq)
OECD-EIT	**115**	**14244**	**6.0**	**11333**	**4.8**	**1.2**	**11333**	**3.1**	**1.7**	**2911**
Coal	42	4736	4.0	3768	3.2		2447	2.2		
Oil	3	309	0.21	246	0.17		246	0.17		
Gas	31	4145	1.8	3298	1.4		1689	0.73		
Nuclear	23	2137		1700			2653			
Hydro	5	1529		1217			1592			
Biomass and Waste	5	405		322			478			
Other Renewables	6	983		782			1274			
Coal - CCS							955			
EIT	**22**	**2468**	**1.2**	**1743**	**0.83**	**0.35**	**1743**	**0.55**	**0.27**	**594**
Coal	4	394	0.42	278	0.29		234	0.26		
Oil	1	61	0.06	43	0.04		42	0.04		
Gas	12	1324	0.70	935	0.49		480	0.25		
Nuclear	3	272		192			436			
Hydro	1	373		263			263			
Biomass and Waste	0	11		8			78			
Other Renewables	0	33		23			122			
Coal - CCS							87			
Non-OECD	**125**	**14944**	**8.6**	**10219**	**5.9**	**2.7**	**10219**	**3.2**	**2.7**	**5109**
Coal	66	6961	6.3	4760	4.3		2239	2.2		
Oil	8	812	0.59	555	0.4		410	0.3		
Gas	30	3860	1.7	2640	1.2		1644	0.74		
Nuclear	6	520		356			1022			
Hydro	8	2346		1604			2044			
Biomass and Waste	4	211		144			1022			
Other Renewables	3	234		160			920			
Coal - CCS							920			
Total	**263**	**31656**	**15.8**	**23295**	**11.5**	**4.3**	**23295**	**6.8**	**4.7**	**8613**

Table 11.A3: *The main results of the emission reductions for the sensitivity cases in GtCO$_2$-eq reduction*

	Default			Change in order[a]			Lowest range	
	Savings		Low-carbon supply	Savings		Low-carbon supply	Savings	Low-carbon supply
	Buildings	Industry		Buildings	Industry			
OECD	0.9	0.3	1.7	0.06	0.03	2.7	1.2	0.9
EIT	0.3	0.1	0.27	0.02	0.02	0.49	0.35	0.18
Non-OECD/EIT	2.3	0.5	2.7	0.25	0.18	4.1	2.7	1.3
Total	3.5	0.8	4.7	0.33	0.24	7.2	4.3	2.4

Note:
[a] For the change in order, the maximum shares of low-carbon technologies were used (the default in Chapter 4)

1.2. Cost distribution

The sector chapters assessed the distribution of the total emission potentials across cost categories. The same cost distribution has been used to present the results in Table 11.3.

2. Sensitivity analysis for potentials in the electricity sector

A sensitivity analysis was carried out to analyse the robustness of the mitigation potential for the electricity sector. The following assumptions were varied:

1) The order of the mitigation option. Instead of assuming that electricity savings occurs before substitution with low-carbon technologies, the potential was also assessed in the reverse order: first substitution, then savings.
2) The value of the 'maximum' shares of low-carbon technologies in the total electricity mix. In Section 4.3 and 4.4 the results are presented for the 'maximum' shares based on various literature sources. Shares differ depending on the different technologies. To assess the sensitivity of these shares, they were varied in the lowest range by 30%, which is consistent with the lowest range in Chapter 4.

The results of each of the sensitivity analyses are presented in Table 11.A3.

Based on the sensitivity analysis it can be concluded that, when assuming the reverse order by allocating emission reductions first to the power sector, followed by the electricity savings, the total emission reduction, i.e. the aggregate of the electricity savings and substitution, would be 1.2 GtCO$_2$-eq lower than the default. This is a consequence of allocating the savings over the total electricity generation mix. The potential is equally sensitive to the 'maximum' shares that are assumed. Reducing these maximum shares by 30% reduces the mitigation potential of the power sector by 50% compared to the default.

12

Sustainable Development and Mitigation

Coordinating Lead Authors:
Jayant Sathaye (USA), Adil Najam (Pakistan)

Lead Authors:
Christopher Cocklin (New Zealand), Thomas Heller (USA), Franck Lecocq (France), Juan Llanes-Regueiro (Cuba), Jiahua Pan (China), Gerhard Petschel-Held † (Germany), Steve Rayner (USA), John Robinson (Canada), Roberto Schaeffer (Brazil), Youba Sokona (Mali), Rob Swart (The Netherlands), Harald Winkler (South Africa)

Contributing Authors:
Sarah Burch (Canada), Jan Corfee Morlot (USA/France), Rutu Dave (The Netherlands), László Pinter (Canada), Andrew Wyatt (Australia)

Review Editors:
Mohan Munasinghe (Sri Lanka), Hans Opschoor (The Netherlands)

This chapter should be cited as:
Sathaye, J., A. Najam, C. Cocklin, T. Heller, F. Lecocq, J. Llanes-Regueiro, J. Pan, G. Petschel-Held , S. Rayner, J. Robinson, R. Schaeffer, Y. Sokona, R. Swart, H. Winkler, 2007: Sustainable Development and Mitigation. In Climate Change 2007: Mitigation. Contribution of Working Group III to the Fourth Assessment Report of the Intergovernmental Panel on Climate Change [B. Metz, O.R. Davidson, P.R. Bosch, R. Dave, L.A. Meyer (eds)], Cambridge University Press, Cambridge, United Kingdom and New York, NY, USA.

Table of Contents

EXECUTIVE SUMMARY

The concept of sustainable development was adopted by the World Commission on Environment and Development, and there is agreement that sustainable development involves a comprehensive and integrated approach to economic, social, and environmental processes. Discourses on sustainable development, however, have focused primarily on the environmental and economic dimensions. The importance of social, political, and cultural factors is only now getting more recognition. Integration is essential in order to articulate development trajectories that are sustainable, including addressing the climate change problem.

There is growing emphasis in the literature on the two-way relationship between climate change mitigation and sustainable development. The relationship may not always be mutually beneficial. In most instances, mitigation can have ancillary benefits or co-benefits that contribute to other sustainable development goals (climate first). Development that is sustainable in many other respects can create conditions in which mitigation can be effectively pursued (development first) (*high agreement, much evidence*).

Although still in early stages, there is growing use of indicators to manage and measure the sustainability of development at the macro and sectoral levels. This is driven in part by the increasing emphasis on accountability in the context of governance and strategy initiatives. At the sectoral level, progress towards sustainable development is beginning to be measured and reported by industry and governments using, for instance, green certification, monitoring tools, and emissions registries. Review of the indicators illustrates, however, that few macro-indicators include measures of progress with respect to climate change (*high agreement, much evidence*).

Climate change is influenced not only by the climate-specific policies but also by the mix of development choices and the resulting development trajectories - a point reinforced by global scenario analyses published since the Third Assessment Report (TAR). Making development more sustainable by changing development paths can thus make a significant contribution to climate goals. But changing development pathways is not about choosing a mapped-out path, but rather about navigating through an uncharted and evolving landscape (*high agreement, much evidence*).

Making decisions about sustainable development and climate change mitigation is no longer the sole purview of governments. There is increasing recognition in the literature of a shift to a more inclusive concept of governance, which includes the contributions of various levels of government, private sector, non-governmental actors, and civil society. The more climate change issues are mainstreamed as part of the planning perspective at the appropriate level of implementation, and the more all relevant parties are involved in the decision-making process in a meaningful way, the more likely they are to achieve the desired goals (*high agreement, medium evidence*).

Regarding governments, a substantial body of political theory identifies and explains the existence of national policy styles or political cultures. The underlying assumption of this work is that individual countries tend to process problems in a specific manner, regardless of the distinctiveness or specific features of any problem; a national 'way of doing things'. Furthermore, the choice of policy instruments is affected by the institutional capacity of governments to implement the instrument. This implies that the preferred mix of policy decisions and their effectiveness in terms of sustainable development and climate change mitigation strongly depend on national characteristics (*high agreement, much evidence*).

The private sector is a central player in ecological and sustainability stewardship. Over the past 25 years, there has been a progressive increase in the number of companies taking steps to address sustainability issues at either the company or industry level. Although there has been progress, the private sector has the capacity to play a much greater role in making development more sustainable in the future, because such a shift is likely to benefit its performance (*medium agreement, medium evidence*).

Citizen groups have been major demanders of sustainable development and are critical actors in implementing sustainable development policy. Apart from implementing sustainable development projects themselves, they can push policy reform through awareness-raising, advocacy, and agitation. They can also pull policy action by filling the gaps and providing policy services, including in the areas of policy innovation, monitoring, and research. Interactions can take the form of partnerships or stakeholder dialogues that can provide citizens' groups with a lever for increasing pressure on both governments and industry (*high agreement, medium evidence*).

Deliberative public-private partnerships work most effectively when investors, local governments and citizen groups are willing to work together to implement new technologies, and produce arenas to discuss these technologies that are locally inclusive (*high agreement, medium evidence*).

Region- and country-specific case studies demonstrate that different paths and policies can achieve noticeable emissions reductions, depending on the capacity to realise sustainability and climate change objectives. These capacities are determined by the same set of conditions that are closely linked to the state of development. The mitigative capacity to realise low emissions can be low due to differentiated national endowments and barriers, even when significant abatement opportunities exist. The challenge of implementing sustainable development exists in both developing and industrialized countries. The nature of the challenge, however, tends to be different in the industrialized countries. (*high agreement, much evidence*).

Some general conclusions emerging from the case studies of how changes in development pathways at the sectoral level have or could lower emissions are reviewed in this chapter (*high agreement, medium evidence*):

- Greenhouse gas (GHG) emissions are influenced by but not rigidly linked to economic growth: policy choices make a difference.
- Sectors where effective production is far below the maximum feasible with the same amount of inputs - sectors far from their production frontier - have opportunities to adopt 'win-win-win' policies. These policies free up resources and bolster growth, meet other sustainable development goals, and also reduce GHG emissions relative to baseline.
- Sectors where production is close to optimal given available inputs – sectors that are closer to the production frontier - also have opportunities to reduce emissions by meeting other sustainable development goals. However, the closer to the production frontier, the more trade-offs are likely to appear.
- To truly have an effect, what matters is that not only a 'good' choice is made at a certain point, but also that the initial policy is sustained for a long period - sometimes several decades.
- It is often not one policy decision, but an array of decisions that are necessary to influence emissions. This raises the issue of coordination between policies in several sectors, and at various scales.

Mainstreaming requires that non-climate policies, programmes, and/or individual actions take climate change mitigation into consideration, in both developing and developed countries. However, merely piggybacking climate change onto an existing political agenda is unlikely to succeed. The ease or difficulty with which mainstreaming is accomplished will depend on both mitigation technologies or practices, and the underlying development path. Weighing other development benefits against climate benefits will be a key basis for choosing development sectors for mainstreaming. Decisions about fiscal policy, multilateral development bank lending, insurance practices, electricity markets, petroleum imports security, forest conservation, for example, which may seem unrelated to climate policy, can have profound impacts on emissions,

the extent of mitigation required, and the resulting costs and benefits. However, in some cases, such as a shift from biomass cooking to LPG in rural areas of developing countries, it may be rational to disregard climate change considerations because of the small increase in emissions compared with its development benefits (*high agreement, medium evidence*).

There is a growing understanding of the possibilities to choose mitigation options and their implementation such that there is no conflict with other dimensions of sustainable development; or, where trade-offs are inevitable, to allow a rational choice to be made. The sustainable development benefits of mitigation options vary within a sector and over regions (*high agreement, much evidence*):

- Generally, mitigation options that improve productivity of resource use, whether energy, water, or land, yield positive benefits across all three dimensions of sustainable development. Other categories of mitigation options have a more uncertain impact and depend on the wider socioeconomic context within which the option is implemented.
- Climate-related policies, such as energy efficiency, are often economically beneficial, improve energy security, and reduce local pollutant emissions. Many energy supply mitigation options can also be designed to achieve other sustainable development benefits, such as avoided displacement of local populations, job creation, and rationalized human settlements design.
- Reducing deforestation can have significant biodiversity, soil, and water conservation benefits, but may result in loss of economic welfare for some stakeholders. Appropriately designed forestation and bioenergy plantations can lead to reclamation of degraded land, manage water runoff, retain soil carbon and benefit rural economies, but could compete with land for agriculture and may be negative for biodiversity.
- There are good possibilities for reinforcing sustainable development though mitigation actions in most sectors, but particularly in waste management, transportation, and building sectors, notably through decreased energy use and reduced pollution.

12.1   Introduction

The concept of sustainable development had its roots in the idea of a sustainable society (Brown, 1981) and in the management of renewable and non-renewable resources. The concept was introduced in the World Conservation Strategy by the International Union for the Conservation of Nature (IUCN, 1980). The World Commission on Environment and Development adopted the concept and launched sustainability into political, public and academic discourses. The concept was defined as "development that meets the needs of the present without compromising the ability of future generations to meet their own needs" (WCED, 1987; Bojo *et al.*, 1992). While this definition is commonly cited, there are divergent views in academic and policy circles on the concept and how to apply it in practice (Banuri *et al.*, 2001; Cocklin, 1995; Pezzoli, 1997; Robinson and Herbert, 2001).

The discussion on sustainable development in the IPCC process has evolved since the First Assessment Report which focused on the technology and cost-effectiveness of mitigation activities. This focus was broadened in the Second Assessment Report (SAR) to include issues related to equity, both procedural and consequential, and across countries and generations, and to environmental (Hourcade *et al.*, 2001) and social considerations (IPCC, 1996). The Third Assessment Report (TAR) further broadened the treatment of sustainable development by addressing issues related to global sustainability (IPCC, 2001b, Chapter 1). The report noted three broad classes of analyses or perspectives: efficiency and cost-effectiveness; equity and sustainable development; and global sustainability and societal learning. The preparation of TAR was supported by IPCC Expert Group Meetings specially targeted at sustainable development and social dimensions of climate change. These groups noted the various ways that the TAR treatment of sustainable development could be improved (Munasinghe and Swart, 2000; Jochem *et al.*, 2001).

In light of this evolution, each chapter of this Fourth Assessment Report focuses to some extent on the links to sustainable development practices. Chapter 1 introduces the concept, Chapter 2 provides a framework for understanding the economic, environmental, and social dimensions, and Chapter 3 addresses the issue of development choices for climate change mitigation in a modelling context. The sector Chapters 4 to 10 and the cross-sectoral Chapter 11 examine the impacts of mitigation options on sustainable development goals; and Chapter 13 describes the extent to which sustainable development is addressed in international policies. Further, IPCC (2007) devotes two chapters that are linked to the mitigation discussion in this report. Chapter 17 in IPCC (2007) considers adaptation practices, options, constraints and capacity, while Chapter 18 examines the inter-relationships between adaptation and mitigation. Finally, Chapter 20 contains discussions of adaptation and sustainable development.

As in the aforementioned chapters, climate change policies can be considered in their own right ('climate first'). Most policy literature about climate change mitigation, and necessarily most of this assessment, focuses on government-driven, climate-specific measures that, through different mechanisms, directly constrain GHG emissions. Such measures will compose an essential element for managing the risks of climate change.

Nevertheless, the greater emphasis in Section 12.2 is on other approaches that may be necessary to go beyond the scope of climate specific actions. Climate change mitigation is treated as an integral element of sustainable development policies ('development first'). Decisions that may seem unrelated to climate policy can have profound impacts on emissions. This analysis does not suggest or imply that non-climate actions can displace climate-specific measures. It emphasizes what more developed and developing countries can do to alter emissions paths in the absence of direct constraints on emissions. Such indirect approaches to climate mitigation are especially relevant in developing countries where mandatory, climate-specific measures are controversial and, at best, prospective.

The relationship between economic development and climate change is of particular importance to developing countries because of where they are in their development process and also because of the particular climate challenges that many of them face. This chapter, therefore, gives particular emphasis to the notion of "making development more sustainable". Making development more sustainable recognizes that there are many ways in which societies balance the economic, social, and environmental, including climate change, dimensions of sustainable development. It also admits the possibility of conflict and trade-offs between measures that advance one aspect of sustainable development while harming another (Munasinghe, 2000).

This chapter (1) describes the evolution of the concept of sustainable development with emphasis on its two-way linkage to climate change mitigation (Section 12.1); (2) explores ways to make development more sustainable, - the role of development paths, how these can be changed, and the role that state, market, and civil society could play in mainstreaming climate change mitigation into development choices (Section 12.2); and (3) summarizes the impacts of climate mitigation on attributes of sustainable development (Section 12.3).

12.1.1 The two-way relationship between sustainable development and climate change

The growing literature on the two-way nature of the relationship between climate change and sustainable development is introduced in Chapter 2 (Metwalli *et al.*, 1998; Rayner and Malone, 1998; Munasinghe and Swart, 2000; Schneider *et al.*, 2000; Banuri *et al.*, 2001; Morita *et al.*, 2001; Smit *et al.*, 2001; Beg *et al.*, 2002; Markandya and Halsnaes, 2002; Metz *et al.*, 2002; Najam and Cleveland, 2003; Swart *et al.*, 2003; Wilbanks, 2003). The notion is that policies pursuing sustainable development and climate change mitigation can be mutually reinforcing. Much of this literature, as elaborated upon in Chapters 4 to 11, emphasizes the degree to which climate change mitigation can have effects. Sometimes called ancillary benefits or co-benefits, these effects will contribute to the sustainable development goals of the jurisdiction in question. This amounts to viewing sustainable development through a climate change lens. It leads to a strong focus on integrating sustainable development goals and consequences into the climate mitigation policy framework, and on assessing the scope for such ancillary benefits. For instance, reductions in GHG emissions might reduce the incidence of death and illness due to air pollution and benefit ecosystem integrity, both elements of sustainable development (Beg *et al.*, 2002). The challenge then becomes ensuring that actions taken to address global environmental problems help to address regional and local development (Beg *et al.*, 2002). Section 12.3 summarizes the impacts of climate mitigation actions on economic, social and environmental aspects of sustainable development noted in Chapters 3 to 11, and 13.

A key finding of the Third Assessment Report (TAR; IPCC, 2001b) is that through climate mitigation alone, it will be extremely difficult and expensive to achieve low stabilization targets (450 ppmv CO_2) from baseline scenarios that embody high emission development paths (also see Chapter 3). Low emission baseline scenarios, however, may go a long way toward achieving low stabilization levels even before climate policy is included in the scenario (Morita *et al.*, 2001) See Section 3.1.2 for a discussion of the distinction between a baseline and stabilization or mitigation scenario. Achieving low emission baseline scenarios consistent with other principles of sustainable development, that is viewing climate change through a sustainable development lens, would illustrate the significant contribution sustainable development can make to stabilization (Metz *et al.*, 2002; Winkler *et al.*, 2002a; Davidson *et al.*, 2003; Heller and Shukla, 2003; Shukla *et al.*, 2003; Swart *et al.*, 2003; Robinson and Bradley, 2006). Section 12.2 focuses on this critical question of the link between sustainable development and ways to mainstream climate change mitigation into sustainable development actions. This is a central element since this topic is not addressed elsewhere in the Fourth Assessment Report in a similarly comprehensive manner that is accessible to a non-climate readership.

By framing the debate as a sustainable development problem rather than only as climate mitigation, the priority goals of all countries and particularly developing countries are better addressed, while acknowledging that the driving forces for emissions are linked to the underlying development path (IPCC, 2007, Chapter 17 and 18; Yohe, 2001; Metz *et al.*, 2002; Winkler *et al.*, 2002a).

Development paths underpin the baseline and stabilization emissions scenarios discussed in Chapter 3 and are used to estimate emissions, climate change and associated climate change impacts[1]. For a development path[2] to be sustainable over a long period, wealth, resources, and opportunity must be shared so that all citizens have access to minimum standards of security, human rights, and social benefits, such as food, health, education, shelter, and opportunity for self-development (Reed, 1996). This was also emphasized by the World Summit on Sustainable Development (WSSD) in Johannesburg in 2002 which introduced the Water, Energy, Health, Agriculture, and Biodiversity (WEHAB) framework.

Several strategies and measures that would advance sustainable development would also enhance adaptive and mitigative capacities. Winkler *et al. (*2006) have suggested that mitigative capacity be defined as "a country's ability to reduce anthropogenic greenhouse gases or enhance natural sinks." There is a close connection between mitigative and adaptive capacities and the underlying socio-economic and technological development paths that give rise to those capacities. In important respects, the determinants of these capacities are critical characteristics of such development paths. For instance, mitigative and adaptive capacities arise out of the more general pool of resources called response capacity, which is strongly affected by the nature of the development path in which it exists.

Prior to exploring these issues further, the evolution of the sustainable development concept is discussed in Section 12.1.2, and the growing use of indicators to measure sustainable development progress at the macro and sectoral levels is described in Section 12.1.3. This review concludes that while the use of quantitative indicators is helping to better define sustainable development, few macro sustainable development indicators explicitly take GHG emissions and climate change impacts into consideration.

12.1.2 Evolution and articulation of the concept of sustainable development

Since the 1992 Earth Summit in Rio de Janeiro, there is general agreement that sustainable development requires the adoption of a comprehensive and integrated approach to economic, social and

---

1 The climate change and climate change impact scenarios assessed in the Fourth Assessment Report are primarily based on the SRES family of emission scenarios. These define a spectrum of development paths, each with associated socio-economic and technological conditions and driving forces. Each family of emission scenarios will, therefore, give rise to a different set of response capacities.

2 Development paths are defined here as a complex array of technological, economic, social, institutional, cultural, and biophysical characteristics that determines the interactions between human and natural systems, including consumption and production patterns in all countries, over time at a particular scale. In the TAR, "alternative development paths" referred to a variety of possible development paths, including a continuation of current trends, but also a variety of other paths. To avoid confusion, the word 'alternative' is avoided in the current report. Development paths will be different in scope and timing in different countries, and can be different for different regions within countries with large differences in internal regional characteristics.

environmental processes (Munasinghe, 1992; Banuri *et al.*, 1994; Najam *et al.*, 2003). The environment-poverty nexus is now well recognized and the linkage between sustainable development and achievement of the Millennium Development Goals (MDGs) has been clearly articulated (Jahan and Umana, 2003). While the challenge of sustainable development is a common one, countries have to adopt different strategies to advance sustainable development goals – especially in the context of achieving the MDGs (Dalal-Clayton, 2003). The paths they adopt will have important implications for the mitigation of climate change (for a more extensive discussion of MDGs, see Section 2.1.6). As noted in Section 4.5.4.4 and Section 6.6, consideration of clean energy services, even though not explicitly mentioned in the MDGs, will be a vital factor in achieving both sustainable development and climate mitigation goals.

However, discourses of sustainable development have historically focused primarily on the environmental and economic dimensions (Barnett, 2001), while overlooking the need for social, political and/or cultural change (Barnett, 2001; Lehtonen, 2004; Robinson, 2004). As Lehtonen (2004) explains, however, most models of sustainable development conceive of social, environmental (and economic) issues as 'independent elements that can be treated, at least analytically, as separate from each-other' (p. 201). The importance of social, political and cultural factors, for example, poverty, social equity, governance, is only now getting more recognition. In particular, there is a growing recognition of the importance of the institutional and governance dimensions (Banuri and Najam, 2002). From a climate change perspective, this integration is essential in order to define sustainable development paths. Moreover, as discussed in this chapter, understanding the institutional context in which policies are made and implemented is critical.

As noted in Chapter 2, the term 'sustainable development,' has given rise to considerable debate and concerns (Robinson, 2004). First, the variety of definitions of sustainable development (Meadowcroft, 1997; Pezzoli, 1997; Mebratu, 1998) has raised concerns about definitional ambiguity or vagueness. In response, it has been argued that this vagueness may constitute a form of constructive ambiguity that allows different interests to engage in the debate, and the concept to be further refined through implementation (Banuri and Najam, 2002; Robinson, 2004). The concept of sustainable development is not unique in this respect, since its conceptual vagueness bears similarities to other norm-based meta-objectives such as 'democracy,' 'freedom,' and 'justice' (Lafferty, 1996; Meadowcroft, 2000).

Second, the term 'sustainable development' can be used to support cosmetic environmentalism, sometimes called greenwashing, or simply hypocrisy (Athanasiou, 1996; Najam, 1999). One response to such practices has been the development of greatly improved monitoring, analytical techniques, and standards, in order to verify claims about sustainable practices (Hardi and Zdan, 1997; OECD, 1998; Bell and Morse, 1999; Parris and Kates, 2003). See Section 12.1.3.

Finally, the most serious concern about sustainable development is that it is inherently delusory. Some critics have argued that because biophysical limits constrain the amount of future development that is sustainable, the term 'sustainable development' is itself an oxymoron (Dovers and Handmer, 1993; Mebratu, 1998; Sachs, 1999). This leads some to argue for a 'strong sustainability' approach in which natural capital must be preserved since it cannot be substituted by any other form of capital (Pearce *et al.*, 1989; Cabeza Gutes, 1996). Others point out that the concept of sustainable development is anthropocentric, thereby avoiding reformulation of values that may be required to pursue true sustainability (Suzuki and McConnell, 1997). While very different in approach and focus, both these criticisms raise fundamental value questions that go to the heart of present debates about environmental and social issues.

Despite these criticisms, basic principles are emerging from the international sustainability discourse, which could help to establish commonly held principles of sustainable development. These include, for instance, the welfare of future generations, the maintenance of essential biophysical life support systems, ecosystem wellbeing, more universal participation in development processes and decision-making, and the achievement of an acceptable standard of human well-being (WCED, 1987; Meadowcroft, 1997; Swart *et al.*, 2003; MA, 2005).

The principles of sustainable development have progressively been internalized in various national and international legal instruments (Boyle and Freestone, 1999; Decleris, 2000). Law contributes to the process of defining the concept of sustainable development through both international (treaty) law and national law. At a national level, principles of sustainable development are being implemented in various regions and countries, including New Zealand and the European Union. For example, New Zealand's Resource Management Act 1991 requires all decisions under the Act to consider and provide for sustainable management of natural and physical resources (Furuseth and Cocklin, 1995). South Africa's National Environmental Management Act provides for the development of assessment procedures that aim to ensure that environmental consequences of policies, plans and programmes are considered (RSA, 1998). India's Planning Commission makes sustainability part of the approach to providing 'Clean Water for All', noting that this requires a shift from groundwater to surface water where possible, or groundwater recharge (Government of India, 2006). Similarly, the 2000 EC Water Framework Directive is seeking to operationalize principles of sustainable use in the management of EU waters (Rieu-Clarke, 2004).

International environmental treaties generally cite sustainable development as a fundamental principle by which they must be interpreted, but rarely provide any further specification of content. The UN Framework Convention on Climate Change, for example, includes in its principles the right to promote sustainable development, but does not elaborate modalities for doing so. In response to the necessity to build a framework of equitable, strong, and effective laws needed to manage humanity's interaction with the Earth and build a fair and sustainable society (Zaelke *et al.*, 2005), the International Network for Environmental Compliance and Enforcement (INECE) launched an initiative at the 2002 WSSD aimed at making a law work for environmental compliance and sustainable development.

Since the 1980s, sustainable development has moved from being an interesting but sometimes contested ideal, to now being the acknowledged goal of much of international policy, including climate change policy. It is no longer a question of whether climate change policy should be understood in the context of sustainable development goals; it is a question of how.

12.1.3 Measurement of progress towards sustainable development

As what is managed needs to be measured, managing the sustainable development process requires a much strengthened evidence base and the development and systematic use of robust sets of indicators and new ways of measuring progress. Measurement not only gauges but also spurs the implementation of sustainable development and can have a pervasive effect on decision-making (Meadows, 1998; Bossel, 1999). In the climate change context, measurement plays an essential role in setting and monitoring progress towards specific climate change related commitments both in the mitigation and adaptation context (CIESIN, 1996-2001).

Agenda 21 (Chapter 40) explicitly recognizes the need for quantitative indicators at various levels (local, provincial, national and international) of the status and trends of the planet's ecosystems, economic activities and social wellbeing (United Nations, 1993). The need for further work on indicators at national and other levels was confirmed by the Johannesburg Plan of Implementation (UNEP, 2002).

As pointed out by Meadows (1998), indicators are ubiquitous, but when poorly chosen create serious malfunctions in socio-economic and ecological systems. Recognizing the shortcomings of mainstream measures, such as GDP, in managing the sustainable development process, alternative indicator systems have been developed and used by an increasing number of entities in various spatial, thematic and organizational contexts (Moldan *et al.*, 1997; IISD, 2006).

Indicator development is also driven by the increasing emphasis on accountability in the context of sustainable development governance and strategy initiatives. In their compilation and analysis of national sustainable development strategies, Swanson *et al.* (2004) emphasize that indicators need to be tied to expected outcomes, policy priorities and

implementation mechanisms. As such, the development of indicators may best be integrated with a process for setting sustainable development objectives and targets, but have an important role in all stages of the strategic policy cycle. Once priority issues are identified, SMART indicators need to be developed - indicators that are Specific, Measurable, Achievable, Relevant/Realistic and Time-bound.

Boulanger (2004) observes that indicators can be classified according to four main approaches: (1) the socio-natural sectors (or systems) approach, which focuses on sustainability as an equilibrium between the three pillars of sustainable development but which overlooks development aspects: (2) the resources approach, which concentrates on sustainable use of natural resources and ignores development issues: (3) a human approach based on human wellbeing, basic needs; and (4) the norms approach, which foresees sustainable development in normative terms. Each approach has its own merits and weaknesses. Despite these efforts at measuring sustainability, few offer an integrated approach to measuring environmental, economic and social parameters (Corson, 1996; Farsari and Prastacos, 2002; Swanson *et al.*, 2004). This review of indicators illustrates a significant gap in macro-indicators in that few include measures of progress with respect to climate change.

Indicator system development typically builds on a conceptual framework serving as a link between relevant world views, sustainability issues and specific indicators. Some of the more common ones include the pressure-state-impact framework and capital-based frameworks covering social, environmental and economic domains. Given the ambiguity of the concept of sustainable development and differences in socio-economic and ecological context, even the use of comparable indicator frameworks usually results in non-identical indicator sets (Parris and Kates, 2003; Pintér *et al.*, 2005).

Various alternative approaches to estimate macro progress towards sustainable development have been developed. Many of these approaches integrate, though not necessarily focus on, aspects of climate change. One approach to indicator development focused on monetary measures and involves adjustment to the GDP. These include, for example, calculation of genuine savings (Hamilton and *et al.*, 1997; Pearce, 2000), Sustainable National Income (Hueting, 1993), and efforts to develop a measure of sustainability (Yohe and Moss, 2000). In an attempt to aggregate and express resource consumption and human impact in the context of a finite earth, a number of indices based on non-monetary, physical measures were created. These indices may be based on the concepts of environmental space or ecospace, and ecological footprint (Wackernagel and Rees, 1996; Venetoulis *et al.*, 2004; Buitenkamp *et al.*, 1993; Opschoor, 1995; Rees, 1996). Vitousek *et al.* (1986) proposed the index of Human Appropriation of Net Primary Production (HANPP). This approach specifies the amount of energy that humans divert for their own use in competition with other species.

In trying to avoid shortcomings from the concept of carrying capacity applied to human societies the formula I = PAT, where I is the human impact on the environment, P the human population, A the affluence (presumably per capita income), and T the effect of technology on the environment, has been commonly used in decomposing the impact of population, economic activity, and fuel use on the environment in general and on historical and future carbon emissions in particular (IEA, 2004c; Kaya, 1990; Schipper *et al.*, 1997; Schumacher and Sathaye, 2000). Other approaches include the development of a 'global entropy model' that inspects the conditions for sustainability (Ruebbelke, 1998). This is done by employing available entropy data to demonstrate the extent to which improvements in entropy efficiency should be accomplished to compensate the effects of increasing economic activity and population growth. Other sets of metrics have less precise ambitions but aim to explain to the larger public the risks of environmental change, such as the notion 'ecological footprint' [see above] used by some NGOs. In this, the aggregate indicators are noted as the number of planets Earth needed to sustain the present way of living of some regions of the World.

As Bartelmus (2001) observes, many of the aggregate indices are yet to be accepted in decision-making due, among others, to measurement, weighting and indicator selection challenges. However, besides efforts to develop aggregate indices either on a monetary or physical basis, many efforts are aimed at developing heterogeneous indicator sets. One of the commonly accepted frameworks uses a classification scheme that groups sustainability issues and indicators according to social, ecological, economic, and in some cases, also institutional categories. Several indicator systems developed at international and national level have adopted a capital-based framework following the above categories. They link indicators more closely to the System of Integrated Environmental and Economic Accounts System of National Accounts (SNA), including its environmental component, (Pintér *et al.*, 2005). At the United Nations, the Division for Sustainable Development led the work on developing a menu and methodology sheets for sustainability indicators that integrate several relevant for climate change from the mitigation and adaptation point of view (UNDSD, 2006). Also, the UNECE/Eurostat/OECD Working Group on Statistics for Sustainable Development is developing a conceptual framework for measuring sustainable development and recommendations for indicator sets. A set of climate change mitigation input and outcome indicators should be included.

While not necessarily focused on climate change per se, many of these indicator efforts include climate change as one of the key issues, on the mitigation or adaptation side. Keeping a broader perspective is essential, as climate change, including its drivers, impacts and related responses, transcend many sectors and issue categories. Indicators are needed in all in order to identify and analyze systemic risks and opportunities. In the mitigation context, quantifying emissions and their underlying driving forces is an essential component of management and accountability mechanisms. GHG emissions accounting is a major new field and is guided by increasingly detailed methodology standards and protocols in both the public and private sector (WBCSD, 2004).

Whether part of integrated indicator systems or developed separately, climate change indicators on the mitigation side may focus on absolute or efficiency measures (Herzog and Baumert, 2006). Absolute measures help track aggregate emissions, thus quantify the direct pressure of human activities on the climate system. Efficiency measures indicate the amount of energy or materials used or GHG emitted in order to produce a unit of economic output, or more generally, to achieve a degree of change in human wellbeing. Depending on the policy context, both absolute measures and efficiency measures may be useful. But from the climate system perspective, it is ultimately indicators of absolute emission levels that matter.

At the sectoral level, several initiatives are being implemented to measure and monitor progress towards sustainable development, including the reduction of greenhouse gas emissions. In the buildings sector, for instance, the US Green Buildings Council, has established Leadership in Energy and Environmental Design (LEED) that sets a voluntary, consensus-based national standard for developing high-performance, sustainable buildings. About 2000 large buildings have received certificates. The Global Reporting Initiative (GRI) is a multi-stakeholder process whose mission is to develop and disseminate globally applicable Sustainability Reporting Guidelines. These Guidelines are for voluntary use by organizations for reporting on the economic, environmental, and social dimensions of their activities, products, and services. Over 700 large industrial corporations are annually reporting their sustainable development progress using these guidelines. Industry sectors, such as cement and aluminium, which are among the most intensive energy users, have their own initiatives to track progress (For more information on sectoral indicators, see Section 12.3.1).

In essence, while tools for measuring progress towards sustainable development are still far from perfect, considerable progress in the development of such tools and considerable uptake in their use has occurred. The trend is clearly towards more refinement in the tools and an increase in their use by governments, business and civil society.

12.2 Implications of development choices for climate change mitigation

The roadmap for this section starts with the concept of development paths. National development paths do not result from integrated policy programmes. They emerge from fragmented decisions made by numerous private actors and public agencies within varied institutional frameworks of state, markets, and civil society. Decisions about the development of the most significant sectors that shape emission profiles - energy, industry, transportation and land use - are made by ministries and companies that do not regularly attend to climate risks. The same is true for even more indirect influences on these sectoral pathways, including financial, macro-economic, and trade practices and policies. The focus on development paths places new emphasis on development's impact on climate and on indirect rather than direct actions that affect climate mitigation. Section 12.2.1 reviews scenario and other literature indicating that in different nations and regions, contingent development paths are plausible and can be associated with widely disparate economic, environmental and social consequences. Section 12.2.2 provides historical evidence that lower emissions pathways are not necessarily associated with lower economic growth.

The second segment of the road map suggests the importance of better understanding in climate policy of how nations organize sectoral and other emissions-determining policies and behaviour. Section 12.2.3 assesses literature that analyze: (1) the particular institutions, organizations, and political cultures that form the installed systems of decision-making and priority-setting from which decisions about key sectors or contexts emerge; and (2) the broader trans-national trends that are reshaping established governance processes. The description of these installed systems and the ways in which they are changing is drawn from an assessment of the social science literature on relationships between states, markets and civil society. Thus, Section 12.2.3 broadens the discourse beyond the economics and technological literature now familiar in climate analysis by incorporating history, political economy, and organization theory. The emphasis moves from government to governance. Rather than focusing on action by governments or states alone, the social science literature suggests more attention on decisions by multiple actors (Rayner and Malone, 1998; Jochem *et al.*, 2001). In some systems, change occurs primarily through actions initiated by either central governments or more federalized local jurisdictions. In others, it proceeds more through initiatives by private organizations that are then complemented by supportive governmental policies.

The final segment of road map relates in Section 12.2.4 to strategies and actions for changing development paths. It builds from the insight that changes in development paths emerge from the interactions of varied, centralized and decentralized public and private decision processes, many of which are not traditionally considered as 'climate policy'. It emphasizes that national circumstances, including endowments in primary energy resources, and the strengths of institutions matter in determining how development policies ultimately impact GHG emissions. Ensuring that key sectors evolve in a more sustainable manner depends on capability to coordinate decentralized choices and decision processes. The literature emphasizes the importance of partnerships between public, private and civil society in actions that contribute to shifts in the direction of development. However, it does not assume that the lead coordinating agency will always be the state. In different societies with different cultures of social change, the lead agent with a strong motivation, whether political or commercial, to bear the costs of organizing change may emerge from states, markets or civil societies.

In sum, Section 12.2 shows that to expand the focus of effective climate action to include development activities involves less emphasis on the search for ideal and general instruments, and involves much more attention on local and fragmented processes for more marginal changes in key sectoral decisions. When added up over time, these decisions could lead to more sustainable development paths and lower emissions.

Clearly, the reformed focus of a broadened scope for climate action raises many questions that have not been highlighted in the research agenda. These are reflected in the agenda for future research in Section 12.4.

12.2.1 Multiplicity of plausible development pathways ahead, with different economic, social and environmental content

Climate policy alone will not solve the climate problem. Making development more sustainable by changing development paths can make a major contribution to climate goals. One of the major findings of TAR in terms of sustainable development was that development choices matter (Banuri *et al.*, 2001). The literature on long-term climate scenarios (Metz *et al.*, 2002; Nakicenovic *et al.*, 2000; Swart *et al.*, 2003), and especially the SRES Report (Morita *et al.*, 2000), points to the same conclusion. Climate outcomes are influenced not only by climate specific policies but also by the mix of development choices made and the development paths that these policies lead to. There are always going to be a variety of development pathways[3] that could possibly be followed and they might lead to future outcomes at global, national, and local levels. The choice of development policies can, therefore, be as consequential to future climate stabilization as the choice of climate-specific policies.

Development pathways can be useful ways to think about possible, even plausible, future states of the world. Over the last century, for example, human health has been improved significantly in most of the world under very different socio-

---

3   Development paths are defined here as a complex array of technological, economic, social, institutional, cultural, and biophysical characteristics that determines the interactions between human and natural systems, including consumption and production patterns in all countries, over time at a particular scale. In the TAR, "alternative development paths" referred to a variety of possible development paths, including a continuation of current trends, but also a variety of other paths. To avoid confusion, the word 'alternative' is avoided in the current report. Development paths will be different in scope and timing in different countries, and can be different for different regions within countries with large differences in internal regional characteristics.

economic pathways and health care systems (e.g., see CGD, 2004; OECD, 2005). Countries have made different decisions with respect to health care, leading to a wide variety of different systems, with still a large divide between industrialized and developing countries (Redclift and Benton, 2006). But in general, the chosen strategies have in common that they have contributed to marked health improvements in almost all regions. Advances have been uneven and improvements are under constant pressure from new developments (e.g., AIDS, new infectious diseases). In general, the health example suggests that human choice can make a positive contribution towards reaching a common goal (Frenk *et al.*, 1993; Smith, 1997). The same could be true for sustainable development in general, and reduced GHG emissions in particular. But changing a development pathway is not about choosing a mapped out path, but rather about navigating through an uncharted and evolving landscape.

Developing scenarios depicting possible development pathways can falsely suggest that these are in some sense latent pathways or routes through the future that have been uncovered through insight or research. In reality, well-defined development pathways are not waiting to be selected. Even understanding the much smaller set of current development paths can be difficult. These are not simply the result of previous policies or decisions of governments, although these certainly affect the outcomes. As Shove *et al.* (1998) argue with respect to energy usage, the present is the result of myriad small activities and practices adopted or developed in the course of everyday life.

In reviewing the literature on development pathways, and in respecting the caveats described above, three key lessons emerge:
- Development paths as well as climate policy determine GHG emissions;
- New global scenario analyses confirm the importance of development pathways for climate change mitigation;
- Development paths can vary by regions and countries because of different priorities and conditions.

These three findings are discussed in the following section.

12.2.1.1 *Development paths as well as climate policies determine GHG emissions*

For much of the last century, the dominant path to industrialization was characterized by high concurrent GHG emissions. The IPCC Third Assessment Report concluded that committing to alternative development paths can result in very different future GHG emissions. Development paths leading to lower emissions will require major policy changes in areas other than climate change. The development pathway pursued is an important determinant of mitigation costs and can be as important as the emissions target in determining overall costs (Hourcade *et al.*, 2001) These findings were based on an extensive analysis of model-based emissions scenarios (Morita

and Lee, 1998), a survey of more qualitative studies (Robinson and Herbert, 2001), and a comparison of stabilization scenarios (Morita *et al.*, 2000) based on the IPCC SRES scenarios (Nakicenovic *et al.*, 2000).

Developing countries do not have to follow the example of developed countries in terms of energy use (UNCSD, 2006), since the early stages of infrastructure development offer opportunities to satisfy their populations' needs in different ways. Many factors that determine a country's or region's development pathway, and, closely related, its energy and GHG emissions are subject to human intervention. Such factors include economic structure, technology, geographical distribution of activities, consumption patterns, urban design and transport infrastructure, demography, institutional arrangements and trade patterns. The later choices with respect to these factors are made, the fewer opportunities there will be to change development paths, because of lock-in effects (e.g., Arthur, 1989). For detailed discussion, see Section 2.7.1 and Section 3.1.3. An assessment of mitigation options should not be limited to technology, although this is certainly a key factor, but should also cover the broader policy agenda. Climate change mitigation can be pursued by specific policies, by coordinating such policies with other policies and integrating them into these other policies. Also, climate mitigation objectives can be mainstreamed into general development choices, by taking climate mitigation objectives routinely into consideration in the pursuance of particular development pathways.

Development policies not explicitly targeting GHG emissions can influence these emissions in a major way. For example, six developing countries (Brazil, China, India, Mexico, South Africa, and Turkey) have avoided through development policy decisions approximately 300 million tons a year of carbon emissions over the past three decades. Many of these efforts were motivated by common drivers, such as economic development and poverty alleviation, energy security, and local environmental protection (Chandler *et al.*, 2002). The current state of knowledge does not allow easy quantitative attribution to specific policies with accuracy, given that other factors (as in any country) also influence these emissions. For example, autonomous technological modernization certainly has played a role. Chandler *et al.* (2002), however, also clearly identify policies that have made a definite contribution. In Brazil, these included production and use of ethanol and sugarcane bagasse, development of the natural gas industrial market, use of alternative energy sources for power generation and a set of demand-side programmes promoting conservation and efficiency in the electricity and transportation sectors (See also Box 12.1).

In China, growth in GHG emissions has been slowed to almost half the economic growth rate over the past two decades through economic reform, energy efficiency improvements, switching from coal to natural gas, renewable energy development, afforestation, and slowing population growth. In India, key factors in GHG emission reductions have been economic restructuring, local environmental protection, and technological change, mediated through economic reform, enforcement of clean air laws by the nation's highest court, renewable energy incentives and development programmes funded by the national government and foreign donors. In Mexico, expanding use of natural gas in place of more carbon-intensive fuels, promoting energy efficiency and fuel substitution by means that included energy pricing mechanisms, and abating some deforestation have played a major role. The policies in South Africa that contribute to lower growth in GHG emissions include restructuring the energy sector, stimulating economic development, increasing access to affordable energy services, managing energy-related environmental impacts, and securing energy supply through diversification. Finally, in Turkey, economic restructuring and price reform resulting from government moves to more market-oriented policies and the expectation of European integration, fuel switching, and energy efficiency measures have contributed to avoided GHG emissions (Chandler et al., 2002).

There are multiple drivers for actions that reduce emissions, and they can produce multiple benefits. The most promising policy approaches are those that capitalize on natural synergies between climate protection and development priorities to simultaneously advance both objectives. Many of these synergies are in energy demand (e.g., efficiency and conservation, education and awareness) and some in energy supply (e.g., renewable options).

Capturing these potential benefits is not always easy, since there are many conflicts and trade-offs. From the perspective of energy security, for example, it can be politically and/or economically attractive to give priority to domestic coal and oil resources over more environmentally friendly imported gas (e.g., SSEB 2006). The adverse economic impact of higher oil prices on oil-importing developing countries is generally more severe than for OECD countries. This is because their economies are more dependent on imported oil and more energy-intensive, and because energy is used less efficiently. On average, oil-importing developing countries use more than twice as much oil to produce a unit of economic output as do OECD countries. Developing countries are also less able to weather the financial turmoil wrought by higher oil-import costs (IEA, 2004a). For a discussion of the role of energy security for development paths, see Section 3.3.6. Some studies have shown that, depending on how priorities are set, some conflict between local atmospheric pollution problems and global climate change issues may arise. This is because some of the most cost-effective, environmentally-friendly power generation technologies for the global environment available in developing countries, such as biomass-fired or even some hydroelectric power plants, may not be sound for the local environment (due to NO_x and particulate emissions in the former case, and flooding in the latter). Conversely, abating local air pollution generally is beneficial from a global perspective. Still, there are a few exceptions. Decreasing sulphur and aerosol emissions (with the exception of black carbon) to address local air pollution problems can increase overall radiative forcing, because these aerosols have a negative radiative forcing. Thus, exploring development paths requires careful assessment of both local environmental priorities and global environmental concerns (Schaeffer and Szklo, 2001).

In developed countries too, development choices made today can lead to very different energy futures. In the TAR, Banuri et al, (2001) distinguished between strategies decoupling growth from resource flows (e.g., resource light infrastructure, eco-intelligent production systems, 'appropriate' technologies and full-cost pricing), and strategies decoupling wellbeing from production (intermediate performance levels, regionalization avoiding long-distance transport, low-resource lifestyles). Technological mitigation options at the sectoral level are mainly

---

Box 12.1: Greenhouse gas emissions avoided by non-climate drivers: a Brazilian example

In the field of energy, experience with policies advancing energy efficiency and renewable energy use confirm that, although developing countries need to increase their energy consumption in order to fuel their social and economic development, it is possible to do so in a cleaner and more sustainable manner. These policy choices can have a significant impact on energy trends, social progress and environmental quality in developing countries (Holliday et al., 2002; Anderson, 2004; Geller et al., 2004). In Brazil, programmes and measures have been undertaken over the past two or three decades in order to mitigate economic and environmental problems. These have included not only improvements in the energy supply and demand side management, but also specific tax incentive policies encouraging the production of cheap, small-engine automobiles (<1000 cc) to allow industry to increase production (and create more jobs while increasing profits) and to make cars more accessible to lower-income sectors of the population. These policies have led to lower carbon dioxide emissions than would otherwise have been the case. Results of these programmes and measures show that, in 2000 alone, some 11% in CO_2 emissions from energy use in Brazil have been reduced compared to what would have been emitted that year, had previous policy decisions not been implemented. Interestingly, although these actions were not motivated by a desire to curb global climate change, if the inherent benefits related to carbon emissions are not fully appraised in the near future, there is a chance that such 'win-win' policies may not be pursued and may even be discontinued (Anderson, 2004; Szklo et al., 2005).

discussed in Chapter 4 to 11 which also cover to some extent non-technological options that relate to different development priorities, as far as the literature allows.

The connections between development pathways and international trade are often left unexplored. International trade allows a country to partially 'de-link' its domestic economic systems from its domestic ecological systems, as some goods can be produced by other economic systems. In such cases, the impacts of producing goods impact the ecological systems of the exporting country (where production takes place) rather than the ecological system of the importing country (where consumption occurs). One popular way of showing that the impacts of economic activities in many nations affect an area much larger than within their national boundaries is the ecological footprint (see Section 12.1.3). For example, the environmental effects of soya and hardwood production for export as fodder and construction material, respectively, are well-known examples. As a consequence, in discussing the implications of development choices for climate change mitigation, it is not enough to discuss development pathways for individual countries. To fully address global emission reductions, an integrated multi-country perspective is needed (Machado *et al.*, 2001).

12.2.1.2   New global scenario analyses confirm the importance of development paths for mitigation

Section 3.1.5 discusses some factors that determine development paths, such as structural changes in production systems, technological patterns in sectors, such as energy, transportation, building, agriculture and forestry, geographical distribution of activities, consumption patterns and trade patterns. After publication of IPCC TAR, several new scenarios relating to climate change or global sustainability were published, making different assumptions for these factors. Most of them confirm the main findings of SRES (see also Chapter 3). It is important, however, to translate the lessons derived from scenarios (which are often global in scale) to national and even local level policy choices that can lead to the desired outcomes.

For the Millennium Ecosystems Assessment (MEA), four scenarios explored implications of development pathways for global and regional ecosystem services, loosely based on the SRES but developed and enriched further (Alcamo *et al.*, 2005; Carpenter and Pingali, 2005; Cork *et al.*, 2005). For the next 50 years, all scenarios find that pressures on ecosystem services increase with the extent of the pressure being determined by the particular development path. The MEA scenarios identify climate change next to land-use change as a major driver of biodiversity loss in the coming century. Quality of the services differs strongly by scenario - with the most positive scenarios finding a clear improvement in some services and the most negative scenario, finding a general decrease. The MEA scenario analysis, thus, emphasizes that development

of ecosystem services, biodiversity, human wellbeing and the capacity of the population to deal with these developments is largely determined by the choice of development pathway.

The United Nations Environment Programme (UNEP, 2002), used SRES scenarios as well as the scenarios of the World Water Vision (Gallopin and Rijsberman, 2000) and the Global Scenario Group (Raskin *et al.*, 1998) as inspiration for the development of four development pathways for the third Global Environmental Outlook (UNEP/RIVM, 2004): Markets First, Security First, Policy First and Sustainability First. Again, the different development pathways reflected by these scenarios are associated with a wide range of GHG emissions similar to the range captured by the SRES scenarios.

Shell's Low Trust Globalization, Open Doors and Flags scenarios explore how different future development pathways could affect the company's business environment. In the Open Doors scenario, CO_2 emissions increase most rapidly as a result of higher economic growth and the absence of security-driven investment in indigenous renewable energy sources, even if people may be more concerned about climate change than in other scenarios. The Low Trust Globalization scenario is characterized by larger barriers to international trade and cooperation. Paradoxically, there could be faster progress towards carbon efficiency as a result of a different set of policies aimed at energy efficiency, conservation and development of renewables, notably wind and, possibly, nuclear power. Finally, the Flags scenario with a patchwork of national approaches could show positive responses to climate change because of factors such as the pursuit of self-reliance (Shell, 2005).

Several scenarios developed since the TAR have explored different development pathways, but without explicitly addressing climate change or GHG emissions. The characteristics of these pathways in terms of the rate and structure of geopolitical, economic, social and technological development, however, would result in large variations in GHG emissions. Four scenarios developed by the US National Intelligence Council (Davos World, Pax Americana, A New Caliphate and Cycle of Fear) explore how the world may evolve until 2020 and what the implications for US policy might be, focusing on security concerns (NIC, 2004). The National Intelligence Council scenarios show the possible impacts of particular development pathways in some regions for other regions. Also, in several developing countries, different future development pathways have been explored in systematic scenario exercises, for example, China (Ogilvy and Schwartz, 2000); the Mont Fleur scenarios for South Africa (Kahane, 2002); the Guatemala Vision (Kahane, 2002); Destino Colombia (Cowan *et al.*, 2000); Kenya at the crossroads (SID/IEA, (Society for International Development and the Institute of Economic Affairs), 2000). Taking global climate change explicitly into account would strengthen and enrich development-oriented scenarios as the ones mentioned above.

Case studies in Tanzania (Agrawala *et al.*, 2003a), Fiji (Agrawala *et al.*, 2003c), Bangladesh (Agrawala *et al.*, 2003b), Nepal (Agrawala *et al.*, 2003a), Egypt (Agrawala *et al.*, 2004b) and Uruguay (Agrawala *et al.*, 2004a) show how climate-change adaptation can be integrated with national and local development policies, often as a no-regrets strategy. Implementation of no-regrets strategies is, however, not without challenges. A study of the Baltic region explores a sustainable development pathway addressing broad environmental, economic and social development goals, including low GHG emissions. It points out that a majority of the population could favour - or at least tolerate - a set of measures that change individual and corporate behaviour to align with local and global sustainability (Raskin *et al.*, 1998). Kaivo-oja *et al.* (2004) conclude that climate change as such may not be a major direct threat to Finland. However, the effects of climate change on the world's socio-economic system and the related consequences for the Finnish system may be considerable. The Finnish scenario analysis, which is based on intensive expert and stakeholder involvement, suggests that such indirect consequences have to be taken into account in developing strategic views of possible future development paths for administrative and business sectors.

Netherlands Environmental Assessment Agency (MNP, 2005) has developed the four IPCC SRES scenarios for a sustainability outlook for the Netherlands. The four scenarios represent four world perspectives with four different views on future priorities for action to make development more sustainable. This outlook points at several dilemmas. Surveys showed that 90% of the Dutch population prefer a future which would be different from the globalizing, market-oriented A1 scenario. Yet, A1 appears to be the future they are heading for. A majority of the population also thinks that something has to be done about unsustainable production and consumption patterns, and suggest that the government should do more. The study suggests that the regional (European) level may be the most appropriate level to address sustainability issues. Global political, economic and cultural differences make effective global policy difficult, while many sustainability issues go beyond local or national capacity to develop and implement effective policies.

Scenarios describe different states of the world that could come about by different developments in the driving forces that are often of a geopolitical nature and are largely unaffected by national or local policy-making. These scenarios studies reveal that different pathways are possible, but also that pursuing them involves many complex challenges. Such challenges include consideration of indirect effects, and difficulties in translating the often positive attitude of the population towards sustainable futures into concrete changes. Decision-makers have to consider the robustness of alternative development pathways they pursue through their policy choices, in the face of global developments they will be confronted with.

12.2.1.3 Development paths can vary by regions and countries because of different priorities and conditions

An understanding of different regional conditions and priorities is essential for mainstreaming climate change policies into sustainable development strategies (See Section 12.2.3). Since regions and countries differ in many dimensions, it is impossible to group them in a way consistent across all dimensions. There is a diversity of regional groupings in the literature using many criteria that are specific to their purpose within the underlying context. (For regional groupings, see Section 2.8).

As noted in Section 12.1.1, the mitigative capacity of a nation is closely related to its underlying development path, which depends on the general pool of resources that may be referred to as response capacity. The response capacity including mitigative capacity of countries varies, amongst other factors, with their ability to pay for abatement costs. Winkler *et al.* (2007) analysed the mitigative capacity of different countries as shaped by two economic factors: namely average abatement cost (or mitigation potential; high cost means low potential); and ability to pay, as approximated by GDP per capita. Ability to pay, measured by GDP per capita, is an important factor in mitigative capacity, since more wealth gives countries greater capacity to reduce emissions. The cost of abatement can act as a barrier in turning mitigative capacity into actual mitigation. Examining these factors together, Winkler *et al.* (2007) found that the abatement costs are not linearly correlated with level of income. Some countries have high mitigative capacity (income) and are also able to translate this into actual mitigation due to low costs. For others, mitigative capacity is clearly low. Relatively high average abatement costs mean that this capacity can be turned into even less actual mitigation. Interestingly, there are some poorer countries with low abatement costs. Conversely, there are also countries with high mitigative capacity, as approximated by income, but high average abatement costs. However, this group of countries still has higher mitigative capacity, simply by virtue of their higher ability to pay. Low-income countries do not spend on mitigation even if they have low-cost mitigation opportunities, simply because the opportunity cost in terms of basic development needs is too high.

Developed economies: Developed economies are included in Annex I to the UNFCCC and are members of the OECD. CO_2 emissions from fossil fuel combustion accounted for over 80% of their total emissions in 2000 with negligible amounts from land-use change (Table 12.1). These countries are also largely responsible for GHG emissions with high radiative forcing. Their population growth is projected to be low or negative (UNDP, 2004), income and level of human development are in the upper middle and high end of the spectrum (UNDP, 2004), and energy consumption and GHG emissions per capita are above the world average (IEA, 2005). These developed countries are assessed to be least vulnerable when compared

Table 12.1: *Profiles of emissions and human development at different levels of development*

	Units	Developed/industrialized/Annex I countries[c]		Developing/Non-Annex I countries[d]	
		OECD	EIT	Developing	Least developed
Emissions profiles by gases, 2000[a]		100		100	100
CO_2 (fossil fuel)	%	81		41	4
CH_4	%	11		16	22
N_2O	%	6		10	12
LUC	%	*0*		33	62
High GWP gases	%	*2*		0	0
Human development profiles[b]					
HDI, 2003		0.892	0.802	0.694	0.518
Life expectancy at birth	years	77.7	68.1	65.0	52.2
Adult literacy	%	100.0	99.2	76.6	54.2
GDP_{ppp}/capita, 2003	US$/capita	25915	7930	4359	1328
Population growth rate (2003-2015)	%/yr	0.5	-0.2	1.3	2.3
GDP/capita growth rate (1990-2003)	%/yr	1.8	0.3	2.9	2.0
Electricity consumption per capita, 2002	kWh/capita	8615	3328	1155	106
CO_2 emissions per capita, 2002	tonnes/capita	11.2	5.9	2.0	0.2
Vulnerability assessment[e]					
Vulnerability scores		10-15	14-22	18->40	

Notes:

a) Source: Baumert et .al., 2004, p. 6. FF: fossil fuel combustion; High GWP (global warming potential) gases: sulphur hexafluoride (SF6), perfluorocarbons (PFCs), and hydrofluorocartbons (HFCs).

b) Source: UNDP, 2005. HDI range: 0.00<HDI<1.00; PPP: purchasing power parity. PPP normally deflates the income level of the developed nations while inflating those in the developing world as one dollar would have larger purchasing power that it has in the developed world.

c) Annex I countries include both developed OECD and EIT countries. However, a few newly admitted OECD countries are not in Annex I list, including South Korea, Singapore, and Mexico. The group of economies in transition (EIT) countries contains several sub-groups: those that are part of the enlarged EU, central Asian Republics, and other members of the CIS. In UNDP (2005) categorization, the coverage is larger, including Central and Eastern Europe and the Commonwealth of Independent Sates (CIS).

d) In emissions profiles, these two subgroups were counted separately while in the UNDP human development profiles, least developed is a subgroup of the developing world.

e) Source: Adger et al., 2004b. Vulnerability scores range from 10 to 50, with 10 the least vulnerable and 50 the most vulnerable. These scores are derived from a series of proxy variables for vulnerability including food security, ecosystem sensitivity, settlement/infrastructure sensitivity, human health sensitivity, economic capacity, human resource capacity, governance capacity and environmental capacity. See, Baumert et al., 2004, p.17.

to other groups of countries (Adger *et al.*, 2004), with vulnerability scores lower than 15, close to the lower end of the spectrum (Table 12.1). In general, mitigative capacity in these economies is high but cost can be high. As well as marginal cost of mitigation increases with the rate of energy efficiency. Nevertheless, there are large mitigation potentials in these countries. For example, passenger vehicle economy in North America and Australia is well below that in EU and Japan, even lower than some developing countries such as China (An and Sauer, 2004). Barring a few newly industrialized countries, most are highly industrialized with limited scope or need for large-scale expansion of the physical infrastructure, such as public utilities, physical transport infrastructure, and buildings (Pan, 2003).

Notwithstanding this limited scope or need for infrastructure expansion and economic growth figures often much lower than in many developing countries, the future will look different from today and low-carbon development pathways are possible. Improving energy efficiency, modernizing production and changing consumption patterns would have a large impact on future GHG emissions (Kotov, 2002). Developed countries possess comparative advantages in technological and financial capabilities in mitigation of climate change. Priority mitigation areas for countries in this group may lie in improving energy efficiency, building new and renewable energy, and carbon capture and storage facilities, and fostering a mutually remunerative low-emissions global development path through technological and financial transfer of resources to the developing world.

In many industrialized countries (e.g., Japan and in Europe), implications of energy systems with very low carbon emissions have been explored, often jointly by governments, energy specialists and stakeholders (e.g., Kok *et al.*, 2000). However, a fundamental and broad discussion in society on the implications of development pathways for climate change in general and climate change mitigation in particular in the industrialized countries has not seriously been initiated. Low-emission pathways apply not only to energy choices. For example, in North-America and Europe, UNEP (2002) identifies land-use development, particularly infrastructure expansion, as a key variable determining future environmental stresses, including GHG emissions. Pathways that capitalize on advances in information technologies to provide a diverse range of lifestyle and spatial planning choices will also affect energy use and GHG emissions.

Economies in Transition: With EU enlargement, economies in transition as a single group no longer exist[4]. Nevertheless, Central and Eastern Europe and Commonwealth of Independent States share some common features in socioeconomic development (UNDP, 2005), and in climate change mitigation and sustainable development (IPCC, 2001b; Adger *et al.*, 2004). With respect to social and economic development, countries in this group fall between the developed and developing countries (Table 12.1). In terms of level of human development and vulnerability, for instance, these countries fall behind the developed countries but are well ahead of the developing countries. In certain key areas, however, they are closer to the developed countries in terms of population growth, levels of industrialization, energy consumption, and GHG emissions.. In other areas, including income levels and distribution, institutions and governance, they can show features similar to the developing world. GDP per capita level in some of these EIT countries is as low as that in the lower middle income developing countries (World Bank, 2003), and energy intensity is in general high (IEA, 2003a).

Although the 0.3 % per annum rate of economic growth in the past 15 years has been low, it is expected that in many countries, future rates could be high, which would contribute to an upward trend in GHG emissions. Measures to decouple economic and emissions growth might be especially important for this group through restructuring the economy (Kotov, 2002). Mitigative capacities are high as compared developing economies, but lower than those for developed economies due to a weaker financial basis. These capacities can be further enlarged through institutional reform, such as liberalization of the energy market and political determination to increase energy efficiency.

Developing Economies: Recently, interest at regional level in exploring development pathways which are consistent with lower GHG emissions has increased (Kok and de Coninck, 2004). This appears to be valid primarily for developing countries. Case studies focus on the future in the priority areas of energy supply, food security and fresh water availability in South Africa (Davidson *et al.*, 2003), Senegal (Sokona *et al.*,

2003), Bangladesh (Rahman *et al.*, 2003), Brazil (La Rovere and Romeiro, 2003), China (Jiang *et al.*, 2003) and India (Shukla *et al.*, 2003) A common finding of these studies is that it is possible to develop pathways that combine low GHG emissions with effective responses to pressing regional problems. In the energy sector, energy security and reduced health risks can be effectively combined with low GHG emissions, even without explicit climate policies. Enhancing soil management, avoiding deforestation, and encouraging reforestation and afforestation can increase carbon storage, while also serving the primary goals of food security and ecosystem protection.

Although the developing economies are highly diverse, their general features contrast to those of the industrialized world. Levels of human development and consumption of energy per capita are much lower than those in the developed countries and in the economies in transition (Table 12.1). GHG emissions from land-use change and agriculture are a significant proportion of their total emissions (Ravindranath and Sathaye, 2002; Baumert *et al.*, 2004).

Given the fact that energy consumption and emission per capita are low in the developing world, focus on climate mitigation alone may have large opportunity cost in terms of fiscal and human capitals, and therefore not be compatible with meeting sustainable development goals. With respect to levels of human development, UNDP (2005) projects that by 2015 almost all developing regions will not be able to meet their Millennium Development Goals. With respect to access to clean water, for example, the 2015 MDG goal will be missed by 210 million people who will not have access, with 50% in South Asia, 40% in Sub-Saharan Africa, 7% in East Asia and the Pacific. Non-climate policies for sustainable development goals can be more effective in addressing climate change, such as population control, poverty eradication, pollution reductions, and energy security, as demonstrated in the People's Republic of China (Winkler *et al.*, 2002b; PRC, 2004). In order to realize the promise of leapfrogging, improvements are needed to the institutional capabilities of the recipient developing country and its energy and environmental policies in order to foster sustainable industrial development (Gallagher, 2006; Lewis and Wiser, 2007).

In aggregate terms, some large developing countries are included in the list of top 25 emitters (Baumert *et al.*, 2004). These few developing countries are projected to increase their emissions at a faster rate than the industrialized world and the rest of developing countries as they are in the stage of rapid industrialization (Pan, 2004b). For these countries, climate change mitigation and sustainable development policies can reinforce one another, however, financial and technological assistance can be help these countries to pursue a low carbon

---

4   EITs are still recognized in international agreements, such as UNFCCC and its Kyoto Protocol.

path of development (Ott *et al.*, 2004). Emissions per capita for some developing countries, however, will continue to be lower than the industrialized countries for many decades.

For most other developing countries, adaptation to climate change takes priority over mitigation as they are more vulnerable to climate change and less carbon dependent (Hasselmann *et al.*, 2003). However, both adaptive and mitigative capacities tend to be low (Huq *et al.*, 2003). OPEC countries are unique in a sense that they may be hurt by development paths that reduce the demand for fossil fuels. Diversification of their economy is high on their agenda. Although climate change mitigation can be one consideration in evaluating poverty alleviation options, poverty has to be alleviated regardless of GHG emissions. Improved access to energy can lead to increasing GHG emissions, for example, where kerosene and propane use is more appropriate than biomass renewables. However, in absolute terms this is a minor increase in global GHG emissions (see also Section 12.2.4).

For most Small Island States, the key issue to sustainable development is the adoption of a comprehensive adaptation and vulnerability assessment and implementing framework with several priorities: sea level rise (high percentage of the population located in coastal areas); coastal zone management (including specially coral reefs and mangroves); water supply (including fresh water catchments);: management of upland forest ecosystem; and food and energy security. For some islands, extreme events, such as tropical hurricanes and El Niño and La Niña events, are an important threat.

In summary, different regions and types of countries have different contextual conditions to respond to, and therefore, their attempts to move towards a development path leading to sustainable development while also mitigating climate change, will vary considerably. Policy decisions will be most effective where made while recognizing these contextual conditions and where they relate and adapt to the existing regional and country realities.

12.2.2 Lower emissions pathways are not necessarily associated with lower economic growth

Section 12.2.1 has demonstrated that business-as-usual futures in countries with similar characteristics can result in very different emission profiles, depending on the development path adopted. Since economic growth figures prominently among the objectives of policy-makers worldwide, the relationship between economic growth and emissions at the national level is reviewed in Section 12.2.2. Consideration is given to whether lower emissions pathways are necessarily associated with lower economic growth The conclusion that there are degrees of freedom between economic growth and GHG emissions is further explored in Section 12.2.3 and Section 12.2.4.

Economic activity is a key driver of CO_2 emissions. How economic growth translates into new emissions, however, is ambiguous. On one hand, as the economy expands, demand for and supply of energy and of energy-intensive goods also increases, pushing up CO_2 emissions.. On the other hand, economic growth may drive technological change, increase efficiency and foster the development of institutions and preferences more conducive to environmental protection and emissions mitigation (see Chapter 3). Also, economic growth may be associated with specialization in sectors high) emissions per unit of output, such as services (manufacturing and heavy industries, respectively), thus resulting in a faster strong or weak relationship between domestic emissions and GDP. Unlike technological change or efficiency, however, specialization does not affect the level of global emissions: it only modifies the distribution of emissions across countries.

The balance between the scale effect of growth and the mitigating factors outlined above has generated intense scrutiny since the early 1990s. Much of the literature focuses on the 'environmental Kuznets curve' (EKC) hypothesis, which posits that at early stages of development, pollution per capita and GDP per capita move in the same direction. Beyond a certain income level, emissions per capita will decrease as GDP per capita increases, thus generating an inverted-U shaped relationship between GDP per capita and pollution. The EKC hypothesis is compatible with several, and possibly joint, explanations: structural shift towards low carbon-intensity sectors; increased environmental awareness with income, policy or technology thresholds; and increasing returns to abatement (Copeland and Taylor, 2004). The EKC hypothesis was initially formulated for local pollutants in the seminal analysis of Grossman and Krueger (1991) but was quickly expanded to CO_2 emissions. Even so, it recognized that some of the theoretical explanations for local pollutants, namely that higher income individuals would be more sensitive to environmental concerns, are less relevant for GHGs that do not have local environmental or health impacts. The EKC hypothesis has generated considerable research, and the field is still very active. Recent summaries can be found in Stern (2004), Copeland and Taylor (2004) or Dasgupta *et al.* (2004). With regard to carbon dioxide, three conclusions can be drawn, as discussed below.

First, using GDP and emissions data over multiple countries and time periods, studies consistently find that GDP per capita and emissions per capita move in the same direction among most or all of the sample (Schmalensee *et al.*, 1998; Ravallion *et al.*, 2000; Heil and Selden, 2001; Wagner and Müller-Fürstenberg, 2004). A 1% increase in GDP per capita is found to lead to an increase in CO_2 emissions per capita of 0.5% to 1.5%, depending on the study. All studies also find evidence that this coefficient, elasticity of per capita CO_2 emissions relative to per capita GDP, is not constant but decreases as per capita income rises. Until recently, empirical studies consistently found a relationship between per capita GDP and per capita CO_2 emissions such that, beyond a certain level of GDP per capita, per capita CO_2 emissions would decrease as income increases - thus confirming the EKC hypothesis for carbon dioxide. However, the reliability of these estimates has been challenged recently on technical grounds. For a general discussion, see Harbaugh *et al.* (2002) and Millimet *et al.* (2003); and for a critical review focusing on carbon dioxide, see Wagner and Müller-Fürstenberg (2004). Two main points emerge from the most recent reviews: (1) they cast doubt on the idea that the EKC hypothesis could be validated based on existing data; (2) they conclude that the relationship between GDP and emissions data is less robust than previously thought.

Second, studies using time series at the country level find less robust relationships between GDP per capita and CO_2 emissions per capita. For example, Moomaw and Unruh (1997) show that international oil price shocks, and not per capita GDP growth, explain most of the variations in per capita emissions in OECD countries. Similarly, Coondoo and Dinda (2002) find a strong correlation between emissions and income in developed countries and in Latin America, but a weaker correlation in Africa and Asia. Recent work on the EKC (Dasgupta *et al.*, 2004) also shows that the relationship between GDP per capita and pollution is not as rigid as it seems, and in fact, mostly disappears when other explanatory variables, notably governance, are introduced.

Third, including trade among the explanatory variables of CO_2 emissions usually yield EKC curves peaking farther in the future (Frankel and Rose, 2002), although there are methodological issues associated with this approach (Heil and Selden, 2001). Using trade-corrected emissions data for USA, Aldy (2005) also shows that taking trade into accounts leads to curves that peak much later. Neither taking trade into account as a new explanatory variable nor correcting emissions for trade effects, however, significantly increases the robustness of the correlation between observed levels of GDP per capita and observed emission levels.

To sum up, the econometric literature on the relationship between GDP per capita and CO_2 emissions per capita does not support an optimistic interpretation of the EKC hypothesis that "the problem will take care of itself" with economic growth. The monotonically increasing relationship between economic activity and CO_2 emissions emerging from the data does not appear to be econometrically very robust, especially at country level and at higher GDP per capita level. The pessimistic interpretation of the literature findings that growth and CO_2 emissions are irrevocably linked is not supported by the data. There is apparently some degree of flexibility between economic growth and CO_2 emissions. For example, CO_2 emissions from fossil-fuel combustion in China remained essentially constant between 1997 and 2001. This was despite a +30% growth in GDP (IEA, 2004a) due to the combination of closing small-scale, inefficient power plants, shift in industry ownership away from the public sector, and introduction of energy efficiency and environmental regulation (Streets *et al.*, 2001; Wu *et al.*, 2005).

However, these econometric studies do not distinguish between structural emissions and emissions that result from policy decisions. Thus, limited information is provided about how future policy choices may or may not influence CO_2 emissions paths. To explore these choices, a more disaggregated approach is necessary, as discussed in the following section.

12.2.3 Changing development pathway requires working with multiple actors, at multiple scales

Over the past two decades, social scientists have observed significant changes in the role of government in relation to social and economic change. These include a shift from government defined strictly by the nation state to a more inclusive concept of governance that recognizes the contributions of various levels of government (global, trans-national, regional, local) as well as the roles of the private sector, non-governmental actors, and civil society (Rhodes, 1996; Goodwin, 1998). The emergence of these new forms of governance has been attributed to the need for new institutions to address the more complex problems of present-day society, among which global environmental risks figure prominently (Beck, 1992; Giddens, 1998; Howes, 2005). Ideology and economic globalization have also played a role in the shifting focus from government to governance. Command-and-control strategies are losing favour while market-based mechanisms, voluntary initiatives, and partnerships with non-governmental organizations have gained wider acceptance (Lewis *et al.*, 2002). However, the shift to discussions of governance does not imply a reduction in the role of government. Governments remain central actors in environmental policy. They ensure the delivery of environmental protection to citizens, and help create the rules, norms, and many organizations that ensure environmental protection (Haas *et al.*, 1993; OECD, 2001; Ostrom *et al.*, 2002).

Recognizing the difficulty and limitations of trying to directly control their domestic economies in an increasingly open and globalized economy, governments now try to pursue economic growth through strategic policies. These policies are designed to increase access to foreign markets, encourage inward foreign investment, maintain national competitiveness, and obtain favourable outcomes from trade agreements (Jessop, 1997). While some believe that globalization has made national governments less powerful, others argue that rather than simply eroding government power, globalization has changed the ways in which governments operate and influence situations (Levi-Faur, 2005). On environmental issues, a strong case has been made for the need for government policy to ensure delivery of environmental protection as a public good (e.g., Liverman, 1999; Haas *et al.*, 1993; OECD, 2001; Ostrom *et al.*, 2002).

The three key institutional sectors– government, market and civil society – have begun to work in closer collaboration, partnering with each other in multiple and diverse ways when their goals are common and their comparative advantages are differentiated (Najam, 1996; Hulme and Edwards, 1997; Davis, 1999). This is not to imply that they always or even mostly work in partnership or have synchronous priorities: it mean that they now do so more often than they did, including in terms of global climate change mitigation (Najam, 2000). The nature of global governance on a range of issues, including on climate change, is today best understood not only as what states do but as a combination of what the state, civil society and markets do or not do (Najam *et al.*, 2004).

The more prominent roles businesses and civil society groups have played in governance has not been without controversy. Some believe that only the state can act in the public interest, while industry and citizens are motivated by self-interest. Others see all actors as motivated by self-interest and, in this context, believe competition and the market ensure the best outcomes – public and private. In this view, civil society, consumers and industry bear greater responsibility and share the risks, while the state maintains a role in setting standards and auditing performance (Dryzek, 1990; Dryzek, 1997; Howes, 2005).

While the roles, responsibilities, and powers assigned to the respective actors remains a hotly contested subject, it is widely acknowledged that responsibility for the environment and sustainability has become a much broader project. It is no longer primarily the preserve of governments, but involves civil society, private sector, and the state (Rayner and Malone, 2000; Najam *et al.*, 2004).

12.2.3.1 *State*

The transition from government to governance recognizes the changing trends among political constitutions in developed and developing countries. While varying in speed and scope in individual states, these institutional reforms broadly span the domains of government and market activity, the powers of public executive administration relative to that of legislatures and courts, the degree of federalism within nation states, the organization of the financial system and capital markets, the demands of corporate governance and corporate social responsibility, the structure of industrial organization and public utilities, the strength and engagement of civil society organizations, and the delegation of national sovereignty to multinational and regional law and regimes (Berger and Dore, 1996; Hollingworth and Boyer, 1997; Schmidt, 2002; Heller and Shukla, 2003).

The specific constellation of these reforms depends on the pre-existing institutions in a country, the local politics of reform and resistant domestic interests. Yet in almost all cases, the re-organization of governance institutions will have important implications for the choice of potential national development paths in key input sectors. For example, a recent study of electricity sector reforms in five leading emerging nations - China, India, South Africa, Brazil and Mexico - found that in no cases did the changes away from power provision through state monopolies correspond closely to the orthodox designs of electricity market reforms (Box 12.2).

All five electricity sectors separate ownership of generation from transmission and distribution and allow participation in the generation markets by independent, often foreign, power producers. Nowhere have competitive generation markets flourished or has the state withdrawn substantially from system planning, tariff setting based on social and political criteria, infrastructure financing, or predominant ownership of major power sector firms (Victor and Heller, 2007). Yet, the consequences for climate friendly energy development have varied across these emerging markets because of nationally specific characteristics. Social goals, including increasing access and renewable power development, have not been interrupted. In some cases, such as the Indian State of Gujarat, the substitution of public grid power by privately developed stand-alone power plants has increased the rate of substitution of coal-fired generation by natural gas (Shukla *et al.*, 2005). In Mexico, complex, financially problematic, government guarantees of tariffs have also encouraged gas fuel diversification from oil to gas. In other cases, including China, the ongoing flux in institutional reforms creates both risks of intensive coal-based power development and the opportunities of more climate friendly energy growth.

The choice of policies that governments seek and are able to pursue is influenced by the political culture and regulatory policy style of a country or region, and the extent of public expectations that their governments will take a strong or weak lead in pursuing policy responses. Earlier efforts to address the issues of institutional capacity for mitigation include a compendium of policy instruments (DOE, 1989); two collections of country studies (Grubb, 1991; Rayner, 1993) and a review of the relevant social science literature on institutions (O'Riordan *et al.*, 1998).

A substantial body of political theory identifies and explains national policy styles or political cultures. The underlying assumption is that individual countries tend to process problems in a specific manner, regardless of the distinctiveness or specific features of any specific problem; a national 'way of doing things.' The key features of prevailing 'policy styles' in various countries and regions of the world are highlighted.

Richardson *et al.* (1982) identified national policy style as deriving from the interaction of two components "(a) the government's approach to problem solving and (b) the relationship between government and other actors in the policy process." Using a basic typology of styles, countries are subdivided according to whether national decision-making is anticipatory or reactive, and whether the political context is consensus-based or impositional. Many studies of national differences in institutional arrangements for making and implementing environment and technology policy emphasize the essentially cooperative approach to environmental

Box 12.2: Poverty tariff in South Africa

The extent to which the policy alleviates poverty depends on the energy burden (percentage of the total household budget spent on energy). The energy burden of poor households in remote rural villages can be up to 18% of the total household budget, according to data from a case study reported in Table 12.2; see also UCT (2002). The 50 kWh provided by the poverty tariff would reduce the energy burden by two-thirds (6 percentage points). Monthly expenditure on electricity and other fuels decline by 18% and 16% respectively, due to the poverty tariff.

Table 12.2: *Mean household expenditure on electricity and other fuels and energy as a percentage of total household expenditure*

Expenditure on	Before subsidy	After subsidy	Difference	
Electricity (Rand/month)	38	31	7	18%
Fuels excluding electricity (Rand/month)	70	59	11	16%
Energy share in household expenditure (%)	18	12	6	

Source: Prasad and Ranninger, 2003.

A recent study in the poor areas of Cape Town showed that monthly electricity consumption has risen by 30-35 kWh/month per customer since the introduction of the poverty tariff, a substantial rise against an average consumption ranging from 100 to 150 kWh per month (Borchers *et al.*, 2001; Holliday *et al.*, 2002). This rise is less than the full 50 kWh/month, suggesting that households make greater use of electricity, but also value some saving on their energy bills (Cowan and Mohlakoana, 2005)

The impacts on climate change mitigation have been broadly scoped. If extended to all customers in a broad-based approach, the poverty tariff might at most increase emissions by 0.146 $MtCO_2$ under the assumption that all the free electricity would be additional to existing energy use (UCT, 2000; Hawken, 1999; Anderson, 1998; Holliday *et al.*, 2002). In practice, it is likely that electricity might displace existing use of paraffin, coal, wood, candles, batteries and other fuels to some extent. This upper-bound estimate represents 0.04% of total GHG emissions, but about 2% of residential sector emissions in 1994. This example from South Africa shows that poverty-alleviation and environmental objectives can be addressed simultaneously. To the extent electricity use displaces indoor fuel use, it may also provide a benefit to public health.

---

protection in Europe and the more confrontational approach that predominates in the United States (Lindquist, 1980; Kelman, 1981; Kunreuther *et al.*, 1982). Jasanoff (1986) shows how information about established technologies, such as formaldehyde use, is interpreted differently by scientific advisory bodies in different countries. In particular, Brickman *et al.* (1985) argue that decentralization of decision-making in the USA both increases the demand for scientific details of technological and environmental hazards and engenders competition between different explanations. Europeans generally expect national government, and increasingly the European Union, to take the lead in all matters pertaining to environmental safety and health, as well as economic and social welfare.

Recent empirical studies confirm the view that only detailed and case-specific analyses of government institutions and policies can illuminate national differences in the pursuit of environmental and other regulatory objectives. Weiner (2002) finds that, contrary to common assertions, the USA and Europe have not differed substantially in their use or implementation of the precautionary principle. Stewart (2001) finds that the USA has successively moved between alternative forms of environmental policies, beginning with command and control,

before switching toward market instruments (permits and taxes), and later experimentation with flexible negotiated regulation and information based instruments.

In these cases, national political and regulatory cultures are distinguished by institutional factors, such as the judicial doctrines of administrative review and regulatory standards of general treatment, more than cultural predilections that support or restrict government action. Finally, governments appear to have varied traditions of policy preferences and authority. European governments and populations appear more comfortable with lifestyle (demand) regulation than do North American governments, which often tend to look to longer-run technology development support in collaboration with market actors (Nelson, 1993).

An important, though often neglected, issue in the choice of policy instruments is the institutional capacity of governments to implement the instrument on the ground (Rayner, 1993). This is often a matter of what countries with highly constrained resources think that they can afford. However, even industrialized nations exhibit significant variation with respect to the characteristics that would be considered ideal for the successful application of the complete suite of policy instruments listed above. These attributes include (O'Riordan *et al.*, 1998):

- a well developed institutional infrastructure to implement regulation;
- an economy that is likely to respond well to fiscal policy instruments because it possesses certain characteristics of the economic models of the free market;
- a highly developed information industry and mass communications infrastructure for educating, advertising, and public opinion formulation;
- a vast combined public and private annual RD&D budget for reducing uncertainties and establishing pilot programmes.

To the extent that these close to ideal conditions for conventional policy instruments are missing, policy-makers are likely to encounter obstacles to their effectiveness. For example, both Brazil and Indonesia (Petrich, 1993) have carefully crafted forest protection laws that could be used to secure forest preservation and carbon management. However, neither country is able to allocate sufficient resources to monitoring and compliance with those laws to ensure that they are effective. Even in industrialized countries, competition for resources among state agencies responsible for promoting economic development and those responsible for environmental protection are almost universally resolved in favour of the former. In much of the developing world, the shortage of programmes resources is exacerbated by pressures to utilize natural resources to earn foreign income. This increases demands of population for energy, and pressures to convert forest land to human habitation. As a result, legislative initiatives often seem to "leave more marks on paper than on the landscape" (Rayner and Richards, 1994).

Less industrialized countries often have poor infrastructures, exacerbated by lack of human, financial, and technological resources. In addition, these countries are likely to focus on more basic considerations of nation building and economic development. The economic conditions of less-industrialized countries also present opportunities to achieve both sustainable development goals and emissions reductions measures at lower cost than in the industrialized countries.

The notions of adaptive and mitigative capacity advanced in the IPCC TAR appear to reinforce the idea that the capacity to develop and implement climate response strategies are essentially the same as those required to develop and implement policies across a wide variety of domains. They are largely synonymous with those of sustainable development. The issues and cases discussed here suggest that the challenges of capacity building for sustainable development is not confined to the less industrialized countries, but that industrialized countries also fall short of the capacity to respond to climate mitigation challenges in a sustainable fashion.

As O'Riordan et al. (1998) note "the more that climate change issues are routinized as part of the planning perspective at the appropriate level of implementation, the national and local government, the firm, the community, the more likely they are to achieve desired goals. Climate policies per se are hard to implement meaningfully. However, merely piggybacking climate change onto an existing political agenda is unlikely to succeed."

12.2.3.2 Market

Industry is a central player in ecological and sustainability stewardship. Accordingly, over the past 25 years or so, there has been a progressive increase in the number of companies taking steps to address sustainability issues (Holliday *et al.*, 2002; Lyon, 2003) at either the company or industry level (see Box 12.3). A number of companies have, as part of their corporate strategy, voluntarily defined goals that reflect social responsibilities and environmental concerns that go beyond traditional company obligations. Following this line of thinking, an increasing number of companies are defining targets for GHG emissions and sinks. Some of the more widely acknowledged corporate sustainability drivers include regulatory compliance, market opportunities, and reputational value. Lyon (2003) hypothesizes that voluntary action on the environment might be explained by either a recognition by companies that pollution is a symptom of production inefficiencies, or a perception that consumers are willing to pay more for products with better environmental credentials. Either explanation would signal that markets are more important than regulation as an incentive for improved environmental performance. Lyon (2003) suggests instead that "it is the opportunity to influence regulation that makes corporate environmentalism profitable".

Some companies have recognized that pursuing sustainability offers potential cost savings (Thompson, 2002; Dunphy *et al.*, 2003). For example, by increasing energy and material efficiency in production and by reducing wastes, companies can reduce costs per unit of production and thereby gain a competitive market advantage (Hawken *et al.*, 1999; Schaltegger *et al.*, 2003). This concept of 'eco-efficiency' further acknowledges that businesses which constantly work to evaluate their environmental performance will be more innovative and responsive businesses. DuPont, for example, has sought to elevate sustainability to the strategic level, using a three-pronged strategy involving integrated science, knowledge intensity and productivity improvements (Holliday, 2001). The company has achieved financial savings in excess of US$1 billion per annum, partly through reduced energy and raw material use and less waste (Holliday, 2001).

Lyon (2003) suggests that the influence of 'green marketing' is modest in terms of shifting industry behaviour with respect to the environment. Senge and Carstedt (2001) position consumers as a key influence in shaping the 'next industrial revolution', founded on an economic system that genuinely connects industry, society and the environment. Their view is that a shift in consumer attitudes and values is an essential prerequisite to building sustainable societies. Schaefer and

Box 12.3: Role of Business

One well-known example of a corporation which has embraced sustainability is Interface Inc., a USA. manufacturer of carpets and upholstery. Since embracing the sustainability goal in 1994, Interface has reduced the carbon intensity of its products by 36% (Hawken et al., 1999; Anderson, 2004) Many of these reductions came through investments in energy efficiency and renewable energy (Holliday et al., 2002). However, Interface has also substantially reduced GHG emissions through other elements of its sustainability strategy, including reduction in raw material use and recycling materials not directly related to energy consumption (Hawken *et al.*, 1999; Anderson, 2004). As most of the materials used by Interface in its production are derived from petrochemicals (Anderson, 1998; Hawken *et al.*, 1999), these strategies have led to substantial reductions in the company's carbon footprint.

CEMEX, a Mexican-based cement manufacturer, was able to achieve similar emissions results through adoption of sustainability-oriented business model. One of the major environmental issues facing cement manufacturers is energy use (Wilson and Change, 2003). As part of its sustainability strategy, Cemex has focused intently on its energy use in an effort to reduce its ecological burden. For example, in 1994 CEMEX embarked on an eco-efficiency programmes to "optimize its consumption of raw materials and energy" (Wilson and Change, 2003), p.29). Through this and other measures, CEMEX reduced CO_2 emissions 2.7 million tons between 1994 and 2003 (Wilson and Change, 2003, p.32).

ITC Ltd, an Indian conglomerate and third largest company in terms of net profits in the country, reportedly sequestered almost a third of its CO_2 emissions in 2003-04, and plans to become a carbon positive corporation through a programmes of energy savings and CO_2 sequestration through farm and social forestry initiatives. Through programmes for rainwater harvesting, the company plans to become a water-positive corporation as well. Its 'e-Choupal' intervention has eliminated the need for brokers and helped 2.4 million farmers across six Indian states participate in global sourcing and marketing of products (Das and Dutta, 2004).

Crane (2005) conclude that a change in behaviour by the majority of consumers is not imminent. They suggest that it will require a sense of crisis to bring about a sea change in consumption patterns.

Managing stakeholder relations has also been identified as a corporate environmental driver. Many companies seek to improve relations with government, NGOs and local communities, because this can offer benefits, such as faster approvals for projects or products (Thompson, 2002), a continuing 'licence to operate', and greater scope for self-regulation. In regard to NGOs, improved relations can reduce or eliminate protests, such as consumer boycotts and direct lobbying (Thompson, 2002). Companies are also improving their environmental and social performance in response to demands from their corporate clients. Many large corporations, in particular, have introduced purchasing guidelines that place demands on suppliers to meet environmental performance standards (Thompson, 2002). The role of trade associations is another factor - including at the international negotiations (Hamilton *et al.*, 2003).

Demands of investors, insurers and other financial institutions are providing further incentives in relation to sustainability. Through improved sustainability performance, companies can potentially increase the attractiveness of their shares in the market, reduce insurance premiums and obtain better loan terms (Thompson, 2002). For example, the rapid growth of socially responsible investment funds (SRIs) in

the last decade is providing an incentive for greater corporate sustainability (Thompson, 2002; Borsky *et al.*, 2006). The role of institutional investors, and the growing concern in some business circles about liability due to inaction on climate change should also be acknowledged. This has led to a growing number of stakeholder initiatives to have publicly owned companies become proactive on climate change, and a growing number of initiatives to monitor and manage GHG emissions, even in the absence of domestic legislation and mandatory requirements (see Innovest, 2005; Cogan, 2006). The Carbon Disclosure Project has emerged as an important framework internationally for company reporting on their carbon footprint. Disclosure of environmental impact is increasingly seen as a crucial element of a company's risk profile for legal liability as well as competitive position in the face of possible future regulation. For example, re-insurers, companies providing insurance to insurance companies, have shown considerable concern about how climate change could impact insurance claims. Zanetti *et al.* (2005) suggest that climate change should be a core element in a company's long term-risk management strategy. Risk and return, demand, compliance and enforcement regimes, amongst other factors, are also likely to have an impact on investment.

Notwithstanding these achievements, there is widespread debate as to whether industry's responses to environmental decline and sustainability issues more generally are sufficient (Elkington, 2001; Sharma, 2002; Doppelt, 2003; Dunphy *et al.*, 2003).

All the same, notions of corporate social responsibility (CSR) have gained a wider hold. The essence of the CSR perspective is that there is a clear basis for businesses to widen their focus from simply profit maximization to include other economic, social, and environmental concerns. The arguments in support of CSR include competitive advantage (Porter and van der Linde, 1995; Porter and Kramer, 2002), notions of corporate citizenship (Marsden, 2000; Andriof and McIntosh, 2001), and stakeholder theory (Post *et al.*, 2002; Driscoll and Starik, 2004; Windsor, 2004). Drawing on the experience of DuPont, Holliday (2001) acknowledges the importance of shareholder value, but adds that business practices focused on sustainability outcomes can generate financial gains.

Colman (2002) reported that 45% of the Fortune Global Top 250 companies have issued environmental, social or sustainability reports. Similarly, CSR would seem to have become a more serious concern to European companies, though Pharaoh (2003) suggests it is primarily sales driven. In the UK, socially responsible investment (Srivastava and Heller, 2003) grew from US$ 46 billion in 1997 to US$ 450 billion in 2001 (Sparkes, 2002). Borsky *et al.* (2006) report that the US$ 2.16 trillion of socially responsible investments held in the USA accounted for approximately 11% of the total investment assets under management in 2003. The standards used by SRI funds to evaluate firms vary widely in the issues they address (with many simply staying away from weapons, tobacco, alcohol, and gambling) and how rigorously these standards are applied. Some SRI companies emphasize diversity and labour relations, while others focus on environment. There is no set of common criteria, and thus not all companies on SRI lists can be considered sustainable. However, growing public interest in SRI has led more companies to be concerned about a variety of social and environmental issues.

In considering the role of business, a distinction between multinationals and smaller, entrepreneurial enterprises is useful. A recent UK report identifies a difference in perspectives and approaches to global climate change in these two groups of businesses, with multinationals taking a long-term view, positioning for the future based on broad policy directions (Hamilton and Kenber, 2006). By contrast, smaller businesses, entrepreneurs or venture capitalists are more sensitive to the details of immediate or shorter term policy reforms. Similarly, there may be a difference even within the multinational sector between the energy suppliers (e.g., electricity producers/ distributors, oil companies, or even coal companies) and energy intensive industries (e.g., chemical or aluminium companies). The former takes a longer term, market development or pro-active view and the later a more reactive view (e.g., BIAC/ OECD/IEA, 1999). Finally, some companies are likely to be 'winners' with any effort to advance sustainable development through clean energy policies (e.g., insulation industry, window manufacturers, energy service companies) and some are likely to be 'losers' (e.g., producers of energy inefficient products). It is therefore difficult to speak about 'market' sector preferences

because there are different types of businesses with significantly different perspectives in different places.

In summary, although there has been progress, the private sector can play a much greater role in making development more sustainable. As the number of companies that operate both profitably and more sustainably increases, the view that addressing social and environmental issues is incompatible with shareholder maximization may loose ground. Opinions vary on the extent to which business can be relied upon to meet sustainability objectives. These range from business being inherently self-interested and exclusively profit-driven, to socially responsible businesses going 'beyond compliance' are on the forefront of the sustainability curve. Although the issues are complicated, there can be no question that the shift towards improved sustainability is fundamentally connected to the social, economic and environmental performance of the private sector. This is especially true in relation to the issue of climate change.

12.2.3.3   Civil society

Civil society refers to the arena of uncoerced collective action around shared interests, purposes and values (Rayner and Malone, 2000). In theory, its institutional forms are distinct from those of the state, family and market, although in practice, the boundaries between state, civil society, family and market are often complex, blurred and negotiated. Civil society commonly embraces a diversity of spaces, actors and institutional forms, varying in their degree of formality, autonomy and power. Civil societies are often populated by organizations such as registered charities, development non-governmental organizations, community groups, women's organizations, faith-based organizations, professional associations, trades unions, self-help groups, social movements, business associations, coalitions and advocacy groups (Najam, 1996). As this definition emphasizes, civil society is closely related to the more recent concept of 'social capital'. As described by Putnam (1993), social capital describes the overlapping networks of associational ties that bind a society together.

During the past three decades, the mantle of civil society has been increasingly claimed by non-governmental organizations (NGOs). The NGO sector has experienced an explosion in numbers worldwide as well as a proliferation of types and functions. There is considerable debate about the extent to which NGOs claim to be or even represent civil society in the traditional sense can be maintained. Certainly, their dependence on either government or business raises questions about the extent to which they are truly independent of the state and the market. According to The Economist (2000), a quarter of Oxfam's US$ 162 million income in 1998 was given by the British Government and the EU. World Vision US, which claims to be the world's largest privately funded Christian relief and development organization, receives millions of dollars worth of resources from the US Government. The role of governments in supporting NGOs is not limited to financial support. At least one UK-based NGO has advised various small governments in climate negotiations and has even drafted text. Other NGOs are closely associated with the market sector, known as BINGOs (Business and Industry NGOs). A question frequently raised about NGOs is of accountability (Jordan and van Tuijl, 2006). Relatively few NGOs are directly accountable to members in the same way that governments are to voters or businesses are to shareholders, raising further questions about the extent to which their claims to the mantle of civil society are justified (Najam, 1996).

Whether they are truly 'civil society' or not, there is little doubt that NGOs can be effective in shaping development and environment. A multitude of interest groups, including civil society in its various manifestations, seek to influence the direction of national and global climate change mitigation policy (Michaelowa, 1998). Non-governmental organizations have been particularly active and often influential in shaping societal debate and policy directions on this issue (Corell and Betsill, 2001; Gough and Shackley, 2001; Newell, 2000). The literature on the various ways in which civil society, and especially NGOs, influence global environmental policy in general and climate policy in particular, points out that civil society employs 'civic will' to the policy discourse and that it can motivate policy in three distinct but related ways (Banuri and Najam, 2002). First, it can push policy reform through awareness-raising, advocacy and agitation. Second, it can pull policy action by filling the gaps and providing policy services such as policy research, policy advice and, in a few cases, actual policy development. Third, it can create spaces for champions of reform within policy systems so that they can assume a salience and create constituencies for change that could not be mobilized otherwise.

The image of civil society 'pushing' for environmental protection and climate change mitigation policies is the most familiar one. There are numerous examples of civil society organizations and movements seeking to push policy reform at the global, national and even local levels. The reform desired by various interest groups within civil society can differ (Michaelowa, 1998). But common to all is the legitimate role civil society has in articulating and seeking their visions of change through a multitude of mechanisms that include public advocacy, voter education, lobbying decision-makers, research, and public protests. Given the nature of the issue, civil society includes not only NGOs but also academic and other non-governmental research institutions, business groups, and broadly stated the 'epistemic' or knowledge communities that work on better understanding of the climate change problematic. Some have argued that civil society has been the critical element in putting global climate change into the policy arena and relentlessly advocating its importance. Governments have eventually began responding to these calls from civil society for systematic environmental protection and global climate change mitigation policies (Gough and Shackley, 2001; Najam et al.,

2004). In particular, studies on the negotiation processes of global climate change policy (Levy and Newell, 2000; Corell and Betsill, 2001) highlight the role of non-governmental and civil society actors in advancing the cause of global climate change mitigation.

The role of civil society in 'pulling' climate change mitigation policy is no less important. In fact, the IPCC assessment process itself is a voluntary knowledge community seeking to organize the state of knowledge on climate change for policy-makers. It is an example of how civil society, and particularly how 'epistemic' or knowledge communities can directly add to or 'pull' the global climate policy debate (Siebenhuner, 2002; Najam and Cleveland, 2003). In addition, the knowledge communities as well as NGOs have been extremely active and instrumental in servicing the needs of national and sub-national climate policy. This is done in various ways: by universities and research institutions writing local and national climate change plans; by NGOs helping in the preparation of national climate change positions for international negotiations and increasingly being part of the national negotiation delegations (Corell and Betsill, 2001); by civil society and epistemic actors playing key roles in climate change policy assessments at all levels from the local to the global.

Finally, civil society plays a very significant role by 'creating spaces for champions of policy reform' and providing platforms where these champions can advance these ideas. The Pew Climate Initiative and the Millennium Ecosystem Assessment are two examples of how civil society has created forums and space for discourse by different actors, and not just civil society actors, to interact and advance the discussion on where climate change mitigation and sustainable development policy should be heading. Increasingly, civil society forums such as these are very cognisant of the need to broaden the participation in these forums to other institutional sectors of society.

12.2.3.4 Interactions

The shift from 'government to governance' has been accompanied by both theoretical and a practical interest in how the three main arenas of actors – state, market and civil society – interact, including how they might work in concert to achieve improved outcomes from a sustainability perspective. A variety of perspectives are offered that cast light on these questions including 'partnerships' (Najam et al., 2003; Hale, 2004; Forsyth, 2005), 'deliberative democracy' (Levine, 2000; O'Riordan and Stoll-Kleeman, 2002; Gutmann and Thompson, 2004), and 'transition theory' (Geels, 2004; Elzen and Wieczorek, 2005)

Each of these studies considers issues of governance in the context of sustainable development and climate change mitigation. Partnerships considers forms of cooperative governance and action, deliberative democracy deals with issues of representation in decision-making, Transition theory seeks to explain how technological innovation occurs and how these processes might be channelled towards changing the technological composition of development pathways, for example, in support of de-carbonization.

Partnerships: Partnerships between public and private actors can maximize impact by taking advantage of each partner's unique strengths and skill sets. Partnership programmes can provide citizens groups with a lever for increasing pressure on both governments and industry to change in support of improved sustainability. From an economic development perspective, one of the potentially fruitful styles of partnership has been between governments and industry through BOT projects - Build, Operate, Transfer. Despite their promise as a means of financing large-scale capital intensive projects, there have been significant difficulties in practice (see Box 12.4).

Cooperative environmental governance models offer advantages such as a more structured framework for pluralist contributions to policy, consensus-building, more stable policy outcomes, and social learning. Although these cooperative models allow for more stakeholder participation, it is also suggested that they fail to fully address exclusion of minority and less powerful groups, non-representative outcomes, and a failure to integrate local knowledge. An analysis of waste-to-energy projects in the Philippines and India confirms that such problems will be encountered (Forsyth, 2005).

The notion that partnerships between sectors is the wave of the future was given particular salience by the World Summit on Sustainable Development in Johannesburg, South Africa, in 2002. There, several 'Type II' partnerships were launched involving various combinations of governments, business and civil society actors (Najam and Cleveland, 2003; Hale, 2004; Bäckstrand and Lövbrand, 2006). Although too early to evaluate the impacts, these particular partnerships, represent a larger trend in the last decade with a far greater level of partnership activities between governments and NGOs, and between government and business, and now increasingly all three. Such multi-sector forums and partnerships are no longer limited to a few industrialized countries or to particular sectoral mixes. There are now cross-sectoral partnerships and the search for meaningful cross-sectoral partnerships in developing and industrialized countries alike and initiated equally by governments, business and civil society.

Box 12.4: Public Private Partnerships

Globally, public private partnerships (PPPs) are an increasingly popular tool governments use to fund large-scale infrastructure projects. Broadly, PPPs involve the investment of private capital and the use of private sector expertise to deliver public infrastructure and services. There are various forms of PPPs. In the power generation sector, popular examples of PPPs are Build-Operate-Transfer (BOT) projects. Private partners (investors) provide the financing and technology, they build, and they operate the power generation facility for a concessionary period of up to 35 years. During the concession, a government partner provides the investor with ownership rights and gradually buys back the project by providing the developer with the right to charge consumers a fee for its product. At the end of the concession period, the facility is transferred to government ownership at no further cost to the government.

BOT projects have enabled developing country governments with growing energy needs to access new financial capital for green or intermediate fuel technologies for power generation. For example, Vietnam is utilizing such investments for natural-gas fired turbines, and Laos is engaging in a large programme of hydropower construction to supply electricity to a regional power grid in the Greater Mekong Sub region. However, BOT projects have also enabled governments to bring on-line more conventional fossil-fuel powered generating capacity in regions where alternative fuels are not available - heavy oils in some regions of China and coal in Thailand.

While PPPs have assisted governments with access to new financial capital and expertise to invest in cleaner power generating capacity, care needs to be taken in evaluating their costs, benefits and risks to governments and consumers. In uncertain investment environments such as that in developing countries, private partners require a range of onerous guarantees from governments to reduce their investment risks over the life of the projects. These include take-or-pay guarantees where governments commit to purchase a minimum level of production, guarantees to cover currency exchange risks, fuel supply price guarantees, political risk guarantees to protect against government regulatory change. In the aftermath of the East Asian financial crisis that began in 1997, governments such as the Philippines and Indonesia, paid a high price for guaranteed power purchases that were denominated in US dollars as their currencies devalued respectively and power demand from industry dropped.

Sources: Estache and Strong, 2000; Handley, 1997; Irwin et al., 1999; Tam, 1999; Wyatt, 2002.

Deliberative Democracy: According to Pimbert and Wakeford (2001), various social and political factors have brought support to the use of deliberative processes in policy-making, planning and technology assessments. According to Levine (2000), public debate over issues such as global warming provides the public opportunity to form opinions, where otherwise such an opportunity might not exist. Additionally, deliberative processes provide decision-makers with insight into the public mood and, public deliberation provides the opportunity for the public to justify their views on matters of concern.

Notions of deliberative democracy emerge from the observed shift from 'government to governance', in that they refer to shared responsibility for the design of policy. O'Riordan and Stoll-Kleeman (2002) suggest that policy spaces are no longer characterized by hierarchical orders; opportunities have been opened for a variety of forms of public-private cooperation, policy networks, formal and informal consultation, working across scales from multinational to local. The drivers, they suggest, include a need for new approaches to decision-making, occasioned by new mixes of private, public and civic actors.

There are at least five issues that continually challenge social scientists engaged in the design and implementation of participatory mechanisms, such as consensus conferences, focus groups, citizens' juries, and community advisory boards. These are:

* *Representation* – Who and how to select. The challenge is achieving representativeness of a community and establishing the legitimacy of those participating to speak on behalf of others;
* *Resources* – Participatory decision-making requires substantial investment by all parties, chiefly, funding and logistical support on the part of governments and business and time on the part of citizens;
* *Agenda framing* – Too narrow a framing prejudges the issues, but overly broad framing frustrates closure;
* *Effectiveness* – Does citizen involvement have impact on decisions. Disaffection deepens when citizen deliberations are not seen to have traction and people think their time has been wasted;
* *Evaluation* – This is seldom done, and when done, is usually self-evaluation of process rather than of outcomes.

Transition Theory: What can loosely be referred to as 'transition theory' (Elzen and Wieczorek, 2005) offers another perspective on 'society – market – state' relations, but importantly also presents some insights into how societies can shift onto more sustainable paths. Berkhout (2002) observed that energy and climate change policy communities are confronted with a major challenge in the form of shaping a substantially de-carbonized future. This necessitates a better understanding of the links between technologies and the institutions in which they are imbedded (Geels, 2004).

The important questions refer to the factors that impede transitions and, of particular interest to policy communities, how transitions could be induced. Socio-technical systems are often characterized by technological lock-in and path dependency. Actors and organizations become imbedded in interdependent networks and mutual dependencies (Walker, 2000; Berkhout, 2002; Geels, 2004). Elzen and Wieczorek (2005) outline options for inducing innovation under different governance paradigms – the top-down, command-and-control approach (state) a market model, or through policy networks (processes, interactions, networks). Geels (2004) and Smith (2003) approach the same question from a different perspective, both concluding that radical innovations are nurtured in 'niches'. Thus: "Climate change, for instance, is currently putting pressure on energy and transport sectors, triggering changes in technical search heuristics and public policies" (Geels, 2004). Berkhout (2002) offers that substantial commitment is required from governments and businesses to invoke transitions.

While the literature on transition theory is vague on how to induce innovations, such as those that might bring about a shift onto a more sustainable development path. It usefully emphasizes the importance of interactions among actors/organizations, technology, and institutions. For a shift to a more sustainable path, Smith (2003) provides an important reminder that technical change has traditionally occurred in the context of economic growth. Sustainable development, he suggests, implies that "the problem ordering shifts subtly yet profoundly", which will establish new challenges in achieving "publicly managed transitions towards environmentally sustainable technological regimes" (Smith, 2003). In the context of climate change, acknowledged in the literature on transition theory as an impetus for technological innovation, this challenge needs to be addressed; this will require new approaches to the governance of technological change and innovation (Berkhout, 2002; Elzen and Wieczorek, 2005).

12.2.3.5 Policy implications

The discussion above implies that actors and actor coalitions are important and that there is increasing evidence of multi-level patterns of governance and transnational networks of influence on climate change and other global environmental issues. These networks join actors across organizational boundaries; business representatives and environmental non-governmental organization activists may join shareholders, government policy communities and scientists to promote (or stall) action (Haas, 1990; Levy and Newell, 2000; Fairhead and Leach, 2003; Paterson *et al.*, 2003; Biermann and Dingwerth, 2004; Haas, 2004; Levy and Newell, 2005). Also, local and regional governments are increasingly active and may provide an invaluable testing ground and experience with mitigation policy in key areas, such as transportation (Betsill and Bulkeley, 2004; Lindseth, 2004; Bulkeley and Betsill, 2005). This suggests that policy-makers could do a number of things differently to promote understanding of climate change and agreement on policy responses to climate change:

- Create 'policy spaces' for non-state actors, scientists, and experts to interact with government actors; actively facilitate interactions between experts and other stakeholders to build trust, understanding and support for action across a wide range of actors (Ostrom, 1990; Ostrom, 2000; Stern, 2000; Banuri and Najam, 2002; Ostrom *et al.*, 2002). Such activity would provide benefits if built from the bottom-up (building on experience and viewpoints from an increasingly active municipal and community level set of response) and from the top-down (working across elites in government or in scientific/expert and other NGO circles).

- Institutionalize opportunities for public debate and wider interactions within the public sphere on environmental issues (Renn, 2001; Bulkeley and Mol, 2003; De Marchi, 2003; Liberatore and Funtowicz, 2003). By creating the means for dialogue and collaboration to construct understanding about global environmental change, participants have the opportunity to formulate views – talk leads to value formation – which can ultimately generate public support for political action (Dietz and Stern, 1995; Dietz, 2003).

- Encourage and facilitate local action and experimentation – where local communities have the potential to work more closely with affected stakeholders and tailor response strategies to the community's values and norms (Cash and Moser, 2000). Local action on climate change interacts with governance and action taken at different scales (e.g., at national and international levels; Bulkeley and Betsill, 2005).

Domestic policy processes influence international policy opportunities and constraints on climate change (Fisher, 2004). Any domestic policy process will necessarily be working to develop a position with input across the range of actors, for example, market, state, civil society and science/expert communities (Hajer, 1995; SLG, 2001; Fisher, 2004). How this plays out will, to some extent, be influenced by different cultural and social biases in governance at the domestic level (e.g., whether science and business have a privileged role in the policy process; the access and influence of environmental organizations; how coalitions of actors across these groups interact with the policy process). On issues of global environmental change, scientists and other experts necessarily play a privileged role to advise governments (Jasanoff, 1990; Giddens, 1991; Beck, 1992; Yearley, 1994; Jasanoff and Wynne, 1998), forming what Haas (1990; 2004) has referred to as transnational epistemic communities or networks of influence. Given large uncertainties, global environmental change science argues for policy processes that give a central role to public deliberation about the issues – to facilitate common framings about the problem and eventual agreement on responses (Funtowicz and Ravetz, 1993; Hajer and Wagenaar, 2003; Stern and Fineberg, 1996).

Ultimately, devising effective climate change mitigation strategies depends on good governance practices, which is

the essence of sustainable development, for example, whole-of-government decision-making; synergies among economic, environmental and social policies; coalition-building; political leadership; integrated approaches; and policy coherence.

12.2.4 Opportunities at the sectoral level to change development pathways towards lower emissions through development policies

The multiplicity of plausible development paths ahead are underlined in Section 12.2.1, in which low emissions are not necessarily associated with low economic growth (Section 12.2.2). However, the vast literature on governance indicates that changing development pathways can rarely be imposed from the top: it requires the coordination of multiple actors, at multiple scales (Section 12.2.3).

On this basis, examples of opportunities to change development pathways towards lower emissions at the sectoral level are presented in Section 12.2.4. Firstly, opportunities in major sectors are reviewed: energy (Section 12.2.4.1); transportation and urban planning (Section 12.2.4.2); rural development (Section 12.2.4.3); and macro-economy and trade (Section 12.2.4.4). Some general lessons are drawn in Section 12.2.4.5. The potential for action on non-climate policies in major sectors is summarized in Section 12.2.4.4, and some insights on how climate considerations could be mainstreamed into non-climate policies in Section 12.2.4.7.

In reviewing how individual policies not intended for climate mitigation impact GHG emissions, examples are drawn from policies already adopted and implemented, and from forward-looking analysis to estimate the impact of future non-climate policies on emissions. However, few case studies directly analyze the link between a given policy and GHG emissions, and these are mostly in the energy sector.

In fact, assessing the impact of specific policies on GHG emissions, even ex post, is difficult for at least four reasons. First, policy packages usually encompass a wide range of measures, making it difficult to disentangle their individual effects. Second, absent command-and-control policies, or cases in which the emission-producing sectors are directly controlled by governments, public policies are only one of many incentives that decision-makers react to (see also Section 12.2.3). Third, indirect effects of policies on emissions, such as increased demand induced by energy efficiency programmes, are even more difficult to evaluate. And last, there is rarely a control group on the basis of which carbon savings can be evaluated.

To make up for the scarce literature on the relationships between policies and emissions, studies of the relationships between policies and proxies and/or key determinants of GHG emissions are also included in the review, for example, studies linking land-use policies with deforestation rate. This allows examples to be drawn from a wider range of sectors, namely energy, transportation and construction, rural development, as well as from macro-economic and trade policies. The depth of the literature, however, is variable across sectors. Finally, the examples below are intended to discuss the relationships between given policies and GHG emissions, and not the pros and cons of each policy.

12.2.4.1 Energy

The implications of four broad categories of energy policies on emissions are discussed: provision of affordable energy services to the poor; liberalization; energy efficiency; and energy security. Policies that support the penetration of renewable energy - which are often introduced for non-climate reasons, but are also obvious tools for climate mitigation in the energy sector - are discussed in Section 4.5.

Access to Energy: Access to energy is critical for the provision of basic services such as lighting, cooking, refrigeration, telecommunication, education, transportation or mechanical power (Najam *et al.*, 2003). Yet, an estimated 2.4 billion people rely on wood, charcoal or dung for cooking, and 1.6 billion are without access to electricity (IEA, 2004c). Providing access to commercial fuel and efficient stoves would have highly positive impacts on human development by reducing child mortality, improving maternal health, and freeing up time used to collect fuel wood, especially for women and girls (Najam and Cleveland, 2003; Modi *et al.*, 2006). For example, indoor air pollution, mainly from cooking and heating from solid fuels, is responsible for 36% of all lower respiratory infections and 22% of chronic obstructive pulmonary disease (WHO, 2002). See also Chapter 4 and Chapter 6. It is estimated that a shift from crop residues to LPG, kerosene, ethanol gel, or biogas could decrease indoor air pollution by approximately 95% (Smith *et al.*, 2000). The impact on GHG emissions depends on the nature of the biomass resources and the carbon intensity of the replacement. Providing reliable access to electricity would also have highly positive impacts on human development, by providing preconditions for the development of new economic and social activities, for example, allowing for education activities at night and employment generating business initiatives (World Bank, 1994; Karekezi and Majoro, 2002; Spalding-Fecher *et al.*, 2002; Toman and Jemelkova, 2003).

The implications of improved access to commercial fuels for cooking on GHG emissions are ambiguous. On the one hand, emissions increase, albeit by a small amount globally. Smith (2002) estimates that providing LPG as fuel for roughly two billion households would increase global GHG emissions by about 2%. On the other hand, unsustainable use of fuelwood and related deforestation decreases. For example, the 'butanization' programmes adopted in Senegal in 1974 to support LPG use through a combination of subsidies to LPG, support for the development for stoves suitable for local conditions and removal of tax on imported equipment, is estimated to have resulted in a 33-fold increase in LPG use, and in a 15% drop of charcoal consumption (Davidson and Sokona, 2002). Similarly, the implications of electrification programmes for GHG emissions are ambiguous. Energy demand is likely to increase as a result of easier access and induced economic benefits. However, emissions per unit of energy consumed might decrease, depending on the relative carbon content of the fuel used in the baseline (typically kerosene) and of the electricity newly provided (de Gouvello and Maigne, 2000). Public policies have a strong influence on this technology choice. In some cases, the technology is set directly by public decision-makers. But even where left to private entities, public policies, such as the choice between centralized or decentralized models of electrification, or the nature of the fiscal system, strongly constrain technology choices.

One example of such indirect impact is documented by Colombier and Hourcade (1989). They found that the "equal price of electricity for all" principle embedded in French law has generated vast implicit subsidies from urban to rural areas and discouraged, over time, the development of cost-effective decentralized electrification alternatives to grid expansion. The expanded grid the country is locked into, however, is the source of very high maintenance and upgrading costs to accommodate increased demand from rural households and companies – much higher than would have occurred had decentralized solutions been implemented at the onset. The implications for GHG emissions (not studied in the paper) are probably limited given the share of nuclear power in France. But similar dynamics could have more important GHG emissions implications in countries with fossil-fuel dominated power grids.

Liberalization: Many countries have embarked on liberalization of their energy sector over the past two decades. These programmes with the objective to reduce costs and improve efficiency of energy services include privatization of the energy producers, separation between production and transmission activities, liberalization of energy markets, and lifting restrictions on capital flows in the sector. Overall, liberalization programmes aim at improving the efficiency of the energy sector, and should, therefore, lead to reduced emissions per unit of output. Effective privatization programmes, however, differ markedly from country to country (Kessides, 2004), depending on prior institutional arrangements. In addition, privatization programmes are often sequentia. See, for example, Jannuzzi (2005) for a discussion on how the Brazilian regulator progressively adapted policies to elicit sufficient resources for energy efficiency and R&D from private utilities. These policies are often 'incomplete', in the sense that former public power generators remain dominant by combining features from both the public and private sector, an outcome very different from the ideal private energy markets (Victor and Heller, 2007: see also Section 12.2.3.1). It may, therefore, not be surprising that there is little literature drawing general lessons on the implications of privatization programmes on GHG emissions.

A great deal of literature, however, deals with the emission implications of some components of privatization programmes, particularly removal of energy subsidies. Energy subsidies removal may also be adopted as a stand-alone policy, independent from privatization. Conversely, subsidies may remain even within competitive markets. Government subsidies in the global energy sector are in the order of US$ 250-300 billion per year, of which around 2-3% support renewable energy (De Moor, 2001). Removing subsidies on energy has well-documented economic benefits. It frees up financial resources for other uses and discourages overuse of natural resources (UNEP, 2004). But, reducing energy subsidies might have important distributional effects, notably on the poor, if not accompanied by appropriate compensation mechanisms. The impact of policies to reduce energy subsidies on CO_2 emissions is expected to be positive in most cases, as higher prices trigger lower demand for energy and induce energy conservation. For example, econometric analyses have shown that price liberalization in Eastern Europe during the period 1992-1999 was an important driver of the decrease in energy intensity in the industrial sector (Cornillie and Fankhauser, 2002). Similarly, removal of energy subsidies has been identified as instrumental in reducing GHG emissions compared with the baseline in China and India over the past 20 years (Chandler et al., 2002). Overall, an OECD study showed global CO_2 emissions could be reduced by more than 6% and real income increased by 0.1% by 2010, if support mechanisms on fossil fuels used by industry and the power generation sector were removed (OECD, 2002). Yet subsidies removal may actually result in increased emissions in cases where poor consumers are forced off-grid and back to highly carbon intensive fuels, such as non-sustainable charcoal or diesel generators. For example, removal of the subsidies for LPG in Senegal under the 'butanization' programmes discussed above is expected to increase charcoal and unsustainable fuelwood use (Deme, 2003). For additional discussion on energy subsidies, see Section 4.5.1 and Section 6.8.3.2 and Section 13.2.1.5.

Energy Efficiency: Policies that increase energy efficiency – both on the demand and on the supply side – are pursued to reduce demand for energy without affecting, or while increasing, output at very low costs. This is the case even though some of the direct efficiency gains might be offset by increased demand due to lower energy costs per unit of output. Efficiency also increases competitiveness, relaxes supply constraints and, therefore, enhances the range of policy options and space, and lowers expenditure on energy thereby freeing up more resources for other development goals. The impact on CO_2 emissions, in turn, tends to be positive, but depends heavily on the carbon content of the energy supply. For example, Brazil National Electricity Conservation Program (PROCEL), created in 1985, has saved an estimated 12.9 TWh and an estimated R$ 2.6 billion from 1986 to 1997. This is 25 times as much

as the amount invested in the programmes, while reducing emissions by an estimated 3.6 Mt CO_2 over the same period of time (La Rovere and Americano, 1999; Szklo et al., 2005). Similarly, Palmer et al. (2004) estimate that the annual energy savings generated by all current Demand-Side Management programmes (DSM) in the USA represent about 6% of the country's non-transportation energy consumption. This leads to reductions in CO_2 emissions equivalent to (at most) 3.5% of the country's total. DSM programmes are also discussed in Section 6.8.3.1 and Section 5.5.1.Over the period 1973-1998, the International Energy Agency (IEA, 2004b) estimates that energy efficiency - driven both by policies and by autonomous technical improvements - have resulted in energy savings corresponding to almost 50% of 1998 energy consumption levels. Without these savings, energy use (and CO_2 emissions) in 1998 would have been almost 50% higher than observed.

Energy Security: Energy security is broadly defined as ensuring long-term security of energy supply at reasonable prices to support the domestic economy. This is a major concern for Governments worldwide, and it has taken new prominence in recent years with the political instability in the Middle East, increased oil prices, and tensions over gas in Europe (Dorian et al., 2006; Turton and Barreto, 2006). Energy security concerns, however, can translate into very different policies depending on national and historical circumstances (Helm, 2002). Their impact on carbon emissions is ambiguous, depending on the nature of the policies and, in particular, on the fuel sources being favoured. For example, in response to the first oil shock, Brazil launched in 1975 the National Alcohol Fuel Program (PRO-ALCOOL) to increase the production of sugarcane ethanol as a substitute for oil, at a time when Brazil was importing about 80% of its oil supply[5]. The programmes resulted in reduction of oil imports and expenditure of foreign currency and job creation, as well as in large emission reductions, estimated at 1.5 Mt CO_2/yr (Szklo et al., 2005). Brazil also provides an example where emissions actually increased as a result of energy security considerations. During the 1990s, Brazil faced lack of public and private investment in the expansion of the power system (both generation and transmission) and a growing supply-demand imbalance, which culminated in electricity shortage and rationing in 2001. This situation forced the country to install and run emergency fossil-fuel plants, which led to a substantial increase in GHG emissions from the power sector in 2001 (Geller et al., 2004). Hourcade and Kostopoulo (1994) show how reactions to the first oil shock by France, Italy, Germany, and Japan led to very different emissions with relatively similar economic outcomes (see Box 12.5).

---

5   PRO-ALCOOL was also a way of assisting the domestic sugar industry at times of low international sugar prices.

Box 12.5: Differentiated reactions to the first oil shock in France, Italy, Germany and Japan

An example of how different development paths can unfold in relatively similar countries is given by Hourcade and Kostopou-lou (1994) for France, Italy, Germany, and Japan - countries with similar levels of GDP per capita in 1973 – in their response to the first oil shock. France moved aggressively to develop domestic supply of nuclear energy and a new building code. Japan made an aggressive shift of its industry towards less energy-intensive activities and simultaneously used its exchange-rate policies to alleviate the burden of oil purchases. Germany built industrial exports to compensate the trade balance deficit in the energy sector. Much of the variations of CO_2 emissions per unit of GDP from 1971 to 1990 can be attributed to these choices (Figure 12.1 left). Yet, while this indicator diminished by half in France, by a third in Japan, and 'only' by a quarter in Germany (IEA 2004b).

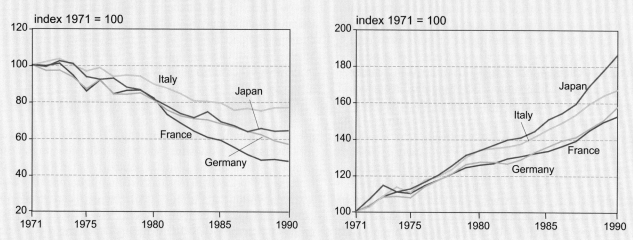

Figure 12.1: *(left) CO_2 emissions from fossil-fuel combustion per unit of GDP; (right) Evolution of GDP per capita*
Note: GDP in US$ in constant prices at market exchange rates.
Source: IEA, 2004a

Hourcade and Kostopoulou (1994) also observe that the macro-economic performance of these countries was relatively comparable between 1973 and 1990 (Figure 12.1 right), suggesting that widely different environmental outcome can be obtained at similar welfare costs in the long-run. In addition, they observe that the responses were for a large part driven by the country's pre-existing technologies and institutions (thus providing an illustration of the general observations about deci-sion-making made in Section 12.2.3.1).

12.2.4.2 Transportation and urban planning

Transportation is a key development issue. Transportation is also one of the fastest growing end-use sectors in terms of CO_2 emissions in both the developed and the developing world. The level of these emissions results from the combination of the distances travelled by goods and people, and the set of technologies used to make these journeys. Demand for and supply of transportation are largely inelastic in the short-term, but become elastic in the longer run as people and activities change location, as new infrastructure is developed and as preferences evolve. A very wide array of policies affects these long-term dynamics. The set of transportation technologies available at time, and their relative costs, are also influenced by public policies.

Three examples of how public policy choices affect transportation supply, transportation demand, technology, and

ultimately emissions from the transport sector are discussed in this section: one of congestion policy, one of urban planning at city level, and the other of national policy driving urban planning. The first example is from the City of London, where a congestion charge was introduced in February 2003 to reduce congestion. Simultaneously, investment in public transport was increased. Early results suggest that congestion in the charging zone has reduced by 30% during the charging hours, that CO_2 emissions have been reduced by 20%, and that primary emissions of NO_x and PM_{10} have been reduced by 16% (Transport for London, 2005). However, the cost-benefit ratio of the operation is questioned (Prud'homme and Bocarejo, 2005; Santos and Fraser, 2006). Other examples of how non-climate transportation policies can impact on emissions are given in Section 5.5.

The second example is the development and steady implementation of an integrated urban planning programme in the city of Curitiba (Brazil) from 1965 onwards. This has allowed the city to grow eight-fold from 1950 to 1990, while maintaining 75% of commute travel by bus – a much higher public transport modal share than in other big Brazilian cities (57% in Rio, 45% in São Paulo) – as well as little congestion. As a result, Curitiba uses 25% less fuel than cities of similar population and socio-economic characteristics. Two characteristics of the programmes seem to have contributed particularly to its success: (i) integration of infrastructure and land-use planning; and (ii) the consistency with which successive municipal administrations have implemented the plan over nearly three decades (Rabinovitch and Leitman, 1993).

The third example concerns urban planning in the United States and Europe (and Japan), the latter being on average rather compact while the former exhibit important sprawl. Nivola (1999) notes that this difference cannot be explained only by differences in demography, geography, technology or income. He argues that the combination of public choices is responsible for most of the differences in urban sprawl between the USA and Europe. Such policies include: (1) a bias towards public financing of roads to the detriment of other modes of transportation in the USA - against a more balanced approach in Europe; (2) dedicated revenues for highway construction in the USA - against funds drawn from general revenues in Europe; (3) lower taxes on gasoline in the USA than in Europe; (4) housing policies more geared towards supporting new homes: (5) a tax system more in favour of homeowners in the USA; (6) lower support from the federal government to local governments in the USA than in Europe; and (7) the quasi-absence of regulations favouring small in-city outlets against shopping malls in the USA. In turn, this difference in urban planning generates widely different transport demand, energy consumption (Newman and Kenworthy, 1991), and CO_2 emissions. Per capita CO_2 emissions from travel in the USA are nearly three times as high as in major European countries, due mostly to a larger number of journeys per capita and a higher energy intensity (Schipper et al., 2001). A key point in the analysis made by Nivola (1999) is that most of these consequences were totally unintended, as these policies were adopted for non-transportation reasons (let alone for emissions reasons).

12.2.4.3   Agriculture and forestry

Vast arrays of policies affect the emissions of the agriculture and forestry sectors, and the emissions or the sequestration rate from biomass and soils. An extensive list of non-climate policies that impact on CO_2, CH_4 and N_2O emissions from the agriculture sector are presented in Chapter 8 (Tables 8.10 and 8.11). The list includes sectoral policies designed to reach environmental goals other than climate change, such as biodiversity conservation or watershed protections; agricultural policies designed to reach non-environmental goals, such as increasing exports of agricultural products or securing farmers' income; and non-agricultural policies with impact on the agriculture sector, such as energy price reforms. For example, the 2003 EU Common Agricultural Policies reform, by decoupling subsidies from production targets, is likely to lead to reduced on-farm CO_2 and N_2O emissions (see Table 8.10). In fact, changes in the Common Agricultural Policy from 1997 to 2001 (in intervention prices, in per-hectare support to grains and oilseeds, as in milk quotas and livestock subsidies) are estimated to have resulted in a 4% decline of agricultural sector emissions in Europe over that period (De Cara et al., 2005).

If the direct emissions of the forestry sector are small, the emissions/uptake related to land conversion from/to forests are extremely large (see Chapter 9). In addition, emissions/uptake related to changes in the quality of existing forests, to the use of forest products in carbon stocks, and to bioenergy are very large. Policies affecting land use and land-use change, policies affecting the substitution between wood-based and other products, and policies related to bioenergy are thus likely to have strong implications for the net emissions from forests and forest products.

The causes of deforestation have been studied specifically. They differ from regions to regions and depend on the interaction of cultural, demographic, economic, technological, political and institutional issues (e.g., Angelsen and Kaimowitz, 1999; Geist and Lambim, 2002). In all cases, the drivers of deforestation are strongly affected by policy decisions. For example, rural road construction or improvement tend to encourage future deforestation (Chomitz and Gray, 1996; Chomitz, 2007), yet may have positive economic implications by providing better access to markets and basic services for remote population in developing countries (Jacoby, 2000). Similarly, agriculture intensification policies have potentially important but ambiguous effects on deforestation. On the one hand, intensification increases the productivity of existing agricultural land and lowers the pressure on forests. On the other, it could also trigger migration and it might, in fact, increase deforestation. Careful design of agriculture intensification policies is thus necessary to avoid unintended outcome on deforestation (Angelsen and Kaimowitz, 2001).

A third example concerns a macro-economic policy decision: the devaluation of Brazil currency in 1999, which fell by 50% against the US dollar. Coupled with an increase of soybean prices on the international market, increased the value of soybean and beef production in the country - notably in the state of Mato Grosso – triggered massive increase in production and massive deforestation of cerrado forests. In fact, a third of total deforestation in the Brazilian Amazon between 1999 and 2003 occurred in Mato Grosso (Chomitz, 2007).

12.2.4.4   Macro-economy and trade

Macro-economic policies such as exchange rate policies, fiscal policies, government budget deficits, or trade policies may have profound impacts on the environment, even though they are designed for other purposes. This link has been extensively studied in the past decades, notably in the context of the evaluation of structural adjustment programmes in developing countries. A key finding from this literature is that the relationship between macro-economic policies and the environment are often complex and country-specific, and depend on whether or not other market or institutional imperfections persist (Munasinghe and Cruz, 1995; Gueorguieva and Bolt, 2003). No case studies discuss the impact of structural adjustment on GHG emissions, but some discuss the relationship between structural adjustment and deforestation and thus, by extension, GHG emissions from land-use change. Again, the effects depend on the mix of policies adopted as part of the structural adjustment programmes, and of country-specific characteristics. For example, Kaimowitz et al. (1999) report that the structural adjustment programmes implemented in Bolivia in 1985 strongly increased profitability of soybean production, and led to massive deforestation in soybean producing areas. Symmetrically, Benhin and Barbier (2004) find that a structural adjustment programmes implemented in Ghana in 1983 led to a reduction of deforestation linked to extension of cocoa culture because, among others, of increased producer price for cocoa, higher availability of inputs, and other measures aimed at rehabilitating existing cocoa farms. Another channel through which structural adjustment programmes could impact on deforestation is through the timber market. Pandey and Wheeler (2001) analyse cross-country data on the markets for wood products in countries where World Bank supported adjustment programmes were implemented. They find that these programmes greatly affect imports, exports, consumption and production in many forest product sectors, but that the impacts on deforestation tend to cancel out. If domestic deforestation does not increase, however, imports of wood products do, suggesting increased pressures on forest in other countries. Finally, as also noted above, Pandey and Wheeler (2001) find that currency devaluation strongly increases the exploitation of forest resources.

Among macro-economic policies, trade policies have attracted particular attention in recent years, due to the fact that international trade has increased dramatically over the past decades. There is a general consensus that, in the long-run, openness to trade is beneficial for economic growth. However, the pace of openness, and how to cope with social consequences of trade policies are subject to much controversy (Winters et al., 2004). Trade has multiple implications for GHG emissions. First, increased demand for transportation of goods and people generates emissions. For example, freight transport now represents more than a third of the total energy use in the transportation sector (see Section 5.2.1). Secondly, trade allows countries to partially 'de-link' consumption from emissions, since some goods and services are produced abroad, with opposite implications for the importing and exporting countries. For example, Welsch (2001) shows that foreign demand for German goods accounts for nearly a third of the observed structural changes in the composition of output and decrease in emissions intensity of West Germany over the period 1985-

1990. At the other end, Machado et al. (2001) report that inflows and outflows of carbon embodied in the international trade of non-energy goods in Brazil accounted for some 10% and 14%, respectively, of the total carbon emissions from energy use of the Brazilian economy in 1995. And the game is often not zero-sum, when production technologies are less carbon-efficient in the exporting country than in the importing one. For example, Shui and Harriss (2006) estimate that USA-China trade represents between 7% and 14% of China's total CO_2 emissions, and that USA-China trade increases world emissions by an average of 100 $MtCO_2$-eq per year over the period 1997-2003 because of higher emissions per kWh and less efficient manufacturing technologies in China. Finally, policies favourable to trade have been accused of favouring the relocation of companies to 'pollution heavens' where environmental constraint would be lower. Empirical analysis, however, do not confirm the 'race to the bottom' hypothesis (Wheeler, 2001). See also Section 11.7.

12.2.4.5 Some general insights on the opportunities to change development pathways at the sectoral level

Although the examples discussed above are very diverse, some general patterns emerge. First, in any given country, sectors where effective production is far below the maximum feasible production with the same amount of inputs – sectors that are far away from their production frontier – have opportunities to adopt 'win-win-win' policies. Such policies free up resources and bolster growth, meet other sustainable development goals, and also, incidentally, reduce GHG emissions relative to baseline. Among the examples discussed above, the removal of energy subsidies in economies in transition, or the mitigation of urban pollution in highly polluted cities in the developing world pertain to the 'win-win-win' category. Of course, these policies may have winners and losers, but compensation mechanisms can be designed to make no-one worse off in the process.

Conversely, sectors where production is close to the optimal given available inputs – sectors that are closer to the production frontier – also have opportunities to reduce emissions by meeting other sustainable development goals. However, the closer to the production frontier, the more trade-offs are likely to appear. For example, as discussed above, diversifying energy supply sources in a country where the energy system is already cost-efficient might be desirable for energy security reasons and/or for local or global environmental reasons. But it might come at a cost to the country if, for example, diversification involves more expensive technologies or more risky investments (Dorian et al., 2006).

Third, in many of the examples reviewed above, what matters is not only that a 'good' choice is made at a certain time, but also that the initial policy has persisted for a long period – sometimes several decades – to truly have effects. The comparison between the development of European and USA cities since the end of World War II is a case in point. The reason is that some of the key dynamics for GHG emissions, such as technological development or land-use patterns, present a lot of inertia, and thus need sustained effort to be re-oriented. This raises deep institutional questions about the possibility of governments to make credible long-term commitments, particularly in democratic societies where policy-makers are in place only for short spans of time (Stiglitz, 1998).

A fourth element that stems from some of the examples outlined above is that often not one policy decision but an array of decisions are necessary to influence emissions. This is especially true when considering large-scale and complex dynamics such as the structure of cities or the dynamics of land-use. This raises, in turn, important issues of coordination between policies in several sectors, and at various scales.

Fifth, as already emphasized in Section 12.2.3, institutions are significant in determining how a given policy or a given set of policies ultimately impact on GHG emissions (World Bank, 2003). For example, the differentiated reactions of Japan, Italy, Germany and France to the first oil shock can be traced to differences in institutions, relative power of different influence groups, and political cultures (Hourcade and Kostopoulou, 1994).

12.2.4.6   Mainstreaming climate change into development choices: Setting priorities

As highlighted in Sections 12.2.4.1 to 12.2.4.5, development policies in various sectors can have strong impacts on GHG emissions. The operational question is how to harness that potential. How can climate change mitigation considerations be mainstreamed into development policies.

Mainstreaming means that development policies, programmes and/or individual actions that otherwise would not have taken climate change mitigation into consideration explicitly include these when making development choices. This makes development more sustainable.

The ease or difficulty with which mainstreaming is accomplished will depend on both the mitigation technology or practice, and the underlying development path. No-regrets energy efficiency options, for instance, are likely to be easier to implement (and labelled as climate change mitigation actions) than others that have higher direct cost, require coordination among stakeholders, and/or require a trade-off against other environmental, and social and economic benefits. Weighing other development benefits against climate benefits will be a key basis for choosing development sectors for mainstreaming climate change considerations. In some cases, it may even be rational to disregard climate change considerations because of an action's other development benefits (Smith, 2002).

Development policies, such as electricity privatization, can increase emissions if they result in construction of natural gas power plants in place of hydroelectric power for instance, but they can reduce emissions if coal power plants are not built. Judicious and informed choices will be needed when pursuing development policies in order to ensure that GHG emissions are reduced and not increased (see above). This section considers which sectors should receive priority for mainstreaming climate change mitigation into development choices; what sectors are better off not pursuing mainstreaming; and which stakeholders might have a bigger stake and voice in mainstreaming. The next section considers concrete ways to mainstream mitigation considerations into development choices.

Prioritizing requires that the current and future associated emissions of the targeted sector and the mitigation potential of the non-climate sustainable development action be estimated. Policy-makers can then weigh the emissions reduction potential against other sustainability aspects of the action in choosing the appropriate policy to implement. In order to implement such an approach, empirical analyses are needed to estimate future associated emissions and current and future mitigation potential of development actions. Few, if any, global analyses provide complete guidance of this type. In light of the lack of empirical analyses, associated emissions for selected sectors in which development actions may be pursued are presented. This provides an initial guide in ranking sustainable development actions. A more complete analysis is needed, however, which would require the estimation of future associated emissions, and current and future mitigation potential of sustainable development actions.

Selected examples of CO_2 emissions associated with sectors where sustainable development actions could be implemented are presented in Table 12.3. These are described below:

Emissions associated with selected sectors:
- *Macro-economy:* Through fiscal tax and subsidy policies, public finance can play an important role in reducing emissions. Rational energy pricing based on long-run-marginal-cost principle can level the playing field for renewables, increase the spread of energy efficient and renewable energy technologies, improve the economic viability of utility companies, and can reduce GHG emissions. Non-climate taxes/subsidies and other fiscal instruments can impact the entire global fossil fuel emissions of CO_2, which amounted to about 51 GtCO$_2$-eq in 2004. Those that directly reduce fossil fuel use could be easily relabelled and mainstreamed as climate taxes, but others, for example a tax on water use, would need to be evaluated for their fossil fuel impacts and climate benefits.
- *Forestry:* Adoption of forest conservation and sustainable forest management practices can contribute to conservation of biodiversity, watershed protection, rural employment generation, increased incomes to forest dwellers and carbon sink enhancement. The forestry sector emissions show a high and low range to signal the uncertainty in estimates of deforestation. A best estimate value is about 7% of global emissions in 2004 (see Table 12.3). There are many country-

Table: 12.3: *Mainstreaming climate change into development choices - selected examples*

Selected sectors	Non-climate policy instruments and actions that are candidates for mainstreaming	Primary decision-makers and actors	Global greenhouse gas emissions by sector that could be addressed by non-climate policies (% of global GHG emissions)[a,d]	Comments
Macro-economy	Implement non-climate taxes/subsidies and/or other fiscal and regulatory policies that promote sustainable development	State (governments at all levels)	100	Total global GHG emissions	Combination of economic, regulatory, and infrastructure non-climate policies could be used to address total global emissions
Forestry	Adoption of forest conservation and sustainable management practices	State (governments at all levels) and civil society (NGOs)	7	GHG emissions from deforestation	Legislation/regulations to halt deforestation, improve forest management, and provide alternative livelihoods can reduce GHG emissions and provide other environmental benefits
Electricity	Adoption of cost-effective renewables, demand-side management programmes, and transmission and distribution loss reduction	State (regulatory commissions), market (utility companies) and, civil society (NGOs, consumer groups)	20[b]	Electricity sector CO_2 emissions (excluding auto producers)	Rising share of GHG-intensive electricity generation is a global concern that can be addressed through non-climate policies
Petroleum imports	Diversifying imported and domestic fuel mix and reducing economy's energy intensity to improve energy security	State and market (fossil fuel industry)	20[b]	CO_2 emissions associated with global crude oil and product imports	Diversification of energy sources to address oil security concerns could be achieved such that GHG emissions are not increased
Rural energy in developing countries	Policies to promote rural LPG, kerosene and electricity for cooking	State and market (utilities and petroleum companies), civil society (NGOs)	<2[c]	GHG emissions from biomass fuel use, not including aerosols	Biomass used for rural cooking causes health impacts due to indoor air pollution, and releases aerosols that add to global warming. Displacing all biomass used for rural cooking in developing countries with LPG would emit 0.70 $GtCO_2$-eq., a relatively modest amount compared to 2004 total global GHG emissions
Insurance for building and transport sectors	Differentiated premiums, liability insurance exclusions, improved terms for green products	State and market (insurance companies)	20	Transport and building sector GHG emissions	Escalating damages due to climate change are a source of concern to insurance industry. Insurance industry could address these through the types of policies noted here
International finance	Country and sector strategies and project lending that reduces emissions	State (international Financial Institutions) and market (commercial banks)	25[b]	CO_2 emissions from developing countries (non-Annex 1)	IFIs can adopt practices so that loans for GHG-intensive projects in developing countries that lock-in future emissions are avoided

Notes:
a. Data from Chapter 1 unless noted otherwise.
b. CO_2 emissions from fossil fuel combustion only; source: IEA, 2006.
c. CO_2 emissions only. Authors estimate, see text.
d. Emissions indicate the relative importance of sectors in 2004. Sectoral emissions are not mutually exclusive and may overlap.

specific studies of the potential to reduce deforestation (Chapter 9).

- *Electricity:* Adoption of cost-effective energy efficiency technologies in electricity generation, transmission distribution, and end-use reduce costs and local pollution in addition to reduction of greenhouse gas emissions. Electricity deregulation or privatization can be practised in any country and can impact the global electricity-related emissions which amounted to about 20% of global emissions.

- *Oil import security:* Oil import security is important to ensure reliable supply of fuels and electricity. Diversification of oil imports, through increasing imported and domestic sources oil and other energy carriers is an approach adopted by countries concerned about energy security. The percentage of net oil imports serves as one indicator of a country's energy security. The CO_2 emissions associated with net oil imports amounted to about 20% of global emissions (see Table 12.3). Reducing oil imports as a strategy to improve energy security thus offers a significant global opportunity to reduce emissions. Minimizing the use of coal as a substitute, and increasing use of less-carbon-intensive energy sources and reducing energy intensity of the economy are options that could be pursued to achieve this goal (IEA, 2004b). However, heavy use of biomass as a fossil fuel substitute may compete with other societal goals such as food security, alleviation of hunger and conservation of biodiversity.

Example of a sector where other benefits outweigh mainstreaming:

- *Rural household energy use:* Development of rural regions, better irrigation and water management, rural schools, better cook stoves in developing countries can promote sustainable development. The emissions associated with rural household activities, mostly derived from energy needed for cooking and some heating, are relatively small, however. These emissions are estimated to be between 10% and 15% of developing-country residential sector emissions or less than 0.5% of global emissions. Rural areas of developing countries rely primarily on traditional bioenergy[6] and consume comparatively small amounts of fossil fuels. The use of improved cook stoves is one way to reduce biomass and fossil fuel use. The worldwide amount estimated by Smith (2002) for provision of LPG as fuel for roughly two billion households is about 2% of global GHG emissions. From a global perspective, Table 12.3 suggests that smaller sectors with significant other welfare benefits need not be burdened with having to reduce CO_2 emissions since larger gains from sustainable development actions that address climate change mitigation are to be had elsewhere.

Emissions that key stakeholders can influence:

- *International finance:* While climate change mitigation is an important component of the multilateral bank (MDB) strategies, in practice climate change issues are not systematically incorporated into lending for all sectors. MDBs could explicitly integrate climate change considerations into their guidelines for country and sector strategies, and apply a greenhouse gas accounting framework in their operations (Sohn *et al.*, 2005). MDBs can directly influence their own lending and indirectly influence the emissions of borrowing countries. The annual emissions from World Bank-funded energy activities alone, for instance, were estimated to range from 0.27 to 0.32 $GtCO_2$ (World Bank, 1999). MDBs could directly influence more than the aforementioned amounts once emissions associated with all lending activities of all MDBs are counted. Indirectly, through policy dialogue and conditionality, MDBs could influence additional emissions from developing countries, which amounted to about 25% of global emissions in 2004 (Table 12.3).

- *Insurance:* Buildings and transport vehicles form the bulk of the insured activities. Emissions from these sectors and from all international marine vessels and aircraft are estimated to be about 20% of global emissions, giving insurers a significant potential role in controlling emissions. Some insurers are beginning to recognize climate-change risks to their business (Vellinga *et al.*, 2001; Mills, 2005). Examples of actions may include premiums differentiated to reflect vehicle fuel economy (this is not unique to the buildings and/or transport sector or distance driven); liability insurance exclusions for large emitters; improved terms to recognize the lower risks associated with green buildings; or new insurance products to help manage technical, regulatory, and financial risks associated with emissions trading (Mills, 2003).

12.2.4.7 Operationalization of mainstreaming

Though there is a considerable amount of literature on how development policies are made (see Section 12.2.3), there is currently very limited literature on how climate mitigation considerations could be mainstreamed into development policies. Based on a number of Indian case studies on integrating climate change mitigation in local development, Heller and Shukla (2003) note operational guidelines which can integrate development and climate policies into the future development pathways of developing countries. In developing countries, which by and large have not yet enacted domestic GHG legislation, the Clean Development Mechanism can play a role as one component of national GHG reduction strategies and sustainable development.

---

6   Bioenergy use is assumed to be GHG emissions neutral.

Based on a United States Environmental Protection Agency (US EPA, 2006) report on best practices for implementation of clean energy policies and programmes, Sathaye *et al.* (2006) conclude that following best practices would benefit the operationalization process: (a) commitment of publicly elected and/or regulatory bodies; (b) involvement and support of key stakeholders; (c) sound economic and environmental analyses conducted using simple and transparent tools; (d) longer time frames for programmes so that they can overcome market and funding cycles; (e) setting annual and cumulative targets to gage progress of mainstreaming; (f) ensuring additionality over and above existing and other planned programmes; (g) selection of an effective entity for implementation; (h) education and regular training of key participants; (i) monitoring and evaluation of mainstreaming results; and (j) maintenance of a functional database on a project's or programme's sustainable development performance.

A study of the Baltic region explores a sustainable development pathway addressing broad environmental, economic and social development goals, including low GHG emissions. A majority of the population could favour - or at least tolerate - a set of measures that change individual and corporate behaviours to align with local and global sustainability (Raskin *et al.*, 1998). Kaivo-oja (2004) concludes that climate change as such may not be a major direct threat to Finland, but the effects of climate change on the world's socio-economic system and the related consequences for the Finnish system may be considerable. The Finnish scenario analysis, which is based on intensive expert and stakeholder involvement, suggests that such indirect consequences have to be taken into account in developing strategic views of possible future development paths for administrative and business sectors.

12.3 Implications of mitigation choices for sustainable development goals

The evolution of the concept of sustainable development with emphasis on its two-way linkage to climate change mitigation is discussed in Section 12.1, and the link between the role of development paths and actors or stakeholders that could make development more sustainable by taking climate change into consideration is explored in Section 12.2. The reverse linkages are summarized in Section 12.3, and the literature on impacts of climate mitigation on attributes of sustainable development is assessed.

The sectoral chapters (Chapters 4–11) provide an overview of the impacts of the implementation of many mitigation technologies and practices that are being or may be deployed at various scales in the world. In this section, the information from the sectoral chapters is summarized and supplemented with findings from the sustainable development literature. Synergies with local sustainable development goals, conditions for their successful implementation, and trade-offs where the climate mitigation and local sustainable development may be at odds with each other are discussed (see overview Table 12.4). In addition, the implications of policy instruments on sustainable development goals are described in Section 12.3.5, with the focus on the Clean Development Mechanism (CDM).

As documented in the sectoral chapters, mitigation options often have positive effects on aspects of sustainability, but may not always be sustainable with respect to all three dimensions of sustainable development - economic, environmental and social. For example, removing subsidies for coal increases its price and creates unemployment of coal mine workers, independently of the actual mitigation (IPCC, 2001). In some cases, the positive effects on sustainability are more indirect, because they are the results of side-effects of reducing GHG emissions. Therefore, it is not always possible to assess the net outcome of the various effects.

The sustainable development benefits of mitigation options vary over sectors and regions. Generally, mitigation options that improve productivity of resource use, whether it is energy, water, or land, yield positive benefits across all three dimensions of sustainable development. In the agricultural sector (Table 8.8), for instance, improved management practices for rice cultivation and grazing land, and use of bioenergy and efficient cooking stoves enhance productivity, and promote social harmony and gender equality. Other categories of mitigation options have a more uncertain impact and depend on the wider socio-economic context within which the option is being implemented.

Some mitigation activities, particularly in the land use sector, have GHG benefits that may be of limited duration. A finite amount of land area is available for forestation, for instance, which limits the amount of carbon that a region can sequester. And, certain practices are carried out in rotation over years and/ or across landscapes, which too limit the equilibrium amount of carbon that can be sequestered. Thus, the incremental sustainable development gains would reach an equilibrium condition after some decades, unless the land yields biofuel that is used as a substitute for fossil fuels.

The sectoral discussion below focuses on the three aspects of sustainable development - economic, environmental, and social. Economic implications include costs and overall welfare. Sectoral costs of various mitigation policies have been widely studied and a range of cost estimates are reported for each sector at both the global and country-specific levels in the sectoral Chapters 4 to 10. Yet, mitigation costs are just one part of the broader economic impacts of sustainable development . Other impacts include growth and distribution of income, employment and availability of jobs, government fiscal budgets, and competitiveness of the economy or sector within a globalizing market.

Table 12.4: *Sectoral mitigation options and sustainable development (economic, local environmental and social) considerations: synergies and trade-offs [a)]*

Sector and mitigation options	Potential sustainable development synergies and conditions for implementation	Potential sustainable development trade-offs
Energy Supply and Use: Chapters 4-7		
Energy efficiency improvement in all sectors (buildings, transportation, industry, and energy supply: Chapters 4-7)	- Almost always cost-effective, reduces or eliminates local pollutant emissions and consequent health impacts, improves indoor comfort and reduces indoor noise level, creates business opportunity and jobs, and improves energy security - Government and industry programmes can help overcome lack of information and principal agent problems - Programmes can be implemented at all levels of government and industry - Important to ensure that low-income household energy needs are given due consideration, and that the process and consequences of implementing mitigation options are, or the result is, gender-neutral	- Indoor air pollution and health impacts of improving biomass cook stove thermal efficiency in developing country rural areas are uncertain
Fuel switching and other options in the transportation and buildings sectors (Chapters 5 and 6)	- CO_2 reduction costs may be offset by increased health benefits - Promotion of public transport and non-motorized transport has large and consistent social benefits - Switching from solid fuels to modern fuels for cooking and heating indoors can reduce indoor air pollution and increase free time for women in developing countries - Institutionalizing planning systems for CO_2 reduction through coordination between national and local governments is important for drawing up common strategies for sustainable transportation systems	- Diesel engines are generally more fuel-efficient than gasoline engines and thus have lower CO_2 emissions, but increase particle emissions - Other measures (CNG buses, hybrid diesel-electric buses and taxi renovation) may provide little climate benefits
Replacing imported fossil fuel with domestic alternative energy sources (DAES: Chapter 4)	- Important to ensure that DAES is cost-effective - Reduces local air pollutant emissions. - Can create new indigenous industries (e.g., Brazil ethanol programme) and hence generate employment	- Balance of trade improvement is traded off against increased capital required for investment - Fossil-fuel-exporting countries may face reduced exports - Hydropower plants may displace local populations and cause environmental damages to water bodies and biodiversity
Replacing domestic fossil fuel with imported alternative energy sources (IAES: Chapter 4)	- Almost always reduces local pollutant emissions - Implementation may be more rapid than DAES - Important to ensure that IAES is cost-effective - Economies and societies of energy-exporting countries would benefit	- Could reduce energy security - Balance of trade may worsen but capital needs may decline
Forestry Sector: Chapter 9		
Afforestation	- Can reduce wasteland, arrest soil degradation, and manage water runoff - Can retain soil carbon stocks if soil disturbance at planting and harvesting is minimized - Can be implemented as agro-forestry plantations that enhance food production - Can generate rural employment and create rural industry - Clear delineation of property rights would expedite implementation of forestation programmes	- Use of scarce land could compete with agricultural land and diminish food security while increasing food costs - Monoculture plantations can reduce biodiversity and are more vulnerable to diseases - Conversion of floodplain and wetland could hamper ecological functions
Avoided deforestation	- Can retain biodiversity, water and soil management benefits, and local rainfall patterns - Reduce local haze and air pollution from forest fires - If suitably managed, it can bring revenue from ecotourism and from sustainably harvested timber sales - Successful implementation requires involving local dwellers in land management and/or providing them alternative livelihoods, enforcing laws to prevent migrants from encroaching on forest land	- Can result in loss of economic welfare for certain stakeholders in forest exploitation (land owners, migrant workers) - Reduced timber supply may lead to reduced timber exports and increased use of GHG-intensive construction materials - Can result in deforestation with consequent sustainable development implications elsewhere
Forest Management	- See afforestation	- Fertilizer application can increase N_2O production and nitrate runoff degrading local (ground)water quality - Prevention of fires and pests has short term benefits but can increase fuel stock for later fires unless managed properly

Table 12.4. Continued.

Sector and mitigation options	Potential sustainable development synergies and conditions for implementation	Potential sustainable development trade-offs
Bioenergy (Chapter 8 and 9)		
Bioenergy production	- Mostly positive when practised with crop residues (shells, husks, bagasse, and/or tree trimmings) - Creates rural employment - Planting crops/trees exclusively for bioenergy requires that adequate agricultural land and labour is available to avoid competition with food production	- Can have negative environmental consequences if practised unsustainably - biodiversity loss, water resource competition, increased use of fertilizer and pesticides - Potential problem with food security (location specific) and increased food costs
Agriculture: Chapter 8		
Cropland management (management of nutrients, tillage, residues, and agro-forestry) Cropland management (water, rice, and set-aside)	- Improved nutrient management can improve ground water quality and environmental health of the cultivated ecosystem	- Changes in water policies could lead to clash of interests and threaten social cohesiveness - Could lead to water overuse
Grazing land management	- Improves livestock productivity, reduces desertification, and provide social security to the poor - Requires laws and enforcement to ban free grazing	
Livestock management	- Mix of traditional rice cultivation and livestock management would enhance incomes even in semi arid and arid regions	
Waste Management: Chapter 10		
Engineered sanitary landfilling with landfill gas recovery	- Can eliminate uncontrolled dumping and open burning of waste, improving health and safety for workers and residents - Sites can provide local energy benefits and public spaces for recreation and other social purposes within the urban infrastructure	- When done unsustainably can cause leaching that leads to soil and groundwater contamination with potentially negative health impacts
Biological processes for waste and wastewater (composting, anaerobic digestion, aerobic and anaerobic wastewater processes)	- Can destroy pathogens and provide useful soil amendments if properly implemented using source-separated organic waste or collected wastewater - Can generate employment - Anaerobic processes can provide energy benefits from CH_4 recovery and use	- A source of odours and water pollution if not properly controlled and monitored
Incineration and other thermal processes	- Obtain the most energy benefit from waste	- Expensive relative to controlled landfilling and composting - Unsustainable in developing countries if technical infrastructure not present - Additional investment for air pollution controls and source separation needed to prevent emissions of heavy metals and other air toxics
Recycling, reuse, and waste minimization	- Provide local employment as well as reductions in energy and raw materials for recycled products - Can be aided by NGO efforts, private capital for recycling industries, enforcement of environmental regulations, and urban planning to segregate waste treatment and disposal activities from community life.	- Uncontrolled waste scavenging results in severe health and safety problems for those who make their living from waste - Development of local recycling industries requires capital.

Note:
a) Material drawn from Chapters 4 to 11. New material is referenced in the accompanying text below that describes the sustainable development implications of mitigation options in each sector.

Environmental impacts include those occurring in local areas on air, water, and land, including the loss of biodiversity. Virtually all forms of energy supply and use, and land-use change activity cause some level of environmental damage. GHG emissions are often directly related to the emissions of other pollutants, either airborne, for example, sulphur dioxide from burning coal which causes local or indoor air pollution, or waterborne, for example, from leaching of nitrates from fertilizer application in intensive agriculture.

The social dimension includes issues such as gender equality, governance, equitable income distribution, housing and education opportunity, health impacts, and corruption. Most mitigation options will impact one or more of these issues, and both benefits and trade-offs are likely.

12.3.1 Energy supply and use

Mitigation options in the energy sector may be classified into those that improve energy efficiency and those that reduce the use of carbon-intensive fuels. The latter may be further classified into domestic and imported fuels. The synergies and trade-offs of these options with economic, local environmental, and social sustainable development goals are presented in Table 12.4. In the case of energy efficiency, it is generally thought to be cost effective and its use reduces or eliminates local pollutant emissions. Improving energy efficiency is thus a desirable option in every energy demand and supply sector.

As noted in Section 12.1.3, over the last decade, quantification of progress towards sustainable development has gained ground. In the industrial sector, several trade associations provide platforms for organizing and implementing GHG mitigation programmes. Chapter 7 notes that performance indicators are being used by the aluminium, semiconductor, and cement industry to measure and report progress towards sustainable development. The Global Reporting Initiative (GRI), a UNEP Collaborating Centre initiative, for example, reports that over 700 companies worldwide make voluntary use of its Sustainability Reporting Guidelines for reporting their sustainable development achievements. Industrial sectors with high environmental impacts lead in reporting and 85% of the reports address progress on climate change (GRI, 2005), and (KPMG Global Sustainability Services, 2005). Another example is in the buildings sector. Several thousand commercial buildings have been certified by the USA Green Building Council's programme on Leadership in Energy and Environmental Design (LEED), which uses 69 criteria to award certificates at various levels of achievement. The certification ensures that a building meets largely quantitative criteria related to energy use, indoor air quality, materials and resource use, water efficiency, and innovation and design process (USGBC, 2005). Economic and ethical considerations are the most cited reasons by businesses in the use of these two guidelines.

12.3.1.1 Energy demand sectors – Transport, Buildings and Industry

In the buildings sector, energy efficiency options may be characterized as integrated and efficient designs and siting, including passive solar technologies and designs and urban planning to limit heat island effect. Considering energy efficiency as the guiding principle during the construction of new homes results in both reduced energy bills -enhancing the affordability of increased energy services- and GHG abatement (see Section 6.6). Policies that actively promote integrated building solutions for both mitigating and adapting to climate change are especially important for the buildings sector. Good

urban planning, including increasing green areas as well as cool roofs in cities, has proven to be an efficient way to limit the heat island effect, which also reduces cooling needs. Mitigation and adaptation can, therefore, be addressed simultaneously by these energy efficiency measures.

In developing countries, efficient cooking stoves that use clean biomass fuels are an important option. These can have significant health benefits including reduction in eye diseases. The incident is disproportionately high amongst rural women in many developing countries where fuelwood and other biomass materials are a principal source of energy (Porritt, 2005). It has also been shown, for example, that the availability of cleaner burning cookers and solar cookers in developing countries not only has important health benefits but also significant social benefit in the lives of women in particular (Dow and Dow, 1998). A move to a more reliable and cleaner fuel not only has benefits in terms of carbon emission and health, it has also the effect of freeing up significant amount of time for women and children, which can be applied to more socially beneficial activities, including going to schools in the case of children. The air pollution benefit of improved stoves, however, is controversial; other studies have noted that efficiency was improved at the expense of higher emissions of harmful pollutants (see Section 4.5.4.1).

In the transport sector, the energy efficiency measures may be categorized into those that are vehicle specific and those that address transportation planning. Vehicle-specific programmes focus on improvement to the technology and vehicle operations. Planning programmes are targeted to street layouts, pavement improvements, lane segregation, and infrastructural measures that improve vehicle movement and facilitate walking, biking and the use of mass transport. Cost-effective mitigation measures of both types have been identified that result in higher vehicle and/or trip fuel economy and reduce local air pollution. Institutionalizing planning systems for CO_2 reduction through coordinated interaction between national and local governments is important for drawing up common strategies for sustainable transportation systems (see Section 5.5.1). While there are many synergies in emission controls for air pollution and climate change, there are also trade-offs (see Section 5.5.4). Promotion of bicycling, walking, and other non-motorized modes of transportation has large and consistent co-benefits of GHG reduction, air quality, and people health improvement (see Section 5.2.1 and 5.5.4). Diesel engines are generally more fuel efficient than gasoline engines and thus have lower CO_2 emissions, but increase particle emissions. Air quality driven measures, such as obligatory particle matter and NO_x filters and in-engine measures, mostly result in higher fuel use and consequently, higher GHG emissions.

In the industrial sector, energy efficiency options may be classified as those aimed at mass-produced products and systems, and those that are process-specific. The potential for cost-effective measures is significant in this sector. Measures in both categories would have a positive impact on the environment. To the extent the measures improve productivity, they would increase economic output and hence add to government tax revenue. Higher tax revenue would benefit national, state and local government fiscal balance sheets (see Section 7.7; Nadel *et al.*, 1997; Barrett *et al.*, 2002; Phadke *et al.*, 2005).

Since energy efficiency improvement reduces reliance on energy supply, it is likely to improve a nation's energy security. Using prices as an instrument to promote energy efficiency mitigation options is often difficult due to the many barriers that impede their progress. Lack of information about such mitigation options and the principal agent problem have been documented to be particularly significant barriers in the residential sector, but these also prevail in the small and medium scale industries sectors (Sathaye and Murtishaw, 2005). Programmes that can overcome such barriers would increase energy efficiency penetration.

12.3.1.2   Energy supply[7]

Switching to low carbon energy supply sources is the other mitigation category in the energy sector with significant GHG benefits. This can be achieved through either increased reliance on imported or indigenous alternative fuels. Using a higher proportion of low carbon imported fuels will almost always reduce local air pollution. Its direct impact will be to increase payment for fuel imports that may result in worsened balance of payments, unless these are utilized to increase a nation's exports (Sathaye *et al.*, 1996). The higher fuel imports will increase dependence on international fuel supply that may result in reduced energy security unless diversification of supply mitigates concerns about increased dependence. Economies and societies of low carbon fuel exporting countries would benefit from the higher trade.

Increased reliance on most indigenous low carbon energy sources[8] would also reduce local air pollution, but the local environmental benefits in certain solid bioenergy applications appear to be uncertain (see Section 4.5.4.1). While indigenous low carbon fuels can reduce fuel imports, these have to be balanced against higher capital requirements for investment in fuel extraction, processing and delivery (Sathaye *et al.*, 1996). The development of large hydro sources can displace local populations and put their livelihood in jeopardy, and in reservoirs with large surface area, the resulting methane emissions may reduce their net GHG benefit substantially. For example, although hydroelectric plants have the potential to reduce GHG emissions significantly, a large amount of literature points to important environmental costs (McCully, 2001; Dudhani *et al.*, 2005), highlights the social disruptions and dislocations (Sarkar and Karagoz 1995; Kaygusuz, 2002), and questions the long-

term economic benefits of major hydropower development. Increased use of indigenous low-carbon fuels can reduce export of fuels from other countries to the extent the latter are substituted away. These may adversely affect the trade balance of exporting countries (Sathaye *et al.*, 1996).

At the same time, low carbon fuels can have other environmental benefits. For example, a move away from coal to cleaner fuels will reduce ecosystem pressures that often accompany mining operations (Azapagic, 2004). Similarly, a move away from charcoal and fuelwood as a source of energy will have the attendant environmental benefits of reducing the pressures of deforestation (Masera *et al.*, 2000; Najam and Cleveland, 2003). This points towards the need to optimize technology choice decisions not only along the dimension of carbon emissions but also other environmental costs.

Wind power can cause harm to bird populations, and may not be aesthetically appealing. Increased use of biomass is viewed as a renewable alternative, but indoor air pollution from solid fuels has been ranked as the fourth most important health risk factor in least developed countries (see Chapter 4). Trade-offs among pollutants are inevitable in the use of some mitigation options, and need to be resolved in the specific context in which the option is to be implemented.

Several examples of corruption that either increases the price of electricity and/or prevent the proceeds from extracted resources to meet development needs are provided in Section 4.5.4.3. This suggests that corruption may reduce the sustainable development benefits of new mitigation technologies and/or low carbon fuels that require a significant modification of social systems.

12.3.1.3   Cross-sectoral sustainable development impacts

Implementation of mitigation options often creates new industries, for example, for energy efficient products such as cooking stoves, efficient lamps, insulation materials, heat pumps, and efficient motors, or for solar panels, windmills, and biogas installations. The success of these new industries depends on various factors, such as the degree of information, costs, the image of the product and its traditional competitors or its attributes other energy efficient. New industries can create new jobs and income, and might be pioneers in new market with significant competitive advantage. Ethanol production from sugar waste has created a new industry and generated employment opportunities and tax revenue for the Government of Brazil. However, the older, outpaced industry may lose jobs. Besides the uncertainty on the overall net effect, this may lead to regional loss of employment. For example, the increased production of biofuels for transportation, or energy production

---

7   Carbon capture and storage (CCS) is an emerging GHG mitigation option that is described in Chapter 4. Its sustainable development impacts would be similar to those described in this section for the siting of power plants.

8   Low carbon energy sources include hydro, biomass, wind, natural gas and other similar energy carriers.

in rural areas, is expected to protect existing employment and to create new jobs in rural areas (Sims, 2003). Renewable energy systems are more labour intensive than fossil fuel systems and a higher proportion of jobs are relatively highly skilled. Thus, an increase in employment of the rural people can only be achieved, if corresponding learning opportunities are created. If, however, labour intensity decreases over time, the long-term effect on jobs might be less pronounced than originally anticipated.

12.3.2 Forestry sector

Mitigation options in the forestry sector may be categorized as those that (1) avoid emissions from deforestation or forest degradation; (2) sequester carbon through forestation; and (3) substitute for energy intensive materials or fossil fuels.

Reducing or avoiding deforestation has considerable environmental benefits. It can retain biodiversity, ecosystem functions, and in cases of large land areas, affect local weather patterns (see Section 9.7.2). Reduction of forest fires improves local air quality. Many deforesting countries have laws that promote conservation of forest areas. The lack of enforcement of laws that ban or limit deforestation or timber extraction has allowed illegal extraction of logs and the burning of forests in Indonesia (Boer, 2001) and Brazil (Boer, 2001; Fearnside, 2001). Avoiding deforestation is relatively expensive, since the opportunity cost of deforested land is high due to its high timber and land values. Stakeholders such as land owners, migrant workers, and local saw mills would be negatively affected.

Transparency and participatory approaches have played a key role in reducing communal tensions and allowed communities to reap the same or larger benefits within an organized legal framework. The Joint Forest Management Programme in India has created a community-based approach to manage forest fringe areas to reduce forest logging for fuelwood and encroachment on forest lands for agriculture (Behera and Engel, 2005). Successful implementation requires that alternative livelihood be provided to the deforesters, programmes to promote forest management jointly with the local population be pursued, and that enforcement be stricter.

Afforestation can provide carbon benefits by increasing carbon stocks on land and in products. Trees planted on wasteland can arrest soil degradation and help manage water runoff. Soil carbon can be increased to the extent soil disturbance during planting and harvesting is minimized. Planting in conjunction with agricultural crops (agro-forestry) enhances economic benefits while increasing food security. Afforestation activities are generally undertaken in rural areas and benefit the rural economy and generate employment for rural dwellers. Clear delineation of property rights would expedite the implementation of forestation programmes. A major concern is that forestation may diminish food security if it were to occur primarily on rich agricultural land, and that monoculture plantations would

reduce biodiversity and increase the risk of catastrophic failure due to diseases. Conversion of floodplains and wetlands to forest plantations could hamper ecological functions.

Afforestation activities can also yield biomass fuel that may be used as a fossil fuel substitute in power plants or as a liquid fuel substitute. Palm-tree plantations are also a rich source of bio-diesel fuel. These sustainable development benefits and potential trade-offs also apply to bioenergy plantations. In regions, where crop residues (rice husks, sugarcane bagasse, nut shells, and/or tree trimmings) are available, these can be harvested synergistically with the crops and pose less potential sustainable development trade-offs.

Forest management activities include sustainable management of native forests, prevention of fires and pests, longer rotation periods, minimizing soil disturbance, reduced harvesting, promoting understory diversity, fertilizer application, and selective and reduced logging. Most of these activities bring positive social and environmental benefits. Minimizing soil disturbance may result in less use of fossil fuels, less emissions from biomass burning, and more employment if less machinery is used. The prevention of fires may result in larger fire events later due to excessive accumulation of fuel. Therefore, such practice should be linked to other practices such as sustainable wood fuel production. Theoretically, N fertilizer application increases net primary productivity (NPP) (and CO_2 removals), but there is a trade-off since at the same time it increases N_2O emissions and may contaminate waters with nitrates.

Some of the social benefits of mitigation policies come through education, training, participation as an integral part of a policy. Participatory approaches to forest management can be more successful than traditional, hierarchical programmes (Stoll, 2003). These participatory programmes can also help to strengthen civil society and democratization. Participatory approaches can create social capital (Dasgupta, 1993): networks and social relations which allow humans to cope better with their livelihoods.

12.3.3 Agriculture sector

Table 12.4 also summarizes the impact of different mitigation activities in agriculture sector on the constituents and determinants of sustainable development (see also Section 8.4.5 and Table 8.8). The table provides a description and tentative direction of impact but the exact magnitude of impact would depend upon the scale and intensity of the activities in the context where they are undertaken.

Several mitigation activities are explored in Chapter 8, ranging from crop, tillage/residue, nutrient, rice, water, manure/biosolid, grazing lands, organic soils, livestock and manure management practice, to land cover change, agro-forestry, land restoration, bioenergy, enhanced energy efficiency and increased carbon storage in agricultural products. It is shown that appropriate adoption of these mitigation measures is likely to help achieve social, economic and environmental goals, although sometimes trade-offs may also occur. Interesting enough, these trade-offs, when and if they occur, seem to be most visible in the short term, as in the long-term synergy amongst the aspects of sustainable development seems to be dominant.

An appropriate and optimal mix of rice cultivation with livestock known as integrated annual crop-animal system and traditionally found in West Africa, India and Indonesia and Vietnam would enhance the net income, improve the condition of cultivated ecosystems and over all human well being (MA, 2005). Such combinations of livestock and crop farming especially for rice would prove effective in income generation even in semi arid and arid areas of the world.

Ground water quality may be enhanced and the loss of biodiversity slowed by greater use of farmyard manure and more targeted pesticides. The impact on social and economic aspects of this mitigation measure remains uncertain. Better nutrient management can improve environmental sustainability.

Controlling overgrazing through pasture improvement has a favourable impact on livestock productivity (greater income from the same number of livestock) and slows/halts desertification (environmental aspect). It also provides social security to the poorest people during extreme events such as drought and other crisis (especially in Sub-Saharan Africa). One effective strategy to control overgrazing is the prohibition of free grazing, as was done in China (Rao, 1994).

This critical sector of the world economy is the biggest, user of the water. In low-income countries, agriculture uses almost 90% of the total extracted water (World Bank, 2000). Policies on free or very cheap energy (electricity, petroleum) as present is some areas for political reasons, contribute to misuse of water as the true economic cost inclusive of environmental and social costs are not reflected in the pricing and other incentive structures. Rationalization of electricity tariffs would aid in improving water allocation across users and over time. Through proper institutions and effective functioning of markets, water management can be operationalized with favourable impact on environmental and economic goals. In the short term, social cohesiveness might come under stress due to a clash of divergent interests.

Land cover and tillage management could encourage favourable impacts on environmental goals. A mix of horticulture with optimal crop rotations would promote carbon sequestration and could also improve agro-ecosystem function. Societal well-being would also be enhanced through provisioning of water and enhanced productivity. Whilst the environmental benefits of tillage/residue management are clear, other impacts are less certain. Land restoration will have positive environmental impacts, but conversion of floodplains

and wetlands to agriculture could hamper ecological function (reduced water recharge, bioremediation, and nutrient cycling) and therefore, could have an adverse impact on sustainable development goals (Kumar, 2001).

Livestock management and manure management mitigation measures are context and location specific in there influence on sustainable development. Appropriate adoption of mitigation measures is likely to help achieve environmental goals, but farmers may incur additional costs, reducing their returns and income.

12.3.4 Waste and wastewater management sector

Better waste and wastewater management is an important sustainable development goal because it can lead directly to improved health, productivity of human resources, and better living conditions. It can also have direct economic benefits in terms of higher value of property due to improved living conditions. The 2002 Johannesburg World Summit on Sustainable Development added a new goal on sanitation, calling for the reduction by 50% of the number of people living without access to safe sanitation by 2015.

Chapter 10 emphasizes that environmentally-responsible waste management to reduce GHG emissions at an appropriate level of technology can promote sustainable development. In many developing countries, uncontrolled open dumpsites, open burning of waste, and poor sewerage practices result in major public health hazards due to vermin, pathogens, safety concerns, air pollution, and contamination of water resources. Often, waste in rural areas is neither collected nor properly managed.

The challenge is to develop improved waste and wastewater management using low to medium-technology strategies that can provide significant public health benefits and GHG mitigation at affordable cost. Some of these strategies include small-scale wastewater management such as septic tanks and recycling of grey water, construction of medium-technology landfills with controlled waste placement and use of daily cover, composting of organic waste, and implementation of landfill bio covers to optimize microbial CH_4 oxidation.

The major impediment in developing countries is the lack of capital. Another challenge is the lack of urban planning so that waste treatment and disposal activities are segregated from community life. A third challenge is often the lack of environmental regulations enforced within urban infrastructure. In many developing countries, waste recycling occurs through the scavenging activities of informal recycling networks. Sustainable development includes a higher standard for these recycling activities so that safety and health concerns are reduced via lower technology solutions that are effective, affordable, and sustainable.

In some cases, landfill gas might be used to provide heating fuel for a factory or commercial venture that can be an alternative source of local employment. Also, compost can be used for agriculture or horticulture applications, and closed re-vegetated landfills can become public parks or recreational areas.

12.3.5 Implications of climate policies for sustainable development

A major policy development since the TAR is the implementation of a large range of climate policies at the international level (e.g., Kyoto Protocol), regional level (e.g., EU Emissions Trading Scheme), national and sub-national level (see the review in Section 13.3.3.4).

The implications of these policies for sustainable development are not assessed in the literature, except for those of the Clean Development Mechanism (CDM) (Michaelowa, 2003; Spalding-Fecher and Simmonds, 2005; Sutter, 2003; UNEP, 2004; Winkler, 2004; Winkler and Thorne, 2002). For extensive discussion, see Section 13.3.3.4.2. The sustainable development implications of particular mitigation activities that can be implemented under the CDM are discussed further in Section 12.3. This section focuses on the sustainable development implications of CDM as a policy. Key findings from this literature that relate to the implications of climate policies on sustainable development are as follows:

- The CDM channels non-trivial amounts of money towards developing countries. In 2005, the CDM channelled about US$2.5 billion to purchase carbon credits in developing countries (Capoor and Ambrosi, 2006), or 0.75% of the (record) net foreign direct investment (FDI) inflow in developing countries for that year (UNCTAD, 2006). In addition, it can be argued that the CDM leverages new private capital to developing countries.
- Since carbon payments are payable in strong currencies, and usually originate from buyers with strong credit ratings, they provide the seller with additional opportunities to raise additional capital and debt from banks and other finance institutions (Mathy *et al.*, 2001; Lecocq and Capoor, 2005).
- The geographical distribution of CDM projects tends to follow FDI flows with most of the financial flows towards large middle-income countries (Fenhann, 2006), and very little financial flows towards least developed countries, notably in Sub-Saharan Africa (Capoor and Ambrosi, 2006).
- Projects mitigating non-CO_2 gases (HFC23, N_2O and CH_4) represent the bulk of the volume of emission reductions exchanged under the CDM. However, projects with the highest direct benefits for local communities deliver fewer emission reductions and are in general accompanied by higher transaction costs. Resolving the tension between global emission reductions and local benefits is a key challenge for the future of climate change regime (Ellisa *et al.*, 2007).

12.4 Gaps in knowledge and future research needs

As noted in Section 12.1, changing development paths will be critical to addressing mitigation and the scale of effort required is unlikely to be forthcoming from the environmental sector on its own. If climate policy on its own will not solve the climate problem, future research on climate change mitigation and sustainable development will need to focus increasingly on development sectors. A better understanding is needed of how countries might get from current development trajectories onto lower-carbon development paths – how to make development more sustainable.

The global GHG emissions reduction potential of such actions varies from a few tens to million tons of carbon, and empirical research is needed to identify and quantify actions that will yield the most emissions savings.

A fundamental yet important step would be to identify relevant non-climate policies affecting GHG emissions/sinks, including trade, finance, rural and urban development, water, energy, health, agriculture, forestry, insurance, and transport among others. Future research will also need to access and use local knowledge. More case studies would help illustrate the link between sustainable development and climate mitigation in developed, developing and transition countries. A particular challenge in this regard is that such policies will necessarily be context specific and will work only when structured within local and national realities. This means that a lot of the research required is at the local and national levels to identify policy options and choices that might best work within the contexts of specific regions, countries and localities.

This chapter has noted that development-oriented scenarios could be enriched by taking global climate change explicitly into account. Future research might develop and analyse scenarios for development paths at different scales and their implications for reducing or avoiding GHG emissions. This may require broadening and deepening the current set of models to better analyse the GHG implications of non-climate scenarios. This also applies to industrialized countries on their development paths and choices.

This chapter has suggested that the capacity to mitigate is rooted in development paths. Considerable research must be carried out to further investigate how mitigation capacity can be turned into actual mitigation, and its connection with components of the underlying development path. Paradoxically, the reviewed literature suggests that a fundamental discussion on the implications of development pathways for climate change in general and climate change mitigation in particular has been and is being explored more extensively for the developing countries than for the industrialized countries. Although the adaptive and mitigative capacity literature does not claim that building capacity will necessarily lead to improved responses to the climate change risk, little work has been done to explicate the widely noted variation in response to climate change among communities and nations with similar capacities. It is apparent, therefore, that capacity is a necessary, but not sufficient, condition for mitigative action. Phenomena such as risk perception, science/policy interactions, and relationships between industry and regulators, for instance, may play some role in determining whether or not capacity is turned into action in response to the climate change risk.

Section 12.1.3 cites several macro-indicators of sustainable development that are being used to track its progress at the national and international level. Few of these take climate change mitigation directly into consideration. Inclusion of this aspect in the use of macro-indicators is identified as an important area of research.

Changing development pathways involves multiple actors, at multiple scales. The roles of different actors and joint actions in changing development pathways need further research, particularly the private sector and civil society (and how they relate to government). A key question revolves around the complex process of decision-making, theories of which need to be applied to sustainable development and mitigation. A particular focus in this area might be identifying patterns of investment and their implications for GHG emissions. Again, much of this research will have to be contextually specific and related to specific local and national contexts.

While future research must focus on multiple sectors, actors and scales, a key area of investigation will remain the role for international agreements. Reconciling the role for international coordination mechanisms with decentralized policy approaches is challenging and requires further evaluation. An area of particular importance in this context is international agreements that are not specific to climate change but whose structure and implementation can affect development paths. These include voluntary international agreements, such as those on the implementation of the Millennium Development Goals (MDGs) to specific Multilateral Environmental Agreements (MEAs), such as those on desertification, on biodiversity, to the related provisions of international policy instruments within the World Trade Organization (WTO). All these agreements, including the WTO, now claim sustainable development as their ultimate goal.

Future research will continue to examine the implications of climate change mitigation for sustainable development. Understanding of the sustainable development implications in each of many sectors is growing, but further analysis will be needed for key sectors and where least information is available. Synergies beyond those in air pollution require more attention, including water, soil management; forest management and others. Apart from investigating synergies, the question of

trade-offs between sustainable development and mitigation (and also adaptation) requires further analysis.

REFERENCES

Adger, W.N., N. Brooks, M. Kelly, S. Bentham, S. Eriksen (eds.), 2004: New indicators of vulnerability and adaptive capacity. Centre for Climate Research.

Agrawala, S., A. Moehner, A. Hemp, M. van Aalst, S. Hitz, J. Smith, H. Meena, S.M. Mwakifwamba, T. Hyera and O.U. Mwaipopo, 2003a: *Development and Climate Change in Tanzania: focus on Mount Kilimanjaro.* OECD, Paris.

Agrawala, S., A. Moehner, M. El Raey, D. Conway, M. van Aalst, M. Hagenstad and J. Smith, 2004b: *Development and Climate Change in Egypt: Focus on Coastal resources and the Nile.* OECD, Paris.

Agrawala, S., T. Ota, T., Ahmed, A.U., Smith, J. and M. van Aalst, 2003b: *Development and Climate Change in Bangladesh: Focus on coastal flooding and the Sundarbans.* OECD, Paris.

Agrawala, S., T. Ota, J. Risbey, M. Hagenstad, J. Smith, M. van Aalst, K. Koshy and B. Prasad, 2003c: *Development and Climate Change in Fiji: Focus on coastal mangroves.* OECD, Paris.

Alcamo, J., D. van Vuuren, C. Ringler, W. Cramer, T. Masui, J. Alder, and K. Schulze, *2005:* Changes in nature's balance sheet: Model-based estimates of future worldwide ecosystem services. *Ecology and Society,* **10**(2), pp.19.

Aldy, J.E., 2005: *An environmental Kuznets Curve analysis of U.S. state-level carbon dioxide emissions. Journal of Environment and Development,* pp. 48-72.

An, F. and A. Sauer, 2004. *Comparison of passenger vehicles fuel economy and greenhouse gas emission standards around the world.* Pew Center on Global Climate Change, Washington D.C., US, pp. 36.

Anderson, R., 1998: *Mid-Course Correction.* The Peregrinzilla Press, Atlanta.

Anderson, R., 2004: Climbing Mount Sustainability. *Quality Progress,* February 2004, pp. 32.

Andriof, J. and M. McIntosh, 2001: *Perspectives on Corporate Citizenship.* Greenleaf Publishing Ltd, Sheffield.

Angelsen, A. and D. Kaimowitz, 1999: Rethinking the causes of deforestation: lessons from economic models. *The World Bank Research Observer,* **14**(1), pp. 73-98. <http://www.worldbank.org/research/journals/wbro/obsfeb99/pdf/article4.pdf> accessed 06/07/07.

Angelsen, A. and D. Kaimowitz (eds.), 2001: *Agricultural technologies and tropical deforestation.* CABI Publishing, New York, USA, 422 pp.

Arthur, W.B., 1989: Competing technologies, increasing returns, and lock-in by historical events. *The Economic Journal,* **99**, pp. 116-131.

Athanasiou, T., 1996: The age of greenwashing. In *Divided planet: The ecology of rich and poor.* University of Georgia Press, Athens, GA, pp. 227-297.

Azapagic, A., 2004: Developing a framework for sustainable development indicators for the mining and minerals industry. *Journal of Cleaner Production,* **12**(6), pp. 639-662.

Bäckstrand, K. and E. Lövbrand, 2006: Planting trees to mitigate climate change: Contested discourses of ecological modernisation, green governmentality and civic environmentalism. *Global Environmental Politics,* **6**(1), pp. 50-75.

Banuri, T., G. Hyden, Banuri, T., C. Juma, M. Rivera, 1994: Sustainable human development: from concept to operation. United Nations Development Programme, New York.

Banuri, T. and A. Najam, 2002: *Civic Entrepreneurship - A Civil Society Perspective on Sustainable Development.* Gandhara Academy Press, Islamabad, <http://www.tellus.org/seib/publications/civic/VOLUME1.pdf> accessed 06/07/07.

Banuri, T., J. Weyant, G. Akumu, A.Najam, L. Pinguelli Rosa, S. Rayner, W. Sachs, R. Sharma, G. Yohe, 2001: Setting the stage: Climate change and sustainable development. In *Climate Change 2001: Mitigation, Report of working group III, Intergovernmental Panel On Climate Change (IPCC)* [Metz, B., O. Davidson, R. Swart, and J. Pan (ed.)], Cambridge University Press, Cambridge.

Barnett, J., 2001: *The meaning of environmental security, ecological politics and policy in the new security era.* Zed Books, London.

Barrett, J., J.A. Hoerner, S. Bernow, B.Dougherty, 2002: *Clean energy and jobs: A comprehensive approach to climate change and energy policy.* Economic Policy Institute and Center for a Sustainable Economy, Washington D.C.

Bartelmus, P., 2001: Accounting for sustainability: greening the national accounts. In *Our Fragile World, Forerunner to the Encyclopaedia of Life Support System.* Tolba, M.K. (ed.), pp. 1721-1735.

Baumert, K. and J. Pershing, *T. Herzog, M. Markoff,* 2004: *Climate data: insights and observations.* Pew Centre on Climate Change, Washington, D.C.

Beck, U., 1992: *Risk Society: Towards a New Modernity.* Sage, London.

Beg, N., J.C. Morlot, O. Davidson, Y. Afrane-Okesse, L. Tyani, F. Denton, Y. Sokona, J.P. Thomasc, E.L. La Rovere, J.K. Parikh, K. Parikh, A.A. Rahman, 2002: Linkages between climate change and sustainable development. *Climate Policy,* **2,** pp. 129-144.

Behera, B. and S. Engel, 2005: Institutional analysis of evolution of joint forest management in India: A new institutional economics approach. *Forest Policy and Economics,* **(8)**4, pp. 350-362.

Bell, S. and S. Morse, 1999: *Sustainability Indicators: Measuring the Immeasurable.* Earthscan, London.

Benhin, J.K.A. and E.B. Barbier, 2004: Structural Adjustment Programme, deforestation and biodiversity loss in Ghana. *Environment and Resources Economic,* **27,** pp. 337-366.

Berger, S. and R. Dore, 1996: *National diversity and global capitalism.* Cornell Press.

Berkhout, F., 2002: Technological regimes, path dependency and the environment. *Global Environmental Change,* **12,** pp. 1-4.

Betsill, M.M. and H. Bulkeley, 2004: Transnational networks and global environmental governance: The cities for climate protection program. *International Studies Quarterly,* **48,** pp. 471-493.

BIAC/OECD/IEA, 1999: *Workshop on Climate Change: Industry view on the Climate Change challenge with special emphasis on the Kyoto Mechanisms - Industry Sector Reports.* OECD/IEA, Paris, 54 pp.

Biermann, F. and K. Dingwerth, 2004: Global environmental change and the nation state. *Global Environmental Politics,* **4,** pp. 1-22.

Boer, R., 2001: Economic assessment of mitigation options for enhancing and maintaining carbon sink capacity in Indonesia. *Mitigation and Adaptation Strategies for Global Change,* **6,** pp. 257-290.

Bojo, J., K.-G. Mäler, L. Unemo, 1992: *Environment and development: An economic approach.* Second Revised Edition. Kluwer Academic Publishers, 211 pp.

Borchers, M., N. Qase, *C.T. Gaunt, J. Mavhungu, H. Winkler, Y. Afrane-Okese, C. Thom,* 2001: *National Electrification Programme evaluation: Summary report.* Evaluation commissioned by the Department of Minerals & Energy and the Development Bank of Southern Africa. Energy & Development Research Centre, University of Cape Town, Cape Town.

Borsky, S., D. Arbelaez-Ruiz, C. Cocklin, D. Holmes, 2006: International trends in socially-responsible investment: implications for corporate managers. In *International Handbook on Environmental Technology Management.* D. Annadale, D. Marinova, and J. Phillimore (ed.), Edward Elgar, London.

Bossel, H., 1999: Indicators for sustainable development: Theory, methods, applications. International Institute for Sustainable Development, Winnipeg, MB, <http://www.iisd.org/pdf/balatonreport.pdf> accessed 06/07/07.

Boulanger, P.-M., 2004: *Les indicateurs du développement durable: un défi scientifique, un enjeu démocratique.* IDDRI, Paris.

Boyle, A. and D. Freestone, 1999: *International law and sustainable development: past achievements and future challenges.* Oxford University Press, Oxford.

Brickman, R., S. Jasanoff, T. Ilgen, 1985: Controlling chemicals: the *politics of regulation in Europe and the United States.* Cornell University Press, Ithaca, New York.

Brown, L., 1981: *Building a sustainable society.* Worldwatch Institute, Washington, D.C.

Buitenkamp, M., H. Venner, T. Wams, 1993: *Action Plan Sustainable Netherlands.* Dutch Friends of the Earth, Amsterdam.

Bulkeley, H. and M.M. Betsill, 2005: Rethinking sustainable cities: Multilevel governance and the 'urban' politics of Climate Change. *Environmental Politics,* **14,** pp. 42-63.

Bulkeley, H. and A.P.J. Mol, 2003: Participation and environmental governance: Consensus, ambivalence and debate. *Environmental Values,* **12,** pp. 143-154.

Cabeza Gutes, M., 1996: The concept of weak sustainability. *Ecological Economics,* **17**(3), pp. 147-156.

Capoor, K. and P. Ambrosi, 2006: *State and trends of the carbon market 2006.* Prototype Carbon Fund and International Emissions Trading Association, Washington D.C.

Carpenter, S., P.L. Pingali, E.M. Bennett, M.B. Zurek, 2005: *Ecosystems and human well-being: Scenarios. Findings of the Scenarios Working Group v. 2* (Millennium Ecosystem Assessment S.), Island Press.

Cash, D.W. and S.C. Moser, 2000: Linking global and local scales: designing dynamic assessment and management processes. *Global Environmental Change,* **10,** pp. 109-120.

CGD, 2004: *Millions saved: Proven successes in global health.* Centre for Global Development.

Chandler, W., R. Schaeffer, D. Zhou, P.R. Shukla, F. Tudela, O. Davidson, S. Alpan-Atamer, 2002: *Climate change mitigation in developing countries: Brazil, China, India, Mexico, South Africa, and Turkey.* Pew Center on Global Climate Change, Washington D.C.

Chomitz, K., 2007: *At Loggerheads? Agricultural expansion, poverty reduction, and environment in the tropical forests.* World Bank: Washington D.C.

Chomitz, K. and D. Gray, 1996: Roads, land use, and deforestation: A spatial model applied to Belize. *The World Bank Economic Review,* **10,** pp. 487-512.

CIESIN, 1996-2001: *Environmental Treaties and Resource Indicators.* Center for International Earth Science Information Network (CIESIN), Palisades, NY, <http://sedac.ciesin.columbia.edu/entri/> accessed 06/07/07.

Cocklin, C., 1995: Agriculture, society, and environment: discourses on sustainability. *International Journal of Sustainable Development and World Ecology,* **2,** pp. 240-256.

Cogan, 2006: *Corporate Governance and Climate Change: Making the Connection.* CERES, Boston, MA, 300 pp.

Metwalli, A.A.M., H.H.J de Jongh, M.A.J.S. van Boekel, S. Cohen, D. Demeritt, J. Robinson, D. Rothman, 1998: Climate Change and Sustainable Development: Towards Dialogue. *Global Environment Change,* **8**(4), pp. 341-371.

Colman, R., 2002: Corporate sustainability a hot commodity. *CMA Management,* **76**(Sep), 9 pp.

Colombier, M. and J.C. Hourcade, 1989: Développement des réseaux et modulations spatio-temporelles des tarifs: l'équité territoriale revisitée. *Revue Economique,* **40**(July), pp. 649-676.

Coondoo, D. and S. Dinda, 2002: Causality between income and emission: a country group-specific econometric analysis. *Ecological Economics,* **40,** pp. 351-367.

Copeland, B.R. and M.S. Taylor, 2004: Trade, growth, and the environment. *Journal of Economic Literature,* **42**(March), pp. 7-71.

Corell, E. and M. Betsill, 2001: A comparative look at NGO influence on international environmental negotiations: Desertification and Climate Change. *Global Environmental Politics,* **2**(1), pp. 86-107.

Cork, S., G. Peterson, G. Petschel-Held, J. Alcamo, J. Alder, E. M. Bennett, E. Carr, D. Deane, G. Nelson, and T. Ribeiro, 2005: Four scenarios. In *Ecosystems and Human Well-being: Scenarios, Volume 2. Findings of the Scenarios Working Group of the Millennium Ecosystems Assessment.* S.R. Carpenter, P.L. Pingali, E.M. Bennett, M.B. Zurek, (eds.), Island Press.

Cornillie, J. and S. Fankhauser, 2002: *The energy intensity of transition countries.* Working Paper N.72 EBRD, London, U.K., 25 pp.

Corson, W. (ed.), 1996: *Measuring sustainability: indicators, trends, and performance.* M.E. Sharpe, Armonk, NY, pp. 325-352.

Cowan, B. and N. Mohlakoana, 2005: *Barriers to access modern fuels in low-income households: Khayelitsha.* Energy Research Centre, University of Cape Town, Cape Town.

Cowan, J., E. Eidinow, *Laura Likely,* 2000: A scenario-planning process for the new millennium. *Deeper News,* **9**(1).

Dalal-Clayton, B. (ed.), 2003: *The MDGs and sustainable development: the need for a strategic approach.* IIED, London, pp. 73-91.

Das, A.K and C. Dutta, 2004: Strengthening rural information infrastructure through E-Choupals. In *Proceedings 21st IASLIC National Seminar,* S.B. Ghosh (ed) Kolkata, India.

Dasgupta, P., 1993: *An Inquiry into Well-being and Destitution.* Clarendon Press, Oxford.

Dasgupta, S., K. Hamilton, K. Pandey, D. Wheeler, 2004: *Air pollution during growth: accounting for governance and vulnerability.* Policy Research Working Paper No.3383, World Bank, Washington D.C, USA, 30 pp.

Davidson, O., K. Halsnaes, S. *Huq, M. Kok, B. Metz, Y. Sokona, J. Verhagen,* 2003: The development and climate nexus: the case of sub-Saharan Africa. *Climate Policy,* **3**(S1), pp. S97-S113.

Davidson, O. and Y. Sokona, 2002: *Think bigger, act faster: a new sustainable energy path for African development.* Energy & Development Research Centre, University of Cape Town & ENDA Tiers Monde, Cape Town and Dakar.

Davis, T.S., 1999: Reflecting on voluntary environmental partnerships: Lessons for the next century. *Corporate Environmental Strategy,* **6**(1), pp. 55-59.

De Cara, S., M. Houzé, P.-A. Jayet, 2005: Methane and nitrous oxide emissions from agriculture in the EU: A spatial assessment of sources and abatement costs. *Environmental and Resource Economics,* **32**, pp. 551-583.

De Gouvello, C. and Y. Maigne (eds.), 2000: *L□Electrification Rurale Décentralisée à l□heure des négociations sur le changement climatique.* Systèmes Solaires, Paris, France, pp. 35-60.

De Marchi, B., 2003: Public participation and risk governance. *Science and Public Policy,* **30**(171-176).

De Moor, A., 2001: Towards a grand deal on subsidies and climate change. *Journal Natural Resources Forum,* **25**(2).

Decleris, M., 2000: *The law of sustainable development - General principles.* European Commission.

Deme, P.A., 2003: Le cas du GPL au Sénégal. *In ESMAP: Énergies modernes et réduction de la pauvreté: Un atelier multi-sectoriel. Actes de l□atelier Régional,* Dakar, Sénégal, 4-6 February 2003, 22 pp.

Dietz, T., 2003: The Darwinian trope in the drama of the commons: Variations on some themes by the Ostroms. In *Academic Conference in Honor of the Work of Elinor and Vincent Ostrom,* George Mason University, Arlington, Virginia.

Dietz, T. and P.C. Stern, 1995: Toward a theory of choice: Socially embedded preference construction. *The Journal of Socio-Economics,* **24**, pp. 261-279.

DOE,1989: *Compendium of options for government policy to encourage private sector responses to potential climate change.* US Department of Energy, Washington D.C.

Doppelt, B., 2003: *Leading change toward sustainability: A change-management guide for business, government and civil society.* Greenleaf Publishing Ltd., Sheffield.

Dorian, J.P., H.T. Franssen, *D.R. Simbeck,* 2006: Global challenges in energy. *Energy Policy,* **34**(15), pp. 1984-1991.

Dovers, S. and J. Handmer, 1993: Contradictions in sustainability. *Environmental Conservation,* **20**(3), pp. 217-222.

Dow, R.M. and C.R. Dow, 1998: using solar cookers and gardens to improve health in urban and rural areas. *Journal of American Dietetic Association,* **99**(9), pp. A58.

Driscoll, C. and M. Starik, 2004: The primordial stakeholder: Advancing the conceptual consideration of stakeholder status for the natural environment. *Journal of Business Ethics,* **49**(1), pp. 55-73.

Dryzek, J., 1990: Designs for environmental discourse: The greening of the administrative state. In *Managing Leviathan: Environmental Politics and the Administrative State.* R. Paehlke and D. Torgerson (ed.), Belhaven Press, London, pp. 97-111.

Dryzek, J., 1997: *The politics of the earth: Environmental discourse.* Oxford University Press, Oxford.

Dudhani, S., A.K. Sinha, *S.S. Inamdar,* 2006: Assessment of small hydropower potential using remote sensing data for sustainable development in India. *Energy Policy* **34**(17), pp. 3195-3205.

Dunphy, D., A.Griffiths, S. Benn, 2003: *Organizational change for corporate sustainability: A guide for leaders and change agents of the future.* Routledge, London.

Elkington, J., 2001: *The Chrysalis Economy: How citizen CEOs and corporations can fuse values and value creation.* Capstone, Oxford.

Ellisa, J., H. Winkler, *J. Corfee-Morlot, F, Gagnon-Lebrun,* 2007: CDM: Taking stock and looking forward. *Energy Policy,* **35**(1), pp. 15-28.

Elzen, B. and A. Wieczorek, 2005: Transitions towards sustainability through system innovation. *Technological Forecasting and Social Change,* **72**, pp. 651-661.

Estache, A. and J. Strong, 2000: The rise, the fall, and the emerging recovery of project finance in transport, private sector development. Working Paper 2385, Washington: World Bank Institute.

ESMAP, 2002: *Energy strategies for rural India: Evidence from six states.* World Bank, Washington D.C. Report 258/02.

Fairhead,J. and M. Leach, 2003: *Science, society and power: Environmental knowledge and policy in West Africa and the Carribbean.* Cambridge, Cambridge University Press.

Farsari, Y. and P. Prastacos, 2002: Sustainable development indicators: an overview. Foundation for the Research and Technology Hellas.

Fearnside, P., 2001: The potential for Brazil's forest sector for mitigating global warming under the Kyoto Protocol. *Mitigation and Adaptation Strategies for Global Change,* **6**, pp. 355-372.

Fenhann, J., 2006: *Overview of the CDM Pipeline,* version 09-08-2006. CD4CDM, UNEP-Risoe Centre, Denmark.

Fisher, D.R., 2004: *National governance and the global Climate Change regime.* Rowman and Littlefield Publishers, Inc., Lanham MD.

Forsyth, T., 2005: Building deliberative public-private partnerships for waste management in Asia. *Geoforum,* **36**(4), pp. 429-439.

Frankel, J.A. and A.K. Rose, 2002: *Is trade good or bad for the environment? Sorting out the causality.* National Bureau of Economic Research Working Paper 9201, Cambridge, MA.

Frenk, J., J.L. Bobadilla, *C. Stern, T. Frejka, R. Lozano,* 1993: Elements for a theory of the health transition. In *Health and social change in international perspective.* L.C. Chen, A. Kleinman, and N.C. Ware (eds.), Harvard University Press, Boston.

Funtowicz, S. and J. Ravetz, 1993: Science for a post-normal age. *Futures,* **25**(739-755).

Furuseth, O. and C. Cocklin, 1995: An institutional framework for sustainable resource management: the New Zealand model. *Natural Resources Journal,* **35**(2), pp. 243-273.

Gallagher, K.S., 2006: Limits to leapfrogging in energy technologies? Evidence from the Chinese automobile industry. *Energy Policy,* **34**(383-394).

Gallopin, G.C. and F. Rijsberman, 2000: Three global water scenarios. *International Journal of Water,* **1**(1), pp. 16-40.

Geels, F., 2004: From sectoral systems of innovation to socio-technical systems: Insights about dynamics and change from sociology and institutional theory. *Research Policy,* **33**, pp. 897-920.

Geist, H.J. and E.F. Lambim, 2002: Proximate causes and underlying driving forces of tropical deforestation. *Bioscience,* **52**(2), pp. 142-150.

Geller, H., R. Schaeffer, *A. Szklo, M. Tolmasquim,* 2004: Policies for advancing energy efficiency and renewable energy use in Brazil. *Energy Policy,* **32,** pp. 1437-1450.

Giddens, A., 1991: *Modernity and self-identity: Self and society in the late modern age.* Stanford, Stanford University Press.

Giddens, A., 1998: Risk society: The context of British politics. In *The Politics of Risk Society,* J. Franklin, ed., Polity Press, Cambridge, pp. 23-34.

Goklany, I.M., 2006: *Integrated Strategies to Reduce Vulnerability and Advance Adaptation, Mitigation, and Sustainable Development.* <http://members.cox.net/igoklany/Goklany-Integrating_A&M_preprint.pdf> accessed 09/07/07.

Goodwin, M., 1998: The governance of rural areas: some emerging research issues and agendas. *Journal of Rural Studies,* **14**(1), pp. 5-12.

Gough, C. and S. Shackley, 2001: The respectable politics of Climate Change: The Epistemic Communities and NGOs. *International Affairs,* **77**(2).

Government of India, 2006: *Towards faster and more inclusive growth: An approach to the 11th Five Year Plan.* Planning Commission, Government of India.

GRI, 2005: Global Reporting Initiative. <http://www.globalreporting.org> accessed 06/07/07.

Grossman, G.M. and A.B. Krueger, 1991: *Environmental impacts of a North American Free Trade Agreement.* National Bureau of Economic Research Working Paper 3914, NBER, Cambridge, MA.

Grubb, M., 1991: *Energy policies and the greenhouse effect.* Energy and environmental programme, Dartmouth for The Royal Institute of International Affairs, Aldershot.

Gueorguieva, A. and K. Bolt, 2003: *A critical review of the literature on structural adjustment and the environment.* World Bank Environmental Economics Series Paper 90, Washington, D.C.

Gutmann, A. and D. Thompson, 2004: *Why Deliberative Democracy?* Princeton University Press, Princeton.

Haas, P., 1990: *Saving the Mediterranean: the politics of international environmental cooperation.* Columbia University Press, New York City.

Haas, P., R.O. Keohane, *M.A. Levy* (eds.), 1993: *Institutions for the earth.* The MIT Press, Cambridge, MA.

Haas, P.M., 2004: When does power listen to truth? A constructivist approach to the policy process. *Journal of European Public Policy,* **(11)**, pp. 569-592.

Hajer, M., 1995: *The politics of environmental discourse.* Oxford University Press, Oxford.

Hajer, M. and H. Wagenaar (eds.), 2003: *Deliberative policy analysis: Understanding governance in the network society.* Cambridge University Press, Cambridge.

Hale, T., 2004: Thinking globally and acting locally: Can the Johannesburg Partnerships coordinate action on sustainable development? *Journal of Environment and Development,* **3**, pp. 220-239.

Hamilton, K., G. Anderson, *D. Pearce,* 1997: *Genuine savings as an indicator of sustainability.* CSERGE Working Paper GEC 97-03, Centre for Social and Economic Research and the World Bank

Hamilton, K., T.L Brewer, T. Aiba, T. Sugiyama, J. Drexhage, 2003: Module 2: Corporate engagement in US, Canada, the EU and Japan and the influence on domestic and international policy. In *The Kyoto-Marrakech System: A Strategic Assessment.* M. Grubb, T. Brewer, B. Müller, J. Drexhage, K. Hamilton, T. Sugiyama, T. Aiba, <http://www.iccept.ic.ac.uk/a5-1.html> accessed 06/07/07.

Hamilton, K. and M. Kenber, 2006: *Business views on international climate and energy.* The Climate Group, UK, 56 pp.

Handley, P., 1997: A critical view of the Build-Operate-Transfer privatisation process in Asia. *Asian Journal of Public Administration,* **19**(2), 203-243.

Harbaugh, W.T., A. Levinson, D. Wilson, 2002: Reexamining the empirical evidence for an environmental Kuznets curve. *The Review of Economics and Statistics,* **84**(3), pp. 541-551.

Hardi, P. and T. Zdan (eds.), 1997: *Assessing Sustainable Development: Principles in practice.* International Institute for Sustainable Development, Winnipeg.

Hasselmann, K., M. Latif, G. Hooss, C. Azar, O. Edenhofer, C. C. Jaeger, 2003: The challenge of long-term climate change. *Science,* **302**, pp. 1923-1925.

Hawken, P., A. Lovins, L. Hunter, 1999: *Natural capitalism: The next industrial revolution.* Earthscan, London.

Heil, M.T. and T.M. Selden, 2001: Carbon emissions and economic development: future trajectories based on historical experience. *Environment and Development Economics,* **6**(1), pp. 63-83.

Heller, T.C. and P.R. Shukla (eds.), 2003: *Development and climate: Engaging developing countries.* Pew Center on Global Climate Change, Arlington, 111 pp.

Helm, D., 2002: Energy policy: security of supply, sustainability and competition. *Energy Policy,* **30**, pp. 173-184.

Herzog, T., K.A. Baumert, J. Pershing, 2006: *Target: Intensity. An analysis of greenhouse gas intensity targets.* World Resources Institute, Washington D.C. <http://www.wri.org/climate/pubs_description.cfm?pid=4195> accessed 06/07/07.

Holliday, C., 2001: Sustainable growth, the DuPont Way. *Harvard Business Review,* September, pp. 129-134.

Holliday, C., S. Schmidheiny, P. Watts, 2002: *Walking the talk: The business case for sustainable development.* Greenleaf Publishing, San Francisco.

Hollingworth, J.R. and R. Boyer, 1997: *Contemporary Capitalism: The embeddedness of institutions,* Cambridge Press.

Hourcade, J.C. and M. Kostopoulou, 1994: Quelles politiques face aux chocs énergétiques. France, Italie, Japon, RFA: quatre modes de résorption des déséquilibres. *Futuribles,* **189**, pp. 7-27.

Hourcade, J.C., P. Shukla, L. Cifuentes, D. Davis, J. Edmonds, B. Fisher, E. Fortin, A. Golub, O. Hohmeyer, A. Krupnick, S. Kverndokk, R. Loulou, R. Richels, H. Segenovic, K.Yamaji , 2001: Global, Regional and National Costs and Ancillary Benefits. In *2001 Climate Change 2001: Mitigation,* B. Metz, O. Davidson, R. Swart, and J. Pan (ed.), Cambridge University Press, Cambridge.

Howes, M., 2005: *Politics and the environment: Risk and the role of government and industry.* Allen and Unwin, Sydney.

Hueting, R. (ed.), 1993: *Calculating a sustainable national income. A practical solution for a theoretical dilemma.* Nagard, Milan.

Hulme, D. and M. Edwards, 1997: NGOs, states and donors: An overview. In *NGOs, States and Donors: Too Close for Comfort?* D. Hulme and M. Edwards (ed.), St. Martin's Press, New York.

Huq, S., H. Reid, L. Murray, 2003: *Mainstreaming adaptation to Climate Change in least developed countries.* Working Paper 1: Country by Country Vulnerability to Climate Change, IIED, <http://www.iied.org/pubs/pdf/full/10002IIED.pdf> accessed 09/07/07.

IEA, 2003a: Energy balance of OECD Countries. International Energy Agency/OECD, Paris. <www.iea.org> accessed 06/07/07.

IEA, 2003b: *Key World Energy Statistics from the IEA.* International Energy Agency/OECD, Paris.

IEA, 2004a: *Analysis of the impact of high oil prices on the global economy.* International Energy Agency, Paris, 15 pp.

IEA, 2004b: *CO_2 emissions from fuel combustion 1971-2002.* International Energy Agency, Paris, France.

IEA, 2004c: *Oil Crises and Climate Challenges: 30 Years of Energy Use in IEA Countries.* International Energy Agency, Paris.

IEA, 2005: *World Energy Outlook. International Energy Agency, Paris.*

IEA, 2006: CO_2 emissions from Fuel Combustion, 1971-2004. *International Energy Agency, Paris.*

IISD, 2006: *Compendium of Sustainable Development Indicator Initiatives.* International Institute for Sustainable Development, Winnipeg, MB, Canada, <http://www.iisd.org/measure/compendium/> accessed 06/07/07.

Innovest, 2005: *Carbon Disclosure Project 2005.* Carbon Disclosure Project, London, 154 pp.

737

IPCC, 1996: *Estimating the costs of mitigating greenhouse gases.* Intergovernmental Panel on Climate Change (IPCC), Cambridge University Press, Cambridge, UK., pp. 263-296.

IPCC, 2000: *Special Report on Emissions Scenarios.* Intergovernmental Panel on Climate Change (IPCC), Cambridge University Press, Cambridge, UK.

IPCC, 2001: *Setting the stage: climate change and sustainable development.* Intergovernmental Panel on Climate Change (IPCC), Cambridge University Press, Cambridge, UK.

IPCC, 2001b: *Climate Change 2001: Mitigation* - Contribution of Working Group III to the Third Assessment Report of the Intergovernmental Panel on Climate Change (IPCC) [Metz, B., O. Davidson, R. Swart, and J. Pan (eds.)]. Cambridge University Press, Cambridge, 700 pp.

IPCC, 2007: *Climate Change 2007: Impacts, Adaptation and Vulnerability.* Contribution of Working Group II to the Fourth Assessment Report of the Intergovernmental Panel on Climate Change [M.L. Parry, O.F. Canziani, J.P. Palutikof, P.J. van der Linden, C.E. Hanson (eds.)]. Cambridge University Press, Cambridge, United Kingdom and New York, NY, USA.

Irwin, T., M. Klein, G.E. Perry, M. Thobani, 1999: Managing government exposure to private infrastructure risks. World Bank Research Observer., **14,** pp. 229-245.

IUCN, 1980: *World conservation strategy: Living resources conservation for sustainable development.* International Union for the Conservation of Nature, United Nations Environment Programme, Worldwide Fund for Nature, Switzerland.

Jacoby, H.G., 2000: Access to markets and the benefits of rural roads. *Economic Journal,* **110,** pp. 713-737. <http://europa.sim.ucm.es/compludoc/GetSumario?r=/W/10001/00130133_7.htm&zfr=0> accessed 06/07/07.

Jahan, S. and A. Umana, 2003: The environment-poverty nexus. *Development Policy Journal,* **3,** pp. 53-70.

Jannuzzi, G.M., 2005: Power sector reforms in Brazil and its impacts on energy efficiency and research and development activities. *Energy Policy,* **33,** pp. 1753-1762.

Jasanoff, S., 1986: *Risk management and political culture: a comparative study of science in the policy context.* Russell Sage Foundation, New York.

Jasanoff, S., 1990: *The fifth branch: Science advisers as policymakers.* Cambridge, Mass.

Jasanoff, S. and B. Wynne, 1998: Science and decisionmaking. In *Human Choice and Climate Change,* S. Rayner and E.L. Malone (ed.), Battelle Press., Columbus, OH, pp. 1-87.

Jessop, R., 1997: Capitalism and its future: Remarks on regulation, government and governance. *Review of International Political Economy,* **4,** pp. 561-581.

Jiang, K., Z. Dadajie, J. Hui, L. Eda, 2003: *Country Study: China.* Energy Research Institute and Academy for Agriculture Science, Beijing, China.

Jochem, E., J. Sathaye, D. Bouille, 2001: *Society, behaviour, and climate change mitigation.* Kluwer Academic Publishers, The Netherlands.

Jordan, L. and P. van Tuijl (eds.), 2006: *NGO Accountability: politics, principles and innovations.* Earthscan, London.

Kahane, A., 2002: *An overview of multi-stakeholder civic scenario work. Global Business Network,* <http://www.generonconsulting.com/publications/> accessed 06/07/07.

Kaimowitz, D., G. Thiele, P. Pacheco, 1999: The effects of structural adjustment on deforestation and forest degradation in lowland Bolivia. *World Development,* **27**(3), pp. 505-520.

Kaivo-oja, J., J. Luukkanen, M. Wilenius, 2004: Defining alternative national-scale socio-economic and technological futures up to 2100: SRES scenarios for the case of Finland. *Boreal Environment Research,* **9,** pp. 109-125.

Karekezi, S. and L. Majoro, 2002: Improving modern energy services for Africa's urban poor. *Energy Policy,* **30**(Special Issue), pp. 1015-1028.

Kaya, Y., 1990: Impact of Carbon Dioxide Emission Control on GNP Growth: Interpretation of Proposed Scenarios. Paper presented to the IPCC Energy and Industry Subgroup, Response Strategies Working Group, Paris.

Kaygusuz, K., 2002 Sustainable development of hydropower and biomass energy in Turkey. *Energy Conversion and Management,* **43**(8), pp. 1099-1120.

Kelman, S., 1981: *Regulating America, regulating Sweden: a comparative study of occupational safety and health policy.* MIT Press, Cambridge, Mass.

Kessides, I.N., 2004: *Reforming infrastructure: privatization, regulation and competition.* Washington D.C, World Bank, 306 pp.

Kok, M., W. Vermeulen, W.J.V. Vermeulen, A.P.C. Faaij, D. de Jager, 2000: *Global warming and social innovation: The challenge of a climate-neutral society.* Earthscan, United Kingdom, 242 pp.

Kok, M.J.T. and H.C. de Coninck, 2004: *Beyond climate: Options for broadening climate policy.* RIVM report 500019001, Bilthoven, The Netherlands.

Kotov, V., 2002: *Policy in transition: New framework for Russia's climate policy.* Fondazione Eni Enrico Mattei (FEEM), Milano, Italy, Report-58.2002.

KPMG Global Sustainability Services, 2005: *KPMG International Survey of Corporate Responsibility Reporting 2005.*

Kumar, P., 2001: *Valuation of ecological services of wetland ecosystems: A case study of Yamuna flood plains in the corridors of Delhi.* Mimeograph, Institute of Economic Growth, Delhi.

Kunreuther, H., J. Linnerooth, R. Starnes, (eds.), 1982: *Liquified energy gas facility siting: International comparisons.* IIASA, Laxenburg, Austria.

La Rovere, E.L. and B. Americano, 1999: *Assessment of global environmental impacts of PROCEL: GHG emissions avoided by PROCEL 1990-2020.* Final Report to Electrobrás, Rio de Janeiro, Brazil.

La Rovere, E.L. and A.R. Romeiro, 2003: *Country study: Brazil.* Centroclima - COPPE/UFRJ and UNICAMP/EMBRAPA, Rio de Janeiro, Brazil.

Lafferty, W., 1996: The politics of sustainable development: global norms for national development. *Environmental Politics,* **5,** pp. 185-208.

Lecocq, F. and K. Capoor, 2005: *State and trends of the carbon market 2005.* Prototype Carbon Fund and International Emissions Trading Association, Washington D.C.

Lehtonen, M., 2004: The environmental-social interface of sustainable development: capabilities, social capital, institutions. *Ecological Economics,* **49**(2), pp. 199-214.

Levi-Faur, D., 2005: The global diffusion of regulatory capitalism. *Annals of the American Academy of Political and Social Sciences,* **598,** pp. 12-32.

Levine, P., 2000: The new progressive era: Toward a fair and deliberative democracy. Rowman & Littlefield Pub. Inc.

Levy, D. and P. Newell, 2000: Oceans apart? Business responses to global environmental issues in Europe and the United States. *Environment,* **42**(9), pp. 9-20.

Levy, D.L. and P.J. Newell, 2005: *The Business of Global Environmental Governance.* MIT Press, Cambridge, MA and London.

Lewis, J.L. and R.H. Wiser, 2007: Fostering a renewable energy technology industry: An international comparison of wind industry policy support mechanism. *Energy Policy* **35,** pp. 1844-1857.

Lewis, N., W. Moran, C. Cocklin, 2002: Restructuring, regulation and sustainability. In *The Sustainability of Rural Systems: Geographical Interpretations.* I. Bowler, C. Bryant, and C. Cocklin (ed.), Kluwer Academic, Dordrecht, pp. 97-121.

Liberatore, A., and S. Funtowicz, 2003: 'Democratising' expertise, 'expertising' democracy: what does this mean, and why bother? *Science and Public Policy,* **30,** pp. 146-155.

Lindquist, L., 1980: *The hare and the tortoise: Clean air policies in the United States and Sweden.* University of Michigan Press, Ann Arbor.

Lindseth, G., 2004: The cities for climate protection campaign (CCPC) and the framing of local climate policy. *Local Environment,* **9**, pp. 325-336.

Liverman, D.M., 1999: Geography and the global environment. *Annals of the Association of American Geographers*, **89**.

Lyon, T., 2003: 'Green' firms bearing gifts. *Regulation*, Fall, pp. 36-40.

MA, 2005: *Current state and trends.* Millenium Ecosystems Assessment, Island Press, Washington D.C.

Machado, G., R. Schaeffer, E. Worrell, 2001: Energy and carbon embodied in the international trade of Brazil: an input-output approach. *Ecological Economics*, **39**, pp. 409-424.

Markandya, A. and K. Halsnaes (eds.), 2002: *Climate change and sustainable development: prospects for developing countries.* Earthscan Publications, London.

Marsden, C., 2000: The new corporate citizenship of big business: Part of the solution to sustainability? *Business and Society Review*, **105**(1), pp. 9.

Masera, O.R., B.D. Saatkamp, D.M. Kammen, 2000: From linear fuel switching to multiple cooking strategies: A critique and alternative to the energy ladder model. *World Development*, **28**(12), pp. 2083-2103.

Mathy, S., J.-C. Hourcade, C. de Gouvello, 2001: Clean development mechanism: leverage for development? *Climate Policy*, **1**, pp. 251-268.

McCully, P., 2001: *Silenced rivers: The ecology and politics of large dams.* Zed Books, London.

Meadowcroft, J., 1997: Planning for sustainable development: Insights from the literatures of political science. *European Journal of Political Research,* **31**, pp. 427-454.

Meadowcroft, J., 2000: Sustainable development: a new(ish) idea for a new century. *Political Studies*, **48**(2), pp. 370-387.

Meadows, D.H., 1998: *Indicators and information systems for sustainable development.* A Report to the Balaton Group. The Sustainability Institute, Hartland Four Corners, VT, <http://sustainabilityinstitute. org/pubs/Indicators&Information.pdf> accessed 06/07/07.

Mebratu, D., 1998: Sustainability and sustainable development: historical and conceptual review. *Environmental Impact Assessment Review*, **18**, pp. 493-520.

Metz, B., M. Berk, M. den Elzen, B. de Vries, D. van Vuuren, 2002: Towards an equitable climate change regime: compatibility with Article 2 of the climate change convention and the link with sustainable development. *Climate Policy*, **2**, pp. 211-230.

Metz, B., O. Davidson, H. de Coninck, M. Loos, L.A. Meyer, (eds.), 2005: *Special Report on Carbon Dioxide Capture and Storage Special Report of the Intergovernmental Panel on Climate Change (IPCC),* Cambridge University Press, Cambridge.

Michaelowa, A., 1998: Climate policy and interest groups: A public choice analysis. *Intereconomics*, **33**(6), pp. 251-259.

Michaelowa, A., 2003: CDM host country institution building. *Mitigation and Adaptation Strategies for Global Change*, **8**, pp. 201-220.

Millimet, D.L., J.A. List, T. Stengos, 2003: The environmental Kuznets curve: Real progress or misspecified models? *Review of Economics and Statistics*, **85**(4), pp. 1038-1047.

Mills, E., 2003: The insurance and risk management industries: New players in the delivery of energy-efficient products and services. *Energy Policy*, **31**, pp. 1257-1272.

Mills, E., 2005: Insurance in a climate of change. In *Science*, **12**(August), pp. 1040-1044.

MNP, 2005: *Quality and the future. Sustainability outlook.* Netherlands Environmental Assessment Agency, Bilthoven, Netherlands, 500013010, 28 pp.

Modi, V., S. McDade, L. Lallement, J. Saghir, 2006: *Energy Services for the Millenium Development Goals.* Energy Sector Management Assistance Program, United Nations Development Programme, UN Millennium Project, World Bank, New York, USA, 109 pp.

Moldan, B., S. Billharz, R. Matravers (eds.), 1997: *Sustainability indicators: A report on the project on indicators of sustainable development.* SCOPE Report no. 58. Wiley, Chicester, UK.

Moomaw, W.R. and G.C. Unruh, 1997: Are environmental Kuznets curves misleading us? The case of CO_2 emissions. *Environment and Development Economics*, **2**, pp. 451-463, <http://ase.tufts.edu/gdae/publications/archives/moomawpaper.pdf> accessed 06/07/07.

Morita, T. and H. Lee, 1998: IPCC Scenario Database. *Mitigation and Adaptation Strategies for Global Change*, **3**(2-4), pp. 121-131.

Morita, T., N. Nakicenovic, J. Robinson, 2000: Overview of Mitigation Scenarios for Global Climate Stabilization Scenarios based on the new IPCC Emissions Scenarios (SRES). *Environmental Economics and Policy Studies*, **3**(2).

Morita, T., J. Robinson, A. Adegbulugbe, J. Alcamo, D. Herbert, E.L. La Rovere, N. Nakicenovic, H. Pitcher, P. Raskin, K. Riahi, A. Sankovski, V. Sokolov, B. de Vries, D. Zhou, 2001: Greenhouse gas emission mitigation scenarios and implications (Chapter 2). In *Climate Change 2001 - Mitigation.* Report of working group III of the Intergovernmental Panel on Climate Change (IPCC). B. Metz, O. Davidson, R. Swart, and J. Pan (ed.), Cambridge University Press, Cambridge, pp. 115-166.

Munasinghe, M., 1992: *Environmental economics and sustainable development.* The World Bank, Washington D.C.

Munasinghe, M. and W. Cruz (eds.), 1995: *Economy wide policies and the environment: Lessons from experience.* World Bank Environment Paper, Washington D.C, USA, 86 pp.

Munasinghe, M. and R. Swart (eds.), 2000: *Climate change and its linkages with development, equity and sustainability.* Intergovernmental Panel on Climate Change, Geneva.

Munasinghe, M., 2000, Development, equity and sustainability (DES) in the context of Climate Change. In *Climate Change and its linkages with development, equity and sustainability,* M. Munasinghe and R. Swart (eds.), Intergovernmental Panel on Climate Change, Geneva.

Nadel, S., 1997: *Energy efficiency and economic development in New York, New Jersey, and Pennsylvania.* American Council for an Energy Efficient Economy, Washington D.C.

Najam, A., 1996: Understanding the Third Sector: Revisiting the prince, the merchant and the citizen. *Nonprofit Management and Leadership,* **7**(2), pp. 203-19.

Najam, A., 1999: World business council for sustainable development: The greening of business or a greenwash? In *Yearbook of International Co-operation on Environment and Development 1999/2000,* H.O. Bergesen, G. Parmann, and Ø.B. Thommessen ed. Earthscan Publications, London, pp. 65-75.

Najam, A., 2000: The four C's of third sector-government relations: Cooperation, confrontation, complementarity, and co-optation. *Nonprofit Management and Leadership,* **10**(4), pp. 375-396.

Najam, A., I. Christopoulou, W.R. Moomaw, 2004: The emergent 'system' of global environmental governance. *Global Environmental Politics*, **4**(23-35).

Najam, A. and C. Cleveland, 2003: Energy and sustainable development at global environmental summits: An evolving agenda. *Environment, Development and Sustainability*, **5**(2), pp. 117-138.

Najam, A. and T. Page, 1998: The Climate Convention: Deciphering the Kyoto Convention. *Environmental Conservation*, **25**(3), pp. 187-194.

Najam, A., A. Rahman, S. Huq, Y. Sokona, 2003: Integrating sustainable development into the fourth IPCC assessment. *Climate Policy*, Special Issue on Climate Change and Sustainable Development, **3**(Supplement 1), pp. S9-S17.

Nakicenovic, N. and R. Swart (eds), 2000: *Emissions Scenarios. Special Report of the Intergovernmental Panel on Climate Change (IPCC).* Cambridge University Press, London.

Nelson, R.R. (ed.), 1993: *National innovation systems. A comparative analysis.* Oxford, New York.

Newell, P., 2000: *Climate for change: Nonstate actors and the global politics of the greenhouse.* Cambridge University Press, Cambridge, UK.

Newman, P.W.G. and J.R. Kenworthy, 1991: *Cities and automobile dependence, an international sourcebook.* Avebury Technical Publishing, Aldershot, UK.

NIC, 2004: *Mapping the global future.* National Intelligence Council, Washington, D.C.

Nivola, P.S., 1999: *Laws of the landscape: How policies shape cities in Europe and America.* Brookings Institution Press, Washington, D.C., USA, 126 pp.

O'Riordan, T., C.L. Cooper, A. Jordan, S. Rayner, K. Richards, P. Runci, and S. Yoffe, 1998: *Institutional frameworks for political action. Human Choice and Climate Change.* Battelle Press, Ohio, 345-370 pp.

O'Riordan, T. and S. Stoll-Kleeman, 2002: Deliberative democracy and participatory biodiversity. In *Biodiversity, Sustainability and Human Communities: Protecting Beyond the Protected.* T. O'Riordan, and S.S.-K. (eds.) ed. Cambridge University Press, Cambridge, pp. 87-112.

OECD, 1998: Sustainable development indicators. In *OECD Expert Workshop*, Organization of Economic Co-operation and Development (OECD), Paris.

OECD, 2001: *Sustainable development: Critical issues.* Organization of Economic Co-operation and Development (OECD), Paris.

OECD, 2005: *Health at a Glance, OECD Indicators.* Organization of Economic Co-operation and Development (OECD), Paris.

OECD, 2002: *Reforming energy subsidies. UN Environmental Programme and Organisation for Economic Cooperation and Development.* OECD/IEA, Oxford, UK, 31 pp.

Ogilvy, J. and P. Schwartz, 2000: *China's futures: Scenarios for the world's fastest growing economy, ecology, and society.* Jossey-Bass, San Francisco.

Opschoor, J.B., 1995: Ecospace and the fall and rise of throughput intensity. *Ecological Economics,* **15**, pp. 137-141.

Ostrom, E., 1990: *Governing the commons: The evolution of institutions for collective action.* Cambridge University Press, Cambridge.

Ostrom, E., 2000: Collective actions and the evolution of norms. *Journal of Economic Perspectives,* **14**, pp. 137-158.

Ostrom, E., T. Dietz, N. Dolsak, P.C. Stern, S. Stonich, E.U. Weber (eds.), 2002: *The drama of the commons.* National Academy Press, Washington, D.C.

Ott, H.E., H. Winkler, B. Brouns, S. Kartha, M. Mace, S. Huq, Y. Kameyama, A.P. Sari, J. Pan, Y. Sokona, P.M. Bhandari, A. Kassenberg, E.L. La Rovere, A. Rahman, 2004: *South-North dialogue on equity in the greenhouse. A proposal for an adequate and equitable global climate agreement.* Eschborn, GTZ.

Palmer, K. and R. Newell, K. Gillingham, 2004: *Retrospective Examination of Demand-side Energy-efficiency Policies*, Discussion Papers dp-04-19, Resources for the Future. <http://www.rff.org/documents/RFF-DP-04-19REV.pdf> accessed 06/07/07.

Pan, J., 2003: Emissions rights and their transferability: equity concerns over climate change mitigation, International Environmental Agreements. *Politics, Law and Economics,* **3**(1), pp. 1-16.

Pan, J., 2004a: China's industrialization and reduction of greenhouse emissions. *China and the World Economy* 12, no.3. <old.iwep.org.cn/wec/**2004**_5-6/**china's**%20**industrialization**.pdf> accessed 31/05/07.

Pan, J., 2004b: Meeting human development needs: emissions reductions from a developing country perspective. *IDS Bulletin on Climate Change and Development,* July 2004, **35**(3), pp. 90-97.

Pan, J. 2004: Industrialisation and emissions reductions: challenges and opportunities. *China and World Economy,* **3**, pp. 1-16.

Pandey, K.D. and D. Wheeler, 2001: *Structural adjustment and forest resources: The impact of World Bank operations.* World Bank Policy Research Working Paper 2584, Washington, D.C.

Parris, T. and R. Kates, 2003: Characterizing and measuring sustainable development. *Annual Review of Energy & the Environment,* **28**, pp. 559-586.

Paterson, M., D. Humphreys, l. Pettiford, 2003: Conceptualizing global environmental governance: from interstate regimes to counter-hegemonic struggles. *Global Environmental Politics,* **3**, pp. 1-10.

Pearce, D., 2000: The policy relevance and use of aggregate indicators: Genuine savings. In *OECD Proceedings, Frameworks to Measure Sustainable Development. An OECD Expert Workshop.* OECD, Paris.

Pearce, D., A. Markandya, E.B. Barbier, 1989: *Blueprint for a green economy.* Earthscan, London.

Petrich, C.H., 1993: Indonesia and global climate change negotiations: potential opportunities and constraints for participation, leadership, and commitment. *Global and Environmental Change,* **3**(1), pp. 53-77.

Pezzoli, K., 1997: Sustainable development: A transdisciplinary overview of the literature. *Journal of Environmental Planning and Management,* **40**(5), pp. 549-574.

Phadke, A., J. Sathaye, S. Padmanabhan, 2005: *Economic benefits of reducing Maharashtra's electricity shortage through end-use efficiency improvement.* LBNL-57053.

Pharaoh, A., 2003: Corporate reputation: the boardroom challenge. Corporate Governance, *International Journal of Business in Society,* **3**(4), pp. 46-51.

Pimbert, M. and T. Wakeford, 2001: Overview - deliberative democracy and citizen empowerment. *PLA Notes,* **40**(February), pp. 23-28.

Pintér, L., P. Hardi, B. Bartelmus, 2005: *Sustainable Development Indicators - The way forward. December 15-17, 2005* Expert Group Meeting of the United Nations Division for Sustainable Development, Winnipeg, IISD for the UN-DSD, <http://www.iisd.org/pdf/2005/measure_indicators_sd_way_forward.pdf> accessed 06/07/07.

Porritt, J., 2005: Healthy environment-healthy people: The links between sustainable development and health. *Public Health,* **119**(11), pp. 952-953.

Porter, M. and M. Kramer, 2002: The competitive advantage of corporate philanthropy. *Harvard Business Review,* **80**(12), pp. 56-68.

Porter, M. and C. van der Linde, 1995: Green and competitive: Ending the stalemate. *Harvard Business Review,* **73**(5), pp. 120-133.

Post, J., L.E. Preston S. Sauter-Sachs, 2002: *Redefining the corporation: Stakeholder management and organizational wealth.* Stanford Business Books, Stanford.

Prasad, G. and H. Ranninger, 2003: *The social impact of the basic electricity support tariff (BEST). Domestic use of energy.* Cape Technikon, pp. 17-22, Cape Town.

PRC, 2004: *Initial National Communication on Climate Change.* China Planning Press, Beijing, People's Republic of China, 80 pp.

Prud'homme, R. and J.P. Bocarejo, 2005: The London congestion charge: a tentative economic appraisal. *Transport Policy,* **12**, pp. 279-287.

Putnam, R., 1993: *Making democracy work: Civic traditions in modern Italy.* Princeton University Press, Princeton.

Rabinovitch, J. and J. Leitman, 1993: *Environmental innovation and management in Curitiba, Brasil.* UNDP/UNCHS/World Bank, Washington, D.C, 62 pp.

Rahman, A., M. Alam, Z. Karim, K. Islam, 2003: *Bangladesh country case study.* Bangladesh Centre for Advanced Studies, Dhaka, Bangladesh.

Rao, C.H., 1994: *Agricultural growth, rural poverty and environmental degradation in India.* Oxford University Press, Delhi.

Raskin, P., G. Gallopin, P. Gutman, A. Hammond, R. Swart, 1998: *Bending the curve: Toward global sustainability. A* Report of the Global Scenario Group. Stockholm Environment Institute, Boston.

Ravallion, M., M. Heil, J. Jalan, 2000: Carbon emissions and income inequality. *Oxford Economic Papers,* **52**, pp. 651-669.

Ravindranath, N.H. and J.A. Sathaye, 2002: *Climate Change and Developing Countries.* Kluwer, 300 pp.

Rayner, S. and E. Malone (eds.), 1998*: Human choice and climate change* (4 volumes). Batelle Press, Columbus, Ohio.

Rayner, S. and E. Malone, 2000: Security, governance and the environment. In: *Environment and Security: Discourses and Practices.* Lowi, M. and B. Shaw (ed.) Macmillan, Basingstoke.

Rayner, S. and K.R. Richards, 1994: *I think that I shall never see ... a lovely forestry policy. Land use programs for conservation of forests.* In Workshop of IPCC WGIII, Tsukuba.

Rayner, S., 1993: National case studies of institutional capabilities to implement greenhouse gas reductions. *Global Environmental Change,* **3**(Special issue), pp. 7-11.

Redclift, M. and T. Benton (eds.), 2006: *Social movements and environmental change.* WHO (World Health Organization), 2006. London and New York: Routledge, pp. 150-168.

Reed, D. (ed.), 1996: *Structural adjustment, the environment, and sustainable development.* Earthscan, London.

Rees, W.E., 1996: Revisiting carrying capacity: Area based indicators of sustainability. Population and Environment. *A Journal of interdisciplinary Studies,* **17**(39).

Renn, O., 2001: The role of social science in environmental policy making: experiences and outlook. *Science and Public Policy* **28**, pp. 427-437.

Rhodes, R., 1996: The new governance: Governing without government. *Political Studies,* **XLIV**, pp. 652-657.

Richardson, J., G. Gustafsson, G. Jordan (eds.), 1982: *The concept of policy style.* George Allen & Unwin, London.

Rieu-Clarke, A.S., 2004: Sustainable use and the EC Water Framework Directive: From principle to practice? In *International Law and Sustainable Development.* N. Schrijver and F. Weiss (ed.), Martinus Nijhoff Publishers, The Netherlands, pp. 714.

Robinson, J., M. Bradley, P. Busby, D. Connor, A. Murray, B. Sampson, W. Soper, 2006: Climate change and sustainable development: Realizing the opportunity. *Ambio,* **35**(1), pp. 2-8.

Robinson, J., 2004: Squaring the circle? Some thoughts on the idea of sustainable development. *Ecological Economics,* **48**, pp. 369-384.

Robinson, J. and D. Herbert, 2001: Integrating climate change and sustainable development. *International Journal of Global Environmental Issues,* **1**(2), pp. 130-148.

RSA, 1998: National Environmental Management Act (Act 107 of 1998). Republic of South Africa, Department of Environmental Affairs and Tourism, Pretoria.

Ruebbelke, D., 1998: Entropic limits of irreversible processes and possible adaptation mechanism for a sustinble development. *World Resources Review,* **10**(2).

Sachs, W., 1999: Sustainable development and the crisis of nature: On the political anatomy of an oxymoron. In *Living with Nature.* F. Fischer and M. Hajer (ed.), Oxford Scholarship Online, Oxford, pp. 23-42.

Santos, G. and G. Fraser, 2006: Road pricing: Lessons from London. *Economic Policy* (April), pp. 263-310.

Sarkar, A.U. and S. Karagoz, 1995: Sustainable development of Hydroelectric Power. *Energy* **20**(10), pp. 977-981.

Sathaye, J., S. de la Rue du Can, S. Kumar, M. Iyer, C. Galitsky, A. Phadke, M. M'cNeill, L. Price, R. Bharvirkar, 2006: *Implementing end-use efficiency improvements in India: Drawing from experience in the US and other countries.* LBNL-60035.

Sathaye, J., P. Monahan, A. Sanstad, 1996: Costs of reducing carbon emissions from the energy sector: China, India, and Brazil, *AMBIO,* **25**(4).

Sathaye, J. and S. Murtishaw, 2005: *Market failures, consumer preferences, and transaction costs in energy efficiency purchase decisions. Lawrence Berkeley National Laboratory for the California Energy Commission,* PIER Energy-Related Environmental Research. CEC-500-2005-020/LBNL-57318.

Schaefer, A. and A. Crane, 2005: Addressing sustainability and consumption. *Journal of Macromarketing,* **1**, pp 76-92.

Schaeffer, R. and A. Szklo, 2001: Future electric power technology choices of Brazil: a possible conflict between local pollution and global climate change. *Energy Policy,* **29**(5), pp. 355-369.

Schaltegger, S., R. Burritt, H. Petersen, 2003: *An introduction to corporate environmental management: Striving for sustainability.* Greenleaf Publishing, UK.

Schipper, L., M. Ting, M. Khrushch, W. Golove, 1997: The evolution of carbon dioxide emissions from energy use in industrialized countries: An end-use analysis. *Energy Policy,* **25**, pp. 651-672.

Schipper, L., S. Murtishaw, F. Unander, 2001: International comparisons of sectoral carbon dioxide emissions using a cross-country decomposition technique. *The Energy Journal,* **22**(2), pp. 35-75.

Schmalensee, R., T.M. Stoker, R.A. Judson, 1998: World carbon dioxide emissions 1950-2050. *Review of Economics and Statistics,* **80**(1), pp. 15-27.

Schmidt, V., 2002: *The Futures of European Capitalism.* Oxford University Press.

Schneider, S.H., W. Easterlig, L.O. Mearns, 2000: Adaptation: Sensitivity to natural variability, agent assumptions and dynamic climate changes. *Climatic Change,* **45**(1), pp. 203-221.

Schumacher, K. and J. Sathaye, 2000: Carbon emissions trends for developing countries and countries with economies in transition. LBNL-44546. *Quarterly Journal of Economic Research,* **4**(Special Issue: Energy Structures Past 2000).

Senge, P. and G. Carstedt, 2001: Innovating our way to the next industrial revolution. *MIT Sloan Management Review,* **Winter**, pp. 24-38.

Sharma, S., 2002: Research in corporate sustainability: What really matters? In *Research in Corporate Sustainability: The Evolving Theory and Practice of Organisations in the Natural Environment.* S. Sharma and M. Starik (ed.), Edward Elgar, Cheltenham, pp. 1-22.

Shell, 2005: *The Shell Global Scenarios to 2025: The business environment: trends, trade-offs and choices.* Shell, London.

Shove, E., L. Lutzenhiser, S. Guy, B. Hackett, H. Wilhite, 1998: Energy and social systems. In *Human Choice and Climate Change, Vol. 2 Resources and Technology.* S. Rayner and E.L. Malone (ed.), Battelle Press, Columbus, OH.

Shui, B. and R.C. Harriss, 2006: The role of CO_2 embodiment in US-China trade. <http://www.globalchange.umd.edu/publications/94/> accessed 06/07/07.

Shukla, P., R. Nair, M. Kapshe, A. Garg, S. Balasubramaniam, D. Menon, K.K. Sharma, 2003: *Development and climate: An assessment for India.* Indian Institute of Management, Ahmedabad.

Shukla, P.K., T. Heller, D.G. Victor, D. Biswas, T. Nag, A. Yajnik, 2005: *Electricity reforms in India: Firm choices and emerging generation markets.* Tata McGraw Hill, New Delhi.

SID/IEA, 2000: *Kenya at the crossroads: scenarios for our future.* Society for International Development and the Institute of Economic Affairs, Nairobi, Kenya.

Siebenhuner, B., 2002: How do scientific assessments learn? Conceptual Framework and Case Study of the IPCC. *Environmental Science and Policy,* **5**(5), pp. 411-420.

Sims, R.E.H., 2003: Bioenergy to mitigate for climate change and meet the needs of the society, the economy and the environment. *Adaptation and Mitigation Strategies for Global Change,* **8**(4), pp. 349-370. <http://www.springerlink.com/app/home/contribution.asp?wasp=23tl pnyvqm1j502g9m7j&referrer=parent&backto=issue,2,7;journal,5,31; searchpublicationsresults,1,2;> acessed 06/07/07.

SLG, 2001: *Learning to manage global environmental risks: A comparative history of climate change, ozone depletion and acid precipitation.* Social Learning Group, MIT Press, Cambridge, MA.

Smit, B., O. Pilifosova, I. Burton, B. Challenger, S. Huq, R.J.T. Klein, G. Yohe, 2001: Adaptation to climate change in the context of sustainable development and equity. I *Climate Change 2001: Impacts, Adaptation and Vulnerability.* J. McCarthy, O. Canziani, N. Leary, D. Dokken, and K. White (ed.), Cambridge University Press, Cambridge.

Smith, A., 2003: Transforming technological regimes for sustainable development: a role for alternative technology niches? *Science and Public Policy,* **2**, pp. 127-135.

Smith, K., 2002: In praise of petroleum? *Science,* **298**, pp. 1847.

Smith, K.R., 1997: Development, health, and the environmental risk transition. In *International Perspectives in Environment, Development, and Health.* G. Shahi, B.S. Levy, A. Binger, T. Kjellstrom, and R. Lawrence (ed.), Springer, New York, pp. 51-62.

Smith, K.R., J.M. Samet, I. Romieu, N. Bruce, 2000: Indoor air pollution in developing countries and acute lower respiratory infections in children. *Thorax,* **55**, pp. 518-532.

Sohn, J., S. Nakhooda, K. Baumert, 2005: *Mainstreaming climate change considerations at the multilateral development banks.* World Resources Institute, Washington, D.C.

Sohngen, B. and R. Sedjo, 2001: Forest set-asides and carbon sequestration. In *World Forests, Markets and Policies.* M. Palo, J. Uusivuori, and G. Mery (ed.), Kluwer Academic.

Sokona, Y., J.-P. Thomas, O. Touré, 2003: *Development and climate: Sengal country study.* Environnement et Développement du Tiers Monde (ENDA-TM), Dakar, Senegal.

Spalding-Fecher, R., A. Clark, M. Davis, G. Simmonds, 2002: The economics of energy efficiency for the poor - a South African case study. Energy: *The International Journal,* **27**(12), pp. 1099-1117.

Spalding-Fecher, R. and G. Simmonds (eds.), 2005: *Sustainable energy development and the Clean Development Mechanism: African Priorities.* Cambridge University Press, Cambridge, 124-135 pp.

Sparkes, R., 2002: *Socially responsible investment: A global revolution.* John Wiley, Chichester.

Srivastava, L. and T. Heller, 2003: *Integrating sustainable development and climate change in AR4 (AR4 SCOP-2/Doc. 8, 12.VIII.2003).* Intergovernmental Panel on Climate Change (IPCC).

SSEB, 2006: *American energy security: Building a bridge to energy independence and to a sustainable energy future.* Southern States Energy Board, Northcross, USA.

Stern, D.I., 2004: The rise and fall of the environmental Kuznets Curve. *World Development,* **32**(8), pp. 1419-1439.

Stern, P.C., 2000: Toward a coherent theory of environmentally significant behavior. *Journal of Social Issues,* **56,** pp. 407-424.

Stern, P.C. and H.V. Fineberg (eds.), 1996: *Understanding risk: Informing decisions in a democratic society.* National Academy Press, Washington D.C.

Stewart, R., 2001: A New generation of environmental regulation? *Capital Law Review,* **21,** pp. 182.

Stiglitz, J.E., 1998: Distinguished lecture on economics in government: the private uses of public interests: incentives and institutions. *Journal of Economic Perspective,* **12**(2), pp. 3-22.

Stoll, S., 2003: Participatory Research.

Streets, D.G., K.J. Jiang, X. Hu, J.E. Sinton, X.-Q. Zhang, D. Xu, M.Z. Jacobson, J.E. Hansen, 2001: Recent reductions in China's greenhouse gas emissions. *Science,* **294,** pp. 1835-1837.

Sutter, C., 2003: *Sustainability check-up for CDM projects: How to assess the sustainability of international projects under the Kyoto Protocol.* Wissenschaftlicher Verlag, Berlin.

Suzuki, D. and A. McConnell, 1997: The *sacred balance: Rediscovering our place in nature.* David Suzuki Foundation and Greystone Books, Vancouver.

Swanson, D.A., L. Pinter, F. Bregha, A. Volkery, K. Jacob, 2004: *National strategies for sustainable development: challenges, approaches and innovations in strategic and co-ordinated action.* IIED-GTZ.

Swart, R., J. Robinson, S. Cohen, 2003: Climate change and sustainable development: expanding the options. *Climate Policy,* Special Issue on Climate Change and Sustainable Development, **3**(S1), pp. S19-S40.

Szklo, A.S., R. Schaeffer, M.E. Schuller, W. Chandler, W., 2005: Brazilian energy policies side-effects on CO_2 emissions reduction. *Energy Policy,* **33**(3), pp. 349-364.

Tam, C.M., 1999: Build-Operate-Transfer model of infrastructure developments in Asia: Reasons for successes and failures. *International Journal of Project Management,* December 1999, pp. 377-382.

The Economist, 2000: Sins of the secular missionaries. January 29, 2000.

Thompson, D., 2002: *Tools for environmental management: A practical introduction and guide.* New Society Publishers, Gabriola, British Columbia.

Toman, M.A. and B. Jemelkova, 2003: Energy and economic development: an assessment of the state of knowledge. *Energy Journal,* **24**(4), pp. 93-112.

Transport for London, 2005: *Central London congestion charging, Impacts monitoring.* Third 45 annual report, April 2005.

Turton, H. and L. Barreto, 2006: Long-term security of energy supply and climate change. *Energy Policy,* **34**(15).

UCT, 2002: *Options for a basic electricity support tariff: Analysis, issues and recommendations for the Department of Minerals & Energy and Eskom.* Cape Town.

UNCSD, 2006: *Johannesburg Plan of Implementation.* United Nations Commission for Sustainable Development, New York.

UNCTAD, 2006: *World Investment Report 2006: FDI from developing and transition economies: Implications for development.* United Nations Conference on Trade and Development, UN, New York and Geneva.

UNDP, 2004: *Human Development Report,* 2004. United Nations Development Programme, Oxford University Press.

UNDP, 2005: *Human Development Report* 2005. United Nations Development Programme Oxford, University Press Oxford.

UNDSD, 2006: United Nations Division for Sustainable Development. <http://www.un.org/esa/sustdev/natlinfo/indicators/egmIndicators/egm.htm> accessed 09/07/07.

UNEP, 2002: *Global Environmental Outlook 3.* United Nations Environment Programme (UNEP), Nairobi, Earthscan, London.

UNEP, 2004: *Energy subsidies: Lessons learned in assessing their impact and designing policy reforms.* Economic and Trade Branch, United Nations Environment Programme (UNEP), Geneva, Switzerland.

UNEP/RIVM, 2004: *The GEO-3 scenarios 2020-2032: quantification and analysis of environmental impacts.* UNEP/DEWA/RS.03-4, RIVM Report 402001022.

United Nations, 1993: *Agenda 21: Earth Summit- The United Nations Programme of Action from Rio.* New York, United Nations, <http://www.springerlink.com/app/home/contribution.asp?wasp=23tlpnyvqm1j502g9m7j&referrer=parent&backto=issue,2,7;journal,5,31;search publicationsresults,1,2;> accessed 06/07/07.

US EPA, 2006: *Clean energy-environment guide to action: Policies, best practices, and action steps for states.* United States Environmental Protection Agency, EPA430-R-06-001.

USGBC, 2005: U.S. Green Building Council, <http://www.usgbc.org/> accessed 10/07/07.

Vellinga, P.V., E. Mills, G. Berz, L. Bouwer, S. Huq, L.A. Kozak, J. Palutikof, B. Schanzenbächer, G. Soler, 2001: Insurance and other financial services. Chapter 8: *Climate Change 2001: Impacts, Vulnerability, and Adaptation.* Contribution of Working Group II to the Third Assessment Report of the Intergovernmental Panel on Climate Change. Cambridge University Press, Cambridge, UK.

Venetoulis, J., D. Chazan, C. Gaudet, 2004: The ecological footprint of nations. Redefining Progress, <http://www.redefiningprogress.org/> accessed 06/07/07.

Vennemo, H., K. Aunan, J. Fang, P. Holtedahl, T. Hu, H.M. Seip, 2006: Domestic environmental benefits of China's energy related CDM potential. *Climatic Change,* **75**(1-2), pp. 215-239.

Victor, D. and T.C. Heller, 2007: *The political economy of power sector reform: The experience of five major developing countries,* Cambridge University Press, Cambridge.

Vitousek, P., P.R. Ehrlich, A.H. Ehrlich, P. Matson, 1986: Human appropriation of the products of photosynthesis. *Bioscience,* **34**(6), pp. 368-373.

Wackernagel, M. and W.E. Rees, 1996: *Our ecological footprint: reducing human impact on the earth.* New Society Publishers, Gabriola Island, BC. Philadelphia, PA. ISBN 086571312X.

Wagner, M. and G. Müller-Fürstenberg, 2004: *The Carbon Kuznets Curve: A cloudy picture emitted by lousy econometrics?* Discussion Paper 04-18. University of Bern, 36 pp.

Walker, W., 2000: Entrapment in large technology systems: institutional commitment and power relations. *Research Policy,* **29,** pp. 833-846.

WBCSD, 2004: *The Greenhouse Gas Protocol: A corporate accounting and reporting standard.* (Revised edition), World Business Council for Sustainable Development (WBCSD)/WRI.

WCED, 1987: *Our common future.* World Commission on Environment and Development. Oxford University Press, Oxford.

Weiner, J., 2002: Comparing precaution in the United States and Europe. *Journal of Risk Research,* **5**(4), pp. 317-349.

Welsch, H., 2001: The determinants of production-related carbon emissions in West Germany 1985-1990: assessing the role of technology and trade. *Structural Change and Economic Dynamics*, **12**, pp. 425-455.

Wheeler, D., 2001: *Racing to the bottom? Foreign investment and air pollution in developing countries.* World Bank, Policy Research Working Paper 2524, Washington D.C, USA, 24 pp.

WHO, 2002: The world health report 2002 - reducing risks, promoting healthy life. World Health Organisation, Geneva, Switzerland, 248 pp.

Wilbanks, T., 2003: Integrating climate change and sustainable development in a place-based context. *Climate Policy*, **3**(S1), pp. S147-S154.

Wilson, D. and C. Change, 2003: CEMEX promotes a sustainable approach with manufacturing excellence. *Environmental Quality Management*, **12**(4).

Windsor, D., 2004: Stakeholder influence strategies for smarter growth. In *Research in Corporate Sustainability: The Evolving Theory and Practice of Organisations in the Natural Environment.* S. Sharma and M. Starik, *et al.* (eds), Edward Elgar, Cheltenham, pp. 93-116.

Winkler, H., 2004: National policies and the CDM: Avoiding perverse incentives. *Journal of Energy in Southern Africa*, **15**(4), pp. 118-122.

Winkler, H., K.A. Baumert, O. Blanchard, S. Burch, J. Robinson, 2007: What factors influence mitigative capacity? *Energy Policy*, **35**(1), pp. 692-703.

Winkler, H., B. Brouns, S. Kartha, 2006: Future mitigation commitments: Differentiating among non-Annex I countries. *Climate Policy*, **5**(5), pp. 469-486.

Winkler, H., R. Spalding-Fecher, S. Mwakasonda, O. Davidson, 2002a: *Sustainable development policies and measures: starting from development to tackle climate change.* World Resources Institute, Washington, D.C, pp. 61-87.

Winkler, H., R. Spalding-Fecher, L.Tyani, 2002b: Comparing developing countries under potential carbon allocation schemes. *Climate Policy*, **2**(4), pp. 303-318.

Winkler, H. and S. Thorne, 2003: Baselines for suppressed demand: CDM projects contribution to poverty alleviation. *South African Journal of Economic and Management Sciences,* **5**(2), pp. 413-429.

Winters, L.A., N. McCullough, A. McKay, 2004: Trade liberalization and poverty, the evidence so far. *Journal of Economic Literature,* **42**(March), pp. 72-115.

World Bank, 1994: *World development report 1994, infrastructures for development.* Washington D.C., USA.

World Bank, 1999: *The effect of a shadow price on carbon emission in the energy portfolio of the World Bank: A carbon backcasting exercise.* Report No. ESM 212/99.

World Bank, 2000: *World Development Report.* World Bank.

World Bank, 2003: *World Development Report. Sustainable development in a dynamic world: Transforming institutions, growth, and quality of life.* World Bank, Washington, D.C, 250 pp.

World Bank, 2004: *World Development Report.* Oxford University Press, Oxford.

Wu, L., S. Kaneko, S. Matsuoka, 2005: Driving forces behind the stagnancy of China's energy-related CO_2 emissions from 1996 to 1999: the relative importance of structural change, intensity change and scale change. *Energy Policy*, **33**, pp. 319-335.

Wyatt, A.B., 2002: The privatisation of public infrastructure in transitional Southeast Asian economies: The case of build-own-operate-transfer projects in Vietnam and Laos. In *Collective Goods, Collective Futures in Asia,* S. Sargeson (ed.), London: Routledge.

Yamin, F. and J. Depledge, 2004: *The international Climate Change regime: A guide to rules, institutions and procedures.* Cambridge University Press.

Yearley, S., 1994: Social movements and environmental change. In *Social Theory and the Global Environment.* M. Redclift, and T. Benton (eds.), Routledge, London and New York, pp. 150-168.

Yohe, G., 2001: Mitigative capacity - the mirror image of adaptive capacity on the emissions side: An editorial. *Climatic Change*, **49**, pp. 247-262.

Yohe, G. and R. Moss, 2000: *Economic sustainability, indicators and climate change. In Climate Change and its linkages with development, equity and sustainability.* Proceedings of the IPCC Expert Meeting [Munasinghe, M. and R. Swart (eds.)]. IPCC and World Meteorological Organization, Geneva (2000), Colombo, Sri Lanka (27-29 April, 1999).

Zaelke, D., D. Kaniaru, E. Kružíková (eds), 2005: *Making Law Work: Environmental Compliance & Sustainable Development. Vol. 1 & 2.* INECE.

Zanetti, A., S. Schwartz, R. Enz, 2005: *Natural catastrophes and man-made disasters in 2004: more than 300,000 fatalities, record insured losses.* Swiss Reinsurance Company, Vol. sigma 1/2005.

13

Policies, Instruments and Co-operative Arrangements

Coordinating Lead Authors:
Sujata Gupta (India) and Dennis A. Tirpak (USA)

Lead Authors:
Nicholas Burger (USA), Joyeeta Gupta (The Netherlands), Niklas Höhne (Germany), Antonina Ivanova Boncheva (Mexico),
Gorashi Mohammed Kanoan (Sudan), Charles Kolstad (USA), Joseph A. Kruger (USA), Axel Michaelowa (Germany),
Shinya Murase (Japan), Jonathan Pershing (USA), Tatsuyoshi Saijo (Japan), Agus Sari (Indonesia)

Contributing Authors:
Michel den Elzen (The Netherlands), Hongwei Yang (PR China)

Review Editors:
Erik Haites (Canada), Ramon Pichs (Cuba)

This chapter should be cited as:
Gupta, S., D. A. Tirpak, N. Burger, J. Gupta, N. Höhne, A. I. Boncheva, G. M. Kanoan, C. Kolstad, J. A. Kruger, A. Michaelowa,
S. Murase, J. Pershing, T. Saijo, A. Sari, 2007: Policies, Instruments and Co-operative Arrangements. In Climate Change 2007:
Mitigation. Contribution of Working Group III to the Fourth Assessment Report of the Intergovernmental Panel on Climate Change
[B. Metz, O.R. Davidson, P.R. Bosch, R. Dave, L.A. Meyer (eds)], Cambridge University Press, Cambridge, United Kingdom and
New York, NY, USA.

Table of Contents

EXECUTIVE SUMMARY

This chapter synthesizes information from the relevant literature on policies, instruments and co-operative arrangements, focusing mainly on new information that has emerged since the Third Assessment Report (TAR). It reviews national policies, international agreements and initiatives of sub-national governments, corporations and non-governmental organizations (NGOs).

National policies

The literature on climate change continues to reflect the wide variety of national policies and measures that are available to governments to limit or reduce greenhouse gas (GHG) emissions. These include regulations and standards, taxes and charges, tradable permits, voluntary agreements, subsidies, financial incentives, research and development programmes and information instruments. Other policies, such as those affecting trade, foreign direct investment, consumption and social development goals, can also affect GHG emissions. Climate change policies, if integrated with other government polices, can contribute to sustainable development in developed and developing countries alike.

Reducing emissions across all sectors and gases requires a portfolio of policies tailored to fit specific national circumstances. While the advantages and disadvantages of any one given instrument can be found in the literature, four main criteria are widely used by policymakers to select and evaluate policies: environmental effectiveness, cost-effectiveness, distributional effects (including equity) and institutional feasibility. Other more specific criteria, such as effects on competitiveness and administrative feasibility, are generally subsumed within these four.

The literature provides a great deal of information for assessing how well different instruments meet these criteria, although it should be kept in mind that all instruments can be designed well or poorly and to be stringent or lax and politically attractive or unattractive. In addition, all instruments must be monitored and enforced to be effective. The general conclusions that can be drawn from the literature are that:
- Regulatory measures and standards generally provide some certainty of emissions levels, but their environmental effectiveness depends on their stringency. They may be preferable when information or other barriers prevent firms and consumers from responding to price signals (*high agreement, much evidence*).
- Taxes and charges are generally cost-effective, but they cannot guarantee a particular level of emissions, and they may be politically difficult to implement and, if necessary, adjust. As with regulations, their environmental effectiveness depends on stringency (*high agreement, much evidence*).
- Tradable permits can establish a carbon price. The volume of allowed emissions determines the carbon price and the

environmental effectiveness of this instrument, while the distribution of allowances can affect cost-effectiveness and competitiveness. Experience has shown that banking provisions can provide significant temporal flexibility (*high agreement, much evidence*). Uncertainty in the price of carbon makes it difficult to estimate the total cost of meeting emission reduction targets.

Voluntary agreements (VAs) between industry and governments, which vary considerably in scope and stringency, are politically attractive, raise awareness among stakeholders and have played a role in the evolution of many national policies. A few have accelerated the application of best available technology and led to measurable reductions of emissions compared to the baseline, particularly in countries with traditions of close cooperation between government and industry. However, there is little evidence that VAs have achieved significant reductions in emissions beyond business as usual (*high agreement, much evidence*). The successful programmes all include clear targets, a baseline scenario, third party involvement in design and review and formal provisions for monitoring.

- Financial incentives are frequently used by governments to stimulate the diffusion of new, less GHG-emitting technologies. While economic costs are generally higher for these than for other instruments, financial incentives are often critical to overcoming the barriers to the penetration of new technologies (*high agreement, much evidence*). Direct and indirect subsidies for fossil fuel use and agriculture remain common practice, although those for coal have declined over the past decade in many Organization for Economic Co-operation and Development (OECD) and in some developing countries.
- Government support through financial contributions, taxation measures, standard setting and market creation is important to the promotion of technology development, innovations and transfer. However, government funding for many energy research programmes has fallen off since the oil shock in the 1970s and stayed constant at this lower level, even after the United Nations Framework Convention on Climate Change (UNFCCC) was ratified. Substantial additional investments in – and policies for – Research and Development (R&D) are needed to ensure that technologies are ready for commercialization in order to arrive at a stabilization of GHGs in the atmosphere (see Chapter 3), as are economic and regulatory instruments to promote their deployment and diffusion (*high agreement, much evidence*).
- Information instruments, including public disclosure requirements, may affect environmental quality by promoting better-informed choices and lead to support for government policy. There is only limited evidence that the provision of information can achieve emissions reductions, but it can improve the effectiveness of other policies (*high agreement, medium evidence*).

In practice, climate-related policies are seldom applied in complete isolation, as they overlap with other national polices relating to the environment, forestry, agriculture, waste management, transport and energy and, therefore, in many cases require more than one instrument. For an environmentally effective and cost-effective instrument mix to be applied, there must be a good understanding of the environmental issue to be addressed, the links with other policy areas and the interactions between the different instruments in the mix. Applicability in specific countries, sectors and circumstances – particularly developing countries and economies in transition – can vary greatly, but may be enhanced when instruments are adapted to local circumstances (*high agreement, much evidence*).

International agreements

As precedents, the UNFCCC and Kyoto Protocol have been significant in providing a means to solve a long-term international environmental problem, but they are only first steps towards the implementation of an international response strategy to combat climate change. The Kyoto Protocol's most notable achievements are the stimulation of an array of national policies, the creation of a carbon market and the establishment of new institutional mechanisms. Its economic impacts on the participating countries are yet to be demonstrated. The Clean Development Mechanism (CDM), in particular, has created a large project pipeline and mobilized substantial financial resources, but it has faced methodological challenges in terms of determining baselines and additionality. The Protocol has also stimulated the development of emissions trading systems, but a fully global system has not been implemented. The Kyoto Protocol is currently constrained by the modest emission limits. It would be more effective if the first commitment period is followed-up by measures to achieve deeper reductions and the implementation of policy instruments covering a higher share of global emissions (*high agreement, much evidence*).

New literature highlights the options for achieving emission reductions both under and outside of the Convention and its Kyoto Protocol by, for example, revising the form and stringency of emission targets, expanding the scope of sectoral and sub-national agreements, developing and adopting common policies, enhancing international Research, Development and Demonstration (RD&D) technology programmes, implementing development-oriented actions and expanding financing instruments (*high agreement, much evidence*). An integration of diverse elements, such as international R&D co-operation and cap and trade programmes, within an agreement is possible, but any comparison of the efforts made by different countries would be complex and resource-intensive (*medium agreement, medium evidence*).

Recent publications examining future international agreements in terms of potential structure and substance report that because climate change is a global problem, any approach that does not include a larger share of global emissions will have a higher global cost or be less environmentally effective (*high agreement, much evidence*). The design of a future regime will have significant implications for global costs and the distribution of cost among regions at different points in time There is a broad consensus in the literature that a successful agreement will have to be environmentally effective and cost-effective, incorporate distributional considerations and equity and be institutionally feasible (*high agreement, much evidence*). Agreements are more likely to be effective if they include goals, specific actions, timetables, participation and institutional arrangements and provisions for reporting and compliance (*high agreement, much evidence*).

Goals determine the extent of participation, the stringency of the measures and the timing of the actions. For example, to limit the temperature increase to 2°C above pre-industrial levels, developed countries would need to reduce emissions in 2020 by 10–40% below 1990 levels and in 2050 by approximately 40–95%. Emissions in developing countries would need to deviate below their current path by 2020, and emissions in all countries would need to deviate substantially below their current path by 2050. A temperature goal of less than 2°C requires earlier reductions and greater participation (and vice versa) (*high agreement, much evidence*). Abatement costs depend on the goal, vary by region and depend on the allocation of emission allowances among regions and the level of participation.

Initiatives of local and regional authorities, corporations, and non-governmental organizations

Corporations, local and regional authorities and NGOs are adopting a variety of actions to reduce GHG emissions. Corporate actions range from voluntary initiatives to emissions targets and, in a few cases, internal trading systems. The reasons corporations undertake independent actions include the desire to influence or pre-empt government action, to create financial value, and to differentiate a company and its products. Actions by regional, state, provincial and local governments include renewable energy portfolio standards, energy efficiency programmes, emission registries and sectoral cap and trade mechanisms. These actions are undertaken to influence national policies, address stakeholder concerns, create incentives for new industries and/or to create environmental co-benefits. Non-government organizations promote programmes that reduce emissions through public advocacy, litigation and stakeholder dialogue. Many of the above actions may limit GHG emissions, stimulate innovative policies, encourage the deployment of new technologies and spur experimentation with new institutions, but they generally have limited impact on their own. To achieve significant emission reductions, these actions must lead to changes in national policies (*high agreement, medium evidence*).

Implications for global climate change policy

Climate change mitigation policies and actions taken by national governments, the private sector and other areas of civil society are inherently interlinked. For example, significant emissions reductions have occurred as a result of actions by governments to address energy security or other national needs (e.g. the switch in the UK to gas, the energy efficiency programmes of China and India, the Brazilian development of a transport fleet driven by bio-fuel or the trend in the 1970s and 1980s toward nuclear power). However, non-climate policy priorities can overwhelm climate mitigation efforts (e.g. decisions in Canada to develop the tar sands reserves, those in Brazil to clear forests for agriculture and in the USA to promote coal power to enhance energy security) and lead to increased emissions. New research to assess the interlinkages between climate change and other national policies and actions might lead to more politically feasible, economically attractive and environmentally beneficial outcomes and international agreements.

13.1 Introduction

Article 4 of the United Nations Framework Convention on climate change (UNFCCC) commits all Parties – taking into account their common but differentiated responsibilities and their specific national and regional priorities, objectives and circumstances – to formulate, implement, publish and regularly update national and, where appropriate, regional programmes containing measures that will result in the mitigation of climate change by addressing anthropogenic emissions of greenhouse gases (GHGs) by sources and removals by sinks. The main purpose of this chapter is to discuss national policy instruments and their implementation, international agreements and other arrangements and initiatives of the private sector, local governments and non-governmental organizations (NGOs). This chapter expands on the literature that has emerged since the Third Assessment Report (TAR) – in particular, on aspects covered in Chapters 6 and 10 of the TAR. There is a relatively heavier focus given to publications proposing new approaches to possible future international agreements, alternative options for international cooperation and initiatives of local governments and the private sector. Wherever feasible, these agreements and arrangements are discussed in the context of criteria such as environmental effectiveness, cost-effectiveness, distributional considerations, institutional feasibility, among others. This chapter does not discuss in detail either sectoral policies, which can be found in other chapters of this report, or adaptation policies, as those may be found in IPCC (2007b).

13.1.1 Types of policies, measures, instruments and co-operative arrangements

A variety of policies, measures, instruments and approaches are available to national governments to limit the emission of GHGs; these include regulations and standards, taxes and charges, tradable permits, voluntary agreements (VAs), informational instruments, subsidies and incentives, research and development and trade and development assistance. Box 13.1 provides a brief definition of each instrument (Hahn, 2001; Sterner, 2003). Depending on the legal framework within which each individual country must operate, these may be implemented at the national level, sub-national level or through bi-lateral or multi-lateral arrangements, and they may be either legally binding or voluntary and either fixed or changeable (dynamic).

Box 13.1 Definitions of selected GHGs abatement policy instruments

Note: The instruments defined below to directly control GHG emissions; instruments may also be used to manage activities that indirectly lead to GHG emissions, such as energy consumption.

Regulations and Standards: These specify the abatement technologies (technology standard) or minimum requirements for pollution output (performance standard) that are necessary for reducing emissions.

Taxes and Charges: A levy imposed on each unit of undesirable activity by a source.

Tradable Permits: These are also known as marketable permits or cap-and-trade systems. This instrument establishes a limit on aggregate emissions by specified sources, requires each source to hold permits equal to its actual emissions and allows permits to be traded among sources.

Voluntary Agreements: An agreement between a government authority and one or more private parties with the aim of achieving environmental objectives or improving environmental performance beyond compliance to regulated obligations. Not all VAs are truly voluntary; some include rewards and/or penalties associated with participating in the agreement or achieving the commitments.[1]

Subsidies and Incentives: Direct payments, tax reductions, price supports or the equivalent thereof from a government to an entity for implementing a practice or performing a specified action.

Information Instruments: Required public disclosure of environmentally related information, generally by industry to consumers. These include labelling programmes and rating and certification systems.

Research and Development (R&D): Activities that involve direct government funding and investment aimed at generating innovative approaches to mitigation and/or the physical and social infrastructure to reduce emissions. Examples of these are prizes and incentives for technological advances.

Non-Climate Policies: Other policies not specifically directed at emissions reduction but which may have significant climate-related effects.

---

1 Voluntary Agreements (VAs) should not be confused with voluntary actions which are undertaken by govern-ment agencies at the sub-national level, corporations, NGOs and other organizations independent of national government authorities. See Section 13.4.

13.1.2   Criteria for policy choice

Four principal criteria for evaluating environmental policy instruments are reported in the literature; these are:

- Environmental effectiveness – the extent to which a policy meets its intended environmental objective or realizes positive environmental outcomes.
- Cost-effectiveness – the extent to which the policy can achieve its objectives at a minimum cost to society.
- Distributional considerations – the incidence or distributional consequences of a policy, which includes dimensions such as fairness and equity, although there are others.
- Institutional feasibility – the extent to which a policy instrument is likely to be viewed as legitimate, gain acceptance, adopted and implemented.

It has to be mentioned, however, that literature in the fields of economics and political science does not provide much guidance in terms of determing which evaluative criteria are the most appropriate for an analysis of environmental policy. However, many authors employ criteria similar to the ones listed above, and although other criteria may also be important in evaluating policies, the analysis presented in this chapter is limited to these four criteria. Criteria may be applied by governments in making ex ante choices among instruments and in ex post evaluation of the performance of instruments.

13.1.2.1   Environmental effectiveness

The main goal of environmental policy instruments and international agreements is to reduce the negative impact of human action on the environment. Policies that achieve specific environmental quality goals better than alternative policies can be said to have a higher degree of environmental effectiveness. It should be noted that although climate protection is the ostensible environmental goal for any climate policy, there may be ancillary environmental benefits (for example, those demonstrated by Burtraw *et al.* (2001a) for air pollution benefits; see also Section 4.5.2. for air quality co-benefits).

The environmental effectiveness of any policy is contingent on its design, implementation, participation, stringency and compliance. For example, a policy that seeks to fully address the climate problem while dealing with only some of the GHGs or some of the sectors will be relatively less effective than one that aims at addressing all gases and all sectors.

The environmental effectiveness of an instrument can only be determined by estimating how well it is likely to perform. Harrington *et al.* (2004) distinguish between estimating how effective an environmental instrument will be ex ante and evaluating its performance ex post. These researchers were able to find or recreate ex ante estimates of expected emissions reductions in a series of U.S. and European case studies. Their comparison of the ex ante and ex post observations suggests a reasonable degree of accuracy in the estimates, with those cases in which emissions reductions were greater than expected involving incentive-based instruments, while the cases in which reductions fell short of expectations involved regulatory approaches.

There are situations in which standards are proven to be effective. Regulators may be unduly pessimistic about the environmental performance of incentive-based instruments or unduly optimistic about the performance of regulatory approaches, or perhaps both. Recent evidence suggests that market-based approaches can provide equal if not superior environmental quality improvements over regulatory approaches (see Ellerman, 2006). As we discuss below, however, institutional constraints may alter the relative efficacy of market- and standards-based instruments.

13.1.2.2   Cost-effectiveness

The cost-effectiveness of a policy is a key decision parameter in a world with scarce resources. Given a particular environmental quality goal, the most cost-effective policy is the one which achieves the desired goal at the least cost. There are many components of cost, and these include both the direct costs of administering and implementing the policy as well as indirect costs, such as how the policy drives cost-reducing technological change.

Cost-effectiveness is distinct from general economic efficiency. Whereas cost-effectiveness takes an environmental goal as given, efficiency involves the process of selecting a specific goal according to economic criteria (Sterner, 2003). Consequently, the choice of a particular environmental goal will likely have dramatic impacts on the overall cost of a policy, even if that policy is implemented using the most cost-effective instrument.

Policies are likely to vary considerably in terms of cost-effectiveness, and any estimation of the costs involved can be challenging (Michaelowa, 2003b). While cost-effectiveness estimates traditionally include the direct expenditures incurred as a result of implementing any specific policy, the policy may also impose indirect social costs, which are more difficult to measure (Davies and Mazurek, 1998). Moreover, costs for which data are limited are often ignored. Harrington *et al.* (2000) provide a summary of commonly excluded costs as well as examples of efforts to estimate these.

Cost-effectiveness can be enhanced with low transaction costs for compliance. This implies limiting the creation of new institutions and keeping implementation procedures as simple as possible while still ensuring system integrity. Studies reported in the literature can be divided into two categories in terms of the economic impacts of the timing of reductions. While some researchers argue that reductions should be postponed until low-cost technologies are available, others argue that necessary decisions have to be made today to avoid a 'lock-in' to an emission intensive pathway that would be expensive to leave at a later time point (see also Chapter 11).

A common concern is that ex ante cost estimates may not reflect the actual costs of a policy when it is assessed from an ex post perspective. Harrington *et al.* (2000) show that the discrepancy between the actual and estimated total costs of 28 environmental regulations in the USA is relatively low and, if anything, that ex ante estimates tend to overstate total costs. While these authors do not systematically evaluate specific environmental instruments, they do find that estimates for market-based instruments tend to overstate unit costs, while unit-costs estimates for other instruments are neither under- nor overestimates.

13.1.2.3   *Distributional considerations*

Policies rarely apportion environmental benefits and costs evenly across stakeholders. Even if a policy meets an environmental goal at least cost, it may face political opposition if it disproportionately impacts – or benefits – certain groups within a society, across societies or across generations. From an economic perspective, a policy is considered to be beneficial if it improves social welfare overall. However, this criterion does not require that the implementation of that policy actually improves the specific situation of any one individual. Consequently, as Keohane *et al.* (1998) argue, distributional considerations may be more important than aggregate cost effectiveness when policymakers evaluate an instrument.

The distributional considerations of climate change policies relate largely to equity. Equity can be defined in a number of ways within the climate context (see IPCC, 2001). Equity and fairness may be perceived differently by different people, depending on the cultural background of the observer. For example, Ringius *et al.* (2002) view responsibility, capacity and need as the basic principles of fairness that seem to be sufficiently widely recognized to serve as a normative basis for a climate policy regime. These three principles have been used in the evaluation of potential international climate agreements (e.g. Torvanger *et al.*, 2004).

A regulation that is perceived as being unfair or for which the incidence is unbalanced may have a difficult time making it through the political process.[2] However, distributional considerations are fundamentally subjective, and the most equitable policy may not be the most politically popular one. For example, a policy that focuses the regulatory burden on a low-income subpopulation or country but directs the benefits to a wealthy interest group may sail with ease through the political process. While highly inequitable in costs and benefits, such an instrument is occasionally attractive to politicians. Bulkeley

(2001) describes the different interests in the Australian climate policy debate and suggests that industrial emitters managed to steer the country away from ambitious reduction target – and toward an emissions increase – at the third Conference of the Parties in Kyoto.

Due to the fact that there is little consensus as to what constitutes optimal distribution, it can be difficult to compare – let alone rank – environmental policies based on distributional criteria (Revesz and Stavins, 2006). One exception is provided by Asheim *et al.* (2001), who construct an axiom of equity which, they argue, can be used to evaluate sustainability.[3] However, while sustainability may be important when evaluating environmental policies, it only captures the inter-generational dimension of distribution and is imperfectly related to political acceptability.

13.1.2.4   *Institutional feasibility*

Institutional realities inevitably constrain environmental policy decisions. Environmental policies that are well adapted to existing institutional constraints have a high degree of institutional feasibility. Economists traditionally evaluate instruments for environmental policy under ideal theoretical conditions; however, those conditions are rarely met in practice, and instrument design and implementation must take political realities into account. In reality, policy choices must be both acceptable to a wide range of stakeholders and supported by institutions, notably the legal system. Other important considerations include human capital and infrastructure as well as the dominant culture and traditions. The decision-making style of each nation is therefore a function of its unique political heritage. Box 13.2 provides an example for one country, taken largely from OECD (2005c).

Certain policies may also be popular due to institutional familiarity. Although market-based instruments are becoming more common, they have often met with resistance from environmental groups. Market-based instruments continue to face strong political opposition, even in the developed world, as demonstrated by environmental taxes in the USA or Europe. Regulatory policies that are outside of the norm of society will always be more difficult to put into effect (e.g. speed limits in Germany, or private sector participation in water services in Bolivia).

Another important dimension of institutional feasibility deals with implementing policies once they have been designed and adopted. Even if a policy receives political support, it may be difficult to implement under certain bureaucratic structures.

---

2   The United States has acknowledged the role of distribution explicitly through Executive Order 12878 (1994), which requires federal agencies to address environmental justice in their missions and activities.

3   For a summary of the economic literature on sustainability and intergenerational equity, see Pezzey and Toman (2002).

Box 13.2 The UK climate change levy: a study in political economy

The UK has a tradition of action on climate change that dates from the early acceptance of the problem by the Conservative Prime Minister Margaret Thatcher in 1988. The Labour government in 1997 reaffirmed the commitment to act and to use market-based instruments wherever possible; however, it voiced concerns on two aspects of this commitment: Firstly, that such measures might have a disproportionate effect on the poor which, in turn, might affect the coal mining communities (an important constituency) and, secondly, that this commitment might perpetuate a perception that the Labour government was committed to high taxes.

A key element of the UK's climate policy is a climate levy. The levy is paid by energy users – not extractors or generators – is levied on industry only and aims to encourage renewable energy. An 80% discount can be secured if the industry in question participates in a negotiated 'climate change agreement' to reduce emissions relative to an established baseline. Any one company over-complying with its agreement can trade the resulting credits in the UK emissions trading scheme, along with renewable energy certificates under a separate renewable energy constraint on generators. However, a number of industrial emitters wanted a heavier discount and, through lobbying, they managed to have a voluntary emissions trading scheme established that enables companies with annual emissions above 10,000 tCO_2-eq to bid for allocation of subsidies. The "auction" offered payments of 360 million and yielded a de-facto payment of 27 € per tonne of CO_2. Thus, the trading part of the scheme has design elements that strongly reflect the interest groups involved (Michaelowa, 2004). The levy itself has limited coverage and, consequently, households, and energy extractors and generators have no incentive to switch to low carbon fuels. However, its design does take household vulnerability, competitiveness concerns and the sensitivity of some sectoral interests into account. Thus, while the levy has contributed to emission reduction, it has not been as effective as a pure tax; a pure tax may not have been institutionally feasible.

13.2 National policy instruments, their implementation and interactions

The policy-making process of almost all governments consists of complex choices involving many stakeholders, including the potential regulated industry, suppliers, producers of complementary products, labour organizations, consumer groups and environmental organizations. The choice and design of virtually any instrument has the potential to benefit some of these stakeholders and to harm others. For example, permits allocated free to existing firms represent a valuable asset transferred from the government to industry, while auctioned permits and taxes generally impose heavier burdens on polluters. As a result, it is likely that a candidate instrument will likely face both support and opposition from the stakeholders. Voluntary measures are often favoured by industry because of their flexibility and potentially lower costs, but these are often opposed by environment groups because of their lack of accountability and enforcement. In practice, policies may be complementary or opposing; moreover, the political calculus used to choose a particular instrument differs for each government.[4]

In formulating a domestic climate policy programme, a combination of policy instruments may work better in practice than reliance on a single instrument. Furthermore, an instrument that works well in one country may not work well in another country with different social norms and institutions. When instruments are to be compared, it is important that the different levels of stringency be taken into consideration and adjusted, for all of the instruments described herein may be set at different levels of stringency. Regulations will also undoubtedly need to be adjusted over time. All instruments must be supplemented with a workable system of monitoring and enforcement. Furthermore, instruments may interact with existing institutions and regulations in other sectors of society.

13.2.1 Climate change and other related policies

In this section we consider a number of instruments that have been used to manage environmental problems in different parts of the world. Some of these tools have been used for climate policy, while others have not; however, experience from dealing with other pollutants suggests their applicability to climate. Mitigation options can range from the purely technological (such as fuel switching) to the purely behavioural (such as reducing vehicle kilometres travelled) as well as innumerable combinations of both technological and behavioural options. Policies, measures and instruments are tools to trigger the implementation of these options.

13.2.1.1 Regulations and standards

Regulatory standards are the most common form of environmental regulation, and they cover a wide variety of

---

4 The design of most instruments assumes effective compliance and penalty provisions.

approaches. A regulatory standard specifies with a certain degree of precision the action(s) that a firm or individual must undertake to achieve environmental objectives and can consist of such actions as specifying technologies or products to use or not use and/or more general standards of performance as well as proclaiming dictates on acceptable and unacceptable behaviour. Two broad classes of regulatory standards are technology and performance standards. Technology standards mandate specific pollution abatement technologies or production methods, while performance standards mandate specific environmental outcomes per unit of product. In this context, where a technology standard might mandate specific CO_2 capture and storage methods on a power plant, a performance standard would limit emissions to a certain number of grams of CO_2 per kilowatt-hour of electricity generated. A product standard would, for example, be the requirement that refrigerators operate minimally at a specified level of efficiency, while a technology-forcing standard would involve setting the refrigerator efficiency requirement slightly beyond present-day technological feasibility but announcing that the efficiency requirement will not go into effect until a number of years following the announcement.

The primary advantage of a regulatory standard is that it may be tailored to an industry or firm, taking into account the specific circumstances of that industry or firm. There is also a more direct connection between the regulatory requirement and the environmental outcome, which can provide some degree of certainty.

Technology standards involve the regulator stipulating the specific technology or equipment that the polluter must use. Technology standards are best used when there are few options open to the polluter for controlling emissions; in this case, the regulator is able to specify the technological steps that a firm should take to control pollution. The information requirements for technology standards are high: the regulator must have good and reliable information on the abatement costs and options open to each firm. Losses in cost effectiveness arise when regulators are less well informed; technology standards may then be applied uniformly to a variety of firms, rather than tailoring the standard to the actual circumstance of the firm. This raises costs without improving environmental effectiveness and is one of the main drawbacks to regulatory standards.

Performance standards can reduce these potential problems with technology standards by providing more flexibility (IPCC, 2001). Costs can generally be lower whenever a firm is given some discretion in how it meets an environmental target. Performance standards expand compliance options beyond a single mandated technology and may include process changes, reduction in output, changes in fuels or other inputs and alternative technologies. Despite this increased flexibility,

performance standards also require well-informed and responsive regulators.

One problem with regulatory standards is that they do not provide polluters with the incentive(s) to search for better approaches to reducing pollution. Thus, they may not perform well in inducing innovation and technological change (Jaffe et al., 2003; Sterner, 2003). If a government mandates a certain technology, there is no economic incentive for firms to develop more effective technologies. Moreover, there may be a 'regulatory ratchet' whereby firms are discouraged from developing more effective technologies out of fear that standards will be tightened yet again (Harrington et al., 2004). Finally, although it may be possible to force some technological change through technology mandates, it is difficult for regulators to determine the amount of change that is possible at a reasonable economic cost. This raises the possibility of implementing either costly, overly stringent requirements or, alternatively, weak, unambitious requirements (Jaffe et al., 2003). Nevertheless, there are examples in the literature of technology innovations spurred by regulatory standards. For example, Wätzold (2004) reported innovative responses from pollution control vendors in Germany in response to standards for SO_2 control.

Although relatively few regulatory standards have been adopted with the sole aim of reducing GHG emissions, standards have been adopted that reduce these gases as a co-benefit. For example, there has been extensive use of standards to increase energy efficiency in over 50 nations (IPCC, 2001). Energy efficiency applications include fuel economy standards for automobiles, appliance standards, and building codes.[5] These types of policies are discussed in more detail in Chapters 5 and 6 of this report. Standards to reduce methane and other emissions from solid waste landfills have been adopted in Europe, the USA and other countries (see Chapter 10) and are often driven by multiple factors, including the reduction of volatile organic compound (VOC) emissions, improved safety by reducing the potential for explosions and reduced odours for local communities (Hershkowitz, 1998).

There are a number of documented situations in which regulatory standards have worked well (see Freeman and Kolstad, 2006; Sterner, 2003). Sterner (2003) reports several cases of such situations, including those in which firms are not responsive to price signals (e.g. in non-competitive settings or with state enterprises) and where monitoring emissions is difficult but tracking the installation of technology is easy. In situations where there is imperfect monitoring and homogeneous abatement costs between firms, Montero (2005) finds that standards may lead to lower emissions and may be economically more efficient than market-based instruments. Based on an analysis of the German SO_2 abatement programme, Wätzold (2004) concludes that a technology standard may be acceptable

---

5   For example, the Green Building Council in the United States of America.

Box 13.3 China mandates energy efficiency standard in urban construction

Approximately 2 billion m² of floor space is being built annually in China, or one half of the world's total. Based on the growing pace of its needs, China will see another 20–30 billion m² of floor space built between the present and 2020. Buildings consume more than one third of all final energy in China, including biomass fuels (IEA, 2006). China's recognition of the need for energy efficiency in the building sector started as early as the 1980s but was impeded due to the lack of feasible technology and funding. Boosted by a nationwide real estate boom, huge investment has flowed into the building construction sector in recent years.

On 1 January, 2006, China introduced a new building construction statute that includes clauses on a mandatory energy efficiency standard for buildings. The Designing Standard for Energy Conservation in Civil Building requires construction contractors to use energy efficient building materials and to adopt energy-saving technology in heating, air conditioning, ventilation and lighting systems in civil buildings. Energy efficiency in building construction has also been written into China's 11th Five-Year National Development Programme (2006–2010), which aims for a 50% reduction in energy use (compared with the current level) and a 65% decrease for municipalities such as Beijing, Shanghai, Tianjin and Chongqing as well as other major cities in the northern parts of the country. Whether future buildings will be able to comply with the requirements in the new statute will be a significant factor in determining whether the country will be able to realize the ambitious energy conservation target of a 20% reduction in energy per gross domestic product (GDP) intensity during the 11th Five-Year Plan of 2005–2010.

when only one technology exists to achieve an environmental result and, therefore, firms do not face differential abatement costs. Finally, standards may be desirable where there are informational barriers that prevent firms or individuals from responding solely to price signals. This may be particularly relevant for energy efficiency standards for household appliances and other similar applications (OECD, 2003d). Chapter 6 provides additional information on this subject.

A growing body of literature is focusing on whether regulatory standards or market-based instruments are preferable for developing countries. One common view is that technology standards may be more appropriate for building the initial capacity for emissions reduction because economic incentive programmes require more specific and greater institutional capacity, have more stringent monitoring requirements and may require fully developed market economies to be effective (IPCC, 2001; Bell and Russell, 2002). Willems and Baumert (2003) support this approach but also note that technology approaches, policies and measures may have greater applicability to the general capacity needs of developing countries interested in pursuing sustainable development strategies (See Box 13.3). Russell and Vaughan (2003) suggest that a transitional strategy is the appropriate approach for developing countries, whereby technology standards are introduced first, followed by performance standards and finally by experimentation with market-based instruments. An alternative view is that, in some cases, a performance standard at the facility level and an overall

emissions cap could provide a more a more effective structure (Ellerman, 2002; Kruger et al., 2003). This type of approach could also facilitate a transition to a tradable permits programme as the institutions and economies develop over time.

13.2.1.2 Taxes and charges

An emission tax on GHG emissions requires individual emitters to pay a fee, charge or tax[6] for every tonne of GHG released into the atmosphere.[7] An emitter must pay this per-unit tax or fee regardless of how much emission reduction is being undertaken.[8] Each emitter weighs the cost of emissions control against the cost of emitting and paying the tax; the end result is that polluters undertake to implement those emission reductions that are cheaper than paying the tax, but they do not implement those that are more expensive, (IPCC, 1996, Section 11.5.1; IPCC, 2001, Section 6.2.2.2; Kolstad, 2000). Since every emitter faces a uniform tax on emissions per tonne of GHG (if energy, equipment and product markets are perfectly competitive), emitters will undertake the least expensive reductions throughout the economy, thereby equalizing the marginal cost of abatement (a condition for cost-effectiveness). Taxes and charges are commonly levelled on commodities that are closely related to emissions, such as energy or road use.

An emissions tax provides some assurance in terms of the marginal cost of pollution control, but it does not ensure a particular level of emissions. Therefore, it may be necessary to

---

6   No distinction is made here among the terms taxes, fees or charges. In actuality, the revenue from taxes may go into the general government coffers, whereas the revenue from fees or charges may be earmarked for specific purposes.

7   Because GHGs have different effects on atmospheric warming per unit of emissions, the use of carbon dioxide equivalents (CO_2-eq) is one way of measuring relative impact.

8   An alternative is the idea of threshold taxes, where the tax per unit of emissions is only assessed on emissions greater than a set threshold (Pezzey 2003). In other words, infra-marginal emissions would be tax-exempt. This type of tax would generate less revenue but could be more politically acceptable.

adjust the tax level to meet an internationally agreed emissions commitment (depending on the structure of the international agreement). Over time, an emissions tax needs to be adjusted for changes in external circumstances, such as inflation, technological progress and new emissions sources (Tietenberg, 2000). Fixed emissions charges in the transition economies of Eastern Europe, for example, have been significantly eroded by the high inflation of the past decade (Bluffstone and Larson, 1997). Innovation and invention generally have the opposite effect by reducing the cost of emissions reductions and increasing the level of reductions implemented. If the tax is intended to achieve a given overall emissions limit, the tax rate will need to be increased to offset the impact of new sources (Tietenberg, 2000).

Most environmentally related taxes with implications for GHG emissions in OECD countries are levied on energy products (150 taxes) and on motor vehicles (125 taxes), rather than on CO_2 emission directly. There is also a significant number of waste-related taxes in OECD countries (about 50 taxes in all), levied either on particular products that can cause particular problems for waste management (about 35 taxes) or on various forms of final waste disposal, including those on incineration and/or land-filling (15 taxes in all). A very significant share of all the revenues from environmentally related taxes originates from taxes on motor fuels. Such taxes were introduced in all member countries many decades ago – primarily as a means to raise revenue. Irregardless of the underlying reasoning for their implementation, however, they do impact on the prices (potential) car users are confronted with and thus have important environmental impacts.

However, there is some experience with the direct taxation of CO_2 emissions. The Nordic Council of Ministers (2002) notes that CO_2 emissions in Denmark decreased by 6% during the period 1988–1997 while the economy grew by 20%, but that they also decreased by 5% in a single year – between 1996 and 1997 – when the tax rate was raised. Bruvoll and Larsen (2004) analysed the specific effect of carbon taxes in Norway. Although total emissions did increase, these researchers found a significant reduction in emissions per unit of GDP over the period due to reduced energy intensity, changes in the energy mix and reduced process emissions. The overall effect of the carbon tax was, however, modest, which may be explained by the extensive tax exemptions and relatively inelastic demand in those sectors in which the tax was actually implemented. Cambridge Econometrics (2005) analysed the impacts of the Climate Change Levy in the UK and found that total CO_2 emissions were reduced by 3.1 MtC – or 2.0% – in 2002 and by 3.6 MtC in 2003 compared to the reference case. The reduction is estimated to grow to 3.7 MtC – or 2.3% – in 2010.

To implement a domestic emissions tax, governments must consider a number of issues, such as the level at which the tax should be set, particularly in the case of pre-existing taxes (e.g. taxes which already exist on energy), or other potential distortions (e.g. subsidies to certain industries or fuels). Consideration must also be given to how the tax is used, with such options as whether it goes directly into general government coffers, is used to offset other taxes (i.e. the double-dividend effect), is transferred across national boundaries to an international body, is earmarked for specific abatement projects, such as renewable energy, or is allocated to those most adversely impacted by either the costs of emission reduction or damage from climate change. Another important issue is the point at which the tax is is should be levied. A tax on gasoline may be levied at the pump and collected directly from consumers or it may be levied on wholesale gasoline production and collected from oil companies. In either case, the final consumer ultimately pays most of this cost, but the administrative and monitoring costs may differ dramatically in the two cases.

Emission taxes do well in both cost effectiveness and environmental effectiveness. The real obstacles facing the use of emission taxes and charges are distributional and, in some countries, institutional. At the best of times, new taxes are not politically popular. Furthermore, emissions or energy taxes often fall disproportionately on lower income classes, thereby creating negative distributional consequences. In developing countries, institutions may be insufficiently developed for the collection of emission fees from a wide variety of dispersed sources. In many countries, state enterprises play a significant role; such public or quasi-private entities may not respond adequately to the incentive effects of a tax or charge.

13.2.1.3   Tradable permits

A steadily increasing amount of research is focusing on tradable permits in terms of, among others, efficiency and equity issues associated with the distribution of permits, implications of economy-wide versus sectoral programmes, mechanisms for handling price uncertainties, different forms of targets and compliance and enforcement issues.

Tradable permit systems can be designed to cover either emissions from a few sectors of the economy or those from virtually the entire economy.[9] A number of analyses have found that economy-wide approaches are superior to sectoral coverage because they equalize marginal costs across the entire economy. Using a variety of models, Pizer et al. (2006) report that in the USA significant cost savings are linked to an economy-wide programme when compared to a sectoral programme coupled with non-market-based policies.[10] Researchers have found similar results for the European Union

---

9   Thus far, emissions trading programmes, such as those for SO_2 and NO_x in the USA and that of the EU Emis-sions Trading System (EU ETS) for CO_2 have only covered certain sectors. In the case of the EU ETS, Chris-tiansen and Wettestad (2003) write that the EU restricted the sectors involved to ease implementation during the first phase of the programme.

10 However, they also find that the exclusion of certain sectors, such as residential and commercial direct use of fossil fuels, does not noticeably affect the cost of an otherwise economy-wide tradable permit system covering electricity production, industry and transportation.

Box 13.4 The EU Emission Trading System

The EU Emissions Trading System (EU ETS) is the world's largest tradable permits programme. The programme was initiated on January 1, 2005, and it applies to approximately 11,500 installations across the EU's 25 Member States. The system covers about 45% of the EU's total CO_2 emissions and includes facilities from the electric power sector and other major industrial sectors.

The first phase of the EU ETS runs from 2005 until 2007. The second phase will begin in 2008 and continue through to 2012, coinciding with the 5-year Kyoto compliance period. Member States develop National Allocation Plans, which describe in detail how allowances will be distributed to different sectors and installations. During the first phase, Member States may auction off up to 5% of their allowances; during the second phase, up to 10% of allowances may be auctioned off.

Market development and prices: A number of factors affect allowance prices in the EU ETS, including the overall size of the allocation, relative fuel prices, weather and the availability of certified emission reductions (CERs) from the Clean Development Mechanism (CDM) (Christiansen *et al*., 2005). The EU ETS experienced significant price volatility during its start-up period, and for a brief period in April 2006 prices rose to nearly 30 per tonne; however, prices subsequently dropped dramatically when the first plant-level emissions data from Member States were released. The sharp decline in prices focused attention on the size of the initial Phase I allocation. Analysts have concluded that this initial allocation was a small reduction from business as usual emissions (Grubb *et al*., 2005; Betz *et al*., 2004).

Consistency in national allocation plans: Several studies have documented differences in the allocation plans and methodologies of Member States (Betz *et al*., 2004; Zetterberg *et al*. 2004; Baron and Philibert, 2005; DEHSt, 2005). Researchers have looked at the impact on innovation and investment incentives of different aspects of allocation rules (Matthes *et al*., 2005; Schleich and Betz, 2005) and have found that these rules can affect technology choices and investment decisions. Ahman *et al*. (2006), Neuhoff *et al*. (2006) and Betz *et al*., (2004) find that when Member States' policies require the confiscation of allowances following the closure of facilities, this creates a subsidy for continued operation of older facilities and a disincentive to build new facilities. They further find that different formulas for new entrants can impact on the market.

Implications of free allocation on electricity prices: Sijm *et al*. (2006) report that a significant percentage of the value of allowances allocated to the power sector was passed on to consumers in the price of electricity and that this pass-through of costs could result in substantially increased profits by some companies. The authors suggest that auctioning a larger share of allowances could address these distributional issues. In a report for the UK government, IPA Energy Consulting found a similar cost pass-through for the UK and other EU Member States (IPA Energy Consulting, 2005).

---

and the EU ETS. (Babiker *et al*., 2003; Betz *et al*., 2004; Klepper and Peterson, 2004; Bohringer and Löschel, 2005).

Not only the coverage of sectors may vary in a tradable permits programme, but also the point of obligation. The responsibility for holding permits may be assigned directly to emitters, such as energy-using industrial facilities (downstream), to producers or processors of fuels (upstream) or to some combination of the two (a 'hybrid system').[11] The upstream system would require permits to be held at the level of fossil fuel wholesalers and importers (Cramton and Kerr, 2002).[12]

There are two basic options for the initial distribution of permits: (1) free distribution of permits to existing polluters or (2) auctions. Cramton and Kerr (2002) describe a number of equity benefits of auctions, including providing a source of revenue that could potentially address inequities brought about by a carbon policy, creating equal opportunity for new entrants and avoiding the potential for "windfall profits" that might accrue to emissions sources if allowances are allocated at no charge.[13] (See Box 13.4 for a discussion of this issue).

Goulder *et al*. (1999) and Dinan and Rogers (2002) find that recycling revenues from auctioned allowances can have economy-wide efficiency benefits if they are used to reduce certain types of taxes. Dinan and Rogers (2002) and Parry (2004) argue that free allocation of tradable permits may be regressive because this type of allowance distribution leads to income

---

11 See IPCC (2001b), Baron and Bygrave (2002), UNEP/UNCTAD (2002), and Baron and Philibert (2005) for a discussion of the advantages and disadvantages of these different approaches.

12 As the discussion below notes, the point of obligation is not necessarily the point at which all permits need be allocated.

13 A hybrid of free allocation and auctioning or emissions taxes is also possible (Pezzey 2003). Bovenburg and Goulder (2001) and Burtraw *et al*. (2002) find that allocating only a small portion of permits at no cost while auctioning the remainder can compensate industry for losses due to a carbon policy.

transfers towards higher income groups (i.e. shareholders) at the expense of households. In contrast, these authors find that government revenues from auctions may be used to address equity issues through reductions in taxes or other distributions to low-income households. Ahman *et al.* (2006) argue that a gradual transition from free allocation to auctioning might be a politically feasible manner to develop a fairer distribution of allowances.

To date, most emissions trading programmes have distributed emissions allowances almost entirely through free allocations.[14] Experience with the US SO_2 programme shows that the no-cost allocation of allowances was critical for gaining political acceptance for the emissions trading concept (Ellerman, 2005). Christiansen and Wettestad (2003) and Markussen and Svendsen (2005) discuss how interest group pressures led to a largely free allocation of allowances in the EU ETS. In a broader sense, the rationale for a policy allowing some free allocation of allowances based on historic emissions is based on the desire to compensate incumbent installations that are affected by the regulation (Tietenberg, 2003; Harrison and Radov, 2002, Ahman *et al.* 2006).

The number of publications exploring the efficiency, equity and competitiveness implications of allowance allocation approaches is continuing to grow. For example, Burtraw *et al.* (2001b) and Fischer (2001) found that periodic updates of allocations on the basis of production are economically inefficient. In an analysis of a potential emissions trading programme in Alberta, Canada, Haites (2003b) found that this type of periodic updating of allocations based on each source's output may reduce the decline in production for some sectors that may arise from an emissions cap but that it may also reduce profits and raise overall costs when compared to a fixed allocation. Demailly and Quirion (2006) find that under certain assumptions, an output-based allocation in the European cement industry would reduce leakage with limited impacts on production. See Chapter 11, Section 11.7.4 for a more extensive discussion on competitiveness issues.

A final issue associated with the distribution of allowances is whether excessive market power can distort prices. Maeda (2003) examines how the initial distribution of permits affects the potential emergence of firms with market power. Tietenberg (2006) summarizes research on market power, including studies on whether different auction designs or initial permit allocation can lead to price manipulation by dominant firms. He concludes that in practice, market power 'typically has not been a problem in emissions trading.' There has yet to be an overall assessment of market power in the EU ETS.

Several authors have compared the advantages and disadvantages of absolute targets (i.e., mass emissions limits on a sector or economy) to those of intensity targets (i.e. limits on emission per unit of GDP).[15] Ellerman and Wing (2003) and Kolstad (2006) find that intensity targets can reduce the uncertainties associated with the cost of emission reduction under uncertain economic growth levels. Pizer (2005b) finds that intensity targets may be more appropriate if the short-term objective is to slow, rather than halt, emissions growth, while Ellerman and Wing (2003) show that an intensity target may be set so stringently that it can halt or reverse growth. Dudek and Golub (2003) argue that absolute targets have more certain environmental results and lower transaction costs for emissions trading, thereby creating stronger incentives for technological change. Kuik and Mulder (2004) find that, for the EU, an intensity or relative target would avoid negative effects on competitiveness but would not reduce emissions at the lowest costs. In contrast, an absolute target combined with permit trading leads to efficient emissions reduction, but its overall macroeconomic costs may be significant. Finally, Quirion (2005) argues that, in the most plausible cases, an emissions tax and an absolute target are superior to an intensity target and that the welfare gaps between the two types of targets are very small. Overall, intensity targets are less effective than absolute targets if the goal is to achieve a certain level of emissions reduction, but they may be more effective at addressing costs when economic growth is uncertain.

Although a tradable permits approach can ensure that a certain quantity of emissions will be reduced, it does not provide any certainty of price. Price uncertainty may be addressed by a 'price cap' or 'safety valve' mechanism, which guarantees that the government will sell additional permits if the market price of allowances hits a certain price (Pizer, 2002; McKibbon and Wilcoxen; 2002, Jacoby and Ellerman; 2004).[16] The underlying reasoning is that GHGs become the focus of concern as they accumulate over an extended period in the atmosphere. There may therefore be less concern about short-term increases in CO_2 as long as the overall trajectory of CO_2 emissions is downward over an extended period (Newell and Pizer, 2003). While the safety valve mechanism shares some advantages with price-based mechanisms, such as a tax, the former may have the added political advantage of providing emitters with an additional allocation of allowances (Pizer, 2005a). A safety valve mechanism does not provide any certainty that a particular emissions level will be met, and it requires additional administrative complexity to link a domestic programme with a safety valve to a programme without a safety valve or with a different safety valve price.

---

14 The US SO_2 trading programme contains a small reserve auction, which was valuable for price discovery during the early years of the programme (Ellerman *et al.*, 2000). Revenue from this auction was returned to the companies affected in the programme. Only four EU Member States (Denmark, 5%; Hungary, 2.5%; Ireland, 0.75%; Lithuania, 1.5%) decided to auction off parts of their ET budget in the first phase of the EU ETS scheme (Betz *et al.*, 2004).

15 Intensity targets are also known as "rate-based", "dynamic," "indexed," and "relative" targets.

16 It is also possible to have a "price floor" to ensure that prices don't go below a certain level. For example, Hepburn *et al.* (2006) discuss how a coordinated auction measure for the EU ETS could be used to support a minimum price.

Experience with trading programmes in the USA has shown significant benefits can be derived from the temporal flexibility provided by banking provisions in cases where the exact timing of emission reductions is not critical to environmental effectiveness (Ellerman *et al.*, 2000; Stavins, 2003). Allowance banking can create a cushion that will prevent price spikes and can hedge uncertainty in allowance prices (Jacoby and Ellerman, 2004).[17] A banking provision allows the arbitrage between actual marginal abatement costs in one phase of a programme and the expected abatement costs in a future phase of a programme. The temporal flexibility of banking is particularly useful for companies facing large capital expenditures because it provides some flexibility in the timing of those expenditures (Tietenberg, 2003). In some emission markets in the USA, banking has been restricted where there was concern about short-term increases in emissions (Tietenberg, 2006). Banking was also restricted between Phase I and Phase 2 in the EU ETS to avoid a large bank that would make it more difficult to meet Kyoto targets.

Several critical elements of an effective enforcement regime for emissions trading have been described in the literature. First, if the goal is strict adherence to the emission limits implied by the number of permits, then excess emissions penalties should be set at levels substantially higher than the prevailing permit price in order to create the appropriate incentives for compliance (Swift, 2001; Stranland *et al.*, 2002).[18] A second component of an enforcement regime is reasonably accurate emissions monitoring (Stranland *et al.*, 2002; Stavins, 2003). San Martin (2003) and Montero (2005) report that incomplete monitoring can undermine the efficiency of trading programmes. Tietenberg (2003) and Kruger *et al.* (2000) emphasize that public access to emissions and trading data provides an additional incentive for compliance.

Finally, there have been several experiments with tradable permits for conventional pollution control in developing countries and economies in transition (Bygrave, 2004; US EPA, 2004). For example, Montero *et al.* (2002) evaluate an experiment with tradable permits for total suspended particulates (TSP) in Santiago, Chile and find that permit markets are underdeveloped due to high transaction costs, uncertainty and poor enforcement. However, they also find an improved documentation of historic emissions inventories and an increased flexibility to address changing market conditions. S. Gupta (2003b) and Wang *et al.* (2004) suggest strengthening the monitoring and enforcement capacity that would be required to implement conventional pollution trading programmes in India and China, respectively. Several authors have concluded that tradable permit programmes may be less appropriate for developing countries due to their lack of appropriate market or enforcement institutions. (Blackman and Harrington, 2000, Bell and Russell, 2002)

13.2.1.4　*Voluntary agreements*

Voluntary agreements are agreements between a government authority and one or more private parties to achieve environmental objectives or to improve environmental performance beyond compliance to regulated obligations. Voluntary agreements are playing an increasingly important role in many countries as a means to achieve environmental and social objectives. They tend to be popular with those directly affected and can be used when other instruments face strong political opposition (Thalmann and Baranzini, 2005). Box 13.5 provides examples of VAs. See Chapter 7, Section 7.9.2 for additional information.

Voluntary agreements can take on many forms with varying levels of stringency. While all VAs are 'voluntary' insofar as firms are not compelled to join, some may involve incentives (rewards or penalties) for participation. Firms may agree to direct emissions reductions or to indirect reductions through changes in product design (see Chapter 6, Section 6.8.2.2.). Agreements may be stand-alone, but they are often used in conjunction with other policy instruments. Voluntary agreements are also a subset of a larger set of 'voluntary approaches' in which industry may first negotiate standards of behaviour with other firms or private groups and then allow third parties to monitor compliance. This larger set also includes unilateral voluntary actions by industry. See Section 13.4, Box 13.5, and Chapter 7, Section 7.9.2 for more information on voluntary actions.

The benefits of VAs for individual companies and for society may be significant. Firms may enjoy lower legal costs, enhance their reputation and improve their relationships with society on a whole and shareholders in particular. Societies gain to the extent that firms translate goals into concrete business practices and persuade other firms to follow their example. The negotiations involved to develop VAs raise awareness of climate change issues and potential mitigative actions within industry (Kågeström *et al.*, 2000), establish a dialogue between industry and government and help shift industries towards best practices.

Evaluating the effectiveness of VAs is not easy. The standard approach is first to measure the environmental performance of a group of firms participating in a VA and then to compare the performance to that of a typical non-participating firm or firms. One problem with this approach is selection bias: it is often the best-performing firms that enter into a VA. A second and related problem is the counterfactual: it is difficult to know what a firm might have done had they not entered into the VA. Very few studies have attempted to evaluate VAs by taking into account both of these issues. Studies which do not take these factors into account can produce an overly optimistic view of the performance of a VA.

17　Price uncertainty may also be addressed by "borrowing" of allowances, i.e. using allowance allocations from future years.
18　The addition of a "make good" provision – that is, the requirement stating that allowances from a subsequent compliance year or period are surrendered for any excess emissions – is a further design element used to ensure that an absolute emissions target is met (Betz and MacGill, 2005).

The environmental effectiveness of VAs is the subject of much discussion. Some governments – as well as industry – believe that VAs are effective in reducing GHG emissions (IAI, 2002; OECD, 2003c). Rietbergen *et al.* (2002) investigated whether the voluntary agreements in The Netherlands have resulted in improvements in energy efficiency beyond what would have occurred in the absence of such agreements. They estimate that, on average, between 25% and 50% of the energy savings in the Dutch manufacturing industry can be attributed to the policy mix of the agreements and supporting measures.

Others are more sceptical about the efficacy of VAs in reducing emissions. Independent assessments of VAs – while acknowledging that investments in cleaner technologies have resulted in absolute emission improvements – indicate that there is little improvement over business-as-usual (BAU) scenarios, as these investments would have probably happened anyway (Harrison, 1999; King and Lenox, 2000; Rietbergen and Blok, 2000; OECD, 2003e; Rivera and deLeon, 2004). The economic efficiency of VAs can also be low, as they seldom incorporate mechanisms to equalize marginal abatement costs between different emitters (Braathen, 2005).

There are a limited, although increasing, number of comprehensive reviews of the effectiveness of VAs, but any comparison of these reviews and assessments is difficult because of the different metrics and evaluative criteria employed (Price, 2005). In general, studies of the design and efficacy of VAs assess only a single programme (e.g. Arora and Cason, 1996; Khanna and Damon, 1999; King and Lenox, 2000; Welch *et al.*, 2000; Rivera, 2002; Croci, 2005). Based on her evaluation of the French experience, Chidiak (2002) suggests that the reductions in GHG emissions cannot necessarily be seen as a direct consequence of the commitments within the agreements and argues that, in actual fact, these improvements have been triggered largely as a result of other environmental regulations and cost reduction efforts. Johannsen (2002) and Helby (2002) present similar results for programmes in Denmark and Sweden, respectively. They note that reductions in specific emissions correspond with industry's BAU behaviour, thereby suggesting that the stated objectives in the agreements were not sufficiently ambitious. In particular, Helby concludes that EKO-Energi, which sought to highlight a new level of best practice and thus pose a challenge to other firms, was 'at best a very modest success,' resulting in a small overall direct effect on total industrial energy consumption. Interestingly, Chidiak also finds that the agreements did not foster intra-industry networking and information exchange on energy management and suggests that their failure to achieve more ambitious goals is a result of the lack of a well-articulated policy-mix. Other analyses indicate that VAs work best as part of a policy package, rather than as a stand-alone instrument (Krarup and Ramesohl, 2002; Torvanger and Skodvin, 2002). OECD (2003e) and Braathen (2005) note that many of the current VAs would perform better if there were a real threat of other instruments being used if targets are not met.

The US Government Accountability Office (2006), in its review of the US Climate Vision and Leaders Programmes, which are supported by the Environmental Protection Agency (EPA) and Department of Energy (DOE), finds that emission reduction goals were set for only 38 of 74 participants, that some goals are intensity-based and others emission-based and that programmes vary in terms of how they are measured, the time periods covered, the requirements for reporting and the means of tracking progress. Brouhle *et al.* (2005) note that the difficulties in evaluating US programmes is associated to the many different programmes and their goals that need be sorted, the availability of adequate data and the measuring of achievement relative to a baseline. Jaccard *et al.* (2006) review various Canadian voluntary programmes that have been in existence for 15 years and report that during that period emissions have grown by 25%.

Darnall and Carmin (2003) review 61 governmental, industry and third-party general environmental agreements, mainly in the USA (see also Lyon and Maxwell, 2000). Overall, their results demonstrate that the voluntary programmes had low programme rigour in that they had limited levels of administrative, environmental and performance requirements. For example, two thirds did not require participants to create environmental targets and to demonstrate that the targets were met. Similarly, almost 50% of the programmes had no monitoring requirements. Compared to government programmes, industry programmes had stronger administrative requirements and third party programmes had yet even slightly stronger requirements. According to Hanks (2002) and OECD (2003e), the best VAs include: a clear goal and baseline scenario; third party participation in the design of the agreement; a description of the parties and their obligations; a defined relationship within the legal and regulatory framework; formal provisions for monitoring, reporting and independent verification of results at the plant level; a clear statement of the responsibilities expected to be self-financed by industry; commitments in terms of individual companies, rather than as sectoral commitments; references to sanctions or incentives in the case of non-compliance.

It must be acknowledged that VAs fit into the cultural traditions of some countries better than others. Japan, for example, has a history of co-operation between government and industry that facilitates the operation of "voluntary" programmes. Some examples of VAs in various countries are provided in Box 13.5.

13.2.1.5 Subsidies and incentives

Direct and indirect subsidies can be important environmental policy instruments, but they have strong market implications and may increase or decrease emissions, depending on their nature. Subsidies aimed at reducing emissions can take different forms, ranging from support for Research and Development

Box 13.5 Examples of national voluntary agreements

- **The Netherlands Voluntary Agreement on Energy Efficiency:** A series of legally binding long-term agreements based on annual improvement targets and benchmarking covenants between 30 industrial sectors and the government with the objective to improve energy efficiency.
- **Australia "Greenhouse Challenge Plus" programme:** An agreement between the government and an enterprise/ industry association to reduce GHG emissions, accelerate the uptake of energy efficiency, integrate GHG issues into business decision making and provide consistent reporting.[19] See http://www.greenhouse.gov.au/challenge.
- **European Automobile Agreement:** An agreement between the European Commission and European, Korean and Japanese car manufacturing associations to reduce average emissions from new cars to 140 gCO_2/km by 2008–2009. See http://ec.europa.eu/environment/CO2/CO2_agreements.htm.
- **Canadian Automobile Agreement:** An agreement between the Canadian government and representatives of the domestic automobile industry to a reduce emissions from cars and light-duty trucks by 5.3 $MtCO_2$-eq by 2010. The agreement also contains provisions relating to research and development and interim reduction goals.
- **Climate Leaders:** An agreement between US companies and the government to develop GHG inventories, set corporate emission reduction targets and report emissions annually to the US EPA. See: http://www.epa.gov/climateleaders/.
- **Keidaren Voluntary Action Plan:** An agreement between the Japanese government and 34 industrial and energy-converting sectors to reduce GHG emissions. A third party evaluation committee reviews the results annually and makes recommendations for adjustments.[20] See http://www.keidanren.or.jp

(R&D), investment tax credit, and price supports (such as feed-in tariffs for renewable electricity).[21] Subsidies that increase emissions typically involve support for fossil fuel production and consumption. They tend to expand the subsidized industry, relative to the non-subsidy case. If the subsidized industry is a source of GHG emissions, subsidies may result in higher emissions. Subsidies to the fossil fuel sector result in over-use of these fuels with resulting higher emissions; subsidies to agriculture can result in the expansion of agriculture into marginal lands and corresponding increases in emissions. Conversely, incentives to encourage the diffusion of new technologies, such as those for renewables or nuclear power, may promote emissions reductions.

One of the significant advantages of subsidies is that they have politically positive distributional consequences. The costs of subsidies are often spread broadly through an economy, whereas the benefits are more concentrated. This means that subsidies may be easier to implement politically than many other forms of regulatory instruments. Subsidies do tend to take on a life of their own, which makes it difficult to eliminate or reduce them, should that be desired.

The International Energy Agency (IEA) estimates that in 2001 energy subsidies in OECD countries alone were

approximately 20–80 billion US$ (IEA, 2001). The level of subsidies in developing and transition economy countries is generally considered to be much higher. One example is low domestic energy prices that are intended to benefit the poor, but which often benefit high users of energy. The result is increased consumption and delayed investments in energy-efficient technologies. In India, kerosene and liquefied petroleum gas (LPG) subsidies are generally intended to shift consumption from biomass to modern fuels, reduce deforestation and improve indoor air quality, particularly in poor rural areas. In reality, these subsidies are largely used by higher expenditure groups in urban areas, thus having little effect on the use of biomass. Nevertheless, removal of subsidies would need to be done cautiously, in the absence of substitutes, as some rural households use kerosene for lighting (Gangopadhyay et al., 2005).

OECD countries are slowly reducing their subsidies to energy production or fuel (such as coal) or changing the structure of their support to reduce the negative effects on trade, the economy and the environment. Coal subsidies in OECD countries fell by 55% between 1991 and 2000 (IEA, 2001).[22] (See Chapter 7 for additional information.)[23] About 460 billion US$ is spent on agricultural subsidies, excluding water and fisheries (Humphreys et al., 2003), with OECD

19 As of 1 July 2006, participation in the programme is a requirement for Australian companies receiving fuel tax credits of more than 3 million US$.
20 This programme is a cross between a mandatory and voluntary programme; see Saito (2001), Yamaguchi (2003) and Tanikawa (2004) for additional information. The special relationship between the government and industry as well as unique societal norms make this voluntary initiative unique. In the context of Japan there is de facto enforcement.
21 One way of promoting the use of renewable sources of electricity is for the government to require electric power producers to purchase such electricity at favourable prices. The US Public Utility Regulatory Policy Act of 1978 required electric utilities to buy renewable energy at "avoided cost". In Europe, specific prices have been set at which utilities must purchase renewable electricity – these are referred to as "feed-in tariffs." These tariffs have been effective at promoting the development of renewable sources of electricity (Ackermann et al., 2001; Menanteau et al., 2003).
22 Calculated using producer subsidy equivalents.
23 It should be noted that a comprehensive analysis of subsidies requires the net effect of subsidies and taxes, including their point of allocation, to be considered.

countries accounting for about 318 billion US$ or 1.2% of the GDP. These subsidies result in the expansion of this sector with associated GHG implications (OECD, 2001, 2002).

Many countries provide financial incentives, such as tax credits for energy-efficient equipment and price supports for renewable energy, to stimulate the diffusion of technologies. In the USA, for example, the Energy Policy Act of 2005 contains an array of financial incentives for various advanced technologies; these financial incentives have been estimated at 11.4 billion US$ over a 10-year period.

One of the most effective incentives for fostering GHG reductions are the price supports associated with the production of renewable electricity, which tend to be set at attractive levels. These price supports have resulted in the significant expansion of the renewable energy sector in OECD countries due to the requirement that electric power producers purchase such electricity at favourable prices. The US Public Utility Regulatory Policy Act of 1978 requires electric utilities to buy renewable energy at "avoided cost". In Europe, specific prices have been set at which utilities must purchase renewable electricity – these are referred to as 'feed-in tariffs'. These tariffs have been effective at promoting the development of renewable sources of electricity (Ackermann *et al.*, 2001; Menanteau *et al.*, 2003). As long as renewables remain a relatively small portion of overall electricity production, consumers see only a small increase in their electricity rates. Incentives therefore have attractive properties in terms of environmental effectiveness, distributional implications and institutional feasibility. The main problem with them is cost-effectiveness: They are costly instruments, particularly in the long-run as interests and industries grow to expect the continuation of subsidy programmes. See Chapter 4.5 for a more extensive discussion.

13.2.1.6   Research and Development[24]

The role of R&D in changing the trajectory of energy economy is unquestionable – new technologies have played a large role in the evolution of the energy sector over the last century. Moreover, the rate at which low emission technologies will improve during the next 20–30 years will be an important determinant of whether low emission paths can be achieved in the long term.

Policy uncertainties, however, often hinder investment in R&D and the dissemination of new technology, although different types of polices may be needed to address different types of investment. Hamilton (2005) notes that investors prefer a policy environment which is 'loud, long and legal'. A number of authors note that long-term policy targets or clear foresight on carbon taxes can overcome social inertia and reduce uncertainty for investors in R&D (Blyth and Yang, 2006; Edenhofer *et al.*, 2006; Reedman, Graham and Coombes, 2006).

Nearly 600 billion US$ was expended worldwide on R&D in all sectors in 2000, with approximately 85% of that amount being spent in only seven countries.[25] Over the last 20 years, the percentage of government-funded R&D has generally declined, while industry-funded R&D has increased in these countries. In a historic context, R&D expenditures as a percentage of GDP have gone up and down in cycles as government priorities have changed over the last 50 years, although in some instances comparisons over time are difficult (US-NSF, 2003; OECD, 2005a; US-GAO, 2005).

Total public funding for energy technologies in IEA countries during the period 1987–2002 was 291 billion US$, with 50% of this allocated to nuclear fission and fusion, 12.3% to fossil fuels and 7.7% to renewable energy technologies (IEA, 2004; see Figure 13.1).[26] Funding has dropped after the initial interest created through the oil shock in the 1970s and has stayed constant, even after the UNFCCC was ratified. Nemet and Kammen (2006) suggest that for the USA a change in direction is warranted and that a five- to tenfold increase in public funding is feasible.

The USA and Japan, the two largest investors in energy R&D, spent on average of 3.38 and 2.45 billion US$, respectively, between 1975 and 1999. However, such figures mask important underlying trends. For example, a large percentage of the funding designated for energy R&D has gone into nuclear power – nearly 75% in the case of Japan (Sagar and van der Zwaan 2006). The support of the US government for R&D declined by 1 billion US$ from 1994 to 2003, with reductions implemented in nearly all energy technologies, while R&D investments in other areas grew by 6% per year. Between the 1980s and 2003, private sector energy R&D declined from nearly 50% of that of government funding to about 25% (Nemet and Kammen, 2006).

Many countries pursue technological (R&D) advancements as a national policy for a variety of different reasons: for example, to foster the development of innovative technologies or to assist domestic industries in being competitive. Countries also chose to co-operate with each other in order to share costs, spread risks, avoid duplication, access facilities, enhance domestic capabilities, support specific economic and political objectives, harmonize standards, accelerate market learning and create goodwill. Cooperation, however, may increase

---

24 As used in this section, the term R&D generally refers to research, development and demonstration.
25 Canada, France, Germany, Italy, Japan, the UK and the USA.
26 In year-2000 US$ and exchange rates.

13.1 (a). RD&D budgets for energy

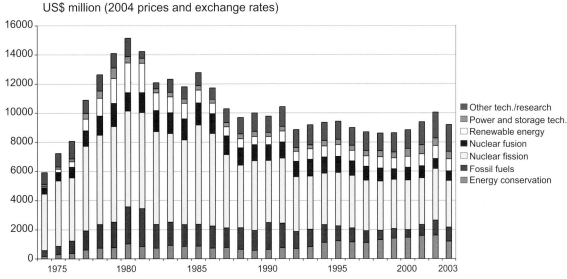

13.1(b). RD&D budgets for renewable energy

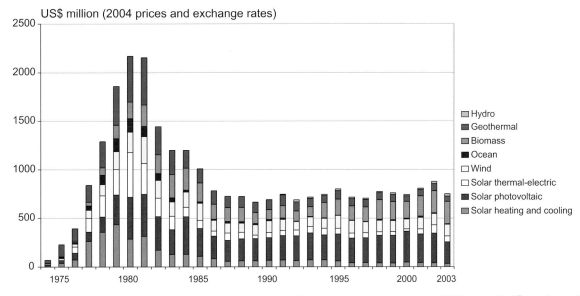

Figure 13.1: *Public funded Research and Development (R&D) expenditures for energy (A) and renewable energy technologies (B) by International Energy Agency (IEA) member countries.*
Source: IEA, 2004, 2005.

transaction costs, require extensive coordination, raise concerns over intellectual property rights and foreclose other technology pathways (Fritsche and Lukas, 2001; Sakakibara, 2001; Ekboir, 2003; Justice and Philibert, 2005). Governments use a number of tools to support R&D, such as grants, contracts, tax credits and allowances and public/private partnerships. The effect of these tools on public budgets and their effectiveness in stimulating innovation will vary as a function of how they are structured and targeted. For example, in the USA, R&D tax credits to industry totalled an estimated 6.4 billion US$ in 2001;

however, industries associated with high GHG emissions did not take advantage of this opportunity in that the utility industry received only 23 million US$.[27]

There are different views on the role of R&D, its links to the overall energy innovation system and processes underlying effective learning. Sagar and van der Zwaan (2006) examined the trends in major industrialized countries and report that public R&D spending does not correlate with changes in national energy intensity or carbon emissions per unit of energy

27 http://www.nsf.gov/statistics/inbrief/nsf/nst05316

consumption. For a more extensive discussion of technological learning, energy supply models and the link to R&D, see Chapter 3, Section 3.4.2 and Chapter 11, Section 11.3.3. Watanabe (1999) argues that government R&D can play a role in achieving breakthroughs in some areas, induce investments by industry in R&D and generate trans-sectoral spill over effects. Others have noted, however, that the benefits of R&D may not be realized for two to three decades, which is beyond the planning horizons of even the most forward-looking companies (Anderson and Bird, 1992) and that, for a variety of reasons, industry can only appropriate a fraction of the benefits of R&D investments (Margolis and Kammen ,1999). In the energy sector in particular, technology 'spill over' to competitors is large; as a result, firms under-invest in R&D (Azar and Dowlatabadi 1999) and face difficulties in evaluating intangible R&D outputs (Alic *et al.* 2003).[28] In addition, regulatory interventions can cap profits in the case of path-breaking research success (Foxon and Kemp, 2004; Grubb, 2004).[29] Goulder and Schneider (1999) argue that increasing R&D expenditures in carbon-free technologies could crowd out R&D in the rest of the economy and therefore reduce overall growth rates. However, Azar and Dowlatabadi (1999) counter that radical technological change will trigger more research overall and therefore increase economy-wide productivity rates.

The OECD (2005b) finds that obligations/quotas, price guarantees and tax preferences have had the most influence on innovation and patent activities in the renewable energy sector and that while public subsidies for R&D have not played a role, the overall level of investment in R&D within the economy of a country has been important. Sathaye *et al.* (2005) observe that government-funded research at government-owned facilities, private companies and universities may help identify patentable technologies and processes. They reviewed the process of allocating patent rights in four OECD countries and found that intellectual property rights (IPR) regimes have changed since the ratification of the UNFCCC, with diffusion typically taking place along a pathway of licensing or royalty payments rather than unrestricted use in the public domain. Popp (2002) also examined patent citations and found that the level of energy-saving R&D depends not only on energy prices, but also on the quality of the accumulated knowledge available to inventors. He finds evidence for diminishing returns to research inputs – both across time and within a given year – and notes that government patents filed in or after 1981 are more likely to be cited. Popp (2004) notes that when in terms of the potential for technology to help solve the climate problem, two market failures lead to underinvestment in climate-friendly R&D: environmental externalities and the public goods nature of the new knowledge. As a result, government subsidies to climate-friendly R&D projects are often proposed as part of a policy solution.

Policies that directly affect the environmental externality have a much larger impact on both atmospheric temperature and economic welfare. Fischer and Newell (2004) examine several policy options to promote renewables and indicate that research subsidies are the most expensive approach to achieve emission reductions – in the absence of higher prices. They note that the process of technological change is less important than the implementation of direct incentives to reduce emission intensity or overall energy use. A more specific example arises from the Danish experience with wind technologies. Meyer (2004) notes that despite significant support for wind energy R&D during the 1980s, wind power only boomed in Denmark when favourable feed-in tariffs were introduced, procedures for construction allowances were simplified and priority was given for green electricity. This is supported by Nemet (2005), who found that the ability to raise capital and take risks has played a much larger role in the recent expansion of the photovoltaic industry than other factors, such as learning by experience.

In summary, national programmes and international cooperation relating to R&D are essential long-term measures to stimulate technological advances. Substantial additional investments in and policies for R&D are needed, depending on the specific goals: for example, if high stabilization levels are desired (e.g. 750 ppmv CO_2-eq, which is scenario category D of Chapter 3 of this report), a technology-focused approach that defers emissions reduction to the future would be sufficient; for low stabilization goals (e.g. 450 ppmv CO_2-eq, which is category A1, or 550 ppmv CO_2-eq, which is category B), strong incentives for short-term emission reductions would be necessary in addition to technological development and deployment programmes. See Section 13.3 for a discussion of goals.

13.2.1.7 *Information instruments*

Information instruments – such as public disclosure requirements and awareness/education campaigns – may positively affect environmental quality by allowing consumers to make better-informed choices. When firms or consumers lack the necessary information about the environmental consequences of their actions, they may act inefficiently. While some research indicates an information provision can be an effective environmental policy instrument, we know less about its efficacy in the context of climate change. Examples of information instruments include labelling programmes for consumer products, information disclosure programmes for firms and public awareness campaigns.

Article 6 of the UNFCCC on Education, Training and Public Awareness calls on governments to promote the development

---

28 An assessment of private public research partnership under the Advanced Technology Programme in the USA indicates that 'Time lags, along with the difficulty inherent in retrospective evaluation of factors affecting the timing and character of innovations, make it difficult to attribute specific commercial advantages to funding awarded much earlier.' In general, companies shift funds away from basic research towards product modifications and extensions.

29 Renewable energy technologies compete in electricity wholesale markets that are frequently exposed to regulations, such as price caps, which reduce incentives for private investment in long-term R&D.

and implementation of educational and public awareness programmes, promote public access to information and public participation and promote the training of scientific, technical and managerial personnel. With decision 11/CP.8, the Conference of the Parties (COP) launched a 5-year country-driven work programme to engage stakeholders in information/education activities. The UNFCCC secretariat notes that there is a general lack of resources, limited technical skill and poor regional coordination relating to information and education campaigns (UNFCCC 2006a).

Information instruments can often be used to improve the effectiveness of other instruments. Another feature common to all information instruments that makes them unique from other environmental policies is that they do not impose penalties for environmentally harmful behaviour per se. A disclosure programme, such as the Toxics Release Inventory (TRI), requires only that firms document and report their emissions; it does not place limits on how much pollution they can emit.

Kennedy *et al.* (1994) demonstrate that environmental externalities can be at least partially corrected through information provision. However, they also point out that when other corrective instruments, such as taxes, are available, these measures are preferable to information policies. Based on a recent theoretical study, Petrtakis *et al.* (2005) reports that information provision can be more effective than tax instruments, especially when the information can be provided at low cost. Osgood (2002) provides limited empirical support in the context of weather information programmes in Mexico and California.

Evidence-to-date suggests that while disclosure mandates may be effective at changing a firm's environmental practices, other information instruments, such as advisory programmes, have less effect on consumer behaviour (Konar and Cohen, 1997). Firms whose stock price declined significantly when pollution data became publicly available reduced their emissions more than other firms in the same industry. Firms may view information policies as overly burdensome and argue that voluntarily provided information is sufficient (Sterner, 2003). Certainly, there is a cost to disclosure and labelling policies, and costs depend on the level of information required by a policy (Beierle, 2004). A firm may have to collect and disseminate information they would not otherwise have gathered, and government agencies must be able to verify that the information is accurate.

13.2.1.8   *Non-climate policies*

There are a number of non-climate national policies that can have an important influence on GHG emissions. These include policies focused on poverty, land use and land use change, energy supply and security; international trade, air pollution, structural reforms and population policies. Only a few types of 'non-climate policies' are touched upon in this section.

The literature available on this topic indicates that poverty reduces the resilience of vulnerable populations and makes them more at risk to the potential impacts of climate change, but it also leads communities to take measures that may increase emissions. Heemst and Bayangos (2004) note that if poverty can be reduced without raising emissions, then a strategy to reduce poverty can be seen as a way to reduce emissions as well as enhance resilience. Typical areas of synergy include small-scale renewables (Richards, 2003) and community forestry (Smith and Scherr, 2002), both of which may benefit the poor.

Land use policies (or the lack thereof), whether terrestrial (agriculture, forestry, nature), aquatic (wetlands) or urban, can lead to enhanced emissions. Verhagen *et al.* (2004) note that policies aimed at integrating climate change concerns with the specific concerns of local people may yield major synergies. For example, within the Netherlands, a major programme is currently underway to understand how spatial planning and climate change policy can be effectively linked. Regional (acid rain abatement), local and indoor air pollution policies can also have climate change co-benefits (Bakker *et al.*, 2004).

The consumption of natural resources varies significantly between developed and developing countries and is ultimately one of the major drivers of global emissions. The global population and income levels affect the consumption of natural resources, particularly those of energy, food and fibre, and hence can also affect GHG emissions. Policies that increase consumption of natural resources have implications for GHG emissions.

13.2.2   Linking national policies

13.2.2.1   *National policy interactions/linkages and packages*

Single instruments are unlikely to be sufficient for climate change mitigation, and it is more likely that a portfolio of policies will be required (see IPCC, 2001). Examples of areas where there are potential synergies include water management strategies, farm practices, forest management strategies and residential building standards. Instruments that maximize potential synergies could become socially and economically efficient and may offer opportunities for countries to achieve sustainable development targets, even in the face of uncertainties. This is especially important given the limited financial and human resources in developing countries (Dang *et al.*, 2003). Climate change considerations also provide both developing and developed countries with an opportunity to look closely at their respective development strategies from a new perspective. Fulfilling development goals through policy reforms in such areas as energy efficiency, renewable energy, sustainable land use and/or agriculture will often also generate benefits related to climate change objectives.

A key synergy is that between adaptation and mitigation policies. Climate policy options can include both mitigation and adaptation (see Chapter 17 of IPCC (2007b) for a discussion on adaptation policies and Chapter 18 for a detailed analysis of interaction between mitigation and adaptation). Many adaptation options are consistent with pathways towards effective and long-term mitigation and, in turn, several mitigation options can facilitate planned adaptation.

In theory, a perfectly functioning market would need only one instrument (e.g. a tax) to address a single environmental problem, such as climate change. In such a situation, the application of two or more overlapping instruments could diminish economic efficiency while increasing administrative costs. In practice, however, there are market failures that may make a mix of instruments desirable. This section describes some of these cases and addresses situations in which multiple or overlapping objectives might justify a mix of policies.

Climate-related policies are seldom applied in complete isolation: in a large number of cases one or more instruments will be applied. The mere existence of an instrument mix, however, is clearly not 'proof' of its environmental effectiveness and economic efficiency. A rather obvious first requirement for applying an environmentally and economically effective instrument mix is to have a good understanding of the environmental issue to be addressed. In practice, many environmental issues can be complex. While a tax can affect the total demand for a product and the choice between different product varieties, it is less suited to address, for example, how a given product is used and when it is used. Hence, other instruments could be needed. A second requirement for designing efficient and effective policies is to have a good understanding of the links with other policy areas: not only do different environmental policies need to be co-ordinated, but co-ordination with other related policies is also necessary. A third requirement is to have a good understanding of the interactions between the different instruments in the mix.

Several authors describe situations in which a combination of policies might be desirable. Johnstone (2002) argues that the price signal from a tradable permits or tax system may not be sufficient to overcome barriers to technological development and diffusion and that additional policies may be warranted. These barriers include: (1) credit market failures that discourage lenders from providing capital to firms for high-risk investments associated with R&D and even the implementation of new technologies and (2) reduced incentives for private investment in R&D if firms can not prevent other firms from benefiting from their investments (i.e. 'spill-over' effects).[30] Fischer and Newell (2004) find that the combination of a technology policy, such as government sponsored R&D, with a tax or tradable permit instrument could help overcome this type of market imperfection.

A second market failure that may require more than one instrument is the lack of information among consumers on the environmental or economic attributes of a technology. In such a case, a price signal alone may not sufficiently spur the diffusion of these types of technologies. One solution to this type of barrier is an eco-labelling system, which can help increase the effectiveness of a price instrument by providing better information on relevant characteristics of the product (OECD, 2003b; Braathen, 2005). Sijm (2005) notes that this type of market failure may exist for households who may lack the relevant information to invest in energy efficiency measures and may not respond to a price signal. Another market failure in the residential sector may be caused by split incentives where neither the landlord nor tenant has an incentive to invest in energy efficiency measures (Sorrell and Sijm, 2003).

With the implementation of the EU ETS, particular attention has been given to the interaction between a tradable permits mechanism and other policies. Sijm (2005) and Sorrell and Sijm (2003) argue that an emissions trading scheme can co-exist with other instruments as long as these other instruments improve the efficiency of the trading mechanism by addressing market failures or contributing to some other policy objective. However, they argue that the combination of an emissions trading scheme with other instruments could also lead to "double regulation", reduced efficiency and increased costs if policies are not designed carefully. NERA (2005) and Morthorst (2001) assess the interaction of renewable energy policies with tradable permits programmes and conclude that if not designed properly, these policies can lower allowance prices but raise the overall costs of the programme.

There may be cases where a package of CO_2 mitigation policies is justified if these policies serve multiple policy objectives. Sijm (2005) gives several examples of policies and objectives that may be compatible with the EU ETS, including direct regulation that also reduces local environmental effects from other pollutants. Renewable energy policies can be used to expand energy supply, increase rural income and reduce conventional pollutants. Policies that encourage bio-fuel production and automobile fuel efficiency have also been advocated for their advantages in encouraging energy security and fuel diversity as well as GHG mitigation. In the USA, these types of energy policies have been proposed in conjunction with a tradable permits system as part of a package to address energy, security and environmental objectives (NCEP, 2004).

13.2.2.2 Criteria assessment

Any evaluation of the instruments based on the criteria discussed herein is challenging for two reasons. First, practitioners must be able to compare potential instruments based on each of the evaluative criteria. Unfortunately, in many cases it can be difficult if not impossible to rank instruments in an objective manner. For example, Fischer et al. (2003) conclude that it is not possible to rank environmental policy

---

30 For a more extensive discussion of these issues, see Jaffe et al., 2003.

Table 13.1: *National environmental policy instruments and evaluative criteria* [a]

Instrument	Criteria			
	Environmental effectiveness	**Cost-effectiveness**	**Meets distributional considerations**	**Institutional feasibility**
Regulations and standards	Emissions level set directly, though subject to exceptions. Depends on deferrals and compliance.	Depends on design; uniform application often leads to higher overall compliance costs.	Depends on level playing field. Small/new actors may be disadvantaged.	Depends on technical capacity; popular with regulators in countries with weakly functioning markets.
Taxes and charges	Depends on ability to set tax at a level that induces behavioural change.	Better with broad application; higher administrative costs where institutions are weak.	Regressive; can be ameliorated with revenue recycling.	Often politically unpopular; may be difficult to enforce with underdeveloped institutions.
Tradable permits	Depends on emissions cap, participation and compliance.	Decreases with limited participation and fewer sectors.	Depends on initial permit allocation. May pose difficulties for small emitters.	Requires well functioning markets and complementary institutions.
Voluntary agreements	Depends on programme design, including clear targets, a baseline scenario, third party involvement in design and review and monitoring provisions.	Depends on flexibility and extent of government incentives, rewards and penalties.	Benefits accrue only to participants.	Often politically popular; requires significant number of administrative staff.
Subsidies and other incentives	Depends on programme design; less certain than regulations/standards.	Depends on level and programme design; can be market distorting.	Benefits selected participants, possibly some that do not need it.	Popular with recipients; potential resistance from vested interests. Can be difficult to phase out.
Research and development	Depends on consistent funding; when technologies are developed and polices for diffusion. May have high benefits in the long term.	Depends on programme design and the degree of risk.	Benefits initially selected participants; potentially easy for funds to be misallocated.	Requires many separate decisions. Depends on research capacity and long-term funding.
Information policies	Depends on how consumers use the information; most effective in combination with other policies.	Potentially low cost, but depends on programme design.	May be less effective for groups (e.g. low-income) that lack access to information.	Depends on cooperation from special interest groups.

Note:
[a] Evaluations are predicated on assumptions that instruments are representative of best practice rather than theoretically perfect. This assessment is based primarily on experiences and published reports from developed countries, as the number of peer reviewed articles on the effectiveness of instruments in other countries is limited. Applicability in specific countries, sectors and circumstances – particularly developing countries and economies in transition – may differ greatly. Environmental and cost effectiveness may be enhanced when instruments are strategically combined and adapted to local circumstances.

instruments based on their technology-stimulating effects. Consequently, it will be difficult to determine which of the available instruments is the most cost-effective. Distributional considerations are also particularly difficult to evaluate. Revesz and Stavins (2006) provide a discussion of the difficulties involved in evaluating instruments based on distribution or equity. They also cite a number of authors that propose different approaches to evaluating policies.

Nevertheless, it is possible to make general statements about each instrument according to the criteria we have selected. For example, it is generally believed that market-based instruments will be more cost effective than regulations and standards (Wiener, 1999). However, this belief implicitly assumes that a country has well-functioning institutions, the lack of which can result in a market-based instrument being more costly to implement (Blackman and Harrington, 2000). Table 13.1 summarizes the seven instruments presented in this chapter for each of the four criteria we discuss. Sterner (2003) and

Harrington *et al.* (2004) provide similar summaries for other instruments.

Second, policymakers must determine how much weight to assign each of the evaluative criteria. Consider two instruments that are equally environmentally effective and both institutionally feasible, but one has unfavourable distributional implications while the other is less cost-effective. In order to choose one instrument over the other, one must assess the relative importance of distribution versus cost-effectiveness. However, the determination of just what weight should be given to each evaluative criterium is a subjective question and one left to policymakers to decide. Some authors do provide some guidelines on how policymakers can determine which evaluative criteria are the most important. Sterner (2003) argues that distributional considerations will likely be less important in an economy with relatively less inequality than in countries with large income disparities and also provides additional guidance on other criteria, including institutional flexibility and incentive compatibility. Bell (2003) and Bell and Russell (2002) argue that institutional feasibility is of critical importance in developing countries, where environmental effectiveness and cost-effectiveness may be determined in large part by a government's institutional capacity. In general, the criteria that receive the most weight will be those that are assessed to be the most important in terms of each country's specific circumstances.

13.3 International climate change agreements and other arrangements

The context of and reasons for an international agreement were relatively well covered in IPCC (2001). The authors of more recent reports cite the reasons presented in older publications for the necessity of agreements – namely, the global nature of the problem and the fact that no single country emits more than approximately 20% of global emissions. This situation means that successful solutions will need to engage multiple countries. Similarly, the fact that no one sector is responsible for more than about 25% of global emissions (the largest sector is that of electricity generation and heat production at 24% of the global, six-gas total; see Baumert et al., 2005a) implies that no single sector will be uniquely required to act.

13.3.1 Evaluations of existing climate change agreements

In contrast, the more recent publications have devoted considerable attention to the limitations of existing international agreements in addressing the climate change. In fact, there are no authoritative assessments of the UNFCCC or its Kyoto Protocol that assert that these agreements have succeeded – or will succeed without changes – in fully solving the climate problem. As its name implies, the UNFCCC was designed as a broad framework, and the Kyoto Protocol's first commitment period for 2008–2012 has been its first detailed step. Both the Convention and the Kyoto Protocol include provisions for further steps as necessary.

A number of limitations and gaps in existing agreements are cited in the literature, namely:
- The lack of an explicit long-term goal means countries do not have a clear direction for national and international policy (see, for example, Corfee-Morlot and Höhne, 2003);
- The targets are inadequately stringent (Den Elzen and Meinshausen, 2005, who argue for more stringent targets);
- The agreements do not engage an adequate complement of countries (see Baumert et al. 1999, who suggest a need to engage developing as well as developed countries, or Bohringer and Welsch 2006, who suggest that with the US withdrawal, the Kyoto Protocol's effect is reduced to zero);
- The agreements are too expensive (Pizer, 1999, 2002);
- The agreements do not have adequately robust compliance provisions (Victor, 2001; Aldy et al., 2003);
- The agreements do not adequately promote the development and/or transfer of technology (Barrett, 2003);
- The agreements, as one consequence of failing to solve the problem, do not adequately propose solutions that will facilitate adaption to the forthcoming changes (Muller, 2002).

Reviews of the current agreements take several forms. Some (e.g. Depledge, 2000) provide detailed article-by-article reviews of the existing agreements, seeking to interpret the legal language as well as to provide a better understanding of their historical derivations. In this manner, they offer insight into how future agreements might be developed. Other studies assess the effect of the emission reductions required by the Kyoto Protocol on global GHG concentrations and conclude that although the effect is currently small (Manne and Richels, 1999), it may be large in the future as present-day emission reductions set the stage for future reduction efforts, which would not have happened otherwise (Höhne and Blok, 2006). Some researchers (e.g. Cooper, 2001; Michaelowa et al., 2005a) evaluate the basic underpinnings of the two climate agreements, looking at problems associated with establishing binding targets and differentiating between countries as well as difficulties in operationalizing the concept of emissions markets. These kinds of assessments – by far the most common – not only assess current limitations but usually proceed to put forth counter-arguments, outline improvements that should be made and propose alternative mechanisms for addressing the climate problem. See the following sections for responses and alternatives to solving the climate problem.

13.3.2 Elements of international agreements and related instruments

The majority of elements identified in the literature draw on existing multilateral agreements, in particular, the UNFCCC and its Kyoto Protocol. Agreements related to climate change, but not specifically focused on GHG mitigation, are less extensively analysed in the climate literature. These include energy policy and technology agreements (see, for example, publications the IEA evaluating their "Implementing Agreements") and the evaluation of VAs with the auto sector (see, for example, Sauer et al., 2005 on the European Automobile Manufacturers Association (ACEA) agreement between the European, Japanese and Korean auto manufacturers). Based on the literature in Table 13.2, it is possible to derive some common elements of international climate change agreements. These are listed in Box 13.6, and expanded upon in the section below.

13.3.3 Proposals for climate change agreements

The literature on climate change contains a large number of proposals on possible future international agreements. Table 13.2 provides a summary review of recent proposals for international climate agreements as reported in the literature (see also Bodansky, 2004; Kameyama, 2004; Philibert, 2005a), although not all proposals cover all elements that are necessary

Box 13.6 Elements for climate change agreements[31]

A number of elements are commonly incorporated in existing – and proposals for new – international climate change agreements. These include:

Goals: Most agreements establish objectives that implementation is supposed to achieve. In the climate context, a variety of goals have been proposed, including those related to emissions reductions, stabilization of GHG concentrations, avoiding "dangerous" interference with climate, technology transfer and sustainable development. Goals can be set at varying degrees of specificity.

Participation: All agreements are undertaken between specific groups of participants. Some have a global scope while others focus on a more limited set of parties (e.g. regional in nature or limited to arrangements between private sector partners). Obligations can be uniform across participants, or differentiated among them.

Actions: All agreements call for some form of action. Actions vary widely and can include national caps or targets on emissions, standards for certain sectors of the economy, financial payments and transfers, technology development, specific programmes for adaptation and reporting and monitoring. The actions can be implicitly or explicitly designed to support sustainable development. The timing for actions varies considerably, from those taking effect immediately, to ones that may take effect only over the longer term; actions may be taken internally (within contracting Parties) or with others (both with non-Parties as well as non-State actors).

Institutions and compliance provisions: Many agreements contain provisions for establishing and maintaining supporting institutions. These perform tasks as varied as serving as repositories for specific, agreement-related data, facilitating or adjudicating compliance, serving as clearing houses for market transactions or information flows, to managing financial arrangements. In addition, most agreements have provisions in case of non-compliance. These include binding and non-binding consequences and may be facilitative or more coercive in nature.

Other elements: Many (although not all) agreements contain additional elements, including, for example, "principles" and other preambular language. These can serve to provide context and guidance for operational elements, although they may be points of contention during negotiations. In addition, many agreements contain provisions for evaluating progress – with a timetable for reviewing the adequacy of efforts and evaluating whether they need to be augmented or modified.

to describe a full regime. The list of proposals is grouped around the following themes: national emission targets and emission trading, sectoral approaches, policies and measures, technology, development-oriented actions, adaptation, financing and proposals focusing on negotiation process and treaty structure.

13.3.3.1 Goals

Most agreements (including those on climate change such as the UNFCCC and the Kyoto Protocol), include specific goals to guide the selection of actions and timing as well as the selection of institutions. Goals can provide a common vision about both the near-term direction and the longer term certainty that is called for by business. In this discussion, goals are distinguished from targets: the former are long-term and systemic, while the latter relate to actions that are near-term and specific. Targets are described under the 'Targets' section (13.3.3.4.1) below.

The choice of the long-term ambition level significantly influences the necessary short-term action and, therefore, the design of the international regime. For example, if the goal is set at high stabilization levels (e.g. stabilizing concentrations at 750 ppmv CO_2-eq, scenario category D of Chapter 3 of this report), a technology-focused approach that defers emissions reduction to the future would be sufficient for the time being. For low stabilization goals (e.g. 450 ppmv CO_2-eq, category A1, or 550 ppmv CO_2-eq, category B), short-term emission reductions would be necessary in addition to technological development programmes.

International regimes can incorporate goals for the short and medium term and for the stabilization of GHG concentrations. One option is to set a goal for long-term GHG concentrations or a maximal temperature rise (such as the 2°C goal proposed by the EU). Such levels might be set based on an agreement of impacts to be avoided (see Den Elzen and Meinshausen, 2005) or on the basis of a cost-benefit analysis (see Nordhaus, 2001). A number of authors have commented on the advantages and disadvantages of setting long-term goals. Pershing and Tudela (2003) suggest that it may be difficult to gain a global agreement on any 'dangerous' level due to political and technical difficulties. Conversely, Corfee-Morlot and Höhne

---

31 While not an element, agreements often contain specific information as to the time for initiating actions and, often, a date by which actions are to be completed. In addition, many agreements contain provisions for evaluating progress – with a timetable for reviewing the adequacy of efforts and evaluating whether they need to be augmented or modified.

Table 13.2: *Overview of recent proposals for international climate agreements.*

Name (reference)	Description
National emission targets and emission trading	
Staged systems	
Multistage with differentiated reductions: Gupta, 1998; Berk and den Elzen, 2001; Blanchard et al., 2003; Criqui *et al.*, 2003; Gupta, 2003a; Höhne et al., 2003; Höhne *et al.*, 2005; Michaelowa *et al.*, 2005b; den Elzen and Meinshausen, 2006, den Elzen *et al.*, 2006a	Countries participate in the system with different stages and stage-specific types of targets; countries transition between stages as a function of indicators; proposal specify stringency of the different stages
Differentiating groups of countries: Storey, 2002; Ott *et al.*, 2004	Countries participate in the system with different stages and stage-specific types of targets
Converging markets: Tangen and Hasselknippe, 2005	Scenario with regional emission trading systems converging to a full global post 2012 market system
Three-part policy architecture: Stavins, 2001	All nations with income above agreed threshold take on different targets (fixed or growth); long-term targets (flexible but stringent); short-term (firm, but moderate); and market-based policy instruments, e.g., emissions trading.
Allocation methods	
Equal per capita allocation: Baer *et al*, 2000; Wicke, 2005	All countries are allocated emission entitlements based on their population.
Contraction and convergence: GCI, 2005	Agreement on a global emission path that leads to an agreed long-term stabilization level for greenhouse gas concentrations ('Contraction'). Emission targets for all individual countries set so per-capita emissions converge ('Convergence').
Basic needs or survival emissions: Aslam, 2002; Pan, 2005	Emission entitlements based on an assessment of emissions to satisfy basic human needs.
Adjusted per capita allocation: Gupta and Bhandari, 1999	Allocation of equal per capita emissions with adjustments using emissions per GDP relative to Annex I average.
Equal per capita emissions over time: Bode, 2004	Allocation based on (1) converging per capita emissions and (2) average per capita emissions for the convergence period that are equal for all countries.
Common but differentiated convergence: Höhne *et al.*, 2006	Annex I countries' per capita emissions converge to low levels within a fixed period. Non-Annex I countries converge to the same level in the same timeframe, but starting when their per capita emissions reach an agreed percentage of the global average. Other countries voluntarily take on "no lose" targets.
Grandfathering: Rose *et al.*, 1998	Reduction obligations based on current emissions.
Global preference score compromise: Müller, 1999	Countries voice preference for either per capita allocation or allocation based on current national emissions.
Historical responsibility – the Brazilian proposal: UNFCCC, 1997b; Rose *et al.*, 1998; Meira Filho and Gonzales Miguez, 2000; Pinguelli Rosa *et al.*, 2001; den Elzen and Schaeffer, 2002; La Rovere *et al.*, 2002; Andronova and Schlesinger, 2004; Pinguelli *et al.*, 2004; Trudinger and Enting, 2005; den Elzen and Lucas, 2005; den Elzen *et al.*, 2005c; Höhne and Blok, 2005; Rive *et al.*, 2006	Reduction obligations between countries are differentiated in proportion to those countries' relative share of responsibility for climate change – i.e. their contribution to the increase of global-average surface temperature over a certain period of time.
Ability to pay: Jacoby *et al.*, 1998; Lecoq and Crassous, 2003	Participation above welfare threshold. Emission reductions as a function of ability to pay (welfare).
Equal mitigation costs: Rose et al., 1998; Babiker and Eckhaus, 2002	Reduction obligations between countries are differentiated so that all participating countries have the same welfare loss.
Triptych: Blok et al., 1997; den Elzen and Berk, 2004; Höhne *et al.*, 2005	National emission targets based on sectoral considerations: Electricity production and industrial production grow with equal efficiency improvements across all countries. "Domestic" sectors converge to an equal per-capita level. National sectoral aggregate levels are then adopted.
Multi-sector convergence: Sijm *et al.*, 2001	Per-capita emission allowances of seven sectors converge to equal levels based on reduction opportunities in these sectors. Countries participate only when they exceed per capita threshold.
Multi-criteria: Ringius *et al.*, 1998; Helm and Simonis, 2001; Ringius *et al.*, 2002	Emission reduction obligations based on a formula that includes several variables, such as population, GDP and others.

Table 13.2: *Continued.*

Name (reference)	Description
National emission targets and emission trading	

<table>
<tr><th colspan="3">Alternative types of emission targets for some countries</th></tr>
<tr>
<td>Dynamic targets:
Hargrave et al., 1998; Lutter, 2000; Müller et al., 2001; Bouille and Girardin, 2002; Chan-Woo, 2002; Lisowski, 2002; Ellerman and Wing, 2003; Höhne et al., 2003; Müller and Müller-Fürstenberger, 2003; Jotzo and Pezzey, 2005; Philibert, 2005b; Pizer, 2005b; Kolstad, 2006</td>
<td colspan="2">Targets are expressed as dynamic variables – including as a function of the GDP ("intensity targets") or variables of physical production (e.g. emissions per tonne of steel produced).</td>
</tr>
<tr>
<td>Dual targets, target range or target corridor:
Philibert and Pershing, 2001; Kim and Baumert, 2002</td>
<td colspan="2">Two emission targets are defined: (1) a lower "selling target" that allows allowance sales if national emissions fall below a certain level; (2) a higher "buying target" that requires the purchase of allowances if a certain level is exceeded.</td>
</tr>
<tr>
<td>Dual intensity targets:
Kim and Baumert, 2002</td>
<td colspan="2">A combination of intensity targets and dual targets.</td>
</tr>
<tr>
<td>"No lose", "non-binding", one-way targets:
Philibert, 2000</td>
<td colspan="2">Emission rights can be sold if the target is reached, while no additional emission rights would have to be bought if target is not met. Allocations are made at a BAU level or at a level below BAU. Structure offers incentives to participate for countries not prepared to take on full commitments but still interested in joining the global trading regime.</td>
</tr>
<tr>
<td>Growth targets, headroom allowances, premium allocation:
Frankel, 1999; Stewart and Wiener, 2001; Viguier, 2004</td>
<td colspan="2">Participation of major developing countries is encouraged by unambitious allocations relative to their likely BAU emissions. To ensure benefit to the atmosphere, a fraction of each permit sold can be banked and definitely removed.</td>
</tr>
<tr>
<td>Action targets:
Goldberg and Baumert, 2004</td>
<td colspan="2">A commitment to reduce GHG emission levels below projected emissions by an agreed date through "actions" taken domestically, or through the purchases of allowances.</td>
</tr>
<tr>
<td>Flexible binding targets:
Murase, 2005</td>
<td colspan="2">A framework for reaching emission targets modelled after the WTO/GATT (World Trade Organization/General Agreement on Tariffs and Trade) scheme for tariff and non-tariff barriers; targets negotiated through rounds of negotiations.</td>
</tr>
<tr><th colspan="3">Modifications to the emission trading system or alternative emission trading system</th></tr>
<tr>
<td>Price cap, safety valve or hybrid trading system:
Pizer, 1999; Pizer, 2002; Jacoby and Ellerman, 2004.</td>
<td colspan="2">Hybrid between a tax and emission trading: after the initial allocation, an unlimited amount of additional allowances are sold at a fixed price.</td>
</tr>
<tr>
<td>Buyer liability:
Victor, 2001b</td>
<td colspan="2">If the seller of a permit did not reduce its emissions as promised, the buyer could not claim the emission credit. Enforcement is more reliable as buyers deal with developed countries with more robust legal procedures.</td>
</tr>
<tr>
<td>Domestic hybrid trading schemes:
McKibbin and Wilcoxen, 1997; McKibbin and Wilcoxen, 2002</td>
<td colspan="2">Two kinds of emissions permits valid only within the country of origin. (1) long-term permits entitle the permit owner to emit 1 tC every year for a long period; permits are distributed once. (2) Annual permits allow 1 tC to be emitted in a single year. An unlimited number of these permits are given out at a fixed price (price cap). Compliance is based on either unit.</td>
</tr>
<tr>
<td>Allowance purchase fund:
Bradford, 2004</td>
<td colspan="2">Countries contribute to an international fund that buys/retires emission reduction units. Countries can sell reductions below their BAU levels.</td>
</tr>
<tr>
<td>Long-term permits:
Peck and Teisberg, 2003</td>
<td colspan="2">Long-term permits could be used once at any time between 2010 and 2070. Depending on the time of emission they are depreciated 1% annually for atmospheric decay of CO_2. The permit would allow the emission of 1 tC in 2070, 1.01 tC in 2069 and 1.0160 (1.71) tons in 2010.</td>
</tr>
</table>

Row group labels (vertical, left margin): National emission targets and emissions trading; Alternative types of emission targets for some countries; Modifications to the emission trading system or alternative emission trading system

Table 13.2: *Continued.*

Name (reference)		Description
Sectoral approaches	Sectoral approaches	
	Sector Clean Development Mechanism, sector Crediting Mechanism : Philibert and Pershing, 2001; Samaniego and Figueres, 2002; Bosi and Ellis, 2005; Ellis and Baron, 2005; Sterk and Wittneben, 2005	Sectoral crediting schemes based on emission reductions below a baseline. Excess allowances can be sold.
	Sector pledge approach: Schmidt et al., 2006	Annex I countries have emission targets, with the ten highest-emitting developing countries pledging to meet voluntary, "no-lose" GHG emissions targets in the electricity and major industrial sectors. Targets are differentiated, based upon national circumstances, and sector-specific energy-intensity benchmarks are developed by experts and supported through a Technology Finance and Assistance Package.
	Caps for multinational cooperation: Sussman et al., 2004	A cap/and trade system associated with the operations of associated enterprises in developing and developed countries.
	Carbon stock protocol: WBGU, 2003	A protocol for the protection of carbon stocks based on a worldwide system of "non-utilization obligations" to share the costs of the non-degrading use of carbon stocks among all states.
	"Non-binding" targets for tropical deforestation[a]: Persson and Azar, 2004	Non-binding commitments for emissions from deforestation under which reduced rates of deforestation could generate emissions allowances.
Policies and measures	Policies and measures	
	Carbon emission tax: Cooper, 1998; Nordhaus, 1998; Cooper, 2001; Nordhaus, 2001; Newell and Pizer, 2003	All countries agree to a common, international GHG emission tax; several of the proposals suggest beginning with a carbon tax limited to emissions from fossil fuel combustion.
	Dual track: Kameyama, 2003	Countries choose either non-legally binding emission targets based on a list of policies and measures or legally-binding emission caps allowing international emissions trading.
	Climate "Marshall Plan": Schelling, 1997, 2002	Financial contributions from developed countries support climate friendly development; similar in scale and oversight to the Marshall Plan.
Technology	Technology	
	Technology research and development: Edmonds and Wise, 1999; Barrett, 2003	Enhanced coordinated technology research and development.
	Energy efficiency standards: Barrett, 2003; Ninomiya, 2003	International agreement on energy efficiency standards for energy-intensive industries.
	Backstop technology protocol: Edmonds and Wise, 1998	New power plants installed after 2020 must be carbon neutral. New synthetic fuels plants must capture CO_2. Non-Annex I countries participate upon reaching Annex I average GDP in 2020.
	Technology prizes for climate change mitigation: Newell and Wilson, 2005	Incentive or inducement prizes targeted at applied research, development and demonstration.
Development-oriented actions	Development-oriented actions	
	Sustainable development policies and measures: Winkler et al., 2002b; Baumert et al., 2005b	Countries integrate policies and measures to reduce GHG emissions into development plans (e.g. developing rural electrification programmes based on renewable energy, or mass transit systems in placed of individual cars).
	Human development goals with low emissions: Pan, 2005	Elements include: identification of development goals/basic human needs; voluntary commitments to low carbon paths via no-regret emission reductions in developing countries conditional to financing and obligatory discouragement of luxurious emissions; reviews of goals and commitments; an international tax on carbon.

Table 13.2: *Continued.*

Name (reference)		Description
Adaptation	Adaptation	
	UNFCCC impact response instrument: Müller, 2002	A new "impact response instrument" under the auspices of the UNFCCC for disaster relief, rehabilitation and recovery.
	Insurance for adaptation; funded by emission trading surcharge: Jaeger, 2003	A portion of the receipts from sales of emissions permits would be used to finance insurance pools.
Financing	Financing	
	Greening investment flows: Sussman and Helme, 2004	Investments through Export Credit Agencies are conditional on projects that are "climate friendly".
	Quantitative finance commitments: Dasgupta and Kelkar, 2003	Annex I countries take on quantitative financial commitments – e.g. expressed as a percentage of the GDP – in addition to emission reduction targets.
Negotiation process and treaty structure	Negotiation process and treaty structure	
	Bottom-up or multi-facet approach, pledge (with review) and review: Reinstein, 2004; Yamaguchi and Sekine, 2006	Each country creates its own initial proposal relating to what it might be able to commit to. Individual actions accumulate one by one. The collective effect of proposals is periodically reviewed for adequacy and – if necessary – additional rounds of proposals are undertaken.
	Portfolio approach: Benedick, 2001	A portfolio including: emission reduction policies, government research/development, technology standards and technology transfer.
	A flexible framework: PEW, 2005	A portfolio including: aspirational long-term goals, adaptation, targets, trading, policies, and technology cooperation.
	Orchestra of treaties: Sugiyama et al., 2003	A system of separate treaties among like-minded countries (emission markets, zero emission technology, climate-wise development) and among all parties to UNFCCC (monitoring, information, funding).
	Case study approach: Hahn, 1998	Multiple case studies of coordinated measures, emissions tax, tradable emission permits and a hybrid system in industrialized countries to learn by doing.

Note:

a There is some potential conflict with the terminology here: "non-binding" targets may be interpreted by some as restricting the capacity of countries to trade as they do not necessarily set up caps that impose prices and thus established tradable commodities.

Source: Earlier overviews by Bodansky, 2004; Kameyama, 2004; Philibert, 2005a

(2003) believe such goal-setting is desirable as it helps structure commitments and institutions, provides an incentive to stimulate action and helps establish criteria against which to measure the success of implementing measures.

An alternative to agreeing on specific CO_2 concentration or temperature levels is an agreement on specific long-term actions (such as a technology-oriented target, such as 'eliminating carbon emissions from the energy sector by 2060'). An advantage of such a goal is that it might be linked to specific actions. While links between such actions, GHG concentrations and climate impacts can be made, there are uncertainties in the precise correlation between them. Additionally, several different targets would have to be set to cover all climate-relevant activities (Schelling, 1997; Pershing and Tudela, 2003).

Another option would be to adopt a 'hedging strategy' (IPCC, 2001, chapter 10), which is defined as a shorter term goal on global emissions, from which it is still possible to reach a range of desirable long-term goals. One example of such a strategy is the California goal of reducing emissions to 1990 levels by 2020, and then reducing them to 80% below 1990 levels by 2050. Once the short-term goal is reached, decisions on subsequent

steps can be made in light of new knowledge and decreased levels of uncertainty. To implement this option, the international community could agree on a maximum quantity of permissible GHG emissions in, for example, 2020 (Corfee-Morlot and Höhne, 2003; Pershing and Tudela, 2003; Yohe *et al.*, 2004).

Another proposal would be to aim at formulating reductions step by step, based on the willingness of countries to act, without explicitly considering a long-term perspective. While such an approach does meet political acceptability criteria, it poses the risk that the individual reductions may not add up to the level required for certain stabilization levels. Some stabilization options may then be out of reach in the near future (see Chapter 3.3, Figure 3.19).

13.3.3.2 Participation

The participation of states in international agreements can vary. At one extreme, participation can be universal; at the other extreme, participation can be limited to just two countries. Many studies propose that participation can be differentiated in different tiers (see Staged systems in Table 13.2). States participating in the same tier would have the same

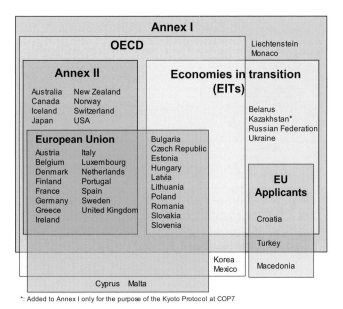

Figure 13.2: *Current country groupings under the UNFCCC, OECD and EU.*
Source: Höhne et al. (2005).

Some have argued that an international agreement needs to include at least the major emitters to be effective, since the largest 15 countries (the EU25 is considered here to be one country) produce as much as 80% of global GHG emissions (Baumert *et al.*, 2005a; PEW, 2005; Stewart and Weiner, 2003; Torvanger *et al.*, 2005; Schmidt *et al.*, 2006). A similar approach has been taken by authors comparing climate change agreements to other multilateral instruments, including disarmament treaties and the Antarctic Treaty (see Murase, 2002a). In these analyses, the authors assert that success can only be achieved if the major stakeholders act. Thus, for example, a nuclear disarmament treaty would be meaningless if it was not ratified by those States with nuclear weapons, even if it was ratified by the 180 non-nuclear States. By analogy, a climate change treaty is meaningful only if commitments are adopted and implemented by the major emitters – noting that the benefits of participation accrue to all countries, including those not taking part in the agreement. Murase (2002a) suggests that a future regime after 2012 thus needs to include key countries or groups such as the USA, EU, Japan, China, India, Korea, Mexico, Brazil, Indonesia, South Africa and Nigeria.

type of commitments (i.e. in the UNFCCC regime). The most important tiers are Annex I and non-Annex I, but there are also special arrangements for economies in transition as well as for least developed countries. Figure 13.2 shows the groupings of countries under the UNFCCC, OECD and EU. The allocation of states into tiers can be made according to quantitative or qualitative criteria or 'ad hoc' (see Table 13.2). According to the principle of sovereignty, states may also choose the tier in which they want to be grouped, provided their choice is accepted by other countries (see Kameyama, 2003; Reinstein, 2004).

Participation in the agreement can be static,[32] or it may change dynamically over time. In the latter case, states can "graduate" from one tier of commitments to the next. Graduation can be linked to the meeting of quantitative thresholds for certain parameters (or combinations of parameters) that have been predefined in the agreement, such as emissions, cumulative emissions, GDP per capita, relative contribution to temperature increase or other measures of development, such as the human development index (see Berk and Den Elzen (2001), Gupta (1998, 2003a) and Höhne *et al.* (2003) for a review of per-capita emissions thresholds; Criqui *et al.* (2003) and Michaelowa *et al.* (2005b) for discussion of a composite index using the sum of per-capita emissions and per-capita GDP and Torvanger *et al.* (2005) for further composite indices). Qualitative thresholds such as adherence to certain country groupings (OECD, Economies in Transition) are already in use. Ott *et al.* (2004) combine quantitative and qualitative thresholds. Thresholds can be derived from agreed-upon GHG concentration targets or global emissions paths or be based on other parameters, such as willingness or capacity to pay.

Much of the literature on game theory suggests that the conditions necessary for achieving large-scale stable coalitions mean that relatively modest emissions reductions will be achieved (e.g. Carraro and Siniscalco, 1993; Hoel and Schneider, 1997). Cooperative game theory emphasizes the prospect of building stable coalitions if a transfer scheme (e.g. by emissions trading) can allocate the gains from cooperation in proportion to the benefits from reduced climate impacts (e.g. Chander and Tulkens, 1995; Germain *et al.*, 1998; Germain *et al.*, 2003). Eykmans and Finus (2003) note that much of the literature focuses on a 'grand (all party) coalition, analyses stability in terms of the aggregate payoff to coalitions and rests on very strong assumptions about implicit punishment of any free-riding countries.' A more extensive discussion of the issues of free-riding is contained in Chapter 10 of the TAR.

Alternative assumptions can provide a richer understanding of possible factors relevant to an agreement by relating relate to the response to payoffs from cooperation, including spillover and trade effects, allowing for the development of multiple coalitions and recognizing trade and the role of technology transfer as well as the potential for other transfer schemes (Tol *et al.*, 2000; Finus, 2002; Kemfert *et al.*, 2004). They also increase the possibility that partial cooperation (including involving more than one coalition) can close the gap between the global optimum (full cooperation) and "no cooperation" by a substantial amount. While this is essentially a theoretical conclusion (based in some cases on modelling reflecting some empirical evidence), it provides some basis for suggesting that it is too restrictive to assume that a single, all-encompassing global intergovernmental agreement is a *necessary* condition

---

32 For example, participation in the tiers of commitments of the Kyoto Protocol can only be changed by an amendment which has to be ratified by all parties. As this is
    extraordinarily difficult, membership in the tiers is essentially fixed.

for effective mitigation action.

Some authors (see, for example, Muller, 2002; Jaeger, 2003) suggest that a climate regime is not exclusively about mitigation but that it also encompasses adaptation and, as such, far wider arrays of countries are vulnerable to climate and must be included in any agreement. Further, several authors (e.g. Meira Filho and Gonzales, 2000; Pan, 2005) argue that even if the majority of emissions are the responsibility of only a few nations, all countries must share the commitments to reduce these for reasons of equity and fairness (recognizing that such actions should be differentiated according to responsibility and capability). Other rationales for global engagement are also used, including that if only some major countries participate, the emissions of non-participating countries could increase by the migration of emission-intensive industries. Therefore, most proposals aim to provide incentives for countries to participate. Some aim at pull incentives, such as temporary over-allocation or no regret structures; others mention push incentives, such as trade sanctions or border tax adjustments (Kuik, 2003; Biermann and Brohm, 2005).

Other authors argue that countries have differentiated historical responsibility and that such a sub-global participation can be effective: Grubb et al. (2002) argue that under some scenarios one can expect that technology development driven by the international climate regime in Annex I countries could offset some or all emissions leakage in non-Annex I countries. Sijm (2004) notes that a number of policies could promote this spillover effect in the longer term. These types of policies include international cooperation on Research, Development and Demonstration (RD&D), promoting open trade or using the Clean Development Mechanism. Others argue that with the participation of some large countries, other countries cannot lag behind and that the climate regime should look for that 'tipping point' (Barrett, 2003).

In general, the literature suggests that actions can occur in parallel and that international agreements could have multiple components, since national circumstances are so diverse. However, the suggestion is also made that care should be taken, particularly for countries with limited institutional capacity, to avoid creating too many simultaneous international activities.

13.3.3.3 *Implications of regime stringency: linking goals, participation and timing*

Several studies have analysed the regional emission allocations or requirements on emission reductions and time of participation in the international climate change regime with the aim of being able to ensure different concentration or temperature stabilization targets (Berk and den Elzen, 2001; Blanchard, 2002; Winkler et al., 2002a; Criqui et al., 2003; WBGU, 2003; Bollen et al., 2004; Groenenberg et al., 2004; Böhringer and Löschel, 2005; den Elzen and Meinshausen, 2005; den Elzen and Lucas, 2005, den Elzen et al., 2005c;

Höhne et al., 2005; Michaelowa et al., 2005a; Böhringer and Welsch, 2006; Höhne, 2006; Persson et al., 2006). A large variety of system designs for allocating emission allowances/ permits were analysed, including contraction and convergence, multistage, Triptych and intensity targets. The studies cover a broad spectrum of parameters and assumptions that influence these results, such as population, GDP development of individual countries or regions, global emission pathways that lead to climate stabilization (including overshooting the desired concentration level), parameters for the thresholds for participation and ways to share emission allowances. For example, the studies include very stringent requirements for developed countries with more lenient requirements for developing countries as well as less stringent requirements for developed countries and more ambitious constraints for developing countries within a plausible range. The conclusions of these studies and their implications for international regimes can be summarized as follows:

- Under regime designs for low and medium concentration stabilization levels (i.e. 450 and 550 ppm CO_2-eq, category A and B; see Chapter 3, Table 3.10) GHG emissions from developed countries would need to be reduced substantially during this century. For low and medium stabilization levels, developed countries as a group would need to reduce their emissions to below 1990 levels in 2020 (on the order of –10% to 40% below 1990 levels for most of the considered regimes) and to still lower levels by 2050 (40% to 95% below 1990 levels), even if developing countries make substantial reductions. The reduction percentages for individual countries vary between different regime designs and parameter settings and may be outside of this range. For high stabilization levels, reductions would have to occur, but at a later date (see Box 13.7).

- Under most of the considered regime designs for low and medium stabilization levels, the emissions from developing countries need to deviate – as soon as possible – from what we believe today would be their baseline emissions, even if developed countries make substantial reductions. For the advanced developing countries, this occurs by 2020 (mostly Latin America, Middle East and East Asia). For high stabilization levels, deviations from the reference level are necessary only at a later date.

- Reaching lower levels of GHG concentrations requires earlier reductions and faster participation compared to higher concentrations.

- For many countries, the overall target set is critical; it dictates the emissions reduction requirements more specifically than does the approach chosen to meet that target.

- The wide diversity of approaches means that not all countries participate under all regimes – even if an identical concentration target is achieved. Obviously, required national actions differ enormously, depending on whether a country participates in a system. However, the difference in reductions required between the various approaches is small for participating countries.

Box 13.7 The range of the difference between emissions in 1990 and emission allowances in 2020/2050 for various GHG concentration levels for Annex I and non-Annex I countries as a group[a]

Scenario category	Region	2020	2050
A-450 ppm CO$_2$-eq[b]	Annex I	–25% to –40%	–80% to –95%
	Non-Annex I	Substantial deviation from baseline in Latin America, Middle East, East Asia and Centrally-Planned Asia	Substantial deviation from baseline in all regions
B-550 ppm CO$_2$-eq	Annex I	-10% to -30%	-40% to -90%
	Non-Annex I	Deviation from baseline in Latin America and Middle East, East Asia	Deviation from baseline in most regions, especially in Latin America and Middle East
C-650 ppm CO$_2$-eq	Annex I	0% to -25%	-30% to -80%
	Non-Annex I	Baseline	Deviation from baseline in Latin America and Middle East, East Asia

Notes:

[a] The aggregate range is based on multiple approaches to apportion emissions between regions (contraction and convergence, multistage, Triptych and intensity targets, among others). Each approach makes different assumptions about the pathway, specific national efforts and other variables. Additional extreme cases – in which Annex I undertakes all reductions, or non-Annex I undertakes all reductions – are not included. The ranges presented here do not imply political feasibility, nor do the results reflect cost variances.

[b] Only the studies aiming at stabilization at 450 ppm CO$_2$-eq assume a (temporary) overshoot of about 50 ppm (See Den Elzen and Meinshausen, 2006).

Source: See references listed in first paragraph of Section 13.3.3.3

Several studies have gone one step further and have, based on emission allocations, calculated emission reduction costs and possible trades of emission allowances at a regional level for different concentration or temperature stabilization targets (Criqui *et al.*, 2003; WBGU, 2003; Bollen *et al.*, 2004; Böhringer and Welsch, 2004, 2006; Böhringer and Löschel, 2005; den Elzen and Lucas, 2005; den Elzen *et al.*, 2005c; Persson *et al.*, 2006). Researchers have also analysed a large variety of system designs. With cost analysis even more assumptions are relevant, such as detailed assumptions on emission reduction costs per sector and region. Costs have been calculated using a variety of models, ranging from those with detailed sectoral representation focussing on the technological aspects to macroeconomic models focussing on the economy as a whole. How (and what) costs are calculated plays a role. Some studies present annual direct mitigation costs (only direct abatement costs) or energy costs, such as mitigation costs and costs of losses of fossil fuel exports or gains from increased exports of biofuels. Other studies present full macro-economic costs, calculated as (cumulative) GDP losses in a specific target year. The cumulative impact of climate policies on GDP may be lower than expected from the annual abatement costs levels due to the fact that climate policy leads mostly to the substitution of investments and activities and much less to an overall reduction of the GDP. The conclusions of these studies on costs can be summarized as follows:

Global costs

- The total global costs are highly dependent on the baseline scenario, marginal abatement costs estimates, the participation level in emission trading and the assumed concentration stabilization level (see also Chapter 11).

- The total global costs does not vary significantly for the same global emission level; however, costs will vary with the degree of participation in emission trading (how and when allowances are allocated). If, for example, some major emitting regions do not participate in the reductions and in emission trading immediately, the global costs of the participating regions may be higher (see also Chapter 3, e.g. Bollen *et al.*, 2004; den Elzen *et al.*, 2005c).

Regional costs

- Regional abatement costs are largely dependent on the assumed stabilization level and baseline scenario. The allocation regime is also an important factor, although in most countries the extent of its effect is less than that of the stabilization level (see Criqui *et al.*, 2003; den Elzen and Lucas, 2005; den Elzen *et al.*, 2006b). The allocation parameter having the largest effect is the timing of participation. Under a staged approach, whether a region participates early or late is of great importance. If, for example, convergence of the per capita emissions were to occur by the end of this century, developing regions would incur high costs relative to what might occur in the reference or baseline cases. Conversely, if convergence were to occur by the middle of the century, developed countries would incur higher costs relative to what they might incur in a reference or baseline case (see Nakicenovic and Riahi, 2003; den Elzen *et al.*, 2005a; Persson *et al.*, 2006).

- Abatement costs (only costs from reducing emissions) as a percentage of GDP vary significantly by region for allocation schemes that ultimately lead to convergence in per capita emissions by the middle of this century. The costs are above the global average for the Middle East and the Russian Federation, including surrounding countries, and – to a lesser extent – for Latin America. The costs are near the world average for the OECD regions and below the world average for China. The other developing regions, such as Africa and South-Asia (India), experience low costs or even gains as a result of financial transfers from emission trading. (Criqui *et al.*, 2003; den Elzen and Lucas, 2005).

- In addition to the abatement costs of reducing emissions, other costs arise from changes in international trade. Fossil fuel-exporting regions are also likely to be affected by losses in coal and oil exports compared to the baseline, while some regions could experience increased bio-energy exports (i.e. the Russian Federation and South America) (see Nakicenovic and Riahi, 2003; van Vuuren *et al.*, 2003; Persson *et al.*, 2006; and also Chapter 11).

- The economic impacts in terms of welfare changes show a similar pattern for different allocation schemes. For example, allocation schemes based on current emissions (sovereignty) lead to welfare losses for the developing countries. Allocation schemes based on a per capita convergence lead to welfare gains for developing countries, without leading to excessive burdens for industrialized countries. (Böhringer and Welsch, 2004)

13.3.3.4 *Actions*

13.3.3.4.1 Targets

While many types of commitments are identified in the literature on climate change, the most frequently evaluated commitment is that of the binding absolute emission reduction target as included in the Kyoto Protocol for Annex I countries. The broad conclusion that can be drawn from the literature is that such targets provide certainty about future emission levels of the participating countries (assuming targets will be met). These targets can also be reached in a flexible manner across GHGs and sectors as well as across borders through emission trading and/or project-based mechanisms (in the Kyoto Protocol case, this is referred to as Joint Implementation (JI) and as the Clean Development Mechanism (CDM).

One crucial element is defining and agreeing on the level of the emission targets. Examples of processes to agree on a target include:

- Participating countries make proposals (pledges) for individual reductions on a bottom-up basis. This approach has the risk that proposed reductions may not be adequate to lead to the desired stabilization levels.
- A common formula can be agreed upon for determining the emission targets. This rule could lead to reduction percentages for each individual country (which could subsequently be modified by negotiations).
- An overall target can be given to a group of countries, with the group deciding internally on how to share the target amongst the participants. This approach has been applied to the EU for the purpose of the Kyoto Protocol. It could, in principle, also be applied to any other group of countries.

Many authors have raised concerns that the absolute or fixed targets may be too rigid and cap economic growth (Philibert and Pershing, 2001; Höhne *et al.*, 2003; Bodansky, 2004). To address these concerns, a number of more flexible national emission targets have been proposed (see alternative types of emission targets in Table 13.2). These options aim at maintaining the advantages of international emissions trading while providing more flexibility to countries to avoid extremely high costs and, thereby, potentially allowing for the adoption of more stringent targets. However, this flexibility reduces the certainty that a given emission level will be reached. Thus, there is a trade-off between costs and certainty in achieving an emissions level (see Jotzo and Pezzey, 2005). Other disadvantages that have been mentioned are adding to the complexity of the system or, in the case of intensity targets, the difficulty in coping with economic recession as well as the potential for creating ambiguity for market investors.

Additional understanding comes from the political science literature which emphasizes the importance of analysing the full range of factors bearing on decisions by nation states, including domestic pressures from the public and affected interest groups, the role of norms and the contribution of NGOs (environment, business and labour) to the negotiation processes. Studies of the European Acid Rain Regime have revealed, for example, that although agreements on an ambitious target can serve as a driver for policy implementation, they may not necessarily result in a good environmental consequence if the countries involved do not have the capacity to comply with what they have committed themselves to in good faith (Victor, 1998). While such case study-based analyses yield conclusions that are dependent on the choice of cases and the manner in which the analysis is carried out, they can provide insights which are more accessible to policymakers than more quantitative economic analyses.

13.3.3.4.2 Flexibility provisions

Many environment agreements seek to address complex issues by allowing for additional flexibility as a means to achieve their goals. Flexibility has been suggested as to 'how', 'when', 'where' and 'what' emissions are to be reduced. In the climate change context, emission reductions under an international agreement can conceptually be achieved any 'where' on the globe. It is also possible to shift the timing ('when') of emission reductions (depending on the emission pathway), the 'how' (i.e. choice of policy instrument) and the 'what' in terms of the specific emission source or sink that is the target of the policy.

The Kyoto Protocol incorporates three articles that provide flexibility as to 'where' emission reductions occur, namely, through provisions on international emission trading, JI and the CDM. Under Kyoto's international ETS, emission allowances may be traded between governments of Annex B parties if a surplus occurs in one country. Emission reductions achieved through projects between Annex I countries are called JI, while emission reduction projects located in non-Annex I countries are called CDM projects. Extensive rules have been agreed upon to ensure that credits created under these project mechanisms actually represent the emissions reduced.

International Emissions Trading

Emissions' trading has become an important implementation mechanism for addressing climate change in many countries. The overall value of the global carbon market was over 10 billion US$ in 2005, and in the first quarter of 2006 the transaction level reached 7.5 billion US$ (World Bank and IETA, 2006). The most advanced ETS is that developed by the EU. While this system is an international one, it bears many of the characteristics of a national programme, with oversight by the European Commission and a centralized regulatory and review mechanism (see Box 13.4 for details, including those on trading prices and volumes). A larger global system of international trading is slowly developing through emission credits generated by the project-based mechanisms.[33] Theoretically, a fully global ETS would provide market players and policymakers with information thus far absent from decision-making: the actual, unfettered, global cost of GHG mitigation in a range of economic activities. In this context, at the international level, such a regime would mirror the information provided by national trading programmes at a global scale.

Lecocq and Capoor (2005) note that while the international GHG emissions market remains fragmented, trading activity has increased substantially during the last 5 years. According to their analysis, regional, national and sub-national trading programmes are all operating under different rules, which could inhibit 'market convergence' and increase the costs of trading. Others indicate that a global market can incorporate diverse domestic and regional systems despite differences in design; they reiterate the point made by others that such a system may be significantly less efficient that a single globally optimized regime (Baron and Philibert, 2005).

A full assessment of the elements required to link multiple regimes is provided by Haites (2003a), who identifies only a few situations that might prevent linkages (a formal prohibition in one system to allow links, and circumstances where a single firm's membership in multiple programmes creates the potential for double counting). However, issues that could complicate links between two or more emissions trading programmes include concerns on the effectiveness of compliance enforcement and on whether the linked regimes provide adequate protection of either system's environmental objectives. As Bygrave and Bosi (2004a,b) note, links do not need to be formal; market arbitrage can provide opportunities for purchasing allowances in multiple markets even if there is no specific recognition of one system's permits under another's structure.

Various authors have analysed the size of the allowance surplus of the countries in transition, barriers to accessing allowances, the potential market power of cartels and links to energy security. Such surpluses can alter the overall costs of compliance with the Kyoto commitments – but only if trade in such surplus allowances is undertaken. Victor et al. (2001a) estimated the joint Russian and Ukrainian surplus at 3.7 billion tCO_2 for the entire commitment period 2008–2012. Berkhout and Smith (2003) estimate the surplus level of the former Soviet Union through to 2030 and state that it could only cover half of an assumed 30% reduction target for a 28-member state EU. Golub and Strukova (2004) see the Russian surplus as being up to 3 billion tCO_2, arguing that due to barriers in the Russian capital market, forward trading with OECD countries represents the only opportunity to raise initial capital to mobilize no-regret and low-cost GHG reductions. Maeda (2003) shows that permits for surplus emissions in the international emissions trading regime may affect the economic efficiency of the Kyoto mechanism and suggests that considerable market power exerted by sellers could affect the price (e.g. if all of the economies in transition form a cartel, if Ukraine forms a cartel with Russia or even if Russia acts alone). Kuik (2003) sees a trade-off between economic efficiency, energy security and carbon dependency with respect to the EU acquisition of Russian and Ukrainian assigned amount units. One proposal for reducing concerns over trading in surplus allowances is that of the 'Green Investment Scheme', in which revenues from sales of surplus allowances are spent on national policies, programmes and projects to further reduce emissions; this option is explained further below.

Project-based mechanisms (Joint Implementation and the Clean Development Mechanism)

The earliest project-based mechanism of the UN Climate Convention process was the pilot phase of 'Activities Implemented Jointly' (AIJ). Most of the 150 AIJ projects were small, and many were only partially implemented due to the lack of financing that resulted from the lack of emissions credits. Only half a dozen investor countries and even fewer host countries developed real, national AIJ programmes. Selection criteria for AIJ programmes often delayed the acceptance of

---

33 The EU ETS has also an international component as it involves cross-border trades and transactions between national allowance registries.

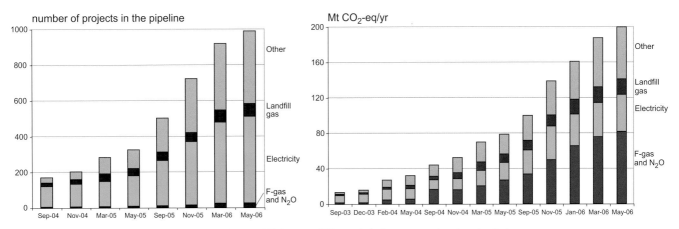

Figure 13.3: *Evolution of the Clean Development Mechanism portfolio in terms of CO_2 -equivalents per year and number of projects.*
Source: Ellis and Karousakis (2006).

projects, and most that were undertaken were commercially viable only if additional financing was provided by a separate investment subsidy (Michaelowa, 2002).

Since 2000, the CDM has allowed crediting of project-based emission reductions in developing countries; this is the first of the Kyoto Protocol's market mechanisms to be implemented. A number of analysts have estimated CDM volume and price. Chen (2003) derived prices of 2.6–4.9 US$/tCO$_2$ and annual volumes of approximately 600–1000 million certified emissions reductions (CERs). Jotzo and Michaelowa (2002) and Michaelowa and Jotzo (2005) model an annual CER demand of 360 million tCO$_2$ and a price of 3.6 €/tCO$_2$. Springer and Varilek (2004) predict a likely CER price of less than 10 US$/tCO$_2$ in 2010. CER prices increased from approximately 3 €/tCO$_2$ in 2003 to more than 20 €/ton in early 2006 (at the time of peak prices in the EU ETS); as of October 2006, they had declined to about 13 €/tCO$_2$. CER prices have been relatively closely tied to EU ETS prices over time.

As of May 2006, the volume of CERs estimated from nearly 1000 proposed projects in 69 countries was 200 MtCO$_2$-eq/ year in 2008–2012 and 330 Mt MtCO$_2$-eq/year in the pre-2008 period (Ellis and Karousakis, 2006; specific project information can be found at http://cdm.unfccc.int; recent updates on the CDM/JI pipeline can also be found at the UNEP/RISO site, www.cd4cdm.org/publications/CDMpipeline.xls) (See Figure 13.3). While not all projects will be implemented, the UNFCCC cites 491 registered projects and estimates CERs equal to 740 MtCO$_2$-eq from those projects through to the end of 2012.[34] Ellis and Karousakis (2006) also indicate that almost half of the proposed CDM projects are in the electricity sector and that many are small renewable projects occurring in 40 countries. However, the majority of credits have come from CDM projects reducing nitrous oxide (N$_2$O), trifluoromethane (HFC-23) and,

to a lesser extent, methane (CH$_4$). Projects that have not yet had methodologies approved will be under-represented in the project mix – even if they represent opportunities for significant emissions reductions at the national or global level. Publicly committed budgets for CER acquisition stood at approximately 7.5 billion US$ (World Bank, 2006) (See Figure 13.4). At such a scale, the CDM begins to reach the same order of magnitude as Global Environment Facility (GEF) and Official Development Assistance (ODA) resources.

It was initially assumed that CDM projects would be undertaken as bilateral arrangements between Annex I and non-Annex I convention Parties (and private sector companies in those countries). As of October 2006, 56% of registered projects were being undertaken unilaterally, indicating that companies in developing countries are procuring the financing to implement projects and sell the CERs to industrialized countries.[35]

A CDM project has to go through an elaborate project cycle that includes external validation and which has been defined by a decision of the 7th Conference of the Parties to the UNFCCC (2001) and is in keeping with the decisions of the CDM Executive Board that is overseeing the project cycle (see, for example, UNFCCC, 2003a–c). As CDM projects are implemented in countries without emissions targets, project 'additionality' becomes important to avoid generating fictitious emission reduction credits through 'business as usual' activities. Several tests of additionality have been discussed in the literature; these include investment additionality (see Greiner and Michaelowa, 2003) and environmental additionality (see Shrestha and Timilsina, 2002). The CDM Executive Board has developed an additionality tool that project proponents can use to test and demonstrate the additionality of a CDM project (http://cdm.unfccc.int/methodologies/PAmethodologies/ Additionality_tool.pdf).

---

34 As of January 22, 2007. See: http://cdm.unfccc.int
35 The CDM Executive Board at its 18th meeting decided that registration can take place without an Annex I Party being involved at the time of registration. An Annex I partner would need to issue a letter of approval after registration in order to receive the CERs.

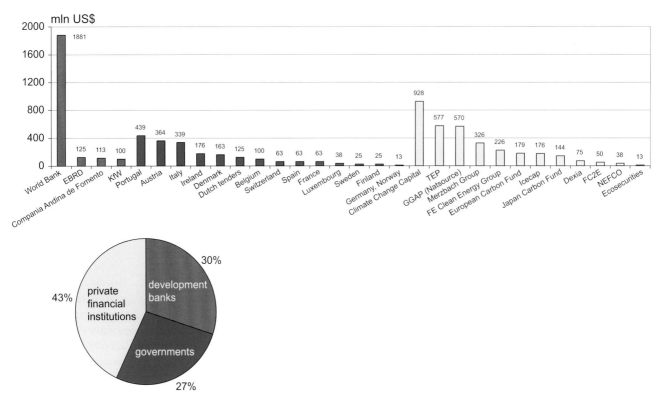

Figure 13.4: *Budgets for the acquisition of certified emissions reductions (CERs) and emission reductions units (ERUs).*
Note: Status as of October 10, 2006 at which time the total budget was almost € 6 billion.

If a project is additional, the next step is to determine a 'baseline' – the emissions that would have occurred if the project had not taken place. One potential risk is the overestimation of baseline emissions, which is a major problem as all participants profit from an overestimate as there is then no incentive to correct it. Stringent rules and modalities are required for determining baselines affecting the efficient processing of the CDM (Bailey *et al.*, 2001). Fischer (2006) argues that due to pressure from industry, rules for standard emission rates are likely to be systematically biased to over-allocation and also risk creating inefficient investment incentives. Alternatively, Broekhoff (2004) focuses on costs and efficiency, arguing that the availability of data and the level of data aggregation determine to a large extent the cost of deriving multi-project baselines. Other authors examine specific baseline issues in the energy sector, particularly the use of models, the need to consider size, vintage, generation type and operational characteristics and issues relating to technology and sectoral approaches (see Fichtner *et al.*, 2001; Zhang *et al.*, 2001; Spalding-Fecher *et al.*, 2002; Begg and Van der Horst, 2004; Illum and Meyer, 2004; Kartha *et al.*, 2004; Rosen *et al.*, 2004; Sathaye *et al.*, 2004).

In order to account for any emissions that occur outside of the CDM project boundary but which are a consequence of the CDM project – emissions referred to a 'carbon leakage' – a CDM project should also include a leakage estimate. According to the UNFCCC CDM glossary of terms, leakage is defined as

the net change of anthropogenic emissions by sources of GHGs that occur outside the project boundary and which is measurable and attributable to the activity of the CDM project. Leakage issues have been discussed by a number of authors (see, for example, Geres and Michaelowa (2002) and Kartha *et al.* (2002) for the electricity sector and the Working Group on Baseline for CDM/JI Project (2001)). There is a general consensus that the determination of project boundaries is critical to any evaluation of leakage.

The coverage of forestry and forest-related projects is a contentious issue under the CDM. The problems primarily relate to the impermanence of the forest and to leakage to other regions. Dutschke (2002) suggests leasing CDM credits to address the non-permanence of forestry sinks. The CDM has addressed the issue of non-permanence through the creation of separate CDM credits, which are called temporary CERs. According to Nelson and de Jong (2003), development priorities can be lost. This is illustrated by the case of a forestry project in Chiapas in which Mexico shifted from a development emphasis with multiple species to two species when the focus changed to carbon sales by individual farmers. Data (or its scarcity) as well as price uncertainty also pose problems. Vöhringer (2004) notes that establishing historical deforestation rates is a major problem in Costa Rica. Van Vliet *et al.* (2003) analysed six proposed plantation forestry projects in Brazil for uncertainty and, based on their results, they suggest that fluctuations in product prices cause variations of up to 200% in CERs and net present value, leading to difficulties in determining the additionality of such projects, thereby making five of the six projects ineligible for CDM.

Perhaps the most critical issue in the context of the viability of the CDM over the longer term is whether there will be an ongoing price signal that encourages both emission reduction commitments and a market demand – over the longer term. This will clearly depend on the shape of both international agreements and evolving national programmes that might support project offsets. Independent of the market demand issues, an important suggestion to enhance the CDM relates to improving the sustainable development benefits of a CDM. One proposal[36] for doing this is the 'Gold Standard', which calls for enhanced environmental assessment, stakeholder consultations and the use of a qualitative sustainability matrix, expanding the CDM regime to allow programmes and policies to be credited – a concept elaborated on in a decision by the first meeting of the Kyoto Parties in 2005, and analysed by Ellis (2006) – and extending CDM project incentives beyond 2012.

Joint Implementation has been much less extensively researched than the CDM. Its later start date and unclear international rules (for example, the 'second track' rules were only agreed upon in October 2006) have generated considerable uncertainty with regard to implementation. Transactions under JI are seen as both cumbersome and beset with institutional obstacles (Korppoo, 2005). In addition, several authors have argued that JI projects will potentially be 'double counted' – given credit under both the project mechanism as well as under the rules for EU ETS. A number of proposals have been made to address this issue. Koch and Michaelowa (1999) and Moe *et al.* (2003) have suggested a 'Green Investment Scheme' (GIS) in which revenues from sales of Assigned Amount Units (AAU) are allocated to projects that reduce GHG emissions. Blyth and Baron (2003) suggest that the scale of a GIS in Russia could reach as much as € 1.25–3.5 billion per annum. This is a very approximate figure and depends on the balance of supply and demand and the prevailing allowance price. Fernandez and Michaelowa (2003) discuss the impact of defining the 'acquis communautaire' as the baseline for JI projects in the new EU Member States and stress the need to establish a predictable legal framework in the host countries, while Van der Gaast (2002) sees a reduced scope for JI in Eastern Europe due to the 'acquis' which could also be increased by using a GIS.

National institutions for project-based mechanisms have been slow to develop. The institutional problem is often exacerbated in countries with unstable economies and institutions and by project developers who often have very short time horizons, are unwilling to wait for the revenues and who cannot provide regular and ongoing monitoring and verification reports of emission reductions (see Michaelowa (2003a) for an overview

of such issues in CDM host countries, Korppoo (2005) for specific issues related to the Russian Federation and Figueres (2004) for issues specific to Latin America).

Sectoral approaches

A number of researchers have suggested that sectoral approaches may provide an appropriate framework for post-Kyoto agreements (see sectoral approaches in Table 13.2). Under such a system, specified targets could be set, starting with specific sectors or industries that are particularly important, politically easier to address, globally homogeneous and/or relatively insulated from competition with other sectors. Such an approach may be binding (e.g. such as an agreement in the International Civil Aviation Organization) or voluntary (such as an agreement through the International Standardization Organization). Targets may be fixed or dynamic, and 'no-lose', binding or non-binding (Philibert and Pershing, 2001; Samaniego and Figures, 2002; Bodansky, 2004). Bosi and Ellis (2005) and Baron and Ellis (2006) have explored different design options for sectoral crediting, including policy, rate-based and fixed limit approaches, and Ellis and Baron (2005) have assessed how these options could be applied to the aluminium and electricity sectors.

Sectoral commitments have the advantage of being able to be specified on a narrower basis than total national emissions. Baumert *et al.* (2005b) consider specific options in aluminium, cement, iron and steel, transportation and electricity generation and conclude that while not all sectors are amenable to such approaches, considerable precedent already exists for agreement both between companies and by governments. Sectoral approaches provide an additional degree of policy flexibility and make the comparison of efforts between countries within a sector a relatively easy process – although comparing efforts across sectors may be difficult (see Philibert, 2005a). An additional disadvantage to sectoral approaches is that they may create economic inefficiency. Trading across all sectors will inherently be at a lower cost than trading only within a single sector.

13.3.3.4.3 Coordination/harmonization of policies

As an alternative to or complementary to internationally agreed caps on emissions, it has been proposed that countries agree to coordinated policies and measures that reduce the emission of GHGs. A number of policies that would achieve this goal have been discussed in the literature, including taxes (such as carbon or energy taxes), trade coordination/liberalization, R&D, sectoral policies and policies that modify foreign direct investment (FDI). Sectoral policies have been discussed above, R&D is discussed in Section 13.2.1.6 and FDI is discussed below on financing. This discussion focuses on harmonized taxes as well on as trade and other policies.

---

36 This is already being applied for some projects on a voluntary basis. See: http://www.cdmgoldstandard.org.

Box 13.7 Climate change and the World Trade Organization (WTO)

There is a history of international cooperation between environmental agreements and the WTO (see, for example, Frankel and Rose, 2003). However, there is also literature pointing to potential conflicts. To date, disputes between climate and trade agreements have not been legally tested. Should a complaint arise, the attitude of a WTO panel may depend on whether the disputed trade measure stems from a treaty obligation or a national policy. Neither the UNFCCC nor the Kyoto Protocol has been formulated in language that can reasonably be interpreted to require or authorize a trade measure as a strategy to promote membership, make the climate regime more effective or enforce the treaty. Thus, any use of a climate trade measure would be considered to be a national-level action (see Fischer et al., 2002).

Two examples help demonstrate the range of possible pitfalls:
- In 1998, Japan introduced the 'top-runner' programme as part of its domestic efforts to implement the Kyoto Protocol. This legislation was intended to ensure that automobiles and other manufactured products would be more energy efficient; it required new appliance and manufactured goods be as efficient as the 'top-runner' in the same category. The legislation raised concern among other automobile-exporting countries, most notably the USA and the EU, which feared that the measures might have adverse effects on their exports; consequently, the latter suggested that the legislation was not compatible with WTO rules on free trade. Conversely, according to Yamaguchi (2004), the Japanese legislation provides for objective standards that would be applied equally to domestic and imported cars and, accordingly, there would be no discriminatory treatment as a matter in law. After discussions between all parties over several years, no formal appeal was ever submitted under the General Agreement on Tariffs and Trade (GATT) or the Technical Barriers to Trade (TBT) Agreement (see Murase, 2004).
- Murase (2002b) considers potential conflicts between the use of the Kyoto Protocol's project-based flexibility mechanisms (CDM and JI) and various trade agreements. Inasmuch as project-based offsets represent foreign direct investment (FDI), they may run counter to both the GATT and Subsidies and Countervailing Measures Agreement as well as the common practice application of the Trade Related Investment Measures (TRIMs) and Agriculture Agreements. Adding an additional point of complexity, Werksman et al. (2001) suggest that the effective functioning of the CDM may require investor discrimination in a manner prohibited by the Most Favored Nation (MFN) clause of international investment agreements.

Assunção and Zhang (2002) explore other areas of interaction between domestic climate policies and the WTO, such as the setting of energy efficiency standards, the requirement for eco-labels and the implementation of targeted government procurement programmes. They suggest that an early process of consultation between WTO members and the Parties to the UNFCCC may be necessary to enhance synergies between the trade and climate regimes. To this end, they recommend the establishment of a joint WTO/Framework Convention on Climate Change (FCCC) working group that would specifically focus on greater coherence between trade, climate change and development policy.

One of the leading proponents of a harmonized tax has been Cooper (1998, 2001). Under his proposals, all participating nations – industrialized and developing alike – would tax their domestic carbon usage at a common rate, thereby achieving cost-effectiveness. Aldy et al. (2003) have suggested a number of problems with Cooper's proposals, including issues of fairness (whether developed and developing countries should have identical tax rates given the relative welfare and relative responsibilities), whether any incentive exists for developed countries to adopt a tax and how to manage gaming behaviour (in which a government may change tax codes to neutralize its effects or to benefit certain economic sectors). Additional criticism of a common tax structure comes from the modelling community: Babiker et al. (2003) note that while an equal marginal abatement cost across countries is economically efficient, it may not be politically feasible in the context of existing tax distortions. They also note that many countries which currently apply such taxes have exempted certain industries, thereby significantly increasing the overall costs of the tax regime. In addition, competitive concerns can arise if one country adopts a tax and a trading partner does not. Several solutions have been proposed, including the use of trade bans or tariffs to induce action. Governments may also seek to use border tax adjustments under such circumstances (Charnovitz, 2003). However, it has been argued that such a measure could be as disadvantageous to a target foreign country as a trade measure. To date, World Trade Organization (WTO) case law has not provided specific rulings on climate-related taxes. Any proposed border adjustments would need careful design and also take WTO law into account (Biermann and Brohm, 2005) (see Box 13.7).

The importance of harmonizing environmental standards – including those related to climate change – has been evaluated by Esty and Ivanova (2002), who conclude that both economic and ecological interdependence demand coordinated national policies and international collective action. To this end, they propose the creation of a Global Environmental Mechanism to help manage the environmental components of a globalizing world, primarily through information and analysis and the creation of a policy space for environmental negotiation and bargaining.

Other fora, in addition to the WTO, also offer opportunities to exchange information and coordinate climate-related policies and activities. For example, the Convention on International Trade in Endangered Species of Wild Fauna and Flora (CITES) offers an opportunity to unite efforts in a common cause to both protect endangered species and the climate. Similarly, meetings of Asia Pacific Economic Cooperation (APEC) provide a platform for regional economies to take steps that meaningfully address the adverse impact of climate change (Ivanova and Angeles, 2005). The APEC Virtual Center (APEC-VC) for region-wide Environmental Technology Exchange launched by the Asia-Pacific economies provides information on environmental technology gathered by regional and local governmental authorities as well as by companies and environment-related organizations. The North American Commission on Environmental Cooperation (the NACEC or CEC), which was created within the North America Free Trade agreement (NAFTA), offers another model: Canada, Mexico and the USA signed an agreement to cooperate on reducing the threat of global change. The trilateral agreement is the basis for public-private partnerships to reduce GHG emissions in North America and to boost investment in green technology. It should be acknowledged that the NACEC could not prevent the detrimental decline in the Mexican environment during their participation in NAFTA (Gallagher, 2004); therefore, some caution must be exercised with regard to the environment when engaging in trade agreements.

13.3.3.4.4 Technology

A number of issues related to technology research, development and deployment (including transfers and investment) have been explored in the literature on climate change. Many authors have asserted that a key element of a successful climate change agreement will be its ability to stimulate the development and transfer of technology – without which it may be difficult or impossible to achieve emissions reductions at a significant scale (Edmonds and Wise, 1999; Barrett, 2003; Pacala and Socolow, 2004).

Technology agreements

The studies reported in the literature make it very clear that R&D support, price signals and other arrangements can all contribute to technology development and diffusion. Financial and human resources, often scarce in developing countries, will be needed to promote R&D, while monetary and political incentives as well as institutional arrangements will be required to promote diffusion (see IPCC (2000) which contains a comprehensive review of technology transfer issues, including proposals for improving international agreements.) Technology agreements may also seek to address barriers in technology

research, development and diffusion. (For additional details on specific sectors and technologies, see Chapters 4–10).

One variant of a technology agreement is formulated by Barrett (2001, 2003) in a proposal which emphasizes both common incentives for climate-friendly technology research and development (R&D) and technology protocols (common standards) rather than targets and timetables. While this proposal could potentially be environmentally effective, depending on the payoffs to the cooperative R&D efforts and the rate of technology deployment, Barrett notes that the system would neither be efficient nor cost-effective, not least because the technology standards would not apply to every sector of the global economy and may entail some technological lock-in. However, Barrett assumes that if standards are set in enough key countries, a 'tipping effect' is created which ultimately would lead to widespread global adoption. In reviewing Barrett's assessment, Philibert (2004) expresses doubts as to whether such a tipping effect would be applicable and suggests, alternatively, that for some technologies (e.g. CO_2 capture and storage), cost constraints may be more critical than acceptability in determining market penetration.

The concept of regional technology-specific agreements has also been explored by Sugiyama and Sinton (2005), who suggest that they may offer an interim path to promote cooperation and develop new, lower cost options to mitigation climate change – allowing any future negotiations on emission caps to proceed more smoothly. Box 13.8 lists some examples of existing international technology coordination programmes.

Technology transfer

One mechanism for technology transfer is through the establishment of – and subsequent contributions to – special funding agencies that disburse money to finance emissions reduction projects or adaptation activities. The UNFCCC and the Kyoto Protocol already include provisions for establishing and funding project activities, although contributions to and participation in these are mostly voluntary. UNFCCC also includes provisions for technology transfer under Article 4.5. The CDM could also be a vehicle for technology transfer, but the effects are unclear at this point.

As part of the Marrakesh Accords, at the seventh Conference of the Parties (COP 7), Parties were able to reach an agreement to work together on a set of technology transfer activities, which were grouped under a framework for meaningful and effective actions to enhance the implementation of Article 4.5 of the Convention. This framework[37] has five main themes:
1. Technology needs and needs assessments;
2. Technology information;
3. Enabling environments;
4. Capacity building;
5. Mechanisms for technology transfer.

---

37 See UNFCCC decision 4/COP 7 on the Development and Transfer of Technologies

Box 13.8 Examples of coordinated international R&D and technology promotion activities

- **International Partnership for a Hydrogen Economy:** Announced in April 2003, the partnership consists of 15 countries and the EU, working together to advance the global transition to the hydrogen economy, with the goal of making fuel cell vehicles commercially available by 2020. The Partnership will work to advance the research, development and deployment of hydrogen and fuel cell technologies and to develop common codes and standards for hydrogen use. See: www.iphe.net.

- **Carbon Sequestration Leadership Forum:** This international partnership was initiated in 2003 and has the aim of advancing technologies for pollution-free and GHG -free coal-fired power plants that can also produce hydrogen for transportation and electricity generation. See: www.cslforum.org.

- **Generation IV International Forum:** This is a multilateral partnership fostering international cooperation in research and development for the next generation of safer, more affordable and more proliferation-resistant nuclear energy systems. This new generation of nuclear power plants could produce electricity and hydrogen with substantially less waste and without emitting any air pollutants or GHG emissions. See: http://nuclear.energy.gov/genIV/neGenIV1.html.

- **Renewable Energy and Energy Efficiency Partnership:** Formed at the World Summit on Sustainable Development in Johannesburg, South Africa, in August 2002, the partnership seeks to accelerate and expand the global market for renewable energy and energy-efficiency technologies. See : http://www.reeep.org

- **Asia-Pacific Partnership on Clean Development and Climate:** Inaugurated in January 2006, the aim of this partnership between Australia, China, India, Japan, Republic of Korea and USA is to focus on technology development related to climate change, energy security and air pollution. Eight public/private task forces are to consider (1) fossil energy, (2) renewable energy and distributed generation, (3) power generation and transmission, (4) steel, (5) aluminium, (6) cement, (7) coal mining and (8) buildings and appliances. See: http://www.asiapacificpartnership.org.

Actions to implement the framework include the organization of meetings and workshops, the development of methodologies to undertake technology needs assessment plans, the development of a technology transfer information clearinghouse, including a network of technology information centres, actions by governments to create enabling environments that will improve the effectiveness of the transfer of environmentally sound technologies and capacity building activities for the enhancement of technology transfer under the Convention. Funding for technology needs assessments has been provided, and further funds for technology may become available from the UNFCCC's Special Climate Change Fund.

Other international efforts have also been undertaken to promote technology transfer in support of climate change mitigation efforts, including those by the UN Industrial Development Organization (UNIDO) and by the Climate Technology Initiative (CTI) of the IEA. As noted by the US National Research Council, additional work is particularly needed to assist poor countries as these lack scientific resources and economic infrastructure as well as the appropriate technologies to reduce their vulnerabilities to potential climate changes (NRC, 2003).

The distinction between public financing for climate change mitigation and private financing for technology investment is often blurred: Clean energy projects are frequently a blend of the two, with public financing used to leverage private investment. For example, the International Finance Corporation (IFC) clean energy financing projects in Eastern Europe, Russia, China and the Philippines use technical assistance funds to train commercial banks in energy efficiency while concurrently lending partial risk guarantees and offering credit lines to encourage banks to provide loans. In this manner public funds are heavily leveraged and provide a source financing for clean energy investments.[38]

Development oriented actions

A 'Sustainable Development Policies and Measures' (SDPAMS) approach proposed by Winkler *et al*. (2002b) and further elaborated by Bradley *et al*. (2005) focuses on linking climate mitigation and adaptation to priority development needs. In its standard form, such an approach would be domestic and unilateral and – with its focus on developmental needs – would also bring GHG benefits. However, the authors also suggest that simultaneous SDPAMS pledges (and possibly harmonized pledges) could be made by both developing and developed countries. However, Bradley *et al*. (2005) do note several limits to this approach and suggest that it may not be suitable for developed countries, nor for every technology or policy. Finally, they note that SDPAMS may not attract the necessary funding for it to be implemented on the scale required for global climate change mitigation.

13.3.3.5 *Financing*

Funding sources for GHG mitigation in developed and developing countries is a crucial issue in the international debate on tackling climate change. Financing is categorized in the literature in terms of public flows (including Development

---

38 See www.ifc.org/CEEF.

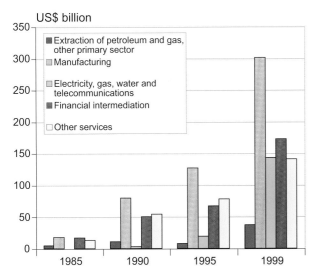

Figure 13.5: *Total OECD foreign direct investment (FDI) outflows to selected sectors.*
Source: OECD (1999)

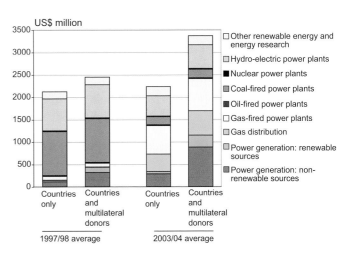

Figure 13.6: *Development assistance for energy*
Source: OECD.

Assistance and government loan guarantees through export credit agencies), private flows or foreign direct investment (FDI) and financing from multilateral institutions, including the Global Environment Facility (GEF) and international financial institutions. Public financing is the main form of assistance for developing country climate change mitigation, while the private sector provides the technology investments. CDM resources are significant when compared with GEF funding, but small in comparison to FDI resources (Ellis *et al.*, 2007). In addition to these instruments, a World Bank survey of contingent financing and risk mitigation instruments for clean infrastructure projects describes the characteristics and potential use of other instruments, such as insurance, reinsurance, loan guarantees, leases and credit derivatives[39] (IPCC, 2000; World Bank, 2003). A small percentage of public funds are used to leverage private investment in clean energy projects.

13.3.3.5.1 Foreign direct investments

OECD trade and FDI have grown strongly in relation to GDP during the past decade: cumulative net FDI outflows between 1995 and 2005 amounted to 1.02 trillion US$. As a share of GDP, outward FDI grew from 1.15% of the GDP in 1994 to 2.02% in 2004. However, while the total sums grew, only 35% went to non-Annex I countries – and of that, nearly 70% went to five countries, namely China (including Hong Kong), Brazil, Mexico, Singapore and South Korea.[40] See also OECD (2005 d) for trends in FDI relative to ODA.

One common assertion in international environmental negotiations is that FDI promotes sustainable development as multinational corporations (MNCs) transfer both cleaner

technology and better environmental management practices. However, empirical studies find little evidence that MNCs transfer either significant cleaner technology or better practices. In statistical studies of Mexico (manufacturing) and Asia (pulp and paper), foreign firms and plants performed no better than domestic companies (Zarsky and Gallagher, 2003). According to Jordaan (2004) the externalities from the presence of foreign-owned firms do not occur automatically, but are dependant on underlying characteristics of the industries and manufacturing firms.

Most FDI in developing countries is targeted to activities such as the extraction of oil and gas, manufacturing and electricity, gas and water, which have the aim to improve economic development but also to increase GHG emissions (Figure 13.5). Maurer and Bhandari (2000) report that during the mid- to late-1990s the major developed countries co-financed energy-intensive projects and exports valued at over 103 billion US$ through their export credit agencies (ECAs). These projects and exports included oil and gas development, fossil fuel power generation, energy-intensive manufacturing, transportation infrastructure and civilian aircraft sales. These countries accounted for 90% of the co-financing provided by ECAs to these energy-intensive exports and projects. By comparison, industrialized countries have directed just a fraction of their ECA financing to renewable energy projects. Between 1994 and 1999 ECAs supported a total of 2 billion US$ in renewable energy projects.

13.3.3.5.2 Direct international transfers

Official development assistance (ODA) remains an important source of financing for those parts of the world and sectors where private flows are comparatively low, although this is a modest financial resource relative to global private direct investment, which was 106 billion US$ in 2005. Data from the OECD suggest that development assistance for energy projects (approximately 3.2 billion US$ in 2004) from bilateral sources has remained relatively flat over the last 6 years.. There has been a shift in support away from coal technologies to those of gas and some extent renewables[41] (see Figure 13.6).

The effectiveness of ODA depends on various factors, the most important of which are good governance, policy and institutional frameworks that encourage private investment (macroeconomic and political stability, respect for human rights and the rule of law), minimum levels of investment in human capital (education, good health, nutrition, social safety nets) and policies and institutions for sound environmental management.

13.3.3.5.3 GEF and the multilateral development banks (MDBs)

The GEF, established in 1991, provides support to developing countries for projects and programmes that protect the global environment. Jointly implemented by the United Nations Development Programme (UNDP), the United Nations Environment Programme (UNEP) and the World Bank, GEF provides grants to fund projects related to biodiversity, climate change, international waters, land degradation, the ozone layer and persistent organic pollutants.[42]

Compared to the magnitude of the environmental challenges facing recipient countries, GEF efforts are relatively modest in scope. From 1991 to 2004, GEF allocated 1.74 billion US$ to climate change projects and activities; even when this amount is matched by the more than 9.29 billion US$ in co-financing, the overall scale of the GEF is small.[43] Funding is given to five project types, namely renewable energy, energy efficiency, sustainable transportation, adaptation, low GHG energy technologies and enabling activities. Hall (2002) analysed the GEF portfolio and noted the focus on incremental, one-time investments in mitigation projects that test and demonstrate a variety of financing and institutional models for promoting technology diffusion. He suggests that this approach should help contribute to a host country's ability to understand, absorb and diffuse technologies.

According to a review of the GEF by the World Bank (2006), 'the GEF's track record in reducing the long-term cost of new low GHG-emitting technologies has not been encouraging'. The continued effectiveness of GEF project funding for technology project types will depend on factors such as the duplication of successful technology transfer models, enhanced links with multilateral banks and co-ordination with other activities that support national systems of innovation and international technology partnerships. It has been suggested that GEF reform will be needed to enhance its effectiveness and transparency, particularly with respect to determining contributions and for evaluating priorities for disbursements (Grafton et al., 2004).

The World Bank (2004a) review of its investments in extractive industries determined that in the future it would be more selective, with a greater focus on the needs of poor people and a stronger emphasis on good governance and on the promotion of environmentally and socially sustainable development. The IFC has revised its performance standards in 2006 to require the reporting of GHG emissions for projects with both direct and indirect emissions of greater than 100,000 tonnes annually. The standards also require the consideration of alternatives or improvements to the energy efficiency of energy intensive projects (see http://www.ifc.org/ifcext/enviro.nsf/Content/ENvSocStandards). However, Sohn et al. (2005) note that the World Bank has continued to both support traditional CO_2-intensive fossil fuels projects and provide relatively limited resources to renewable and low CO_2-emitting energy alternatives. They suggest that Governments may use their leverage to direct the activities of multilateral development banks through their respective Boards and Councils in order to strengthen MDB programmes to account for the environmental consequences of their lending; develop programmatic approaches to lending that remove institutional barriers and create enabling environments for private technology transfers.

The higher perceived risk in developing countries, as reflected in sovereign credit ratings, can be compounded further by including new and emerging technologies. International or regional financing institutions can play a critical role in lowering the risk and leveraging private finance into the sector. MDBs have responded to this challenge by establishing several new initiatives. For example, the European Bank for Reconstruction and Development's (EBRD) new Sustainable Energy Initiative was launched in May 2006 to address the wasteful and polluting use of energy. The EBRD plans to invest up to € 1.5 billion in energy efficiency, renewables and clean energy projects over the next 3 years, which could lead to up to € 5 billion of total investment. The Asian Development Bank (ADB) launched the Energy Efficiency Initiative (EEI) in July 2005, the core objective of which is to expand ADB's investments in energy efficiency projects (including renewable energy), with an indicative annual lending target of 1 billion US$ between 2008 and 2010. The World Bank has announced the establishment of the Clean Energy Fund Vehicle with a capitalization of 10 billion US$ and an annual disbursement of 2 billion US$ to accelerate the transition to a low carbon economy.

---

41 See OECD website for information on development activities, including statistics, data, indicators and methods for accessing data: http://www.oecd.org/topicstatsportal/0,2647,en_2825_495602_1_1_1_1_1,00.html
42 See the website of the Global Environment Facility for additional information: http://www.gefweb.org/
43 http://www.gefweb.org/Projects/focal_areas/focal_areas.html#cc

13.3.3.6 *Capacity building*

The literature on climate change has not addressed capacity building to any extent, despite its critical relevance to the climate change issue. Part of the solution to the climate change problem has been cast in terms of helping developing countries with technology transfer and assistance. The importance of this is recognized in the text of the UNFCCC and Kyoto Protocol as well as in the more detailed implementing framework of the Marrakech Accords.

The capacity building framework within the climate change regime focuses on developing the capacity in developing countries to implement decisions. Capacity building has been defined historically as the formal training of employees, technological gate-keeping and learning-by-doing, with the recognition that this is a slow and complex process. According to Yamin and Depledge (2004), the Marrakesh Accords have been partially successful in bringing some additional coherence, coordination and prioritization into the process of capacity building. These authors argue that the effort to promote country-driven and contextually tailored efforts that are both iterative and involve learning-by-doing are appropriate.

Other ideas on capacity building also abound. Sagar (2000) argues that it may be more relevant to strengthen the domestic capacity for undertaking policy research and innovation as well as for managing technological and institutional change rather than merely creating the capacity for implementing policies developed elsewhere. This proposal is based on the idea that only context-relevant policy instruments are likely to work within the specific domestic circumstances of the relevant countries.

A number of recent analyses carried out on this subject have questioned whether capacity building can be initiated from outside a country. Since capacity issues are embedded in local contexts, the OECD has argued that it may be a mistake to assume that capacity building can be easily accomplished from outside this context.

Najam *et al.* (2003) note the importance of capacity building for developing countries and require that it be an integral part of any future agreement if it is to have wide support from this group. In particular, they argue that inasmuch as efforts to combat climate change and promote sustainable development are 'two sides of the same coin' enhancing the capacities of communities and countries to fight climate change will have multiple benefits. They also make the case that the most pressing need in this context is to strengthen the social, economic and technical resilience of the poorest and most vulnerable countries against extreme climatic events.

13.3.3.7 *Compliance*

Using game theory, Hovi and Areklett (2004) argue that a compliance system has to meet several criteria: (1) consequences of non-compliance have to be more than proportionate; (2) punishment needs to take place when behaviour is suboptimal; (3) an effective enforcement system must be able to curb collective as well as individual incentives to cheat. The compliance system agreed under Kyoto is viewed as only partially fulfilling these criteria. For example, Nentjes and Klaassen (2004) note that the obligation to fully restore any excess emissions in subsequent periods does not exclude the option of postponing restoration forever. If such an outcome occurs, the trading mechanisms under the Protocol may be substantially weakened. However, it is pointed out that introducing adversarial elements (such as sanctions) into the system are highly undesirable in view of the fact that the Kyoto Protocol currently covers only one third of the total GHG emissions of the world (Murase, 2005).

There are two schools of thought regarding the appropriate response to non-compliance contemplated under the Kyoto Protocol (see Murase, 2002b). One view advocates 'soft' compliance-management, which favours primarily facilitative and promotional approaches by rendering assistance to non-compliant States; those holding this view often refer to 'the non-compliance procedure' used under the Montreal Protocol. The other view takes a 'hard' enforcement approach in order to coerce compliance by imposing penalties or sanctions on non-complying parties. Financial penalties and economic or trade sanctions have been proposed along these lines. However, it has been suggested that such measures could be in conflict with WTO/GATT rules on trade liberalization (Mitchell, 2005).

A more nuanced view is provided by Wettestad (2005), who concludes that there are eight lessons to be learnt from other regimes. These include the need for an institutional warm-up period, wise institutional engineering, moderate expectations from the verification process, increased transparency, efforts to maintain close cooperation between the Facilitative and Enforcement Branch of the Compliance Committee, the search for opportunities to engage civil society in the process and a focus on assistance and compliance facilitation using the enforcement mechanism as an important but 'hidden' stick.

In his review of the Kyoto Protocol's compliance mechanism, Barrett (2003) argues that failure to comply over two compliance periods can essentially be equivalent to indefinitely postponing action: A country that is found in non-compliance in the first period has to make up the difference plus 30% in the next period. If it fails to achieve the latter target as well, it will have to make up the difference in the period thereafter – a process that can continue indefinitely. Perhaps the most important point in his proposal is that if countries feel that they cannot easily meet their commitments, they will negotiate for higher allowances in the period thereafter – or even withdraw from the agreement entirely. He also notes that the Protocol does not have any procedures to deal with countries that decide not to cooperate with the rules.

There is a significant body of research that compares various dispute settlement procedures. A number of these assessments examine environmental agreements (see, for example, Werksman, 2005), while others more specifically focus on possible conflicts between climate agreements and trade agreements (see, for example, Murase, 2002b). With respect to the latter, Murase notes the need for a coordinating authority to be established between a multilateral environmental agreement (MEA) and the WTO. Given that MEAs and the WTO are independent treaties on equal footing, neither can automatically be given the right to make a decision in the case of a conflict. As a result, a number of authors (e.g. Esty, 2001; Murase, 2002b) have called for the establishment of a new institution, such as a World Environment Organization (WEO), that would embody its own dispute settlement mechanism. This institution would function as a counterpart of WTO by attaining an equal footing between the two regimes.

13.3.3.8 Adaptation

The element of adaptation in international climate agreements has been far less explored to date than mitigation.[44] While most authors agree that adaptation is a vital part of a future agreement (although Schipper (2006) suggests that it was not a key focus of the initial UNFCCC negotiators), there is little mention in climate change literature of concrete proposals detailing the actions or obligations that should be undertaken by countries. Most proposals focus on leveraging funding for adaptation activities with an additional set of proposals addressing more specifically the links between adaptation, vulnerability and development agendas (see, for example, Najam et al., 2003).

Parry et al. (2005) develop an assessment of how adaptation may be incorporated into a future climate change architecture. They begin by noting that much of the adaptive response is likely to be local and, consequently, it is less conducive to a common international approach. Instead, they argue that a key need will be for efforts to incorporate adaptation into development policies and practices, including local, sectoral and national decision-making – a process they refer to as 'climate-proofing'. At the local level, this would incorporate strategies for municipal planning, including developing and maintaining seed banks, emergency preparedness services and community social services. At the sectoral level, it would include efforts to build climate into infrastructure design and maintenance codes and standards. At the national level, it would include integration into national planning and budget processes – for example, by examining whether planned expenditures will increase exposure to the impacts of climate change – and by doing so, minimize the financial risk, promote macro-economic stability and set aside sufficient funds to manage the consequences of climate shocks. Finally, at the international level, they suggest that key opportunities exist for integrating adaptation into the Millennium Development Goals and into lending practices of international institutions and bilateral aid agencies.

Three funds have been created under the UNFCCC and the Kyoto Protocol to manage adaptation issues: the Least Developed Countries Fund, the Special Climate Change Fund (both under the UNFCCC) and the Adaptation Fund (under the Protocol). In addition, the GEF has been requested to consider adopting more flexible approaches to funding adaptation (though this may not happen with core GEF funds, but with new money from these other funds that would be disbursed by the GEF).

Corfee-Morlot et al. (2002) suggest that it would be unrealistic to expect the GEF to cover the full cost of adaptation as such expenses would quickly exhaust their resources. Huq and Burton (2003) propose integrating adaptation into the mainstream work of development agencies, thereby allowing for more cost-effective and wider ranging support. However, as noted by Huq and Reid (2004), doing so runs the risk of diluting other existing aid efforts – which often have considerably higher priorities in-country than climate change adaptation.

The potential role for private (and public) insurance has also been suggested as a possible mechanism to pay for adaptation (e.g. Bals et al., 2005). Parry et al. (2005) list possible insurance schemes and risk transfer instruments, including:

- An international insurance pool (a collective loss-sharing fund to compensate victims of climate change damages);
- Public-private insurance partnerships (where the insurer is the government, but policies are developed and managed by the private sector);
- Regional catastrophic insurance schemes (regional cash reserves are pooled through mandatory contributions from member governments, and reserves are used for weather-related catastrophes);
- Micro-insurance (risk pooling for low-income individuals affected by specific risks);
- Catastrophe bonds (giving private insurers protection against extreme events; capital is provided by large institutional investors);
- Weather derivatives (financial mechanisms to hedge financial risk from catastrophic weather events)
- Weather hedges (providing protection for farmers; currently sold by banks, farm cooperatives and micro-finance institutions).

13.3.3.9 Negotiating process

It is important that several technical issues be taken into consideration when an agreement is negotiated and implemented. Since the international negotiation process under the UNFCCC is based on decisions by consensus, an approach

---

44 See IPCC(2007b), Chapter 17 and 18 for a broad review of adaptation issues.

Table 13.3: *Assessment of international agreements on climate change.*[45]

Approach	Environmental effectiveness	Cost-effectiveness	Meets distributional considerations	Institutional feasibility
National emission targets and international emission trading (including offsets)	Depends on participation and compliance.	Decreases with limited participation and reduced gas and sector coverage.	Depends on initial allocation.	Depends on capacity to prepare inventories and compliance. Defections weaken regime stability.
Sectoral agreements	Not all sectors amenable to such agreements, thereby limiting overall effectiveness. Effectivenss depends on whether agreement is binding or non-binding.	Lack of trading across sectors increases overall costs, although they may be cost-effective within individual sectors. Competitive concerns reduced within each sector.	Depends on participation. Within-sector competitiveness concerns are alleviated if treated equally at global level.	Requires many separate decisions and technical capacity. Each sector may require cross-country institutions to manage agreements.
Coordinated policies and measures	Individual measures can be effective; emission levels may be uncertain; success will be a function of compliance.	Depends on policy design.	Extent of coordination could limit national flexibility, but may increase equity.	Depends on the number of countries (easier among smaller groups of countries than at the global level).
Cooperation on Technology RD&D[a]	Depends on funding, when technologies are developed and policies for diffusion.	Varies with degree of R&D risk. Cooperation reduces individual national risk.	Intellectual property concerns may negate the benefits of cooperation.	Requires many separate decisions. Depends on research capacity and long-term funding.
Development-oriented actions	Depends on national policies and design to create synergies.	Depends on the extent of synergies with other development objectives.	Depends on distributional effects of development policies.	Depends on priority given to sustainable development in national policies and goals of national institutions.
Financial mechanisms	Depends on funding selection criteria.	Depends on country and project type.	Depends on project and country.	Depends on national institutions.
Capacity building	Varies over time and depends on critical mass.	Depends on programme design.	Depends on selection of recipient group.	Depends on country and institutional frameworks.

[a] Research, Development and Demonstration.

that is simple and requires a small number of separate decisions by international bodies most likely has a higher chance of being agreed upon. This may be true of any agreement that engages multiple countries.

It has been reported in the literature that ownership of an instrument – and hence its commitment and effectiveness – is linked to the manner in which the agreement was negotiated, and that the leadership (directional, instrumental and structural) demonstrated in a regime may stimulate its effectiveness. Kanie (2003) concludes that in the EU, the introduction of policies and measures and institution building changed the dynamics of the climate change negotiation process by enhancing leadership capacity.

The role and influence of non-State actors in the process of negotiation also increase the legitimacy and compliance-pull of a regime, both because such participation promotes the broader acceptability of the agreement and because it may increase knowledge about the regime. Agreements are also more likely

to be effective when they are negotiated in accordance with established rules of procedure, when the negotiators of key countries have been able to adequately prepare themselves for the negotiation and when the subject matter of the negotiations is designed to address the problem and has not been artificially limited to make the solutions more attractive to the more powerful countries (Andresen and Wettestad, 1992; Benedick, 1993; Sebenius, 1993; Greene, 1996; Gupta and Grubb, 2000; Gupta and Ringius, 2001). The attention of the regular media to climate negotiations can also mobilize awareness of the issue which then increases pressure on the negotiators to achieve a result (Newell, 2000).

13.3.4 Evaluating international climate change agreements

This section reviews the literature using the same criteria as in Section 13.2: environmental effectiveness, cost-effectiveness, distributional considerations and institutional feasibility. The discussion is summarized in Table 13.3, and then discussed in

45 The table examines each approach based on its capacity to meet its internal goals – not in relation to achieving a global environmental goal. If such targets are to be achieved, a combination of instruments needs to be adopted. Not all approaches have received an equivalent evaluation in the literature; evidence for individual elements of the matrix varies.

greater depth in the text. As is the case with national policies, international agreements are instruments that can be designed well or poorly and be stringent or lax, binding or non-binding, or politically attractive or unattractive.

13.3.4.1 Environmental effectiveness

Environmentally effective international agreements lead to reductions in global GHG emissions and/or concentrations or to decreased climate impacts. The literature suggests that to achieve such success, agreements must provide incentives or deterrents to both State and individual behaviour in order to achieve a specific outcome. However, at the international level, there is some dispute as to whether agreements change trends, or merely codify actions already underway.

An additional critical element in the effectiveness of an international agreement is that of the implementation context: The relevant literature shows that agreements tend to be more successful in countries with both a high level of domestic awareness and resources and a strong institutional and legal framework and where there is clear political will. Where global agreements are designed using only blue-print approaches to instruments, these instruments may ultimately ignore the specific cultural and institutional contexts within which they are designed to function and may actually not work as well (see conclusions of the Millennium Ecosystem Assessment, 2005). Agreements that promote ancillary objectives, such as reductions in ordinary air pollution levels, also have a higher chance of success.

An agreement that includes a limited group of countries (particularly if they are not major emitters) may be less effective – and this weakness may be exaggerated when emissions of non-participating countries increase by the migration of emission-intensive industries. Conversely, additional benefits may accrue due to technology spillover that may enhance environmental effectiveness (see Section 13.3.3.2).

The timing of an agreement's provisions may also affect its effectiveness: Focusing only on longer term emission reductions (as suggested under some forms of technology agreements) may preclude the possibility of reaching low climate stabilization levels, as many lower levels require immediate emission reductions.

13.3.4.2 Cost-effectiveness

A cost-effective international agreement would minimize global and national costs and provide participating sovereign nations with sufficient flexibility to reach their commitments in a fashion tailored to their national needs and priorities. To achieve this, agreements would need to avoid being prescriptive in its actions but, instead, leave room for the implementation of the target, (e.g. while reducing emissions in different sectors or reducing the emissions of different gases, they should not create significant distortions in competitiveness between countries).

Many analysts argue that the most cost-effect system would be one which enables emission trading with the broadest possible participation of countries. Such a system would allow the emission reductions to occur in those countries, sectors and gases where they can be achieved at the lowest cost. An approach based on specific policies and measures would have to be designed carefully to be as efficient as an emission trading system. The flexibility provided to private actors in a trading regime also increases the system's cost-effectiveness.

13.3.4.3 Distributional considerations, including equity

Perhaps the most politically charged issue in international negotiations is that of equity. Whether a system of national emission targets within an international agreement can be conducive to social development and equity depends on participation and the initial allocation of emission rights. For example, Pan (2005) suggests that all countries should participate – but that emissions associated with basic needs should be exempt from limits, while emissions associated with luxury activities should be constrained. Conversely, Gupta and Bhandari (2003) suggest that in the initial stages of an agreement, obligations should only be assigned to a limited set of (wealthier) parties. Exemptions to sectors or countries and modifications to the allocation of obligations can help address equity issues.

13.3.4.4 Institutional feasibility

Two aspects of institutional feasibility are critical in reaching successful international agreements: (1) negotiating and adopting an agreement and (2) the subsequent (usually national) implementation of that agreement.

Since international agreements are usually adopted by consensus, successful agreements are often relatively simple and require only a limited number of separate decisions by international bodies. In addition, global agreements usually require that all data and tools necessary for enforcement be widely available and verifiable (or if not, that they become available in the future). While there has been no comprehensive critique of the proposals in Table 13.3 in terms of their institutional feasibility, the latter clearly varies widely – for example, in terms of the extent to which they try to accommodate national circumstances and different levels of technical sophistication. Hence, the feasibility of reaching agreements will also vary accordingly.

A sectoral or technology approach would require multiple decisions: which sectors, which types of technologies, and how to regulate or support them. Choosing the sectors (and determining sectoral boundaries) or technologies for agreement may be difficult – unless participation were voluntary (e.g. the current suite of IEA implementing agreements, or the bilateral and multilateral efforts on specific technologies). This may require compromises on environmental effectiveness and equity. In addition, the assessment of whether a country had fulfilled its obligations would be complex. Philibert (2005a) notes that determining the effectiveness of technology or sectoral agreements could be difficult. In the case of a technology approach, definitive conclusions would likely be delayed until the technologies began to diffuse – and that could mean concomitant requirements for establishing long-lived institutions. The establishment of international institutions to manage coordinated policies and measures or development-oriented approaches may also be complex. While some private sector international institutions exist (e.g., the Aluminium Institute, which has set targets for GHG reductions in aluminium processing among its member companies), most sectors do not have such institutional arrangements. Similarly, while there are institutions designed to promote development (e.g., the Bretton Woods institutions), few have integrated climate change into their portfolios (see Maurer and Bhandari, 2000). Kanie (2006) argues that while the Kyoto Protocol will remain the core of the institutional system, a network will ultimately be both necessary – and increase effectiveness. The creation of a web of institutions tackling climate change and related issues not only ensures that any shortcoming in one institution does not lead to the collapse of the whole system, but it also enhances collective strength.

13.4 Insights from and interactions with private, local and non-governmental initiatives

This section addresses voluntary actions taken by sub-national governments, corporations, NGO's and others that are independent of national government programs or policies. See Box 13.9. Note that in contrast, section 13.2 addresses voluntary agreements between national governments and private parties.[46]

13.4.1 Sub-national initiatives

Local, state, provincial or regional governments have developed GHG policies and programmes that are either synergistic with national policies or are independent of these policies. Several reasons are given in the literature as to why sub-national entities undertake independent policies on GHGs or other environmental issues. Oates (2001) and Vogel *et al.* (2005) highlight the influence that State governments in the USA have had on national policy by experimenting with innovative initiatives. Rabe (2004) argues that some US states have enacted GHG policies to create incentives for new emission reduction technologies or to facilitate the recognition of emission reductions by companies in the event of future national regulations. Regional or local GHG reductions may also

be motivated by the desire to achieve additional environmental co-benefits, such as reductions in air pollution.

On the other hand, sub-national actions to address climate change may be viewed as a 'free rider' problem because non-participating regions may benefit from the actions of the participating areas without paying the costs (Kousky and Schneider, 2003). Regional or local initiatives may also cause 'leakage' if mandatory requirements in one jurisdiction cause a shift in economic activity and emissions to other jurisdictions without mandatory requirements (Kruger, 2006).

Sub-national governments in the USA and Australia, two countries that are not Parties to the Kyoto Protocol, have been among the most active on GHG policy, with a number of US states having adopted or proposed a variety of programmes to address GHGs, including renewable energy portfolio standards, energy efficiency programmes, automobile emissions standards and emissions registries. Perhaps the most notable examples of such an initiative are those of eight states in north-eastern and mid-Atlantic USA announcing their intent to adopt a regional cap-and-trade programme, known as the Regional Greenhouse Gas Initiative (RGGI); three western states – California, Washington and Oregon – may explore a similar initiative (McKinstry, 2004; Peterson, 2004; Pew Center, 2004; Rabe, 2004). Australian states have developed a broad array of programmes to reduce, sequester or measure GHG emissions (see http://www.epa.qld.gov.au/environmental_management/ sustainability/greenhouse/greenhouse_policy/other_states_ and_territories/). For example, the Australian state of Victoria has adopted a series of programmes to support renewable energy projects and the development of a 'green power' market (Northrop, 2004), while that of New South Wales has developed a credit-based emissions trading scheme for electricity retailers, generators and some electricity users. (Fowler, 2004; Baron and Philibert, 2005; MacGill, *et al.*, 2006). Finally, the Australian states have announced their intention to explore the development of a multi-jurisdictional emissions trading system (see http:// www.cabinet. nsw.gov. au/ greenhouse/report.pdf).

Northrop (2004) reports that more than 600 cities worldwide have participated in programmes to implement measures aimed at reducing local GHG emissions.[47] These include cities in developing countries. In total, 18 cities in South America,[48] 12 cities in South Africa[49] and 17 cities in India[50] are becoming more active in developing environmental measures at the local level. Kousky and Schneider (2003) find that cities have primarily adopted GHG policies with co-benefits, including more efficient energy use. Fleming and Webber (2004) describe a variety of GHG measurement and energy efficiency measures undertaken at the regional and local level in the UK, and Pizer and

---

46 See Higley *et al.* (2001), OECD (2003e) and Lyon and Maxwell (2004) for typologies of different types of approaches and initiatives.
47 These cities participate in the International Council for Local Environmental Initiatives (ICLEI), Cities for Climate Protection (CCP) programme. See http://www.iclei.org.
48 http://www.iclei.org/index.php?id=528.
49 http://www.iclei.org/index.php?id=700.
50 http://www.iclei.org/index.php?id=1089.

Tamura (2004) summarize measures undertaken by the Tokyo city government to reduce GHGs and control the 'heat island' effect. These types of initiatives may influence sub-national and national government policies and serve as incubators for new approaches to achieve GHG emission reductions.

13.4.2 Corporate and NGO actions

Corporations and NGOs, including industry associations and environmental advocacy groups, have started a variety of programmes and initiatives to address GHG emissions. The various factors leading corporations to adopt voluntary environmental action have been explored in the literature (Lyon and Maxwell, 2004; Thalmann and Baranzini, 2005). While some companies have attributed these actions to sustainable development goals or environmental stewardship policies (Margolick and Russell, 2001), it is often difficult to separate these goals from economic motives (Kolk and Pinske, 2004). Less controversial is the notion that companies adopt voluntary initiatives to create financial value in one form or another (Lyon and Maxwell, 2004).

There are both political and non-political drivers of corporate voluntary environmental action. Political drivers include a desire to pre-empt or influence future regulation. For example, trade associations in 30 countries have sponsored codes of management practices, the objectives of which are partly intended to forestall the imposition of government mandates (Nash and Ehrenfeld, 1996). Alternatively, corporations may adopt voluntary initiatives to influence future regulation in ways that improve their strategic positions. By adopting environmental technologies or other strategies ahead of regulatory mandates, corporations can signal to regulators that these alternatives are practical or relatively cost-effective (Reinhardt, 1999). Hoffman (2005) finds that some companies have adopted internal emissions trading schemes or GHG measurement programmes to gain expertise that will help them influence future national or international policies. A related motivation for voluntary action is the desire to manage the risks of future regulations by taking action that would increase profitability or protect a company's competitive position in the event of future regulatory mandates (Margolick and Russell, 2001).

Non-political drivers of voluntary corporate environ-mentalism include the desire to reduce costs through practices that also have environmental benefits (sometimes known as 'eco-efficiency'). Esty and Porter (1998) discuss how the desire to reduce energy or material costs drives corporate voluntary action, although this point of view is subject to some debate (Palmer et al., 1995; Porter and van der Linde, 1995). Hoffman (2005) and Margolick and Russell (2001) describe a variety of actions taken by US and Canadian companies to reduce GHG emissions while also reducing energy and operational costs.

Companies may also adopt environmental initiatives to appeal to green consumers, environmentally conscious stakeholders or even their own employees. Reinhardt (1998) discusses how this approach can take the form of companies differentiating their products by their environmental performance. Other companies have identified market opportunities for new products from potential GHG gas regimes (Reinhardt and Packard, 2001; Kolk and Pinske, 2005). In terms of the composition of the stakeholders, Maxwell et al. (2000) find that firms located in US states with a higher per capita membership in environmental organizations had more rapid reductions of toxic emissions. Margolick and Russell (2001) and Reinhardt (2000) report that corporate managers cited employee retention and recruitment as reasons for taking voluntary action.

Voluntary corporate-wide emissions targets for GHGs have become particularly popular. For example, Hoffman (2005) finds that as many as 60 US corporations have adopted corporate GHG emissions reduction targets and that some of these companies have participated in one of several partnership programmes run by NGOs (see Box 13.9). Under many of these programmes, companies develop a corporate GHG inventory and adopt an emission target. These targets take different forms, including absolute targets and intensity targets based on emissions or energy use per unit of production or sales (Margolick and Russell, 2001; King et al., 2004). Corporate targets have also been implemented with internal trading systems, such as those operated by British Petroleum (Margolick and Russell, 2001; Akhurst et al., 2003) and Petroleos Mexicana (PEMEX) (Bygrave, 2004).

Levy and Newell (2005) describe how the business sector, sometimes in partnership with NGOs, has initiated environmental certification or standardization regimes to fulfill a quasi-governmental role or to augment the role of governments. One of the most widely-used examples of this type of standard setting is the Greenhouse Gas Protocol, an initiative organized by the World Business Council for Sustainable Development (WBCSD) and the World Resources Institute (WRI) to develop an internationally accepted accounting and reporting standard for GHGs (WRI/WBCSD, 2004). The WRI/WBCSD reporting standard has been used by corporations, NGOs and government voluntary programmes. The International Standards Organization (ISO), based on the WRI/WBCSD, has adopted standards for the reporting of GHGs at the company and project level.[51]

Other standardization or certification efforts have been formed to support markets for project-based mechanisms or emissions trading. For example, the International Financial Reporting Interpretations Committee (IFRIC), which is the interpretive arm of the International Accounting Standards Board (IASB), has issued guidelines on financial accounting for emission allowances.[52] The International Emissions Trading

---

51 The relevant ISO standards are ISO 14064 Part 1. This may be found at: http://www.iso.org/iso/en/CatalogueDetailPage.CatalogueDetail?CSNUMBER=38381&scopelist=PROGRAMME
52 See http://www.iasb.org/news/index.asp?showPageContent=no&xml=10_262_25_02122004_31122009.htm

Box 13.9 Examples of private partnerships and programmes

Business Leader Initiative on Climate Change (BLICC): Under this initiative, five European companies monitor and report their GHG emissions and set a reduction target. See http://www.respecteurope.com/rt2/BLICC/

Carbon Disclosure Project: Under this project, 940 companies report their GHG emissions. The project is supported by institutional investors controlling about 25% of the global stock markets. See http://www.cdproject.net

Carbon Trust: The Carbon Trust is a not-for-profit company set up by the UK government to reduce carbon emissions. The Trust provides technical assistance, investment funds and other services to companies on emission reduction strategies and for the development of new technologies. See http://www.thecarbontrust.co.uk/default.ct

Cement Sustainability Initiative: Ten companies have developed 'The Cement Sustainability Initiative' for 2002–2007 under the umbrella of the World Business Council for Sustainable Development. This initiative out-lines individual or joint actions to set emissions targets and monitor and report emissions.

Chicago Climate Exchange: The Chicago Climate Exchange is a GHG emission reduction and trading pilot programme for emission sources and offset projects in the USA, Canada and Mexico. It is a self-regulatory, rules-based exchange designed and governed by the members who have made a voluntary commitment to reduce their GHG emissions by 4% below the average of their 1998–2001 baseline by 2006. See http://www.chicagoclimatex.com

Offset Programmes: Braun and Stute (2004) identified 35 organizations that offer services to offset the emissions of companies, communities and private individuals. These organizations first calculate the emissions of their participants and then undertake emission reduction or carbon sequestration projects or acquire and retire emission reduction units or emission allowances.

Pew Center on Climate Change Business Environmental Leadership Council: Under this initiative, 41 companies establish emissions reduction objectives, invest in new, more efficient products, practices, and technologies and support actions to achieve cost-effective emissions reductions. See: http://www.pewclimate.org/companies_leading_the_way_belc/

Top ten consumer information system: This NGO-sponsored programme provides consumers with information on the most efficient consumer products and services available in local markets. The service is available in ten EU countries, with plans to expand to China and Latin America. See http://www.topten.info

WWF Climate Savers: The NGO World Wide Fund of Nature (WWF) has build partnerships with individual leading corporations that pledge to reduce their global warming emissions worldwide by 7% below 1990 levels by the year 2010. Six companies have entered this programme. See http://www.panda.org/about_wwf/what_we_do/climate_change/our_solutions/business_industry/climate_savers/ index.cfm

Association, together with the World Bank Carbon Finance Group/Prototype Carbon Fund have developed a validation and verification manual to be used by stakeholders involved in developing, financing, validating and verifying CDM and JI projects.

13.4.3 Litigation related to climate change

The authors of many technical articles point out that litigation is likely to be used increasingly as countries and citizens become dissatisfied with the pace of international and national decision-making on climate change (Penalver, 1998; Marburg, 2001; Weisslitz, 2002; Allen, 2003; Grossman, 2003; Verheyen, 2003; Gillespie, 2004; Thackeray, 2004; Dlugolecki, 2005; Hancock, 2005; Jacobs, 2005; Lipanovich, 2005; Mank, 2005). These authors argue that the possible causes of action in litigation include (1) customary law principle of state responsibility,

(2) nuisance and the no-harm principle, (3) violation of international agreements including the WTO and the United National Convention on the Law of the Sea (UNCLOS) and the violation of human rights and (4) the abdication of authority by states to legislate on environmental issues based on the existing environmental legislation in the country concerned. However, they also emphasize that although there are often strong legal grounds for taking action, there may also be reasons for a strong defence.

Gillespie (2004) argues that if the international process is arguably not taking place in good faith, there is sound reason for requesting the International Court of Justice for an Advisory Opinion in this area, especially when the significant (potential) harm faced by small island states are taken into account. Jacobs (2005) and Verheyen (2003) analysed the potential case for a small island state actually suing the USA before the International

Court of Justice. Burns (2004) and Doelle (2004) point out that non-ratification of the Kyoto Protocol could imply illegal subsidies to national industries under the WTO and pollution of the seas under UNCLOS. Hancock (2005) sees the potential for liability suits increasing and advises companies to disclose their emissions to the Securities and Exchange Commission as a step to limit liability. Issues of causality are being dealt with in the literature (Allen, 2003) and through precedent (Lipanovich, 2005).

There are currently a number of court cases in Kyoto Party countries, both developed (Germany) and developing (Nigeria), and non-Parties (Australia and the USA). For example, in Germany, NGOs have sued the export credit support agencies for not disclosing information on the GHG emissions of the projects they support in developing countries. (See www.climatelaw.org/media/german.suit). A similar case was filed in the US District Court for the Northern District of California, on August 26, 2002 by Friends of the Earth, Greenpeace and the city of Boulder, Colorado, which have sued the Export-Import Bank and the Overseas Private Investment Corporation under the National Environmental Policy Act, alleging that these two US government agencies had provided 32 billion US$ for supporting the finance and insurance of oil fields, pipelines and coal-fired plants in developing countries over the previous 10 years without assessing the impacts on the environment including global warming. A Federal Judge in California has ruled in favour of the plaintiffs.[53]

In a case filed in Argentina, the plaintiffs allege a violation of Article 6 of the Climate Convention. In Nigeria, NGOs have sued the major oil companies and the state for continuing gas flaring, an industrial process which contributes about 70 million tonnes of CO_2 annually to global GHG emissions (Climate Justice Programme, 2005) and which is viewed as a violation of the Convention and the human rights of the local people.[54] In Australia, NGOs have filed a suit against a minister for permitting a mine expansion project without examining the GHG emissions. See www.austlii.edu.au/au/cases/vic/VCAT/2004/2029.html.

There are two law cases in the USA where a coalition of states[55] and environmental NGOs argue that the US EPA has

the authority to regulate CO_2 and other GHGs as air pollutants under the Clean Air Act.[56] In addition, eight US States, New York City and two land conservation trusts initiated a lawsuit in July 2004 against the five US power companies with the largest CO_2 emissions, on the grounds that these companies contribute to a public nuisance (global warming). That case, though dismissed by the trial court, is on appeal.[57] Non-government organizations in Australia have also given notice to the major GHG emitters in the USA about their obligations under national and international law to reduce their emissions (http://www.cana.net.au/documents/legal/aus_fin_rev.doc). In July 2005, a wildlife organization sued the Australian Government for failing to protect the Great Barrier Reef (http://www.climatelaw.org/media/Australia.emissions.suit). A court case was filed in December 2005 by the Inuit people before the Inter-American Commission of Human Rights against the US government for human rights violations of the Inuit people's way of life.[58] There have also been cases that have challenged the allocation of emission allowances. With the entry into force of the EU Emissions Trading Directive,[59] there has been some litigation in Germany that has challenged the manner in which the German Government has interpreted and transposed the directive into its National Allocation Plan in 2004.[60] The courts have thus far decided that the Emission Allocation Law is in conformity with German law and with European rules on property rights.[61]

While many of the these legal cases have not yet led to interim judgments in favour of the plaintiff, they do reveal there is a decided interest in pursuing the legal route as the means to pushing for action on climate change. These cases are based on a number of different legal grounds for doing so, but it may take some years before courts decide which, if any, of these grounds are valid.

13.4.4 Interactions between private, local and non-governmental initiatives and national/international efforts

The preceding sections have touched on a number of the interactions that take place between private, sub-national and non-governmental initiatives and national and international climate change efforts. As discussed, some of these efforts have been designed, at least in part, to influence the development of

---

53 Order Denying Defendants' Motion for Summary Judgment, in the case of Friends of the earth, Greenpeace, Inc. and City of Boulder Colorado versus Peter Watson (Overseas Private Investment Corporation) and Phillip Lerrill (Export-Import Bank of the United States), No. C 02-4106 JSW.

54 Suit No. FHC/CS/B/126/2005; filed in the Federal High Court of Nigeria, in the Benin Judicial Division, Holden at Benin City.

55 California, Connecticut, Illinois, Maine, Massachusetts, New Jersey, New Mexico, New York, Oregon, Rhode Island, Vermont and Washington together with New York City, Baltimore, and Washington, DC.

56 Massachusetts vs. Environmental Protection Agency, 415 F.3d 50 (D.C. Cir. 2005). A petition for Supreme Court review is pending. This case concerns motor vehicle emissions. Another case has been filed in the US Court of Appeals for the District of Columbia Circuit by a coalition of states and NGOs led by New York over an EPA decision not to regulate CO_2 from power plants.

57 Connecticut, *et al.* vs. American Electric Power Company Inc., *et al.*; 406 F.Supp.2d 265 (S.D.N.Y. 2005), appeal pending in the Court of Appeals for the Second Circuit.

58 Petition to the Inter-American Commission on Human Rights Seeking Relief From Violations Resulting From Global Warming Caused by Acts and Omissions of the United States, December 7, 2005.

59 Directive 2003/87/EC of the European Parliament and the Council of 13 October 2003 (OJ L 275, 25-10-2003), establishing a scheme for GHG allowance trading within the community and amending Council Directive 96/61/EC (OJ L257, 10-10-1996); available at < http://europa.eu.int/eur-lex/pri/en/oj/dat/2003/l_275/l_27520031025en00320046.pdf>

60 Gesetz über den nationalen Zuteilungsplan für Treibhausgasemissionsberechtigungen in der Zuteilungsperiode 2005-2007 (Zuteilungsgesetz 2007 - ZuG2007), Bundesgesetzblatt Jahrgang 2004, Teil I, Nr. 45, 30. August 2004.

61 Beschluss vom 1.9.2004, NVwZ2004, S.1389 ff; Beschluss vom 18.10.2004, NVwZ2005, S.112 ff; BverwG, Urteil vom 309.6.2005, NVwZ2005, S. 1178ff.

national programmes or the international climate regime. Other programmes have been designed to fill roles in these regimes that may be appropriate for private or non-governmental entities. Finally, other legal or programmatic initiatives have been launched because of the perceived inadequacy of national or international efforts.

One of the most important drivers of these interactions is the development of a global GHG emission trading market. Many of the standardization and certification efforts described above have been designed to build institutions for the emerging GHG market which in turn may also facilitate interactions between sub-national initiatives and national or international climate regimes. For example, the eight north-eastern and mid-Atlantic states in the US Regional Greenhouse Gas Initiative (RGGI) cap and trade programme will allow the use of CDM credits and EU ETS allowances under certain circumstances (RGGI, 2005). Similarly, there has been an exploration of a possible linkage between the NSW Greenhouse Gas abatement scheme and the EU ETS and Kyoto mechanisms (Fowler, 2004; Betz and MacGill, 2005).

In addition to international carbon markets, there are other frameworks that facilitate interactions between private, sub-national, and non-governmental initiatives and national and international climate change efforts. For example, NGOs, private companies and governments have formed partnerships to help implement the World Summit on Sustainable Development (WSSD). These partnerships, known as 'type II agreements' are self-organized and are formed as voluntary cooperative initiatives and have the common goal of integrating the economic, social and environmental dimensions of sustainable development. To date, more than 300 partnerships are registered. A significant number of these partnerships are climate change-related (see http://www.un.org/esa/sustdev/partnerships/partnerships.htm).

13.5 Implications for global climate change policy

This chapter has provided information on the national and international policy options available to governments and the global community to address global climate change. We note that there are many tools available and that each has its own unique advantages and disadvantages. While further studies are likely to yield additional insights, particularly with respect to the implementation of policy choices, it is unlikely that the suite of policies available to governments will grow substantially in the future.

With this in mind, it is useful to consider several questions in the light of the following background information. Since the IPCC was formed nearly 20 years ago atmospheric GHG concentrations have gone up from 354 to 385 ppm (or approximately 25% of the total increase since the pre-industrial

level of 270 ppm) as the emissions of GHG have risen (see http://cdiac.ornl.gov/ftp/trends/CO$_2$/maunaloa. CO$_2$). We have measurement data that indicates that the world is warming, and we can calculate, given the data on past and current emissions, that there is at the present time approximately 0.6 degrees of additional warming 'in the bank' (See IPCC, 2007a). Therefore:

- Why has the application of policies been so modest?
- Why is the global community not on a faster implementation track?
- Why have – at the very least – hedging strategies not emerged in many more countries?
- Is the scale of the problem too large for current institutions?
- Is there a lack of information on potential impacts or on low-cost options?
- Has policy-making been influenced by the special interests of a few?

Assuming that policies have been carefully designed, there appears to be no need to delay their implementation – indeed, there is an abundance of information in climate change literature that continues to suggest the non-climate benefits of many of these policies and the potential climate benefits of many non-climate policies. Moreover, as outlined in other chapters of this report, with a few exceptions, these policies would have only a very small impact on national economic growth – albeit the impact would be large in absolute terms.

One answer to these questions may lie in the complex nature of the policy-making processes – both for climate change policy and, even more importantly, in other areas at the national and sub-national level. For example, some of the most significant emissions reductions in both developed and developing countries have occurred at this intersection of policies (e.g. the switch to gas in the UK, the Chinese energy efficiency programmes for energy security, the Brazilian development of a bio-fuel-driven transport fleet, or the trend in the 1970s and 1980s toward nuclear power). Conversely, some of the most significant increases in emissions have been the result of non-climate policy priorities which have overwhelmed climate mitigation efforts (e.g. decisions in Canada to exploit the tar sands reserves, those in Brazil to clear forests for agriculture and those in the USA to promote coal-powered electricity generation to enhance energy security). Assessing how these mega-decisions are made and how they can be linked with climate change policies is the topic of chapter 12 and may be crucial to the future.

A second answer may be linked to the over-riding drive by all governments (reflecting both corporate and individual desires) for cheap and secure energy and for economic growth, to the competitive nature of the global economy and to the perception that any step, however modest, will disadvantage some special interest. Finding a way to mitigate the impacts on the losers – as well as create new winners – may be a key to accelerating the pace of policy implementation. Most importantly perhaps, finding ways to eliminate the climate of 'fear' that prevents actions (or more aggressive actions) and to promote a climate of 'opportunity' may be crucial to moving beyond modest steps. As outlined in other chapters of this report, the impact of mitigation efforts on national economic growth is relatively small, although the economic impacts differ among countries and may be larger than the impacts of other environmental problems. Mitigation is also more complicated as it involves more political actors and greater levels of cooperation and/or coordination. In this respect, better estimates of the risks, costs and benefits of climate policies in terms of market and no-market terms as well as ethical terms may enable governments to make informed decisions.

From the literature reviewed in this chapter, it is clear that governments, companies and civil society have been actively grappling with these questions. The very diversity of the policy mix, the activism of NGOs and the wealth of modelling, research and analysis (even if, to date, these have yielded only modest changes in emissions) collectively provide a framework for taking additional steps.

New research might provide further insight into why some policies have succeeded – and why others have not. In particular, additional work is needed to bolster the currently sparse body of research addressing the concerns of developing countries. Understanding how to accelerate policy adoption may be the most important research topic for the immediate future. As this chapter and others have noted, technology and policy tools do exist for taking that significant first step in addressing climate change. Potential future agreements can take advantage of this learning to encourage economically prudent and politically feasible actions.

REFERENCES

Ackermann, T., G. Andersson, and L. Söder, 2001: Overview of government and market driven programs for the promotion of renewable power generation. *Renewable Energy*, **22**(1-3), January-March 2001, pp. 197-204.

Ahman, M., D. Burtraw, J. Kruger, and L. Zetterberg, 2006: A ten year rule to guide the allocation of EU emission allowances. *Energy Policy*.

Akhurst, M., J. Morgheim, and R. Lewis, 2003: Greenhouse gas emissions trading in BP. *Energy Policy*, **31**(7), June 2003, pp. 657-663.

Aldy, J.E., S. Barrett, and R.N. Stavins, 2003: Thirteen plus one: a comparison of global climate policy architectures. *Climate Policy*, **3**(4), December 2003, pp. 373-397.

Alic, J.A., D.C. Mowery, E.S. Rubin, 2003: U.S. Technology and Innovation Policies: Lessons for Climate Change. Pew Center on Global Climate Change, Washington, D.C.

Allen, M., 2003: Liability for Climate Change: Will it ever be possible to sue anyone for damaging the climate? Commentary in *Nature*, **421**, 27 February 2003, pp. 891-89.

Anderson, D., C.D. Bird, 1992: Carbon accumulations and technical progress - A simulation study of costs. *Bulletin of Economics and Statistics*, **54**(1), pp. 1-27, Oxford.

Andresen, S. and J. Wettestad, 1992: International resource cooperation and the greenhouse problem. *Global Environmental Change, Human and Policy Dimensions*, **2**(4), December 1992, Butterworth-Heinemann, Oxford, pp. 277-291.

Andronova, N.G. and M.E. Schlesinger, 2004: Importance of sulfate aerosol in evaluating the relative contributions of regional emissions to the historical global temperature change. *Adaptation and Mitigation Strategies for Global Change*, **9**, pp. 383-390.

Arora, S. and T. Cason, 1996: Why do firms volunteer to exceed environmental regulations? Understanding Participation in EPA's 33/50 Program. *Land Economics*, **72**, pp. 413-32.

Asheim, G.B., W. Buchholz, and B. Tungodden, 2001: Justifying sustainability. In *Journal of Environmental Economics and Management*, **41**, pp. 252-268.

Aslam, M.A., 2002: Equal per capita entitlements: a key to global participation on climate change? K.A. Baumert, O. Blanchard, S. Llosa, and J.F. Perkaus (eds), Options for protecting the climate (pp. 175-201). Washington, WRI.

Assunção, L., and Z.X. Zhang, 2002: Domestic climate policies and the WTO. United Nations Conference on Trade and Development, Discussion Paper No. 164.

Azar, C. and H. Dowlatabadi, 1999: A review of technical change in assessment of climate policy. *Annual Review Energy Economics*, **24**, pp.513-544.

Babiker, M.H., R.S. Eckhaus 2002: Rethinking the Kyoto targets. *Climatic Change*, **54**, pp. 99-114.

Babiker, M., L., P. Criqui, A.D. Ellerman, J. Reilly and L. Viguier, 2003: Assessing the impact of carbon tax differentiation in the European Union. *Environmental Modeling and Assessment*, **8**(3), pp. 187-197.

Baer, P., J. Harte, B. Haya, A.V. Herzog, J. Holdren, N.E. Hultman, D.M. Kammen, R.B. Norgaard, and L. Raymond, 2000: Equity and greenhouse gas responsibility. *Science*, **289** (2287.12 Discussion paper 2003-2).

Bailey, P., T. Jackson, S. Parkinson, and K. Begg, 2001: Searching for baselines constructing joint implementation project emission reductions. *Global Environmental Change*, **11**, pp.185-192.

Bakker, S.J.A., H.C. de Coninck, and J.C. Jansen, 2004: Air pollution, health and climate change. M.T.J. Kok and H.C. de Coninck (ed.), *Beyond Climate: Options for Broadening Climate Policy*, RIVM Report 500036 001, Bilthoven, pp. 139-171.

Bals, C., I. Burton, S. Butzengeiger, A. Dlugolecki, E.Gurenko, E. Hoekstra, P. Höppe, R. Kumar, J. Linnerooth-Bayer, R. Mechler, and K. Warner, 2005: Insurance-related options for adaptation to climate change, the Munich Climate Insurance Initiative (MCII). <http://www.germanwatch.org/rio/c11insur.pdf>, accessed 03/07/07.

Baron, R. and C. Philibert, 2005: Act Locally, Trade Globally: Emissions Trading for Climate. International Energy Agency, 2005, Paris.

Baron, R. and J. Ellis, 2006: Sectoral crediting mechanisms for greenhouse gas mitigation: institutional and operational issues. OECD/IEA Annex I Group, Paris.

Baron, R. and S. Bygrave (2002). "Towards International Emissions Trading: Design Implication for Linkages". Paper presented at the 3rd CATEP Workshop on Global Trading, Kiel Institute for World Economics, 30 September - 1 October 2002.

Barrett, S., 2001: Towards a better climate treaty. Policy matters. *World Economics*, **3**(2), pp. 35-45.

Barrett, S., 2003: Environment and Statecraft. Oxford University Press, Oxford.

Baumert, K., A.R. Bhandari, N. Kete, 1999: What might a developing country climate commitment look like? World Resources Institute, Washington D.C.

Baumert, K., T. Herzog, and J. Pershing, 2005a: Navigating the numbers: Greenhouse gases and international climate change agreements. World Resources Institute, Washington D.C., USA, ISBN: 1-56973-599-9.

Baumert, K., R. Bradley, N.K. Dubach, J.R. Moreira, S. Mwakasonda, W.-S. Ng, L.A. Horta Nogueira, V. Parente, J. Pershing, L. Schipper, and H. Winkler, 2005b: Growing in the greenhouse: Policies and measures for sustainable development while protecting the climate. Washington, USA, World Resources Institute.

Begg, K. and D. van der Horst, 2004: Preserving environmental integrity in standardised baselines: The role of additionality and uncertainty. *Mitigation and Adaptation Strategies for Global Change*, **9**, pp. 180-200.

Beierle, T.C., 2004: The benefits and costs of environmental information disclosure: What do we know about right-to-know? RFF Discussion Paper 03-05, March.

Bell, R. and C. Russell, 2002: Environmental policy for developing countries. *Issues in Science and Technology*. Spring, pp. 63-70.

Bell, R.G., 2003: Choosing environmental policy instruments in the real world. OECD: Paris.

Benedick, R.E., 1993: Perspectives of a negotiation practioner. Sjostedt, G. (ed.) *International Environment Negotiation*, IIASA, pp. 219-243.

Benedick, R., 2001: Striking a new deal on climate change. *Science and Technology Online*, Fall 2001.

Berk, M.M. and M.G.J. den Elzen, 2001: Options for differentiation of future commitments in climate policy: how to realise timely participation to meet stringent climate goals. *Climate Policy*, **1**, pp. 465-480.

Berkhout, F. and A. Smith, 2003: Carbon flows between the EU and Eastern Europe: Baselines, scenarios and policy options. *International Environmental Agreements: Politics, Law and Economics*, **3**, pp. 199-219.

Betz, R., W. Eichhammer, and J. Schleich, 2004: Designing national allocation plans for EU emissions trading - A first analysis of the outcomes. *Energy & Environment*, **15**, pp. 375-425.

Betz, R. and I. MacGill, 2005: Emissions trading for Australia: Design, transition and linking options. CEEM Discussion Paper, DP_050815.

Biermann, F. and R. Brohm, 2005: Border adjustments on energy taxes: A possible tool for european policymakers in implementing the Kyoto Protocol? In *Vierteljahrshefte zur Wirtschaftsforschung*. **74**, pp. 249-258.

Blackman, A. and W. Harrington, 2000: The use of economic incentives in developing countries: Lessons from international experience with industrial air pollution. *Journal of Environment and Development*, **9**(1), pp. 5-44.

Blanchard, O., 2002: Scenarios for differentiating commitments. K.A. Baumert, O. Blanchard, S. Llosa and J.F. Perkaus (eds), *Options for protecting the climate*. WRI, Washington D.C.

Blanchard, O, C. Criqui, A. Kitous, and L. Vinguier, 2003: Efficiency with equity: A pragmatic Approach. I. Kaul, P. Conceição, K. Le Goulven, R.U. Mendoza (eds), *Providing public goods: managing globalization*. Oxford, Oxford University Press, Office of Development Studies, United Nations Development Program, Grenoble, France.

Blok, K., G.J.M. Phylipsen, and J.W. Bode, 1997: The Triptych Approach, burden sharing differentiation of CO_2 emissions reduction among EU Member States. Discussion paper for the informal workshop for the European Union Ad Hoc Group on Climate, Zeist, the Netherlands, January 16-17, 1997, Dept. of Science, Technology and Society, Utrecht University, Utrecht, 1997 (9740).

Bluffstone, R. and B.A. Larson (eds) 1997: Controlling Pollution in Transition economies: Theories and Methods. London, UK: Edward Elgar.

Blyth, W. and R. Baron, 2003: Green investment schemes: Options and issues. COM/ENV/EPOC/IEA/SLT(2003)9, OECD/IEA, Paris.

Blyth, W. and M. Yang, 2006: Impact of climate change policy uncertainty on power investment. Document no. IEA/SLT(2006)11. IEA, Paris.

Bodansky, D., 2004: International climate efforts beyond 2012: A survey of approaches. Pew Climate Center, Washington D.C., <www.pewclimate.org>, accessed 03/07/07.

Bode, S., 2004: Equal emissions per capita over time - a proposal to combine responsibility and equity of rights for post-2012 GHG emission entitlement allocation. *European Environment*, **14**, pp. 300-316.

Böhringer, C. and H. Welsch, 2004: Contraction and convergence of carbon emissions: an intertemporal multi-region CGE analysis. *Journal of Policy Modeling*, **26**, pp. 21-39.

Böhringer, C. and A. Löschel, 2005: Climate policy beyond Kyoto: Quo vadis? A computable general equilibrium analysis based on expert judgements. KYKLOS 58(4), pp. 467-493.

Böhringer, C. and H. Welsch, 2006: Burden Sharing in a greenhouse: egalitarianism and sovereignty reconciled. *Applied Economics*, **38**, pp. 981-996.

Bollen, J.C., A.J.G. Manders, and P.J.J. Veenendaal, 2004: How much does a 30% emission reduction cost? Macroeconomic effects of post-Kyoto climate policy in 2020. CPB Document no 64, Netherlands Bureau for Economic Policy Analysis, The Hague.

Bosi, M. and J. Ellis, 2005: Exploring options for sectoral crediting mechanisms. OECD/IEA Annex 1 Expert Group, Paris.

Bouille, D., O. Girardin, 2002: Learning from the Argentine Voluntary Commitment. K.A. Baumert (ed), *Options for Protecting the Climate*. World Resource Institute, Washington D.C.

Bovenburg, A. and L. Goulder, 2001: Neutralizing the adverse industry impacts of CO_2 abatement policies: What does it cost? Behavioral and distributional effects of environmental policy (eds, C. Carraro and G. Metcalf), University of Chicago Press.

Braathen, N.A., 2005: Environmental agreements in combination with other policy instruments. E. Croci, *the Handbook of Environmental Voluntary Agreements*, Springer, 2005.

Bradford, D.F., 2004: Improving on Kyoto: Greenhouse gas control as the purchase of a global public good. Bradnee Chambers *et al.* (2005). Typology of Response Options, Millennium Ecosystem Assessment.

Bradley, R., K. Baumert, N.K. Dubash, J.R. Moreira, S. Mwakasonda, W.-S. Ng, L.A. Horta Nogueira, V. Parente, J. Pershing, L. Schipper, and H. Winkler, 2005: Growing in the greenhouse: Policies and measures for sustainable development while protecting the climate. Washington, USA, World Resources Institute.

Braun, M. and Stute, E. 2004: Anbieter von Dienstleistungen fur den Ausgleich von Treibhaus-gasemissionen. Germanwatch-Hintergrundpapier, Bonn/Berlin, <www.germanwatch.org/rio/thg-ad03.htm> accessed 03/07/07.

Broekhoff, M., 2004: Institutional and cost implications of multi-project baselines. *Mitigation and Adaptation Strategies for Global Change*, **9**, pp. 201-216, Centre for Economic Policy Research, London.

Brouhle, K., C. Griffiths, and A. Wolverton, 2005: The use of voluntary approaches for environmental policymaking in the US. E. Croci, *the Handbook of Environmental Voluntary Agreements*, Springer, 2005.

Bruvoll, A. and B.M. Larsen, 2004: Greenhouse gas emissions in Norway: Do carbon taxes work? Energy Policy, 32. A previous version is available as Discussion Papers 337 by Statistics Norway, <www.ssb.no/cgi-bin/publsoek?job=forside&id=dp-337&kode=dp&lang=en>, accessed 03/07/07.

Bulkeley, H., 2001: No regrets? Economy and environment in Australia's domestic climate change policy process. *Global Environmental Change*, **11**, pp. 155-169.

Burns, W.C.G., 2004: Climate justice: The prospect for climate change litigation. The exigencies that drive potential causes of action for climate change damages at the international level. American Society of International Law Proceedings, **98**, 223 pp.

Burtraw, D., A. Krupnick, K. Palmer, A. Paul, M. Toman, and C. Bloyd, 2001a: Ancillary benefits of reduced air pollution in the United States from moderate greenhouse gas mitigation policies in the electricity sector. RFF Discussion Paper 01-61.

Burtraw, D., K. Palmer, A. Paul, R. Bharvirkar, 2001b: The effect of allowance allocation on the cost of carbon emissions trading. RFF Discussion Paper 01-30.

Burtraw, D., K. Palmer, R. Bharvirkar, and A. Paul, 2002: The effect on asset values of the allocation of carbon dioxide emission allowances. *The Electricity Journal*, June 2002, **15**(5), pp. 51-62.

Bygrave, S., 2004: Experience with emissions trading and project-based mechanisms in OECD and non-OECD countries. *Greenhouse Gas Emissions Trading and Project-Based Mechanisms*, Proceedings of OECD Global Forum on Sustainable Development: Emissions Trading, OECD, Paris.

Bygrave, S. and M. Bosi, 2004a: Linking non-EU domestic emisisons trading schemes with the EU emissions trading scheme. OECD/IEA Annex I Expert Group, Paris.

Bygrave, S. and M. Bosi, 2004b: Linking project-based mechanisms with domestic greenhouse gas emissions trading schemes. OECD/IEA Annex I Expert Group, Paris.

Cambridge Econometrics, 2005: Modelling the initial effects of the climate change levy. Report submitted to HM Customs and Excise by Cambridge Econometrics, Department of Applied Economics, University of Cambridge and the Policy Studies Institute, <http://customs.hmrc.gov.uk/channelsPortalWebApp/channelsPortalWebApp.portal?_nfpb=true&_pageLabel=pageLibrary_MiscellaneousReports&propertyType=document&columns=1&id=HMCE_PROD1_023971>, accessed 03/07/07.

Carraro, C. and D. Siniscalco, 1993: Strategies for the international protection of the environment. *Journal of Public Economics*, **52**, pp. 309-328.

Chander, P. and H. Tulkens, 1995: A core-theoretic solution for the design of cooperative agreements on transfrontier pollution. *International Tax and Public Finance*, **2**, pp. 279-293.

Chan-Woo, K., 2002: Negotiations on Climate Change: Debates on commitment of developing countries and possible responses. *East Asian Review*, **14**, pp. 42-60.

Charnovitz, S., 2003: Trade and climate: Potential conflicts and synergies. In *Beyond Kyoto: Advancing the International Effort Against Climate Change*, Aldy *et al.*, Report prepared for the Pew Center on Global Climate Change, 2003.

Chen, W., 2003: Carbon quota price and CDM potential after Marrakesh. *Energy Policy*, **31**, pp. 709-719.

Chidiak, M., 2002: Lessons from the French experience with voluntary agreements for greenhouse-gas reduction. *Journal of Cleaner Production*, **10**, pp 121-128.

Christiansen, A.C. and J. Wettestad, 2003: The EU as a frontrunner on emissions trading: how did it happen and will the EU succeed? *Climate Policy*, **3**, pp. 3-13.

Christiansen, A.C., A. Arvanitakis, K. Tangen, H. Hasselkippe, 2005: Price determinants in the EU emissions trading scheme. *Climate Policy*, **5**, pp. 1-17.

Climate Justice Programme, 2005: Gas flaring in Nigeria: A human rights, environmental change. *Natural Resources Journal*, **38**, pp. 563-569.

Cooper, R., 1998: Towards a real treaty on global warming. *Foreign Affairs*, **77**, pp. 66-79.

Cooper, R., 2001: The Kyoto Protocol: A flawed concept. *Environmental Law Reporter*, **31**(11), pp. 484-411, 492 pp.

Corfee-Morlot, J., M. Berg, and G. Caspary, 2002: Exploring linkages between natural resource management and climate adaptation strategies. OECD, Paris, COM/ENV/EPOC/DCD/DAC(2002)3/Final.

Corfee-Morlot, J., N. Höhne, 2003: Climate change: long-term targets and short-term commitments. *Global Environmental Change*, **13**, 2003, pp. 277-293.

Cramton, P. and S. Kerr, 2002: Tradable carbon permit auctions: How and why to auction not grandfather. *Energy Policy*, **30**, pp. 333-345.

Criqui, P., A. Kitous, M. Berk, M. Den Elzen, B. Eickhout, P. Lucas, D. van Vuuren, N. Kouvaritakis, and D. Vanregemorter, 2003: Greenhouse gas reduction pathways in the UNFCCC process up to 2025 - Technical Report. Study Contract: B4-3040/2001/325703/MAR/E.1 for the European Commission, DG Environment, Brussels.

Croci, E. (ed.), 2005: The handbook of environmental voluntary agreements: Design, implementation and evaluation issues. Springer, The Netherlands.

Dang, H., A. Michaelowa, and D. Tuan, 2003: Synergy of adaptation and mitigation strategies in the context of sustainable development: the case of Vietnam. *Climate Policy*, **3**, Supplement 1, pp. S81-S96.

Darnall, N. and J. Carmin, 2003: The design and rigor of U.S. voluntary environmental programs: results from the survey. Raleigh, North Carolina State University.

Dasgupta, C. and U. Kelkar, 2003: Indian perspectives on beyond-2012. Presentation at the open symposium 'International Climate Regime beyond 2012: Issues and Challenges', Tokyo.

Davies, J.C. and J. Mazurek, 1998: Pollution control in the United States: Evaluating the system.

DEHSt, 2005: Implementation of the emissions trading in the EU: National allocation plans of all EU states. German Emissions Trading Authority (DEHSt) at the Federal Environmental Agency (UBA), Berlin, November 2005.

Demailly, D. and P. Quirion, 2006: CO_2 abatement, competitiveness and leakage in the European cement industry under the EU ETS: Grandfathering vs. output-based allocation. *Climate Policy*.

Den Elzen, M.G.J. and M. Schaeffer, 2002: Responsibility for past and future global warming: uncertainties in attributing anthropogenic climate change. *Climatic change*, **54**, pp. 29-73.

Den Elzen, M.G.J. and M.M. Berk, 2004: Bottom-up approaches for defining future climate mitigation commitments. MNP Report 728001029/2005, Netherlands Environmental Assessment Agency (MNP), Bilthoven, the Netherlands.

Den Elzen, M.G.J. and M. Meinshausen, 2005: Meeting the EU 2°C climate target: global and regional emission implications. MNP Report 728001031/2005, Netherlands Environmental Assessment Agency (MNP), Bilthoven, the Netherlands.

Den Elzen, M.G.J. and M. Meinshausen, 2006: Multi-gas emission pathways for meeting the EU 2°C climate target. *Avoiding Dangerous Climate Change*, H.J. Schellnhuber, W. Cramer, N. Nakicenovic, T. Wigley, and G. Yohe (eds), Cambridge University Press, Cambridge, UK.

Den Elzen, M.G.J. and P. Lucas, 2005: The FAIR model: a tool to analyse environmental and costs implications of climate regimes. *Environmental Modeling & Assessment*, **10**(2) pp. 115-134.

Den Elzen, M.G.J., F. Fuglestvedt, N. Höhne, C. Trudinger, J. Lowe, B. Matthews, B. Romstad, C. Pires de Campos, and N. Andronova, 2005a: Analysing countries' contribution to climate change: Scientific and policy-related choices. *Environmental Science & Policy*, **8**, pp. 614-636.

Den Elzen, M.G.J., M. Schaeffer, and P. Lucas, 2005b: Differentiation of future commitments based on Parties' contribution to climate change. Climate Change: uncertainties in the 'Brazilian Proposal' in the context of a policy implementation. *Climatic Change*, **71**(3), pp. 277-301.

Den Elzen, M.G.J., P. Lucas, and D.P. van Vuuren, 2005c: Abatement costs of post-Kyoto climate regimes. *Energy Policy*, **33**(16), pp. 2138-2151.

Den Elzen, M.G.J., M.M. Berk, P. Lucas, C. Criqui, and A. Kitous, 2006a: Multi-Stage: a rule-based evolution of future commitments under the Climate Change Convention. *International Environmental Agreements: Politics, Law and Economics*, **6**(2006), pp. 1-28.

Den Elzen, M.G.J., P. Lucas, and D.P. van Vuuren, 2006b: Exploring regional abatement costs and options under allocation schemes for emission allowances. MNP report, Bilthoven, the Netherlands.

Depledge, J., 2000: Tracing the origins of the Kyoto Protocol: An article-by-article textual history. UNFCCC document FCCC/TP/2000/2, <www.unfccc.int>, accessed 16/04/2007.

Dinan, T. and D.L. Rogers, 2002: Distributional effects of carbon allowances trading: How government decisions determine winners and losers. *National Tax Journal*, **55**(2).

Dlugolecki, A. and S. Lafeld, 2005: Climate change - agenda for action: the financial sector's perspective. Allianz Group and WWF, Munich.

Doelle, M., 2004: Climate Change and the WTO: Opportunities to motivate state action on climate change through the world trade organization. *Review of European Community and International Environmental Law.* **13**(1), pp. 85-103.

Dudek, D. and A. Golub, 2003: 'Intensity' targets: Pathway or roadblock to preventing climate change while enhancing economic growth? Climate Policy, 3(S2), pp. S21-28.

Dutschke, M., 2002: Fractions of permanence - Squaring the cycle of sink carbon accounting. *Mitigation and Adaptation Strategies for Global Change*, **7**(4), pp.381-402.

Edenhofer, O., K. Lessmann, C. Kemfert, M. Grubb, and J. Köhler, 2006: Induced technological change: Exploring its implication for the ecnomics of atmospheric stabilization. *The Energy Journal Special Issue*, **57**, pp. 107.

Edmonds, J. and M. Wise, 1998: Building backstop technologies and policies to implement the framework convention on climate change. Washington, D.C., Pacific Northwest National Laboratory.

Edmonds, J. and M.A. Wise, 1999: Exploring a technology strategy for stabilising atmospheric CO_2. *International Environmental Agreements on Climate Change*, C. Carraro, Dordrecht, Netherlands, Kluwer Academic Publishers, 19.

Ekboir, J.M., 2003: Research and technology policies in innovation systems: zero tillage in Brazil, *Research Policy*, **32**, pp. 573-586.

Ellerman, A.D., P. Joskow, R. Schmalensee, J.P. Montero, and E. Bailey, 2000: Markets for clean air. The U.S. Acid Rain Program, Cambridge University Press.

Ellerman, A.D., 2002: Designing a tradeable permit system to control SO_2 emissions in China: Principles and practice. *The Energy Journal*, **23**(2), pp. 1-26.

Ellerman, A.D. and I. Sue Wing, 2003: Absolute v. intensity-based emission caps. Climate Policy **3**(Supplement 2), pp. S7-S20.

Ellerman, A.D., 2005: U.S. experience with emissions trading: Lessons for CO_2. Climate Policy and Emissions Trading After Kyoto, edited by Bernd Hansjurgen. Cambridge, England, Cambridge University Press.

Ellerman, A.D., 2006: Are cap-and-trade programs more environmentally effective than conventional regulation? In *Moving to Markets: Lessons from Thirty Years of Experience*, J. Freeman and C. Kolstad (eds), Oxford University Press, New York.

Ellis, J. and R. Baron, 2005: Sectoral Crediting Mechanisms: An initial assessment of electricity and aluminium. OECD/IEA Annex I Group, Paris.

Ellis, J. and K. Karousakis, 2006: The developing CDM market: May 2006 Update. OECD/IEA Annex I Expert Group, Paris.

Ellis, J., 2006: Issues related to a programme of activities under the CDM. OECD/IEA Annex I Group, Paris.

Ellis, J., H. Winkler, J. Corfee-Morlot, and F. Gagnon-Lebrun, 2007: CDM: Taking stock, Looking forward. *Energy Policy*, **35**, pp.15-28.

Esty, D.C. and M.E. Porter, 1998: Industrial ecology and competitiveness, Strategic implications for the firm. *Journal of Industrial Ecology*, **2**(1), pp. 35-43.

Esty, D.C., 2001: Bridging the trade-environment divide. *Journal of Economic Perspectives*, **15**(3), Summer 2001, pp. 113-130.

Esty, D. and M. Ivanova (eds), 2002: Global environmental governance: Options and opportunities. Yale School of Forestry and Environmental Studies, New Haven.

Eyckmans, J. and M. Finus, 2003: Coalition formation in global warming game: How the design of protocols affects the success of environmental treaty-making. Pre-publication draft.

Fernandez, M. and A. Michaelowa, 2003: Joint implementation and EU accession countries. *Global Environmental Change*, **13**, pp. 269-275.

Fichtner, W., M. Goebelt, and O. Rentz, 2001: The efficiency of international cooperation in mitigating climate change: analysis of joint implementation, the clean development mechanism and emission trading for the Federal Republic of Germany, the Russian Federation and Indonesia. *Energy Policy*, **29**, pp. 817-830.

Figures, C., 2004: Institutional capacity to integrate institutional development and climate change consideration: An assessment of DNAs in LAC. Interamerican Development Bank, October 2004, Washington, D.C.

Finus, M., 2002: New developments in coalition theory: An application to the case of global pollution. *The International Dimension of Environmental Policy*. Kluwer, Dordrecht, Holland.

Fischer, C., 2001: Rebating environmental policy revenues: Output-based allocations and tradable performance standards. RFF Discussion Paper, 01-22.

Fischer, C., S. Hoffman, and Y. Yoshino, 2002: Multilateral trade agreements and market-based environmental policies. RFF Discussion Paper, May.

Fischer, C., 2006: Project-based mechanisms for emissions reductions: balancing trade-offs with baselines. *Energy Policy*.

Fisher, C., I.W.H. Parry, and W.A. Pizer, 2003: Instrument choice for environmental protection when technological innovation is endogenous. *Journal of Environmental Economics and Management*, **45**, pp.523-545.

Fisher, C. and R. Newell, 2004: Environmental and technology policies for climate change and renewable energy, resources for the future. Discussion paper 04-05, April 2004.

Fleming, P.D. and P.H. Webber, 2004: Local and regional greenhouse gas management. *Energy Policy*, **32**(2004), pp. 761-771.

Fowler, R., 2004: Lessons from the implementation of a mandatory project-based greenhouse gas abatement scheme in the state of New South Wales, Australia. *Greenhouse Gas Market 2004: Ready for Takeoff*, International Emissions Trading Association, 2004.

Foxon, T. and R. Kemp, 2004: Innovation impacts of environmental policies. International Handbook on Environment and Technology Management (ETM).

Frankel, J.A., 1999: Greenhouse Gas Emissions. Policy Brief, Brookings Institution, Washington D.C., **52**.

Frankel, J.A. and A.K. Rose, 2003: Is trade good or bad for the environment? Sorting out the causality. National Bureau of Economic Research (NBER).

Freeman, J. and C. Kolstad (eds), 2006: Moving to markets: Lessons from thirty years of experience. (Oxford University Press, New York.

Fritsche, M. and R. Lukas, 2001: Who cooperates on R&D? *Research Policy*, **30**, pp. 297-312.

Gallagher, K.P., 2004: *Free Trade and the Environment. Mexico, NAFTA and Beyond*. Palo Alto, CA: Stanford University Press.

Gangopadhyay, S., B. Ramaswami, and W. Wadhwa, 2005: Reducing subsidies on household fuels in India: How will it affect the poor? *Energy Policy*, **33**(2005), pp. 2326-2336.

GCI, 2005: GCI Briefing: Contraction & Convergence. <http://www.gci.org.uk/briefings/ICE.pdf>, accessed April 2006.

Geres, R. and A. Michaelowa, 2002: A qualitative method to consider leakage effects from CDM and JI projects. *Energy Policy*, **30**(2002), pp. 461-463.

Germain, M., P.L. Toint, and H. Tulkens, 1998: Financial transfers to sustain cooperative international optimality in stock pollutant abatement. *Sustainability and Firms: Technological Change and the Changing Regulatory Environment*. Edward Elgar, Cheltenham, UK, pp. 205-219.

Germain, M., P.L. Toint, H. Tulkens, and A.d. Zeeuw, 2003: Transfers to sustain core theoretic cooperation in international stock pollutant control. *Journal of Economic Dynamics & Control*, **28**, pp. 79-99.

Gillespie, A., 2004: Small Island States in the face of Climate Change: The end of the line in international environmental responsibility. *UCLA Journal of Environmental Law and Policy*, **22**, 107 pp.

Goldberg, D. and K. Baumert, 2004: Action targets: A new form of GHG commitment. *Joint Implementation Quarterly*, **10**(3), pp. 8-9.

Golub, A. and E. Strukova, 2004: Russia and the GHG market. *Climatic Change*, **63**, pp.223-243.

Goulder, L., I. Parry, R. Williams, and D. Burtraw, 1999: The cost effectiveness of alternative instruments for environmental effectiveness in a second best setting. *Journal of Public Economics*, **72**(3), pp. 329-360.

Goulder, L.H. and S.H. Schneider, 1999: Induced technological change and the attractiveness of CO_2 emissions abatement policies. *Resource and Energy Economics*, **21**, pp. 211-253.

Grafton, Q., F. Jotzo, and M. Wasson, 2004: Financing sustainable development: country undertakings and rights for environmental sustainability (CURES). Ecological Economics, 51, pp. 65-78.

Greene, O., 1996: Lessons from other international environmental agreements. In *Sharing the effort - Options for differentiating commitments on climate change*, M. Paterson and M. Grubb (eds) The Royal Institute of International Affairs, London, pp. 23-44.

Greiner, S. and A. Michaelowa, 2003: Defining investment additionality for CDM projects - practical approaches. *Energy Policy*, **31**(2003), pp. 1007-1015.

Groenenberg, H., K. Blok, and J. van der Sluijs, 2004: Global Triptych: a bottom-up approach for the differentiation of commitments under the Climate Convention. *Climate Policy*, **4**(4), pp. 153-175.

Grossman, D.A., 2003: Warming up to a not-so-radical idea: Tort based climate change litigation. *Colombia Journal of Environmental Law*, **28**, pp. 1-61.

Grubb, M., C. Hope, and R. Fouquet, 2002: Climatic implications of the Kyoto Protocol: The contribution of international spillover. *Climatic Change*, **54**, pp. 11-28.

Grubb, M., 2004: Technology innovation and climate change policy: an overview of issues and options. Keio Economic Studies, **41**(2), pp.103-132.

Grubb, M., C. Azar, and U.M. Persson, 2005: Allowance allocation in the European emissions trading scheme: a commentary. *Climate Policy*, **5**, pp. 127-136.

Gupta, J., 1998: Encouraging developing country participation in the climate change regime. Institute for Environmental Studies (IVM), Vrije Universiteit, Amsterdam, The Netherlands.

Gupta, J. and M. Grubb (eds), 2000: Climate Change and European leadership: A sustainable role for Europe. Environment and Policy Series, Kluwer Academic Publishers, Dordrecht, 344 pp.

Gupta, J. and L. Ringius, 2001: The EU's climate leadership: Between ambition and reality. International Environmental Agreements. *Politics, law and economics*, **1**(2), pp. 281-299, London, 178 pp.

Gupta, J., 2003a: Engaging developing countries in climate change: (KISS and Make-up!). *Climate Policy for the 21st Century, Meeting the Long-Term Challenge of Global Warming*, D. Michel (ed.), Washington D.C., Center for Transatlantic Relations, <http://transatlantic.sais-jhu.edu>, accessed 03/07/07.

Gupta, S., 2003b: Incentive-based approaches for mitigating greenhouse gas emissions: Issues and prospects for India. *India and Global Climate Change Perspectives on Economics and Policy from a Developing Country*, M.A. Toman, U. Chakravorty, and S. Gupta (eds), Washington, Resources for the Future Press.

Gupta, S. and P.M. Bhandari, 1999: An effective allocation criterion for CO_2 emissions. *Energy Policy*, **27**(12), November.

Gupta, S. and P.M. Bhandari, 2003: Allocation of GHG emissions - An example of short term and long term criteria. *India and Global Climate Change, Perspectives on Economics and Policy from a Developing Country, Resources for the Future*, M.A. Toman, U. Chakravorty, S. Gupta (eds).

Hahn, R., 1998: The economic and politics of Climate Change. Washington D.C., American Enterprise Institute Press.

Hahn, R.W., 2001: A primer on environmental policy design. London, Routledge.

Haites, E., 2003a: Harmonization between national and International Tradeable Permit Schemes. CATEP Synthesis Paper, OECD, Paris.

Haites, E., 2003b: Output-based allocation as a form of protection for internationally competitive industries. *Climate Policy*, **3**(Supplement 2), December 2003, pp. S29-S41.

Hall, B.H., 2002: The financing of research and development NBER Working Papers 8773. International Bureau of Economic Research.

Hamilton, K., 2005: The finance-policy gap: Policy conditions for attracting long-term investment. *The Finance of Climate Change*, K. Tang (ed.), Risk Books, London.

Hancock, E.E., 2005: Red dawn, blue thunder, purple rain: Corporate risk of liability for global climate change and the sec disclosure dilemma. *Georgetown International Environmental Law Review*, **17**, 223 pp.

Hanks, J., 2002: Voluntary agreements, climate change and industrial energy efficiency. *Journal of Cleaner Production*, **10**, pp. 103-107.

Hargrave, T., N. Helme, and C. Vanderlan, 1998: Growth baselines. Washington, D.C., USA, Center for Clean Air Policy.

Harrington, W., R.D. Morgenstern, and P. Nelson, 2000: On the accuracy of regulatory cost estimates. *Journal of Policy Analysis and Management*, **19**(2), pp.297-322.

Harrington, W., R.D. Morgenstern, and T. Sterner, 2004: Overview: Comparing instrument choices. *Choosing Environmental Policy*, W. Harrington, R.D. Morgenstern, and T. Sterner (eds), Washington, D.C., Resources for the Future Press.

Harrison, K., 1999: Talking with the donkey: Cooperative approaches to environmental protection. *Journal of Industrial Ecology*, **2**(3), pp. 51-72.

Harrison, D. and D. Radov, 2002: Evaluation of alternative initial allocation mechanisms in a European Union greenhouse gas emissions allowance trading scheme. National Economic Research Associates, prepared for DG Environment, European Commission.

Heemst, J. van, and V. Bayangos, 2004: Poverty and Climate Change. *Beyond Climate: Options for Broadening Climate Policy*, M.T.J. Kok and H.C. de Coninck (ed.), RIVM Report 500036 001, Bilthoven, pp. 21-51.

Helby, P., 2002: EKO-Energi - a public voluntary programme: targeted at Swedish firms with ambitious environmental goals. *Journal of Cleaner Production*, **10**, pp 143-151.

Helm, C. and U.E. Simonis, 2001: Distributive justice in international environmental policy: Axiomatic foundation and exemplary formulation. Environmental Values, 10, pp. 5-18.

Hepburn, C., M. Grubb, K. Heuhoff, F. Matthes, and M. Tse, 2006: Auctioning of EU ETS Phase II allowances: how and why? *Climate Policy*.

Hershkowitz, A., 1998: In defense of recycling. *Social research*, **65**(1), Spring.

Herzog, H., K. Caldeira, and J. Reilly, 2003: An issue of permanence: Assessing the effectiveness of temporary carbon storage. *Climatic Change*, **59**, pp. 293-310.

Higley, C.J., F. Convery, and F. Leveque, 2001: Voluntary approaches: An introduction. *Environmental Voluntary Approaches: Research Insights for Policy Makers*, C.J. Higley and F. Leveque (eds), CERNA, 2001.

Hoel, M. and K. Schneider, 1997: Incentives to participate in an international environmental agreement. *Environmental and Resource Economics*, **9**, pp. 153-170.

Hoffman, A., 2005: Climate change strategy: The business logic behind voluntary greenhouse gas reductions. California Management Review, **47**(3).

Höhne, N., J. Harnisch, G.J.M. Phylipsen, K. Blok, and C. Galleguillos, 2003: Evolution of commitments under the UNFCCC: Involving newly industrialized economies and developing countries. Research Report 201 41 255. UBA-FB 000412, <http://www.umweltbundesamt.de/uba-info-medien-e/index.htm >, accessed 03/07/07.

Höhne, N. and K. Blok, 2005: Calculating historical contributions to climate change - discussing the 'Brazilian Proposal'. *Climatic Change*, **71**, pp. 141-173.

Höhne, N., D. Phylipsen, S. Ullrich, and K. Blok, 2005: Options for the second commitment period of the Kyoto Protocol. For the German Federal Environmental Agency, 02/05, ISSN 1611-8855, <http://www.fiacc.net/data/Climate_Change_02.05.pdf>, accessed 03/07/07.

Höhne, N. and K. Blok, 2006: The impact of the Kyoto Protocol on climate stabilization. What is next after the Kyoto Protocol? Assessment of options for international climate policy post 2012. Amsterdam, The Netherlands, Techne Press.

Höhne, N., 2006: *What is next after the Kyoto Protocol. Assessment of options for international climate policy post 2012*. Techne Press, Amsterdam, the Netherlands, ISBN 908594005-2.

Höhne, N., M.G.J. den Elzen, and M. Weiss, 2006: Common but differentiated convergence (CDC), a new conceptual approach to long-term climate policy. *Climate Policy* (accepted).

Hovi, J. and I. Areklett, 2004: Enforcing the climate regime: Game theory and the Marrakesh Accords. In *International Environmental Agreements: Politics, Law and Economics*, **4**, pp. 1-26.

Humphreys, J., M. van Bueren, and A. Stoeckel, 2003: Greening farm subsidies, rural industries research and development corporation, Barton ACT ISBN 064258608.

Huq, S. and I. Burton, 2003: Funding adaptation to climate change: What, who and how to fund? Sustainable Development opinion paper, IIED, London.

Huq, S. and H. Reid, 2004: Mainstreaming adaptation in development. *IDS Bulletin*. **35**, pp. 15-21.

IAI, 2002: Aluminium: Industry as a partner for sustainable development. International Aluminium Institute, Paris: ICCA and UNEP, <http://www.uneptie.org/outreach/wssd/docs/sectors/final/aluminium.pdf>, accessed 03/07/07.

IEA, 2001: Energy subsidy reform and sustainable development challenges for policymakers. International Energy Agency, Paris.

IEA, 2003: World Energy Investment Outlook. International Energy Agency, Paris.

IEA, 2004: Renewable Energy: Market and Policy Trends in IEA countries. International Energy Agency, Paris.

IEA, 2005: Renewables Information. International Energy Agency, Paris.

IEA, 2006: Energy Statistics of Non-OECD Countries. International Energy Agency, Paris.

Illum, K. and N. Meyer, 2004: Joint implementation: methodology and policy considerations. *Energy Policy*, **32**, pp. 1013-1023.

IPA Energy Consulting (IPA), 2005: Implications of the EU Emissions Trading Scheme for the U.K. power Generation Sector. Report to the Department of Trade and Industry, November 11, 2005.

IPCC, 1996: Climate Change 1995 - Economic and social dimensions of climate change. Contribution of Working Group III to the Second Assessment Report of the Intergovernmental Panel on Climate Change. J.P. Bruce, H. Lee, E.F. Haites [eds]. Cambridge University Press, Cambridge, United Kingdom and New York, NY, USA, 448 pp.

IPCC, 2001: Climate Change 2001: Mitigation. Contribution of Working Group III to the third assessment report of the Intergovernmental Panel on Climate Change. B. Metz, O. Davidson, R. Swart, J. Pan [eds.]. Cambridge University Press, Cambridge, United Kingdom and New York, NY, USA, 752 pp.

IPCC, 2007a: Climate Change 2007: The Physical Science Basis. Contribution of Working Group I to the Fourth Assessment Report of the Intergovernmental Panel on Climate Change [Solomon, S., D. Qin, M. Manning, Z. Chen, M. Marquis, K.B.M.Tignor and H.L. Miller (eds.)]. Cambridge University Press, Cambridge, United Kingdom and New York, NY, USA, 996 pp.

IPCC, 2007b: Climate Change 2007: Impacts, Adaptation and Vulnerability. Contribution of Working Group II to the Fourth Assessment Report of the Intergovernmental Panel on Climate Change [Parry, M.L., O.F. Canziani, J.P. Palutikof, P.J. van der Linden, C.E. Hanson (eds.)]. Cambridge University Press, Cambridge, United Kingdom and New York, NY, USA.

Ivanova, A. and M. Angeles, 2005: Trade and environmental issues in APEC. *Globalization: Opportunities and Challenges for East Asia*, K. Fatemi, (ed), The Haworth Press, Binhampton, N.Y.

Jaccard, M., N. Rivers, C. Bataille, R. Murphy, J. Nyboer, and B. Sadownik, 2006: Burning our money. C.D. Howe Institute, ISSN 0824-8001, **204**, May 2006.

Jacobs, R.W., 2005: Treading deep waters: Substantive law issues in Tuvalu's threat to sue the United States in the international court of justice. *Pacific Rim Law and Policy Journal*, **14**, 103 pp.

Jacoby, H.D., R. Prinn, and R. Schmalensee, 1998: Kyoto's Unfinished Business. Foreign Affairs.

Jacoby, H.D. and A.D. Ellerman, 2004: The safety valve and climate policy. *Energy Policy*, **32**(4), pp. 481-491.

Jaeger, C.C., 2003: Climate Change: Combining mitigation and adaptation. Climate Policy for the 21st Century: Meeting the Long-Term Challenge of Global Warming. D. Michel,Washington, D.C., Center for Transatlantic Relations.

Jaffe, A.B., R.G. Newell, and R.N. Stavins, 2003: Technological change and the environment. In *Handbook of Environmental Economics*. K.-G. Mäler and J. Vincent (eds), Amsterdam, The Netherlands, Elsevier Science, 2003.

Johannsen, K.S., 2002: Combining voluntary agreements and taxes: an evaluation of the Danish agreement scheme on energy efficiency in industry. *Journal of Cleaner Production*, **10**, pp 129-141.

Johnstone, N., 2002: The use of tradable permits in combination with other policy instruments: A scoping paper. ENV/EPOC/WPNEP(2002)28, Working Party on National Environmental Policy, OECD, Paris.

Jordaan, J., 2004: Competition, geographical concentration and FDI-induced externalities: New evidence for Mexican manufacturing industries. Working Paper 32, Technological Institute of Monterrey, Mexico.

Jotzo, F. and A. Michaelowa, 2002: Estimating the CDM market under the Marrakech Accords. *Climate Policy*, **2**, pp. 179-196.

Jotzo, F. and J.C.V. Pezzey, 2005: Optimal intensity targets for emissions trading under uncertainty. Draft. Canberra, AU: Centre for Resource and Environmental Studies - Australian National University.

Justice, D. and C. Philibert, 2005: International energy technology collaboration and climate change mitigation. Synthesis Report, OECD/IEA Annex I Expert Group on the UNFCCC, Paris.

Kågeström, J., K. Astrand, and P. Helby, 2000: Voluntary Agreements-Implementation and Efficiency: Swedish Country Report,

Kameyama, Y., 2003: Dual track approach: an optimal climate architecture for beyond 2012? COM/ENV/EPOC/IEA/SLT(2005)6 30. Tsukuba, JP, National Institute for Environmental Studies.

Kameyama, Y., 2004: The future climate regime: a regional comparison of proposals. *International Environmental Agreements: Politics, Law and Economics*, **4**, pp. 307-326.

Kanie, N., 2003: Leadership and domestic policy in multilateral diplomacy: The case of the Netherlands. *International Negotiation*, **8**(2), pp. 339-365.

Kartha, S., M. Lazarus, and M. Bosi, 2002: Practical baseline recommendations for greenhouse gas projects in the electric power sector. OECD, Paris.

Kartha, S., M. Lazarus, and M. Bosi, 2004: Baseline recommendations for greenhouse gas mitigation projects in the electric power sector. *Energy Policy*, **32**, pp.545-566.

Kemfert, C., W. Liseb, and R.S.J. Tol, 2004: Games of Climate Change with international trade. Kluwer Academic Publishers. Printed in the Netherlands, *Environmental and Resource Economics*, **28**, pp. 209-232

Kennedy, P.W., B. Laplante, and J. Maxwell, 1994: Pollution policy: The role for publicly provided information. *Journal of Environmental Economics and Management*, **26**, pp. 31-43.

Keohane, N., R.L. Revesz, and R.N. Stavins, 1998: The choice of regulatory instruments in environmental policy. *Harvard Environmental Law Review*, **22**, pp.313-367.

Khanna, M. and L.A. Damon, 1999: EPA's voluntary 33/50 Program: Impact on toxic releases and economic performance of firms. *Journal of Environmental Economics and Management*, **37**(1), pp. 1-25.

Kim, Y. and K. Baumert, 2002: Reducing uncertainty through dual intensity targets. K. Baumert (ed.), *Building on the Kyoto Protocol, Options for protecting the climate*, World resources Institute, Washington, USA, <http://pubs.wri.org/pubs_pdf.cfm?PubID=3762>, accessed 03/07/07.

King, A. and M. Lenox, 2000: Industry self-regulation without sanctions: The chemical industry's responsible care program. *Academy of Management Journal*, **43**(4),pp. 698-716.

King, M.D., P. Sarria, D.J. Moss, and N.J. Numark, 2004: U.S. business actions to address climate change: case studies of five industry sectors. Sustainable Energy Institute, November, 2004.

Klepper, G. and S. Peterson, 2004: The EU emissions trading scheme - allowance prices, trade flows and competitiveness effects. *European Environment*, **14**, pp.201-218.

Koch, T. and A. Michaelowa, 1999: 'Hot air' reduction through non-quantifiable measures and early JI. *Joint Implementation Quarterly*, **5**(2), pp. 9-10.

Kolk, A. and J. Pinske, 2004: Market strategies for Climate Change. *European Management Journal*, **22**(3), pp.304-314.

Kolk, A. and J. Pinske, 2005: Business responses to Climate Change: Identifying emergent strategies. *California Management Review*, Spring, 2005.

Kolstad, C., 2000: *Environmental Economics*. Oxford University Press, New York.

Kolstad, C.D. 2006: The simple analytics of greenhouse gas emission intensity reduction targets. *Energy Policy*, **33**(17), pp. 2231-2236.

Konar, S. and M.A. Cohen, 1997: Information as regulation: The effect of community right-to-know laws on toxic emissions. *Journal of Environmental Economics and Management*, **32**, pp.109-124.

Korppoo, A., 2005: Russian energy efficiency projects: lessons learnt from activities implemented jointly. Pilot phase. *Energy Policy*, **33**, pp. 113-126.

Kousky, C. and S. Schneider, 2003: Global climate policy: will cities lead the way? **3**(4), December 2003, pp. 359-372.

Krarup, S. and S. Ramesohl, 2002: Voluntary agreements: key to higher energy efficiency in industry? *Voluntary Environmental Agreements: Process, Practice and Future Use*, P. ten Brink (ed.), Sheffield, UK.

Kruger, J., B. McLean, and R. Chen, 2000: A tale of two revolutions: Administration of the SO_2 trading program. *Emissions Trading: Environment al Policy's New Approach*, R. Kosobud (ed.), New York, John Wiley & Sons.

Kruger, J., K. Grover, and J. Schreifels, 2003: Building institutions to address air pollution in developing countries: The Cap-and-trade approach. *Greenhouse Gas Emissions Trading and Project-Based Mechanisms*, Proceedings of OECD Global Forum on Sustainable Development: Emissions Trading, OECD, Paris.

Kruger, J., 2006: From SO_2 to greenhouse gases: Trends and events shaping future emissions trading programs in the United States. In *Acid in the Environment: Lessons Learned and Future Prospects*, G.R. Visgilio and D.M. Whitelaw (ed.), Springer Science+Business Media, Inc. 2006.

Kuik, O., 2003: Climate change policies, energy security and carbon dependency. Trade-offs for the European Union in the Longer Term. *International Environmental Agreements: Politics, Law and Economics*, **3**, pp.221-242.

Kuik, O., and M. Mulder, 2004: Emissions trading and competitiveness: pros and cons of relative and absolute schemes. *Energy Policy*, **32**, pp. 737-745.

La Rovere, E.L., S.K. Ribeiro, and K.A. Baumert, 2002: The Brazilian proposal on relative responsibility for global warming. Options for protecting the climate. K.A. Baumert, O. Blanchard, S. Llosa, and J. F. Perkaus. Washington, World Resource Institute (WRI).

Lecoq, F. and R. Crassous, 2003: International climate regime beyond 2012 - Are quota allocation rules robust to uncertainty? Policy Research Working Paper. Washington D.C.

Lecoq, F. and K. Capoor, 2005: State and trends of the carbon market 2005. World Bank, Washington D.C.

Levy, D.L. and P.J. Newell, 2005. The business of global environmental governance (Global Environmental Accord: Strategies for sustainability and institutional innovation). Cambridge, MIT Press.

Lipanovich, A., 2005: Smoke before oil: Modeling a suit against the auto and oil industry on the tobacco tort litigation is feasible. *Golden Gate University Law Review*, **35**, 429 pp.

Lisowski, M., 2002: The emperor's new clothes: redressing the Kyoto Protocol. *Climate Policy* **2**(2-3).

Lutter, R., 2000: Developing countries' greenhouse emissions: Uncertainty and implications for participation in the kyoto protocol. The Energy Journal, 21(4).

Lyon, T. P. and J.W. Maxwell, 2000: Voluntary approaches to environmental regulation: A survey. Economic Institutions and Environmental Policy, M. Franzini and A. Nicita (eds), Chapter 7, pp. 142-174.

Lyon, T.P. and J.W. Maxwell, 2004: Corporate environmentalism and public policy. Cambridge University Press, Cambridge, UK, 2004.

MacGill, I., H. Outhred, and K. Nolles, 2006: Some design lessons from market-based greenhouse gas regulation in the restructured Australian electricity industry. *Energy Policy*, **34**, pp.11-25.

Maeda, A., 2003: The emergence of market power in emission rights markets: The role of initial permit distribution. *Journal of Regulatory Economics*, **24**(3), pp. 293-314.

Mank, B.C., 2005: Standing and global warming: Is injury to all, injury to none? *Environmental Law*, **35**, pp.1-83.

Manne, A.S. and R.G. Richels, 1999: The Kyoto Protocol: A cost-effective strategy for meeting environmental objectives? *Energy Journal, Kyoto Special Issue*, pp. 1-25.

Marburg, K.L., 2001: Combating the impacts of global warming: A novel legal strategy. *Colorado Journal of International Environmental Law and Policy*, 171 pp.

Margolick, M. and D. Russell, 2001: Corporate greenhouse gas reduction targets. Pew Center on Global Climate Change, Washington, D.C.

Margolis, R.M. and D.M. Kammen, 1999: Evidence of under-investment in energy R&D in the United States and the impact of federal policy. *Energy Policy*, **27**, pp. 575-584.

Markussen, P. and G.T. Svendsen, 2005: Industry lobbying and the political economy of GHG trade in the European Union. *Energy Policy*, **33**, pp. 245-255.

Matthes, F. *et al*., 2005: The environmental effectiveness and economic efficiency of the European Union Emissions Trading Scheme (Öko Institut).

Maurer, C. and R. Bhandari, 2000: The climate of export credit agencies. World Resources Institute, Washington D.C.

Maxwell, J.W., T.P. Lyon, and S.C. Hackett, 2000: Self-regulation and social welfare: The political economy of corporate environmentalism. *Journal of Law and Economics*, **XLIII**, October, pp. 583-618.

McKibbin, W. and P. Wilcoxen, 1997: Salvaging the Kyoto climate negotiations. *Policy Brief, Brookings Institution*, **27**.

McKibbin, W.J. and P.J. Wilcoxen, 2002: Climate change policy after Kyoto: A blueprint for a realistic approach. Washington D.C., Brookings Institution.

McKinstry, R., 2004: Laboratories for local solutions for global problems: State, local, and private leadership in developing strategies to mitigate the causes and effects of climate change. *Penn State Environmental Law Review*, **12**(1), pp. 15-82.

Meira Filho, L.G. and J.D. Gonzales Miguez, 2000: Note on the time-dependant relationship between emissions of greenhouse gases and climate change. Technical note, Ministry of Science and Technology Federal Republic of Brazil, <http://www.mct.gov.br/clima>, accessed 03/07/07.

Menanteau, P., D. Finon, and M-L. Lamy, 2003: Prices versus quantities: choosing policies for promoting the development of renewable energy. *Energy Policy*, **31**(8), pp.799-812.

Meyer, N.I., 2004: Development of Danish wind power market. *Energy &Environment*, **15**(4), pp.657-672.

Michaelowa, A., 2002: The AIJ pilot phase as laboratory for CDM and JI. *International Journal of Global Environmental Issues*, **2**(3-4), pp. 267-280.

Michaelowa, A., 2003a: CDM host country institution building. *Mitigation and Adaptation Strategies for Global Change*, **8**, pp.201-220.

Michaelowa, A., 2003b: Germany - a pioneer on earthen feet? *Climate Policy*, **3**(1), pp. 31-44.

Michaelowa, A., 2004: *International review for Environmental Strategies*, **5**, 217-231.

Michaelowa, A. and F. Jotzo, 2005: Transaction costs, institutional rigidities and the size of the clean development mechanism. *Energy Policy*, **33**, pp. 511-523.

Michaelowa, A., K. Tangen, and H. Hasselknippe, 2005a: Issues and options for the post-2012 climate architecture -- An overview. *International Environmental Agreements*, **5**, pp. 5-24.

Michaelowa, A., S. Butzengeiger, and M. Jung, 2005b: Graduation and deepening: An ambitious post-2012 climate policy scenario. *International Environmental Agreements: Politics, Law and Economics*. **5**, pp. 25-46.

Millennium Ecosystem Assessment, 2005: Ecosystems and human well-being: Synthesis. Island Press, Washington, D.C.

Mitchell, R.B., 2005: Flexibility, compliance and norm development in the climate regime. *Implementing Climate Regime: International Compliance*, O.S. Stokke, J. Hovi, and G. Ulfstein, Earthscan, London, pp. 65-83.

Moe, A., K. Tangen, V. Berdin, and O. Pluzhnikov, 2003: emissions trading and green investments in Russia. *Energy & Environment*, **14**(6), pp. 841-858.

Montero, J-P., J.M. Sanchez, and R. Katz, 2002: A market-based environmental policy experiment in Chile. *Journal of Law and Economics*, **XLV**, pp. 267-287 (April 2002).

Montero, J-P., 2005: Pollution markets with imperfectly observed emissions. *RAND Journal of Economics*, Autumn 2005, **36**(3), pp. 645-660.

Montero. J-P., 2007: Tradeable Permits with incomplete monitoring: Evidence from Santiago's Particulate Permits Program. Ch 6 in *Moving to Markets in Environmental Regulation*, J. Freeman and C.D. Kolstad (eds), New York, Oxford University Press.

Morthorst, P.E., 2001: Interactions of a tradable green certificate market with a tradable permits market. *Energy Policy*, **29**(5), pp. 345-353.

Müller, B., 1999: Justice in global warming negotiations - How to achieve a procedurally fair compromise. Oxford, Osford Institute for Energy Studies.

Müller, B., M. Michealowa, and C. Vrolijk, 2001: Rejecting Kyoto: A study of proposed alternatives to the Kyoto Protocol. Oxford, Climate Strategies International Network for Climate Policy Analysis, Oxford Institute for Energy Analysis.

Müller, B., 2002: Equity in global climate change: the great divide. Oxford, UK, Oxford University Press.

Müller, B. and G. Müller-Fürstenberger, 2003: Price-related sensitivities of greenhouse gas intensity targets. *Mimeo*, Oxford Energy Institute.

Murase, S., 2002a: Implementation of international environmental law: Its international and domestic aspects: Case of the Kyoto Protocol. *Jurisuto*,**1232**, pp.71-78 (in Japanese).

Murase, S., 2002b: Conflict of international regimes: Trade and the environment, Institute of International Public Law and International Relations of Thessaloniki. *Thesaurus Acroasium*, **XXXI**, pp. 297-340.

Murase, S., 2003: Problem on compliance of the Kyoto Protocol and building a new international regime: Possibility of an alternative regime that includes the United States and major developing countries. Keio University, *Mita Gakkai Zasshi*, **96**(2), pp. 5-18 (in Japanese).

Murase, S., 2004: Japan's measures on global warming and the TBT agreement. *Survey and Research on Coordination of Trade and Environment*. GISPRI, 2004 (in Japanese), pp. 81-92.

Murase, S., 2005: Trade and the environment: With particular reference to climate change issues. *Manchester Journal of International Economic Law*, 2(2), pp. 18-38.

Najam, A., S. Huq, and Y. Sokona, 2003: Climate negotiations beyond Kyoto: developing countries concerns and interests. *Climate Policy*, **3**, pp. 221-231.

Nakicenovic, N. and K. Riahi, 2003. Model runs with MESSAGE in the context of the further development of the kyoto-protocol. Technical report, Wissenschaftlicher Beirat der Bundesregierung Globale Umweltveränderungen, Berlin.

Nash, J. and J. Ehrenfeld, 1996: Code Green: Business adopts voluntary environmental standards. *Environment*, **38**(1), pp. 16-20, 36-45.

National Commission on Energy Policy (NCEP), 2004: Ending the energy stalemate: A bipartisan strategy to address America's energy challenges. Washington, D.C.

National Research Council (NRC), 2003: Planning climate and global change research: A review of the draft U.S. Climate Change Science Program Strategic Plan. Washington, D.C.

Nelson, K. and B. de Jong, 2003: Making global initiatives local realities: carbon mitigation projects in Chiapas, Mexico. *Global Environmental Change*, **13**, pp.19-30.

Nemet, G.F., 2005: Beyond the learning curve: factors influencing cost reductions in photovoltaics. *Energy Policy*, August 2005.

Nemet, G.F. and D.M. Kammen, 2006: U.S. energy research and development: Declining investment, increasing need, and feasibility of expansion. *Energy Policy*, February 2006.

Nentjes, A., and G. Klaassen, 2004: On the quality of compliance mechanisms in the Kyoto Protocol. *Energy Policy*, **32**, pp. 531-544.

NERA, 2005: Interactions of the EU ETS with Green and White Certificate Schemes: Summary Report for Policy Makers. <http://europa.eu.int/comm/environment/climat/pdf/ec_green_summary_report051117.pdf>, accessed 03/07/07.

Neuhoff, K., K. Keats, and M. Sata, 2006: Allocation and incentives: Impacts of CO_2 emission allowance allocations to the electricity sector. *Climate Policy* (forthcoming).

Newell, P., 2000: Climate for change. Non-state actors and the global politics of the greenhouse. Cambridge University Press, Cambridge, 2000, pp. 222,

Newell, R. and W.A. Pizer, 2003: Regulating stock externalities under uncertainty. *Journal of Environmental Economics and Management*, **45**, pp. 416-432.

Newell, R. and W. Pizer, 2004: Uncertain discount rates in climate policy analysis. *Energy Policy*, **32**, pp. 519-529.

Newell, R. and N. Wilson, 2005: Technology prizes for Climate Change Mitigation, Resources for the future. Discussion paper 05-33, June, 2005, Washington, D.C.

Ninomiya, Y., 2003: Prospects for energy efficiency improvement through an international agreement. *Climate Regime Beyond 2012: Incentives for Global Participation*, National Institute for Environmental Studies and the Institute for Global Environmental Strategies.

Nordhaus, W., 1998: Is the Kyoto Protocol a dead duck? Are there any live ducks around? Comparison of Alternative Global Tradable Emissions Regimes.

Nordhaus, W., 2001: After Kyoto: Alternative mechanisms to control global warming. Paper prepared for a joint session of the American Economic Association and Association of Environmental and Resource Economists.

Nordic Council of Ministers, 2002: The use of economic instruments in Nordic Environmental Policy 1999-2001. TemaNord 2002:581, Nordic Council of Ministers, Copenhagen.

Northrop, M., 2004: Leading by example: Profitable corporate strategies and successful public policies for reducing greenhouse gas emissions. *Widener Law Journal*, **14**(1), pp. 21-80.

Oates, W.E., 2001: A Reconsideration of Environmental Federalism, Resources for the Future. Discussion Paper 01-54, November, 2001.

OECD, 1999: Conference on foreign direct investment & environment, The Hague, 28-29 January 1999, BIAC Discussion Paper.

OECD, 2001: Environmental indicators for Agriculture: methods and Results. Volume **3**, Paris.

OECD, 2002: Agricultural practices that reduce greenhouse gas emissions: Overview and results of survey instruments. Paris.

OECD, 2003(b): Choosing environmental policy instruments in the real world. Global forum on Sustainable Development: Emission Trading, 17-18 March, 2003, Paris.

OECD, 2003(c): Policies to reduce greenhouse gas emissions in industry - successful approaches and lessons learned. Workshop Report, Paris.

OECD, 2003(d): Technology innovation, development and diffusion. OECD and IEA Information Paper, COM/ENV/EPOC/IEA/SLT(2003)4, Paris.

OECD, 2003(e): Voluntary approaches for environmental policy: Effectiveness, efficiency and usage in policy mixes. Paris.

OECD, 2005(a): Main science and technology indicators. 2005/2, Paris.

OECD, 2005(b): An empirical study of environmental R&D: What encourages facilities to be environmentally innovative. Paris.

OECD, 2005(c): The United Kingdom Climate Change Levy - A study in political economy. Paris.

OECD, 2005(d): Mobilising private investment for development: Policy lessons on the role of ODA. *The DAC Journal*, **6**(2) - ISSN: 15633152, Paris.

Osgood, D.E., 2002: Environmental improvement with economic development through public information provision. Cambridge University Press, *Environment and Development Economics*, **7**, pp. 751-768

Ott, H.E., H. Winkler, B. Brouns, S. Kartha, M. Mace, S. Huq, Y. Kameyama, A.P. Sari, J. Pan, Y. Sokona, P.M. Bhandari, A. Kassenberg, E.L. La Rovere, and A. Rahman, 2004: South-North dialogue on equity in the greenhouse. A proposal for an adequate and equitable global climate agreement. GTZ, Eschborn, Germany.

Pacala, S. and R. Socolow, 2004: Stabilization wedges: Solving the climate problem for the next 50 years with current technologies. *Science*, **305**(5686), pp. 968-972.

Palmer, K., W.E. Oates, and P.R. Portney, 1995: Tightening environmental wtandards: The benefit-cost or the no-cost paradigm? *Journal of Economic Perspectives*, **9**(4), pp. 119-132.

Pan, J., 2005: Meeting human development goals with low emissions: An alternative to emissions caps for post-Kyoto from a developing country perspective. *International Environmental Agreements: Politics, Law and Economics*, **5**, pp. 89-104.

Parry, I., 2004: Are emissions permits regressive? *Journal of Environmental Economics and Management*, **47**, pp. 364-387.

Parry, J., A. Hammill, and J. Drexhage, 2005: Climate change and adaptation: A summary. International Institute for Sustainable Development (IISD), Winnipeg, <http://www.iisd.org/pdf/2005/climate_adaptation_sum.pdf>, accessed 03/07/07.

Peck, S.C. and T.J. Teisberg, 2003: An innovative policy approach to solve the global climate issue.

Penalver, E.M., 1998: Acts of God or toxic torts? Applying tort principles to the problem of climate change. *Natural Resources Journal*, **38**, pp. 563-569.

Pershing, J. and F. Tudela, 2003: A long-term target: Framing the climate effort. *Beyond Kyoto: advancing the international effort against climate change*. Pew Climate Center, Washington D.C.

Persson, M. and C. Azar, 2004: Brazil beyond Kyoto - Prospects and problems in handling tropical deforestation in a second commitment period.

Persson, T.A., C. Azar, and K. Lindgren, 2006: Allocation of CO_2 emission permits - economic incentives for emission reductions in developing countries. *Energy Policy*.

Peterson, T.D., 2004: The evolution of state climate change policy in the United States: Lessons learned and new directions. *Widener Law Journal*, **14**(1), pp. 81-120.

Petrtakis, E., E.S. Sartzetakis, and A. Xepapadeas, 2005: Environmental information provision as a public policy instrument. Contributions to *Economic Analysis & Policy*, **4**(1), Article 14. <http://www.bepress.com/bejeap/contributions/vol4/iss1/art14>, accessed 03/07/07.

Pew Center Report, 2004: Learning from State Action on Climate Change. In *Brief*, **8**.

Pew Center Report, 2005: International climate efforts beyond 2012: Report of the climate dialogue at Pocantico. Arlington, VA, USA, Pew Center on Global Climate Change.

Pezzey, J.C.V. and M.A. Toman, 2002: The economics of sustainability: A review of journal articles. Discussion Paper 02-03, Resources for the Future, Washington, D.C.

Pezzey, J.C.V., 2003: Emission taxes and tradable permits: A comparison of views on long run efficiency. *Environmental and Resource Economics*, **26**(2), pp. 329-342.

Philibert, C., 2000: How could emissions trading benefit developing countries. *Energy Policy*, **28**(13), pp. 947-956.

Philibert, C. and J. Pershing, 2001: Considering the options: climate targets for all countries. *Climate Policy*, **1**(2), pp. 211-227.

Philibert, C., 2004: International energy technology collaboration and climate change mitigation. COM/ENV/EPOC/IEA/SLT(2004)1, OECD/IEA, Paris.

Philibert, C., 2005a: Climate mitigation: Integrating approaches for future international co-operation. OECD/IEA Annex I Expert Group, Paris.

Philibert, C., 2005b: New commitment options: Compatibility with emissions trading. No. COM/ENV/EPOC/IEA/SLT(2005)9. Paris, France: Organisation for Economic Co-operation and Development/International Energy Agency, <http://www.oecd.org/dataoecd/62/40/35798709.pdf>, accessed 03/07/07.

Phillips, G., 2004: GHG assessors - A new career option? Greenhouse Gas Market 2004: Ready for Takeoff, International Emissions Trading Association, 2004.

Pinguelli, R., S. Luiz, and S. Kahn Ribeiro, 2001: The present, past, and future contributions to global warming of CO_2 emissions from fuels. *Climatic Change*, **48**, pp. 289-308.

Pinguelli, R., S. Luiz, K. Ribeiro, M.S. Muylaert, and C. Pires de Campos, 2004: Comments on the Brazilian Proposal and contributions to global temperature increase with different climate responses - CO_2 emissions due to fossil fuels, CO_2 emissions due to land use change. *Energy Policy*, **32**(13), pp. 1499-1510.

Pizer, W., 1999: Choosing price or quantity controls for greenhouse gases. Climate Issues Brief, *Resources for the Future*, **17**.

Pizer, W.A., 2002: Combining price and quantity controls to mitigate global climate change. Journal of Public Economics, 85(3), pp. 409-434.

Pizer, W.A. and K. Tamura, 2004: Climate Policy in the U.S. and Japan: A Workshop Summary, Resources for the Future Discussion Paper. 04-22, March, 2004.

Pizer, W.A., 2005a: Climate policy design under uncertainty. Discussion Paper 05-44, Resources for the Future.

Pizer, W.A., 2005b: The case for intensity targets. *Climate Policy*, **5**(4), pp. 455-462.

Pizer, W., D. Burtraw, W. Harrington, R. Newell, and J. Sanchirico, 2006: Modeling Economy-wide vs. Sectoral Climate Policies. *The Energy Journal*, **27**(3).

Popp, D., 2002: They don't invent them like they used to: an examination of energy patent citation over time. National Bureau of Economic Research, Working Paper No. 11415, Cambridge, Massachusetts.

Popp, D., 2004: R&D subsidies and climate policy: Is there a free lunch? National Bureau of Economic Research, Working Paper No. 10880, Cambridge, Massachusetts.

Porter, M.E. and C. van der Linde, 1995: Toward a new conception of the environment-competitiveness relationship. *The Journal of Economic Perspectives*, **9**(4), pp. 97-118.

Price, L., 2005: Voluntary agreements for energy efficiency or GHG emission reduction in industry: An assessment of programs around the world. Lawrence Berkeley National Laboratory, LBNL-58138, April 2005.

Quirion, P., 2005: Does uncertainty justify intensity emission caps? *Resource & Energy Economics*, **27**(4), November 2005, pp. 343-353.

Rabe, B.G., 2004: Statehouse and greenhouse, the emerging politics of American climate change policy. Washington, Brookings Institution Press, 2004.

Reedman, L., P. Graham, and P. Coombes, 2006: Using a real-options approach to model technology adoption under carbon price uncertainty: An application to the Australian electricity sector. *The Economic Record*, **82**, Special Issue, pp. 864-873.

Regional Greenhouse Gas Initiative (RGGI), 2005: Regional greenhouse gas initiative, Memorandum of Understanding. December, 2005, accessed at <http://www.rggi.org/agreement.htm>, accessed 03/07/07.

Reinhardt, F., 1998: Environmental product differentiation: Implications for corporate strategy. *California Management Review*, **40**(1), pp. 43-73.

Reinhardt, F., 1999: Market failure and the environmental policies of firms, economic rationales for 'beyond compliance' behavior. *Journal of Industrial Ecology*, **3**(1), pp. 9-21.

Reinhardt, F., 2000: What every executive needs to know about global warming. *Harvard Business Review*, July/August, 2000.

Reinhardt, F. and K.O. Packard, 2001: A business manager's approach to climate change. *Climate Change: Science, Strategies, and Solutions*, E. Claussen (ed.), Leiden, Boston, Koln, Brill, 2001.

Reinstein, R., 2004: A possible way forward on climate change. *Mitigation and Adaptation Strategies for Global Change*, **9**, 2004, pp. 295-309.

Revesz, R.L. and R.N. Stavins, 2006: Environmental Law and Policy. Handbook of Law and Economics, A.M. Polinsky and S. Shavell (eds). Amsterdam: Elsevier Science.

Richards, M., 2003: Poverty reduction, equity and climate change: Global governance synergies or contradictions? Overseas Development Institute, Globalisation and Poverty Programme.

Rietbergen, M. and K. Blok, 2000: Voluntary agreements- implementation and efficiency. <http://www.akf.dk/VAIE/>, accessed 03/07/07.

Rietbergen, M.G., J.C.M. Farla, and K. Blok, 2002: Do agreements enhance energy efficiency improvement? Analysing the actual outcome of long-term agreements on industrial energy efficiency improvement. *The Netherlands Journal of Cleaner Production*, **10**, pp. 153-163.

Ringius, L., A. Torvanger, and B. Holtsmark, 1998: Can multi-criteria rules fairly distribute climate burdens? - OECD results from three burden sharing rules. *Energy Policy*, **26**(10), pp. 777-793.

Ringius, L., A. Torvanger, and A. Underdal, 2002: Burden sharing and fairness principles in international climate policy. *International Environmental Agreements: Politics, Law and Economics*, **2**, pp. 1-22.

Rive, N., A. Torvanger and J.S. Fuglestvedt, 2006: Climate agreements based on responsibility for global warming: periodic updating, policy choices, and regional costs. *Global Environmental Change*.

Rivera, J., 2002: Assessing a voluntary environmental initiative in the developing world: The Costa Rican certification for sustainable tourism. *Policy Sciences*, **35**, pp. 333-360.

Rivera, J. and P. deLeon, 2004: Is greener whiter? The Sustainable Slopes Program and the voluntary environmental performance of western ski areas. *Policy Studies Journal*, **32**(3), pp. 417-437.

Rose, A., B. Stevens, J. Edmonds, and M. Wise, 1998: International equity and differentiation in global warming policy. *Environmental & Resource Economics*, **12**(1), pp. 25-51.

Rosen, J., W. Fichtner, and O. Rentz, 2004: Baseline standardization with optimising energy system models. *Mitigation and Adaptation Strategies for Global Change*, **9**, pp.121-146.

Russell, C. and W. Vaughan, 2003: The choice of pollution control policy instruments in developing countries: Arguments, evidence and suggestions. H. Folmer and T. Tietenberg (eds), The International Yearbook of Environmental and Resource Economics 2003/2004, Cheltenham, UK, Edward Elgar.

Sagar, A., 2000: Capacity building for the environment: A view for the South, A view for the North. *Annual Review of Energy and the Environment*, **25**, pp. 377-439.

Sagar, A.D. and B.C.C. van der Zwaan, 2006: Technological innovation in the energy sector: R&D deployment, and learning-by-doing. *Energy Policy*, November 2006.

Saito, K., 2001: Observation on voluntary initiatives as environmental policy. Keizai Kenkyuu Nenpo (Yearbook of Economic Studies), Tokyo University, **26**, pp. 91-108.

Sakakibara, M., 2001: Cooperative research and development: who participates and in which industries do projects take place? *Research Policy*, **30**, pp. 993-1018.

Samaniego, J. and C. Figueres, 2002: Evolving to a sector-based Clean Development Mechanism. Options for protecting the climate. K.A. Baumert, O. Blanchard, S. Llosa, and J.F. Perkaus. Washington, WRI.

San Martin, R., 2003: Marketable emissions permits with imperfect monitoring. *Energy Policy*, **31**, pp. 1369-1378.

Sathaye, J., S. Murtishaw, L. Price, M. Lefranc, J. Roy, H. Winkler, and R. Spalding-Fecher, 2004: Multiproject baselines for evaluation of electric power projects. *Energy Policy*, **32**, pp. 1303-1317.

Sathaye, J., E. Holt, and S. De La Rue du Can, 2005: Overview of IPR practices for publicly-funded technologies. Lawrence Berkeley Laboratory Report # 59072, Berkeley.

Sauer, A., P. Mettler, F. Wellington, and G.G. Hartmann, 2005: Transparency issues with ACEA agreement: Are investors driving blindly? World Resources Institute, Washington, D.C.

Schelling, T.C., 1997: The cost of combating global warming, facing the tradeoffs. Foreign Affairs, November/December.

Schelling, T.C., 2002: What makes greenhouse sense? Foreign Affairs, May/June COM/ENV/EPOC/IEA/SLT(2005)6 32.

Schipper, L., 2006: Conceptual history of adaptation in the UNFCCC Process. Review of European Community & International Environmental Law, **15**(1), pp. 82 - April 2006.

Schleich, J. and R. Betz, 2005: Incentives for energy efficiency and innovation in the European Emission Trading System. Proceedings of the 2005 ECEEE Summer Study - What works and who delivers? Mandelieu, France, pp.1495.

Schmidt, J., N. Helme, J. Lee, M. Houdashelt, and N. Höhne, 2006: Sector-based approach to the post-2012 climate change policy architecture. *Climate Policy*.

Sebenius, J.K., 1993: The Law of the Sea Conference: Lessons for negotiations to control global warming. G. Sjostedt (ed.), International Environment Negotiations, IIASA, pp. 189-216.

Shrestha, R. and G. Timilsina, 2002: The additionality criterion for identifying clean development mechanism projects under the Kyoto Protocol. *Energy Policy*, **30**, pp.73-79.

Sijm, J., J.C. Jansen, and A. Torvanger, 2001: Differentiation of mitigation commitments: the multi-sector convergence approach. *Climate Policy*, **1**(4), pp. 481-497.

Sijm, J., 2004: Induced technological change and spillovers in climate policy modeling. Energy Research Centre, The Netherlands, ECN-C-04-073.

Sijm, J., 2005: The interaction between the EU emissions trading scheme and national energy policies. *Climate Policy*, **5**(1), pp. 79-96.

Sijm, J., K. Neuhoff, and Y. Chen, 2006: CO_2 cost pass through and windfall profits in the power sector. *Climate Policy*, **6**(1), pp. 49-72.

Smith, J. and S.J. Scherr, 2002: Forest carbon and local livelihoods: assessment of opportunities and policy recommendations. CIFOR Occasional Paper No. 37. CIFOR (Center for International Forestry Research), Bogor, Indonesia.

Sohn, J., S. Nakhouda, and K. Baumert, 2005: Mainstreaming climate change at the multilateral development banks. World Resources Institute, Washington, D.C.

Sorrell, S. and J. Sijm, 2003: Carbon trading in the policy mix. *Oxford Review of Economic Policy*, **19**(3), pp. 420-437.

Spalding-Fecher, R., S. Thorne, and N. Wamukonya, 2002: Residential solar water heating as a potential Clean Development Mechanism project: A South African case study. *Mitigation and Adaptation Strategies for Global Change*, **7**, pp. 135-153.

Springer, U. and M. Varilek, 2004: Estimating the price of tradable permits for greenhouse gas emissions in 2008-12. *Energy Policy*, **32**, pp. 611-621.

Stavins, R.N., 2001: Economic analysis of global climate change policy: A primer. *Climate Change: Science, Strategies, and Solutions*. E. Claussen, V.A. Cochran, and D.P. Davis. Boston. Brill 18 Discussion paper 2003-2: draft ver. 1 August 2003 Publishing.

Stavins, 2003: Experience with market-based environmental policy instruments. *Handbook of Environmental Economics*, I, K.-G. Mäler and J. Vincent (eds), Chapter 9, pp. 355-435. Amsterdam: Elsevier Science.

Sterk, W. and B. Wittneben, 2005: Addressing opportunities and challenges of a sectoral approach to the clean development mechanism. JIKO Policy Paper 1/2005, August 2005, Wuppertal Institute, Wuppertal. Available at <http://www.wupperinst.org/jiko>, accessed 03/07/07.

Sterner, T., 2003: Policy instruments for environmental and natural resource management. Washington, D.C., Resources for the Future Press.

Stewart, R.B. and J.B. Wiener, 2001: Reconstructing climate policy: The paths ahead. *Policy Matters*, AEI-Brookings Joint Center for Regulatory Studies, Washington, DC August: 01-23.

Stewart, R. and J. Wiener, 2003: Practical climate change policy. *Issues on Line in Science and Technology*. National Academy of Science, Washington, D.C.

Storey, M., 2002: Kyoto and beyond, issues and options in the global response to climate change. Stockholm, Sweden: Swedish Environmental Protection Agency. <http://www.internat.naturvardsverket.se/documents/issues/climate/report/Kyoto.pdf>, accessed 03/07/07.

Stranland, J., C. Chavez, and B. Field, 2002: Enforcing emissions trading programs: Theory, practice, and performance. *Policy Studies Journal*, **303**(3), pp. 343-361.

Sugiyama, T., J. Sinton, O. Kimura, and T. Ueno, 2003: Orchestra of Treaties. CRIEPI, Tokyo, Japan.

Sugiyama, T. and J. Sinton, 2005: Orchestra of treaties: A future climate regime scenario with multiple treaties among like-minded countries. *International Environmental Agreements: Politics, Law and Economics*, **5**, pp. 65-88.

Sussman, F. and N. Helme, 2004: Harnessing financial flows from export credit agencies for climate protection. Washington D.C., USA, Center for Clean Air Policy.

Sussman, F., N. Helme, and C. Kelly, 2004: *Establishing Greenhouse Gas Emission Caps for Multinational Corporations*. Washington D.C., USA, Center for Clean Air Policy.

Swift, B., 2001: How environmental laws work: An analysis of the utility sector's response to regulation of nitrogen oxides and sulfur dioxide under the clean air act. *Tulane Environmental Law Journal*, **14**, 309 pp.

Tangen, K. and H. Hasselknippe, 2005: Converging markets. *International Environmental Agreements: Politics, Law and Economics*. **5**, pp. 47-64.

Tanikawa, H., 2004: Incentive structure of the voluntary environmental responses of the japanese industrial enterprises. RIETI Discussion Paper 04-J-30, pp. 1-55, 2004, Summary, *RIETI Journal*, pp.1-4.

Thackeray, R.W., 2004: Struggling for air: The Kyoto Protocol, citizen's suits under the Clean Air Act, and the United States Options for Addressing Global Climate Change. Indiana International and Comparative Law Review, 14, 855 pp.

Thalmann, P. and A. Baranzini 2005: An overview of the economics of voluntary approaches in climate policies. Edward Elgar, United Kingdom.

Tietenberg, T., 2000: Environmental and natural resource economics, 4th ed. New York: Harper-Collins.

Tietenberg, T., 2003: The tradable permits approach to protecting the commons: Lessons for Climate Change. Oxford Review of Economic Policy, 19(3), pp. 400-419.

Tietenberg, T., 2006: Emissions trading: Principles and practice, Second Edition. Washington: RFF Press.

Tol, R.S.J., W. Lise, and B.C.C. Van der Zwaan, 2000: Technology diffusion and the stability of climate coalitions. February 2000.

Torvanger, A. and T. Skodvi, 2002: Environmental agreements in climate politics. *Voluntary Environmental Agreements- Process, Practice and Future Use*, P. Brink, Greenleaf Publishing, Sheffield, UK.

Torvanger, A. and G. Odd, 2004: An evaluation of pre-kyoto differentiation proposals for national greenhouse gas abatement targets. *International Environmental Agreements: Politics, Law and Economics*, **4**, pp. 65-91.

Torvanger, A., G. Bang, H.H. Kolshus, and J. Vevatne, 2005: Broadening the climate regime: Design and feasibility of multi-stage climate agreements. Oslo, Norway, Center for International Climate and Environmental Research.

Trudinger, C.M. and I.G. Enting, 2005: Comparison of formalisms for attributing responsibility for climate change: Non-linearities in the Brazilian Proposal. *Climatic Change*, **68**(1-2), pp. 67-99.

U.S. Environmental Protection Agency (US EPA), 2004: International experiences with economic incentives for protecting the environment. National Center for Environmental Economics, EPA-236-R-04-001.

U.S. Government Accountability Office, 2006: Climate Change: EPA and DOE should do more to encourage progress under two voluntary programs. GAO 06 97, April 2006, Washington, D.C. <http://www.gao.gov/new.items/d0697.pdf>, accessed 03/07/07.

U.S. National Science Foundation, 2003: National patterns of research development resources: 2003. <http://www.nsf.gov/statistics/nsf05308/sectd.htm>, accessed 03/07/07.

UNFCCC, 1997: Implementation of the Berlin Mandate, Additional proposal by Parties. Document FCCC/AGBM/1997/MISC.1/Add.3, <http://www.unfccc.int>, accessed 03/07/07.

UNFCCC, 1997b: Paper no.1: Brazil, proposed elements of a protocol to the United Nations Framework Convention on Climate Change, Bonn.

UNFCCC, 2001: Report of the Conference of the Parties on its seventh session, Addendum, Part 2: Action taken by the Conference of the Parties. Vol II, FCCC/CP/2001/13/Add.2, Bonn.

UNFCCC, 2003a: Indicative simplified baseline and monitoring methodologies for selected small-scale CDM project activity categories. Appendix B1 of the simplified modalities and procedures for small-scale CDM project activities, Annex 6, Report of the 7th meeting of the Executive Board, Bonn.

UNFCCC, 2003b: Clarifications on issues relating to baselines and monitoring methodologies. Annex 1, Report of the 8th meeting of the Executive Board, Bonn.

UNFCCC, 2003c: Procedures for submission and consideration of a proposed new methodology. Annex 2, Report of the 8th meeting of the Executive Board, Bonn.

UNFCCC, 2006a: Synthesis report on regional workshops on Article 6 of the Convention. FCCC/2006/SBI/17, Bonn.

Van der Gaast, W., 2002: The scope for joint implementation in the EU candidate countries. *International Environmental Agreements: Politics, Law and Economics*. **2**, pp. 277-292.

Van Vliet, O., A. Faaj, and C. Dieperink, 2003: Forestry projects under the Clean Development Mechanism? *Climatic Change*, **61**, pp. 123-156.

Van Vuuren, D.P., M.G.J. den Elzen, M.M. Berk, P. Lucas, B. Eickhout, H. Eerens, and R. Oostenrijk, 2003: Regional costs and benefits of alternative post-Kyoto climate regimes. Netherlands Environmental Assessment Agency (MNP), Bilthoven, the Netherlands. RIVM-report 728001025, <www.mnp.nl/en>, accessed 02/07/07.

Verhagen, A., G.J. Nabuurs, and J. Veraart, 2004: The role of land use in sustainable development: Options and constraints under climate change. M.T.J. Kok and H.C. de Coninck (ed.) *Beyond Climate: Options for Broadening Climate Policy*, RIVM Report 500036 001, Bilthoven, pp. 53-75.

Verheyen, R., 2003: Climate Change Damage in International Law, Universität Hamburg, Fachbereich Rechtswissenschaften, Dissertation.

Victor, D.G., 1998: The operation and effectiveness of the Montreal Protocol's non-compliance procedure. *The Implementation and effectiveness of International Environmental Commitments: Theory and Practice*. D.G. Victor, K. Raustiala, and E.B. Skolnikoff (eds), Cambridge, MA, MIT Press, Ch.4, pp. 137-176.

Victor, D.G., 2001a: The collapse of the Kyoto Protocol and the struggle to slow global warming. Princeton, University Press.

Victor, D., 2001b: International agreements and the struggle to tame carbon. *Global Climate Change*, pp. 204-229.

Viguier, L., 2004: A proposal to increase developing country participation in international climate policy. *Environmental Science & Policy*, **7**, pp. 195-204.

Vogel, D., Toffel, M., and D. Post, 2005: Environmental federalism in the European Union and the United States. A handbook of globalization and environmental policy: National Government Interventions in a Global Arena. F. Wiken, K. Zoeteman, and J. Pieters (eds), E. Elgar, United Kingdom.

Vöhringer, F. 2004: Forest conservation and the Clean Development Mechanism: Lessons from the Costa Rican protected areas project. *Mitigation and Adaptation Strategies for Global Change*, **9**, pp. 217-240.

Wang, J., J. Yang, C. Ge, D. Cao, and J. Schreifels, 2004: Controlling sulfur dioxide in China: Will emission trading work? *Environment*, June. Washington, D.C., RFF Press.

Watanabe, C., 1999: Systems options for sustainable development-effect and limit of the Ministry of International Trade and Industry's efforts to substitute technology for energy. *Research Policy*, **28**, pp. 719-749.

Wätzold, F., 2004: SO_2 emissions in Germany, Regulations to fight Waldsterben. *Choosing Environmental Policy*, W. Harrington, R.D. Morgenstern, and T. Sterner, eds, Washington, D.C. Resources for the Future Press.

WBGU, 2003: Climate Protection Strategies for the 21st Century. Kyoto and Beyond. German Advisory Council on Global Change, Berlin.

Weisslitz, M., 2002: Rethinking the equitable principle of common but differentiates responsibility: differential versus absolute norms of compliance and contribution in the global climate change context. *Colorado Journal of International Environmental Law and Policy*, **13**, pp. 473.

Welch, E.W., A. Mazur, and S. Bretschneider, 2000: Voluntary behavior by electric utilities: Levels of adoption and contribution of the climate challenge program to the reduction of carbon dioxide. *Journal of Public Policy Analysis and Management*, **19**(3), pp. 407-426.

Werksman, J., 2001: Greenhouse gas emissions trading and the WTO: 153. In *Inter-linkages. The Kyoto Protocol and the International Trade and Investment Regime*. W. Bradnee, Chambers (ed.), United Nations University Press.

Werksman, J., 2005: The negotiations of the Kyoto Compliance System: Towards hard enforcement. In *Implementing Climate Regime: International Compliance*, O.S. Stokke, J. Hovi, and G. Ulfstein, *Earthscan*, London, pp. 17-37.

Wettestad, J., 1996: Acid Lessons? Assessing and explaining LRTAP implementation and effectiveness IIASA. Working Paper, March 1996.

Wettestad, J., 2005: Enhancing climate compliance - What are the lessons to learn from environmental regimes and the EU? In *Implementing Climate Regime: International Compliance*, O.S. Stokke, J. Hovi and G. Ulfstein, *Earthscan*, London, pp. 209-231.

Wicke, L., 2005: Beyond Kyoto - A new global climate certificate system. Heidelberg, Germany, Springer Verlag.

Wiener, J.B., 1999: Global environmental regulation: Instrument choice in legal context. In *The Yale Law Journal*, **108**(4), pp. 677-800.

Willems, S. and K. Baumert, 2003: Institutional capacity and climate change, OECD document COM/ENV/EPOC/IEA/SLT(2003)5, Paris.

Winkler, H., R. Spalding-Fecher, and L. Tyani, 2002a: Comparing developing countries under potential carbon allocation schemes. *Climate Policy*, **2**, pp. 303-318.

Winkler, H., R. Spalding-Fecher, S. Mwakasonda, and O. Davidson, 2002b: Sustainable development policies and measures: Starting from development to tackle climate change. Options for protecting the climate. K.A. Baumert, O. Blanchard, S. Llosa, and J.F. Perkaus. Washington DC, WRI: 61-87.

Working Group on Baseline for CDM/JI Project, 2001: *Technical Procedures for CDM/JI Projects at the Planning Stage, Interim Report to Ministry of the Environment, Government of Japan*, Tokyo, Ministry of the Environment, pp. 70.

World Bank, 2003: Survey of contingent financing and risk mitigation instruments for clean infrastructure projects. Nov 3003, Washington, D.C. <http://carbonfinance.org/Router.cfm?Page=DocLib&ht=23&dl=1>, accessed 02/07/07.

World Bank, 2004a: Striking a better balance - the Final Report of the Extractive Industries Review. The World Bank Group on Extractive Industries Group, Washington D.C.

World Bank, 2006: An investment framework for clean energy and development: A progress report. World Bank Group Development Committee, September 5, 2006 (DC2006-0012).

World Bank and International Emission Trading Association, 2006: State and trends of the Carbon Market 2006. Washington, D.C.

World Resources Institute/World Business Council for Sustainable Development (WRI/WBCSD), 2004: The Greenhouse Gas Protocol, a corporate accounting and reporting standard. Revised Edition, ISBN 1-56973-568-9.

Yamaguchi, M., 2003: Environmental effectiveness of voluntary agreement to cope with climate change: An evaluation methodology, *Mita Journal of Economics*, **96**(2), pp. 19-47.

Yamaguchi, M., 2004: Implementing the Kyoto Protocol commitment and their impacts on trade: Focusing on Japanese automobile fuel efficiency standards. *Keio Economic Studies*, **41**(1).

Yamaguchi, M. and T. Sekine, 2006: A proposal for the Post-Kyoto framework. *Keio Economic Studies*, **43**, pp. 85-112.

Yamin, F. and J. Depledge, 2004: The International Climate Change Regime: A guide to rules, institutions and procedures. Cambridge University Press.

Yohe, G., N. Andronova and M. Schlesinger, 2004: To hedge or not against an uncertain climate future? *Science*, **306**, pp. 416-417.

Zarsky, L. and K. Gallagher, 2003: Searching for the Holy Grail? Making FDI Work for Sustainable Development. Analytical Paper, World Wildlife Fund (WWF), Switzerland.

Zetterberg, L., K. Nilsson, M. Ahman, A.S. Kumlin, and L. Birgirsdotter, 2004: Analysis of national allocation plans of the EU ETS. Research Report, IVL, B-1591.

Zhang, C., T. Heller, M. May, 2001: Impact on global warming of development and structural changes in the electricity sector of Guangdong Province, China. In *Energy Policy*, **29**, pp. 179-203.

Annex I

Glossary

Editor: Aviel Verbruggen (Belgium)

Notes: Glossary entries (highlighted in **bold**) are by preference subjects; a main entry can contain **subentries**, also in bold, e.g. **Final Energy** is defined under the entry **Energy**. Some definitions are adapted from Cleveland C.J. and C. Morris, 2006: Dictionary of Energy, Elsevier, Amsterdam. The Glossary is followed by a list of Acronyms/Abbreviations and by a list of Chemical Compounds (Annex II).

Activities Implemented Jointly (AIJ)

The pilot phase for Joint Implementation, as defined in Article 4.2(a) of the UNFCCC, which allows for project activity among developed countries (and their companies) and between developed and developing countries (and their companies). AIJ is intended to allow parties to the UNFCCC to gain experience in jointly implemented projects. AIJ under the pilot phase do not lead to any credits. Decisions remain about the future of AIJ projects and how they may relate to the Kyoto Mechanisms. As a simple form of tradable permits, AIJ and other market-based schemes represent potential mechanisms for stimulating additional resource flows for reducing emissions. See also Clean Development Mechanism, and Emissions Trading.

Actual net greenhouse gas removals by sinks

The sum of the verifiable changes in carbon stocks in the carbon pools within the project boundary of an afforestation or reforestation project, minus the increase in GHG emissions as a result of the implementation of the project activity. The term stems from the Clean Development Mechanism (CDM) afforestation and reforestation modalities and procedures.

Adaptation

Initiatives and measures to reduce the vulnerability of natural and human systems against actual or expected climate change effects. Various types of adaptation exist, e.g. anticipatory and reactive, private and public, and autonomous and planned. Examples are raising river or coastal dikes, the substitution of more temperature-shock resistant plants for sensitive ones, etc.

Adaptive capacity

The whole of capabilities, resources and institutions of a country or region to implement effective adaptation measures.

Additionality

Reduction in emissions by sources or enhancement of removals by sinks that is additional to any that would occur in the absence of a Joint Implementation (JI) or a Clean Development Mechanism (CDM) project activity as defined in the Kyoto Protocol Articles on JI and CDM. This definition may be further broadened to include financial, investment, technology, and environmental additionality. Under **financial additionality**, the project activity funding is additional to existing Global Environmental Facility, other financial commitments of parties included in Annex I, Official Development Assistance, and other systems of cooperation. Under **investment additionality**, the value of the Emissions Reduction Unit/Certified Emission Reduction Unit shall significantly improve the financial or commercial viability of the project activity. Under **technology additionality**, the technology used for the project activity shall be the best available for the circumstances of the host party. **Environmental additionality** refers to the environmental integrity of the claimed amount by which greenhouse gas emissions are reduced due to a project relative to its baseline. A project activity is further additional, if the incentive from the sale of emission allowances helps to overcome barriers to its implementation.

Aerosols

A collection of airborne solid or liquid particles, typically between 0.01 and 10 μm in size and residing in the atmosphere for at least several hours. Aerosols may be of either natural or anthropogenic origin. Aerosols may influence climate in several ways: directly through scattering and absorbing radiation, and indirectly through acting as condensation nuclei for cloud formation or modifying the optical properties and lifetime of clouds.

Afforestation

Direct human-induced conversion of land that has not been forested for a period of at least 50 years to forested land through planting, seeding and/or the human-induced promotion of natural seed sources.[1] See also Re- and Deforestation.

Agreement

In this Report, the degree of agreement is the relative level of convergence of the literature as assessed by the authors.

Alliance of Small Island States (AOSIS)

Formed at the Second World Climate Conference (1990). AOSIS comprises small-island and low-lying coastal developing countries that are particularly vulnerable to the adverse consequences of climate change, such as sea-level rise, coral bleaching, and the increased frequency and intensity of tropical storms. With more than 35 states from the Atlantic, Caribbean, Indian Ocean, Mediterranean, and Pacific, AOSIS share common objectives on environmental and sustainable development matters in the UNFCCC process.

Ancillary benefits

Policies aimed at some target, e.g. climate change mitigation, may be paired with positive side effects, such as increased resource-use efficiency, reduced emissions of air pollutants associated with fossil fuel use, improved transportation, agriculture, land-use practices, employment, and fuel security. **Ancillary impacts** is also used when the effects may be negative. Policies directed at abating air pollution may consider greenhouse-gas mitigation an ancillary benefit, but this perspective is not considered in this assessment. See also **co-benefits**.

Annex I countries

The group of countries included in Annex I (as amended in 1998) to the UNFCCC, including all the OECD countries and economies in transition. Under Articles 4.2 (a) and 4.2 (b) of the Convention, Annex I countries committed themselves specifically to the aim of returning individually or jointly to their 1990 levels of greenhouse-gas emissions by the year 2000. By default, the other countries are referred to as Non-Annex I countries.

---

1 For a discussion of the term *forest* and related terms such as *afforestation*, *reforestation*, and *deforestation* (ARD), see the IPCC Special Report on Land Use, Land-Use Change and Forestry, Cambridge University Press, 2000.

Annex II countries

The group of countries included in Annex II to the UNFCCC, including all OECD countries. Under Article 4.2 (g) of the Convention, these countries are expected to provide financial resources to assist developing countries to comply with their obligations, such as preparing national reports. Annex II countries are also expected to promote the transfer of environmentally sound technologies to developing countries.

Annex B countries

The countries included in Annex B to the Kyoto Protocol that have agreed to a target for their greenhouse-gas emissions, including all the Annex I countries (as amended in 1998) except for Turkey and Belarus.

Anthropogenic emissions

Emissions of greenhouse gases, greenhouse-gas precursors, and aerosols associated with human activities. These include the burning of fossil fuels, deforestation, land-use changes, livestock, fertilization, etc. that result in a net increase in emissions.

Assigned Amount (AA)

Under the Kyoto Protocol, the assigned amount is the quantity of greenhouse-gas emissions that an Annex B country has agreed to as its ceiling for its emissions in the first commitment period (2008 to 2012). The AA is the country's total greenhouse-gas emissions in 1990 multiplied by five (for the five-year commitment period) and by the percentage it agreed to as listed in Annex B of the Kyoto Protocol (e.g. 92% for the EU; 93% for the USA).

Assigned Amount Unit (AAU)

An AAU equals 1 tonne (metric ton) of CO_2-equivalent emissions calculated using the Global Warming Potential.

Backstop technology

Models estimating mitigation often characterize an arbitrary carbon-free technology (often for power generation) that becomes available in the future in unlimited supply over the horizon of the model. This allows models to explore the consequences and importance of a generic solution technology without becoming enmeshed in picking the technology. This "backstop" technology might be a nuclear technology, fossil technology with capture and sequestration, solar, or something as yet unimagined. The backstop technology is typically assumed either not to currently exist, or to exist only at higher costs relative to conventional alternatives.

Banking

According to the Kyoto Protocol [Article 3 (13)], parties included in Annex I to the UNFCCC may save excess AAUs from the first commitment period for compliance with their respective cap in subsequent commitment periods (post-2012).

Barrier

Any obstacle to reaching a goal, adaptation or mitigation potential that can be overcome or attenuated by a policy, programme, or measure. **Barrier removal** includes correcting market failures directly or reducing the transactions costs in the public and private sectors by e.g. improving institutional capacity, reducing risk and uncertainty, facilitating market transactions, and enforcing regulatory policies.

Baseline

The reference for measurable quantities from which an alternative outcome can be measured, e.g. a non-intervention scenario is used as a reference in the analysis of intervention scenarios.

Benchmark

A measurable variable used as a baseline or reference in evaluating the performance of an organization. Benchmarks may he drawn from internal experience, that of other organizations or from legal requirement and are often used to gauge changes in performance over time.

Benefit transfer

An application of monetary values from one particular analysis to another policy-decision setting, often in a geographic area other than the one in which the original study was performed.

Biochemical Oxygen Demand (BOD)

The amount of dissolved oxygen consumed by micro-organisms (bacteria) in the bio-chemical oxidation of organic and inorganic matter in waste water.

Biocovers

Layers placed on top of landfills that are biologically active in oxidizing methane into CO_2.

Biofilters

Filters using biological material to filter or chemically process pollutants like oxidizing methane into CO_2.

Biodiversity

The variability among living organisms from all sources including, inter alia, terrestrial, marine and other aquatic ecosystems and the ecological complexes of which they are part; this includes diversity within species, between species and of ecosystems.

Bioenergy

Energy derived from biomass.

Biofuel

Any liquid, gaseous, or solid fuel produced from plant or animal organic matter. E.g. soybean oil, alcohol from fermented sugar, black liquor from the paper manufacturing process, wood as fuel, etc. **Second-generation biofuels** are products such as ethanol and biodiesel derived from ligno-cellulosic biomass by chemical or biological processes.

Biological options

Biological options for mitigation of climate change involve one or more of the three strategies: conservation - conserving an existing carbon pool, thereby preventing CO_2 emissions to the atmosphere; sequestration - increasing the size of existing carbon pools, thereby extracting CO_2 from the atmosphere; substitution - substituting biomass for fossil fuels or energy-intensive products, thereby reducing CO_2 emissions.

Biomass

The total mass of living organisms in a given area or of a given species usually expressed as dry weight. Organic matter consisting of, or recently derived from, living organisms (especially regarded as fuel) excluding peat. Biomass includes products, by-products and waste derived from such material. **Cellulosic biomass** is biomass from cellulose, the primary structural component of plants and trees

Black Carbon

Particle matter in the atmosphere that consists of soot, charcoal and/or possible light-absorbing refractory organic material. Black carbon is operationally defined matter based on measurement of light absorption and chemical reactivity and/or thermal stability.

Bottom-up models

Models represent reality by aggregating characteristics of specific activities and processes, considering technological, engineering and cost details. See also **top-down models**.

Bubble

Policy instrument for pollution abatement named for its treatment of multiple emission points as if they were contained in an imaginary bubble. Article 4 of the Kyoto Protocol allows a group of countries to meet their target listed in Annex B jointly by aggregating their total emissions under one 'bubble' and sharing the burden (e.g. the EU).

Carbon Capture and Storage (CCS)

A process consisting of separation of CO_2 from industrial and energy-related sources, transport to a storage location, and long-term isolation from the atmosphere.

Carbon cycle

The set of processes such as photosynthesis, respiration, decomposition, and air-sea exchange, by which carbon continuously cycles through various reservoirs, such as the atmosphere, living organisms, soils, and oceans.

Carbon dioxide (CO_2)

CO_2 is a naturally occurring gas, and a by-product of burning fossil fuels or biomass, of land-use changes and of industrial processes. It is the principal anthropogenic greenhouse gas that affects Earth's radiative balance. It is the reference gas against which other greenhouse gases are measured and therefore it has a Global Warming Potential of 1.

Carbon dioxide fertilization

The enhancement of the growth of plants because of increased atmospheric CO_2 concentration. Depending on their mechanism of photosynthesis, certain types of plants are more sensitive to changes in atmospheric CO_2 concentration than others.

Carbon intensity

The amount of emissions of CO_2 per unit of GDP.

Carbon leakage

The part of emissions reductions in Annex B countries that may be offset by an increase of the emissions in the non-constrained countries above their baseline levels. This can occur through (1) relocation of energy-intensive production in non-constrained regions; (2) increased consumption of fossil fuels in these regions through decline in the international price of oil and gas triggered by lower demand for these energies; and (3) changes in incomes (thus in energy demand) because of better terms of trade. Leakage also refers to GHG-related effects of GHG-emission reduction or CO_2-sequestration project activities that occur outside the project boundaries and that are measurable and attributable to the activity. On most occasions, leakage is understood as counteracting the initial activity. Nevertheless, there may be situations where effects attributable to the activity outside the project area lead to GHG-emission reductions. These are commonly called spill-over. While (negative) leakage leads to a discount of emission reductions as verified, positive spill-over may not in all cases be accounted for.

Carbon pool

Carbon pools are: above-ground biomass, belowground biomass, litter, dead wood and soil organic carbon. CDM project participants may choose not to account one or more carbon pools if they provide transparent and verifiable information showing that the choice will not increase the expected net anthropogenic GHG removals by sinks.

Carbon price

What has to be paid (to some public authority as a tax rate, or on some emission permit exchange) for the emission of 1 tonne of CO_2 into the atmosphere. In the models and this Report, the carbon price is the social cost of avoiding an additional unit of CO_2 equivalent emission. In some models it is represented by the shadow price of an additional unit of CO_2 emitted, in others by the rate of carbon tax, or the price of emission-permit allowances. It has also been used in this Report as a cut-off rate for marginal abatement costs in the assessment of economic mitigation potentials.

Cap

Mandated restraint as an upper limit on emissions. The Kyoto Protocol mandates emissions caps in a scheduled timeframe on the anthropogenic GHG emissions released by Annex B countries. By 2008-2012 the EU e.g. must reduce its CO_2-equivalent emissions of six greenhouse gases to a level 8% lower than the 1990-level.

Capacity building

In the context of climate change, capacity building is developing technical skills and institutional capabilities in developing countries and economies in transition to enable their participation in all aspects of adaptation to, mitigation of, and research on climate change, and in the implementation of the Kyoto Mechanisms, etc.

CCS-ready

If rapid deployment of CCS is desired, new power plants could be designed and located to be 'CCS-ready' by reserving space for the capture installation, designing the unit for optimal performance when capture is added and siting the plant to enable access to storage reservoirs.

Certified Emission Reduction Unit (CER)

Equal to one metric tonne of CO_2-equivalent emissions reduced or sequestered through a Clean Development Mechanism project, calculated using Global Warming Potentials. In order to reflect potential non-permanence of afforestation and reforestation project activities, the use of temporary certificates for Net Anthropogenic Greenhouse Gas Removal was decided by COP 9. See also **Emissions Reduction Units.**

Chemical oxygen demand (COD)

The quantity of oxygen required for the complete oxidation of organic chemical compounds in water; used as a measure of the level of organic pollutants in natural and waste waters.

Chlorofluorocarbons (CFCs)

Greenhouse gases covered under the 1987 Montreal Protocol and used for refrigeration, air conditioning, packaging, insulation, solvents, or aerosol propellants. Because they are not destroyed in the lower atmosphere, CFCs drift into the upper atmosphere where, given suitable conditions, they break down ozone. These gases are being replaced by other compounds, including hydrochlorofluorocarbons and hydrofluorocarbons, which are greenhouse gases covered under the Kyoto Protocol.

Clean Development Mechanism (CDM)

Defined in Article 12 of the Kyoto Protocol, the CDM is intended to meet two objectives: (1) to assist parties not included in Annex I in achieving sustainable development and in contributing to the ultimate objective of the convention; and (2) to assist parties included in Annex I in achieving compliance with their quantified emission limitation and reduction commitments. Certified Emission Reduction Units from CDM projects undertaken in Non-Annex I countries that limit or reduce GHG emissions, when certified by operational entities designated by Conference of the Parties/ Meeting of the Parties, can be accrued to the investor (government or industry) from parties in Annex B. A share of the proceeds from certified project activities is used to cover administrative expenses as well as to assist developing country parties that are particularly vulnerable to the adverse effects of climate change to meet the costs of adaptation.

Climate Change (CC)

Climate change refers to a change in the state of the climate that can be identified (e.g. using statistical tests) by changes in the mean and/or the variability of its properties, and that persists for an extended period, typically decades or longer. Climate change may be due to natural internal processes or external forcings, or to persistent anthropogenic changes in the composition of the atmosphere or in land use.

Note that UNFCCC, in its Article 1, defines "climate change" as "a change of climate which is attributed directly or indirectly to human activity that alters the composition of the global atmosphere and which is in addition to natural climate variability observed over comparable time periods". The UNFCCC thus makes a distinction between "climate change" attributable to human activities altering the atmospheric composition, and "climate variability" attributable to natural causes.

Climate feedback

An interaction mechanism between processes in the climate system is a climate feedback when the result of an initial process triggers changes in secondary processes that in turn influence the initial one. A positive feedback intensifies the initial process; a negative feedback reduces the initial process. Example of a **positive climate feedback**: higher temperatures as initial process cause melting of the arctic ice leading to less reflection of solar radiation, what leads to higher temperatures. Example of a **negative feedback**: higher temperatures increase the amount of cloud cover (thickness or extent) that could reduce incoming solar radiation and so limit the increase in temperature.

Climate sensitivity

In IPCC Reports, equilibrium climate sensitivity refers to the equilibrium change in annual mean global surface temperature following a doubling of the atmospheric CO_2-equivalent concentration. The evaluation of the equilibrium climate sensitivity is expensive and often hampered by computational constraints.

The **effective climate sensitivity** is a related measure that circumvents the computational problem by avoiding the requirement of equilibrium. It is evaluated from model output for evolving non-equilibrium conditions. It is a measure of the strengths of the feedbacks at a particular time and may vary with forcing history and climate state. The climate sensitivity parameter refers to the equilibrium change in the annual mean global surface temperature following a unit change in radiative forcing ($K/W/m^2$)

The **transient climate response** is the change in the global surface temperature, averaged over a 20-year period, centred at the time of CO_2 doubling, i.e., at year 70 in a 1% per year compound CO_2 increase experiment with a global coupled climate model. It is a measure of the strength and rapidity of the surface temperature response to greenhouse gas forcing.

Climate threshold

The point at which the atmospheric concentration of greenhouse gases triggers a significant climatic or environmental event, which is considered unalterable, such as widespread bleaching of corals or a collapse of oceanic circulation systems.

CO_2-equivalent concentration

The concentration of carbon dioxide that would cause the same amount of radiative forcing as a given mixture of carbon dioxide and other greenhouse gases.

CO_2-equivalent emission

The amount of CO_2 emission that would cause the same radiative forcing as an emitted amount of a well mixed greenhouse gas, or a mixture of well mixed greenhouse gases, all multiplied with their respective Global Warming Potentials to take into account the differing times they remain in the atmosphere.

Co-benefits

The benefits of policies implemented for various reasons at the same time, acknowledging that most policies designed to address greenhouse gas mitigation have other, often at least equally important, rationales (e.g., related to objectives of development, sustainability, and equity). The term co-impact is also used in a more generic sense to cover both positive and negative side of the benefits. See also **ancillary benefits**.

Co-generation

The use of waste heat from thermal electricity-generation plants. The heat is e.g. condensing heat from steam turbines or hot flue gases exhausted from gas turbines, for industrial use, buildings or district heating. Synonym for **Combined Heat and Power (CHP)** generation.

Combined-cycle Gas Turbine (CCGT)

Power plant that combines two processes for generating electricity. First, gas or light fuel oil feeds a gas turbine that inevitably exhausts hot flue gases (>800°C). Second, heat recovered from these gases, with additional firing, is the source for producing steam that drives a steam turbine. The turbines rotate separate alternators.

It becomes an **integrated CCGT** when the fuel is syngas from a coal or biomass gasification reactor with exchange of energy flows between the gasification and CCGT plants.

Compliance

Compliance is whether and to what extent countries do adhere to the provisions of an accord. Compliance depends on implementing policies ordered, and on whether measures follow up the policies. Compliance is the degree to which the actors whose behaviour is targeted by the agreement, local government units, corporations, organizations or individuals, conform to the implementing obligations. See also **implementation**.

Conference of the Parties (COP)

The supreme body of the UNFCCC, comprising countries with right to vote that have ratified or acceded to the convention. The first session of the Conference of the Parties (COP-1) was held in Berlin (1995), followed by 2.Geneva (1996), 3.Kyoto (1997), 4.Buenos Aires (1998), 5.Bonn (1999), 6.The Hague/Bonn (2000, 2001), 7.Marrakech (2001), 8.Delhi (2002), 9.Milan (2003), 10.Buenos Aires (2004), 11.Montreal (2005), 12.Nairobi (2006). See also **Meeting of the Parties (MOP).**

Contingent Valuation Method (CVM)

CVM is an approach to quantitatively assess values assigned by people in monetary (willingness to pay) and non monetary (willingness to contribute with time, resources etc.) terms. It is a direct method to estimate economic values for ecosystem and environmental services. A survey of people are asked their willingness to pay for access to, or their willingness to accept compensation for removal of, a specific environmental service, based on a hypothetical scenario and description of the environmental service. See also **values**.

Cost

The consumption of resources such as labor time, capital, materials, fuels and so on as the consequence of an action. In economics all resources are valued at their **opportunity cost**, being the value of the most valuable alternative use of the resources. Costs are defined in a variety of ways and under a variety of assumptions that affect their value

Cost types include: **administrative costs** of planning, management, monitoring, audits, accounting, reporting, clerical activities, etc. associated with a project or programme; **damage costs** to ecosystems, economies and people due to negative effects from climate change; **implementation costs** of changing existing rules and regulation, capacity building efforts, information, training and education, etc. to put a policy into place; **private costs** are carried by individuals, companies or other private entities that undertake the action, where **social costs** include additionally the external costs on the environment and on society as a whole.

Costs can be expressed as **total**, **average** (**unit, specific**) being the total divided by the number of units of the item for which the cost is being assessed, and **marginal** or **incremental** costs as the cost of the last additional unit.

The perspectives adopted in this report are: **Project level** considers a "standalone" activity that is assumed not to have significant indirect economic impacts on markets and prices (both demand and supply) beyond the activity itself. The activity can be the implementation of specific technical facilities, infrastructure, demand-side regulations, information efforts, technical standards, etc. **Technology level** considers a specific greenhouse-gas mitigation technology, usually with several applications in different projects and sectors. The literature on technologies covers their technical characteristics, especially evidence on learning curves as the technology diffuses and matures. **Sector level** considers sector policies in a "partial-equilibrium" context, for which other sectors and the macroeconomic variables are assumed to be as given. The policies can include economic instruments related to prices, taxes, trade, and financing, specific large-scale investment projects, and demand-side regulation efforts. **Macroeconomic level** considers the impacts of policies on real income and output, employment and economic welfare across all sectors and markets. The policies include all sorts of economic policies, such as taxes, subsidies, monetary policies, specific investment programmes, and technology and innovation policies.

The negative of costs are benefits, and often both are considered together.

Cost-benefit analysis

Monetary measurement of all negative and positive impacts associated with a given action. Costs and benefits are compared in terms of their difference and/or ratio as an indicator of how a given investment or other policy effort pays off seen from the society's point of view.

Cost-effectiveness analysis

A special case of cost-benefit analysis in which all the costs of a portfolio of projects are assessed in relation to a fixed policy goal. The policy goal in this case represents the benefits of the projects and all the other impacts are measured as costs or as negative costs (co-benefits). The policy goal can be, for example, a specified goal of emissions reductions of greenhouse gases.

Crediting period

The CDM crediting period is the time during which a project activity is able to generate GHG-emission reduction or CO_2 removal certificates. Under certain conditions, the crediting period can be renewed up to two times.

Deforestation

The natural or anthropogenic process that converts forest land to non-forest. See **afforestation** and **reforestation.**

Demand-side management (DSM)

Policies and programmes for influencing the demand for goods and/ or services. In the energy sector, DSM aims at reducing the demand for electricity and energy sources. DSM helps to reduce greenhouse gas emissions.

Dematerialization

The process by which economic activity is decoupled from matter–energy throughput, through processes such as eco-efficient production or industrial ecology, allowing environmental impact to fall per unit of economic activity.

Deposit-refund system

A deposit or fee (tax) is paid when acquiring a commodity and a refund or rebate is received for implementation of a specified action (mostly delivering the commodity at a particular place).

Desertification

This refers to land degradation in arid, semi-arid, and dry sub-humid areas resulting from various factors, including climatic variations and human activities. The United Nations Convention to Combat Desertification defines land degradation as a reduction or loss, in arid, semi-arid, and dry sub-humid areas, of the biological or economic productivity and complexity of rain-fed cropland, irrigated cropland, or range, pasture, forest and woodlands resulting from land uses or from a process or combination of processes, including processes arising from human activities and habitation patterns, such as soil erosion caused by wind and/or water, deterioration of the physical, chemical and biological or economic properties of soil and long-term loss of natural vegetation.

Devegetation

This is loss of vegetation density within one land-cover class.

Development path

An evolution based on an array of technological, economic, social, institutional, cultural and biophysical characteristics that determine the interactions between human and natural systems, including production and consumption patterns in all countries, over time at a particular scale. **Alternative development paths** refer to different possible trajectories of development, the continuation of current trends being just one of the many paths.

Discounting

A mathematical operation making monetary (or other) amounts received or expended at different points in time (years) comparable across time. The operator uses a fixed or possibly time-varying discount rate (>0) from year to year that makes future value worth less today. In a **descriptive discounting approach** one accepts the discount rates people (savers and investors) actually apply in their day-to-day decisions (**private discount rate**). In a **prescriptive (ethical or normative) discounting approach** the discount rate is fixed from a social perspective, e.g. based on an ethical judgement about the interests of future generations (**social discount rate**).

District heating

Hot water (steam in old systems) is distributed from central stations to buildings and industries in a densely occupied area (a district, a city or an industrialized area such as the Ruhr or Saar in Germany). The insulated two-pipe network functions like a water-based central heating system in a building. The central heat sources can be waste-heat recovery at industrial processes, waste-incineration plants, cogeneration power plants or stand-alone boilers burning fossil fuels or biomass.

Double dividend

The extent to which revenue-generating instruments, such as carbon taxes or auctioned (tradable) carbon emission permits can (1) limit or reduce GHG emissions and (2) offset at least part of the potential welfare losses of climate policies through recycling the revenue in the economy to reduce other taxes likely to cause distortions. In a world with involuntary unemployment, the climate change policy adopted may have an effect (a positive or negative 'third dividend') on employment. **Weak double dividend** occurs as long as there is a revenue-recycling effect. That is, revenues are recycled through reductions in the marginal rates of distorting taxes. **Strong double dividend** requires that the (beneficial) revenue-recycling effect more than offsets the combination of the primary cost and in this case, the net cost of abatement is negative. See also **interaction effect**.

Economies in Transition (EITs)

Countries with their economies changing from a planned economic system to a market economy.

Economies of scale (scale economies)

The unit cost of an activity declines when the activity is extended (e.g., more units are produced).

Ecosystem

A system of living organisms interacting with each other and their physical environment. The boundaries of what could be called an ecosystem are somewhat arbitrary, depending on the focus of interest or study. Thus, the extent of an ecosystem may range from very small spatial scales to the entire planet Earth ultimately.

Emissions Direct / Indirect

Direct emissions or "point of emission" are defined at the point in the energy chain where they are released and are attributed to that point in the energy chain, whether a sector, a technology or an activity. E.g. emissions from coal-fired power plants are considered direct emissions from the energy supply sector. **Indirect emissions** or emissions "allocated to the end-use sector" refer to the energy use in end-use sectors and account for the emissions associated with the upstream production of the end-use energy. E.g. some emissions associated with electricity generation can be attributed to the buildings sector corresponding to the building sector's use of electricity.

Emission factor

An emission factor is the rate of emission per unit of activity, output or input. E.g. a particular fossil fuel power plant has a CO_2 emission factor of 0.765 kg/kWh generated.

Emission permit

An emission permit is a non-transferable or tradable entitlement allocated by a government to a legal entity (company or other emitter) to emit a specified amount of a substance. A **tradable permit** is an economic policy instrument under which rights to discharge pollution - in this case an amount of greenhouse gas emissions - can be exchanged through either a free or a controlled permit-market.

Emission quota

The portion of total allowable emissions assigned to a country or group of countries within a framework of maximum total emissions.

Emissions Reduction Unit (ERU)

Equal to one metric tonne of CO_2-equivalent emissions reduced or sequestered arising from a Joint Implementation (defined in Article 6 of the Kyoto Protocol) project. See also **Certified Emission Reduction Unit** and **emissions trading**.

Emission standard

A level of emission that by law or by voluntary agreement may not be exceeded. Many standards use emission factors in their prescription and therefore do not impose absolute limits on the emissions.

Emissions trading

A market-based approach to achieving environmental objectives. It allows those reducing GHG emissions below their emission cap to use or trade the excess reductions to offset emissions at another source inside or outside the country. In general, trading can occur at the intra-company, domestic, and international levels. The Second Assessment Report by the IPCC adopted the convention of using permits for domestic trading systems and quotas for international trading systems. Emissions trading under Article 17 of the Kyoto Protocol is a tradable quota system based on the assigned amounts calculated from the emission reduction and limitation commitments listed in Annex B of the Protocol.

Emission trajectories

These are projections of future emission pathways, or observed emission patterns.

Energy

The amount of work or heat delivered. Energy is classified in a variety of types and becomes useful to human ends when it flows from one place to another or is converted from one type into another. **Primary energy** (also referred to as energy sources) is the energy embodied in natural resources (e.g., coal, crude oil, natural gas, uranium) that has not undergone any anthropogenic conversion. It is transformed into **secondary energy** by cleaning (natural gas), refining (oil in oil products) or by conversion into electricity or heat. When the secondary energy is delivered at the end-use facilities it is called **final energy** (e.g., electricity at the wall outlet), where it becomes **usable energy** (e.g., light). Daily, the sun supplies large quantities of energy as rainfall, winds, radiation, etc. Some share is stored in biomass or rivers that can be harvested by men. Some share is directly usable such as daylight, ventilation or ambient heat. **Renewable energy** is obtained from the continuing or repetitive currents of energy occurring in the natural environment and includes non-carbon technologies such as solar energy, hydropower, wind, tide and waves and geothermal heat, as well as carbon-neutral technologies such as biomass. **Embodied energy** is the energy used to produce a material substance (such as processed metals or building materials), taking into account energy used at the manufacturing facility (zero order), energy used in producing the materials that are used in the manufacturing facility (first order), and so on.

Energy efficiency

The ratio of useful energy output of a system, conversion process or activity to its energy input.

Energy intensity

The ratio of energy use to economic output. At the national level, energy intensity is the ratio of total domestic primary energy use or final energy use to Gross Domestic Product. See also **specific energy use**

Energy security

The various security measures that a given nation, or the global community as a whole, must carry out to maintain an adequate energy supply.

Energy Service Company (ESCO)

A company that offers energy services to end-users, guarantees the energy savings to be achieved tying them directly to its remuneration, as well as finances or assists in acquiring financing for the operation of the energy system, and retains an on-going role in monitoring the savings over the financing term.

Environmental effectiveness

The extent to which a measure, policy or instrument produces a decided, decisive or desired environmental effect.

Environmentally sustainable technologies

Technologies that are less polluting, use resources in a more sustainable manner, recycle more of their wastes and products, and handle residual wastes in a more acceptable manner than the technologies that they substitute. They are also more compatible with nationally determined socio-economic, cultural and environmental priorities.

Evidence

Information or signs indicating whether a belief or proposition is true or valid. In this Report, the degree of evidence reflects the amount of scientific/technical information on which the Lead Authors are basing their findings.

Externality / External cost / External benefit

Externalities arise from a human activity, when agents responsible for the activity do not take full account of the activity's impact on others' production and consumption possibilities, while there exists no compensation for such impact. When the impact is negative, so are external costs. When positive they are referred to as external benefits.

Feed-in tariff

The price per unit of electricity that a utility or power supplier has to pay for distributed or renewable electricity fed into the grid by non-utility generators. A public authority regulates the tariff.

Flaring

Open air burning of waste gases and volatile liquids, through a chimney, at oil wells or rigs, in refineries or chemical plants and at landfills.

Forecast

Projected outcome from established physical, technological, economic, social, behavioral, etc. patterns.

Forest

Defined under the Kyoto Protocol as a minimum area of land of 0.05-1.0 ha with tree-crown cover (or equivalent stocking level) of more than 10-30 % with trees with the potential to reach a minimum height of 2-5 m at maturity in situ. A forest may consist either of closed forest formations where trees of various storey and undergrowth cover a high proportion of the ground or of open forest. Young natural stands and all plantations that have yet to reach a crown density of 10-30 % or tree height of 2-5 m are included under forest, as are areas normally forming part of the forest area that are temporarily un-stocked as a result of human intervention such as harvesting or natural causes but which are expected to revert to forest. See also **Afforestation, Deforestation** and **Reforestation**.

Fossil fuels

Carbon-based fuels from fossil hydrocarbon deposits, including coal, peat, oil and natural gas.

Free Rider

One who benefits from a common good without contributing to its creation or preservation.

Fuel cell

A fuel cell generates electricity in a direct and continuous way from the controlled electrochemical reaction of hydrogen or another fuel and oxygen. With hydrogen as fuel it emits only water and heat (no CO_2) and the heat can be utilized (see **cogeneration**).

Fuel switching

In general, this is substituting fuel A for fuel B. In the climate-change discussion it is implicit that fuel A has lower carbon content than fuel B, e.g., natural gas for coal.

Full-cost pricing

Setting the final prices of goods and services to include both the private costs of inputs and the external costs created by their production and use.

G77/China. See Group of 77 and China.

General circulation (climate) model (GCM)

A numerical representation of the climate system based on the physical, chemical and biological properties of its components, their interactions and feedback processes, and accounting for all or some of its known properties. The climate system can be represented by models of varying complexity, i.e. for any one component or combination of components a hierarchy of models can be identified, differing in such aspects as the number of spatial dimensions, the extent to which physical, chemical or biological processes are explicitly represented, or the level at which the parameters are assessed empirically. Coupled atmosphere/ocean/sea-ice General Circulation Models provide a comprehensive representation of the climate system. There is an evolution towards more complex models with active chemistry and biology.

General equilibrium analysis

General equilibrium analysis considers simultaneously all the markets and feedback effects among these markets in an economy leading to market clearance. See also **market equilibrium**.

Geo-engineering

Technological efforts to stabilize the climate system by direct intervention in the energy balance of the Earth for reducing global warming

Global Environmental Facility (GEF)

The Global Environment Facility (GEF), established in 1991, helps developing countries fund projects and programmes that protect the global environment. GEF grants support projects related to biodiversity, climate change, international waters, land degradation, the ozone layer, and persistent organic pollutants

Global warming

Global warming refers to the gradual increase, observed or projected, in global surface temperature, as one of the consequences of radiative forcing caused by anthropogenic emissions.

Global Warming Potential (GWP)

An index, based upon radiative properties of well mixed greenhouse gases, measuring the radiative forcing of a unit mass of a given well mixed greenhouse gas in today's atmosphere integrated over a chosen time horizon, relative to that of CO_2. The GWP represents the combined effect of the differing lengths of time that these gases remain in the atmosphere and their relative effectiveness in absorbing outgoing infrared radiation. The Kyoto Protocol is based on GWPs from pulse emissions over a 100-year time frame.

Green accounting

Attempts to integrate into macroeconomic studies a broader set of social welfare measures, covering e.g., social, environmental, and development oriented policy aspects. Green accounting includes both monetary valuations that attempt to calculate a 'green national product' with the economic damage by pollutants subtracted from the national product, and accounting systems that include quantitative non-monetary pollution, depletion and other data.

Greenhouse effect

Greenhouse gases effectively absorb infrared radiation, emitted by the Earth's surface, by the atmosphere itself due to the same gases and by clouds. Atmospheric radiation is emitted to all sides, including downward to the Earth's surface. Thus, greenhouse gases trap heat within the surface-troposphere system. This is called the greenhouse effect.

Thermal infrared radiation in the troposphere is strongly coupled to the temperature at the altitude at which it is emitted. In the troposphere, the temperature generally decreases with height. Effectively, infrared radiation emitted to space originates from an altitude with a temperature of, on average, $-19°C$, in balance with the net incoming solar radiation, whereas the Earth's surface is kept at a much higher temperature of, on average, $+14°C$.

An increase in the concentration of greenhouse gases leads to an increased infrared opacity of the atmosphere and therefore to an effective radiation into space from a higher altitude at a lower temperature. This causes a radiative forcing that leads to an enhancement of the greenhouse effect, the so-called **enhanced greenhouse effect**.

815

Greenhouse gases (GHGs)

Greenhouse gases are those gaseous constituents of the atmosphere, both natural and anthropogenic, that absorb and emit radiation at specific wavelengths within the spectrum of infrared radiation emitted by the Earth's surface, the atmosphere and clouds. This property causes the greenhouse effect. Water vapour (H_2O), carbon dioxide (CO_2), nitrous oxide (N_2O), methane (CH_4) and ozone (O_3) are the primary greenhouse gases in the earth's atmosphere. Moreover, there are a number of entirely human-made greenhouse gases in the atmosphere, such as the halocarbons and other chlorine- and bromine-containing substances, dealt with under the Montreal Protocol. Besides carbon dioxide, nitrous oxide and methane, the Kyoto Protocol deals with the greenhouse gases sulphur hexafluoride, hydrofluorocarbons, and perfluorocarbons.

Gross Domestic Product (GDP)

The sum of gross value added, at purchasers' prices, by all resident and non-resident producers in the economy, plus any taxes and minus any subsidies not included in the value of the products in a country or a geographic region for a given period, normally one year. It is calculated without deducting for depreciation of fabricated assets or depletion and degradation of natural resources.

Gross National Product (GNP)

GNP is a measure of national income. It measures value added from domestic and foreign sources claimed by residents. GNP comprises Gross Domestic Product plus net receipts of primary income from non-resident income.

Gross World Product

An aggregation of the individual country's Gross Domestic Products to obtain the sum for the world.

Group of 77 and China (G77/China)

Originally 77, now more than 130, developing countries that act as a major negotiating bloc in the UNFCCC process. G77/China is also referred to as Non-Annex I countries in the context of the UNFCCC.

Governance

The way government is understood has changed in response to social, economic and technological changes over recent decades. There is a corresponding shift from govern*ment* defined strictly by the nation-state to a more inclusive concept of govern*ance*, recognizing the contributions of various levels of government (global, international, regional, local) and the roles of the private sector, of non-governmental actors and of civil society.

Hot air

Under the terms of the 1997 Kyoto Protocol, national emission targets in Annex B are expressed relative to emissions in the year 1990. For countries in the former Soviet Union and Eastern Europe this target has proven to be higher than their current and projected emissions for reasons unrelated to climate-change mitigation activities. Russia and Ukraine, in particular, are expected to have a substantial volume of excess emission allowances over the period 2008-2012 relative to their forecast emissions. These allowances are sometimes referred to as hot air because, while they can be traded under the Kyoto Protocol's flexibility mechanisms, they did not result from mitigation activities.

Hybrid vehicle

Any vehicle that employs two sources of propulsion, especially a vehicle that combines an internal combustion engine with an electric motor.

Hydrofluorocarbons (HFCs)

One of the six gases or groups of gases to be curbed under the Kyoto Protocol. They are produced commercially as a substitute for chlorofluorocarbons. HFCs are largely used in refrigeration and semiconductor manufacturing. Their Global Warming Potentials range from 1,300 to 11,700.

Implementation

Implementation describes the actions taken to meet commitments under a treaty and encompasses legal and effective phases. **Legal implementation** refers to legislation, regulations, judicial decrees, including other actions such as efforts to administer progress, which governments take to translate international accords into domestic law and policy. **Effective implementation** needs policies and programmes that induce changes in the behaviour and decisions of target groups. Target groups then take effective measures of mitigation and adaptation.

Income elasticity (of demand)

This is the ratio of the percentage change in quantity of demand for a good or service to a one percentage change in income. For most goods and services, demand goes up when income grows, making income elasticity positive. When the elasticity is less than one, goods and services are called necessities.

Industrial ecology

The relationship of a particular industry with its environment. It often refers to the conscious planning of industrial processes to minimize their negative externalities (e.g., by heat and materials cascading).

Inertia

In the context of climate-change mitigation, inertia relates to the difficulty of change resulting from pre-existing conditions within society such as physical man-made capital, natural capital and social non-physical capital, including institutions, regulations and norms. Existing structures lock in societies, making change more difficult.

Integrated assessment

A method of analysis that combines results and models from the physical, biological, economic and social sciences, and the interactions between these components in a consistent framework to evaluate the status and the consequences of environmental change and the policy responses to it.

Integrated Design Process (IDP) of buildings

Optimizing the orientation and shape of buildings and providing high-performance envelopes for minimizing heating and cooling loads. Passive techniques for heat transfer control, ventilation and daylight access reduce energy loads further. Properly sized and controlled, efficient mechanical systems address the left-over loads. IDP requires an iterative design process involving all the major stakeholders from building users to equipment suppliers, and can achieve 30-75% savings in energy use in new buildings at little or no additional investment cost.

Intelligent controls

In this report, the notion of 'intelligent control' refers to the application of information technology in buildings to control heating, ventilation, air-conditioning, and electricity use effectively. It requires effective monitoring of parameters such as temperature, convection, moisture, etc., with appropriate control measurements ('smart metering').

Interaction effect

The consequence of the interaction of climate-change policy instruments with existing domestic tax systems, including both cost-increasing tax interaction and cost-reducing revenue-recycling effect. The former reflects the impact that greenhouse gas policies can have on labour and capital markets through their effects on real wages and the real return to capital. Restricting allowable GHG emissions, raises the carbon price and so the costs of production and the prices of output, thus reducing the real return to labour and capital. With policies that raise revenue for the government, carbon taxes and auctioned permits, the revenues can be recycled to reduce existing distortional taxes. See also **double dividend**.

Intergovernmental Organization (IGO)
Organizations constituted of governments. Examples include the World Bank, the Organization of Economic Co-operation and Development (OECD), the International Civil Aviation Organization (ICAO), the Intergovernmental Panel on Climate Change (IPCC), and other UN and regional organizations. The Climate Convention allows accreditation of these IGOs to attend negotiating sessions.

International Energy Agency (IEA)
Established in 1974, the agency is linked with the OECD. It enables OECD member countries to take joint measures to meet oil supply emergencies, to share energy information, to coordinate their energy policies, and to cooperate in developing rational energy use programmes.

Joint Implementation (JI)
A market-based implementation mechanism defined in Article 6 of the Kyoto Protocol, allowing Annex I countries or companies from these countries to implement projects jointly that limit or reduce emissions or enhance sinks, and to share the Emissions Reduction Units. JI activity is also permitted in Article 4.2(a) of the UNFCCC. See also **Activities Implemented Jointly** and **Kyoto Mechanisms**.

Kyoto Mechanisms (also called Flexibility Mechanisms)
Economic mechanisms based on market principles that parties to the Kyoto Protocol can use in an attempt to lessen the potential economic impacts of greenhouse gas emission-reduction requirements. They include **Joint Implementation** (Article 6), **Clean Development Mechanism** (Article 12), and **Emissions trading** (Article 17).

Kyoto Protocol
The Kyoto Protocol to the UNFCCC was adopted at the Third Session of the Conference of the Parties (COP) in 1997 in Kyoto. It contains legally binding commitments, in addition to those included in the FCCC. Annex B countries agreed to reduce their anthropogenic GHG emissions (carbon dioxide, methane, nitrous oxide, hydrofluorocarbons, perfluorocarbons and sulphur hexafluoride) by at least 5% below 1990 levels in the commitment period 2008-2012. The Kyoto Protocol came into force on 16 February 2005.

Landfill
A landfill is a solid waste disposal site where waste is deposited below, at or above ground level. Limited to engineered sites with cover materials, controlled placement of waste and management of liquids and gases. It excludes uncontrolled waste disposal.

Land-use
The total of arrangements, activities and inputs undertaken in a certain land-cover type (a set of human actions). The social and economic purposes for which land is managed (e.g., grazing, timber extraction, and conservation). Land-use change occurs when, e.g., forest is converted to agricultural land or to urban areas.

Leapfrogging
The ability of developing countries to bypass intermediate technologies and jump straight to advanced clean technologies. Leapfrogging can enable developing countries to move to a low-emissions development trajectory.

Learning by doing
As researchers and firms gain familiarity with a new technological process, or acquire experience through expanded production they can discover ways to improve processes and reduce cost. Learning by doing is a type of experience-based technological change.

Levelized cost price
The unique price of the outputs of a project that makes the present value of the revenues (benefits) equal to the present value of the costs over the lifetime of the project. See also **discounting** and **present value**.

Likelihood
The likelihood of an occurrence, outcome or result, where this can be estimated probabilistically, is expressed in IPCC reports using a standard terminology:

Particular, or a range of, occurrences/outcomes of an uncertain event owning a probability of		are said to be:	
	>99%		Virtually certain
	>90%		Very likely
	>66%		Likely
	33 to 66%		About as likely as not
	<33%		Unlikely
	<10%		Very unlikely
	<1%		Exceptionally unlikely

Lock-in effect
Technologies that cover large market shares continue to be used due to factors such as sunk investment costs, related infrastructure development, use of complementary technologies and associated social and institutional habits and structures.

Low-carbon technology
A technology that over its life cycle causes less CO_2-eq. emissions than other technological options do. See also **Environmentally sustainable technologies**.

Macroeconomic costs
These costs are usually measured as changes in Gross Domestic Product or changes in the growth of Gross Domestic Product, or as loss of welfare or consumption.

Marginal cost pricing
The pricing of goods and services such that the price equals the additional cost arising when production is expanded by one unit. Economic theory shows that this way of pricing maximizes social welfare in a first-best economy.

Market barriers
In the context of climate change mitigation, market barriers are conditions that prevent or impede the diffusion of cost-effective technologies or practices that would mitigate GHG emissions.

Market-based regulation
Regulatory approaches using price mechanisms (e.g., taxes and auctioned tradable permits), among other instruments, to reduce GHG emissions.

Market distortions and imperfections
In practice, markets will always exhibit distortions and imperfections such as lack of information, distorted price signals, lack of competition, and/or institutional failures related to regulation, inadequate delineation of property rights, distortion-inducing fiscal systems, and limited financial markets

Market equilibrium
The point at which the demand for goods and services equals the supply; often described in terms of price levels, determined in a competitive market, 'clearing' the market.

Market Exchange Rate (MER)
This is the rate at which foreign currencies are exchanged. Most economies post such rates daily and they vary little across all the exchanges. For some developing economies official rates and black-market rates may differ significantly and the MER is difficult to pin down.

Material efficiency options
In this report, options to reduce GHG emissions by decreasing the volume of materials needed for a certain product or service

Measures
Measures are technologies, processes, and practices that reduce GHG emissions or effects below anticipated future levels. Examples of measures are renewable energy technologies, waste minimization processes and public transport commuting practices, etc. See also **policies**.

Methane (CH$_4$)
Methane is one of the six greenhouse gases to be mitigated under the Kyoto Protocol. It is the major component of natural gas and associated with all hydrocarbon fuels, animal husbandry and agriculture. **Coal-bed methane** is the gas found in coal seams.

Methane recovery
Methane emissions, e.g., from oil or gas wells, coal beds, peat bogs, gas transmission pipelines, landfills, or anaerobic digesters, are captured and used as a fuel or for some other economic purpose (e.g., chemical feedstock).

Meeting of the Parties (to the Kyoto Protocol) (MOP)
The Conference of the Parties (COP) of the UNFCCC serves as the Meeting of the Parties (MOP), the supreme body of the Kyoto Protocol, since the latter entered into force on 16 February 2005. Only parties to the Kyoto Protocol may participate in deliberations and make decisions.

Millennium Development Goals (MDG)
A set of time-bound and measurable goals for combating poverty, hunger, disease, illiteracy, discrimination against women and environmental degradation, agreed at the UN Millennium Summit in 2000.

Mitigation
Technological change and substitution that reduce resource inputs and emissions per unit of output. Although several social, economic and technological policies would produce an emission reduction, with respect to climate change, mitigation means implementing policies to reduce GHG emissions and enhance sinks.

Mitigative capacity
This is a country's ability to reduce anthropogenic GHG emissions or to enhance natural sinks, where ability refers to skills, competencies, fitness and proficiencies that a country has attained and depends on technology, institutions, wealth, equity, infrastructure and information. Mitigative capacity is rooted in a country's sustainable development path.

Montreal Protocol
The Montreal Protocol on Substances that Deplete the Ozone Layer was adopted in Montreal in 1987, and subsequently adjusted and amended in London (1990), Copenhagen (1992), Vienna (1995), Montreal (1997) and Beijing (1999). It controls the consumption and production of chlorine- and bromine-containing chemicals that destroy stratospheric ozone, such as chlorofluorocarbons, methyl chloroform, carbon tetrachloride, and many others.

Multi-attribute analysis.
Integrates different decision parameters and values in a quantitative analysis without assigning monetary values to all parameters. Multi-attribute analysis can combine quantitative and qualitative information.

Multi-gas
Next to CO$_2$ also the other greenhouse gases (methane, nitrous oxide and fluorinated gases) are taken into account in e.g. achieving reduction of emissions (**multi-gas reduction**) or stabilization of concentrations (**multi-gas stabilization**).

National Action Plans
Plans submitted to the COP by parties outlining the steps that they have adopted to limit their anthropogenic GHG emissions. Countries must submit these plans as a condition of participating in the UNFCCC and, subsequently, must communicate their progress to the COP regularly. The National Action Plans form part of the National Communications, which include the national inventory of GHG sources and sinks.

Net anthropogenic greenhouse gas removals by sinks
For CDM afforestation and reforestation projects, 'net anthropogenic GHG removals by sinks' equals the actual net GHG removals by sinks minus the baseline net GHG removals by sinks minus leakage.

Nitrous oxide (N$_2$O)
One of the six types of greenhouse gases to be curbed under the Kyoto Protocol.

Non-Annex I Countries/Parties
The countries that have ratified or acceded to the UNFCCC but are not included in Annex I.

Non-Annex B Countries/Parties
The countries not included in Annex B of the Kyoto Protocol.

No-regret policy (options / potential)
Such policy would generate net social benefits whether or not there is climate change associated with anthropogenic emissions of greenhouse gases. **No-regret options** for GHG emissions reduction refer to options whose benefits (such as reduced energy costs and reduced emissions of local/regional pollutants) equal or exceed their costs to society, excluding the benefits of avoided climate change.

Normative analysis
Economic analysis in which judgments about the desirability of various policies are made. The conclusions rest on value judgments as well as on facts and theories.

Oil sands and oil shale
Unconsolidated porous sands, sandstone rock and shales containing bituminous material that can be mined and converted to a liquid fuel.

Opportunities
Circumstances to decrease the gap between the market potential of any technology or practice and the economic potential or technical potential.

Ozone (O$_3$)
Ozone, the tri-atomic form of oxygen, is a gaseous atmospheric constituent. In the troposphere, ozone is created both naturally and by photochemical reactions involving gases resulting from human activities. Troposphere ozone acts as a greenhouse gas. In the stratosphere, ozone is created by the interaction between solar ultraviolet radiation and molecular oxygen (O$_2$). Stratospheric ozone plays a dominant role in the stratospheric radiative balance. Its concentration is highest in the ozone layer.

Pareto criterion
A criterion testing whether an individual's welfare can be increased without making others in the society worse off. A **Pareto improvement** occurs when an individual's welfare is improved without making the welfare of the rest of society worse off. A

Pareto optimum is reached when no one's welfare can be increased without making the welfare of the rest of society worse off, given a particular distribution of income. Different income distributions lead to different Pareto optima.

Passive solar design

Structural design and construction techniques that enable a building to utilize solar energy for heating, cooling, and lighting by non-mechanical means.

Perfluorocarbons (PFCs)

Among the six greenhouse gases to be abated under the Kyoto Protocol. These are by-products of aluminium smelting and uranium enrichment. They also replace chlorofluorocarbons in manufacturing semiconductors. The Global Warming Potential of PFCs is 6500–9200.

Policies

In UNFCCC parlance, policies are taken and/or mandated by a government - often in conjunction with business and industry within its own country, or with other countries - to accelerate mitigation and adaptation measures. Examples of policies are carbon or other energy taxes, fuel efficiency standards for automobiles, etc. **Common and co-ordinated or harmonised policies** refer to those adopted jointly by parties. See also **measures.**

Portfolio analysis

Deals with a portfolio of assets or policies that are characterized by different risks and pay-offs. The objective function is built up around the variability of returns and their risks, leading up to the decision rule to choose the portfolio with highest expected return.

Post-consumer waste

Waste from consumption activities, e.g. packaging materials, paper, glass, rests from fruits and vegetables, etc.

Potential

In the context of climate change, potential is the amount of mitigation or adaptation that could be - but is not yet – realized over time. As potential levels are identified: market, economic, technical and physical.

- **Market potential** indicates the amount of GHG mitigation that might be expected to occur under forecast market conditions including policies and measures in place at the time. It is based on private unit costs and discount rates, as they appear in the base year and as they are expected to change in the absence of any additional policies and measures.
- **Economic potential** is in most studies used as the amount of GHG mitigation that is cost-effective for a given carbon price, based on social cost pricing and discount rates, including energy savings, but without most externalities. Theoretically, it is defined as the potential for cost-effective GHG mitigation when non-market social costs and benefits are included with market costs and benefits in assessing the options for particular levels of carbon prices (as affected by mitigation policies) and when using social discount rates instead of private ones. This includes externalities, i.e., non-market costs and benefits such as environmental co-benefits
- **Technical potential** is the amount by which it is possible to reduce GHG emissions or improve energy efficiency by implementing a technology or practice that has already been demonstrated. No explicit reference to costs is made but adopting 'practical constraints' may take into account implicit economic considerations.
- **Physical potential** is the theoretical (thermodynamic) and sometimes, in practice, rather uncertain upper limit to mitigation.

Precautionary Principle

A provision under Article 3 of the UNFCCC, stipulating that the parties should take precautionary measures to anticipate, prevent or minimize the causes of climate change and mitigate its adverse effects. Where there are threats of serious or irreversible damage, lack of full scientific certainty should not be used as a reason to postpone such measures, taking into account that policies and measures to deal with climate change should be cost-effective in order to ensure global benefits at the lowest possible cost.

Precursors

Atmospheric compounds which themselves are not greenhouse gases or aerosols, but which have an effect on greenhouse gas or aerosol concentrations by taking part in physical or chemical processes regulating their production or destruction rates.

Pre-industrial

The era before the industrial revolution of the late 18th and 19th centuries, after which the use of fossil fuel for mechanization started to increase.

Present value

The value of a money amount differs when the amount is available at different moments in time (years). To make amounts at differing times comparable and additive, a date is fixed as the 'present'. Amounts available at different dates in the future are discounted back to a present value, and summed to get the present value of a series of future cash flows. **Net present value** is the difference between the present value of the revenues (benefits) with the present value of the costs. See also **discounting**.

Price elasticity of demand

The ratio of the percentage change in the quantity of demand for a good or service to one percentage change in the price of that good or service. When the absolute value of the elasticity is between 0 and 1, demand is called inelastic; when it is greater than one, demand is called elastic.

'Primary market' and 'secondary market' trading

In commodities and financial exchanges, buyers and sellers who trade directly with each other constitute the 'primary market', while buying and selling through exchange facilities represent the 'secondary market'.

Production frontier

The maximum outputs attainable with the optimal uses of available inputs (natural resources, labour, capital, information).

Public sector leadership programmes in energy efficiency

Government purchasing and procurement of energy-efficient products and services. Government agencies are responsible for a wide range of energy-consuming facilities and services such as government office buildings, schools, and health care facilities. The government is often a country's largest consumer of energy and largest buyer of energy-using equipment. Indirect beneficial impacts occur when governments act effectively as market leaders. First, government buying power can create or expand demand for energy-efficient products and services. Second, visible government energy-saving actions can serve as an example for others.

Purchasing Power Parity (PPP)

The purchasing power of a currency is expressed using a basket of goods and services that can be bought with a given amount in the home country. International comparison of, e.g., Gross Domestic Products of countries can be based on the purchasing power of currencies rather than on current exchange rates. PPP estimates tend to lower per capita GDPs in industrialized countries and raise per capita GDPs in developing countries. (**PPP** is also an acronym for polluter-pays-principle).

Radiative forcing

Radiative forcing is the change in the net vertical irradiance (expressed in Watts per square metre: W/m^2) at the tropopause due to an internal change or a change in the external forcing of the climate system, such as, for example, a change in the concentration of CO_2 or in the output of the sun.

Rebound effect

After implementation of efficient technologies and practices, part of the savings is taken back for more intensive or other consumption, e.g., improvements in car-engine efficiency lower the cost per kilometre driven, encouraging more car trips or the purchase of a more powerful vehicle.

Reforestation

Direct human-induced conversion of non-forested land to forested land through planting, seeding and/or the human-induced promotion of natural seed sources, on land that was previously forested but converted to non-forested land. For the first commitment period of the Kyoto Protocol, reforestation activities will be limited to reforestation occurring on those lands that did not contain forest on 31 December 1989. See also **afforestation** and **deforestation.**

Reservoir

A component of the climate system, other than the atmosphere, which has the capacity to store, accumulate or release a substance of concern, e.g., carbon, a greenhouse gas or a precursor. Oceans, soils, and forests are examples of reservoirs of carbon. **Stock** is the absolute quantity of substance of concerns, held within a reservoir at a specified time. See also **Carbon pool**.

Safe landing approach. See **tolerable windows approach**.

Scenario

A plausible description of how the future may develop based on a coherent and internally consistent set of assumptions about key driving forces (e.g., rate of technological change, prices) and relationships. Note that scenarios are neither predictions nor forecasts, but are useful to provide a view of the implications of developments and actions.

Sequestration

Carbon storage in terrestrial or marine reservoirs. **Biological sequestration** includes direct removal of CO_2 from the atmosphere through land-use change, afforestation, reforestation, carbon storage in landfills and practices that enhance soil carbon in agriculture.

Shadow pricing

Setting prices of goods and services that are not, or incompletely, priced by market forces or by administrative regulation, at the height of their social marginal value. This technique is used in cost-benefit analysis.

Sinks

Any process, activity or mechanism that removes a greenhouse gas or aerosol, or a precursor of a greenhouse gas or aerosol from the atmosphere.

Smart metering. See **Intelligent control**.

Social cost of carbon (SCC)

The discounted monetized sum (e.g. expressed as a price of carbon in $\$/tCO_2$) of the annual net losses from impacts triggered by an additional ton of carbon emitted today. According to usage in economic theory, the social cost of carbon establishes an economically optimal price of carbon at which the associated marginal costs of mitigation would equal the marginal benefits of mitigation.

Social unit costs of mitigation

Carbon prices in $US\$/tCO_2$ and $US\$/tC$-eq (as affected by mitigation policies and using social discount rates) required to achieve a particular level of mitigation (**economic potential**) in the form of a reduction below a baseline for GHG emissions. The reduction is usually associated with a policy target, such as a cap in an emissions trading scheme or a given level of stabilization of GHG concentrations in the atmosphere.

Source

Source mostly refers to any process, activity or mechanism that releases a greenhouse gas, aerosol or a precursor of a greenhouse gas or aerosol into the atmosphere. Source can also refer to, e.g., an energy source.

Specific energy use

The energy used in the production of a unit material, product or service.

Spill-over effect

The effects of domestic or sector mitigation measures on other countries or sectors. Spill-over effects can be positive or negative and include effects on trade, carbon leakage, transfer of innovations, and diffusion of environmentally sound technology and other issues.

Stabilization

Keeping constant the atmospheric concentrations of one or more GHG (e.g., CO_2) or of a CO_2-equivalent basket of GHG. Stabilization analyses or scenarios address the stabilization of the concentration of GHG in the atmosphere.

Standards

Set of rules or codes mandating or defining product performance (e.g., grades, dimensions, characteristics, test methods, and rules for use). **Product, technology or performance standards** establish minimum requirements for affected products or technologies. Standards impose reductions in GHG emissions associated with the manufacture or use of the products and/or application of the technology.

Storyline

A narrative description of a scenario (or a family of scenarios) that highlights the scenario's main characteristics, relationships between key driving forces, and the dynamics of the scenarios.

Structural change

Changes, for example, in the relative share of Gross Domestic Product produced by the industrial, agricultural, or services sectors of an economy; or more generally, systems transformations whereby some components are either replaced or potentially substituted by other ones.

Subsidy

Direct payment from the government or a tax reduction to a private party for implementing a practice the government wishes to encourage. The reduction of GHG emissions is stimulated by lowering existing subsidies that have the effect of raising emissions (such as subsidies to fossil fuel use) or by providing subsidies for practices that reduce emissions or enhance sinks (e.g. for insulation of buildings or for planting trees).

Sulphur hexafluoride (SF$_6$)

One of the six greenhouse gases to be curbed under the Kyoto Protocol. It is largely used in heavy industry to insulate high-voltage equipment and to assist in the manufacturing of cable-cooling systems and semi-conductors. Its Global Warming Potential is 23,900.

Supplementarity

The Kyoto Protocol states that emissions trading and Joint Implementation activities are to be supplemental to domestic policies (e.g. energy taxes, fuel efficiency standards) taken by developed countries to reduce their GHG emissions. Under some proposed definitions of supplementarity (e.g., a concrete ceiling on level of use), developed countries could be restricted in their use of the Kyoto Mechanisms to achieve their reduction targets. This is a subject for further negotiation and clarification by the parties.

Sustainable Development (SD)

The concept of sustainable development was introduced in the World Conservation Strategy (IUCN 1980) and had its roots in the concept of a sustainable society and in the management of renewable resources. Adopted by the WCED in 1987 and by the Rio Conference in 1992 as a process of change in which the exploitation of resources, the direction of investments, the orientation of technological development and institutional change are all in harmony and enhance both current and future potential to meet human needs and aspirations. SD integrates the political, social, economic and environmental dimensions.

Targets and timetables

A target is the reduction of a specific percentage of GHG emissions from a baseline date (e.g., below 1990 levels) to be achieved by a set date or timetable (e.g., 2008-2012). Under the Kyoto Protocol the EU agreed to reduce its GHG emissions by 8% below 1990 levels by the 2008-2012 commitment period. Targets and timetables are an emissions cap on the total amount of GHG emissions that can be emitted by a country or region in a given time period.

Tax

A **carbon tax** is a levy on the carbon content of fossil fuels. Because virtually all of the carbon in fossil fuels is ultimately emitted as CO_2, a carbon tax is equivalent to an **emission tax** on each unit of CO_2-equivalent emissions. An **energy tax** - a levy on the energy content of fuels - reduces demand for energy and so reduces CO_2 emissions from fossil fuel use. An **eco-tax** is designed to influence human behaviour (specifically economic behaviour) to follow an ecologically benign path.

An **international** carbon/emission/energy tax is a tax imposed on specified sources in participating countries by an international authority. The revenue is distributed or used as specified by this authority or by participating countries. A **harmonized tax** commits participating countries to impose a tax at a common rate on the same sources, because imposing different rates across countries would not be cost-effective. A **tax credit** is a reduction of tax in order to stimulate purchasing of or investment in a certain product, like GHG emission reducing technologies. A **carbon charge** is the same as a carbon tax. See also **Interaction effect**

Technological change

Mostly considered as technological *improvement*, i.e., more or better goods and services can be provided from a given amount of resources (production factors). Economic models distinguish autonomous (exogenous), endogenous and induced technological change.

Autonomous (exogenous) technological change is imposed from outside the model, usually in the form of a time trend affecting energy demand or world output growth. **Endogenous technological change** is the outcome of economic activity *within* the model, i.e., the choice of technologies is included within the model and affects energy demand and/or economic growth. **Induced technological change** implies endogenous technological change but adds further changes *induced* by policies and measures, such as carbon taxes triggering R&D efforts.

Technology

The practical application of knowledge to achieve particular tasks that employs both technical artefacts (hardware, equipment) and (social) information ('software', know-how for production and use of artefacts).

Technology transfer

The exchange of knowledge, hardware and associated software, money and goods among stakeholders, which leads to the spreading of technology for adaptation or mitigation The term encompasses both diffusion of technologies and technological cooperation across and within countries.

Tolerable windows approach (TWA)

This approach seeks to identify the set of all climate-protection strategies that are simultaneously compatible with 1) prescribed long-term climate-protection goals, and 2) normative restrictions on the emissions mitigation burden. The constraints may include limits on the magnitude and rate of global mean temperature change, on the weakening of the thermohaline circulation, on ecosystem losses and on economic welfare losses resulting from selected climate damages, adaptation costs and mitigation efforts. For a given set of constraints, and given a solution exists, the TWA delineates an emission corridor of complying emission paths.

Top-down models

Models applying macroeconomic theory, econometric and optimization techniques to aggregate economic variables. Using historical data on consumption, prices, incomes, and factor costs, top-down models assess final demand for goods and services, and supply from main sectors, such as the energy sector, transportation, agriculture, and industry. Some top-down models incorporate technology data, narrowing the gap to **bottom-up models**.

Trace gas

A minor constituent of the atmosphere, next to nitrogen and oxygen that together make up 99% of all volume. The most important trace gases contributing to the greenhouse effect are carbon dioxide, ozone, methane, nitrous oxide, perfluorocarbons, chlorofluorocarbons, hydrofluorocarbons, sulphur hexafluoride and water vapour.

Tradable permit. See **emission permit**

Tradable quota system. See **emissions trading**.

Uncertainty

An expression of the degree to which a value is unknown (e.g. the future state of the climate system). Uncertainty can result from lack of information or from disagreement about what is known or even knowable. It may have many types of sources, from quantifiable errors in the data to ambiguously defined concepts or terminology, or uncertain projections of human behavior. Uncertainty can therefore be represented by quantitative measures (e.g., a range of values calculated by various models) or by qualitative statements (e.g., reflecting the judgment of a team of experts). See also **likelihood.**

United Nations Framework Convention on Climate Change (UNFCCC)

The Convention was adopted on 9 May 1992 in New York and signed at the 1992 Earth Summit in Rio de Janeiro by more than 150 countries and the European Economic Community. Its ultimate objective is the 'stabilization of greenhouse gas concentrations in the atmosphere at a level that would prevent dangerous anthropogenic interference with the climate system'. It contains commitments for all parties. Under the Convention parties included in Annex I aimed to return greenhouse gas emission not controlled by the Montreal Protocol to 1990 levels by the year 2000. The convention came into force in March 1994.

Value added

The net output of a sector or activity after adding up all outputs and subtracting intermediate inputs.

Values

Worth, desirability or utility based on individual preferences. Most social science disciplines use several definitions of value. Related to nature and environment, there is a distinction between intrinsic and instrumental values, the latter assigned by humans. Within instrumental values, there is an unsettled catalogue of different values, such as (direct and indirect) use, option, conservation, serendipity, bequest, existence, etc.

Mainstream economics define the total value of any resource as the sum of the values of the different individuals involved in the use of the resource. The economic values, which are the foundation of the estimation of costs, are measured in terms of the willingness to pay by individuals to receive the resource or by the willingness of individuals to accept payment to part with the resource. See also **contingent valuation method**.

Voluntary action

Informal programmes, self-commitments and declarations, where the parties (individual companies or groups of companies) entering into the action set their own targets and often do their own monitoring and reporting.

Voluntary agreement

An agreement between a government authority and one or more private parties to achieve environmental objectives or to improve environmental performance beyond compliance to regulated obligations. Not all voluntary agreements are truly voluntary; some include rewards and/or penalties associated with joining or achieving commitments.

Annex II

Abbreviations & Acronyms

A	
ADB	Asian Development Bank
AIJ	Activities Implemented Jointly
ALM	Africa, Latin America, Middle East Region (SRES and post-SRES scenarios)
A/R	Afforestation and Reforestation
B	
BAU	Business As Usual
BECS	Bioenergy with CCS
BOD	Biochemical Oxygen Demand / Biological Oxygen Demand
BRT	Bus Rapid transport
C	
CAA	Clean Air Act
CAFÉ	Corporate Average Fuel Economy
CANZ	Canada, Australia and New Zealand
CBA	Cost Benefit Analysis
CCGT	Combined Cycle Gas Turbine
CCS	Carbon Capture and Storage
CDM	Clean Development Mechanism
CER	Certified Emission Reduction
CFCs	Chlorofluorocarbons
CFL	Compact Fluorescent Lamp
CGE	Computable General Equilibrium
CHP	Combined Heat and Power
CONCAWE	European Oil Company Organisation for Environment, Health, and Safety
COD	Chemical Oxygen Demand
COP	Conference of the Parties / Coefficient of Performance
CSD	Commission for Sustainable Development
CSP	Concentrating Solar Power
D	
DAES	Domestic Alternative Energy Sources
DES	Development, Equity and Sustainability
DSM	Demand Side Management
E	
EBRD	European Bank for Reconstruction and Development
EC	European Commission
EEA	European Environmental Agency
EECCA	Countries of Eastern Europe, the Caucasus and Central Asia
EIT	Economy In Transition
EMAS	ECO-Management and Audit Scheme (EU)
EMF	Energy Modeling Forum (Stanford University)
EPR	Extended Producer Responsibility
EPRI	Electric Power Research Institute
ESCO	Energy Service Company
ETS	Emission Trading Scheme (EU)
EU	European Union
F	
FAO	Food & Agriculture Organization
FC	Fuel Cell
FCCC	Framework Convention on Climate Change (UN)
FDI	Foreign Direct Investment
FOB	Free on Board
FOD	First-Order Decay

G	
GATT	General Agreement on Trade and Tariffs
GCM	Global Climate Model
GDP	Gross Domestic Product
GEF	Global Environment Facility
GHG	Green House Gas
GMT	Global Mean Temperature
GNP	Gross National Product
GPP	Gross Primary Production
GWP	Global Warming Potential
H	
HDI	Human Development Index
HFC	Hydrofluorocarbons
HSS	High Strength Steels
HVAC	Heating, Ventilation and Air Conditioning
I	
IA	Integrated Assessment
IAEA	International Atomic Energy Agency
IAES	Imported Alternative Energy Sources
ICAO	International Civil Aviation Organization
ICE	Internal Combustion Engine
IDP	Integrated Design Process (for buildings)
IEA	International Energy Agency
IET	International Emission Trading
IGO	Intergovernmental Organization
IIASA	International Institute for Applied System Analysis
IMO	International Maritime Organization
IPCC	Intergovernmental Panel on Climate Change
ISIC	International Standard Industrial Classification
ISO	International Standardization Organization
IT	Information Technology
ITC	Induced Technological Change
ITER	International Thermonuclear Experimental Reactor
IUCN	International Union for the Conservation of Nature and natural resources
J	
J	Joule =Newton x meter (International Standard unit of energy)
JI	Joint Implementation
JRC	Joint Research Centre (EU)
L	
LCA	Life Cycle Assessment
LHV	Lower Heating Value
LNG	Liquefied Natural Gas
LPG	Liquid Petroleum Gas
M	
MAC	Marginal Abatement Cost
MBT	Mechanical Biological Treatment
MDG	Millennium Development Goals
MEA	Multilateral Environmental Agreements
MOP	Meeting Of the Parties (Kyoto Protocol)
N	
NAFTA	North American Free Trade Agreement
NEDC	New European Driving Cycle
NGO	Non-Governmental Organization
NIC	Newly Industrialized Country
NMHC	Non-Methane Hydro Carbon
NMT	Non-Motorised Transport
NMVOC	Non-Methane Volatile Organic Compounds

O	
O&M	Operation and Maintenance
ODA	Official Development Assistance
ODS	Ozone Depleting Substances
OECD	Organization for Economic Co-operation and Development
OPEC	Organization of Petroleum Exporting Countries
P	
PAYT	Pay As You Throw
PFC	Perfluorocarbon
ppm [v/w]	parts per million [by volume / weight]
PPP	Purchasing Power Parity
PV	Photovoltaïc
Q	
QELRCs	Quantified Emission Limitation or Reduction Commitments
R	
RD&D	Research,Development and Demonstration
S	
SAR	Second Assessment Report (IPCC)
SBSTA	Subsidiary Body for Scientific and Technological Advice
SCC	Social Cost of Carbon
SD	Sustainable Development
SMEs	Small and Medium Enterprises
SPM	Summary for Policy Makers
SRCCS	Special Report on Carbon Capture and Storage (IPCC)
SRES	Special Report on Emissions Scenarios (IPCC)
SRLULUCF	Special Report on Land-Use, Land-Use Change and Forestry (IPCC)
SROC	Special Report on Ozone and Climate (IPCC)
SRTT	Special Report on methodological and technological issues in Technology Transfer (IPCC)
T	
TAR	Third Assessment Report (IPCC)
TPES	Total Primary Energy Supply
TWA	Tolerable Windows Approach
U	
UK	United Kingdom
UN	United Nations
UNCED	UN Conference on Environment and Development
UNDP	UN Development Programme
UNEP	UN Environment Programme
UNFCCC	UN Framework Convention on Climate Change
USA	United States of America
V	
VAT	Value Added Tax
VOC	Volatile organic compound
W	
W	Watt = Joule/second (International Standard unit of power)
WBCSD	World Business Council for Sustainable Development
WCED	World Commission on Environment and Development
WEC	World Energy Council
WG I	Working Group One of the IPCC
WG II	Working Group Two of the IPCC
WG III	Working Group Three of the IPCC
WHO	World Health Organization (UN)
WTO	World Trade Organization
WWF	World Wide Fund for nature

Chemical Symbols

C	Carbon
C_2F_6	Perfluoroethane / Hexafluoroethane
CF_4	Perfluoromethane / Tetrafluoromethane
CFCs	Chlorofluorocarbons
CH_4	Methane
CO	Carbon monoxide
CO_2	Carbon dioxide
H / H_2	Hydrogen (element / gas)
H_2O	Water / Water vapor
HFCs	Hydrofluorocarbons
N / N_2	Nitrogen (element /gas)
N_2O	Nitrous Oxide
NF_3	Nitrogen trifluoride
NH_3 / NH_4^+	Ammonia / Ammonium ion
NMVOC	Non-methane Volatile Organic Compounds
NO	Nitric oxide
NO_2	Nitrogen dioxide
NO_x	The sum of NO and NO_2 expressed in NO_2 mass equivalent
O / O_2	Oxygen (element / gas)
O_3	Ozone
P	Phosphorus
SF_6	Sulphur hexafluoride
SO_2	Sulphur dioxide
SO_x	Sulphur oxides expressed in SO_2 mass equivalent
VOCs	Volatile Organic Compounds

Prefixes for basic physical units

Enlargement			Reduction		
name	symbol	factor	name	symbol	factor
deca	da	10^{+1}	deci	d	10^{-1}
hecto	h	10^{+2}	centi	c	10^{-2}
kilo	k	10^{+3}	milli	m	10^{-3}
mega	M	10^{+6}	micro	μ	10^{-6}
giga	G	10^{+9}	nano	n	10^{-9}
tera	T	10^{+12}	pico	p	10^{-12}
peta	P	10^{+15}	femto	f	10^{-15}
exa	E	10^{+18}	atto	a	10^{-18}

Annex III

Contributors to the IPCC WGIII Fourth Assessment Report

ACHARD, Frédéric
Joint Research Centre of the EC
Italy, France

ADEGBULUGBE, Anthony
Centre for Energy Research
and Development
Nigeria

AHMED, Mohammed Abdelrafie
University of Khartoum, Faculty
of Engineering & Architecture
Sudan

AHUJA, Dilip
National Institute of Advanced Studies
India

AKUMU, Grace
Climate Network Africa
Kenya

ALFSEN, Knut H.
Statistics Norway
Norway

ALHARTHI, Awwad
Saudi Aramco
Saudi Arabia

AMANN, Marcus
International Institute for
Applied Systems Analysis
Austria

AMBROSI, Philippe
World Bank
France

ANAYA, Carlos A.
Universidad Nacional Autónoma de México
México

ANDRASKO, Kenneth
USEPA
United States of America

APPS, Mike
Natural Resources Canada,
Canadian Forest Service
Canada

ASMUSSEN, Arne
Joanneum Research, Graz
Austria, Germany

AWERBUCH†, Shimon
United States of America

BABIKER, Mustafa
ARAMCO
Saudi Arabia, Sudan

BARKER, Terry
University of Cambridge, Dep.
of Land Economy, 4CMR
United Kingdom

BASHMAKOV, Igor Alexeyevich
Center for Energy Efficiency (CENEf)
Russia

BEALE, Roger
Allen Consulting Group
Australia

BENITEZ, Pablo
British Columbia Ministry of Environment
Canada, Equador

BERNSTEIN, Lenny
L. S. Bernstein & Associates, L.L.C.
United States of America

BERTOLDI, Paolo
Joint Research Centre of the EC
Italy

BEUTHE, Michel
Facultés Universitaires Catholique de Mons
Belgium

BIROL, Fatih
International Energy Agency
France, Turkey

BLOK, Kornelis
Ecofys
The Netherlands

BOARDMAN, Brenda
Oxford ECI
United Kingdom

BOER, Rizaldi
Climatology Laboratory
Indonesia

BOGNER, Jean E.
Landfills +, Inc, and University of Illiois
United States of America

BOSCH, Peter
Ecofys
The Netherlands

BOSE, Ranjan
TERI
India

BOUILLE, Daniel
Bariloche Foundation
Argentina

BRADLEY, Richard
International Energy Agency
United States of America

BRINKMAN, Sander
Brinkman Climate Change, Consultancy
The Netherlands

BROWN, Marilyn A.
Georgia Institute of Technology and
Oak Ridge National Laboratory
United States of America

BURCH, Sarah
University of British Columbia
Canada

BURGER, Nicholas
University of California, Santa Barbara
United States of America

CAI, Zucong
Institute of Soil Science, Chinese
Academy of Sciences
P.R. China

CALVO BUENDIA, Eduardo
Universidad Nacional Mayor de San Marcos
Peru

CHEN, Wenjun
Canada Centre for Remote Sensing
Canada

CIFUENTES, Luis
P. Universidad Catolica de Chile
Chile

CLAPP, Christa
U.S. Environmental Protection Agency
United States of America

CLARKE, Leon
Pacific Northwest National Laboratory
United States of America

COCKLIN, Christopher R
James Cook University
Australia, New Zealand

CORFEE-MORLOT, Jan
OECD
France, United States of America

CRABBÉ, Philippe J.
University of Ottawa
Canada, Belgium

DAVE, Rutu
Netherlands Environmental
Assessment Agency
The Netherlands

DE LA CHESNAYE, Francisco
U.S. Environmental Protection Agency
United States of America

Coordinating lead authors, lead authors, contributing authors and review editors are listed alphabetically by surname; when nationality is different from country of residence it is mentioned second.

DE LA RUE DU CAN, Stephane
Lawrence Berkeley National Laboratory
United States of America, France

DELHOTAL, Casey
Research Triangle Institute
(RTI) International
United States of America

DEN ELZEN, Michel
Netherlands Environmental
Assessment Agency
The Netherlands

DIAZ, Luis F.
CalRecovery, Inc.
United States of America

DIAZ MOREJON, Cristobal Felix
Ministry of Science, Technology
and the Environment
Cuba

DREXHAGE, John
International Institute for
Sustainable Development
Canada

DUAN, Maosheng
Tsinghua University
P.R. China

DUTSCHKE, Michael
Biocarbon Consult
Germany

EDENHOFER, Ottmar
Potsdam Institute for Climate
Impact Research
Germany

EDMONDS, Jae
Pacific Northwest National Laboratory
United States of America

EICKHOUT, Bas
Netherlands Environmental
Assessment Agency
The Netherlands

ELGIZOULI, Ismail
High Council for Environment and
Natural Resources (HCENR)
Sudan

ELSIDDIG, Elnour Abdalla
University of Khartoum
Sudan

FAAIJ, Andre P.C.
Copernicus Institute for Sustainable
Development - Utrecht University
The Netherlands

FENHANN, Jørgen
Risø National Laboratory, Technical
University of Denmark
Denmark

FISHER, Brian S.
CRA International
Australia

FLANNERY, Brian
Exxon Mobil Corporation
United States of America

FORD-ROBERTSON, Justin
Ford-Robertson Initiatives (FRI) Ltd
New Zealand

FRUMHOFF, Peter C.
Union of Concerned Scientists
United States of America

GAO, Qingxian
Chinese Research Academy of
Environmental Science
P.R. China

GARG, Amit
Government of India/UNEP Risø Centre
Denmark, India

GASCA, Jorge
Mexican Petroleum Institute
México

GEHL, Stephen
EPRI
United States of America

GENG, Luis J.
Consultant
Peru

GIELEN, Dolf
International Energy Agency
France, Netherlands

GOLAY, Michael
Massachusetts Institute of Technology
United States of America

GOLLIER, Christian
University of Toulouse
France, Belgium

GREENE, David L.
Oak Ridge National Laboratory
National Transport Research Center
United States of America

GREGORY, Robert
Golder Associates (UK) Ltd.
United Kingdom

GRUBB, Michael
Carbon Trust /Cambridge University/
Imperial College London
United Kingdom

GRÜBLER, Arnulf
International Institute for Applied
Systems Analysis/Yale University
Austria

GULLISON, Raymond E.
Hardner & Gullison Associates, LLC
Canada

GUPTA, Joyeeta
Institute for Environmental
Studies, Free University
The Netherlands

GUPTA, Sujata
Asian Development Bank
Philippines, India

GWARY, Daniel
University of Maiduguri
Nigeria

HA DUONG, Minh
Cired, Centre National de la
Recherche Scientifique
France

HAITES, Erik
Margaree Consultants Inc.
Canada

HALSNÆS, Kirsten
Risø National Laboratory, Technical
University of Denmark
Denmark

HANAOKA, Tatsuya
National Institute for Environmental Studies
Japan

HARE, Bill
Potsdam Institute for Climate
Impact Research
Germany, Australia

HARNISCH, Jochen
Ecofys
Germany

HARRIS, Jeffrey
Alliance to Save Energy
United States of America

HARVEY, Danny
University of Toronto
Canada

HASHIMOTO, Seiji
National Institute for Environmental Studies
Japan

HATA, Hiroshi
Railway Technical Research Institute
Japan

HAVES, Philip
Lawrence Berkeley National Laboratory
United States of America, United Kingdom

HAWKINS, David
Natural Resources Defence Council
United States of America

HAYAMI, Hitoshi
Keio University
Japan

HEGGEDAL, Tom-Reiel
Statistics Norway
Norway

HEIJ, BertJan
The Netherlands

HELLER, Thomas C.
Stanford University
United States of America

HIGUCHI, Niro
National Institute for Research
in the Amazon
Brazil

HOHMEYER, Olav
University of Flensburg
Germany

HÖHNE, Niklas
Ecofys
Germany

HOOGWIJK, Monique
Ecofys
The Netherlands

HOURCADE, Jean-Charles
Cired, Centre National de la
Recherche Scientifique
France

HOWDEN, S. Mark
CSIRO Sustainable Ecosystems
Australia

IBITOYE, Francis I.
Centre for Energy Research
and Development
Nigeria

IVANOVA BONCHEVA, Antonina
Autonomous University of
Southern Baja California
México

JANZEN, H. Henry
Agriculture and Agri-Food Canada
Canada

JEPMA, Catrinus J.
University of Groningen
The Netherlands

JIANG, Kejun
Energy Research Institute
P.R. China

JOBBAGY, Esteban
Universidad Nacional de San
Luis & CONICET
Argentina

JOCHEM, Eberhard
Fraunhofer Institute for Systems
and Innovation Research
Germany

JOOSEN, Suzanne
Ecofys
The Netherlands

KAHN RIBEIRO, Suzana
Federal University of Rio de Janeiro
Brazil

KAINUMA, Mikiko
National Institute for Environmental Studies
Japan

KANOAN, Gorashi Mohammed
University of Nizwa
Sultanate of Oman, Sudan

KARJALAINEN, Timo
Finnish Forest Research Insititue
Finland

KHESHGI, Haroon
ExxonMobil Research and
Engineering Company
United States of America

KJELDSEN, Peter
Technical University of Denmark
Denmark

KOBAYASHI, Shigeki
Toyota Central R&D Labs, Inc
Japan

KOLSTAD, Charles
University of California, Santa Barbara,
Dept of Economics & Bren School
United States of America

KONAR, Manaswita
Global Change Programme,
Jadavpur University
India

KONSTANTINAVICIUTE, Inga
Lithuanian Energy Institute
Lithuania

KRANKINA, Olga
Oregon State University
Department of Forest Science
United States of America, Russia

KRUGER, Joseph A.
National Commission on Energy Policy
United States of America

KUIJPERS, Lambert
Technical University Eindhoven
The Netherlands

KUMAR, Pushpam
University of Liverpool
UK, India

KURZ, Werner A.
Natural Resources Canada,
Canadian Forest Service
Canada

KVERNDOKK, Snorre
Ragnar Frisch Centre for
Economic Research
Norway

LA ROVERE, Emilio Lèbre
Federal University of Rio de Janeiro
Brazil

LANG, Siwei
China Academy of Building Research
P.R. China

LARSEN, Hans
Risø National Laboratory, Technical
University of Denmark
Denmark

LATHAM, John
University Corporation for
Atmospheric Research
United States of America

LECOCQ, Franck
Laboratory of Forestry Economics,
INRA-AgroParisTech (ENGREF)
France

LEE, David S.
Manchester Metropolitan University
United Kingdom

LEE, Hoesung
Council on Energy and Environment
Korea/Keimyung University
Korea

LEFEVRE, Nicolas
International Energy Agency
France, United States of America

LEVERMORE, Geoffrey J.
The Univerity of Manchester
United Kingdom

LEVINE, Mark D.
Lawrence Berkeley National Laboratory
United States of America

LLANES-REGUEIRO, Juan F.
Havana University, Centre for
Environmental Studies
Cuba

MALONE, Elizabeth L.
Joint Global Change Research Institute
United States of America

MARECKOVA, Katarina
Umweltbundesambt
Austria, Slovakia

MARKANDYA, Anil
University of Bath/FEEM
Italy, United Kingdom

MARTINO, Daniel L.
Carbosur
Uruguay

MARTINOT, Eric
Tsinghua University
P.R. China, United States of America

MASERA, Omar
Centro de Investigaciones en
Ecosistemas, UNAM
México, Argentina

MASTRANDREA, Michael D.
Stanford University
United States of America

MATSUHASI, Ryuji
Tokyo University
Japan

MATSUMOTO, Mitsuo
Forestry and Forest Products
Research Institute
Japan

MATSUOKA, Yuzuru
Kyoto University
Japan

MATYSEK, Anna
CRA International
Australia

McALLISTER, Tim
Agriculture and Agri-Food Canada
Canada

McCARL, Bruce
Texas A&M University
United States of America

McFARLAND, Mack
DuPont Fluoroproducts
United States of America

MEHLWANA, Monga
CSIR, Natural Resources
and the Environment
South Africa

METZ, Bert
Netherlands Environmental
Assessment Agency
The Netherlands

MEYER, Leo
Netherlands Environmental
Assessment Agency
The Netherlands

MICHAELOWA, Axel
University of Zurich
Switzerland, Germany

MINER, Reid
NCASI
United States of America

MIRASGEDIS, Sevastian
National Observatory of Athens
Greece

MOEZZI, Mithra
HELIO International
United States of America

MONNI, Suvi
Benviroc Ltd
Finland

MOOMAW, William
US The Fletcher School of Law and
Diplomacy Tufts University
United States of America

MOREIRA, Jose Roberto
National Reference Center on
Biomass, Institute of Electr.&
Energy, Univ. of Sao Paulo
Brazil

MUNASINGHE, Mohan
Munasinghe Institute for Development
Sri Lanka

MURASE, Shinya
Faculty of Law, Sophia University, Tokyo
Japan

MUROMACHI, Yasunori
Tokyo Institute of Technology
Japan

NABUURS, Gert-Jan
Alterra
The Netherlands

NAJAM, Adil
Boston University Center for Energy
and Environmental Studies
United States of America, Pakistan

NAKICENOVIC, Nebojsa
International Institute for Applied Systems
Analysis/Vienna University of Technology
Austria, Montenegro

NEWTON, Peter J.
Dept. for Business, Enterprise
and Regulatory Reform
United Kingdom

NIKITINA, Elena
Russian Academy of Science Inst.of
World Economy and Int. Relations
Russia

NIMIR, Hassan B.
University of Khartoum, Dept.
of Petroleum Engineering
Sudan

NOVIKOVA, Aleksandra
Central European University
Hungary, Russia

ODINGO, Richard
University of Nairobi
Kenya

OGLE, Stephen
Colorado State University
United States of America

OKAZAKI, Teruo
Nippon Steel Corporation
Japan

OLIVIER, Jos G.J.
Netherlands Environmental
Assessment Agency
The Netherlands

O'MARA, Frank P.
University College Dublin
Ireland

O'NEILL, Brian
International Institute for
Applied Systems Analysis
Austria, United States of America

OPSCHOOR, Hans (J.B.)
Insititute of Social Studies/Free University
The Netherlands

OYHANTÇABAL, Walter
Minstry of Livestock, Agriculture
and Fisheries. Uruguay
Uruguay

PAN, Genxing
Nanjing Agricultural University
P.R. China

PAN, Jiahua
USDA Forest Service
P.R. China

PERSHING, Jonathan
World Resources Institute
United States of America

PETSCHEL-HELD†, Gerhard
Germany

PICHS, Ramon
CIEM
Cuba

PIPATTI, Riitta
Statistics Finland
Finland

PITCHER, Hugh M.
Joint Global Change Research Institute
United States of America

PIZER, William A.
Resources for the Future
United States of America

PLOTKIN, Steven
Argonne National Laboratory
United States of America

POPP, David
The Maxwell School, Syracuse University
United States of America

PRICE, Lynn K.
Lawrence Berkeley National Laboratory
United States of America

RANA, Ashish
Reliance Industries Ltd.
India

RAO, Shilpa
International Institute of
Applied Systems Analysis
Austria, India

RAVINDRANATH, N. H.
Indian Institute of Science
India

RAYNER, Steve
James Martin Institute,
University of Oxford
United Kingdom, United States of America

READ, Peter
Massey University
New Zealand

RIAHI, Keywan
International Institute for
Applied Systems Analysis
Austria

RICE, Charles
Kansas State University
United States of America

RICHELS, Richard G.
Electric Power Research Institute
United States of America

RILLING, Jacques
CSTB Building Research Center
France

ROBINSON, John
UBC
Canada

ROGNER, H-Holger
International Atomic Energy Agency
Austria, Germany

ROMANENKOV, Vladimir
All-Russian Institute of Agrochemistry
named after D.Pryanishnikov
Russia

ROSE, Steven K.
United States Environmental
Protection Agency
United States of America

ROY, Joyashree
Jadavpur University
India

RYPDAL, Kristin
CICERO
Norway

SAIJO, Tatsuyoshi
Osaka University
Japan

SANDERS, Johan
Wageningen University and Research Centre
The Netherlands,

SÁNZ SÁNCHEZ, María José
Fundación CEAM
Spain

SARI, Agus
PT Ecosecurities
Indonesia

SATHAYE, Jayant
Lawrence Berkeley National Laboratory
United States of America

SCHAEFFER, Roberto
Federal University of Rio de Janeiro
Brazil

SCHLAMADINGER, Bernhard
Joanneum Research, Graz
Austria

SCHLEICHER, Stefan P.
University of Graz
Austria

SCHNEIDER, Uwe A.
Hamburg University
Germany

SCHOCK, Robert
Lawrence Livermore National
Laboratory/World Energy Council
United States of America/United
Kingdom, United States of America

SCHOLES, Robert (Bob) J.
Council for Scientific and
Industrial Research
South Africa

SEKI, Shigetaka
Ministry of Economy Trade and Industry
Japan

SHEINBAUM PARDO, Claudia
Universidad Nacional Autónoma de México
México

SHUKLA, Priyadarshi
Indian Institute of Management, Ahmedabad
India

SILVEIRA SOARES FILHO, Britaldo
Centro de Sensoriamento Remoto
Brazil

SIMS, Ralph E. H.
Massey University/International
Energy Agency
New Zealand

SIROTENKO, Oleg D.
All-Russian Research Ins. For
Agricultural Meteorology
Russia

SKJOLSVIK, Kjell Olav
Marintek
Norway

SMITH, Kirk R.
University of California / Woods
Hole Research Center
United States of America

SMITH, Michael H.
University of Leeds
United Kingdom

SMITH, Peter
University of Aberdeen
United Kingdom

SOHNGEN, Brent
The Ohio State University
United States of America

SOKONA, Youba
Sahara and Sahel Observatory (OSS) Tunis
Tunisia, Mali

SPERLING, Daniel
University of California, Davis
United States of America

STRENGERS, Bart
Netherlands Environmental
Assessment Agency
The Netherlands

SUGIYAMA, Taishi
Central Research Institute of
Electric Power Industry
Japan

SUTAMIHARDJA, R.T.M.
Ministry of Environment
Indonesia

SWART, Rob J.
Netherlands Environmental
Assessment Agency
The Netherlands

TANAKA, Kanako
International Energy Agency
France, Japan

TIRPAK, Dennis A.
UNFCCC Secretariat
France, United States of America

TORRES-MARTÍNEZ, Julio
Cuban Observatory for
Science and Technology
Cuba

TOTH, Ferenc L.
International Atomic Energy Agency/
Corvinus University of Budapest
Austria, Hungary

TOWPRAYOON, Sirintornthep
KM University
Thailand

TRINES, Eveline
Treeness Consult
The Netherlands

TURNER, Clive
Eskom
South Africa

UCHIYAMA, Yohji
The University of Tsukuba
Japan

URGE-VORSATZ, Diana
Central European University, Dept. of
Environmental Sciences and Policy
Hungary

VAN KOOTEN, G. Cornelis
University of Victoria
Canada

VAN VUUREN, Detlef P.
Netherlands Environmental
Assessment Agency
The Netherlands

VERBRUGGEN, Aviel
University of Antwerp
Belgium

VILLAVICENCIO, Arturo
Equador

VUORI, Seppo
Technical Research Centre of Finland
Finland

WA GITHENDU, Mukiri
Ministry of Science and Technology
Kenya

WAMUKONYA, Njeri
Division of Policy Development and Law
Kenya

WARREN, Rachel
Tyndall Centre for Climate Change
Research, School of Environmental
Sciences, University of East Anglia
United Kingdom

WATTENBACH, Martin
University of Aberdeen
United Kingdom, Germany

WEYANT, John
Stanford University
United States of America

WINKLER, Harald
University of Cape Town
South Africa

WIT, Ron
Stichting Natuur en Milieu/CE
The Netherlands

WORRELL, Ernst
Lawrence Berkeley National
Laboratory/Ecofys
The Netherlands

WYATT, Andrew B.
University of Sydney
Australia

XU, Huaging
Chinese Acedemy of Sciences
P.R. China

YAMAGUCHI, Mitsutsune
The University of Tokyo
Japan

YAMAJI, Kenji
The University of Tokyo
Japan

YAMBA, Francis
University of Zambia
Zambia

YANG, Hongwei
Energy Research Institute
P.R. China

YOSHINO, Hiroshi
Tohoku University
Japan

ZHANG, Tianzhu
Dept. of Environmental Science and
Engineering, Tsinghua University
P.R. China

ZHANG, Xiaoquan
Chinese Academy of Forestry
P.R. China

ZHANG, Xiliang
Tsinghua University
P.R. China

ZHOU, Dadi
Energy Research Institute
P.R. China

ZHOU, Fengqi
Energy Research Institute of NDRC
P.R. China

ZHOU, Peter P.
EECG Consultants (Pty) Ltd.
Botswana, Zimbabwe

ZOU, Ji
Renmin University of China
P.R. China

Annex IV

Reviewers of the IPCC WGIII Fourth Assessment Report

Argentina

Ginzo, Hector
Ministerio de Relaciones Exteriores,
Comercio Internacional y Culto

Pedace, Alberto
Buenos Aires Univertsity

Taboada, Miguel Angel
Facultad de Agronomía

Australia

Betz, Regina Annette
University of New South Wales (UNSW)

Denniss, Tom
Energetech Australia Pty Ltd

Edwards, Spencer
Australian Greenhouse Office

Enting, Ian
MASCOS

Gifford, Roger
CSIRO

Jotzo, Frank
Australian National University

MacGill, Iain
University of NSW

Pezzey, Jack
Australian National University

Quiggin, John
University of Queensland

Raupach, Michael
CSIRO Marine and Atmospheric Research

Trewin, Dennis
Australian Bureau of Statistics

Uddin, NOIM
Macquarie University, Sydney

Austria

Amon, Barbara
Institute of Agricultural Engineering

Clemens, Torsten
OMV E&P

Furukawa, Michinobu
International Institute for
Applied Systems Analysis

Hojesky, Helmut

Klaus, Radunsky
Umweltbundesamt

Kolp, Peter
IIASA

Ma, Tieju
International Institute for
Applied Systems Analysis

Manfred, Lexer
University of Natural Resources and
Applied Life Sciences, Vienna

Miketa, Asami
International Atomic Energy Agency

Rogner, H-Holger
IAEA

Schrattenholzer, Leo
IIASA

Shihab-Eldin, Adnan
OPEC

Belgium

Botschek, Peter
European Chemical Industry Council (Cefic)

Claude, Lorea
CEMBUREAU

Debruxelles, Jean-Pierre
EUROFER

Friberg, Lars
Climate Action Network (CAN) Europe

Herbst, Stephan
Toyota Motor Europe

Kjaer, Christian
European Wind Energy Association

Kris, Piessens
Royal Belgian Institute of Naturals Sciences

Lorea, Claude
CEMBUREAU, The European
Cement Industry

Macey, Kirsten
Climate Action Network Europe

Martchek, Kenneth
Alcoa

Tulkens, Henry
Center for Operations Research
and Econoetrics (CORE)

Van Houte, Frédéric
Comité Permanent des Industries
du Verre Européennes

Van Wesemael, Bas
Université catholique de Louvain

Van Ypersele, Jean-Pascal
Université catholique de Louvain

Vanderstraeten, Martine

Verbruggen, Aviel
University of Antwerp

Wittoeck, Peter
Belgian Federal Administration

Bénin

Guendehou, G. H. Sabin
Benin Centre for Scientific
and Technical Research

Ibouraïma, Yabi
LECREDE/DGAT/UAC

Vissin, Expédit Wilfrid
FLASH/Abomey-Calavi University

Bolivia

Halloy, Stephan
Universidad Mayor de San Andrés

Trujillo Blanco, Ramiro Juan
National Programme on Climate Changes

Brazil

Barbosa da Silva, Demóstenes
AES Brazil

Lima, Magda Aparecida
Brazilian Agricultural Research
Corporation - Embrapa

Machado, Pedro
Embrapa Rice and Beans

Moreira, Jose Roberto
Institute of Electrotechnology and Energy,
University of Sao Paulo-IEE-USP

Volpi, Giulio
WWF International

Canada

Drexhage, John
International Institute for
Sustainable Development

Edwards, Patti
Environment Canada

Etcheverry, Jose
David Suzuki Foundation

Gagnon, Luc
Hydro-Quebec

Gilbert, Richard
Centre for Sustainable Transportation

Green, Christopher
McGill University

Harvey, Danny
University of Toronto

Hupe, Jane
ICAO - International Civil
Aviation Organization (UN)

Hyndman, Richard
Canadian Association of
Petroleum Producers

Jackson, David
McMaster University

Meadowcroft, James
Carleton University

Mirza, Monirul
Adaptation and Impacts Research
Division (AIRD), Environment Canada

Nyboer, John
Simon Fraser University

Sheppard, Stephen
University of British Columbia

Stewart, Cohen
Environment Canada

Turk, Eli
Canadian Electricity Association (CEA)

Chile

Fernando, Farias
CONAMA (IPCC Chilean focal point)

Meza, Francisco
Facultad de Agronomia. Pontificia
Universidad Catolica de Chile

Costa Rica

Locatelli, Bruno
CIRAD-CATIE

Cuba

Diaz Morejon, Cristobal Felix
Ministry of Science, Technology
and the Environment

Hernández, Gladys Cecilia
Centre for World Economy Research

Llanes-Regueiro, Juan F.
Havana University

Somoza, José
National Institute of Economic Research

Torres-Martinez, Julio
Cuban Observatory for
Science and Technology

Czech Republic

Jan, Pretel
Climate Change

Novotny, Vladimir
International Lime Association ILA

Denmark

Fenhann, Jørgen
Risø

Jesper, Jensen
J-Consulting ApS

Juhler-Kristoffersen, Helle
Confederation of Danish Industries

Estonia

Ahas, Rein
University of Tartu

European Community

Lars, Müller
Directorate General Environment

Finland

Hamalainen, Raimo
Helsinki University of Technology

Käyhkö, Jukka
University of Turku

Melanen, Matti
Finnish Environment Institute

Saari, Mikko
Technical Research Centre of Finland VTT

Savolainen, Ilkka
Technical Research Centre of Finland VTT

Syri, Sanna
VTT

France

Birol, Fatih
International Energy Agency

Bonduelle, Antoine
E&E_Consultant

Braathen, Nils-Axel
OECD

Campbell, Nick
ARKEMA SA

Caneill, Jean-Yves
Electricité de France

Cazzola, Pierpaolo
IEA

Ceron, Jean-Paul
CRIDEAU/Université de Limoges

Corfee-Morlot, Jan
University College London / OECD

Czernichowski-Lauriol, Isabelle
BRGM

De T'Serclaes, Philippine
International Energy Agency

Dellero, Nicole
Corporate Strategy AREVA

Ghersi, Frédéric
CNRS

Gielen, Dolf
International Energy Agency

Gordelier, Stanley
Nuclear Energy Agency of the OECD

Gundogdu, Niyazi
La Farge Group

Hallegatte, Stephane
Météo-France

Jacques, Leonardi
Institut National de Recherche sur
les Transports et leur Sécurité

Karousakis, Katia
OECD

Ladoux, Norbert
University of Toulouse and IDEI

Leonardi, Jacques
INRETS - Institut National de Recherche
sur les Transports et leur Sécurité

Levina, Ellina
OECD

Marc, Gillet
ONERC

Mocilnikar, Antoine-Tristan
Délégué Interministériel au
Développement Durable

Muirheid, Ben
International Fertilizer Industry
Association (IFA)

Paillard, Michel
IFREMER

Perkins, Stephen
European Conference of Ministers
of Transport (ECMT)

Petit, Michel
CGTI

Philibert, Cédric
International Energy Agency

Poot, Brigitte
Total s.a.

Puig, Daniel
United Nations Environment Programme

Ricketts, Brian
International Energy Agency

Riedacker, Arthur
INRA

Rilling, Jacques
CSTB Building Research Center

Ruffing, Kenneth
Non-affiliated

Sinton, Jonathan
International Energy Agency

Stevens, Candice
OECD

Sukalac, Kristen Elizabeth
International Fertilizer Industry
Association (IFA)

Taylor, Michael
International Energy Agency

Taylor, Peter
International Energy Agency

Varet, Jacques
BRGM-French Geological Survey

Wickstrom, Johanna
World LP Gas Association

Germany

Brockhagen, Dietrich
Atmosfair gGmbH

Bruckner, Thomas
Technical University of Berlin

Fuentes, Ursula
German Federal Environment Ministry

Glauner, Reinhold
Institute for World Forestry

Grassl, Hartmut
Max Planck Institute for Meteorology

Hein, Joachim
BDI - Federation of German Industries

Jung, Martina
Ecofys

Kemfert, Claudia
German Institute for Economic Research

Kohlhaas, Michael
German Institute for Economic Research

Kriegler, Elmar
Potsdam Institute for Climate
Impact Research

Kuckshinrichs, Wilhelm
Forschungszentrum Juelich GmbH

Kulessa, Margareta
Mainz University of Applied Sciences

Martin, Janicke
Forschungsstelle für Umweltpolitik

Meyer, Bernd
University of Osnabrueck and GWS mbH

Ott, Hermann E.
Wuppertal Institute for Climate,
Environment and Energy

Peterson, Sonja
Kiel Institute for World Economics

Pitz-Paal, Robert
German Aerospace Centre (DLR)

Plöger, Friedrich
Siemens AG

Renn, Ortwin Wolfgang
Institute for Social Science,
University of Stuttgart

Reusswig, Fritz
Potsdam Institute for Climate
Impact Research

Rothermel, Joerg
German Chemical Industry Association

Schleich, Joachim
Fraunhofer Institute Systems
and Innovation Research

Schwarze, Reimund
DIW Berlin

Supersberger, Nikolaus
Wuppertal Institute for Climate
Environment and Energy

Tol, Richard
Hamburg University

Treber, Manfred
Germanwatch

Von Goerne, Gabriela
Greenpeace

Walz, Rainer
Fraunhofer Institute Systems
and Innovation Research

Weimer-Jehle,
Institute for Social Science,
University of Stuttgart

Greece

Georgopoulou, Elena
National Observatory of Athens

Hungary

Urge-Vorsatz, Diana
Central European University

Zoltán, Somogyi
Forest Research Institute

India

Chatterjee, Anish
Development Alternatives

Grover, Ravi B
Department of Atomic Energy

Roy, Joyashree
Jadavpur University

Tulkens, Philippe
TERI School of Advanced Studies

Subodh, Sharma
Ministry of Environment and Forests

Ireland

Finnegan, Pat
Grian

McGovern, Frank
Environmental Protection Agency

Israel

Deutscher, Guy
Tel Aviv University

Italy

Caserini, Stefano
Politecnico di Milano

Dragoni, Walter
Perugia University

Romeri, Mario Valentino

Japan

Akimoto, Keigo
Resaerch Institute of Innovative
Technology for the Earth (RITE)

Amano, Akihiro
University of Hyogo

Chiba, Yukihiro
Forestry and Forest Products
Research Institute

Fujino, Junichi
National Institute for Environmental Studies

Hashimoto, Seiji
National Institute for Environmental Studies

Hirata, Kimiko
Kiko Network

Hori, Masahiko
Japan Automobile Research Institute

Ichinose, Toshiaki
National Institute for Environmental Studies

Inoue, Takashi
Tokyo University of Science

Ishii, Atsushi
Tohoku University

Itoh, Kiminori
Yokohama National University

Izumi, Yoshito
Taiheiyo Cement Corporation

Kadono, Koji
Global Industrial and Social
Progress Research Institute

Kaibara, Makoto
Matsushita Electric Industrial Co,. Ltd.

Kainuma, Mikiko
National Institute for Environmental Studies

Kanie, Norichika
Tokyo Institute of Technology

Kawase, Reina
Kyoto University

Kiyono, Yoshiyuki
Forestry and Forest Products
Research Institute

Kobayashi, Noriyuki
Nihon University

Koide, Hitoshi
Waseda University

Komiyama, Ryoichi
The Institute of Energy
Economics, Japan (IEEJ)

Kosugi, Takanobu
Ritsumeikan University

Maeda, Akira
Kyoto University

Masui, Toshihiko
National Institute for Environmental Studies

Miura, Shuichi
Tohoku university of art & design

Mizuno, Koichi
National Institute of Advanced
Industrial Science and Technology

Mori, Shunsuke
Tokyo University of Science

Murayama, Shigeo
The Federation of Electric
Power Companies

Nakakuki, Shinichi
Tokyo Electric Power Company

Nakamaru, Susumu
Sun Management Instutute

Nakamura, Takuya
Mizuho Information & Research Institute

Naoto, Hisajima
Climate Change Division,
International Cooperation Bureau

Nishio, Masahiro
Ministry of Economy, Trade and Industry

Nishioka, Shuzo
National Institute for Environmental Studies

Noriyuki, Kobayashi

Okayama, Takayuki
Tokyo University of Agriculture
and Technology

Omori, Ryota
Japan Science and Technology Agency

Onuma, Eiichi

Oshima, Kenichi
Ritsumeikan University

Sano, Fuminori
Research Institute of Innovative
Technology for the Earth

Seki, Shigetaka
METI

Shimoda, Yoshiyuki
Osaka University

Shirato, Yasuhito
Agriculture, Forestry and
Fisheries Research Council

Sugiyama, Taishi
CRIEPI

Takenaka
Japan Gas Association

Takeshita, Takayuki
The University of Tokyo

Tonooka, Yutaka
Saitama University

Tonosaki, Mario
Forestry and Forest Products
Research Institute

Uezono, Masatake
Citizens' Alliance for saving the
Atmosphere and the Earth

Xue, Ziqiu
Research Institute of Innovative
Technology for the Earth

Yagi, Kazuyuki
National Institute for Agro-
Environmental Sciences

Yamaguchi, Mitsutsune
Teikyo University

Yanagisawa, Yukio
The University of Tokyo

Yaoita, Yasuhisa
Global Industrial and Social
Progress Research Institute

Yoshida, Satoshi
Tokyo Gas Company

Kenya

Demkine, Volodymyr
UNEP

Korea

Kang, Yoon-Young
Korea Energy Economics Institute

Noh, Dong-Woon
Korea Energy Economics Institute

Malawi

Donald Reuben, Kamdonyo
Meteorological Services

Kainja, Samuel
Malawi Water Partnership

Malaysia

Chow, Kok Kee
Malaysian Meteorological Deparment

México

Aguayo, Francisco
El Colegio de México

Gasca, Jorge
Mexican Petroleum Institute

Romero Lankao, Patricia
Metropolitan Autonomous
University, campus Xochimilco

Morocco

Senhaji, Faouzi
I.A.V. Hassan II   (GERERE)

New Zealand

Baisden, W. Troy
Landcare Research

Beard, Catherine
Greenhouse Policy Coalition (NGO
representing energy intensive sector)

Becken, Susanne
Landcare Research

Chapman, Ralph
Victoria University of Wellington

Gray, Vincent
Climate Consultant

Isaacs, Nigel
Branz Ltd

Law, Cliff
National Institute of Water and
Atmospheric Research

Maclaren, Piers
Piers Maclaren & Associates

Read, Peter
Massey University

Walton, Darren
Opus Central Laboratories

Nicaragua

Ebinger, Michael
Atmosphere, Climate, &
Environmental Dynamics (EES-2)

Norway

Aunan, Kristin
Center for International Climate and
Environmental Research-Oslo

Berntsen, Terje
CICERO

Fuglestvedt, Jan
CICERO

Gerlagh, Reyer
Centre for Advanced Study

Jenssen, Tore
Yara Hesq

Kjell, Oren
Norsk Hydro ASA

Oren, Kjell
Norsk Hydro ASA

Øyvind, Christophersen
Norwegian Pollution Control Authority

Schei, Tormod André
Statkraft AS

Torvanger, Asbjørn
CICERO

Vennemo, Haakon
ECON

P. R. China

Chen, Wenying
Tsinghua University

Dahe, Qin
China Meteorological Administration

Guo, Yuan
Energy Research Institute, National
Development and Reform Commission

Hongfa, Di
Tsinghua University

Junfeng, Li
Energy Research Institute of
National Development and
Reform Committee
and Secretary General of China Renewable
Energy Industry Association.

Wang, Yanjia
Tsinghua University

Wu, Shaohong
Institute of Geographical Sciences
and Natural Resources Research,
Chinese Academy of Sciences

Zhao, Yong
China Huaneng Technical and
Economic Research Institute

Pakistan

Athar, Ghulam Rasul
Pakistan Atomic Energy Commission

Latif, Muhammad
Pakistan Atomic Energy Commission

Shahzad, Iqbal
National Conservation Strategy (NCS) Unit

Peru

Bartra, Valentin
Instituto Andino y Amazónico
de Derecho Ambiental

Philippines

Guha, Ajay
Asian Development Bank

Poland

Zyrina, Olga
MCPFE, LU Warsaw

Russia

Shmakin, Andrey
Institute of Geography, Russian
Academy of Sciences.

Saudi Arabia

Al-Fehaid, Mohammed
Ministry of Petroleum

Senegal

Thiam, Nogoye
ENDA-TM

South Africa

Asamoah, Joe
International Energy Foundation

Winkler, Harald
University of Cape Town

Spain

Abanades-García, Juan Carlos
Instituto Nacional del Carbon-CSIC

Catalan, Jordi

Ciscar, Juan Carlos
IPTS, European Commission

Hernández, Félix
Economía y Geografía. Consejo Superior
de Investigaciones Científicas (IEG-CSIC)

Löschel, Andreas
European Commission, DG Joint
Research Centre, Institute for
Pospective Technological Studies

Pardo Buendía, Mercedes
Political Science and Sociology Department

Pedro, Fernández
Ordenación del Territorio,
Urbanismo y Medio Ambiente. ETSI
Caminos, Canales y Puertos

Rafael, Martínez Carrasco
Instituto de Recursos Naturales y
Agrobiologia de Salamanca

Russ, Peter
IPTS, Joint Research Centre,
European Commission

Sanz Sanchez, M.J.
Fundación CEAM

Yábar Sterling, Ana
Institute of Environmental Studies

Sri Lanka

Munasinghe, Mohan
Munasinghe Institute for
Development (MIND)

Sweden

Berndes, Göran
Chalmers University of Technology

Berntsson, Thore
Chalmers

Mattias, Lundblad
Climate Unit

Möllersten, Kenneth
Swedish Energy Agency

Sterner, Thomas
Univ of Göteborg

Switzerland

Aebischer, Bernard
CEPE/ETHZ

Baranzini, Andrea
Geneva School of Business Administration

Barreto, Leonardo
Paul Scherrer Institute

Hardeman, Andreas
International Air Transport Association

Koch, Lorenz
World Business Council for Sustainable
Development (WBCSD)

Kreuzer, Michael
Institute of Animal Science, Swiss
Federal Institute of Technology (ETH)

Marcu, Andrei
IETA

Mazzotti, Marco
Institute of Process Engineering

Michael, Kreuzer
Institute of Animal Science

Romero, Jose
Federal Office for the Environment

The Netherlands

Annema, Jan-Anne
Ministerie V & W

Beerkens, Ruud
TNO Science & Industry

Berk, Marcel
Netherlands Environmental
Assessment Agency

Bosch, Peter
IPCC TSU WGIII

Brinkman, Sander
TSU WG III / Brinkman Climate Change

Brok, Paul
National Aerospace Laboratory NLR

Bruggink, Jos J.C.
Energy Research Centre of
the Netherlands ECN

Dave, Rutu
IPCC WGIII TSU

De Coninck, Heleen
Energy Research Centre of
the Netherlands ECN

De Gans, Gert
Kerkinactie

De Lange, Theo J.
Energy Research Centre of
the Netherlands ECN

De Wilde, Hein
Energy Research Centre of
the Netherlands ECN

Den Elzen, Michel
Netherlands Environmental
Assessment Agency

Eerens, Hans
Netherlands Environmental
Assessment Agency

Elbersen, Berien
WUR-Alterra

Elbersen, Wolter
Agrotechnolgy and Food
Sciences Groep of WUR

Faber, Albert
Netherlands Environmental
Assessment Agency

Fiechter, Wilco
Nederlandse Spoorwegen (Dutch Railways)

Groenenberg, Heleen
Energy Research Centre of
the Netherlands ECN

Gunning, Jan Willem
Free University, Amsterdam

Hoogwijk, Monique
Ecofys

Hutjes, Ronald
Alterra

Jansen, Daniel
Energy Research Institute of
the Netherlands ECN

Jeeninga, Harm
Energy Research Institute of
the Netherlands ECN

Kemp, Rene
Erasmus University

Kessels, John
Energy Research Institute of
the Netherlands ECN

Kok, Marcel T.J.
Netherlands Environmental
Assessment Agency

Kornelis, Blok
Ecofys

Kram, Tom
Netherlands Environmental
Assessment Agency

Kruijt, Bart
Alterra

Kuikman, Peter
Alterra

Labohm, Hans H.J.
Consultant

Metz, Bert
IPCC WGIII

Misdorp, Robbert
Ministry of Transport and Public Works

Olivier, Jos
Netherlands Environmental
Assessment Agency

Oonk, Hans
TNO

Pols, Donald
Friends of the Earth Netherlands/
Milieudefensie

Ren, Tao
Utrecht University

Ronald, Flipphi
Dutch Ministry of the
Environment (VROM)

Rothman, Jan
Erasmus University

Ruijgrok, Walter
EnergieNed

Sawyer, Steve
Greenpeace International

Schyns, Vianney
DSM and SABIC

Seebregts, Ad
Energy Research Institute of
the Netherlands ECN

Sijm, Jos
Energy Research Institute of
the Netherlands ECN

Sinke, Wim
Energy Research Institute of
the Netherlands ECN

Skutsch, Margaret
University of Twente

Smulders, Sjak
Tilburg University

Stead, Dominic
Delft University of Technology

Stevens, Arjette
Natuur en Milieu

Storm van Leeuwen, Jan Willem
Ceedata Consultants

Swart, Rob
Netherlands Environmental
Assessment Agency

Teske, Sven
Greenpeace International

Trines, Eveline
Treeness Consult

Van Bussel, Gerard
Section Wind Energy, Faculty LR, TU Delft.

Van der Meer, Peter
Alterra

Van der Schoor, Tineke
Sustainability Centre Lauwersoog/
RUG-Bedrijfskunde

van der Zwaan, Bob
Energy Research Institute of
the Netherlands ECN

Van Soest, Jan Paul
Advies voor Duurzaamheid on request of International Gas Union

Van Wee, Bert
Delft University of Technology

Verhagen, Jan
Plant Research International, Wageningen UR

Tunisia

Osman, Néjib
CIEDE/Agence Nationale pour la Maîtrise de l'Energie

United Kingdom

Ashford, Paul
Calebgroup

Awerbuch, Shimon
SPRU - University of Sussex

Barker, Terry
4CMR Centre for Climate Change Mitigation Research, University of Cambridge

Chase, Robert
International Aluminium Institute

Cinzia, Losenno
Global Atmosphere Division

Cobb, Jonathan
World Nuclear Association

Cook, Ian
United Kingdom Atomic Energy Authority

Copley, Christine
World Coal Institute

Curran, James
Scottish Environmental Protection Agency

Dlugolecki, Andrew
University of east anglia

Gardiner, Ann
AEA Technology

Green, Richard
University of Birmingham

Grubb, Michael
Carbon Trust/Cambridge University/ Imperial College London

Hamilton, Kirsty
Chatham House/UK Business Council for Sustainable Energy

Hedger, Merylyn
Environment Agency

Jefferson, Michael
World Renewable Energy Network & Congresses

Köhler, Jonathan
Tyndall Centre, University of Cambridge

Lawrence, Martin
Earth & Ocean Carbon Ltd

McCulloch, Archie
Marbury Technical Consulting

Mottershead, Chris
BP

Parker, Claire
Environmental Policy Consultant

Pearce, Catherine
Friends of the Earth International

Phylipsen, Dian
Ecofys

Reisinger, Andy
TSU IPCC Synthesis Report

Tanner, Thomas
Institute of Development Studies

Viner, David
University of East Anglia

Warren, Luke
IPIECA

Warren, Rachel
University of East Anglia

United States of America

Ackerman, Frank
Global Development and Environment Institute, Tufts University

Anderson, Peter
Center for a Competitive Waste Industry / RecycleWorlds Consulting Corp

Arquit Niederberger, Anne
Policy Solutions

Baer, Paul
Stanford University

Bahor, Brian
Covanta Energy Corporation

Baughcum, Steven
Boeing Company

Bernstein, Lenny
L. S. Bernstein & Associates, L.L.C.

Bero, James
BASF Corporation

Blanford, Geoffrey
Stanford University

Bogner, Jean
Landfills +, Inc.

Bowman, Michael
GE Global Research

Boyden, James
Vulcan Inc.

Brieno Rankin, Veronica
CH2M HILL

Brown, Donald
Pennsylvania Consortium for Interdisciplinary Environmental Policy

Brown, Sandra
Winrock International

Bull, Stan
National Renewable Energy Laboratory

Clemmer, Steve
Union of Concerned Scientists

Cointreau, Sandra
World Bank

Crawford, James
Trane/American Standard

Danilin, Michael
The Boeing Company

De la Chesnaye, Francisco
USEPA

Delhotal, Katherine Casey
Research Trinagle Institute

Diaz, Luis
CalRecovery, Inc.

Doctor, Richard
Argonne National Laboratory

Dooley, James
Battelle

Emerson, Sarah
Energy Security Analysis, Inc

Epstein, Paul
Harvard Medical School

Freedman, Steven
Energy Consultant

Fuessel, Hans-Martin
Stanford University

Gallagher, Kelly Sims
Harvard University

Ghosh, Debyani
Global Energy Partners, LLC

Golay, Michael
MIT

Goldston, Robert
Princeton Plasma Physics Laboratory

Golub, Alexander
Environmental Defense

Heil, Mark
U.S. Environmental Protection Agency

John Oliver, Niles
Tropical Forest Group

Jones, Russell
American Petroleum Institute

Kheshgi, Haroon
ExoonMobil Research and
Engineering Company

Kirilenko, Andrei
University of North Dakota

Klages-Buechner, Sabine
E.I. du Pont de Nemours and Company

Kruger, Dina
Environmental Protection Agency

Larson, Robert
US Environmental Protection Agency

Lee, Arthur
Chevron Corporation

Lewis, Joanna
Pew Center on Global Climate Change

Lynd, Lee
Dartmouth College

Malone, Elizabeth L
Pacific Northwest National Laboratory

Mark, Jason
Union of Concerned Scientists

Marks, Jerry
International Aluminium Institute/J
Marks & Associates

Maurice, Lourdes
US Government

Michaels, Patrick
University of Virginia and Cato Insitutute

Miller, Alan
International Finance Corporation

Miner, Reid
NCASI

Montgomery, David
CRA International

Murray, Brian
Center for Regulatory Economics
and Policy Research

Niles, John
Coalition for Rainforest Nations

Nissen, David
Columbia University

Park, Jacob
Green Mountain College

Pizer, William
Resources for the Future

Price, Jeff
California State University, Chico

Price, Lynn
Lawrence Berkeley National Laboratory

Ragland, Jim
Aramco Services Company

Raskin, Paul
Tellus Institute

Rose, Steven
US-EPA

Rosenberg, Norman

Rubin, Edward
Carnegie Mellon University

Sprinz, Detlef
The University of Michigan

Trigg, Talley
U.S. Department of State

Widge, Vikram
International Finance Corporation

Zambia

Yamba, Francis
University of Zambia

Zimbabwe

Nziramasanga, Norbert
Southern centre for Energy and Environment

Index

Note: * indicates the term also appears in the Glossary. Bold page numbers indicate page spans for entire chapters. Italicized page numbers denote tables, figures and boxed material.

Indexer: Marilyn Anderson (United States)

policies, 225, 748-749, 753-765, 795-796 (*See also* Policies)
'public good' and, 127
sustainable development and, 32-34, 81-82, 119, 121-122, 121-127, 693, 695-696
technology and, 148-152
Climate feedback*, 101, 173
Climate sensitivity*, 41-42, 173, 229-231
Climate threshold*, 173
Cloud generation/enhancement, 625
Co-benefits*, 81, 98, 138, 623, 669-677, 696
agriculture sector, 66, 500, 522, 526-530, 673
air quality, 213-214, 623, 669-673
buildings sector, 58, 389-390, 416-418
ecosystems, 673
energy sector, 47, 310-311
forestry sector, 70-71, 543, 549, 574-576
health, 98, 416-417, 623, 670-672
industry sector, 62, 484-485
transport sector, 52, 378-379
waste sector, 75, 587, 588, 610, *612,* 613
Co-generation*, 459, 465, 469, 470-471
Co-operative arrangements. *See* Policies
CO₂ equivalents*, 27, 38, 139, 171, *229,* 621
Combined-cycle gas turbine (CCCGT)*, 274, 278, *284*
Combined heat and power (CHP), 284, 305, 459, 465, 469, 469-470
Competitiveness, 81, 666-667, 709
Compliance*, *769,* 787-788
Conference of the Parties (COP)*, 113, 783
Confidence, 34
Cooking stoves, 54, 55, 390, 403, *417,* 670, 729
Cost-benefit analysis*, 42, 130, 134, 144, 173, 231-232, 235
Cost-effectiveness analysis*, 130
Cost-effectiveness of policies, 747, 751-752, 790
Costs*, 35, 79-80, 93, 134-140, 139-140, 625-641
agriculture sector, 64-65, 499-500, 516-519, 531, *632, 636*
baseline, 622
buildings sector, 55-56, 389, 409-416, *632, 636*
carbon capture and storage, 286
decarbonization, 664
definitions, 35, 134-136
determinants of, 136
discounting, 136-138, 633
energy sector, 44-46, 286, 289-305, *632, 636*
forestry sector, 68-69, 543, 560-563, *632, 636*

industry sector, 61, 472-475, *632, 636*
irreversible, 102
macro-economic, 79-80, 135, 172, 621, 642, 776-777
mitigation, medium- to long-term, 658-664
risk and uncertainty and, 134
social cost of carbon, 42, 173, 232-233
social costs, 35, 36, 135, 136
stabilization, 40, 172, 203-206
technology, 148, 650-658
transaction and implementation, 138-139
transport sector, 50-51, 359-366, *632, 636*
waste sector, 73-74, 589-590, 603-606, *632, 636*
Crops/cropland. *See* Agriculture

D

Dangerous anthropogenic interference, 32, 97, 99-100
Decarbonization, 172, 219-222, 325, 391, 659, 660-661
co-benefits, 669-673
costs, 664
Decision-making, 34-35, 42, 119, 127-131, 173
assumptions in, 128, 130, 131, 235
hedging, 43, 173, 234, 663
irreversibility and, 127-128
risks and uncertainty, 33, 42, 119, 128-129, 142, 173
sequential, 128-129
support tools, 34, 130-131, 587
See also Trade-offs
Deforestation*, 504, 543, 546, 552, 574
in Amazon basin, 552, *553*
avoided, 552, 574, *727,* 731
carbon price and, 621, 660
causes of, 566, 721
CO₂ emissions from, *104*
reduction of, 87, 550, 551-552, 565, 566-568, 574, 694, 731
Demand-side management (DSM)*, 425-426
Dematerialization*, 609
Desertification*, 525
Developed and developing countries, 82-83, 704-707
terminology use, 160-161
Development, 176, 693, 699-726
adaptation-mitigation interactions, 225-226
equity and, 143-144
forestry sector trends, 67
goals, 123, 124-125 (*See also* Millennium Development Goals)
industry sector trends, 58-59, 62, 451-452
paradigms, 123-124
waste sector trends, 71-72, 593-595

See also Development paths; Sustainable development
Development paths*, 33, 82-83, 173, 226, 693, 696, 699-726
GHG emissions and, 177, 701-703
importance of, 82-83, 700, 701, 703-704
multiplicity of, 700-707
opportunities to change, 82, 717-726
regional conditions/priorities and, 704-707
technological change and, 151
Diesel fuel/vehicles, 50, 325, 328, *336,* 337, 675
Direct emissions*, 55, *104,* 338, 389, 449, 587, *639*
Disclosure requirements, 747
Discounting*, 136-138, 633
Distribution and equity. *See* Equity
District heating*, 265, 276-277, 279, 283, 284, 308
Double dividend*, 138

E

Ecological footprint, 698, 699
Economic efficiency, 147
Economic potentials, 35, 77-78, 140, 621, 631-635
aggregate, 626-640
agriculture sector, 499-500, 516-519
carbon prices and, 621
definition, 35, 140
forestry sector, 69, 543, 551, *561-563*
overview, *632*
waste sector, 73-74, 587, 605, *606*
Economic scenarios, 171, 176-177, 180
Economies in transition (EITs)*, 83, 113, 429-430, 609, 706, *774*
Economies of scale*, *154,* 661
Ecosystem* co-benefits, 673
Education, 307
EITs. *See* Economies in transition
Electric vehicles, 50, 347
Electricity, 254, 260, 282, 299-305, *724*
buildings sector use/emissions, 391-392, *393*
carbon content, 637
governance/ownership, 709
mitigation potential, 627-628, *632*
projections, 301, 659
savings, 457, 473, *475, 477,* 621, *632,* 687-690
Embodied energy*, 389, 391, 405
Emission permits*, 307, 658
Emission quotas*, 147, 306-307
Emission standards*, 754, *769,* 791
Emission trajectories*, 171, 175, 199, 235, 647-649
Emissions, 27-31, 41, 97, *103,* 121
baseline scenarios, 37-38, 186-194
cap*, 755, 758, *767, 769, 772*
co-control, 674
CO₂-equivalents, 27, 38, 139, 171, *229,* 621